Roark's Formulas for Stress and Strain

About the Authors

Richard G. Budynas (deceased) was professor emeritus of mechanical engineering at the Rochester Institute of Technology. He was the author of *Advanced Strength and Applied Stress Analysis*, Second Edition (McGraw-Hill, 1999), and coauthor of *Shigley's Mechanical Engineering Design*, Ninth Edition (McGraw-Hill, 2011).

Ali M. Sadegh is professor of mechanical engineering, former chairman of the department, and the founder and director of the Center for Advanced Engineering Design and Development at the City College of the City University of New York. He is a Fellow of ASME and SME. He is a PE and CMfgE with over 40 years of experience in academia and practicing mechanical engineering design. He led the development of the 11th and 12th editions of *Mark's Standard Handbook for Mechanical Engineers* (McGraw-Hill, 2007, 2018).

Roark's Formulas for Stress and Strain

RICHARD G. BUDYNAS

ALI M. SADEGH

Ninth Edition

McGraw Hill

New York Chicago San Francisco
Athens London Madrid
Mexico City Milan New Delhi
Singapore Sydney Toronto

Library of Congress Control Number: 2019945013

Roark's Formulas for Stress and Strain, Ninth Edition

3 4 5 6 7 8 9 LWI 24 23

ISBN 978-1-260-45375-1
MHID 1-260-45375-8

The pages within this book were printed on acid-free paper.

Sponsoring Editor
Robert Argentieri

Editorial Supervisor
Donna M. Martone

Acquisitions Coordinator
Elizabeth Houde

Project Manager
Sarika Gupta,
Cenveo® Publisher Services

Copy Editor
Kevin Campbell

Proofreader
Cenveo Publisher Services

Indexer
ARC Indexing

Production Supervisor
Lynn M. Messina

Composition
Cenveo Publisher Services

Art Director, Cover
Jeff Weeks

Contents

Preface to the Ninth Edition

It is recognized that recent high-powered computers incorporated in computational mechanics such as finite element methods have advanced and facilitated design and stress and strain analyses of mechanical and structural systems. While numerical methods are widely used in engineering practice, the very essence of engineering is rooted in the power of classical closed-form solutions and established analytical methods. In addition, for optimization, parametric design, and for understanding the mechanical behavior of a system, a closed-form analytical solution is advantageous to a numerical solution.

Since the publication of the 8th edition of *Roark's Formulas for Stress and Strain*, 10 years ago, we have witnessed significant advances in engineering methodology in solving stress analysis problems. This has motivated the authors to embark on an improved edition of this book.

Thus in preparation of this 9th edition, the authors had three continuing objectives: first, to modernize with newly designed artwork and update the contents as required, second, to introduce new topics and chapters that will maintain the high standard of this book, and finally, to improve upon the material retained from the 8th edition. The 9th edition of Roark's is intended to make available a compact, comprehensive summary of the formulas and principles pertaining to strength of materials for both practicing engineers and engineering students.

This book is intended primarily to be a reference book that is authoritative and covers the field of stress and strain analyses in a comprehensive manner. Similar to the 8th edition, the tabular format is continued in this edition. This format has been particularly successful when implementing problem solutions on user-friendly computer software such as MATLAB, MathCAD, TK Solver, and Excel. Commercial packages are available which integrate the abovementioned software with Roark's. Tabulated coefficients are in dimensionless form for convenience in using either system of units. Design formulas drawn from works published in the past remain in the system units originally published or quoted.

The authors are mindful of the competing requirements to offer the user a broad spectrum of information that has made this book so useful for over 80 years. Therefore, in this edition, the authors have included a number of new topics in the chapters. The main organizational change in the 9th edition is that majority of tables are published in portrait format for ease of reading. Other changes/additions included in the 9th edition are as follows:

- Chapter 2, Stress and Strain: Important Relationships: A new section on three-dimensional Mohr's circle analysis for simple configurations is added.
- Chapter 3, The Behavior of Bodies under Stress: Mechanical properties of some constructional steel including modulus of elasticity of materials and yield strength are added. In addition, a section on factor of safety is also added.
- Chapter 4, Principles and Analytical Methods: The energy method "Castigliano's theorem" for deflections calculation is revised, and more examples are added.
- Chapter 5, Numerical Methods: The numerical methods "finite element method" and "boundary element method" with more references are updated and revised.

- Chapter 6, Experimental Methods: A new section "Nondestructive Testing" for quantitatively characterizing mechanical properties is added.
- Chapter 16, Dynamic and Temperature Stresses: Tables of natural frequencies of cylindrical shells and springs are added.
- Chapter 17, Stress Concentration: The table for stress concentration factor is expanded.
- Chapter 18, Fatigue and Fracture: A new section on fatigue with a table of fatigue limits of materials and a new section on creep with a table of creep of materials at high temperature are added.
- Chapter 19, Stresses in Fasteners, Joints, and Gears: A new section "Bolt Strength and Design" and a new section Gearing and Gear Stress are added.
- Chapter 20, Composite Materials: Two new sections "Composite Sandwich Structures" and "Composite Cellular Structures" are added. In addition, a table of Mechanical Properties of Graphite-Polymer Composite Material with Different Volume Fraction and several new references are added.
- Chapter 21, Solid Biomechanics: Biomechanics is updated and revised, and artificial intervertebral discs are discussed.
- The references and publications of most chapters are expanded and updated.
- All the chapters are redesigned to make a much better user experience including: adding color to the figures and tables; newly designed artwork and headings; and portrait format of majority of tables for ease of reading.

The authors would especially like to thank those individuals, publishers, institutions, and corporations who have generously given permission to use material in this and previous editions, and the many dedicated readers and users of *Roark's Formulas for Stress and Strain.*

Meticulous care has been exercised to avoid errors. However, if any are inadvertently included in this newly designed printing, the authors will appreciate being informed so that these errors can be eliminated in subsequent printings of this edition.

Richard G. Budynas
Ali M. Sadegh

Preface to the First Edition

This book was written for the purpose of making available a compact, adequate summary of the formulas, facts, and principles pertaining to strength of materials. It is intended primarily as a reference book and represents an attempt to meet what is believed to be a present need of the designing engineer.

This need results from the necessity for more accurate methods of stress analysis imposed by the trend of engineering practice. That trend is toward greater speed and complexity of machinery, greater size and diversity of structures, and greater economy and refinement of design. In consequence of such developments, familiar problems, for which approximate solutions were formerly considered adequate, are now frequently found to require more precise treatment, and many less familiar problems, once of academic interest only, have become of great practical importance. The solutions and data desired are often to be found only in advanced treatises or scattered through an extensive literature, and the results are not always presented in such form as to be suited to the requirements of the engineer. To bring together as much of this material as is likely to prove generally useful and to present it in convenient form have been the author's aim.

The scope and management of the book are indicated by the contents. In Part 1 are defined all terms whose exact meanings might otherwise not be clear. In Part 2 certain useful general principles are stated; analytical and experimental methods of stress analysis are briefly described; and information concerning the behavior of material under stress is given. In Part 3 the behavior of structural elements under various conditions of loading is discussed, and extensive tables of formulas for the calculation of stress, strain, and strength are given.

Because they are not believed to serve the purpose of this book, derivations of formulas and detailed explanations, such as are appropriate in a textbook, are omitted, but a sufficient number of examples are included to illustrate the application of the various formulas and methods. Numerous references to more detailed discussions are given, but for the most part these are limited to sources that are generally available, and no attempt has been made to compile an exhaustive bibliography.

That such a book as this derives almost wholly from the work of others is self-evident, and it is the author's hope that due acknowledgment has been made of the immediate sources of all material presented here. To the publishers and others who have generously permitted the use of material, he wishes to express his thanks. The helpful criticisms and suggestions of his colleagues, Professors E. R. Maurer, M. O. Withey, J. B. Kommers, and K. F. Wendt, are gratefully acknowledged. A considerable number of the tables of formulas have been published from time to time in *Product Engineering*, and the opportunity thus afforded for criticism and study of arrangement has been of great advantage.

Finally, it should be said that, although every care has been taken to avoid errors, it would be oversanguine to hope that none had escaped detection; for any suggestions that readers may make concerning needed corrections, the author will be grateful.

Raymond J. Roark

List of Tables

Roark's Formulas for Stress and Strain

CHAPTER 1

Introduction

The widespread use of personal computers, which have the power to solve problems solvable in the past only on mainframe computers, has influenced the tabulated format of this book. Computer programs for structural analysis, employing techniques such as the finite element method and the boundary element method, are also available for general use. These programs are very powerful; however, in many cases, elements of structural systems can be analyzed quite effectively independently without the need for an elaborate finite element model. In some instances, finite element models or programs are verified by comparing their solutions with the results given in a book such as this. Contained within this book are simple, accurate, and thorough tabulated formulations that can be applied to the stress analysis of a comprehensive range of structural components.

This chapter serves to introduce the reader to the terminology, state property units and conversions, and contents of the book.

1.1 TERMINOLOGY

Definitions of terms used throughout the book can be found in the glossary in App. C.

1.2 STATE PROPERTIES, UNITS, AND CONVERSIONS

The basic state properties associated with stress analysis include the following: geometrical properties such as length, area, volume, centroid, center of gravity, and second-area moment (area moment of inertia); material properties such as mass density, modulus of elasticity, Poisson's ratio, and thermal expansion coefficient; loading properties such as force, moment, and force distributions (e.g., force per unit length, force per unit area, and force per unit volume); other properties associated with loading, including energy, work, and power; and stress analysis properties such as deformation, strain, and stress.

Two basic systems of units are employed in the field of stress analysis: SI units and USCU units.* SI units are mass-based units using the kilogram (kg), meter (m), second (s), and Kelvin (K) or degree Celsius (°C) as the fundamental units of mass, length, time, and temperature, respectively. Other SI units, such as that used for force, the Newton (kg-m/s^2), are derived quantities. USCU units are force-based units using the pound force (lbf),

*SI and USCU are abbreviations for the International System of Units (from the French Systéme International d'Unités) and the United States Customary Units, respectively.

TABLE 1.1 Units Appropriate to Structural Analysis*

Quantity	International Metric (SI)	U.S. Customary
Length	(meter) m	(foot) ft
Force and weight, W	(newton) N(kg-m/s^2)	(pound) lbf
Time	s	s
Angle	rad	rad
Second area moment	m^4	ft^4
Mass	kg	lbf-s^2/ft (slug)
Area	m^2	ft^2
Mass moment of inertia	kg-m^2	lbf-s^2-ft
Moment	N-m	lbf-ft
Volume	m^3	ft^3
Mass density	kg/m^3	lbf-s^2/ft^4
Stiffness of linear spring	N/m	lbf/ft
Stiffness of rotary spring	N-m/rad	lbf-ft/rad
Temperature	K (Kelvin)	°F (degrees Fahrenheit)
Torque, work, energy	N-m (Joule)	lbf-ft
Stiffness of torsional spring	N-m/rad	lbf-ft/rad
Stress or pressure	N/m^2 (pascal)	lbf/ft^2 (psi)

*In stress anlaysis, the unit of length used most often is the inch.

inch (in) or foot (ft), second (s), and degree Fahrenheit (°F) as the fundamental units of force, length, time, and temperature, respectively. Other USCU units, such as that used for mass, the slug (lbf-s^2/ft) or the nameless lbf-s^2/in, are derived quantities. Table 1.1 gives a listing of the primary SI and USCU units used for structural analysis. Other SI units are given in Table 1.2. Certain prefixes may be appropriate, depending on the size of the quantity. Common prefixes are given in Table 1.3. For example, the modulus of elasticity of carbon steel is approximately 207 GPa = 207×10^9 Pa = 207×10^9 N/m^2. Prefixes are normally used with SI units. However, there are cases where prefixes are also used with USCU units. Some examples are the kpsi (1 kpsi = 10^3 psi = 10^3 lbf/in^2), kip (1 kip = 1 kilopound = 1,000 lbf), and Mpsi (1 Mpsi = 10^6 psi).

Depending on the application, different units may be specified. It is important that the analyst be aware of all the implications of the units and make consistent use of them. For example, if you are building a model from a CAD file in which the design dimensional units are given in mm, it is unnecessary to change the system of units or to scale the model to units of m. However, if in this example the input forces are in Newtons, then the output stresses will be in N/mm^2, which is correctly expressed as MPa. If in this example applied moments are to be specified, the units should be N-mm. For deflections in this example, the modulus of elasticity E should also be specified in MPa and the output deflections will be in mm.

Tables 1.4 and 1.5 present the conversions from USCU units to SI units and vice versa for some common state property units. The more detailed conversion units are given in Table 1.6.

TABLE 1.2 SI Units

Quantity	Unit (SI)	Formula
Base Units		
Length	meter (m)	
Mass	kilogram (kg)	
Time	second (s)	
Thermodynamic temperature	Kelvin (K)	
Supplementary Units		
Plane angle	radian (rad)	
Solid angle	steradian (sr)	
Derived Units		
Acceleration	meter per second square	m/s^2
Angular acceleration	radian per second square	rad/s^2
Angular velocity	radian per second	rad/s
Area	square meter	m^2
Density	kilogram per cubic meter	kg/m^3
Energy	joule (J)	N-m
Force	Newton (N)	$kg\text{-}m/s^2$
Frequency	hertz (Hz)	1/s
Power	watt (W)	J/s
Pressure	Pascal (Pa)	N/m^2
Quantity of heat	joule (J)	N-m
Stress	Pascal (Pa)	N/m^2
Thermal conductivity	watt per meter-Kelvin	W/(m-K)
Velocity	meter per second	m/s
Viscosity dynamic	Pascal-second	Pa-s
Viscosity kinematic	square meter per second	m^2/s
Work	joule (J)	N-m

1.3 CONTENTS

Following the introduction, the state of stress and the important relationships associated with stress and strain and their transformations including Mohr's circle is described in Chap. 2. The behavior of bodies under stress is presented in Chap. 3. Chapter 4 describes equation of motion and equilibrium of solid and analytical methods of solving for the stresses and deflections in an elastic body, including the energy methods. Numerical methods such as Finite Element Method and Boundary Element Method are presented in Chap. 5. The experimental methods for stress and strain measurements are presented in Chap. 6. Many topics associated with the stress analysis of structural components, including direct tension, compression, shear, and combined stresses; bending of straight and curved beams; torsion; bending of flat plates; columns and other compression members; shells of revolution, pressure vessels, and pipes; direct bearing and shear stress; elastic stability;

TABLE 1.3 Multiples and Submultiples of SI Units

Prefix	Symbol		Multiplying Factor
exa	E	10^{18}	1 000 000 000 000 000 000
peta	P	10^{15}	1 000 000 000 000 000
tera	T	10^{12}	1 000 000 000 000
giga	G	10^{9}	1 000 000 000
mega	M	10^{6}	1 000 000
kilo	k	10^{3}	1 000
hecto	h	10^{2}	100
deca	da	10	10
deci	d	10^{-1}	0.1
centi	c	10^{-2}	0.01
milli	m	10^{-3}	0.001
micro	µ	10^{-6}	0.000 001
nano	n	10^{-9}	0.000 000 001
pico	p	10^{-12}	0.000 000 000 001
femto	f	10^{-15}	0.000 000 000 000 001
atto	a	10^{-18}	0.000 000 000 000 000 001

TABLE 1.4 SI Conversion Table

SI Units	From SI to English	From English to SI
Length		
kilometer (km) = 1000 m	1 km = 0.621 mi	1 mi = 1.609 km
meter (m) = 100 cm	1 m = 3.281 ft	1 ft = 0.305 m
centimeter (cm) = 0.01 m	1 cm = 0.394 in	1 in = 2.540 cm
millimeter (mm) = 0.001 m	1 mm = 0.039 in	1 in = 25.4 mm
micrometer (µm) = 0.000 001 m	1 µm = 3.93×10^{-5} in	1 in = 25,400 mm
nanometer (nm) = 0.000 000 001 m	1 nm = 3.93×10^{-8} in	1 in = 25,400,000 mm
Area		
square kilometer (km^2) = 100 hectares	1 km^2 = 0.386 mi^2	1 mi^2 = 2.590 km^2
hectare (ha) = 10,000 m^2	1 ha = 2.471 acres	1 acre = 0.405 ha
square meter (m^2) = 10,000 cm^2	1 m^2 = 10.765 ft^2	1 ft^2 = 0.093 m^2
square centimeter (cm^2) = 100 mm^2	1 cm^2 = 0.155 in^2	1 in^2 = 6.452 cm^2
Volume		
liter (L) = 1000 mL = 1 dm^3	1 L = 1.057 fl qt	1 fl qt = 0.946 L
milliliter (mL) = 0.001 L = 1 cm^3	1 mL = 0.034 fl oz	1 fl oz = 29.575 mL
microliter (µL) = 0.000 001 L	1 µL = 3.381×10^{-5} fl oz	1 fl oz = 29,575 µL

(Continued)

TABLE 1.4 SI Conversion Table (*Continued*)

SI Units	From SI to English	From English to SI
Mass		
kilogram (kg) = 1000 g	1 kg = 2.205 lb	1 lb = 0.454 kg
gram (g) = 1000 mg	1 g = 0.035 oz	1 oz = 28.349 g
milligram (mg) = 0.001 g	1 mg = 3.52×10^{-5} oz	1 oz = 28,349 mg
microgram (µg) = 0.000 001 g	1 µg = 3.52×10^{-8} oz	1 oz = 28,349,523 µg

TABLE 1.5 Multiplication Factors to Convert from USCU Units to SI Units

To Convert from	To	Multiply by
Mass		
ounce (avoirdupois)	kilogram (kg)	2.834952×10^{-2}
pound (avoirdupois)	kilogram (kg)	4.535924×10^{-1}
ton (short, 2000 lb)	kilogram (kg)	9.071847×10^{2}
ton (long, 2240 lb)	kilogram (kg)	1.016047×10^{3}
kilogram (kg)	ounce (avoirdupois)	3.527396×10^{1}
kilogram (kg)	pound (avoirdupois)	2.204622
kilogram (kg)	ton (short, 2000 lb)	1.102311×10^{-3}
kilogram (kg)	ton (long, 2240 lb)	9.842064×10^{-4}
Mass Per Unit Length		
pound per foot (lb/ft)	kilogram per meter (kg/m)	1.488164
pound per inch (lb/in)	kilogram per meter (kg/m)	1.785797×10^{1}
kilogram per meter (kg/m)	pound per foot (lb/ft)	6.719689×10^{-1}
kilogram per meter (kg/m)	pound per inch (lb/in)	5.599741×10^{-2}
Mass Per Unit Area		
pound per square foot (lb/ft^2)	kilogram per square meter (kg/m^2)	4.882428
kilogram per square meter (kg/m^2)	pound per square foot (lb/ft^2)	2.048161×10^{-1}
Mass Per Unit Volume		
pound per cubic foot (lb/ft^3)	kilogram per cubic meter (kg/m^3)	1.601846×10^{1}
pound per cubic inch (lb/in^3)	kilogram per cubic meter (kg/m^3)	2.767990×10^{4}
kilogram per cubic meter (kg/m^3)	pound per cubic foot (lb/ft^3)	6.242797×10^{-2}
kilogram per cubic meter (kg/m^3)	pound per cubic inch (lb/in^3)	3.612730×10^{-5}
pound per cubic foot (lb/ft^3)	pound per cubic inch (lb/in^3)	1.728000×10^{3}
Length		
foot (ft)	meter (m)	3.048000×10^{-1}
inch (in)	meter (m)	2.540000×10^{-2}
mil	meter (m)	2.540000×10^{-5}
inch (in)	micrometer (µm)	2.540000×10^{4}
meter (m)	foot (ft)	3.28084
meter (m)	inch (in)	3.937008×10^{1}
meter (m)	mil	3.937008×10^{4}
micrometer (µm)	inch (in)	3.937008×10^{-5}

(*Continued*)

TABLE 1.5 Multiplication Factors to Convert from USCU Units to SI Units (*Continued*)

To Convert from	To	Multiply by
Area		
$foot^2$	square meter (m^2)	9.290304×10^{-2}
$inch^2$	square meter (m^2)	6.451600×10^{-4}
circular mil	square meter (m^2)	5.067075×10^{-10}
square centimeter (cm^2)	square inch (in^2)	1.550003×10^{-1}
square meter (m^2)	$foot^2$	1.076391×10^{1}
square meter (m^2)	$inch^2$	1.550003×10^{3}
square meter (m^2)	circular mil	1.973525×10^{9}
Volume		
$foot^3$	cubic meter (m^3)	2.831685×10^{-2}
$inch^3$	cubic meter (m^3)	1.638706×10^{-5}
cubic centimeter (cm^3)	cubic inch (in^3)	6.102374×10^{-2}
cubic meter (m^3)	$foot^3$	3.531466×10^{1}
cubic meter (m^3)	$inch^3$	6.102376×10^{4}
gallon (U.S. liquid)	cubic meter (m^3)	3.785412×10^{-3}
Force		
pounds-force (lbf)	newtons (N)	4.448222
Pressure or Stress		
pound force per square inch (lbf/in^2)(psi)	pascal (Pa)	6.894757×10^{3}
kip per square inch (kip/in^2)(ksi)	pascal (Pa)	6.894757×10^{6}
pound force per square inch (lbf/in^2)(psi)	megapascal (MPa)	6.894757×10^{-3}
pascal (Pa)	pound force per square inch (psi)	1.450377×10^{-4}
pascal (Pa)	kip per square inch (ksi)	1.450377×10^{-7}
megapascals (MPa)	pound force per square inch (lbf/in^2) (psi)	1.450377×10^{2}
Section Properties		
section modulus S (in^3)	S (m^3)	1.638706×10^{-5}
moment of inertia I (in^4)	I (m^4)	4.162314×10^{-7}
modulus of elasticity E (psi)	E (Pa)	6.894757×10^{3}
section modulus S (m^3)	S (in^3)	6.102374×10^{4}
moment of inertia I (m^4)	I (in^4)	2.402510×10^{6}
modulus of elasticity E (Pa)	E (psi)	1.450377×10^{-4}
Temperature		
degree Fahrenheit	degree Celsius	$t°C = (t°F - 32)/1.8$
degree Celsius	degree Fahrenheit	$t°F = 1.8\,t°C + 32$
Angle		
degree	radian (rad)	1.745329×10^{-2}
radian (rad)	degree	5.729578×10^{1}

TABLE 1.6 Conversion Factors

To Convert from	To	Multiply by
acceleration of free fall, standard (g_n)	meter per second square (m/s^2)	9.80665
acre (based on U.S. survey foot)	square meter (m^2)	4.046873×10^3
acre foot (based on U.S. survey foot)	cubic meter (m^3)	1.233489×10^3
atmosphere, standard (atm)	pascal (Pa)	1.01325×10^5
atmosphere, standard (atm)	kilopascal (kPa)	1.01325×10^2
atmosphere, technical (at)	pascal (Pa)	9.80665×10^4
atmosphere, technical (at)	kilopascal (kPa)	9.80665×10^1
bar (bar)	pascal (Pa)	1.0×10^5
bar (bar)	kilopascal (kPa)	1.0×10^2
barn (b)	square meter (m^2)	1.0×10^{-28}
barrel [for petroleum, 42 gallons (U.S.)] (bbl)	cubic meter (m^3)	1.589873×10^{-1}
barrel [for petroleum, 42 gallons (U.S.)] (bbl)	liter (L)	1.589873×10^2
$calorie_{IT}$ (cal_{IT})	joule (J)	4.1868
$calorie_{th}$ (cal_{th})	joule (J)	4.184
calorie (cal) (mean)	joule (J)	4.19002
calorie (15°C) (cal_{15})	joule (J)	4.18580
calorie (20°C) (cal_{20})	joule (J)	4.18190
centimeter of mercury (0°C)	pascal (Pa)	1.33322×10^3
centimeter of mercury (0°C)	kilopascal (kPa)	1.33322
centimeter of mercury, conventional (cmHg)	pascal (Pa)	1.333224×10^3
centimeter of mercury, conventional (cmHg)	kilopascal (kPa)	1.333224
centimeter of water (4°C)	pascal (Pa)	9.80638×10^1
centimeter of water, conventional (cmH_2O)	pascal (Pa)	9.80665×10^1
centipoise (cP)	pascal second (Pa-s)	1.0×10^{-3}
centistokes (cSt)	meter square per second (m^2/s)	1.0×10^{-6}
chain (based on U.S. survey foot) (ch)	meter (m)	2.011684×10^1
circular mil	square meter (m^2)	5.067075×10^{-10}
cubic foot (ft^3)	cubic meter (m^3)	2.831685×10^{-2}
cubic foot per minute (ft^3/min)	cubic meter per second (m^3/s)	4.719474×10^{-4}
cubic foot per minute (ft^3/min)	liter per second (L/s)	4.719474×10^{-1}
cubic foot per second (ft^3/s)	cubic meter per second (m^3/s)	2.831685×10^{-2}
cubic inch (in^3)	cubic meter (m^3)	1.638706×10^{-5}

(Continued)

TABLE 1.6 Conversion Factors (*Continued*)

To Convert from	To	Multiply by
cubic inch per minute (in^3/min)	cubic meter per second (m^3/s)	2.731177×10^{-7}
cubic mile (mi^3)	cubic meter (m^3)	4.168182×10^{9}
cubic yard (yd^3)	cubic meter (m^3)	7.645549×10^{-1}
cubic yard per minute (yd^3/min)	cubic meter per second (m^3/s)	1.274258×10^{-2}
cup (U.S.)	cubic meter (m^3)	2.365882×10^{-4}
cup (U.S.)	liter (L)	2.365882×10^{-1}
cup (U.S.)	milliliter (mL)	2.365882×10^{2}
day (d)	second (s)	8.64×10^{4}
day (sidereal)	second (s)	8.616409×10^{4}
degree (angle) (°)	radian (rad)	1.745329×10^{-2}
degree Celsius (temperature) (°C)	kelvin (K)	$T/\text{K} = t/°\text{C} + 273.15$
degree Celsius (temperature interval) (°C)	kelvin (K)	1.0
degree centigrade (temperature)	degree Celsius (°C)	$t/°\text{C} \approx t/\text{deg. cent.}$
degree centigrade (temperature interval)	degree Celsius (°C)	1.0
degree Fahrenheit (temperature) (°F)	degree Celsius (°C)	$t/°\text{C} = (t/°\text{F} - 32)/1.8$
degree Fahrenheit (temperature) (°F)	kelvin (K)	$T/\text{K} = (t/°\text{F} + 459.67)/1.8$
dyne (dyn)	newton (N)	1.0×10^{-5}
dyne centimeter (dyn-cm)	newton meter (N-m)	1.0×10^{-7}
dyne per square centimeter (dyn/cm^2)	pascal (Pa)	1.0×10^{-1}
fermi	femtometer (fm)	1.0
fluid ounce (U.S.) (fl oz)	cubic meter (m^3)	2.957353×10^{-5}
fluid ounce (U.S.) (fl oz)	milliliter (mL)	2.957353×10^{1}
foot (ft)	meter (m)	3.048×10^{-1}
foot (U.S. survey) (ft)	meter (m)	3.048006×10^{-1}
foot of mercury, conventional (ftHg)	pascal (Pa)	4.063666×10^{4}
foot of mercury, conventional (ftHg)	kilopascal (kPa)	4.063666×10^{1}
foot of water (39.2°F)	pascal (Pa)	2.98898×10^{3}
foot of water (39.2°F)	kilopascal (kPa)	2.98898
foot of water, conventional (ftH_2O)	pascal (Pa)	2.989067×10^{3}
foot of water, conventional (ftH_2O)	kilopascal (kPa)	2.989067
foot per hour (ft/h)	meter per second (m/s)	8.466667×10^{-5}
foot per minute (ft/min)	meter per second (m/s)	5.08×10^{-3}
foot per second (ft/s)	meter per second (m/s)	3.048×10^{-1}
foot per second square (ft/s^2)	meter per second square (m/s^2)	3.048×10^{-1}

(Continued)

TABLE 1.6 Conversion Factors (*Continued*)

To Convert from	To	Multiply by
foot poundal	joule (J)	4.214011×10^{-2}
foot pound-force (ft-lbf)	joule (J)	1.355818
foot pound-force per hour (ft-lbf/h)	watt (W)	3.766161×10^{-4}
foot pound-force per minute (ft-lbf/min)	watt (W)	2.259697×10^{-2}
foot pound-force per second (ft-lbf/s)	watt (W)	1.355818
foot to the fourth power (ft^4)	meter to the fourth power (m^4)	8.630975×10^{-3}
gal (gal)	meter per second square (m/s^2)	1.0×10^{-2}
gallon [Canadian and U.K. (Imperial)] (gal)	cubic meter (m^3)	4.54609×10^{-3}
gallon [Canadian and U.K. (Imperial)] (gal)	liter (L)	4.54609
gallon (U.S.) (gal)	cubic meter (m^3)	3.785412×10^{-3}
gallon (U.S.) (gal)	liter (L)	3.785412
gallon (U.S.) per day (gal/d)	cubic meter per second (m^3/s)	4.381264×10^{-8}
gallon (U.S.) per day (gal/d)	liter per second (L/s)	4.381264×10^{-5}
gallon (U.S.) per horsepower hour [gal/(hp-h)]	cubic meter per joule (m^3/J)	1.410089×10^{-9}
gallon (U.S.) per horsepower hour [gal/(hp-h)]	liter per joule (L/J)	1.410089×10^{-6}
gallon (U.S.) per minute (gpm)(gal/min)	cubic meter per second (m^3/s)	6.309020×10^{-5}
gallon (U.S.) per minute (gpm)(gal/min)	liter per second (L/s)	6.309020×10^{-2}
gon (also called grade) (gon)	radian (rad)	1.570796×10^{-2}
gon (also called grade) (gon)	degree (angle) (°)	9.0×10^{-1}
grain (gr)	kilogram (kg)	6.479891×10^{-5}
grain (gr)	milligram (mg)	6.479891×10^{1}
grain per gallon (U.S.) (gr/gal)	kilogram per cubic meter (kg/m^3)	1.711806×10^{-2}
grain per gallon (U.S.) (gr/gal)	milligram per liter (mg/L)	1.711806×10^{1}
gram-force per square centimeter (gf/cm^2)	pascal (Pa)	9.80665×10^{1}
gram per cubic centimeter (g/cm^3)	kilogram per cubic meter (kg/m^3)	1.0×10^{3}
hectare (ha)	square meter (m^2)	1.0×10^{4}
horsepower (550 ft-lbf/s) (hp)	watt (W)	7.456999×10^{2}
horsepower (boiler)	watt (W)	9.80950×10^{3}
horsepower (electric)	watt (W)	7.46×10^{2}
horsepower (metric)	watt (W)	7.354988×10^{2}

(*Continued*)

TABLE 1.6 Conversion Factors (*Continued*)

To Convert from	To	Multiply by
horsepower (U.K.)	watt (W)	7.4570×10^2
horsepower (water)	watt (W)	7.46043×10^2
hour (h)	second (s)	3.6×10^3
inch (in)	meter (m)	2.54×10^{-2}
inch (in)	centimeter (cm)	2.54
inch of mercury (32°F)	pascal (Pa)	3.38638×10^3
inch of mercury (32°F)	kilopascal (kPa)	3.38638
inch of mercury (60°F)	pascal (Pa)	3.37685×10^3
inch of mercury (60°F)	kilopascal (kPa)	3.37685
inch of mercury, conventional (inHg)	pascal (Pa)	3.386389×10^3
inch of mercury, conventional (inHg)	kilopascal (kPa)	3.386389
inch of water (39.2°F)	pascal (Pa)	2.49082×10^2
inch of water (60°F)	pascal (Pa)	2.4884×10^2
inch of water, conventional (inH_2O)	pascal (Pa)	2.490889×10^2
inch per second (in/s)	meter per second (m/s)	2.54×10^{-2}
kelvin (K)	degree Celsius (°C)	$t/°C = T/K - 273.15$
kilogram-force (kgf)	newton (N)	9.80665
kilogram-force meter (kgf-m)	newton meter (N-m)	9.80665
kilogram-force per square centimeter (kgf/cm^2)	pascal (Pa)	9.80665×10^4
kilogram-force per square centimeter (kgf/cm^2)	kilopascal (kPa)	9.80665×10^1
kilogram-force per square meter (kgf/m^2)	pascal (Pa)	9.80665
kilogram-force per square millimeter (kgf/mm^2)	pascal (Pa)	9.80665×10^6
kilogram-force per square millimeter (kgf/mm^2)	megapascal (MPa)	9.80665
kilogram-force second square per meter ($kgf\text{-}s^2/m$)	kilogram (kg)	9.80665
light year (l.y.)	meter (m)	9.46073×10^{15}
liter (L)	cubic meter (m^3)	1.0×10^{-3}
microinch	meter (m)	2.54×10^{-8}
microinch	micrometer (μm)	2.54×10^{-2}
micron (μ)	meter (m)	1.0×10^{-6}
micron (μ)	micrometer (μm)	1.0
mil (0.001 in)	meter (m)	2.54×10^{-5}
mil (0.001 in)	millimeter (mm)	2.54×10^{-2}

(*Continued*)

TABLE 1.6 Conversion Factors (*Continued*)

To Convert from	To	Multiply by
mil (angle)	radian (rad)	9.817477×10^{-4}
mil (angle)	degree (°)	5.625×10^{-2}
mile (mi)	meter (m)	1.609344×10^{3}
mile (mi)	kilometer (km)	1.609344
mile (based on U.S. survey foot) (mi)	meter (m)	1.609347×10^{3}
mile (based on U.S. survey foot) (mi)	kilometer (km)	1.609347×10^{0}
mile, nautical	meter (m)	1.852×10^{3}
mile per gallon (U.S.) (mpg) (mi/gal)	meter per cubic meter (m/m^3)	4.251437×10^{5}
mile per gallon (U.S.) (mpg) (mi/gal)	kilometer per liter (km/L)	4.251437×10^{-1}
mile per gallon (U.S.) (mpg) (mi/gal)	liter per 100 kilometer (L/100 km)	divide 235.215 by number of miles per gallon
mile per hour (mi/h)	meter per second (m/s)	4.4704×10^{-1}
mile per hour (mi/h)	kilometer per hour (km/h)	1.609344
mile per minute (mi/min)	meter per second (m/s)	2.68224×10^{1}
mile per second (mi/s)	meter per second (m/s)	1.609344×10^{3}
millibar (mbar)	pascal (Pa)	1.0×10^{2}
millibar (mbar)	kilopascal (kPa)	1.0×10^{-1}
millimeter of mercury, conventional (mmHg)	pascal (Pa)	1.333224×10^{2}
millimeter of water, conventional (mmH_2O)	pascal (Pa)	9.80665
minute (angle) (')	radian (rad)	2.908882×10^{-4}
minute (min)	second (s)	6.0×10^{1}
minute (sidereal)	second (s)	5.983617×10^{1}
ounce [Canadian and U.K. fluid (Imperial)] (fl oz)	cubic meter (m^3)	2.841306×10^{-5}
ounce [Canadian and U.K. fluid (Imperial)] (fl oz)	milliliter (mL)	2.841306×10^{1}
ounce (U.S. fluid) (fl oz)	cubic meter (m^3)	2.957353×10^{-5}
ounce (U.S. fluid) (fl oz)	milliliter (mL)	2.957353×10^{1}
ounce (avoirdupois)-force (ozf)	newton (N)	2.780139×10^{-1}
ounce (avoirdupois)-force inch (ozf-in)	newton meter (N-m)	7.061552×10^{-3}
ounce (avoirdupois)-force inch (ozf-in)	millinewton meter (mN-m)	7.061552×10
ounce (avoirdupois) per cubic inch (oz/in^3)	kilogram per cubic meter (kg/m^3)	1.729994×10^{3}
ounce (avoirdupois) per gallon [Canadian and U.K. (Imperial)] (oz/gal)	kilogram per cubic meter (kg/m^3)	6.236023

(*Continued*)

TABLE 1.6 Conversion Factors (*Continued*)

To Convert from	To	Multiply by
ounce (avoirdupois) per gallon [Canadian and U.K. (Imperial)] (oz/gal)	gram per liter (g/L)	6.236023
poise (P)	pascal second (Pa-s)	1.0×10^{-1}
pound (avoirdupois) (lb)	kilogram (kg)	4.535924×10^{-1}
pound (troy or apothecary) (lb)	kilogram (kg)	3.732417×10^{-1}
poundal	newton (N)	1.382550×10^{-1}
poundal per square foot	pascal (Pa)	1.488164
poundal second per square foot	pascal second (Pa-s)	1.488164
pound foot square (lb-ft^2)	kilogram meter square (kg-m^2)	4.214011×10^{-2}
pound-force (lbf)	newton (N)	4.448222
pound-force foot (lbf-ft)	newton meter (N-m)	1.355818
pound-force foot per inch (lbf-ft/in)	newton meter per meter (N-m/m)	5.337866×10^{1}
pound-force inch (lbf-in)	newton meter (N-m)	1.129848×10^{-1}
pound-force inch per inch (lbf-in/in)	newton meter per meter (N-m/m)	4.448222
pound-force per foot (lbf/ft)	newton per meter (N/m)	1.459390×10^{1}
pound-force per inch (lbf/in)	newton per meter (N/m)	1.751268×10^{2}
pound-force per pound (lbf/lb) (thrust to mass ratio)	newton per kilogram (N/kg)	9.80665
pound-force per square foot (lbf/ft^2)	pascal (Pa)	4.788026×10^{1}
pound-force per square inch (psi) (lbf/in^2)	pascal (Pa)	6.894757×10^{3}
pound-force per square inch (psi) (lbf/in^2)	kilopascal (kPa)	6.894757
revolution (r)	radian (rad)	6.283185
revolution per minute (rpm) (r/min)	radian per second (rad/s)	1.047198×10^{-1}
rhe	reciprocal pascal second (Pa-s)	1.0×10^{1}
rpm (revolution per minute) (r/min)	radian per second (rad/s)	1.047198×10^{-1}
square foot (ft^2)	square meter (m^2)	9.290304×10^{-2}
square foot per hour (ft^2/h)	square meter per second (m^2/s)	2.58064×10^{-5}
square foot per second (ft^2/s)	square meter per second (m^2/s)	9.290304×10^{-2}
square inch (in^2)	square meter (m^2)	6.4516×10^{-4}
square inch (in^2)	square centimeter (cm^2)	6.4516
square mile (mi^2)	square meter (m^2)	2.589988×10^{6}
square mile (mi^2)	square kilometer (km^2)	2.589988

(*Continued*)

TABLE 1.6 Conversion Factors (*Continued*)

To Convert from	To	Multiply by
square mile (based on U.S. survey foot) (mi^2)	square meter (m^2)	2.589998×10^6
ton-force (2,000 lbf)	newton (N)	8.896443×10^3
ton-force (2,000 lbf)	kilonewton (kN)	8.896443
ton, long, per cubic yard	kilogram per cubic meter (kg/m^3)	1.328939×10^3
ton, metric (t)	kilogram (kg)	1.0×10^3
tonne (called "metric ton" in U.S.) (t)	kilogram (kg)	1.0×10^3
watt hour (W-h)	joule (J)	3.6×10^3
watt per square centimeter (W/cm^2)	watt per square meter (W/m^2)	1.0×10^4
watt per square inch (W/in^2)	watt per square meter (W/m^2)	1.550003×10^3
watt second (W-s)	joule (J)	1.0
yard (yd)	meter (m)	9.144×10^{-1}
year (365 days)	second (s)	3.1536×10^7

stress concentrations; and dynamic and temperature stresses and vibration, are presented in Chaps. 7 to 16. Each chapter contains many tables associated with most conditions of geometry, loading, and boundary conditions for a given element type. The definition of each term used in a table is completely described in the introduction of the table. Tables for stress concentration factors are given in Chap. 17. Fatigue and fracture mechanics are presented in Chap. 18. In Chap. 19, stresses in fasteners and joints including welding are described. Stresses and strains in fiber reinforced composite materials and laminates are presented in Chap. 20. Finally, biomechanics including the stresses in hard and soft tissues and biomaterials are discussed in Chap. 21.

Appendices

The first appendix deals with the properties of a plane area. The second appendix presents mathematical formulas and matrices. The third appendix provides a glossary of the terminology employed in the field of stress analysis.

The references given in a particular chapter are always referred to by number, and are listed at the end of each chapter.

1.4 REFERENCES

1. Thompson, A., and B. N. Taylor (eds.), *NIST Guide for the Use of the International System of Units (SI)*, NIST, 2008.
2. Avallone, E. A., T. Baumeister, and A. M. Sadegh, *Marks' Standard Handbook for Mechanical Engineers*, 11th ed., McGraw-Hill, 2007.
3. Zwillinger, D., *CRC Standard Mathematical Tables and Formulae*, 31st ed., Chapman and Hall/CRC, 2003.
4. Jeffrey, A., and Hui-Hui Dai, *Handbook of Mathematical Formulas and Integrals*, 4th ed., Academic Press, 2008.
5. Spiegel, M. R., *Schaum's Mathematical Handbook of Formulas and Tables*, McGraw-Hill, 1998.
6. Adams, R. A., *Calculus: A Complete Course*, 5th ed., Pearson Education Ltd., 2003.

CHAPTER 2

Stress and Strain: Important Relationships

Understanding the physical properties of stress and strain is a prerequisite to utilizing the many methods and results of structural analysis in design. This chapter provides the definitions and important relationships of stress and strain.

2.1 STRESS

Stress is simply a distributed force on an external or internal surface of a body. To obtain a physical feeling of this idea, consider being submerged in water at a particular depth. The "force" of the water one feels at this depth is a pressure, which is a compressive *stress*, and not a finite number of "concentrated" forces. Other types of force distributions (stress) can occur in a liquid or solid. Tensile (pulling rather than pushing) and shear (rubbing or sliding) force distributions can also exist.

Consider a general solid body loaded as shown in Fig. 2.1(a). P_i and p_i are applied concentrated forces and applied surface force distributions, respectively; and R_i and r_i are possible support reaction force and surface force distributions, respectively. To determine the state of stress at point Q in the body, it is necessary to expose a surface containing the point Q. This is done by making a planar slice, or break, through the body intersecting the point Q. The orientation of this slice is arbitrary, but it is generally made in a convenient plane where the state of stress can be determined easily or where certain geometric relations can be utilized. The first slice, illustrated in Fig. 2.1(b), is arbitrarily oriented by the surface normal x. This establishes the yz plane. The external forces on the remaining body are shown, as well as the internal force (stress) distribution across the exposed internal surface containing Q. In the general case, this distribution will not be uniform along the surface, and will be neither normal nor tangential to the surface at Q. However, the force distribution at Q will have components in the normal and tangential directions. These components will be tensile or compressive and shear stresses, respectively.

Following a right-handed rectangular coordinate system, the y and z axes are defined perpendicular to x, and tangential to the surface. Examine an infinitesimal area $\Delta A_x = \Delta y \Delta z$ surrounding Q, as shown in Fig. 2.2(a). The equivalent concentrated force due to the force distribution across this area is ΔF_x, which in general is neither normal nor tangential to the surface (the subscript x is used to designate the *normal* to the area). The force ΔF_x has components in the x, y, and z directions, which are labeled ΔF_{xx}, ΔF_{xy}, and ΔF_{xz}, respectively, as shown in Fig. 2.2(b). Note that the first subscript denotes the direction normal to

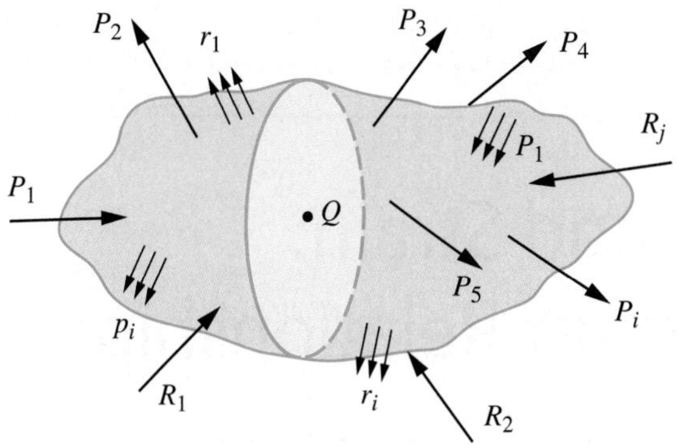

(*a*) Structural member

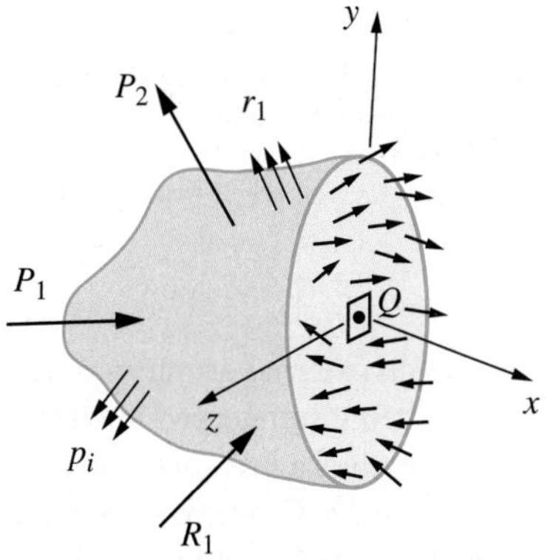

(*b*) Isolated section

Figure 2.1

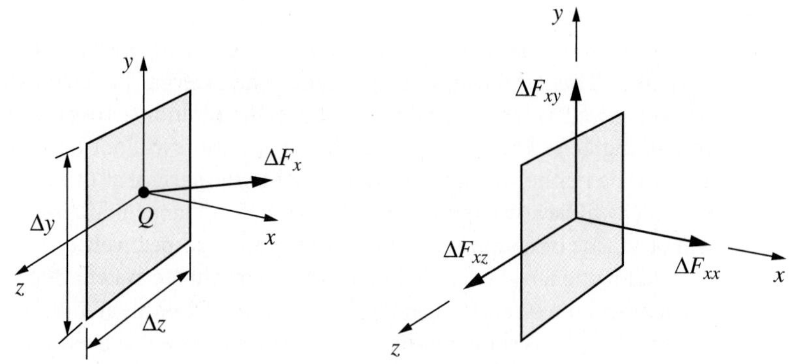

(*a*) Force on the ΔA surface (*b*) Force components

Figure 2.2

the surface and the second gives the actual direction of the force component. The average distributed force per unit area (average stress) in the x direction is*

$$\bar{\sigma}_{xx} = \frac{\Delta F_{xx}}{\Delta A_x}$$

Recalling that stress is actually a point function, we obtain the exact stress in the x direction at point Q by allowing ΔA_x to approach zero.
Thus,

$$\sigma_{xx} = \lim_{\Delta A_x \to 0} \frac{\Delta F_{xx}}{\Delta A_x}$$

or,

$$\sigma_{xx} = \frac{dF_{xx}}{dA_x} \tag{2.1-1}$$

Stresses arise from the tangential forces ΔF_{xy} and ΔF_{xz} as well, and since these forces are tangential, the stresses are shear stresses. Similar to Eq. (2.1-1),

$$\tau_{xy} = \frac{dF_{xy}}{dA_x} \tag{2.1-2}$$

$$\tau_{xz} = \frac{dF_{xz}}{dA_x} \tag{2.1-3}$$

Since, by definition, σ represents a normal stress acting in the same direction as the corresponding surface normal, double subscripts are redundant, and standard practice is to drop one of the subscripts and write σ_{xx} as σ_x. The three stresses existing on the exposed surface at the point are illustrated together using a single arrow vector for each stress as shown in Fig. 2.3. However, it is important to realize that the stress arrow represents a force distribution (stress, force per unit area), and *not* a concentrated force.

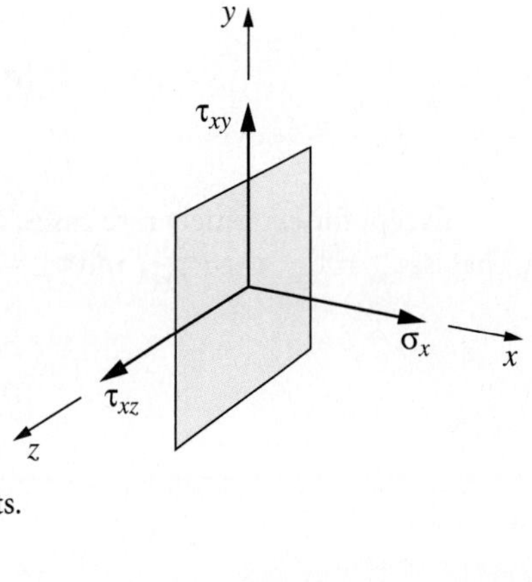

Figure 2.3 Stress components.

*Standard engineering practice is to use the Greek symbols σ and τ for normal (tensile or compressive) and shear stresses, respectively.

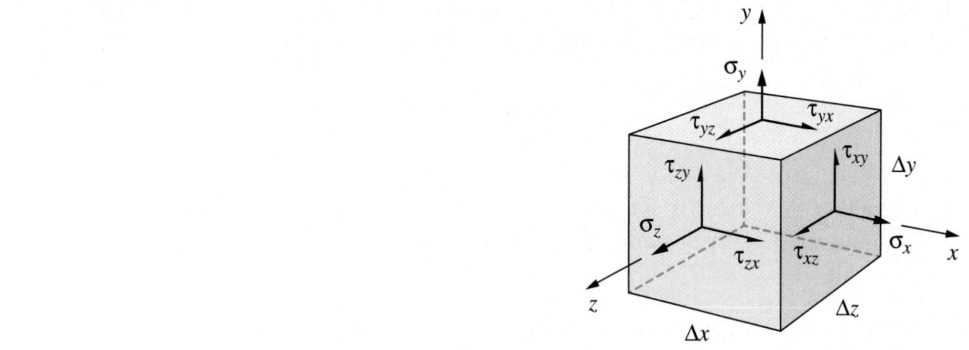

Figure 2.4 Stresses on three orthogonal surfaces.

The shear stresses τ_{xy} and τ_{xz} are the components of the net shear stress acting on the surface, where the net shear stress is given by*

$$(\tau_x)_{\text{net}} = \sqrt{\tau_{xy}^2 + \tau_{xz}^2} \tag{2.1-4}$$

To describe the complete state of stress at point Q completely, it would be necessary to examine other surfaces by making different planar slices. Since different planar slices would necessitate different coordinates and different free-body diagrams, the stresses on each planar surface would be, in general, quite different. As a matter of fact, in general, an infinite variety of conditions of normal and shear stress exist at a given point within a stressed body. So, it would take an infinitesimal spherical surface surrounding the point Q to understand and describe the complete state of stress at the point. Fortunately, through the use of the method of *coordinate transformation*, it is only necessary to know the state of stress on *three* different surfaces to describe the state of stress on *any* surface. This method is described in Sec. 2.3.

The three surfaces are generally selected to be mutually perpendicular, and are illustrated in Fig. 2.4 using the stress subscript notation as earlier defined. This state of stress can be written in matrix form, where the stress matrix $[\sigma]$ is given by

$$[\sigma] = \begin{bmatrix} \sigma_x & \tau_{xy} & \tau_{xz} \\ \tau_{yx} & \sigma_y & \tau_{yz} \\ \tau_{zx} & \tau_{zy} & \sigma_z \end{bmatrix} \tag{2.1-5}$$

Except for extremely rare cases, it can be shown that adjacent shear stresses are equal. That is, $\tau_{yx} = \tau_{xy}$, $\tau_{zy} = \tau_{yz}$, and $\tau_{xz} = \tau_{zx}$, and the stress matrix is symmetric and written as

$$[\sigma] = \begin{bmatrix} \sigma_x & \tau_{xy} & \tau_{zx} \\ \tau_{xy} & \sigma_y & \tau_{yz} \\ \tau_{zx} & \tau_{yz} & \sigma_z \end{bmatrix} \tag{2.1-6}$$

*Stresses can only be added as vectors if they exist on a common surface.

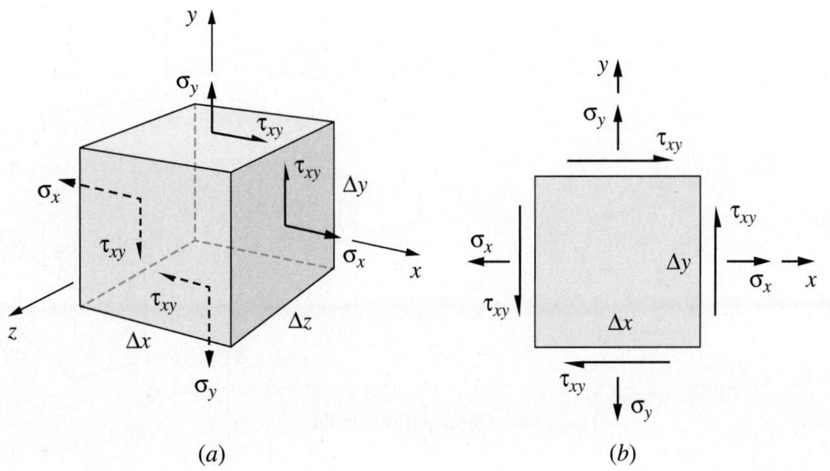

Figure 2.5 Plane stress.

Plane Stress

There are many practical problems where the stresses in one direction are zero. This situation is referred to as a case of *plane stress*. Arbitrarily selecting the *z* direction to be stress-free with $\sigma_z = \tau_{yz} = \tau_{yz} = 0$, the last row and column of the stress matrix can be eliminated, and the stress matrix is written as

$$[\sigma] = \begin{bmatrix} \sigma_x & \tau_{xy} \\ \tau_{xy} & \sigma_y \end{bmatrix} \tag{2.1-7}$$

and the corresponding stress element, viewed three-dimensionally and down the *z* axis, is shown in Fig. 2.5.

2.2 STRAIN AND THE STRESS–STRAIN RELATIONS

As with stresses, two types of strains exist: normal and shear strains, which are denoted by ε and γ, respectively. Normal strain is the rate of change of the length of the stressed element in a particular direction. Shear strain is a measure of the distortion of the stressed element, and has two definitions: the *engineering shear strain* and the *elasticity shear strain*. Here, we will use the former, more popular, definition. However, a discussion of the relation of the two definitions will be provided in Sec. 2.4. The engineering shear strain is defined as the change in the corner angle of the stress cube, in radians.

Normal Strain

Initially, consider only one normal stress σ_x applied to the element as shown in Fig. 2.6. We see that the element increases in length in the *x* direction and decreases in length in the *y* and *z* directions. The dimensionless rate of increase in length is defined as the *normal strain*, where ε_x, ε_y, and ε_z represent the normal strains in the *x*, *y*, and *z* directions,

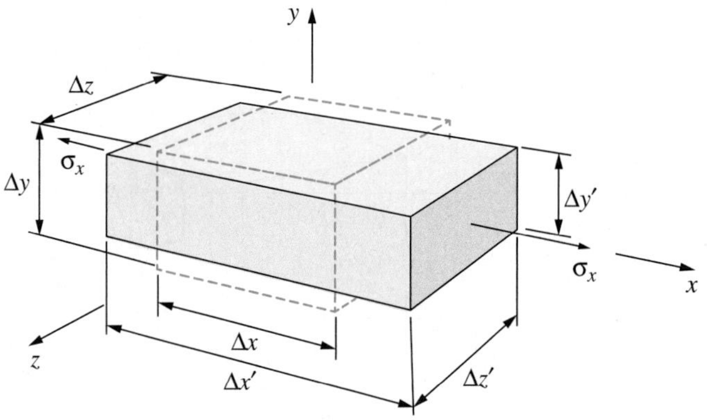

Figure 2.6 Deformation attributed to σ_x.

respectively. Thus, the new length in any direction is equal to its original length plus the rate of increase (normal strain) times its original length. That is,

$$\Delta x' = \Delta x + \varepsilon_x \Delta x, \qquad \Delta y' = \Delta y + \varepsilon_y \Delta y, \qquad \Delta z' = \Delta z + \varepsilon_z \Delta z \tag{2.2-1}$$

There is a direct relationship between strain and stress. *Hooke's law* for a linear, homogeneous, isotropic material is simply that the normal strain is directly proportional to the normal stress, and is given by

$$\varepsilon_x = \frac{1}{E}[\sigma_x - \nu(\sigma_y + \sigma_z)] \tag{2.2-2a}$$

$$\varepsilon_y = \frac{1}{E}[\sigma_y - \nu(\sigma_z + \sigma_x)] \tag{2.2-2b}$$

$$\varepsilon_z = \frac{1}{E}[\sigma_z - \nu(\sigma_x + \sigma_y)] \tag{2.2-2c}$$

where the material constants, E and ν, are the *modulus of elasticity* (also referred to as *Young's modulus*) and *Poisson's ratio*, respectively. Typical values of E and ν for some materials are given in Table 2.1 at the end of this chapter.

If the strains in Eqs. (2.2-2) are known, the stresses can be solved for simultaneously to obtain

$$\sigma_x = \frac{E}{(1+\nu)(1-2\nu)}[(1-\nu)\varepsilon_x + \nu(\varepsilon_y + \varepsilon_z)] \tag{2.2-3a}$$

$$\sigma_y = \frac{E}{(1+\nu)(1-2\nu)}[(1-\nu)\varepsilon_y + \nu(\varepsilon_z + \varepsilon_x)] \tag{2.2-3b}$$

$$\sigma_z = \frac{E}{(1+\nu)(1-2\nu)}[(1-\nu)\varepsilon_z + \nu(\varepsilon_x + \varepsilon_y)] \tag{2.2-3c}$$

For *plane stress*, with $\sigma_z = 0$, Eqs. (2.2-2) and (2.2-3) become

$$\varepsilon_x = \frac{1}{E}(\sigma_x - \nu\sigma_y) \tag{2.2-4a}$$

$$\varepsilon_y = \frac{1}{E}(\sigma_y - \nu\sigma_x) \tag{2.2-4b}$$

$$\varepsilon_z = -\frac{\nu}{E}(\sigma_x + \sigma_y) \tag{2.2-4c}$$

and

$$\sigma_x = \frac{E}{1-\nu^2}(\varepsilon_x + \nu\varepsilon_y) \tag{2.2-5a}$$

$$\sigma_y = \frac{E}{1-\nu^2}(\varepsilon_y + \nu\varepsilon_x) \tag{2.2-5b}$$

Shear Strain

The change in shape of the element caused by the shear stresses can be first illustrated by examining the effect of τ_{xy} alone as shown in Fig. 2.7. The engineering shear strain γ_{xy} is a measure of the skewing of the stressed element from a rectangular parallelepiped. In Fig. 2.7(*b*), the shear strain is defined as the change in the angle *BAD*. That is,

$$\gamma_{xy} = \angle BAD - \angle B'A'D'$$

where γ_{xy} is in dimensionless radians.

For a linear, homogeneous, isotropic material, the shear strains in the *xy*, *yz*, and *zx* planes are directly related to the shear stresses by

$$\gamma_{xy} = \frac{\tau_{xy}}{G} \tag{2.2-6a}$$

$$\gamma_{yz} = \frac{\tau_{yz}}{G} \tag{2.2-6b}$$

$$\gamma_{zx} = \frac{\tau_{zx}}{G} \tag{2.2-6c}$$

where the material constant, *G*, is called the *shear modulus.*

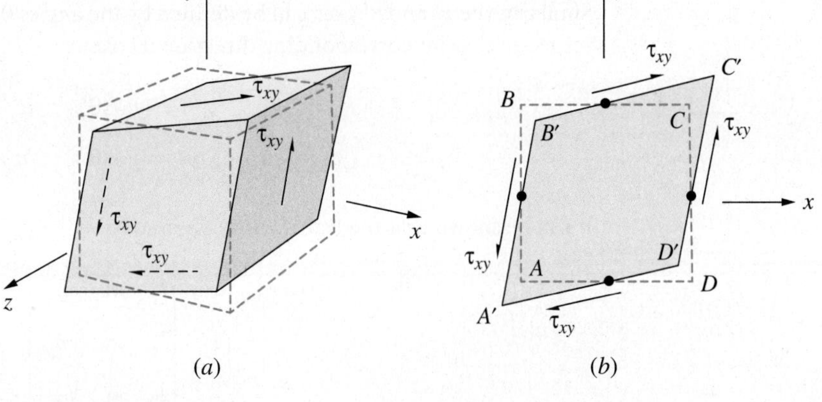

Figure 2.7 Shear deformation.

It can be shown that for a linear, homogeneous, isotropic material the shear modulus is related to Poisson's ratio by (Ref. 1)

$$G = \frac{E}{2(1+\nu)} \tag{2.2-7}$$

2.3 STRESS TRANSFORMATIONS

As was stated in Sec. 2.1, knowing the state of stress on three mutually orthogonal surfaces at a point in a structure is sufficient to generate the state of stress for *any* surface at the point. This is accomplished through the use of *coordinate transformations*. The development of the transformation equations is quite lengthy and is not provided here (see Ref. 1). Consider the element shown in Fig. 2.8(*a*), where the stresses on surfaces with normals in the *x*, *y*, and *z* directions are known and are represented by the stress matrix

$$[\sigma]_{xyz} = \begin{bmatrix} \sigma_x & \tau_{xy} & \tau_{zx} \\ \tau_{xy} & \sigma_y & \tau_{yz} \\ \tau_{zx} & \tau_{yz} & \sigma_z \end{bmatrix} \tag{2.3-1}$$

Now consider the element, shown in Fig. 2.8(*b*), to correspond to the state of stress *at the same point* but defined relative to a different set of surfaces with normals in the x', y', and z' directions. The stress matrix corresponding to this element is given by

$$[\sigma]_{x'y'z'} = \begin{bmatrix} \sigma_{x'} & \tau_{x'y'} & \tau_{z'x'} \\ \tau_{x'y'} & \sigma_{y'} & \tau_{y'z'} \\ \tau_{z'x'} & \tau_{y'z'} & \sigma_{z'} \end{bmatrix} \tag{2.3-2}$$

To determine $[\sigma]_{x'y'z'}$ by coordinate transformation, we need to establish the relationship between the $x'y'z'$ and the xyz coordinate systems. This is normally done using *directional cosines*. First, let us consider the relationship between the x' axis and the xyz coordinate system. The orientation of the x' axis can be established by the angles $\theta_{x'x}$, $\theta_{x'y}$, and $\theta_{x'z}$, as shown in Fig. 2.9. The directional cosines for x' are given by

$$l_{x'} = \cos\theta_{x'x} \quad m_{x'} = \cos\theta_{x'y} \quad n_{x'} = \cos\theta_{x'z} \tag{2.3-3}$$

Similarly, the y' and z' axes can be defined by the angles $\theta_{y'x}$, $\theta_{y'y}$, $\theta_{y'z}$ and $\theta_{z'x}$, $\theta_{z'y}$, $\theta_{z'z}$, respectively, with corresponding directional cosines

$$l_{y'} = \cos\theta_{y'x} \quad m_{y'} = \cos\theta_{y'y} \quad n_{y'} = \cos\theta_{y'z} \tag{2.3-4}$$

$$l_{z'} = \cos\theta_{z'x} \quad m_{z'} = \cos\theta_{z'y} \quad n_{z'} = \cos\theta_{z'z} \tag{2.3-5}$$

It can be shown that the transformation matrix

$$[\mathbf{T}] = \begin{bmatrix} l_{x'} & m_{x'} & n_{x'} \\ l_{y'} & m_{y'} & n_{y'} \\ l_{z'} & m_{z'} & n_{z'} \end{bmatrix} \tag{2.3-6}$$

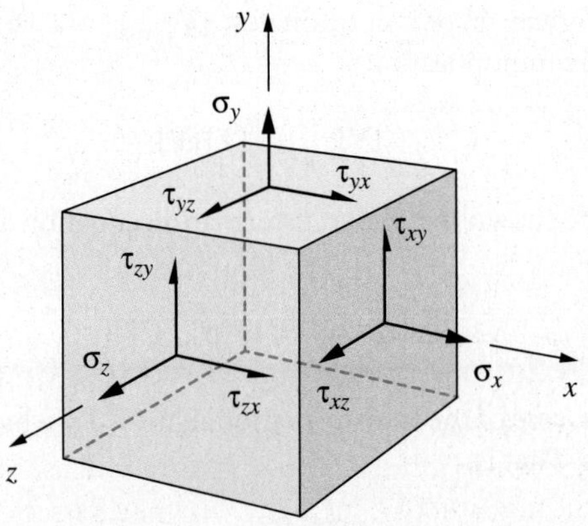

(*a*) Stress element relative to *xyz* axes

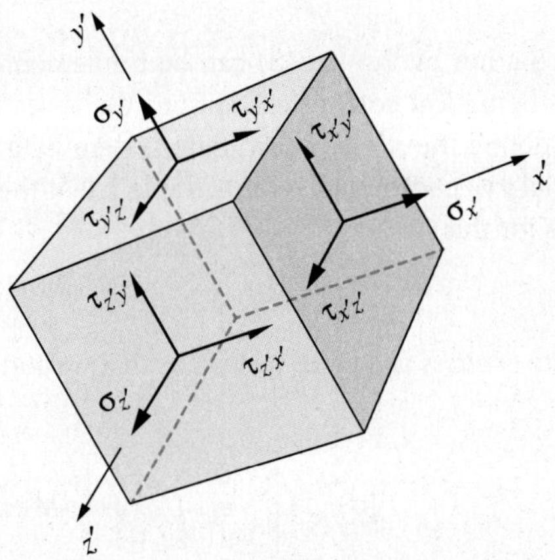

(*b*) Stress element relative to *x'y'z'* axes

Figure 2.8 The stress at a point using different coordinate systems.

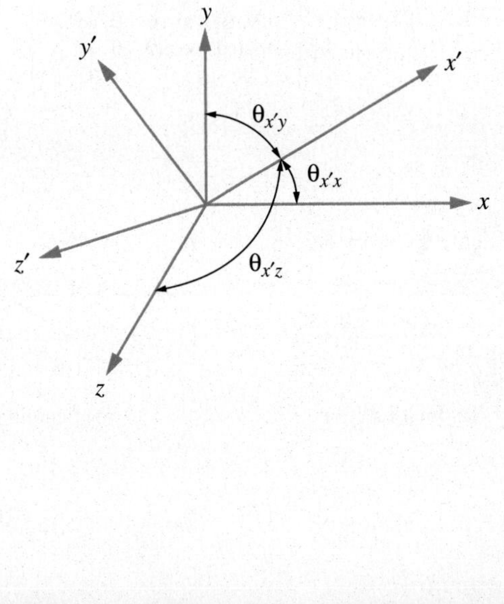

Figure 2.9 Coordinate transformation.

transforms a vector given in *xyz* coordinates, $\{\mathbf{V}\}_{xyz}$, to a vector in $x'y'z'$ coordinates, $\{\mathbf{V}\}_{x'y'z'}$, by the matrix multiplication

$$\{\mathbf{V}\}_{x'y'z'} = [\mathbf{T}]\{\mathbf{V}\}_{xyz} \tag{2.3-7}$$

Furthermore, it can be shown that the transformation equation for the stress matrix is given by (see Ref. 1)

$$[\sigma]_{x'y'z'} = [\mathbf{T}][\sigma]_{xyz}[\mathbf{T}]^T \tag{2.3-8}$$

where $[\mathbf{T}]^T$ is the *transpose* of the transformation matrix $[\mathbf{T}]$, which is simply an interchange of rows and columns. That is,

$$[\mathbf{T}]^T = \begin{bmatrix} l_{x'} & l_{y'} & l_{z'} \\ m_{x'} & m_{y'} & m_{z'} \\ n_{x'} & n_{y'} & n_{z'} \end{bmatrix} \tag{2.3-9}$$

The stress transformation by Eq. (2.3-8) can be implemented very easily using a computer spreadsheet or mathematical software. Defining the directional cosines is another matter. One method is to define the $x'y'z'$ coordinate system by a series of two-dimensional rotations from the initial *xyz* coordinate system. Table 2.2 at the end of this chapter gives transformation matrices for this.

EXAMPLE The state of stress at a point relative to an *xyz* coordinate system is given by the stress matrix

$$[\sigma]_{xyz} = \begin{bmatrix} -8 & 6 & -2 \\ 6 & 4 & 2 \\ -2 & 2 & -5 \end{bmatrix} \text{ MPa}$$

Determine the state of stress on an element that is oriented by first rotating the *xyz* axes 45° about the *z* axis, and then rotating the resulting axes 30° about the new *x* axis.

SOLUTION The surface normals can be found by a series of coordinate transformations for each rotation. From Fig. 2.10(*a*), the vector components for the first rotation can be represented by

$$\begin{Bmatrix} x_1 \\ y_1 \\ z_1 \end{Bmatrix} = \begin{bmatrix} \cos\theta & \sin\theta & 0 \\ -\sin\theta & \cos\theta & 0 \\ 0 & 0 & 1 \end{bmatrix} \begin{Bmatrix} x \\ y \\ z \end{Bmatrix} \tag{a}$$

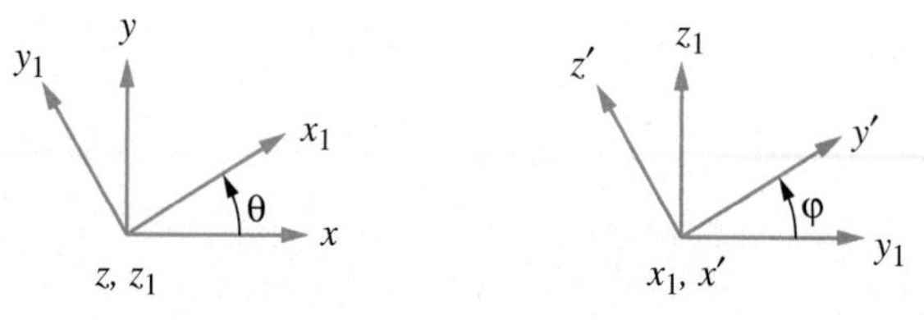

(*a*) First rotation (*b*) Second rotation

Figure 2.10

The last rotation establishes the $x'y'z'$ coordinates as shown in Fig. 2.10(b), and they are related to the $x_1y_1z_1$ coordinates by

$$\begin{Bmatrix} x' \\ y' \\ z' \end{Bmatrix} = \begin{bmatrix} 1 & 0 & 0 \\ 0 & \cos\varphi & \sin\varphi \\ 0 & -\sin\varphi & \cos\varphi \end{bmatrix} \begin{Bmatrix} x_1 \\ y_1 \\ z_1 \end{Bmatrix} \tag{b}$$

Substituting Eq. (a) in (b) gives

$$\begin{Bmatrix} x' \\ y' \\ z' \end{Bmatrix} = \begin{bmatrix} 1 & 0 & 0 \\ 0 & \cos\varphi & \sin\varphi \\ 0 & -\sin\varphi & \cos\varphi \end{bmatrix} \begin{bmatrix} \cos\theta & \sin\theta & 0 \\ -\sin\theta & \cos\theta & 0 \\ 0 & 0 & 1 \end{bmatrix} \begin{Bmatrix} x \\ y \\ z \end{Bmatrix} = \begin{bmatrix} \cos\theta & \sin\theta & 0 \\ -\sin\theta\cos\varphi & \cos\theta\cos\varphi & \sin\varphi \\ \sin\theta\sin\varphi & -\cos\theta\sin\varphi & \cos\varphi \end{bmatrix} \begin{Bmatrix} x \\ y \\ z \end{Bmatrix} \tag{c}$$

Equation (c) is of the form of Eq. (2.3-7). Thus, the transformation matrix is

$$[\mathbf{T}] = \begin{bmatrix} \cos\theta & \sin\theta & 0 \\ -\sin\theta\cos\varphi & \cos\theta\cos\varphi & \sin\varphi \\ \sin\theta\sin\varphi & -\cos\theta\sin\varphi & \cos\varphi \end{bmatrix} \tag{d}$$

Substituting $\theta = 45°$ and $\varphi = 30°$ gives

$$[\mathbf{T}] = \frac{1}{4}\begin{bmatrix} 2\sqrt{2} & 2\sqrt{2} & 0 \\ -\sqrt{6} & \sqrt{6} & 2 \\ \sqrt{2} & -\sqrt{2} & 2\sqrt{3} \end{bmatrix} \tag{e}$$

The transpose of [T] is

$$[\mathbf{T}]^T = \frac{1}{4}\begin{bmatrix} 2\sqrt{2} & -\sqrt{6} & \sqrt{2} \\ 2\sqrt{2} & \sqrt{6} & -\sqrt{2} \\ 0 & 2 & 2\sqrt{3} \end{bmatrix} \tag{f}$$

From Eq. (2.3-8),

$$[\sigma]_{x'y'z'} = \frac{1}{4}\begin{bmatrix} 2\sqrt{2} & 2\sqrt{2} & 0 \\ -\sqrt{6} & \sqrt{6} & 2 \\ \sqrt{2} & -\sqrt{2} & 2\sqrt{3} \end{bmatrix} \begin{bmatrix} -8 & 6 & -2 \\ 6 & 4 & 2 \\ -2 & 2 & -5 \end{bmatrix} \frac{1}{4}\begin{bmatrix} 2\sqrt{2} & -\sqrt{6} & \sqrt{2} \\ 2\sqrt{2} & \sqrt{6} & -\sqrt{2} \\ 0 & 2 & 2\sqrt{3} \end{bmatrix}$$

This matrix multiplication can be performed by simply using either a computer spreadsheet or mathematical software, resulting in

$$[\sigma]_{x'y'z'} = \begin{bmatrix} 4 & 5.196 & -3 \\ 5.196 & -4.801 & 2.714 \\ -3 & 2.714 & -8.199 \end{bmatrix} \text{MPa}$$

Stresses on a Single Surface

If one was concerned about the state of stress on one particular surface, a complete stress transformation would be unnecessary. Let the directional cosines for the normal of the surface be given by l, m, and n. It can be shown that the normal stress on the surface is given by

$$\sigma = \sigma_x l^2 + \sigma_y m^2 + \sigma_z n^2 + 2\tau_{xy} lm + 2\tau_{yz} mn + 2\tau_{zx} nl \tag{2.3-10}$$

and the net shear stress on the surface is

$$\tau=[(\sigma_x l+\tau_{xy}m+\tau_{zx}n)^2+(\tau_{xy}l+\sigma_y m+\tau_{yz}n)^2+(\tau_{zx}l+\tau_{yz}m+\sigma_z n)^2-\sigma^2]^{1/2} \tag{2.3-11}$$

The direction of τ is established by the directional cosines

$$\begin{aligned} l_\tau &= \frac{1}{\tau}[(\sigma_x-\sigma)l+\tau_{xy}m+\tau_{zx}n] \\ m_\tau &= \frac{1}{\tau}[\tau_{xy}l+(\sigma_y-\sigma)m+\tau_{yz}n] \\ n_\tau &= \frac{1}{\tau}[\tau_{zx}l+\tau_{yz}m+(\sigma_z-\sigma)n] \end{aligned} \tag{2.3-12}$$

EXAMPLE The state of stress at a particular point relative to the *xyz* coordinate system is

$$[\sigma]_{xyz}=\begin{bmatrix} 14 & 7 & -7 \\ 7 & 10 & 0 \\ -7 & 0 & 35 \end{bmatrix} \text{ kpsi}$$

Determine the normal and shear stress on a surface at the point where the surface is parallel to the plane given by the equation

$$2x-y+3z=9$$

SOLUTION The normal to the surface is established by the *directional numbers* of the plane and are simply the coefficients of *x*, *y*, and *z* terms of the equation of the plane. Thus, the directional numbers are 2, –1, and 3. The directional cosines of the normal to the surface are simply the normalized values of the directional numbers, which are the directional numbers divided by $\sqrt{2^2+(-1)^2+3^2}=\sqrt{14}$. Thus,

$$l=2/\sqrt{14}, \quad m=-1/\sqrt{14}, \quad n=3/\sqrt{14}$$

From the stress matrix, $\sigma_x=14$, $\tau_{xy}=7$, $\tau_{zx}=-7$, $\sigma_y=10$, $\tau_{yz}=0$, and $\sigma_z=35$ kpsi. Substituting the stresses and directional cosines into Eq. (2.3-10) gives

$$\begin{aligned} \sigma &= 14(2/\sqrt{14})^2+10(-1/\sqrt{14})^2+35(3/\sqrt{14})^2+2(7)(2/\sqrt{14})(-1/\sqrt{14}) \\ &\quad +2(0)(-1/\sqrt{14})(3/\sqrt{14})+2(-7)(3/\sqrt{14})(2/\sqrt{14})=19.21\text{ kpsi} \end{aligned}$$

The shear stress is determined from Eq. (2.3-11), and is

$$\begin{aligned} \tau &= \{[14(2/\sqrt{14})+7(-1/\sqrt{14})+(-7)(3/\sqrt{14})]^2 \\ &\quad +[7(2/\sqrt{14})+10(-1/\sqrt{14})+(0)(3/\sqrt{14})]^2 \\ &\quad +[(-7)(2/\sqrt{14})+(0)(-1/\sqrt{14})+35(3/\sqrt{14})]^2-(19.21)^2\}^{1/2}=14.95\text{ kpsi} \end{aligned}$$

From Eq. (2.3-12), the directional cosines for the direction of τ are

$$l_\tau=\frac{1}{14.95}[(14-19.21)(2/\sqrt{14})+7(-1/\sqrt{14})+(-7)(3/\sqrt{14})]=-0.687$$

$$m_\tau=\frac{1}{14.95}[7(2/\sqrt{14})+(10-19.21)(-1/\sqrt{14})+(0)(3/\sqrt{14})]=0.415$$

$$n_\tau=\frac{1}{14.95}[(-7)(2/\sqrt{14})+(0)(-1/\sqrt{14})+(35-19.21)(3/\sqrt{14})]=0.596$$

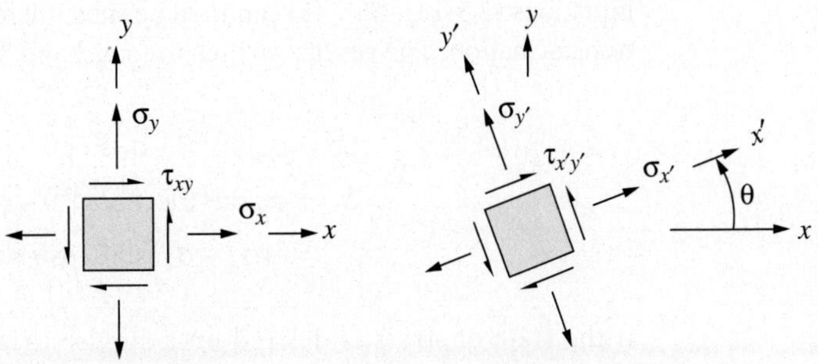

(*a*) Initial element (*b*) Transformed element

Figure 2.11 Plane stress transformations.

Plane Stress

For the state of plane stress shown in Fig. 2.11(*a*), $\sigma_z = \tau_{yz} = \tau_{zx} = 0$. Plane stress transformations are normally performed in the *xy* plane, as shown in Fig. 2.11(*b*). The angles relating the $x'y'z'$ axes to the *xyz* axes are

$$\begin{array}{lll} \theta_{x'x} = \theta, & \theta_{x'y} = 90^\circ - \theta, & \theta_{x'z} = 90^\circ \\ \theta_{y'x} = \theta + 90^\circ, & \theta_{y'y} = \theta, & \theta_{y'z} = 90^\circ \\ \theta_{z'x} = 90^\circ, & \theta_{z'y} = 90^\circ, & \theta_{z'z} = 0 \end{array}$$

Thus, the directional cosines are

$$\begin{array}{lll} l_{x'} = \cos\theta & m_{x'} = \sin\theta & n_{x'} = 0 \\ l_{y'} = -\sin\theta & m_{y'} = \cos\theta & n_{y'} = 0 \\ l_{z'} = 0 & m_{z'} = 0 & n_{z'} = 1 \end{array}$$

The last rows and columns of the stress matrices are zero so the stress matrices can be written as

$$[\sigma]_{xy} = \begin{bmatrix} \sigma_x & \tau_{xy} \\ \tau_{xy} & \sigma_y \end{bmatrix} \tag{2.3-13}$$

and

$$[\sigma]_{x'y'} = \begin{bmatrix} \sigma_{x'} & \tau_{x'y'} \\ \tau_{x'y'} & \sigma_{y'} \end{bmatrix} \tag{2.3-14}$$

Since the plane stress matrices are 2 × 2, the transformation matrix and its transpose are written as

$$[\mathbf{T}] = \begin{bmatrix} \cos\theta & \sin\theta \\ -\sin\theta & \cos\theta \end{bmatrix}, \quad [\mathbf{T}]^T = \begin{bmatrix} \cos\theta & -\sin\theta \\ \sin\theta & \cos\theta \end{bmatrix} \tag{2.3-15}$$

Equations (2.3-13)–(2.3-15) can then be substituted into Eq. (2.3-8) to perform the desired transformation. The results, written in long-hand form, would be

$$\begin{aligned}
\sigma_{x'} &= \sigma_x \cos^2\theta + \sigma_y \sin^2\theta + 2\tau_{xy}\cos\theta\sin\theta \\
\sigma_{y'} &= \sigma_x \sin^2\theta + \sigma_y \cos^2\theta - 2\tau_{xy}\cos\theta\sin\theta \\
\tau_{x'y'} &= -(\sigma_x - \sigma_y)\sin\theta\cos\theta + \tau_{xy}(\cos^2\theta - \sin^2\theta)
\end{aligned} \tag{2.3-16}$$

If the state of stress is desired on a single surface with a normal rotated θ counterclockwise from the x axis, the first and third equations of Eqs. (2.3-16) can be used as given. However, using trigonometric identities, the equations can be written in slightly different form. Letting σ and τ represent the desired normal and shear stresses on the surface, the equations are

$$\begin{aligned}
\sigma &= \frac{\sigma_x + \sigma_y}{2} + \frac{\sigma_x - \sigma_y}{2}\cos 2\theta + \tau_{xy}\sin 2\theta \\
\tau &= -\frac{\sigma_x - \sigma_y}{2}\sin 2\theta + \tau_{xy}\cos 2\theta
\end{aligned} \tag{2.3-17}$$

Equations (2.3-17) represent a set of parametric equations of a circle in the $\sigma\tau$ plane. This circle is commonly referred to as *Mohr's circle* and is discussed in Sec. 2.5.

Principal Stresses

In general, maximum and minimum values of the normal stresses occur on surfaces where the shear stresses are zero. These stresses, which are actually the eigenvalues of the stress matrix, are called the *principal stresses*. Three principal stresses exist, σ_1, σ_2, and σ_3, where they are commonly ordered as $\sigma_1 \geq \sigma_2 \geq \sigma_3$.

Considering the stress state given by the matrix of Eq. (2.3-1) to be known, the principal stresses σ_p are related to the given stresses by

$$\begin{aligned}
(\sigma_x - \sigma_p)l_p + \tau_{xy}m_p + \tau_{zx}n_p &= 0 \\
\tau_{xy}l_p + (\sigma_y - \sigma_p)m_p + \tau_{yz}n_p &= 0 \\
\tau_{zx}l_p + \tau_{yz}m_p + (\sigma_z - \sigma_p)n_p &= 0
\end{aligned} \tag{2.3-18}$$

where l_p, m_p, and n_p are the directional cosines of the normals to the surfaces containing the principal stresses. One possible solution to Eqs. (2.3-18) is $l_p = m_p = n_p = 0$. However, this cannot occur, since

$$l_p^2 + m_p^2 + n_p^2 = 1. \tag{2.3-19}$$

To avoid the zero solution of the directional cosines of Eqs. (2.3-18), the determinant of the coefficients of l_p, m_p, and n_p in the equation is set to zero. This makes the solution of the directional cosines indeterminate from Eqs. (2.3-18). Thus,

$$\begin{vmatrix}
(\sigma_x - \sigma_p) & \tau_{xy} & \tau_{zx} \\
\tau_{xy} & (\sigma_y - \sigma_p) & \tau_{yz} \\
\tau_{zx} & \tau_{yz} & (\sigma_z - \sigma_p)
\end{vmatrix} = 0$$

Expanding the determinant yields

$$\sigma_p^3 - (\sigma_x + \sigma_y + \sigma_z)\sigma_p^2 + (\sigma_x\sigma_y + \sigma_y\sigma_z + \sigma_z\sigma_x - \tau_{xy}^2 - \tau_{yz}^2 - \tau_{zx}^2)\sigma_p$$
$$- (\sigma_x\sigma_y\sigma_z + 2\tau_{xy}\tau_{yz}\tau_{zx} - \sigma_x\tau_{yz}^2 - \sigma_y\tau_{zx}^2 - \sigma_z\tau_{xy}^2) = 0 \qquad (2.3\text{-}20)$$

where Eq. (2.3-20) is a cubic equation yielding the three principal stresses σ_1, σ_2, and σ_3.

To determine the directional cosines for a specific principal stress, the stress is substituted into Eqs. (2.3-18). The three resulting equations in the unknowns l_p, m_p, and n_p will not be independent since they were used to obtain the principal stress. Thus, only two of Eqs. (2.3-18) can be used. However, the second-order Eq. (2.3-19) can be used as the third equation for the three directional cosines. Instead of solving one second-order and two linear equations simultaneously, a simplified method is demonstrated in the following example.*

EXAMPLE For the following stress matrix, determine the principal stresses and the directional cosines associated with the normals to the surfaces of each principal stress.

$$[\sigma] = \begin{bmatrix} 3 & 1 & 1 \\ 1 & 0 & 2 \\ 1 & 2 & 0 \end{bmatrix} \text{ MPa}$$

SOLUTION Substituting $\sigma_x = 3, \tau_{xy} = 1, \tau_{zx} = 1, \sigma_y = 0, \tau_{yz} = 2$, and $\sigma_z = 0$ into Eq. (2.3-20) gives

$$\sigma_p^3 - (3+0+0)\sigma_p^2 + [(3)(0)+(0)(0)+(0)(3)-2^2-1^2-1^2]\sigma_p$$
$$-[(3)(0)(0)+(2)(2)(1)(1)-(3)(2^2)-(0)(1^2)-(0)(1^2)] = 0$$

which simplifies to

$$\sigma_p^3 - 3\sigma_p^2 - 6\sigma_p + 8 = 0 \qquad (a)$$

The solutions to the cubic equation are $\sigma_p = 4, 1$, and -2 MPa. Following the conventional ordering,

$$\sigma_1 = 4 \text{ MPa}, \qquad \sigma_2 = 1 \text{ MPa}, \qquad \sigma_3 = -2 \text{ MPa}$$

The directional cosines associated with each principal stress are determined independently. First, consider σ_1 and substitute $\sigma_p = 4$ MPa into Eqs. (2.3-18). This results in

$$-l_1 + m_1 + n_1 = 0 \qquad (b)$$

$$l_1 - 4m_1 + 2n_1 = 0 \qquad (c)$$

$$l_1 + 2m_1 - 4n_1 = 0 \qquad (d)$$

where the subscript agrees with that of σ_1.

*Mathematical software packages can be used quite easily to extract the eigenvalues (σ_p) and the corresponding eigenvectors (l_p, m_p, and n_p) of a stress matrix. The reader is urged to explore software such as *Mathcad, Matlab, Maple, and Mathematica.*

Equations (*b*), (*c*), and (*d*) are no longer independent since they were used to determine the values of σ_p. Only two independent equations can be used, and in this example, any two of the above can be used. Consider Eqs. (*b*) and (*c*), which are independent. A third equation comes from Eq. (2.3-19), which is nonlinear in l_1, m_1, and n_1. Rather than solving the three equations simultaneously, consider the following approach.

Arbitrarily, let $l_1 = 1$ in Eqs. (*b*) and (*c*). Rearranging gives

$$m_1 + n_1 = 1$$
$$4m_1 - 2n_1 = 1$$

solving these simultaneously gives $m_1 = n_1 = ½$. These values of l_1, m_1, and n_1 do not satisfy Eq. (2.3-19). However, all that remains is to normalize their values by dividing by $\sqrt{1^2 + (½)^2 + (½)^2} = \sqrt{6}/2$. Thus,*

$$l_1 = (1)(2/\sqrt{6}) = \sqrt{6}/3$$
$$m_1 = (½)(2/\sqrt{6}) = \sqrt{6}/6$$
$$n_1 = (½)(2/\sqrt{6}) = \sqrt{6}/6$$

Repeating the same procedure for σ_2 = 1 MPa results in

$$l_2 = \sqrt{3}/3, \quad m_2 = -\sqrt{3}/3, \quad n_2 = -\sqrt{3}/3$$

and for $\sigma_3 = -2$ MPa

$$l_3 = 0, \quad m_3 = \sqrt{2}/2, \quad n_3 = -\sqrt{2}/2$$

If two of the principal stresses are equal, there will exist an infinite set of surfaces containing these principal stresses, where the normals of these surfaces are perpendicular to the direction of the third principal stress. If all three principal stresses are equal, a *hydrostatic* state of stress exists, and regardless of orientation, all surfaces contain the same principal stress with no shear stress.

Principal Stresses, Plane Stress

Considering the stress element shown in Fig. 2.11(*a*), the shear stresses on the surface with a normal in the *z* direction are zero. Thus, the normal stress $\sigma_z = 0$ is a principal stress. The directions of the remaining two principal stresses will be in the *xy* plane. If $\tau_{x'y'} = 0$ in Fig. 2.11(*b*), then $\sigma_{x'}$ would be a principal stress, σ_p with $l_p = \cos\theta$, $m_p = \sin\theta$, and $n_p = 0$. For this case, only the first two of Eqs. (2.3-18) apply, and are

$$\begin{aligned} (\sigma_x - \sigma_p)\cos\theta + \tau_{xy}\sin\theta = 0 \\ \tau_{xy}\cos\theta + (\sigma_y - \sigma_p)\sin\theta = 0 \end{aligned} \qquad (2.3\text{-}21)$$

As before, we eliminate the trivial solution of Eqs. (2.3-21) by setting the determinant of the coefficients of the directional cosines to zero. That is,

$$\begin{vmatrix} (\sigma_x - \sigma_p) & \tau_{xy} \\ \tau_{xy} & (\sigma_y - \sigma_p) \end{vmatrix} = (\sigma_x - \sigma_p)(\sigma_y - \sigma_p) - \tau_{xy}^2$$
$$= \sigma_p^2 - (\sigma_x + \sigma_y)\sigma_p + (\sigma_x\sigma_y - \tau_{xy}^2) = 0 \qquad (2.3\text{-}22)$$

---

*This method has one potential flaw. If l_1 is actually zero, then a solution would not result. If this happens, simply repeat the approach letting either m_1 or n_1 equal unity.

Equation (2.3-22) is a quadratic equation in σ_p for which the two solutions are

$$\sigma_p = \frac{1}{2}\left[(\sigma_x + \sigma_y) \pm \sqrt{(\sigma_x - \sigma_y)^2 + 4\tau_{xy}^2}\right] \tag{2.3-23}$$

Since for plane stress, one of the principal stresses (σ_z) is always zero, numbering of the stresses ($\sigma_1 \geq \sigma_2 \geq \sigma_3$) cannot be performed until Eq. (2.3-23) is solved.

Each solution of Eq. (2.3-23) can then be substituted into *one* of Eqs. (2.3-21) to determine the direction of the principal stress. Note that if $\sigma_x = \sigma_y$ and $\tau_{xy} = 0$, then σ_x and σ_y are principal stresses and Eqs. (2.3-21) are satisfied for all values of θ. This means that all stresses in the plane of analysis are equal and the state of stress at the point is *isotropic* in the plane. Summary of the formulas are given in Table 2.3.

EXAMPLE Determine the principal stresses for a case of plane stress given by the stress matrix

$$[\sigma] = \begin{bmatrix} 5 & -4 \\ -4 & 11 \end{bmatrix} \text{ kpsi}$$

Show the element containing the principal stresses properly oriented with respect to the initial *xyz* coordinate system.

SOLUTION From the stress matrix, $\sigma_x = 5, \sigma_y = 11$, and $\tau_{xy} = -4$ kpsi and Eq. (2.3-23) gives

$$\sigma_p = \frac{1}{2}\left[(5+11) \pm \sqrt{(5-11)^2 + 4(-4)^2}\right] = 13,\ 3\,\text{kpsi}$$

Thus, the three principal stresses (σ_1, σ_2, σ_3), are (13, 3, 0) kpsi, respectively. For directions, first substitute $\sigma_1 = 13$ kpsi into either one of Eqs. (2.3-21). Using the first equation with $\theta = \theta_1$

$$(\sigma_x - \sigma_1)\cos\theta_1 + \tau_{xy}\sin\theta_1 = (5-13)\cos\theta_1 + (-4)\sin\theta_1 = 0$$

or

$$\theta_1 = \tan^{-1}\left(-\frac{8}{4}\right) = -63.4°$$

Now for the other principal stress, $\sigma_2 = 3$ kpsi, the first of Eqs. (2.3-21) gives

$$(\sigma_x - \sigma_2)\cos\theta_2 + \tau_{xy}\sin\theta_2 = (5-3)\cos\theta_2 + (-4)\sin\theta_2 = 0$$

or

$$\theta_2 = \tan^{-1}\left(\frac{2}{4}\right) = 26.6°$$

Figure 2.12(*a*) illustrates the initial state of stress, whereas the orientation of the element containing the in-plane principal stresses is shown in Fig. 2.12(*b*).

Maximum Shear Stresses

Consider that the principal stresses for a general stress state have been determined using the methods just described and are illustrated by Fig. 2.13. The 123 axes represent the normals for the principal surfaces with directional cosines determined by Eqs. (2.3-18) and (2.3-19).

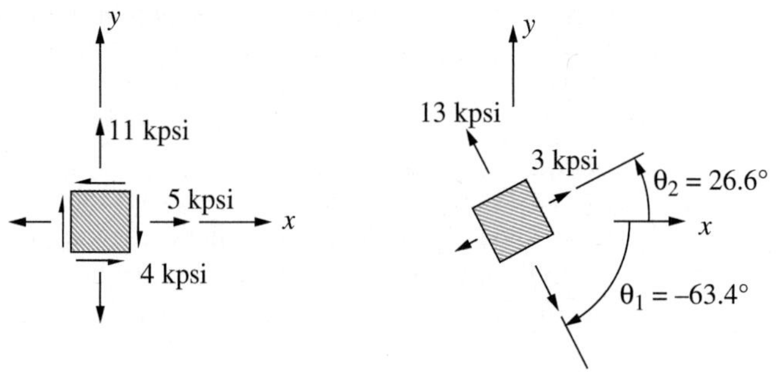

(*a*) Initial element

(*b*) Transformed element containing the principal stresses

Figure 2.12 Plane stress example.

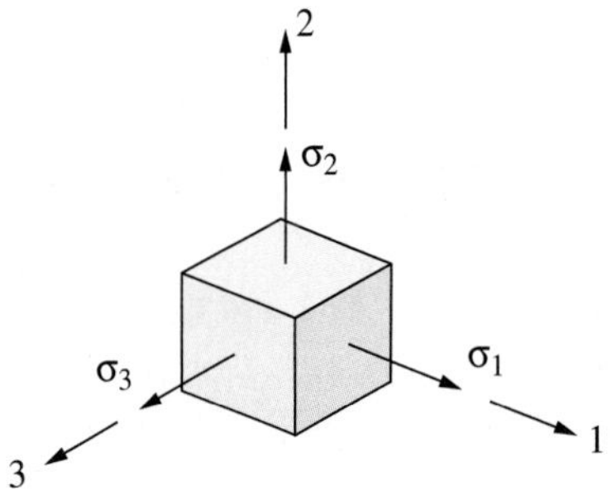

Figure 2.13 Principal stress state.

Viewing down a principal stress axis (e.g., the 3 axis) and performing a plane stress transformation in the plane normal to that axis (e.g., the 12 plane), one would find that the shear stress is a maximum on surfaces ±45° from the two principal stresses in that plane (e.g., σ_1, σ_2). On these surfaces, the maximum shear stress would be one-half the difference of the principal stresses [e.g., $\tau_{max} = (\sigma_1 - \sigma_2)/2$] and will also have a normal stress equal to the average of the principal stresses [e.g., $\sigma_{ave} = (\sigma_1 + \sigma_2)/2$]. Viewing along the three principal axes would result in three shear stress maxima, sometimes referred to as the *principal shear stresses*. These stresses together with their accompanying normal stresses are

$$\begin{array}{lll} \textbf{Plane 1, 2:} & (\tau_{max})_{1,2} = (\sigma_1 - \sigma_2)/2, & (\sigma_{ave})_{1,2} = (\sigma_1 + \sigma_2)/2 \\ \textbf{Plane 2, 3:} & (\tau_{max})_{2,3} = (\sigma_2 - \sigma_3)/2, & (\sigma_{ave})_{2,3} = (\sigma_2 + \sigma_3)/2 \\ \textbf{Plane 1, 3:} & (\tau_{max})_{1,3} = (\sigma_1 - \sigma_3)/2, & (\sigma_{ave})_{1,3} = (\sigma_1 + \sigma_3)/2 \end{array} \qquad (2.3\text{-}24)$$

Since conventional practice is to order the principal stresses by $\sigma_1 \geq \sigma_2 \geq \sigma_3$, the largest shear stress of all is given by the third of Eqs. (2.3-24) and will be repeated here for emphasis:

$$\tau_{max} = (\sigma_1 - \sigma_3)/2 \qquad (2.3\text{-}25)$$

EXAMPLE In the previous example, the principal stresses for the stress matrix

$$[\sigma] = \begin{bmatrix} 5 & -4 \\ -4 & 11 \end{bmatrix} \text{ kpsi}$$

were found to be $(\sigma_1, \sigma_2, \sigma_3) = (13, 3, 0)$ kpsi. The orientation of the element containing the principal stresses was shown in Fig. 2.12(*b*), where axis 3 was the *z* axis and normal to the page. Determine the maximum shear stress and show the orientation and complete state of stress of the element containing this stress.

SOLUTION The initial element and the transformed element containing the principal stresses are repeated in Fig. 2.14(*a*) and (*b*), respectively. The maximum shear stress will exist in the 1, 3 plane and is determined by substituting $\sigma_1 = 13$ and $\sigma_3 = 0$ into Eqs. (2.3-24). This results in

$$(\tau_{max})_{1,3} = (13-0)/2 = 6.5 \text{ kpsi}, \quad (\sigma_{ave})_{1,3} = (13+0)/2 = 6.5 \text{ kpsi}$$

To establish the orientation of these stresses, view the element along the axis containing $\sigma_2 = 3$ kpsi [View A, Fig. 2.14(*c*)] and rotate the surfaces ±45° as shown in Fig. 2.14(*d*).

The directional cosines associated with the surfaces are found through successive rotations. Rotating the *xyz* axes to the 123 axes yields

$$\begin{Bmatrix} 1 \\ 2 \\ 3 \end{Bmatrix} = \begin{bmatrix} \cos 63.4^\circ & -\sin 63.4^\circ & 0 \\ \sin 63.4^\circ & \cos 63.4^\circ & 0 \\ 0 & 0 & 1 \end{bmatrix} \begin{Bmatrix} x \\ y \\ z \end{Bmatrix} \tag{a}$$

$$= \begin{bmatrix} 0.4472 & -0.8944 & 0 \\ 0.8944 & 0.4472 & 0 \\ 0 & 0 & 1 \end{bmatrix} \begin{Bmatrix} x \\ y \\ z \end{Bmatrix}$$

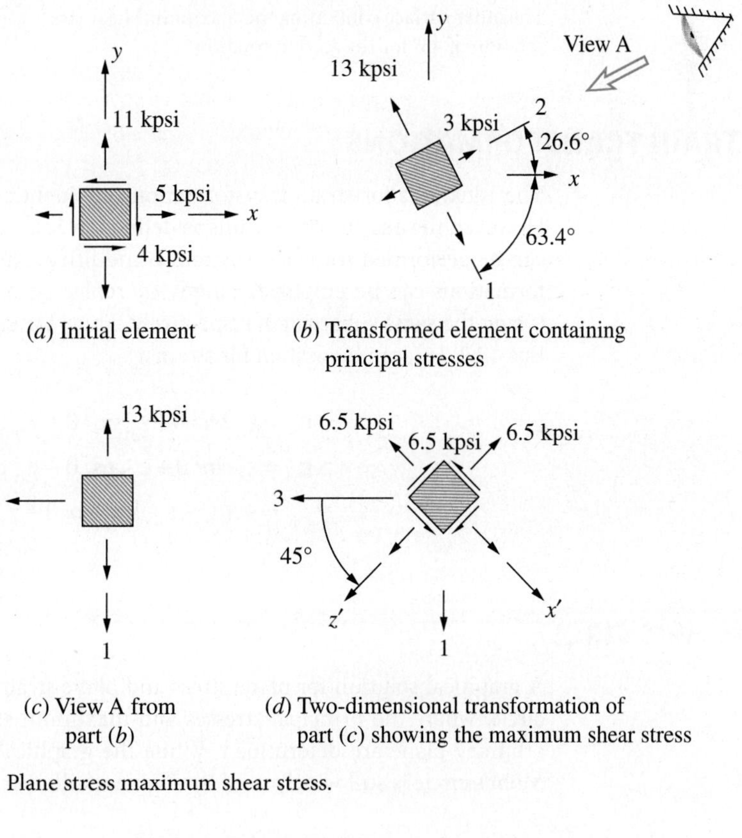

Figure 2.14 Plane stress maximum shear stress.

A counterclockwise rotation of 45° of the normal in the 3 direction about axis 2 is represented by

$$\begin{Bmatrix} x' \\ y' \\ z' \end{Bmatrix} = \begin{bmatrix} \cos 45^\circ & 0 & -\sin 45^\circ \\ 0 & 1 & 0 \\ \sin 45^\circ & 0 & \cos 45^\circ \end{bmatrix} \begin{Bmatrix} 1 \\ 2 \\ 3 \end{Bmatrix} = \begin{bmatrix} 0.7071 & 0 & -0.7071 \\ 0 & 1 & 0 \\ 0.7071 & 0 & 0.7071 \end{bmatrix} \begin{Bmatrix} 1 \\ 2 \\ 3 \end{Bmatrix} \tag{b}$$

Thus,

$$\begin{Bmatrix} x' \\ y' \\ z' \end{Bmatrix} = \begin{bmatrix} 0.7071 & 0 & -0.7071 \\ 0 & 1 & 0 \\ 0.7071 & 0 & 0.7071 \end{bmatrix} \begin{bmatrix} 0.4472 & -0.8944 & 0 \\ 0.8944 & 0.4472 & 0 \\ 0 & 0 & 1 \end{bmatrix} \begin{Bmatrix} x \\ y \\ z \end{Bmatrix}$$

$$= \begin{bmatrix} 0.3162 & -0.6325 & -0.7071 \\ 0.8944 & 0.4472 & 0 \\ 0.3162 & -0.6325 & 0.7071 \end{bmatrix} \begin{Bmatrix} x \\ y \\ z \end{Bmatrix}$$

The directional cosines for Eq. (2.1-14*d*) are therefore

$$\begin{bmatrix} n_{x'x} & n_{x'y} & n_{x'z} \\ n_{y'x} & n_{y'y} & n_{y'z} \\ n_{z'x} & n_{z'y} & n_{z'z} \end{bmatrix} = \begin{bmatrix} 0.3162 & -0.6325 & -0.7071 \\ 0.8944 & 0.4472 & 0 \\ 0.3162 & -0.6325 & 0.7071 \end{bmatrix}$$

The other surface containing the maximum shear stress can be found similarly except for a clockwise rotation of 45° for the second rotation.

2.4 STRAIN TRANSFORMATIONS

The equations for strain transformations are identical to those for stress transformations. However, the engineering strains as defined in Sec. 2.2 *will not* transform. Transformations can be performed if the shear strain is modified. All of the equations for the stress transformations can be employed simply by replacing σ and τ in the equations by ε and $\gamma/2$ (using the same subscripts), respectively. Thus, for example, the equations for plane stress, Eqs. (2.3-16), can be written for strain as

$$\begin{aligned} \varepsilon_{x'} &= \varepsilon_x \cos^2\theta + \varepsilon_y \sin^2\theta + \gamma_{xy} \cos\theta \sin\theta \\ \varepsilon_{y'} &= \varepsilon_x \sin^2\theta + \varepsilon_y \cos^2\theta - \gamma_{xy} \cos\theta \sin\theta \\ \gamma_{x'y'} &= -2(\varepsilon_x - \varepsilon_y)\sin\theta\cos\theta + \gamma_{xy}(\cos^2\theta - \sin^2\theta) \end{aligned} \tag{2.4-1}$$

2.5 MOHR'S CIRCLE

A graphical solution for plane stress and plane strain transformations is known as Mohr's circle, where the principal stresses and maximum shear stresses as well as stresses in an arbitrary plane are determined. While the graphical solution is an old classical method, Mohr's circle is still widely used by engineers all over the world.

The stress transformation formulas for plane stress at a given location, Eqs. 2.3-17 can be rewritten as

$$\sigma_{x'} = \frac{\sigma_x + \sigma_y}{2} + \frac{\sigma_x - \sigma_y}{2}\cos 2\theta + \tau_{xy}\sin 2\theta$$
$$\tau_{x'y'} = -\frac{\sigma_x - \sigma_y}{2}\sin 2\theta + \tau_{xy}\cos 2\theta \tag{2.5-1}$$

where $x'y'$ represent arbitrary axes rotated counterclockwise in the xy plane by the angle θ from the xy axes. By squaring each equation and adding the equations together we have

$$\left(\sigma_{x'} - \frac{\sigma_x + \sigma_y}{2}\right)^2 + \tau_{x'y'}^2 = \left(\frac{\sigma_x - \sigma_y}{2}\right)^2 + \tau_{xy}^2 \tag{2.5-2}$$

The above equation can be written in a more compact form as

$$(\sigma_{x'} - \sigma_{ave})^2 + \tau_{x'y'}^2 = R^2 \tag{2.5-3}$$

where

$$\sigma_{ave} = \frac{\sigma_x + \sigma_y}{2}$$

$$R = \sqrt{\left(\frac{\sigma_x - \sigma_y}{2}\right)^2 + \tau_{xy}^2}$$

Equation (2.5-3) is the equation of a circle where the abscissa is the normal stress and the ordinate is the shear stress, as shown in Fig. 2.15.

In Fig. 2.15, σ_{min} and σ_{max} are the minimum and maximum values of normal stress. τ_{min} and τ_{max} are the minimum and maximum values of shear stress.

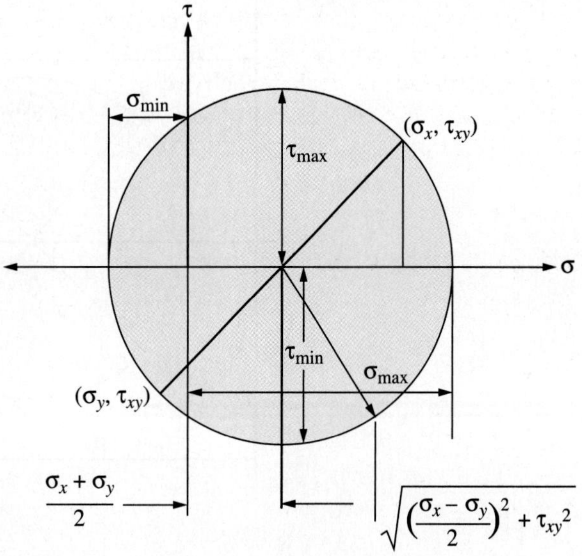

Figure 2.15 Mohr's circle.

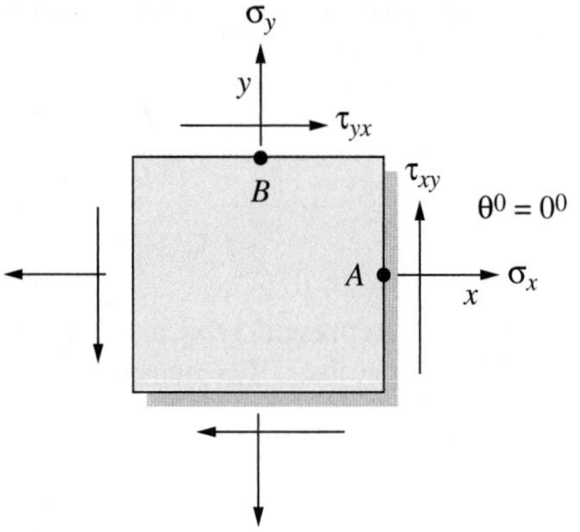

Figure 2.16

Construction of Mohr's Circle

Mohr's circle can be used to determine the normal and shear stress components $\sigma_{x'}$ and $\tau_{x'y'}$ acting on any arbitrary plane. Consider an element shown in Fig. 2.16, having normal and shear stresses. Draw a coordinate system of σ (abscissa) and τ (ordinate). Denote points A and B on the two perpendicular sides of the element as shown, where their coordinates are $A(\sigma_x, \tau_{xy})$ and $B(\sigma_y, \tau_{xy})$. Note that the direction of τ_{xy} at points A and B is positive for this element. Locate points A and B on the coordinate system (σ, τ) as follows: For a positive shear at point A (i.e., CCW direction of the shear with respect to the center of the element) use downward τ. For a positive shear at point B (i.e., CW direction of the shear with respect to the center of the element) use upward τ, as shown in Fig. 2.17. That is, the upper half of the vertical ordinate is for positive τ_{xy} of point B and the lower half of the vertical ordinate is for positive τ_{xy} of point A.

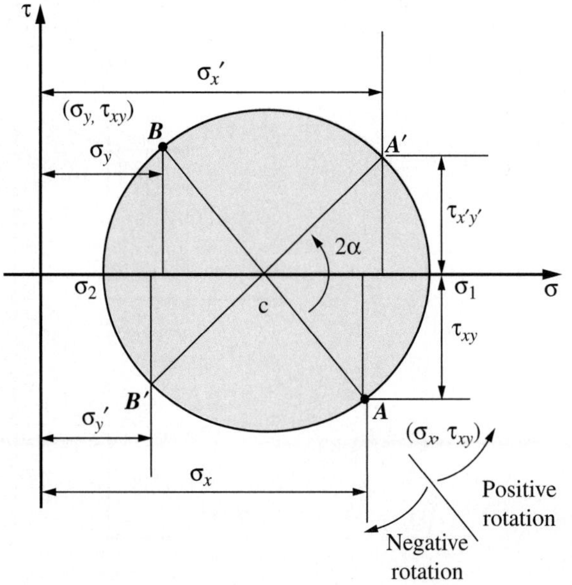

Figure 2.17

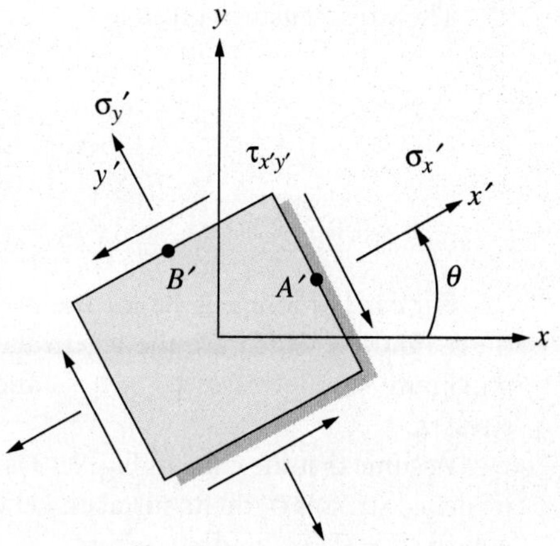

Figure 2.18

Once the two points A and B are located, connect the two points and draw a circle that passes through points A and B with its center on the horizontal axis, (where the line AB intersect with the horizontal axis). This is the Mohr's circle of the element.

The intersections of the circle with the x axis are the maximum (σ_1) and minimum (σ_2) principal stresses only in the plane of analysis and the radius of the circle is the maximum shear stress, τ_{max}. To find the orientation of the element in which the maximum stresses occur, the element must be rotated half of the angle between CA and the horizontal axis. That is, the orientation of the principal stresses is determined by rotating the element θ degrees, as shown in Fig. 2.17.

To determine the normal and shear stresses in an arbitrary plane, for example θ degrees from the x axis, one can rotate the diameter AB the amount 2θ in the same direction as in the element. The new coordinates of A and B are the stress components on the new plane; see Fig. 2.18. The stresses in the new orientation can be determined from Mohr's circle, as shown in Fig. 2.17. Mohr's circle for some common state of stress are given in Table 2.4.

2.6 MOHR'S CIRCLES FOR 3D STRESS ANALYSIS

Consider a three-dimensional (3D) state of stress where there are six stress components, as shown in Fig. 2.19.

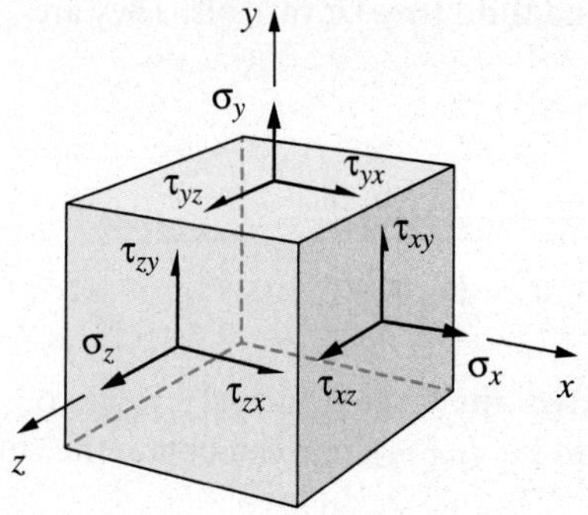

Figure 2.19 3D stress state relative to xyz axes.

The stress tensor is given as

$$[\sigma]_{xyz} = \begin{bmatrix} \sigma_x & \tau_{xy} & \tau_{zx} \\ \tau_{xy} & \sigma_y & \tau_{yz} \\ \tau_{zx} & \tau_{yz} & \sigma_z \end{bmatrix} \tag{2.6-1}$$

Similar to Mohr's circle for the 2D, we are interested in finding the principal stresses σ_1, σ_2, and σ_3, which are the eigenvalues of the stress tensor [Eq. (2.6-1)], and the three maximum shear stresses $\tau_{\max1}$, $\tau_{\max2}$, and $\tau_{\max3}$, which can be calculated from the principal stresses.

Assume that the cube in Fig. 2.19 is rotated such that it has only normal stresses, that is, principal stresses σ_p on its surfaces. Let the directional cosines of the new orientation of the cube be, $l_i = (l_p, m_p,$ and $n_p)$, where

$$l_p^2 + m_p^2 + n_p^2 = 1 \tag{2.6-2}$$

Solving the following eigenvalue problem,

$$(\sigma_p \delta_{ij} - \sigma_{ij}) l_i = 0 \tag{2.6-3}$$

where δ_{ij} is the Kronecker delta, leads to

$$\begin{aligned} (\sigma_x - \sigma_p) l_p + \tau_{xy} m_p + \tau_{zx} n_p &= 0 \\ \tau_{xy} l_p + (\sigma_y - \sigma_p) m_p + \tau_{yz} n_p &= 0 \\ \tau_{zx} l_p + \tau_{yz} m_p + (\sigma_z - \sigma_p) n_p &= 0 \end{aligned} \tag{2.6-4}$$

By equating the determinant of this equation to zero, we have

$$\begin{vmatrix} (\sigma_x - \sigma_p) & \tau_{xy} & \tau_{zx} \\ \tau_{xy} & (\sigma_y - \sigma_p) & \tau_{yz} \\ \tau_{zx} & \tau_{yz} & (\sigma_z - \sigma_p) \end{vmatrix} = 0 \tag{2.6-5}$$

which leads to

$$\sigma_p^3 - I_1 \sigma_p^2 + I_2 \sigma_p - I_3 = 0 \tag{2.6-6}$$

where I's are first, second, and third stress invariants. They are

$$\begin{aligned} I_1 &= \sigma_x + \sigma_y + \sigma_z \\ I_2 &= \sigma_x \sigma_y + \sigma_y \sigma_z + \sigma_x \sigma_z - \tau_{xy}^2 - \tau_{yz}^2 - \tau_{xz}^2 \\ I_3 &= \sigma_x \sigma_y \sigma_z + 2\tau_{xy}\tau_{yz}\tau_{xz} - \sigma_x \tau_{yz}^2 - \sigma_y \tau_{xz}^2 - \sigma_z \tau_{xy}^2 \end{aligned} \tag{2.6-7}$$

Once Eq. (2.6-6) is solved, the three principal stresses σ_1, σ_2, σ_3 will be determined. Substituting σ_1, σ_2, or σ_3 into Eq. (2.6-4), one can obtain the corresponding principal axes

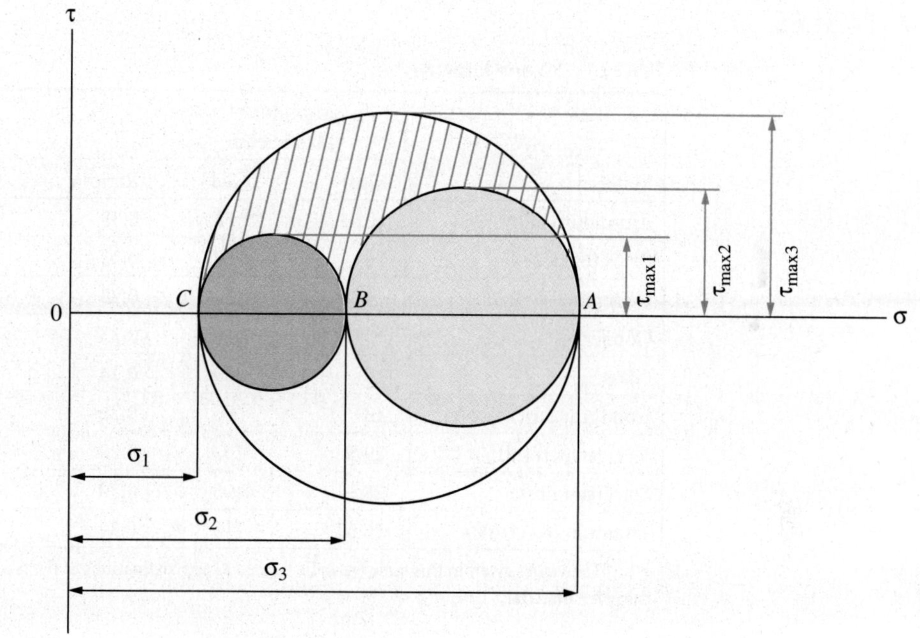

Figure 2.20 The 3D representation of Mohr's circle.

l_p, m_p, and n_p, respectively. If one rotates the axes of the cube of Fig. 2.19 such that the new cube faces are normal to l_p, m_p, and n_p, the stresses on that plane are principal normal stresses σ_1, σ_2, and σ_3, respectively, and there will be no shear stress on those planes. Considering $\sigma_1 > \sigma_2 > \sigma_3$, the corresponding 3D Mohr's circle can be constructed by three circles shown in Fig. 2.20.

If the new cube is rotated about the axis n_p, the stresses on the rotating four surfaces can be analyzed through the 2D transformation by means of 2D Mohr's circle. Note that during this rotation the shear stresses on the faces normal to the n_p axis remain equal to zero, and the normal stress σ_3 is perpendicular to that plane, where the transformation takes place, remains the same and does not affect this transformation.

The circle of diameter *AB* is used to determine the normal and shear stresses or the rotating faces of the new cube about the n_p axis. Similarly, the circles of diameters *BC* and *AC* may be used to determine the stresses on the cube as it is rotated about the l_p, m_p axes, respectively. The maximum shear stresses are then determined by the radii of the three circles, that is,

$$\tau_{max1} = \frac{\sigma_1 - \sigma_3}{2} \tag{2.6-8}$$

Similarly one can find: $\tau_{max2} = \dfrac{\sigma_1 - \sigma_2}{2}$, $\tau_{max1} = \dfrac{\sigma_2 - \sigma_3}{2}$.

If the axes of Fig. 2.19 are rotated in random directions other than principal axes, it can be shown that the normal and shear stresses on any cutoff plane would be located within the hatched area of Fig. 2.20, that is bounded by the AC biggest circle and the other two AB and BC circles. Therefore, one can determine the maximum and minimum normal and shear stresses from a state of stress, using 3D Mohr's circles, as shown in Fig. 2.20.

2.7 TABLES

TABLE 2.1 Material Properties*

	Modulus of Elasticity, E		Poisson's	Thermal Expansion Coefficient, α	
Material	Mpsi	GPa	Ratio, ν	μ/°F	μ/°C
Aluminum alloys	10.5	72	0.33	13.1	23.5
Brass (65/35)	16	110	0.32	11.6	20.9
Concrete	4	34	0.20	5.5	9.9
Copper	17	118	0.33	9.4	16.9
Glass	10	69	0.24	5.1	9.2
Iron (gray cast)	13	90	0.26	6.7	12.1
Steel (structural)	29.5	207	0.29	6.5	11.7
Steel (stainless)	28	193	0.30	9.6	17.3
Titanium (6 A1/4V)	16.5	115	0.34	5.2	9.5

*The values given in this table are to be treated as approximations of the true behavior of an actual batch of the given material.

TABLE 2.2 Transformation Matrices for Positive Rotations About an Axis*

Axis	Transformation Matrix
x axis: 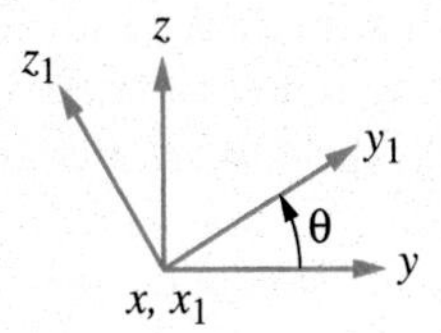	$\begin{Bmatrix} x_1 \\ y_1 \\ z_1 \end{Bmatrix} = \begin{bmatrix} 1 & 0 & 0 \\ 0 & \cos\theta & \sin\theta \\ 0 & -\sin\theta & \cos\theta \end{bmatrix} \begin{Bmatrix} x \\ y \\ z \end{Bmatrix}$
y axis: 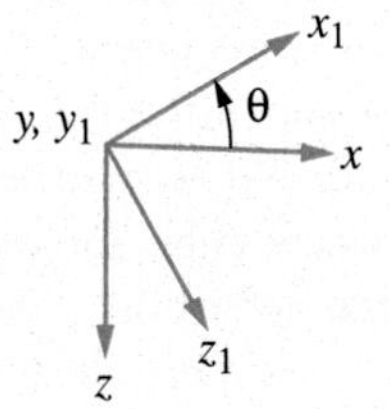	$\begin{Bmatrix} x_1 \\ y_1 \\ z_1 \end{Bmatrix} = \begin{bmatrix} \cos\theta & 0 & -\sin\theta \\ 0 & 1 & 0 \\ \sin\theta & 0 & \cos\theta \end{bmatrix} \begin{Bmatrix} x \\ y \\ z \end{Bmatrix}$
z axis: 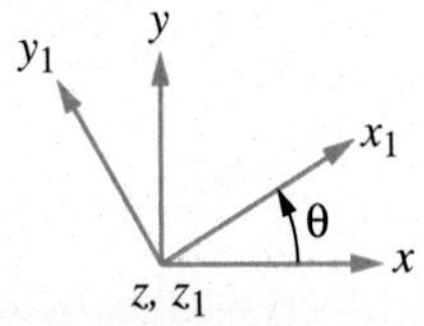	$\begin{Bmatrix} x_1 \\ y_1 \\ z_1 \end{Bmatrix} = \begin{bmatrix} \cos\theta & \sin\theta & 0 \\ -\sin\theta & \cos\theta & 0 \\ 0 & 0 & 1 \end{bmatrix} \begin{Bmatrix} x \\ y \\ z \end{Bmatrix}$

*A positive rotation about a given axis is counterclockwise about the axis (as viewed from the positive axis direction).

TABLE 2.3 Transformation Equations and Principle Stress Formulas

General state of stress

$$[\sigma]_{x'y'z'} = [\mathbf{T}][\sigma]_{xyz}[\mathbf{T}]^T$$

where

$$[\sigma]_{x'y'z'} = \begin{bmatrix} \sigma_{x'} & \tau_{x'y'} & \tau_{z'x'} \\ \tau_{x'y'} & \sigma_{y'} & \tau_{y'z'} \\ \tau_{z'x'} & \tau_{y'z'} & \sigma_{z'} \end{bmatrix}, \quad [\mathbf{T}] = \begin{bmatrix} l_{x'} & m_{x'} & n_{x'} \\ l_{y'} & m_{y'} & n_{y'} \\ l_{z'} & m_{z'} & n_{z'} \end{bmatrix}, \quad [\sigma]_{xyz} = \begin{bmatrix} \sigma_x & \tau_{xy} & \tau_{zx} \\ \tau_{xy} & \sigma_y & \tau_{yz} \\ \tau_{zx} & \tau_{yz} & \sigma_z \end{bmatrix}$$

Stresses on a single surface (l, m, n are directional cosines of surface normal)

$$\sigma = \sigma_x l^2 + \sigma_y m^2 + \sigma_z n^2 + 2\tau_{xy} lm + 2\tau_{yz} mn + 2\tau_{zx} nl$$

$$\tau = [(\sigma_x l + \tau_{xy} m + \tau_{zx} n)^2 + (\tau_{xy} l + \sigma_y m + \tau_{yz} n)^2 + (\tau_{zx} l + \tau_{yz} m + \sigma_z n)^2 - \sigma^2]^{1/2}$$

$$l_\tau = \frac{1}{\tau}[(\sigma_x - \sigma)l + \tau_{xy} m + \tau_{zx} n]$$

$$m_\tau = \frac{1}{\tau}[\tau_{xy} l + (\sigma_y - \sigma)m + \tau_{yz} n]$$

$$n_\tau = \frac{1}{\tau}[\tau_{zx} l + \tau_{yz} m + (\sigma_z - \sigma)n]$$

l_τ, m_τ, and n_τ are directional cosines for the direction of τ.

Plane stress (θ is counterclockwise from x axis to surface normal, x')

$$\sigma = \frac{1}{2}(\sigma_x + \sigma_y) + \frac{1}{2}(\sigma_x - \sigma_y)\cos 2\theta + \tau_{xy} \sin 2\theta$$

$$\tau = -\frac{1}{2}(\sigma_x - \sigma_y)\sin 2\theta + \tau_{xy} \cos 2\theta$$

Principal stresses (general case)

$$\sigma_p^3 - (\sigma_x + \sigma_y + \sigma_z)\sigma_p^2 + (\sigma_x\sigma_y + \sigma_y\sigma_z + \sigma_z\sigma_x - \tau_{xy}^2 - \tau_{yz}^2 - \tau_{zx}^2)\sigma_p$$
$$- (\sigma_x\sigma_y\sigma_z + 2\tau_{xy}\tau_{yz}\tau_{zx} - \sigma_x\tau_{yz}^2 - \sigma_y\tau_{zx}^2 - \sigma_z\tau_{xy}^2) = 0$$

Directional cosines (l_p, m_p, n_p) are found from three of the following equations:

$$\left.\begin{aligned} (\sigma_x - \sigma_p)l_p + \tau_{xy} m_p + \tau_{zx} n_p &= 0 \\ \tau_{xy} l_p + (\sigma_y - \sigma_p)m_p + \tau_{yz} n_p &= 0 \\ \tau_{zx} l_p + \tau_{yz} m_p + (\sigma_z - \sigma_p)n_p &= 0 \end{aligned}\right\} \text{ select two independent equations}$$

$$l_p^2 + m_p^2 + n_p^2 = 1$$

Principal stresses (plane stress) One principal stress is zero and the remaining two are given by

$$\sigma_p = \frac{1}{2}\left[(\sigma_x + \sigma_y) \pm \sqrt{(\sigma_x - \sigma_y)^2 + 4\tau_{xy}^2}\right]$$

Angle of surface normal relative to the x axis is given by

$$\theta_p = \tan^{-1}\left(\frac{\sigma_p - \sigma_x}{\tau_{xy}}\right)$$

TABLE 2.4 Mohr's Circle for Some Common States of Stress

A: Compression

1 Uniaxial compression

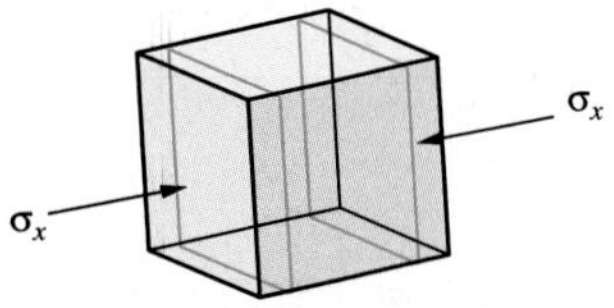

$(\sigma_x = \sigma_3)$,
$(\sigma_1 = \sigma_2 = 0)$

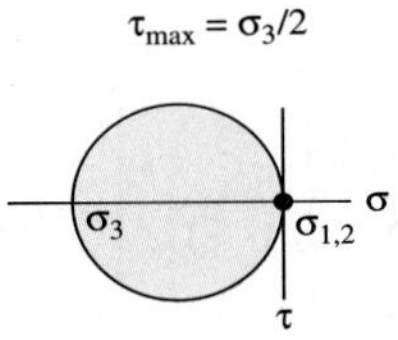

2 Equal biaxial compression

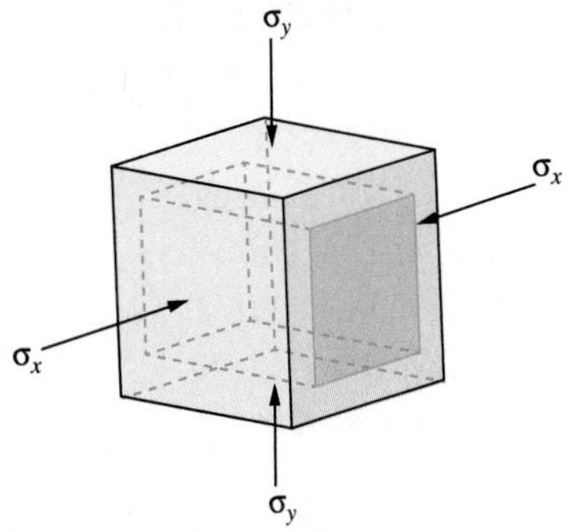

$(\sigma_x = \sigma_y = \sigma_2 = \sigma_3)$

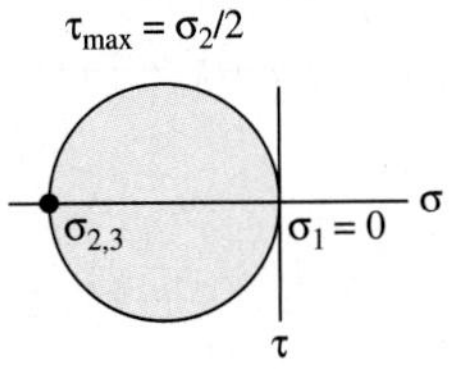

3 Equal triaxial compression

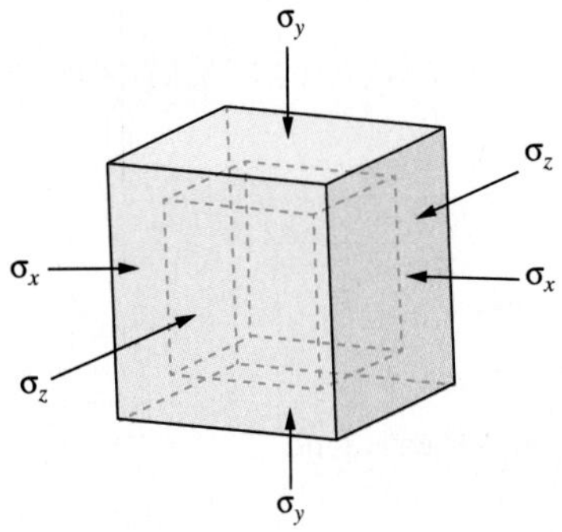

$(\sigma_x = \sigma_y = \sigma_z = \sigma_1 = \sigma_2 = \sigma_3)$

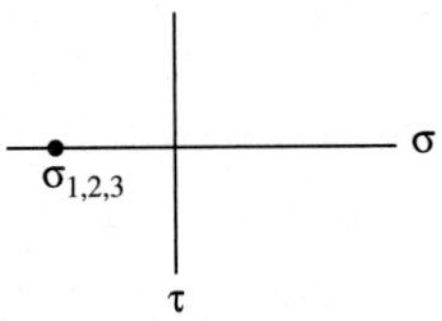

B: Tension

1 Uniaxial tension

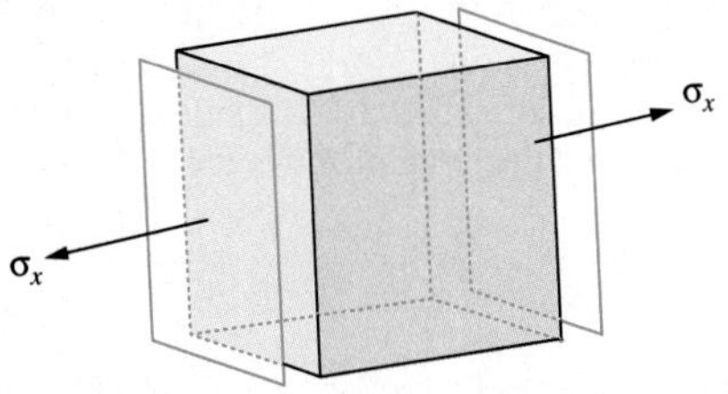

$(\sigma_x = \sigma_1)$,
$(\sigma_2 = \sigma_3 = 0)$

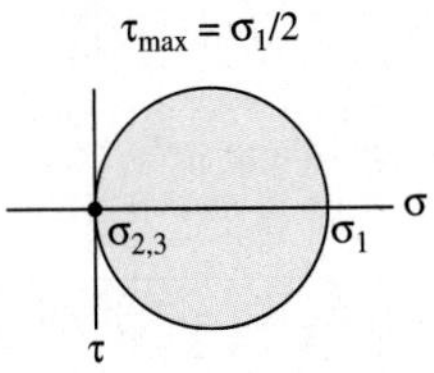

(Continued)

TABLE 2.4 Mohr's Circle for Some Common States of Stress (*Continued*)

B: Tension

2 Equal biaxial tension

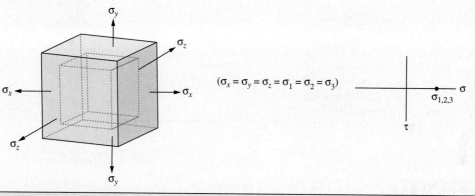

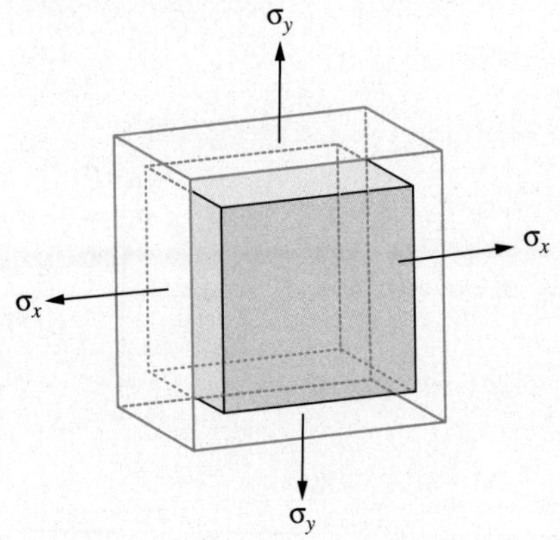

$(\sigma_x = \sigma_y = \sigma_1 = \sigma_2)$

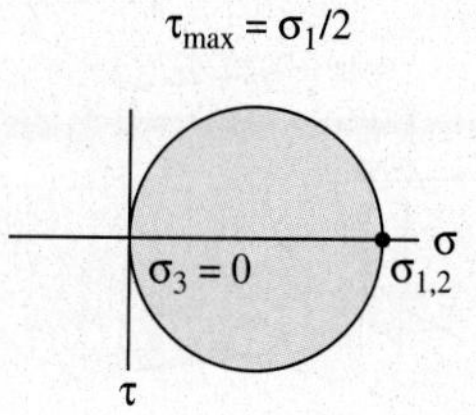

3 Equal triaxial tension

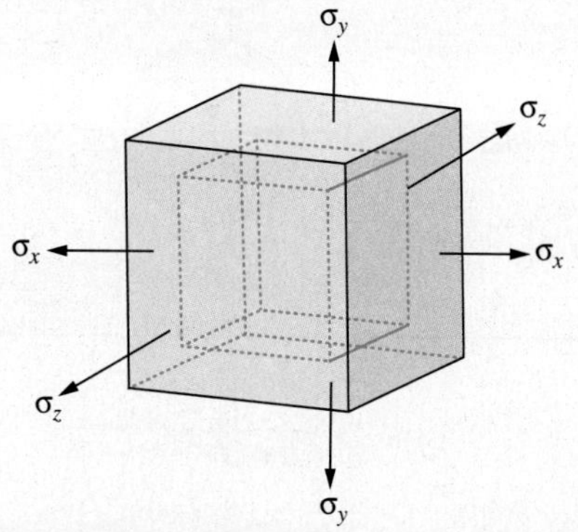

$(\sigma_x = \sigma_y = \sigma_z = \sigma_1 = \sigma_2 = \sigma_3)$

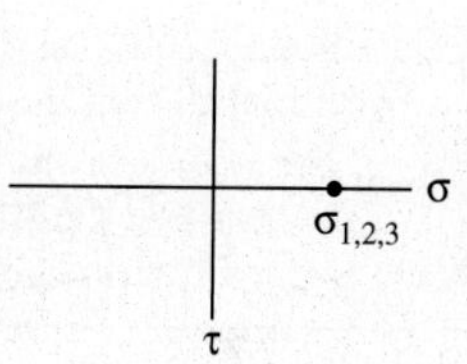

C: Tension and Compression

1 Equal tension (2 planes *x* and *y*) with lateral compression (plane z)

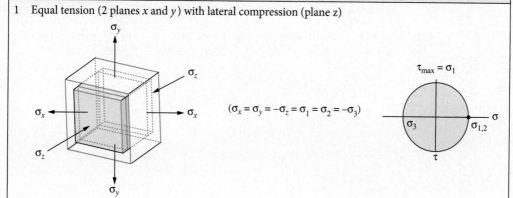

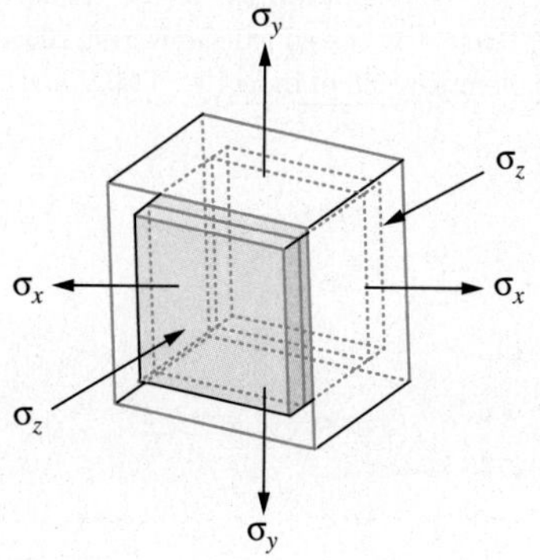

$(\sigma_x = \sigma_y = -\sigma_z = \sigma_1 = \sigma_2 = -\sigma_3)$

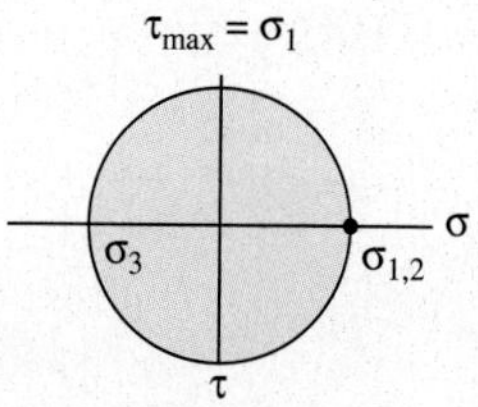

(Continued)

TABLE 2.4 Mohr's Circle for Some Common States of Stress (*Continued*)

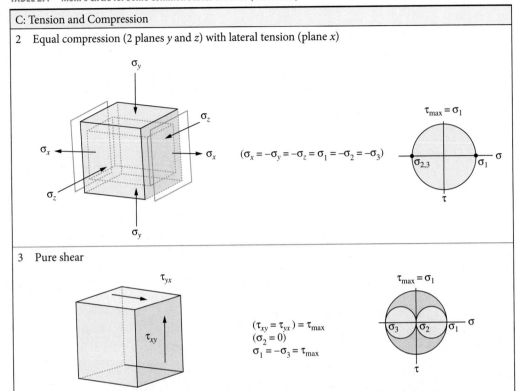

C: Tension and Compression
2 Equal compression (2 planes *y* and *z*) with lateral tension (plane *x*)
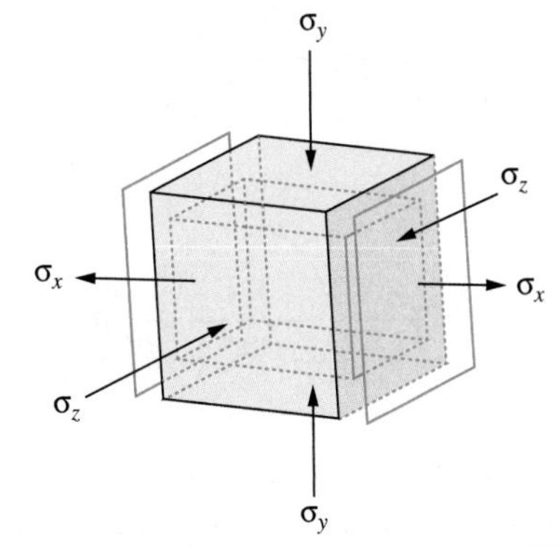 $(\sigma_x = -\sigma_y = -\sigma_z = \sigma_1 = -\sigma_2 = -\sigma_3)$ 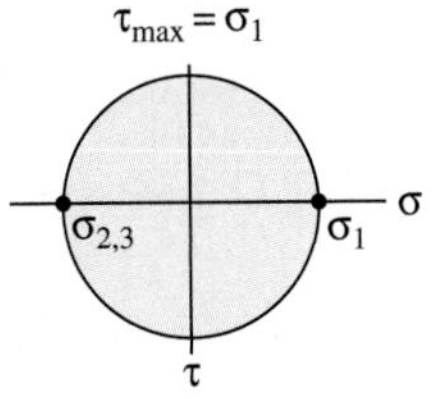
3 Pure shear
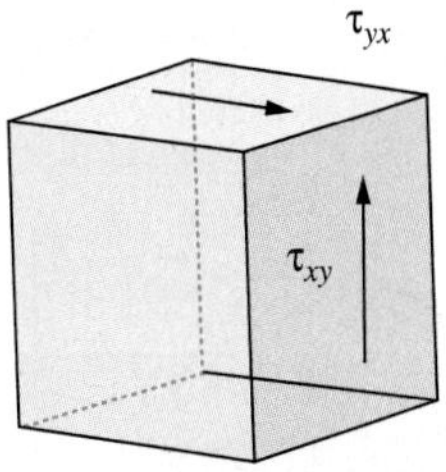 $(\tau_{xy} = \tau_{yx}) = \tau_{max}$ $(\sigma_2 = 0)$ $\sigma_1 = -\sigma_3 = \tau_{max}$ 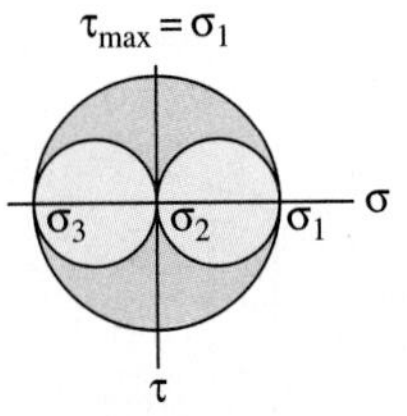

2.8 REFERENCES

1. Budynas, R. G., *Advanced Strength and Applied Stress Analysis*, 2nd ed., McGraw-Hill, 1999.
2. Avallone, E. A., T. Baumeister, and A. M. Sadegh, *Marks' Standard Handbook for Mechanical Engineers*, 11th ed., McGraw-Hill, 2007.
3. Rothbart, H., and T. H. Brown, *Mechanical Design Handbook*, 2nd ed., McGraw-Hill, 2006.
4. Gambhir, M. L., *Fundamentals of Solid Mechanics*, Prentice-Hall of India Pvt., Ltd., 2009.

CHAPTER 3

The Behavior of Bodies under Stress

This discussion pertains to the behavior of what are commonly designated as *structural materials.* That is, materials suitable for structures and members that must sustain loads without suffering damage. Included in this category are most of the metals, concrete, wood, composite materials, some plastics, etc. It is beyond the scope of this book to give more than a mere statement of a few important facts concerning the behavior of a stressed material. Extensive literature is available on every phase of the subject, and the articles contained herein will serve as an introduction only.

3.1 METHODS OF LOADING

The mechanical properties of a material are usually determined by laboratory tests, and the commonly accepted values of ultimate strength, elastic limit, etc., are those found by testing a specimen of a certain form in a certain manner. To apply results so obtained in engineering design requires an understanding of the effects of many different variables, such as form and scale, temperature and other conditions of service, and method of loading.

The method of loading, in particular, affects the behavior of bodies under stress. There are an infinite number of ways in which stress may be applied to a body, but for most purposes it is sufficient to distinguish the types of loading now to be defined.

1. *Short-time static loading.* The load is applied so gradually that at any instant all parts are essentially in equilibrium. In testing, the load is increased progressively until failure occurs, and the total time required to produce failure is not more than a few minutes. In service, the load is increased progressively up to its maximum value, is maintained at that maximum value for only a limited time, and is not reapplied often enough to make fatigue a consideration. The ultimate strength, elastic limit, yield point, yield strength, and modulus of elasticity of a material are usually determined by short-time static testing at room temperature.
2. *Long-time static loading.* The maximum load is applied gradually and maintained. In testing, it is maintained for a sufficient time to enable its probable final effect to be predicted; in service, it is maintained continuously or intermittently during the life of the structure. The creep, or flow characteristics, of a material and its probable permanent strength are determined by long-time static testing at the temperatures prevailing under service conditions. (See Sec. 3.6.)

3. *Repeated loading.* Typically, a load or stress is applied and wholly or partially removed or reversed repeatedly. This type of loading is important if high stresses are repeated for a few cycles or if relatively lower stresses are repeated many times; it is discussed under *Fatigue.* (See Sec. 3.8.)
4. *Dynamic loading.* The circumstances are such that the rate of change of momentum of the parts must be taken into account. One such condition may be that the parts are given definite *accelerations* corresponding to a controlled motion, such as the constant acceleration of a part of a rotating member or the repeated accelerations suffered by a portion of a connecting rod. As far as stress effects are concerned, these loadings are treated as virtually static and the *inertia forces* (Sec. 16.2) are treated exactly as though they were ordinary static loads.

A second type of *quasi-static* loading, *quick static loading,* can be typified by the rapid burning of a powder charge in a gun barrel. Neither the powder, gas, nor any part of the barrel acquires appreciable radial momentum; therefore, equilibrium may be considered to exist at any instant and the maximum stress produced in the gun barrel is the same as though the powder pressure had developed gradually.

In static loading and the two types of dynamic loading just described, the loaded member is required to resist a definite *force*. It is important to distinguish this from *impact loading,* where the loaded member is usually required to absorb a definite amount of *energy*.

Impact loading can be divided into two general categories. In the first case a relatively large slow-moving mass strikes a less massive beam or bar and the *kinetic energy* of the moving mass is assumed to be converted into *strain energy* in the beam. All portions of the beam and the moving mass are assumed to stop moving simultaneously. The shape of the elastic axis of the deflected beam or bar is thus the same as in static loading. A special case of this loading, generally called *sudden loading,* occurs when a mass that is not moving is released when in contact with a beam and falls through the distance the beam deflects. This produces approximately twice the stress and deflection that would have been produced had the mass been "eased" onto the beam (see Sec. 16.4). The second case of impact loading involves the mass of the member being struck. *Stress waves* travel through the member during the impact and continue even after the impacting mass has rebounded (see Sec. 16.3).

On consideration, it is obvious that methods of loading really differ only in degree. As the time required for the load to be applied increases, short-time static loading changes imperceptibly into longtime static loading; impact may be produced by a body moving so slowly that the resulting stress conditions are practically the same as though equal deflection had been produced by static loading; the number of stress repetitions at which fatigue becomes involved is not altogether definite. Furthermore, all these methods of loading may be combined or superimposed in various ways. Nonetheless, the classification presented is convenient because most structural and machine parts function under loading that may be classified definitely as one of the types described.

3.2 ELASTICITY; PROPORTIONALITY OF STRESS AND STRAIN

In determining stress by mathematical analysis, it is customary to assume that material is elastic, isotropic, homogeneous, and infinitely divisible without change in properties and that it conforms to Hooke's law, which states that strain is proportional to stress. Actually, none of these assumptions is strictly true. A structural material is usually an aggregate of crystals, fibers, or cemented particles, the arrangement of which may be either random or systematic. When the arrangement is random, the material is essentially isotropic if the part considered is large in comparison with the constituent units; when the arrangement is systematic, the elastic properties and strength are usually different in different directions and the material is anisotropic. Again, when subdivision is carried to the point where the part under consideration comprises only a portion of a single crystal, fiber, or other unit, in all probability its properties will differ from those of a larger part that is an aggregate of such units. Finally, very careful experiments show that for all materials there is probably some set and some deviation from Hooke's law for any stress, however small.

These facts impose certain limitations upon the conventional methods of stress analysis and must often be taken into account, but formulas for stress and strain, mathematically derived and based on the assumptions stated, give satisfactory results for nearly all problems of engineering design. In particular, Hooke's law may be regarded as practically true up to a proportional limit, which, though often not sharply defined, can be established for most materials with sufficient definiteness. So, too, a fairly definite elastic limit is determinable; in most cases it is so nearly equal to the proportional limit that no distinction need be made between the two.

3.3 FACTORS AFFECTING ELASTIC PROPERTIES

For ordinary purposes it may be assumed that the elastic properties of most metals, when stressed below a nominal proportional limit, are constant with respect to stress, unaffected by ordinary atmospheric variations of temperature, unaffected by prior applications of moderate stress, and independent of the rate of loading. When precise relations between stress and strain are important, as in the design or calibration of instruments, these assumptions cannot always be made. The fourth edition of this book (Ref. 1) discussed in detail the effects of strain rate, temperature, etc., on the elastic properties of many metals and gave references for the experiments performed. The relationships between atomic and molecular structure and the elastic properties are discussed in texts on materials science.

Wood exhibits a higher modulus of elasticity and much higher proportional limit when tested rapidly than when tested slowly. The standard impact test on a beam indicates a fiber stress at the proportional limit approximately twice as great as that found by the standard static bending test. Absorption of moisture up to the fiber saturation point greatly lowers both the modulus of elasticity and the proportional limit (Ref. 2).

Both concrete and cast iron have stress-strain curves more or less curved throughout, and neither has a definite proportional limit. For these materials, it is customary to define E as the ratio of some definite stress (e.g., the allowable stress or one-fourth the ultimate strength) to the corresponding unit strain; the quantity so determined is called the *secant* modulus since it represents the slope of the secant of the stress-strain diagram drawn from the origin to the point representing the stress chosen. The moduli of elasticity of cast iron are much more variable than those of steel, and the stronger grades are stiffer than the weaker ones. Cast iron suffers a distinct set from the first application of even a moderate stress; but after several repetitions of that stress, the material exhibits perfect elasticity up to, but not beyond, that stress. The modulus of elasticity is slightly less in tension than in compression.

Concrete also shows considerable variation in modulus of elasticity, and in general its stiffness increases with its strength. Like cast iron, concrete can be made to exhibit perfect elasticity up to a moderate stress by repeated loading up to that stress. Because of its tendency to yield under continuous loading, the modulus of elasticity indicated by long-time loading is much less than that obtained by progressive loading at ordinary speeds.

3.4 LOAD DEFORMATION RELATION FOR A BODY

If Hooke's law holds for the material of which a member or structure is composed, the member or structure will usually conform to a similar law of load-deformation proportionality and the deflection of a beam or truss, the twisting of a shaft, the dilation of a pressure container, etc., may in most instances be assumed proportional to the magnitude of the applied load or loads.

There are two important exceptions to this rule. One is to be found in any case where the stresses due to the loading are appreciably affected by the deformation. Examples of this are a beam subjected to axial and transverse loads; a flexible wire or cable held at the ends and loaded transversely; a thin diaphragm held at the edges and loaded normal to its plane; a ball pressed against a plate or against another ball; and a helical spring under severe extension.

The second exception is represented by any case in which failure occurs through elastic instability, as in the compressive loading of a long, slender column. Here, for compression loads less than a specific *critical* (*Euler*) *load*, elastic instability plays no part and the axial deformation is linear with load. At the critical load, the type of deformation changes, and the column bends instead of merely shortening axially. For any load beyond the critical load, high bending stresses and failure occurs through excessive deflection (see Sec. 3.13).

3.5 PLASTICITY

Elastic deformation represents an actual change in the distance between atoms or molecules; plastic deformation represents a permanent change in their relative positions. In crystalline materials, this permanent rearrangement consists largely of group displacements of the atoms in the crystal lattice brought about by slip on planes of least resistance, parts of a crystal sliding past one another and in some instances suffering angular displacement. In amorphous materials, the rearrangement appears to take place through the individual shifting from positions of equilibrium of many atoms or molecules, the cause being thermal agitation due to external work and the result appearing as a more or less uniform flow like that of a viscous liquid. It should be noted that plastic deformation before rupture is much less for biaxial or triaxial tension than for one-way stress; for this reason metals that are ordinarily ductile may prove brittle when thus stressed.

3.6 CREEP AND RUPTURE UNDER LONG-TIME LOADING

More materials will creep or flow to some extent and eventually fail under a sustained stress less than the short-time ultimate strength. After a short time at load, the initial creep related to stress redistribution in the structure and strain hardening ceases and the *steady state*, or *viscous creep*, predominates. The viscous creep will continue until fracture unless the load is reduced sufficiently, but it is seldom important in materials at temperatures less than 40 to 50% of their absolute melting temperatures. Thus, creep and long-time strength at atmospheric temperatures must sometimes be taken into account in designing members of nonferrous metals and in selecting allowable stresses for wood, plastics, and concrete.

Metals

Creep is an important consideration in high-pressure steam and distillation equipment, gas turbines, nuclear reactors, supersonic vehicles, etc. Marin, Odqvist, and Finnie, in Ref. 3, give excellent surveys and list references on creep in metals and structures. Conway (Refs. 4 and 5) discusses the effectiveness of various parametric equations, and Conway and Flagella (Ref. 6) present extensive creep-rupture data for the refractory metals. Odqvist (Ref. 7) discusses the theory of creep and its application to large deformation and stability problems in plates, shells, membranes, and beams and tabulates creep constants for 15 common metals and alloys. Hult (Ref. 8) also discusses creep theory and its application to many structural problems. Penny and Marriott (Ref. 9) discuss creep theories and the design of experiments to verify them. They also discuss the development of several metals for increased resistance to creep at high temperatures as well as polymeric and composite materials at lower temperatures. Reference 10 is a series of papers with extensive references covering creep theory, material properties, and structural problems.

Plastics

The literature on the behavior of the many plastics being used for structural or machine applications is too extensive to list here.

Concrete

Under sustained compressive stress, concrete suffers considerable plastic deformation and may flow for a very long time at stresses less than the ordinary working stress. Continuous flow has been observed over a period of 10 years, though ordinarily it ceases or becomes imperceptible within 1 or 2 years. The rate of flow is greater for air than for water storage, greater for small than for large specimens, and for moderate stresses increases approximately as the applied stress. On removal of stress, some elastic recovery occurs. Concrete also shows creep under tensile stress, the early creep rate being greater than the flow rate under compression (Refs. 11 and 16).

Under very gradually applied loading concrete exhibits an ultimate strength considerably less than that found under short-time loading; in certain compression tests it was found that increasing the time of testing from 1 s to 4 h decreased the unit stress at failure about 30%, most of this decrease occurring between the extremely quick (1 or 2 s) and the conventional (several minutes) testing. This indicates that the compressive stress that concrete can sustain indefinitely may be considerably less than the ultimate strength as determined by a conventional test. On the other hand, the long-time imposition of a moderate loading appears to have no harmful effect; certain tests show that after 10 years of constant loading equal to one-fourth the ultimate strength, the compressive strength of concrete cylinders is practically the same and the modulus of elasticity is considerably greater than for similar cylinders that were not kept under load (Ref. 15).

The modulus of rupture of plain concrete also decreases with the time of loading, and some tests indicate that the long-time strength in cross-breaking may be only 55 to 75% of the short-time strength (Ref. 12).

Reference 17 is a compilation of 12 papers, each with extensive references, dealing with the effect of volumetric changes on concrete structures. Design modifications to accommodate these volumetric changes are the main thrust of the papers.

Wood

Wood also yields under sustained stress; the long-time (several years) strength is about 55% of the short-time (several minutes) strength in bending; for direct compression parallel to the grain the corresponding ratio is about 75% (Ref. 2).

3.7 CRITERIA OF ELASTIC FAILURE AND OF RUPTURE

For the purpose of this discussion it is convenient to divide metals into two classes: (1) *ductile* metals, in which marked plastic deformation commences at a fairly definite stress (yield point, yield strength, or possibly elastic limit) and which exhibit considerable ultimate elongation; and (2) *brittle* metals, for which the beginning of plastic deformation is not clearly defined and which exhibit little ultimate elongation. Mild steel is typical of the first class, and cast iron is typical of the second; an ultimate elongation of 5% has been suggested as the arbitrary dividing line between the two classes of metals.

A ductile metal is usually considered to have failed when it has suffered *elastic failure*, that is, when marked plastic deformation has begun. Under simple uniaxial tension this occurs when the stress reaches a value we will denote by σ_{ys}, which represents the yield strength, yield point, or elastic limit, according to which one of these is the most satisfactory indication of elastic failure for the material in question. The question arises: when does elastic failure occur under other conditions of stress, such as compression, shear, or a combination of tension, compression, and shear?

There are many theories of elastic failure that can be postulated for which the consequences can be seen in the tensile test. When the tensile specimen begins to yield at a tensile stress of σ_{ys}, the following events occur:

1. *The maximum-principal-stress theory.* The maximum principal stress reaches the tensile yield strength, σ_{ys}.
2. *The maximum-shear-stress theory* (also called the *Tresca theory*). The maximum shear stress reaches the shear yield strength, $0.5\,\sigma_{ys}$.
3. *The maximum-principal-strain theory.* The maximum principal strain reaches the yield strain, σ_{ys}/E.
4. *The maximum-strain-energy theory.* The strain energy per unit volume reaches a maximum of $0.5\,\sigma_{ys}^2/E$.
5. *The maximum-distortion-energy theory* (also called the *von Mises theory and the Maxwell-Huber-Hencky-von Mises theory*). The energy causing a change in shape (distortion) reaches $[(1+\nu)/(3E)]\sigma_{ys}^2$.
6. *The maximum-octahedral-shear-stress theory.* The shear stress acting on each of eight (octahedral) surfaces containing a hydrostatic normal stress, $\sigma_{\text{ave}} = (\sigma_1+\sigma_2+\sigma_3)/3$, reaches a value of $\sqrt{2}\sigma_{ys}/3$. It can be shown that this theory yields identical conditions as that provided by the maximum-distortion-energy theory.

Of these six theories, for ductile materials, the fifth and sixth are the ones that agree best with experimental evidence. However, the second leads to results so nearly the same and is simpler and more conservative for design applications. Thus, it is more widely used as a basis for design.

Failure theories for yield of ductile materials are based on shear or distortion. The maximum-distortion-energy theory equates the distortion energy for a general case of stress to the distortion energy when a simple tensile specimen yields. In terms of the principle stresses the distortion energy for the general case can be shown to be (see Ref. 59)

$$u_d = \frac{1+\nu}{6E}[(\sigma_1-\sigma_2)^2+(\sigma_2-\sigma_3)^2+(\sigma_3-\sigma_1)^2] \tag{3.7-1}$$

For the simple tensile test, yielding occurs when $\sigma_1 = \sigma_{ys}$ and $\sigma_2 = \sigma_3 = 0$. From Eq. (3.7-1), this gives a distortion energy at yield of

$$(u_d)_y = \frac{1+\nu}{3E}\sigma_{ys}^2 \tag{3.7-2}$$

Equating the energy for the general case, Eq. (3.7-1), to that for yield, Eq. (3.7-2), gives

$$\sqrt{0.5[(\sigma_1-\sigma_2)^2+(\sigma_2-\sigma_3)^2+(\sigma_3-\sigma_1)^2]}=\sigma_{ys} \tag{3.7-3}$$

For yield under a single, uniaxial state of stress, the stress would be equated to σ_{ys}. Thus, for yield, a single, uniaxial stress *equivalent* to the general state of stress is equated to the left-hand side of Eq. (3.7-3). This equivalent stress is called the *von Mises stress*, σ_{vM}, and is given by

$$\sigma_{vM}=\sqrt{0.5[(\sigma_1-\sigma_2)^2+(\sigma_2-\sigma_3)^2+(\sigma_3-\sigma_1)^2]} \tag{3.7-4}$$

Therefore, the maximum-distortion-energy theory predicts elastic failure when the von Mises stress reaches the yield strength.

The maximum-octahedral-shear-stress theory yields identical results to that of the maximum-distortion-energy theory (see Ref. 59). Through stress transformation, a stress element can be isolated in which all normal stresses on it are equal. These normal stresses are the averages of the normal stresses of the stress matrix, which are also the averages of the principal stresses and are given by

$$\sigma_{\text{ave}}=\frac{1}{3}(\sigma_x+\sigma_y+\sigma_z)=\frac{1}{3}(\sigma_1+\sigma_2+\sigma_3) \tag{3.7-5}$$

The element with these normal stresses in an octahedron where the eight surfaces are symmetric with respect to the principal axes. The directional cosines of the normal of these surfaces, relative to the principal axes, are eight combinations of $\pm 1/\sqrt{3}$ (e.g., one set is $1/\sqrt{3}, 1/\sqrt{3}, 1/\sqrt{3}$; another is $1/\sqrt{3}, -1/\sqrt{3}, 1/\sqrt{3}$, etc.). The octahedron is as shown in Fig. 3.1. The shear stresses on these surfaces are also equal, called the *octahedral shear stresses*, and are given by

$$\tau_{\text{oct}}=\frac{1}{3}\sqrt{(\sigma_1-\sigma_2)^2+(\sigma_2-\sigma_3)^2+(\sigma_3-\sigma_1)^2} \tag{3.7-6}$$

Again, for the simple tensile test, yield occurs when $\sigma_1=\sigma_{ys}$ and $\sigma_2=\sigma_3=0$. From Eq. (3.7-6), this gives an octahedral shear stress at yield of

$$(\tau_{\text{oct}})_y=\frac{\sqrt{2}}{3}\sigma_{ys} \tag{3.7-7}$$

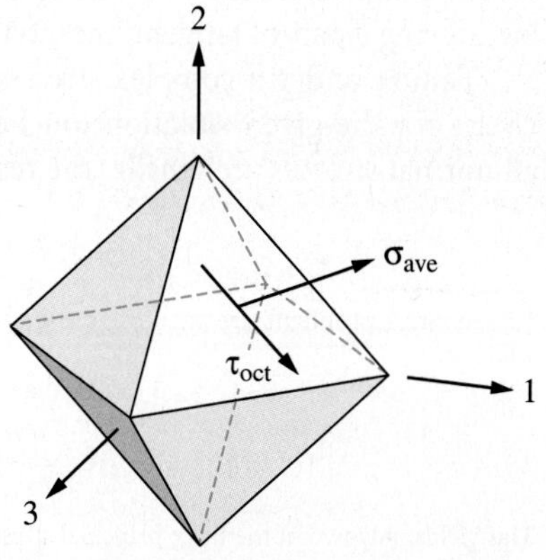

Figure 3.1 Octahedral surfaces containing octahedral shear stresses (shown relative to the principal axes, with only one set of stresses displayed).

Equating Eqs. (3.7-6) and (3.7-7) results in Eq. (3.7-3) again, proving that the maximum-octahedral-shear-stress theory is identical to the maximum-distortion-energy theory.

The maximum-shear-stress theory equates the maximum shear stress for a general state of stress to the maximum shear stress obtained when the tensile specimen yields. If the principal stresses are ordered such that $\sigma_1 \geq \sigma_2 \geq \sigma_3$, the maximum shear stress is given by $0.5(\sigma_1 - \sigma_3)$ (see Sec. 2.3, Eq. 2.3-25). The maximum shear stress obtained when the tensile specimen yields is 0.5 σ_{ys}. Thus, the condition for elastic failure for the maximum-shear-stress theory is*

$$\sigma_1 - \sigma_3 = \sigma_{ys} \tag{3.7-8}$$

The criteria just discussed concern the elastic failure of *material.* Such failure may occur locally in a *member* and may do no real damage if the volume of material affected is so small or so located as to have only negligible influence on the form and strength of the member as a whole. Whether or not such local overstressing is significant depends upon the properties of the material and the conditions of service. Fatigue properties, resistance to impact, and mechanical functioning are much more likely to be affected than static strength, and a degree of local overstressing that would constitute failure in a high-speed machine part might be of no consequence whatever in a bridge member.

A brittle material cannot be considered to have definitely failed until it has broken, which can occur either through a *tensile fracture*, when the maximum tensile stress reaches the ultimate strength, or through what appears to be a *shear fracture,* when the maximum compressive stress reaches a certain value. The fracture occurs on a plane oblique to the maximum compressive stress but not, as a rule, on the plane of maximum shear stress, and so it cannot be considered to be purely a shear failure (see Ref. 14). The results of some tests on glass and Bakelite (Ref. 26) indicate that for these brittle materials either the maximum stress or the maximum strain theory affords a satisfactory criterion of rupture while neither the maximum shear stress nor the constant energy of distortion theory does. These tests also indicate that strength increases with rate of stress application and that the increase is more marked when the location of the most stressed zone changes during the loading (pressure of a sphere on a flat surface) than when this zone is fixed (axial tension).

Another failure theory that is applicable to brittle materials is the *Coulomb-Mohr theory of failure.* Brittle materials have ultimate compressive strengths σ_{uc} greater than their ultimate tensile strengths σ_{ut}, and therefore both a uniaxial tensile test and a uniaxial compressive test must be run to use the Coulomb–Mohr theory. First we draw on a single plot both Mohr's stress circle for the tensile test at the instant of failure and Mohr's stress circle for the compressive test at the instant of failure; then we complete a failure envelope simply by drawing a pair of tangent lines to the two circles, as shown in Fig. 3.2.

Failure under a complex stress situation is expected if the largest of the three Mohr circles for the given situation touches or extends outside the envelope just described. If all normal stresses are tensile, the results coincide with the maximum stress theory. For a

*Plane stress problems are encountered quite often where the principal stresses are found from Eq. (2.3-23), which is

$$\sigma_p = \left(\frac{\sigma_x + \sigma_y}{2}\right) \pm \sqrt{\left(\frac{\sigma_x - \sigma_y}{2}\right)^2 + \tau_{xy}^2}$$

This yields only two of the three principal stresses. The third principal stress for plane stress is zero. Once the *three* principal stresses are determined, they can be ordered according to $\sigma_1 \geq \sigma_2 \geq \sigma_3$ and then Eq. (3.7-8) can be employed.

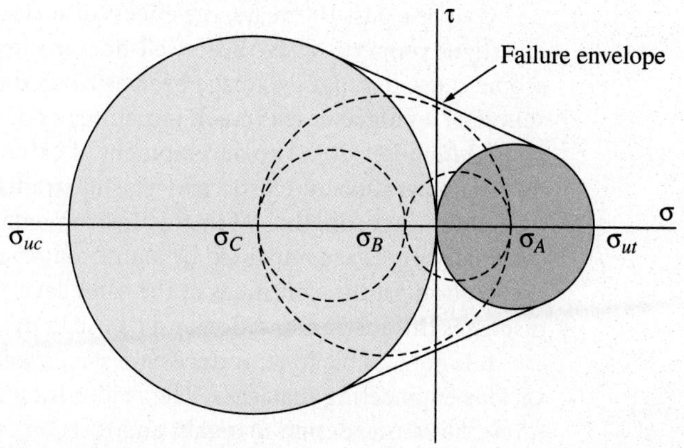

Figure 3.2

condition where the three principal stresses are σ_A, σ_B, and σ_C, as shown in Fig. 3.2, failure is being approached but will not take place unless the dashed circle passing through σ_A and σ_C reaches the failure envelope.

The accurate prediction of the breaking strength of a member composed of brittle metal requires a knowledge of the effect of *form and scale*, and these effects are expressed by the *rupture factor* (see Sec. 3.11). In addition, what has been said here concerning brittle metals applies also to any essentially isotropic brittle material.

Thus far, our discussion of failure has been limited to *isotropic* materials. For wood, which is distinctly *anisotropic*, the possibility of failure in each of several ways and directions must be taken into account, viz.: (1) by tension parallel to the grain, which causes fracture; (2) by tension transverse to the grain, which causes fracture; (3) by shear parallel to the grain, which causes fracture; (4) by compression parallel to the grain, which causes gradual buckling of the fibers usually accompanied by a shear displacement on an oblique plane; and (5) by compression transverse to the grain, which causes sufficient deformation to make the part unfit for service. The unit stress producing each of these types of failure must be ascertained by suitable tests (Ref. 2).

Another anisotropic class of material of consequence is that of the composites. It is well-known that composite members (see Secs. 7.3, 8.2, and definitions), such as steel reinforced concrete beams, more effectively utilize the more expensive, higher-strength materials in high-stress areas and the less expensive, lower-strength materials in the low-stress areas. Composite materials accomplish the same effect at microstructural and macrostructural levels. Composite materials come in many forms, but are generally formulated by embedding a reinforcement material in the form of fibers, flakes, particles, or laminations, in a randomly or orderly oriented fashion within a base matrix of polymeric, metallic, or ceramic material. For more detail properties of composites, see Ref. 60.

3.8 FATIGUE*

Practically all materials will break under numerous repetitions of a stress that is not as great as the stress required to produce immediate rupture. This phenomenon is known as *fatigue*.

*For further discussion on fatigue see Chap. 18.

Over the past 100 years, the effects of surface condition, corrosion, temperature, etc., on fatigue properties have been well-documented, but only in recent years has the microscopic cause of fatigue damage been attributed to *cyclic plastic flow* in the material at the source of a fatigue crack (crack initiation) or at the tip of an existing fatigue crack (*crack propagation*; Ref. 20). The development of extremely sensitive extensometers has permitted the separation of elastic and plastic strains when testing axially loaded specimens over short gage lengths. With this instrumentation it is possible to determine whether cyclic loading is accompanied by significant cyclic plastic strain and, if it is, whether the cyclic plastic strain continues at the same level, increases, or decreases. Sandor (Ref. 44) discusses this instrumentation and its use in detail.

It is not feasible to reproduce here even a small portion of the fatigue data available for various engineering materials. The reader should consult materials handbooks, manufacturers' literature, design manuals, and texts on fatigue. See Refs. 44–48. Some of the more important factors governing fatigue behavior in general will be outlined in the following material:

Number of Cycles to Failure

Most data concerning the number of cycles to failure are presented in the form of an S–N curve where the cyclic stress amplitude is plotted versus the number of cycles to failure. This generally leads to a straight-line log-log plot if we account for the scatter in the data. For ferrous metals a lower limit exists on the stress amplitude and is called the *fatigue limit*, or *endurance limit*. This generally occurs at a life of from 10^5–10^7 cycles of reversed stress, and we assume that stresses below this limit will not cause failure regardless of the number of repetitions. With the ability to separate elastic and plastic strains accurately, there are instances when a plot of plastic-strain amplitudes versus N and elastic-strain amplitudes versus N will reveal more useful information (Refs. 44 and 45).

Method of Loading and Size of Specimen

Uniaxial stress can be produced by axial load, bending, or a combination of both. In flat-plate bending, only the upper and lower surfaces are subjected to the full range of cyclic stress. In rotating bending, all surface layers are similarly stressed, but in axial loading, the entire cross section is subjected to the same average stress. Since fatigue properties of a material depend upon the statistical distribution of defects throughout the specimen, it is apparent that the three methods of loading will produce different results.

In a similar way, the size of a bending specimen will affect the fatigue behavior while it will have little effect on an axially loaded specimen. Several empirical formulas have been proposed to represent the influence of size on a machine part or test specimen in bending. For steel, Moore (Ref. 38) suggests the equation

$$\sigma'_e\left(1-\frac{0.016}{d'}\right)=\sigma''_e\left(1-\frac{0.016}{d''}\right)$$

where σ'_e is the endurance limit for a specimen of diameter d' and σ''_e is the endurance limit for a specimen of diameter d''. This formula was based on test results obtained with specimens from 0.125–1.875 inches in diameter and shows good agreement within that size range. Obviously it cannot be used for predicting the endurance limit of very small specimens. The few relevant test results available indicate a considerable decrease in endurance limit for very large diameters (Refs. 22–24).

Stress Concentrations

Fatigue failures occur at stress levels less than those necessary to produce the gross yielding which would blunt the sharp rise in stress at a stress concentration. It is necessary, therefore, to apply the fatigue strengths of a smooth specimen to the peak stresses expected at the stress concentrations unless the size of the stress-concentrating notch or fillet approaches the grain size or the size of an anticipated defect in the material itself (see *Factor of stress concentration in fatigue* in Sec. 3.10). References 40 and 41 discuss the effect of notches on low-cycle fatigue. Table 17.1 provides stress concentration factor for many load and geometric conditions.

Surface Conditions

Surface roughness constitutes a form of stress raiser. Discussion of the effect of surface coatings and platings is beyond the scope of this book (see Refs. 28 and 36).

Corrosion Fatigue

Under the simultaneous action of *corrosion* and repeated stress, the fatigue strength of most metals is drastically reduced, sometimes to a small fraction of the strength in air, and a true endurance limit can no longer be said to exist. Liquids and gases not ordinarily thought of as especially conducive to corrosion will often have a very deleterious effect on fatigue properties, and resistance to corrosion is more important than normal fatigue strength in determining the relative rating of different metals (Refs. 24, 25, and 31).

Range of Stress

Stressing a ductile material beyond the elastic limit or yield point in tension will raise the elastic limit for subsequent cycles but lower the elastic limit for compression. The consequence of this Bauschinger effect on fatigue is apparent if one accepts the statement that fatigue damage is a result of cyclic plastic flow; that is, if the range of cyclic stress is reduced sufficiently, higher peak stresses can be accepted without suffering continuing damage.

Various empirical formulas for the endurance limit corresponding to any given range of stress variation have been suggested, the most generally accepted of which is expressed by the *Goodman diagram* or some modification thereof. Figure 3.3 shows one method of constructing this diagram. In each cycle, the stress varies from a maximum value σ_{max} to a

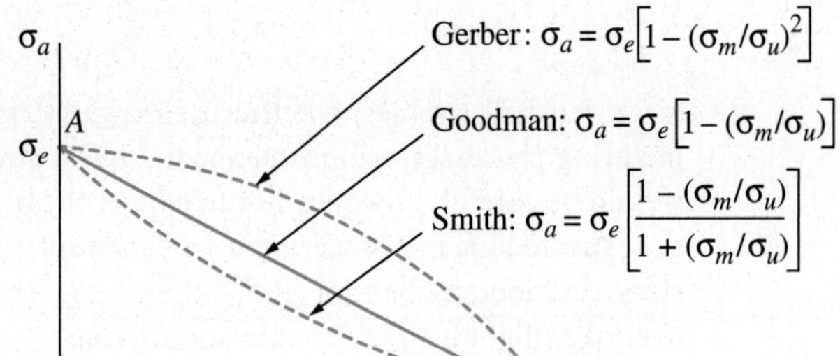

Figure 3.3

minimum value σ_{min}, either of which is plus or minus according to whether it is tensile or compressive. The *mean stress* is

$$\sigma_m = \frac{1}{2}(\sigma_{max} + \sigma_{min})$$

and the *alternating stress* is

$$\sigma_a = \frac{1}{2}(\sigma_{max} - \sigma_{min})$$

the addition and subtraction being algebraic. With reference to rectangular axes, σ_m is measured horizontally and σ_a vertically. Obviously when $\sigma_m = 0$, the limiting value of σ_a is the endurance limit for fully reversed stress, denoted here by σ_e. When $\sigma_a = 0$, the limiting value of σ_m is the ultimate tensile strength, denoted here by σ_u. Points *A* and *B* on the axes are thus located.

According to the Goodman theory, the ordinate to a point on the straight line *AB* represents the maximum alternating stress σ_a that can be imposed in conjunction with the corresponding mean stress σ_m. Any point above *AB* represents a stress condition that would eventually cause failure; any point below *AB* represents a stress condition with more or less margin of safety. A more conservative construction, suggested by Soderberg (Ref. 13), is to move point *B* back to σ_{ys}, the yield strength. A less conservative but sometimes preferred construction, proposed by Gerber, is to replace the straight line by the parabola.

The Goodman diagrams described can be used for steel and for aluminum and titanium alloys, but for cast iron many test results fall below the straight line *AB* and the lower curved line, suggested by Smith (Ref. 21), is preferred. Test results for magnesium alloys also sometimes fall below the straight line.

Figure 3.3 represents conditions where σ_m is tensile. If σ_m is compressive, σ_a is increased; and for values of σ_m less than the compression yield strength, the relationship is represented approximately by the straight line *AB* extended to the left with the same slope. When the mean stress and alternating stress are both torsional, σ_a is practically constant until σ_m exceeds the yield strength in shear; and for alternating bending combined with mean torsion, the same thing is true. But when σ_m is tensile and σ_a is torsional, σ_a diminishes as σ_m increases in almost the manner represented by the Goodman line. When stress concentration is to be taken into account, the accepted practice is to apply K_f (or K_t if K_f is not known) to σ_a only, not to σ_m (for K_t and K_f, see Sec. 3.10).

Residual Stress

Since residual stresses, whether deliberately introduced or merely left over from manufacturing processes, will influence the *mean* stress, their effects can be accounted for. One should be careful, however, not to expect the beneficial effects of a residual stress if during the expected life of a structure it will encounter overloads sufficient to change the residual-stress distribution. Sandor (Ref. 44) discusses this in detail and points out that an occasional overload might be beneficial in some cases.

The several modified forms of the Goodman diagram are used for predicting the stress levels which will form cracks, but other more extensive plots such as the Haigh diagram (Ref. 45) can be used to predict in addition the stress levels for which cracks, once formed, will travel, fatigue lives, etc.

Combined Stress

No single theory of failure in Sec. 3.7 can be applied to all fatigue loading conditions. The *maximum-distortion-energy theory* seems to be conservative in most cases, however. Reference 18 gives a detailed description of an acceptable procedure for designing for fatigue under conditions of combined stress. The procedure described also considers the effect of mean stress on the cyclic stress range. Three criteria for failure are discussed: *gross yielding, crack initiation,* and *crack propagation.* An extensive discussion of fatigue under combined stress is found in Refs. 27, 31, and 45.

Stress History

A very important question and one that has been given much attention is the influence of previous stressing on fatigue strength. One theory that has had considerable acceptance is the *linear damage law* (Miner in Ref. 27); here the assumption is made that the damage produced by repeated stressing at any level is directly proportional to the number of cycles. Thus, if the number of cycles producing failure (100% damage) at a stress range σ_1 is N_1, then the proportional damage produced by N cycles of the stress is N/N_1 and stressing at various stress levels for various numbers of cycles causes cumulative damage equal to the summation of such fractional values. Failure occurs, therefore, when $\Sigma N/N_1 = 1$. The formula implies that the effect of a given number of cycles is the same, whether they are applied continuously or intermittently, and does not take into account the fact that for some metals understressing (stressing below the endurance limit) raises the endurance limit. The linear damage law is not reliable for all stress conditions, and various modifications have been proposed, such as replacing 1 in the formula by a quantity x whose numerical value, either more or less than unity, must be determined experimentally. Attempts have been made to develop a better theory (e.g., Corten and Dolan, Freudenthal and Gumbel, in Ref. 32). Though all the several theories are of value when used knowledgeably, it does not appear that as yet any generally reliable method is available for predicting the life of a stressed part under variable or random loading. (See Refs. 19 and 39.) See Refs. 44 and 45 for a more detailed discussion.

A modification of the foil strain gage called an *S–N fatigue life gage* (Refs. 33 and 34) measures accumulated plastic deformation in the form of a permanent change in resistance. A given total change in resistance can be correlated with the damage necessary to cause a fatigue failure in a given material.

3.9 BRITTLE FRACTURE

Brittle fracture is a term applied to an unexpected brittle failure of a material such as low-carbon steel where large plastic strains are usually noted before actual separation of the part. Major studies of brittle fracture started when failures such as those of welded ships operating in cold seas led to a search for the effect of temperature on the mode of failure. For a brittle fracture to take place, the material must be subjected to a tensile stress at a location where a crack or other very sharp notch or defect is present and the temperature must be lower than the so-called *transition temperature.* To determine a transition temperature for a given material, a series of notched specimens is tested under impact loading, each at a different temperature, and the ductility or the energy required to cause fracture is noted. There will be a limited range of temperatures over which the ductility or fracture energy will drop significantly. Careful examination of the fractured specimens will show that the material at the root of the notch has tried to contract laterally.

Where the fracture energy is large, there is evidence of a large lateral contraction; and where the fracture energy is small, the lateral contraction is essentially zero. In all cases the lateral contraction is resisted by the adjacent less stressed material. The deeper and sharper cracks have relatively more material to resist lateral contraction. Thicker specimens have a greater distance over which to build up the necessary *triaxial* tensile stresses that lead to a tensile failure without producing enough shear stress to cause yielding. Thus, the term *transition temperature* is somewhat relative since it depends upon notch geometry as well as specimen size and shape. Since yielding is a flow phenomenon, it is apparent that rate of loading is also important. Static loading of sufficient intensity may start a brittle fracture, but it can continue under much lower stress levels owing to the higher rate of loading.

The ensuing research in the field of *fracture mechanics* has led to the development of both acceptable theories and experimental techniques, the discussion of which is beyond the scope of this book. Users should examine Refs. 49–58 for information and for extensive bibliographies.

3.10 STRESS CONCENTRATION

The distribution of elastic stress across the section of a member may be nominally uniform or may vary in some regular manner, as illustrated by the linear distribution of stress in flexure. When the variation is abrupt so that within a very short distance the intensity of stress increases greatly, the condition is described as *stress concentration*. It is usually due to local irregularities of form such as small holes, screw threads, scratches, and similar *stress raisers*. There is obviously no hard and fast line of demarcation between the rapid variation of stress brought about by a stress raiser and the variation that occurs in such members as sharply curved beams, but in general the term *stress concentration* implies some form of irregularity not inherent in the member as such but accidental (tool marks) or introduced for some special purpose (screw thread).

The maximum intensity of elastic stress produced by many of the common kinds of stress raisers can be ascertained by mathematical analysis, photoelastic analysis, or direct strain measurement and is usually expressed by the *stress concentration factor*. This term is defined in App. C, but its meaning may be made clearer by an example. Consider a straight rectangular beam, originally of uniform breadth b and depth D, which has had cut across the lower face a fairly sharp transverse V-notch of uniform depth h, making the net depth of the beam section at that point $D - h$. If now the beam is subjected to a uniform bending moment M, the *nominal* fiber stress at the root of the notch may be calculated by ordinary flexure formula $\sigma = Mc/I$, which here reduces to $\sigma = 6M/[b(D-h)^2]$. But the actual stress σ' is very much greater than this because of the stress concentration that occurs at the root of the notch. The ratio σ'/σ, actual stress divided by nominal stress, is the stress concentration factor K_t for this particular case. Values of K_t for a number of common stress raisers are given in Table 17.1. The most complete single source for numerical values of stress concentration factors is Peterson (Ref. 42). It also contains an extensive bibliography.

The abrupt variation and high local intensity of stress produced by stress raisers are characteristics of *elastic behavior*. The plastic yielding that occurs on overstressing greatly mitigates stress concentration even in relatively brittle materials and causes it to have much less influence on breaking strength than might be expected from a consideration of the elastic stresses only. The practical significance of stress concentration therefore depends on circumstances. For ductile metal under static loading it is usually (though not always) of little or no importance; for example, the high stresses that occur at the edges of rivet holes in structural steel members are safely ignored, the stress due to a tensile load being assumed uniform on the net section. (In the case of eyebars and similar pin-connected members, however, a reduction of 25% in allowable stress on the net section is recommended.) For brittle material under static loading, stress concentration is often a serious consideration, but its effect varies widely and cannot be predicted either from K_t or from the brittleness of the material (see Ref. 35).

What may be termed the *stress concentration factor at rupture*, or the *strength reduction factor*, represents the significance of stress concentration for static loading. This factor, which will be denoted by K_r is the ratio of the computed stress at rupture for a plain specimen to the computed stress at rupture for the specimen containing the stress raiser. For the case just described, it would be the ratio of the modulus of rupture of the plain beam to that of the notched beam, the latter being calculated for the net section. K_r is therefore a ratio of stresses, one or both of which may be fictitious, but is nonetheless a measure of the strength-reducing effect of stress concentration. Some values of K_r are given in Table 17 of Ref. 1.

It is for conditions involving fatigue that stress concentration is most important. Even the most highly localized stresses, such as those produced by small surface scratches, may greatly lower the apparent endurance limit, but materials vary greatly in *notch sensitivity,* as susceptibility to this effect is sometimes called. Contrary to what might be expected, ductility (as ordinarily determined by axial testing) is not a measure of immunity to stress concentration in fatigue; for example, steel is much more susceptible than cast iron. What may be termed the *fatigue stress concentration factor* K_f is the practical measure of notch sensitivity. It is the ratio of the endurance limit of a plain specimen to the nominal stress at the endurance limit of a specimen containing the stress raiser.

A study of available experimental data shows that K_f is almost always less, and often significantly less, than K_t, and various methods for estimating K_f from K_t have been proposed. Neuber (Ref. 37) proposes the formula

$$K_f = 1 + \frac{K_t - 1}{1 + \pi\sqrt{\rho'/\rho}/(\pi - \omega)} \tag{3.10-1}$$

where ω is the flank angle of the notch (called θ in Table 17.1), ρ is the radius of curvature (in inches) at the root of the notch (called r in Table 17.1), and ρ' is a dimension related to the grain size, or size of some type of basic building block, of the material and may be taken as 0.0189 in for steel.

All the methods described are valuable and applicable within certain limitations, but none can be applied with confidence to all situations (Ref. 29). Probably none of them gives sufficient weight to the effect of scale in the larger size range. There is abundant evidence to show that the significance of stress concentration increases with size for both static and repeated loading, especially the latter.

An important fact concerning stress concentration is that a single isolated notch or hole has a worse effect than have a number of similar stress raisers placed close together; thus, a single V-groove reduces the strength of a part more than does a continuous screw thread of almost identical form. The deleterious effect of an unavoidable stress raiser can, therefore, be mitigated sometimes by juxtaposing additional form irregularities of like nature, but the actual superposition of stress raisers, such as the introduction of a small notch in a fillet, may result in a stress concentration factor equal to or even exceeding the product of the factors for the individual stress raisers (Refs. 30 and 43).

3.11 EFFECT OF FORM AND SCALE ON STRENGTH; RUPTURE FACTOR

It has been pointed out (Sec. 3.7) that a member composed of brittle material breaks in tension when the maximum tensile stress reaches the ultimate strength or in shear when the maximum compressive stress reaches a certain value. In calculating the stress at rupture in such a member it is customary to employ an *elastic-stress formula*; thus the ultimate fiber stress in a beam is usually calculated by the ordinary flexure formula. It is known that the result (modulus of rupture) is not a true stress, but it can be used to predict the strength of a similar beam of the same material. However, if another beam of the same material but of different cross section, span/depth ratio, size, or manner of loading and support is tested, the modulus of rupture will be found to be different. (The effect of the shape of the section is often taken into account by the *form factor*, and the effects of the span/depth ratio and manner of loading are recognized in the testing procedure.) Similarly, the *calculated* maximum stress at rupture in a curved beam, flat plate, or torsion member is not equal to the ultimate strength of the material, and the magnitude of the disparity will vary greatly with the material, form of the member, manner of loading, and absolute scale. In order to predict accurately the breaking load for such a member, it is necessary to take this variation into account, and the *rupture factor* (defined in App. C) provides a convenient means of doing so. Values of the rupture factor for a number of materials and types of members are given in Table 18 of Ref. 1.

On the basis of many experimental determinations of the rupture factor (Ref. 35), the following generalizations may be made:

1. The smaller the proportional part of the member subjected to high stress, the larger the rupture factor. This is exemplified by the facts that a beam of circular section exhibits a higher modulus of rupture than a rectangular beam and that a flat plate under a concentrated center load fails at a higher computed stress than one uniformly loaded. The extremes in this respect are, on the one hand, a uniform bar under axial tension for which the rupture factor is unity and, on the other hand, a case of severe stress concentration such as a sharply notched bar for which the rupture factor may be indefinitely large.
2. In the flexure of statically indeterminate members, the redistribution of bending moments that occurs when plastic yielding starts at the most highly stressed section increases the rupture factor. For this reason a flat plate gives a higher value than a simple beam, and a circular ring gives a higher value than a portion of it tested as a statically determinate curved beam.
3. The rupture factor seems to vary inversely with the absolute scale for conditions involving abrupt stress variation, which is consistent with the fact (already noted) that for cases of stress concentration both K_r and K_f diminish with the absolute scale.
4. As a rule, the more brittle the material, the more nearly all rupture factors approach unity. There are, however, many exceptions to this rule. It has been pointed out (Sec. 3.10) that immunity to notch effect even under static loading is not always proportional to ductility.

The practical significance of these facts is that for a given material and given factor of safety, some members may be designed with a much higher allowable stress than others. This fact is often recognized in design; for example, the allowable stress for wooden airplane spars varies according to the form factor and the proportion of the stress that is flexural.

What has been said here pertains especially to comparatively brittle materials, that is, materials for which failure consists in fracture rather than in the beginning of plastic deformation. The effect of form on the ultimate strength of ductile members is less important, although even for steel the allowable unit stress is often chosen with regard to circumstances such as those discussed previously. For instance, in gun design the maximum stress is allowed to approach and even exceed the nominal elastic limit, the volume of material affected being very small, and in structural design extreme fiber stresses in bending are permitted to exceed the value allowed for axial loading. In testing, account must be taken of the fact that some ductile metals exhibit a higher ultimate strength when fracture occurs at a reduced section such as would be formed in a tensile specimen by a concentric groove or notch. Whatever effect of stress concentration may remain during plastic deformation is more than offset by the supporting action of the shoulders, which tends to prevent the normal "necking down."

3.12 PRESTRESSING

Parts of an elastic system, by accident or design, may have introduced into them stresses that cause and are balanced by opposing stresses in other parts, so that the system reaches a state of stress without the imposition of any external load. Examples of such initial, or locked-up, stresses are the temperature stresses in welded members, stresses in a statically indeterminate truss due to tightening or "rigging" some of the members by turnbuckles, and stresses in the flange couplings of a pipeline caused by screwing down the nuts. The effects of such prestressing upon the rigidity and strength of a system will now be considered, the assumption being made that prestressing is not so severe as to affect the properties of the *material.*

In discussing this subject it is necessary to distinguish two types of systems, viz., one in which the component parts can sustain reversal of stress and one in which at least some of the component parts cannot sustain reversal of stress. Examples of the first type are furnished by a solid bar and by a truss, all members of which can sustain either tension or compression. Examples of the second type are furnished by the bolt-flange combination mentioned and by a truss with wire diagonals that can take tension only.

For the first type of system, prestressing has no effect on initial rigidity. Thus a plain bar with locked-up temperature stresses will exhibit the same modulus of elasticity as a similar bar from which these stresses have been removed by annealing; two prestressed helical springs arranged in parallel, the tension in one balancing the compression in the other, will deflect neither more nor less than the same two springs similarly placed without prestressing.

Prestressing will lower the elastic limit (or allowable load, or ultimate strength) provided that in the absence of prestressing all parts of the system reach their respective elastic limits (or allowable loads, or ultimate strengths) simultaneously. But if this relation between the parts does not exist, then prestressing may raise any or all of these quantities. One or two examples illustrating each condition may make this clear.

Consider first a plain bar that is to be loaded in axial tension. If there are no locked-up stresses, then (practically speaking) all parts of the bar reach their allowable stress, elastic limit, and ultimate strength simultaneously. But if there are locked-up stresses present, then the parts in which the initial tension is highest reach their elastic limit before other parts, and the elastic limit of the bar as a whole is thus lowered. The load at which the allowable unit stress is first reached is similarly lowered, and the ultimate strength may also be reduced; although if the material is ductile, the equalization of stress that occurs during elongation will largely prevent this.

As an example of the second condition (all parts do not simultaneously reach the elastic limit or allowable stress) consider a thick cylinder under internal pressure. If the cylinder is not prestressed, the stress at the interior surface reaches the elastic limit first and so governs the pressure that may be applied. But if the cylinder is prestressed by shrinking on a jacket or wrapping with wire under tension, as is done in gun construction, then the walls are put into an initial state of compression. This compressive stress also is greatest at the inner surface, and the pressure required to reverse it and produce a tensile stress equal to the elastic limit is much greater than before. As another example, consider a composite member comprising two rods of equal length, one aluminum and the other steel, that are placed side by side to jointly carry a tensile load. For simplicity, it will be assumed that the allowable unit stresses for the materials are the same. Because the modulus of elasticity of the steel is about three times that of the aluminum, it will reach the allowable stress first and at a total load less than the sum of the allowable loads for the bars acting separately. But if the composite bar is properly prestressed, the steel being put into initial compression and the aluminum into initial tension (the ends being in some way rigidly connected to permit this), then on the application of a tensile load the two bars will reach the allowable stress simultaneously, and the load-carrying capacity of the combination is thus greater than before. Similarly the elastic limit and sometimes the ultimate strength of a composite member may be raised by prestressing.

In a system of the second type (in which all parts *cannot* sustain stress reversal) prestressing increases the rigidity for any load less than that required to produce stress reversal. The effect of prestressing up to that point is to make the rigidity of the system the same as though all parts were effective. Thus in the case of the truss with wire diagonals, it is as though the counterwires were taking compression; in the case of the flange-bolt combination, it is as though the flanges were taking tension. (If the flanges are practically rigid in comparison with the bolts, there is no deformation until the applied load exceeds the bolt tension and so the system is rigid.) When the applied load becomes large enough to cause stress reversal (to make the counterwires go slack or to separate the flanges), the effect of prestressing disappears and the system is neither more nor less rigid than a similar one not prestressed provided, of course, none of the parts has been overstressed.

The elastic limit (or allowable load, or ultimate strength) of a system of this type is not affected by prestressing unless the elastic limit (or allowable load, or ultimate strength) of one or more of the parts is reached before the stress reversal occurs. In effect, a system of this type is exactly like a system of the first type until stress reversal occurs, after which all effects of prestressing vanish.

The effects of prestressing are often taken advantage of, notably in bolted joints (flanges, cylinder heads, etc.), where high initial tension in the bolts prevents stress fluctuation and consequent fatigue, and in prestressed reinforced-concrete members, where the initially compressed concrete is enabled, in effect, to act in tension without cracking up to the point of stress reversal. The example of the prestressed thick cylinder has already been mentioned.

3.13 ELASTIC STABILITY

Under certain circumstances the maximum load a member will sustain is determined not by the strength of the material but by the stiffness of the member. This condition arises when the load produces a bending or a twisting moment that is proportional to the corresponding deformation. The most familiar example is the *Euler column*. When a straight slender column is loaded axially, it remains straight and suffers only axial compressive deformation under small loads. If while thus loaded it is slightly deflected by a transverse force, it will straighten after removal of that force. But there is obviously some axial load that will just hold the column in the deflected position, and since both the bending moment due to the load and the resisting moment due to the stresses are directly proportional to the deflection, the load required thus to hold the column is independent of the amount of the deflection. If this condition of balance exists at stresses less than the elastic limit, the condition is called *elastic stability,* and the load that produces this condition is called the critical load. Any increase of the load beyond this critical value is usually attended by immediate collapse of the member.

It is the compressive stresses within long, thin sections of a structure that can cause instabilities. The compressive stress can be elastic or inelastic, and the instability can be global or local. Global instabilities can cause catastrophic failure, whereas local instabilities may cause permanent deformation but not necessarily a catastrophic failure. For the Euler column, when instability occurs, it is global since the entire cross section is involved in the deformation. Localized buckling of the edges of the flange in compression of a wide-flange I-beam in bending can occur. Likewise, the center of the web of a transversely loaded I-beam or plate girder in bending undergoes pure shear where along the diagonal (45°) compressive stresses are present and localized buckling is possible.

Other examples of elastic stability are afforded by a thin cylinder under external pressure, a thin plate under edge compression or edge shear, and a deep thin cantilever beam under a transverse end load applied at the top surface. Some such elements, unlike the simple column described previously, do not fail under the load that initiates elastic buckling but demonstrate increasing resistance as the buckling progresses. Such postbuckling behavior is important in many problems of shell design. Elastic stability is discussed further in Chap. 15, and formulas for the critical loads for various members and types of loadings are given in Tables 15.1 and 15.2.

Factor of Safety

Due to uncertainty in the design process, including calculations, variability of material properties, and quality of manufacturing processes, a design margin over the theoretical design, a factor of safety also known as safety factor (SF), is used. In other words, considering uncertainty and variability in the design process, SF expresses how much sturdier a device is than it needs to be for an intended load. The value of the SF is related to the lack of confidence in the design process or the safety issue of the device. Factor of safety of a component of a system or device, n, is expressed as

$$n = \frac{\text{Strength of component}}{\text{Stress (load) on component}}$$

Or it can be written as $n = \frac{S}{\sigma}$, where σ is allowable stress for the design of the component, that is,

$$\sigma_{\text{allowable}} = \frac{S}{n}$$

Depending on the structure and the loading condition, S could be yield stress, ultimate stress, or failure stress such as critical buckling in columns or stability stress in a structure.

In the design process, to reduce probability of component failure, designers have to select a design SF. When selecting a design SF, each of the following factors should be considered:

1. *Variations in material properties:* Note that no two batches of raw materials are exactly alike and there is a variability of strength and mechanical properties of materials.
2. *Effect of size in material strength properties:* Some materials may fail at lower stress when the size is larger than the laboratory test specimen.
3. *Class of materials:* SF varies for different materials where their probability of failure are different. For example, ductile materials, brittle materials, ceramic materials, and plastic materials each has a different range of SFs for different loading conditions.
4. *Type of loading:* Depending on the loading condition SF varies significantly. The four class of loadings are static load, dynamic load, cyclic load, and impact/shock load.

5. *Manufacturing process:* The process of production of materials usually introduces stress concentration and residual stresses. For example, hot or cold rolling, forging, forming, tempering, and machining could introduce undesirable residual stress in materials.
6. *Environmental effect:* The environment in which the device is expected to operate such as salty, acidic, and radioactive environment could affect the properties of the materials over long period of time.
7. *Specific requirement for life expectancy and reliability:* Some machines are expected to have a prolong service life with high degree of reliability.
8. *Overall concern for human safety:* All designs must consider safety and welfare of the operator and other persons who may be near or in contact with the machine or device. This concern should be taken seriously when safety of public is of concern.

The following table contains reasonable SFs for different type of loading and materials, as suggested by Joseph P. Vidosic (Ref. 65).

Item	Material and Loading Conditions	Safety Factor
1	When exceptionally reliable known materials are used under controllable conditions and subjected to loads and stresses that can be determined with certainty, and where low weight is a particularly important consideration	$n = 1.25–1.5$
2	When well-known materials, under reasonably constant environmental conditions, subjected to loads and stresses that can be determined using qualified design procedures	$n = 1.5–2.0$
3	When average materials are used and are operated in ordinary environments and subjected to loads and stresses that can be determined	$n = 2–2.5$
4	When brittle materials under average conditions of environment, load, and stress are used. Also when less tried or uncommon materials are used	$n = 2.5–3$
5	When untested materials are used under average conditions of environment, load, and stress	$n = 3–4$
6	When better known materials are used in uncertain environments or subjected to uncertain stresses such as dynamic load	$n = 3–4$
7	When repeated or cyclic loads are used, the endurance limit (rather than the yield strength) should be used. Depending on the environment and materials' SF of 1 to 6 are acceptable	$n = 1–6$
8	When impact forces are used, depending on the severity of impact SF given in item 3 to 6 are acceptable	$n = 2–4$
9	When brittle materials are used, ultimate strength should be used as the theoretical maximum and depending on the probability of failure of the material the factors presented in items 1 to 6 should be approximately doubled	$n = 2–12$
10	When impact forces are used, the factors given in items 3 to 6 are acceptable, but an impact factor should be included	$n = 2–4$

SF values can be thought of as a standardized way for comparing strength and reliability between systems. However, the use of a factor of safety does not imply that an item, structure, or design is "safe." Many quality assurances, engineering design, manufacturing, installation, and end-use factors may influence whether or not something is safe in any particular situation.

Depending on designer's experience and accumulated experience of specific industry empirical, reliable SFs could be established. A designer of a specific device or produce may depend upon such empirical SFs where the product or device has a long history of use and the factors based upon such a history are reliable. Many specific industries have codes and standards where specific SFs are recommended. For example, the ASME Pressure Vessel Codes, various building codes, and specific values that are stipulated in contracts for both civilian and governmental designs are reliable sources. Other codes and standards for design of vessels, pumps, valves, piping systems as well as BS 5950-1:2000 for structural steel work in building should be followed for related designs. Finally, for a specific industry, statistical methods have also been employed in establishing a factor of safety.

Reliability

Reliability is the ability of a machine, or a system to consistently perform its intended function without degradation or failure. In simple term, reliability is the ratio of how many times a system performs as intended without failure over how many times it is used. The reliability R can be expressed by a number as

$$R = \frac{\text{Number times the device worked satisfactory}}{\text{Number of total times the device was used}}$$

The range of reliability is $0 \leq R < 1$.

For example, a reliability of $R = 0.95$ means that there is a 95% chance that the part will perform proper function without failure. The failure of 5 parts per 1000 parts is considered an acceptable failure rate for certain products. Then the reliability is

$$R = 1 - \frac{5N}{1000N} = 0.995 \text{ or } 99.5\%$$

If a system has more than one part and each part has a reliability of R_i, where $i = 1, 2, 3, \ldots.$ Then the total reliability of the system is

$$R = R_1 \cdot R_2 \cdot R_3 \cdot R_4 \cdots$$

For example, a system has three parts and each part's reliability is 98%, then the total system reliability is

$$R = (0.98)^3 = 0.941 \text{ or } 94.1\%$$

Proper selection of SF could lead to higher percentage of reliability.

3.14 TABLES: MECHANICAL PROPERTIES OF MATERIALS

Elastic Constants—Bulk, Shear, and Lame Modulus

Young's modulus and Poisson's ratio are the most common properties used to characterize elastic solids, but other measures are also used. For example, we define the **shear modulus, bulk modulus** and **Lame modulus** of an elastic solid as follows:

$$\text{Bulk modulus } K = \frac{E}{3(1-2\nu)}$$

$$\text{Shear modulus } \mu = G = \frac{E}{2(1+\nu)}$$

$$\text{Lame' modulus } \lambda = \frac{\nu E}{(1+\nu)(1-2\nu)}$$

A table relating all the possible combinations of moduli to all other possible combinations is given below.

TABLE 3.1 Modulus of Elasticity Relationships

	Lame Modulus λ	Shear Modulus μ	Young's Modulus E	Poisson's Ratio ν	Bulk Modulus K
λ, μ			$\frac{\mu(3\lambda+2\mu)}{\lambda+\mu}$	$\frac{\lambda}{2(\lambda+\mu)}$	$\frac{3\lambda+2\mu}{3}$
λ, E		Irrational		Irrational	Irrational
λ, ν		$\frac{\lambda(1-2\nu)}{2\nu}$	$\frac{\lambda(1+\nu)(1-2\nu)}{\nu}$		$\frac{\lambda(1+\nu)}{3\nu}$
λ, K		$\frac{3(K-\lambda)}{2}$	$\frac{9K(K-\lambda)}{3K-\lambda}$	$\frac{\lambda}{3K-\lambda}$	
μ, E	$\frac{\mu(2\mu-E)}{E-3\mu}$			$\frac{E-2\mu}{2\mu}$	$\frac{\mu E}{3(3\mu-E)}$
μ, ν	$\frac{2\mu\nu}{1-2\nu}$		$2\mu(1+\nu)$		$\frac{2\mu(1+\nu)}{3(1-2\nu)}$
μ, K	$\frac{3K-2\mu}{3}$		$\frac{9K\mu}{3K+\mu}$	$\frac{3K-2\mu}{2(3K+\mu)}$	
E, ν	$\frac{\nu E}{(1+\nu)(1-2\nu)}$	$\frac{E}{2(1+\nu)}$			$\frac{E}{3(1-2\nu)}$
E, K	$\frac{3K(3K-E)}{9K-E}$	$\frac{3EK}{9K-E}$		$\frac{3K-E}{6K}$	
ν, K	$\frac{3K\nu}{(1+\nu)}$	$\frac{3K(1-2\nu)}{2(1+\nu)}$	$3K(1-2\nu)$		

TABLE 3.2 Material Classification, Names, and Abbreviations

Metals (the metals and alloys of engineering)	Aluminum alloys	Al alloys
	Copper alloys	Cu alloys
	Lead alloys	Lead alloys
	Magnesium alloys	Mg alloys
	Nickel alloys	Ni alloys
	Carbon steels	Steels
	Stainless steels	Stainless steels
	Tin alloys	Tin alloys
	Titanium alloys	Ti alloys
	Tungsten alloys	W alloys
	Lead alloys	Pb alloys
	Zinc alloys	Zn alloys

(Continued)

TABLE 3.2 Material Classification, Names, and Abbreviations (*Continued*)

Ceramics	Alumina	Al_2O_3
Technical ceramics (fine ceramics capable of load-bearing application)	Aluminum nitride	AIN
	Boron carbide	B_4C
	Silicon carbide	SiC
	Silicon nitride	Si_3N_4
	Tungsten carbide	WC
Nontechnical ceramics (porous ceramics of construction)	Brick	Brick
	Concrete	Concrete
	Stone	Stone
Glasses	Soda-lime glass	Soda-lime glass
	Borosilicate glass	Borosilicate glass
	Silica glass	Silica glass
	Glass ceramic	Glass ceramic
Polymers (the thermoplastics and thermosets of engineering)	Acrylonitrile butadiene styrene	ABS
	Cellulose polymers	CA
	lonomers	lonomers
	Epoxies	Epoxy
	Phenolics	Phenolics
	Polyamides (nylons)	PA
	Polycarbonate	PC
	Polyesters	Polyesters
	Polyetheretherkeytone	PEEK
	Polyethylene	PE
	Polyethylene terephalate	PET or PETE
	Polymethylmethacrylate	PMMA
	Polyoxymethylene (acetal)	POM
	Polypropylene	PP
	Polystyrene	PS
	Polytetrafluorethylene	PTFE
	Polyvinylchloride	PVC
Elastomers (engineering rubbers, natural and synthetic)	Butyl rubber	Butyl rubber
	EVA	EVA
	Isoprene	Isoprene
	Natural rubber	Natural rubber
	Polychloroprene (Neoprene)	Neoprene
	Polyurethane	PU
	Silicon elastomers	Silicones
Hybrids	Carbon-fiber reinforced polymers	CFRP
Composites	Glass-fiber reinforced polymers	GFRP
	SIC reinforced aluminum	AI-SiC
Foams	Flexible polymer foams	Flexible foams
	Rigid polymer foams	Rigid foams
Natural materials	Cork	Cork
	Bamboo	Bamboo
	Wood	Wood

TABLE 3.3 Moduli and Strength of Materials

Material	Young's Modulus (Modulus of Elasticity) –E– (10^6 psi)	Young's Modulus (Modulus of Elasticity) –E– (10^9 N/m^2, GPa)	Ultimate Tensile Strength –S_u– (10^6 N/m^2, MPa)	Yield Strength –S_y– (10^6 N/m^2, MPa)
ABS Plastics		2.3	40	
Acrylic		3.2	70	
Aluminum	10	69	110	95
Antimony	11.3			
Beryllium	42			
Bismuth	4.6			
Bone		9	170 (compression)	
Boron				3100
Brasses		100–125	250	
Bronzes		100–125		
Cadmium	4.6			
Carbon Fiber Reinforced Plastic		150		
Cast Iron 4.5% C, ASTM A-48			170	
Chromium	36			
Cobalt	30			
Concrete, High Strength (compression)		30	40	
Copper	17		220	70
Diamond		1050–1200		
Douglas Fir Wood		13	50 (compression)	
Glass		50–90	50 (compression)	
Gold	10.8			
Iridium	75			
Iron	28.5			
Lead	2			
Magnesium	6.4	45		
Manganese	23			
Marble			15	
Molybdenum	40			
Nickel	31			

Material	Young's Modulus (Modulus of Elasticity) –E– (10^6 psi)	Young's Modulus (Modulus of Elasticity) –E– (10^9 N/m^2, GPa)	Ultimate Tensile Strength –S_u– (10^6 N/m^2, MPa)	Yield Strength –S_y– (10^6 N/m^2, MPa)
Niobium (Columbium)	15			
Nylon			75	45
Oak Wood (along grain)		11		
Osmium	80			
Pine Wood			40	
Platinum	21.3			
Plutonium	14			
Polycarbonate		2.6	70	
Polyethylene HDPE		0.8	15	
Polyethylene Terephthalate (PET)		2–2.7	55	
Polyimide		2.5	85	
Polypropylene		1.5–2	40	
Polystyrene		3–3.5	40	
Rhodium	42			
Rubber		0.01–0.1		
Selenium	8.4			
Silicon Carbide		450		3440
Silver	10.5			
Stainless Steel, AISI 302			860	502
Steel, Structural ASTM-A36		200	400	250
Steel, High Strength Alloy ASTM A-514			760	690
Tantalum	27			
Thorium	8.5			
Titanium	16			
Titanium Alloy		105–120	900	730
Tungsten		400–410		
Tungsten Carbide		450–650		
Uranium	24			
Vanadium	19			
Wrought Iron		190–210		
Zinc	12			

TABLE 3.4 Temperature Effects of Elastic Modulus

Young Modulus of Elasticity –E– (106 psi)															
	Temperature (°C)														
	–200	–129	–73	21	93	149	204	260	316	371	427	482	538	593	649
	Temperature (°F)														
Metal	–325	–200	–100	70	200	300	400	500	600	700	800	900	1000	1100	1200
Cast Iron															
Gray cast iron				13.4	13.2	12.9	12.6	12.2	11.7	11	10.2				
Steel															
Carbon steel C ≤ 0.3%	31.4	30.8	30.2	29.5	28.8	28.3	27.7	27.3	26.7	25.5	24.2	22.4	20.4	18	
Carbon steel C ≥ 0.3%	31.2	30.6	30	29.3	28.6	28.1	27.5	27.1	26.5	25.3	24	22.2	20.2	17.9	15.4
Carbon-moly steels	31.1	30.5	29.9	29.2	28.5	28	27.4	27	26.4	25.3	23.9	22.2	20.1	17.8	15.3
Nickel steels Ni 2%–9%	29.6	29.1	28.5	27.8	27.1	26.7	26.1	25.7	25.2	24.6	23				
Cr-Mo steels Cr ½%–2%	31.6	31	30.4	29.7	29	28.5	27.9	27.5	26.9	26.3	25.5	24.8	23.9	23	21.8
Cr-Mo steels Cr 2¼%–3%	32.6	32	31.4	30.6	29.8	29.4	28.8	28.3	27.7	27.1	26.3	25.6	24.6	23.7	22.5
Cr-Mo steels Cr 5%–9%	32.9	32.3	31.7	30.9	30.1	29.7	29	28.6	28	27.3	26.1	24.7	22.7	20.4	18.2
Chromium steels Cr 12%, 17%, 27%	31.2	30.7	30.1	29.2	28.5	27.9	27.3	26.7	26.1	25.6	24.7	23.2	21.5	19.1	16.6
Austenitic steels (TP304, 310, 316, 321, 347)	30.3	29.7	29.1	28.3	27.6	27	26.5	25.8	25.3	24.8	24.1	23.5	22.8	22.1	21.2
Copper and Copper Alloys															
Comp. and leaded-Sn bronze (C83600, C92200)	14.8	14.6	14.4	14	13.7	13.4	13.2	12.9	12.5	12					
Naval brass Si & A1 bronze (C46400, C65500, C95200, C95400)	15.9	15.6	15.4	15	14.6	14.4	14.1	13.8	13.4	12.8					
Copper (C11000)	16.9	16.6	16.5	16	15.6	15.4	15	14.7	14.2	13.7					
Copper red brass A1-bronze (C10200, C12000, C12200, C125000, C142000, C23000, C61400)	18	17.7	17.5	17	16.6	16.3	16	15.6	15.1	14.5					
Nickel and Nickel Alloys															
Monel 400 (N04400)	27.8	27.3	26.8	26	25.4	25	24.7	24.3	24.1	23.7	23.1	22.6	22.1	21.7	21.2
Titanium															
Unalloyed titanium grades 1, 2, 3, and 7				15.5	15	14.6	14	13.3	12.6	11.9	11.2				
Aluminum and Aluminum Alloys															
Grades 443, 1060, 1100, 3003, 3004, 6063	11.1	10.8	10.5	10	9.6	9.2	8.7								

TABLE 3.5 Extended Mechanical Properties at Room Temperature

Typical Mechanical Properties at Room Temperature					
Metal	Tensile Strength, 1000 lb/in²	Yield Strength 1000 lb/in²	Ultimate Elongation, %	Reduction of Area, %	Brinell No.
Cast iron	18–60	8–40	0	0	100–300
Wrought iron	45–55	25–35	35–25	55–30	100
Commercially pure iron, annealed	42	19	48	85	70
Hot-rolled	48	30	30	75	90
Cold-rolled	100	95			200
Structural steel, ordinary	50–65	30–40	40–30		120
Low-alloy, high-strength	65–90	40–80	30–15	70–40	150
Steel, SAE 1300, annealed	70	40	26	70	150
Quenched, drawn 1300°F	100	80	24	65	200
Drawn 1000°F	130	110	20	60	260
Drawn 700°F	200	180	14	45	400
Drawn 400°F	240	210	10	30	480
Steel, SAE 4340, annealed	80	45	25	70	170
Quenched, drawn 1300°F	130	110	20	60	270
Drawn 1000°F	190	170	14	50	395
Drawn 700°F	240	215	12	48	480
Drawn 400°F	290	260	10	44	580
Cold-rolled steel, SAE 1112	84	76	18	45	160
Stainless steel, 18-S	85–95	30–35	60–65	75–65	145–160
Steel castings, heat-treated	60–125	30–90	33–14	65–20	120–250
Aluminum, pure, rolled	13–24	5–21	35–5		23–44
Aluminum-copper alloys, cast	19–23	12–16	4–0		50–80
Wrought, heat-treated	30–60	10–50	33–15		50–120
Aluminum die castings	30		2		
Aluminum alloy 17ST	56	34	26	39	100
Aluminum alloy 51ST	48	40	20	35	105
Copper, annealed	32	5	58	73	45
Copper, hard-drawn	68	60	4	55	100
Brasses, various	40–120	8–80	60–3		50–170
Phosphor bronze	40–130		55–5		50–200
Tobin bronze, rolled	63	41	40	52	120
Magnesium alloys, various	21–45	11–30	17–0.5		47–78
Monel 400, Ni-Cu alloy	79	30	48	75	125
Molybdenum, rolled	100	75	30		250
Silver, cast, annealed	18	8	54		27
Titanium 6–4 alloy, annealed	130	120	10	25	352
Ductile iron, grade 80-55-06	80	55	6		225–255

TABLE 3.6 Hardness Test Indenters

Test	Indenter	Shape of Indentation: Side View	Shape of Indentation: Top View	Load	Formula for Hardness Number[a]
Brinell	10-mm sphere of steel or tungsten carbide	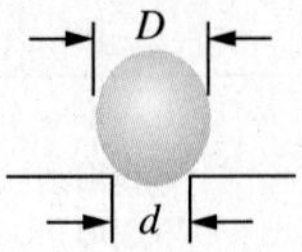	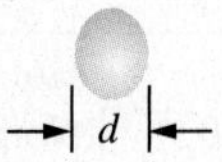	P	$HB = \frac{2P}{\pi D[D - \sqrt{D^2 - d^2}]}$
Vickers microhardness	Diamond pyramid	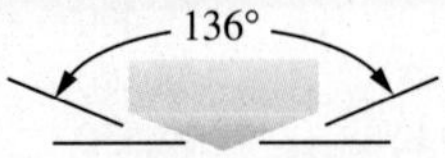	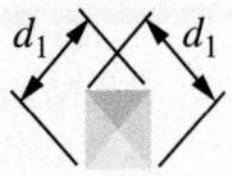	P	$HV = 1.854P/d_1^2$
Knoop microhardness	Diamond pyramid	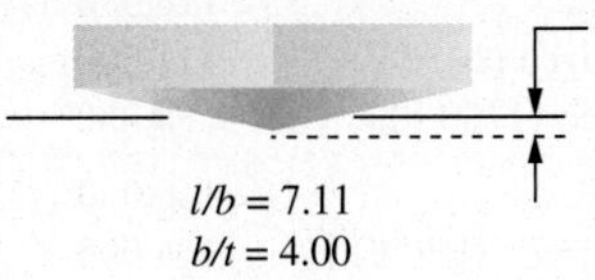	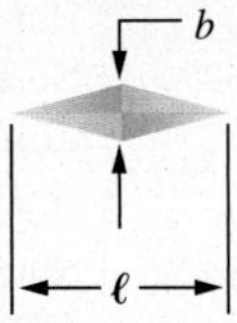	P	$HK = 14.2P/l^2$
Rockwell and superficial Rockwell	Diamond cone	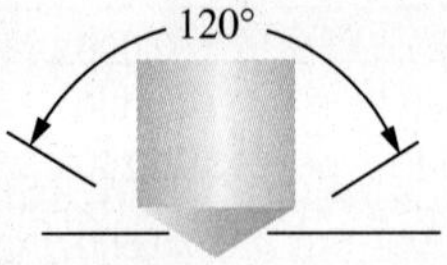		60 kg, 100 kg, 150 kg	Rockwell
	$\frac{1}{16}, \frac{1}{8}, \frac{1}{4}, \frac{1}{2}$ in. diameter steel spheres			15 kg, 30 kg, 45 kg	Superficial Rockwell

[a]For more information refer to Hayden, H. W., W. G. Moffatt, and J. Wulff, *The Structure and Properties of Materials*, Wiley, New York, 1965.

TABLE 3.7 ANSI Carbon Steel Mechanical Characteristics

AISI Reference	Manufactured	Strength: Tensile psi (lb/in.²)	Strength: Yield psi (lb/in.²)	Elongation (%)	Hardness, Bhn
1015	Hot Rolled	61,000	45,500	39.0	126
	Normalized (1700 F)	61,500	47,000	37.0	121
	Annealed (1600 F)	56,000	41,250	37.0	111
1020	Hot Rolled	65,000	48,000	36.0	143
	Normalized (1600 F)	64,000	50,250	35.8	131
	Annealed (1600 F)	57,250	42,750	36.5	111
1022	Hot Rolled	73,000	52,000	35.0	149
	Normalized (1700 F)	70,000	52,000	34.0	143
	Annealed (1600 F)	65,250	46,000	35.0	137
1030	Hot Rolled	80,000	50,000	32.0	179
	Normalized (1700 F)	75,000	50,000	32.0	149
	Annealed (1550 F)	67,250	49,500	31.2	126

(Continued)

TABLE 3.7 ANSI Carbon Steel Mechanical Characteristics (*Continued*)

AISI Reference	Manufactured	Strength: Tensile	Strength: Yield	Elongation (%)	Hardness, Bhn
		psi (lb/in.^{2})			
1040	Hot Rolled	90,000	60,000	25.0	201
	Normalized (1650 F)	85,500	54,250	28.0	170
	Annealed (1450 F)	75,250	51,250	30.2	149
1050	Hot Rolled	105,000	60,000	20.0	229
	Normalized (1650 F)	108,500	62,000	20.0	217
	Annealed (1450 F)	92,250	53,000	23.7	187
1060	Hot Rolled	118,000	70,000	17.0	241
	Normalized (1650 F)	112,500	61,000	18.0	229
	Annealed (1450 F)	90,750	54,000	22.5	179
1080	Hot Rolled	140,000	85,000	12.0	293
	Normalized (1650 F)	146,500	76,000	11.0	293
	Annealed (1450 F)	89,250	54,500	24.7	174
1095	Hot Rolled	140,000	83,000	9.0	293
	Normalized (1650 F)	147,000	72,500	9.5	293
	Annealed (1450 F)	95,250	55,000	13.0	192
1117	Hot Rolled	70,600	44,300	33.0	143
	Normalized (1650 F)	67,750	44,000	33.5	137
	Annealed (1575 F)	62,250	40,500	32.8	121
1118	Hot Rolled	75,600	45,900	32.0	149
	Normalized (1700 F)	69,250	46,250	33.5	143
	Annealed (1450 F)	65,250	41,250	34.5	131
1137	Hot Rolled	91,000	55,000	28.0	192
	Normalized (1650 F)	97,000	57,500	22.5	197
	Annealed (1450 F)	84,750	50,000	26.8	174
1141	Hot Rolled	98,000	52,000	22.0	192
	Normalized (1650 F)	102,500	58,750	22.7	201
	Annealed (1500 F)	86,800	51,200	25.5	163
1144	Hot Rolled	102,000	61,000	21.0	212
	Normalized (1650 F)	96,750	58,000	21.0	197
	Annealed (1450 F)	84,750	50,250	24.8	167
1340	Normalized (1600 F)	121,250	81,000	22.0	248
	Annealed (1475 F)	102,000	63,250	25.5	207
3140	Normalized (1600 F)	129,250	87,000	19.7	262
	Annealed (1500 F)	100,000	61,250	24.5	197
4130	Normalized (1600 F)	97,000	63,250	25.5	197
	Annealed (1585 F)	81,250	52,250	28.2	156
4140	Normalized (1600 F)	148,000	95,000	17.7	302
	Annealed (1500 F)	95,000	60,500	25.7	197
4150	Normalized (1600 F)	167,500	106,500	11.7	321
	Annealed (1500 F)	105,750	55,000	20.2	197
4320	Normalized (1640 F)	115,000	67,250	20.8	235
	Annealed (1560 F)	84,000	61,625	29.0	163

(*Continued*)

TABLE 3.7 ANSI Carbon Steel Mechanical Characteristics (*Continued*)

AISI Reference	Manufactured	Strength Tensile psi (lb/in.²)	Strength Yield psi (lb/in.²)	Elongation (%)	Hardness, Bhn
4340	Normalized (1600 F)	185,500	125,000	12.2	363
	Annealed (1490 F)	108,000	68,500	22.0	217
4620	Normalized (1650 F)	83,250	53,125	29.0	174
	Annealed (1575 F)	74,250	54,000	31.3	149
4820	Normalized (1580 F)	109,500	70,250	24.0	229
	Annealed (1500 F)	98,750	67,250	22.3	197
5140	Normalized (1600 F)	115,000	68,500	22.7	229
	Annealed (1525 F)	83,000	42,500	28.6	167
5150	Normalized (1600 F)	126,250	76,750	20.7	255
	Annealed (1520 F)	98,000	51,750	22.0	197
5160	Normalized (1575 F)	138,750	77,000	17.5	269
	Annealed (1495 F)	104,750	40,000	17.2	197
8630	Normalized (1600 F)	94,250	62,250	23.5	187
	Annealed (1550 F)	81,750	54,000	29.0	156
8650	Normalized (1600 F)	148,500	99,750	14.0	302
	Annealed (1465 F)	103,750	56,000	22.5	212
8740	Normalized (1600 F)	134,750	88,000	16.0	269
	Annealed (1500 F)	100,750	60,250	22.2	201
9255	Normalized (1650 F)	135,250	84,000	19.7	269
	Annealed (1550 F)	112,250	70,500	21.7	229
9310	Normalized (1630 F)	131,500	82,750	18.8	269
	Annealed (1550 F)	119,000	63,750	17.3	241

TABLE 3.8 Coefficients of Thermal Expansion

Product	Linear Temperature Expansion Coefficient -α- (10^{-6} m/m K)	(10^{-6} in/in °F)	Product	Linear Temperature Expansion Coefficient -α- (10^{-6} m/m K)	(10^{-6} in/in °F)	Product	Linear Temperature Expansion Coefficient -α- (10^{-6} m/m K)	(10^{-6} in/in °F)	Product	Linear Temperature Expansion Coefficient -α- (10^{-6} m/m K)	(10^{-6} in/in °F)
ABS (Acrylonitrile butadiene styrene) thermoplastic	73.8	41	Invar	1.5	0.8	Polyester-glass fiber-reinforced	25	14	Ethylene vinyl acetate (EVA)	180	100
ABS-glass fiber-reinforced	30.4	17	Iridium	6.4	3.6	Polyethylene (PE)	200	111	Europium	35	19.4
Acetal	106.5	59.2	Iron, pure	12	6.7	Polyethylene (PE)-high molecular weight		60	Fluoroethylene propylene (FEP)	135	75

(*Continued*)

TABLE 3.8 Coefficients of Thermal Expansion (*Continued*)

Product	Linear Temperature Expansion Coefficient -α-		Product	Linear Temperature Expansion Coefficient -α-		Product	Linear Temperature Expansion Coefficient -α-		Product	Linear Temperature Expansion Coefficient -α-	
	(10^{-6} m/m K)	(10^{-6} in/in °F)		(10^{-6} m/m K)	(10^{-6} in/in °F)		(10^{-6} m/m K)	(10^{-6} in/in °F)		(10^{-6} m/m K)	(10^{-6} in/in °F)
Acetal-glass fiber-reinforced	39.4	22	Iron, cast	10.4	5.9	Polyethylene terephthalate (PET)	59.4	33	Gadolinium	9	5
Acrylic, sheet, cast	81	45	Iron, forged	11.3	6.3	Polyphenylene-glass fiber-reinforced	35.8	20	Germanium	6.1	3.4
Acrylic, extruded	234	130	Lanthanum	12.1	6.7	Polypropylene (PP), unfilled	90.5	50.3	Glass, hard	5.9	3.3
Alumina	5.4	3	Lead	28	15.1	Polypropylene-glass fiber-reinforced	32	18	Glass, Pyrex	4	2.2
Aluminum	22.2	12.3	Limestone	8	4.4	Polystyrene (PS)	70	38.9	Glass, plate	9	5
Antimony	10.4	5.8	Lithium	46	25.6	Polysulfone (PSO)	55.8	31	Gold	14.2	8.2
Arsenic	4.7	2.6	Lutetium	9.9	5.5				Granite	7.9	4.4
Barium	20.6	11.4	Magnesium	25	14	Polyurethane (PUR), rigid	57.6	32	Graphite, pure	7.9	4.4
Beryllium	11.5	6.4	Manganese	22	12.3				Ice	51	28.3
Bismuth	13	7.3	Marble	5.5–14.1	3.1–7.9	Polyvinyl chloride (PVC)	50.4	28	Inconel	12.6	7
Brass	18.7	10.4	Masonry	4.7–9.0	2.6–5.0	Polyvinylidene fluoride (PVDF)	127.8	71	Indium	33	18.3
Brick masonry	5.5	3.1	Mica	3	1.7	Porcelain	4.5	2.5	Wood, oak parallel to grain	4.9	2.7
Bronze	18	10	Molybdenum	5	2.8	Potassium	83	46.1	Wood, oak across to grain	5.4	3
Cadmium	30	16.8	Monel	13.5	7.5	Praseodymium	6.7	3.7	Wood, pine	5	2.8
Calcium	22.3	12.4	Mortar	7.3–13.5	4.1–7.5	Promethium	11	6.1	Ytterbium	26.3	14.6
Carbon-diamond	1.2	0.67	Neodymium	9.6	5.3	Quartz	0.77–1.4	0.43–0.79	Yttrium	10.6	5.9
Cast Iron Gray	10.8	6	Nickel	13	7.2	Rhenium	6.7	3.7	Zinc	29.7	16.5
Cellulose acetate (CA)	130	72.2	Niobium (Columbium)	7	3.9	Rhodium	8	4.5	Zirconium	5.7	3.2
Cellulose acetate butynate (CAB)		80–95	Nylon, general purpose	72	40	Rubber, hard	77	42.8	Terne	11.6	6.5
Cellulose nitrate (CN)	100	55.6	Nylon, Type 11, molding and extruding compound	100	55.6	Ruthenium	9.1	5.1	Thallium	29.9	16.6
Cement	10	6	Nylon, Type 12, molding and extruding compound	80.5	44.7	Samarium	12.7	7.1	Thorium	12	6.7

(*Continued*)

TABLE 3.8 Coefficients of Thermal Expansion (*Continued*)

Product	Linear Temperature Expansion Coefficient -α- (10^{-6} m/m K)	(10^{-6} in/in °F)	Product	Linear Temperature Expansion Coefficient -α- (10^{-6} m/m K)	(10^{-6} in/in °F)	Product	Linear Temperature Expansion Coefficient -α- (10^{-6} m/m K)	(10^{-6} in/in °F)	Product	Linear Temperature Expansion Coefficient -α- (10^{-6} m/m K)	(10^{-6} in/in °F)
Cerium	5.2	2.9	Nylon, Type 6, cast	85	47.2	Sandstone	11.6	6.5	Thulium	13.3	7.4
Chlorinated polyvinylchloride (CPVC)	66.6	37	Nylon, Type 6/6, molding compound	80	44.4	Scandium	10.2	5.7	Tin	23.4	13
Chromium	6.2	3.4	Osmium	5	2.8	Selenium	3.8	2.1	Titanium	8.6	4.8
Clay tile structure	5.9	3.3	Palladium	11.8	6.6	Silicon	5.1	2.8	Tungsten	4.3	2.4
Cobalt	12	6.7	Phenolic resin without fillers	80	44.4	Silver	19.5	10.7	Uranium	13.9	7.7
Concrete	14.5	8	Plaster	16.4	9.2	Slate	10.4	5.8	Vanadium	8	4.5
Concrete structure	9.8	5.5	Platinum	9	5	Sodium	70	39.1	Vinyl ester	16–22	8.7–12
Constantan	18.8	10.4	Plutonium	54	30.2	Solder 50–50	24	13.4	Wood, fir	3.7	2.1
Copper	16.6	9.3	Polyallomer	91.5	50.8	Steatite	8.5	4.7	Epoxy, castings, resins, and compounds, unfilled	55	31
Copper, Beryllium 25	17.8	9.9	Polyamide (PA)	110	61.1	Steel	13	7.3	Erbium	12.2	6.8
Corundum, sintered	6.5	3.6	Polybutylene (PB)		72	Steel Stainless Austenitic (304)	17.3	9.6	Ethylene ethyl acrylate (EEA)	205	113.9
Cupronickel 30%	16.2	9	Polycarbonate (PC)	70.2	39	Steel Stainless Austenitic (310)	14.4	8	Hard alloy K20	6	3.3
Diamond	1.1	0.6	Polycarbonate-glass fiber-reinforced	21.5	12	Steel Stainless Austenitic (316)	16	8.9	Hastelloy C	11.3	6.3
Dysprosium	9.9	5.5	Polyester	123.5	69	Steel Stainless Ferritic (410)	9.9	5.5	Holmium	11.2	6.2
Ebonite	76.6	42.8	Hafnium	5.9	3.3	Strontium	22.5	12.5	Tellurium	36.9	20.5

TABLE 3.9 Elastic Constants of Selected Polycrystalline Ceramics

Elastic Constants of Selected Polycrystalline Ceramics (20°C)					
Material	Crystal Type	μ (GPa)	B (GPa)	ν	E (GPa)
Carbides					
C	Cubic	468	416	0.092	1022
SiC	Cubic	170	210	0.181	402
TaC	Cubic	118	217	0.270	300
TiC	Cubic	182	242	0.199	437
ZrC	Cubic	170	223	0.196	407

(*Continued*)

TABLE 3.9 Elastic Constants of Selected Polycrystalline Ceramics (*Continued*)

Elastic Constants of Selected Polycrystalline Ceramics (20°C)					
Material	Crystal Type	μ (GPa)	B (GPa)	ν	E (GPa)
Oxides					
Al_2O_3	Trigonal	163	251	0.233	402
$Al_2O_3 \cdot MgO$	Cubic	107	195	0.268	271
$BaO \cdot TiO_2$	Tetragonal	67	177	0.332	178
BeO	Tetragonal	165	224	0.204	397
CoO	Cubic	70	185	0.332	186
$FeO \cdot Fe_2O_3$	Cubic	91	162	0.263	230
Fe_2O_3	Trigonal	93	98	0.140	212
MgO	Cubic	128	154	0.175	300
$2MgO \cdot SiO_2$	Orthorhombic	81	128	0.239	201
MnO	Cubic	66	154	0.313	173
SrO	Cubic	59	82	0.210	143
$SrO \cdot TiO_2$	Cubic	266	183	0.010	538
TiO_2	Tetragonal	113	206	0.268	287
UO_2	Cubic	87	212	0.319	230
ZnO	Hexagonal	45	143	0.358	122
ZrO_2–$12Y_2O_3$	Cubic	89	204	0.310	233
SiO_2	Trigonal	44	38	0.082	95
Chalcogenides					
CdS	Hexagonal	15	59	0.38	42
PbS	Cubic	33	62	0.27	84
ZnS	Cubic	33	78	0.31	87
PbTe	Cubic	22	41	0.27	56
Fluorides					
BaF_2	Cubic	25	57	0.31	65
CaF_2	Cubic	42	88	0.29	108
SrF_2	Cubic	35	70	0.29	90
LiF	Cubic	48	67	0.21	116
NaF	Cubic	31	49	0.24	77
Other Halides					
CsBr	Cubic	8.8	16	0.26	23
CsCl	Cubic	10	18	0.27	25
Csl	Cubic	7.1	13	0.27	18
KCl	Cubic	10	18	0.27	25
NaBr	Cubic	11	19	0.26	28
NaCl	Cubic	15	25	0.25	38
Nal	Cubic	8.5	15	0.27	20
RbCl	Cubic	7.5	16	0.29	21

TABLE 3.10 Mechanical Properties of Some Constructional Steels*

ASTM Designation	Thickness Range, mm (in)	Yield Point, min		Tensile Strength		Elongation in 200 mm (8 in) min, %	Suitable for Welding?
		MPa	1000 lb/in²	MPa	1000 lb/in²		
Structural Carbon-Steel Plates							
ASTM A36	To 100 mm (4 in), incl.	248	36	400–552	58–80	20	Yes
Low- and Intermediate-Tensile-Strength Carbon-Steel Plates							
ASTM A283	(structural quality)						
Grade A	All thicknesses	165	24	310	45	28	Yes
Grade B	All thicknesses	186	27	345	50	25	Yes
GradeC	All thicknesses	207	30	379	55	22	Yes
Grade D	All thicknesses	228	33	414	60	20	Yes
Carbon-Silicon Steel Plates for Machine Parts and General Construction							
ASTM A284							
Grade A	To 305 mm (12 in)	172	25	345	50	25	Yes
Grade B	To 305 mm (12 in)	159	23	379	55	23	Yes
Grade C	To 305 mm (12 in)	145	21	414	60	21	Yes
Grade D	To 200 mm (8 in)	145	21	414	60	21	Yes
Carbon-Steel Pressure-Vessel Plates							
ASTM A285							
Grade A	To 50 mm (2 in)	165	24	303–379	44–55	27	Yes
Grade B	To 50 mm (2 in)	186	27	345–414	50–60	25	Yes
Grade C	To 50 mm (2 in)	207	30	379–448	55–65	23	Yes
Structural Steel for Locomotives and Railcars							
ASTM A113							
Grade A	All thicknesses	228	33	414–496	60–72	21	No
Grade B	All thicknesses	186	27	345–427	50–62	24	No
Grade C	All thicknesses	179	26	331–400	48–58	26	No
Structural Steel for Ships							
ASTM A131 (all grades)		221	32	400–490	58–71	21	No
Heat-Treated Constructional Alloy-Steel Plates							
ASTM A514	To 64 mm (2 in), incl.	700	100	800–950	115–135	18†	Yes
	Over 64 to 102 mm (2 to 4 in), incl.	650	90	750–950	105–135	17†	Yes

*See appropriate ASTM documents for properties of other plate steels, shapes, bars, wire, tubing, etc.
†Elongation in 50 mm (2 in), min.

TABLE 3.11 Representative Average Mechanical Properties of Cold-Drawn Steel

SAE No.	Tensile Strength MPa	Tensile Strength 1000 lb/in²	Yield Strength MPa	Yield Strength 1000 lb/in²	Elongation in 50 mm (2 in), %	Reduction of Area, %	Brinell Hardness
1010	462	67	379	55.0	25.0	57	137
1015	490	71	416	60.3	22.0	55	149
1020	517	75	439	63.7	20.0	52	156
1025	552	80	469	68.0	18.5	50	163
1030	600	87	509	73.9	17.5	48	179
1035	634	92	539	78.2	17.0	45	187
1040	669	97	568	82.4	16.0	40	197
1045	703	102	598	86.7	15.0	35	207
1117	552	80	469	68.0	19.0	51	163
1118	569	82.5	483	70.1	18.5	50	167
1137	724	105	615	89.2	16.0	35	217
1141	772	112	656	95.2	14.0	30	223

Sizes 16 to 50 mm (% to 2 in) diameter, test specimens 50 × 13 mm (2 × 0.505 in).
Source: ASM "Metals Hand book."

TABLE 3.12 Material Properties of Fibers Used in Composites

	Graphite Fiber	Glass Fiber*	Polymer Matrix*
E_1 (GPa)	233	73.1	4.62
E_2 (GPa)	23.1	73.1	4.62
G_{12} (GPa)	8.96	30.0	1.699
ν_{12}	0.200	0.22	0.36
α_1(/°C)	-0.540×10^{-6}	5.04	41.4×10^{-6}
α_2(/°C)	10.10×10^{-6}	5.04	41.4×10^{-6}

*Assumed to be isotropic.

3.15 REFERENCES

1. Roark, R. J., *Formulas for Stress and Strain*, 4th ed., McGraw-Hill, 1965.
2. U.S. Dept. of Agriculture, *Wood Handbook*, Forest Products Laboratory, 1987.
3. Abramson, H. N., H. Leibowitz, J. M. Crowley, and S. Juhasz (eds.), *Applied Mechanics Surveys*, Spartan Books, 1966.
4. Conway, J. B., *Stress-Rupture Parameters: Origin, Calculation, and Use*, Gordon and Breach Science Publishers, 1969.
5. Conway, J. B., *Numerical Methods for Creep and Rupture Analyses*, Gordon and Breach Science Publishers, 1967.
6. Conway, J. B., and P. N. Flagella, *Creep-Rupture Data for the Refractory Metals to High Temperatures*, Gordon and Breach Science Publishers, 1971.
7. Odqvist, F. K. G., *Mathematical Theory of Creep and Creep Rupture*, Oxford University Press, 1966.
8. Hult, J. A. H., *Creep in Engineering Structures*, Blaisdell, 1966.

9. Penny, R. K., and D. L. Marriott, *Design for Creep*, McGraw-Hill, 1971.
10. Smith, A. I., and A. M. Nicolson (eds.), *Advances in Creep Design: The A. E. Johnson Memorial Volume*, Applied Science Publishers, 1971.
11. Davis, R. E., H. E. Davis, and J. S. Hamilton, "Plastic Flow of Concrete under Sustained Stress," *Proc. ASTM*, 34(II): 854, 1934.
12. Report of Committee on Materials of Construction, *Bull. Assoc. State Eng. Soc.*, July 1934.
13. Soderberg, R., "Working Stresses," ASME Paper A-106, *J. Appl. Mech.*, 2(3): 1935.
14. Nadai, A., "Theories of Strength," ASME Paper APM 55-15, *J. Appl. Mech.*, 1(3): 1933.
15. Washa, G. W., and P. G. Fluck, "Effect of Sustained Loading on Compressive Strength and Modulus of Elasticity of Concrete," *J. Am. Concr. Inst.*, 46, May 1950.
16. Neville, A. M., *Creep of Concrete: Plain, Reinforced, and Prestressed*, North-Holland, 1970.
17. Designing for Effects of Creep, Shrinkage, Temperature in Concrete Structures, *Am. Concr. Inst. Publ.* SP-27, 1971.
18. *Fatigue Design Handbook*, Society of Automotive Engineers, Inc., 1968.
19. Structural Fatigue in Aircraft, *ASTM Spec. Tech. Publ.* 404, 1965.
20. Fatigue Crack Propagation, *ASTM Spec. Tech. Publ.* 415, 1966.
21. Smith. J. O., "The Effect of Range of Stress on Fatigue Strength," *Univ. Ill., Eng. Exp. Sta. Bull.* 334, 1942.
22. Horger, O. J., and H. R. Neifert, "Fatigue Strength of Machined Forgings 6 to 7 Inches in Diameter," *Proc. ASTM*, 39, 1939.
23. Eaton, F. C., "Fatigue Tests of Large Alloy Steel Shafts," *Symposium on Large Fatigue Testing Machines and their Results*, ASTM Spec. Tech. Publ. 216, 1957.
24. Jiro, H., and A. Junich, "Studies on Rotating Beam Fatigue of Large Mild Steel Specimens," *Proc. 9th Japan National Congress Appl. Mech.*, 1959.
25. Gould, A. J., Corrosion Fatigue (in Ref. 32).
26. Weibull, W., "Investigations into Strength Properties of Brittle Materials," *Proc. B. Swed. Inst. Eng. Res.*, 149, 1938.
27. Sines, G., and J. L. Waisman (eds.), "*Metal Fatigue*," McGraw-Hill, 1959.
28. Heywood, R. B., *Designing against Fatigue of Metals*, Reinhold, 1962.
29. Yen, C. S., and T. J. Dolan, "A Critical Review of the Criteria for Notch Sensitivity in Fatigue of Metals," *Univ. Ill., Exp. Sta. Bull.* 398, 1952.
30. Mowbray, A. Q., Jr., "The Effect of Superposition of Stress Raisers on Members Subjected to Static or Repeated Loads," *Proc. Soc. Exp. Stress Anal*, 10(2): 1953.
31. Forrest, P. G., *Fatigue of Metals*, Pergamon Press, Addison-Wesley Series in Metallurgy and Materials, 1962.
32. International Conference on Fatigue of Metals, Institution of Mechanical Engineers, London, and American Society of Mechanical Engineers, New York, 1956.
33. Harting, D. R., "The S–N Fatigue Life Gage: A Direct Means of Measuring Cumulative Fatigue Damage," *Exp. Mech.*, 6(2), February 1966.
34. Descriptive literature, Micro-Measurements, Inc., Romulus, Michigan.
35. Roark, R. J., R. S. Hartenberg, and R. Z. Williams, "The Influence of Form and Scale on Strength," *Univ. Wis. Exp. Ste. Bull.*, 84, 1938.
36. Battelle Memorial Institute, *Prevention of Fatigue of Metals*, John Wiley & Sons, 1941.
37. Neuber, H., *Theory of Notch Stresses*," J. W. Edwards, Publisher, Incorporated, 1946.
38. Moore, H. F., "A Study of Size Effect and Notch Sensitivity in Fatigue Tests of Steel," *Proc. Am. Soc. Test. Mater.*, 45, 1945.
39. Metal Fatigue Damage: Mechanism, Detection, Avoidance, and Repair, *ASTM Spec. Tech. Publ.*, 495, 1971.
40. Cyclic Stress-Strain Behavior: Analysis, Experimentation, and Failure Prediction, *ASTM Spec. Tech. Publ.*, 519, 1973.
41. Effect of Notches on Low-Cycle Fatigue: A Literature Survey, *ASTM Spec. Tech. Publ.*, 490, 1972.
42. Pilkey, W. D., and D. F. Pilkey, *Peterson's Stress Concentration Factors*, 3rd ed., John Wiley & Sons, 2008.
43. Vicentini, V., "Stress-Concentration Factors for Superposed Notches," *Exp. Mech.*, 7(3), March 1967.
44. Sandor, B. I., *Fundamentals of Cyclic Stress and Strain*, The University of Wisconsin Press, 1972.
45. Fuchs, H. O., and R. I. Stephens, *Metal Fatigue in Engineering*, John Wiley & Sons, 1980.
46. "Fatigue and Microstructure," papers presented at the 1978 ASM Materials Science Seminar, American Society for Metals, 1979.
47. Pook, L. P., *The Role of Crack Growth in Metal Fatigue*, The Metals Society, London, 1983.
48. Ritchie, R. O., and J. Larkford, (eds.), "Small Fatigue Cracks," *Proceedings of the Second Engineering Foundation International Conference/Workshop*, The Metallurgical Society, Inc., Santa Barbara, California, Jan. 5–10, 1986.
49. *Int. J. Fracture*, Martinus Nijhoff.
50. *Eng. Fracture Mech.*, Pergamon Journals.
51. *Journal of Reinforced Plastics and Composites*, Technomic Publishing.

52. Liebowitz, H. (ed.), *Fracture*, Academic Press, 1968.
53. Sih, G. C., *Handbook of Stress Intensity Factors*, Institute of Fracture and Solid Mechanics, Lehigh University, 1973.
54. Kobayashi, A. S. (ed.), *Experimental Techniques in Fracture Mechanics, 1 and 2*, Iowa State University Press, 1973 and 1975.
55. Broek, D., *Elementary Engineering Fracture Mechanics*, 3rd ed., Martinus Nijhoff, 1982.
56. Atluri, S. N. (ed.), *Computational Methods in the Mechanics of Fracture*, North-Holland, 1984.
57. Sih, G. C., E. Sommer, and W. Dahl (eds.), *Application of Fracture Mechanics to Materials and Structures*, Martinus Nijhoff, 1984.
58. Kobayashi, A. S. (ed.), *Handbook on Experimental Mechanics*, Prentice-Hall, 1987.
59. Budynas, R. G., *Advanced Strength and Applied Stress Analysis*, 2nd ed., McGraw-Hill, 1999.
60. Schwartz, M. M., *Composite Materials Handbook*, 2nd ed., McGraw-Hill, 1992.
61. Budynas, R. G., and J. K. Nisbett, *Shigley's Mechanical Engineering Design*, 8th ed., McGraw-Hill, 2010.
62. Sadegh, A. M., and W. M. Worek, *Marks' Standard Handbook for Mechanical Engineers*, 12th ed., McGraw-Hill, 2018.
63. www.engineersedge.com, Tables for mechanical properties of materials.
64. *ASM Handbook 20: Materials Selection and Design*, 1997.
65. Vidosic, J. P., *Machine Design Project*, The Ronald Press, New York, 1957.

CHAPTER 4

Principles and Analytical Methods

Most of the formulas of mechanics of materials express the relations among the form and dimensions of a member, the loads applied thereto, and the resulting stress or deformation. Any such formula is valid only within certain limitations and is applicable only to certain problems. An understanding of these limitations and of the way in which formulas may be combined and extended for the solution of problems to which they do not immediately apply requires a knowledge of certain principles and methods that are stated briefly in this chapter. The significance and use of these principles and methods are illustrated in Chaps. 7 to 16 by examples that accompany the discussion of specific problems.

4.1 EQUATIONS OF MOTION AND OF EQUILIBRIUM

The relations that exist at any instant between the motion of a body and the forces acting on it may be expressed by these two equations: (1) F_x (the component along any line x of all forces acting on a body) $= m\bar{a}_x$ (the product of the mass of the body and the x component of the acceleration of its mass center); (2) T_x (torque about any line x of all forces acting on the body) $= dH_x/dt$ (the time rate at which its angular momentum about that line is changing). If the body in question is in equilibrium, these equations reduce to (1) $F_x = 0$ and (2) $T_x = 0$.

These equations, Hooke's law, and experimentally determined values of the elastic constants E, G, and ν constitute the basis for the mathematical analysis of most problems of mechanics of materials. The majority of the common formulas for stress are derived by considering a portion of the loaded member as a body in equilibrium under the action of forces that include the stresses sought and then solving for these stresses by applying the equations of equilibrium.

4.2 PRINCIPLE OF SUPERPOSITION

With certain exceptions, the effect (stress, strain, or deflection) produced on an elastic system by any final state of loading is the same whether the forces that constitute that loading are applied simultaneously or in any given sequence and is the result of the effects that the several forces would produce if each acted singly.

An exception to this principle is afforded by any case in which some of the forces cause a deformation that enables other forces to produce an effect they would not have otherwise.

A beam subjected to transverse and axial loading is an example; the transverse loads cause a deflection that enables the longitudinal load to produce a bending effect it would not produce if acting alone. In no case does the principle apply if the deformations are so large as to alter appreciably the geometrical relations of the parts of the system.

The principle of superposition is important and has many applications. It often makes it possible to resolve or break down a complex problem into a number of simple ones, each of which can be solved separately for like stresses, deformations, etc., which are then algebraically added to yield the solution of the original problem.

4.3 PRINCIPLE OF RECIPROCAL DEFLECTIONS

Let A and B be any two points of an elastic system. Let the displacement of B in any direction U due to force P acting in any direction V at A be u; and let the displacement of A in the direction V due to a force Q acting in the direction U at B be v. Then $Pv = Qu$.

This is the general statement of the *principle of reciprocal deflections.* If P and Q are equal and parallel and u and v are parallel, the statement can be simplified greatly. Thus, for a horizontal beam with vertical loading and deflection understood, the principle expresses the following relation: A load applied at any point A produces the same deflection at any other point B as it would produce at A if applied at B.

The principle of reciprocal deflections is a corollary of the principle of superposition and so can be applied only to cases for which that principle is valid. It can be used to advantage in many problems involving deformation. Examples of the application of the principle are given in Chaps. 8 and 11.

4.4 METHOD OF CONSISTENT DEFORMATIONS (STRAIN COMPATIBILITY)

Many statically indeterminate problems are easily solved by utilizing the obvious relations among the deformations of the several parts or among the deformations produced by the several loads. Thus, the division of load between the parts of a composite member is readily ascertained by expressing the deformation or deflection of each part in terms of the load it carries and then equating these deformations or deflections. For example, the reaction at the supported end of a beam with one end fixed and the other supported can be found by regarding the beam as a cantilever, acted on by the actual loads and an upward end load (the reaction), and setting the resultant deflection at the support end equal to zero.

The method of consistent deformations is based on the principle of superposition; it can be applied only to cases for which that principle is valid.

4.5 ENERGY METHODS

In addition to the differential equations method the analysis of stress and deformation can be accomplished through the use of energy methods. These methods are based on minimization of energy associated with deformation and stress. Energy methods are effective in problems involving complex shapes and cross sections. In particular, strain energy methods offer simple approaches for computation of displacements of structural and machine elements subjected to combined loading.

Principles and Methods Involving Strain Energy

Strain energy is defined as the mechanical energy stored up in an elastically stressed system; formulas for the amount of strain energy developed in members under various conditions of loading are given in Chaps. 7 to 15. It is the purpose of this article to state certain relations between strain energy and external forces that are useful in the analysis of stress and deformation. For convenience, external forces with points of application that do not move will here be called *reactions*, and external forces with points of application that move will be called *loads*.

External Work Equal to Strain Energy

When an *elastic* system is subjected to static loading, the external work done by the loads as they increase from zero to their maximum value is equal to the strain energy acquired by the system.

This relation may be used directly to determine the deflection of a system under a single load; for such a case, assuming a linear material, it shows that the deflection at the point of loading in the direction of the load is equal to twice the strain energy divided by the load. The relationship also furnishes a means of determining the critical load that produces elastic instability in a member. A reasonable form of curvature, compatible with the boundary conditions, is assumed, and the corresponding critical load found by equating the work of the load to the strain energy developed, both quantities being calculated for the curvature assumed. For each such assumed curvature, a corresponding approximate critical load will be found and the least load so found represents the closest approximation to the true critical load (see Refs. 3–5).

4.6 CASTIGLIANO'S THEOREM

In 1873 Alberto Castigliano, an Italian engineer, employed Betti's reciprocal theorem and developed a formula for linearly elastic structures that is known as Castigliano's Theorem. The theorem states that for a linear structure, the partial derivative of the strain energy, U, with respect to external load, P_i, is equal to the deflection, δ_i, of the structure at the point of the application and in the direction of that load. In equation form, the theorem is

$$\frac{\partial U}{\partial P_i} = \delta_i$$

Similarly, in the case of an applied moment, Castigliano's theorem can also be written as

$$\frac{\partial U}{\partial C_i} = \theta_i$$

where, C_i is the bending moment or twisting torque and θ_i is the associated rotation, slope, or angle of twist at the point of applied moment (see Refs. 8–9).

In applying Castigliano's theorem, the strain energy must be expressed as a function of a load. If the structure is subjected to a combination of loads, for example, axial load N, bending moment M, shearing force V, torsional moment T, the strain energy is

$$U = \int \frac{N^2 dx}{2AE} + \int \frac{M^2 dx}{2EI} + \int \frac{\alpha V^2 dx}{2AG} + \int \frac{T^2 dx}{2JG}$$

where α is the shape factor and is equal to 10/9, 2, and 6/5 for circle, thin walled circular and rectangular respectively. For beams with I-section, box section, or channels the value of α is equal to A/A_{web} where A is the entire area of the cross section and $A_{web} = ht$ is the area of the web, assuming that h is the beam depth and t is the web thickness.

Note that the last integral is valid only for a circular cross section.* Using Castigliano's theorem, the displacement δ_i is given by

$$\delta_i = \frac{1}{AE}\int N\frac{\partial N}{\partial P_i}dx + \frac{1}{EI}\int M\frac{\partial M}{\partial P_i}dx + \frac{1}{AG}\int \alpha V\frac{\partial V}{\partial P_i}dx + \frac{1}{JG}\int T\frac{\partial T}{\partial P_i}dx$$

Similarly, the angle of rotation may be written as:

$$\theta_i = \frac{1}{AE}\int N\frac{\partial N}{\partial C_i}dx + \frac{1}{EI}\int M\frac{\partial M}{\partial C_i}dx + \frac{1}{AG}\int \alpha V\frac{\partial V}{\partial C_i}dx + \frac{1}{JG}\int T\frac{\partial T}{\partial C_i}dx$$

EXAMPLE A cantilever beam of constant flexural rigidity, EI, and area A is designed to carry P_v and P_h loads as shown in Fig. 4.1. Use Castigliano's theorem and determine the angular rotation of the free end, point B. Also, find the horizontal and vertical deflections at point B. Neglect the shear effect.

SOLUTION Apply a moment C at point B. Denote vertical and horizontal parts of the beam as sections 1 and 2, respectively, as shown in Fig. 4.2. The moments M_1 and M_2, and the loads N_1 and N_2 at sections 1 and 2, respectively, are

$$M_1 = P_h x_1 + C \quad \text{and} \quad N_1 = 0$$

$$M_2 = P_v x_2 + P_h h + C \quad \text{and} \quad N_2 = -P_h$$

Use Castigliano's theorem,

$$\frac{\partial U}{\partial C_i} = \theta_i$$

Thus,

$$\theta_i = \frac{\partial U}{\partial C_i} = \frac{1}{EI}\left(\int M\frac{\partial M}{\partial C}dx\right) + \frac{1}{EA}\left(\int N\frac{\partial N}{\partial C}dx\right)$$

$$\frac{\partial N}{\partial C} = 0, \quad \frac{\partial M_1}{\partial C} = 1, \quad \frac{\partial M_2}{\partial C} = 1$$

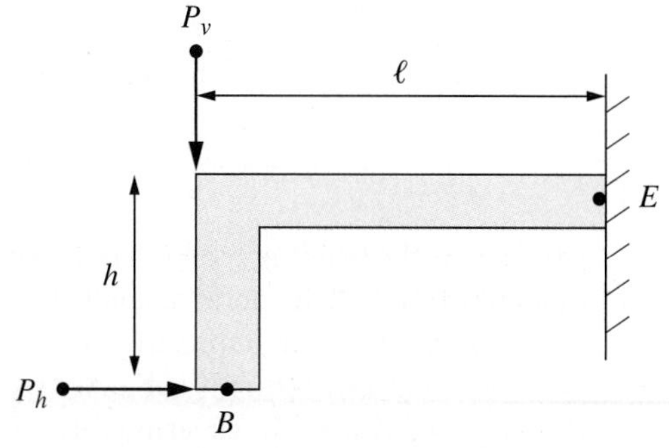

Figure 4.1

*For noncircular cross-sections, see Sec. 10.2 and Table 10.7 for equivalent J terms (called K).

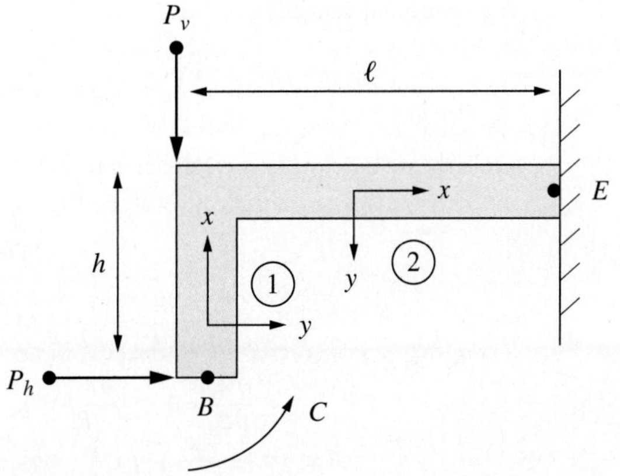

Figure 4.2

Then using the theorem, we solve for θ

$$\theta = \frac{1}{EI}\left(\int_0^h M_1 \frac{\partial M_1}{\partial C} dx_1 + \int_0^l M_2 \frac{\partial M_2}{\partial C} dx_2\right)$$

$$= \frac{1}{EI}\left[\int_0^h (P_h x_1 + C)(1) dx_1 + \int_0^l (P_v x_2 + P_h h + C)(1) dx_2\right]$$

$$= \frac{1}{EI}\left(\frac{1}{2}P_h h^2 + Ch + \frac{1}{2}P_v l^2 + P_h hl + Cl\right)$$

Since $C = 0$, then

$$\theta = \frac{1}{EI}\left(\frac{1}{2}P_h h^2 + \frac{1}{2}P_v l^2 + P_h hl\right) = \frac{1}{2EI}\left[P_h h[h + 2l] + P_v l^2\right]$$

Using Castigliano's theorem,

$$\frac{\partial U}{\partial P_h} = \delta_h$$

Thus,

$$\delta_h => = \frac{\partial U}{\partial P_h} = \frac{1}{EI}\left(\int M \frac{\partial M}{\partial P_h} dx\right) + \frac{1}{EA}\left(\int N \frac{\partial N}{\partial P_h} dx\right)$$

$$\frac{\partial N_1}{\partial P_h} = 0, \quad \frac{\partial M_1}{\partial P_h} = \mathrm{X}_1, \quad \frac{\partial M_2}{\partial P_h} = h, \quad \frac{\partial N_2}{\partial P_h} = -P_h$$

For the vertical and horizontal deflections of B, the moment C is unnecessary, so, set $C = 0$. Then using the theorem, we solve for δ_h

$$\delta_h = \frac{1}{EI}\left(\int_0^h M_1 \frac{\partial M_1}{\partial P_h} dx_1 + \int_0^l M_2 \frac{\partial M_2}{\partial P_h} dx_2\right) + \frac{1}{EA}\left(\int_0^h N_1 \frac{\partial N_1}{\partial P_h} dx_1 + \int_0^l N_2 \frac{\partial N_2}{\partial P_h} dx_2\right)$$

$$= \frac{1}{EI}\left[\int_0^h (P_h x_1) x_1 \, dx_1 + \int_0^l (P_v x_2 + P_h h) h \, dx_2\right] + \frac{1}{EA}\left[\int_0^l (-P_h)(-1) dx_2\right]$$

$$= \frac{1}{EI}\left(\frac{1}{3}P_h h^3 + \frac{1}{2}P_v hl^2 + P_h h^2 l\right) + \frac{P_h l}{EA}$$

The horizontal deflection is thus

$$\delta_h = \frac{1}{6EI}[2P_h h^2(h+3l)+3P_v hl^2]+\frac{P_h l}{EA}$$

Similarly, we have the vertical deflection as

$$\frac{\partial U}{\partial P_v} = \delta_v$$

$$\delta_v = \frac{\partial U}{\partial P_v} = \frac{1}{EI}\left(\int M\frac{\partial M}{\partial P_v}dx\right)+\frac{1}{EA}\left(\int N\frac{\partial N}{\partial P_v}dx\right)$$

$$\frac{\partial N_1}{\partial P_v}=0, \quad \frac{\partial M_1}{\partial P_v}=0, \quad \frac{\partial M_2}{\partial P_v}=x_2, \quad \frac{\partial N_2}{\partial P_v}=0;$$

$$\delta_v = \frac{1}{EI}\left(\frac{1}{3}P_v l^3+\frac{1}{2}P_h hl^2\right)$$

EXAMPLE Determine the vertical deflection of point *B* of the truss as shown in Fig. 4.3.

SOLUTION We apply a force *F* at point *B* in the vertical direction. The forces on each element of the truss can be calculated and are shown in Fig. 4.4.

Using Castigliano's theorem, the strain energy is

$$U=\int\frac{N^2dx}{2AE}$$

$$(\sigma_v)_B=(U/F)_{F=0}=\sum_{i=1}^{N}\frac{N_i}{EA}\frac{dN_i}{dF}L_i$$

Substituting the force of each member, N_i in the equation, the vertical displacement of *B* is

$$(\delta_v)_B=\frac{1}{EA}\left[-\left(P\sqrt{2}+F\frac{\sqrt{2}}{2}\right)\left(-\frac{\sqrt{2}}{2}\right)(L\sqrt{2})2+\left(P+\frac{F}{2}\right)\left(\frac{1}{2}\right)(L)2+F(1)(L)\right]$$

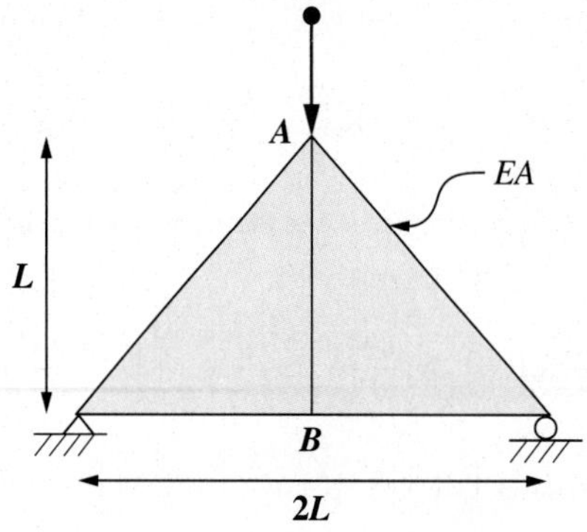

Figure 4.3

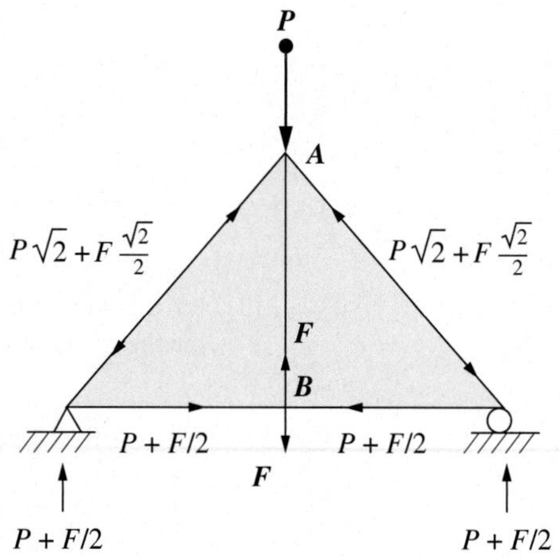

Figure 4.4

Since in this example, $F = 0$, the vertical deflection of B is

$$(\delta_v)_B = \frac{PL}{EA}[2\sqrt{2}+1]$$

EXAMPLE A cylindrical rod in the form of a ring of radius R that is supported at point A and is subjected to load P at point B, as shown in Fig. 4.5. Assume bending stiffness of EI for the ring and neglect normal and shear stresses in the rod. Determine the vertical deflection at point B.

SOLUTION

$$M = QR\sin\theta$$

$$U = \int_0^l \frac{M^2}{2EI}dx \qquad dx = Rd\theta$$

$$e_B = -2\frac{\partial u}{\partial Q} = \frac{1}{EI}\int_0^\pi M\frac{\partial M}{\partial Q}Rd\theta$$

$$= 2\left(\frac{1}{EI}\int_0^\pi (QR\sin\theta)(R\sin\theta)Rd\theta\right)$$

$$= 2\left(\frac{QR^3}{EI}\int_0^\pi \sin^2\theta\, d\theta\right) = \frac{QR^3}{EI}\int_0^\pi \left(\frac{1-\cos 2\theta}{2}\right)d\theta = \frac{QR^3}{EI}\left(\frac{\theta}{2}\Big|_0^\pi - \frac{1}{4}\sin 2\theta\Big|_0^\pi\right)$$

$$= 2\left(\frac{QR^3}{EI}\left(\left(\frac{\pi}{2}\right) - (0)\right)\right)$$

$$= 2\left(\frac{QR^3\pi}{2EI}\right)$$

$$e_B = \frac{\pi PR^3}{EI}$$

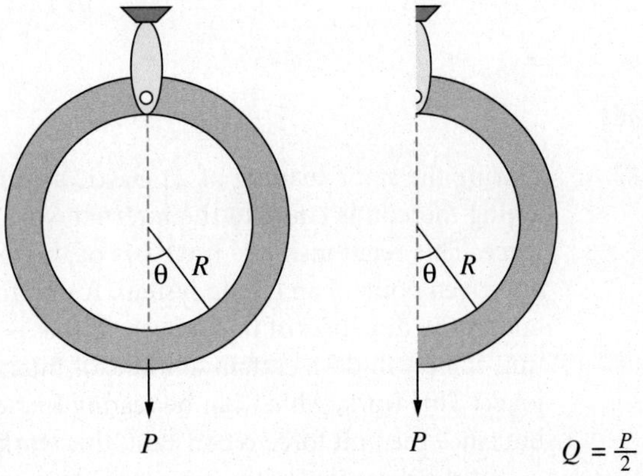

Figure 4.5

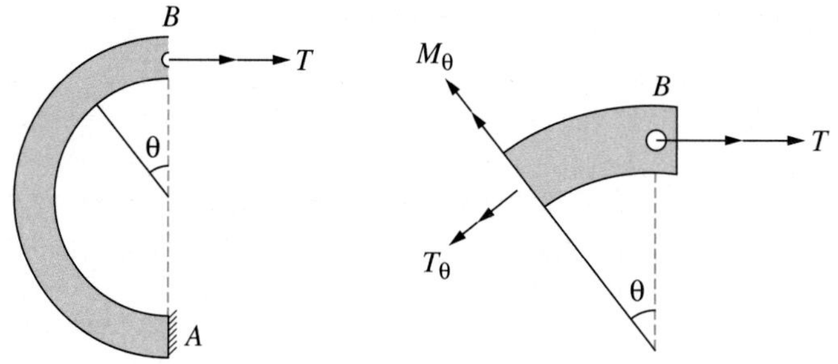

Figure 4.6

EXAMPLE The free end of a cylindrical rod in the form of a half circle is fixed and the other end is subjected to a torsional couple T, as shown in Fig. 4.6. Assume bending stiffness of EI and torsional stiffness of JG for the rod and neglect normal and shear stresses. Determine the angle of twist at the free end.

SOLUTION

$$M_\theta = T\sin\theta$$

$$T_\theta = T\cos\theta$$

$$U = \frac{1}{EI}\int_0^\pi M^2 R\,d\theta + \frac{1}{JG}\int_0^\pi T^2 R\,d\theta$$

Using Castigliano's theorem rotation at B is

$$\phi = \frac{\partial u}{\partial T} = \frac{1}{2EI}\int_0^\pi M_\theta \frac{\partial M_\theta}{\partial T} R\,d\theta + \frac{1}{2JG}\int_0^\pi T_\theta \frac{\partial T_\theta}{\partial T} R\,d\theta$$

$$\phi = \frac{1}{2EI}\int_0^\pi T\sin^2\theta\, R\,d\theta + \frac{1}{2JG}\int_0^\pi T\cos^2\theta\, R\,d\theta$$

$$\phi = \frac{\pi RT}{2}\left(\frac{1}{EI} + \frac{1}{JG}\right)$$

Method of Unit Loads

During the static loading of an elastic system the external work done by a *constant* force acting thereon is equal to the internal work done by the stresses caused by that constant force. This relationship is the basis of the following method for finding the deflection of any given point of an elastic system: A unit force is imagined to act at the point in question and in the direction of the deflection that is to be found. The stresses produced by such a unit force will do a certain amount of internal work during the application of the actual loads. This work, which can be readily found, is equal to the work done by the unit force; but since the unit force is constant, this work is equal to the deflection sought.

If the direction of the deflection cannot be ascertained in advance, its horizontal and vertical components can be determined separately in the way described and the resultant deflection found therefrom. Examples of application of the method are given in Sec. 7.4.

Deflection, the Partial Derivative of Strain Energy

When a linear *elastic* system is statically loaded, the partial derivative of the strain energy with respect to any one of the applied forces is equal to the movement of the point of application of that force in the direction of that force. This relationship provides a means of finding the deflection of a beam or truss under several loads (see Refs. 3, 5, and 7).

Theorem of Least Work

When an elastic system is statically loaded, the distribution of stress is such as to make the strain energy a minimum consistent with equilibrium and the imposed boundary conditions. This principle is used extensively in the solution of statically indeterminate problems. In the simpler type of problem (beams with redundant supports or trusses with redundant members), the first step in the solution consists in arbitrarily selecting certain reactions or members to be considered redundant, the number and identity of these being such that the remaining system is just determinate. The strain energy of the entire system is then expressed in terms of the unknown redundant reactions or stresses. The partial derivative of the strain energy with respect to each of the redundant reactions or stresses is then set equal to zero and the resulting equations solved for the redundant reactions or stresses. The remaining reactions or stresses are then found by the equations of equilibrium. An example of the application of this method is given in Sec. 7.4.

As defined by this procedure, the *theorem of least work* is implicit in Castigliano's theorem: It furnishes a method of solution identical with the method of consistent deflections, the deflection used being zero and expressed as a partial derivative of the strain energy. In a more general type of problem, it is necessary to determine which of an infinite number of possible stress distributions or configurations satisfies the condition of minimum strain energy. Since the development of software based on the finite-element method of analysis the electronic computer has made practicable the solution of many problems of this kind—shell analysis, elastic and plastic buckling, etc.—that formerly were relatively intractable.

4.7 DIMENSIONAL ANALYSIS

Most physical quantities can be expressed in terms of mass, length, and time conveniently represented by the symbols M, L, and t, respectively. Thus velocity is Lt^{-1}, acceleration is Lt^{-2}, force is MLt^{-2}, unit stress is $ML^{-1}t^{-2}$, etc. A formula in which the several quantities are thus expressed is a dimensional formula, and the various applications of this system of representation constitute *dimensional analysis*.

Dimensional analysis may be used to check formulas for homogeneity, check or change units, derive formulas, and establish the relationships between similar physical systems that differ in scale (e.g., a model and its prototype). In strength of materials, dimensional analysis is especially useful in checking formulas for homogeneity. To do this, it is not always necessary to express *all* quantities dimensionally since it may be possible to cancel some terms. Thus it is often convenient to express force by some symbol, as F, until it is ascertained whether or not all terms representing force can be canceled.

For example, consider the formula for the deflection y at the free end of a cantilever beam of length l carrying a uniform load per unit length, w. This formula (Table 8.1) is

$$y = -\frac{1}{8}\frac{wl^4}{EI}$$

To test for homogeneity, omit the negative sign and the coefficient 1/8 (which is dimensionless) and write the formula

$$L = \frac{(F/L)L^4}{(F/L^2)L^4}$$

It is seen that F cancels and the equation reduces at once to $L = L$, showing that the original equation was homogeneous.

Instead of the symbols M, L, t, and F, we can use the names of the *units* in which the quantities are to be expressed. Thus the above equation may be written

$$\text{inches} = \frac{(\text{pounds/inch})(\text{inches}^4)}{(\text{pounds/inch}^2)(\text{inches}^4)} = \text{inches}$$

This practice is especially convenient if it is desired to change units. Thus it might be desired to write the above formula so that y is given in inches when l is expressed in feet. It is only necessary to write

$$\text{inches} = \frac{1}{8}\frac{(\text{pounds/inch})(\text{feet} \times 12)^4}{(\text{pounds/inches}^2)\text{inches}^4}$$

and the coefficient is thus found to be 2592 instead of 1/8.

By what amounts to a reversal of the checking process described, it is often possible to determine the way in which a certain term or terms should appear in a formula provided the other terms involved are known. For example, consider the formula for the critical load of the Euler column. Familiarity with the theory of flexure suggests that this load will be directly proportional to E and I. It is evident that the length l will be involved in some way as yet unknown. It is also reasonable to assume that the load is independent of the deflection since both the bending moment and the resisting moment would be expected to vary in direct proportion to the deflection. We can then write $P = kEIl^a$, where k is a dimensionless constant that must be found in some other way and the exponent a shows how l enters the expression. Writing the equation dimensionally and omitting k, we have

$$F = \frac{F}{L^2}L^4L^a \qquad \text{or} \qquad L^2 = L^{4+a}$$

Equating the exponents of L (as required for homogeneity) we find $a = -2$, showing that the original formula should be $P = kEI/l^2$. Note that the derivation of a formula in this way requires at least a partial knowledge of the relationship that is to be expressed.

A much more detailed discussion of similitude, modeling, and dimensional analysis can be found in Chaps. 15 and 8 of Refs. 6 and 7, respectively. Reference 6 includes a section where the effect of Poisson's ratio on the stresses in two- and three-dimensional problems is discussed. Since Poisson's ratio is dimensionless, it would have to be the same in model and prototype for perfect modeling and this generally is not possible. References to work on this problem are included and will be helpful.

4.8 REMARKS ON THE USE OF FORMULAS

No calculated value of stress, strength, or deformation can be regarded as exact. The formulas used are based on certain assumptions as to properties of materials, regularity of form, and boundary conditions that are only approximately true, and they are derived by mathematical procedures that often involve further approximations. In general, therefore, great precision in numerical work is not justified. Each individual problem requires the exercise of judgment, and it is impossible to lay down rigid rules of procedure; but the following suggestions concerning the use of formulas may be of value:

1. For most cases, calculations giving results to three significant figures are sufficiently precise. An exception is afforded by any calculation that involves the algebraic addition of quantities that are large in comparison with the final result (e.g., some of the formulas for beams under axial and transverse loading, some of the formulas for circular rings, and any case of superposition in which the effects of several loads tend to counteract each other). For such cases more significant figures should be carried throughout the calculations.
2. In view of uncertainties as to actual conditions, many of the formulas may appear to be unnecessarily elaborate and include constants given to more significant figures than is warranted. For this reason, we may often be inclined to simplify a formula by dropping unimportant terms, "rounding off" constants, etc. It is sometimes advantageous to do this, but it is usually better to use the formula as it stands, bearing in mind that the result is at best only a close approximation. The only disadvantage of using an allegedly "precise" formula is the possibility of being misled into thinking that the result it yields corresponds exactly to a real condition. So far as the time required for calculation is concerned, little is saved by simplification.
3. When using an unfamiliar formula, we may be uncertain as to the correctness of the numerical substitutions made and mistrustful of the result. It is nearly always possible to effect some sort of check by analogy, superposition, reciprocal deflections, comparison, or merely by judgment and common sense. Thus the membrane analogy (Sec. 6.4) shows that the torsional stiffness of any irregular section is greater than that of the largest inscribed circular section and less than that of the smallest circumscribed section. Superposition shows that the deflection and bending moment at the center of a beam under triangular loading (Table 8.1, case 2e) is the same as under an equal load uniformly distributed. The principle of reciprocal deflections shows that the stress and deflection at the center of a circular flat plate under eccentric concentrated load (Table 11.2, case 18) are the same as for an equal load uniformly distributed along a concentric circle with radius equal to the eccentricity (case 9a). Comparison shows that the critical unit compressive stress is greater for a thin plate under edge loading than for a strip of that plate regarded as a Euler column. Common sense and judgment should generally serve to prevent the acceptance of grossly erroneous calculations.
4. A difficulty frequently encountered is uncertainty as to boundary conditions—whether a beam or flat plate should be calculated as freely supported or fixed, whether a load should be assumed uniformly or otherwise distributed, etc. In any such case it is a good plan to make *bracketing assumptions*, i.e., to calculate the desired quantity on the basis of each of two assumptions representing limits between which the actual conditions must lie. Thus for a beam with ends having an unknown degree of fixity, the bending moment at the center cannot be more than if the ends were freely supported and the bending moments at the ends cannot be more than if the ends were truly fixed. If so designed as to be safe for either extreme condition, the beam will be safe for any intermediate degree of fixity.
5. The stress and deflections predicted by most formulas do not account for localized effects of the loads. For example, the stresses and deflections given for a straight, simply-supported beam with a centered, concentrated lateral force only account for that due to bending. Additional compressive bearing stresses and deflections exist depending on the exact nature of the interaction of the applied and reaction forces with the beam. Normally, the state of stress and deformation at distances

greater than the dimensions of the loaded regions only depend on the *net* effect of the localized applied and reaction forces and are independent of the form of these forces. This is an application of *Saint Venant's principle* (defined in App. C). This principle may not be reliable for thin-walled structures or for some orthotropic materials.

6. Formulas concerning the validity of which there is a reason for doubt, especially empirical formulas, should be checked dimensionally. If such a formula expresses the results of some intermediate condition, it should be checked for extreme or terminal conditions; thus an expression for the deflection of a beam carrying a uniform load over a portion of its length should agree with the corresponding expression for a fully loaded beam when the loaded portion becomes equal to the full length and should vanish when the loaded portion becomes zero.

4.9 REFERENCES

1. Love, A. E. H., *Mathematical Theory of Elasticity*, 2nd ed., Cambridge University Press, 1906.
2. Morley, A., *Theory of Structures*, 5th ed., Longmans, Green, 1948.
3. Langhaar, H. L., *Energy Methods in Applied Mechanics*, John Wiley & Sons, 1962.
4. Timoshenko, S., and J. M. Gere, *Theory of Elastic Stability*, 2nd ed., McGraw-Hill, 1961.
5. Cook, R. D., and W. C. Young, *Advanced Mechanics of Materials*, 2nd ed., Prentice-Hall, 1999.
6. Kobayashi, A. S. (ed.), *Handbook on Experimental Mechanics*, 2nd ed., Society for Experimental Mechanics, VCH, 1993.
7. Budynas, R. G., *Advanced Strength and Applied Stress Analysis*, 2nd ed., McGraw- Hill, 1999.
8. Srivastava, A. K., and P. C. Gope, *Strength of Materials*, Prentice-Hall, 2007.
9. Avallone, E. A., T. Baumeister, and A. M. Sadegh, *Marks' Standard Handbook for Mechanical Engineers*, 11th ed., McGraw-Hill, 2007.

CHAPTER 5

Numerical Methods

The analysis of stress and deformation of the loading of simple geometric structures can usually be accomplished by closed-form techniques. As the structures become more complex, the analyst is forced to approximations of closed-form solutions, experimentation, or numerical methods. There are a great many numerical techniques used in engineering applications for which digital computers are very useful. In the field of structural analysis, the numerical techniques generally employ a method which discretizes the continuum of the structural system into a finite collection of points (or nodes) whereby mathematical relations from elasticity are formed. The most popular techniques used currently are the finite element method (FEM) and boundary element method (BEM). For this reason, most of this chapter is dedicated to a general description of these methods. A great abundance of papers and textbooks have been presented on the finite element and the boundary element methods, and a complete listing is beyond the scope of this book. However, some textbooks and historical papers are included for introductory purposes. Other methods, some of which FEM is based upon, include *trial functions* via *variational methods* and *weighted residuals*, the *finite difference method* (FDM), *structural analogues*, and the *boundary element method* (BEM). In this chapter, other numerical methods such as finding zeroes of polynomials, solution of differential equations and numerical integrations are also presented.

5.1 THE FINITE DIFFERENCE METHOD

In the field of structural analysis, one of the earliest procedures for the numerical solutions of the governing differential equations of stressed continuous solid bodies was the *finite difference method*. In the finite difference approximation of differential equations, the derivatives in the equations are replaced by difference quotients of the values of the dependent variables at discrete mesh points of the domain. Finite difference method generally employs three types of quotations known as forward, backward, and central differences.

A **forward difference** is the difference between the function values at x and $x + h$ as

$$\Delta f(x) = f(x+h) - f(x)$$

where h is the spacing between two adjacent points, which is a constant or variable. However, a **backward difference** uses the function values at x and $x - h$, and is expressed as

$$\nabla f(x) = f(x) - f(x-h)$$

Finally, the **central difference** is given by

$$\delta f(x) = f\left(x + \frac{h}{2}\right) - f\left(x - \frac{h}{2}\right)$$

Using the forward quotation, for example, the derivative of a function *f* at a point *x* which is defined by

$$f'(x) = \lim_{h \to 0} \frac{f(x+h) - f(x)}{h}$$

can be written as $[f(x+h) - f(x)]/h = \Delta f(x)/h$ for a non-zero and small *h*. This is forward difference derivative of function *f*.

Assuming that f is continuous and differentiable, and using the Taylor's theorem, the error in the forward and backward approximation can be derived, and it is in the order of h, that is, as h goes to zero, the error approaches to zero. However, the error in the central difference is in the order of h^2, that is, it yields a more accurate approximation.

To solve a problem, the domain of the differential equation is discretized and the derivatives in the differential equation are replaced by one of the quotations of the finite difference approximations. After imposing the appropriate boundary conditions on the structure, the discrete equations are solved obtaining the values of the variables at the mesh points. The technique has many disadvantages, including inaccuracies of the derivatives of the approximated solution, difficulties in imposing boundary conditions along curved boundaries, difficulties in accurately representing complex geometric domains, and the inability to utilize non-uniform and non-rectangular meshes.

5.2 THE FINITE ELEMENT METHOD

The finite element method (FEM) evolved from the use of trial functions via variational methods and weighted residuals, the finite difference method, and structural analogues (see Table 1.1 of Ref. 1). FEM overcomes the difficulties encountered by the finite difference method in that the solution of the differential equations of the structural problem is obtained by utilizing an integral formulation to generate a system of algebraic equations with continuous piecewise-smooth (trial) functions that approximate the unknown quantities. A geometrically complex domain of the structural problem can be systematically represented by a large, but finite, collection of simpler subdomains, called *finite elements.* For structural problems, the displacement field of each element is approximated by polynomials, which are interpolated with respect to preselected points *(nodes)* on, and possibly within, the element. The polynomials are referred to as *interpolation functions,* where variational or weighted residual methods (e.g., Rayleigh–Ritz, Galerkin, etc.) are applied to determine the unknown nodal values. Boundary conditions can easily be applied along curved boundaries, complex geometric domains can be modeled, and non-uniform and non-rectangular meshes can be employed.

The modern development of FEM began in the 1940s in the field of structural mechanics with the work of Hrennikoff, McHenry, and Newmark, who used a lattice of line elements (rods and beams) for the solution of stresses in continuous solids (see Refs. 2–4). In 1943, from a 1941 lecture, Courant suggested piecewise-polynomial interpolation over triangular subregions as a method to model torsional problems (see Ref. 5).

With the advent of digital computers in the 1950s, it became practical for engineers to write and solve the stiffness equations in matrix form (see Refs. 6–8). A classic paper by Turner, Clough, Martin, and Topp published in 1956 presented the matrix stiffness equations for the truss, beam, and other elements (see Ref. 9). The expression *finite element* is first attributed to Clough (see Ref. 10).

Since these early beginnings, a great deal of effort has been expended in the development of FEM in the areas of element formulations and computer implementation of the entire solution process. The major advances in computer technology include the rapidly expanding computer hardware capabilities, efficient and accurate matrix solver routines, and computer graphics for ease in the preprocessing stages of model building, including automatic adaptive mesh generation, and in the postprocessing stages of reviewing the solution results. A great abundance of literature has been presented on the subject, including many textbooks. A partial list of some textbooks, introductory and more comprehensive, is given at the end of this chapter. For a brief introduction to FEM and modeling techniques, see Chaps. 9 and 10, respectively, of Ref. 11.

Formulation of the Finite Element Method

The finite element method relies on the minimization of the total potential energy of the system, expressed in terms of displacement functions. The principle of potential energy may be written as

$$\Delta\Pi = \Delta(U - W) = 0$$

where U is the strain energy, i.e., $\left[\int_V (\sigma_x \varepsilon_x + \cdots + \sigma_z \varepsilon_z)\,dV\right]$, and W is the work done by a body force F per unit volume and a surface force (traction) P per unit area, as

$$\Delta W = \int_V (F_x \Delta u + F_y \Delta \upsilon + F_z \Delta w)\,dV + \int_A (p_x \Delta u + p_y \Delta \upsilon + p_z \Delta w)\,dA$$

Consider an elastic body of Fig. 5.1 that is divided into n number of finite elements. Therefore, the principle of potential energy of the body is:

$$\Delta\Pi = \sum_1^n \int_V (\sigma_x \Delta\varepsilon_x + \cdots + \sigma_z \Delta\varepsilon_z)\,dV - \sum_1^n \int_V (F_x \Delta u + F_y \Delta\upsilon + F_z \Delta w)\,dV$$
$$- \sum_1^n \int_s (p_x \Delta u + p_y \Delta\upsilon + p_z \Delta w)\,ds = 0 \qquad \text{(a)}$$

where n = number of elements comprising the body
V = volume of a discrete element
s = portion of the boundary surface area over which forces are prescribed
F = body forces per unit volume
p = prescribed boundary force or surface traction per unit area

The equation (a) can be written in a matrix form as

$$\sum_1^n \int_V \left(\{\Delta\varepsilon\}_e^T \{\sigma\}_e - \{\Delta f\}_e^T \{F\}_e\right) dV - \sum_1^n \int_s \{\Delta f\}_e^T \{p\}_e\,ds = 0 \qquad \text{(b)}$$

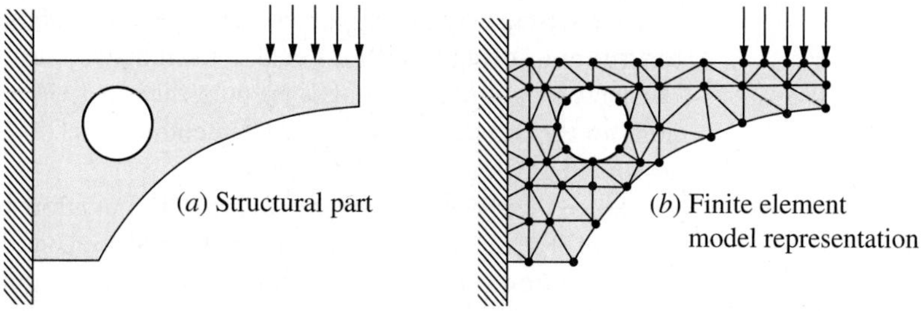

Figure 5.1 Discretization of a continuous structure.

where T denotes the transpose of a matrix. The strain ε and displacement δ relationship, for each element can be written in a simplified matrix form as, $\{\varepsilon\}_e = [B]\{\delta\}_e$ and similarly the stress-strain relationship in a simplified matrix form is

$$\{\sigma\}_e = [D]\{\varepsilon\}_e$$

Therefore, the equation (b) can be written as

$$\sum_1^n \{\Delta\delta\}_e^T([k]_e\{\delta\}_e - \{Q\}_e) = 0 \tag{c}$$

where

$$[k]_e = \int_V [B]^T[D][B]dV \tag{d}$$

$$\{Q\}_e = \int_V [N]^T\{F\}dV + \int_V [B]^T[D]\{\varepsilon_0\}dV + \int_s [N]^T\{p\}ds \tag{e}$$

Considering the entire continuum body, equation (c) can be assembled as

$$(\Delta\delta)^T([K]\{\delta\} - \{Q\}) = 0$$

Since this equation must be satisfied for all nodal displacements $\{\Delta\delta\}$ then we have

$$[K]\{\delta\} = \{Q\}$$

where

$$[K] = \sum_1^n [k]_e, \quad \{Q\} = \sum_1^n \{Q\}_e$$

$[K]$ is known as coefficient matrix and $\{Q\}$ is the nodal force matrix. Finally the nodal displacement is calculated by:

$$\{\delta\} = [K]^{-1}\{Q\} \tag{f}$$

To summarize the general procedure for solving an elasticity problem using the finite element method, use the following steps:

a. Calculate stiffness matrix $[K]_e$ using equation (d) and assemble them for the whole body as $[K]$ matrix.

b. Calcuate the nodal force matrix $\{Q\}_e$ using equation (e) and assemble them for the whole body as $\{Q\}$.

c. Solve the matrix equation (f) for the nodal displacement by satisfying the boundary conditions, and determine $\{\delta\}$.

d. Once the nodal displacements are found, the strain and stress for each element is found by $\{\varepsilon\}_e = [B]\{\delta\}_e$ and $\{\sigma\}_e = [D]\{\varepsilon\}_e$ respectively.

Implementation of the Finite Element Method

FEM is ideally suited to digital computers, in which a continuous elastic structure (continuum) is divided (discretized) into small but finite well-defined substructures *(elements)*. Using matrices, the continuous elastic behavior of each element is categorized in terms of the element's material and geometric properties, the distribution of loading (static, dynamic, and thermal) within the element, and the loads and displacements at the *nodes* of the element. The element's nodes are the fundamental governing entities of the element, since it is the node where the element connects to other elements, where elastic properties of the element are established, where boundary conditions are assigned, and where forces (contact or body) are ultimately applied. A node possesses *degrees of freedom* (dof's). Degrees of freedom are the translational and rotational motion that can exist at a node. At most, a node can possess three translational and three rotational degrees of freedom. Once each element within a structure is defined locally in matrix form, the elements are then globally assembled (attached) through their common nodes (dof's) into an overall system matrix. Applied loads and boundary conditions are then specified, and through matrix operations the values of all unknown displacement degrees of freedom are determined. Once this is done, it is a simple matter to use these displacements to determine strains and stresses through the constitutive equations of elasticity.

Many geometric shapes of elements are used in finite element analysis for specific applications. The various elements used in a general-purpose commercial FEM software code constitute what is referred to as the *element library* of the code. Elements can be placed in the following categories: *line elements, surface elements, solid elements,* and *special purpose elements.* Table 5.1 provides some, but not all, of the types of elements available for finite element analysis.

Since FEM is a numerical technique that discretizes the domain of a continuous structure, errors are inevitable. These errors are:

1. *Computational Errors* These are due to round-off errors from the computer floating-point calculations and the formulations of the numerical integration schemes that are employed. Most commercial finite element codes concentrate on reducing these errors and consequently the analyst generally is concerned with discretization factors.

2. *Discretization Errors* The geometry and the displacement distribution of a true structure vary continuously. Using a finite number of elements to model the structure introduces errors in matching geometry and the displacement distribution due to the inherent limitations of the elements. For example, consider the thin-plate structure shown in Fig. 5.1(*a*). Figure 5.1(*b*) shows a finite element model of the structure where three-noded, plane-stress, triangular elements are employed. The plane-stress triangular element has a flaw, which creates two basic problems. The element has straight sides, which remain straight after deformation. The strains throughout the plane-stress triangular element are constant. The first problem, a geometric one, is the modeling of curved edges. Note that the surface

TABLE 5.1 Sample Finite Element Library

Element Type	Name	Shape	Number of Nodes	Applications
Line	Truss		2	Pin-ended bar in tension or compression
	Beam	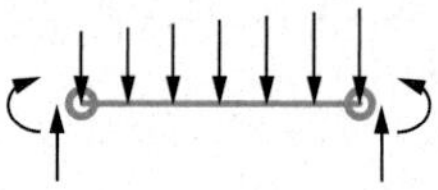	2	Bending
	Frame	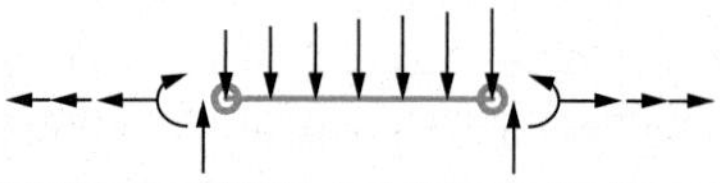	2	Axial, torsional, and bending. With or without load stiffening
Surface	4-noded quadrilateral	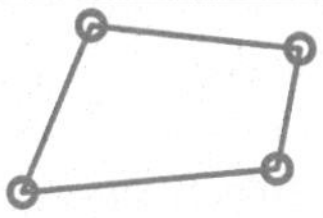	4	Plane stress or strain, axisymmetry, shear panel, thin flat plate in bending
	8-noded quadrilateral	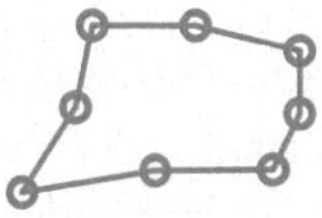	8	Plane stress or strain, thin plate or shell in bending
	3-noded triangle	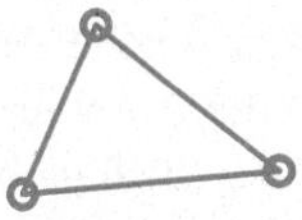	3	Plane stress or strain, axisymmetry, shear panel, thin flat plate in bending. Prefer quad where possible. Used for transitions of quads
	6-noded triangle	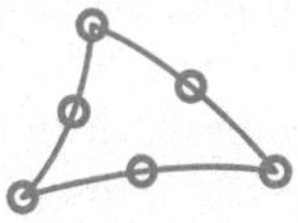	6	Plane stress or strain, axisymmetry, thin plate or shell in bending. Prefer quad where possible. Used for transitions of quads
Solid*	8-noded hexagon (brick)	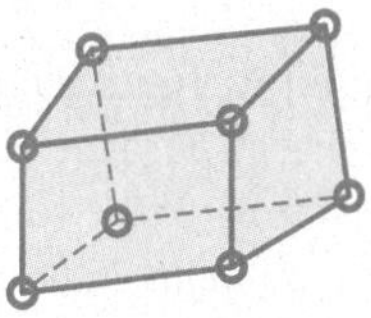	8	Solid, thick plate (using mid-size nodes)
	6-noded pentagon (wedge)	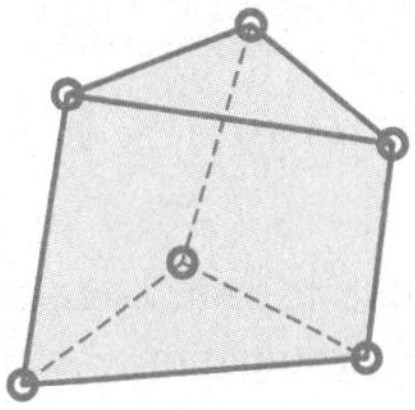	6	Solid, thick plate (using mid-size nodes). Used for transitions
	4-noded tetrahedron (tet)	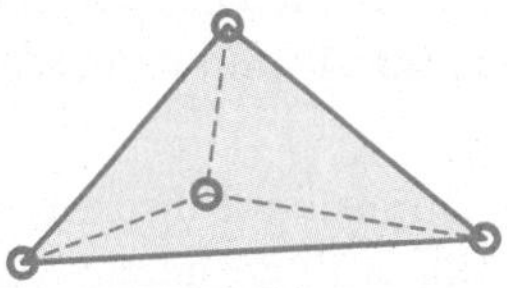	4	Solid, thick plate (using mid-size nodes). Used for transitions
Special purpose	Gap		2	Free displacement for prescribed compressive gap
	Hook		2	Free displacement for prescribed extension gap
	Rigid	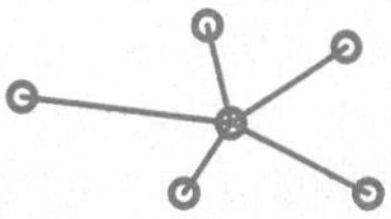	Variable	Rigid constraints between nodes

*These element are also available with mid-size nodes.

of the model with a large curvature appears reasonably modeled, whereas the surface of the hole is very poorly modeled. The second problem, which is much more severe, is that the strains in various regions of the actual structure are changing rapidly, and the constant-strain element will only provide an approximation of the average strain at the center of the element. So, in a nutshell, the results predicted using this model will be relatively poor. The results can be improved by significantly increasing the number of elements used (increased mesh density). Alternatively, using a better element, such as an eight-noded quadrilateral, which is more suited to the application, will provide the improved results. Due to higher-order interpolation functions, the eight-noded quadrilateral element can model curved edges and provide for a higher-order function for the strain distribution.

Advances in Finite Element Methods

With the advent of computational speed, mesh generation techniques, and new algorithm for solving engineering problems, finite element methods have become ever more important to engineers as tools for design, stress analysis, and optimization. In particular, nonlinear finite element methods have been developed for solving nonlinear and plastic deformation of technological problems. Use of linear finite element is often inadequate for some structures, as well as new materials such as composites, rubber or plastic, that exhibit nonlinear behavior. Theoretical foundations of these nonlinear problems are prerequisite to their finite-element discretization as well as algorithms for solving the nonlinear equations.

5.3 THE BOUNDARY ELEMENT METHOD

The idea of *boundary element method* (BEM), also known as *bound integral equation method* (BIEM), developed more recently than FEM, is that the governing partial differential equation (or PDE) of a domain is transformed into an integral equation on the boundary and then uses the boundary solution to find the solution of the problem inside the domain, (see Refs. 12–16). This transformation is commonly accomplished through the *Green's Integral Equations.* Since differential equation of the domain is transferred to an integral equation on the boundary, the dimension of the problem is reduced by one, that is, a two-dimensional problem is reduced to a line integer (one dimension) and a three dimensional problem is reduced to a surface integral (two dimension). The dimension reduction feature of the BEM makes it very attractive for problems involving very large domains where FEM solution requires large number of elements. That is, unlike the finite element method, in the boundary element method the domain does not need to be discretized into finite elements, rather only the boundary is discretized (Fig. 5.2). However, in the boundary

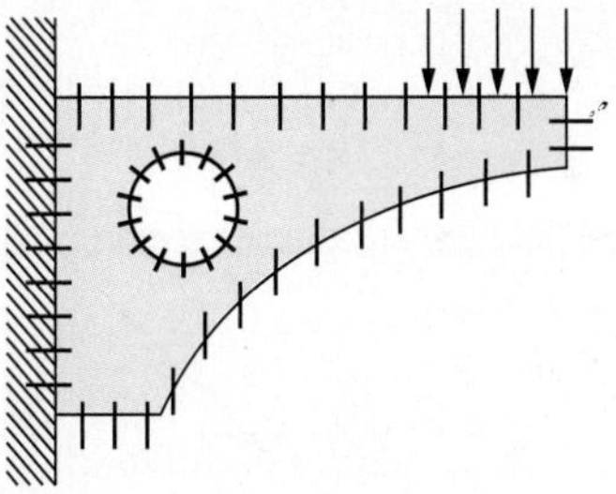

Figure 5.2 Structural part.

element method a *fundamental solution* or *the Green's function* is needed. A comparison of the two methods is shown in the table below.

BEM	FEM
Discretization of boundary	Discretization of whole domain
Good on infinite or semi-infinite domains	Good on finite domains
Approximate stress and then displacement Approximation of stress is more accurate	Approximate displacement and then stress Approximation of stress may be less accurate
The coefficient matrix is small but fully populated	The coefficient matrix is large but sparsely populated
Requires a fundamental solution to the PDE	Requires no prior knowledge of solution
Can be difficult to solve inhomogeneous or nonlinear problems	Solves most linear second-order PDEs

BEM Solution to the Laplacian Equation

Consider a class of problems having the Laplacian Equation as their governing differential equations, such as torsion, heat conduction, perfect laminar flow, electro statics, etc.

The differential equation is

$$\nabla^2 u = 0 \quad \text{in } \Omega \tag{5.3-1}$$

The domain Ω, shown in Fig. 5.3, is subjected to the boundary conditions:

$$u = f(x, y, z) \quad \text{on } \Gamma_1 \text{ (Dirichlet B.C.)} \tag{5.3-2}$$

$$\frac{\partial u}{\partial n} = g(x, y, z) \quad \text{on } \Gamma_2 \text{ (Neumann B.C.)} \tag{5.3-3}$$

where $\Gamma = \Gamma_1 + \Gamma_2$ is the boundary of Ω and n is the direction of the outward-directed normal to Γ at the boundary points (x, y, z). The functions f and g are prescribed known functions on Γ_1 and Γ_2, respectively.

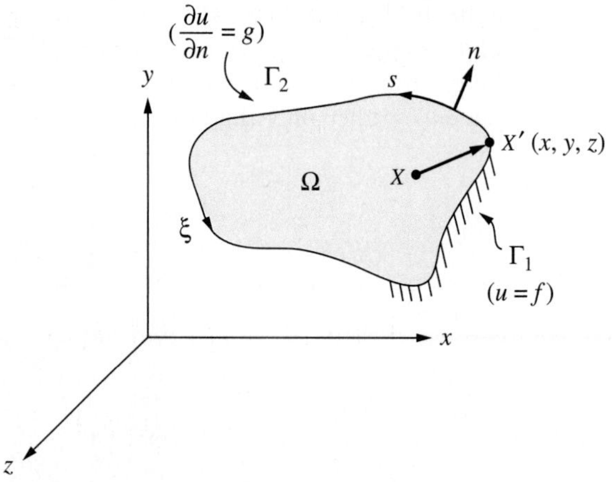

Figure 5.3 An elastic domain Ω with boundary conditions.

To solve the Laplacian we employ fundamental solution, also called *Green's function* or *Influence function*, as

$$\nabla^2 G = \delta(x-\xi) \qquad \text{in } \Omega' \tag{5.3-4}$$

where Ω' is an infinite region and denoting the points (x, y, z) and (ξ, η, ς) by x and ξ, respectively. The solution of Eq. (5.3-4) is well known and it is

$$G(x,\xi) = -\frac{1}{4\pi r(x,\xi)} \qquad \text{for 3D problems} \tag{5.3-5}$$

and

$$G(x,\xi) = +\frac{1}{2\pi}\ln r(x,\xi) \qquad \text{for 2D problems} \tag{5.3-6}$$

where r is the distance between x and ξ and

$$r(x,\xi) = r = [(x-\xi)^2 + (y-\eta)^2 + (z-\varsigma)^2]^{1/2}$$

There are two approaches to the solution of the Laplacian: *direct* and *indirect methods.*

Direct Method

In the direct BEM, the Green's Identity is used as

$$\int_\Omega (u\nabla^2 G - G\nabla^2 u)\,d\Omega = \int_\Gamma \left(u\frac{\partial G}{\partial n} - G\frac{\partial u}{\partial n}\right) d\Gamma \tag{5.3-7}$$

Substituting the boundary conditions into Eq. (5.3-7), then the solution of the Eq. (5.3-1) is

$$u(x) = \int_\Gamma \left[f(\xi)\frac{\partial G}{\partial n}(x,\xi) - g(\xi)G(x,\xi) \right] d\Gamma(\xi) \tag{5.3-8}$$

where $x(x, y, z)$ is an interior point of Ω.

Letting x approach to a boundary point x' and extracting the singularity we arrive at

$$\frac{1}{2}f(x') + \int_\Gamma g(\xi)G(x',\xi)\,d\Gamma(\xi) = \int_\Gamma f(\xi)\frac{\partial G}{\partial n}(x',\xi)\,d\Gamma(\xi) \tag{5.3-9}$$

This is the boundary-integral equation relating the boundary values of f and g. If f is prescribed on the boundary, then this equation leads to a Fredholm equation of the first kind, and if g is prescribed on the boundary, the equation leads to a Fredholm equation of the second kind. In either case, f and g can be determined numerically by subdividing the boundary into boundary elements. Once f and g are determined, the solution of Laplacian at any point in the region can be calculated using Eq. (5.3-8).

Indirect Method

In the indirect approach, the region Ω is embedded into an infinite region Ω' for which the fundamental solution, or the Green's function is known (see Fig. 5.4). We then subjected the boundary to a layer of unknown source points $P^*(x')$. Using the superposition principle, the solution is

$$u(x) = \int_\Gamma P^*(\xi)G(x,\xi)\,d\Gamma(\xi) \tag{5.3-10}$$

Letting x approach to a boundary point x' and extracting the singularity, we arrive at

$$f(x') = \int_\Gamma P^*(\xi)G(x',\xi)\,d\Gamma(\xi) \qquad \text{for } x' \in \Gamma_1 \tag{5.3-11}$$

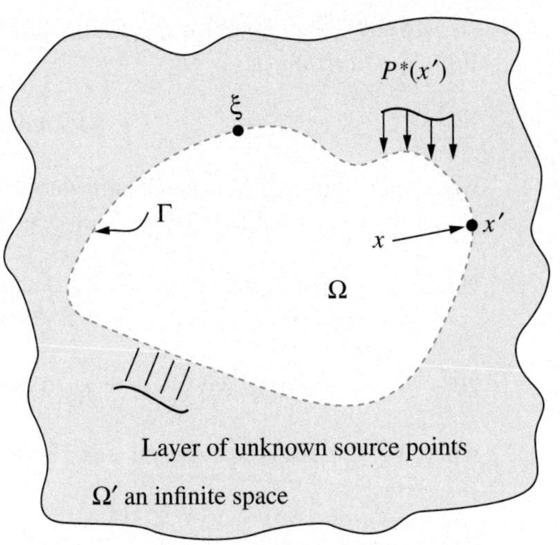

Figure 5.4 Imbedded domain Ω into an infinite region Ω′ of the same material.

and

$$\frac{1}{2}P^*(x')+\int_{\Gamma}P^*(\xi)\frac{\partial}{\partial n}G(x',\xi)d\Gamma(\xi)=g(x') \qquad \text{for } x'\in\Gamma_2 \qquad (5.3\text{-}12)$$

$P^*(x')$ can be calculated through either Eq. (5.3-11) or (5.3-12), and the final solution of the Laplacian could be found using Eq. (5.3-10).

BEM in Elasticity

Direct Method

In the direct approach, which is due Rizzo (Ref. 12), consider an elastic region R with boundary B subjected to a traction and displacement boundary as shown in Fig. 5.5.

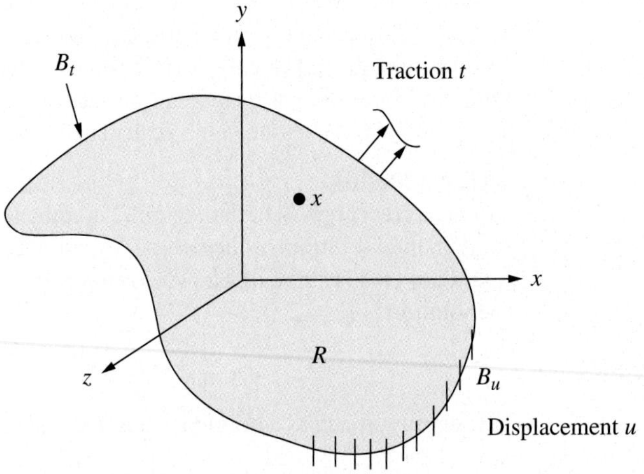

Figure 5.5 An elastic region R and the boundary conditions.

Two states of stresses are considered: One is the traction influence function, T^* and the second one is displacement influence function, U^*. Both T^* and U^* are known as influence functions or the "*Green's Function,*" and their solution for an infinite elastic region is known as *Love's* solution. Rizzo employed Betti's reciprocal theorem and arrived at

$$u_j(x) = \int_B t_i(\xi) U^*_{ij}(\xi, x) ds(\xi) - \int_B u_i(\xi) T^*_{ij}(\xi, x) ds(\xi) \tag{5.3-13}$$

where $x \in R$ (the domain) and $\xi \in B$ (the boundary), and where t_i, u_i the boundary traction and displacement data, respectively, are unknown.

To obtain the boundary integral equation, let x approach the boundary point x'. Similar to Laplacian solution, after extracting the singularity Eq. (5.3-13) becomes

$$\frac{1}{2} u_j(x') + \int_B u_i(\xi) T^*_{ij}(\xi, x') ds(\xi) = \int_B t_i(\xi) U^*_{ij}(\xi, x') ds(\xi) \tag{5.3-14}$$

where the integral on the right side of the equation is nonsingular and is considered the Chausy principal value. Equation (5.3-14) is an integral equation which can be solved for u_i or t_i. Once these values are determined the solution of the problem is given by Eq. (5.3-13).

Indirect Method

In the indirect method, similar to Laplacian we embed the region R into an infinite region of the same materials. Then, we subject a fictitious layer of body force $P^*(x)$ on the boundary points, Fig. 5.6.

Assume the two fundamental solutions as $H_{ij;q}(x, \xi)$, which is the ij^{th} stress component at a field point x due to a unit load in the q direction at a source point ξ, and $I_{i;q}(x, \xi)$, which is the ith displacement component at a field point x due to a unit load in the q direction at a source point ξ. Then, using the superposition principal, the stress and displacement at any point is

$$\sigma_{ij}(x) = \int_B H_{ij;q}(x, \xi) P^*_q(\xi) ds(\xi) \qquad \xi \in B$$

$$u_i(x) = \int_B I_{i;q}(x, \xi) P^*_q(\xi) ds(\xi) \qquad \xi \in B \tag{5.3-15}$$

In order to solve the boundary value problem of interest, the boundary conditions on B are yet to be satisfied. These are:

$$\sigma_{ij} n_j = p_i^s \qquad \text{on } B_t$$

$$U_i = u_i^s \qquad \text{on } B_u \tag{5.3-16}$$

Substituting Eq. (5.3-15) into (5.3-16) leads to the integral equation. By subdividing the boundary into elements one can solve the integral equations numerically. Note that boundary elements for a general three-dimensional solid are quadrilateral or triangular surface elements covering the surface area of the component. Also for two-dimensional and axisymmetric problems, only line elements tracing the outline of the component are used.

Advances in the BEM

In the last decade, BEM has been further developed in theoretical formulations of the Green's function, which are the key ingredients in the BEM as well as computational method of solving engineering problems. As for the formulation, these advances in the

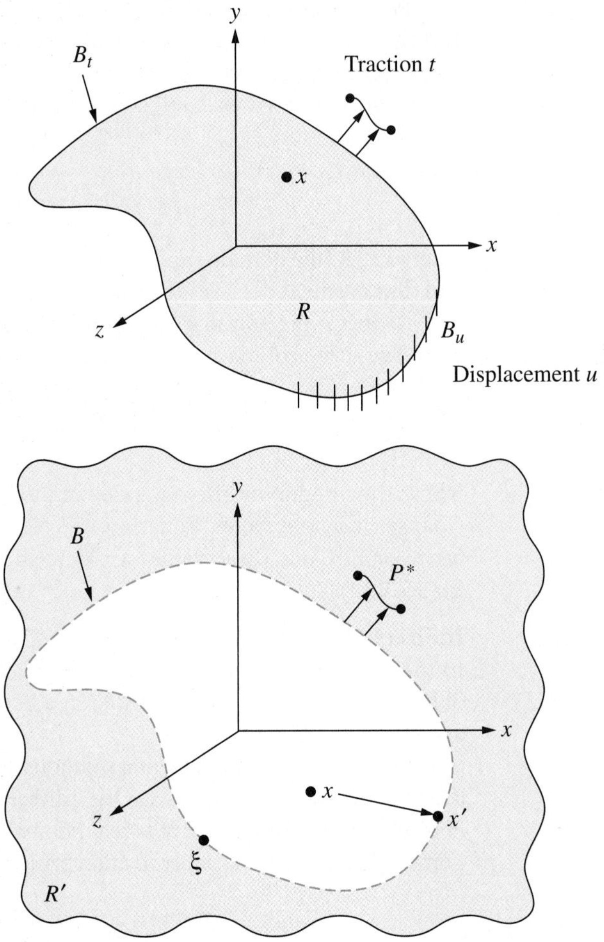

Figure 5.6 (a) Region R and the boundary conditions, (b) Embedded region R into an infinite region R'.

boundary integral equation are associated with boundary mesh free methods, the symmetric Galerkin BEM formulation, the BIE-related boundary mesh-free methods, and the BEM formulations with vibrational base. Green's function has also been developed for transversely isotropic, general anisotropic, and layered elastic materials. Based on Green's function, BEM has also been developed for layered structures used in advanced composite materials. This requires derivation of Green's function with the application of Fourier transform and inverse transform which requires numerical integration.

In reference to the solution methods, new algorithms such as adaptive cross-approximation method, fast multipole method, and the pre-corrected fast Fourier transformation method have been developed to enhance the accuracy, reduction of computational time, and speed of the analysis.

Theoretical formulations of combined boundary element method and FEM have also been developed. For large domains that BEM is computationally advantageous, one might need a more detailed analysis of a smaller region with FEM. Consider a large two-dimensional or three-dimensional domain where a fictitious small boundary is defined around a small region of interest, within or adjacent to the large domain. The solution is that first subdivide the boundary of the large domain and the fictitious boundary into boundary elements.

Second, use BEM to solve the larger domain (excluding the smaller region) and determine tractions around the fictitious boundary. Third, discretize the smaller domain with finite elements. Finally, use FE and solve the smaller region using the known traction boundary conditions from BE analysis. In this process, the need for discretizing the large domain with finite elements is eliminated and the results of the smaller region with high accuracy are achieved. One can interchange the order of the process, that is, first solve the small region with FE and then solve the larger domain with BEM.

5.4 ZEROES OF POLYNOMIALS

Newton-Raphson Method

Let x_0 be an approximate root of the equation $f(x) = 0$, and let $x_1 = x_0 + h$ be the exact root, then $f(x_1) = 0$.

Expanding $f(x_0 + h)$ in a *Taylor's series*, we have

$$f(x_0) + hf'(x_0) + h^2 f''(x_0)/2! + \cdots \cdots \cdots \cdots \cdots = 0$$

Since h is small, neglecting h^2 and higher powers of h, we get

$$f(x_0) + hf'(x_0) = 0$$

or

$$h = -f(x_0)/f'(x_0)$$

A closer approximation to the root is given by

$$x_1 = x_0 - f(x_0)/f'(x_0)$$

Similarly starting with x_1, a still better approximation x_2 is given by

$$x_2 = x_1 - f(x_0)/f'(x_0)$$

In general,

$$x_{n+1} = x_n - f(x_0)/f'(x_0) \qquad (n = 0, 1, 2, 3 \ldots \ldots)$$

This method is known as the *Newton-Raphson formula* or *Newton's iteration formula.*

Gauss-Elimination Method

To solve a system of linear equations, n equations, n unknown, the Gauss-Elimination method is employed. In this method, the unknowns are eliminated successively and the system of equations are reduced to an upper triangular system from which the unknowns are found by back substitution.

Consider the system of equations

$$\begin{aligned} a_1x + b_1y + c_1z &= d_1 \\ a_2x + b_2y + c_2z &= d_2 \\ a_3x + b_3y + c_3z &= d_3 \end{aligned} \tag{1}$$

Using simple subtraction techniques and eliminating x from second and third equations we get

$$\begin{aligned} a_1x + b_1y + c_1z &= d_1 \\ b_2'y + c_2'z &= d_2' \\ b_3'y + c_3'z &= d_3' \end{aligned} \tag{2}$$

In the next step, eliminate y from third equation in (2) we have

$$\begin{aligned} a_1x + b_1y + c_1z &= d_1 \\ b_2'y + c_2'z &= d_2' \\ c_3''\, z &= d_3'' \end{aligned} \tag{3}$$

The values of x, y, z are found from the reduced system (3) by *back substitution.* Thus the unknowns are found by Gauss-Elimination method.

5.5 SOLUTION OF DIFFERENTIAL EQUATIONS

Euler's Method

Assume that $f(x, y)$ is continuous in the variable y, and consider the initial value problem

$$y' = f(x, y) \text{ with } y(a) = x_o = \alpha, \qquad \text{over the interval} \qquad a \le x \le b$$

Let $x_{k+1} = x_k + h$ where h is the specified interval, then the iterative solution is

$$y_{k+1} = y_k + hf(x_k, y_k) \qquad \text{for } k = 0, 1, 2 \ldots\ldots. n-1$$

The approximate solution at the discrete set of points is $\{(x_k, y_k)\}_k^n = 0$

Runge-Kutta Method

Let $x_{k+1} = x_k + h$ where h is the specified interval, then the iterative solution is

$$y_{j+1} = y_j + \frac{1}{6}(k_1 + 2k_2 + 2k_3 + k_4) \qquad \text{for} \qquad k = 0, 1, 2, \ldots, m-1$$

where

$$k_1 = hf(x_j, y_j)$$

$$k_2 = hf\left(x_j + \frac{h}{2}, y_j + \frac{k_1}{2}\right)$$

$$k_3 = hf\left(x_j + \frac{h}{2}, y_j + \frac{k_2}{2}\right)$$

$$k_4 = hf(x_j + h, y_j + k_3)$$

The approximate solution at the discrete set of points is $\{(x_k, y_k)\}_{k=0}^n$.

5.6 NUMERICAL INTEGRATION

There exists a unique polynomial $P_m(x)$ of degree $\le m$, passing through the $m + 1$ equally spaced points $\{(x_k, f(x_k))\}_{k=0}^m$. When this polynomial is used to approximate $f(x)$ over $[a, b]$, and the integral of $f(x)$ is approximated by the integral of $P_m(x)$, the resulting formula is called a Newton-Cotes quadrature formula. When the sample points $x_0 = a$ and $x_m = b$ are used, it is called a closed Newton-Cotes formula.

Newton-Cotes Quadrature Formula in generic form is written as

$$\int_{x_0}^{x_0+nh} f(x)dx = y_0 + \frac{n}{2}\Delta y_0 + \frac{n}{12}(2n-3)\Delta^2 y_0 \cdots\cdots$$

Case 1

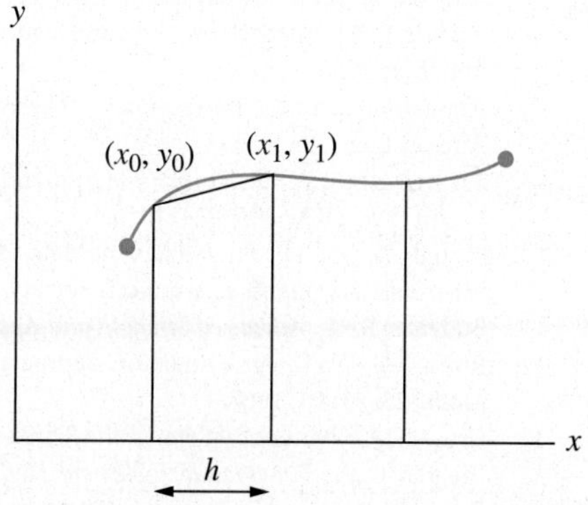

Initially assuming $n = 1$ in above and taking the curve through (x_0, y_0) *and* (x_1, y_1) as a straight line, we get

$$\int_{x_0}^{x_0+nh} f(x)dx = \frac{h}{2}[(y_0 + y_n) + 2(y_1 + y_2 + \cdots\cdots + y_{n-1})]$$

This is known as the ***Trapezoidal rule.***

Case 2

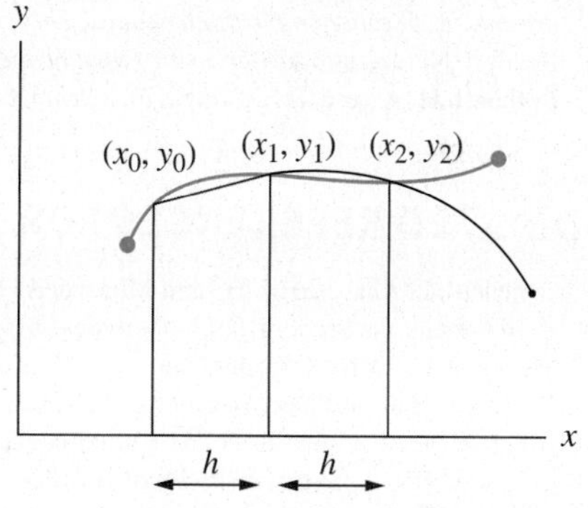

Considering $n = 2$ in above and taking the curve through (x_0, y_0), (x_1, y_1), and (x_2, y_2) as a parabola, we get

$$\int_{x_0}^{x_0+nh} f(x)dx = \frac{h}{3}[(y_0 + y_n) + 4(y_1 + y_3 + \cdots\cdots + y_{n-1}) + 2(y_2 + y_4 + \cdots\cdots + y_{n-2})]$$

This is known as ***Simpson's rule.***

5.7 REFERENCES

1. Zienkiewicz, O. C., and R. L. Taylor, *The Finite Element Method, Vol. 1, Basic Formulations and Linear Problems*, 4th ed., McGraw-Hill, 1989.
2. Hrennikoff, A., "Solution of Problems in Elasticity by the Frame Work Method," *J. Appl. Mech.*, 8(4): 169–175, 1941.
3. McHenry, D., A Lattice Analogy for the Solution of Plane Stress Problems, *J. Inst. Civil Eng.*, 21, 59–82, 1943.
4. Newmark, N. M., Numerical Methods of Analysis in Bars, Plates, and Elastic Bodies, in L. E. Grinter (ed.), *Numerical Methods in Analysis in Engineering*, Macmillan, 1949.

5. Courant, R., "Variational Methods for the Solution of Problems of Equilibrium and Vibrations," *Bull. Am. Math. Soc.*, 49, 1–23, 1943.
6. Levy, S., "Structural Analysis and Influence Coefficients for Delta Wings," *J. Aero. Sci.*, 20(7): 449–454, 1953.
7. Argyris, J. H., "Energy Theorems and Structural Analysis," *Aircraft Eng.*, Oct., Nov., Dec. 1954 and Feb., Mar., Apr., May 1955.
8. Argyris, J. H., and S. Kelsey, *Energy Theorems and Structural Analysis*, Butterworths, 1960 (reprinted from Aircraft Eng., 1954–55).
9. Turner, M. J., R. W. Clough, H. C. Martin, and L. J. Topp, "Stiffness and Deflection Analysis of Complex Structures," *J. Aero. Sci.*, 23(9): 805–824, 1956.
10. Clough, R. W., "The Finite Element Method in Plane Stress Analysis," *Proceedings of the Second Conference on Electronic Computation*, American Society of Civil Engineers, Pittsburgh, PA, pp. 345–378, September 1960.
11. Budynas, R. G., *Advanced Strength and Applied Stress Analysis*, 2nd ed., McGraw-Hill, 1999.
12. Rizzo, F. J., "An Integral Equation Approach to Boundary Value Problems of Classical Elastostatics," *Q. Appl. Math.*, 25, 83–95, 1967.
13. Cruse, T. A., "Numerical Solutions in Three-Dimensional Elastostatics," *Int. J. Solids Struct.*, 5, 1258–1274, 1969.
14. Brebbia, C. A., *The Boundary Element Method for Engineers*, Pentech Press, 1978.
15. Banerjee, P. K., and R. Butterfield, *Boundary Element Methods in Engineering Science*, McGraw-Hill, 1981.
16. Trevelyan, J., *Boundary Elements for Engineers*, Computational Mechanics Publications, 1994.

5.8 ADDITIONAL UNCITED REFERENCES FOR FINITE ELEMENTS

Bathe, K. J., *Finite Element Procedures*, Prentice-Hall, 1996.

Chandrupatla, T. R., and A. D. Belegundu, *Introduction to Finite Elements in Engineering*, 2nd ed., Prentice-Hall, 1997.

Cook, R. D., *Concepts and Applications of Finite Element Analysis*, 2nd ed., John Wiley & Sons, 1981.

Cook, R. D., *Finite Element Modeling for Stress Analysis*, John Wiley & Sons, 1995.

Cook, R. D., D. S. Malkus, and M. E. Plesha, *Concepts and Applications of Finite Element Analysis*, 3rd ed., John Wiley & Sons, 1989.

Hoffman, J. D., *Numerical Methods for Engineers and Scientists*, 1st ed., McGraw-Hill, 2001.

Love, A. E., *Treatise on the Mathematical Theory of Elasticity*, Dover Publications, 1927.

Reddy, J. N., *An Introduction to the Finite Element Method*, 2nd ed., McGraw-Hill, 1984.

Rothbart, H. A., and T. H. Brown, Jr., *Mechanical Design Handbook*, 2nd ed., McGraw-Hill, 2006.

5.9 ADDITIONAL UNCITED REFERENCES FOR BOUNDARY ELEMENTS

Benedetti, I., Aliabadi, M. H., and Milazzo, A., "A Fast BEM for the Analysis of Damaged Structures with Bonded Piezoelectric Sensors, *Comput. Methods Appl. Mech. Eng.*, 199, 490–501, 2010.

Buroni, F. C., Ortiz, J. E., and Sáez, A., "Multiple Pole Residue Approach for 3D BEM Analysis of Mathematical Degenerate and Non-Degenerate Materials," *Int. J. Numer. Methods Eng.*, 86, 1125–1143, 2011.

Chati, M. K., Paulino, G. H., and Mukherjee, S., "The Meshless Standard and Hypersingular Boundary Node Methods—Applications to Error Estimation and Adaptivity in Three-Dimensional Problems," *Int. J. Numer. Methods Eng.*, 50, 2233–2269, 2001.

Liu, Y. J., Fast Multipole Boundary Element Method—Theory and Applications in Engineering, Cambridge University Press, Cambridge, 2009.

Mukherjee, S., "Boundary Element Methods in Solid Mechanics—A Tribute to Frank Rizzo," *Electron. J. Boundary Elem.*, 1(1): 47–55, 2003.

Pan, E., "Anisotropic Green's Functions and BEMs (Editor)," *Eng. Anal. Boundary Elem.*, 29, 161, 2005.

Benedetti, I., Aliabadi, M. H., and Milazzo, A., "A Fast BEM for the Analysis of Damaged Structures with Bonded Piezoelectric Sensors, *Comput. Methods Appl. Mech. Eng.*, 199, 490–501, 2010.

Buroni, F. C., Ortiz, J. E., and Sáez, A., "Multiple Pole Residue Approach for 3D BEM Analysis of Mathematical Degenerate and Non-Degenerate Materials," *Int. J. Numer. Methods Eng.*, 86, 1125–1143, 2011.

Chati, M. K., Paulino, G. H., and Mukherjee, S., "The Meshless Standard and Hypersingular Boundary Node Methods—Applications to Error Estimation and Adaptivity in Three-Dimensional Problems," *Int. J. Numer. Methods Eng.*, 50, 2233–2269, 2001.

Liu, Y. J., *Fast Multipole Boundary Element Method—Theory and Applications in Engineering*, Cambridge University Press, Cambridge, 2009.

Mukherjee, S., "Boundary Element Methods in Solid Mechanics—A Tribute to Frank Rizzo," *Electron. J. Boundary Elem.*, 1(1): 47–55, 2003.

Pan, E., "Anisotropic Green's Functions and BEMs (Editor)," *Eng. Anal. Boundary Elem.*, 29, 161, 2005.

CHAPTER 6

Experimental Methods

A structural member may be of such a form or may be loaded in such a way that the direct use of formulas for the calculation of stress and strain produced in it is ineffective. One then must resort either to numerical techniques such as that presented in Chap. 5, finite element method or to experimental methods. Experimental methods can be applied to the actual member in some cases, or to a model thereof. Which choice is made depends upon the results desired, the accuracy needed, the practicality of size, and the cost associated with the experimental method. There has been a tremendous increase in the use of numerical methods over the years, but the use of experimental methods is still very effective. Many investigations make use of both numerical and experimental results to cross feed information from one to the other for increased accuracy and cost effectiveness. Some of the more important experimental methods are described briefly in Sec. 6.1 of this chapter. Of these methods, the most popular method employs electrical resistance strain gages and is described in more detail in Sec. 6.2. Only textbooks, reference books, handbooks, and lists of journals are referenced, since there are several organizations (see Refs. 1, 25, and 26) devoted either partially or totally to experimental methods, and a reasonable listing of papers would be excessive and soon out of date. The most useful reference for users wanting information on experimental methods is Ref. 27, the *Handbook on Experimental Mechanics*, edited by A. S. Kobayashi and dedicated to the late Dr. M. Hetenyi, who edited Ref. 2. Reference 27 contains 22 chapters contributed by 27 authors under the sponsorship of the Society for Experimental Mechanics. Experimental methods applied specifically to the field of fracture mechanics are treated extensively in Refs. 13, 15, 17, 19, 22, and Chaps. 14 and 20 of Ref. 27.

6.1 MEASUREMENT TECHNIQUES

The determination of stresses produced under a given loading of a structural system by means of experimental techniques is based on the measurement of deflections. Since strain is directly related to (the rate of change of) deflection, it is common practice to say that the measurements made are that of strain. Stresses are then determined implicitly using the stress–strain relations. Deflections in a structural system can be measured through changes in resistance, capacitance, or inductance of electrical elements; optical effects of interference, diffraction, or refraction; or thermal emissions. Measurement is comparatively easy when the stress is fairly uniform over a considerable length of the part in question, but becomes more difficult when the stress is localized or varies greatly with position. Short gage lengths and great precision require stable gage elements and stable electronic amplification if used. If dynamic strains are to be measured, a suitable high-frequency response is also necessary. In an isotropic material undergoing uniaxial stress, one normal strain measurement is all that is necessary. On a free surface under biaxial stress conditions, two measured orthogonal normal strains will provide the stresses in the same directions of the measured strains. On a free surface under a general state of plane stress, three measured normal strains in different directions will allow the determination of the stresses in directions at that position (see Sec. 6.2). At a free edge in a member that is thin perpendicular to the free edge, the state of stress is uniaxial and, as stated earlier, can be determined from one normal strain tangent to the edge. Another tactic might be to measure the change in thickness or the through-thickness strain at the edge. This might be more practical, such as measuring the strain at the bottom of a groove in a thin plate. For example, assume an orthogonal *xyz* coordinate system where *x* is parallel to the edge and *z* is in the direction of the thickness at the edge. Considering a linear, isotropic material, from Hooke's law, $\varepsilon_z = -\nu\sigma_x/E$. Thus, $\sigma_x = -E\varepsilon_z/\nu$.

The following descriptions provide many of the successful instruments and techniques used for strain measurement. They are listed in a general order of mechanical, electrical, optical, and thermal methods. Optical and thermal techniques have been greatly enhanced by advances in digital image processing technology for computers (see Chap. 21 of Ref. 27).

Mechanical Measurement

A direct measurement of strain can be made with an Invar tape over a gage length of several meters or with a pair of dividers over a reasonable fraction of a meter. For shorter gage lengths, mechanical amplification can be used, but friction is a problem and vibration can make them difficult to mount and to read. Optical magnification using mirrors still requires mechanical levers or rollers and is an improvement but still not satisfactory for most applications. In a laboratory setting, however, such mechanical and optical magnification can be used successfully. See Ref. 3 for more detailed descriptions. A scratch gage uses scratches on a polished target to determine strain amplitudes, and while the scratches are in general not strictly related to time, they are usually related to events in such a way as to be extremely useful in measuring some dynamic events. The scratched target is viewed with a microscope to obtain peak-to-peak strains per event, and a zero strain line can also be scratched on the target if desired (Ref. 3). The use of lasers and/or optical telescopes with electronic detectors to evaluate the motion of fiduciary marks on distant structures makes remote-displacement measurements possible, and when two such detectors are used, strains can be measured. While the technique is valuable when needed for remote measurement, generally for environmental reasons, it is an expensive technique for obtaining the strain at a single location.

Brittle Coatings

Surface coatings formulated to crack at strain levels well within the elastic limit of most structural materials provide a means of locating points of maximum strain and the directions of principal strains. Under well-controlled environmental conditions and with suitable calibration, such coatings can yield quantitative results (Refs. 2, 3, 7, 9, 20, 21, and 27). This technique, however, is not universally applicable, since the coatings may not be readily available due to environmental problems with the coating materials.

Electrical Strain and Displacement Gages

The evolution of electrical gages has led to a variety of configurations where changes in resistance, capacitance, or inductance can be related to strain and displacement with proper instrumentation (Refs. 2–5, 20, 21, 23, 24, and 27).

(a) Resistance Strain Gage

For the electrical resistance strain gages, the gage lengths vary from less than 0.01 in. to several inches. The gage grid material can be metallic or a semiconductor. The gages can be obtained in alloys that are designed to provide minimum output due to temperature strains alone and comparatively large outputs due to stress-induced strains. Metallic bonded-foil gages are manufactured by a photoetching process that allows for a wide range of configurations of the grid(s). The semiconductor strain gages provide the largest resistance change for a given strain, but are generally very sensitive to temperature changes. They are used in transducers where proper design can provide temperature compensation. The use of electrical resistance strain gages for stress analysis purposes constitutes the majority of experimental applications. For this reason, Sec. 6.2 provides further information on the use of these gages.

(b) Capacitance Strain Gage

Capacitance strain gages are larger and more massive than bonded electric resistance strain gages and are more widely used for applications beyond the upper temperature limits of the bonded resistance strain gages.

(c) Inductance Strain Gages

The change in air gap in a magnetic circuit can create a large change in *inductance* depending upon the design of the rest of the magnetic circuit. The large change in inductance is accompanied by a large change in *force* across the gap, and so the very sensitive inductance strain gages can be used only on more massive structures. They have been used as overload indicators on presses with no electronic amplification necessary. The linear relationship between core motion and output voltage of a *linear differential transformer* makes possible accurate measurement of displacements over a wide range of gage lengths and under a wide variety of conditions. The use of displacement data as input for work in experimental modal analysis is discussed in Chap. 16 of Ref. 27 and in many of the technical papers in Ref. 24.

Interferometric Strain Gages

Whole-field interferometric techniques will be discussed later, but a simple strain gage with a short length and high sensitivity can be created by several methods. In one, a diffraction grating is deposited at the desired location and in the desired direction, and the change in grating pitch under strain is measured. With a metallic grid, these strain gages can be used at elevated temperatures. Another method, also useable at high temperatures, makes use of the interference of light reflected from the inclined surfaces of two very closely spaced indentations in the surface of a metallic specimen. Both of these methods are discussed and referenced in Ref. 27.

Photoelastic Analysis

When a beam of polarized light passes through an elastically stressed transparent isotropic material, the beam may be treated as having been decomposed into two rays polarized in the planes of the principal stresses in the material. In birefringent materials the indexes of refraction of the material encountered by these two rays will depend upon the principal stresses. Therefore, interference patterns will develop which are proportional to the differences in the principal stresses.

(a) Two-Dimensional Analysis

With suitable optical elements—polarizers and wave plates of specific relative retardation—both the principal stress differences and the directions of principal stresses may be determined at every point in a two-dimensional specimen (Refs. 2–6, 10, 14, 18, 27, and 28). Many suitable photoelastic plastics are available. The material properties that must be considered are transparency, sensitivity (relative index of refraction change with stress), optical and mechanical creep, modulus of elasticity, ease of machining, cost, and stability (freedom from stresses developing with time). Materials with appropriate creep properties may be used for *photoplasticity* studies (Ref. 16).

(b) Three-Dimensional Analysis

Several photoelastic techniques are used to determine stresses in three-dimensional specimens. If information is desired at a single point only, the optical polarizers, wave plates, and photoelastically sensitive material can be *embedded* in a transparent model (Ref. 2) and two-dimensional techniques used. A modification of this technique, *stress freezing*, is possible in some biphase materials. By heating, loading, cooling, and unloading, it is possible to lock permanently into the specimen, on a *molecular* level, strains proportional to those present under load. Since equilibrium exists at a molecular level, the specimen can be cut into two-dimensional slices and all *secondary principal stress differences* determined. The secondary principal stresses at a point are defined as the largest and smallest normal stresses in the plane of the slice; these in general will not correspond with the principal stresses at that same point in the three-dimensional structure. If desired, the specimen can be cut into cubes and the three principal stress differences determined. The individual principal stresses at a given point cannot be determined from photoelastic data taken at that point alone since the addition of a hydrostatic stress to any cube of material would not be revealed by differences in the indexes of refraction. Mathematical integration techniques, which start at a point where the hydrostatic stress component is known, can be used with photoelastic data to determine all individual principal stresses.

A third method, *scattered light photoelasticity*, uses a laser beam of intense monochromatic polarized light or a similar thin sheet of light passing through photoelastically sensitive transparent models that have the additional property of being able to scatter uniformly a small portion of the light from any point on the beam or sheet. The same general restrictions apply to this analysis as applied to the stress-frozen three-dimensional analysis except that the specimen does not have to be cut. However, the amount of light available for analysis is much less, the specimen must be immersed in a fluid with an index of refraction that very closely matches that of the specimen, and in general the data are much more difficult to analyze.

(c) Photoelastic Coating

Photoelastic coatings have been sprayed, bonded in the form of thin sheets, or cast directly in place on the surface of models or structures to determine the two-dimensional surface strains. The surface is made reflective before bonding the plastics in place so the effective thickness of the photoelastic plastic is doubled and all two-dimensional techniques can be applied with suitable instrumentation.

Moiré Techniques

All moiré techniques can be explained by optical interference, but the course-grid techniques can also be evaluated on the basis of obstructive or mechanical interference.

(a) Geometric Moiré

Geometric moiré techniques use grids of alternate equally wide bands of relatively transparent or light-colored material and opaque or dark-colored material in order to observe the relative motion of two such grids. The most common technique (Refs. 2, 5, 8, and 11) uses an alternate transparent and opaque grid to produce photographically a matching grid on the flat surface of the specimen. Then the full-field relative motion is observed between the reproduction and the original when the specimen is loaded. Similarly, the original may be used with a projector to produce the photographic image on the specimen and then produce interference with the projected image after loading. These methods can use ordinary white light, and the interference is due merely to *geometric* blocking of the light as it passes through or is reflected from the grids.

Another similar technique, *shadow moiré*, produces interference patterns due to motion of the specimen at right angles to its surface between an alternately transparent and opaque grid and the shadow of the grid on the specimen.

(b) Moiré Interferometry

Interferometry provides a means of producing both specimen gratings and reference gratings. Virtual reference gratings of more than 100,000 lines per inch have been utilized. Moiré interferometry provides contour maps of in-plane displacements, and, with the fine pitches attainable, differentiation to obtain strains from this experimental process is comparable to that used in the finite-element method of numerical analysis where displacement fields are generally the initial output. See Chap. 7 in Ref. 27.

Holographic and Laser Speckle Interferometry

The rapid evolution of holographic and laser speckle interferometry is related to the development of high-power lasers and to the development of digital computer enhancement of the resulting images. Various techniques are used to measure the several displacement components of diffuse reflecting surfaces. Details are beyond the scope of this book and are best reviewed in Chap. 8 of Ref. 27.

Shadow Optical Method of Caustics

The very simple images created by the reflection or refraction of light from the surface contours of high-gradient stress concentrations such as those at the tips of cracks make the use of the shadow optical method of caustics very useful for dynamic studies of crack growth or arrest. Chapter 9 of Ref. 27 gives a detailed discussion of this technique and a comparison to photoelastic studies for the same loadings.

X-ray Diffraction

X-ray diffraction makes possible the determination of changes in interatomic distance and thus the measurement of elastic strain. The method has the particular advantages that it can be used at points of high stress concentration and to determine residual stresses without cutting the object of investigation.

Stress-Pattern Analysis by Thermal Emission

This technique uses computer enhancement of infrared detection of very small temperature changes in order to produce digital output related to stress at a point on the surface of a structure, a stress graph along a line on the surface, or a full-field isopachic stress map of the surface. Under cyclic loading, at a frequency high enough to assure that any heat transfer due to stress gradients is insignificant, the thermoelastic effect produces a temperature change proportional to the change in the sum of the principal stresses. Although calibration corrections must be made for use at widely differing ambient temperatures, the technique works over a wide range of temperatures and on a variety of structural materials including metals, wood, concrete, and plain and reinforced plastics. Tests have been made on some metals at temperatures above 700°C. Chapter 14 of Ref. 27 describes and surveys work on this technique.

6.2 ELECTRICAL RESISTANCE STRAIN GAGES

General

The use of electrical resistance strain gages is probably the most common method of measurement in experimental stress analysis. In addition, strain gage technology is quite important in the design of transducer instrumentation for the measurement of force, torque, pressure, etc.

Electrical resistance strain gages are based on the principal that the resistance R of a conductor changes as a function of normal strain ε. The resistance of a conductor can be expressed as

$$R=\rho\frac{L}{A} \tag{6.2-1}$$

where ρ is the resistivity of the conductor (ohms-length), and L and A are the length and cross-sectional area of the conductor, respectively. It can be shown that a change in R due to changes in ρ, L, and A is given by

$$\frac{\Delta R}{R}=(1+2\nu)\varepsilon+\frac{\Delta\rho}{\rho} \tag{6.2-2}$$

where ν is Poisson's ratio, and assuming small strain on the conductor, ε, which is given by $\Delta L/L$. If the change in the resistance of the conductor is considered to be only due to the applied strain, then Eq. (6.2-2) can be written as

$$\frac{\Delta R}{R}=S_a\varepsilon \tag{6.2-3}$$

where

$$S_a=1+2\nu+\frac{\Delta\rho/\rho}{\varepsilon} \tag{6.2-4}$$

S_a is the *sensitivity* of the conductor to strain*. The first two terms come directly from changes in dimension of the conductor where for most metals the quantity $1+2\nu$ varies from 1.4 to 1.7. The last term in Eq. (6.2-4) is called the change in specific resistance relative to strain, and for some metals can account for much of the sensitivity to strain. The most commonly used material for strain gages is a copper–nickel alloy called Constantan, which has a strain sensitivity of 2.1. Other alloys used for strain gage applications are modified

*When using a commercial strain indicator, one must enter the sensitivity provided by the gage manufacturer. This sensitivity is referred to the gage factor of the gage, S_g. This is defined slightly differently than S_a and will be discussed shortly.

Karma, Nichrome V, and Isoelastic, which have sensitivities of 2.0, 2.2, and 3.6, respectively. The primary advantages of Constantan are:

1. The strain sensitivity S_a is linear over a wide range of strain and does not change significantly as the material goes plastic.
2. The thermal stability of the material is excellent and is not greatly influenced by temperature changes when used on common structural materials.
3. The metallurgical properties of Constantan are such that they can be processed to minimize the error induced due to the mismatch in the thermal expansion coefficients of the gage and the structure to which it is adhered over a wide range of temperature.

Isoelastic, with a higher sensitivity, is used for dynamic applications. Semiconductor gages are also available, and can reach sensitivities as high as 175. However, care must be exercised with respect to the poor thermal stability of these piezoresistive gages.

Most gages have a nominal resistance of 120 ohm or 350 ohm. Considering a 120-ohm Constantan gage, to obtain a measurement of strain within an accuracy of $\pm 5\ \mu$, it would be necessary to measure a change in resistance within ±1.2 Mohm. To measure these small changes in resistance accurately, commercial versions of the Wheatstone bridge, called *strain gage indicators*, are available.

Metallic alloy electrical resistance strain gages used in experimental stress analysis come in two basic types: bonded-wire and bonded-foil (see Fig. 6.1). Today, bonded-foil gages are by far the more prevalent. The resistivity of Constantan is approximately 49 μohm·cm. Thus if a strain gage is to be fabricated using a wire 0.025 mm in diameter and is to have a resistance of 120 ohm, the gage would require a wire approximately 120 mm long. To make the gage more compact over a shorter active length, the gage is constructed with many loops as shown in Fig. 6.1. Typical commercially available bonded-foil gage lengths vary from 0.20 mm (0.008 in) to 101.6 mm (4.000 in). For normal applications, bonded-foil gages either come mounted on a very thin polyimide film carrier (backing) or are encapsulated between two thin films of polyimide. Other carrier materials are available for special cases such as high-temperature applications.

The most widely used adhesive for bonding a strain gage to a test structure is the pressure-curing methyl 2-cyanoacrylate cement. Other adhesives include epoxy, polyester, and ceramic cements.

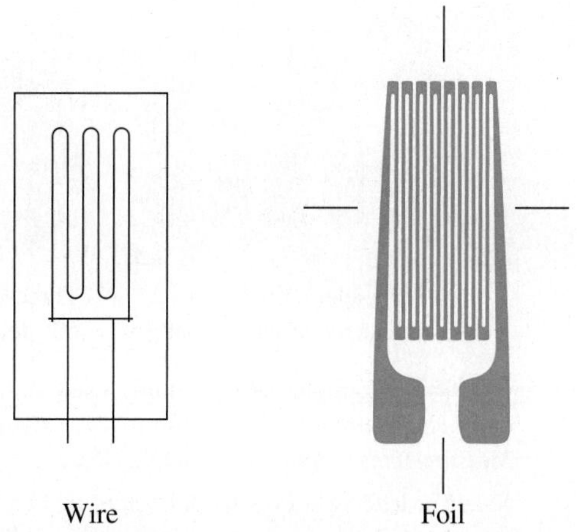

Figure 6.1 Forms of electrical resistance strain gages.

Extreme care must be exercised when installing a gage, since a good bond and an electrically insulated gage are necessary. The installation procedures can be obtained from technical instruction bulletins supplied by the manufacturer. Once a gage is correctly mounted, wired, resistance tested for continuity and insulation from the test structure, and waterproofed (if appropriate), it is ready for instrumentation and testing.

Strain Gage Configurations

In both wire or foil gages, many configurations and sizes are available. Strain gages come in many forms for transducer or stress-analysis applications. The fundamental configurations for stress-analysis work are shown in Fig. 6.2.

A strain gage is mounted on a free surface, which in general, is in a state of plane stress where the state of stress with regards to a specific *xy* rectangular coordinate system can be unknown up to the three stresses, σ_x, σ_y, and τ_{xy}. Thus, if the state of stress is completely unknown on a free surface, it is necessary to use a three-element rectangular or delta rosette since each gage element provides only one piece of information, the indicated normal strain at the point in the direction of the gage.

To understand how the rosettes are used, consider the three-element rectangular rosette shown in Fig. 6.3(*a*), which provides normal strain components in three directions spaced at angles of 45°.

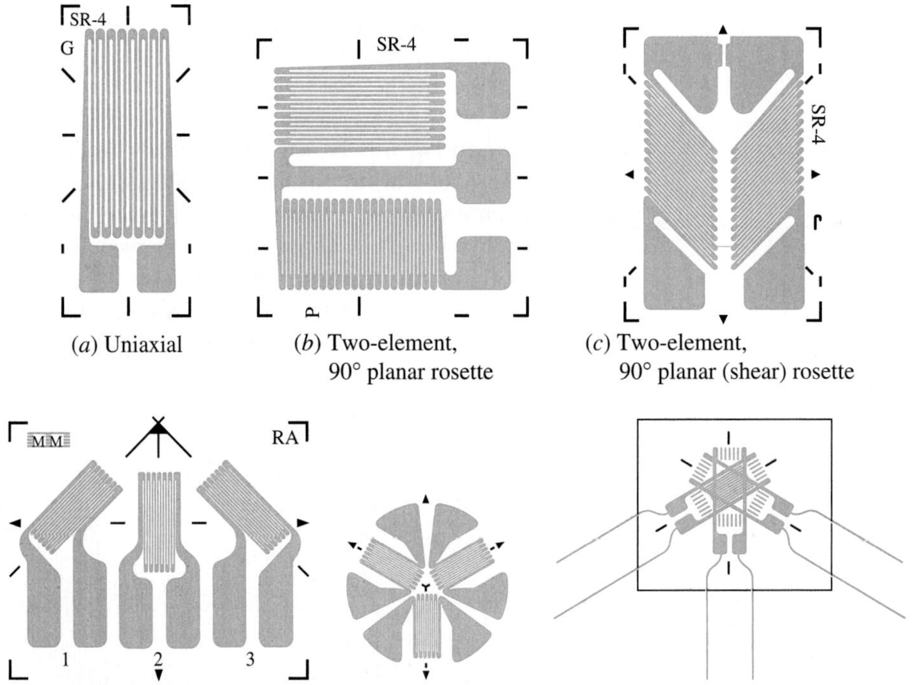

(*a*) Uniaxial (*b*) Two-element, 90° planar rosette (*c*) Two-element, 90° planar (shear) rosette

(*d*) Three-element, 45° planar rectangular rosette (*e*) Three-element, 60° delta rosette (*f*) Three-element, 60° stacked delta rosette

Figure 6.2 Examples of commonly used strain gage configurations. (*Source*: Figures *a-c* courtesy of *BLH Electronics, Inc.*, Canton, MA. Figures *d-f* courtesy of *Micro-Measurements Division of Measurements Group, Inc.*, Raleigh, NC.)

Note: The letters *SR-4* on the *BLH* gages are in honor of E. E. Simmons and Arthur C. Ruge and their two assistants (a total of four individuals), who, in 1937–1938, independently produced the first bonded-wire resistance strain gage.

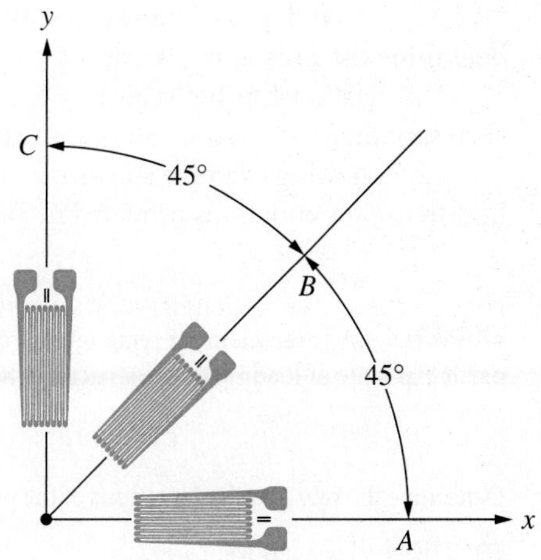

(*a*) Rectangular rosette

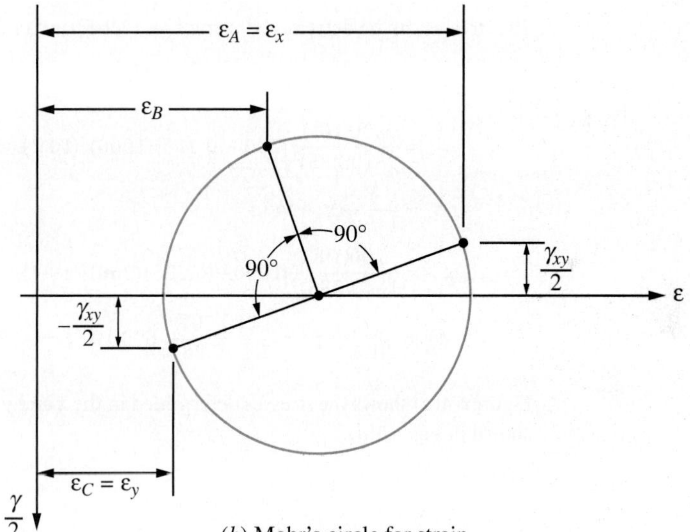

(*b*) Mohr's circle for strain

Figure 6.3 Three-element strain gage rosette.

If an *xy* coordinate system is assumed to coincide with gages *A* and *C*, then $\varepsilon_x = \varepsilon_A$ and $\varepsilon_y = \varepsilon_C$. Gage *B* in conjunction with gages *A* and *C* provides information necessary to determine γ_{xy}. Recalling the first of Eqs. (2.4-1), $\varepsilon_{x'} = \varepsilon_x \cos^2\theta + \varepsilon_y \sin^2\theta + \gamma_{xy}\cos\theta\sin\theta$, with $\theta = 45°$

$$\varepsilon_B = \varepsilon_x \cos^2 45° + \varepsilon_y \sin^2 45° + \gamma_{xy} \cos 45° \sin 45°$$
$$= \frac{1}{2}(\varepsilon_x + \varepsilon_y + \gamma_{xy}) = \frac{1}{2}(\varepsilon_A + \varepsilon_C + \gamma_{xy})$$

Solving for γ_{xy} yields

$$\gamma_{xy} = 2\varepsilon_B - \varepsilon_A - \varepsilon_C$$

Once ε_x, ε_y, and γ_{xy} are known, Hooke's law [Eqs. (2.2-5) and (2.2-6*a*)] can be used to determine the stresses σ_x, σ_y, and τ_{xy}.

The relationship between ε_A, ε_B, and ε_C can be seen from Mohr's circle of strain corresponding to the strain state at the point under investigation [see Fig. 6.3(*b*)].

The following example shows how to use the above equations for an analysis as well as how to use the equations provided in Table 6.3.

EXAMPLE A three-element rectangular rosette strain gage is installed on a steel specimen. For a particular state of loading of the structure the strain gage readings are*

$$\varepsilon_A = 200\,\mu, \qquad \varepsilon_B = 900\,\mu, \qquad \varepsilon_C = 1000\,\mu$$

Determine the values and orientations of the principal stresses at the point. Let $E = 200$ GPa and $\nu = 0.285$.

SOLUTION From above,

$$\varepsilon_x = \varepsilon_A = 200\,\mu, \qquad \varepsilon_y = \varepsilon_C = 1000\,\mu$$

$$\gamma_{xy} = 2\varepsilon_B - \varepsilon_A - \varepsilon_C = (2)(900) - 200 - 1000 = 600\,\mu$$

The stresses can be determined using Eqs. (2.2-5) and (2.2-6*a*):

$$\sigma_x = \frac{E}{1-\nu^2}(\varepsilon_x + \nu\varepsilon_y)$$

$$= \frac{200(10^9)}{1-(0.285)^2}[200 + (0.285)(1000)](10^{-6}) = 105.58(10^6)\,\text{N/m}^2 = 105.58\,\text{MPa}$$

$$\sigma_y = \frac{E}{1-\nu^2}(\varepsilon_y + \nu\varepsilon_x)$$

$$= \frac{200(10^9)}{1-(0.285)^2}[1000 + (0.285)(200)](10^{-6}) = 230.09(10^6)\,\text{N/m}^2 = 230.09\,\text{MPa}$$

$$\tau_{xy} = \frac{E}{2(1+\nu)}\gamma_{xy} = \frac{200(10^9)}{2(1+0.285)}600(10^{-6}) = 46.69(10^6)\,\text{N/m}^2 = 46.69\,\text{MPa}$$

Figure 6.4(*a*) shows the stresses determined in the *x* and *y* directions as related to the gage orientation shown in Fig. 6.3(*a*).

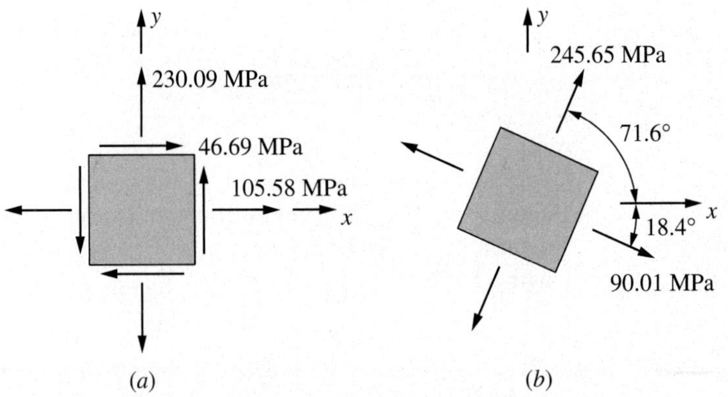

Figure 6.4 (*a*) Stresses in the *x* and *y* directions. (*b*) Principal stresses.

*The strain gage readings are typically corrected due to the effect of transverse strains on each gage. This will be discussed shortly.

For the principal stress axes, we use Eq. (2.3-23) given by

$$\sigma_p = \frac{1}{2}\left[(\sigma_x + \sigma_y) \pm \sqrt{(\sigma_x - \sigma_y)^2 + 4\tau_{xy}^2}\right]$$
$$= \frac{1}{2}\left[105.58 + 230.09 \pm \sqrt{(105.58 - 230.09)^2 + 4(46.67)^2}\right]$$
$$= 245.65, \quad 90.01\,\text{MPa}$$

For the orientation of the principal stress axes, using the first of Eqs. (2.3-21) gives

$$\theta_p = \tan^{-1}\left(\frac{\sigma_p - \sigma_x}{\tau_{xy}}\right) \tag{a}$$

For the principal stress, $\sigma_1 = 245.65$ MPa, Eq. (*a*) gives

$$\theta_{p1} = \tan^{-1}\left(\frac{245.65 - 105.58}{46.69}\right) = 71.6°$$

For the other principal stress, $\sigma_2 = 90.01$ MPa

$$\theta_{p2} = \tan^{-1}\left(\frac{90.01 - 105.58}{46.69}\right) = -18.4°$$

Recalling that θ_p is defined positive in the counterclockwise direction, the principal stress state at the point relative to the *xy* axes of the strain gage rosette corresponds to that shown in Fig. 6.4(*b*).

Using the equations given in Table 6.3 at the end of the chapter,

$$\frac{\varepsilon_A + \varepsilon_C}{1 - \nu} = \frac{200 + 1000}{1 - 0.285} = 1678.3\,\mu$$

$$\frac{1}{1+\nu}\sqrt{(\varepsilon_A - \varepsilon_C)^2 + (2\varepsilon_B - \varepsilon_A - \varepsilon_C)^2} = \frac{1}{1 + 0.285}\sqrt{(200 - 1000)^2 + [2(900) - 200 - 1000]^2} = 778.2\,\mu$$

Thus,

$$\sigma_{p1} = \frac{200(10)^9}{2}(1678.3 + 778.2) = 245.65\,\text{MPa}$$

$$\sigma_{p2} = \frac{200(10)^9}{2}(1678.3 - 778.2) = 90.01\,\text{MPa}$$

The principal angle is

$$\theta_p = \frac{1}{2}\tan^{-1}\left(\frac{2(900) - 200 - 1000}{200 - 1000}\right) = \frac{1}{2}\tan^{-1}\left(\frac{+600}{-800}\right) = \frac{1}{2}(143.13°) = 71.6°$$

counterclockwise from the *x* axis (*A* gage) to $\sigma_{p1} = 245.65$ MPa.* Note that this agrees with Fig. 6.4(*b*).

*When calculating θ_p, do not change the signs of the numerator and denominator in the equation. The $\tan^{-1}$ is defined from 0° to 360°. For example, $\tan^{-1}(+/+)$ is in the range 0°–90°, $\tan^{-1}(+/-)$ is in the range 90°–180°, $\tan^{-1}(-/-)$ is in the range 180°–270°, and $\tan^{-1}(-/+)$ is in the range 270°–360°. Using this definition for θ_p, the calculation will yield the counterclockwise angle from the *x* axis to σ_{p1}, the greater of the two principal stresses in the plane of the gages.

Strain Gage Corrections

There are many types of corrections that may be necessary to obtain accurate strain gage results (see Refs. 27 and 28). Two fundamental corrections that are necessary correct the indicated strain errors due to strains on the specimen perpendicular (transverse) to the longitudinal axis of the gage and changes in temperature of the gage installation. With each strain gage, the manufacturer provides much information on the performance of the gage, such as its sensitivity to longitudinal and transverse strain and how the sensitivity of the gage behaves relative to temperature changes.

(a) Transverse Sensitivity Corrections

The strain sensitivity of a single straight uniform length of conductor in a uniform *uniaxial strain field* ε in the longitudinal direction of the conductor is given by Eq. (6.2-3), which is $S_a = (\Delta R/R)/\varepsilon$. In a general strain field, there will be strains perpendicular to the longitudinal axis of the conductor (*transverse strains*). Due to the width of the conductor elements and the geometric configuration of the conductor in the gage pattern, the transverse strains will also effect a change in resistance in the conductor. This is not desirable, since only the effect of the strain in the direction of the gage length is being sought.

To further complicate things, the sensitivity of the strain gage provided by the gage manufacturer is *not* based on a uniaxial strain field, but that of a uniaxial *stress* field in a tensile test specimen. For a uniaxial stress field let the axial and transverse strains be ε_a and ε_t respectively. The sensitivity provided by the gage manufacturer, called the *gage factor* S_g, is defined as $S_g = (\Delta R/R)\varepsilon_a$, where under a uniaxial stress field, $\varepsilon_t = -\nu_0 \varepsilon_a$. Thus

$$\frac{\Delta R}{R} = S_g \varepsilon_a \qquad \text{with} \qquad \varepsilon_t = -\nu_0 \varepsilon_a \tag{6.2-5}$$

The term ν_0 is Poisson's ratio of the material on which the manufacturer's gage factor was measured and is normally taken to be 0.285. If the gage is used under conditions where the transverse strain is $\varepsilon_t = -\nu\varepsilon_a$, then the equation $\Delta R/R = S_g \varepsilon_a$ would yield exact results. If $\varepsilon_t \neq -\nu_0 \varepsilon_a$, then some error will occur. This error depends on the sensitivity of the gage to transverse strain and the deviation of the ratio of $\varepsilon_t/\varepsilon_a$ from $-\nu_0$. The strain gage manufacturer generally supplies a transverse sensitivity coefficient, K_t, defined as S_t/S_a, where S_t is the transverse sensitivity factor. One cannot correct the indicated strain from a single strain reading. Thus it is necessary to have multiple strain readings from that of a strain gage rosette. Table 6.2 at the end of the chapter gives equations for the corrected strain values of the three most widely used strain gage rosettes. Corrected strain readings are given by ε whereas uncorrected strains from the strain gage indicator are given by $\hat{\varepsilon}$.

EXAMPLE In the previous example the indicated strains are $\hat{\varepsilon}_A = 200\,\mu$, $\hat{\varepsilon}_B = 900\,\mu$, and $\hat{\varepsilon}_C = 1000\,\mu$. Determine the principal stresses and directions if the transverse sensitivity coefficients of the gages are $K_{tA} = K_{tC} = 0.05$ and $K_{tB} = 0.06$.

SOLUTION From Table 6.2, let

$$\begin{aligned}\varepsilon_A &= \frac{(1-\nu_0 K_{tA})\hat{\varepsilon}_A - K_{tA}(1-\nu_0 K_{tC})\hat{\varepsilon}_C}{1-K_{tA}K_{tC}} \\ &= \frac{[1-(0.285)(0.05)](200)-(0.05)[1-(0.285)(0.05)](1000)}{1-(0.05)(0.05)} \\ &= 148.23\,\mu\end{aligned}$$

$$\varepsilon_B = \frac{(1-\nu_0 K_{tB})\hat{\varepsilon}_B - \dfrac{K_{tB}}{1-K_{tA}K_{tC}}[(1-\nu_0 K_{tA})(1-K_{tC})\hat{\varepsilon}_A + (1-\nu_0 K_{tC})(1-K_{tA})\hat{\varepsilon}_C]}{1-K_{tB}}$$

$$= \left([1-0.285(0.06)](900) - \frac{0.06}{1-0.05(0.05)} \right.$$

$$\left. \times \{[1-0.285(0.05)](1-0.05)(200) + [1-0.285(0.05)](1-0.05)(1000)\} \right)$$

$$\times (1-0.06)^{-1}$$

$$= 869.17\,\mu$$

and

$$\varepsilon_C = \frac{(1-\nu_0 K_{tC})\hat{\varepsilon}_C - K_{tC}(1-\nu_0 K_{tA})\hat{\varepsilon}_A}{1-K_{tA}K_{tC}}$$

$$= \frac{[1-(0.285)(0.05)](1000) - (0.05)[1-(0.285)(0.05)](200)}{1-(0.05)(0.05)}$$

$$= 978.34\,\mu$$

From Table 6.3,*

$$\frac{\varepsilon_A + \varepsilon_C}{1-\nu} = \frac{148.23+978.34}{1-0.285} = 1575.62\,\mu$$

$$\frac{1}{1+\nu}\sqrt{(\varepsilon_A - \varepsilon_C)^2 + (2\varepsilon_B - \varepsilon_A - \varepsilon_C)^2}$$

$$= \frac{1}{1+0.285}\sqrt{(148.23-978.34)^2 + [2(869.17)-148.23-978.34]^2} = 802.48\,\mu$$

and

$$\sigma_{p1} = \frac{200(10)^9}{2}(1575.62+802.48)(10^{-6}) = 237.8(10^6)\,\text{N/m}^2 = 237.81\,\text{MPa}$$

$$\sigma_{p2} = \frac{200(10)^9}{2}(1575.62-802.48)(10^{-6}) = 90.01(10^6)\,\text{N/m}^2 = 77.31\,\text{MPa}$$

$$\theta_p = \frac{1}{2}\tan^{-1}\left(\frac{2(869.17)-148.23-978.34}{148.23-978.34}\right) = \frac{1}{2}\tan^{-1}\left(\frac{611.77}{-830.11}\right)$$

$$= \frac{1}{2}(143.61^\circ) = 71.8^\circ$$

The principal stress element is shown in Fig. 6.5 relative to the *xy* coordinate system of the gage rosette as shown in Fig. 6.3(*a*).

(b) Corrections Due to Temperature Changes

Temperature changes on an installed strain gage cause a change in resistance, which is due to a mismatch in the thermal expansion coefficients of the gage and the specimen, a change in the resistivity of the gage material, and a change in the gage factor, S_g. This effect can be compensated for by two different methods. The first method of temperature compensation is achieved using an additional compensating gage on an adjacent arm of the Wheatstone bridge circuit. This compensating gage must be identical to the active gage, mounted on the

*Note that if ν_s for the specimen was different from $\nu_0 = 0.285$, ν_s would be used in the equations of Table 6.3 but *not* for Table 6.4.

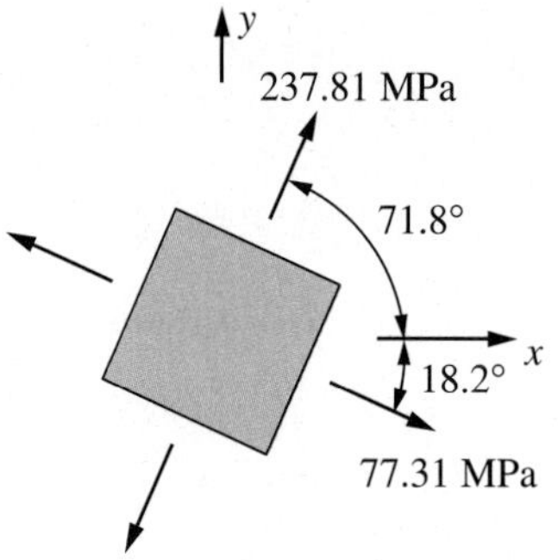

Figure 6.5 Principal stress element corrected for transverse sensitivity.

same material as the active gage, and undergoing an identical temperature change as that of the active gage.

The second method involves calibration of the gage relative to temperature changes. The gage can be manufactured and calibrated for the application on a specific specimen material. The metallurgical properties of alloys such as Constantan and modified Karma can be processed to minimize the effect of temperature change over a limited range of temperatures, somewhat centered about room temperature. Gages processed in this manner are called *self-temperature-compensated* strain gages. An example of the characteristics of a BLH self-temperature-compensated gage specifically processed for use on a low-carbon steel is shown in Fig. 6.6. Note that the apparent strain is zero at 22°C and 45°C and approximately

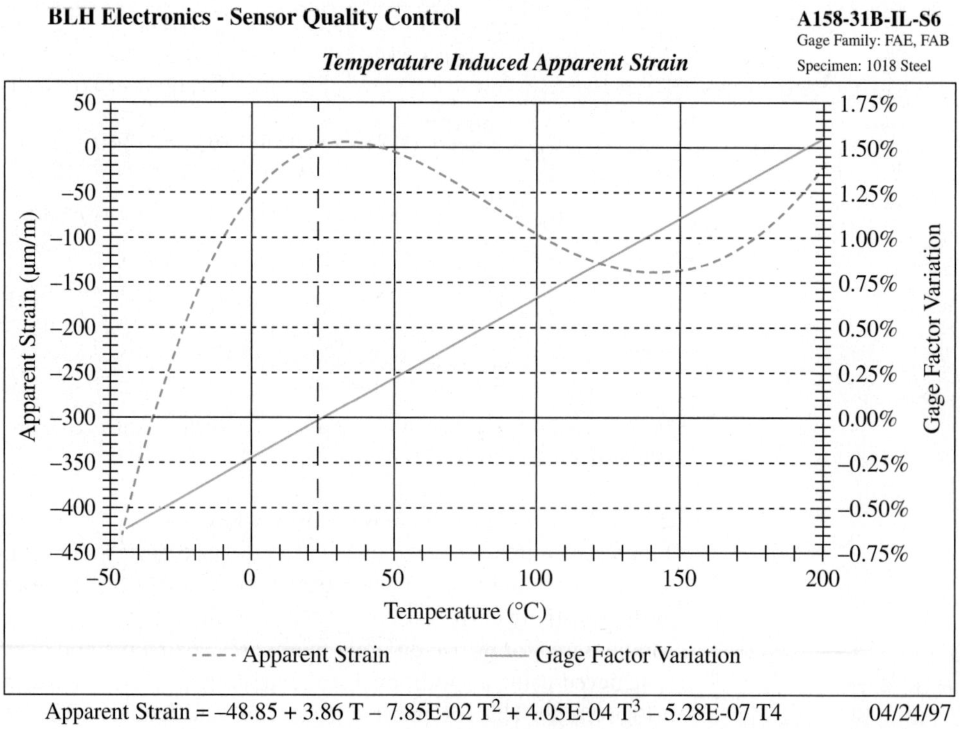

Figure 6.6 Strain gage temperature characteristics. (*Source*: Data sheet courtesy *BLH Electronics*, Inc., Canton, MA.)

zero in the vicinity of these temperatures. For temperatures beyond this region, compensation can be achieved by monitoring the temperature at the strain gage site. Then, using either the curve from the data sheet or the fitted polynomial equation, the strain readings can be corrected numerically. Note, however, that the curve and the polynomial equation given on the data sheet are based on a gage factor of 2.0. If corrections are anticipated, the gage factor adjustment of the strain indicator should be set to 2.0. An example that demonstrates this correction is given at the end of this section.

The gage factor variation with temperature is also presented in the data sheet of Fig. 6.6. If the strain gage indicator is initially set at $(S_g)_i$, the actual gage factor at temperature T is $(S_g)_T$, and the indicator registers a strain measurement of $\varepsilon_{\text{reading}}$, the corrected strain is

$$\varepsilon_{\text{actual}} = \frac{(S_g)_i}{(S_g)_T}\varepsilon_{\text{reading}} \tag{6.2-6}$$

where

$$(S_g)_T = \left(1 + \frac{\Delta S_g(\%)}{100}\right)(S_g)_i \tag{6.2-7}$$

and $\Delta S_g(\%)$ being the percent variation in gage factor given in Fig. 6.6.

If a simultaneous correction for apparent strain and gage factor variation is necessary, the corrected strain is given by

$$\varepsilon_{\text{actual}} = \frac{(S_g)_i}{(S_g)_T}(\varepsilon_{\text{reading}} - \varepsilon_{\text{apparent}}) \tag{6.2-8}$$

EXAMPLE A strain gage with the characteristics of Fig. 6.6 has a room-temperature gage factor of 2.1 and is mounted on a 1018 steel specimen. A strain measurement of −1800 μ is recorded during the test when the temperature is 150°C. Determine the value of actual test strain if:

(*a*) the gage is in a half-bridge circuit with a dummy temperature compensating gage and prior to testing, the indicator is zeroed with the gage factor set at 2.1.

(*b*) the gage is the only gage in a quarter-bridge circuit and prior to testing, the indicator is zeroed with the gage factor set at 2.0.

SOLUTION From Fig. 6.6, the gage factor variation at 150°C is $\Delta S_g(\%) = 1.13\%$. Thus, from Eq. (6.2-7), the gage factor at the test temperature is

$$(S_g)_T = \left(1 + \frac{1.13}{100}\right)(2.1) = 2.124$$

(*a*) Since in this part, a dummy gage is present that cancels the apparent strain, the only correction that is necessary is due to the change in the gage factor. From Eq. (6.2-6),

$$\varepsilon_{\text{actual}} = \left(\frac{2.1}{2.124}\right)(-1800) = -1780\,\mu$$

which we see is a minor correction.

(*b*) In this part, we must use Eq. (6.2-8). Using the equation given in Fig. 6.6, the apparent strain at the test temperature is

$$\begin{aligned}\varepsilon_{\text{apparent}} &= -48.85 + (3.86)(150) - (7.85E-02)(150)^2 \\ &\quad + (4.05E-04)(150)^3 - (5.28E-07)(150)^4 = -136.5\,\mu\end{aligned}$$

Substituting this into Eq. (6.2-8), with $(S_g)_i = 2.0$, gives

$$\varepsilon_{\text{actual}} = \left(\frac{2.0}{2.124}\right)[-1800-(-136.5)] = -1566\,\mu$$

which is *not* a minor correction.

6.3 DETECTION OF PLASTIC YIELDING

In parts made of ductile metal, sometimes a great deal can be learned concerning the location of the most highly stressed region and the load that produces elastic failure by noting the first signs of plastic yielding. Such yielding may be detected in the following ways.

Observation of Slip Lines

If yielding occurs first at some point on the surface, it can be detected by the appearance of slip lines if the surface is suitably polished.

Brittle Coating

If a member is coated with some material that will flake off easily, this flaking will indicate local yielding of the member. A coating of rosin or a wash of lime or white portland cement, applied and allowed to dry, is best for this purpose, but chalk or mill scale will often suffice. By this method zones of high stress such as those that occur in pressure vessels around openings and projections can be located and the load required to produce local yielding can be determined approximately.

Photoelastic Coatings

Thin photoelastic coatings show very characteristic patterns analogous to slip lines when the material beneath the coating yields.

6.4 ANALOGIES

Certain problems in elasticity involve equations that cannot be solved but that happen to be mathematically identical with the equations that describe some other physical phenomenon which can be investigated experimentally. Among the more useful of such analogies are the following.

Membrane Analogy

This is especially useful in determining the torsion properties of bars having noncircular sections. If in a thin flat plate holes are cut having the outlines of various sections and over each of these holes a soap film (or other membrane) is stretched and slightly distended by pressure from one side, the volumes of the bubbles thus formed are proportional to the torsional rigidities of the corresponding sections and the slope of a bubble surface at any point is proportional to the stress caused at that point of the corresponding section by a given twist per unit length of bar. By cutting in the plate one hole the shape of the section to be studied and another hole that is circular, the torsional properties of the irregular section can be determined by comparing the bubble formed on the hole of that shape with the bubble formed on the circular hole since the torsional properties of the circular section are known.

Electrical Analogy for Isopachic Lines

Isopachic lines are lines along which the sums of the principal stresses are equal in a two-dimensional plane stress problem. The voltage at any point on a uniform two-dimensional conducting surface is governed by the same form of equation as is the principal stress sum. Teledeltos paper is a uniform layer of graphite particles on a paper backing and makes an excellent material from which to construct the electrical analog. The paper is cut to a geometric outline corresponding to the shape of the two-dimensional structure or part, and boundary potentials are applied by an adjustable power supply. The required boundary potentials are obtained from a photoelastic study of the part where the principal stress sums can be found from the principal stress differences on the boundaries (Refs. 2 and 3). A similar membrane analogy has the height of a nonpressurized membrane proportional to the principal stress sum (Refs. 2 and 3).

6.5 WHEATSTONE BRIDGE

To measure the strain with strain gages an electric circuit is needed. The best known circuit is the Wheatstone bridge and is used to measure unknown electrical resistances by balancing legs of a bridge circuit. Its operation is similar to a potentiometer. The basic bridge is constructed from four resistances R_i ($i = 1$ to 4) as shown in Fig. 6.7. Initially the bridge is balanced such that $V_2 = 0$. To achieve this, $R_1R_3 = R_2R_4$. If one of the resistances, say R_4, is a strain gage, the resistance will change when the gage is strained. This will cause a change in the galvanometer voltage V_2. With one unknown active resistance, the bridge is said to be a *quarter bridge*. The bridge can be used for more than one active resistance as shown in Table 6.5. Consider the possibility that all four resistances can change from R_i to $R_i + \Delta R_i$ ($i = 1$ to 4). With the bridge initially balanced, it can be shown that for very small changes in ΔR_i the voltage change is

$$\Delta V_2 \approx r\left(\frac{\Delta R_4}{R_4} - \frac{\Delta R_3}{R_3} + \frac{\Delta R_2}{R_2} - \frac{\Delta R_1}{R_1}\right)V_1 \tag{6.2-9}$$

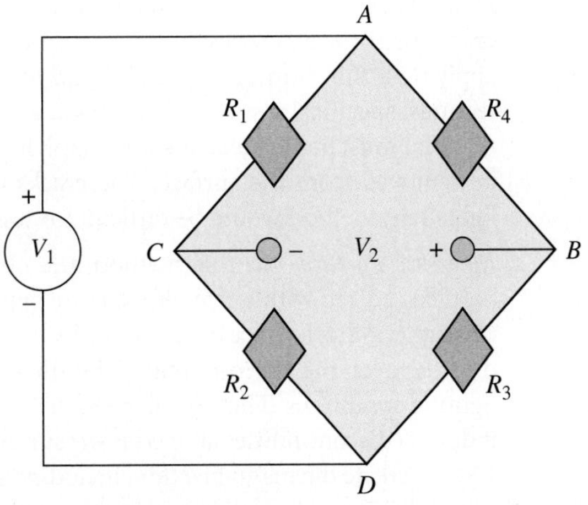

Figure 6.7 Basic Wheatstone bridge.

where

$$r = \frac{R_1 R_2}{(R_1 + R_2)^2} = \frac{R_3 R_4}{(R_3 + R_4)^2}$$

As discussed in Sec. 6.2, temperature correction can be achieved by using an additional compensating gage on an adjacent arm of the Wheatstone bridge circuit. This compensating (dummy) gage must be identical to the active gage, mounted on the same material as the active gage, and undergoing an identical temperature change as that of the active gage. Assuming the R_4 gage is the active gage and the R_3 gage is the dummy gage we see from Eq. (6.2-9) that a change in R_4 due to temperature is equal to the change in R_3 and is consequently cancelled. The second bridge in Table 6.5 shows this arrangement which is a *half bridge* arrangement.

6.6 NONDESTRUCTIVE TESTING

Nondestructive testing (NDT) is noninvasive techniques to determine the integrity of a material, or to quantitatively characterize mechanical properties of a part. The NDT does not affect or change the part's properties. While, there are many NDT methods, the most common methods are visual inspection, liquid penetrant methods, magnetic particle testing, ultrasonic testing, acoustic emission testing, X-ray radiography, eddy current methods, microwave inspection, and infrared thermography. These methods of NDT are briefly described. For more information, see Sec. 3.4 of Ref. 35. The popular three-dimensional film-based measurement hologram interferometry (HI) technique is being succeeded by its electronic analog, digital holographic interferometry (DHI).

Visual Inspection In this method in addition to use of naked-eye observation by skilled operators, more sophisticated implementations of this method include the use of fiberscopes and borescopes to inspect hard-to-access locations such as the interiors of tubes and vessels. Robotic devices have also been used in hazardous environments that extend beyond the reach of fiberscopes. Subjective observation and human error is a drawback of visual inspection. For this reason, proper training of the inspectors is a critical requirement of visual inspection.

Liquid Penetrant Tests This method is widely used in industry for the detection of surface cracks or surface porosity in various materials. The liquid penetrant technique consists of applying a thin film of a special liquid to the test part to highlight the images of surface features, specifically surface-breaking cracks. Before applying the liquid, the surface of the material must be cleaned; and after applying the liquid, the excessive penetrant liquid must be removed from the surface. The cracks can be visually observed under normal visual light. This method would be difficult for testing porous materials.

Magnetic Particle In this method, the object is first magnetized and then milled iron particles coated with a dye pigment are applied to the object. These particles are attracted to magnetic flux leakage fields caused by the discontinuities and cluster to form an indication directly over the discontinuity. This indication can be visually detected under proper lighting conditions. The advantage of this method over the liquid penetrant methods is that it detects discontinuities at or near the surface of ferromagnetic materials. There are several ways to create the magnetization, including solenoids, yokes, threaded cables, etc. There are two types of magnetic particles, namely, the "dry" type (dyed, usually red) and the "wet" type (fluorescent). The "dry" type can be viewed under visible light, whereas the fluorescent "wet" type requires illumination from a black light in a darkened environment.

Ultrasonic Method This is a popular class of NDT methods that uses ultrasonic waves (defined here as transient vibrations at frequencies above the audible range, i.e., >20 kHz) to probe the test object in order to detect the presence of defects or characterize geometrical, compositional, or mechanical properties. There are several wave modes that can be utilized in an ultrasonic test of a solid object, including longitudinal shear waves and Rayleigh waves. The discontinuities in the material affect the ultrasonic waves. When the size of the defect is equal to or larger than the ultrasonic wavelength, the wave undergoes a reflection (echo) due to the mismatch in acoustic impedances between the base material and the discontinuity. When, instead, the size of the defect is smaller than the ultrasonic wavelength, the wave undergoes amplitude scattering that macroscopically manifests itself as an attenuation (damping) during propagation.

Acoustic Emission This method uses the same propagating wave modes as the ultrasonic method; however, it relies on the test object itself to excite such waves. Therefore, acoustic emission is a method that requires only receiving transducers on the test object with the necessary excitation being provided only through mechanically or thermally stressing the object. The method is based on the fact that growing cracks, moving dislocations, and similar transient events within the test object release mechanical energy in the form of an acoustic or ultrasonic wave that can then be detected by appropriately spaced transducers. This method is primarily used for active and growing defects such as proof-testing of pressure vessels and curve-shaped structures.

X-Ray Radiography In this method, high-energy electromagnetic waves with short wavelengths ranging between 0.01 and 10 nanometers are used to penetrate into a material and they attenuate depending on the material composition and density. Changes in material composition or density cause changes in attenuation of the traveling radiation. Radiation detectors convert the incident X-rays into an optical or electrical signal that can then be converted into a negative image of the part that is impressed on the detector plate. This method allows to reconstruct three-dimensional images of the test object through computed tomography (CT), where the X-ray projection angles are changed to obtain a complete image of the object. The various projection angles can be generated by either rotating the test object with fixed positions of the X-ray source and detector, or by rotating the positions of the X-ray source and detector for a fixed orientation of the test object. The most common implementation of X-ray radiography in medicine and NDT is the projection radiography, where a collimated X-ray beam is sent through the test part and detected, in transmission, by an appropriate X-ray detector.

Eddy Current This method is used for inspection of electrically conductive materials. Currents in the test object are induced in response to a changing electromagnetic field that is generated by a changing current in a conductor that is placed in close proximity to the material. The magnetic fields are then detected either by electromagnetic induction in a coil or system of coils, or by other sensors such as the Hall effect devices. The presence of discontinuities or material variations alters the eddy currents, thus changing the apparent impedance of the inducing coil or of a detection coil.

Typical applications of eddy current testing include the measurement of internal or external diameter of tubes and pipes, the detection of corrosion or cracking defects, and the inspection of coatings. A particularly successful application is the thickness measurement of metallic and nonmetallic coatings on metals. Coating thicknesses measured typically range from 0.0001 to 0.100 in (0.00025 to 0.25 cm). For these measurements to be possible, the coating conductivity must differ from that of the base metal.

Microwaves Microwave inspection generally consists of measuring various properties of the electromagnetic wave scattered by, or transmitted through, the test object. Microwaves are electromagnetic radiations with wavelengths lying between 1 mm and 1 m. The operating frequency can be selected to maximize the interaction with dielectric layers, voids, inclusions, and surface flaws in the test object. The signal strength into a dielectric material is dictated by the incident power level, the loss factor (or absorption) of the material, and the wave frequency. In NDT, the waves are usually polarized to allow for selective interaction with an elongated anomaly. Microwaves generated in a test instrument are transmitted by a waveguide through air to the test object. Analysis of the scattered (reflected) or transmitted energy indicates certain material characteristics, such as moisture content, composition, structure, density, degree of cure, aging, and presence of flaws.

Infrared Thermography This test is a thermography method that relies on the disruption of the heat conduction by an object's discontinuities (e.g., internal defects). The thermal diffusion equation governs the conduction of heat in a solid. An important material property governing this process is the thermal diffusivity. When an object has an internal discontinuity (e.g., a defect), the difference in thermal diffusivities between the base material and the discontinuity causes a disruption in the heat flow that generally slows down the heat conduction through the discontinuity, revealing the presence of the subsurface defect as a "hot spot" on the temperature profile of the object's surface. That is, the defects produce discontinuities in the temperature distribution of the object's surface. An infrared thermographic test involves a heat source (e.g., flashlamps) and an infrared camera. Because of the nature of heat diffusion, infrared thermography works best on defects that are close to the surface (subsurface features), for example, delamination in composite panels or disbonds in adhesively bonded joints.

MEMS Application in NDT

Recently application of microelectromechanical systems (MEMS) in nondestructive testing equipment significantly improved the detection and evaluation of the micro-cracks that are too small to find. Particular application of MEMS is in the electromagnetic nondestructive testing (EM NDT) techniques that plays an important role in many industry fields (Ref. 36). However, there are challenges in the use of EM NDT. For example, micro-cracks are too small to find and evaluate and complex components such as average sensors are difficult to be embedded inside structures and they are hardly to get close the micro-cracks. The advent of MEMS fabrication has brought an advantage to the EM sensors, such as high special resolution, high sensitivity, flexible substrate, be easily embedded, and on-chip integration. In addition, MEMS technique is capable of fabricating the coils of a solenoid with complex three-dimensional structure. Because the EM signal by micro-crack is weak, the low-noise preamplifier is necessary in the measure circuit. It is possible to integrate the sensor and preamplifier in one chip by means of MEMS fabrication technologies. Therefore, MEMS-based eddy current sensors are used as NDT equipment due to their higher resolution, higher sensitivity, complex structure, flexible substrate, be easily embedded and on-chip integration (Ref. 37).

6.7 TABLES

TABLE 6.1 Change in Resistance with Strain for Various Strain Gage Element Materials

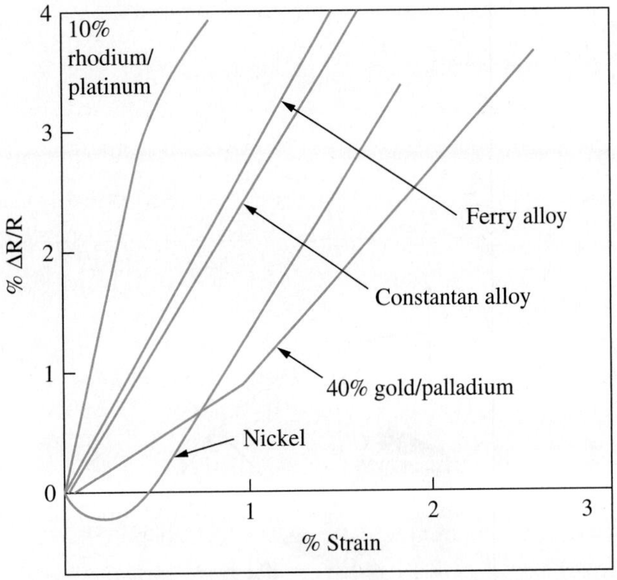

TABLE 6.2 Properties of Various Strain Gage Grid Materials

Material	Composition	Use	GF	Resistivity (Ohm/mil-ft)	Temp. Coef. of Resistance (ppm/F)	Temp. Coef. of Expansion (ppm/F)	Max Operating Temp. (F)
Constantan	45% NI, 55% Cu	Strain gage	2.0	290	6	8	900
Isoelastic	36% Ni, 8% Cr, 0.5% Mo, 55.5% Fe	Strain gage (dynamic)	3.5	680	260		800
Manganin	84% Cu, 12% Mn, 4% Ni	Strain gage (shock)	0.5	260	6		
Nichrome	80% Ni, 20% Cu	Thermometer	2.0	640	220	5	2000
Iridium-Platinum	95% Pt, 5% Ir	Thermometer	5.1	135	700	5	2000
Monel	67% Ni, 33% Cu		1.9	240	1100		
Nickel			−12	45	2400	8	
Karma	74% Ni, 20% Cr, 3% Al, 3% Fe	Strain gage (hi temp)	2.4	800	10		1500

TABLE 6.3 Strain Gage Rosette Equations Applied to a Specimen of a Linear, Isotropic Material

The principal strains and stresses are given relative to the *xy* coordinate axes as shown.

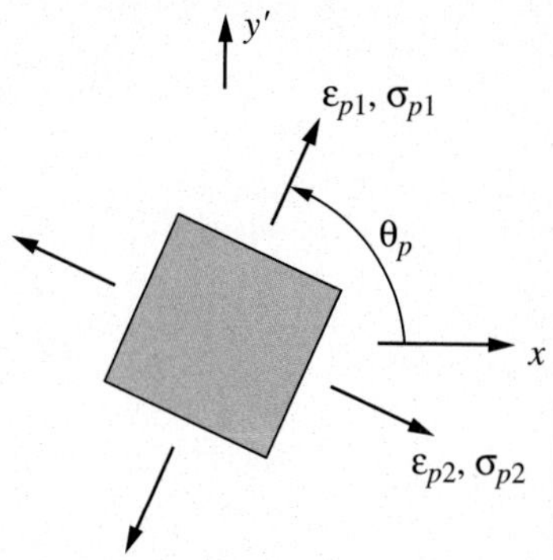

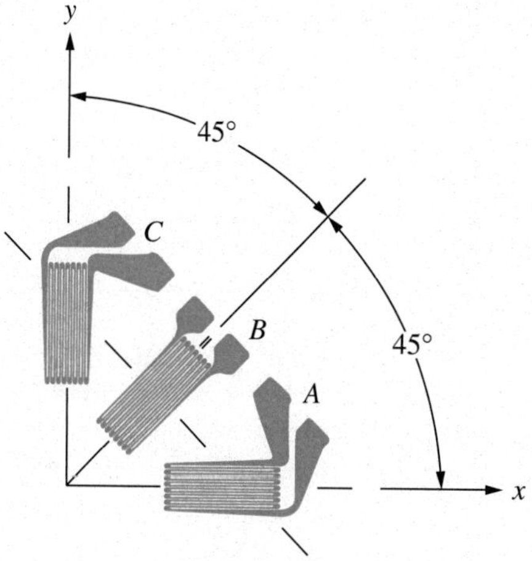

Three-element rectangular rosette

Principal strains

$$\varepsilon_{p1} = \frac{\varepsilon_A + \varepsilon_C}{2} + \frac{1}{2}\sqrt{(\varepsilon_A - \varepsilon_C)^2 + (2\varepsilon_B - \varepsilon_A - \varepsilon_C)^2}$$

$$\varepsilon_{p2} = \frac{\varepsilon_A + \varepsilon_C}{2} - \frac{1}{2}\sqrt{(\varepsilon_A - \varepsilon_C)^2 + (2\varepsilon_B - \varepsilon_A - \varepsilon_C)^2}$$

Principal stresses

$$\sigma_{p1} = \frac{E}{2}\left[\frac{\varepsilon_A + \varepsilon_C}{1-\nu} + \frac{1}{1+\nu}\sqrt{(\varepsilon_A - \varepsilon_C)^2 + (2\varepsilon_B - \varepsilon_A - \varepsilon_C)^2}\right]$$

$$\sigma_{p2} = \frac{E}{2}\left[\frac{\varepsilon_A + \varepsilon_C}{1-\nu} - \frac{1}{1+\nu}\sqrt{(\varepsilon_A - \varepsilon_C)^2 + (2\varepsilon_B - \varepsilon_A - \varepsilon_C)^2}\right]$$

Principal angle

Treating the $\tan^{-1}$ as a *single-valued function,** the angle *counterclockwise* from gage *A* to the axis containing ε_{p1} or σ_{p1} is given by

$$\theta_p = \frac{1}{2}\tan^{-1}\left(\frac{2\varepsilon_B - \varepsilon_A - \varepsilon_C}{\varepsilon_A - \varepsilon_C}\right)$$

*See Example in Sec. 6.2.

(*Continued*)

TABLE 6.3 Strain Gage Rosette Equations Applied to a Specimen of a Linear, Isotropic Material (*Continued*)

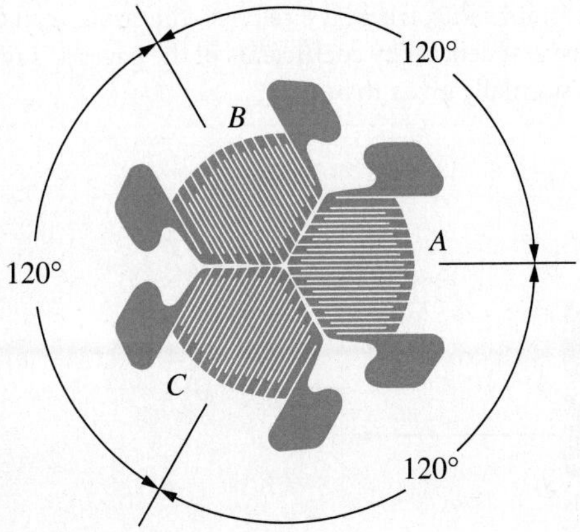

Three-element delta rosette

Principal strains

$$\varepsilon_{p1} = \frac{\varepsilon_A + \varepsilon_B + \varepsilon_C}{3} + \frac{\sqrt{2}}{3}\sqrt{(\varepsilon_A - \varepsilon_B)^2 + (\varepsilon_B - \varepsilon_C)^2 + (\varepsilon_C - \varepsilon_A)^2}$$

$$\varepsilon_{p2} = \frac{\varepsilon_A + \varepsilon_B + \varepsilon_C}{3} - \frac{\sqrt{2}}{3}\sqrt{(\varepsilon_A - \varepsilon_B)^2 + (\varepsilon_B - \varepsilon_C)^2 + (\varepsilon_C - \varepsilon_A)^2}$$

Principal stresses

$$\sigma_{p1} = \frac{E}{3}\left[\frac{\varepsilon_A + \varepsilon_B + \varepsilon_C}{1-\nu} + \frac{\sqrt{2}}{1+\nu}\sqrt{(\varepsilon_A - \varepsilon_B)^2 + (\varepsilon_B - \varepsilon_C)^2 + (\varepsilon_C - \varepsilon_A)^2}\right]$$

$$\sigma_{p2} = \frac{E}{3}\left[\frac{\varepsilon_A + \varepsilon_B + \varepsilon_C}{1-\nu} - \frac{\sqrt{2}}{1+\nu}\sqrt{(\varepsilon_A - \varepsilon_B)^2 + (\varepsilon_B - \varepsilon_C)^2 + (\varepsilon_C - \varepsilon_A)^2}\right]$$

Principal angle

Treating the $\tan^{-1}$ as a *single-valued function** the angle *counterclockwise* from gage A to the axis containing ε_{p1} or σ_{p1} is given by

$$\theta_p = \frac{1}{2}\tan^{-1}\left[\frac{\sqrt{3}(\varepsilon_C - \varepsilon_B)}{2\varepsilon_A - \varepsilon_B - \varepsilon_C}\right]$$

*See Example (as applied to a rectangular rosette) in Sec. 6.2.

TABLE 6.4 Corrections for the Transverse Sensitivity of Electrical Resistance Strain Gages

ε refers to corrected strain value, whereas $\hat{\varepsilon}$ refers to the strain read from the strain indicator. The K_t terms are the transverse sensitivity coefficients of the gages as supplied by the manufacturer. Poisson's ratio, ν_0, is normally given to be 0.285.

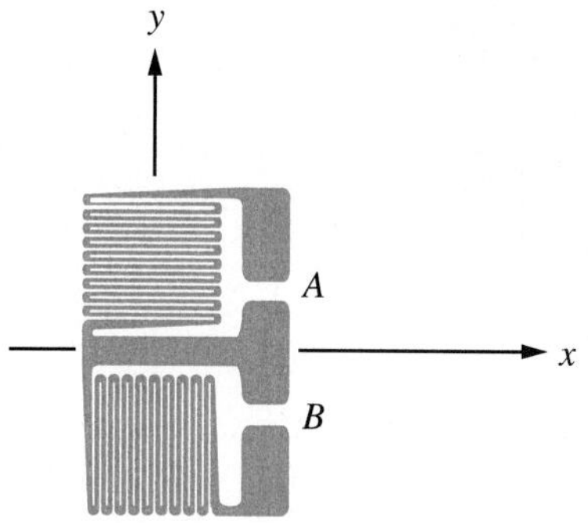

Two-element rectangular rosette

$$\varepsilon_x = \frac{(1-\nu_0 K_{tA})\hat{\varepsilon}_A - K_{tA}(1-\nu_0 K_{tB})\hat{\varepsilon}_B}{1-K_{tA}K_{tB}}$$

$$\varepsilon_y = \frac{(1-\nu_0 K_{tB})\hat{\varepsilon}_B - K_{tB}(1-\nu_0 K_{tA})\hat{\varepsilon}_A}{1-K_{tA}K_{tB}}$$

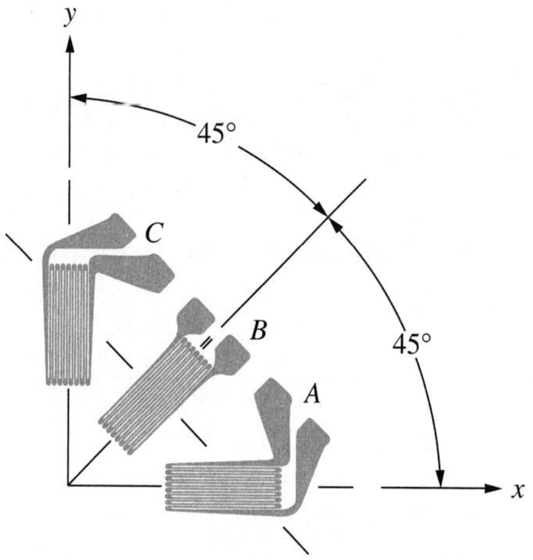

Three-element rectangular rosette

$$\varepsilon_A = \frac{(1-\nu_0 K_{tA})\hat{\varepsilon}_A - K_{tA}(1-\nu_0 K_{tC})\hat{\varepsilon}_C}{1-K_{tA}K_{tC}}$$

$$\varepsilon_B = \frac{(1-\nu_0 K_{tB})\hat{\varepsilon}_B - \dfrac{K_{tB}}{1-K_{tA}K_{tC}}[(1-\nu_0 K_{tA})(1-K_{tC})\hat{\varepsilon}_A + (1-\nu_0 K_{tC})(1-K_{tA})\hat{\varepsilon}_C]}{1-K_{tB}}$$

$$\varepsilon_C = \frac{(1-\nu_0 K_{tC})\hat{\varepsilon}_C - K_{tC}(1-\nu_0 K_{tA})\hat{\varepsilon}_A}{1-K_{tA}K_{tC}}$$

(Continued)

TABLE 6.4 Corrections for the Transverse Sensitivity of Electrical Resistance Strain Gages (*Continued*)

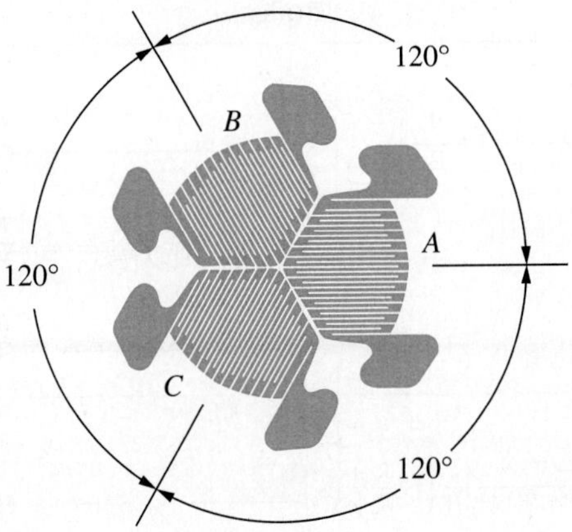

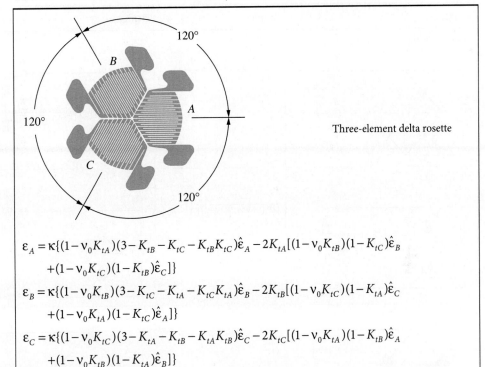

Three-element delta rosette

$$\varepsilon_A = \kappa\{(1-\nu_0 K_{tA})(3-K_{tB}-K_{tC}-K_{tB}K_{tC})\hat{\varepsilon}_A - 2K_{tA}[(1-\nu_0 K_{tB})(1-K_{tC})\hat{\varepsilon}_B + (1-\nu_0 K_{tC})(1-K_{tB})\hat{\varepsilon}_C]\}$$

$$\varepsilon_B = \kappa\{(1-\nu_0 K_{tB})(3-K_{tC}-K_{tA}-K_{tC}K_{tA})\hat{\varepsilon}_B - 2K_{tB}[(1-\nu_0 K_{tC})(1-K_{tA})\hat{\varepsilon}_C + (1-\nu_0 K_{tA})(1-K_{tC})\hat{\varepsilon}_A]\}$$

$$\varepsilon_C = \kappa\{(1-\nu_0 K_{tC})(3-K_{tA}-K_{tB}-K_{tA}K_{tB})\hat{\varepsilon}_C - 2K_{tC}[(1-\nu_0 K_{tA})(1-K_{tB})\hat{\varepsilon}_A + (1-\nu_0 K_{tB})(1-K_{tA})\hat{\varepsilon}_B]\}$$

where

$$\kappa = (3K_{tA}K_{tB}K_{tC} - K_{tA}K_{tB} - K_{tB}K_{tC} - K_{tA}K_{tC} - K_{tA} - K_{tB} - K_{tC} + 3)^{-1}$$

TABLE 6.5 Strain Gauge Equations for Several Types of Bridge Configurations

Bridge	Strain (ε)	Bridge	Strain (ε)
Quarter bridge 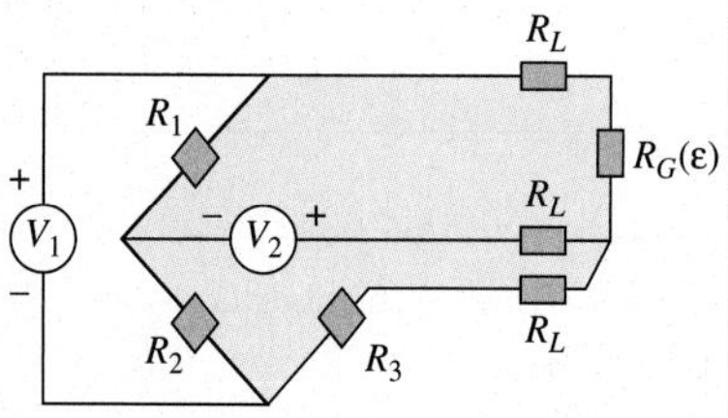	$\frac{-4V_r}{GF(1+2V_r)}\cdot\left(1+\frac{R_L}{R_G}\right)$	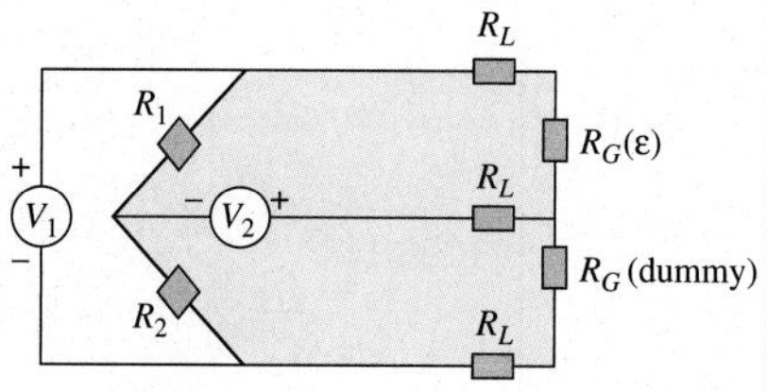	$\frac{-4V_r}{GF(1+2V_r)}\cdot\left(1+\frac{R_L}{R_G}\right)$
Half bridge 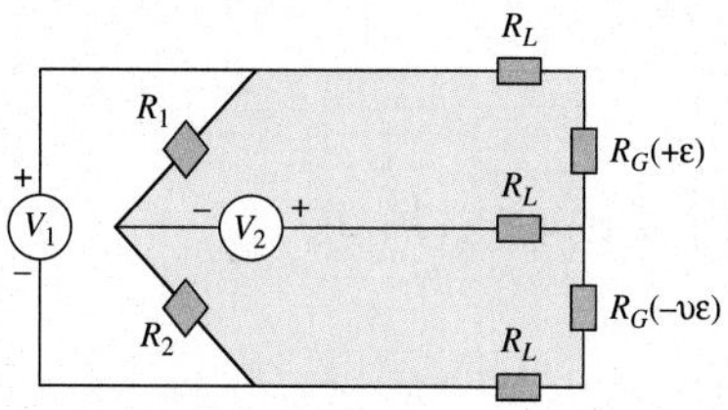	$\frac{-4V_r}{GF[(1+\nu)-2V_r(\nu-1)]}\cdot\left(1+\frac{R_L}{R_G}\right)$	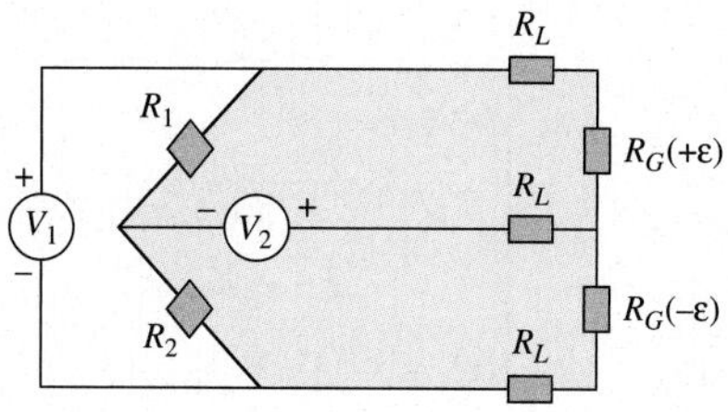	$\frac{-2V_r}{GF}\cdot\left(1+\frac{R_L}{R_G}\right)$
Full bridge 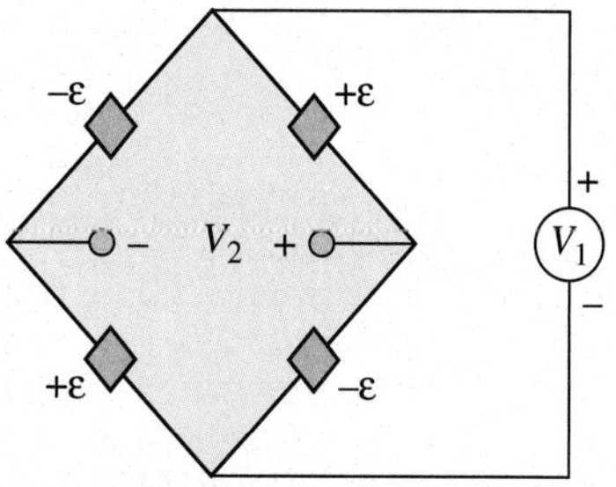	$\frac{-V_r}{GF}$	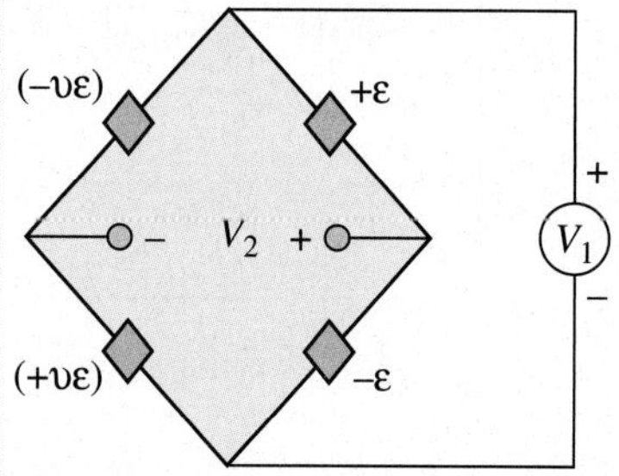	$\frac{-2V_r}{GF(\nu+1)}$
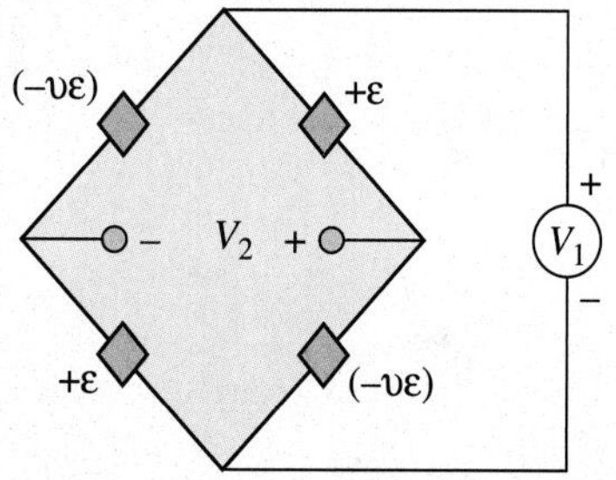	$\frac{-2V_r}{GF[(\nu+1)-V_r(\nu-1)]}$		

Note: The strain gauge equations for several types configurations of bridge are given in this table. To account for unbalanced bridges in the nonstrained state and to simplify the bridge equations, we introduce the ration V_r as:

$$V_r = (V_{2(\text{strained})} - V_{2(\text{unstrained})})/V_1$$

where $V_{2(\text{strained})}$ is the measured output when strained, and $V_{2(\text{unstrained})}$ is the initial, unstrained output voltage. V_1 is the excitation voltage.
Also, the designation (+ε) and (−ε) indicates active strain gages mounted in tension and compression, respectively.
The designation (−νε) indicates that the strain gage is mounted in the transversal direction, so that its resistance change is primarily due to the Poisson's strain, whose magnitude is given as (−νε).
Other nomenclature used in the equations includes:

R_G = nominal resistance value of strain gage
GF = gage factor of strain gage
R_L = lead resistance

6.8 REFERENCES

1. *Proc. Soc. Exp. Stress Anal.*, 1943–1960; *Exp. Mech., J. Soc. Exp. Stress Anal.*, from Jan., 1961 to June 1984; and *Exp. Mech., J. Soc. Exp.* Mechanics after June 1984.
2. Hetényi, M. (ed.), *Handbook of Experimental Stress Analysis*, John Wiley & Sons, 1950.
3. Dally, J. W., and W. F. Riley, *Experimental Stress Analysis*, 2nd ed., McGraw-Hill, 1978.
4. Dove, R. C., and P. H. Adams, *Experimental Stress Analysis and Motion Measurement*, Charles E. Merrill Books, 1955.
5. Holister, G. S., *Experimental Stress Analysis: Principles and Methods*, Cambridge University Press, 1967.
6. Frocht, M. M., *Photoelasticity*, vols. 1 and 2, John Wiley & Sons, 1941, 1948.
7. Durelli, A. J., *Applied Stress Analysis*, Prentice-Hall, 1967.
8. Durelli, A. J., and V. J. Parks, *Moiré Analysis of Strain*, Prentice-Hall, 1970.
9. Durelli, A. J., E. A. Phillips, and C. H. Tsao, Introduction to the Theoretical and Experimental Analysis of Stress and Strain, Prentice-Hall, 1958.
10. Durelli, A. J., and W. F. Riley, *Introduction to Photomechanics*, Prentice-Hall, 1965.
11. Theocaris, P. S., *Moiré Fringes in Strain Analysis*, Pergamon Press, 1969.
12. Savin, G. N., *Stress Concentration around Holes*, Pergamon Press, 1961.
13. Liebowitz, H. (ed.): *Fracture*, Academic Press, 1968.
14. Heywood, R. B.: *Photoelasticity for Designers*, Pergamon Press, 1969.
15. Sih, G. C., *Handbook of Stress Intensity Factors, Institute of Fracture and Solid Mechanics*, Lehigh University, 1973.
16. Javornicky, J., *Photoplasticity*, Elsevier Scientific, 1974.
17. Kobayashi, A. S. (ed.), *Experimental Techniques in Fracture Mechanics, 1 and 2*, Iowa State University Press, 1973 and 1975.
18. Aben, H., Integrated Photoelasticity, McGraw-Hill, 1979.
19. Broek, D., Elementary Engineering Fracture Mechanics, Martinus Nijhoff, 1982.
20. Kobayashi, A. S. (ed.), *Manual of Engineering Stress Analysis*, 3rd ed., Prentice-Hall, 1982.
21. Kobayashi, A. S. (ed.), *Manual on Experimental Stress Analysis*, 4th ed., Society of Experimental Mechanics, 1983.
22. Atluri, S. N. (ed.), *Computational Methods in the Mechanics of Fracture*, North-Holland, 1984.
23. Dally, J. W., W. F. Riley, and K. G. McConnell, *Instrumentation for Engineering Measurements*, John Wiley & Sons, 1984.
24. *Int. J. Anal. Exp. Modal Anal., Journal of the Society of Experimental Mechanics.*
25. *J. Strain Analysis, Journal of the Institution of Mechanical Engineers.*
26. Strain, *Journal of the British Society for Strain Measurement.*
27. Kobayashi, A. S., (ed.), *Handbook on Experimental Mechanics*, 2nd ed., Society for Experimental Mechanics, VCH, 1993.
28. Budynas, R. G., *Advanced Strength and Applied Stress Analysis*, 2nd ed., McGraw-Hill, 1999.
29. Dally, J. W., and W. F. Riley, *Experimental Stress Analysis*, McGraw-Hill, 1991.
30. Shukla, A., and J. W. Dally, *Experimental Solid Mechanics*, 4th ed., College House Enterprises, 2010.
31. http://www.omega.com/Literature/Transactions/volume3/strain.html#sendes
32. http://www.eidactics.com/Downloads/Refs-Methods/NI_Strain_Gauge_tutorial.pdf
33. Liptak, B., *Instrument Engineers' Handbook*, CRC Press LLC, 1995.
34. Hoffman, K., *Applying the Wheatstone Bridge Circuit*, HBM Publication, VD 72001e, Germany, 1976.
35. Sadegh A., and W. Worek, *Marks Standard Handbook for Mechanical Engineers*, 12th ed., McGraw-Hill Publication, November 2017.
36. He Y., F. Luo, M. Pan, F. Weng, X. Hu, J. Gao, and B. Liu, "Pulsed Eddy Current Technique for Defect Detection in Aircraft Riveted Structures," NDT&E Int., 43, 176–181, 2010.
37. Pan, M., Y. He, R. Xie, and W. Zhou, "The Influence of MEMS on Electromagnetic NDT," IEEE Conference Paper, June 2014, DOI: 10.1109/FENDT.2014.6928297.

CHAPTER 7

Tension, Compression, Shear, and Combined Stress

7.1 BAR UNDER AXIAL TENSION (OR COMPRESSION); COMMON CASE

The bar is straight, of any uniform cross section, of homogeneous material, and (if under compression) short or constrained against lateral buckling. The loads are applied at the ends, centrally, and in such a manner as to avoid nonuniform stress distribution at any section of the part under consideration. The stress does not exceed the proportional limit.

Behavior

Parallel to the load the bar elongates (under tension) or shortens (under compression), the unit longitudinal strain being ε and the total longitudinal deflection in the length l being δ. At right angles to the load the bar contracts (under tension) or expands (under compression); the unit lateral strain ε' is the same in all transverse directions, and the total lateral deflection δ' in any direction is proportional to the lateral dimension d measured in that direction. Both longitudinal and lateral strains are proportional to the applied load. On any right section there is a uniform tensile (or compressive) stress σ; on any oblique section there is a uniform tensile (or compressive) normal stress σ_θ and a uniform shear stress τ_θ. The deformed bar under tension is represented in Fig. 7.1(*a*), and the stresses in Fig. 7.1(*b*).

Formulas

Let

P = applied load
A = cross-sectional area (before loading)
l = length (before loading)
E = modulus of elasticity
ν = Poisson's ratio

Then

$$\sigma = \frac{P}{A} \tag{7.1-1}$$

$$\sigma_\theta = \frac{P}{A}\cos^2\theta, \qquad \max\,\sigma_\theta = \sigma\ (\text{when } \theta = 0°)$$

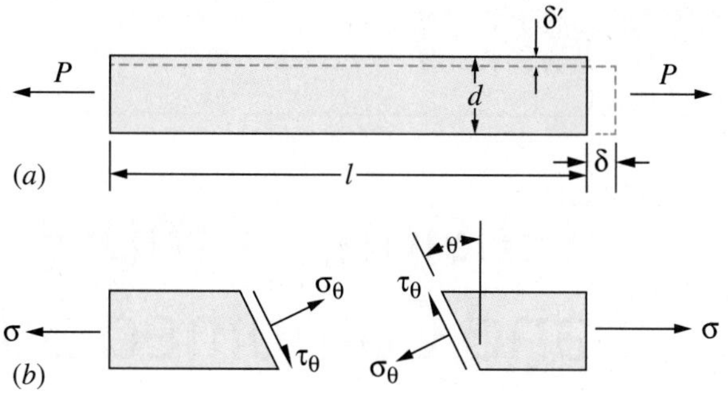

Figure 7.1

$$\tau = \frac{P}{2A}\sin 2\theta, \qquad \max \tau_\theta = \frac{1}{2}\sigma \text{ (when } \theta = 45 \text{ or } 135°)$$

$$\varepsilon = \frac{\sigma}{E} \tag{7.1-2}$$

$$\delta = l\varepsilon = \frac{Pl}{AE} \tag{7.1-3}$$

$$\varepsilon' = -\nu\varepsilon \tag{7.1-4}$$

$$\delta' = \varepsilon'\delta \tag{7.1-5}$$

$$\text{Strain energy per unit volume } U = \frac{1}{2}\frac{\sigma^2}{E} \tag{7.1-6}$$

$$\text{Total strain energy } U = \frac{1}{2}\frac{\sigma^2}{E}Al = \frac{1}{2}P\delta \tag{7.1-7}$$

For small strain, each unit area of cross section changes by $(-2\nu\varepsilon)$ under load, and each unit of volume changes by $(1-2\nu)\varepsilon$ under load.

In some discussions it is convenient to refer to the *stiffness* of a member, which is a measure of the resistance it offers to being deformed. The stiffness of a uniform bar under axial load is shown by Eq. (7.1-3) to be proportional to A and E directly and to l inversely, i.e., proportional to AE/l.

EXAMPLE A cylindrical rod of steel 4 in long and 1.5 in diameter has an axial compressive load of 20,000 lb applied to it. For this steel $\nu = 0.285$ and $E = 30{,}000{,}000$ lb/in^2. Determine (*a*) the unit compressive stress, σ; (*b*) the total longitudinal deformation, δ; (*c*) the total transverse deformation, δ'; (*d*) the change in volume, ΔV; and (*e*) the total energy, or work done in applying the load.

SOLUTION

(*a*) $$\sigma = \frac{P}{A} = \frac{4P}{\pi d^2} = \frac{4(-20{,}000)}{\pi(1.5)^2} = -11{,}320 \text{ lb/in}^2$$

(*b*) $$\varepsilon = \frac{\sigma}{E} = \frac{-11{,}320}{30{,}000{,}000} = -377(10^{-6})$$

$$\delta = \varepsilon l = (-377)(10^{-6})(4) = -1.509(10^{-3}) \text{ in ("}-\text{" means shortening)}$$

(*c*) $\varepsilon' = -\nu\varepsilon = -0.285(-377)(10^{-6}) = 107.5(10^{-6})$

$$\delta' = \varepsilon' d = (107.5)(10^{-6})(1.5) = 1.613(10^{-4}) \text{ in ("+" means expansion)}$$

(*d*) $\Delta V/V = (1-2\nu)\varepsilon = [1-2(0.285)](-377)(10^{-6}) = -162.2(10^{-6})$

$$\Delta V = -162.2(10^{-6})V = -162.2(10^{-6})\frac{\pi}{4}d^2 l = -162.2(10^{-6})\frac{\pi}{4}(1.5)^2(4)$$

$$= -1.147(10^{-3}) \text{ in}^3 \text{ ("–" means decrease)}$$

(*e*) Increase in strain energy,

$$U = \frac{1}{2}P\delta = \frac{1}{2}(-20{,}000)(-1.509)(10^{-3}) = 15.09 \text{ in-lb}$$

7.2 BAR UNDER TENSION (OR COMPRESSION); SPECIAL CASES

If the bar is not straight, it is subject to bending; formulas for this case are given in Sec. 12.4.

If the load is applied eccentrically, the bar is subject to bending; formulas for this case are given in Secs. 8.7 and 12.4. If the load is compressive and the bar is long and not laterally constrained, it must be analyzed as a column by the methods of Chaps. 12 and 15.

If the stress exceeds the proportional limit, the formulas for stress given in Sec. 7.1 still hold, but the deformation and work done in producing it can be determined only from experimental data relating unit strain to unit stress.

If the section is not uniform but changes *gradually*, the stress at any section can be found by dividing the load by the area of that section; the total longitudinal deformation over a length l is given by $\int_0^l (P/AE)\,dx$, and the strain energy is given by $\int_0^l \frac{1}{2}(P^2/AE)\,dx$, where dx is an infinitesimal length in the longitudinal direction. If the change in section is *abrupt* stress concentration may have to be taken into account, values of K_t being used to find elastic stresses and values of K_r being used to predict the breaking load. Stress concentration may also have to be considered if the end attachments for loading involve pinholes, screw threads, or other stress raisers (see Sec. 3.10 and Chap. 17).

If instead of being applied at the ends of a uniform bar the load is applied at an intermediate point, both ends being held, the *method of consistent deformations* shows that the load is apportioned to the two parts of the bar in inverse proportion to their respective lengths.

If a uniform bar is supported at one end in a vertical position and loaded only by its own weight, the maximum stress occurs at the supported end and is equal to the weight divided by the cross-sectional area. The total elongation is *half* as great and the total strain energy *one-third* as great as if a load equal to the weight were applied at the unsupported end. A bar supported at one end and loaded by its own weight and an axial downward load P (force) applied at the unsupported end will have the same unit stress σ (force per unit area) at all sections if it is tapered so that all sections are similar in form but vary in scale according to the formula

$$y = \frac{\sigma}{w}\log_e \frac{A\sigma}{P} \tag{7.2-1}$$

where y is the distance from the free end of the bar to any section, A is the area of that section, and w is the density of the material (force per unit volume).

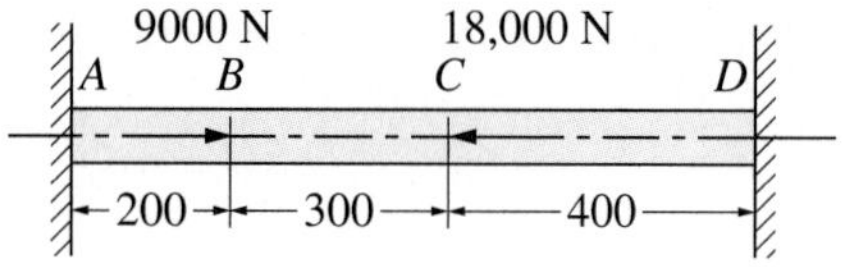

Figure 7.2

If a bar is stressed by having both ends rigidly held while a change in temperature is imposed, the resulting stress is found by calculating the longitudinal expansion (or contraction) that the change in temperature would produce if the bar were not held and then calculating the load necessary to shorten (or lengthen) it by that amount (principle of superposition). If the bar is uniform, the unit stress produced is independent of the length of the bar if restraint against buckling is provided. If a bar is stressed by being struck an axial blow at one end, the case is one of *impact* loading, discussed in Sec. 16.3.

EXAMPLE 1 Figure 7.2 represents a uniform bar rigidly held at the ends A and D and axially loaded at the intermediate points B and C. It is required to determine the total force in each portion of the bar AB, BC, CD. The loads are in newtons and the lengths in centimeters.

SOLUTION Each load is divided between the portions of the bar to right and left in inverse proportion to the lengths of these parts (consistent deformations), and the total force sustained by each part is the algebraic sum of the forces imposed by the individual loads (superposition). Of the 9000 N load therefore 7/9, or 7000 N, is carried in tension by segment AB, and 2/9, or 2000 N, is carried in compression by the segment BD. Of the 18,000 N load, 4/9, or 8000 N, is carried in compression by segment AC, and 5/9, or 10,000 N, is carried in tension by segment CD. Denoting tension by the plus sign and compression by the minus sign, and adding algebraically, the actual stresses in each segment are found to be

$$AB\text{: } 7000 - 8000 = -1000 \text{ N}$$

$$BC\text{: } -2000 - 8000 = -10{,}000 \text{ N}$$

$$CD\text{: } -2000 + 10{,}000 = +8000 \text{ N}$$

The results are quite independent of the diameter of the bar and of E provided the bar is completely uniform.

If instead of being *held* at the ends, the bar is prestressed by wedging it between rigid walls under an initial compression of, say, 10,000 N and the loads at B and C are then applied, the results secured above would represent the *changes* in force the several parts would undergo. The final forces in the bar would therefore be 11,000 N compression in AB, 20,000 N compression in BC, and 2000 N compression in CD. But if the initial compression were less than 8000 N, the bar would break contact with the wall at D (no tension possible); there would be no force at all in CD, and the forces in AB and BC, now statically determinate, would be 9000 and 18,000 N compression, respectively.

EXAMPLE 2 A steel bar 24 in long has the form of a truncated cone, being circular in section with a diameter at one end of 1 in and at the other of 3 in. For this steel, $E = 30{,}000{,}000$ lb/in^2 and the coefficient of thermal expansion is 0.0000065/°F. This bar is rigidly held at both ends and subjected to a drop in temperature of 50°F. It is required to determine the maximum tensile stress thus caused.

SOLUTION Using the principle of superposition, the solution is effected in three steps: (*a*) the shortening δ due to the drop in temperature is found, assuming the bar is free to contract; (*b*) the force P required to produce an elongation equal to δ, that is, to stretch the bar back to its original length, is calculated; (*c*) the maximum tensile stress produced by this force P is calculated.

(*a*) $\delta = 50(0.0000065)(24) = 0.00780$ in.

(*b*) Let d denote the diameter and A the area of any section a distance x in from the small end of the bar. Then

$$d = 1 + \frac{x}{12}, \qquad A = \frac{\pi}{4}\left(1 + \frac{x}{12}\right)^2$$

and

$$\delta = \int_0^l \frac{P}{EA}dx = \int_0^{24} \frac{4P}{(\pi E)(1+x/12)^2}dx = \frac{4P}{\pi(30)(10^6)}\frac{(-12)}{(1+x/12)}\bigg|_0^{24} = 3.395(10^{-7})P$$

Equating this to the thermal contraction of 0.00780 in yields

$$P = 22{,}970 \text{ lb}$$

(*c*) The maximum stress occurs at the smallest section, and is

$$\sigma = \frac{4P}{\pi d_{\min}^2} = \frac{4(22{,}970)}{\pi(1)^2} = 29{,}250 \text{ lb/in}^2$$

The result can be accepted as correct only if the proportional limit of the steel is known to be as great as or greater than the maximum stress and if the concept of a rigid support can be accepted. (See cases 8, 9, and 10 in Table 14.1.)

7.3 COMPOSITE MEMBERS

A tension or compression member may be made up of parallel elements or parts which jointly carry the applied load. The essential problem is to determine how the load is apportioned among the several parts, and this is easily done by the method of consistent deformations. If the parts are so arranged that all undergo the same total elongation or shortening, then each will carry a portion of the load proportional to its stiffness, i.e., proportional to AE/l if each is a uniform bar and proportional to AE if all these uniform bars are of equal length. It follows that if there are n bars, with section areas $A_1, A_2, \ldots, A_n$, lengths $l_1, l_2, \ldots, l_n$, and moduli $E_1, E_2, \ldots, E_n$, then the loads on the several bars $P_1, P_2, \ldots, P_n$ are given by

$$P_1 = P\frac{\dfrac{A_1E_1}{l_1}}{\dfrac{A_1E_1}{l_1} + \dfrac{A_2E_2}{l_2} + \cdots + \dfrac{A_nE_n}{l_n}} \tag{7.3-1}$$

$$P_2 = P\frac{\dfrac{A_2E_2}{l_2}}{\dfrac{A_1E_1}{l_1} + \dfrac{A_2E_2}{l_2} + \cdots + \dfrac{A_nE_n}{l_n}} \tag{7.3-2}$$

A composite member of this kind can be *prestressed*. P_1, P_2, etc., then represent the *increments* of force in each member due to the applied load, and can be found by Eqs. (7.3-1) and (7.3-2), provided all bars can sustain reversal of stress, or provided the applied load is not great enough to cause such reversal in any bar which cannot sustain it. As explained in Sec. 3.12, by proper prestressing, all parts of a composite member can be made to reach their allowable loads, elastic limits, or ultimate strengths simultaneously (Example 2).

EXAMPLE 1 A ring is suspended by three vertical bars, *A*, *B*, and *C* of unequal lengths. The upper ends of the bars are held at different levels, so that as assembled none of the bars is stressed. *A* is 4 ft long, has a section area of 0.3 in², and is of steel for which $E = 30{,}000{,}000$ lb/in²; *B* is 3 ft long, has a section area of 0.2 in², and is of copper for which $E = 17{,}000{,}000$ lb/in²; *C* is 2 ft long, has a section area of 0.4 in², and is of aluminum for which $E = 10{,}000{,}000$ lb/in². A load of 10,000 lb is hung on the ring. It is required to determine how much of this load is carried by each bar.

SOLUTION Denoting by P_A, P_B, and P_C the loads carried by A, B, and C, respectively, and expressing the moduli of elasticity in millions of pounds per square inch and the lengths in feet, we substitute in Eq. (7.3-1) and find

$$P_A = 10{,}000\left[\frac{\dfrac{(0.3)(30)}{4}}{\dfrac{(0.3)(30)}{4}+\dfrac{(0.2)(17)}{3}+\dfrac{(0.4)(10)}{2}}\right] = 4180 \text{ lb}$$

Similarly

$$P_B = 2100 \text{ lb} \qquad \text{and} \qquad P_C = 3720 \text{ lb}$$

EXAMPLE 2 A composite member is formed by passing a steel rod through an aluminum tube of the same length and fastening the two parts together at both ends. The fastening is accomplished by adjustable nuts, which make it possible to assemble the rod and tube so that one is under initial tension and the other is under an equal initial compression. For the steel rod the section area is 1.5 in², the modulus of elasticity is 30,000,000 lb/in², and the allowable stress is 15,000 lb/in². For the aluminum tube the section area is 2 in², the modulus of elasticity is 10,000,000 lb/in² and the allowable stress is 10,000 lb/in². It is desired to prestress the composite member so that under a tensile load both parts will reach their allowable stresses simultaneously.

SOLUTION When the allowable stresses are reached, the force in the steel rod will be 1.5(15,000) = 22,500 lb, the force in the aluminum tube will be 2(10,000) = 20,000 lb, and the total load on the member will be 22,500 + 20,000 = 42,500 lb. Let P_i denote the initial tension or compression in the members, and, as before, let tension be considered positive and compression negative. Then, since Eq. (7.3-1) gives the *increment* in force, we have for the aluminum tube

$$P_i + 42{,}500\frac{(2)(10)}{(2)(10)+(1.5)(30)} = 20{,}000$$

or

$$P_i = +6920 \text{ lb} \qquad \text{(initial tension)}$$

For the steel rod, we have

$$P_i + 42{,}500\frac{(1.5)(30)}{(2)(10)+(1.5)(30)} = 22{,}500$$

or

$$P_i = -6920 \text{ lb} \qquad \text{(initial compression)}$$

If the member were not prestressed, the unit stress in the steel would always be just three times as great as that in the aluminum because it would sustain the same unit deformation and its modulus of elasticity is three times as great. Therefore, when the steel reached its allowable stress of 15,000 lb/in^2, the aluminum would be stressed to only 5000 lb/in^2 and the allowable load on the composite member would be only 32,500 lb instead of 42,500 lb.

7.4 TRUSSES

A conventional truss is essentially an assemblage of straight uniform bars that are subjected to axial tension or compression when the truss is loaded at the joints. The deflection of any joint of a truss is easily found by the *method of unit loads* (Sec. 4.5). Let p_1, p_2, p_3, etc., denote the forces produced in the several members by an *assumed unit load* acting in the direction x at the joint whose deflection is to be found, and let δ_1, δ_2, δ_3, etc., denote the longitudinal deformations produced in the several members by the *actual applied loads*. The deflection Δ_x in the direction x of the joint in question is given by

$$\Delta_x = p_1\delta_1 + p_2\delta_2 + p_3\delta_3 + \cdots = \sum_{i=1}^{n} p_i\delta_i \qquad (7.4\text{-}1)$$

The deflection in the direction y, at right angles to x, can be found similarly by assuming the unit load to act in the y direction; the resultant deflection is then determined by combining the x and y deflections. Attention must be given to the *signs* of p and δ: p is positive if a member is subjected to tension and negative if under compression, and δ is positive if it represents an elongation and negative if it represents a shortening. A positive value for $\Sigma p\delta$ means that the deflection is in the direction of the assumed unit load, and a negative value means that it is in the opposite direction. (This procedure is illustrated in Example 1 below.)

A statically indeterminate truss can be solved by the *method of least work* (Sec. 4.5). To do this, it is necessary to write down the expression for the total strain energy in the structure, which, being simply the sum of the strain energies of the constituent bars, is given by

$$\frac{1}{2}P_1\delta_1 + \frac{1}{2}P_2\delta_2 + \frac{1}{2}P_3\delta_3 + \cdots = \sum_{i=1}^{n}\frac{1}{2}P_i\delta_i = \sum_{i=1}^{n}\frac{1}{2}\left(\frac{P^2l}{AE}\right)_i \qquad (7.4\text{-}2)$$

Here P_1, P_2, etc., denote the forces in the individual members due to the applied loads and δ has the same meaning as above. It is necessary to express each force P_i as the sum of the two forces; one of these is the force the applied loads would produce with the redundant member removed, and the other is the force due to the unknown force (say, F) exerted by this redundant member on the rest of the structure. The total strain energy is thus expressed as a function of the known applied forces and F, the force in the redundant member. The partial derivative with respect to F of this expression for strain energy is then set equal to zero and solved for F. If there are two or more redundant members, the expression for strain energy with all the redundant forces, F_1, F_2, etc., represented is differentiated once with respect to each. The equations thus obtained are then solved simultaneously for the unknown forces. (The procedure is illustrated in Example 2.)

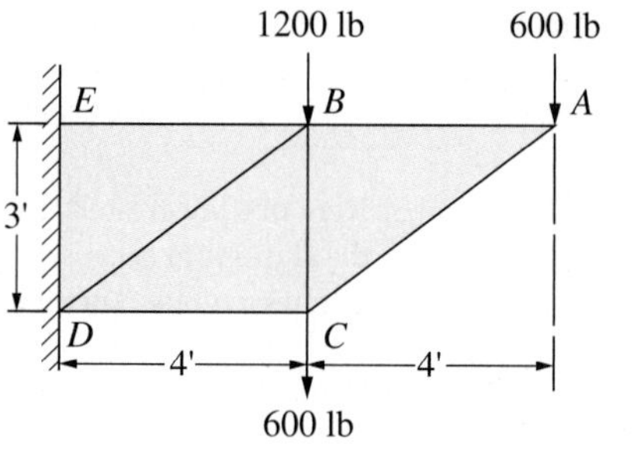

Figure 7.3

EXAMPLE 1 The truss shown in Fig. 7.3 is composed of tubular steel members, for which $E = 30{,}000{,}000$ lb/in². The section areas of the members are given in the table below. It is required to determine Δ_x and Δ_y, the horizontal and vertical components of the displacement of joint A produced by the indicated loading.

SOLUTION The method of unit loads is used. The force P in each member due to the applied loads is found, and the resulting elongation or shortening δ is calculated. The force p_x in each member due to a load of 1 lb acting to the right at A, and the force p_y in each member due to a load of 1 lb acting down at A are calculated. By Eq. (7.4-1), $\Sigma p_x\delta$ then gives the horizontal and $\Sigma p_y\delta$ gives the vertical displacement or deflection of A. Tensile forces and elongations are denoted by +, compressive forces and shortenings by −. The work is conveniently tabulated as follows:

Member	Area, A_i, in²	Length, l_i, in	P_i, lb	$\delta_i = \left(\frac{Pl}{AE}\right)_i$, in	$(p_x)_i$	$(p_x\delta)_i$, in (*a*)	$(p_y)_i$	$(p_y\delta)_i$, in (*b*)
(1) *AB*	0.07862	48	800	0.01628	1.000	0.01628	1.333	0.02171
(2) *AC*	0.07862	60	−1000	−0.02544	0	0	−1.667	0.04240
(3) *BC*	0.1464	36	1200	0.00984	0	0	1.000	0.00984
(4) *BE*	0.4142	48	4000	0.01545	1.000	0.01545	2.667	0.04120
(5) *BD*	0.3318	60	−4000	−0.02411	0	0	−1.667	0.04018
(6) *CD*	0.07862	48	−800	−0.01628	0	0	−1.333	0.02171
						$\Delta_x = 0.03173$ in		$\Delta_y = 0.17704$ in

Δ_x and Δ_y are both found to be positive, which means that the displacements are in the directions of the assumed unit loads—to the right and down. Had either been found to be negative, it would have meant that the displacement was in a direction opposite to that of the corresponding unit load.

EXAMPLE 2 Assume a diagonal member, running from A to D and having a section area 0.3318 in² and length 8.544 ft, is to be added to the truss of Example 1; the structure is now statically indeterminate. It is required to determine the force in each member of the altered truss due to the loads shown.

SOLUTION We use the method of least work. The truss has one redundant member; any member except BE may be regarded as redundant, since if any one were removed, the remaining structure would be stable and statically determinate. We select AD to be regarded as redundant, denote the unknown force in AD by F, and assume F to be tension. We find the force in each member assuming AD to be removed, then find the force in each member due to a pull F exerted at A by AD, and then add these forces, thus getting an expression for the force in each member of the actual truss in terms of F. The expression for the strain energy can then be written out, differentiated with respect to F, equated to zero, and solved for F. F being known, the force in each member of the truss is easily found. The computations are conveniently tabulated as follows:

	Forces in Members†			
	Due to Applied Loads without *AD*	Due to Pull, F, Exerted by *AD*	Total Forces, P_i. Superposition of (*a*) and (*b*)	Actual Total Values with $F = -1050$ lb in (*c*)
Member	(*a*) (lb)	(*b*)	(*c*)	(*d*) (lb)
(1) *AB*	800	−0.470 *F*	800 − 0.470 *F*	1290
(2) *AC*	−1000	−0.584 *F*	−1000 − 0.584 *F*	−390
(3) *BC*	1200	0.351 *F*	1200 + 0.351 *F*	830
(4) *BE*	4000	0	4000	4000
(5) *BD*	−4000	−0.584 *F*	−4000 − 0.584 *F*	−3390
(6) *CD*	−800	−0.470 *F*	−800 − 0.470 *F*	−306
(7) *AD*	0	*F*	*F*	−1050

†+ for tension and − for compression.

$$U = \sum_{i=1}^{7} \frac{1}{2}\left(\frac{P^2 l}{AE}\right)_i = \frac{1}{2E}\left[\frac{(800-0.470F)^2(48)}{0.07862} + \frac{(-1000-0.584F)^2(60)}{0.07862}\right.$$
$$+\frac{(1200+0.351F)^2(36)}{0.1464} + \frac{(4000)^2(48)}{0.4142}$$
$$+\frac{(-4000-0.584F)^2(60)}{0.3318} + \frac{(-800-0.470F)^2(48)}{0.07862}$$
$$\left.+\frac{F^2(102.5)}{0.3318}\right]$$

Setting the partial derivative of U relative to F to zero,

$$\frac{\partial U}{\partial F} = \frac{1}{2E}\left[\frac{2(800-0.470F)(-0.470)(48)}{0.07862} + \frac{2(-1000-0.584F)(-0.584)(60)}{0.07862} + \cdots\right] = 0$$

and solving for F gives $F = -1050$ lb.

The negative sign here simply means that *AD* is in compression. A positive value of F would have indicated tension. Substituting the value of F into the terms of column (*c*) yields the actual total forces in each member as tabulated in column (*d*).

7.5 BODY UNDER PURE SHEAR STRESS

A condition of pure shear may be produced by any one of the methods of loading shown in Fig. 7.4. In Fig. 7.4(*a*), a rectangular block of length a, height b, and uniform thickness t is shown loaded by forces P_1 and P_2, uniformly distributed over the surfaces to which they are applied and satisfying the equilibrium equation $P_1 b = P_2 a$. There are equal shear stresses on all vertical and horizontal planes, so that any contained cube oriented like *ABCD* has on each of four faces the shear stress $\tau = P_1/at = P_2/bt$ and no other stress.

In Fig. 7.4(*b*), a rectangular block is shown under equal and opposite biaxial stresses σ_t and σ_c. There are equal shear stresses on all planes inclined at 45° to the top and bottom faces, so that a contained cube oriented like *ABCD* has on each of four faces the shear stress $\tau = \sigma_t = \sigma_c$ and no other stress.

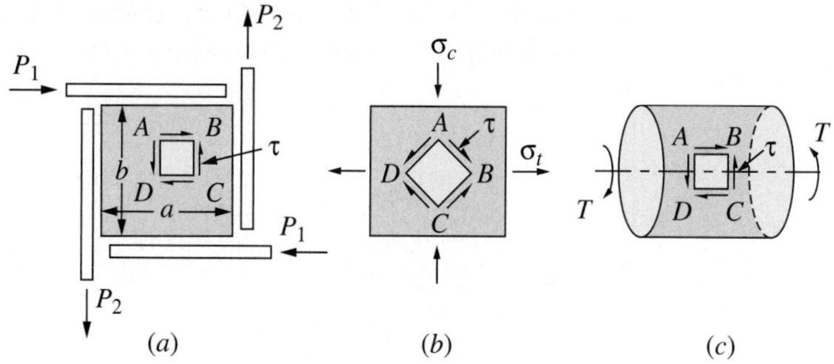

Figure 7.4

In Fig. 7.4(*c*), a circular shaft is shown under a twisting moment *T*; a cube of infinitesimal dimensions, a distance *z* from the axis and oriented like *ABCD* has on each of four faces an essentially uniform shear stress $\tau = Tz/J$ (Sec. 10.1) and no other stress.

In whatever way the loading is accomplished, the result is to impose on an elementary cube of the loaded body the condition of stress represented in Fig. 7.5, that is, shearing stress alone on each of four faces, these stresses being equal and so directed as to satisfy the equilibrium condition $T_x = 0$ (Sec. 4.1).

The stresses, σ_θ and τ_θ on a transformed surface rotated counterclockwise through the angle θ can be determined from the transformation equations given by Eq. (2.3-17). They are given by

$$\sigma_\theta = \tau \sin 2\theta, \qquad \tau_\theta = \tau \cos 2\theta \tag{7.5-1}$$

where $(\sigma_\theta)_{\max,\min} = \pm\tau$ at $\theta = \pm 45°$.

The strains produced by pure shear are shown in Fig. 7.5(*b*), where the cube *ABCD* is deformed into a rhombohedron *A′B′C′D′*. The unit shear strain, γ, referred to as the *engineering shear strain*, is the reduction of angles $\angle ABC$ and $\angle ADC$, and the increase in angles $\angle DAB$ and $\angle BCD$ in radians. Letting *G* denote the modulus of rigidity, the shear strain is related to the shear stress as

$$\gamma = \frac{\tau}{G} \tag{7.5-2}$$

Assuming a linear material, the strain energy per unit volume for pure shear, u_s, within the elastic range is given by

$$u_s = \frac{1}{2}\frac{\tau^2}{G} \tag{7.5-3}$$

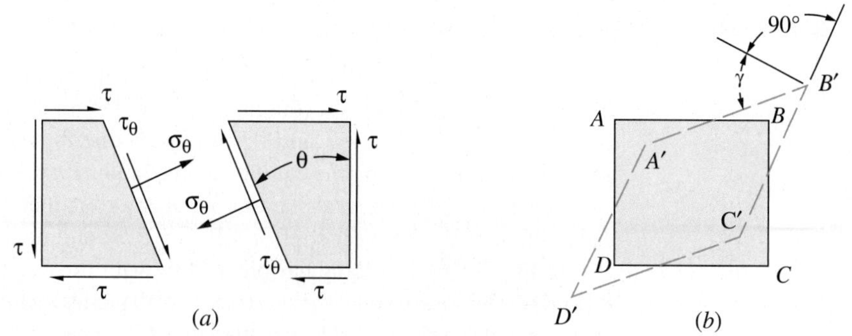

Figure 7.5 (*a*) Shear stress and transformation. (*b*) Shear strain.

The relations between τ, σ, and the strains represented in Fig. 7.5(b) make it possible to express G in terms of E and Poisson's ratio, ν, for a linear, homogeneous, isotropic material. The relationship is

$$G = \frac{F}{2(1+\nu)} \qquad (7.5\text{-}4)$$

From known values of E (determined by a tensile test) and G (determined by a torsion test) it is thus possible to calculate ν.

7.6 CASES OF DIRECT SHEAR LOADING

By *direct shear loading* is meant any case in which a member is acted on by equal, parallel, and opposite forces so nearly colinear that the material between them is subjected primarily to shear stress, with negligible bending. Examples of this are provided by rivets, bolts, and pins, shaft splines and keys, screw threads, short lugs, etc. These are not really cases of pure shear; the actual stress distribution is complex and usually indeterminate because of the influence of fit and other factors. In designing such parts, however, it is usually assumed that the shear is uniformly distributed on the critical section, and since working stresses are selected with due allowance for the approximate nature of this assumption, the practice is usually permissible. In *beams* subject to transverse shear, this assumption cannot be made as a rule.

Shear and other stresses in rivets, pins, keys, etc., are discussed more fully in Chap. 14, shear stresses in beams in Chap. 8, and shear stresses in torsion members in Chap. 10.

7.7 COMBINED STRESS

Under certain circumstances of loading, a body is subjected to a combination of tensile and compressive stresses (usually designated as *biaxial* or *triaxial* stress) or to a combination of tensile, compressive, and shear stresses (usually designated as *combined stress*). For example, the material at the inner surface of a thick cylindrical pressure vessel is subjected to triaxial stress (radial compression, longitudinal tension, and circumferential tension), and a shaft simultaneously bent and twisted is subjected to combined stress (longitudinal tension or compression, and torsional shear).

In most instances the normal and shear stresses on each of three mutually perpendicular planes are due to flexure, axial loading, torsion, beam shear, or some combination of these which separately can be calculated readily by the appropriate formulas. Normal stresses arising from different load conditions acting on the same plane can be combined simply by algebraic addition considering tensile stresses positive and compressive stresses negative. Similarly, shear stresses can be combined by algebraic addition following a consistent sign convention. Further analysis of the combined states of normal and shear stresses must be performed using the transformation techniques outlined in Sec. 2.3. The principal stresses, the maximum shear stress, and the normal and shear stresses on any given plane can be found by the equations given in Sec. 2.3.

The strains produced by any combination of stresses not exceeding the proportional limit can also be found using Hooke's law for each stress and then combined by superposition. Consideration of the strains caused by equal triaxial stresses leads to an expression for the bulk modulus of elasticity given by

$$K = \frac{E}{3(1-2\nu)} \qquad (7.7\text{-}1)$$

EXAMPLE 1 A rectangular block 12 in long, 4 in high, and 2 in thick is subjected to a longitudinal tensile stress σ_x = 12,000 lb/in², a vertical compressive stress σ_y = 15,000 lb/in², and a lateral compressive stress σ_z = 9000 lb/in². The material is steel, for which E = 30,000,000 lb/in² and ν = 0.30. It is required to find the total change in length.

SOLUTION The longitudinal deformation is found by superposition. The unit strain due to each stress is computed separately by Eqs. (7.1-2) and (7.1-4); these results are added to give the resultant longitudinal unit strain, which is multiplied by the length to give the total elongation. Denoting unit longitudinal strain by ε_x and total longitudinal deflection by δ_x, we have

$$\varepsilon_x = \frac{12{,}000}{E} - \nu\frac{-15{,}000}{E} - \nu\frac{-9000}{E}$$

$$= 0.000400 + 0.000150 + 0.000090 = +0.00064$$

$$\delta_x = 12(0.00064) = 0.00768 \text{ in}$$

The lateral dimensions have nothing to do with the result since the lateral stresses, not the lateral loads, are given.

EXAMPLE 2 A piece of "standard extra-strong" pipe, 2 in nominal diameter, is simultaneously subjected to an internal pressure of p = 2000 lb/in² and to a twisting moment of T = 5000 lb-in caused by tightening a cap screwed on at one end. Determine the maximum tensile stress and the maximum shear stress thus produced in the pipe.

SOLUTION The calculations will be made, first, for a point at the outer surface and, second, for a point at the inner surface. The dimensions of the pipe and properties of the cross section are as follows: inner radius r_i = 0.9695 in, outer radius r_o = 1.1875 in, cross-sectional area of bore A_b = 2.955 in², cross-sectional area of pipe wall A_w = 1.475 in², and polar moment of inertial J = 1.735 in⁴.

We take axis x along the axis of the pipe, axis y tangent to the cross section, and axis z radial in the plane of the cross section. For a point at the outer surface of the pipe, σ_x is the longitudinal tensile stress due to pressure and σ_y is the circumferential (hoop) stress due to pressure, the radial stress $\sigma_z = 0$ (since the pressure is zero on the outer surface of the pipe), and τ_{xy} is the shear stress due to torsion. Equation (7.1-1) can be used for σ_x, where $P = pA_b$ and $A = A_\omega$. To calculate σ_y, we use the formula for stress in thick cylinders (Table 13.5, case 1b). Finally, for τ_{xy}, we use the formula for torsional stress [Eq. (10.1-2)]. Thus,

$$\sigma_x = \frac{pA_b}{A_\omega} = \frac{(2000)(2.955)}{1.475} = 4007 \text{ lb/in}^2$$

$$\sigma_y = p\frac{r_i^2(r_o^2 + r_o^2)}{r_o^2(r_o^2 - r_i^2)} = 2000\frac{(0.9695^2)(1.1875^2 + 1.1875^2)}{(1.1875^2)(1.1875^2 - 0.9695^2)} = 7996 \text{ lb/in}^2$$

$$\tau_{xy} = \frac{Tr_o}{J} = \frac{(5000)(1.1875)}{1.735} = 3422 \text{ lb/in}^2$$

This is a case of plane stress where Eq. (2.3-23) applies. The principal stresses are thus

$$\sigma_p = \frac{1}{2}[(\sigma_x + \sigma_y) \pm \sqrt{(\sigma_x - \sigma_y)^2 + 4\tau_{xy}^2}]$$

$$= \frac{1}{2}[(4007 + 7996) \pm \sqrt{(4007 - 7996)^2 + 4(3422^2)}] = 9962,\ 2041 \text{ lb/in}^2$$

Thus, $\sigma_{max} = 9962 \text{ lb/in}^2$.

In order to determine the maximum shear stress, we order the *three* principal stresses such that $\sigma_1 \geq \sigma_2 \geq \sigma_3$. For plane stress, the out-of-plane principal stresses are zero. Thus, $\sigma_1 = 9962 \text{ lb/in}^2$, $\sigma_2 = 2041 \text{ lb/in}^2$, and $\sigma_3 = 0$. From Eq. (2.3-25), the maximum shear stress is

$$\tau_{max} = \frac{1}{2}(\sigma_1 - \sigma_3) = \frac{1}{2}(9962 - 0) = 4981 \text{ lb/in}^2$$

For a point on the inner surface, the stress conditions are three-dimensional since a radial stress due to the internal pressure is present. The longitudinal stress is the same; however, the circumferential stress and torsional shear stress change. For the inner surface,

$$\sigma_x = 4007 \text{ lb/in}^2$$

$$\sigma_y = p\frac{r_o^2 + r_i^2}{r_o^2 - r_i^2} = 2000\frac{1.1875^2 + 0.9695^2}{1.1875^2 - 0.9695^2} = 9996 \text{ lb/in}^2$$

$$\sigma_z = -p = -2000 \text{ lb/in}^2$$

$$\tau_{xy} = \frac{Tr_i}{J} = \frac{(5000)(0.9695)}{1.735} = 2794 \text{ lb/in}^2$$

$$\tau_{yz} = \tau_{zx} = 0$$

The principal stresses are found using Eq. (2.3-20):*

$$\begin{aligned}\sigma_p^3 &- (4007 + 9996 - 2000)\sigma_p^2 + [(4007)(9996) + (9996)(-2000) \\ &+ (-2000)(4007) - 2794^2 - 0 - 0]\sigma_p - [(4007)(9996)(-2000) + 2(2794)(0)(0) \\ &- (4007)(0^2) - (9996)(0^2) - (-2000)(2794^2)] = 0\end{aligned}$$

or

$$\sigma_p^3 - 12.003(10^3)\sigma_p^2 + 4{,}2415(10^6)\sigma_p + 64.495(10^9) = 0$$

Solving this gives $\sigma_p = 11{,}100, 2906,$ and -2000 lb/in^2, which are the principal stresses σ_1, σ_2, and σ_3, respectively. Obviously, the maximum tensile stress is 11,100 lb/in². Again, the maximum shear stress comes from Eq. (2.3-25), and is ½ [11,100 – (–2000)] = 6550 lb/in².

Note that for this problem, if the pipe is a ductile material, and one were looking at failure due to shear stress (see Sec. 3.7), the stress conditions for the pipe are more severe at the inner surface compared with the outer surface.

**Note:* Since $\tau_{yz} = \tau_{zx} = 0$, σ_z is one of the principal stresses and the other two can be found from the plane stress equations. Consequently, the other two principal stresses are in the xy plane.

CHAPTER 8

Beams; Flexure of Straight Bars

8.1 STRAIGHT BEAMS (COMMON CASE) ELASTICALLY STRESSED

The formulas in this section are based on the following assumptions: (1) The beam is of homogeneous material that has the same modulus of elasticity in tension and compression. (2) The beam is straight or nearly so; if it is slightly curved, the curvature is in the plane of bending and the radius of curvature is at least 10 times the depth. (3) The cross section is uniform. (4) The beam has at least one longitudinal plane of symmetry. (5) All loads and reactions are perpendicular to the axis of the beam and lie in the same plane, which is a longitudinal plane of symmetry. (6) The beam is long in proportion to its depth, the span/depth ratio being 8 or more for metal beams of compact section, 15 or more for beams with relatively thin webs, and 24 or more for rectangular timber beams. (7) The beam is not disproportionately wide (see Sec. 8.11 for a discussion on the effect of beam width). (8) The maximum stress does not exceed the proportional limit.

Applied to any case for which these assumptions are not valid, the formulas given yield results that at best are approximate and that may be grossly in error; such cases are discussed in subsequent sections. The limitations stated here with respect to straightness and proportions of the beam correspond to a maximum error in calculated results of about 5%.

In the following discussion, it is assumed for convenience that the beam is horizontal and the loads and reactions vertical.

Behavior

As the beam bends, fibers on the convex side lengthen, and fibers on the concave side shorten. The neutral surface is normal to the plane of the loads and contains the centroids of all sections, hence the neutral axis of any section is the horizontal central axis. Plane sections remain plane, and hence unit fiber strains and stresses are proportional to distance from the neutral surface. Longitudinal displacements of points on the neutral surface are negligible. Vertical deflection is largely due to bending, that due to shear being usually negligible under the conditions stated.

There is at any point a longitudinal fiber stress σ, which is tensile if the point lies between the neutral and convex surfaces of the beam and compressive if the point lies between the neutral and concave surfaces of the beam. This fiber stress σ usually may be assumed uniform across the width of the beam (see Secs. 8.11 and 8.12).

There is at any point a longitudinal shear stress τ on the horizontal plane and an equal vertical shear stress on the transverse plane. These shear stresses, due to the transverse beam forces, may be assumed uniform across the width of the beam (see sections 8.11 and 8.12).

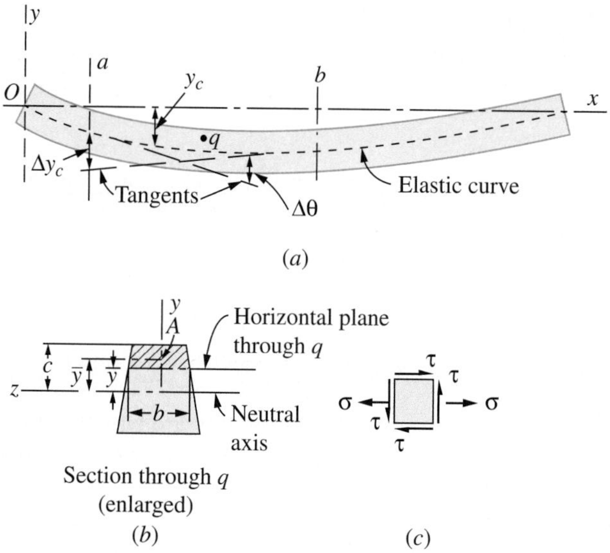

Figure 8.1

Figure 8.1(a,b) represents a beam under load and shows the various dimensions that appear in the formulas; Fig. 8.1(c) shows a small stress element at a point q acted on by the stresses σ and τ.

Formulas

Let I = the moment of inertia of the section of the beam with respect to the neutral axis and E = modulus of elasticity of the material.

The fiber stress σ at any point q is

$$\sigma = -\frac{My}{I} \tag{8.1-1}$$

where M is the bending moment at the section containing q, and y is the vertical distance from the neutral axis to q.

The shear stress τ at any point q is

$$\tau = \frac{VA'\bar{y}}{Ib} \tag{8.1-2}$$

where V is the vertical shear at the section containing q, A' is the area of that part of the section above (or below) q, $\bar{y}$ is the distance from the neutral axis to the centroid of A', and b is the net breadth of the section measured through q.

The complementary energy of flexure U_f, is

$$U_f = \int \frac{M^2}{2EI} dx \tag{8.1-3}$$

where M represents the bending moment equation in terms of x, the distance from the left end of the beam to any section.

The radius of curvature ρ of the elastic curve at any section is

$$\rho = \frac{EI}{M} \tag{8.1-4}$$

where M is the bending moment at the section in question.

The general differential equation of the elastic curve is

$$EI\frac{d^2 y_c}{dx^2} = M \tag{8.1-5}$$

where M has the same meaning as in Eq. (8.1-3) and y_c represents the vertical deflection of the centroidal axis of the beam. Solution of this equation for the vertical deflection y_c is effected by writing out the expression for M, integrating twice, and determining the constants of integration by the boundary conditions.

By the method of unit loads the vertical deflection at any point is found to be

$$y_c = \int \frac{Mm}{EI}\,dx \tag{8.1-6}$$

where M has the same meaning as in Eq. (8.1-3) and m is the equation of the bending moment due to a unit load acting vertically at the section where y_c is to be found. The integration indicated must be performed over each portion of the beam for which either M or m is expressed by a different equation. A positive result for y_c means that the deflection is in the direction of the assumed unit load; a negative result means it is in the opposite direction (see Example 2 at the end of this section).

Using Castigliano's second theorem the deflection is found to be

$$y_c = \frac{\partial U}{\partial P} \tag{8.1-7}$$

where U is given by Eq. (8.1-3) and P is a vertical load, real or imaginary, applied at the section where y_c is to be found. It is most convenient to perform the differentiation within the integral sign; as with Eq. (8.1-6), the integration must extend over the entire length of the beam, and the sign of the result is interpreted as before.

The change in slope of elastic curve Δθ (radians) between any two sections a and b is

$$\Delta\theta = \int_a^b \frac{M}{EI}\,dx \tag{8.1-8}$$

where M has the same meaning as in Eq. (8.1-3).

The deflection Δy_c at any section a, measured vertically from a tangent drawn to the elastic curve at any section b, is

$$\Delta y_c = \int_a^b \frac{M}{EI}\,x\,dx \tag{8.1-9}$$

where x is the distance from a to any section dx between a and b.

Important relations between the bending moment and shear equations are

$$V = \frac{dM}{dx} \tag{8.1-10}$$

$$M = \int V\,dx \tag{8.1-11}$$

These relations are useful in constructing shear and moment diagrams and locating the section or sections of maximum bending moment, since Eq. (8.1-10) shows that the maximum moment occurs when V, its first derivative, passes through zero, and Eq. (8.1-11) shows that the increment in bending moment that occurs between any two sections is equal to the area under the shear diagram between those sections.

Maximum Fiber Stress

The maximum fiber stress at any section occurs at the point or points most remote from the neutral axis and is given by Eq. (8.1-1) when $y = c$; hence

$$\sigma_{\max} = \frac{Mc}{I} = \frac{M}{I/c} = \frac{M}{S} \tag{8.1-12}$$

where S is the *section modulus*, I/c. The maximum fiber stress in the beam occurs at the section of greatest bending moment. If the section is not symmetrical about the neutral axis, the stresses should be investigated at both the section of greatest positive moment and the section of greatest negative moment.

Maximum Transverse Shear Stress†

The maximum transverse shear stress in the beam occurs at the section of greatest vertical shear. The maximum transverse shear stress at any section occurs at the neutral axis, provided the net width b is as small there as anywhere else; if the section is narrower elsewhere, the maximum shear stress may not occur at the neutral axis. This maximum transverse shear stress can be expressed conveniently by the formula

$$(\tau_{\max})_V = \alpha \frac{V}{A} \tag{8.1-13}$$

where V/A is the *average* shear stress on the section and α is a factor that depends on the form of the section. For a rectangular section, $\alpha = 3/2$ and the maximum stress is at the neutral axis; for a solid circular section, $\alpha = 4/3$ and the maximum stress is at the neutral axis; for a triangular section, $\alpha = 3/2$ and the maximum stress is halfway between the top and bottom of the section; for a diamond-shaped section of depth h, $\alpha = 9/8$ and the maximum stress is at points that are a distance $h/8$ above and below the neutral axis.

In the derivation of Eq. (8.1-2) and in the preceding discussion, it is assumed that the shear stress is uniform across the width of the beam; that is, it is the same at all points on any transverse line parallel to the neutral axis. Actually this is not the case; exact analysis (Ref. 1) shows that the shear stress varies across the width and that for a rectangle the maximum intensity occurs at the ends of the neutral axis where, for a wide beam, it is twice the average intensity. Photoelastic investigation of beams under concentrated loading shows that localized shearing stresses about four times as great as the maximum stress given by Eq. (8.1-2) occur near the points of loading and support (Ref. 2), but experience shows that this variation may be ignored and design based on the average value as determined by Eq. (8.1-2).

---

†Note that the transverse shear stress denoted here is the shear stress due to the vertical transverse force, V. The maximum transverse shear stress in a beam is not necessarily the maximum shear stress in the beam. One needs to look at the overall state of stress in light of stress transformations. For long beams, the maximum shear stress is normally due to the maximum fiber stress, and, using Eq. (2.3-25), the maximum shear stress is $\tau_{\max} = \frac{1}{2}\sigma_{\max} = \frac{1}{2} M/S$. For this reason, we will denote the maximum transverse shear stress in a beam as $(\tau_{\max})_V$.

For some sections the greatest horizontal shear stress at a given point occurs, not on a horizontal plane, but on an inclined longitudinal plane which cuts the section so as to make b a minimum. Thus, for a circular tube or pipe the greatest horizontal shear stress at any point occurs on a *radial* plane; the corresponding shear stress in the plane of the section is not vertical but tangential, and in computing τ by Eq. (8.1-2) b should be taken as twice the thickness of the tube instead of the net horizontal breadth of the member. (See Table 9.2, case 20 for an example of where this shear stress in a tube is of importance.)

In an I-, T-, or box section there is a horizontal shear stress on any vertical longitudinal plane through the flange, and this stress is usually a maximum at the juncture of flange and web. It may be calculated by Eq. (8.1-2), taking A' as the area outside of the vertical plane (for outstanding flanges) or between the vertical plane and the center of the beam section (for box girders), and b as the flange thickness (see the solution to Example l*b*). The other terms have the same meanings as explained previously.

Shear stresses are not often of controlling importance except in wood or metal beams that have thin webs or a small span/depth ratio. For beams that conform to the assumptions stated previously, strength will practically always be governed by fiber stress.

Change in Projected Length due to Bending

The apparent shortening of a beam due to bending, that is, the difference between its original length and the horizontal projection of the elastic curve, is given by

$$\Delta l = -\frac{1}{2}\int_0^l \left(\frac{dy}{dx}\right)^2 dx \tag{8.1-14}$$

To evaluate Δl, dy/dx is expressed in terms of x [Eq. (8.1-5)] and the square of this is integrated as indicated.

The extreme fibers of the beam undergo a change in actual length due to stress given by

$$e = \int_0^l \frac{Mc}{EI} dx \tag{8.1-15}$$

By means of these equations the actual relative horizontal displacement of points on the upper or lower surface of the beam can be predicted and the necessary allowances made in the design of rocker bearings, clearance for the corners, etc.

Tabulated Formulas

Table 8.1 gives formulas for the reactions, moments, slopes and deflections at each end of single-span beams supported and loaded in various ways. The table also gives formulas for the vertical shears, bending moments, slopes, and deflections at any distance x from the left end of the span.

In these formulas, the unit step function is used by itself and in combination with ordinary functions.

The unit step function is denoted by $\langle x-a\rangle^0$ where the use of the angle brackets $\langle\,\rangle$ is defined as follows: If $x<a$, $\langle x-a\rangle^0=0$; if $x>a$, $\langle x-a\rangle^0=1$. At $x=a$ the unit step function is undefined just as vertical shear is undefined directly beneath a concentrated load. The use of the angle brackets $\langle\,\rangle$ is extended to other cases involving powers of the unit step function and the ordinary function $(x-a)^n$. Thus the quantity $(x-a)^n\langle x-a\rangle^0$ is shortened to $\langle x-a\rangle^n$ and again is given a value of zero if $x<a$ and is $(x-a)^n$ if $x>a$.

In addition to the usual concentrated vertical loads, concentrated couples, and distributed loads, Table 8.1 also presents cases where the loading is essentially a development of reactions due to deformations created within the span. These cases include the concentrated angular displacement, concentrated transverse deflection, and linear temperature differential across the beam from top to bottom. *In all three cases it can be assumed that initially the beam was deformed but free of stress and then is forced to conform to the end conditions.* (In many cases no forces are created, and the formulas give only the deformed shape.)

Hetényi (Ref. 29) discusses in detail the Maclaurin series form of the equations used in this article and suggests (Ref. 53) that the deformation type of loading might be useful in solving beam problems. Thomson (Ref. 65) describes the use of the unit step function in the determination of beam deflections. By superposition, the formulas can be made to apply to almost any type of loading and support. The use of the tabulated and fundamental formulas given in this article is illustrated in the following examples:

EXAMPLE 1 For a beam supported and loaded as shown in Fig. 8.2, it is required to determine the maximum tensile stress, maximum shear stress, and maximum compressive stress, assuming, (a) that the beam is made of wood with section as shown in Fig. 8.2(*a*); (b) that the beam is made of wood with section as shown in Fig. 8.2(*b*); and (c) that the beam is a 4-in, 7.7-lb steel I-beam, not shown.

SOLUTION By using the equations of equilibrium (Sec. 4.1), the left and right reactions are found to be 900 and 1500 lb, respectively. The shear and moment equations are therefore

$$(x=0 \text{ to } x=160)\text{: } V=900-12x$$

$$M=900x-12x\left(\frac{1}{2}x\right)$$

$$(x=160 \text{ to } x=200)\text{: } V=900-12x+1500$$

$$M=900x-12x\left(\frac{1}{2}x\right)+1500(x-160)$$

Using the step function described previously, these equations can be reduced to

$$V=900-12x+1500\langle x-160\rangle^0$$

$$M=900x-6x^2+1500\langle x-160\rangle$$

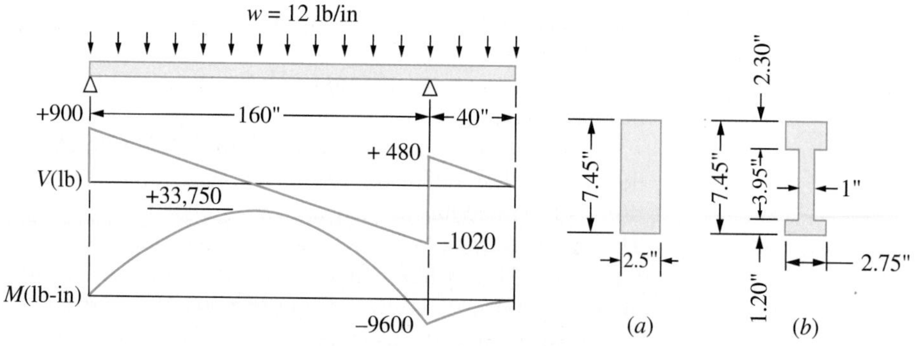

Figure 8.2

These equations are plotted, giving the shear and moment diagrams shown in Fig. 8.2. The maximum positive moment evidently occurs between the supports; the exact location is found by setting the first shear equation equal to zero and solving for x, which gives $x = 75$ in. Substitution of this value of x in the first moment equation gives $M = 33{,}750$ lb-in. The maximum negative moment occurs at the right support where the shear diagram again passes through zero and is 9600 lb-in.

The results obtained so far are independent of the cross section of the beam. The stresses will now be calculated for each of the sections (*a*), (*b*), and (*c*).

(*a*) For the rectangular section: $I = \frac{1}{12}bd^3 = 86.2$ in^4; $I/c = \frac{1}{6}bd^2 = 23.1$ in^3; and $A = bd = 18.60$ in^2. Therefore,

$$\sigma_{max} = \frac{M_{max}}{I/c} = \frac{33{,}750}{23.1} = 1460 \text{ lb/in}^2 \quad \text{[by Eq. (8.1-12)]}$$

This stress occurs at $x = 75$ in and is tension in the bottom fibers of the beam and compression in the top. The maximum *transverse* shear stress is†

$$\tau_{max} = \frac{3}{2}\frac{V_{max}}{A} = \frac{3}{2}\frac{1020}{18.60} = 82 \text{ lb/in}^2 \quad \text{[by Eq. (8.1-13)]}$$

which is the horizontal and vertical shear stress at the neutral axis of the section just to the left of the right support.

(*b*) For the routed section, it is found (App. A) that the neutral axis is 4 in from the base of the section and $I = 82.6$ in^4. The maximum transverse shear stress on a horizontal plane occurs at the neutral axis since b is as small there as anywhere, and so in Eq. (8.1-2) the product $A'\bar{y}$ represents the statical moment about the neutral axis of all that part of the section above the neutral axis. Taking the moment of the flange and web portions separately, we find $A'\bar{y} = (2.75)(2.3)(2.30) + (1)(1.15)(0.575) = 15.2$ in^3. Also, $b = 1.00$ in.

Since the section is not symmetrical about the neutral axis, the fiber stresses will be calculated at the section of maximum positive moment and at the section of maximum negative moment. We have

$$\text{At } x = 75 \text{ in:} \quad \sigma = \begin{cases} \dfrac{(33{,}750)(4)}{82.6} = 1630 \text{ lb/in}^2 & \text{(tension in bottom fiber)} \\ \dfrac{(33{,}750)(3.45)}{82.6} = 1410 \text{ lb/in}^2 & \text{(compression in top fiber)} \end{cases}$$

$$\text{At } x = 160 \text{ in:} \quad \sigma = \begin{cases} \dfrac{(9600)(4)}{82.6} = 456 \text{ lb/in}^2 & \text{(compression in bottom fiber)} \\ \dfrac{(9600)(3.45)}{82.6} = 400 \text{ lb/in}^2 & \text{(tension in top fibers)} \end{cases}$$

It is seen that for this beam the maximum fiber stresses in both tension and compression occur at the section of maximum positive bending moment. The maximum transverse shear stress is

$$(\tau_{max})_V = \frac{(1020)(15.2)}{(82.6)(1)} = 188 \text{ lb/in}^2 \quad \text{[by Eq. (8.1-2)]}$$

†The actual maximum shear stress for this example is found from a stress transformation of the maximum bending stress. Thus, at the outer fibers of the beam at $x = 75$, $\tau_{max} = \frac{1}{2}\sigma_{max} = \frac{1}{2}(1460) = 730$ lb/in^2.

This is the maximum shear stress on a horizontal plane and occurs at the neutral axis of the section just to the left of the right support. The actual maximum shear stress is $\tau_{max} = ½\,\sigma_{max} = ½ \times (1630) = 815$ lb/in^2.

(*c*) For the steel I-beam, the structural steel handbook gives $I/c = 3.00$ in^3 and $t = 0.190$ in. Therefore,

$$\sigma_{max} = \frac{33{,}750}{3} = 11{,}250 \text{ lb/in}^2$$

This stress occurs at $x = 75$ in, where the beam is subject to tension in the bottom fibers and compression in the top. The maximum transverse shear stress is approximated by

$$(\tau_{max})_V \approx \frac{1020}{(4)(0.19)} = 1340 \text{ lb/in}^2$$

Although this method of calculating τ (where the shear force is assumed to be carried entirely by the web) is only approximate, it is usually sufficiently accurate to show whether or not this shear stress is important. If it indicates that this shear stress may govern, then the stress at the neutral axis may be calculated by Eq. (8.1-2). For standard I-beams, the allowable vertical shear is given by the structural-steel handbooks, making computation unnecessary.

EXAMPLE 2 The beam shown in Fig. 8.3 has a rectangular section 2 in wide and 4 in deep and is made of spruce, where $E = 1{,}300{,}000$ lb/in^2. It is required to determine the deflection of the left end.

SOLUTION The solution will be effected first by superposition, using the formulas of Table 8.1. The deflection y of the left end is the sum of the deflection y_1 produced by the distributed load and the deflection y_2 produced by the concentrated load. Each of these is computed independently of the other and added using superposition. Thus

$$y_1 = -40\theta = (-40)\left[-\frac{1}{24}\frac{(2.14)(140^3)}{EI}\right] = +\frac{9{,}800{,}000}{EI}$$

by formula for θ at A, case 2e, where $a = 0$, $l = 140$ in and $w_a = w_l = 2.14$ lb/in. y_2 is calculated as the sum of the deflection the 150-lb load would produce if the beam were *fixed* at the left support and the deflection produced by the fact that it actually assumes a slope there. The first part of the deflection is given by the formula for max y (case 1a), and the second part is found by multiplying the overhang (40 in) by the slope produced at the left end of the 140-in span by a counterclockwise couple equal to $150(40) = 6000$ lb-in applied at that point (formula for θ at A, case 3e, where $a = 0$).

$$y_2 = -\frac{1}{3}\frac{(150)(40^3)}{EI} + (-40)\left[-\frac{1}{3}\frac{(-6000)(140)}{EI}\right] = -\frac{14{,}400{,}000}{EI}$$

Adding algebraically, the deflection of the left end is

$$y = y_1 + y_2 = -\frac{4{,}600{,}000}{EI} = -0.33 \text{ in} \quad \text{(deflection is downward)}$$

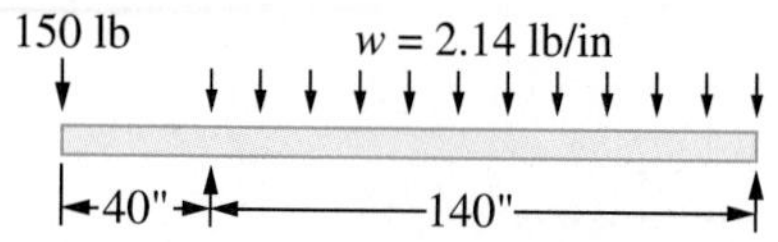

Figure 8.3

The solution of this problem can also be effected readily by using Eq. (8.1-6). The reaction at the left support due to the actual loads is 343 lb and the reaction due to a unit load acting down at the left support is 1.286. If x is measured from the extreme left end of the beam,

$$M=-150x+343\langle x-40\rangle-2.14\frac{\langle x-40\rangle^2}{2} \quad \text{and} \quad m=-x+1.286\langle x-40\rangle$$

Simplifying the equations, we have

$$\begin{aligned} y &= \int \frac{Mm}{EI}dx \\ &= \frac{1}{EI}\left[\int_0^{40}(-150x)(-x)dx+\int_{40}^{180}(-1.071x^2+278.8x-15{,}430)(0.286x-51.6)dx\right] \\ &= +0.33 \text{ in} \end{aligned}$$

(Here the plus sign means that y is in the direction of the assumed unit load, i.e., downward.)

This second solution involves much more labor than the first, and the calculations must be carried out with great accuracy to avoid the possibility of making a large error in the final result.

EXAMPLE 3 A cast iron beam is simply supported at the left end and fixed at the right end on a span of 100 cm. The cross section is 4 cm wide and 6 cm deep ($I = 72$ cm^4). The modulus of elasticity of cast iron is 10^7 N/cm^2, and the coefficient of expansion is 0.000012 cm/cm/°C. It is desired to determine the locations and magnitudes of the maximum vertical deflection and the maximum bending stress in the beam. The loading consists of a uniformly increasing distributed load starting at 0 at 40 cm from the left end and increasing to 200 N/cm at the right end. In addition, the beam, which was originally 20°C, is heated to 50°C on the top and 100°C on the bottom with the temperature assumed to vary linearly from top to bottom.

SOLUTION Superimposing cases 2c and 6c of Table 8.1, the following reactions are determined. (*Note*: For case 2c, $w_a = 0$, $a = 40$ cm, $w_l = 200$ N/cm, and $l = 100$ cm; for case 6c, $T_1 = 50$°C, $T_2 = 100$°C, $\gamma = 0.000012$ cm/cm/°C, $t = 6$ cm, and $a = 0$.)

$$R_A = \frac{200(100-40)^3(4\cdot 100+40)}{40(100^3)} - \frac{3(10^7)(72)(0.000012)(100-50)}{2(6)(100)} = -604.8 \text{ N}$$

$$\text{M}_A = 0 \qquad y_A = 0$$

$$\begin{aligned} \theta_A &= \frac{-200(100-40)^3(2\cdot 100+3\cdot 40)}{240(10^7)(72)(100)} + \frac{0.000012(100-50)(-100)}{4(6)} \\ &= -0.0033 \text{ rad} \end{aligned}$$

Therefore,

$$\begin{aligned} y &= -0.0033x - \frac{604.8x^3}{6EI} - \frac{200\langle x-40\rangle^5}{(100-40)(120)EI} + \frac{0.000012(100-50)x^2}{2(6)} \\ &= -0.0033x - 1.4(10^{-7})x^3 - 3.86(10^{-11})\langle x-40\rangle^5 + 5.0(10^{-5})x^2 \end{aligned}$$

and

$$\frac{dy}{dx} = -0.0033 - 4.2(10^{-7})x^2 - 19.3(10^{-11})\langle x-40\rangle^4 + 10(10^{-5})x$$

The maximum deflection will occur at a position x_1 where the slope dy/dx is zero. At this time an assumption must be made as to which span, $x_1 < 40$ or $x_1 > 40$, will contain the maximum deflection. Assuming that x_1 is less than 40 cm and setting the slope equal to zero,

$$0 = -0.0033 - 4.2(10^{-7})x_1^2 + 10(10^{-5})x_1$$

Of the two solutions for x_1, 39.7 cm and 198 cm, only the value of 39.7 cm is valid since it is less than 40 cm. Substituting $x = 39.7$ cm into the deflection equation gives the maximum deflection of -0.061 cm. Similarly,

$$M = -604.8x - \frac{200}{6(100-40)}\langle x-40\rangle^3$$

which has a maximum negative value where x is a maximum, that is, at the right end:

$$M_{max} = -604.8(100) - \frac{200}{6(100-40)}(100-40)^3 = -180{,}480 \text{ N-cm}$$

$$\sigma_{max} = \frac{Mc}{I} = \frac{180{,}480(3)}{72} = 7520 \text{ N/cm}^2$$

EXAMPLE 4 The cast iron beam in Example 3 is to be simply supported at both ends and carry a concentrated load of 10,000 N at a point 30 cm from the right end. It is desired to determine the relative displacement of the lower edges of the end section.

SOLUTION For this example, case 1e can be used with $a = 70$ cm:

$$R_A = \frac{10{,}000(100-70)}{100} = 3000 \text{ N} \qquad M_A = 0$$

$$\theta_A = \frac{-10{,}000(70)}{6(10^7)(72)(100)}(200-70)(100-70) = -0.00632 \text{ rad}, \qquad y_A = 0$$

$$\theta_B = \frac{10{,}000(70)}{6(10^7)(72)(100)}(100^2-70^2) = 0.00826 \text{ rad}$$

Then

$$\frac{dy}{dx} = -0.00632 + 2.083(10^{-6})x^2 - \frac{10{,}000}{2(10^7)(72)}\langle x-70\rangle^2$$
$$= -0.00632 + 2.083(10^{-6})x^2 - 6.94(10^{-6})\langle x-70\rangle^2$$

The shortening of the neutral surface of the beam is given in Eq. (8.1-14) to be

$$\Delta l = \frac{1}{2}\int_0^l \left(\frac{dy}{dx}\right)^2 dx$$
$$= \frac{1}{2}\int_0^{70} [-0.00632 + 2.083(10^{-6})x^2]^2\, dx$$
$$+ \frac{1}{2}\int_{70}^{100} [-0.04033 + 9.716(10^{-4})x - 4.857(10^{-6})x^2]^2\, dx$$

or

$$\Delta l = 0.00135 \text{ cm} \qquad \text{(a shortening)}$$

In addition to the shortening of the neutral surface, the lower edge of the left end moves to the left by an amount $\theta_A c$ or $0.00632(3) = 0.01896$ cm. Similarly, the lower edge of the right end moves to the right by an amount $\theta_B c$ or $0.00826(3) = 0.02478$ cm. Evaluating the motion of the lower edges in this manner is equivalent to solving Eq. (8.1-15) for the total strain in the lower fibers of the beam.

The total relative motion of the lower edges of the end sections is therefore a moving apart by an amount $0.01896 + 0.02478 - 0.00135 = 0.0424$ cm.

8.2 COMPOSITE BEAMS AND BIMETALLIC STRIPS

Beams that are constructed of more than one material can be treated by using an equivalent width technique if the maximum stresses in each of the several materials remain within the proportional limit. An equivalent cross section is developed in which the width of each component parallel to the principal axis of bending is increased in the same proportion that the modulus of elasticity of that component makes with the modulus of the assumed material of the equivalent beam.

EXAMPLE The beam cross section shown in Fig. 8.4(*a*) is composed of three portions of equal width and depth. The top portion is made of aluminum for which $E_A = 10 \cdot 10^6$ lb/in²; the center is made of brass for which $E_B = 15 \cdot 10^6$ lb/in²; and the bottom is made of steel for which $E_S = 30 \cdot 10^6$ lb/in². Figure 8.4(*b*) shows the equivalent cross section, which is assumed to be made of aluminum. For this equivalent cross section the centroid must be located and the moment of inertia determined for the centroidal axis.

SOLUTION

$$\bar{y} = \frac{3(2)(5) + 4.5(2)(3) + 9(2)(1)}{6 + 9 + 18} = 2.27 \text{ in}$$

$$I_x = \frac{3(2^3)}{12} + 6(5-2.27)^2 + \frac{4.5(2^3)}{12} + 9(3-2.27)^2 + \frac{9(2^3)}{12} + 18(2.27-1)^2$$

$$= 89.5 \text{ in}^4$$

The equivalent stiffness EI of this beam is therefore $10 \cdot 10^6(89.5)$, or $895 \cdot 10^6$ lb-in².

A flexure stress computed by $\sigma = Mc/I_x$ will give a stress in the equivalent beam which can thereby be converted into the stress in the actual composite beam by multiplying by the modulus ratio. If a bending moment of 300,000 lb-in were applied to a beam with the cross section shown, the stress at the top surface of the equivalent beam would be $\sigma = 300{,}000(6 - 2.27)/89.5$, or 12,500 lb/in². Since the material at the top is the same in both the actual and equivalent beams, this is also the maximum stress in the aluminum portion of the actual beam. The stress at the bottom of the equivalent beam would be $\sigma = 300{,}000(2.27)/89.5 = 7620$ lb/in². Multiplying this stress by the modulus ratio, the actual stress at the bottom of the steel portion of the beam would be $\sigma = 7620(30/10) = 22{,}900$ lb/in².

Bimetallic strips are widely used in instruments to sense or control temperatures. The following formula gives the equivalent properties of the strip for which the cross section is shown in Fig. 8.5:

$$\text{Equivalent } EI = \frac{w t_b^3 t_a E_b E_a}{12(t_a E_a + t_b E_b)} K_1 \tag{8.2-1}$$

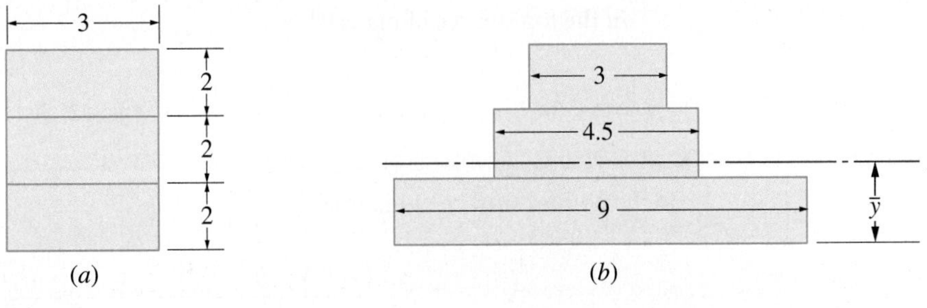

Figure 8.4

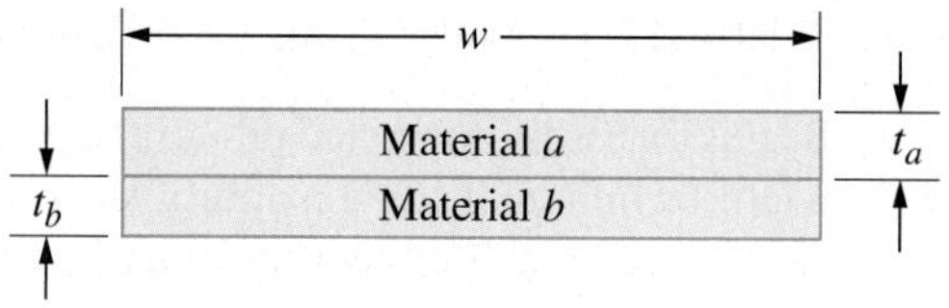

Figure 8.5

or

$$K_1 = 4 + 6\frac{t_a}{t_b} + 4\left(\frac{t_a}{t_b}\right)^2 + \frac{E_a}{E_b}\left(\frac{t_a}{t_b}\right)^3 + \frac{E_b}{E_a}\frac{t_b}{t_a} \tag{8.2-2}$$

All the formulas in Table 8.1, cases 1–5, can be applied to the bimetallic beam by using this equivalent value of *EI*. Since a bimetallic strip is designed to deform when its temperature differs from T_o, the temperature at which the strip is straight, Table 8.1, case 6, can be used to solve for reaction forces and moments as well as deformations of the bimetallic strip under a uniform temperature *T*. To do this, the term $\gamma(T_2 - T_1)/t$ is replaced by the term $6(\gamma_b - \gamma_a)(T - T_o)(t_a + t_b)/(t_b^2 K_1)$ and *EI* is replaced by the equivalent *EI* given by Eq. (8.2-1).

After the moments and deformations have been determined, the flexure stresses can be computed. The stresses due to the bending moments caused by the restraints and any applied loads are given by the following expressions:

In the top surface of material *a*:

$$\sigma = \frac{-6M}{wt_b^2K_1}\left(2 + \frac{t_b}{t_a} + \frac{E_at_a}{E_bt_b}\right) \tag{8.2-3}$$

In the bottom surface of material *b*:

$$\sigma = \frac{6M}{wt_b^2K_1}\left(2 + \frac{t_a}{t_b} + \frac{E_bt_b}{E_at_a}\right) \tag{8.2-4}$$

If there are no restraints imposed, the distortion of a bimetallic strip due to a temperature change is accompanied by flexure stresses in the two materials. This differs from a beam made of a single material which deforms free of stress when subjected to a linear temperature variation through the thickness if there are no restraints. Therefore, the following stresses must be added algebraically to the stresses caused by the bending moments, if any:

In the top surface of material *a*:

$$\sigma = \frac{-(\gamma_b - \gamma_a)(T - T_o)E_a}{K_1}\left[3\frac{t_a}{t_b} + 2\left(\frac{t_a}{t_b}\right)^2 - \frac{E_bt_b}{E_at_a}\right] \tag{8.2-5}$$

In the bottom surface of material *b*:

$$\sigma = \frac{(\gamma_b - \gamma_a)(T - T_o)E_b}{K_1}\left[3\frac{t_a}{t_b} + 2 - \frac{E_a}{E_b}\left(\frac{t_a}{t_b}\right)^3\right] \tag{8.2-6}$$

EXAMPLE A bimetallic strip is made by bonding a piece of titanium alloy ¼ in wide by 0.030 in thick to a piece of stainless steel ¼ in wide by 0.060 in thick. For titanium, $E = 17 \cdot 10^6$ lb/in^2 and $\gamma = 5.7 \cdot 10^{-6}$ in/in/°F; for stainless steel, $E = 28 \cdot 10^6$ lb/in^2 and $\gamma = 9.6 \cdot 10^{-6}$ in/in/°F. It is desired to find the length of bimetal required to develop a reaction force of 5 oz at a simply supported left end when the right end is fixed and the temperature is raised 50°F; also the maximum stresses must be determined.

SOLUTION First find the value of K_1 from Eq. (8.2-2) and then evaluate the equivalent stiffness from Eq. (8.2-1):

$$K_1 = 4 + 6\frac{0.03}{0.06} + 4\left(\frac{0.03}{0.06}\right)^2 + \frac{17}{28}\left(\frac{0.03}{0.06}\right)^3 + \frac{28}{17}\frac{0.06}{0.03} = 11.37$$

$$\text{Equivalent } EI = \frac{0.25(0.06^3)(0.03)(28\cdot 10^6)(17\cdot 10^6)}{12[0.03(17\cdot 10^6)+0.06(28\cdot 10^6)]}11.37 = 333 \text{ lb-in}^2$$

Under a temperature rise over the entire length, the bimetallic strip curves just as a single strip would curve under a temperature differential. To use case 6c in Table 8.1, the equivalent to $\gamma(T_2 - T_1)/t$ must be found. This equivalent value is given by

$$\frac{6(9.6\cdot 10^{-6} - 5.7\cdot 10^{-6})(50)(0.03+0.06)}{(0.06^2)(11.37)} = 0.00257 \text{ in}^{-1}$$

The expression for R_A can now be obtained from case 6c in Table 8.1 and, noting that $a = 0$, the value of the length l can be determined:

$$R_A = \frac{-3(l^2 - a^2)}{2l^3} EI \frac{\gamma}{t}(T_2 - T_1) = \frac{-3}{2l}(333)(0.00257) = \frac{-5}{16} \text{ lb}$$

Therefore, $l = 4.11$ in.

The maximum bending moment is found at the fixed end and is equal to $R_A l$:

$$\max M = -\frac{5}{16}(4.11) = -1.285 \text{ lb-in}$$

Combining Eqs. (8.2-3) and (8.2-5), the flexure stress on the top of the titanium is

$$\sigma = \frac{-6(-1.285)}{0.25(0.06)^2(11.37)}\left(2 + \frac{0.06}{0.03} + \frac{17}{28}\frac{0.03}{0.06}\right)$$
$$-\frac{(9.6\cdot 10^{-6} - 5.7\cdot 10^{-6})(50)(17\cdot 10^6)}{11.37}\left[3\frac{0.03}{0.06} + 2\left(\frac{0.03}{0.06}\right)^2 - \frac{28}{17}\frac{0.06}{0.03}\right]$$
$$= 3242 + 378 = 3620 \text{ lb/in}^2$$

Likewise, the flexure stress on the bottom of the stainless steel is

$$\sigma = \frac{6(-1.285)}{0.25(0.06)^2(11.37)}\left(2 + \frac{0.03}{0.06} + \frac{28}{17}\frac{0.06}{0.03}\right)$$
$$+\frac{(9.6\cdot 10^{-6} - 5.7\cdot 10^{-6})(50)(28\cdot 10^6)}{11.37}\left[3\frac{0.03}{0.06} + 2 - \frac{17}{28}\left(\frac{0.03}{0.06}\right)^3\right]$$
$$= -4365 + 1644 = -2720 \text{ lb/in}^2$$

8.3 THREE-MOMENT EQUATION

The *three-moment equation*, which expresses the relationship between the bending moments found at three *consecutive* supports in a *continuous* beam, can be readily derived for any loading shown in Table 8.1. This is accomplished by taking any two consecutive spans and evaluating the slope for each span at the end where the two spans join. These slopes, which are expressed in terms of the three moments and the loads on the spans, are then equated and the equation reduced to its usual form.

EXAMPLE Consider two contiguous spans loaded as shown in Fig. 8.6. In addition to the loading shown, it is known that the left end of span 1 had settled an amount $y_2 - y_1$ relative to the right end of the span, and similarly that the left end of span 2 has settled an amount $y_3 - y_2$ relative to the right end. (Note that y_1, y_2, and y_3 are considered positive upward as usual.) The individual spans with their loadings are shown in Fig. 8.7(*a*,*b*). Determine the relationship between the applied loads and the moment at the intermediate support.

SOLUTION Using cases 2e and 3e from Table 8.1 and noting the relative deflections mentioned above, the expression for the slope at the right end of span 1 is

$$\theta_2 = \frac{w_1(l_1^2 - a_1^2)^2}{24E_1I_1l_1} - \frac{w_1(l_1 - a_1)^2}{360E_1I_1l_1}(8l_1^2 + 9a_1l_1 + 3a_1^2) + \frac{M_1l_1^2}{6E_1I_1l_1} + \frac{-M_2(l_1^2 - 3a_1^2)}{6E_1I_1l_1} + \frac{y_2 - y_1}{l_1}$$

Similarly, using cases 1e and 3e from Table 8.1, the expression for the slope at the left end of span 2 is

$$\theta_2 = \frac{-W_2a_2}{6E_2I_2l_2}(2l_2 - a_2)(l_2 - a_2) - \frac{M_2}{6E_2I_2l_2}(2l_2^2) + \frac{-M_3}{6E_2I_2l_2}(2l_2^2 - 6l_2^2 + 3l_2^2) + \frac{y_3 - y_2}{l_2}$$

Equating these slopes gives

$$\frac{M_1l_1}{6E_1I_1} + \frac{M_2l_1}{3E_1I_1} + \frac{M_2l_2}{3E_2I_2} + \frac{M_3l_2}{6E_2I_2} = \frac{-w_1(l_1 - a_1)^2}{360E_1I_1l_1}(7l_1^2 + 21a_1l_1 + 12a_1^2) - \frac{y_2 - y_1}{l_1} - \frac{W_2a_2}{6E_2I_2l_2}(2l_2 - a_2)(l_2 - a_2) + \frac{y_3 - y_2}{l_2}$$

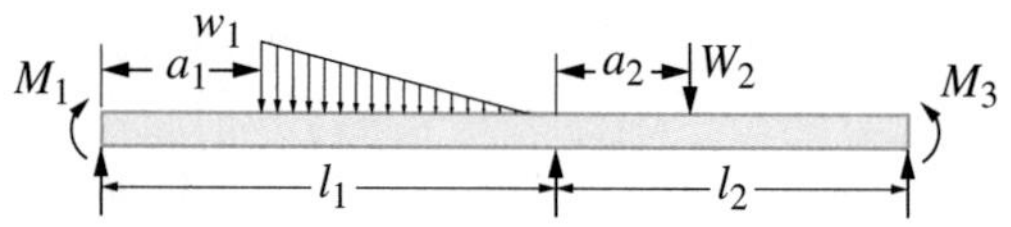

Figure 8.6

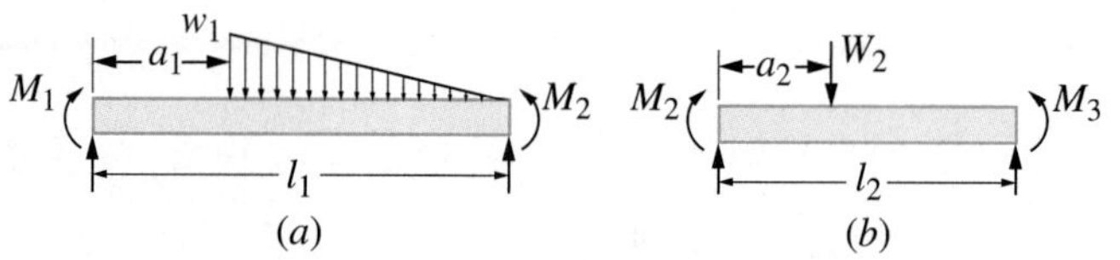

Figure 8.7

If M_1 and M_3 are known, this expression can be solved for M_2; if not, similar expressions for the adjacent spans must be written and the set of equations solved for the moments.

The three-moment equation can also be applied to beams carrying axial tension or compression in addition to transverse loading. The procedure is exactly the same as that described above except the slope formulas to be used are those given in Tables 8.8 and 8.9.

8.4 RIGID FRAMES

By superposition and the matching of slopes and deflections, the formulas in Table 8.1 can be used to solve for the indeterminate reactions in rigid frames or to determine the deformations where the support conditions permit such deformations to take place. The term *rigid* in this section simply means that any deformations are small enough to have negligible effect on bending moments.

In Table 8.2 formulas are given for the indeterminate reactions and end deformations for rigid frames consisting of three members. Only in-plane deformations and only those due to bending moments have been included in the expressions found in this table. Since deformations due to transverse shear and axial loading are not included, the results are limited to those frames made up of members which are long in proportion to their depths (see assumptions 6–8 in Sec. 8.1). Each member must be straight and of uniform cross section having a principal axis lying in the plane of the frame. The elastic stability of the frame, either in the plane of the frame or out of this plane, has not been treated in developing Table 8.2. The effects of axial load on the bending deformations, as documented in Tables 8.7-8.9, have not been considered. A final check on a solution must verify that indeed these effects can be neglected. Very extensive compilations of formulas for rigid frames are available, notably those of Kleinlogel (Ref. 56) and Leontovich (Ref. 57).

While Table 8.2 is obviously designed for frames with three members where the vertical members both lie on the same side of the horizontal member, its use is not limited to this configuration. One can set the lengths of either of the vertical members, members 1 and 2, equal to zero and solve for reactions and deformations of two-member frames. The length of the horizontal member, member 3, should not be set to zero for two reasons: (1) It does not generally represent a real problem; and (2) the lengths of members 1 and 2 are assumed not to change, and this adds a restraint to member 3 that would force it to have zero slope if its length was very short. Another very useful application of the expressions in Table 8.2 is to apply them to frames where one of the two vertical members lies on the opposite side of the horizontal member. Instead of forming a U-shape in the side view, it forms a Z-shape. To do this one must change the signs of three variables associated with the reversed member: (1) the sign of the length of the reversed member, (2) the sign of the distance a which locates any load on the reversed member, and (3) the sign of the product EI of the reversed member. All the reactions and end-point deflections retain their directions as given in the figures in Table 8.2; that is, if member 1 is reversed and extends upward from the left end of member 3, H_A now acts at the upper end of member 1 and is positive to the left as is δ_{HA}. Example 3 illustrates this application as well as showing how the results of using Tables 8.1 and 8.2 together can be used to determine the deformations anywhere in a given frame.

When the number of members is large, as in a framed building, a relaxation method such as moment distribution might be used or a digital computer could be programmed to solve the large number of equations. In all rigid frames, corner or knee design is important;

much information and experimental data relating to this problem are to be found in the reports published by the Fritz Engineering Laboratories of Lehigh University. The frames given in Table 8.2 are assumed to have rigid corners; however, corrections can be made easily once the rigidity of a given corner design is known by making use of the concentrated angular displacement loading with the displacement positioned at the corner. This application is illustrated in Example 2.

EXAMPLE 1 The frame shown in Fig. 8.8(a) is fixed at the lower end of the right-hand member and guided at the lower end of the left-hand member in such a way as to prevent any rotation of this end but permitting horizontal motion if any is produced by the loading. The configuration could represent the upper half of the frame shown in Fig. 8.8(b); for this frame the material properties and physical dimensions are given as $l_1 = 40$ in, $l_2 = 20$ in, $l_3 = 15$ in, $E_1 = E_2 = E_3 = 30\cdot 10^6$ lb/in², $I_1 = 8$ in⁴, $I_2 = 10$ in⁴, and $I_3 = 4$ in⁴. In addition to the load P of 1000 lb, the frame has been subjected to a temperature rise of 50°F since it was assembled in a stress-free condition. The coefficient of expansion of the material used in all three portions is 0.0000065 in/in/°F.

SOLUTION An examination of Table 8.2 shows the required end or support conditions in case 7 with the loading cases f and p listed under case 5. For cases 5–12, the frame constants are evaluated as follows:

$$C_{HH} = \frac{l_1^3}{3E_1I_1} + \frac{l_1^3 - (l_1 - l_2)^3}{3E_2I_2} + \frac{l_1^2 l_3}{E_3I_3} = \frac{\dfrac{40^3}{3(8)} + \dfrac{40^3 - (40-20)^3}{3(10)} + \dfrac{(40^2)(15)}{4}}{30(10^6)}$$

$$= \frac{2666.7 + 1866.7 + 6000}{30(10^6)} = 0.0003511 \text{ in/lb}$$

Similarly

$$C_{HV} = C_{VH} = 0.0000675 \text{ in/lb}$$

$$C_{HM} = C_{MH} = 0.00001033 \text{ lb}^{-1}$$

$$C_{VV} = 0.0000244 \text{ in/lb}$$

$$C_{VM} = C_{MV} = 0.00000194 \text{ lb}^{-1}$$

$$C_{MM} = 0.000000359 \text{ (lb-in)}^{-1}$$

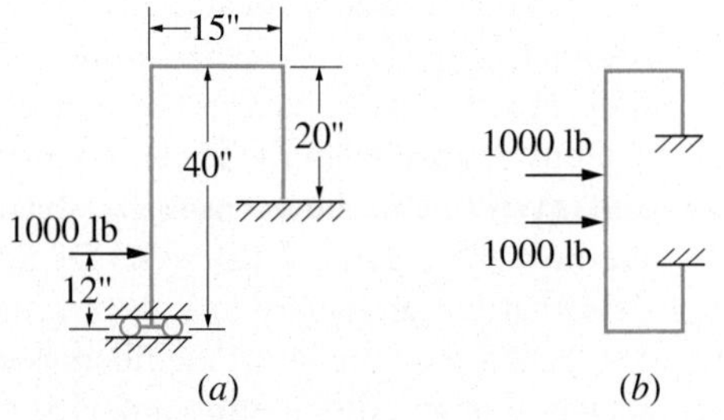

Figure 8.8

For case 5f

$$LF_H = W\left(C_{HH} - aC_{HM} + \frac{a^3}{6E_1I_1}\right)$$

$$= 1000\left[0.0003511 - 12(0.00001033) + \frac{12^3}{6(30\cdot 10^6)(8)}\right]$$

$$= 1000(0.0003511 - 0.000124 + 0.0000012) = 0.2282 \text{ in}$$

Similarly

$$LF_V = 0.0442 \text{ in} \quad \text{and} \quad LF_M = 0.00632 \text{ rad}$$

For case 5p

$$LF_H = -(T - T_o)\gamma_3 l_3 = -50(0.0000065)(15) = -0.004875 \text{ in}$$

$$LF_V = 0.0065 \text{ in} \quad \text{and} \quad LF_M = 0 \text{ rad}$$

For the combined loading

$$LF_H = 0.2282 - 0.004875 = 0.2233 \text{ in}$$

$$LF_V = 0.0507 \text{ in}$$

$$LF_M = 0.00632 \text{ in}$$

Now the left end force, moment, and displacement can be evaluated:

$$V_A = \frac{LF_V C_{MM} - LF_M C_{VM}}{C_{VV}C_{MM} - C_{VM}^2} = \frac{0.0507(0.359\cdot 10^{-6}) - 0.00632(1.94\cdot 10^{-6})}{(24.4\cdot 10^{-6})(0.359\cdot 10^{-6}) - (1.94\cdot 10^{-6})^2}$$

$$= 1189 \text{ lb}$$

$$M_A = 11{,}179 \text{ lb-in}$$

$$\delta_{HA} = 0.0274 \text{ in}$$

Figure 8.9 shows the moment diagram for the entire frame.

EXAMPLE 2 If the joint at the top of the left vertical member in Example 1 had not been rigid but instead had been found to be deformable by 10^{-7} rad for every inch-pound of bending moment applied to it, the solution can be modified as follows.

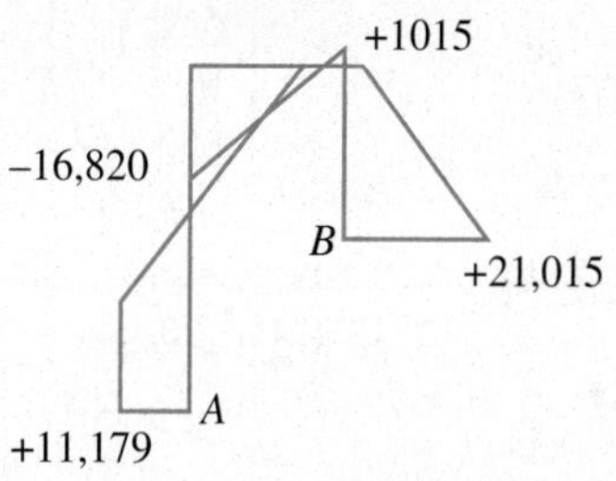

Figure 8.9 (units are in lb-in)

SOLUTION The bending moment as the corner in question would be given by $M_A - 28(1000)$, and so the corner rotation would be $10^{-7}(M_A - 28{,}000)$ rad in a direction opposite to that shown by θ_o in case 5*a*. Note that the position of θ_o is at the corner, and so *A* would be 40 in. Therefore, the following load terms due to the corner deformation can be added to the previously determined load terms:

$$LF_H = -10^{-7}(M_A - 28{,}000)(40)$$

$$LF_V = 0$$

$$LF_M = -10^{-7}(M_A - 28{,}000)$$

Thus the resultant load terms become

$$LF_H = 0.2233 - 4\cdot10^{-6}M_A + 0.112 = 0.3353 - 4\cdot10^{-6}M_A$$

$$LF_V = 0.0507 \text{ in}$$

$$LF_M = 0.00632 - 10^{-7}M_A + 0.0028 = 0.00912 - 10^{-7}M_A$$

Again, the left end force, moment and displacement are evaluated:

$$V_A = \frac{0.0507(0.359\cdot10^{-6}) - (0.00912 - 10^{-7}M_A)(1.94\cdot10^{-6})}{4.996\cdot10^{-12}}$$

$$= 100 + 0.0388M_A$$

$$M_A = \frac{(0.00912 - 10^{-7}M_A)(24.4\cdot10^{-6}) - 0.0507(1.94\cdot10^{-6})}{4.996\cdot10^{-12}}$$

$$= 24{,}800 - 0.488M_A$$

or

$$M_A = 16{,}670 \text{ lb-in}$$

$$\delta_{HA} = -0.0460 \text{ in}$$

$$V_A = 747 \text{ lb}$$

EXAMPLE 3 Find the reactions and deformations at the four positions *A* to *D* in the pinned-end frame shown in Fig. 8.10. All lengths are given in millimeters, $M_o = 2000$ N-mm, and all members are made of aluminum for which $E = 7(10^4)\text{N/mm}^2$ with a constant cross section for which $I = 100 \text{ mm}^4$.

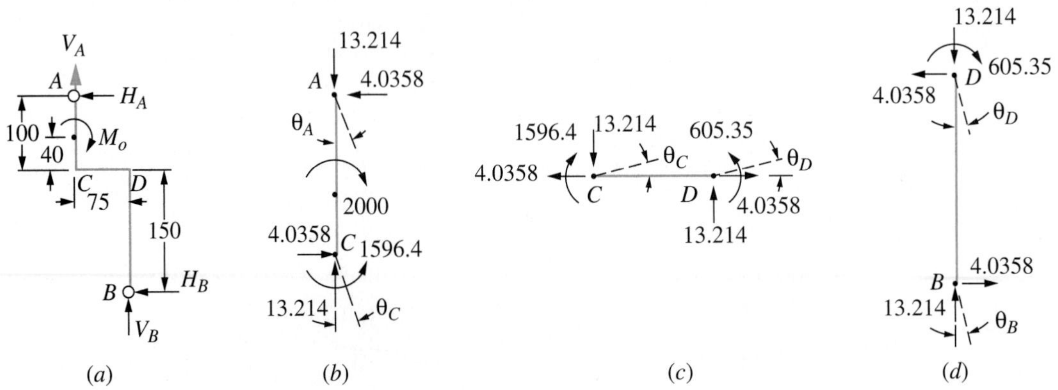

Figure 8.10

SOLUTION Case 1h of Table 8.2 covers this combination of loading and end supports, if, due to the upward reach of member 1, appropriate negative values are used. The need for negative values is described in the introduction to Sec. 8.4. Substitute the following material properties and physical dimensions into case 1h: $l_1 = -100$ mm, $l_2 = 150$ mm, $l_3 = 75$ mm, $a = -(100 - 40) = -60$ mm, $E_1I_1 = -7(10^6)$ N-mm², and $E_2I_2 = E_3I_3 = 7(10^6)$ N-mm². Using these data gives the frame and load constants as

$$A_{HH} = 0.2708 \text{ mm/N}, \quad A_{HM} = A_{MH} = -0.0008036 \text{ N}^{-1}$$

$$A_{MM} = 0.00001786 \text{ (N-mm)}^{-1}, \quad LP_H = 0.0005464(2000) = 1.0928 \text{ mm}$$

$$LP_M = -0.000009286(2000) = -0.01857 \text{ rad}$$

Using these frames and load terms in case 1, the pinned-end case, the reaction and deformations are found to be

$$\delta_{HA} = 0 \quad M_A = 0 \quad H_A = \frac{LP_H}{A_{HH}} = 4.0352 \text{ N} \quad \text{(to left)}$$

and

$$\psi_A = A_{MH}H_A - LP_M = 0.01533 \text{ rad} \qquad \text{(clockwise)}$$

Applying the three independent statics equations to Fig. 8.10(*a*) results in

$$V_A = -13.214 \text{ N} \quad H_B = -4.0358 \text{ N} \quad \text{and} \quad V_B = 13.214 \text{ N}$$

Now treat each of the three members as separate bodies in equilibrium, and find the deformations as pinned-end beams using equations from Table 8.1 as appropriate.

For member 1 as shown in Fig. 8.10(*b*), using case 3e twice, once for each of the two moment loadings, gives

$$\theta_A = \frac{-2000}{6(7 \cdot 10^6)(100)}[2(100)^2 - 6(60)(100) + 3(60)^2]$$
$$+ \frac{1596.4}{6(7 \cdot 10^6)(100)}[2(100)^2 - 6(100)(100) + 3(100)^2]$$
$$= -0.001325 \text{ rad}$$

$$\theta_C = \frac{2000}{6(7 \cdot 10^6)(100)}[(100)^2 - 3(60)^2] - \frac{1596.4}{6(7 \cdot 10^6)(100)}[(100)^2 - 3(100)^2]$$
$$= 0.007221 \text{ rad}$$

To obtain the angle $\psi_A = 0.01533$ rad at position A, member 1 must be given an additional rigid-body clockwise rotation of 0.01533 – 0.001325 = 0.01401 rad. This rigid-body motion moves position C to the left a distance of 1.401 mm and makes the slope at position C equal to 0.007221 – 0.01401 = –0.006784 rad (clockwise).

For member 3 as shown in Fig. 8.10(*c*), again use case 3e from Table 8.1 twice to get

$$\theta_C = \frac{-1596.4}{6(7 \cdot 10^6)(75)}[2(75)^2 - 0 + 0] - \frac{-605.35}{6(7 \cdot 10^6)(75)}[2(75)^2 - 6(75)(75) + 3(75)^2]$$
$$= -0.006782 \text{ rad}$$

$$\theta_D = \frac{1596.4}{6(7 \cdot 10^6)(75)}[(75)^2 - 0] + \frac{-605.35}{6(7 \cdot 10^6)(75)}[(75)^2 - 3(75)^2]$$
$$= 0.005013 \text{ rad}$$

No rigid-body rotation is necessary for member 3 since the left end has the same slope as the lower end of member 1, which is as it should be.

For member 2 as shown in Fig. 8.10*(d)*, use case 3e from Table 8.1 to get

$$\theta_D = \frac{-605.35}{6(7\cdot 10^6)(150)}[2(150)^2 - 0 + 0] = -0.004324 \text{ rad}$$

$$\theta_B = \frac{605.35}{6(7\cdot 10^6)(150)}[(150)^2 - 0] = 0.002162 \text{ rad}$$

To match the slope at the right end of member 3, a rigid-body counterclockwise rotation of $0.005013 + 0.00432 = 0.009337$ rad must be given to member 2. This creates a slope $\psi_B = 0.009337 + 0.002162 = 0.01150$ rad counterclock-wise and a horizontal deflection at the top end of $0.009337(150) = 1.401$ mm to the left. This matches the horizontal deflection of the lower end of member 1 as a final check on the calculations.

To verify that the effect of axial load on the bending deformations of the members is negligible, the Euler load on the longest member is found to be more than 100 times the actual load. Using the formulas from Table 8.8 would not produce significantly different results from those in Table 8.1.

8.5 BEAMS ON ELASTIC FOUNDATIONS

There are cases in which beams are supported on foundations which develop essentially continuous reactions that are proportional at each position along the beam to the deflection of the beam at that position. This is the reason for the name *elastic foundation*. Solutions are available (Refs. 41 and 42) which consider that the foundation transmits shear forces within the foundation such that the reaction force is not directly proportional to the deflection at a given location but instead is proportional to a function of the deflections near the given location; these solutions are much more difficult to use and are not justified in many cases since the linearity of most foundations is open to question anyway.

It is not necessary, in fact, that a foundation be continuous. If a discontinuous foundation, such as is encountered in the support provided a rail by the cross ties, is found to have at least three concentrated reaction forces in every half-wavelength of the deflected beam, then the solutions provided in this section are adequate.

Table 8.5 provides formulas for the reactions and deflections at the left end of a finite-length beam on an elastic foundation as well as formulas for the shear, moment, slope, and deflection at any point x along the length. The format used in presenting the formulas is designed to facilitate programming for use on a digital computer or programmable calculator.

In theory the equations in Table 8.5 are correct for any finite-length beam or for any finite foundation modulus, but for practical purposes they should not be used when βl exceeds a value of 6 because the roundoff errors that are created where two very nearly equal large numbers are subtracted will make the accuracy of the answer questionable. For this reason, Table 8.6 has been provided. Table 8.6 contains formulas for semi-infinite- and infinite-length beams on elastic foundations. These formulas are of a much simpler form since the far end of the beam is assumed to be far enough away so as to have no effect on the response of the left end to the loading. If $\beta l > 6$ and the load is nearer the left end, this is the case.

Hetényi (Ref. 53) discusses this problem of a beam supported on an elastic foundation extensively and shows how the solutions can be adapted to other elements such as hollow cylinders. Hetényi (Ref. 51) has also developed a series solution for beams supported on elastic foundations in which the stiffness parameters of the beam and foundation are not incorporated in the arguments of trigonometric or hyperbolic functions. He gives tables of coefficients derived for specific boundary conditions from which deformation, moments, or shears can be found at any specific point along the beam. Any degree of accuracy can be obtained by using enough terms in the series.

Tables of numerical values, Tables 8.3 and 8.4, are provided to assist in the solution of the formulas in Table 8.5. Interpolation is possible for values that are included but should be used with caution if it is noted that differences of large and nearly equal numbers are being encountered. A far better method of interpolation for a beam with a single load is to solve the problem twice. For the first solution move the load to the left until $\beta(l-a)$ is a value found in Table 8.3, and for the second solution move the load similarly to the right. A linear interpolation from these solutions should be very accurate.

Presenting the formulas for end reactions and displacements in Table 8.5 in terms of the constants C_i and C_{ai} is advantageous since it permits one to solve directly for loads anywhere on the span. If the loads are at the left end such that $C_i = C_{ai}$, then the formulas can be presented in a simpler form as is done in Ref. 6 of Chap. 13 for cylindrical shells. To facilitate the use of Table 8.5 when a concentrated load, moment, angular rotation, or lateral displacement is at the left end (i.e., $a = 0$), the following equations are presented to simplify the numerators:

$$C_1C_2 + C_3C_4 = C_{12}, \qquad 2C_1^2 + C_2C_4 = 2 + C_{11}$$

$$C_2C_3 - C_1C_4 = C_{13}, \qquad C_2^2 - 2C_1C_3 = C_{14}$$

$$C_1^2 + C_3^2 = 1 + C_{11}, \qquad 2C_3^2 - C_2C_4 = C_{11}$$

$$C_2^2 + C_4^2 = 2C_{14}, \qquad 2C_1C_3 + C_4^2 = C_{14}$$

EXAMPLE 1 A 6-in, 12.5-lb I-beam 20 ft long is used as a rail for an overhead crane and is in turn being supported every 2 ft of its length by being bolted to the bottom of a 5-in, 10-lb I-beam at midlength. The supporting beams are each 21.5 ft long and are considered to be simply supported at the ends. This is a case of a discontinuous foundation being analyzed as a continuous foundation. It is desired to determine the maximum bending stresses in the 6-in beam as well as in the supporting beams when a load of 1 ton is supported at one end of the crane.

SOLUTION The spring constant for each supporting beam is $48EI/l^3$, or $(48)(30 \cdot 10^6)(12.1)/(21.5 \cdot 12)^3 =$ 1013 lb/in. If this is assumed to be distributed over a 2-ft length of the rail, the equivalent value of $b_o k_o$ is 1.013/24 = 42.2 lb/in per inch of deflection. Therefore,

$$\beta = \left(\frac{b_o k_o}{4EI}\right)^{1/4} = \left[\frac{42.2}{4(30 \cdot 10^6)(21.8)}\right]^{1/4} = 0.01127 \text{ in}^{-1}$$

and

$$\beta l = (0.01127)(240) = 2.70$$

An examination of the deflection of a beam on an elastic foundation shows that it varies cyclicly in amplitude with the sine and cosine of βx. A half-wavelength of this cyclic variation would occur over a span l_1, where $\beta l_1 = \pi$, or $l_1 = \pi/0.01127 = 279$ in. There is no question about there being at least three supporting forces over this length, and so the use of the solution for a continuous foundation is entirely adequate.

Since βl is less than 6, Table 8.5 will be used. Refer to case 1 where both ends are free. It must be pointed out that a simple support refers to a reaction force, developed by a support other than the foundation, which is large enough to prevent any vertical deflection of the end of the beam. From the table we find that $R_A = 0$ and $M_A = 0$; and since the load is at the left end, $a = 0$. When $a = 0$, the C_a terms are equal to the C terms, and so the four terms C_1, C_2, C_3, and C_4 are calculated:

$$C_1 = \cosh\beta l\cos\beta l = 7.47(-0.904) = -6.76$$

$$C_2 = \cosh\beta l\sin\beta l + \sinh\beta l\cos\beta l = 7.47(0.427) + 7.41(-0.904) = -3.50$$

Similarly $C_3 = 3.17$, $C_4 = 9.89$, and $C_{11} = 54.7$. (see Tables 8.3 and 8.4.) Therefore,

$$\theta_A = \frac{2000}{2(30\cdot 10^6)(21.8)(0.01127^2)}\frac{(-3.50^2)-(2)(3.17)(-6.76)}{54.7} = 0.01216 \text{ rad}$$

$$y_A = \frac{2000}{2(30\cdot 10^6)(21.8)(0.01127^3)}\frac{(9.89)(-6.76)-(3.17)(-3.50)}{54.7} = -1.092 \text{ in}$$

With the deformations at the left end known, the expression for the bending moment can be written:

$$M = -y_A 2EI\beta^2 F_3 - \theta_A EI\beta F_4 - \frac{W}{2\beta}F_{a2}$$

$$= 1.092(2)(30\cdot 10^6)(21.8)(0.01127^2)F_3 - 0.01216(30\cdot 10^6)(21.8)(0.01127)F_4$$

$$-\frac{2000}{2(0.01127)}F_{a2}$$

$$= 181{,}400F_3 - 89{,}600F_4 - 88{,}700F_{a2}$$

Now substituting the expressions for F_{a2}, F_3, and F_4 gives

$$M = 181{,}400\sinh\beta x\sin\beta x - 89{,}600(\cosh\beta x\sin\beta x - \sinh\beta x\cos\beta x)$$

$$-88{,}700(\cosh\beta x\sin\beta x + \sinh\beta x\cos\beta x)$$

or

$$M = 181{,}400\sinh\beta x\sin\beta x - 178{,}300\cosh\beta x\sin\beta x + 900\sinh\beta x\cos\beta x$$

The maximum value of M can be found by trying values of x in the neighborhood of $x = \pi/4\beta = \pi/4(0.01127) = 69.7$ in, which would be the location of the maximum moment if the beam were infinitely long (see Table 8.6). This procedure reveals that the maximum moment occurs at $x = 66.5$ in and has a value of −55,400 lb-in.

The maximum stress in the 6-in I-beam is therefore 55,400(3)/21.8 = 7620 lb/in^2. The maximum stress in the supporting 5-in I-beams is found at the midspan of the beam directly above the load. The deflection of this beam is known to be 1.092 in, and the spring constant is 1013 lb/in, so that the center load on the beam is 1.092(1013) = 1107 lb. Therefore, the maximum bending moment is $Pl/4$ = 1107(21.5)(12)/4 = 71,400 lb-in and the maximum stress is 71,400(2.5)/12.1 = 14,780 lb/in^2.

EXAMPLE 2 If the 6-in I-beam in Example 1 had been much longer but supported in the same manner, Table 8.6 could have been used. Case 8 reveals that for an end load the end deflection is $-W/2EI\beta^3 = -2000/2(30\cdot 10^6)(21.8)(0.01127^3) = -1.070$ in and the maximum moment would have equaled $-0.3225W/\beta = -0.3225(2000)/0.01127 = -57{,}200$ in-lb at 69.7 in from the left end. We should not construe from this example that increasing the length will always increase the stresses; if the load had been placed elsewhere on the span, the longer beam could have had the lower maximum stress.

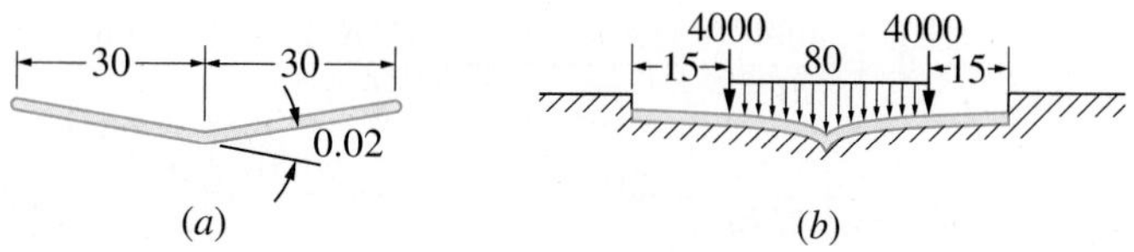

Figure 8.11

EXAMPLE 3 An aluminum alloy beam 3 in wide, 2 in deep, and 60 in long is manufactured with an initial concentrated angular deformation of 0.02 rad at midlength; this initial shape is shown in Fig. 8.11(*a*). In use, the beam is placed on an elastic foundation which develops 500 lb/in² vertical upward pressure for every 1 in it is depressed. The beam is loaded by two concentrated loads of 4000 lb each and a uniformly distributed load of 80 lb/in over the portion between the concentrated loads. The loading is shown in Fig. 8.11(*b*). It is desired to determine the maximum bending stress in the aluminum beam.

SOLUTION First determine the beam and foundation parameters:

$$E = 9.5 \cdot 10^6 \text{ lb/in}^2, \quad I = \frac{1}{12}(3)(2^3) = 2 \text{ in}^4, \quad k_0 = 500 \text{ lb/in}^2\text{/in}, \quad b_o = 3 \text{ in}$$

$$\beta = \left[\frac{3(500)}{4(9.5 \cdot 10^6)(2)}\right]^{1/4} = 0.0666, \quad l = 60 \text{ in}, \quad \beta l = 4.0$$

$$C_1 = -17.85, \quad C_2 = -38.50, \quad C_3 = -20.65, \quad C_4 = -2.83, \quad C_{11} = 744$$

An examination of Table 8.5 shows the loading conditions covered by the superposition of three cases in which both ends are free: case 1 used twice with $W_1 = 4000$ lb and $a_1 = 15$ in, and $W_2 = 4000$ lb and $a_2 = 45$ in; case 2 used twice with $w_3 = 80$ lb/in and $a_3 = 15$ in, and $w_4 = -80$ lb/in and $a_4 = 45$ in; case 5 used once with $\theta_o = 0.02$ and $a = 30$ in.

The loads and deformations at the left end are now evaluated by summing the values for the five different loads, which is done in the order in which the loads are mentioned. But before actually summing the end values, a set of constants involving the load positions must be determined for each case. For case 1, load 1:

$$C_{a1} = \cosh\beta(60-15)\cos\beta(60-15) = 10.068(-0.99) = -9.967$$

$$C_{a2} = -8.497$$

For case 1, load 2:

$$C_{a1} = 0.834, \quad C_{a2} = 1.933$$

For case 2, load 3:

$$C_{a2} = -8.497, \quad C_{a3} = 1.414$$

For case 2, load 4:

$$C_{a2} = 1.933, \quad C_{a3} = 0.989$$

For case 5:

$$C_{a3} = 3.298, \quad C_{a4} = 4.930$$

Therefore, $R_A = 0$ and $M_A = 0$.

$$\theta_A = \frac{4000}{2EI\beta^2}\frac{(-38.50)(-8.497)-(2)(-20.65)(-9.967)}{744} + \frac{4000}{2EI\beta^2}\frac{(-38.50)(1.933)-(2)(-20.65)(0.834)}{744}$$

$$+\frac{80}{2EI\beta^3}\frac{(-38.50)(1.414)-(-20.65)(-8.497)}{744} + \frac{-80}{2EI\beta^3}\frac{(-38.50)(0.989)-(-20.65)(1.933)}{744}$$

$$+0.02\frac{(-38.50)(4.93)-(2)(-20.65)(3.298)}{744}$$

$$=\frac{400}{EI\beta^2}(-0.0568)+\frac{4000}{EI\beta^2}(-0.02688)+\frac{80}{EI\beta^3}(-0.1545)-\frac{80}{EI\beta^3}(0.00125)+0.02(-0.0721)$$

$$=-0.007582 \text{ rad}$$

Similarly,

$$y_A = -0.01172 \text{ in}$$

An examination of the equation for the transverse shear V shows that the value of the shear passes through zero at $x = 15$, 30, and 45 in. The maximum positive bending moment occurs at $x = 15$ in and is evaluated as follows, noting again that R_A and M_A are zero:

$$M_{15} = -(-0.01172)(2)(9.5\cdot 10^6)(2)(0.06666^2)[\sinh(0.06666)(15)\sin 1]$$

$$-(-0.007582)(9.5\cdot 10^6)(2)(0.06666)(\cosh 1 \sin 1 - \sinh 1 \cos 1)$$

$$= 8330 \text{ lb-in}$$

Similarly, the maximum negative moment at $x = 30$ in is evaluated, making sure that the terms for the concentrated load at 15 in and the uniformly distributed load from 15 to 30 in are included:

$$M_{30} = -13{,}000 \text{ lb-in}$$

The maximum bending stress is given by $\sigma = Mc/I$ and is found to be 6500 lb/in^2.

8.6 DEFORMATION DUE TO THE ELASTICITY OF FIXED SUPPORTS

The formulas in Tables 8.1, 8.2, 8.5, 8.6, and 8.8–8.10 that apply to those cases where fixed or guided end supports are specified are based on the assumption that the support is rigid and holds the fixed or guided end truly horizontal or vertical. The slight deformation that actually occurs at the support permits the beam to assume there a slope $\Delta\theta$, which for the conditions represented in Fig. 8.12, that is, a beam integral with a semi-infinite supporting foundation, is given by

$$\Delta\theta = \frac{16.67M}{\pi E h_1^2} + \frac{(1-\nu)V}{Eh_1}$$

Here M is the bending moment per unit width and V is the shear force per unit width of the beam at the support; E is the modulus of elasticity, and ν is Poisson's ratio for the foundation material; and $h_1 = h + 1.5r$ (Ref. 54). The effect of this deformation is to increase the deflections of the beam. For a cantilever, this increase is simply $x\,\Delta\theta$, but for other support conditions the concept of the externally created angular deformation may be utilized (see Example 2 on page 167).

For the effect of many moment-loaded cantilever beams spaced closely one above the next, see Ref. 67.

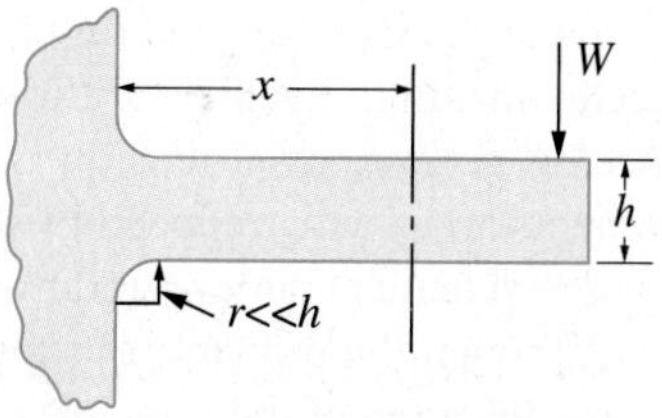

Figure. 8.12

8.7 BEAMS UNDER SIMULTANEOUS AXIAL AND TRANSVERSE LOADING

Under certain conditions a beam may be subjected to axial tension or compression in addition to the transverse loads. Axial tension tends to straighten the beam and thus reduce the bending moments produced by the transverse loads, but axial compression has the opposite effect and may greatly increase the maximum bending moment and deflection and obviously must be less than the critical or buckling load. See Chap. 15. In either case a solution cannot be effected by simple superposition but must be arrived at by methods that take into account the change in deflection produced by the axial load.

For any condition of loading, the maximum normal stress in an extreme fiber is given by

$$\sigma_{max} = \frac{P}{A} \pm \frac{Mc}{I} \tag{8.7-1}$$

where P is the axial load (positive if tensile and negative if compressive), A is the cross-sectional area of the beam, I/c is the section modulus, and M is the maximum bending moment due to the combined effect of axial and transverse loads. (Use the plus sign if M causes tension at the point in question and the minus sign if M causes compression.)

It is the determination of M that offers difficulty. For some cases, especially if P is small or tensile, it is permissible to ignore the small additional moment caused by P and to take M equal to M', the bending moment due to transverse loads only. Approximate formulas of the type (Ref. 33)

$$y_{max} = \frac{y'_{max}}{1 \pm \alpha_y P^2/EI}, \quad \theta_{max} = \frac{\theta'_{max}}{1 \pm \alpha_\theta P^2/EI}, \quad M_{max} = \frac{M'_{max}}{1 \pm \alpha_M P^2/EI} \tag{8.7-2}$$

have been used, but the values of α_y, α_θ, and α_M are different for each loading and each manner of supporting the beam.

Instead of tabulating the values of α, which give answers with increasing error as P increases, Table 8.7(a–d) gives values of the coefficient C_p which can be used in the expressions

$$y_A = C_P y'_A, \quad \theta_A = C_P \theta'_A, \quad M_A = C_P M'_A, \quad \text{etc.} \tag{8.7-3}$$

where the primed values refer to the laterally loaded beam without the axial load and can be evaluated from expressions found in Table 8.1. For those cases listed where the reactions are statically indeterminate, the reaction coefficients given will enable the remaining reactions to be evaluated by applying the principles of static equilibrium. The given values of C_p are exact, based on the assumption that deflections due to transverse shear are negligible. This same assumption was used in developing the equations for transverse shear, bending moment, slope, and deflection shown in Tables 8.8 and 8.9.

Table 8.8 lists the general equations just mentioned as well as boundary values and selected maximum values for the case of axial compressive loading plus transverse loading. Since, in general, axial tension is a less critical condition, where deflections, slopes, and moments are usually reduced by the axial load, Table 8.9 is much more compact and gives only the general equations and the left-end boundary values.

Although the principle of superposition does not apply to the problem considered here, this modification of the principle can be used: the moment (or deflection) for a combination of transverse loads can be found by adding the moments (or deflections) for each transverse load combined with the entire axial load. Thus a beam supported at the ends and subjected to a uniform load, a center load, and an axial compression would have a maximum bending moment (or deflection) given by the sum of the maximum moments (or deflections) for Table 8.8, cases 1e and 2e, the end load being included once for each transverse load.

A problem closely related to the beam under combined axial and lateral loading occurs when the ends of a beam are axially restrained from motion along the axis of the beam (held) and a lateral load is applied. A solution is effected by equating the increase in length of the neutral surface of the beam Pl/AE to the decrease in length due to the curvature of the neutral surface $1/2\int_0^1 \theta^2 dx$ [Eq. (8.1-14)]. In general, solving the resulting equation for P is difficult owing to the presence of the hyperbolic functions and the several powers of the load P in the equation. If the beam is long, slender, and heavily loaded, this will be necessary for good accuracy; but if the deflections are small, the deflection curve can be approximated with a sine or cosine curve, obtaining the results given in Table 8.10. The following examples will illustrate the use of the formulas in Tables 8.7-8.10:

EXAMPLE 1 A 4-in, 7.7-lb steel I-beam 20 ft long is simply supported at both ends and simultaneously subjected to a transverse load of 50 lb/ft (including its own weight), a concentrated lateral load of 600 lb acting vertically downward at a position 8 ft from the left end, and an axial compression of 3000 lb. It is required to determine the maximum fiber stress and the deflection at midlength.

SOLUTION Here $P = 3000$ lb; $l = 240$ in; $I = 6$ in^4; $I/c = 3$ in^3; $A = 2.21$ in^2; $w_a = w_l = 50/12 = 4.17$ lb/in; and $a = 0$ for case 2e; $W = 600$ lb and $a = 96$ in for case 1e; $k = \sqrt{P/EI} = 0.00408$ in^{-1}; $kl = 0.98$. The solution will be carried out (*a*) ignoring deflection, (*b*) using coefficients from Table 8.7 and (*c*) using precise formulas from Table 8.8.

(*a*) $R_A = 860$ lb, and max $M_{8'} = 860(8) - 8(50)(4) = 5280$ lb-ft:

$$\text{max compressive stress} = -\frac{P}{A} - \frac{M}{I/c} = -\frac{3000}{2.21} - \frac{5280(12)}{3} = -22{,}475 \text{ lb/in}^2$$

For the uniform load (Table 8.1, case 2e):

$$y_{l/2} = \frac{-5}{384}\frac{w_a l^4}{EI} = \frac{-5(4.17)(240^4)}{384(30\cdot 10^6)(6)} = -1.00 \text{ in}$$

For the concentrated load (Table 8.1, case 1e):

$$R_A = 360 \text{ lb}$$

$$\theta_A = \frac{-600(96)[2(240)-96](240-96)}{6(30\cdot 10^6)(6)(240)} = -0.123 \text{ rad}$$

$$y_{l/2} = -0.123(120) + \frac{360(120^3)}{6(30\cdot 10^6)(6)} - \frac{600(120-96)^3}{6(30\cdot 10^6)(6)} = -0.907 \text{ in}$$

Thus

$$\text{Total midlength deflection} = -1.907 \text{ in}$$

(*b*) From Table 8.7(*b*) (simply supported ends), coefficients are given for concentrated loads at $l/4$ and $l/2$. Plotting curves of the coefficients versus kl and using linear interpolation to correct for $a = 0.4l$ give a value of $C_p = 1.112$ for the midlength deflection and 1.083 for the moment under the load. Similarly, for a uniform load on the entire span ($a = 0$), the values of C_p are found to be 1.111 for the midlength deflection and 1.115 for the moment at midlength. If it is assumed that this last coefficient is also satisfactory for the moment at $x = 0.4l$, the following deflections and moments are calculated:

$$\text{Max } M_{8'} = 360(8)(1.083) + [500(8) - 8(50)(4)](1.115) = 3120 + 2680 = 5800 \text{ lb-ft}$$

$$\text{Max compressive stress} = -\frac{P}{A} - \frac{M}{I/c} = -\frac{3000}{2.21} - \frac{5800(12)}{3} = -24{,}560 \text{ lb/in}^2$$

$$\text{Midlength deflection} = -0.907(1.112) - 1.00(1.111) = -2.12 \text{ in}$$

(*c*) From Table 8.8 cases 1e and 2e, $R_A = 860$ lb and

$$\theta_A = \frac{-600}{3000}\left[\frac{\sin 0.00408(240-96)}{\sin 0.98} - \frac{240-96}{240}\right] + \frac{-4.17}{0.00408(3000)}\left(\tan\frac{0.98}{2} - \frac{0.98}{2}\right)$$

$$= \frac{-600}{3000}\left(\frac{0.5547}{0.8305} - 0.6\right) - 0.341(0.533 - 0.49)$$

$$= -0.0283 \text{ rad}$$

$$\text{Max } M_{8'} = \frac{860}{0.00408}\sin 0.00408(96) - \frac{-0.0283(3000)}{0.00408}\sin 0.392 - \frac{4.17}{0.00408^2}(1 - \cos 0.392)$$

$$= 80{,}500 + 7950 - 19{,}000 = 69{,}450 \text{ lb-in}$$

$$\text{Max compressive stress} = -\frac{3000}{2.21} - \frac{69{,}450}{3} = -24{,}500 \text{ lb/in}^2$$

$$\text{Midlength deflection} = \frac{-0.0283}{0.00408}\sin 0.49 + \frac{860}{0.00408(3000)}(0.49 - \sin 0.49)$$

$$- \frac{600}{0.00408(3000)}[0.00408(120-96) - \sin 0.00408(120-96)]$$

$$- \frac{4.17}{0.00408^2(3000)}\left[\frac{0.00408^2(120^2)}{2} - 1 + \cos 0.00408(120)\right]$$

$$= -3.27 + 1.36 - 0.00785 - 0.192 = -2.11 \text{ in}$$

The ease with which the coefficients C_p can be obtained from Tables 8.7(*a*–*d*) makes this a very desirable way to solve problems of axially loaded beams. Some caution must be observed, however, when interpolating for the position of the load. For example, the concentrated moment in Tables 8.7*c* and 8.7*d* shows a large variation in C_p for the end moments when the load position is changed from 0.25 to 0.50, especially under axial tension. Note that there are some cases in which C_p either changes sign or increases and then decreases when kl is increased; in these cases the loading produces both positive and negative moments and deflections in the span.

EXAMPLE 2 A solid round brass bar 1 cm in diameter and 120 cm long is rigidly fixed at both ends by supports assumed to be sufficiently rigid to preclude any relative horizontal motion of the two supports. If this bar is subjected to a transverse center load of 300 N at midlength, what is the center deflection and what is the maximum tensile stress in the bar?

SOLUTION Here P is an unknown tensile load; $W = 300$ N and $a = 60$ cm; $l = 120$ cm; $A = 0.785$ cm^2, $I = 0.0491$ cm^4; and $E = 10 \cdot 10^6$ N/cm^2. (This situation is described in Table 8.10, case 2.) The first equation is solved for y_{max}:

$$y_{\max} + \frac{0.785}{16(0.0491)} y_{\max}^3 = \frac{300(120^3)}{2(\pi^4)(10 \cdot 10^6)(0.0491)}$$

$$y_{\max} + y_{\max}^3 = 5.44$$

Therefore, $y_{max} = 1.57$ cm. The second equation is now solved for P:

$$P = \frac{\pi^2(10 \cdot 10^6)(0.785)}{4(120^2)} 1.57^2 = 3315 \text{ N}$$

$$k = \sqrt{\frac{P}{EI}} = \left[\frac{3315}{(10 \cdot 10^6)(0.0491)}\right]^{1/2} = 0.0822 \text{ cm}^{-1}$$

$$kl = 9.86$$

From Table 8.7, case 1d, the values of R_A and M_A can be calculated. (Note that θ_A and y_A are zero.) First evaluate the necessary constants:

$$C_2 = \sinh 9.86 = 9574.4$$

$$C_3 = \cosh 9.86 - 1 = 9574.4 - 1 = 9573.4$$

$$C_4 = \sinh 9.86 - 9.86 = 9564.5$$

$$C_{a3} = \cosh \frac{9.86}{2} - 1 = 69.193 - 1 = 68.193$$

$$C_{a4} = \sinh 4.93 - 4.93 = 69.186 - 4.93 = 64.256$$

$$R_A = W\frac{C_3C_{a3} - C_2C_{a4}}{C_3^2 - C_2C_4} = 300\frac{9573.4(68.193) - 9574.4(64.256)}{9573.4^2 - 9574.4(9564.5)} = 300(0.5)$$

$$= 150 \text{ N}$$

$$M_A = \frac{-W}{k}\frac{C_4C_{a3} - C_3C_{a4}}{C_3^2 - C_2C_4} = \frac{-300}{0.0822}\frac{9564.5(68.193) - 9573.4(64.256)}{74{,}900}$$

$$= \frac{-300}{0.0822} 0.493 = -1800 \text{ N-cm}$$

$$\text{Max tensile stress} = \frac{P}{A} + \frac{Mc}{I} = \frac{3315}{0.785} + \frac{1800(0.5)}{0.0491} = 4220 + 18{,}330$$

$$= 22{,}550 \text{ N/cm}^2$$

$$\text{Midlength deflection} = \frac{-1800}{3315}\left(\cosh\frac{9.86}{2} - 1\right) + \frac{150}{3315(0.0822)}(\sinh 4.93 - 4.93)$$

$$= -37.0 + 35.4 = -1.6 \text{ cm}$$

This compares favorably with the value $y_{max} = 1.57$ cm obtained from the equation which was based on the assumption of a cosine curve for the deflection.

An alternative to working with the large numerical values of the hyperbolic sines and cosines as shown in the preceding calculations would be to simplify the equations for this case where the load is at the center by using the double-angle identities for hyperbolic functions. If this is done here, the expressions simplify to

$$R_A = \frac{W}{2} \quad M_A = \frac{-W}{2k}\tanh\frac{kl}{4} \quad y_{l/2} = \frac{-W}{kP}\left(\frac{kl}{4} - \tanh\frac{kl}{4}\right)$$

Using these expressions gives $R_A = 150$ N, $M_A = -1800$ N-cm, and $y_{l/2} = -1.63$ cm. Table 8.6 for axial compression gives the formulas for these special cases, but when the lateral loads are not placed at midlength or any of the other common locations, a desk calculator or digital computer must be used. If tables of hyperbolic functions are employed, it should be kept in mind that adequate solutions can be made using values of kl close to the desired values if such values are given in the table and the desired ones are not. For example, if the values for the arguments 9.86 and 4.93 are not available but values for 10 and 5 are (note that it is necessary to maintain the correct ratio a/l), these values could be used with no noticeable change in the results. Finally, an energy approach, using Rayleigh's technique, is outlined in Chap. 6, Sec. 13, of Ref. 72. The method works well with simple, axially constrained, and unconstrained beams.

8.8 BEAMS OF VARIABLE SECTION

Stress

For a beam whose cross section changes gradually, Eqs. (8.1-1), (8.1-4), and (8.1-10)–(8.1-12) apply with sufficient accuracy; Eqs. (8.1-3) and (8.1-5)-(8.1-7) apply if I is treated as a variable, as in the examples that follow. All the formulas given in Table 8.1 for vertical shear and bending moments in *statically determinate* beams apply, but the formulas given for statically indeterminate beams and for deflection and slope are inapplicable to beams of nonuniform section unless the section varies in such a way that I is constant.

Accurate analysis (Ref. 3) shows that in an end-loaded cantilever beam of rectangular section which is symmetrically tapered in the plane of bending, the maximum fiber stress is somewhat less than is indicated by Eq. (8.1-12), the error amounting to about 5% for a surface slope of 15° (wedge angle 30°) and about 10% for a surface slope of 20°. See also Prob. 2.35 in Ref. 66. The maximum horizontal and vertical shear stress is shown to occur at the upper and lower surfaces instead of at the neutral axis and to be approximately three times as great as the average shear stress on the section for slopes up to 20°. It is very doubtful, however, if this shear stress is often critical even in wood beams, although it may possibly start failure in short, heavily reinforced concrete beams that are deepened or "haunched" at the ends. Such a failure, if observed, would probably be ascribed to compression since it would occur at a point of high compressive stress. It is also conceivable, of course, that this shear stress might be of importance in certain metal parts subject to repeated stress.

Abrupt changes in the section of a beam cause high local stresses, the effect of which is taken into account by using the proper factor of stress concentration (Sec. 3.10 and Table 17.1).

Deflection

Determining deflections or statically indeterminate reactions for beams of variable section can be considered in two categories: where the beam has a continuously varying cross section from one end to the other, and where the cross section varies in a stepwise fashion.

Considering the first category, where the section varies continuously, we sometimes find a variation where Eq. (8.1-5) can be integrated directly, with the moment of inertia treated as a variable. This has been accomplished in Ref. 20 for tapered beams of circular section, but using the expressions presented, one must carry more than the usual number of digits to get accurate results. In most instances, however, this is not easy, if possible, and a more productive approach is to integrate Eq. (8.1-6) numerically using small incremental lengths Δx. This has been done for a limited number of cases, and the results are tabulated in Tables 8.11(a)–(d).

These tables give coefficients by which the stated reaction forces or moments or the stated deformations for uniform beams, as given in Table 8.1, must be multiplied to obtain the comparable reactions or deformations for the tapered beams. The coefficients are dependent upon the ratio of the moment of inertia at the right end of the beam I_B to the moment of inertia at the left end I_A, assuming that the uniform beam has a moment of inertia I_A. The coefficients are also dependent upon the manner of variation between the two end values. This variation is of the form $I_x = I_A(1+Kx/l)^n$, where x is measured from the left end and $K=(I_B/I_A)^{1/n}-1$. Thus if the beam is uniform, $n = 0$; if the width of a rectangular cross section varies linearly, $n = 1$; if the width of a rectangular cross section varies parabolically, $n = 2$; if the depth of a rectangular cross section varies linearly, $n = 3$; and if the lateral dimensions of any cross section vary linearly and proportionately, $n = 4$. Beams having similar variations in cross section can be analyzed approximately by comparing the given variations to those found in Table 8.11.

Coefficients are given for only a few values of a/l, so it is not desirable to interpolate to determine coefficients for other values of a/l. Instead it is advisable to determine the corrected deformations or reactions with the loads at the tabulated values of a/l and then interpolate. This allows the use of additional known values as shown in the second example below. For beams with symmetric end conditions, such as both ends fixed or both ends simply supported, the data given for any value of $a/l < 0.5$ can be used twice by reversing the beam end for end.

EXAMPLE 1 A tapered beam 30 in long with a depth varying linearly from 2 in at the left end to 4 in at the right end and with a constant width of 1.5 in is fixed on the right end and simply supported on the left end. A concentrated clockwise couple of 5000 lb-in is applied at midlength and it is desired to know the maximum bending stress in the beam.

SOLUTION First determine the left-end reaction force for a uniform cross section. From Table 8.1, case 3c, the left reaction is

$$R_A = \frac{-3M_o(l^2-a^2)}{2l^3} = \frac{-3(5000)(30^2-15^2)}{2(30^3)} = -187.5 \text{ lb}$$

For the tapered beam

$$I_A = \frac{1.5(2^3)}{12} = 1 \text{ in}^4, \quad I_B = \frac{1.5(4^3)}{12} = 8 \text{ in}^4$$

In Table 8.11(c) for $n = 3$, $I_B/I_A = 8$; and for case 3c with the loading at $l/2$, the coefficient is listed as 0.906. Therefore, the left-end reaction is $-187.5(0.906) = -170$ lb.

The maximum negative moment will occur just left of midlength and will equal $-170(15) = -2550$ lb-in. The maximum positive moment will occur just right of midlength and will equal $-2550 + 5000 = 2450$ lb-in. At midlength the moment of inertia $I = 1.5(3^3)/12 = 3.37$ in^4, and so the maximum stress is given by $\sigma = Mc/I = 2550(1.5)/3.37 = 1135$ lb/in^2 just left of midlength.

EXAMPLE 2 A machine part is an 800-mm-long straight beam with a variable wide-flange cross section. The single central web has a constant thickness of 1.5 mm but a linearly varying depth from 6 mm at the left end A to 10 mm at the right end B. The web and flanges are welded together continuously over the entire length and are also welded to supporting structures at each end to provide fixed ends. A concentrated lateral load of 100 N acts normal to the central axis of the beam parallel to the web at a distance of 300 mm from the left end. The maximum bending stress and the deflection under the load are desired. The modulus of elasticity is 70GPa, or 70,000 N/mm^2.

SOLUTION First determine the left-end reaction force and moment for a beam of constant cross section. From Table 8.1, case 1d,

$$R_A = \frac{W}{l^3}(l-a)^2(l+2a) = \frac{100}{800^3}(800-300)^2(800+600) = 68.36 \text{ N}$$

$$M_A = \frac{-Wa}{l^2}(l-a)^2 = \frac{-100(300)}{800^2}(800-300)^2 = -11{,}720 \text{ N-mm}$$

For the tapered beam,

$$I_A = \frac{4(10^3)-2.5(6^3)}{12} = 288.3 \text{ mm}^4$$

$$I_B = \frac{8(14^3)-6.5(10^3)}{12} = 1287.7 \text{ mm}^4$$

and at midlength where $x = l/2$, the moment of inertia is given by

$$I_{l/2} = \frac{6(12^3)-4.5(8^3)}{12} = 672.0 \text{ mm}^4$$

Using the formula for the variation of I with x and these three values for the moment of inertia, approximate values for K and n can be found.

$$I_B = I_A(1+K)^n, \quad I_{l/2} = I_A\left(1+\frac{K}{2}\right)^n$$

$$\frac{1287.7}{288.3} = 4.466 = (1+K)^n, \quad \frac{672}{288.3} = 2.331 = \left(1+\frac{K}{2}\right)^n$$

$$4.466^{1/n} - 2(2.331)^{1/n} + 1 = 0$$

Solving this last expression gives $1/n = 0.35$, $n = 2.86$, and $K = 0.689$.

An examination of Tables 8.11(a–d) shows that for a fixed-ended beam with a concentrated load, which is case 1d in Table 8.1, values of coefficients are given only for $a/l = 0.25$ and 0.50. For this problem $a/l = 0.375$. Simple linear interpolation is not sufficiently accurate. However, if one imagines the load at $a/l = 0.25$ the values for R_A and M_A can be found. This procedure can be repeated for $a/l = 0.50$. Two other sets of data are also available. If the load were placed at the left end, $a/l = 0$, $M_A = 0$, $R_A =$ 100 N, and $dM_A/da = -100$ N. If the load were placed at the right end, $a/l = 1$, $R_A = 0$, $M_A = 0$, and $dM_A/da = 0$. The variations of the tabulated coefficients with I_B/I_A and with n do not pose a comparable problem since many data are available. Plotting curves for the variation with I_B/I_A and interpolating linearly between $n = 2$ and $n = 3$ for $n = 2.86$ give the coefficients used below to find the values for R_A and M_A at $a/l = 0.25$ and 0.50:

	Untapered beam		Tapered beam where $n = 2.86$ and $I_B/I_A = 4.466$	
a/l	0.25	0.50	0.25	0.50
R_A (N)	84.38	50.00	84.38(0.922) = 77.80	50(0.0805) = 40.25
M_A (N-mm)	−11,250	−10,000	−11,250(0.788) = −8865	−10,000(0.648) = −6480

Plotting these values of R_A and M_A versus a/l for the four positions allows one to pick from the graphs at $a/l = 0.375$, $R_A = 60$ N, and $M_A = -8800$ N-mm. The use of static equilibrium now gives $M_B = -10{,}800$ N-mm and the moment at the load of 9200 N-mm. The bending stress at the left end is found to be the largest.

$$\sigma_A = \frac{M_A c_A}{I_A} = \frac{8800(5)}{288.3} = 152.6 \text{ MPa}$$

No correction coefficients for deflections are included for case ld in Table 8.11. The deflection at the load position can be found from the information in this table, however, if either end of the beam is isolated and treated as a cantilever with an end load and an end moment. The moment of inertia at the load position C is given by

$$I_C = \frac{5.5(11.5^3) - 4(7.5^3)}{12} = 556.4 \text{ mm}^4$$

Treat the left portion of the beam as a 300-mm-long cantilever, using case 1a with an end load of 60 N and case 3a with an end moment of 9200 N-mm. Determine the correction coefficients for $a/l = 0$, $n = 3$, and the moment of inertia ratio of $288.33/556.44 = 0.518$. Interpolation between data for $n = 2$ and $n = 3$ is not justified when considering the approximations already made from plotted curves. Noting that all correction coefficients in Table 8.11 are unity for $I_B/I_A = 1$ and using data points for $I_B/I_A = 0.25$ and 0.50, the correction coefficients used below were found.

$$y_C = -\frac{60(300^3)(1.645)}{3(70{,}000)(556.4)} + \frac{9200(300^2)(1.560)}{2(70{,}000)(556.4)} = -22.8 + 16.6 = -6.2 \text{ mm}$$

This deflection at the load can be checked by repeating the above procedure by using the right-hand portion of the beam. The slope of the beam at the load can also be used as a check.

Alternative Solution

The solution just presented was intended to illustrate appropriate methods of interpolation with the limited load positions shown in the tables. There is also an alternative solution involving superposition of cases. Remove the fixity at end A and treat the 500-mm-long right portion as a cantilever with an end load of 100 N. Use $n = 3$ as being close enough to $n = 2.86$ and $I_B/I_C = 1287.7/556.4 = 2.314$. Interpolate between $I_B/I_C = 2$ and 4 to obtain from case 1a in Table 8.11(c) the correction coefficients used below to calculate the slope and deflection at the load.

$$y_C = -\frac{100(500^3)(0.555)}{3(70{,}000)(556.4)} = -59.37 \text{ mm} \quad \theta_C = \frac{100(500^2)(0.593)}{2(70{,}000)(556.4)} = 0.1903 \text{ rad}$$

Since the left portion is unloaded and remains straight, the end deflection and slope are $y_A = -59.37 - 300(0.1903) = -116.5$ mm, and $\theta_A = 0.1903$ rad. Next treat the complete beam as a cantilever under an end load R_A and an end moment M_A. Let $I_A = 228.3$ mm^4, $I_B/I_A = 4.466$, and again let $n = 3$. From cases 1a and 3a in Table 8.11(c),

$$y_A = \frac{R_A(800^3)(0.332)}{3(70{,}000)(288.3)} + \frac{M_A(800^2)(0.380)}{2(70{,}000)(288.3)} = 2.808R_A + 0.00602M_A$$

$$\theta_A = \frac{R_A(800^2)(0.380)}{2(70{,}000)(288.3)} - \frac{M_A(800)(0.497)}{(70{,}000)(288.3)} = -0.006024R_A - 19.7(10^{-6})M_A$$

Adding the slopes and deflections from the load of 100 N to those above and equating each to zero to represent the fixed end gives $R_A = 60.3$ N and $M_A = -8790$ N-mm. This is a reasonable check on those values from the first solution.

The second category of determining deflections, where the cross section varies in steps from one uniform section to another, can be solved in several ways. Equation (8.1-5) can be integrated, matching slopes and deflections at the transition sections, or Eq. (8.1-6) can be integrated over the separate portions and summed to obtain the desired deflections. A third method utilizes the advantages of the step function and its application to beam deflections as given in Table 8.1. In a given portion of the span where the cross section is uniform it is apparent that the shape of the elastic curve will remain the same if the internal bending moments and the moments of inertia are increased or decreased in proportion. By this means, a modified moment diagram can be constructed which could be applied to a beam with a single constant cross section and thereby produce an elastic curve identical to the one produced by the actual moments and the several moments of inertia present in the actual span. It is also apparent that this modified moment diagram could be produced by adding appropriate loads to the beam. (See Refs. 29 and 65.) In summary, then, a new loading is constructed which will produce the required elastic curve, and the solution for this loading is carried out by using the formulas in Table 8.1. This procedure will be illustrated by the following example:

EXAMPLE The beam shown in Fig. 8.13 has a constant depth of 4 in and a step increase in the width from 2 to 5 in at a point 5 ft from the left end. The left end is simply supported, and the right end is fixed; the loading is a uniform 200 lb/ft from $x = 3$ ft to the right end. Find the value of the reaction at the left end and the maximum stress.

SOLUTION For the left 5 ft, $I_1 = 2(4^3)/12 = 10.67\ \text{in}^4$. For the right 5 ft, $I_2 = 5(4^3)/12 = 26.67\ \text{in}^4$, or 2.5 times I_1.

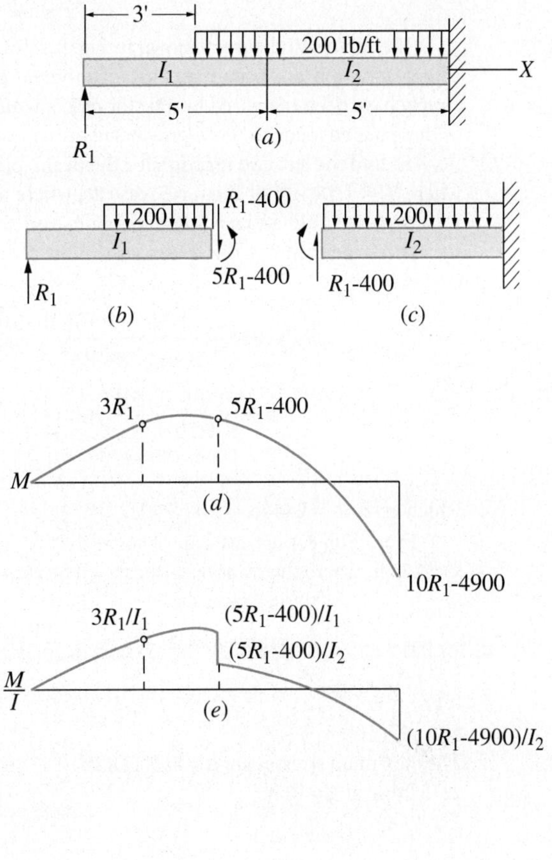

Figure 8.13

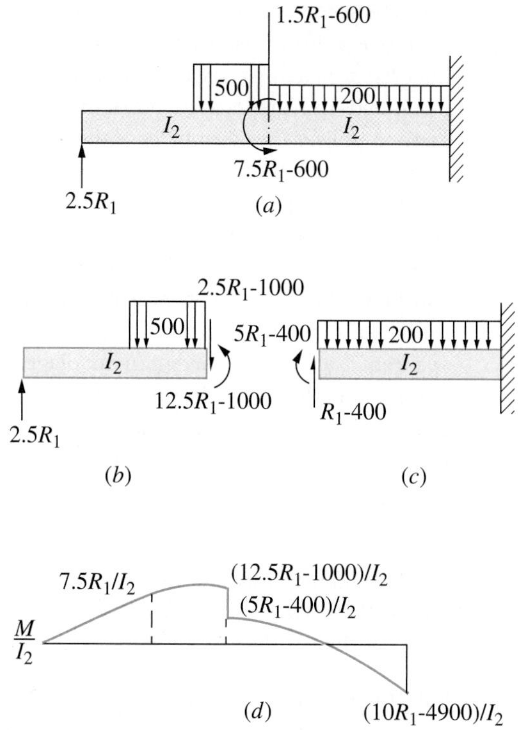

Figure 8.14

The same M/I diagram shown in Fig. 8.13(e) can be produced by the loading shown in Fig. 8.14 acting upon a beam having a constant moment of inertia I_2. Note that all loads on the left portion simply have been increased by a factor of 2.5, while added loads at the 5-ft position reduce the effects of these added loads to those previously present on the right portion.

To find the left-end reaction for the beam loaded as shown in Fig. 7.14(a), use Table 8.1, case 1c, where $W = 1.5R_1 - 600$ and $a = 5$; case 2c, where $w_a = w_l = 500$ lb/ft and $a = 3$ ft; case 2c, again, where $w_a = w_l = -300$ lb/ft and $a = 5$ ft; and finally case 3c, where $M_o = -(7.5R_1 - 600)$ and $a = 5$ ft. Summing the expressions for R_A from these cases in the order above, we obtain

$$R_A = 2.5R_1 = \frac{(1.5R_1 - 600)(10-5)^2}{2(10^3)}[2(10)+5] + \frac{500(10-3)^3}{8(10^3)}[3(10)+3]$$
$$+\frac{(-300)(10-5)^3}{8(10^3)}[3(10)+5] - \frac{3[-(7.5R_1 - 600)]}{2(10^3)}(10^2 - 5^2)$$

which gives $R_1 = 244$ lb.

From Fig. 8.13(a) we can observe that the maximum positive bending moment will occur at $x = 4.22$ ft, where the transverse shear will be zero. The maximum moments are therefore

$$\text{Max} + M = 244(4.22) - \frac{200}{2}(1.22^2) = 881 \text{ lb-ft}$$

$$\text{Max} - M = 244(10) - 4900 = -2460 \text{ lb-ft} \qquad \text{at the right end}$$

The maximum stresses are $\sigma = 881(12)(2)/10.67 = 1982$ lb/in^2 at $x = 4.22$ ft and $\sigma = 2460(12)(2)/26.67 = 2215$ lb/in^2 at $x = 10$ ft.

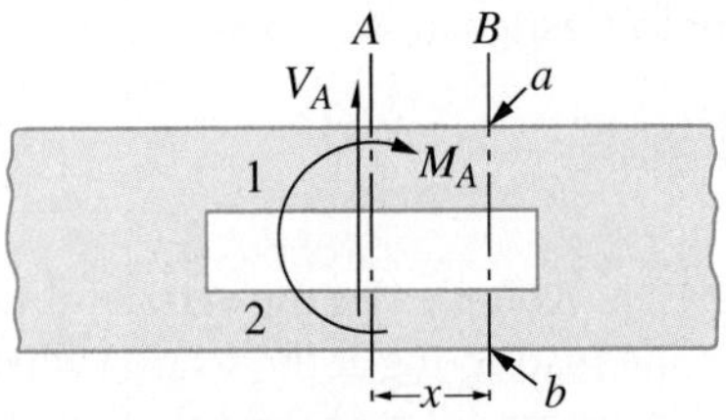

Figure 8.15

8.9 SLOTTED BEAMS

If the web of a beam is pierced by a hole or slot (Fig. 8.15), the stresses in the extreme fibers a and b at any section B are given by

$$\sigma_a = -\frac{M_A}{I/c} - \frac{V_A x I_1/(I_1+I_2)}{(I/c)_1} \quad \text{(compression)}$$

$$\sigma_b = \frac{M_A}{I/c} + \frac{V_A x I_2/(I_1+I_2)}{(I/c)_2} \quad \text{(tension)}$$

Here M_A is the bending moment at A (midlength of the slot), V_A is the vertical shear at A, I/c is the section modulus of the net beam section at B, I_1 and I_2 are the moments of inertia, and $(I/c)_1$ and $(I/c)_2$ are the section moduli of the cross sections of parts 1 and 2 about their own central axes. M and V are positive or negative according to the usual convention, and x is positive when measured to the right.

The preceding formulas are derived by replacing all forces acting on the beam to the left of A by an equivalent couple M_A and shear V_A at A. The couple produces a bending stress given by the first term of the formula. The shear divides between parts 1 and 2 in proportion to their respective I's and produces in each part an additional bending stress given by the second term of the formula. The stress at any other point in the cross section can be found similarly by adding the stresses due to M_A and those due to this secondary bending caused by the shear. (At the ends of the slot there is a stress concentration at the corners which is not taken into account here.)

The above analysis applies also to a beam with multiple slots of equal length; all that is necessary is to modify the term $(I_1 + I_2)$. The numerator is still the I of the part in question and the denominator is the sum of the I's of all the parts 1, 2, 3, etc. The formulas can also be used for a rigid frame consisting of beams of equal length joined at their ends by rigid members; thus in Fig. 8.15 parts 1 and 2 might equally well be two separate beams joined at their ends by rigid crosspieces.

8.10 BEAMS OF RELATIVELY GREAT DEPTH

In beams of small span/depth ratio, the transverse shear stresses are likely to be high and the resulting deflection due to shear may not be negligible. For span/depth ratios of 3 or more, the deflection y_s due to shear is found by the method of unit loads to be

$$y_s = F\int \frac{Vv}{AG}dx \qquad (8.10\text{-}1)$$

or by Castigliano's theorem to be

$$y_s = \frac{\partial U_s}{\partial P} \tag{8.10-2}$$

In Eq. (8.10-1), V is the vertical shear due to the actual loads, v is the vertical shear due to a unit load acting at the section where the deflection is desired, A is the area of the section, G is the modulus of rigidity, F is a factor depending on the form of the cross section, and the integration extends over the entire length of the beam, with due regard to the signs of V and v. For three solid sections, a rectangle, a triangle with base either up or down, and a trapezoid with parallel sides top and bottom, $F = 6/5$; for a diamond-shaped section, $F = 31/30$; for a solid circular section, $F = 10/9$; for a thin-walled hollow circular section, $F = 2$; for an I- or box section with flanges and webs of uniform thickness,

$$F = \left[1 + \frac{3(D_2^2 - D_1^2)D_1}{2D_2^3}\left(\frac{t_2}{t_1} - 1\right)\right]\frac{4D_2^2}{10r^2}$$

where

D_1 = distance from neutral axis to the nearest surface of the flange
D_2 = distance from neutral axis to extreme fiber
t_1 = thickness of web (or webs in box beams)
t_2 = width of flange
r = radius of gyration of section with respect to the neutral axis

If the I- or box beam has flanges of nonuniform thickness, it may be replaced by an "equivalent" section whose flanges, of uniform thickness, have the same width and areas as those of the actual section (Ref. 19). Approximate results may be obtained for I-beams using $F = 1$ and taking for A the area of the web.

Application of Eq. (8.10-1) to several common cases of loading yields the following results:

End support, center load P $\quad y_s = \frac{1}{4}F\frac{Pl}{AG}$

End support, uniform load w $\quad y_s = \frac{1}{8}F\frac{wl^2}{AG}$

Cantilever, end load P $\quad y_s = F\frac{Pl}{AG}$

Cantilever, uniform load w $\quad y_s = \frac{1}{2}F\frac{wl^2}{AG}$

In Eq. (8.10-2), $U_s = F\int(V^2/2AG)dx$, P is a vertical load, real or imaginary, applied at the section where y_s is to be found, and the other terms have the same meaning as in Eq. (8.10-1).

The deflection due to shear will usually be negligible in metal beams unless the span/depth ratio is extremely small; in wood beams, because of the small value of G compared with E, deflection due to shear is much more important. In computing deflections it may be allowed for by using for E a value obtained from bending tests (shear deflection ignored) on beams of similar proportions or a value about 10% less than that found by testing in direct compression if the span/depth ratio is between 12 and 24. For larger ratios the effect of shear is negligible, and for lower ratios it should be calculated by the preceding method.

For extremely short deep beams, the assumption of linear stress distribution, on which the simple theory of flexure is based, is no longer valid. Equation (8.1-1) gives sufficiently accurate results for span/depth ratios down to about 3; for still smaller ratios it was believed formerly that the actual stresses were smaller than the formula indicates (Refs. 1 and 2), but more recent analyses by numerical methods (Refs. 43 and 44) indicate that the contrary is true. These analyses show that at s/d between 1.5 and 1, depending on the manner of loading and support, the stress distribution changes radically and the ratio of maximum stress to Mc/I becomes greater than 1 and increases rapidly as s/d becomes still smaller. In the following table, the influence of s/d on both maximum fiber stress and maximum horizontal shear stress is shown in accordance with the solution given in Ref. 43. Reference 44 gives comparable results, and both strain-gage measurements (Ref. 45) and photoelastic studies (Ref. 46) support the conclusions reached in these analyses.

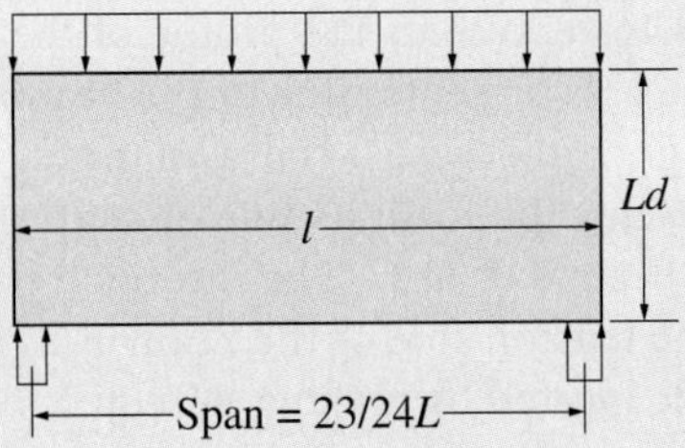

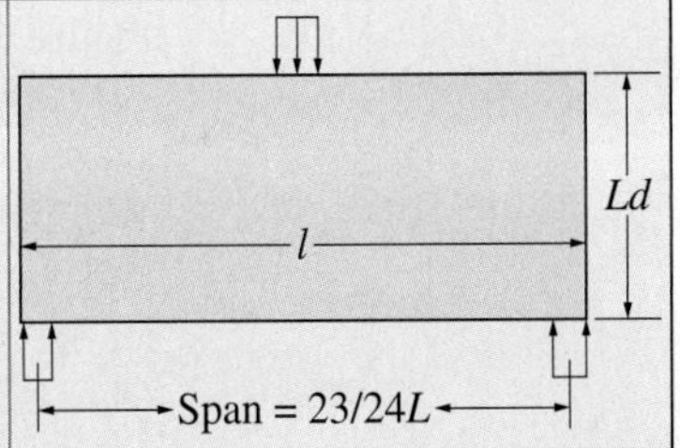

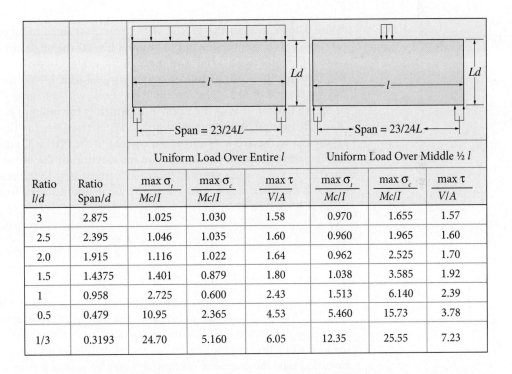

		Uniform Load Over Entire l			Uniform Load Over Middle ½ l		
Ratio l/d	Ratio Span/d	$\frac{\max \sigma_t}{Mc/I}$	$\frac{\max \sigma_c}{Mc/I}$	$\frac{\max \tau}{V/A}$	$\frac{\max \sigma_t}{Mc/I}$	$\frac{\max \sigma_c}{Mc/I}$	$\frac{\max \tau}{V/A}$
3	2.875	1.025	1.030	1.58	0.970	1.655	1.57
2.5	2.395	1.046	1.035	1.60	0.960	1.965	1.60
2.0	1.915	1.116	1.022	1.64	0.962	2.525	1.70
1.5	1.4375	1.401	0.879	1.80	1.038	3.585	1.92
1	0.958	2.725	0.600	2.43	1.513	6.140	2.39
0.5	0.479	10.95	2.365	4.53	5.460	15.73	3.78
1/3	0.3193	24.70	5.160	6.05	12.35	25.55	7.23

These established facts concerning elastic stresses in short beams seem incompatible with the contrary influence of s/d on modulus of rupture, discussed in Sec. 8.15, unless it is assumed that there is a very radical redistribution of stress as soon as plastic action sets in.

The stress produced by a concentrated load acting on a very short *cantilever* beam or projection (gear tooth, sawtooth, screw thread) can be found by the following formula, due to Heywood (Chap. 2, Ref. 28) and modified by Kelley and Pedersen (Ref. 59). As given here, the formula follows this modification, with some changes in notation.

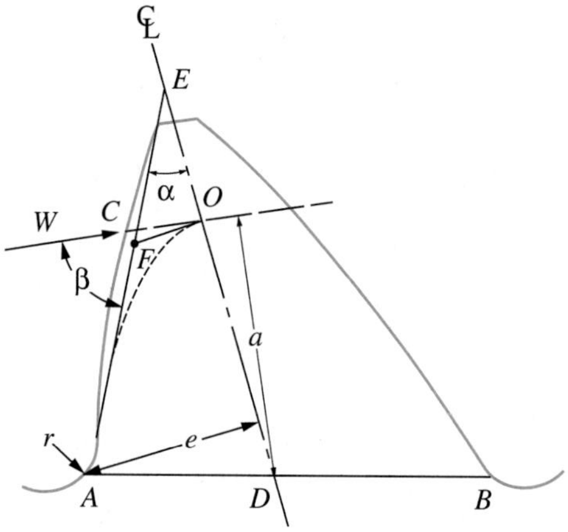

Figure 8.16

Figure 8.16 represents the profile of the beam, assumed to be of uniform thickness t. ED is the axis or center line of the beam; it bisects the angle between the sides if these are straight; otherwise it is drawn through the centers of two unequal inscribed circles. W represents the load; its line of action, or load line, intersects the beam profile at C and the beam axis at O. The inscribed parabola, with vertex at O, is tangent to the fillet on the tension side of the beam at A, which is the point of maximum tensile stress. (A can be located by making AF equal to FE by trial, F being the intersection of a perpendicular to the axis at O and a trial tangent to the fillet.) B is the corresponding point on the compression side, and D is the intersection of the beam axis with section AB. The dimensions a and e are perpendicular, respectively, to the load line and to the beam axis, r is the fillet radius, and b is the straight-line distance from A to C. The tensile stress at A is given by

$$\sigma = \frac{W}{t}\left[1 + 0.26\left(\frac{e}{r}\right)^{0.7}\right]\left[\frac{1.5a}{e^2} + \frac{\cos\beta}{2e} + \frac{0.45}{(be)^{1/2}}\right]$$

Here the quantity in the first pair of brackets is the factor of stress concentration for the fillet. In the second pair of brackets, the first term represents the bending moment divided by the section modulus, the second term represents the effect of the component of the load along the tangent line, positive when tensile, and the third term represents what Heywood calls the *proximity effect*, which may be regarded as an adjustment for the very small span/depth ratio.

Kelley and Pedersen have suggested a further refinement in locating the point of maximum stress, putting it at an angular distance equal to $25° - ½\alpha$, positive toward the root of the fillet. Heywood suggests locating this point at 30° from the outer end of the fillet, reducing this to 12° as the ratio of b to e increases; also, Heywood locates the moment of W about a point halfway between A and B instead of about D. For most cases the slightly different procedures seem to give comparable results and agree well with photoelastic analysis. However, more recent experimental studies (1963), including fatigue tests, indicate that actual stresses may considerably exceed those computed by the formula (Ref. 63).

8.11 BEAMS OF RELATIVELY GREAT WIDTH

Because of prevention of the lateral deformation that would normally accompany the fiber stresses, wide beams, such as thin metallic strips, are more rigid than the formulas of Sec. 8.1 indicate. This stiffening effect is taken into account by using $E/(1 - \nu^2)$ instead of E in the formulas for deflection and curvature if the beams are very wide (Ref. 21). The anticlastic curvature that exists on narrow rectangular beams is still present at the extreme edges of very wide beams, but the central region remains flat in a transverse direction and transverse bending stresses equal to Poisson's ratio times the longitudinal bending stresses are present. For rectangular beams of moderate width, Ashwell (Ref. 10) shows that the stiffness depends not only upon the ratio of depth to width of the beam but also upon the radius of curvature to which the beam is bent. For a rectangular beam of width b and depth h bent to a radius of curvature ρ by a bending moment M, these variables are related by the expression $1/\rho = M/KEI$, where $I = bh^3/12$, and the following table of values for K is given for several values of Poisson's ratio and for the quantity $b^2/\rho h$.

	$b^2/\rho h$						
Value of ν	0.25	1.00	4.00	16.0	50.0	200	800
0.1000	1.0000	1.0003	1.0033	1.0073	1.0085	1.0093	1.0097
0.2000	1.0001	1.0013	1.0135	1.0300	1.0349	1.0383	1.0400
0.3000	1.0002	1.0029	1.0311	1.0710	1.0826	1.0907	1.0948
0.3333	1.0002	1.0036	1.0387	1.0895	1.1042	1.1146	1.1198
0.4000	1.0003	1.0052	1.0569	1.1357	1.1584	1.1744	1.1825
0.5000	1.0005	1.0081	1.0923	1.2351	1.2755	1.3045	1.3189

In very short wide beams, such as the concrete slabs used as highway-bridge flooring, the deflection and fiber-stress distribution cannot be regarded as uniform across the width. In calculating the strength of such a slab, it is convenient to make use of the concept of *effective width,* that is, the width of a spanwise strip which, acting as a beam with uniform extreme fiber stress equal to the maximum stress in the slab, develops the same resisting moment as does the slab. The effective width depends on the manner of support, manner of loading, and ratio of breadth to span b/a. It has been determined by Holl (Ref. 22) for a number of assumed conditions, and the results are given in the following table for a slab that is freely supported at each of two opposite edges (Fig. 8.17). Two kinds of loading are considered, viz. uniform load over the entire slab and load uniformly distributed over a central circular area of radius c. The ratio of the effective width e to the span a is given for each of a number of ratios of c to slab thickness h and each of a number of b/a values.

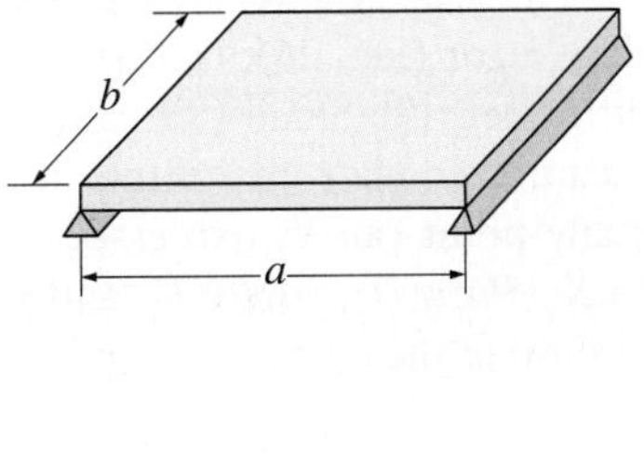

Figure 8.17

Loading	Values of e/a for				
	$b/a=1$	$b/a=1.2$	$b/a=1.6$	$b/a=2$	$b/a=\infty$
Uniform	0.960	1.145	1.519	1.900	
Central, $c=0$	0.568	0.599	0.633	0.648	0.656
Central, $c=0.125h$	0.581	0.614	0.649	0.665	0.673
Central, $c=0.250h$	0.599	0.634	0.672	0.689	0.697
Central, $c=0.500h$	0.652	0.694	0.740	0.761	0.770

For the same case (a slab that is supported at opposite edges and loaded on a central circular area) Westergaard (Ref. 23) gives $e=0.58a+4c$ as an approximate expression for effective width. Morris (Ref. 24) gives $e=\frac{1}{2}e_c+d$ as an approximate expression for the effective width for midspan *off-center* loading, where e_c is the effective width for central loading and d is the distance from the load to the nearer unsupported edge.

For a slab that is *fixed* at two opposite edges and uniformly loaded, the stresses and deflections may be calculated with sufficient accuracy by the ordinary beam formulas, replacing E by $E/(1-\nu^2)$. For a slab thus supported and loaded at the center, the maximum stresses occur under the load, except for relatively large values of c, where they occur at the midpoints of the fixed edges. The effective widths are approximately as given in the following table (values from the curves of Ref. 22). Here b/a and c have the same meaning as in the preceding table, but it should be noted that values of e/b are given instead of e/a.

Values of c	Values of e/b for				Max Stress at
	$b/a=1$	$b/a=1.2$	$b/a=1.6$	$b/a=2.0$	
0	0.51	0.52	0.53	0.53	Load
$0.01a$	0.52	0.54	0.55	0.55	Load
$0.03a$	0.58	0.59	0.60	0.60	Load
$0.10a$	0.69	0.73	0.81	0.86	Fixed edges

Holl (Ref. 22) discusses the deflections of a wide beam with two edges supported and the distribution of pressure under the supported edges. The problem of determining the effective width in concrete slabs and tests made for that purpose are discussed by Kelley (Ref. 25), who also gives a brief bibliography on the subject.

The case of a very wide *cantilever* slab under a concentrated load is discussed by MacGregor (Ref. 26), Holl (Ref. 27), Jaramillo (Ref. 47), Wellauer and Seireg (Ref. 48), Little (Ref. 49), Small (Ref. 50), and others. For the conditions represented in Fig. 8.18, a cantilever plate of infinite length with a concentrated load, the bending stress σ at any point can be expressed by $\sigma=K_m(6P/t^2)$, and the deflection y at any point by $y=K_y(Pa^2/\pi D)$, where K_m and K_y are dimensionless coefficients that depend upon the location of the load and the point, and D is as defined in Table 11.2. For the load at $x=c$,

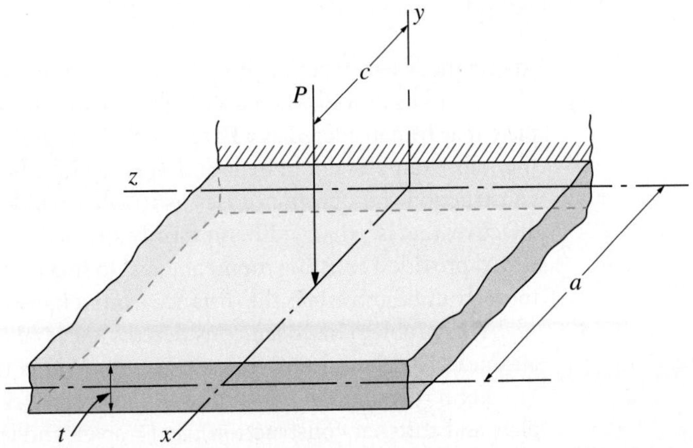

Figure 8.18

$z = 0$, the stress at any point on the fixed edge $x = 0$, $z = z$, and the deflection at any point on the free edge $x = a$, $z = z$, can be found by using the following values of K_m and K_y:

c/a \ z/a		0	0.25	0.50	1.0	1.5	2	∞
1.0	K_m	0.509	0.474	0.390	0.205	0.091	0.037	0
	K_y	0.524	0.470	0.380	0.215	0.108	0.049	0
0.75	K_m	0.428	0.387	0.284	0.140	0.059	0.023	0
	K_y	0.318	0.294	0.243	0.138	0.069	0.031	0
0.50	K_m	0.370	0.302	0.196	0.076	0.029	0.011	0
0.25	K_m	0.332	0.172	0.073	0.022	0.007	0.003	0

These values are based on the analysis of Jaramillo (Ref. 47), who assumes an infinite length for the plate, and are in good agreement, so far as comparable, with coefficients given by MacGregor (Ref. 26). They differ only slightly from results obtained by Holl (Ref. 27) for a length/span ratio of 4 and by Little (Ref. 49) for a length/span ratio of 5 and are in good agreement with available test data.

Wellauer and Seireg (Ref. 48) discuss the results of tests on beams of various proportions and explain and illustrate an empirical method by which the K_m values obtained by Jaramillo (Ref. 47) for the infinite plate under concentrated loading can be used to determine approximately the stress in a finite plate under an arbitrary transverse loading.

The stresses corresponding to the tabulated values of K_m are *span-wise* (x direction) stresses; the maximum *crosswise* (z direction) stress occurs under the load when the load is applied at the midpoint of the free edge and is approximately equal to the maximum span-wise stress for that loading.

Although the previous formulas are based on the assumption of infinite width of a slab, tests (Ref. 26) on a plate with a width of 8½ in and span a of 1¼ in showed close agreement between calculated and measured deflections, and Holl's analysis (Ref. 27), based on the assumption of a plate width four times the span, gives results that differ only slightly from MacGregor's (Ref. 26). The formulas given should therefore be applicable to slabs of breadth as small as four times the span.

8.12 BEAMS WITH WIDE FLANGES; SHEAR LAG

In thin metal construction, box, T-, or I-beams with very wide thin cover plates or flanges are sometimes used, and when a thin plate is stiffened by an attached member, a portion of the plate may be considered as a flange, acting integrally with the attached member which forms the web; examples are to be found in ship hulls, floors, tanks, and aircraft. In either type of construction the question arises as to what width of flange or plate would be considered effective; that is, what width, uniformly stressed to the maximum stress that actually occurs, would provide a resisting moment equal to that of the actual stresses, which are greatest near the web and diminish as the distance from it increases.

This problem has been considered by several investigators; the results tabulated below are due to Hildebrand and Reissner (Ref. 38), Winter (Ref. 39), and Miller (Ref. 28).

Let b = actual net width of the flange (or clear distance between webs in continuous plate and stiffener construction), let l = span, and let b' = effective width of the flange at the section of maximum bending moment. Then the approximate value of b'/b, which varies with the loading and with the ratio l/b, can be found for beams of uniform section in the table below. (In this table the case numbers refer to the manner of loading and support represented in Table 8.1.) See also Ref. 37.

Case No. and Load Positions (from Table 8.1)	Reference No.	l/b													
		1	1.25	1.50	1.75	2	2.5	3	4	5	6	8	10	15	20
1a. $a = 0$	38	0.571	0.638	0.690	0.730	0.757	0.801	0.830	0.870	0.895	0.913	0.934	0.946		
2a. $w_a = w_l$, $a = 0$	38			0.550	0.600	0.632	0.685	0.724	0.780	0.815	0.842	0.876	0.899		
2a. $w_a = 0$, $a = 0$	38						0.609	0.650	0.710	0.751	0.784	0.826	0.858		
1e. $a = l/2$	38				0.530	0.571	0.638	0.686	0.757	0.801	0.830	0.870	0.895	0.936	0.946
1e. $a = l/2$	39							0.550	0.670	0.732	0.779	0.850	0.894	0.945	
1e. $a = l/2$	28						0.525			0.750					
1e. $a = l/4$	38				0.455	0.495	0.560	0.610	0.686	0.740	0.788	0.826	0.855	0.910	0.930
2e. $w_a = w_l$, $a = 0$	38				0.640	0.690	0.772	0.830	0.897	0.936	0.957	0.977	0.985	0.991	0.995

Ratio of effective width to total width b′/b for wide flanges

Some of the more important conclusions stated in Ref. 38 can be summarized as follows:

The amount of shear lag depends not only on the method of loading and support and the ratio of span to flange width but also on the ratio of G to E and on the ratio $m = (3I_w + I_s)/(I_w + I_s)$, where I_w and I_s are the moments of inertia about the neutral axis of the beam of the side plates and cover plates, respectively. (The values tabulated from Ref. 38 are for $G/E = 0.375$ and $m = 2$.) The value of b'/b increases with increasing m, but for values of m between 1.5 and 2.5 the variation is small enough to be disregarded. Shear lag at the critical section does not seem to be affected appreciably by the taper of the beam in width, but the taper in cover-plate thickness may have an important effect. In beams with fixed ends the effect of shear lag at the end sections is the same as for a cantilever of span equal to the distance from the point of inflection to the adjacent end.

In Ref. 39 it is stated that for a given l/b ratio the effect of shear lag is practically the same for box, I-, T-, and U-beams.

Flange in Compression

The preceding discussion and tabulated factors apply to any case in which the flange is subjected to tension or to compression less than that required to produce elastic instability (see Chap. 15). When a thin flange or sheet is on the compression side, however, it may be stressed beyond the stability limit. For this condition, the effective width decreases with the actual stress. A formula for effective width used in aircraft design is

$$b' = Kt\sqrt{\frac{E}{s}}$$

where s is the maximum compressive stress (adjacent to the supporting web or webs) and K is a coefficient which may be conservatively taken as 0.85 for box beams and 0.60 for a T- or an I-beam having flanges with unsupported outer edges.

A theoretical analysis that takes into account both compressive buckling and shear lag is described in Ref. 40. Problems involving shear lag and buckling are most frequently encountered in design with thin-gage metal; good guides to such design are the books "Cold-Formed Steel Design Manual" in 5 parts including commentary, published in 1982 by the American Iron and Steel Institute, and "Aluminum Construction Manual," 4th ed., published in 1981 by the Aluminum Association. See also Ref. 68.

8.13 BEAMS WITH VERY THIN WEBS

In beams with extremely thin webs, such as are used in airplane construction, buckling due to shear will occur at stresses well below the elastic limit. This can be prevented if the web is made *shear-resistant* by the addition of stiffeners such as those used in plate girders, but the number of these required may be excessive. Instead of making the web shear-resistant, it may be permitted to buckle elastically without damage, the shear being carried wholly in *diagonal tension.* This tension tends to pull the upper and lower flanges together, and to prevent this, vertical struts are provided which carry the vertical component of the diagonal web tension. A girder so designed is, in effect, a Pratt truss, the web replacing the diagonal-tension members and the vertical struts constituting the compression members. In appearance, these struts resemble the stiffeners of an ordinary plate girder, but their function is obviously quite different.

A beam of this kind is called a *diagonal-tension field beam,* or *Wagner beam,* after Professor Herbert Wagner of Danzig, who is largely responsible for developing the theory. Because of its rather limited field of application, only one example of the Wagner beam will be considered here, viz. a cantilever under end load.

Let P = end load, h = depth of the beam, t = thickness of the web, d = spacing of the vertical struts, x = distance from the loaded end to the section in question, H_t and H_c = total stresses in the tension and compression flanges, respectively, at the given section, C = total compression on a vertical strut, and f = unit diagonal tensile stress in the web. Then

$$H_t = \frac{Px}{h} - \frac{1}{2}P, \quad H_c = \frac{Px}{h} + \frac{1}{2}P, \quad C = \frac{Pd}{h}, \quad f = \frac{2P}{ht}$$

The vertical component of the web tension constitutes a beam loading on each individual flange between struts; the maximum value of the resulting bending moment occurs at the struts and is given by $M_f = 1/12\ Pd^2/h$. The flexural stresses due to M_f must be added to the stresses due to H_t or H_c, which may be found simply by dividing H_t or H_c by the area of the corresponding flange.

The horizontal component of the web tension causes a bending moment $M = 1/8\ Ph$ in the vertical strut at the end of the beam unless bending there is prevented by some system of bracing. This end strut must also distribute the load to the web, and should be designed to carry the load as a pin-ended column of length ½ h as well as to resist the moment imposed by the web tension.

The intermediate struts are designed as pin-ended columns with lengths somewhat less than h. An adjacent portion of the web is included in the area of the column, the width of the strip considered effective being $30t$ in the case of aluminum and $60t$ in the case of steel.

Obviously the preceding formulas will apply also to a beam with end supports and center load if P is replaced by the reaction ½ P. Because of various simplifying assumptions made in the analysis, these formulas are conservative; in particular the formula for stress in the vertical struts or stiffeners gives results much larger than actual stresses that have been discovered experimentally. More accurate analyses, together with experimental data from various sources, will be found in Refs. 30, 34, 35, 62, and 69–71.

8.14 BEAMS NOT LOADED IN PLANE OF SYMMETRY; FLEXURAL CENTER

The formulas for stress and deflection given in Sec. 8.1 are valid if the beam is loaded in a plane of symmetry; they are also valid if the applied loads are parallel to either principal central axis of the beam section, but unless the loads also pass through the *elastic axis*, the beam will be subjected to torsion as well as bending.

For the general case of a beam of any section loaded by a transverse load P in any plane therefore the solution comprises the following steps: (1) The load P is resolved into an equal and parallel force P' passing through the flexural center Q of the section, and a twisting couple T equal to the moment of P about Q; (2) P' is resolved at Q into rectangular components P'_u and P'_v, each parallel to a principal central axis of the section; (3) the flexural stresses and deflections due to P'_u and P'_v are calculated independently by the formulas of Sec. 8.1 and superimposed to find the effect of P'; and (4) the stresses due to T are computed independently and superimposed on the stresses due to P', giving the stresses due to the actual loading. (It is to be noted that T may cause longitudinal fiber stresses as well as shear stresses. See Sec. 10.3 and the example at the end of this section.) If there are several loads, the effect of each is calculated separately and these effects added. For a distributed load the same procedure is followed as for a concentrated load.

The above procedure requires the determination of the position of the *flexural center* Q. For any section having two or more axes of symmetry (rectangle, I-beam, etc.) and for any section having a point of symmetry (equilateral triangle, Z-bar, etc.), Q is at the centroid. For any section having only one axis of symmetry, Q is on that axis but in general not at the centroid. For such sections and for unsymmetrical sections in general, the position of Q must be determined by calculation, direct experiment, or the soap-film method (Sec. 6.4).

Table 8.12 gives the position of the flexural center for each of a number of sections.

Neutral Axis

When a beam is bent by one or more loads that lie in a plane not parallel to either principal central axis of the section, the neutral axis passes through the centroid but is not perpendicular to the plane of the loads. Let axes 1 and 2 be the principal central axes of the section, and let I_1 and I_2 represent the corresponding moments of inertia. Then, if the plane of the loads makes with axis 1 an angle α, the neutral axis makes with axis 2 an angle β such that $\tan\beta = (I_2/I_1)\tan\alpha$. It can be seen from this equation that the neutral axis tends to approach the principal central axis about which the moment of inertia is least.

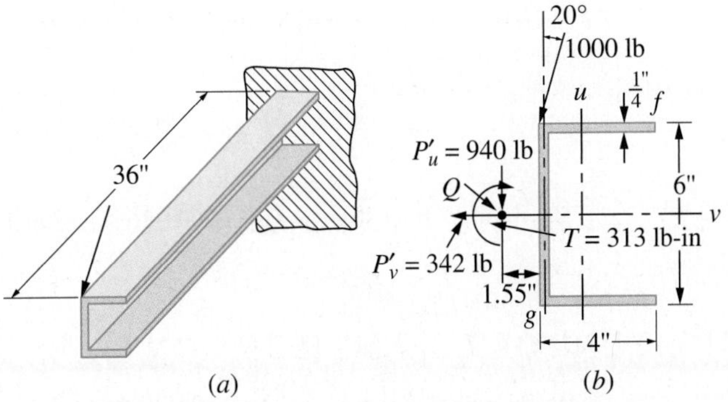

Figure 8.19

EXAMPLE Figure 8.19(*a*) represents a cantilever beam of channel section under a diagonal end load applied at one corner. It is required to determine the maximum resulting fiber stress.

SOLUTION For the section (Fig. 8.19*b*): $I_u = 5.61$ in⁴, $I_v = 19.9$ in⁴; $b = 3.875$ in, $h = 5.75$ in, and $t = \frac{1}{4}$ in. By the formula from Table 8.12, $e = b^2h^2t/4I_v = 1.55$ in; therefore the flexural center is at Q, as shown. When the load is resolved into vertical and horizontal components at Q and a couple, the results are as shown in Fig. 8.19(*b*). (Vertical and horizontal components are used because the *principal central axes u* and *v* are vertical and horizontal.)

The maximum fiber stress will occur at the corner where the stresses due to the vertical and horizontal bending moments are of the same kind; at the upper-right corner *f* both stresses are tensile, and since *f* is farther from the *u* axis than the lower-left corner *g* where both stresses are compressive, it will sustain the greater stress. This stress will be simply the sum of the stresses due to the vertical and horizontal components of the load, or

$$\sigma = \frac{940(36)(3)}{19.9} + \frac{342(36)(2.765)}{5.61} = 5100 + 6070 = 11{,}200 \text{ lb/in}^2$$

The effect of the couple depends on the way in which the inner end of the beam is supported. If it is simply constrained against rotation in the planes of bending and twisting, the twisting moment will be resisted wholly by shear stress on the cross section, and these stresses can be found by the appropriate torsion formula of Table 10.1. If, however, the beam is built in so that the flanges are fixed in the *horizontal* plane, then part of the torque is resisted by the bending rigidity of the flanges and the corresponding moment causes a further fiber stress. This can be found by using the formulas of Sec. 10.3.

For the channel section, K is given with sufficient accuracy by the formula $K = (t^3/3)(h + 2b)$ (Table 10.2, case 1), which gives $K = 0.073$ in⁴. Taking $G = 12{,}000{,}000$ lb/in², and $E = 30{,}000{,}000$ lb/in², and the formula for C_w as

$$C_w = \frac{h^2b^3t}{12}\frac{2h + 3b}{h + 6b} = 38.4 \text{ in}^6$$

the value for β can be found. From Table 10.3, the formula for β is given as

$$\beta = \left(\frac{KG}{C_wE}\right)^{1/2} = \left[\frac{0.073(12)}{38.4(30)}\right]^{1/2} = 0.0276$$

From Table 10.3, case 1b, the value of θ″ at the wall is given as

$$\theta'' = \frac{T_o}{C_wE\beta}\tanh\beta l = \frac{313}{38.4(30\cdot10^6)(0.0276)}\tanh 0.0276(36) = 7.47(10^{-6}) \text{ in}^{-2}$$

Therefore, the longitudinal stress at f can be found from the expression for σ_x in Table 10.2, case 1, as

$$\sigma_x = \frac{hb}{2}\frac{h+3b}{h+6b}E\theta'' = \frac{6(4)}{2}\frac{6+3(4)}{6+6(4)}(30)(10^6)(7.47)(10^{-6}) = 1610 \text{ lb/in}^2$$

The resultant fiber stress at f is $11{,}200 - 1610 = 9590 \text{ lb/in}^2$.

8.15 STRAIGHT UNIFORM BEAMS (COMMON CASE); ULTIMATE STRENGTH

When a beam is stressed beyond the elastic limit, plane sections remain plane or nearly so but unit stresses are no longer proportional to strains and hence no longer proportional to distance from the neutral surface. If the material has similar stress-strain curves in tension and compression, the stress distribution above and below the neutral surface will be similar and the neutral axis of any section which is symmetric about a horizontal axis will still pass through the centroid; if the material has different properties in tension and compression, then the neutral axis will shift away from the side on which the fibers yield the most; this shift causes an additional departure from the stress distribution assumed by the theory outlined in Sec. 8.1.

Failure in Bending

The strength of a beam of ordinary proportions is determined by the maximum bending moment it can sustain. For beams of nonductile material (cast iron, concrete, or seasoned wood) this moment may be calculated by the formula $M_m = \sigma'(I/c)$ if σ', the *modulus of rupture*, is known. The modulus of rupture depends on the material and other factors (see Sec. 3.11), and attempts have been made to calculate it for a given material and section from the form of the complete stress-strain diagram. Thus for cast iron an approximate value of σ' may be found by the formula $\sigma' = K\sqrt{c/z'}\sigma_t$, where c is the distance to the extreme fiber, z' is the distance from the neutral axis to the centroid of the tensile part of the section, and K is an experimental coefficient equal to 6/5 for sections that are flat at the top and bottom (rectangle, I, T, etc.) and 4/3 for sections that are pointed or convex at the top and bottom (circle, diamond, etc.) (Ref. 4). Some tests indicate that this method of calculating the breaking strength of cast iron is sometimes inaccurate but generally errs on the side of safety (Ref. 5).

In general, the breaking strength of a beam can be predicted best from experimentally determined values of the rupture factor and ultimate strength or the form factor and modulus of rupture. The rupture factors are based on the ultimate tensile strength for all materials except wood, for which it is based on compressive strength. Form factors are based on a rectangular section. For structural steel, wrought aluminum, and other ductile metals, where beams do not actually break, the modulus of rupture means the computed fiber stress at the maximum bending moment (Refs. 6–9).

When the maximum bending moment occurs at but one section, as for a single concentrated load, the modulus of rupture is higher than when the maximum moment extends over a considerable part of the span. For instance, the modulus of rupture of short beams of brittle material is about 20% higher when determined by center loading than when determined by third-point loading. The disparity decreases as the span/depth ratio increases.

Beams of ductile material (structural steel or aluminum) do not ordinarily fracture under static loading but fail through excessive deflection. For such beams, if they are of relatively thick section so as to preclude local buckling, the maximum bending moment is that which corresponds to plastic yielding throughout the section. This maximum moment, or "plastic" moment, is usually denoted by M_p and can be calculated by the formula $M_p = \sigma_y Z$, where σ_y is the lower yield point of the material and Z, called the *plastic section modulus,* is the arithmetical sum of the statical moments about the neutral axis of the parts of the cross section above and below that axis. Thus, for a rectangular section of depth d and width b,

$$Z = \left(\frac{1}{2}bd\right)\left(\frac{1}{4}d\right) + \left(\frac{1}{2}bd\right)\left(\frac{1}{4}d\right) = \frac{1}{4}bd^2$$

This method of calculating the maximum resisting moment of a ductile-material beam is widely used in "plastic design" and is discussed further in Sec. 8.16. It is important to note that when the plastic moment has been developed, the neutral axis divides the cross-sectional area into halves and so is not always a centroidal axis. It is also important to note that the plastic moment is always greater than the moment required to just stress the extreme fiber to the lower yield point. This moment, which may be denoted by M_y, is equal to $\sigma_y I/c$, and so

$$\frac{M_p}{M_y} = \frac{Z}{I/c}$$

This ratio $Z/(I/c)$, called the *shape factor*, depends on the form of the cross section. For a solid rectangle it would be $\frac{1}{4}bd^2/\frac{1}{6}bd^2$, or 1.5; for an I-section it is usually about 1.15. Table A.1 gives formulas or numerical values for the plastic section modulus Z and for the shape factor for most of the cross sections listed.

In tubes and beams of thin open section, local buckling or crippling will sometimes occur before the full plastic resisting moment is realized, and the length of the member will have an influence. Tubes of steel or aluminum alloy generally will develop a modulus of rupture exceeding the ultimate tensile strength when the ratio of diameter to wall thickness is less than 50 for steel or 35 for aluminum. Wide-flanged steel beams will develop the full plastic resisting moment when the outstanding width/thickness ratio is less than 8.7 for σ_y = 33,000 lb/in^2 or 8.3 for σ_y = 36,000 lb/in^2. Charts giving the effective modulus of rupture of steel, aluminum, and magnesium tubes of various proportions may be found in Ref. 55.

Failure in Shear

Failure by an actual shear fracture is likely to occur only in wood beams, where the shear strength parallel to the grain is, of course, small.

In I-beams and similar thin-webbed sections the diagonal compression that accompanies shear (Sec. 7.5) may lead to a buckling failure (see the discussion of *web buckling* that follows), and in beams of cast iron and concrete the diagonal tension that similarly accompanies shear may cause rupture. The formula for shear stress [Eq. (8.1-2)] may be considered valid as long as the *fiber* stresses do not exceed the proportional limit, and therefore it may be used to calculate the vertical shear necessary to produce failure in any case where the ultimate shearing strength of the beam is reached while the fiber stresses, at the section of maximum shear, are still within the proportional limit.

Web Buckling; Local Failure

An I-beam or similar thin-webbed member may fail by buckling of the web owing to diagonal compression when the shear stress reaches a certain value. Ketchum and Draffin (Ref. 11) and Wendt and Withey (Ref. 12) found that in light I-beams this type of buckling occurs when the shear stress, calculated by $\tau = 1.25\ V/\text{web area}$ (Ref. 11) or $\tau = V/\text{web area}$ (Ref. 12), reaches a value equal to the unit load that can be carried by a vertical strip of the beam as a round-ended column. For the thin webs of the beams tested, such a thin strip would be computed as a Euler column; for heavier beams an appropriate parabolic or other formula should be used (Chap. 12).

In plate girders, web buckling may be prevented by vertical or diagonal stiffeners, usually consisting of double angles that are riveted or welded, one on each side of the web. Steel-construction specifications (Ref. 13) require that such stiffeners be provided when h/t exceeds 70 and v exceeds $64{,}000{,}000/(h/t)^2$. Such stiffeners should have a moment of inertia (figured for an axis at the center line of the web) equal to at least $0.00000016H^4$ and should be spaced so that the clear distance between successive stiffeners is not more than $11{,}000t/\sqrt{v}$ or 84 in, whichever is least. Here h is the clear depth of the web between flanges, t is the web thickness, v is the shear stress V/ht, and H is the total depth of the web. In light-metal airplane construction, the stiffeners are sometimes designed to have a moment of inertia about an axis parallel to the web given by $I = (2.29d/t)(Vh/33E)^{4/3}$, where V = the (total) vertical shear and d = the stiffener spacing center to center (Ref. 14).

Buckling failure may occur also as a result of vertical compression at a support or concentrated load, which is caused by either column-type buckling of the web (Refs. 11 and 12) or crippling of the web at the toe of the fillet (Ref. 15). To guard against this latter type of failure, present specifications provide that for interior loads $R/t(N+2k) \le 24{,}000$ and for end reactions $R/t(N+k) \le 24{,}000$, where R is the concentrated load or end reaction, t the web thickness, N the length of bearing, and k the distance from the outer face of the flange to the web toe of the fillet. Here R is in pounds and all linear dimensions are in inches.

Wood beams will crush locally if the supports are too narrow or if a load is applied over too small a bearing area. The unit bearing stress in either case is calculated by dividing the force by the nominal bearing area, no allowance being made for the nonuniform distribution of pressure consequent upon bending (Ref. 9). Metal beams also may be subjected to high local pressure stresses; these are discussed in Chap. 14.

Lateral Buckling

The compression flange of an I-beam or similar member may fail as a column as a result of lateral buckling if it is unsupported. Such buckling may be *elastic or plastic*; that is, it may occur at a maximum fiber stress below or above the elastic limit. In the first case the buckling is an example of elastic instability, for which relevant formulas are given in Table 15.1. For buckling above the elastic range analytical solutions are difficult to obtain, and empirical expressions based on experiment are used (as will be shown to be true also of the columns discussed in Chap. 12).

Moore (Ref. 16) found that standard I-beams fail by lateral buckling when

$$s' = 40{,}000 - 60\frac{ml}{r}$$

where s' is the compressive stress in the extreme fiber [computed by Eq. (8.1-1)], l is the span (in inches), r is the radius of gyration (in inches) of the beam section about a central axis parallel to the web, and m is a coefficient which depends on the manner of loading and support and has the following values:

Loading and support	Value of m
End supports, uniform load	0.667
End supports, midpoint load	0.500
End supports, single load at any point	0.500
End supports, loads at third points	0.667
End supports, loads at quarter points	0.750
End supports, loads at sixth points	0.833
Cantilever beam, uniform load	0.667
Cantilever beam, end load	1.000
Fixed-ended beam, uniform load	0.281
Fixed-ended beam, midpoint load	0.250

For very light I-beam, Ketchum and Draffin (Ref. 11) found that the lower limit of test results is given by

$$s' = 24{,}000 - 40\frac{ml}{r}$$

where the terms have the same meaning and m the same values as given previously.

The beams tested by Moore generally failed at stresses below but very close to the yield point and so probably could be regarded as representing plastic buckling. The lighter beams tested by Ketchum and Draffin, however, failed at stresses below the limit of proportionality and are examples of elastic buckling.

In Ref. 13 rules are given for the reduction in allowable compressive stress according to the unbraced length of the compression flange. A review of the literature on this subject of the lateral buckling of structural members and a bibliography through 1959 are to be found in Ref. 58.

Narrow rectangular beams may fail also as a result of buckling of the compression edge. When this buckling occurs below the elastic limit, the strength is determined by elastic stability; formulas for this case are given in Table 15.1. For buckling at stresses beyond the elastic limit, no simple formula for the critical stress can be given, but methods for calculating this critical stress are given for aluminum beams by Dumont and Hill (Ref. 17) and for wood beams by Trayer and March (Ref. 18).

8.16 PLASTIC, OR ULTIMATE STRENGTH, DESIGN

The foregoing discussion of beams and frames is based for the most part on the assumption of purely elastic action and on the acceptance of maximum fiber stress as the primary criterion of safety. These constitute the basis of *elastic* analysis and design. An alternative and often preferred method of design, applicable to rigid frames and statically indeterminate beams made of materials capable of plastic action, is the method of *plastic,* or *ultimate strength,* design. It is based on the fact that such a frame or beam cannot deflect indefinitely or collapse until the full plastic moment M_p (see Sec. 8.15) has been developed at each of several critical sections. If it is assumed that the plastic moment—a determinable couple—does indeed act at each such section, then the problem becomes a statically determinate one and the load corresponding to the collapse condition can be readily calculated.

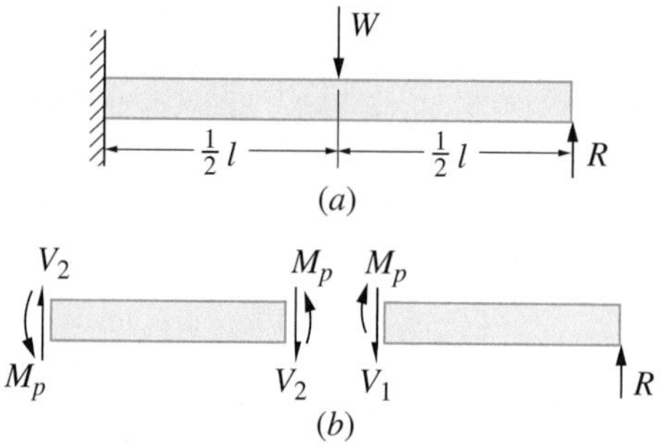

Figure 8.20

A simple illustration of the procedure is afforded by the beam of Fig. 8.20(*a*), corresponding to case lc of Table 8.1. Suppose it is desired to determine the maximum value of the load W that the beam can support. It is shown by elastic analysis, and is indeed apparent from inspection, that the maximum bending moments occur at the load and at the left end of the beam. The maximum possible value of each such moment is M_p. It is evident that the beam cannot collapse until the moment at each of these points reaches this value. Therefore, when W has reached its maximum value and collapse is imminent, the beam is acted on by the force system represented in Fig. 8.20(*b*); there is a *plastic hinge* and a known couple M_p at each of the critical sections and the problem is statically determinate. For equilibrium of the right half, $R = M_p/(l/2)$ and $V_1 = R$; and for equilibrium of the left half, $V_2 = W - R$ and $[W - M_p/(l/2)]l/2 = 2M_p$ or $W = 6M_p/l$.

In attempting to predict the collapse load on the basis of elastic analysis, it is easy to fall into the error of equating the maximum elastic moment 3/16 Wl at the wall (Table 8.1) to M_p, thus obtaining $W = 16/3\ M_p/l$. This erroneous procedure fails to take into account the fact that as W increases and yielding commences and progresses at the wall section, there is a redistribution of moments; the moment at the wall becomes less than 3/16 Wl, and the moment at the load becomes greater than 5/32 Wl until finally each moment becomes equal to M_p. An important point to note is that although the elastic moments are affected by even a very slight departure from the assumed conditions—perfect fixity at one end and rigid support at the other—the collapse load is not thus affected. So long as the constraints are rigid enough to develop the plastic hinges as indicated, the ultimate load will be the same. Similarly, the method does not require that the beam be uniform in section, although a local reduction in section leading to the formation of a hinge at some point other than those assumed, of course, would alter the solution.

For a beam with multiple concentrated transverse loads or with distributed transverse loads, the locations of hinges are not known and must be assumed and verified. A virtual work approach to plastic collapse may permit a more rapid analysis than does the use of equilibrium equations, see Ref. 66. Verification consists of using the equilibrium conditions to construct a moment diagram and determine that no moments larger than the locally permitted values of the fully plastic moment are present.

Since nonlinear behavior does not permit superposition of results, one must consider all loads which are acting at a given time. If any of the several loads on a beam tend to cancel the maximum moments due to other loads, one must also consider the order in which the loads are applied in service to assure that the partially loaded beam has not collapsed before all loads have been applied.

Column 4 of Table A.1 contains an expression or a numerical value for the plastic section modulus Z and for the shape factor $SF = Zc/I$ for many of the cross sections. Using the plastic section modulus and the value of the yield strength of the material, one can find the full plastic moment M_p. Table 8.13 contains expressions for the loadings which will cause plastic collapse and the locations of the plastic hinges associated with each such loading.

The following example problems illustrate (1) the direct use of the tabulated material and (2) the use of the virtual work approach to a problem where two loads are applied simultaneously and where one plastic hinge location is not obvious.

EXAMPLE 1 A hollow aluminum cylinder is used as a transversely loaded beam 6 ft long with a 3-in outer diameter and a 1-in inner diameter. It is fixed at both ends and subjected to a distributed loading which increases linearly from zero at midspan to a maximum value w_l at the right end. The yield strength of this material is 27,000 psi, and the value of W_{lc} at plastic collapse is desired.

SOLUTION From Table A.1 case 15, the expression for the plastic section modulus is given as $Z = 1.333(R^3 - R_i^3)$, which gives

$$Z = 1.333(1.5^3 - 0.5^3) = 4.33 \text{ in}^3$$

and

$$M_p = 4.33(27{,}000) = 117{,}000 \text{ lb-in}$$

From Table 8.13 case 2d, with $w_a = 0$ for a uniformly increasing load, the locations of the fully developed plastic hinges are at the two ends and at a distance x_{h2} from the left end, where

$$x_{h2} = a + \left(a^2 - al + \frac{l^2}{3} - \frac{a^3}{3l}\right)^{1/2}$$

Since $l = 72$ in and $a = 86$ in, the third hinge is found at $x_{h2} = 50.70$ in. The expression for the collapse load w_{lc} is given as

$$w_{lc} = \frac{12M_p(l-a)}{(l - x_{h2})(x_{h2}^2 - 3ax_{h2} + lx_{h2} + a^3/l)} = 1703 \text{ lb/in}$$

EXAMPLE 2 A steel beam of trapezoidal section is shown in Fig. 8.21. It is 1500 mm long, fixed at the right end and simply supported at the left end. The factor of safety of the loading shown is to be determined based on plastic collapse under a proportionately increased set of similar loads. The yield strength of the material is 200 N/mm^2 in both tension and compression. All dimensions are in millimeters.

SOLUTION First evaluate M_p. An examination of Table A.1 shows that the plastic section modulus for this cross section is not given. The yield strength in tension and compression is the same, so half the cross-sectional area of 2100 mm^2 will be in tension and half in compression under a fully developed plastic hinge. Calculations show that a horizontal axis 16.066 mm above the base will divide this

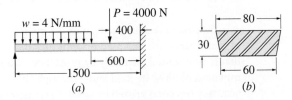

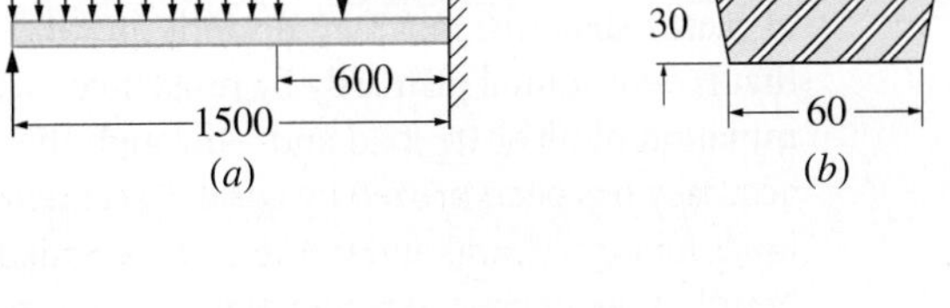

Figure 8.21

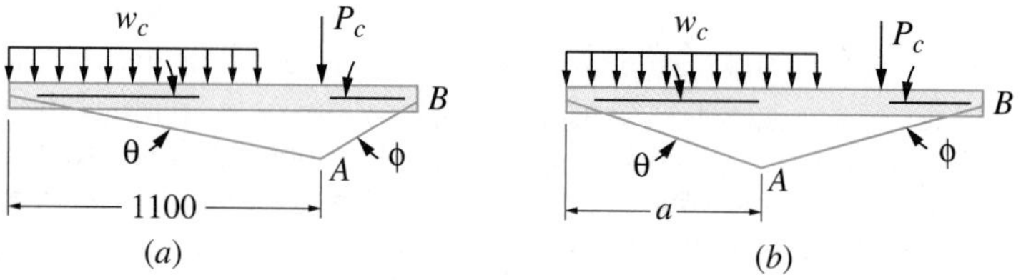

Figure 8.22

area into equal parts. The centroid of the upper portion is 7.110 mm above this axis, and the centroid of the lower portion is 7.814 mm below. Therefore,

$$M_p = 200(1050)(7.110+7.814) = 3.134(10^6) \text{ N-mm}$$

Let the concentrated load at collapse P_c be accompanied by a uniformly distributed load w_c, which equals $P_c/1000$. During a virtual displacement of the beam when plastic rotation is taking place about the fully developed plastic hinges, the elastic deformations and any deformations due to the development of the plastic hinges remain constant and can be neglected in computing the work done by the loading. The angles shown in Fig. 8.22 are not true slopes at these locations but merely represent the virtual rotations of the fully developed plastic hinges. The location of hinge A is not known at this point in the solution, but it is either under the concentrated load or somewhere in the portion of the beam under the distributed load.

Trial 1. Assume that hinge A is under the concentrated load and the virtual displacements are represented by Fig. 8.22(a). The work performed by the loads during their vertical displacements is absorbed by the two plastic hinges. The hinge at A rotates through the angle $\theta+\phi$ and the hinge at B through the angle ϕ. Thus

$$M_p(\theta+\phi+\phi) = P_c 1100\theta + w_c(900)(450)\theta$$

where $w_c = P_c/1000$ and from geometry $400\phi = 1100\theta$ so that $P_c = 4.319(-10^{-3})M_p$, and from the equilibrium of the entire beam one obtains $R_1 = 3.206(10^{-3})M_p$. Using these two values and constructing a moment diagram, one finds that a maximum moment of $1.190M_p$ will be present at a distance of 742 mm from the left end.

Thus the assumption that hinge A was under the concentrated load was incorrect. A second trial solution will be carried out by assuming that hinge A is a distance a from the left end.

Trial 2. Figure 8.22(b) shows the virtual displacements for this second assumption. Again set the virtual work done by the loading equal to the energy absorbed by the two plastic hinges, or

$$M_p(\theta+2\phi) = P_c 400\phi + \frac{w_c\theta a^2}{2} + w_c(900-a)\left(1500-a-\frac{900-a}{2}\right)\phi$$

Note that $w_c = P_c/1000$ and from geometery $\phi(1500-a) = \theta a$ so that $P_c = M_p[(1500+a)/(1345a-0.75a^2)]$. A minimum value of P_c is desired so this expression is differentiated with respect to a and the derivative set equal to zero. This leads to $a =$ 722.6 mm and $P_c = 3.830(10^{-3})M_p$, a significantly smaller value than before and one which leads to a moment diagram with a maximum positive moment of M_p at $a = 722.6$ mm and a maximum negative moment of $-M_p$ at the right end. This then is the correct solution, and substituting the numerical value for M_p one gets $P_c = 12{,}000$ N and $w_c = 12$ N/mm. The applied loads were $P = 4000$ N and $w = 4$ N/mm, so the factor of safety is 3.0.

Because of the simplicity, these examples may give an exaggerated impression of the ease of plastic analysis, but they do indicate that for any indeterminate structure with strength that is determined primarily by resistance to bending, the method is well-suited to the determination of ultimate load and—through the use of a suitable factor of safety—to design. Its accuracy has been proved by good agreement between computed and experimental ultimate loads for both beams and frames. An extended discussion of plastic analysis is not appropriate here, but the interested reader will find an extensive literature on the subject (Refs. 60 and 61).

8.17 TABLES

TABLE 8.1 Shear, Moment, Slope, and Deflection Formulas for Elastic Straight Beams

Notation: W = load (force); w = unit load (force per unit length); M_o = applied couple (force-length); θ_o externally created concentrated angular displacement (radians); Δ_o = externally created concentrated lateral displacement; T_1 and T_2 = temperatures on the top and bottom surfaces, respectively (degrees). R_A and R_B are the vertical end reactions at the left and right, respectively, and are positive upward. M_A and M_B are the reaction end moments at the left and right, respectively. All moments are positive when producing compression on the upper portion of the beam cross section. The transverse shear force V is positive when acting upward on the left end of a portion of the beam. All applied loads, couples, and displacements are positive as shown. All deflections are positive upward, and all slopes are positive when up and to the right. E is the modulus of elasticity of the beam material, and I is the area moment of inertia about the centroidal axis of the beam cross section. γ is the temperature coefficient of expansion (unit strain per degree).

1. Concentrated intermediate load

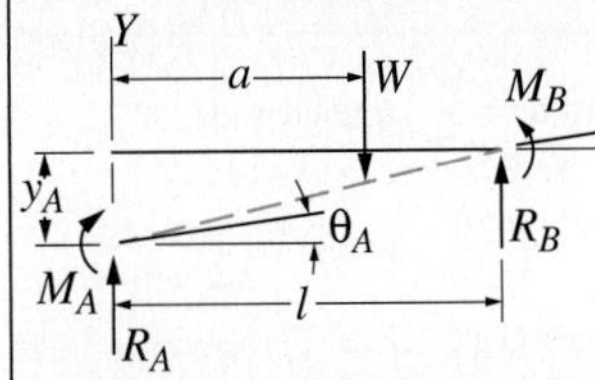

Transverse shear $= V = R_A - W\langle x-a\rangle^0$

Bending moment $= M = M_A + R_A x - W\langle x-a\rangle$

$$\text{Slope} = \theta = \theta_A + \frac{M_A x}{EI} + \frac{R_A x^2}{2EI} - \frac{W}{2EI}\langle x-a\rangle^2$$

$$\text{Deflection} = y = y_A + \theta_A x + \frac{M_A x^2}{2EI} + \frac{R_A x^3}{6EI} - \frac{W}{6EI}\langle x-a\rangle^3$$

(*Note:* See page 155 for a definition of the term $\langle x-a\rangle^n$.)

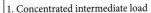

End Restraints, Reference No.	Boundary Values	Selected Maximum Values of Moments and Deformations
1a. Left end free, right end fixed (cantilever) 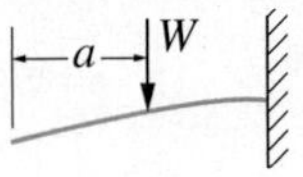	$R_A = 0 \quad M_A = 0 \quad \theta_A = \frac{W(l-a)^2}{2EI}$ $y_A = \frac{-W}{6EI}(2l^3 - 3l^2 a + a^3)$ $R_B = W \quad M_B = -W(l-a)$ $\theta_B = 0 \quad y_B = 0$	Max $M = M_B$; max possible value $= -Wl$ when $a = 0$ Max $\theta = \theta_A$; max possible value $= \frac{Wl^2}{2EI}$ when $a = 0$ Max $y = y_A$; max possible value $= \frac{-Wl^3}{3EI}$ when $a = 0$
1b. Left end guided, right end fixed 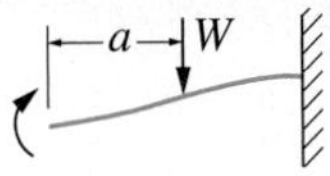	$R_A = 0 \quad M_A = \frac{W(l-a)^2}{2l} \quad \theta_A = 0$ $y_A = \frac{-W}{12EI}(l-a)^2(l+2a)$ $R_B = W \quad M_B = \frac{-W(l^2-a^2)}{2l}$ $\theta_B = 0 \quad y_B = 0$	Max $+M = M_A$; max possible value $= \frac{Wl}{2}$ when $a = 0$ Max $-M = M_B$; max possible value $= -\frac{Wl}{2}$ when $a = 0$ Max $y = y_A$; max possible value $= \frac{-Wl^3}{12EI}$ when $a = 0$
1c. Left end simply supported right end fixed 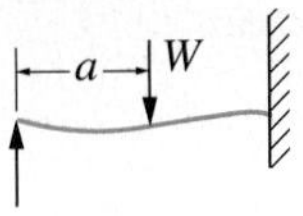	$R_A = \frac{W}{2l^3}(l-a)^2(2l+a) \quad M_A = 0$ $\theta_A = \frac{-Wa}{4EIl}(l-a)^2 \quad y_A = 0$ $R_B = \frac{Wa}{2l^3}(3l^2 - a^2) \quad \theta_B = 0$ $M_B = \frac{-Wa}{2l^2}(l^2 - a^2) \quad y_B = 0$	$\text{Max} + M = \frac{Wa}{2l^3}(l-a)^2(2l+a)$ at $x = a$; max possible value $= 0.174Wl$ when $a = 0.366l$ Max $-M = M_B$; max possible value $= -0.1924Wl$ when $a = 0.5773l$ Max $y = \frac{-Wa}{6EI}(l-a)^2\left(\frac{a}{2l+a}\right)^{1/2}$ at $x = l\left(\frac{a}{2l+a}\right)^{1/2}$ when $a > 0.414l$ Max $y = \frac{-Wa(l^2-a^2)^3}{3EI(3l^2-a^2)^2}$ at $x = \frac{l(l^2+a^2)}{3l^2-a^2}$ when $a < 0.414l$; max possible $y = -0.0098\frac{Wl^3}{EI}$ when $x = a = 0.414l$

(*Continued*)

TABLE 8.1 Shear, Moment, Slope, and Deflection Formulas for Elastic Straight Beams (*Continued*)

End Restraints, Reference No.	Boundary Values	Selected Maximum Values of Moments and Deformations
1d. Left end fixed, right end fixed 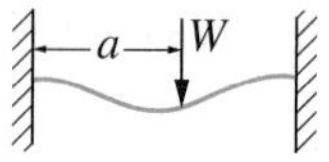	$R_A = \frac{W}{l^3}(l-a)^2(l+2a)$ $M_A = \frac{-Wa}{l^2}(l-a)^2$ $\theta_A = 0 \quad y_A = 0$ $R_B = \frac{Wa^2}{l^3}(3l-2a)$ $M_B = \frac{-Wa^2}{l^2}(l-a)$ $\theta_B = 0 \quad y_B = 0$	$\text{Max} + M = \frac{2Wa^2}{l^3}(l-a)^2$ at $x=a$; max possible value $= \frac{Wl}{8}$ when $a = \frac{l}{2}$ $\text{Max} - M = M_A$ if $a < \frac{l}{2}$; max possible value $= -0.1481Wl$ when $a = \frac{l}{3}$ $\text{Max } y = \frac{-2W(l-a)^2a^3}{3EI(l+2a)^2}$ at $x = \frac{2al}{l+2a}$ if $a > \frac{l}{2}$; max possible value $= \frac{-Wl^3}{192EI}$ when $x = a = \frac{l}{2}$
1e. Left end simply supported, right end simply supported 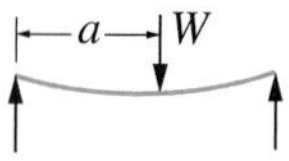	$R_A = \frac{W}{l}(l-a) \quad M_A = 0$ $\theta_A = \frac{-Wa}{6EIl}(2l-a)(l-a) \quad y_A = 0$ $R_B = \frac{Wa}{l} \quad M_B = 0$ $\theta_B = \frac{Wa}{6EIl}(l^2-a^2) \quad y_B = 0$	$\text{Max } M = R_A a$ at $x=a$; max possible value $= \frac{Wl}{4}$ when $a = \frac{l}{2}$ $\text{Max } y = \frac{-Wa}{3EIl}\left(\frac{l^2-a^2}{3}\right)^{3/2}$ at $x = l - \left(\frac{l^2-a^2}{3}\right)^{1/2}$ when $a < \frac{l}{2}$; max possible value $= \frac{-Wl^3}{48EI}$ at $x = \frac{l}{2}$ when $a = \frac{l}{2}$ $\text{Max } \theta = \theta_A$ when $a < \frac{l}{2}$; max possible value $= -0.0642\frac{Wl^2}{EI}$ when $a = 0.423l$
1f. Left end guided, right end simply supported 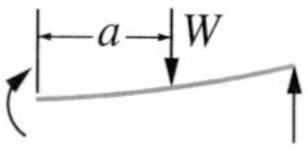	$R_A = 0 \quad M_A = W(l-a) \quad \theta_A = 0$ $y_A = \frac{-W(l-a)}{6EI}(2l^2 + 2al - a^2)$ $R_B = W \quad M_B = 0$ $\theta_B = \frac{W}{2EI}(l^2 - a^2) \quad y_B = 0$	$\text{Max } M = M_A$ for $0 < x < a$; max possible value $= Wl$ when $a = 0$ $\text{Max } \theta = \theta_B$; max possible value $= \frac{Wl^2}{2EI}$ when $a = 0$ $\text{Max } y = y_A$; max possible value $= \frac{-Wl^3}{3EI}$ when $a = 0$
2. Partial distributed load 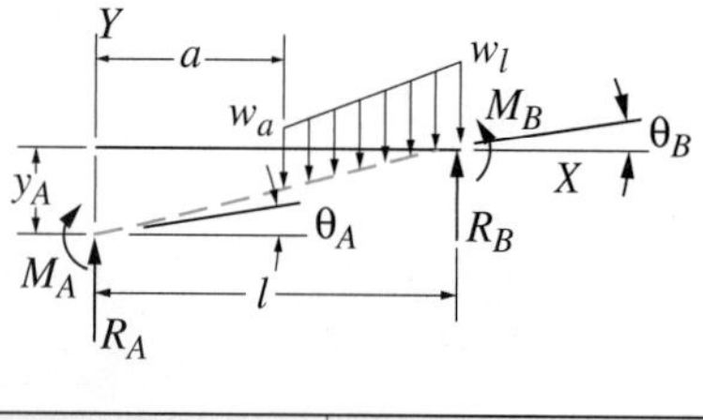		Transverse shear $= V = R_A - w_a\langle x-a\rangle - \frac{w_l - w_a}{2(l-a)}\langle x-a\rangle^2$ Bending moment $= M = M_A + R_A x - \frac{w_a}{2}\langle x-a\rangle^2 - \frac{w_l - w_a}{6(l-a)}\langle x-a\rangle^3$ Slope $= \theta = \theta_A + \frac{M_A x}{EI} + \frac{R_A x^2}{2EI} - \frac{w_a}{6EI}\langle x-a\rangle^3 - \frac{w_l - w_a}{24EI(l-a)}\langle x-a\rangle^4$ Deflection $= y = y_A + \theta_A + \theta_A x + \frac{M_A x^2}{2EI} + \frac{R_A x^3}{6EI} - \frac{w_a}{24EI}\langle x-a\rangle^4 - \frac{(w_l - w_a)}{120EI(l-a)}\langle x-a\rangle^5$
End Restraints, Reference No.	Boundary Values	Selected Maximum Values of Moments and Deformations

2a. Left end free, right end fixed (cantilever) 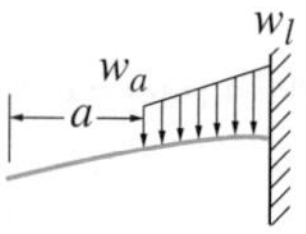	$R_A = 0 \quad M_A = 0$
	$\theta_A = \frac{w_a}{6EI}(l-a)^3 + \frac{w_l - w_a}{24EI}(l-a)^3$
	$y_A = \frac{-w_a}{24EI}(l-a)^3(3l+a) - \frac{w_l - w_a}{120EI}(l-a)^3(4l+a)$
	$R_B = \frac{w_a - w_l}{2}(l-a)$
	$M_B = \frac{-w_a}{2}(l-a)^2 - \frac{w_l - w_a}{6}(l-a)^2$
	$\theta_B = 0 \quad y_B = 0$
		If $a = 0$ and $w_l = w_a$ (uniform load on entire span), then
		$\text{Max } M = M_B = \frac{-w_a l^2}{2} \quad \text{Max } \theta = \theta_A = \frac{w_a l^3}{6EI}$
		$\text{Max } y = y_A = \frac{-w_a l^4}{8EI}$
		If $a = 0$ and $w_a = 0$ (uniformly increasing load), then
		$\text{Max } M = M_B = \frac{-w_l l^2}{6} \quad \text{Max } \theta = \theta_A = \frac{w_l l^3}{24EI}$
		$\text{Max } y = y_A = \frac{-w_l l^4}{30EI}$

(*Continued*)

TABLE 8.1 Shear, Moment, Slope, and Deflection Formulas for Elastic Straight Beams (*Continued*)

End Restraints, Reference No.	Boundary Values	Selected Maximum Values of Moments and Deformations
		If $a = 0$ and $w_l = 0$ (uniformly decreasing load), then $\text{Max } M = M_B = \frac{-w_a l^2}{3} \qquad \text{Max } \theta = \theta_A \frac{w_a l^3}{8EI}$ $\text{Max } y = y_A = \frac{-11 w_a l^4}{120EI}$

2b. Left end guided, right end fixed
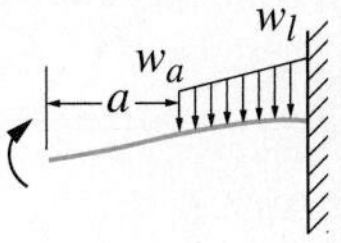
 	
	$R_A = 0 \qquad \theta_A = 0$
	$M_A = \frac{w_a}{6l}(l-a)^3 + \frac{w_l - w_a}{24l}(l-a)^3$
	$y_A = \frac{-w_a}{24EI}(l-a)^3(l+a) - \frac{w_l - w_a}{240EI}(l-a)^3(3l+2a)$
	$R_B = \frac{w_a + w_l}{2}(l-a)$
	$M_B = \frac{-w_a}{6l}(l-a)^2(2l+a) - \frac{w_l - w_a}{24l}(l-a)^2(3l+a)$
	$\theta_B = 0 \qquad y_B = 0$
		If $a = 0$ and $w_l = w_a$ (uniform load on entire span), then
		$\text{Max} - M = M_B = \frac{-w_a l^2}{3} \qquad \text{Max} + M = M_A = \frac{w_a l^2}{6}$
		$\text{Max } y = y_A = \frac{-w_a l^4}{24EI}$
		If $a = 0$ and $w_a = 0$ (uniformly increasing load), then
		$\text{Max} - M = M_B = \frac{-w_l l^2}{8} \qquad \text{Max} + M = M_A = \frac{w_l l^2}{24}$
		$\text{Max } y = y_A = \frac{-w_l l^4}{80EI}$
		If $a = 0$ and $w_l = 0$ (uniformly decreasing load), then
		$\text{Max} - M = M_B = \frac{-5 w_a l^2}{24} \qquad \text{Max} + M = M_A = \frac{w_a l^2}{8}$
		$\text{Max } y = y_A = \frac{-7 w_a l^4}{240EI}$

2c. Left end simply supported, right end fixed
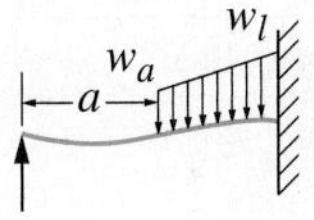
 	
	$R_A = \frac{w_a}{8l^3}(l-a)^3(3l+a) + \frac{w_l - w_a}{40l^3}(l-a)^3(4l+a)$
	$\theta_A = \frac{-w_a}{48EIl}(l-a)^3(l+3a) - \frac{w_l - w_a}{240EIl}(l-a)^3(2l+3a)$
	$M_A = 0 \qquad y_A = 0$
	$R_B = \frac{w_a + w_l}{2}(l-a) - R_A$
	$M_B = R_A l - \frac{w_a}{2}(l-a)^2 - \frac{w_l - w_a}{6}(l-a)^2$
	$\theta_B = 0 \qquad y_B = 0$
		If $a = 0$ and $w_l = w_a$ (uniform load on entire span), then
		$R_A = \frac{3}{8} w_a l \qquad R_B = \frac{5}{8} w_a l \qquad \text{Max} - M = M_B = \frac{-w_a l^2}{8}$
		$\text{Max} + M = \frac{9 w_a l^2}{128}$ at $x = \frac{3}{8} l \quad \text{Max } \theta = \theta_A = \frac{-w_a l^3}{48EI}$
		$\text{Max } y = -0.0054 \frac{w_a l^4}{EI}$ at $x = 0.4215l$
		If $a = 0$ and $w_a = 0$ (uniformly increasing load), then
		$R_A = \frac{w_l l}{10} \qquad R_B = \frac{2 w_l l}{5} \qquad \text{Max} - M = M_B = \frac{-w_l l^2}{15}$
		$\text{Max} + M = 0.0298 w_l l^2$ at $x = 0.4472l \quad \text{Max } \theta = \theta_A = \frac{-w_l l^3}{120EI}$
		$\text{Max } y = -0.00239 \frac{w_l l^4}{EI}$ at $x = 0.4472l$
		If $a = 0$ and $w_l = 0$ (uniformly decreasing load), then
		$R_A = \frac{11}{40} w_a l \qquad R_B = \frac{9}{40} w_a l \qquad \text{Max} - M = M_B = \frac{-7}{120} w_a l^2$
		$\text{Max} + M = 0.0422 w_a l^2$ at $x = 0.329l$
		$\text{Max } \theta = \theta_A = \frac{-w_a l^3}{80EI} \qquad \text{Max } y = -0.00304 \frac{w_a l^4}{EI}$; at $x = 0.4025l$

(*Continued*)

TABLE 8.1 Shear, Moment, Slope, and Deflection Formulas for Elastic Straight Beams (*Continued*)

End Restraints, Reference No.	Boundary Values	Selected Maximum Values of Moments and Deformations

2d. Left end fixed, right end fixed 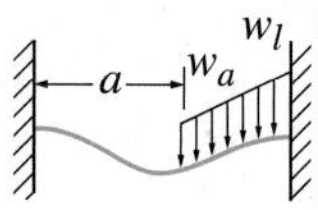	$R_A = \frac{w_a}{2l^3}(l-a)^3(l+a) + \frac{w_l - w_a}{20l^3}(l-a)^3(3l+2a)$
	$M_A = \frac{-w_a}{12l^2}(l-a)^3(l+3a) - \frac{w_l - w_a}{60l^2}(l-a)^3(2l+3a)$
	$\theta_A = 0 \quad y_A = 0$
	$R_B = \frac{w_a + w_l}{2}(l-a) - R_A$
	$M_B = R_A l + M_A - \frac{w_a}{2}(l-a)^2 - \frac{w_l - w_a}{6}(l-a)^2$
	$\theta_B = 0 \quad y_B = 0$
		If $a=0$ and $w_l = w_a$ (uniform load on entire span), then
		$\text{Max} - M = M_A = M_B = \frac{-w_a l^2}{12} \quad \text{Max} + M = \frac{w_a l^2}{24} \text{ at } x = \frac{l}{2}$
		$\text{Max } y = \frac{-w_a l^4}{384EI} \text{ at } x = \frac{l}{2}$
		If $a=0$ and $w_a = 0$ (uniformly increasing load), then
		$R_A = \frac{3w_l l}{20} \quad M_A = \frac{-w_l l^2}{30} \quad R_B = \frac{7w_l l}{20} \quad \text{Max} - M = M_B = \frac{-w_l l^2}{20}$
		$\text{Max} + M = 0.0215 w_l l^2 \text{ at } x = 0.548l$
		$\text{Max } y = -0.001309\frac{w_l l^4}{EI} \text{ at } x = 0.525l$

2e. Left end simply supported, right end simply supported 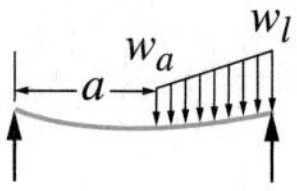	$R_A = \frac{w_a}{2l}(l-a)^2 + \frac{w_l - w_a}{6l}(l-a)^2$
	$M_A = 0 \quad y_A = 0$
	$\theta_A = \frac{-w_a}{24EIl}(l-a)^2(l^2 + 2al - a^2)$
	$- \frac{w_l - w_a}{360EIl}(l-a)^2(7l^2 + 6al - 3a^2)$
	$R_B = \frac{w_a + w_l}{2}(l-a) - R_A$
	$\theta_B = \frac{w_a}{24EIl}(l^2 - a^2)^2$
	$+ \frac{w_l - w_a}{360EIl}(l-a)^2(8l^2 + 9al + 3a^2)$
	$M_B = 0 \quad y_B = 0$
		If $a=0$ and $w_l = w_a$ (uniform load on entire span), then
		$R_A = R_B = \frac{w_a l}{2} \quad \text{Max } M = \frac{w_a l^2}{8} \text{ at } x = \frac{l}{2}$
		$\text{Max } \theta = \theta_B = \frac{w_a l^3}{24EI} \quad \text{Max } y = \frac{-5w_a l^4}{384EI} \text{ at } x = \frac{l}{2}$
		If $a=0$ and $w_a = 0$ (uniformly increasing load), then
		$R_A = \frac{w_l l}{6} \quad R_B = \frac{w_l l}{3} \quad \text{Max } M = 0.0641 w_l l^2 \text{ at } x = 0.5773l$
		$\theta_A = \frac{-7w_l l^3}{360EI} \quad \theta_B = \frac{w_l l^3}{45EI}$
		$\text{Max } y = -0.00653\frac{w_l l^4}{EI} \text{ at } x = 0.5195l$

2f. Left end guided, right end simply supported 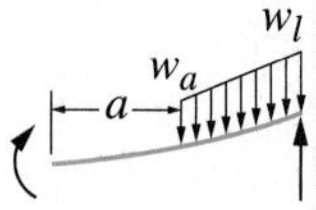	$R_A = 0 \quad \theta_A = 0$
	$M_A = \frac{w_a}{2}(l-a)^2 + \frac{w_l - w_a}{6}(l-a)^2$
	$y_A = \frac{-w_a}{24EI}(l-a)^2(5l^2 + 2al - a^2)$
	$- \frac{w_l - w_a}{120EI}(l-a)^2(9l^2 + 2al - a^2)$
	$R_B = \frac{w_a + w_l}{2}(l-a)$
	$\theta_B = \frac{w_a}{6EI}(l-a)^2(2l+a) + \frac{w_l - w_a}{24EI}(l-a)^2(3l+a)$
	$M_B = 0 \quad y_B = 0$
		If $a=0$ and $w_l = w_a$ (uniform load on entire span), then
		$\text{Max } M = M_A = \frac{w_a l^2}{2} \quad \text{Max } \theta = \theta_B = \frac{w_a l^3}{3EI}$
		$\text{Max } y = y_A = \frac{-5w_a l^4}{24EI}$
		If $a=0$ and $w_a = 0$ (uniformly increasing load), then
		$\text{Max } M = M_A = \frac{w_l l^2}{6} \quad \text{Max } \theta = \theta_B = \frac{w_l l^3}{8EI}$
		$\text{Max } y = y_A = \frac{-3w_l l^4}{40EI}$
		If $a=0$ and $w_l = 0$ (uniformly decreasing load), then
		$\text{Max } M = M_A = \frac{w_a l^2}{3} \quad \text{Max } \theta = \theta_B = \frac{5w_a l^3}{24EI}$
		$\text{Max } y = y_A = \frac{-2w_a l^4}{15EI}$

3. Concentrated intermediate moment 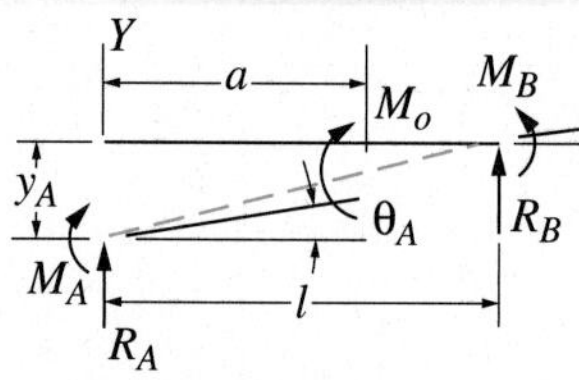		Transverse shear $= V = R_A$ Bending moment $= M = M_A + R_A x + M_o\langle x-a \rangle^0$ Slope $= \theta = \theta_A + \frac{M_A x}{EI} + \frac{R_A x^2}{2EI} + \frac{M_o}{EI}\langle x-a \rangle$ Deflection $= y = y_A + \theta_A x + \frac{M_A x^2}{2EI} + \frac{R_A x^3}{6EI} + \frac{M_o}{2EI}\langle x-a \rangle^2$

(*Continued*)

TABLE 8.1 Shear, Moment, Slope, and Deflection Formulas for Elastic Straight Beams (*Continued*)

End Restraints, Reference No.	Boundary Values	Selected Maximum Values of Moments and Deformations
3a. Left end free, right end fixed (cantilever) 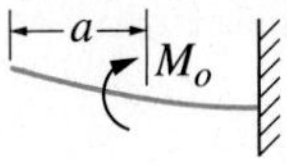 	$R_A = 0 \quad M_A = 0$ $\theta_A = \frac{-M_o(l-a)}{EI}$ $y_A = \frac{M_o(l^2 - a^2)}{2EI}$ $R_B = 0 \quad M_B = M_o$ $\theta_B = 0 \quad y_B = 0$	Max $M = M_o$ Max $\theta = \theta_A$; max possible value $= \frac{-M_o l}{EI}$ when $a = 0$ Max $y = y_A$; max possible value $= \frac{M_o l^2}{2EI}$ when $a = 0$
3b. Left end guided, right end fixed 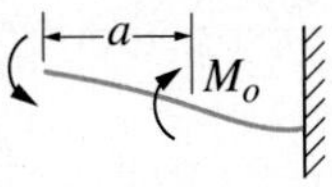 	$R_A = 0 \quad \theta_A = 0$ $M_A = \frac{-M_o(l-a)}{l}$ $y_A = \frac{M_o a(l-a)}{2EI}$ $R_B = 0 \quad \theta_B = 0$ $M_B = \frac{M_o a}{l} \quad y_B = 0$	Max $+M = M_B$; max possible value $= M_o$ when $a = l$ Max $-M = M_A$; max possible value $= -M_o$ when $a = 0$ Max $y = y_A$; max possible value $= \frac{M_o l^2}{8EI}$ when $a = \frac{l}{2}$
3c. Left end simply, supported right end fixed 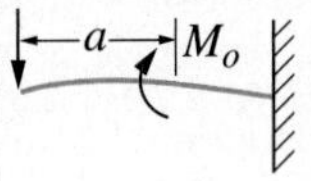 	$R_A = \frac{-3M_o}{2l^3}(l^2 - a^2)$ $\theta_A = \frac{M_o}{4EIl}(l-a)(3a-l)$ $M_A = 0 \quad y_A = 0$ $R_B = \frac{3M_o}{2l^3}(l^2 - a^2)$ $M_B = \frac{M_o}{2l^2}(3a^2 - l^2)$ $\theta_B = 0 \quad y_B = 0$	Max $+M = M_o + R_A a$ just right of $x = a$; max possible value $= M_o$ when $a = 0$ or $a = l$; min possible value $= 0.423M_o$ when $a = 0.577l$ Max $-M = \frac{-3M_o a}{2l^3}(l^2 - a^2)$ just left of $x = a$ if $a > 0.282l$; max possible value $= -0.577M_o$ when $a = 0.577l$ Max $-M = \frac{-M_o}{2l^2}(l^2 - 3a^2)$ at B if $a < 0.282l$; max possible value $= -0.5M_o$ when $a = 0$ Max $+y = \frac{M_o(l-a)}{6\sqrt{3(l+a)}EI}(3a-l)^{3/2}$ at $x = l\sqrt{\frac{3a-l}{3l+3a}}$; max possible value $= 0.0257\frac{M_o l^2}{EI}$ at $x = 0.474l$ when $a = 0.721l$ (*Note*: There is no positive deflection if $a < \frac{l}{3}$.) Max $-y$ occurs at $x = \frac{2l^3}{3(l^2 - a^2)}\left[1 - \frac{1}{2}\sqrt{1 - 6\left(\frac{a}{l}\right)^2 + 9\left(\frac{a}{l}\right)^4}\right]$; max possible value $= \frac{-M_o l^2}{27EI}$ at $x = \frac{l}{3}$ when $a = 0$
3d. Left end fixed, right end fixed 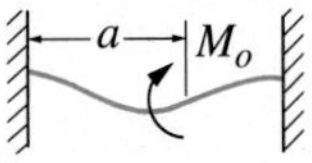 	$R_A = \frac{-6M_o a}{l^3}(l-a)$ $M_A = \frac{-M_o}{l^2}(l^2 - 4al + 3a^2)$ $\theta_A = 0 \quad y_A = 0$ $R_B = -R_A$ $M_B = \frac{M_o}{l^2}(3a^2 - 2al)$ $\theta_B = 0 \quad y_B = 0$	Max $+M = \frac{M_o}{l^3}(4al^2 - 9a^2l + 6a^3)$ just right of $x = a$; max possible value $= M_o$ when $a = l$ Max $-M = \frac{M_o}{l^3}(4al^2 - 9a^2l + 6a^3 - l^3)$ just left of $x = a$; max possible value $= -M_o$ when $a = 0$ Max $+y = \frac{2M_A^3}{3R_A^2 EI}$ at $x = \frac{l}{3a}(3a - l)$; max possible value $= 0.01617\frac{M_o l^2}{EI}$ at $x = 0.565l$ when $a = 0.767l$ (*Note*: There is no positive deflection if $a < \frac{l}{3}$.)

(*Continued*)

TABLE 8.1 Shear, Moment, Slope, and Deflection Formulas for Elastic Straight Beams (*Continued*)

End Restraints, Reference No.	Boundary Values	Selected Maximum Values of Moments and Deformations
3e. Left end simply supported, right end simply supported 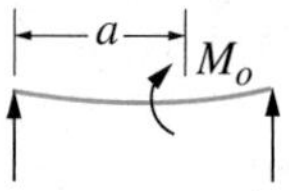	$R_A = \frac{-M_o}{l}$ $\theta_A = \frac{-M_o}{6EIl}(2l^2 - 6al + 3a^2)$ $M_A = 0 \quad y_A = 0$ $R_B = \frac{M_o}{l}$ $\theta_B = \frac{M_o}{6EIl}(l^2 - 3a^2)$ $M_B = 0 \quad y_B = 0$	Max $+M = \frac{M_o}{l}(l-a)$ just right of $x = a$; max possible value $= M_o$ when $a = 0$ Max $-M = \frac{-M_o a}{l}$ just left of $x = a$; max possible value $= -M_o$ when $a = l$ Max $+y = \frac{M_o(6al - 3a^2 - 2l^2)^{3/2}}{9\sqrt{3}EIl}$ at $x = (2al - a^2 - \frac{2}{3}l^2)^{1/2}$ when $a > 0.423l$; max possible value $= 0.0642\frac{M_o l^2}{EI}$ at $x = 0.577l$ when $a = l$ (*Note:* There is no positive deflection if $a < 0.423l$.)
3f. Left end guided, right end simply supported 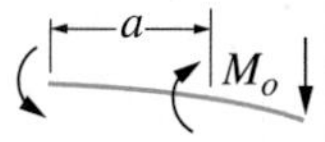	$R_A = 0 \quad \theta_A = 0$ $M_A = -M_o$ $y_A = \frac{M_o a}{2EI}(2l - a)$ $R_B = 0 \quad M_B = 0 \quad y_B = 0$ $\theta_B = \frac{-M_o a}{EI}$	Max $M = -M_o$ for $0 < x < a$ Max $\theta = \theta_B$; max possible value $= \frac{-M_o l}{EI}$ when $a = l$ Max $y = y_A$; max possible value $= \frac{M_o l^2}{2EI}$ when $a = l$
4. Intermediate externally created angular deformation 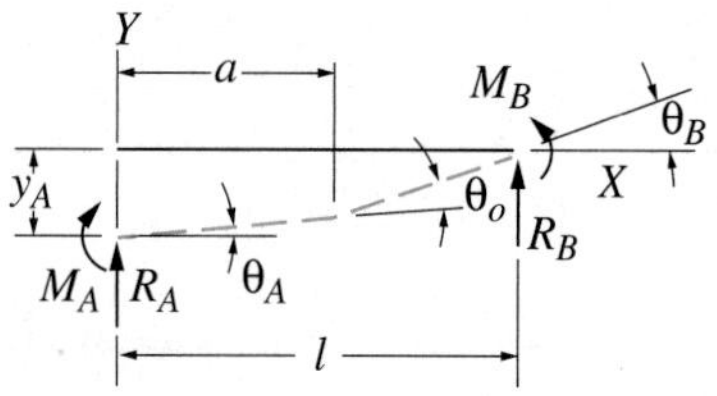		Transverse shear $= V = R_A$ Bending moment $= M = M_A + R_A x$ Slope $= \theta = \theta_A + \frac{M_A x}{EI} + \frac{R_A x^2}{2EI} + \theta_o \langle x - a \rangle^0$ Deflection $= y = y_A + \theta_A x + \frac{M_A x^2}{2EI} + \frac{R_A x^3}{6EI} + \theta_o \langle x - a \rangle$
End Restraints, Reference No.	**Boundary Values**	**Selected Maximum Values of Moments and Deformations**
4a. Left end free, right end fixed 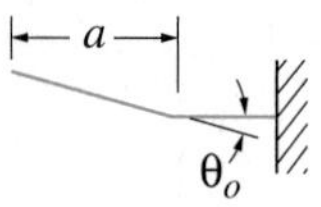	$R_A = 0 \quad M_A = 0$ $\theta_A = -\theta_o \quad y_A = \theta_o a$ $R_B = 0 \quad M_B = 0$ $\theta_B = 0 \quad y_B = 0$	Max $y = y_A$, max possible value $= \theta_o l$ when $a = l$
4b. Left end guided, right end fixed 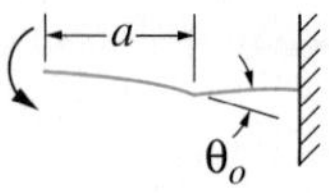	$R_A = 0 \quad M_A = \frac{-EI\theta_o}{l}$ $\theta_A = 0 \quad y_A = \theta_o\left(a - \frac{l}{2}\right)$ $R_B = 0 \quad M_B = \frac{-EI\theta_o}{l}$ $\theta_B = 0 \quad y_B = 0$	Max $M = M_A$ Max $+y = y_A$ when $a > \frac{l}{2}$; max possible value $= \frac{\theta_o l}{2}$ when $a = l$ Max $-y = \frac{-\theta_o}{2l}(l-a)^2$ at $x = a$; max possible value $= \frac{-\theta_o l}{2}$ when $a = 0$

(*Continued*)

TABLE 8.1 Shear, Moment, Slope, and Deflection Formulas for Elastic Straight Beams (*Continued*)

End Restraints, Reference No.	Boundary Values	Selected Maximum Values of Moments and Deformations
4c. Left end simply supported, right end fixed 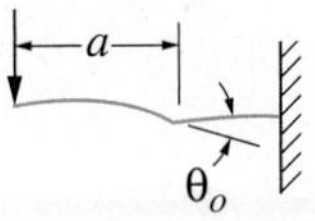 	$M_A=0 \quad y_A=0$ $R_A=\frac{-3EIa\theta_o}{l^3}$ $\theta_A=-\theta_o\left(1-\frac{3a}{2l}\right)$ $R_B=-R_A$ $M_B=\frac{-3EIa\theta_o}{l^2}$ $\theta_B=0 \quad y_B=0$	Max $M=M_B$; max possible value $=\frac{-3EI\theta_o}{l}$ when $a=l$ Max $+y=\theta_o a\left(1-\frac{2l}{3a}\right)^{3/2}$ at $x=l\left(1-\frac{2l}{3a}\right)^{1/2}$ When $a\ge\frac{2}{3}l$; max possible value $=0.1926\theta_o l$ at $x=0.577l$ when $a=l$ (*Note:* There is no positive deflection if $a<\frac{2}{3}l$.) Max $-y=-\theta_o a\left(1-\frac{3a}{2l}+\frac{a^3}{2l^3}\right)$ at $x=a$; max possible value $=-0.232\theta_o l$ at $x=0.366l$ when $a=0.366l$
4d. Left end fixed, right end fixed 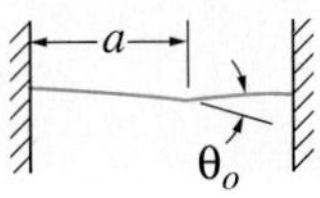 	$R_A=\frac{6EI\theta_o}{l^3}(l-2a)$ $M_A=\frac{2EI\theta}{l^2}(3a-2l)$ $\theta_A=0 \quad y_A=0$ $R_B=-R_A$ $M_B=\frac{2EI\theta_o}{l^2}(l-3a)$ $\theta_B=0 \quad y_B=0$	Max $+M=M_B$ when $a<\frac{l}{2}$; max possible value $=\frac{2EI\theta_o}{l}$ when $a=0$ Max $-M=M_A$ when $a<\frac{l}{2}$; max possible value $=\frac{-4EI\theta_o}{l}$ when $a=0$ Max $+y$ occurs at $x=\frac{l^2}{3(l-2a)}$ if $a<\frac{l}{3}$; max possible value $=\frac{4}{27}l\theta_o$ when $a=0$ (*Note:* There is no positive deflection if $\frac{l}{3}<a<\frac{2}{3}l$.) Max $-y=\frac{-2\theta_o a^2}{l^3}(l-a)^2$ at $x=a$; max possible value $=\frac{-\theta_o l}{8}$ when $a=\frac{l}{2}$
4e. Left end simply supported, right end simply supported 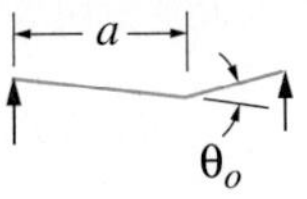 	$R_A=0 \quad M_A=0$ $\theta_A=\frac{-\theta}{l}(l-a) \quad y_A=0$ $R_B=0 \quad M_B=0$ $y_B=0 \quad \theta_B=\frac{\theta_o}{l}$	Max $y=\frac{-\theta_o a}{l}(l-a)$ at $x=a$; max possible value $=\frac{-\theta_o l}{4}$ when $a=\frac{l}{2}$
4f. Left end guided, right end simply supported 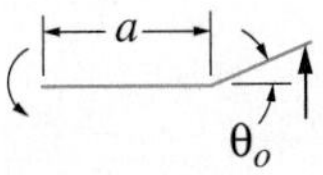 	$R_A=0 \quad M_A=0$ $\theta_A=0 \quad y_A=-\theta_o(l-a)$ $R_B=0 \quad M_B=0$ $y_B=0 \quad \theta_B=\theta_o$	Max $y=y_A$; max possible value $=-\theta_o l$ when $a=0$

(*Continued*)

TABLE 8.1 Shear, Moment, Slope, and Deflection Formulas for Elastic Straight Beams (*Continued*)

End Restraints, Reference No.	Boundary Values	Selected Maximum Values of Moments and Deformations
5. Intermediate externally created lateral displacement 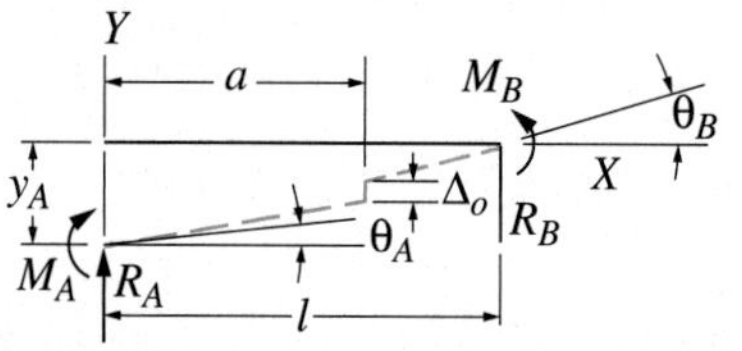		Transverse shear $= V = R_A$ Bending moment $= M = M_A + R_A x$ Slope $= \theta = \theta_A + \frac{M_A x}{EI} + \frac{R_A x^2}{2EI}$ Deflection $= y = y_A + \theta_A x + \frac{M_A x^2}{2EI} + \frac{R_A x^3}{6EI} + \Delta_o \langle x-a \rangle^0$

End Restraints, Reference No.	Boundary Values	Selected Maximum Values of Moments and Deformations
5a. Left end free, right end fixed 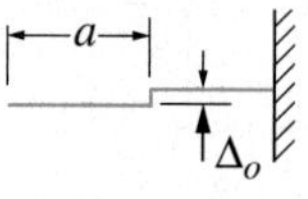	$R_A = 0 \quad M_A = 0$ $\theta_A = 0 \quad y_A = -\Delta_O$ $R_B = 0 \quad M_B = 0$ $\theta_B = 0 \quad y_B = 0$	Max $y = y_A$ when $x < a$
5b. Left end guided, right end fixed 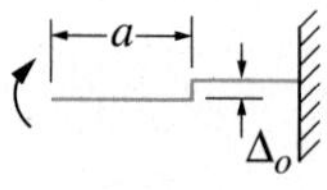	$R_A = 0 \quad M_A = 0$ $\theta_A = 0 \quad y_A = -\Delta_O$ $R_B = 0 \quad M_B = 0$ $\theta_B = 0 \quad y_B = 0$	Max $y = y_A$ when $x < a$
5c. Left end simply supported, right end fixed 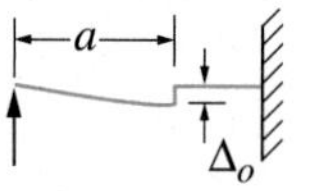	$R_A = \frac{3EI\Delta_o}{l^3} \quad M_A = 0$ $\theta_A = \frac{-3\Delta_o}{2l} \quad y_A = 0$ $R_B = -R_A \quad M_B = \frac{3EI\Delta_o}{l^2}$ $\theta_B = 0 \quad y_B = 0$	Max $M = M_B$ Max $\theta = \theta_A$ Max $+ y = \frac{\Delta_o}{2l^3}(2l^3 + a^3 - 3l^2 a)$ just right of $x = a$; max possible value $= \Delta_o$ when $a = 0$ Max $- y = \frac{-\Delta_o a}{2l^3}(3l^2 - a^2)$ just left of $x = a$; max possible value $= -\Delta_o$ when $a = l$
5d. Left end fixed, right end fixed 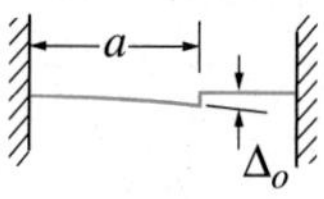	$R_A = \frac{12EI\Delta_o}{l^3} \quad \theta_A = 0$ $M_A = \frac{-6EI\Delta_o}{l^2} \quad y_A = 0$ $R_B = -R_A \quad M_B = -M_A$ $\theta_B = 0 \quad y_B = 0$	Max $+ M = M_B$ Max $- M = M_A$ Max $\theta = \frac{-3\Delta_o}{2l}$ at $x = \frac{l}{2}$ Max $+ y = \frac{\Delta_o}{l^3}(l^3 + 2a^3 - 3a^2 l)$ just right of $x = a$; max possible value $= \Delta_o$ when $a = 0$ Max $- y = \frac{-\Delta_o a^2}{l^3}(3l - 2a)$ just left of $x = a$; max possible value $= -\Delta_o$ when $a = l$
5e. Left end simply supported, right end simply supported 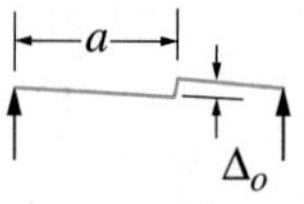	$R_A = 0 \quad M_A = 0$ $y_A = 0 \quad \theta_A = \frac{-\Delta_o}{l}$ $R_B = 0 \quad M_B = 0$ $y_B = 0 \quad \theta_B = \frac{-\Delta_o}{l}$	Max $+ y = \frac{\Delta_o}{l}(l - a)$ just right of $x = a$; max possible value $= \Delta_o$ when $a = 0$ Max $- y = \frac{-\Delta_o a}{l}$ just left of $x = a$; max possible value $= -\Delta_o$ when $a = l$

(*Continued*)

TABLE 8.1 Shear, Moment, Slope, and Deflection Formulas for Elastic Straight Beams (*Continued*)

End Restraints, Reference No.	Boundary Values	Selected Maximum Values of Moments and Deformations
5f. Left end guided, right end simply supported 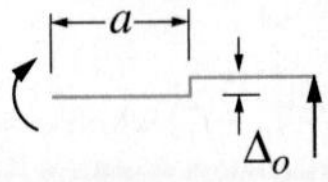	$R_A = 0 \quad M_A = 0$ $\theta_A = 0 \quad y_A = -\Delta_O$ $R_B = 0 \quad M_B = 0$ $\theta_B = 0 \quad y_B = 0$	Max $y = y_A$ when $x < a$

6. Uniform temperature variation from top to bottom from a to l

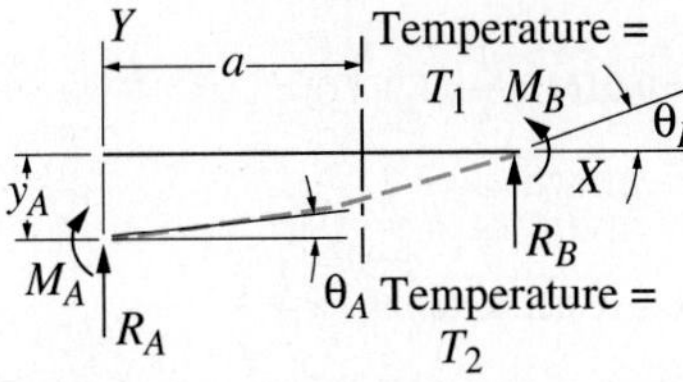

Transverse shear $= V = R_A$

Bending moment $= M = M_A + R_A x$

$$\text{Slope} = \theta = \theta_A + \frac{M_A x}{EI} + \frac{R_A x^2}{2EI} + \frac{\gamma}{t}(T_2 - T_1)\langle x - a\rangle$$

$$\text{Deflection} = y = y_A + \theta_A x + \frac{M_A x^2}{2EI} + \frac{R_A x^3}{6EI} + \frac{\gamma}{2t}(T_2 - T_1)\langle x - a\rangle^2$$

γ = temperature coefficient of expansion (unit strain/°) t = depth of beam

End Restraints, Reference No.	Boundary Values	Selected Maximum Values of Moments and Deformations
6a. Left end free, right end fixed 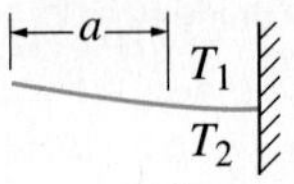	$R_A = 0 \quad M_A = 0$ $\theta_A = \frac{-\gamma}{t}(T_2 - T_1)(l - a)$ $y_A = \frac{\gamma}{2t}(T_2 - T_1)(l^2 - a^2)$ $R_B = 0 \quad M_B = 0$ $\theta_B = 0 \quad y_B = 0$	$M = 0$ everywhere Max $\theta = \theta_A$; max possible value $= \frac{-\gamma l}{t}(T_2 - T_1)$ when $a = 0$ Max $y = y_A$; max possible value $= \frac{-\gamma l^2}{2t}(T_2 - T_1)$ when $a = 0$
6b. Left end guided, right end fixed 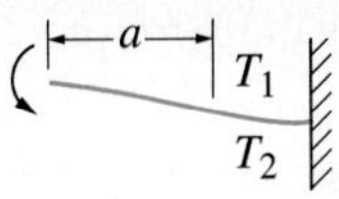	$R_A = 0 \quad \theta_A = 0$ $M_A = \frac{-EI\gamma}{lt}(T_2 - T_1)(l - a)$ $y_A = \frac{a\gamma}{2t}(T_2 - T_1)(l - a)$ $R_B = 0 \quad M_B = M_A$ $\theta_B = 0 \quad y_B = 0$	$M = M_A$ everywhere; max possible value $= \frac{-EI\gamma}{t}(T_2 - T_1)$ where $a = 0$ Max $\theta = \frac{-a\gamma}{lt}(T_2 - T_1)(l - a)$ *at* $x = a$; max possible value $=$ $\frac{-\gamma l}{4t}(T_2 - T_1)$ when $a = \frac{l}{2}$ Max $y = y_A$; max possible value $= \frac{\gamma l^2}{8t}(T_2 - T_1)$ when $a = \frac{l}{2}$
6c. Left end simply supported, right end fixed 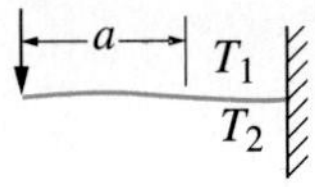	$M_A = 0 \quad y_A = 0$ $R_A = \frac{-3EI\gamma}{2tl^3}(T_2 - T_1)(l^2 - a^2)$ $\theta_A = \frac{\gamma}{4tl}(T_2 - T_1)(l - a)(3a - l)$ $R_B = -R_A \quad M_B = R_A l$ $\theta_B = 0 \quad y_B = 0$	Max $M = M_B$; max possible value $= \frac{-3EI\gamma}{2t}(T_2 - T_1)$ when $a = 0$ Max $+ y = \frac{\gamma(T_2 - T_1)(l - a)}{6t\sqrt{3(l + a)}}(3a - l)^{3/2}$ at $x = l\left(\frac{3a - l}{3l + 3a}\right)^{1/2}$; max possible value $= 0.0257\frac{\gamma l^2}{t}(T_2 - T_1)$ at $x = 0.474l$ when $a = 0.721l$ (*Note:* There is no positive deflection if $a < l/3$.) Max $- y$ occurs at $x = \frac{2l^3}{3(l^2 - a^2)}\left[1 - \frac{1}{2}\sqrt{1 - 6\left(\frac{a}{l}\right)^2 + 9\left(\frac{a}{l}\right)^4}\right]$; max possible value $= \frac{-\gamma l^2}{27t}(T_2 - T_1)$ at $x = \frac{l}{3}$ when $a = 0$

(*Continued*)

TABLE 8.1 Shear, Moment, Slope, and Deflection Formulas for Elastic Straight Beams (*Continued*)

End Restraints, Reference No.	Boundary Values	Selected Maximum Values of Moments and Deformations
6d. Left end fixed, right end fixed 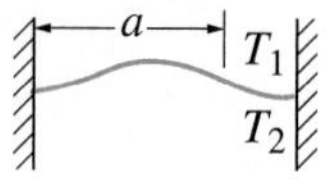 	$R_A = \frac{-6EIa\gamma}{tl^3}(T_2 - T_1)(l-a)$ $M_A = \frac{EI\gamma}{tl^2}(T_2 - T_1)(l-a)(3a-l)$ $\theta_A = 0 \quad y_A = 0$ $R_B = -R_A$ $M_B = \frac{-EI\gamma}{tl^2}(T_2 - T_1)(l-a)(3a+l)$ $\theta_B = 0 \quad y_B = 0$	$\text{Max} + M = M_A$; max possible value $= \frac{EI\gamma}{3t}(T_2 - T_1)$ when $a = \frac{2}{3}l$ (*Note:* There is no positive moment if $a < \frac{l}{3}$.) $\text{Max} - M = M_B$; max possible value $= \frac{-4EI\gamma}{3t}(T_2 - T_1)$ when $a = \frac{l}{3}$ $\text{Max} + y = \frac{2M_A^2}{3R_A^2 EI}$ at $x = \frac{l}{3a}(3l - a)$; max possible value $= 0.01617\frac{\gamma l^2}{t}(T_2 - T_1)$ at $x = 0.565l$ when $a = 0.767l$ (*Note:* There is no positive deflection if $a < \frac{l}{3}$.)
6e. Left end simply supported, right end simply supported 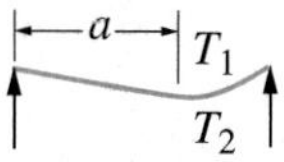 	$R_A = 0 \quad M_A = 0 \quad y_A = 0$ $\theta_A = \frac{-\gamma}{2tl}(T_2 - T_1)(l-a)^2$ $R_B = 0 \quad M_B = 0 \quad y_B = 0$ $\theta_B = \frac{\gamma}{2tl}(T_2 - T_1)(l^2 - a^2)$	$M = 0$ everywhere $\text{Max} + \theta = \theta_B$; max possible value $= \frac{\gamma l}{2t}(T_2 - T_1)$ when $a = 0$ $\text{Max} - \theta = \theta_A$; max possible value $= \frac{-\gamma l}{2t}(T_2 - T_1)$ when $a = 0$ $\text{Max } y = \frac{-\gamma}{8tl^2}(T_2 - T_1)(l^2 - a^2)^2$; max possible value $= \frac{-\gamma l^2}{8t}(T_2 - T_1)$ at $x = \frac{l}{2}$ when $a = 0$
6f. Left end guided, right end simply supported 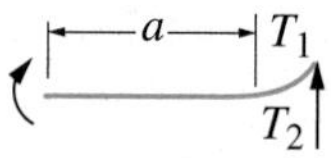 	$R_A = 0 \quad M_A = 0 \quad \theta_A = 0$ $y_A = \frac{-\gamma}{2t}(T_2 - T_1)(l-a)^2$ $R_B = 0 \quad M_B = 0 \quad y_B = 0$ $\theta_B = \frac{\gamma}{l}(T_2 - T_1)(l-a)$	$M = 0$ everywhere $\text{Max } \theta = \theta_B$; max possible value $= \frac{\gamma l}{t}(T_2 - T_1)$ when $a = 0$ $\text{Max } y = y_A$; max possible value $= \frac{-\gamma l^2}{2t}(T_2 - T_1)$ when $a = 0$

TABLE 8.2 Reaction and Deflection Formulas for In-Plane Loading of Elastic Frames

Notation: W = load (force); w = unit load (force per unit length); M_o = applied couple (force-length); θ_o = externally created concentrated angular displacement (radians); Δ_o = externally created concentrated lateral displacement (length); $T - T_o$ = uniform temperature rise (degrees); T_1 and T_2 = temperature on outside and inside, respectively (degrees). H_A and H_B are the horizontal end reactions at the left and right, respectively, and are positive to the left; V_A and V_B are the vertical end reactions at the left and right, respectively, and are positive upwards; M_A and M_B are the reaction moments at the left and right, respectively, and are positive clockwise. I_1, I_2, and I_3 are the respective area moments of inertia for bending in the plane of the frame for the three members (length to the fourth); E_1, E_2, and E_3 are the respective moduli of elasticity (force per unit area); γ_1, γ_2, and γ_3 are the respective temperature coefficients of expansions (unit strain per degree).

General reaction and deformation expressions for cases 1–4, right end pinned in all four cases

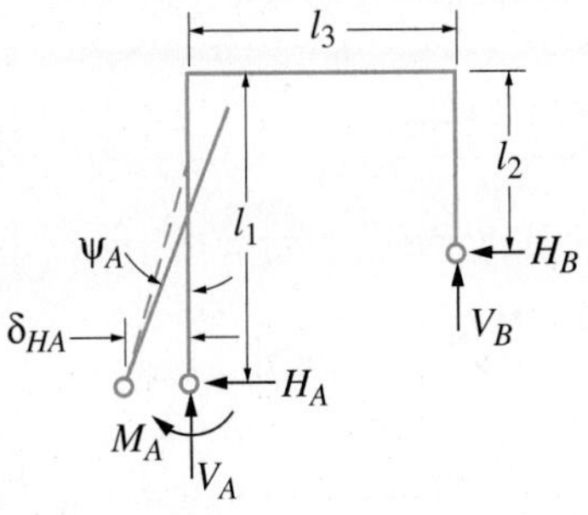

Deformation equations:

Horizontal deflection at $A = \delta_{HA} = A_{HH}H_A + A_{HM}M_A - LP_H$

Angular rotation at $A = \psi_A = A_{MH}H_A + A_{MM}M_A - LP_M$

where $$A_{HH} = \frac{l_1^3}{3E_1I_1} + \frac{l_2^3}{3E_2I_2} + \frac{l_3}{3E_3I_3}(l_1^2 + l_1l_2 + l_2^2)$$

$$A_{HM} = A_{MH} = \frac{l_1^2}{2E_1I_1} + \frac{l_3}{6E_3I_3}(2l_1 + l_2)$$

$$A_{MM} = \frac{l_1}{E_1I_1} + \frac{l_3}{3E_3I_3}$$

and where LP_H and LP_M are loading terms given below for several types of load

(*Note:* V_A, V_B, and H_B are to be evaluated from equilibrium equations after calculating H_A and M_A.)

1. Left end pinned, right end pinned

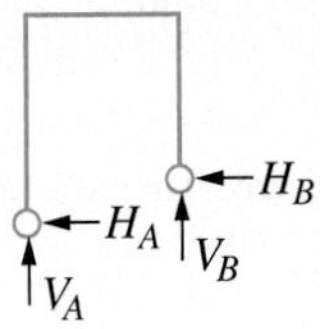

Since $\delta_{HA} = 0$ and $M_A = 0$,

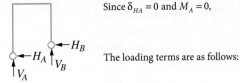

$$H_A = \frac{LP_H}{A_{HH}} \text{ and } \psi_A = A_{MH}H_A - LP_M$$

The loading terms are as follows:

Reference No., Loading		Loading Terms
1a. Concentrated load on the horizontal member	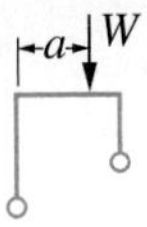	$LP_H = \frac{Wa}{6E_3I_3}\left[3l_1a - 2l_1l_3 - l_2l_3 - \frac{a^2}{l_3}(l_1 - l_2)\right]$ $LP_M = \frac{Wa}{6E_3I_3}\left(3a - 2l_3 - \frac{a^2}{l_3}\right)$
1b. Distributed load on the horizontal member	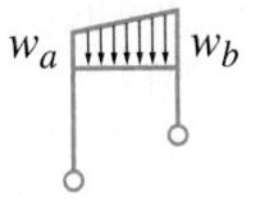	$LP_H = \frac{-w_al_3^3}{24E_3I_3}(l_1 + l_2)\frac{(w_b - w_a)l_3^3}{360E_3I_3}(7l_1 + 8l_2)$ $LP_M = \frac{-7w_al_3^3}{72E_3I_3} - \frac{11(w_b - w_a)l_3^3}{180E_3I_3}$
1c. Concentrated moment on the horizontal member	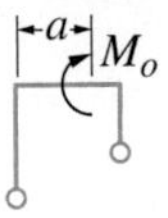	$LP_H = \frac{M_o}{6E_3I_3}\left[6l_1a - 2l_1l_3 - l_2l_3 - \frac{3a^2}{l_3}(l_1 - l_2)\right]$ $LP_M = \frac{M_o}{6E_3I_3}\left(4a - 2l_3 - \frac{3a^2}{l_3}\right)$
1d. Concentrated angular displacement on the horizontal member	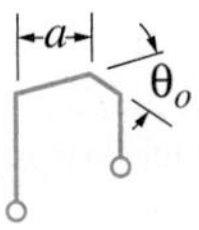	$LP_H = \theta_o\left[l_1 - \frac{a}{l_3}(l_1 - l_2)\right]$ $LP_M = \theta_o\left(\frac{2}{3} - \frac{a}{l_3}\right)$

(*Continued*)

TABLE 8.2 Reaction and Deflection Formulas for In-Plane Loading of Elastic Frames (*Continued*)

Reference No., Loading	Loading Terms
1e. Concentrated lateral displacement on the horizontal member 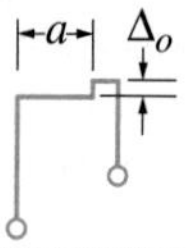	$LP_H = \frac{-\Delta_o(l_1 - l_2)}{l_3}$ (*Note:* Δ_o could also be an increase in the length l_1 or a decrease in the length l_2.) $LP_M = \frac{-\Delta_o}{l_3}$
1f. Concentrated load on the left vertical member 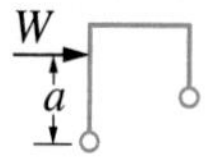	$LP_H = W\left(A_{HH} - aA_{HM} + \frac{a^3}{6E_1I_1}\right)$ $LP_M = W\left(A_{MH} - aA_{MM} + \frac{a^2}{2E_1I_1}\right)$
1g. Distributed load on the left vertical member 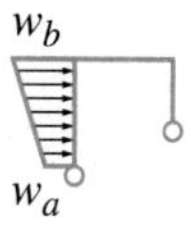	$LP_H = w_a\left(A_{HH}l_1 - A_{HM}\frac{l_1^2}{2} + \frac{l_1^4}{24E_1I_1}\right) + (w_b - w_a)\left(A_{HH}\frac{l_1}{2} - A_{HM}\frac{l_1^2}{3} + \frac{l_1^4}{30E_1I_1}\right)$ $LP_M = w_a\left[\frac{l_1^3}{6E_1I_1} + \frac{l_1l_3}{6E_3I_3}(l_1 + l_2)\right] + (w_b - w_a)\left[\frac{l_1^3}{24E_1I_1} + \frac{l_1l_3}{36E_3I_3}(2l_1 + 3l_2)\right]$
1h. Concentrated moment on the left vertical member 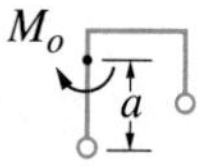	$LP_H = M_o\left(\frac{a^2}{2E_1I_1} - A_{HM}\right)$ $LP_M = M_o\left(\frac{a}{E_1I_1} - A_{MM}\right)$
1i. Concentrated angular displacement on the left vertical member 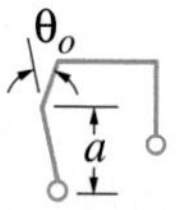	$LP_H = \theta_o(a)$ $LP_M = \theta_o(1)$
1j. Concentrated lateral displacement on the left vertical member 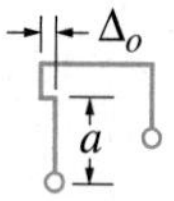	$LP_H = \Delta_o(1)$ (*Note:* Δ_o could also be a decrease in the length l_3.) $LP_M = 0$
1k. Concentrated load on the right vertical member 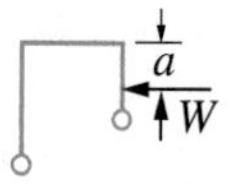	$LP_H = W\left[\frac{1}{6E_2I_2}(3l_2a^2 - 2l_2^3 - a^3) - \frac{l_3}{6E_3I_3}(l_2 - a)(l_1 + 2l_2)\right]$ $LP_M = W\left[\frac{-l_3}{6E_3I_3}(l_2 - a)\right]$
1l. Distributed load on the right vertical member 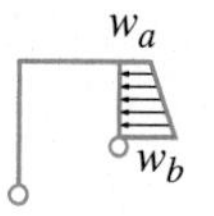	$LP_H = w_a\left[\frac{-5l_2^4}{24E_2I_2} - \frac{l_2^2l_3}{12E_3I_3}(l_1 + 2l_2)\right] + (w_b - w_a)\left[\frac{-3l_2^4}{40E_2I_2} - \frac{l_2^2l_3}{36E_3I_3}(l_1 + 2l_2)\right]$ $LP_M = w_a\left(\frac{-l_2^2l_3}{12E_3I_3}\right) + (w_b - w_a)\left(\frac{-l_2^2l_3}{36E_3I_3}\right)$
1m. Concentrated moment on the right vertical member 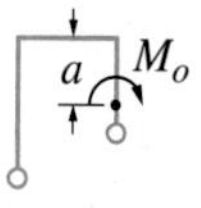	$LP_H = M_o\left[\frac{a}{2E_2I_2}(2l_2 - a) + \frac{l_3}{6E_3I_3}(l_1 + 2l_2)\right]$ $LP_M = M_o\frac{l_3}{6E_3I_3}$
1n. Concentrated angular displacement on the right vertical member 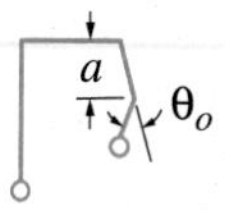	$LP_H = \theta_o(l_2 - a)$ $LP_M = 0$
1p. Concentrated lateral displacement on the right vertical member 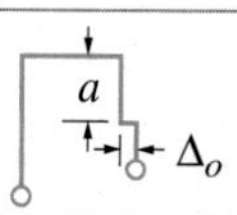	$LP_H = \Delta_o(-1)$ (*Note:* Δ_o could also be an increase in the length l_3.) $LP_M = 0$

(*Continued*)

TABLE 8.2 Reaction and Deflection Formulas for In-Plane Loading of Elastic Frames (*Continued*)

Reference No., Loading	Loading Terms
1q. Uniform temperature rise: T = uniform temperature T_o = unloaded temperature 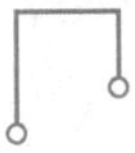	$LP_H = (T-T_o)\left[\gamma_3 l_3 - \frac{(l_1-l_2)}{l_3}(\gamma_1 l_1 - \gamma_2 l_2)\right]$ γ = temperature coefficient of expansion (unit strain/degree) $LP_M = (T-T_o)\left[\frac{-1}{l_3}(\gamma_1 l_1 - \gamma_2 l_2)\right]$
1r. Uniform temperature differential from outside to inside. Average temperature is T_o 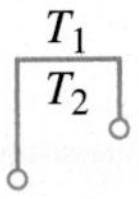	$LP_H = \frac{(T_1-T_2)}{2}\left[\frac{\gamma_1 l_1^2}{t_1}+\frac{\gamma_2 l_2^2}{t_2}+\frac{\gamma_3 l_3(l_1+l_2)}{t_3}\right]$ t_1, t_2, and t_3 are beam thicknesses from inside to outside $LP_M = \frac{(T_1-T_2)}{2}\left(\frac{4\gamma_1 l_1}{t_1}+\frac{\gamma_3 l_3}{3t_3}\right)$
2. Left end guided horizontally, right end pinned 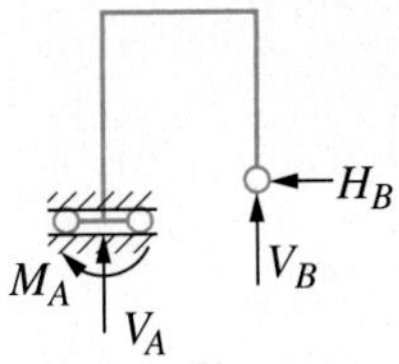	Since $\psi_A = 0$ and $H_A = 0$ $M_A = \frac{LP_M}{A_{MM}}$ and $\delta_{HA} = A_{HM}M_A - LP_H$ Use the loading terms for cases 1a to 1r.
3. Left end roller supported along the horizontal, right end pinned 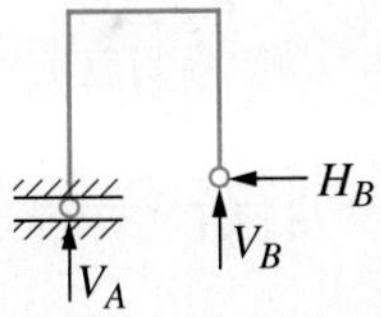	Since H_A and M_A are both zero, this is a statically determinate case. $\delta_{HA} = -LP_H$ and $\psi_A = -LP_M$ Use the loading terms for cases 1a to 1r.
4. Left end fixed, right end pinned 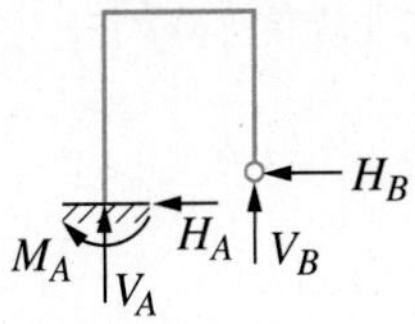	Since $\delta_{HA} = 0$ and $\psi_A = 0$, $H_A = \frac{A_{MM}LP_H - A_{HM}LP_M}{A_{HH}A_{MM} - (A_{HM})^2}$ and $M_A = \frac{A_{HH}LP_M - A_{HM}LP_H}{A_{HH}A_{MM} - (A_{HM})^2}$ Use the loading terms for cases 1a–1r.

General reaction and deformation expressions for cases 5–12, right end fixed in all eight cases:

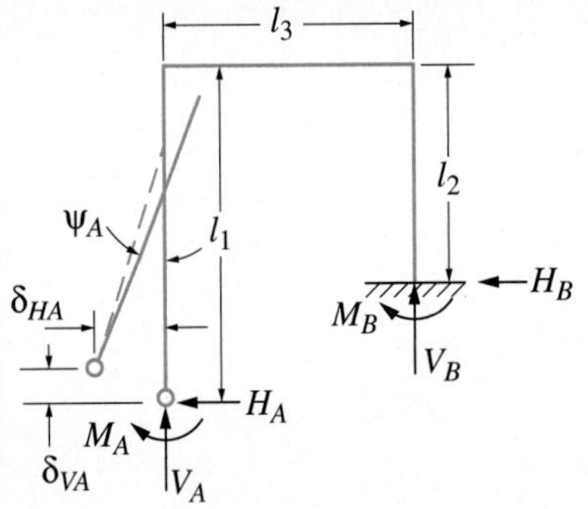

Deformation equations:

Horizontal deflection at $A = \delta_{HA} = C_{HH}H_A + C_{HV}V_A + C_{HM}M_A - LF_H$

Vertical deflection at $A = \delta_{VA} = C_{VH}H_A + C_{VV}V_A + C_{VM}M_A - LF_V$

Angular rotation at $A = \psi_A = C_{MH}H_A + C_{MV}V_A + C_{MM}M_A - LF_M$

where $C_{HH} = \frac{l_1^3}{3E_1I_1} + \frac{l_1^3 - (l_1-l_2)^3}{3E_2I_2} + \frac{l_1^2 l_3}{E_3I_3}$

$$C_{HV} = C_{VH} = \frac{l_2 l_3}{2E_2I_2}(2l_1 - l_2) + \frac{l_1 l_3^2}{2E_3I_3}$$

$$C_{HM} = C_{MH} = \frac{l_1^2}{2E_1I_1} + \frac{l_2}{2E_2I_2}(2l_1 - l_2) + \frac{l_1 l_3}{E_3I_3}$$

$$C_{VV} = \frac{l_2 l_3^2}{E_2I_2} + \frac{l_3^3}{3E_3I_3}$$

$$C_{VM} = C_{MV} = \frac{l_2 l_3}{E_2I_2} + \frac{l_3^2}{2E_3I_3}$$

$$C_{MM} = \frac{l_1}{E_1I_1} + \frac{l_2}{E_2I_2} + \frac{l_3}{E_3I_3}$$

and where LF_H, LF_V, and LF_M are loading terms given below for several types of load

(*Note:* If desired, H_B, V_B, and M_B are to be evaluated from equilibrium equations after calculating H_A, V_A, and M_A.)

(*Continued*)

TABLE 8.2 Reaction and Deflection Formulas for In-Plane Loading of Elastic Frames (*Continued*)

5. Left end fixed, right end fixed	Since $\delta_{HA}=0$, $\delta_{VA}=0$, and $\psi_A=0$, these three equations are solved simultaneously for H_A, V_A, and M_A:
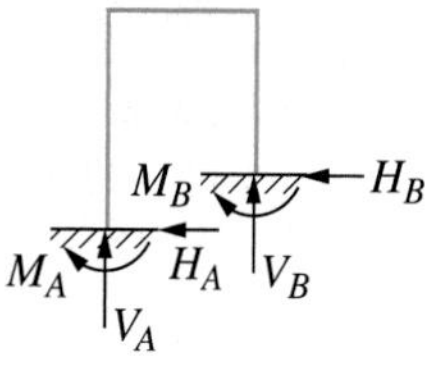	$C_{HH}H_A + C_{HV}V_A + C_{HM}M_A = LF_H$ $C_{VH}H_A + C_{VV}V_A + C_{VM}M_A = LF_V$ $C_{MH}H_A + C_{MV}V_A + C_{MM}M_A = LF_M$ The loading terms are given below.

Reference No., Loading	Loading Terms
5a. Concentrated load on the horizontal member 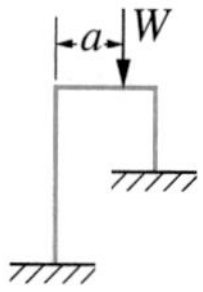	$LF_H = W\left[\frac{l_2}{2E_2I_2}(2l_1 - l_2)(l_3 - a) + \frac{l_1}{2E_3I_3}(l_3 - a)^2\right]$ $LF_V = W\left(C_{VV} - aC_{VM} + \frac{a^3}{6E_3I_3}\right)$ $LF_M = W\left[\frac{l_2}{E_2I_2}(l_3 - a) + \frac{1}{2E_3I_3}(l_3 - a)^2\right]$
5b. Distributed load on the horizontal member 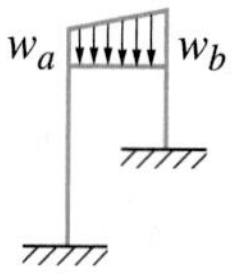	$LF_H = w_a\left[\frac{l_2 l_3^2}{4E_2I_2}(2l_1 - l_2) + \frac{l_1 l_3^3}{6E_3I_3}\right] + (w_b - w_a)\left[\frac{l_2 l_3^2}{12E_2I_2}(2l_1 - l_2) + \frac{l_1 l_3^3}{24E_3I_3}\right]$ $LF_V = w_a\left(\frac{l_2 l_3^3}{2E_2I_2} + \frac{l_3^4}{8E_3I_3}\right) + (w_b - w_a)\left(\frac{l_2 l_3^3}{6E_2I_2} + \frac{l_3^4}{30E_3I_3}\right)$ $LF_M = w_a\left(\frac{l_2 l_3^2}{2E_2I_2} + \frac{l_3^3}{6E_3I_3}\right) + (w_b - w_a)\left(\frac{l_2 l_3^2}{6E_2I_2} + \frac{l_3^3}{24E_3I_3}\right)$
5c. Concentrated moment on the horizontal member 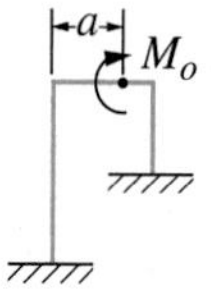	$LF_H = M_o\left[\frac{-l_2}{2E_2I_2}(2l_1 - l_2) - \frac{l_1}{E_3I_3}(l_3 - a)\right]$ $LF_V = M_o\left(-C_{VM} + \frac{a^3}{2E_3I_3}\right)$ $LF_M = M_o\left[\frac{-l_2}{E_2I_2} - \frac{1}{E_3I_3}(l_3 - a)\right]$
5d. Concentrated angular displacement on the horizontal member 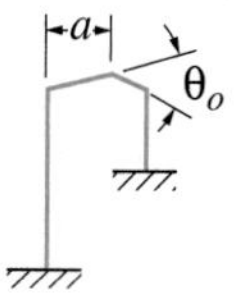	$LF_H = \theta_o(l_1)$ $LF_V = \theta_o(a)$ $LF_M = \theta_o(1)$
5e. Concentrated lateral displacement on the horizontal member 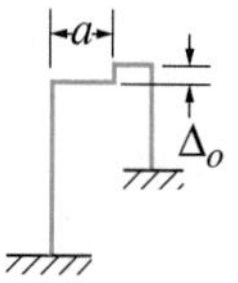	$LF_H = 0$ $LF_V = \Delta_o(1)$ $LF_M = 0$
5f. Concentrated load on the left vertical member 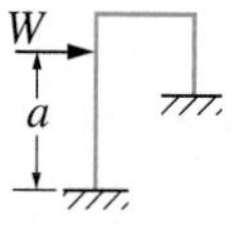	$LF_H = W\left(C_{HH} - aC_{HM} + \frac{a^3}{6E_1I_1}\right)$ $LF_V = W(C_{VH} - aC_{VM})$ $LF_M = W\left(C_{MH} - aC_{MM} + \frac{a^2}{2E_1I_1}\right)$

(*Continued*)

TABLE 8.2 Reaction and Deflection Formulas for In-Plane Loading of Elastic Frames (*Continued*)

Reference No., Loading		Loading Terms
5g. Distributed load on the left vertical member	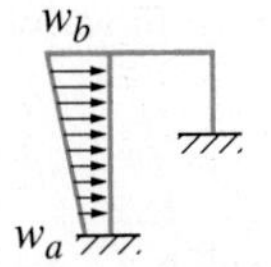	$LF_H = w_a\left(C_{HH}l_1 - C_{HM}\frac{l_1^2}{2} + \frac{l_1^4}{24E_1I_1}\right) + (w_b - w_a)\left(C_{HH}\frac{l_1}{2} - C_{HM}\frac{l_1^2}{3} + \frac{l_1^4}{30E_1I_1}\right)$ $LF_V = w_a\left(C_{VH}l_1 - C_{VM}\frac{l_1^2}{2}\right) + (w_b - w_a)\left(C_{VH}\frac{l_1}{2} - C_{VM}\frac{l_1^2}{3}\right)$ $LF_M = w_a\left(C_{MH}l_1 - C_{MM}\frac{l_1^2}{2} + \frac{l_1^3}{6E_1I_1}\right) + (w_b - w_a)\left(C_{MH}\frac{l_1}{2} - C_{MM}\frac{l_1^2}{3} + \frac{l_1^3}{8E_1I_1}\right)$
5h. Concentrated moment on the left vertical member	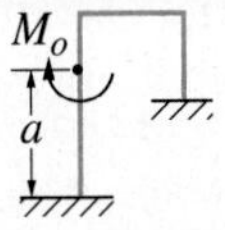	$LF_H = M_o\left(-C_{HM} + \frac{a^2}{2E_1I_1}\right)$ $LF_V = M_o(-C_{VM})$ $LF_M = M_o\left(-C_{MM} + \frac{a}{E_1I_1}\right)$
5i. Concentrated angular displacement on the left vertical member	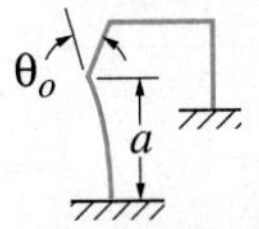	$LF_H = \theta_o(a)$ $LF_V = 0$ $LF_M = \theta_o(1)$
5j. Concentrated lateral displacement on the left vertical member	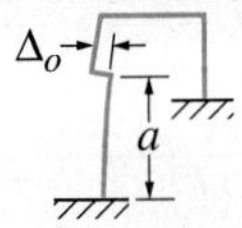	$LF_H = \Delta_o(1)$ $LF_V = 0$ $LF_M = 0$
5k. Concentrated load on the right vertical member	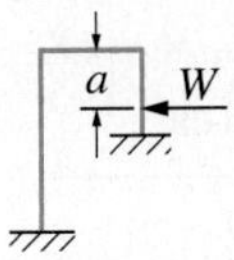	$LF_H = \frac{W}{6E_2I_2}[3l_1(l_2 - a)^2 - 2l_2^3 - a^3 + 3al_2^2]$ $LF_V = \frac{W}{2E_2I_2}[l_3(l_2 - a)^2]$ $LF_M = \frac{W}{2E_2I_2}(l_2 - a)^2$
5l. Distributed load on the right vertical member	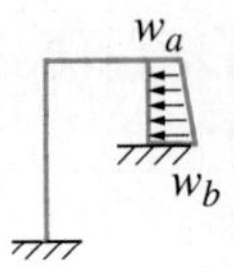	$LF_H = w_a\left[\frac{l_2^3}{24E_2I_2}(4l_1 - 3l_2)\right] + (w_b - w_a)\left[\frac{l_2^3}{120E_2I_2}(5l_1 - 4l_2)\right]$ $LF_V = w_a\frac{l_2^3 l_3}{6E_2I_2} + (w_b - w_a)\frac{l_2^3 l_3}{24E_2I_2}$ $LF_M = w_a\frac{l_2^3}{6E_2I_2} + (w_b - w_a)\frac{l_2^3}{24E_2I_2}$
5m. Concentrated moment on the right vertical member	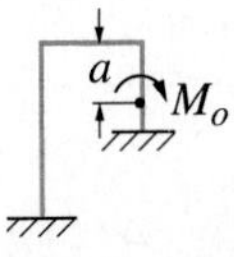	$LF_H = \frac{M_o}{2E_2I_2}[-2l_1(l_2 - a) - a^2 + l_2^2]$ $LF_V = \frac{M_o}{E_2I_2}[-l_3(l_2 - a)]$ $LF_M = \frac{M_o}{E_2I_2}[-(l_2 - a)]$
5n. Concentrated angular displacement on the right vertical member	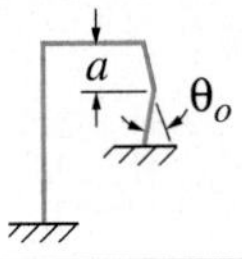	$LF_H = \theta_o(l_1 - a)$ $LF_V = \theta_o(l_3)$ $LF_M = \theta_o(1)$
5o. Concentrated lateral displacement on the right vertical member	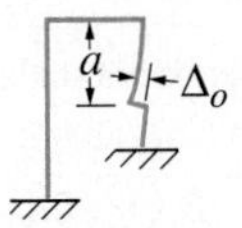	$LF_H = \Delta_o(-1)$ (*Note*: Δ_o could also be an increase in the length l_3.) $LF_V = 0$ $LF_M = 0$

(*Continued*)

TABLE 8.2 Reaction and Deflection Formulas for In-Plane Loading of Elastic Frames (*Continued*)

Reference No., Loading	Loading Terms
5p. Uniform temperature rise: T = uniform temperature T_o = unloaded temperature 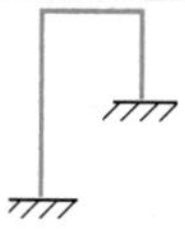	$LF_H=(T-T_o)(-\gamma_3 l_3)$ γ = temperature coefficient of expansion (inches/inch/degree) $LF_V=(T-T_o)(\gamma_1 l_1-\gamma_2 l_2)$ $LF_M=0$
5q. Uniform temperature differential from outside to inside; average temperature is T_o 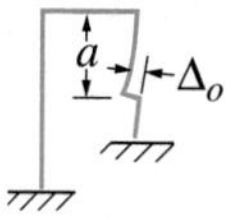	$LF_H=(T_1-T_2)\left[\frac{l_1^2\gamma_1}{2t_1}+\frac{l_2\gamma_2}{2t_2}(2l_1-l_2)+\frac{l_1 l_3\gamma_3}{t_3}\right]$ $LF_V=(T_1-T_2)\left(\frac{l_2 l_3\gamma_2}{t_2}+\frac{l_3^2\gamma_3}{2t_3}\right)$ t_1, t_2, and t_3 are beam thicknesses from inside to outside $LF_M=(T_1-T_2)\left(\frac{l_1\gamma_1}{t_1}+\frac{l_2\gamma_2}{t_2}+\frac{l_3\gamma_3}{t_3}\right)$
6. Left end pinned, right end fixed 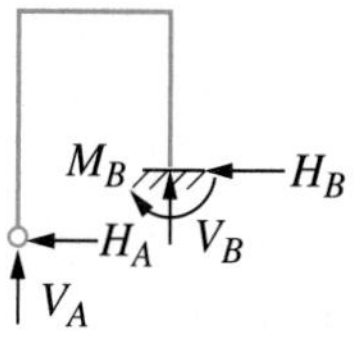	Since $\delta_{HA}=0$, $\delta_{VA}=0$ and $M_A=0$, $H_A=\frac{LF_H C_{VV}-LF_V C_{HV}}{C_{HH}C_{VV}-(C_{HV})^2}$ $V_A=\frac{LF_V C_{HH}-LF_H C_{HV}}{C_{HH}C_{VV}-(C_{HV})^2}$ $\psi_A=C_{MH}H_A+C_{MV}V_A-LF_M$ Use the loading terms for cases 5a–5r.
7. Left end guided horizontally, right end fixed 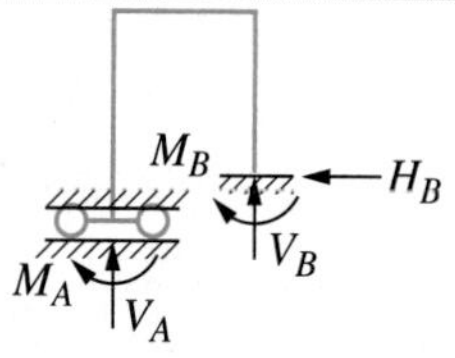	Since $\delta_{VA}=0$, $\psi_A=0$ and $H_A=0$, $V_A=\frac{LF_V C_{MM}-LF_M C_{VM}}{C_{VV}C_{MM}-(C_{VM})^2}$ $M_A=\frac{LF_M C_{VV}-LF_V C_{VM}}{C_{VV}C_{MM}-(C_{VM})^2}$ $\delta_{HA}=C_{HV}V_A+C_{HM}M_A-LF_H$ Use the loading terms for cases 5a–5r.
8. Left end guided vertically, right end fixed 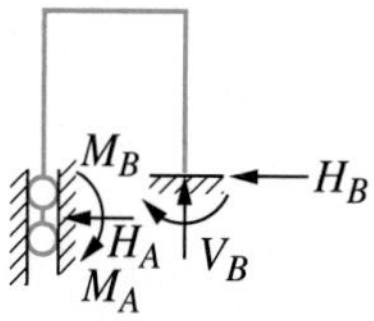	Since $\delta_{HA}=0$, $\psi_A=0$, and $V_A=0$, $H_A=\frac{LF_H C_{MM}-LF_M C_{HM}}{C_{HH}C_{MM}-(C_{HM})^2}$ $M_A=\frac{LF_M C_{HH}-LF_H C_{HM}}{C_{HH}C_{MM}-(C_{HM})^2}$ $\delta_{VA}=C_{VH}H_A+C_{VM}M_A-LF_V$ Use the loading terms for cases 5a–5r.
9. Left end roller supported along the horizontal, right end fixed 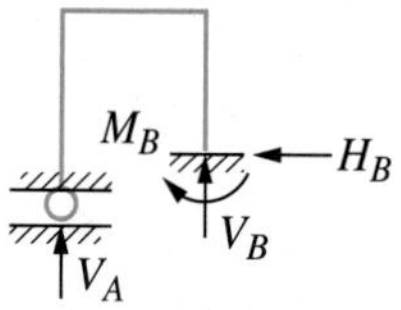	Since $\delta_{VA}=0$, $H_A=0$, and $M_A=0$, $V_A=\frac{LF_V}{C_{VV}}$ $\delta_{HA}=C_{HV}V_A-LF_H$ and $\psi_A=C_{MV}V_A-LF_M$ Use the loading terms for cases 5a–5r.
10. Left end roller supported along the vertical, right end fixed 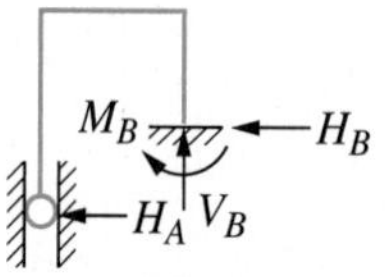	Since $\delta_{HA}=0$, $V_A=0$, $M_A=0$, $H_A=\frac{LF_H}{C_{HH}}$ $\delta_{VA}=C_{VH}H_A-LF_V$ and $\psi_A=C_{MH}H_A-LF_M$ Use the loading terms for cases 5a–5r.
11. Left end guided by moment only (zero slope at the left end), right end fixed 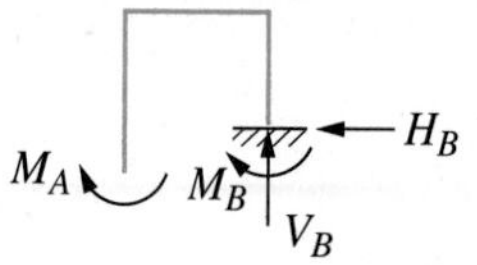	Since $\psi_A=0$, $H_A=0$, and $V_A=0$, $M_A=\frac{LF_M}{C_{MM}}$ $\delta_{HA}=C_{HM}M_A-LF_H$ and $\delta_{VA}=C_{VM}M_A-LF_V$ Use the loading terms for cases 5a–5r.
12. Left end free, right end fixed 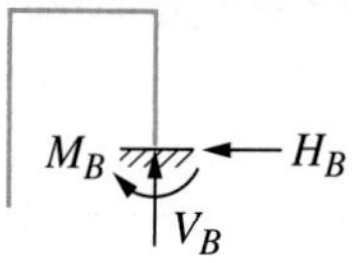	Since $H_A=0$, $V_A=0$ and $M_A=0$, this is a statically determinate case. The deflections are given by $\delta_{HA}=-LF_H$ $\delta_{VA}=-LF_V$ and $\psi_A=-LF_M$ Use the loading terms for cases 5a–5r.

TABLE 8.3 Numerical Values for Functions Used in Table 8.5

βx	F_1	F_2	F_3	F_4
0.00	1.00000	0.00000	0.00000	0.00000
0.10	0.99998	0.20000	0.01000	0.00067
0.20	0.99973	0.39998	0.04000	0.00533
0.30	0.99865	0.59984	0.08999	0.01800
0.40	0.99573	0.79932	0.15995	0.04266
0.50	0.98958	0.99792	0.24983	0.08331
0.60	0.97841	1.19482	0.35948	0.14391
0.70	0.96001	1.38880	0.48869	0.22841
0.80	0.93180	1.57817	0.63709	0.34067
0.90	0.89082	1.76067	0.80410	0.48448
1.00	0.83373	1.93342	0.98890	0.66349
1.10	0.75683	2.09284	1.19034	0.88115
1.20	0.65611	2.23457	1.40688	1.14064
1.30	0.52722	2.35341	1.63649	1.44478
1.40	0.36558	2.44327	1.87659	1.79593
1.50	0.16640	2.49714	2.12395	2.19590
1.60	−0.07526	2.50700	2.37456	2.64573
1.70	−0.36441	2.46387	2.62358	3.14562
1.80	−0.70602	2.35774	2.86523	3.69467
1.90	−1.10492	2.17764	3.09266	4.29076
2.00	−1.56563	1.91165	3.29789	4.93026
2.10	−2.09224	1.54699	3.47170	5.60783
2.20	−2.68822	1.07013	3.60355	6.31615
2.30	−3.35618	−0.46690	3.68152	7.04566
2.40	−4.09766	−0.27725	3.69224	7.78428
2.50	−4.91284	−1.17708	3.62088	8.51709
2.60	−5.80028	−2.24721	3.45114	9.22607
2.70	−6.75655	−3.50179	3.16529	9.88981
2.80	−7.77591	−4.95404	2.74420	10.48317
2.90	−8.84988	−6.61580	2.16749	10.97711
3.00	−9.96691	−8.49687	1.41372	11.33837
3.20	−12.26569	−12.94222	−0.71484	11.50778
3.40	−14.50075	−18.30128	−3.82427	10.63569

(Continued)

TABLE 8.3 Numerical Values for Functions Used in Table 8.5 (*Continued*)

βx	F_1	F_2	F_3	F_4
3.60	−16.42214	−24.50142	−8.09169	8.29386
3.80	−17.68744	−31.35198	−13.66854	3.98752
4.00	−17.84985	−38.50482	−20.65308	−2.82906
4.20	−16.35052	−45.41080	−29.05456	−12.72446
4.40	−12.51815	−51.27463	−38.74857	−26.24587
4.60	−5.57927	−55.01147	−49.42334	−43.85518
4.80	5.31638	−55.21063	−60.51809	−65.84195
5.00	21.05056	−50.11308	−71.15526	−92.21037
5.20	42.46583	−37.61210	−80.07047	−122.53858
5.40	70.26397	−15.28815	−85.54576	−155.81036
5.60	104.86818	19.50856	−85.35442	−190.22206
5.80	146.24469	69.51236	−76.72824	−222.97166
6.00	193.68136	137.31651	−56.36178	−250.04146

TABLE 8.4 Numerical Values for Denominators Used in Table 8.5

βl	C_{11}	C_{12}	C_{13}	C_{14}
0.00	0.00000	0.00000	0.00000	0.00000
0.10	0.00007	0.20000	0.00133	0.02000
0.20	0.00107	0.40009	0.01067	0.08001
0.30	0.00540	0.60065	0.03601	0.18006
0.40	0.01707	0.80273	0.08538	0.32036
0.50	0.04169	1.00834	0.16687	0.50139
0.60	0.08651	1.22075	0.28871	0.72415
0.70	0.16043	1.44488	0.45943	0.99047
0.80	0.27413	1.68757	0.68800	1.30333
0.90	0.44014	1.95801	0.98416	1.66734
1.00	0.67302	2.26808	1.35878	2.08917
1.10	0.98970	2.63280	1.82430	2.57820
1.20	1.40978	3.07085	2.39538	3.14717
1.30	1.95606	3.60512	3.08962	3.81295
1.40	2.65525	4.26345	3.92847	4.59748
1.50	3.53884	5.07950	4.93838	5.52883
1.60	4.64418	6.09376	6.15213	6.64247
1.70	6.01597	7.35491	7.61045	7.98277

(Continued)

TABLE 8.4 Numerical Values for Denominators Used in Table 8.5 (*Continued*)

βl	C_{11}	C_{12}	C_{13}	C_{14}
1.80	7.70801	8.92147	9.36399	9.60477
1.90	9.78541	10.86378	11.47563	11.57637
2.00	12.32730	13.26656	14.02336	13.98094
2.10	15.43020	16.23205	17.10362	16.92046
2.20	19.21212	19.88385	20.83545	20.51946
2.30	23.81752	24.37172	25.36541	24.92967
2.40	29.42341	29.87747	30.87363	30.33592
2.50	36.24681	36.62215	37.58107	36.96315
2.60	44.55370	44.87496	45.75841	45.08519
2.70	54.67008	54.96410	55.73686	55.03539
2.80	66.99532	67.29005	67.92132	67.21975
2.90	82.01842	82.34184	82.80645	82.13290
3.00	100.33792	100.71688	100.99630	100.37775
3.20	149.95828	150.51913	150.40258	149.96510
3.40	223.89682	224.70862	224.21451	224.02742
3.60	334.16210	335.25438	334.46072	334.55375
3.80	498.67478	500.03286	499.06494	499.42352
4.00	744.16690	745.73416	744.74480	745.31240
4.20	1110.50726	1112.19410	1111.33950	1112.02655
4.40	1657.15569	1658.85362	1658.26871	1658.96679
4.60	2472.79511	2474.39393	2474.17104	2474.76996
4.80	3689.70336	3691.10851	3691.28284	3691.68805
5.00	5505.19766	5506.34516	5506.88918	5507.03673
5.20	8213.62683	8214.49339	8215.32122	8215.18781
5.40	12254.10422	12254.71090	12255.69184	12255.29854
5.60	18281.71463	18282.12354	18283.10271	18282.51163
5.80	27273.73722	27274.04166	27274.86449	27274.16893
6.00	40688.12376	40688.43354	40688.97011	40688.27990

TABLE 8.5 Shear, Moment, Slope, and Deflection Formulas for Finite-Length Beams on Elastic Foundations

Notation: W = load (force); w = unit load (force per unit length); M_o = applied couple (force-length); θ_o = externally created concentrated angular displacement (radians); Δ_o = externally created concentrated lateral displacement (length); γ = temperature coefficient of expansion (unit strain per degree); T_1 and T_2 = temperatures on top and bottom surfaces, respectively (degrees). R_A and R_B are the vertical end reactions at the left and right, respectively, and are positive upward. M_A and M_B are the reaction end moments at the left and right, respectively, and all moments are positive when producing compression on the upper portion of the beam cross section. The transverse shear force V is positive when acting upward on the left end of a portion of the beam. All applied loads, couples, and displacements are positive as shown. All slopes are in radians, and all temperatures are in degrees. All deflections are positive upward and slopes positive when up and to the right. Note that M_A and R_A are reactions, not applied loads. They exist only when necessary end restraints are provided.

The following constants and functions, involving both beam constants and foundation constants, are hereby defined in order to permit condensing the tabulated formulas which follow.

k_o = foundation modulus (unit stress per unit deflection); b_o = beam width; and $\beta = (b_o k_o/4EI)^{1/4}$. (*Note:* See page 155 for a definition of $\langle x-a\rangle^n$.) The functions $\cosh\beta\langle x-a\rangle$, $\sinh\beta\langle x-a\rangle$, $\cos\beta\langle x-a\rangle$, and $\sin\beta\langle x-a\rangle$ are also defined as having a value of zero if $x<a$.

$F_1 = \cosh\beta x\cos\beta x$

$F_2 = \cosh\beta x\sin\beta x + \sinh\beta x\cos\beta x$

$F_3 = \sinh\beta x\sin\beta_x$

$F_4 = \cosh\beta x\sin\beta x - \sinh\beta x\cos\beta x$

$F_{a1} = \langle x-a\rangle^0\cosh\beta\langle x-a\rangle\cos\beta\langle x-a\rangle$

$F_{a2} = \cosh\beta\langle x-a\rangle\sin\beta\langle x-a\rangle + \sinh\beta\langle x-a\rangle\cos\beta\langle x-a\rangle$

$F_{a3} = \sinh\beta\langle x-a\rangle\sin\beta\langle x-a\rangle$

$F_{a4} = \cosh\beta\langle x-a\rangle\sin\beta\langle x-a\rangle - \sinh\beta\langle x-a\rangle\cos\beta\langle x-a\rangle$

$F_{a5} = \langle x-a\rangle^0 - F_{a1}$

$F_{a6} = 2\beta(x-a)\langle x-a\rangle^0 - F_{a2}$

$C_1 = \cosh\beta l\cos\beta l$

$C_2 = \cosh\beta l\sin\beta l + \sinh\beta l\cos\beta l$

$C_3 = \sinh\beta l\sin\beta l$

$C_4 = \cosh\beta l\sin\beta l - \sinh\beta l\cos\beta l$

$C_{a1} = \cosh\beta(l-a)\cos\beta(l-a)$

$C_{a2} = \cosh\beta(l-a)\sin\beta(l-a) + \sinh\beta(l-a)\cos\beta(l-a)$

$C_{a3} = \sinh\beta(l-a)\sin\beta(l-a)$

$C_{a4} = \cosh\beta(l-a)\sin\beta(l-a) - \sinh\beta(l-a)\cos\beta(l-a)$

$C_{a5} = 1 - C_{a1}$

$C_{a6} = 2\beta(l-a) - C_{a2}$

$C_{11} = \sinh^2\beta l - \sin^2\beta l$

$C_{12} = \cosh\beta l\sinh\beta l + \cos\beta l\sin\beta l$

$C_{13} = \cosh\beta l\sinh\beta l - \cos\beta l\sin\beta l$

$C_{14} = \sinh^2\beta l + \sin^2\beta l$

1. Concentrated intermediate load

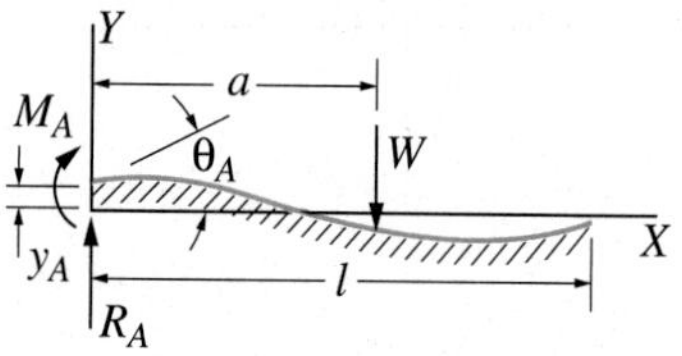

Transverse shear $= V = R_A F_1 - y_A 2EI\beta^3 F_2 - \theta_A 2EI\beta^2 F_3 - M_A\beta F_4 - WF_{a1}$

Bending moment $= M = M_A F_1 + \dfrac{R_A}{2\beta}F_2 - y_A 2EI\beta^2 F_3 - \theta_A EI\beta F_4 - \dfrac{W}{2\beta}F_{a2}$

Slope $= \theta = \theta_A F_1 + \dfrac{M_A}{2EI\beta}F_2 + \dfrac{R_A}{2EI\beta^2}F_3 - y_A\beta F_4 - \dfrac{W}{2EI\beta^2}F_{a3}$

Deflection $= y = y_A F_1 + \dfrac{\theta_A}{2\beta}F_2 + \dfrac{M_A}{2EI\beta^2}F_3 + \dfrac{R_A}{4EI\beta^3}F_4 - \dfrac{W}{4EI\beta^3}F_{a4}$

If $\beta l > 6$, see Table 8.6.

Expressions for R_A, M_A, θ_A, and y_A are found below for several combinations of end restraints.

Left End \ Right End	Free	Guided	Simply Supported	Fixed
Free	$R_A = 0 \quad M_A = 0$ $\theta_A = \dfrac{W}{2EI\beta^2}\dfrac{C_2C_{a2} - 2C_3C_{a1}}{C_{11}}$ $y_A = \dfrac{W}{2EI\beta^3}\dfrac{C_4C_{a1} - C_3C_{a2}}{C_{11}}$	$R_A = 0 \quad M_A = 0$ $\theta_A = \dfrac{W}{2EI\beta^2}\dfrac{C_2C_{a3} - C_4C_{a1}}{C_{12}}$ $y_A = \dfrac{-W}{2EI\beta^3}\dfrac{C_1C_{a1} + C_3C_{a3}}{C_{12}}$	$R_A = 0 \quad M_A = 0$ $\theta_A = \dfrac{W}{2EI\beta^2}\dfrac{C_1C_{a2} + C_3C_{a4}}{C_{13}}$ $y_A = \dfrac{-W}{4EI\beta^3}\dfrac{C_4C_{a4} + C_2C_{a2}}{C_{13}}$	$R_A = 0 \quad M_A = 0$ $\theta_A = \dfrac{W}{2EI\beta^2}\dfrac{2C_1C_{a3} + C_4C_{a4}}{2 + C_{11}}$ $y_A = \dfrac{W}{2EI\beta^3}\dfrac{C_1C_{a4} - C_2C_{a3}}{2 + C_{11}}$
Guided	$R_A = 0 \quad \theta_A = 0$ $M_A = \dfrac{W}{2\beta}\dfrac{C_2C_{a2} - 2C_3C_{a1}}{C_{12}}$ $y_A = \dfrac{-W}{4EI\beta^3}\dfrac{2C_1C_{a1} + C_4C_{a2}}{C_{12}}$	$R_A = 0 \quad \theta_A = 0$ $M_A = \dfrac{W}{2\beta}\dfrac{C_2C_{a3} - C_4C_{a1}}{C_{14}}$ $y_A = \dfrac{-W}{4EI\beta^3}\dfrac{C_2C_{a1} + C_4C_{a3}}{C_{14}}$	$R_A = 0 \quad \theta_A = 0$ $M_A = \dfrac{W}{2\beta}\dfrac{C_1C_{a2} + C_3C_{a4}}{1 + C_{11}}$ $y_A = \dfrac{W}{4EI\beta^3}\dfrac{C_1C_{a4} - C_3C_{a2}}{1 + C_{11}}$	$R_A = 0 \quad \theta_A = 0$ $M_A = \dfrac{W}{2\beta}\dfrac{2C_1C_{a3} + C_4C_{a4}}{C_{12}}$ $y_A = \dfrac{W}{4EI\beta^3}\dfrac{C_2C_{a4} - 2C_3C_{a3}}{C_{12}}$

(*Continued*)

TABLE 8.5 Shear, Moment, Slope, and Deflection Formulas for Finite-Length Beams on Elastic Foundations (*Continued*)

Left End \ Right End	Free	Guided	Simply Supported	Fixed
Simply Supported	$M_A=0 \quad y_A=0$ $R_A=W\frac{C_3C_{a2}-C_4C_{a1}}{C_{13}}$ $\theta_A=\frac{W}{2EI\beta^2}\frac{C_1C_{a2}-C_2C_{a1}}{C_{13}}$	$M_A=0 \quad y_A=0$ $R_A=W\frac{C_1C_{a1}+C_3C_{a3}}{1+C_{11}}$ $\theta_A=\frac{W}{2EI\beta^2}\frac{C_1C_{a3}-C_3C_{a1}}{1+C_{11}}$	$M_A=0 \quad y_A=0$ $R_A=\frac{W}{2}\frac{C_2C_{a2}+C_4C_{a4}}{C_{14}}$ $\theta_A=\frac{W}{4EI\beta^2}\frac{C_2C_{a4}-C_4C_{a2}}{C_{14}}$	$M_A=0 \quad y_A=0$ $R_A=W\frac{C_2C_{a3}-C_1C_{a4}}{C_{13}}$ $\theta_A=\frac{W}{2EI\beta^2}\frac{C_3C_{a4}-C_4C_{a3}}{C_{13}}$
Fixed	$\theta_A=0 \quad y_A=0$ $R_A=W\frac{2C_1C_{a1}+C_4C_{a2}}{2+C_{11}}$ $M_A=\frac{W}{\beta}\frac{C_1C_{a2}-C_2C_{a1}}{2+C_{11}}$	$\theta_A=0 \quad y_A=0$ $R_A=W\frac{C_4C_{a3}+C_2C_{a1}}{C_{12}}$ $M_A=\frac{W}{\beta}\frac{C_1C_{a3}-C_3C_{a1}}{C_{12}}$	$\theta_A=0 \quad y_A=0$ $R_A=W\frac{C_3C_{a2}-C_1C_{a4}}{C_{13}}$ $M_A=\frac{W}{2\beta}\frac{C_2C_{a4}-C_4C_{a2}}{C_{13}}$	$\theta_A=0 \quad y_A=0$ $R_A=W\frac{2C_3C_{a3}-C_2C_{a4}}{C_{11}}$ $M_A=\frac{W}{\beta}\frac{C_3C_{a4}-C_4C_{a3}}{C_{11}}$

2. Partial uniformly distributed load

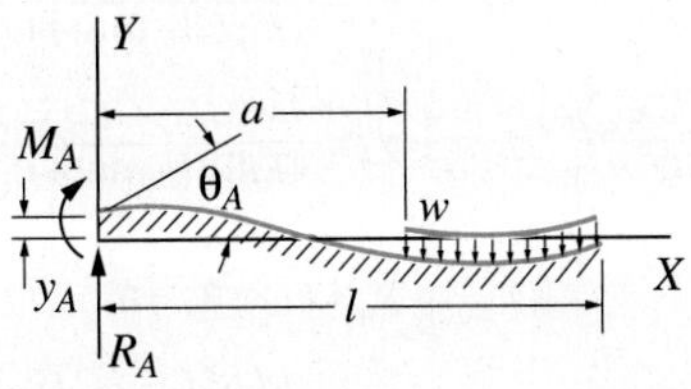

$$\text{Transverse shear} = V = R_AF_1 - y_A2EI\beta^3F_2 - \theta_A2EI\beta^2F_3 - M_A\beta F_4 - \frac{w}{2\beta}F_{a2}$$

$$\text{Bending moment} = M = M_AF_1 + \frac{R_A}{2\beta}F_2 - y_A2EI\beta^2F_3 - \theta_AEI\beta F_4 - \frac{w}{2\beta^2}F_{a3}$$

$$\text{Slope} = \theta = \theta_AF_1 + \frac{M_A}{2EI\beta}F_2 + \frac{R_A}{2EI\beta^2}F_3 - y_A\beta F_4 - \frac{w}{4EI\beta^3}F_{a4}$$

$$\text{Deflection} = y = y_AF_1 + \frac{\theta_A}{2\beta}F_2 + \frac{M_A}{2EI\beta^2}F_3 + \frac{R_A}{4EI\beta^3}F_4 - \frac{w}{4EI\beta^4}F_{a5}$$

If $\beta l > 6$, see Table 8.6.

Expressions for R_A, M_A, θ_A, and y_A are found below for several combinations of end restraints.

Left End \ Right End	Free	Guided	Simply Supported	Fixed
Free	$R_A=0 \quad M_A=0$ $\theta_A=\frac{w}{2EI\beta^3}\frac{C_2C_{a3}-C_3C_{a2}}{C_{11}}$ $y_A=\frac{w}{4EI\beta^4}\frac{C_4C_{a2}-2C_3C_{a3}}{C_{11}}$	$R_A=0 \quad M_A=0$ $\theta_A=\frac{w}{4EI\beta^3}\frac{C_2C_{a4}-C_4C_{a2}}{C_{12}}$ $y_A=\frac{-w}{4EI\beta^4}\frac{C_1C_{a2}+C_3C_{a4}}{C_{12}}$	$R_A=0 \quad M_A=0$ $\theta_A=\frac{w}{2EI\beta^3}\frac{C_1C_{a3}+C_3C_{a5}}{C_{13}}$ $y_A=\frac{-w}{4EI\beta^4}\frac{C_4C_{a5}+C_2C_{a3}}{C_{13}}$	$R_A=0 \quad M_A=0$ $\theta_A=\frac{w}{2EI\beta^3}\frac{C_1C_{a4}+C_4C_{a5}}{2+C_{11}}$ $y_A=\frac{w}{4EI\beta^4}\frac{2C_1C_{a5}-C_2C_{a4}}{2+C_{11}}$
Guided	$R_A=0 \quad \theta_A=0$ $M_A=\frac{w}{2\beta^2}\frac{C_2C_{a3}-C_3C_{a2}}{C_{12}}$ $y_A=\frac{-w}{4EI\beta^4}\frac{C_1C_{a2}+C_4C_{a3}}{C_{12}}$	$R_A=0 \quad \theta_A=0$ $M_A=\frac{w}{4\beta^2}\frac{C_2C_{a4}-C_4C_{a2}}{C_{14}}$ $y_A=\frac{-w}{8EI\beta^4}\frac{C_2C_{a2}+C_4C_{a4}}{C_{14}}$	$R_A=0 \quad \theta_A=0$ $M_A=\frac{w}{2\beta^2}\frac{C_1C_{a3}+C_3C_{a5}}{1+C_{11}}$ $y_A=\frac{w}{4EI\beta^4}\frac{C_1C_{a5}-C_3C_{a3}}{1+C_{11}}$	$R_A=0 \quad \theta_A=0$ $M_A=\frac{w}{2\beta^2}\frac{C_1C_{a4}+C_4C_{a5}}{C_{12}}$ $y_A=\frac{w}{4EI\beta^4}\frac{C_2C_{a5}-C_3C_{a4}}{C_{12}}$
Simply Supported	$M_A=0 \quad y_A=0$ $R_A=\frac{w}{2\beta}\frac{2C_2C_{a3}-C_4C_{a2}}{C_{13}}$ $\theta_A=\frac{w}{4EI\beta^3}\frac{2C_1C_{a3}-C_2C_{a2}}{C_{13}}$	$M_A=0 \quad y_A=0$ $R_A=\frac{w}{2\beta}\frac{C_1C_{a2}+C_3C_{a4}}{1+C_{11}}$ $\theta_A=\frac{w}{4EI\beta^3}\frac{C_1C_{a4}-C_3C_{a2}}{1+C_{11}}$	$M_A=0 \quad y_A=0$ $R_A=\frac{w}{2\beta}\frac{C_2C_{a3}+C_4C_{a5}}{C_{14}}$ $\theta_A=\frac{w}{4EI\beta^3}\frac{C_2C_{a5}-C_4C_{a3}}{C_{14}}$	$M_A=0 \quad y_A=0$ $R_A=\frac{w}{2\beta}\frac{C_2C_{a4}-2C_1C_{a5}}{C_{13}}$ $\theta_A=\frac{w}{4EI\beta^3}\frac{2C_3C_{a5}-C_4C_{a4}}{C_{13}}$
Fixed	$\theta_A=0 \quad y_A=0$ $R_A=\frac{w}{\beta}\frac{C_1C_{a2}+C_4C_{a3}}{2+C_{11}}$ $M_A=\frac{w}{2\beta^2}\frac{2C_1C_{a3}-C_2C_{a2}}{2+C_{11}}$	$\theta_A=0 \quad y_A=0$ $R_A=\frac{w}{2\beta}\frac{C_4C_{a4}+C_2C_{a2}}{C_{12}}$ $M_A=\frac{w}{2\beta^2}\frac{C_1C_{a4}-C_3C_{a2}}{C_{12}}$	$\theta_A=0 \quad y_A=0$ $R_A=\frac{w}{\beta}\frac{C_3C_{a3}-C_1C_{a5}}{C_{13}}$ $M_A=\frac{w}{2\beta^2}\frac{C_2C_{a5}-C_4C_{a3}}{C_{13}}$	$\theta_A=0 \quad y_A=0$ $R_A=\frac{w}{\beta}\frac{C_3C_{a4}-C_2C_{a5}}{C_{11}}$ $M_A=\frac{w}{2\beta^2}\frac{2C_3C_{a5}-C_4C_{a4}}{C_{11}}$

(*Continued*)

TABLE 8.5 Shear, Moment, Slope, and Deflection Formulas for Finite-Length Beams on Elastic Foundations (*Continued*)

3. Partial uniformly increasing load

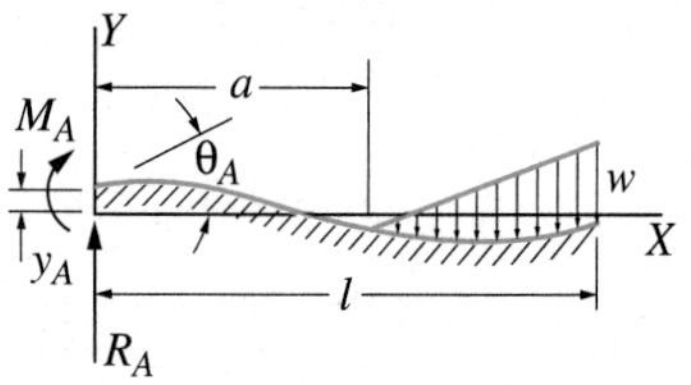

$$\text{Transverse shear} = V = R_A F_1 - y_A 2EI\beta^3 F_2 - \theta_A 2EI\beta^2 F_3 - M_A \beta F_4 - \frac{wF_{a3}}{2\beta^2(l-a)}$$

$$\text{Bending moment} = M = M_A F_1 + \frac{R_A}{2\beta} F_2 - y_A 2EI\beta^2 F_3 - \theta_A EI\beta F_4 - \frac{wF_{a4}}{4\beta^3(l-a)}$$

$$\text{Slope} = \theta = \theta_A F_1 + \frac{M_A}{2EI\beta} F_2 + \frac{R_A}{2EI\beta^2} F_3 - y_A \beta F_4 - \frac{wF_{a5}}{4EI\beta^4(l-a)}$$

$$\text{Deflection} = y = y_A F_1 + \frac{\theta_A}{2\beta} F_2 + \frac{M_A}{2EI\beta^2} F_3 + \frac{R_A}{4EI\beta^3} F_4 - \frac{wF_{a6}}{8EI\beta^5(l-a)}$$

If $\beta l > 6$, see Table 8.6.
Expressions for R_A, M_A, θ_A, and y_A are found below for several combinations of end restraints.

Left End \ Right End	Free	Guided	Simply Supported	Fixed
Free	$R_A = 0 \quad M_A = 0$ $\theta_A = \frac{w(C_2C_{a4} - 2C_3C_{a3})}{4EI\beta^4(l-a)C_{11}}$ $y_A = \frac{w(C_4C_{a3} - C_3C_{a4})}{4EI\beta^5(l-a)C_{11}}$	$R_A = 0 \quad M_A = 0$ $\theta_A = \frac{w(C_2C_{a5} - C_4C_{a3})}{4EI\beta^4(l-a)C_{12}}$ $y_A = \frac{-w(C_1C_{a3} + C_3C_{a5})}{4EI\beta^5(l-a)C_{12}}$	$R_A = 0 \quad M_A = 0$ $\theta_A = \frac{w(C_1C_{a4} + C_3C_{a6})}{4EI\beta^4(l-a)C_{13}}$ $y_A = \frac{-w(C_2C_{a4} + C_4C_{a6})}{8EI\beta^5(l-a)C_{13}}$	$R_A = 0 \quad M_A = 0$ $\theta_A = \frac{w(2C_1C_{a5} + C_4C_{a6})}{4EI\beta^4(l-a)(2+C_{11})}$ $y_A = \frac{w(C_1C_{a6} - C_2C_{a5})}{4EI\beta^5(l-a)(2+C_{11})}$
Guided	$R_A = 0 \quad \theta_A = 0$ $M_A = \frac{w(C_2C_{a4} - 2C_3C_{a3})}{4\beta^3(l-a)C_{12}}$ $y_A = \frac{-w(2C_1C_{a3} + C_4C_{a4})}{8EI\beta^5(l-a)C_{12}}$	$R_A = 0 \quad \theta_A = 0$ $M_A = \frac{w(C_2C_{a5} - C_4C_{a3})}{4\beta^3(l-a)C_{14}}$ $y_A = \frac{-w(C_2C_{a3} + C_4C_{a5})}{8EI\beta^5(l-a)C_{14}}$	$R_A = 0 \quad \theta_A = 0$ $M_A = \frac{w(C_1C_{a4} + C_3C_{a6})}{4\beta^3(l-a)(1+C_{11})}$ $y_A = \frac{w(C_1C_{a6} - C_3C_{a4})}{8EI\beta^5(l-a)(1+C_{11})}$	$R_A = 0 \quad \theta_A = 0$ $M_A = \frac{w(2C_1C_{a5} + C_4C_{a6})}{4\beta^3(l-a)C_{12}}$ $y_A = \frac{w(C_2C_{a6} - 2C_3C_{a5})}{8EI\beta^5(l-a)C_{12}}$
Simply supported	$M_A = 0 \quad y_A = 0$ $R_A = \frac{w(C_3C_{a4} - C_4C_{a3})}{2\beta^2(l-a)C_{13}}$ $\theta_A = \frac{w(C_1C_{a4} - C_2C_{a3})}{4EI\beta^4(l-a)C_{13}}$	$M_A = 0 \quad y_A = 0$ $R_A = \frac{w(C_1C_{a3} + C_3C_{a5})}{2\beta^2(l-a)(1+C_{11})}$ $\theta_A = \frac{w(C_1C_{a5} - C_3C_{a3})}{4EI\beta^4(l-a)(1+C_{11})}$	$M_A = 0 \quad y_A = 0$ $R_A = \frac{w(C_2C_{a4} + C_4C_{a6})}{4\beta^2(l-a)C_{14}}$ $\theta_A = \frac{w(C_2C_{a6} - C_4C_{a4})}{8EI\beta^4(l-a)C_{14}}$	$M_A = 0 \quad y_A = 0$ $R_A = \frac{w(C_2C_{a5} - C_1C_{a6})}{2\beta^2(l-a)C_{13}}$ $\theta_A = \frac{w(C_3C_{a6} - C_4C_{a5})}{4EI\beta^4(l-a)C_{13}}$
Fixed	$\theta_A = 0 \quad y_A = 0$ $R_A = \frac{w(2C_1C_{a3} + C_4C_{a4})}{2\beta^2(l-a)(2+C_{11})}$ $M_A = \frac{w(C_1C_{a4} - C_2C_{a3})}{2\beta^3(l-a)(2+C_{11})}$	$\theta_A = 0 \quad y_A = 0$ $R_A = \frac{w(C_4C_{a5} + C_2C_{a3})}{2\beta^2(l-a)C_{12}}$ $M_A = \frac{w(C_1C_{a5} - C_3C_{a3})}{2\beta^3(l-a)C_{12}}$	$\theta_A = 0 \quad y_A = 0$ $R_A = \frac{w(C_3C_{a4} - C_1C_{a6})}{2\beta^2(l-a)C_{13}}$ $M_A = \frac{w(C_2C_{a6} - C_4C_{a4})}{4\beta^3(l-a)C_{13}}$	$\theta_A = 0 \quad y_A = 0$ $R_A = \frac{w(2C_3C_{a5} - C_2C_{a6})}{2\beta^2(l-a)C_{11}}$ $M_A = \frac{w(C_3C_{a6} - C_4C_{a5})}{2\beta^3(l-a)C_{11}}$

(*Continued*)

TABLE 8.5 Shear, Moment, Slope, and Deflection Formulas for Finite-Length Beams on Elastic Foundations (*Continued*)

4. Concentrated intermediate moment

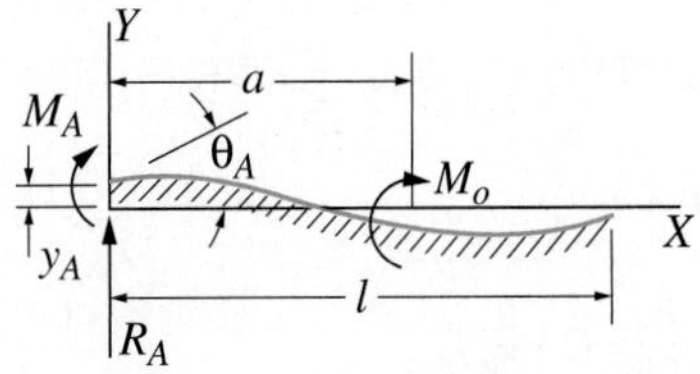

$$\text{Transverse shear} = V = R_A F_1 - y_A 2EI\beta^3 F_2 - \theta_A 2EI\beta^2 F_3 - M_A \beta F_4 - M_o \beta F_{a4}$$

$$\text{Bending moment} = M = M_A F_1 + \frac{R_A}{2\beta} F_2 - y_A 2EI\beta^2 F_3 - \theta_A EI\beta F_4 + M_o F_{a1}$$

$$\text{Slope} = \theta = \theta_A F_1 + \frac{M_A}{2EI\beta} F_2 + \frac{R_A}{2EI\beta^2} F_3 - y_A \beta F_4 + \frac{M_o}{2EI\beta} F_{a2}$$

$$\text{Deflection} = y = y_A F_1 + \frac{\theta_A}{2\beta} F_2 + \frac{M_A}{2EI\beta^2} F_3 + \frac{R_A}{4EI\beta^3} F_4 + \frac{M_o}{2EI\beta^2} F_{a3}$$

If $\beta l > 6$, see Table 8.6.

Expressions for R_A, M_A, θ_A, and y_A are found below for several combinations of end restraints.

Left End \ Right End	Free	Guided	Simply Supported	Fixed
Free	$R_A = 0 \quad M_A = 0$ $\theta_A = \frac{-M_o}{EI\beta} \frac{C_3 C_{a4} + C_2 C_{a1}}{C_{11}}$ $y_A = \frac{M_o}{2EI\beta^2} \frac{2C_3 C_{a1} + C_4 C_{a4}}{C_{11}}$	$R_A = 0 \quad M_A = 0$ $\theta_A = \frac{-M_o}{2EI\beta} \frac{C_2 C_{a2} + C_4 C_{a4}}{C_{12}}$ $y_A = \frac{M_o}{2EI\beta^2} \frac{C_3 C_{a2} - C_1 C_{a4}}{C_{12}}$	$R_A = 0 \quad M_A = 0$ $\theta_A = \frac{-M_o}{EI\beta} \frac{C_1 C_{a1} + C_3 C_{a3}}{C_{13}}$ $y_A = \frac{M_o}{2EI\beta^2} \frac{C_4 C_{a3} + C_2 C_{a1}}{C_{13}}$	$R_A = 0 \quad M_A = 0$ $\theta_A = \frac{-M_o}{EI\beta} \frac{C_1 C_{a2} + C_4 C_{a3}}{2 + C_{11}}$ $y_A = \frac{-M_o}{2EI\beta^2} \frac{2C_1 C_{a3} - C_2 C_{a2}}{2 + C_{11}}$
Guided	$R_A = 0 \quad \theta_A = 0$ $M_A = -M_o \frac{C_2 C_{a1} + C_3 C_{a4}}{C_{12}}$ $y_A = \frac{-M_o}{2EI\beta^2} \frac{C_1 C_{a4} - C_4 C_{a1}}{C_{12}}$	$R_A = 0 \quad \theta_A = 0$ $M_A = \frac{-M_o}{2} \frac{C_2 C_{a2} + C_4 C_{a4}}{C_{14}}$ $y_A = \frac{M_o}{4EI\beta^2} \frac{C_4 C_{a2} - C_2 C_{a4}}{C_{14}}$	$R_A = 0 \quad \theta_A = 0$ $M_A = -M_o \frac{C_1 C_{a1} + C_3 C_{a3}}{1 + C_{11}}$ $y_A = \frac{M_o}{2EI\beta^2} \frac{C_3 C_{a1} - C_1 C_{a3}}{1 + C_{11}}$	$R_A = 0 \quad \theta_A = 0$ $M_A = -M_o \frac{C_1 C_{a2} + C_4 C_{a3}}{C_{12}}$ $y_A = \frac{M_o}{2EI\beta^2} \frac{C_3 C_{a2} - C_2 C_{a3}}{C_{12}}$
Simply Supported	$M_A = 0 \quad y_A = 0$ $R_A = -M_o \beta \frac{2C_3 C_{a1} + C_4 C_{a4}}{C_{13}}$ $\theta_A = \frac{-M_o}{2EI\beta} \frac{2C_1 C_{a1} + C_2 C_{a4}}{C_{13}}$	$M_A = 0 \quad y_A = 0$ $R_A = -M_o \beta \frac{C_3 C_{a2} - C_1 C_{a4}}{1 + C_{11}}$ $\theta_A = \frac{-M_o}{2EI\beta} \frac{C_1 C_{a2} + C_3 C_{a4}}{1 + C_{11}}$	$M_A = 0 \quad y_A = 0$ $R_A = -M_o \beta \frac{C_2 C_{a1} + C_4 C_{a3}}{C_{14}}$ $\theta_A = \frac{-M_o}{2EI\beta} \frac{C_2 C_{a3} - C_4 C_{a1}}{C_{14}}$	$M_A = 0 \quad y_A = 0$ $R_A = -M_o \beta \frac{C_2 C_{a2} - 2C_1 C_{a3}}{C_{13}}$ $\theta_A = \frac{-M_o}{2EI\beta} \frac{2C_3 C_{a3} - C_4 C_{a2}}{C_{13}}$
Fixed	$\theta_A = 0 \quad y_A = 0$ $R_A = -M_o 2\beta \frac{C_4 C_{a1} - C_1 C_{a4}}{2 + C_{11}}$ $M_A = -M_o \frac{2C_1 C_{a1} + C_2 C_{a4}}{2 + C_{11}}$	$\theta_A = 0 \quad y_A = 0$ $R_A = -M_o \beta \frac{C_4 C_{a2} - C_2 C_{a4}}{C_{12}}$ $M_A = -M_o \frac{C_1 C_{a2} + C_3 C_{a4}}{C_{12}}$	$\theta_A = 0 \quad y_A = 0$ $R_A = -M_o 2\beta \frac{C_3 C_{a1} - C_1 C_{a3}}{C_{13}}$ $M_A = -M_o \frac{C_2 C_{a3} - C_4 C_{a1}}{C_{13}}$	$\theta_A = 0 \quad y_A = 0$ $R_A = -M_o 2\beta \frac{C_3 C_{a2} - C_2 C_{a3}}{C_{11}}$ $M_A = -M_o \frac{2C_3 C_{a3} - C_4 C_{a2}}{C_{11}}$

(*Continued*)

TABLE 8.5 Shear, Moment, Slope, and Deflection Formulas for Finite-Length Beams on Elastic Foundations (*Continued*)

5. Externally created concentrated angular displacement

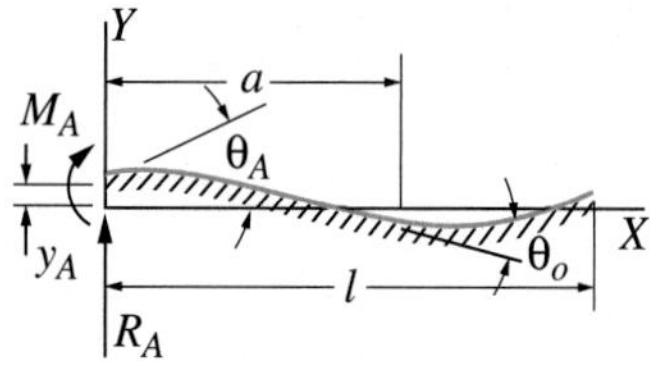

$$\text{Transverse shear} = V = R_A F_1 - y_A 2EI\beta^3 F_2 - \theta_A 2EI\beta^2 F_3 - M_A \beta F_4 - \theta_o 2EI\beta^2 F_{a3}$$

$$\text{Bending moment} = M = M_A F_1 + \frac{R_A}{2\beta} F_2 - y_A 2EI\beta^2 F_3 - \theta_A EI\beta F_4 - \theta_o EI\beta F_{a4}$$

$$\text{Slope} = \theta = \theta_A F_1 + \frac{M_A}{2EI\beta} F_2 + \frac{R_A}{2EI\beta^2} F_3 - y_A \beta F_4 + \theta_o F_{a1}$$

$$\text{Deflection} = y = y_A F_1 + \frac{\theta_A}{2\beta} F_2 + \frac{M_A}{2EI\beta^2} F_3 + \frac{R_A}{4EI\beta^3} F_4 + \frac{\theta_o}{2\beta} F_{a2}$$

If $\beta l > 6$, see Table 8.6.

Expressions for R_A, M_A, θ_A, and y_A are found below for several combinations of end restraints.

Left End \ Right End	Free	Guided	Simply Supported	Fixed
Free	$R_A = 0 \quad M_A = 0$ $\theta_A = \theta_o \frac{C_2 C_{a4} - 2C_3 C_{a3}}{C_{11}}$ $y_A = \frac{\theta_o}{\beta} \frac{C_4 C_{a3} - C_3 C_{a4}}{C_{11}}$	$R_A = 0 \quad M_A = 0$ $\theta_A = -\theta_o \frac{C_2 C_{a1} + C_4 C_{a3}}{C_{12}}$ $y_A = \frac{\theta_o}{\beta} \frac{C_3 C_{a1} - C_1 C_{a3}}{C_{12}}$	$R_A = 0 \quad M_A = 0$ $\theta_A = \theta_o \frac{C_1 C_{a4} - C_3 C_{a2}}{C_{13}}$ $y_A = \frac{\theta_o}{2\beta} \frac{C_4 C_{a2} - C_2 C_{a4}}{C_{13}}$	$R_A = 0 \quad M_A = 0$ $\theta_A = -\theta_o \frac{2C_1 C_{a1} + C_4 C_{a2}}{2 + C_{11}}$ $y_A = \frac{-\theta_o}{\beta} \frac{C_1 C_{a2} - C_2 C_{a1}}{2 + C_{11}}$
Guided	$R_A = 0 \quad \theta_A = 0$ $M_A = \theta_o EI\beta \frac{C_2 C_{a4} - 2C_3 C_{a3}}{C_{12}}$ $y_A = \frac{-\theta_o}{2\beta} \frac{2C_1 C_{a3} + C_4 C_{a4}}{C_{12}}$	$R_A = 0 \quad \theta_A = 0$ $M_A = -\theta_o EI\beta \frac{C_2 C_{a1} + C_4 C_{a3}}{C_{14}}$ $y_A = \frac{\theta_o}{2\beta} \frac{C_4 C_{a1} - C_2 C_{a3}}{C_{14}}$	$R_A = 0 \quad \theta_A = 0$ $M_A = \theta_o EI\beta \frac{C_1 C_{a4} - C_3 C_{a2}}{1 + C_{11}}$ $y_A = \frac{-\theta_o}{2\beta} \frac{C_1 C_{a2} + C_3 C_{a4}}{1 + C_{11}}$	$R_A = 0 \quad \theta_A = 0$ $M_A = -\theta_o EI\beta \frac{2C_1 C_{a1} + C_4 C_{a2}}{C_{12}}$ $y_A = \frac{\theta_o}{2\beta} \frac{2C_3 C_{a1} - C_2 C_{a2}}{C_{12}}$
Simply Supported	$M_A = 0 \quad y_A = 0$ $R_A = \theta_o 2EI\beta^2 \frac{C_3 C_{a4} - C_4 C_{a3}}{C_{13}}$ $\theta_A = \theta_o \frac{C_1 C_{a4} - C_2 C_{a3}}{C_{13}}$	$M_A = 0 \quad y_A = 0$ $R_A = \theta_o 2EI\beta^2 \frac{C_1 C_{a3} - C_3 C_{a1}}{1 + C_{11}}$ $\theta_A = -\theta_o \frac{C_1 C_{a1} + C_3 C_{a3}}{1 + C_{11}}$	$M_A = 0 \quad y_A = 0$ $R_A = \theta_o EI\beta^2 \frac{C_2 C_{a4} - C_4 C_{a2}}{C_{14}}$ $\theta_A = \frac{-\theta_o}{2} \frac{C_2 C_{a2} + C_4 C_{a4}}{C_{14}}$	$M_A = 0 \quad y_A = 0$ $R_A = \theta_o 2EI\beta^2 \frac{C_1 C_{a2} - C_2 C_{a1}}{C_{13}}$ $\theta_A = \theta_o \frac{C_4 C_{a1} - C_3 C_{a2}}{C_{13}}$
Fixed	$\theta_A = 0 \quad y_A = 0$ $R_A = \theta_o 2EI\beta^2 \frac{2C_1 C_{a3} + C_4 C_{a4}}{2 + C_{11}}$ $M_A = \theta_o 2EI\beta \frac{C_1 C_{a4} - C_2 C_{a3}}{2 + C_{11}}$	$\theta_A = 0 \quad y_A = 0$ $R_A = \theta_o 2EI\beta^2 \frac{C_2 C_{a3} - C_4 C_{a1}}{C_{12}}$ $M_A = -\theta_o 2EI\beta \frac{C_1 C_{a1} + C_3 C_{a3}}{C_{12}}$	$\theta_A = 0 \quad y_A = 0$ $R_A = \theta_o 2EI\beta^2 \frac{C_1 C_{a2} + C_3 C_{a4}}{C_{13}}$ $M_A = -\theta_o EI\beta \frac{C_2 C_{a2} + C_4 C_{a4}}{C_{13}}$	$\theta_A = 0 \quad y_A = 0$ $R_A = \theta_o 2EI\beta^2 \frac{C_2 C_{a2} - 2C_3 C_{a1}}{C_{11}}$ $M_A = \theta_o 2EI\beta \frac{C_4 C_{a1} - C_3 C_{a2}}{C_{11}}$

(*Continued*)

TABLE 8.5 Shear, Moment, Slope, and Deflection Formulas for Finite-Length Beams on Elastic Foundations (*Continued*)

6. Externally created concentrated lateral displacement

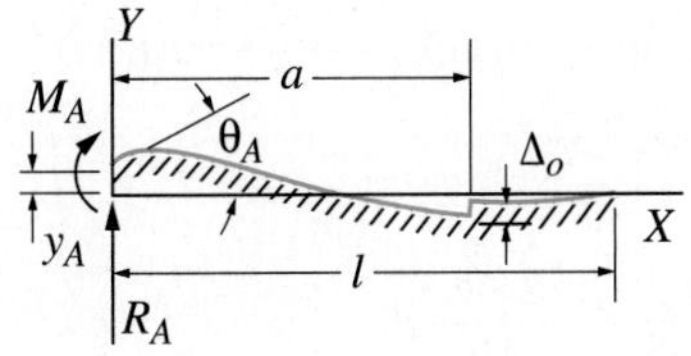

Transverse shear $= V = R_A F_1 - y_A 2EI\beta^3 F_2 - \theta_A 2EI\beta^2 F_3 - M_A \beta F_4 - \Delta_o 2EI\beta^3 F_{a2}$

Bending moment $= M = M_A F_1 + \frac{R_A}{2\beta} F_2 - y_A 2EI\beta^2 F_3 - \theta_A EI\beta F_4 - \Delta_o 2EI\beta^2 F_{a3}$

Slope $= \theta = \theta_A F_1 + \frac{M_A}{2EI\beta} F_2 + \frac{R_A}{2EI\beta^2} F_3 - y_A \beta F_4 - \Delta_o \beta F_{a4}$

Deflection $= y = y_A F_1 + \frac{\theta_A}{2\beta} F_2 + \frac{M_A}{2EI\beta^2} F_3 + \frac{R_A}{4EI\beta^3} F_4 + \Delta_o F_{a1}$

If $\beta l > 6$, see Table 8.6.

Expressions for R_A, M_A, θ_A, and y_A are found below for several combinations of end restraints.

Left End \ Right End	Free	Guided	Simply Supported	Fixed
Free	$R_A = 0 \quad M_A = 0$ $\theta_A = \Delta_o 2\beta \frac{C_2 C_{a3} - C_3 C_{a2}}{C_{11}}$ $y_A = \Delta_o \frac{C_4 C_{a2} - 2C_3 C_{a3}}{C_{11}}$	$R_A = 0 \quad M_A = 0$ $\theta_A = \Delta_o \beta \frac{C_2 C_{a4} - C_4 C_{a2}}{C_{12}}$ $y_A = -\Delta_o \frac{C_1 C_{a2} + C_3 C_{a4}}{C_{12}}$	$R_A = 0 \quad M_A = 0$ $\theta_A = \Delta_o 2\beta \frac{C_1 C_{a3} - C_3 C_{a1}}{C_{13}}$ $y_A = \Delta_o \frac{C_4 C_{a1} - C_2 C_{a3}}{C_{13}}$	$R_A = 0 \quad M_A = 0$ $\theta_A = \Delta_o 2\beta \frac{C_1 C_{a4} - C_4 C_{a1}}{2 + C_{11}}$ $y_A = -\Delta_o \frac{2C_1 C_{a1} + C_2 C_{a4}}{2 + C_{11}}$
Guided	$R_A = 0 \quad \theta_A = 0$ $M_A = \Delta_o 2EI\beta^2 \frac{C_2 C_{a3} - C_3 C_{a2}}{C_{12}}$ $y_A = -\Delta_o \frac{C_1 C_{a2} + C_4 C_{a3}}{C_{12}}$	$R_A = 0 \quad \theta_A = 0$ $M_A = \Delta_o EI\beta^2 \frac{C_2 C_{a4} - C_4 C_{a2}}{C_{14}}$ $y_A = \frac{-\Delta_o}{2} \frac{C_2 C_{a2} + C_4 C_{a4}}{C_{14}}$	$R_A = 0 \quad \theta_A = 0$ $M_A = \Delta_o 2EI\beta^2 \frac{C_1 C_{a3} - C_3 C_{a1}}{1 + C_{11}}$ $y_A = -\Delta_o \frac{C_1 C_{a1} + C_3 C_{a3}}{1 + C_{11}}$	$R_A = 0 \quad \theta_A = 0$ $M_A = \Delta_o 2EI\beta^2 \frac{C_1 C_{a4} - C_4 C_{a1}}{C_{12}}$ $y_A = -\Delta_o \frac{C_2 C_{a1} + C_3 C_{a4}}{C_{12}}$
Simply Supported	$M_A = 0 \quad y_A = 0$ $R_A = \Delta_o 2EI\beta^3 \frac{2C_3 C_{a3} - C_4 C_{a2}}{C_{13}}$ $\theta_A = \Delta_o \beta \frac{2C_1 C_{a3} - C_2 C_{a2}}{C_{13}}$	$M_A = 0 \quad y_A = 0$ $R_A = \Delta_o 2EI\beta^3 \frac{C_1 C_{a2} + C_3 C_{a4}}{1 + C_{11}}$ $\theta_A = \Delta_o \beta \frac{C_1 C_{a4} - C_3 C_{a2}}{1 + C_{11}}$	$M_A = 0 \quad y_A = 0$ $R_A = \Delta_o 2EI\beta^3 \frac{C_2 C_{a3} - C_4 C_{a1}}{C_{14}}$ $\theta_A = -\Delta_o \beta \frac{C_2 C_{a1} + C_4 C_{a3}}{C_{14}}$	$M_A = 0 \quad y_A = 0$ $R_A = \Delta_o 2EI\beta^3 \frac{C_2 C_{a4} + 2C_1 C_{a1}}{C_{13}}$ $\theta_A = -\Delta_o \beta \frac{2C_3 C_{a1} + C_4 C_{a4}}{C_{13}}$
Fixed	$\theta_A = 0 \quad y_A = 0$ $R_A = \Delta_o 4EI\beta^3 \frac{C_1 C_{a2} + C_4 C_{a3}}{2 + C_{11}}$ $M_A = \Delta_o 2EI\beta^2 \frac{2C_1 C_{a3} - C_2 C_{a2}}{2 + C_{11}}$	$\theta_A = 0 \quad y_A = 0$ $R_A = \Delta_o 2EI\beta^3 \frac{C_4 C_{a4} + C_2 C_{a2}}{C_{12}}$ $M_A = \Delta_o 2EI\beta^2 \frac{C_1 C_{a4} - C_3 C_{a2}}{C_{12}}$	$\theta_A = 0 \quad y_A = 0$ $R_A = \Delta_o 4EI\beta^3 \frac{C_3 C_{a3} + C_1 C_{a1}}{C_{13}}$ $M_A = -\Delta_o 2EI\beta^2 \frac{C_2 C_{a1} + C_4 C_{a3}}{C_{13}}$	$\theta_A = 0 \quad y_A = 0$ $R_A = \Delta_o 4EI\beta^3 \frac{C_3 C_{a4} + C_2 C_{a1}}{C_{11}}$ $M_A = -\Delta_o 2EI\beta^2 \frac{2C_3 C_{a1} + C_4 C_{a4}}{C_{11}}$

(*Continued*)

TABLE 8.5 Shear, Moment, Slope, and Deflection Formulas for Finite-Length Beams on Elastic Foundations (*Continued*)

7. Uniform temperature differential from top to bottom

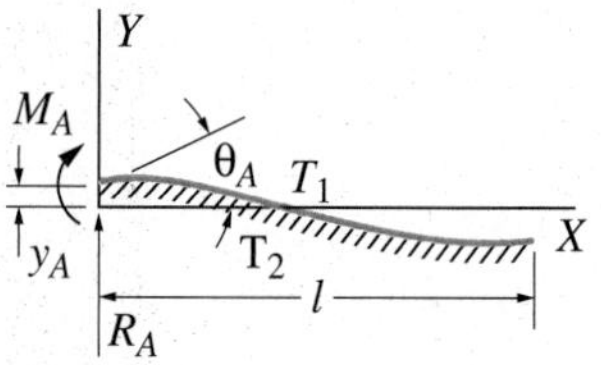

$$\text{Transverse shear} = V = R_A F_1 - y_A 2EI\beta^3 F_2 - \theta_A 2EI\beta^2 F_3 - M_A \beta F_4 + \frac{T_1 - T_2}{t}\gamma EI\beta F_4$$

$$\text{Bending moment} = M = M_A F_1 + \frac{R_A}{2\beta} F_2 - y_A 2EI\beta^2 F_3 - \theta_A EI\beta F_4 - \frac{T_1 - T_2}{t}\gamma EI(F_1 - 1)$$

$$\text{Slope} = \theta = \theta_A F_1 + \frac{M_A}{2EI\beta} F_2 + \frac{R_A}{2EI\beta^2} F_3 - y_A \beta F_4 - \frac{T_1 - T_2}{2t\beta}\gamma F_2$$

$$\text{Deflection} = y = y_A F_1 + \frac{\theta_A}{2\beta} F_2 + \frac{M_A}{2EI\beta^2} F_3 + \frac{R_A}{4EI\beta^3} F_4 - \frac{T_1 - T_2}{2t\beta^2}\gamma F_3$$

If $\beta l > 6$, see Table 8.6.

Expressions for R_A, M_A, θ_A, and y_A are found below for several combinations of end restraints.

Left End \ Right End	Free	Guided	Simply Supported	Fixed
Free	$R_A = 0 \quad M_A = 0$ $\theta_A = \frac{(T_1 - T_2)\gamma}{\beta t}\frac{C_1C_2 + C_3C_4 - C_2}{C_{11}}$ $y_A = \frac{-(T_1 - T_2)\gamma}{2\beta^2 t}\frac{C_4^2 + 2C_1C_3 - 2C_3}{C_{11}}$	$R_A = 0 \quad M_A = 0$ $\theta_A = \frac{(T_1 - T_2)\gamma}{2\beta t}\frac{C_2^2 + C_4^2}{C_{12}}$ $y_A = \frac{-(T_1 - T_2)\gamma}{2\beta^2 t}\frac{C_2C_3 - C_1C_4}{C_{12}}$	$R_A = 0 \quad M_A = 0$ $\theta_A = \frac{(T_1 - T_2)\gamma}{\beta t}\frac{C_1^2 + C_3 - C_4}{C_{13}}$ $y_A = \frac{-(T_1 - T_2)\gamma}{2\beta^2 t}\frac{C_1C_2 + C_3C_4 - C_2}{C_{13}}$	$R_A = 0 \quad M_A = 0$ $\theta_A = \frac{(T_1 - T_2)\gamma}{\beta t}\frac{C_1C_2 + C_3C_1}{2 + C_{11}}$ $y_A = \frac{(T_1 - T_2)\gamma}{2\beta^2 t}\frac{2C_1C_3 - C_2^2}{2 + C_{11}}$
Guided	$R_A = 0 \quad \theta_A = 0$ $M_A = \frac{(T_1 - T_2)\gamma EI}{t}\frac{C_1C_2 + C_3C_4 - C_2}{C_{12}}$ $y_A = \frac{(T_1 - T_2)\gamma}{2\beta^2 t}\frac{C_4}{C_{12}}$	$R_A = 0 \quad \theta_A = 0$ $M_A = \frac{(T_1 - T_2)\gamma EI}{t}$ $y_A = 0$	$R_A = 0 \quad \theta_A = 0$ $M_A = \frac{(T_1 - T_2)\gamma EI}{t}\frac{C_1^2 + C_3^2 - C_1}{1 + C_{11}}$ $y_A = \frac{(T_1 - T_2)\gamma}{2\beta^2 t}\frac{C_3}{1 + C_{11}}$	$R_A = 0 \quad \theta_A = 0$ $M_A = \frac{(T_1 - T_2)\gamma EI}{t}$ $y_A = 0$
Simply Supported	$M_A = 0 \quad y_A = 0$ $R_A = \frac{(T_1 - T_2)\gamma\beta EI}{t}\frac{2C_1C_3 + C_4^2 - 2C_3}{C_{13}}$ $\theta_A = \frac{(T_1 - T_2)\gamma}{2\beta t}\frac{2C_1^2 + C_2C_4 - 2C_1}{C_{13}}$	$M_A = 0 \quad y_A = 0$ $R_A = \frac{(T_1 - T_2)\gamma\beta EI}{t}\frac{C_2C_3 - C_1C_4}{1 + C_{11}}$ $\theta_A = \frac{(T_1 - T_2)\gamma}{2\beta t}\frac{C_1C_2 + C_3C_4}{1 + C_{11}}$	$M_A = 0 \quad y_A = 0$ $R_A = \frac{(T_1 - T_2)\gamma\beta EI}{t}\frac{C_1C_2 + C_3C_4 - C_2}{C_{14}}$ $\theta_A = \frac{(T_1 - T_2)\gamma}{2\beta t}\frac{C_2C_3 - C_1C_4 + C_4}{C_{14}}$	$M_A = 0 \quad y_A = 0$ $R_A = \frac{(T_1 - T_2)\gamma\beta EI}{t}\frac{C_2^2 - 2C_1C_3}{C_{13}}$ $\theta_A = \frac{(T_1 - T_2)\gamma}{2\beta t}\frac{2C_3^2 - C_2C_4}{C_{13}}$
Fixed	$\theta_A = 0 \quad y_A = 0$ $R_A = \frac{(T_1 - T_2)\gamma 2\beta EI}{t}\frac{-C_4}{2 + C_{11}}$ $M_A = \frac{(T_1 - T_2)\gamma EI}{t}\frac{2C_1^2 + C_2C_4 - 2C_1}{2 + C_{11}}$	$\theta_A = 0 \quad y_A = 0$ $R_A = 0$ $M_A = \frac{(T_1 - T_2)\gamma EI}{t}$	$\theta_A = 0 \quad y_A = 0$ $R_A = \frac{(T_1 - T_2)\gamma\beta EI}{t}\frac{-2C_3}{C_{13}}$ $M_A = \frac{(T_1 - T_2)\gamma EI}{t}\frac{C_2C_3 - C_1C_4 + C_4}{C_{13}}$	$\theta_A = 0 \quad y_A = 0$ $R_A = 0$ $M_A = \frac{(T_1 - T_2)\gamma EI}{t}$

TABLE 8.6 Shear, Moment, Slope, and Deflection Formulas for Semi-Infinite Beams on Elastic Foundations

Notation: All notation is the same as that for Table 8.5. No length is defined since these beams are assumed to extend from the left end, for which restraints are defined, to a length beyond that portion affected by the loading. Note that M_A and R_A are reactions, not applied loads.

The following constants and functions, involving both beam constants and foundation constants, are hereby defined in order to permit condensing the tabulated formulas which follow.

k_o = foundation modulus (unit stress per unit deflection); b_o = beam width, and $\beta = (b_o k_o / 4EI)^{1/4}$. (*Note:* See page 155 for a definition of $\langle x-a \rangle^n$.)

$F_1 = \cosh \beta x \cos \beta x$ $\qquad A_1 = 0.5e^{-\beta a} \cos \beta a$ $\qquad B_1 = 0.5e^{-\beta b} \cos \beta b$

$F_2 = \cosh \beta x \sin \beta x + \sinh \beta x \cos \beta x$ $\qquad A_2 = 0.5e^{-\beta a}(\sin \beta a - \cos \beta a)$ $\qquad B_2 = 0.5e^{-\beta b}(\sin \beta b - \cos \beta b)$

$F_3 = \sinh \beta x \sin \beta x$ $\qquad A_3 = -0.5e^{-\beta a} \sin \beta a$ $\qquad B_3 = -0.5e^{-\beta b} \sin \beta b$

$F_4 = \cosh \beta x \sin \beta x - \sinh \beta x \cos \beta x$ $\qquad A_4 = 0.5e^{-\beta a}(\sin \beta a + \cos \beta a)$ $\qquad B_4 = 0.5e^{-\beta b}(\sin \beta b + \cos \beta b)$

$F_{a1} = \langle x-a \rangle^0 \cosh \beta \langle x-a \rangle \cos \beta \langle x-a \rangle$

$F_{a2} = \cosh \beta \langle x-a \rangle \sin \beta \langle x-a \rangle + \sinh \beta \langle x-a \rangle \cos \beta \langle x-a \rangle$

$F_{a3} = \sinh \beta \langle x-a \rangle \sin \beta \langle x-a \rangle$

$F_{a4} = \cosh \beta \langle x-a \rangle \sin \beta \langle x-a \rangle - \sinh \beta \langle x-a \rangle \cos \beta \langle x-a \rangle$

$F_{a5} = \langle x-a \rangle^0 - F_{a1}$

$F_{a6} = 2\beta(x-a)\langle x-a \rangle^0 - F_{a2}$

$F_{b1} = \langle x-b \rangle^0 \cosh \beta \langle x-b \rangle \cos \beta \langle x-b \rangle$

$F_{b2} = \cosh \beta \langle x-b \rangle \sin \beta \langle x-b \rangle + \sinh \langle \beta x-b \rangle \cos \beta \langle x-b \rangle$

$F_{b3} = \sinh \beta \langle x-b \rangle \sin \beta \langle x-b \rangle$

$F_{b4} = \cosh \beta \langle x-b \rangle \sin \beta \langle x-b \rangle - \sinh \beta \langle x-b \rangle \cos \beta \langle x-b \rangle$

$F_{b5} = \langle x-b \rangle^0 - F_{b1}$

$F_{b6} = 2\beta\langle x-b \rangle\langle x-b \rangle^0 - F_{b2}$

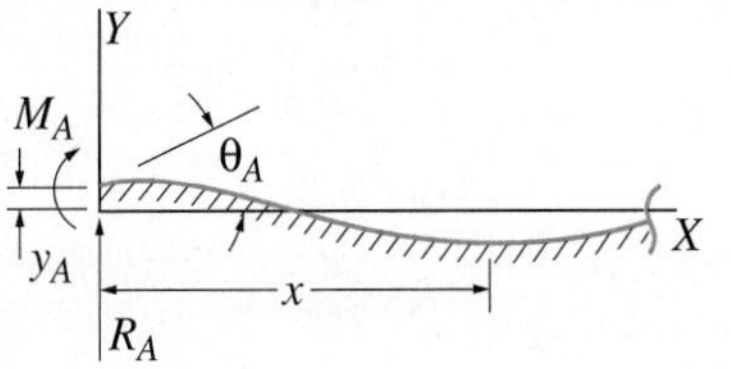

$$\text{Transverse shear} = V = R_A F_1 - y_A 2EI\beta^3 F_2 - \theta_A 2EI\beta^2 F_3 - M_A \beta F_4 + LT_V$$

$$\text{Bending moment} = M = M_A F_1 + \frac{R_A}{2\beta} F_2 - y_A 2EI\beta^2 F_3 - \theta_A EI\beta F_4 + LT_M$$

$$\text{Slope} = \theta = \theta_A F_1 + \frac{M_A}{2EI\beta} F_2 + \frac{R_A}{2EI\beta^2} F_3 - y_A \beta F_4 + LT_\theta$$

$$\text{Deflection} = y = y_A F_1 + \frac{\theta_A}{2\beta} F_2 + \frac{M_A}{2EI\beta^2} F_3 + \frac{R_A}{4EI\beta^3} F_4 + LT_y$$

Expressions for R_A, M_A, θ_A, and y_A are found below for several combinations of loading and left end restraints. The loading terms LT_V, LT_M, LT_θ, and LT_y are given for each loading condition.

(Continued)

TABLE 8.6 Shear, Moment, Slope, and Deflection Formulas for Semi-Infinite Beams on Elastic Foundations (*Continued*)

Loading, Reference No. \ Left End Restraint	Free	Guided	Simply Supported	Fixed	Loading Terms
1. Concentrated intermediate load (if $\beta a > 3$, see case 10) 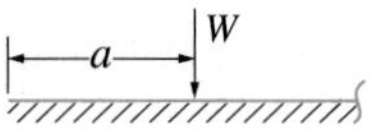	$R_A=0 \quad M_A=0$ $\theta_A=\frac{-W}{EI\beta^2}A_2$ $y_A=\frac{-W}{EI\beta^3}A_1$ (if $a = 0$, see case 8)	$R_A=0 \quad \theta_A=0$ $M_A=\frac{-W}{\beta}A_2$ $y_A=\frac{-W}{2EI\beta^3}A_4$	$M_A=0 \quad y_A=0$ $R_A=2WA_1$ $\theta_A=\frac{W}{EI\beta^2}A_3$	$\theta_A=0 \quad y_A=0$ $R_A=2WA_4$ $M_A=\frac{2W}{\beta}A_3$	$LT_V=-WF_{a1}$ $LT_M=\frac{-W}{2\beta}F_{a2}$ $LT_\theta=\frac{-W}{2EI\beta^2}F_{a3}$ $LT_y=\frac{-W}{4EI\beta^3}F_{a4}$
2. Uniformly distributed load from a to b 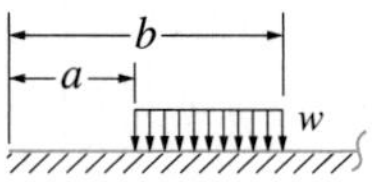	$R_A=0 \quad M_A=0$ $\theta_A=\frac{-w}{EI\beta^3}(B_3-A_3)$ $y_A=\frac{-w}{2EI\beta^4}(B_2-A_2)$	$R_A=0 \quad \theta_A=0$ $M_A=\frac{-w}{\beta^2}(B_3-A_3)$ $y_A=\frac{w}{2EI\beta^4}(B_1-A_1)$	$M_A=0 \quad y_A=0$ $R_A=\frac{w}{\beta}(B_2-A_2)$ $\theta_A=\frac{w}{2EI\beta^3}(B_4-A_4)$	$\theta_A=0 \quad y_A=0$ $R_A=\frac{-2w}{\beta}(B_1-A_1)$ $M_A=\frac{w}{\beta^2}(B_4-A_4)$	$LT_V=\frac{-w}{2\beta}(F_{a2}-F_{b2})$ $LT_M=\frac{-w}{2\beta^2}(F_{a3}-F_{b3})$ $LT_\theta=\frac{-w}{4EI\beta^3}(F_{a4}-F_{b4})$ $LT_y=\frac{-w}{4EI\beta^4}(F_{a5}-F_{b5})$
3. Uniform increasing load from a to b 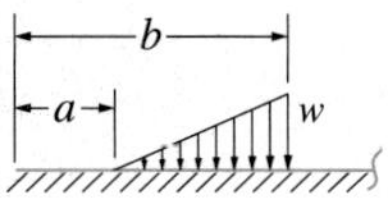	$R_A=0 \quad M_A=0$ $\theta_A=\frac{w}{2EI\beta^4}\left(\frac{B_4-A_4}{b-a}-2\beta B_3\right)$ $y_A=\frac{w}{2EI\beta^5}\left(\frac{B_3-A_3}{b-a}-\beta B_2\right)$	$R_A=0 \quad \theta_A=0$ $M_A=\frac{w}{2\beta^3}\left(\frac{B_4-A_4}{b-a}-2\beta B_3\right)$ $y_A=\frac{-w}{4EI\beta^5}\left(\frac{B_2-A_2}{b-a}-2\beta B_1\right)$	$M_A=0 \quad y_A=0$ $R_A=\frac{-w}{\beta^2}\left(\frac{B_3-A_3}{b-a}-\beta B_2\right)$ $\theta_A=\frac{w}{2EI\beta^4}\left(\frac{B_1-A_1}{b-a}+\beta B_4\right)$	$\theta_A=0 \quad y_A=0$ $R_A=\frac{w}{\beta^2}\left(\frac{B_2-A_2}{b-a}-2\beta B_1\right)$ $M_A=\frac{w}{\beta^3}\left(\frac{B_1-A_1}{b-a}+\beta B_4\right)$	$LT_V=\frac{-w}{2\beta^2}\left(\frac{F_{a3}-F_{b3}}{b-a}-\beta F_{b2}\right)$ $LT_M=\frac{-w}{4\beta^3}\left(\frac{F_{a4}-F_{b4}}{b-a}-2\beta F_{b3}\right)$ $LT_\theta=\frac{-w}{4EI\beta^4}\left(\frac{F_{a5}-F_{b5}}{b-a}-\beta F_{b4}\right)$ $LT_y=\frac{-w}{8EI\beta^5}\left(\frac{F_{a6}-F_{b6}}{b-a}-2\beta F_{b5}\right)$
4. Concentrated intermediate moment (if $\beta a > 3$, see case 11) 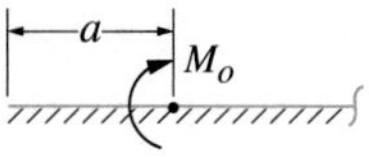	$R_A=0 \quad M_A=0$ $\theta_A=\frac{-2M_o}{EI\beta}A_1$ $y_A=\frac{M_o}{EI\beta^2}A_4$ (If $a = 0$, see case 9.)	$R_A=0 \quad \theta_A=0$ $M_A=-2M_oA_1$ $y_A=\frac{-M_o}{EI\beta^2}A_3$	$M_A=0 \quad y_A=0$ $R_A=-2M_o\beta A_4$ $\theta_A=\frac{M_o}{EI\beta}A_2$	$\theta_A=0 \quad y_A=0$ $R_A=4M_o\beta A_3$ $M_A=2M_oA_2$	$LT_V=-M_o\beta F_{a4}$ $LT_M=M_oF_{a1}$ $LT_\theta=\frac{M_o}{2EI\beta}F_{a2}$ $LT_y=\frac{M_o}{2EI\beta^2}F_{a3}$
5. Externally created concentrated angular displacement 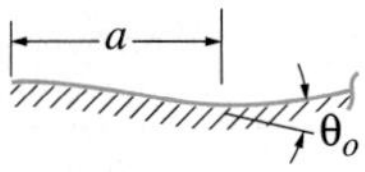	$R_A=0 \quad M_A=0$ $\theta_A=-2\theta_oA_4$ $y_A=\frac{-2\theta_o}{\beta}A_3$	$R_A=0 \quad \theta_A=0$ $M_A=-2\theta_oEI\beta A_4$ $y_A=\frac{\theta_o}{\beta}A_2$	$M_A=0 \quad y_A=0$ $R_A=4\theta_oEI\beta^2A_3$ $\theta_A=-2\theta_oA_1$	$\theta_A=0 \quad y_A=0$ $R_A=-4\theta_oEI\beta^2A_2$ $M_A=-4\theta_oEI\beta A_1$	$LT_V=-2\theta_oEI\beta^2F_{a3}$ $LT_M=-\theta_oEI\beta F_{a4}$ $LT_\theta=\theta_oF_{a1}$ $LT_y=\frac{\theta_o}{2\beta}F_{a2}$
6. Externally created concentrated lateral displacement 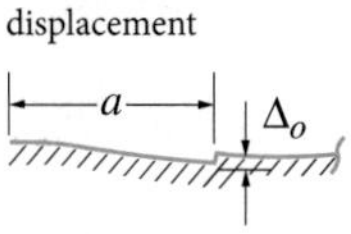	$R_A=0 \quad M_A=0$ $\theta_A=4\Delta_o\beta A_3$ $y_A=2\Delta_oA_2$	$R_A=0 \quad \theta_A=0$ $M_A=4\Delta_oEI\beta^2A_3$ $y_A=-2\Delta_oA_1$	$M_A=0 \quad y_A=0$ $R_A=-4\Delta_oEI\beta^3A_2$ $\theta_A=-2\Delta_o\beta A_4$	$\theta_A=0 \quad y_A=0$ $R_A=8\Delta_oEI\beta^3A_1$ $M_A=-4\Delta_oEI\beta^2A_4$	$LT_V=-2\Delta_oEI\beta^3F_{a2}$ $LT_M=-2\Delta_oEI\beta^2F_{a3}$ $LT_\theta=-\Delta_o\beta F_{a4}$ $LT_y=\Delta_oF_{a1}$
7. Uniform temperature differential from top to bottom 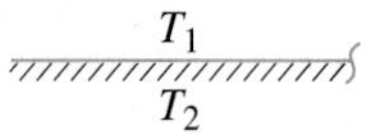	$R_A=0 \quad M_A=0$ $\theta_A=\frac{T_1-T_2}{t\beta}\gamma$ $y_A=-\frac{T_1-T_2}{2t\beta^2}\gamma$	$R_A=0 \quad \theta_A=0$ $M_A=\frac{T_1-T_2}{t}\gamma EI$ $y_A=0$	$M_A=0 \quad y_A=0$ $R_A=\frac{T_1-T_2}{t}\gamma EI\beta$ $\theta_A=\frac{T_1-T_2}{2t\beta}\gamma$	$\theta_A=0 \quad y_A=0$ $R_A=0$ $M_A=\frac{T_1-T_2}{t}\gamma EI$	$LT_V=\frac{T_1-T_2}{t}\gamma EI\beta F_4$ $LT_M=\frac{T_1-T_2}{t}\gamma EI(1-F_1)$ $LT_\theta=-\frac{T_1-T_2}{2t\beta}\gamma F_2$ $LT_y=-\frac{T_1-T_2}{2t\beta^2}\gamma F_3$

(Continued)

TABLE 8.6 Shear, Moment, Slope, and Deflection Formulas for Semi-Infinite Beams on Elastic Foundations (*Continued*)

Simple Loads on Semi-Infinite and on Infinite Beams on Elastic Foundations		
Loading, Reference No.	Shear, Moment, and Deformation Equations	Selected Maximum Values
8. Concentrated end load on a semi-infinite beam, left end free 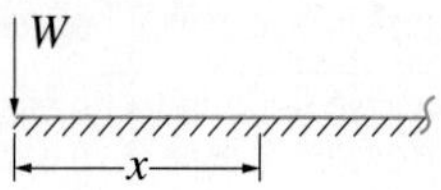	$V = -We^{-\beta x}(\cos\beta x - \sin\beta x)$ $M = -\frac{W}{\beta}e^{-\beta x}\sin\beta x$ $\theta = \frac{W}{2EI\beta^2}e^{-\beta x}(\cos\beta x + \sin\beta x)$ $y = -\frac{W}{2EI\beta^3}e^{-\beta x}\cos\beta x$	Max $V = -W$ at $x = 0$ Max $M = -0.3224\frac{W}{\beta}$ at $x = \frac{\pi}{4\beta}$ Max $\theta = \frac{W}{2EI\beta^2}$ at $x = 0$ Max $y = \frac{-W}{2EI\beta^3}$ at $x = 0$
9. Concentrated end moment on a semi-infinite beam, left end free 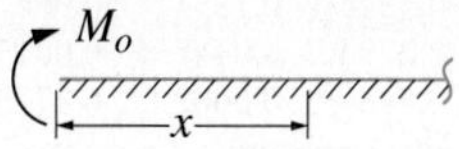	$V = -2M_o\beta e^{-\beta x}\sin\beta x$ $M = M_o e^{-\beta x}(\cos\beta x + \sin\beta x)$ $\theta = -\frac{M_o}{EI\beta}e^{-\beta x}\cos\beta x$ $y = -\frac{M_o}{2EI\beta^2}e^{-\beta x}(\sin\beta x - \cos\beta x)$	Max $V = -0.6448M_o\beta$ at $x = \frac{\pi}{4\beta}$ Max $M = M_o$ at $x = 0$ Max $\theta = -\frac{M_o}{EI\beta}$ at $x = 0$ Max $y = \frac{M_o}{2EI\beta^2}$ at $x = 0$
10. Concentrated load on an infinite beam 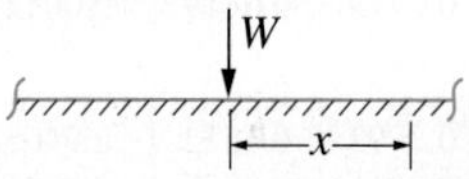	$V = -\frac{W}{2}e^{-\beta x}\cos\beta x$ $M = \frac{W}{4\beta}e^{-\beta x}(\cos\beta x - \sin\beta x)$ $\theta = \frac{W}{4EI\beta^2}e^{-\beta x}\sin\beta x$ $y = -\frac{W}{8EI\beta^3}e^{-\beta x}(\cos\beta x + \sin\beta x)$	Max $V = -\frac{W}{2}$ at $x = 0$ Max $M = \frac{W}{4\beta}$ at $x = 0$ Max $\theta = 0.0806\frac{W}{EI\beta^2}$ at $x = \frac{\pi}{4\beta}$ Max $y = -\frac{W}{8EI\beta^3}$ at $x = 0$
11. Concentrated moment on an infinite beam 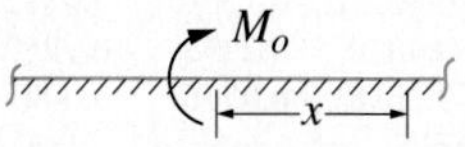	$V = -\frac{M_o\beta}{2}e^{-\beta x}(\cos\beta x + \sin\beta x)$ $M = \frac{M_o}{2}e^{-\beta x}\cos\beta x$ $\theta = -\frac{M_o}{4EI\beta}e^{-\beta x}(\cos\beta x - \sin\beta x)$ $y = -\frac{M_o}{4EI\beta^2}e^{-\beta x}\sin\beta x$	Max $V = -\frac{M_o\beta}{2}$ at $x = 0$ Max $M = \frac{M_o}{2}$ at $x = 0$ Max $\theta = -\frac{M_o}{4EI\beta}$ at $x = 0$ Max $y = -0.0806\frac{M_o}{EI\beta^2}$ at $x = \frac{\pi}{4\beta}$

TABLE 8.7(*a*) Reaction and Deflection Coefficients for Beams Under Simultaneous Axial and Transverse Loading: Cantilver End Support

Case No. in Table 8.1	Load Location a/l	Coefficient Listed for	Axial Compressive Load, $kl=\sqrt{Pl^2/EI}$					Axial Tensile Load, $kl=\sqrt{Pl^2/EI}$				
			0.2	0.4	0.6	0.8	1.0	0.5	1.0	2.0	4.0	8.0
1a. Conc. load	0	y_A	1.0163	1.0684	1.1686	1.3455	1.6722	0.9092	0.7152	0.3885	0.1407	0.0410
		θ_A	1.0169	1.0713	1.1757	1.3604	1.7016	0.9054	0.7039	0.3671	0.1204	0.0312
		M_B	1.0136	1.0570	1.1402	1.2870	1.5574	0.9242	0.7616	0.4820	0.2498	0.1250
	0.5	y_A	1.0153	1.0646	1.1589	1.3256	1.6328	0.9142	0.7306	0.4180	0.1700	0.0566
		θ_A	1.0195	1.0821	1.2026	1.4163	1.8126	0.8914	0.6617	0.2887	0.0506	0.0022
		M_B	1.0085	1.0355	1.0869	1.1767	1.3402	0.9524	0.8478	0.6517	0.4333	0.2454
2a. Uniform load	0	y_A	1.0158	1.0665	1.1638	1.3357	1.6527	0.9117	0.7228	0.4031	0.1552	0.0488
		θ_A	1.0183	1.0771	1.1900	1.3901	1.7604	0.8980	0.6812	0.3243	0.0800	0.0117
		M_B	1.0102	1.0427	1.1047	1.2137	1.4132	0.9430	0.8193	0.5969	0.3792	0.2188
	0.5	y_A	1.1050	1.0629	1.1548	1.3171	1.6161	0.9164	0.7373	0.4314	0.1851	0.0667
		θ_A	1.0198	1.0835	1.2062	1.4239	1.8278	0.8896	0.6562	0.2794	0.0447	0.0015
		M_B	1.0059	1.0248	1.0606	1.1229	1.2357	0.9666	0.8925	0.7484	0.5682	0.3773
2a. Uniformly increasing	0	y_A	1.0155	1.0652	1.1604	1.3287	1.6389	0.9135	0.7283	0.4137	0.1662	0.0552
		θ_A	1.0190	1.0799	1.1972	1.4051	1.7902	0.8942	0.6700	0.3039	0.0629	0.0057
		M_B	1.0081	1.0341	1.0836	1.1701	1.3278	0.9543	0.8543	0.6691	0.4682	0.2930
	0.5	y_A	1.0147	1.0619	1.1523	1.3118	1.6056	0.9178	0.7415	0.4400	0.1951	0.0740
		θ_A	1.0200	1.0843	1.2080	1.4277	1.8355	0.8887	0.6535	0.2748	0.0419	0.0012
		M_B	1.0046	1.0191	1.0467	1.0944	1.1806	0.9742	0.9166	0.8020	0.6489	0.4670
2a. Uniformly decreasing	0	y_A	1.0159	1.0670	1.1650	1.3382	1.6578	0.9110	0.7208	0.3992	0.1512	0.0465
		θ_A	1.0181	1.0761	1.1876	1.3851	1.7505	0.8992	0.6850	0.3311	0.0857	0.0136
		M_B	1.0112	1.0469	1.1153	1.2355	1.4559	0.9374	0.8018	0.5609	0.3348	0.1817
	0.5	y_A	1.0150	1.0633	1.1557	1.3189	1.6197	0.9159	0.7358	0.4284	0.1816	0.0642
		θ_A	1.0198	1.0833	1.2056	1.4226	1.8253	0.8899	0.6571	0.2809	0.0456	0.0016
		M_B	1.0066	1.0276	1.0676	1.1372	1.2632	0.9628	0.8804	0.7215	0.5279	0.3324
3a. Conc. moment	0	y_A	1.0169	1.0713	1.1757	1.3604	1.7016	0.9054	0.7039	0.3671	0.1204	0.0312
		θ_A	1.0136	1.0570	1.1402	1.2870	1.5574	0.9242	0.7616	0.4820	0.2498	0.1250
		M_B	1.0203	1.0857	1.2116	1.4353	1.8508	0.8868	0.6481	0.2658	0.0366	0.0007
	0.5	y_A	1.0161	1.0677	1.1668	1.3418	1.6646	0.9101	0.7180	0.3932	0.1437	0.0409
		θ_A	1.0186	1.0785	1.1935	1.3974	1.7747	0.8961	0.6754	0.3124	0.0664	0.0046
		M_B	1.0152	1.0641	1.1575	1.3220	1.6242	0.9147	0.7308	0.4102	0.1378	0.0183

TABLE 8.7(*b*) Reaction and Deflection Coefficients for Beams Under Simultaneous Axial and Transverse Loading: Simply Supported Ends

Case No. in Table 8.1	Load Location a/l	Coefficient Listed for	Axial Compressive Load, $kl=\sqrt{Pl^2/EI}$					Axial Tensile Load, $kl=\sqrt{Pl^2/EI}$				
			0.4	0.8	1.2	1.6	2.0	1.0	2.0	4.0	8.0	12.0
1e. Conc. load	0.25	$y_{l/2}$	1.0167	1.0702	1.1729	1.3546	1.6902	0.9069	0.7082	0.3751	0.1273	0.0596
		θ_A	1.0144	1.0605	1.1485	1.3031	1.5863	0.9193	0.7447	0.4376	0.1756	0.0889
		θ_B	1.0185	1.0779	1.1923	1.3958	1.7744	0.8972	0.6805	0.3311	0.0990	0.0444
		$M_{l/4}$	1.0101	1.0425	1.1039	1.2104	1.4025	0.9427	0.8158	0.5752	0.3272	0.2217
	0.50	$y_{l/2}$	1.0163	1.0684	1.1686	1.3455	1.6722	0.9092	0.7152	0.3885	0.1407	0.0694
		θ_A	1.0169	1.0713	1.1757	1.3604	1.7016	0.9054	0.7039	0.3671	0.1204	0.0553
		$M_{l/2}$	1.0136	1.0570	1.1402	1.2870	1.5574	0.9242	0.7616	0.4820	0.2498	0.1667

(Continued)

TABLE 8.7(*b*) Reaction and Deflection Coefficients for Beams Under Simultaneous Axial and Transverse Loading: Simply Supported Ends (*Continued*)

Case No. in Table 8.1	Load Location a/l	Coefficient Listed for	Axial Compressive Load, $kl = \sqrt{Pl^2/EI}$					Axial Tensile Load, $kl = \sqrt{Pl^2/EI}$				
			0.4	0.8	1.2	1.6	2.0	1.0	2.0	4.0	8.0	12.0
2e. Uniform load	0	$y_{l/2}$	1.0165	1.0696	1.1714	1.3515	1.6839	0.9077	0.7107	0.3797	0.1319	0.0630
		θ_A	1.0163	1.0684	1.1686	1.3455	1.6722	0.9092	0.7152	0.3885	0.1407	0.0694
		$M_{l/2}$	1.0169	1.0713	1.1757	1.3604	1.7016	0.9054	0.7039	0.3671	0.1204	0.0553
	0.50	$y_{l/2}$	1.0165	1.0696	1.1714	1.3515	1.6839	0.9077	0.7107	0.3797	0.1319	0.0630
		θ_A	1.0180	1.0759	1.1873	1.3851	1.7524	0.8997	0.6875	0.3418	0.1053	0.0475
		θ_B	1.0149	1.0626	1.1540	1.3147	1.6099	0.9166	0.7368	0.4248	0.1682	0.0865
		$M_{l/2}$	1.0169	1.0713	1.1757	1.3604	1.7016	0.9054	0.7039	0.3671	0.1204	0.0553
2e. Uniformly increasing	0	$y_{l/2}$	1.0165	1.0696	1.1714	1.3515	1.6839	0.9077	0.7107	0.3797	0.1319	0.0630
		θ_A	1.0172	1.0722	1.1781	1.3656	1.7127	0.9044	0.7011	0.3643	0.1214	0.0570
		θ_B	1.0155	1.0651	1.1603	1.3280	1.6368	0.9134	0.7276	0.4097	0.1575	0.0803
		$M_{l/2}$	1.0169	1.0713	1.1757	1.3604	1.7016	0.9054	0.7039	0.3671	0.1204	0.0553
	0.50	$y_{l/2}$	1.0167	1.0702	1.1729	1.3545	1.6899	0.9069	0.7084	0.3754	0.1278	0.0601
		θ_A	1.0184	1.0776	1.1915	1.3942	1.7710	0.8976	0.6816	0.3329	0.1002	0.0450
		θ_B	1.0140	1.0588	1.1445	1.2948	1.5702	0.9215	0.7516	0.4521	0.1936	0.1048
		$M_{l/2}$	1.0183	1.0771	1.1900	1.3901	1.7604	0.8980	0.6812	0.3243	0.0800	0.0270
3e. Conc. moment	0	$y_{l/2}$	1.0169	1.0713	1.1757	1.3604	1.7016	0.9054	0.7039	0.3671	0.1204	0.0553
		θ_A	1.0108	1.0454	1.1114	1.2266	1.4365	0.9391	0.8060	0.5630	0.3281	0.2292
		θ_B	1.0190	1.0801	1.1979	1.4078	1.7993	0.8945	0.6728	0.3200	0.0932	0.0417
	0.25	$y_{l/2}$	1.0161	1.0677	1.1668	1.3418	1.6646	0.9101	0.7180	0.3932	0.1437	0.0704
		θ_A	1.0202	1.0852	1.2102	1.4318	1.8424	0.8873	0.6485	0.2595	0.0113	−0.0244
		θ_B	1.0173	1.0728	1.1795	1.3682	1.7174	0.9035	0.6982	0.3571	0.1131	0.0512

TABLE 8.7(*c*) Reaction and Deflection Coefficients for Beams Under Simultaneous Axial and Transverse Loading: Left End Simply Supported, Right End Fixed

Case No. in Table 8.1	Load Location a/l	Coefficient Listed for	Axial Compressive Load, $kl = \sqrt{Pl^2/EI}$					Axial Tensile Load, $kl = \sqrt{Pl^2/EI}$				
			0.6	1.2	1.8	2.4	3.0	1.0	2.0	4.0	8.0	12.0
1c. Conc. load	0.25	$y_{l/2}$	1.0190	1.0804	1.2005	1.4195	1.8478	0.9507	0.8275	0.5417	0.2225	0.1108
		θ_A	1.0172	1.0726	1.1803	1.3753	1.7530	0.9553	0.8429	0.5762	0.2576	0.1338
		M_B	1.0172	1.0728	1.1818	1.3812	1.7729	0.9554	0.8443	0.5881	0.3018	0.1940
	0.50	$y_{l/2}$	1.0170	1.0719	1.1786	1.3718	1.7458	0.9557	0.8444	0.5802	0.2647	0.1416
		θ_A	1.0199	1.0842	1.2101	1.4406	1.8933	0.9485	0.8202	0.5255	0.2066	0.1005
		M_B	1.1037	1.0579	1.1432	1.2963	1.5890	0.9642	0.8733	0.6520	0.3670	0.2412
2c. Uniform load	0	$y_{l/2}$	1.0176	1.0742	1.1846	1.3848	1.7736	0.9543	0.8397	0.5694	0.2524	0.1323
		θ_A	1.0183	1.0776	1.1933	1.4042	1.8162	0.9524	0.8334	0.5561	0.2413	0.1263
		M_B	1.0122	1.0515	1.1273	1.2635	1.5243	0.9681	0.8874	0.6900	0.4287	0.3033
	0.50	$y_{l/2}$	1.0163	1.0689	1.1709	1.3549	1.7094	9.9575	0.8502	0.5932	0.2778	0.1505
		θ_A	1.0202	1.0856	1.2139	1.4496	1.9147	0.9477	0.8179	0.5224	0.2087	0.1048
		M_B	1.0091	1.0383	1.0940	1.1920	1.3744	0.9760	0.9141	0.7545	0.5126	0.3774
2c. Uniformly increasing	0	$y_{l/2}$	1.0170	1.0719	1.1785	1.3716	1.7453	0.9557	0.8444	0.5799	0.2637	0.1405
		θ_A	1.0192	1.0814	1.2030	1.4255	1.8619	0.9502	0.8259	0.5394	0.2237	0.1136
		M_B	1.0105	1.0440	1.1084	1.2230	1.4399	0.9726	0.9028	0.7277	0.4799	0.3504
	0.50	$y_{l/2}$	1.0160	1.0674	1.1669	1.3463	1.6911	0.9584	0.8533	0.6003	0.2860	0.1571
		θ_A	1.0202	1.0855	1.2138	1.4499	1.9165	0.9478	0.8183	0.5245	0.2141	0.1105
		M_B	1.0071	1.0298	1.0726	1.1473	1.2843	0.9813	0.9325	0.8029	0.5900	0.4573

(*Continued*)

TABLE 8.7(*c*) Reaction and Deflection Coefficients for Beams Under Simultaneous Axial and Transverse Loading: Left End Simply Supported, Right End Fixed (*Continued*)

Case No. in Table 8.1	Load Location a/l	Coefficient Listed for	Axial Compressive Load, $kl=\sqrt{Pl^2/EI}$					Axial Tensile Load, $kl=\sqrt{Pl^2/EI}$				
			0.6	1.2	1.8	2.4	3.0	1.0	2.0	4.0	8.0	12.0
2c. Uniformly decreasing	0	$y_{l/2}$	1.0180	1.0762	1.1895	1.3957	1.7968	0.9532	0.8359	0.5608	0.2431	0.1256
		θ_A	1.0177	1.0751	1.1868	1.3900	1.7857	0.9539	0.8383	0.5673	0.2531	0.1347
		M_B	1.0142	1.0600	1.1489	1.3098	1.6207	0.9630	0.8698	0.6470	0.3701	0.2495
	0.50	$y_{l/2}$	1.0165	1.0695	1.1725	1.3584	1.7169	0.9571	0.8490	0.5902	0.2743	0.1477
		θ_A	1.0202	1.0856	1.2139	1.4495	1.9140	0.9477	0.8177	0.5216	0.2066	0.1026
		M_B	1.0104	1.0439	1.1078	1.2208	1.4327	0.9726	0.9023	0.7232	0.4625	0.3257
3c. Conc. moment	0	$y_{l/2}$	1.0199	1.0842	1.2101	1.4406	1.8933	0.9485	0.8202	0.5255	0.2066	0.1005
		θ_A	1.0122	1.0515	1.1273	1.2635	1.5243	0.9681	0.8874	0.6900	0.4287	0.3030
		M_B	1.0183	1.0779	1.1949	1.4105	1.8379	0.9525	0.8348	0.5684	0.2842	0.1704
	0.50	$y_{l/2}$	1.0245	1.1041	1.2613	1.5528	2.1347	0.9368	0.7812	0.4387	0.1175	0.0390
		θ_A	1.0168	1.0707	1.1750	1.3618	1.7186	0.9562	0.8452	0.5760	0.2437	0.1176
		M_B	0.9861	0.9391	0.8392	0.6354	0.1828	1.0346	1.1098	1.1951	0.9753	0.7055

TABLE 8.7(*d*) Reaction and Deflection Coefficient for Beams Under Simultaneous Axial and Transverse Loading: Fixed Ends

Case No. in Table 8.1	Load Location a/l	Coefficient Listed for	Axial Compressive Load, $kl=\sqrt{Pl^2/EI}$					Axial Tensile Load, $kl=\sqrt{Pl^2/EI}$				
			0.8	1.6	2.4	3.2	4.0	1.0	2.0	4.0	8.0	12.0
1d. Conc. load	0.25	$y_{l/2}$	1.0163	1.0684	1.1686	1.3455	1.6722	0.9756	0.9092	0.7152	0.3885	0.2228
		R_A	1.0007	1.0027	1.0064	1.0121	1.0205	0.9990	0.9960	0.9859	0.9613	0.9423
		M_A	1.0088	1.0366	1.0885	1.1766	1.3298	0.9867	0.9499	0.8350	0.6008	0.4416
		M_B	1.0143	1.0603	1.1498	1.3117	1.6204	0.9787	0.9213	0.7583	0.4984	0.3645
	0.50	$y_{l/2}$	1.0163	1.0684	1.1686	1.3455	1.6722	0.9756	0.9092	0.7152	0.3885	0.2228
		M_A	1.0136	1.0570	1.1402	1.2870	1.5574	0.9797	0.9242	0.7616	0.4820	0.3317
		$M_{l/2}$	1.0136	1.0570	1.1402	1.2870	1.5574	0.9797	0.9242	0.7616	0.4820	0.3317
2d. Uniform load	0	$y_{l/2}$	1.0163	1.0684	1.1686	1.3455	1.6722	0.9756	0.9092	0.7152	0.3885	0.2228
		M_A	1.0108	1.0454	1.1114	1.2266	1.4365	0.9837	0.9391	0.8060	0.5630	0.4167
		$M_{l/2}$	1.0190	1.0801	1.1979	1.4078	1.7993	0.9716	0.8945	0.6728	0.3200	0.1617
	0.50	$y_{l/2}$	1.0146	1.0667	1.1667	1.3434	1.6696	0.9741	0.9077	0.7141	0.3879	0.2224
		R_A	0.9982	0.9927	0.9828	0.9677	0.9453	1.0027	1.0106	1.0375	1.1033	1.1551
		M_A	1.0141	1.0595	1.1473	1.3045	1.5999	0.9789	0.9217	0.7571	0.4868	0.3459
		M_B	1.0093	1.0390	1.0950	1.1913	1.3622	0.9859	0.9470	0.8282	0.5976	0.4488
2d. Uniformly increasing	0	$y_{l/2}$	1.0163	1.0684	1.1686	1.3455	1.6722	0.9756	0.9092	0.7152	0.3885	0.2228
		R_A	0.9995	0.9979	0.9951	0.9908	0.9845	1.0008	1.0030	1.0108	1.0303	1.0463
		M_A	1.0124	1.0521	1.1282	1.2627	1.5107	0.9814	0.9307	0.7818	0.5218	0.3750
		M_B	1.0098	1.0410	1.1001	1.2026	1.3870	0.9853	0.9447	0.8221	0.5904	0.4445
	0.50	$y_{l/2}$	1.0161	1.0679	1.1672	1.3427	1.6667	0.9758	0.9099	0.7174	0.3927	0.2274
		R_A	0.9969	0.9875	0.9707	0.9449	0.9070	1.0047	1.0182	1.0648	1.1815	1.2778
		M_A	1.0141	1.0595	1.1476	1.3063	1.6076	0.9790	0.9222	0.7602	0.4995	0.3647
		M_B	1.0075	1.0312	1.0755	1.1507	1.2819	0.9887	0.9573	0.8594	0.6582	0.5168
3d. Conc. moment	0.25	$y_{l/2}$	1.0169	1.0713	1.1757	1.3604	1.7016	0.9746	0.9054	0.7039	0.3671	0.2001
		R_A	0.9993	0.9972	0.9932	0.9867	0.9763	1.0010	1.0038	1.0122	1.0217	1.0134
		M_A	1.0291	1.1227	1.3025	1.6203	2.2055	0.9563	0.8376	0.4941	−0.0440	−0.2412
		M_B	1.0151	1.0635	1.1571	1.3244	1.6380	0.9775	0.9164	0.7404	0.4517	0.3035
	0.50	$\theta_{l/2}$	1.0054	1.0220	1.0515	1.0969	1.1641	0.9918	0.9681	0.8874	0.6900	0.5346
		R_A	1.0027	1.0110	1.0260	1.0492	1.0842	0.9959	0.9842	0.9449	0.8561	0.7960
		M_A	1.0081	1.0331	1.0779	1.1477	1.2525	0.9877	0.9525	0.8348	0.5684	0.3881

TABLE 8.8 Shear, Moment, Slope, and Deflection Formulas for Beams Under Simultaneous Axial Compression and Transverse Loading

Notation: P = axial compressive load (force); all other notation is the same as that for Table 8.1. P must be less than P_{cr} where $P_{cr} = K_1\pi^2 EI/l^2$ and where, for cases 1a–6a and 1f–6f, $K_1 = 0.25$; for cases 1b–6b and 1e–6e; $K_1 = 1$; for cases 1c–6c, $K_1 = 2.046$; and for cases 1d–6d, $K_1 = 4$.

The following constants and functions are hereby defined in order to permit condensing the tabulated formulas which follow. $k = (P/EI)^{1/2}$. (*Note:* See page 155 for a definition of $\langle x-a\rangle^n$.) The function $\sin k\langle x-a\rangle$ is also defined as having a value of zero if $x < a$.

$F_1 = \cos kx$	$F_{a1} = \langle x-a\rangle^0 \cos k\langle x-a\rangle$	$C_1 = \cos kl$	$C_{a1} = \cos k(l-a)$	(*Note:* M_A and R_A as well as M_B and R_B are reactions, not applied loads. They exist only when the necessary end restraints are provided.)
$F_2 = \sin kx$	$F_{a2} = \sin k\langle x-a\rangle$	$C_2 = \sin kl$	$C_{a2} = \sin k(l-a)$	
$F_3 = 1-\cos kx$	$F_{a3} = \langle x-a\rangle^0[1-\cos k\langle x-a\rangle]$	$C_3 = 1-\cos kl$	$C_{a3} = 1-\cos k(l-a)$	
$F_4 = kx-\sin kx$	$F_{a4} = k\langle x-a\rangle - \sin k\langle x-a\rangle$	$C_4 = kl-\sin kl$	$C_{a4} = k(l-a)-\sin k(l-a)$	
	$F_{a5} = \frac{k^2}{2}\langle x-a\rangle^2 - F_{a3}$		$C_{a5} = \frac{k^2}{2}(l-a)^2 - C_{a3}$	
	$F_{a6} = \frac{k^3}{6}\langle x-a\rangle^3 - F_{a4}$		$C_{a6} = \frac{k^3}{6}(l-a)^3 - C_{a4}$	

1. Axial compressive load plus concentrated intermediate lateral load

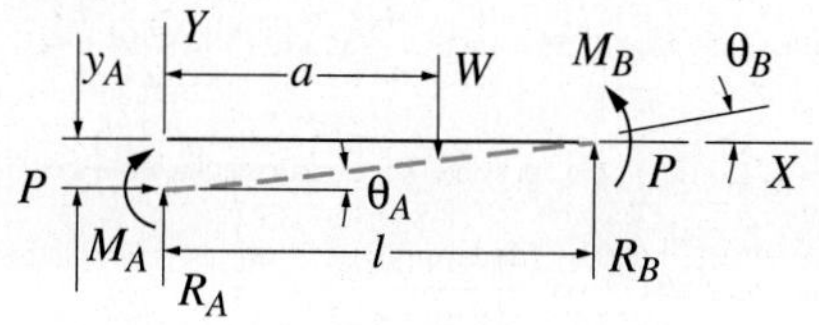

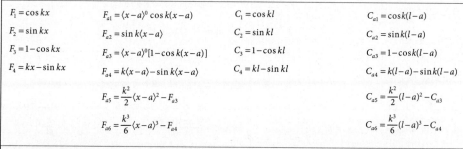

Transverse shear $= V = R_A F_1 - M_A k F_2 - \theta_A P F_1 - W F_{a1}$

Bending moment $= M = M_A F_1 + \frac{R_A}{k} F_2 - \frac{\theta_A P}{k} F_2 - \frac{W}{k} F_{a2}$

Slope $= \theta = \theta_A F_1 + \frac{M_A k}{P} F_2 + \frac{R_A}{P} F_3 - \frac{W}{P} F_{a3}$

Deflection $= y = y_A + \frac{\theta_A}{k} F_2 + \frac{M_A}{P} F_3 + \frac{R_A}{kP} F_4 - \frac{W}{kP} F_{a4}$

End Restraints, Reference No.	Boundary Values	Selected Maximum Values of Moments and Deformations
1a. Left end free, right end fixed (cantilever) 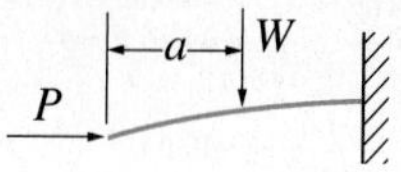 	$R_A = 0 \quad M_A = 0 \quad \theta_A = \frac{W}{P}\frac{C_{a3}}{C_1}$ $y_A = \frac{-W}{kP}\frac{C_2 C_{a3} - C_1 C_{a4}}{C_1}$ $R_B = W \quad \theta_B = 0 \quad y_B = 0$ $M_B = \frac{-W}{k}\frac{C_2 C_{a3} + C_1 C_{a2}}{C_1}$	Max $M = M_B$; max possible value $= \frac{-W}{k}\tan kl$ when $a = 0$ Max $\theta = \theta_A$; max possible value $= \frac{W}{P}\frac{1-\cos kl}{\cos kl}$ when $a = 0$ Max $y = y_A$; max possible value $= \frac{-W}{kP}(\tan kl - kl)$ when $a = 0$
1b. Left end guided, right end fixed 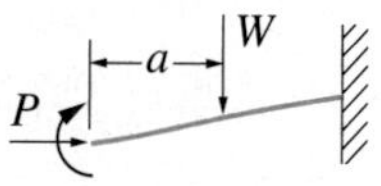 	$R_A = 0 \quad M_A = \frac{W}{k}\frac{C_{a3}}{C_2} \quad \theta_A = 0$ $y_A = \frac{-W}{kP}\frac{C_3 C_{a3} - C_2 C_{a4}}{C_2}$ $R_B = W \quad \theta_B = 0 \quad y_B = 0$ $M_B = \frac{-W}{k}\frac{\cos ka - \cos kl}{\sin kl}$	Max + $M = M_A$; max possible value $= \frac{W}{k}\tan\frac{kl}{2}$ when $a = 0$ Max $-M = M_B$; max possible value $= \frac{-W}{k}\tan\frac{kl}{2}$ when $a = 0$ Max $y = y_A$; max possible value $= \frac{-W}{kP}\left(2\tan\frac{kl}{2} - kl\right)$ when $a = 0$
1c. Left end simply supported, right end fixed 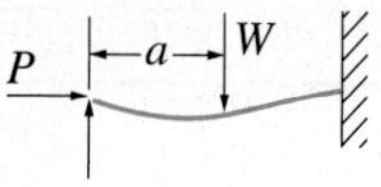 	$R_A = W\frac{C_2 C_{a3} - C_1 C_{a4}}{C_2 C_3 - C_1 C_4} \quad M_A = 0$ $\theta_A = \frac{-W}{P}\frac{C_4 C_{a3} - C_3 C_{a4}}{C_2 C_3 - C_1 C_4} \quad y_A = 0$ $R_B = W - R_A \quad \theta_B = 0 \quad y_B = 0$ $M_B = \frac{-W}{k}\frac{kl\sin ka - ka\sin kl}{\sin kl - kl\cos kl}$	Max $-M = M_B$; max possible value occurs when $a = \frac{1}{k}\cos^{-1}\frac{\sin kl}{kl}$ If $a = l/2$ (transverse center load), then $R_A = W\frac{\sin kl - \sin\frac{kl}{2} - \frac{kl}{2}\cos kl}{\sin kl - kl\cos kl}$ $M_B = -Wl\frac{\sin\frac{kl}{2}\left(1-\cos\frac{kl}{2}\right)}{\sin kl - kl\cos kl}$

(*Continued*)

TABLE 8.8 Shear, Moment, Slope, and Deflection Formulas for Beams Under Simultaneous Axial Compression and Transverse Loading (*Continued*)

End Restraints, Reference No.	Boundary Values	Selected Maximum Values of Moments and Deformations
1d. Left end fixed, right end fixed 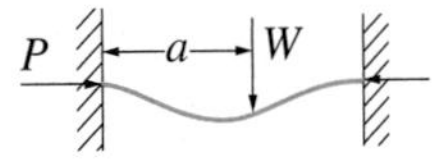 	$R_A = W\frac{C_3C_{a3} - C_2C_{a4}}{C_3^2 - C_2C_4}$ $M_A = \frac{-W}{k}\frac{C_4C_{a3} - C_3C_{a4}}{C_3^2 - C_2C_4}$ $\theta_A = 0 \quad y_A = 0$ $R_B = W - R_A \quad \theta_B = 0 \quad y_B = 0$ $M_B = M_A + R_A l - W(l-a)$	$\text{Max} - M = M_A$ if $a < \frac{l}{2}$ If $a = \frac{l}{2}$ (transverse center load), then $R_A = R_B = \frac{W}{2} \quad M_B = M_A = \frac{-W}{2k}\tan\frac{kl}{4}$ $\text{Max} + M = \frac{W}{2k}\tan\frac{kl}{4} \quad$ at $x = \frac{l}{2}$ $\text{Max } y = \frac{-W}{kP}\left(\tan\frac{kl}{4} - \frac{kl}{4}\right) \quad$ at $x = \frac{l}{2}$
1e. Left end simply supported, right end simply supported 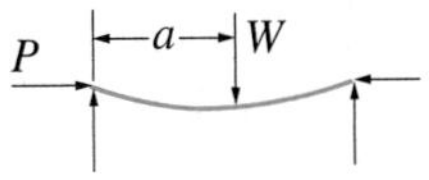 	$R_A = \frac{W}{l}(l-a) \quad M_A = 0 \quad y_A = 0$ $\theta_A = \frac{-W}{P}\left[\frac{\sin k(l-a)}{\sin kl} - \frac{l-a}{l}\right]$ $R_B = W\frac{a}{l} \quad M_B = 0 \quad y_B = 0$ $\theta_B = \frac{W}{P}\left(\frac{\sin ka}{\sin kl} - \frac{a}{l}\right)$	$\text{Max } M = \frac{W \sin k(l-a)}{k \sin kl}\sin ka \quad$ at $x = a$ if $\frac{l}{2} < a < \frac{\pi}{2k}$ $\text{Max } M = \frac{W \sin k(l-a)}{k \sin kl} \quad$ at $x = \frac{\pi}{2k}$ if $a > \frac{\pi}{2k}$ and $a > \frac{l}{2}$; max possible value of $M = \frac{W}{2k}\tan\frac{kl}{2}$ at $x = a$ when $a = \frac{l}{2}$ $\text{Max } \theta = \theta_B$ if $a > \frac{l}{2}$; max possible value occurs when $a = \frac{1}{k}\cos^{-1}\frac{\sin kl}{kl}$ Max y occurs at $x = \frac{1}{k}\cos^{-1}\frac{(l-a)\sin kl}{l \sin(l-a)}$ if $a > \frac{l}{2}$; max possible value $= \frac{-W}{2kP}\left(\tan\frac{kl}{2} - \frac{kl}{2}\right)$ at $x = \frac{l}{2}$ when $a = \frac{l}{2}$
1f. Left end guided, right end simply supported 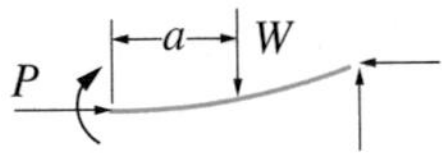 	$R_A = 0 \quad \theta_A = 0$ $M_A = \frac{W}{k}\frac{\sin k(l-a)}{\cos kl}$ $y_A = \frac{-W}{kP}\left[\frac{\sin k(l-a)}{\cos kl} - k(l-a)\right]$ $R_B = W \quad M_B = 0 \quad y_B = 0$ $\theta_B = \frac{W}{P}\left(\frac{\cos ka}{\cos kl} - 1\right)$	$\text{Max } M = M_A$; max possible value $= \frac{W}{k}\tan kl \quad$ when $a = 0$ $\text{Max } \theta = \theta_B$; max possible value $= \frac{W}{P}\frac{1 - \cos kl}{\cos kl} \quad$ when $a = 0$ $\text{Max } y = y_A$; max possible value $= \frac{-W}{kP}(\tan kl - kl)$ when $a = 0$
2. Axial compressive load plus distributed lateral load 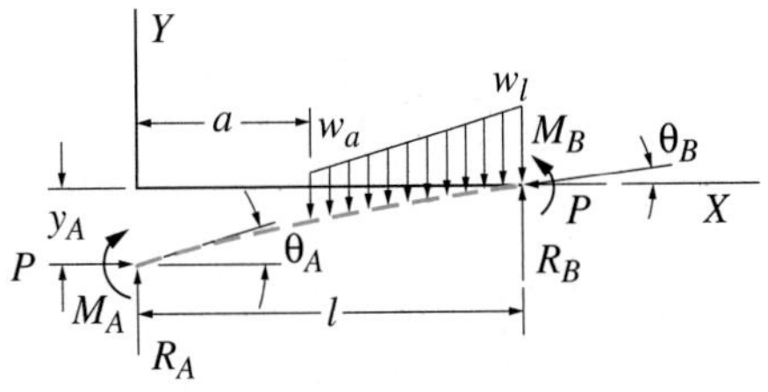 		Transverse shear $= V = R_A F_1 - M_A k F_2 - \theta_A P F_1 - \frac{w_a}{k} F_{a2} - \frac{w_l - w_a}{k^2(l-a)} F_{a3}$ Bending moment $= M = M_A F_1 + \frac{R_A}{k} F_2 - \frac{\theta_A P}{k} F_2 - \frac{w_a}{k^2} F_{a3} - \frac{w_l - w_a}{k^3(l-a)} F_{a4}$ Slope $= \theta = \theta_A F_1 + \frac{M_A k}{P} F_2 + \frac{R_A}{P} F_3 - \frac{w_a}{kP} F_{a4} - \frac{w_l - w_a}{k^2 P(l-a)} F_{a5}$ Deflection $= y = y_A + \frac{\theta_A}{k} F_2 + \frac{M_A}{P} F_3 + \frac{R_A}{kP} F_4 - \frac{w_a}{k^2 P} F_{a5} - \frac{w_l - w_a}{k^3 P(l-a)} F_{a6}$

(*Continued*)

TABLE 8.8 Shear, Moment, Slope, and Deflection Formulas for Beams Under Simultaneous Axial Compression and Transverse Loading (*Continued*)

End Restraints, Reference No.	Boundary Values	Selected Maximum Values of Moments and Deformations

2a. Left end free, right end fixed (cantilever)
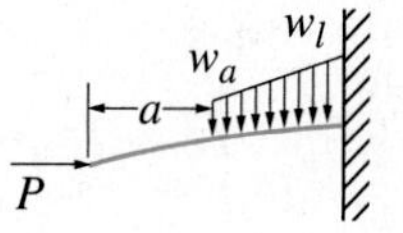
 	
	$R_A = 0 \quad M_A = 0$
	$\theta_A = \frac{w_a}{kP}\frac{C_{a4}}{C_1} + \frac{w_l - w_a}{k^2P(l-a)}\frac{C_{a5}}{C_1}$
	$y_A = \frac{-w_a}{k^2P}\frac{C_2C_{a4} - C_1C_{a5}}{C_1} - \frac{w_l - w_a}{k^3P(l-a)}\frac{C_2C_{a5} - C_1C_{a6}}{C_1}$
	$R_B = \frac{w_a + w_l}{2}(l-a) \quad \theta_B = 0$
	$M_B = \frac{-w_a}{k^2}\frac{C_2C_{a4} + C_1C_{a3}}{C_1} - \frac{w_l - w_a}{k^3(l-a)}\frac{C_2C_{a5} + C_1C_{a4}}{C_1}$
	$y_B = 0$
		If $a = 0$ and $w_a = w_l$ (uniform load on entire span), then
		$\text{Max } M = M_B = \frac{-w_a}{k^2}\left(1 + kl\tan kl - \frac{1}{\cos kl}\right)$
		$\text{Max } \theta = \theta_A = \frac{w_a}{kP}\left(\frac{kl}{\cos kl} - \tan kl\right)$
		$\text{Max } y = y_A = \frac{-w_a}{k^2P}\left(1 + kl\tan kl - \frac{k^2l^2}{2} - \frac{1}{\cos kl}\right)$
		If $a = 0$ and $w_a = 0$ (uniformly increasing load), then
		$\text{Max } M = M_B = \frac{-w_l}{k^2}\left(1 + \frac{kl}{2}\tan kl - \frac{\tan kl}{kl}\right)$
		$\text{Max } \theta = \theta_A = \frac{w_l}{kP}\left(\frac{1}{kl} + \frac{kl}{2\cos kl} - \frac{1}{kl\cos kl}\right)$
		$\text{Max } y = y_A = \frac{-w_l}{k^2P}\left(1 + \frac{kl}{2}\tan kl - \frac{k^2l^2}{6} - \frac{\tan kl}{kl}\right)$

2b. Left end guided, right end fixed
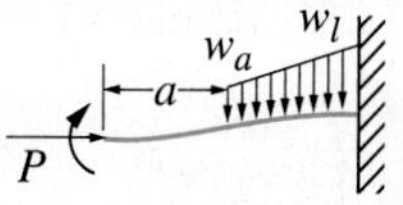
 	
	$R_A = 0 \quad \theta_A = 0$
	$M_A = \frac{w_a}{k^2}\frac{C_{a4}}{C_2} + \frac{w_l - w_a}{k^3(l-a)}\frac{C_{a5}}{C_2}$
	$y_A = \frac{-w_a}{k^2P}\frac{C_3C_{a4} - C_2C_{a5}}{C_2} - \frac{w_l - w_a}{k^3P(l-a)}\frac{C_3C_{a5} - C_2C_{a6}}{C_2}$
	$R_B = \frac{w_a + w_l}{2}(l-a) \quad \theta_B = 0$
	$M_B = \frac{-w_a}{k^2}\frac{C_2C_{a3} - C_1C_{a4}}{C_2} - \frac{(w_1 - w_a)}{k^3(l-a)}\frac{C_2C_{a4} - C_1C_{a5}}{C_2}$
	$y_B = 0$
		If $a = 0$ and $w_a = w_l$ (uniform load on entire span), then
		$\text{Max} + M = M_A = \frac{w_a}{k^2}\left(\frac{kl}{\sin kl} - 1\right)$
		$\text{Max} - M = M_B = \frac{-w_a}{k^2}\left(1 - \frac{kl}{\tan kl}\right)$
		$\text{Max } y = y_A = \frac{-w_al}{kP}\left(\tan\frac{kl}{2} - \frac{kl}{2}\right)$
		If $a = 0$ and $w_a = 0$ (uniformly increasing load), then
		$\text{Max} + M = M_A = \frac{w_l}{k^2}\left(\frac{kl}{2\sin kl} - \frac{\tan(kl/2)}{kl}\right)$
		$\text{Max} - M = M_B = \frac{-w_l}{k^2}\left(1 - \frac{kl}{2\tan kl} - \frac{1 - \cos kl}{kl\sin kl}\right)$
		$\text{Max } y = y_A = \frac{-w_l}{k^2P}\left[\left(\frac{kl}{2} - \frac{2}{kl}\right)\tan\frac{kl}{2} - \frac{k^2l^2}{6} + 1\right]$

2c. Left end simply supported, right end fixed
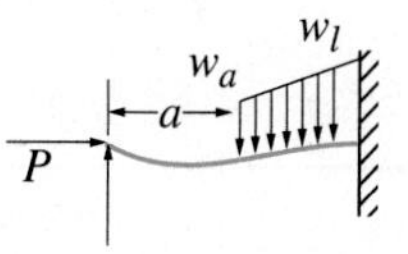
 	
	$M_A = 0 \quad y_A = 0$
	$R_A = \frac{w_a}{k}\frac{C_2C_{a4} - C_1C_{a5}}{C_2C_3 - C_1C_4} + \frac{w_l - w_a}{k^2(l-a)}\frac{C_2C_{a5} - C_1C_{a6}}{C_2C_3 - C_1C_4}$
	$\theta_A = \frac{-w_a}{kP}\left[\frac{C_4C_{a4} - C_3C_{a5}}{C_2C_3 - C_1C_4}\right] - \frac{w_l - w_a}{k^2P(l-a)}\frac{C_4C_{a5} - C_3C_{a6}}{C_2C_3 - C_1C_4}$
	$R_B = \frac{w_a + w_l}{2}(l-a) - R_A$
	$M_B = \frac{-w_a}{k^2}\left(\frac{C_2C_{a5} - klC_2C_{a4}}{C_2C_3 - C_1C_4} + C_{a3}\right)$
	$\quad - \frac{(w_l - w_a)}{k^3(l-a)}\left(\frac{C_2C_{a6} - klC_2C_{a5}}{C_2C_3 - C_1C_4} + C_{a4}\right)$
	$\theta_B = 0 \quad y_B = 0$
		If $a = 0$ and $w_a = w_l$ (uniform load on entire span), then
		$\text{Max } \theta = \theta_A = \frac{-w_a}{kP}\frac{4 - 2kl\sin kl - (2 - k^2l^2/2)(1 + \cos kl)}{\sin kl - kl\cos kl}$
		$\text{Max} - M = M_B = \frac{-w_al}{k}\frac{\tan kl[\tan(kl/2) - kl/2]}{\tan kl - kl}$
		$R_A = \frac{w_a}{k}\frac{kl\sin kl - 1 + (1 - k^2l^2/2)\cos kl}{\sin kl - kl\cos kl}$
		If $a = 0$ and $w_a = 0$ (uniformly increasing load), then
		$\text{Max } \theta = \theta_A = \frac{-w_ll}{6P}\frac{2kl + kl\cos kl - 3\sin kl}{\sin kl - kl\cos kl}$
		$\text{Max} - M = M_B = \frac{-w_l}{k^2}\frac{(l - k^2l^2/3)\tan kl - kl}{\tan kl - kl}$
		$R_A = \frac{w_ll}{6}\left(\frac{2\tan kl}{\tan kl - kl} - \frac{6}{k^2l^2} + 1\right)$

(Continued)

TABLE 8.8 Shear, Moment, Slope, and Deflection Formulas for Beams Under Simultaneous Axial Compression and Transverse Loading (*Continued*)

End Restraints, Reference No.	Boundary Values	Selected Maximum Values of Moments and Deformations

2d. Left end fixed, right end fixed
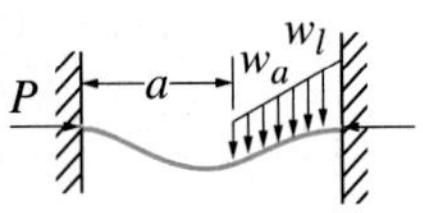
 	
	$\theta_A = 0 \quad y_A = 0$
	$R_A = \frac{w_a}{k}\frac{C_3C_{a4} - C_2C_{a5}}{C_3^2 - C_2C_4} + \frac{w_l - w_a}{k^2(l-a)}\frac{C_3C_{a5} - C_2C_{a6}}{C_3^2 - C_2C_4}$
	$M_A = \frac{-w_a}{k^2}\frac{C_4C_{a4} - C_3C_{a5}}{C_3^2 - C_2C_4} - \frac{w_l - w_a}{k^3(l-a)}\frac{C_4C_{a5} - C_3C_{a6}}{C_3^2 - C_2C_4}$
	$R_B = \frac{w_a + w_l}{2}(l-a) - R_A$
	$M_B = M_A + R_A l - \frac{w_a}{2}(l-a)^2 - \frac{w_l - w_a}{6}(l-a)^2$
	$\theta_B = 0 \quad y_B = 0$
		If $a = 0$ and $w_a = w_l$ (uniform load on entire span), then
		$\text{Max} - M = M_A = M_B = \frac{-w_a}{k^2}\left[1 - \frac{kl/2}{\tan(kl/2)}\right]$
		$\text{Max} + M = \frac{w_a}{k^2}\left[\frac{kl/2}{\sin(kl/2)} - 1\right]$ at $x = \frac{l}{2}$
		$\text{Max } y = \frac{-w_a l}{2kP}\left(\tan\frac{kl}{4} - \frac{kl}{4}\right)$ at $x = \frac{l}{2}$
		$R_A = R_B = \frac{w_a l}{2}$
		If $a = 0$ and $w_a = 0$ (uniformly increasing load), then
		$\text{Max} - M = M_B = \frac{-w_l}{k^2}\left(1 - \frac{(kl/2)\sin kl - k^2l^2/6 - (k^2l^2/3)\cos kl}{2 - 2\cos kl - kl\sin kl}\right)$
		$M_A = \frac{w_l l}{6k}\frac{3\sin kl - kl(2 + \cos kl)}{2 - 2\cos kl - kl\sin kl}$
		$R_A = \frac{w_l}{k^2 l}\left(\frac{k^2l^2}{6}\frac{3 - 3\cos kl - kl\sin kl}{2 - 2\cos kl - kl\sin kl} - 1\right)$

2e. Left end simply supported, right end simply supported
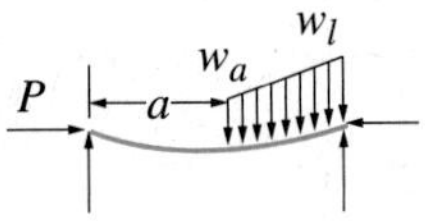
 	
	$M_A = 0 \quad y_A = 0$
	$R_A = \frac{w_a}{2l}(l-a)^2 + \frac{w_l - w_a}{6l}(l-a)^2$
	$\theta_A = \frac{-w_a}{kP}\left[\frac{1 - \cos k(l-a)}{\sin kl} - \frac{k}{2l}(l-a)^2\right]$
	$- \frac{w_l - w_a}{kP}\left[\frac{k(l-a) - \sin k(l-a)}{k(l-a)\sin kl} - \frac{k}{6l}(l-a)^2\right]$
	$R_B = \frac{w_a + w_1}{2}(l-a) - R_A$
	$M_B = 0 \quad y_B = 0$
	$\theta_B = \frac{w_a}{kP}\left[\frac{\cos ka - \cos kl}{\sin kl} - \frac{k(l^2 - a^2)}{2l}\right] + \frac{w_l - w_a}{k^2P(l-a)}$
	$\left[\frac{k^2}{6l}(3al^2 - 2l^3 - a^3) + 1 - \frac{\sin ka + k(l-a)\cos kl}{\sin kl}\right]$
		If $a = 0$ and $w_a = w_l$ (uniform load on entire span), then
		$\text{Max} + M = \frac{w_a}{k^2}\left[\frac{1}{\cos(kl/2)} - 1\right]$ at $x = \frac{l}{2}$
		$\text{Max } \theta = \theta_B = -\theta_A = \frac{w_a}{kP}\left(\tan\frac{kl}{2} - \frac{kl}{2}\right)$
		$\text{Max } y = \frac{-w_a}{k^2P}\left[\frac{1}{\cos(kl/2)} - \frac{k^2l^2}{8} - 1\right]$ at $x = \frac{l}{2}$
		If $a = 0$ and $w_a = 0$ (uniformly increasing load), then
		$M = \frac{w_l}{k^2}\left(\frac{\sin kx}{\sin kl} - \frac{x}{l}\right)$; max M occurs at $x = \frac{1}{k}\cos^{-1}\frac{\sin kl}{kl}$
		$\theta_A = \frac{-w_l}{kP}\left(\frac{1}{\sin kl} - \frac{1}{kl} - \frac{kl}{6}\right)$
		$\text{Max } \theta = \theta_B = \frac{w_l}{kP}\left(\frac{1}{kl} - \frac{kl}{3} - \frac{1}{\tan kl}\right)$
		$y = \frac{-w_l}{k^2P}\left(\frac{\sin kx}{\sin kl} - \frac{x}{l} - \frac{k^2x}{6l}(l^2 - x^2)\right)$

(*Continued*)

TABLE 8.8 Shear, Moment, Slope, and Deflection Formulas for Beams Under Simultaneous Axial Compression and Transverse Loading (*Continued*)

End Restraints, Reference No.	Boundary Values	Selected Maximum Values of Moments and Deformations

2f. Left end guided, right end simply supported
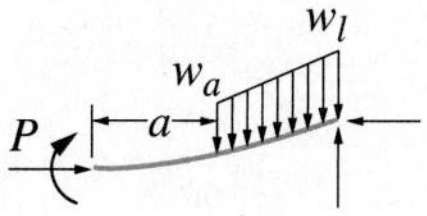
	$R_A=0 \quad \theta_A=0$
	$M_A=\frac{w_a}{k^2}\frac{C_{a3}}{C_1}+\frac{w_l-w_a}{k^3(l-a)}\frac{C_{a4}}{C_1}$
	$y_A=\frac{-w_a}{k^2P}\left[\frac{C_{a3}}{C_1}-\frac{k^2}{2}(l-a)^2\right]-\frac{w_l-w_a}{k^3P(l-a)}\left[\frac{C_{a4}}{C_1}-\frac{k^3}{6}(l-a)^3\right]$
	$R_B=\frac{w_a+w_l}{2}(l-a) \quad M_B=0$
	$\theta_B=\frac{w_a}{kP}\left[\frac{\sin kl-\sin ka}{\cos kl}-k(l-a)\right]+\frac{w_l-w_a}{k^2P(l-a)}\left[\frac{k(l-a)\sin kl-\cos ka}{\cos kl}-\frac{k^2(l-a)^2}{2}+1\right]$
	$y_B=0$
		If $a=0$ and $w_a=w_l$ (uniform load on entire span), then
		$\text{Max } M=M_A=\frac{w_a}{k^2}\left(\frac{1}{\cos kl}-1\right)$
		$\text{Max } \theta=\theta_B=\frac{w_a}{kP}(\tan kl-kl)$
		$\text{Max } y=y_A=\frac{-w_a}{k^2P}\left(\frac{1}{\cos kl}-1-\frac{k^2l^2}{2}\right)$
		If $a=0$ and $w_a=0$ (uniformly increasing load), then
		$\text{Max } M=M_A=\frac{w_l}{k^3l}\frac{kl-\sin kl}{\cos kl}$
		$\text{Max } \theta=\theta_B=\frac{w_l}{k^2Pl}\left(1-\frac{k^2l^2}{2}-\frac{1-kl\sin kl}{\cos kl}\right)$
		$\text{Max } y=y_A=\frac{-w_l}{k^2P}\left(\frac{kl-\sin kl}{kl\cos kl}-\frac{k^2l^2}{6}\right)$

3. Axial compressive load plus concentrated intermediate moment 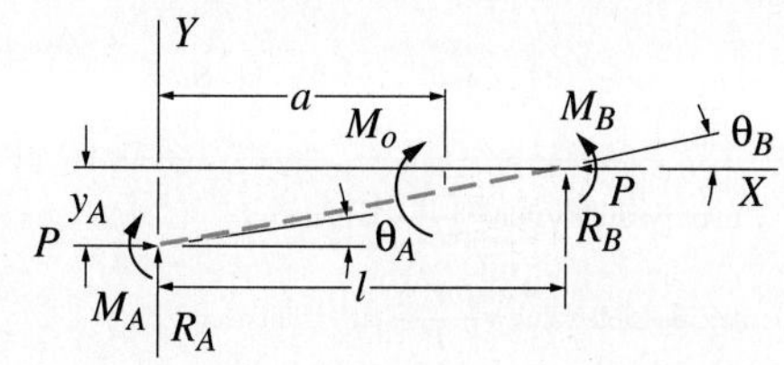		Transverse shear $=V=R_AF_1-M_AkF_2-\theta_APF_1-M_okF_{a2}$ Bending moment $=M=M_AF_1+\frac{R_A}{k}F_2-\frac{\theta_AP}{k}F_2+M_oF_{a1}$ Slope $=\theta=\theta_AF_1+\frac{M_Ak}{P}F_2+\frac{R_A}{P}F_3+\frac{M_ok}{P}F_{a2}$ Deflection $=y=y_A+\frac{\theta_A}{k}F_2+\frac{M_A}{P}F_3+\frac{R_A}{kP}F_4+\frac{M_o}{P}F_{a3}$
3a. Left end free, right end fixed (cantilever) 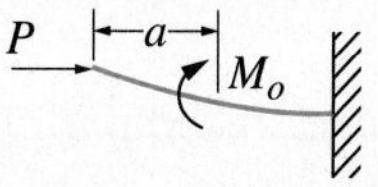	$R_A=0 \quad M_A=0$ $\theta_A=\frac{-M_ok}{P}\frac{\sin k(l-a)}{\cos kl}$ $y_A=\frac{M_o}{P}\left(\frac{\cos ka}{\cos kl}-1\right)$ $R_B=0 \quad \theta_B=0 \quad y_B=0$ $M_B=M_o\frac{\cos ka}{\cos kl}$	$\text{Max } M=M_B$; max possible value $=\frac{M_o}{\cos kl}$ when $a=0$ $\text{Max } \theta=\theta_A$; max possible value $=\frac{-M_ok}{P}\tan kl$ when $a=0$ $\text{Max } y=y_A$; max possible value $=\frac{M_o}{P}\left(\frac{1}{\cos kl}-1\right)$ when $a=0$
3b. Left end guided, right end fixed 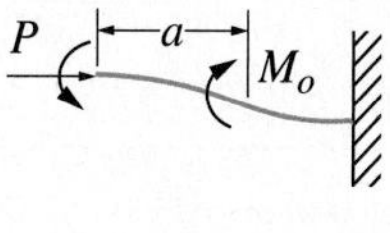	$R_A=0 \quad \theta_A=0$ $M_A=-M_o\frac{\sin k(l-a)}{\sin kl}$ $y_A=\frac{M_o}{P}\left[\frac{\sin k(l-a)+\sin ka}{\sin kl}-1\right]$ $R_B=0 \quad \theta_B=0 \quad y_B=0$ $M_B=M_o\frac{\sin ka}{\sin kl}$	$\text{Max} + M=M_B$; max possible value $=M_o$ when $a=l$ $\text{Max} - M=M_A$; max possible value $=-M_o$ when $a=0$ $\text{Max } \theta=\frac{-M_ok}{P}\frac{\sin k(l-a)}{\sin kl}$ at $x=a$; max possible value $=\frac{-M_ok}{2P}\tan\frac{kl}{2}$ when $a=\frac{l}{2}$ $\text{Max } y=y_A$; max possible value $=\frac{M_o}{P}\left[\frac{1}{\cos(kl/2)}-1\right]$ when $a=\frac{l}{2}$
3c. Left end simply supported, right end fixed 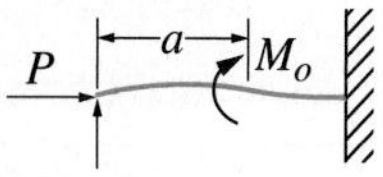	$M_A=0 \quad y_A=0$ $R_A=-M_ok\frac{\cos ka-\cos kl}{\sin kl-kl\cos kl}$ $\theta_A=\frac{-M_ok}{P}\frac{C_3C_{a3}-C_4C_{a2}}{C_2C_3-C_1C_4}$ $R_B=-R_A \quad \theta_B=0 \quad y_B=0$ $M_B=M_o\frac{\sin kl-kl\cos ka}{\sin kl-kl\cos kl}$	If $a=0$ (concentrated end moment), then $R_A=-M_ok\frac{1-\cos kl}{\sin kl-kl\cos kl}$ $\theta_A=\frac{-M_ok}{P}\frac{2-2\cos kl-kl\sin kl}{\sin kl-kl\cos kl}$ $M_B=-M_o\frac{kl-\sin kl}{\sin kl-kl\cos kl}$

(*Continued*)

TABLE 8.8 Shear, Moment, Slope, and Deflection Formulas for Beams Under Simultaneous Axial Compression and Transverse Loading (*Continued*)

End Restraints, Reference No.	Boundary Values	Selected Maximum Values of Moments and Deformations
3d. Left end fixed, right end fixed 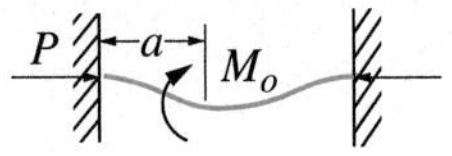	$\theta_A = 0 \quad y_A = 0$ $R_A = -M_o k \dfrac{C_3C_{a2} - C_2C_{a3}}{C_3^2 - C_2C_4}$ $M_A = -M_o \dfrac{C_3C_{a3} - C_4C_{a2}}{C_3^2 - C_2C_4}$ $R_B = -R_A \quad \theta_B = 0 \quad y_B = 0$ $M_B = R_A l + M_A + M_o$	If $a = l/2$ (concentrated center moment), then $R_A = -M_o k \dfrac{[1/\cos(kl/2)] - 1}{2\tan(kl/2) - kl}$ $M_A = -M_o \dfrac{1 - \cos kl - kl\sin(kl/2)}{2 - 2\cos kl - kl\sin kl}$ At the center, $y = 0$ and $\theta = \left(\dfrac{-M_o k}{2P}\right)\left(\dfrac{2 - 2\cos\frac{kl}{2} - \frac{kl}{2}\sin\frac{kl}{2}}{\sin\frac{kl}{2} - \frac{kl}{2}\cos\frac{kl}{2}}\right)$

3e. Left end simply supported, right end simply supported 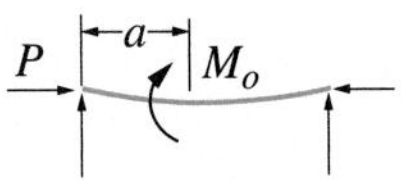	$M_A = 0 \quad y_A = 0$
	$R_A = \dfrac{-M_o}{l}$
	$\theta_A = \dfrac{M_o}{Pl}\left[\dfrac{kl\cos k(l-a)}{\sin kl} - 1\right]$
	$R_B = -R_A \quad M_B = 0 \quad y_B = 0$
	$\theta_B = \dfrac{M_o}{Pl}\left(\dfrac{kl\cos ka}{\sin kl} - 1\right)$
		If $a = 0$ (concentrated moment at the left end), then
		$\theta_A = \dfrac{-M_o}{Pl}\left(1 - \dfrac{kl}{\tan kl}\right)$
		$\theta_B = \dfrac{M_o}{Pl}\left(\dfrac{kl}{\sin kl} - 1\right)$
		$M = M_o \cos kx \left(1 - \dfrac{\tan kx}{\tan kl}\right)$
		If $a = l/2$ (concentrated moment at the center), then
		$\theta_A = \theta_B = \dfrac{M_o}{Pl}\left[\dfrac{kl}{2\sin(kl/2)} - 1\right]$ and $y = 0$ at the center

3f. Left end guided, right end simply supported 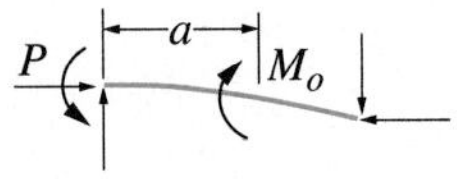	$R_A = 0 \quad \theta_A = 0$ $M_A = -M_o \dfrac{\cos k(l-a)}{\cos kl}$ $y_A = \dfrac{M_o}{P}\left[\dfrac{\cos k(l-a)}{\cos kl} - 1\right]$ $R_B = 0 \quad M_B = 0 \quad y_B = 0$ $\theta_B = \dfrac{-M_o k}{P} \dfrac{\sin ka}{\cos kl}$	Max $M = M_A$; max possible value $- \dfrac{-M_o}{\cos kl}$ when $a - l$ Max $\theta = \theta_B$; max possible value $= \dfrac{-M_o k}{P} \tan kl$ when $a = l$ Max $y = y_a$; max possible value $= \dfrac{M_o}{P}\left(\dfrac{1}{\cos kl} - 1\right)$ when $a = l$
4. Axial compressive load plus externally created concentrated angular displacement 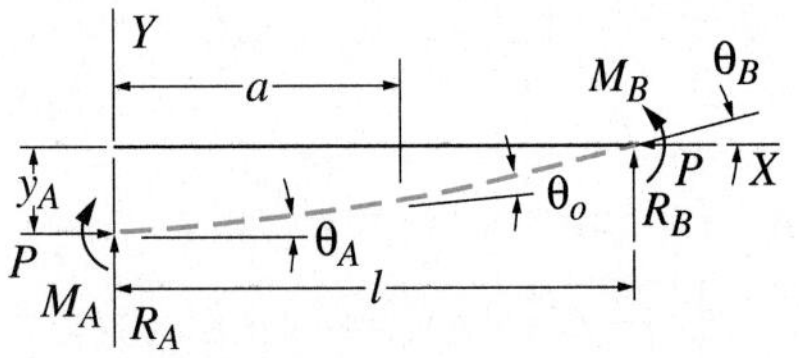		Transverse shear $= V = R_A F_1 - M_A k F_2 - \theta_A P F_1 - \theta_o P F_{a1}$ Bending moment $= M = M_A F_1 + \dfrac{R_A}{k} F_2 - \dfrac{\theta_A P}{k} F_2 - \dfrac{\theta_o P}{k} F_{a2}$ Slope $= \theta = \theta_A F_1 + \dfrac{M_A k}{P} F_2 + \dfrac{R_A}{P} F_3 + \theta_o F_{a1}$ Deflection $= y = y_A + \dfrac{\theta_A}{k} F_2 + \dfrac{M_A}{P} F_3 + \dfrac{R_A}{kP} F_4 + \dfrac{\theta_o}{k} F_{a2}$
4a. Left end free, right end fixed 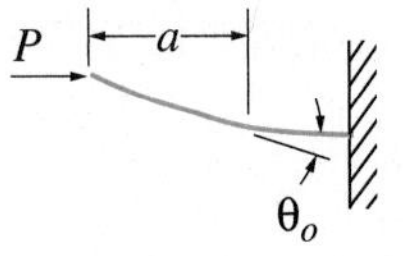	$R_A = 0 \quad M_A = 0$ $\theta_A = -\theta_o \dfrac{\cos k(l-a)}{\cos kl}$ $y_A = \dfrac{\theta_0}{k} \dfrac{\sin ka}{\cos kl}$ $R_B = 0 \quad \theta_B = 0 \quad y_B = 0$ $M_B = \dfrac{\theta_o P}{k} \dfrac{\sin ka}{\cos kl}$	Max $M = M_B$; max possible value $= \dfrac{\theta_o P}{k} \tan kl$ when $a = l$ Max $\theta = \theta_A$; max possible value $= \dfrac{-\theta_o}{\cos kl}$ when $a = l$ Max $y = y_A$; max possible value $= \dfrac{\theta_o}{k} \tan kl$ when $a = l$

(*Continued*)

TABLE 8.8 Shear, Moment, Slope, and Deflection Formulas for Beams Under Simultaneous Axial Compression and Transverse Loading (*Continued*)

End Restraints, Reference No.	Boundary Values	Selected Maximum Values of Moments and Deformations
4b. Left end guided, right end fixed 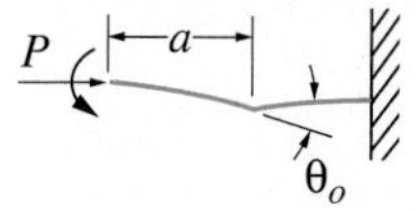	$R_A=0 \quad \theta_A=0$ $M_A=\frac{-\theta_o P}{k}\frac{\cos k(l-a)}{\sin kl}$ $y_A=\frac{\theta_o}{k}\frac{\cos k(l-a)-\cos ka}{\sin kl}$ $R_B=0 \quad \theta_B=0 \quad y_B=0$ $M_B=\frac{-\theta_o P}{k}\frac{\cos ka}{\sin kl}$	Max $-M=M_B$ if $a<\frac{l}{2}$; max possible value $=\frac{-\theta_o P}{k\sin kl}$ when $a=0$ Max $-M=M_A$ if $a>\frac{l}{2}$; max possible value $=\frac{-\theta_o P}{k\sin kl}$ when $a=l$ Max $+y=y_A$; max possible value $=\frac{\theta_o}{k}\tan\frac{kl}{2}$ when $a=l$ Max $-y$ occurs at $x=a$; max possible value $=\frac{-\theta_o}{k}\tan\frac{kl}{2}$ at $x=0$ when $a=0$
4c. Left end simply supported, right end fixed 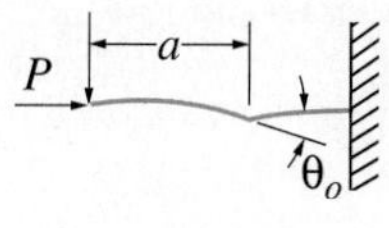	$M_A=0 \quad y_A=0$ $R_A=-\theta_o P\frac{\sin ka}{\sin kl-kl\cos kl}$ $\theta_A=-\theta_o\frac{C_3C_{a2}-C_4C_{a1}}{C_2C_3-C_1C_4}$ $R_B=-R_A \quad \theta_B=0 \quad y_B=0 \quad M_B=R_Al$	
4d. Left end fixed, right end fixed 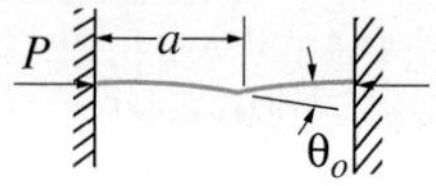	$\theta_A=0 \quad y_A=0$ $R_A=-\theta_o P\frac{C_3C_{a1}-C_2C_{a2}}{C_3^2-C_2C_4}$ $M_A=\frac{-\theta_o P}{k}\frac{C_3C_{a2}-C_4C_{a1}}{C_3^2-C_2C_4}$ $R_B=-R_A \quad \theta_B=0 \quad y_B=0 \quad M_B=M_A+R_Al$	
4e. Left end simply supported, right end simply supported 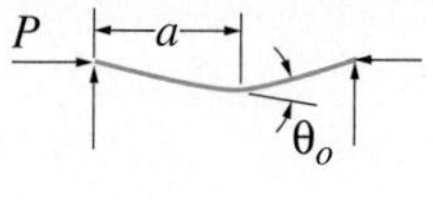	$M_A=0 \quad y_A=0 \quad R_A=0$ $\theta_A=-\theta_o\frac{\sin k(l-a)}{\sin kl}$ $M_B=0 \quad y_B=0 \quad R_B=0$ $\theta_B=\theta_o\frac{\sin ka}{\sin kl}$	Max $M=\frac{\theta_o P}{k}\frac{\sin k(l-a)\sin ka}{\sin kl}$ at $x=a$; max possible value $=\frac{\theta_o P}{k\cos(kl/2)}$ when $a=\frac{l}{2}$ Max $\theta=\theta_A$ if $a<l/2$; max possible value $=-\theta_o$ when $a=0$ Max $y=\frac{-\theta_o}{k}\frac{\sin k(l-a)\sin ka}{\sin kl}$ at; $x=a$; max possible value $=\frac{-\theta_o}{k\cos(kl/2)}$ when $a=\frac{l}{2}$
4f. Left end guided, right end simply supported 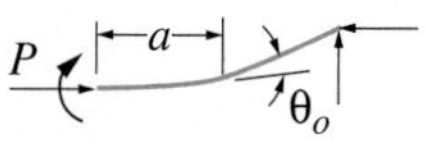	$R_A=0 \quad \theta_A=0$ $M_A=\frac{\theta_o P}{k}\frac{\sin k(l-a)}{\cos kl}$ $y_A=\frac{-\theta_o}{k}\frac{\sin k(l-a)}{\cos kl}$ $R_B=0 \quad M_B=0 \quad y_B=0$ $\theta_B=\theta_o\frac{\cos ka}{\cos kl}$	Max $M=M_A$; max possible value $=\frac{\theta_o P}{k}\tan kl$ when $a=0$ Max $\theta=\theta_B$; max possible value $=\frac{\theta_o}{\cos kl}$ when $a=0$ Max $y=y_A$; max possible value $=\frac{-\theta_o}{k}\tan kl$ when $a=0$

(*Continued*)

TABLE 8.8 Shear, Moment, Slope, and Deflection Formulas for Beams Under Simultaneous Axial Compression and Transverse Loading (*Continued*)

End Restraints, Reference No.	Boundary Values	Selected Maximum Values of Moments and Deformations
5. Axial compressive load plus externally created concentrated lateral displacement 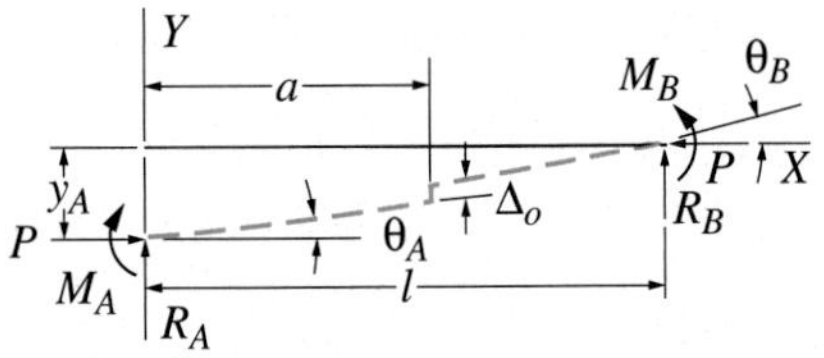 	Transverse shear $= V = R_A F_1 - M_A k F_2 - \theta_A P F_1 + \Delta_o P k F_{a2}$ Bending moment $= M = M_A F_1 + \frac{R_A}{k} F_2 - \frac{\theta_A P}{k} F_2 - \Delta_o P F_{a1}$ Slope $= \theta = \theta_A F_1 + \frac{M_A k}{P} F_2 + \frac{R_A}{P} F_3 - \Delta_o k F_{a2}$ Deflection $= y = y_A + \frac{\theta_A}{k} F_2 + \frac{M_A}{P} F_3 + \frac{R_A}{kP} F_4 + \Delta_o F_{a1}$	
5a. Left end free, right end fixed (cantilever) 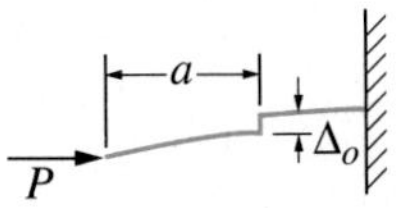 	$R_A = 0 \quad M_A = 0$ $\theta_A = \Delta_o k \frac{\sin k(l-a)}{\cos kl}$ $y_A = -\Delta_o \frac{\cos ka}{\cos kl}$ $R_B = 0 \quad \theta_B = 0 \quad y_B = 0$ $M_B = -\Delta_o P \frac{\cos ka}{\cos kl}$	Max $M = M_B$; max possible value $= \frac{-\Delta_o P}{\cos kl}$ when $a = 0$ Max $\theta = \theta_A$; max possible value $= \Delta_o k \tan kl$ when $a = 0$ Max $y = y_A$; max possible value $= \frac{-\Delta_o}{\cos kl}$ when $a = 0$
5b. Left end guided, right end fixed 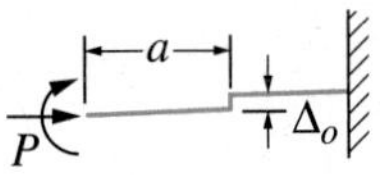 	$R_A = 0 \quad \theta_A = 0$ $M_A = \Delta_o P \frac{\sin k(l-a)}{\sin kl}$ $y_A = -\Delta_o \frac{\sin k(l-a) + \sin ka}{\sin kl}$ $R_B = 0 \quad \theta_B = 0 \quad y_B = 0 \quad M_B = -\Delta_o P \frac{\sin ka}{\sin kl}$	Max $+M = M_A$; max possible value $= \Delta_o P$ when $a = 0$ Max $-M = M_B$; max possible value $= -\Delta_o P$ when $a = l$ Max $y = y_A$; max possible value $= \frac{-\Delta_o}{\cos(kl/2)}$ when $a = \frac{l}{2}$
5c. Left end simply supported, right end fixed 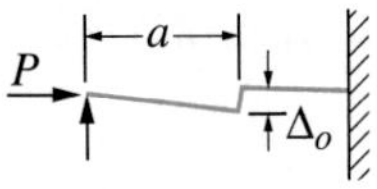 	$M_A = 0 \quad y_A = 0$ $R_A = \Delta_o P k \frac{\cos ka}{\sin kl - kl \cos kl}$ $\theta_A = -\Delta_o k \frac{C_3 C_{a1} + C_4 C_{a2}}{C_2 C_3 - C_1 C_4}$ $R_B = -R_A \quad \theta_B = 0 \quad y_B = 0 \quad M_B = R_A l$	
5d. Left end fixed, right end fixed 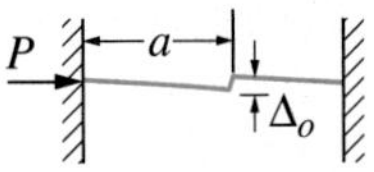 	$\theta_A = 0 \quad y_A = 0$ $R_A = \Delta_o P k \frac{C_3 C_{a2} + C_2 C_{a1}}{C_3^2 - C_2 C_4}$ $M_A = -\Delta_o P \frac{C_3 C_{a1} + C_4 C_{a2}}{C_3^2 - C_2 C_4}$ $R_B = -R_A \quad \theta_B = 0 \quad y_B = 0 \quad M_B = M_A + R_A l$	

(*Continued*)

TABLE 8.8 Shear, Moment, Slope, and Deflection Formulas for Beams Under Simultaneous Axial Compression and Transverse Loading (*Continued*)

End Restraints, Reference No.	Boundary Values	Selected Maximum Values of Moments and Deformations
5e. Left end simply supported, right end simply supported 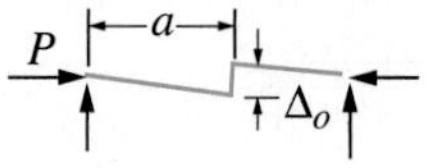	$R_A=0$ $M_A=0$ $y_A=0$ $\theta_A=-\Delta_o k\frac{\cos k(l-a)}{\sin kl}$ $R_B=0$ $M_B=0$ $y_B=0$ $\theta_B=-\Delta_o k\frac{\cos ka}{\sin kl}$	Max $+M=\Delta_o P\frac{\sin ka}{\sin kl}\cos k(l-a)$ at x just left of a; max possible value $=\Delta_o P$ when $a=l$ Max $-M=-\Delta_o P\frac{\cos ka}{\sin kl}\sin k(l-a)$ at x just right of a; max possible value $=-\Delta_o P$ when $a=0$ Max $+y=\Delta_o\frac{\cos ka}{\sin kl}\sin k(l-a)$ at x just right of a; max possible value $=\Delta_o$ when $a=0$ Max $-y=-\Delta_o\frac{\sin ka}{\sin kl}\cos k(l-a)$ at x just left of a; max possible value $=-\Delta_o$ when $a=l$
5f. Left end guided, right end simply supported 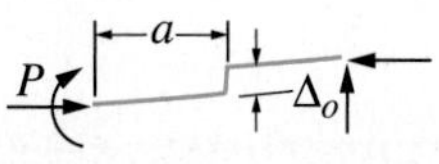	$R_A=0$ $\theta_A=0$ $M_A=\Delta_0 P\frac{\cos k(l-a)}{\cos kl}$ $y_A=-\Delta_o\frac{\cos k(l-a)}{\cos kl}$ $R_B=0$ $M_B=0$ $y_B=0$ $\theta_B=\Delta_o k\frac{\sin ka}{\cos kl}$	Max $M=M_A$; max possible value $=\frac{\Delta_o P}{\cos kl}$ when $a=l$ Max $\theta=\theta_B$; max possible value $=\Delta_o k\tan kl$ when $a=l$ Max $y=y_A$; max possible value $=\frac{-\Delta_o}{\cos kl}$ when $a=l$
6. Axial compressive load plus a uniform temperature variation from top to bottom in the portion from a to l; t is the thickness of the beam 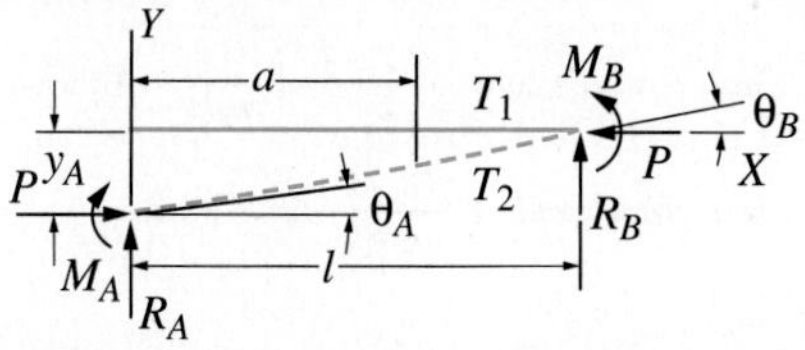	Transverse shear $=V=R_AF_1-M_AkF_2-\theta_APF_1-\frac{\gamma(T_2-T_1)P}{kt}F_{a2}$ Bending moment $=M=M_AF_1+\frac{R_A}{k}F_2-\frac{\theta_AP}{k}F_2-\frac{\gamma(T_2-T_1)EI}{t}F_{a3}$ Slope $=\theta=\theta_AF_1+\frac{M_Ak}{P}F_2+\frac{R_A}{P}F_3+\frac{\gamma(T_2-T_1)}{kt}F_{a2}$ Deflection $=y=y_A+\frac{\theta_A}{k}F_2+\frac{M_A}{P}F_3+\frac{R_A}{kP}F_4+\frac{\gamma(T_2-T_1)}{k^2t}F_{a3}$	
6a. Left end free, right end fixed 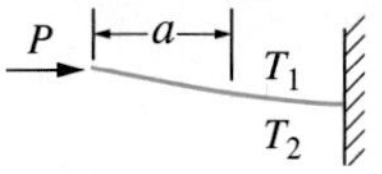	$R_A=0$ $M_A=0$ $\theta_A=\frac{-\gamma(T_2-T_1)}{kt}\frac{\sin k(l-a)}{\cos kl}$ $y_A=\frac{\gamma(T_2-T_1)}{k^2t}\left(\frac{\cos ka}{\cos kl}-1\right)$ $R_B=0$ $\theta_B=0$ $y_B=0$ $M_B=Py_A$	Max $M=M_B$ max possible value $=\frac{\gamma(T_2-T_1)EI}{t}\left(\frac{1}{\cos kl}-1\right)$ when $a=0$ Max $\theta=\theta_A$; max possible value $=\frac{-\gamma(T_2-T_1)}{kt}\tan kl$ when $a=0$ Max $y=y_A$; max possible value $=\frac{\gamma(T_2-T_1)}{k^2t}\left(\frac{1}{\cos kl}-1\right)$ when $a=0$
6b. Left end guided, right end fixed 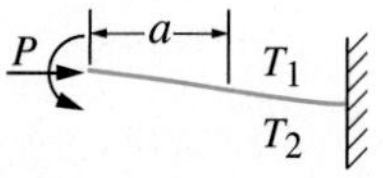	$R_A=0$ $\theta_A=0$ $M_A=\frac{-\gamma(T_2-T_1)EI}{t}\frac{\sin k(l-a)}{\sin kl}$ $y_A=-\frac{\gamma(T_2-T_1)}{k^2t}\frac{C_3C_{a2}-C_2C_{a3}}{C_2}$ $R_B=0$ $\theta_B=0$ $y_B=0$ $M_B=\frac{\gamma(T_2-T_1)EI}{t}\left(\frac{\sin ka}{\sin kl}-1\right)$	Max $-M=M_A$; max possible value $=\frac{-\gamma(T_2-T_1)EI}{t}$ when $a=l$ (*Note*: There is no positive moment in the beam.) Max $\theta=\frac{-\gamma(T_2-T_1)}{kt}\frac{\sin ka}{\sin kl}\sin k(l-a)$ at $x=a$; max possible value $=\frac{-\gamma(T_2-T_1)}{2kt}\tan\frac{kl}{2}$ when $a=\frac{l}{2}$ Max $y=y_A$; max possible value $=\frac{\gamma(T_2-T_1)}{k^2t}\left[\frac{1}{\cos(kl/2)}-1\right]$ when $a=\frac{l}{2}$

(*Continued*)

TABLE 8.8 Shear, Moment, Slope, and Deflection Formulas for Beams Under Simultaneous Axial Compression and Transverse Loading (*Continued*)

End Restraints, Reference No.	Boundary Values	Selected Maximum Values of Moments and Deformations
6c. Left end simply supported, right end fixed 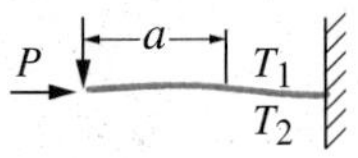	$M_A = 0 \quad y_A = 0$ $R_A = \frac{-\gamma(T_2-T_1)P}{kt}\frac{\cos ka - \cos kl}{\sin kl - kl\cos kl}$ $\theta_A = \frac{-\gamma(T_2-T_1)}{kt}\frac{C_3C_{a3}-C_4C_{a2}}{C_2C_3-C_1C_4}$ $R_B = -R_A \quad \theta_B = 0 \quad y_B = 0 \quad M_B = R_A l$	If $a = 0$ (temperature variation over entire span), then $\text{Max} - M = M_B = \frac{-\gamma(T_2-T_1)Pl}{kt}\frac{1-\cos kl}{\sin kl - kl\cos kl}$ $\text{Max}\,\theta = \theta_A = \frac{-\gamma(T_2-T_1)}{kt}\frac{2-2\cos kl - kl\sin kl}{\sin kl - kl\cos kl}$
6d. Left end fixed, right end fixed 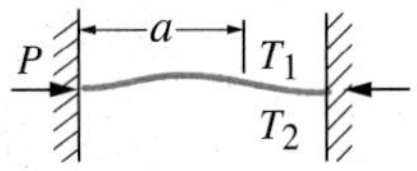	$\theta_A = 0 \quad y_A = 0$ $R_A = \frac{-\gamma(T_2-T_1)P}{kt}\frac{C_3C_{a2}-C_2C_{a3}}{C_3^2-C_2C_4}$ $M_A = \frac{-\gamma(T_2-T_1)EI}{t}\frac{C_3C_{a3}-C_4C_{a2}}{C_3^2-C_2C_4}$ $R_B = -R_A \quad \theta_B = 0 \quad y_B = 0 \quad M_B = M_A + R_A l$	If $a = 0$ (temperature variation over entire span), then $R_A = R_B = 0$ $M = \frac{-\gamma(T_2-T_1)EI}{t}$ everywhere in the span $\theta = 0$ and $y = 0$ everywhere in the span
6e. Left end simply supported, right end simply supported 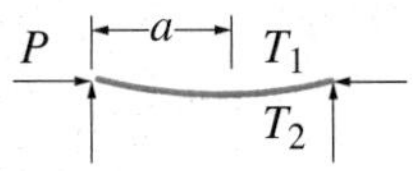	$R_A = 0 \quad M_A = 0 \quad y_A = 0$ $\theta_A = \frac{-\gamma(T_2-T_1)}{kt}\frac{1-\cos k(l-a)}{\sin kl}$ $R_B = 0 \quad M_B = 0 \quad y_B = 0$ $\theta_B = \frac{\gamma(T_2-T_1)}{kt}\frac{\cos ka - \cos kl}{\sin kl}$	Max y occurs at $x = \frac{l}{k}\tan^{-1}\frac{1-\cos kl\cos ka}{\sin kl}$; max possible value $= \frac{-\gamma(T_2-T_1)}{k^2t}\left[\frac{1}{\cos(kl/2)}-1\right]$ at $x = \frac{l}{2}$ when $a = 0$ Max $M = P(\max y)$ Max $\theta = \theta_B$; max possible value $= \frac{\gamma(T_2-T_1)}{kt}\tan\frac{kl}{2}$ when $a = 0$
6f. Left end guided, right end simply supported 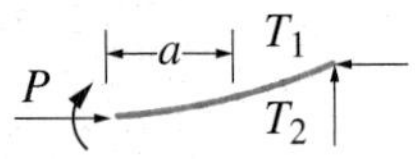	$R_A = 0 \quad \theta_A = 0$ $M_A = \frac{\gamma(T_2-T_1)EI}{t}\frac{1-\cos k(l-a)}{\cos kl}$ $y_A = \frac{-\gamma(T_2-T_1)}{k^2t}\frac{1-\cos k(l-a)}{\cos kl}$ $R_B = 0 \quad M_B = 0 \quad y_B = 0$ $\theta_B = \frac{\gamma(T_2-T_1)}{kt}\frac{\sin kl - \sin ka}{\cos kl}$	Max $M = M_A$; max possible value $= \frac{\gamma(T_2-T_1)EI}{t}\left(\frac{1}{\cos kl}-1\right)$ when $a = 0$ Max $y = y_A$; max possible value $= \frac{-\gamma(T_2-T_1)}{k^2t}\left(\frac{1}{\cos kl}-1\right)$ when $a = 0$ Max $\theta = \theta_B$; max possible value $= \frac{\gamma(T_2-T_1)}{kt}\tan kl$ when $a = 0$

TABLE 8.9 Shear, Moment, Slope, and Deflection Formulas for Beams Under Simultaneous Axial Tension and Transverse Loading

Notation: P = axial tensile load (force); all other notation is the same as that for Table 8.1; see Table 8.8 for loading details.

The following constants and functions are hereby defined in order to permit condensing the tabulated formulas which follow. $k = (P/EI)^{1/2}$. (*Note:* See page 155 for a definition of $\langle x-a\rangle^n$.) The function $\sinh k\langle x-a\rangle$ is also defined as having a value of zero if $x < a$.

$F_1 = \cosh kx$	$F_{a1} = \langle x-a\rangle^0 \cosh k(x-a)$	$C_1 = \cosh kl$	$C_{a1} = \cosh k(l-a)$	(*Note:* Load terms LT_V, LT_M, LT_θ, and LT_y are found at the end of the table for each of the several loadings.)
$F_2 = \sinh kx$	$F_{a2} = \sinh k\langle x-a\rangle$	$C_2 = \sinh kl$	$C_{a2} = \sinh k(l-a)$	
$F_3 = \cosh kx - 1$	$F_{a3} = \langle x-a\rangle^0[\cosh k(x-a) - 1]$	$C_3 = \cosh kl - 1$	$C_{a3} = \cosh k(l-a) - 1$	
$F_4 = \sinh hx - kx$	$F_{a4} = \sinh k\langle x-a\rangle - k\langle x-a\rangle$	$C_4 = \sinh kl - kl$	$C_{a4} = \sinh k(l-a) - k(l-a)$	
	$F_{a5} = F_{a3} - \frac{k^2}{2}\langle x-a\rangle^2$		$C_{a5} = C_{a3} - \frac{k^2}{2}(l-a)^2$	
	$F_{a6} = F_{a4} - \frac{k^3}{6}\langle x-a\rangle^3$		$C_{a6} = C_{a4} - \frac{k^3}{6}(l-a)^3$	

Axial tensile load plus lateral loading

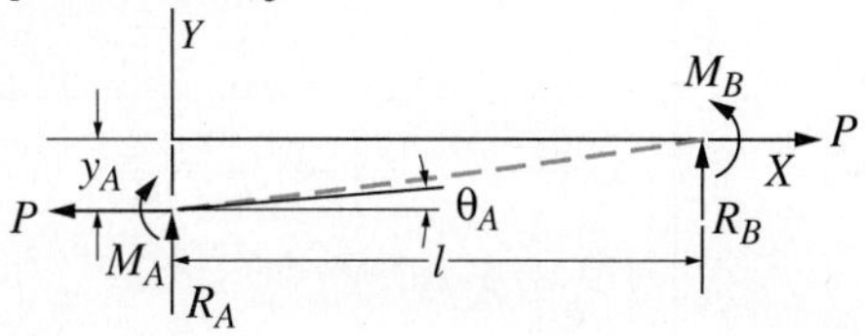

Transverse shear = $V = R_A F_1 + M_A k F_2 + \theta_A P F_1 + LT_V$

Bending moment = $M = M_A F_1 + \frac{R_A}{k} F_2 + \frac{\theta_A P}{k} F_2 + LT_M$

Slope = $\theta = \theta_A F_1 + \frac{M_A k}{P} F_2 + \frac{R_A}{P} F_3 + LT_\theta$

Deflection = $y = y_A + \frac{\theta_A}{k} F_2 + \frac{M_A}{P} F_3 + \frac{R_A}{Pk} F_4 + LT_y$

(*Note:* For each set of end restraints the two initial parameters not listed are zero. For example, with the left end free and the right end fixed, the values of R_A and M_A are zero.)

End Restraints \ Lateral Load		Case 1, Concentrated Lateral Load	Case 2, Distributed Lateral Load	Case 3, Concentrated Moment	Case 4, Concentrated Angular Displacement	Case 5, Concentrated Lateral Displacement	Case 6, Uniform Temperature Variation
Left end free, right end fixed (a)	θ_A	$\frac{W}{P}\frac{C_{a3}}{C_1}$	$\frac{w_a}{kP}\frac{C_{a4}}{C_1} + \frac{(w_l - w_a)C_{a5}}{k^2 P(l-a)C_1}$	$\frac{-M_o k}{P}\frac{C_{a2}}{C_1}$	$-\theta_o \frac{C_{a1}}{C_1}$	$-\Delta_o k \frac{C_{a2}}{C_1}$	$\frac{-\gamma(T_2 - T_1)}{kt}\frac{C_{a2}}{C_1}$
	y_A	$\frac{-W}{kP}\left(\frac{C_2 C_{a3}}{C_1} - C_{a4}\right)$	$\frac{-w_a}{k^2 P}\left(\frac{C_2 C_{a4}}{C_1} - C_{a5}\right) + \frac{-(w_l - w_a)}{k^3 P(l-a)}\left(\frac{C_2 C_{a5}}{C_1} - C_{a6}\right)$	$\frac{M_o}{P}\left(\frac{C_2 C_{a2}}{C_1} - C_{a3}\right)$	$\frac{\theta_o}{k}\left(\frac{C_2 C_{a1}}{C_1} - C_{a2}\right)$	$\Delta_o\left(\frac{C_2 C_{a2}}{C_1} - C_{a1}\right)$	$\frac{\gamma(T_2 - T_1)}{k^2 t}\left(\frac{C_2 C_{a2}}{C_1} - C_{a3}\right)$
Left end guided, right end fixed (b)	M_A	$\frac{W}{k}\frac{C_{a3}}{C_2}$	$\frac{w_a}{k^2}\frac{C_{a4}}{C_2} + \frac{w_l - w_a}{k^3(l-a)}\frac{C_{a5}}{C_2}$	$-M_o \frac{C_{a2}}{C_2}$	$\frac{-\theta_o P}{k}\frac{C_{a1}}{C_2}$	$-\Delta_o P \frac{C_{a2}}{C_2}$	$\frac{-\gamma(T_2 - T_1)P}{k^2 t}\frac{C_{a2}}{C_2}$
	y_A	$\frac{-W}{kP}\left(\frac{C_3 C_{a3}}{C_2} - C_{a4}\right)$	$\frac{-w_a}{k^2 P}\left(\frac{C_3 C_{a4}}{C_2} - C_{a5}\right) + \frac{-(w_l - w_a)}{k^3 P(l-a)}\left(\frac{C_3 C_{a5}}{C_2} - C_{a6}\right)$	$\frac{M_o}{P}\left(\frac{C_3 C_{a2}}{C_2} - C_{a3}\right)$	$\frac{\theta_o}{k}\left(\frac{C_3 C_{a1}}{C_2} - C_{a2}\right)$	$\Delta_o\left(\frac{C_3 C_{a2}}{C_2} - C_{a1}\right)$	$\frac{\gamma(T_2 - T_1)}{k^2 t}\left(\frac{C_3 C_{a2}}{C_2} - C_{a3}\right)$
Left end simply supported, right end fixed (c)	R_A	$W\frac{C_2 C_{a3} - C_1 C_{a4}}{C_2 C_3 - C_1 C_4}$	$\frac{w_a}{k}\frac{C_2 C_{a4} - C_1 C_{a5}}{C_2 C_3 - C_1 C_4} + \frac{w_l - w_a}{k^2(l-a)}\frac{C_2 C_{a5} - C_1 C_{a6}}{C_2 C_3 - C_1 C_4}$	$-M_o k \frac{C_2 C_{a2} - C_1 C_{a3}}{C_2 C_3 - C_1 C_4}$	$-\theta_o P \frac{C_2 C_{a1} - C_1 C_{a2}}{C_2 C_3 - C_1 C_4}$	$\Delta_o kP \frac{C_1 C_{a1} - C_2 C_{a2}}{C_2 C_3 - C_1 C_4}$	$\frac{-\gamma(T_2 - T_1)P}{kt}\frac{C_2 C_{a2} - C_1 C_{a3}}{C_2 C_3 - C_1 C_4}$
	θ_A	$\frac{-W}{P}\frac{C_4 C_{a3} - C_3 C_{a4}}{C_2 C_3 - C_1 C_4}$	$\frac{-w_a}{kP}\frac{C_4 C_{a4} - C_3 C_{a5}}{C_2 C_3 - C_1 C_4} + \frac{-(w_l - w_a)}{k^2 P(l-a)}\frac{C_4 C_{a5} - C_3 C_{a6}}{C_2 C_3 - C_1 C_4}$	$\frac{-M_o k}{P}\frac{C_3 C_{a3} - C_4 C_{a2}}{C_2 C_3 - C_1 C_4}$	$-\theta_o \frac{C_3 C_{a2} - C_4 C_{a1}}{C_2 C_3 - C_1 C_4}$	$\Delta_o k \frac{C_4 C_{a2} - C_3 C_{a1}}{C_2 C_3 - C_1 C_4}$	$\frac{-\gamma(T_2 - T_1)}{kt}\frac{C_3 C_{a3} - C_4 C_{a2}}{C_2 C_3 - C_1 C_4}$

(*Continued*)

TABLE 8.9 Shear, Moment, Slope, and Deflection Formulas for Beams Under Simultaneous Axial Tension and Transverse Loading (*Continued*)

End Restraints \ Lateral Load		Case 1, Concentrated Lateral Load	Case 2, Distributed Lateral Load	Case 3, Concentrated Moment	Case 4, Concentrated Angular Displacement	Case 5, Concentrated Lateral Displacement	Case 6, Uniform Temperature Variation
Left end fixed, right end fixed (d)	R_A	$W\frac{C_3C_{a3}-C_2C_{a4}}{C_3^2-C_2C_4}$	$\frac{w_a}{k}\frac{C_3C_{a4}-C_2C_{a5}}{C_3^2-C_2C_4}+\frac{w_l-w_a}{k^2(l-a)}\frac{C_3C_{a5}-C_2C_{a6}}{C_3^2-C_2C_4}$	$-M_ok\frac{C_3C_{a2}-C_2C_{a3}}{C_3^2-C_2C_4}$	$-\theta_oP\frac{C_3C_{a1}-C_2C_{a2}}{C_3^2-C_2C_4}$	$\Delta_oPk\frac{C_2C_{a1}-C_3C_{a2}}{C_3^2-C_2C_4}$	$\frac{-\gamma(T_2-T_1)P}{kt}\frac{C_3C_{a2}-C_2C_{a3}}{C_3^2-C_2C_4}$
	M_A	$\frac{-W}{k}\frac{C_4C_{a3}-C_3C_{a4}}{C_3^2-C_2C_4}$	$\frac{-w_a}{k^2}\frac{C_4C_{a4}-C_3C_{a5}}{C_3^2-C_2C_4}+\frac{-(w_l-w_a)}{k^3(l-a)}\frac{C_4C_{a5}-C_3C_{a6}}{C_3^2-C_2C_4}$	$-M_o\frac{C_3C_{a3}-C_4C_{a2}}{C_3^2-C_2C_4}$	$\frac{-\theta_oP}{k}\frac{C_3C_{a2}-C_4C_{a1}}{C_3^2-C_2C_4}$	$\Delta_oP\frac{C_4C_{a2}-C_3C_{a1}}{C_3^2-C_2C_4}$	$\frac{-\gamma(T_2-T_1)P}{k^2t}\frac{C_3C_{a3}-C_4C_{a2}}{C_3^2-C_2C_4}$
Left end simply supported, right end simply supported (e)	R_A	$\frac{W}{l}(l-a)$	$\frac{w_a}{2l}(l-a)^2+\frac{w_l-w_a}{6l}(l-a)^2$	$\frac{-M_o}{l}$	0	0	0
	θ_A	$\frac{-W}{Pkl}\left(\frac{C_4C_{a2}}{C_2}-C_{a4}\right)$	$\frac{-w_a}{Pk}\left[\frac{k(l-a)^2}{2l}-\frac{C_{a3}}{C_2}\right]+\frac{-(w_l-w_a)}{Pk^2(l-a)}\left[\frac{k^2(l-a)^3}{6l}-\frac{C_{a4}}{C_2}\right]$	$\frac{M_ok}{P}\left(\frac{1}{kl}-\frac{C_{a1}}{C_2}\right)$	$-\theta_o\frac{C_{a2}}{C_2}$	$-\Delta_ok\frac{C_{a1}}{C_2}$	$\frac{-\gamma(T_2-T_1)}{kt}\frac{C_{a3}}{C_2}$
Left end guided, right end simply supported (f)	M_A	$\frac{W}{k}\frac{C_{a2}}{C_1}$	$\frac{w_a}{k^2}\frac{C_{a3}}{C_1}+\frac{w_l-w_a}{k^3(l-a)}\frac{C_{a4}}{C_1}$	$-M_o\frac{C_{a1}}{C_1}$	$\frac{-\theta_oP}{k}\frac{C_{a2}}{C_1}$	$-\Delta_oP\frac{C_{a1}}{C_1}$	$\frac{-\gamma(T_2-T_1)P}{k^2t}\frac{C_{a3}}{C_1}$
	y_A	$\frac{-W}{Pk}\left(\frac{C_3C_{a2}}{C_1}-C_{a4}\right)$	$\frac{-w_a}{k^2P}\left[\frac{k^2(l-a)^2}{2}-\frac{C_{a3}}{C_1}\right]+\frac{-(w_l-w_a)}{k^3P(l-a)}\left[\frac{k^3(l-a)^3}{6}-\frac{C_{a4}}{C_1}\right]$	$\frac{m_o}{P}\left(1-\frac{C_{a1}}{C_1}\right)$	$\frac{-\theta_o}{k}\frac{C_{a2}}{C_1}$	$-\Delta_o\frac{C_{a1}}{C_1}$	$\frac{-\gamma(T_2-T_1)}{k^2t}\frac{C_{a3}}{C_1}$
Load terms for all end restraints (a)-(f)	LT_V	$-WF_{a1}$	$\frac{-w_a}{k}F_{a2}-\frac{w_l-w_a}{k^2(l-a)}F_{a3}$	M_okF_{a2}	θ_oPF_{a1}	ΔPkF_{a2}	$\frac{\gamma(T_2-T_1)P}{kt}F_{a2}$
	LT_M	$\frac{-W}{k}F_{a2}$	$\frac{-w_a}{k^2}F_{a3}-\frac{w_l-w_a}{k^3(l-a)}F_{a4}$	M_oF_{a1}	$\frac{\theta_oP}{k}F_{a2}$	Δ_oPF_{a1}	$\frac{\gamma(T_2-T_1)P}{k^2t}F_{a3}$
	LT_θ	$\frac{-W}{P}F_{a3}$	$\frac{-w_a}{Pk}F_{a4}-\frac{(w_l-w_a)}{Pk^2(l-a)}F_{a5}$	$\frac{M_ok}{P}F_{a2}$	θ_oF_{a1}	Δ_okF_{a2}	$\frac{\gamma(T_2-T_1)}{kt}F_{a2}$
	LT_y	$\frac{-W}{Pk}F_{a4}$	$\frac{-w_a}{Pk^2}F_{a5}-\frac{w_l-w_a}{Pk^3(l-a)}F_{a6}$	$\frac{M_o}{P}F_{a3}$	$\frac{\theta_o}{k}F_{a2}$	Δ_oF_{a1}	$\frac{\gamma(T_2-T_1)}{k^2t}F_{a3}$

TABLE 8.10 Beams Restrained Against Horizontal Displacement at the Ends

Case No., Manner of Loading and Support	Formulas to Solve for y_{max} and P
1. Ends pinned to rigid supports, concentrated center load W	$y_{max} + \frac{A}{4I} y_{max}^3 = \frac{2Wl^3}{\pi^4 EI}$ (Solve for y_{max}) $P = \frac{\pi^2 EA}{4l^2} y_{max}^2$ Use case 1e from Table 8.7(*b*) or Table 8.9 to determine maximum slopes and moments after solving for *P*.
2. Ends fixed to rigid supports, concentrated center load W	$y_{max} + \frac{A}{16I} y_{max}^3 = \frac{Wl^3}{2\pi^4 EI}$ (Solve for y_{max}) $P = \frac{\pi^2 EA}{4l^2} y_{max}^2$ Use case 1d from Table 8.7(*d*) or Table 8.9 to determine maximum slopes and moments after solving for *P*.
3. Ends pinned to rigid supports, uniformly distributed transverse load w on entire span	$y_{max} + \frac{A}{4I} y_{max}^3 = \frac{5wl^4}{4\pi^4 EI}$ (Solve for y_{max}) $P = \frac{\pi^2 EA}{4l^2} y_{max}^2$ Use case 2e from Table 8.7(*b*) or Table 8.9 to determine maximum slopes and moments after solving for *P*.
4. Ends fixed to rigid supports, uniformly distributed transverse load w on entire span	$y_{max} + \frac{A}{16I} y_{max}^3 = \frac{wl^4}{4\pi^4 EI}$ (Solve for y_{max}) $P = \frac{\pi^2 EA}{4l^2} y_{max}^2$ Use case 2d from Table 8.7(*d*) or Table 8.9 to determine maximum slopes and moments after solving for *P*.
5. Same as case 1, except beam is perfectly flexible like a cable or chain and has an unstretched length l	$\tan\theta - \sin\theta = \frac{W}{2EA}$ or if $\theta < 12°$, $\theta = \left(\frac{W}{EA}\right)^{1/3}$ $P = \frac{W}{2\tan\theta}$ 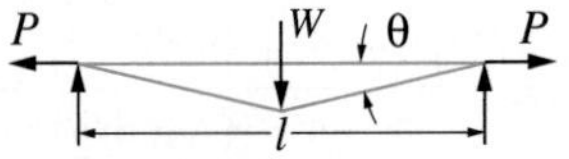
6. Same as case 3, except beam is perfectly flexible like a cable or chain and has an unstretched length l	$y_{max} = l\left(\frac{3wl}{64EA}\right)^{1/3}$ $P = \frac{wl^2}{8y_{max}}$ 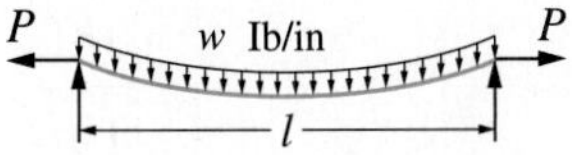

TABLE 8.11(*a*) Reaction and Deflection Coefficients for Tapered Beams

Case No. in Table 8.1	Load Location a/l	Multiplier Listed for	I_B/I_A				
			0.25	0.50	2.0	4.0	8.0
1a	0	y_A	2.525	1.636	0.579	0.321	0.171
		θ_A	2.262	1.545	0.614	0.359	0.201
	0.25	y_A	2.663	1.682	0.563	0.303	0.159
		θ_A	2.498	1.631	0.578	0.317	0.168
	0.50	y_A	2.898	1.755	0.543	0.284	0.146
		θ_A	2.811	1.731	0.548	0.289	0.149
	0.75	y_A	3.289	1.858	0.521	0.266	0.135
		θ_A	3.261	1.851	0.522	0.267	0.135
1c	0.25	R_A	1.055	1.028	0.972	0.946	0.926
		θ_A	1.492	1.256	0.744	0.514	0.330
	0.50	R_A	1.148	1.073	0.936	0.887	0.852
		θ_A	1.740	1.365	0.682	0.435	0.261
1d	0.25	R_A	1.046	1.026	0.968	0.932	0.895
		M_A	1.137	1.077	0.905	0.797	0.686
	0.50	R_A	1.163	1.085	0.915	0.837	0.771
		M_A	1.326	1.171	0.829	0.674	0.542
1e	0.25	θ_A	1.396	1.220	0.760	0.531	0.342
		$y_{l/2}$	1.563	1.301	0.703	0.452	0.268
	0.50	θ_A	1.524	1.282	0.718	0.476	0.293
		$y_{l/2}$	1.665	1.349	0.674	0.416	0.239
2a. Uniform load	0	y_A	2.711	1.695	0.561	0.302	0.158
		θ_A	2.525	1.636	0.579	0.321	0.171
	0.25	y_A	2.864	1.742	0.547	0.289	0.149
		θ_A	2.745	1.708	0.556	0.296	0.154
	0.50	y_A	3.091	1.806	0.532	0.275	0.140
		θ_A	3.029	1.790	0.535	0.278	0.142
	0.75	y_A	3.435	1.890	0.516	0.262	0.132
		θ_A	3.415	1.886	0.516	0.263	0.133
2c. Uniform load	0	R_A	1.074	1.036	0.968	0.941	0.922
		θ_A	1.663	1.326	0.710	0.473	0.296
	0.50	R_A	1.224	1.104	0.917	0.858	0.818
		θ_A	1.942	1.438	0.653	0.403	0.237

Moments of inertia vary as $(1 + Kx/l)^n$, where $n = 1.0$

(*Continued*)

TABLE 8.11(*a*) Reaction and Deflection Coefficients for Tapered Beams (*Continued*)

Case No. in Table 8.1	Load Location a/l	Multiplier Listed for	I_B/I_A 0.25	0.50	2.0	4.0	8.0
2d. Uniform load	0	R_A	1.089	1.046	0.954	0.911	0.872
		M_A	1.267	1.137	0.863	0.733	0.615
	0.50	R_A	1.267	1.130	0.886	0.791	0.717
		M_A	1.481	1.234	0.794	0.625	0.491
2e. Uniform load	0	θ_A	1.508	1.271	0.729	0.492	0.309
		$y_{l/2}$	1.678	1.352	0.676	0.420	0.243
	0.50	θ_A	1.616	1.320	0.700	0.454	0.275
		$y_{l/2}$	1.765	1.389	0.658	0.398	0.225
2a. Uniformly increasing load	0	y_A	2.851	1.737	0.549	0.291	0.150
		θ_A	2.711	1.695	0.561	0.302	0.158
	0.25	y_A	3.005	1.781	0.538	0.280	0.143
		θ_A	2.915	1.757	0.543	0.285	0.147
	0.50	y_A	3.220	1.839	0.525	0.270	0.137
		θ_A	3.172	1.827	0.527	0.272	0.138
	0.75	y_A	3.526	1.910	0.513	0.260	0.131
		θ_A	3.511	1.907	0.513	0.260	0.131
2c. Uniformly increasing load	0	R_A	1.129	1.062	0.948	0.907	0.878
		θ_A	1.775	1.372	0.686	0.442	0.269
	0.50	R_A	1.275	1.124	0.907	0.842	0.799
		θ_A	2.063	1.479	0.639	0.388	0.225
2d. Uniformly increasing load	0	R_A	1.157	1.079	0.926	0.860	0.804
		M_A	1.353	1.177	0.833	0.685	0.559
	0.50	R_A	1.334	1.157	0.870	0.767	0.690
		M_A	1.573	1.269	0.777	0.601	0.468
2e. Uniformly increasing load	0	θ_A	1.561	1.295	0.714	0.472	0.291
		$y_{l/2}$	1.722	1.370	0.667	0.409	0.234
	0.50	θ_A	1.654	1.335	0.693	0.447	0.269
		$y_{l/2}$	1.806	1.404	0.651	0.392	0.221

(*Continued*)

TABLE 8.11(*a*) Reaction and Deflection Coefficients for Tapered Beams (*Continued*)

Case No. in Table 8.1	Load Location a/l	Multiplier Listed for	I_B/I_A				
			0.25	0.50	2.0	4.0	8.0
3a	0	y_A	2.262	1.545	0.614	0.359	0.201
		θ_A	1.848	1.386	0.693	0.462	0.297
	0.25	y_A	2.337	1.575	0.597	0.337	0.182
		θ_A	2.095	1.492	0.627	0.367	0.203
	0.50	y_A	2.566	1.658	0.566	0.305	0.159
		θ_A	2.443	1.622	0.575	0.313	0.164
	0.75	y_A	3.024	1.795	0.532	0.275	0.140
		θ_A	2.985	1.785	0.534	0.277	0.141
3c	0	R_A	0.896	0.945	1.059	1.118	1.173
		θ_A	1.312	1.166	0.823	0.645	0.482
	0.50	R_A	1.016	1.014	0.977	0.952	0.929
		θ_A	1.148	1.125	0.794	0.565	0.365
3d	0.25	R_A	0.796	0.890	1.116	1.220	1.298
		M_A	1.614	1.331	0.653	0.340	0.106
	0.50	R_A	0.958	0.988	0.988	0.958	0.919
		M_A	0.875	0.965	0.965	0.875	0.758
3e	0	θ_A	1.283	1.159	0.818	0.631	0.460
		$y_{l/2}$	1.524	1.282	0.718	0.476	0.293
	0.25	θ_A	1.628	1.338	0.666	0.393	0.208
		$y_{l/2}$	1.651	1.345	0.671	0.408	0.229

TABLE 8.11(*b*) Reaction and Deflection Coefficients for Tapered Beams

Case No. in Table 8.1	Load Location a/l	Multiplier Listed for	I_B/I_A				
			0.25	0.50	2.0	4.0	8.0
1a	0	y_A	2.729	1.667	0.589	0.341	0.194
		θ_A	2.455	1.577	0.626	0.386	0.235
	0.25	y_A	2.872	1.713	0.572	0.320	0.176
		θ_A	2.708	1.663	0.588	0.338	0.190
	0.50	y_A	3.105	1.783	0.549	0.296	0.157
		θ_A	3.025	1.761	0.555	0.301	0.161
	0.75	y_A	3.460	1.877	0.525	0.272	0.140
		θ_A	3.437	1.872	0.526	0.273	0.140
1c	0.25	R_A	1.052	1.028	0.970	0.938	0.905
		θ_A	1.588	1.278	0.759	0.559	0.398
	0.50	R_A	1.138	1.070	0.932	0.867	0.807
		θ_A	1.867	1.390	0.695	0.468	0.306
1d	0.25	R_A	1.049	1.027	0.969	0.934	0.895
		M_A	1.155	1.082	0.909	0.813	0.713
	0.50	R_A	1.169	1.086	0.914	0.831	0.753
		M_A	1.358	1.177	0.833	0.681	0.548
1e	0.25	θ_A	1.509	1.246	0.778	0.586	0.428
		$y_{l/2}$	1.716	1.334	0.721	0.501	0.334
	0.50	θ_A	1.668	1.313	0.737	0.525	0.363
		$y_{l/2}$	1.840	1.385	0.692	0.460	0.294
2a. Uniform load	0	y_A	2.916	1.724	0.569	0.318	0.174
		θ_A	2.729	1.667	0.589	0.341	0.194
	0.25	y_A	3.067	1.770	0.554	0.301	0.161
		θ_A	2.954	1.737	0.563	0.311	0.169
	0.50	y_A	3.282	1.830	0.537	0.283	0.148
		θ_A	3.226	1.816	0.540	0.287	0.150
	0.75	y_A	3.580	1.906	0.518	0.266	0.136
		θ_A	3.564	1.902	0.519	0.267	0.136
2c. Uniform load	0	R_A	1.068	1.035	0.965	0.932	0.899
		θ_A	1.774	1.349	0.723	0.510	0.351
	0.50	R_A	1.203	1.098	0.910	0.831	0.761
		θ_A	2.076	1.463	0.664	0.430	0.271

Moments of inertia vary as $(1 + Kx/l)^n$, where $n = 2.0$

(*Continued*)

TABLE 8.11(*b*) Reaction and Deflection Coefficients for Tapered Beams (*Continued*)

Case No. in Table 8.1	Load Location a/l	Multiplier Listed for	I_B/I_A				
			0.25	0.50	2.0	4.0	8.0
2d. Uniform load	0	R_A	1.091	1.046	0.954	0.909	0.865
		M_A	1.290	1.142	0.866	0.741	0.628
	0.50	R_A	1.267	1.129	0.833	0.779	0.689
		M_A	1.509	1.239	0.795	0.625	0.486
2e. Uniform load	0	θ_A	1.645	1.301	0.747	0.542	0.382
		$y_{l/2}$	1.853	1.387	0.694	0.463	0.298
	0.50	θ_A	1.774	1.352	0.718	0.500	0.339
		$y_{l/2}$	1.955	1.426	0.675	0.438	0.274
2a. Uniformly increasing load	0	y_A	3.052	1.765	0.556	0.304	0.163
		θ_A	2.916	1.724	0.569	0.318	0.174
	0.25	y_A	3.199	1.807	0.543	0.290	0.153
		θ_A	3.116	1.784	0.550	0.297	0.158
	0.50	y_A	3.395	1.860	0.529	0.276	0.143
		θ_A	3.354	1.849	0.532	0.279	0.144
	0.75	y_A	3.653	1.923	0.515	0.263	0.134
		θ_A	3.641	1.921	0.515	0.263	0.134
2c. Uniformly increasing load	0	R_A	1.119	1.059	0.944	0.890	0.841
		θ_A	1.896	1.396	0.698	0.475	0.315
	0.50	R_A	1.244	1.116	0.898	0.810	0.736
		θ_A	2.196	1.503	0.649	0.411	0.255
2d. Uniformly increasing load	0	R_A	1.159	1.079	0.925	0.854	0.789
		M_A	1.379	1.182	0.836	0.691	0.565
	0.50	R_A	1.328	1.154	0.866	0.752	0.656
		M_A	1.596	1.272	0.777	0.598	0.457
2e. Uniformly increasing load	0	θ_A	1.708	1.326	0.732	0.521	0.360
		$y_{l/2}$	1.904	1.407	0.684	0.451	0.286
	0.50	θ_A	1.817	1.368	0.711	0.491	0.331
		$y_{l/2}$	2.001	1.442	0.668	0.430	0.268

(Continued)

TABLE 8.11(*b*) Reaction and Deflection Coefficients for Tapered Beams (*Continued*)

Case No. in Table 8.1	Load Location a/l	Multiplier Listed for	I_B/I_A				
			0.25	0.50	2.0	4.0	8.0
3a	0	y_A	2.455	1.577	0.626	0.386	0.235
		θ_A	2.000	1.414	0.707	0.500	0.354
	0.25	y_A	2.539	1.608	0.609	0.363	0.211
		θ_A	2.286	1.526	0.641	0.400	0.243
	0.50	y_A	2.786	1.691	0.575	0.323	0.177
		θ_A	2.667	1.657	0.586	0.333	0.185
	0.75	y_A	3.234	1.821	0.538	0.284	0.148
		θ_A	3.200	1.812	0.540	0.286	0.149
3c	0	R_A	0.900	0.946	1.062	1.132	1.212
		θ_A	1.375	1.181	0.835	0.688	0.558
	0.50	R_A	1.021	1.015	0.977	0.946	0.911
		θ_A	1.223	1.148	0.814	0.622	0.451
3d	0.25	R_A	0.785	0.888	1.117	1.230	1.333
		M_A	1.682	1.347	0.660	0.348	0.083
	0.50	R_A	0.966	0.991	0.991	0.966	0.928
		M_A	0.890	0.972	0.974	0.905	0.807
3e	0	θ_A	1.364	1.179	0.833	0.682	0.549
		$y_{l/2}$	1.668	1.313	0.737	0.525	0.363
	0.25	θ_A	1.801	1.376	0.686	0.441	0.263
		$y_{l/2}$	1.826	1.382	0.690	0.454	0.284

TABLE 8.11(*c*) Reaction and Deflection Coefficients for Tapered Beams

Case No. in Table 8.1	Load Location a/l	Multiplier Listed for	I_B/I_A				
			0.25	0.50	2.0	4.0	8.0
1a	0	y_A	2.796	1.677	0.593	0.349	0.204
1a	0	θ_A	2.520	1.587	0.630	0.397	0.250
1a	0.25	y_A	2.939	1.722	0.575	0.327	0.184
1a	0.25	θ_A	2.777	1.674	0.592	0.346	0.200
1a	0.50	y_A	3.169	1.791	0.551	0.300	0.162
1a	0.50	θ_A	3.092	1.770	0.558	0.307	0.167
1a	0.75	y_A	3.509	1.883	0.526	0.274	0.142
1a	0.75	θ_A	3.488	1.878	0.527	0.275	0.143
1c	0.25	R_A	1.051	1.027	0.969	0.936	0.899
1c	0.25	θ_A	1.626	1.286	0.764	0.573	0.422
1c	0.50	R_A	1.134	1.068	0.930	0.860	0.791
1c	0.50	θ_A	1.916	1.399	0.700	0.480	0.322
1d	0.25	R_A	1.050	1.027	0.969	0.934	0.895
1d	0.25	M_A	1.161	1.084	0.911	0.818	0.724
1d	0.50	R_A	1.171	1.086	0.914	0.829	0.748
1d	0.50	M_A	1.378	1.179	0.834	0.684	0.553
1e	0.25	θ_A	1.554	1.256	0.784	0.605	0.460
1e	0.25	$y_{l/2}$	1.774	1.346	0.728	0.519	0.362
1e	0.50	θ_A	1.723	1.324	0.743	0.543	0.391
1e	0.50	$y_{l/2}$	1.907	1.397	0.699	0.477	0.318
2a. Uniform load	0	y_A	2.981	1.734	0.572	0.324	0.182
2a. Uniform load	0	θ_A	2.796	1.677	0.593	0.349	0.204
2a. Uniform load	0.25	y_A	3.130	1.779	0.556	0.306	0.167
2a. Uniform load	0.25	θ_A	3.020	1.747	0.566	0.317	0.176
2a. Uniform load	0.50	y_A	3.338	1.837	0.538	0.287	0.151
2a. Uniform load	0.50	θ_A	3.285	1.823	0.542	0.291	0.154
2a. Uniform load	0.75	y_A	3.620	1.911	0.519	0.268	0.137
2a. Uniform load	0.75	θ_A	3.606	1.097	0.520	0.269	0.138
2c. Uniform load	0	R_A	1.066	1.034	0.965	0.928	0.891
2c. Uniform load	0	θ_A	1.817	1.357	0.727	0.522	0.370
2c. Uniform load	0.50	R_A	1.194	1.096	0.908	0.821	0.741
2c. Uniform load	0.50	θ_A	2.125	1.471	0.668	0.439	0.284

Moments of inertia vary as $(1 + Kx/l)^n$, where $n = 3.0$

(Continued)

TABLE 8.11(*c*) Reaction and Deflection Coefficients for Tapered Beams (*Continued*)

Case No. in Table 8.1	Load Location a/l	Multiplier Listed for	I_B/I_A				
			0.25	0.50	2.0	4.0	8.0
2d. Uniform load	0	R_A	1.092	1.046	0.954	0.908	0.863
		M_A	1.297	1.144	0.867	0.745	0.635
	0.50	R_A	1.266	1.128	0.882	0.776	0.680
		M_A	1.517	1.240	0.796	0.626	0.487
2e. Uniform load	0	θ_A	1.697	1.311	0.753	0.560	0.411
		$y_{l/2}$	1.919	1.400	0.700	0.480	0.322
	0.50	θ_A	1.833	1.363	0.724	0.517	0.365
		$y_{l/2}$	2.025	1.438	0.680	0.453	0.296
2a. Uniformly increasing load	0	y_A	3.115	1.773	0.559	0.309	0.169
		θ_A	2.981	1.734	0.572	0.324	0.182
	0.25	y_A	3.258	1.815	0.545	0.294	0.157
		θ_A	3.178	1.792	0.552	0.301	0.163
	0.50	y_A	3.446	1.866	0.531	0.279	0.146
		θ_A	3.407	1.856	0.533	0.282	0.148
	0.75	y_A	3.687	1.927	0.516	0.264	0.135
		θ_A	3.676	1.925	0.516	0.265	0.135
2c. Uniformly increasing load	0	R_A	1.114	1.058	0.942	0.885	0.829
		θ_A	1.942	1.404	0.702	0.486	0.332
	0.50	R_A	1.233	1.113	0.895	0.800	0.713
		θ_A	2.244	1.511	0.652	0.419	0.266
2d. Uniformly increasing load	0	R_A	1.159	1.078	0.925	0.853	0.785
		M_A	1.386	1.183	0.837	0.694	0.596
	0.50	R_A	1.325	1.153	0.865	0.747	0.645
		M_A	1.602	1.273	0.777	0.598	0.456
2e. Uniformly increasing load	0	θ_A	1.764	1.337	0.738	0.538	0.387
		$y_{l/2}$	1.972	1.419	0.690	0.466	0.309
	0.50	θ_A	1.878	1.379	0.717	0.508	0.356
		$y_{l/2}$	2.072	1.454	0.674	0.445	0.288

(*Continued*)

TABLE 8.11(*c*) Reaction and Deflection Coefficients for Tapered Beams (*Continued*)

Case No. in Table 8.1	Load Location a/l	Multiplier Listed for	I_B/I_A				
			0.25	0.50	2.0	4.0	8.0
3a	0	y_A	2.520	1.587	0.630	0.397	0.250
		θ_A	2.054	1.424	0.712	0.513	0.375
	0.25	y_A	2.607	1.619	0.613	0.373	0.224
		θ_A	2.352	1.537	0.646	0.412	0.260
	0.50	y_A	2.858	1.702	0.579	0.330	0.185
		θ_A	2.741	1.668	0.590	0.342	0.194
	0.75	y_A	3.296	1.829	0.539	0.288	0.152
		θ_A	3.264	1.821	0.542	0.290	0.153
3c	0	R_A	0.901	0.947	1.063	1.136	1.223
		θ_A	1.401	1.186	0.839	0.701	0.583
	0.50	R_A	1.022	1.015	0.977	0.945	0.906
		θ_A	1.257	1.157	0.820	0.642	0.483
3d	0.25	R_A	0.781	0.887	1.117	1.233	1.343
		M_A	1.705	1.352	0.663	0.355	0.088
	0.50	R_A	0.969	0.992	0.992	0.969	0.932
		M_A	0.897	0.975	0.977	0.916	0.828
3e	0	θ_A	1.397	1.186	0.838	0.699	0.579
		$y_{l/2}$	1.723	1.324	0.743	0.543	0.391
	0.25	θ_A	1.868	1.389	0.693	0.460	0.289
		$y_{l/2}$	1.892	1.394	0.697	0.471	0.308

TABLE 8.11(*d*) Reaction and Deflection Coefficients for Tapered Beams

Case No. in Table 8.1	Load Location a/l	Multiplier Listed for	I_B/I_A				
			0.25	0.50	2.0	4.0	8.0
1a	0	y_A θ_A	2.828 2.552	1.682 1.593	0.595 0.632	0.354 0.402	0.210 0.258
	0.25	y_A θ_A	2.971 2.811	1.727 1.679	0.576 0.593	0.330 0.350	0.188 0.206
	0.50	y_A θ_A	3.200 3.124	1.796 1.774	0.553 0.559	0.303 0.310	0.165 0.170
	0.75	y_A θ_A	3.532 3.511	1.886 1.881	0.527 0.528	0.276 0.277	0.143 0.144
1c	0.25	R_A θ_A	1.051 1.646	1.027 1.290	0.969 0.767	0.935 0.581	0.896 0.434
	0.50	R_A θ_A	1.131 1.941	1.068 1.404	0.929 0.702	0.857 0.485	0.784 0.331
1d	0.25	R_A M_A	1.051 1.164	1.027 1.085	0.969 0.912	0.935 0.821	0.896 0.730
	0.50	R_A M_A	1.172 1.373	1.086 1.180	0.914 0.835	0.828 0.686	0.746 0.556
1e	0.25	θ_A $y_{l/2}$	1.578 1.805	1.260 1.351	0.787 0.731	0.615 0.528	0.476 0.376
	0.50	θ_A $y_{l/2}$	1.752 1.941	1.329 1.404	0.746 0.702	0.552 0.485	0.406 0.331
2a. Uniform load	0	y_A θ_A	3.013 2.828	1.738 1.682	0.573 0.595	0.328 0.354	0.187 0.210
	0.25	y_A θ_A	3.161 3.052	1.783 1.751	0.558 0.568	0.309 0.320	0.170 0.180
	0.50	y_A θ_A	3.365 3.314	1.841 1.827	0.539 0.543	0.289 0.293	0.154 0.157
	0.75	y_A θ_A	3.639 3.625	1.913 1.910	0.520 0.521	0.269 0.270	0.138 0.139
2c. Uniform load	0	R_A θ_A	1.065 1.839	1.034 1.361	0.964 0.729	0.927 0.528	0.888 0.380
	0.50	R_A θ_A	1.190 2.151	1.095 1.476	0.907 0.670	0.817 0.443	0.731 0.290

Moments of inertia vary as $(1 + Kx/l)^n$, where $n = 4.0$

(Continued)

TABLE 8.11(*d*) Reaction and Deflection Coefficients for Tapered Beams (*Continued*)

Case No. in Table 8.1	Load Location a/l	Multiplier Listed for	I_B/I_A				
			0.25	0.50	2.0	4.0	8.0
2d. Uniform load	0	R_A	1.092	1.046	0.954	0.908	0.862
		M_A	1.301	1.145	0.867	0.747	0.639
	0.50	R_A	1.266	1.128	0.882	0.774	0.676
		M_A	1.521	1.241	0.796	0.627	0.488
2e. Uniform load	0	θ_A	1.724	1.316	0.756	0.569	0.426
		$y_{l/2}$	1.953	1.406	0.703	0.488	0.335
	0.50	θ_A	1.864	1.396	0.727	0.526	0.379
		$y_{l/2}$	2.061	1.445	0.683	0.461	0.307
2a. Uniformly increasing load	0	y_A	3.145	1.778	0.560	0.312	0.173
		θ_A	3.013	1.738	0.573	0.328	0.187
	0.25	y_A	3.287	1.819	0.546	0.297	0.160
		θ_A	3.207	1.796	0.553	0.304	0.166
	0.50	y_A	3.470	1.869	0.532	0.281	0.147
		θ_A	3.432	1.859	0.534	0.284	0.150
	0.75	y_A	3.703	1.929	0.516	0.265	0.136
		θ_A	3.692	1.927	0.517	0.266	0.136
2c. Uniformly increasing load	0	R_A	1.112	1.057	0.942	0.882	0.823
		θ_A	1.966	1.408	0.704	0.492	0.340
	0.50	R_A	1.227	1.111	0.894	0.794	0.701
		θ_A	2.269	1.515	0.653	0.423	0.271
2d. Uniformly increasing load	0	R_A	1.159	1.078	0.924	0.852	0.783
		M_A	1.390	1.184	0.837	0.695	0.572
	0.50	R_A	1.323	1.153	0.864	0.744	0.639
		M_A	1.605	1.274	0.777	0.598	0.456
2e. Uniformly increasing load	0	θ_A	1.793	1.343	0.741	0.547	0.402
		$y_{l/2}$	2.007	1.425	0.693	0.475	0.321
	0.50	θ_A	1.909	1.385	0.719	0.516	0.369
		$y_{l/2}$	2.108	1.461	0.677	0.453	0.299

(*Continued*)

TABLE 8.11(*d*) Reaction and Deflection Coefficients for Tapered Beams (*Continued*)

Case No. in Table 8.1	Load Location a/l	Multiplier Listed for	I_B/I_A				
			0.25	0.50	2.0	4.0	8.0
3a	0	y_A	2.552	1.593	0.632	0.402	0.258
		θ_A	2.081	1.428	0.714	0.520	0.386
	0.25	y_A	2.641	1.624	0.615	0.378	0.231
		θ_A	2.386	1.543	0.648	0.419	0.270
	0.50	y_A	2.893	1.707	0.581	0.334	0.190
		θ_A	2.778	1.674	0.592	0.346	0.200
	0.75	y_A	3.326	1.833	0.540	0.290	0.154
		θ_A	3.295	1.825	0.543	0.292	0.155
3c	0	R_A	0.902	0.947	1.063	1.138	1.227
		θ_A	1.414	1.189	0.841	0.707	0.595
	0.50	R_A	1.023	1.015	0.976	0.944	0.904
		θ_A	1.275	1.161	0.823	0.652	0.499
3d	0.25	R_A	0.780	0.887	1.117	1.234	1.347
		M_A	1.716	1.354	0.665	0.359	0.092
	0.50	R_A	0.971	0.993	0.993	0.971	0.935
		M_A	0.902	0.976	0.979	0.922	0.839
3e	0	θ_A	1.414	1.189	0.841	0.707	0.595
		$y_{l/2}$	1.752	1.329	0.746	0.552	0.406
	0.25	θ_A	1.903	1.396	0.697	0.470	0.304
		$y_{l/2}$	1.927	1.401	0.700	0.480	0.321

TABLE 8.12 Position of Flexural Center *Q* for Different Sections

Form of Section	Position of Q
1. Any narrow section symmetrical about the x axis; centroid at $x = 0$ $y = 0$ 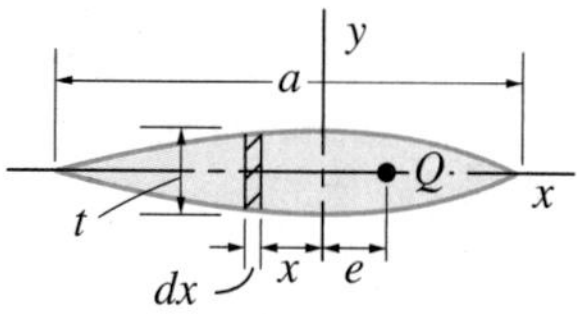	$e = \frac{1+3\nu}{1+\nu} \frac{\int xt^3 dx}{\int t^3 dx}$ For narrow triangle (with $\nu = 0.25$), $e = 0.187a$ (Refs. 32 and 52).
2. Beam composed of n elements of any form, connected of separate, with common neutral axis (e.g., multiple-spar airplane wing) 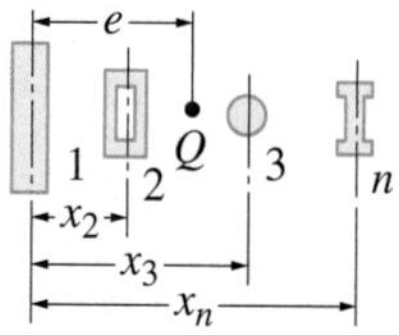	$e = \frac{E_2 I_2 x_2 + E_3 I_3 x_3 + \cdots + E_n I_n x_n}{E_1 I_1 + E_2 I_2 + E_3 I_3 + \cdots + E_n I_n}$ where I_1, I_2, etc., are moments of inertia of the several elements about the X axis (i.e., Q is at the centroid of the products EI for the several elements).
3. Semicircular area 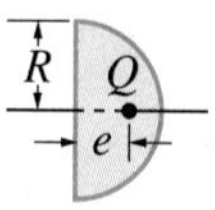	$e = \frac{8}{15\pi} \frac{3+4\nu}{1+\nu} R$ (Q is to right of centroid) (Refs. 1 and 64) For any sector of solid or hollow circular area, see Ref. 32.
4. Angle 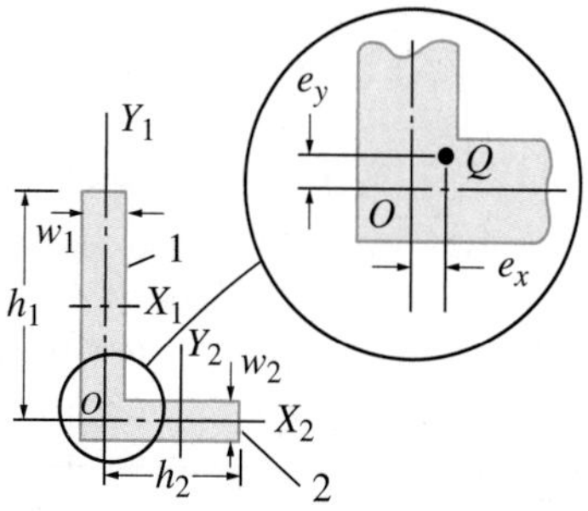	Leg 1 = rectangle $w_1 h_1$; leg 2 = rectangle $w_2 h_2$ I_1 = moment of inertia of leg 1 about Y_1 (central axis) I_2 = moment of inertia of leg about Y_2 (central axis) $e_y = \frac{h_1}{2} \frac{I_1}{I_1 + I_2}$ (for e_x, use X_1 and X_2 central axes) (Ref. 31) If w_1 and w_2 are small, $e_x = e_y = 0$ (practically) and Q is at 0.
5. Channel 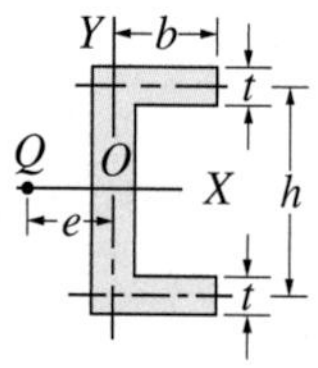	$e = h\frac{I_{xy}}{I_x}$ where I_{xy} = product of inertia of the half section (above X) with respect to axes X and Y, and I_x = moment of inertia of whole section with respect to axis X If t is uniform, $e = (b^2 - t^2/4)h^2 t/4I_x$.
6. T 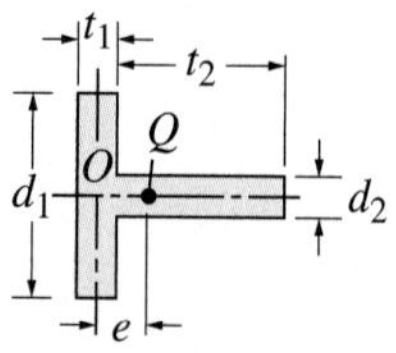	$e = \frac{1}{2}(t_1 + t_2)\frac{1}{1 + d_1^3 t_1 / d_2^3 t_2}$ For a T-beam of ordinary proportions, Q may be assumed to be at 0.

(Continued)

TABLE 8.12 Position of Flexural Center *Q* for Different Sections (*Continued*)

Form of Section	Position of Q
7. *I* with unequal flanges and thin web 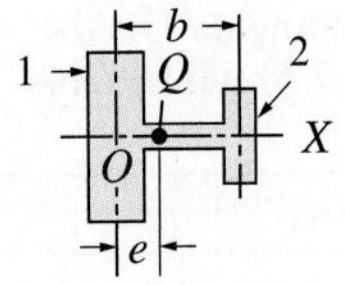	$e = b\dfrac{I_2}{I_1 + I_2}$ where I_1 and I_2, respectively, denote moments of inertia about X axis of flanges 1 and 2
8. Hollow thin-walled triangular section 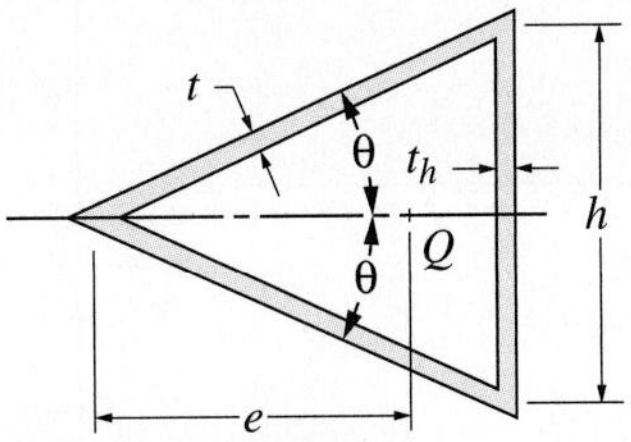	$\dfrac{e}{h} = \dfrac{1}{2\tan\theta(1 + t\sin\theta/t_h)}$
9. Hollow thin-walled section bounded by a circular arc and a straight edge 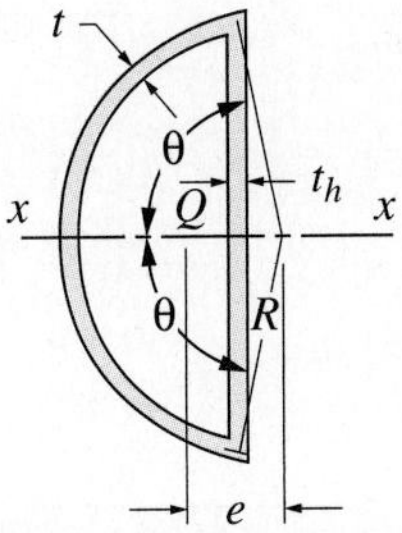 *Note:* Expressions are valid for $0 < \theta < \pi$	$I_x = tR^3\left(\theta - \sin\theta\cos\theta + \dfrac{2t_h}{3t}\sin^3\theta\right)$ $\dfrac{e}{R} = \dfrac{2t_h R^3 \sin\theta(\cos\theta + t/t_h)}{I_x(\sin\theta + t_h\theta/t)}\left(\sin\theta - \theta\cos\theta + \dfrac{t_h\theta}{3t}\sin^2\theta\right)$ If $\theta = \pi/2$ $I_x = \dfrac{tR^3}{2}\left(\pi + \dfrac{4t_h}{3t}\right)$ $\dfrac{e}{R} = \dfrac{4(6 + \pi t_h/t)}{(2 + \pi t_h/t)(3\pi + 4t_h/t)}$
10. Hollow thin-walled rectangular section 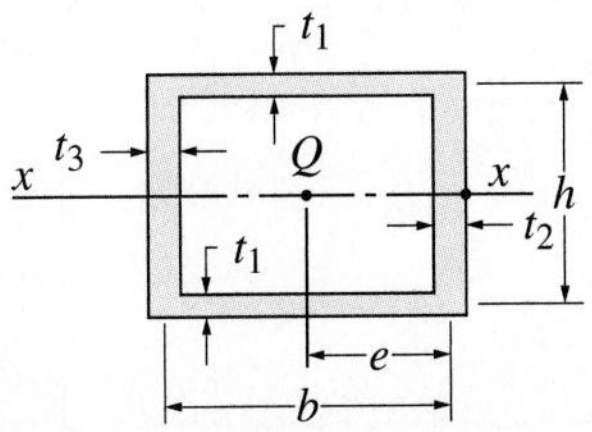	$I_x = \dfrac{h^3}{12}(t_2 + t_3) + \dfrac{t_1 b h^2}{2}$ $\dfrac{e}{b} = \dfrac{bh^2}{12I_x}\left[9t_1 + \dfrac{t_3 h}{b} - \dfrac{12(b + t_1 h/t_3)}{2b/t_1 + h/t_3 + h/t_2}\right]$
11. For thin-walled sections, such as lipped channels, hat sections, and sectors of circular tubes, see Table 9.2. The position of the flexural centers and shear centers coincide.	

TABLE 8.13 Collapse Loads with Plastic Hinge Locations for Straight Beams

Notation: M_p = fully plastic bending moment (force-length); x_h = position of a plastic hinge (length); W_c = concentrated load necessary to produce plastic collapse of the beam (force); w_c = unit load necessary to produce plastic collapse of the beam (force per unit length); M_{oc} = applied couple necessary to produce plastic collapse (force-length). The fully plastic bending moment M_p is the product of the yield strength of the material σ_{ys} and the plastic section modulus Z found in Table A.1 for the given cross sections.

Reference No., End Restraints	Collapse Loads with Plastic Hinge Locations
1a. Left end free, right end fixed (cantilever) 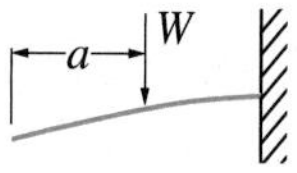	$W_c = \frac{M_p}{l-a}$ $x_h = l$
1b. Left end guided, right end fixed 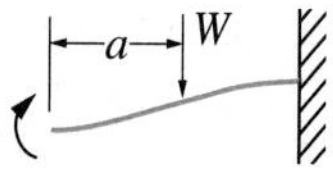	$W_c = \frac{2M_p}{l-a}$ $0 \leq x_{h1} \leq a \qquad x_{h2} = l$
1c. Left end simply supported, right end fixed 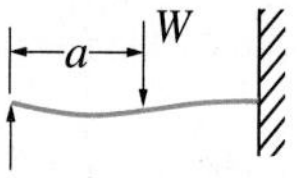	$W_c = \frac{M_p(l+a)}{a(l-a)}$ $x_{h1} = a \qquad x_{h2} = l$
1d. Left end fixed, right end fixed 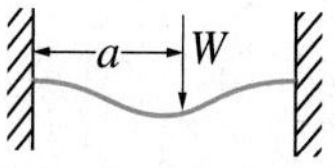	$W_c = \frac{2M_p l}{a(l-a)}$ $x_{h1} = 0 \qquad x_{h2} = a \qquad x_{h3} = l$
1e. Left end simply supported, right end simply supported 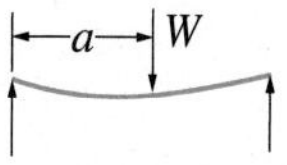	$W_c = \frac{M_p l}{a(l-a)}$ $x_h = a$
1f. Left end guided, right end simply supported 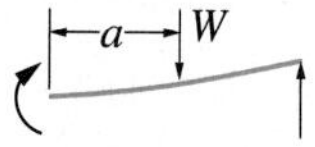	$W_c = \frac{M_p}{l-a}$ $0 < x_h < a$

2a. Left end free, right end fixed (cantilever) 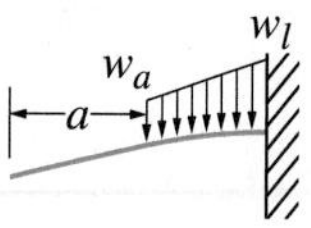	If $w_l = w_a$ (uniform load), then
	 $w_{ac} = \frac{2M_p}{(l-a)^2} \qquad x_h = l$
	 If $w_a = 0$ (uniformly increasing load), then
	 $w_{lc} = \frac{6M_p}{(l-a)^2} \qquad x_h = l$
	 If $w_l = 0$ (uniformly decreasing load), then
	 $w_{ac} = \frac{3M_p}{(l-a)^2} \qquad x_h = l$

(*Continued*)

TABLE 8.13 Collapse Loads with Plastic Hinge Locations for Straight Beams (*Continued*)

<table>
<tr><th>Reference No., End Restraints</th><th>Collapse Loads with Plastic Hinge Locations</th></tr>
<tr><td>2b. Left end guided, right end fixed
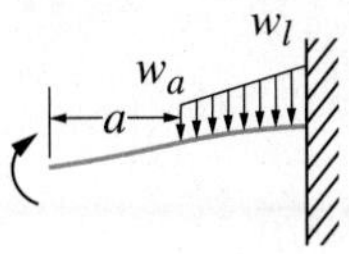</td><td>

If $w_l = w_a$ (uniform load), then

$$w_{ac} = \frac{4M_p}{(l-a)^2} \quad x_{h1} = a \quad x_{h2} = l$$

If $w_a = 0$ (uniformly increasing load), then

$$w_{lc} = \frac{12M_p}{(l-a)^2} \quad x_{h1} = a \quad x_{h2} = l$$

If $w_l = 0$ (uniformly decreasing load), then

$$w_{ac} = \frac{6M_p}{(l-a)^2} \quad x_{h1} = a \quad x_{h2} = l$$

</td></tr>
<tr><td>2c. Left end simply supported, right end fixed
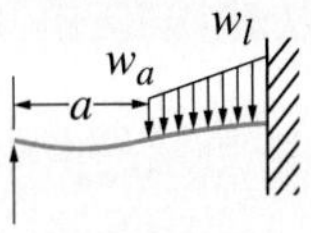</td><td>

If $w_l = w_a$ (uniformly load), then

$$w_{ac} = \frac{2M_p(l+x_{h1})}{(l-x_{h1})(lx_{h1}-a^2)}$$

where $x_{h1} = [2(l^2+a^2)]^{1/2} - l \quad x_{h2} = l$

If $w_a = 0$ (uniformly increasing load), then

$$w_{lc} = \frac{6K_1M_p}{(l-a)^2} \quad x_{h1} = K_2 l \quad x_{h2} = l$$

<table>
<tr><td>a/l</td><td>0</td><td>0.2</td><td>0.4</td><td>0.6</td><td>0.8</td></tr>
<tr><td>K_1</td><td>4.000</td><td>3.324</td><td>2.838</td><td>2.481</td><td>2.211</td></tr>
<tr><td>K_2</td><td>0.500</td><td>0.545</td><td>0.616</td><td>0.713</td><td>0.838</td></tr>
</table>

If $w_l = 0$ (uniformly decreasing load), then

$$w_{ac} = \frac{6K_3M_p}{(l-a)^2} \quad x_{h1} = K_4 l \quad x_{h2} = l$$

<table>
<tr><td>a/l</td><td>0</td><td>0.2</td><td>0.4</td><td>0.6</td><td>0.8</td></tr>
<tr><td>K_3</td><td>3.596</td><td>2.227</td><td>1.627</td><td>1.310</td><td>1.122</td></tr>
<tr><td>K_4</td><td>0.347</td><td>0.387</td><td>0.490</td><td>0.634</td><td>0.808</td></tr>
</table>

</td></tr>
</table>

(*Continued*)

TABLE 8.13 Collapse Loads with Plastic Hinge Locations for Straight Beams (*Continued*)

Reference No., End Restraints	Collapse Loads with Plastic Hinge Locations

2d. Left end fixed, right end fixed
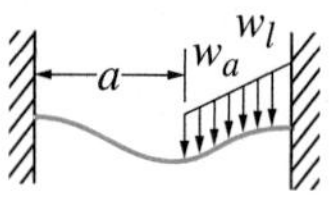
 	
	If $w_l = w_a$ (uniform load), then
	$w_{ac} = \dfrac{16M_p l^2}{(l^2-a^2)^2} \quad x_{h1}=0 \quad x_{h2}=\dfrac{l^2+a^2}{2l} \quad x_{h3}=l$
	If $w_a = 0$ (uniformly increasing load), then
	$w_{lc} = \dfrac{12M_p(l-a)}{(l-x_{h2})(x_{h2}^2-3ax_{h2}+lx_{h2}+a^3/l)}$
	$x_{h1}=0 \quad x_{h2}=a+\left(a^2-al+\dfrac{l^2}{3}-\dfrac{a^3}{3l}\right)^{1/2} \quad x_{h3}=l$
	If $w_l = 0$ (uniformly decreasing load), then
	$w_{ac} = \dfrac{12M_p(l-a)}{(l-x_{h2})(2lx_{h2}-3a^2-x_{h2}^2+2a^3/l)}$
	$x_{h1}=0 \quad x_{h2}=l-\left(\dfrac{l^2}{3}-a^2+\dfrac{2a^3}{3l}\right)^{1/2} \quad x_{h3}=l$

2e. Left end simply supported, right end simply supported
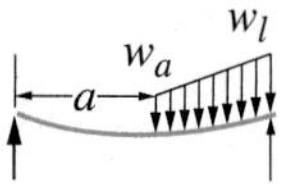
 	
	If $w_l = w_a$ (uniform load), then
	$w_{ac} = \dfrac{8M_p l^2}{(l^2-a^2)^2} \quad x_h=\dfrac{l^2+a^2}{2l}$
	If $w_a = 0$ (uniformly increasing load), then
	$w_{lc} = \dfrac{6M_p(l-a)}{(l-x_h)(x_h^2-3ax_h+lx_h+a^3/l)}$
	$x_h=a+\left(a^2-al+\dfrac{l^2}{3}-\dfrac{a^3}{3l}\right)^{1/2}$
	If $w_l = 0$ (uniformly decreasing load), then
	$w_{ac} = \dfrac{6M_p(l-a)}{(l-x_h)(2lx_h-3a^2-x_h^2+2a^3/l)}$
	$x_h=l-\left(\dfrac{l^2}{3}-a^2+\dfrac{2a^3}{3l}\right)^{1/2}$

2f. Left end guided, right end simply supported
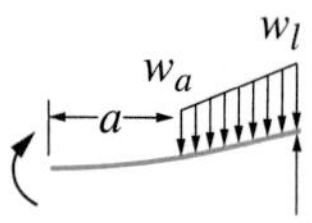
 	
	If $w_l = w_a$ (uniformly load), then
	$w_{ac} = \dfrac{2M_p}{(l-a)^2} \quad 0 \le x_h \le a$
	If $w_a = 0$ (uniformly increasing load), then
	$w_{lc} = \dfrac{6M_p}{(l-a)^2} \quad 0 \le x_h \le a$
	If $w_l = 0$ (uniformly decreasing load), then
	$w_{ac} = \dfrac{3M_p}{(l-a)^2} \quad 0 \le x_h \le a$

(*Continued*)

TABLE 8.13 Collapse Loads with Plastic Hinge Locations for Straight Beams (*Continued*)

Reference No., End Restraints	Collapse Loads with Plastic Hinge Locations
3a. Left end free, right end fixed (cantilever) 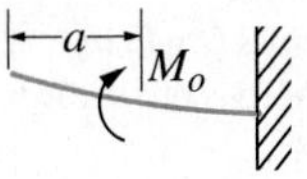	$M_{oc} = M_p$ $a < x_h < l$
3b. Left end guided, right end fixed 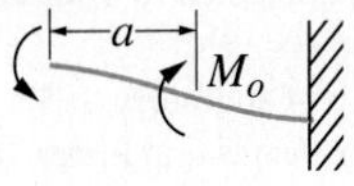	$M_{oc} = 2M_p$ $0 < x_{h1} < a \quad a < x_{h2} < l$
3c. Left end simply supported, right end fixed 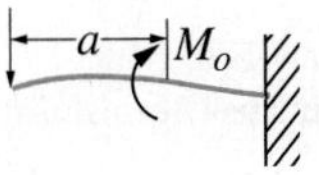	If $l/3 \le a \le l$, then $M_{oc} = 2M_p$ and two plastic hinges form, one on each side of and adjacent to the loading M_o If $0 \le a \le l/3$, then $M_{oc} = \frac{M_p(l+a)}{l-a}$ $x_{h1} = a$ just to the right of the loading M_o $\quad x_{h2} = l$
3d. Left end fixed, right end fixed 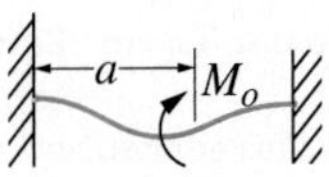	$M_{oc} = 2M_p$ and two plastic hinges form, one on each side of and adjacent to the loading M_o If $0 < a < l/2$, then a third hinge forms at the right end If $l/2 < a < l$, then the third hinge forms at the left end If $a = l/2$, two hinges form at any two locations on one side of the load and one at any location on the other side

3e. Left end simply supported, right end simply supported 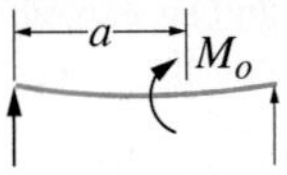	If $0 \le a < l/2$, then
	 $M_{oc} = \frac{M_p l}{l-a}$
	 $x_h = a$ just to the right of the loading M_o
	 If $l/2 < a \le l$, then
	 $M_{oc} = \frac{M_p l}{a}$
	 $x_h = a$ just to the right of the loading M_o
	 If $a = l/2$, then
	 $M_{oc} = 2M_p$
	 and two plastic hinges form, one on each side of and adjacent to the loading M_o

3f. Left end guided, right end simply supported 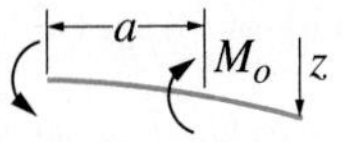	$M_{oc} = M_p$ $0 < x_h < a$

8.18 REFERENCES

1. Timoshenko, S. P., and J. N. Goodier, *Theory of Elasticity*, 3rd ed., McGraw-Hill, 1970.
2. Frocht, M. M., "A Photoelastic Investigation of Shear and Bending Stresses in Centrally Loaded Simple Beams," *Eng. Bull., Carnegie Inst. Technol.*, 1937.
3. Timoshenko, S., *Strength of Materials*, D. Van Nostrand, 1930.
4. Bach, C., "Zur Beigungsfestigkeit des Gusseisens," *Z. Vereines Dtsch. Ing.*, 32: 1089, 1888.
5. Schlick, W. J., and B. A. Moore, "Strength and Elastic Properties of Cast Iron," *Iowa Eng. Exp. Sta.*, Iowa State College, Bull. 127, 1930.
6. Symposium on Cast Iron, *Proc. ASTM*, 33(II): 115, 1933.
7. Roark, R. J., R. S. Hartenberg, and R. Z. Williams, "The Effect of Form and Scale on Strength," *Eng. Exp. Sta.*, Univ. Wis., Bull. 82, 1938.
8. Newlin, J. A., and G. W. Trayer, "Form Factors of Beams Subjected to Transverse Loading Only," *Natl. Adv. Comm. Aeron*, Rept. 181, 1924.
9. U.S. Dept. of Agriculture, *Wood Handbook*, Forest Products Laboratory, 1987.
10. Ashwell, D. G., "The Anticlastic Curvature of Rectangular Beams and Plates," *J. R. Aeron. Soc.*, 54, 1950.
11. Ketchum, M. S., and J. O. Draffin, "Strength of Light I-beams," *Eng. Exp. Sta.*, Univ. Ill., Bull. 241, 1932.
12. Wendt, K. F., and M. O. Withey, "The Strength of Light Steel Joists," *Eng. Exp. Sta.*, Univ. Wis., Bull. 79, 1934.
13. American Institute of Steel Construction, *Specifications for the Design, Fabrication, and Erection of Structural Steel for Buildings*, 1978.
14. Younger, J. E., *Structural Design of Metal Airplanes*, McGraw-Hill, 1935.
15. Lyse, I., and H. J. Godfrey, "Investigation of Web Buckling in Steel Beams," *Trans. Am. Soc. Civil Eng.*, 100: 675, 1935.
16. Moore, H. F., "The Strength of I-beams in Flexure," *Eng. Exp. Sta., Univ. Ill.*, Bull. 68, 1913.
17. Dumont, C., and H. N. Hill, "The Lateral Instability of Deep Rectangular Beams," *Nat. Adv. Comm. Aeron.*, Tech. Note 601, 1937.
18. Trayer, G. W., and H. W. March, "Elastic Instability of Members having Sections Common in Aircraft Construction," *Natl. Adv. Comm. Aeron.*, Rept. 382, 1931.
19. Newlin, J. A., and G, W. Trayer, "Deflection of Beams with Special Reference to Shear Deformation," *Natl. Adv. Comm. Aeron.*, Rept. 180, 1924.
20. McCutcheon, William J., "Deflections and Stresses in Circular Tapered Beams and Poles," *Civ. Eng. Pract. Des. Eng.*, 2: 1983.
21. Timoshenko. S., "Mathematical Determination of the Modulus of Elasticity," *Mech. Eng.*, 45: p. 259, 1923.
22. Holl, D. L., "Analysis of Thin Rectangular Plates Supported on Opposite Edges," *Iowa Eng. Exp. Sta.*, Iowa State College, Bull. 129, 1936.
23. Westergaard, H. M., *Computation of Stress Due to Wheel Loads, Public Roads*, U.S. Dept. of Agriculture, Bureau of Public Roads, 11: 9, 1930.
24. Morris, C. T., "Concentrated Loads on Slabs," *Ohio State Univ. Eng. Exp. Sta.*, Bull. 80, 1933.
25. Kelley, E. F., "Effective Width of Concrete Bridge Slabs Supporting Concentrated Loads," *Pub. Roads*, U.S. Dept. of Agriculture, Bureau of Public Roads, 7(1): 1926.
26. MacGregor, C. W., Deflection of Long Helical Gear Tooth," *Mech. Eng.*, 57: 225, 1935.
27. Holl, D. L., Cantilever Plate with Concentrated Edge Load, ASME Paper A-8, *J. Appl. Mech.*, 4(1): 1937.
28. Miller, A. B., "Die mittragende Breite, and Über die mittragende Breite, *Luftfahrtforsch.*, 4(1): 1929.
29. Hetényi, M., "Application of Maclaurin Series to the Analysis of Beams in Bending," *J. Franklin Inst.*, 254: 1952.
30. Kuhn, P., J. P. Peterson, and L. R. Levin, "A Summary of Diagonal Tension, Parts I and II," *Natl. Adv. Comm. Aeron.*, Tech. Notes 2661 and 2662, 1952.
31. Schwalbe, W. L. S., "The Center of Torsion for Angle and Channel Sections," *Trans. ASME*, 54(11): 125, 1932.
32. Young, A. W., E. M. Elderton, and K. Pearson, "On the Torsion Resulting from Flexure in Prisms with Cross-Sections of Uniaxial Symmetry," Drapers' Co. Research Memoirs, Tech. Ser. 7, 1918.
33. Maurer, E. R., and M. O. Withey, *Strength of Materials*, John Wiley & Sons, 1935.
34. Peery, D. J., *Aircraft Structures*, McGraw-Hill, 1950.
35. Sechler, E. E., and L. G. Dunn, *Airplane Structural Analysis and Design*, John Wiley & Sons, 1942.
36. Griffel, W., *Handbook of Formulas for Stress and Strain*, Frederick Ungar, 1966.
37. Reissner, E., "Least Work Solutions of Shear Lag Problems," *J. Aeron. Sci.*, 8(7): 284, 1941.
38. Hildebrand, F. B., and E. Reissner, "Least-Work Analysis of the Problem of Shear Lag in Box Beams," *Natl. Adv. Comm. Aeron.*, Tech. Note 893, 1943.
39. Winter, G., "Stress Distribution in and Equivalent Width of Flanges of Wide, Thinwall Steel Beams," *Natl. Adv. Comm. Aeron.*, Tech. Note 784, 1940.
40. Tate, M. B., "Shear Lag in Tension Panels and Box Beams," *Iowa Eng. Exp. Sta. Iowa State College*, Eng. Rept. 3, 1950.

41. Vlasov, V. Z., and U. N. Leontév, *Beams, Plates and Shells on Elastic Foundations*, transl. from Russian, Israel Program for Scientific Translations, Jerusalem, NASA TT F-357, US, 1966.
42. Kameswara Rao, N. S. V., Y. C. Das, and M. Anandakrishnan, "Variational Approach to Beams on Elastic Foundations," *Proc. Am. Soc. Civil Eng., J. Eng. Mech. Div.*, 97(2), 1971.
43. White, Richard N., *Rectangular Plates Subjected to Partial Edge Loads: Their Elastic Stability and Stress Distribution*, Doctoral Dissertation, University of Wisconsin, 1961.
44. Chow, L., Harry D. Conway, and George Winter, "Stresses in Deep Beams," *Trans. Am. Soc. Civil Eng.*, 118: 686, 1963.
45. Kaar, P. H., "Stress in Centrally Loaded Deep Beams," *Proc. Soc. Exp. Stress Anal.*, 15(1): 77, 1957.
46. Saad, S., and A. W. Hendry, "Stresses in a Deep Beam with a Central Concentrated Load," *Exp. Mech., J. Soc. Exp. Stress Anal.*, 18(1): 192, 1961.
47. Jaramillo, T. J., "Deflections and Moments due to a Concentrated Load on a Cantilever Plate of Infinite Length," *ASME J. Appl. Mech.*, 17(1): 1950.
48. Wellauer, E. J., and A. Seireg, Bending Strength of Gear Teeth by Cantilever-Plate Theory, *ASME J. Eng. Ind.*, 82: August 1960.
49. Little, Robert W., *Bending of a Cantilever Plate*, Master's Thesis, University of Wisconsin, 1959.
50. Small, N. C., "Bending of a Cantilever Plate Supported from an Elastic Half Space," *ASME J. Appl. Mech.*, 28(3): 1961.
51. Hetényi, M., "Series Solutions for Beams on Elastic Foundations," *ASME J. Appl. Mech.*, 38(2): 1971.
52. Duncan, W. J., "The Flexural Center or Center of Shear," *J. R. Aeron. Soc.*, 57: September 1953.
53. Hetényi, Miklos, *Beams on Elastic Foundation*, The University of Michigan Press, 1946.
54. O' Donpell, W. J., "The Additional Deflection of a Cantilever Due to the Elasticity of the Support," *ASME J. Appl. Mech.*, 27(3): 1960.
55. *ANC Mil-Hdbk-5, Strength of Metal Aircraft Elements*, Armed Forces Supply Support Center, March 1959.
56. Kleinlogel, A., *Rigid Frame Formulas*, Frederick Ungar, 1958.
57. Leontovich, Valerian, *Frames and Arches*, McGraw-Hill, 1959.
58. Lee, G. C., "A Survey of Literature on the Lateral Instability of Beams," Bull. 63 Weld. Res. Counc, August 1960.
59. Kelley, B. W., and R. Pedersen, "The Beam Strength of Modern Gear Tooth Design," *Trans. SAE*, 66: 1950.
60. Beedle, Lynn S., *Plastic Design of Steel Frames*, John Wiley & Sons, 1958.
61. The Steel Skeleton, vol. II, *Plastic Behaviour and Design*, Cambridge University Press, 1956.
62. Johnston, B. G., F. J. Lin, and T. V. Galambos, *Basic Steel Design*, 3rd ed., Prentice-Hall, 1986.
63. Weigle, R. E., R. R. Lasselle, and J. P. Purtell, "Experimental Investigation of the Fatigue Behavior of Thread-Type Projections," *Exp. Mech.*, 3(5): 1963.
64. Leko, T., "On the Bending Problem of Prismatical Beam by Terminal Transverse Load," *ASME J. Appl. Mech.*, 32(1): 1965.
65. Thomson, W. T., "Deflection of Beams by the Operational Method," *J. Franklin Inst.*, 247(6): 1949.
66. Cook, R. D., and W. C. Young, *Advanced Mechanics of Materials*, 2nd ed., Prentice-Hall, 1998.
67. Cook, R. D., "Deflections of a Series of Cantilevers Due to Elasticity of Support," *ASME J. Appl. Mech.*, 34(3): 1967.
68. Yu, Wei-Wen, *Cold-Formed Steel Design*, John Wiley & Sons, 1985.
69. White, R. N., and C. G. Salmon (eds.), *Building Structural Design Handbook*, John Wiley & Sons, 1987.
70. American Institute of Steel Construction, *Manual of Steel Construction-Load and Resistance Factor Design*, 1st ed., 1986.
71. Salmon, C. G., and J. E. Johnson, *Steel Structures: Design and Behavior*, 2nd ed., Harper & Row, 1980.
72. Budynas, R. G., *Advanced Strength and Applied Stress Analysis*, 2nd ed., McGraw-Hill, 1999.

CHAPTER 9

Curved Beams

9.1 BENDING IN THE PLANE OF THE CURVE

In a straight beam having either a constant cross section or a cross section which changes gradually along the length of the beam, the neutral surface is defined as the longitudinal surface of zero fiber stress when the member is subjected to pure bending. It contains the neutral axis of every section, and these neutral axes pass through the centroids of the respective sections. In this section on bending in the plane of the curve, the use of the many formulas is restricted to those members for which that axis passing through the centroid of a given section and directed normal to the plane of bending of the member is a principal axis. The one exception to this requirement is for a condition equivalent to the beam being constrained to remain in its original plane of curvature such as by frictionless external guides.

To determine the stresses and deformations in curved beams satisfying the restrictions given above, one first identifies several cross sections and then locates the centroids of each. From these centroidal locations the curved centroidal surface can be defined. For bending in the plane of the curve there will be at each section (1) a force N normal to the cross section and taken to act through the centroid, (2) a shear force V parallel to the cross section in a radial direction, and (3) a bending couple M in the plane of the curve. In addition there will be radial stresses σ_r in the curved beam to establish equilibrium. These internal loadings are shown in Fig. 9.1(a), and the stresses and deformations due to each will be evaluated.

Circumferential Normal Stresses Due to Pure Bending

When a curved beam is bent in the plane of initial curvature, plane sections remain plane, but because of the different lengths of fibers on the inner and outer portions of the beam, the distribution of unit strain, and therefore stress, is not linear. The neutral axis does not pass through the centroid of the section and Eqs. (8.1-1) and (8.1-2) do not apply. The error involved in their use is slight as long as the radius of curvature is more than about eight times the depth of the beam. At that curvature the errors in the maximum stresses are in the range of 4–5%. The errors created by using the straight-beam formulas become large for sharp curvatures as shown in Table 9.1, which gives formulas and selected numerical data for curved beams of several cross sections and for varying degrees of curvature. In part the formulas and tabulated coefficients are taken from the University of Illinois Circular by Wilson and Quereau (Ref. 1) with modifications suggested by Neugebauer (Ref. 28). For cross sections not included in Table 9.1 and for determining circumferential stresses at locations other than the extreme fibers, one can find formulas in texts on advanced mechanics of materials, for example, Refs. 29 and 36.

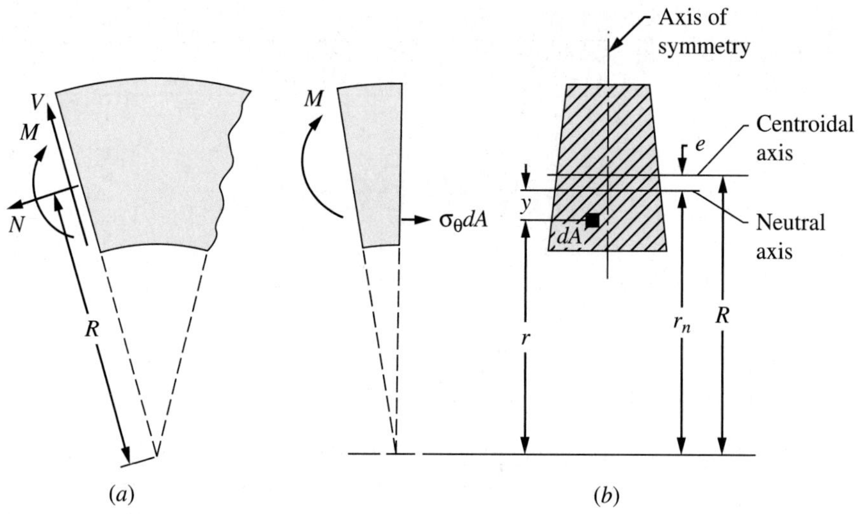

Figure 9.1

The circumferential normal stress σ_θ is given as

$$\sigma_\theta = \frac{My}{Aer} \tag{9.1-1}$$

where M is the applied bending moment, A is the area of the cross section, e is the distance from the centroidal axis to the neutral axis, and y and r locate the radial position of the desired stress from the neutral axis and the center of the curvature, respectively. See Fig. 9.1(b).

$$e = R - r_n = R - \frac{A}{\int_{\text{area}} dA/r} \quad \text{for } \frac{R}{d} < 8 \tag{9.1-2}$$

Equations (9.1-1) and (9.1-2) are based on derivations that neglect the contribution of radial normal stress to the circumferential strain. This assumption does not cause appreciable error for curved beams of compact cross section for which the radial normal stresses are small, and it leads to acceptable answers for beams having thin webs where, although the radial stresses are higher, they occur in regions of the cross section where the circumferential bending stresses are small. The use of the equations in Table 9.1 and of Eqs. (9.1-1) and (9.1-2) is limited to values of $R/d > 0.6$ where, for a rectangular cross section, a comparison of this mechanics-of-materials solution [Eq. (9.1-1)] to the solution using the theory of elasticity shows the mechanics of materials solution to indicate stresses approximately 10% too large.

While in theory the curved-beam formula for circumferential bending stress, Eq. (9.1-1), could be used for beams of very large radii of curvature, one should not use the expression for e from Eq. (9.1-2) for cases where R/d, the ratio of the radius of the curvature R to the depth of the cross section, exceeds 8. The calculation for e would have to be done with careful attention to precision on a computer or calculator to get an accurate answer. Instead one should use the following approximate expression for e which becomes very accurate for large values of R/d. See Ref. 29.

$$e \approx \frac{I_c}{RA} \quad \text{for } \frac{R}{d} > 8 \tag{9.1-3}$$

where I_c is the area moment of inertia of the cross section about the centroidal axis. Using this expression for e and letting R approach infinity leads to the usual straight-beam formula for bending stress.

For complex sections where the table or Eq. (9.1-3) are inappropriate, a numerical technique that provides excellent accuracy can be employed. This technique is illustrated on pp. 318–321 of Ref. 36.

In summary, use Eq. (9.1-1) with e from Eq. (9.1-2) for $0.6 < R/d < 8$. Use Eq. (9.1-1) with e from Eq. (9.1-3) for those curved beams for which $R/d > 8$ and where errors of less than 4–5% are desired, or use straight-beam formulas if larger errors are acceptable or if $R/d \gg 8$.

Circumferential Normal Stresses Due to Hoop Tension *N* (*M* = 0)

The normal force N was chosen to act through the centroid of the cross section, so a constant normal stress N/A would satisfy equilibrium. Solutions carried out for rectangular cross sections using the theory of elasticity show essentially a constant normal stress with higher values on a thin layer of material on the inside of the curved section and lower values on a thin layer of material on the outside of the section. In most engineering applications the stresses due to the moment M are much larger than those due to N, so the assumption of uniform stress due to N is reasonable.

Shear Stress Due to The Radial Shear Force *V*

Although Eq. (8.1-2) does not apply to curved beams, Eq. (8.1-13), used as for a straight beam, gives the *maximum* shear stress with sufficient accuracy in most instances. Again an analysis for a rectangular cross section carried out using the theory of elasticity shows that the peak shear stress in a curved beam occurs not at the centroidal axis as it does for a straight beam but toward the inside surface of the beam. For a very sharply curved beam, $R/d = 0.7$, the peak shear stress was $2.04V/A$ at a position one-third of the way from the inner surface to the centroid. For a sharply curved beam, $R/d = 1.5$, the peak shear stress was $1.56V/A$ at a position 80% of the way from the inner surface to the centroid. These values can be compared to a peak shear stress of $1.5V/A$ at the centroid for a straight beam of rectangular cross section.

If a mechanics-of-materials solution for the shear stress in a curved beam is desired, the element in Fig. 9.2(*b*) can be used and moments taken about the center of curvature. Using the normal stress distribution, $\sigma_\theta = N/A + My/AeR$, one can find the shear stress expression to be

$$\tau_{r\theta} = \frac{V(R-e)}{t_r A e r^2}(RA_r - Q_r) \qquad (9.1\text{-}4)$$

where t_r is the thickness of the section normal to the plane of curvature at the radial position r and

$$A_r = \int_b^r dA_1 \quad \text{and} \quad Q_r = \int_b^r r_1\, dA_1 \qquad (9.1\text{-}5)$$

Equation (9.1-4) gives conservative answers for the peak values of shear stress in rectangular sections when compared to elasticity solutions. The locations of peak shear stress are the same in both analyses, and the error in magnitude is about 1%.

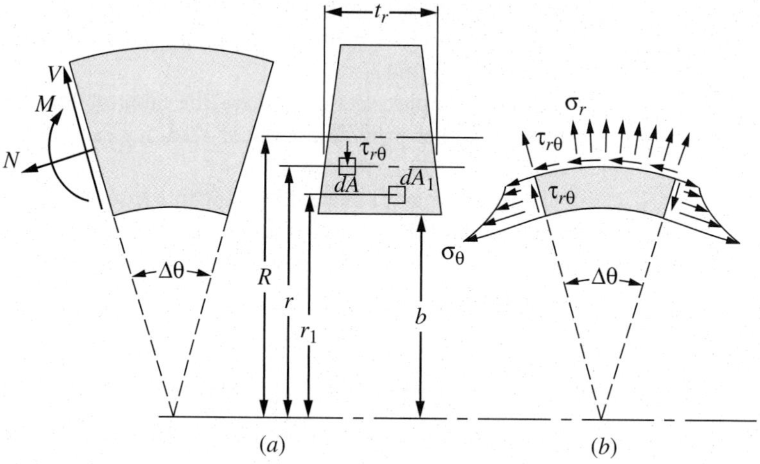

Figure 9.2

Radial Stresses Due to Moment *M* and Normal Force *N*

Owing to the radial components of the fiber stresses, radial stresses are present in a curved beam; these are tensile when the bending moment tends to straighten the beam and compressive under the reverse condition. A mechanics-of-materials solution may be developed by summing radial forces and summing forces perpendicular to the radius using the element in Fig. 9.2(*a*).

$$\sigma_r = \frac{R-e}{t_r Aer}\left[(M-NR)\left(\int_b^r \frac{dA_1}{r_1} - \frac{A_r}{R-e}\right) + \frac{N}{r}(RA_r - Q_r)\right] \tag{9.1-6}$$

Equation (9.1-6) is as accurate for radial stress as is Eq. (9.1-4) for shear stress when used for a rectangular cross section and compared to an elasticity solution. However, the complexity of Eq. (9.1-6) coupled with the fact that the stresses due to N are generally smaller than those due to M leads to the usual practice of omitting the terms involving N. This leads to the equation for radial stress found in many texts, such as Refs. 29 and 36.

$$\sigma_r = \frac{R-e}{t_r Aer} M\left(\int_b^r \frac{dA_1}{r_1} - \frac{A_r}{R-e}\right) \tag{9.1-7}$$

Again care must be taken when using Eqs. (9.1-4), (9.1-6), and (9.1-7) to use an accurate value for e as explained above in the discussion following Eq. (9.1-3).

Radial stress is usually not a major consideration in compact sections for it is smaller than the circumferential stress and is low where the circumferential stresses are large. However, in flanged sections with thin webs, the radial stress may be large at the junction of the flange and web, and the circumferential stress is also large at this position. This can lead to excessive shear stress and the possible yielding if the radial and circumferential stresses are of opposite sign. A large compressive radial stress in a thin web may also lead to a buckling of the web. Corrections for curved-beam formulas for sections having thin flanges are discussed in the next paragraph but corrections are also needed if a section has a *thin* web and *very thick* flanges. Under these conditions the individual flanges tend to rotate about their own neutral axes and larger radial and shear stresses are developed. Broughton et al. discuss this configuration in Ref. 31.

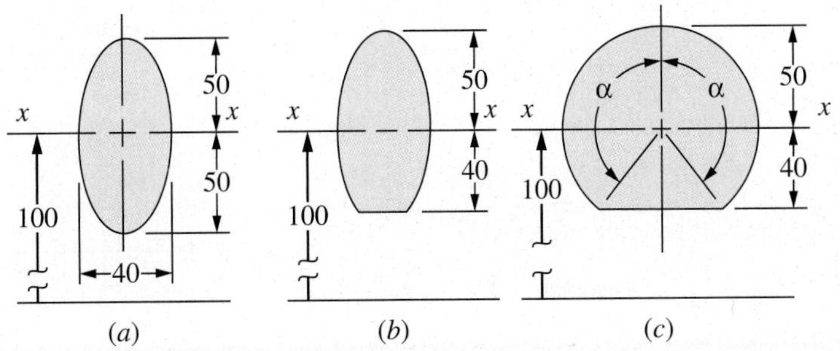

Figure 9.3

EXAMPLE 1 The sharply curved beam with an elliptical cross section shown in Fig. 9.3(*a*) has been used in a machine and has carried satisfactorily a bending moment of $2(10^6)$ N-mm. All dimensions on the figures and in the calculations are given in millimeters. A redesign does not provide as much space for this part, and a decision has been made to salvage the existing stock of this part by machining 10 mm from the inside. The question has been asked as to what maximum moment the modified part can carry without exceeding the peak stress in the original installation.

SOLUTION First compute the maximum stress in the original section by using case 6 of Table 9.1. $R=100$, $c=50$, $R/c=2$, $A=\pi(50)(20)=3142$, $e/c=0.5[2-(2^2-1)^{1/2}]=0.1340$, $e=6.70$, and $r_n=100-6.7=93.3$. Using these values, the stress σ_i can be found as

$$\sigma_i=\frac{My}{Aer}=\frac{2(10^6)(93.3-50)}{3142(6.7)(50)}=82.3\ \text{N/mm}^2$$

Alternatively one can find σ_i from $\sigma_i=k_iMc/I_x$, where k_i is found to be 1.616 in the table of values from case 6

$$\sigma_i=\frac{(1.616)(2)(10^6)(50)}{\pi(20)(50)^3/4}=82.3\ \text{N/mm}^2$$

Next consider the same section with 10 mm machined from the inner edge as shown in Fig. 9.3(*b*). Use case 9 of Table 9.1 with the initial calculations based on the equivalent modified circular section shown in Fig. 9.3(*c*). For this configuration $\alpha=\cos^{-1}(-40/50)=2.498$ rad (143.1°), $\sin\alpha=0.6$, $\cos\alpha=-0.8$, $R_x=100$, $a=50$, $a/c=1.179$, $c=42.418$, $R=102.418$, and $R/c=2.415$. In this problem $R_x>a$, so by using the appropriate expression from case 9 one obtains $e/c=0.131$ and $e=5.548$. R, c, and e have the same values for the machined ellipse, Fig. 9.3(*b*), and from case 18 of Table A.1 the area is found to be $A=20(50)(\alpha-\sin\alpha\cos\alpha)=2978$. Now the maximum stress on the inner surface can be found and set equal to 82.3 N/mm².

$$\sigma_i=82.3=\frac{My}{Aer}=\frac{M(102.42-5.548-60)}{2978(5.548)(60)}$$

$$82.3=37.19(10^6)M,\quad M=2.21(10^6)\ \text{N-mm}$$

One might not expect this increase in M unless consideration is given to the machining away of a stress concentration. Be careful, however, to note that, although removing the material reduced the peak stress in this case, the part will undergo greater deflections under the same moment that it carried before.

EXAMPLE 2 A curved beam with a cross section shown in Fig. 9.4 is subjected to a bending moment of 10^7 N-mm in the plane of the curve and to a normal load of 80,000 N in tension. The center of the circular portion of the cross section has a radius of curvature of 160 mm. All dimensions are given and used in the formulas in millimeters. The circumferential stresses in the inner and outer fibers are desired.

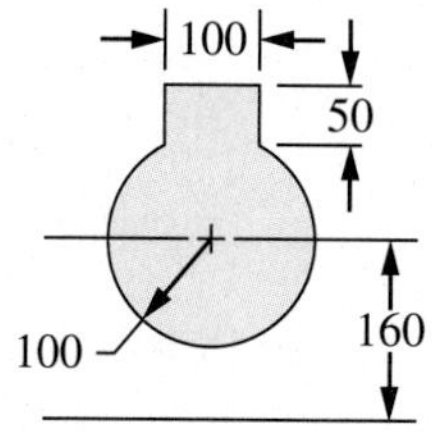

Figure 9.4

SOLUTION This section can be modeled as a summation of three sections: (1) a solid circular section, (2) a negative (materials removed) segment of a circle, and (3) a solid rectangular section. The section properties are evaluated in the order listed above and the results summed for the composite section.

Section 1. Use case 6 of Table 9.1. $R=160$, $b=200$, $c=100$, $R/c=1.6$, $\int dA/r=200[1.6-(1.6^2-1)^{1/2}]=220.54$, and $A=\pi(100^2)=31{,}416$.
Section 2. Use case 9 of Table 9.1. $a=\pi/6(30°)$, $R_x=160$, $a=100$, $R_x/a=1.6$, $a/c=18.55$, $c=5.391$, $R=252.0$, $\int dA/r=3.595$, and from case 20 of Table A.1, $A=905.9$.
Section 3. Use case 1 of Table 9.1. $R=160+100\cos 30°+25=271.6$, $b=100$, $c=25$, $R/c=10.864$, $A=5000$, $\int dA/r=100\ln(11.864/9.864)=18.462$.

For the composite section, $A=31{,}416-905.9+5000=35{,}510$, $R=[31{,}416(160)-905.9(252)+5000(272.6)]/35{,}510=173.37$, $c=113.37$, $\int dA/r=220.54-3.595+18.462=235.4$, $r_n=A/(\int dA/r)=35{,}510/235.4=150.85$, and $e=R-r_n=22.52$.

Using these data the stresses on the inside and outside are found to be

$$\sigma_i=\frac{My}{Aer}+\frac{N}{A}=\frac{10^7(150.85-60)}{35{,}510(22.52)(60)}+\frac{80{,}000}{35{,}510}$$

$$=18.93+2.25=21.18\ \text{N/mm}^2$$

$$\sigma_o=\frac{10^7(150.85-296.6)}{35{,}510(22.52)(296.6)}+\frac{80{,}000}{35{,}510}$$

$$=-6.14+2.25=-3.89\ \text{N/mm}^2$$

Curved Beams with Wide Flanges

In reinforcing rings for large pipes, airplane fuselages, and ship hulls, the combination of a curved sheet and attached web or stiffener forms a curved beam with wide flanges. Formulas for the effective width of a flange in such a curved beam are given in Ref. 9 and are as follows.

When the flange is indefinitely wide (e.g., the inner flange of a pipe-stiffener ring), the effective width is

$$b'=1.56\sqrt{Rt}$$

where b' is the total width assumed effective, R is the mean radius of curvature of the flange, and t is the thickness of the flange.

When the flange has a definite unsupported width b (gross width less web thickness), the ratio of effective to actual width b'/b is a function of qb, where

$$q=\sqrt[4]{\frac{3(1-\nu^2)}{R^2t^2}}$$

Corresponding values of qb and b'/b are as follows:

qb	1	2	3	4	5	6	7	8	9	10	11
b'/b	0.980	0.850	0.610	0.470	0.380	0.328	0.273	0.244	0.217	0.200	0.182

For the curved beam each flange should be considered as replaced by one of corresponding effective width b', and all calculations for direct, bending, and shear stresses, including corrections for curvature, should be based on this transformed section.

Bleich (Ref. 10) has shown that under a *straightening* moment where the curvature is decreased, the radial components of the fiber stresses in the flanges bend both flanges radially away from the web, thus producing tension in the fillet between flange and web in a direction normal to both the circumferential and radial normal stresses discussed in the previous section. Similarly, a moment which increases the curvature causes both flanges to bend radially toward the web and produce compressive stresses in the fillet between flange and web. The *nominal* values of these transverse bending stresses σ' in the fillet, without any correction for the stress concentration at the fillet, are given by $|\sigma'| = |\beta\sigma_m|$, where σ_m is the circumferential bending stress at the *midthickness* of the flange. This is less than the maximum value found in Table 9.1 and can be calculated by using Eq. (9.1-1). See the first example problem. The value of the coefficient β depends upon the ratio c^2/Rt, where c is the actual unsupported projecting width of the flange to either side of the web and R and t have the same meaning they did in the expressions for b' and q. Values of β may be found from the following table; they were taken from Ref. 10, where values of b' are also tabulated.

$c^2/Rt = 0$ $\beta = 0$	0.1 0.297	0.2 0.580	0.3 0.836	0.4 1.056	0.5 1.238	0.6 1.382	0.8 1.577
$c^2/Rt = 1$ $\beta = 1.677$	1.2 1.721	1.4 1.732	1.5 1.732	2 1.707	3 1.671	4 1.680	5 1.700

Derivations of expressions for b'/b and for β are also found in Ref. 29. Small differences in the values given in various references are due to compensations for secondary effects. The values given here are conservative.

In a similar way, the radial components of the circumferential normal stresses distort thin tubular cross sections of curved beams. This distortion affects both the stresses and deformations and is discussed in the next section.

U-Shaped Members

A U-shaped member having a semicircular inner boundary and a rectangular outer boundary is sometimes used as a punch or riveter frame. Such a member can usually be analyzed as a curved beam having a concentric outer boundary, but when the back thickness is large, a more accurate analysis may be necessary. In Ref. 11 are presented the results of a photoelastic stress analysis of such members in which the effects of variations in the several dimensions were determined. See case 23, Table 17.1.

Deflections

If a sharply curved beam is only a small portion of a larger structure, the contribution to deflection made by the curved portion can best be calculated by using the stresses at the inner and outer surfaces to calculate strains and the strains then used to determine the rotations of the plane sections. If the structure is made up primarily of a sharply curved beam or a combination of such beams, then refer to the next section.

9.2 DEFLECTION OF CURVED BEAMS

Deflections of curved beams can generally be found most easily by applying an energy method such as Castigliano's second theorem. One such expression is given by Eq. (8.1-7). The proper expression to use for the complementary energy depends upon the degree of curvature in the beam.

Deflection of Curved Beams of Large Radius

If for a curved beam, the radius of curvature is large enough such that Eqs. (8.1-1) and (8.1-2) are acceptable, that is, the radius of curvature is greater than 10 times the depth, then the stress distribution across the depth of the beam is very nearly linear and the complementary energy of flexure is given with sufficient accuracy by Eq. (8.1-3). If, in addition, the angular span is great enough such that deformations due to axial stress from the normal force N and the shear stresses due to transverse shear V can be neglected, deflections can be obtained by applying Eqs. (8.1-3) and (8.1-7) and rotations by Eq. (8.1-8). The following example shows how this is done.

EXAMPLE Figure 9.5 represents a slender uniform bar curved to form the quadrant of a circle; it is fixed at the lower end and at the upper end is loaded by a vertical force V, a horizontal force H, and a couple M_0. It is desired to find the vertical deflection δ_y, the horizontal deflection δ_x, and the rotation θ of the upper end.

SOLUTION According to Castigliano's second theorem, $\delta_y = \partial U/\partial V$, $\delta_x = \partial U/\partial H$, and $\theta = \partial U/\partial M_0$. Denoting the angular position of any section by x, it is evident that the moment there is $M = VR\sin x + HR(1-\cos x) + M_0$. Disregarding shear and axial stress, and replacing ds by $R\,dx$, we have [Eq. (8.1-3)]

$$U = U_f = \int_0^{\pi/2} \frac{[VR\sin x + HR(1-\cos x) + M_0]^2 R\,dx}{2EI}$$

Instead of integrating this and then carrying out the partial differentiations, we will differentiate first and then integrate, and for convenience suppress the constant term EI until all computations are completed. Thus

$$\begin{aligned}
\delta_y &= \frac{\partial U}{\partial V} \\
&= \int_0^{\pi/2} [VR\sin x + HR(1-\cos x) + M_0](R\sin x)R\,dx \\
&= VR^3\left(\frac{1}{2}x - \frac{1}{2}\sin x\cos x\right) - HR^3\left(\cos x + \frac{1}{2}\sin^2 x\right) - M_0R^2\cos x\Big|_0^{\pi/2} \\
&= \frac{(\pi/4)VR^3 + \frac{1}{2}HR^3 + M_0R^2}{EI}
\end{aligned}$$

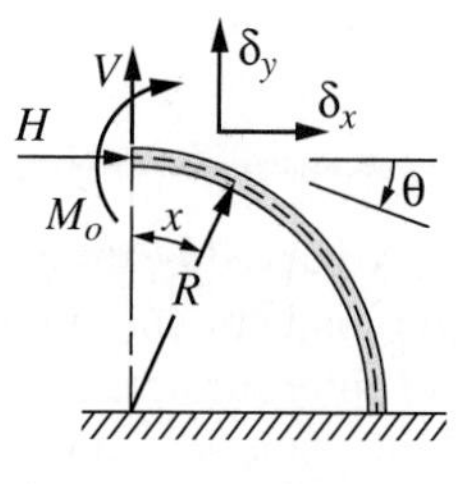

Figure 9.5

$$\delta_x = \frac{\partial U}{\partial H}$$

$$= \int_0^{\pi/2} [VR\sin x + HR(1-\cos x) + M_0]R(1-\cos x)R\,dx$$

$$= VR^3\left(-\cos x - \frac{1}{2}\sin^2 x\right) + HR^3\left(\frac{3}{2}x - 2\sin x + \frac{1}{2}\sin x\cos x\right) + M_0R^2(x - \sin x)\Big|_0^{\pi/2}$$

$$= \frac{\frac{1}{2}VR^3 + \left(\frac{3}{4}\pi - 2\right)HR^3 + (\pi/2 - 1)M_0R^2}{EI}$$

$$\theta = \frac{\partial U}{\partial M_0}$$

$$= \int_0^{\pi/2} [VR\sin x + HR(1-\cos x) + M_0]R\,dx$$

$$= -VR^2\cos x + HR^2(x - \sin x) + M_0Rx\Big|_0^{\pi/2}$$

$$= \frac{VR^2 + (\pi/2 - 1)HR^2 + (\pi/2)M_0R}{EI}$$

The deflection produced by any one load or any combination of two loads is found by setting the other load or loads equal to zero; thus, V alone would produce $\delta_x = \frac{1}{2}VR^3/EI$, and M alone would produce $\delta_y = M_0R^2/EI$. In this example all results are positive, indicating that δ_x is in the direction of H, δ_y in the direction of V, and θ in the direction of M_0.

Distortion of Tubular Sections

In curved beams of thin tubular section, the distortion of the cross section produced by the radial components of the fiber stresses reduces both the strength and stiffness. If the beam curvature is not so sharp as to make Eqs. (8.1-1) and (8.1-4) inapplicable, the effect of this distortion of the section can be taken into account as follows.

In calculating deflection of curved beams of hollow circular section, replace I by KI, where

$$K = 1 - \frac{9}{10 + 12(tR/a^2)^2}$$

(Here R = the radius of curvature of the beam axis, a = the outer radius of tube section, and t = the thickness of tube wall.) In calculating the maximum bending stress in curved beams of hollow circular section, use the formulas

$$\sigma_{max} = \frac{Ma}{I}\frac{2}{3K\sqrt{3\beta}} \quad \text{at } y = \frac{\alpha}{\sqrt{3\beta}} \quad \text{if } \frac{tR}{a^2} < 1.472$$

or

$$\sigma_{max} = \frac{Ma}{I}\frac{1-\beta}{K} \quad \text{at } y = a \quad \text{if } \frac{tR}{a^2} > 1.472$$

where

$$\beta = \frac{6}{5 + 6(tR/a^2)^2}$$

and y is measured from the neutral axis. Torsional stresses and deflections are unchanged.

In calculating deflection or stress in curved beams of hollow square section and uniform wall thickness, replace I by

$$\frac{1+0.0270n}{1+0.0656n}I$$

where $n = b^4/R^2t^2$. (Here R = the radius of curvature of the beam axis, b = the length of the side of the square section, and t = the thickness of the section wall.)

The preceding formulas for circular sections are from von Kármán (Ref. 4); the formulas for square sections are from Timoshenko (Ref. 5), who also gives formulas for rectangular sections.

Extensive analyses have been made for thin-walled pipe elbows with sharp curvatures for which the equations given above do not apply directly. Loadings may be *in-plane*, *out-of-plane*, or in various combinations (Ref. 8). Internal pressure increases and external pressure decreases pipe-bend stiffness. To determine ultimate load capacities of pipe bends or similar thin shells, elastic-plastic analyses, studies of the several modes of instability, and the stabilizing effects of flanges and the piping attached to the elbows are some of the many subjects presented in published works. Bushnell (Ref. 7) included an extensive list of references. Using numerical results from computer codes, graphs of stress indices and flexibility factors provide design data (Refs. 7, 19, and 34).

Deflection of Curved Beams of Small Radius

For a sharply curved beam, that is, the radius of curvature is less than 10 times the depth, the stress distribution is not linear across the depth. The expression for the complementary energy of flexure is given by

$$U_f = \int \frac{M^2}{2AEeR} R\,dx = \int \frac{M^2}{2AEe} dx \qquad (9.2\text{-}1)$$

where A is the cross-sectional area, E is the modulus of elasticity, and e is the distance from the centroidal axis to the neutral axis as given in Table 9.1. The differential change in angle dx is the same as is used in the previous example. See Fig. 9.1. Also keep clearly in mind that the bending in the plane of the curvature must be about a *principal axis* or the restraints described in the first paragraph of Sec. 9.1 must be present.

For all cross sections the value of the product AeR approaches the value of the moment of inertia I when the radius of curvature becomes greater than 10 times the depth. This is seen clearly in the following table where values of the ratio AeR/I are given for several sections and curvatures.

		R/d			
Case No.	Section	1	3	5	10
1	Solid rectangle	1.077	1.008	1.003	1.001
2	Solid circle	1.072	1.007	1.003	1.001
5	Triangle (base inward)	0.927	0.950	0.976	0.988
6	Triangle (base outward)	1.268	1.054	1.030	1.014

For curved beams of large radius the effect on deflections of the shear stresses due to V and the circumferential normal stresses due to N were small unless the length was small. For sharply curved beams the effects of these stresses must be considered. Only the effects of the radial stresses σ_r will be neglected. The expression for the complementary energy including all but the radial stresses is given by

$$U_f = \int \frac{M^2}{2AEe} dx + \int \frac{FV^2R}{2AG} dx + \int \frac{N^2R}{2AE} dx - \int \frac{MN}{AE} dx \tag{9.2-2}$$

where all the quantities are defined in the notation at the beginning of Table 9.2.

The last term, hereafter referred to as the *coupling* term, involves the complementary energy developed from coupling the strains from the bending moment M and the normal force N. A positive bending moment M produces a negative strain at the position of the *centroidal* axis in a curved beam, and the resultant normal force N passes through the centroid. Reasons have been given for and against including the coupling term in attempts to improve the accuracy of calculated deformations (see Refs. 3 and 29). Ken Tepper, Ref. 30, called attention to the importance of the coupling term for sharply curved beams. The equations in Tables 9.2 and 9.3 have been modified and now include the effect of the coupling term. With this change, the formulas given in Tables 9.2 and 9.3 for the indeterminate reactions and for the deformations are no longer limited to thin rings and arches but can be used as well for thick rings and arches. As before, for thin rings and arches α and β can be set to zero with little error.

To summarize this discussion and its application to the formulas in Tables 9.2 and 9.3, one can place a given curved beam into one of three categories: a thin ring, a moderately thick ring, and a very thick or sharply curved ring. The boundaries between these categories depend upon the R/d ratio and the shape of the cross section. Reference to the preceding tabulation of the ratio AeR/I will be helpful.

For thin rings the effect of normal stress due to N and shear stress due to V can be neglected, that is, set α and β equal to zero. For moderately thick rings and arches use the equations as they are given in Tables 9.2 and 9.3. For thick rings and arches replace the moment of inertia I with the product AeR in all equations including those for α and β. To illustrate the accuracy of this approach, the previous example problem will be repeated but for a thick ring of rectangular cross section. The rectangular cross section was chosen because a solution can be obtained by using the theory of elasticity with which to compare and evaluate the results.

EXAMPLE Figure 9.6(a) represents a thick uniform bar of rectangular cross section having a curved centroidal surface of radius R. It is fixed at the lower end, and the upper end is loaded by a vertical force V, a horizontal force H, and a couple M_o. It is desired to find the vertical deflection δ_y, the horizontal deflection δ_x, and the rotation θ of the upper end. Note that the deflections δ_y and δ_x are the values at the free end and at the radial position R at which the load H is applied.

FIRST SOLUTION Again Castigliano's theorem will be used. First find the moment, shear, and axial force at the angular position x.

$$M_x = VR\sin x + HR(1-\cos x) + M_o$$

$$V_x = V\cos x + H\sin x$$

$$N_x = -H\cos x + V\sin x$$

Since the beam is to be treated as a thick beam, the expression for complementary energy is given by

$$U = \int \frac{M_x^2}{2AEe} dx + \int \frac{FV_x^2R}{2AG} dx + \int \frac{N_x^2R}{2AE} dx - \int \frac{M_xN_x}{AE} dx$$

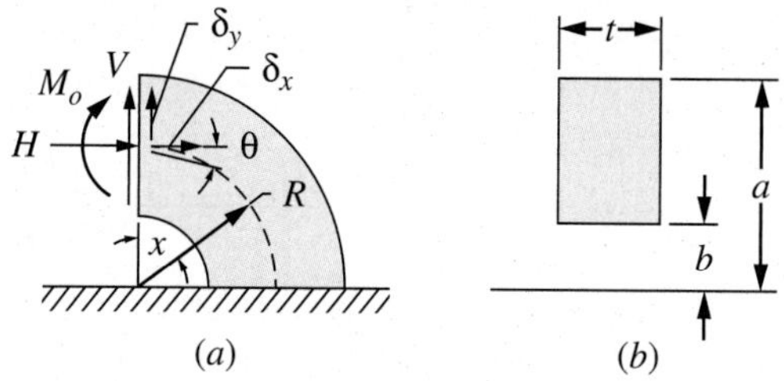

Figure 9.6

The deflection can now be calculated

$$\delta_y = \frac{\partial U}{\partial V} = \int_0^{\pi/2} \frac{M_x}{AEe}(R\sin x)dx + \int_0^{\pi/2} \frac{FV_xR}{AG}(\cos x)dx + \int_0^{\pi/2} \frac{N_xR}{AE}(\sin x)dx$$

$$- \int_0^{\pi/2} \frac{M_x}{AE}(\sin x)dx - \int_0^{\pi/2} \frac{N_x}{AE}(R\sin x)dx$$

$$= \frac{(\pi/4)VR^3 + 0.5HR^3 + M_oR^2}{EAeR} + \frac{0.5R(\pi V/2 + H)[2F(1+v) - 1] - M_o}{AE}$$

$$\delta_x = \frac{\partial U}{\partial H} = \int_0^{\pi/2} \frac{M_xR}{AEe}(1-\cos x)dx + \int_0^{\pi/2} \frac{FV_xR}{AG}(\sin x)dx + \int_0^{\pi/2} \frac{N_xR}{AE}(-\cos x)dx$$

$$- \int_0^{\pi/2} \frac{M_x}{AE}(-\cos x)dx - \int_0^{\pi/2} \frac{N_xR}{AE}(1-\cos x)dx$$

$$= \frac{0.5VR^3 + (3\pi/4 - 2)HR^3 + (\pi/2 - 1)M_oR^2}{EAeR}$$

$$+ \frac{0.5VR[2F(1+v) - 1] + (\pi/4)HR[2F(1+v) + 8/\pi - 1] + M_o}{EA}$$

$$\theta = \frac{\partial U}{\partial M_o} = \int_0^{\pi/2} \frac{M_x}{AEe}(1)dx + \int_0^{\pi/2} \frac{FV_xR}{AG}(0)dx + \int_0^{\pi/2} \frac{N_xR}{AE}(0)dx - \int_0^{\pi/2} \frac{M_x}{AE}(0)dx$$

$$- \int_0^{\pi/2} \frac{N_x}{AE}(1)dx$$

$$= \frac{VR^2 + (\pi/2 - 1)HR^2 + (\pi/2)M_oR}{EAeR} + \frac{H - V}{AE}$$

There is no need to reduce these expressions further in order to make a numerical calculation, but it is of interest here to compare to the solutions in the previous example. Therefore, let $\alpha = e/R$ and $\beta = FEe/GR = 2F(1+v)/R$ as defined previously.

$$\delta_y = \frac{(\pi/4)VR^3(1 - \alpha + \beta) + 0.5HR^3(1 - \alpha + \beta) + M_oR^2(1 - \alpha)}{EAeR}$$

$$\delta_x = \frac{0.5VR^3(1 - \alpha + \beta) + HR^3[(3\pi/4 - 2) + (2 - \pi/4)\alpha + (\pi/4)\beta]}{EAeR}$$

$$+ \frac{M_oR^2(\pi/2 - 1 + \alpha)}{EAeR}$$

$$\theta = \frac{VR^2(1 - \alpha) + HR^2(\pi/2 - 1 + \alpha) + (\pi/2)M_oR}{EAeR}$$

Up to this point in the derivation, the cross section has not been specified. For a rectangular cross section having an outer radius a and an inner radius b and of thickness t normal to the surface shown in Fig. 9.6(b), the following substitutions can be made in the deformation equations. Let $\nu = 0.3$.

$$R = \frac{a+b}{2}, \qquad A = (a-b)t, \qquad F = 1.2 \qquad \text{(see Sec. 8.10)}$$

$$\alpha = \frac{e}{R} = 1 - \frac{2(a-b)}{(a+b)\ln(a/b)}, \qquad \beta = 3.12\alpha$$

In the following table the value of a/b is varied from 1.1, where $R/d = 10.5$, a thin beam, to $a/b = 5.0$, where $R/d = 0.75$, a very thick beam. Three sets of numerical values are compared. The first set consists of the three deformations δ_y, δ_x, and θ evaluated from the equations just derived and due to the vertical load V. The second set consists of the same deformations due to the same loading but evaluated by applying the equations for a thin curved beam from the first example. The third set consists of the same deformations due to the same loading but evaluated by applying the theory of elasticity. See Ref. 2. The abbreviation MM in parentheses identifies the values from the *mechanics-of materials* solutions and the abbreviation EL similarly identifies those found from the *theory of elasticity*.

		From Thick Beam Theory			From Thin Beam Theory		
a/b	R/d	$\frac{\delta_y(\text{MM})}{\delta_y(\text{EL})}$	$\frac{\delta_x(\text{MM})}{\delta_x(\text{EL})}$	$\frac{\theta(\text{MM})}{\theta(\text{EL})}$	$\frac{\delta_y(\text{MM})}{\delta_y(\text{EL})}$	$\frac{\delta_x(\text{MM})}{\delta_x(\text{EL})}$	$\frac{\theta(\text{MM})}{\theta(\text{EL})}$
1.1	10.5	0.9996	0.9990	0.9999	0.9986	0.9980	1.0012
1.3	3.83	0.9974	0.9925	0.9991	0.9900	0.9852	1.0094
1.5	2.50	0.9944	0.9836	0.9976	0.9773	0.9967	1.0223
1.8	1.75	0.9903	0.9703	0.9944	0.9564	0.9371	1.0462
2.0	1.50	0.9884	0.9630	0.9916	0.9431	0.9189	1.0635
3.0	1.00	0.9900	0.9485	0.9729	0.8958	0.8583	1.1513
4.0	0.83	1.0083	0.9575	0.9511	0.8749	0.8345	1.2304
5.0	0.75	1.0230	0.9763	0.9298	0.8687	0.8290	1.2997

If reasonable errors can be tolerated, the strength-of-materials solutions are very acceptable when proper recognition of thick and thin beams is given.

SECOND SOLUTION Table 9.3 is designed to enable one to take any angular span 2θ and any single load or combination of loads and find the necessary indeterminate reactions and the desired deflections. To demonstrate this use of Table 9.3 in this example the deflection δ_x will be found due to a load H. Use case 12, with load terms from case 5d and with $\theta = \pi/4$ and $\phi = \pi/4$. Both load terms LF_H and LF_V are needed since the desired deflection δ_x is not in the direction of either of the deflections given in the table. Let $c = m = s = n = 0.7071$.

$$LF_H = H\left\{\frac{\pi}{2}0.7071 + \frac{k_1}{2}\left[\frac{\pi}{2}0.7071 - 0.7071^3(2)\right] - k_2 2(0.7071)\right\}$$

$$LF_V = H\left\{-\frac{\pi}{2}0.7071 - \frac{k_1}{2}\left[\frac{\pi}{2}0.7071 + 0.7071^3(2)\right] + k_2 4(0.7071^3)\right\}$$

$$\delta_x = (\delta_{VA} - \delta_{HA})0.7071 = \frac{-R^3}{EAeR}(LF_V - LF_H)0.7071$$

$$= \frac{-R^3 H}{EAeR}\left(-\frac{\pi}{2} - \frac{k_1}{2}\frac{\pi}{2} + 2k_2\right) = \frac{R^3 H}{EAeR}\left[\frac{3\pi}{4} - 2 + \left(2 - \frac{\pi}{4}\right)\alpha + \frac{\pi}{4}\beta\right]$$

This expression for δ_x is the same as the one derived directly from Castigliano's theorem. For angular spans of 90° or 180° the direct derivation is not difficult, but for odd-angle spans the use of the equations in Table 9.3 is recommended.

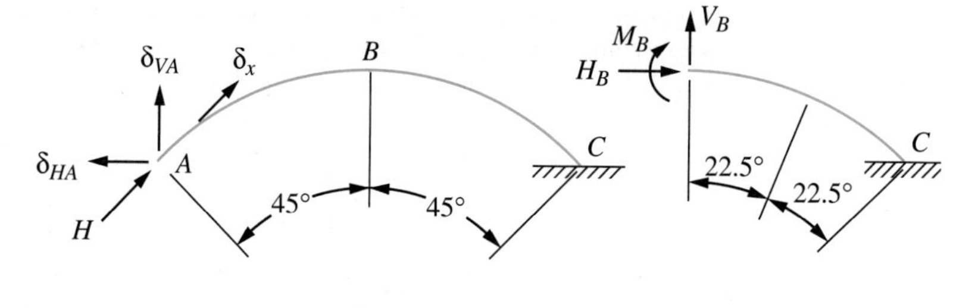

Figure 9.7

The use of the equations in Table 9.3 is also recommended when deflections are desired at positions other than the load point. For example, assume the deflections of the midspan of this arch are desired when it is loaded with the end load H as shown in Fig. 9.7(a). To do this, isolate the span from B to C and find the loads H_B, V_B, and M_B which act at point B. This gives $H_B = V_B = 0.7071H$ and $M_B = HR(1-0.7071)$. Now, superpose cases 12c, 12d, and 12n using these loads and $\theta = \phi = \pi/8$. In a problem with neither end fixed, a rigid-body motion may have to be superposed to satisfy the boundary conditions.

Deflection of Curved Beams of Variable Cross Section and/or Radius

None of the tabulated formulas applies when either the cross section or the radius of curvature varies along the span. The use of Eqs. (9.2-1) and (9.2-2), or of comparable expressions for thin curved beams, with numerical integration carried out for a finite number of elements along the span, provides an effective means of solving these problems. This is best shown by example.

EXAMPLE A rectangular beam of constant thickness and a depth varying linearly along the length is bent such that the centroidal surface follows the curve $x = 0.25y^2$ as shown in Fig. 9.8. The vertical deflection at the loaded end is desired. To keep the use of specific dimensions to a minimum let the depth of the curved beam at the fixed end $= 1.0$, the thickness $= 0.5$, and the horizontal location of the load $P = 1.0$. The beam will be subdivided into eight segments, each spanning 0.25 units in the y direction. Normally a constant length along the span is used, but using constant Δy gives shorter spans where moments are larger and curvatures are sharper. The numerical calculations are also easier. Use will be made of the following expressions in order to provide the tabulated information

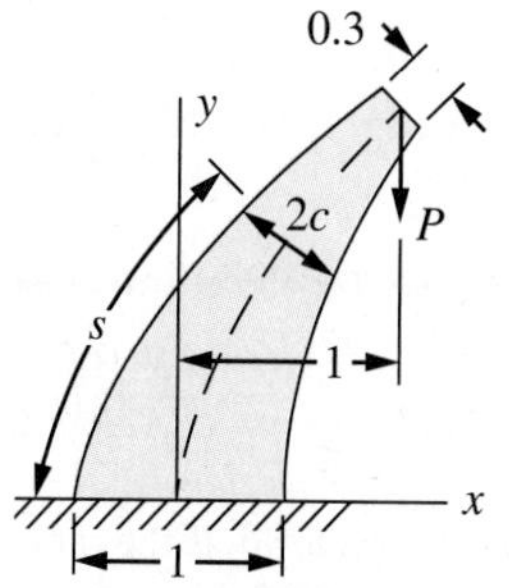

Figure 9.8

from which the needed summation can be found. Note that y_i and x_i are used here as the y and x positions of the midlength of each segment.

$$x = 0.25y^2, \quad \frac{dx}{dy} = 0.5y, \quad \frac{d^2x}{d^2y} = 0.5, \quad \Delta l = \Delta y(1+x_i)^{1/2}$$

$$R = \frac{[1+(dx/dy)^2]^{3/2}}{d^2x/d^2y}$$

$$\frac{e}{c} = \frac{R}{c} - \frac{2}{\ln[(R/c+1)/(R/c-1)]} \quad \text{for } \frac{R}{2c} < 8$$

[see Eq. (9.1-1) and case 1 of Table 9.1] or

$$\frac{e}{c} = \frac{I_c}{RAc} = \frac{t(2c)^3}{12(Rt2c^2)} = \frac{c}{3R} \quad \text{for } \frac{R}{2c} > 8$$

[See Eq. (9.1-3).]

The desired vertical deflection of the loaded end can be determined from Castigliano's theorem, using Eq. (9.2-2) for U_f in summation form rather than integral form. This reduces to

$$\delta = \frac{\delta U}{\delta P} = \frac{P}{E}\Sigma\left[\frac{(M/P)^2}{eR} + F\left(\frac{V}{P}\right)^2 2(1+\nu) + \left(\frac{N}{P}\right)^2 - 2\frac{M}{P}\frac{N}{P}\right]\frac{\Delta l}{A}$$

$$= \frac{P}{E}\Sigma\frac{\Delta l}{A}[B]$$

where $[B]$ and $[B]\Delta l/A$ are the last two columns in the following table. The internal forces and moments can be determined from equilibrium equations as

$$\frac{M}{P} = -(1-x_i), \quad \theta_i = \tan^{-1}\frac{dx}{dy}, \quad V = P\sin\theta_i, \quad \text{and} \quad N = -P\cos\theta_i$$

In the evaluation of the above equations for this problem, $F = 1.2$ and $\nu = 0.3$. In the table below one must fill in the first five columns in order to find the total length of the beam before the midsegment depth $2c$ can be found and the table completed.

Element No.	y_i	x_i	R	Δl	c	R/c
1	0.125	0.004	2.012	0.251	0.481	4.183
2	0.375	0.035	2.106	0.254	0.442	4.761
3	0.625	0.098	2.300	0.262	0.403	5.707
4	0.875	0.191	2.601	0.273	0.362	7.180
5	1.125	0.316	3.020	0.287	0.320	9.451
6	1.375	0.473	3.574	0.303	0.275	13.019
7	1.625	0.660	4.278	0.322	0.227	18.860
8	1.875	0.879	5.151	0.343	0.176	29.243
				2.295		

Element No.	e/c	M/P	V/P	N/P	$[B]$	$[B]\frac{\Delta l}{A}$
1	0.0809	−0.996	0.062	−0.998	11.695	6.092
2	0.0709	−0.965	0.184	−0.983	13.269	7.627
3	0.0589	−0.902	0.298	−0.954	14.370	9.431
4	0.0467	−0.809	0.401	−0.916	14.737	11.101
5	0.0354	−0.684	0.490	−0.872	14.007	12.569
6	0.0256	−0.527	0.567	−0.824	11.856	13.105
7	0.0177	−0.340	0.631	−0.776	8.049	11.431
8	0.0114	−0.121	0.684	−0.730	3.232	6.290
						77.555

Therefore, the deflection at the load and in the direction of the load is $77.56P/E$ in whatever units are chosen as long as the depth at the fixed end is unity. If one maintains the same length-to-depth ratio and the same shape, the deflection can be expressed as $\delta = 77.56P/(E2t_o)$, where t_o is the constant thickness of the beam. Michael Plesha (Ref. 33) provided a finite-element solution for this configuration and obtained for the load point a vertically downward deflection of 72.4 units and a horizontal deflection of 88.3 units. The 22 elements he used were nine-node, quadratic displacement, Lagrange elements. The reader is invited to apply a horizontal dummy load and verify the horizontal deflection.

9.3 CIRCULAR RINGS AND ARCHES

In large pipelines, tanks, aircraft, and submarines, the circular ring is an important structural element, and for correct design it is often necessary to calculate the stresses and deflections produced in such a ring under various conditions of loading and support. The circular arch of uniform section is often employed in buildings, bridges, and machinery.

Rings

A closed circular ring may be regarded as a *statically indeterminate beam* and analyzed as such by the use of Castigliano's second theorem. In Table 9.2 are given formulas thus derived for the bending moments, tensions, shears, horizontal and vertical deflections, and rotations of the load point in the plane of the ring for various loads and supports. By superposition, these formulas can be combined so as to cover almost any condition of loading and support likely to occur.

The ring formulas are based on the following assumptions: (1) The ring is of uniform cross section and has symmetry about the plane of curvature. An exception to this requirement of symmetry can be made if moment restraints are provided to prevent rotation of each cross section out of its plane of curvature. Lacking the plane of symmetry and any external constraints, out-of-plane deformations will accompany in-plane loading. Meck, in Ref. 21, derives expressions concerning the coupling of in-plane and out-of-plane deformations of circular rings of arbitrary compact cross section and resulting instabilities. (2) All loadings are applied at the radial position of the centroid of the cross section. For thin rings this is of little concern, but for radially thick rings a concentrated load acting in other than a radial direction and not at the centroidal radius must be replaced by a statically equivalent load at the centroidal radius and a couple. For case 15, where the loading is due to gravity or a constant linear acceleration, and for case 21, where the loading is due to rotation around an axis normal to the plane of the ring, the proper distribution of loading through the cross section is accounted for in the formulas. (3) It is nowhere stressed beyond the elastic limit. (4) It is not so severely deformed as to lose its essentially circular shape. (5) Its deflection is due primarily to bending, but for thicker rings the deflections due to deformations caused by axial tension or compression in the ring and/or by transverse shear stresses in the ring may be included. To include these effects, we can evaluate first the coefficients α and β, the *axial stress deformation factor* and the *transverse shear deformation factor*, and then the constants k_1 and k_2. Such corrections are more often necessary when composite or sandwich construction is employed. If no axial or shear stress corrections are desired, α and β are set equal to zero and the values of k are set equal to unity. (6) In the case of pipes acting as beams between widely spaced supports, the distribution of shear stress across the section of the pipe is in accordance with Eq. (8.1-2), and the direction of the resultant shear stress at any point of the cross section is tangential.

Note carefully the deformations given regarding the point or points of loading as compared with the deformations of the horizontal and vertical diameters. For many of the cases listed, the numerical values of load and deflection coefficients have been given for several positions of the loading. These coefficients do not include the effect of axial and shear deformation.

No account has been taken in Table 9.2 of the effect of radial stresses in the vicinity of the concentrated loads. These stresses and the local deformations they create can have a significant effect on overall ring deformations and peak stresses. In case 1, a reference is made to Sec. 14.3 in which thick-walled rollers or rings are loaded on the outer ends of a diameter. The stresses and deflections given here are different from those predicted by the equations in case 1. If a concentrated load is used only for purposes of superposition, as is often the case, there is no cause for concern, but if an actual applied load is concentrated over a small region and the ring is sharply curved with thick walls, then one must be aware of the possible errors.

EXAMPLE 1 A pipe with a diameter of 13 ft and thickness of ½ in is supported at intervals of 44 ft by rings, each ring being supported at the extremities of its horizontal diameter by vertical reactions acting at the centroids of the ring sections. It is required to determine the bending moments in a ring at the bottom, sides, and top, and the maximum bending moment when the pipe is filled with water.

SOLUTION We use the formulas for cases 4 and 20 of Table 9.2. Taking the weight of the water as 62.4 lb/ft^3 and the weight of the shell as 20.4 lb/ft^2, the total weight W of 44 ft of pipe carried by one ring is found to be 401,100 lb. Therefore, for case 20, $W = 401{,}100$ lb, and for case 4, $W = 250{,}550$ lb and $\theta = \pi/2$. Assume a thin ring, $\alpha = \beta = 0$.

At bottom:

$$M = M_C = 0.2387(401{,}100)(6.5)(12) - 0.50(200{,}550)(78)$$

$$= 7.468(10^6) - 7.822(10^6) = -354{,}000 \text{ lb-in}$$

At top:

$$M = M_A = 0.0796(401{,}100)(78) - 0.1366(200{,}550)(78) = 354{,}000 \text{ lb-in}$$

$$N = N_A = 0.2387(401{,}100) - 0.3183(200{,}500) = 31{,}900 \text{ lb}$$

$$V = V_A = 0$$

At sides:

$$M = M_A - N_A R(1-u) + V_A Rz + LT_M$$

where for $x = \pi/2$, $u = 0$, $z = 1$, and $LT_M = (WR/\pi)(1-u-xz/2) = [401{,}100(78)/\pi](1-\pi/4) = 2.137(10^6)$ for case 20, and $LT_M = 0$ for case 4 for since $z - s = 0$. Therefore,

$$M = 354{,}000 - 31{,}900(78)(1-0) + 0 + 2.137(10^6) = 2800 \text{ lb-in}$$

The value of 2800 lb-in is due to the small differences in large numbers used in the superposition. An exact solution would give zero for this value. It is apparent that at least four digits must be carried.

To determine the location of maximum bending moment let $0 < x < \pi/2$ and examine the expression for M:

$$M = M_A - N_A R(1-\cos x) + \frac{WR}{\pi}\left(1 - \cos x - \frac{x \sin x}{2}\right)$$

$$\frac{dM}{dx} = -N_A R \sin x + \frac{WR}{\pi}\sin x - \frac{WR}{2\pi}\sin x - \frac{WRx}{2\pi}\cos x$$

$$= 31{,}950 R \sin x - 63{,}800 Rx \cos x$$

At $x = x_1$ let $dM/dx = 0$ or $\sin x_1 = 2x_1 \cos x_1$, which yields $x_1 = 66.8°$ (1.166 rad). At $x = x_1 = 66.8°$,

$$M = 354{,}00 - 31{,}900(78)(1-0.394) + \frac{401{,}100(78)}{\pi}\left[1 - 0.394 - \frac{1.166(0.919)}{2}\right]$$

$$= -455{,}000 \text{ lb-in} \quad \text{(max negative moment)}$$

Similarly, at $x = 113.2°$, $M = 455{,}000$ lb-in (max positive moment).

By applying the supporting reactions outside the center line of the ring at a distance a from the centroid of the section, side couples that are each equal to $Wa/2$ would be introduced. The effect of these, found by the formulas for case 3, would be to reduce the maximum moments, and it can be shown that the optimum condition obtains when $a = 0.04R$.

EXAMPLE 2 The pipe of Example 1 rests on soft ground, with which it is in contact over 150° of its circumference at the bottom. The supporting pressure of the soil may be assumed to be radial and uniform. It is required to determine the bending moment at the top and bottom and at the surface of the soil. Also the bending stresses at these locations and the change in the horizontal diameter must be determined.

SOLUTION A section of pipe 1 in long is considered. The loading may be considered as a combination of cases 12, 15, and 16. Owing to the weight of the pipe (case 15, $w = 0.1416$ lb-in), and letting $K_T = k_1 = k_2 = 1$, and $\alpha = \beta = 0$,

$$M_A = \frac{0.1416(78)^2}{2} = 430 \text{ lb-in}$$

$$N_A = \frac{0.1416(78)}{2} = 5.52 \text{ lb}$$

$$V_A = 0$$

and at $x = 180 - \frac{150}{2} = 105° = 1.833$ rad,

$$LT_M = -0.1416(78^2)[1.833(0.966) - 0.259 - 1] = -440 \text{ lb-in}$$

Therefore,

$$M_{105°} = 430 - 5.52(78)(1 + 0.259) - 440 = -552 \text{ lb-in}$$

$$M_C = 1.5(0.1416)(78) = 1292 \text{ lb-in}$$

Owing to the weight of contained water (case 16, $\rho = 0.0361$ lb/in^3),

$$M_A = \frac{0.0361(78^3)}{4} = 4283 \text{ lb-in/in}$$

$$N_A = \frac{0.0361(78^2)(3)}{4} = 164.7 \text{ lb/in}$$

$$V_A = 0$$

and at $x = 105°$,

$$LT_M = 0.0361(78^3)\left[1 + 0.259 - \frac{1.833(0.966)}{2}\right] = 6400 \text{ lb-in/in}$$

Therefore,

$$M_{105^\circ} = 4283 - 164.7(78)(1 + 0.259) + 6400 = -5490 \text{ lb-in/in}$$

$$M_C = \frac{0.0361(78^3)(3)}{4} = 12{,}850 \text{ lb-in/in}$$

Owing to earth pressure and the reversed reaction (case 12, $\theta = 105°$),

$$2wR\sin\theta = 2\pi R(0.1416) + 0.0361\pi R^2 = 759 \text{ lb} \qquad (w = 5.04 \text{ lb/in})$$

$$M_A = \frac{-5.04(78^2)}{\pi}[0.966 + (\pi - 1.833)(-0.259) - 1(\pi - 1.833 - 0.966)]$$

$$= -2777 \text{ in-lb}$$

$$N_A = \frac{-5.04(78)}{\pi}[0.966 + (\pi - 1.833)(-0.259)] = -78.5 \text{ lb}$$

$$V_A = 0$$

$$LT_M = 0$$

$$M_{105^\circ} = -2777 + 78.5(78)(1.259) = 4930 \text{ lb-in}$$

$$M_C = -5.04(78^2)\frac{1.833(1 - 0.259)}{\pi} = -13{,}260 \text{ lb-in}$$

Therefore, for the 1 in section of pipe

$$M_A = 430 + 4283 - 2777 = 1936 \text{ lb-in}$$

$$\sigma_A = \frac{6M_A}{t^2} = 46{,}500 \text{ lb/in}^2$$

$$M_{105^\circ} = -552 - 5490 + 4930 = -1112 \text{ lb-in}$$

$$\sigma_{105^\circ} = 26{,}700 \text{ lb/in}^2$$

$$M_C = 1292 + 12{,}850 - 13{,}260 = 882 \text{ lb-in}$$

$$\sigma_C = 21{,}200 \text{ lb/in}^2$$

The change in the horizontal diameter is found similarly by superimposing the three cases. For E use $30(10^6)/(1 - 0.285^2) = 32.65(10^6) \text{ lb/in}^2$, since a plate is being bent instead of a narrow beam (see page 189). For I use the moment of inertia of a 1-in-wide piece, 0.5 in thick:

$$I = \frac{1}{12}(1)(0.5^3) = 0.0104 \text{ in}^4, \qquad EI = 340{,}000 \text{ lb-in}^2$$

From case 12:

$$\Delta D_H = \frac{-5.04(78^4)}{340{,}000}\left[\frac{(\pi - 1.833)(-0.259) + 0.966}{2} - \frac{2}{\pi}(\pi - 1.833 - 0.966)\right]$$

$$= \frac{-5.04(78)^4}{340{,}000}(0.0954) = -52.37 \text{ in}$$

From case 15:

$$\Delta D_H = \frac{0.4292(0.1416)78^4}{340{,}000} = 6.616 \text{ in}$$

From case 16:

$$\Delta D_H = \frac{0.2146(0.0361)78^5}{340{,}000} = 65.79 \text{ in}$$

The total change in the horizontal diameter is 20 in. It must be understood at this point that the answers are somewhat in error since this large a deflection does violate the assumption that the loaded ring is very nearly circular. This was expected when the stresses were found to be so large in such a thin pipe.

Arches

Table 9.3 gives formulas for end reactions and end deformations for circular arches of constant radius of curvature and constant cross section under 18 different loadings and with 14 combinations of end conditions. The corrections for axial stress and transverse shear are accomplished as they were in Table 9.2 by the use of the constants α and β. Once the indeterminate reactions are known, the bending moments, axial loads, and transverse shear forces can be found from equilibrium equations. If deformations are desired for points away from the ends, the unit-load method [Eq. (8.1-6)] can be used or the arch can be divided at the position where the deformations are desired and either portion analyzed again by the formulas in Table 9.3. Several examples illustrate this last approach. Note that in many instances the answer depends upon the difference of similar large terms, and so appropriate attention to accuracy must be given.

EXAMPLE 1 A $WT4 \times 6.5$ structural steel T-beam is formed in the plane of its web into a circular arch of 50-in radius spanning a total angle of 120°. The right end is fixed, and the left end has a pin which is constrained to follow a horizontal slot in the support. The load is applied through a vertical bar welded to the beam, as shown in Fig. 9.9. Calculate the movement of the pin at the left end, the maximum bending stress, and the rotation of the bar at the point of attachment to the arch.

SOLUTION The following material and cross-sectional properties may be used for this beam. $E = 30(10^6)$ lb/in^2, $G = 12(10^6)$ lb/in^2, $I_x = 2.90$ in^4, $A = 1.92$ in^2, flange thickness $= 0.254$ in, and web thickness $= 0.230$ in. The loading on the arch can be replaced by a concentrated moment of 8000 lb-in and a horizontal force of 1000 lb at a position indicated by $\phi = 20°$ (0.349 rad). $R = 50$ in. and $\theta = 60°$ (1.047 rad). For these loads and boundary conditions, cases 9b and 9n of Table 9.3 can be used.

Since the radius of 50 in is only a little more than 10 times the depth of 4 in, corrections for axial load and shear will be considered. The axial-stress deformation factor $\alpha = I/AR^2 = 2.9/1.92(50^2) = 0.0006$. The transverse-shear deformation factor $\beta = FEI/GAR^2$, where F will be approximated here by using

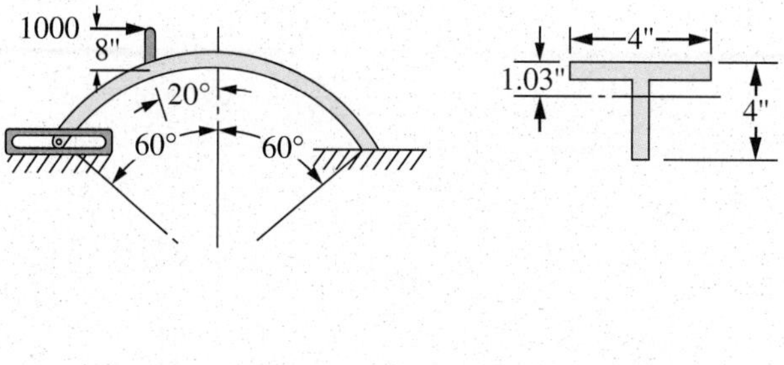

Figure 9.9

$F = 1$ and A = web area = $4(0.23) = 0.92$. This gives $\beta = 1(30)(10^6)(2.90)/12(10^6)(0.92)(50^2) = 0.003$. The small values of α and β indicate that bending governs the deformations, and so the effect of axial load and transverse shear will be neglected. Note that $s = \sin 60°$, $c = \cos 60°$, $n = \sin 20°$, and $m = \cos 20°$.

For case 9b,

$$LF_H = 1000\left[\frac{1.0472+0.3491}{2}(1+2\cos 20°\cos 60°) - \frac{\sin 60°\cos 60°}{2}\right.$$
$$\left. - \frac{\sin 20°\cos 20°}{2} - \cos 20°\sin 60° - \sin 20°\cos 60°\right]$$
$$= 1000(-0.00785) = -7.85\text{ lb}$$

Similarly,

$$LF_V = 1000(-0.1867) = -186.7\text{ lb and } LF_M = 1000(-0.1040) = -104.0\text{ lb}$$

For the case 9n,

$$LF_H = \frac{8000}{50}(-0.5099) = -81.59\text{ lb}$$
$$LF_V = \frac{8000}{50}(-1.6489) = -263.8\text{ lb}$$
$$LF_M = \frac{8000}{50}(-1.396) = -223.4\text{ lb}$$

Also,

$$B_{VV} = 1.0472 + 2(1.0472)\sin^2 60° - \sin 60°\cos 60° = 2.1850\text{ lb}$$
$$B_{HV} = 0.5931\text{ lb}$$
$$B_{MV} = 1.8138\text{ lb}$$

Therefore,

$$V_A = -\frac{186.7}{2.1850} - \frac{263.8}{2.1850} = -85.47 - 120.74 = -206.2\text{ lb}$$

$$\delta_{HA} = \frac{50^3}{30(10^6)(2.9)}[0.5931(-206.2) + 7.85 + 81.59] = -0.0472\text{ in}$$

The expression for the bending moment can now be obtained by an equilibrium equation for a position located by an angle x measured from the left end:

$$M_x = V_A R[\sin\theta - \sin(\theta - x)] + 8000\langle x - (\theta - \phi)\rangle^0$$
$$-1000R[\cos(\theta - x) - \cos\phi]\langle x - (\theta - \phi)\rangle^0$$

At $x = 40°-$ $\quad M_x = -206.2(50)[\sin 60° - \sin(60° - 40°)] = -5403\text{ lb-in}$

At $x = 40°+$ $\quad M_x = -5403 + 8000 = 2597\text{ lb-in}$

At $x = 60°$ $\quad M_x = -206.2(50)(0.866) + 8000 - 1000(50)(1 - 0.940)$

$\quad = -3944\text{ lb-in}$

At $x = 120°$ $\quad M_x = 12{,}130\text{ lb-in}$

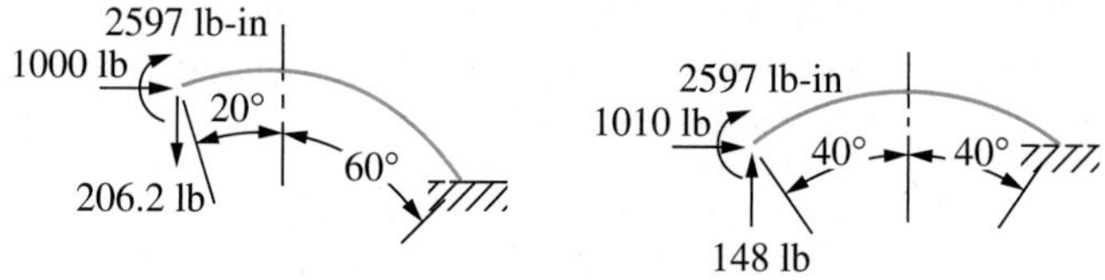

Figure 9.10

The maximum bending stress is therefore

$$\sigma = \frac{12{,}130(4-1.03)}{2.9} = 12{,}420\ \text{lb/in}^2$$

To obtain the rotation of the arch at the point of attachment of the bar, we first calculate the loads on the portion to the right of the loading and then establish an equivalent symmetric arch (see Fig. 9.10). Now from cases 12a, 12b, and 12n, where $\theta = \phi = 40°\ (0.698\ \text{rad})$, we can determine the load terms as follows:

$$\text{For case 12a} \qquad LF_M = -148[2(0.698)(0.643)] = -133\ \text{lb}$$

$$\text{For case 12b} \qquad LF_M = 1010[0.643 + 0.643 - 2(0.698)(0.766)] = 218\ \text{lb}$$

$$\text{For case 12n} \qquad LF_M = \frac{2597}{50}(-0.698 - 0.698) = -72.5\ \text{lb}$$

Therefore, the rotation at the load is

$$\psi_A = \frac{-50^2}{30(10^6)(2.9)}(-133 + 218 - 72.5) = -0.00036\ \text{rad}$$

We would not expect the rotation to be in the opposite direction to the applied moment, but a careful examination of the problem shows that the point on the arch where the bar is fastened moves to the right 0.0128 in. Therefore, the net motion in the direction of the 1000-lb load on the end of the 8-in bar is 0.0099 in, and so the applied load does indeed do positive work on the system.

EXAMPLE 2 The deep circular arch of titanium alloy has a triangular cross section and spans 120° as shown in Fig. 9.11. It is driven by the central load P to produce an acceleration of 40 g. The tensile stress at A and the deformations at the extreme ends are required. All dimensions given and used in the formulas are in centimeters.

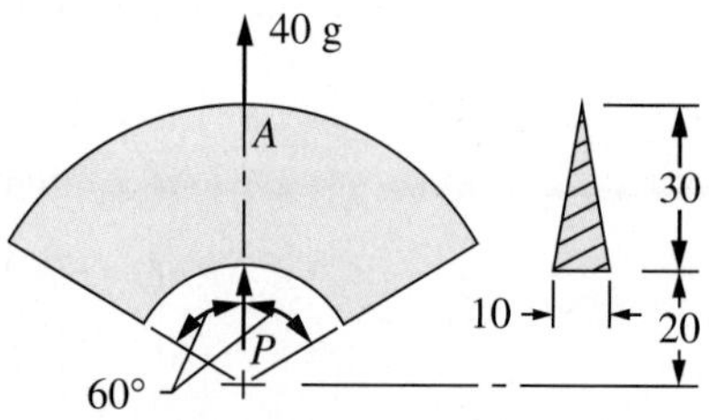

Figure 9.11

SOLUTION This is a statically determinate problem, so the use of information from Table 9.3 is needed only to obtain the deformations. Superposing the central load and the forces needed to produce the acceleration on the total span can be accomplished readily by using cases 3a and 3h. This solution, however, will provide only the horizontal and rotational deformations of the ends. Using the symmetry one can also superpose the loadings from cases 12h and 12i on the left half of the arch and obtain all three deformations. Performing both calculations provides a useful check. All dimensions are given in centimeters and used with expressions from Table 9.1, case 5, to obtain the needed factors for this section. Thus, $b = 10$, $d=30$, $A=150$, $c=10$, $R=30$, $R/c=3$, $e/c=0.155$, $e=1.55$ and, for the peak stresses, $k_i=1.368$ and $k_o=0.697$. The titanium alloy has a modulus of elasticity of 117 GPa [$11.7(10^6)\text{N/cm}^2$], a Poisson's ratio of 0.33, and a mass density of 4470 kg/m^3, or 0.00447 kg/cm^3. One g of acceleration is 9.81 m/s^2, and 1 cm of arc length at the centroidal radius of 30 cm will have a volume of 150 cm^3 and a mass of 0.6705 kg. This gives a loading parallel to the driving force P of 0.6705(40)(9.81) = 263 N/cm of centroidal arc length. Since this is a very sharply curved beam, $R/d = 1$, one must recognize that the resultant load of 263 N/cm does not act through the centroid of the cross-sectional area but instead acts through the mass center of the differential length. The radius to this point is given as R_{cg} and is found from the expression $R_{cg}/R=1+I/AR^2$, where I is the area moment of inertia about the centroidal axis of the cross section. Therefore, $R_{cg}/R=1+(bd^3/36)/(bd/2)R^2=1.056$. Again due to the sharp curvature the axial- and shear-stress contributions to deformation must be considered. From the introduction to Table 9.3 we find that $\alpha=h/R=0.0517$ and $\beta=2F(1+\nu)h/R=0.1650$, where $F=1.2$ for a triangular cross section as given in Sec. 8.10. Therefore, $k_1=1-\alpha+\beta=1.1133$, and $k_2=1-\alpha=0.9483$.

For a first solution use the full span and superpose cases 3a and 3h. To obtain the load terms LP_H and LP_M use cases 1a and 1h.

For case 1a, $W=-263(30)(2\pi/3)=-16{,}525\text{ N}$, $\theta=60°$, $\phi=0°$, $s=0.866$, $c=0.500$, $n=0$, and $m=1.000$.

$$LP_H=-16{,}525\left[\frac{\pi}{3}(0.866)(0.5)-0+\frac{1.1133}{2}(0.5^2-1.0^2)+0.9483(0.5)(0.5-1.0)\right]$$

$$=-16{,}525(-0.2011)=3323\text{ N}$$

Similarly, $LP_M=3575\text{ N}$.

For case 1h, $w=263$ N/cm, $R=30$, $R_{cg}/R=1.056$, $\theta=60°$, $s=0.866$, and $c=0.5000$.

$$LP_H=263(30)(-0.2365)=-1866\text{ N}$$

and

$$LP_M=263(30)(-0.2634)=-2078\text{ N}$$

Returning now to case 3 where M_A and H_A are zero, one finds that $V_A=0$ by superposing the loadings. To obtain δ_{HA} and ψ_A we superpose cases a and h and substitute AhR for I because of the sharp curvature

$$\delta_{HA1}=-30^3\frac{3323-1866}{11.7(10^6)(150)(1.55)(30)}=-482(10^{-6})\text{ cm}$$

$$\psi_{A1}=-30^2\frac{3575-2078}{8.161(10^{10})}=-16.5(10^{-6})\text{ rad}$$

Now for the second and more complete solution, use will be made of cases 12h and 12i. The left half spans 60°, so $\theta = 30°$, $s = 0.5000$, and $c = 0.8660$. In this solution the central symmetry axis of the left half being used is inclined at 30° to the gravitational loading of 263 N/cm. Therefore, for case 5h, $w = 263\cos 30° = 227.8\ \text{N/cm}$

$$LF_H = 227.8(30)\left\{\frac{1.1133}{2}\left[2\left(\frac{\pi}{6}\right)(0.866^2) - \frac{\pi}{6} - 0.5(0.866)\right]\right.$$

$$\left. + 0.9483(1.056+1)\left[\frac{\pi}{6} - 0.5(0.866)\right] + 1.056(2)(0.866)\left(\frac{\pi}{6}0.866 - 0.5\right)\right\}$$

$$= 227.8(30)(-0.00383) = -26.2\ \text{N}$$

Similarly,

$$LF_V = 227.8(30)(0.2209) = 1510\ \text{N} \quad \text{and} \quad LF_M = 227.8(30)(0.01867) = 1276\ \text{N}$$

For case 5i, $w = -263 \sin 30° = -131.5\ \text{N/cm}$ and again $\theta = 30°$

$$LF_H = -131.5(30)(0.0310) = -122.3\ \text{N}$$

$$LF_V = -131.5(30)(-0.05185) = 204.5\ \text{N}$$

$$LF_M = -131.5(30)(-0.05639) = 222.5\ \text{N}$$

Using case 12 and superposition of the loadings gives

$$\delta_{HA2} = -30^3\frac{-26.2-122.3}{8.161(10^{10})} = 49.1(10^{-6})\ \text{cm}$$

$$\delta_{VA2} = -30^3\frac{1510+204.5}{8.161(10^{10})} = -567(10^{-6})\ \text{cm}$$

$$\psi_{A2} = -30^2\frac{1276+222.5}{8.161(10^{10})} = -16.5(10^{-6})\ \text{rad}$$

Although the values of ψ_A from the two solutions check, one further step is needed to check the horizontal and vertical deflections of the free ends. In the last solution the reference axes are tilted at 30°. Therefore, the horizontal and vertical deflections of the left end are given by

$$\delta_{HA} = \delta_{HA2}(0.866) + \delta_{VA2}(0.5) = -241(10^{-6})\ \text{cm}$$

$$\delta_{VA} = \delta_{HA2}(-0.5) + \delta_{VA2}(0.866) = -516(10^{-6})\ \text{cm}$$

Again the horizontal deflection of –0.000241 cm for the left half of the arch checks well with the value of –0.000482 cm for the entire arch. With the two displacements of the centroid and the rotation of the end cross section now known, one can easily find the displacements of any other point on the end cross section.

To find the tensile stress at point A we need the bending moment at the center of the arch. This can be found by integration as

$$M = \int_{\pi/6}^{\pi/2} -263Rd\theta(R_{cg}\cos\theta) = -263RR_{cg}\sin\theta\Big|_{\pi/6}^{\pi/2} = -125{,}000\ \text{N-cm}$$

Using the data from Table 9.1, the stress in the outer fiber at the point A is given by

$$\sigma_A = \frac{k_o M_C}{I} = \frac{0.697(125{,}000)(20)}{10(30^3)/36} = 232\ \text{N/cm}^2$$

9.4 ELLIPTICAL RINGS

For an elliptical ring of semiaxes a and b, under equal and opposite forces W (Fig. 9.12), the bending moment M_1 at the extremities of the major axis is given by $M_1 = K_1Wa$, and for equal and opposite outward forces applied at the ends of the minor axis, the moment M_1 at the ends of the major axis is given by $M_1 = -K_2Wa$, where K_1 and K_2 are coefficients which depend on the ratio a/b and have the following values:

a/b	1	1.1	1.2	1.3	1.4	1.5	1.6	1.7
K_1	0.318	0.295	0.274	0.255	0.240	0.227	0.216	0.205
K_2	0.182	0.186	0.191	0.195	0.199	0.203	0.206	0.208

a/b	1.8	1.9	2.0	2.1	2.2	2.3	2.4	2.5
K_1	0.195	0.185	0.175	0.167	0.161	0.155	0.150	0.145
K_2	0.211	0.213	0.215	0.217	0.219	0.220	0.222	0.223

Burke (Ref. 6) gives charts by which the moments and tensions in elliptical rings under various conditions of concentrated loading can be found; the preceding values of K were taken from these charts.

Timoshenko (Ref. 13) gives an analysis of an elliptical ring (or other ring with two axes of symmetry) under the action of a uniform outward pressure, which would apply to

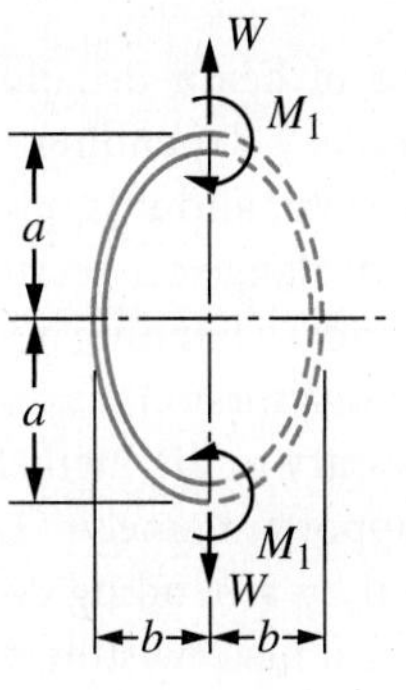

Figure 9.12

a tube of elliptical section under internal pressure. For this case $M = Kpa^2$, where M is the bending moment at a section a distance x along the ring from the end of the minor axis, p is the outward normal pressure per linear inch, and K is a coefficient that depends on the ratios b/a and x/S, where S is one-quarter of the perimeter of the ring. Values of K are given in the following table; M is positive when it produces tension at the inner surface of the ring:

x/S \ b/a	0.3	0.5	0.6	0.7	0.8	0.9
0	−0.172	−0.156	−0.140	−0.115	−0.085	−0.045
0.1	−0.167	−0.152	−0.135	−0.112	−0.082	−0.044
0.2	−0.150	−0.136	−0.120	−0.098	−0.070	−0.038
0.4	−0.085	−0.073	−0.060	−0.046	−0.030	−0.015
0.6	0.020	0.030	0.030	0.028	0.022	0.015
0.7	0.086	0.090	0.082	0.068	0.050	0.022
0.8	0.160	0.150	0.130	0.105	0.075	0.038
0.9	0.240	0.198	0.167	0.130	0.090	0.046
1.0	0.282	0.218	0.180	0.140	0.095	0.050

Values of M calculated by the preceding coefficients are correct only for a ring of uniform moment of inertia I; if I is not uniform, then a correction ΔM must be added. This correction is given by

$$\Delta M = \frac{-\int_0^x \frac{M}{I}\,dx}{\int_0^x \frac{dx}{I}}$$

The integrals can be evaluated graphically. Reference 12 gives charts for the calculation of moments in elliptical rings under uniform radial loading; the preceding values of K were taken from these charts.

9.5 CURVED BEAMS LOADED NORMAL TO PLANE OF CURVATURE

This type of beam usually presents a statically indeterminate problem, the degree of indeterminacy depending upon the manner of loading and support. Both bending and twisting occur, and it is necessary to distinguish between an analysis that is applicable to compact or flangeless sections (circular, rectangular, etc.) in which torsion does not produce secondary bending and one that is applicable to flanged sections (I-beams, channels, etc.) in which torsion may be accompanied by such secondary bending (see Sec. 10.3). It is also necessary to distinguish among three types of constraints that may or may not occur at the supports, namely: (1) the beam is prevented from *sloping*, its horizontal axis held horizontal by a bending couple; (2) the beam is prevented from *rolling*, its vertical axis held vertical by a twisting couple; and (3) in the case of a flanged section, the flanges are prevented from turning about their vertical axes by horizontal secondary bending couples.

These types of constraints will be designated here as (1) fixed as to slope, (2) fixed as to roll, and (3) flanges fixed.

Compact Sections

Table 9.4 treats the curved beam of uniform cross section under concentrated and distributed loads normal to the plane of curvature, out-of-plane concentrated bending moments, and concentrated and distributed torques. Expressions are given for transverse shear, bending moment, twisting moment, deflection, bending slope, and roll slope for 10 combinations of end conditions. To keep the presentation to a reasonable size, use is made of the singularity functions discussed in detail previously, and an extensive list of constants and functions is given. In previous tables the representative functional values have been given, but in Table 9.4 the value of β depends upon both bending and torsional properties, and so a useful set of tabular values would be too large to present. The curved beam or ring of circular cross section is so common, however, that numerical coefficients are given in the table for $\beta = 1.3$ which will apply to a solid or hollow circular cross section of material for which Poisson's ratio is 0.3.

Levy (Ref. 14) has treated the closed circular ring of arbitrary compact cross section for six loading cases. These cases have been chosen to permit apropriate superposition in order to solve a large number of problems, and both isolated and distributed out-of-plane loads are discussed. Hogan (Ref. 18) presents similar loadings and supports. In a similar way the information in Table 9.4 can be used by appropriate superposition to solve most out-of-plane loading problems on closed rings of compact cross section if strict attention is given to the symmetry and boundary conditions involved. Several simple examples of this reasoning are described in the following three cases:

1. If a closed circular ring is supported on any number of equally spaced simple supports (two or more) and if identical loading on each span is symmetrically placed relative to the center of the span, then each span can be treated by boundary condition f of Table 9.4, case 1. This boundary condition has both ends with no deflection or slope, although they are free to roll as needed.
2. If a closed circular ring is supported on any even number of equally spaced simple supports and if the loading on any span is antisymmetrically placed relative to the center line of each span and symmetrically placed relative to each support, then boundary condition f can be applied to each full span. This problem can also be solved by applying boundary condition g to each half span. Boundary condition g has one end simply supported and slope-guided and the other end simply supported and roll-guided.
3. If a closed circular ring is supported on any even number of equally spaced simple supports (four or more) and if each span is symmetrically loaded relative to the center of the span with adjacent spans similarly loaded in opposite directions, then boundary condition i can be applied to each span. This boundary condition has both ends simply supported and roll-guided.

Once any indeterminate reaction forces and moments have been found and the indeterminate internal reactions found at at least one location in the ring, all desired internal bending moment, torques, and transverse shears can be found by equilibrium equations. If a large number of such calculations need be made, one should consider using a theorem published in 1922 by Biezeno. For details of this theorem see Ref. 32. A brief illustration of this work for loads normal to the plane of the ring is given in Ref. 29.

A treatment of curved beams on elastic foundations is beyond the scope of this book. See Ref. 20.

The following examples illustrate the applications of the formulas in Table 9.4 to both curved beams and closed rings with out-of-plane loads.

EXAMPLE 1 A piece of 8-in standard pipe is used to carry water across a passageway 40 ft wide. The pipe must come out of a wall normal to the surface and enter normal to a parallel wall at a position 16.56 ft down the passageway at the same elevation. To accomplish this a decision was made to bend the pipe into two opposite arcs of 28.28-ft radius with a total angle of 45° in each arc. If it is assumed that both ends are rigidly held by the walls, determine the maximum combined stress in the pipe due to its own weight and the weight of a full pipe of water.

SOLUTION An 8-in standard pipe has the following properties: $A = 8.4\text{ in}^2$, $I = 72.5\text{ in}^4$, $w = 2.38\text{ lb/in}$, $E = 30(10^6)\text{ lb/in}^2$, $\nu = 0.3$, $J = 145\text{ in}^4$, $\text{OD} = 8.625\text{ in}$, $\text{ID} = 7.981\text{ in}$, and $t = 0.322\text{ in}$. The weight of water in a 1-in length of pipe is 1.81 lb. Owing to the symmetry of loading it is apparent that at the center of the span where the two arcs meet there is neither slope nor roll. An examination of Table 9.4 reveals that a curved beam that is fixed at the right end and roll- and slope-guided at the left end is not included among the 10 cases. Therefore, a solution will be carried out by considering a beam that is fixed at the right end and free at the left end with a uniformly distributed load over the entire span and both a concentrated moment and a concentrated torque on the left end. (These conditions are covered in cases 2a, 3a, and 4a.)

Since the pipe is round, $J = 2I$; and since $G = E/2(1+\nu)$, $\beta = 1.3$. Also note that for all three cases $\phi = 45°$ and $\theta = 0°$. For these conditions, numerical values of the coefficients are tabulated and the following expressions for the deformations and moments can be written directly from superposition of the three cases:

$$y_A = 0.3058\frac{M_o R^2}{EI} - 0.0590\frac{T_o R^2}{EI} - 0.0469\frac{(2.38+1.81)R^4}{EI}$$

$$\Theta_A = -0.8282\frac{M_o R}{EI} - 0.0750\frac{T_o R}{EI} + 0.0762\frac{4.19R^3}{EI}$$

$$\psi_A = 0.0750\frac{M_o R}{EI} + 0.9782\frac{T_o R}{EI} + 0.0267\frac{4.19R^3}{EI}$$

$$V_B = 0 + 0 - 4.19R(0.7854)$$

$$M_B = 0.7071M_o - 0.7071T_o - 0.2929(4.19)R^2$$

$$T_B = 0.7071M_o + 0.7071T_o - 0.0783(4.19)R^2$$

Since both Θ_A and ψ_A are zero and $R = 28.28(12) = 339.4\text{ in}$,

$$0 = -0.8282M_o - 0.0750T_o + 36{,}780$$

$$0 = 0.0750M_o + 0.9782T_o + 12{,}888$$

Solving these two equations gives $M_o = 45{,}920\text{ lb-in}$ and $T_o = -16{,}700\text{ lb-in}$. Therefore,

$$y_A = -0.40\text{ in}, \qquad M_B = -97{,}100\text{ lb-in}$$

$$T_B = -17{,}000\text{ lb-in}, \qquad V_B = -1120\text{ lb}$$

The maximum combined stress would be at the top of the pipe at the wall where $\sigma = Mc/I = 97{,}100(4.3125)/72.5 = 5575\ \text{lb/in}^2$ and $\tau = Tr/J = 17{,}100(4.3125)/145 = 509\ \text{lb/in}^2$

$$\sigma_{max} = \frac{5775}{2} + \sqrt{\left(\frac{5775}{2}\right)^2 + 509^2} = 5819\ \text{lb/in}^2$$

EXAMPLE 2 A hollow steel rectangular beam 4 in wide, 8 in deep, and with 0.1-in wall thickness extends over a loading dock to be used as a crane rail. It is fixed to a warehouse wall at one end and is simply supported on a post at the other. The beam is curved in a horizontal plane with a radius of 15 ft and covers a total angular span of 60°. Calculate the torsional and bending stresses at the wall when a load of 3000 lb is 20° out from the wall. Neglect the weight of the beam.

SOLUTION The beam has the following properties: $R = 180$ in; $\phi = 60°(\pi/3\ \text{rad})$; $\theta = 40°$; $\phi - \theta = 20°$ $(\pi/9\ \text{rad})$; $I = 1/12[4(8^3) - 3.8(7.8^3)] = 20.39\ \text{in}^4$; $K = 2(0.1^2)(7.9^2)(3.9^2)/[8(0.1) + 4(0.1) - 2(0.1^2)] = 16.09\ \text{in}^4$ (see Table 10.1, case 16); $E = 30(10^6)$; $G = 12(10^6)$; and $\beta = 30(10^6)(20.39)/12(10^6)(16.09) = 3.168$. Equations for a curved beam that is fixed at one end and simply supported at the other with a concentrated load are found in Table 9.4, case 1b. To obtain the bending and twisting moments at the wall requires first the evaluation of the end reaction V_A, which, in turn, requires the following constants:

$$C_3 = -3.168\left(\frac{\pi}{3} - \sin 60°\right) - \frac{1+3.168}{2}\left(\frac{\pi}{3}\cos 60° - \sin 60°\right) = 0.1397$$

$$C_{a3} = -3.168\left(\frac{\pi}{9} - \sin 20°\right) - C_{a2} = 0.006867$$

Similarly,

$$C_6 = C_1 = 0.3060, \qquad C_{a6} = C_{a1} = 0.05775$$

$$C_9 = C_2 = -0.7136, \qquad C_{a9} = C_{a2} = -0.02919$$

Therefore,

$$V_A = 3000\frac{-0.02919(1-\cos 60°) - 0.05775\sin 60° + 0.006867}{-0.7136(1-\cos 60°) - 0.3060\sin 60° + 0.1397} = 359.3\ \text{lb}$$

$$M_B = 359.3(180)(\sin 60°) - 3000(180)(\sin 20°) = -128{,}700\ \text{lb-in}$$

$$T_B = 359.3(180)(1-\cos 60°) - 3000(180)(1-\cos 20°) = -230\ \text{lb-in}$$

At the wall,

$$\sigma = \frac{Mc}{I} = \frac{128{,}700(4)}{20.39} = 25{,}240\ \text{lb/in}^2$$

$$\tau = \begin{cases} \dfrac{VA'\bar{y}}{Ib} = \dfrac{(3000-359.3)[4(4)(2) - 3.9(3.8)(1.95)]}{20.39(0.2)} = 2008\ \text{lb/in}^2 & \text{(due to transverse shear)} \\ \dfrac{T}{2t(a-t)(b-t)} = \dfrac{230}{2(0.1)(7.9)(3.9)} = 37.31\ \text{lb/in}^2 & \text{(due to torsion)} \end{cases}$$

EXAMPLE 3 A solid round aluminum bar is in the form of a horizontal closed circular ring of 100-in radius resting on three equally spaced simple supports. A load of 1000 lb is placed midway between two supports, as shown in Fig. 9.13(a). Calculate the deflection under this load if the bar is of such diameter as to make the maximum normal stress due to combined bending and torsion equal to 20,000 lb/in Let $E = 10(10^6)$ lb/in^2 and $\nu =$ (

SOLUTION The reactions R_B, R_C, and R_D are statically determinate, and a solution yields $R_B = -333.3$ lb and $R_C = R_D = 666.7$ lb. The internal bending and twisting moments are statically indeterminate, and so an energy solution would be appropriate. However, there are several ways that Table 9.4 can be used by superimposing various loadings. The method to be described here is probably the most straightforward.

Consider the equivalent loading shown in Fig. 9.13(b), where $R_B = -333.3$ lb and $R_A = -1000$ lb. The only difference is in the point of zero deflection. Owing to the symmetry of loading, one-half of the ring can be considered slope-guided at both ends, points A and B. Case 1f gives tabulated values of the necessary coefficients for $\phi = 180°$ and $\theta = 60°$. We can now solve for the following values:

$$V_A = -666.7(0.75) = -500 \text{ lb}$$

$$M_A = -666.7(100)(-0.5774) = 38,490 \text{ lb-in}$$

$$\psi_A = \frac{-666.7(100^2)}{EI}(-0.2722) = \frac{1.815(10^6)}{EI}$$

$$T_A = 0 \quad y_A = 0 \quad \Theta_A = 0$$

$$M_B = -666.7(100)(-0.2887) = 19,250 \text{ lb-in}$$

$$M_{60°} = -666.7(100)(0.3608) = -24,050 \text{ lb-in}$$

The equations for M and T can now be examined to determine the location of the maximum combined stress:

$$M_x = -50,000 \sin x + 38,490 \cos x + 66,667 \sin(x - 60°)\langle x - 60°\rangle^0$$

$$T_x = -50,000(1 - \cos x) + 38,490 \sin x + 66,667[1 - \cos(x - 60°)]\langle x - 60°\rangle^0$$

A careful examination of the expression for M shows no maximum values except at the ends and at the position of the load. The torque, however, has a maximum value of 13,100 in-lb at $x = 37.59°$ and a minimum value of -8790 in-lb at $x = 130.9°$. At these same locations the bending moments are zero. At the position of the load, the torque $T = 8330$ lb-in. Nowhere is the combined stress larger than the bending stress at point A. Therefore,

$$\sigma_A = 20,000 = \frac{M_A c}{I} = \frac{38,490 d/2}{(\pi/64)d^4} = \frac{392,000}{d^3}$$

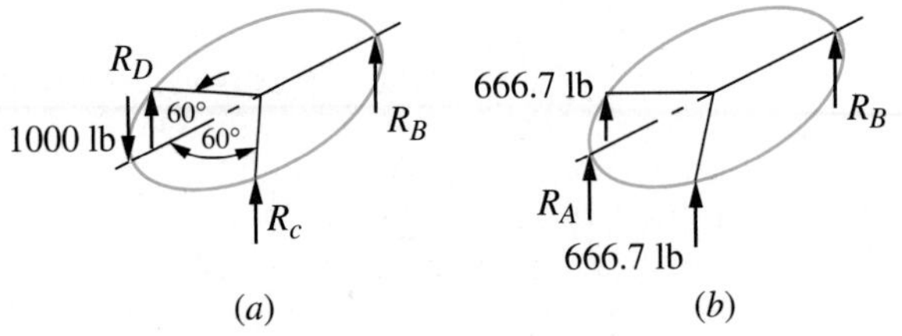

Figure 9.13

which gives

$$d = 2.70\,\text{in} \qquad \text{and} \qquad I = 2.609\,\text{in}^4$$

To obtain the deflection under the 1000-lb load in the original problem, first we must find the deflection at the position of the load of 666.7 lb in Fig. 9.13(*b*). At $x = 60°$,

$$y_x = 0 + 0 + \frac{1.815(10^6)(100)}{10(10^6)2.609}(1 - \cos 60°) + \frac{38{,}490(100^2)}{10(10^6)(2.609)}F_1$$
$$+ 0 + \frac{-500(100^3)}{10(10^6)(2.609)}F_3$$

where

$$F_1 = \frac{1 + 1.3}{2}\frac{\pi}{3}\sin 60° - 1.3(1 - \cos 60°) = 0.3029 \qquad \text{and} \qquad F_3 = 0.1583$$

Therefore,

$$y_{60} = 3.478 + 5.796 - 3.033 = 6.24\,\text{in}$$

If the entire ring were now rotated as a rigid body about point *B* in order to lower points *C* and *D* by 6.24 in, point *A* would be lowered a distance of $6.24(2)/(1 + \cos 60°) = 8.32\,\text{in}$, which is the downward deflection of the 1000-lb load.

The use of a fictitious support, as was done in this problem at point *A*, is generalized for asymmetric loadings, both in-plane and out-of-plane, by Barber in Ref. 35.

Flanged Sections

The formulas in Table 9.4 for flangeless or compact sections apply also to flanged sections when the ends are fixed as to slope only or when fixed as to slope and roll but not as to flange bending and if the loads are distributed or applied only at the ends. If the flanges are fixed or if concentrated loads are applied within the span, the additional torsional stiffness contributed by the bending resistance of the flanges [*warping restraint* (see Sec. 10.3)] may appreciably affect the value and distribution of twisting and bending moments. The flange stresses caused by the secondary bending or warping may exceed the primary bending stresses. References 15–17 and 22 show methods of solution and give some numerical solutions for simple concentrated loads on curved I-beams with both ends fixed completely. Brookhart (Ref. 22) also includes results for additional boundary conditions and uniformly distributed loads. Results are compared with cases where the warping restraint was not considered.

Dabrowski (Ref. 23) gives a thorough presentation of the theory of curved thin-walled beams and works out many examples including multispan beams and beams with open cross sections, closed cross sections, and cross sections which contain both open and closed elements; an extensive bibliography is included. Vlasov (Ref. 27) also gives a very thorough derivation and discusses, among many other topics, vibrations, stability, laterally braced beams of open cross section, and thermal stresses. He also examines the corrections necessary to account for shear deformation in flanges being warped. Verden (Ref. 24) is primarily concerned with multispan curved beams and works out many examples. Sawko and Cope (Ref. 25) and Meyer (Ref. 26) apply finite-element analysis to curved box girder bridges.

9.6 TABLES

TABLE 9.1 Formulas for Curved Beams Subjected to Bending in the Plane of the Curve

Notation: R = radius of curvature measured to centroid of section; c = distance from centroidal axis to extreme fiber on concave side of beam; A = area of section; e = distance from centroidal axis to neutral axis measured toward center of curvature; I = moment of inertia of cross section about centroidal axis perpendicular to plane of curvature; and $k_i = \sigma_i/\sigma$ and $k_o = \sigma_o/\sigma$ where σ_i = actual stress in exteme fiber on concave side, σ_o = actual stress in extreme fiber on convex side, and σ = fictitious unit stress in corresponding fiber as computed by ordinary flexure formula for a straight beam.

Form and Dimensions of Cross Section, Reference No.	Formulas	Values of $\frac{e}{c}$, k_i, and k_o for Various Values of $\frac{R}{c}$
1. Solid rectangular section 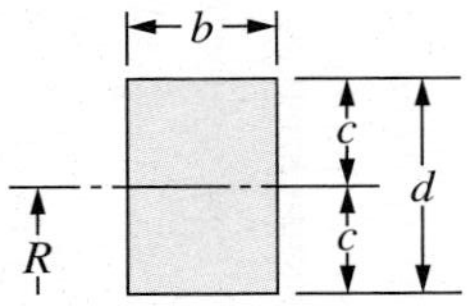	$\frac{e}{c}=\frac{R}{c}-\frac{2}{\ln\frac{R/c+1}{R/c-1}}$ (*Note:* e/c, k_i, and k_o are independent of the width b.) $k_i=\frac{1}{3e/c}\frac{1-e/c}{R/c-1}$ $k_o=\frac{1}{3e/c}\frac{1+e/c}{R/c+1}$ $\int_{\text{area}}\frac{dA}{r}=b\ln\frac{R/c+1}{R/c-1}$	$\frac{R}{c}$ = 1.20 1.40 1.60 1.80 2.00 3.00 4.00 6.00 8.00 10.00 $\frac{e}{c}$ = 0.366 0.284 0.236 0.204 0.180 0.115 0.085 0.056 0.042 0.033 k_i = 2.888 2.103 1.798 1.631 1.523 1.288 1.200 1.124 1.090 1.071 k_o = 0.566 0.628 0.671 0.704 0.730 0.810 0.853 0.898 0.922 0.937
2. Trapezoidal section 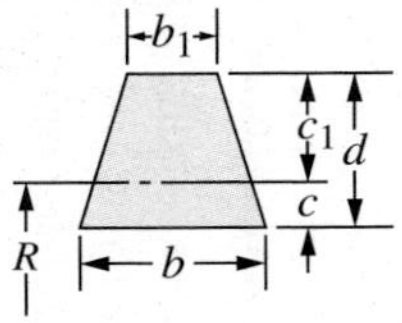	$\frac{d}{c}=\frac{3(1+b_1/b)}{1+2b_1/b}$, $\frac{c_1}{c}=\frac{d}{c}-1$ $\frac{e}{c}=\frac{R}{c}-\frac{\frac{1}{2}(1+b_1/b)(d/c)^2}{\left[\frac{R}{c}+\frac{c_1}{c}-\frac{b_1}{b}\left(\frac{R}{c}-1\right)\right]\ln\left(\frac{R/c+c_1/c}{R/c-1}\right)-\left(1-\frac{b_1}{b}\right)\frac{d}{c}}$ $k_i=\frac{1}{2e/c}\frac{1-e/c}{R/c-1}\frac{1+4b_1/b+(b_1/b)^2}{(1+2b_1/b)^2}$ $k_o=\frac{c_1/c}{2e/c}\frac{c_1/c+e/c}{R/c+c_1/c}\frac{1+4b_1/b+(b_1/b)^2}{(2+b_1/b)^2}$ (*Note:* While e/c, k_i, k_o depend upon the width ratio b_1/b, they are independent of the width b.)	(When $b_1/b = ½$) $\frac{R}{c}$ = 1.20 1.40 1.60 1.80 2.00 3.00 4.00 6.00 8.00 10.00 $\frac{e}{c}$ = 0.403 0.318 0.267 0.232 0.206 0.134 0.100 0.067 0.050 0.040 k_i = 3.011 2.183 1.859 1.681 1.567 1.314 1.219 1.137 1.100 1.078 k_o = 0.544 0.605 0.648 0.681 0.707 0.790 0.836 0.885 0.911 0.927
3. Triangular section, base inward 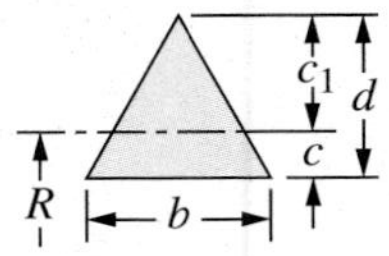	$\frac{e}{c}=\frac{R}{c}-\frac{4.5}{\left(\frac{R}{c}+2\right)\ln\left(\frac{R/c+2}{R/c-1}\right)-3}$, $c=\frac{d}{3}$ $k_i=\frac{1}{2e/c}\frac{1-e/c}{R/c-1}$ $k_o=\frac{1}{4e/c}\frac{2+e/c}{R/c+2}$ $\int_{\text{area}}\frac{dA}{r}=b\left[\left(\frac{R}{3c}+\frac{2}{3}\right)\ln\left(\frac{R/c+2}{R/c-1}\right)-1\right]$ (*Note:* e/c, k_i, and k_o are independent of the width b.)	$\frac{R}{c}$ = 1.20 1.40 1.60 1.80 2.00 3.00 4.00 6.00 8.00 10.00 $\frac{e}{c}$ = 0.434 0.348 0.296 0.259 0.232 0.155 0.117 0.079 0.060 0.048 k_i = 3.265 2.345 1.984 1.784 1.656 1.368 1.258 1.163 1.120 1.095 k_o = 0.438 0.497 0.539 0.573 0.601 0.697 0.754 0.821 0.859 0.883

4. Triangular section, base outward 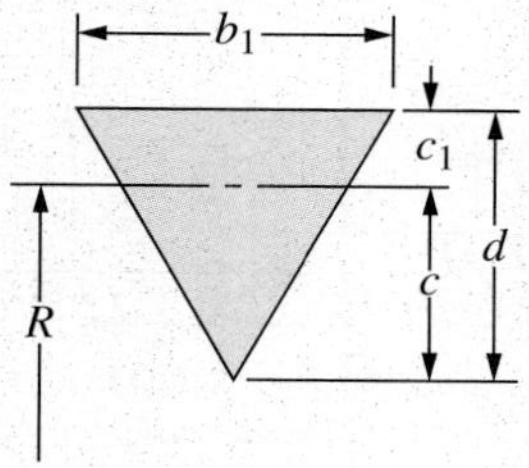	$\frac{e}{c}=\frac{R}{c}-\frac{1.125}{1.5-\left(\frac{R}{c}-1\right)\ln\frac{R/c+0.5}{R/c-1}}, \quad c=\frac{2d}{3}$ $k_i=\frac{1}{8e/c}\frac{1-e/c}{R/c-1}$ $k_o=\frac{1}{4e/c}\frac{2e/c+1}{2R/c+1}$ $\int_{\text{area}}\frac{dA}{r}=b_1\left[1-\frac{2}{3}\left(\frac{R}{c}-1\right)\ln\frac{R/c+0.5}{R/c-1}\right]$ (*Note:* e/c, k_i, and k_o are independent of the width b_1.)	$\frac{R}{c}=1.20$ 1.40 1.60 1.80 2.00 3.00 4.00 6.00 8.00 10.00 $\frac{e}{c}=0.151$ 0.117 0.097 0.083 0.073 0.045 0.033 0.022 0.016 0.013 $k_i=3.527$ 2.362 1.947 1.730 1.595 1.313 1.213 1.130 1.094 1.074 $k_o=0.636$ 0.695 0.735 0.765 0.788 0.857 0.892 0.927 0.945 0.956
5. Diamond 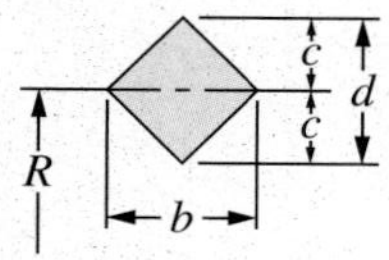	$\frac{e}{c}=\frac{R}{c}-\frac{1}{\frac{R}{c}\ln\left[1-\left(\frac{c}{R}\right)^2\right]+\ln\frac{R/c+1}{R/c-1}}$ $k_i=\frac{1}{6e/c}\frac{1-e/c}{R/c-1}$ $k_o=\frac{1}{6e/c}\frac{1+e/c}{R/c+1}$ $\int_{\text{area}}\frac{dA}{r}=b\left[\frac{R}{c}\ln\left[1-\left(\frac{c}{R}\right)^2\right]+\ln\frac{R/c+1}{R/c-1}\right]$ (*Note:* e/c, k_i, and k_o are independent of the width b.)	$\frac{R}{c}=1.200$ 1.400 1.600 1.800 2.000 3.000 4.000 6.000 8.000 10.000 $\frac{e}{c}=0.175$ 0.138 0.116 0.100 0.089 0.057 0.042 0.028 0.021 0.017 $k_i=3.942$ 2.599 2.118 1.866 1.709 1.377 1.258 1.159 1.115 1.090 $k_o=0.510$ 0.572 0.617 0.652 0.681 0.772 0.822 0.875 0.904 0.922
6. Solid circular or elliptical section 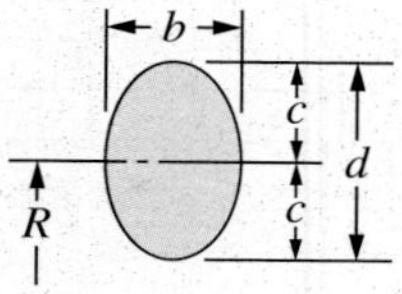	$\frac{e}{c}=\frac{1}{2}\left[\frac{R}{c}-\sqrt{\left(\frac{R}{c}\right)^2-1}\right]$ $k_i=\frac{1}{4e/c}\frac{1-e/c}{R/c-1}$ $k_o=\frac{1}{4e/c}\frac{1+e/c}{R/c+1}, \quad \int_{\text{area}}\frac{dA}{r}=\pi b\left[\frac{R}{c}-\sqrt{\left(\frac{R}{c}\right)^2-1}\right]$ (*Note:* e/c, k_i, and k_o are independent of the width b.)	$\frac{R}{c}=1.20$ 1.40 1.60 1.80 2.00 3.00 4.00 6.00 8.00 10.00 $\frac{e}{c}=0.268$ 0.210 0.176 0.152 0.134 0.086 0.064 0.042 0.031 0.025 $k_i=3.408$ 2.350 1.957 1.748 1.616 1.332 1.229 1.142 1.103 1.080 $k_o=0.537$ 0.600 0.644 0.678 0.705 0.791 0.837 0.887 0.913 0.929

(*Continued*)

TABLE 9.1 Formulas for Curved Beams Subjected to Bending in the Plane of the Curve (*Continued*)

Form and Dimensions of Cross Section, Reference No.	Formulas	Values of $\frac{e}{c}$, k_i, and k_o for Various Values of $\frac{R}{c}$
7. Solid semicircle or semiellipse, base inward 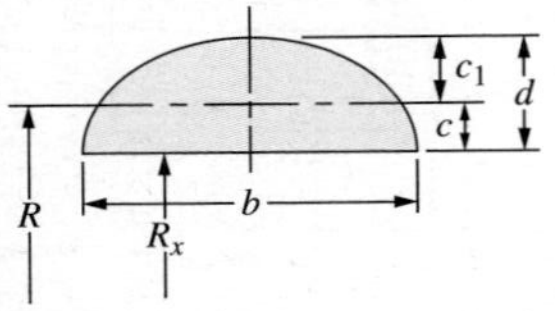 (*Note:* For a semicircle, $b/2 = d$.)	$R = R_x + c, \quad \frac{d}{c} = \frac{3\pi}{4}$ (*Note:* e/c, k_i, and k_o are independent of the width b.) $k_i = \frac{0.3879}{e/c} \frac{1-e/c}{R/c-1}$ $k_o = \frac{0.2860}{e/c} \frac{e/c+1.3562}{R/c+1.3562}$ For $R_x \geq d : R/c \geq 3.356$ and $\int_{\text{area}} \frac{dA}{r} = \frac{\pi R_x b}{2d} - b - \frac{b}{d}\sqrt{R_x^2 - d^2}\left(\frac{\pi}{2} - \sin^{-1}\frac{d}{R_x}\right)$ $\frac{e}{c} = \frac{R}{c} - \frac{(d/c)^2/2}{\frac{R}{c} - 2.5 - \sqrt{\left(\frac{R}{c}-1\right)^2 - \left(\frac{d}{c}\right)^2}\left(1 - \frac{2}{\pi}\sin^{-1}\frac{d/c}{R/c-1}\right)}$ For $R_x < d : R/c < 3.356$ and $\int_{\text{area}} \frac{dA}{r} = \frac{\pi R_x b}{2d} - b + \frac{b}{d}\sqrt{d^2 - R_x^2}\ln\frac{d+\sqrt{d^2 - R_x^2}}{R_x}$ $\frac{e}{c} = \frac{R}{c} - \frac{(d/c)^2/2}{\frac{R}{c} - 2.5 + \frac{2}{\pi}\sqrt{\left(\frac{d}{c}\right)^2 - \left(\frac{R}{c}-1\right)^2}\ln\frac{d/c+\sqrt{(d/c)^2-(R/c-1)^2}}{R/c-1}}$	$\frac{R}{c} = 1.200$ 1.400 1.600 1.800 2.000 3.000 4.000 6.000 8.000 10.000 $\frac{e}{c} = 0.388$ 0.305 0.256 0.222 0.197 0.128 0.096 0.064 0.048 0.038 $k_i = 3.056$ 2.209 1.878 1.696 1.579 1.321 1.224 1.140 1.102 1.080 $k_o = 0.503$ 0.565 0.609 0.643 0.671 0.761 0.811 0.867 0.897 0.916
8. Solid semicircle or semiellipse, base outward 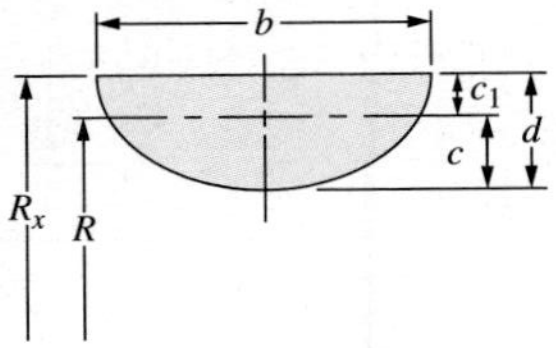 (*Note:* for a semicircle, $b/2 = d$.)	$R = R_x - c_1, \quad \frac{d}{c} = \frac{3\pi}{3\pi - 4}, \quad \frac{c_1}{c} = \frac{4}{3\pi - 4}$ (*Note:* e/c, k_i, and k_o are independent of the width b.) $\int_{\text{area}} \frac{dA}{c} = \frac{\pi R_x b}{2d} + b - \frac{b}{d}\sqrt{R_x^2 - d^2}\left(\frac{\pi}{2} + \sin^{-1}\frac{d}{R_x}\right)$ $\frac{e}{c} = \frac{R}{c} - \frac{(d/c)^2/2}{\frac{R}{c} + \frac{10}{3\pi - 4} - \sqrt{\left(\frac{R}{c}+\frac{c_1}{c}\right)^2 - \left(\frac{d}{c}\right)^2}\left(1 + \frac{2}{\pi}\sin^{-1}\frac{d/c}{R/c+c_1/c}\right)}$ $k_i = \frac{0.2109}{e/c} \frac{1-e/c}{R/c-1}$ $k_o = \frac{0.2860}{e/c} \frac{e/c+0.7374}{R/c+0.7374}$	$\frac{R}{c} = 1.200$ 1.400 1.600 1.800 2.000 3.000 4.000 6.000 8.000 10.000 $\frac{e}{c} = 0.244$ 0.189 0.157 0.135 0.118 0.075 0.055 0.036 0.027 0.021 $k_i = 3.264$ 2.262 1.892 1.695 1.571 1.306 1.210 1.130 1.094 1.073 $k_o = 0.593$ 0.656 0.698 0.730 0.755 0.832 0.871 0.912 0.933 0.946

9. Segment of a solid circle, base inward

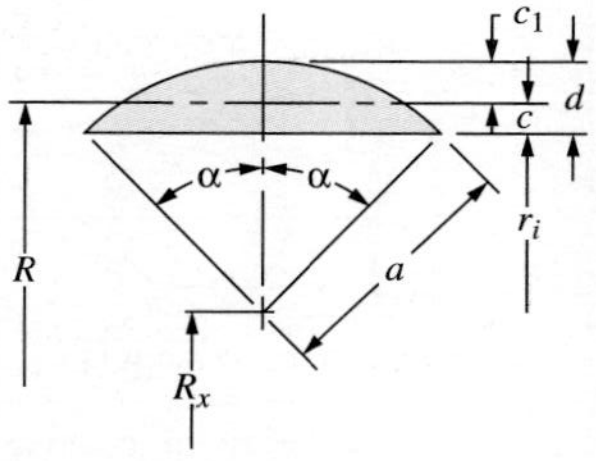

Note: R, c, e/c, k_i, and k_o are all independent of the width of the segment provided all horizontal elements of the segment change width proportionately. Therefore to use these expressions for a segment of an ellipse, find the circle which has the same radial dimensions R_x, r_i, and d and evaluate e/c, k_i, and k_o which have the same values for both the circle and ellipse. To find dA/r for the ellipse, multiply the value of dA/r for the circle by the ratio of the horizontal to vertical semiaxes for the ellipse. See the example.

$$R = R_x + c + a\cos\alpha$$

$$\frac{a}{c} = \frac{3\alpha - 3\sin\alpha\cos\alpha}{3\sin\alpha - 3\alpha\cos\alpha - \sin^3\alpha}, \qquad \frac{c_1}{c} = \frac{3\alpha - 3\sin\alpha\cos\alpha - 2\sin^3\alpha}{3\sin\alpha - 3\alpha\cos\alpha - \sin^3\alpha}$$

$$k_i = \frac{I}{Ac^2}\frac{1}{e/c}\frac{1-e/c}{R/c-1}$$

where expression for I and A are found in Table A.1, case 19

$$k_o = \frac{I}{Ac^2}\frac{1}{(e/c)(c_1/c)}\frac{e/c+c_1/c}{R/c+c_1c}$$

For $R_x \geq a$: $R/c \geq (a/c)(1+\cos\alpha) + 1$ and

$$\int_{\text{area}} \frac{dA}{r} = 2R_x\alpha - 2a\sin\alpha - 2\sqrt{R_x^2 - a^2}\left(\frac{\pi}{2} - \sin^{-1}\frac{a + R_x\cos\alpha}{R_x + a\cos\alpha}\right)$$

$$\frac{e}{c} = \frac{R}{c} - \frac{(\alpha - \sin\alpha\cos\alpha)a/c}{\dfrac{2\alpha R_x}{a} - 2\sin\alpha - 2\sqrt{\left(\dfrac{R_x}{a}\right)^2 - 1}\left(\dfrac{\pi}{2} - \sin^{-1}\dfrac{1 + (R_x/a)\cos\alpha}{R_x/a + \cos\alpha}\right)}$$

(*Note:* Values of $\sin^{-1}$ between $-\pi/2$ and $\pi/2$ are to be taken in above expressions.)

For $R_x < a$: $R/c < (a/c)(1+\cos\alpha) + 1$ and

$$\int_{\text{area}} \frac{dA}{r} = 2R_x\alpha - 2a\sin\alpha + 2\sqrt{a^2 - R_x^2} \times \ln\frac{\sqrt{a^2 - R_x^2}\sin\alpha + a + R_x\cos\alpha}{R_x + a\cos\alpha}$$

$$\frac{e}{c} = \frac{R}{c} - \frac{(\alpha - \sin\alpha\cos\alpha)a/c}{\dfrac{2\alpha R_x}{a} - 2\sin\alpha + 2\sqrt{1 - \left(\dfrac{R_x}{a}\right)^2}\ln\dfrac{\sqrt{1-(R_x/a)^2}\sin\alpha + 1 + (R_x/a)\cos\alpha}{R_x/a + \cos\alpha}}$$

For $\alpha = 60°$:

R/c	1.200	1.400	1.600	1.800	2.000	3.000	4.000	6.000	8.000	10.000
e/c	0.401	0.317	0.266	0.232	0.206	0.134	0.101	0.067	0.051	0.041
k_i	3.079	2.225	1.891	1.707	1.589	1.327	1.228	1.143	1.104	1.082
k_o	0.498	0.560	0.603	0.638	0.665	0.755	0.806	0.862	0.893	0.913

For $\alpha = 30°$:

R/c	1.200	1.400	1.600	1.800	2.000	3.000	4.000	6.000	8.000	10.000
e/c	0.407	0.322	0.271	0.236	0.210	0.138	0.103	0.069	0.052	0.042
k_i	3.096	2.237	1.900	1.715	1.596	1.331	1.231	1.145	1.106	1.083
k_o	0.495	0.556	0.600	0.634	0.662	0.752	0.803	0.860	0.891	0.911

(Continued)

TABLE 9.1 Formulas for Curved Beams Subjected to Bending in the Plane of the Curve (*Continued*)

Form and Dimensions of Cross Section, Reference No.	Formulas	Values of $\frac{e}{c}$, k_i, and k_o for Various Values of $\frac{R}{c}$
10. Segment of a solid circle, base outward 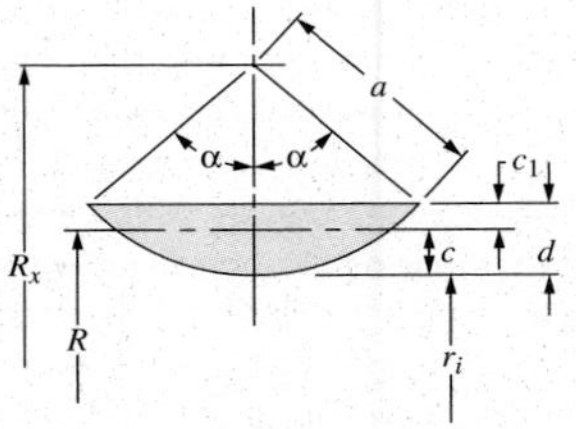 *Note:* R, c, e/c, k_i and k_o are all independent of the width of the segment provided all horizontal elements of the segment change width proportionately. To use these expressions for a segment of an ellipse, refer to the explanation in case 9.	$R = R_x + c - a$ $\frac{a}{c} = \frac{3\alpha - 3\sin\alpha\cos\alpha}{3\alpha - 3\sin\alpha\cos\alpha - 2\sin^3\alpha}, \quad \frac{c_1}{c} = \frac{3\sin\alpha - 3\alpha\cos\alpha - \sin^3\alpha}{3\alpha - 3\sin\alpha\cos\alpha - 2\sin^3\alpha}$ $k_i = \frac{I}{Ac^2}\frac{1}{e/c}\frac{1-e/c}{R/c-1}$ where expression for I and A are found in Table A.1, case 19 $k_o = \frac{I}{Ac^2}\frac{1}{(e/c)(c_1/c)}\frac{e/c + c_1/c}{R/c + c_1c}$ $\int_{\text{area}} \frac{dA}{r} = 2R_x\alpha + 2a\sin\alpha - 2\sqrt{R_x^2 - a^2}\left(\frac{\pi}{2} + \sin^{-1}\frac{a - R_x\cos\alpha}{R_x - a\cos\alpha}\right)$ $\frac{e}{c} = \frac{R}{c} - \frac{(\alpha - \sin\alpha\cos\alpha)a/c}{\frac{2\alpha R_x}{a} + 2\sin\alpha - 2\sqrt{\left(\frac{R_x}{a}\right)^2 - 1}\left(\frac{\pi}{2} - \sin^{-1}\frac{1-(R_x/a)\cos\alpha}{R_x/a - \cos\alpha}\right)}$ (*Note*: Values of $\sin^{-1}$ between $-\pi/2$ and $\pi/2$ are to be taken in above expressions.)	For $\alpha = 60°$ $\frac{R}{c}$ = 1.200 1.400 1.600 1.800 2.000 3.000 4.000 6.000 8.000 10.000 $\frac{e}{c}$ = 0.235 0.181 0.150 0.129 0.113 0.071 0.052 0.034 0.025 0.020 k_i = 3.241 2.247 1.881 1.686 1.563 1.301 1.207 1.127 1.092 1.072 k_o = 0.598 0.661 0.703 0.735 0.760 0.836 0.874 0.914 0.935 0.948 For $\alpha = 30°$ $\frac{R}{c}$ = 1.200 1.400 1.600 1.800 2.000 3.000 4.000 6.000 8.000 10.000 $\frac{e}{c}$ = 0.230 0.177 0.146 0.125 0.110 0.069 0.051 0.033 0.025 0.020 k_i = 3.232 2.241 1.876 1.682 1.560 1.299 1.205 1.126 1.091 1.072 k_o = 0.601 0.663 0.706 0.737 0.763 0.838 0.876 0.916 0.936 0.948
11. Hollow circular section 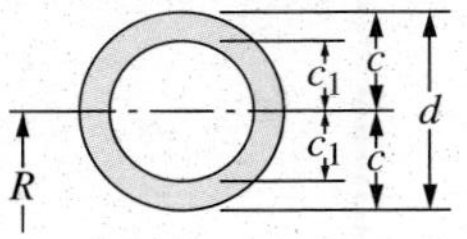	$\frac{e}{c} = \frac{1}{2}\left[\frac{2R}{c} - \sqrt{\left(\frac{R}{c}\right)^2 - 1} - \sqrt{\left(\frac{R}{c}\right)^2 - \left(\frac{c_1}{c}\right)^2}\right]$ $k_i = \frac{1}{4e/c}\frac{1-e/c}{R/c-1}\left[1 + \left(\frac{c_1}{c}\right)^2\right]$ $k_o = \frac{1}{4e/c}\frac{1+e/c}{R/c+1}\left[1 + \left(\frac{c_1}{c}\right)^2\right]$ (*Note:* For thin-walled tubes the discussion on page 277 should be considered.)	(When $c_1/c = ½$) $\frac{R}{c}$ = 1.20 1.40 1.60 1.80 2.00 3.00 4.00 6.00 8.00 10.00 k_i = 3.276 2.267 1.895 1.697 1.573 1.307 1.211 1.130 1.094 1.074 $\frac{e}{c}$ = 0.323 0.256 0.216 0.187 0.166 0.107 0.079 0.052 0.039 0.031 k_o = 0.582 0.638 0.678 0.708 0.733 0.810 0.852 0.897 0.921 0.936

12. Hollow elliptical section

12a. Inner and outer perimeters are ellipses, wall thickness is not constant

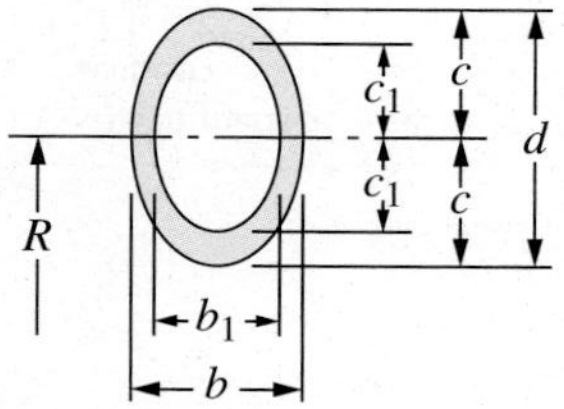

$$\frac{e}{c} = \frac{R}{c} - \frac{\frac{1}{2}[1-(b_1/b)(c_1/c)]}{\frac{R}{c} - \sqrt{\left(\frac{R}{c}\right)^2 - 1} - \frac{b_1/b}{c_1/c}\left[\frac{R}{c} - \sqrt{\left(\frac{R}{c}\right)^2 - \left(\frac{c_1}{c}\right)^2}\right]}$$

$$k_i = \frac{1}{4e/c}\,\frac{1-e/c}{R/c-1}\,\frac{1-(b_1/b)(c_1/c)^3}{1-(b_1/b)(c_1/c)}$$

$$k_o = \frac{1}{4e/c}\,\frac{1+e/c}{R/c+1}\,\frac{1-(b_1/b)(c_1/c)^3}{1-(b_1/b)(c_1/c)}$$

(*Note:* While e/c, k_i, and k_o depend upon the width ratio b_1/b, they are independent of the width b.)

(When $b_1/b = 3/5$, $c_1/c = 4/5$)

$R/c =$	1.20	1.40	1.60	1.80	2.00	3.00	4.00	6.00	8.00	10.00
$e/c =$	0.345	0.279	0.233	0.202	0.178	0.114	0.085	0.056	0.042	0.034
$k_i =$	3.033	2.154	1.825	1.648	1.535	1.291	1.202	1.125	1.083	1.063
$k_o =$	0.579	0.637	0.677	0.709	0.734	0.812	0.854	0.899	0.916	0.930

12b. Constant wall thickness, midthickness perimeter is an ellipse (shown dashed)

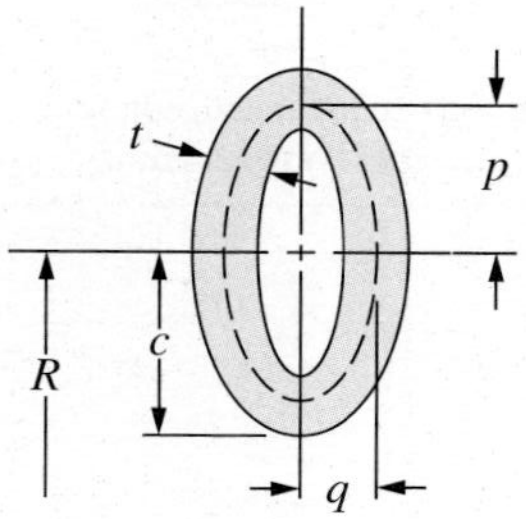

Note: There is a limit on the maximum wall thickness allowed in this case. Cusps will form on the inner perimeter at the ends of the major axis if this maximum is exceeded. If $p/q \le 1$, then $t_{max} = 2p^2/q$. If $p/q \ge 1$, then $t_{max} = 2q^2/p$.

There is no closed-form solution for this case, so numerical solutions were run for the ranges $1.2 < R/c < 5$; $0 < t < t_{max}$; $0.2 < p/q < 5$. Results are expressed below in terms of the solutions for case 12a for which $c = p + t/2$, $c_1 = p - t/2$, $b = 2q + t$, and $b_1 = 2q - t$.

$$\frac{e}{c} = K_1\left(\frac{e}{c} \text{ from case 12a}\right), \quad k_i = K_2(k_i \text{ from case 12a})$$

$$k_o = K_3(k_o \text{ from case 12a})$$

where K_1, K_2, and K_3, are given in the following table and are essentially independent of t and R/c.

p/q	0.200	0.333	0.500	0.625	1.000	1.600	2.000	3.000	4.000	5.000
K_1	0.965	0.985	0.995	0.998	1.000	1.002	1.007	1.027	1.051	1.073
K_2	1.017	1.005	1.002	1.001	1.000	1.000	1.000	0.998	0.992	0.985
K_3	0.982	0.992	0.998	0.999	1.000	1.002	1.004	1.014	1.024	1.031

(*Continued*)

TABLE 9.1 Formulas for Curved Beams Subjected to Bending in the Plane of the Curve (*Continued*)

Form and Dimensions of Cross Section, Reference No.	Formulas	Values of $\frac{e}{c}$, k_i, and k_o for Various Values of $\frac{R}{c}$
13. T-beam or channel section 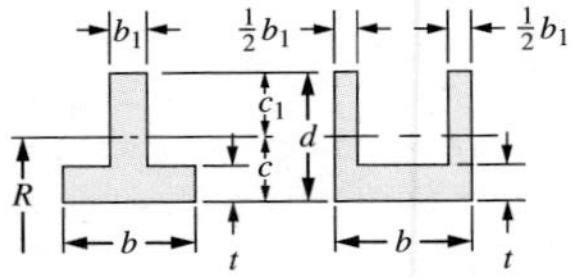	$\frac{d}{c}=\frac{2[b_1/b+(1-b_1/b)(t/d)]}{b_1/b+(1-b_1/b)(t/d)^2}, \quad \frac{c_1}{c}=\frac{d}{c}-1$ $\frac{e}{c}=\frac{R}{c}-\frac{(d/c)[b_1/b+(1-b_1/b)(t/d)]}{\frac{b_1}{b}\ln\frac{d/c+R/c-1}{(d/c)(t/d)+R/c-1}+\ln\frac{(d/c)(t/d)+R/c-1}{R/c-1}}$ $k_i=\frac{I_c}{Ac^2(R/c-1)}\frac{1-e/c}{e/c}$ where $\frac{I_c}{Ac^2}=\frac{1}{3}\left(\frac{d}{c}\right)^2\left[\frac{b_1/b+(1-b_1/b)(t/d)^3}{b_1/b+(1-b_1/b)(t/d)}\right]-1$ $k_o=\frac{I_c}{Ac^2(e/c)}\frac{d/c+e/c-1}{R/c+d/c-1}\frac{1}{d/c-1}$ (*Note:* While e/c, k_i, and k_o depend upon the width ratio b_1/b, they are independent of the width b.)	(When $b_1/b=$ ¼, $t/d=$ ¼) $\frac{R}{c}=$ 1.200 1.400 1.600 1.800 2.000 3.000 4.000 6.000 8.000 10.000 $\frac{e}{c}=$ 0.502 0.419 0.366 0.328 0.297 0.207 0.160 0.111 0.085 0.069 $k_i=$ 3.633 2.538 2.112 1.879 1.731 1.403 1.281 1.176 1.128 1.101 $k_o=$ 0.583 0.634 0.670 0.697 0.719 0.791 0.832 0.879 0.905 0.922
14. Symmetrical I-beam or hollow rectangular section 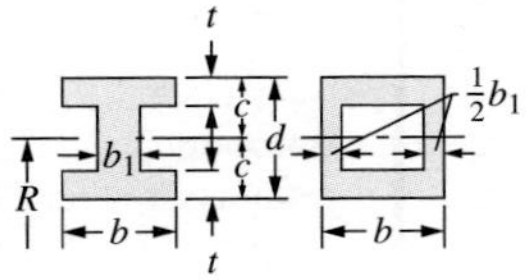	$\frac{e}{c}=\frac{R}{c}-\frac{2[t/c+(1-t/c)(b_1/b)]}{\ln\frac{R/c^2+(R/c-1)(t/c)-1}{(R/c)^2-(R/c-1)(t/c)-1}+\frac{b_1}{b}\ln\frac{R/c-t/c+1}{R/c+t/c-1}}$ $k_i=\frac{I_c}{Ac^2(R/c-1)}\frac{1-e/c}{e/c}$ where $\frac{I_c}{Ac^2}=\frac{1}{3}\frac{1-(1-b_1/b)(1-t/c)^3}{1-(1-b_1/b)(1-t/c)}$ $k_o=\frac{I_c}{Ac^2(R/c+1)}\frac{1+e/c}{e/c}$ (*Note:* While e/c, k_i, and k_o depend upon the width ratio b_1/b, they are independent of the width b.)	(When $b_1/b=1/3$, $t/d=1/6$) $\frac{R}{c}=$ 1.20 1.40 1.60 1.80 2.00 3.00 4.00 6.00 8.00 10.00 $\frac{e}{c}=$ 0.489 0.391 0.330 0.287 0.254 0.164 0.122 0.081 0.060 0.048 $k_i=$ 2.156 1.876 1.630 1.496 1.411 1.225 1.156 1.097 1.071 1.055 $k_o=$ 0.666 0.714 0.747 0.771 0.791 0.853 0.886 0.921 0.940 0.951
15. Unsymmetrical I-beam section 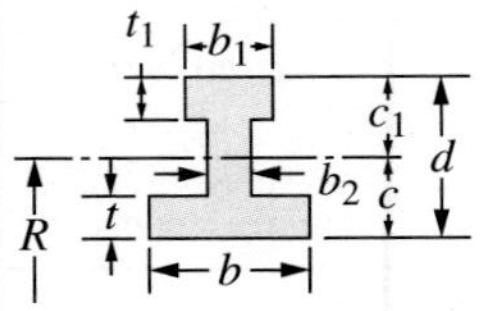	$A=bd[b_1/b+(1-b_2/b)(t/d)-(b_1/b-b_2/b)(1-t_1/d)]$ $\frac{d}{c}=\frac{2A/bd}{(b_1/b-b_2/b)(2-t_1/d)(t_1/d)+(1-b_2/b)(t/d)^2+b_2/b}$ $\frac{e}{c}=\frac{R}{c}-\frac{(A/bd)(d/c)}{\ln\frac{R/c+t/c-1}{R/c-1}+\frac{b_2}{b}\ln\frac{R/c+c_1/c-t_1/c}{R/c+t/c-1}+\frac{b_1}{b}\ln\frac{R/c+c_1/c}{R/c+c_1/c-t_1/c}}$ $k_i=\frac{I_c}{Ac^2(R/c-1)}\frac{1-e/c}{e/c}$ where $\frac{I_c}{Ac^2}=\frac{1}{3}\left(\frac{d}{c}\right)^2\left[\frac{b_1/b+(1-b_2/b)(t/d)^3-(b_1/b-b_2/b)(1-t_1/d)^3}{b_1/b+(1-b_2/b)(t/d)-(b_1/b-b_2/b)(1-t_1/d)}\right]-1$ $k_o=\frac{I_c}{Ac^2(e/c)}\frac{d/c+e/c-1}{R/c+d/c-1}\frac{1}{d/c-1}$ (*Note:* While e/c, k_i, and k_o depend upon the width ratio b_1/b and b_2/b, they are independent of the width b.)	(When $b_1/b=2/3$, $b_2/b=1/6$, $t_1/d=1/6$, $t/d=1/3$) $\frac{R}{c}=$ 1.20 1.40 1.60 1.80 2.00 3.00 4.00 6.00 8.00 10.00 $\frac{e}{c}=$ 0.491 0.409 0.356 0.318 0.288 0.200 0.154 0.106 0.081 0.066 $k_i=$ 3.589 2.504 2.083 1.853 1.706 1.385 1.266 1.165 1.120 1.094 $k_o=$ 0.671 0.721 0.754 0.779 0.798 0.856 0.887 0.921 0.938 0.950

TABLE 9.2 Formulas for Circular Rings

Notation: W = load (force); w and v = unit loads (force per unit of circumferential length); ρ = unit weight of contained liquid (force per unit volume); M_o = applied couple (force-length). M_A, M_B, M_C, and M are internal moments at A, B, C, and x, respectively, positive as shown. N_A, N, V_A, and V are internal forces, positive as shown. E = modulus of elasticity (force per unit area); ν = Poisson's ratio; A = cross-sectional area (length squared); R = radius to the centroid of the cross section (length); I = area moment of inertia of ring cross section about the principal axis perpendicular to the plane of the ring (length⁴). [Note that for a pipe or cylinder, a representative segment of unit axial length may be used by replacing EI by $Et^3/12(1-\nu^2)$.] e = positive distance measured radially inward from the centroidal axis of the cross section to the neutral axis of pure bending (see Sec. 9.1). θ, x, and ϕ are angles (radians) and are limited to the range zero to π for all cases except 18 and 19; $s=\sin\theta$, $c=\cos\theta$, $z=\sin x$, $u=\cos x$, $n=\sin\phi$, and $m=\cos\phi$.

ΔD_V and ΔD_H are changes in the vertical and horizontal diameters, respectively, and an increase is positive. ΔL is the change in the lower half of the vertical diameter or the vertical motion relative to point C of a line connecting points B and D on the ring. Similarly ΔL_W is the vertical motion relative to point C of a horizontal line connecting the load points on the ring. ΔL_{WH} is the change in length of a horizontal line connecting the load points on the ring. ψ is the angular rotation (radians) of the load point in the plane of the ring and is positive in the direction of positive θ. For the distributed loadings the load points just referred to are the points where the distributed loading starts, i.e., the position located by the angle θ. The reference to points A, B, and C and to the diameters refer to positions on a circle of radius R passing through the centroids of the several sections, i.e., diameter = $2R$. It is important to consider this when dealing with thick rings. Similarly, all concentrated and distributed loadings are assumed to be applied at the radial position of the centroid with the exception of the cases where the ring is loaded by its own weight or by dynamic loading, cases 15 and 21. In these two cases the actual radial distribution of load is considered. If the loading is on the outer or inner surfaces of thick rings, an equivalent loading at the centroidal radius R must be used. See the examples to determine how this might be accomplished.

The hoop-stress deformation factor is $\alpha = I/AR^2$ for thin rings or $\alpha = e/R$ for thick rings. The transverse (radial) shear deformation factor is $\beta = FEI/GAR^2$ for thin rings or $\beta = 2F(1+\nu)e/R$ for thick rings, where G is the shear modulus of elasticity and F is a shape factor for the cross section (see Sec. 8.10). The following constants are defined to simplify the expressions which follow. Note that these constants are unity if no correction for hoop stress or shear stress is necessary or desired for use with thin rings. $k_1 = 1-\alpha+\beta$, $k_2 = 1-\alpha$.

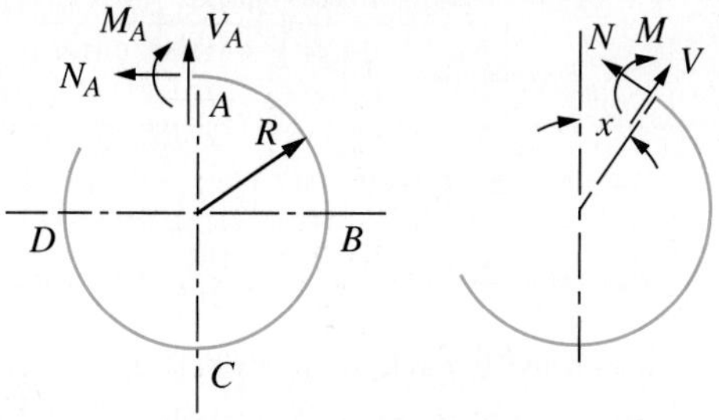

General formulas for moment, hoop load, and radial shear

$$M = M_A - N_A R(1-u) + V_A Rz + LT_M$$

$$N = N_A u + V_A z + LT_N$$

$$V = -N_A z + V_A u + LT_V$$

where LT_M, LT_N, and LT_V are load terms given below for several types of load.

Note: Due to symmetry in most of the cases presented, the loads beyond 180° are not included in the load terms. Only for cases 16, 17, and 19 should the equations for M, N, and V be used beyond 180°.

Note: The use of the bracket $\langle x-\theta\rangle^0$ is explained on page 159 and has a value of zero unless $x > \theta$.

Reference No., Loading, and Load Terms	Formulas for Moments, Loads, and Deformations and Some Selected Numerical Values	
1. 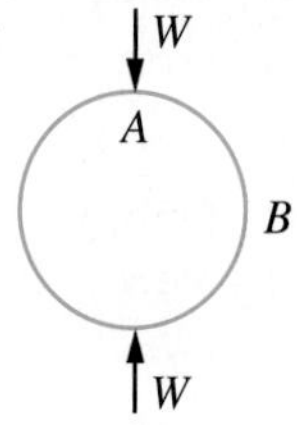 $LT_M = \frac{-WRz}{2}$ $LT_N = \frac{-Wz}{2}$ $LT_V = \frac{-Wu}{2}$	$M_A = \frac{WRk_2}{\pi}$ $N_A = 0$ $V_A = 0$ $\Delta D_H = \frac{WR^3}{EI}\left(\frac{k_1}{2} - k_2 + \frac{2k_2^2}{\pi}\right)$ $\Delta D_V = \frac{-WR^3}{EI}\left(\frac{\pi k_1}{4} - \frac{2k_2^2}{\pi}\right)$	$\text{Max} + M = M_A = 0.3183WRk_2$ $\text{Max} - M = M_B = -(0.5 - 0.3183k_2)WR$ If $\alpha = \beta = 0$, $\Delta D_H = 0.1366\frac{WR^3}{EI}$ and $\Delta D_V = -0.1488\frac{WR^3}{EI}$ *Note:* For concentrated loads on thick-walled rings, study the material in Sec. 14.3 on hollow pins and rollers. Radial stresses under the concentrated loads have a significant effect not considered here.

(Continued)

TABLE 9.2 Formulas for Circular Rings (*Continued*)

Reference No., Loading, and Load Terms	Formulas for Moments, Loads, and Deformations and Some Selected Numerical Values

2.

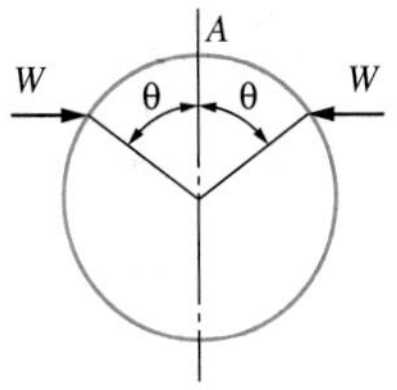

$LT_M = -WR(c-u)\langle x-\theta\rangle^0$

$LT_N = Wu\langle x-\theta\rangle^0$

$LT_V = -Wz\langle x-\theta\rangle^0$

$$M_A = \frac{-WR}{\pi}[(\pi-\theta)(1-c)-s(k_2-c)]$$

$$M_C = \frac{-WR}{\pi}[\theta(1+c)-s(k_2+c)]$$

$$N_A = \frac{-W}{\pi}[\pi-\theta+sc]$$

$$V_A = 0$$

$$\Delta D_H = \begin{cases} \dfrac{-WR^3}{EI\pi}[0.5\pi k_1(\theta-sc)+2k_2\theta c-2k_2^2 s] & \text{if } \theta \le \dfrac{\pi}{2} \\ \dfrac{-WR^3}{EI\pi}[0.5\pi k_1(\pi-\theta+sc)-2k_2(\pi-\theta)c-2k_2^2 s] & \text{if } \theta \ge \dfrac{\pi}{2} \end{cases}$$

$$\Delta D_V = \frac{WR^3}{EI}\left[\frac{k_1 s^2}{2}-k_2\left(1-c+\frac{2\theta c}{\pi}\right)+\frac{2k_2^2 s}{\pi}\right]$$

$$\Delta L = \begin{cases} \dfrac{WR^3}{EI}\left[\dfrac{\theta c}{2}+\dfrac{k_1(\theta-sc)}{2\pi}-k_2\left(\dfrac{\theta c}{\pi}+\dfrac{s}{2}\right)+\dfrac{k_2^2 s}{\pi}\right] & \text{if } \theta \le \pi/2 \\ \dfrac{WR^3}{EI}\left[\dfrac{(\pi-\theta)c}{2}+\dfrac{k_1(\theta-sc-\pi c^2)}{2\pi}-k_2\left(1+\dfrac{\theta c}{\pi}-\dfrac{s}{2}\right)+k_2^2 s/\pi\right] & \text{if } \theta \ge \pi/2 \end{cases}$$

$$\Delta L_W = \frac{WR^3}{EI\pi}[(\pi-\theta)\theta sc+0.5k_1 s^2(\theta-sc)+k_2(2\theta s^2-\pi s^2-\theta c-\theta)+k_2^2 s(1+c)]$$

$$\Delta L_{WH} = \frac{-WR^3}{EI\pi}[(\pi-\theta)2\theta c^2-k_1(\pi sc+s^2c^2-2\theta sc-\pi\theta+\theta^2)-2k_2 sc(\pi-2\theta)-2k_2^2 s^2]$$

$$\Delta\psi = \frac{-WR^2}{EI\pi}[(\pi-\theta)\theta c-k_2 s(sc+\pi-2\theta)]$$

$$\text{Max}+M = \frac{WRs(k_2-c^2)}{\pi} \quad \text{at } x=\theta$$

$$\text{Max}-M = \begin{cases} M_A & \text{if } \theta \le \dfrac{\pi}{2} \\ M_C & \text{if } \theta \ge \dfrac{\pi}{2} \end{cases}$$

If $\alpha=\beta=0$, $M=K_M WR$, $N=K_N W$,

$\Delta D = K_{\Delta D}WR^3/EI$, $\Delta\psi = K_{\Delta\psi}WR^2/EI$, etc.

θ	30°	45°	60°
K_{M_A}	−0.0903	−0.1538	−0.1955
K_{M_θ}	0.0398	0.1125	0.2068
K_{N_A}	−0.9712	−0.9092	−0.8045
$K_{\Delta D_H}$	−0.0157	−0.0461	−0.0891
$K_{\Delta D_V}$	0.0207	0.0537	0.0930
$K_{\Delta L}$	0.0060	0.0179	0.0355
$K_{\Delta L_W}$	0.0119	0.0247	0.0391
$K_{\Delta L_{WH}}$	−0.0060	−0.0302	−0.0770
$K_{\Delta\psi}$	0.0244	0.0496	0.0590

3.

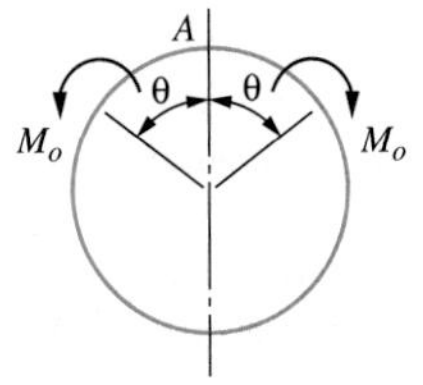

$LT_M = M_o\langle x-\theta\rangle^0$

$LT_N = 0$

$LT_V = 0$

$$M_A = \frac{-M_o}{\pi}\left(\pi-\theta-\frac{2sk_2}{k_1}\right)$$

$$M_C = \frac{M_o}{\pi}\left(\theta-\frac{2sk_2}{k_1}\right)$$

$$N_A = \frac{M_o}{R\pi}\left(\frac{2sk_2}{k_1}\right)$$

$$V_A = 0$$

$$\Delta D_H = \begin{cases} \dfrac{M_o R^2}{EI}k_2\left(\dfrac{2\theta}{\pi}-s\right) & \text{if } \theta \le \dfrac{\pi}{2} \\ \dfrac{M_o R^2}{EI}k_2\left(\dfrac{2\theta}{\pi}-2+s\right) & \text{if } \theta \ge \dfrac{\pi}{2} \end{cases}$$

$$\Delta D_V = \frac{M_o R^2}{EI}k_2\left(\frac{2\theta}{\pi}-1+c\right)$$

$$\Delta L = \begin{cases} \dfrac{-M_o R^2}{EI}\left[\dfrac{\theta}{2}-\dfrac{k_2(\theta+s)}{\pi}\right] & \text{if } \theta \le \dfrac{\pi}{2} \\ \dfrac{-M_o R^2}{EI}\left[\dfrac{\pi-\theta}{2}-\dfrac{k_2(\theta+s-\pi c)}{\pi}\right] & \text{if } \theta \ge \dfrac{\pi}{2} \end{cases}$$

$$\Delta L_W = \frac{-M_o R^2}{EI\pi}[(\pi-\theta)\theta s-k_2(s^3+\theta+\theta c)]$$

$$\Delta L_{WH} = \frac{M_o R^2}{EI\pi}[2\theta c(\pi-\theta)+2k_2 s(2\theta-\pi-sc)]$$

$$\Delta\psi = \frac{M_o R}{EI\pi}\left[\theta(\pi-\theta)-\frac{2s^2k_2^2}{k_1}\right]$$

$$\text{Max}+M = \frac{M_o}{\pi}\left(\theta+\frac{2sck_2}{k_1}\right) \quad \text{at } x \text{ just greater than } \theta$$

$$\text{Max}-M = \frac{-M_o}{\pi}\left(\pi-\theta-\frac{2sck_2}{k_1}\right) \quad \text{at } x \text{ just less than } \theta$$

If $\alpha=\beta=0$, $M=K_M M_o$, $N=K_N M_o/R$,

$\Delta D = K_{\Delta D}M_o R^2/EI$, $\Delta\psi = K_{\Delta\psi}M_o R/EI$, etc.

θ	30°	45°	60°	90°
K_{M_A}	−0.5150	−0.2998	−0.1153	0.1366
K_{N_A}	0.3183	0.4502	0.5513	0.6366
K_{M_θ}	−0.5577	−0.4317	−0.3910	−0.5000
$K_{\Delta D_H}$	−0.1667	−0.2071	−0.1994	0.0000
$K_{\Delta D_V}$	0.1994	0.2071	0.1667	0.0000
$K_{\Delta L}$	0.0640	0.0824	0.0854	0.0329
$K_{\Delta L_W}$	0.1326	0.1228	0.1022	0.0329
$K_{\Delta L_{WH}}$	−0.0488	−0.0992	−0.1180	0.0000
$K_{\Delta\psi}$	0.2772	0.2707	0.2207	0.1488

(*Continued*)

TABLE 9.2 Formulas for Circular Rings (*Continued*)

Reference No., Loading, and Load Terms	Formulas for Moments, Loads, and Deformations and Some Selected Numerical Values

4.

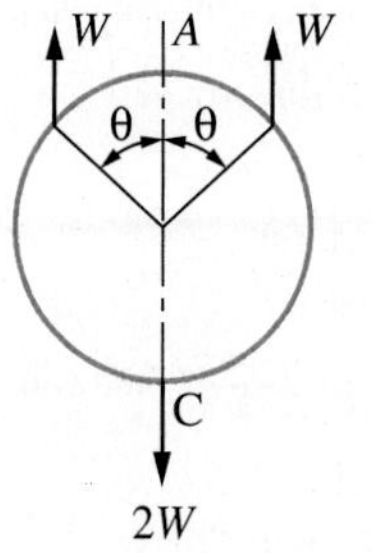

$LT_M = WR(z-s)\langle x-\theta\rangle^0$

$LT_N = Wz\langle x-\theta\rangle^0$

$LT_V = Wu\langle x-\theta\rangle^0$

$$M_A = \frac{-WR}{\pi}[s(s-\pi+\theta)+k_2(1+c)]$$

$$M_C = \frac{-WR}{\pi}\left[s\theta - s^2 + k_2(1+c)\right]$$

$$N_A = \frac{-W}{\pi}s^2$$

$$V_A = 0$$

$$\Delta D_H = \begin{cases} \dfrac{-WR^3}{EI\pi}\left[\pi k_1\left(1-\dfrac{s^2}{2}\right) - 2k_2(\pi-\theta s) + 2k_2^2(1+c)\right] & \text{if } \theta \le \dfrac{\pi}{2} \\ \dfrac{-WR^3}{EI\pi}\left[\dfrac{\pi k_1 s^2}{2} - 2sk_2(\pi-\theta) + 2k_2^2(1+c)\right] & \text{if } \theta \ge \dfrac{\pi}{2} \end{cases}$$

$$\Delta D_V = \frac{WR^3}{EI\pi}\left[\frac{\pi k_1(\pi-\theta-sc)}{2} + k_2 s(\pi-2\theta) - 2k_2^2(1+c)\right]$$

$$\Delta L = \begin{cases} \dfrac{WR^3}{2EI}\left[\theta s + k_1\left(\dfrac{\pi}{2}-\dfrac{s^2}{\pi}\right) - k_2\left(1-c+\dfrac{2\theta s}{\pi}\right) - \dfrac{2k_2^2(1+c)}{\pi}\right] & \text{if } \theta \le \dfrac{\pi}{2} \\ \dfrac{WR^3}{2EI}\left[s(\pi-\theta) + k_1\left(\pi-\theta-sc-\dfrac{s^2}{\pi}\right) - k_2\left(1+c+\dfrac{2\theta s}{\pi}\right) - \dfrac{2k_2^2(1+c)}{\pi}\right] & \text{if } \theta \ge \dfrac{\pi}{2} \end{cases}$$

$$\Delta L_W = \frac{WR^3}{EI\pi}\left[\theta s^2(\pi-\theta) + \frac{\pi k_1(\pi-\theta-sc-s^4/\pi)}{2} + k_2 s(\pi c - 2\theta - 2\theta c) - k_2^2(1+c)^2\right]$$

$$\Delta L_{WH} = \frac{-WR^3}{EI\pi}\left[2\theta sc(\pi-\theta) + k_1 s^2(\theta - sc) - 2k_2\pi s^2 - \theta s^2 - \theta c + \theta c^2) + 2k_2^2 s(1+c)\right]$$

$$\Delta\psi = \frac{WR^2}{EI\pi}\left[-\theta s(\pi-\theta) + k_2(\theta+\theta c+s^3)\right]$$

Max $+M$ occurs at an angular position

$x_1 = \tan^{-1}\dfrac{-\pi}{s^2}$ if $\theta < 106.3°$

Max $+M$ occurs at the load if $\theta \ge 106.3°$

Max $-M = M_c$

If $\alpha = \beta = 0$, $M = K_M WR$, $N = K_M W$,

$\Delta D = K_{\Delta D} WR^3/EI$, $\Delta\psi = K_{\Delta\psi} WR^2/EI$, etc.

θ	30°	60°	90°	120°	150°
K_{M_A}	−0.2569	−0.1389	−0.1366	−0.1092	−0.0389
K_{N_A}	−0.0796	−0.2387	−0.3183	−0.2387	−0.0796
K_{M_C}	−0.5977	−0.5274	−0.5000	−0.4978	−0.3797
K_{M_θ}	−0.2462	−0.0195	0.1817	0.2489	0.1096
$K_{\Delta D_H}$	−0.2296	−0.1573	−0.1366	−0.1160	−0.0436
$K_{\Delta D_V}$	0.2379	0.1644	0.1488	0.1331	0.0597
$K_{\Delta L}$	0.1322	0.1033	0.0933	0.0877	0.0431
$K_{\Delta L_W}$	0.2053	0.1156	0.0933	0.0842	0.0271
$K_{\Delta L_{WH}}$	−0.0237	−0.0782	−0.1366	−0.1078	−0.0176
$K_{\Delta\psi}$	0.1326	0.1022	0.0329	−0.0645	−0.0667

(*Continued*)

TABLE 9.2 Formulas for Circular Rings (*Continued*)

Reference No., Loading, and Load Terms	Formulas for Moments, Loads, and Deformations and Some Selected Numerical Values

5.

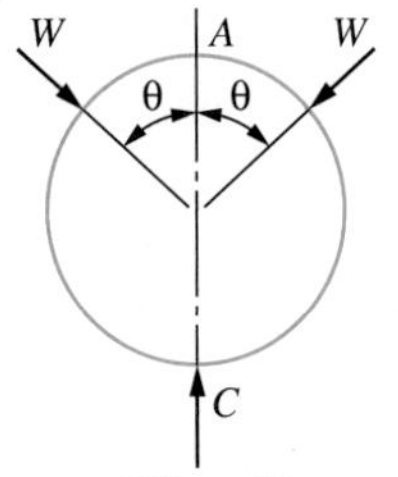

$2W\cos\theta$

$LT_M = -WR\sin(x-\theta)\langle x-\theta\rangle^0$

$LT_N = -W\sin(x-\theta)\langle x-\theta\rangle^0$

$LT_V = -W\cos(x-\theta)\langle x-\theta\rangle^0$

$$M_A = \frac{-WR}{\pi}[s(\pi-\theta) - k_2(1+c)]$$

$$M_C = \frac{-WR}{\pi}[s\theta - k_2(1+c)]$$

$$N_A = \frac{-W}{\pi}s(\pi-\theta)$$

$$V_A = 0$$

$$\Delta D_H = \begin{cases} \dfrac{-WR^3}{EI}\left[k_1\left(\dfrac{\theta s}{2} - c\right) + 2k_2 c - \dfrac{2k_2^2(1+c)}{\pi}\right] & \text{if } \theta \le \dfrac{\pi}{2} \\ \dfrac{-WR^3}{EI}\left[\dfrac{k_1 s(\pi-\theta)}{2} - \dfrac{2k_2^2(1+c)}{\pi}\right] & \text{if } \theta \ge \dfrac{\pi}{2} \end{cases}$$

$$\Delta D_V = \frac{WR^3}{EI}\left[\frac{k_1(s-\pi c+\theta c)}{2} - k_2 s + \frac{2k_2^2(1+c)}{\pi}\right]$$

$$\Delta L = \begin{cases} \dfrac{WR^3}{2EI}\left[k_1\left(\dfrac{\theta s}{\pi} - \dfrac{\pi c}{2}\right) - k_2(1-c) + \dfrac{2k_2^2(1+c)}{\pi}\right] & \text{if } \theta \le \dfrac{\pi}{2} \\ \dfrac{WR^3}{2EI}\left[k_1\left(\dfrac{\theta s}{\pi} - \pi c + \theta c\right) + k_2(1+c-2s) + \dfrac{2k_2^2(1+c)}{\pi}\right] & \text{if } \theta \ge \dfrac{\pi}{2} \end{cases}$$

$$\Delta L_W = \frac{WR^3}{EI}\left\{k_1\frac{s - s^3(1-\theta/\pi) - c(\pi-\theta)}{2} + k_2\left[\frac{\theta s(1+c)}{\pi} - s\right] + \frac{k_2^2(1+c)^2}{\pi}\right\}$$

$$\Delta L_{WH} = \frac{-WR^3}{EI\pi}[k_1 s(\pi-\theta)(\theta - sc) + 2\theta c k_2(1+c) - 2sk_2^2(1+c)]$$

$$\Delta\psi = \frac{WR^2}{EI\pi}[\pi s^2 - \theta(1+c+s^2)]k_2$$

Max $+M = M_C$ if $\theta \le 60°$

Max $+M$ occurs at the load if $\theta > 60°$ where

$$M_\theta = \frac{WR}{\pi}[k_2(1+c) - sc(\pi-\theta)]$$

$$\text{Max} - M = \begin{cases} M_C & \text{if } \theta \ge 90° \\ M_A & \text{if } 60° \le \theta \le 90° \end{cases}$$

Max $-M$ occurs at an angular position $x_1 = \tan^{-1}\dfrac{-\pi c}{\theta s}$ if $\theta \le 60°$

If $\alpha = \beta = 0$, $M = K_M WR$, $N = K_N W$,

$\Delta D = K_{\Delta D} WR^3/EI$, $\Delta\psi = K_{\Delta\psi} WR^2/EI$, etc.

θ	30°	60°	90°	120°	150°
K_{M_A}	0.1773	−0.0999	−0.1817	−0.1295	−0.0407
K_{N_A}	−0.4167	−0.5774	−0.5000	−0.2887	−0.0833
K_{M_C}	0.5106	0.1888	−0.1817	−0.4182	−0.3740
K_{M_θ}	0.2331	0.1888	0.3183	0.3035	0.1148
$K_{\Delta D_H}$	0.1910	0.0015	−0.1488	−0.1351	−0.0456
$K_{\Delta D_V}$	−0.1957	−0.0017	0.1366	0.1471	0.0620
$K_{\Delta L}$	−0.1115	−0.0209	0.0683	0.0936	0.0447
$K_{\Delta L_W}$	−0.1718	−0.0239	0.0683	0.0888	0.0278
$K_{\Delta L_{WH}}$	0.0176	−0.0276	−0.1488	−0.1206	−0.0182
$K_{\Delta\psi}$	−0.1027	0.0000	0.0000	0.0833	−0.0700

6.

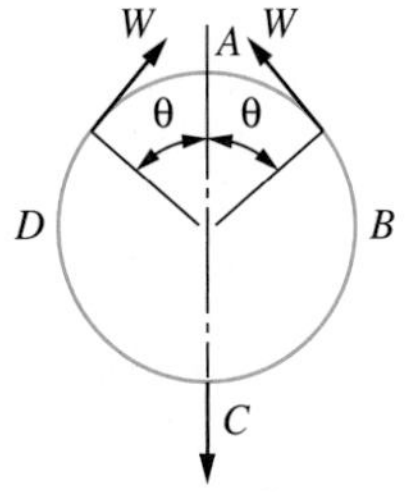

$2W\sin\theta$

$LT_M = -WR[1-\cos(x-\theta)]\langle x-\theta\rangle^0$

$LT_N = W\cos(x-\theta)\langle x-\theta\rangle^0$

$LT_V = -W\sin(x-\theta)\langle x-\theta\rangle^0$

$$M_A = \frac{-WR}{\pi}[s(1+k_2) - (\pi-\theta)(1-c)]$$

$$M_C = \frac{-WR}{\pi}[s(k_2-1) + \theta(1+c)]$$

$$N_A = \frac{-W}{\pi}[s + (\pi-\theta)c]$$

$$V_A = 0$$

$$\Delta D_H = \begin{cases} \dfrac{-WR^3}{EI}\left[\dfrac{k_1(s+\theta c)}{2} - 2k_2\left(s - \dfrac{\theta}{\pi}\right) + \dfrac{2k_2^2 s}{\pi}\right] & \text{if } \theta \le \dfrac{\pi}{2} \\ \dfrac{-WR^3}{EI}\left[\dfrac{k_1(s+\pi c-\theta c)}{2} - 2k_2\left(1 - \dfrac{\theta}{\pi}\right) + \dfrac{2k_2^2 s}{\pi}\right] & \text{if } \theta \ge \dfrac{\pi}{2} \end{cases}$$

$$\Delta D_V = \frac{WR^3}{EI}\left[\frac{k_1 s(\pi-\theta)}{2} + k_2\left(1 - c - \frac{2\theta}{\pi}\right) - \frac{2k_2^2 s}{\pi}\right]$$

$$\Delta L = \begin{cases} \dfrac{WR^3}{EI}\left[\dfrac{\theta}{2} + \dfrac{k_1(\pi^2 s + 2\theta c - 2s)}{4\pi} - k_2\left(\dfrac{s}{2} + \dfrac{\theta}{\pi}\right) - \dfrac{k_2^2 s}{\pi}\right] & \text{if } \theta \le \dfrac{\pi}{2} \\ \dfrac{WR^3}{EI}\left[\dfrac{\pi}{2} - \dfrac{\theta}{2} + \dfrac{k_1(\pi s - \theta s + \theta c/\pi - s/\pi - c)}{2} - k_2\left(\dfrac{\theta}{\pi} + \dfrac{s}{2} + c\right) - \dfrac{k_2^2 s}{\pi}\right] & \text{if } \theta \ge \dfrac{\pi}{2} \end{cases}$$

$$\Delta L_W = \frac{WR^3}{EI\pi}\left[\theta s(\pi-\theta) + \frac{k_1 s(\theta sc - s^2 - sc\pi + \pi^2 - \theta\pi)}{2} - k_2\theta(1+s^2+c) - k_2^2 s(1+c)\right]$$

$$\Delta L_{WH} = \frac{-WR^3}{EI\pi}\left[2\theta c(\pi-\theta) - k_1(sc^2\pi - 2\theta sc^2 + s^2 c - \theta c\pi + \theta^2 c - \theta s^3) - 2k_2 s(\pi - \theta + \theta c) + 2k_2^2 s^2\right]$$

$$\Delta\psi = \frac{-WR^2}{EI\pi}[\theta(\pi-\theta) - k_2 s(\theta + s + \pi c - \theta c)]$$

$$\text{Max} + M = \frac{WR}{\pi}[\pi s\sin x_1 - (s - \theta c)\cos x_1 - k_2 s - \theta]$$

at an angular position $x_1 = \tan^{-1}\dfrac{-\pi s}{s - \theta c}$

(*Note:* $x_1 > \theta$ and $x_1 > \pi/2$)

Max $-M = M_C$

If $\alpha = \beta = 0$, $M = K_M WR$, $N = K_N W$,

$\Delta D = K_{\Delta D} WR^3/EI$, $\Delta\psi = K_{\Delta\psi} WR^2/EI$, etc.

θ	30°	60°	90°	120°	150°
K_{M_A}	−0.2067	−0.2180	−0.1366	−0.0513	−0.0073
K_{N_A}	−0.8808	−0.6090	−0.3183	−0.1090	−0.0148
K_{M_C}	−0.3110	−0.5000	−0.5000	−0.3333	−0.1117
$K_{\Delta D_H}$	−0.1284	−0.1808	−0.1366	−0.0559	−0.0083
$K_{\Delta D_V}$	0.1368	0.1889	0.1488	0.0688	0.0120
$K_{\Delta L}$	0.0713	0.1073	0.0933	0.0472	0.0088
$K_{\Delta L_W}$	0.1129	0.1196	0.0933	0.0460	0.0059
$K_{\Delta L_{WH}}$	−0.0170	−0.1063	−0.1366	−0.0548	−0.0036
$K_{\Delta\psi}$	0.0874	0.1180	0.0329	−0.0264	−0.0123

(*Continued*)

TABLE 9.2 Formulas for Circular Rings (*Continued*)

Reference No., Loading, and Load Terms	Formulas for Moments, Loads, and Deformations and Some Selected Numerical Values

7. Ring under any number of equal radial forces equally spaced

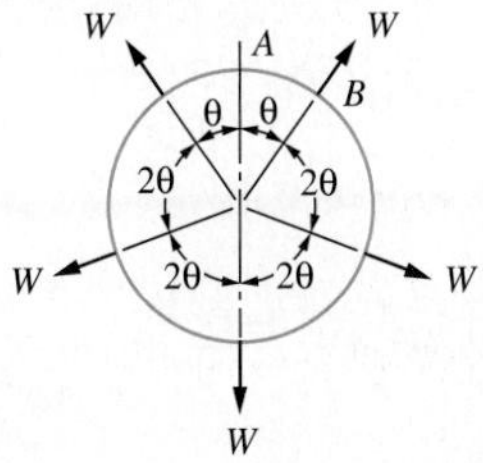

For $0<x<\theta$ $\quad M=\dfrac{WR(u/s-k_2/\theta)}{2} \quad N=\dfrac{Wu}{2s} \quad V=\dfrac{-Wz}{2s}$

$\text{Max}+M=M_A=\dfrac{WR(1/s-k_2/\theta)}{2} \qquad \text{Max}-M=\dfrac{-WR}{2}\left(\dfrac{k_2}{\theta}-\dfrac{c}{s}\right)$ at each load position

Radial displacement at each load point $=\Delta R_B=\dfrac{WR^3}{EI}\left[\dfrac{k_1(\theta-sc)}{4s^2}+\dfrac{k_2c}{2s}-\dfrac{k_2^2}{2\theta}\right]$

Radial displacement at $x=0, 2\theta, \ldots=\Delta R_A=\dfrac{-WR^3}{EI}\left[\dfrac{k_1(s-\theta c)}{4s^2}-\dfrac{k_2}{2s}+\dfrac{k_2^2}{2\theta}\right]$

If $\alpha=\beta=0$, $M=K_MWR$, $\Delta R=K_{\Delta R}WR^3/EI$

θ	15°	30°	45°	60°	90°
K_{M_A}	0.02199	0.04507	0.07049	0.09989	0.18169
K_{M_B}	−0.04383	−0.08890	−0.13662	−0.18879	−0.31831
$K_{\Delta R_B}$	0.00020	0.00168	0.00608	0.01594	0.07439
$K_{\Delta R_A}$	−0.00018	−0.00148	−0.00539	−0.01426	−0.06831

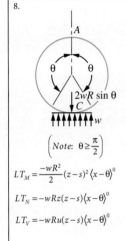

8.

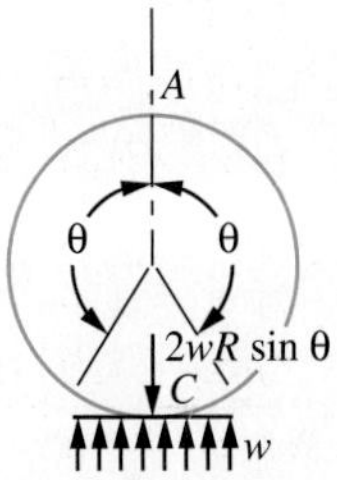

$\left(\textit{Note: } \theta \ge \dfrac{\pi}{2}\right)$

$LT_M=\dfrac{-wR^2}{2}(z-s)^2\langle x-\theta\rangle^0$

$LT_N=-wRz(z-s)\langle x-\theta\rangle^0$

$LT_V=-wRu(z-s)\langle x-\theta\rangle^0$

$$M_A=\frac{wR^2}{2\pi}\left[\pi(s^2-0.5)-\frac{sc-\theta}{2}-s^2\left(\theta+\frac{2s}{3}\right)-k_2(2s+sc-\pi+\theta)\right]$$

$$M_C=\frac{-wR^2}{2\pi}\left[\frac{\pi}{2}+\frac{sc}{2}-\frac{\theta}{2}+\theta s^2-\frac{2s^3}{3}+k_2(2s+sc-\pi+\theta)\right]$$

$$N_A=\frac{-wRs^3}{3\pi}$$

$$V_A=0$$

$$\Delta D_H=\frac{-wR^4}{2EI\pi}\left[\frac{k_1\pi s^3}{3}+k_2(\pi-2\pi s^2-\theta+2\theta s^2+sc)+2k_2^2(2s+sc-\pi+\theta)\right]$$

$$\Delta D_V=\frac{wR^4}{2EI\pi}\left[k_1\pi\left(\pi s-\theta s-\frac{2}{3}-c+\frac{c^3}{3}\right)-k_2(\pi c^2+sc-\theta+2\theta s^2)\right.$$
$$\left.-2k_2^2(2s+sc-\pi+\theta)\right]$$

$$\Delta L=\frac{wR^4}{4EI\pi}\left[\pi(\pi-\theta)\frac{2s^2-1}{2}-\frac{\pi sc}{2}-2\pi k_1\left(\frac{2}{3}-\pi s+c+\theta s-\frac{c^3}{3}+\frac{s^3}{3\pi}\right)\right.$$
$$\left.-k_2(sc+\pi-\theta+2\theta s^2+2s\pi-\pi^2+\pi\theta+\pi sc)-2k_2^2(2s+sc-\pi+\theta)\right]$$

$Max+M$ occurs at an angular position x_1 where $x_1>\theta, x_1>123.1°$, and

$$\tan x_1+\frac{3\pi(s-\sin x_1)}{s^3}=0$$

$\text{Max}-M=M_C$

If $\alpha=\beta=0$, $M=K_MwR^2$, $N=K_NwR$, $\Delta D=K_{\Delta D}wR^4/EI$, etc.

θ	90°	120°	135°	150°
K_{M_A}	−0.0494	−0.0329	−0.0182	−0.0065
K_{N_A}	−0.1061	−0.0689	−0.0375	−0.0133
K_{M_C}	−0.3372	−0.2700	−0.1932	−0.1050
$K_{\Delta D_H}$	−0.0533	−0.0362	−0.0204	−0.0074
$K_{\Delta D_V}$	0.0655	0.0464	0.0276	0.0108
$K_{\Delta L}$	0.0448	0.0325	0.0198	0.0080

(*Continued*)

TABLE 9.2 Formulas for Circular Rings (*Continued*)

Reference No., Loading, and Load Terms	Formulas for Moments, Loads, and Deformations and Some Selected Numerical Values

9.

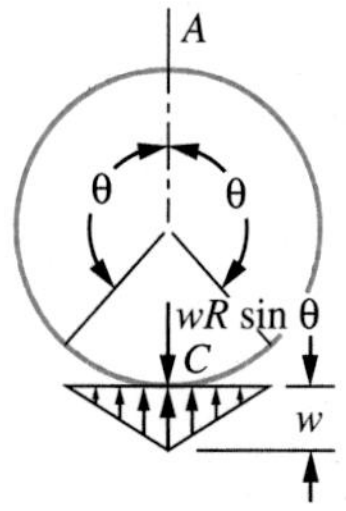

$\left(Note:\ \theta \ge \frac{\pi}{2}\right)$

$LT_M = \frac{wR^2}{6s}(z-s)^3\langle x-\theta\rangle^0$

$LT_N = \frac{wRz}{s^2}(z-s)^2\langle x-\theta\rangle^0$

$LT_V = \frac{wRu}{2s}(z-s)^2\langle x-\theta\rangle^0$

$$M_A = \frac{wR^2}{36\pi s}\{(\pi-\theta)(6s^3-9s)-3s^4+8+8c-5s^2c-6k_2[3s(s-\pi+\theta)+s^2c+2+2c]\}$$

$$M_C = \frac{-wR^2}{36\pi s}\{9s(\pi-\theta)+6\theta s^3-3s^4-8-8c+5s^2c+6k_2[3s(s-\pi+\theta)+s^2c+2+2c]\}$$

$$N_A = \frac{-wRs^3}{12\pi}$$

$$V_A = 0$$

$$\Delta D_H = \frac{-wR^4}{18EI\pi}\left\{\frac{3k_1\pi s^3}{4}-k_2\left[(\pi-\theta)(6s^2-9)+\frac{8(1+c)}{s}-5sc\right]+6k_2^2\left[sc+\frac{2(1+c)}{s}-3(\pi-\theta-s)\right]\right\}$$

$$\Delta D_V = \frac{wR^4}{18EI\pi}\left\{18\pi k_1\left[\left(\frac{s}{4}+\frac{1}{16s}\right)(\pi-\theta)-\frac{13c}{48}-\frac{s^2c}{24}-\frac{1}{3}\right]+k_2\left[(\pi-\theta)(3s^2-9)-3s^2\theta+\frac{8(1+c)}{s}-5sc\right]-6k_2^2\left[sc+\frac{2(1+c)}{s}-3(\pi-\theta-s)\right]\right\}$$

$$\Delta L = \frac{wR^4}{EI}\left[(\pi-\theta)\left(\frac{s^2}{12}-\frac{1}{8}\right)+\frac{1+c}{9s}-\frac{5sc}{72}+k_1\frac{(\pi-\theta)(12s+3/s)-13c-2s^2c-16-2s^3/\pi}{48}\right.$$
$$\left.-k_2\frac{(1+c)(2\pi-\frac{8}{3})/s-3(\pi-\theta)(\pi-1)+2\theta s^2+3\pi s+sc\left(\pi+\frac{5}{3}\right)}{12\pi}-k_2^2\frac{3(s-\pi+\theta)+2(1+c)s+sc}{6\pi}\right]$$

Max $+M$ occurs at an angular position x_1 where $x_1 > \theta, x_1 > 131.1°$, and $\tan x_1 + \frac{6\pi(s-\sin x_1)^2}{s^4} = 0$

Max $-M = M_C$

If $\alpha = \beta = 0$, $M = K_M wR^2$, $N = K_N wR$, $\Delta D = K_{\Delta D} wR^4/EI$, etc.

θ	90°	120°	135°	150°
K_{M_A}	−0.0127	−0.0084	−0.0046	−0.0016
K_{N_A}	−0.0265	−0.0172	−0.0094	−0.0033
K_{M_C}	−0.1263	−0.0989	−0.0692	−0.0367
$K_{\Delta D_H}$	−0.0141	−0.0093	−0.0052	−0.0019
$K_{\Delta D_V}$	0.0185	0.0127	0.0074	0.0028
$K_{\Delta L}$	0.0131	0.0092	0.0054	0.0021

10.

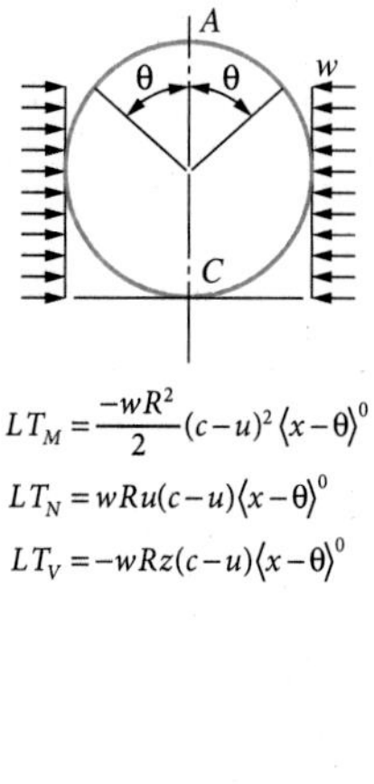

$LT_M = \frac{-wR^2}{2}(c-u)^2\langle x-\theta\rangle^0$

$LT_N = wRu(c-u)\langle x-\theta\rangle^0$

$LT_V = -wRz(c-u)\langle x-\theta\rangle^0$

$$M_A = \frac{-wR^2}{4\pi}\left[(\pi-\theta)(4c+2s^2-1)+s\left(4-\frac{4s^2}{3}-c\right)-2k_2(\pi-\theta+sc)\right]$$

$$M_C = \frac{-wR^2}{4\pi}\left[3\pi+\theta+4\theta c-2\theta s^2-4s-sc+\frac{4s^3}{3}-2k_2(\pi-\theta+sc)\right]$$

$$N_A = \frac{-wR}{\pi}\left(\pi c+s-\theta c-\frac{s^3}{3}\right)$$

$$V_A = 0$$

$$\Delta D_H = \begin{cases} \frac{-wR^4}{6EI\pi}[\pi k_1(s^3+3\theta c+4-3s)+3k_2(\pi-\theta+2\theta c^2-sc)-6k_2^2(\pi-\theta+sc)] & \text{for } \theta \le \frac{\pi}{2} \\ \frac{-wR^4}{2EI\pi}\left\{\pi k_1\left[c(\pi-\theta)+s-\frac{s^3}{3}\right]+k_2[(\pi-\theta)(2s^2-1)-sc]-2k_2^2(\pi-\theta+sc)\right\} & \text{for } \theta \ge \frac{\pi}{2} \end{cases}$$

$$\Delta D_V = \frac{wR^4}{3EI\pi}\{\pi k_1(2-c^3+3c)+3k_2[2\theta s^2-\theta+sc-\pi(1+2c+s^2)]+6k_2^2(\pi-\theta+sc)\}$$

$$\Delta L = \begin{cases} \frac{wR^4}{12EI\pi}[1.5\pi(\theta-2\theta s^2-sc)+2k_1(2\pi+s^3+3\theta c-3s)+3k_2(sc+\theta\pi+2\theta s^2-3\pi-\theta-\pi sc)+6k_2^2(\pi-\theta+sc)] & \text{for } \theta \le \frac{\pi}{2} \\ \frac{wR^4}{12EI\pi}\{1.5\pi[(\pi-\theta)(1-2s^2)+sc]+2k_1(2\pi+s^3+3\theta c-3s-\pi c^3)+3k_2[(\pi+1)(\pi-\theta+sc)+2\theta s^2-4\pi(1+c)]+6k_2^2(\pi-\theta+sc)\} & \text{for } \theta \ge \frac{\pi}{2} \end{cases}$$

Max $+M$ occurs at an angular position x_1 where $x_1 > \theta, x_1 > 90°$, and $x_1 = \cos^{-1}\frac{s^3/3+\theta c-s}{\pi}$

Max $-M = M_C$

If $\alpha = \beta = 0, M = K_M wR^2, N = K_N wR, \Delta D = K_{\Delta D} wR^4/EI$, etc.

θ	0°	30°	45°	60°	90°	120°	135°	150°
K_{M_A}	−0.2500	−0.2434	−0.2235	−0.1867	−0.0872	−0.0185	−0.0052	−0.00076
K_{NA}	−1.0000	−0.8676	−0.7179	−0.5401	−0.2122	−0.0401	−0.0108	−0.00155
K_{M_C}	−0.2500	−0.2492	−0.2448	−0.2315	−0.1628	−0.0633	−0.0265	−0.00663
$K_{\Delta D_H}$	−0.1667	−0.1658	−0.1610	−0.1470	−0.0833	−0.0197	−0.0057	−0.00086
$K_{\Delta D_V}$	0.1667	0.1655	0.1596	0.1443	0.0833	0.0224	0.0071	0.00118
$K_{\Delta L}$	0.0833	0.0830	0.0812	0.0756	0.0486	0.0147	0.0049	0.00086

(*Continued*)

TABLE 9.2 Formulas for Circular Rings (*Continued*)

<table>
<tr><th>Reference No., Loading, and Load Terms</th><th>Formulas for Moments, Loads, and Deformations and Some Selected Numerical Values</th></tr>
<tr><td>

11.

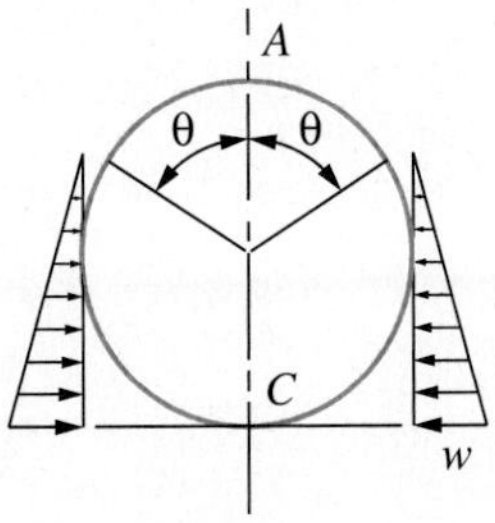

$LT_M = \frac{-wR^2}{6(1+c)}(c-u)^3\langle x-\theta\rangle^0$

$LT_N = \frac{wRu}{2(1+c)}(c-u)^2\langle x-\theta\rangle^0$

$LT_V = \frac{-wRz}{2(1+c)}(c-u)^2\langle x-\theta\rangle^0$

</td><td>

$$M_A = \frac{-wR^2}{(1+c)\pi}\left[(\pi-\theta)\frac{3+12c^2+2c+4cs^2}{24} - \frac{3s^3c-3s-5s^3}{36} + \frac{5sc}{8} - k_2\left(\frac{\pi c}{2} - \frac{\theta c}{2} + \frac{s^3}{3} + \frac{sc^2}{2}\right)\right]$$

$$M_C = \frac{-wR^2}{(1+c)\pi}\left[(\pi-\theta)\frac{-3-12c^2+2c+4cs^2}{24} + \frac{\pi(1+c)^3}{6} + \frac{3s^3c+3s+5s^3}{36} - \frac{5sc}{8} - k_2\left(\frac{\pi c}{2} - \frac{\theta c}{2} + \frac{s^3}{3} + \frac{sc^2}{2}\right)\right]$$

$$N_A = \frac{-wR}{(1+c)\pi}\left[(\pi-\theta)\frac{1+4c^2}{8} + \frac{5sc}{8} - \frac{s^3c}{12}\right]$$

$$V_A = 0$$

$$\Delta D_H = \begin{cases} \frac{-wR^4}{EI(1+c)\pi}\left\{\pi k_1\left(\frac{\theta+4\theta c^2-5sc}{16} + \frac{s^3c+16c}{24}\right) - k_2\left(\frac{5sc^2+3\theta c+6\theta s^2c-8s}{18} - \frac{\pi c}{2}\right) - k_2^2\left[c(\pi-\theta) + \frac{2s^3}{3} + sc^2\right]\right\} & \text{for } \theta \le \frac{\pi}{2} \\ \frac{-wR^4}{EI(1+c)\pi}\left\{\pi k_1\left[(\pi-\theta)\frac{1+4c^2}{16} + \frac{5sc}{16} - \frac{s^3c}{24}\right] - k_2\left[\frac{5sc^2-8s}{18} - (\pi-\theta)\frac{c+2s^2c}{6}\right] - k_2^2\left[c(\pi-\theta) + s - \frac{s^3}{3}\right]\right\} & \text{for } \theta \ge \frac{\pi}{2} \end{cases}$$

$$\Delta D_V = \frac{wR^4}{EI(1+c)}\left\{k_1\left[\frac{(1+c)^2}{6} - \frac{s^4}{24}\right] + k_2\left(\frac{5sc^2+3\theta c+6\theta s^2c-8s}{18\pi} + \frac{s^2}{2} + \frac{c^3}{6} - c - \frac{2}{3}\right) + k_2^2\frac{c(\pi-\theta)+s-s^3/3}{\pi}\right\}$$

$$\Delta L = \begin{cases} \frac{wR^4}{EI(1+c)}\left\{\frac{3s+5s^3+6\theta c^3-9\theta c-16}{72} + k_1\left(\frac{c}{3} + \frac{1}{16} + \frac{12\theta c^2+3\theta+2s^3c-15sc}{48\pi}\right)\right. \\ \qquad \left. + k_2\left[\frac{1-s^3}{6} - \frac{c(3+sc-\theta)}{4} + \frac{3\theta c+6\theta s^2c-3s-5s^3}{36\pi}\right] + k_2^2\frac{c(\pi-\theta)/2+sc^2/2+s^3/3}{\pi}\right\} & \text{for } \theta \le \frac{\pi}{2} \\ \frac{wR^4}{EI(1+c)}\left\{\frac{-(\pi-\theta)c(1+2s^2)}{24} - \frac{sc^2}{24} - \frac{s^3}{9} + k_1\left(\frac{c}{3} + \frac{1}{16} - \frac{c^4}{24} + \frac{12\theta c^2+3\theta+2s^3c-15sc}{48\pi}\right)\right. \\ \qquad \left. + k_2\left[\frac{2s-2+sc^2}{12} + \frac{c(\pi-\theta-3-2c)}{4} + \frac{3\theta c+6\theta s^2c-3s-5s^3}{36\pi}\right] + k_2^2\frac{3c(\pi-\theta)+2s+sc^2}{6\pi}\right\} & \text{for } \theta \ge \frac{\pi}{2} \end{cases}$$

Max $+M$ occurs at an angular position x_1

where $x_1 > \theta$, $x_1 > 96.8°$, and $x_1 = \arccos\left\{c - \left[(c^2+0.25)\left(1-\frac{\theta}{\pi}\right) + \frac{sc(5-2s^2/3)}{4\pi}\right]^{1/2}\right\}$

Max $-M = M_C$

If $\alpha = \beta = 0$, $M = K_M wR^2$, $N = K_N wR$, $\Delta D = K_{\Delta D} wR^4/EI$, etc.

θ	0°	30°	45°	60°	90°	120°	135°	150°
K_{M_A}	−0.1042	−0.0939	−0.0808	−0.0635	−0.0271	−0.0055	−0.0015	−0.00022
K_{N_A}	−0.3125	−0.2679	−0.2191	−0.1628	−0.0625	−0.0116	−0.0031	−0.00045
K_{M_C}	−0.1458	−0.1384	−0.1282	−0.1129	−0.0688	−0.0239	−0.0096	−0.00232
$K_{\Delta D_H}$	−0.8333	−0.0774	−0.0693	−0.0575	−0.0274	−0.0059	−0.0017	−0.00025
$K_{\Delta D_V}$	0.0833	0.0774	0.0694	0.0579	0.0291	0.0071	0.0022	0.00035
$K_{\Delta L}$	0.0451	0.0424	0.0387	0.0332	0.0180	0.0048	0.0015	0.00026

</td></tr>
</table>

(Continued)

TABLE 9.2 Formulas for Circular Rings (*Continued*)

Reference No., Loading, and Load Terms	Formulas for Moments, Loads, and Deformations and Some Selected Numerical Values

12.

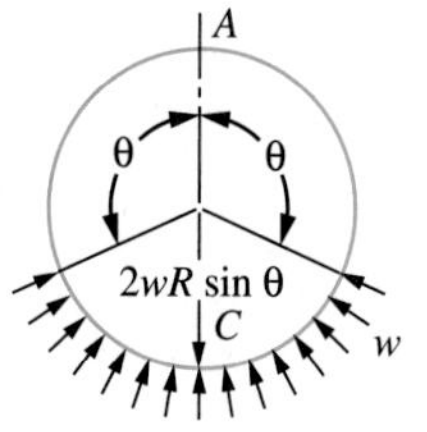

$LT_M = -wR^2[1-\cos(x-\theta)]\langle x-\theta\rangle^0$

$LT_N = -wR[1-\cos(x-\theta)]\langle x-\theta\rangle^0$

$LT_V = -wR\sin(x-\theta)\langle x-\theta\rangle^0$

$$M_A = \frac{-wR^2}{\pi}[s+\pi c-\theta c-k_2(\pi-\theta-s)]$$

$$M_C = \frac{-wR^2}{\pi}[\pi-s+\theta c-k_2(\pi-\theta-s)]$$

$$N_A = \frac{-wR}{\pi}(s+\pi c-\theta c)$$

$$V_A = 0$$

$$\Delta D_H = \begin{cases} \dfrac{-wR^4}{EI}\left[\dfrac{k_1(s+\theta c)}{2}+2k_2(1-s)-\dfrac{2k_2^2(\pi-\theta-s)}{\pi}\right] & \text{for } \theta \le \dfrac{\pi}{2} \\ \dfrac{-wR^4}{EI}\left[\dfrac{k_1(s+\pi c-\theta c)}{2}-\dfrac{2k_2^2(\pi-\theta-s)}{\pi}\right] & \text{for } \theta \ge \dfrac{\pi}{2} \end{cases}$$

$$\Delta D_V = \frac{wR^4}{EI}\left[\frac{k_1 s(\pi-\theta)}{2}-k_2(1+c)+\frac{2k_2^2(\pi-\theta-s)}{\pi}\right]$$

$$\Delta L = \begin{cases} \dfrac{wR^4}{2EI\pi}\left[k_1\left(\dfrac{\pi^2 s}{2}-s+\theta c\right)+k_2\pi(\theta-s-2)+2k_2^2(\pi-\theta-s)\right] & \text{for } \theta \le \dfrac{\pi}{2} \\ \dfrac{wR^4}{2EI\pi}[k_1(\pi^2 s-\pi\theta s-\pi c-s+\theta c)+k_2\pi(\pi-\theta-s-2-2c)+2k_2^2(\pi-\theta-s)] & \text{for } \theta \ge \dfrac{\pi}{2} \end{cases}$$

Max $+M$ occurs at an angular position x_1 where $x_1 > \theta, x_1 > 90°$, and $x_1 = \tan^{-1}\dfrac{-\pi s}{s-\theta c}$

Max $-M = M_C$

If $\alpha = \beta = 0$, $M = K_M wR^2$, $N = K_N wR$, $\Delta D = K_{\Delta D} wR^4/EI$, etc.

θ	30°	60°	90°	120°	150°
K_{M_A}	−0.2067	−0.2180	−0.1366	−0.0513	−0.0073
K_{N_A}	−0.8808	−0.6090	−0.3183	−0.1090	−0.0148
K_{M_C}	−0.3110	−0.5000	−0.5000	−0.3333	−0.1117
$K_{\Delta D_H}$	−0.1284	−0.1808	−0.1366	−0.0559	−0.0083
$K_{\Delta D_V}$	0.1368	0.1889	0.1488	0.0688	0.0120
$K_{\Delta L}$	0.0713	0.1073	0.0933	0.0472	0.0088

13.

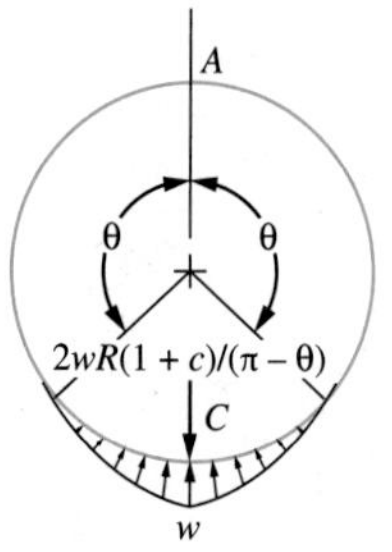

The radial pressure w_x varies linearly with x from 0 at $x = \theta$ to w at $x = \pi$.

$LT_M = \dfrac{-wR^2}{\pi-\theta}(x-\theta-zc+us)\langle x-\theta\rangle^0$

$LT_N = \dfrac{-wR}{\pi-\theta}(x-\theta-zc+us)\langle x-\theta\rangle^0$

$LT_V = \dfrac{-wR}{\pi-\theta}(1-uc-zs)\langle x-\theta\rangle^0$

$$M_A = \frac{-wR^2}{\pi(\pi-\theta)}\left\{2+2c-s(\pi-\theta)+k_2\left[1+c-\frac{(\pi-\theta)^2}{2}\right]\right\}$$

$$M_C = \frac{-wR^2}{\pi(\pi-\theta)}\left\{\pi(\pi-\theta)-2-2c-s\theta+k_2\left[1+c-\frac{(\pi-\theta)^2}{2}\right]\right\}$$

$$N_A = \frac{-wR}{\pi(\pi-\theta)}[2+2c-s(\pi-\theta)]$$

$$V_A = 0$$

$$\Delta D_H = \begin{cases} \dfrac{-wR^4}{EI(\pi-\theta)}\left\{k_1\left(1-\dfrac{s\theta}{2}\right)+k_2(\pi-2\theta-2c)+k_2^2\,\dfrac{2+2c-(\pi-\theta)^2}{\pi}\right\} & \text{for } \theta \le \dfrac{\pi}{2} \\ \dfrac{-wR^4}{EI(\pi-\theta)}\left\{k_1\left[1+c-\dfrac{s(\pi-\theta)}{2}\right]+k_2^2\,\dfrac{2+2c-(\pi-\theta)^2}{\pi}\right\} & \text{for } \theta \ge \dfrac{\pi}{2} \end{cases}$$

$$\Delta D_V = \frac{wR^4}{EI(\pi-\theta)}\left\{k_1\,\frac{s+c(\pi-\theta)}{2}-k_2(\pi-\theta-s)-k_2^2\,\frac{2+2c-(\pi-\theta)^2}{\pi}\right\}$$

$$\Delta L = \begin{cases} \dfrac{wR^4}{2EI\pi(\pi-\theta)}\left\{k_1\left(\dfrac{\pi^2 c}{2}-2c+2\pi-2-\theta s\right)-k_2\pi\left[2(\pi-\theta)-1+c-\dfrac{\pi^2}{4}+\dfrac{\theta^2}{2}\right]-k_2^2[2+2c-(\pi-\theta)^2]\right\} & \text{for } \theta \le \dfrac{\pi}{2} \\ \dfrac{wR^4}{2EI\pi(\pi-\theta)}\left\{k_1[\pi c(\pi-\theta)+2\pi s-2c-2-\theta s]-k_2\pi\left[2(\pi-\theta)+1+c-2s-\dfrac{(\pi-\theta)^2}{2}\right]-k_2^2[2+2c-(\pi-\theta)^2]\right\} & \text{for } \theta \ge \dfrac{\pi}{2} \end{cases}$$

Max $+M$ occurs at an angular position x_1 where $x_1 > \theta, x_1 > 103.7°$, and x_1 is found from

$$\left(1+c+\frac{s\theta}{2}\right)\sin x_1 + c\cos x_1 - 1 = 0$$

Max $-M = M_C$

(*Continued*)

TABLE 9.2 Formulas for Circular Rings (*Continued*)

Reference No., Loading, and Load Terms	Formulas for Moments, Loads, and Deformations and Some Selected Numerical Values
14. 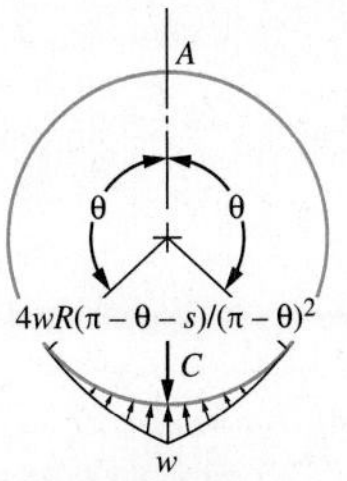 The radial pressure w_x varies with $(x=\theta)^2$ from 0 at $x=\theta$ to w at $x=\pi$ $LT_M=\frac{-wR^2}{(\pi-\theta)^2}[(x-\theta)^2-2+2uc+2zs]\langle x-\theta\rangle^0$ $LT_N=\frac{-wR}{(\pi-\theta)^2}[(x-\theta)^2-2+2uc+2zs]\langle x-\theta\rangle^0$ $LT_V=\frac{-2wR}{(\pi-\theta)^2}(x-\theta-zc+us)\times\langle x-\theta\rangle^0$	$M_A=\frac{-wR^2}{\pi(\pi-\theta)^2}\left\{2(\pi-\theta)(2-c)-6s+k_2\left[2(\pi-\theta-s)-\frac{(\pi-\theta)^3}{3}\right]\right\}$ $M_C=\frac{-wR^2}{\pi(\pi-\theta)^2}\left\{2\theta(2-c)+6s-6\pi+\pi(\pi-\theta)^2+k_2\left[2(\pi-\theta-s)-\frac{(\pi-\theta)^3}{3}\right]\right\}$ $N_A=\frac{-wR}{\pi(\pi-\theta)^2}[2(\pi-\theta)(2-c)-6s]$ $V_A=0$ $\Delta D_H=\begin{cases}\frac{-wR^4}{EI(\pi-\theta)^2}\left[k_1\left(\frac{\pi^2}{4}-6-\theta^2 s+3s-3\theta c+\frac{3\pi}{2}+c+\theta s-2\theta\right)+k_2\left(\frac{\pi^2}{2}-4+4s-2\theta\pi+2\theta^2\right)-2k_2^2\frac{(\pi-\theta)^3-6(\pi-\theta-s)}{3\pi}\right] & \text{for } \theta\le\frac{\pi}{2}\\ \frac{-wR^4}{EI(\pi-\theta)^2}\left\{k_1[(2-c)(\pi-\theta)-3s]-2k_2^2\frac{(\pi-\theta)^3-6(\pi-\theta-s)}{3\pi}\right\} & \text{for } \theta\ge\frac{\pi}{2}\end{cases}$ $\Delta D_V=\frac{wR^4}{EI(\pi-\theta)^2}\left\{k_1[2+2c-s(\pi-\theta)]+k_2[(2+2c-(\pi-\theta)^2]+2k_2^2\frac{(\pi-\theta)^3-6(\pi-\theta-s)}{3\pi}\right\}$ $\Delta L=\begin{cases}\frac{wR^4}{EI(\pi-\theta)^2}\left\{\left(k_1\frac{3s+2\theta-\theta c}{\pi}+\pi-2\theta-\frac{\pi s}{2}\right)+k_2\left[2+s-\theta+\frac{\pi^3}{8}+\frac{\theta^3}{6}-\frac{\theta\pi^2}{4}-(\pi-\theta)^2\right]+k_2^2\frac{(\pi-\theta)^3-6(\pi-\theta-s)}{3\pi}\right\} & \text{for } \theta\le\frac{\pi}{2}\\ \frac{wR^4}{EI(\pi-\theta)^2}\left\{k_1\left[\frac{3s+2\theta-\theta c}{\pi}-s(\pi-\theta)+3c\right]+k_2\left[\frac{(\pi-\theta)^3}{6}-(\pi-\theta)^2-\pi+\theta+s+2+2c\right]+k_2^2\frac{(\pi-\theta)^3-6(\pi-\theta-s)}{3\pi}\right\} & \text{for } \theta\ge\frac{\pi}{2}\end{cases}$ Max $+M$ occurs at an angular position x_1 where $x_1>\theta$, $x_1>108.6°$, and x_1 is found from $(x_1-\theta+s\cos x_1)+(3s-2\pi+2\theta-\theta c)\sin x_1=0$ Max $-M=M_C$
15. Ring supported at base and loaded by own weight per unit length of circumference w 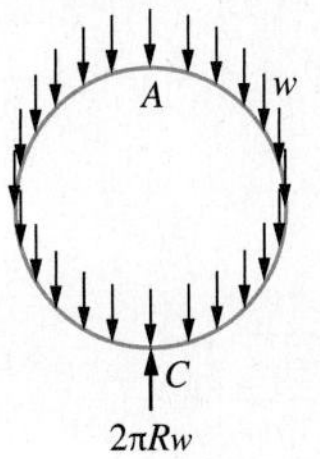 $LT_M=-wR^2[xz+K_T(u-1)]$ $LT_N=-wRxz$ $LT_V=-wRxu$	$M_A=wR^2\left[k_2-0.5-\frac{(K_T-1)\beta}{k_1}\right]$ where $K_T=1+\frac{I}{AR^2}$ $M_C=wR^2\left[k_2+0.5+\frac{(K_T-1)\beta}{k_1}\right]$ $N_A=wR\left[0.5+\frac{(K_T-1)k_2}{k_1}\right]$ $V_A=0$ $\Delta D_H=\frac{wR^3}{EAe}\left(\frac{k_1\pi}{2}-k_2\pi+2k_2^2\right)$ $\Delta D_V=\frac{-wR^3}{EAe}\left(\frac{k_1\pi^2}{4}-2k_2^2\right)$ $\Delta L=\frac{-wR^3}{EAe}\left[1+\frac{3k_1\pi^2}{16}-\frac{k_2\pi}{2}-k_2^2+(K_T-1)\alpha\right]$ *Note:* The constant K_T accounts for the radial distribution of mass in the ring. Max $+M=M_C$ Max $-M$ occurs at an angular position x_1 where $\frac{x_1}{\tan x_1}=-0.5+\frac{(K_T-1)\beta}{k_1}$ For a thin ring where $K_T\approx 1$, Max $-M=wR^2(1.6408-k_2)$ at $x=105.23°$ If $\alpha=\beta=0$, $M_A=\frac{wR^2}{2}$ $N_A=\frac{wR}{2}$ $\Delta D_H=0.4292\frac{wR^4}{EI}$ $\Delta D_V=-0.4674\frac{wR^4}{EI}$ $\Delta L=-0.2798\frac{wR^4}{EI}$ Max $+M=\frac{3}{2}wR^2$ at C
16. Unit axial segment of pipe filled with liquid of weight per unit volume ρ and supported at the base 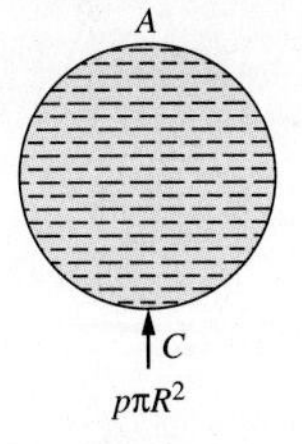 $LT_M=\rho R^3\left(1-u-\frac{xz}{2}\right)$ $LT_N=\rho R^2\left(1-u-\frac{xz}{2}\right)$ $LT_V=\rho R^2\left(\frac{z}{2}-\frac{xu}{2}\right)$	$M_A=\rho R^3\left(0.75-\frac{k_2}{2}\right)$ $M_C=\rho R^3\left(1.25-\frac{k_2}{2}\right)$ $N_A=0.75\rho R^2$ $V_A=0$ $\Delta D_H=\frac{\rho R^5 12(1-\nu^2)}{Et^3}\left[\frac{k_1\pi}{4}+k_2\left(2-\frac{\pi}{2}\right)-k_2^2\right]$ $\Delta D_V=\frac{-\rho R^5 12(1-\nu^2)}{Et^3}\left(\frac{k_1\pi^2}{8}-2k_2+k_2^2\right)$ $\Delta L=\frac{-\rho R^5 12(1-\nu^2)}{Et^3}\left[\frac{k_1 3\pi^2}{32}-k_2\left(0.5+\frac{\pi}{4}\right)+\frac{k_2^2}{2}\right]$ *Note:* For this case and case 17, $\alpha=\frac{t^2}{12R^2(1-\nu^2)}$ $\beta=\frac{t^2}{6R^2(1-\nu)}$ where $t=$ pipe wall thickness Max $+M=M_C$ Max $-M=-\rho R^3\left(\frac{k_2}{2}-0.1796\right)$ at $x=105.23°$ If $\alpha=\beta=0$, $\Delta D_H=0.2146\frac{\rho R^5 12(1-\nu^2)}{Et^3}$ $\Delta D_V=-0.2337\frac{\rho R^5 12(1-\nu^2)}{Et^3}$ $\Delta L=-0.1399\frac{\rho R^5 12(1-\nu^2)}{Et^3}$

(*Continued*)

TABLE 9.2 Formulas for Circular Rings (*Continued*)

Reference No., Loading, and Load Terms	Formulas for Moments, Loads, and Deformations and Some Selected Numerical Values

17. Unit axial segment of pipe partly filled with liquid of weight per unit volume ρ and supported at the base

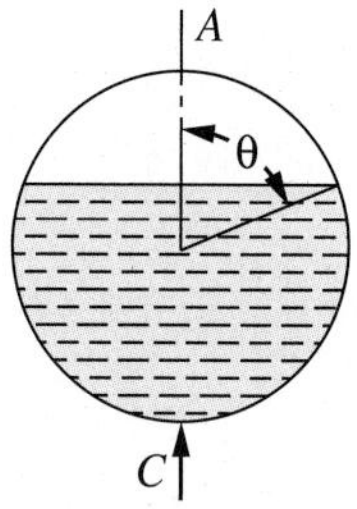

$\rho R^2(\pi - \theta + sc)$

$$LT_M = \frac{\rho R^3}{2}[2c - z(x-\theta+sc) - u(1+c^2)]\langle x-\theta\rangle^0$$

$$LT_N = \frac{\rho R^2}{2}[2c - z(x-\theta+sc) - u(1+c^2)]\langle x-\theta\rangle^0$$

$$LT_V = \frac{\rho R^2}{2}[zc^2 - u(x-\theta+sc)] \times \langle x-\theta\rangle^0$$

Note: see case 16 for expression for α and β.

$$M_A = \frac{\rho R^3}{4\pi}\{2\theta s^2 + 3sc - 3\theta + \pi + 2\pi c^2 + 2k_2[sc - 2s + (\pi-\theta)(1-2c)]\}$$

$$N_A = \frac{\rho R^2}{4\pi}[3sc + (\pi-\theta)(1+2c^2)]$$

$$V_A = 0$$

$$\Delta D_H = \begin{cases} \dfrac{\rho R^5 3(1-\nu^2)}{2Et^3\pi}\left\{k_1\pi(sc+2\pi-3\theta+2\theta c^2) + 8k_2\pi\left(2c - sc - \dfrac{\pi}{2} + \theta\right) + 8k_2^2[(\pi-\theta)(1-2c)+sc-2s]\right\} & \text{for } \theta \le \dfrac{\pi}{2} \\ \dfrac{\rho R^5 3(1-\nu^2)}{2Et^3\pi}\left\{k_1\pi[(\pi-\theta)(1+2c^2)+3sc] + 8k_2^2[(\pi-\theta)(1-2c)+sc-2s]\right\} & \text{for } \theta \ge \dfrac{\pi}{2} \end{cases}$$

$$\Delta D_V = \frac{-\rho R^5 3(1-\nu^2)}{2Et^3\pi}\left\{k_1\pi[s^2 + (\pi-\theta)(\pi-\theta+2sc)] - 4k_2\pi(1+c)^2 - 8k_2^2[(\pi-\theta)(1-2c)+sc-2s]\right\}$$

$$\Delta L = \begin{cases} \dfrac{-\rho R^5 3(1-\nu^2)}{2Et^3\pi}\left\{k_1\left[2\theta c^2 + \theta - 3sc + \pi^2\left(sc - \theta + \dfrac{3\pi}{4}\right)\right] + 2k_2\pi(2 + 2\theta c - 2s - 4c - \pi + \theta - sc) - 4k_2^2[(\pi-\theta)(1-2c)+sc-2s]\right\} \\ \qquad \text{for } \theta \le \dfrac{\pi}{2} \\ \dfrac{-\rho R^5 3(1-\nu^2)}{2Et^3\pi}\left\{k_1[2\theta c^2 + \theta - 3sc + \pi(\pi-\theta)(\pi-\theta+2sc) - 3\pi c^2] + 2k_2\pi[2s - 2(1+c)^2 - sc - (\pi-\theta) - (1-2c)] - 4k_2^2[(\pi-\theta)(1-2c)+sc-2s]\right\} \\ \qquad \text{for } \theta \ge \dfrac{\pi}{2} \end{cases}$$

$$\text{Max} + M = M_C = \frac{\rho R^3}{4\pi}\{4\pi c + \pi + 2\theta c^2 + \theta - 3sc + 2k_2[(\pi-\theta)(1-2c)+sc-2s]\}$$

Max $-M$ occurs at an angular position where $x_1 > \theta, x_1 > 105.23°$, and x_1 is found from

$$(\theta + 2\theta c^2 - 3sc - \pi)\tan x_1 + 2\pi(\theta - sc - x_1) = 0$$

If $\alpha = \beta = 0$, $M = K_M \rho R^3$, $N = K_N \rho R^2$, $\Delta D = K_{\Delta D}\rho R^5 12(1-\nu^2)/Et^3$, etc.

θ	0°	30°	45°	60°	90°	120°	135°	150°
K_{M_A}	0.2500	0.2290	0.1935	0.1466	0.0567	0.0104	0.0027	0.00039
K_{N_A}	0.7500	0.6242	0.4944	0.3534	0.1250	0.0216	0.0056	0.00079
K_{M_C}	0.7500	0.7216	0.6619	0.5649	0.3067	0.0921	0.0344	0.00778
$K_{\Delta D_H}$	0.2146	0.2027	0.1787	0.1422	0.0597	0.0115	0.0031	0.00044
$K_{\Delta D_V}$	−0.2337	−0.2209	−0.1955	−0.1573	−0.0700	−0.0150	−0.0043	−0.00066
$K_{\Delta L}$	−0.1399	−0.1333	−0.1198	−0.0986	−0.0465	−0.0106	−0.0031	−0.00050

(Continued)

TABLE 9.2 Formulas for Circular Rings (*Continued*)

Reference No., Loading, and Load Terms	Formulas for Moments, Loads, and Deformations and Some Selected Numerical Values

18.

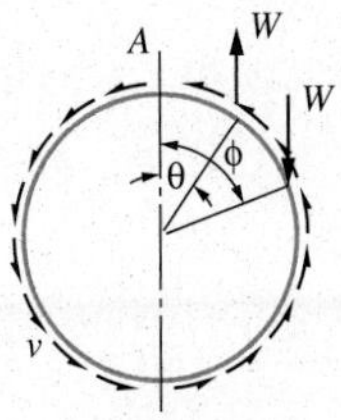

$$\upsilon = \frac{W}{2\pi R}(\sin\phi - \sin\theta)$$

$$LT_M = -\frac{WR}{2\pi}(n-s)(x-z) + WR(z-s)\langle x-\phi\rangle^0 - WR(z-n)\langle x-\phi\rangle^0$$

$$LT_N = \frac{W}{2\pi}(n-s)z + Wz\langle x-\theta\rangle^0 - Wz\langle x-\phi\rangle^0$$

$$LT_V = \frac{-W}{2\pi}(n-s)(1-u) + Wu\langle x-\phi\rangle^0 - Wu\langle x-\phi\rangle^0$$

$$M_A = \frac{WR}{2\pi}[n^2 - s^2 - (\pi-\phi)n + (\pi-\theta)s - k_2(c-m)]$$

$$N_A = \frac{W}{2\pi}(n^2 - s^2)$$

$$V_A = \frac{W}{2\pi}(\theta - \phi + s - n + sc - nm)$$

If $\alpha = \beta = 0$, $M = K_M WR$, $N = K_N W$, $V = K_V W$

θ	ϕ − θ	30°	45°	60°	90°	120°	135°	150°	180°
0°	K_{M_A}	−0.1899	−0.2322	−0.2489	−0.2500	−0.2637	−0.2805	−0.2989	−0.3183
	K_{N_A}	0.0398	0.0796	0.1194	0.1592	0.1194	0.0796	0.0398	0.0000
	K_{V_A}	−0.2318	−0.3171	−0.3734	−0.4092	−0.4022	−0.4080	−0.4273	−0.5000
30°	K_{M_A}	−0.0590	−0.0613	−0.0601	−0.0738	−0.1090	−0.1231	−0.1284	−0.1090
	K_{N_A}	0.0796	0.1087	0.1194	0.0796	−0.0000	−0.0291	−0.0398	−0.0000
	K_{V_A}	−0.1416	−0.1700	−0.1773	−0.1704	−0.1955	−0.2279	−0.2682	−0.3408
45°	K_{M_A}	−0.0190	−0.0178	−0.0209	−0.0483	−0.0808	−0.0861	−0.0808	−0.0483
	K_{N_A}	0.0689	0.0796	0.0689	0.0000	−0.0689	−0.0796	−0.0689	−0.0000
	K_{V_A}	−0.0847	−0.0920	−0.0885	−0.0908	−0.1426	−0.1829	−0.2231	−0.2749
60°	K_{M_A}	−0.0011	−0.0042	−0.0148	−0.0500	−0.0694	−0.0641	−0.0500	−0.0148
	K_{N_A}	0.0398	0.0291	−0.0000	−0.0796	−0.1194	−0.1087	−0.0796	0.0000
	K_{V_A}	−0.0357	−0.0322	−0.0288	−0.0539	−0.1266	−0.1668	−0.1993	−0.2243
90°	K_{M_A}	−0.0137	−0.0305	−0.0489	−0.0683	−0.0489	−0.0305	−0.0137	0.0000
	K_{N_A}	−0.0398	−0.0796	−0.1194	−0.1592	−0.1194	−0.0796	−0.0398	0.0000
	K_{V_A}	0.0069	0.0012	−0.0182	−0.0908	−0.1635	−0.1829	−0.1886	−0.1817

19.

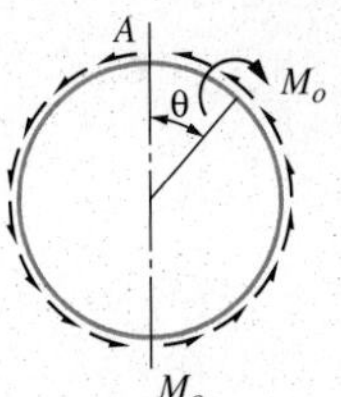

$$v = \frac{M_o}{2\pi R^2}$$

$$LT_M = \frac{-M_o}{2\pi}(x-z) + M_o\langle x-\theta\rangle^0$$

$$LT_N = \frac{M_o z}{2\pi R}$$

$$LT_V = \frac{-M_o}{2\pi R}(1-u)$$

$$M_A = \frac{-M_o}{2\pi}\left(\pi - \theta - \frac{2k_2 s}{k_1}\right)$$

$$N_A = \frac{M_o}{\pi R}\left(\frac{k_2 s}{k_1}\right)$$

$$V_A = \frac{-M_o}{2\pi R}\left(1 + \frac{2k_2 c}{k_1}\right)$$

$\text{Max} + M = \frac{M_o}{2}$ for x just greater than θ

$\text{Max} - M = \frac{-M_o}{2}$ for x just less than θ

At $x = \theta + 180°$, $M = 0$

Other maxima are, for $\alpha = \beta = 0$

$$M \begin{cases} -0.1090 M_o & \text{at } x = \theta + 120° \\ 0.1090 M_o & \text{at } x = \theta + 240° \end{cases}$$

If $\alpha = \beta = 0$, $M = k_M M_o$, $N = K_N M_o / R$, $V = K_V M_o / R$

θ	0°	30°	45°	60°	90°	120°	135°	150°	180°
K_{M_A}	−0.5000	−0.2575	−0.1499	−0.0577	0.0683	0.1090	0.1001	0.0758	0.0000
K_{N_A}	0.0000	0.1592	0.2251	0.2757	0.3183	0.2757	0.2250	0.1592	0.0000
K_{V_c}	−0.4775	−0.4348	−0.3842	−0.3183	−0.1592	0.0000	0.0659	0.1165	0.1592

(Continued)

TABLE 9.2 Formulas for Circular Rings (*Continued*)

Reference No., Loading, and Load Terms	Formulas for Moments, Loads, and Deformations and Some Selected Numerical Values
20. Bulkhead or supporting ring in pipe, supported at bottom and carrying total load W transferred by tangential shear υ distributed as shown 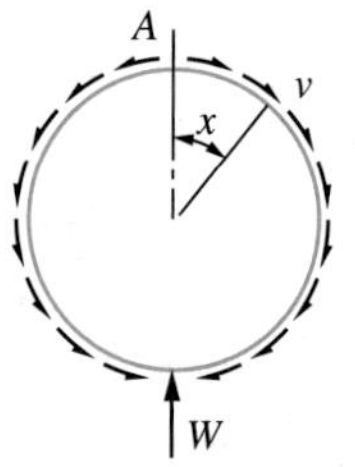 $v=\frac{W\sin x}{\pi R}$ $LT_M=\frac{WR}{\pi}\left(1-u-\frac{xz}{2}\right)$ $LT_N=\frac{-W}{2\pi}xz$ $LT_V=\frac{W}{2\pi}(z-xu)$ for $0<x<180°$	$M_A=\frac{WR}{2\pi}(k_2-0.5)$ $M_C=\frac{WR}{2\pi}(k_2+0.5)$ $N_A=\frac{0.75W}{\pi}$ $V_A=0$ $\Delta D_H=\frac{WR^3}{EI}\left(\frac{k_1}{4}-\frac{k_2}{2}+\frac{k_2^2}{\pi}\right)$ $\Delta D_V=\frac{-WR^3}{EI}\left(\frac{k_1\pi}{8}-\frac{k_2^2}{\pi}\right)$ $\Delta L=\frac{-WR^3}{4EI\pi}\left[4+k_1\frac{3\pi^2}{8}-k_2(\pi+2)-2k_2^2\right]$ $\text{Max}+M=M_C$ $\text{Max}-M=\frac{-WR}{4\pi}(3.2815-2k_2)$ at $x=105.2°$ If $\alpha=\beta=0$, $M_A=0.0796WR$ $N_A=0.2387W$ $V_A=0$ $\Delta D_H=0.0683\frac{WR^3}{EI}$ $\Delta D_V=-0.0744\frac{WR^3}{EI}$ $\Delta R=-0.0445\frac{WR^3}{EI}$
21. Ring rotating at angular rate ω rad/s about an axis perpendicular to the plane of the ring. Note the requirement of symmetry of the cross section in Sec. 9.3. 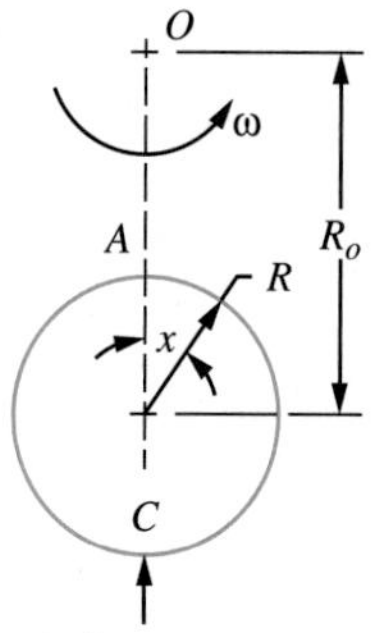 $\delta\omega^2 2\pi RR_oA$ δ = mass density of ring material $LT_M=\delta\omega^2AR^3\left\{K_T(1-u)-\frac{R_o}{R}[xz-K_T(1-u)]\right\}$ $LT_N=\delta\omega^2AR^2\left[K_T(1-u)-\frac{R_o}{R}xz\right]$ $LT_V=\delta\omega^2AR^2\left[zK_T(2u-1)-\frac{R_o}{R}xu\right]$	$M_A=\delta\omega^2AR^3\left\{K_T\alpha+\frac{R_o}{R}\left[k_2-0.5-\frac{(K_T-1)\beta}{k_1}\right]\right\}$ where $K_T=1+\frac{I}{AR^2}$ $M_C=\delta\omega^2AR^3\left\{K_T\alpha+\frac{R_o}{R}\left[k_2+0.5+\frac{(K_T-1)\beta}{k_1}\right]\right\}$ $N_A=\delta\omega^2AR^2\left\{K_T+\frac{R_o}{R}\left[0.5+(K_T-1)\frac{k_2}{k_1}\right]\right\}$ $V_A=0$ $\Delta D_H=\frac{\delta\omega^2R^4}{Ee}\left[2K_Tk_2\alpha+\frac{R_o}{R}\left(\frac{k_1\pi}{2}-k_2\pi+2k_2^2\right)\right]$ $\Delta D_V=\frac{\delta\omega^2R^4}{Ee}\left[2K_Tk_2\alpha-\frac{R_o}{R}\left(\frac{k_1\pi^2}{4}-2k_2^2\right)\right]$ $\Delta L=\frac{\delta\omega^2R^4}{Ee}\left[K_Tk_2\alpha-\frac{R_o}{R}\left(\frac{k_1 3\pi^2}{16}+k_2-\frac{k_2\pi}{2}-k_2^2+K_T\alpha\right)\right]$ *Note:* The constant K_T accounts for the radial distribution of the mass in the ring. $\text{Max}+M=M_C$ $\text{Max}-M$ occurs at the angular position x_1 where $\frac{x_1}{\tan x_1}=-0.5+\frac{(K_T-1)\beta}{k_1}$ For a thin ring where $K_T\approx 1$, $\text{Max}-M=-\delta\omega^2AR^3\left[\frac{R_o}{R}(1.6408-k_2)-\alpha\right]$ at $x=105.23°$

TABLE 9.3 Reaction and Deformation Formulas for Circular Arches

Notation: W = load (force); w = unit load (force per unit of circumferential length); M_o = applied couple (force-length). θ_o = externally created concentrated angular displacement (radians); Δ_o = externally created concentrated radial displacement; $T - T_o$ = uniform temperature rise (degrees); T_1 and T_2 = temperatures on outside and inside, respectively (degrees). H_A and H_B are the horizontal end reactions at the left and right, respectively, and are positive to the left; V_A and V_B are the vertical end reactions at the left and right ends, respectively, and are positive upward; M_A and M_B are the reaction moments at the left and right, respectively, and are positive clockwise. E = modulus of elasticity (force per unit area); ν = Poisson's ratio; A is the cross-sectional area; R is the radius at the centroid at the cross section; I = area moment of inertia of arch cross section about the principal axis perpendicular to the plane of the arch. [Note that for a wide curved plate or a sector of a cylinder, a representative segment of unit axial length may be used by replacing EI by $Et^3/12(1-\nu^2)$.] e is the positive distance measured radially inward from the centroidal axis of the cross section to the neutral axis of pure bending (see Sec. 9.1). θ (radians) is one-half of the total subtended angle of the arch and is limited to the range zero to π. For an angle θ close to zero, round-off errors may cause troubles; for an angle θ close to π, the possibility of static or elastic instability must be considered. Deformations have been assumed small enough so as to not affect the expressions for the internal bending moments, radial shear, and circumferential normal forces. Answers should be examined to be sure that such is the case before accepting them. ϕ (radians) is the angle measured counterclockwise from the midspan of the arch to the position of a concentrated load or the start of a distributed load. $s = \sin\theta$, $c = \cos\theta$, $n = \sin\phi$, and $m = \cos\phi$. γ = temperature coefficient of expansion.

The references to end points A and B refer to positions on a circle of radius R passing through the centroids of the several sections. It is important to note this carefully when dealing with thick rings. Similarly, all concentrated and distributed loadings are assumed to be applied at the radial position of the centroid with the exception of cases 1 and 5, subcases h and i where the ring is loaded by its own weight or by a constant linear acceleration. In these two cases the actual radial distribution of load is considered. If the loading is on the outer or inner surfaces of thick rings, a statically equivalent loading at the centroidal radius R must be used. See examples to determine how this might be accomplished.

The hoop-stress deformation factor is $\alpha = I/AR^2$ for thin rings or $\alpha = e/R$ for thick rings. The transverse- (radial-) shear deformation factor is $\beta = FEI/GAR^2$ for thin rings or $\beta = 2F(1+\nu)e/R$ for thick rings, where G is the shear modulus of elasticity and F is a shape factor for the cross section (see Sec. 8.10). The following constants are defined to simplify the expressions which follow. Note that these constants are unity if no correction for hoop stress or shear stress is necessary or desired for use with thin rings. $k_1 = 1 - \alpha + \beta$, $k_2 = 1 - \alpha$.

General reaction and expressions for cases 1–4; right end pinned in all four cases, no vertical motion at the left end

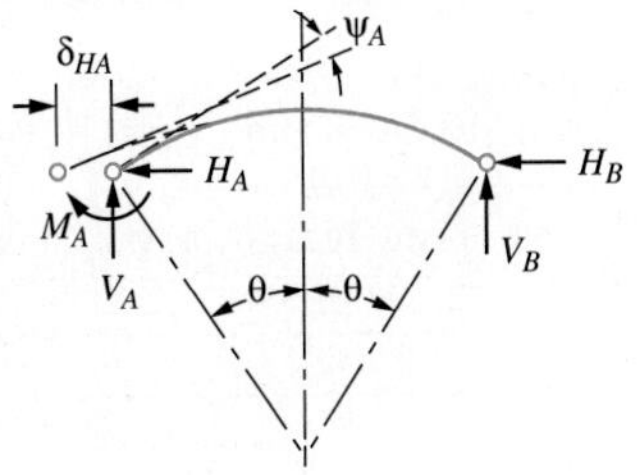

Deformation equations:

Horizontal deflection at $A = \delta_{HA} = \dfrac{R^3}{EI}\left(A_{HH}H_A + A_{HM}\dfrac{M_A}{R} - LP_H\right)$

Angular rotation at $A = \psi_A = \dfrac{R^2}{EI}\left(A_{MH}H_A + A_{MM}\dfrac{M_A}{R} - LP_M\right)$

where $A_{HH} = 2\theta c^2 + k_1(\theta - sc) - k_2 2sc$

$A_{MH} = A_{HM} = k_2 s - \theta c$

$A_{MM} = \dfrac{1}{4s^2}[2\theta s^2 + k_1(\theta + sc) - k_2 2sc]$

and where LP_H and LP_M are loading terms given below for several types of load.

(*Note:* If desired, V_A, V_B, and H_B can be evaluated from equilibrium equations after calculating H_A and M_A.)

1. Left end pinned, right end pinned

Since $\delta_{HA} = 0$ and $M_A = 0$,

$$H_A = \frac{LP_H}{A_{HH}} \quad \text{and} \quad \psi_A = \frac{R^2}{EI}(A_{MH}H_A - LP_M)$$

The loading terms are given below.

Reference No., Loading	Loading Terms and Some Selected Numerical Values
1a. Concentrated vertical load 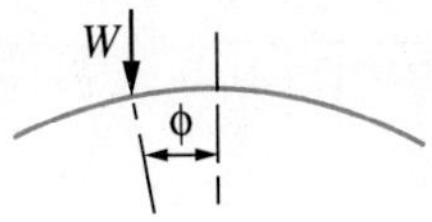	$LP_H = W\left[\theta sc - \phi nc + \dfrac{k_1}{2}(c^2 - m^2) + k_2 c(c - m)\right]$ $LP_M = \dfrac{W}{2}\left[\phi n - \theta s + \dfrac{k_1}{2s^2}(\theta n - \phi s + snc - snm) - k_2(c - m)\right]$

For $\alpha = \beta = 0$

θ	30°		60°		90°	
ϕ	0°	15°	0°	30°	0°	45°
LP_H/W	−0.0143	−0.0100	−0.1715	−0.1105	−0.5000	−0.2500
LP_M/W	−0.0639	−0.0554	−0.2034	−0.1690	−0.2854	−0.1978

(*Continued*)

TABLE 9.3 Reaction and Deformation Formulas for Circular Arches (*Continued*)

Reference No., Loading	Loading Terms and Some Selected Numerical Values

1b. Concentrated horizontal load

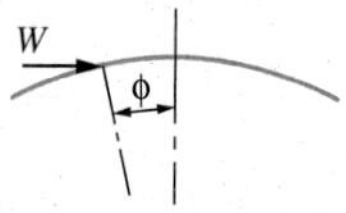

$$LP_H = W\left[\theta c^2 + \phi mc + \frac{k_1}{2}(\theta + \phi - sc - nm) - k_2 c(s+n)\right]$$

$$LP_M = \frac{W}{2}\left[-\theta c - \phi m + \frac{k_1}{2s^2}(\theta c - \theta m + sm^2 - scm) + k_2(s+n)\right]$$

For $\alpha = \beta = 0$

θ	30°		60°		90°	
ϕ	0°	15°	0°	30°	0°	45°
$\frac{LP_H}{W}$	0.0050	0.0057	0.1359	0.1579	0.7854	0.9281
$\frac{LP_M}{W}$	0.0201	0.0222	0.1410	0.1582	0.3573	0.4232

1c. Concentrated radial load

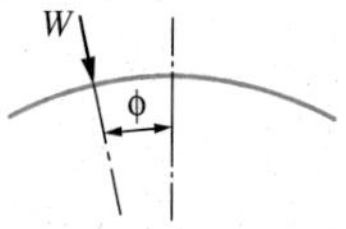

$$LP_H = W\left[\theta c(cn+sm) + \frac{k_1}{2}(\theta n + \phi n - scn - s^2 m) - k_2 c(1 + sn - cm)\right]$$

$$LP_M = \frac{W}{2}\left[-\theta(cn+sm) + \frac{k_1}{2s^2}(\theta cn - \phi sm) + k_2(1 + sn - cm)\right]$$

For $\alpha = \beta = 0$

θ	30°		60°		90°	
ϕ	0°	15°	0°	30°	0°	45°
$\frac{LP_H}{W}$	−0.0143	−0.0082	−0.1715	−0.0167	−0.5000	0.4795
$\frac{LP_M}{W}$	−0.0639	−0.0478	−0.2034	−0.0672	−0.2854	0.1594

1d. Concentrated tangential load

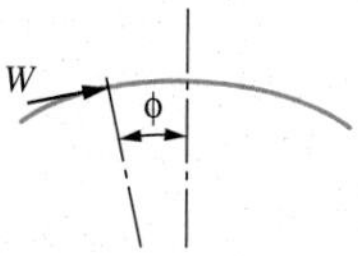

$$LP_H = W\left[\theta c(cm - sn) + \phi c + \frac{k_1}{2}(\theta m + \phi m - scm - c^2 n) - k_2 c(sm + cn)\right]$$

$$LP_M = \frac{W}{2}\left[\theta(sn - cm) - \phi + \frac{k_1}{2s^2}(\theta cm - \theta + \phi sn - sc + sm) + k_2(sm + cn)\right]$$

For $\alpha = \beta = 0$

θ	30°		60°		90°	
ϕ	0°	15°	0°	30°	0°	45°
$\frac{LP_H}{W}$	0.0050	0.0081	0.1359	0.1920	0.7854	0.8330
$\frac{LP_M}{W}$	0.0201	0.0358	0.1410	0.2215	0.3573	0.4391

1e. Uniform vertical load on partial span

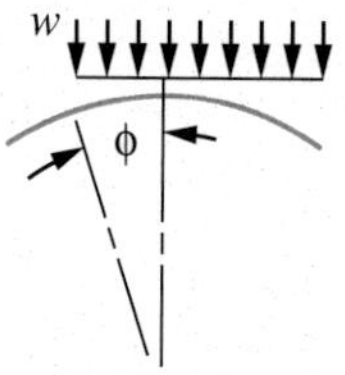

$$LP_H = \frac{wR}{4}\left[\theta c(1 + 4sn + 2s^2) + \phi c(m^2 - n^2) - c(sc + mn) + \frac{2k_1}{3}(n^3 - 3ns^2 - 2s^3) + 2k_2 c(2cn + cs - \theta - \phi - mn)\right]$$

$$LP_M = \frac{wR}{8}\left\{mn + sc - \theta(4sn + 2s^2 + 1) - \phi(m^2 - n^2) + \frac{k_1}{s}\left[\frac{\theta}{s}(n^2 + s^2) + 2(c - m) - \frac{2}{3}(c^3 - m^3) + c(n^2 - s^2) - 2\phi n\right] + 2k_2(\theta + \phi + mn - sc - 2cn)\right\}$$

If $\phi = 0$ (the full span is loaded),

$$LP_H = \frac{wR}{6}[3c(2\theta s^2 + \theta - sc) - 4k_1 s^3 + 6k_2 c(sc - \theta)]$$

$$LP_M = \frac{wR}{4}[sc - \theta - 2\theta s^2 + 2k_2(\theta - sc)]$$

For $\alpha = \beta = 0$

θ	30°		60°		90°	
ϕ	0°	15°	0°	30°	0°	45°
$\frac{LP_H}{wR}$	−0.0046	−0.0079	−0.0969	−0.1724	−0.3333	−0.6280
$\frac{LP_M}{wR}$	−0.0187	−0.0350	−0.1029	−0.2031	−0.1667	−0.3595

1f. Uniform horizontal load on left side only

$$LP_H = \frac{wR}{12}[3\theta c(1 - 6c^2 + 4c) + 3sc^2 + k_1(6\theta - 6sc - 12\theta c + 12c - 8s^3) + 6k_2 c(3sc - 2s - \theta)]$$

$$LP_M = \frac{wR}{8}\left\{6\theta c^2 - \theta - 4\theta c - sc + \frac{k_1}{3s^2}[s(2 - 3c + c^3) - 3\theta(1-c)^2] + 2k_2(\theta + 2s - 3sc)\right\}$$

For $\alpha = \beta = 0$

θ	30°	60°	90°
$\frac{LP_H}{wR}$	0.0010	0.0969	1.1187
$\frac{LP_M}{wR}$	0.0040	0.1060	0.5833

(*Continued*)

TABLE 9.3 Reaction and Deformation Formulas for Circular Arches (*Continued*)

Reference No., Loading	Loading Terms and Some Selected Numerical Values

1g. Uniform horizontal load on right side only

$$LP_H = \frac{wR}{12}[3\theta c(1+2c^2-4c)+3sc^2+2k_1(2s^3-3\theta+3sc)+6k_2c(2s-sc-\theta)]$$

$$LP_M = \frac{wR}{8}\left\{4\theta c-2\theta c^2-\theta-sc-\frac{k_1}{3s^2}[s(2-3c+c^3)-3\theta(1-c)^2] + 2k_2(\theta-2s+sc)\right\}$$

For $\alpha = \beta = 0$			
θ	30°	60°	90°
$\frac{LP_H}{wR}$	−0.0004	−0.0389	−0.4521
$\frac{LP_M}{wR}$	−0.0015	−0.0381	−0.1906

1h. Vertical loading uniformly distributed along the circumference (by gravity or linear acceleration)

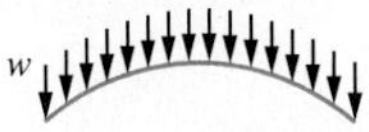

(*Note:* The full span is loaded.)

$$LP_H = wR\left\{2\theta^2 sc+\left(\frac{k_1}{2}+k_2\right)(2\theta c^2-\theta-sc)+\frac{R_{cg}}{R}[k_2(\theta-sc)-2c(s-\theta c)]\right\}$$

$$LP_M = wR\left[\left(\frac{R_{cg}}{R}+k_2\right)(s-\theta c)-\theta^2 s\right]$$

where R_{cg} is the radial distance to the center of mass for a differential length of the circumference for radially thicker arches. $R_{cg}/R=1+I_c/(AR^2)$. I_c is the area moment of inertia about the centroidal axis of the cross section.

For radially thin arches let $R_{cg} = R$. See the discussion on page 355.

For $\alpha = \beta = 0$ and for $R_{cg} = R$			
θ	30°	60°	90°
$\frac{LP_H}{wR}$	−0.0094	−0.2135	−0.7854
$\frac{LP_M}{wR}$	−0.0440	−0.2648	−0.4674

1i. Horizontal loading uniformly distributed along the circumference (by gravity or linear acceleration)

(*Note:* The full span is loaded.)

$$LP_H = wR\theta[2\theta c^2+k_1(\theta-sc)-2k_2sc]$$

$$LP_M = \frac{wR}{2s}\left[-2\theta^2 sc+\frac{k_1}{2s}(2\theta^2c+\theta s+s^2c)+k_2(2\theta s^2-\theta-sc) - \frac{R_{cg}}{R}(k_1-k_2)(\theta+sc)\right]$$

See case 1h for a definition of the radius R_{cg}.

For $\alpha = \beta = 0$ and for $R_{cg} = R$			
θ	30°	60°	90°
$\frac{LP_H}{wR}$	0.0052	0.2846	2.4674
$\frac{LP_M}{wR}$	0.0209	0.2968	1.1781

1j. Partial uniformly distributed radial loading

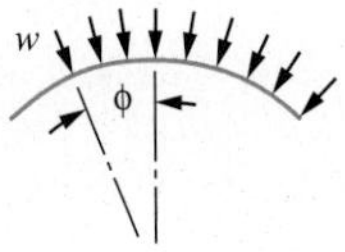

$$LP_H = wRc\left[\theta(1-cm+sn)+\frac{k_1}{2c}(scm+c^2n-\theta m-\phi m)+k_2(sm+cn-\theta-\phi)\right]$$

$$LP_M = \frac{wR}{2}\left[\theta(cm-1-sn)+\frac{k_1}{2s^2}[\theta-\theta cm-\phi sn+sc-sm]+k_2(\theta+\phi-sm-cn)\right]$$

If $\phi=\theta$ (the full span is loaded),

$$LP_H = wRc[2\theta s^2-k_1(\theta-sc)-2k_2(\theta-sc)]$$

$$LP_M = wR[-\theta s^2+k_2(\theta-sc)]$$

For $\alpha = \beta = 0$						
θ	30°		60°		90°	
ϕ	0°	15°	0°	30°	0°	45°
$\frac{LP_H}{wR}$	−0.0050	−0.0081	−0.1359	−0.1920	−0.7854	−0.8330
$\frac{LP_M}{wR}$	−0.0201	−0.0358	−0.1410	−0.2215	−0.3573	−0.4391

1k. Partial uniformly increasing distributed radial loading

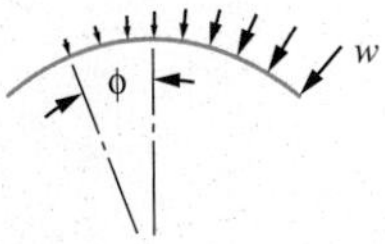

$$LP_H = \frac{wR}{\theta+\phi}\left\{\theta c(\theta+\phi-cn-sm)+\frac{k_1}{2}[n(sc-\theta-\phi)+s^2m+2c-2m] + k_2\left[scn+s^2m-\frac{c}{2}(\theta+\phi)^2+c-m\right]\right\}$$

$$LP_M = \frac{wR}{\theta+\phi}\left\{\frac{\theta}{2}(cn+sm-\theta-\phi)+\frac{k_1}{4s^2}[(\theta+\phi)(\theta+sc)+\phi sm-\theta cn -2s^2-2sn]+\frac{k_2}{4}[(\theta+\phi)^2+2cm-2sn-2]\right\}$$

If $\phi = \theta$ (the full span is loaded),

$$LP_H = \frac{wRc}{\theta}\left[\theta(\theta-cs)-\frac{k_1 s}{2c}(\theta-sc)-k_2(\theta^2-s^2)\right]$$

$$LP_M = \frac{wR}{2\theta}\left[\theta sc-\theta^2+\frac{k_1}{2s^2}(\theta^2+\theta sc-2s^2)+k_2(\theta^2-s^2)\right]$$

For $\alpha = \beta = 0$						
θ	30°		60°		90°	
ϕ	0°	15°	0°	30°	0°	45°
$\frac{LP_H}{wR}$	−0.0018	−0.0035	−0.0518	−0.0915	−0.3183	−0.5036
$\frac{LP_M}{wR}$	−0.0072	−0.0142	−0.0516	−0.0968	−0.1366	−0.2335

(*Continued*)

TABLE 9.3 Reaction and Deformation Formulas for Circular Arches (*Continued*)

Reference No., Loading	Loading Terms and Some Selected Numerical Values

1l. Partial second-order increase in distributed radial loading

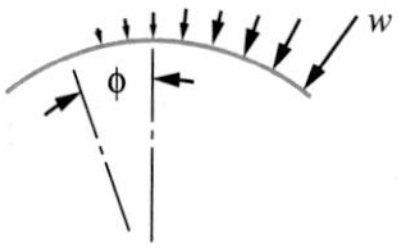

$$LP_H = \frac{wRc}{(\theta+\phi)^2}\left\{\theta(\theta+\phi)^2 - 2\theta(1-cm+sn) + \frac{k_1}{c}[(\theta+\phi)(2c+m) - c(sm+cn) - 2n - 2s] + \frac{k_2}{3}[6(\theta+\phi-sm-cn)-(\theta+\phi)^3]\right\}$$

$$LP_M = \frac{wR}{(\theta+\phi)^2}\left\{\theta(1-cm+sn) - \frac{\theta}{2}(\theta+\phi)^2 + \frac{k_1}{4s^2}[(\theta+\phi)^2(\theta+sc)+2s(\phi n+3m-3c)-4s^2(\theta+\phi)-2\theta(1-cm)] + k_2\left[sm+cn-\theta-\phi+\frac{(\theta+\phi)^3}{6}\right]\right\}$$

If $\phi = \theta$ (the full span is loaded),

$$LP_H = \frac{wR}{6\theta^2}[6\theta^3 c - 6\theta s^2 c + 3k_1(3\theta c - 3s + s^3) + 2k_2(3\theta c - 2\theta^3 c - 3sc^2)]$$

$$LP_M = \frac{wR}{2\theta^2}\left[\theta s^2 - \theta^3 + k_1\frac{\theta}{2s^2}(\theta^2 + \theta sc - 2s^2) + k_2\left(sc - \theta + \frac{2\theta^3}{3}\right)\right]$$

For $\alpha = \beta = 0$

θ	30°		60°		90°	
ϕ	0°	15°	0°	30°	0°	45°
$\frac{LP_H}{wR}$	−0.0010	−0.0019	−0.0276	−0.0532	−0.1736	−0.3149
$\frac{LP_M}{wR}$	−0.0037	−0.0077	−0.0269	−0.0542	−0.0726	−0.1388

1m. Partial uniformly distributed tangential loading

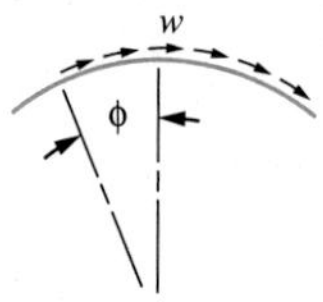

$$LP_H = \frac{wR}{2}\left\{2\theta c(cn+sm) - c(\theta^2-\phi^2) + k_1[(n\theta+\phi)+c(cm-sn-2)+e] + k_2 2c(cm-sn-1)\right\}$$

$$LP_M = \frac{wR}{4}\left\{\theta^2 - \phi^2 - 2\theta(cn+sm) + \frac{k_1}{s^2}[\theta(cn-cs-\theta-\phi)-\phi s(c+m)+2s(s+n)] + k_2 2(1+sn-cm)\right\}$$

If $\phi = \theta$ (the full span is loaded),

$$LP_H = wR[2\theta c^2 s + k_1 s(\theta - sc) - 2k_2 cs^2]$$

$$LP_M = wR\left[-\theta sc + \frac{k_1}{2s^2}(2s^2 - \theta sc - \theta^2) + k_2 s^2\right]$$

For $\alpha = \beta = 0$

θ	30°		60°		90°	
ϕ	0°	15°	0°	30°	0°	45°
$\frac{LP_H}{wR}$	0.0010	0.0027	0.0543	0.1437	0.5000	1.1866
$\frac{LP_M}{wR}$	0.0037	0.0112	0.0540	0.1520	0.2146	0.5503

1n. Concentrated couple

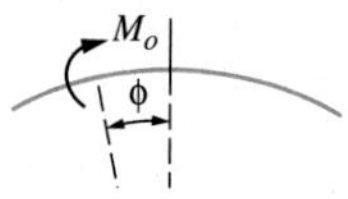

$$LP_H = \frac{M_o}{R}(\phi c - k_2 n)$$

$$LP_M = \frac{M_o}{4s^2R}[-2s^2\phi - k_1(\theta + sc) + k_2 2sm]$$

For $\alpha = \beta = 0$

θ	30°		60°		90°	
ϕ	0°	15°	0°	30°	0°	45°
$\frac{LP_H R}{M_o}$	0.0000	−0.0321	0.0000	−0.2382	0.0000	−0.7071
$\frac{LP_M R}{M_o}$	0.0434	−0.1216	0.0839	−0.2552	0.1073	−0.4318

1p. Concentrated angular displacement

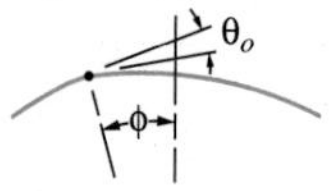

$$LP_H = \frac{\theta_o EI}{R^2}(m - c)$$

$$LP_M = \frac{\theta_o EI}{R^2}\left(\frac{1}{2} + \frac{n}{2s}\right)$$

1q. Concentrated radial displacement

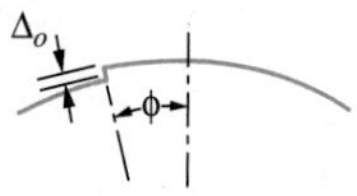

$$LP_H = \frac{\Delta_o EI}{R^3} n$$

$$LP_M = \frac{\Delta_o EI}{R^3}\left(-\frac{m}{2s}\right)$$

(*Continued*)

TABLE 9.3 Reaction and Deformation Formulas for Circular Arches (*Continued*)

Reference No., Loading	Loading Terms and Some Selected Numerical Values
1r. Uniform temperature rise over that span to the right of point Q 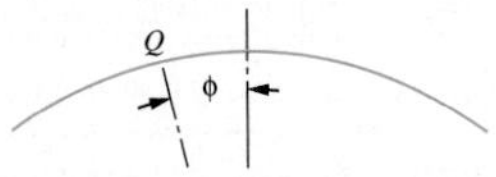	$LP_H = -(T - T_o)\frac{\gamma EI}{R^2}(s+n)$ $LP_M = (T - T_o)\frac{\gamma EI}{2R^2 s}(m-c)$ T = uniform temperature T_o = unload temperature
1s. Linear temperature differential through the thickness t for that span to the right of point Q 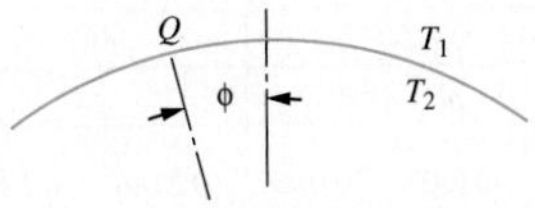	$LP_H = (T_1 - T_2)\frac{\gamma EI}{Rt}(n + s - \theta c - \phi c)$ $LP_M = (T_1 - T_2)\frac{\gamma EI}{2Rts}(\theta s + \phi s - m + c)$ where t is the radial thickness and T_o, the unloaded temperature, is the temperature at the radius of the centroid
2. Left end guided horizontally, right end pinned 	Since $\psi_A = 0$ and $H_A = 0$, $M_A = \frac{LP_M}{A_{MM}}R$ and $\delta_{HA} = \frac{R^3}{EI}\left(A_{HM}\frac{M_A}{R} - LP_H\right)$ Use load terms given above for cases 1a – 1s.
3. Left end roller supported in vertical direction only, right end pinned 	Since both M_A and H_A are zero, this is a statically determinate case: $\delta_{HA} = \frac{-R^3}{EI}LP_H$ and $\psi_A = \frac{-R^2}{EI}LP_M$ Use load terms given above for cases 1a – 1s.
4. Left end fixed, right end pinned 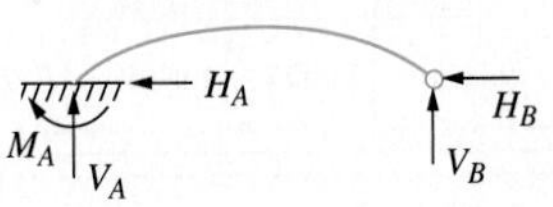	Since $\delta_{HA} = 0\,and\,\psi_A = 0$, $H_A = \frac{A_{MM}LP_H - A_{HM}LP_M}{A_{HH}A_{MM} - A_{HM}^2}$ and $\frac{M_A}{R} = \frac{A_{HH}LP_M - A_{HM}LP_H}{A_{HH}A_{MM} - A_{HM}^2}$ Use load terms given above for cases 1a – 1s.
General reaction and deformation expressions for cases 5–14, right end fixed in all 10 cases 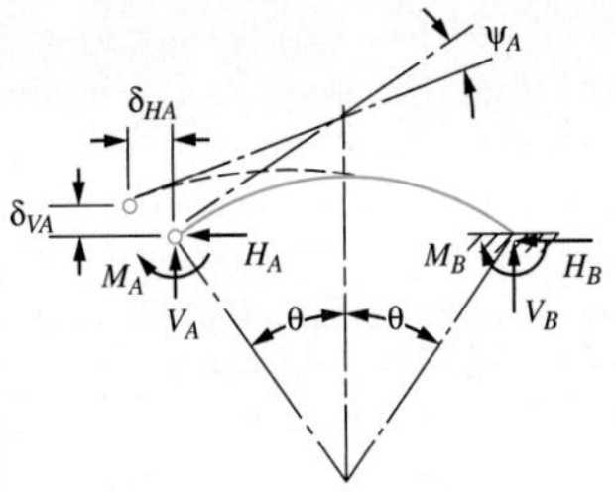	Deformation equations: Horizontal deflection at $A = \delta_{HA} = \frac{R^3}{EI}\left(B_{HH}H_A + B_{HV}V_A + B_{HM}\frac{M_A}{R} - LF_H\right)$ Vertical deflection at $A = \delta_{VA} = \frac{R^3}{EI}\left(B_{VH}H_A + B_{VV}V_A + B_{VM}\frac{M_A}{R} - LF_V\right)$ Angular rotation at $A = \psi_A = \frac{R^2}{EI}\left(B_{MH}H_A + B_{MV}V_A + B_{MM}\frac{M_A}{R} - LF_M\right)$ where $B_{HH} = 2\theta c^2 + k_1(\theta - sc) - k_2 2sc$ $B_{HV} = B_{VH} = -2\theta sc + k_2 2s^2$ $B_{HM} = B_{MH} = -2\theta c + k_2 2s$ $B_{VV} = 2\theta s^2 + k_1(\theta + sc) - k_2 2sc$ $B_{VM} = B_{MV} = 2\theta s$ $B_{MM} = 2\theta$ and where LF_H, LF_V, and LF_M are loading terms given below for several types of load
(*Note:* If desired, H_B, V_B, and M_B can be evaluated from equilibrium equations after calculating H_A, V_A, and M_A.)	
5. Left end fixed, right end fixed 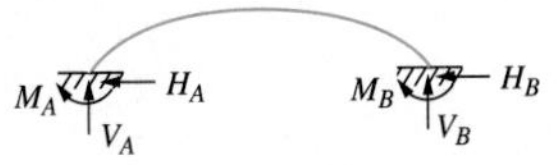	Since $\delta_{HA} = 0$, $\delta_{VA} = 0$, $\psi_A = 0$, these equations must be solved simultaneously for H_A, V_A, and M_A/R. The loading terms are given in cases 5a–5s. $B_{HH}H_A + B_{HV}V_A + B_{HM}M_A/R = LF_H$ $B_{VH}H_A + B_{VV}V_A + B_{VM}M_A/R = LF_V$ $B_{MH}H_A + B_{MV}V_A + B_{MM}M_A/R = LF_M$

(*Continued*)

TABLE 9.3 Reaction and Deformation Formulas for Circular Arches (*Continued*)

Reference No., Loading	Loading Terms and Some Selected Numerical Values

5a. Concentrated vertical load

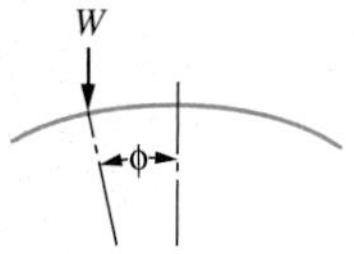

$$LF_H = W\left[-(\theta+\phi)cn+\frac{k_1}{2}(c^2-m^2)+k_2(1+sn-cm)\right]$$

$$LF_V = W\left[(\theta+\phi)sn+\frac{k_1}{2}(\theta+\phi+sc+nm)-k_2(2sc-sm+cn)\right]$$

$$LF_M = W[(\theta+\phi)n+k_2(m-c)]$$

For $\alpha=\beta=0$

θ	30°		60°		90°	
ϕ	0°	15°	0°	30°	0°	45°
LF_H/W	0.0090	0.0253	0.1250	0.3573	0.5000	1.4571
LF_V/W	0.1123	0.2286	0.7401	1.5326	1.7854	3.8013
LF_M/W	0.1340	0.3032	0.5000	1.1514	1.0000	2.3732

5b. Concentrated horizontal load

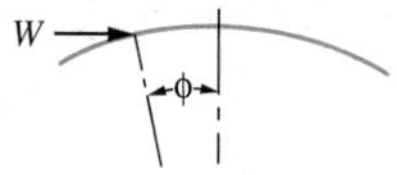

$$LF_H = W\left[(\theta+\phi)mc+\frac{k_1}{2}(\theta+\phi-sc-nm)-k_2(sm+cn)\right]$$

$$LF_V = W\left[-(\theta+\phi)sm+\frac{k_1}{2}(c^2-m^2)+k_2(1-2c^2+cm+sn)\right]$$

$$LF_M = W[-(\theta+\phi)m+k_2(s+n)]$$

For $\alpha=\beta=0$

θ	30°		60°		90°	
ϕ	0°	15°	0°	30°	0°	45°
LF_H/W	−0.0013	0.0011	−0.0353	0.0326	−0.2146	0.2210
LF_V/W	−0.0208	−0.0049	−0.2819	−0.0621	−1.0708	−0.2090
LF_M/W	−0.0236	0.0002	−0.1812	0.0057	−0.5708	0.0410

5c. Concentrated radial load

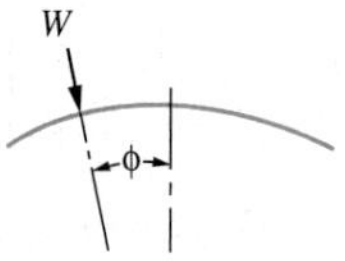

$$LF_H = W\left[\frac{k_1}{2}(\theta n+\phi n-scn-s^2m)+k_2(m-c)\right]$$

$$LF_V = W\left[\frac{k_1}{2}(\theta m+\phi m+scm+c^2n)+k_2(s+n-2scm-2c^2n)\right]$$

$$LF_M = W[k_2(1+sn-cm)]$$

For $\alpha=\beta=0$

θ	30°		60°		90°	
ϕ	0°	15°	0°	30°	0°	45°
LF_H/W	0.0090	0.0248	0.1250	0.3257	0.5000	1.1866
LF_V/W	0.1123	0.2196	0.7401	1.2962	1.7854	2.5401
LF_M/W	0.1340	0.2929	0.5000	1.0000	1.0000	1.7071

5d. Concentrated tangential load

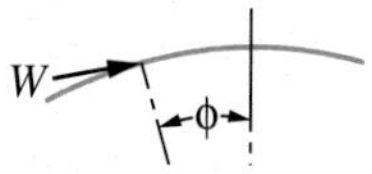

$$LF_H = W\left[(\theta+\phi)c+\frac{k_1}{2}(\theta m+\phi m-scm-c^2n)-k_2(s+n)\right]$$

$$LF_V = W\left[-(\theta+\phi)s-\frac{k_1}{2}(\theta n+\phi n+scn+s^2m)+k_2(2s^2m+2scn+c-m)\right]$$

$$LF_M = W[-\theta-\phi+k_2(sm+cn)]$$

For $\alpha=\beta=0$

θ	30°		60°		90°	
ϕ	0°	15°	0°	30°	0°	45°
LF_H/W	−0.0013	−0.0055	−0.0353	−0.1505	−0.2146	−0.8741
LF_V/W	−0.0208	−0.0639	−0.2819	−0.8200	−1.0708	−2.8357
LF_M/W	−0.0236	−0.0783	−0.1812	−0.5708	−0.5708	−1.6491

5e. Uniform vertical load on partial span

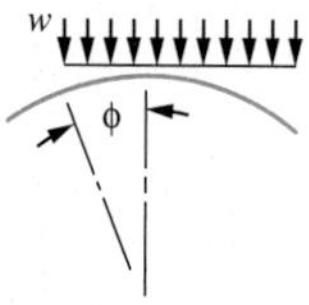

$$LF_H = \frac{wR}{4}\left\{c[(1-2n^2)(\theta+\phi)-sc-mn]-\frac{2k_1}{3}(2s^3+3s^2n-n^3)+2k_2[s+2n+sn^2-c(\theta+\phi+mn)]\right\}$$

$$LF_V = \frac{wR}{4}\left\{s[(1-2m^2)(\theta+\phi)+sc+mn]+\frac{2k_1}{3}[3n(\theta+\phi+sc)+3m-m^3-2c^3]+2k_2[s(\theta+\phi-2sc+nm-4cn)-cn^2]\right\}$$

$$LF_M = \frac{wR}{4}[(1-2m^2)(\theta+\phi)+nm+sc+2k_2(\theta+\phi+nm-sc-2cn)]$$

If $\phi=\theta$ (the full span is loaded),

$$LF_H = \frac{wR}{2}\left[\theta c(1-2s^2)-sc^2-\frac{k_1 4s^3}{3}+k_2 2(s^3+s-c\theta)\right]$$

$$LF_V = wR\left[\frac{s}{2}(\theta s^2-\theta c^2+sc)+k_1 s(\theta+sc)+k_2 s(\theta-3sc)\right]$$

$$LF_M = wR\left[\frac{1}{2}(\theta s^2-\theta c^2+sc)+k_2(\theta-sc)\right]$$

For $\alpha=\beta=0$

θ	30°		60°		90°	
ϕ	0°	15°	0°	30°	0°	45°
LF_H/wR	0.0012	0.0055	0.0315	0.1471	0.1667	0.8291
LF_V/wR	0.0199	0.0635	0.2371	0.7987	0.7260	2.6808
LF_M/wR	0.0226	0.0778	0.1535	0.5556	0.3927	1.5531

(*Continued*)

TABLE 9.3 Reaction and Deformation Formulas for Circular Arches (*Continued*)

Reference No., Loading	Loading Terms and Some Selected Numerical Values

5f. Uniform horizontal load on left side only

$$LF_H = \frac{wR}{4}\left[\theta c(4s^2-1)+sc^2+2k_1\left(\theta-2\theta c-sc+2sc^2+\frac{2s^2}{3}\right)+2k_2(sc^2-s^3-\theta c)\right]$$

$$LF_V = \frac{wR}{4}\left\{-\theta s(4s^2-1)-s^2c-\frac{2k_1}{3}(1-3c^2+2c^3)+2k_2[\theta s+2(1-c)(1-2c^2)]\right\}$$

$$LF_M = \frac{wR}{4}\left[-\theta(4s^2-1)-sc+2k_2(2s-3sc+\theta)\right]$$

For $\alpha = \beta = 0$

θ	30°	60°	90°
$\frac{LF_H}{wR}$	0.0005	0.0541	0.6187
$\frac{LF_V}{wR}$	0.0016	0.0729	0.4406
$\frac{LF_M}{wR}$	0.0040	0.1083	0.6073

5g. Uniform horizontal load on right side only

$$LF_H = \frac{wR}{4}\left[sc^2-\theta c+\frac{2k_1}{3}(2s^3+3sc-3\theta)+2k_2(s-\theta c)\right]$$

$$LF_V = \frac{wR}{4}\left[\theta s-s^2c+\frac{2k_1}{3}(1-3c^2+2c^3)-2k_2(2-4c^2+2c^3-\theta s)\right]$$

$$LF_M = \frac{wR}{4}[\theta-sc+2k_2(\theta-2s+sc)]$$

For $\alpha = \beta = 0$

θ	30°	60°	90°
$\frac{LF_H}{wR}$	0.0000	0.0039	0.0479
$\frac{LF_V}{wR}$	0.0009	0.0448	0.3448
$\frac{LF_M}{wR}$	0.0010	0.0276	0.1781

5h. Vertical loading uniformly distributed along the circumference (by gravity or linear acceleration)

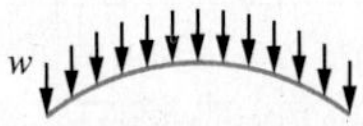

(*Note:* The full span is loaded.)

$$LF_H = wR\left[\frac{k_1}{2}(2\theta c^2-\theta-sc)+k_2\left(\frac{R_{cg}}{R}+1\right)(\theta-sc)+\frac{R_{cg}}{R}2c(\theta c-s)\right]$$

$$LF_V = wR\left[k_1\theta(\theta+sc)+2k_2 s(s-2\theta c)-\frac{R_{cg}}{R}2s(\theta c-s)\right]$$

$$LF_M = 2wR\left(\frac{R_{cg}}{R}+k_2\right)(s-\theta c)$$

See case 1h for a definition of the radius R_{cg}.

For $\alpha = \beta = 0$ and $R_{cg} = R$

θ	30°	60°	90°
$\frac{LF_H}{wR}$	0.0149	0.4076	2.3562
$\frac{LF_V}{wR}$	0.1405	1.8294	6.4674
$\frac{LF_M}{wR}$	0.1862	1.3697	4.0000

5i. Horizontal loading uniformly distributed along the circumference (by gravity or linear acceleration)

(*Note:* The full span is loaded.)

$$LF_H = wR\left[k_1\theta(\theta-sc)+\frac{R_{cg}}{R}2s(\theta c-k_2 s)\right]$$

$$LF_V = wR\left[\frac{k_1}{2}(2\theta c^2-\theta-sc)+\left(\frac{R_{cg}}{R}+1\right)k_2(sc-\theta)+\left(2k_2-\frac{R_{cg}}{R}\right)2\theta s^2\right]$$

$$LF_M = wR\left(k_2-\frac{R_{cg}}{R}\right)2\theta s$$

See case 1h for a definition of the radius R_{cg}.

For $\alpha = \beta = 0$ and $R_{cg} = R$

θ	30°	60°	90°
$\frac{LF_H}{wR}$	0.0009	0.0501	0.4674
$\frac{LF_V}{wR}$	−0.0050	−0.1359	−0.7854
$\frac{LF_M}{wR}$	0.0000	0.0000	0.0000

5j. Partial uniformly distributed radial loading

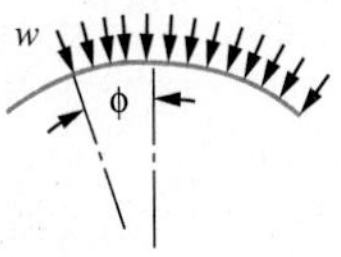

$$LF_H = wR\left[\frac{k_1}{2}(scm+c^2n-\theta m-\phi m)+k_2(s+n-\theta c-\phi c)\right]$$

$$LF_V = wR\left[\frac{k_1}{2}(\theta n+\phi n+scn+s^2m)+k_2(\theta s+\phi s-2scn+2c^2m-c-m)\right]$$

$$LF_M = wR[k_2(\theta+\phi-sm-cn)]$$

If $\phi = \theta$ (the full span is loaded),

$$LF_H = wR[k_1c(sc-\theta)+2k_2(s-\theta c)]$$

$$LF_V = wR[k_1 s(\theta+sc)+2k_2 s(\theta-2sc)]$$

$$LF_M = wR[2k_2(\theta-sc)]$$

For $\alpha = \beta = 0$

θ	30°		60°		90°	
ϕ	0°	15°	0°	30°	0°	45°
$\frac{LF_H}{wR}$	0.0013	0.0055	0.0353	0.1505	0.2146	0.8741
$\frac{LF_V}{wR}$	0.0208	0.0639	0.2819	0.8200	1.0708	2.8357
$\frac{LF_M}{wR}$	0.0236	0.0783	0.1812	0.5708	0.5708	1.6491

(*Continued*)

TABLE 9.3 Reaction and Deformation Formulas for Circular Arches (*Continued*)

Reference No., Loading	Loading Terms and Some Selected Numerical Values

5k. Partial uniformly increasing distributed radial loading

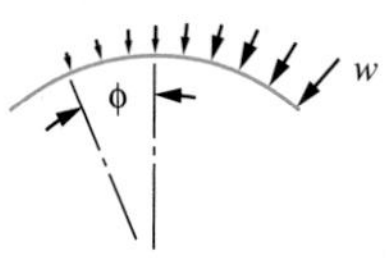

$$LF_H = \frac{wR}{\theta+\phi}\left\{\frac{k_1}{2}[scn-(\theta+\phi)n+2c-m-c^2m] + \frac{k_2}{2}[(\theta+\phi)(2s-\theta c-\phi c)+2c-2m]\right\}$$

$$LF_V = \frac{wR}{\theta+\phi}\left\{\frac{k_1}{2}[2s+2n-(\theta+\phi)m-smc-c^2n] + \frac{k_2}{2}[(\theta+\phi)(\theta s+\phi s-2c)-2s-2n+4smc+4c^2n]\right\}$$

$$LF_M = \frac{wR}{\theta+\phi}\left\{\frac{k_2}{2}[(\theta+\phi)^2+2(cm-sn-1)]\right\}$$

If $\phi = \theta$ (the full span is loaded),

$$LF_H = \frac{wR}{2\theta}[k_1 s(sc-\theta)+2k_2\theta(s-\theta c)]$$

$$LF_V = \frac{wR}{2\theta}[k_1(2s-sc^2-\theta c)+2k_2(2sc^2+s\theta^2-s-\theta c)]$$

$$LF_M = \frac{wR}{2\theta}[k_2(\theta^2-s^2)]$$

For $\alpha = \beta = 0$

θ	30°		60°		90°	
ϕ	0°	15°	0°	30°	0°	45°
$\frac{LF_H}{wR}$	0.0003	0.0012	0.0074	0.0330	0.0451	0.1963
$\frac{LF_V}{wR}$	0.0054	0.0169	0.0737	0.2246	0.2854	0.8245
$\frac{LF_M}{wR}$	0.0059	0.0198	0.0461	0.1488	0.1488	0.4536

5l. Partial second-order increase in distributed radial loading

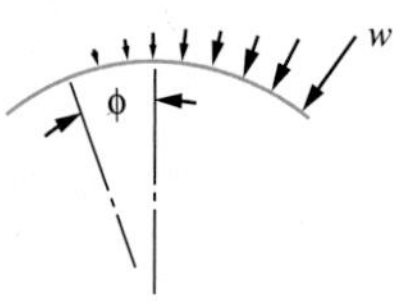

$$LF_H = \frac{wR}{(\theta+\phi)^2}\left\{k_1[(\theta+\phi)(2c+m)-2s-2n-c^2n-scm] + \frac{k_2}{3}[3(\theta+\phi)(\theta s+\phi s+2c)-6s-6n-c(\theta+\phi)^3]\right\}$$

$$LF_V = \frac{wR}{(\theta+\phi)^2}\left\{k_1[(\theta+\phi)(2s-n)+mc^2-3m-scn+2c] - \frac{k_2}{3}[3(\theta+\phi)(\theta c+\phi c+2s)-6c-6m+12c(mc-sn)-s(\theta+\phi)^3]\right\}$$

$$LF_M = \frac{wR}{(\theta+\phi)^2}\left\{\frac{k_2}{3}[6(sm+cn-\theta-\phi)+(\theta+\phi)^3]\right\}$$

If $\phi = \theta$ (the full span is loaded),

$$LF_H = \frac{wR}{2\theta^2}\left[k_1(3\theta c-3s+s^3)+2k_2\left(\theta c-s+s\theta^2-\frac{2c\theta^3}{3}\right)\right]$$

$$LF_V = \frac{wR}{2\theta^2}\left[k_1 s(\theta-sc)+2k_2\left(2s^2c-\theta s-c\theta^2+\frac{2s\theta^3}{3}\right)\right]$$

$$LF_M = \frac{wR}{\theta^2}\left[k_2\left(sc-\theta+\frac{2\theta^3}{3}\right)\right]$$

For $\alpha = \beta = 0$

θ	30°		60°		90°	
ϕ	0°	15°	0°	30°	0°	45°
$\frac{LF_H}{wR}$	0.0001	0.0004	0.0025	0.0116	0.0155	0.0701
$\frac{LF_V}{wR}$	0.0022	0.0070	0.0303	0.0947	0.1183	0.3579
$\frac{LF_M}{wR}$	0.0024	0.0080	0.0186	0.0609	0.0609	0.1913

5m. Partial uniformly distributed tangential loading

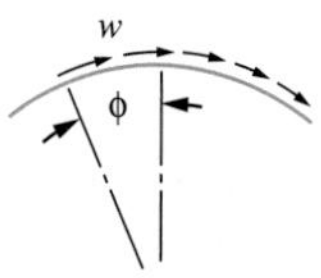

$$LF_H = \frac{wR}{2}[(\theta+\phi)^2c+k_1(\theta n+\phi n-scn-s^2m+2m-2c) + 2k_2(m-c-\theta s-\phi s)]$$

$$LF_V = \frac{wR}{2}[-(\theta+\phi)^2s+k_1(\theta m+\phi m+c^2n+scm-2s-2n) + 2k_2(\theta c+\phi c+2s^2n-n-2scm+s)]$$

$$LF_M = \frac{wR}{2}[-(\theta+\phi)^2+2k_2(1+sn-cm)]$$

If $\phi = \theta$ (the full span is loaded),

$$LF_H = wR[2\theta^2c+k_1 s(\theta-sc)-k_2 2\theta s]$$

$$LF_V = wR[-2\theta^2s+k_1(\theta c-s-s^3)+2k_2(\theta c+s-2sc^2)]$$

$$LF_M = wR(-2\theta^2+k_2 2s^2)$$

For $\alpha = \beta = 0$

θ	30°		60°		90°	
ϕ	0°	15°	0°	30°	0°	45°
$\frac{LF_H}{wR}$	−0.0001	−0.0009	−0.0077	−0.0518	−0.0708	−0.4624
$\frac{LF_V}{wR}$	−0.0028	−0.0133	−0.0772	−0.3528	−0.4483	−1.9428
$\frac{LF_M}{wR}$	−0.0031	−0.0155	−0.0483	−0.2337	−0.2337	−1.0687

(Continued)

TABLE 9.3 Reaction and Deformation Formulas for Circular Arches (*Continued*)

Reference No., Loading	Loading Terms and Some Selected Numerical Values
5n. Concentrated couple 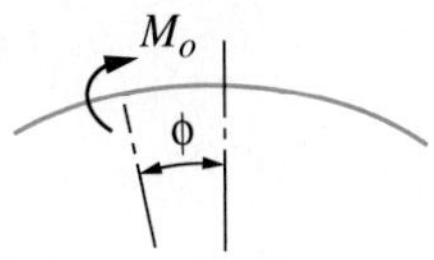 	$LF_H = \frac{M_o}{R}[(\theta+\phi)c - k_2(s+n)]$ $LF_V = \frac{M_o}{R}[-(\theta+\phi)s + k_2(c-m)]$ $LF_M = \frac{M_o}{R}(-\theta-\phi)$

For $\alpha = \beta = 0$

θ	30°		60°		90°	
ϕ	0°	15°	0°	30°	0°	45°
$\frac{LF_H R}{M_o}$	−0.0466	−0.0786	−0.3424	−0.5806	−1.0000	−1.7071
$\frac{LF_V R}{M_o}$	−0.3958	−0.4926	−1.4069	−1.7264	−2.5708	−3.0633
$\frac{LF_M R}{M_o}$	−0.5236	−0.7854	−1.0472	−1.5708	−1.5708	−2.3562

Reference No., Loading	Loading Terms and Some Selected Numerical Values
5p. Concentrated angular displacement 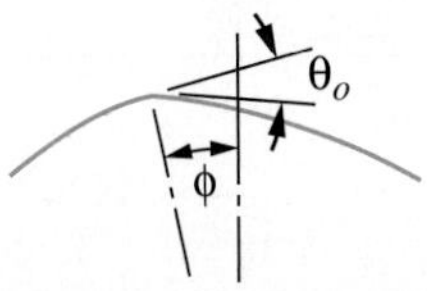 	$LF_H = \frac{\theta_o EI}{R^2}(m-c)$ $LF_V = \frac{\theta_o EI}{R^2}(s-n)$ $LF_M = \frac{\theta_o EI}{R^2}(1)$
5q. Concentrated radial displacement 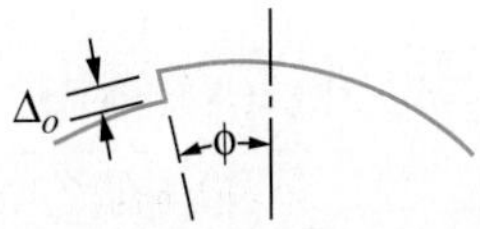 	$LF_H = \frac{\Delta_o EI}{R^3}(n)$ $LF_V = \frac{\Delta_o EI}{R^3}(m)$ $LF_M = 0$
5r. Uniform temperature rise over that span to the right of point Q 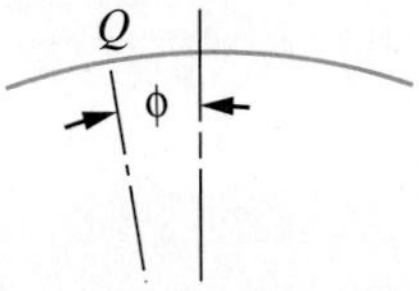 	$LF_H = -(T-T_o)\frac{\gamma EI}{R^2}(n+s)$ $LF_V = (T-T_o)\frac{\gamma EI}{R^2}(c-m)$ $LF_M = 0$ T = uniform temperature T_o = unloaded temperature
5s. Linear temperature differential through the thickness t for that span to the right of point Q 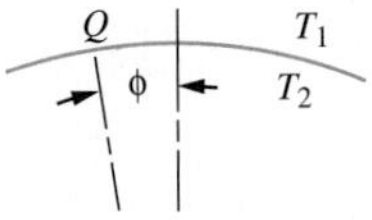 	$LF_H = (T_1-T_2)\frac{\gamma EI}{Rt}(n+s-\theta c-\phi c)$ $LF_V = (T_1-T_2)\frac{\gamma EI}{Rt}(m-c+\theta s+\phi s)$ $LF_M = (T_1-T_2)\frac{\gamma EI}{Rt}(\theta+\phi)$ *Note:* The temperature at the centroidal axis is the initial unloaded temperature.
6. Left end pinned, right end fixed 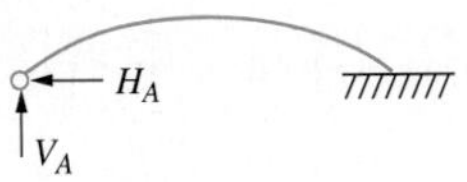 	Since $\delta_{HA} = 0$, $\delta_{VA} = 0$, and $M_A = 0$, $H_A = \frac{B_{VV}LF_H - B_{HV}LF_V}{B_{HH}B_{VV} - B_{HV}^2}$ $V_A = \frac{B_{HH}LF_V - B_{HV}LF_H}{B_{HH}B_{VV} - B_{HV}^2}$ Use load terms given above for cases 5a–5s. $\psi_A = \frac{R^2}{EI}(B_{MH}H_A + B_{MV}V_A - LF_M)$
7. Left end guided in horizontal direction, right end fixed  	Since $\delta_{VA} = 0$, $\psi_A = 0$, and $H_A = 0$, $V_A = \frac{B_{MM}LF_V - B_{MV}LF_M}{B_{VV}B_{MM} - B_{MV}^2}$ $\frac{M_A}{R} = \frac{B_{VV}LF_M - B_{MV}LF_V}{B_{VV}B_{MM} - B_{MV}^2}$ Use load terms given above for cases 5a–5s. $\delta_{HA} = \frac{R^3}{EI}\left(B_{HV}V_A + B_{HM}\frac{M_A}{R} - LF_H\right)$

(*Continued*)

TABLE 9.3 Reaction and Deformation Formulas for Circular Arches (*Continued*)

Reference No., Loading	Loading Terms and Some Selected Numerical Values
8. Left end guided in vertical direction, right end fixed 	Since $\delta_{HA}=0$, $\psi_A=0$, and $V_A=0$, $$H_A=\frac{B_{MM}LF_H-B_{HM}LF_M}{B_{HH}B_{MM}-B_{HM}^2}$$ $$\frac{M_A}{R}=\frac{B_{HH}LF_M-B_{HM}LF_H}{B_{HH}B_{MM}-B_{HM}^2}$$ $$\delta_{VA}=\frac{R^3}{EI}\left(B_{VH}H_A+B_{VM}\frac{M_A}{R}-LF_V\right)$$ Use load terms given above for cases 5a–5s.
9. Left end roller supported in vertical direction only, right end fixed 	Since $\delta_{VA}=0$, $H_A=0$, and $M_A=0$, $$V_A=\frac{LF_V}{B_{VV}},\quad \delta_{HA}=\frac{R^3}{EI}(B_{HV}V_A-LF_H)$$ $$\psi_A=\frac{R^2}{EI}(B_{MV}V_A-LF_M)$$ Use load terms given above for cases 5a–5s.
10. Left end roller supported in horizontal direction only, right end fixed 	Since $\delta_{HA}=0$, $V_A=0$, and $M_A=0$, $$H_A=\frac{LF_H}{B_{HH}},\quad \delta_{VA}=\frac{R^3}{EI}(B_{VH}H_A-LF_V)$$ $$\psi_A=\frac{R^2}{EI}(B_{MH}H_A-LF_M)$$ Use load terms given above for cases 5a–5s.
11. Left end restrained against rotation only, right end fixed 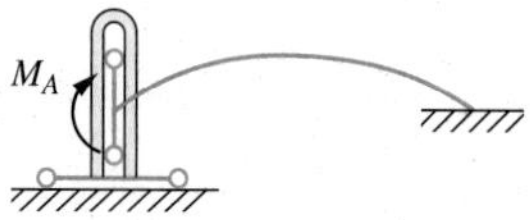	Since $\psi_A=0$, $H_A=0$, and $V_A=0$, $$\frac{M_A}{R}=\frac{LF_M}{B_{MM}},\quad \delta_{HA}=\frac{R^3}{EI}\left(B_{HM}\frac{M_A}{R}-LF_H\right)$$ $$\delta_{VA}=\frac{R^3}{EI}\left(B_{VM}\frac{M_A}{R}-LF_V\right)$$ Use load terms given above for cases 5a–5s.
12. Left end free, right end fixed 	Since $H_A=0$, $V_A=0$, and $M_A=0$, this is a statically determinate problem. The deflections at the free end are given by $$\delta_{HA}=\frac{-R^3}{EI}LF_H,\quad \delta_{VA}=\frac{-R^3}{EI}LF_V$$ $$\psi_A=\frac{-R^2}{EI}LF_M$$ Use load terms given above for cases 5a–5s.
13. Left end guided along an inclined surface at angle ζ, right end fixed 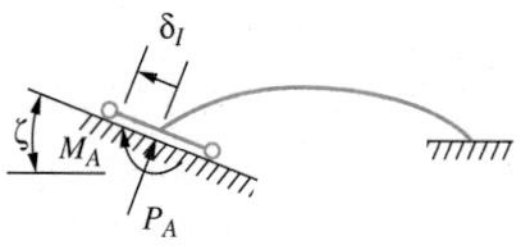	Since there is no deflection perpendicular to the incline, the following three equations must be solved for M_A, P_A, and δ_I: $$\delta_I\frac{EI\cos\zeta}{R^3}=P_A(B_{HV}\cos\zeta-B_{HH}\sin\zeta)+B_{HM}\frac{M_A}{R}-LF_H$$ $$\delta_I\frac{EI\sin\zeta}{R^3}=P_A(B_{VV}\cos\zeta-B_{VH}\sin\zeta)+B_{VM}\frac{M_A}{R}-LF_V$$ $$0=P_A(B_{MV}\cos\zeta-B_{MH}\sin\zeta)+B_{MM}\frac{M_A}{R}-LF_M$$ Use load terms given above for cases 5a–5s.
14. Left end roller supported along an inclined surface at angle ζ, right end fixed 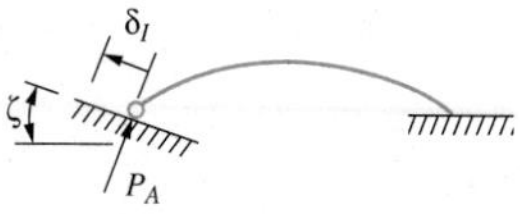	Since there is no deflection perpendicular to the incline and $M_A=0$, the following equations give P_A, δ_I, and ψ_A: $$P_A=\frac{LF_V\cos\zeta-LF_H\sin\zeta}{B_{HH}\sin^2\zeta-2B_{HV}\sin\zeta\cos\zeta+B_{VV}\cos^2\zeta}$$ $$\delta_I=\frac{R^3}{EI}\{P_A[B_{HV}(\cos^2\zeta-\sin^2\zeta)+(B_{VV}-B_{HH})\sin\zeta\cos\zeta]-LF_H\cos\zeta-LF_V\sin\zeta\}$$ $$\psi_A=\frac{R^2}{EI}[P_A(B_{MV}\cos\zeta-B_{MH}\sin\zeta)-LF_M]$$ Use load terms given above for cases 5a–5s.

TABLE 9.4 Formulas for Curved Beams of Compact Cross Section Loaded Normal to the Plane of Curvature

Notation: W = applied load normal to the plane of curvature (force); M_o = applied bending moment in a plane tangent to the curved axis of the beam (force-length); T_o = applied twisting moment in a plane normal to the curved axis of the beam (force-length); w = distributed load (force per unit length); t_o = distributed twisting moment (force-length per unit length); V_A = reaction force, M_A = reaction bending moment, T_A = reaction twisting moment, y_A = deflection normal to the plane of curvature, Θ_A = slope of the beam axis in the plane of the moment M_A, and ψ_A = roll of the beam cross section in the plane of the twisting moment T_A, all at the left end of the beam. Similarly, V_B, M_B, T_B, y_B, Θ_B, and ψ_B are the reactions and displacements at the right end; V, M, T, y, Θ, and ψ are internal shear forces, moments, and displacements at an angular position x rad from the left end. All loads and reactions are positive as shown in the diagram; y is positive upward; Θ is positive when y increases as x increases; and ψ is positive in the direction of T.

R = radius of curvature of the beam axis (length); E = modulus of elasticity (force per unit area); I = area moment of inertia about the bending axis (length to the fourth power) (note that this must be a principal axis of the beam cross section); G = modulus of rigidity (force per unit area); ν = Poisson's ratio; K = torsional stiffness constant of the cross section (length to the fourth power) (see page 402); θ = angle in radians from the left end to the position of the loading; ϕ = angle (radians) subtended by the entire span of the curved beam. See page 159 for a definition of the term $\langle x-\theta\rangle^n$.

The following constants and functions are hereby defined to permit condensing the tabulated formulas which follow, $\beta = EI/GK$.

$F_1 = \frac{1+\beta}{2} x\sin x - \beta(1-\cos x)$	$C_1 = \frac{1+\beta}{2}\phi\sin\phi - \beta(1-\cos\phi)$
$F_2 = \frac{1+\beta}{2}(x\cos x - \sin x)$	$C_2 = \frac{1+\beta}{2}(\phi\cos\phi - \sin\phi)$
$F_3 = -\beta(x-\sin x) - \frac{1+\beta}{2}(x\cos x - \sin x)$	$C_3 = -\beta(\phi-\sin\phi) - \frac{1+\beta}{2}(\phi\cos\phi - \sin\phi)$
$F_4 = \frac{1+\beta}{2}x\cos x + \frac{1-\beta}{2}\sin x$	$C_4 = \frac{1+\beta}{2}\phi\cos\phi + \frac{1-\beta}{2}\sin\phi$
$F_5 = -\frac{1+\beta}{2}x\sin x$	$C_5 = -\frac{1+\beta}{2}\phi\sin\phi$
$F_6 = F_1$	$C_6 = C_1$
$F_7 = F_5$	$C_7 = C_5$
$F_8 = \frac{1-\beta}{2}\sin x - \frac{1+\beta}{2}x\cos x$	$C_8 = \frac{1-\beta}{2}\sin\phi - \frac{1+\beta}{2}\phi\cos\phi$
$F_9 = F_2$	$C_9 = C_2$
$F_{a1} = \left\{\frac{1+\beta}{2}(x-\theta)\sin(x-\theta) - \beta[1-\cos(x-\theta)]\right\}\langle x-\theta\rangle^0$	$C_{a1} = \frac{1+\beta}{2}(\phi-\theta)\sin(\phi-\theta) - \beta[1-\cos(\phi-\theta)]$
$F_{a2} = \frac{1+\beta}{2}[(x-\theta)\cos(x-\theta) - \sin(x-\theta)]\langle x-\theta\rangle^0$	$C_{a2} = \frac{1+\beta}{2}[(\phi-\theta)\cos(\phi-\theta) - \sin(\phi-\theta)]$
$F_{a3} = \{-\beta[x-\theta-\sin(x-\theta)] - F_{a2}\}\langle x-\theta\rangle^0$	$C_{a3} = -\beta[\phi-\theta-\sin(\phi-\theta)] - C_{a2}$
$F_{a4} = \left[\frac{1+\beta}{2}(x-\theta)\cos(x-\theta) + \frac{1-\beta}{2}\sin(x-\theta)\right]\langle x-\theta\rangle^0$	$C_{a4} = \frac{1+\beta}{2}(\phi-\theta)\cos(\phi-\theta) + \frac{1-\beta}{2}\sin(\phi-\theta)$
$F_{a5} = -\frac{1+\beta}{2}(x-\theta)\sin(x-\theta)\langle x-\theta\rangle^0$	$C_{a5} = -\frac{1+\beta}{2}(\phi-\theta)\sin(\phi-\theta)$
$F_{a6} = F_{a1}$	$C_{a6} = C_{a1}$
$F_{a7} = F_{a5}$	$C_{a7} = C_{a5}$
$F_{a8} = \left[\frac{1-\beta}{2}\sin(x-\theta) - \frac{1+\beta}{2}(x-\theta)\cos(x-\theta)\right]\langle x-\theta\rangle^0$	$C_{a8} = \frac{1-\beta}{2}\sin(\phi-\theta) - \frac{1+\beta}{2}(\phi-\theta)\cos(\phi-\theta)$
$F_{a9} = F_{a2}$	$C_{a9} = C_{a2}$
$F_{a12} = \frac{1+\beta}{2}[(x-\theta)\sin(x-\theta) - 2 + 2\cos(x-\theta)]\langle x-\theta\rangle^0$	$C_{a12} = \frac{1+\beta}{2}[(\phi-\theta)\sin(\phi-\theta) - 2 + 2\cos(\phi-\theta)]$
$F_{a13} = \left\{\beta\left[1-\cos(x-\theta) - \frac{(x-\theta)^2}{2}\right] - F_{a12}\right\}\langle x-\theta\rangle^0$	$C_{a13} = \beta\left[1-\cos(\phi-\theta) - \frac{(\phi-\theta)^2}{2}\right] - C_{a12}$
$F_{a15} = F_{a2}$	$C_{a15} = C_{a2}$
$F_{a16} = F_{a3}$	$C_{a16} = C_{a3}$
$F_{a18} = \left[1-\cos(x-\theta) - \frac{1+\beta}{2}(x-\theta)\sin(x-\theta)\right]\langle x-\theta\rangle^0$	$C_{a18} = 1-\cos(\phi-\theta) - \frac{1+\beta}{2}(\phi-\theta)\sin(\phi-\theta)$
$F_{a19} = F_{a12}$	$C_{a19} = C_{a12}$

(Continued)

TABLE 9.4 Formulas for Curved Beams of Compact Cross Section Loaded Normal to the Plane of Curvature (*Continued*)

1. Concentrated intermediate lateral load

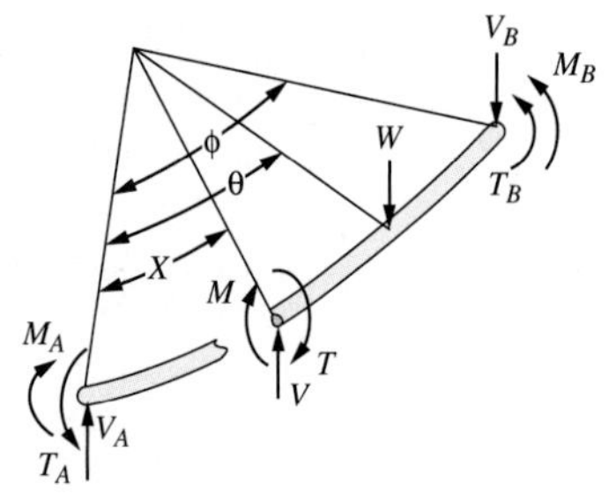

Transverse shear $= V = V_A - W\langle x-\theta\rangle^0$

Bending moment $= M = V_A R\sin x + M_A\cos x - T_A\sin x - WR\sin(x-\theta)\langle x-\theta\rangle^0$

Twisting moment $= T = V_A R(1-\cos x) + M_A\sin x + T_A\cos x - WR[1-\cos(x-\theta)]\langle x-\theta\rangle^0$

Deflection $= y = y_A + \Theta_A R\sin x + \psi_A R(1-\cos x) + \dfrac{M_A R^2}{EI}F_1 + \dfrac{T_A R^2}{EI}F_2 + \dfrac{V_A R^3}{EI}F_3 - \dfrac{WR^3}{EI}F_{a3}$

Bending slope $= \Theta = \Theta_A\cos x + \psi_A\sin x + \dfrac{M_A R}{EI}F_4 + \dfrac{T_A R}{EI}F_5 + \dfrac{V_A R^2}{EI}F_6 - \dfrac{WR^2}{EI}F_{a6}$

Roll slope $= \psi = \psi_A\cos x - \Theta_A\sin x + \dfrac{M_A R}{EI}F_7 + \dfrac{T_A R}{EI}F_8 + \dfrac{V_A R^2}{EI}F_9 - \dfrac{WR^2}{EI}F_{a9}$

For tabulated values: $V = K_V W,\ M = K_M WR,\ T = K_T WR,\ y = K_y\dfrac{WR^3}{EI},\ \Theta = K_\Theta\dfrac{WR^2}{EI},\ \psi = K_\psi\dfrac{WR^2}{EI}$

End Restraints, Reference No.	Formulas for Boundary Values and Selected Numerical Values

1a. Right end fixed, left end free

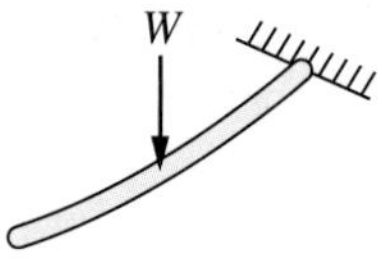

$V_A = 0\quad M_A = 0\quad T_A = 0$

$y_B = 0\quad \Theta_B = 0\quad \psi_B = 0$

$y_A = \dfrac{-WR^3}{EI}[C_{a6}\sin\phi - C_{a9}(1-\cos\phi) - C_{a3}],\quad \Theta_A = \dfrac{WR^2}{EI}(C_{a6}\cos\phi - C_{a9}\sin\phi)$

$\psi_A = \dfrac{WR^2}{EI}(C_{a9}\cos\phi + C_{a6}\sin\phi)$

$V_B = -W$

$M_B = -WR\sin(\phi-\theta)$

$T_B = -WR[1-\cos(\phi-\theta)]$

If $\beta = 1.3$ (solid or hollow round cross section, $\nu = 0.3$)

ϕ	45°	90°			180°		
θ	0°	0°	30°	60°	0°	60°	120°
K_{yA}	−0.1607	−1.2485	−0.6285	−0.1576	−7.6969	−3.7971	−0.6293
$K_{\Theta A}$	0.3058	1.1500	0.3938	0.0535	2.6000	−0.1359	−0.3929
$K_{\psi A}$	0.0590	0.5064	0.3929	0.1269	3.6128	2.2002	0.3938
K_{VB}	−1.0000	−1.0000	−1.0000	−1.0000	−1.0000	−1.0000	−1.0000
K_{MB}	−0.7071	−1.0000	−0.8660	−0.5000	−0.0000	−0.8660	−0.8660
K_{TB}	−0.2929	−1.0000	−0.5000	−0.1340	−2.0000	−1.5000	−0.5000

1b. Right end fixed, left end simply supported

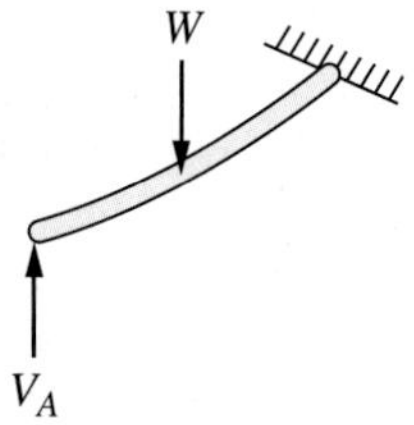

$M_A = 0\quad T_A = 0\quad y_A = 0$

$y_B = 0\quad \Theta_B = 0\quad \psi_B = 0$

$V_A = W\dfrac{C_{a9}(1-\cos\phi) - C_{a6}\sin\phi + C_{a3}}{C_9(1-\cos\phi) - C_6\sin\phi + C_3}$

$\Theta_A = \dfrac{WR^2}{EI}\dfrac{(C_{a3}C_9 - C_{a9}C_3)\sin\phi + (C_{a9}C_6 - C_{a6}C_9)(1-\cos\phi) + (C_{a6}C_3 - C_{a3}C_3)\cos\phi}{C_9(1-\cos\phi) - C_6\sin\phi + C_3}$

$\psi_A = \dfrac{WR^2}{EI}\dfrac{[C_{a6}(C_3 + C_9) - C_6(C_{a3} + C_{a9})]\sin\phi + (C_{a9}C_3 - C_{a3}C_9)\cos\phi}{C_9(1-\cos\phi) - C_6\sin\phi + C_3}$

$V_B = V_A - W$

$M_B = V_A R\sin\phi - WR\sin(\phi-\theta)$

$T_B = V_A R(1-\cos\phi) - WR[1-\cos(\phi-\theta)]$

If $\beta = 1.3$ (solid or hollow round cross section, $\nu = 0.3$)

ϕ	45°		90°		180°	
θ	15°	30°	30°	60°	60°	120°
K_{VA}	0.5136	0.1420	0.5034	0.1262	0.4933	0.0818
$K_{\Theta A}$	−0.0294	−0.0148	−0.1851	−0.0916	−1.4185	−0.6055
$K_{\psi A}$	0.0216	0.0106	0.1380	0.0630	0.4179	0.0984
K_{MB}	−0.1368	−0.1584	−0.3626	−0.3738	−0.8660	−0.8660
K_{TB}	0.0165	0.0075	0.0034	−0.0078	−0.5133	−0.3365
$K_{M\theta}$	0.1329	0.0710	0.2517	0.1093	0.4272	0.0708

(*Continued*)

TABLE 9.4 Formulas for Curved Beams of Compact Cross Section Loaded Normal to the Plane of Curvature (*Continued*)

End Restraints, Reference No.	Formulas for Boundary Values and Selected Numerical Values

1c. Right end fixed, left end supported and slope-guided

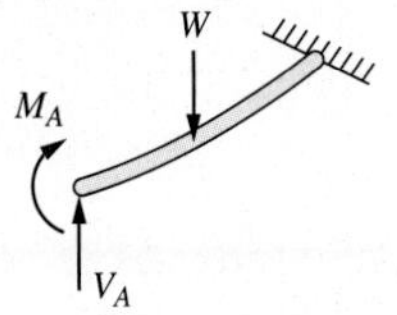

$T_A = 0 \quad y_A = 0 \quad \Theta_A = 0$

$y_B = 0 \quad \Theta_B = 0 \quad \psi_B = 0$

$$V_A = W\frac{(C_{a9}C_4 - C_{a6}C_7)(1-\cos\phi) + (C_{a6}C_1 - C_{a3}C_4)\cos\phi + (C_{a3}C_7 - C_{a9}C_1)\sin\phi}{(C_4C_9 - C_6C_7)(1-\cos\phi) + (C_1C_6 - C_3C_4)\cos\phi + (C_3C_7 - C_1C_9)\sin\phi}$$

$$M_A = WR\frac{(C_{a6}C_9 - C_{a9}C_6)(1-\cos\phi) + (C_{a3}C_6 - C_{a6}C_3)\cos\phi + (C_{a9}C_3) - C_{a3}C_9)\sin\phi}{(C_4C_9 - C_6C_7)(1-\cos\phi) + (C_1C_6 - C_3C_4)\cos\phi + (C_3C_7 - C_1C_9)\sin\phi}$$

$$\psi_A = \frac{WR^2}{EI}\frac{C_{a3}(C_4C_9 - C_6C_7) + C_{a6}(C_3C_7 - C_1C_9) + C_{a9}(C_1C_6 - C_3C_4)}{(C_4C_9 - C_6C_7)(1-\cos\phi)(C_1C_6 - C_3C_4)\cos\phi + (C_3C_7 - C_1C_9)\sin\phi}$$

$$V_B = V_A - W$$

$$M_B = V_A R\sin\phi + M_A\cos\phi - WR\sin(\phi - \theta)$$

$$T_B = V_A R(1-\cos\phi) + M_A\sin\phi - WR[1-\cos(\phi-\theta)]$$

If $\beta = 1.3$ (solid or hollow round cross section, $\nu = 0.3$)

ϕ	45°		90°		180°	
θ	15°	30°	30°	60°	60°	120°
K_{VA}	0.7407	0.2561	0.7316	0.2392	0.6686	0.1566
K_{MA}	−0.1194	−0.0600	−0.2478	−0.1226	−0.5187	−0.2214
$K_{\psi A}$	−0.0008	−0.0007	−0.0147	−0.0126	−0.2152	−0.1718
K_{MB}	−0.0607	−0.1201	−0.1344	−0.2608	−0.3473	−0.6446
K_{TB}	−0.0015	−0.0015	−0.0161	−0.0174	−0.1629	−0.1869
$K_{M\theta}$	0.0764	0.0761	0.1512	0.1458	0.3196	0.2463

1d. Right end fixed, left end supported and roll-guided

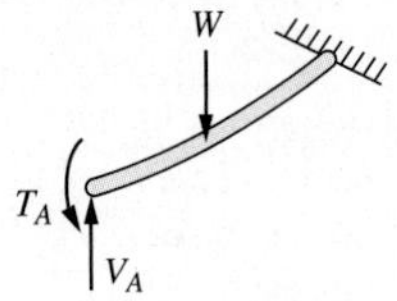

$M_A = 0 \quad y_A = 0 \quad \psi_A = 0$

$y_B = 0 \quad \Theta_B = 0 \quad \psi_B = 0$

$$V_A = W\frac{[(C_{a3} + C_{a9})C_5 - C_{a6}(C_2 + C_8)]\sin\phi + (C_{a3}C_8 - C_{a9}C_2)\cos\phi}{[C_5(C_3 + C_9) - C_6(C_2 + C_8)]\sin\phi + (C_3C_8 - C_2C_9)\cos\phi}$$

$$T_A = WR\frac{[C_{a6}(C_3 + C_9) - C_6(C_{a3} + C_{a9})]\sin\phi + (C_{a9}C_3 - C_{a3}C_9)\cos\phi}{[C_5(C_3 + C_9) - C_6(C_2 + C_8)]\sin\phi + (C_3C_8 - C_2C_9)\cos\phi}$$

$$\Theta_A = \frac{WR^2}{EI}\frac{C_{a3}(C_5C_9 - C_6C_8) + C_{a6}(C_3C_8 - C_2C_9) + C_{a9}(C_2C_6 - C_3C_5)}{[C_5(C_3 + C_9) - C_6(C_2 + C_8)]\sin\phi + (C_3C_8 - C_2C_9)\cos\phi}$$

$$V_B = V_A - W$$

$$M_B = V_A R\sin\phi - T_A\sin\phi - WR\sin(\phi - \theta)$$

$$T_B = V_A R(1-\cos\phi) + T_A\cos\phi - WR[1-\cos(\phi-\theta)]$$

If $\beta = 1.3$ (solid or hollow round cross section, $\nu = 0.3$)

ϕ	45°		90°		180°	
θ	15°	30°	30°	60°	60°	120°
K_{VA}	0.5053	0.1379	0.4684	0.1103	0.3910	0.0577
K_{TA}	−0.0226	−0.0111	−0.0862	−0.0393	−0.2180	−0.0513
$K_{\Theta A}$	−0.0252	−0.0127	−0.1320	−0.0674	−1.1525	−0.5429
K_{MB}	−0.1267	−0.1535	−0.3114	−0.3504	−0.8660	−0.8660
K_{TB}	−0.0019	−0.0015	−0.0316	−0.0237	−0.5000	−0.3333
$K_{M\theta}$	0.1366	0.0745	0.2773	0.1296	0.5274	0.0944

(*Continued*)

TABLE 9.4 Formulas for Curved Beams of Compact Cross Section Loaded Normal to the Plane of Curvature (*Continued*)

End Restraints, Reference No.	Formulas for Boundary Values and Selected Numerical Values

1e. Right end fixed, left end fixed

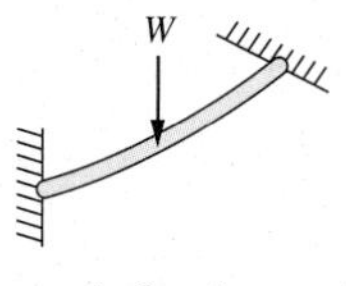

$y_A = 0 \quad \Theta_A = 0 \quad \psi_A = 0$

$y_B = 0 \quad \Theta_B = 0 \quad \psi_B = 0$

$$V_A = W\frac{C_{a3}(C_4C_8 - C_5C_7) + C_{a6}(C_2C_7 - C_1C_8) + C_{a9}(C_1C_5 - C_2C_4)}{C_1(C_5C_9 - C_6C_8) + C_4(C_3C_8 - C_2C_9) + C_7(C_2C_6 - C_3C_5)}$$

$$M_A = WR\frac{C_{a3}(C_5C_9 - C_6C_8) + C_{a6}(C_3C_8 - C_2C_9) + C_{a9}(C_2C_6 - C_3C_5)}{C_1(C_5C_9 - C_6C_8) + C_4(C_3C_8 - C_2C_9) + C_7(C_2C_6 - C_3C_5)}$$

$$T_A = WR\frac{C_{a3}(C_6C_7 - C_4C_9) + C_{a6}(C_1C_9 - C_3C_7) + C_{a9}(C_3C_4 - C_1C_6)}{C_1(C_5C_9 - C_6C_8) + C_4(C_3C_8 - C_2C_9) + C_7(C_2C_6 - C_3C_5)}$$

$$V_B = V_A - W$$

$$M_B = V_A R\sin\phi + M_A\cos\phi - T_A\sin\phi - WR\sin(\phi - \theta)$$

$$T_B = V_A R(1 - \cos\phi) + M_A\sin\phi + T_A\cos\phi - WR[1 - \cos(\phi - \theta)]$$

If $\beta = 1.3$ (solid or hollow round cross section, $\nu = 0.3$)

ϕ	45°	90°	180°	270°	360°	
θ	15°	30°	60°	90°	90°	180°
K_{VA}	0.7424	0.7473	0.7658	0.7902	0.9092	0.5000
K_{MA}	−0.1201	−0.2589	−0.5887	−0.8488	−0.9299	−0.3598
K_{TA}	0.0009	0.0135	0.1568	0.5235	0.7500	1.0000
K_{MB}	−0.0606	−0.1322	−0.2773	−0.2667	0.0701	−0.3598
K_{TB}	−0.0008	−0.0116	−0.1252	−0.3610	−0.2500	−1.0000
$K_{M\theta}$	0.0759	0.1427	0.2331	0.2667	0.1592	0.3598

1f. Right end supported and slope-guided, left end supported and slope-guided

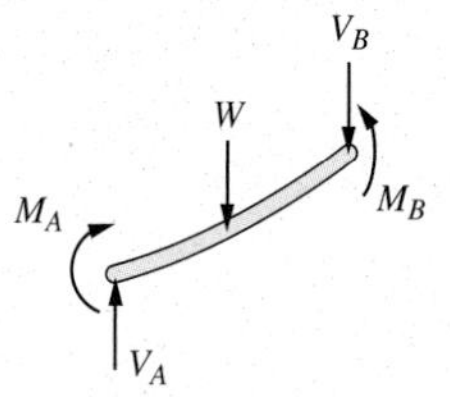

$T_A = 0 \quad y_A = 0 \quad \Theta_A = 0$

$T_B = 0 \quad y_B = 0 \quad \Theta_B = 0$

$$V_A = W\frac{[-C_1\sin\phi + C_4(1 - \cos\phi)][1 - \cos(\phi - \theta)] + C_{a3}\sin^2\phi - C_{a6}\sin\phi(1 - \cos\phi)}{C_4(1 - \cos\phi)^2 + C_3\sin^2\phi - (C_1 + C_6)(1 - \cos\phi)\sin\phi}$$

$$M_A = WR\frac{C_{a6}(1 - \cos\phi)^2 - C_{a3}(1 - \cos\phi)\sin\phi + [C_3\sin\phi - C_6(1 - \cos\phi)][1 - \cos(\phi - \theta)]}{C_4(1 - \cos\phi)^2 + C_3\sin^2\phi - (C_1 + C_6)(1 - \cos\phi)\sin\phi}$$

$$\psi_A = \frac{WR^2}{EI}\frac{(C_{a3}C_4 - C_{a6}C_1)(1 - \cos\phi) - (C_{a3}C_6 - C_{a6}C_3)\sin\phi - (C_3C_4 - C_1C_6)[1 - \cos(\phi - \theta)]}{C_4(1 - \cos\phi)^2 + C_3\sin^2\phi - (C_1 + C_6)(1 - \cos\phi)\sin\phi}$$

$$V_B = V_A - W$$

$$M_B = V_A R\sin\phi + M_A\cos\phi - WR\sin(\phi - \theta)$$

$$\psi_B = \psi_A\cos\phi + \frac{M_A R}{EI}C_7 + \frac{V_A R^2}{EI}C_9 - \frac{WR^2}{EI}C_{a9}$$

If $\beta = 1.3$ (solid or hollow round cross section, $\nu = 0.3$)

ϕ	45°	90°	180°	270°
θ	15°	30°	60°	90°
K_{VA}	0.7423	0.7457	0.7500	0.7414
K_{MA}	−0.1180	−0.2457	−0.5774	−1.2586
$K_{\psi A}$	−0.0024	−0.0215	−0.2722	−2.5702
K_{MB}	−0.0586	−0.1204	−0.2887	−0.7414
$K_{\psi B}$	−0.0023	−0.0200	−0.2372	−2.3554
$K_{M\theta}$	0.0781	0.1601	0.3608	0.7414

1g. Right end supported and slope-guided, left end supported and roll-guided

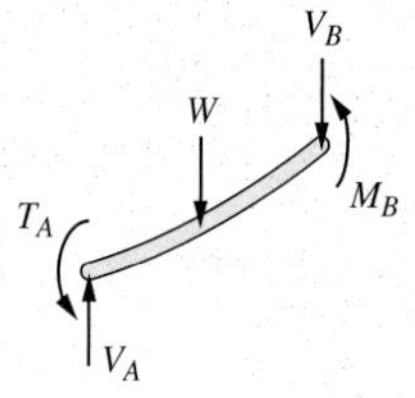

$M_A = 0 \quad y_A = 0 \quad \psi_A = 0$

$T_B = 0 \quad y_B = 0 \quad \Theta_B = 0$

$$V_A = W\frac{(C_5\sin\phi - C_2\cos\phi)[1 - \cos(\phi - \theta)] + C_{a3}\cos^2\phi - C_{a6}\sin\phi\cos\phi}{(C_5\sin\phi - C_2\cos\phi)(1 - \cos\phi) + C_3\cos^2\phi - C_6\sin\phi\cos\phi}$$

$$T_A = WR\frac{(C_3\cos\phi - C_6\sin\phi)[1 - \cos(\phi - \theta)] - (C_{a3}\cos\phi - C_{a6}\sin\phi)(1 - \cos\phi)}{(C_5\sin\phi - C_2\cos\phi)(1 - \cos\phi) + C_3\cos^2\phi - C_6\sin\phi\cos\phi}$$

$$\Theta_A = \frac{WR^2}{EI}\frac{(C_2C_6 - C_3C_5)[1 - \cos(\phi - \theta)] + (C_{a3}C_5 - C_{a6}C_2)(1 - \cos\phi) + (C_{a6}C_3 - C_{a3}C_6)\cos\phi}{(C_5\sin\phi - C_2\cos\phi)(1 - \cos\phi) + C_3\cos^2\phi - C_6\sin\phi\cos\phi}$$

$$V_B = V_A - W$$

$$M_B = V_A R\sin\phi - T_A\sin\phi - WR\sin(\phi - \theta)$$

$$\psi_B = -\Theta_A\sin\phi + \frac{T_A R}{EI}C_8 + \frac{V_A R^2}{EI}C_9 - \frac{WR^2}{EI}C_{a9}$$

If $\beta = 1.3$ (solid or hollow round cross section, $\nu = 0.3$)

ϕ	45°		90°		180°	
θ	15°	30°	30°	60°	60°	120°
K_{VA}	0.5087	0.1405	0.5000	0.1340	0.6257	0.2141
K_{TA}	−0.0212	−0.0100	−0.0774	−0.0327	−0.2486	−0.0717
$K_{\Theta A}$	−0.0252	−0.0127	−0.1347	−0.0694	−1.7627	−0.9497
K_{MB}	−0.1253	−0.1524	−0.2887	−0.3333	−0.8660	−0.8660
$K_{\psi B}$	−0.0016	−0.0012	−0.0349	−0.0262	−0.9585	−0.6390
$K_{M\theta}$	0.1372	0.0753	0.2887	0.1443	0.7572	0.2476

(*Continued*)

TABLE 9.4 Formulas for Curved Beams of Compact Cross Section Loaded Normal to the Plane of Curvature (*Continued*)

End Restraints, Reference No.	Formulas for Boundary Values and Selected Numerical Values

1h. Right end supported and slope-guided, left end simply supported

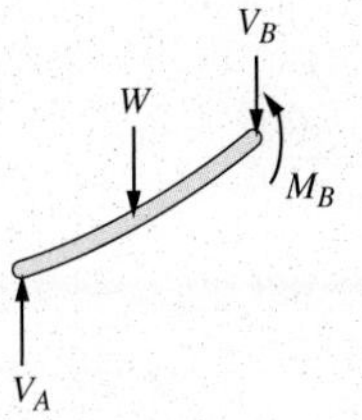

$M_A = 0 \quad T_A = 0 \quad y_A = 0$

$T_B = 0 \quad y_B = 0 \quad \Theta_B = 0$

$$V_A = W\frac{1-\cos(\phi-\theta)}{1-\cos\phi}$$

$$\Theta_A = \frac{WR^2}{EI}\left\{\frac{C_{a3}\sin\phi + C_6[1-\cos(\phi-\theta)]}{1-\cos\phi} - \frac{C_3\sin\phi[1-\cos(\phi-\theta)]}{(1-\cos\phi)^2} - C_{a6}\right\}$$

$$\psi_A = \frac{WR^2}{EI}\left[\frac{C_{a6}\sin\phi - C_{a3}\cos\phi}{1-\cos\phi} - (C_6\sin\phi - C_3\cos\phi)\frac{1-\cos(\phi-\theta)}{(1-\cos\phi)^2}\right]$$

$$V_B = V_A - W$$

$$M_B = V_A R\sin\phi - WR\sin(\phi-\theta)$$

$$\psi_B = \psi_A\cos\phi - \Theta_A\sin\phi + \frac{V_A R^2}{EI}C_9 - \frac{WR^2}{EI}C_{a9}$$

If $\beta = 1.3$ (solid or hollow round cross section, $\nu = 0.3$)

ϕ	45°		90°		180°	
θ	15°	30°	30°	60°	60°	120°
K_{VA}	0.4574	0.1163	0.5000	0.1340	0.7500	0.2500
$K_{\Theta A}$	−0.0341	−0.0169	−0.1854	−0.0909	−2.0859	−1.0429
$K_{\psi A}$	0.0467	0.0220	0.1397	0.0591	0.4784	0.1380
K_{MB}	−0.1766	−0.1766	−0.3660	−0.3660	−0.8660	−0.8660
$K_{\psi B}$	0.0308	0.0141	0.0042	−0.0097	−0.9878	−0.6475
$K_{M\theta}$	0.1184	0.0582	0.2500	0.1160	0.6495	0.2165

1i. Right end supported and roll-guided, left end supported and roll-guided

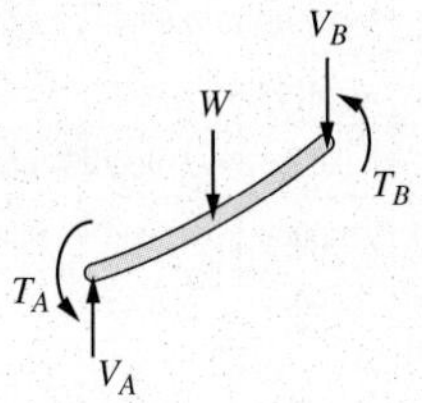

$M_A = 0 \quad y_A = 0 \quad \psi_A = 0$

$M_B = 0 \quad y_B = 0 \quad \psi_B = 0$

$$V_A = W\frac{(C_{a3}+C_{a9})\sin\phi + (C_2+C_8)\sin(\phi-\theta)}{(C_2+C_3+C_8+C_9)\sin\phi}$$

$$T_A = WR\frac{(C_{a3}+C_{a9})\sin\phi - (C_3+C_9)\sin(\phi-\theta)}{(C_2+C_3+C_8+C_9)\sin\phi}$$

$$\Theta_A = \frac{WR^2}{EI}\frac{C_{a3}(C_8+C_9) - C_{a9}(C_2+C_3) + (C_2C_9 - C_3C_8)\sin(\phi-\theta)/\sin\phi}{(C_2+C_3+C_8+C_9)\sin\phi}$$

$$V_B = V_A - W$$

$$T_B = V_A R(1-\cos\phi) + T_A\cos\phi - WR[1-\cos(\phi-\theta)]$$

$$\Theta_B = \Theta_A\cos\phi + \frac{V_A R^2}{EI}C_6 + \frac{T_A R}{EI}C_5 - \frac{WR^2}{EI}C_{a6}$$

If $\beta = 1.3$ (solid or hollow round cross section, $\nu = 0.3$)

ϕ	45°	90°	270°
θ	15°	30°	90°
K_{VA}	0.6667	0.6667	0.6667
K_{TA}	−0.0404	−0.1994	0.0667
$K_{\Theta A}$	−0.0462	−0.3430	−4.4795
K_{TB}	0.0327	0.1667	−1.3333
$K_{\Theta B}$	0.0382	0.3048	1.7333
$K_{M\theta}$	0.1830	0.4330	0.0000

1j. Right end supported and roll-guided, left end simply supported

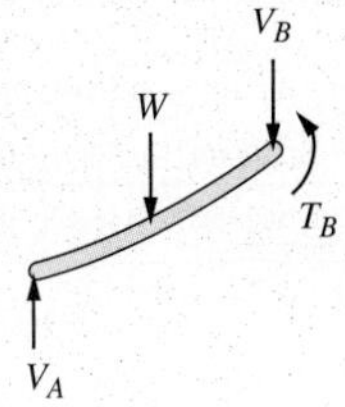

$M_A = 0 \quad T_A = 0 \quad y_A = 0$

$M_B = 0 \quad y_B = 0 \quad \psi_B = 0$

$$V_A = W\frac{\sin(\phi-\theta)}{\sin\phi}$$

$$\Theta_A = \frac{WR^2}{EI}\left\{\frac{C_{a3}\cos\phi - C_{a9}(1-\cos\phi)}{\sin\phi} - [C_3\cos\phi - C_9(1-\cos\phi)]\frac{\sin(\phi-\theta)}{\sin^2\phi}\right\}$$

$$\psi_A = \frac{WR^2}{EI}\left[C_{a3} + C_{a9} - (C_3+C_9)\frac{\sin(\phi-\theta)}{\sin\phi}\right]$$

$$V_B = V_A - W$$

$$T_B = V_A R(1-\cos\phi) - WR[1-\cos(\phi-\theta)]$$

$$\Theta_B = \Theta_A\cos\phi + \psi_A\sin\phi + \frac{V_A R^2}{EI}C_6 - \frac{WR^2}{EI}C_{a6}$$

If $\beta = 1.3$ (solid or hollow round cross section, $\nu = 0.3$)

ϕ	45°		90°		270°	
θ	15°	30°	30°	60°	90°	180°
K_{VA}	0.7071	0.3660	0.8660	0.5000	0.0000	−1.0000
$K_{\Theta A}$	−0.0575	−0.0473	−0.6021	−0.5215	−3.6128	0.0000
$K_{\psi A}$	0.0413	0.0334	0.4071	0.3403	−4.0841	−8.1681
K_{TB}	0.0731	0.0731	0.3660	0.3660	−2.0000	−2.0000
$K_{\Theta B}$	0.0440	0.0509	0.4527	0.4666	6.6841	14.3810
$K_{M\theta}$	0.1830	0.1830	0.4330	0.4330	0.0000	0.0000

(*Continued*)

TABLE 9.4 Formulas for Curved Beams of Compact Cross Section Loaded Normal to the Plane of Curvature (*Continued*)

End Restraints, Reference No.	Formulas for Boundary Values and Selected Numerical Values

2. Concentrated in intermediate bending moment

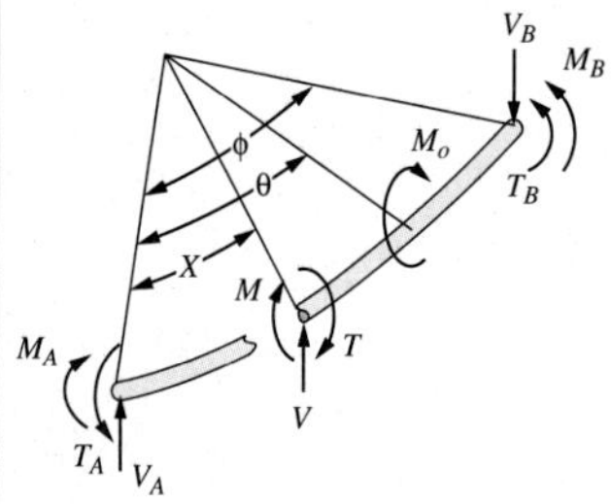

Transverse shear $= V = V_A$

Bending moment $= M = V_A R \sin x + M_A \cos x - T_A \sin x + M_o \cos(x-\theta)\langle x-\theta\rangle^0$

Twisting moment $= T = V_A R(1-\cos x) + M_A \sin x + T_A \cos x + M_o \sin(x-\theta)\langle x-\theta\rangle^0$

Vertical deflection $y = y_A + \Theta_A R \sin x + \psi_A R(1-\cos x) + \frac{M_A R^2}{EI}F_1 + \frac{T_A R^2}{EI}F_2 + \frac{V_A R^3}{EI}F_3 + \frac{M_o R^2}{EI}F_{a1}$

Bending slope $= \Theta = \Theta_A \cos x + \psi_A \sin x + \frac{M_A R}{EI}F_4 + \frac{T_A R}{EI}F_5 + \frac{V_A R^2}{EI}F_6 + \frac{M_o R}{EI}F_{a4}$

Roll slope $= \psi = \psi_A \cos x - \Theta_A \sin x + \frac{M_A R}{EI}F_7 + \frac{T_A R}{EI}F_8 + \frac{V_A R^2}{EI}F_9 + \frac{M_o R}{EI}F_{a7}$

For tabulated values $V = K_V \frac{M_o}{R}$, $M = K_M M_o$, $T = K_T M_o$, $y = K_y \frac{M_o R^2}{EI}$, $\Theta = K_\Theta \frac{M_o R}{EI}$, $\psi = K_\psi \frac{M_o R}{EI}$

2a. Right end fixed, left end free

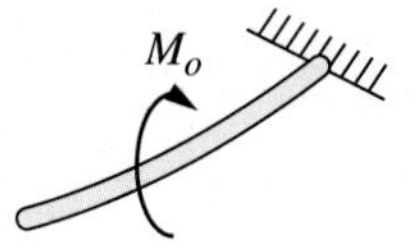

$V_A = 0 \quad M_A = 0 \quad T_A = 0$

$y_B = 0 \quad \Theta_B = 0 \quad \psi_B = 0$

$$y_A = \frac{M_o R^2}{EI}[C_{a4}\sin\phi - C_{a7}(1-\cos\phi) - C_{a1}]$$

$$\Theta_A = \frac{M_o R}{EI}(C_{a7}\sin\phi - C_{a4}\cos\phi)$$

$$\psi_A = -\frac{M_o R}{EI}(C_{a4}\sin\phi + C_{a7}\cos\phi)$$

$V_B = 0, \quad M_B = M_o\cos(\phi-\theta)$

$T_B = M_o \sin(\phi-\theta)$

If $\beta = 1.3$ (solid or hollow round cross section, $\nu = 0.3$)

ϕ	45°	90°			180°		
θ	0°	0°	30°	60°	0°	60°	120°
K_{yA}	0.3058	1.1500	1.1222	0.6206	2.6000	4.0359	1.6929
$K_{\Theta A}$	−0.8282	−1.8064	−1.0429	−0.3011	−3.6128	−1.3342	0.4722
$K_{\psi A}$	0.0750	0.1500	−0.4722	−0.4465	0.0000	−2.0859	−1.0429
K_{MB}	0.7071	0.0000	0.5000	0.8660	−1.0000	−0.5000	0.5000
K_{TB}	0.7071	1.0000	0.8660	0.5000	0.0000	0.8660	0.8660

2b. Right end fixed, left end simply supported

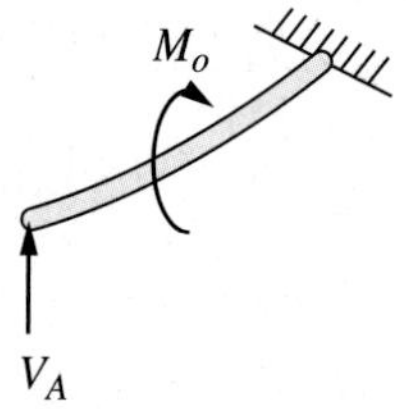

$M_A = 0 \quad T_A = 0 \quad y_A = 0$

$y_B = 0 \quad \Theta_B = 0 \quad \psi_B = 0$

$$V_A = \frac{-M_o}{R}\frac{C_{a7}(1-\cos\phi) - C_{a4}\sin\phi + C_{a1}}{C_9(1-\cos\phi) - C_6\sin\phi + C_3}$$

$$\Theta_A = -\frac{M_o R}{EI}\frac{(C_{a1}C_9 - C_{a7}C_3)\sin\phi + (C_{a7}C_6 - C_{a4}C_9)(1-\cos\phi) + (C_{a4}C_3 - C_{a1}C_6)\cos\phi}{C_9(1-\cos\phi) - C_6\sin\phi + C_3}$$

$$\psi_A = -\frac{M_o R}{EI}\frac{[(C_{a4}(C_9+C_3) - C_6(C_{a1}+C_{a7})]\sin\phi + (C_{a7}C_3 - C_{a1}C_9)\cos\phi}{C_9(1-\cos\phi) - C_6\sin\phi + C_3}$$

$V_B = V_A$

$M_B = V_A R\sin\phi + M_o\cos(\phi-\theta)$

$T_B = V_A R(1-\cos\phi) + M_o\sin(\phi-\theta)$

If $\beta = 1.3$ (solid or hollow round cross section, $\nu = 0.3$)

ϕ	45°	90°			180°		
θ	0°	0°	30°	60°	0°	60°	120°
K_{VA}	−1.9021	−0.9211	−0.8989	−0.4971	−0.3378	−0.5244	−0.2200
$K_{\Theta A}$	−0.2466	−0.7471	−0.0092	0.2706	−2.7346	0.0291	1.0441
$K_{\psi A}$	0.1872	0.6165	−0.0170	−0.1947	1.2204	−0.1915	−0.2483
K_{MB}	−0.6379	−0.9211	−0.3989	0.3689	−1.0000	−0.5000	0.5000
K_{TB}	0.1500	0.0789	−0.0329	0.0029	−0.6756	−0.1827	0.4261

(*Continued*)

TABLE 9.4 Formulas for Curved Beams of Compact Cross Section Loaded Normal to the Plane of Curvature (*Continued*)

End Restraints, Reference No.	Formulas for Boundary Values and Selected Numerical Values

2c. Right end fixed, left end supported and slope-guided

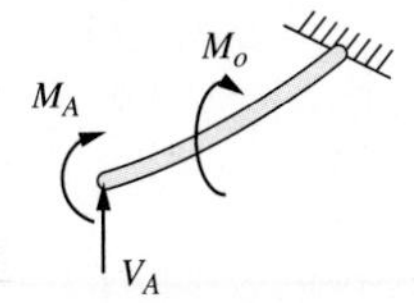

$T_A = 0 \quad y_A = 0 \quad \Theta_A = 0$

$y_B = 0 \quad \Theta_B = 0 \quad \psi_B = 0$

$$V_A = -\frac{M_o}{R}\frac{(C_{a7}C_4 - C_{a4}C_7)(1-\cos\phi) + (C_{a4}C_1 - C_{a1}C_4)\cos\phi + (C_{a1}C_7 - C_{a7}C_1)\sin\phi}{(C_4C_9 - C_6C_7)(1-\cos\phi) + (C_1C_6 - C_3C_4)\cos\phi + (C_3C_7 - C_1C_9)\sin\phi}$$

$$M_A = -M_o\frac{(C_{a4}C_9 - C_{a7}C_6)(1-\cos\phi) + (C_{a1}C_6 - C_{a4}C_3)\cos\phi + (C_{a7}C_3 - C_{a1}C_9)\sin\phi}{(C_4C_9 - C_6C_7)(1-\cos\phi) + (C_1C_6 - C_3C_4)\cos\phi + (C_3C_7 - C_1C_9)\sin\phi}$$

$$\psi_A = -\frac{M_oR}{EI}\frac{C_{a1}(C_4C_9 - C_6C_7) + C_{a4}(C_3C_7 - C_1C_9) + C_{a7}(C_1C_6 - C_3C_4)}{(C_4C_9 - C_6C_7)(1-\cos\phi) + (C_1C_6 - C_3C_4)\cos\phi + (C_3C_7 - C_1C_9)\sin\phi}$$

$$V_B = V_A$$

$$M_B = V_AR\sin\phi + M_A\cos\phi + M_o\cos(\phi - \theta)$$

$$T_B = V_AR(1-\cos\phi) + M_A\sin\phi + M_o\sin(\phi - \theta)$$

If β = 1.3 (solid or hollow round cross section, ν = 0.3)

ϕ	45°		90°		180°	
θ	15°	30°	30°	60°	60°	120°
K_{VA}	−1.7096	−1.6976	−0.8876	−0.8308	−0.5279	−0.3489
K_{MA}	−0.0071	0.3450	−0.0123	0.3622	0.0107	0.3818
$K_{\psi A}$	−0.0025	0.0029	−0.0246	0.0286	−0.1785	0.2177
K_{MB}	−0.3478	0.0095	−0.3876	0.0352	−0.5107	0.1182
K_{TB}	−0.0057	0.0056	−0.0338	0.0314	−0.1899	0.1682

2d. Right end fixed, left end supported and roll-guided

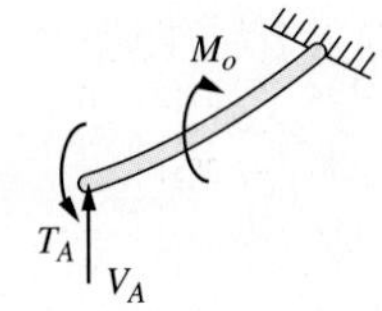

$M_A = 0 \quad y_A = 0 \quad \psi_A = 0$

$y_B = 0 \quad \Theta_B = 0 \quad \psi_B = 0$

$$V_A = -\frac{M_o}{R}\frac{[(C_{a1} + C_{a7})C_5 - C_{a4}(C_2 + C_8)]\sin\phi + (C_{a1}C_8 - C_{a7}C_2)\cos\phi}{[C_5(C_3 + C_9) - C_6(C_2 + C_8)]\sin\phi + (C_3C_8 - C_2C_9)\cos\phi}$$

$$T_A = -M_o\frac{[C_{a4}(C_3 + C_9) - (C_{a1} + C_{a7})C_6]\sin\phi + (C_{a7}C_3 - C_{a1}C_9)\cos\phi}{[C_5(C_3 + C_9) - C_6(C_2 + C_8)]\sin\phi + (C_3C_8 - C_2C_9)\cos\phi}$$

$$\Theta_A = -\frac{M_oR}{EI}\frac{C_{a1}(C_5C_9 - C_6C_8) + C_{a4}(C_3C_8 - C_2C_9) + C_{a7}(C_2C_6 - C_3C_5)}{[C_5(C_3 + C_9) - C_6(C_2 + C_8)]\sin\phi + (C_3C_8 - C_2C_9)\cos\phi}$$

$$V_B = V_A$$

$$M_B = V_AR\sin\phi - T_A\sin\phi + M_o\cos(\phi - \theta)$$

$$T_B = V_AR(1-\cos\phi) + T_A\cos\phi + M_o\sin(\phi - \theta)$$

If β = 1.3 (solid or hollow round cross section, ν = 0.3)

ϕ	45°	90°			180°		
θ	0°	0°	30°	60°	0°	30°	60°
K_{VA}	−1.9739	−1.0773	−0.8946	−0.4478	−0.6366	−0.4775	−0.1592
K_{TA}	−0.1957	−0.3851	0.0106	0.1216	−0.6366	0.0999	0.1295
$K_{\Theta A}$	−0.2100	−0.5097	−0.0158	0.1956	−1.9576	−0.0928	0.8860
K_{MB}	−0.5503	−0.6923	−0.4052	0.2966	−1.0000	−0.5000	0.5000
K_{TB}	−0.0094	−0.0773	−0.0286	0.0522	−0.6366	−0.1888	0.4182

2e. Right end fixed, left end fixed

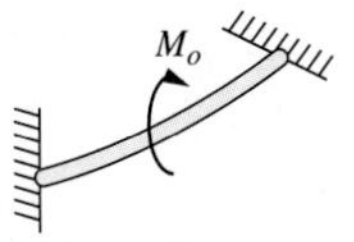

$y_A = 0 \quad \Theta_A = 0 \quad \psi_A = 0$

$y_B = 0 \quad \Theta_B = 0 \quad \psi_B = 0$

$$V_A = -\frac{M_o}{R}\frac{C_{a1}(C_4C_8 - C_5C_7) + C_{a4}(C_2C_7 - C_1C_8) + C_{a7}(C_1C_5 - C_2C_4)}{C_1(C_5C_9 - C_6C_8) + C_4(C_3C_8 - C_2C_9) + C_7(C_2C_6 - C_3C_5)}$$

$$M_A = -M_o\frac{C_{a1}(C_5C_9 - C_6C_8) + C_{a4}(C_3C_8 - C_2C_9) + C_{a7}(C_2C_6 - C_3C_5)}{C_1(C_5C_9 - C_6C_8) + C_4(C_3C_8 - C_2C_9) + C_7(C_2C_6 - C_3C_5)}$$

$$T_A = -M_o\frac{C_{a1}(C_6C_7 - C_4C_9) + C_{a4}(C_1C_9 - C_3C_7) + C_{a7}(C_3C_4 - C_1C_6)}{C_1(C_5C_9 - C_6C_8) + C_4(C_3C_8 - C_2C_9) + C_7(C_2C_6 - C_3C_5)}$$

$$V_B = V_A$$

$$M_B = V_AR\sin\phi + M_A\cos\phi - T_A\sin\phi + M_o\cos(\phi - \theta)$$

$$T_B = V_AR(1-\cos\phi) + M_A\sin\phi + T_A\cos\phi + M_o\sin(\phi - \theta)$$

If β = 1.3 (solid or hollow round cross section, ν = 0.3)

ϕ	45°	90°	180°	270°	360°	
θ	15°	30°	60°	90°	90°	180°
K_{VA}	−1.7040	−0.8613	−0.4473	−0.3115	−0.1592	−0.3183
K_{MA}	−0.0094	−0.0309	−0.0474	0.0584	−0.0208	0.5000
K_{TA}	0.0031	0.0225	0.1301	0.2788	0.5908	−0.3183
K_{MB}	−0.3477	−0.3838	−0.4526	−0.4097	−0.0208	−0.5000
K_{TB}	−0.0036	−0.0262	−0.1586	−0.3699	−0.4092	−0.3183

(*Continued*)

TABLE 9.4 Formulas for Curved Beams of Compact Cross Section Loaded Normal to the Plane of Curvature (*Continued*)

End Restraints, Reference No.	Formulas for Boundary Values and Selected Numerical Values

2f. Right end supported and slope-guided, left end supported and slope-guided

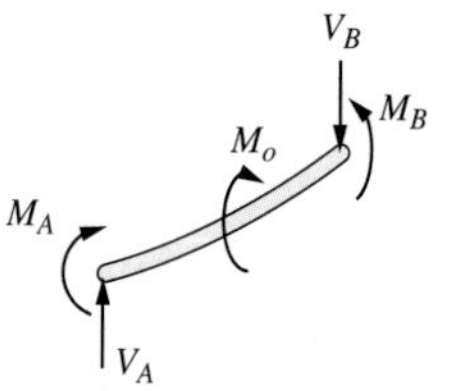

$T_A=0 \quad y_A=0 \quad \Theta_A=0$

$T_B=0 \quad y_B=0 \quad \Theta_B=0$

$$V_A=+\frac{M_o}{R}\frac{[C_1\sin\phi-C_4(1-\cos\phi)]\sin(\phi-\theta)-C_{a1}\sin^2\phi+C_{a4}\sin\phi(1-\cos\phi)}{C_4(1-\cos\phi)^2+C_3\sin^2\phi-(C_1+C_6)(1-\cos\phi)\sin\phi}$$

$$M_A=-M_o\frac{[C_3\sin\phi-C_6(1-\cos\phi)]\sin(\phi-\theta)-C_{a1}(1-\cos\phi)\sin\phi+C_{a4}(1-\cos\phi)^2}{C_4(1-\cos\phi)^2+C_3\sin^2\phi-(C_1+C_6)(1-\cos\phi)\sin\phi}$$

$$\psi_A=\frac{M_oR}{EI}\frac{(C_3C_4-C_1C_6)\sin(\phi-\theta)+(C_{a1}C_6-C_{a4}C_3)\sin\phi-(C_{a1}C_4-C_{a4}C_1)(1-\cos\phi)}{C_4(1-\cos\phi)^2+C_3\sin^2\phi-(C_1+C_6)(1-\cos\phi)\sin\phi}$$

$$V_B=V_A$$

$$M_B=V_AR\sin\phi+M_A\cos\phi+M_o\cos(\phi-\theta)$$

$$\psi_B=\psi_A\cos\phi+\frac{M_AR}{EI}C_7+\frac{V_AR^2}{EI}C_9+\frac{M_oR}{EI}C_{a7}$$

If $\beta = 1.3$ (solid or hollow round cross section, $\nu = 0.3$)

ϕ	45°	90°	180°	270°
θ	15°	30°	60°	90°
K_{VA}	−1.7035	−0.8582	−0.4330	−0.2842
K_{MA}	−0.0015	−0.0079	−0.0577	−0.2842
$K_{\psi A}$	−0.0090	−0.0388	−0.2449	−1.7462
K_{MB}	−0.3396	−0.3581	−0.4423	−0.7159
$K_{\psi B}$	−0.0092	−0.0418	−0.2765	−1.8667

2g. Right end supported and slope-guided, left end supported and roll-guided

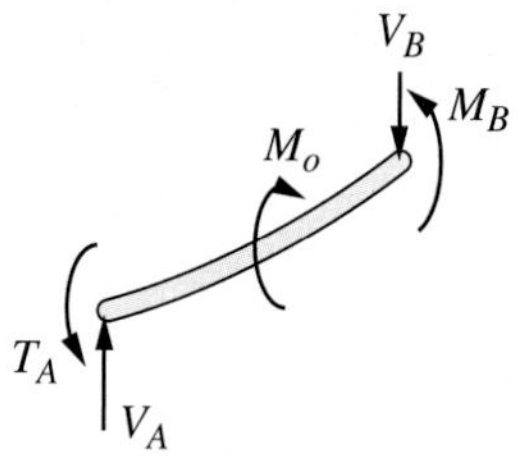

$M_A=0 \quad y_A=0 \quad \psi_A=0$

$T_B=0 \quad y_B=0 \quad \Theta_B=0$

$$V_A=-\frac{M_o}{R}\frac{C_{a1}\cos^2\phi-C_{a4}\sin\phi\cos\phi+(C_5\sin\phi-C_2\cos\phi)\sin(\phi-\theta)}{(C_5\sin\phi-C_2\cos\phi)(1-\cos\phi)+C_3\cos^2\phi-C_6\sin\phi\cos\phi}$$

$$T_A=-M_o\frac{(C_{a4}\sin\phi-C_{a1}\cos\phi)(1-\cos\phi)+(C_3\cos\phi-C_6\sin\phi)\sin(\phi-\theta)}{(C_5\sin\phi-C_2\cos\phi)(1-\cos\phi)+C_3\cos^2\phi-C_6\sin\phi\cos\phi}$$

$$\Theta_A=\frac{-M_oR}{EI}\frac{(C_{a1}C_5-C_{a4}C_2)(1-\cos\phi)+(C_{a4}C_3-C_{a1}C_6)\cos\phi+(C_2C_6-C_3C_5)\sin(\phi-\theta)}{(C_5\sin\phi-C_2\cos\phi)(1-\cos\phi)+C_3\cos^2\phi-C_6\sin\phi\cos\phi}$$

$$V_B=V_A$$

$$M_B=V_AR\sin\phi-T_A\sin\phi+M_o\cos(\phi-\theta)$$

$$\psi_B=-\Theta_A\sin\phi+\frac{T_AR}{EI}C_8+\frac{V_AR^2}{EI}C_9+\frac{M_oR}{EI}C_{a7}$$

If $\beta = 1.3$ (solid or hollow round cross section, $\nu = 0.3$)

ϕ	45°	90°			180°		
θ	0°	0°	30°	60°	0°	60°	120°
K_{VA}	−1.9576	−1.0000	−0.8660	−0.5000	−0.3378	−0.3888	−0.3555
K_{TA}	−0.1891	−0.3634	0.0186	0.1070	−0.6756	0.0883	0.1551
$K_{\Theta A}$	−0.2101	−0.5163	−0.0182	0.2001	−2.7346	−0.3232	1.3964
K_{MB}	−0.5434	−0.6366	−0.3847	0.2590	−1.0000	−0.5000	0.5000
$K_{\psi B}$	−0.0076	−0.0856	−0.0316	0.0578	−1.2204	−0.3619	0.8017

2h. Right end supported and slope-guided, left end simply supported

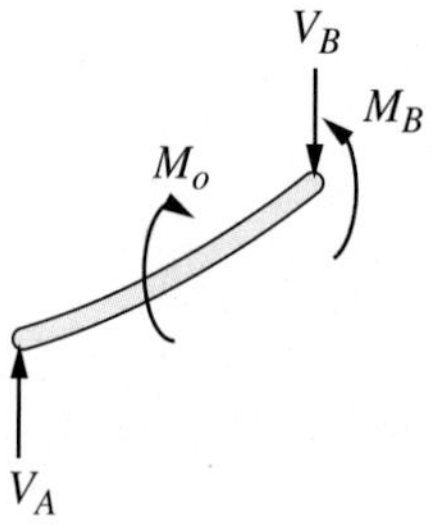

$M_A=0 \quad T_A=0 \quad y_A=0$

$T_B=0 \quad y_B=0 \quad \Theta_B=0$

$$V_A=-\frac{M_o}{R}\frac{\sin(\phi-\theta)}{1-\cos\phi}$$

$$\Theta_A=-\frac{M_oR}{EI}\left[\frac{C_{a1}\sin\phi+C_6\sin(\phi-\theta)}{1-\cos\phi}-\frac{C_3\sin\phi\sin(\phi-\theta)}{(1-\cos\phi)^2}-C_{a4}\right]$$

$$\psi_A=-\frac{M_oR}{EI}\left[\frac{C_{a4}\sin\phi-C_{a1}\cos\phi}{1-\cos\phi}+\frac{(C_3\cos\phi-C_6\sin\phi)\sin(\phi-\theta)}{(1-\cos\phi)^2}\right]$$

$$V_B=V_A$$

$$M_B=V_AR\sin\phi+M_o\cos(\phi-\theta)$$

$$\psi_B=\psi_A\cos\phi-\Theta_A\sin\phi+\frac{V_AR^2}{EI}C_9+\frac{M_oR}{EI}C_{a7}$$

If $\beta = 1.3$ (solid or hollow round cross section, $\nu = 0.3$)

ϕ	45°	90°			180°		
θ	0°	0°	30°	60°	0°	60°	120°
K_{VA}	−2.4142	−1.0000	−0.8660	−0.5000	0.0000	−0.4330	−0.4330
$K_{\Theta A}$	−0.2888	−0.7549	−0.0060	0.2703	−3.6128	−0.2083	1.5981
$K_{\psi A}$	0.4161	0.6564	−0.0337	−0.1933	1.3000	−0.1700	−0.2985
K_{MB}	−1.0000	−1.0000	−0.3660	0.3660	−1.0000	−0.5000	0.5000
$K_{\psi B}$	0.2811	0.0985	−0.0410	0.0036	−1.3000	−0.3515	0.8200

(*Continued*)

TABLE 9.4 Formulas for Curved Beams of Compact Cross Section Loaded Normal to the Plane of Curvature (*Continued*)

End Restraints, Reference No.	Formulas for Boundary Values and Selected Numerical Values

2i. Right end supported and roll-guided, left end supported and roll-guided

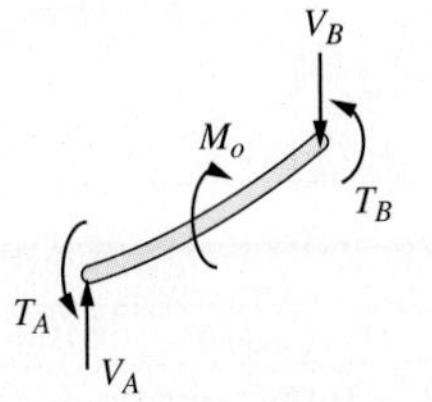

$M_A = 0 \quad y_A = 0 \quad \psi_A = 0$

$M_B = 0 \quad y_B = 0 \quad \psi_B = 0$

$$V_A = -\frac{M_o}{R}\frac{(C_{a1}+C_{a7})\sin^2\phi + (C_2+C_8)\cos(\phi-\theta)\sin\phi}{(C_2+C_3+C_8+C_9)\sin^2\phi}$$

$$T_A = -M_o\frac{(C_{a1}+C_{a7})\sin^2\phi - (C_3+C_9)\cos(\phi-\theta)\sin\phi}{(C_2+C_3+C_8+C_9)\sin^2\phi}$$

$$\Theta_A = -\frac{M_oR}{EI}\frac{[C_{a1}(C_8+C_9) - C_{a7}(C_2+C_3)]\sin\phi + (C_2C_9 - C_3C_8)\cos(\phi-\theta)}{(C_2+C_3+C_8+C_9)\sin^2\phi}$$

$$V_B = V_A$$

$$T_B = V_AR(1-\cos\phi) + T_A\cos\phi + M_o\sin(\phi-\theta)$$

$$\Theta_B = \Theta_A\cos\phi + \frac{T_AR}{EI}C_5 + \frac{V_AR^2}{EI}C_6 + \frac{M_oR}{EI}C_{a4}$$

If $\beta = 1.3$ (solid or hollow round cross section, $\nu = 0.3$)

ϕ	45°		90°		270°	
θ	0°	15°	0°	30°	0°	90°
K_{VA}	−1.2732	−1.2732	−0.6366	−0.6366	−0.2122	−0.2122
K_{TA}	−0.2732	−0.0485	−0.6366	−0.1366	−0.2122	0.7878
$K_{\Theta A}$	−0.3012	−0.0605	−0.9788	−0.2903	−5.1434	0.1259
K_{TB}	0.1410	0.0928	0.3634	0.2294	−1.2122	−0.2122
$K_{\Theta B}$	0.1658	0.1063	0.6776	0.3966	0.4259	2.0823

2j. Right end supported and roll-guided, left end simply supported

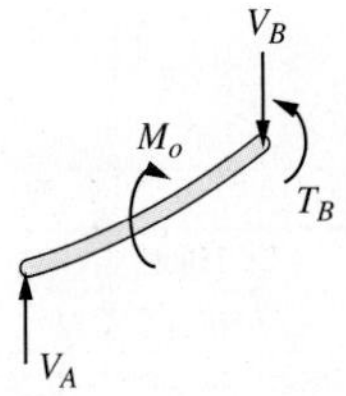

$M_A = 0 \quad T_A = 0 \quad y_A = 0$

$M_B = 0 \quad y_B = 0 \quad \psi_B = 0$

$$V_A = -\frac{M_o\cos(\phi-\theta)}{R\sin\phi}$$

$$\Theta_A = -\frac{M_oR}{EI}\left\{\frac{C_{a1}\cos\phi - C_{a7}(1-\cos\phi)}{\sin\phi} - \frac{[C_3\cos\phi - C_9(1-\cos\phi)]\cos(\phi-\theta)}{\sin^2\phi}\right\}$$

$$\psi_A = -\frac{M_oR}{EI}\left[C_{a1} + C_{a7} - \frac{(C_3+C_9)\cos(\phi-\theta)}{\sin\phi}\right]$$

$$V_B = V_A$$

$$T_B = V_AR(1-\cos\phi) + M_o\sin(\phi-\theta)$$

$$\Theta_B = \Theta_A\cos\phi + \psi_A\sin\phi + \frac{V_AR^2}{EI}C_6 + \frac{M_oR}{EI}C_{a4}$$

If $\beta = 1.3$ (solid or hollow round cross section, $\nu = 0.3$)

ϕ	45°			90°		
θ	0°	15°	30°	0°	30°	60°
K_{VA}	−1.0000	−1.2247	−1.3660	0.0000	−0.5000	−0.8660
$K_{\Theta A}$	−0.3774	−0.0740	0.1322	−1.8064	−0.4679	0.6949
$K_{\psi A}$	0.2790	0.0495	−0.0947	1.3000	0.2790	−0.4684
K_{TB}	0.4142	0.1413	−0.1413	1.0000	0.3660	−0.3660
$K_{\Theta B}$	0.2051	0.1133	−0.0738	1.1500	0.4980	−0.4606

3. Concentrated intermediate twisting moment (torque)

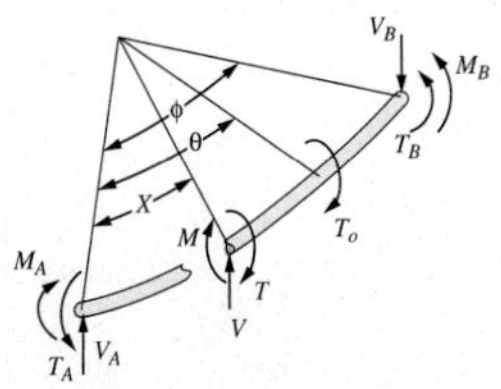

Transverse shear $= V = V_A$

Bending moment $= M = V_AR\sin x + M_A\cos x - T_A\sin x - T_o\sin(x-\theta)\langle x-\theta\rangle^0$

Twisting moment $= T = V_AR(1-\cos x) + M_A\sin x + T_A\cos x + T_o\cos(x-\theta)\langle x-\theta\rangle^0$

Vertical deflection $= y = y_A + \Theta_AR\sin x + \psi_AR(1-\cos x) + \frac{M_AR^2}{EI}F_1 + \frac{T_AR^2}{EI}F_2 + \frac{V_AR^3}{EI}F_3 + \frac{T_oR^2}{EI}F_{a2}$

Bending slope $= \Theta = \Theta_A\cos x + \psi_A\sin x + \frac{M_AR}{EI}F_4 + \frac{T_AR}{EI}F_5 + \frac{V_AR^2}{EI}F_6 + \frac{T_oR}{EI}F_{a5}$

Roll slope $= \psi = \psi_A\cos x - \Theta_A\sin x + \frac{M_AR}{EI}F_7 + \frac{T_AR}{EI}F_8 + \frac{V_AR^2}{EI}F_9 + \frac{T_oR}{EI}F_{a8}$

For tabulated values: $V = K_V\frac{T_o}{R}, \quad M = K_MT_o, \quad T = K_TT_o, \quad y = K_y\frac{T_oR^2}{EI}, \quad \Theta = K_\Theta\frac{T_oR}{EI}, \quad \psi = K_\psi\frac{T_oR}{EI}$

(*Continued*)

TABLE 9.4 Formulas for Curved Beams of Compact Cross Section Loaded Normal to the Plane of Curvature (*Continued*)

End Restraints, Reference No.	Formulas for Boundary Values and Selected Numerical Values

3a. Right end fixed, left end free

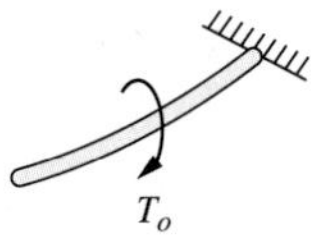

$V_A=0 \quad M_A=0 \quad T_A=0$

$y_B=0 \quad \Theta_B=0 \quad \psi_B=0$

$$y_A=\frac{T_oR^2}{EI}[C_{a5}\sin\phi-C_{a8}(1-\cos\phi)-C_{a2}]$$

$$\Theta_A=-\frac{T_oR}{EI}(C_{a5}\cos\phi-C_{a8}\sin\phi)$$

$$\psi_A=-\frac{T_oR}{EI}(C_{a8}\cos\phi+C_{a5}\sin\phi)$$

$$V_B=0$$

$$M_B=-T_o\sin(\phi-\theta)$$

$$T_B=T_o\cos(\phi-\theta)$$

If $\beta=1.3$ (solid or hollow round cross section, $\nu=0.3$)

ϕ	45°	90°			180°		
θ	0°	0°	30°	60°	0°	60°	120°
K_{yA}	−0.0590	−0.5064	0.0829	0.3489	−3.6128	0.0515	1.8579
$K_{\Theta A}$	−0.0750	−0.1500	−0.7320	−0.5965	0.0000	−2.0859	−1.0429
$K_{\psi A}$	0.9782	1.8064	1.0429	0.3011	3.6128	1.0744	−0.7320
K_{MB}	−0.7071	−1.0000	−0.8660	−0.5000	0.0000	−0.8660	−0.8660
K_{TB}	0.7071	0.0000	0.5000	0.8660	−1.0000	−0.5000	0.5000

3b. Right end fixed, left end simply supported

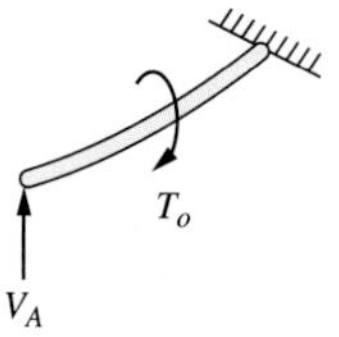

$M_A=0 \quad T_A=0 \quad y_A=0$

$y_B=0 \quad \Theta_B=0 \quad \psi_B=0$

$$V_A=-\frac{T_o}{R}\frac{C_{a8}(1-\cos\phi)-C_{a5}\sin\phi+C_{a2}}{C_9(1-\cos\phi)-C_6\sin\phi+C_3}$$

$$\Theta_A=-\frac{T_oR}{EI}\frac{(C_{a2}C_9-C_{a8}C_3)\sin\phi+(C_{a8}C_6-C_{a5}C_9)(1-\cos\phi)+(C_{a5}C_3-C_{a2}C_6)\cos\phi}{C_9(1-\cos\phi)-C_6\sin\phi+C_3}$$

$$\psi_A=-\frac{T_oR}{EI}\frac{[C_{a5}(C_9+C_3)-C_6(C_{a2}+C_{a8})]\sin\phi+(C_{a8}C_3-C_{a2}C_9)\cos\phi}{C_9(1-\cos\phi)-C_6\sin\phi+C_3}$$

$$V_B=V_A$$

$$M_B=V_AR\sin\phi-T_o\sin(\phi-\theta)$$

$$T_B=V_AR(1-\cos\phi)+T_o\cos(\phi-\theta)$$

If $\beta=1.3$ (solid or hollow round cross section, $\nu=0.3$)

ϕ	45°	90°			180°		
θ	0°	0°	30°	60°	0°	60°	120°
K_{VA}	0.3668	0.4056	−0.0664	−0.2795	0.4694	−0.0067	−0.2414
$K_{\Theta A}$	−0.1872	−0.6165	−0.6557	−0.2751	−1.2204	−2.0685	−0.4153
$K_{\psi A}$	0.9566	1.6010	1.0766	0.4426	1.9170	1.0985	0.1400
K_{MB}	−0.4477	−0.5944	−0.9324	−0.7795	0.0000	−0.8660	−0.8660
K_{TB}	0.8146	0.4056	0.4336	0.5865	−0.0612	−0.5134	0.0172

3c. Right end fixed, left end supported and slope-guided

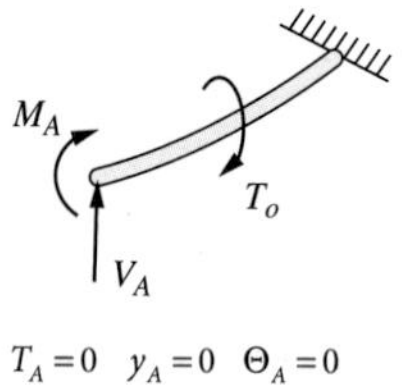

$T_A=0 \quad y_A=0 \quad \Theta_A=0$

$y_B=0 \quad \Theta_B=0 \quad \psi_B=0$

$$V_A=-\frac{T_o}{R}\frac{(C_{a8}C_4-C_{a5}C_7)(1-\cos\phi)+(C_{a5}C_1-C_{a2}C_4)\cos\phi+(C_{a2}C_7-C_{a8}C_1)\sin\phi}{(C_4C_9-C_6C_7)(1-\cos\phi)+(C_1C_6-C_3C_4)\cos\phi+(C_3C_7-C_1C_9)\sin\phi}$$

$$M_A=-T_o\frac{(C_{a5}C_9-C_{a8}C_6)(1-\cos\phi)+(C_{a2}C_6-C_{a5}C_3)\cos\phi+(C_{a8}C_3-C_{a2}C_9)\sin\phi}{(C_4C_9-C_6C_7)(1-\cos\phi)+(C_1C_6-C_3C_4)\cos\phi+(C_3C_7-C_1C_9)\sin\phi}$$

$$\psi_A=-\frac{T_oR}{EI}\frac{C_{a2}(C_4C_9-C_6C_7)+C_{a5}(C_3C_7-C_1C_9)+C_{a8}(C_1C_6-C_3C_4)}{(C_4C_9-C_6C_7)(1-\cos\phi)+(C_1C_6-C_3C_4)\cos\phi+(C_3C_7-C_1C_9)\sin\phi}$$

$$V_B=V_A$$

$$M_B=V_AR\sin\phi+M_A\cos\phi-T_o\sin(\phi-\theta)$$

$$T_B=V_AR(1-\cos\phi)+M_A\sin\phi+T_o\cos(\phi-\theta)$$

If $\beta=1.3$ (solid or hollow round cross section, $\nu=0.3$)

ϕ	45°	90°			180°		
θ	0°	0°	30°	60°	0°	60°	120°
K_{VA}	1.8104	1.1657	0.7420	0.0596	0.6201	0.2488	−0.1901
K_{MA}	−0.7589	−0.8252	−0.8776	−0.3682	−0.4463	−0.7564	−0.1519
$K_{\psi A}$	0.8145	1.0923	0.5355	0.2156	1.3724	0.1754	−0.0453
K_{MB}	0.0364	0.1657	−0.1240	−0.4404	0.4463	−0.1096	−0.7141
K_{TB}	0.7007	0.3406	0.3644	0.5575	0.2403	−0.0023	0.1199

(*Continued*)

TABLE 9.4 Formulas for Curved Beams of Compact Cross Section Loaded Normal to the Plane of Curvature (*Continued*)

End Restraints, Reference No.	Formulas for Boundary Values and Selected Numerical Values

3d. Right end fixed, left end supported and roll-guided

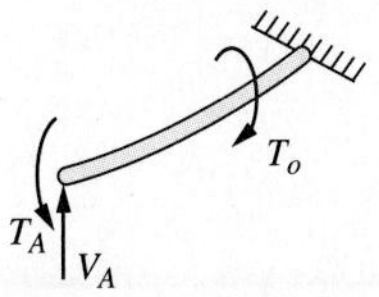

$M_A = 0 \quad y_A = 0 \quad \psi_A = 0$

$y_B = 0 \quad \Theta_B = 0 \quad \psi_B = 0$

$$V_A = -\frac{T_o}{R}\frac{[(C_{a2}+C_{a8})C_5 - C_{a5}(C_2+C_8)]\sin\phi + (C_{a2}C_8 - C_{a8}C_2)\cos\phi}{[C_5(C_3+C_9)-C_6(C_2+C_8)]\sin\phi + (C_3C_8 - C_2C_9)\cos\phi}$$

$$T_A = -T_o\frac{[C_{a5}(C_3+C_9)-C_6(C_{a2}+C_{a8})]\sin\phi + (C_{a8}C_3 - C_{a2}C_9)\cos\phi}{[C_5(C_3+C_9)-C_6(C_2+C_8)]\sin\phi + (C_3C_8 - C_2C_9)\cos\phi}$$

$$\Theta_A = -\frac{T_oR}{EI}\frac{C_{a2}(C_5C_9 - C_6C_8) + C_{a5}(C_3C_8 - C_2C_9) + C_{a8}(C_2C_6 - C_3C_5)}{[C_5(C_3+C_9)-C_6(C_2+C_8)]\sin\phi + (C_3C_8 - C_2C_9)\cos\phi}$$

$$V_B = V_A$$

$$M_B = V_AR\sin\phi - T_A\sin\phi - T_o\sin(\phi-\theta)$$

$$T_B = V_AR(1-\cos\phi) + T_A\cos\phi + T_o\cos(\phi-\theta)$$

If $\beta = 1.3$ (solid or hollow round cross section, $\nu = 0.3$)

ϕ	45°		90°		180°	
θ	15°	30°	30°	60°	60°	120°
K_{VA}	−0.3410	−0.4177	−0.3392	−0.3916	−0.2757	−0.2757
K_{TA}	−0.6694	−0.3198	−0.6724	−0.2765	−0.5730	−0.0730
$K_{\Theta A}$	−0.0544	−0.0263	−0.2411	−0.1046	−1.3691	−0.3262
K_{MB}	−0.2678	−0.3280	−0.5328	−0.6152	−0.8660	−0.8660
K_{TB}	0.2928	0.6175	0.1608	0.4744	−0.4783	0.0217

3e. Right end fixed, left end fixed

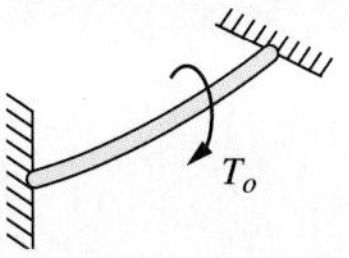

$y_A = 0 \quad \Theta_A = 0 \quad \psi_A = 0$

$y_B = 0 \quad \Theta_B = 0 \quad \psi_B = 0$

$$V_A = -\frac{T_o}{R}\frac{C_{a2}(C_4C_8 - C_5C_7) + C_{a5}(C_2C_7 - C_1C_8) + C_{a8}(C_1C_5 - C_2C_4)}{C_1(C_5C_9 - C_6C_8) + C_4(C_3C_8 - C_2C_9) + C_7(C_2C_6 - C_3C_5)}$$

$$M_A = -T_o\frac{C_{a2}(C_5C_9 - C_6C_8) + C_{a5}(C_3C_8 - C_2C_9) + C_{a8}(C_2C_6 - C_3C_5)}{C_1(C_5C_9 - C_6C_8) + C_4(C_3C_8 - C_2C_9) + C_7(C_2C_6 - C_3C_5)}$$

$$T_A = -T_o\frac{C_{a2}(C_6C_7 - C_4C_9) + C_{a5}(C_1C_9 - C_3C_7) + C_{a8}(C_3C_4 - C_1C_6)}{C_1(C_5C_9 - C_6C_8) + C_4(C_3C_8 - C_2C_9) + C_7(C_2C_6 - C_3C_5)}$$

$$V_B = V_A$$

$$M_B = V_AR\sin\phi + M_A\cos\phi - T_A\sin\phi - T_o\sin(\phi-\theta)$$

$$T_B = V_AR(1-\cos\phi) + M_A\sin\phi + T_A\cos\phi + T_o\cos(\phi-\theta)$$

If $\beta = 1.3$ (solid or hollow round cross section, $\nu = 0.3$)

ϕ	45°	90°	180°	270°	360°	
θ	15°	30°	60°	90°	90°	180°
K_{VA}	0.1704	0.1705	0.1696	0.1625	0.1592	0.0000
K_{MA}	−0.2591	−0.4731	−0.6994	−0.7073	−0.7500	0.0000
K_{TA}	−0.6187	−0.4903	−0.1278	0.2211	0.1799	0.5000
K_{MB}	−0.1252	−0.2053	−0.1666	0.0586	0.2500	0.0000
K_{TB}	0.2953	0.1974	−0.0330	−0.1302	0.1799	−0.5000

3f. Right end supported and slope-guided, left end supported and slope-guided

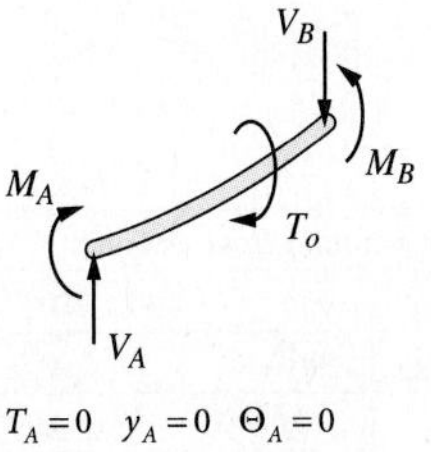

$T_A = 0 \quad y_A = 0 \quad \Theta_A = 0$

$T_B = 0 \quad y_B = 0 \quad \Theta_B = 0$

$$V_A = \frac{T_o}{R}\frac{[C_1\sin\phi - C_4(1-\cos\phi)]\cos(\phi-\theta) - C_{a2}\sin^2\phi + C_{a5}(1-\cos\phi)\sin\phi}{C_4(1-\cos\phi)^2 + C_3\sin^2\phi - (C_1+C_6)(1-\cos\phi)\sin\phi}$$

$$M_A = -T_o\frac{[C_3\sin\phi - C_6(1-\cos\phi)]\cos(\phi-\theta) - C_{a2}(1-\cos\phi)\sin\phi + C_{a5}(1-\cos\phi)^2}{C_4(1-\cos\phi)^2 + C_3\sin^2\phi - (C_1+C_6)(1-\cos\phi)\sin\phi}$$

$$\psi_A = \frac{T_oR}{EI}\frac{(C_3C_4 - C_1C_6)\cos(\phi-\theta) + (C_{a2}C_6 - C_{a5}C_3)\sin\phi - (C_{a2}C_4 - C_{a5}C_1)(1-\cos\phi)}{C_4(1-\cos\phi)^2 + C_3\sin^2\phi - (C_1+C_6)(1-\cos\phi)\sin\phi}$$

$$V_B = V_A$$

$$M_B = V_AR\sin\phi + M_A\cos\phi - T_o\sin(\phi-\theta)$$

$$\psi_B = \psi_A\cos\phi + \frac{M_AR}{EI}C_7 + \frac{V_AR^2}{EI}C_9 + \frac{T_oR}{EI}C_{a8}$$

If $\beta = 1.3$ (solid or hollow round cross section, $\nu = 0.3$)

ϕ	45°		90°		180°	
θ	0°	15°	0°	30°	0°	60°
K_{VA}	1.0645	0.5147	0.8696	0.4252	0.5000	0.2500
K_{MA}	−1.4409	−1.4379	−0.8696	−0.9252	−0.3598	−0.7573
$K_{\psi A}$	1.6003	1.3211	1.2356	0.6889	1.4564	0.1746
K_{MB}	−0.9733	−1.1528	−0.1304	−0.4409	0.3598	−0.1088
$K_{\psi B}$	1.1213	1.1662	0.4208	0.4502	0.3500	−0.0034

(*Continued*)

TABLE 9.4 Formulas for Curved Beams of Compact Cross Section Loaded Normal to the Plane of Curvature (*Continued*)

End Restraints, Reference No.	Formulas for Boundary Values and Selected Numerical Values

3g. Right end supported and slope-guided, left end supported and roll-guided

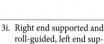

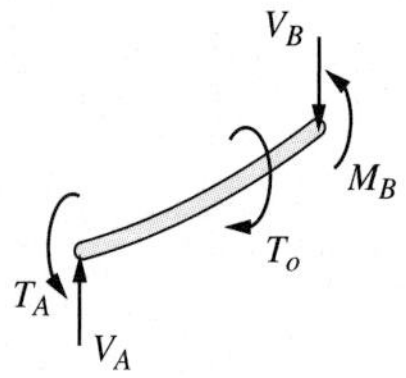

$M_A = 0 \quad y_A = 0 \quad \psi_A = 0$

$T_B = 0 \quad y_B = 0 \quad \Theta_B = 0$

$$V_A = -\frac{T_o}{R}\,\frac{C_{a2}\cos^2\phi - C_{a5}\sin\phi\cos\phi + (C_5\sin\phi - C_2\cos\phi)\cos(\phi-\theta)}{(C_5\sin\phi - C_2\cos\phi)(1-\cos\phi) + C_3\cos^2\phi - C_6\sin\phi\cos\phi}$$

$$T_A = -T_o\,\frac{(C_{a5}\sin\phi - C_{a2}\cos\phi)(1-\cos\phi) + (C_3\cos\phi - C_6\sin\phi)\cos(\phi-\theta)}{(C_5\sin\phi - C_2\cos\phi)(1-\cos\phi) + C_3\cos^2\phi - C_6\sin\phi\cos\phi}$$

$$\Theta_A = -\frac{T_oR}{EI}\,\frac{(C_{a2}C_5 - C_{a5}C_2)(1-\cos\phi) + (C_{a5}C_3 - C_{a2}C_6)\cos\phi + (C_2C_6 - C_3C_5)\cos(\phi-\theta)}{(C_5\sin\phi - C_2\cos\phi)(1-\cos\phi) + C_3\cos^2\phi - C_6\sin\phi\cos\phi}$$

$$V_B = V_A$$

$$M_B = V_A R\sin\phi - T_A\sin\phi - T_o\sin(\phi-\theta)$$

$$\psi_B = -\Theta_A\sin\phi + \frac{T_AR}{EI}C_8 + \frac{V_AR^2}{EI}C_9 + \frac{T_oR}{EI}C_{a8}$$

If $\beta = 1.3$ (solid or hollow round cross section, $\nu = 0.3$)

ϕ	45°		90°		180°	
θ	15°	30°	30°	60°	60°	120°
K_{VA}	−0.8503	−1.4915	−0.5000	−0.8660	−0.0512	−0.2859
K_{TA}	−0.8725	−0.7482	−0.7175	−0.4095	−0.6023	−0.0717
$K_{\Theta A}$	−0.0522	−0.0216	−0.2274	−0.0640	−1.9528	−0.2997
K_{MB}	−0.4843	−0.7844	−0.6485	−0.9566	−0.8660	−0.8660
$K_{\psi B}$	0.2386	0.5031	0.1780	0.5249	−0.9169	0.0416

3h. Right end supported and slope-guided, left end simply supported

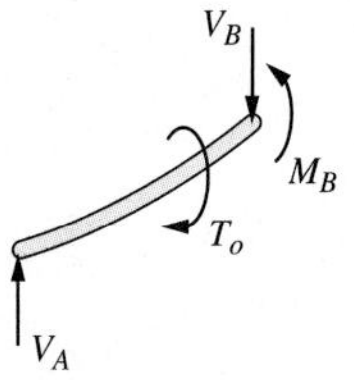

$M_A = 0 \quad T_A = 0 \quad y_A = 0$

$T_B = 0 \quad y_B = 0 \quad \Theta_B = 0$

$$V_A = -\frac{T_o\cos(\phi-\theta)}{R(1-\cos\phi)}$$

$$\Theta_A = -\frac{T_oR}{EI}\left[\frac{C_{a2}\sin\phi + C_6\cos(\phi-\theta)}{1-\cos\phi} - \frac{C_3\sin\phi\cos(\phi-\theta)}{(1-\cos\phi)^2} - C_{a5}\right]$$

$$\psi_A = -\frac{T_oR}{EI}\left[\frac{C_{a5}\sin\phi - C_{a2}\cos\phi}{1-\cos\phi} + (C_3\cos\phi - C_6\sin\phi)\frac{\cos(\phi-\theta)}{(1-\cos\phi)^2}\right]$$

$$V_B = V_A$$

$$M_B = V_A R\sin\phi - T_o\sin(\phi-\theta)$$

$$\psi_B = \psi_A\cos\phi - \Theta_A\sin\phi + \frac{V_AR^2}{EI}C_9 + \frac{T_oR}{EI}C_{a8}$$

If $\beta = 1.3$ (solid or hollow round cross section, $\nu = 0.3$)

ϕ	45°	90°			180°		
θ	0°	0°	30°	60°	0°	60°	120°
K_{VA}	−2.4142	0.0000	−0.5000	−0.8660	0.5000	0.2500	−0.2500
$K_{\Theta A}$	−0.4161	−0.6564	−0.6984	−0.3328	−1.3000	−2.7359	−0.3929
$K_{\psi A}$	2.1998	1.8064	1.2961	0.7396	1.9242	1.1590	0.1380
K_{MB}	−2.4142	−1.0000	−1.3660	−1.3660	0.0000	−0.8660	−0.8660
$K_{\psi B}$	1.5263	0.5064	0.5413	0.7323	−0.1178	−0.9878	0.0332

3i. Right end supported and roll-guided, left end supported and roll-guided

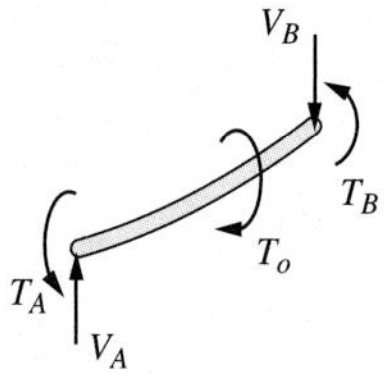

$M_A = 0 \quad y_A = 0 \quad \psi_A = 0$

$M_B = 0 \quad y_B = 0 \quad \psi_B = 0$

$$V_A = 0$$

$$T_A = -T_o\,\frac{(C_{a2}+C_{a8})\sin^2\phi + (C_3+C_9)\sin(\phi-\theta)\sin\phi}{(C_2+C_3+C_8+C_9)\sin^2\phi}$$

$$\Theta_A = -\frac{T_oR}{EI}\,\frac{[C_{a2}(C_8+C_9) - C_{a8}(C_2+C_3)]\sin\phi - (C_2C_9 - C_3C_8)\sin(\phi-\theta)}{(C_2+C_3+C_8+C_9)\sin^2\phi}$$

$$V_B = 0$$

$$T_B = V_AR(1-\cos\phi) + T_A\cos\phi + T_o\cos(\phi-\theta)$$

$$\Theta_B = \Theta_A\cos\phi + \frac{T_AR}{EI}C_5 + \frac{V_AR^2}{EI}C_6 + \frac{T_oR}{EI}C_{a5}$$

If $\beta = 1.3$ (solid or hollow round cross section, $\nu = 0.3$)

ϕ	45°	90°	270°
θ	15°	30°	90°
K_{VA}	0.0000	0.0000	0.0000
K_{TA}	−0.7071	−0.8660	0.0000
$K_{\Theta A}$	−0.0988	−0.6021	−3.6128
K_{TB}	0.3660	0.5000	−1.0000
$K_{\Theta B}$	0.0807	0.5215	0.0000

(*Continued*)

TABLE 9.4 Formulas for Curved Beams of Compact Cross Section Loaded Normal to the Plane of Curvature (*Continued*)

End Restraints, Reference No.	Formulas for Boundary Values and Selected Numerical Values

3j. Right end supported and roll-guided, left end simply supported

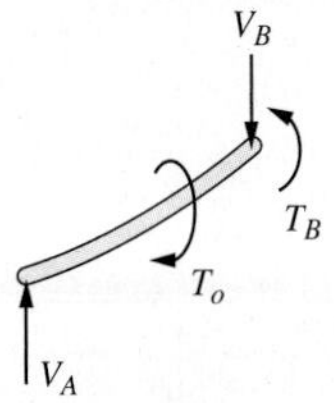

$M_A = 0 \quad T_A = 0 \quad y_A = 0$

$M_B = 0 \quad y_B = 0 \quad \psi_B = 0$

$$V_A = \frac{T_o \sin(\phi-\theta)}{R\sin\phi}$$

$$\Theta_A = -\frac{T_o R}{EI}\left\{\frac{C_{a2}\cos\phi - C_{a8}(1-\cos\phi)}{\sin\phi} + [C_3\cos\phi - C_9(1-\cos\phi)]\frac{\sin(\phi-\theta)}{\sin^2\phi}\right\}$$

$$\psi_A = -\frac{T_o R}{EI}\left[C_{a2} + C_{a8} + (C_3 + C_9)\frac{\sin(\phi-\theta)}{\sin\phi}\right]$$

$$V_B = V_A$$

$$T_B = V_A R(1-\cos\phi) + T_o\cos(\phi-\theta)$$

$$\Theta_B = \Theta_A\cos\phi + \psi_A\sin\phi + \frac{V_A R^2}{EI}C_6 + \frac{T_o R}{EI}C_{a5}$$

If $\beta = 1.3$ (solid or hollow round cross section, $\nu = 0.3$)

ϕ	45°			90°		
θ	0°	15°	30°	0°	30°	60°
K_{VA}	1.0000	0.7071	0.3660	1.0000	0.8660	0.5000
$K_{\Theta A}$	−0.2790	−0.2961	−0.1828	−1.3000	−1.7280	−1.1715
$K_{\psi A}$	1.0210	0.7220	0.3737	2.0420	1.7685	1.0210
K_{TB}	1.0000	1.0731	1.0731	1.0000	1.3660	1.3660
$K_{\Theta B}$	0.1439	0.1825	0.1515	0.7420	1.1641	0.9732

4. Uniformly distributed lateral load

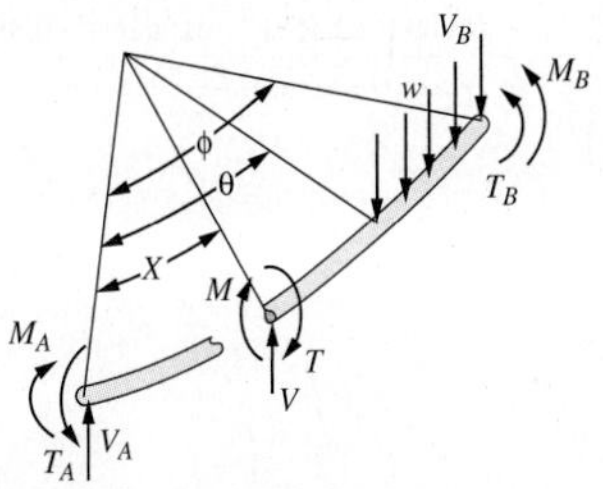

Transverse shear $= V = V_A - wR\langle x-\theta\rangle^1$

Bending moment $= M = V_A R\sin x + M_A\cos x - T_A\sin x - wR^2[1-\cos(x-\theta)]\langle x-\theta\rangle^0$

Twisting moment $= T = V_A R(1-\cos x) + M_A\sin x + T_A\cos x - wR^2[x-\theta-\sin(x-\theta)]\langle x-\theta\rangle^0$

Vertical deflection $= y = y_A + \Theta_A R\sin x + \psi_A R(1-\cos x) + \frac{M_A R^2}{EI}F_1 + \frac{T_A R^2}{EI}F_2 + \frac{V_A R^3}{EI}F_3 - \frac{wR^4}{EI}F_{a13}$

Bending slope $= \Theta = \Theta_A\cos x + \psi_A\sin x + \frac{M_A R}{EI}F_4 + \frac{T_A R}{EI}F_5 + \frac{V_A R^2}{EI}F_6 - \frac{wR^3}{EI}F_{a16}$

Roll slope $\psi = \psi_A\cos x - \Theta_A\sin x + \frac{M_A R}{EI}F_7 + \frac{T_A R}{EI}F_8 + \frac{V_A R^2}{EI}F_9 - \frac{wR^3}{EI}F_{a19}$

For tabulated values: $V = K_V wR,\quad M = K_M wR^2,\quad T = K_T wR^2,\quad y = K_y\frac{wR^4}{EI},\quad \Theta = K_\Theta\frac{wR^3}{EI},\quad \psi = K_\psi\frac{wR^3}{EI}$

4a. Right end fixed, left end free

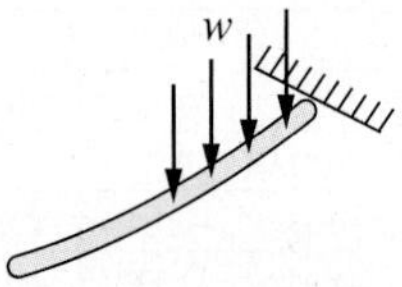

$V_A = 0 \quad M_A = 0 \quad T_A = 0$

$y_B = 0 \quad \Theta_B = 0 \quad \psi_B = 0$

$$y_A = -\frac{wR^4}{EI}[C_{a16}\sin\phi - C_{a19}(1-\cos\phi) - C_{a13}]$$

$$\Theta_A = \frac{wR^3}{EI}(C_{a16}\cos\phi - C_{a19}\sin\phi)$$

$$\psi_A = \frac{wR^3}{EI}(C_{a19}\cos\phi + C_{a16}\sin\phi)$$

$$V_B = -wR(\phi-\theta)$$

$$M_B = -wR^2[1-\cos(\phi-\theta)]$$

$$T_B = -wR^2[\phi-\theta-\sin(\phi-\theta)]$$

If $\beta = 1.3$ (solid or hollow round cross section, $\nu = 0.3$)

ϕ	45°	90°			180°		
θ	0°	0°	30°	60°	0°	60°	120°
K_{yA}	−0.0469	−0.7118	−0.2211	−0.0269	−8.4152	−2.2654	−0.1699
$K_{\Theta A}$	0.0762	0.4936	0.1071	0.0071	0.4712	−0.6033	−0.1583
$K_{\psi A}$	0.0267	0.4080	0.1583	0.0229	4.6000	1.3641	0.1071
K_{MB}	−0.2929	−1.0000	−0.5000	−0.1340	−2.0000	−1.5000	−0.5000
K_{TB}	−0.0783	−0.5708	−0.1812	−0.0236	−3.1416	−1.2284	−0.1812

4b. Right end fixed, left end simply supported

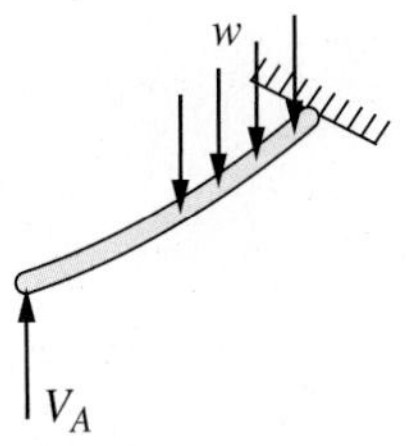

$M_A = 0 \quad T_A = 0 \quad y_A = 0$

$y_B = 0 \quad \Theta_B = 0 \quad \psi_B = 0$

$$V_A = wR\frac{C_{a19}(1-\cos\phi) - C_{a16}\sin\phi + C_{a13}}{C_9(1-\cos\phi) - C_6\sin\phi + C_3}$$

$$\Theta_A = \frac{wR^3}{EI}\,\frac{(C_{a13}C_9 - C_{a19}C_3)\sin\phi + (C_{a19}C_6 - C_{a16}C_9)(1-\cos\phi) + (C_{a16}C_3 - C_{a13}C_6)\cos\phi}{C_9(1-\cos\phi) - C_6\sin\phi + C_3}$$

$$\psi_A = \frac{wR^3}{EI}\,\frac{[C_{a16}(C_3+C_9) - C_6(C_{a13}+C_{a19})]\sin\phi + (C_{a19}C_3 - C_{a13}C_9)\cos\phi}{C_9(1-\cos\phi) - C_6\sin\phi + C_3}$$

$$V_B = V_A - wR(\phi-\theta)$$

$$M_B = V_A R\sin\phi - wR^2[1-\cos(\phi-\theta)]$$

$$T_B = V_A R(1-\cos\phi) - wR^2[\phi-\theta-\sin(\phi-\theta)]$$

If $\beta = 1.3$ (solid or hollow round cross section, $\nu = 0.3$)

ϕ	45°	90°			180°		
θ	0°	0°	30°	60°	0°	60°	120°
K_{VA}	0.2916	0.5701	0.1771	0.0215	1.0933	0.2943	0.0221
$K_{\Theta A}$	−0.1300	−0.1621	−0.0966	−0.0177	−2.3714	−1.3686	−0.2156
$K_{\psi A}$	0.0095	0.1192	0.0686	0.0119	0.6500	0.3008	0.0273
K_{MB}	−0.0867	−0.4299	−0.3229	−0.1124	−2.0000	−1.5000	−0.5000
K_{TB}	0.0071	−0.0007	−0.0041	−0.0021	−0.9549	−0.6397	−0.1370

(*Continued*)

TABLE 9.4 Formulas for Curved Beams of Compact Cross Section Loaded Normal to the Plane of Curvature (*Continued*)

End Restraints, Reference No.	Formulas for Boundary Values and Selected Numerical Values

4c. Right end fixed, left end supported and slope-guided

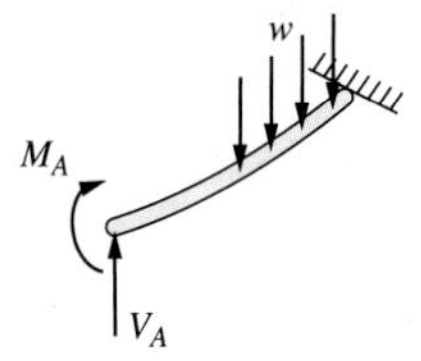

$T_A = 0 \quad y_A = 0 \quad \Theta_A = 0$

$y_B = 0 \quad \Theta_B = 0 \quad \psi_B = 0$

$$V_A = wR\frac{(C_{a19}C_4 - C_{a16}C_7)(1-\cos\phi) + (C_{a16}C_1 - C_{a13}C_4)\cos\phi + (C_{a13}C_7 - C_{a19}C_1)\sin\phi}{(C_4C_9 - C_6C_7)(1-\cos\phi) + (C_1C_6 - C_3C_4)\cos\phi + (C_3C_7 - C_1C_9)\sin\phi}$$

$$M_A = wR^2\frac{(C_{a16}C_9 - C_{a19}C_6)(1-\cos\phi) + (C_{a13}C_6 - C_{a16}C_3)\cos\phi + (C_{a19}C_3 - C_{a13}C_9)\sin\phi}{(C_4C_9 - C_6C_7)(1-\cos\phi) + (C_1C_6 - C_3C_4)\cos\phi + (C_3C_7 - C_1C_9)\sin\phi}$$

$$\psi_A = \frac{wR^3}{EI}\frac{C_{a13}(C_4C_9 - C_6C_7) + C_{a16}(C_3C_7 - C_1C_9) + C_{a19}(C_1C_6 - C_3C_4)}{(C_4C_9 - C_6C_7)(1-\cos\phi) + (C_1C_6 - C_3C_4)\cos\phi + (C_3C_7 - C_1C_9)\sin\phi}$$

$$V_B = V_A - wR(\phi - \theta)$$

$$M_B = V_A R\sin\phi + M_A\cos\phi - wR^2[1 - \cos(\phi - \theta)]$$

$$T_B = V_A R(1 - \cos\phi) + M_A\sin\phi - wR^2[\phi - \theta - \sin(\phi - \theta)]$$

If β = 1.3 (solid or hollow round cross section, ν = 0.3)

ϕ	45°	90°			180°		
θ	0°	0°	30°	60°	0°	60°	120°
K_{VA}	0.3919	0.7700	0.2961	0.0434	1.3863	0.4634	0.0487
K_{MA}	−0.0527	−0.2169	−0.1293	−0.0237	−0.8672	−0.5005	−0.0789
$K_{\psi A}$	−0.0004	−0.0145	−0.0111	−0.0027	−0.4084	−0.3100	−0.0689
K_{MB}	−0.0531	−0.2301	−0.2039	−0.0906	−1.1328	−0.9995	−0.4211
K_{TB}	−0.0008	−0.0178	−0.0143	−0.0039	−0.3691	−0.3016	−0.0838

4d. Right end fixed, left end supported and roll-guided

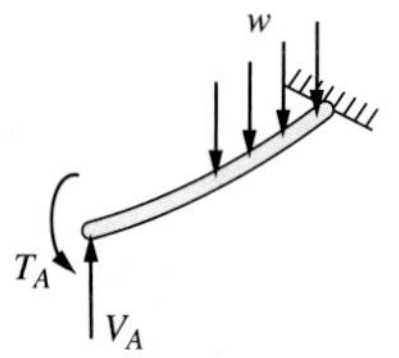

$M_A = 0 \quad y_A = 0 \quad \psi_A = 0$

$y_B = 0 \quad \Theta_B = 0 \quad \psi_B = 0$

$$V_A = wR\frac{[(C_{a13} + C_{a19})C_5 - C_{a16}(C_2 + C_8)]\sin\phi + (C_{a13}C_8 - C_{a19}C_2)\cos\phi}{[C_5(C_3 + C_9) - C_6(C_2 + C_8)]\sin\phi + (C_3C_8 - C_2C_9)\cos\phi}$$

$$T_A = wR^2\frac{[C_{a16}(C_3 + C_9) - C_6(C_{a13} + C_{a19})]\sin\phi + (C_{a19}C_3 - C_{a13}C_9)\cos\phi}{[C_5(C_3 + C_9) - C_6(C_2 + C_8)]\sin\phi + (C_3C_8 - C_2C_9)\cos\phi}$$

$$\Theta_A = \frac{wR^3}{EI}\frac{C_{a13}(C_5C_9 - C_6C_8) + C_{a16}(C_3C_8 - C_2C_9) + C_{a19}(C_2C_6 - C_3C_5)}{[C_5(C_3 + C_9) - C_6(C_2 + C_8)]\sin\phi + (C_3C_8 - C_2C_9)\cos\phi}$$

$$V_B = V_A - wR(\phi - \theta)$$

$$M_B = V_A R\sin\phi - T_A\sin\phi - wR^2[1 - \cos(\phi - \theta)]$$

$$T_B = V_A R(1 - \cos\phi) + T_A\cos\phi - wR^2[\phi - \theta - \sin(\phi - \theta)]$$

If β = 1.3 (solid or hollow round cross section, ν = 0.3)

ϕ	45°	90°			180°		
θ	0°	0°	30°	60°	0°	60°	120°
K_{VA}	0.2880	0.5399	0.1597	0.0185	0.9342	0.2207	0.0154
K_{TA}	−0.0099	−0.0745	−0.0428	−0.0075	−0.3391	−0.1569	−0.0143
$K_{\Theta A}$	−0.0111	−0.1161	−0.0702	−0.0131	−1.9576	−1.1171	−0.1983
K_{MB}	−0.0822	−0.3856	−0.2975	−0.1080	−2.0000	−1.5000	−0.5000
K_{TB}	−0.0010	−0.0309	−0.0215	−0.0051	−0.9342	−0.6301	−0.1362

4e. Right end fixed, left end fixed

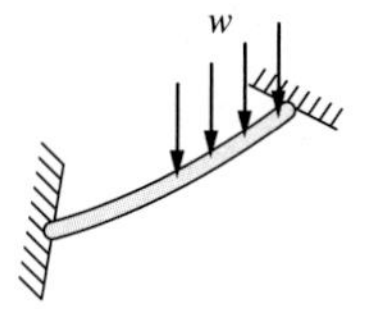

$y_A = 0 \quad \Theta_A = 0 \quad \psi_A = 0$

$y_B = 0 \quad \Theta_B = 0 \quad \psi_B = 0$

$$V_A = wR\frac{C_{a13}(C_4C_8 - C_5C_7) + C_{a16}(C_2C_7 - C_1C_8) + C_{a19}(C_1C_5 - C_2C_4)}{C_1(C_5C_9 - C_6C_8) + C_4(C_3C_8 - C_2C_9) + C_7(C_2C_6 - C_3C_5)}$$

$$M_A = wR^2\frac{C_{a13}(C_5C_9 - C_6C_8) + C_{a16}(C_3C_8 - C_2C_9) + C_{a19}(C_2C_6 - C_3C_5)}{C_1(C_5C_9 - C_6C_8) + C_4(C_3C_8 - C_2C_9) + C_7(C_2C_6 - C_3C_5)}$$

$$T_A = wR^2\frac{C_{a13}(C_6C_7 - C_4C_9) + C_{a16}(C_1C_9 - C_3C_7) + C_{a19}(C_3C_4 - C_1C_6)}{C_1(C_5C_9 - C_6C_8) + C_4(C_3C_8 - C_2C_9) + C_7(C_2C_6 - C_3C_5)}$$

$$V_B = V_A - wR(\phi - \theta)$$

$$M_B = V_A R\sin\phi + M_A\cos\phi - T_A\sin\phi - wR^2[1 - \cos(\phi - \theta)]$$

$$T_B = V_A R(1 - \cos\phi) + M_A\sin\phi + T_A\cos\phi - wR^2[\phi - \theta - \sin(\phi - \theta)]$$

If β = 1.3 (solid or hollow round cross section, ν = 0.3)

ϕ	45°		90°		180°		360°
θ	0°	15°	0°	30°	0°	60°	0°
K_{VA}	0.3927	0.1548	0.7854	0.3080	1.5708	0.6034	3.1416
K_{MA}	−0.0531	−0.0316	−0.2279	−0.1376	−1.0000	−0.6013	−2.1304
K_{TA}	0.0005	0.0004	0.0133	0.0102	0.2976	0.2259	3.1416
K_{MB}	−0.0531	−0.0471	−0.2279	−0.2022	−1.0000	−0.8987	−2.1304
K_{TB}	−0.0005	−0.0004	−0.0133	−0.0108	−0.2976	−0.2473	−3.1416

(*Continued*)

TABLE 9.4 Formulas for Curved Beams of Compact Cross Section Loaded Normal to the Plane of Curvature (*Continued*)

End Restraints, Reference No.	Formulas for Boundary Values and Selected Numerical Values

4f. Right end supported and slope-guided, left end supported and slope-guided

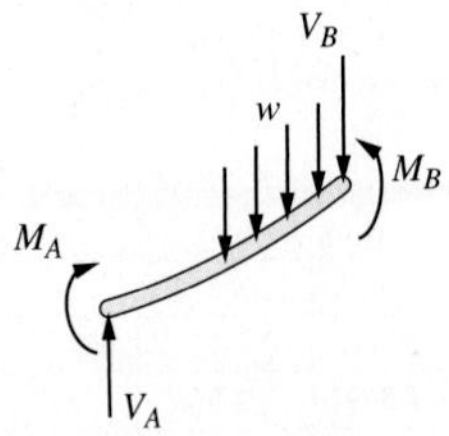

$T_A = 0 \quad y_A = 0 \quad \Theta_A = 0$

$T_B = 0 \quad y_B = 0 \quad \Theta_B = 0$

$$V_A = wR\frac{[C_4(1-\cos\phi)-C_1\sin\phi][\phi-\theta-\sin(\phi-\theta)]+C_{a13}\sin^2\phi-C_{a16}\sin\phi(1-\cos\phi)}{C_4(1-\cos\phi)^2+C_3\sin^2\phi-(C_1+C_6)(1-\cos\phi)\sin\phi}$$

$$M_A = wR^2\frac{[C_3\sin\phi-C_6(1-\cos\phi)][\phi-\theta-\sin(\phi-\theta)]+C_{a16}(1-\cos\phi)^2-C_{a13}\sin\phi(1-\cos\phi)}{C_4(1-\cos\phi)^2+C_3\sin^2\phi-(C_1+C_6)(1-\cos\phi)\sin\phi}$$

$$\psi_A = \frac{wR^3}{EI}\frac{(C_{a13}C_4-C_{a16}C_1)(1-\cos\phi)-(C_{a13}C_6-C_{a16}C_3)\sin\phi-(C_3C_4-C_1C_6)[\phi-\theta-\sin(\phi-\theta)]}{C_4(1-\cos\phi)^2+C_3\sin^2\phi-(C_1+C_6)(1-\cos\phi)\sin\phi}$$

$$V_B = V_A - wR(\phi-\theta)$$

$$M_B = V_A R\sin\phi + M_A\cos\phi - wR^2[1-\cos(\phi-\theta)]$$

$$\psi_B = \psi_A\cos\phi + \frac{M_A R}{EI}C_7 + \frac{V_A R^2}{EI}C_9 - \frac{wR^3}{EI}C_{a19}$$

If $\beta = 1.3$ (solid or hollow round cross section, $\nu = 0.3$)

ϕ	45°		90°		180°	
θ	0°	15°	0°	30°	0°	60°
K_{VA}	0.3927	0.1549	0.7854	0.3086	1.5708	0.6142
K_{MA}	−0.0519	−0.0308	−0.2146	−0.1274	−1.0000	−0.6090
$K_{\psi A}$	−0.0013	−0.0010	−0.0220	−0.0171	−0.5375	−0.4155
K_{MB}	−0.0519	−0.0462	−0.2146	−0.1914	−1.0000	−0.8910
$K_{\psi B}$	−0.0013	−0.0010	−0.0220	−0.0177	−0.5375	−0.4393

4g. Right end supported and slope-guided, left end supported and roll-guided

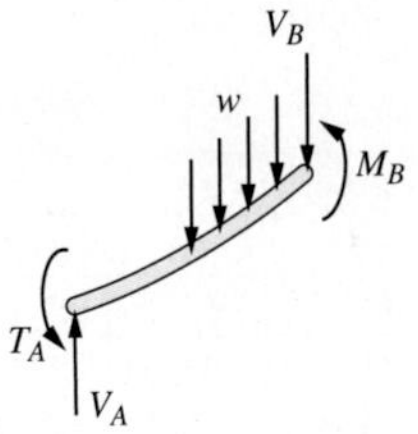

$M_A = 0 \quad y_A = 0 \quad \psi_A = 0$

$T_B = 0 \quad y_B = 0 \quad \Theta_B = 0$

$$V_A = wR\frac{(C_5\sin\phi-C_2\cos\phi)[\phi-\theta-\sin(\phi-\theta)]+C_{a13}\cos^2\phi-C_{a16}\sin\phi\cos\phi}{(C_5\sin\phi-C_2\cos\phi)(1-\cos\phi)+C_3\cos^2\phi-C_6\sin\phi\cos\phi}$$

$$T_A = wR^2\frac{(C_3\cos\phi-C_6\sin\phi)[\phi-\theta-\sin(\phi-\theta)]-(C_{a13}\cos\phi-C_{a16}\sin\phi)(1-\cos\phi)}{(C_5\sin\phi-C_2\cos\phi)(1-\cos\phi)+C_3\cos^2\phi-C_6\sin\phi\cos\phi}$$

$$\Theta_A = \frac{wR^3}{EI}\frac{(C_2C_6-C_3C_5)[\phi-\theta-\sin(\phi-\theta)]+(C_{a13}C_5-C_{a16}C_2)(1-\cos\phi)+(C_{a16}C_3-C_{a13}C_6)\cos\phi}{(C_5\sin\phi-C_2\cos\phi)(1-\cos\phi)+C_3\cos^2\phi-C_6\sin\phi\cos\phi}$$

$$V_B = V_A - wR(\phi-\theta)$$

$$M_B = V_A R\sin\phi - T_A\sin\phi - wR^2[1-\cos(\phi-\theta)]$$

$$\psi = -\Theta_A\sin\phi + \frac{T_A R}{EI}C_8 + \frac{V_A R^2}{EI}C_9 - \frac{wR^3}{EI}C_{a19}$$

If $\beta = 1.3$ (solid or hollow round cross section, $\nu = 0.3$)

ϕ	45°	90°			180°		
θ	0°	0°	30°	60°	0°	60°	120°
K_{VA}	0.2896	0.5708	0.1812	0.0236	1.3727	0.5164	0.0793
K_{TA}	−0.0093	−0.0658	−0.0368	−0.0060	−0.3963	−0.1955	−0.0226
$K_{\Theta A}$	−0.0111	−0.1188	−0.0720	−0.0135	−3.0977	−1.9461	−0.3644
K_{MB}	−0.0815	−0.3634	−0.2820	−0.1043	−2.0000	−1.5000	−0.5000
$K_{\psi B}$	−0.0008	−0.0342	−0.0238	−0.0056	−1.7908	−1.2080	−0.2610

4h. Right end supported and slope-guided, left end simply supported

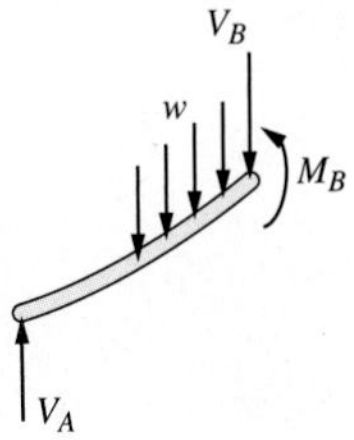

$M_A = 0 \quad T_A = 0 \quad y_A = 0$

$T_B = 0 \quad y_B = 0 \quad \Theta_B = 0$

$$V_A = wR\frac{\phi-\theta-\sin(\phi-\theta)}{1-\cos\phi}$$

$$\Theta_A = \frac{wR^3}{EI}\left\{\frac{C_{a13}\sin\phi+C_6[\phi-\theta-\sin(\phi-\theta)]}{1-\cos\phi}-\frac{C_3\sin\phi[\phi-\theta-\sin(\phi-\theta)]}{(1-\cos\phi)^2}-C_{a16}\right\}$$

$$\psi_A = \frac{wR^3}{EI}\left[\frac{C_{a16}\sin\phi-C_{a13}\cos\phi}{1-\cos\phi}-(C_6\sin\phi-C_3\cos\phi)\frac{\phi-\theta-\sin(\phi-\theta)}{(1-\cos\phi)^2}\right]$$

$$V_B = V_A - wR(\phi-\theta)$$

$$M_B = V_A R\sin\phi - wR^2[1-\cos(\phi-\theta)]$$

$$\psi_B = \psi_A\cos\phi - \Theta_A\sin\phi + \frac{V_A R^2}{EI}C_9 - \frac{wR^3}{EI}C_{a19}$$

If $\beta = 1.3$ (solid or hollow round cross section, $\nu = 0.3$)

ϕ	45°	90°			180°		
θ	0°	0°	30°	60°	0°	60°	120°
K_{VA}	0.2673	0.5708	0.1812	0.0236	1.5708	0.6142	0.0906
$K_{\Theta A}$	−0.0150	−0.1620	−0.0962	−0.0175	−3.6128	−2.2002	−0.3938
$K_{\psi A}$	0.0204	0.1189	0.0665	0.0109	0.7625	0.3762	0.0435
K_{MB}	−0.1039	−0.4292	−0.3188	−0.1104	−2.0000	−1.5000	−0.5000
$K_{\psi B}$	−0.0133	−0.0008	−0.0051	−0.0026	−1.8375	−1.2310	−0.2637

(*Continued*)

TABLE 9.4 Formulas for Curved Beams of Compact Cross Section Loaded Normal to the Plane of Curvature (*Continued*)

End Restraints, Reference No.	Formulas for Boundary Values and Selected Numerical Values

4i. Right end supported and roll-guided, left end supported and roll-guided

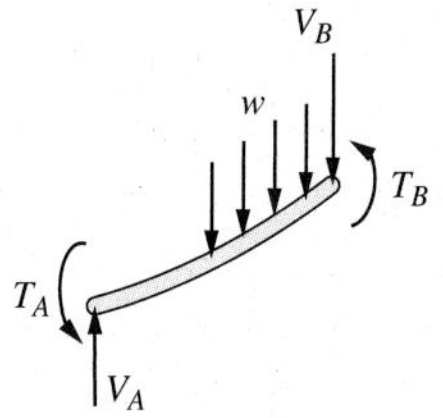

$M_A = 0 \quad y_A = 0 \quad \psi_A = 0$

$M_B = 0 \quad y_B = 0 \quad \psi_B = 0$

$$V_A = wR\frac{(C_{a13}+C_{a19})\sin\phi + (C_2+C_8)[1-\cos(\phi-\theta)]}{(C_2+C_3+C_8+C_9)\sin\phi}$$

$$T_A = wR^2\frac{(C_{a13}+C_{a19})\sin\phi - (C_3+C_9)[1-\cos(\phi-\theta)]}{(C_2+C_3+C_8+C_9)\sin\phi}$$

$$\Theta_A = \frac{wR^3}{EI}\,\frac{C_{a13}(C_8+C_9) - C_{a19}(C_2+C_3) + (C_2C_9 - C_3C_8)[1-\cos(\phi-\theta)]/\sin\phi}{(C_2+C_3+C_8+C_9)\sin\phi}$$

$$V_B = V_A - wR(\phi-\theta)$$

$$T_B = V_A R(1-\cos\phi) + T_A\cos\phi - wR^2[\phi-\theta-\sin(\phi-\theta)]$$

$$\Theta_B = \Theta_A\cos\phi + \frac{T_A R}{EI}C_5 + \frac{V_A R^2}{EI}C_6 - \frac{wR^3}{EI}C_{a16}$$

If $\beta = 1.3$ (solid or hollow round cross section, $\nu = 0.3$)

ϕ	45°		90°		270°	
θ	0°	15°	0°	30°	0°	90°
K_{VA}	0.3927	0.1745	0.7854	0.3491	2.3562	1.0472
K_{TA}	−0.0215	−0.0149	−0.2146	−0.1509	3.3562	3.0472
$K_{\Theta A}$	−0.0248	−0.0173	−0.3774	−0.2717	−10.9323	−6.2614
K_{MB}	0.0215	0.0170	0.2146	0.1679	−3.3562	−2.0944
$K_{\Theta B}$	0.0248	0.0194	0.3774	0.2912	10.9323	9.9484

4j. Right end supported and roll-guided, left end simply supported

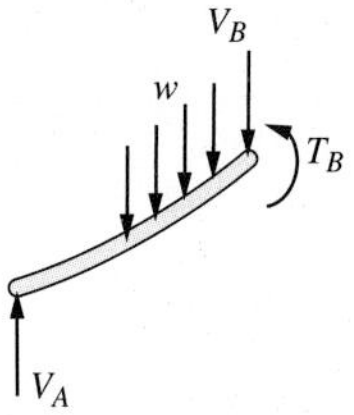

$M_A = 0 \quad T_A = 0 \quad y_A = 0$

$M_B = 0 \quad y_B = 0 \quad \psi_B = 0$

$$V_A = wR\frac{1-\cos(\phi-\theta)}{\sin\phi}$$

$$\Theta_A = \frac{wR^3}{EI}\left\{\frac{C_{a13}\cos\phi - C_{a19}(1-\cos\phi)}{\sin\phi} - [C_3\cos\phi - C_9(1-\cos\phi)]\frac{1-\cos(\phi-\theta)}{\sin^2\phi}\right\}$$

$$\psi_A = \frac{wR^3}{EI}\left[C_{a13} + C_{a19} - (C_3+C_9)\frac{1-\cos(\phi-\theta)}{\sin\phi}\right]$$

$$V_B = V_A - wR(\phi-\theta)$$

$$T_B = V_A R(1-\cos\phi) - wR^2[\phi-\theta-\sin(\phi-\theta)]$$

$$\Theta_B = \Theta_A\cos\phi + \psi_A\sin\phi + \frac{V_A R^2}{EI}C_6 - \frac{wR^3}{EI}C_{a16}$$

If $\beta = 1.3$ (solid or hollow round cross section, $\nu = 0.3$)

ϕ	45°			90°		
θ°	0°	15°	30°	0°	30°	60°
K_{VA}	0.4142	0.1895	0.0482	1.0000	0.5000	0.1340
$K_{\Theta A}$	−0.0308	−0.0215	−0.0066	−0.6564	−0.4679	−0.1479
$K_{\psi A}$	0.0220	0.0153	0.0047	0.4382	0.3082	0.0954
K_{TB}	0.0430	0.0319	0.0111	0.4292	0.3188	0.1104
$K_{\Theta B}$	0.0279	0.0216	0.0081	0.5367	0.4032	0.1404

5. Uniformly distributed torque

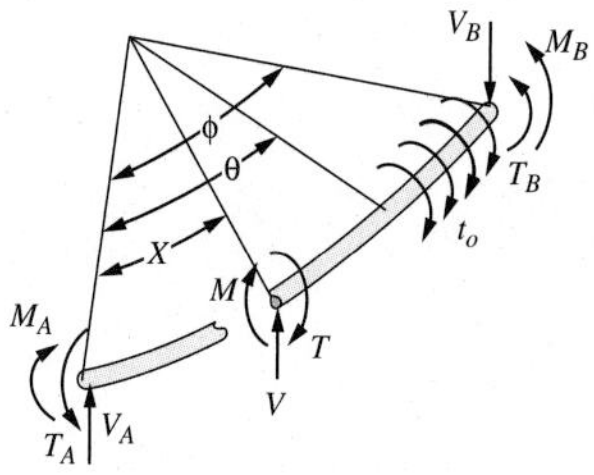

Transverse shear $= V = V_A$

Bending moment $= M = V_A R\sin x + M_A\cos x - T_A\sin x - t_o R[1-\cos(x-\theta)]\langle x-\theta\rangle^0$

Twisting moment $= T = V_A R(1-\cos x) + M_A\sin x + T_A\cos x + t_o R\sin(x-\theta)\langle x-\theta\rangle^0$

Vertical deflection $= y = y_A + \Theta_A R\sin x + \psi_A R(1-\cos x) + \frac{M_A R^2}{EI}F_1 + \frac{T_A R^2}{EI}F_2 + \frac{V_A R^3}{EI}F_3 + \frac{t_o R^3}{EI}F_{a12}$

Bending slope $= \Theta = \Theta_A\cos x + \psi_A\sin x + \frac{M_A R}{EI}F_4 + \frac{T_A R}{EI}F_5 + \frac{V_A R^2}{EI}F_6 + \frac{t_o R^2}{EI}F_{a15}$

Roll slope $= \psi = \psi_A\cos x - \Theta_A\sin x + \frac{M_A R}{EI}F_7 + \frac{T_A R}{EI}F_8 + \frac{V_A R^2}{EI}F_9 + \frac{t_o R^2}{EI}F_{a18}$

For tabulated values: $V = K_V t_o$, $M = K_M t_o R$, $T = K_T t_o R$, $y = K_y\frac{t_o R^3}{EI}$, $\Theta = K_\Theta\frac{t_o R^2}{EI}$, $\psi = K_\psi\frac{t_o R^2}{EI}$

(*Continued*)

TABLE 9.4 Formulas for Curved Beams of Compact Cross Section Loaded Normal to the Plane of Curvature (*Continued*)

End Restraints, Reference No.	Formulas for Boundary Values and Selected Numerical Values

5a. Right end fixed, left end free

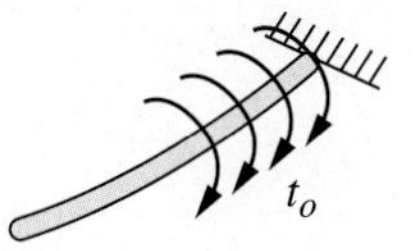

$V_A = 0 \quad M_A = 0 \quad T_A = 0$

$y_B = 0 \quad \Theta_B = 0 \quad \psi_B = 0$

$$y_A = \frac{t_o R^3}{EI}[C_{a15}\sin\phi - C_{a18}(1-\cos\phi) - C_{a12}]$$

$$\Theta_A = -\frac{t_o R^2}{EI}(C_{a15}\cos\phi - C_{a18}\sin\phi)$$

$$\psi_A = -\frac{t_o R^2}{EI}(C_{a18}\cos\phi + C_{a15}\sin\phi)$$

$$V_B = 0$$

$$M_B = -t_0 R[1-\cos(\phi-\theta)]$$

$$T_B = t_o R\sin(\phi-\theta)$$

If $\beta = 1.3$ (solid or hollow round cross section, $\nu = 0.3$)

ϕ	45°	90°			180°		
θ	0°	0°	30°	60°	0°	60°	120°
K_{yA}	0.0129	0.1500	0.2562	0.1206	0.6000	2.5359	1.1929
$K_{\Theta A}$	−0.1211	−0.8064	−0.5429	−0.1671	−3.6128	−2.2002	−0.3938
$K_{\psi A}$	0.3679	1.1500	0.3938	0.0535	2.0000	−0.5859	−0.5429
K_{MB}	−0.2929	−1.0000	−0.5000	−0.1340	−2.0000	−1.5000	−0.5000
K_{TB}	0.7071	1.0000	0.8660	0.5000	0.0000	0.8660	0.8660

5b. Right end fixed, left end simply supported

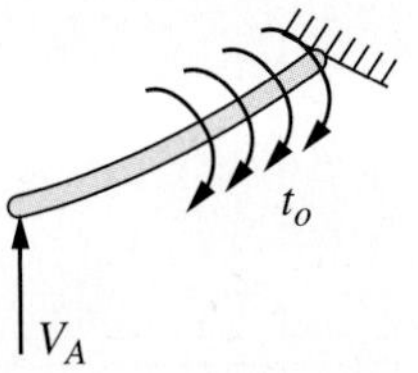

$M_A = 0 \quad T_A = 0 \quad y_A = 0$

$y_B = 0 \quad \Theta_B = 0 \quad \psi_B = 0$

$$V_A = -t_o \frac{C_{a18}(1-\cos\phi) - C_{a15}\sin\phi + C_{a12}}{C_9(1-\cos\phi) - C_6\sin\phi + C_3}$$

$$\Theta_A = -\frac{t_o R^2}{EI}\,\frac{(C_{a12}C_9 - C_{a18}C_3)\sin\phi + (C_{a18}C_6 - C_{a15}C_9)(1-\cos\phi) + (C_{a15}C_3 - C_{a12}C_6)\cos\phi}{C_9(1-\cos\phi) - C_6\sin\phi + C_3}$$

$$\psi_A = -\frac{t_o R^2}{EI}\,\frac{[C_{a15}(C_9 + C_3) - C_6(C_{a12} + C_{a18})]\sin\phi + (C_{a18}C_3 - C_{a12}C_9)\cos\phi}{C_9(1-\cos\phi) - C_6\sin\phi + C_3}$$

$$V_B = V_A$$

$$M_B = V_A R\sin\phi - t_o R[1-\cos(\phi-\theta)]$$

$$T_B = V_A R(1-\cos\phi) + t_o R\sin(\phi-\theta)$$

If $\beta = 1.3$ (solid or hollow round cross section, $\nu = 0.3$)

ϕ	45°	90°			180°		
θ	0°	0°	30°	60°	0°	60°	120°
K_{VA}	−0.0801	−0.1201	−0.2052	−0.0966	−0.0780	−0.3295	−0.1550
$K_{\Theta A}$	−0.0966	−0.6682	−0.3069	−0.0560	−3.4102	−1.3436	0.0092
$K_{\psi A}$	0.3726	1.2108	0.4977	0.1025	2.2816	0.6044	0.0170
K_{MB}	−0.3495	−1.1201	−0.7052	−0.2306	−2.0000	−1.5000	−0.5000
K_{TB}	0.6837	0.8799	0.6608	0.4034	−0.1559	0.2071	0.5560

5c. Right end fixed, left end supported and slope-guided

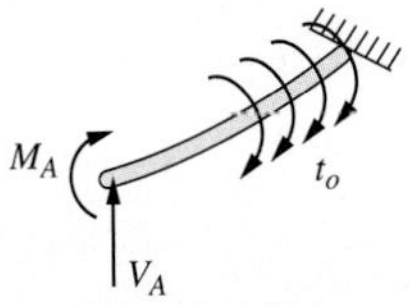

$T_A = 0 \quad y_A = 0 \quad \Theta_A = 0$

$y_B = 0 \quad \Theta_B = 0 \quad \psi_B = 0$

$$V_A = -t_o \frac{(C_{a18}C_4 - C_{a15}C_7)(1-\cos\phi) + (C_{a15}C_1 - C_{a12}C_4)\cos\phi + (C_{a12}C_7 - C_{a18}C_1)\sin\phi}{(C_4C_9 - C_6C_7)(1-\cos\phi) + (C_1C_6 - C_3C_4)\cos\phi + (C_3C_7 - C_1C_9)\sin\phi}$$

$$M_A = -t_0 R\frac{(C_{a15}C_9 - C_{a18}C_6)(1-\cos\phi) + (C_{a12}C_6 - C_{a15}C_3)\cos\phi + (C_{a18}C_3 - C_{a12}C_9)\sin\phi}{(C_4C_9 - C_6C_7)(1-\cos\phi) + (C_1C_6 - C_3C_4)\cos\phi + (C_3C_7 - C_1C_9)\sin\phi}$$

$$\psi_A = -\frac{t_o R^2}{EI}\,\frac{C_{a12}(C_4C_9 - C_6C_7) + C_{a15}(C_3C_7 - C_1C_9) + C_{a18}(C_1C_6 - C_3C_4)}{(C_4C_9 - C_6C_7)(1-\cos\phi) + (C_1C_6 - C_3C_4)\cos\phi + (C_3C_7 - C_1C_9)\sin\phi}$$

$$V_B = V_A$$

$$M_B = V_A R\sin\phi + M_A\cos\phi - t_o R[1-\cos(\phi-\theta)]$$

$$T_B = V_A R(1-\cos\phi) + M_A\sin\phi + t_o R\sin(\phi-\theta)$$

If $\beta = 1.3$ (solid or hollow round cross section, $\nu = 0.3$)

ϕ	45°	90°			180°		
θ	0°	0°	30°	60°	0°	60°	120°
K_{VA}	0.6652	0.7038	0.1732	−0.0276	0.3433	−0.1635	−0.1561
K_{MA}	−0.3918	−0.8944	−0.4108	−0.0749	−1.2471	−0.4913	0.0034
$K_{\psi A}$	0.2993	0.6594	0.2445	0.0563	0.7597	0.0048	0.0211
K_{MB}	−0.0996	−0.2962	−0.3268	−0.1616	−0.7529	−1.0087	−0.5034
K_{TB}	0.6249	0.8093	0.6284	0.3975	0.6866	0.5390	0.5538

(*Continued*)

TABLE 9.4 Formulas for Curved Beams of Compact Cross Section Loaded Normal to the Plane of Curvature (*Continued*)

End Restraints, Reference No.	Formulas for Boundary Values and Selected Numerical Values

5d. Right end fixed, left end supported and roll-guided

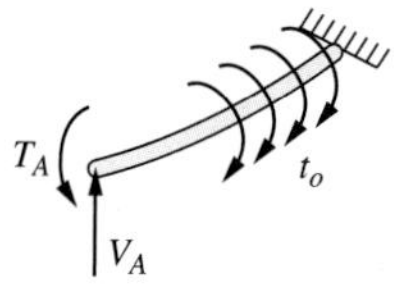

$M_A = 0 \quad y_A = 0 \quad \psi_A = 0$

$y_B = 0 \quad \Theta_B = 0 \quad \psi_B = 0$

$$V_A = -t_o \frac{[(C_{a12}+C_{a18})C_5 - C_{a15}(C_2+C_8)]\sin\phi + (C_{a12}C_8 - C_{a18}C_2)\cos\phi}{[C_5(C_3+C_9) - C_6(C_2+C_8)]\sin\phi + (C_3C_8 - C_2C_9)\cos\phi}$$

$$T_A = -t_o R \frac{[C_{a15}(C_3+C_9) - C_6(C_{a12}+C_{a18})]\sin\phi + (C_{a18}C_3 - C_{a12}C_9)\cos\phi}{[C_5(C_3+C_9) - C_6(C_2+C_8)]\sin\phi + (C_3C_8 - C_2C_9)\cos\phi}$$

$$\Theta_A = -\frac{t_o R^2}{EI} \frac{C_{a12}(C_5C_9 - C_6C_8) + C_{a15}(C_3C_8 - C_2C_9) + C_{a18}(C_2C_6 - C_3C_5)}{[C_5(C_3+C_9) - C_6(C_2+C_8)]\sin\phi + (C_3C_8 - C_2C_9)\cos\phi}$$

$$V_B = V_A$$

$$M_B = V_A R\sin\phi - T_A \sin\phi - t_o R[1-\cos(\phi-\theta)]$$

$$T_B = V_A R(1-\cos\phi) + T_A \cos\phi + t_o R\sin(\phi-\theta)$$

If β = 1.3 (solid or hollow round cross section, ν = 0.3)

ϕ	45°	90°			180°		
θ	0°	0°	30°	60°	0°	60°	120°
K_{VA}	−0.2229	−0.4269	−0.3313	−0.1226	−0.6366	−0.4775	−0.1592
K_{TA}	−0.3895	−0.7563	−0.3109	−0.0640	−1.1902	−0.3153	−0.0089
$K_{\Theta A}$	−0.0237	−0.2020	−0.1153	−0.0165	−1.9576	−0.9588	0.0200
K_{MB}	−0.1751	−0.6706	−0.5204	−0.1926	−2.0000	−1.5000	−0.5000
K_{TB}	0.3664	0.5731	0.5347	0.3774	−0.0830	0.2264	0.5566

5e. Right end fixed, left end fixed

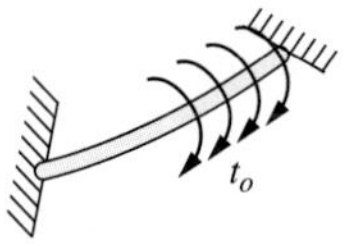

$y_A = 0 \quad \Theta_A = 0 \quad \psi_A = 0$

$y_B = 0 \quad \Theta_B = 0 \quad \psi_B = 0$

$$V_A = -t_o \frac{C_{a12}(C_4C_8 - C_5C_7) + C_{a15}(C_2C_7 - C_1C_8) + C_{a18}(C_1C_5 - C_2C_4)}{C_1(C_5C_9 - C_6C_8) + C_4(C_3C_8 - C_2C_9) + C_7(C_2C_6 - C_3C_5)}$$

$$M_A = -t_o R \frac{C_{a12}(C_5C_9 - C_6C_8) + C_{a15}(C_3C_8 - C_2C_9) + C_{a18}(C_2C_6 - C_3C_5)}{C_1(C_5C_9 - C_6C_8) + C_4(C_3C_8 - C_2C_9) + C_7(C_2C_6 - C_3C_5)}$$

$$T_A = -t_o R \frac{C_{a12}(C_6C_7 - C_4C_9) + C_{a15}(C_1C_9 - C_3C_7) + C_{a18}(C_3C_4 - C_1C_6)}{C_1(C_5C_9 - C_6C_8) + C_4(C_3C_8 - C_2C_9) + C_7(C_2C_6 - C_3C_5)}$$

$$V_B = V_A$$

$$M_B = V_A R\sin\phi + M_A\cos\phi - T_A\sin\phi - t_o R[1-\cos(\phi-\theta)]$$

$$T_B = V_A R(1-\cos\phi) + M_A\sin\phi + T_A\cos\phi + t_o R\sin(\phi-\theta)$$

If β = 1.3 (solid or hollow round cross section, ν = 0.3)

ϕ	45°		90°		180°	
θ	0°	15°	0°	30°	0°	60°
K_{VA}	0.0000	−0.0444	0.0000	−0.0877	0.0000	−0.1657
K_{MA}	−0.1129	−0.0663	−0.3963	−0.2262	−1.0000	−0.4898
K_{TA}	−0.3674	−0.1571	−0.6037	−0.2238	−0.5536	−0.0035
K_{MB}	−0.1129	−0.1012	−0.3963	−0.3639	−1.0000	−1.0102
K_{TB}	0.3674	0.3290	0.6037	0.5522	0.5536	0.5382

5f. Right end supported and slope-guided, left end supported and slope-guided

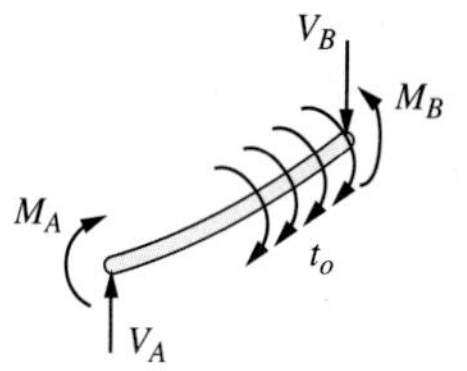

$T_A = 0 \quad y_A = 0 \quad \Theta_A = 0$

$T_B = 0 \quad y_B = 0 \quad \Theta_B = 0$

$$V_A = t_o \frac{[C_1\sin\phi - C_4(1-\cos\phi)]\sin(\phi-\theta) - C_{a12}\sin^2\phi + C_{a15}(1-\cos\phi)\sin\phi}{C_4(1-\cos\phi)^2 + C_3\sin^2\phi - (C_1+C_6)(1-\cos\phi)\sin\phi}$$

$$M_A = -t_o R \frac{[C_3\sin\phi - C_6(1-\cos\phi)]\sin(\phi-\theta) - C_{a12}(1-\cos\phi)\sin\phi + C_{a15}(1-\cos\phi)^2}{C_4(1-\cos\phi)^2 + C_3\sin^2\phi - (C_1+C_6)(1-\cos\phi)\sin\phi}$$

$$\psi_A = \frac{t_o R^2}{EI} \frac{(C_3C_4 - C_1C_6)\sin(\phi-\theta) + (C_{a12}C_6 - C_{a15}C_3)\sin\phi - (C_{a12}C_4 - C_{a15}C_1)(1-\cos\phi)}{C_4(1-\cos\phi)^2 + C_3\sin^2\phi - (C_1+C_6)(1-\cos\phi)\sin\phi}$$

$$V_B = V_A$$

$$M_B = V_A R\sin\phi + M_A\cos\phi - t_o R[1-\cos(\phi-\theta)]$$

$$\psi_B = \psi_A\cos\phi + \frac{M_A R}{EI}C_7 + \frac{V_A R^2}{EI}C_9 + \frac{t_o R^2}{EI}C_{a18}$$

If β = 1.3 (solid or hollow round cross section, ν = 0.3)

ϕ	45°		90°		180°	
θ	0°	15°	0°	30°	0°	60°
K_{VA}	0.0000	−0.2275	0.0000	−0.3732	0.0000	−0.4330
K_{MA}	−1.0000	−0.6129	−1.0000	−0.4928	−1.0000	−0.2974
$K_{\psi A}$	1.0000	0.6203	1.0000	0.5089	1.0000	0.1934
K_{MB}	−1.0000	−0.7282	−1.0000	−0.8732	−1.0000	−1.2026
$K_{\psi B}$	1.0000	0.7027	1.0000	0.7765	1.0000	0.7851

(*Continued*)

TABLE 9.4 Formulas for Curved Beams of Compact Cross Section Loaded Normal to the Plane of Curvature (*Continued*)

End Restraints, Reference No.	Formulas for Boundary Values and Selected Numerical Values

5g. Right end supported and slope-guided, left end supported and roll-guided

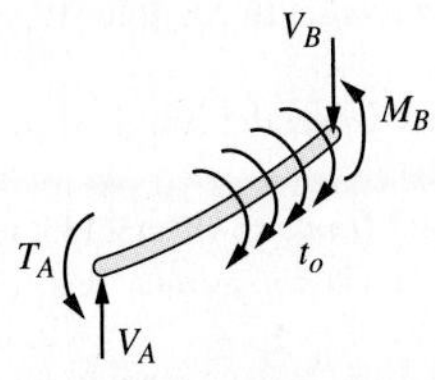

$M_A = 0 \quad y_A = 0 \quad \psi_A = 0$
$T_B = 0 \quad y_B = 0 \quad \Theta_B = 0$

$$V_A = -t_o \frac{C_{a12}\cos^2\phi - C_{a15}\sin\phi\cos\phi + (C_5\sin\phi - C_2\cos\phi)\sin(\phi-\theta)}{(C_5\sin\phi - C_2\cos\phi)(1-\cos\phi) + C_3\cos^2\phi - C_6\sin\phi\cos\phi}$$

$$T_A = -t_o R \frac{(C_{a15}\sin\phi - C_{a12}\cos\phi)(1-\cos\phi) + (C_3\cos\phi - C_6\sin\phi)\sin(\phi-\theta)}{(C_5\sin\phi - C_2\cos\phi)(1-\cos\phi) + C_3\cos^2\phi - C_6\sin\phi\cos\phi}$$

$$\Theta_A = -\frac{t_o R^2}{EI} \frac{(C_{a12}C_5 - C_{a15}C_2)(1-\cos\phi) + (C_{a15}C_3 - C_{a12}C_6)\cos\phi + (C_2C_6 - C_3C_5)\sin(\phi-\theta)}{(C_5\sin\phi - C_2\cos\phi)(1-\cos\phi) + C_3\cos^2\phi - C_6\sin\phi\cos\phi}$$

$$V_B = V_A$$

$$M_B = V_A R\sin\phi - T_A\sin\phi - t_o R[1-\cos(\phi-\theta)]$$

$$\psi_B = -\Theta_A\sin\phi + \frac{T_A R}{EI}C_8 + \frac{V_A R^2}{EI}C_9 + \frac{t_o R^2}{EI}C_{a18}$$

If $\beta = 1.3$ (solid or hollow round cross section, $\nu = 0.3$)

ϕ	45°	90°			180°		
θ	0°	0°	30°	60°	0°	60°	120°
K_{VA}	−0.8601	−1.0000	−0.8660	−0.5000	−0.5976	−0.5837	−0.4204
K_{TA}	−0.6437	−0.9170	−0.4608	−0.1698	−1.1953	−0.3014	0.0252
$K_{\Theta A}$	−0.0209	−0.1530	−0.0695	0.0158	−2.0590	−0.6825	0.6993
K_{MB}	−0.4459	−1.0830	−0.9052	−0.4642	−2.0000	−1.5000	−0.5000
$K_{\psi B}$	0.2985	0.6341	0.5916	0.4716	−0.1592	0.4340	1.0670

5h. Right end supported and slope-guided, left end simply supported

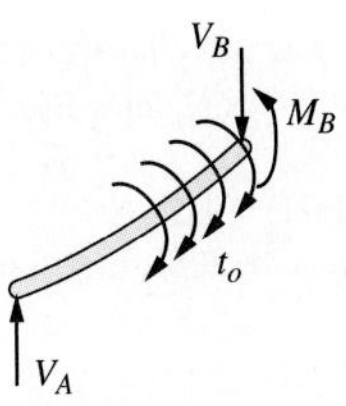

$M_A = 0 \quad T_A = 0 \quad y_A = 0$
$T_B = 0 \quad y_B = 0 \quad \Theta_B = 0$

$$V_A = -\frac{t_o\sin(\phi-\theta)}{1-\cos\phi}$$

$$\Theta_A = -\frac{t_o R^2}{EI}\left[\frac{C_{a12}\sin\phi + C_6\sin(\phi-\theta)}{1-\cos\phi} - \frac{C_3\sin\phi\sin(\phi-\theta)}{(1-\cos\phi)^2} - C_{a15}\right]$$

$$\psi_A = -\frac{t_o R^2}{EI}\left[\frac{C_{a15}\sin\phi - C_{a12}\cos\phi}{1-\cos\phi} + (C_3\cos\phi - C_6\sin\phi)\frac{\sin(\phi-\theta)}{(1-\cos\phi)^2}\right]$$

$$V_B = V_A$$

$$M_B = V_A R\sin\phi - t_o R[1-\cos(\phi-\theta)]$$

$$\psi_B = \psi_A\cos\phi - \Theta_A\sin\phi + \frac{V_A R^2}{EI}C_9 + \frac{t_o R^2}{EI}C_{a18}$$

If $\beta = 1.3$ (solid or hollow round cross section, $\nu = 0.3$)

ϕ	45°	90°			180°		
θ	0°	0°	30°	60°	0°	60°	120°
K_{VA}	−2.4142	−1.0000	−0.8660	−0.5000	0.0000	−0.4330	−0.4330
$K_{\Theta A}$	−0.2888	−0.7579	−0.3720	−0.0957	−3.6128	−1.0744	0.7320
$K_{\psi A}$	1.4161	1.6564	0.8324	0.3067	2.3000	0.5800	−0.0485
K_{MB}	−2.0000	−2.0000	−1.3660	−0.6340	−2.0000	−1.5000	−0.5000
$K_{\psi B}$	1.2811	1.0985	0.8250	0.5036	−0.3000	0.3985	1.0700

5i. Right end supported and roll-guided, left end supported and roll-guided

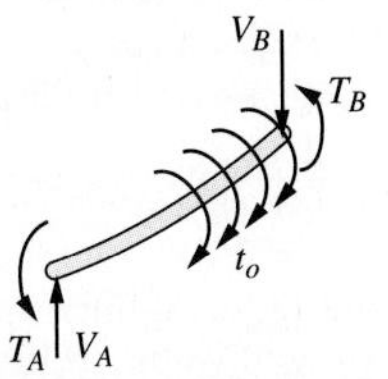

$M_A = 0 \quad y_A = 0 \quad \psi_A = 0$
$M_B = 0 \quad y_B = 0 \quad \psi_B = 0$

$$V_A = 0$$

$$T_A = -t_o R\frac{(C_{a12} + C_{a18})\sin^2\phi + (C_3 + C_9)\sin\phi[1-\cos(\phi-\theta)]}{(C_2 + C_3 + C_8 + C_9)\sin^2\phi}$$

$$\Theta_A = -\frac{t_o R^2}{EI}\frac{[C_{a12}(C_8 + C_9) - C_{a18}(C_2 + C_3)]\sin\phi - (C_2C_9 - C_3C_8)[1-\cos(\phi-\theta)]}{(C_2 + C_3 + C_8 + C_9)\sin^2\phi}$$

$$V_B = 0$$

$$T_B = T_A\cos\phi + t_o R\sin(\phi-\theta)$$

$$\Theta_B = \Theta_A\cos\phi + \frac{T_A R}{EI}C_5 + \frac{t_o R^2}{EI}C_{a15}$$

If $\beta = 1.3$ (solid or hollow round cross section, $\nu = 0.3$)

ϕ	45°		90°		270°	
θ	0°	15°	0°	30°	0°	90°
K_{VA}	0.0000	0.0000	0.0000	0.0000	0.0000	0.0000
K_{TA}	−0.4142	−0.1895	−1.0000	−0.5000	1.0000	2.0000
$K_{\Theta A}$	−0.0527	−0.0368	−0.6564	−0.4679	−6.5692	−2.3000
K_{TB}	0.4142	0.3660	1.0000	0.8660	−1.0000	0.0000
$K_{\Theta B}$	0.0527	0.0415	0.6564	0.5094	6.5692	7.2257

5j. Right end supported and roll-guided, left end simply supported

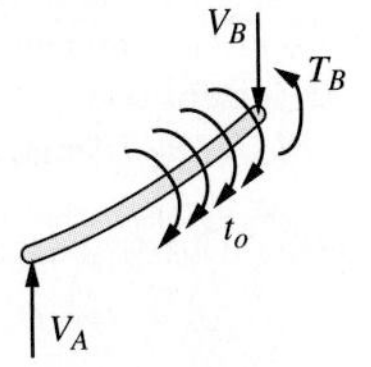

$M_A = 0 \quad T_A = 0 \quad y_A = 0$
$M_B = 0 \quad y_B = 0 \quad \psi_B = 0$

$$V_A = \frac{t_o[1-\cos(\phi-\theta)]}{\sin\phi}$$

$$\Theta_A = -\frac{t_o R^2}{EI}\left\{\frac{C_{a12}\cos\phi - C_{a18}(1-\cos\phi)}{\sin\phi} + \frac{[C_3\cos\phi - C_9(1-\cos\phi)][1-\cos(\phi-\theta)]}{\sin^2\phi}\right\}$$

$$\psi_A = -\frac{t_o R^2}{EI}\left\{C_{a12} + C_{a18} + \frac{(C_3 + C_9)[1-\cos(\phi-\theta)]}{\sin\phi}\right\}$$

$$V_B = V_A$$

$$T_B = V_A R(1-\cos\phi) + t_o R\sin(\phi-\theta)$$

$$\Theta_B = \Theta_A\cos\phi + \phi_A\sin\phi + \frac{V_A R^2}{EI}C_6 + \frac{t_o R^2}{EI}C_{a15}$$

If $\beta = 1.3$ (solid or hollow round cross section, $\nu = 0.3$)

ϕ	45°			90°		
θ	0°	15°	30°	0°	30°	60°
K_{VA}	0.4142	0.1895	0.0482	1.0000	0.5000	0.1340
$K_{\Theta A}$	−0.1683	−0.0896	−0.0247	−1.9564	−1.1179	−0.3212
$K_{\psi A}$	0.4229	0.1935	0.0492	2.0420	1.0210	0.2736
K_{TB}	0.8284	0.5555	0.2729	2.0000	1.3660	0.6340
$K_{\Theta B}$	0.1124	0.0688	0.0229	1.3985	0.8804	0.2878

9.7 REFERENCES

1. Wilson, B. J., and J. F. Quereau, "A Simple Method of Determining Stress in Curved Flexural Members," *Univ. Ill. Eng. Exp. Sta.*, Circ. 16, 1927.
2. Timoshenko, S., and J. N. Goodier, *Theory of Elasticity*, 2nd ed., Engineering Society Monograph, McGraw-Hill, 1951.
3. Boresi, A. P., R. J. Schmidt, and O. M. Sidebottom, *Advanced Mechanics of Materials*, 5th ed., John Wiley & Sons, 1993.
4. von Kármán, Th., "Über die Formänderung dünnwandiger Rohre, insbesondere federnder Ausgleichrohre," *Z. Vereines Dtsch. Ing.*, 55: 1889, 1911.
5. Timoshenko, S., "Bending Stresses in Curved Tubes of Rectangular Cross-Section," *Trans. ASME*, 45: 135, 1923.
6. Burke, W. F., "Working Charts for the Stress Analysis of Elliptic Rings," *Nat. Adv. Comm. Aeron.*, Tech. Note 444, 1933.
7. Bushnell, D., "Elastic-Plastic Bending and Buckling of Pipes and Elbows," *Comp. Struct.*, 13: 1981.
8. Utecht, E. A., "Stresses in Curved, Circular Thin-Wall Tubes," *AME J. Appl. Mech.*, 30(1): 1963.
9. U.S. Dept. of Agriculture, Bureau of Reclamation, "Penstock Analysis and Stiffener Design," Boulder Canyon Project, Final Reports, Pt. V, Bull. 5, 1940.
10. Bleich, H., "Stress Distribution in the Flanges of Curved T- and I-Beams," U.S. Dept. of Navy, David W. Taylor Model Basin, transl. 228, 1950.
11. Mantle, J. B., and T. J. Dolan, "A Photoelastic Study of Stresses in U-Shaped Members," *Proc. Soc. Exp. Stress Anal.*, 6(1): 1948.
12. Stressed Skin Structures, Royal Aeronautical Society, data sheets.
13. Timoshenko, S., *Strength of Materials*, D. Van Nostrand, 1930.
14. Levy, R., "Displacements of Circular Rings with Normal Loads," *Proc. Am. Soc. Civil Eng., J. Struct. Div.*, 88(1): 1962.
15. Moorman, R. B. B., "Stresses in a Curved Beam under Loads Normal to the Plane of Its Axis," *Iowa Eng. Exp. Sta.*, Iowa State College, Bull. 145, 1940.
16. Fisher, G. P., "Design Charts for Symmetrical Ring Girders," *ASME J. Appl. Mech.*, 24(1): 1957.
17. Moorman, R. B. B., "Stresses in a Uniformly Loaded Circular-Arc I-Beam," *Univ. Missouri Bull.*, Eng. Ser. 36, 1947.
18. Hogan, M. B., Utah Eng. Exp. Sta., Bulls. 21, 27, and 31.
19. Karabin, M. E., E. C. Rodabaugh, and J. F. Whatham, "Stress Component Indices for Elbow-Straight Pipe Junctions Subjected to In-Plane Bending," *Trans. ASME J. Pressure Vessel Tech.*, 108, February 1986.
20. Volterra, E., and T. Chung, "Constrained Circular Beam on Elastic Foundations," Trans. Am. Soc. Civil Eng., vol. 120, 1955 (paper 2740).
21. Meck, H. R., "Three-Dimensional Deformation and Buckling of a Circular Ring of Arbitrary Section," *ASME J. Eng. Ind.*, 91(1): 1969.
22. Brookhart, G. C., "Circular-Arc I-Type Girders," *Proc. Am. Soc. Civil Eng., J. Struct. Div.*, 93(6): 1967.
23. Dabrowski, R., *Curved Thin-Walled Girders, Theory and Analyses*, Cement and Concrete Association, 1972.
24. Verden, W., *Curved Continuous Beams for Highway Bridges*, Frederick Ungar, 1969 (English transl.).
25. Sawko, F., and R. J. Cope, "Analysis of Multi-Cell Bridges Without Transverse Diaphragms—A Finite Element Approach," *Struct. Eng.*, 47(11): 1969.
26. Meyer, C., "Analysis and Design of Curved Box Girder Bridges," Univ. California, Berkeley, Struct, Eng. & Struct. Mech. Rept. SESM-70–22, December 1970.
27. Vlasov, V. Z., *Thin-Walled Elastic Beams*, Clearing House for Federal Scientific and Technical Information, U.S. Dept. Of Commerce, 1961.
28. Neugebauer, G. H., Private communication.
29. Cook, R. D., and W. C. Young, *Advanced Mechanics of Materials*, 2nd ed., Prentice Hall, 1998.
30. Tepper, K., Private communication.
31. Broughton, D. C., M. E. Clark, and H. T. Corten, "Tests and Theory of Elastic Stresses in Curved Beams Having I- and T-Sections," *Exp. Mech.*, 8(1): 1950.
32. Biezeno, C. B., and R. Grammel, *Engineering Dynamics*, vol. II (Elastic Problems of Single Machine Elements), Blackie & Son, 1956 (English translation of 1939 edition in German).
33. Plesha, M. E., "Department of Engineering Mechanics," University of Wisconsin–Madison, private communication.
34. Whatham, J. F., "Pipe Bend Analysis by Thin Shell Theory," *ASME J. Appl. Mech.*, 53: March 1986.
35. Barber, J. R., "Force and Displacement Influence Functions for the Circular Ring," *Inst. Mech. Eng. J. Strain Anal.*, 13(2): 1978.
36. Budynas, R. G., *Advanced Strength and Applied Analysis*, 2nd ed., McGraw-Hill, 1999.

CHAPTER 10

Torsion

10.1 STRAIGHT BARS OF UNIFORM CIRCULAR SECTION UNDER PURE TORSION

The formulas in this section are based on the following assumptions: (1) The bar is straight, of uniform circular section (solid or concentrically hollow), and of homogeneous isotropic material; (2) the bar is loaded only by equal and opposite twisting couples, which are applied at its ends in planes normal to its axis; and (3) the bar is not stressed beyond the elastic limit.

Behavior

The bar twists, each section rotating about the longitudinal axis. Plane sections remain plane, and radii remain straight. There is at any point a shear stress τ on the plane of the section; the magnitude of this stress is proportional to the distance from the center of the section, and its direction is perpendicular to the radius drawn through the point. Accompanying this shear stress there is an equal longitudinal shear stress on a radial plane and equal tensile and compressive stresses σ_t and σ_c at 45° (see Sec. 7.5). The deformation and stresses described are represented in Fig. 10.1.

In addition to these deformations and stresses, there is some longitudinal strain and stress. A solid circular cylinder wants to lengthen under twist, as shown experimentally by Poynting (Ref. 26). In any event, for elastic loading of metallic circular bars, neither longitudinal deformation nor stress is likely to be large enough to have engineering significance.

Formulas

Let T = twisting moment, l = length of the member, r = outer radius of the section, J = polar moment of inertia of the section, ρ = radial distance from the center of the section to any point q, τ = the shear stress, θ = angle of twist (radians), G = modulus of rigidity of the material, and U = strain energy. Then

$$\theta = \frac{Tl}{JG} \qquad (10.1\text{-}1)$$

$$\tau = \frac{T\rho}{J} \text{ (at point } q) \qquad (10.1\text{-}2)$$

$$\tau_{max} = \frac{Tr}{J} \text{ (at outer surface)} \qquad (10.1\text{-}3)$$

$$U = \frac{1}{2}\frac{T^2 l}{JG} \qquad (10.1\text{-}4)$$

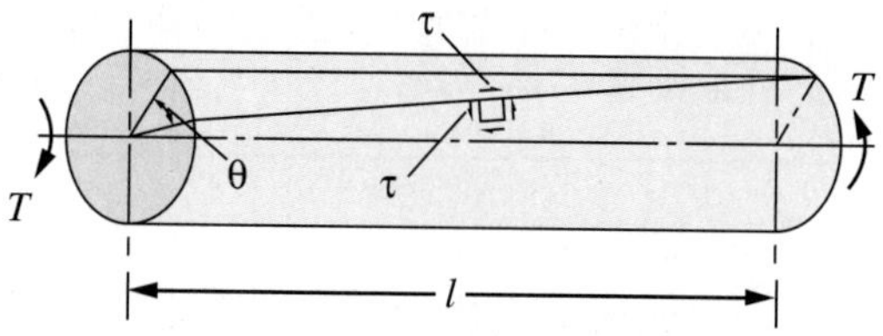

Figure 10.1

By substituting for J in Eqs. (10.1-1) and (10.1-3), its value $2I$ from Table A.1, the formulas for cases 1 and 10 in Table 10.1 are readily obtained. If a solid or hollow circular shaft has a slight taper, the formulas above for shear stress are sufficiently accurate and the expressions for θ and U can be modified to apply to a differential length by replacing l by dl. If the change in section is abrupt, as at a shoulder with a small fillet, the maximum stress should be found by the use of a suitable factor of stress concentration K_t. Values of K_t are given in Table 17.1.

10.2 BARS OF NONCIRCULAR UNIFORM SECTION UNDER PURE TORSION

The formulas of this section are based on the same assumptions as those of Sec. 10.1 except that the cross section of the bar is not circular. It is important to note that the condition of loading implies that the end sections of the bar are free to warp, there being no constraining forces to hold them in their respective planes.

Behavior

The bar twists, each section rotating about its torsional center. Sections do not remain plane, but warp, and some radial lines through the torsional center do not remain straight. The distribution of shear stress on the section is not necessarily linear, and the direction of the shear stress is not necessarily normal to a radius.

Formulas

The torsional stiffness of the bar can be expressed by the general equations

$$T = \frac{\theta}{l} KG \quad \text{or} \quad \theta = \frac{Tl}{KG} \tag{10.2-1}$$

where K is a factor dependent on the form and dimensions of the cross section. For a *circular* section K is the polar moment of inertia J [Eq. (10.1-1)] for other sections K is less than J and may be only a very small fraction of J. The maximum stress is a function of the twisting moment and of the form and dimensions of the cross section. In Table 10.1, formulas are given for K and for max τ for a variety of sections. The formulas for cases 1–3, 5, 10, and 12 are based on rigorous mathematical analysis. The equations for case 4 are given in a simplified form involving an approximation, with a resulting error not greater than 4%. The K formulas for cases 13–21 and the stress formulas for cases 13–18 are based on mathematical analysis but are approximate (Ref. 2); their accuracy depends upon how nearly the actual section conforms to the assumptions indicated as to form. The K formulas for cases 22–26 and the stress formulas for cases 19–26 are based on the membrane analogy and are to be regarded as reasonably close approximations giving results that are rarely as much as 10% in error (Refs. 2–4 and 11).

It will be noted that, formulas for K in cases 23–26 are based on the assumption of uniform flange thickness. For slightly tapering flanges, D should be taken as the diameter of the largest circle that can be inscribed in the actual section, and b as the average flange thickness. For sharply tapering flanges the method described by Griffith (Ref. 3) may be used. Charts relating especially to structural H- and I-sections are in Ref. 11.

Cases 7, 9, and 27–35 present the results of curve fitting to data from Isakower, Refs. 12 and 13. These data were obtained from running a computer code CLYDE (Ref. 14) based on a finite-difference solution using central differences with a constant-size square mesh. Reference 12 also suggests an extension of this work to include sections containing hollows. For some simple concentric hollows the results of solutions in Table 10.1 can be superposed to obtain closely approximate solutions if certain limitations are observed. See the examples at the end of this section.

The formulas of Table 10.1 make possible the calculation of the strength and stiffness of a bar of almost any form, but an understanding of the membrane analogy (Sec. 6.4) makes it possible to draw certain conclusions as to the *comparative* torsional properties of different sections by simply visualizing the bubbles that would be formed over holes of corresponding size and shape. From the volume relationship, it can be seen that of two sections having the same area, the one more nearly circular is the stiffer, and that although any extension whatever of the section increases its torsional stiffness, narrow outstanding flanges and similar protrusions have little effect. It is also apparent that any member having a narrow section, such as a thin plate, has practically the same torsional stiffness when flat as when bent into the form of an open tube or into channel or angle section.

From the slope relationship it can be seen that the greatest stresses (slopes) in a given section occur at the boundary adjacent to the thicker portions, and that the stresses are very low at the ends of outstanding flanges or protruding corners and very high at points where the boundary is sharply concave. Therefore, a longitudinal slot or groove that is sharp at the bottom or narrow will cause high local stresses, and if it is deep will greatly reduce the torsional stiffness of the member. The direction of the shear stresses at any point is along the contour of the bubble surface at the corresponding point, and at points corresponding to local maximum and minimum elevations of the bubble having zero slopes in all directions the shear stress is zero. Therefore, there may be several points of zero shear stress in a section. Thus for an I-section, there are high points of zero slope at the center of the largest inscribed circles (at the junction of web and flanges) and a low point of zero slope at the center of the web, and eight points of zero slope at the external corners. At these points in the section the shear stress is zero.

The preceding generalizations apply to solid sections, but it is possible to make somewhat similar generalizations concerning hollow or tubular sections from the formulas given for cases 10–16. These formulas show that the strength and stiffness of a hollow section depend largely upon the area inclosed by the median boundary. For this reason a circular tube is stiffer and stronger than one of any other form, and the more nearly the form of any hollow section approaches the circular, the greater will be its strength and stiffness. It is also apparent from the formulas for strength that even a local reduction in the thickness of the wall of a tube, such as would be caused by a longitudinal groove, may greatly increase the maximum shear stress, though if the groove is narrow the effect on stiffness will be small.

The torsional strengths and stiffnesses of thin-walled multicelled structures such as airplane wings and boat hulls can be calculated by the same procedures as for single-celled sections. The added relationships needed are developed from the fact that all cells twist at the same angular rate at a given section (Ref. 1).

EXAMPLE 1 It is required to compare the strength and stiffness of a circular steel tube, 4 in outside diameter and $\frac{5}{32}$ in thick, with the strength and stiffness of the same tube after it has been split by cutting full length along an element. No warping restraint is provided.

SOLUTION The strengths will be compared by comparing the twisting moments required to produce the same stress; the stiffnesses will be compared by comparing the values of K.

(a) For the tube (Table 10.1, case 10), $K=\frac{1}{2}\pi\left(r_o^4-r_i^4\right)=\frac{1}{2}\pi\left[2^4-\left(1\frac{27}{32}\right)^4\right]=6.98\text{ in}^4$,

$$T=\tau\frac{\pi\left(r_o^4-r_i^4\right)}{2r_o}=3.49\tau\text{ lb-in}$$

(b) For the split tube (Table 10.1, case 17), $K=\frac{2}{3}\pi rt^3=\frac{2}{3}\pi\left(1\frac{59}{64}\right)\left(\frac{5}{32}\right)^3=0.0154\text{ in}^4$,

$$T=\tau\frac{4\pi^2r^2t^2}{6\pi r+1.8t}=0.097\tau\text{ lb-in}$$

The closed section is therefore more than 400 times as stiff as the open section and more than 30 times as strong.

EXAMPLE 2 It is required to determine the angle through which an airplane-wing spar of spruce, 8 ft long and having the section shown in Fig. 10.2, would be twisted by end torques of 500 lb-in, and to find the maximum resulting stress. For the material in question, $G=100{,}000\text{ lb/in}^2$ and $E=1{,}500{,}000\text{ lb/in}^2$.

SOLUTION All relevant dimensions are shown in Fig. 10.2, with notation corresponding to that used in the formulas. The first step is to compute K by the formulas given for case 26 (Table 10.1), and we have

$$K=2K_1+K_2+2\alpha D^4$$

$$K_1=2.75(1.045^3)\left\{\frac{1}{3}-\frac{0.21(1.045)}{2.75}\left[1-\frac{1.045^4}{12(2.75^4)}\right]\right\}=0.796\text{ in}^4$$

$$K_2=\tfrac{1}{3}(2.40)(0.507^3)=0.104\text{ in}^4$$

$$\alpha=\frac{0.507}{1.045}\left[0.150+\frac{0.1(0.875)}{1.045}\right]=0.1133$$

Thus,

$$K=2(0.796)+0.104+2(0.1133)(1.502^4)=2.85\text{ in}^4$$

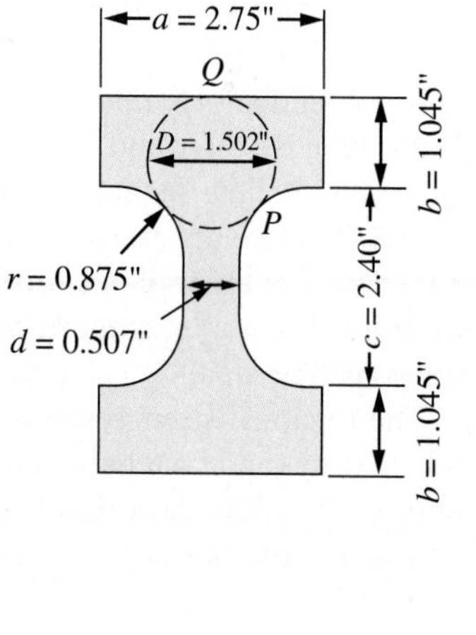

Figure 10.2

Therefore,

$$\theta = \frac{Tl}{KG} = \frac{500(96)}{2.85(100,000)} = 0.168 \text{ rad} = 9.64°$$

The maximum stress will probably be at *P*, the point where the largest inscribed circle touches the section boundary at a fillet. The formula is

$$\tau_{\max} = \frac{T}{K}C$$

where

$$C = \frac{1.502}{1 + \frac{\pi^2(1.502^4)}{16(7.63^2)}}\left\{1 + \left[0.118\ln\left(1 - \frac{1.502}{2(-0.875)}\right) - 0.238\frac{1.502}{2(-0.875)}\right] \times \tanh\frac{2(\pi/2)}{\pi}\right\} = 1.73 \text{ in}$$

Substituting the values of *T*, *C*, and *K*, it is found that

$$\tau_{\max} = \frac{500}{2.85}(1.73) = 303 \text{ lb/in}^2$$

It will be of interest to compare this stress with the stress at *Q*, the other point where the maximum inscribed circle touches the boundary. Here the formula that applies is

$$\tau = \frac{T}{K}C$$

where

$$C = \frac{1.502}{1 + \pi^2 \frac{1.502^4}{16(7.63^2)}}\left[1 + 0.15\left(\frac{\pi^2(1.502^4)}{16(7.63^2)} - \frac{1.502}{\infty}\right)\right] = 1.437 \text{ in}$$

(Here r = infinity because the boundary is straight.)

Substituting the values of *T*, *C*, and *K* as before, it is found that $\tau = 252 \text{ lb/in}^2$.

EXAMPLE 3 For each of the three cross sections shown in Fig. 10.3, determine the numerical relationships between torque and maximum shear stress and between torque and rate of twist.

SOLUTION To illustrate the method of solution, superposition will be used for section *A* despite the availability of a solution from case 10 of Table 10.1. For a shaft, torque and angle of twist are related in the same way that a soap-film volume under the film is related to the pressure which inflates the film, provided the same cross section is used for each. See the discussion of the membrane analogy in Sec. 6.4. One can imagine then a soap film blown over a circular hole of radius R_o and then imagine the removal of that portion of the volume extending above the level of the soap film for which the

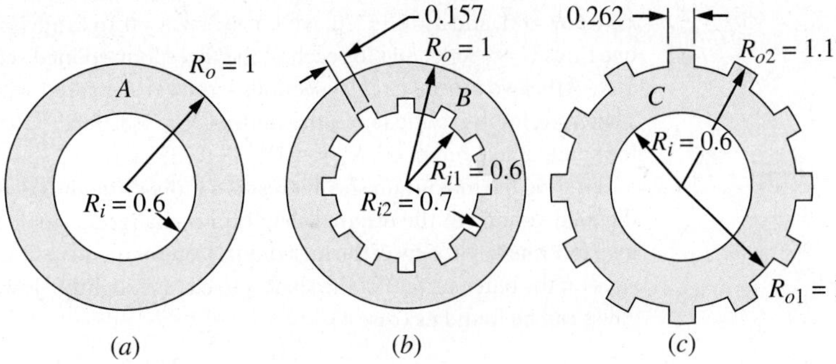

Figure 10.3

radius is R_i. Doing the equivalent operation with the shaft assumes that the rate of twist is θ/L and that it applies to both the outer round shaft of radius R_o and the material removed with radius R_i. The resulting torque T_R is then the difference of the two torques or

$$T_R = T_o - T_i = \frac{K_o G\theta}{L} - \frac{K_i G\theta}{L}$$

where from case 1

$$K_o = \frac{\pi R_o^4}{2} \text{ and } K_i = \frac{\pi R_i^4}{2}$$

or

$$T_R = \frac{\pi/2\left(R_o^4 - R_i^4\right)G\theta}{L}$$

Case 10 for the hollow round section gives this same relationship. Slicing off the top of the soap film would not change the maximum slope of the bubble which is present along the outer radius. Thus the maximum shear stress on the hollow shaft with torque T_R is the same as the shear stress produced on a solid round section by the torque T_o. From case 1

$$\pi = \frac{2T_o}{\pi R_o^3} = \frac{2T_R}{\pi R_o^3}\frac{T_o}{T_R} = \frac{2T_R}{\pi R_o^3}\frac{R_o^4}{R_o^4 - R_i^4} = \frac{2T_R R_o}{\pi\left(R_o^4 - R_i^4\right)}$$

which again checks the expression for shear stress in case 10. Inserting the numerical values for R_o and R_i gives

$$T_R = \frac{1.3672G\theta}{L} \text{ and } \tau = 0.7314T_R$$

It is important to note that the exact answer was obtained because there was a concentric circular contour line on the soap film blown over the large hole. Had the hole in the shaft been slightly off center, none of the equivalent contour lines on the soap film would have been absolutely circular and the answers just obtained would be slightly in error.

Now apply this same technique to the hollow shaft with section B having a 12-spline internal hole. To solve this case exactly, one would need to find a contour line on the equivalent soap film blown over the circular hole of radius R_o. Such a contour does not exist, but it can be created as discussed in Sec. 6.4. Imagine a massless fine wire bent into the shape of the internal spline in a single plane and allow this wire to float at a constant elevation in the soap film in equilibrium with the surface tension. The volume which is to be removed from the soap film is now well-defined and will be worked out using cases 1 and 33, but the unanswered question is what the addition of the wire did to the total volume of the original soap film. It is changed by a small amount. This information comes from Ref. 12, where Isakower shows numerical solutions to some problems with internal noncircular holes. The amount of the error depends upon the number and shape of the splines and upon how close they come to the outer boundary of the shaft. Ignoring these errors, the solution can be carried out as before.

The torque carried by the solid round shaft is the same as for section A. For the 12-point internal spline, one needs to use case 33 three times since case 33 carries only four splines and then remove the extra material added by using case 1 twice for the material internal to the splines. For case 33 let $r = 0.6$, $b = 0.1$, and $a = 0.157/2$, which gives $b/r = 0.167$ and $a/b = 0.785$. Using the equations in case 33 one finds $C = 0.8098$, and for each of the three four-splined sections, $K = 2(0.8098)(0.6)^4 = 0.2099$. For each of the two central circular sections removed, use case 1 with $r = 0.6$, getting $K = \pi(0.6^4)/2 = 0.2036$. Therefore, for the splined hole, the value of $K_i = 3(0.2099) - 2(0.2036) = 0.2225$. For the solid shaft with the splined hole removed, $K_R = \pi(1)^4/2 - 0.2255 = 1.3483$ so that $T_R = 1.3483G\theta/L$.

Finding the maximum shear stress is a more difficult task for this cross section. If, as stated before, the total volume of the original soap film is changed a small amount when the spline-shaped wire is inserted, one might expect the meridional slope of the soap film at the outer edge and the corresponding stress at the outer surface of the shaft A to change slightly. However, if one ignores this effect, this shear stress can be found as (case 1)

$$\tau_o = \frac{2T_o}{\pi R_o^3} = \frac{2T_R}{\pi R_o^3}\frac{T_o}{T_R} = \frac{2T_R}{\pi R_o^3}\frac{K_o}{K_R} = \frac{2T_R}{\pi(1)^3}\frac{\pi/2}{1.3378} = 0.7475T_R$$

For this section, however, there is a possibility that the maximum shear stress and maximum slope of soap film will be bound at the outer edge of an internal spline. No value for this is known, but it would be close to the maximum shear stress on the inner edge of the spline for the material removed. This is given in case 33 as $\tau_i = TB/r^3$, where B can be found from the equations to be $B = 0.6264$, so that $\tau_i = T(0.6264)/0.6^3 = 2.8998T$. Since the torque here is the torque necessary to give a four-splined shaft a rate of twist θ/L, which is common to all the elements used, both positive and negative,

$$\tau_i = 2.8998\frac{KG\theta}{L} = 2.8998(0.2099)\frac{G\theta}{L} = 0.6087\frac{G\theta}{L}$$

$$= 0.6087\frac{T_R}{1.3483} = 0.454T_R$$

Any errors in this calculation would not be of consequence unless the stress concentrations in the corners of the splines raise the peak shear stresses above $\tau_o = 0.7475T_R$. Since this is possible, one would want to carry out a more complete analysis if considerations of fatigue or brittle fracture were necessary.

Using the same arguments already presented for the first two sections, one can find K_R for section C by using three of the 4-splined sections from case 33 and removing twice the solid round material with radius 1.0 and once the solid round material with radius 0.6. For case 33, $r = 1.0, b = 0.17, a = 0.262/2$, $b/r = 0.17$, and $a/b = 0.771$. Using these data, one finds that $C = 0.8100$, $B = 0.6262$, and $K = 1.6200$. This gives then for the hollow section C,

$$K_R = 1.6200(3) - \frac{2\pi(1^4)}{2} - \frac{\pi(0.6^4)}{2} = 1.5148 \quad \text{and} \quad T_R = \frac{1.5148G\theta}{L}$$

Similarly,

$$\tau_{max} = \frac{T(0.6262)}{1^3} = \frac{0.6262(1.6200)G\theta}{L} = \frac{1.0144G\theta}{L}$$

$$= 1.0144\frac{T_R}{1.5148} = 0.6697T_R$$

Again one would expect the maximum shear stress to be somewhat larger with twelve splines than with four and again the stress concentrations in the corners of the splines must be considered.

10.3 EFFECT OF END CONSTRAINT

It was pointed out in Sec. 10.2 that when noncircular bars are twisted, the sections do not remain plane but warp, and that the formulas of Table 10.1 are based on the assumption that this warping is not prevented. If one or both ends of a bar are so fixed that warping is prevented, or if the torque is applied to a section other than at the ends of a bar, the stresses and the angle of twist produced by the given torque are affected. In compact sections the effect is slight, but in the case of open thin-walled sections the effect may be considerable.

Behavior

To visualize the additional support created by warping restraint, consider a very thin rectangular cross section and an I-beam having the same thickness and the same total cross-sectional area as the rectangle. With no warping restraint the two sections will have essentially the same stiffness factor K (see Table 10.1, cases 4 and 26). With warping restraint provided at one end of the bar, the rectangle will be stiffened very little but the built-in flanges of the I-beam act as cantilever beams. The shear forces developed in the flanges as a result of the bending of these cantilevers will assist the torsional shear stresses in carrying the applied torque and greatly increase the stiffness of the bar unless the length is very great.

Formulas

Table 10.2 gives formulas for the warping stiffness factor C_w, the torsional stiffness factor K, the location of the shear center, the magnitudes and locations within the cross section of the maximum shear stresses due to simple torsion, the maximum shear stresses due to warping, and the maximum bending stresses due to warping. All the cross sections listed are assumed to have thin walls and the same thickness throughout the section unless otherwise indicated.

Table 10.3 provides the expressions necessary to evaluate the angle of rotation θ and the first three derivatives of θ along the span for a variety of loadings and boundary restraints. The formulas in this table are based on deformations from bending stresses in the thin-walled open cross sections due to warping restraint and consequently, since the transverse shear deformations of the beam action are neglected, are not applicable to cases where the torsion member is short or where the torsional loading is applied close to a support which provides warping restraint.

In a study of the effect on seven cross sections, all of which were approximately 4 in deep and had walls approximately 0.1 in thick, Schwabenlender (Ref. 28) tested them with one end fixed and the other end free to twist but not warp with the torsional loading applied to the latter end. He found that the effect of the transverse shear stress noticeably reduced the torsional stiffness of cross sections such as those shown in Table 10.2, cases 1 and 6–8, when the length was less than six times the depth; for sections such as those in cases 2–5 (Table 10.2), the effect became appreciable at even greater lengths. To establish an absolute maximum torsional stiffness constant we note that for any cross section, when the length approaches zero, the effective torsional stiffness constant K' cannot exceed J, the polar moment of inertia, where $J = I_x + I_y$ for axes through the centroid of the cross section. (Example 1 illustrates this last condition.)

Reference 19 gives formulas and graphs for the angle of rotation and the first three derivatives for 12 cases of torsional loading of open cross sections. Payne (Ref. 15) gives the solution for a box girder that is fixed at one end and has a torque applied to the other. (This solution was also presented in detail in the fourth edition of this book.) Chu (Ref. 29) and Vlasov (Ref. 30) discuss solutions for cross sections with both open and closed parts. Kollbrunner and Basler (Ref. 31) discuss the warping of continuous beams and consider the multicellular box section, among other cross sections.

EXAMPLE 1 A steel torsion member has a cross section in the form of a twin channel with flanges inward as dimensioned in Fig. 10.4. Both ends of this channel are rigidly welded to massive steel blocks to provide full warping restraint. A torsional load is to be applied to one end block while the other is fixed. Determine the angle of twist at the loaded end for an applied torque of 1000 lb-in for lengths of 100, 50, 25, and 10 in. Assume $E = 30(10^6)$ lb/in^2 and $\nu = 0.285$.

SOLUTION First determine cross-sectional constants, noting that $b = 4 - 0.1 = 3.9$ in, $b_1 = 1.95 - 0.05 = 1.9$ in, $h = 4 - 0.1 = 3.9$ in, and $t = 0.1$ in.

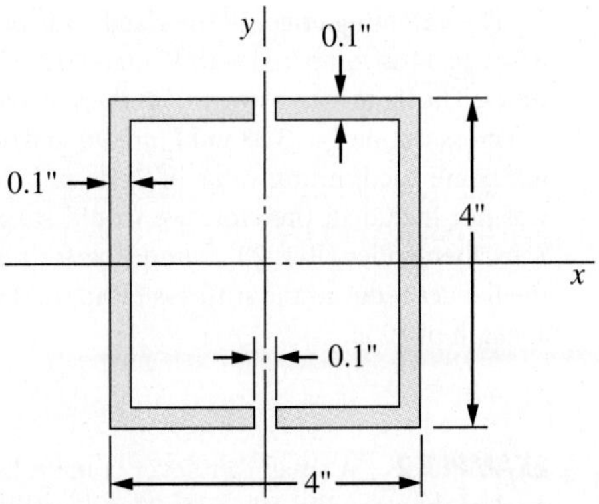

Figure 10.4

From Table 10.2, case 4,

$$K = \frac{t^3}{3}(2b + 4b_1) = \frac{0.1^3}{3}[2(3.9) + 4(1.9)] = 0.005133\ \text{in}^4$$

$$C_w = \frac{tb^2}{24}\left(8b_1^3 + 6h^2b_1 + h^2b + 12b_1^2h\right) = 28.93\ \text{in}^6$$

$$G = \frac{E}{2(1+\nu)} = \frac{30(10^6)}{2(1+0.285)} = 11.67(10^6)\ \text{lb/in}^2$$

$$\beta = \left(\frac{KG}{C_wE}\right)^{1/2} = \left[\frac{0.005133(11.67)(10^6)}{28.93(30)(10^6)}\right]^{1/2} = 0.00831\ \text{in}^{-1}$$

From Table 10.3, case 1d, when $a = 0$, the angular rotation at the loaded end is given as

$$\theta = \frac{T_o}{C_wE\beta^3}\left(\beta l - 2\tanh\frac{\beta l}{2}\right)$$

If we were to describe the total angle of twist at the loaded end in terms of an equivalent torsional stiffness constant K' in the expression

$$\theta = \frac{T_o l}{K'G}$$

then

$$K' = \frac{T_o l}{G\theta} \quad \text{or} \quad K' = K\frac{\beta l}{\beta l - 2\tanh\dfrac{\beta l}{2}}$$

The following table gives both K' and θ for the several lengths:

l	βl	$\tanh\left(\frac{\beta l}{2}\right)$	K'	θ
200	1.662	0.6810	0.0284	34.58°
100	0.831	0.3931	0.0954	5.15°
50	0.416	0.2048	0.3630	0.68°
25	0.208	0.1035	1.4333	0.09°
10	0.083	0.0415	8.926	0.006°

The stiffening effect of the fixed ends of the flanges is obvious even at a length of 200 in where $K' = 0.0284$ as compared with $K = 0.00513$. The warping restraint increases the stiffness more than five times. The large increases in K' at the shorter lengths of 25 and 10 in must be examined carefully. For this cross section $I_x = 3.88$ and $I_y = 3.96$, and so $J = 7.84$ in^4. The calculated stiffness $K' = 8.926$ at $l = 10$ in is beyond the limiting value of 7.84, and so it is known to be in error because shear deformation was not included; therefore we would suspect the value of $K' = 1.433$ at $l = 25$ in as well. Indeed, Schwabenlender (Ref. 28) found that for a similar cross section the effect of shear deformation in the flanges reduced the stiffness by approximately 25% at a length of 25 in and by more than 60% at a length of 10 in.

EXAMPLE 2 A small cantilever crane rolls along a track welded from three pieces of 0.3-in-thick steel, as shown in Fig. 10.5. The 20-ft-long track is solidly welded to a rigid foundation at the right end and simply supported 4 ft from the left end, which is free. The simple support also provides resistance to rotation about the beam axis but provides no restraint against warping or rotation in horizontal and vertical planes containing the axis.

The crane weighs 300 lb and has a center of gravity which is 20 in out from the web of the track. A load of 200 lb is carried at the end of the crane 60 in from the web. It is desired to determine the maximum flexure stress and the maximum shear stress in the track and also the angle of inclination of the crane when it is located 8 ft from the welded end of the track.

SOLUTION The loading will be considered in two stages. First consider a vertical load of 500 lb acting 8 ft from the fixed end of a 16-ft beam fixed at one end and simply supported at the other. The following constants are needed:

$$\bar{y} = \frac{4(0.3)(10) + 9.7(0.3)(5)}{(4 + 9.7 + 8)(0.3)} = 4.08 \text{ in}$$

$$E = 30(10^6) \text{ lb/in}^2$$

$$I_x = \frac{4(0.3^3)}{12} + 4(0.3)(10 - 4.08)^2 + \frac{0.3(9.7^3)}{12} + 9.7(0.3)(5 - 4.08)^2 + \frac{8(0.3^3)}{12} + 8(0.3)(4.08^2)$$

$$= 107.3 \text{ in}^4$$

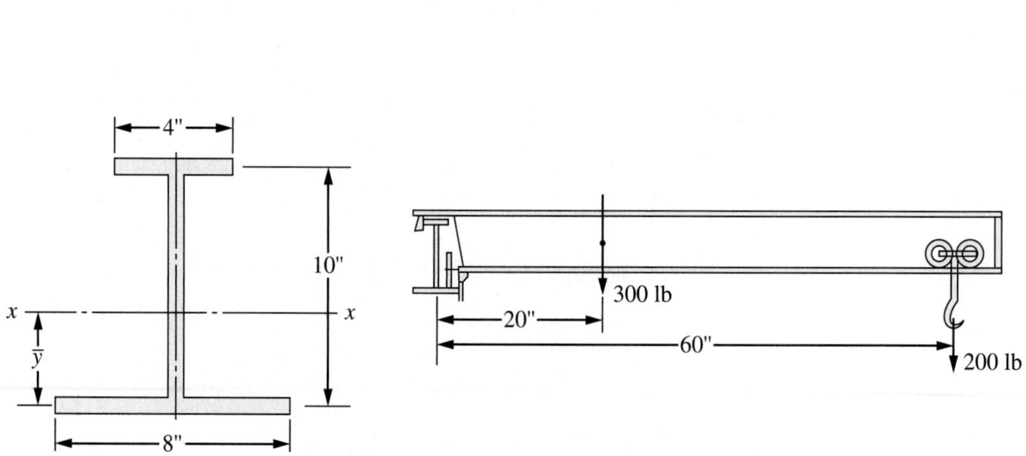

Figure 10.5

From Table 8.1, case 1c, $a = 8$ ft, $l = 16$ ft, $W = 500$ lb, $M_A = 0$, and $R_A = [500/2(16^3)](16 - 8)^2 [2(16) + 8] = 156.2$ lb. Now construct the shear and moment diagrams.

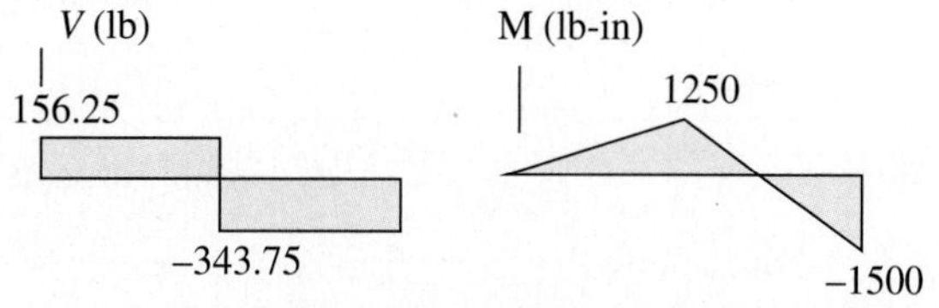

The second portion of the solution considers a beam fixed at the right end against both rotation and warping and free at the left end. It is loaded at a point 4 ft from the left end with an unknown torque T_c, which makes the angle of rotation zero at that point, and a known torque of 300(20) + 200(60) = 18,000 lb-in at a point 12 ft from the left end. Again evaluate the following constants; assume $G = 12(10^6)$ lb/in^2:

$$K = \tfrac{1}{3}(4 + 8 + 10)(0.3^3) = 0.198 \text{ in}^4 \quad \text{(Table 10.2, case 7)}$$

$$C_w = \frac{(10^2)(0.3)(4^3)(8^3)}{(12)(4^3 + 8^3)} = 142.2 \text{ in}^6 \quad \text{(Table 10.2, case 7)}$$

$$\beta = \left(\frac{KG}{C_w E}\right)^{1/2} = \left[\frac{(0.198)(12)(10^6)}{(142.2)(30)(10^6)}\right]^{1/2} = 0.0236 \text{ in}^{-1}$$

Therefore, $\beta l = 0.0236(20)(12) = 5.664$, $\beta(l - a) = 0.0236(20 - 12)(12) = 2.2656$ for $a = 12$ ft, and $\beta(l - a) = 0.0236(20 - 4)(12) = 4.5312$ for $a = 4$ ft.

From Table 10.3, case 1b, consider two torsional loads: an unknown torque T_c at $a = 4$ ft and a torque of 18,000 lb-in at $a = 12$ ft. The following constants are needed:

$$C_1 = \cosh \beta l = \cosh 5.664 = 144.1515$$

$$C_2 = \sinh \beta l = \sinh 5.664 = 144.1480$$

For $a = 4$ ft,

$$C_{a3} = \cosh \beta(l - a) - 1 = \cosh 4.5312 - 1 = 45.4404$$

$$C_{a4} = \sinh \beta(l - a) - \beta(l - a) = \sinh 4.5312 - 4.5312 = 41.8984$$

For $a = 12$ ft,

$$C_{a3} = \cosh 2.2656 - 1 = 3.8703$$

$$C_{a4} = \sinh 2.2656 - 2.2656 = 2.5010$$

At the left end $T_A = 0$ and $\theta''_A = 0$:

$$\theta_A = \frac{18{,}000}{(142.2)(30)(10^6)(0.0236^3)}\left[\frac{(144.1480)(3.8703)}{144.1515} - 2.5010\right]$$

$$+ \frac{T_c}{(142.2)(30)(10^6)(0.0236^3)}\left[\frac{(144.1480)(45.4404)}{144.1515} - 41.8984\right]$$

$$= 0.43953 + 6.3148(10^{-5})\,T_c$$

Similarly, $\theta'_A = -0.0002034 - 1.327(10^{-7})\,T_c$.

To evaluate T_c the angle of rotation at $x = 4$ ft is set equal to zero:

$$\theta_c = 0 = \theta_A + \frac{\theta'_A}{\beta} F_{2(c)}$$

where

$$F_{2(c)} = \sinh[0.0236(48)] = \sinh 1.1328 = 1.3911$$

or

$$0 = 0.43953 + 6.3148(10^{-5})\,T_c - \frac{(0.0002034)(1.3911)}{0.0236} - \frac{(1.327)(10^{-7})(1.3911)}{0.0236} T_c$$

This gives $T_c = -7728$ lb-in, $\theta_A = -0.04847$ rad, and $\theta'_A = 0.0008221$ rad/in.

To locate positions of maximum stress it is desirable to sketch curves of θ', θ'', and θ''' versus the position x:

$$\theta' = \theta'_A F_1 + \frac{T_c}{C_w E\beta^2} F_{a3(c)} + \frac{T_o}{C_w E\beta^2} F_{a3}$$

$$= 0.0008221 \cosh \beta x - \frac{7728}{(142.2)(30)(10^6)(0.0236^2)} [\cosh \beta(x-48) - 1] \langle x-48 \rangle^0$$

$$+ \frac{18{,}000}{(142.2)(30)(10^6)(0.0236^2)} [\cosh \beta(x-144) - 1] \langle x-144 \rangle^0$$

This gives

$$\theta' = 0.0008221 \cosh \beta x - 0.003253[\cosh \beta(x-48) - 1] \langle x-48 \rangle^0$$
$$+ 0.0007575[\cosh \beta(x-144) - 1] \langle x-144 \rangle^0$$

Similarly,

$$\theta'' = 0.00001940 \sinh \beta x - 0.00007676 \sinh \beta \langle x-48 \rangle + 0.00001788 \sinh \beta \langle x-144 \rangle$$

$$\theta''' = 10^{-6}[0.458 \cosh \beta x - 1.812 \cosh \beta(x-48) \langle x-48 \rangle^0 + 422 \cosh \beta(x-144) \langle x-144 \rangle^0]$$

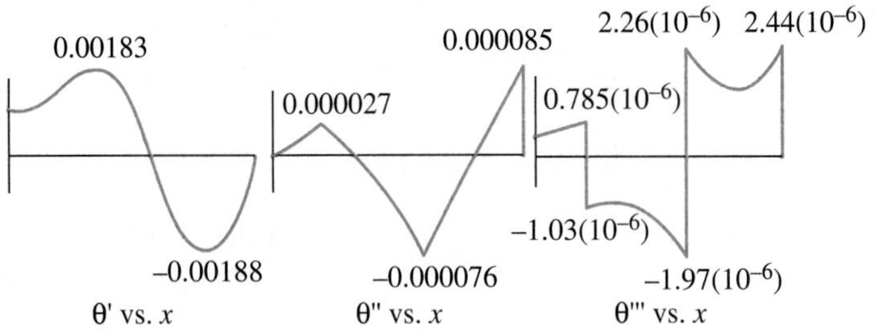

Maximum bending stresses are produced by the beam bending moments of +1250 lb-ft at $x = 12$ ft and −1500 lb-ft at $x = 20$ ft and by maximum values of θ'' of −0.000076 at $x = 12$ ft and +0.000085 at $x = 20$ ft. Since the largest magnitudes of both M_x and θ'' occur at $x = 20$ ft, the maximum bending stress will be at the wall. Therefore, at the wall,

$$\sigma_A = \frac{1500(12)(10-4.08+0.15)}{107.26} - \frac{10(4)}{2}\frac{0.3(8^3)}{0.3(4^3)+0.3(8^3)}(30)(10^6)(0.000085)$$

$$= 970 - 45{,}300 = -44{,}300 \text{ lb/in}^2$$

$$\sigma_B = 970 + 45{,}300 = 46{,}300 \text{ lb/in}^2$$

$$\sigma_C = \frac{-1500(12)(4.08+0.15)}{107.26} + \frac{10(8)}{2}\frac{0.3(4^3)}{0.3(4^3)+0.3(8^3)}(30)(10^6)(0.000085)$$

$$= -700 + 11{,}300 = 10{,}600 \text{ lb/in}^2$$

$$\sigma_D = -700 - 11{,}300 = -12{,}000 \text{ lb/in}^2$$

Maximum shear stresses are produced by θ', θ''', and beam shear V. The shear stress due to θ''' is maximum at the top of the web, that due to θ' is maximum on the surface anywhere, and that due to V is maximum at the neutral axis but is not much smaller at the top of the web. The largest shear stress in a given cross section is therefore found at the top of the web and is the sum of the absolute values of the three components at one of four possible locations at the top of the web. This gives

$$|\tau_{\max}| = \left| \frac{1}{8}\frac{10(0.3)(8^3)(4^2)}{0.3(4^3+8^3)}30(10^6)\theta''' \right| + |0.3(12)(10^6)\theta'| + \left| \frac{(2-0.15)(0.3)(10-4.08)}{107.26(0.3)} V \right|$$

$$= |533.3(10^6)\theta'''| + |3.6(10^6)\theta'| + |0.1021\, V|$$

The following maximum values of θ''', θ', V, and $\tau_{\max}$ are found at the given values of the position x:

x	θ'''	θ'	V	$\lvert\tau_{\max}\rvert$, lb/in^2
48+	$-1.03(10^{-6})$	$1.41(10^{-3})$	156.3	5633
79.2	$-0.80(10^{-6})$	$1.83(10^{-3})$	156.3	7014
144−	$-1.97(10^{-6})$	$-0.28(10^{-3})$	156.3	2073
144+	$2.26(10^{-6})$	$-0.28(10^{-3})$	−343.7	2247
191.8	$1.37(10^{-6})$	$-1.88(10^{-3})$	−343.7	7522
240	$2.44(10^{-6})$	0	−343.7	1355

To obtain the rotation of the crane at the load, substitute $x = 144$ into

$$\theta = \theta_A + \frac{\theta'_A}{\beta}F_2 + \frac{T_c}{C_w E\beta^3}F_{a4(c)}$$

$$= -0.04847 + \frac{0.0008221}{0.0236}14.9414 - \frac{7728(4.7666-2.2656)}{142.2(30)(10^6)(0.0236^3)} = 0.1273$$

$$= 7.295°$$

10.4 EFFECT OF LONGITUDINAL STRESSES

It was pointed out in Sec. 10.1 that the elongation of the outer fibers consequent upon twist caused longitudinal stresses, but that in a bar of circular section these stresses were negligible. In a flexible bar, the section of which comprises one or more narrow rectangles, the stresses in the longitudinal fibers may become large; and since after twisting these fibers are inclined, the stresses in them have components, normal to the axis of twist, which contribute to the torsional resistance of the member.

The stress in the longitudinal fibers of a thin twisted strip and the effect of these stresses on torsional stiffness have been considered by Timoshenko (Ref. 5), Green (Ref. 6), Cook and Young (Ref. 1), and others. The following formulas apply to this case: Let $2a$ = width of strip; $2b$ = thickness of strip; τ, σ_t, and σ_c = maximum shear, maximum tensile, and maximum compressive stress due to twisting, respectively; T = applied twisting moment; and θ/l = angle of twist per unit length. Then

$$\sigma_t = \frac{E\tau^2}{12G^2}\left(\frac{a}{b}\right)^2 \tag{10.4-1}$$

$$\sigma_c = \tfrac{1}{2}\sigma_t \tag{10.4-2}$$

$$T = KG\frac{\theta}{l} + \frac{8}{45}E\left(\frac{\theta}{l}\right)^3 ba^5 \tag{10.4-3}$$

The first term on the right side of Eq. (10.4-3), $KG\theta/l$, represents the part of the total applied torque T that is resisted by torsional shear; the second term represents the part that is resisted by the tensile stresses in the (helical) longitudinal fibers. It can be seen that this second part is small for small angles of twist but increases rapidly as θ/l increases.

To find the stresses produced by a given torque T, first the value of θ/l is found by Eq. (10.4-3), taking K as given for Table 10.1, case 4. Then τ is found by the stress formula for case 4, taking $KG\theta/l$ for the twisting moment. Finally σ_t and σ_c can be found by Eqs. (10.4-1) and (10.4-2).

This stiffening and strengthening effect of induced longitudinal stress will manifest itself in any bar having a section composed of narrow rectangles, such as a I-, T-, or channel, provided that the parts are so thin as to permit a large unit twist without overstressing. At the same time the accompanying longitudinal compression [Eq. (10.4-2)] may cause failure through elastic instability (see Table 15.1).

If a thin strip of width a and maximum thickness b is *initially* twisted (as by cold working) to a helical angle β, then there is an initial stiffening effect in torsion that can be expressed by the ratio of effective K to nominal K (as given in Table 10.3):

$$\frac{\text{Effective K}}{\text{Nominal K}} = 1 + C(1+\nu)\beta^2\left(\frac{a}{b}\right)^2$$

where C is a numerical coefficient that depends on the shape of the cross section and is $\frac{2}{15}$ for a rectangle, $\frac{1}{8}$ for an ellipse, $\frac{1}{10}$ for a lenticular form, and $\frac{7}{60}$ for a double wedge (Ref. 22).

If a bar of any cross section is independently loaded in tension, then the corresponding longitudinal tensile stress σ_t similarly will provide a resisting torque that again depends on the angle of twist, and the total applied torque corresponding to any angle of twist θ is $T = (KG + \sigma_t J)\theta/l$, where J is the centroidal polar moment of inertia of the cross section. If the longitudinal loading causes a compressive stress σ_c, the equation becomes

$$T = (KG - \sigma_c J)\frac{\theta}{l}$$

Bending also influences the torsional stiffness of a rod unless the cross section has (1) two axes of symmetry, (2) point symmetry, or (3) one axis of symmetry that is normal to the plane of bending. (The influences of longitudinal loading and bending are discussed in Ref. 23.)

10.5 ULTIMATE STRENGTH OF BARS IN TORSION

When twisted to failure, bars of ductile material usually break in shear, the surface of fracture being normal to the axis and practically flat. Bars of brittle material usually break in tension, the surface of fracture being helicoidal.

Circular Sections

The formulas of Sec. 10.1 apply only when the maximum stress does not exceed the elastic limit. If Eq. (10.1-3) is used with T equal to the twisting moment at failure, a fictitious value of τ is obtained, which is called the *modulus of rupture in torsion* and which for convenience will be denoted here by τ'. For solid bars of steel, τ' slightly exceeds the ultimate tensile strength when the length is only about twice the diameter but drops to about 80% of the tensile strength when the length becomes 25 times the diameter. For solid bars of aluminum, τ' is about 90% of the tensile strength.

For tubes, the modulus of rupture decreases with the ratio of diameter D to wall thickness t. Younger (Ref. 7) gives the following approximate formula, applicable to tubes of steel and aluminum:

$$\tau' = \frac{1600\tau'_0}{(D/t - 2)^2 + 1600}$$

where τ' is the modulus of rupture in torsion of the tube and τ'_0 is the modulus of rupture in torsion of a solid circular bar of the same material. (Curves giving τ' as a function of D/t for various steels and light alloys may be found in Ref. 18.)

10.6 TORSION OF CURVED BARS; HELICAL SPRINGS

The formulas of Secs. 10.1 and 10.2 can be applied to slightly curved bars without significant error, but for sharply curved bars, such as helical springs, account must be taken of the influence of curvature and slope. Among others, Wahl (Ref. 8) and Ancker and Goodier (Ref. 24) have discussed this problem, and the former presents charts which greatly facilitate the calculation of stress and deflection for springs of non-circular section. Of the following formulas cited, those for round wire were taken from Ref. 24, and those for square and rectangular wire from Ref. 8 (with some changes of notation).

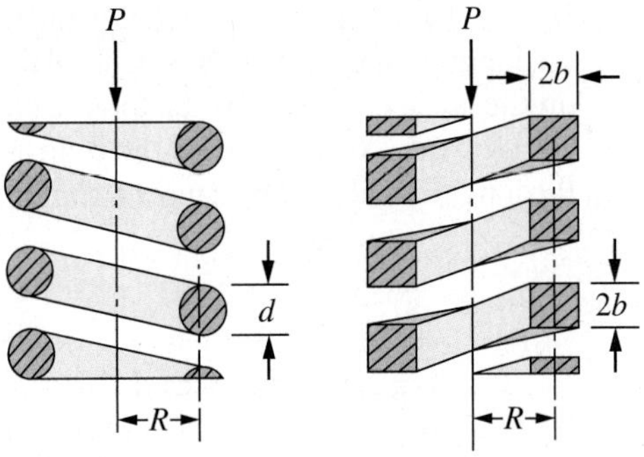

Figure 10.6

Let R = radius of coil measured from spring axis to center of section (Fig. 10.6), d = diameter of circular section, $2b$ = thickness of square section, P = load (either tensile or compressive), n = number of active turns in spring, α = pitch angle of spring, f = total stretch or shortening of spring, and τ = maximum shear stress produced. Then for a spring of *circular* wire,

$$f = \frac{64PR^3n}{Gd^4}\left[1 - \frac{3}{64}\left(\frac{d}{R}\right)^2 + \frac{3+\nu}{2(1+\nu)}(\tan\alpha)^2\right] \tag{10.6-1}$$

$$\tau = \frac{16PR}{\pi d^3}\left[1 + \frac{5}{8}\frac{d}{R} + \frac{7}{32}\left(\frac{d}{R}\right)^2\right] \tag{10.6-2}$$

For a spring of *square* wire,

$$f = \frac{2.789PR^3n}{Gb^4} \quad \text{for} \quad c > 3 \tag{10.6-3}$$

$$\tau = \frac{4.8PR}{8b^3}\left(1 + \frac{1.2}{c} + \frac{0.56}{c^2} + \frac{0.5}{c^3}\right) \tag{10.6-4}$$

where $c = R/b$.

For a spring of *rectangular* wire, section $2a \times 2b$ where $a > b$,

$$f = \frac{3\pi PR^3n}{8Gb^4}\,\frac{1}{a/b - 0.627[\tanh(\pi b/2a) + 0.004]} \tag{10.6-5}$$

for $c > 3$ if the long dimension $2a$ is parallel to the spring axis or for $c > 5$ if the long dimension $2a$ is perpendicular to the spring axis,

$$\tau = \frac{PR(3b + 1.8a)}{8b^2a^2}\left(1 + \frac{1.2}{c} + \frac{0.56}{c^2} + \frac{0.5}{c^3}\right) \tag{10.6-6}$$

It should be noted that in each of these cases the maximum stress is given by the ordinary formula for the section in question (from Table 10.1) multiplied by a corrective factor that takes account of curvature, and these corrective factors can be used for any curved bar of the corresponding cross section. Also, for compression springs with the end turns ground down for even bearing, n, should be taken as the actual number of turns (including the tapered end turns) less 2. For tension springs n should be taken as the actual number of turns or slightly more.

Unless laterally supported, compression springs that are relatively long will buckle when compressed beyond a certain critical deflection. This critical deflection depends on the ratio of L, the free length, to D, the mean diameter, and is indicated approximately by the following tabulation, based on Ref. 27. Consideration of coil closing before reaching the critical deflection is necessary.

L/D	1	2	3	4	5	6	7	8
Critical deflection/L	0.72	0.71	0.68	0.63	0.53	0.39	0.27	0.17

Precise Formula

For very accurate calculation of the extension of a spring, as is necessary in designing precision spring scales, account must be taken of the change in slope and radius of the coils caused by stretching. Sayre (Ref. 9) gives a formula which takes into account not only the effect of this change in form but also the deformation due to direct transverse shear and flexure. This formula can be written as

$$f = P\left\{\left[\frac{R_0^2 L}{GK} - \frac{R_0^2 H_0^2}{GKL}\left(1 - \frac{GK}{EI}\right) + \frac{FL}{AG}\right] - \left[\frac{R_0^2}{3GKL}\left(3 - \frac{2GK}{EI}\right)\left(H^2 + HH_0 - 2H_0^2\right)\right]\right\} \quad (10.6\text{-}7)$$

where f = stretch of the spring; P = load; R_0 = initial radius of the coil; H = variable length of the effective portion of the stretched spring; H_0 = initial value of H; L = actual developed length of the wire of which the spring is made; A = cross-sectional area of this wire; K = the torsional-stiffness factor for the wire section, as given in Table 10.1 ($K = \frac{1}{2}\pi r^4$ for a circle; $K = 2.25a^4$ for a square; etc.); F = the section factor for shear deformation [Eq. (8.10-1); $F = \frac{10}{9}$ for a circle or ellipse, $F = \frac{6}{5}$ for a square or rectangle]; and I = moment of inertia of the wire section about a central axis parallel to the spring axis. The first term in brackets represents the initial rate of stretch, and the second term in brackets represents the change in this rate due to change in form consequent upon stretch. The final expression shows that f is not a linear function of P.

10.7 TABLES

TABLE 10.1 Formulas for Torsional Deformation and Stress

General Formulas: $\theta = TL/KG$ and $\tau = T/Q$, where θ = angle of twist (radians); T = twisting moment (force-length); L = length, τ = unit shear stress (force per unit area); G = modulus of rigidity (force per unit area); K (length to the fourth) and Q (length cubed) are functions of the cross section.

Form and Dimensions of Cross Sections, Other Quantities Involved, and Case No.	Formula for K in $\theta = \frac{TL}{KG}$	Formula for Shear Stress
1. Solid circular section 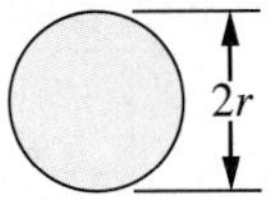	$K = \frac{1}{2}\pi r^4$	$\tau_{max} = \frac{2T}{\pi r^3}$ at boundary
2. Solid elliptical section 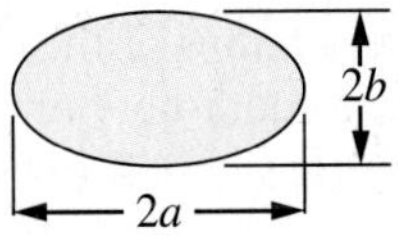	$K = \frac{\pi a^3 b^3}{a^2 + b^2}$	$\tau_{max} = \frac{2T}{\pi a b^2}$ at ends of minor axis
3. Solid square section 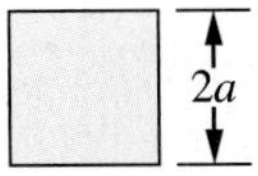	$K = 2.25a^4$	$\tau_{max} = \frac{0.601T}{a^3}$ at midpoint of each side
4. Solid rectangular section 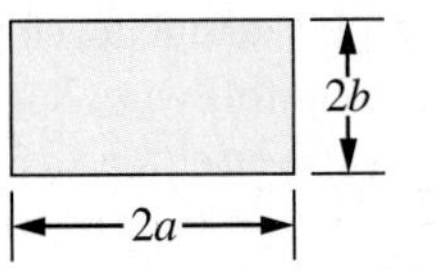	$K = ab^3\left[\frac{16}{3} - 3.36\frac{b}{a}\left(1 - \frac{b^4}{12a^4}\right)\right]$ for $a \geq b$	$\tau_{max} = \frac{3T}{8ab^2}\left[1 + 0.6095\frac{b}{a} + 0.8865\left(\frac{b}{a}\right)^2 - 1.8023\left(\frac{b}{a}\right)^3 + 0.9100\left(\frac{b}{a}\right)^4\right]$ at the midpoint of each longer side for $a \geq b$
5. Solid triangular section (equilaterial) 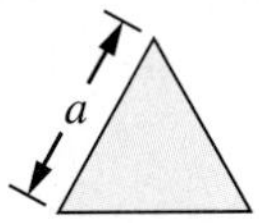	$K = \frac{a^4\sqrt{3}}{80}$	$\tau_{max} = \frac{20T}{a^3}$ at midpoint of each side
6. Isosceles triangle 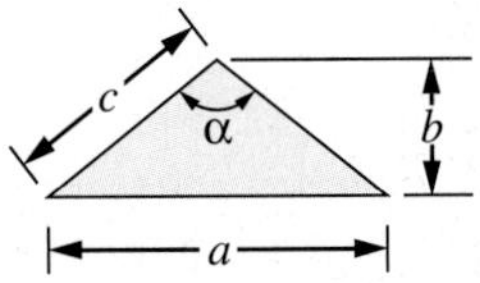 (*Note:* See also Ref. 21 for graphs of stress magnitudes and locations and stiffness factors.)	For $\frac{2}{3} < a/b < \sqrt{3}$ $(39° < \alpha < 82°)$ For $K = \frac{a^3 b^3}{15a^2 + 20b^2}$ approximate formula which is exact at $\alpha = 60°$ where $K = 0.02165c^4$. For $\sqrt{3} < a/b < 2\sqrt{3}$ $(82° < \alpha < 120°)$ $K = 0.0915b^4\left(\frac{a}{b} - 0.8592\right)$ approximate formula which is exact at $\alpha = 90°$ where $K = 0.0261c^4$ (errors < 4%) (Ref. 20)	For $39° < \alpha < 120°$ $Q = \frac{K}{b[0.200 + 0.309a/b - 0.0418(a/b)^2]}$ approximate formula which is exact at $\alpha = 60°$ and $\alpha = 90°$ For $\alpha = 60°$ $Q = 0.0768b^3 = 0.0500c^3$ For $\alpha = 90°$ $Q = 0.1604b^3 = 0.0567c^3$ τ_{max} at center of longest side

(*Continued*)

TABLE 10.1 Formulas for Torsional Deformation and Stress (*Continued*)

Form and Dimensions of Cross Sections, Other Quantities Involved, and Case No.	Formula for K in $\theta = \frac{TL}{KG}$	Formula for Shear Stress
7. Circular segmental section 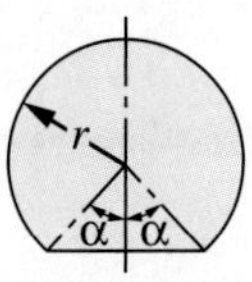 [*Note*: $h = r(1-\cos\alpha)$]	$K = 2Cr^4$ where C varies with $\frac{h}{r}$ as follows: For $0 \le \frac{h}{r} \le 1.0$, $C = 0.7854 - 0.0333\frac{h}{r} - 2.6183\left(\frac{h}{r}\right)^2 + 4.1595\left(\frac{h}{r}\right)^3 - 3.0769\left(\frac{h}{r}\right)^4 + 0.9299\left(\frac{h}{r}\right)^5$	$\tau_{max} = \frac{TB}{r^3}$ where B varies with $\frac{h}{r}$ as follows: For $0 \le \frac{h}{r} \le 1.0$, $B = 0.6366 + 1.7598\frac{h}{r} - 5.4897\left(\frac{h}{r}\right)^2 + 14.062\left(\frac{h}{r}\right)^3 - 14.510\left(\frac{h}{r}\right)^4 + 6.434\left(\frac{h}{r}\right)^5$ (Data from Refs. 12 and 13)
8. Circular sector 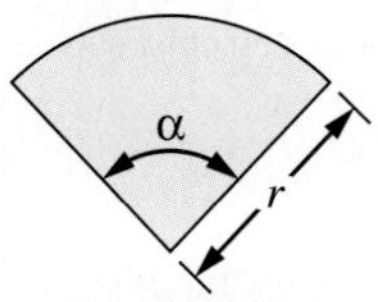 (*Note*: See also Ref. 21.)	$K = Cr^4$ where C varies with $\frac{\alpha}{\pi}$ as follows: For $0.1 \le \frac{\alpha}{\pi} \le 2.0$, $C = 0.0034 - 0.0697\frac{\alpha}{\pi} + 0.5825\left(\frac{\alpha}{\pi}\right)^2 - 0.2950\left(\frac{\alpha}{\pi}\right)^3 + 0.0874\left(\frac{\alpha}{\pi}\right)^4 - 0.0111\left(\frac{\alpha}{\pi}\right)^5$	$\tau_{max} = \frac{T}{Br^3}$ on a radial boundary. B varies with $\frac{\alpha}{\pi}$ as follows: For $0.1 \le \frac{\alpha}{\pi} \le 1.0$, $B = 0.0117 - 0.2137\frac{\alpha}{\pi} + 2.2475\left(\frac{\alpha}{\pi}\right)^2 - 4.6709\left(\frac{\alpha}{\pi}\right)^3 + 5.1764\left(\frac{\alpha}{\pi}\right)^4 - 2.2000\left(\frac{\alpha}{\pi}\right)^5$ (Data from Ref. 17)
9. Circular shaft with opposite sides flattened 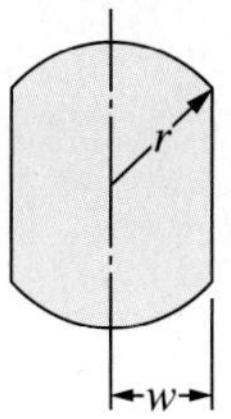 (*Note*: $h = r - w$)	$K = 2Cr^4$ where C varies with $\frac{h}{r}$ as follows: For two flat sides where $0 \le \frac{h}{r} \le 0.8$, $C = 0.7854 - 0.4053\frac{h}{r} - 3.5810\left(\frac{h}{r}\right)^2 + 5.2708\left(\frac{h}{r}\right)^3 - 2.0772\left(\frac{h}{r}\right)^4$ For four flat sides where $0 \le \frac{h}{r} \le 0.293$, $C = 0.7854 - 0.7000\frac{h}{r} - 7.7982\left(\frac{h}{r}\right)^2 + 14.578\left(\frac{h}{r}\right)^3$	$\tau_{max} = \frac{BT}{r^3}$ where B varies with $\frac{h}{r}$ as follows: For two flat sides where $0 \le \frac{h}{r} \le 0.6$, $B = 0.6366 + 2.5303\frac{h}{r} - 11.157\left(\frac{h}{r}\right)^2 + 49.568\left(\frac{h}{r}\right)^3 - 85.886\left(\frac{h}{r}\right)^4 + 69.849\left(\frac{h}{r}\right)^5$ For four flat sides where $0 \le \frac{h}{r} \le 0.293$, $B = 0.6366 + 2.6298\frac{h}{r} - 5.6147\left(\frac{h}{r}\right)^2 + 30.853\left(\frac{h}{r}\right)^3$ (Data from Refs. 12 and 13)
10. Hollow concentric circular section 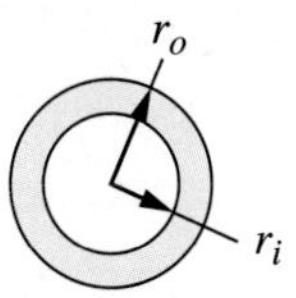 	$K = \frac{1}{2}\pi\left(r_o^4 - r_i^4\right)$	$\tau_{max} = \frac{2Tr_o}{\pi\left(r_o^4 - r_i^4\right)}$ at outer boundary

(*Continued*)

TABLE 10.1 Formulas for Torsional Deformation and Stress (*Continued*)

Form and Dimensions of Cross Sections, Other Quantities Involved, and Case No.	Formula for K in $\theta = \frac{TL}{KG}$	Formula for Shear Stress
11. Eccentric hollow circular section 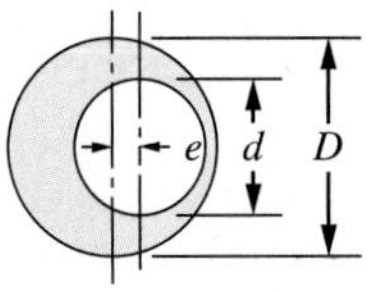 $\frac{e}{D} = \lambda$ $\frac{d}{D} = n$	$K = \frac{\pi(D^4 - d^4)}{32C}$ where $C = 1 + \frac{16n^2}{(1-n^2)(1-n^4)}\lambda^2 + \frac{384n^4}{(1-n^2)^2(1-n^4)^4}\lambda^4$	$\tau_{max} = \frac{16TDF}{\pi(D^4 - d^4)}$ $F = 1 + \frac{4n^2}{1-n^2}\lambda + \frac{32n^2}{(1-n^2)(1-n^4)}\lambda^2 + \frac{48n^2(1+2n^2+3n^4+2n^6)}{(1-n^2)(1-n^4)(1-n^6)}\lambda^3 + \frac{64n^2(2+12n^2+19n^4+28n^6+18n^8+14n^{10}+3n^{12})}{(1-n^2)(1-n^4)(1-n^6)(1-n^8)}\lambda^4$ (Ref. 10)
12. Hollow elliptical section, outer and inner boundaries similar ellipses 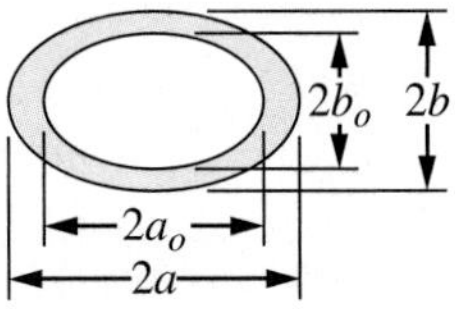	$K = \frac{\pi a^3 b^3}{a^2 + b^2}(1 - q^4)$ where $q = \frac{a_o}{a} = \frac{b_o}{b}$ (*Note:* The wall thickness is not constant.)	$\tau_{max} = \frac{2T}{\pi a b^2 (1 - q^4)}$ at ends of minor axis on outer surface
13. Hollow, thin-walled section of uniform thickness; U = length of elliptical median boundary, shown dashed $U = \pi(a + b - t)\times\left[1 + 0.258\frac{(a-b)^2}{(a+b-t)^2}\right]$ 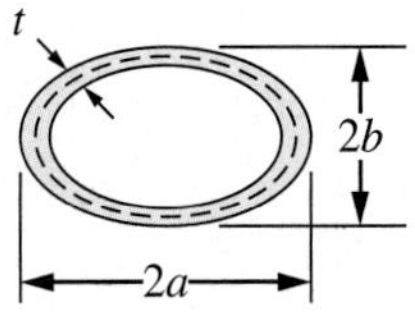 (approximately)	$K = \frac{4\pi^2 t\left[\left(a - \frac{1}{2}t\right)^2\left(b - \frac{1}{2}t\right)^2\right]}{U}$	$\tau_{average} = \frac{T}{2\pi t(a - \frac{1}{2}t)(b - \frac{1}{2}t)}$ (Stress is nearly uniform if t is small.)
14. Any thin tube of uniform thickness; U = length of median boundary; A = mean of areas enclosed by outer and inner boundaries, or (approximate) area within median boundary 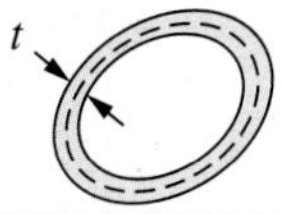	$K = \frac{4A^2 t}{U}$	$\tau_{average} = \frac{T}{2tA}$ (Stress is nearly uniform if t is small.)

(*Continued*)

TABLE 10.1 Formulas for Torsional Deformation and Stress (*Continued*)

Form and Dimensions of Cross Sections, Other Quantities Involved, and Case No.	Formula for K in $\theta = \frac{TL}{KG}$	Formula for Shear Stress
15. Any thin tube. U and A as for case 14; t = thickness at any point 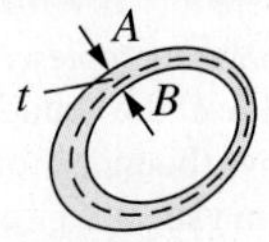	$K = \frac{4A^2}{\int dU/t}$	$\tau_{average}$ on any thickness $AB = \frac{T}{2tA}$ (τ_{max} where t is a minimum)
16. Hollow rectangle, thin-walled 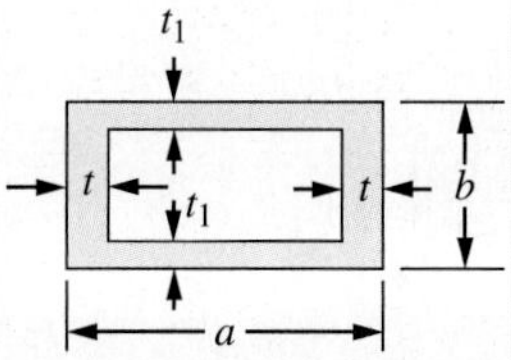 (*Note:* For thick-walled hollow rectangles see Refs. 16 and 25. Reference 25 illustrates how to extend the work presented to cases with more than one enclosed region.)	$K = \frac{2tt_1(a-t)^2(b-t_1)^2}{at + bt_1 - t^2 - t_1^2}$	$\tau_{average} = \begin{cases} \frac{T}{2t(a-t)(b-t_1)} & \text{near midlength of short sides} \\ \frac{T}{2t_1(a-t)(b-t_1)} & \text{near midlength of long sides} \end{cases}$ (There will be higher stresses at inner corners unless fillets of fairly large radius are provided.)
17. Thin circular open tube of uniform thickness; r = mean radius 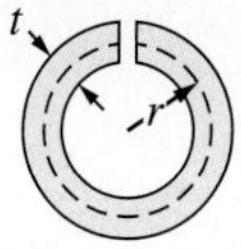	$K = \frac{2}{3}\pi rt^3$	$\tau_{max} = \frac{T(6\pi r + 1.8t)}{4\pi^2 r^2 t^2}$ along both edges remote from ends (This assumes t is small compared with mean radius.)
18. Any thin open tube of uniform thickness; U = length of median line, shown dashed 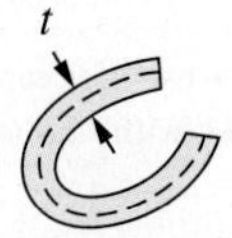	$K = \frac{1}{3}Ut^3$	$\tau_{max} = \frac{T(3U + 1.8t)}{U^2 t^2}$ along both edges remote from ends (This assumes t is small compared with least radius of curvature of median line; otherwise use the formulas given for cases 19–26.)
19. Any elongated section with axis of symmetry OX; U = length, A = area of section, I_x = moment of inertia about axis of symmetry 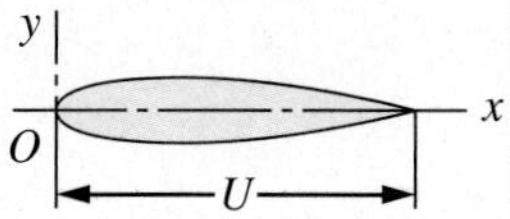	$K = \frac{4I_x}{1 + 16I_x/AU^2}$	

(*Continued*)

TABLE 10.1 Formulas for Torsional Deformation and Stress (*Continued*)

Form and Dimensions of Cross Sections, Other Quantities Involved, and Case No.	Formula for K in $\theta = \frac{TL}{KG}$	Formula for Shear Stress
20. Any elongated section or thin open tube; dU = elementary length along median line, t = thickness normal to median line, A = area of section 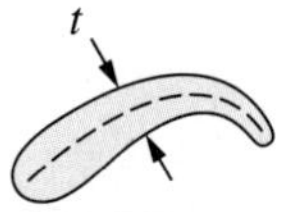	$K = \frac{F}{3+4F/AU^2}$ where $F = \int_0^U t^3\,dU$	For all solid sections of irregular form (cases 19–26 inclusive) the maximum shear stress occurs at or very near one of the points where the largest inscribed circle touches the boundary,* and of these, at the one where the curvature of the boundary is algebraically least. (Convexity represents positive and concavity negative curvature of the boundary.) At a point where the curvature is positive (boundary of section straight or convex) this maximum stress is given approximately by $\tau_{max} = G\frac{\theta}{L}C$ or $\tau_{max} = \frac{T}{K}C$
21. Any solid, fairly compact section without reentrant angles, j = polar moment of inertia about centroid axis, A = area of section 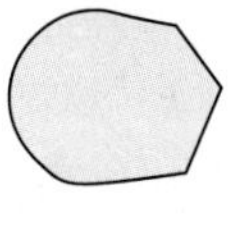	$K = \frac{A^4}{40J}$	where $C = \frac{D}{1+\frac{\pi^2 D^4}{16A^2}}\left[1+0.15\left(\frac{\pi^2 D^4}{16A^2} - \frac{D}{2r}\right)\right]$ D = diameter of largest inscribed circle r = radius of curvature of boundary at the point (positive for this case) A = area of the section *Unless at some point on the boundary there is a sharp reentant angle, causing high local stress.
22. Trapezoid 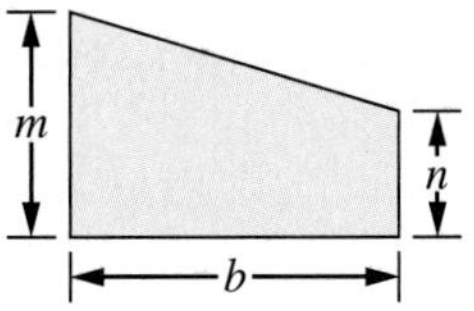	$K = \frac{1}{12}b(m+n)(m^2+n^2) - V_L m^4 - V_s n^4$ where $V_L = 0.10504 - 0.10s + 0.0848s^2 - 0.06746s^3 + 0.0515s^4$ $V_s = 0.10504 + 0.10s + 0.0848s^2 + 0.06746s^3 + 0.0515s^4$ $s = \frac{m-n}{b}$ (Ref. 11)	At a point where the curvature is negative (boundary of section concave or reentrant), this maximum stress is given approximately by $\tau_{max} = G\frac{\theta}{L}C$ or $\tau_{max} = \frac{T}{K}C$ where $C = \frac{D}{1+\frac{\pi^2 D^4}{16A^2}}\left\{1+\left[0.118\ln\left(1-\frac{D}{2r}\right) - 0.238\frac{D}{2r}\right]\tanh\frac{2\phi}{\pi}\right\}$ and D, A, and r have the same meaning as before and ϕ = a positive angle through which a tangent to the boundary rotates in turning or traveling around the reentrant portion, measured in radians (here r is *negative*).
23. T-section, flange thickness uniform. For definitions of r, D, t, and t_1, see case 26. 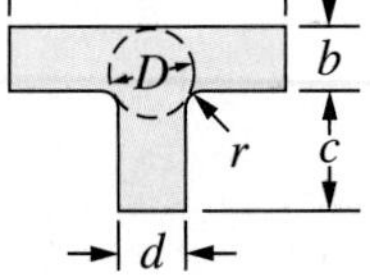	$K = K_1 + K_2 + \alpha D^4$ where $K_1 = ab^3\left[\frac{1}{3} - 0.21\frac{b}{a}\left(1-\frac{b^4}{12a^4}\right)\right]$ $K_2 = cd^3\left[\frac{1}{3} - 0.105\frac{d}{c}\left(1-\frac{d^4}{192c^4}\right)\right]$ $\alpha = \frac{t}{t_1}\left(0.15 + 0.10\frac{r}{b}\right)$ $D = \frac{(b+r)^2 + rd + d^2/4}{(2r+b)}$ for $d < 2(b+r)$	The preceding formulas should also be used for cases 17 and 18 when t is relatively large compared with radius of median line.

(Continued)

TABLE 10.1 Formulas for Torsional Deformation and Stress (*Continued*)

Form and Dimensions of Cross Sections, Other Quantities Involved, and Case No.	Formula for K in $\theta = \frac{TL}{KG}$	Formula for Shear Stress
24. L-section; $b \ge d$. For definitions of r and D, see case 26. 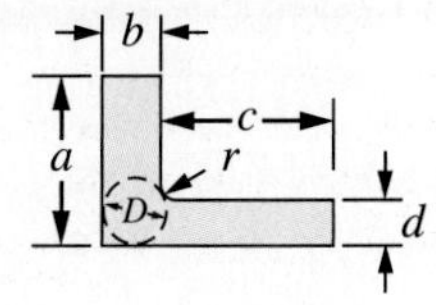	$K = K_1 + K_2 + \alpha D^4$ where $K_1 = ab^3\left[\frac{1}{3} - 0.21\frac{b}{a}\left(1 - \frac{b^4}{12a^4}\right)\right]$ $K_2 = cd^3\left[\frac{1}{3} - 0.105\frac{d}{c}\left(1 - \frac{d^4}{192c^4}\right)\right]$ $\alpha = \frac{d}{b_1}\left(0.07 + 0.076\frac{r}{b}\right)$ $D = 2[d + b + 3r - \sqrt{2(2r+b)(2r+d)}]$ for $b < 2(d+r)$	
25. U- or Z-section 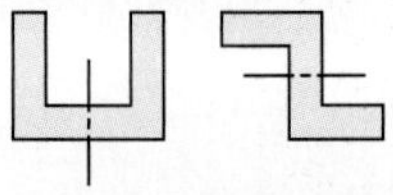	K = sum of K's of constituent L-sections computed as for case 24	
26. I-section, flange thickness uniform; r = fillet radius, D = diameter largest inscribed circle, $t = b$ if $b < d$; $t = d$ if $d < b$; $t_1 = b$ if $b > d$; $t_1 = d$ if $d > b$ 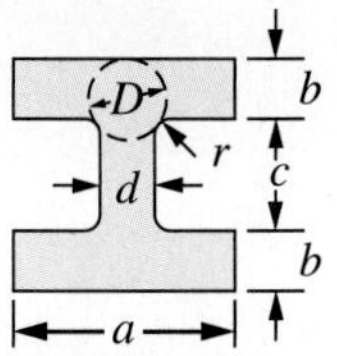	$K = 2K_1 + K_2 + 2\alpha D^4$ where $K_1 = ab^3\left[\frac{1}{3} - 0.21\frac{b}{a}\left(1 - \frac{b^4}{12a^4}\right)\right]$ $K_2 = \frac{1}{3}cd^3$ $\alpha = \frac{t}{t_1}\left(0.15 + 0.1\frac{r}{b}\right)$ Use expression for D from case 23.	
27. Split hollow shift 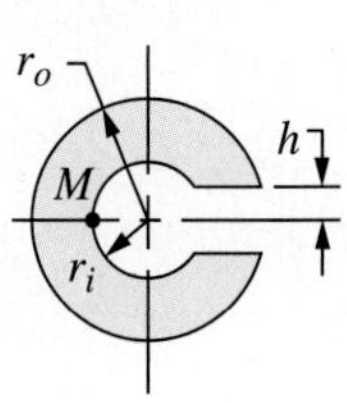	$K = 2Cr_o^4$ where C varies with $\frac{r_i}{r_o}$ as follows: For $0.2 \le \frac{r_i}{r_o} \le 0.6$, $C = K_1 + K_2\frac{r_i}{r_o} + K_3\left(\frac{r_i}{r_o}\right)^2 + K_4\left(\frac{r_i}{r_o}\right)^3$ where for $0.1 \le h/r_i \le 1.0$, $K_1 = 0.4427 + 0.0064\frac{h}{r_i} - 0.0201\left(\frac{h}{r_i}\right)^2$ $K_2 = -0.8071 - 0.4047\frac{h}{r_i} + 0.1051\left(\frac{h}{r_i}\right)^2$ $K_3 = -0.0469 + 1.2063\frac{h}{r_i} - 0.3538\left(\frac{h}{r_i}\right)^2$ $K_4 = 0.5023 - 0.9618\frac{h}{r_i} + 0.3639\left(\frac{h}{r_i}\right)^2$	At M, $\tau = \frac{TB}{r_o^3}$ where B varies with $\frac{r_i}{r_o}$ as follows: For $0.2 \le \frac{r_i}{r_o} \le 0.6$, $B = K_1 + K_2\frac{r_i}{r_o} + K_3\left(\frac{r_i}{r_o}\right)^2 + K_4\left(\frac{r_i}{r_o}\right)^3$ where for $0.1 \le h/r_i \le 1.0$, $K_1 = 2.0014 - 0.1400\frac{h}{r_i} - 0.3231\left(\frac{h}{r_i}\right)^2$ $K_2 = 2.9047 + 3.0069\frac{h}{r_i} + 4.0500\left(\frac{h}{r_i}\right)^2$ $K_3 = -15.721 - 6.5077\frac{h}{r_i} - 12.496\left(\frac{h}{r_i}\right)^2$ $K_4 = 29.553 + 4.1115\frac{h}{r_i} + 18.845\left(\frac{h}{r_i}\right)^2$ (Data from Refs. 12 and 13)

(*Continued*)

TABLE 10.1 Formulas for Torsional Deformation and Stress (*Continued*)

Form and Dimensions of Cross Sections, Other Quantities Involved, and Case No.	Formula for K in $\theta = \frac{TL}{KG}$	Formula for Shear Stress
28. Shaft with one keyway 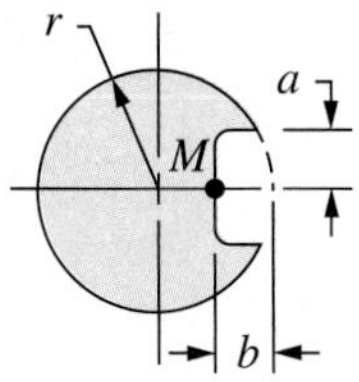	$K = 2Cr^4$ where C varies with $\frac{b}{r}$ as follows: For $0 \le \frac{b}{r} \le 0.5$, $C = K_1 + K_2\frac{b}{r} + K_3\left(\frac{b}{r}\right)^2 + K_4\left(\frac{b}{r}\right)^3$ where for $0.3 \le a/b \le 1.5$, $K_1 = 0.7854$ $K_2 = -0.0848 + 0.1234\frac{a}{b} - 0.0847\left(\frac{a}{b}\right)^2$ $K_3 = -0.4318 - 2.2000\frac{a}{b} + 0.7633\left(\frac{a}{b}\right)^2$ $K_4 = -0.0780 + 2.0618\frac{a}{b} - 0.5234\left(\frac{a}{b}\right)^2$	At M, $\tau = \frac{TB}{r^3}$ where B varies with $\frac{b}{r}$ as follows: For $0.2 \le \frac{b}{r} \le 0.5$, $B = K_1 + K_2\frac{b}{r} + K_3\left(\frac{b}{r}\right)^2 + K_4\left(\frac{b}{r}\right)^3$ where for $0.5 \le a/b \le 1.5$, $K_1 = 1.1690 - 0.3168\frac{a}{b} + 0.0490\left(\frac{a}{b}\right)^2$ $K_2 = 0.43490 - 1.5096\frac{a}{b} + 0.8677\left(\frac{a}{b}\right)^2$ $K_3 = -1.1830 + 4.2764\frac{a}{b} - 1.7024\left(\frac{a}{b}\right)^2$ $K_4 = 0.8812 - 0.2627\frac{a}{b} - 0.1897\left(\frac{a}{b}\right)^2$ (Data from Refs. 12 and 13)
29. Shaft with two keyways 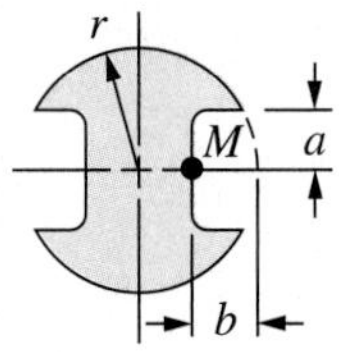	$K = 2Cr^4$ where C varies with $\frac{b}{r}$ sas follows: For $0 \le \frac{b}{r} \le 0.5$, $C = K_1 + K_2\frac{b}{r} + K_3\left(\frac{b}{r}\right)^2 + K_4\left(\frac{b}{r}\right)^3$ where for $0.3 \le a/b \le 1.5$, $K_1 = 0.7854$ $K_2 = -0.0795 + 0.1286\frac{a}{b} - 0.1169\left(\frac{a}{b}\right)^2$ $K_3 = -1.4126 - 3.8589\frac{a}{b} + 1.3292\left(\frac{a}{b}\right)^2$ $K_4 = 0.7098 + 4.1936\frac{a}{b} - 1.1053\left(\frac{a}{b}\right)^2$	At M, $\tau = \frac{TB}{r^3}$ where B varies with $\frac{b}{r}$ as follows: For $0.2 \le \frac{b}{r} \le 0.5$, $B = K_1 + K_2\frac{b}{r} + K_3\left(\frac{b}{r}\right)^2 + K_4\left(\frac{b}{r}\right)^3$ where for $0.5 \le a/b \le 1.5$, $K_1 = 1.2512 - 0.5406\frac{a}{b} + 0.0387\left(\frac{a}{b}\right)^2$ $K_2 = -0.9385 + 2.3450\frac{a}{b} + 0.3256\left(\frac{a}{b}\right)^2$ $K_3 = 7.2650 - 15.338\frac{a}{b} + 3.1138\left(\frac{a}{b}\right)^2$ $K_4 = -11.152 + 33.710\frac{a}{b} - 10.007\left(\frac{a}{b}\right)^2$ (Data from Refs. 12 and 13)

(Continued)

TABLE 10.1 Formulas for Torsional Deformation and Stress (*Continued*)

Form and Dimensions of Cross Sections, Other Quantities Involved, and Case No.	Formula for K in $\theta = \frac{TL}{KG}$	Formula for Shear Stress
30. Shaft with four keyways 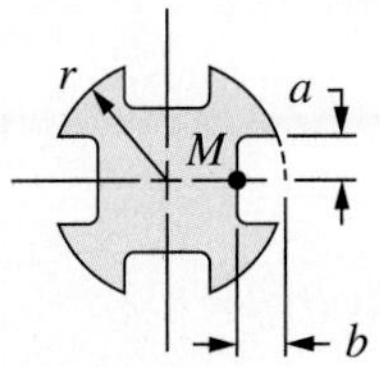	$K = 2Cr^4$ where C varies with $\frac{b}{r}$ as follows: For $0 \le \frac{b}{r} \le 0.4$, $C = K_1 + K_2\frac{b}{r} + K_3\left(\frac{b}{r}\right)^2 + K_4\left(\frac{b}{r}\right)^3$ where for $0.3 \le a/b \le 1.2$, $K_1 = 0.7854$ $K_2 = -0.1496 + 0.2773\frac{a}{b} - 0.2110\left(\frac{a}{b}\right)^2$ $K_3 = -2.9138 - 8.2354\frac{a}{b} + 2.5782\left(\frac{a}{b}\right)^2$ $K_4 = 2.2991 + 12.097\frac{a}{b} - 2.2838\left(\frac{a}{b}\right)^2$	At M, $\tau = \frac{TB}{r^3}$ where B varies with $\frac{b}{r}$ as follows: For $0.2 \le \frac{b}{r} \le 0.4$, $B = K_1 + K_2\frac{b}{r} + K_3\left(\frac{b}{r}\right)^2 + K_4\left(\frac{b}{r}\right)^3$ where for $0.5 \le a/b \le 1.2$, $K_1 = 1.0434 + 1.0449\frac{a}{b} - 0.2977\left(\frac{a}{b}\right)^2$ $K_2 = 0.0958 - 9.8401\frac{a}{b} + 1.6847\left(\frac{a}{b}\right)^2$ $K_3 = 15.749 - 6.9650\frac{a}{b} + 14.222\left(\frac{a}{b}\right)^2$ $K_4 = -35.878 + 88.696\frac{a}{b} - 47.545\left(\frac{a}{b}\right)^2$ (Data from Refs. 12 and 13)
31. Shaft with one spline 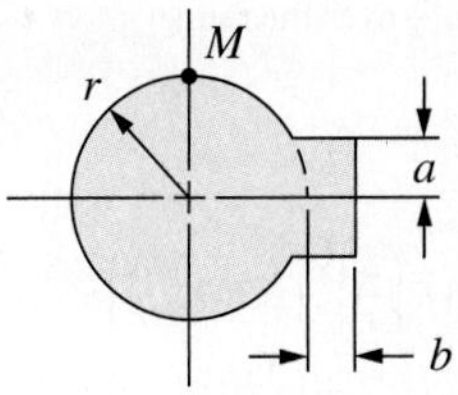	$K = 2Cr^4$ where C varies with $\frac{b}{r}$ as follows: For $0 \le \frac{b}{r} \le 0.5$, $C = K_1 + K_2\frac{b}{r} + K_3\left(\frac{b}{r}\right)^2 + K_4\left(\frac{b}{r}\right)^3$ where for $0.2 \le a/b \le 1.4$, $K_1 = 0.7854$ $K_2 = 0.0264 - 0.1187\frac{a}{b} + 0.0868\left(\frac{a}{b}\right)^2$ $K_3 = -0.2017 + 0.9019\frac{a}{b} - 0.4947\left(\frac{a}{b}\right)^2$ $K_4 = 0.2911 - 1.4875\frac{a}{b} + 2.0651\left(\frac{a}{b}\right)^2$	At M, $\tau = \frac{TB}{r^3}$ where B varies with $\frac{b}{r}$ as follows: For $0 \le \frac{b}{r} \le 0.5$, $B = K_1 + K_2\frac{b}{r} + K_3\left(\frac{b}{r}\right)^2 + K_4\left(\frac{b}{r}\right)^3$ where for $0.2 \le a/b \le 1.4$, $K_1 = 0.6366$ $K_2 = -0.0023 + 0.0168\frac{a}{b} + 0.0093\left(\frac{a}{b}\right)^2$ $K_3 = 0.0052 + 0.0225\frac{a}{b} - 0.3300\left(\frac{a}{b}\right)^2$ $K_4 = 0.0984 - 0.4936\frac{a}{b} + 0.2179\left(\frac{a}{b}\right)^2$ (Data from Refs. 12 and 13)
32. Shaft with two splines 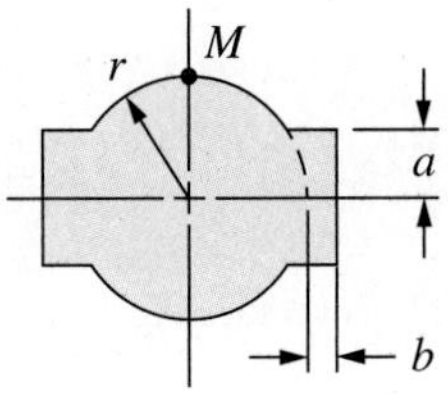	$K = 2Cr^4$ where C varies with $\frac{b}{r}$ as follows: For $0 \le \frac{b}{r} \le 0.5$, $C = K_1 + K_2\frac{b}{r} + K_3\left(\frac{b}{r}\right)^2 + K_4\left(\frac{b}{r}\right)^3$ where for $0.2 \le a/b \le 1.4$, $K_1 = 0.7854$ $K_2 = 0.0204 - 0.1307\frac{a}{b} + 0.1157\left(\frac{a}{b}\right)^2$ $K_3 = -0.2075 + 1.1544\frac{a}{b} - 0.5937\left(\frac{a}{b}\right)^2$ $K_4 = 0.3608 - 2.2582\frac{a}{b} + 3.7336\left(\frac{a}{b}\right)^2$	At M, $\tau = \frac{TB}{r^3}$ where B varies with $\frac{b}{r}$ as follows: For $0 \le \frac{b}{r} \le 0.5$, $B = K_1 + K_2\frac{b}{r} + K_3\left(\frac{b}{r}\right)^2 + K_4\left(\frac{b}{r}\right)^3$ where for $0.2 \le a/b \le 1.4$, $K_1 = 0.6366$ $K_2 = 0.0069 - 0.0229\frac{a}{b} + 0.0637\left(\frac{a}{b}\right)^2$ $K_3 = -0.0675 + 0.3996\frac{a}{b} - 1.0514\left(\frac{a}{b}\right)^2$ $K_4 = 0.3582 - 1.8324\frac{a}{b} + 1.5393\left(\frac{a}{b}\right)^2$ (Data from Refs. 12 and 13)

(*Continued*)

TABLE 10.1 Formulas for Torsional Deformation and Stress (*Continued*)

Form and Dimensions of Cross Sections, Other Quantities Involved, and Case No.	Formula for K in $\theta = \frac{TL}{KG}$	Formula for Shear Stress
33. Shaft with four splines 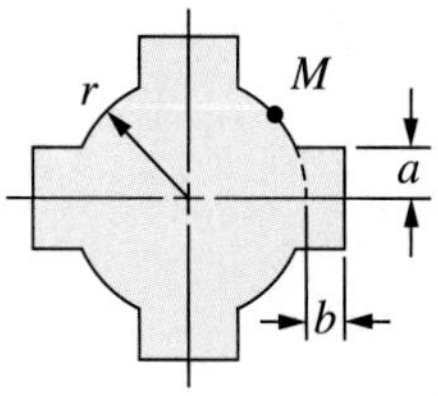 	$K = 2Cr^4$ where C varies with $\frac{b}{r}$ as follows: For $0 \le \frac{b}{r} \le 0.5$, $C = K_1 + K_2\frac{b}{r} + K_3\left(\frac{b}{r}\right)^2 + K_4\left(\frac{b}{r}\right)^3$ where for $0.2 \le a/b \le 1.0$, $K_1 = 0.7854$ $K_2 = 0.0595 - 0.3397\frac{a}{b} + 0.3239\left(\frac{a}{b}\right)^2$ $K_3 = -0.6008 + 3.1396\frac{a}{b} - 2.0693\left(\frac{a}{b}\right)^2$ $K_4 = 1.0869 - 6.2451\frac{a}{b} + 9.4190\left(\frac{a}{b}\right)^2$	At M, $\tau = \frac{TB}{r^3}$ where B varies with $\frac{b}{r}$ as follows: For $0 \le \frac{b}{r} \le 0.5$, $B = K_1 + K_2\frac{b}{r} + K_3\left(\frac{b}{r}\right)^2 + K_4\left(\frac{b}{r}\right)^3$ where for $0.2 \le a/b \le 1.0$, $K_1 = 0.6366$ $K_2 = 0.0114 - 0.0789\frac{a}{b} + 0.1767\left(\frac{a}{b}\right)^2$ $K_3 = -0.1207 + 1.0291\frac{a}{b} - 2.3589\left(\frac{a}{b}\right)^2$ $K_4 = 0.5132 - 3.4300\frac{a}{b} + 4.0226\left(\frac{a}{b}\right)^2$ (Data from Refs. 12 and 13)
34. Pinned shaft with one, two, or four grooves 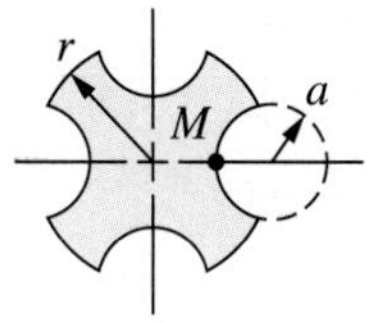 	$K = 2Cr^4$ where C varies with $\frac{a}{r}$ over the range $0 \le \frac{a}{r} \le 0.5$ as follows: For one groove, $C = 0.7854 - 0.0225\frac{a}{r} - 1.4154\left(\frac{a}{r}\right)^2 + 0.9167\left(\frac{a}{r}\right)^3$ For two grooves, $C = 0.7854 - 0.0147\frac{a}{r} - 3.0649\left(\frac{a}{r}\right)^2 + 2.5453\left(\frac{a}{r}\right)^3$ For four grooves, $C = 0.7854 - 0.0409\frac{a}{r} - 6.2371\left(\frac{a}{r}\right)^2 + 7.2538\left(\frac{a}{r}\right)^3$	At M, $\tau = \frac{TB}{r^3}$ where B varies with $\frac{a}{r}$ over the range $0.1 \le \frac{a}{r} \le 0.5$ as follows: For one groove, $B = 1.0259 + 1.1802\frac{a}{r} - 2.7897\left(\frac{a}{r}\right)^2 + 3.7092\left(\frac{a}{r}\right)^3$ For two grooves, $B = 1.0055 + 1.5427\frac{a}{r} - 2.9501\left(\frac{a}{r}\right)^2 + 7.0534\left(\frac{a}{r}\right)^3$ For four grooves, $B = 1.2135 - 2.9697\frac{a}{r} + 33.713\left(\frac{a}{r}\right)^2 - 99.506\left(\frac{a}{r}\right)^3 + 130.49\left(\frac{a}{r}\right)^4$ (Data from Refs. 12 and 13)
35. Cross shaft 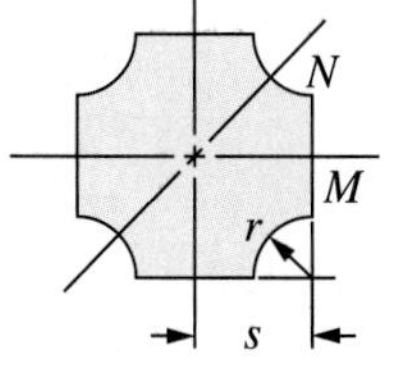 	$K = 2Cs^4$ where C varies with $\frac{r}{s}$ over the range $0 \le \frac{r}{s} \le 0.9$ as follows: $C = 1.1266 - 0.3210\frac{r}{s} + 3.1519\left(\frac{r}{s}\right)^2 - 14.347\left(\frac{r}{s}\right)^3$ $+ 15.223\left(\frac{r}{s}\right)^4 - 4.7767\left(\frac{r}{s}\right)^5$	At M, $\tau = \frac{B_M T}{s^3}$ where B_M varies with $\frac{r}{s}$ over the range $0 \le \frac{r}{s} \le 0.5$ as follows: $B_M = 0.6010 + 0.1059\frac{r}{s} - 0.9180\left(\frac{r}{s}\right)^2 + 3.7335\left(\frac{r}{s}\right)^3 - 2.8686\left(\frac{r}{s}\right)^4$ At N, $\tau = \frac{B_N T}{s^3}$ where B_N varies with $\frac{r}{s}$ over the range $0.3 \le \frac{r}{s} \le 0.9$ as follows: $B_N = -0.3281 + 9.1405\frac{r}{s} - 42.520\left(\frac{r}{s}\right)^2$ $+ 109.04\left(\frac{r}{s}\right)^3 - 133.95\left(\frac{r}{s}\right)^4 + 66.054\left(\frac{r}{s}\right)^5$ (*Note:* $B_N > B_M$ for $r/s > 0.32$) (Data from Refs. 12 and 13)

TABLE 10.2 Formulas for Torsional Properties and Stresses in Thin-Walled Open Cross Sections

Notation: Point 0 indicates the shear center. e = distance from a reference to the shear center; K = torsional stiffness constant (length to the fourth power); C_w = warping constant (length to the sixth power); τ_1 = shear stress due to torsional rigidity of the cross section (force per unit area); τ_2 = shear stress due to warping rigidity of the cross section (force per unit area); σ_x = bending stress due to warping rigidity of the cross section (force per unit area); E = modulus of elasticity of the material (force per unit area); and G = modulus of rigidity (shear modulus) of the material (force per unit area).

The appropriate values of θ', θ'', and θ''' are found in Table 10.3 for the loading and boundary restraints desired.

Cross Section, Reference No.	Constants	Selected Maximum Values
1. Channel 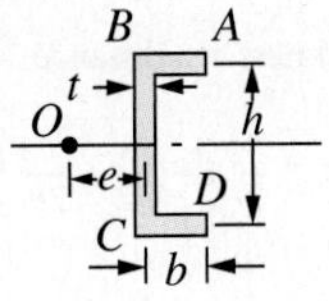	$e=\frac{3b^2}{h+6b}$ $K=\frac{t^3}{3}(h+2b)$ $C_w=\frac{h^2b^3t}{12}\frac{2h+3b}{h+6b}$	$(\sigma_x)_{max}=\frac{hb}{2}\frac{h+3b}{h+6b}E\theta''$ throughout the thickness at corners A and D $(\tau_2)_{max}=\frac{hb^2}{4}\left(\frac{h+3b}{h+6b}\right)^2E\theta'''$ throughout the thickness at a distance $b\frac{h+3b}{h+6b}$ from corners A and D $(\tau_1)_{max}=tG\theta'$ at the surface everywhere
2. C-section 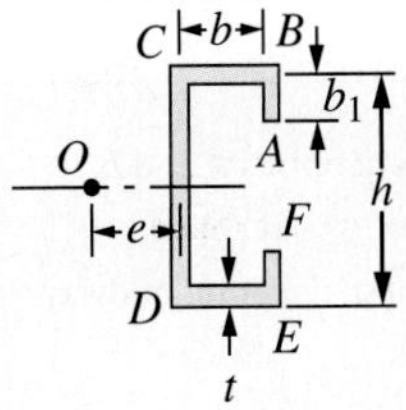	$e=b\frac{3h^2b+6h^2b_1-8b_1^3}{h^3+6h^2b+6h^2b_1+8b_1^3-12hb_1^2}$ $K=\frac{t^3}{3}(h+2b+2b_1)$ $C_w=t\left[\frac{h^2b^2}{2}\left(b_1+\frac{b}{3}-e-\frac{2eb_1}{b}+\frac{2b_1^2}{h}\right)+\frac{h^2e^2}{2}\left(b+b_1+\frac{h}{6}-\frac{2b_1^2}{h}\right)+\frac{2b_1^3}{3}(b+e)^2\right]$	$(\sigma_x)_{max}=\left[\frac{h}{2}(b-e)+b_1(b+e)\right]E\theta''$ throughout the thickness at corners A and F $(\tau_2)_{max}=\left[\frac{h}{4}(b-e)(2b_1+b-e)+\frac{b_1^2}{2}(b+e)\right]E\theta'''$ throughout the thickness on the top and bottom flanges at a distance e from corners C and D $(\tau_1)_{max}=tG\theta'$ at the surface everywhere
3. Hat section 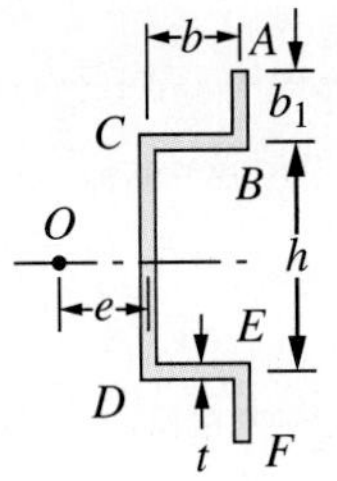	$e=b\frac{3h^2b+6h^2b_1-8b_1^3}{h^3+6h^2b+6h^2b_1+8b_1^3-12hb_1^2}$ $K=\frac{t^3}{3}(h+2b+2b_1)$ $C_w=t\left[\frac{h^2b^2}{2}\left(b_1+\frac{b}{3}-e-\frac{2eb_1}{b}-\frac{2b_1^2}{h}\right)+\frac{h^2e^2}{2}\left(b+b_1+\frac{h}{6}+\frac{2b_1^2}{h}\right)+\frac{2b_1^3}{3}(b+e)^2\right]$	$\sigma_x=\left[\frac{h}{2}(b-e)-b_1(b+e)\right]E\theta''$ throughout the thickness at corners A and F $\sigma_x=\frac{h}{2}(b-e)E\theta''$ throughout the thickness at corners B and E $\tau_2=\left[\frac{h^2(b-e)^2}{8(b+e)}+\frac{b_1^2}{2}(b+e)-\frac{hb_1}{2}(b-e)\right]E\theta'''$ throughout the thickness at a distance $\frac{h(b-e)}{2(b+e)}$ from corner B toward corner A $\tau_2=\left[\frac{b_1^2}{2}(b+e)-\frac{hb_1}{2}(b-e)-\frac{h}{4}(b-e)^2\right]E\theta'''$ throughout the thickness at a distance e from corner C toward corner B $\tau_1=tG\theta'$ at the surface everywhere
4. Twin channel with flanges inward 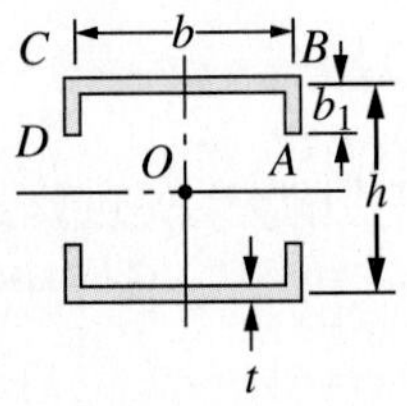	$K=\frac{t^3}{3}(2b+4b_1)$ $C_w=\frac{tb^2}{24}(8b_1^3+6h^2b_1+h^2b+12b_1^2h)$	$(\sigma_x)_{max}=\frac{b}{2}\left(b_1+\frac{h}{2}\right)E\theta''$ throughout the thickness at points A and D $(\tau_2)_{max}=\frac{-b}{16}(4b_1^2+4b_1h+hb)E\theta'''$ throughout the thickness midway between corners B and C $(\tau_1)_{max}=tG\theta'$ at the surface everywhere

(Continued)

TABLE 10.2 Formulas for Torsional Properties and Stresses in Thin-Walled Open Cross Sections (*Continued*)

Cross Section, Reference No.	Constants	Selected Maximum Values
5. Twin channel with flanges outward 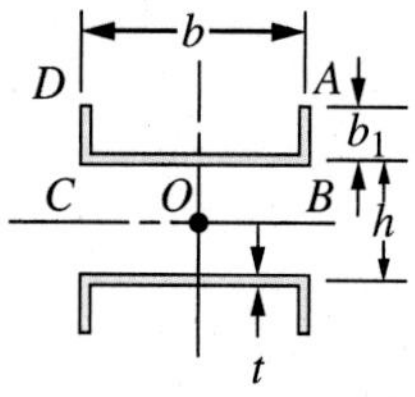 	$K=\frac{t^3}{3}(2b+4b_1)$ $C_w=\frac{tb^2}{24}(8b_1^3+6h^2b_1+h^2b-12b_1^2h)$	$(\sigma_x)_{max}=\frac{hb}{4}E\theta''$ throughout the thickness at points B and C if $h>b_1$ $(\sigma_x)_{max}=\left(\frac{hb}{4}-\frac{bb_1}{2}\right)E\theta''$ throughout the thickness at points A and D if $h<b_1$ $(\tau_2)_{max}=\frac{b}{4}\left(\frac{h}{2}-b_1\right)^2E\theta'''$ throughout the thickness at a distance $\frac{h}{2}$ from corner B toward point A if $b_1>\frac{h}{2}\left(1+\sqrt{\frac{1}{2}+\frac{b}{2h}}\right)$ $(\tau_2)_{max}=\frac{b}{4}\left(b_1^2-\frac{hb}{4}-hb_1\right)E\theta'''$ throughout the thickness at a point midway between corners B and C if $b_1<\frac{h}{2}\left(1+\sqrt{\frac{1}{2}+\frac{b}{2h}}\right)$ $(\tau_1)_{max}=tG\theta'$ at the surface everywhere
6. Wide flanged beam with equal flanges 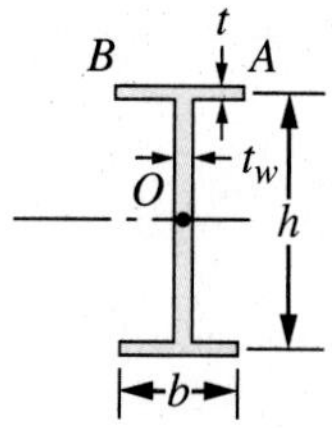 	$K=\frac{1}{2}(2t^3b+t_w^3h)$ $C_w=\frac{h^2tb^3}{24}$	$(\sigma_x)_{max}=\frac{hb}{4}E\theta''$ throughout the thickness at points A and B $(\tau_2)_{max}=-\frac{hb^2}{16}E\theta'''$ throughout the thickness at a point midway between A and B $(\tau_1)_{max}=tG\theta'$ at the surface everywhere
7. Wide flanged beam with unequal flanges 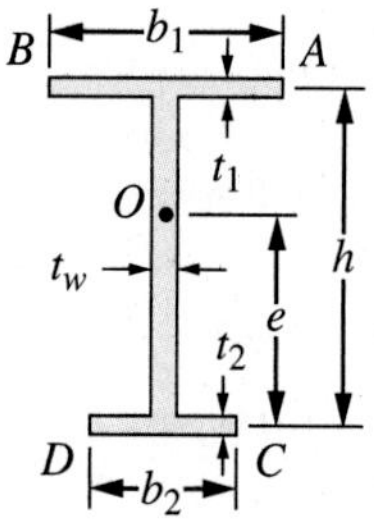 	$e=\frac{t_1b_1^3h}{t_1b_1^3+t_2b_2^3}$ $K=\frac{1}{3}(t_1^3b_1+t_2^3b_2+t_w^3h)$ $C_w=\frac{h^2t_1t_2b_1^3b_2^3}{12(t_1b_1^3+t_2b_2^3)}$	$(\sigma_x)_{max}=\frac{hb_1}{2}\frac{t_2b_2^3}{t_1b_1^3+t_2b_2^3}E\theta''$ throughout the thickness at points A and B if $t_2b_2^2>t_1b_1^2$ $(\sigma_x)_{max}=\frac{hb_2}{2}\frac{t_1b_1^3}{t_1b_1^3+t_2b_2^3}E\theta''$ throughout the thickness at points C and D if $t_2b_2^2<t_1b_1^2$ $(\tau_2)_{max}=\frac{-1}{8}\frac{ht_2b_2^3b_1^2}{t_1b_1^3+t_2b_2^3}E\theta'''$ throughout the thickness at a point midway between A and B if $t_2b_2>t_1b_1$ $(\tau_2)_{max}=\frac{-1}{8}\frac{ht_1b_1^3b_2^2}{t_1b_1^3+t_2b_2^3}E\theta'''$ throughout the thickness at a point midway between C and D if $t_2b_2<t_1b_1$ $(\tau_1)_{max}t_{max}=G\theta'$ at the surface on the thickest portion

(*Continued*)

TABLE 10.2 Formulas for Torsional Properties and Stresses in Thin-Walled Open Cross Sections (*Continued*)

Cross Section, Reference No.	Constants	Selected Maximum Values
8. *Z*-section 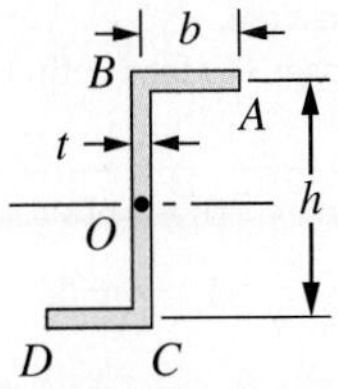	$K = \frac{t^3}{3}(2b+h)$ $C_w = \frac{th^2b^3}{12}\left(\frac{b+2h}{2b+h}\right)$	$(\sigma_x)_{max} = \frac{hb}{2}\frac{b+h}{2b+h}E\theta''$ throughout the thickness at points *A* and *D* $(\tau_2)_{max} = \frac{-hb^2}{4}\left(\frac{b+h}{2b+h}\right)^2 E\theta'''$ throughout the thickness at a distance $\frac{b(b+h)}{2b+h}$ from point *A* $(\tau_1)_{max} = tG\theta'$ at the surface everywhere
9. Segment of a circular tube 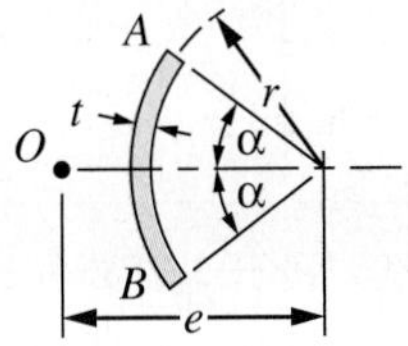 (*Note:* If t/r is small, α can be larger than π to evaluate constants for the case when the walls overlap.)	$e = 2r\frac{\sin\alpha - \alpha\cos\alpha}{\alpha - \sin\alpha\cos\alpha}$ $K = \frac{2}{3}t^3 r\alpha$ $C_w = \frac{2tr^5}{3}\left[\alpha^3 - 6\frac{(\sin\alpha - \alpha\cos\alpha)^2}{\alpha - \sin\alpha\cos\alpha}\right]$	$(\sigma_x) = (r^2\alpha - re\sin\alpha)E\theta''$ throughout the thickness at points *A* and *B* $(\tau_2)_{max} = r^2\left[e(1-\cos\alpha) - \frac{r\alpha^2}{2}\right]E\theta'''$ throughout the thickness at midlength $(\tau_1)_{max} = tG\theta'$ at the surface everywhere
10. 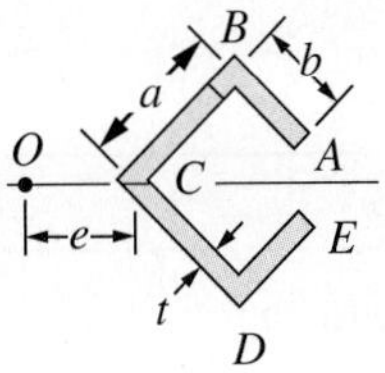	$e = 0.707ab^2\frac{3a-2b}{2a^3-(a-b)^3}$ $K = \frac{2}{3}t^3(a+b)$ $C_w = \frac{ta^4b^3}{6}\frac{4a+3b}{2a^3-(a-b)^3}$	$(\sigma_x)_{max} = \frac{a^2b}{2}\frac{2a^2+3ab-b^2}{2a^3-(a-b)^3}E\theta''$ throughout the thickness at points *A* and *E* $\tau_2 = \frac{a^2b^2}{4}\frac{a^2-2ab-b^2}{2a^3-(a-b)^3}E\theta'''$ throughout the thickness at point *C* $(\tau_1)_{max} = tG\theta'$ at the surface everywhere
11. 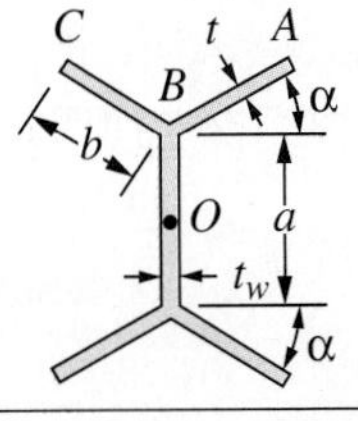	$K = \frac{1}{3}(4t^3b + t_w^3a)$ $C_w = \frac{a^2b^3t}{3}\cos^2\alpha$ (*Note:* Expressions are equally valid for + and −α.)	$(\sigma_x)_{max} = \frac{ab}{2}\cos\alpha E\theta''$ throughout the thickness at points *A* and *C* $(\tau_2)_{max} = \frac{-ab^2}{4}\cos\alpha E\theta'''$ throughout the thickness at point *B* $(\tau_1)_{max} = tG\theta'$ at the surface everywhere

TABLE 10.3 Formulas for the Elastic Deformations of Uniform Thin-Walled Open Members Under Torsional Loading

Notation: T_o = applied torsional load (force-length); t_o = applied distributed torsional load (force-length per unit length); T_A and T_B are the reaction end torques at the left and right ends, respectively. θ = angle of rotation at a distance x from the left end (radians). θ', θ'', and θ''' are the successive derivatives of θ with respect to the distance x. All rotations, applied torsional loads, and reaction end torques are positive as shown (*CCW* when viewed from the right end of the member). E is the modulus of elasticity of the material; C_w is the warping constant for the cross section; K is the torsional constant (see Table 10.2 for expressions for C_w and K); and G is the modulus of rigidity (shear modulus) of the material.

The constants and functions are hereby defined in order to permit condensing the tabulated formulas which follow. See page 160 for a definition of $\langle x-a\rangle^n$. The function $\sinh\beta\langle x-a\rangle$ is also defined as zero if x is less than a. $\beta=(KG/C_wE)^{1/2}$.

$F_1=\cosh\beta x$	$F_{a1}=\langle x-a\rangle^0\cosh\beta(x-a)$	$C_1=\cosh\beta l$	$C_{a1}=\cosh\beta(l-a)$	$A_1=\cosh\beta a$
$F_2=\sinh\beta x$	$F_{a2}=\sinh\beta\langle x-a\rangle$	$C_2=\sinh\beta l$	$C_{a2}=\sinh\beta(l-a)$	$A_2=\sinh\beta a$
$F_3=\cosh\beta x-1$	$F_{a3}=\langle x-a\rangle^0[\cosh\beta(x-a)-1]$	$C_3=\cosh\beta l-1$	$C_{a3}=\cosh\beta(l-a)-1$	
$F_4=\sinh\beta x-\beta x$	$F_{a4}=\sinh\beta\langle x-a\rangle-\beta\langle x-\alpha\rangle$	$C_4=\sinh\beta l-\beta l$	$C_{a4}=\sinh\beta(l-a)-\beta(l-a)$	
	$F_{a5}=F_{a3}-\dfrac{\beta^2\langle x-a\rangle^2}{2}$		$C_{a5}=C_{a3}-\dfrac{\beta^2(l-a)^2}{2}$	
	$F_{a6}=F_{a4}-\dfrac{\beta^3\langle x-a\rangle^3}{6}$		$C_{a6}=C_{a4}-\dfrac{\beta^3(l-a)^3}{6}$	

1. Concentrated intermediate torque

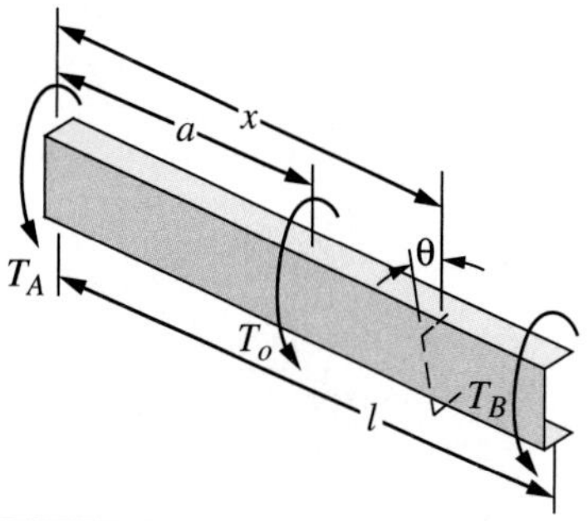

$$\theta=\theta_A+\frac{\theta'_A}{\beta}F_2+\frac{\theta''_A}{\beta^2}F_3+\frac{T_A}{C_wE\beta^3}F_4+\frac{T_o}{C_wE\beta^3}F_{a4}$$

$$\theta'=\theta'_AF_1+\frac{\theta''_A}{\beta}F_2+\frac{T_A}{C_wE\beta^2}F_3+\frac{T_o}{C_wE\beta^2}F_{a3}$$

$$\theta''=\theta''_AF_1+\theta'_A\beta F_2+\frac{T_A}{C_wE\beta}F_2+\frac{T_o}{C_wE\beta}F_{a2}$$

$$\theta'''=\theta'_A\beta^2F_1+\theta''_A\beta F_2+\frac{T_A}{C_wE}F_1+\frac{T_o}{C_wE}F_{a1}$$

$$T=T_A+T_o\langle x-a\rangle^0$$

End Restraints, Reference No.	Boundary Value	Selected Special Cases and Maximum Values
1a. Left end free to twist and warp, right end free to warp but not twist 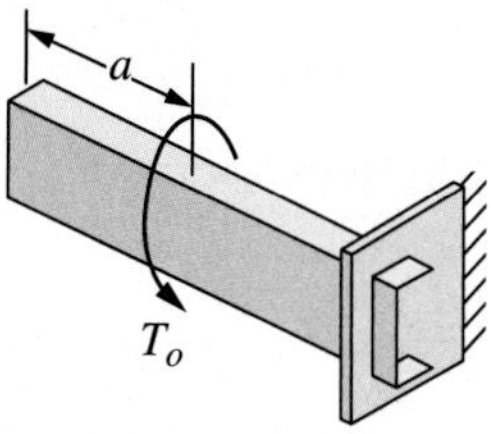	$T_A=0\quad \theta''_A=0$ $\theta_A=\dfrac{T_o}{C_wE\beta^2}(l-a)$ $\theta'_A=\dfrac{-T_o}{C_wE\beta^2}\dfrac{C_{a2}}{C_2}$ $\theta_B=0,\quad \theta''_B=0,\quad T_B=-T_o$ $\theta'_B=\dfrac{-T_o}{C_wE\beta^2}\left(1-\dfrac{A_2}{C_2}\right)$ $\theta'''_B=\dfrac{T_o}{C_wE}\dfrac{A_2}{C_2}$	$\theta_{max}=\theta_A$; max possible value $=\dfrac{T_ol}{C_wE\beta^2}$ when $a=0$ $\theta'_{max}=\theta'_B$; max possible value $=\dfrac{-T_o}{C_wE\beta^2}$ when $a=0$ $\theta''_{max}=\dfrac{-T_o}{C_wE\beta}\dfrac{C_{a2}}{C_2}A_2$ at $x=a$; max possible value $=\dfrac{-T_o}{2C_wE\beta}\tanh\dfrac{\beta l}{2}$ when $a=l/2$ $(-\theta''')_{max}=\dfrac{-T_o}{C_wE}\dfrac{C_{a2}}{C_2}A_1$ just left of $x=a$ $(+\theta''')_{max}=\dfrac{T_o}{C_wE}\left(1-\dfrac{C_{a2}}{C_2}A_1\right)$ just right of $x=a$; max possible value $=\dfrac{T_o}{C_wE}$ when a approaches l If $a=0$ (torque applied at the left end), $\theta=\dfrac{T_o}{KG}(l-x)$, $\theta'=\dfrac{-T_o}{KG}$, $\theta''=0$, $\theta'''=0$

(*Continued*)

TABLE 10.3 Formulas for the Elastic Deformations of Uniform Thin-Walled Open Members Under Torsional Loading (*Continued*)

End Restraints, Reference No.	Boundary Value	Selected Special Cases and Maximum Values
1b. Left end free to twist and warp right end fixed (no twist or warp) 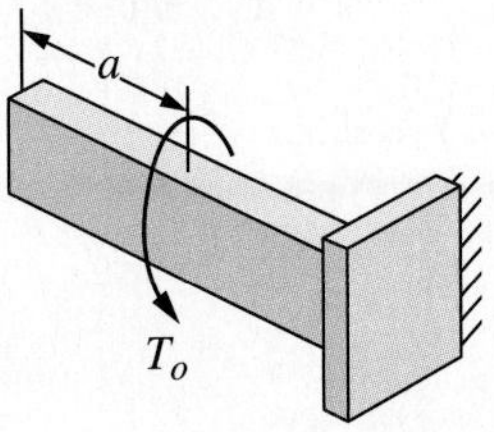	$T_A = 0 \quad \theta''_A = 0$ $\theta_A = \frac{T_o}{C_w E\beta^3}\left(\frac{C_2 C_{a3}}{C_1} - C_{a4}\right)$ $\theta'_A = \frac{-T_o}{C_w E\beta^2}\frac{C_{a3}}{C_1}$ $\theta_B = 0, \quad \theta'_B = 0, \quad T_B = -T_o$ $\theta''_B = \frac{-T_o}{C_w E\beta}\frac{A_2 - C_2}{C_1}$ $\theta'''_B = \frac{T_o}{C_w E}$	If $a = 0$ (torque applied at the left end), $\theta = \frac{T_o}{C_w E\beta^3}\left[\beta(l-x) - \tanh\beta l + \frac{\sinh\beta x}{\cosh\beta l}\right]$ $\theta' = \frac{T_o}{C_w E\beta^2}\left(1 - \frac{\cosh\beta x}{\cosh\beta l}\right), \quad \theta'' = \frac{T_o}{C_w E\beta}\frac{\sinh\beta x}{\cosh\beta}$ $\theta''' = \frac{T_o}{C_w E}\frac{\cosh\beta x}{\cosh\beta l}$ $\theta_{max} = \frac{T_o}{C_w E\beta^3}(\beta l - \tanh\beta l)$ at $x = 0$ $\theta'_{max} = \frac{-T_o}{C_w E\beta^2}\left(1 - \frac{1}{\cosh\beta l}\right)$ at $x = 0$ $\theta''_{max} = \frac{T_o}{C_w E\beta}\tanh\beta l$ at $x = l$ $\theta'''_{max} = \frac{T_o}{C_w E}$ at $x = l$
1c. Left end free to twist but not warp, right end free to warp but not twist 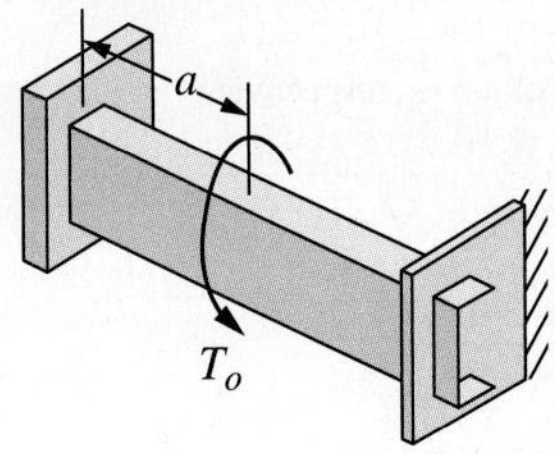	$T_A = 0, \quad \theta'_A = 0$ $\theta_A = \frac{T_o}{C_w E\beta^3}\left(\frac{C_3 C_{a2}}{C_1} - C_{a4}\right)$ $\theta''_A = \frac{-T_o}{C_w E\beta}\frac{C_{a2}}{C_1}$ $\theta_B = 0, \quad \theta''_B = 0, \quad T_B = -T_o$ $\theta'_B = \frac{-T_o}{C_w E\beta^2}\left(\frac{C_2 C_{a2}}{C_1} - C_{a3}\right)$ $\theta'''_B = \frac{-T_o}{C_w E}\left(\frac{C_2 C_{a2}}{C_1} - C_{a1}\right)$	If $a = 0$ (torque applied at the left end), $\theta = \frac{T_o}{C_w E\beta^3}[\sinh\beta x - \tanh\beta l\cosh\beta x + \beta(l-x)]$ $\theta' = \frac{T_o}{C_w E\beta^2}(\cosh\beta x - \tanh\beta l\sinh\beta x - 1)$ $\theta'' = \frac{T_o}{C_w E\beta}(\sinh\beta x - \tanh\beta l\cosh\beta x),$ $\theta''' = \frac{T_o}{C_w E}(\cosh\beta x - \tanh\beta l\sinh\beta x)$ $\theta_{max} = \frac{T_o}{C_w E\beta^3}(\beta l - \tanh\beta l)$ at $x = 0$ $\theta'_{max} = \frac{-T_o}{C_w E\beta^2}\left(\frac{-1}{\cosh\beta l} + 1\right)$ at $x = l$ $\theta''_{max} = \frac{-T_0}{C_w E\beta}\tanh\beta l$ at $x = 0$ $\theta'''_{max} = \frac{T_o}{C_w E}$ at $x = 0$
1d. Left end free to twist but not warp, right end fixed (no twist or warp) 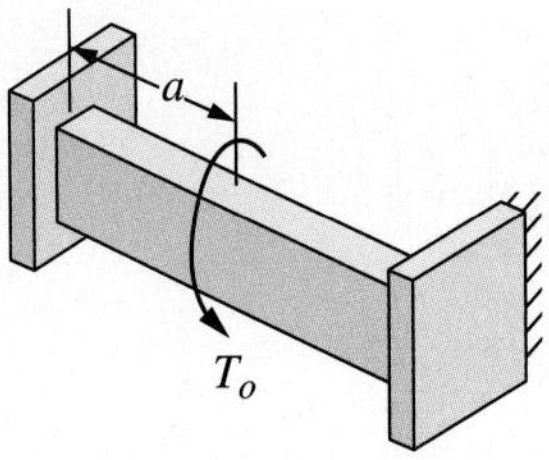	$T_A = 0, \quad \theta'_A = 0$ $\theta_A = \frac{T_o}{C_w E\beta^3}\left(\frac{C_3 C_{a3}}{C_2} - C_{a4}\right)$ $\theta''_A = \frac{-T_o}{C_w E\beta}\frac{C_{a3}}{C_2}$ $\theta_B = 0, \quad \theta'_B = 0, \quad T_B = -T_o$ $\theta''_B = \frac{-T_o}{C_w E\beta}\left(\frac{C_1 C_{a3}}{C_2} - C_{a2}\right)$ $\theta'''_B = \frac{T_o}{C_w E}$	If $a = 0$ (torque applied at the left end), $\theta = \frac{T_o}{C_w E\beta^3}\left[\sinh\beta x + \beta(l-x) - \tanh\frac{\beta l}{2}(1 + \cosh\beta x)\right]$ $\theta' = \frac{T_o}{C_w E\beta^2}\left(\cosh\beta x - 1 - \tanh\frac{\beta l}{2}\sinh\beta x\right)$ $\theta'' = \frac{T_o}{C_w E\beta}\left(\sinh\beta x - \tanh\frac{\beta l}{2}\cosh\beta x\right),$ $\theta''' = \frac{T_o}{C_w E}\left(\cosh\beta x - \tanh\frac{\beta l}{2}\sinh\beta x\right)$ $\theta_{max} = \frac{T_o}{C_w E\beta^3}\left(\beta l - 2\tanh\frac{\beta l}{2}\right)$ at $x = 0$ $\theta'_{max} = \frac{T_o}{C_w E\beta^2}\left[\frac{1}{\cosh(\beta l/2)} - 1\right]$ at $x = \frac{l}{2}$ $\theta''_{max} = \frac{\mp T_o}{C_w E\beta}\tanh\frac{\beta l}{2}$ at $x = 0$ and $x = l$, respectively $\theta'''_{max}\ \frac{T_o}{C_w E}$ at $x = 0$ and $x = l$

(*Continued*)

TABLE 10.3 Formulas for the Elastic Deformations of Uniform Thin-Walled Open Members Under Torsional Loading (*Continued*)

End Restraints, Reference No.	Boundary Value	Selected Special Cases and Maximum Values
1e. Both ends free to warp but not twist 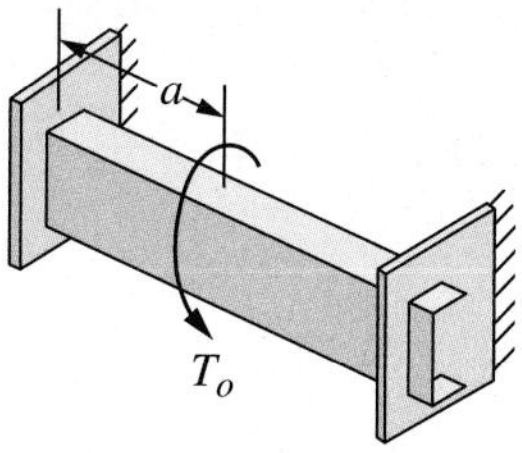	$\theta_A=0,\ \theta''_A=0$ $T_A=-T_o\left(1-\frac{a}{l}\right)$ $\theta'_A=\frac{T_o}{C_wE\beta^2}\left(1-\frac{a}{l}-\frac{C_{a2}}{C_2}\right)$ $\theta_B=0,\ \theta''_B=0$ $\theta'_B=\frac{-T_o}{C_wE\beta^2}\left(\frac{a}{l}-\frac{A_2}{C_2}\right)$ $\theta'''_B=\frac{T_o}{C_wE}\frac{A_2}{C_2}$ $T_B=-T_o\frac{a}{l}$	If $a=l/2$ (torque applied at midlength), $\theta=\frac{T_o}{C_wE\beta^3}\left[\frac{\beta x}{2}-\frac{\sinh\beta x}{2\cosh(\beta l/2)}=\sinh\beta\left\langle x-\frac{1}{2}\right\rangle-\beta\left\langle x-\frac{l}{2}\right\rangle\right]$ $\theta'=\frac{T_o}{C_wE\beta^2}\left\{\frac{1}{2}-\frac{\cosh\beta x}{2\cosh(\beta l/2)}+\left\langle x-\frac{l}{2}\right\rangle^0\left[\cosh\beta\left(x-\frac{l}{2}\right)-1\right]\right\}$ $\theta''=\frac{-T_o}{C_wE\beta}\left[\frac{\sinh\beta x}{2\cosh(\beta l/2)}-\sinh\beta\left\langle x-\frac{l}{2}\right\rangle\right]$ $\theta'''=\frac{-T_o}{C_wE}\left[\frac{\cosh\beta x}{2\cosh(\beta l/2)}-\left\langle x-\frac{l}{2}\right\rangle^0\cosh\beta\left(x-\frac{l}{2}\right)\right]$ $\theta_{max}=\frac{T_o}{2C_wE\beta^3}\left(\frac{\beta l}{2}-\tanh\frac{\beta l}{2}\right)$ at $x=\frac{l}{2}$ $\theta'_{max}=\frac{\pm T_o}{2C_wE\beta^2}\left[1-\frac{1}{\cosh(\beta l/2)}\right]$ at $x=0$ and $x=l$, respectively $\theta''_{max}=\frac{-T_o}{2C_wE\beta}\tanh\frac{\beta l}{2}$ at $x=\frac{l}{2}$ $\theta'''_{max}=\frac{\mp T_o}{2C_wE}$ just left and just right of $x=\frac{l}{2}$, respectively
1f. Left end free to warp but not twist, right end fixed (no twist or warp) 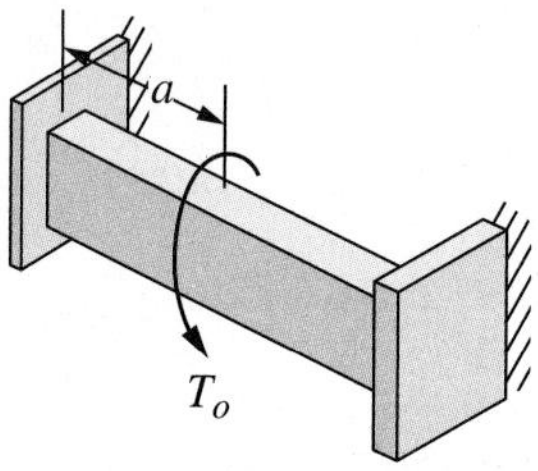	$\theta_A=0,\ \ \theta''_A=0$ $T_A=-T_o\frac{C_1C_{a4}-C_2C_{a3}}{C_1C_4-C_2C_3}$ $\theta'_A=\frac{T_o}{C_wE\beta^2}\frac{C_3C_{a4}-C_4C_{a3}}{C_1C_4-C_2C_3}$ $\theta_B=0,\ \ \theta'_B=0$ $\theta''_B=\frac{T_o}{C_wE\beta}\frac{\beta lA_2-\beta aC_2}{C_1C_4-C_2C_3}$ $\theta'''_B=\frac{T_o}{C_wE}\frac{A_2-\beta aC_1}{C_1C_4-C_2C_3}$ $T_B=-T_o-T_A$	If $a=l/2$ (torque applied at midlength), $T_A=-T_o\frac{\sinh\beta l-(\beta l/2)\cosh\beta l-\sinh(\beta l/2)}{\sinh\beta l-\beta l\cosh\beta l}$ $\theta'_A=\frac{T_o}{C_wE\beta^2}\frac{2\sinh(\beta l/2)-\beta l\cosh(\beta l/2)}{\sinh\beta l-\beta l\cosh\beta l}\left(\cosh\frac{\beta l}{2}-1\right)$
1g. Both ends fixed (no twist or warp) 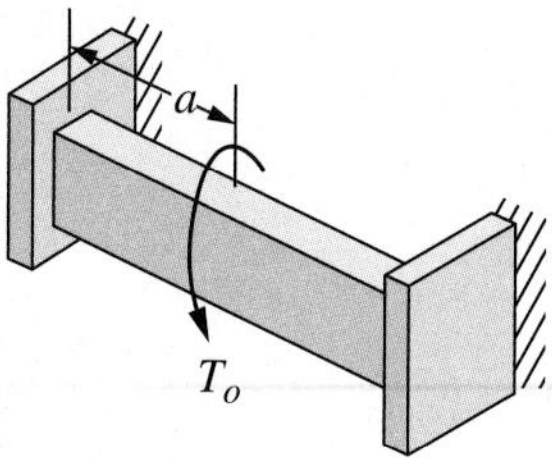	$\theta_A=0,\ \ \theta'_A=0$ $\theta''_A=\frac{T_o}{C_wE\beta}\frac{C_3C_{a4}-C_4C_{a3}}{C_2C_4-C_3^2}$ $T_A=-T_o\frac{C_2C_{a4}-C_3C_{a3}}{C_2C_4-C_3^2}$ $\theta_B=0,\ \ \theta'_B=0$ $\theta''_B=\theta''_AC_1+\frac{T_A}{C_wE\beta}C_2+\frac{T_o}{C_wE\beta}C_{a2}$ $\theta'''_B=\theta''_A\beta C_2+\frac{T_A}{C_wE}C_1+\frac{T_o}{C_wE}C_{a1}$ $T_B=-T_o-T_A$	If $a=l/2$ (torque applied at midlength), $T_A=T_B=\frac{-T_o}{2}$ $\theta_{max}=\frac{T_o}{C_wE\beta^3}\left(\frac{\beta l}{4}-\tanh\frac{\beta l}{4}\right)$ at $x=\frac{l}{2}$ $\theta'_{max}=\frac{T_o}{2C_wE\beta^2}\left[1-\frac{1}{\cosh(\beta l/4)}\right]$ at $x=\frac{l}{4}$ $\theta''_{max}=\overset{+}{\underset{+}{-}}\frac{T_o}{2C_wE\beta}\tanh\frac{\beta l}{4}$ at $x=0$, $x=\frac{l}{2}$, and $x=l$, respectively $(-\theta''')_{max}=\frac{-T_o}{2C_wE}$ at $x=0$ and just left of $x=\frac{l}{2}$ $(+\theta''')_{max}=\frac{T_o}{2C_wE}$ at $x=l$ and just right of $x=\frac{l}{2}$

(*Continued*)

TABLE 10.3 Formulas for the Elastic Deformations of Uniform Thin-Walled Open Members Under Torsional Loading (*Continued*)

2. Uniformly distributed torque from a to l

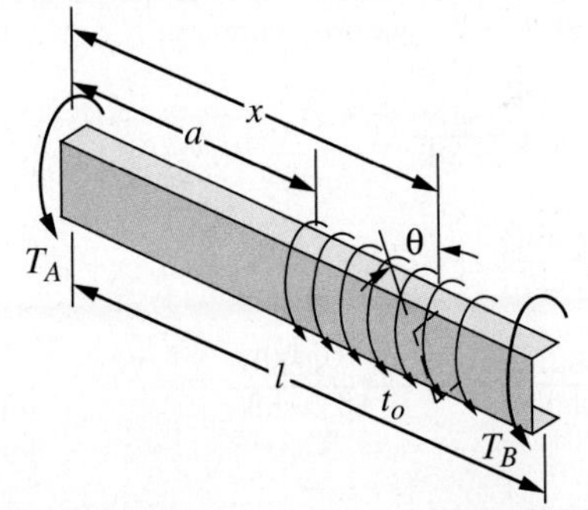

$$\theta = \theta_A + \frac{\theta'_A}{\beta}F_2 + \frac{\theta''_A}{\beta^2}F_3 + \frac{T_A}{C_w E\beta^3}F_4 + \frac{t_o}{C_w E\beta^4}F_{a5}$$

$$\theta' = \theta'_A F_1 + \frac{\theta''_A}{\beta}F_2 + \frac{T_A}{C_w E\beta^2}F_3 + \frac{t_o}{C_w E\beta^3}F_{a4}$$

$$\theta'' = \theta''_A F_1 + \theta'_A\beta F_2 + \frac{T_A}{C_w E\beta}F_2 + \frac{t_o}{C_w E\beta^2}F_{a3}$$

$$\theta''' = \theta'_A\beta^2 F_1 + \theta''_A\beta F_2 + \frac{T_A}{C_w E}F_1 + \frac{t_o}{C_w E\beta}F_{a2}$$

$$T = T_A + t_o\langle x-a\rangle$$

End Restraints, Reference No.	Boundary Value	Selected Special Cases and Maximum Values
2a. Left end free to twist and warp, right end free to warp but not twist 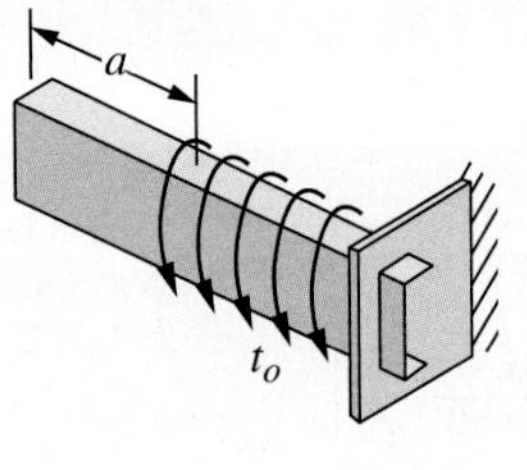	$T_A = 0, \quad \theta''_A = 0$ $\theta_A = \frac{t_o}{2C_w E\beta^2}(l-a)^2$ $\theta'_A = \frac{-t_o}{C_w E\beta^3}\frac{C_{a3}}{C_2}$ $\theta_B = 0, \qquad \theta''_B = 0$ $\theta'_B = \frac{-t_o}{C_w E\beta^3}\left[\frac{A_1 - C_1}{C_2} + \beta(l-a)\right]$ $\theta'''_B = \frac{-t_o}{C_w E\beta}\frac{A_1 - C_1}{C_2}$ $T_B = -t_o(l-a)$	If $a = 0$ (uniformly distributed torque over entire span), $\theta = \frac{t_o}{C_w E\beta^4}\left[\frac{\beta^2(l^2-x^2)}{2} + \frac{\sinh\beta(l-x) + \sinh\beta x}{\sinh\beta l} - 1\right]$ $\theta' = \frac{-t_o}{C_w E\beta^3}\left[\beta x + \frac{\cosh\beta(l-x) + \cosh\beta x}{\sinh\beta l}\right]$ $\theta'' = \frac{-t_o}{C_w E\beta^2}\left(1 - \frac{\sinh\beta(l-x) + \sinh\beta x}{\sinh\beta l}\right)$ $\theta''' = \frac{-t_o}{C_w E\beta}\frac{\cosh\beta(l-x) - \cosh\beta x}{\sinh\beta l}$ $\theta_{max} = \frac{t_o l^2}{2C_w E\beta^3}$ at $x = 0$ $\theta'_{max} = \frac{t_o}{C_w E\beta^3}\left(\beta l - \tanh\frac{\beta l}{2}\right)$ at $x = l$ $\theta''_{max} = \frac{-t_o}{C_w E\beta^2}\left[1 - \frac{1}{\cosh(\beta l/2)}\right]$ at $x = \frac{l}{2}$ $\theta'''_{max} = \frac{\mp t_o}{C_w E\beta}\tanh\frac{\beta l}{2}$ at $x = 0$ and $x = l$, respectively
2b. Left end free to twist and warp, right end fixed (no twist or warp) 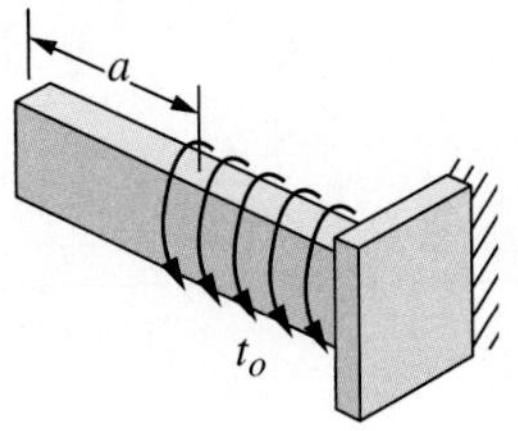	$T_A = 0, \qquad \theta''_A = 0$ $\theta_A = \frac{t_o}{C_w E\beta^4}\left(\frac{C_2 C_{a4}}{C_1} - C_{a5}\right)$ $\theta'_A = \frac{-t_o}{C_w E\beta^3}\frac{C_{a4}}{C_1}$ $\theta_B = 0, \qquad \theta'_B = 0$ $\theta''_B = \frac{-t_o}{C_w E\beta^2}\left(\frac{C_2 C_{a4}}{C_1} - C_{a3}\right)$ $\theta'''_B = \frac{t_o}{C_w E}(l-a)$ $T_B = -t_o(l-a)$	If $a = 0$ (uniformly distributed torque over entire span), $\theta = \frac{t_o}{C_w E\beta^4}\left[\frac{1 - \cosh\beta(l-x) + \beta(\sinh\beta l - \sinh\beta x)}{\cosh\beta l} - \frac{\beta^2(l^2-x^2)}{2}\right]$ $\theta' = \frac{-t_o}{C_w E\beta^3}\left[\frac{\sinh\beta(l-x) - \beta l\cosh\beta x}{\cosh\beta l} + \beta x\right]$ $\theta'' = \frac{-t_o}{C_w E\beta^2}\left[1 - \frac{\cosh\beta(l-x) + \beta l\sinh\beta x}{\cosh\beta l}\right]$ $\theta''' = \frac{-t_o}{C_w E\beta}\frac{\sinh\beta(l-x) - \beta l\cosh\beta x}{\cosh\beta l}$ $\theta_{max} = \frac{t_o}{C_w E\beta^4}\left(1 + \frac{\beta^2 l^2}{2} - \frac{1 + \beta l\sinh\beta l}{\cosh\beta l}\right)$ at $x = 0$ The location of max θ' depends upon βl $\theta''_{max} = \frac{t_o}{C_w E\beta^2}\left(\frac{1 + \beta l\sinh\beta l}{\cosh\beta l} - 1\right)$ at $x = l$ $\theta'''_{max} = \frac{t_o l}{C_w E}$ at $x = l$

(*Continued*)

TABLE 10.3 Formulas for the Elastic Deformations of Uniform Thin-Walled Open Members Under Torsional Loading (*Continued*)

End Restraints, Reference No.	Boundary Value	Selected Special Cases and Maximum Values
2c. Left end free to twist but not warp, right end free to warp but not twist 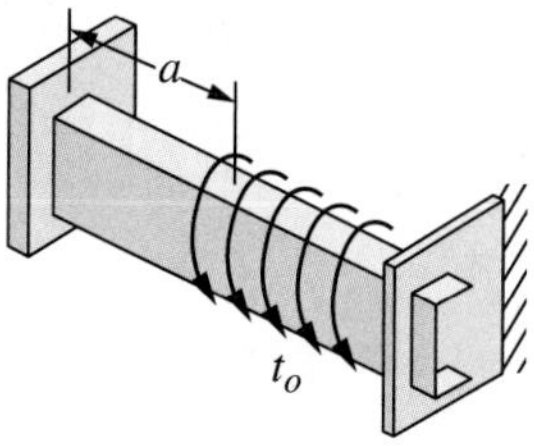 	$T_A = 0, \quad \theta'_A = 0$ $\theta_A = \frac{t_o}{C_w E\beta^4}\left(\frac{C_3 C_{a3}}{C_1} - C_{a5}\right)$ $\theta''_A = \frac{-t_o}{C_w E\beta^2}\frac{C_{a3}}{C_1}$ $\theta_B = 0, \quad \theta''_B = 0$ $\theta'_B = \frac{-t_o}{C_w E\beta^3}\left(\frac{C_2 C_{a3}}{C_1} - C_{a4}\right)$ $\theta'''_B = \frac{-t_o}{C_w E\beta}\left(\frac{C_2 C_{a3}}{C_1} - C_{a2}\right)$ $T_B = -t_o(l-a)$	If $a = 0$ (uniformly distributed torque over entire span), $\theta = \frac{t_o}{C_w E\beta^4}\left[\frac{\beta^2(l^2 - x^2)}{2} + \frac{\cosh\beta x}{\cosh\beta l} - 1\right]$ $\theta' = \frac{-t_o}{C_w E\beta^3}\left(\beta x - \frac{\sinh\beta x}{\cosh\beta l}\right)$ $\theta'' = \frac{-t_o}{C_w E\beta^2}\left(1 - \frac{\cosh\beta x}{\cosh\beta l}\right), \quad \theta''' = \frac{t_o}{C_w E\beta}\frac{\sinh\beta x}{\cosh\beta l}$ $\theta_{max} = \frac{t_o}{C_w E\beta^4}\left(\frac{\beta^2 l^2}{2} + \frac{1}{\cosh\beta l} - 1\right)$ at $x = 0$ $\theta'_{max} = \frac{-t_o}{C_w E\beta^3}(\beta l - \tanh\beta l)$ at $x = l$ $\theta''_{max} = \frac{-t_o}{C_w E\beta^2}\left(1 - \frac{1}{\cosh\beta l}\right)$ at $x = 0$ $\theta'''_{max} = \frac{t_o}{C_w E\beta}\tanh\beta l$ at $x = l$
2d. Left end free to twist but not warp, right end fixed (no twist or warp) 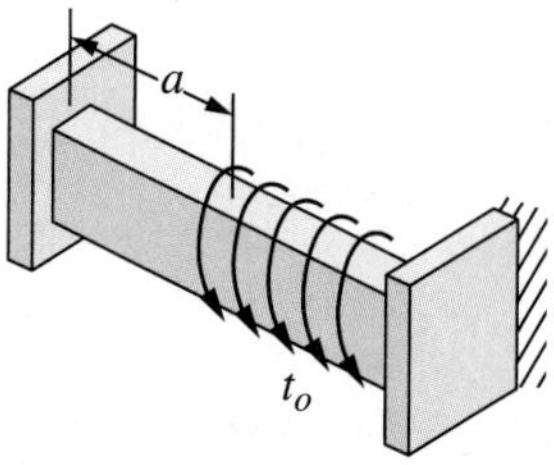 	$T_A = 0, \quad \theta'_A = 0$ $\theta_A = \frac{t_o}{C_w E\beta^4}\left(\frac{C_3 C_{a3}}{C_2} - C_{a5}\right)$ $\theta''_A = \frac{-t_o}{C_w E\beta^2}\frac{C_{a4}}{C_2}$ $\theta_B = 0, \quad \theta'_B = 0$ $\theta''_B = \frac{-t_o}{C_w E\beta^2}\left(\frac{C_1 C_{a4}}{C_2} - C_{a3}\right)$ $\theta'''_B = \frac{t_o}{C_w E}(l-a)$ $T_B = -t_o(l-a)$	If $a = 0$ (uniformly distributed torque over entire span), $\theta = \frac{t_o}{C_w E\beta^4}\left[\frac{\beta^2}{2}(l^2 - x^2) + \beta l\frac{\cosh\beta x - \cosh\beta l}{\sinh\beta l}\right]$ $\theta' = \frac{-t_o}{C_w E\beta^3}\left(\beta x - \frac{\beta l \sinh\beta x}{\sinh\beta l}\right)$ $\theta'' = \frac{-t_o}{C_w E\beta^2}\left(1 - \frac{\beta l\cosh\beta x}{\sinh\beta l}\right), \quad \theta''' = \frac{t_o l}{C_w E}\frac{\sinh\beta x}{\sinh\beta l}$ $\theta_{max} = \frac{t_o l}{C_w E\beta^3}\left(\frac{\beta l}{2} - \tanh\frac{\beta l}{2}\right)$ at $x = 0$ $(+\theta'')_{max} = \frac{t_o}{C_w E\beta^2}\left(\frac{\beta l}{\tanh\beta l} - 1\right)$ at $x = l$ $(-\theta'')_{max} = \frac{-t_o}{C_w E\beta^2}\left(1 - \frac{\beta l}{\sinh\beta l}\right)$ at $x = 0$ $\theta'''_{max} = \frac{t_o l}{C_w E}$ at $x = l$

(*Continued*)

TABLE 10.3 Formulas for the Elastic Deformations of Uniform Thin-Walled Open Members Under Torsional Loading (*Continued*)

End Restraints, Reference No.	Boundary Value	Selected Special Cases and Maximum Values
2e. Both ends free to warp but not twist 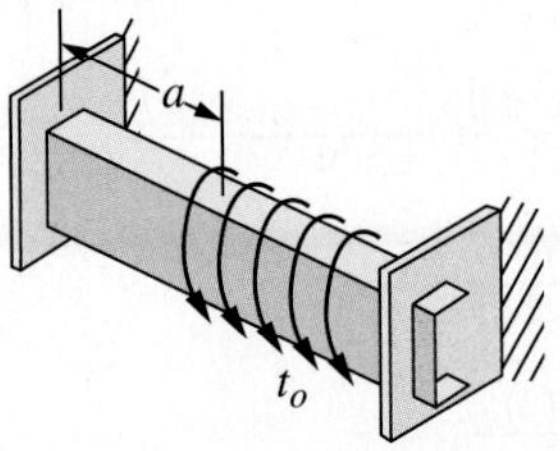 	$\theta_A = 0, \quad \theta''_A = 0$ $T_A = \frac{-t_o}{2l}(l-a)^2$ $\theta'_A = \frac{t_o}{C_w E\beta^3}\left[\frac{\beta}{2l}(l-a)^2 - \frac{C_{a3}}{C_2}\right]$ $\theta_B = 0, \quad \theta''_B = 0$ $\theta'_B = \frac{t_o}{C_w E\beta^3}\left[\frac{\beta}{2l}(l-a)^2 - \frac{C_1 C_{a3}}{C_2} + C_{a4}\right]$ $\theta'''_B = \frac{-t_o}{C_w E\beta}\frac{\cosh\beta a - \cosh\beta l}{\sinh\beta l}$ $T_B = \frac{-t_o}{2l}(l^2 - a^2)$	If $a = 0$ (uniformly distributed torque over entire span), $\theta = \frac{t_o}{C_w E\beta^4}\left[\frac{\beta^2 x(l-x)}{2} + \frac{\cosh\beta(x-l/2)}{\cosh(\beta/2)} - 1\right]$ $\theta' = \frac{t_o}{C_w E\beta^3}\left[\frac{\sinh\beta(x-l/2)}{\cosh(\beta l/2)} - \beta(x-l/2)\right]$ $\theta'' = \frac{t_o}{C_w E\beta^2}\left[\frac{\cosh(x-l/2)}{\cosh(\beta l/2)} - 1\right]$ $\theta''' = \frac{t_o}{C_w E\beta}\frac{\sinh\beta(x-l/2)}{\cosh(\beta l/2)}$ $\theta_{max} = \frac{t_o}{C_w E\beta^4}\left[\frac{\beta^2 l^2}{8} + \frac{1}{\cosh(\beta l/2)} - 1\right]$ at $x = \frac{l}{2}$ $\theta'_{max} = \frac{\pm t_o}{C_w E\beta^3}\left(\frac{\beta l}{2} - \tanh\frac{\beta l}{2}\right)$ at $x = 0$ and $x = l$, respectively $\theta''_{max} = \frac{t_o}{C_w E\beta^2}\left[\frac{1}{\cosh(\beta l/2)} - 1\right]$ at $x = \frac{l}{2}$ $\theta'''_{max} = \frac{\mp t_o}{C_w E\beta}\tanh\frac{\beta l}{2}$ at $x = 0$ and $x = l$, respectively
2f. Left end free to warp but not twist, right end fixed (no twist or warp) 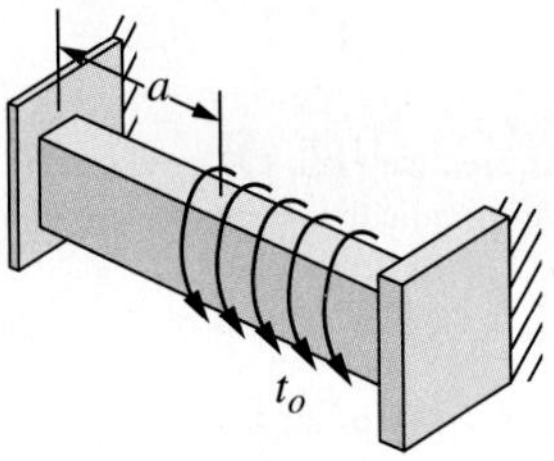 	$\theta_A = 0, \quad \theta''_A = 0$ $T_A = \frac{-t_o}{\beta}\frac{C_1 C_{a5} - C_2 C_{a4}}{C_1 C_4 - C_2 C_3}$ $\theta'_A = \frac{t_o}{C_w E\beta^3}\frac{C_3 C_{a5} - C_4 C_{a4}}{C_1 C_4 - C_2 C_3}$	If $a = 0$ (uniformly distributed torque over entire span), θ_{max} occurs very close to $x = 0.425l$ $\theta'_{max} = \frac{t_o}{C_w E\beta^3}\frac{2 - \beta^2 l^2/2 + 2\beta l \sinh\beta l - (2 + \beta^2 l^2/2)\cosh\beta l}{\sinh\beta l - \beta l\cosh\beta l}$ at $x = 0$ $\theta''_{max} = \frac{-t_o}{C_w E\beta^2}\frac{(\beta^2 l^2/2)\sinh\beta l + \beta l(1 - \cosh\beta l)}{\sinh\beta l - \beta l\cosh\beta l}$ at $x = l$ $(-\theta''')_{max} = \frac{t_o}{C_w E\beta}\frac{1 - \beta^2 l^2/2 - \cosh\beta l + \beta l\sinh\beta l}{\sinh\beta l - \beta l\cosh\beta l}$ at $x = 0$ $(+\theta''')_{max} = \frac{t_o}{C_w E\beta}\frac{(1 - \beta^2 l^2/2)\cosh\beta l - 1}{\sinh\beta l - \beta l\cosh\beta l}$ at $x = l$

(*Continued*)

TABLE 10.3 Formulas for the Elastic Deformations of Uniform Thin-Walled Open Members Under Torsional Loading (*Continued*)

End Restraints, Reference No.	Boundary Value	Selected Special Cases and Maximum Values
2g. Both ends fixed (no twist or warp) 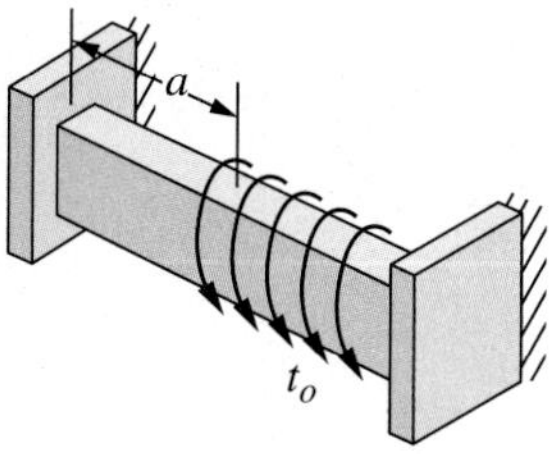 	$\theta_A = 0, \quad \theta'_A = 0$ $\theta''_A = \dfrac{t_o}{C_w E \beta^2} \dfrac{C_3 C_{a5} - C_4 C_{a4}}{C_2 C_4 - C_3^2}$ $T_A = \dfrac{-t_o}{\beta} \dfrac{C_2 C_{a5} - C_3 C_{a4}}{C_2 C_4 - C_3^2}$ $\theta'''_A = \dfrac{T_A}{C_w E}$	If $a = 0$ (uniformly distributed torque over entire span), $T_A = \dfrac{-t_o l}{2}$ $\theta = \dfrac{t_o l}{2 C_w E \beta^3}\left[\dfrac{\beta x}{l}(l - x) + \dfrac{\cosh\beta(x - l/2) - \cosh(\beta l/2)}{\sinh(\beta l/2)}\right]$ $\theta' = \dfrac{t_o l}{2 C_w E \beta^2}\left[1 - \dfrac{2x}{l} + \dfrac{\sinh\beta(x - l/2)}{\sinh(\beta l/2)}\right]$ $\theta'' = \dfrac{t_o}{C_w E \beta^2}\left[\dfrac{\beta l \cosh\beta(x - l/2)}{2\sinh(\beta l/2)} - 1\right]$ $\theta''' = \dfrac{t_o}{C_w E \beta} \dfrac{\beta l \sinh\beta(x - l/2)}{2\sinh(\beta l/2)}$ $\theta_{max} = \dfrac{t_o l}{2 C_w E \beta^3}\left(\dfrac{\beta l}{4} - \tanh\dfrac{\beta l}{4}\right)$ at $x = \dfrac{l}{2}$ $(-\theta'')_{max} = \dfrac{-t_o}{C_w E \beta^2}\left[1 - \dfrac{\beta l}{2\sinh(\beta l/2)}\right]$ at $x = \dfrac{l}{2}$ $(+\theta'')_{max} = \dfrac{t_o}{C_w E \beta^2}\left(\dfrac{\beta l}{2\tanh(\beta l/2)} - 1\right)$ at $x = 0$ and $x = l$ $\theta'''_{max} = \dfrac{\mp t_o l}{2 C_w E}$ at $x = 0$ and $x = l$, respectively

10.8 REFERENCES

1. Cook, R. D., and W. C. Young, *Advanced Mechanices of Materials*, 2nd ed., Prentice-Hall, 1998.
2. Trayer, G. W., and H. W. March, "The Torsion of Members having Sections Common in Aircraft Construction," *Natl. Adv. Comm. Aeron.*, Rept. 334, 1929.
3. Griffith, A. A., "The Determination of the Torsional Stiffness and Strength of Cylindrical Bars of any Shape," Repts. Memo. 334, *Adv. Comm. Aeron. (British)*, 1917.
4. Taylor, G. I., and A. A. Griffith, "The Use of Soap Films in Solving Torsion Problems, Reports and Memoranda 333," *Adv. Comm. Aeron. (British)*, 1917.
5. Timoshenko, S., *Strength of Materials*, pt. II, D. Van Nostrand, 1930.
6. Green, A. E., "The Equilibrium and Elastic Stability of a Thin Twisted Strip," *Proc. R. Soc. Lond. Ser. A*, vol. 154, 1936.
7. Younger, J. E., *Structural Design of Metal Airplanes*, McGraw-Hill, 1935.
8. Wahl, A. M., *Mechanical Springs*, 2nd ed., McGraw-Hill, 1963. See also Wahl, A. M., "Helical Compression and Tension Springs," ASME Paper A-38, *J. Appl. Mech.*, 2(1): 1935.
9. Sayre, M. E., "New Spring Formulas and New Materials for Precision Spring Scales," *Trans. ASME*, 58: 379, 1936.
10. Wilson, T. S., "The Eccentric Circular Tube," *Aircr. Eng.*, 14(157), March 1942.
11. Lyse, I., and B. G. Johnston, "Structural Beams in Torsion," *Inst. Res.*, Lehigh Univ., Circ. 113, 1935.
12. Isakower, R. I., *The Shaft Book (Design Charts for Torsional Properties of Non-Circular Shafts)*, U.S. Army ARRADCOM, MISD Users' Manual 80-5, March 1980.
13. Isakower, R. I., "Don't Guess on Non-Circular Shafts," *Des. Eng.*, November 1980.
14. Isakower, R. I., and R. E. Barnas, *The Book of CLYDE—With a Torquing Chapter*, U.S. Army ARRADCOM Users' Manual MISD UM 77-3, October 1977.
15. Payne, J. H., "Torsion in Box Beams," *Aircr. Eng.*, January 1942.
16. Abramyan, B. L., "Torsion and Bending of Prismatic Rods of Hollow Rectangular Section," *NACA Tech. Memo. 1319*, November 1951.

17. Timoshenko, S. P., and J. N. Goodier, *Theory of Elasticity*, 2nd ed., McGraw-Hill, 1951.
18. "ANC Mil-Hdbk-5, Strength of Metal Aircraft Elements," Armed Forces Supply Support Center, March 1959.
19. Bethlehem Steel Co., *Torsion Analysis of Rolled Steel Sections*.
20. Nuttall, Henry, "Torsion of Uniform Rods with Particular Reference to Rods of Triangular Cross Section," *ASME J. Appl. Mech.*, 19(4): 1952.
21. Gloumakoff, N. A., and Yi-Yuan Yu, "Torsion of Bars With Isosceles Triangular and Diamond Sections," *ASME J. Appl. Mech.*, 31(2): 1964.
22. Chu, Chen, "The Effect of Initial Twist on the Torsional Rigidity of Thin Prismatical Bars and Tubular Members," *Proc. 1st U.S. Nat. Congr. Appl. Mech.*, p. 265, 1951.
23. Engel, H. L., and J. N. Goodier, "Measurements of Torsional Stiffness Changes and Instability Due to Tension, Compression, and Bending," *ASME J. Appl. Mech.*, 20(4): 1953.
24. Ancker, C. J., Jr., and J. N. Goodier, "Pitch and Curvature Correction for Helical Springs," *ASME J. Appl. Mech.*, 25(4): 1958.
25. Marshall, J., "Derivation of Torsion Formulas for Multiply Connected Thick-Walled Rectangular Sections," *ASME J. Appl. Mech.*, 37(2): 1970.
26. Poynting, J. H., *Proc. R. Soc. Lond.*, Ser. A, vol. 32, 1909 and vol. 36, 1912.
27. *Mechanical Springs: Their Engineering and Design*, 1st ed., William D. Gibson Co., Division of Associated Spring Corp., 1944.
28. Schwabenlender, C. W., "Torsion of Members of Open Cross Section," masters thesis, University of Wisconsin, 1965.
29. Chu, Kuang-Han, and A. Longinow, "Torsion in Sections with Open and Closed Parts," *Proc. Am. Soc. Civil Eng., J. Struct. Div.*, 93(6): 1967.
30. Vlasov, V. Z., *Thin-Walled Elastic Beams*, Clearing House for Federal Scientific and Technical Information, U.S. Department of Commerce, 1961.
31. Kollbrunner, C. F., and K. Basler, *Torsion in Structures*, Springer-Verlag, 1969.

CHAPTER 11

Flat Plates

11.1 COMMON CASE

The formulas of this section are based on the following assumptions: (1) The plate is flat, of uniform thickness, and of homogeneous isotropic material; (2) the thickness is not more than about one-quarter of the least transverse dimension, and the maximum deflection is not more than about one-half the thickness; (3) all forces—loads and reactions—are normal to the plane of the plate; and (4) the plate is nowhere stressed beyond the elastic limit. For convenience in discussion, it will be assumed further that the plane of the plate is horizontal.

Behavior

The plate deflects. The middle surface (halfway between top and bottom surfaces) remains unstressed; at other points there are biaxial stresses in the plane of the plate. Straight lines in the plate that were originally vertical remain straight but become inclined; therefore, the intensity of either principal stress at points on any such line is proportional to the distance from the middle surface, and the maximum stresses occur at the outer surfaces of the plate.

Formulas

Unless otherwise indicated, the formulas given in Tables 11.2* to 11.4 are based on very closely approximate mathematical analysis and may be accepted as sufficiently accurate so long as the assumptions stated hold true. Certain additional facts of importance in relation to these formulas are as follows.

Concentrated Loading

It will be noted that all formulas for maximum stress due to a load applied over a small area give very high values when the radius of the loaded area approaches zero. Analysis by a more precise method (Ref. 12) shows that the actual maximum stress produced by a load concentrated on a very small area of radius r_o can be found by replacing the actual r_o by a so-called *equivalent radius* r'_o, which depends largely upon the thickness of the plate t and to a lesser degree on its least transverse dimension. Holl (Ref. 13) shows how r'_o varies with the width of a flat plate. Westergaard (Ref. 14) gives an approximate expression for this equivalent radius:

$$r'_o = \sqrt{1.6r_o^2 + t^2} - 0.675t \qquad (11.1\text{-}1)$$

**Note:* Table 11.1 contains numerical values for functions used in Table 11.2.

This formula, which applies to a plate of any form, may be used for all values of r_o less than $0.5t$; for larger values, the actual r_o may be used.

Use of the equivalent radius makes possible the calculation of the finite maximum stresses produced by a (nominal) point loading whereas the ordinary formula would indicate that these stresses were infinite.

Edge Conditions

The formulas of Tables 11.2–11.4 are given for various combinations of edge support: free, guided (zero slope but free to move vertically), and simply supported or fixed. No exact edge condition is likely to be realized in ordinary construction, and a condition of true edge fixity is especially difficult to obtain. Even a small horizontal force at the line of contact may appreciably reduce the stress and deflection in a simply supported plate; however, a very slight yielding at nominally fixed edges will greatly relieve the stresses there while increasing the deflection and center stresses. For this reason, it is usually advisable to design a fixed-edged plate that is to carry uniform load for somewhat higher center stresses than are indicated by theory.

11.2 BENDING OF UNIFORM-THICKNESS PLATES WITH CIRCULAR BOUNDARIES

In Table 11.2, cases 1–5 consider annular and solid circular plates of constant thickness under *axisymmetric* loading for several combinations of boundary conditions. In addition to the formulas, tabulated values of deformation and moment coefficients are given for many common loading cases. The remaining cases include concentrated loading and plates with some circular and straight boundaries. Only the deflections due to bending strains are included; in Sec. 11.3, the additional deflections due to shear strains are considered.

Formulas

For cases 1–15 (Table 11.2), expressions are given for deformations and reactions at the edges of the plates as well as general equations which allow the evaluation of *deflections, slopes, moments*, and *shears* at any point in the plate. The several axisymmetric loadings include uniform, uniformly increasing, and parabolically increasing normal pressure over a portion of the plate. This permits the approximation of any reasonable axisymmetric distributed loading by fitting an approximate second-order curve to the variation in loading and solving the problem by superposition. (See the Examples at the end of this section.)

In addition to the usual loadings, Table 11.2 also includes loading cases that may be described best as *externally applied conditions which force a lack of flatness into the plate*. For example, in cases 6 and 14, expressions are given for a manufactured concentrated change in slope in a plate, which could also be used if a plastic hinge were to develop in a plate and the change in slope at the plastic hinge is known or assumed. Similarly, case 7 treats a plate with a small step manufactured into the otherwise flat surface and gives the reactions which develop when this plate is forced to conform to the specified boundary conditions. These cases are also useful when considering known boundary rotations or lateral displacements. (References 46, 47, 57, and 58 present tables and graphs for many of the loadings given in these cases.)

The use of the constants C_1 to C_9 and the functions F_1 to F_9, L_1 to L_{19}, and G_1 to G_{19} in Table 11.2 appears to be a formidable task at first. However, when we consider the large number of cases it is possible to present in a limited space, the reason for this method of presentation becomes clear. With careful inspection, we find that the constants and functions with *like subscripts* are the same except for the change in variable. We also note the use of the *singularity function* $\langle r - r_o \rangle^0$, which is given a value of 0 for $r < r_o$ and a value of 1 for $r > r_o$. In Table 11.1, values are listed for all the preceding functions for several values of the variables b/r, b/a, r_o/a, and r_o/r; also listed are five of the most used denominators for the several values of b/a. (Note that these values are for $\nu = 0.30$.)

EXAMPLE 1 A solid circular steel plate, 0.2 in thick and 20 in in diameter, is simply supported along the edge and loaded with a uniformly distributed load of 3 lb/in². It is required to determine the center deflection, the maximum stress, and the deflection equation. Given: $E = 30(10^6)$ lb/in² and $\nu = 0.285$.

SOLUTION This plate and loading are covered in Table 11.2, case 10a. The following constants are obtained:

$$D = \frac{30(10^6)(0.2^3)}{12(1-0.285^2)} = 21{,}800, \quad q = 3, \quad a = 10, \quad r_o = 0$$

Since $r_o = 0$,

$$y_c = \frac{-qa^4}{64D}\frac{5+\nu}{1+\nu} = \frac{-3(10^4)(5.285)}{64(21{,}800)(1.285)} = -0.0833 \text{ in}$$

and
$$M_{\max} = M_c = \frac{qa^2}{16}(3+\nu) = \frac{3(10^2)(3.285)}{16} = 61.5 \text{ lb-in/in}$$

Therefore,
$$\sigma_{\max} = \frac{6M_c}{t^2} = \frac{6(61.5)}{0.2^2} = 9240 \text{ lb/in}^2$$

The general deflection equation for these several cases is

$$y = y_c + \frac{M_c r^2}{2D(1+\nu)} + LT_y$$

where for this case $LT_y = (-qr^4/D)G_{11}$. For $r_o = 0$, $G_{11} = \frac{1}{64}$ (note that $r > r_o$ everywhere in the plate, so that $\langle r - r_o \rangle^0 = 1$); therefore,

$$y = -0.0833 + \frac{61.5r^2}{2(21{,}800)(1.285)} - \frac{3r^4}{21{,}800(64)}$$
$$= -0.0883 + 0.001098r^2 - 0.00000215r^4$$

As a check, the deflection at the outer edge can be evaluated as

$$y_a = -0.0883 + 0.001098(10^2) - 0.00000215(10^4)$$
$$= -0.0883 + 0.1098 - 0.0215 = 0$$

EXAMPLE 2 An annular aluminum plate with an outer radius of 20 in and an inner radius of 5 in is to be loaded with an annular line load of 40 lb/in at a radius of 10 in. Both the inner and outer edges are simply supported, and it is required to determine the maximum deflection and maximum stress as a function of the plate thickness. Given: $E = 10(10^6)$ lb/in² and $\nu = 0.30$.

SOLUTION The solution to this loading and support condition is found in Table 11.2, case 1c, where $b/a = 0.25$, $r_o/a = 0.50$, $a = 20$ in, and $w = 40$ lb/in. No numerical solutions are presented for this combination of b/a and r_o/a, and so either the equations for C_1, C_3, C_7, C_9, L_3, and L_9 must be evaluated or values for these coefficients must be found in Table 11.1. Since the values of C are found for the variable b/a, from Table 11.1, under the column headed 0.250, the following coefficients are determined:

$$C_1 = 0.881523, \quad C_3 = 0.033465, \quad C_7 = 1.70625$$

$$C_9 = 0.266288, \quad C_1C_9 - C_3C_7 = 0.177640$$

The values of L are found for the variable r_o/a, and so from Table 11.1, under the column headed 0.500, the following coefficients are determined:

$$L_3 = 0.014554 \quad \text{and} \quad L_9 = 0.290898$$

Whether the numbers in Table 11.1 can be interpolated and used successfully depends upon the individual problem. In some instances, where lesser degrees of accuracy are required, interpolation can be used; in other instances, requiring greater degrees of accuracy, it would be better to solve the problem for values of b and r_o that do fit Table 11.1 and then interpolate between the values of the final deflections or stresses.

Using the preceding coefficients, the reaction force and slope can be determined at the inside edge and the deflection equation developed (note that $y_b = 0$ and $M_{rb} = 0$):

$$\theta_b = \frac{-wa^2}{D}\frac{C_3L_9 - C_9L_3}{C_1C_9 - C_3C_7} = \frac{-40(20)^2}{D}\frac{0.033465(0.290898) - 0.266288(0.014554)}{0.177640}$$

$$= \frac{-527.8}{D}\text{rad}$$

$$Q_b = w\frac{C_1L_9 - C_7L_3}{C_1C_9 - C_3C_7} = 40\frac{0.881523(0.290898) - 1.70625(0.014554)}{0.177640}$$

$$= 52.15 \text{ lb/in}$$

Therefore,
$$y = 0 - \frac{527.8r}{D}F_1 + 0 + \frac{52.15r^3}{D}F_3 - \frac{40r^3}{D}G_3$$

Substituting the appropriate expressions for F_1, F_3, and G_3 would produce an equation for y as a function of r, but a reduction of this equation to simple form and an evaluation to determine the location and magnitude of maximum deflection would be extremely time-consuming. Table 11.1 can be used again to good advantage to evalute y at specific values of r, and an excellent approximation to the maximum deflection can be obtained.

b/r	r	F_1	$-527.8rF_1$	F_3	$52.15r^3F_3$	r_o/r	G_3	$-40r^3G_3$	$y(D)$
1.00	5.000	0.000	0.0	0.000	0.0		0.000	0.0	0.0
0.90	5.555	0.09858	−289.0	0.000158	1.4		0.000	0.0	−287.6
0.80	6.250	0.194785	−642.5	0.001191	15.2		0.000	0.0	−627.3
0.70	7.143	0.289787	−1092.0	0.003753	71.3		0.000	0.0	−1020.7
0.60	8.333	0.385889	−1697.1	0.008208	247.7		0.000	0.0	−1449.4
0.50	10.000	0.487773	−2574.2	0.014554	759.0	1.00	0.000	0.0	−1815.2
0.40	12.500	0.605736	−3996.0	0.022290	2270.4	0.80	0.001191	−93.0	−1818.6
0.33	15.000	0.704699	−5578.6	0.027649	4866.4	0.67	0.005019	−677.6	−1389.8
0.30	16.667	0.765608	−6734.2	0.030175	7285.4	0.60	0.008208	−1520.0	−968.8
0.25	20.000	0.881523	−9304.5	0.033465	13961.7	0.50	0.014554	−4657.3	−0.1

An examination of the last column on the right shows the deflection at the outer edge to be approximately zero and indicates that the maximum deflection is located at a radius near 11.25 in and has a value of approximately

$$\frac{-1900}{D}=\frac{-1900(12)(1-0.3^2)}{10(10^6)t^3}=\frac{-0.00207}{t^3}\text{in}$$

The maximum bending moment will be either a tangential moment at the inside edge or a radial moment at the load line:

$$M_{tb}=\frac{\theta_b D(1-\nu^2)}{b}=\frac{-527.8(1-0.3^2)}{5}=-96.2\text{ lb-in/in}$$

$$M_{r(r_o)}=\theta_b\frac{D}{r}F_{7(r_o)}+Q_b r F_{9(r_o)}$$

Where at $r=r_o$, $b/r=0.5$. Therefore,

$$F_{7(r_o)}=0.6825$$

$$F_{9(r_o)}=0.290898$$

$$M_{r(r_o)}=\frac{-527.8}{10}(0.6825)+52.15(10)(0.290898)$$

$$=-36.05+151.5=115.45\text{ lb-in/in}$$

The maximum bending stress in the plate is

$$\sigma=\frac{6(115.45)}{t^2}=\frac{693}{t^2}\text{ lb/in}^2$$

EXAMPLE 3 A flat phosphor bronze disk with thickness of 0.020 in and a diameter of 4 in is upset locally in a die to produce an abrupt change in slope in the radial direction of 0.05 rad at a radius of 3/4 in. It is then clamped between two flat dies as shown in Fig. 11.1. It is required to determine the maximum bending stress due to the clamping. Given: $E=16(10^6)\text{ lb/in}^2$ and $\nu=0.30$.

SOLUTION This example of forcing a known change in slope into a plate clamped at both inner and outer edges is covered in Table 11.2, case 6h, where $\theta_o=0.05$, $b/a=0.10$, and $r_o/a=0.50$. These dimensions were chosen to fit the tabulated data for a case where $\nu=0.30$. For this case $M_{rb}=-2.054(0.05)(11.72)/1.5=-0.803$ lb-in/in, $Q_b=-0.0915(0.05)(11.72)/1.5^2=-0.0238$ lb/in, $y_b=0$, and $\theta_b=0$. The expression for M_r then becomes

$$M_r=-0.803F_8-0.0238rF_9+\frac{0.05(11.72)}{r}G_7$$

An examination of the numerical values of F_8 and F_9 shows that F_8 decreases slightly less than F_9 increases as r increases, but the larger coefficient of the first term indicates that M_{rb} is indeed the maximum moment. The maximum stress is therefore $\sigma=0.803(6)/0.02^2=12{,}050\text{ lb/in}^2$ in tension on the top surface at the inner edge. The maximum deflection is at $r_o=0.75$ in and equals $-0.1071(0.05)(1.5)=-0.00803$ in.

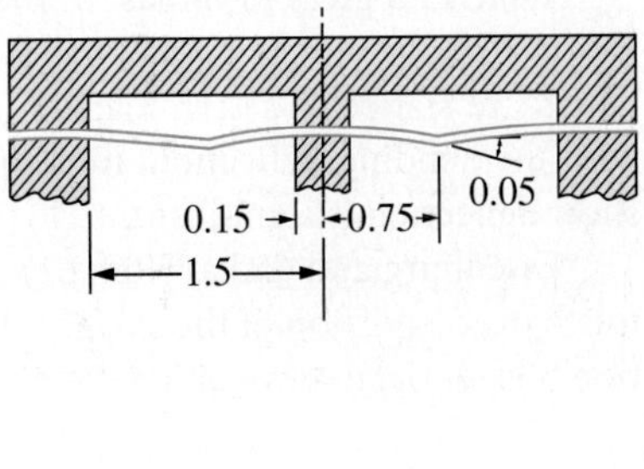

Figure 11.1

EXAMPLE 4 A circular steel plate 2 in thick and 20 ft in diameter is simply supported at the outer edge and supported on a center support which can be considered to provide uniform pressure over a diameter of 1.8 in. The plate is loaded in an axisymmetric manner with a distributed load which increases linearly with radius from a value of 0 at $r = 4$ ft to a value of 2000 lb/ft^2 at the outer edge. Determine the maximum bending stress. Given: $E = 30(10^6)$ lb/in^2 and $\nu = 0.30$.

SOLUTION Table 11.2, case 11a, deals with this loading and a simply supported outer edge. For this problem $q = 2000/144 = 13.9$ lb/in^2, $a = 120$ in, and $r_o = 48$ in, and so $r_o/a = 0.4$. From the tabulated data for these quantities, $K_{y_c} = -0.01646, K_{\theta_a} = 0.02788$, and $K_{M_c} = 0.04494$. Therefore,

$$y_c = \frac{-0.01646(13.9)(120^4)}{D} = \frac{-0.475(10^8)}{D} \text{ in}$$

$$M_c = 0.04494\,(13.9)(120^2) = 9000 \text{ lb-in/in}$$

Case 16 (Table 11.2) considers the center load over a small circular area. It is desired to determine W such that the max $y = 0.475(10^8)/D$. Therefore,

$$-\frac{W120^2}{16\pi D}\frac{3+0.3}{1+0.3} = \frac{0.475(10^8)}{D}$$

which gives $W = -65{,}000$ lb. The maximum moment is at the center of the plate where

$$M_r = \frac{W}{4\pi}\left[(1+\nu)\ln\frac{a}{b}+1\right]$$

The equivalent radius r_o' is given by [Eq. (11.1-1)]

$$r_o' = \sqrt{1.6r_o^2 + t^2} - 0.675t = \sqrt{1.6(0.9^2)+2^2} - 0.675\,(2) = 0.95 \text{ in}$$

Therefore, $$M_{\max} = \frac{-65{,}000}{4\pi}\left(1.3\ln\frac{120}{0.95}+1\right) = -37{,}500 \text{ lb-in/in}$$

The maximum stress is at the center of the plate where

$$\sigma = \frac{6M}{t^2} = \frac{6(-37{,}500+9000)}{2^2} = -43.200 \text{ lb/in}^2 \quad \text{(tension on the top surface)}$$

11.3 CIRCULAR-PLATE DEFLECTION DUE TO SHEAR

The formulas for deflection given in Table 11.2 take into account bending stresses only; there is, in every case, some additional deflection due to shear. Usually this is so slight as to be negligible, but in circular pierced plates with large openings the deflection due to shear may constitute a considerable portion of the total deflection. Wahl (Ref. 19) suggests that this is the case when the thickness is greater than one-third the difference in inner and outer diameters for plates with simply supported edges, or greater than one-sixth this difference for plates with one or both edges fixed.

Table 11.3 gives formulas for the additional deflection due to shear in which the form factor F has been taken equal to 1.2, as in Sec. 8.10. All the cases listed have shear forces which are statically determinate. For the *indeterminate* cases, the shear deflection, along with the bending deflection, must be considered in the determination of the reactions if shear deflection is significant.

Essenburg and Gulati (Ref. 61) discuss the problem in which two plates when loaded touch over a portion of the surface. They indicate that the consideration of shear deformation is essential in developing the necessary expressions. Two examples are worked out.

EXAMPLE An annular plate with an inner radius of 1.4 in, an outer radius of 2 in, and a thickness of 0.50 in is simply supported at the inner edge and loaded with an annular line load of 800 lb/in at a radius of 1.8 in. The deflection of the free outer edge is desired. Given: $E = 18(10^6)$ lb/in^2 and $\nu = 0.30$.

SOLUTION To evaluate the deflection due to bending one can refer to Table 11.2, case 1k. Since $b/a = 0.7$, in Table 11.1, under the column headed 0.700, we obtain the following constants

$$C_1 = 0.2898,\quad C_3 = 0.003753,\quad C_7 = 0.3315,\quad C_9 = 0.2248$$

Similarly, $r_o/a = 0.9$, and again in Table 11.1, under the column headed 0.900, we obtain the additional constants $L_3 = 0.0001581$ and $L_9 = 0.09156$.

The plate constant $D = Et^3/12(1-\nu^2) = 18(10^6)(0.5)^3/12(1-0.3^2) = 206{,}000$ lb-in. and the shear modulus $G = E/2(1+\nu) = 18(10^6)/2(1+0.3) = 6.92(10^6)$ lb/in^2. The bending deflection of the outer edge is given by

$$y_a = \frac{-wa^3}{D}\left[\frac{C_1}{C_7}\left(\frac{r_o C_9}{b} - L_9\right) - \frac{r_o C_3}{b} + L_3\right]$$

$$= \frac{-800(2)^3}{206{,}000}\left\{\frac{0.2898}{0.3315}\left[\frac{1.8(0.2248)}{1.4} - 0.09156\right] - \frac{1.8(0.003753)}{1.4} + 0.0001581\right\}$$

$$= \frac{-800(2)^3}{206{,}000}(0.16774) = -0.00521 \text{ in}$$

For the deflection due to shear we refer to Table 11.3, case 1k, and obtain

$$y_{sa} = \frac{-wa}{tG}\left(1.2\frac{r_o}{a}\ln\frac{r_o}{b}\right) = \frac{800(2)}{0.5(6.92)(10^6)}\left[1.2(0.9)\ln\frac{1.8}{1.4}\right] = -0.000125 \text{ in}$$

Thus, the total deflection of the outer edge is $-0.00521 - 0.000125 = -0.00534$ in. Note that the thickness 0.50 is somewhat more than one-third the difference in inner and outer diameters 1.2, and the shear deflection is only 2.4% of the bending deflection.

11.4 BIMETALLIC PLATES

A very wide beam of rectangular cross section can be treated as a beam if E is replaced by $E/(1-\nu^2)$ and I by $t^3/12$ (see Sec. 8.11). It can also be treated as a plate with two opposite edges free as shown in Figs. 8.16 and 11.2. For details see Ref. 88.

To use the beam equations in Tables 8.1, 8.5, 8.6, 8.8, and 8.9 for plates like that shown in Fig. 11.2 with two opposite edges free, the loadings must be uniformly distributed across the plate parallel to side b as shown. At every position in such a plate, except close to the free edges a, there will be bending moments $M_z = \nu M_x$. If the plate is isotropic and homogeneous, and in the absence of any in-plane loading, there will be no change in

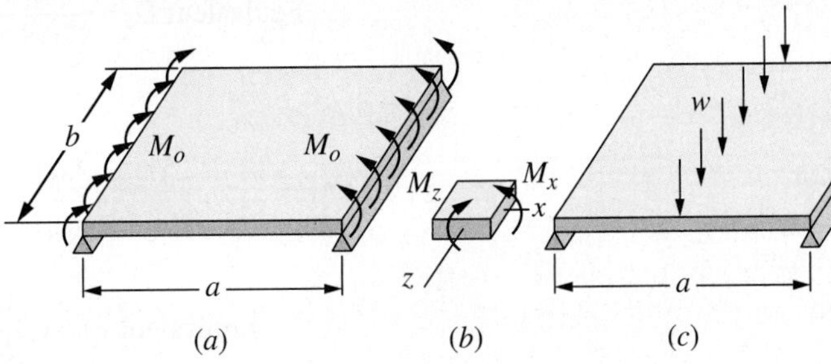

Figure 11.2

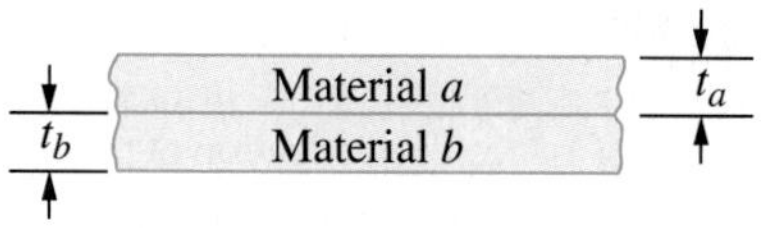

Figure 11.3

length of any line parallel to side b. The response of a *bimetallic plate* differs from that of the homogeneous plate in one important respect. If the values of Poisson's ratio differ for the two materials, there will be a change in length of those lines parallel to side b due to an in-plane strain ε_z developed from the loadings shown in Fig. 11.2. Using the notations from Figs. 11.2 and 11.3 and from the expression for K_{2p} on the next page, Eq. (11.3-2),

$$\varepsilon_z = \frac{6M_x(1-\nu_a^2)}{E_b t_b^2 K_{2p}} \frac{(t_b/t_a)(1+t_b/t_a)(\nu_a-\nu_b)}{(1+E_a t_a/E_b t_b)^2-(\nu_a+\nu_b E_a t_a/E_b t_b)^2}$$

For the moment loading M_o in Fig. 11.2(a), the value of ε_z, will be everywhere the same and the plate will merely expand or contract in the z direction. For the line loading shown in Fig. 11.2(c), however, the unit strains ε_z, will differ from place to place depending upon the value M_x, and consequently in-plane stresses will be developed. For more general analyses of this type of problem see Refs. 89 and 90.

Bimetallic Circular Plates

Applying this same reasoning to a bimetallic circular plate leads to the following conclusions:

1. If the Poisson's ratios for the two materials are equal, any of the cases in Table 11.2 can be used if the following equivalent value of the plate stiffness constant D_e is substituted for D.
2. If the Poisson's ratios differ by a significant amount, the equivalent values of D_e and ν_e may be used for any combination of loading and edge restraints which deform the plate into a spherical surface providing the edge restraints do not prevent motion parallel to the surface of the plate. This restriction assures that bending moments are constant in magnitude at all locations in the plate and in all directions. Thus one can use cases 8a, 8f, 8h, and 15 with either a uniform temperature rise or a temperature variation through the thickness which is the same everywhere in the plate. Obviously one needs also an equivalent temperature coefficient of expansion or an equivalent loading expression for each such temperature loading as well as the equivalent material constants D_e and ν_e.

$$\text{Equivalent } D_e = \frac{E_a t_a^3}{12(1-\nu_a^2)} K_{2p} \tag{11.3-1}$$

where

$$K_{2p} = 1 + \frac{E_b t_b^3(1-\nu_a^2)}{E_a t_a^3(1-\nu_b^2)} + \frac{3(1-\nu_a^2)(1+t_b/t_a)^2(1+E_a t_a/E_b t_b)}{(1+E_a t_a/E_b t_b)^2-(\nu_a+\nu_b E_a t_a/E_b t_b)^2} \tag{11.3-2}$$

$$\text{Equivalent } \nu_e = \nu_a \frac{K_{3p}}{K_{2p}} \tag{11.3-3}$$

where

$$K_{3p}=1+\frac{\nu_b E_b t_b^3(1-\nu_a^2)}{\nu_a E_a t_a^3(1-\nu_b^2)}+\frac{3(1-\nu_a^2)(1+t_b/t_a)^2(1+\nu_b E_a t_a/\nu_a E_b t_b)}{(1+E_a t_a/E_b t_b)^2-(\nu_a+\nu_b E_a t_a/E_b t_b)^2} \tag{11.3-4}$$

A bimetallic plate deforms laterally into a spherical surface when its uniform temperature differs from T_o, the temperature at which the plate is flat. Cases 8 and 15 (Table 11.2) can be used to solve for reaction moments and forces as well as the deformations of a bimetallic plate subjected to a uniform temperature T provided that any guided and/or fixed edges are not capable of developing in-plane resisting forces but instead allow the plate to expand or contract in its plane as necessary. To use these cases we need only to replace the term $\gamma(1+\nu)\Delta T/t$ by an equivalent expression

$$\left[\frac{\gamma(1+\nu)\Delta T}{t}\right]_e=\frac{6(\gamma_b-\gamma_a)(T-T_o)(t_a+t_b)(1+\nu_e)}{t_b^2 K_{1p}} \tag{11.3-5}$$

where

$$K_{1p}=4+6\frac{t_a}{t_b}+4\left(\frac{t_a}{t_b}\right)^2+\frac{E_a t_a^3(1-\nu_b)}{E_b t_b^3(1-\nu_a)}+\frac{E_b t_b(1-\nu_a)}{E_a t_a(1-\nu_b)} \tag{11.3-6}$$

and replace D by the equivalent stiffness D_e given previously.

After the moments and deformations have been determined, the flexural stresses can be evaluated. The stresses due to the bending moments caused by restraints and any applied loads are given by the following expressions. In the top surface of material a, in the direction of any moment M

$$\sigma=\frac{-6M}{t_a^2 K_{2p}}\left[1+\frac{(1-\nu_a^2)(1+t_b/t_a)(1+E_a t_a/E_b t_b)}{(1+E_a t_a/E_b t_b)^2-(\nu_a+\nu_b E_a t_a/E_b t_b)^2}\right] \tag{11.3-7}$$

In the bottom surface of material b,

$$\sigma=\frac{6M}{t_a^2 K_{2p}}\left[\frac{E_b t_b(1-\nu_a^2)}{E_a t_a(1-\nu_b^2)}+\frac{t_a}{t_b}\frac{(1-\nu_a^2)(1+t_b/t_a)(1+E_a t_a/E_b t_b)}{(1+E_a t_a/E_b t_b)^2-(\nu_a+\nu_b E_a t_a/E_b t_b)^2}\right] \tag{11.3-8}$$

Even when no restraints are imposed, the distortion of a bimetallic plate due to a temperature change is accompanied by flexural stresses in the two materials. This differs from the plate made of a single material, which deforms free of stress when subjected to a linear temperature variation through the thickness when there are no restraints. Therefore, the following stresses must be added algebraically to the preceding stresses due to bending moments, if any. In the top surface of material a, in all directions

$$\sigma=\frac{-(\gamma_b-\gamma_a)(T-T_o)E_a}{(1-\nu_a)K_{1p}}\left[3\frac{t_a}{t_b}+2\left(\frac{t_a}{t_b}\right)^2-\frac{E_b t_b(1-\nu_a)}{E_a t_a(1-\nu_b)}\right] \tag{11.3-9}$$

In the bottom surface of material b,

$$\sigma=\frac{(\gamma_b-\gamma_a)(T-T_o)E_b}{(1-\nu_b)K_{1p}}\left[3\frac{t_a}{t_b}+2-\frac{E_a t_a^3(1-\nu_b)}{E_b t_b^3(1-\nu_a)}\right] \tag{11.3-10}$$

EXAMPLE An annular bimetallic plate has a 3-in outer diameter and a 2.1-in inner diameter; the top portion is 0.020-in-thick stainless steel, and the bottom is 0.030-in-thick titanium (see Fig. 11.4). For the stainless steel $E = 28(10^6)$ lb/in^2, $\nu = 0.3$, and $\gamma = 9.6(10^{-6})$ in/in/°F; for the titanium $E = 17(10^6)$ lb/in^2, $\nu = 0.3$, and $\gamma = 5.7\,(10^{-6})$ in/in/°F. The outer edge is simply supported, and the inner edge is elastically supported by a spring which develops 500 lb of load for each inch of deflection. It is necessary to determine the center deflection and the maximum stress for a temperature rise of 50°F.

SOLUTION First evaluate the constants K_{1p}, K_{2p}, and K_{3p}, the equivalent stiffness D_e, and the equivalent Poisson's ratio ν_e. From Eq. (11.3-6),

$$K_{1p} = 4 + 6\frac{0.02}{0.03} + 4\left(\frac{2}{3}\right)^2 + \frac{28}{17}\left(\frac{2}{3}\right)^3\left(\frac{1-0.3}{1-0.3}\right) + \frac{17}{28}\left(\frac{3}{2}\right)\left(\frac{1-0.3}{1-0.3}\right)$$
$$= 11.117$$

Since $\nu_a = \nu_b$ for this example, $K_{3p} = k_{2p} = 11.986$ and the equivalent Poisson's ratio $\nu_e = 0.3$. From Eq. (11.3-1),

$$D_e = \frac{28(10^6)(0.02^3)}{12(1-0.3^2)}(11.986) = 246 \text{ lb-in}$$

Table 11.2, case 8a, treats an annular plate with the inner edge free and the outer edge simply supported. As in Eq. (11.3-5), the term $\gamma\Delta T/t$ must be replaced by

$$\frac{6(\gamma_b - \gamma_a)(T - T_0)(t_a + t_b)}{t_b^2 K_{1p}} = \frac{6(5.7 - 9.6)(10^{-6})(50)(0.02 + 0.03)}{(0.03^2)(11.177)} = -0.00582$$

Since $b/a = 1.05/1.5 = 0.7$ and $\nu_e = 0.3$, the tabulated data can be used, and $K_{yb} = -0.255$ and $K_{\theta b} = 0.700$. Therefore, $y_b = -0.255\,(-0.00582)\,(1.5^2) = 0.00334$ in, and $\theta_b = 0.7(-0.00582)\,(1.5) = -0.0061$ rad. There are no moments or edge loads in the plate, and so $M_{rb} = 0$ and $Q_b = 0$. Case 1a treats an annular plate with an annular line load. For $r_o = b$ and $b/a = 0.7$, $K_{yb} = -0.1927$ and $K_{\theta b} = 0.6780$. Therefore, $y_b = -0.1927w(1.5^3)/246 = -0.002645w$, $\theta_b = -0.678w(1.5^2)/246 = 0.0062w$ rad, $M_{rb} = 0$, and $Q_b = 0$.

Equating the deflection of the inner edge of the plate to the deflection of the elastic support gives $y_b = 0.00334 - 0.002645w = 2\pi(1.05)w/500 = 0.0132w$. Solving for w, we obtain $w = 0.211$ lb/in for a total center load of 1.39 lb. The deflection of the inner edge is $y_b = 0.0132(0.211) = 0.00279$ in. The maximum moment developed in the plate is the tangential moment at the inner edge.

$M_{tb} = 0.8814\,(0.211)\,(1.5) = 0.279$ lb-in. The stresses can now be computed. On the top surface of the stainless steel combining Eqs. (11.3-7) and (11.3-9) yields

$$\sigma = \frac{-6(0.279)}{0.02^2(11.986)}\left\{1 + \frac{(1-0.3^2)(1+3/2)[1+28(2)/17(3)]}{[1+28(2)/17(3)]^2 - [0.3+0.3\,(28)(2)/17(3)]^2}\right\}$$
$$-\frac{(5.7-9.6)(10^{-6})(50)(28)(10^6)}{(1-0.3)(11.177)}\left[3\left(\frac{2}{3}\right) + 2\left(\frac{2}{3}\right)^2 - \frac{17}{28}\left(\frac{3}{2}\right)\right]$$
$$= -765 + 1381 = 616 \text{ lb/in}^2$$

Similarly, on the bottom surface of the titanium, Eqs. (11.3-8) and (11.3-10) give

$$\sigma = 595 - 1488 = -893 \text{ lb/in}^2$$

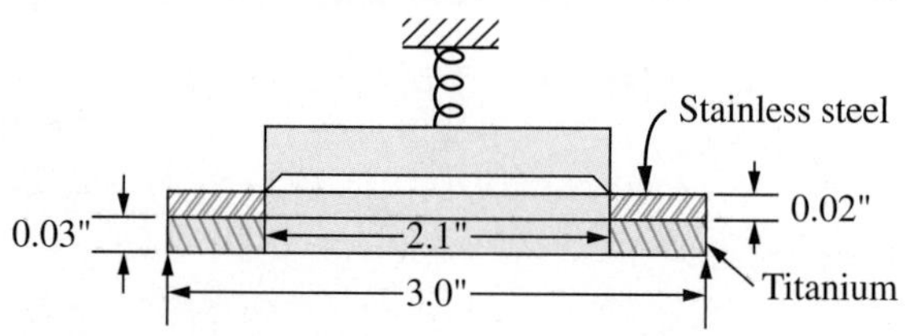

Figure 11.4

11.5 NONUNIFORM LOADING OF CIRCULAR PLATES

The case of a circular plate under a nonuniformly distributed loading symmetrical about the center can be solved by treating the load as a series of elementary ring loadings and summing the stresses and deflections produced by such loadings. The number of ring loadings into which the actual load should be resolved depends upon the rate at which the distributed load varies along the radius and the accuracy desired. In general, a division of the load into rings each having a width equal to one-fifth the loaded length of the radius should be sufficient.

If the nonuniformly distributed loading can be reasonably approximated by a second-order curve, the loadings in Table 11.2, cases 2–4, can be superimposed in the necessary proportions. (This technique is illustrated in Sec. 11.6.) Heap (Ref. 48) gives tabular data for circular plates loaded with a lateral pressure varying inversely with the square of the radius.

Concentrated Loads

In Refs. 60 and 75–79 similar numerical techniques are discussed for concentrated loads on either of two concentric annular plates in combination with edge beams in some cases. The numerical data presented are limited but are enough to enable the reader to approximate many other cases.

11.6 CIRCULAR PLATES ON ELASTIC FOUNDATIONS

Discussions of the theory of bending of circular plates on elastic foundations can be found in Refs. 21 and 46, and in Ref. 41 of Chap.8. The complexity of these solutions prohibits their inclusion in this handbook, but a simple *iterative* approach to this problem is possible.

The procedure consists in evaluating the deflection of the loaded plate without the elastic foundation and then superimposing a given fraction of the foundation reaction resulting from this deflection until finally the given fraction increases to 1 and the assumed and calculated foundation reactions are equal.

EXAMPLE Given the same problem stated in Example 1 of Sec. 11.2, but in addition to the simply supported edge, an elastic foundation with a modulus of 20 lb/in²/in is present under the entire plate.

SOLUTION An examination of the deflection equation resulting from the uniform load shows that the term involving r^4 is significant only near the outer edge where the effect of foundation pressure would not be very large. We must also account for the fact that the foundation reactions will reduce the plate deflections or the procedure described may not converge. Therefore, for a first trial let us assume that the foundation pressure is given by

$$q_f = 20(-0.0883+0.001098r^2)(0.50) = -0.883+0.01098r^2$$

The total loading on the plate then consists of a uniform load of $3-0.883 = 2.117$ lb/in² and a parabolically increasing load of 1.098 lb/in² maximum value. From Table 11.2, case 10a,

$$y_c = \frac{-qa^4(5+\nu)}{64D(1+\nu)} = \frac{-2.117(10^4)(5.285)}{64(21{,}800)(1.285)} = -0.063 \text{ in}$$

$$M_c = \frac{qa^2}{16}(3+\nu) = \frac{2.117(10^2)(3.285)}{16} = 43.5 \text{ lb-in/in}$$

$$LT_y = \frac{-qr^4}{D}G_{11} = \frac{-2.117r^4}{21{,}800}\frac{1}{64} = -1.517(10^{-6})r^4$$

From Table 11.2, case 12a,

$$y_c = \frac{-qa^4(7+\nu)}{288D(1+\nu)} = \frac{-1.098(10^4)(7.285)}{288(21{,}800)(1.285)} = -0.00992 \text{ in}$$

$$M_c = \frac{qa^2(5+\nu)}{96} = \frac{1.098(10^2)(5.285)}{96} = 6.05 \text{ lb-in/in}$$

$$LT_y = \frac{-qr^6}{Da^2}G_{13} = \frac{-1.098r^6}{21{,}800(10^2)}\frac{25}{14{,}400} = -8.75(10^{-10})r^6$$

Using these values, the deflection equation can be written

$$y = -0.0623 - 0.00992 + \frac{(43.5+6.05)r^2}{2(21{,}800)1.285} - 1.517(10^{-6})r^4 - 8.75(10^{-10})r^6$$
$$= -0.0722 + 0.000885r^2 - 1.517(10^{-6})r^4 - 8.75(10^{-10})r^6$$

This deflection would create a foundation reaction

$$q_f = 20(-0.0722 + 0.000885r^2) = -1.445 + 0.0177r^2$$

if the higher-order terms were neglected. Again applying a 50% factor to the difference between the assumed and calculated foundation pressure gives an improved loading from the foundation

$$q_f = -1.164 + 0.01434r^2$$

Repeating the previous steps again, we obtain

$$y_c = -0.0623\frac{3-1.164}{2.117} - 0.00992\frac{0.01434}{0.01098} = -0.0671 \text{ in}$$

$$M_c = 43.5\frac{3-1.164}{2.117} + 6.05\frac{0.01434}{0.01098} = 45.61 \text{ lb-in/in}$$

$$y = -0.0671 + 0.000813r^2$$

$$q_f = -1.342 + 0.01626r^2$$

Successive repetitions of the previous steps give improved values for q_f:

$$q_f = -1.306 + 0.1584r^2, \quad q_f = -1.296 + 0.1566r^2, \quad q_f = -1.290 + 0.1566r^2$$

Using values from the last iteration, the final answers are

$$y_c = -0.0645 \text{ in}, \quad M_c = 43.8 \text{ lb-in/in}, \quad \text{and } \sigma_{max} = 6580 \text{ psi}$$

An exact analysis using expressions from Ref. 46 gives

$$y_c = -0.0637 \text{ in} \quad \text{and} \quad M_c = 43.3 \text{ lb-in/in}$$

11.7 CIRCULAR PLATES OF VARIABLE THICKNESS

For any circular plate of variable thickness, loaded symmetrically with respect to the center, the stresses and deflections can be found as follows: The plate is divided into an arbitrary number of concentric rings, each of which is assumed to have a uniform thickness equal to its mean thickness. Each such ring is loaded by radial moments M_a and M_b at its outer and inner circumferences, respectively, by vertical shears at its inner and outer circumferences, and by whatever load is distributed over its surface. The shears are known, each being equal to the total load on the plate within the corresponding circumference. The problem is to determine the edge moments, and this is done by making use of the fact that the slope of each ring at its inner circumference is equal to the slope of the next inner ring at its outer circumference. This condition, together with the known slope (or moment) at the outer edge of the plate and the known slope (or moment) at the inside edge or center of the plate, enables as many equations to be written as there are unknown quantities M. Having found all the edge moments, stresses and deflections can be calculated for each ring by the appropriate formulas of Table 11.2 and the deflections added to find the deflection of the plate.

A more direct solution (Ref. 21) is available if the plate is of such form that the variation in thickness can be expressed fairly closely by the equation $t = t_o e^{-nx^2/6}$, where t is the thickness at any point a distance r from the center, t_o is the thickness at the center, e is the base for the napierian system of logarithms (2.178), x is the ratio r/a, and n is a number chosen so as to make the equation agree with the actual variation in thickness. The constant n is positive for a plate that decreases in thickness toward the edge and negative for a plate that increases in thickness toward the edge. For a plate of uniform thickness, $n = 0$; and for a plate twice as thick at the center as at the edge, $n = +4.16$. The maximum stress and deflection for a uniformly loaded circular plate are given by $\sigma_{max} = \beta qa^2/t_0^2$ and $y_{max} = \alpha qa^4/Et_o^3$ respectively, where β and α depend on n, where $\nu = 0.3$, and for values of n from 4 to −4 can be found by interpolation from the following table:

		n								
Edge Conditions		+4	+3	+2	+1	0	−1	−2	−3	−4
Edges supported Case 10a, $r_o = 0$	β	1.63	1.55	1.45	1.39	1.24	1.16	1.04	0.945	0.855
	α	1.220	1.060	0.924	0.804	0.695	0.600	0.511	0.432	0.361
Edges fixed Case 10b, $r_o = 0$	β	2.14	1.63	1.31	0.985	0.75	0.55	0.43	0.32	0.26
	α	0.4375	0.3490	0.276	0.217	0.1707	0.1343	0.1048	0.0830	0.0653

For the loadings in the preceding table as well as for a simply supported plate with an edge moment, Ref. 46 gives graphs and tables which permit the evaluation of radial and tangential stresses throughout the plate. This same reference gives extensive tables of moment and shear coefficients for a variety of loadings and support conditions for plates in which the thickness varies as $t = t_a(r/a)^{-n/3}$, where t_a is the thickness at the outer edge: Values are tabulated for $n = 0, 1, 1.5$, and 2 and for $\nu = 1/6$.

Stresses and deflections for plates with thicknesses varying linearly with radius are tabulated in Refs. 46 and 57. Annular plates with the outer edges fixed and the inner edges guided and with thicknesses increasing linearly with the radii from zero at the center are discussed in Ref. 36 and tabulated in previous editions of this handbook. A uniformly loaded circular plate with a fixed edge and a thickness varying linearly along a diameter is discussed by Strock and Yu (Ref. 65). Conway (Ref. 66) considers the necessary proportions for a rib along the diameter of a uniformly loaded, clamped circular plate to affect a minimum weight design for a given maximum stress.

Perforated Plates

Slot and O'Donnell (Ref. 62) present the relationship between the effective elastic constants for *thick perforated plates* under bending and *thin perforated plates* under in-plane loading. Numerical results are presented in the form of tables and graphs, and many references are listed.

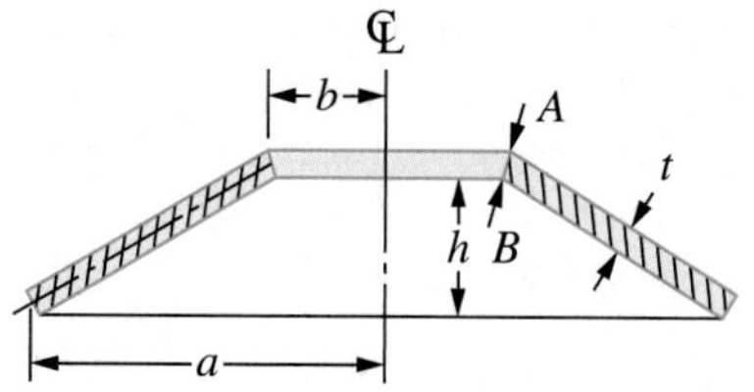

Figure 11.5

11.8 DISK SPRINGS

The *conical disk*, or *Belleville spring* (Fig. 11.5), is not a flat plate, of course, but it may appropriately be considered in this chapter because it bears a superficial resemblance to a flat ring and is sometimes erroneously analyzed by the formulas for case 1a. The stress and deflection produced in a spring of this type are not proportional to the applied load because the change in form consequent upon deflection markedly changes the load-deflection and load-stress relationships. This is indeed the peculiar advantage of this form of spring because it makes it possible to secure almost any desired variation of "spring rate" and also possible to obtain a considerable range of deflection under almost constant load. The largest stresses occur at the inner edge.

Formulas for deflection and stress at points A and B are (Ref. 27)

$$P = \frac{E\delta}{(1-\nu^2)Ma^2}\left[(h-\delta)\left(h-\frac{\delta}{2}\right)t + t^3\right]$$

$$\sigma_A = \frac{-E\delta}{(1-\nu^2)Ma^2}\left[C_1\left(h-\frac{\delta}{2}\right)+C_2 t\right]$$

$$\sigma_B = \frac{-E\delta}{(1-\nu^2)Ma^2}\left[C_1\left(h-\frac{\delta}{2}\right)-C_2 t\right]$$

where P = total applied load; E = modulus of elasticity; δ = deflection; h = cone height of either inner or outer surface; t = thickness; a and b are the outer and inner radii of the middle surface; and M, C_1, and C_2 are constants whose values are functions of a/b and are given in the following table:

a/b	M	C_1	C_2
1.0	0		
1.2	0.31	1.02	1.05
1.4	0.46	1.07	1.14
1.6	0.57	1.14	1.23
1.8	0.64	1.18	1.30
2.0	0.70	1.23	1.39
2.2	0.73	1.27	1.46
2.6	0.76	1.35	1.60
3.0	0.78	1.43	1.74
3.4	0.80	1.50	1.88
3.8	0.80	1.57	2.00
4.2	0.80	1.64	2.14
4.6	0.80	1.71	2.26
5.0	0.79	1.77	2.38

The formulas for stress may give either positive or negative results, depending upon δ; a negative result indicates compressive stress, and a positive result a tensile stress. It is to be noted that P also may become negative.

Wempner (Refs. 67 and 68) derives more exacting expressions for the conical spring. Unless the center hole is small or the cone angle is outside the range normally used for disk springs, however, the differences are slight. Reference 69 presents useful design curves based on Ref. 27.

Conical spring washers can be stacked to operate in either series or parallel. One must be careful to consider the effect of friction, however, when using them in the parallel configuration.

11.9 NARROW RING UNDER DISTRIBUTED TORQUE ABOUT ITS AXIS

When the inner radius b is almost as great as the outer radius a, the loading for cases 1a, 1k, 2a, 2k, and so on, becomes almost equivalent to that shown in Fig. 11.6, which represents a ring subjected to a uniformly distributed torque of M (force-length/unit length) about that circumference passing through the centroids at the radius R. An approximation to this type of loading also occurs in clamping, or "follower," rings used for joining pipe; here the bolt forces and the balancing gasket or flange pressure produce the distributed torque, which obviously tends to "roll" the ring, or turn it inside out, so to speak.

Under this loading the ring, whatever the shape of its cross section (as long as it is reasonably compact) is subjected to a bending moment at every section equal to MR, the neutral axis being the central axis of the cross section in the plane of the ring. The maximum resulting stress occurs at the extreme fiber and is given by Eq. (8.1-12); that is,

$$\sigma = \frac{MR}{I/c} \tag{11.9-1}$$

The ring does not bend, and there is no twisting, but every section rotates in its own plane about its centroid through an angle

$$\theta = \frac{MR^2}{EI} = \frac{\sigma R}{Ec} \tag{11.9-2}$$

These formulas may be used to obtain approximate results for the cases of flat-plate loading listed previously when the difference between a and b is small, as well as for pipe flanges, etc. Paul (Ref. 70) discusses the collapse or inversion of rings due to plastic action.

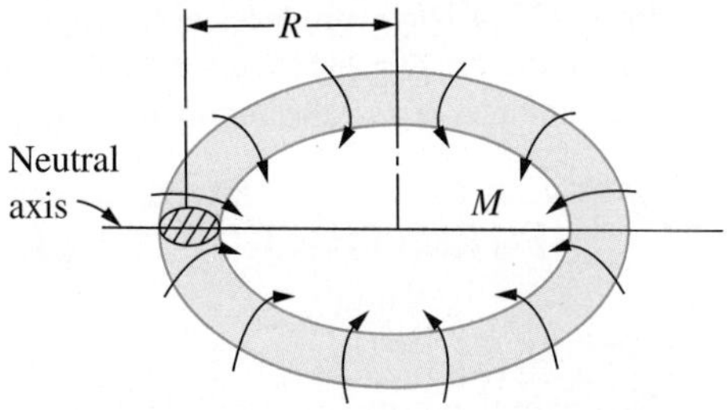

Figure 11.6

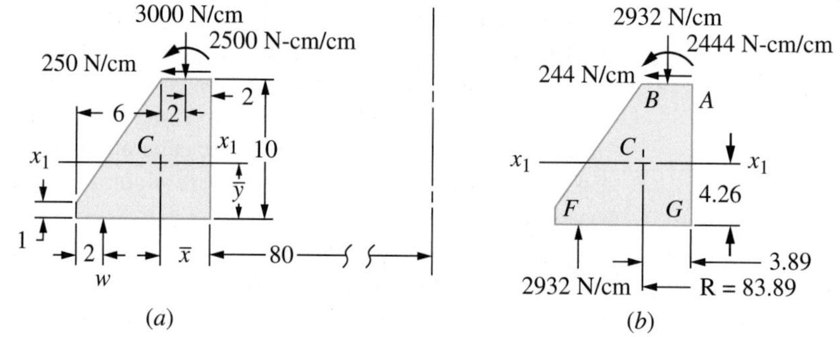

Figure 11.7 (All dimensions in centimeters)

EXAMPLE The cross section shown in Fig. 11.7 is from a roll-forged ring of steel used to support the bottom of a large shell. The modulus of elasticity is 207 GPa, or $20.7(10^6)$N/cm^2, and Poisson's ratio is 0.285. The loadings from the shell are shown in Fig. 11.7(*a*) and are unit loads at a radius of 82 cm where they are applied.

SOLUTION In the equations for stress and angular rotation the moment distributed around the ring must be evaluated as that moment acting upon a segment of the ring having a unit length along the circumference at the radius of the centroid of the cross section. In Fig. 11.7(*b*) these appropriate loadings are shown. Before they could be found, however, the centroid and the moment of inertia about the *x* axis through this centroid must have been evaluated. This was done as follows:

$$A = 10(10) - \frac{9(6)}{2} = 73\ \text{cm}^2$$

$$\bar{y} = \frac{100(5) - 27(7)}{73} = 4.26\ \text{cm}, \quad \bar{x} = \frac{100(5) - 27(8)}{73} = 3.89\ \text{cm}$$

$$I_{x1} = \frac{10^4}{12} + 100(5 - 4.26)^2 - \frac{6(9^3)}{36} - 27(7 - 4.26)^2 = 563.9\ \text{cm}^4$$

First calculate the value of w which will put into equilibrium at a radius of 88 cm the vertical load of 3000 N/cm at a radius of 82 cm. This is 2795 N/cm. Next convert all these loads to the values they will have when applied to a free-body diagram consisting of a segment that is 1 cm long at the centroidal radius of 83.89 cm. For the loads on the top of the free-body diagram the length upon which they act is 82/83.89 = 0.9775 cm so that the desired couple is then 2500(0.9775) = 2444 N-cm/cm. All the remaining forces were computed in a similar manner.

Using the loads shown in Fig. 11.7(*b*), the clockwise moment about the centroid is found to be $M = 2932(6) - 2444 - 244(10 - 4.26) = 13{,}747$ N-cm. This gives the section a clockwise rotation of $\theta = 13{,}747(83.89^2)/20.7(10^6)(563.9) = 0.00829$ rad. All material in the section lying above the x_1 axis will then move toward the central axis and be in compression. The stresses at positions *A* and *B* will then be $\sigma = -13{,}747(83.89)(5.74)/563.9 = -11{,}739$ N/cm^2. Similarly, the stresses at positions *F* and *G* are $\sigma = 13{,}747(83.89)(4.26)/563.9 = 8712$ N/cm^2.

In addition to the stresses caused by the rotation of the cross section, the radially outward shear force of 244 N/cm produces everywhere in the cross section a circumferential tensile stress of $\sigma = 244(83.89)/73 = 280$ N/cm^2. Note that a tacit assumption has been made that no radially directed friction forces exist at the bottom of the ring.

11.10 BENDING OF UNIFORM-THICKNESS PLATES WITH STRAIGHT BOUNDARIES

Formulas

No general expression for deflection as a function of position in a plate is given since solutions for plates with straight boundaries are generally obtained numerically for specific ratios of plate dimensions, load location, and boundary conditions. In a few instances

Poisson's ratio is included in the expressions given, but in most cases a specific value of Poisson's ratio has been used in obtaining the tabulated numerical results and the value used is indicated. Reference 47 includes results obtained using several values of Poisson's ratio and shows the range of values that can be expected as this ratio is changed. Errors in deflection should not exceed 7 or 8% and in maximum stress 15% for values of Poisson's ratio in the range from 0.15 to 0.30. Since much of the data are obtained using finite-difference approximations for the plate differential equations and a limited number of elements have been used, it is not always possible to identify maximum values if they occur at points between the chosen grid points.

Table 11.4 presents maximum values where possible and the significant values otherwise for deflections normal to the plate surface, bending stresses, and in many cases the boundary reaction forces *R*. For rectangular plates with simply supported edges the maximum stresses are shown to be near the center of the plate. There are, however, stresses of similar magnitude near the corners if the corners are held down as has been assumed for all cases presented. Reference 21 discusses the increase in stress at the center of the plate when the corners are permitted to rise. For a uniformly loaded square plate this increase in stress is approximately 35%.

It is impractical to include plates of all possible shapes and loadings, but many more cases can be found in the literature. Bareš (Ref. 47) presents tabulated values of bending moments and deflections for a series of plates in the form of *isosceles triangles* and *symmetric trapezoids* for linearly varying lateral pressures and for values of Poisson's ratio of 0.0 and 0.16. Tabulated values are given for *skew* plates with uniform lateral loading and concentrated lateral loads for the support conditions where two opposite edges are simply supported and two edges are free; the value of Poisson's ratio used was zero. In addition to many cases also included in Table 11.4, Marguerre and Woernle (Ref. 50) give results for line loading and uniform loading on a narrow strip across a rectangular plate. They also discuss the case of a rectangular plate supported within the span by elastic cross beams. Morley (Ref. 51) discusses solutions of problems involving *parallelogram*, or *skew*, plates and box structures. A few graphs and tables of results are given.

For plates with boundary shapes or restraints not discussed in the literature, we can only approximate an answer or resort to a direct numerical solution of the problem at hand. All numerical methods are approximate but can be carried to any degree of accuracy desired at the expense of time and computer costs. There are many numerical techniques used to solve plate problems, and the choice of a method for a given problem can be difficult. Leissa et al. (Ref. 56) have done a very complete and competent job of comparing and rating 9 approximate numerical methods on the basis of 11 different criteria. Szilard (Ref. 84) discusses both classical and numerical methods and tabulates many solutions.

Variable Thickness

Petrina and Conway (Ref. 63) give numerical data for two sets of boundary conditions, three aspect ratios and two nearly linear tapers in plate thickness. The loading was uniform and they found that the center deflection and center moment differed little from the same uniform-thickness case using the average thickness; the location and magnitude of maximum stress, however, did vary.

11.11 EFFECT OF LARGE DEFLECTION; DIAPHRAGM STRESSES

When the deflection becomes larger than about one-half the thickness, as may occur in thin plates, the middle surface becomes appreciably strained and the stress in it cannot be ignored. This stress, called *diaphragm* stress, or *direct* stress, enables the plate to carry part of the load as a diaphragm in direct tension. This tension may be balanced by radial tension at the edges if the edges are *held* or by circumferential compression if the edges are not horizontally restrained. In thin plates this circumferential compression may cause buckling.

When this condition of large deflection exists, the plate is stiffer than indicated by the ordinary theory and the load-deflection and load-stress relations are nonlinear. Stresses for a given load are less and stresses for a given deflection are generally greater than the ordinary theory indicates.

Circular Plates

Formulas for stress and deflection when middle surface stresses are taken into account are given below. These formulas should be used whenever the maximum deflection exceeds half the thickness if accurate results are desired. The table following gives the necessary constants for the several loadings and support conditions listed.

Let t = thickness of plate; a = outer radius of plate; q = unit lateral pressure; y = maximum deflection; σ_b = bending stress; σ_d = diaphragm stress; $\sigma = \sigma_b + \sigma_d$ = maximum stress due to flexure and diaphragm tension combined. Then the following formulas apply:

$$\frac{qa^4}{Et^4} = K_1 \frac{y}{t} + K_2 \left(\frac{y}{t}\right)^3 \tag{11.11-1}$$

$$\frac{\sigma a^2}{Et^2} = K_3 \frac{y}{t} + K_4 \left(\frac{y}{t}\right)^2 \tag{11.11-2}$$

First solve for y in Eq. (11.11-1) and then obtain the stresses from Eq. (11.11-2).

EXAMPLE For the plate of Example 1 of Sec. 11.2, it is desired to determine the maximum deflection and maximum stress under a load of 10 lb/in^2.

SOLUTION If the linear theory held, the stresses and deflections would be directly proportional to the load, which would indicate a maximum stress of 9240(10)/3 = 30,800 lb/in^2 and a maximum deflection of 0.0883(10)/3 = 0.294 in. Since this deflection is much more than half the thickness, Eqs. (11.11-3) and (11.11-2) with the constants from case 1 in the table will be used to solve for the deflection and stress. From Eq. (11.11-1), we obtain

$$\frac{10(10^4)}{30(10^6)(0.2^4)} = \frac{1.016}{1-0.3}\frac{y}{t} + 0.376\left(\frac{y}{t}\right)^3$$

$$2.0833 = 1.4514\frac{y}{t} + 0.376\left(\frac{y}{t}\right)^3$$

Starting with a trial value for y somewhat less than 0.294 in, a solution is found when y = 0.219 in. From Eq. (11.11-2) the maximum stress is found to be 27,500 lb/in^2.

Warshawsky (Ref. 3) fitted Eqs. (11.11-1) and (11.11-2) to the data presented by Mah in Ref. 71, and cases 5–9 in the following table give these results. Chia in Ref. 91 has a chapter on nonlinear bending of isotropic nonrectangular plates in which he covers in great detail the derivations, plotted results, and formulas similar to Eqs. (11.11-1) and (11.11-2) for distributed loadings, concentrated center loads, applied edge moments, and combined loadings for circular plates with various boundary conditions. The uniformly loaded circular plate on an elastic foundation is discussed and results presented for several boundary conditions. He also treats annular plates, elliptical plates, and skew plates under uniform loading. Reference 54 presents the results of a study of the large deflections of clamped annular sector plates for sector angles from 30–90° in steps of 30° and for ratios of inner to outer radii from 0–0.6 in steps of 0.2.

Circular Plates under Distributed Load Producing Large Deflections

<table>
<tr><th>Case No., Edge Condition</th><th>Constants</th></tr>
<tr><td>1. Simply supported (neither fixed nor held). Uniform pressure q over entire plate.</td><td>$K_1 = \frac{1.016}{1-\nu}$ $K_2 = 0.376$
$K_3 = \frac{1.238}{1-\nu}$ $K_4 = 0.294$
(Ref. 5)</td></tr>
<tr><td>2. Fixed but not held (no edge tension). Uniform pressure q over entire plate.</td><td>$K_1 = \frac{5.33}{1-\nu^2}$ $K_2 = 0.857$
(At center) $K_3 = \frac{2}{1-\nu}$ $K_4 = 0.50$
(At edge) $K_3 = \frac{4}{1-\nu^2}$ $K_4 = 0.0$
(Ref. 5)</td></tr>
<tr><td>3. Fixed and held. Uniform pressure q over entire plate.</td><td>$K_1 = \frac{5.33}{1-\nu^2}$ $K_2 = \frac{2.6}{1-\nu^2}$
(At center) $K_3 = \frac{2}{1-\nu}$ $K_4 = 0.976$
(At edge) $K_3 = \frac{4}{1-\nu^2}$ $K_4 = 0.476$
(Refs. 15 and 16)</td></tr>
<tr><td>4. Diaphragm without flexural stiffness, edge held. Uniform pressure q over entire plate.</td><td>$K_1 = 0.0$ $K_2 = 3.44$
(At center) $K_3 = 0.0$ $K_4 = 0.965$
(At edge) $K_3 = 0.0$ $K_4 = 0.748$
(At r from the center) $y = y_{max}\left(1 - 0.9\frac{r^2}{a^2} - 0.1\frac{r^5}{a^5}\right)$
(Refs. 18 and 29)</td></tr>
<tr><td>5. Fixed and held. Uniform pressure q over a central area of radius r_o. $\nu = 0.3$</td><td>
<table>
<tr><th></th><th></th><th></th><th colspan="2">At edge</th><th colspan="2">At center</th></tr>
<tr><th>r_o/a</th><th>K_1</th><th>K_2</th><th>K_3</th><th>K_4</th><th>K_3</th><th>K_4</th></tr>
<tr><td>1.00</td><td>5.86</td><td>3.32</td><td>4.40</td><td>1.73</td><td></td><td></td></tr>
<tr><td>0.75</td><td>6.26</td><td>3.45</td><td>3.80</td><td>1.32</td><td></td><td></td></tr>
<tr><td>0.50</td><td>9.17</td><td>5.50</td><td></td><td></td><td>3.38</td><td>0.76</td></tr>
<tr><td>0.25</td><td>27.1</td><td>13.9</td><td></td><td></td><td>4.62</td><td>1.18</td></tr>
</table>
(Ref. 3)</td></tr>
<tr><td>6. Simply supported and held radially. Uniform pressure q over a central area of radius r_o. $\nu = 0.3$</td><td>
<table>
<tr><th></th><th></th><th></th><th colspan="2">At center</th></tr>
<tr><th>r_o/a</th><th>K_1</th><th>K_2</th><th>K_3</th><th>K_4</th></tr>
<tr><td>0.75</td><td>1.71</td><td>3.21</td><td>1.84</td><td>0.81</td></tr>
<tr><td>0.50</td><td>2.95</td><td>5.07</td><td>2.06</td><td>0.95</td></tr>
<tr><td>0.25</td><td>9.95</td><td>13.8</td><td>2.60</td><td>1.31</td></tr>
</table>
(Ref. 3)</td></tr>
<tr><td>7. Fixed and held with a central support. Uniform pressure q over entire plate. $\nu = 0.3$</td><td>y_{max} at $r = 0.45a$
$K_1 = 36.4$ $K_2 = 20.0$
(Ref. 3)</td></tr>
</table>

(Continued)

Circular Plates under Distributed Load Producing Large Deflections (*Continued*)

Case No., Edge Condition	Constants
8. Annular plate fixed and held at both inner and outer edges. Uniform pressure q over entire annular plate. $\nu = 0.3$	For inner edge radius $= 0.2a$, max deflection y at $r = 0.576a$ $K_1 = 84.0 \quad K_2 = 63.5$ For stress at $r = 0.2a$, $K_3 = 36.0 \quad K_4 = 25.8$ (Ref. 3)
9. Annular plate simply supported and held radially at both inner and outer edges. Uniform pressure q over entire annular plate. $\nu = 0.3$	For inner edge radius $= 0.2a$, max deflection y at $r = 0.576a$ $K_1 = 20.3 \quad K_2 = 51.8$ For stress at $r = 0.2a$, $K_3 = 12.14 \quad K_4 = 2.41$
	For inner edge radius $= 0.4a$, max deflection y at $r = 0.688a$ $K_1 = 57.0 \quad K_2 = 159$ For stress at $r = 0.664a$, $K_3 = 14.52 \quad K_4 = 6.89$ (Ref. 3)

Elliptical Plates

Nash and Cooley (Ref. 72) present graphically the results of a uniform pressure on a clamped elliptical plate for $a/b = 2$. Their method of solution is presented in detail, and the numerical solution is compared with experimental results and with previous solutions they have referenced. Ng (Ref. 73) has tabulated the values of center deflection for clamped elliptical plates on elastic foundations for ratios of a/b from 1 to 2 and for a wide range of foundation moduli. Large deflections are also graphed for two ratios a/b (1.5 and 2) for the same range of foundation moduli.

Rectangular Plates

Analytical solutions for uniformly loaded rectangular plates with large deflections are given in Refs. 30–34, where the relations among load, deflection, and stress are expressed by numerical values of the dimensionless coefficients y/t, qb^4/Et^4, and $\sigma b^2/Et^2$. The values of these coefficients given in the table following are taken from these references and are for $\nu = 0.316$. In this table, a, b, q, E, y, and t have the same meaning as in Table 11.4, σ_d is the diaphragm stress, and σ is the total stress found by adding the diaphragm stress and the bending stress. See also Ref. 17.

In Ref. 35 experimentally determined deflections are given and compared with those predicted by theory. In Ref. 74 a numerical solution for uniformly loaded rectangular plates with simply supported edges is discussed, and the results for a square plate are compared with previous approximate solutions. Graphs are presented to show how stresses and deflections vary across a square plate.

Chia in Ref. 91 includes a chapter on moderately large deflections of isotropic rectangular plates. Not only are the derivations presented but the results of most cases are presented in the form of graphs usable for engineering calculations. Cases of initially deflected plates are included, and the comprehensive list of references is useful. Aalami and Williams in Ref. 92 present 42 tables of large-deflection reduction coefficients over a range of length ratios a/b and for a variety—three bending and four membrane—of symmetric and nonsymmetric boundary conditions. Loadings include overall uniform and linearly varying pressures as well as pressures over limited areas centered on the plates.

Parallelogram Plates

Kennedy and Ng (Ref. 53) present several graphs of large elastic deflections and the accompanying stresses for uniformly loaded skew plates with clamped edges. Several aspect ratios and skew angles are represented.

11.12 PLASTIC ANALYSIS OF PLATES

The onset of yielding in plates may occur before the development of appreciable diaphragm stress if the plate is relatively thick. For thinner plates, the nonlinear increase in stiffness due to diaphragm stresses is counteracted by the decrease in stiffness which occurs when the material starts to yield (Refs. 52 and 80). Save and Massonnet (Ref. 81) discuss the effect of the several yield criteria on the response of circular and rectangular plates under various loadings and give an extensive list of references. They also compare the results of theory with referenced experiments which have been performed. *Orthotropy* in plates can be caused by cold-forming the material or by the positioning of stiffeners. The effect of this orthotropic behavior on the yielding of circular plates is discussed by Save and Massonnet (Ref. 81) as well as by Markowitz and Hu (Ref. 82).

Crose and Ang (Ref. 83) describe an iterative solution scheme which first solves the elastic case and then increments the loading upward to allow a slow expansion of the yielded volume after it forms. The results of a test on a clamped plate are compared favorably with a theoretical solution.

11.13 ULTIMATE STRENGTH

Plates of brittle material fracture when the actual maximum tensile stress reaches the ultimate tensile strength of the material. A flat-plate modulus of rupture, analogous to the modulus of rupture of a beam, may be determined by calculating the (fictitious) maximum stress corresponding to the breaking load, using for this purpose the appropriate formula for elastic stress. This flat-plate modulus of rupture is usually greater than the modulus of rupture determined by testing a beam of rectangular section.

Rectangular Plates Under Uniform Load Producing Large Deflection

a/b	Edges and Point of Max σ	Coef.	qb^4/Et^4										
			0	12.5	25	50	75	100	125	150	175	200	250
1	Held, not fixed At center of plate	y/t	0	0.430	0.650	0.930	1.13	1.26	1.37	1.47	1.56	1.63	1.77
		$\sigma_d b^2/Et^2$	0	0.70	1.60	3.00	4.00	5.00	6.10	7.00	7.95	8.60	10.20
		$\sigma b^2/Et^2$	0	3.80	5.80	8.70	10.90	12.80	14.30	15.60	17.00	18.20	20.50
1	Held and riveted At center of plate	y/t	0	0.406	0.600	0.840	1.00	1.13	1.23	1.31	1.40	1.46	1.58
		$\sigma_d b^2/Et^2$	0	0.609	1.380	2.68	3.80	4.78	5.75	6.54	7.55	8.10	9.53
		$\sigma b^2/Et^2$	0	3.19	5.18	7.77	9.72	11.34	12.80	14.10	15.40	16.40	18.40
1	Held and fixed At center of long edges	y/t	0	0.165	0.32	0.59	0.80	0.95	1.08	1.19	1.28	1.38	1.54
		$\sigma_d b^2/Et^2$	0	0.070	0.22	0.75	1.35	2.00	2.70	3.30	4.00	4.60	5.90
		$\sigma b^2/Et^2$	0	3.80	6.90	14.70	21.0	26.50	31.50	36.20	40.70	45.00	53.50
	At center of plate	$\sigma_d b^2/Et^2$	0	0.075	0.30	0.95	1.65	2.40	3.10	3.80	4.50	5.20	6.50
		$\sigma b^2/Et^2$	0	1.80	3.50	6.60	9.20	11.60	13.0	14.50	15.80	17.10	19.40
1.5	Held, not fixed At center of plate	y/t	0	0.625	0.879	1.18	1.37	1.53	1.68	1.77	1.88	1.96	2.12
		$\sigma_d b^2/Et^2$	0	1.06	2.11	3.78	5.18	6.41	7.65	8.60	9.55	10.60	12.30
		$\sigma b^2/Et^2$	0	4.48	6.81	9.92	12.25	14.22	16.0	17.50	18.90	20.30	22.80
2 to ∞	Held, not fixed At center of plate	y/t	0	0.696	0.946	1.24	1.44	1.60	1.72	1.84	1.94	2.03	2.20
		$\sigma_d b^2/Et^2$	0	1.29	2.40	4.15	5.61	6.91	8.10	9.21	10.10	10.90	12.20
		$\sigma b^2/Et^2$	0	4.87	7.16	10.30	12.60	14.60	16.40	18.00	19.40	20.90	23.60
1.5 to ∞	Held and fixed At center of long edges	y/t	0	0.28	0.51	0.825	1.07	1.24	1.40	1.50	1.63	1.72	1.86
		$\sigma_d b^2/Et^2$	0	0.20	0.66	1.90	3.20	4.35	5.40	6.50	7.50	8.50	10.30
		$\sigma b^2/Et^2$	0	5.75	11.12	20.30	27.8	35.0	41.0	47.0	52.50	57.60	67.00

Plates of ductile material fail by excessive plastic deflection, as do beams of similar material. For a number of cases the load required to produce collapse has been determined analytically, and the results for some of the simple loadings are summarized as follows.

1. *Circular plate; uniform load, edges simply supported*

$$W_u = \sigma_y\left(\tfrac{3}{2}\pi t^2\right) \qquad \text{(Ref. 43)}$$

2. *Circular plate; uniform load, fixed edges*

$$W_u = \sigma_y(2.814\pi t^2) \qquad \text{(Ref. 43)}$$

(For collapse loads on partially loaded orthotropic annular plates see Refs. 81 and 82.)

3. *Rectangular plate, length a, width b; uniform load, edges supported*

$$W_u = \beta\sigma_y t^2$$

where β depends on the ratio of b to a and has the following values (Ref. 44):

b/a	1	0.9	0.8	0.7	0.6	0.5	0.4	0.3	0.2
β	5.48	5.50	5.58	5.64	5.89	6.15	6.70	7.68	9.69

4. *Plate of any shape and size, any type of edge support, concentrated load at any point*

$$W_u = \sigma_y\left(\tfrac{1}{2}\pi t^2\right) \qquad \text{(Ref. 45)}$$

In each of the above cases W_u denotes the total load required to collapse the plate, t the thickness of the plate, and σ_y the yield point of the material. Accurate prediction of W_u is hardly to be expected; the theoretical error in some of the formulas may range up to 30%, and few experimental data seem to be available.

11.14 TABLES

TABLE 11.1 Numerical Values for Functions Used in Table 11.2

r_o/r	1.000	0.900	0.800	0.750	0.700	2/3	0.600	0.500
G_1	0.000	0.098580346	0.19478465	0.2423283	0.2897871	0.3215349	0.3858887	0.487773
G_2	0.000	0.004828991	0.01859406	0.0284644	0.0401146	0.0487855	0.0680514	0.100857
G_3	0.000	0.000158070	0.00119108	0.0022506	0.0037530	0.0050194	0.0082084	0.014554
G_4	1.000	0.973888889	0.95750000	0.9541667	0.9550000	0.9583333	0.9733333	1.025000
G_5	0.000	0.095000000	0.18000000	0.2187500	0.2550000	0.2777778	0.3200000	0.375000
G_6	0.000	0.004662232	0.01725742	0.0258495	0.0355862	0.0425624	0.0572477	0.079537
G_7	0.000	0.096055556	0.20475000	0.2654167	0.3315000	0.3791667	0.4853333	0.682500
G_8	1.000	0.933500000	0.87400000	0.8468750	0.8215000	0.8055556	0.7760000	0.737500
G_9	0.000	0.091560902	0.16643465	0.1976669	0.2247621	0.2405164	0.2664220	0.290898
G_{11}	0.000	0.000003996	0.00006104	0.0001453	0.0002935	0.0004391	0.0008752	0.001999
G_{12}	0.000	0.000000805	0.00001240	0.0000297	0.0000603	0.0000905	0.0001820	0.000422
G_{13}	0.000	0.000000270	0.00000418	0.0000100	0.0000205	0.0000308	0.0000623	0.000146
G_{14}	0.000	0.000158246	0.00119703	0.0022693	0.0038011	0.0051026	0.0084257	0.015272
G_{15}	0.000	0.000039985	0.00030618	0.0005844	0.0009861	0.0013307	0.0022227	0.004111
G_{16}	0.000	0.000016107	0.00012431	0.0002383	0.0004039	0.0005468	0.0009196	0.001721
G_{17}	0.000	0.004718219	0.01775614	0.0268759	0.0374539	0.0452137	0.0621534	0.090166
G_{18}	0.000	0.001596148	0.00610470	0.0093209	0.0131094	0.0159275	0.0221962	0.032948
G_{19}	0.000	0.000805106	0.00310827	0.0047694	0.0067426	0.0082212	0.0115422	0.017341
$C_1C_6 - C_3C_4$	0.000	0.000305662	0.00222102	0.0041166	0.0067283	0.0088751	0.0141017	0.023878
$C_1C_9 - C_3C_7$	0.000	0.009010922	0.03217504	0.0473029	0.0638890	0.0754312	0.0988254	0.131959
$C_2C_6 - C_3C_5$	0.000	0.000007497	0.00010649	0.0002435	0.0004705	0.0006822	0.0012691	0.002564
$C_2C_9 - C_3C_8$	0.000	0.000294588	0.00205369	0.0037205	0.0059332	0.0076903	0.0117606	0.018605
$C_4C_9 - C_6C_7$	0.000	0.088722311	0.15582772	0.1817463	0.2028510	0.2143566	0.2315332	0.243886

Numerical values for the plate coefficients F, C, L, and G for values of b/r, b/a, r_o/a, and r_o/r, respectively, from 0.05 to 1.0. Poisson's ratio is 0.30. The table headings are given for G_1 to G_{19} for the various values of r_o/r.† Also listed in the last five lines are values for the most used denominators for the ratios b/a.

(Continued)

TABLE 11.1 Numerical Values for Functions Used in Table 11.2 (*Continued*)

r_o/r	0.400	1/3	0.300	0.250	0.200	0.125	0.100	0.050
G_1	0.605736	0.704699	0.765608	0.881523	1.049227	1.547080	1.882168	3.588611
G_2	0.136697	0.161188	0.173321	0.191053	0.207811	0.229848	0.235987	0.245630
G_3	0.022290	0.027649	0.030175	0.033465	0.035691	0.035236	0.033390	0.025072
G_4	1.135000	1.266667	1.361667	1.562500	1.880000	2.881250	3.565000	7.032500
G_5	0.420000	0.444444	0.455000	0.468750	0.480000	0.492187	0.495000	0.498750
G_6	0.099258	0.109028	0.112346	0.114693	0.112944	0.099203	0.090379	0.062425
G_7	0.955500	1.213333	1.380167	1.706250	2.184000	3.583125	4.504500	9.077250
G_8	0.706000	0.688889	0.681500	0.671875	0.664000	0.655469	0.653500	0.650875
G_9	0.297036	0.289885	0.282550	0.266288	0.242827	0.190488	0.166993	0.106089
G_{11}	0.003833	0.005499	0.006463	0.008057	0.009792	0.012489	0.013350	0.014843
G_{12}	0.000827	0.001208	0.001435	0.001822	0.002266	0.003027	0.003302	0.003872
G_{13}	0.000289	0.000427	0.000510	0.000654	0.000822	0.001121	0.001233	0.001474
G_{14}	0.024248	0.031211	0.034904	0.040595	0.046306	0.054362	0.056737	0.060627
G_{15}	0.006691	0.008790	0.009945	0.011798	0.013777	0.016917	0.017991	0.020139
G_{16}	0.002840	0.003770	0.004290	0.005138	0.006065	0.007589	0.008130	0.009252
G_{17}	0.119723	0.139340	0.148888	0.162637	0.175397	0.191795	0.196271	0.203191
G_{18}	0.044939	0.053402	0.057723	0.064263	0.070816	0.080511	0.083666	0.089788
G_{19}	0.023971	0.028769	0.031261	0.035098	0.039031	0.045057	0.047086	0.051154
$C_1C_6 - C_3C_4$	0.034825	0.041810	0.044925	0.048816	0.051405	0.051951	0.051073	0.047702
$C_1C_9 - C_3C_7$	0.158627	0.170734	0.174676	0.177640	0.176832	0.168444	0.163902	0.153133
$C_2C_6 - C_3C_5$	0.004207	0.005285	0.005742	0.006226	0.006339	0.005459	0.004800	0.002829
$C_2C_9 - C_3C_8$	0.024867	0.027679	0.028408	0.028391	0.026763	0.020687	0.017588	0.009740
$C_4C_9 - C_6C_7$	0.242294	0.234900	0.229682	0.220381	0.209845	0.193385	0.188217	0.179431

† To obtain a value of either C_i, L_i, or F_i for a corresponding value of either b/a, r_o/a, or b/r, respectively, use the tabulated value of G_i for the corresponding value of r_o/r.

TABLE 11.2 Formulas for Flat Circular Plates of Constant Thickness

Notation: W = total applied load (force); w = unit line load (force per unit of circumferential length); q = load per unit area; M_o = unit applied line moment loading (force-length per unit of circumferential length); θ_o = externally applied change in radial slope (radians); y_o = externally applied radial step in the vertical deflection (length); y = vertical deflection of plate (length) θ = radial slope of plate; M_r = unit radial bending moment; M_t = unit tangential bending moment; Q = unit shear force (force per unit of circumferential length); E = modulus of elasticity (force per unit area); ν = Poisson's ratio; γ = temperature coefficient of expansion (unit strain per degree); a = outer radius; b = inner radius for annular plate; t = plate thickness; r = radial location of quantity being evaluated; r_o = radial location of unit line loading or start of a distributed load. F_1 to F_9 and G_1 to G_{19} are the several functions of the radial location r. C_1 to C_9 are plate constants dependent upon the ratio a/b. L_1 to L_{19} are loading constants dependent upon the ratio a/r_o. When used as subscripts, r and t refer to radial and tangential directions, respectively. When used as subscripts, a, b, and o refer to an evaluation of the quantity subscripted at the outer edge, inner edge, and the position of the loading or start of distributed loading, respectively. When used as a subscript, c refers to an evaluation of the quantity subscripted at the center of the plate.

Positive signs are associated with the several quantities in the following manner: Deflections y and y_o are positive upward; slopes θ and θ_o are positive when the deflection y increases positively as r increases; moments M_r, M_t, and M_o are positive when creating compression on the top surface; and the shear force Q is positive when acting upward on the inner edge of a given annular section.

Note: Bending stresses can be found from the moments M_r and M_t by the expression $\sigma = 6M/t^2$. The plate constant $D = Et^3/12(1-\nu^2)$. The singularity function brackets $\langle\ \rangle$ indicate that the expression contained within the brackets must be equated to zero unless $r > r_o$, after which they are treated as any other brackets. Note that Q_b, Q_a, M_{rb}, and M_{ra} are reactions, not loads. They exist only when necessary edge restraints are provided.

General Plate Functions and Constants for Solid and Annular Circular Plates

$F_1 = \frac{1+\nu}{2}\frac{b}{r}\ln\frac{r}{b} + \frac{1-\nu}{4}\left(\frac{r}{b}-\frac{b}{r}\right)$	$C_1 = \frac{1+\nu}{2}\frac{b}{a}\ln\frac{a}{b} + \frac{1-\nu}{4}\left(\frac{a}{b}-\frac{b}{a}\right)$
$F_2 = \frac{1}{4}\left[1-\left(\frac{b}{r}\right)^2\left(1+2\ln\frac{r}{b}\right)\right]$	$C_2 = \frac{1}{4}\left[1-\left(\frac{b}{a}\right)^2\left(1+2\ln\frac{a}{b}\right)\right]$
$F_3 = \frac{b}{4r}\left\{\left[\left(\frac{b}{r}\right)^2+1\right]\ln\frac{r}{b}+\left(\frac{b}{r}\right)^2-1\right\}$	$C_3 = \frac{b}{4a}\left\{\left[\left(\frac{b}{a}\right)^2+1\right]\ln\frac{a}{b}+\left(\frac{b}{a}\right)^2-1\right\}$
$F_4 = \frac{1}{2}\left[(1+\nu)\frac{b}{r}+(1-\nu)\frac{r}{b}\right]$	$C_4 = \frac{1}{2}\left[(1+\nu)\frac{b}{a}+(1-\nu)\frac{a}{b}\right]$
$F_5 = \frac{1}{2}\left[1-\left(\frac{b}{r}\right)^2\right]$	$C_5 = \frac{1}{2}\left[1-\left(\frac{b}{a}\right)^2\right]$
$F_6 = \frac{b}{4r}\left[\left(\frac{b}{r}\right)^2-1+2\ln\frac{r}{b}\right]$	$C_6 = \frac{b}{4a}\left[\left(\frac{b}{a}\right)^2-1+2\ln\frac{a}{b}\right]$
$F_7 = \frac{1}{2}(1-\nu^2)\left(\frac{r}{b}-\frac{b}{r}\right)$	$C_7 = \frac{1}{2}(1-\nu^2)\left(\frac{a}{b}-\frac{b}{a}\right)$
$F_8 = \frac{1}{2}\left[1+\nu+(1-\nu)\left(\frac{b}{r}\right)^2\right]$	$C_8 = \frac{1}{2}\left[1+\nu+(1-\nu)\left(\frac{b}{a}\right)^2\right]$
$F_9 = \frac{b}{r}\left\{\frac{1+\nu}{2}\ln\frac{r}{b}+\frac{1-\nu}{4}\left[1-\left(\frac{b}{r}\right)^2\right]\right\}$	$C_9 = \frac{b}{a}\left\{\frac{1+\nu}{2}\ln\frac{a}{b}+\frac{1-\nu}{4}\left[1-\left(\frac{b}{a}\right)^2\right]\right\}$

(*Continued*)

TABLE 11.2 Formulas for Flat Circular Plates of Constant Thickness (*Continued*)

General Plate Functions and Constants for Solid and Annular Circular Plates	
$L_1=\frac{1+\nu}{2}\frac{r_o}{a}\ln\frac{r_o}{a}+\frac{1-\nu}{4}\left(\frac{a}{r_o}-\frac{r_o}{a}\right)$	$G_1=\left[\frac{1+\nu}{2}\frac{a}{r_o}\ln\frac{r}{r_o}+\frac{1-\nu}{4}\left(\frac{r}{r_o}-\frac{r_o}{r}\right)\right]\langle r-r_o\rangle^0$
$L_2=\frac{1}{4}\left[1-\left(\frac{r_o}{a}\right)^2\left(1+2\ln\frac{a}{r_o}\right)\right]$	$G_2=\frac{1}{4}\left[1-\left(\frac{r_o}{r}\right)^2\left(1+2\ln\frac{r}{r_o}\right)\right]\langle r-r_o\rangle^0$
$L_3=\frac{r_o}{4a}\left\{\left[\left(\frac{r_o}{a}\right)^2+1\right]\ln\frac{a}{r_o}+\left(\frac{r_o}{a}\right)^2-1\right\}$	$G_3=\frac{r_o}{4r}\left\{\left[\left(\frac{r_o}{r}\right)^2+1\right]\ln\frac{r}{r_o}+\left(\frac{r_o}{r}\right)^2-1\right\}\langle r-r_o\rangle^0$
$L_4=\frac{1}{2}\left[(1+\nu)\frac{r_o}{a}+(1-\nu)\frac{a}{r_o}\right]$	$G_4=\frac{1}{2}\left[(1+\nu)\frac{r_o}{r}+(1-\nu)\frac{r}{r_o}\right]\langle r-r_o\rangle^0$
$L_5=\frac{1}{2}\left[1-\left(\frac{r_o}{a}\right)^2\right]$	$G_5=\frac{1}{2}\left[1-\left(\frac{r_o}{r}\right)^2\right]\langle r-r_o\rangle^0$
$L_6=\frac{r_o}{4a}\left[\left(\frac{r_o}{a}\right)^2-1+2\ln\frac{a}{r_o}\right]$	$G_6=\frac{r_o}{4r}\left[\left(\frac{r_o}{r}\right)^2-1+2\ln\frac{r}{r_o}\right]\langle r-r_o\rangle^0$
$L_7=\frac{1}{2}(1-\nu^2)\left(\frac{a}{r_o}-\frac{r_o}{a}\right)$	$G_7=\frac{1}{2}(1-\nu^2)\left(\frac{r}{r_o}-\frac{r_o}{r}\right)\langle r-r_o\rangle^0$
$L_8=\frac{1}{2}\left[1+\nu+(1-\nu)\left(\frac{r_o}{a}\right)^2\right]$	$G_8=\frac{1}{2}\left[1+\nu+(1-\nu)\left(\frac{r_o}{r}\right)^2\right]\langle r-r_o\rangle^0$
$L_9=\frac{r_o}{a}\left\{\frac{1+\nu}{2}\ln\frac{a}{r_o}+\frac{1-\nu}{4}\left[1-\left(\frac{r_o}{a}\right)^2\right]\right\}$	$G_9=\frac{r_o}{r}\left\{\frac{1+\nu}{2}\ln\frac{r}{r_o}+\frac{1-\nu}{4}\left[1-\left(\frac{r_o}{r}\right)^2\right]\right\}\langle r-r_o\rangle^0$
$L_{11}=\frac{1}{64}\left\{1+4\left(\frac{r_o}{a}\right)^2-5\left(\frac{r_o}{a}\right)^4-4\left(\frac{r_o}{a}\right)^2\left[2+\left(\frac{r_o}{a}\right)^2\right]\ln\frac{a}{r_o}\right\}$	$G_{11}=\frac{1}{64}\left\{1+4\left(\frac{r_o}{r}\right)^2-5\left(\frac{r_o}{r}\right)^4-4\left(\frac{r_o}{r}\right)^2\left[2+\left(\frac{r_o}{r}\right)^2\right]\ln\frac{r}{r_o}\right\}\langle r-r_o\rangle^0$
$L_{12}=\frac{a}{14{,}400(a-r_o)}\left\{64-225\frac{r_o}{a}-100\left(\frac{r_o}{a}\right)^3+261\left(\frac{r_o}{a}\right)^5+60\left(\frac{r_o}{a}\right)^3\left[3\left(\frac{r_o}{a}\right)^2+10\right]\ln\frac{a}{r_o}\right\}$	$G_{12}=\frac{r\langle r-r_o\rangle^0}{14{,}400(r-r_o)}\left\{64-225\frac{r_o}{r}-100\left(\frac{r_o}{r}\right)^3+261\left(\frac{r_o}{r}\right)^5+60\left(\frac{r_o}{r}\right)^3\left[3\left(\frac{r_o}{r}\right)^2+10\right]\ln\frac{r}{r_o}\right\}$
$L_{13}=\frac{a^2}{14{,}400(a-r_o)^2}\left\{25-128\frac{r_o}{a}+225\left(\frac{r_o}{a}\right)^2-25\left(\frac{r_o}{a}\right)^4-97\left(\frac{r_o}{a}\right)^6-60\left(\frac{r_o}{a}\right)^4\left[5+\left(\frac{r_o}{a}\right)^2\right]\ln\frac{a}{r_o}\right\}$	$G_{13}=\frac{r^2\langle r-r_o\rangle^0}{14{,}400(r-r_o)^2}\left\{25-128\frac{r_o}{r}+225\left(\frac{r_o}{r}\right)^2-25\left(\frac{r_o}{r}\right)^4-97\left(\frac{r_o}{r}\right)^6-60\left(\frac{r_o}{r}\right)^4\left[5+\left(\frac{r_o}{r}\right)^2\right]\ln\frac{r}{r_o}\right\}$

(*Continued*)

TABLE 11.2 Formulas for Flat Circular Plates of Constant Thickness (*Continued*)

General Plate Functions and Constants for Solid and Annular Circular Plates	
$L_{14}=\frac{1}{16}\left[1-\left(\frac{r_o}{a}\right)^4-4\left(\frac{r_o}{a}\right)^2\ln\frac{a}{r_o}\right]$	$G_{14}=\frac{1}{16}\left[1-\left(\frac{r_o}{r}\right)^4-4\left(\frac{r_o}{r}\right)^2\ln\frac{r}{r_o}\right]\langle r-r_o\rangle^0$
$L_{15}=\frac{a}{720(a-r_o)}\left[16-45\frac{r_o}{a}+9\left(\frac{r_o}{a}\right)^5+20\left(\frac{r_o}{a}\right)^3\left(1+3\ln\frac{a}{r_o}\right)\right]$	$G_{15}=\frac{r\langle r-r_o\rangle^0}{720(r-r_o)}\left[16-45\frac{r_o}{a}+9\left(\frac{r_o}{a}\right)^5+20\left(\frac{r_o}{a}\right)^3\left(1+3\ln\frac{r}{r_o}\right)\right]$
$L_{16}=\frac{a^2}{1440(a-r_o)^2}\left[15-64\frac{r_o}{a}+90\left(\frac{r_o}{a}\right)^2-6\left(\frac{r_o}{a}\right)^6-5\left(\frac{r_o}{a}\right)^4\left(7+12\ln\frac{a}{r_o}\right)\right]$	$G_{16}=\frac{r^2\langle r-r_o\rangle^0}{1440(r-r_o)^2}\left[15-64\frac{r_o}{a}+90\left(\frac{r_o}{r}\right)^2-6\left(\frac{r_o}{r}\right)^6-5\left(\frac{r_o}{r}\right)^4\left(7+12\ln\frac{r}{r_o}\right)\right]$
$L_{17}=\frac{1}{4}\left\{1-\frac{1-\nu}{4}\left[1-\left(\frac{r_o}{a}\right)^4\right]-\left(\frac{r_o}{a}\right)^2\left[1+(1+\nu)\ln\frac{a}{r_o}\right]\right\}$	$G_{17}=\frac{1}{4}\left\{1-\frac{1-\nu}{4}\left[1-\left(\frac{r_o}{r}\right)^4\right]-\left(\frac{r_o}{r}\right)^2\left[1+(1+\nu)\ln\frac{r}{r_o}\right]\right\}\langle r-r_o\rangle^0$
$L_{18}=\frac{a}{720(a-r_o)}\left\{\left[20\left(\frac{r_o}{a}\right)^3+16\right](4+\nu)-45\frac{r_o}{a}(3+\nu)-9\left(\frac{r_o}{a}\right)^5(1-\nu)+60\left(\frac{r_o}{a}\right)^3(1+\nu)\ln\frac{a}{r_o}\right\}$	$G_{18}=\frac{r\langle r-r_o\rangle^0}{720(r-r_o)}\left\{\left[20\left(\frac{r_o}{r}\right)^3+16\right](4+\nu)-45\frac{r_o}{r}(3+\nu)-9\left(\frac{r_o}{r}\right)^5(1-\nu)+60\left(\frac{r_o}{r}\right)^3(1+\nu)\ln\frac{r}{r_o}\right\}$
$L_{19}=\frac{a^2}{1440(a-r_o)^2}\left[15(5+\nu)-64\frac{r_o}{a}(4+\nu)+90\left(\frac{r_o}{a}\right)^2(3+\nu)-5\left(\frac{r_o}{a}\right)^4(19+7\nu)+6\left(\frac{r_o}{a}\right)^6(1-\nu)-60\left(\frac{r_o}{a}\right)^4(1+\nu)\ln\frac{a}{r_o}\right]$	$G_{19}=\frac{r^2\langle r-r_o\rangle^0}{1440(r-r_o)^2}\left[15(5+\nu)-64\frac{r_o}{r}(4+\nu)+90\left(\frac{r_o}{r}\right)^2(3+\nu)-5\left(\frac{r_o}{r}\right)^4(19+7\nu)+6\left(\frac{r_o}{r}\right)^6(1-\nu)-60\left(\frac{r_o}{r}\right)^4(1+\nu)\ln\frac{r}{r_o}\right]$

(Continued)

TABLE 11.2 Formulas for Flat Circular Plates of Constant Thickness (*Continued*)

Case 1. Annular plate with a uniform annular line load w at a radius r_o

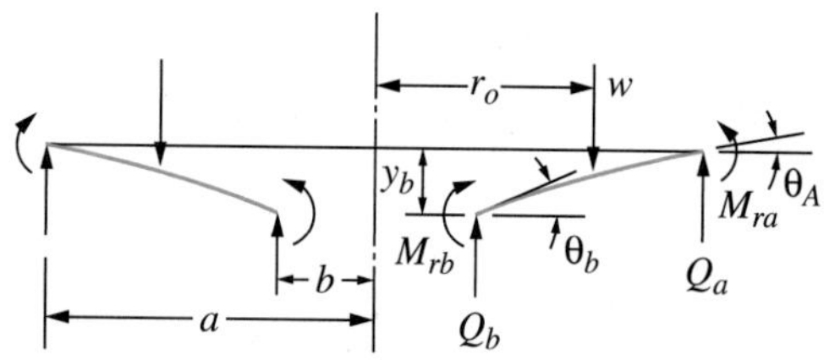

General expressions for deformations, moments, and shears:

$$y = y_b + \theta_b r F_1 + M_{rb}\frac{r^2}{D}F_2 + Q_b\frac{r^3}{D}F_3 - w\frac{r^3}{D}G_3$$

$$\theta = \theta_b F_4 + M_{rb}\frac{r}{D}F_5 + Q_b\frac{r^2}{D}F_6 - w\frac{r^2}{D}G_6$$

$$M_r = \theta_b\frac{D}{r}F_7 + M_{rb}F_8 + Q_b r F_9 - wrG_9$$

$$M_t = \frac{\theta D(1-\nu^2)}{r} + \nu M_r$$

$$Q = Q_b\frac{b}{r} - w\frac{r_o}{r}\langle r - r_o\rangle^0$$

For the numerical data given below, $\nu = 0.3$

$$y = K_y\frac{wa^3}{D}, \quad \theta = K_\theta\frac{wa^2}{D}, \quad M = K_M wa, \quad Q = K_Q w$$

Case No., Edge Restraints	Boundary Values	Special Cases

1a. Outer edge simply supported, inner edge free

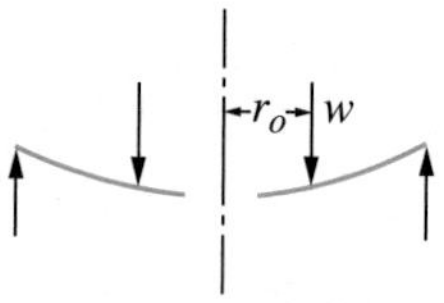

Boundary values: $M_{rb} = 0, \; Q_b = 0, \; y_a = 0, \; M_{ra} = 0$

$$y_b = \frac{-wa^3}{D}\left(\frac{C_1 L_9}{C_7} - L_3\right)$$

$$\theta_b = \frac{wa^2}{DC_7}L_9$$

$$\theta_a = \frac{wa^2}{D}\left(\frac{C_4 L_9}{C_7} - L_6\right)$$

$$Q_b = -w\frac{r_o}{a}$$

Special cases: $y_{max} = y_b, \quad M_{max} = M_{tb}$

If $r_o = b$ (load at inner edge),

b/a	0.1	0.3	0.5	0.7	0.9
K_{y_b}	−0.0364	−0.1266	−0.1934	−0.1927	−0.0938
K_{θ_b}	0.0371	0.2047	0.4262	0.6780	0.9532
K_{θ_a}	0.0418	0.1664	0.3573	0.6119	0.9237
$K_{M_{tb}}$	0.3374	0.6210	0.7757	0.8814	0.9638

1b. Outer edge simply supported, inner edge guided

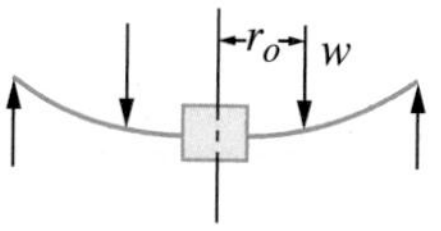

Boundary values: $\theta_b = 0, \; Q_b = 0, \; y_a = 0, \; M_{ra} = 0$

$$y_b = \frac{-wa^3}{D}\left(\frac{C_2 L_9}{C_8} - L_3\right)$$

$$M_{rb} = \frac{wa}{C_8}L_9$$

$$\theta_a = \frac{wa^2}{D}\left(\frac{C_5 L_9}{C_8} - L_6\right)$$

$$Q_a = -w\frac{r_o}{a}$$

Special cases: $y_{max} = y_b, \quad M_{max} = M_{rb}$

If $r_o = b$ (load at inner edge),

b/a	0.1	0.3	0.5	0.7	0.9
K_{y_b}	−0.0269	−0.0417	−0.0252	−0.0072	−0.0003
K_{θ_a}	0.0361	0.0763	0.0684	0.0342	0.0047
$K_{M_{rb}}$	0.2555	0.4146	0.3944	0.2736	0.0981

1c. Outer edge simply supported, inner edge simply supported

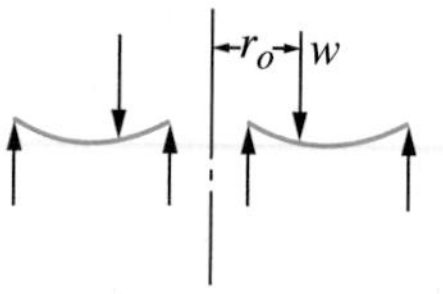

Boundary values: $y_b = 0, \; M_{rb} = 0, \; y_a = 0, \; M_{ra} = 0$

$$\theta_b = \frac{-wa^2}{D}\frac{C_3 L_9 - C_9 L_3}{C_1 C_9 - C_3 C_7}$$

$$Q_b = w\frac{C_1 L_9 - C_7 L_3}{C_1 C_9 - C_3 C_7}$$

$$\theta_a = \theta_b C_4 + Q_b\frac{a^2}{D}C_6 - \frac{wa^2}{D}L_6$$

$$Q_a = Q_b\frac{b}{a} - \frac{wr_o}{a}$$

Special cases:

b/a	0.1		0.5		0.7
r_o/a	0.5	0.7	0.7	0.9	0.9
$K_{y_{max}}$	−0.0102	−0.0113	−0.0023	−0.0017	−0.0005
K_{θ_a}	0.0278	0.0388	0.0120	0.0122	0.0055
K_{θ_b}	−0.0444	−0.0420	−0.0165	−0.0098	−0.0048
$K_{M_{tb}}$	−0.4043	−0.3819	−0.0301	−0.0178	−0.0063
$K_{M_{ro}}$	0.1629	0.1689	0.1161	0.0788	0.0662
K_{Q_b}	2.9405	2.4779	0.8114	0.3376	0.4145

(*Continued*)

TABLE 11.2 Formulas for Flat Circular Plates of Constant Thickness (*Continued*)

Case No., Edge Restraints	Boundary Values	Special Cases

1d. Outer edge simply supported, inner edge fixed

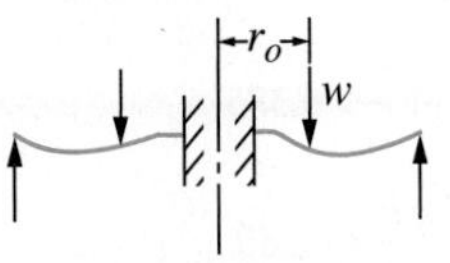

Boundary Values:

$y_b = 0,\ \theta_b = 0,\ y_a = 0,\ M_{ra} = 0$

$$M_{rb} = -wa\frac{C_3L_9 - C_9L_3}{C_2C_9 - C_3C_8}$$

$$Q_b = w\frac{C_2L_9 - C_8L_3}{C_2C_9 - C_3C_8}$$

$$\theta_a = M_{rb}\frac{a}{D}C_5 + Q_b\frac{a^2}{D}C_6 - \frac{wa^2}{D}L_6$$

$$Q_a = Q_b\frac{b}{a} - \frac{wr_o}{a}$$

Special Cases:

b/a	0.1		0.5		0.7
r_o/a	0.5	0.7	0.7	0.9	0.9
$K_{y_{max}}$	−0.0066	−0.0082	−0.0010	−0.0010	−0.0003
K_{θ_a}	0.0194	0.0308	0.0056	0.0084	0.0034
$K_{M_{rb}}$	−0.4141	−0.3911	−0.1172	−0.0692	−0.0519
K_{Q_b}	3.3624	2.8764	1.0696	0.4901	0.5972

1e. Outer edge fixed, inner edge free

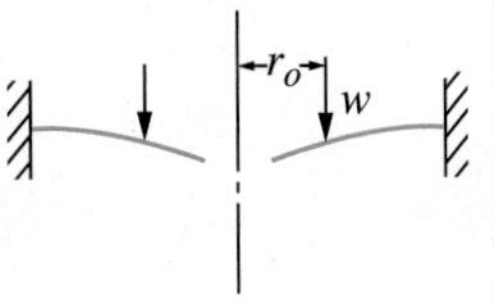

Boundary Values:

$M_{rb} = 0,\ Q_b = 0,\ y_a = 0,\ \theta_a = 0$

$$y_b = \frac{-wa^3}{D}\left(\frac{C_1L_6}{C_4} - L_3\right)$$

$$\theta_b = \frac{wa^2}{DC_4}L_6$$

$$M_{ra} = -wa\left(L_9 - \frac{C_7L_6}{C_4}\right)$$

$$Q_a = \frac{-wr_o}{a}$$

Special Cases:

If $r_o = b$ (load at inner edge),

b/a	0.1	0.3	0.5	0.7	0.9
K_{y_b}	−0.0143	−0.0330	−0.0233	−0.0071	−0.0003
K_{θ_b}	0.0254	0.0825	0.0776	0.0373	0.0048
$K_{M_{ra}}$	−0.0528	−0.1687	−0.2379	−0.2124	−0.0911
$K_{M_{tb}}$	0.2307	0.2503	0.1412	0.0484	0.0048

(*Note:* $|M_{ra}| > |M_{tb}|$ if $b/a > 0.385$)

1f. Outer edge fixed, inner edge guided

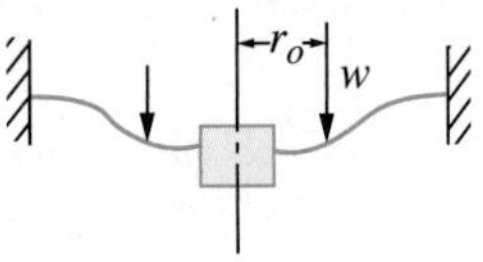

Boundary Values:

$\theta_b = 0,\ Q_b = 0,\ y_a = 0,\ \theta_a = 0$

$$y_b = \frac{-wa^3}{D}\left(\frac{C_2L_6}{C_5} - L_3\right)$$

$$M_{rb} = \frac{wa}{C_5}L_6$$

$$M_{ra} = -wa\left(L_9 - \frac{C_8L_6}{C_5}\right)$$

$$Q_a = \frac{-wr_o}{a}$$

Special Cases:

If $r_o = b$ (load at inner edge),

b/a	0.1	0.3	0.5	0.7	0.9
K_{y_b}	−0.0097	−0.0126	−0.0068	−0.0019	−0.0001
$K_{M_{rb}}$	0.1826	0.2469	0.2121	0.1396	0.0491
$K_{M_{ra}}$	−0.0477	−0.1143	−0.1345	−0.1101	−0.0458

1g. Outer edge fixed, inner edge simply supported

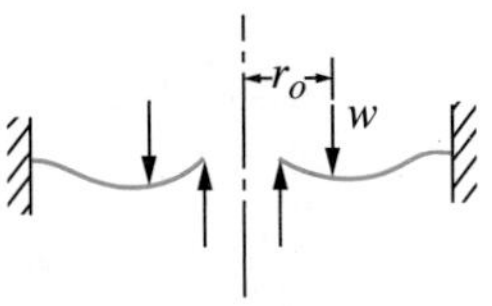

Boundary Values:

$y_b = 0,\ M_{rb} = 0,\ y_a = 0,\ \theta_a = 0$

$$\theta_b = \frac{-wa^2}{D}\frac{C_3L_6 - C_6L_3}{C_1C_6 - C_3C_4}$$

$$Q_b = w\frac{C_1L_6 - C_4L_3}{C_1C_6 - C_3C_4}$$

$$M_{ra} = \theta_b\frac{D}{a}C_7 + Q_b aC_9 - waL_9$$

$$Q_a = Q_b\frac{b}{a} - \frac{wr_o}{a}$$

Special Cases:

b/a	0.1		0.5		0.7
r_o/a	0.5	0.7	0.7	0.9	0.9
$K_{y_{max}}$	−0.0053	−0.0041	−0.0012	−0.0004	−0.0002
K_{θ_b}	−0.0262	−0.0166	−0.0092	−0.0023	−0.0018
$K_{M_{tb}}$	−0.2388	−0.1513	−0.0167	−0.0042	−0.0023
$K_{M_{ro}}$	0.1179	0.0766	0.0820	0.0208	0.0286
$K_{M_{ra}}$	−0.0893	−0.1244	−0.0664	−0.0674	−0.0521
K_{Q_b}	1.9152	1.0495	0.5658	0.0885	0.1784

1h. Outer edge fixed, inner edge fixed

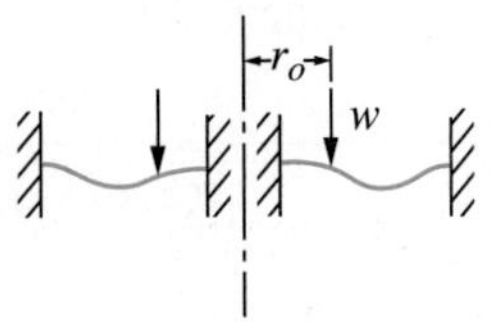

Boundary Values:

$y_b = 0,\ \theta_b = 0,\ y_a = 0,\ \theta_a = 0$

$$M_{rb} = -wa\frac{C_3L_6 - C_6L_3}{C_2C_6 - C_3C_5}$$

$$Q_b = w\frac{C_2L_6 - C_5L_3}{C_2C_6 - C_3C_5}$$

$$M_{ra} = M_{rb}C_8 + Q_b aC_9 - waL_9$$

$$Q_a = Q_b\frac{b}{a} - \frac{wr_o}{a}$$

Special Cases:

b/a	0.1		0.5		0.7
r_o/a	0.5	0.7	0.7	0.9	0.9
$K_{y_{max}}$	−0.0038	−0.0033	−0.0006	−0.0003	−0.0001
$K_{M_{rb}}$	−0.2792	−0.1769	−0.0856	−0.0216	−0.0252
$K_{M_{ra}}$	−0.0710	−0.1128	−0.0404	−0.0608	−0.0422
$K_{M_{ro}}$	0.1071	0.0795	0.0586	0.0240	0.0290
K_{Q_b}	2.4094	1.3625	0.8509	0.1603	0.3118

(*Continued*)

TABLE 11.2 Formulas for Flat Circular Plates of Constant Thickness (*Continued*)

Case No., Edge Restraints	Boundary Values	Special Cases
1i. Outer edge guided, inner edge simply supported 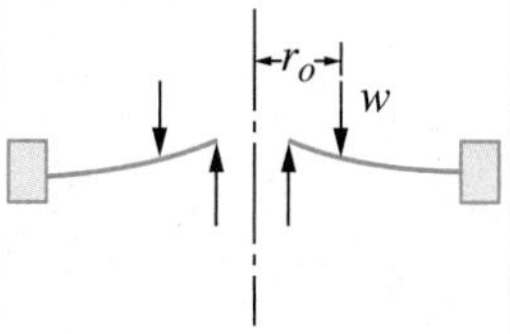	$y_b = 0,\ M_{rb} = 0,\ \theta_a = 0,\ Q_a = 0$ $\theta_b = \frac{-wa^2}{DC_4}\left(\frac{r_oC_6}{b} - L_6\right)$ $Q_b = \frac{wr_o}{b}$ $y_a = \frac{-wa^3}{D}\left[\frac{C_1}{C_4}\left(\frac{r_oC_6}{b} - L_6\right) - \frac{r_oC_3}{b} + L_3\right]$ $M_{ra} = wa\left[\frac{C_7}{C_4}\left(L_6 - \frac{r_oC_6}{b}\right) + \frac{r_oC_9}{b} - L_9\right]$	If $r_o = a$ (load at outer edge), $y_{\max} = y_a = \frac{-wa^4}{bD}\left(\frac{C_1C_6}{C_4} - C_3\right)$ $M_{\max} = M_{ra} = \frac{wa^2}{b}\left(C_9 - \frac{C_6C_7}{C_4}\right)$ if $\frac{b}{a} > 0.385$ $M_{\max} = M_{tb} = \frac{-wa^3}{b^2}(1-\nu^2)\frac{C_6}{C_4}$ if $\frac{b}{a} < 0.385$ (For numerical values, see case 1e after computing the loading at the inner edge.)
1j. Outer edge guided, inner edge fixed 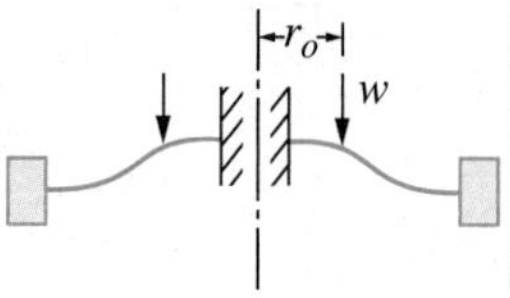	$y_b = 0,\ \theta_b = 0,\ \theta_a = 0,\ Q_a = 0$ $M_{rb} = \frac{-wa}{C_5}\left(\frac{r_oC_6}{b} - L_6\right)$ $Q_b = \frac{wr_o}{b}$ $y_a = \frac{-wa^3}{D}\left[\frac{C_2}{C_5}\left(\frac{r_oC_6}{b} - L_6\right) - \frac{r_oC_3}{b} + L_3\right]$ $M_{ra} = wa\left[\frac{C_8}{C_5}\left(L_6 - \frac{r_oC_6}{b}\right) + \frac{r_oC_9}{b} - L_9\right]$	If $r_o = a$ (load at outer edge), $y_{\max} = y_a = \frac{-wa^4}{bD}\left(\frac{C_2C_6}{C_5} - C_3\right)$ $M_{\max} = M_{rb} = \frac{-wa^2C_6}{bC_5}$ (For numerical values, see case 1f after computing the loading at the inner edge.)
1k. Outer edge free, inner edge simply supported 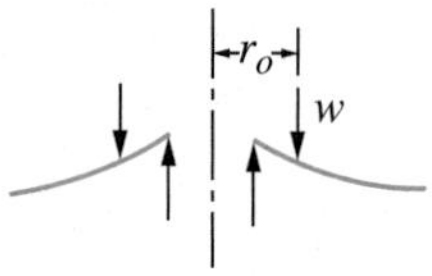	$y_b = 0,\ M_{rb} = 0,\ M_{ra} = 0,\ Q_a = 0$ $\theta_b = \frac{-wa^2}{DC_7}\left(\frac{r_oC_9}{b} - L_9\right)$ $Q_b = \frac{wr_o}{b}$ $y_a = \frac{-wa^3}{D}\left[\frac{C_1}{C_7}\left(\frac{r_oC_9}{b} - L_9\right) - \frac{r_oC_3}{b} + L_3\right]$ $\theta_a = \frac{-wa^2}{D}\left[\frac{C_4}{C_7}\left(\frac{r_oC_9}{b} - L_9\right) - \frac{r_oC_6}{b} + L_6\right]$	If $r_o = a$ (load at outer edge), $y_{\max} = y_a = \frac{-wa^4}{bD}\left(\frac{C_1C_9}{C_7} - C_3\right)$ $M_{\max} = M_{tb} = \frac{-wa^3}{b^2}(1-\nu^2)\frac{C_9}{C_7}$ (For numerical values, see case 1a after computing the loading at the inner edge.)
1l. Outer edge free, inner edge fixed 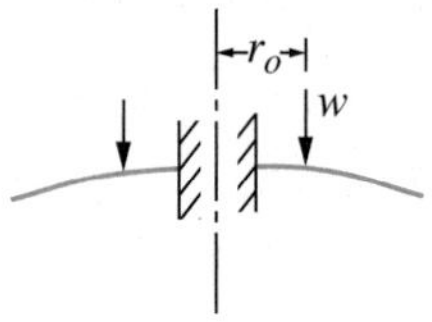	$y_b = 0,\ \theta_b = 0,\ M_{ra} = 0,\ Q_a = 0$ $M_{rb} = \frac{-wa}{C_8}\left(\frac{r_oC_9}{b} - L_9\right)$ $Q_b = \frac{wr_o}{b}$ $y_a = \frac{-wa^3}{D}\left[\frac{C_2}{C_8}\left(\frac{r_oC_9}{b} - L_9\right) - \frac{r_oC_3}{b} + L_3\right]$ $\theta_a = \frac{-wa^2}{D}\left[\frac{C_5}{C_8}\left(\frac{r_oC_9}{b} - L_9\right) - \frac{r_oC_6}{b} + L_6\right]$	If $r_o = a$ (load at outer edge), $y_{\max} = y_a = \frac{-wa^4}{bD}\left(\frac{C_2C_9}{C_8} - C_3\right)$ $M_{\max} = M_{rb} = \frac{-wa^2}{b}\frac{C_9}{C_8}$ (For numerical values, see case 1b after computing the loading at the inner edge.)

(*Continued*)

TABLE 11.2 Formulas for Flat Circular Plates of Constant Thickness (*Continued*)

Case 2. Annular plate with a uniformly distributed pressure q over the portion from r_o to a

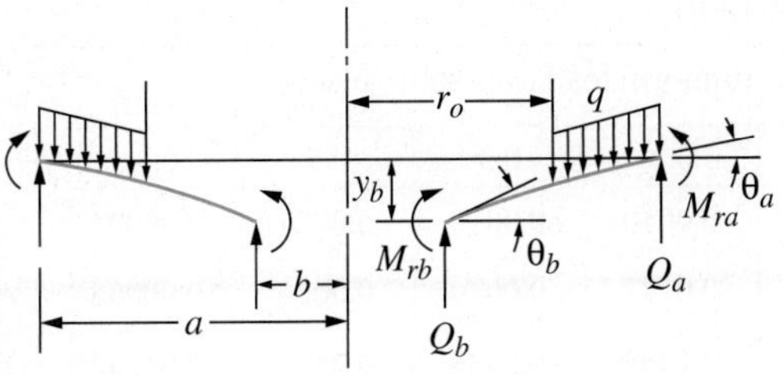

General expressions for deformations, moments, and shears:

$$y = y_b + \theta_b r F_1 + M_{rb}\frac{r^2}{D}F_2 + Q_b\frac{r^3}{D}F_3 - q\frac{r^4}{D}G_{11}$$

$$\theta = \theta_b F_4 + M_{rb}\frac{r}{D}F_5 + Q_b\frac{r^2}{D}F_6 - q\frac{r^3}{D}G_{14}$$

$$M_r = \theta_b\frac{D}{r}F_7 + M_{rb}F_8 + Q_b r F_9 - qr^2 G_{17}$$

$$M_t = \frac{\theta D(1-\nu^2)}{r} + \nu M_r$$

$$Q = Q_b\frac{b}{r} - \frac{q}{2r}(r^2 - r_o^2)^0\langle r - r_o\rangle^0$$

For the numerical data given below, $\nu = 0.3$

$$y = K_y\frac{qa^4}{D}, \quad \theta = K_\theta\frac{qa^3}{D}, \quad M = K_M qa^2, \quad Q = K_Q qa$$

Case No., Edge Restraints	Boundary Values	Special Cases

2a. Outer edge simply supported, inner edge free

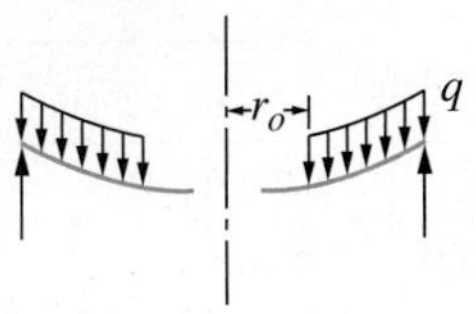

Boundary values: $M_{rb} = 0, \quad Q_b = 0, \quad y_a = 0, \quad M_{ra} = 0$

$$y_b = \frac{-qa^4}{D}\left(\frac{C_1 L_{17}}{C_7} - L_{11}\right)$$

$$\theta_b = \frac{qa^3}{DC_7}L_{17}$$

$$\theta_a = \frac{qa^3}{D}\left(\frac{C_4 L_{17}}{C_7} - L_{14}\right)$$

$$Q_a = \frac{-q}{2a}(a^2 - r_o^2)$$

Special cases: $y_{max} = y_b, \quad M_{max} = M_{tb}$

If $r_o = b$ (uniform load over entire plate),

b/a	0.1	0.3	0.5	0.7	0.9
K_{y_b}	−0.0687	−0.0761	−0.0624	−0.0325	−0.0048
K_{θ_a}	0.0986	0.1120	0.1201	0.1041	0.0477
K_{θ_b}	0.0436	0.1079	0.1321	0.1130	0.0491
$K_{M_{tb}}$	0.3965	0.3272	0.2404	0.1469	0.0497

2b. Outer edge simply supported, inner edge guided

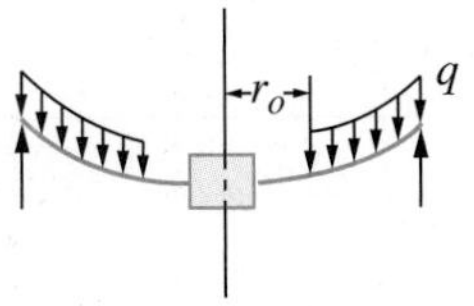

Boundary values: $\theta_b = 0, \quad Q_b = 0, \quad y_a = 0, \quad M_{ra} = 0$

$$y_b = \frac{-qa^4}{D}\left(\frac{C_2 L_{17}}{C_8} - L_{11}\right)$$

$$M_{rb} = \frac{qa^2}{C_8}L_{17}$$

$$\theta_a = \frac{qa^3}{D}\left(\frac{C_5 L_{17}}{C_8} - L_{14}\right)$$

$$Q_a = \frac{-q}{2a}(a^2 - r_o^2)$$

Special cases: $y_{max} = y_b, \quad M_{max} = M_{rb}$

If $r_o = b$ (uniform load over entire plate),

b/a	0.1	0.3	0.5	0.7	0.9
K_{y_b}	−0.0575	−0.0314	−0.0103	−0.0015	−0.00002
K_{θ_a}	0.0919	0.0645	0.0306	0.0078	0.00032
$K_{M_{rb}}$	0.3003	0.2185	0.1223	0.0456	0.00505

2c. Outer edge simply supported, inner edge simply supported

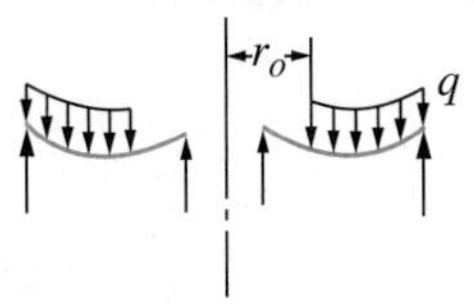

Boundary values: $y_b = 0, \quad M_{rb} = 0, \quad y_a = 0, \quad M_{ra} = 0$

$$\theta_b = \frac{-qa^3}{D}\frac{C_3 L_{17} - C_9 L_{11}}{C_1 C_9 - C_3 C_7}$$

$$Q_b = qa\frac{C_1 L_{17} - C_7 L_{11}}{C_1 C_9 - C_3 C_7}$$

$$\theta_a = \theta_b C_4 + Q_b\frac{a^2}{D}C_6 - \frac{qa^3}{D}L_{14}$$

$$Q_a = Q_b\frac{b}{a} - \frac{q}{2a}(a^2 - r_o^2)$$

Special cases: If $r_o = b$ (uniform load over entire plate),

b/a	0.1	0.3	0.5	0.7
$K_{y_{max}}$	−0.0060	−0.0029	−0.0008	−0.0001
K_{θ_b}	−0.0264	−0.0153	−0.0055	−0.0012
K_{θ_a}	0.0198	0.0119	0.0047	0.0011
$K_{M_{tb}}$	−0.2401	−0.0463	−0.0101	−0.0015
$K_{M_{r\,max}}$	0.0708	0.0552	0.0300	0.0110
K_{Q_b}	1.8870	0.6015	0.3230	0.1684

(*Continued*)

TABLE 11.2 Formulas for Flat Circular Plates of Constant Thickness (*Continued*)

Case No., Edge Restraints	Boundary Values	Special Cases

2d. Outer edge simply supported, inner edge fixed

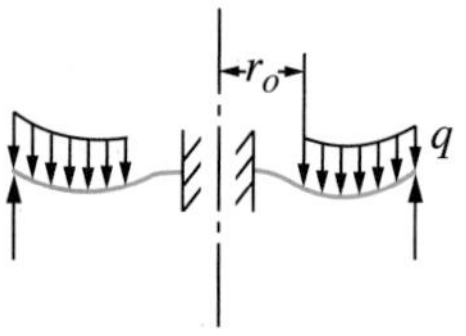

$y_b = 0,\ \theta_b = 0,\ y_a = 0,\ M_{ra} = 0$

$$M_{rb} = -qa^2 \frac{C_3 L_{17} - C_9 L_{11}}{C_2 C_9 - C_3 C_8}$$

$$Q_b = qa \frac{C_2 L_{17} - C_8 L_{11}}{C_2 C_9 - C_3 C_8}$$

$$\theta_a = M_{rb} \frac{a}{D} C_5 + Q_b \frac{a^2}{D} C_6 - \frac{qa^3}{D} L_{14}$$

$$Q_a = Q_b \frac{b}{a} - \frac{q}{2a}(a^2 - r_o^2)$$

If $r_o = b$ (uniform load over entire plate),

b/a	0.1	0.3	0.5	0.7
$K_{y_{\max}}$	−0.0040	−0.0014	−0.0004	−0.00004
K_{θ_a}	0.0147	0.0070	0.0026	0.00056
$K_{M_{tb}}$	−0.2459	−0.0939	−0.0393	−0.01257
K_{Q_b}	2.1375	0.7533	0.4096	0.21259

2e. Outer edge fixed, inner edge free

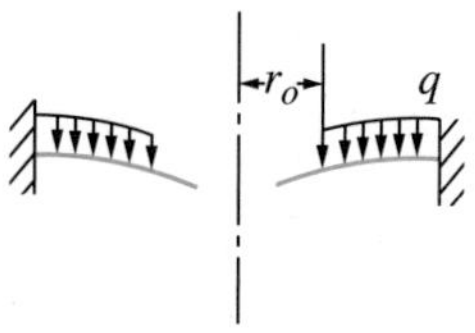

$M_{rb} = 0,\ Q_b = 0,\ y_a = 0,\ \theta_a = 0$

$$y_b = \frac{-qa^4}{D}\left(\frac{C_1 L_{14}}{C_4} - L_{11}\right)$$

$$\theta_b = \frac{qa^3 L_{14}}{DC_4}$$

$$M_{ra} = -qa^2\left(L_{17} - \frac{C_7}{C_4} L_{14}\right)$$

$$Q_a = \frac{-q}{2a}(a^2 - r_o^2)$$

If $r_o = b$ (uniform load over entire plate),

b/a	0.1	0.3	0.5	0.7	0.9
K_{y_b}	−0.0166	−0.0132	−0.0053	−0.0009	−0.00001
K_{θ_b}	0.0159	0.0256	0.0149	0.0040	0.00016
$K_{M_{ra}}$	−0.1246	−0.1135	−0.0800	−0.0361	−0.00470
$K_{M_{tb}}$	0.1448	0.0778	0.0271	0.0052	0.00016

2f. Outer edge fixed, inner edge guided

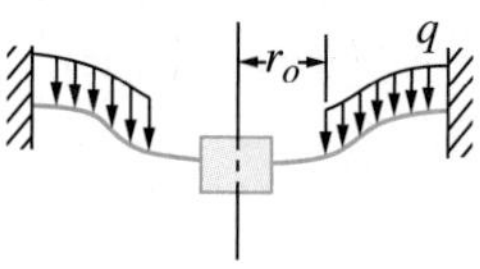

$\theta_b = 0,\ Q_b = 0,\ y_a = 0,\ \theta_a = 0$

$$y_b = \frac{-qa^4}{D}\left(\frac{C_2 L_{14}}{C_5} - L_{11}\right)$$

$$M_{rb} = \frac{qa^2 L_{14}}{C_5}$$

$$M_{ra} = -qa^2\left(L_{17} - \frac{C_8}{C_5} L_{14}\right)$$

$$Q_a = \frac{-q}{2a}(a^2 - r_o^2)$$

If $r_o = b$ (uniform load over entire plate),

b/a	0.1	0.3	0.5	0.7	0.9
K_{y_b}	−0.0137	−0.0068	−0.0021	−0.0003	
$K_{M_{rb}}$	0.1146	0.0767	0.0407	0.0149	0.00167
$K_{M_{ra}}$	−0.1214	−0.0966	−0.0601	−0.0252	−0.00316

2g. Outer edge fixed, inner edge simply supported

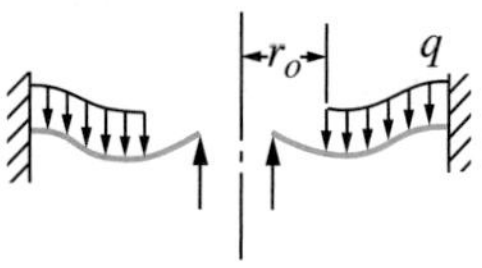

$y_b = 0,\ M_{rb} = 0,\ y_a = 0,\ \theta_a = 0$

$$\theta_B = \frac{-qa^3}{D} \frac{C_3 L_{14} - C_6 L_{11}}{C_1 C_6 - C_3 C_4}$$

$$Q_b = qa \frac{C_1 L_{14} - C_4 L_{11}}{C_1 C_6 - C_3 C_4}$$

$$M_{ra} = \theta_b \frac{D}{a} C_7 + Q_b a C_9 - qa^2 L_{17}$$

$$Q_a = Q_b \frac{b}{a} - \frac{q}{2a}(a^2 - r_o^2)$$

If $r_o = b$ (uniform load over entire plate),

b/a	0.1	0.3	0.5	0.7	0.9
$K_{y_{\max}}$	−0.0025	−0.0012	−0.0003		
K_{θ_b}	−0.0135	−0.0073	−0.0027	−0.0006	
$K_{M_{tb}}$	−0.1226	−0.0221	−0.0048	−0.0007	
$K_{M_{ra}}$	−0.0634	−0.0462	−0.0262	−0.0102	−0.0012
K_{Q_b}	1.1591	0.3989	0.2262	0.1221	0.0383

(*Continued*)

TABLE 11.2 Formulas for Flat Circular Plates of Constant Thickness (*Continued*)

<table>
<tr><th>Case No., Edge Restraints</th><th>Boundary Values</th><th>Special Cases</th></tr>
<tr>
<td>2h. Outer edge fixed, inner edge fixed
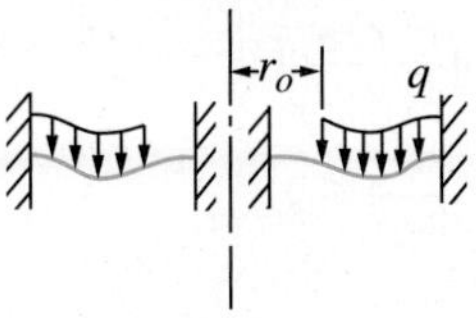</td>
<td>$y_b = 0,\ \theta_b = 0,\ y_a = 0,\ \theta_a = 0$
$M_{rb} = -qa^2 \dfrac{C_3 L_{14} - C_6 L_{11}}{C_2 C_6 - C_3 C_5}$
$Q_b = qa \dfrac{C_2 L_{14} - C_5 L_{11}}{C_2 C_6 - C_3 C_5}$
$M_{ra} = M_{rb} C_8 + Q_b a C_9 - qa^2 L_{17}$
$Q_a = Q_b \dfrac{b}{a} - \dfrac{q}{2a}(a^2 - r_o^2)$</td>
<td>If $r_o = b$ (uniform load over entire plate),
<table>
<tr><th>b/a</th><th>0.1</th><th>0.3</th><th>0.5</th><th>0.7</th></tr>
<tr><td>$K_{y_{max}}$</td><td>−0.0018</td><td>−0.0006</td><td>−0.0002</td><td></td></tr>
<tr><td>$K_{M_{rb}}$</td><td>−0.1433</td><td>−0.0570</td><td>−0.0247</td><td>−0.0081</td></tr>
<tr><td>$K_{M_{ra}}$</td><td>−0.0540</td><td>−0.0347</td><td>−0.0187</td><td>−0.0070</td></tr>
<tr><td>K_{Q_b}</td><td>1.4127</td><td>0.5414</td><td>0.3084</td><td>0.1650</td></tr>
</table></td>
</tr>
<tr>
<td>2i. Outer edge guided, inner edge simply supported
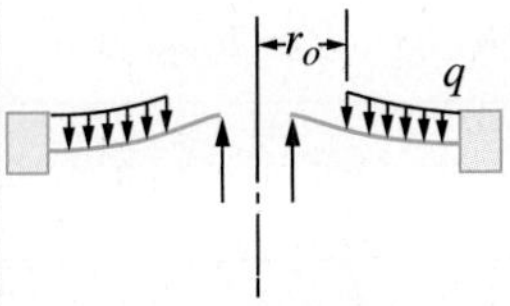</td>
<td>$y_b = 0,\ M_{rb} = 0,\ \theta_a = 0,\ Q_a = 0$
$\theta_b = \dfrac{-qa^3}{DC_4}\left[\dfrac{C_6}{2ab}(a^2 - r_o^2) - L_{14}\right]$
$Q_b = \dfrac{q}{2b}(a^2 - r_o^2)$
$y_a = \theta_b a C_1 + Q_b \dfrac{a^3}{D} C_3 - \dfrac{qa^4}{D} L_{11}$
$M_{ra} = \theta_b \dfrac{D}{a} C_7 + Q_b a C_9 - qa^2 L_{17}$</td>
<td>If $r_o = b$ (uniform load over entire plate),
<table>
<tr><th>b/a</th><th>0.1</th><th>0.3</th><th>0.5</th><th>0.7</th><th>0.9</th></tr>
<tr><td>K_{y_a}</td><td>−0.0543</td><td>−0.0369</td><td>−0.0122</td><td>−0.0017</td><td>−0.00002</td></tr>
<tr><td>K_{θ_b}</td><td>−0.1096</td><td>−0.0995</td><td>−0.0433</td><td>−0.0096</td><td>−0.00034</td></tr>
<tr><td>$K_{M_{ra}}$</td><td>0.1368</td><td>0.1423</td><td>0.0985</td><td>0.0412</td><td>0.00491</td></tr>
<tr><td>$K_{M_{tb}}$</td><td>−0.9971</td><td>−0.3018</td><td>−0.0788</td><td>−0.0125</td><td>−0.00035</td></tr>
</table></td>
</tr>
<tr>
<td>2j. Outer edge guided, inner edge fixed
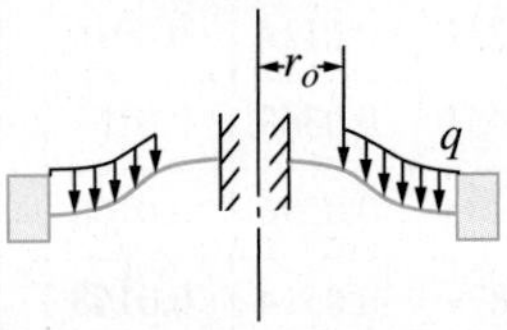</td>
<td>$y_b = 0,\ \theta_b = 0,\ \theta_a = 0,\ Q_a = 0$
$M_{rb} = \dfrac{-qa^2}{C_5}\left[\dfrac{C_6}{2ab}(a^2 - r_o^2) - L_{14}\right]$
$Q_b = \dfrac{q}{2b}(a^2 - r_o^2)$
$y_a = M_{rb} \dfrac{a^2}{D} C_2 + Q_b \dfrac{a^3}{D} C_3 - \dfrac{qa^4}{D} L_{11}$
$M_{ra} = M_{rb} C_8 + Q_b a C_9 - qa^2 L_{17}$</td>
<td>If $r_o = b$ (uniform load over entire plate),
<table>
<tr><th>b/a</th><th>0.1</th><th>0.3</th><th>0.5</th><th>0.7</th><th>0.9</th></tr>
<tr><td>K_{y_a}</td><td>−0.0343</td><td>−0.0123</td><td>−0.0030</td><td>−0.0004</td><td></td></tr>
<tr><td>$K_{M_{rb}}$</td><td>−0.7892</td><td>−0.2978</td><td>−0.1184</td><td>−0.0359</td><td>−0.00351</td></tr>
<tr><td>$K_{M_{ra}}$</td><td>0.1146</td><td>0.0767</td><td>0.0407</td><td>0.0149</td><td>0.00167</td></tr>
</table></td>
</tr>
<tr>
<td>2k. Outer edge free, inner edge simply supported
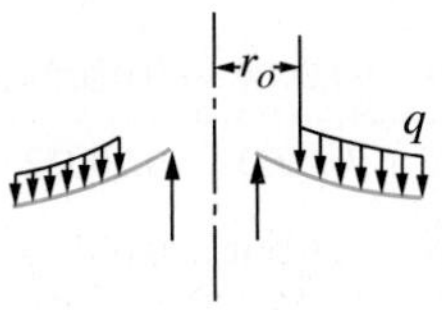</td>
<td>$y_b = 0,\ M_{rb} = 0,\ M_{ra} = 0,\ Q_a = 0$
$\theta_b = \dfrac{-qa^3}{DC_7}\left[\dfrac{C_9}{2ab}(a^2 - r_o^2) - L_{17}\right]$
$Q_b = \dfrac{q}{2b}(a^2 - r_o^2)$
$y_a = \theta_b a C_1 + Q_b \dfrac{a^3}{D} C_3 - \dfrac{qa^4}{D} L_{11}$
$\theta_a = \theta_b C_4 + Q_b \dfrac{a^2}{D} C_6 - \dfrac{qa^3}{D} L_{14}$</td>
<td>If $r_o = b$ (uniform load over entire plate),
<table>
<tr><th>b/a</th><th>0.1</th><th>0.3</th><th>0.5</th><th>0.7</th><th>0.9</th></tr>
<tr><td>K_{y_a}</td><td>−0.1115</td><td>−0.1158</td><td>−0.0826</td><td>−0.0378</td><td>−0.0051</td></tr>
<tr><td>K_{θ_b}</td><td>−0.1400</td><td>−0.2026</td><td>−0.1876</td><td>−0.1340</td><td>−0.0515</td></tr>
<tr><td>K_{θ_a}</td><td>−0.1082</td><td>−0.1404</td><td>−0.1479</td><td>−0.1188</td><td>−0.0498</td></tr>
<tr><td>$K_{M_{tb}}$</td><td>−1.2734</td><td>−0.6146</td><td>−0.3414</td><td>−0.1742</td><td>−0.0521</td></tr>
</table></td>
</tr>
<tr>
<td>2l. Outer edge free, inner edge fixed
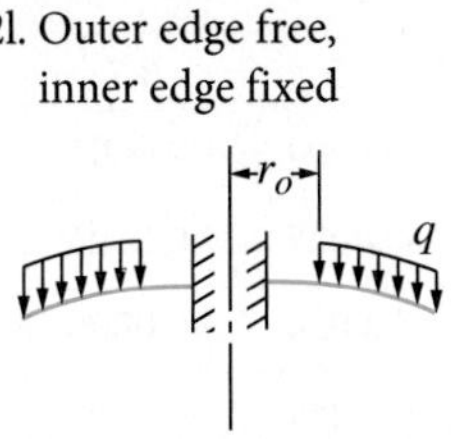</td>
<td>$y_b = 0,\ \theta_b = 0,\ M_{ra} = 0,\ Q_a = 0$
$M_{rb} = \dfrac{-qa^2}{C_8}\left[\dfrac{C_9}{2ab}(a^2 - r_o^2) - L_{17}\right]$
$Q_b = \dfrac{q}{2b}(a^2 - r_o^2)$
$y_a = M_{rb} \dfrac{a^2}{D} C_2 + Q_b \dfrac{a^3}{D} C_3 - \dfrac{qa^4}{D} L_{11}$
$\theta_a = M_{rb} \dfrac{a}{D} C_5 + Q_b \dfrac{a^2}{D} C_6 - \dfrac{qa^3}{D} L_{14}$</td>
<td>If $r_o = b$ (uniform load over entire plate),
<table>
<tr><th>b/a</th><th>0.1</th><th>0.3</th><th>0.5</th><th>0.7</th><th>0.9</th></tr>
<tr><td>K_{y_a}</td><td>−0.0757</td><td>−0.0318</td><td>−0.0086</td><td>−0.0011</td><td></td></tr>
<tr><td>K_{θ_a}</td><td>−0.0868</td><td>−0.0512</td><td>−0.0207</td><td>−0.0046</td><td>−0.00017</td></tr>
<tr><td>$K_{M_{rb}}$</td><td>−0.9646</td><td>−0.4103</td><td>−0.1736</td><td>−0.0541</td><td>−0.00530</td></tr>
</table></td>
</tr>
</table>

(*Continued*)

TABLE 11.2 Formulas for Flat Circular Plates of Constant Thickness (*Continued*)

Case 3. Annular plate with a distributed pressure increasing linearly from zero at r_o to q at a

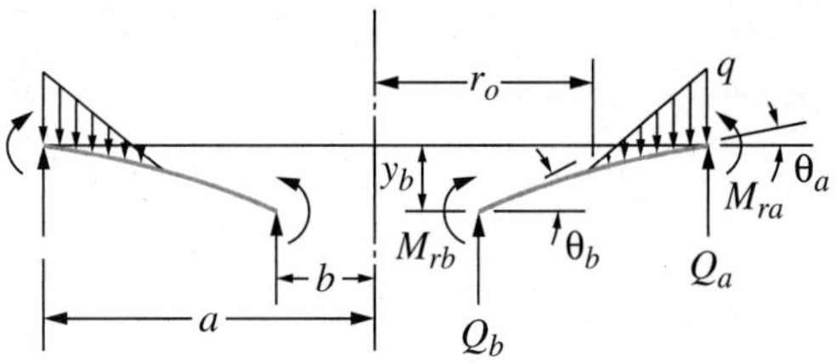

General expressions for deformations, moments, and shears:

$$y = y_b + \theta_b r F_1 + M_{rb}\frac{r^2}{D}F_2 + Q_b\frac{r^3}{D}F_3 - q\frac{r^4}{D}\frac{r-r_o}{a-r_o}G_{12}$$

$$\theta = \theta_b F_4 + M_{rb}\frac{r}{D}F_5 + Q_b\frac{r^2}{D}F_6 - q\frac{r^3}{D}\frac{r-r_o}{a-r_o}G_{15}$$

$$M_r = \theta_b\frac{D}{r}F_7 + M_{rb}F_8 + Q_b r F_9 - qr^2\frac{r-r_o}{a-r_o}G_{18}$$

$$M_t = \frac{\theta D(1-\nu^2)}{r} + \nu M_r$$

$$Q = Q_b\frac{b}{r} - \frac{q}{6r(a-r_o)}(2r^3 - 3r_o r^2 + r_o^3)\langle r-r_o\rangle^0$$

For the numerical data given below, $\nu = 0.3$

$$y = K_y\frac{qa^4}{D},\quad \theta = K_\theta\frac{qa^3}{D},\quad M = K_M qa^2,\quad Q = K_Q qa$$

Case No., Edge Restraints	Boundary Values	Special Cases

3a. Outer edge simply supported, inner edge free

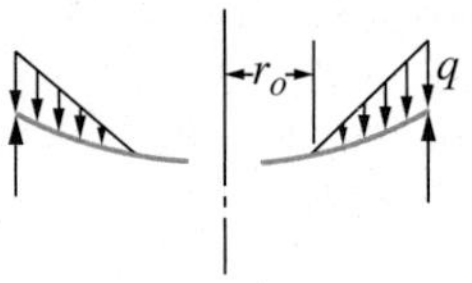

Boundary Values: $M_{rb} = 0,\ Q_b = 0,\ y_a = 0,\ M_{ra} = 0$

$$y_b = \frac{-qa^4}{D}\left(\frac{C_1 L_{18}}{C_7} - L_{12}\right)$$

$$\theta_b = \frac{qa^3}{DC_7}L_{18}$$

$$\theta_a = \frac{qa^3}{D}\left(\frac{C_4 L_{18}}{C_7} - L_{15}\right)$$

$$Q_a = \frac{-q}{6a}(2a^2 - r_o a - r_o^2)$$

Special Cases: $y_{max} = y_b$, $M_{max} = M_{tb}$

If $r_o = b$ (linearly increasing load from b to a),

b/a	0.1	0.3	0.5	0.7	0.9
K_{y_b}	−0.0317	−0.0306	−0.0231	−0.0114	−0.0016
K_{θ_a}	0.0482	0.0470	0.0454	0.0368	0.0161
K_{θ_b}	0.0186	0.0418	0.0483	0.0396	0.0166
$K_{M_{tb}}$	0.1690	0.1269	0.0879	0.0514	0.0168

3b. Outer edge simply supported, inner edge guided

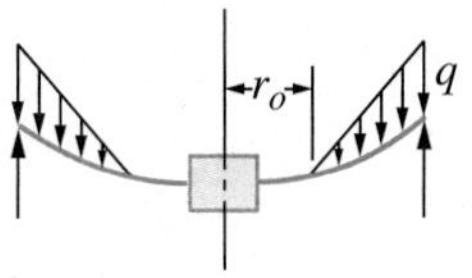

Boundary Values: $\theta_b = 0,\ Q_b = 0,\ y_a = 0,\ M_{ra} = 0$

$$y_b = \frac{-qa^4}{D}\left(\frac{C_2 L_{18}}{C_8} - L_{12}\right)$$

$$M_{rb} = \frac{qa^2 L_{18}}{C_8}$$

$$\theta_a = \frac{qa^3}{D}\left(\frac{C_5 L_{18}}{C_8} - L_{15}\right)$$

$$Q_a = \frac{-q}{6a}(2a^2 - r_o a - r_o^2)$$

Special Cases: $y_{max} = y_b$, $M_{max} = M_{rb}$

If $r_o = b$ (linearly increasing load from b to a),

b/a	0.1	0.3	0.5	0.7	0.9
K_{y_b}	−0.0269	−0.0133	−0.0041	−0.0006	−0.00001
K_{θ_a}	0.0454	0.0286	0.0126	0.0031	0.00012
$K_{M_{rb}}$	0.1280	0.0847	0.0447	0.0160	0.00171

3c. Outer edge simply supported, inner edge simply supported

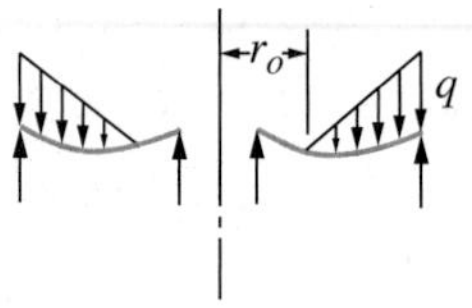

Boundary Values: $y_b = 0,\ M_{rb} = 0,\ y_a = 0,\ M_{ra} = 0$

$$\theta_b = \frac{-qa^3}{D}\frac{C_3 L_{18} - C_9 L_{12}}{C_1 C_9 - C_3 C_7}$$

$$Q_b = qa\frac{C_1 L_{18} - C_7 L_{12}}{C_1 C_9 - C_3 C_7}$$

$$\theta_a = \theta_b C_4 + Q_b\frac{a^2}{D}C_6 - \frac{qa^3}{D}L_{15}$$

$$Q_a = Q_b\frac{b}{a} - \frac{q}{6a}(2a^2 - r_o a - r_o^2)$$

Special Cases: If $r_o = b$ (linearly increasing load from b to a),

b/a	0.1	0.3	0.5	0.7
$K_{y_{max}}$	−0.0034	−0.0015	−0.0004	−0.0001
K_{θ_b}	−0.0137	−0.0077	−0.0027	−0.0006
K_{θ_a}	0.0119	0.0068	0.0026	0.0006
$K_{M_{tb}}$	−0.1245	−0.0232	−0.0049	−0.0007
$K_{M_r\,max}$	0.0407	0.0296	0.0159	0.0057
K_{Q_b}	0.8700	0.2417	0.1196	0.0591

(*Continued*)

TABLE 11.2 Formulas for Flat Circular Plates of Constant Thickness (*Continued*)

<table>
<tr><th>Case No., Edge Restraints</th><th>Boundary Values</th><th>Special Cases</th></tr>
<tr><td>3d. Outer edge simply supported, inner edge fixed
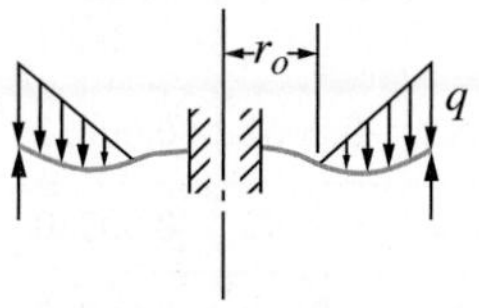</td>
<td>$y_b = 0,\ \theta_b = 0,\ y_a = 0,\ M_{ra} = 0$
$M_{rb} = -qa^2 \dfrac{C_3 L_{18} - C_9 L_{12}}{C_2 C_9 - C_3 C_8}$
$Q_b = qa \dfrac{C_2 L_{18} - C_8 L_{12}}{C_2 C_9 - C_3 C_8}$
$\theta_a = M_{rb} \dfrac{a}{D} C_5 + Q_b \dfrac{a^2}{D} C_6 - \dfrac{qa^3}{D} L_{15}$
$Q_a = Q_b \dfrac{b}{a} - \dfrac{q}{6a}(2a^2 - r_o a - r_o^2)$</td>
<td>If $r_o = b$ (linearly increasing load from b to a),
<table>
<tr><th>b/a</th><th>0.1</th><th>0.3</th><th>0.5</th><th>0.7</th></tr>
<tr><td>$K_{y_{max}}$</td><td>−0.0024</td><td>−0.0008</td><td>−0.0002</td><td>−0.00002</td></tr>
<tr><td>K_{θ_a}</td><td>0.0093</td><td>0.0044</td><td>0.0016</td><td>0.00034</td></tr>
<tr><td>$K_{M_{rb}}$</td><td>−0.1275</td><td>−0.0470</td><td>−0.0192</td><td>−0.00601</td></tr>
<tr><td>K_{Q_b}</td><td>0.9999</td><td>0.3178</td><td>0.1619</td><td>0.08029</td></tr>
</table></td></tr>
<tr><td>3e. Outer edge fixed, inner edge free
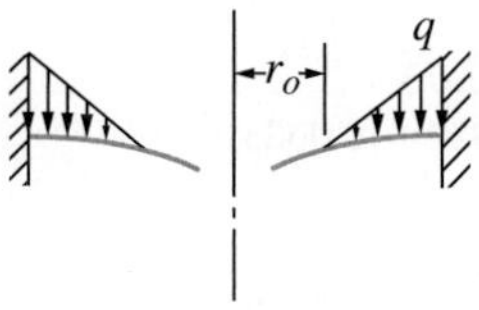</td>
<td>$M_{rb} = 0,\ Q_b = 0,\ y_a = 0,\ \theta_a = 0$
$y_b = \dfrac{-qa^4}{D}\left(\dfrac{C_1 L_{15}}{C_4} - L_{12}\right)$
$\theta_b = \dfrac{qa^3 L_{15}}{D C_4}$
$M_{ra} = -qa^2\left(L_{18} - \dfrac{C_7}{C_4} L_{15}\right)$
$Q_a = \dfrac{-q}{6a}(2a^2 - r_o a - r_o^2)$</td>
<td>If $r_o = b$ (linearly increasing load from b to a),
<table>
<tr><th>b/a</th><th>0.1</th><th>0.3</th><th>0.5</th><th>0.7</th><th>0.9</th></tr>
<tr><td>K_{y_b}</td><td>−0.0062</td><td>−0.0042</td><td>−0.0015</td><td>−0.00024</td><td></td></tr>
<tr><td>K_{θ_b}</td><td>0.0051</td><td>0.0073</td><td>0.0040</td><td>0.00103</td><td>0.00004</td></tr>
<tr><td>$K_{M_{ra}}$</td><td>−0.0609</td><td>−0.0476</td><td>−0.0302</td><td>−0.01277</td><td>−0.00159</td></tr>
<tr><td>$K_{M_{tb}}$</td><td>0.0459</td><td>0.0222</td><td>0.0073</td><td>0.00134</td><td>0.00004</td></tr>
</table></td></tr>
<tr><td>3f. Outer edge fixed, inner edge guided
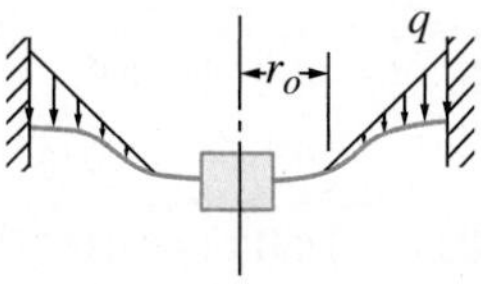</td>
<td>$\theta_b = 0,\ Q_b = 0,\ y_a = 0,\ \theta_a = 0$
$y_b = \dfrac{-qa^4}{D}\left(\dfrac{C_2 L_{15}}{C_5} - L_{12}\right)$
$M_{rb} = \dfrac{qa^2 L_{15}}{C_5}$
$M_{ra} = -qa^2\left(L_{18} - \dfrac{C_8}{C_5} L_{15}\right)$
$Q_a = \dfrac{-q}{6a}(2a^2 - r_o a - r_o^2)$</td>
<td>If $r_o = b$ (linearly increasing load from b to a),
<table>
<tr><th>b/a</th><th>0.1</th><th>0.3</th><th>0.5</th><th>0.7</th><th>0.9</th></tr>
<tr><td>K_{y_b}</td><td>−0.0053</td><td>−0.0024</td><td>−0.0007</td><td>−0.0001</td><td></td></tr>
<tr><td>$K_{M_{rb}}$</td><td>0.0364</td><td>0.0219</td><td>0.0110</td><td>0.0039</td><td>0.00042</td></tr>
<tr><td>$K_{M_{tb}}$</td><td>−0.0599</td><td>−0.0428</td><td>−0.0249</td><td>−0.0099</td><td>−0.00120</td></tr>
</table></td></tr>
<tr><td>3g. Outer edge fixed, inner edge simply supported
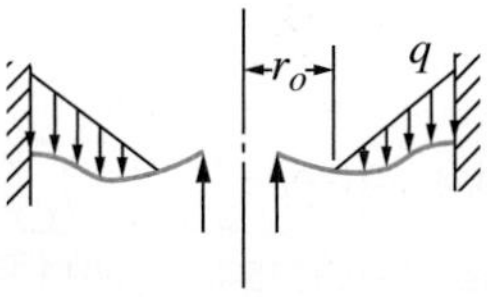</td>
<td>$y_b = 0,\ M_{rb} = 0,\ y_a = 0,\ \theta_a = 0$
$\theta_b = \dfrac{-qa^3}{D} \dfrac{C_3 L_{15} - C_6 L_{12}}{C_1 C_6 - C_3 C_4}$
$Q_b = qa \dfrac{C_1 L_{15} - C_4 L_{12}}{C_1 C_6 - C_3 C_4}$
$M_{ra} = \theta_b \dfrac{D}{a} C_7 + Q_b a C_9 - qa^2 L_{18}$
$Q_a = Q_b \dfrac{b}{a} - \dfrac{q}{6a}(2a^2 - r_o a - r_o^2)$</td>
<td>If $r_o = b$ (linearly increasing load from b to a),
<table>
<tr><th>b/a</th><th>0.1</th><th>0.3</th><th>0.5</th><th>0.7</th><th>0.9</th></tr>
<tr><td>$K_{y_{max}}$</td><td>−0.0013</td><td>−0.0005</td><td>−0.0002</td><td></td><td></td></tr>
<tr><td>K_{θ_b}</td><td>−0.0059</td><td>−0.0031</td><td>−0.0011</td><td>−0.0002</td><td></td></tr>
<tr><td>$K_{M_{tb}}$</td><td>−0.0539</td><td>−0.0094</td><td>−0.0020</td><td>−0.0003</td><td></td></tr>
<tr><td>$K_{M_{ra}}$</td><td>−0.0381</td><td>−0.0264</td><td>−0.0145</td><td>−0.0056</td><td>−0.0006</td></tr>
<tr><td>K_{Q_b}</td><td>0.4326</td><td>0.1260</td><td>0.0658</td><td>0.0339</td><td>0.0104</td></tr>
</table></td></tr>
<tr><td>3h. Outer edge fixed, inner edge fixed
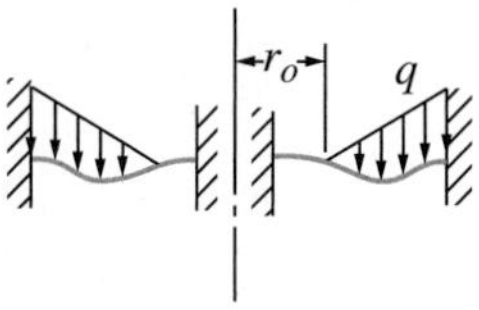</td>
<td>$y_b = 0,\ \theta_b = 0,\ y_a = 0,\ \theta_a = 0$
$M_{rb} = -qa^2 \dfrac{C_3 L_{15} - C_6 L_{12}}{C_2 C_6 - C_3 C_5}$
$Q_b = qa \dfrac{C_2 L_{15} - C_5 L_{12}}{C_2 C_6 - C_3 C_5}$
$M_{ra} = M_{rb} C_8 + Q_b a C_9 - qa^2 L_{18}$
$Q_a = Q_b \dfrac{b}{a} - \dfrac{q}{6a}(2a^2 - r_o a - r_o^2)$</td>
<td>If $r_o = b$ (linearly increasing load from b to a),
<table>
<tr><th>b/a</th><th>0.1</th><th>0.3</th><th>0.5</th><th>0.7</th><th>0.9</th></tr>
<tr><td>$K_{y_{max}}$</td><td>−0.0009</td><td>−0.0003</td><td>−0.0001</td><td></td><td></td></tr>
<tr><td>$K_{M_{rb}}$</td><td>−0.0630</td><td>−0.0242</td><td>−0.0102</td><td>−0.0033</td><td>−0.00035</td></tr>
<tr><td>$K_{M_{rn}}$</td><td>−0.0340</td><td>−0.0215</td><td>−0.0114</td><td>−0.0043</td><td>−0.00048</td></tr>
<tr><td>K_{Q_b}</td><td>0.5440</td><td>0.1865</td><td>0.0999</td><td>0.0514</td><td>0.01575</td></tr>
</table></td></tr>
</table>

(*Continued*)

TABLE 11.2 Formulas for Flat Circular Plates of Constant Thickness (*Continued*)

<table>
<tr><th>Case No.,
Edge Restraints</th><th>Boundary Values</th><th>Special Cases</th></tr>
<tr>
<td>3i. Outer edge guided, inner edge simply supported
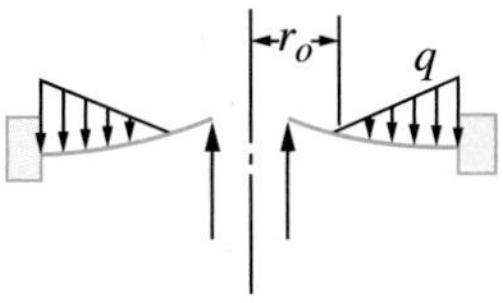</td>
<td>$y_b = 0, \; M_{rb} = 0, \; \theta_a = 0, \; Q_a = 0$
$\theta_b = \dfrac{-qa^3}{DC_4}\left[\dfrac{C_6}{6ab}(2a^2 - r_o a - r_o^2) - L_{15}\right]$
$Q_b = \dfrac{q}{6b}(2a^2 - r_o a - r_o^2)$
$y_a = \theta_b a C_1 + Q_b \dfrac{a^3}{D} C_3 - \dfrac{qa^4}{D} L_{12}$
$M_{ra} = \theta_b \dfrac{D}{a} C_7 + Q_b a C_9 - qa^2 L_{18}$</td>
<td>If $r_o = b$ (linearly increasing load from b to a),
<table>
<tr><th>b/a</th><th>0.1</th><th>0.3</th><th>0.5</th><th>0.7</th><th>0.9</th></tr>
<tr><td>K_{y_a}</td><td>−0.0389</td><td>−0.0254</td><td>−0.0082</td><td>−0.0011</td><td>−0.00001</td></tr>
<tr><td>K_{θ_b}</td><td>−0.0748</td><td>−0.0665</td><td>−0.0283</td><td>−0.0062</td><td>−0.00022</td></tr>
<tr><td>$K_{M_{ra}}$</td><td>0.1054</td><td>0.1032</td><td>0.0689</td><td>0.0282</td><td>0.00330</td></tr>
<tr><td>$K_{M_{tb}}$</td><td>−0.6808</td><td>−0.2017</td><td>−0.0516</td><td>−0.0080</td><td>−0.00022</td></tr>
</table></td>
</tr>
<tr>
<td>3j. Outer edge guided, inner edge fixed
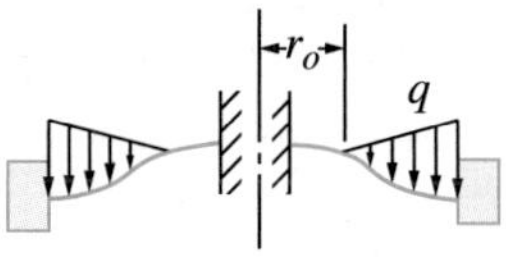</td>
<td>$y_b = 0, \; \theta_b = 0, \; \theta_a = 0, \; Q_a = 0$
$M_{rb} = \dfrac{-qa^2}{C_5}\left[\dfrac{C_6}{6ab}(2a^2 - r_o a - r_o^2) - L_{15}\right]$
$Q_b = \dfrac{q}{6b}(2a^2 - r_o a - r_o^2)$
$y_a = M_{rb} \dfrac{a^2}{D} C_2 + Q_b \dfrac{a^3}{D} C_3 - \dfrac{qa^4}{D} L_{12}$
$M_{ra} = M_{rb} C_8 + Q_b a C_9 - qa^2 L_{18}$</td>
<td>If $r_o = b$ (linearly increasing load from b to a),
<table>
<tr><th>b/a</th><th>0.1</th><th>0.3</th><th>0.5</th><th>0.7</th><th>0.9</th></tr>
<tr><td>K_{y_a}</td><td>−0.0253</td><td>−0.0089</td><td>−0.0022</td><td>−0.0003</td><td></td></tr>
<tr><td>$K_{M_{rb}}$</td><td>−0.5388</td><td>−0.1990</td><td>−0.0774</td><td>−0.0231</td><td>−0.00221</td></tr>
<tr><td>$K_{M_{ra}}$</td><td>0.0903</td><td>0.0594</td><td>0.0312</td><td>0.0113</td><td>0.00125</td></tr>
</table></td>
</tr>
<tr>
<td>3k. Outer edge free, inner edge simply supported
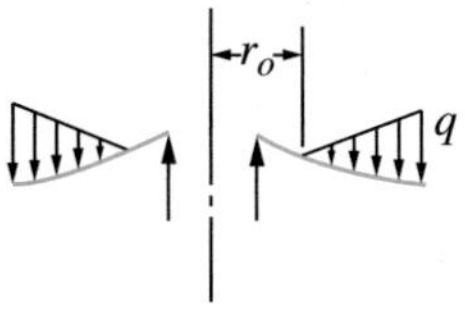</td>
<td>$y_b = 0, \; M_{rb} = 0, \; M_{ra} = 0, \; Q_a = 0$
$\theta_b = \dfrac{-qa^3}{DC_7}\left[\dfrac{C_9}{6ab}(2a^2 - r_o a - r_o^2) - L_{18}\right]$
$Q_b = \dfrac{q}{6b}(2a^2 - r_o a - r_o^2)$
$y_a = \theta_b a C_1 + Q_b \dfrac{a^3}{D} C_3 - \dfrac{qa^4}{D} L_{12}$
$\theta_a = \theta_b C_4 + Q_b \dfrac{a^2}{D} C_6 - \dfrac{qa^3}{D} L_{15}$</td>
<td>If $r_o = b$ (linearly increasing load from b to a),
<table>
<tr><th>b/a</th><th>0.1</th><th>0.3</th><th>0.5</th><th>0.7</th><th>0.9</th></tr>
<tr><td>K_{y_a}</td><td>−0.0830</td><td>−0.0826</td><td>−0.0574</td><td>−0.0258</td><td>−0.0034</td></tr>
<tr><td>K_{θ_b}</td><td>−0.0982</td><td>−0.1413</td><td>−0.1293</td><td>−0.0912</td><td>−0.0346</td></tr>
<tr><td>K_{θ_a}</td><td>−0.0834</td><td>−0.1019</td><td>−0.1035</td><td>−0.0812</td><td>−0.0335</td></tr>
<tr><td>$K_{M_{tb}}$</td><td>−0.8937</td><td>−0.4286</td><td>−0.2354</td><td>−0.1186</td><td>−0.0350</td></tr>
</table></td>
</tr>
<tr>
<td>3l. Outer edge free, inner edge fixed
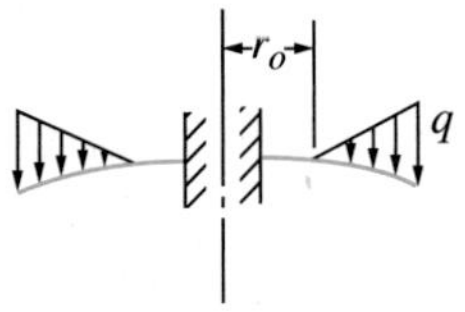</td>
<td>$y_b = 0, \; \theta_b = 0, \; M_{ra} = 0, \; Q_a = 0$
$M_{rb} = \dfrac{-qa^2}{C_8}\left[\dfrac{C_9}{6ab}(2a^2 - r_o a - r_o^2) - L_{18}\right]$
$Q_b = \dfrac{q}{6b}(2a^2 - r_o a - r_o^2)$
$y_a = M_{rb} \dfrac{a^2}{D} C_2 + Q_b \dfrac{a^3}{D} C_3 - \dfrac{qa^4}{D} L_{12}$
$\theta_a = M_{rb} \dfrac{a}{D} C_5 + Q_b \dfrac{a^2}{D} C_6 - \dfrac{qa^3}{D} L_{15}$</td>
<td>If $r_o = b$ (linearly increasing load from b to a),
<table>
<tr><th>b/a</th><th>0.1</th><th>0.3</th><th>0.5</th><th>0.7</th><th>0.9</th></tr>
<tr><td>K_{y_a}</td><td>−0.0579</td><td>−0.0240</td><td>−0.0064</td><td>−0.0008</td><td></td></tr>
<tr><td>K_{θ_a}</td><td>−0.0684</td><td>−0.0397</td><td>−0.0159</td><td>−0.0035</td><td>−0.00013</td></tr>
<tr><td>$K_{M_{rb}}$</td><td>−0.6769</td><td>−0.2861</td><td>−0.1197</td><td>−0.0368</td><td>−0.00356</td></tr>
</table></td>
</tr>
</table>

(*Continued*)

TABLE 11.2 Formulas for Flat Circular Plates of Constant Thickness (*Continued*)

Case 4. Annular plate with a distributed pressure increasing parabolically from zero at r_o to q at a

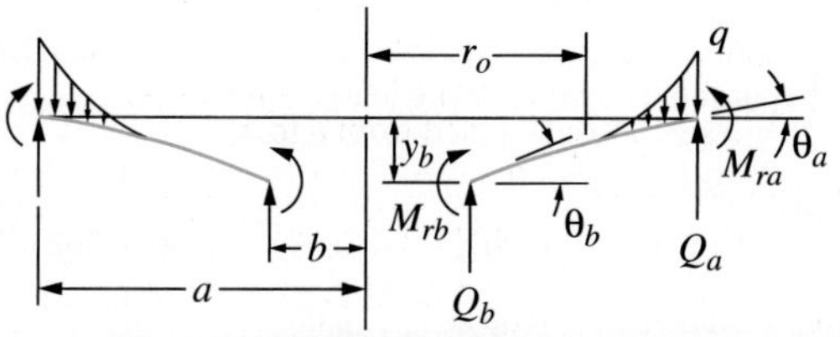

General expressions for deformations, moments, and shears:

$$y = y_b + \theta_b r F_1 + M_{rb}\frac{r^2}{D}F_2 + Q_b\frac{r^3}{D}F_3 - a\frac{r^4}{D}\left(\frac{r-r_o}{a-r_o}\right)^2 G_{13}$$

$$\theta = \theta_b F_4 + M_{rb}\frac{r}{D}F_5 + Q_b\frac{r^2}{D}F_6 - q\frac{r^3}{D}\left(\frac{r-r_o}{a-r_o}\right)^2 G_{16}$$

$$M_r = \theta_b\frac{D}{r}F_7 + M_{rb}F_8 + Q_b r F_9 - qr^2\left(\frac{r-r_o}{a-r_o}\right)^2 G_{19}$$

$$M_t = \frac{\theta D(1-\nu^2)}{r} + \nu M_r$$

$$Q = Q_b\frac{b}{r} - \frac{q}{12r(a-r_o)^2}(3r^4 - 8r_o r^3 + 6r_o^2 r^2 - r_o^4)\langle r - r_o\rangle^0$$

For the numerical data given below, $\nu = 0.3$

$$y = K_y\frac{qa^4}{D}, \quad \theta = K_\theta\frac{qa^3}{D}, \quad M = K_M qa^2, \quad Q = K_Q qa$$

Case No., Edge Restraints	Boundary Values	Special Cases

4a. Outer edge simply supported, inner edge free

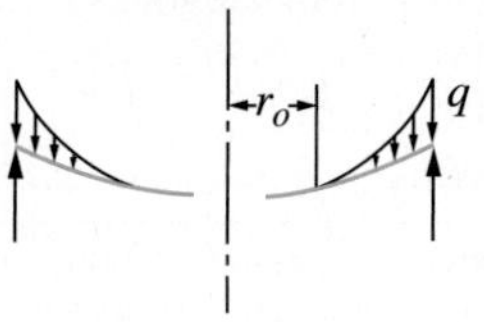

Boundary Values:

$M_{rb} = 0, \quad Q_b = 0, \quad y_a = 0, \quad M_{ra} = 0$

$$y_b = \frac{-qa^4}{D}\left(\frac{C_1 L_{19}}{C_7} - L_{13}\right)$$

$$\theta_b = \frac{qa^3}{DC_7}L_{19}$$

$$\theta_a = \frac{qa^3}{D}\left(\frac{C_4 L_{19}}{C_7} - L_{16}\right)$$

$$Q_a = \frac{-q}{12a}(3a^2 - 2ar_o - r_o^2)$$

Special Cases:

$y_{max} = y_b \quad M_{max} = M_{tb}$

If $r_o = b$ (parabolically increasing load from b to a),

b/a	0.1	0.3	0.5	0.7	0.9
K_{y_b}	−0.0184	−0.0168	−0.0122	−0.0059	−0.0008
K_{θ_a}	0.0291	0.0266	0.0243	0.0190	0.0082
K_{θ_b}	0.0105	0.0227	0.0254	0.0203	0.0084
$K_{M_{tb}}$	0.0951	0.0687	0.0462	0.0264	0.0085

4b. Outer edge simply supported, inner edge guided

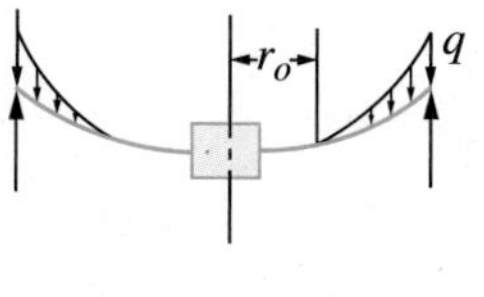

Boundary Values:

$\theta_b = 0, \quad Q_b = 0, \quad y_a = 0, \quad M_{ra} = 0$

$$y_b = \frac{-qa^4}{D}\left(\frac{C_2 L_{19}}{C_8} - L_{13}\right)$$

$$M_{rb} = \frac{qa^2 L_{19}}{C_8}$$

$$\theta_a = \frac{qa^3}{D}\left(\frac{C_5 L_{19}}{C_8} - L_{16}\right)$$

$$Q_a = \frac{-q}{12a}(3a^2 - 2ar_o - r_o^2)$$

Special Cases:

$y_{max} = y_b, \quad M_{max} = M_{rb}$

If $r_o = b$ (parabolically increasing load from b to a),

b/a	0.1	0.3	0.5	0.7	0.9
K_{y_b}	−0.0158	−0.0074	−0.0022	−0.0003	
K_{θ_a}	0.0275	0.0166	0.0071	0.0017	0.00007
$K_{M_{rb}}$	0.0721	0.0459	0.0235	0.0082	0.00086

4c. Outer edge simply supported, inner edge simply supported

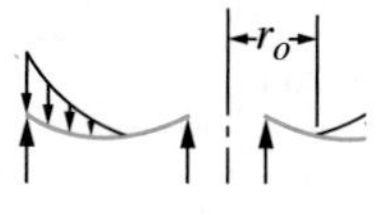

Boundary Values:

$y_b = 0, \quad M_{rb} = 0, \quad y_a = 0, \quad M_{ra} = 0$

$$\theta_b = \frac{-qa^3}{D}\frac{C_3 L_{19} - C_9 L_{13}}{C_1 C_9 - C_3 C_7}$$

$$Q_b = qa\frac{C_1 L_{19} - C_7 L_{13}}{C_1 C_9 - C_3 C_7}$$

$$\theta_a = \theta_b C_4 + Q_b\frac{a^2}{D}C_6 - \frac{qa^3}{D}L_{16}$$

$$Q_a = Q_b\frac{b}{a} - \frac{q}{12a}(3a^2 - 2ar_o - r_o^2)$$

Special Cases:

If $r_o = b$ (parabolically increasing load from b to a),

b/a	0.1	0.3	0.5	0.7
$K_{y_{max}}$	−0.0022	−0.0009	−0.0003	
K_{θ_b}	−0.0083	−0.0046	−0.0016	−0.0003
K_{θ_a}	0.0080	0.0044	0.0017	0.0004
$K_{M_{tb}}$	−0.0759	−0.0139	−0.0029	−0.0004
$K_{M_{r\,max}}$	0.0267	0.0185	0.0098	0.0035
K_{Q_b}	0.5068	0.1330	0.0633	0.0305

(*Continued*)

TABLE 11.2 Formulas for Flat Circular Plates of Constant Thickness (*Continued*)

Case No., Edge Restraints	Boundary Values	Special Cases

4d. Outer edge simply supported, inner edge fixed

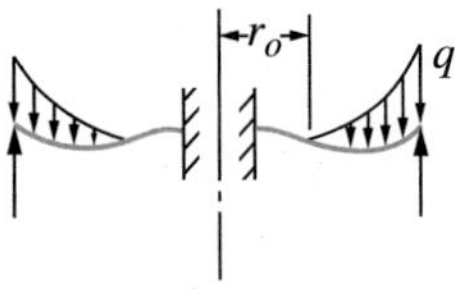

Boundary Values:

$y_b = 0,\ \theta_b = 0,\ y_a = 0,\ M_{ra} = 0$

$$M_{rb} = -qa^2 \frac{C_3 L_{19} - C_9 L_{13}}{C_2 C_9 - C_3 C_8}$$

$$Q_b = qa \frac{C_2 L_{19} - C_8 L_{13}}{C_2 C_9 - C_3 C_8}$$

$$\theta_a = M_{rb} \frac{a}{D} C_5 + Q_b \frac{a^2}{D} C_6 - \frac{qa^3}{D} L_{16}$$

$$Q_a = Q_b \frac{b}{a} - \frac{q}{12a}(3a^2 - 2ar_o - r_o^2)$$

Special Cases: If $r_o = b$ (parabolically increasing load from b to a),

b/a	0.1	0.3	0.5	0.7
$K_{y_{max}}$	−0.0016	−0.0005	−0.0001	−0.00002
K_{θ_a}	0.0064	0.0030	0.0011	0.00023
$K_{M_{rb}}$	−0.0777	−0.0281	−0.0113	−0.00349
K_{Q_b}	0.5860	0.1785	0.0882	0.04276

4e. Outer edge fixed, inner edge free

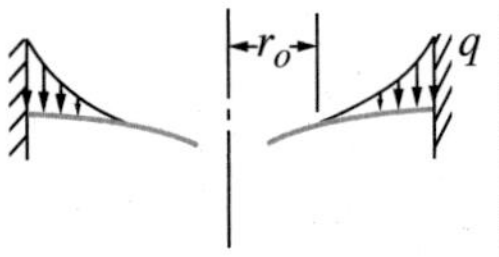

Boundary Values:

$M_{rb} = 0,\ Q_b = 0,\ y_a = 0,\ \theta_a = 0$

$$y_b = \frac{-qa^4}{D}\left(\frac{C_1 L_{16}}{C_4} - L_{13}\right)$$

$$\theta_b = \frac{qa^3 L_{16}}{DC_4}$$

$$M_{ra} = -qa^2\left(L_{19} - \frac{C_7}{C_4} L_{16}\right)$$

$$Q_a = \frac{-q}{12a}(3a^2 - 2ar_o - r_o^2)$$

Special Cases: If $r_o = b$ (parabolically increasing load from b to a),

b/a	0.1	0.3	0.5	0.7	0.9
K_{y_b}	−0.0031	−0.0019	−0.0007	−0.0001	
K_{θ_b}	0.0023	0.0032	0.0017	0.0004	0.00002
$K_{M_{ra}}$	−0.0368	−0.0269	−0.0162	−0.0066	−0.00081
$K_{M_{tb}}$	0.0208	0.0096	0.0031	0.0006	0.00002

4f. Outer edge fixed, inner edge guided

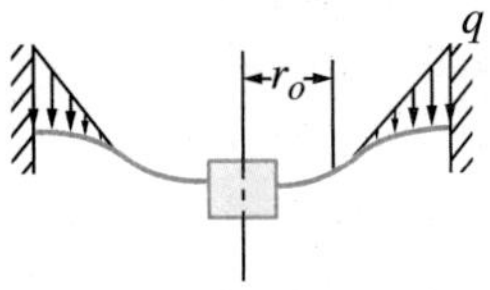

Boundary Values:

$\theta_b = 0,\ Q_b = 0,\ y_a = 0,\ \theta_a = 0$

$$y_b = \frac{-qa^4}{D}\left(\frac{C_2 L_{16}}{C_5} - L_{13}\right)$$

$$M_{rb} = \frac{qa^2 L_{16}}{C_5}, \quad M_{ra} = -qa^2\left(L_{19} - \frac{C_8}{C_5} L_{16}\right)$$

$$Q_a = \frac{-q}{12a}(3a^2 - 2ar_o - r_o^2)$$

Special Cases: If $r_o = b$ (parabolically increasing load from b to a),

b/a	0.1	0.3	0.5	0.7	0.9
K_{y_b}	−0.0026	−0.0011	−0.0003		
$K_{M_{rb}}$	0.0164	0.0094	0.0046	0.0016	0.00016
$K_{M_{ra}}$	−0.0364	−0.0248	−0.0140	−0.0054	−0.00066

4g. Outer edge fixed, inner edge simply supported

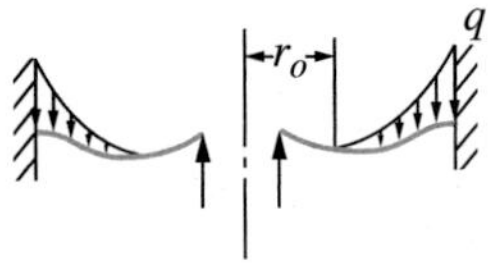

Boundary Values:

$y_b = 0,\ M_{rb} = 0,\ y_a = 0,\ \theta_a = 0$

$$\theta_b = \frac{-qa^3}{D} \frac{C_3 L_{16} - C_6 L_{13}}{C_1 C_6 - C_3 C_4}$$

$$Q_b = qa \frac{C_1 L_{16} - C_4 L_{13}}{C_1 C_6 - C_3 C_4}$$

$$M_{ra} = \theta_b \frac{D}{a} C_7 + Q_b a C_9 - qa^2 L_{19}$$

$$Q_a = Q_b \frac{b}{a} - \frac{q}{12a}(3a^2 - 2ar_o - r_o^2)$$

Special Cases: If $r_o = b$ (parabolically increasing load from b to a),

b/a	0.1	0.3	0.5	0.7
$K_{y_{max}}$	−0.0007	−0.0003	−0.0001	
K_{θ_b}	−0.0031	−0.0016	−0.0006	−0.00012
$K_{M_{tb}}$	−0.0285	−0.0049	−0.0010	−0.00015
$K_{M_{ra}}$	−0.0255	−0.0172	−0.0093	−0.00352
K_{Q_b}	0.2136	0.0577	0.0289	0.01450

4h. Outer edge fixed, inner edge fixed

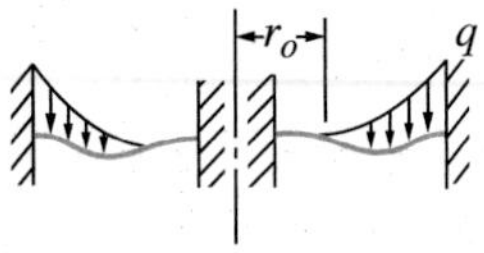

Boundary Values:

$y_b = 0,\ \theta_b = 0,\ y_a = 0,\ \theta_a = 0$

$$M_{rb} = -qa^2 \frac{C_3 L_{16} - C_6 L_{13}}{C_2 C_6 - C_3 C_5}$$

$$Q_b = qa \frac{C_2 L_{16} - C_5 L_{13}}{C_2 C_6 - C_3 C_5}$$

$$M_{ra} = M_{rb} C_8 + Q_b a C_9 - qa^2 L_{19}$$

$$Q_a = Q_b \frac{b}{a} - \frac{q}{12a}(3a^2 - 2ar_o - r_o^2)$$

Special Cases: If $r_o = b$ (parabolically increasing load from b to a),

b/a	0.1	0.3	0.5	0.7
$K_{y_{max}}$	−0.0005	−0.0002	−0.00005	
$K_{M_{rb}}$	−0.0333	−0.0126	−0.00524	−0.00168
$K_{M_{ra}}$	−0.0234	−0.0147	−0.00773	−0.00287
K_{Q_b}	0.2726	0.0891	0.04633	0.02335

(*Continued*)

TABLE 11.2 Formulas for Flat Circular Plates of Constant Thickness (*Continued*)

Case No., Edge Restraints | Boundary Values | Special Cases

4i. Outer edge guided, inner edge simply supported

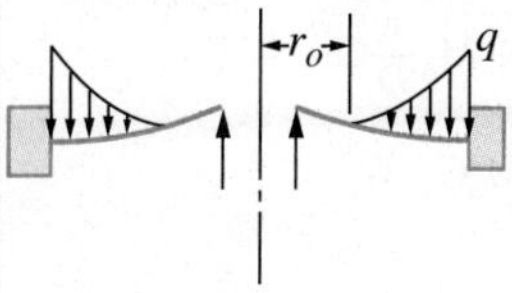

$y_b = 0,\quad M_{rb} = 0,\quad \theta_a = 0,\quad Q_a = 0$

$$\theta_b = \frac{-qa^3}{DC_4}\left[\frac{C_6}{12ab}(3a^2 - 2ar_o - r_o^2) - L_{16}\right]$$

$$Q_b = \frac{q}{12b}(3a^2 - 2ar_o - r_o^2)$$

$$y_a = \theta_b a C_1 + Q_b \frac{a^3}{D} C_3 - \frac{qa^4}{D} L_{13}$$

$$M_{ra} = \theta_b \frac{D}{a} C_7 + Q_b a C_9 - qa^2 L_{19}$$

If $r_o = b$ (parabolically increasing load from b to a),

b/a	0.1	0.3	0.5	0.7	0.9
K_{y_a}	−0.0302	−0.0193	−0.0061	−0.0008	−0.00001
K_{θ_b}	−0.0567	−0.0498	−0.0210	−0.0045	−0.00016
$K_{M_{ra}}$	0.0859	0.0813	0.0532	0.0215	0.00249
$K_{M_{tb}}$	−0.5156	−0.1510	−0.0381	−0.0059	−0.00016

4j. Outer edge guided, inner edge fixed

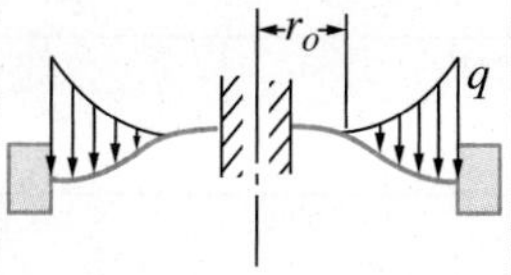

$y_b = 0,\quad \theta_b = 0,\quad \theta_a = 0,\quad Q_a = 0$

$$M_{rb} = \frac{-qa^2}{C_5}\left[\frac{C_6}{12ab}(3a^2 - 2ar_o - r_o^2) - L_{16}\right]$$

$$Q_b = \frac{q}{12b}(3a^2 - 2ar_o - r_o^2)$$

$$y_a = M_{rb}\frac{a^2}{D} C_2 + Q_b \frac{a^3}{D} C_3 - \frac{qa^4}{D} L_{13}$$

$$M_{ra} = M_{rb} C_8 + Q_b a C_9 - qa^2 L_{19}$$

If $r_o = b$ (parabolically increasing load from b to a),

b/a	0.1	0.3	0.5	0.7	0.9
K_{y_a}	−0.0199	−0.0070	−0.0017	−0.0002	
$K_{M_{rb}}$	−0.4081	−0.1490	−0.0573	−0.0169	−0.00161
$K_{M_{rn}}$	0.0745	0.0485	0.0253	0.0091	0.00100

4k. Outer edge free, inner edge simply supported

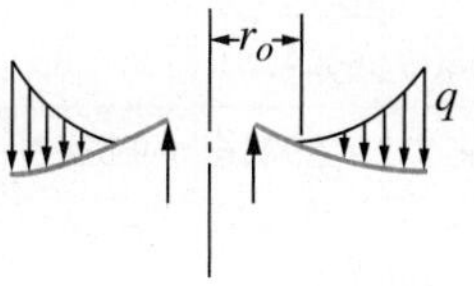

$y_b = 0,\quad M_{rb} = 0,\quad M_{ra} = 0,\quad Q_a = 0$

$$\theta_a = \frac{-qa^3}{DC_7}\left[\frac{C_9}{12ab}(3a^2 - 2ar_o - r_o^2) - L_{19}\right]$$

$$Q_b = \frac{q}{12b}(3a^2 - 2ar_o - r_o^2)$$

$$y_a = \theta_b a C_1 + Q_b \frac{a^3}{D} C_3 - \frac{qa^4}{D} L_{13}$$

$$\theta_a = \theta_b C_4 + Q_b \frac{a^2}{D} C_6 - \frac{qa^3}{D} L_{16}$$

If $r_o = b$ (parabolically increasing load from b to a),

b/a	0.1	0.3	0.5	0.7	0.9
K_{y_b}	−0.0662	−0.0644	−0.0441	−0.0196	−0.0026
K_{θ_b}	−0.0757	−0.1087	−0.0989	−0.0693	−0.0260
K_{θ_a}	−0.0680	−0.0802	−0.0799	−0.0618	−0.0252
$K_{M_{tb}}$	−0.6892	−0.3298	−0.1800	−0.0900	−0.0263

4l. Outer edge free, inner edge fixed

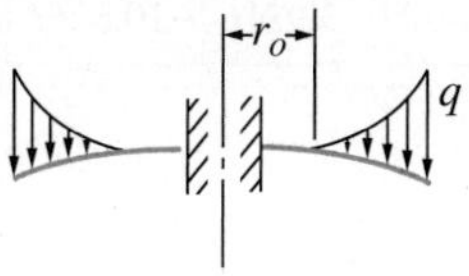

$y_b = 0,\quad \theta_b = 0,\quad M_{ra} = 0,\quad Q_a = 0$

$$M_{rb} = \frac{-qa^2}{C_8}\left[\frac{C_9}{12ab}(3a^2 - 2ar_o - r_o^2) - L_{19}\right]$$

$$Q_b = \frac{q}{12b}(3a^2 - 2ar_o - r_o^2)$$

$$y_a = M_{rb}\frac{a^2}{D} C_2 + Q_b \frac{a^3}{D} C_3 - \frac{qa^4}{D} L_{13}$$

$$\theta_a = M_{rb}\frac{a}{D} C_5 + Q_b \frac{a^2}{D} C_6 - \frac{qa^3}{D} L_{16}$$

If $r_o = b$ (parabolically increasing load from b to a),

b/a	0.1	0.3	0.5	0.7	0.9
K_{y_a}	−0.0468	−0.0193	−0.0051	−0.0006	−0.00001
K_{θ_a}	−0.0564	−0.0324	−0.0128	−0.0028	−0.00010
$K_{M_{rb}}$	−0.5221	−0.2202	−0.0915	−0.0279	−0.00268

(*Continued*)

TABLE 11.2 Formulas for Flat Circular Plates of Constant Thickness (*Continued*)

Case 5. Annular plate with a uniform line moment M_o at a radius r_o

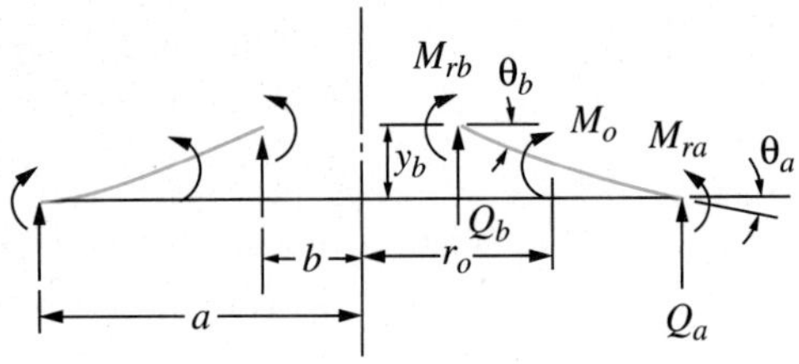

Note: If the loading M_o is on the inside edge, $r > r_o$ everywhere, so $\langle r - r_o \rangle^0 = 1$ everywhere.

General expressions for deformations, moments, and shears:

$$y = y_b + \theta_b r F_1 + M_{rb}\frac{r^2}{D}F_2 + Q_b\frac{r^3}{D}F_3 - M_o\frac{r^2}{D}G_2$$

$$\theta = \theta_b F_4 + M_{rb}\frac{r}{D}F_5 + Q_b\frac{r^2}{D}F_6 + M_o\frac{r}{D}G_5$$

$$M_r = \theta_b\frac{D}{r}F_7 + M_{rb}F_8 + Q_b r F_9 + M_o G_8$$

$$M_t = \frac{\theta D(1-\nu^2)}{r} + \nu M_r$$

$$Q = Q_b\frac{b}{r}$$

For the numerical data given below, $\nu = 0.3$

$$y = K_y\frac{M_o a^2}{D}, \quad \theta = K_\theta\frac{M_o a}{D}, \quad M = K_M M_o, \quad Q = K_Q\frac{M_o}{a}$$

Case No., Edge Restraints	Boundary Values	Special Cases

5a. Outer edge simply supported, inner edge free

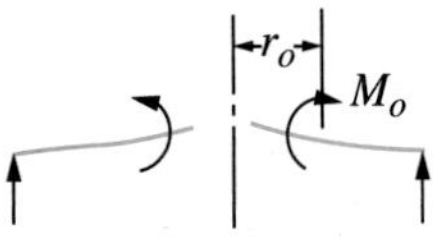

Boundary Values:

$M_{rb} = 0, \quad Q_b = 0, \quad y_a = 0, \quad M_{ra} = 0$

$$y_b = \frac{M_o a^2}{D}\left(\frac{C_1 L_8}{C_7} - L_2\right)$$

$$\theta_b = \frac{-M_o a}{DC_7}L_8$$

$$\theta_a = \frac{-M_o a}{D}\left(\frac{C_4 L_8}{C_7} - L_5\right)$$

$$Q_a = 0$$

Special Cases:

$y_{max} = y_b, \quad M_{max} = M_{tb}$

If $r_o = b$ (moment M_o at the inner edge),

b/a	0.1	0.3	0.5	0.7	0.9
K_{y_b}	0.0371	0.2047	0.4262	0.6780	0.9532
K_{θ_a}	−0.0222	−0.2174	−0.7326	−2.1116	−9.3696
K_{θ_b}	−0.1451	−0.4938	−1.0806	−2.4781	−9.7183
$K_{M_{tb}}$	−1.0202	−1.1978	−1.6667	−2.9216	−9.5263

If $r_o = a$ (moment M_o at the outer edge),

b/a	0.1	0.3	0.5	0.7	0.9
K_{y_b}	0.4178	0.5547	0.7147	0.8742	1.0263
K_{θ_a}	−0.7914	−0.9866	−1.5018	−2.8808	−10.1388
K_{θ_b}	−0.2220	−0.7246	−1.4652	−3.0166	−10.4107
$K_{M_{tb}}$	−2.0202	−2.1978	−2.6667	−3.9216	−10.5263

5b. Outer edge simply supported, inner edge guided

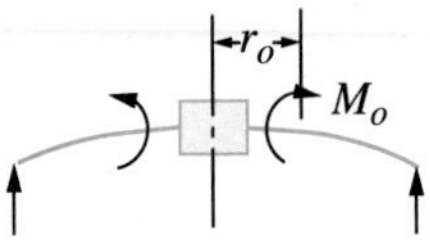

Boundary Values:

$\theta_b = 0, \quad Q_b = 0, \quad y_a = 0, \quad M_{ra} = 0$

$$y_b = \frac{M_o a^2}{D}\left(\frac{C_2 L_8}{C_8} - L_2\right)$$

$$M_{rb} = \frac{-M_o L_{18}}{C_8}$$

$$\theta_a = \frac{-M_o a}{D}\left(\frac{C_5 L_8}{C_8} - L_5\right)$$

$$Q_a = 0$$

Special Cases:

$y_{max} = y_b, \quad M_{max} = M_{rb}$

If $r_o = a$ (moment M_o at the outer edge),

b/a	0.1	0.3	0.5	0.7	0.9
K_{y_b}	0.3611	0.2543	0.1368	0.0488	0.0052
K_{θ_a}	−0.7575	−0.6676	−0.5085	−0.3104	−0.1018
$K_{M_{rb}}$	−1.5302	−1.4674	−1.3559	−1.2173	−1.0712

(*Continued*)

TABLE 11.2 Formulas for Flat Circular Plates of Constant Thickness (*Continued*)

Columns: Case No., Edge Restraints | Boundary Values | Special Cases

5c. Outer edge simply supported, inner edge simply supported

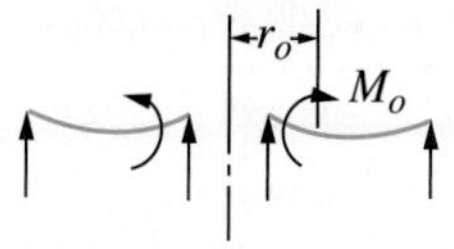

Boundary Values:

$y_b = 0,\quad M_{rb} = 0,\quad y_a = 0,\quad M_{ra} = 0$

$$\theta_b = \frac{M_o a}{D}\frac{C_3 L_8 - C_9 L_2}{C_1 C_9 - C_3 C_7}$$

$$Q_b = \frac{-M_o}{a}\frac{C_1 L_8 - C_7 L_2}{C_1 C_9 - C_3 C_7}$$

$$\theta_a = \theta_b C_4 + Q_b \frac{a^2}{D} C_6 + \frac{M_o a}{D} L_5$$

$$Q_a = Q_b \frac{b}{a}$$

Special Cases:

If $r_o = b$ (moment M_o at the inner edge),

b/a	0.1	0.3	0.5	0.7	0.9
$K_{y_{max}}$	−0.0095	−0.0167	−0.0118	−0.0050	−0.0005
K_{θ_a}	0.0204	0.0518	0.0552	0.0411	0.0158
K_{θ_b}	−0.1073	−0.1626	−0.1410	−0.0929	−0.0327
$K_{M_{tb}}$	−0.6765	−0.1933	0.0434	0.1793	0.2669
K_{Q_b}	−1.0189	−1.6176	−2.2045	−3.5180	−10.1611

If $r_o = a$ (moment M_o at the outer edge),

b/a	0.1	0.3	0.5	0.7	0.9
$K_{y_{max}}$	0.0587	0.0390	0.0190	0.0063	0.0004
K_{θ_a}	−0.3116	−0.2572	−0.1810	−0.1053	−0.0339
K_{θ_b}	0.2037	0.1728	0.1103	0.0587	0.0175
$K_{M_{tb}}$	1.8539	0.5240	0.2007	0.0764	0.0177
K_{Q_b}	−11.4835	−4.3830	−3.6964	−4.5358	−10.9401

5d. Outer edge simply supported, inner edge fixed

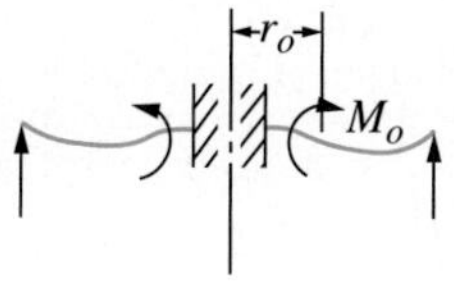

Boundary Values:

$y_b = 0,\quad \theta_b = 0,\quad y_a = 0,\quad M_{ra} = 0$

$$M_{rb} = M_o \frac{C_3 L_8 - C_9 L_2}{C_2 C_9 - C_3 C_8}$$

$$Q_b = \frac{-M_o}{a}\frac{C_2 L_8 - C_8 L_2}{C_2 C_9 - C_3 C_8}$$

$$\theta_a = M_{rb}\frac{a}{D} C_5 + Q_b \frac{a^2}{D} C_6 + \frac{M_o a}{D} L_5$$

$$Q_a = Q_b \frac{b}{a}$$

Special Cases:

If $r_o = a$ (moment M_o at the outer edge),

b/a	0.1	0.3	0.5	0.7	0.9
$K_{y_{max}}$	0.0449	0.0245	0.0112	0.0038	0.0002
K_{θ_a}	−0.2729	−0.2021	−0.1378	−0.0793	−0.0255
$K_{M_{rb}}$	1.8985	1.0622	0.7823	0.6325	0.5366
K_{Q_b}	−13.4178	−6.1012	−5.4209	−6.7611	−16.3923

5e. Outer edge fixed, inner edge free

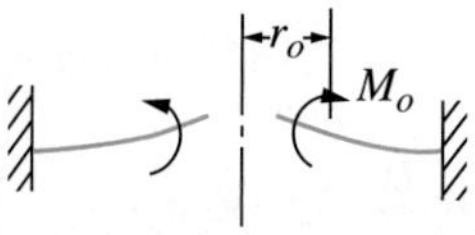

Boundary Values:

$M_{rb} = 0,\quad Q_b = 0,\quad y_a = 0,\quad \theta_a = 0$

$$y_b = \frac{M_o a^2}{D}\left(\frac{C_1 L_5}{C_4} - L_2\right)$$

$$\theta_b = \frac{-M_o a}{D C_4} L_5$$

$$M_{ra} = M_o\left(L_8 - \frac{C_7}{C_4} L_5\right)$$

$$Q_a = 0$$

Special Cases:

If $r_o = b$ (moment M_o at the inner edge),

b/a	0.1	0.3	0.5	0.7	0.9
K_{y_b}	0.0254	0.0825	0.0776	0.0373	0.0048
K_{θ_b}	−0.1389	−0.3342	−0.3659	−0.2670	−0.0976
$K_{M_{tb}}$	−0.9635	−0.7136	−0.3659	−0.0471	0.2014

5f. Outer edge fixed, inner edge guided

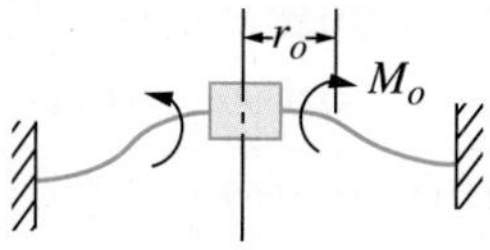

Boundary Values:

$\theta_b = 0,\quad Q_b = 0,\quad y_a = 0,\quad \theta_a = 0$

$$y_b = \frac{M_o a^2}{D}\left(\frac{C_2 L_5}{C_5} - L_2\right)$$

$$M_{rb} = \frac{-M_o}{C_5} L_5$$

$$M_{ra} = M_o\left(L_8 - \frac{C_8}{C_5} L_5\right)$$

$$Q_a = 0$$

Special Cases:

b/a	0.1		0.5		0.7
r_o/a	0.5	0.7	0.7	0.9	0.9
K_{y_b}	0.0779	0.0815	0.0285	0.0207	0.0101
$K_{M_{rb}}$	−0.7576	−0.5151	−0.6800	−0.2533	−0.3726
$K_{M_{ra}}$	0.2424	0.4849	0.3200	0.7467	0.6274

(*Continued*)

TABLE 11.2 Formulas for Flat Circular Plates of Constant Thickness (*Continued*)

Case No., Edge Restraints	Boundary Values	Special Cases

5g. Outer edge fixed, inner edge simply supported

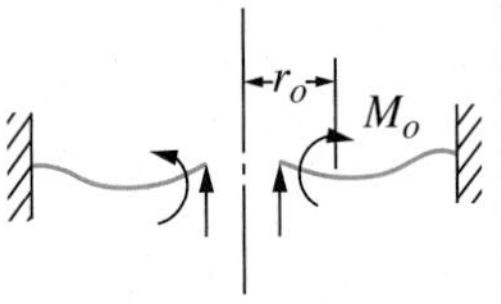

Boundary Values:

$y_b = 0, \quad M_{rb} = 0, \quad y_a = 0, \quad \theta_a = 0$

$$\theta_b = \frac{M_o a}{D} \frac{C_3 L_5 - C_6 L_2}{C_1 C_6 - C_3 C_4}$$

$$Q_b = \frac{-M_o}{a} \frac{C_1 L_5 - C_4 L_2}{C_1 C_6 - C_3 C_4}$$

$$M_{ra} = \theta_b \frac{D}{a} C_7 + Q_b a C_9 + M_o L_8$$

$$Q_a = Q_b \frac{b}{a}$$

Special Cases: If $r_o = b$ (moment M_o at the inner edge),

b/a	0.1	0.3	0.5	0.7	0.9
$K_{y_{max}}$	−0.0067	−0.0102	−0.0066	−0.0029	−0.0002
K_{θ_b}	−0.0940	−0.1278	−0.1074	−0.0699	−0.0245
K_{Q_b}	−1.7696	−2.5007	−3.3310	−5.2890	−15.2529

5h. Outer edge fixed, inner edge fixed

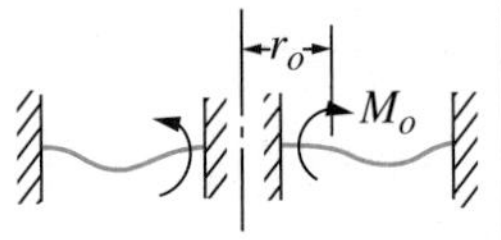

Boundary Values:

$y_b = 0, \quad \theta_b = 0, \quad y_a = 0, \quad \theta_a = 0$

$$M_{rb} = M_o \frac{C_3 L_5 - C_6 L_2}{C_2 C_6 - C_3 C_5}$$

$$Q_b = \frac{-M_o}{a} \frac{C_2 L_5 - C_5 L_2}{C_2 C_6 - C_3 C_5}$$

$$M_{ra} = M_{rb} C_8 + Q_b a C_9 + M_o L_8$$

$$Q_a = Q_b \frac{b}{a}$$

Special Cases:

b/a	0.1		0.5		0.7
r_o/a	0.5	0.7	0.7	0.9	0.9
$K_{M_{rb}}$	0.7096	1.0185	0.2031	0.3895	0.3925
$K_{M_{ra}}$	−0.1407	0.0844	−0.2399	0.3391	0.0238
$K_{M_{ro}}$	−0.5045	−0.5371	−0.4655	−0.4671	−0.5540
$K_{M_{ro}}$	0.4955	0.4629	0.5345	0.5329	0.4460
K_{Q_b}	−8.0354	−8.3997	−4.1636	−3.0307	−5.4823

(*Note:* the two values of $K_{M_{ro}}$ are for positions just before and after the applied moment M_o.)

Case 6. Annular plate with an externally applied change in slope θ_o on an annulus with a radius r_o

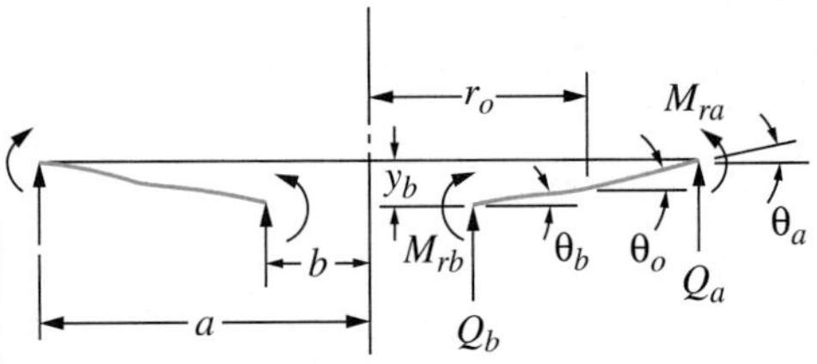

General expressions for deformations, moments, and shears:

$$y = y_b + \theta_b r F_1 + M_{rb} \frac{r^2}{D} F_2 + Q_b \frac{r^3}{D} F_3 + \theta_o r G_1$$

$$\theta = \theta_b F_4 + M_{rb} \frac{r}{D} F_5 + Q_b \frac{r^2}{D} F_6 + \theta_o G_4$$

$$M_r = \theta_b \frac{D}{r} F_7 + M_{rb} F_8 + Q_b r F_9 + \frac{\theta_o D}{r} G_7$$

$$M_t = \frac{\theta D(1 - \nu^2)}{r} + \nu M_r$$

$$Q = Q_b \frac{b}{r}$$

For the numerical data given below, $\nu = 0.3$ all values given for $K_{M_{to}}$ are found just outside r_o

$$y = K_y \theta_o a, \quad \theta = K_\theta \theta_o, \quad M = K_M \theta_o \frac{D}{a}, \quad Q = K_Q \theta_o \frac{D}{a^2}$$

Case No., Edge Restraints	Boundary Values	Special Cases

6a. Outer edge simply supported, inner edge free

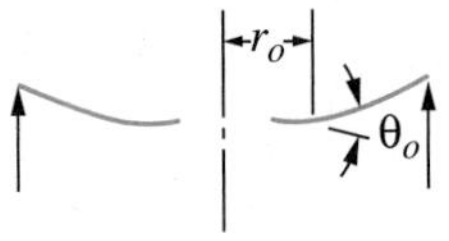

Boundary Values:

$M_{rb} = 0, \quad Q_b = 0, \quad y_a = 0, \quad M_{ra} = 0$

$$y_b = \theta_o a \left(\frac{C_1 L_7}{C_7} - L_1 \right)$$

$$\theta_b = \frac{-\theta_o}{C_7} L_7$$

$$\theta_a = -\theta_o \left(\frac{C_4 L_7}{C_7} - L_4 \right)$$

$$Q_a = 0$$

Special Cases:

b/a	0.1		0.5		0.7
r_o/a	0.5	0.7	0.7	0.9	0.9
K_{y_b}	−0.2026	−0.1513	−0.0529	−0.0299	−0.0146
K_{y_o}	−0.2821	−0.2224	−0.1468	−0.0844	−0.0709
K_{θ_b}	−0.1515	−0.0736	−0.4857	−0.1407	−0.2898
$K_{M_{tb}}$	−1.3788	−0.6697	−0.8840	−0.2562	−0.3767
$K_{M_{to}}$	1.1030	0.9583	0.6325	0.8435	0.7088

(*Continued*)

TABLE 11.2 Formulas for Flat Circular Plates of Constant Thickness (*Continued*)

Case No., Edge Restraints	Boundary Values	Special Cases

6b. Outer edge simply supported, inner edge guided

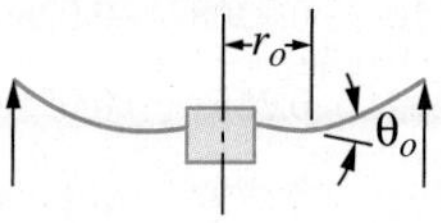

$$\theta_b = 0,\quad Q_b = 0,\quad y_a = 0,\quad M_{ra} = 0$$

$$y_b = \theta_o a\left(\frac{C_2 L_7}{C_8} - L_1\right)$$

$$M_{rb} = \frac{-\theta_o D L_7}{a C_8}$$

$$\theta_a = -\theta_o\left(\frac{C_5 L_7}{C_8} - L_4\right)$$

$$Q_a = 0$$

b/a	0.1		0.5		0.7
r_o/a	0.5	0.7	0.7	0.9	0.9
K_{y_b}	−0.2413	−0.1701	−0.2445	−0.0854	−0.0939
K_{θ_a}	0.5080	0.7039	0.7864	0.9251	0.9441
$K_{M_{rb}}$	−1.0444	−0.5073	−0.4495	−0.1302	−0.1169
$K_{M_{tb}}$	−0.3133	−0.1522	−0.1349	−0.0391	−0.0351

6c. Outer edge simply supported, inner edge simply supported

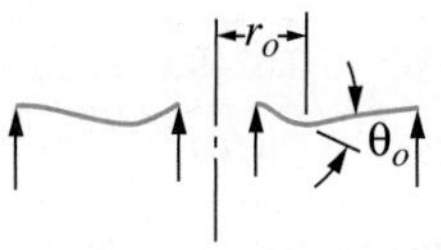

$$y_b = 0,\quad M_{rb} = 0,\quad y_a = 0,\quad M_{ra} = 0$$

$$\theta_b = \theta_o \frac{C_3 L_7 - C_9 L_1}{C_1 C_9 - C_3 C_7}$$

$$Q_b = \frac{-\theta_o D}{a^2}\,\frac{C_1 L_7 - C_7 L_1}{C_1 C_9 - C_3 C_7}$$

$$\theta_a = \theta_b C_4 + Q_b \frac{a^2}{D} C_6 + \theta_o L_4$$

$$Q_a = Q_b \frac{b}{a}$$

b/a	0.1		0.5		0.7
r_o/a	0.5	0.7	0.7	0.9	0.9
K_{y_o}	−0.1629	−0.1689	−0.1161	−0.0788	−0.0662
K_{θ_b}	−0.3579	−0.2277	−0.6023	−0.2067	−0.3412
K_{θ_a}	0.2522	0.5189	0.3594	0.7743	0.6508
$K_{M_{tb}}$	−3.2572	−2.0722	−1.0961	−0.3762	−0.4435
$K_{M_{to}}$	0.6152	0.6973	0.4905	0.7851	0.6602
K_{Q_b}	5.5679	4.1574	0.2734	0.1548	0.0758

6d. Outer edge simply supported, inner edge fixed

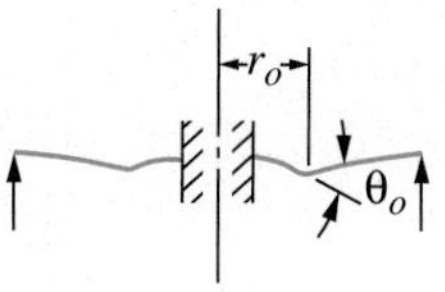

$$y_b = 0,\quad \theta_b = 0,\quad y_a = 0,\quad M_{ra} = 0$$

$$M_{rb} = \frac{\theta_o D}{a}\,\frac{C_3 L_7 - C_9 L_1}{C_2 C_9 - C_3 C_8}$$

$$Q_b = \frac{-\theta_o D}{a^2}\,\frac{C_2 L_7 - C_8 L_1}{C_2 C_9 - C_3 C_8}$$

$$\theta_a = M_{rb}\frac{a}{D} C_5 + Q_b \frac{a^2}{D} C_6 + \theta_o L_4$$

$$Q_a = Q_b \frac{b}{a}$$

b/a	0.1		0.5		0.7
r_o/a	0.5	0.7	0.7	0.9	0.9
K_{y_o}	−0.1333	−0.1561	−0.0658	−0.0709	−0.0524
K_{θ_a}	0.1843	0.4757	0.1239	0.6935	0.4997
$K_{M_{rb}}$	−3.3356	−2.1221	−4.2716	−1.4662	−3.6737
K_{Q_b}	8.9664	6.3196	9.6900	3.3870	12.9999

6e. Outer edge fixed, inner edge free

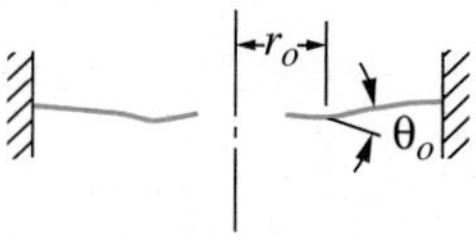

$$M_{rb} = 0,\quad Q_b = 0,\quad y_a = 0,\quad \theta_a = 0$$

$$y_b = \theta_o a\left(\frac{C_1 L_4}{C_4} - L_1\right)$$

$$\theta_b = \frac{-\theta_o L_4}{C_4}$$

$$M_{ra} = \frac{\theta_o D}{a}\left(L_7 - \frac{C_7}{C_4} L_4\right)$$

$$Q_a = 0$$

b/a	0.1		0.5		0.7
r_o/a	0.5	0.7	0.7	0.9	0.9
K_{y_b}	0.0534	0.2144	0.1647	0.3649	0.1969
K_{y_o}	−0.0975	−0.0445	−0.0155	−0.0029	−0.0013
K_{θ_b}	−0.2875	−0.2679	−0.9317	−0.9501	−1.0198
$K_{M_{tb}}$	−2.6164	−2.4377	−1.6957	−1.7293	−1.3257

6f. Outer edge fixed, inner edge guided

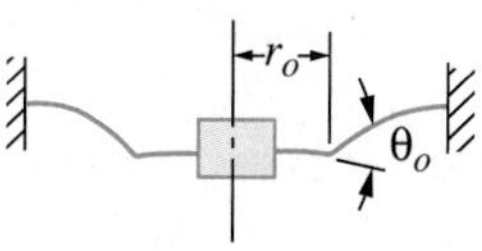

$$\theta_b = 0,\quad Q_b = 0,\quad y_a = 0,\quad \theta_a = 0$$

$$y_b = \theta_o a\left(\frac{C_2 L_4}{C_5} - L_1\right)$$

$$M_{rb} = \frac{-\theta_o D L_4}{a C_5}$$

$$M_{ra} = \frac{\theta_o D}{a}\left(L_7 - \frac{C_8}{C_5} L_4\right)$$

$$Q_a = 0$$

b/a	0.1		0.5		0.7
r_o/a	0.5	0.7	0.7	0.9	0.9
K_{y_b}	0.0009	0.1655	−0.0329	0.1634	0.0546
K_{y_o}	−0.1067	−0.0472	−0.0786	−0.0094	−0.0158
$K_{M_{rb}}$	−2.0707	−1.9293	−2.5467	−2.5970	−3.8192
$K_{M_{ra}}$	−0.6707	−0.9293	−1.5467	−1.8193	−3.0414

(Continued)

TABLE 11.2 Formulas for Flat Circular Plates of Constant Thickness (*Continued*)

Case No., Edge Restraints	Boundary Values	Special Cases

6g. Outer edge fixed, inner edge simply supported

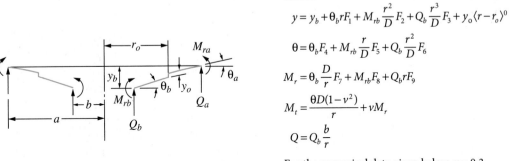

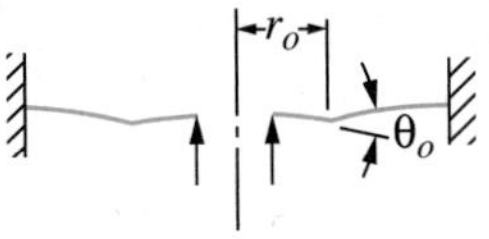

Boundary values:

$y_b = 0, \quad M_{rb} = 0, \quad y_a = 0, \quad \theta_a = 0$

$$\theta_b = \theta_o \frac{C_3 L_4 - C_6 L_1}{C_1 C_6 - C_3 C_4}$$

$$Q_b = \frac{-\theta_o D}{a^2} \frac{C_1 L_4 - C_4 L_1}{C_1 C_6 - C_3 C_4}$$

$$M_{ra} = \theta_b \frac{D}{a} C_7 + Q_b a C_9 + \frac{\theta_o D}{a} L_7$$

$$Q_a = Q_b \frac{b}{a}$$

Special cases:

b/a	0.1		0.5		0.7
r_o/a	0.5	0.7	0.7	0.9	0.9
K_{y_o}	−0.1179	−0.0766	−0.0820	−0.0208	−0.0286
K_{θ_b}	−0.1931	0.1116	−0.3832	0.2653	0.0218
$K_{M_{ra}}$	−0.8094	−1.6653	−1.9864	−4.2792	−6.1794
$K_{M_{tb}}$	−1.7567	1.0151	−0.6974	0.4828	0.0284
K_{Q_b}	−3.7263	−14.9665	−7.0690	−15.6627	−27.9529

6h. Outer edge fixed, inner edge fixed

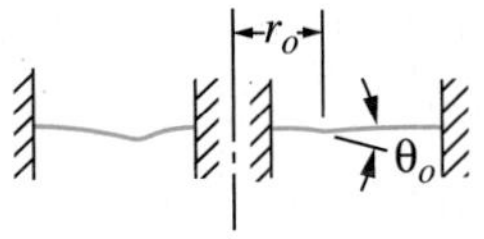

Boundary values:

$y_b = 0, \quad \theta_b = 0, \quad y_a = 0, \quad \theta_a = 0$

$$M_{rb} = \frac{\theta_o D}{a} \frac{C_3 L_4 - C_6 L_1}{C_2 C_6 - C_3 C_5}$$

$$Q_b = \frac{-\theta_o D}{a^2} \frac{C_2 L_4 - C_5 L_1}{C_2 C_6 - C_3 C_5}$$

$$M_{ra} = M_{rb} C_8 + Q_b a C_9 + \theta_o \frac{D}{a} L_7$$

$$Q_a = Q_b \frac{b}{a}$$

Special cases:

b/a	0.1		0.5		0.7
r_o/a	0.5	0.7	0.7	0.9	0.9
K_{y_o}	−0.1071	−0.0795	−0.0586	−0.0240	−0.0290
$K_{M_{rb}}$	−2.0540	1.1868	−3.5685	2.4702	0.3122
$K_{M_{ra}}$	−0.6751	−1.7429	−0.8988	−5.0320	−6.3013
K_{Q_b}	−0.0915	−17.067	4.8176	−23.8910	−29.6041

Case 7. Annular plate with an externally applied vertical deformation y_o at a radius r_o

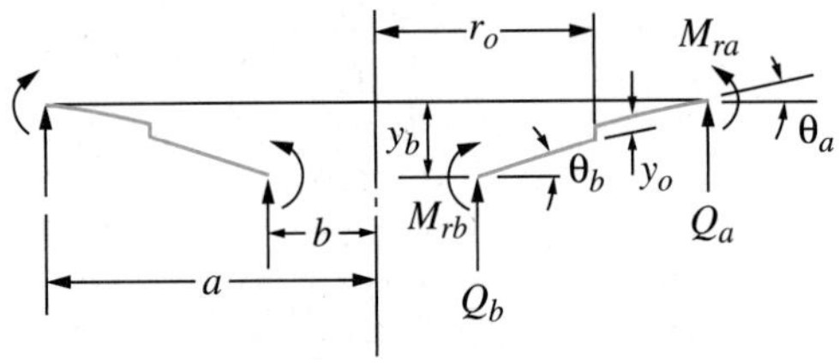

General expressions for deformations, moments, and shears:

$$y = y_b + \theta_b r F_1 + M_{rb} \frac{r^2}{D} F_2 + Q_b \frac{r^3}{D} F_3 + y_o \langle r - r_o \rangle^0$$

$$\theta = \theta_b F_4 + M_{rb} \frac{r}{D} F_5 + Q_b \frac{r^2}{D} F_6$$

$$M_r = \theta_b \frac{D}{r} F_7 + M_{rb} F_8 + Q_b r F_9$$

$$M_t = \frac{\theta D (1 - \nu^2)}{r} + \nu M_r$$

$$Q = Q_b \frac{b}{r}$$

For the numerical data given below, $\nu = 0.3$

$$y = K_y y_o, \quad \theta = K_\theta \frac{y_o}{a}, \quad M = K_M y_o \frac{D}{a^2}, \quad Q = K_Q y_o \frac{D}{a^3}$$

7c. Outer edge simply supported, inner edge simply supported

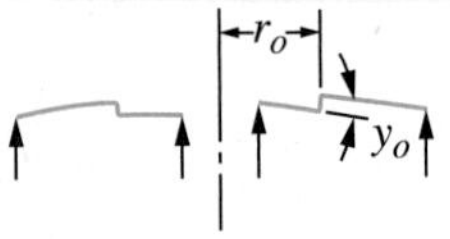

Boundary values:

$y_b = 0, \quad M_{rb} = 0, \quad y_a = 0, \quad M_{ra} = 0$

$$\theta_b = \frac{-y_o C_9}{a(C_1 C_9 - C_3 C_7)}$$

$$Q_b = \frac{y_o D C_7}{a^3 (C_1 C_9 - C_3 C_7)}$$

$$\theta_a = \frac{y_o}{a} \frac{C_7 C_6 - C_9 C_4}{C_1 C_9 - C_3 C_7}$$

$$Q_a = Q_b \frac{b}{a}$$

Special cases:

b/a	0.1	0.3	0.5	0.7	0.9
K_{θ_b}	−1.0189	−1.6176	−2.2045	−3.5180	−10.1611
K_{θ_a}	−1.1484	−1.3149	−1.8482	−3.1751	−9.8461
$K_{M_{tb}}$	−9.2716	−4.9066	−4.0121	−4.5734	−10.2740
K_{Q_b}	27.4828	7.9013	5.1721	5.1887	10.6599

(*Note:* Constants given are valid for all values of $r_o > b$.)

(*Continued*)

TABLE 11.2 Formulas for Flat Circular Plates of Constant Thickness (*Continued*)

Case No., Edge Restraints	Boundary Values	Special Cases

7d. Outer edge simply supported, inner edge fixed

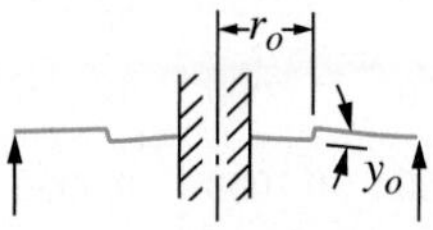

Boundary Values:

$y_b = 0,\quad \theta_b = 0,\quad y_a = 0,\quad M_{ra} = 0$

$$M_{rb} = \frac{-y_o D C_9}{a^2(C_2C_9 - C_3C_8)}$$

$$Q_b = \frac{y_o D C_8}{a^3(C_2C_9 - C_3C_8)}$$

$$\theta_a = \frac{y_o}{a}\,\frac{C_6C_8 - C_5C_9}{C_2C_9 - C_3C_8}$$

$$Q_a = Q_b \frac{b}{a}$$

Special Cases:

b/a	0.1	0.3	0.5	0.7	0.9
K_{θ_a}	−1.3418	−1.8304	−2.7104	−4.7327	−14.7530
$K_{M_{rb}}$	−9.4949	−9.9462	−15.6353	−37.8822	−310.808
K_{Q_b}	37.1567	23.9899	39.6394	138.459	3186.83

(*Note:* Constants given are valid for all values of $r_o > b$.)

7e. Outer edge fixed, inner edge simply supported

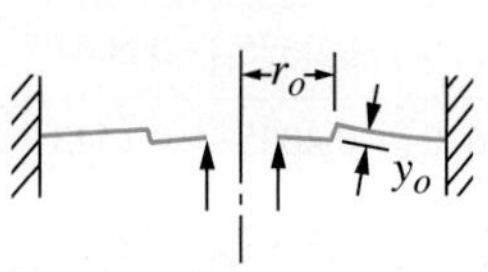

Boundary Values:

$y_b = 0,\quad M_{rb} = 0,\quad y_a = 0,\quad \theta_a = 0$

$$\theta_b = \frac{-y_o C_6}{a(C_1C_6 - C_3C_4)}$$

$$Q_b = \frac{y_o D C_4}{a^3(C_1C_6 - C_3C_4)}$$

$$M_{ra} = \frac{y_o D}{a^2}\,\frac{C_4C_9 - C_6C_7}{C_1C_6 - C_3C_4}$$

$$Q_a = Q_b \frac{b}{a}$$

Special Cases:

b/a	0.1	0.3	0.5	0.7	0.9
K_{θ_b}	−1.7696	−2.5008	−3.3310	−5.2890	−15.2528
$K_{M_{ra}}$	3.6853	5.1126	10.2140	30.1487	290.2615
$K_{M_{tb}}$	−16.1036	−7.5856	−6.0624	−6.8757	−15.4223
K_{Q_b}	69.8026	30.3098	42.9269	141.937	3186.165

(*Note:* Constants given are valid for all values of $r_o > b$.)

7f. Outer edge fixed, inner edge fixed

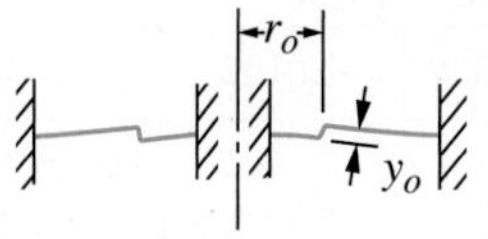

Boundary Values:

$y_b = 0,\quad \theta_b = 0,\quad y_a = 0,\quad \theta_a = 0$

$$M_{rb} = \frac{-y_o D C_6}{a^2(C_2C_6 - C_3C_5)}$$

$$Q_b = \frac{y_o D C_5}{a^3(C_2C_6 - C_3C_5)}$$

$$M_{ra} = \frac{y_o D}{a^2}\,\frac{C_5C_9 - C_6C_8}{C_2C_6 - C_3C_5}$$

$$Q_a = Q_b \frac{b}{a}$$

Special Cases:

b/a	0.1	0.3	0.5	0.7	0.9
$K_{M_{rb}}$	−18.8284	−19.5643	−31.0210	−75.6312	−621.8586
$K_{M_{ra}}$	4.9162	9.0548	19.6681	59.6789	579.6755
K_{Q_b}	103.1218	79.2350	146.258	541.958	12671.35

(*Note:* Constants given are valid for all values of $r_o > b$.)

Case 8. Annular plate with, from r_o to a, a uniform temperature differential ΔT between the bottom and the top surface (the midplane temperature is assumed to be unchanged, and so no in-plane forces develop)

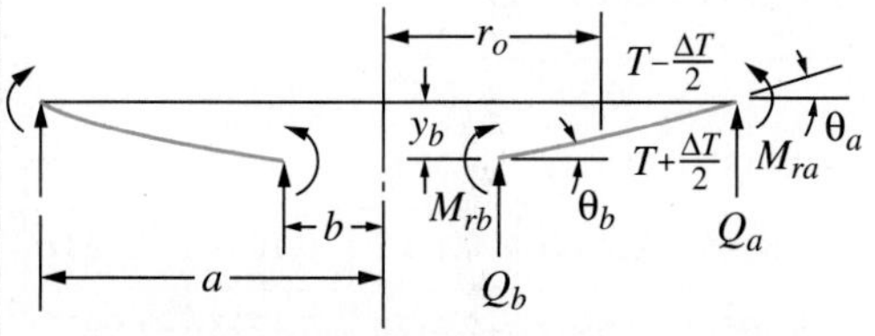

Note: If the temperature difference ΔT occurs over the entire plate, $r > r_o$ everywhere, so $\langle r - r_o\rangle^0 = 1$ everywhere therefore all numerical data for K_{Mtb} are given at a radius just greater than b.

General expressions for deformations, moments, and shears:

$$y = y_b + \theta_b r F_1 + M_{rb}\frac{r^2}{D}F_2 + Q_b\frac{r^3}{D}F_3 + \frac{\gamma(1+\nu)\Delta T}{t}r^2G_2$$

$$\theta = \theta_b F_4 + M_{rb}\frac{r}{D}F_5 + Q_b\frac{r^2}{D}F_6 + \frac{\gamma(1+\nu)\Delta T}{t}rG_5$$

$$M_r = \theta_b\frac{D}{r}F_7 + M_{rb}F_8 + Q_b r F_9 + \frac{\gamma(1+\nu)\Delta T}{t}D(G_8 - \langle r - r_o\rangle^0)$$

$$M_t = \frac{\theta D(1-\nu^2)}{r} + \nu M_r - \frac{\gamma(1-\nu^2)\Delta T D}{t}\langle r - r_o\rangle^0$$

$$Q = Q_b\frac{b}{r}$$

For the numerical data given below, $\nu = 0.3$

$$y = K_y\frac{\gamma\Delta T a^2}{t},\qquad \theta = K_\theta\frac{\gamma\Delta T a}{t},$$

$$M = K_M\frac{\gamma\Delta T D}{t},\qquad Q = K_Q\frac{\gamma\Delta T D}{at}$$

(*Continued*)

TABLE 11.2 Formulas for Flat Circular Plates of Constant Thickness (*Continued*)

Case No., Edge Restraints	Boundary Values	Special Cases
8a. Outer edge simply supported, inner edge free 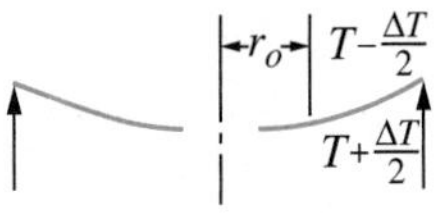	$M_{rb}=0,\quad Q_b=0,\quad y_a=0,\quad M_{ra}=0$ $y_b=\frac{-\gamma(1+\nu)\Delta T a^2}{t}\left[L_2+\frac{C_1}{C_7}(1-L_8)\right]$ $\theta_b=\frac{\gamma(1+\nu)\Delta T a}{tC_7}(1-L_8)$ $Q_a=0$ $\theta_a=\frac{\gamma(1+\nu)\Delta T a}{t}\left[L_5+\frac{C_4}{C_7}(1-L_8)\right]$	If $r_o=b$ (ΔT over entire plate), see table 8a below. (*Note:* There are no moments in the plate.)
8b. Outer edge simply supported, inner edge guided 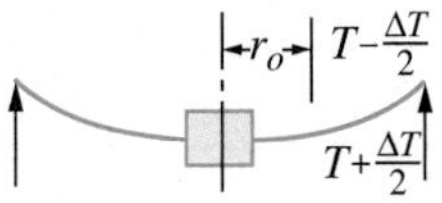	$\theta_b=0,\quad Q_b=0,\quad y_a=0,\quad M_{ra}=0$ $y_b=\frac{-\gamma(1+\nu)\Delta T a^2}{t}\left[L_2+\frac{C_2}{C_8}(1-L_8)\right]$ $M_{rb}=\frac{\gamma(1+\nu)\Delta T D}{tC_8}(1-L_8)$ $\theta_a=\frac{\gamma(1+\nu)\Delta T a}{t}\left[L_5+\frac{C_5}{C_8}(1-L_8)\right]$ $Q_a=0$	If $r_o=b$ (ΔT over entire plate), see table 8b below.
8c. Outer edge simply supported, inner edge simply supported 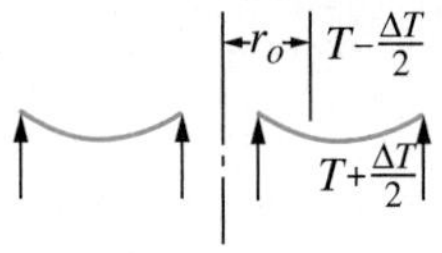	$y_b=0,\quad M_{rb}=0,\quad y_a=0,\quad M_{ra}=0$ $\theta_b=\frac{-\gamma(1+\nu)\Delta T a}{t}\,\frac{C_9L_2+C_3(1-L_8)}{C_1C_9-C_3C_7}$ $Q_b=\frac{\gamma(1+\nu)\Delta T D}{at}\,\frac{C_7L_2+C_1(1-L_8)}{C_1C_9-C_3C_7}$ $\theta_a=\theta_bC_4+Q_b\frac{a^2}{D}C_6+\frac{\gamma(1+\nu)\Delta T a}{t}L_5$ $Q_a=Q_b\frac{b}{a}$	If $r_o=b$ (ΔT over entire plate), see table 8c below.
8d. Outer edge simply supported, inner edge fixed 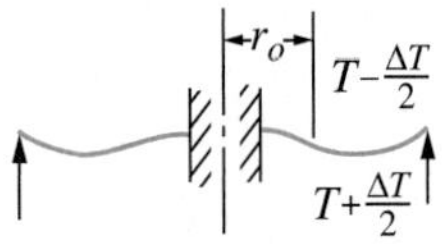	$y_b=0,\quad \theta_b=0,\quad y_a=0,\quad M_{ra}=0$ $M_{rb}=\frac{-\gamma(1+\nu)\Delta T D}{t}\,\frac{C_9L_2+C_3(1-L_8)}{C_2C_9-C_3C_8}$ $Q_b=\frac{\gamma(1+\nu)\Delta T D}{at}\,\frac{C_8L_2+C_2(1-L_8)}{C_2C_9-C_3C_8}$ $\theta_a=M_{rb}\frac{a}{D}C_5+Q_b\frac{a^2}{D}C_6+\frac{\gamma(1+\nu)\Delta T a}{t}L_5$ $Q_a=Q_b\frac{b}{a}$	If $r_o=b$ (ΔT over entire plate), see table 8d below.

Special cases, 8a. If $r_o=b$ (ΔT over entire plate),

b/a	0.1	0.3	0.5	0.7	0.9
K_{y_b}	−0.4950	−0.4550	−0.3750	−0.2550	−0.0950
K_{θ_a}	1.0000	1.0000	1.0000	1.0000	1.0000
K_{θ_b}	0.1000	0.3000	0.5000	0.7000	0.9000

(*Note:* There are no moments in the plate.)

Special cases, 8b. If $r_o=b$ (ΔT over entire plate),

b/a	0.1	0.3	0.5	0.7	0.9
K_{y_b}	−0.4695	−0.3306	−0.1778	−0.0635	−0.0067
K_{θ_a}	0.9847	0.8679	0.6610	0.4035	0.1323
$K_{M_{rb}}$	0.6893	0.6076	0.4627	0.2825	0.0926
$K_{M_{tb}}$	−0.7032	−0.7277	−0.7712	−0.8253	−0.8822

Special cases, 8c. If $r_o=b$ (ΔT over entire plate),

b/a	0.1	0.3	0.5	0.7	0.9
$K_{y_{max}}$	−0.0865	−0.0701	−0.0388	−0.0142	
K_{θ_b}	−0.4043	−0.4360	−0.3267	−0.1971	−0.0653
K_{θ_a}	0.4316	0.4017	0.3069	0.1904	0.0646
$K_{M_{tb}}$	−4.5894	−2.2325	−1.5045	−1.1662	−0.9760
K_{Q_b}	13.6040	3.5951	1.9395	1.3231	1.0127

Special cases, 8d. If $r_o=b$ (ΔT over entire plate),

b/a	0.1	0.3	0.5	0.7	0.9
$K_{y_{max}}$	−0.0583	−0.0318	−0.0147	−0.0049	
K_{θ_a}	0.3548	0.2628	0.1792	0.1031	0.0331
$K_{M_{rb}}$	−3.7681	−2.6809	−2.3170	−2.1223	−1.9975
K_{Q_b}	17.4431	7.9316	7.0471	8.7894	21.3100

(*Continued*)

TABLE 11.2 Formulas for Flat Circular Plates of Constant Thickness (*Continued*)

Case No., Edge Restraints	Boundary Values	Special Cases

8e. Outer edge fixed, inner edge free

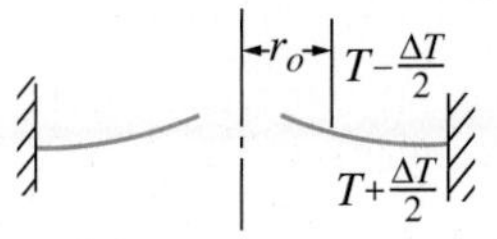

Boundary Values:

$M_{rb}=0,\quad Q_b=0,\quad y_a=0,\quad \theta_a=0$

$$y_b=\frac{-\gamma(1+\nu)\Delta T a^2}{t}\left(L_2-\frac{C_1}{C_4}L_5\right)$$

$$\theta_b=\frac{-\gamma(1+\nu)\Delta T a}{tC_4}L_5$$

$$M_{ra}=\frac{-\gamma(1+\nu)\Delta T D}{t}\left(\frac{C_7}{C_4}L_5+1-L_8\right)$$

$$Q_a=0$$

Special Cases:

If $r_o=b$ (ΔT over entire plate),

b/a	0.1	0.3	0.5	0.7	0.9
K_{y_b}	0.0330	0.1073	0.1009	0.0484	0.0062
K_{θ_b}	−0.1805	−0.4344	−0.4756	−0.3471	−0.1268
$K_{M_{ra}}$	−1.2635	−1.0136	−0.6659	−0.3471	−0.0986
$K_{M_{tb}}$	−2.5526	−2.2277	−1.7756	−1.3613	−1.0382

8f. Outer edge fixed, inner edge guided

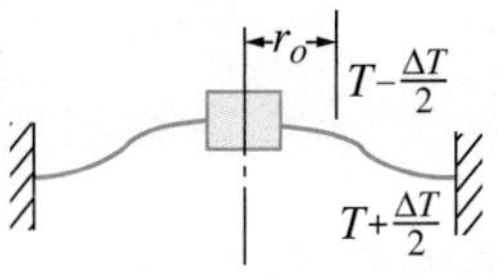

Boundary Values:

$\theta_b=0,\quad Q_b=0,\quad y_a=0,\quad \theta_a=0$

$$y_b=\frac{-\gamma(1+\nu)\Delta T a^2}{t}\left(L_2-\frac{C_2}{C_5}L_5\right)$$

$$M_{rb}=\frac{-\gamma(1+\nu)\Delta T D}{tC_5}L_5$$

$$M_{ra}=\frac{-\gamma(1+\nu)\Delta T D}{t}\left(\frac{C_8}{C_5}L_5+1-L_8\right)$$

$$Q_a=0$$

Special Cases:

If $r_o=b$ (ΔT over entire plate), all deflections are zero and $K_{M_r}=K_{M_t}=-1.30$ everywhere in the plate. If $r_o>b$, the following tabulated values apply:

b/a	0.1		0.5		0.7
r_o/a	0.5	0.7	0.7	0.9	0.9
K_{y_b}	0.1013	0.1059	0.0370	0.0269	0.0132
$K_{M_{rb}}$	−0.9849	−0.6697	−0.8840	−0.3293	−0.4843
$K_{M_{ra}}$	−0.9849	−0.6697	−0.8840	−0.3293	−0.4843
$K_{M_{to}}$	−1.5364	−1.3405	−1.3267	−1.0885	−1.1223

8g. Outer edge fixed, inner edge simply supported

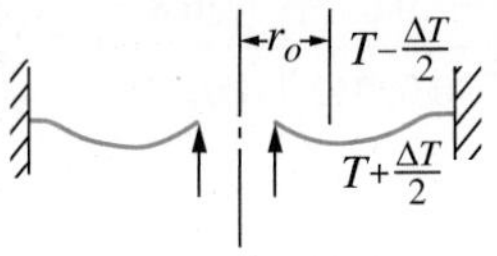

Boundary Values:

$y_b=0,\quad M_{rb}=0,\quad y_a=0,\quad \theta_a=0$

$$\theta_b=\frac{-\gamma(1+\nu)\Delta T a}{t}\,\frac{C_6L_2-C_3L_5}{C_1C_6-C_3C_4}$$

$$Q_b=\frac{\gamma(1+\nu)\Delta T D}{at}\,\frac{C_4L_2-C_1L_5}{C_1C_6-C_3C_4}$$

$$M_{ra}=\theta_B\frac{D}{a}C_7+Q_baC_9-\frac{\gamma(1+\nu)\Delta T D}{t}(1-L_8)$$

$$Q_a=Q_b\frac{b}{a}$$

Special Cases:

If $r_o=b$ (ΔT over entire plate),

b/a	0.1	0.3	0.5	0.7	0.9
$K_{y_{max}}$	−0.0088	−0.0133	−0.0091	−0.0039	
K_{θ_b}	−0.1222	−0.1662	−0.1396	−0.0909	−0.0319
$K_{M_{tb}}$	−2.0219	−1.4141	−1.1641	−1.0282	−0.9422
$K_{M_{ra}}$	−1.3850	−1.5620	−1.6962	−1.8076	−1.9050
K_{Q_b}	−2.3005	−3.2510	−4.3303	−6.8757	−19.8288

8h. Outer edge fixed, inner edge fixed

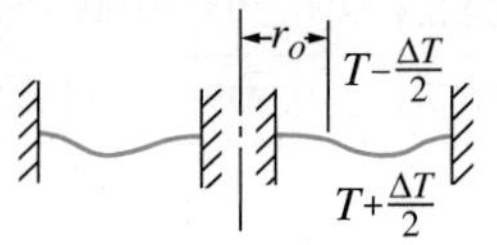

Boundary Values:

$y_b=0,\quad \theta_b=0,\quad y_a=0,\quad \theta_a=0$

$$M_{rb}=\frac{-\gamma(1+\nu)\Delta T D}{t}\,\frac{C_6L_2-C_3L_5}{C_2C_6-C_3C_5}$$

$$Q_b=\frac{\gamma(1+\nu)\Delta T D}{at}\,\frac{C_5L_2-C_2L_5}{C_2C_6-C_3C_5}$$

$$M_{ra}=M_{rb}C_8+Q_baC_9-\frac{\gamma(1+\nu)\Delta T D}{t}(1-L_8)$$

$$Q_a=Q_b\frac{b}{a}$$

Special Cases:

If $r_o=b$ (ΔT over entire plate), all deflections are zero and $K_{M_r}=K_{M_t}=-1.30$ everywhere in the plate. If $r_o>b$, the following tabulated values apply:

b/a	0.1		0.5		0.7
r_o/a	0.5	0.7	0.7	0.9	0.9
$K_{M_{rb}}$	0.9224	1.3241	0.2640	0.5063	0.5103
$K_{M_{ra}}$	−1.4829	−1.1903	−1.6119	−0.8592	−1.2691
$K_{M_{ta}}$	−1.3549	−1.2671	−1.3936	−1.1677	−1.2907
K_{Q_b}	−10.4460	−10.9196	−5.4127	−3.9399	−7.1270

(Continued)

TABLE 11.2 Formulas for Flat Circular Plates of Constant Thickness (*Continued*)

Cases 9 to 15. Solid circular plate under the several indicated loadings

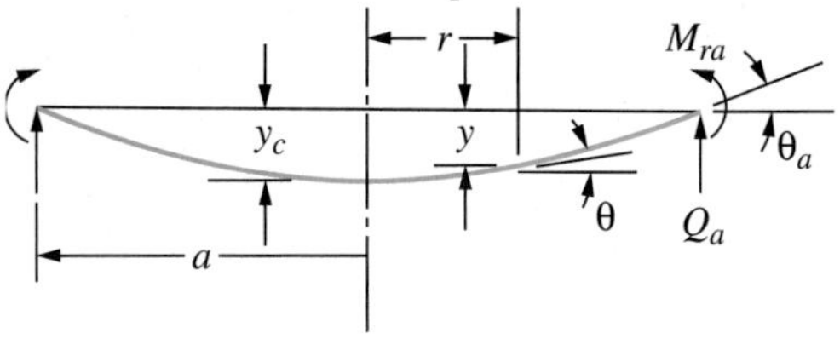

General expressions for deformations, moments, and shears:

$$y = y_c + \frac{M_c r^2}{2D(1+\nu)} + LT_y$$ (*Note:* y_c is the center deflection.)

$$\theta = \frac{M_c r}{D(1+\nu)} + LT_\theta$$ (*Note:* M_c is the moment at the center.)

$$M_r = M_c + LT_M$$

$$M_t = \frac{\theta D(1-\nu^2)}{r} + \nu M_r$$ (*Note:* For $r < r_o$, $M_t = M_r = M_c$.)

$$Q_r = LT_Q$$

For the numerical data given below, $\nu = 0.3$. (*Note:* ln = natural logarithm.)

Case No., Loading, Load Terms	Edge Restraint	Boundary Values	Special Cases
9. Uniform annular line load	9a. Simply supported	$y_a = 0,\ M_{ra} = 0$ $y_c = \frac{-wa^3}{2D}\left(\frac{L_9}{1+\nu} - 2L_3\right)$ $M_c = waL_9$ $Q_a = -w\frac{r_o}{a}$ $\theta_a = \frac{wr_o(a^2 - r_o^2)}{2D(1+\nu)a}$	$y = k_y\frac{wa^3}{D},\ \theta = k_\theta\frac{wa^2}{D},\ M = K_M wa$ (see table 9a below) (*Note:* If r_o approaches 0, see case 16.)
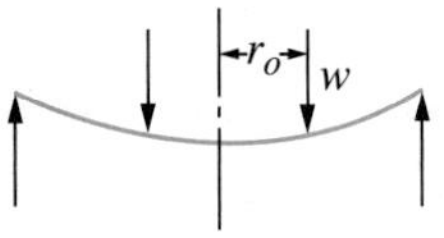 $LT_y = \frac{-wr^3}{D}G_3$ $LT_\theta = \frac{-wr^2}{D}G_6$ $LT_M = -wrG_9$ $LT_Q = \frac{-wr_o}{r}\langle r - r_o\rangle^0$	9b. Fixed	$y_c = \frac{-wa^3}{2D}(L_6 - 2L_3)$ $M_c = wa(1+\nu)L_6$ $M_{ra} = \frac{-wr_o}{2a^2}(a^2 - r_o^2)$ $y_a = 0,\ \theta_a = 0$	(see table 9b below) (*Note:* If r_o approaches 0, see case 17.)
10. Uniformly distributed pressure from r_o to a 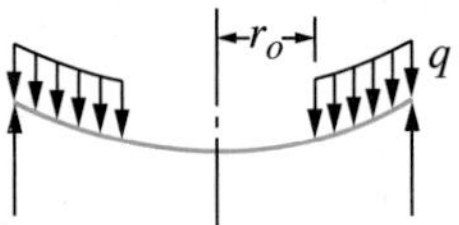 $LT_y = \frac{-qr^4}{D}G_{11}$ $LT_\theta = \frac{-qr^3}{D}G_{14}$ $LT_M = -qr^2G_{17}$ $LT_Q = \frac{-q}{2r}(r^2 - r_o^2)\langle r - r_o\rangle^0$	10a. Simply supported	$y_a = 0,\ M_{ra} = 0$ $y_c = \frac{-qa^4}{2D}\left(\frac{L_{17}}{1+\nu} - 2L_{11}\right)$ $M_c = qa^2L_{17}$ $\theta_a = \frac{q}{8Da(1+\nu)}(a^2 - r_o^2)^2$ $Q_a = \frac{-q}{2a}(a^2 - r_o^2)$	$y = k_y\frac{qa^4}{D},\ \theta = K_\theta\frac{qa^3}{D},\ M = K_M qa^2$ (see table 10a below) *Note:* If $r_o = 0$, $G_{11} = \frac{1}{64}$, $G_{14} = \frac{1}{16}$, $G_{17} = \frac{(3+\nu)}{16}$ $y_c = \frac{-qa^4(5+\nu)}{64D(1+\nu)}$, $M_c = \frac{qa^2(3+\nu)}{16}$, $\theta_a = \frac{qa^3}{8D(1+\nu)}$

Case 9a special cases:

r_o/a	0.2	0.4	0.6	0.8
K_{y_c}	−0.05770	−0.09195	−0.09426	−0.06282
K_{θ_a}	0.07385	0.12923	0.14769	0.11077
K_{M_c}	0.24283	0.29704	0.26642	0.16643

Case 9b special cases:

r_o/a	0.2	0.4	0.6	0.8
K_{y_c}	−0.02078	−0.02734	−0.02042	−0.00744
K_{M_c}	0.14683	0.12904	0.07442	0.02243
$K_{M_{ra}}$	−0.09600	−0.16800	−0.19200	−0.14400

Case 10a special cases:

r_o/a	0.0	0.2	0.4	0.6	0.8
K_{y_c}	−0.06370	−0.05767	−0.04221	−0.02303	−0.00677
K_{θ_a}	0.09615	0.08862	0.06785	0.03939	0.01246
K_{M_c}	0.20625	0.17540	0.11972	0.06215	0.01776

(*Continued*)

TABLE 11.2 Formulas for Flat Circular Plates of Constant Thickness (*Continued*)

<table>
<tr><th>Case No., Loading, Load Terms</th><th>Edge Restraint</th><th>Boundary Values</th><th>Special Cases</th></tr>
<tr><td></td><td>10b. Fixed</td><td>$y_a=0,\ \theta_a=0$
$y_c=\frac{-qa^4}{2D}(L_{14}-2L_{11})$
$M_c=qa^2(1+\nu)L_{14}$
$M_{ra}=\frac{-q}{8a^2}(a^2-r_o^2)^2$</td><td>
<table>
<tr><th>r_o/a</th><th>0.0</th><th>0.2</th><th>0.4</th><th>0.6</th><th>0.8</th></tr>
<tr><td>K_{y_c}</td><td>−0.01563</td><td>−0.01336</td><td>−0.00829</td><td>−0.00344</td><td>−0.00054</td></tr>
<tr><td>K_{M_c}</td><td>0.08125</td><td>0.06020</td><td>0.03152</td><td>0.01095</td><td>0.00156</td></tr>
<tr><td>$K_{M_{ra}}$</td><td>−0.12500</td><td>−0.11520</td><td>−0.08820</td><td>−0.05120</td><td>−0.01620</td></tr>
</table>
Note: If
$r_o=0,\ G_{11}=\frac{1}{64},\ G_{14}=\frac{1}{16},\ G_{17}=\frac{(3+\nu)}{16}$
$y_c=\frac{-qa^4}{64D},\ M_c=\frac{qa^2(1+\nu)}{16},\ M_{ra}=\frac{-qa^2}{8}$</td></tr>
<tr><td rowspan="2">11. Linearly increasing pressure from r_o to a
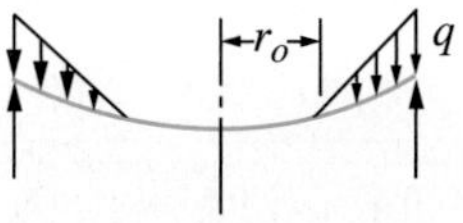

$LT_y=\frac{-qr^4}{D}\frac{r-r_o}{a-r_o}G_{12}$
$LT_\theta=\frac{-qr^3}{D}\frac{r-r_o}{a-r_o}G_{15}$
$LT_M=-qr^2\frac{r-r_o}{a-r_o}G_{18}$
$LT_Q=\frac{-q(2r^3-3r_or^2+r_o^3)}{6r(a-r^o)}\times\langle r-r_o\rangle^0$</td><td>11a. Simply supported</td><td>$M_{ra}=0,\ y_a=0$
$y_c=\frac{-qa^4}{2D}\left(\frac{L_{18}}{1+\nu}-2L_{12}\right)$
$M_c=qa^2L_{18}$
$\theta_a=\frac{qa^3}{D}\left(\frac{L_{18}}{1+\nu}-L_{15}\right)$
$Q_a=\frac{-q}{6a}(2a^2-r_oa-r_o^2)$</td><td>$y=k_y\frac{qa^4}{D},\ \theta=K_\theta\frac{qa^3}{D},\ M=K_Mqa^2$
<table>
<tr><th>r_o/a</th><th>0.0</th><th>0.2</th><th>0.4</th><th>0.6</th><th>0.8</th></tr>
<tr><td>K_{y_c}</td><td>−0.03231</td><td>−0.02497</td><td>−0.01646</td><td>−0.00836</td><td>−0.00234</td></tr>
<tr><td>K_{θ_a}</td><td>0.05128</td><td>0.04070</td><td>0.02788</td><td>0.01485</td><td>0.00439</td></tr>
<tr><td>K_{M_c}</td><td>0.09555</td><td>0.07082</td><td>0.04494</td><td>0.02220</td><td>0.00610</td></tr>
</table>
Note: If
$r_o=0,\ G_{12}=\frac{1}{225},\ G_{15}=\frac{1}{45},\ G_{18}=\frac{(4+\nu)}{45}$
$y_c=\frac{-qa^4(6+\nu)}{150D(1+\nu)},\ M_c=\frac{qa^2(4+\nu)}{45},$
$\theta_a=\frac{qa^3}{15D(1+\nu)}$</td></tr>
<tr><td>11b. Fixed</td><td>$y_a=0,\ \theta_a=0$
$y_c=\frac{-qa^4}{2D}(L_{15}-2L_{12})$
$M_c=qa^2(1+\nu)L_{15}$
$M_{ra}=-qa^2[L_{18}-(1+\nu)L_{15}]$</td><td>
<table>
<tr><th>r_o/a</th><th>0.0</th><th>0.2</th><th>0.4</th><th>0.6</th><th>0.8</th></tr>
<tr><td>K_{y_c}</td><td>−0.00667</td><td>−0.00462</td><td>−0.00252</td><td>−0.00093</td><td>−0.00014</td></tr>
<tr><td>K_{M_c}</td><td>0.02889</td><td>0.01791</td><td>0.00870</td><td>0.00289</td><td>0.00040</td></tr>
<tr><td>$K_{M_{ra}}$</td><td>−0.06667</td><td>−0.05291</td><td>−0.03624</td><td>−0.01931</td><td>−0.00571</td></tr>
</table>
Note: If
$r_o=0,\ G_{12}=\frac{1}{225},\ G_{15}=\frac{1}{45},\ G_{18}=\frac{(4+\nu)}{45}$
$y_c=\frac{-qa^4}{150D},\ M_c=\frac{qa^2(1+\nu)}{45},\ M_{ra}=\frac{-qa^2}{15}$</td></tr>
</table>

(*Continued*)

TABLE 11.2 Formulas for Flat Circular Plates of Constant Thickness (*Continued*)

Case No., Loading, Load Terms	Edge Restraint	Boundary Values	Special Cases

12. Parabolically increasing pressure from r_o to a

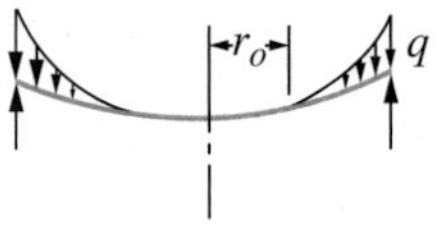

$$LT_y = \frac{-qr^4}{D}\left(\frac{r-r_o}{a-r_o}\right)^2 G_{13}$$

$$LT_\theta = \frac{-qr^3}{D}\left(\frac{r-r_o}{a-r_o}\right)^2 G_{16}$$

$$LT_M = -qr^2\left(\frac{r-r_o}{a-r_o}\right)^2 G_{19}$$

$$LT_Q = \frac{-q(3r^4-8r_or^3+6r_o^2r^2-r_o^4)}{12r(a-r_o)^2} \times \langle r-r_o\rangle^0$$

12a. Simply supported

Boundary Values:

$$y_a = 0, \quad M_{ra} = 0$$

$$y_c = \frac{-qa^4}{2D}\left(\frac{L_{19}}{1+\nu} - 2L_{13}\right)$$

$$M_c = qa^2L_{19}$$

$$\theta_a = \frac{qa^3}{D}\left(\frac{L_{19}}{1+\nu} - L_{16}\right)$$

$$Q_a = \frac{-q}{12a}(3a^2 - 2ar_o - r_o^2)$$

Special Cases:

$$y = K_y\frac{qa^4}{D}, \quad \theta = K_\theta\frac{qa^3}{D}, \quad M = K_Mqa^2$$

r_o/a	0.0	0.2	0.4	0.6	0.8
K_{yc}	−0.01949	−0.01419	−0.00893	−0.00438	−0.00119
$K_{\theta a}$	0.03205	0.02396	0.01560	0.00796	0.00227
K_{Mc}	0.05521	0.03903	0.02397	0.01154	0.00311

Note: If

$$r_o = 0, \quad G_{13} = \frac{1}{576}, \quad G_{16} = \frac{1}{96}, \quad G_{19} = \frac{(5+\nu)}{96}$$

$$y_c = \frac{-qa^4(7+\nu)}{288D(1+\nu)}, \quad M_c = \frac{qa^2(5+\nu)}{96}, \quad \theta_a = \frac{qa^3}{24D(1+\nu)}$$

12b. Fixed

Boundary Values:

$$y_a = 0, \quad \theta_a = 0$$

$$y_c = \frac{-qa^4}{2D}(L_{16} - 2L_{13})$$

$$M_c = qa^2(1+\nu)L_{16}$$

$$M_{ra} = -qa^2[L_{19} - (1+\nu)L_{16}]$$

Special Cases:

r_o/a	0.0	0.2	0.4	0.6	0.8
K_{yc}	−0.00347	−0.00221	−0.00113	−0.00040	−0.000058
K_{Mc}	0.01354	0.00788	0.00369	0.00120	0.000162
K_{Mra}	−0.04167	−0.03115	−0.02028	−0.01035	−0.002947

Note: If

$$r_o = 0, \quad G_{13} = \frac{1}{576}, \quad G_{16} = \frac{1}{96}, \quad G_{19} = \frac{(5+\nu)}{96}$$

$$y_c = \frac{-qa^4}{288D}, \quad M_c = \frac{qa^2(1+\nu)}{96}, \quad M_{ra} = \frac{-qa^2}{24}$$

13. Uniform line moment at r_o

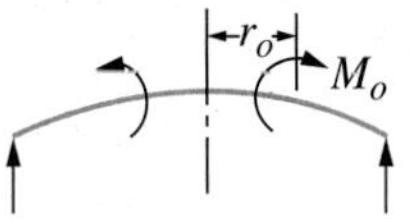

$$LT_y = \frac{M_or^2}{D}G_2$$

$$LT_\theta = \frac{M_or}{D}G_5$$

$$LT_M = M_oG_8$$

$$LT_Q = 0$$

13a. Simply supported

Boundary Values:

$$y_a = 0, \quad M_{ra} = 0, \quad Q_a = 0$$

$$y_c = \frac{M_or_o^2}{2D}\left(\frac{1}{1+\nu} + \ln\frac{a}{r_o}\right)$$

$$M_c = -M_oL_8$$

$$\theta_a = \frac{-M_or_o^2}{Da(1+\nu)}$$

Special Cases:

$$y = k_y\frac{M_oa^2}{D}, \quad \theta = K_\theta\frac{M_oa}{D}, \quad M = K_MM_o$$

r_o/a	0.2	0.4	0.6	0.8	1.0
K_{yc}	0.04757	0.13484	0.23041	0.31756	0.38462
$K_{\theta a}$	−0.03077	−0.12308	−0.27692	−0.49231	−0.76923
K_{Mc}	−0.66400	−0.70600	−0.77600	−0.87400	−1.00000

13b. Fixed

Boundary Values:

$$y_a = 0, \quad \theta_a = 0, \quad Q_a = 0$$

$$y_c = \frac{M_or_o^2}{2D}\ln\frac{a}{r_o}$$

$$M_c = \frac{-M_o(1+\nu)}{2a^2}(a^2 - r_o^2)$$

$$M_{ra} = \frac{M_or_o^2}{a^2}$$

Special Cases:

r_o/a	0.2	0.4	0.6	0.8
K_{yc}	0.03219	0.07330	0.09195	0.07141
K_{Mc}	−0.62400	−0.54600	−0.41600	−0.23400
K_{Mra}	0.04000	0.16000	0.36000	0.64000

(Continued)

TABLE 11.2 Formulas for Flat Circular Plates of Constant Thickness (*Continued*)

<table>
<tr><th>Case No., Loading, Load Terms</th><th>Edge Restraint</th><th>Boundary Values</th><th>Special Cases</th></tr>
<tr><td rowspan="2">14. Externally applied change in slope at a radius r_o
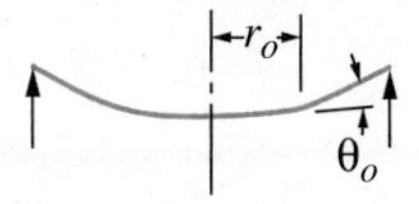
$LT_y = \theta_o r G_1$
$LT_\theta = \theta_o G_4$
$LT_M = \frac{\theta_o D}{r} G_7$
$LT_Q = 0$</td>
<td>14a. Simply supported</td>
<td>$y_a = 0,\ M_{ra} = 0,\ Q_a = 0$
$y_c = \frac{-\theta_o r_o(1+\nu)}{2}\ln\frac{a}{r_o}$
$M_c = \frac{-\theta_o D(1-\nu^2)}{2r_o a^2}(a^2 - r_o^2)$
$\theta_a = \frac{\theta_o r_o}{a}$</td>
<td>$y = K_y \theta_o a,\ \theta = K_\theta \theta_o,\ M = K_M \theta_o \frac{D}{a}$
<table>
<tr><th>r_o/a</th><th>0.2</th><th>0.4</th><th>0.6</th><th>0.8</th></tr>
<tr><td>K_{y_c}</td><td>−0.20923</td><td>−0.23824</td><td>−0.19922</td><td>−0.11603</td></tr>
<tr><td>K_{θ_a}</td><td>0.20000</td><td>0.40000</td><td>0.60000</td><td>0.80000</td></tr>
<tr><td>K_{M_c}</td><td>−2.18400</td><td>−0.95550</td><td>−0.48533</td><td>−0.20475</td></tr>
<tr><td>$K_{M_{to}}$</td><td>2.33600</td><td>1.31950</td><td>1.03133</td><td>0.93275</td></tr>
</table></td></tr>
<tr><td>14b. Fixed</td>
<td>$y_a = 0,\ \theta_a = 0,\ Q_a = 0$
$y_c = \frac{\theta_o r_o}{2}\left[1-(1+\nu)\ln\frac{a}{r_o}\right]$
$M_c = \frac{-\theta_o D(1+\nu)}{a}L_4$
$M_{ra} = \frac{-\theta_o D r_o}{a^2}(1+\nu)$</td>
<td><table>
<tr><th>r_o/a</th><th>0.2</th><th>0.4</th><th>0.6</th><th>0.8</th><th>1.0</th></tr>
<tr><td>K_{y_c}</td><td>−0.10923</td><td>−0.03824</td><td>0.10078</td><td>0.28396</td><td>0.50000</td></tr>
<tr><td>K_{M_c}</td><td>−2.44400</td><td>−1.47550</td><td>−1.26533</td><td>−1.24475</td><td>−1.30000</td></tr>
<tr><td>$K_{M_{ra}}$</td><td>−0.26000</td><td>−0.52000</td><td>−0.78000</td><td>−1.04000</td><td>−1.30000</td></tr>
</table></td></tr>
<tr><td rowspan="2">15. Uniform temperature differential ΔT between the bottom and top surface from r_o to a
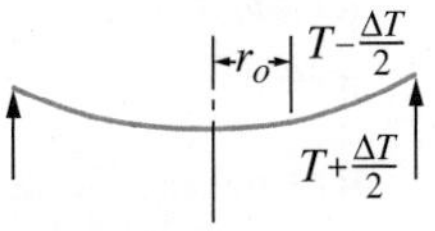
$LT_y = \frac{\gamma(1+\nu)\Delta T}{t} r^2 G_2$
$LT_\theta = \frac{\gamma(1+\nu)\Delta T}{t} r G_5$
$LT_M = \frac{\gamma D(1+\nu)\Delta T}{t}(G_8 - \langle r - r_o\rangle^0)$
$LT_Q = 0$
Note: Values for $K_{M_{to}}$ are given at a radius just greater than r_o.</td>
<td>15a. Simply supported</td>
<td>$y_a = 0,\ M_{ra} = 0,\ Q_a = 0$
$y_c = \frac{-\gamma\Delta T}{2t}\left[a^2 - r_0^2 - r_0^2(1+\nu)\ln\frac{a}{r_o}\right]$
$M_c = \frac{\gamma D(1+\nu)\Delta T}{t}(1 - L_8)$
$\theta_a = \frac{\gamma\Delta T}{ta}(a^2 - r_o^2)$</td>
<td>$y = K_y \frac{\gamma\Delta T a^2}{t},\ \theta = K_\theta \frac{\gamma\Delta T a}{t},\ M = K_M \frac{\gamma\Delta T D}{t}$
<table>
<tr><th>r_o/a</th><th>0.0</th><th>0.2</th><th>0.4</th><th>0.6</th><th>0.8</th></tr>
<tr><td>K_{y_c}</td><td>−0.50000</td><td>−0.43815</td><td>−0.32470</td><td>−0.20047</td><td>−0.08717</td></tr>
<tr><td>K_{θ_a}</td><td>1.00000</td><td>0.96000</td><td>0.84000</td><td>0.64000</td><td>0.36000</td></tr>
<tr><td>K_{M_c}</td><td>0.00000</td><td>0.43680</td><td>0.38220</td><td>0.29120</td><td>0.16380</td></tr>
<tr><td>$K_{M_{to}}$</td><td></td><td>−0.47320</td><td>−0.52780</td><td>−0.61880</td><td>−0.74620</td></tr>
</table>
Note: When the entire plate is subjected to the temperature differential, there is no stress anywhere in the plate.</td></tr>
<tr><td>15b. Fixed</td>
<td>$y_a = 0,\ \theta_a = 0,\ Q_a = 0$
$y_c = \frac{\gamma(1+\nu)\Delta T}{2t} r_o^2 \ln\frac{a}{r_o}$
$M_c = \frac{-\gamma D(1+\nu)^2\Delta T}{2ta^2}(a^2 - r_o^2)$
$M_{ra} = \frac{-\gamma D(1+\nu)\Delta T}{ta^2}(a^2 - r_o^2)$</td>
<td><table>
<tr><th>r_o/a</th><th>0.0</th><th>0.2</th><th>0.4</th><th>0.6</th><th>0.8</th></tr>
<tr><td>K_{y_c}</td><td>0.00000</td><td>0.04185</td><td>0.09530</td><td>0.11953</td><td>0.09283</td></tr>
<tr><td>K_{M_c}</td><td>−1.30000</td><td>−0.81120</td><td>−0.70980</td><td>−0.54080</td><td>−0.30420</td></tr>
<tr><td>$K_{M_{ro}}$</td><td></td><td>−1.24800</td><td>−1.09200</td><td>−0.83200</td><td>−0.46800</td></tr>
<tr><td>$K_{M_{to}}$</td><td></td><td>−1.72120</td><td>−1.61980</td><td>−1.45080</td><td>−1.21420</td></tr>
</table>
Note: When the entire plate is subjected to the temperature differential, the moments are the same everywhere in the plate and there are no deflections.</td></tr>
<tr><td colspan="4">*Note:* The term $\frac{-\gamma(1-\nu^2)\Delta T D}{t}\langle r - r_o\rangle^0$ must be added to M_t for this case 15. Also, if $r_o = 0$, then $G_2 = \frac{1}{4}, G_5 = \frac{1}{2}, G_8 = (1+\nu)/2$, and $\langle r - r_o\rangle^0 = 1$ for all values of r.</td></tr>
</table>

(*Continued*)

TABLE 11.2 Formulas for Flat Circular Plates of Constant Thickness (*Continued*)

Cases 16 to 31. The following cases include loadings on circular plates or plates bounded by some circular boundaries (each case is complete in itself). (*Note:* ln = natural logarithm)

Case No., Loading, Restraints	Formulas	Special Cases
16. Uniform load over a very small central circular area of radius r_o; edge simply supported 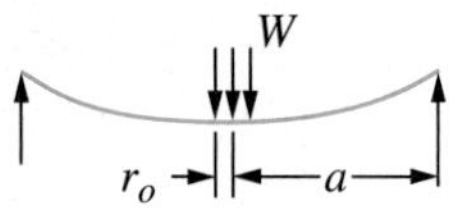 $W=q\pi r_o^2$	For $r>r_o$ $y=\frac{-W}{16\pi D}\left[\frac{3+\nu}{1+\nu}(a^2-r^2)-2r^2\ln\frac{a}{r}\right]$ $\theta=\frac{Wr}{4\pi D}\left(\frac{1}{1+\nu}+\ln\frac{a}{r}\right)$ $M_r=\frac{W}{16\pi}\left[4(1+\nu)\ln\frac{a}{r}+(1-\nu)\left(\frac{a^2-r^2}{a^2}\right)\frac{r_o'^2}{r^2}\right]$ where $r_o'=\sqrt{1.6r_o^2+t^2}-0.675t$ if $r_o<0.5t$ or $r_o'=r_o$ if $r_o>0.5t$ $M_t=\frac{W}{16\pi}\left[4(1+\nu)\ln\frac{a}{r}+(1-\nu)\left(4-\frac{r_o'^2}{r^2}\right)\right]$	$y_{max}=\frac{-Wa^2}{16\pi D}\frac{3+\nu}{1+\nu}$ at $r=0$ $\theta_{max}=\frac{Wa}{4\pi D(1+\nu)}$ at $r=a$ $(M_r)_{max}=\frac{W}{4\pi}\left[(1+\nu)\ln\frac{a}{r_o'}+1\right]$ at $r=0$ $(M_t)_{max}=(M_r)_{max}$ at $r=0$
17. Uniform load over a very small central circular area of radius r_o; edge fixed 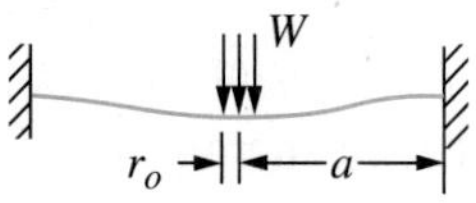 $W=q\pi r_o^2$	For $r>r_o'$ $y=\frac{-W}{16\pi D}\left[a^2-r^2\left(1+2\ln\frac{a}{r}\right)\right]$ $\theta=\frac{Wr}{4\pi D}\ln\frac{a}{r}$ $M_r=\frac{W}{4\pi}\left[(1+\nu)\ln\frac{a}{r}-1+\frac{(1-\nu)r_o'^2}{4r^2}\right]$ where $r_o'=\sqrt{1.6r_o^2+t^2}-0.675t$ if $r_o<0.5t$ or $r_o'=r_o$ if $r_o\ge 0.5t$ $M_t=\frac{W}{4\pi}\left[(1+\nu)\ln\frac{a}{r}-\nu+\frac{\nu(1-\nu)r_o'^2}{4r^2}\right]$	$y_{max}=\frac{-Wa^2}{16\pi D}$ at $r=0$ $\theta_{max}=0.0293\frac{Wa}{D}$ at $r=0.368a$ $(+M_r)_{max}=\frac{W}{4\pi}(1+\nu)\ln\frac{a}{r_o'}$ at $r=0$ $(-M_r)_{max}=\frac{-W}{4\pi}$ at $r=a$ $(+M_t)_{max}=(+M_r)_{max}$ at $r=0$ $(-M_t)_{max}=\frac{-\nu W}{4\pi}$ at $r=a$
18. Uniform load over a small eccentric circular area of radius r_o; edge simply supported 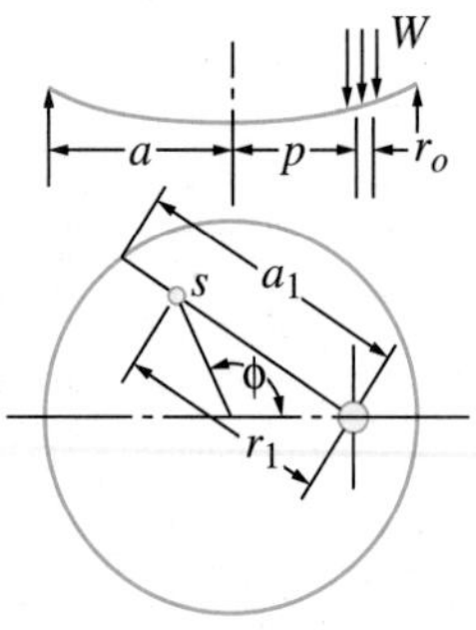 *Note:* $r_o'=\sqrt{1.6r_o^2+t^2}-0.675t$ if $r_o<0.5t$ or $r_o'=r_o$ if $r_o\ge 0.5t$	$(M_r)_{max}=(M_t)_{max}=\frac{W}{4\pi}\left[1+(1+\nu)\ln\left(\frac{a-p}{r_o'}\right)-\frac{(1-\nu)r_o'^2}{4(a-p)^2}\right]$ at the load At any point *s*, $M_r=(M_r)_{max}\frac{(1+\nu)\ln(a_1/r_1)}{1+(1+\nu)\ln(a_1/r_o')}$, $M_t=(M_t)_{max}\frac{(1+\nu)\ln(a_1/r_1)+1-\nu}{1+(1+\nu)\ln(a_1/r_o')}$ $y=-[K_o(r^3-b_oar^2+c_oa^3)+K_1(r^4-b_1ar^3+c_1a^3r)\cos\phi+K_2(r^4-b_2ar^3+c_2a^2r^2)\cos 2\phi]$ where $K_o=\frac{W}{\pi Da^4}\frac{2(1+\nu)}{9(5+\nu)}(p^3-b_oap^2+c_oa^3)$ $K_1=\frac{W}{\pi Da^6}\frac{2(3+\nu)}{3(9+\nu)}(p^4-b_1ap^3+c_1a^3p)$ $K_2=\frac{W}{\pi Da^6}\frac{(4+\nu)^2}{(9+\nu)(5+\nu)}(p^4-b_2ap^3+c_2a^2p^2)$ where $b_o=\frac{3(2+\nu)}{2(1+\nu)}$, $b_1=\frac{3(4+\nu)}{2(3+\nu)}$, $b_2=\frac{2(5+\nu)}{4+\nu}$, $c_o=\frac{4+\nu}{2(1+\nu)}$, $c_1=\frac{6+\nu}{2(3+\nu)}$, $c_2=\frac{6+\nu}{4+\nu}$ (See Ref. 1)	

(*Continued*)

TABLE 11.2 Formulas for Flat Circular Plates of Constant Thickness (*Continued*)

Case No., Loading, Restraints	

19. Uniform load over a small eccentric circular area of radius r_o; edge fixed

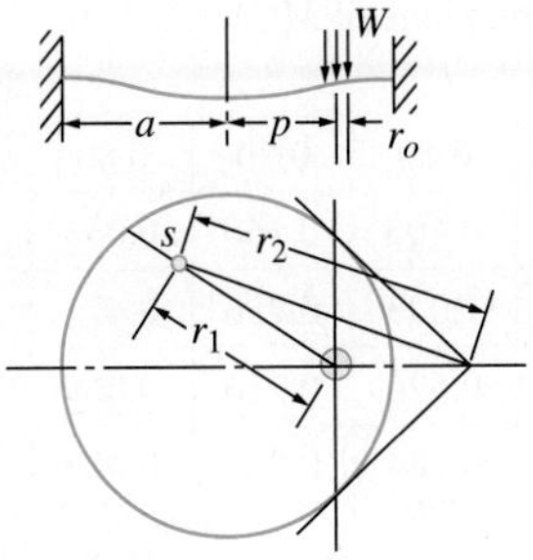

At any point s,

$$y=\frac{-W}{16\pi D}\left[\frac{p^2 r_2^2}{a^2}-r_1^2\left(1+2\ln\frac{pr_2}{ar_1}\right)\right] \quad (\textit{Note}: \text{As } p\to 0,\ pr_2\to a^2)$$

At the load point,

$$y=\frac{-W}{16\pi D}\frac{(a^2-p^2)^2}{a^2}$$

$$M_r=\frac{W(1+\nu)}{16\pi}\left[4\ln\left(\frac{a-p}{r_o'}\right)+\left(\frac{r_o'}{a-p}\right)^2\right]=M_{\max} \quad \text{if } r_o'<0.6(a-p) \qquad (\textit{Note: } r_o' \text{ defined in case 18})$$

At the near edge,

$$M_r=\frac{-W}{8\pi}\left[2-\left(\frac{r_o'}{a-p}\right)^2\right]=M_{\max} \quad \text{if } r_o>0.6(a-p)$$

[Formulas due to Michell (Ref. 2). See Ref. 60 for modified boundary conditions.]

20. Central couple on an annular plate with a simply supported outer edge (trunnion loading)

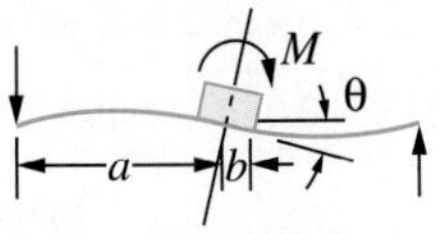

(*Note:* For eccentric trunnions loaded with vertical loads, couples, and pressure on the plate, see Refs. 86 and 87.)

20a. Trunnion simply supported to plate. For $\nu=0.3$

$$\theta=\frac{\alpha M}{Et^3},\quad \tau_{\max}=\tau_{rt}=\frac{\lambda M}{at^2} \text{ at } r=b \text{ at } 90^\circ \text{ to the plane of } M$$

$$\sigma_{\max}=\sigma_t=\frac{\gamma M}{at^2} \text{ at } r=b \text{ in the plane of } M$$

b/a	0.10	0.15	0.20	0.25	0.30	0.40	0.50	0.60	0.70	0.80
λ	9.475	6.256	4.630	3.643	2.976	2.128	1.609	1.260	1.011	0.827
γ	12.317	8.133	6.019	4.735	3.869	2.766	2.092	1.638	1.314	1.075
α	2.624	2.256	1.985	1.766	1.577	1.257	0.984	0.743	0.528	0.333

(Ref. 85)

20b. Trunnion fixed to the plate

$$\theta=\frac{\alpha M}{Et^3},\quad (\sigma_r)_{\max}=\frac{\beta M}{at^2} \text{ at } r=b \text{ in the plane of } M$$

b/a	0.10	0.15	0.20	0.25	0.30	0.40	0.50	0.60	0.70	0.80
β	9.478	6.252	4.621	3.625	2.947	2.062	1.489	1.067	0.731	0.449
α	1.403	1.058	0.820	0.641	0.500	0.301	0.169	0.084	0.035	0.010

(Ref. 21)

(*Continued*)

TABLE 11.2 Formulas for Flat Circular Plates of Constant Thickness (*Continued*)

Case No., Loading, Restraints	

21. Central couple on an annular plate with a fixed outer edge (trunnion loading)

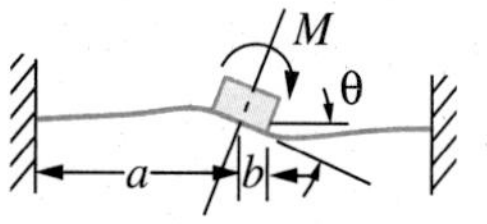

(*Note:* For eccentric trunnions see note in case 20 above.)

21a. Trunnion simply supported to plate. For $\nu = 0.3$

$\theta = \frac{\alpha M}{Et^3}$, $\tau_{max} = \tau_{rt} = \frac{\lambda M}{at^2}$ at $r = b$ at 90° to the plane of M

$(\sigma_r)_{max} = \frac{\beta M}{at^2}$ at $r = a$ in the plane of M $\quad$ $\max \sigma_t = \frac{\gamma M}{at^2}$ at $r = b$ in the plane of M

b/a	0.10	0.15	0.20	0.25	0.30	0.40	0.50	0.60	0.70	0.80
λ	9.355	6.068	4.367	3.296	2.539	1.503	0.830	0.405	0.166	0.053
β	0.989	1.030	1.081	1.138	1.192	1.256	1.205	1.023	0.756	0.471
γ	12.161	7.889	5.678	4.285	3.301	1.954	1.079	0.526	0.216	0.069
α	2.341	1.949	1.645	1.383	1.147	0.733	0.405	0.184	0.064	0.015

(Ref. 85)

21b. Trunnion fixed to the plate. For $\nu = 0.3$

$\theta = \frac{\alpha M}{Et^3}$, $\sigma_{max} = \sigma_r = \frac{\beta M}{at^2}$ at $r = b$ in the plane of M

$\sigma_r = \frac{\beta b M}{a^2 t^2}$ at $r = a$ in the plane of M

b/a	0.10	0.15	0.20	0.25	0.30	0.40	0.50	0.60	0.70	0.80
β	9.36	6.08	4.41	3.37	2.66	1.73	1.146	0.749	0.467	0.262
α	1.149	0.813	0.595	0.439	0.320	0.167	0.081	0.035	0.013	0.003

(Ref. 22)

22. Linearly distributed load symmetrical about a diameter; edge simply supported

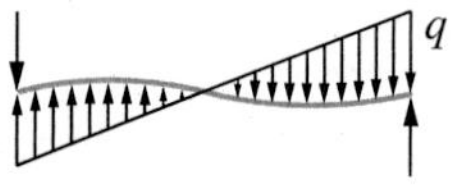

$(M_r)_{max} = \frac{qa^2(5+\nu)}{72\sqrt{3}}$ at $r = 0.577a$

$(M_t)_{max} = \frac{qa^2(5+\nu)(1+3\nu)}{72(3+\nu)}$ at $r = 0.675a$

Max edge reaction per linear inch $= \frac{qa}{4}$

$y_{max} = 0.042\frac{qa^4}{Et^3}$ at $r = 0.503a$ (for $\nu = 0.03$)

(Refs. 20 and 21)

23. Central couple balanced by linearly distributed pressure (footing)

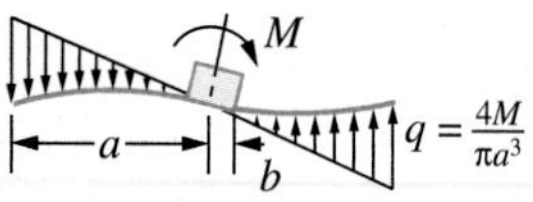

(At inner edge) $(\sigma_r)_{max} = \beta \frac{M}{at^2}$ where β is given in the following table:

a/b	1.25	1.50	2.00	3.00	4.00	5.00
β	0.1625	0.4560	1.105	2.250	3.385	4.470

(Values for $\nu = 0.3$)

(Ref. 21)

(Continued)

TABLE 11.2 Formulas for Flat Circular Plates of Constant Thickness (*Continued*)

Case No., Loading, Restraints

24. Concentrated load applied at the outer edge of an annular plate with a fixed inner edge

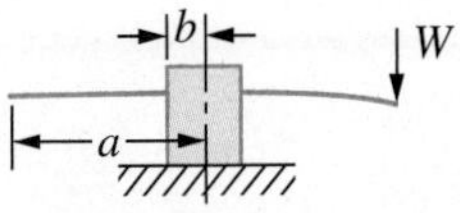

(At inner edge) $(\sigma_r)_{max} = \beta \frac{W}{t^2}$ where β is given in the following table:

a/b	1.25	1.50	2.00	3.00	4.00	5.00
β	3.665	4.223	5.216	6.904	8.358	9.667

(Values for $\nu = 0.3$) (Ref. 93)

(See Ref. 64 for this loading on a plate with radially varying thickness. See graphs in Ref. 59 for the load distributed over an arc at the edge. See Ref. 60 for the load W placed away from the edge.)

25. Solid circular plate with a uniformly distributed load q over the shaded segment

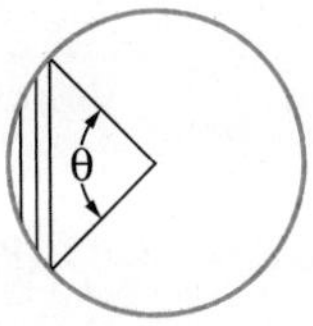

$\sigma_{max} = (\sigma_r)_{max} = \beta \frac{qa^2}{t^2}$

$y_{max} = \alpha \frac{qa^4}{Et^3}$ on the symmetrical diameter at the value of r given in the table

Edge	Coefficient	θ: 90°	120°	180°
Supported	α	0.0244, $r = 0.39a$	0.0844, $r = 0.30a$	0.345, $r = 0.15a$
	β	0.306, $r = 0.60a$		
Fixed	α	0.00368, $r = 0.50a$	0.0173, $r = 0.4a$	0.0905, $r = 0.20a$
	β	0.285, $r = a$		

Values for $\nu = \frac{1}{3}$ (Ref. 39)

26. Solid circular plate, uniform load q over the shaded sector

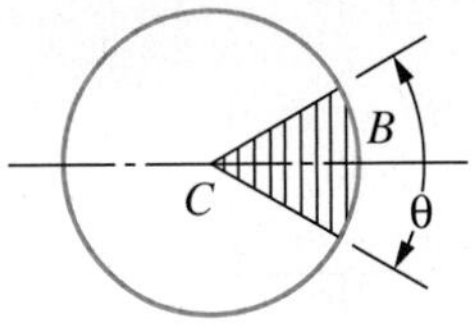

For simply supported edges:

$\sigma_{max} = \sigma_r$ near the center along the loaded radius of symmetry (values not given)

σ_r at the center $= \frac{\theta}{360} \sigma_r$ at the center of a fully loaded plate

$y_{max} = -\alpha_1 \frac{qa^4}{Et^3}$ at approximately $\frac{1}{4}$ the radius from center along the radius of symmetry

(α_1 given in table)

For fixed edges:

$\sigma_{max} = \sigma_r$ at point $B = \beta \frac{qa^2}{t^2}$

$y_{max} = -\alpha_2 \frac{qa^4}{Et^3}$ at approximately $\frac{1}{4}$ the radius from center along the radius of symmetry

(β and α_2 given in table)

Edge condition	Coefficient	θ: 30°	60°	90°	120°	150°	180°
Simply supported	α_1	0.061	0.121	0.179	0.235	0.289	0.343
Fixed	α_2	0.017	0.034	0.050	0.064	0.077	0.089
	β	0.240	0.371	0.457	0.518	0.564	0.602

[*Note:* For either edge condition $y_c = (\theta/360) y_c$ for a fully loaded plate.] (Ref. 38)

(*Continued*)

TABLE 11.2 Formulas for Flat Circular Plates of Constant Thickness (*Continued*)

Case No., Loading, Restraints	

27. Solid circular sector, uniformly distributed load q over the entire surface; edges simply supported

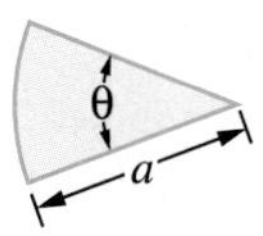

$$(\sigma_r)_{max} = \beta \frac{qa^2}{t^2}, \quad (\sigma_t)_{max} = \beta_1 \frac{qa^2}{t^2}, \quad y_{max} = \alpha \frac{qa^2}{Et^3}$$

θ	45°	60°	90°	180°
β	0.102	0.147	0.240	0.522
β_1	0.114	0.155	0.216	0.312
α	0.0054	0.0105	0.0250	0.0870

(Values for $\nu = 0.3$) (Ref. 21)

28. Solid circular sector, uniformly distributed load q over the entire surface; straight edges simply supported, curved edge fixed

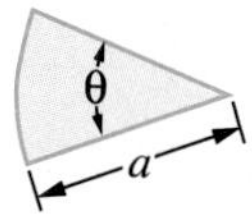

$$\sigma_{max} = \sigma_r \text{ at curved edge} = \beta \frac{qa^2}{t^2}, \quad y_{max} = \alpha \frac{qa^4}{Et^3}$$

θ	45°	60°	90°	180°
β	0.1500	0.2040	0.2928	0.4536
α	0.0035	0.0065	0.0144	0.0380

(Values for $\nu = 0.3$) (Ref. 21)

29. Solid circular sector of infinite radius, uniformly distributed load q over entire surface; straight edges fixed

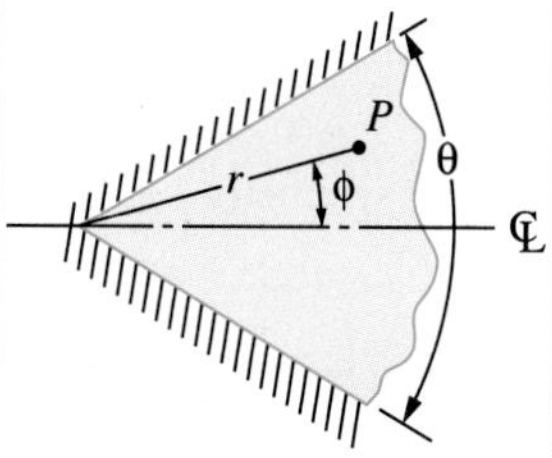

At point P:

$$\sigma_r = \frac{9qr^2}{8t^2}\left[\frac{3+\nu}{3} - \frac{4\cos\theta\cos 2\phi - (1-\nu)\cos 4\phi}{2\cos^2\theta + 1}\right]$$

$$\sigma_t = \frac{9qr^2}{8t^2}\left[\frac{1+3\nu}{3} - \frac{4\nu\cos\theta\cos 2\phi + (1-\nu)\cos 4\phi}{2\cos^2\theta + 1}\right]$$

$$y = \frac{-3(1-\nu^2)qr^4}{16Et^3}\left(1 + \frac{\cos 4\phi - 4\cos\theta\cos 2\phi}{2\cos^2\theta + 1}\right)$$

(*Note:* θ should not exceed 60°.)

At the edge, $\phi = \pm\theta/2$:

$$\sigma_t = \frac{3qr^2}{2t^2}\frac{\sin^2\theta}{1 + 2\cos^2\theta}$$

$$\sigma_r = \nu\sigma_t$$

Along the center line, $\phi = 0$

$$\sigma_r = \frac{3qr^2}{4t^2}\frac{3(1-\cos\theta)^2 - \nu\sin^2\theta}{1 + 2\cos^2\theta}$$

$$\sigma_t = \frac{3qr^2}{4t^2}\frac{3\nu(1-\cos\theta)^2 - \sin^2\theta}{1 + 2\cos^2\theta}$$

$$y = \frac{-3(1-\nu^2)qr^4}{8Et^3}\frac{(1-\cos\theta)^2}{1 + 2\cos^2\theta}$$

(Ref. 37)

30. Solid semicircular plate, uniformly distributed load q over the entire surface; all edges fixed

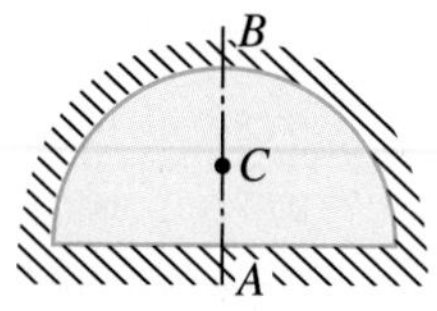

$$\sigma_{max} = \sigma_r \text{ at } A = -0.42\frac{qa^2}{t^2} \text{ (values for } \nu = 0.2)$$

$$\sigma_r \text{ at } B = -0.36\frac{qa^2}{t^2}$$

$$\sigma_r \text{ at } C = 0.21\frac{qa^2}{t^2}$$

(Ref. 40)

(*Continued*)

TABLE 11.2 Formulas for Flat Circular Plates of Constant Thickness (*Continued*)

Case No., Loading, Restraints	

31. Semicircular annular plate, uniformly loaded over entire surface; outer edge supported, all other edges free

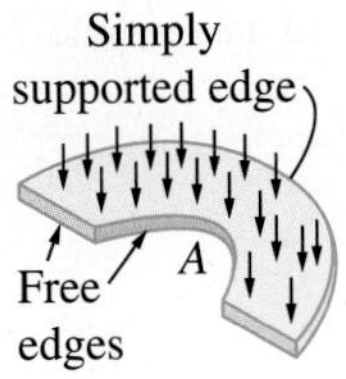

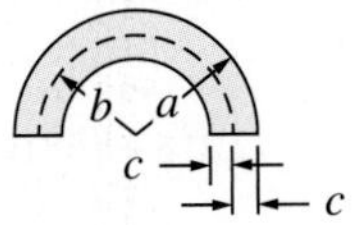

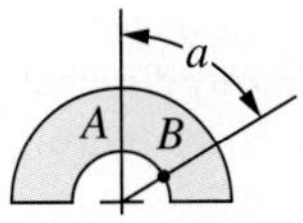

Formulas valid for $b \geq 0.7a$

(At A) $\sigma_t = \frac{6qcb}{t^2}\left(\frac{b}{c}-\frac{1}{3}\right)\left[c_1\left(1-\gamma_1^2\frac{c}{b}\right)+c_2\left(1-\gamma_2^2\frac{c}{b}\right)+\frac{c}{b}\right]K$ max stress occurs on inner edge over central 60°

(At B) $y = \frac{-24qc^2b^2}{Et^3}\left(\frac{b}{c}-\frac{1}{3}\right)\left[c_1\cosh\gamma_1\alpha + c_2\cosh\gamma_2\alpha+\frac{c}{b}\right]$ max deflection occurs when $\alpha = \frac{\pi}{2}$

where $c_1 = \frac{1}{\left(\frac{b}{c}-\gamma_1^2\right)(\lambda-1)\cosh\frac{\gamma_1\pi}{2}}$, $c_2 = \frac{1}{\left(\frac{b}{c}-\gamma_2^2\right)\left(\frac{1}{\lambda}-1\right)\cosh\frac{\gamma_2\pi}{2}}$

$$\gamma_1 = \frac{\gamma}{\sqrt{2}}\sqrt{1+\sqrt{1-\frac{4b^2}{c^2\gamma^4}}},\quad \gamma_2 = \frac{\gamma}{\sqrt{2}}\sqrt{1-\sqrt{1-\frac{4b^2}{c^2\gamma^4}}}$$

$$\gamma = \sqrt{\frac{2b}{c}+4\left(1-\frac{0.625t}{2c}\right)\frac{G}{E}\left(1+\frac{b}{c}\right)^2},$$

$$\lambda = \frac{\gamma_1\left(\frac{b}{c}-y_1^2+\lambda_1\right)\left(\frac{b}{c}-\gamma_2^2\right)\tanh\frac{\gamma_1\pi}{2}}{\gamma_2\left(\frac{b}{c}-\gamma_2^2+\lambda_1\right)\left(\frac{b}{c}-\gamma_1^2\right)\tanh\frac{\gamma_2\pi}{2}},$$

$$\lambda_1 = 4\left(1-\frac{0.625t}{2c}\right)\frac{G}{E}\left(1+\frac{b}{c}\right)^2$$

K = function of $(b-c)/(b+c)$ and has values as follows:

$(b-c)/(b+c)$	0.4	0.5	0.6	0.7	0.8	0.9	1.0
K	1.58	1.44	1.32	1.22	1.13	1.06	1.0

[Formulas due to Wahl (Ref.10).]

(*Continued*)

TABLE 11.2 Formulas for Flat Circular Plates of Constant Thickness (*Continued*)

Case No., Loading, Restraints	Boundary Condition	Special Cases
32. Elliptical plate, uniformly distributed load q over entire surface 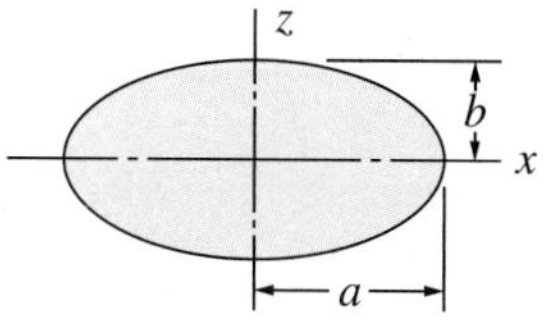 $\alpha = \frac{b}{a}$	32a. Simply supported	At the center: $\sigma_{max} = \sigma_z = -[2.816+1.581\nu-(1.691+1.206\nu)\alpha]\frac{qb^2}{t^2}$ $y_{max} = -[2.649+0.15\nu-(1.711+0.75\nu)\alpha]\frac{qb^4(1-\nu^2)}{Et^3}$ [Approximate formulas for $0.2<\alpha<1.0$ (see numerical data in Refs. 21 and 56)]
	32b. Fixed	At the edge of span b: $\sigma_{max} = \sigma_z = \frac{6qb^2}{t^2(3+2\alpha^2+3\alpha^4)}$ At the edge of span a: $\sigma_x = \frac{6qb^2\alpha^2}{t^2(3+2\alpha^2+3\alpha^4)}$ At the center: $\sigma_z = \frac{-3qb^2(1+\nu\alpha^2)}{t^2(3+2\alpha^2+3\alpha^4)}, \sigma_x = \frac{-3qb^2(\alpha^2+\nu)}{t^2(3+2\alpha^2+3\alpha^4)}, y_{max} = \frac{-3qb^4(1-\nu^2)}{2Et^3(3+2\alpha^2+3\alpha^4)}$ (Ref. 5)
33. Elliptical plate, uniform load over a small concentric circular area of radius r_o (note definition of r'_o in case 18) 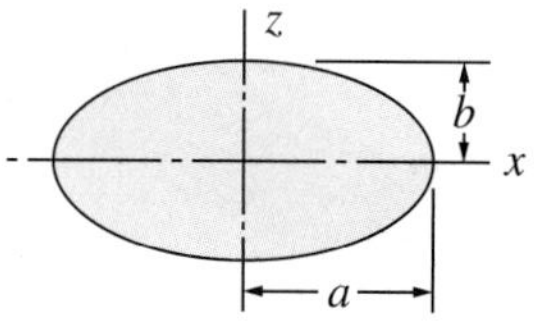 $\alpha = \frac{b}{a}$	33a. Simply supported	At the center: $\sigma_{max} = \sigma_z = \frac{-3W}{2\pi t^2}\left[(1+\nu)\ln\frac{b}{r'_o}+\nu(6.57-2.57\alpha)\right]$ $y_{max} = \frac{-Wb^2}{Et^3}(0.76-0.18\alpha)$ for $\nu=0.25$ [Approximate formulas by interpolation between cases of circular plate and infinitely long narrow strip (Ref. 4)]
	33b. Fixed	At the center: $\sigma_z = \frac{-3W(1+\nu)}{2\pi t^2}\left(\ln\frac{2b}{r'_o}-0.317\alpha-0.376\right)$ $y_{max} = \frac{-Wb^2}{Et^3}(0.326-0.104\alpha)$ for $\nu=0.25$ [Approximate formulas by interpolation between cases of circular plate and infinitely long narrow strip (Ref. 6).]

TABLE 11.3 Shear Deflections for Flat Circular Plates of Constant Thickness

Notation: y_{sb}, y_{sa} and y_{sro} are the deflections at b, a, and r_o, respectively, caused by transverse shear stresses. K_{sb}, K_{sa}, K_{sro} are deflection coefficients defined by the relationships $y_s = K_s wa/tG$ for an annular line load and $y_s = K_s qa^2/tG$ for all distributed loadings (see Table 11.2 for all other notation and for the loading cases referenced).

Case No.	Shear Deflection Coefficients	Tabulated Values for Specific Cases

1a, 1b, 1e, 1f, 9

$$K_{sr_o} = K_{sb} = -1.2\frac{r_o}{a}\ln\frac{a}{r_o} \quad (\textit{Note: } r_o > 0)$$

r_o/a	0.1	0.3	0.5	0.7	0.9
K_{sb}	−0.2763	−0.4334	−0.4159	−0.2996	−0.1138

2a, 2b, 2e, 2f, 10

$$K_{sr_o} = K_{sb} = -0.30\left[1-\left(\frac{r_o}{a}\right)^2\left(1+2\ln\frac{a}{r_o}\right)\right]$$

r_o/a	0.1	0.3	0.5	0.7	0.9
K_{sb}	−0.2832	−0.2080	−0.1210	−0.0481	−0.0058

3a, 3b, 3e, 3f, 11

$$K_{sr_o} = K_{sb} = \frac{-a}{30(a-r_o)}\left[4-9\frac{r_o}{a}+\left(\frac{r_o}{a}\right)^3\left(5+6\ln\frac{a}{r_o}\right)\right]$$

r_o/a	0.1	0.3	0.5	0.7	0.9
K_{sb}	−0.1155	−0.0776	−0.0430	−0.0166	−0.0019

4a, 4b, 4e, 4f, 12

$$K_{sr_o} = K_{sb}$$

$$= \frac{-a^2}{120(a-r_o)^2}\left[9-32\frac{r_o}{a}+36\left(\frac{r_o}{a}\right)^2-\left(\frac{r_o}{a}\right)^4\left(13+12\ln\frac{a}{r_o}\right)\right]$$

r_o/a	0.1	0.3	0.5	0.7	0.9
K_{sb}	−0.0633	−0.0411	−0.0223	−0.0084	−0.00098

1i, 1j, 1k, 1l

$$K_{sr_o} = K_{sb} = -1.2\frac{r_o}{a}\ln\frac{r_o}{b} \quad (\textit{Note: } b > 0)$$

b/a \ r_o/a	0.2	0.4	0.6	0.8	1.0
0.1	−0.1664	−0.6654	−1.2901	−1.9963	−2.7631
0.3		−0.1381	−0.4991	−0.9416	−1.4448
0.5			−0.1313	−0.4512	−0.8318
0.7	Values of K_{sa}			−0.1282	−0.4280
0.9					−0.1264

2i, 2j, 2k, 2l

$$K_{sr_o} = -0.60\left[1-\left(\frac{r_o}{a}\right)^2\right]\ln\frac{r_o}{b} \quad (\textit{Note: } b > 0)$$

b/a \ r_o/a	0.1	0.3	0.5	0.7	0.9
0.1	−0.0000	−0.5998	−0.7242	−0.5955	−0.2505
0.3		−0.0000	−0.2299	−0.2593	−0.1252
0.5			−0.0000	−0.1030	−0.0670
0.7	Values of K_{sr_o}			−0.0000	−0.0287
0.9					−0.0000

$$K_{sa} = -0.30\left[2\ln\frac{a}{b}-1+\left(\frac{r_o}{a}\right)^2\left(1-2\ln\frac{r_o}{b}\right)\right] \quad (\textit{Note: } b > 0)$$

b/a \ r_o/a	0.1	0.3	0.5	0.7	0.9
0.1	−1.0846	−1.0493	−0.9151	−0.6565	−0.2567
0.3		−0.4494	−0.4208	−0.3203	−0.1315
0.5			−0.1909	−0.1640	−0.0732
0.7	Values of K_{sa}			−0.0610	−0.0349
0.9					−0.0062

(*Continued*)

TABLE 11.3 Shear Deflections for Flat Circular Plates of Constant Thickness (*Continued*)

Case No.	Shear Deflection Coefficients	Tabulated Values for Specific Cases

Case No. 3i, 3j, 3k, 3l

$$K_{sr_o} = -0.20\left[2 - \frac{r_o}{a} - \left(\frac{r_o}{a}\right)^2\right]\ln\frac{r_o}{b} \quad (\textit{Note: } b>0)$$

b/a \ r_o/a	0.1	0.3	0.5	0.7	0.9
0.1	−0.0000	−0.3538	−0.4024	−0.3152	−0.1274
0.3		−0.0000	−0.1277	−0.1373	−0.0637
0.5			−0.0000	−0.0545	−0.0341
0.7	Values of K_{sr_o}			−0.0000	−0.0146
0.9					−0.0000

$$K_{sa} = \frac{-a}{30(a-r_o)}\left[6\left(2 - 3\frac{r_o}{a}\right)\ln\frac{a}{b} - 4 + 9\frac{r_o}{a} - \left(\frac{r_o}{a}\right)^3\left(5 - 6\ln\frac{r_o}{b}\right)\right] \quad (\textit{Note: } b>0)$$

b/a \ r_o/a	0.1	0.3	0.5	0.7	0.9
0.1	−0.7549	−0.6638	−0.5327	−0.3565	−0.1316
0.3		−0.3101	−0.2580	−0.1785	−0.0679
0.5			−0.1303	−0.0957	−0.0383
0.7	Values of K_{sa}			−0.0412	−0.0187
0.9					−0.0042

Case No. 4i, 4j, 4k, 4l

$$K_{sr_o} = -0.10\left[3 - 2\frac{r_o}{a} - \left(\frac{r_o}{a}\right)^2\right]\ln\frac{r_o}{b} \quad (\textit{Note: } b>0)$$

b/a \ r_o/a	0.1	0.3	0.5	0.7	0.9
0.1	−0.0000	−0.2538	−0.2817	−0.2160	−0.0857
0.3		−0.0000	−0.0894	−0.0941	−0.0428
0.5			−0.0000	−0.0373	−0.0229
0.7	Values of K_{sr_o}			−0.0000	−0.0098
0.9					−0.0000

$$K_{sa} = \frac{-a^2}{120(a-r_o)^2}\left\{12\left[3 - 8\frac{r_o}{a} + 6\left(\frac{r_o}{a}\right)^2\right]\ln\frac{a}{b} - 9 + 32\frac{r_o}{a} - 36\left(\frac{r_o}{a}\right)^2 + \left(\frac{r_o}{a}\right)^4\left(13 - 12\ln\frac{r_o}{b}\right)\right\} \quad (\textit{Note: } b>0)$$

b/a \ r_o/a	0.1	0.3	0.5	0.7	0.9
0.1	−0.5791	−0.4908	−0.3807	−0.2472	−0.0888
0.3		−0.2370	−0.1884	−0.1252	−0.0460
0.5			−0.0990	−0.0685	−0.0261
0.7	Values of K_{sa}			−0.0312	−0.0129
0.9					−0.0031

TABLE 11.4 Formulas for Flat Plates with Straight Boundaries and Constant Thickness

Notation: The notation for Table 11.2 applies with the following modifications: a and b refer to plate dimensions, and when used as subscripts for stress, they refer to the stresses in directions parallel to the sides a and b, respectively. σ is a bending stress which is positive when tensile on the bottom and compressive on the top if loadings are considered vertically downward. R is the reaction force per unit length normal to the plate surface exerted by the boundary support on the edge of the plate. r'_o is the equivalent radius of contact for a load concentrated on a very small area and is given by $r'_o = \sqrt{1.6r_o^2 + t^2} - 0.675t$ if $r_o < 0.5t$ and $r'_o = r_o$ if $r_o \geq 0.5t$.

Case No., Shape, and Supports	Case No., Loading	Formulas and Tabulated Specific Values

1. Rectangular plate; all edges simply supported

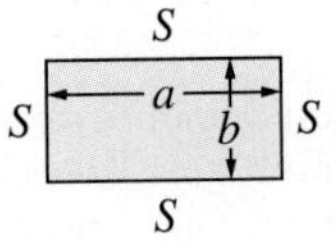

1a. Uniform over entire plate

(At centre) $\sigma_{max} = \sigma_b = \dfrac{\beta q b^2}{t^2}$ and $y_{max} = \dfrac{-\alpha q b^4}{E t^3}$

(At centre of long sides) $R_{max} = \gamma q b$

a/b	1.0	1.2	1.4	1.6	1.8	2.0	3.0	4.0	5.0	∞
β	0.2874	0.3762	0.4530	0.5172	0.5688	0.6102	0.7134	0.7410	0.7476	0.7500
α	0.0444	0.0616	0.0770	0.0906	0.1017	0.1110	0.1335	0.1400	0.1417	0.1421
γ	0.420	0.455	0.478	0.491	0.499	0.503	0.505	0.502	0.501	0.500

(Ref. 21 for $v = 0.3$)

1b. Uniform over small concentric circle of radius r_o (note definition of r'_o)

(At centre) $\sigma_{max} = \dfrac{3W}{2\pi t^2}\left[(1+v)\ln\dfrac{2b}{\pi r'_o} + \beta\right]$

$y_{max} = \dfrac{-\alpha W b^2}{E t^3}$

a/b	1.0	1.2	1.4	1.6	1.8	2.0	∞
β	0.435	0.650	0.789	0.875	0.927	0.958	1.000
α	0.1267	0.1478	0.1621	0.1715	0.1770	0.1805	0.1851

(Ref. 21 for $v = 0.3$)

1c. Uniform over central rectangular area

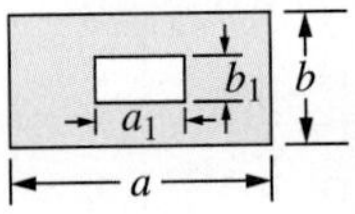

(At center) $\sigma_{max} = \sigma_b = \dfrac{\beta W}{t^2}$ where $W = q a_1 b_1$

a_1/b → / b_1/b ↓	$a = b$						$a = 1.4b$						$a = 2b$					
	0	0.2	0.4	0.6	0.8	1.0	0	0.2	0.4	0.8	1.2	1.4	0	0.4	0.8	1.2	1.6	2.0
0		1.82	1.38	1.12	0.93	0.76		2.0	1.55	1.12	0.84	0.75		1.64	1.20	0.97	0.78	0.64
0.2	1.82	1.28	1.08	0.90	0.76	0.63	1.78	1.43	1.23	0.95	0.74	0.64	1.73	1.31	1.03	0.84	0.68	0.57
0.4	1.39	1.07	0.84	0.72	0.62	0.52	1.39	1.13	1.00	0.80	0.62	0.55	1.32	1.08	0.88	0.74	0.60	0.50
0.6	1.12	0.90	0.72	0.60	0.52	0.43	1.10	0.91	0.82	0.68	0.53	0.47	1.04	0.90	0.76	0.64	0.54	0.44
0.8	0.92	0.76	0.62	0.51	0.42	0.36	0.90	0.76	0.68	0.57	0.45	0.40	0.87	0.76	0.63	0.54	0.44	0.38
1.0	0.76	0.63	0.52	0.42	0.35	0.30	0.75	0.62	0.57	0.47	0.38	0.33	0.71	0.61	0.53	0.45	0.38	0.30

(Values from charts of Ref. 8; $v = 0.3$)

1d. Uniformly increasing along length

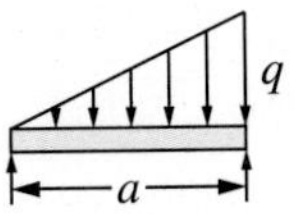

$\sigma_{max} = \dfrac{\beta q b^2}{t^2}$ and $y_{max} = \dfrac{-\alpha q b^4}{E t^3}$

a/b	1	1.5	2.0	2.5	3.0	3.5	4.0
β	0.16	0.26	0.34	0.38	0.43	0.47	0.49
α	0.022	0.043	0.060	0.070	0.078	0.086	0.091

(Values from charts of Ref. 8; $v = 0.3$)

1e. Uniformly increasing along width

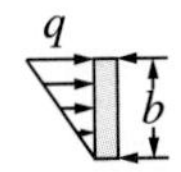

$\sigma_{max} = \dfrac{\beta q b^2}{t^2}$ and $y_{max} = \dfrac{-\alpha q b^4}{E t^3}$

a/b	1	1.5	2.0	2.5	3.0	3.5	4.0
β	0.16	0.26	0.32	0.35	0.37	0.38	0.38
α	0.022	0.042	0.056	0.063	0.067	0.069	0.070

(Values from charts of Ref. 8; $v = 0.3$)

(*Continued*)

TABLE 11.4 Formulas for Flat Plates with Straight Boundaries and Constant Thickness (*Continued*)

Case No., Shape, and Supports	Case No., Loading	Formulas and Tabulated Specific Values
	1f. Uniform over entire plate plus uniform tension or compression P lb/linear in applied to short edges	$y_{max} = \alpha \frac{qb^4}{Et^3}$ $(\sigma_a)_{max} = \beta_x \frac{qb^2}{t^2}$ $(\sigma_b)_{max} = \beta_y \frac{qb^2}{t^2}$. Here α, β_x, and β_y depend on ratios $\frac{a}{b}$ and $\frac{P}{P_E}$, where $P_E = \frac{\pi^2 Et^3}{3(1-\nu^2)b^2}$, and have the following values:

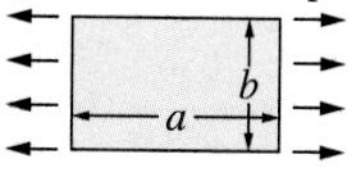

Coef.	a/b \ P/P_E	0	0.15	0.25	0.50	0.75	1	2	3	4	5
					P, Tension						
α	1	0.044	0.039		0.030		0.023	0.015	0.011	0.008	0.0075
	1½	0.084	0.075		0.060		0.045	0.0305	0.024	0.019	0.0170
	2	0.110	0.100		0.084		0.067	0.0475	0.0375	0.030	0.0260
	3	0.133	0.125		0.1135		0.100	0.081	0.066	0.057	0.0490
	4	0.140	0.136		0.1280		0.118	0.102	0.089	0.080	0.072
β_x	1	0.287					0.135	0.096	0.072	0.054	0.045
	1½	0.300					0.150	0.105	0.078	0.066	0.048
	2	0.278					0.162	0.117	0.093	0.075	0.069
	3	0.246					0.180	0.150	0.126	0.105	0.093
	4	0.222					0.192	0.168	0.156	0.138	0.124
β_y	1	0.287					0.132	0.084	0.054	0.036	0.030
	1½	0.487					0.240	0.156	0.114	0.090	0.072
	2	0.610					0.360	0.258	0.198	0.162	0.138
	3	0.713					0.510	0.414	0.348	0.294	0.258
	4	0.741					0.624	0.540	0.480	0.420	0.372
					P, Compression						
α	1	0.044		0.060	0.094	0.180					
	1½	0.084		0.109	0.155	0.237					
	2	0.110		0.139	0.161	0.181					
	3	0.131		0.145	0.150	0.150					
	4	0.140		0.142	0.142	0.138					
β_x	1	0.287		0.372	0.606	1.236					
	1½	0.300		0.372	0.522	0.846					
	2	0.278		0.330	0.390	0.450					
	3	0.246		0.228	0.228	0.210					
	4	0.222		0.225	0.225	0.225					
β_y	1	0.287		0.420	0.600	1.260					
	1½	0.487		0.624	0.786	1.380					
	2	0.610		0.720	0.900	1.020					
	3	0.713		0.750	0.792	0.750					
	4	0.741		0.750	0.750	0.750					

In the above formulas σ_a and σ_b are stresses due to bending only; add direct stress P/t to σ_a. (Ref. 41)

(Continued)

TABLE 11.4 Formulas for Flat Plates with Straight Boundaries and Constant Thickness (*Continued*)

Case No., Shape, and Supports	Case No., Loading	Formulas and Tabulated Specific Values
	1g. Uniform over entire plate plus uniform tension P lb/linear in applied to all edges	$y_{max} = \alpha \frac{qb^4}{Et^3}$ $(\sigma_a)_{max} = \beta_x \frac{qb^2}{t^2}$ $(\sigma_b)_{max} = \beta_y \frac{qb^2}{t^2}$. Here α, β_x, and β_y depend on ratios $\frac{a}{b}$ and $\frac{P}{P_E}$, where $P_E = \frac{\pi^2 Et^3}{3(1-\nu^2)b^2}$, and have the following values:

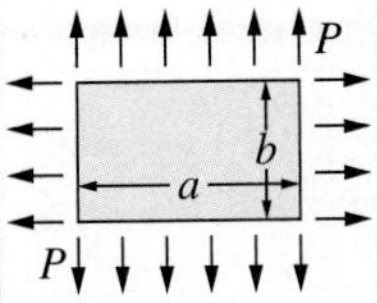

Coef.	a/b \ P/P_E	0	0.15	0.5	1	2	3	4	5
α	1	0.044	0.035	0.022	0.015	0.008	0.006	0.004	0.003
	1½	0.084	0.060	0.035	0.022	0.012	0.008	0.006	0.005
	2	0.110	0.075	0.042	0.025	0.014	0.010	0.007	0.006
	3	0.133	0.085	0.045	0.026	0.016	0.011	0.008	0.007
	4	0.140	0.088	0.046	0.026	0.016	0.011	0.008	0.007
β_x	1	0.287	0.216	0.132	0.084	0.048	0.033	0.026	0.021
	1½	0.300	0.204	0.117	0.075	0.045	0.031	0.024	0.020
	2	0.278	0.189	0.111	0.072	0.044	0.031	0.024	0.020
	3	0.246	0.183	0.108	0.070	0.043	0.031	0.025	0.020
	4	0.222	0.183	0.108	0.074	0.047	0.032	0.027	0.024
β_y	1	0.287	0.222	0.138	0.090	0.051	0.036	0.030	0.024
	1½	0.487	0.342	0.186	0.108	0.066	0.042	0.036	0.030
	2	0.610	0.302	0.216	0.132	0.072	0.051	0.042	0.036
	3	0.713	0.444	0.234	0.141	0.078	0.054	0.042	0.036
	4	0.741	0.456	0.240	0.144	0.078	0.054	0.042	0.036

In the above formulas σ_a and σ_b are stresses due to bending only; add direct stress P/t to σ_a and σ_b. (Ref. 42)

2. Rectangular plate; three edges simply supported, one edge (b) free

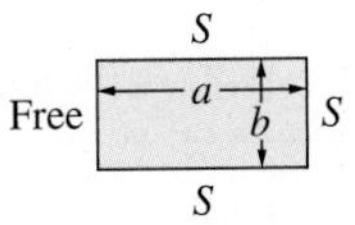

2a. Uniform over entire plate

$\sigma_{max} = \frac{\beta q b^2}{t^2}$ and $y_{max} = \frac{-\alpha q b^4}{Et^3}$

a/b	0.50	0.667	1.0	1.5	2.0	4.0
β	0.36	0.45	0.67	0.77	0.79	0.80
α	0.080	0.106	0.140	0.160	0.165	0.167

(Ref. 8 for $\nu = 0.3$)

2d. Uniformly increasing along the a side

$\sigma_{max} = \frac{\beta q b^2}{t^2}$ and $y_{max} = \frac{-\alpha q b^4}{Et^3}$

a/b	0.50	0.667	1.0	1.5	2.0	2.5	3.0	3.5	4.0
β	0.11	0.16	0.20	0.28	0.32	0.35	0.36	0.37	0.37
α	0.026	0.033	0.040	0.050	0.058	0.064	0.067	0.069	0.070

(Ref. 8 for $\nu = 0.3$)

3. Rectangular plate; three edges simply supported, one short edge (b) fixed

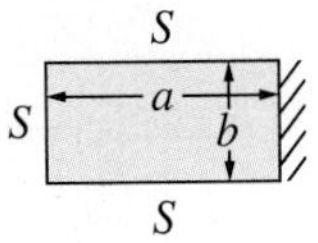

3a. Uniform over entire plate

$\sigma_{max} = \frac{\beta q b^2}{t^2}$ and $y_{max} = \frac{-\alpha q b^4}{Et^3}$

a/b	1	1.5	2.0	2.5	3.0	3.5	4.0
β	0.50	0.67	0.73	0.74	0.75	0.75	0.75
α	0.030	0.071	0.101	0.122	0.132	0.137	0.139

(Values from charts of Ref. 8; $\nu = 0.3$)

4. Rectangular plate; three edges simply supported, one long edge (a) fixed

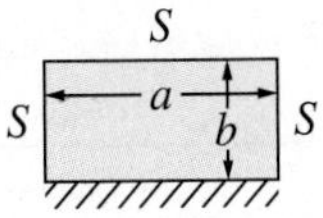

4a. Uniform over entire plate

$\sigma_{max} = \frac{\beta q b^2}{t^2}$ and $y_{max} = \frac{-\alpha q b^4}{Et^3}$

a/b	1	1.5	2.0	2.5	3.0	3.5	4.0
β	0.50	0.66	0.73	0.74	0.74	0.75	0.75
α	0.030	0.046	0.054	0.056	0.057	0.058	0.058

(Values from charts of Ref. 8; $\nu = 0.3$)

(*Continued*)

TABLE 11.4 Formulas for Flat Plates with Straight Boundaries and Constant Thickness (*Continued*)

Case No., Shape, and Supports	Case No., Loading	Formulas and Tabulated Specific Values

5. Rectangular plate; two long edges simply supported, two short edges fixed

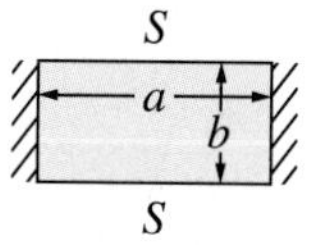

5a. Uniform over entire plate

(At centre of short edges) $\sigma_{max} = \dfrac{-\beta q b^2}{t^2}$

(At centre) $y_{max} = \dfrac{-\alpha q b^4}{E t^3}$

a/b	1	1.2	1.4	1.6	1.8	2	∞
β	0.4182	0.5208	0.5988	0.6540	0.6912	0.7146	0.750
α	0.0210	0.0349	0.0502	0.0658	0.0800	0.0922	

(Ref. 21)

6. Rectangular plate; two long edges fixed, two short edges simply supported

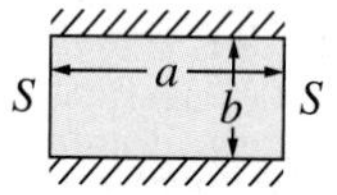

6a. Uniform over entire plate

(At centre of long edges) $\sigma_{max} = \dfrac{-\beta q b^2}{t^2}$

(At centre) $y_{max} = \dfrac{-\alpha q b^4}{E t^3}$

a/b	1	1.2	1.4	1.6	1.8	2	∞
β	0.4182	0.4626	0.4860	0.4968	0.4971	0.4973	0.500
α	0.0210	0.0243	0.0262	0.0273	0.0280	0.0283	0.0285

(Ref. 21)

7. Rectangular plate; one edge fixed, opposite edge free, remaining edges simply supported

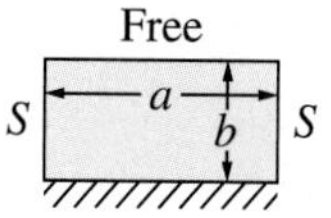

7a. Uniform over entire plate

(At center of fixed edge) $\sigma = \dfrac{-\beta_1 q b^2}{t^2}$ and $R = \gamma_1 q b$

(At center of free edge) $\sigma = \dfrac{\beta_2 q b^2}{t^2}$

(At end of free edge) $R = \gamma_2 q b$

a/b	0.25	0.50	0.75	1.0	1.5	2.0	3.0
β_1	0.044	0.176	0.380	0.665	1.282	1.804	2.450
β_2	0.048	0.190	0.386	0.565	0.730	0.688	0.434
γ_1	0.183	0.368	0.541	0.701	0.919	1.018	1.055
γ_2	0.131	0.295	0.526	0.832	1.491	1.979	2.401

(Ref. 49 for $\nu = 0.2$)

7aa. Uniform over 2/3 of plate from fixed edge

(At center of fixed edge) $\sigma = \dfrac{-\beta q b^2}{t^2}$ and $R = \gamma q b$

a/b	0.25	0.50	0.75	1.0	1.5	2.0	3.0
β	0.044	0.161	0.298	0.454	0.730	0.932	1.158
γ	0.183	0.348	0.466	0.551	0.645	0.681	0.689

(Ref. 49 for $\nu = 0.2$)

7aaa. Uniform over 1/3 of plate from fixed edge

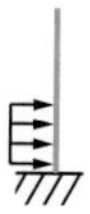

(At center of fixed edge) $\sigma = \dfrac{-\beta q b^2}{t^2}$ and $R = \gamma q b$

a/b	0.25	0.50	0.75	1.0	1.5	2.0	3.0
β	0.040	0.106	0.150	0.190	0.244	0.277	0.310
γ	0.172	0.266	0.302	0.320	0.334	0.338	0.338

(Ref. 49 for $\nu = 0.2$)

7d. Uniformly decreasing from fixed edge to free edge

(At center of fixed edge) $\sigma = \dfrac{-\beta q b^2}{t^2}$ and $R = \gamma q b$

a/b	0.25	0.50	0.75	1.0	1.5	2.0	3.0
β	0.037	0.120	0.212	0.321	0.523	0.677	0.866
γ	0.159	0.275	0.354	0.413	0.482	0.509	0.517

(Ref. 49 for $\nu = 0.2$)

(*Continued*)

TABLE 11.4 Formulas for Flat Plates with Straight Boundaries and Constant Thickness (*Continued*)

Case No., Shape, and Supports	Case No., Loading	Formulas and Tabulated Specific Values

7dd. Uniformly decreasing from fixed edge to zero at 2/3*b*

(At center of fixed edge) $\sigma = \dfrac{-\beta q b^2}{t^2}$ and $R = \gamma q b$

a/b	0.25	0.50	0.75	1.0	1.5	2.0	3.0
β	0.033	0.094	0.146	0.200	0.272	0.339	0.400
γ	0.148	0.233	0.277	0.304	0.330	0.339	0.340

(Ref. 49 for $\nu = 0.2$)

7ddd. Uniformly decreasing from fixed edge to zero at 1/3*b*

(At center of fixed edge) $\sigma = \dfrac{-\beta q b^2}{t^2}$ and $R = \gamma q b$

a/b	0.25	0.50	0.75	1.0	1.5	2.0	3.0
β	0.023	0.048	0.061	0.073	0.088	0.097	0.105
γ	0.115	0.149	0.159	0.164	0.167	0.168	0.168

(Ref. 49 for $\nu = 0.2$)

7f. Distributed line load *w* lb/in along free edge

(At center of fixed edge) $\sigma_b = \dfrac{-\beta_1 w b}{t^2}$ and $R = \gamma_1 w$

(At center of free edge) $\sigma_a = \dfrac{\beta_2 w b}{t^2}$

(At ends of free edge) $R = \gamma_2 w$

a/b	0.25	0.50	0.75	1.0	1.5	2.0	3.0
β_1	0.000	0.024	0.188	0.570	1.726	2.899	4.508
β_2	0.321	0.780	1.204	1.554	1.868	1.747	1.120
γ_1	0.000	0.028	0.160	0.371	0.774	1.004	1.119
γ_2	1.236	2.381	3.458	4.510	6.416	7.772	9.031

(Ref. 49 for $\nu = 0.2$)

8. Rectangular plate, all edges fixed

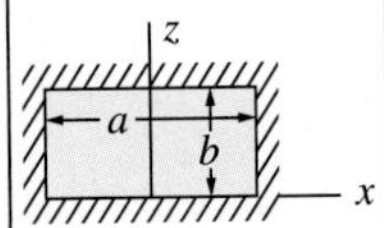

8a. Uniform over entire plate

(At centre of long edge) $\sigma_{max} = \dfrac{-\beta_1 q b^2}{t^2}$

(At centre) $\sigma = \dfrac{\beta_2 q b^2}{t^2}$ and $y_{max} = \dfrac{\alpha q b^4}{E t^3}$

a/b	1.0	1.2	1.4	1.6	1.8	2.0	∞
β_1	0.3078	0.3834	0.4356	0.4680	0.4872	0.4974	0.5000
β_2	0.1386	0.1794	0.2094	0.2286	0.2406	0.2472	0.2500
α	0.0138	0.0188	0.0226	0.0251	0.0267	0.0277	0.0284

(Refs. 7 and 25 and Ref. 21 for $\nu = 0.3$)

8b. Uniform over small concentric circle of radius r_o (note definition of r'_o)

(At center) $\sigma_b = \dfrac{3W}{2\pi t^2}\left[(1+\nu)\ln\dfrac{2b}{\pi r'_o} + \beta_1\right]$ and $y_{max} = \dfrac{\alpha W b^2}{E t^3}$

(At center of long edge) $\sigma_b = \dfrac{-\beta_2 W}{t^2}$

a/b	1.0	1.2	1.4	1.6	1.8	2.0	∞
β_1	−0.238	−0.078	0.011	0.053	0.068	0.067	0.067
β_2	0.7542	0.8940	0.9624	0.9906	1.0000	1.004	1.008
α	0.0611	0.0706	0.0754	0.0777	0.0786	0.0788	0.0791

(Ref. 26 and Ref. 21 for $\nu = 0.3$)

(*Continued*)

TABLE 11.4 Formulas for Flat Plates with Straight Boundaries and Constant Thickness (*Continued*)

Case No., Shape, and Supports	Case No., Loading	Formulas and Tabulated Specific Values

8d. Uniformly decreasing parallel to side b

(At $x=0, z=0$) $(\sigma_b)_{max} = \dfrac{-\beta_1 qb^2}{t^2}$

(At $x=0, z=0.4b$) $\sigma_b = \dfrac{\beta_2 qb^2}{t^2}$ and $\sigma_a = \dfrac{\beta_3 qb^2}{t^2}$

(At $x=0, z=b$) $\sigma_b = \dfrac{-\beta_4 qb^2}{t^2}$

(At $x=\pm\dfrac{a}{2}, z=0.45b$) $(\sigma_a)_{max} = \dfrac{-\beta_5 qb^2}{t^2}$

$y_{max} = \dfrac{-\alpha qb^4}{Et^3}$

a/b	0.6	0.8	1.0	1.2	1.4	1.6	1.8	2.0
β_1	0.1132	0.1778	0.2365	0.2777	0.3004	0.3092	0.3100	0.3068
β_2	0.0410	0.0633	0.0869	0.1038	0.1128	0.1255	0.1157	0.1148
β_3	0.0637	0.0688	0.0762	0.0715	0.0610	0.0509	0.0415	0.0356
β_4	0.0206	0.0497	0.0898	0.1249	0.1482	0.1615	0.1680	0.1709
β_5	0.1304	0.1436	0.1686	0.1800	0.1845	0.1874	0.1902	0.1908
α	0.0016	0.0047	0.0074	0.0097	0.0113	0.0126	0.0133	0.0136

(Ref. 28 for $\nu = 0.3$)

9. Rectangular plate, three edges fixed, one edge (a) simply supported

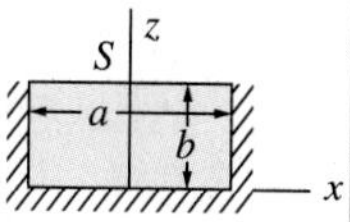

9a. Uniform over entire plate

(At $x=0, z=0$) $(\sigma_b)_{max} = \dfrac{-\beta_1 qb^2}{t^2}$ and $R = \gamma_1 qb$

(At $x=0, z=0.6b$) $\sigma_b = \dfrac{\beta_2 qb^2}{t^2}$ and $\sigma_a = \dfrac{\beta_3 qb^2}{t^2}$

(At $x=0, z=b$) $R = \gamma_2 qb$

(At $x=\pm\dfrac{a}{2}, z=0.6b$) $\sigma_a = \dfrac{-\beta_4 qb^2}{t^2}$ and $R = \gamma_3 qb$

a/b	0.25	0.50	0.75	1.0	1.5	2.0	3.0
β_1	0.020	0.081	0.173	0.307	0.539	0.657	0.718
β_2	0.004	0.018	0.062	0.134	0.284	0.370	0.422
β_3	0.016	0.061	0.118	0.158	0.164	0.135	0.097
β_4	0.031	0.121	0.242	0.343	0.417	0.398	0.318
γ_1	0.115	0.230	0.343	0.453	0.584	0.622	0.625
γ_2	0.123	0.181	0.253	0.319	0.387	0.397	0.386
γ_3	0.125	0.256	0.382	0.471	0.547	0.549	0.530

(Ref. 49 for $\nu = 0.2$)

9aa. Uniform over 2/3 of plate from fixed edge

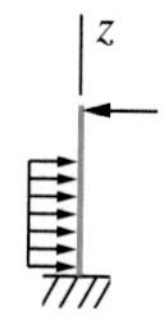

(At $x=0, z=0$) $(\sigma_b)_{max} = \dfrac{-\beta_1 qb^2}{t^2}$ and $R = \gamma_1 qb$

(At $x=0, z=0.6b$) $\sigma_b = \dfrac{\beta_2 qb^2}{t^2}$ and $\sigma_a = \dfrac{\beta_3 qb^2}{t^2}$

(At $x=0, z=b$) $R = \gamma_2 qb$

(At $x=\pm\dfrac{a}{2}, z=0.4b$) $\sigma_a = \dfrac{-\beta_4 qb^2}{t^2}$ and $R = \gamma_3 qb$

a/b	0.25	0.50	0.75	1.0	1.5	2.0	3.0
β_1	0.020	0.080	0.164	0.274	0.445	0.525	0.566
β_2	0.003	0.016	0.044	0.093	0.193	0.252	0.286
β_3	0.012	0.043	0.081	0.108	0.112	0.091	0.066
β_4	0.031	0.111	0.197	0.255	0.284	0.263	0.204
γ_1	0.115	0.230	0.334	0.423	0.517	0.542	0.543
γ_2	0.002	0.015	0.048	0.088	0.132	0.139	0.131
γ_3	0.125	0.250	0.345	0.396	0.422	0.417	0.405

(Ref. 49 for $\nu = 0.2$)

(Continued)

TABLE 11.4 Formulas for Flat Plates with Straight Boundaries and Constant Thickness (*Continued*)

Case No., Shape, and Supports	Case No., Loading	Formulas and Tabulated Specific Values

9aaa. Uniform over 1/3 of plate from fixed edge

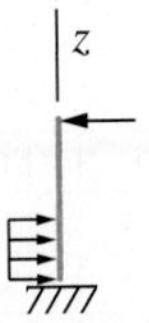

(At $x=0, z=0$) $\quad (\sigma_b)_{max} = \dfrac{-\beta_1 q b^2}{t^2}$ and $R = \gamma_1 q b$

(At $x=0, z=0.2b$) $\quad \sigma_b = \dfrac{\beta_2 q b^2}{t^2}$ and $\sigma_a = \dfrac{\beta_3 q b^2}{t^2}$

(At $x=0, z=b$) $\quad R = \gamma_2 q b$

(At $x=\pm\dfrac{a}{2}, z=0.2b$) $\quad \sigma_a = \dfrac{-\beta_4 q b^2}{t^2}$ and $R = \gamma_3 q b$

a/b	0.25	0.50	0.75	1.0	1.5	2.0	3.0
β_1	0.020	0.068	0.108	0.148	0.194	0.213	0.222
β_2	0.005	0.026	0.044	0.050	0.047	0.041	0.037
β_3	0.013	0.028	0.031	0.026	0.016	0.011	0.008
β_4	0.026	0.063	0.079	0.079	0.068	0.056	0.037
γ_1	0.114	0.210	0.261	0.290	0.312	0.316	0.316
γ_2	0.000	0.000	0.004	0.011	0.020	0.021	0.020
γ_3	0.111	0.170	0.190	0.185	0.176	0.175	0.190

(Ref. 49 for $\nu = 0.2$)

9d. Uniformly decreasing from fixed edge to simply supported edge

(At $x=0, z=0$) $\quad (\sigma_b)_{max} = \dfrac{-\beta_1 q b^2}{t^2}$ and $R = \gamma_1 q b$

$\left(\text{At } x=\pm\dfrac{a}{2}, z=0.4b\right) \quad \sigma_a = \dfrac{-\beta_2 q b^2}{t^2}$ and $R = \gamma_2 q b$

a/b	0.25	0.50	0.75	1.0	1.5	2.0	3.0
β_1	0.018	0.064	0.120	0.192	0.303	0.356	0.382
β_2	0.019	0.068	0.124	0.161	0.181	0.168	0.132
γ_1	0.106	0.195	0.265	0.323	0.383	0.399	0.400
γ_2	0.075	0.152	0.212	0.245	0.262	0.258	0.250

(Ref. 49 for $\nu = 0.2$)

9dd. Uniformly decreasing from fixed edge to zero at 2/3b

(At $x=0, z=0$) $\quad (\sigma_b)_{max} = \dfrac{-\beta_1 q b^2}{t^2}$ and $R = \gamma_1 q b$

$\left(\text{At } x=\pm\dfrac{a}{2}, z=0.4b \text{ if } a \ge b \text{ or } z=0.2b \text{ if } a<b\right) \quad \sigma_a = \dfrac{-\beta_2 q b^2}{t^2}$ and $R = \gamma_2 q b$

a/b	0.25	0.50	0.75	1.0	1.5	2.0	3.0
β_1	0.017	0.056	0.095	0.140	0.201	0.228	0.241
β_2	0.019	0.050	0.068	0.098	0.106	0.097	0.074
γ_1	0.101	0.177	0.227	0.262	0.294	0.301	0.301
γ_2	0.082	0.129	0.146	0.157	0.165	0.162	0.158

(Ref. 49 for $\nu = 0.2$)

9ddd. Uniformly decreasing from fixed edge to zero at 1/3b

(At $x=0, z=0$) $\quad (\sigma_b)_{max} = \dfrac{-\beta_1 q b^2}{t^2}$ and $R = \gamma_1 q b$

$\left(\text{At } x=\pm\dfrac{a}{2}, z=0.2b\right) \quad \sigma_a = \dfrac{-\beta_2 q b^2}{t^2}$ and $R = \gamma_2 q b$

a/b	0.25	0.50	0.75	1.0	1.5	2.0	3.0
β_1	0.014	0.035	0.047	0.061	0.075	0.080	0.082
β_2	0.010	0.024	0.031	0.030	0.025	0.020	0.013
γ_1	0.088	0.130	0.146	0.155	0.161	0.162	0.162
γ_2	0.046	0.069	0.079	0.077	0.074	0.074	0.082

(Ref. 49 for $\nu = 0.2$)

(*Continued*)

TABLE 11.4 Formulas for Flat Plates with Straight Boundaries and Constant Thickness (*Continued*)

Case No., Shape, and Supports	Case No., Loading	Formulas and Tabulated Specific Values
10. Rectangular plate; three edges fixed, one edge (a) free		

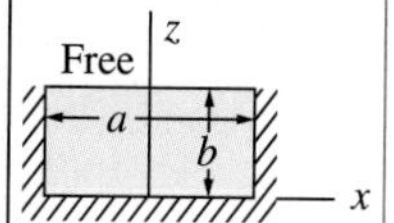

10a. Uniform over entire plate

(At $x = 0, z = 0$) $(\sigma_b)_{max} = \dfrac{-\beta_1 q b^2}{t^2}$ and $R = \gamma_1 q b$

(At $x = 0, z = b$) $\sigma_a = \dfrac{\beta_2 q b^2}{t^2}$

$\left(\text{At } x = \pm\dfrac{a}{2}, z = b\right)$ $\sigma_a = \dfrac{-\beta_3 q b^2}{t^2}$ and $R = \gamma_2 q b$

a/b	0.25	0.50	0.75	1.0	1.5	2.0	3.0
β_1	0.020	0.081	0.173	0.321	0.727	1.226	2.105
β_2	0.016	0.066	0.148	0.259	0.484	0.605	0.519
β_3	0.031	0.126	0.286	0.511	1.073	1.568	1.982
γ_1	0.114	0.230	0.341	0.457	0.673	0.845	1.012
γ_2	0.125	0.248	0.371	0.510	0.859	1.212	1.627

(Ref. 49 for $\nu = 0.2$)

10aa. Uniform over 2/3 of plate from fixed edge

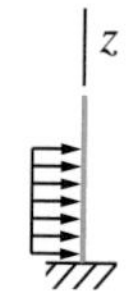

(At $x = 0, z = 0$) $(\sigma_b)_{max} = \dfrac{-\beta_1 q b^2}{t^2}$ and $R = \gamma_1 q b$

$\left(\text{At } x = \pm\dfrac{a}{2}, z = 0.6b \text{ for } a > b \text{ or } z = 0.4b \text{ for } a \le b\right)$ $\sigma_a = \dfrac{-\beta_2 q b^2}{t^2}$ and $r = \gamma_2 q b$

a/b	0.25	0.50	0.75	1.0	1.5	2.0	3.0
β_1	0.020	0.080	0.164	0.277	0.501	0.710	1.031
β_2	0.031	0.110	0.198	0.260	0.370	0.433	0.455
γ_1	0.115	0.230	0.334	0.424	0.544	0.615	0.674
γ_2	0.125	0.250	0.344	0.394	0.399	0.409	0.393

(Ref. 49 for $\nu = 0.2$)

10aaa. Uniform over 1/3 of plate from fixed edge

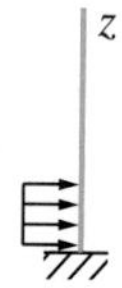

(At $x = 0, z = 0$) $(\sigma_b)_{max} = \dfrac{-\beta_1 q b^2}{t^2}$ and $R = \gamma_1 q b$

$\left(\text{At } x = \pm\dfrac{a}{2}, z = 0.2b\right)$ $\sigma_a = \dfrac{-\beta_2 q b^2}{t^2}$ and $R = \gamma_2 q b$

a/b	0.25	0.50	0.75	1.0	1.5	2.0	3.0
β_1	0.020	0.068	0.110	0.148	0.202	0.240	0.290
β_2	0.026	0.063	0.084	0.079	0.068	0.057	0.040
γ_1	0.115	0.210	0.257	0.291	0.316	0.327	0.335
γ_2	0.111	0.170	0.194	0.185	0.174	0.170	0.180

(Ref. 49 for $\nu = 0.2$)

10d. Uniformly decreasing from fixed edge to zero at free edge

(At $x = 0, z = 0$) $(\sigma_b)_{max} = \dfrac{-\beta_1 q b^2}{t^2}$ and $R = \gamma_1 q b$

$\left(\text{At } x = \pm\dfrac{a}{2}, z = b \text{ if } a > b \text{ or } z = 0.4b \text{ if } a < b\right)$ $\sigma_a = \dfrac{-\beta_2 q b^2}{t^2}$ and $R = \gamma_2 q b$

a/b	0.25	0.50	0.75	1.0	1.5	2.0	3.0
β_1	0.018	0.064	0.120	0.195	0.351	0.507	0.758
β_2	0.019	0.068	0.125	0.166	0.244	0.387	0.514
γ_1	0.106	0.195	0.265	0.324	0.406	0.458	0.505
γ_2	0.075	0.151	0.211	0.242	0.106	0.199	0.313

(Ref. 49 for $\nu = 0.2$)

10dd. Uniformly decreasing from fixed edge to zero at 2/3b

(At $x = 0, z = 0$) $(\sigma_b)_{max} = \dfrac{-\beta_1 q b^2}{t^2}$ and $R = \gamma_1 q b$

$\left(\text{At } x = \pm\dfrac{a}{2}, z = 0.4b \text{ if } a \ge b \text{ or } z = 0.2b \text{ if } a < b\right)$ $\sigma_b = \dfrac{-\beta_2 q b^2}{t^2}$ and $R = \gamma_2 q b$

a/b	0.25	0.50	0.75	1.0	1.5	2.0	3.0
β_1	0.017	0.056	0.095	0.141	0.215	0.277	0.365
β_2	0.019	0.050	0.068	0.099	0.114	0.113	0.101
γ_1	0.102	0.177	0.227	0.263	0.301	0.320	0.336
γ_2	0.082	0.129	0.146	0.157	0.163	0.157	0.146

(Ref. 49 for $\nu = 0.2$)

(*Continued*)

TABLE 11.4 Formulas for Flat Plates with Straight Boundaries and Constant Thickness (*Continued*)

Case No., Shape, and Supports	Case No., Loading	Formulas and Tabulated Specific Values

10ddd. Uniformly decreasing from fixed edge to zero at 1/3b

(At $x=0, z=0$) $(\sigma_b)_{max} = \dfrac{-\beta_1 q b^2}{t^2}$ and $R = \gamma_1 q b$

$\left(\text{At } x = \pm\dfrac{a}{2}, z = 0.2b\right)$ $\sigma_a = \dfrac{-\beta_2 q b^2}{t^2}$ and $R = \gamma_2 q b$

a/b	0.25	0.50	0.75	1.0	1.5	2.0	3.0
β_1	0.014	0.035	0.047	0.061	0.076	0.086	0.100
β_2	0.010	0.024	0.031	0.030	0.025	0.020	0.014
γ_1	0.088	0.130	0.146	0.156	0.162	0.165	0.167
γ_2	0.046	0.069	0.079	0.077	0.073	0.073	0.079

(Ref. 49 for $\nu = 0.2$)

11. Rectangular plate; two adjacent edges fixed, two remaining edges free

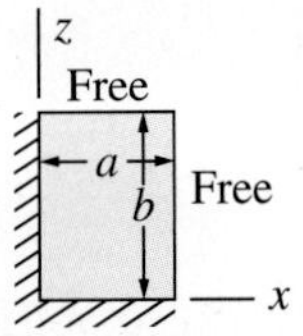

11a. Uniform over entire plate

(At $x=a, z=0$) $(\sigma_b)_{max} = \dfrac{-\beta_1 q b^2}{t^2}$ and $R = \gamma_1 q b$

$\left(\text{At } x = 0, z = b \text{ if } a > \dfrac{b}{2} \text{ or } a = 0.8b \text{ if } a \le \dfrac{b}{2}\right)$ $\sigma_a = \dfrac{-\beta_2 q b^2}{t^2}$ and $R = \gamma_2 q b$

a/b	0.125	0.25	0.375	0.50	0.75	1.0
β_1	0.050	0.182	0.353	0.631	1.246	1.769
β_2	0.047	0.188	0.398	0.632	1.186	1.769
γ_1	0.312	0.572	0.671	0.874	1.129	1.183
γ_2	0.127	0.264	0.413	0.557	0.829	1.183

(Ref. 49 for $\nu = 0.2$)

11aa. Uniform over plate from $z=0$ to $z=2/3b$

(At $x=a, z=0$) $(\sigma_b)_{max} = \dfrac{-\beta_1 q b^2}{t^2}$ and $R = \gamma_1 q b$

$\left(\text{At } x = 0, z = 0.6b \text{ if } a > \dfrac{b}{2} \text{ or } z = 0.4b \text{ if } a \le \dfrac{b}{2}\right)$ $\sigma_a = \dfrac{-\beta_2 q b^2}{t^2}$ and $R = \gamma_2 q b$

a/b	0.125	0.25	0.375	0.50	0.75	1.0
β_1	0.050	0.173	0.297	0.465	0.758	0.963
β_2	0.044	0.143	0.230	0.286	0.396	0.435
γ_1	0.311	0.543	0.563	0.654	0.741	0.748
γ_2	0.126	0.249	0.335	0.377	0.384	0.393

(Ref. 49 for $\nu = 0.2$)

11aaa. Uniform over plate from $z=0$ to $z=1/3\,b$

(At $x=a, z=0$) $(\sigma_b)_{max} = \dfrac{-\beta_1 q b^2}{t^2}$ and $R = \gamma_1 q b$

$\left(\text{At } x = 0, z = 0.4b \text{ if } a > \dfrac{b}{2} \text{ or } z = 0.2b \text{ if } a \le \dfrac{b}{2}\right)$ $\sigma_a = \dfrac{-\beta_2 q b^2}{t^2}$ and $R = \gamma_2 q b$

a/b	0.125	0.25	0.375	0.50	0.75	1.0
β_1	0.034	0.099	0.143	0.186	0.241	0.274
β_2	0.034	0.068	0.081	0.079	0.085	0.081
γ_1	0.222	0.311	0.335	0.343	0.349	0.347
γ_2	0.109	0.162	0.180	0.117	0.109	0.105

(Ref. 49 for $\nu = 0.2$)

11d. Uniformly decreasing from $z=0$ to $z=b$

(At $x=a, z=0$) $(\sigma_b)_{max} = \dfrac{-\beta_1 q b^2}{t^2}$ and $R = \gamma_1 q b$

$\left(\text{At } x = 0, z = b \text{ if } a = b, \text{ or } z = 0.6b \text{ if } \dfrac{b}{2} \le a < b, \text{ or } z = 0.4b \text{ if } a < \dfrac{b}{2}\right)$ $\sigma_a = \dfrac{-\beta_2 q b^2}{t^2}$ and $R = \gamma_2 q b$

a/b	0.125	0.25	0.375	0.50	0.75	1.0
β_1	0.043	0.133	0.212	0.328	0.537	0.695
β_2	0.028	0.090	0.148	0.200	0.276	0.397
γ_1	0.271	0.423	0.419	0.483	0.551	0.559
γ_2	0.076	0.151	0.205	0.195	0.230	0.192

(Ref. 49 for $\nu = 0.2$)

(*Continued*)

TABLE 11.4 Formulas for Flat Plates with Straight Boundaries and Constant Thickness (*Continued*)

Case No., Shape, and Supports	Case No., Loading	Formulas and Tabulated Specific Values
	11dd. Uniformly decreasing from $z=0$ to $z=2/3b$	(At $x=a, z=0$) $(\sigma_b)_{max}=\dfrac{-\beta_1 qb^2}{t^2}$ and $R=\gamma_1 qb$ (At $x=0$, $z=0.4b$ if $a\geq 0.375b$, or $z=0.2b$ if $a<0.375b$) $\sigma_a=\dfrac{-\beta_2 qb^2}{t^2}$ and $R=\gamma_2 qb$

a/b	0.125	0.25	0.375	0.50	0.75	1.0
β_1	0.040	0.109	0.154	0.215	0.304	0.362
β_2	0.026	0.059	0.089	0.107	0.116	0.113
γ_1	0.250	0.354	0.316	0.338	0.357	0.357
γ_2	0.084	0.129	0.135	0.151	0.156	0.152

(Ref. 49 for $\nu=0.2$)

Case No., Shape, and Supports	Case No., Loading	Formulas and Tabulated Specific Values
	11ddd. Uniformly decreasing from $z=0$ to $z=1/3b$	(At $x=a, z=0$) $(\sigma_b)_{max}=\dfrac{-\beta_1 qb^2}{t^2}$ and $R=\gamma_1 qb$ (At $x=0, z=0.2b$) $\sigma_a=\dfrac{-\beta_2 qb^2}{t^2}$ and $R=\gamma_2 qb$

a/b	0.125	0.25	0.375	0.50	0.75	1.0
β_1	0.025	0.052	0.071	0.084	0.100	0.109
β_2	0.014	0.028	0.031	0.029	0.025	0.020
γ_1	0.193	0.217	0.170	0.171	0.171	0.171
γ_2	0.048	0.072	0.076	0.075	0.072	0.072

(Ref. 49 for $\nu=0.2$)

Case No., Shape, and Supports	Case No., Loading	Formulas and Tabulated Specific Values
12. Continuous plate; supported at equal intervals a on circular supports of radius r_o	12a. Uniform over entire surface	(At edge of support) $\sigma_a=\dfrac{0.15q}{t^2}\left(a-\dfrac{4}{3}r_o\right)^2\left(\dfrac{1}{n}+4\right)$ when $0.15\leq n<0.30$ (Ref. 9) or $\sigma_a=\dfrac{3qa^2}{2\pi t^2}\left[(1+\nu)\ln\dfrac{a}{r_o}-21(1-\nu)\dfrac{r_o^2}{a^2}-0.55-1.50\nu\right]$ when $n<0.15$ where $n=\dfrac{2r_o}{a}$ (Ref. 11)
13. Continuous plate; supported continuously on an elastic foundation of modulus k (lb/in²/in)	13b. Uniform over a small circle of radius r_o, remote from edges	(Under the load) $\sigma_{max}=\dfrac{3W(1+\nu)}{2\pi t^2}\left(\ln\dfrac{L_e}{r_o}+0.6159\right)$ where $L_e=\sqrt[4]{\dfrac{Et^3}{12(1-\nu^2)k}}$ Max foundation pressure $q_o=\dfrac{W}{8L_e^2}$ $y_{max}=\dfrac{-W}{8kL_e^2}$ (Ref. 14)
	13bb. Uniform over a small circle of radius r_o, adjacent to edge but remote from corner	(Under the load) $\sigma_{max}=\dfrac{0.863W(1+\nu)}{t^2}\left(\ln\dfrac{L_e}{r_o}+0.207\right)$ $y_{max}=0.408(1+0.4\nu)\dfrac{W}{kL_e^2}$ (Ref. 14)
	13bbb. Uniform over a small circle of radius r_o, adjacent to a corner	(At the corner) $y_{max}=\left(1.1-1.245\dfrac{r_o}{L_e}\right)\dfrac{W}{kL_e^2}$ (At a distance $=2.38\sqrt{r_oL_e}$ from the corner along diagonal) $\sigma_{max}=\dfrac{3W}{t^2}\left[1-1.083\left(\dfrac{r_o}{L_e}\right)^{0.6}\right]$ (Ref. 14)

(*Continued*)

TABLE 11.4 Formulas for Flat Plates with Straight Boundaries and Constant Thickness (*Continued*)

Case No., Shape, and Supports	Case No., Loading	Formulas and Tabulated Specific Values
14. Parallelogram plate (skew slab); all edges simply supported	14a. Uniform over entire plate	(At center of plate) $\sigma_{max} = \dfrac{\beta q b^2}{t^2}$ and $y_{max} = \dfrac{\alpha q b^4}{Et^3}$ For $a/b = 2.0$

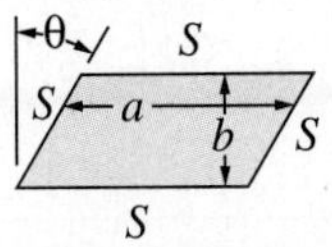

θ	0°	30°	45°	60°	75°
β	0.585	0.570	0.539	0.463	0.201
α	0.119	0.118	0.108	0.092	0.011

(Ref. 24 for $\nu = 0.2$)

Case No., Shape, and Supports	Case No., Loading	Formulas and Tabulated Specific Values
15. Parallelogram plate (skew slab); shorter edges simply supported, longer edges free	15a. Uniform over entire plate	(Along free edge) $\sigma_{max} = \dfrac{\beta_1 q b^2}{t^2}$ and $y_{max} = \dfrac{\alpha_1 q b^4}{Et^3}$ (At center of plate) $\sigma_{max} = \dfrac{\beta_2 q b^2}{t^2}$ and $y_{max} = \dfrac{\alpha_2 q b^4}{Et^3}$ For $a/b = 2.0$

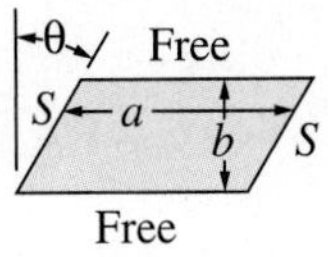

θ	0°	30°	45°	60°
β_1	3.05	2.20	1.78	0.91
β_2	2.97	2.19	1.75	1.00
α_1	2.58	1.50	1.00	0.46
α_2	2.47	1.36	0.82	0.21

(Ref. 24 for $\nu = 0.2$)

Case No., Shape, and Supports	Case No., Loading	Formulas and Tabulated Specific Values
16. Parallelogram plate (skew slab); all edges fixed	16a. Uniform over entire plate	(Along longer edge toward obtuse angle) $\sigma_{max} = \dfrac{\beta_1 q b^2}{t^2}$ (At centre of plate) $\sigma = \dfrac{\beta_2 q b^2}{t^2}$ and $y_{max} = \dfrac{\alpha q b^4}{Et^3}$

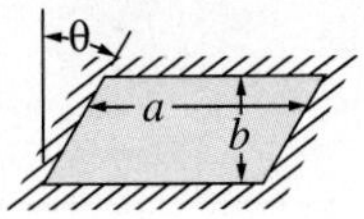

θ	a/b	1.00	1.25	1.50	1.75	2.00	2.25	2.50	3.00
0°	β_1	0.308	0.400	0.454	0.481	0.497			
	β_2	0.138	0.187	0.220	0.239	0.247			
	α	0.0135	0.0195	0.0235	0.0258	0.0273			
15°	β_1	0.320	0.412	0.483	0.531	0.553			
	β_2	0.135	0.200	0.235	0.253	0.261			
	α	0.0127	0.0189	0.0232	0.0257	0.0273			
30°	β_1		0.400	0.495	0.547	0.568	0.580		
	β_2		0.198	0.221	0.235	0.245	0.252		
	α		0.0168	0.0218	0.0249	0.0268	0.0281		
45°	β_1			0.394	0.470	0.531	0.575	0.601	
	β_2			0.218	0.244	0.260	0.265	0.260	
	α			0.0165	0.0208	0.0242	0.0265	0.0284	
60°	β_1					0.310	0.450	0.538	0.613
	β_2					0.188	0.204	0.214	0.224
	α					0.0136	0.0171	0.0198	0.0245

(Ref. 53 for $\nu = 1/3$)

Case No., Shape, and Supports	Case No., Loading	Formulas and Tabulated Specific Values
17. Equilateral triangle; all edges simply supported	17a. Uniform over entire plate	(At $x = 0, z = -0.062a$) $(\sigma_z)_{max} = \dfrac{0.1488qa^2}{t^2}$ (At $x = 0, z = 0.129a$) $(\sigma_x)_{max} = \dfrac{0.1554qa^2}{t^2}$ (At $x = 0, z = 0$) $y_{max} = \dfrac{-qa^4(1-\nu^2)}{81Et^3}$ (Refs. 21 and 23 for $\nu = 0.3$)
	17b. Uniform over small circle of radius r_o at $x = 0, z = 0$	(At $x = 0, z = 0$) $\sigma_{max} = \dfrac{3W}{2\pi t^2}\left[\dfrac{1-\nu}{2} + (1+\nu)\ln\dfrac{0.377a}{r'_o}\right]$ $y_{max} = 0.069W(1-\nu^2)a^2/Et^3$

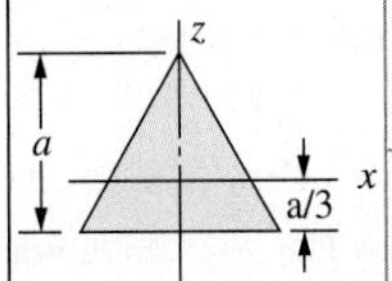

(*Continued*)

TABLE 11.4 Formulas for Flat Plates with Straight Boundaries and Constant Thickness (*Continued*)

<table>
<tr><th>Case No., Shape, and Supports</th><th>Case No., Loading</th><th>Formulas and Tabulated Specific Values</th></tr>
<tr><td>18. Right-angle isosceles triangle; all edges simply supported
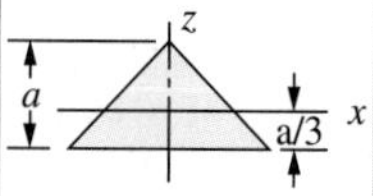</td><td>18a. Uniform over entire plate</td><td>$\sigma_{max} = \sigma_z = \dfrac{0.262qa^2}{t^2}$
$(\sigma_x)_{max} = \dfrac{0.225qa^2}{t^2}$
$y_{max} = \dfrac{0.038qa^4}{Et^3}$
(Ref. 21 for $\nu = 0.3$)</td></tr>
<tr><td>19. Regular polygonal plate; all edges simply supported
Number of sides = n</td><td>19a. Uniform over entire plate</td><td>(At center) $\sigma = \dfrac{\beta qa^2}{t^2}$ and $y_{max} = \dfrac{-\alpha qa^4}{Et^3}$
(At center of straight edge) Max slope $= \dfrac{\xi qa^3}{Et^3}$
<table>
<tr><td>n</td><td>3</td><td>4</td><td>5</td><td>6</td><td>7</td><td>8</td><td>9</td><td>10</td><td>15</td><td>∞</td></tr>
<tr><td>β</td><td>1.302</td><td>1.152</td><td>1.086</td><td>1.056</td><td>1.044</td><td>1.038</td><td>1.038</td><td>1.044</td><td>1.074</td><td>1.236</td></tr>
<tr><td>α</td><td>0.910</td><td>0.710</td><td>0.635</td><td>0.599</td><td>0.581</td><td>0.573</td><td>0.572</td><td>0.572</td><td>0.586</td><td>0.695</td></tr>
<tr><td>ξ</td><td>1.535</td><td>1.176</td><td>1.028</td><td>0.951</td><td>0.910</td><td>0.888</td><td>0.877</td><td>0.871</td><td>0.883</td><td>1.050</td></tr>
</table>
(Ref. 55 for $\nu = 0.3$)</td></tr>
<tr><td>20. Regular polygonal plate; all edges fixed
Number of sides = n</td><td>20a. Uniform over entire plate</td><td>(At center) $\sigma = \dfrac{\beta_1 qa^2}{t^2}$ and $y_{max} = \dfrac{-\alpha qa^4}{Et^3}$
(At center of straight edge) $\sigma_{max} = \dfrac{-\beta_2 qa^2}{t^2}$
<table>
<tr><td>n</td><td>3</td><td>4</td><td>5</td><td>6</td><td>7</td><td>8</td><td>9</td><td>10</td><td>∞</td></tr>
<tr><td>β</td><td>0.589</td><td>0.550</td><td>0.530</td><td>0.518</td><td>0.511</td><td>0.506</td><td>0.503</td><td>0.500</td><td>0.4875</td></tr>
<tr><td>β_2</td><td>1.423</td><td>1.232</td><td>1.132</td><td>1.068</td><td>1.023</td><td>0.990</td><td>0.964</td><td>0.944</td><td>0.750</td></tr>
<tr><td>α</td><td>0.264</td><td>0.221</td><td>0.203</td><td>0.194</td><td>0.188</td><td>0.184</td><td>0.182</td><td>0.180</td><td>0.171</td></tr>
</table>
(Ref. 55 for $\nu = 0.3$)</td></tr>
</table>

11.15 REFERENCES

1. Roark, R. J., "Stresses Produced in a Circular Plate by Eccentric Loading and by a Transverse Couple," *Univ. Wis. Eng. Exp. Sta.*, Bull. 74, 1932. The deflection formulas are due to Föppl. See Die Biegung einer kreisförmigen Platte, Sitzungsber. math.-phys. Kl. K. B. Akad. Wiss. Münch., p. 155, 1912.
2. Michell, J. H., "The Flexure of Circular Plates," *Proc. Math. Soc. Lond*, p. 223, 1901.
3. Warshawsky, I., Private communication.
4. Timoshenko, S., "Über die Biegung der allseitig unterstützten rechteckigen Platte unter Wirkung einer Einzellast," *Der Bauingenieur*, 3: Jan. 31, 1922.
5. Prescott, J., *Applied Elasticity*, Longmans, Green, 1924.
6. Nadai, A., "Über die Spannungsverteilung in einer durch eine Einzelkraft belasteten rechteckigen Platte," *Der Bauingenieur*, 2: Jan. 15, 1921.
7. Timoshenko, S., and J. M. Lessells, *Applied Elasticity*, Westinghouse Technical Night School Press, 1925.
8. Wojtaszak, I. A., "Stress and Deflection of Rectangular Plates," ASME Paper A-71, *J. Appl. Mech.*, 3(2): 1936.
9. Westergaard, H. M., and A. Slater, "Moments and Stresses in Slabs," *Proc. Am. Concr. Inst.*, 17: 1921.
10. Wahl, A. M., "Strength of Semicircular Plates and Rings under Uniform External Pressure," *Trans. ASME*, 54(23): 1932.
11. Nadai, A., "Die Formänderungen und die Spannungen von durchlaufenden Platten," *Der Bauingenieur*, 5: 102, 1924.
12. Nadai, A., *Elastische Platten*, Julius Springer, 1925.
13. Holl, D. L., "Analysis of Thin Rectangular Plates Supported on Opposite Edges," *Iowa Eng. Exp. Sta.*, Iowa State College, Bull. 129, 1936.
14. Westergaard, H. M., "Stresses in Concrete Pavements Computed by Theoretical Analysis," *Public Roads*, U.S. Dept. of Agriculture, Bureau of Public Roads, 7(2): 1926.

15. Timoshenko, S., *Vibration Problems in Engineering*, p. 319, D. Van Nostrand Company, 1928.
16. Way, S., "Bending of Circular Plates with Large Deflection," *Trans. ASME*, 56(8) 1934 (see also discussion by E. O. Waters).
17. Sturm, R. G., and R. L. Moore, "The Behavior of Rectangular Plates under Concentrated Load," ASME Paper A-75, *J. Appl. Mech.*, 4(2): 1937.
18. Hencky, H., "Uber den Spannungszustand in kreisrunder Platten mit verschwindender Biegungssteifigkeit," *Z. Math. Phys.*, 63: 311, 1915.
19. Wahl, A. M., "Stresses and Deflections in Flat Circular Plates with Central Holes," *Trans. ASME Paper APM-52-3*, 52(1): 29, 1930.
20. Flügge, W., "Kreisplatten mit linear veränderlichen Belastungen," *Bauingenieur*, 10(13): 221, 1929.
21. Timoshenko, S., and S. Woinowsky-Krieger, *Theory of Plates and Shells*, 2nd ed., McGraw-Hill, 1959.
22. Reissner, H., "Über die unsymmetrische Biegung dünner Kreisringplatte," *Ing.-Arch.*, 1: 72, 1929.
23. Woinowsky-Krieger, S., "Berechnung der ringsum frei aufliegenden gleichseitigen Dreiecksplatte," *Ing.-Arch.*, 4: 254, 1933.
24. Jensen, V. P., "Analysis of Skew Slabs," *Eng. Exp. Sta. Univ. Ill.*, Bull. 332, 1941.
25. Evans, T. H., "Tables of Moments and Deflections for a Rectangular Plate Fixed at All Edges and Carrying a Uniformly Distributed Load," *ASME J. Appl. Mech.*, 6(1): March 1939.
26. Young, D., "Clamped Rectangular Plates with a Central Concentrated Load," *ASME Paper A-114, J. Appl. Mech.*, 6(3): 1939.
27. Almen, J. O., and A. Laszlo, "The Uniform-Section Disc Spring," *Trans. ASME*, 58: 305, 1936.
28. Odley, E. G., "Deflections and Moments of a Rectangular Plate Clamped on All Edges and under Hydrostatic Pressure," *ASME J. Appl. Mech.*, 14(4): 1947.
29. Stevens, H. H., "Behavior of Circular Membranes Stretched above the Elastic Limit by Air Pressure," *Exp. Stress Anal.*, 2(1): 1944.
30. Levy, S., "Bending of Rectangular Plates with Large Deflections," *Natl. Adv. Comm. Aeron*, Tech. Note 846, 1942.
31. Levy, S., "Square Plate with Clamped Edges under Normal Pressure Producing Large Deflections," *Natl. Adv. Comm. Aeron.*, Tech. Note 847, 1942.
32. Levy, S., and S. Greenman, "Bending with Large Deflection of a Clamped Rectangular Plate with Length-Width Ratio of 1.5 under Normal Pressure," *Natl. Adv. Comm. Aeron.*, Tech. Note 853, 1942.
33. Chi-Teh Wang, "Nonlinear Large Deflection Boundary-Value Problems of Rectangular Plates," *Natl. Adv. Comm. Aeron.*, Tech. Note 1425, 1948.
34. Chi-Teh Wang, "Bending of Rectangular Plates with Large Deflections," *Natl. Adv. Comm. Aeron.*, Tech. Note 1462, 1948.
35. Ramberg, W., A. E. McPherson, and S. Levy, "Normal Pressure Tests of Rectangular Plates," *Natl. Adv. Comm. Aeron.*, Rept. 748, 1942.
36. Conway, H. D., "The Bending of Symmetrically Loaded Circular Plates of Variable Thickness," *ASME J. Appl. Mech.*, 15(1), 1948.
37. Reissmann, Herbert, "Bending of Clamped Wedge Plates," *ASME J. Appl. Mech.*, 20: March 1953.
38. Bassali, W. A., and R. H. Dawoud, "Bending of an Elastically Restrained Circular Plate under Normal Loading on a Sector," *ASME J. Appl. Mech.*, 25(1): 1958.
39. Bassali, W. A., and M. Nassif, "Stresses and Deflections in Circular Plate Loaded over a Segment," *ASME J. Appl. Mech.*, 26(1): 1959.
40. Jurney, W. H., "Displacements and Stresses of a Laterally Loaded Semicircular Plate with Clamped Edges," *ASME J. Appl. Mech.*, 26(2): 1959.
41. Conway, H. D., "Bending of Rectangular Plates Subjected to a Uniformly Distributed Lateral Load and to Tensile or Compressive Forces in the Plane of the Plate," *ASME J. Appl. Mech.*, 16(3): 1949.
42. Morse, R. F., and H. D. Conway, "The Rectangular Plate Subjected to Hydrostatic Tension and to Uniformly Distributed Lateral Load," *ASME J. Appl. Mech.*, 18(2): June 1951.
43. Hodge, P. G., Jr., *Plastic Analysis of Stuctures*, McGraw-Hill, 1959.
44. Shull, H. E., and L. W. Hu, "Load-Carrying Capacities of Simply Supported Rectangular Plates," *ASME J. Appl. Mech.*, 30(4): 1963.
45. Zaid, M., "Carrying Capacity of Plates of Arbitrary Shape," *ASME J. Appl. Mech.*, 25(4) 1958.
46. Márkus, G., *Theorie und Berechnung rotations symmetrischer Bauwerke*, Werner-Verlag, 1967.
47. Bareš, R., *Tables for the Analysis of Plates, Slabs, and Diaphragms Based on the Elastic Theory*, 3rd ed., Bauverlag GmbH (English transl. by Carel van Amerogen), Macdonald and Evans, 1979.
48. Heap, J., "Bending of Circular Plates Under a Variable Symmetrical Load," *Argonne Natl. Lab. Bull. 6882*, 1964.
49. Moody, W., "Moments and Reactions for Rectangular Plates," *Bur. Reclamation Eng. Monogr. 27*, 1960.
50. Marguerre, K., and H. Woernle, *Elastic Plates*, Blaisdell, 1969.

51. Morley, L., *Skew Plates and Structures*, Macmillan, 1963.
52. Hodge, P., *Limit Analysis of Rotationally Symmetric Plates and Shells*, Prentice-Hall, 1963.
53. Kennedy, J., and S. Ng, "Linear and Nonlinear Analyses of Skewed Plates," *ASME J. Appl. Mech.*, 34(2): 1967.
54. Srinivasan, R. S., and V. Thiruvenkatachari, "Large Deflection Analysis of Clamped Annular Sector Plates," *Inst. Mech. Eng. J. Strain Anal.*, 19(1): 1948.
55. Leissa, A., C. Lo, and F. Niedenfuhr, "Uniformly Loaded Plates of Regular Polygonal Shape," *AIAA J.*, 3(3): 1965.
56. Leissa, A., W. Clausen, L. Hulbert, and A. Hopper, "A Comparison of Approximate Methods for the Solution of Plate Bending Problems," *AIAA J.*, 7(5): 1969.
57. Stanek, F. J., *Stress Analysis of Circular Plates and Cylindrical Shells*, Dorrance, 1970.
58. Griffel, W., *Plate Formulas*, Frederick Ungar, 1968.
59. Tuncel, Özcan, "Circular Ring Plates Under Partial Arc Loading," *ASME J. Appl. Mech.*, 31(2): 1964.
60. Lee, T. M., "Flexure of Circular Plate by Concentrated Force," *Proc. Am. Soc. Civil Eng., Eng. Mech. Div.*, 94(3): 1968.
61. Essenburg, F., and S. T. Gulati, "On the Contact of Two Axisymmetric Plates," *ASME J. Appl. Mech.*, 33(2): 1966.
62. Slot, T., and W. J. O'Donnell, "Effective Elastic Constants for Thick Perforated Plates with Square and Triangular Penetration Patterns," *ASME J. Eng. Ind.*, 11(4): 1971.
63. Petrina, P., and H. D. Conway, "Deflection and Moment Data for Rectangular Plates of Variable Thickness," *ASME J. Appl. Mech.*, 39(3): 1972.
64. Conway, H. D., "Nonaxial Bending of Ring Plates of Varying Thickness," *ASME J. Appl. Mech.*, 25(3): 1958.
65. Strock, R. R., and Yi-Yuan Yu, "Bending of a Uniformly Loaded Circular Disk With a Wedge-Shaped Cross Section," *ASME J. Appl. Mech.*, 30(2): 1963.
66. Conway, H. D., "The Ribbed Circular Plate," *ASME J. Appl. Mech.*, 30(3): 1963.
67. Wempner, G. A., "The Conical Disc Spring," *Proc. 3rd Natl. Congr. Appl. Mech., ASME*, 1958.
68. Wempner, G. A., "Axisymmetric Deflections of Shallow Conical Shells," *Proc. Am. Soc. Civil Eng., Eng. Mech. Div.*, 90(2): 1964.
69. Owens, J. H., and D. C. Chang, "Belleville Springs Simplified," *Mach. Des.*, May 14, 1970.
70. Paul, B., "Collapse Loads of Rings and Flanges Under Uniform Twisting Moment and Radial Force," *ASME J. Appl. Mech.*, 26(2): 1959.
71. Mah, G. B. J., "Axisymmetric Finite Deflection of Circular Plates," *Proc. Am. Soc. Civil Eng., Eng. Mech. Div.*, 95(5): 1969.
72. Nash, W. A., and I. D. Cooley, "Large Deflections of a Clamped Elliptical Plate Subjected to Uniform Pressure," *ASME J. Appl. Mech.*, 26(2): 1959.
73. Ng, S. F., "Finite Deflection of Elliptical Plates on Elastic Foundations," *AIAA J.*, 8(7): 1970.
74. Bauer, F., L. Bauer, W. Becker, and E. Reiss, "Bending of Rectangular Plates With Finite Deflections," *ASME J. Appl. Mech.*, 32(4): 1965.
75. Dundurs, J., and Tung-Ming Lee, "Flexure by a Concentrated Force of the Infinite Plate on a Circular Support," *ASME J. Appl. Mech.*, 30(2): 1963.
76. Amon, R., and O. E. Widera, "Clamped Annular Plate under a Concentrated Force," *AIAA J.*, 7(1): 1969.
77. Amon, R., O. E. Widera, and R. G. Ahrens, "Problem of the Annular Plate, Simply Supported and Loaded with an Eccentric Concentrated Force," *AIAA J.*, 8(5): 1970.
78. Widera, O. E., R. Amon, and P. L. Panicali, "On the Problem of the Infinite Elastic Plate with a Circular Insert Reinforced by a Beam," *Nuclear Eng. Des.*, 12(3): 1970.
79. Amon, R., O. E. Widera, and S. M. Angel, "Green's Function of an Edge-Beam Reinforced Circular Plate," *Proc. CANCAM*, Calgary, May 1971.
80. Ohashi, Y., and S. Murakami, "Large Deflection in Elastoplastic Bending of a Simply Supported Circular Plate Under a Uniform Load," *ASME J. Appl. Mech.*, 33(4): 1966.
81. Save, M. A., and C. E. Massonnet, *Plastic Analysis and Design of Plates, Shells and Disks*, North-Holland, 1972.
82. Markowitz, J., and L. W. Hu, "Plastic Analysis of Orthotropic Circular Plates," *Proc. Am. Soc. Civil Eng., Eng. Mech. Div.*, 90(5): 1964.
83. Crose, J. G., and A. H.-S. Ang, "Nonlinear Analysis Method for Circular Plates," *Proc. Am. Soc. Civil Eng., Eng. Mech. Div.*, 95(4): 1969.
84. Szilard, R., *Theory and Analysis of Plates: Classical and Numerical Methods*, Prentice-Hall, 1974.
85. Ollerton, E., "Thin Annular Plates Having an Inner Clamp Carrying Bending Moments," *Inst. Mech. Eng. J. Strain Anal.*, 12(3): 1977.
86. Ollerton, E., "The Deflection of a Thin Circular Plate with an Eccentric Circular Hole," *Inst. Mech. Eng. J. Strain Anal.*, 11(2): 1976.

87. Ollerton, E., "Bending Stresses in Thin Circular Plates with Single Eccentric Circular Holes," *Inst. Mech. Eng. J. Strain Anal.*, 11(4): 1976.
88. Cook, R. D., and W. C. Young, *Advanced Mechanics of Materials*, 2nd ed., Prentice- Hall, 1998.
89. Pister, K. S., and S. B. Dong, "Elastic Bending of Layered Plates," *Proc. Am. Soc. Civ. Eng., Eng. Mech. Div.*, 85(4): 1959.
90. Goldberg, M. A., M. Newman, and M. J. Forray, "Stresses in a Uniformly Loaded Two-Layer Circular Plate," *Proc. Am. Soc. Civ. Eng., Eng. Mech. Div.*, 91(3): 1965.
91. Chia, Chuen-Yuan, *Nonlinear Analysis of Plates*, McGraw-Hill, 1980.
92. Aalami, B., and D. G. Williams, *Thin Plate Design for Transverse Loading*, John Wiley & Sons, 1975.
93. Boedo, S., and V. C. Prantil, "Corrected Solution of Clamped Ring Plate with Edge Point Load," *J. Eng. Mech.*, 124(6): 1998.

CHAPTER 12

Columns and Other Compression Members

12.1 COLUMNS; COMMON CASE

The formulas and discussion of this section are based on the following assumptions: (1) The column is nominally straight and is subjected only to nominally concentric and axial end loads, and such crookedness and eccentricity as may occur are accidental and not greater than is consistent with standard methods of fabrication and ordinary conditions of service; (2) the column is homogeneous and of uniform cross section; (3) if the column is made up of several longitudinal elements, these elements are so connected as to act integrally; (4) there are no parts so thin as to fail by local buckling before the column as a whole has developed its full strength.

End Conditions

The strength of a column is in part dependent on the *end conditions*, that is, the degree of end fixity or constraint. A column with ends that are supported and fixed, so that there can be neither lateral displacement nor change in slope at either end, is called *fixed-ended.* A column with ends that are supported against lateral displacement but not constrained against change in slope is called *round-ended.* A column with one end fixed and the other end neither laterally supported nor otherwise constrained is called *free-ended.* A column with both end surfaces that are flat and normal to the axis and that bear evenly against rigid loading surfaces is called *flat-ended.* A column with ends that bear against transverse pins is called *pin-ended.*

Truly fixed-ended and truly round-ended columns practically never occur in practice; the actual conditions are almost always intermediate. The greatest degree of fixity is found in columns with ends that are riveted or welded to relatively rigid parts that are also fixed. Theoretically, a flat-ended column is equivalent to a fixed-ended column until the load reaches a certain critical value at which the column "kicks out" and bears only on one edge of each end surface instead of on the whole surface. Actually, flat-ended columns have a degree of end constraint considerably less than that required to produce fixity. The nearest approach to round-ended conditions is found in pin-ended columns subject to vibration or other imposed motion. The degree of end fixity may be expressed by the *coefficient of constraint* [explained following Eq. (12.1-1)] or by the *free* or *effective* length, which is the length measured between points of counterflexure or the length of a round-ended column of equal strength.

Behavior

If sufficiently slender, a column will fail by elastic instability (see Chap. 15). In this case the maximum unit stress sustained is less than the proportional limit of the material; it depends on the modulus of elasticity, the slenderness ratio (ratio of the length of the column to the least radius of gyration of the section), and the end conditions and is independent of the strength of the material. Columns which fail in this way are called *long columns*.

Columns that are too short to fail by elastic instability are called *short columns*; such a column will fail when the maximum fiber stress due to direct compression and to the bending that results from accidental crookedness and eccentricity reaches a certain value. For structural steel this value is about equal to the tensile yield point; for light alloys it is about equal to the compressive yield strength; and for wood it lies between the flexural elastic limit and the modulus of rupture.

For a given material and given end conditions, there is a certain slenderness ratio which marks the dividing point between long and short columns called the *critical slenderness ratio*.

Formulas for Long Columns

The unit stress at which a long column fails by elastic instability is given by the Euler formula

$$\frac{P}{A} = \frac{C\pi^2 E}{(L/r)^2} \tag{12.1-1}$$

where P = total load, A = area of section, E = modulus of elasticity, L/r = slenderness ratio, and C is the coefficient of constraint, which depends on end conditions. For round ends, $C = 1$; for fixed ends, $C = 4$; and for the end conditions that occur in practice, C can rarely be assumed greater than 2. It is generally not considered good practice to employ long columns in building and bridge construction, but they are used in aircraft. (Formulas for the loads producing elastic instability of uniform and tapered bars under a wide variety of conditions of loading and support are given in Table 15.1.)

Formulas for Short Columns

It is not possible to calculate with accuracy the maximum stress produced in a short column by a nominally concentric load because of the large influence of the indeterminate crookedness and eccentricity. The maximum unit stress that a column will sustain, however, can be expressed by any of a number of formulas, each of which contains one or more terms that is empirically adjusted to secure conformity with test results. Of such formulas, those given below are the best known and provide the basis for most of the design formulas used in American practice. In these equations P denotes the load at failure, A the cross-sectional area, L the length, and r the least radius of gyration of the section; the meaning of other symbols used is explained in the discussion of each formula.

Secant Formula

$$\frac{P}{A} = \frac{\sigma}{1 + \dfrac{ec}{r^2}\sec\left(\dfrac{KL}{2r}\sqrt{\dfrac{P}{AE}}\right)} \tag{12.1-2}$$

This formula is adapted from the formula for stress due to eccentric loading [Eq. (12.4-1)]. Here σ denotes the maximum fiber stress at failure (usually taken as the yield point for steel and as the yield strength for light alloys); e denotes the *equivalent eccentricity* (that eccentricity which in a perfectly straight column would cause the same amount of bending as the actual eccentricity and crookedness); c denotes the distance from the central axis about which bending occurs to the extreme fiber on the concave or compression side of the bent column; and K is a numerical coefficient, dependent on end conditions, such that KL is the effective length of the column, or distance between points of inflection. The term ec/r^2 is called the *eccentric ratio*, and a value is assumed for this ratio which makes the formula agree with the results of tests on columns of the type under consideration. For example, tests on structural steel columns of conventional design indicate that the average value of the eccentric ratio is 0.25. In using the secant formula, P/A must be solved for by trial or by the use of prepared charts.

Rankine Formula

$$\frac{P}{A} = \frac{\sigma}{1+\phi(L/r)^2} \qquad (12.1\text{-}3)$$

This is a semirational formula. The value of σ is sometimes taken as the ultimate strength of the material and the value of ϕ as $\sigma/C\pi^2E$, thus making the formula agree with the results of tests on short prisms when L/r is very small and with Euler's equation when L/r is very large. More often σ and ϕ are adjusted empirically to make the equation agree with the results of tests through the L/r range of most importance.

Simple Polynomial Formula

$$\frac{P}{A} = \sigma - k\left(\frac{L}{r}\right)^n \qquad (12.1\text{-}4)$$

This is an empirical formula. For most steels the exponent n is chosen as 2 and σ is chosen as the yield point to give the well-known parabolic formula. The constant k is generally chosen to make the parabola intersect tangent to the Euler curve for a long column. See Refs. 1–4.

For cast irons and for many of the aluminum alloys (Refs. 1 and 5) the exponent n is taken as unity to give a *straight-line* formula. If σ were here taken as the maximum fiber stress at failure and the straight line made tangent to the Euler curve, the formula would give values well below experimental values. For this reason the straight-line formula is generally used for *intermediate* lengths with σ and k modified to make the straight line pass through the experimental data of interest. For columns shorter than intermediate in length, a constant value of P/A is given; for those longer than intermediate, the Euler curve is specified.

For timber the exponent n is usually chosen as 4 (Refs. 1 and 6) and then treated in the same manner as was the parabolic formula. The value used for σ is generally the ultimate compressive strength of the timber.

Exponential Formula

$$\frac{P}{A} = C_1^{\lambda^2}\sigma \qquad \text{where } \lambda = \frac{KL}{r\pi}\left(\frac{\sigma}{E}\right)^{1/2} \qquad (12.1\text{-}5)$$

The American Institute of Steel Construction in Ref. 7 suggests the use of a formula of the form shown in Eq. (12.1-5), where K, L, r, and σ are as defined for the secant formula. The constant λ combines the column dimensions L and r, the degree of end fixity indicated by K, and the material properties σ and E. The secant, Rankine, parabolic, exponential, and

Euler formulas can all be expressed in terms of λ and a simple tabulation used to compare them.

Parabolic formula $\qquad \dfrac{P/A}{\sigma} = 1 - \dfrac{\lambda^2}{4} \qquad$ for $\lambda < 1.414$

Exponential formula $\qquad \dfrac{P/A}{\sigma} = 0.6922^{\lambda^2} \qquad$ for $\lambda < 1.649$

Rankine formula $\qquad \dfrac{P/A}{\sigma} = \dfrac{1}{1+\lambda^2} \qquad$ for all λ

Secant formula $\qquad \dfrac{P/A}{\sigma} = \dfrac{1}{1+0.25\sec\left(\dfrac{\pi\lambda}{2}\sqrt{\dfrac{P/A}{\sigma}}\right)} \qquad$ for all λ

The Euler formula becomes $(P/A)/\sigma = 1\lambda^2$, and both the parabolic and exponential formulas given above have been derived to be tangent to the Euler formulas where they intersect. In the table given below are shown the values of $(P/A)/\sigma$ for each of the given equations or sets of equations.

In the case of the secant formula at very short lengths, the lower values can be attributed to the chosen eccentricity ratio of 0.25.

By applying a proper factor of safety, a *safe* load can be calculated. Many codes make use of safety factors which vary with the effective L/r ratios, so the safe-load formulas may differ somewhat in form from the ultimate-load formulas just discussed. See Ref. 1 for extensive listings of applicable codes and design specifications.

Calculation of Stress

The best way to compute the probable value of the maximum fiber stress in a short column, caused by the imposition of a nominally concentric load that is less than the ultimate load, is to use the secant formula [Eq. (12.1-2)] with an assumed value of e or ec/r^2. However, by transposing terms, any one of Eqs. (12.1-3)–(12.1-5) can be written so as to give the maximum stress σ in terms of the load P. Such procedure is logical only when σ is the fiber stress at failure and P the ultimate load; but if the maximum stress due to some load that is less than the ultimate load is thus computed, the result, although probably considerably in error, is almost sure to be greater than the true stress and hence the method errs on the side of safety.

Values of $(p/a)/\sigma$ for Four Different Equations or Sets of Equations

	λ															
Equations	0.000	0.200	0.400	0.600	0.800	1.000	1.200	1.400	1.600	1.800	2.000	2.200	2.400	2.600	2.800	3.000
Parabolic-Euler	1.000	0.990	0.960	0.910	0.840	0.750	0.640	0.510	0.391	0.309	0.250	0.207	0.174	0.148	0.128	0.111
Exponential-Euler	1.000	0.985	0.943	0.876	0.790	0.692	0.589	0.486	0.390	0.309	0.250	0.207	0.174	0.148	0.128	0.111
Rankine	1.000	0.962	0.862	0.735	0.610	0.500	0.410	0.338	0.281	0.236	0.200	0.171	0.148	0.129	0.113	0.100
Secant	0.800	0.794	0.773	0.734	0.673	0.589	0.494	0.405	0.331	0.273	0.227	0.191	0.163	0.140	0.122	0.107

12.2 LOCAL BUCKLING

If a column is composed wholly or partially of thin material, local buckling may occur at a unit load less than that required to cause failure of the column as a whole. When such local buckling occurs at a unit stress less than the proportional limit, it represents elastic instability; the critical stress at which this occurs can be determined by mathematical analysis. Formulas for the critical stress at which bars and thin plates exhibit elastic instability, under various conditions of loading and support, are given in Tables 15.1 and 15.2. All such formulas are based upon assumptions as to homogeneity of material, regularity of form, and boundary conditions that are never realized in practice; the critical stress to be expected under any actual set of circumstances is nearly always less than that indicated by the corresponding theoretical formula and can be determined with certainty only by test. This is also true of the ultimate load that will be carried by such parts as they buckle since elastic buckling is not necessarily attended by failure, and thin flanges and webs, by virtue of the support afforded by attached parts, may carry a load considerably in excess of that at which buckling occurs (see Sec. 12.6).

In the following paragraphs, the more important facts and relations that have been established concerning local buckling are stated, insofar as they apply to columns of more or less conventional design. In the formulas given, b represents the unsupported width of the part under consideration, t its thickness, σ_y the yield point or yield strength, and E and ν have their usual meanings.

Outstanding Flanges

For a long flange having one edge fixed and the other edge free, the theoretical formula for buckling stress is

$$\sigma' = \frac{1.09E}{1-\nu^2}\left(\frac{t}{b}\right)^2 \tag{12.2-1}$$

and for a flange having one edge simply supported and the other edge free, the corresponding formula is (see Table 15.2)

$$\sigma' = \frac{0.416E}{1-\nu^2}\left(\frac{t}{b}\right)^2 \tag{12.2-2}$$

For the outstanding flange of a column, the edge condition is intermediate, with the degree of constraint depending upon the torsional rigidity of the main member and on the way in which the flange is attached. The conclusions of the ASCE Column Research Committee (Ref. 8) on this point may be summed up as follows: For columns of structural steel having a proportional limit of 30,000 lb/in^2, an outstanding flange riveted between two angles, each having a thickness equal to that of the flange, will not fail by elastic buckling if b/t is less than 15, b being measured from the free edge of the flange to the first row of rivets; for wider flanges, the formula for buckling stress is

$$\sigma' = 0.4E\left(\frac{t}{b}\right)^2 \tag{12.2-3}$$

If the thickness of each supporting angle is twice that of the flange, elastic buckling will not occur if b/t is less than 20, b in this case being measured from the free edge of the flange to the toe of the angle; for wider flanges, the formula for buckling stress is

$$\sigma' = 0.6E\left(\frac{t}{b}\right)^2 \tag{12.2-4}$$

The ultimate strength of an outstanding flange is practically equal to the area times the yield point up to a b/t ratio of 15; for wider flanges the ultimate load is not appreciably greater, and so there is no substantial gain in load-carrying capacity when the width of a flange is increased to more than 15 times the thickness. In Ref. 2 are given recommended limiting values of width/thickness ratios in terms of σ_y for webs, flanges, and other parts subject to buckling.

In the case of aluminum, the *allowable* unit stress on an outstanding flange may be found by the formulas

$$(\text{Allowable, lb/in}^2)\quad \sigma = \begin{cases} 15{,}000 - 123k\dfrac{b}{t} & \text{when } k\dfrac{b}{t} < 81 \quad (12.2\text{-}5) \\[2ex] \dfrac{33{,}000{,}000}{\left(k\dfrac{b}{t}\right)^2} & \text{when } k\dfrac{b}{t} > 81 \quad (12.2\text{-}6) \end{cases}$$

Here k is to be taken as 4 when the outstanding flange is one leg of an angle T or other section having relatively little torsional rigidity, and may be taken as 3 when the flange is part of or firmly attached to a heavy web or other part that offers relatively great edge constraint.

A formula (Ref. 13) for the ultimate strength of short compression members consisting of single angles, which takes into account both local and general buckling, is

$$\frac{P}{A} = \sigma \tanh\left[K\left(\frac{t}{b}\right)^2\right] \tag{12.2-7}$$

where $K = 149.1 + 0.1(L/r - 47)^2$ and σ, which depends on L/r, has the following values:

L/r	0	20	40	60	80
σ	40,000	38,000	34,000	27,000	18,000

This formula is for an alloy (24ST) having a yield strength of 43,000 lb/in² and a modulus of elasticity of 10,500,000 lb/in², and is for round-ended columns ($c = 1$). A more general formula for thin sections other than angles is

$$\frac{P}{A} = \sigma \tanh(Kt) \tag{12.2-8}$$

Here $\sigma = \sigma_y(1+B)/(1+B+B^2)$, where $B = \sigma_y(L/r)^2/(c\pi^2E)$, and $K = K_o(\sigma_y/\sigma)^{1/2}$, where K_o is a *shape factor*, the value of which is found from Eq. (12.2-8), P/A being experimentally determined by testing columns of the section in question that have a slenderness ratio of about 20. For a closed box or "hat" section, $K_o = 15.6$; for a section with flat flanges with a width that is not more than 25 times the thickness, $K_o = 10.8$; for a section of oval form or having wholly or partially curved flanges, K_o ranges from 12 to 32 (Ref. 13). (An extensive discussion of design procedures and buckling formulas for aluminum columns and other structural elements is found in Ref. 5.)

For spruce and other wood of similar properties, Trayer and March (Chap. 14, Ref. 3) give as the formula for buckling stress

$$\sigma' = 0.07E\left(\frac{t}{b}\right)^2 \tag{12.2-9}$$

when the edge constraint is as great as normally can be expected in all-wood construction and

$$\sigma' = 0.044E\left(\frac{t}{b}\right)^2 \tag{12.2-10}$$

when conditions are such as to make the edge constraint negligible.

Thin Webs

For a long thin web that is fixed along each edge, the theoretical formula for buckling stress is

$$\sigma' = \frac{5.73E}{1-\nu^2}\left(\frac{t}{b}\right)^2 \tag{12.2-11}$$

and for a web that is simply supported along each edge, the corresponding formula is (See Table 15.2.)

$$\sigma' = \frac{3.29E}{1-\nu^2}\left(\frac{t}{b}\right)^2 \tag{12.2-12}$$

For structural steel columns, the conclusion of the ASCE Column Research Committee (Ref. 8) is that elastic buckling will not occur at b/t ratios less than 30. Tests made by the Bureau of Standards (Ref. 14) on steel members consisting of wide webs riveted between edge angles indicate that this conclusion is conservative and that b/t may be safely as great as 35 if b is taken as the width between rivet lines.

For aluminum columns, the same formulas for allowable stress on a thin web are suggested as are given previously for the outstanding flange [(Eqs. (12.2-5) and (12.2-6)] but with $k = 1.2$. (For discussion of the ultimate strength developed by a thin web, see Sec. 12.6.)

Thin Cylindrical Tubes

For a thin cylindrical tube, the theoretical formula for the critical stress at which buckling occurs is

$$\sigma' = \frac{E}{\sqrt{3}\sqrt{1-\nu^2}}\frac{t}{R} \tag{12.2-13}$$

when R denotes the mean radius of the tube (see Table 15.2). Tests indicate that the critical stress actually developed is usually only 40–60% of this theoretical value.

Much recent work has been concerned with measuring initial imperfections in manufactured cylindrical tubes and correlating these imperfections with measured critical loads. For more detailed discussions and recommendations refer to Refs. 1–5 in this chapter and to Refs. 101–109 in Chap. 15.

Attached Plates

When the flanges or web of a column are formed by riveting a number of plates placed flat against one another, there is a possibility that the outer plate or plates will buckle between points of attachment if the unsupported length is too great compared with the thickness. If the full yield strength σ_y of an outer plate is to be developed, the ratio of unsupported length a to thickness t should not exceed the value indicated by the formula (Ref. 17).

$$\frac{a}{t} = 0.52\sqrt{\frac{E}{\sigma_y}} \qquad (12.2\text{-}14)$$

Some specifications (Ref. 10) guard against the possibility of such buckling by limiting the maximum distance between rivets (in the direction of the stress) to 16 times the thickness of the thinnest outside plate and to 20 times the thickness of the thinnest inside plate; the ratio 16 is in agreement with Eq. (12.2-14).

Local Buckling of Latticed Columns

To guard against the possibility that the longitudinal elements of a latticed column will buckle individually between points of support, some specifications (Ref. 10) limit the slenderness ratio of such parts between points of attachment of lacing bars to 40 or to two-thirds the slenderness ratio of the column as a whole, whichever is less.

Lacing Bars

In a column composed of channels or other structural shapes connected by lacing bars, the function of the lacing bars is to resist the transverse shear due to initial obliquity and that consequent upon such bending as may occur under load. The amount of this shear is conjectural since the obliquity is accidental and indeterminate.

Salmon (Ref. 17) shows that with the imperfections usually to be expected, the transverse shear will be at least 1% of the axial load. Moore and Talbot (Ref. 18) found that for certain experimental columns the shear amounted to from 1–3% of the axial load. Some specifications require that in buildings the lacing be designed to resist a shear equal to 2% of the axial load (Ref. 2), and that in bridges it be designed to resist a shear V given by

$$V = \frac{P}{100}\left[\frac{100}{(L/r)+10} + \frac{L/r}{100}\right]$$

where P is the allowable axial load and r is the radius of gyration of the column section with respect to the central axis perpendicular to the plane of the lacing (Ref. 10).

The strength of individual lacing bars as columns has been investigated experimentally. For a bar of rectangular section with a single rivet at each end, the ultimate strength (in psi) is given by

$$\frac{P}{A} = 25{,}000 - 50\frac{L}{r} \qquad \text{(Ref. 8)}$$

or

$$\frac{P}{A} = 21{,}400 - 45\frac{L}{r} \qquad \text{(Ref. 18)}$$

For bars of angle or channel section, these formulas are conservative. For flat bars used as double lacing, the crossed bars being riveted together, tests show that the effective L is about half the actual distance between end rivets. Some specifications (Refs. 2 and 10) require lacing bars of any section to be designed by the regular column formula, L being taken as the distance between end rivets for single lacing and as 70% of that distance for double lacing. There are additional limitations as to slope of lacing, minimum section, and method of riveting.

12.3 STRENGTH OF LATTICED COLUMNS

Although it is customary to assume that a latticed column acts integrally and develops the full strength of the nominal section, tests show that when bending occurs in the plane of the lacing, the column is less stiff than would be the case if this assumption were valid. For a column so designed that buckling occurs in a plane normal to that of the lacing, this fact is unimportant; but in long open columns laced on all sides, such as are often used for derrick booms and other light construction, it may be necessary to take it into account.

For any assumed transverse loading, it is easy to calculate that part of the deflection of a latticed member which is due to strains in the lacing bars and thus to derive a value for what may be called the *reduced modulus of elasticity KE*. Such calculations agree reasonably well with the results of tests (see Ref. 8), but K, of course, varies with the nature of the assumed transverse loading or with the form of the assumed elastic curve, which amounts to the same thing. For uniformly distributed loading and end support, and for the type of lacing shown in Fig. 12.1(a), K is given by

$$K = \frac{1}{1 + \dfrac{4.8I}{AL^2 \cos^2\theta \sin\theta}} \tag{12.3-1}$$

where L = length of the column, I = moment of inertia of the column cross section about the principal axis which is normal to the plane of battens, and A = cross-sectional area of a single lacing bar. For double lacing, 2.4 should be used in place of 4.8. If KE is used in place of E, the effect of reduced stiffness on the strength of a long column will be approximately allowed for. The method is theoretically inexact mainly because the form of elastic curve assumed is not identical with that taken by the column, but the error due to this is small.

Timoshenko (Ref. 19) gives formulas based upon the assumption that the elastic curve of the column is a sinusoid, from which the following expressions for K may be derived. For the arrangement shown in Fig. 12.1(a),

$$K = \frac{1}{1 + \dfrac{4.93I}{AL^2 \cos^2\theta \sin\theta}} \tag{12.3-2}$$

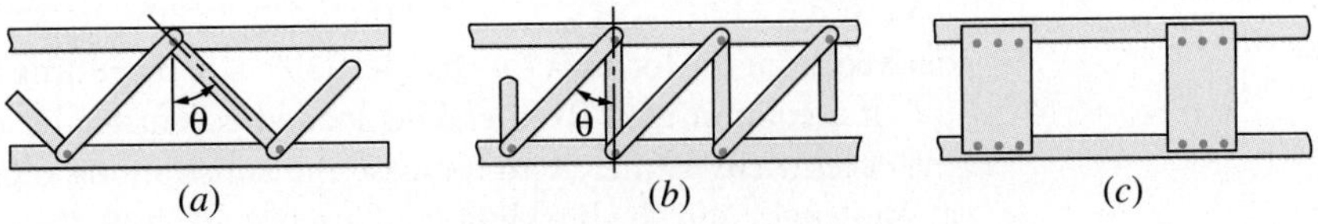

Figure 12.1

For the arrangement shown in Fig. 12.1(*b*),

$$K = \frac{1}{1 + \dfrac{4.93I}{A_1 L^2 \cos^2\theta \sin\theta} + \dfrac{4.93I}{A_2 L^2 \tan\theta}} \tag{12.3-3}$$

where A_1 = cross-sectional area of each diagonal bar and A_2 = cross-sectional area of each transverse bar. For the channel and batten-plate arrangement shown in Fig. 12.1(*c*),

$$K = \frac{1}{1 + \dfrac{\pi^2 I}{L^2}\left(\dfrac{ab}{12I_2} + \dfrac{a^2}{24I_1}\right)} \tag{12.3-4}$$

where a = center-to-center distance between battens; b = length of a batten rivets; I_1 = moment of inertia of a single-channel section (or any similar section being used for each column leg) about its own centroidal axis normal to the plane of the battens; and I_2 = moment of inertia of a pair of batten plates (i.e., $I_2 = 2tc^3/12$, where t is the batten-plate thickness and c is the batten-plate width measured parallel to the length of the column).

In all the preceding expressions for K, it is assumed that all parts have the same modulus of elasticity, and only the additional deflection due to longitudinal strain in the lacing bars and to secondary flexure of channels and batten plates is taken into account. For fairly long columns laced over practically the entire length, the values of K given by Eqs. (12.3-1)–(12.3-3) are probably sufficiently accurate. More elaborate formulas for shear deflection, in which direct shear stress in the channels, bending of the end portions of channels between stay plates, and rivet deformation, as well as longitudinal strains in lacing bars, are taken into account, are given in Ref. 8; these should be used when calculating the deflection of a short latticed column under direct transverse loading.

The use of K as a correction factor for obtaining a reduced value of E is convenient in designing long latticed columns; for short columns the correction is best made by replacing L in whatever column formula is selected by $\sqrt{(1/K)L}$.

Several failure modes and the critical loads associated with these modes are discussed in Refs. 11 and 12.

12.4 ECCENTRIC LOADING; INITIAL CURVATURE

When a round-ended column is loaded eccentrically with respect to one of the principal axes of the section (here called axis 1), the formula for the maximum stress produced is

$$\sigma = \frac{P}{A}\left\{1 + \frac{ec}{r^2}\sec\left[\frac{P}{4EA}\left(\frac{L}{r}\right)^2\right]^{1/2}\right\} \tag{12.4-1}$$

where e = eccentricity, c = distance from axis 1 to the extreme fiber on the side nearest the load, and r = radius of gyration of the section with respect to axis 1. (This equation may be derived from the formula for case 3e, Table 8.8, by putting $M_1 = Pe$.)

If a column with fixed ends is loaded eccentrically, as is assumed here, the effect of the eccentricity is merely to increase the constraining moments at the ends; the moment at midlength and the buckling load are not affected. If the ends are *partially* constrained, as by a frictional moment M, this constraint may be taken into account by considering the actual eccentricity e reduced to $e - M/P$. If a free-ended column is loaded eccentrically, as is assumed here, the formula for the maximum stress is

$$\sigma = \frac{P}{A}\left\{1 + \frac{ec}{r^2}\sec\left[\frac{P}{EA}\left(\frac{L}{r}\right)^2\right]^{1/2}\right\} \tag{12.4-2}$$

where the notation is the same as for Eq. (12.4-1).

When a round-ended column is loaded eccentrically with respect to *both* principal axes of the section (here called axes 1 and 2), the formula for the maximum stress is

$$\sigma = \frac{P}{A}\left\{1 + \frac{e_1c_1}{r_1^2}\sec\left[\frac{P}{4EA}\left(\frac{L}{r_1}\right)^2\right]^{1/2} + \frac{e_2c_2}{r_2^2}\sec\left[\frac{P}{4EA}\left(\frac{L}{r_2}\right)^2\right]^{1/2}\right\} \tag{12.4-3}$$

where the subscripts 1 and 2 have reference to axes 1 and 2 and the notation is otherwise the same as for Eq. (12.4-1). [The use of Eq. (12.4-1) is illustrated in the example below, which also shows the use of Eq. (12.3-1) in obtaining a reduced modulus of elasticity to use with a latticed column.]

If a round-ended column is initially curved in a plane that is perpendicular to principal axis 1 of the section, the formula for the maximum stress produced by concentric end loading is

$$\sigma = \frac{P}{A}\left(1 + \frac{dc}{r^2}\frac{8EA}{P(L/r)^2}\left\{\sec\left[\frac{P}{4EA}\left(\frac{L}{r}\right)^2\right]^{1/2} - 1\right\}\right) \tag{12.4-4}$$

where d = maximum initial deflection, c = distance from axis 1 to the extreme fiber on the concave side of the column, and r = radius of gyration of the section with respect to axis 1. If the column is initially curved in a plane that is not the plane of either of the principal axes 1 and 2 of the section, the formula for the maximum stress is

$$\sigma = \frac{P}{A}\left(1 + \frac{d_1c_1}{r_1^2}\frac{8EA}{P(L/r_1)^2}\left\{\sec\left[\frac{P}{4EA}\left(\frac{L}{r_1}\right)^2\right]^{1/2} - 1\right\}\right.$$
$$\left. + \frac{d_2c_2}{r_2^2}\frac{8EA}{P(L/r_2)^2}\left\{\sec\left[\frac{P}{4EA}\left(\frac{L}{r_2}\right)^2\right]^{1/2} - 1\right\}\right) \tag{12.4-5}$$

where d_1 = the component of the initial deflection perpendicular to the plane of axis 1, d_2 = the component of the initial deflection perpendicular to the plane of axis 2, and c_1, c_2, r_1, and r_2 each has reference to the axis indicated by the subscript.

Eccentrically loaded columns and columns with initial curvature can also be designed by the interaction formulas given in Sec. 12.5.

EXAMPLE Figure 12.2 represents the cross section of a structural steel column composed of two 10-in, 35-lb channels placed 12 in back to back and latticed together. The length of the column is 349.3 in, and it has single lacing, the bars being of rectangular section, 2½ by ¼ in, and inclined at 45°. This column is loaded eccentrically, the load being applied on axis 2 but 2.40 in from axis 1.

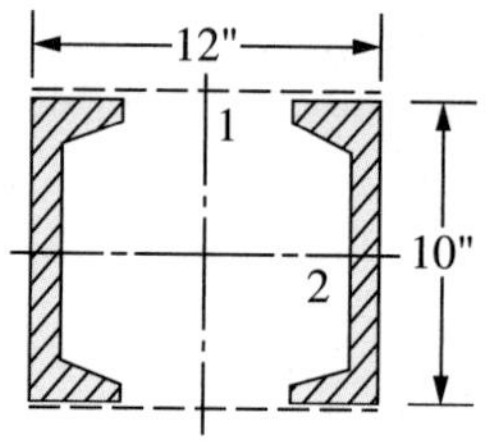

Figure 12.2

With respect to bending in the plane of the eccentricity, the column is round-ended. It is required to calculate the maximum fiber stress in the column when a load of 299,000 lb, or 14,850 lb/in², is thus applied.

SOLUTION For axis 1, $r = 5.38$ in, $c = 6.03$ in (measured), and $e = 2.40$ in. Since the bending due to eccentricity is in the plane of the lacing, a reduced E is used. K is calculated by Eq. (12.3-1), where $I = 583.3$ in^4, $A = 2\frac{1}{2} \times \frac{1}{4} = 0.625$ in^2, $L = 349.3$ in, and $\theta = 45°$. Therefore,

$$K = \frac{1}{1 + \dfrac{(4.8)(583)}{(0.625)(349.3^2)(0.707^2)(0.707)}} = 0.94$$

and using the secant formula [Eq. (12.4-1), (Ref. 23)], we have

$$\sigma = 14{,}850\left\{1 + \frac{(2.40)(6.03)}{5.38^2}\sec\left[\frac{14{,}850}{(4)(0.625)(30{,}000{,}000)}\left(\frac{349.3}{5.38}\right)^2\right]^{1/2}\right\}$$

$$= 25{,}300 \text{ lb/in}^2$$

[This column was actually tested under the loading just described, and the maximum stress (as determined by strain-gage measurements) was found to be 25,250 lb/in². Such close agreement between measured and calculated stress must be regarded as fortuitous, however.]

12.5 COLUMNS UNDER COMBINED COMPRESSION AND BENDING

A column bent by lateral forces or by couples presents essentially the same problem as a beam under axial compression, and the stresses produced can be found by the formulas of Table 8.8 provided the end conditions are determinable. Because these and other uncertainties generally preclude precise solution, it is common practice to rely upon some interaction formula, such as one of those given below. The column may be considered safe for the given loading when the relevant equations are satisfied.

The following notation is common to all the equations; other terms are defined as introduced:

F_a = allowable value of P/A for the member considered as a concentrically loaded column

F_b = allowable value of compressive fiber stress for the member considered as a beam under bending only

$f_a = P/A$ = average compressive stress due to the axial load P

f_b = computed maximum bending stress due to the transverse loads, applied couples, or a combination of these

L = unbraced length in plane of bending

L/r = slenderness ratio for buckling in that plane

For Structural Steel

$$\frac{f_a}{F_a} = + \frac{C_m f_b}{(1 - f_a/F_e)F_b} \le 1 \qquad \text{when } \frac{f_a}{F_a} > 0.15$$

$$\frac{f_a}{F_a} + \frac{f_b}{F_b} \le 1 \qquad \text{when } \frac{f_a}{F_a} < 0.15$$

for sections between braced points, and

$$\frac{f_a}{0.6F_y} + \frac{f_b}{F_b} \le 1$$

for sections at braced points only. Here $F_e = 149{,}000{,}000/(L/r)^2$ and F_y = yield point of steel; $C_m = 0.85$ except that for restrained compression members in frames braced against joint translation and without transverse loading between joints; $C_m = 0.6 + 0.4(M_1/M_2)$, where M_1 is the smaller and M_2 the larger of the moments at the ends of the critical unbraced length of the member. M_1/M_2 is positive when the unbraced length is bent in single curvature and negative when it is bent in reverse curvature. For such members with transverse loading between joints, C_m may be determined by rational analysis, or the appropriate formula from Table 8.8 may be used. (Formulas are adapted from Ref. 2, with F_e given in English units.)

For Structural Aluminum

$$\frac{f_a}{F_a} + \frac{f_b}{F_b(1 - f_a/F_e)} \le 1$$

Here $F_e = 51{,}000{,}000/(L/r)^2$ for building structures and $F_e = 45{,}000{,}000/(L/r)^2$ for bridge structures. (Formulas are taken from Ref. 9 with some changes of notation and with F_e given in English units.)

For Wood (Solid Rectangular)

When $L/d \le \sqrt{0.3E/F_a}$,

$$\frac{f_b}{F_b} + \frac{f_a}{F_a} \le 1$$

When $L/d > \sqrt{0.3E/F_a}$,

1. Concentric end loads plus lateral loads,

$$\frac{f_b}{F_b - f_a} + \frac{f_a}{F_a} \le 1$$

2. Eccentric end load,

$$\frac{1.25 f_b}{F_b - f_a} + \frac{f_a}{F_a} \le 1$$

3. Eccentric end load plus lateral loads,

$$\frac{f_{bl} + 1.25 f_{be}}{F_b - f_a} + \frac{f_a}{F_a} \le 1$$

Here d = dimension of the section in the plane of bending, f_{bl} = computed bending stress due to lateral loads, and f_{be} = computed bending stress due to the eccentric moment. (Formulas are taken from Ref. 16 with some changes of notation.)

12.6 THIN PLATES WITH STIFFENERS

Compression members and compression flanges of flexural members are sometimes made of a very thin sheet reinforced with attached stiffeners; this construction is especially common in airplanes, where both wings and fuselage are often of the "stressed-skin" type.

When a load is applied to such a combination, the portions of the plate not very close to the stiffeners buckle elastically at a very low unit stress, but those portions immediately adjacent to the stiffeners develop the same stress as do the latter, and portions a short distance from the stiffeners develop an intermediate stress. In calculating the part of any applied load that will be carried by the plate or in calculating the strength of the combination, it is convenient to make use of the concept of "effective," or "apparent," width, that is, the width of that portion of the sheet which, if it developed the same stress as the stiffener, would carry the same load as is actually carried by the entire sheet.

For a flat, rectangular plate that is supported but not fixed along each of two opposite edges and subjected to a uniform shortening parallel to those edges, the theoretical expression (Ref. 20) for the effective width is

$$w = \frac{\pi t}{2\sqrt{3(1-\nu^2)}}\sqrt{\frac{E}{\sigma}} \tag{12.6-1}$$

where w = the effective width along each supported edge, t = the thickness of the plate, and σ = the unit stress at the supported edge. Since the maximum value of σ is σ_y (the yield point or yield strength), the maximum load that can be carried by the effective strip or by the whole plate (which amounts to the same thing) is

$$P = \frac{\pi t^2}{\sqrt{3(1-\nu^2)}}\sqrt{E\sigma_y} \tag{12.6-2}$$

This formula can be written

$$P = Ct^2\sqrt{E\sigma_y} \tag{12.6-3}$$

where C is an empirical constant to be determined experimentally for any given material and manner of support. Tests (Ref. 21) made on single plates of various metals, supported at the edges, gave values for C ranging from 1.18 to 1.67; its theoretical value from Eq. (12.6-2) (taking $\nu = 0.25$) is 1.87.

Sechler (Ref. 22) represents C as a function of $\lambda = (t/b)\sqrt{E/\sigma_y}$, where b is the panel width, and gives a curve showing experimentally determined values of C plotted against λ. The following table of corresponding values is taken from Sechler's corrected graph:

λ	0.02	0.05	0.1	0.15	0.2	0.3	0.4	0.5	0.6	0.8
C	2.0	1.76	1.62	1.50	1.40	1.28	1.24	1.20	1.15	1.10

The effective width at failure can be calculated by the relation

$$w = \frac{1}{2}Ct\sqrt{\frac{E}{\sigma_y}} = \frac{1}{2}Cb\lambda$$

In the case of a cylindrical panel loaded parallel to the axis, the effective width at failure can be taken as approximately equal to that for a flat sheet, but the increase in the buckling stress in the central portion of the panel due to curvature must be taken into account. Sechler shows that the contribution of this central portion to the strength of the panel may be allowed for by using for C in the formula $P = Ct^2\sqrt{E\sigma_y}$, a value given by

$$C = C_f - 0.3C_f\lambda\eta + 0.3\eta$$

where $\lambda = (t/b)\sqrt{E/\sigma_y}$, $\eta = (b/r)\sqrt{E/\sigma_y}$, and C_f, is the value of C for a flat sheet, as given by the above table.

The above formulas and experimental data refer to single sheets supported along each edge. In calculating the load carried by a flat sheet with longitudinal stiffeners at any given stiffener stress σ_s, the effective width corresponding to that stress is found by

$$w = b(0.25 + 0.91\lambda^2)$$

where $\lambda = (t/b)\sqrt{E/\sigma_s}$ and b = distance between the stiffeners (Ref. 23). The total load carried by n stiffeners and the supported plate is then

$$P = n(A_s + 2wt)\sigma_s$$

where A_s is the section area of one stiffener. When σ_s is the maximum unit load the stiffener can carry as a column, P becomes the ultimate load for the reinforced sheet.

In calculating the ultimate load on a curved sheet with stiffeners, the strength of each unit or panel may be found by adding to the buckling strength of the central portion of the panel the strength of a column made up of the stiffener and the effective width of the attached sheet, this effective width being found by

$$w = \frac{1}{2}C_f t\sqrt{\frac{E}{\sigma_c}}$$

where C_f is the flat-sheet coefficient corresponding to $\lambda = (t/b)\sqrt{E/\sigma_c}$ and σ_c is the unit load that the stiffener-and-sheet column will carry before failure, determined by an appropriate column formula. [For the type of thin section often used for stiffeners in airplane construction, σ_c may be found by Eq. (12.2-7) or (12.2-8).] Since the unit load σ_c and the effective width w are interdependent (because of the effect of w on the column radius of gyration), it is necessary to assume a value of σ_c, calculate the corresponding w, and then ascertain if the value of σ_c is consistent (according to the column formula used) with this w. (This procedure may have to be repeated several times before agreement is reached.) Then, σ_c and w being known, the strength of the stiffener-and-sheet combination is calculated as

$$P = n[\sigma_c(A_s + 2wt) + (b - 2w)t\sigma']$$

where n is the number of stiffeners, A_s is the section area of one stiffener, b is the distance between stiffeners (rivet line to rivet line) and σ' is the critical buckling stress for the central portion of the sheet, taken as $\sigma' = 0.3Et/r$ (r being the radius of curvature of the sheet).

Methods of calculating the strength of stiffened panels and thin columns subject to local and torsional buckling are being continually modified in the light of current study and experimentation. A more extensive discussion than is appropriate here can be found in books on airplane stress analysis, as well as in Refs. 4 and 5.

12.7 SHORT PRISMS UNDER ECCENTRIC LOADING

When a compressive or tensile load is applied eccentrically to a short prism (i.e., one so short that the effect of deflection is negligible), the resulting stresses are readily found by superposition. The eccentric load P is replaced by an equal axial load P' and by couples Pe_1 and Pe_2, where e_1 and e_2 denote the eccentricities of P with respect to the principal axes 1 and 2, respectively. The stress at any point, or the maximum stress, is then found by superposing the direct stress P'/A due to the axial load and the bending stresses due to the couples Pe_1 and Pe_2, these being found by the ordinary flexure formula (Sec. 8.1).

If, however, the prism is composed of a material that can withstand compression only (masonry) or tension only (very thin shell), this method cannot be employed when the load acts *outside the kern* because the reversal of stress implied by the flexure formula cannot occur. By assuming a linear stress distribution and making use of the facts that the volume of the stress solid must equal the applied load P and that the center of gravity of the stress solid must lie on the line of action of P, formulas can be derived for the position of the neutral axis (line of zero stress) and for maximum fiber stress in a prism of any given cross section. A number of such formulas are given in Table 12.1, together with the dimensions of the kern for each of the sections considered. For any section that is symmetrical about the axis of eccentricity, the maximum stress $K(P/A)$ is just twice the average stress P/A when the load is applied at the edge of the kern and increases as the eccentricity increases, becoming (theoretically) infinite when the load is applied at the extreme fiber. A prism made of material incapable of sustaining both tension and compression will fail completely when the resultant of the loads falls outside the boundary of any cross section, and will crack (under tension) or buckle (under compression) part way across any section through which the resultant of the loads passes at a point lying outside the kern.

For any section not shown in Table 12.1, a chart may be constructed showing the relation between e and x; this is done by assuming successive positions of the neutral axis (parallel to one principal axis) and solving for the corresponding eccentricity by the relation $b = I/M$, where b = distance from the neutral axis to the point of application of the load (assumed to be on the other principal axis), I = moment of inertia, and M = the statical moment about the neutral axis of that part of the section carrying stress. The position of the neutral axis for any given eccentricity being known, the maximum stress can be found by the relation $\sigma_{max} = Px/M$. These equations simply express the facts stated above—that the center of gravity of the stress solid lies on the line of action of P, and that the volume of the stress solid is equal to P. The procedure outlined is simple in principle but rather laborious when applied to any except the simpler type of section since both M and I may have to be determined by graphical integration.

The method of solution just outlined and all the formulas of Table 12.1 are based on the assumption that the load is applied on one of the principal axes of the section. If the load is applied outside the kern and on neither principal axis, solution is more difficult because neither the position nor the direction of the neutral axis corresponding to a given position of the load is known. The following graphical method, which involves successive trials, may be used for a section of any form.

Let a prism of any section (Fig. 12.3) be loaded at any point P. Guess the position of the neutral axis NN. Draw from NN to the most remote fiber q the perpendicular aq. That part of the section on the load side of NN is under compression, and the intensity of stress varies linearly from 0 at NN to σ at q. Divide the stressed area into narrow strips of uniform width dy running parallel to NN. The total stress on any strip acts at the center of that strip and is proportional to the area of the strip $w\,dy$ and to its distance y from NN. Draw the locus of the centers of the strips bcq and mark off a length of strip extending $\frac{1}{2}\, wy/x$ to each side of this locus. This

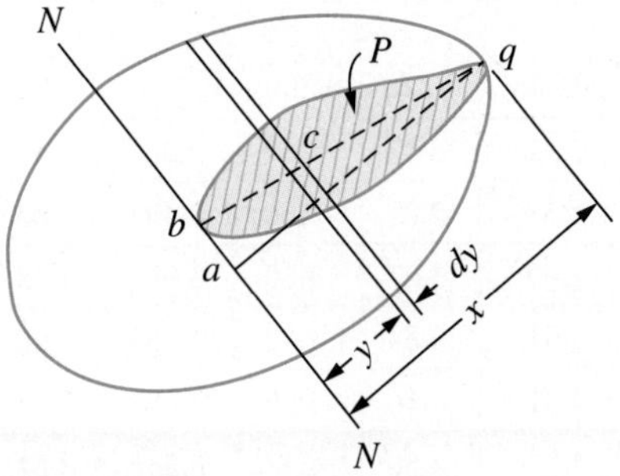

Figure 12.3

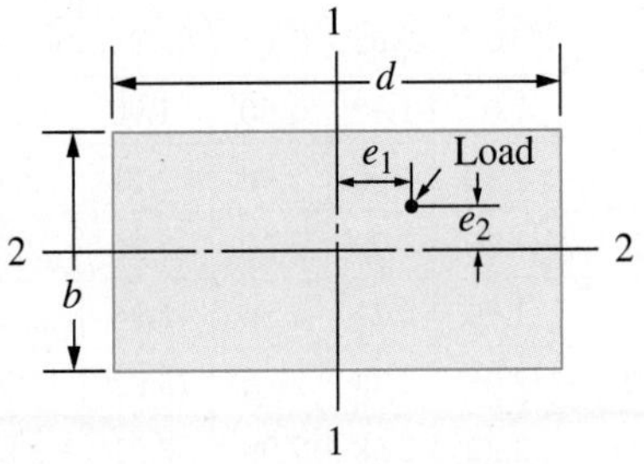

Figure 12.4

portion of the strip, if it sustained a unit stress σ, would carry the same total load as does the whole strip when sustaining the actual unit stress $\sigma y/x$ and may be called the *effective portion* of the strip. The effective portions of all strips combine to form the *effective area*, shown as the shaded portion of Fig. 12.3. Now if the assumed position of *NN* is correct, the centroid of this effective area will coincide with the point *P* and the maximum stress σ will then be equal to the load *P* divided by the effective area.

To ascertain whether or not the centroid of the effective area does coincide with *P*, trace its outline on stiff cardboard; then cut out the piece so outlined and balance it on a pin thrust through at *P*. If the piece balances in any position, *P* is, of course, the centroid. Obviously the chance of guessing the position of *NN* correctly at the first attempt is remote, and a number of trials are likely to be necessary. Each trial, however, enables the position of *NN* to be estimated more closely, and the method is less tedious than might be supposed.

For a solid rectangular section, Esling (Ref. 15) explains a special method of analysis and gives tabulated constants which greatly facilitate solution for this particular case. The coefficient *K*, by which the average stress P/A is multiplied to give the maximum stress σ, is given as a function of the eccentric ratios e_1/d and e_2/b, where the terms have the meaning shown by Fig. 12.4. The values of *K*, taken from Esling's paper, are as shown in the table given on the next page.

EXAMPLE A bridge pier of masonry, 80 ft high, is rectangular in section, measuring at the base 20 by 10 ft, the longer dimension being parallel to the track. This pier is subjected to a vertical load *P* (including its own weight) of 1500 tons, a horizontal braking load (parallel to the track) of 60 tons, and a horizontal wind load P_z (transverse to the track) of 50 tons. It is required to determine the maximum compressive stress at the base of the pier, first assuming that the masonry can sustain tension and, second, that it cannot.

SOLUTION For convenience in numerical work, the ton will be retained as the unit of force and the foot as the unit of distance.

(*a*) *Masonry can sustain tension:* Take $d = 20$ ft, and $b = 10$ ft, and take axes 1 and 2 as shown in Fig. 12.5. Then, with respect to axis 1, the bending moment $M_1 = 60 \times 80 = 4800$ ton-ft, and the section modulus $(I/c)_1 = 1/6(10)(20^2) = 667$ ft^3. With respect to axis 2, the bending moment $M_2 = 50 \times 80 = 4000$ ton-ft, and the section modulus $(I/c)_2 = 1/6(20)(10^2) = 333$ ft^3. The section area is $10 \times 20 = 200$ ft^2. The maximum stress obviously occurs at the corner where both bending moments cause compression and is

$$\sigma = \frac{1500}{200} + \frac{4800}{667} + \frac{4000}{333} = 7.5 + 7.2 + 12 = 26.7 \text{ tons/ft}^2$$

e_2/b \ e_1/d	0	0.05	0.10	0.15	0.175	0.200	0.225	0.250	0.275	0.300	0.325	0.350	0.375	0.400
0	1.0	1.30	1.60	1.90	2.05	2.22	2.43	2.67	2.96	3.33	3.81	4.44	5.33	6.67
0.05	1.30	1.60	1.90	2.21	2.38	2.58	2.81	3.09	3.43	3.87	4.41	5.16	6.17	7.73
0.10	1.60	1.90	2.20	2.56	2.76	2.99	3.27	3.60	3.99	4.48	5.14	5.99	7.16	9.00
0.15	1.90	2.21	2.56	2.96	3.22	3.51	3.84	4.22	4.66	5.28	6.03	7.04	8.45	10.60
0.175	2.05	2.38	2.76	3.22	3.50	3.81	4.16	4.55	5.08	5.73	6.55	7.66	9.17	11.50
0.200	2.22	2.58	2.99	3.51	3.81	4.13	4.50	4.97	5.54	6.24	7.12	8.33	9.98	
0.225	2.43	2.81	3.27	3.84	4.16	4.50	4.93	5.18	6.05	6.83	7.82	9.13	10.90	
0.250	2.67	3.09	3.60	4.22	4.55	4.97	5.48	6.00	6.67	7.50	8.57	10.0	12.00	
0.275	2.96	3.43	3.99	4.66	5.08	5.54	6.05	6.67	7.41	8.37	9.55	11.1		
0.300	3.33	3.87	4.48	5.28	5.73	6.24	6.83	7.50	8.37	9.37	10.80			
0.325	3.81	4.41	5.14	6.03	6.55	7.12	7.82	8.57	9.55	10.80				
0.350	4.44	5.16	5.99	7.04	7.66	8.33	9.13	10.00	11.10					
0.375	5.33	6.17	7.16	8.45	9.17	9.98	10.90	12.00						
0.400	6.67	7.73	9.00	10.60	11.50									

By double linear interpolation, the value of K for any eccentricity within the limits of the table may readily be found.

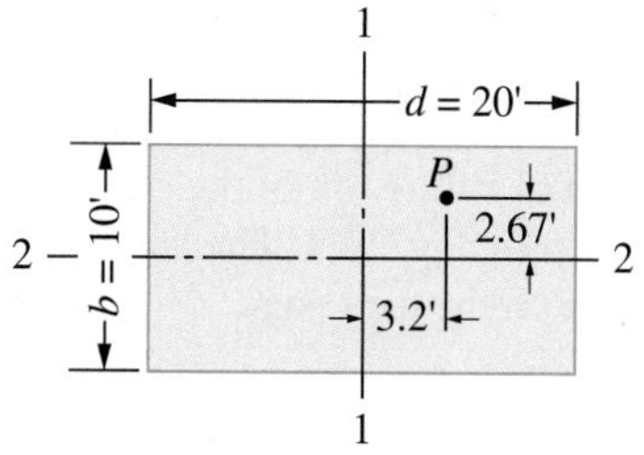

Figure 12.5

(*b*) *Masonry cannot sustain tension:* The resultant of the loads pierces the base section of the pier at point P, at a distance $e_1 = 60 \times 80/1500 = 3.2$ ft from axis 1 at $e_2 = 50 \times 80/1500 = 2.67$ ft from axis 2. This resultant is resolved at point P into rectangular components, the only one of which causing compression is the vertical component, equal to 1500, with eccentricities e_1 and e_2. The eccentric ratios are $e_1/d = 0.16$ and $e_2/b = 0.267$. Referring to the tabulated values of K, linear interpolation between $e_1/d = 0.15$ and 0.175 at $e_2/b = 0.250$ gives $K = 4.35$. Similar interpolation at $e_2/b = 0.275$ gives $K = 4.83$. Linear interpolation between these values gives $K = 4.68$ as the true value at $e_2/b = 0.267$. The maximum stress is therefore

$$\sigma = \frac{KP}{A} = 4.68 \times \frac{1500}{200} = 35.1 \text{ tons/ft}^2$$

12.8 TABLE

TABLE 12.1 Formulas for Short Prisms Loaded Eccentrically; Stress Reversal Impossible

Notation: m and n are dimensions of the kern which is shown shaded in each case; x is the distance from the most stressed fiber to the neutral axis; and A is the net area of the section. Formulas for x and for maximum stress assume the prism to be subjected to longitudinal load P acting outside the kern on one principal axis of the section and at a distance e from the other principal axis.

Form of Section, Form of Kern, and Case No.	Formulas for m, n, x, and σ_{max}
1. Solid rectangular section 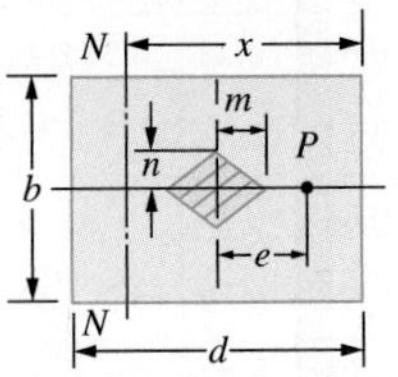	$m=\frac{1}{6}d \quad n=\frac{1}{6}b$ $x=3\left(\frac{1}{2}d-e\right)$ $\sigma_{max}=\frac{P}{A}\frac{4d}{3d-6e}$
2. Hollow rectangular section 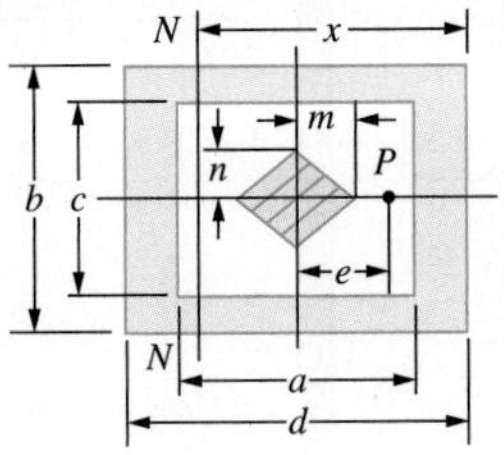	$m=\frac{1}{6}\frac{bd^3-ca^3}{d(db-ac)}, \quad n=\frac{1}{6}\frac{db^3-ac^3}{b(db-ac)}$ x satisfies $\frac{e}{d}=\begin{cases}\frac{1}{2}-\frac{\frac{1}{6}bx^3-\frac{1}{2}a^2c(\frac{1}{2}d-\frac{1}{6}a)-\frac{1}{2}acd(x-\frac{1}{2}a-\frac{1}{2}d)}{d[\frac{1}{2}bx^2-ac(x-\frac{1}{2}d)]} & \text{if } \frac{a+d}{2}<x<d \\ \frac{1}{2}-\frac{\frac{1}{6}bx^3-\frac{1}{2}c(x-\frac{1}{2}d-\frac{1}{2}a)^2(\frac{1}{3}x+\frac{1}{6}d-\frac{1}{6}a)}{d[\frac{1}{2}bx^2-\frac{1}{2}c(x-\frac{1}{2}d-\frac{1}{2}a)^2]} & \text{if } \frac{d-a}{2}<x<\frac{a+d}{2}\end{cases}$ $\sigma_{max}=\begin{cases}\frac{P}{\frac{1}{2}bx-ac(x-\frac{1}{2}d)/x} & \text{if } \frac{a+d}{2}<x<d \\ \frac{P}{\frac{1}{2}bx-c(x-\frac{1}{2}d+\frac{1}{2}a)^2/2x} & \text{if } \frac{d-a}{2}<x<\frac{a+d}{2}\end{cases}$
3. Thin-walled rectangular shell 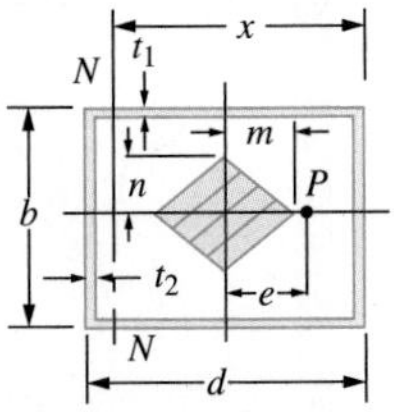	$m=\frac{1}{6}d\left(\frac{dt_1+3bt_2}{dt_1+bt_2}\right), \quad n=\frac{1}{6}b\left(\frac{bt_2+3dt_1}{dt_1+bt_2}\right)$ $x=\frac{1}{2}\left(\frac{3}{2}d-3e\right)+\sqrt{b\left(\frac{3}{2}d\frac{t_2}{t_1}-3e\frac{t_2}{t_1}\right)+\frac{1}{4}\left(\frac{3}{2}d-3e\right)^2}$ $\sigma_{max}=\frac{P}{xt_1+bt_2}$
4. Solid circular section 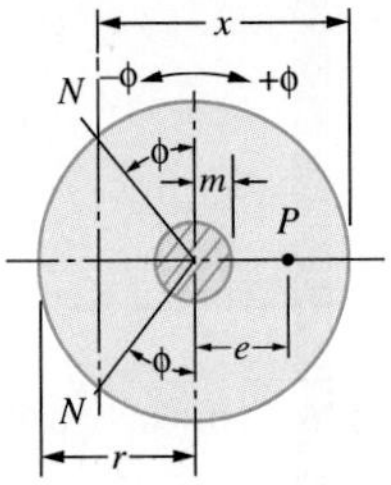	$m=\frac{r}{4}$ $x=r(1-\sin\phi)$ where ϕ satisfies $\frac{e}{r}=\frac{\frac{1}{8}\pi-\frac{1}{4}\phi-\frac{5}{12}\sin\phi\cos\phi+\frac{1}{6}\sin^3\phi\cos\phi}{\cos\phi-\frac{1}{3}\cos^3\phi-\frac{1}{2}\pi\sin\phi+\phi\sin\phi}$ $\sigma_{max}=\frac{P}{A}\frac{\pi(1-\sin\phi)}{A\cos\phi-\frac{1}{3}\cos^3\phi-\frac{1}{2}\pi\sin\phi+\phi\sin\phi}$ or $\sigma_{max}=\frac{P}{A}K$ where K is given by following table:

e/r	0.25	0.30	0.35	0.40	0.45	0.50	0.55	0.60	0.65	0.70	0.75	0.80	0.85	0.90
x/r	2.00	1.82	1.66	1.51	1.37	1.23	1.10	0.97	0.84	0.72	0.60	0.47	0.35	0.24
K	2.00	2.21	2.46	2.75	3.11	3.56	4.14	4.90	5.94	7.43	9.69	13.4	20.5	37.5

(*Continued*)

TABLE 12.1 Formulas for Short Prisms Loaded Eccentrically; Stress Reversal Impossible (*Continued*)

Form of Section, Form of Kern, and Case No.	Formulas for m, n, x, and σ_{max}

5. Hollow circular section

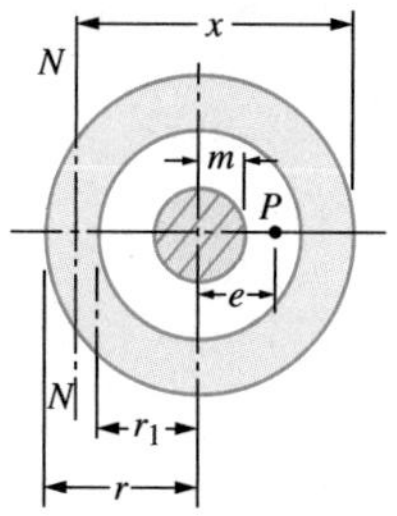

$$m = \frac{\frac{1}{4}(r^2 + r_1^2)}{r}$$

$$\sigma_{max} = \frac{P}{A}K \quad \text{where } K \text{ is given by following table:}$$

r_1/r		0.29	0.30	0.34	0.35	0.40	0.41	0.45	0.50	0.55	0.60	0.65	0.70	0.75	0.80	0.85	0.90
		e/r															
0.4	x/r	2.00	1.97		1.81	1.67		1.53	1.38	1.22	1.05	0.88	0.73	0.60	0.48	0.35	0.24
	K	2.00	2.03		2.22	2.43		2.68	2.99	3.42	4.03	4.90	6.19	8.14	11.3	17.3	31.5
0.6	x/r			2.00	1.97	1.84		1.71	1.56	1.39	1.21	1.02	0.82	0.64	0.48	0.35	0.24
	K			2.00	2.03	2.18		2.36	2.58	2.86	3.24	3.79	4.64	6.04	8.54	13.2	24.0
0.8	x/r						2.00	1.91	1.78	1.62	1.45	1.26	1.05	0.84	0.63	0.41	0.24
	K						2.00	2.10	2.24	2.42	2.65	2.94	3.34	3.95	4.98	7.16	13.4
1.0	x/r								2.00	1.87	1.71	1.54	1.35	1.15	0.94	0.72	0.48
	K								2.00	2.12	2.26	2.43	2.63	2.90	3.27	3.79	4.68

6. Thin-walled circular shell

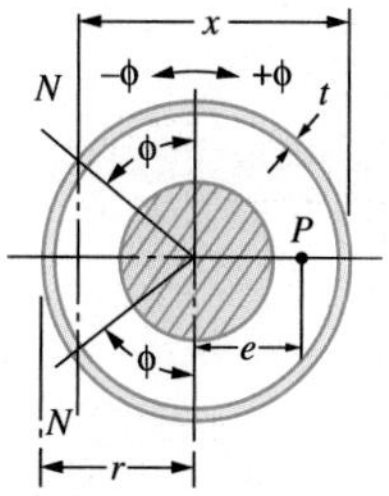

$$m = \frac{1}{2}r$$

$$x = r(1 - \sin\phi) \quad \text{where } \phi \text{ satisfies} \quad \frac{e}{r} = \frac{\frac{1}{2}\pi - \phi - \sin\phi\cos\phi}{2\cos\phi - \pi\sin\phi + 2\phi\sin\phi}$$

$$\sigma_{max} = \frac{P}{tr}\frac{1-\sin\phi}{2\cos\phi - \pi\sin\phi + 2\phi\sin\phi} \quad \text{or} \quad \sigma_{max} = \frac{P}{A}K$$

where K is given by the above table (case 5) for $\frac{r_1}{r} = 1$

7. Solid isosceles triangle

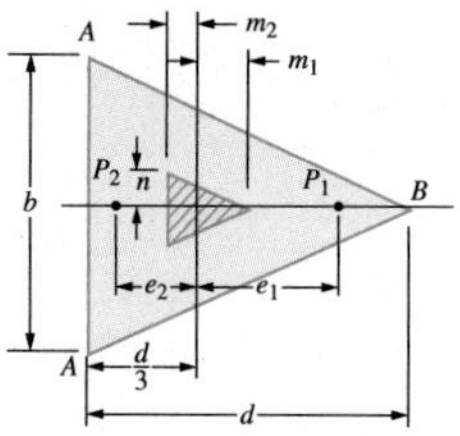

$$m_1 = \frac{d}{6}, \quad m_2 = \frac{d}{12}, \quad n = \frac{b}{8}, \quad A = \frac{bd}{2}$$

For the load right of centroid and kern, i.e., $e_1 > m_1$:

Distance from vertex to neutral axis $= \frac{4d}{3} - 2e_1$

$$\sigma_{max} = \sigma_B = \frac{P_1}{A}\frac{6.75}{(2 - 3e_1/d)^2}$$

For the load left of centroid and kern, i.e., $e_2 > m_2$:

Distance from base to neutral axis $= x_2$ where x_2 satisfies

$$\frac{x_2}{2d}\frac{2 - x_2/d}{3 - x_2/d} = \frac{1}{3} - \frac{e_2}{d}$$

$$\sigma_{max} = \sigma_A = \frac{P_2}{A}\frac{3d}{x_2(3 - x_2/d)} \quad \text{or} \quad \sigma_{max} = \frac{P}{A}K$$

where K is given by the following table:

e_2/d	0.083	0.100	0.125	0.150	0.175	0.200	0.225	0.250	0.275	0.300	0.325
x_2/d	1.000	0.885	0.750	0.635	0.532	0.437	0.348	0.263	0.181	0.102	0.025
K	1.500	1.602	1.778	1.997	2.283	2.677	3.252	4.173	5.886	10.17	40.18

(*Continued*)

TABLE 12.1 Formulas for Short Prisms Loaded Eccentrically; Stress Reversal Impossible (*Continued*)

Form of Section, Form of Kern, and Case No.	Formulas for m, n, x, and σ_{max}
8. Solid trapezoidal section 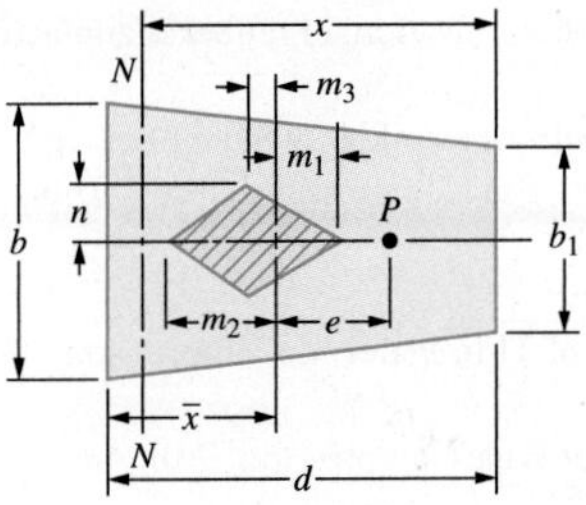	$q=\frac{b_1}{b}$, $\bar{x}=\frac{d}{3}\frac{1+2q}{1+q}$, $n=\frac{b}{8}\frac{1+q+q^2+q^3}{1+q+q^2}$ $m_1=\frac{d}{6}\frac{1+4q+q^2}{1+3q+2q^2}$, $m_2=\frac{d}{6}\frac{1+4q+q^2}{2+3q+q^2}$, $m_3=\frac{d}{12}\frac{1+3q-3q^2-q^3}{1+2q+2q^2+q^3}$ $A=\frac{d(b+b_1)}{2}$ $\sigma_{max}=\frac{P}{A}\frac{3(1+q)}{3qx/d+(1-q)(x/d)^2}$ where x satisfies $\frac{e}{d}=\frac{2+q}{3(1+q)}-\frac{x}{2d}\frac{2q+(1-q)x/d}{3q+(1-q)x/d}$ or $\sigma_{max}=\frac{P}{A}K$ where K is given by following table:

q	$\bar{x}/d$	m_1/d	m_2/d	m_3/d	n/b	x/d	1.000	0.900	0.800	0.700	0.600	0.500	0.400	0.300	0.200
0.40	0.429	0.183	0.137	0.063	0.130	e/d	0.183	0.225	0.267	0.308	0.348	0.388	0.427	0.465	0.502
						K	2.333	2.682	3.125	3.704	4.487	5.600	7.292	10.14	15.91
0.80	0.481	0.172	0.160	0.018	0.151	e/d	0.172	0.208	0.244	0.279	0.314	0.349	0.383	0.417	0.451
						K	2.077	2.326	2.637	3.037	3.571	4.320	5.444	7.317	11.07
1.25	0.519	0.160	0.172	−0.018	0.189	e/d	0.160	0.191	0.222	0.254	0.286	0.318	0.350	0.383	0.415
						K	1.929	2.128	2.377	2.697	3.135	3.724	4.623	6.122	9.122
2.50	0.571	0.137	0.183	−0.063	0.325	e/d	0.137	0.161	0.187	0.214	0.242	0.271	0.301	0.332	0.363
						K	1.750	1.897	2.083	2.326	2.652	3.111	3.804	4.965	7.292

12.9 REFERENCES

1. White, R. N., and C. G. Salmon (eds.), *Building Structural Design Handbook*, John Wiley & Sons, 1987.
2. Specification for the Design, Fabrication, and Erection of Structural Steel for Buildings, with Commentary, American Institute of Steel Construction, 1978.
3. Specification for the Design of Cold-Formed Steel Structural Members, American Iron and Steel Institute, September 1980.
4. *Cold-Formed Steel Design Manual*, Part I, Specification; Part II, Commentary; Part III, Supplementary Information; Part IV, Illustrative Examples; and Part V, Charts and Tables, American Iron and Steel Institute, November 1982.
5. Specifications for Aluminum Structures, Sec. 1, *Aluminum Construction Manual*, 4th ed., The Aluminum Association, Inc., 1982.
6. *Wood Handbook*, Forest Products Laboratory, U.S. Dept. of Agriculture, 1987.
7. American Institute of Steel Construction, *Manual of Steel Construction—Load and Resistance Factor Design*, 1st ed., 1986.
8. "Final Report of the Special Committee on Steel Column Research," *Trans. Am. Soc. Civil Eng.*, 98: 1376, 1933.
9. "Suggested Specifications for Structures of Aluminum Alloys 6061-T6 and 6062-T6," *Proc. Am. Soc. Civil Eng.*, ST6, paper 3341, vol. 88, December 1962.
10. Specifications for Steel Railway Bridges, American Railway Association, 1950.
11. Narayanan, R. (ed.), *Axially Compressed Structures: Stability and Strength*, Elsevier Science, 1982.
12. Thompson, J. M. T., and G. W. Hunt (eds.), *Collapse: The Buckling of Structures in Theory and Practice*, IUTAM, Cambridge University Press, 1983.

13. Kilpatrick, S. A., and O. U. Schaefer, "Stress Calculations for Thin Aluminum Alloy Sections," *Prod. Eng.*, February, March, April, May, 1936.
14. Johnston, R. S., "Compressive Strength of Column Web Plates and Wide Web Columns," *Tech. Paper Bur. Stand.*, 327, 1926.
15. Esling, K. E., "A Problem Relating to Railway-Bridge Piers of Masonry or Brickwork," *Proc. Inst. Civil Eng.*, 165: 219, 1905–1906.
16. National Design Specification for Stress-Grade Lumber and Its Fastenings, National Lumber Manufacturers Association, 1960.
17. Salmon, E. H., *Columns*, Oxford Technical Publications, Henry Frowde and Hodder & Stoughton, 1921.
18. Talbot, A. N., and H. F. Moore, "Tests of Built-Up Steel and Wrought Iron Compression Pieces," *Trans. Am. Soc. Civil Eng.*, 65: 202, 1909. (Also Univ. Ill. Eng. Exp. Sta., Bull. 44.)
19. Timoshenko, S., *Strength of Materials*, D. Van Nostrand Company, Inc., 1930.
20. von Kármán, Th., E. E. Sechler, and L. H. Donnell, "The Strength of Thin Plates in Compression," *Trans. ASME*, 54(2): 53, 1932.
21. Schuman, L., and G. Black, "Strength of Rectangular Flat Plates under Edge Compression," *Nat. Adv. Comm. Aeron.*, Rept. 356, 1930.
22. Sechler, E. E., "A Preliminary Report on the Ultimate Compressive Strength of Curved Sheet Panels," Guggenheim Aeron. Lab., Calif. Inst. Tech., Publ. 36, 1937.
23. Sechler, E. E., "Stress Distribution in Stiffened Panels under Compression," *J. Aeron. Sci.*, 4(8): 320, 1937.

CHAPTER 13

Shells of Revolution; Pressure Vessels; Pipes

13.1 CIRCUMSTANCES AND GENERAL STATE OF STRESS

The discussion and formulas in this section apply to any vessel that is a figure of revolution. For convenience of reference, a line that represents the intersection of the wall and a plane containing the axis of the vessel is called a *meridian*, and a line representing the intersection of the wall and a plane normal to the axis of the vessel is called a *circumference*. Obviously the meridian through any point is perpendicular to the circumference through that point.

When a vessel of the kind under consideration is subjected to a distributed loading, such as internal or external pressure, the predominant stresses are membrane stresses, that is, stresses constant through the thickness of the wall. There is a meridional membrane stress σ_1 acting parallel to the meridian, a circumferential, or hoop, membrane stress σ_2 acting parallel to the circumference, and a generally small radial stress σ_3 varying through the thickness of the wall. In addition, there may be bending and/or shear stresses caused by loadings or physical characteristics of the shell and its supporting structure. These include (1) concentrated loads, (2) line loads along a meridian or circumference, (3) sudden changes in wall thickness or an abrupt change in the slope of the meridian, (4) regions in the vessel where a meridian becomes normal to or approaches being normal to the axis of the vessel, and (5) wall thicknesses greater than those considered thin-walled, resulting in variations of σ_1 and σ_2 through the wall.

In consequence of these stresses, there will be meridional, circumferential, and radial strains leading to axial and radial deflections and changes in meridional slope. If there is axial symmetry of both the loading and the vessel, there will be no tendency for any circumference to depart from the circular form unless buckling occurs.

13.2 THIN SHELLS OF REVOLUTION UNDER DISTRIBUTED LOADINGS PRODUCING MEMBRANE STRESSES ONLY

If the walls of the vessel are relatively thin (less than about one-tenth the smaller principal radius of curvature) and have no abrupt changes in thickness, slope, or curvature, and if the loading is uniformly distributed or smoothly varying and axisymmetric, the stresses σ_1 and σ_2 are practically uniform throughout the thickness of the wall and are the only important ones present; the radial stress σ_3 and such bending stresses as occur are negligibly small. Table 13.1 gives formulas for the stresses and deformations under loadings such as those just described for cylindrical, conical, spherical, and toroidal vessels as well as for general smooth figures of revolution as listed under case 4.

If two thin-walled shells are joined to produce a vessel, and if it is desired to have no bending stresses at the joint under uniformly distributed or smoothly varying loads, then it is necessary to choose shells for which the radial deformations and the rotations of the meridians are the same for each shell at the point of connection. For example, a *cylindrical* shell under uniform internal pressure will have a radial deformation of $qR^2(1-\nu/2)/Et$ while a *hemispherical* head of equal thickness under the same pressure will have a radial deformation of $qR^2(1-\nu)/2Et$; the meridian rotation ψ is zero in both cases. This mismatch in radial deformation will produce bending and shear stresses in the near vicinity of the joint. An examination of case 4a (Table 13.1) shows that if R_1 is infinite at $\theta = 90°$ for a smooth figure of revolution, the radial deformation and the rotation of the meridian will match those of the cylinder.

Flügge (Ref. 5) points out that the family of *cassinian* curves has the property just described. He also discusses in some detail the *ogival* shells, which have a constant radius of curvature R_1 for the meridian but for which R_2 is a variable. If R_2 is everywhere less than R_1, the ogival shell has a pointed top, as shown in Fig. 13.1(a). If R_2 is infinite, as it is at point A in Fig. 13.1(b), the center of the shell must be supported to avoid large bending stresses although some bending stresses are still present in the vicinity of point A. For more details of these deviations from membrane action see Refs. 66 and 74–76.

For very thin shells where bending stresses are negligible, a nonlinear membrane theory can provide more realistic values near the crown, point A. Rossettos and Sanders have carried out such a solution (Ref. 52). Chou and Johnson (Ref. 57) have examined large deflections of elastic toroidal membranes of a type used in some sensitive pressure-measuring devices.

Galletly in Ref. 67 shows that simple membrane theory is not adequate for the stress analysis of most torispherical pressure vessels. Ranjan and Steele in Ref. 68 have worked with asymptotic expansions and give a simple design formula for the maximum stress in the toroidal segment which is in good agreement with experimental and numerical studies. They present a simple condition that gives the optimum knuckle radius for prescribed spherical cap and cylinder geometries and also give expressions leading to a lower limit for critical internal pressure at which wrinkles are formed due to circumferential compression in the toroid.

Baker, Kovalevsky, and Rish (Ref. 6) give formulas for toroidal segments, ogival shells, elliptical shells, and Cassini shells under various loadings; all these cases can be evaluated from case 4 of Table 13.1 once R_1 and R_2 are calculated. In addition to the axisymmetric shells considered in this chapter, Refs. 5, 6, 45, 59, 66, 74, 81, and 82 discuss in some detail the membrane stresses in nonaxisymmetric shells, such as barrel vaults, elliptic cylinders, and hyperbolic paraboloids.

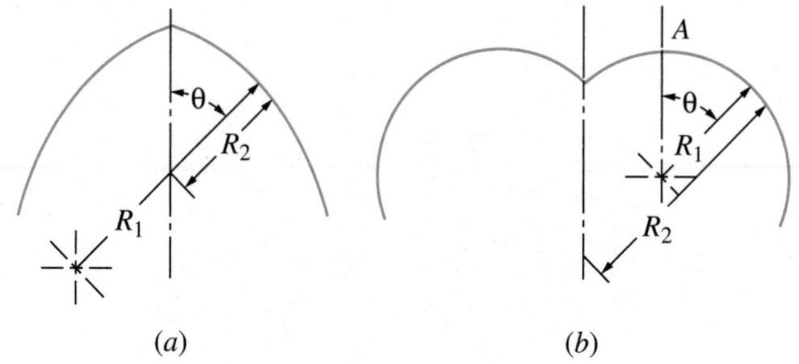

Figure 13.1

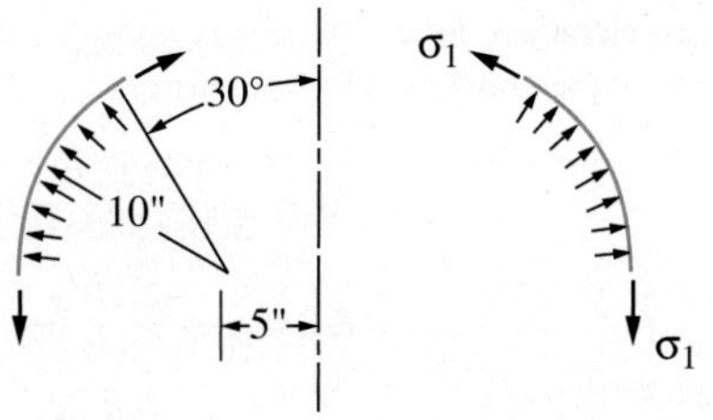

Figure 13.2

EXAMPLE 1 A segment of a toroidal shell shown in Fig. 13.2 is to be used as a transition between a cylinder and a head closure in a thin-walled pressure vessel. To properly match the deformations, it is desired to know the change in radius and the rotation of the meridian at both ends of the toroidal segment under an internal pressure loading of 200 lb/in². Given: $E = 30(10^6)\,\text{lb/in}^2$, $\nu = 0.3$, and the wall thickness $t = 0.1$ in.

SOLUTION Since this particular case is not included in Table 13.1, the general case 4a can be used. At the upper end $\theta = 30°$, $R_1 = 10$ in, and $R_2 = 10 + 5/\sin 30° = 20$ in; therefore,

$$\Delta R_{30°} = \frac{200(20^2)(0.5)}{2(30)(10^6)(0.1)}\left(2 - \frac{20}{10} - 0.3\right) = -0.002\,\text{in}$$

Since R_1 is a constant, $dR_1/d\theta = 0$ throughout the toroidal segment; therefore,

$$\psi_{30°} = \frac{200(20^2)}{2(30)(10^6)(0.1)(10)(0.577)}\left[3\frac{10}{20} - 5 + \frac{20}{10}(2+0)\right] = 0.00116\,\text{rad}$$

At the lower end, $\theta = 90°$, $R_1 = 10$ in, and $R_2 = 15$ in; therefore,

$$\Delta R_{90°} = \frac{200(15^2)(1)}{2(30)(10^6)(0.1)}\left(2 - \frac{15}{10} - 0.3\right) = 0.0015\,\text{in}$$

Since tan 90° = infinity and $dR_1/d\theta = 0$, $\psi_{90°} = 0$. In this problem $R_2/R_1 \le 2$, so the value of σ_2 is never compressive, but this is not always true. One must check for the possibility of circumferential buckling.

EXAMPLE 2 The truncated thin-walled cone shown in Fig. 13.3 is supported at its base by the membrane stress σ_1. The material in the cone weighs 0.10 lb/in³, $t = 0.25$ in, $E = 10(10^6)\,\text{lb/in}^2$, and Poisson's ratio is 0.3. Find the stress σ_1 at the base, the change in radius at the base, and the change in height of the cone if the cone is subjected to an acceleration parallel to its axis of 399g.

SOLUTION Since the formulas for a cone loaded by its own weight are given only for a complete cone, superposition will have to be used. From Table 13.1 cases 2c and 2d will be applicable. First take a complete cone loaded by its own weight with its density multiplied by 400 to account for the

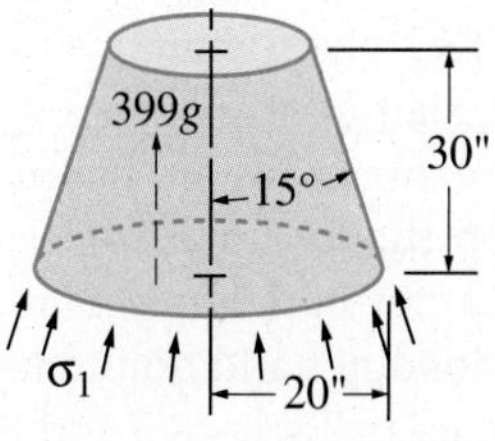

Figure 13.3

acceleration. Since the vertex is up instead of down, a negative value can be used for δ. $R = 20$ in, $\delta = -40.0$, and $\alpha = 15°$; therefore,

$$\sigma_1 = \frac{-40(20)}{2\cos 15°\cos 15°} = -1600 \text{ lb/in}^2$$

$$\Delta R = \frac{-40(20^2)}{10(10^6)\cos 15°}\left(\sin 15° - \frac{0.3}{2\sin 15°}\right) = 0.000531 \text{ in}$$

$$\Delta y = \frac{-40(20^2)}{10(10^6)\cos^2 15°}\left(\frac{1}{4\sin^2 15°} - \sin^2 15°\right) = -0.00628 \text{ in}$$

Next we find the radius of the top as 11.96 in and calculate the change in length and effective weight of the portion of the complete cone to be removed. $R = 11.96$ in, $\delta = -40.0$, and $\alpha = 15°$; therefore,

$$\Delta y = -0.00628\left(\frac{11.96}{20}\right)^2 = -0.00225 \text{ in}$$

The volume of the removed cone is

$$\frac{11.96}{\sin 15°}\frac{11.96(2\pi)}{2}(0.25) = 434 \text{ in}^3$$

and the effective weight of the removed cone is $434(0.1)(400) = 17{,}360$ lb.

Removing the load of 17,360 lb can be accounted for by using case 2d, where $P = 17{,}360$, $R = 20$ in, $r = 11.96$ in, $h = 30$ in, and $\alpha = 15°$:

$$\sigma_1 = \frac{17{,}360}{2\pi(20)(0.25)\cos 15°} = 572 \text{ lb/in}^2$$

$$\Delta R = \frac{-0.3(17{,}360)}{2\pi(10)(10^6)(0.25)\cos 15°} = -0.000343 \text{ in}$$

$$\Delta h = \frac{17{,}360 \ln(20/11.96)}{2\pi(10)(10^6)(0.25)\sin 15°\cos^2 15°} = 0.002353 \text{ in}$$

Therefore, for the truncated cone,

$$\sigma_1 = -1600 + 572 = -1028 \text{ lb/in}^2$$

$$\Delta R = 0.000531 - 0.000343 = 0.000188 \text{ in}$$

$$\Delta h = -0.00628 + 0.00225 + 0.002353 = -0.00168 \text{ in}$$

13.3 THIN SHELLS OF REVOLUTION UNDER CONCENTRATED OR DISCONTINUOUS LOADINGS PRODUCING BENDING AND MEMBRANE STRESSES

Cylindrical Shells

Table 13.2 gives formulas for forces, moments, and displacements for several *axisymmetric* loadings on both *long* and *short* thin-walled *cylindrical* shells having free ends. These expressions are based on differential equations similar in form to those used to develop the formulas for beams on elastic foundations in Chap. 8. To avoid excessive redundancy in the presentation, only the free-end cases are given in this chapter, but all of the loadings and boundary conditions listed in Tables 8.5 and 8.6 as well as the tabulated data in Tables 8.3 and 8.4 are directly applicable to cylindrical shells by substituting the shell parameters λ and D for the beam parameters β and EI, respectively. (This will be demonstrated in the examples which follow.) Since many loadings on cylindrical shells occur at the ends, note carefully on page 171 the modified numerators to be used in the equations in Table 8.5 for the condition when $a = 0$. A special application of this would be the situation where one end of a cylindrical shell is forced to increase a known amount in radius while maintaining zero slope at that same end. This reduces to an application of an externally created concentrated lateral displacement Δ_0 at $a = 0$ (Table 8.5, case 6) with the left end fixed. (See Example 4.)

Pao (Ref. 60) has tabulated influence coefficients for short cylindrical shells under edge loads with wall thicknesses varying according to $t = Cx^n$ for values of $n = ¼ (¼) (2)$ and for values of t_1/t_2 of 2, 3, and 4. Various degrees of taper are considered by representing data for $k = 0.2(0.2)(1.0)$ where $k^4 = 3(1-\nu^2)x_1^4/R^2t_1^2$. Stanek (Ref. 49) has tabulated similar coefficients for constant-thickness cylindrical shells.

A word of caution is in order at this point. The original differential equations used to develop the formulas presented in Table 13.2 were based on the assumption that radial deformations were small. If the magnitude of the radial deflection approaches the wall thickness, the accuracy of the equations declines. In addition, if axial loads are involved on a relatively short shell, the moments of these axial loads might have an appreciable effect if large deflections are encountered. The effects of these moments are not included in the expressions given.

EXAMPLE 1 A steel tube with a 4.180-in outside diameter and a 0.05-in wall thickness is free at both ends and is 6 in long. At a distance of 2 in from the left end a steel ring with a circular cross section is shrunk onto the outside of the tube such as to compress the tube radially inward a distance of 0.001 in. The maximum tensile stress in the tube is desired. Given: $E = 30(10^6)$ lb/in² and $\nu = 0.30$.

SOLUTION We calculate the following constants:

$$R = 2.090 - 0.025 = 2.065$$

$$\lambda = \left[\frac{3(1-0.3^2)}{2.065^2(0.05^2)}\right]^{1/4} = 4.00$$

$$D = \frac{30(10^6)(0.05^3)}{12(1-0.3^2)} = 344$$

Since $6/\lambda = 6/4.0 = 1.5$ in and the closest end of the tube is 2 in from the load, this can be considered a very long tube. From Table 13.2, case 15 indicates that both the maximum deflection and the maximum moment are under the load, so that

$$-0.001 = \frac{-p}{8(344)(4.00^3)} \quad \text{or} \quad p = 176 \text{ lb/in}$$

$$M_{\max} = \frac{176}{4(4)} = 11.0 \text{ in-lb/in}$$

At the cross section under the load and on the inside surface, the following stresses are present:

$$\sigma_1 = 0$$

$$\sigma_1' = \frac{6M}{t^2} = \frac{6(11.0)}{0.05^2} = 26{,}400 \text{ lb/in}^2$$

$$\sigma_2 = \frac{yE}{R} + \nu\sigma_1 = \frac{-0.001(30)(10^6)}{2.065} = -14{,}500 \text{ lb/in}^2$$

$$\sigma_2' = 0.30(26{,}400) = 7920 \text{ lb/in}^2$$

The principal stresses on the inside surface are 26,400 and −6580 lb/in².

EXAMPLE 2 Given the same tube and loading as in Example 1, except the tube is only 1.2 in long and the ring is shrunk in place 0.4 in from the left end, the maximum tensile stress is desired.

SOLUTION Since both ends are closer than $6/\lambda = 1.5$ in from the load, the free ends influence the behavior of the tube under the load. From Table 13.2, case 2 applies in this example, and since the deflection under the load is the given value from which to work, we must evaluate y at $x = a = 0.4$ in. Note that $\lambda l = 4.0(1.2) = 4.8$, $\lambda x = \lambda a = 4.0(0.4) = 1.6$, and $\lambda(l-a) = 4.0(1.2-0.4) = 3.2$. Also,

$$y = y_A F_1 + \frac{\psi_A}{2\lambda} + LT_y$$

$$y_A = \frac{-p}{2D\lambda^3} \frac{C_3 C_{a2} - C_4 C_{a1}}{C_{11}}$$

$$\psi_A = \frac{p}{2D\lambda^2} \frac{C_3 C_{a2} - C_4 C_{a1}}{C_{11}}$$

where

$$\begin{aligned} C_3 &= -60.51809 && \text{(from Table 8.3, under } F_3 \text{ for } \beta x = 4.8) \\ C_{a2} &= -12.94222 && \text{(from Table 8.3, under } F_2 \text{ for } \beta x = 3.2) \\ C_4 &= -65.84195 \\ C_{11} &= 3689.703 \\ C_{a1} &= -12.26569 \\ C_2 &= -55.21063 \end{aligned}$$

Also F_1 (at $x = a$) $= -0.07526$ and F_2 (at $x = a$) $= 2.50700$; therefore,

$$y_A = \frac{-p}{2(344)(4.0^3)} \frac{-60.52(-12.94) - (-65.84)(-12.27)}{3689.7} = 0.154(10^{-6})p$$

$$\psi_A = \frac{-p}{2(344)(4.0^2)} \frac{-55.21(-12.94) - 2(-60.52)(-12.27)}{3689.7} = -19.0(10^{-6})p$$

and $LT_y = 0$ since x is not greater than a. Substituting into the expression for y at $x = a$ gives

$$-0.001 = 0.154(10^{-6})p(-0.07526) - \frac{19.0(10^{-6})p(2.507)}{2(4.0)} = -5.96(10^{-6})p$$

or $p = 168$ lb/in, $y_A = 0.0000259$ in, and $\psi_A = -0.00319$ rad.

Although the position of the maximum moment depends upon the position of the load, the maximum moment in this case would be expected to be under the load since the load is some distance from the free end,

$$M = -y_A 2D\lambda^2 F_3 - \psi_A D\lambda F_4 + LT_M$$

and at $x = a$, $F_3 = 2.37456$, $F_4 = 2.64573$, and $LT_M = 0$ since x is not greater than a. Therefore,

$$\begin{aligned} M_{max} &= -(0.0000259)(2)(344)(4.0^2)(2.375) - (-0.00319)(344)(4.0)(2.646) \\ &= 10.92 \text{ lb-in/in} \end{aligned}$$

At the cross section under the load and on the inside surface, the following stresses are present:

$$\sigma_1 = 0 \qquad \sigma_2 = \frac{-0.001(30)(10^6)}{2.065} = -14{,}500 \text{ lb/in}^2$$

$$\sigma_1' = \frac{6(10.92)}{0.05^2} = 26{,}200 \text{ lb/in}^2 \qquad \sigma_2' = 0.30(26{,}200) = 7{,}860 \text{ lb/in}^2$$

The small change in the maximum stress produced in this shorter tube points out how localized the effect of a load on a shell can be. Had the radial load been the same, however, instead of the radial deflection, a greater difference might have been noted and the stress σ_2 would have increased in magnitude instead of decreased.

EXAMPLE 3 A cylindrical aluminum shell is 10 in long and 15 in in diameter and must be designed to carry an internal pressure of 300 lb/in² without exceeding a maximum tensile stress of 12,000 lb/in². The ends are capped with massive flanges, which are sufficiently clamped to the shell to effectively resist any radial or rotational deformation at the ends. Given: $E = 10(10^6)$ lb/in² and $\nu = 0.3$.

FIRST SOLUTION Case 1c from Table 13.1 and cases 1 and 3 or cases 8 and 10 from Table 13.2 can be superimposed to find the radial end load and the end moment which will make the slopes and deflections at both ends zero. Figure 13.4 shows the loadings applied to the shell. First we evaluate the necessary constants:

$$R = 7.5 \text{ in}, \qquad l = 10 \text{ in}, \qquad D\frac{10(10^6)t^3}{12(1-0.3^2)} = 915{,}800t^3$$

$$\lambda = \left[\frac{3(1-0.3^2)}{7.5^2 t^2}\right]^{1/4} = \frac{0.4694}{t^{0.5}}, \qquad \lambda l = \frac{4.694}{t^{0.5}}$$

Since the thickness is unknown at this step in the calculation, we can only estimate whether the shell must be considered long or short, that is, whether the loads at one end will have any influence on the deformations at the other. To make an estimate of this effect we can calculate the wall thickness necessary for just the internal pressure. From case 1c of Table 13.1, the value of the hoop stress $\sigma_2 = \theta R/t$ can be equated to 12,000 lb/in² and the expression solved for the thickness:

$$t = \frac{300(7.5)}{12{,}000} = 0.1875 \text{ in}$$

Using this value for t gives $\lambda l = 10.84$, which would be a very long shell.

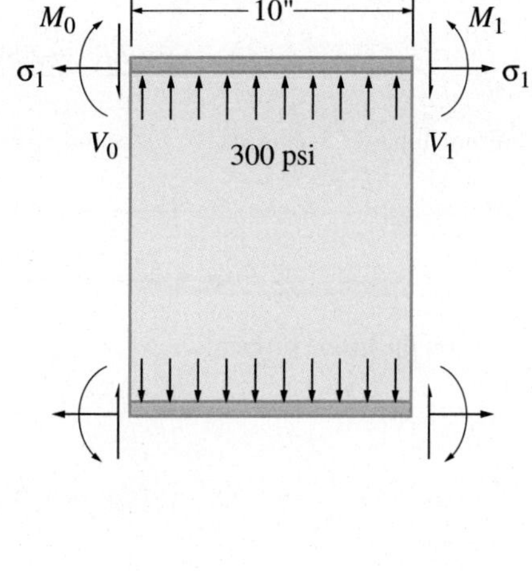

Figure 13.4

For a trial solution the assumption will be made that the radial load and bending moment at the right end do not influence the deformations at the left end. Owing to the rigidity of the end caps, the radial deformation and the angular rotation of the left end will be set equal to zero. From Table 13.1, case 1c,

$$\sigma_1 = \frac{qR}{2t} = \frac{300(7.5)}{2t} = \frac{1125}{t}, \qquad \sigma_2 = \frac{qR}{t} = \frac{2250}{t}, \qquad \psi = 0$$

$$\Delta R = \frac{qR^2}{Et}\left(1 - \frac{\nu}{2}\right) = \frac{300(7.5^2)}{10(10^6)t}\left(1 - \frac{0.3}{2}\right) = \frac{0.001434}{t}$$

From Table 13.2, case 8,

$$y_A = \frac{-V_o}{2D\lambda^3} = \frac{-V_o}{2(915{,}800t^3)(0.4694/t^{1/2})^3} = \frac{-5.279(10^{-6})V_o}{t^{3/2}}$$

$$\psi_A = \frac{V_o}{2D\lambda^2} = \frac{2.478(10^{-6})V_o}{t^{5/2}}$$

From Table 13.2, case 10,

$$y_A = \frac{M_o}{2D\lambda^2} = \frac{2.478(10^{-6})M_o}{t^2}$$

$$\psi_A = \frac{-M_o}{D\lambda} = \frac{-2.326(10^{-6})M_o}{t^{5/2}}$$

Summing the radial deformations to zero gives

$$\frac{0.001434}{t} - \frac{5.279(10^{-6})V_o}{t^{3/2}} + \frac{2.478(10^{-6})M_o}{t^2} = 0$$

Similarly, summing the end rotations to zero gives

$$\frac{2.478(10^{-6})V_o}{t^2} - \frac{2.326(10^{-6})M_o}{t^{5/2}} = 0$$

Solving these two equations gives

$$V_o = 543t^{1/2} \quad \text{and} \quad M_o = 579t$$

A careful examination of the problem reveals that the maximum bending stress will occur at the end, and so the following stresses must be combined. From Table 13.1, case 1c,

$$\sigma_1 = \frac{1125}{t}, \qquad \sigma_2 = \frac{2250}{t}$$

From Table 13.2, case 8,

$$\sigma_1 = 0, \quad \sigma_1' = 0, \quad \sigma_2' = 0$$

$$\sigma_2 = \frac{-2V_o\lambda R}{t} = \frac{-2(543t^{1/2})(0.4694/t^{1/2})(7.5)}{t} = \frac{-3826}{t}$$

From Table 13.2, case 10,

$$\sigma_1 = 0$$

$$\sigma_2 = \frac{2M_o\lambda^2 R}{t} = \frac{2(579t)(0.4694/t^{1/2})^2(7.5)}{t} = \frac{1913}{t}$$

and on the inside surface

$$\sigma_1' = \frac{6M_o}{t^2} = \frac{3473}{t}$$

$$\sigma_2' = \nu\sigma_1' = \frac{1042}{t}$$

Therefore, at the end of the cylinder the maximum longitudinal tensile stress is $1125/t + 3473/t = 4598/t$; similarly the maximum circumferential tensile stress is $2250/t - 3826/t + 1913/t + 1042/t = 1379/t$.

Since the allowable tensile stress was $12{,}000$ lb/in^2, we can evaluate $4598/t = 12{,}000$ to obtain $t = 0.383$ in. This allows λl to be calculated as 7.59, which verifies the assumption that the shell can be considered a long shell for this loading and support.

SECOND SOLUTION This loaded shell represents a case where both ends are fixed and a uniform radial pressure is applied over the entire length. Since the shell is considered long, we can find the expressions for R_A and M_A in Table 8.6, case 2, under the condition of the left end fixed and where the distance $a = 0$ and b can be considered infinite.

$$R_A = \frac{-2w}{\beta}(B_1 - A_1) \quad \text{and} \quad M_A = \frac{w}{\beta^2}(B_4 - A_4)$$

If $-V_o$ is substituted for R_A, M_o for M_A, λ for β, and D for EI, the solution should apply to the problem at hand. Care must be exercised when substituting for the distributed load w. A purely radial pressure would produce a radial deformation $\Delta R = qR^2/Et$, while the effect of the axial pressure on the ends reduces this to $\Delta R = qR^2(1-\nu/2)/Et$. Therefore, for w we must substitute $-300(1-\nu/2) = -255$ lb/in^2. Also note that for $a = 0$, $A_1 = A_4 = 0.5$, and for $b = \infty$, $B_1 = B_4 = 0$. Therefore,

$$V_o = \frac{-2(255)}{\lambda}(0-0.5) = \frac{255}{\lambda} = 543t^{1/2}$$

$$M_o = \frac{-255}{\lambda^2}(0-0.5) = \frac{127.5}{\lambda^2} = 579t$$

which verifies the results of the first solution.

If we examine case 2 of Table 8.5 under the condition of both ends fixed, we find the expression

$$M_o = M_A = \frac{w}{2\lambda^2}\frac{2C_3C_5 - C_4^2}{C_{11}}$$

Substituting for the several constants and reducing the expression to a simple form, we obtain

$$M_o = \frac{-w}{2\lambda^2}\frac{\sinh\lambda l - \sin\lambda l}{\sinh\lambda l + \sin\lambda l}$$

The hyperbolic sine of 7.59 is 989, and so for all practical purposes

$$M_o = \frac{-w}{2\lambda^2} = 579t$$

which, of course, is the justification for the formulas in Table 8.6.

EXAMPLE 4 A 2-in length of steel tube described in Example 1 is heated, and rigid plugs are inserted ½ in into each end. The rigid plugs have a diameter equal to the inside diameter of the tube plus 0.004 in at room temperature. Find the longitudinal and circumferential stresses at the outside of the tube adjacent to the end of the plug and the diameter at midlength after the tube is shrunk into the plugs.

SOLUTION The most straightforward solution would consist of assuming that the portion of the tube outside the rigid plug is, in effect, displaced radially a distance of 0.002 in and owing to symmetry the midlength has zero slope. A steel cylindrical shell ½ in in length, fixed on the left end with a radial displacement of 0.002 in at $a = 0$ and with the right end guided, that is, slope equal to zero, is the case to be solved.

From Example 1, $R = 2.065$ in, $\lambda = 4.00$, and $D = 344$; $\lambda l = 4.0(0.5) = 2.0$. From Table 8.5, case 6, for the left end fixed and the right end guided, we find the following expressions when $a = 0$:

$$R_A = \Delta_o 2EI\beta^3\frac{C_4^2 + C_2^2}{C_{12}} \quad \text{and} \quad M_A = \Delta_o 2EI\beta^2\frac{C_1C_4 - C_3C_2}{C_{12}}$$

Replace EI with D and b with λ; $\Delta_o = 0.002$, and from page 171 note that $C_4^2 + C_2^2 = 2C_{14}$ and $C_1C_4 - C_3C_2 = -C_{13}$. From Table 8.3, for $\lambda l = 2.00$, we find that $2C_{14} = 27.962$, $-C_{13} = -14.023$, and $C_{12} = 13.267$. Therefore,

$$R_A = 0.002(2)(344)(4.0^3)\frac{27.962}{13.267} = 185.6 \text{ lb/in}$$

$$M_A = 0.002(2)(344)(4.0^2)\frac{-14.023}{13.267} = -23.27 \text{ in-lb/in}$$

To find the deflection at the midlength of the shell, which is the right end of the half-shell being used here, we solve for y at $x = 0.5$ in and $\lambda x = 4.0(0.5) = 2.0$. Note that $y_A = 0$ because the deflection of 0.002 in was forced into the shell just beyond the end in the solution being considered here. Therefore,

$$y = \frac{M_A}{2D\lambda^2}F_3 + \frac{R_A}{2D\lambda^3}F_4 + \Delta_o F_{a1}$$

where from Table 8.3 at $\lambda x = 2.0$

$$F_3 = 3.298, \qquad F_4 = 4.930, \qquad F_{a1} = F_1 = -1.566 \text{ since } a = 0$$

$$y_{x=0.5} = \frac{-23.27}{2(344)(4.0^2)}3.298 + \frac{185.6}{4(344)(4.0^3)}4.93 + 0.002(-1.566) = 0.00029 \text{ in}$$

For a partial check on the solution we can calculate the slope at midlength. From Table 8.5, case 6,

$$\theta = \frac{M_A}{2EI\beta}F_2 + \frac{R_A}{2EI\beta^2}F_3 - \Delta_o\beta F_{a4}$$

where $F_2 = 1.912$ and $F_{a4} = F_4$ since $a = 0$. Therefore,

$$\theta = \frac{-23.27}{2(344)(4.0)}1.912 + \frac{185.6}{2(344)(4.0^2)}3.298 - 0.002(4.0)(4.930) = 0.00000$$

Now from Table 13.2,

$$\sigma_1' = \frac{-6M}{t^2} = \frac{-6(-23.27)}{0.05^2} = 55{,}850 \text{ lb/in}^2$$

Since $\sigma_1 = 0$,

$$\sigma_2 = \frac{0.002(30)(10^6)}{2.065} = 29{,}060 \text{ lb/in}^2$$

$$\sigma_2' = 0.3(55{,}800) = 16{,}750 \text{ lb/in}^2$$

On the outside surface at the cross section adjacent to the plug the longitudinal stress is 55,850 lb/in^2 and the circumferential stress is 29,060 + 16,750 = 45,810 lb/in^2. Since a rigid plug is only hypothetical, the actual stresses present would be smaller when a solid but elastic plug is used. External clamping around the shell over the plugs would also be necessary to fulfill the assumed fixed-end condition. The stresses calculated are therefore maximum possible values and would be conservative.

Spherical Shells

The format used to present the formulas for the finite-length cylindrical shells could be adapted for finite portions of open spherical and conical shells with both edge loads and loads applied within the shells if we were to accept the approximate solutions based on equivalent cylinders. Baker, Kovalevsky, and Rish (Ref. 6) present formulas based on this approximation for open spherical and conical shells under edge loads and edge displacements.

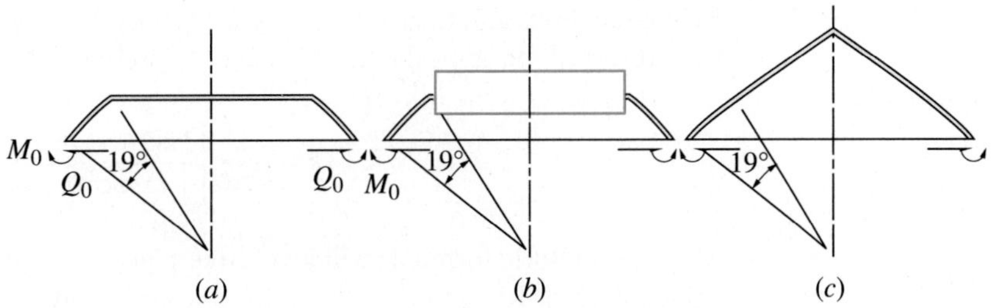

Figure 13.5

For partial spherical shells under axisymmetric loading, Hetényi, in an earlier work (Ref. 14), discusses the errors introduced by this same approximation and compares it with a better approximate solution derived therein. Table 13.3, case 1, gives formulas based on Hetényi's work, and although it is estimated that the calculational effort is twice that of the simpler approximation, the errors in maximum stresses are decreased substantially, especially when the opening angle ϕ is much different from 90°.

Stresses and deformations due to edge loads decrease exponentially away from the loaded edges of axisymmetric shells, and consequently boundary conditions or other restraints are not important if they are far enough from the loaded edge. For example, the exponential term decreases to approximately 1% when the product of the spherical shell parameter β (see Table 13.3, case 1) and the angle ω (in radians) is greater than 4.5; similarly it reduces to approximately 5% at $\beta\omega = 3$. This means that a spherical shell with a radius/thickness ratio of 50, for which $\beta \approx 9$, can have an opening angle ϕ as small as 1/3 rad, or 19°, and still be solved with formulas for cases 1 with very little error. Figure 13.5 shows three shells, for which R/t is approximately 50, which would respond similarly to the edge loads M_o and Q_o. In fact, the conical portion of the shell in Fig. 13.5(*c*) could be extended much closer than 19° to the loaded edge since the conical portion near the junction of the cone and sphere would respond in a similar way to the sphere. (Hetényi discusses this in Ref. 14.)

Similar bounds on *nonspherical* but axisymmetric shells can be approximated by using closely matching equivalent spherical shells (Ref. 6). (We should note that the angle ϕ in Table 13.3, case 1, is not limited to a maximum of 90°, as will be illustrated in the examples at the end of this section.)

For shallow spherical shells where ϕ is small, Gerdeen and Niedenfuhr (Ref. 46) have developed influence coefficients for uniform pressure and for edge loads and moments. Shells with central cutouts are also included as are loads and moments on the edge of the cutouts. Many graphs as well as tabulated data are presented, which permit the solution of a wide variety of problems by superposition.

Cheng and Angsirikul (Ref. 80) present the results of an elasticity solution for edge-loaded spherical domes with thick walls and with thin walls.

Conical Shells

Exact solutions to the differential equations for both long and short thin-walled truncated conical shells are described in Refs. 30, 31, 64, and 65. Verifications of these expressions by tests are described in Ref. 32, and applications to reinforced cones are described in Ref. 33. In Table 13.3, case 4 for long cones, where the loads at one end do not influence the displacements at the other, is based on the solution described by Taylor (Ref. 65) in which the

Kelvin functions and their derivatives are replaced by asymptotic formulas involving negative powers of the conical shell parameter k (presented here in a modified form):

$$k = \frac{2}{\sin\alpha}\left[\frac{12(1-\nu^2)R^2}{t^2\sec^2\alpha}\right]^{1/4}$$

These asymptotic formulas will give three-place accuracy for the Kelvin functions for all values of $k > 5$. To appreciate this fully, one must understand that a truncated thin-walled cone with an R/t ratio of 10 at the small end, a semiapex angle of 80°, and a Poisson's ratio of 0.3 will have a value of $k = 4.86$. For problems where k is much larger than 5, fewer terms can be used in the series, but a few trial calculations will soon indicate the number of terms it is necessary to carry. If only displacements and stresses at the loaded edge are needed, the simpler forms of the expressions can be used. (See the example at the end of Sec. 13.4.)

Baltrukonis (Ref. 64) obtains approximations for the influence coefficients which give the edge displacements for *short* truncated conical shells under axisymmetric edge loads and moments; this is done by using one-term asymptotic expressions for the Kelvin functions. Applying the multiterm asymptotic expressions suggested by Taylor to a short truncated conical shell leads to formulas that are too complicated to present in a reasonable form. Instead, in Table 13.3, case 5 tabulates numerical coefficients based upon this more accurate formulation but evaluated by a computer for the case where Poisson's ratio is 0.3. Because of limited space, only five values of k and six values of the length parameter $\mu_D = |k_A - k_B|/\sqrt{2}$ are presented. If μ_D is greater than 4, the end loads do not interact appreciably and the formulas from case 4 may be used.

Tsui (Ref. 58) derives expressions for deformations of conical shells for which the thickness tapers linearly with distance along the meridian; influence coefficients are tabulated for a limited range of shell parameters. Blythe and Kyser (Ref. 50) give formulas for thinwalled conical shells loaded in torsion.

Toroidal Shells

Simple closed-form solutions for toroidal shells are generally valid for a rather limited range of parameters, so that usually it is necessary to resort to numerical solutions. Osipova and Tumarkin (Ref. 18) present extensive tables of functions for the asymptotic method of solution of the differential equations for toroidal shells; this reference also contains an extensive bibliography of work on toroidal shells. Tsui and Massard (Ref. 43) tabulate the results of numerical solutions in the form of influence coefficients and influence functions for internal pressure and edge loadings on finite portions of segments of toroidal shells. Segments having *positive* and *negative* gaussian curvatures are considered; when both positive and negative curvatures are present in the same shell, the solutions can be obtained by matching slopes and deflections at the junction. References 29, 51, and 61 describe similar solutions.

Stanley and Campbell (Ref. 77) present the principal test results on 17 full-size, production-quality torispherical ends and compare them to theory. Kishida and Ozawa (Ref. 78) compare results arrived at from elasticity, photoelasticity, and shell theory. References 67 and 68 discuss torispherical shells and present design formulas. See the discussion in Sec. 13.2 on this topic.

Jordon (Ref. 53) works with the shell-equilibrium equations of a *deformed* shell to examine the effect of pressure on the stiffness of an axisymmetrically loaded toroidal shell.

Kraus (Ref. 44), in addition to an excellent presentation of the theory of thin elastic shells, devotes one chapter to numerical analysis under static loadings and another to numerical analysis under dynamic loadings. Comparisons are made among results obtained by finite-element methods, finite-difference methods, and analytic solutions. Numerical techniques, element sizes, and techniques of shell subdivision are discussed in detail. It would be impossible to list here all the references describing the finite-element computer programs available for solving shell problems, but Perrone (Ref. 62) has presented an excellent summary and Bushnell (Ref. 63) describes work on shells in great detail.

EXAMPLE 1 Two partial spheres of aluminum are to be welded together as shown in Fig. 13.6 to form a pressure vessel to withstand an internal pressure of 200 lb/in². The mean radius of each sphere is 2 ft, and the wall thickness is 0.5 in. Calculate the stresses at the seam. Given: $E = 10(10^6)$ lb/in² and $\nu = 033$.

SOLUTION The edge loading will be considered in three parts, as shown in Fig. 13.6(b). The tangential edge force T will be applied to balance the internal pressure and, together with the pressure, will cause only membrane stresses and the accompanying change in circumferential radius ΔR; this loading will produce no rotation of the meridian. Owing to the symmetry of the two shells, there is no resultant radial load on the edge, and so Q_o is added to eliminate that component of T. M_o is needed to ensure no edge rotation.

First apply the formulas from Table 13.1, case 3a:

$$\sigma_1 = \sigma_2 = \frac{qR_2}{2t} = \frac{200(24)}{2(0.5)} = 4800 \text{ lb/in}^2$$

$$\Delta R = \frac{qR_2^2(1-\nu)\sin\theta}{2Et} = \frac{200(24^2)(1-0.33)\sin 120°}{2(10)(10^6)(0.5)} = 0.00668 \text{ in}$$

$$T = \sigma_1 t = 4800(0.5) = 2400 \text{ lb/in}$$

$$\psi = 0$$

Next apply case 1a from Table 13.3:

$$Q_o = T\sin 30° = 2400(0.5) = 1200 \text{ lb/in}$$

$$\phi = 120°$$

$$\beta = \left[3(1-\nu^2)\left(\frac{R_2}{t}\right)^2\right]^{1/4} = \left[3(1-0.33^2)\left(\frac{24}{0.5}\right)^2\right]^{1/4} = 8.859$$

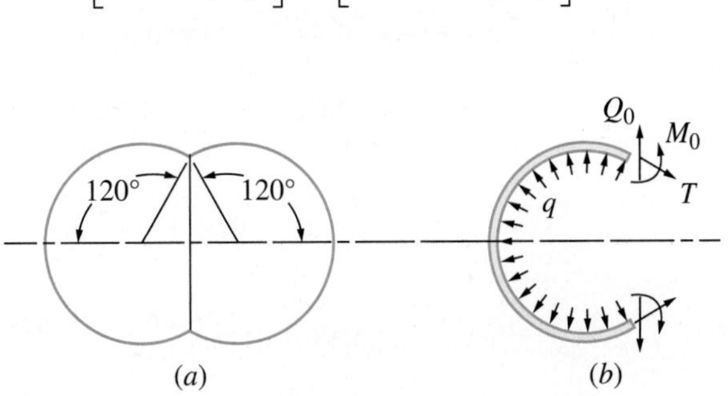

Figure 13.6

At the edge where $\omega = 0$,

$$K_1 = 1 - \frac{1-2\nu}{2\beta}\cot\phi = 1 - \frac{1-2(0.33)}{2(8.859)}\cot 120° = 1.011$$

$$K_2 = 1 - \frac{1+2\nu}{2\beta}\cot\phi = 1.054$$

$$\Delta R = \frac{Q_o R_2 \beta \sin^2\phi}{EtK_1}(1 + K_1 K_2) = \frac{1200(24)(8.859)\sin^2 120°}{10(10^6)(0.5)(1.011)}[1 + 1.011(1.054)]$$

$$= 0.0782 \text{ in}$$

$$\psi = \frac{Q_o 2\beta^2 \sin\phi}{EtK_1} = \frac{1200(2)(8.859^2)\sin 120°}{10(10^6)(0.5)(1.011)} = 0.0323 \text{ rad}$$

$$\sigma_1 = \frac{Q_o \cos\phi}{t} = \frac{1200\cos 120°}{0.5} = -1200 \text{ lb/in}^2$$

$$\sigma_1' = 0$$

$$\sigma_2 = \frac{Q_o \beta \sin\phi}{2t}\left(\frac{2}{K_1} + K_1 + K_2\right) = \frac{1200(8.859)\sin 120°}{2(0.5)}\left(\frac{2}{1.011} + 1.011 + 1.054\right)$$

$$= 37{,}200 \text{ lb/in}^2$$

$$\sigma_2' = \frac{-Q_o \beta^2 \cos\phi}{K_1 R_2} = \frac{-1200(8.859^2)\cos 120°}{1.011(24)} = 1940 \text{ lb/in}^2$$

Now apply case 1b from Table 13.3:

$$\Delta R = \frac{M_o 2\beta^2 \sin\phi}{EtK_1} = 0.00002689 M_o$$

$$\psi = \frac{M_o 4\beta^3}{EtR_2K_1} = \frac{M_o 4(8.859)^3}{10(10^6)(0.5)(24)(1.011)} = 0.00002292 M_o$$

Since the combined edge rotation ψ must be zero,

$$0 = 0 + 0.0323 + 0.00002292 M_o \quad \text{or} \quad M_o = -1409 \text{ lb-in/in}$$

and

$$\Delta R = 0.00668 + 0.0782 + 0.00002689(-1409) = 0.04699 \text{ in}$$

$$\sigma_1 = 0$$

$$\sigma_1' = \frac{-6(-1409)}{0.05^2} = 33{,}800 \text{ lb/in}^2$$

$$\sigma_2 = \frac{M_o 2\beta^2}{R^2 K_1 t} = \frac{-1409(2)(8.859^2)}{24(1.011)(0.5)} = -18{,}200 \text{ lb/in}^2$$

$$M_2 = \frac{M_o}{2\nu K_1}[(1+\nu^2)(K_1 + K_2) - 2K_2]$$

$$= \frac{-1409}{2(0.33)(1.011)}[(1 + 0.33^2)(1.011 + 1.054) - 2(1.054)] = -384 \text{ lb-in/in}$$

$$\sigma_2' = \frac{-6(-384)}{0.5^2} = 9220 \text{ lb/in}^2$$

The superimposed stresses at the joint are therefore

$$\sigma_1 = 4800 - 1200 + 0 = 3600 \text{ lb/in}^2$$

$$\sigma_1' = 0 + 0 + 33{,}800 = 33{,}800 \text{ lb/in}^2$$

$$\sigma_2 = 4800 + 37{,}200 - 18{,}200 = 23{,}800 \text{ lb/in}^2$$

$$\sigma_2' = 0 + 1940 + 9220 = 11{,}160 \text{ lb/in}^2$$

The maximum stress is a tensile meridional stress of 37,400 lb/in^2 on the outside surface at the joint. A further consideration would be given to any stress concentrations due to the shape of the weld cross section.

EXAMPLE 2 To reduce the high stresses in Example 1, it is proposed to add to the joint a reinforcing ring of aluminum having a cross-sectional area A. Calculate the optimum area to use.

SOLUTION If the ring could be designed to expand in circumference by the same amount that the sphere does under membrane loading only, then all bending stresses could be eliminated. Therefore, let a ring be loaded radially with a load of $2Q_o$ and have the radius increase by 0.00668 in. Since $\Delta R/R = 2Q_o R/AE$, then

$$A = \frac{2Q_o R^2}{E\Delta R} = \frac{2(1200)(24^2)\sin^2 60°}{10(10^6)(0.00668)} = 15.5 \text{ in}^2$$

With this large an area required, the simple expression just given for $\Delta R/R$ based on a thin ring is not adequate; furthermore, there is not enough room to place such a ring external to the shell. An internal reinforcement seems more reasonable. If a 6-in-diameter hole is required for passage of the fluid, the internal reinforcing disk can have an outer radius of 20.78 in, an inner radius of 3 in, and a thickness t_1 to be determined. The loading on the disk is shown in Fig. 13.7. The change in the outer radius is desired.

From Table 13.5, case 1a, the effect of the 200 lb/in^2 internal pressure can be evaluated:

$$\Delta a = \frac{q}{E}\frac{2ab^2}{a^2 - b^2} = \frac{200}{10(10^6)}\frac{2(20.78)(3^2)}{20.78^2 - 3^2} = 0.0000177 \text{ in}$$

From Table 13.5, case 1c, the effect of the loads Q_o can be determined if the loading is modeled as an outward pressure of $-2Q_o/t_1$. Therefore,

$$\Delta a = \frac{-qa}{E}\left(\frac{a^2 + b^2}{a^2 - b^2} - \nu\right) = \frac{2(1200)(20.78)}{t_1 10(10^6)}\left(\frac{20.78^2 + 3^2}{20.78^2 - 3^2} - 0.33\right) = \frac{0.00355}{t_1}$$

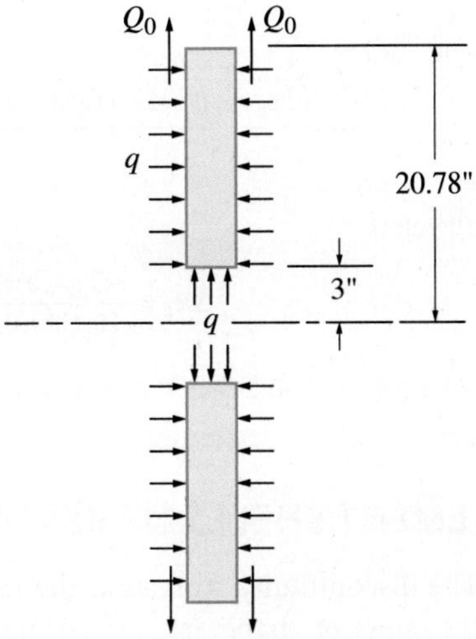

Figure 13.7

The longitudinal pressure of 200 lb/in² will cause a small lateral expansion in the outer radius of

$$\Delta_a = \frac{200(0.33)(20.78)}{10(10^6)} = 0.000137 \text{ in}$$

Summing the changes in the outer radius to the desired value gives

$$0.00668 = 0.0000177 + 0.000137 + \frac{0.00355}{t_1} \quad \text{or} \quad t_1 = 0.545 \text{ in}$$

(Undoubtedly further optimization could be carried out on the volume of material required and the ease of welding the joint by varying the thickness of the disk and the size of the internal hole.)

EXAMPLE 3 A truncated cone of aluminum with a uniform wall thickness of 0.050 in and a semi-apex angle of 55° has a radius of 2 in at the small end and 2.5 in at the large end. It is desired to know the radial loading at the small end which will increase the radius by half the wall thickness. Given: $E = 10(10^6)$ lb/in² and $\nu = 0.33$.

SOLUTION Evaluate the distances from the apex along a meridian to the two ends of the shell and then obtain the shell parameters:

$$R_A = 2.5 \text{ in}$$

$$R_B = 2.0 \text{ in}$$

$$k_A = \frac{2}{\sin 55°}\left[\frac{12(1-0.33^2)(2.5^2)}{0.050^2 \sec^2 55°}\right]^{1/4} = 23.64$$

$$k_B = 21.15$$

$$\mu_D = \frac{23.64 - 21.15}{2} = 1.76$$

$$\beta = [12(1-0.33^2)]^{1/2} = 3.27$$

From Table 13.3, case 6c, tabulated constants for shell forces, moments, and deformations can be found when a radial load is applied to the small end. For the present problem the value of $K_{\Delta R}$ at the small end ($\Omega = 1.0$) is needed when $\mu_D = 1.76$ and $k_A = 23.64$. Interpolation from the following data gives $K_{\Delta R} = 1.27$:

k_A	10.0				20.0				40.0			
μ_n	0.8	1.2	1.6	3.2	0.8	1.2	1.6	3.2	0.8	1.2	1.6	3.2
$K_{\Delta R}$	2.085	1.610	1.343	1.113	2.400	1.696	1.351	1.051	2.491	1.709	1.342	1.025

At $\Omega = 1.0$
Therefore,

$$\Delta R_B = \frac{-Q_B(2.0)\sin 55°}{10(10^6)(0.050)} \frac{21.15}{\sqrt{2}}(1.27) = -0.00006225 Q_B$$

Since $\Delta R_B = 0.050/2$ (half the thickness), $Q_B = -402$ lb/in (outward).

13.4 THIN MULTIELEMENT SHELLS OF REVOLUTION

The discontinuity stresses at the junctions of shells or shell elements due to changes in thickness or shape are not serious under static loading of ductile materials; however, they are serious under conditions of cyclic or fatigue loading. In Ref. 9, discontinuity stresses are discussed with a numerical example; also, allowable levels of the membrane stresses due to internal pressure are established, as well as allowable levels of membrane and bending stresses due to discontinuities under both static and cyclic loadings.

Langer (Ref. 10) discusses four modes of failure of a pressure vessel—bursting due to general yielding, ductile tearing at a discontinuity, brittle fracture, and creep rupture—and the way in which these modes are affected by the choice of material and wall thickness; he also compares pressure-vessel codes of several countries. Zaremba (Ref. 47) and Johns and Orange (Ref. 48) describe in detail the techniques for accurate deformation matching at the intersections of axisymmetric shells. See also Refs. 74 and 75.

The following example illustrates the use of the formulas in Tables 13.1–13.3 to determine discontinuity stresses.

EXAMPLE The vessel shown in quarter longitudinal section in Fig. 13.8(*a*) consists of a cylindrical shell ($R = 24$ in and $t = 0.633$ in) with conical ends ($\alpha = 45°$ and $t = 0.755$ in). The parts are welded together, and the material is steel, for which $E = 30(10^6)$ lb/in² and $\nu = 0.25$. It is required to determine the maximum stresses at the junction of the cylinder and cone due to an internal pressure of 300 lb/in². (This vessel corresponds to one for which the results of a supposedly precise analysis and experimentally determined stress values are available. See Ref. 17.)

SOLUTION For the cone, case 2a in Table 13.1 and cases 4a and 4b in Table 13.3 can be used: $R = 24$ in, $\alpha = 45°$, and $t = 0.755$ in. The following conditions exist at the end of the cone: From Table 13.1, case 2a, for the load T and pressure q,

$$\sigma_1 = \frac{300(24)}{2(0.755)\cos 45°} = 6740 \text{ lb/in}^2, \qquad T = 6740(0.755) = 5091 \text{ lb/in}$$

$$\sigma_2 = 13{,}480 \text{ lb/in}^2, \qquad \sigma_1' = 0, \qquad \sigma_2' = 0$$

$$\Delta R = \frac{300(24^2)}{30(10^6)(0.755)\cos 45°}\left(1 - \frac{0.25}{2}\right) = 0.00944 \text{ in}$$

$$\psi = \frac{3(300)(24)(1)}{2(30)(10^6)(0.755)\cos 45°} = 0.000674 \text{ rad}$$

From Table 13.3, case 4a, for the radial edge load Q_o,

$$R_A = 24 \text{ in}$$

$$k_A = \frac{2}{\sin 45°}\left[\frac{12(1-0.25^2)(24^2)}{0.755^2 \sec^2 45°}\right]^{1/4} = 24.56$$

$$\beta = [12(1-0.25^2)]^{1/2} = 3.354$$

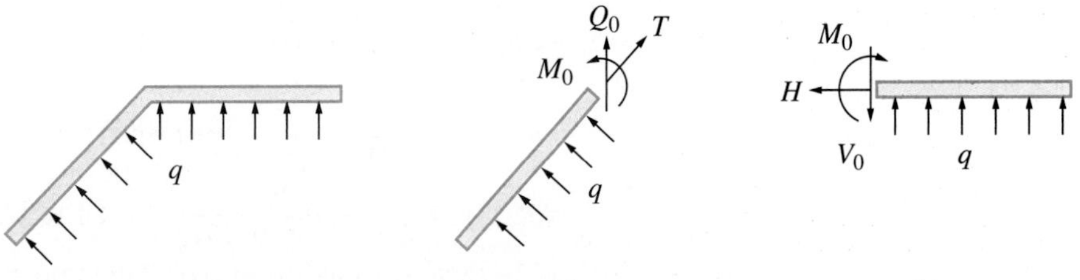

(*a*) (*b*) (*c*)

Figure 13.8

Only values at $R = R_A$ are needed for this solution. Therefore, the series solutions for the constants can be used to give

$$F_{9A} = C_1 = 0.9005, \quad F_{1A} = 0, \quad F_{3A} = 0, \quad F_{2A} = 0.8977$$

$$F_{4A} = 0.8720, \quad F_{5A} = F_{8A} = 0.8746, \quad F_{10A} = F_{7A} = F_{6A} = 0.8947$$

$$\Delta R_A = \frac{Q_o 24(0.7071)(24.56)}{30(10^6)(0.755)(\sqrt{2})(0.9005)}\left[0.8720 - \frac{4(0.25^2)}{24.56^2}0.8977\right] = 12.59(10^{-6})Q_o$$

$$\psi_A = \frac{Q_o 24(3.354)}{30(10^6)(0.755^2)(0.9005)}(0.8947) = 4.677(10^{-6})Q_o$$

$$N_{1A} = 0.7071Q_o, \quad M_{1A} = 0$$

$$N_{2A} = \frac{Q_o(0.7071)(24.56)}{2(0.9005)}\left[0.8720 + \frac{2(0.25)}{24.56}0.8746\right] = 12.063Q_o$$

$$M_{2A} = \frac{Q_o(0.7071)(1-0.25^2)(0.755)}{3.354(0.9005)}0.8947 = 0.1483Q_o$$

From Table 13.3, case 4b, for the edge moment M_A,

$\Delta R_A = 4.677(10^{-6})M_o$ (same coefficient shown for ψ_A for the loading Q_o as would be expected from Maxwell's theorem)

$$\psi_A = \frac{M_o 2\sqrt{2}(3.354^2)(24)}{30(10^6)(0.755^3)(24.56)(0.7071)}\frac{0.8977}{0.9005} = 3.395(10^{-6})M_o$$

$$N_{1A} = 0, \quad N_{2A} = M_o\frac{3.354(0.8947)}{0.755(0.9005)} = 4.402M_o$$

$$M_{1A} = M_o, \quad M_{2A} = M_o\left[0.25 + \frac{2(2)(1-0.25^2)(0.8977)}{24.56(0.9005)}\right] = 0.3576M_o$$

For the cylinder, case 1c in Table 13.1 and cases 8 and 10 in Table 13.2 can be used (it is assumed that the other end of the cylinder is far enough away so as to not affect the deformations and stresses at the cone-cylinder junction):

R = 24 in; t = 0.633 in; $\lambda = [3(1-0.25^2)/24^2/0.633^2]^{1/4} = 0.3323$; and $D = 30(10^6)(0.633^3)/12(1-0.25^2) = 6.76(10^5)$. The following conditions exist at the end of the cylinder:

From Table 13.1, case 1c, for the axial load H and the pressure q,

$$\sigma_1 = \frac{300(24)}{2(0.633)} = 5690 \text{ lb/in}^2, \quad H = 5690(0.633) = 3600 \text{ lb/in}$$

$$\sigma_2 = 11380 \text{ lb/in}^2, \quad \sigma_1' = 0, \quad \sigma_2' = 0$$

$$\Delta R = \frac{300(24^2)}{30(10^6)(0.633)}\left(1 - \frac{0.25}{2}\right) = 0.00796 \text{ in}$$

$$\psi = 0$$

From Table 13.2, case 8, for the radial end load V_o,

$$\psi_A = \frac{V_o}{2(6.76)(10^5)(0.3323^2)} = 6.698(10^{-6})V_o$$

$$\Delta R_A = y_A = \frac{-V_o}{2(6.76)(10^5)(0.3323^2)} = -20.16(10^{-6})V_o$$

$$\sigma_1 = 0, \quad \sigma_2 = \frac{yE}{R} = \frac{-20.16(10^{-6})V_o(30)(10^6)}{24} = -25.20\ V_o$$

$$\sigma_1' = 0, \quad \sigma_2' = 0$$

From Table 13.2, case 10, for the end moment M_o,

$$\psi_A = \frac{-M_o}{6.76(10^5)(0.3323)} = -4.452(10^{-6})M_o$$

$$\Delta R_A = y_A = \frac{M_o}{2(6.76)(10^5)(0.3323^2)} = 6.698(10^{-6})M_o$$

$$\sigma_1 = 0, \qquad \sigma_2 = \frac{2M_o\lambda^2 R}{t} = \frac{2M_o(0.3323^2)(24)}{0.633} = 8.373M_o$$

$$\sigma_1' = \frac{-6M_o}{t^2} = \frac{-6M_o}{0.633^2} = -14.97M_o, \qquad \sigma_2' = \nu\sigma_1' = -3.74M_o$$

Summing the radial deflections for the end of the cone and equating to the sum for the cylinder gives

$$0.00944 + 12.59(10^{-6})Q_o + 4.677(10^{-6})M_o = 0.00796 - 20.16(10^{-6})V_o + 6.698(10^{-6})M_o$$

Doing the same with the meridian rotations gives

$$0.000674 + 4.677(10^{-6})Q_o + 3.395(10^{-6})M_o = 0 + 6.698(10^{-6})V_o - 4.452(10^{-6})M_o$$

Finally, equating the radial forces gives

$$Q_o + 5091 \cos 45° = V_o$$

Solving the three equations simultaneously yields

$$Q = -2110 \text{ lb/in}, \qquad V_o = 1490 \text{ lb/in}, \qquad M_o = 2443 \text{ lb-in/in}$$

In the cylinder,

$$\sigma_1 = 5690 + 0 + 0 = 5690 \text{ lb/in}^2$$

$$\sigma_2 = 11{,}380 - 25.20(1490) + 8.373(2443) = -5712 \text{ lb/in}^2$$

$$\sigma_1' = 0 + 0 - 14.97(2443) = -36{,}570 \text{ lb/in}^2$$

$$\sigma_2' = 0 + 0 - 3.74(2443) = -9140 \text{ lb/in}^2$$

Combined hoop stress on outside $= -5712 - 9140 = -14{,}852 \text{ lb/in}^2$

Combined hoop stress on inside $= -5712 + 9140 = 3428 \text{ lb/in}^2$

Combined meridional stress on outside $= 5690 - 36{,}570$

$= -30{,}880 \text{ lb/in}^2$

Combined meridional stress on inside $= 5690 + 36{,}570$

$= 42{,}260 \text{ lb/in}^2$

Similarly, in the cone,

$$\sigma_1 = 6740 + \frac{0.7071(-2110)}{0.755} + 0 = 4764 \text{ lb/in}^2$$

$$\sigma_2 = 13{,}480 + \frac{12.063(-2110)}{0.755} + \frac{4.402(2443)}{0.755} = -5989 \text{ lb/in}^2$$

$$\sigma_1' = 0 + 0 - \frac{2443(6)}{0.755^2} = -25{,}715 \text{ lb/in}^2$$

$$\sigma_2' = 0 - \frac{0.1483(-2110)(6)}{0.755^2} - \frac{0.3576(2443)(6)}{0.755^2} = -5902 \text{ lb/in}^2$$

$$\text{Combined hoop stress on outside} = -5989 - 5902 = -11{,}891 \text{ lb/in}^2$$

$$\text{Combined hoop stress on inside} = -5989 + 5902 = -87 \text{ lb/in}^2$$

$$\text{Combined meridional stress on outside} = 4764 - 25{,}715 = -20{,}951 \text{ lb/in}^2$$

$$\text{Combined meridional stress on inside} = 4764 + 25{,}715 = 30{,}480 \text{ lb/in}^2$$

These stress values are in substantial agreement with the computed and experimental values cited in Refs. 17 and 26. Note that the radial deflections are much less than the wall thicknesses. See the discussion in the third paragraph of Sec. 13.3.

In the problem just solved by the method of deformation matching only two shells met at their common circumference. The method, however, can be extended to cases where more than two shells meet in this manner. The primary source of difficulty encountered when setting up the equations to carry out such a solution is the rigor needed when labeling the several edge loads and the establishment of the proper signs for the radial and rotational deformations. An additional problem arises when the several shells intersect not at a single circumference but at two or more closely spaced circumferences.

Figure 13.9 illustrates two conical shells and a spherical shell joined together by a length of cylindrical shell. The length of this central cylinder is a critical dimension in determining how the cylinder is treated. If the length is small enough for a given radius and wall thickness, it may be sufficient to treat it as a narrow ring whose cross section deflects radially and rotates with respect to the original meridian but whose cross section does not change shape. For an example as to how these narrow rings are treated see Sec. 11.9. For a longer cylinder the cross section does change shape and it is treated as a short cylinder, using expressions from Table 13.2. Here there are two circumferences where slopes and deflections are to be matched but the loads on each end of the cylinder influence the deformations at the other end. Finally, if the cylinder is long enough, $\lambda l > 6$, for example, the ends are far enough apart so that two separate problems may be solved.

Table 13.3 presents formulas and tabulated data for several combinations of thin shells of revolution and thin circular plates joined two at a time at a common circumference. All shells are assumed long enough so that the end interactions can be neglected. Loadings include axial load, a loading due to a rotation at constant angular velocity about the central axis, and internal or external pressure where the pressure is either constant or varying linearly along the axis of the shell. For the pressure loading the equations represent the case where the junction of the shells carries no axial loading such as when a cylindrical shell carries a frictionless piston which is supported axially by structures other than the cylinder walls. *The decision to present the pressure loadings in this form was based primarily on the ease of presentation. When used for closed pressure vessels, the deformations and stresses for the axial load must be superposed on those for the pressure loading.*

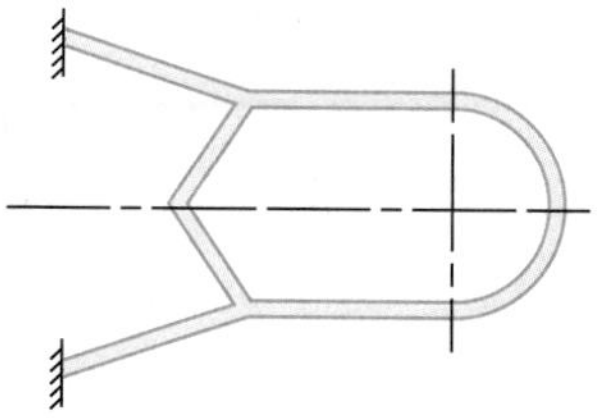

Figure 13.9

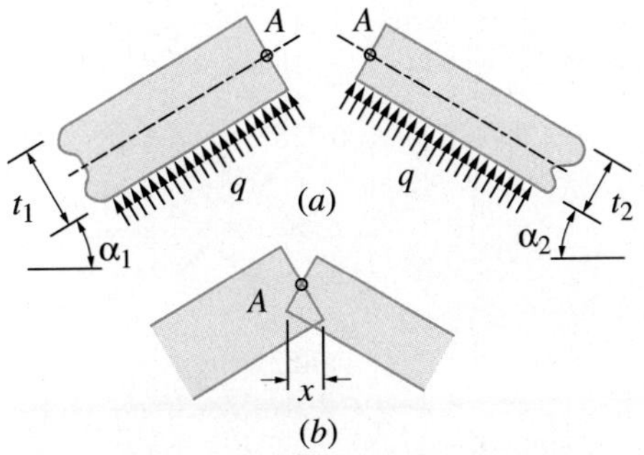

Figure 13.10

The reasons for presenting the tabulated data in this table are several. (1) In many instances one merely needs to know whether the stresses and deformations at such discontinuities are important to the safety and longevity of a structure. Using interpolation one can explore quickly the tables of tabulated data to make such a determination. (2) The tabulated data also allow those who choose to solve the formulas to verify their results.

The basic information in Table 13.4 can be developed as needed from formulas in the several preceding tables, but the work has been extended a few steps further by modifying the expressions in order to make them useful for shells with somewhat thicker walls.

In the sixth edition of this book, correction terms were presented to account for the fact that internal pressure loading acts on the inner surface, not at the mid-thickness. For external pressure, the proper substitutions are indicated by notes for the several cases. This has already been accounted for in the general pressure loadings on the several shell types, but there is an additional factor to account for at the junction of the shells. In Fig. 13.10(*a*), the internal pressure is shown acting all the way to the hypothetical end of the left-hand shell. The general equations in Table 13.1 assume this to be the case, and the use of these equations in Table 13.4 makes this same assumption. The correction terms in the sixth edition of this book added or subtracted, depending upon the signs of α_1, α_1, and q, the pressure loading over the length x shown in Fig. 13.10(*b*). These corrections included the effects of the radial components, the axial components, and the moments about point A of this local change in loading. The complexity of these corrections may seem out of proportion to the benefits derived, and, depending upon their needs, users will have to decide whether or not to include them in their calculations. To assist users in making this decision, the following example will compare results with and without the correction terms and show the relative effect of using only the radial component of the change in the local pressure loading at the junction of a cone and cylinder.

EXAMPLE For this example, the pressurized shell is that of the previous example shown in Fig. 13.8. The calculations for that example were carried out using equations from Tables 13.1 and 13.3. The stresses in the cylinder at the junction are given at the end of the solution, and the radial deflection and the rotation at the junction can be calculated from the expressions given just before the stress calculations. The following results table lists these stresses and deflections in column [1]. As stated above, the equations used in Table 13.4 to solve for the shell junction stresses were those given in Tables 13.1–13.3, but modified somewhat to make them more accurate for shells with thicker walls. Using cases 2a and 2b from Table 13.4 gives the results shown in columns [2]–[7] in the results table. The axial load used for column [2] was $P = \pi(24 - 0.633/2)^2(300) = 528{,}643$ lb. All of the stresses in the results table are those found in the cylinder at the junction. Column [3] gives data for the internal pressure loading with no correction factors and column [4] is the sums for the axial load and internal pressure, columns [2] plus [3]. Column [5] is for the internal pressure corrected for the change in

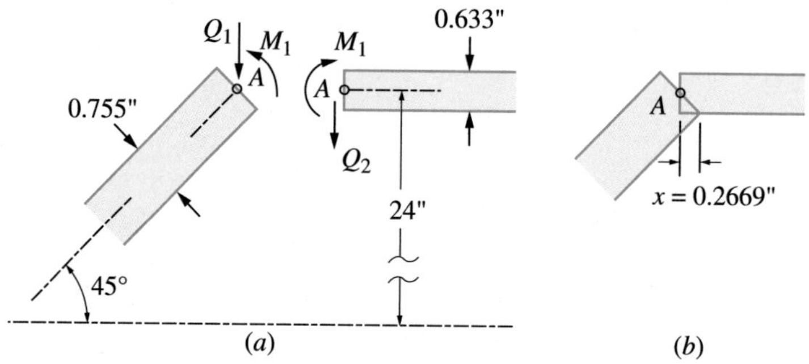

Figure 13.11

loading at the joint. Column [7] is the difference in the numbers of columns [4] and [6], and gives the changes due to the correction factors in Table 13.4.

Column [8] shows the changes due to the radial component of the correction in the joint loading which are calculated as follows. Figure 13.11(*a*) shows the joint being considered, with the dimensions. The value of $x = 0.2669$ in, and when this is multiplied by the internal pressure of 300 lb/in², one obtains $Q_1 + Q_2 = 80.07$ lb/in, the radially inward load needed to compensate for the radial component of the internal pressure *not* acting on the joint.

Using already evaluated expressions from the previous example, the following equations can be written:

For the cone,

$$\Delta R_A = -12.59(10^{-6})Q_1 + 4.677(10^{-6})M_1$$

$$\psi_A = -4.677(10^{-6})Q_1 + 3.395(10^{-6})M_1$$

For the cylinder,

$$\Delta R_A = -20.16(10^{-6})Q_2 + 6.698(10^{-6})M_1$$

$$\psi_A = 6.698(10^{-6})Q_2 - 4.452(10^{-6})M_1$$

Equating the equations for ΔR_A and ψ_A with $Q_1 + Q_2 = 80.07$ lb/in, yields $Q_1 = 48.83$ lb/in, $Q_2 = 31.24$ lb/in, $M_1 = 55.77$ lb-in/in, $\Delta R_A = -0.000354$ in, and $\psi_A = 0.000039$ rad.

As would be expected, the radial component is a major contributor for the joint being discussed and would be for most pressure vessel joints.

From Previous Example		From Table 13.4						Change Due to the Approx. Corrections Given Above
		Case 2b Axial Load	Case 2a Internal Pressure without Corrections	Sum [2] + [3]	Case 2a Internal Pressure with Corrections	Sum [2] + [5]	Change Due to the Use of the Correction Terms	
	[1]	[2]	[3]	[4]	[5]	[6]	[7]	[8]
σ_1	5,690	5,538	0	5,538	0	5,538	0	0
σ_2	-5,712	-17,038	11,647	-5,391	11,252	-5,786	-395	-321
σ'_1	-36,570	-3,530	1,927	-35,604	1,030	-36,500	-896	-835
σ'_2	-9,140	-9,382	482	-8,900	258	-9,124	-224	-209
ΔR_A	-0.005699	-0.01474	0.00932	-0.00542	0.00900	-0.00574	-0.00032	-0.000354
ψ_A	-0.000900	0.001048	-0.000174	0.000874	-0.000146	0.000902	-0.000028	-0.000039

Results Table (stresses in lb/in², deflection in inches, rotation in radians)

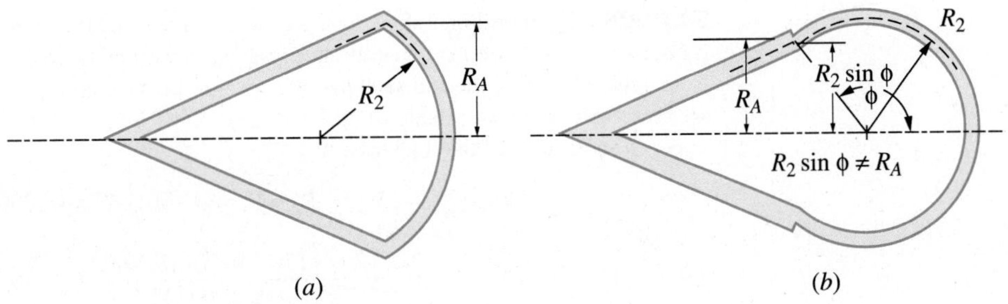

Figure 13.12

Most shell intersections have a common circumference, identified by the radius R_A, and defined as the intersection of the midsurfaces of the shells. If the two shells have meridional slopes which differ substantially at this intersection, the shape of the joint is easily described. See Fig. 13.12(*a*). If, however, these slopes are very nearly the same and the shell thicknesses differ appreciably, the intersection of the two midsurfaces could be far away from an actual joint, and the midthickness radius must be defined for each shell. See Fig. 13.12(*b*).

For this reason there are two sets of correction terms based on these two joint contours. All correction terms are treated as external loads on the right-hand member. The appropriate portion of this loading is transferred back to the left-hand member by small changes in the radial load V_1 and the moment M_1 which are found by equating the deformations in the two shells at the junction. In each case the formulas for the stresses at the junction are given only for the left-hand member. Stresses are computed on the assumption that each member ends abruptly at the joint with the end cross section normal to the meridian. No stress concentrations are considered, and no reduction in stress due to any added weld material or joint reinforcement has been made. The examples show how such corrections can be made for the stresses.

While the discussion above has concentrated primarily on the stresses at or very near the junction of the members, there are cases where stresses at some distance from the junction can be a source of concern. Although a toroidal shell is not included in Table 13.4, the presence of large circumferential compressive stress in the toroidal region of a torispherical head on a pressure vessel is known to create buckling instabilities when such a vessel is subjected to internal pressure. Section 15.4 describes this problem and others of a similar nature such as a truncated spherical shell under axial tension.

EXAMPLE 1 The shell consisting of a cone and a partial sphere shown in Fig. 13.13 is subjected to an internal pressure of 500 N/cm². The junction deformations and the circumferential and meridional stress components at the inside surface of the junction are required. Use $E = 7(10^6)$ N/cm² and $\nu = 0.3$ for the material in both portions of the shell. All the linear dimensions will be given and used in centimeters.

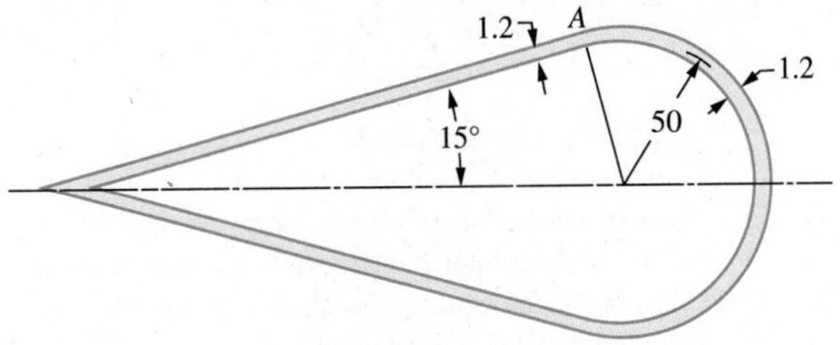

Figure 13.13

SOLUTION The meridional slopes of the cone and sphere are the same at the junction, and the sphere is not truncated nor are any penetrations present at any other location, so, $\theta_2 = \phi_2 = 105°$. Using case 6 from Table 13.4, the cone and shell parameters and the limiting values for which the given equations are acceptable are now evaluated.

For the cone using Table 13.3, case 4:

$$\alpha_1 = 15°, \qquad R_A = R_1 = 50 \sin 105° = 48.296$$

$$k_A = \frac{2}{\sin 15°}\left[\frac{12(1-0.3^2)(48.296^2)}{1.2^2 \sec^2 15°}\right]^{0.25} = 87.58$$

Where $\mu = 4$, the value of $k_B = 87.58 - 4\sqrt{2} = 81.93$ and $R_B = 42.26$. Since both K_A and K_B are greater than 5 and both R_A and R_B are greater than 5 (1.2 cos 15°), the cone parameters are within the acceptable range for use with the equations. $b_1 = 48.296 - 1.2\cos 15°/2 = 47.717$, $\alpha_1 = 48.876$, and $\beta_1 = [12(1-0.3^2)]^{0.5} = 3.305$.

For the sphere using Table 13.3, case 1:

$$\beta_2 = \left[3(1-0.3^2)\left(\frac{50}{1.2}\right)^2\right]^{0.25} = 8.297$$

$$b_2 = 50 - \frac{1.2}{2} = 49.40, \qquad a_2 = 50.60$$

and, at the edge, where $\omega = 0$,

$$K_{12} = 1 - [1 - 2(0.3)]\frac{(\cot 105°)/2}{8.297} = 1.0065 \quad \text{and} \quad K_{22} = 1.0258$$

Since $3/\beta_2 = 0.3616$ and $\pi - 3/\beta_2 = 2.78$, the value of $\phi_2 = 1.833$ rad lies within this range, so the spherical shell parameters are also acceptable.

Next the several junction constants are determined from the shell parameters just found and from any others required. Again from Table 13.3, case 4:

$$F_{2A} = 1 - \frac{2.652}{87.58} + \frac{3.516}{87.58^2} - \frac{2.610}{87.58^3} + \frac{0.038}{87.58^4} = 0.9702$$

Similarly,

$$F_{4A} = 0.9624, \qquad F_{7A} = 0.9699, \qquad \text{and} \qquad C_1 = 0.9720$$

Using these values, $C_{AA1} = 638.71$, $C_{AA2} = 651.39$, $C_{AA} = 1290.1$, $C_{AB1} = -132.72$, $C_{AB2} = 132.14$, $C_{AB} = -0.5736$, $C_{BB1} = 54.736$, $C_{BB2} = 54.485$, and $C_{BB} = 109.22$.

Turning now to the specific loadings needed, one uses case 6a for internal pressure with *no axial load on the junction* and case 6b with an axial load $P = 500\pi(47.72)^2 = 3.577(10^6)$N.

For case 6a: Although the tables of numerical data include α_1 = 15° and ϕ_1 = 105° as a given pair of parameters, the value of R_1/t_1 = 40.25 is not one of the values for which data are given. The load terms are

$$LT_{A1} = \frac{47.72(48.3)}{1.2^2 \cos 15°} = 1656.8, \qquad LT_{A2} = -1637.0, \qquad LT_{B1} = -22.061, \qquad LT_{B2} = 0$$

In this example the junction meridians are tangent and the inside surface is smooth, so there are no correction terms to consider. Had the radii and the thicknesses been such that the welded junction had an internal step, either abrupt or tapered, the internal pressure acting upon this step would be accounted for by the appropriate correction terms (see the next example).

Now combining the shell and load terms, $LT_A = 1656.8 - 1637.0 + 0 = 19.8$, $LT_B = -22.06$, $K_{V1} = 0.0153$, $K_{M1} = -0.2019$, $V_1 = 9.913$, $M_1 = -145.4$, $N_1 = -2.379$, $\Delta R_A = 0.1389$, $\psi_A = 641(10^{-6})$, $\sigma_1 = -1.98$, $\sigma_2 = 20{,}128$, $\sigma_1' = 605.7$, and $\sigma_2' = 196.2$.

For case 6b:

$$LT_{A1} = \frac{-0.3(48.30^2)}{2(1.2^2)\cos 15°} = -251.5, \qquad LT_{A2} = 1090.0$$

$$LT_{B1} = 5.582, \qquad LT_{B2} = 0, \qquad LT_{AC} = 0, \qquad LT_{BC} = 0$$

$$LT_A = 838.5, \qquad LT_B = 5.582$$

Again combine the shell and load terms to get $K_{V1} = 0.6500$, $K_{M1} = 0.0545$, $V_1 = 380.7$, $M_1 = 38.33$, $N_1 = 12{,}105$, $\Delta R_A = -0.0552$, $\psi_A = -0.0062$, $\sigma_1 = 10{,}087$, $\sigma_2 = -4972$, $\sigma_1' = -159.7$, and $\sigma_2' = -188$.

The final step is to sum the deformations and stresses. $\Delta R_A = 0.0837$, $\psi_A = -0.0056$, $\sigma_1 = 10{,}085$, $\sigma_2 = 15{,}156$, $\sigma_1' = 446.0$, $\sigma_2' = 8.2$. A check of these values against the tabulated constants shows that reasonable values could have been obtained by interpolation. At the junction the shell moves outward a distance of 0.0837 cm, and the upper meridian as shown in Fig. 13.13 rotates 0.0056 rad clockwise. On the inside surface of the junction the circumferential stress is 15,164 N/cm^2 and the meridional stress is 10,531 N/cm^2.

As should have been expected by the smooth transition from a conical to a spherical shell of the same thickness, the bending stresses are very small. In the next example the smooth inside surface will be retained but the cone and sphere will be different in thickness, and external pressure will be applied to demonstrate the use of the terms which correct for the pressure loading on the step in the wall thickness at the junction.

EXAMPLE 2 The only changes from Example 1 will be to make the pressure external at 1000 N/cm^2 and to increase the cone thickness to 4 cm and the sphere thickness to 2 cm. The smooth inside surface will be retained. If the correction terms are not used in this example, the external pressure will be presumed to act on the outer surface of the cone up to the junction and on the external spherical surface of the sphere. There will be no consideration given for the external pressure acting upon the 2-cm-wide external shoulder at the junction. The correction terms treat the additional axial and radial loadings and the added moment due to this pressure loading on the shoulder. If a weld fillet were used at the junction, the added pressure loading would be the same, so the correction terms are still applicable but *no* consideration is made for the added stiffness due to the extra material in the weld fillet. If the meridians for the cone and for the sphere intersect at an angle more than about 5°, a different correction term is used. This second correction term assumes that no definite step occurs on either the inner or the outer surface. See the discussion in Sec. 13.4 related to Fig. 13.12.

SOLUTION For the cone using Table 13.3, case 4: $\alpha_1 = 15°$, $R_A = R_1 = 51 \sin 105°$, and $K_A = 48.449$ when the radius and thickness are changed to the values shown in Fig. 13.14. Where $\mu = 4$, the value of $K_B = 42.793$ and $R_B = 38.4$, which is greater than 5 $(4 \cos 15°)$. Thus, again K_A and K_B are greater than 5 and the cone parameters are within the acceptable range for use with the equations. In a similar manner the parameters for the sphere are found to be within the range for which the equations are acceptable. Repeating the calculations as was done for the first example, with and without correction terms, one finds the following stresses:

	Case 6a ($q = -1000$ N/cm^2)		
	Without Correction Terms	With Correction Terms	Case 6b ($P = -8{,}233{,}600$ N)
ΔR_A	−0.1244	−0.1262	0.0566
γ_A	−0.00443	−0.00425	0.00494
σ_1	−110.15	−133.64	−6739.1
σ_2	−17,710	−17,972	6021.3
σ'_1	1317.2	342.3	−1966.2
σ'_2	283.7	−99.0	−454.3

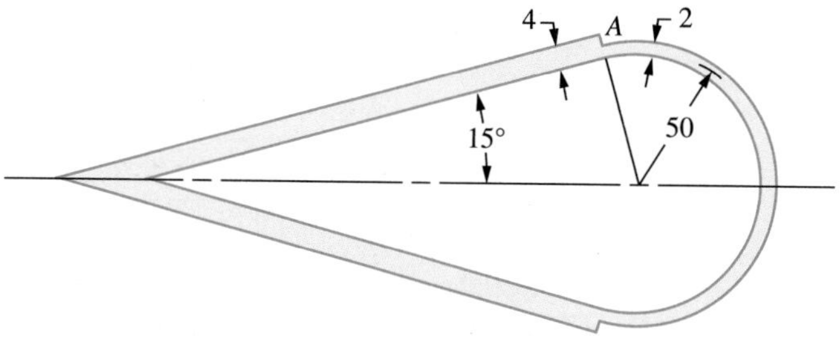

Figure 13.14

The effect of the correction terms is apparent but does not cause large changes in the larger stresses or the deformations. Summing the values for cases 6a with correction terms and for 6b gives the desired results as follows. The radial deflection at the junction is 0.0696 cm inward, the upper meridian rotates 0.00069 rad clockwise, on the inside of the junction the circumferential stress is $-12{,}504\ \text{N/cm}^2$, and the meridional stress is $-8497\ \text{N/cm}^2$.

EXAMPLE 3 The vessel shown in Fig. 13.15 is conical with a flat-plate base supported by an annular line load at a radius of 35 in. The junction deformations and the meridional and circumferential stresses on the outside surface at the junction of the cone and plate are to be found. The only loading to be considered is the hydrostatic loading due to the water contained to a depth of 50 in. Use $E=10(10^6)\ \text{lb/in}^2$ and $\nu=0.3$ for the material in the shell and in the plate. All linear dimensions will be given and used in inches.

SOLUTION The proportions chosen in this example are ones matching the tabulated data in cases 7b and 7c from Table 13.4 in order to demonstrate the use of this tabulated information.

From case 7c with the loading of an internal hydrostatic pressure with *no axial load on the junction* and for $\alpha=-30°$, $R_1=50$, $t_1=1$, $R_1/t_1=50$, $t_2=2$, $t_2/t_1=2$, $x_1=50$, $x_1/R_1=1$, $R_2/R_1=35/50=0.7$, $\nu_1=\nu_2=0.3$; and for $a_2=a_1$ due to the plate extending only to the outer surface of the cone at the junction, we find from the table the following coefficients: $K_{V1}=3.3944$, $K_{M1}=2.9876$, $K_{\Delta RA}=0.1745$, $K_{\psi A}=-7.5439$, and $K_{\sigma 2}=0.1847$. Since water has a weight of 62.4 lb/ft³, the internal pressure q_1 at the junction is $62.4(50)/1728=1.806\ \text{lb/in}^2$. Using these coefficients and the dimensions and the material properties we find that

$$V_1=1.806(1)(3.3944)=6.129$$

$$M_1=1.806(1^2)(2.9876)=5.394$$

$$N_1=-6.129\sin -30°=3.064$$

$$\Delta R_A=\frac{1.806(50^2)}{10(10^6)(1)}0.1745=78.77(10^{-6})$$

$$\psi_A=\frac{1.806(50)}{10(10^6)(1)}(-7.5439)=-68.10(10^{-6})$$

$$\sigma_1=\frac{3.064}{1}=3.064$$

$$\sigma_2=78.77(10^{-6})\frac{10(10^6)}{50}+0.3(3.064)=16.673$$

$$\sigma_1'=\frac{-6(5.394)}{1^2}=-32.364$$

$\sigma_2'=-2.895$ (*Note*: The extensive calculations are not shown.)

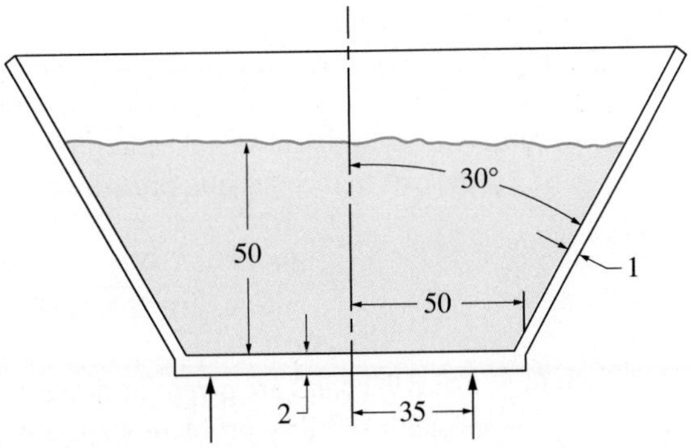

Figure 13.15

In the above calculations no correction terms were used. When the correction terms are included and the many calculations carried out, the deformations and stresses are found to be

$$\Delta R_A = 80.13(10^{-6}), \qquad \psi_A = 68.81(10^{-6})$$

$$\sigma_1 = 3.028, \qquad \sigma_2 = 16.934, \qquad \sigma_1' = -30.595, \qquad \sigma_2' = -5.519$$

There is not a great change due to the correction terms.

For case 7b the axial load to be used must now be calculated. The radius of the fluid at the plate is $50 - 0.5/\cos 30° = 49.423$. The radius of the fluid at the top surface is $49.423 + 50\tan 30° = 78.290$. The vertical distance below the top of the plate down to the tip of the conical inner surface is 49.423/ tan 30° = 85.603. The volume of fluid = $\pi(78.290)^2(85.603 + 50)/3 - \pi(49.423)^2(85.603)/3 = 651{,}423$ in^3. The total weight of the fluid = 651,423(62.4)/1728 = 23,524 lb. The axial load acting on the plate in case 7c = $q_1\pi b_1^2 = 1.806\pi(49.56)^2 = 13{,}940$ lb. Using case 7b with the axial compressive load $P = 103{,}940 - 23{,}524 = -9584$ gives the following results:

$$\Delta R_A = 496.25(10^{-6}), \qquad \psi_A = -606.03(10^{-6})$$

$$\sigma_1 = -57.973, \qquad \sigma_2 = 81.858, \qquad \sigma_1' = 1258.4, \qquad \sigma_2' = 272.04$$

Summing the results from case 7c with correction terms and from case 7b produces $\Delta R_A = 576.38(10^{-6})$, $\psi_A = -674.84(10^{-6})$, $\sigma_1 = -54.945$, $\sigma_2 = 98.792$, $\sigma_1' = 1227.8$, and $\sigma_2' = 269.5$.

The junction moves radially outward a distance of 0.00058 in, and the junction meridian on the right in Fig. 13.15 rotates 0.000675 rad clockwise. On the outside of the cone at the junction, the circumferential stress is 368.3 lb/in^2 and the meridional stress is 1173 lb/in^2.

13.5 THIN SHELLS OF REVOLUTION UNDER EXTERNAL PRESSURE

All formulas given in Tables 13.1 and 13.3 for thin vessels under distributed pressure are for internal pressure, but they will apply equally to cases of external pressure if q is given a negative sign. The formulas in Table 13.2 for distributed pressure are for external pressure in order to correspond to similar loadings for beams on elastic foundations in Chap. 8. It should be noted with care that the application of external pressure may cause an instability failure due to stresses lower than the elastic limit, and in such a case the formulas in this chapter do not apply. This condition is discussed in Chap. 15, and formulas for the critical pressures or stresses producing instability are given in Table 15.2.

A vessel of moderate thickness may collapse under external pressure at stresses just below the yield point, its behavior being comparable to that of a short column. The problem of ascertaining the pressure that produces failure of this kind is of special interest in connection with cylindrical vessels and pipe. For external loading such as that in Table 13.1, case 1c, the external collapsing pressure can be given by

$$q' = \frac{t}{R}\frac{\sigma_y}{1+(4\sigma_y/E)(R/t)^2} \qquad \text{(see Refs. 1, 7, and 8)}$$

In Refs. 8 and 9, charts are given for designing vessels under external pressure.

A special instability problem should be considered when designing long cylindrical vessels or even *relatively short corrugated tubes* under internal pressure. Haringx (Refs. 54 and 55) and Flügge (Ref. 5) have shown that vessels of this type will buckle laterally if the ends are restrained against longitudinal displacement and if the product of the internal pressure and the cross-sectional area reaches the Euler load for the column as a whole. For cylindrical shells this is seldom a critical factor, but for corrugated tubes or bellows this is recognized as a so-called *squirming instability*. To determine the Euler load for a bellows, an equivalent thin-walled circular cross section can be established which will have a radius equal to the mean radius of the bellows and a product Et, for which the equivalent cylinder will have the same axial deflection under end load as would the bellows. The overall bending moment of inertia I of the very thin equivalent cylinder can then be used in the expression $P_u = K\pi^2 EI/l^2$ for the Euler load. In a similar way Seide (Ref. 56) discusses the effect of pressure on the lateral bending of a bellows.

EXAMPLE A corrugated-steel tube has a mean radius of 5 in, a wall thickness of 0.015 in, and 60 semicircular corrugations along its 40-in length. The ends are rigidly fixed, and the internal pressure necessary to produce a squirming instability is to be calculated. Given: $E = 30(10^6)$ lb/in^2 and $\nu = 0.3$.

SOLUTION Refer to Table 13.3, case 6b: $a = 5$ in, length = 40 in, $b = 40/120 = 0.333$ in, and $t = 0.015$ in

$$\mu = \frac{b^2}{at}\sqrt{12(1-\nu^2)} = \frac{0.333^2}{5(0.015)}\sqrt{12(1-0.3^2)} = 4.90$$

$$\text{Axial stretch} = \frac{-0.577Pbn\sqrt{1-\nu^2}}{Et^2} = \frac{0.577P(0.333)(60)\sqrt{0.91}}{30(10^6)(0.015^2)} = -0.00163P$$

If a cylinder with a radius of 5 in and product E_1t_1 were loaded in compression with a load P, the stretch would be

$$\text{Stretch} = \frac{-Pl}{A_1E_1} = \frac{-P(40)}{2\pi 5t_1E_1} = -0.00163P$$

or

$$t_1E_1 = \frac{40}{2(5)(0.00163)} = 780.7 \text{ lb/in}$$

The bending moment of inertia of such a cylinder is $I_1 = \pi R^3 t_1$ (see Table A.1, case 13). The Euler load for fixed ends is

$$P_{cr} = \frac{4\pi^2E_1I_1}{l^2} = \frac{4\pi^2E_1\pi R^3t_1}{l^2} = \frac{4\pi^3 5^3(780.7)}{40^2} = 7565 \text{ lb}$$

The internal pressure is therefore

$$q' = \frac{P_{cr}}{\pi R^2} = \frac{7565}{\pi 5^2} = 96.3 \text{ lb/in}^2$$

From Table 13.3, case 6c, the maximum stresses caused by this pressure are

$$(\sigma_2)_{\max} = 0.955(96.3)(0.91)^{1/6}\left[\frac{5(0.333)}{0.015^2}\right]^{2/3} = 34{,}400 \text{ lb/in}^2$$

$$(\sigma')_{\max} = 0.955(96.3)(0.91)^{-1/3}\left[\frac{5(0.333)}{0.015^2}\right]^{2/3} = 36{,}060 \text{ lb/in}^2$$

If the yield strength is greater than 36,000 lb/in^2, the corrugated tube should buckle laterally, that is, squirm, at an internal pressure of 96.3 lb/in^2.

13.6 THICK SHELLS OF REVOLUTION

If the wall thickness of a vessel is more than about one-tenth the radius, the meridional and hoop stresses cannot be considered uniform throughout the thickness of the wall and the radial stress cannot be considered negligible. These stresses in thick vessels, called *wall stresses*, must be found by formulas that are quite different from those used in finding membrane stresses in thin vessels.

It can be seen from the formulas for cases 1a and 1b of Table 13.5 that the stress σ_2 at the inner surface of a thick cylinder approaches q as the ratio of outer to inner radius approaches infinity. It is apparent therefore that if the stress is to be limited to some specified value σ, the pressure must never exceed $q = \sigma$, no matter how thick the wall is made. To overcome this limitation, the material at and near the inner surface must be put into a state of initial compression. This can be done by shrinking on one or more jackets (as explained in Sec. 3.12 and in the example therein) or by subjecting the vessel to a high internal pressure that stresses the inner part into the plastic range and, when removed, leaves residual compression there and residual tension in the outer part. This procedure is called *autofrettage*, or *self-hooping*. If many successive jackets are superimposed on the original tube by shrinking or wrapping, the resulting structure is called a *multilayer vessel*. Such a construction has certain advantages, but it should be noted that the formulas for hoop stresses are based on the assumption that an isotropic material is used. In a multilayered vessel the effective radial modulus of elasticity is less than the tangential modulus, and in consequence the hoop stress at and near the outer wall is less than the formula would indicate; therefore, the outer layers of material contribute less to the strength of the vessel than might be supposed.

Cases 1e and 1f in Table 13.5 represent radial body-force loading, which can be superimposed to give results for centrifugal loading, etc. (see Sec. 16.2). Case 1f is directly applicable to thick-walled disks with embedded electrical conductors used to generate magnetic fields. In many such cases the magnetic field varies linearly through the wall to zero at the outside. If there is a field at the outer turn, cases 1e and 1f can be superimposed in the necessary proportions.

The tabulated formulas for elastic wall stresses are accurate for both thin and thick vessels, but formulas for predicted yield pressures do not always agree closely with experimental results (Refs. 21, 34–37, and 39). The expression for q_y given in Table 13.5 is based on the minimum strain-energy theory of elastic failure. The expression for bursting pressure

$$q_u = 2\sigma_u \frac{a-b}{a+b} \tag{13.6-1}$$

commonly known as the *mean diameter formula*, is essentially empirical but agrees reasonably well with experiment for both thin and thick cylindrical vessels and is convenient to use. For very thick vessels the formula

$$q_u = \sigma_u \ln\frac{a}{b} \tag{13.6-2}$$

is preferable. Greater accuracy can be obtained by using with this formula a multiplying factor that takes into account the strain-hardening properties of the material (Refs. 10, 20, and 37). With the same objective, Faupel (Ref. 39) proposes (with different notation) the formula

$$q_u = \frac{2\sigma_y}{\sqrt{3}}\left(2 - \frac{\sigma_y}{\sigma_u}\right)\ln\frac{a}{b} \tag{13.6-3}$$

A rather extensive discussion of bursting pressure is given in Ref. 38, which presents a tabulated comparison between bursting pressures as calculated by a number of different formulas and as determined by actual experiment.

EXAMPLE At the powder chamber, the inner radius of a 3-in gun tube is 1.605 in and the outer radius is 2.425 in. It is desired to shrink a jacket on this tube to produce a radial pressure between the tube and jacket of 7600 lb/in^2. The outer radius of this jacket is 3.850 in. It is required to determine the difference between the inner radius of the jacket and the outer radius of the tube in order to produce the desired pressure, calculate the stresses in each part when assembled, and calculate the stresses in each part when the gun is fired, generating a powder pressure of 32,000 lb/in^2.

SOLUTION Using the formulas for Table 13.5, case 1c, it is found that for an external pressure of 7600 lb/in^2, the stress σ_2 at the outer surface of the tube is –19,430 lb/in^2, the stress σ_2 at the inner surface is –27,050 lb/in^2, and the change in outer radius $\Delta a = -0.001385$ in; for an internal pressure of 7600 lb/in^2, the stress σ_2 at the inner surface of the jacket is +17,630 lb/in^2, the stress σ_2 at the outer surface is +10,050 lb/in^2, and the change in inner radius $\Delta b = +0.001615$ in. (In making these calculations the inner radius of the jacket is assumed to be 2.425 in.) The initial difference between the inner radius of the jacket and the outer radius of the tube must be equal to the sum of the radial deformations they suffer, or $0.001385 + 0.001615 = 0.0030$ in; therefore, the initial radius of the jacket should be $2.425 - 0.0030 = 2.422$ in.

The stresses produced by the powder pressure are calculated at the inner surface of the tube, at the common surface of tube and jacket (r = 2.425 in), and at the outer surface of the jacket. These stresses are then superimposed on those found previously. The calculations are as follows:

For the tube at the inner surface,

$$\sigma_2 = +32{,}000\frac{3.85^2 + 1.605^2}{3.85^2 - 1.605^2} = 45{,}450 \text{ lb/in}^2$$

$$\sigma_3 = -32{,}000 \text{ lb/in}^2$$

For tube and jacket at the interface,

$$\sigma_2 = +32{,}000\frac{1.605^2}{2.425^2}\frac{3.85^2 + 2.425^2}{3.85^2 - 1.605^2} = +23{,}500 \text{ lb/in}^2$$

$$\sigma_3 = -32{,}000\frac{1.605^2}{2.425^2}\frac{3.85^2 - 2.425^2}{3.85^2 - 1.605^2} = -10{,}200 \text{ lb/in}^2$$

For the jacket at the outer surface,

$$\sigma_2 = +32{,}000\frac{1.605^2}{3.85^2}\frac{3.85^2 + 3.85^2}{3.85^2 - 1.605^2} = +13{,}500 \text{ lb/in}^2$$

These are the stresses due to the powder pressure. Superimposing the stresses due to the shrinkage, we have as the resultant stresses:

At inner surface of tube,

$$\sigma_2 = -27{,}050 + 45{,}450 = +18{,}400 \text{ lb/in}^2$$

$$\sigma_3 = 0 - 32{,}000 = -32{,}000 \text{ lb/in}^2$$

At outer surface of tube,

$$\sigma_2 = -19{,}430 + 23{,}500 = +4070 \text{ lb/in}^2$$

$$\sigma_3 = -7600 - 10{,}200 = -17{,}800 \text{ lb/in}^2$$

At inner surface of jacket,

$$\sigma_2 = +17{,}630 + 23{,}500 = +41{,}130 \text{ lb/in}^2$$

$$\sigma_3 = -7600 - 10{,}200 = -17{,}800 \text{ lb/in}^2$$

At outer surface of jacket,

$$\sigma_2 = 10{,}050 + 13{,}500 = +23{,}550 \text{ lb/in}^2$$

13.7 PIPE ON SUPPORTS AT INTERVALS

For a pipe or cylindrical tank supported at intervals on saddles or pedestals and filled or partly filled with liquid, the stress analysis is difficult and the results are rendered uncertain by doubtful boundary conditions. Certain conclusions arrived at from a study of tests (Refs. 11 and 12) may be helpful in guiding design. See also Ref. 75.

1. For a circular pipe or tank supported at intervals and held circular at the supports by rings or bulkheads, the ordinary theory of flexure is applicable if the pipe is completely filled.
2. If the pipe is only partially filled, the cross section at points between supports becomes out of round and the distribution of longitudinal fiber stress is neither linear nor symmetrical across the section. The highest stresses occur for the half-full condition; then the maximum longitudinal compressive stress and the maximum circumferential bending stresses occur at the ends of the horizontal diameter, the maximum longitudinal tensile stress occurs at the bottom, and the longitudinal stress at the top is practically zero. According to theory (Ref. 4), the greatest of these stresses is the longitudinal compression, which is equal to the maximum longitudinal stress for the full condition divided by

$$K = \left(\frac{L}{R}\sqrt{\frac{t}{R}} \right)^{1/2}$$

where R = pipe radius, t = thickness, and L = span. The maximum circumferential stress is about one-third of this. Tests (Ref. 11) on a pipe having $K = 1.36$ showed a longitudinal stress that is somewhat less and a circumferential stress that is considerably greater than indicated by this theory.
3. For an unstiffened pipe resting in saddle supports, there are high local stresses, both longitudinal and circumferential, adjacent to the tips of the saddles. These stresses are less for a large saddle angle β (total angle subtended by arc of contact

between pipe and saddle) than for a small angle, and for the ordinary range of dimensions they are practically independent of the thickness of the saddle, that is, its dimension parallel to the pipe axis. For a pipe that fits the saddle well, the maximum value of these localized stresses will probably not exceed that indicated by the formula

$$\sigma_{max} = k\frac{P}{t^2}\ln\frac{R}{t}$$

where P = total saddle reaction, R = pipe radius, t = pipe thickness, and K = coefficient given by

$$k = 0.02 - 0.00012(\beta - 90)$$

where β is in degrees. This stress is almost wholly due to circumferential bending and occurs at points about 15° above the saddle tips.

4. The maximum value of P the pipe can sustain is about 2.25 times the value that will produce a maximum stress equal to the yield point of the pipe material, according to the formula given above.
5. The comments in conclusion 3 above are based on the results of tests performed on very thin-walled pipe. Evces and O'Brien in Ref. 73 describe similar tests on thicker-walled ductile-iron pipe for which R/t does not normally exceed 50. They found that optimum saddle angles lie in the range $90° > \beta > 120°$ and that for $R/t \geq 28$ the formulas for σ_{max} can be used if the value of k is given by

$$k = 0.03 - 0.00017(\beta - 90)$$

The maximum stress will be located within ±15° of the tip if the pipe fits the saddle well.
6. For a pipe supported in flexible slings instead of on rigid saddles, the maximum local stresses occur at the points of tangency of sling and pipe section; in general, they are less than the corresponding stresses in the saddle-supported pipe but are of the same order of magnitude.

A different but closely related support system for horizontal cylindrical tanks consists of a pair of longitudinal line loads running the full length of the vessel. If the tank wall is thin, accounting for the deformations, which are normally ignored in standard stress formulas, it shows that the stresses are significantly lower. Cook in Ref. 79 uses a nonlinear analysis to account for deformations and reports results for various positions of the supports, radius/thickness ratios, and depths of fill in the tank.

13.8 TABLES

TABLE 13.1 Formulas for Membrane Stresses and Deformations in Thin-Walled Pressure Vessels

Notation: P = axial load (force); p = unit load (force per unit length); q and w = unit pressures (force per unit area); δ = density (force per unit volume); σ_1 = meridional stress; σ_2 = circumferential, or hoop, stress; R_1 = radius of curvature of a meridian, a principal radius of curvature of the shell surface; R_2 = length of the normal between the point on the shell and the axis of rotation, the second principal radius of curvature; R = radius of curvature of a circumference; ΔR = radial displacement of a circumference; Δy = change in the height dimension y; y = length of cylindrical or conical shell and is also used as a vertical position coordinate, positive upward, from an indicated origin in some cases; ψ = rotation of a meridian from its unloaded position, positive when that meridional rotation represents an increase in ΔR when y or θ increases; E = modulus of elasticity; and ν = Poisson's ratio.

Case No., Form of Vessel	Manner of Loading	Formulas
1. Cylindrical 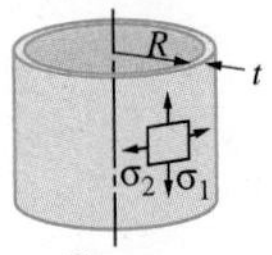 $\frac{R}{t} > 10$	1a. Uniform axial load, p force/unit length 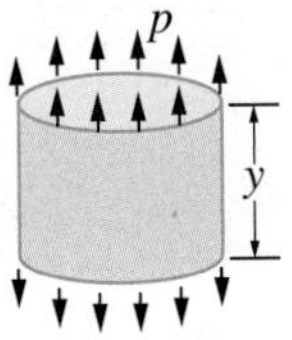	$\sigma_1 = \frac{p}{t}$ $\sigma_2 = 0$ $\Delta R = \frac{-p\nu R}{Et}$ $\Delta y = \frac{py}{Et}$ $\psi = 0$
	1b. Uniform radial pressure, q force/unit area 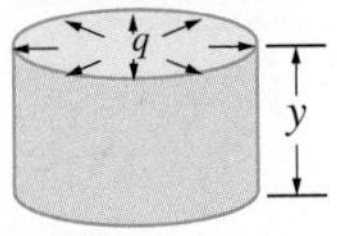	$\sigma_1 = 0$ $\sigma_2 = \frac{qR}{t}$ $\Delta R = \frac{qR^2}{Et}$ $\Delta y = \frac{-qR\nu y}{Et}$ $\psi = 0$
	1c. Uniform internal or external pressure, q force/unit area (ends capped)	At point away from the ends $\sigma_1 = \frac{qR}{2t}$ $\sigma_2 = \frac{qR}{t}$ $\Delta R = \frac{qR^2}{Et}\left(1 - \frac{\nu}{2}\right)$ $\Delta y = \frac{qRy}{Et}(0.5 - \nu)$ $\psi = 0$
	1d. Linearly varying radial pressure, q force/unit area 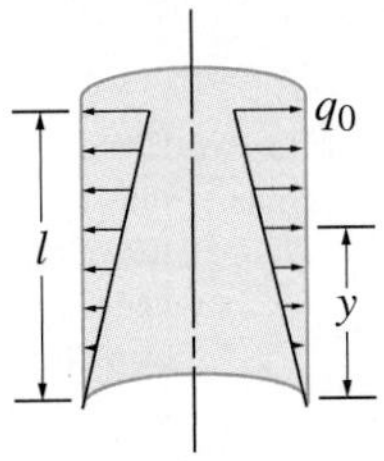	$q = \frac{q_0 y}{l}$ (where y must be measured from a free end. if pressure starts away from the end, see case 6 in Table 13.2) $\sigma_1 = 0$ $\sigma_2 = \frac{qR}{t} = \frac{q_0 Ry}{lt}$ $\Delta R = \frac{qR^2}{Et} = \frac{q_0 R^2 y}{Etl}$ $\Delta y = \frac{-q_0 R\nu y^2}{2Etl}$ $\psi = \frac{q_0 R^2}{Etl}$

(*Continued*)

TABLE 13.1 Formulas for Membrane Stresses and Deformations in Thin-Walled Pressure Vessels (*Continued*)

Case No., Form of Vessel	Manner of Loading	Formulas
	1e. Own weight, δ force/unit volume; top edge support, bottom edge free	$\sigma_1 = \delta y$ $\sigma_2 = 0$ $\Delta R = \dfrac{-\delta \nu R y}{E}$ $\Delta y = \dfrac{\delta y^2}{2E}$ $\psi = \dfrac{-\delta \nu R}{E}$
	1f. Uniform rotation, ω rad/s, about central axis 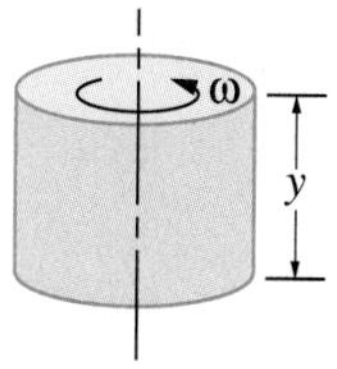 δ_m = mass density	$\sigma_1 = 0$ $\sigma_2 = \delta_m R^2 \omega^2$ $\Delta R = \dfrac{\delta_m R^3 \omega^2}{E}$ $\Delta y = \dfrac{-\nu \delta_m R^2 \omega^2 y}{E}$ $\psi = 0$
2. Cone 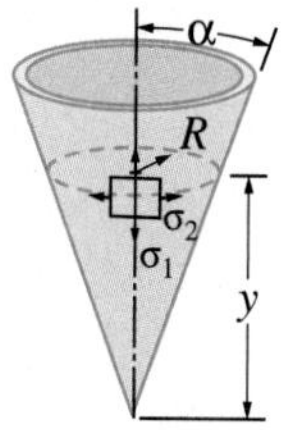 $\dfrac{R}{t} > 10$	2a. Uniform internal or external pressure, q force/unit area; tangential edge support 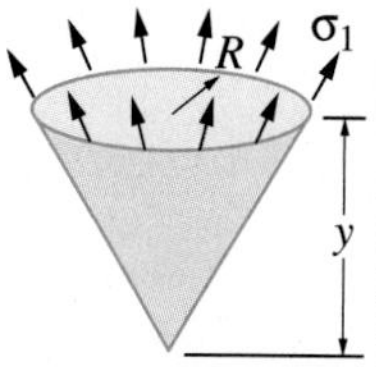	$\sigma_1 = \dfrac{qR}{2t\cos\alpha}$ $\sigma_2 = \dfrac{qR}{t\cos\alpha}$ $\Delta R = \dfrac{qR^2}{Et\cos\alpha}\left(1 - \dfrac{\nu}{2}\right)$ $\Delta y = \dfrac{qR^2}{4Et\sin\alpha}(1 - 2\nu - 3\tan^2\alpha)$ $\psi = \dfrac{3qR\tan\alpha}{2Et\cos\alpha}$
	2b. Filled to depth d with liquid of density δ force/ unit volume; tangential edge support 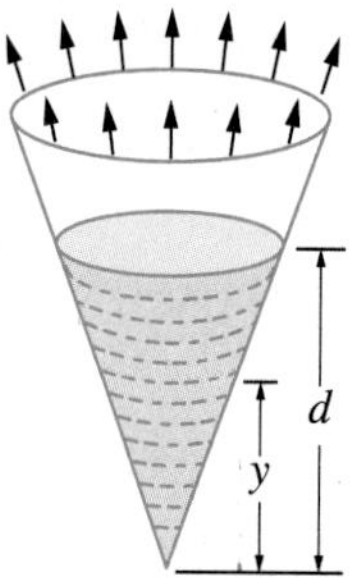	At any level y below the liquid surface $y \le d$ $\sigma_1 = \dfrac{\delta y \tan\alpha}{2t\cos\alpha}\left(d - \dfrac{2y}{3}\right), \quad (\sigma_1)_{max} = \dfrac{3\delta^2 \tan\alpha}{16t\cos\alpha} \quad \text{when } y = \dfrac{3d}{4}$ $\sigma_2 = \dfrac{y(d-y)\delta\tan\alpha}{t\cos\alpha}, \quad (\sigma_2)_{max} = \dfrac{\delta d^2 \tan\alpha}{4t\cos\alpha} \quad \text{when } y = \dfrac{d}{2}$ $\Delta R = \dfrac{\delta y^2 \tan^2\alpha}{Et\cos\alpha}\left[d\left(1 - \dfrac{\nu}{2}\right) - y\left(1 - \dfrac{\nu}{3}\right)\right]$ $\Delta y = \dfrac{\delta y^2 \sin\alpha}{Et\cos^4\alpha}\left\{\dfrac{d}{4}(1-2\nu) - \dfrac{y}{9}(1-3\nu) - \sin^2\alpha\left[\dfrac{d}{2}(2-\nu) - \dfrac{y}{3}(3-\nu)\right]\right\}$ $\psi = \dfrac{\delta y \sin^2\alpha}{6Et\cos^3\alpha}(9d - 16y)$ At any level y above the liquid level $\sigma_1 = \dfrac{\delta d^3 \tan\alpha}{6ty\cos\alpha}, \quad \sigma_2 = 0, \quad \Delta R = \dfrac{-\nu\delta d^3 \tan^2\alpha}{6Et\cos\alpha}$ $\Delta y = \dfrac{\delta d^3 \sin\alpha}{6Et\cos^4\alpha}\left[\dfrac{5}{6} - \nu(1 - \sin^2\alpha) + \ln\dfrac{y}{d}\right]$ $\psi = \dfrac{-\delta d^3 \sin^2\alpha}{6Et\cos^3\alpha}\dfrac{1}{y}$ (*Note:* There is a discontinuity in the rate of increase in fluid pressure at the top of the liquid. This leads to some bending in this region and is indicated by a discrepancy in the two expressions for the meridional slope at $y = d$.

(*Continued*)

TABLE 13.1 Formulas for Membrane Stresses and Deformations in Thin-Walled Pressure Vessels (*Continued*)

Case No., Form of Vessel	Manner of Loading	Formulas
	2c. Own weight, δ force/unit volume; tangential top edge support	$\sigma_1 = \frac{\delta R}{2\sin\alpha\cos\alpha}$ $\sigma_2 = \delta R \tan\alpha$ $\Delta R = \frac{\delta R^2}{E\cos\alpha}\left(\sin\alpha - \frac{\nu}{2\sin\alpha}\right)$ $\Delta y = \frac{\delta R^2}{E\cos^2\alpha}\left(\frac{1}{4\sin^2\alpha} - \sin^2\alpha\right)$ $\psi = \frac{2\delta R}{E\cos^2\alpha}\left[\sin^2\alpha\left(1+\frac{\nu}{2}\right) - \frac{1}{4}(1+2\nu)\right]$
	2d. Tangential loading only; resultant load = P 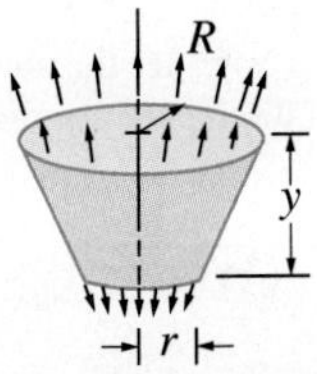 	$\sigma_1 = \frac{P}{2\pi R t\cos\alpha}$ $\sigma_2 = 0$ $\Delta R = \frac{-\nu P}{2\pi E t\cos\alpha}$ $\Delta y = \frac{P}{2\pi E t\sin\alpha\cos^2\alpha}\ln\frac{R}{r}$ $\psi = \frac{-P\sin\alpha}{2\pi E R t\cos^2\alpha}$ r must be finite to avoid infinite stress and $r/t > 10$ to be considered thin-walled
	2e. Uniform loading, force/unit area; on the horizontal projected area; tangential top edge support 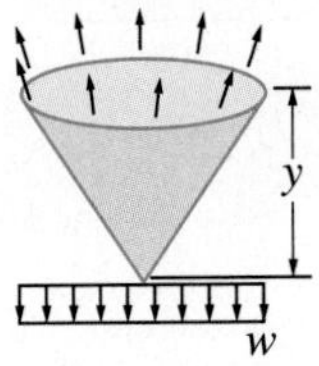 	$\sigma_1 = \frac{wR}{2t\cos\alpha}$ $\sigma_2 = \frac{wR\sin^2\alpha}{t\cos\alpha}$ $\Delta R = \frac{wR^2}{Et\cos\alpha}\left(\sin^2\alpha - \frac{\nu}{2}\right)$ $\Delta y = \frac{wR^2}{2Et\cos^2\alpha}\left[\frac{1}{2\sin\alpha} + \nu(1-\sin\alpha) - 2\sin^2\alpha\right]$ $\psi = \frac{wR\sin\alpha}{2Et\cos^2\alpha}(4\sin^2\alpha - 1 - 2\nu\cos^2\alpha)$
	2f. Uniform rotation, ω rad/s, about central axis 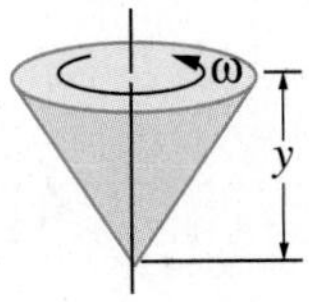 δ_m = mass density	$\sigma_1 = 0$ $\sigma_2 = \delta_m R^2\omega^2$ $\Delta R = \frac{\delta_m R^3\omega^2}{E}$ $\Delta y = \frac{-\delta_m R^3\omega^2}{E\cos\alpha}\left(\sin\alpha + \frac{\nu}{3\sin\alpha}\right)$ $\psi = \frac{\delta_m R^2\omega^2\tan\alpha}{E}(3+\nu)$

(*Continued*)

TABLE 13.1 Formulas for Membrane Stresses and Deformations in Thin-Walled Pressure Vessels (*Continued*)

Case No., Form of Vessel	Manner of Loading	Formulas
3. Spherical 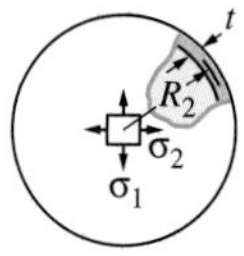 $\frac{R_2}{t} > 10$	3a. Uniform internal or external pressure, q force/unit area; tangential edge support 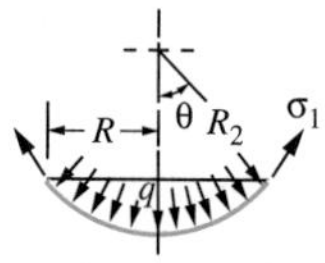	$\sigma_1 = \sigma_2 = \frac{qR_2}{2t}$ $\Delta R = \frac{qR_2^2(1-\nu)\sin\theta}{2Et}$ $\Delta R_2 = \frac{qR_2^2(1-\nu)}{2Et}$ $\Delta y = \frac{qR_2^2(1-\nu)(1-\cos\theta)}{2Et}$ $\psi = 0$
	3b. Filled to depth d with liquid of density δ force/unit volume; tangential edge support 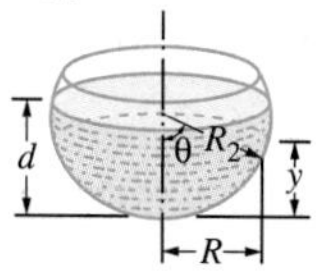	At any level y below the liquid surface, $y < d$ $\sigma_1 = \frac{\delta R_2^2}{6t}\left(3\frac{d}{R_2} - 1 + \frac{2\cos^2\theta}{1+\cos\theta}\right)$ $\sigma_2 = \frac{\delta R_2^2}{6t}\left[3\frac{d}{R_2} - 5 + \frac{(3+2\cos\theta)2\cos\theta}{1+\cos\theta}\right]$ (*Note:* See the note in case 2b regarding the discrepancy in the meridional slope $y = d$.) $\Delta R = \frac{\delta R_2^3 \sin\theta}{6Et}\left[3(1-\nu)\frac{d}{R_2} - 5 + \nu + 2\cos\theta\frac{3+(2-\nu)\cos\theta}{1+\cos\theta}\right]$ $\Delta y = \frac{\delta R_2^3}{6Et}\left\{3(1-\nu)\left[\frac{d}{R_2}(1-\cos\theta) + \cos\theta\right] - (2-\nu)\cos^2\theta + (1+\nu)\left(1 + 2\ln\frac{2}{1+\cos\theta}\right)\right\}$ $\psi = \frac{-\delta R_2^2}{Et}\sin\theta$ Weight of liquid $= P = \delta\pi d^2\left(R_2 - \frac{d}{3}\right)$ At any level y above the liquid level use case 3d with the load equal to the entire weight of the liquid.
	3c. Own weight, δ force/unit volume; tangential top edge support 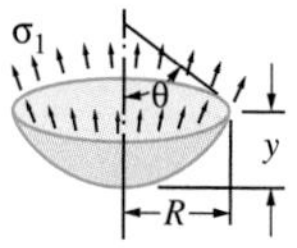	$\sigma_1 = \frac{\delta R_2}{1+\cos\theta}$, $\sigma_2 = -\delta R_2\left(\frac{1}{1+\cos\theta} - \cos\theta\right)$ Max tensile $\sigma_2 = \frac{\delta R_2}{2}$ at $\theta = 0$ $\sigma_2 = 0$ at $\theta = 51.83°$ $\Delta R = \frac{-\delta R_2^2 \sin\theta}{E}\left(\frac{1+\nu}{1+\cos\theta} - \cos\theta\right)$ $\Delta y = \frac{\delta R_2^2}{E}\left[\sin^2\theta + (1+\nu)\ln\frac{2}{1+\cos\theta}\right]$ $\psi = \frac{-\delta R_2}{E}(2+\nu)\sin\theta$
	3d. Tangential loading only; resultant load $= P$ 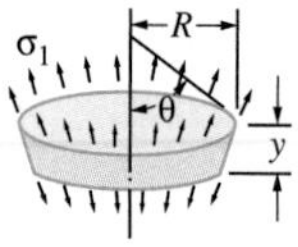	$\sigma_1 = \frac{P}{2\pi R_2 t \sin^2\theta}$, $\sigma_2 = -\sigma_1$ $\Delta R = \frac{-P(1+\nu)}{2\pi Et\sin\theta}$ $\Delta y = \frac{P(1+\nu)}{2\pi Et}\left[\ln\left(\tan\frac{\theta}{2}\right) - \ln\left(\tan\frac{\theta_o}{2}\right)\right]$ $\psi = 0$ (*Note:* θ_o is the angle to the lower edge and cannot go to zero without local bending occurring in the shell.)

(*Continued*)

TABLE 13.1 Formulas for Membrane Stresses and Deformations in Thin-Walled Pressure Vessels (*Continued*)

Case No., Form of Vessel	Manner of Loading	Formulas
	3e. Uniform loading, w force/unit area; on the horizontal projected area; tangential top edge support 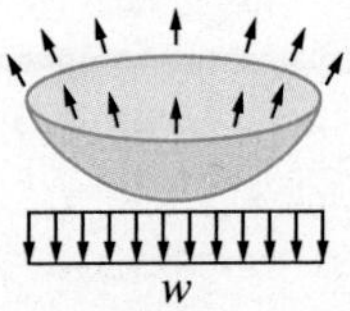	For $\theta \le 90°$ $\sigma_1 = \frac{wR_2}{2t}$ $\sigma_2 = \frac{wR_2}{2t}\cos 2\theta$ $\Delta R = \frac{wR_2^2 \sin\theta}{2Et}(\cos 2\theta - \nu)$ $\Delta y = \frac{wR_2^2}{2Et}[2(1-\cos^3\theta)+(1+\nu)(1-\cos\theta)]$ $\psi = \frac{-wR_2}{Et}(3+\nu)\sin\theta\cos\theta$
	3f. Uniform rotation, ω rad/s, about central axis 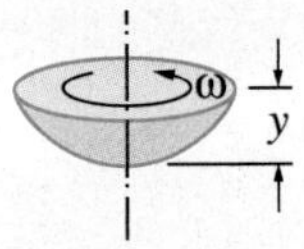 δ_m = mass density	$\sigma_1 = 0$ $\sigma_2 = \delta_m R^2\omega^2$ $\Delta R = \frac{\delta_m R^3\omega^2}{E}$ $\Delta y = \frac{-\delta_m R_2^3\omega^2}{E}(1+\nu-\nu\cos\theta-\cos^3\theta)$ $\psi = \frac{\delta_m R_2^2\omega^2\sin\theta\cos\theta}{E}(3+\nu)$
4. Any smooth figure of revolution if R_2 is less than infinity 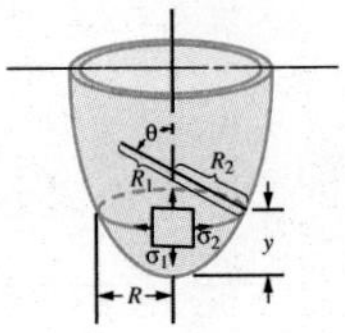 $\frac{R_2}{t} > 10$	4a. Uniform internal or external pressure, q force/unit area; tangential edge support	$\sigma_1 = \frac{qR_2}{2t}$ $\sigma_2 = \frac{qR_2}{2t}\left(2-\frac{R_2}{R_1}\right)$ $\Delta R = \frac{qR_2^2\sin\theta}{2Et}\left(2-\frac{R_2}{R_1}-\nu\right)$ $\psi = \frac{qR_2^2}{2EtR_1\tan\theta}\left[3\frac{R_1}{R_2}-5+\frac{R_2}{R_1}\left(2+\frac{1}{R_1}\frac{dR_1}{d\theta}\tan\theta\right)\right]$
	4b. Filled to depth d with liquid of density δ force/unit volume; tangential edge support. W = weight of liquid contained to a depth y	At any level y below the liquid surface, $y < d$, $\sigma_1 = \frac{W}{2\pi R_2 t\sin^2\theta}+\frac{\delta R_2(d-y)}{2t}$ $\sigma_2 = \frac{-W}{2\pi R_1 t\sin^2\theta}+\frac{\delta R_2(d-y)}{2t}\left(2-\frac{R_2}{R_1}\right)$ $\Delta R = \frac{R_2\sin\theta}{E}(\sigma_2-\nu\sigma_1)$ At any level y above the liquid level use case 4d with the load equal to the entire weight of the liquid

(Continued)

TABLE 13.1 Formulas for Membrane Stresses and Deformations in Thin-Walled Pressure Vessels (*Continued*)

Case No., Form of Vessel	Manner of loading	Formulas
	4c. Own weight, δ force/unit volume; tangential top edge support. W = weight of vessel below the level y	$\sigma_1 = \frac{W}{2\pi R_2 t \sin^2\theta}$ $\sigma_2 = \frac{W}{2\pi R_1 t \sin^2\theta} + \delta R_2 \cos\theta$ $\Delta R = \frac{R_2 \sin\theta}{E}(\sigma_2 - \nu\sigma_1)$
	4d. Tangential loading only, resultant load = P	$\sigma_1 = \frac{P}{2\pi R_2 t \sin^2\theta}$ $\sigma_2 = \frac{-P}{2\pi R_1 t \sin^2\theta}$ $\Delta R = \frac{-P}{2\pi E t \sin\theta}\left(\frac{R_2}{R_1} + \nu\right)$ $\psi = \frac{-P}{2\pi E t R_1 \sin^2\theta}\left[\frac{1}{\tan\theta}\left(1 + \frac{R_1}{R_2} - 2\frac{R_2}{R_1}\right) - \frac{R_2}{R_1^2}\frac{dR_1}{d\theta}\right]$
	4e. Uniform loading, w force/unit area, on the horizontal projected area; tangential top edge support	For $\theta \le 90°$ $\sigma_1 = \frac{wR_2}{2t}$ $\sigma_2 = \frac{wR_2}{2t}\left(2\cos^2\theta - \frac{R_2}{R_1}\right)$ $\Delta R = \frac{wR_2^2 \sin\theta}{2Et}\left(2\cos^2\theta - \frac{R_2}{R_1} - \nu\right)$ $\psi = \frac{w}{2EtR_1 \tan\theta}\left[R_1R_2(4\cos^2\theta - 1 - 2\nu\sin^2\theta) - R_2^2(7 - 2\cos\theta) + \frac{R_2^3}{R_1}\left(2 + \frac{\tan\theta}{R_1}\frac{dR_1}{d\theta}\right)\right]$
	4f. Uniform rotation, ω rad/s, about central axis δ_m = mass density	$\sigma_1 = 0$ $\sigma_2 = \delta_m R^2 \omega^2$ $\Delta R = \frac{\delta_m R^3 \omega^2}{E}$ $\psi = \frac{\delta_m R^2 \omega^2}{E \tan\theta}(3 + \nu)$
5. Toroidal shell 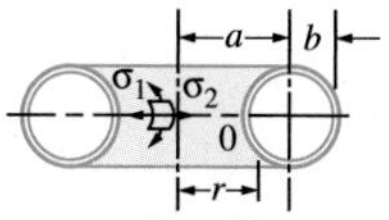 $b/t > 10$ b/t > 10	5a. Uniform internal or external pressure, q force/unit area	$\sigma_1 = \frac{qb}{2t}\frac{r+a}{r}$ $(\sigma_1)_{max} = \frac{qb}{2t}\frac{2a-b}{a-b}$ at point O $\sigma_2 = \frac{qb}{2t}$ (throughout) $\Delta r = \frac{qb}{2Et}[r - \nu(r+a)]$ [*Note:* There are some bending stresses at the top and bottom where R_2 (see case 4) is infinite (see Ref. 42).]

TABLE 13.2 Shear, Moment, Slope, and Deflection Formulas for Long and Short Thin-Walled Cylindrical Shells under Axisymmetric Loading

Notation: V_o, H, and p = unit loads (force per unit length); q = unit pressure (force per unit area); M_o = unit applied couple (force-length per unit length); all loads are positive as shown. At a distance x from the left end, the following quantities are defined: V = meridional radial shear, positive when acting outward on the right hand portion; M = meridional bending moment, positive when compressive on the outside; ψ = meridional slope (radians), positive when the deflection increases with x; y = radial deflection, positive outward. σ_1 and σ_2 = meridional and circumferential membrane stresses; positive when tensile; σ'_1 and σ'_2 = meridional and circumferential bending stresses, positive when tensile on the outside; τ = meridional radial shear stress; E = modulus of elasticity; ν = Poisson's ratio; R = mean radius; t = wall thickness.

The following constants and functions are hereby defined in order to permit condensing the tabulated formulas which follow:

$$\lambda = \left[\frac{3(1-\nu^2)}{R^2t^2}\right]^{1/4} \qquad D = \frac{Et^3}{12(1-\nu^2)}$$

(*Note:* See page 155 for a definition $\langle x-a\rangle^n$; also all hyperbolic and trigonometric functions of the argument $\langle x-a\rangle$ are also defined as zero if $(x<a)$ (Note: when the limitations on maximum deflections discussed in paragraph 3 of Sec 13.3.).

$F_1 = \cosh\lambda x\cos\lambda x$	$C_1 = \cosh\lambda l\cos\lambda l$	$C_{11} = \sinh^2\lambda l - \sin^2\lambda l$
$F_2 = \cosh\lambda x\sin\lambda x + \sinh\lambda x\cos\lambda x$	$C_2 = \cosh\lambda l\sin\lambda l + \sinh\lambda l\cos\lambda l$	$C_{12} = \cosh\lambda l\sinh\lambda l + \cos\lambda l\sin\lambda l$
$F_3 = \sinh\lambda x\sin\lambda x$	$C_3 = \sinh\lambda l\sin\lambda l$	$C_{13} = \cosh\lambda l\sinh\lambda l - \cos\lambda l\sin\lambda l$
$F_4 = \cosh\lambda x\sin\lambda x - \sinh\lambda x\cos\lambda x$	$C_4 = \cosh\lambda l\sin\lambda l - \sinh\lambda l\cos\lambda l$	$C_{14} = \sinh^2\lambda l + \sin^2\lambda l$

$F_{a1} = \langle x-a\rangle^0\cosh\lambda\langle x-a\rangle\cos\lambda\langle x-a\rangle$	$C_{a1} = \cosh\lambda(l-a)\cos\lambda(l-a)$
$F_{a2} = \cosh\lambda\langle x-a\rangle\sin\lambda\langle x-a\rangle + \sinh\lambda\langle x-a\rangle\cos\lambda\langle x-a\rangle$	$C_{a2} = \cosh\lambda(l-a)\sin\lambda(l-a) + \sinh\lambda(l-a)\cos\lambda(l-a)$
$F_{a3} = \sinh\lambda\langle x-a\rangle\sin\lambda\langle x-a\rangle$	$C_{a3} = \sinh\lambda(l-a)\sin\lambda(l-a)$
$F_{a4} = \cosh\lambda\langle x-a\rangle\sin\lambda\langle x-a\rangle - \sinh\lambda\langle x-a\rangle\cos\lambda\langle x-a\rangle$	$C_{a4} = \cosh\lambda(l-a)\sin\lambda(l-a) - \sinh\lambda(l-a)\cos\lambda(l-a)$
$F_{a5} = \langle x-a\rangle^0 - F_{a1}$	$C_{a5} = 1 - C_{a1}$
$F_{a6} = 2\lambda\langle x-a\rangle\langle x-a\rangle^0 - F_{a2}$	$C_{a6} = 2\lambda(l-a) - C_{a2}$

$A_1 = \frac{1}{2}e^{-\lambda a}\cos\lambda a$	$B_1 = \frac{1}{2}e^{-\lambda b}\cos\lambda b$
$A_2 = \frac{1}{2}e^{-\lambda a}(\sin\lambda a - \cos\lambda a)$	$B_2 = \frac{1}{2}e^{-\lambda b}(\sin\lambda b - \cos\lambda b)$
$A_3 = -\frac{1}{2}e^{-\lambda a}\sin\lambda a$	$B_3 = \frac{-1}{2}e^{-\lambda b}\sin\lambda b$
$A_4 = \frac{1}{2}e^{-\lambda a}(\sin\lambda a + \cos\lambda a)$	$B_4 = \frac{1}{2}e^{-\lambda b}(\sin\lambda b + \cos\lambda b)$

Numerical values F_1, F_2, F_3, and F_4 for λx ranging from 0–6 are tabulated in Table 8.3; numerical values of C_{11}, C_{12}, C_{13}, and C_{14} are tabulated in Table 8.4.

Short shell with free ends

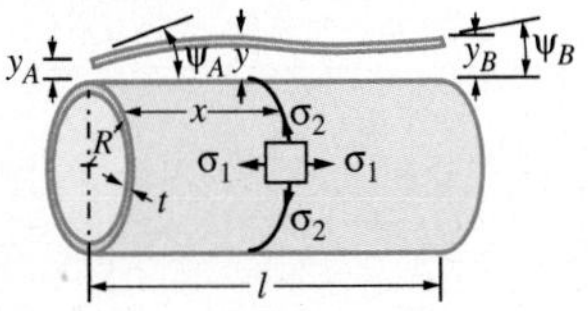

$R/t > 10$

Meridional radial shear $= V = -y_A 2D\lambda^3 F_2 - \psi_A 2D\lambda^2 F_3 + LT_V$

Meridional bending $= M = -y_A 2D\lambda^2 F_3 - \psi_A D\lambda F_4 + LT_M$ (*Note: The* load terms LT_V, LT_M etc., are given for each of the following cases.)

Meridional slope $= \psi = \psi_A F_1 - y_A \lambda F_4 + LT_\psi$

Radial deflection $= y = y_A F_1 + \frac{\psi_A}{2\lambda}F_2 + LT_y$

Circumferential membrane stress $= \sigma_2 = \frac{yE}{R} + \nu\sigma_1$

Meridional bending stress $= \sigma'_1 = \frac{-6M}{t^2}$

Circumferential bending stress $= \sigma'_2 = \nu\sigma'_1$

Meridional radial shear stress $= \tau = \frac{V}{t}$ (average value)

(*Continued*)

TABLE 13.2 Shear, Moment, Slope, and Deflection Formulas for Long and Short Thin-Walled Cylindrical Shells under Axisymmetric Loading (*Continued*)

Loading and Case No.	End Deformations	Load Terms or Load and Deformation Equations	Selected Values
1. Radial end load, V_o lb/in 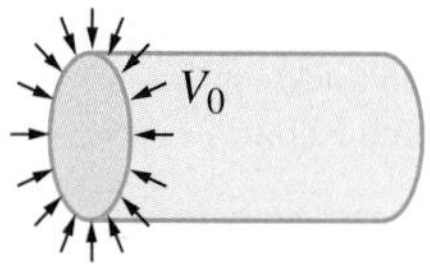 If $\lambda l > 6$, see case 8	$\psi_A = \frac{V_o}{2D\lambda^2}\frac{C_{14}}{C_{11}}$ $y_A = \frac{-V_o}{2D\lambda^3}\frac{C_{13}}{C_{11}}$ $\psi_B = \frac{V_o}{2D\lambda^2}\frac{2C_3}{C_{11}}$ $y_B = \frac{V_o}{2D\lambda^3}\frac{C_4}{C_{11}}$	$LT_V = -V_o F_1$ $LT_M = \frac{-V_o}{2\lambda}F_2$ $LT_\psi = \frac{-V_o}{2D\lambda^2}F_3$ $LT_y = \frac{-V_o}{4D\lambda^3}F_4$	$\sigma_1 = 0 \quad (\sigma_2)_{max} = \frac{y_A E}{R}$ $\psi_{max} = \psi_A$ $y_{max} = y_A$
2. Intermediate radial load, p lb/in 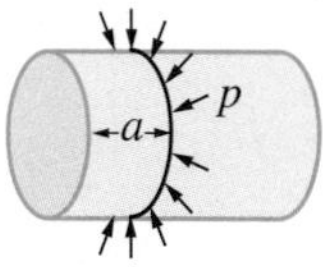 If $\lambda l > 6$, case 9	$\psi_A = \frac{p}{2D\lambda^2}\frac{C_2C_{a2} - 2C_3C_{a1}}{C_{11}}$ $y_A = \frac{-p}{2D\lambda^3}\frac{C_3C_{a2} - C_4C_{a1}}{C_{11}}$ $\psi_B = \psi_A C_1 - y_A\lambda C_4 - \frac{p}{2D\lambda^2}C_{a3}$ $y_B = y_A C_1 + \frac{\psi_A C_2}{2\lambda} - \frac{p}{4D\lambda^3}C_{a4}$	$LT_V = -pF_{a1}$ $LT_M = \frac{-p}{2\lambda}F_{a2}$ $LT_\psi = \frac{-p}{2D\lambda^2}F_{a3}$ $LT_y = \frac{-p}{4D\lambda^3}F_{a4}$	$\sigma_1 = 0$
3. End moment, M_o lb-in/in 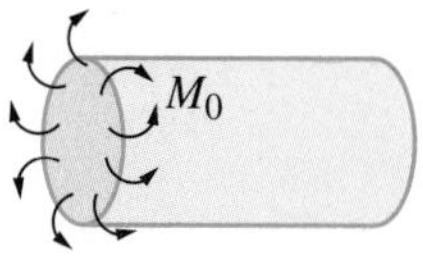 if $\lambda l > 6$, see case 10	$\psi_A = \frac{-M_o}{D\lambda}\frac{C_{12}}{C_{11}}$ $y_A = \frac{M_o}{2D\lambda^2}\frac{C_{14}}{C_{11}}$ $\psi_B = \frac{-M_o}{D\lambda}\frac{C_2}{C_{11}}$ $y_B = \frac{-M_o}{D\lambda^2}\frac{C_3}{C_{11}}$	$LT_V = -M_o\lambda F_4$ $LT_M = M_o F_1$ $LT_\psi = \frac{M_o}{2D\lambda}F_2$ $LT_y = \frac{M_o}{2D\lambda^2}F_3$	$\sigma_1 = 0$ $(\sigma_2)_{max} = \frac{y_A E}{R}$ $M_{max} = M_o \quad (\text{at } x = 0)$ $\psi_{max} = \psi_A$ $y_{max} = y_A$
4. Intermediate applied moment, M_o lb-in/in 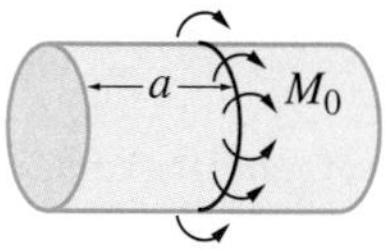 If $\lambda l > 6$, consider case 11	$\psi_A = \frac{-M_o}{D\lambda}\frac{C_2C_{a1} + C_3C_{a4}}{C_{11}}$ $y_A = \frac{M_o}{2D\lambda^2}\frac{2C_3C_{a1} + C_4C_{a4}}{C_{11}}$ $\psi_B = \psi_A C_1 - y_A\lambda C_4 + \frac{M_o}{2D\lambda}C_{a2}$ $y_B = y_A C_1 + \frac{\psi_A}{2\lambda}C_2 + \frac{M_o}{2D\lambda^2}C_{a3}$	$LT_V = -M_o\lambda F_{a4}$ $LT_M = M_o F_{a1}$ $LT_\psi = \frac{M_o}{2D\lambda}F_{a2}$ $LT_y = \frac{M_o}{2D\lambda^2}F_{a3}$	
5. Uniform radial pressure from a to l 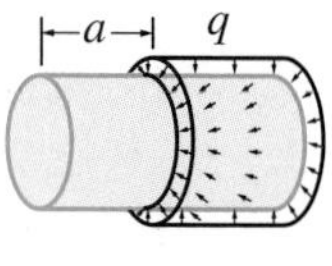 If $\lambda l > 6$, consider case 12	$\psi_A = \frac{q}{2D\lambda^3}\frac{C_2C_{a3} - C_3C_{a2}}{C_{11}}$ $y_A = \frac{-q}{4D\lambda^4}\frac{2C_3C_{a3} - C_4C_{a2}}{C_{11}}$ $\psi_B = \psi_A C_1 - y_A\lambda C_4 - \frac{q}{4D\lambda^3}C_{a4}$ $y_B = y_A C_1 + \frac{\psi_A}{2\lambda}C_2 - \frac{q}{4D\lambda^4}C_{a5}$	$LT_V = \frac{-q}{2\lambda}F_{a2}$ $LT_M = \frac{-q}{2\lambda^2}F_{a3}$ $LT_\psi = \frac{-q}{4D\lambda^3}F_{a4}$ $LT_y = \frac{-q}{4D\lambda^4}F_{a5}$	

(*Continued*)

TABLE 13.2 Shear, Moment, Slope, and Deflection Formulas for Long and Short Thin-Walled Cylindrical Shells under Axisymmetric Loading (*Continued*)

Loading and Case No.	End Deformations	Load Terms or Load and Deformation Equations	Selected Values
6. Uniformly increasing pressure from a to l 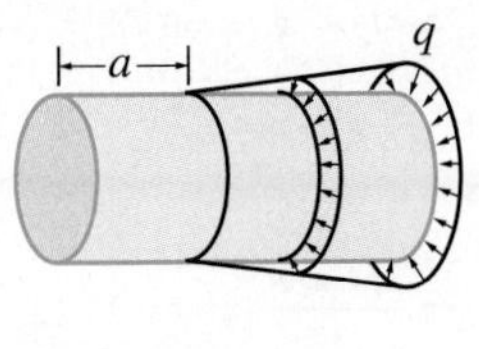	$\psi_A = \frac{-q}{4D\lambda^4(l-a)} \frac{2C_3C_{a3} - C_2C_{a4}}{C_{11}}$ $y_A = \frac{-q}{4D\lambda^5(l-a)} \frac{C_3C_{a4} - C_4C_{a3}}{C_{11}}$ $\psi_B = \psi_A C_1 - y_A \lambda C_4 - \frac{qC_{a5}}{4D\lambda^4(l-a)}$ $y_B = y_A C_1 + \frac{\psi_A}{2\lambda} C_2 - \frac{qC_{a6}}{8D\lambda^5(l-a)}$	$LT_V = \frac{-q}{2\lambda^2(l-a)} F_{a3}$ $LT_M = \frac{-q}{4\lambda^3(l-a)} F_{a4}$ $LT_\psi = \frac{-q}{4D\lambda^4(l-a)} F_{a5}$ $LT_y = \frac{-q}{8D\lambda^5(l-a)} F_{a6}$	$\sigma_1 = 0$ $(\sigma_2)_{max} = \frac{y_B E}{R}$ $y_{max} = y_B$ $\psi_{max} = \psi_B$
7. Axial load along the portion from a to l 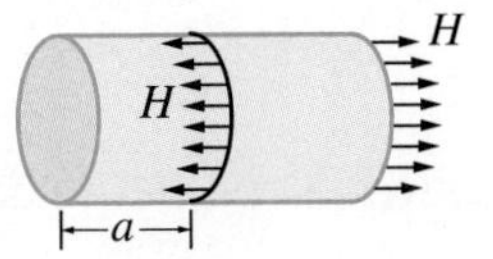	$\psi_A = \frac{\nu H}{2D\lambda^3 R} \frac{C_2C_{a3} - C_3C_{a2}}{C_{11}}$ $y_A = \frac{-\nu HR}{Et} \frac{2C_3C_{a3} - C_3C_{a2}}{C_{11}}$ $\psi_B = \psi_A C_1 - y_A \lambda C_4 - \frac{\nu HR\lambda}{Et} C_{a4}$ $y_B = y_A C_1 + \frac{\psi_A C_2}{2\lambda} - \frac{\nu HR}{Et} C_{a5}$	$LT_V = \frac{-\nu H}{2\lambda R} F_{a2}$ $LT_M = \frac{-\nu H}{2\lambda^2 R} F_{a3}$ $LT_\psi = \frac{-\nu HR\lambda}{Et} F_{a4}$ $LT_y = \frac{-\nu HR}{Et} F_{a5}$	$\sigma_1 = \frac{H}{t}\langle x-a \rangle^0$
Long shells with the left end free (right end more than $6/\lambda$ unit of length from the closest load) 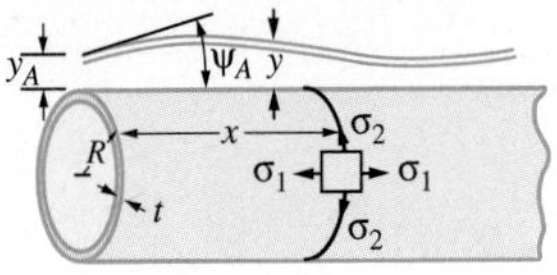 $R/t > 10$	Meridional radial shear $= V = -y_A 2D\lambda^3 F_2 - \psi_A 2D\lambda^2 F_3 + LT_V$ Meridional bending moment $= M = -y_A 2D\lambda^2 F_3 - \psi_A D\lambda F_4 + LT_M$ Meridional slope $= \psi = \psi_A F_1 - y_A \lambda F_4 + LT_\psi$ Radial deflection $= y = y_A F_1 + \frac{\psi_A}{2\lambda} F_2 + LT_y$ Circumferential membrane stress $= \sigma_2 = \frac{yE}{R} + \nu\sigma_1$ Meridional bending stress $= \sigma'_1 = \frac{-6M}{t^2}$ Circumferential bending stress $= \sigma'_2 = \nu\sigma'_1$ Meridional radial shear stress $= \tau = \frac{V}{t}$ (average value)		(*Note: The* load terms LT_V, LT_M, etc., are given where needed in the following cases.)
8. Radial end load, V_o lb/in 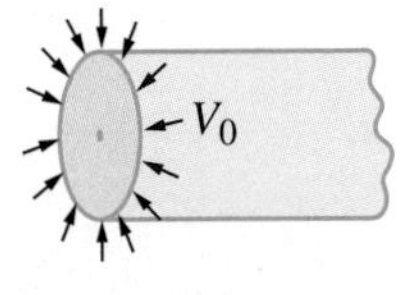	$\psi_A = \frac{V_o}{2D\lambda^2}$ $y_A = \frac{-V_o}{2D\lambda^3}$	$V = -V_o e^{-\lambda x}(\cos\lambda x - \sin\lambda x)$ $M = \frac{-V_o}{\lambda} e^{-\lambda x} \sin\lambda x$ $\psi = \frac{V_o}{2D\lambda^2} e^{-\lambda x}(\cos\lambda x + \sin\lambda x)$ $y = \frac{-V_o}{2D\lambda^3} e^{-\lambda x} \cos\lambda x$	$V_{max} = -V_o$ at $x = 0$ $M_{max} = -0.3224 \frac{V_o}{\lambda}$ at $x = \frac{\pi}{4\lambda}$ $\psi_{max} = \psi_A$ $y_{max} = y_A$ $\sigma_1 = 0$ $(\sigma_2)_{max} = \frac{-2V_o \lambda R}{t}$ at $x = 0$
9. Intermediate radial load, p lb/in 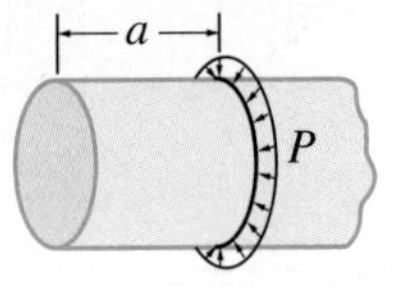 If $\lambda a > 3$, consider case 15	$\psi_A = \frac{-p}{D\lambda^2} A_2$ $y_A = \frac{-p}{D\lambda^3} A_1$	$LT_V = -pF_{a1}$ $LT_M = \frac{-p}{2\lambda} F_{a2}$ $LT_\psi = \frac{-p}{2D\lambda^2} F_{a3}$ $LT_y = \frac{-p}{4D\lambda^3} F_{a4}$	$\sigma_1 = 0$

(Continued)

TABLE 13.2 Shear, Moment, Slope, and Deflection Formulas for Long and Short Thin-Walled Cylindrical Shells under Axisymmetric Loading (*Continued*)

Loading and Case No.	End Deformations	Load Terms or Load and Deformation Equations	Selected Values
10. End moment, M_o lb-in/in 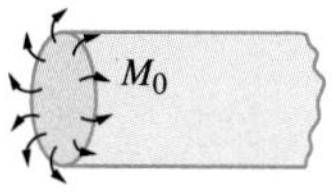	$\psi_A = \frac{-M_o}{D\lambda}$ $y_A = \frac{M_o}{2D\lambda^2}$	$V = -2M_o\lambda e^{-\lambda x}\sin\lambda x$ $M = M_o e^{-\lambda x}(\cos\lambda x + \sin\lambda x)$ $\psi = \frac{-M_o}{D\lambda}e^{-\lambda x}\cos\lambda x$ $y = \frac{-M_o}{2D\lambda^2}e^{-\lambda x}(\sin\lambda x - \cos\lambda x)$	$V_{max} = -0.6448M_o\lambda$ at $x = \frac{\pi}{4\lambda}$ $M_{max} = M_o$ at $x = 0$ $\psi_{max} = \psi_A$, $y_{max} = y_A$ $\sigma_1 = 0$ $(\sigma_2)_{max} = \frac{2M_o\lambda^2 R}{t}$ at $x = 0$
11. Intermediate applied moment, M_o lb-in/in 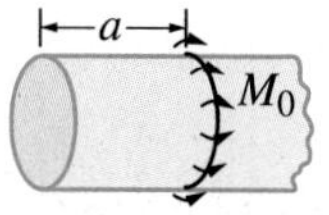 if $\lambda a > 3$, consider case 16	$\psi_A = \frac{-2M_o}{D\lambda}A_1$ $y_A = \frac{M_o}{D\lambda^2}A_4$	$LT_V = -M_o\lambda F_{a4}$ $LT_M = M_o F_{a1}$ $LT_\psi = \frac{M_o}{2D\lambda}F_{a2}$ $LT_y = \frac{M_o}{2D\lambda^2}F_{a3}$	$\sigma_1 = 0$
12. Uniform radial pressure from a to b 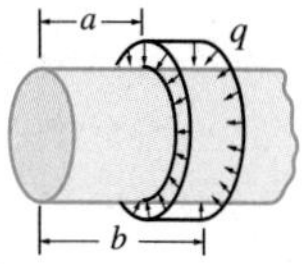	$\psi_A = \frac{-q}{D\lambda^3}(B_3 - A_3)$ $y_A = \frac{-q}{2D\lambda^4}(B_2 - A_2)$	$LT_V = \frac{-q}{2\lambda}(F_{a2} - F_{b2})$ $LT_M = \frac{-q}{2\lambda^2}(F_{a3} - F_{b3})$ $LT_\psi = \frac{-q}{4D\lambda^3}(F_{a4} - F_{b4})$ $LT_y = \frac{-q}{4D\lambda^4}(F_{a5} - F_{b5})$ For values of F_{b1} to F_{b6} substitute b for a in the expressions for F_{a1} to F_{a6}	$\sigma_1 = 0$
13. Uniformly increasing pressure from a to b 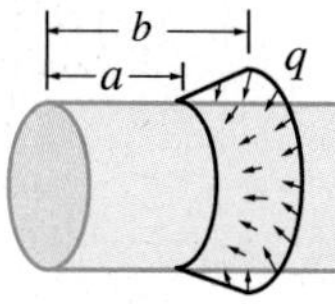	$\psi_A = \frac{q}{D}\left[\frac{B_4 - A_4}{2\lambda^4(b-a)} - \frac{B_3}{\lambda^3}\right]$ $y_A = \frac{q}{2D}\left[\frac{B_3 - A_3}{\lambda^5(b-a)} - \frac{B_2}{\lambda^4}\right]$	$LT_V = \frac{-q}{2}\left[\frac{F_{a3} - F_{b3}}{\lambda^2(b-a)} - \frac{F_{b2}}{\lambda}\right]$ See note in case 12 $LT_M = \frac{-q}{2}\left[\frac{F_{a4} - F_{b4}}{2\lambda^3(b-a)} - \frac{F_{b3}}{\lambda^2}\right]$ $LT_\psi = \frac{-q}{4D}\left[\frac{F_{a5} - F_{b5}}{\lambda^4(b-a)} - \frac{F_{b4}}{\lambda^3}\right]$ $LT_y = \frac{-q}{4D}\left[\frac{F_{a6} - F_{b6}}{2\lambda^5(b-a)} - \frac{F_{b5}}{\lambda^4}\right]$	$\sigma_1 = 0$
14. Axial load along the portion from a to b 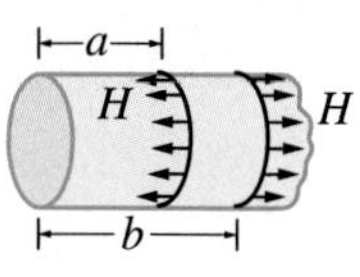	$\psi_A = \frac{-\nu H}{RD\lambda^3}(B_3 - A_3)$ $y_A = \frac{-\nu H}{2RD\lambda^4}(B_2 - A_2)$	$LT_V = \frac{-\nu H}{2R\lambda}(F_{a2} - F_{b2})$ $LT_M = \frac{-\nu H}{2R\lambda^2}(F_{a3} - F_{b3})$ $LT_\psi = \frac{-\nu H}{4RD\lambda^3}(F_{a4} - F_{b4})$ $LT_y = \frac{-\nu H}{4RD\lambda^4}(F_{a5} - F_{b5})$	$\sigma_1 = \frac{H}{t}\langle x-a\rangle^0 - \frac{H}{t}\langle x-b\rangle^0$

(Continued)

TABLE 13.2 Shear, Moment, Slope, and Deflection Formulas for Long and Short Thin-Walled Cylindrical Shells under Axisymmetric Loading (*Continued*)

Very long shells (both ends more than $6/\lambda$ units of length from the nearest loading)

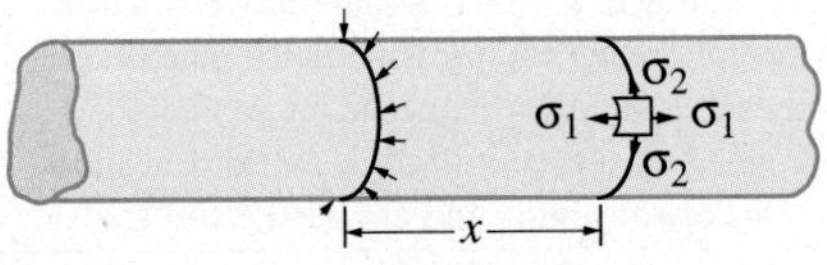

$R/t > 10$

Circumferential membrane stress $= \sigma_2 = \frac{yE}{R} + \nu\sigma_1$

Meridional bending stress $= \sigma'_1 = \frac{-6M}{t^2}$

Circumferential bending stress $= \sigma'_2 = \nu\sigma'_1$

Meridional radial shear stress $= \tau = \frac{V}{t}$

Loading and Case No.	Load and Deformation Equations	Selected Values
15. Concentrated radial load, p (lb/linear in of circumference) 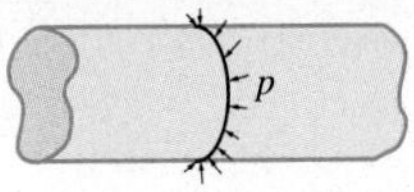	$V = \frac{-p}{2} e^{-\lambda x} \cos\lambda x$ $M = \frac{p}{4\lambda} e^{-\lambda x}(\cos\lambda x - \sin\lambda x)$ $\psi = \frac{p}{4D\lambda^2} e^{-\lambda x} \sin\lambda x$ $y = \frac{-p}{8D\lambda^3} e^{-\lambda x}(\cos\lambda x + \sin\lambda x)$	$V_{max} = \frac{-p}{2}$ at $x = 0$, $\sigma_1 = 0$ $M_{max} = \frac{p}{4\lambda}$ at $x = 0$ $\psi_{max} = 0.0806 \frac{p}{D\lambda^2}$ at $x = \frac{\pi}{4\lambda}$ $y_{max} = \frac{-p}{8D\lambda^3}$ at $x = 0$
16. Applied moment 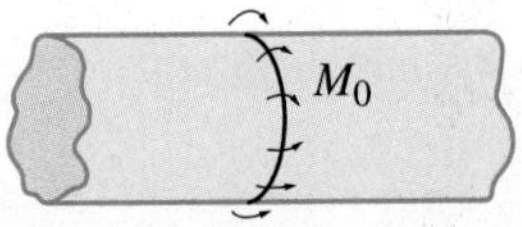	$V = \frac{-M_o\lambda}{2} e^{-\lambda x}(\cos\lambda x + \sin\lambda x)$ $M = \frac{M_o}{2} e^{-\lambda x} \cos\lambda x$ $\psi = \frac{-M_o}{4D\lambda} e^{-\lambda x}(\cos\lambda x - \sin\lambda x)$ $y = \frac{-M_o}{4D\lambda^2} e^{-\lambda x} \sin\lambda x$	$V_{max} = \frac{-M_o\lambda}{2}$ at $x = 0$, $\sigma_1 = 0$ $M_{max} = \frac{M_o}{2}$ at $x = 0$ $\psi_{max} = \frac{-M_o}{4D\lambda}$ at $x = 0$ $y_{max} = -0.0806 \frac{M_o}{D\lambda^2}$ at $x = \frac{\pi}{4\lambda}$
17. Uniform pressure over a band of width $2a$ 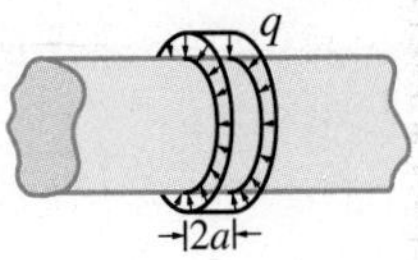	Superimpose cases 10 and 12 to make ψ_A (at $x = 0$) $= 0$ [*Note:* x is measured from the midlength of the loaded band]	$M_{max} = \frac{q}{2\lambda^2} e^{-\lambda a} \sin\lambda a$ at $x = 0$ $y_{max} = \frac{-q}{4D\lambda^4}(1 - e^{-\lambda a} \cos\lambda a)$ at $x = 0$ $\sigma_1 = 0$

TABLE 13.3 Formulas for Bending and Membrane Stresses and Deformations in Thin-Walled Pressure Vessels

Notation: Q_o and p unit loads (force per unit length); q = unit pressure (force per unit area); M_o = unit applied couple (force-length per unit length); ψ_o = applied edge rotation (radians); Δ_o = applied edge displacement, all loads are positive as shown. V = meridional transverse shear, positive as shown; M_1 and M_2 = meridional and circumferential bending moments, respectively, positive when compressive on the outside; ψ = change in meridional slope (radians), positive when the change is in the same direction as a positive M_1; ΔR = change in circumferential radius, positive outward; σ_1 and σ_2 = meridional and circumferential membrane stresses, positive when tensile; σ'_1 and σ'_2 = meridional and circumferential bending stresses, positive when tensile on the outside. E = modulus of elasticity, ν = Poisson's ratio; and $D = Et^3/12(1-\nu^2)$.

1. Partial spherical shells

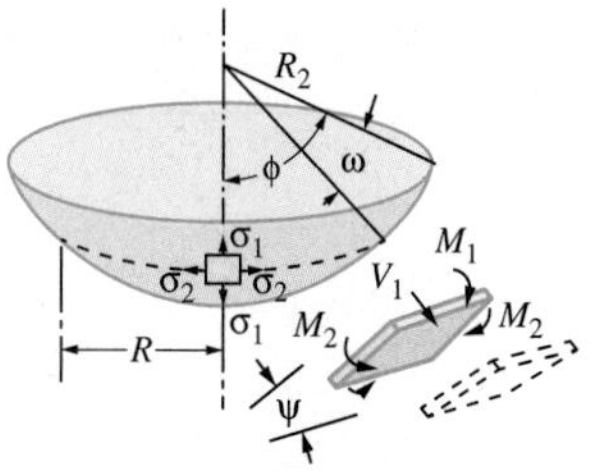

$$\beta = \left[3(1-\nu^2)\left(\frac{R_2}{t}\right)^2\right]^{1/4}$$

$$K_1 = 1 - \frac{1-2\nu}{2\beta}\cot(\phi-\omega)$$

$$K_2 = 1 - \frac{1+2\nu}{2\beta}\cot(\phi-\omega)$$

$$K_3 = \frac{e^{-\beta\omega}}{\sqrt{\sin(\phi-\omega)}}$$

$$\frac{R_2}{t} > 10$$

Meridional radial shear $= V_1 = -CK_3\sin(\beta\omega+\zeta)$ (*Note:* Expressions for C and ζ are given below for the several loads.)

Meridional bending moment $= M_1 = \dfrac{CR_2K_3}{2\beta}[K_1\cos(\beta\omega+\zeta)+\sin(\beta\omega+\zeta)]$

Circumferential bending moment $= M_2 = \dfrac{CR_2K_3}{2\beta}\left\{\nu\sin(\beta\omega+\zeta)+\left[2-\left(1-\dfrac{\nu}{2}\right)(K_1+K_2)\right]\cos(\beta\omega+\zeta)\right\}$

Change in meridional slope $= \psi = \dfrac{C2\beta^2K_3}{Et}\cos(\beta\omega+\zeta)$

Change in radius of circumference $= \Delta R = \dfrac{CR_2\beta K_3\sin(\phi-\omega)}{Et}[\cos(\beta\omega+\zeta)]-K_2\sin(\beta\omega+\zeta)]$

Meridional membrane stress $= \sigma_1 = \dfrac{-CK_3\cot(\phi-\omega)}{t}\sin(\beta\omega+\zeta)$

Meridional bending stress $= \sigma'_1 = \dfrac{-6M_1}{t^2}$

Circumferential membrane stress $= \sigma_2 = \dfrac{C\beta K_3}{2t}[2\cos(\beta\omega+\zeta)-(K_1+K_2)\sin(\beta\omega+\zeta)]$

Circumferential bending stress $= \sigma'_2 = \dfrac{-6M_2}{t^2}$

Meridional radial shear stress $= \tau = \dfrac{V_1}{t}$ (average value)

[*Note:* For reasonable accuracy $3/\beta < \phi < \pi - 3/\beta$. Deformations and stresses due to edge loads and displacements are essentially zero when $\omega > 3/\beta$ (see discussion).]

Case No., Loading	Formulas

1a. Uniform radial force Q_o at the edge

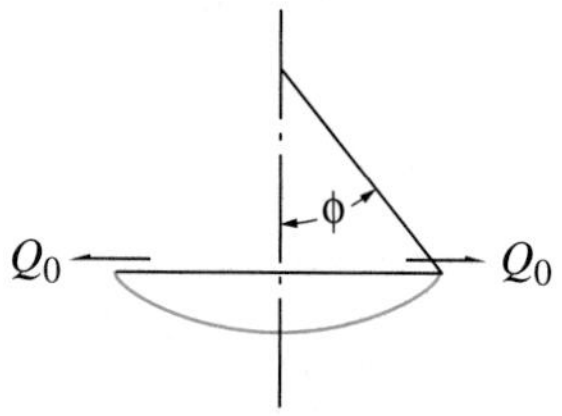

$C = \dfrac{Q_o(\sin\phi)^{3/2}\sqrt{1+K_1^2}}{K_1}$ and $\zeta = \tan^{-1}(-K_1)$ where $-\dfrac{\pi}{2} < \zeta < \dfrac{\pi}{2}$

Max value of M_1 occurs at $\omega = \pi/4\beta$

At the edge where $\omega = 0$,

$V_1 = Q_o\sin\phi,\quad M_1 = 0,\quad \sigma'_1 = 0,\quad \sigma_1 = \dfrac{Q_o\cos\phi}{t}$

$\sigma_2 = \dfrac{Q_o\,\beta\sin\phi}{2t}\left(\dfrac{2}{K_1}+K_1+K_2\right) = (\sigma_2)_{max}$

$M_2 = \dfrac{Q_o t^2 B^2\cos\phi}{6K_1R_2},\quad \sigma'_2 = \dfrac{-Q_o B^2\cos\phi}{K_1R_2}$

$\psi = \dfrac{Q_o 2\beta^2\sin\phi}{EtK_1},\quad \Delta R = \dfrac{Q_o R_2\,\beta\sin^2\phi}{EtK_1}(1+K_1K_2)$ (Refs. 14 and 42)

1b. Uniform edge moment M_o

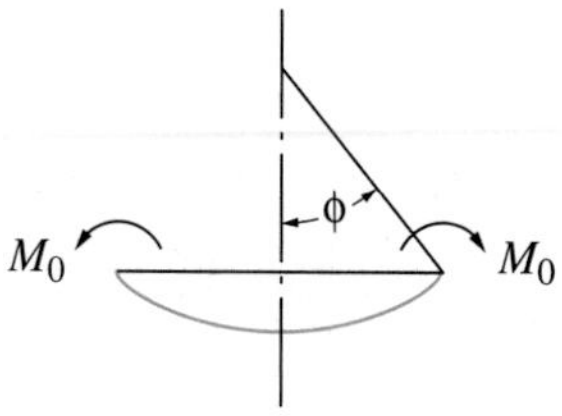

$C = \dfrac{M_o 2\beta\sqrt{\sin\phi}}{R_2K_1},\quad \zeta = 0$

At the edge where $\omega = 0$,

$V_1 = 0,\quad \sigma_1 = 0,\quad M_1 = M_o,\quad \sigma'_1 = \dfrac{-6M_o}{t^2}$

$\sigma_2 = \dfrac{M_o 2\beta^2}{R_2K_1t},\quad M_2 = \dfrac{M_o}{2\nu K_1}[(1+\nu^2)(K_1+K_2)-2K_2],\quad \sigma'_2 = \dfrac{-6M_2}{t^2}$

$\psi = \dfrac{M_o 4\beta^3}{EtR_2K_1},\quad \Delta R = \dfrac{M_o 2\beta^2\sin\phi}{EtK_1}$ (Refs. 14 and 42)

(Continued)

TABLE 13.3 Formulas for Bending and Membrane Stresses and Deformations in Thin-Walled Pressure Vessels (*Continued*)

Case No., Loading	Formulas
1c. Radial displacement Δ_o; no edge rotation 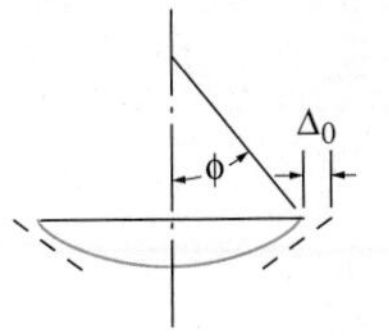 	$C=\frac{-\Delta_o Et}{R_2\beta K_2\sqrt{\sin\phi}}, \quad \zeta=\frac{\pi}{2}=90°$ At the edge where $\omega=0$, $V_1=\frac{\Delta_o Et}{R_2\beta K_2\sin\phi}, \quad \sigma_1=\frac{\Delta_o E\cos\phi}{R_2\beta K_2\sin^2\phi}$ Resultant radial edge force $=\frac{\Delta_o Et}{R_2\beta K_2\sin^2\phi}$ $M_1=\frac{-\Delta_o Et}{2\beta^2 K_2\sin\phi}, \quad \sigma_1'=\frac{3\Delta_o E}{t\beta^2 K_2\sin\phi}$ $\sigma_2=\frac{\Delta_o E(K_1+K_2)}{2R_2K_2\sin\phi}, \quad M_2=\frac{-\Delta_o Et\nu}{2\beta^2K_2\sin\phi}, \quad \sigma'_2=\frac{3\Delta_o E\nu}{t\beta^2K_2\sin\phi}$ $\psi=0, \quad \Delta R=\Delta_o$
1d. Edge rotation ψ_o rad; no edge displacement 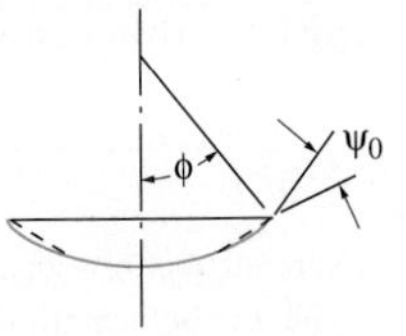 	$C=\frac{\psi_o Et\sqrt{(1+K_2^2)\sin\phi}}{2\beta^2K_2}, \quad \zeta=\tan^{-1}\frac{1}{K_2} \quad$ where $0<\zeta<\pi$ At the edge where $\omega=0$, $V_1=\frac{-\psi_o Et}{2\beta^2K_2}, \quad \sigma_1=\frac{-\psi_o E\cot\phi}{2\beta^2K_2}$ Resultant radial edge force $=\frac{-\psi_o Et}{2\beta^2K_2\sin\phi}$ $M_1=\frac{\psi_o EtR_2}{4\beta^3}\left(K_1+\frac{1}{K_2}\right), \quad \sigma_1'=\frac{-6M_1}{t^2}$ $\sigma_2=\frac{-\psi_o E}{4\beta K_2}(K_1-K_2), \quad M_2=\frac{\psi_o EtR_2}{8\nu\beta^3}\left[(1+\nu^2)(K_1+K_2)-2K_2+\frac{2\nu^2}{K_2}\right], \quad \sigma_2'=\frac{-6M_2}{t^2}$ $\psi=\psi_o \quad \Delta R=0$
2. Partial spherical shell, load P concentrated on small circular area of radius r_o at pole; any edge support 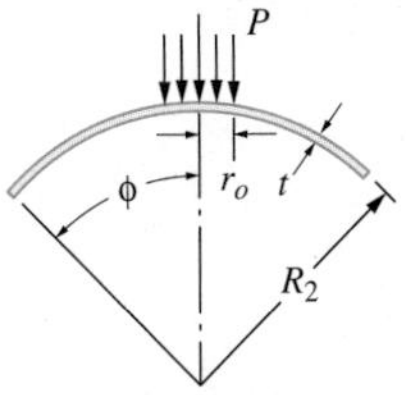 $R_2/t>10$ *Note:* $r'_o=\sqrt{1.6t_o^2+t^2}-0.675t$ if $r_o<0.5t$; $r'_o=r_o$ if $r_o\geq 0.5t$ (see Sec. 11.1). For $\phi>\sin^{-1}(1.65\sqrt{t/R_2})$	*Note:* The deflection for this case is measured locally relative to the undeformed shell. It does not include any deformations due to the edge supports or membrane stresses remote from the loading. The formulas for deflection and stress are applicable also to off-axis loads if no edge is closer than $\phi=\sin^{-1}(1.65\sqrt{t/R_2})$. If an edge were as close as half this angle, the results would be modified very little. Deflection under the center of the load $=-A\frac{PR_2\sqrt{1-\nu^2}}{Et^2}$ Max membrane stress under the center of the load $\sigma_1=\sigma_2=-B\frac{P\sqrt{1-\nu^2}}{t^2}$ *Note*: See also Ref. 72. Max bending stress under the center of the load $\sigma'_1=\sigma'_2=-C\frac{P(1+\nu)}{t^2}$ Here A, B, and C are numerical coefficients that depend upon $\mu=r'_o\left[\frac{12(1-\nu^2)}{R_2^2t^2}\right]^{1/4}$ and have the values tabulated below

μ	0	0.1	0.2	0.4	0.6	0.8	1.0	1.2	1.4
A	0.433	0.431	0.425	0.408	0.386	0.362	0.337	0.311	0.286
B	0.217	0.215	0.212	0.204	0.193	0.181	0.168	0.155	0.143
C	∞	1.394	1.064	0.739	0.554	0.429	0.337	0.266	0.211

(*Continued*)

TABLE 13.3 Formulas for Bending and Membrane Stresses and Deformations in Thin-Walled Pressure Vessels (*Continued*)

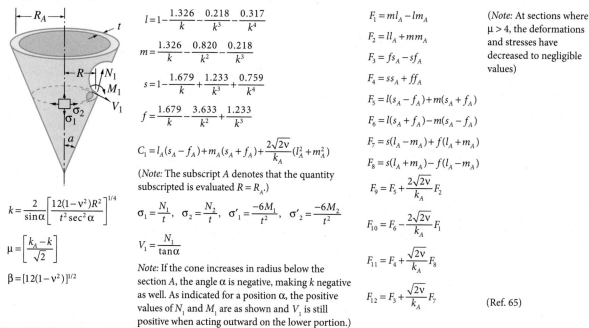

Case No., Loading	Formulas
3. Shallow spherical shell, point load P at the pole	3a. Edge vertically supported and guided: Max deflection $y = -A_1 \dfrac{PR^2}{16\pi D}$; Edge moment $M_o = -B_1 \dfrac{P}{4\pi}$
	3b. Edge fixed and held: Max deflection $y = -A_2 \dfrac{PR^2}{16\pi D}$; Edge moment $M_o = -B_2 \dfrac{P}{4\pi}$

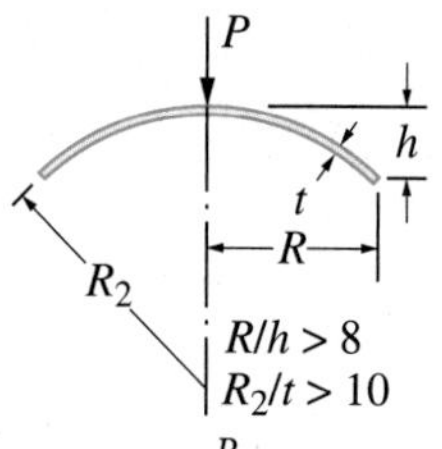

$h < \dfrac{R}{8}$

$t < \dfrac{R_2}{10}$

Here A and B are numerical coefficients that depend upon $\alpha = 2\left[\dfrac{3(1-\nu^2)h^2}{t^2}\right]^{1/4}$ and have the values tabulated below

α	0	1	2	3	4	5	6	7	8	9	10
A_1	1.000	0.996	0.935	0.754	0.406	0.321	0.210	0.148	0.111	0.085	0.069
B_1	1.000	0.995	0.932	0.746	0.498	0.324	0.234	0.192	0.168	0.153	0.140
A_2	1.000	0.985	0.817	0.515	0.320	0.220	0.161	0.122	0.095	0.075	0.061
B_2	1.000	0.975	0.690	0.191	−0.080	−0.140	−0.117	−0.080	−0.059	−0.039	−0.026

4. Long conical shells with edge loads. Expressions are accurate if $R/(t\cos\alpha) > 10$ and $|k| > 5$ everywhere in the region from the loaded end to the position where $\mu > 4$.

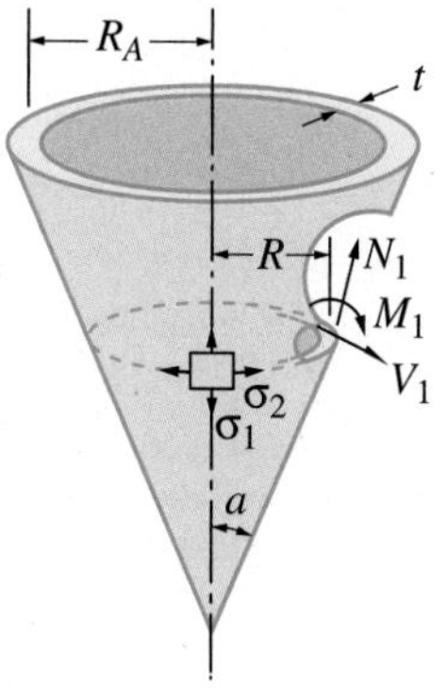

$$k = \frac{2}{\sin\alpha}\left[\frac{12(1-\nu^2)R^2}{t^2\sec^2\alpha}\right]^{1/4}$$

$$\mu = \left[\frac{k_A - k}{\sqrt{2}}\right]$$

$$\beta = [12(1-\nu^2)]^{1/2}$$

$$l = 1 - \frac{1.326}{k} - \frac{0.218}{k^3} - \frac{0.317}{k^4}$$

$$m = \frac{1.326}{k} - \frac{0.820}{k^2} - \frac{0.218}{k^3}$$

$$s = 1 - \frac{1.679}{k} + \frac{1.233}{k^3} + \frac{0.759}{k^4}$$

$$f = \frac{1.679}{k} - \frac{3.633}{k^2} + \frac{1.233}{k^3}$$

$$C_1 = l_A(s_A - f_A) + m_A(s_A + f_A) + \frac{2\sqrt{2}\nu}{k_A}(l_A^2 + m_A^2)$$

(*Note:* The subscript A denotes that the quantity subscripted is evaluated $R = R_A$.)

$$\sigma_1 = \frac{N_1}{t}, \quad \sigma_2 = \frac{N_2}{t}, \quad \sigma'_1 = \frac{-6M_1}{t^2}, \quad \sigma'_2 = \frac{-6M_2}{t^2}$$

$$V_1 = \frac{N_1}{\tan\alpha}$$

Note: If the cone increases in radius below the section A, the angle α is negative, making k negative as well. As indicated for a position α, the positive values of N_1 and M_1 are as shown and V_1 is still positive when acting outward on the lower portion.)

$$F_1 = ml_A - lm_A$$
$$F_2 = ll_A + mm_A$$
$$F_3 = fs_A - sf_A$$
$$F_4 = ss_A + ff_A$$
$$F_5 = l(s_A - f_A) + m(s_A + f_A)$$
$$F_6 = l(s_A + f_A) - m(s_A - f_A)$$
$$F_7 = s(l_A - m_A) + f(l_A + m_A)$$
$$F_8 = s(l_A + m_A) - f(l_A - m_A)$$
$$F_9 = F_5 + \frac{2\sqrt{2}\nu}{k_A}F_2$$
$$F_{10} = F_6 - \frac{2\sqrt{2}\nu}{k_A}F_1$$
$$F_{11} = F_4 + \frac{\sqrt{2}\nu}{k_A}F_8$$
$$F_{12} = F_3 + \frac{\sqrt{2}\nu}{k_A}F_7$$

(*Note:* At sections where $\mu > 4$, the deformations and stresses have decreased to negligible values)

(Ref. 65)

For use at the loaded end where $R = R_A$,

$$F_{1A} = F_{3A} = 0$$

$$F_{2A} = 1 - \frac{2.652}{k_A} + \frac{3.516}{k_A^2} - \frac{2.610}{k_A^3} + \frac{0.038}{k_A^4}$$

$$F_{4A} = 1 - \frac{3.359}{k_A} + \frac{5.641}{k_A^2} - \frac{9.737}{k_A^3} + \frac{14.716}{k_A^4}$$

$$F_{8A} = F_{5A} = 1 - \frac{3.359}{k_A} + \frac{7.266}{k_A^2} - \frac{10.068}{k_A^3} + \frac{5.787}{k_A^4}$$

$$F_{10A} = F_{7A} = F_{6A} = 1 - \frac{2.652}{k_A} + \frac{1.641}{k_A^2} - \frac{0.290}{k_A^3} - \frac{2.211}{k_A^4}$$

$$F_{9A} = C_1 = F_{5A} + \frac{2\sqrt{2}\nu}{k_A}F_{2A}$$

(*Continued*)

TABLE 13.3 Formulas for Bending and Membrane Stresses and Deformations in Thin-Walled Pressure Vessels (*Continued*)

Case No., Loading	Formulas
4a. Uniform radial force Q_A 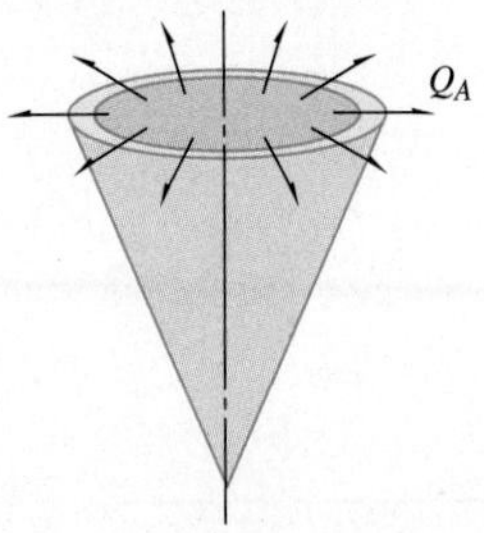 	$N_1 = Q_A \sin\alpha \left(\frac{k_A}{k}\right)^{5/2} \frac{e^{-\mu}}{C_1}(F_9 \cos\mu - F_{10} \sin\mu)$ $N_2 = Q_A \sin\alpha \left(\frac{k_A}{k}\right)^{3/2} \frac{k_A}{\sqrt{2}} \frac{e^{-\mu}}{C_1}(F_{11} \cos\mu + F_{12} \sin\mu)$ $M_1 = Q_A \sin\alpha \left(\frac{k_A}{k}\right)^{3/2} \frac{k_A t}{\sqrt{2}\beta} \frac{e^{-\mu}}{C_1}\left[-\left(F_{12} - \frac{\sqrt{2}\nu}{k} F_{10}\right)\cos\mu + \left(F_{11} + \frac{\sqrt{2}\nu}{k} F_9\right)\sin\mu\right]$ $M_2 = Q_A \nu \sin\alpha \left(\frac{k_A}{k}\right)^{3/2} \frac{k_A t}{\sqrt{2}\beta} \frac{e^{-\mu}}{C_1}\left[-\left(F_{12} - \frac{\sqrt{2}\nu}{\nu k} F_{10}\right)\cos\mu + \left(F_{11} + \frac{\sqrt{2}}{\nu k} F_9\right)\sin\mu\right]$ $\Delta R = \frac{Q_A R}{Et} \sin\alpha \left(\frac{k_A}{k}\right)^{3/2} \frac{k_A}{\sqrt{2}} \frac{e^{-\mu}}{C_1}\left[\left(F_{11} - \frac{\sqrt{2}\nu}{k} F_9\right)\cos\mu + \left(F_{12} + \frac{\sqrt{2}\nu}{k} F_{10}\right)\sin\mu\right]$ $\psi = \frac{Q_A R_A \beta}{Et^2}\left(\frac{k_A}{k}\right)^{1/2} \frac{e^{-\mu}}{C_1}(F_{10} \cos\mu + F_9 \sin\mu)$ At the loaded end where $R = R_A$, $N_{1A} = Q_A \sin\alpha, \quad N_{2A} = Q_A \sin\alpha \frac{k_A}{\sqrt{2}C_1}\left(F_{4A} + \frac{\sqrt{2}\nu}{k_A} F_{8A}\right)$ $M_{1A} = 0, \quad M_{2A} = Q_A \sin\alpha(1-\nu^2)\frac{t}{\beta C_1} F_{10A}$ $\Delta R_A = \frac{Q_A R_A \sin\alpha}{Et} \frac{k_A}{\sqrt{2}C_1}\left(F_{4A} - \frac{4\nu^2}{k_A^2} F_{2A}\right), \quad \psi_A = \frac{Q_A R_A \beta}{Et^2 C_1} F_{10A}$
4b. Uniform edge moment M_A 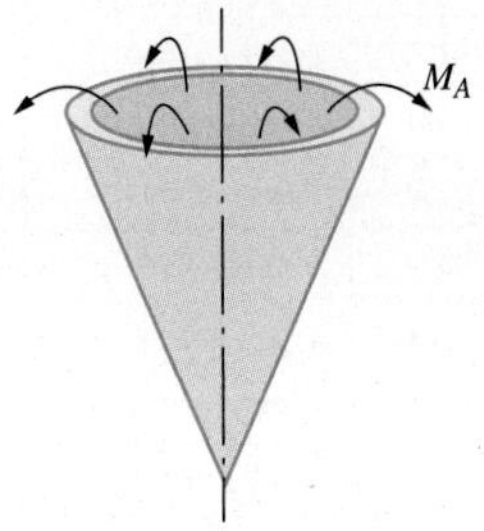 	$N_1 = M_A \left(\frac{k_A}{k}\right)^{5/2} \frac{2\sqrt{2}\beta}{tk_A} \frac{e^{-\mu}}{C_1}(F_1 \cos\mu - F_2 \sin\mu)$ $N_2 = M_A \left(\frac{k_A}{k}\right)^{3/2} \frac{\beta}{t} \frac{e^{-\mu}}{C_1}(F_7 \cos\mu - F_8 \sin\mu)$ $M_1 = M_A \left(\frac{k_A}{k}\right)^{3/2} \frac{e^{-\mu}}{C_1}\left[\left(F_8 + \frac{2\sqrt{2}\nu}{k} F_2\right)\cos\mu + \left(F_7 + \frac{2\sqrt{2}\nu}{k} F_1\right)\sin\mu\right]$ $M_2 = M_A \left(\frac{k_A}{k}\right)^{3/2} \nu \frac{e^{-\mu}}{C_1}\left[\left(F_8 + \frac{\sqrt{2}\nu}{\nu k} F_2\right)\cos\mu + \left(F_7 + \frac{2\sqrt{2}\nu}{\nu k} F_1\right)\sin\mu\right]$ $\Delta R = M_A \left(\frac{k_A}{k}\right)^{3/2} \frac{\beta R}{Et^2} \frac{e^{-\mu}}{C_1}\left[\left(F_7 - \frac{2\sqrt{2}\nu}{k} F_1\right)\cos\mu - \left(F_8 - \frac{2\sqrt{2}\nu}{k} F_2\right)\sin\mu\right]$ $\psi = M_A \left(\frac{k_A}{k}\right)^{1/2} \frac{2\sqrt{2}\beta^2 R_A}{Et^3 k_A \sin\alpha} \frac{e^{-\mu}}{C_1}(F_2 \cos\mu + F_1 \sin\mu)$ At the loaded end where $R = R_A$, $N_{1A} = 0, \quad N_{2A} = M_A \frac{\beta}{tC_1} F_{7A}$ $M_{1A} = M_A, \quad M_{2A} = M_A\left[\nu + \frac{2\sqrt{2}(1-\nu^2)}{k_A C_1} F_{2A}\right]$ $\Delta R_A = M_A \frac{\beta R_A}{Et^2 C_1} F_{7A}, \quad \psi_A = M_A \frac{2\sqrt{2}\beta^2 R_A}{Et^3 k_A C_1 \sin\alpha} F_{2A}$

(*Continued*)

TABLE 13.3 Formulas for Bending and Membrane Stresses and Deformations in Thin-Walled Pressure Vessels (*Continued*)

5. Short conical shells. Expressions are accurate if $R/(t\cos\alpha)>10$ and $|k|>5$ everywhere in the cone.

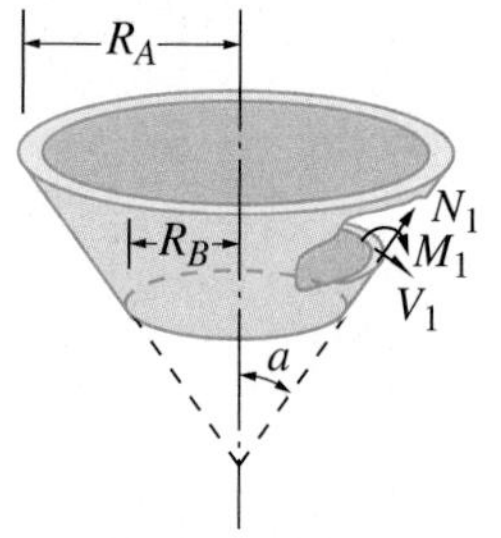

$$k=\frac{2}{\sin\alpha}\left[\frac{12(1-\nu^2)R^2}{t^2\sec^2\alpha}\right]^{1/4},\quad \sigma_1=\frac{N_1}{t},\quad \sigma_2=\frac{N_2}{t}$$

$$\mu_D=\left|\frac{k_A-k_B}{\sqrt{2}}\right|,\quad \sigma'_1=\frac{-6M_1}{t^2},\quad \sigma_2=\frac{-6M_2}{t^2}$$

$$\beta=[12(1-\nu^2)]^{1/2},\quad V_1=\frac{N_1}{\tan\alpha}$$

Case No., Loading	Formulas
5a. Uniform radial force Q_A at the large end	$N_1=Q_A\sin\alpha K_{N1}$, $N_2=Q_A\sin\alpha\frac{k_A}{\sqrt{2}}K_{N2}$

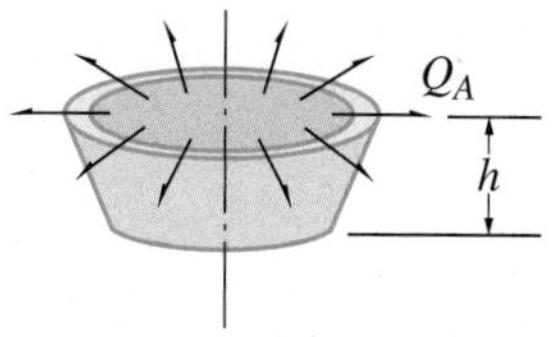

$$N_1=Q_A\sin\alpha K_{N1},\qquad N_2=Q_A\sin\alpha\frac{k_A}{\sqrt{2}}K_{N2}$$

$$M_1=Q_A\sin\alpha\frac{k_At}{\sqrt{2}\beta}K_{M1},\qquad M_2=Q_A\nu\sin\alpha\frac{k_At}{\sqrt{2}\beta}K_{M2},\qquad \Delta h=\frac{Q_AR_A}{Et}K_{\Delta h1}-\frac{Q_AR_A\sin^2\alpha}{Et\cos\alpha}\frac{k_A}{\sqrt{2}}K_{\Delta h2}$$

$$\Delta R=\frac{Q_AR_A\sin\alpha}{Et}\frac{k_A}{\sqrt{2}}K_{\Delta R},\qquad \psi=\frac{Q_AR_A\beta}{Et^2}K_\psi$$

For $R_B/(t\cos\alpha)>10$ and $k_B>5$ and for $\nu=0.3$, the following tables give the values of K at several locations along the shell [$\Omega=(R_A-R)/(R_A-R_B)$]:

k_A	$\mu_D=0.4$		$k_B=9.434$	$K_{\Delta h1}=-0.266$	$K_{\Delta h2}=2.979$		$\mu_D=0.6$	$k_B=9.151$	$K_{\Delta h1}=-0.266$		$K_{\Delta h2}=2.899$	$\mu_D=0.8$	$k_B=8.869$	$K_{\Delta h1}=-0.269$		$K_{\Delta h2}=2.587$
	Ω	0.000	0.250	0.500	0.750	1.000	0.000	0.250	0.500	0.750	1.000	0.000	0.250	0.500	0.750	1.000
10	K_{N1}	1.000	0.548	0.216	0.025	0.000	1.000	0.450	0.068	−0.102	0.000	1.000	0.392	−0.029	−0.192	0.000
	K_{N2}	2.748	2.056	1.319	0.533	−0.307	2.304	1.643	0.919	0.123	−0.761	1.952	1.371	0.720	−0.016	−0.862
	K_{M1}	0.000	0.047	0.054	0.034	0.000	0.000	0.076	0.084	0.047	0.000	0.000	0.106	0.114	0.060	0.000
	K_{M2}	3.729	3.445	3.576	3.690	3.801	2.151	2.340	2.467	2.566	2.675	1.468	1.671	1.780	1.848	1.941
	$K_{\Delta R}$	2.706	1.977	1.238	0.489	−0.274	2.262	1.558	0.842	0.111	−0.637	1.909	1.282	0.644	−0.006	−0.678
	K_ψ	7.644	7.703	7.759	7.819	7.887	5.014	5.064	5.106	5.156	5.223	3.421	3.454	3.470	3.500	3.559
k_A	$\mu_D=1.2$		$k_B=8.303$	$K_{\Delta h1}=-0.278$	$K_{\Delta h2}=1.986$		$\mu_D=1.6$	$k_B=7.737$	$K_{\Delta h1}=-0.287$		$K_{\Delta h2}=1.571$	$\mu_D=3.2$	$k_B=5.475$	$K_{\Delta h1}=-0.323$		$K_{\Delta h2}=0.953$
	Ω	0.000	0.250	0.500	0.750	1.000	0.000	0.250	0.500	0.750	1.000	0.000	0.250	0.500	0.750	1.000
10	K_{N1}	1.000	0.339	−0.133	−0.308	0.000	1.000	0.315	−0.185	−0.377	0.000	1.000	0.156	−0.301	−0.361	0.000
	K_{N2}	1.470	1.023	0.514	−0.080	−0.810	1.197	0.815	0.394	−0.082	−0.696	0.940	0.466	0.088	−0.097	−0.182
	K_{M1}	0.000	0.165	0.176	0.088	0.000	0.000	0.220	0.235	0.117	0.000	0.000	0.347	0.318	0.131	0.000
	K_{M2}	0.810	1.040	1.105	1.098	1.147	0.555	0.813	0.832	0.746	0.743	0.406	0.729	0.538	0.191	0.010
	$K_{\Delta R}$	1.428	0.931	0.439	−0.051	−0.559	1.155	0.721	0.321	−0.046	−0.416	0.898	0.379	0.066	−0.039	−0.055
	K_ψ	1.887	1.880	1.829	1.806	1.843	1.294	1.242	1.112	1.025	1.038	0.948	0.734	0.333	0.066	0.007
k_A	$\mu_D=0.4$		$k_B=19.434$	$K_{\Delta h1}=-0.293$	$K_{\Delta h2}=5.488$		$\mu_D=0.6$	$k_B=9.151$	$K_{\Delta h1}=-0.294$		$K_{\Delta h2}=4.247$	$\mu_D=0.8$	$k_B=18.869$	$K_{\Delta h1}=-0.296$		$K_{\Delta h2}=3.368$
	Ω	0.000	0.250	0.500	0.750	1.000	0.000	0.250	0.500	0.750	1.000	0.000	0.250	0.500	0.750	1.000
20	K_{N1}	1.000	0.346	−0.053	−0.0175	0.000	1.000	0.281	−0.149	−0.256	0.000	1.000	0.255	−0.193	−0.296	0.000
	K_{N2}	4.007	2.674	1.298	−0.123	−1.591	2.975	1.954	0.885	−0.234	−1.410	2.334	1.528	0.677	−0.244	−0.186
	K_{M1}	0.000	0.052	0.052	0.025	0.000	0.000	0.082	0.078	0.035	0.000	0.000	0.111	0.106	0.045	0.000
	K_{M2}	2.974	3.079	3.133	3.163	3.198	1.546	1.668	1.704	1.703	1.717	0.933	1.072	1.093	1.064	1.059
	$K_{\Delta R}$	3.985	2.629	1.263	−0.114	−1.052	2.954	1.908	0.852	−0.214	−1.293	2.313	1.481	0.644	−0.200	−1.056
	K_ψ	13.866	13.918	13.967	14.020	14.079	7.210	7.241	7.263	7.293	7.338	4.349	4.359	4.351	4.360	4.393

(*Continued*)

TABLE 13.3 Formulas for Bending and Membrane Stresses and Deformations in Thin-Walled Pressure Vessels (*Continued*)

k_A	$\mu_D = 1.2$		$k_B = 18.303$	$K_{\Delta h1} = -0.298$	$K_{\Delta h2} = 2.334$		$\mu_D = 1.6$	$k_B = 17.737$	$K_{\Delta h1} = -0.300$		$K_{\Delta h2} = 1.777$	$\mu_D = 3.2$	$k_B = 15.475$	$K_{\Delta h1} = -0.305$		$K_{\Delta h2} = 1.023$
	Ω	0.000	0.250	0.500	0.750	1.000	0.000	0.250	0.500	0.750	1.000	0.000	0.250	0.500	0.750	1.000
20	K_{N1}	1.000	0.236	−0.230	−0.337	0.000	1.000	0.222	−0.247	−0.352	0.000	1.000	0.049	−0.274	−0.215	0.000
	K_{N2}	1.629	1.059	0.460	−0.175	−0.867	1.283	0.808	0.334	−0.139	−0.655	0.977	0.390	0.027	−0.094	−0.112
	K_{M1}	0.000	0.168	0.160	0.066	0.000	0.000	0.222	0.211	0.087	0.000	0.000	0.342	0.253	0.079	0.000
	K_{M2}	0.458	0.639	0.634	1.553	1.514	0.299	0.519	0.489	0.359	0.289	0.211	0.507	0.322	0.089	−0.002
	$K_{\Delta R}$	1.608	1.011	0.427	−0.147	−0.726	1.262	0.761	0.303	−0.111	−0.515	0.956	0.350	0.026	−0.062	−0.067
	K_{ψ}	2.137	2.105	2.031	1.992	2.005	1.392	1.313	1.160	1.065	1.061	0.938	0.692	0.259	0.032	−0.007
k_A	$\mu_D = 0.4$		$k_B = 39.434$	$K_{\Delta h1} = -0.299$	$K_{\Delta h2} = 6.866$		$\mu_D = 0.6$	$k_B = 39.151$	$K_{\Delta h1} = -0.299$		$K_{\Delta h2} = 4.776$	$\mu_D = 0.8$	$k_B = 38.869$	$K_{\Delta h1} = -0.300$		$K_{\Delta h2} = 3.634$
	Ω	0.000	0.250	0.500	0.750	1.000	0.000	0.250	0.500	0.750	1.000	0.000	0.250	0.500	0.750	1.000
40	K_{N1}	1.000	0.238	−0.192	−0.275	0.000	1.000	0.216	−0.225	−0.304	0.000	1.000	0.209	−0.238	−0.317	0.000
	K_{N2}	4.692	2.998	1.277	−0.471	−2.248	3.234	1.060	0.860	−0.366	−1.621	2.459	1.565	0.649	−0.289	−1.255
	K_{M1}	0.000	0.055	0.051	0.021	0.000	0.000	0.084	0.077	0.030	0.000	0.000	0.112	0.103	0.040	0.000
	K_{M2}	1.851	1.922	1.934	1.921	1.917	0.863	0.957	0.959	0.924	0.906	0.498	0.617	0.611	0.555	0.524
	$K_{\Delta R}$	4.681	2.974	1.261	−0.458	−2.185	3.223	2.036	0.844	−0.351	−1.553	2.448	1.541	0.634	−0.273	−1.185
	K_{ψ}	17.256	17.286	17.311	17.341	17.377	8.048	8.058	8.059	8.069	8.092	4.645	4.637	4.612	4.603	4.617
k_A	$\mu_D = 1.2$		$k_B = 38.303$	$K_{\Delta h1} = -0.300$	$K_{\Delta h2} = 2.446$		$\mu_D = 1.6$	$k_B = 37.737$	$K_{\Delta h1} = -0.300$		$K_{\Delta h2} = 1.847$	$\mu_D = 3.2$	$k_B = 35.475$	$K_{\Delta h1} = -0.300$		$K_{\Delta h2} = 1.053$
	Ω	0.000	0.250	0.500	0.750	1.000	0.000	0.250	0.500	0.750	1.000	0.000	0.025	0.500	0.750	1.000
40	K_{N1}	1.000	0.202	−0.249	−0.327	0.000	1.000	0.189	−0.253	−0.324	0.000	1.000	0.012	−0.245	−0.165	0.000
	K_{N2}	1.677	1.055	0.430	−0.200	−0.850	1.309	0.792	0.304	−0.153	−0.616	0.991	0.352	0.007	−0.087	−0.092
	K_{M1}	0.000	0.168	0.154	0.060	0.000	0.000	0.221	0.200	0.077	0.000	0.000	0.333	0.223	0.063	0.000
	K_{M2}	0.237	0.405	0.385	0.291	0.236	0.152	0.367	0.331	0.201	0.126	0.107	0.408	0.250	0.066	−0.001
	$K_{\Delta R}$	1.666	1.031	0.415	−0.184	−0.780	1.299	0.768	0.290	−0.137	−0.548	0.981	0.333	0.009	−0.072	−0.072
	K_{ψ}	2.206	2.159	2.071	2.021	2.022	1.419	1.324	1.160	1.061	1.048	0.994	0.665	0.229	0.023	−0.009
k_A	$\mu_D = 0.4$		$k_B = 79.434$	$K_{\Delta h1} = -0.300$	$K_{\Delta h2} = 7.324$		$\mu_D = 0.6$	$k_B = 79.151$	$K_{\Delta h1} = -0.300$		$K_{\Delta h2} = 4.935$	$\mu_D = 0.8$	$k_B = 78.869$	$K_{\Delta h1} = -0.300$		$K_{\Delta h2} = 3.713$
	Ω	0.000	0.250	0.500	0.750	1.000	0.000	0.250	0.500	0.750	1.000	0.000	0.250	0.500	0.750	1.000
80	K_{N1}	1.000	0.202	−0.235	−0.305	0.000	1.000	0.196	−0.245	−0.313	0.000	1.000	0.194	−0.248	−0.317	0.000
	K_{N2}	4.917	3.098	1.264	−0.584	−2.446	3.309	2.083	0.846	−0.403	−1.667	2.494	1.568	0.635	−0.306	−1.259
	K_{M1}	0.000	0.056	0.050	0.019	0.000	0.000	0.084	0.076	0.029	0.000	0.000	0.112	0.101	0.038	0.000
	K_{M2}	0.985	1.045	1.043	1.017	1.002	0.444	0.531	0.524	0.479	0.454	0.253	0.367	0.355	0.293	0.257
	$K_{\Delta R}$	4.912	3.086	1.256	−0.576	−2.412	3.303	2.071	0.838	−0.395	−1.631	2.489	1.556	0.627	−0.298	−1.224
	K_{ψ}	18.365	18.378	18.386	18.399	18.416	8.287	8.285	8.272	8.270	8.280	4.725	4.707	4.672	4.654	4.658
k_A	$\mu_D = 1.2$		$k_B = 78.303$	$K_{\Delta h1} = -0.300$	$K_{\Delta h2} = 2.483$		$\mu_D = 1.6$	$k_B = 77.737$	$K_{\Delta h1} = -0.300$		$K_{\Delta h2} = 1.873$	$\mu_D = 3.2$	$k_B = 75.475$	$K_{\Delta h1} = -0.298$		$K_{\Delta h2} = 1.067$
	Ω	0.000	0.250	0.500	0.750	1.000	0.000	0.250	0.500	0.750	1.000	0.000	0.250	0.500	0.750	1.000
80	K_{N1}	1.000	0.189	−0.251	−0.318	0.000	1.000	0.176	−0.252	−0.308	0.000	1.000	−0.002	−0.230	−0.145	0.000
	K_{N2}	1.691	1.047	0.415	−0.206	−0.832	1.318	0.781	0.290	−0.157	−0.593	0.997	0.334	0.000	−0.083	−0.084
	K_{M1}	0.000	0.168	0.150	0.057	0.000	0.000	0.220	0.194	0.073	0.000	0.000	0.327	0.208	0.056	0.000
	K_{M2}	0.119	0.285	0.264	0.169	0.113	0.076	0.292	0.258	0.132	0.059	0.053	0.363	0.221	0.057	−0.001
	$K_{\Delta R}$	1.686	1.035	0.408	−0.198	−0.797	1.313	0.769	0.283	−0.149	−0.560	0.992	0.324	0.001	−0.076	−0.075
	K_{ψ}	2.224	2.170	2.075	2.019	2.014	1.426	1.324	1.155	1.053	1.037	0.998	0.651	0.216	0.020	−0.009

(*Continued*)

TABLE 13.3 Formulas for Bending and Membrane Stresses and Deformations in Thin-Walled Pressure Vessels (*Continued*)

k_A	$\mu_D = 0.4$		$k_B = 159.434$	$K_{\Delta h1} = -0.300$	$K_{\Delta h2} = 7.452$		$\mu_D = 0.6$	$k_B = 159.151$	$K_{\Delta h1} = -0.300$		$K_{\Delta h2} = 4.980$	$\mu_D = 0.8$	$k_B = 158.869$	$K_{\Delta h1} = -0.300$		$K_{\Delta h2} = 3.738$
	Ω	0.000	0.250	0.500	0.750	1.000	0.000	0.250	0.500	0.750	1.000	0.000	0.250	0.500	0.750	1.000
	K_{N1}	1.000	0.192	−0.247	−0.312	0.000	1.000	0.190	−0.249	−0.314	0.000	1.000	0.189	−0.250	−0.315	0.000
	K_{N2}	4.979	3.121	1.257	−0.164	−2.493	3.329	2.086	0.839	−0.413	−1.671	2.505	1.566	0.628	−0.311	−1.254
160	K_{M1}	0.000	0.056	0.050	0.019	0.000	0.000	0.084	0.075	0.028	0.000	0.000	0.112	0.100	0.038	0.000
	K_{M2}	0.500	0.558	0.552	0.522	0.504	0.224	0.309	0.300	0.253	0.225	0.127	0.240	0.227	0.164	0.127
	$K_{\Delta R}$	4.977	3.115	1.253	−0.610	−2.475	3.327	2.080	0.835	−0.409	−1.654	2.502	1.560	0.624	−0.307	−1.236
	K_ψ	18.666	18.668	18.667	18.670	18.678	8.350	8.341	8.322	8.313	8.316	4.746	4.724	4.683	4.660	4.660

k_A	$\mu_D = 1.2$		$k_B = 158.303$	$K_{\Delta h1} = -0.300$	$K_{\Delta h2} = 2.497$		$\mu_D = 1.6$	$k_B = 157.737$	$K_{\Delta h1} = -0.300$		$K_{\Delta h2} = 1.884$	$\mu_D = 3.2$	$k_B = 155.475$	$K_{\Delta h1} = -0.298$		$K_{\Delta h2} = 1.073$
	Ω	0.000	0.250	0.500	0.750	1.000	0.000	0.250	0.500	0.750	1.000	0.000	0.250	0.500	0.750	1.000
	K_{N1}	1.000	0.185	−0.251	−0.312	0.000	1.000	0.171	−0.250	−0.300	0.000	1.000	−0.008	−0.223	−0.136	0.000
	K_{N2}	1.696	1.042	0.408	−0.208	−0.821	1.322	0.775	0.283	−0.158	−0.582	1.000	0.325	−0.003	−0.081	−0.080
160	K_{M1}	0.000	0.168	0.149	0.056	0.000	0.000	0.219	0.191	0.071	0.000	0.000	0.324	0.201	0.053	0.000
	K_{M2}	0.060	0.226	0.205	0.111	0.055	0.038	0.255	0.223	0.100	0.028	0.027	0.341	0.207	0.054	−0.000
	$K_{\Delta R}$	1.693	1.036	0.404	−0.204	−0.804	1.319	0.769	0.280	−0.154	−0.565	0.998	0.320	−0.003	−0.077	−0.076
	K_ψ	2.230	2.171	2.074	2.015	2.006	1.429	1.323	1.151	1.048	1.030	0.999	0.644	0.210	0.018	−0.010

Case No., Loading	Formulas
5b. Uniform edge moment M_A at the large end 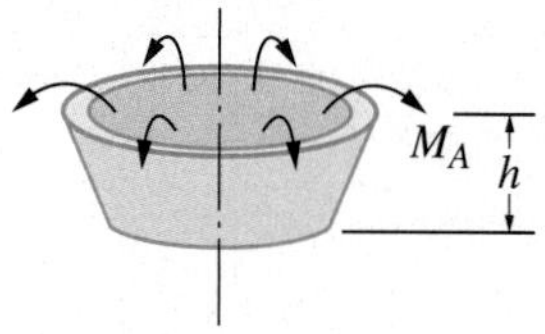	$N_1 = M_A \frac{2\sqrt{2}\beta}{t k_A} K_{N1}$, $N_2 = M_A \frac{\beta}{t} K_{N2}$ $M_1 = M_A K_{M1}$, $M_2 = M_A \nu K_{M2}$ $\Delta R = M_A \frac{\beta R_A}{E t^2} K_{\Delta R}$, $\psi = M_A \frac{2\sqrt{2}\beta^2 R_A}{E t^3 k_A \sin\alpha} K_\psi$, $\Delta h = \frac{M_A R_A \beta}{E t^2 \sin\alpha} K_{\Delta h1} - \frac{M_A \beta R_A \sin\alpha}{E t^2 \cos\alpha} K_{\Delta h2}$ For $R_B/(t\cos\alpha) > 10$ and $k_B > 5$ and for $\nu = 0.3$, the following tables give the values of K at several locations along the shell $[\Omega = (R_A - R)/(R_A - R_B)]$.

k_A	$\mu_D = 0.4$		$k_B = 9.434$	$K_{\Delta h1} = -0.017$	$K_{\Delta h2} = 15.507$		$\mu_D = 0.6$	$k_B = 9.151$	$K_{\Delta h1} = -0.026$		$K_{\Delta h2} = 9.766$	$\mu_D = 0.8$	$k_B = 8.869$	$K_{\Delta h1} = -0.031$		$K_{\Delta h2} = 6.554$
	Ω	0.000	0.250	0.500	0.750	1.000	0.000	0.250	0.500	0.750	1.000	0.000	0.250	0.500	0.750	1.000
	K_{N1}	0.000	−0.584	−0.819	−0.647	0.000	0.000	−0.579	−0.830	−0.672	0.000	0.000	−0.527	−0.772	−0.641	0.000
	K_{N2}	7.644	4.020	0.170	−3.934	−8.328	5.014	2.687	0.163	−2.604	−5.674	3.421	1.851	0.138	−1.780	−3.983
10	K_{M1}	1.000	0.816	0.538	0.238	0.000	1.000	0.844	0.554	0.229	0.000	1.000	0.865	0.568	0.226	0.000
	K_{M2}	17.470	17.815	18.122	18.475	18.961	8.300	8.428	8.481	8.581	8.867	4.850	4.836	4.718	4.644	4.792
	$K_{\Delta R}$	7.644	3.957	0.227	−3.559	−7.413	5.014	2.625	0.215	−2.236	−4.752	3.421	1.795	0.181	−1.449	−3.133
	K_ψ	19.197	19.268	19.368	19.503	19.671	8.509	8.480	8.489	8.548	8.656	4.488	4.381	4.320	4.325	4.393

k_A	$\mu_D = 1.2$		$k_B = 8.303$	$K_{\Delta h1} = -0.038$	$K_{\Delta h2} = 3.401$		$\mu_D = 1.6$	$k_B = 7.737$	$K_{\Delta h1} = -0.042$		$K_{\Delta h2} = 2.086$	$\mu_D = 3.2$	$k_B = 5.475$	$K_{\Delta h1} = -0.050$		$K_{\Delta h2} = 0.953$
	Ω	0.000	0.250	0.500	0.750	1.000	0.000	0.250	0.500	0.750	1.000	0.000	0.250	0.500	0.750	1.000
	K_{N1}	0.000	−0.429	−0.642	−0.554	0.000	0.000	−0.370	−0.547	−0.476	0.000	0.000	−0.375	−0.389	−0.193	0.000
	K_{N2}	1.887	0.993	0.070	−0.951	−2.196	1.294	0.602	−0.004	−0.594	−1.323	0.948	0.127	−0.236	−0.240	−0.018
10	K_{M1}	1.000	0.890	0.590	0.227	0.000	1.000	0.899	0.596	0.225	0.000	1.000	0.816	0.395	0.071	0.000
	K_{M2}	2.623	2.444	2.119	1.834	1.814	2.052	1.772	1.319	0.904	0.784	1.833	1.258	0.479	−0.058	−0.262
	$K_{\Delta R}$	1.887	0.949	0.106	−0.694	−1.514	1.294	0.570	0.034	−0.387	−0.792	0.948	0.131	−0.132	−0.106	−0.005
	K_ψ	1.892	1.670	1.506	1.436	1.458	1.266	0.915	0.674	0.553	0.547	0.971	0.425	0.064	−0.071	−0.091

(*Continued*)

TABLE 13.3 Formulas for Bending and Membrane Stresses and Deformations in Thin-Walled Pressure Vessels (*Continued*)

k_A	$\mu_D = 0.4$		$k_B =$ 19.434	$K_{\Delta h1} =$ −0.008	$K_{\Delta h2} = 27.521$		$\mu_D = 0.6$	$k_B =$ 19.151	$K_{\Delta h1} = -0.009$		$K_{\Delta h2} =$ 14.217	$\mu_D = 0.8$	$k_B =$ 18.869	$K_{\Delta h1} = -0.010$		$K_{\Delta h2} =$ 8.477
	Ω	0.000	0.250	0.500	0.750	1.000	0.000	0.250	0.500	0.750	1.000	0.000	0.250	0.500	0.750	1.000
20	K_{N1}	0.000	−1.050	−1.435	−1.104	0.000	0.000	−0.821	−1.135	−0.883	0.000	0.000	−0.660	−0.919	−0.722	0.000
	K_{N2}	13.866	7.110	0.149	−7.036	−14.461	7.210	3.726	0.107	−3.672	−7.642	4.349	2.246	0.072	−2.210	−4.638
	K_{M1}	1.000	0.832	0.517	0.191	0.000	1.000	0.848	0.524	0.183	0.000	1.000	0.858	0.530	0.181	0.000
	K_{M2}	15.905	15.971	15.910	15.860	15.967	6.219	6.150	5.930	5.721	5.697	3.421	3.286	2.988	2.703	2.617
	$K_{\Delta R}$	13.866	7.055	0.204	−6.696	−13.655	7.210	3.683	0.149	−3.408	−7.007	4.349	2.212	0.105	−1.999	−4.128
	K_ψ	34.745	34.800	34.881	34.998	35.145	12.165	12.103	12.079	12.107	12.178	5.643	5.504	5.415	5.395	5.431
k_A	$\mu_D = 1.2$		$k_B =$ 18.303	$K_{\Delta h1} =$ −0.010	$K_{\Delta h2} = 3.962$		$\mu_D = 1.6$	$k_B =$ 17.737	$K_{\Delta h1} = -0.011$		$K_{\Delta h2} =$ 2.326	$\mu_D = 3.2$	$k_B =$ 15.475	$K_{\Delta h1} = -0.011$		$K_{\Delta h2} =$ 0.978
	Ω	0.000	0.250	0.500	0.750	1.000	0.000	0.250	0.500	0.750	1.000	0.000	0.250	0.500	0.750	1.000
20	K_{N1}	0.000	−0.476	−0.662	−0.524	0.000	1.000	−0.389	−0.519	−0.401	0.000	0.000	−0.357	−0.279	−0.091	0.000
	K_{N2}	2.137	1.051	0.008	−1.045	−2.179	1.392	0.587	−0.055	−0.611	−1.187	0.983	0.048	−0.240	−0.177	0.010
	K_{M1}	1.000	0.867	0.537	0.180	0.000	1.000	0.865	0.529	0.175	0.000	1.000	0.730	0.287	0.040	0.000
	K_{M2}	1.899	1.692	1.314	0.954	0.814	1.552	1.292	0.861	0.466	0.305	1.425	0.906	0.297	−0.011	−0.067
	$K_{\Delta R}$	2.137	1.027	0.033	−0.898	−1.825	1.392	0.571	−0.029	−0.499	−0.934	0.983	0.057	−0.182	−0.121	0.006
	K_ψ	2.096	1.846	1.664	1.585	1.588	1.286	0.943	0.690	0.571	0.559	0.992	0.369	0.019	−0.083	−0.094
k_A	$\mu_D = 0.4$		$k_B =$ 39.434	$K_{\Delta h1} =$ −0.002	$K_{\Delta h2} = 34.371$		$\mu_D = 0.6$	$k_B =$ 39.151	$K_{\Delta h1} = -0.002$		$K_{\Delta h2} =$ 15.957	$\mu_D = 0.8$	$k_B =$ 38.869	$K_{\Delta h1} = -0.002$		$K_{\Delta h2} =$ 9.123
	Ω	0.000	0.250	0.500	0.750	1.000	0.000	0.250	0.500	0.750	1.000	0.000	0.250	0.500	0.750	1.000
40	K_{N1}	0.000	−1.300	−1.755	−1.333	0.000	0.000	−0.910	−1.234	−0.942	0.000	0.000	−0.698	−0.948	−0.726	0.000
	K_{N2}	17.256	8.734	0.089	−8.689	−17.610	8.048	4.080	0.052	−4.054	−8.255	4.645	2.339	0.024	−2.327	−4.743
	K_{M1}	1.000	0.842	0.508	0.168	0.000	1.000	0.849	0.511	0.166	0.000	1.000	0.852	0.514	0.165	0.000
	K_{M2}	10.272	10.179	9.917	9.657	9.577	3.910	3.768	3.449	3.134	3.009	2.290	2.125	1.780	1.441	1.298
	$K_{\Delta R}$	17.256	8.700	0.124	−8.478	−17.115	8.048	4.056	0.076	−3.907	−7.909	4.645	2.321	0.043	−2.214	−4.478
	K_ψ	43.228	43.226	43.250	43.310	43.397	13.566	13.468	13.409	13.403	13.438	6.015	5.852	5.741	5.700	5.714
k_A	$\mu_D = 1.2$		$k_B =$ 38.303	$K_{\Delta h1} =$ −0.003	$K_{\Delta h2} = 4.136$		$\mu_D = 1.6$	$k_B =$ 37.737	$K_{\Delta h1} = -0.003$		$K_{\Delta h2} =$ 2.404	$\mu_D = 3.2$	$k_B =$ 34.475	$K_{\Delta h1} = -0.003$		$K_{\Delta h2} =$ 0.986
	Ω	0.000	0.250	0.500	0.750	1.000	0.000	0.250	0.500	0.750	1.000	0.000	0.250	0.500	0.750	1.000
40	K_{N1}	0.000	−0.484	−0.649	−0.494	0.000	0.000	−3.388	−0.494	−0.362	0.000	0.000	−0.341	−0.233	−0.065	0.000
	K_{N2}	2.206	1.043	−0.026	−1.055	−2.105	1.419	0.565	−0.079	−0.603	−1.107	0.994	0.014	−0.231	−0.150	0.011
	K_{M1}	1.000	0.853	0.513	0.164	0.000	1.000	0.845	0.498	0.156	0.000	1.000	0.684	0.243	0.130	0.000
	K_{M2}	1.462	1.267	0.892	0.532	0.375	1.280	1.053	0.654	0.289	0.133	1.214	0.762	0.244	0.009	−0.025
	$K_{\Delta R}$	2.206	1.032	−0.012	−0.980	−1.930	1.419	0.557	−0.065	−0.547	−0.985	0.994	0.020	−0.201	−0.125	0.009
	K_ψ	2.154	1.888	1.696	1.610	1.603	1.303	0.994	0.687	0.569	0.552	0.999	0.342	0.005	−0.084	−0.091
k_A	$\mu_D = 0.4$		$k_B =$ 79.434	$K_{\Delta h1} =$ −0.001	$K_{\Delta h2} = 36.643$		$\mu_D = 0.6$	$k_B =$ 79.151	$K_{\Delta h1} = -0.001$		$K_{\Delta h2} =$ 16.473	$\mu_D = 0.8$	$k_B =$ 78.869	$K_{\Delta h1} = -0.001$		$K_{\Delta h2} =$ 9.313
	Ω	0.000	0.250	0.500	0.750	1.000	0.000	0.250	0.500	0.750	1.000	0.000	0.250	0.500	0.750	1.000
80	K_{N1}	0.000	−1.380	−1.851	−1.397	0.000	0.000	−0.934	−1.254	−0.948	0.000	0.000	−0.707	−0.948	−0.716	0.000
	K_{N2}	18.365	9.234	0.044	−9.211	−18.539	8.287	4.159	0.019	−4.150	−8.363	4.725	2.348	−0.001	−2.348	−4.720
	K_{M1}	1.000	0.845	0.504	0.160	−0.000	1.000	0.847	0.505	0.160	0.000	1.000	0.848	0.506	0.160	0.000
	K_{M2}	5.934	5.791	5.466	5.142	5.004	2.498	2.340	1.998	1.657	1.507	1.656	1.489	1.138	0.791	0.636
	$K_{\Delta R}$	18.365	9.216	0.063	−9.099	−18.277	8.287	4.147	0.032	−4.075	−8.186	4.725	2.339	0.009	−2.291	−4.588
	K_ψ	46.009	45.963	45.944	45.961	46.004	13.967	13.848	13.768	13.742	13.755	6.117	5.941	5.818	5.765	5.767

(Continued)

TABLE 13.3 Formulas for Bending and Membrane Stresses and Deformations in Thin-Walled Pressure Vessels (*Continued*)

k_A	$\mu_D = 1.2$		$k_B = 78.303$	$K_{\Delta h1} = -0.001$	$K_{\Delta h2} = 4.192$		$\mu_D = 1.6$	$k_B = 77.737$	$K_{\Delta h1} = -0.001$		$K_{\Delta h2} = 2.432$	$\mu_D = 3.2$	$k_B = 75.475$	$K_{\Delta h1} = -0.001$		$K_{\Delta h2} = 0.988$
	Ω	0.000	0.250	0.500	0.750	1.000	0.000	0.250	0.500	0.750	1.000	0.000	0.250	0.500	0.750	1.000
80	K_{N1}	0.000	−0.485	−0.637	−0.476	0.000	0.000	−0.386	−0.479	−0.342	0.000	0.000	−0.331	−0.212	−0.055	0.000
	K_{N2}	2.224	1.031	−0.043	−1.052	−2.054	1.426	0.551	−0.091	−0.596	−1.065	0.998	−0.001	−0.224	−0.138	0.010
	K_{M1}	1.000	0.845	0.501	0.157	0.000	1.000	0.834	0.483	0.148	0.000	1.000	0.660	0.224	0.027	0.000
	K_{M2}	1.233	1.051	0.687	0.335	0.179	1.140	0.937	0.558	0.211	0.062	1.107	0.697	0.223	0.017	−0.011
	$K_{\Delta R}$	2.224	1.025	−0.035	−1.014	−1.968	1.426	0.547	−0.083	−0.568	−1.005	0.998	0.002	−0.210	−0.126	0.009
	K_ψ	2.170	1.896	1.699	1.609	1.598	1.309	0.942	0.683	0.565	0.547	1.002	0.329	−0.001	−0.083	−0.089
k_A	$\mu_D = 0.4$		$k_B = 159.434$	$K_{\Delta h1} = -0.000$	$K_{\Delta h2} = 37.272$		$\mu_D = 0.6$	$k_B = 159.151$	$K_{\Delta h1} = -0.000$		$K_{\Delta h2} = 16.619$	$\mu_D = 0.8$	$k_B = 158.869$	$K_{\Delta h1} = -0.000$		$K_{\Delta h2} = 9.371$
	Ω	0.000	0.250	0.500	0.750	1.000	0.000	0.250	0.500	0.750	1.000	0.000	0.250	0.500	0.750	1.000
160	K_{N1}	0.000	−1.401	−1.873	−1.409	−0.000	0.000	−0.939	−1.225	−0.944	0.000	0.000	−0.708	−0.944	−0.709	0.000
	K_{N2}	18.666	9.353	0.019	−9.342	−18.739	8.350	4.170	0.002	−4.169	−8.357	4.746	2.343	−0.014	−2.350	−4.691
	K_{M1}	1.000	0.845	0.502	0.158	0.000	1.000	0.845	0.502	0.158	0.000	1.000	0.845	0.502	0.157	0.000
	K_{M2}	3.508	3.353	3.012	2.672	2.520	1.755	1.595	1.249	0.904	0.749	1.329	1.166	0.817	0.470	0.314
	$K_{\Delta R}$	18.666	9.344	0.029	−9.285	−18.607	8.350	4.164	0.009	−4.131	−8.269	4.746	2.338	−0.009	−2.321	−4.625
	K_ψ	46.763	46.693	46.650	46.643	46.662	14.074	13.943	13.852	13.815	13.818	6.145	5.962	5.833	5.775	5.770
k_A	$\mu_D = 1.2$		$k_B = 158.303$	$K_{\Delta h1} = -0.000$	$K_{\Delta h2} = 4.123$		$\mu_D = 1.6$	$k_B = 157.737$	$K_{\Delta h1} = -0.000$		$K_{\Delta h2} = 2.443$	$\mu_D = 3.2$	$k_B = 155.475$	$K_{\Delta h1} = -0.000$		$K_{\Delta h2} = 0.990$
	Ω	0.000	0.250	0.500	0.750	1.000	0.000	0.250	0.500	0.750	1.000	0.000	0.250	0.500	0.750	1.000
160	K_{N1}	0.000	−0.484	−0.630	−0.466	0.000	1.000	−0.385	−0.471	−0.332	0.000	0.000	−0.325	−0.202	−0.051	0.000
	K_{N2}	2.230	1.022	−0.051	−1.048	−2.026	1.429	0.543	−0.096	−0.592	−1.044	0.999	−0.008	−0.221	−0.133	0.010
	K_{M1}	1.000	0.841	0.495	0.154	0.000	1.000	0.829	0.475	0.144	0.000	1.000	0.649	0.214	0.025	0.000
	K_{M2}	1.117	0.943	0.587	0.242	0.087	1.070	0.879	0.512	0.175	0.030	1.054	0.666	0.214	0.020	−0.005
	$K_{\Delta R}$	2.230	1.020	−0.047	−1.029	−1.983	1.429	0.542	−0.092	−0.577	−1.014	0.999	−0.007	−0.214	−0.127	0.009
	K_ψ	2.175	1.897	1.698	1.606	1.593	1.311	0.940	0.680	0.562	0.543	1.003	0.323	−0.004	−0.083	−0.088

Case No., Loading | **Formulas**

5c. Uniform radial force Q_B, at the small end

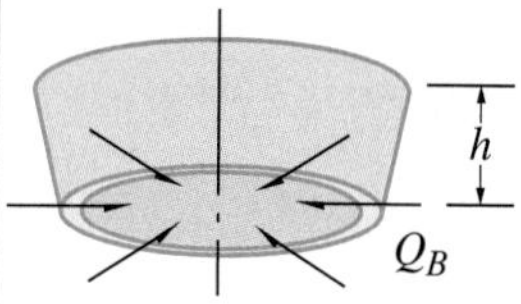

$$N_1 = Q_B \sin\alpha K_{N1}, \qquad N_2 = -Q_B \sin\alpha \frac{k_B}{\sqrt{2}} K_{N2}$$

$$M_1 = -Q_B \sin\alpha \frac{k_B t}{\sqrt{2}\beta} K_{M1}, \quad M_2 = -Q_B \nu \sin\alpha \frac{k_B t}{\sqrt{2}\beta} K_{M2}, \quad \Delta h = \frac{Q_B R_B}{Et} K_{\Delta h1} - \frac{Q_B R_B \sin^2\alpha}{Et\cos\alpha} \frac{k_B}{\sqrt{2}} K_{\Delta h2}$$

$$\Delta R = \frac{-Q_B R_B \sin\alpha}{Et} \frac{k_B}{\sqrt{2}} K_{\Delta R}, \quad \psi = \frac{Q_B R_B \beta}{Et^2} K_\psi$$

For $R_B/(t\cos\alpha) > 10$ and $k_B > 5$ and for $\nu = 0.3$, the following tables give the values of K at several locations along the shell $[\Omega = (R_A - R)/(R_A - R_B)]$.

k_A	$\mu_D = 0.4$		$k_B = 9.434$	$K_{\Delta h1} = 0.335$	$K_{\Delta h2} = 3.162$		$\mu_D = 0.6$	$k_B = 9.151$	$K_{\Delta h1} = 0.036$		$K_{\Delta h2} = 3.129$	$\mu_D = 0.8$	$k_B = 8.869$	$K_{\Delta h1} = 0.336$		$K_{\Delta h2} = 2.850$
	Ω	0.000	0.250	0.500	0.750	1.000	0.000	0.250	0.500	0.750	1.000	0.000	0.250	0.500	0.750	1.000
10	K_{N1}	0.000	0.016	0.172	0.491	1.000	0.000	−0.075	0.033	0.372	1.000	0.000	−0.122	−0.048	0.292	1.000
	K_{N2}	−0.258	0.436	1.178	1.973	2.827	−0.583	0.047	0.744	1.519	2.387	−0.601	−0.077	0.523	1.219	2.037
	K_{M1}	0.000	0.026	0.046	0.043	0.000	0.000	0.032	0.066	0.069	0.000	0.000	0.037	0.085	0.095	0.000
	K_{M2}	−3.000	−3.086	−3.188	−3.321	−3.502	−1.866	−1.939	−2.026	−2.159	−2.376	−1.192	−1.246	−1.310	−1.434	−1.678
	$K_{\Delta R}$	−0.290	0.478	1.259	2.057	2.872	−0.696	0.050	0.818	1.611	2.433	−0.764	−0.100	0.591	1.317	2.085
	K_ψ	7.413	7.478	7.550	7.627	7.702	4.752	4.817	4.895	4.985	5.070	3.133	3.192	3.274	3.376	3.469

(*Continued*)

TABLE 13.3 Formulas for Bending and Membrane Stresses and Deformations in Thin-Walled Pressure Vessels (*Continued*)

k_A	$\mu_D = 1.2$		$k_B =$ 8.303	$K_{\Delta h1} =$ 0.336	$K_{\Delta h2} = 2.283$		$\mu_D = 1.6$	$k_B =$ 7.737	$K_{\Delta h1} = 0.340$		$K_{\Delta h2} =$ 1.882	$\mu_D = 3.2$	$k_B =$ 5.475	$K_{\Delta h1} = -0.431$		$K_{\Delta h2} =$ 1.213
	Ω	0.000	0.250	0.500	0.750	1.000	0.000	0.250	0.500	0.750	1.000	0.000	0.250	0.500	0.750	1.000
	K_{N1}	0.000	−0.148	−0.120	0.197	1.000	0.000	−0.135	−0.138	0.134	1.000	0.000	−0.026	−0.065	−0.039	1.000
	K_{N2}	−0.464	−0.125	0.295	0.837	1.559	−0.322	−0.113	0.172	0.604	1.288	−0.030	−0.034	−0.020	0.137	1.036
10	K_{M1}	0.000	0.043	0.155	0.144	0.000	0.000	0.045	0.134	0.185	0.000	0.000	0.014	0.073	0.190	0.000
	K_{M2}	−0.539	−0.560	−0.583	−0.691	−0.992	−0.263	−0.263	−0.258	−0.358	−0.729	−0.001	0.010	0.034	−0.019	−0.721
	$K_{\Delta R}$	−0.673	−0.177	0.354	0.943	1.610	−0.538	−0.181	0.220	0.713	1.343	−0.100	−0.098	−0.053	0.213	1.113
	K_{ψ}	1.514	1.562	1.657	1.798	1.920	0.792	0.833	0.945	1.142	1.314	0.005	0.016	0.108	0.422	0.920
k_A	$\mu_D = 0.4$		$k_B =$ 19.434	$K_{\Delta h1} =$ 0.308	$K_{\Delta h2} = 5.616$		$\mu_D = 0.6$	$k_B =$ 19.151	$K_{\Delta h1} = 0.306$		$K_{\Delta h2} =$ 4.390	$\mu_D = 0.8$	$k_B =$ 18.869	$K_{\Delta h1} = 0.306$		$K_{\Delta h2} =$ 3.518
	Ω	0.000	0.250	0.500	0.750	1.000	0.000	0.250	0.500	0.750	1.000	0.000	0.250	0.500	0.750	1.000
	K_{N1}	0.000	−0.157	−0.057	0.322	1.000	0.000	−0.216	−0.145	0.247	1.000	0.000	−0.235	−0.181	0.211	1.000
	K_{N2}	−1.460	−0.151	1.201	2.600	4.048	−1.238	−0.255	0.778	1.867	3.018	−0.996	−0.242	0.565	1.435	2.377
20	K_{M1}	0.000	0.022	0.048	0.051	0.000	0.000	0.029	0.071	0.079	0.000	0.000	0.036	0.093	0.107	0.000
	K_{M2}	−2.846	−2.877	−2.906	−2.962	−3.072	−1.439	−1.450	−1.454	−1.495	−1.624	−0.835	−0.831	−0.813	−0.844	−0.996
	$K_{\Delta R}$	−1.546	−0.162	1.235	2.645	4.070	−1.350	−0.278	0.810	1.915	3.040	−1.119	−0.271	0.595	1.484	2.400
	K_{ψ}	13.655	13.714	13.781	13.851	13.918	7.007	7.055	7.116	7.188	7.251	4.128	4.169	4.232	4.314	4.381
k_A	$\mu_D = 1.2$		$k_B =$ 18.303	$K_{\Delta h1} =$ 0.305	$K_{\Delta h2} = 2.490$		$\mu_D = 1.6$	$k_B =$ 17.737	$K_{\Delta h1} = 0.306$		$K_{\Delta h2} =$ 1.932	$\mu_D = 3.2$	$k_B =$ 15.475	$K_{\Delta h1} = -0.313$		$K_{\Delta h2} =$ 1.138
	Ω	0.000	0.250	0.500	0.750	1.000	0.000	0.250	0.500	0.750	1.000	0.000	0.250	0.500	0.750	1.000
	K_{N1}	0.000	−0.236	−0.205	0.171	1.000	0.000	−0.217	−0.208	0.138	1.000	0.000	−0.069	−0.144	−0.046	1.000
	K_{N2}	−0.664	−0.189	0.344	0.959	1.673	−0.457	−0.149	0.221	0.701	1.327	−0.052	−0.059	−0.022	0.234	1.023
20	K_{M1}	0.000	0.048	0.133	0.160	0.000	0.000	0.057	0.165	0.207	0.000	0.000	0.030	0.137	0.277	0.000
	K_{M2}	−0.358	−0.332	−0.281	−0.302	−0.506	−0.178	−0.138	−0.065	−0.080	−0.341	0.001	0.028	0.105	0.145	−0.276
	$K_{\Delta R}$	−0.793	−0.223	0.372	1.010	1.696	−0.581	−0.185	0.246	0.752	1.351	−0.087	−0.092	−0.035	0.272	1.051
	K_{ψ}	1.824	1.859	1.940	2.066	2.160	0.934	0.967	1.078	1.268	1.410	−0.006	0.012	0.155	0.557	0.994
k_A	$\mu_D = 0.4$		$k_B =$ 39.434	$K_{\Delta h1} =$ 0.301	$K_{\Delta h2} = 6.940$		$\mu_D = 0.6$	$k_B =$ 39.151	$K_{\Delta h1} = 0.301$		$K_{\Delta h2} =$ 4.853	$\mu_D = 0.8$	$k_B =$ 38.869	$K_{\Delta h1} = 0.301$		$K_{\Delta h2} =$ 3.711
	Ω	0.000	0.250	0.500	0.750	1.000	0.000	0.250	0.500	0.750	1.000	0.000	0.250	0.500	0.750	1.000
	K_{N1}	0.000	−0.260	−0.189	0.227	1.000	0.000	−0.279	−0.219	0.200	1.000	0.000	−0.282	−0.229	0.188	1.000
	K_{N2}	−2.154	−0.480	1.222	2.952	4.713	−1.520	−0.374	0.803	2.012	3.255	−1.152	−0.296	0.591	1.516	2.480
40	K_{M1}	0.000	0.020	0.049	0.055	0.000	0.000	0.028	0.073	0.083	0.000	0.000	0.036	0.097	0.111	0.000
	K_{M2}	−1.810	−1.807	−1.795	−1.808	−1.881	−0.830	−0.814	−0.782	−0.787	−0.884	−0.467	−0.440	−0.391	−0.391	−0.515
	$K_{\Delta R}$	−2.216	−0.493	1.237	2.976	4.724	−1.587	−0.389	0.818	2.036	3.266	−1.220	−0.312	0.606	1.540	2.491
	K_{ψ}	17.115	17.153	17.196	17.244	17.288	7.909	7.937	7.978	8.029	8.071	4.478	4.502	4.549	4.614	4.662
k_A	$\mu_D = 1.2$		$k_B =$ 38.303	$K_{\Delta h1} =$ 0.301	$K_{\Delta h2} = 2.523$		$\mu_D = 1.6$	$k_B =$ 37.737	$K_{\Delta h1} = 0.301$		$K_{\Delta h2} =$ 1.923	$\mu_D = 3.2$	$k_B =$ 35.475	$K_{\Delta h1} = 0.300$		$K_{\Delta h2} =$ 1.107
	Ω	0.000	0.250	0.500	0.750	1.000	0.000	0.250	0.500	0.750	1.000	0.000	0.250	0.500	0.750	1.000
	K_{N1}	0.000	−0.275	−0.234	0.170	1.000	0.000	−0.256	−0.232	0.149	1.000	0.000	−0.096	−0.182	−0.033	1.000
	K_{N2}	−0.747	−0.206	0.372	1.004	1.698	−0.517	−0.157	0.249	0.739	1.331	−0.064	−0.070	−0.016	0.278	1.013
40	K_{M1}	0.000	0.051	0.141	0.164	0.000	0.000	0.063	0.177	0.214	0.000	0.000	0.040	0.167	0.303	0.000
	K_{M2}	−0.198	−0.154	−0.076	−0.070	−0.248	−0.100	−0.043	0.057	0.070	−0.162	0.001	0.039	0.148	0.235	−0.121
	$K_{\Delta R}$	−0.814	−0.223	0.386	1.029	1.709	−0.581	−0.175	0.261	0.764	1.342	−0.081	−0.086	−0.020	0.296	1.025
	K_{ψ}	1.930	1.954	2.026	2.140	2.218	0.985	1.012	1.120	1.303	1.428	−0.009	0.014	0.180	0.602	1.001

(*Continued*)

TABLE 13.3 Formulas for Bending and Membrane Stresses and Deformations in Thin-Walled Pressure Vessels (*Continued*)

k_A	$\mu_D = 0.4$		$k_B =$ 79.434	$K_{\Delta h1} =$ 0.300	$K_{\Delta h2} = 7.362$		$\mu_D = 0.6$	$k_B =$ 79.151	$K_{\Delta h1} = 0.300$		$K_{\Delta h2} =$ 4.974	$\mu_D = 0.8$	$k_B =$ 78.869	$K_{\Delta h1} = 0.300$		$K_{\Delta h2} =$ 3.752
	Ω	0.000	250	0.500	0.750	1.000	0.000	0.250	0.500	0.750	1.000	0.000	0.250	0.500	0.750	1.000
80	K_{N1}	0.000	−0.296	−0.233	0.197	1.000	0.000	−0.300	−0.241	0.188	1.000	0.000	−0.299	−0.243	0.184	1.000
	K_{N2}	−2.395	−0.588	1.235	3.073	4.928	−1.614	−0.407	0.817	2.059	3.319	−1.207	−0.309	0.606	1.543	2.505
	K_{M1}	0.000	0.019	0.050	0.056	0.000	0.000	0.028	0.074	0.084	0.000	0.000	0.037	0.098	0.112	0.000
	K_{M2}	−0.973	−0.959	−0.933	−0.932	−0.993	−0.434	−0.410	−0.367	−0.362	−0.450	−0.243	−0.208	−0.151	−0.142	−0.257
	$K_{\Delta R}$	−2.429	−0.595	1.243	3.085	4.933	−1.649	−0.415	0.824	2.071	3.325	−1.241	−0.318	0.613	1.556	2.510
	K_ψ	18.277	18.298	18.324	18.355	18.382	8.186	8.202	8.231	8.270	8.299	4.588	4.603	4.640	4.695	4.734
k_A	$\mu_D = 1.2$		$k_B =$ 78.303	$K_{\Delta h1} =$ 0.300	$K_{\Delta h2} = 2.522$		$\mu_D = 1.6$	$k_B =$ 77.737	$K_{\Delta h1} = 0.300$		$K_{\Delta h2} =$ 1.911	$\mu_D = 3.2$	$k_B =$ 75.475	$K_{\Delta h1} = 0.298$		$K_{\Delta h2} =$ 1.093
	Ω	0.000	0.250	0.500	0.750	1.000	0.000	0.250	0.500	0.750	1.000	0.000	0.250	0.500	0.750	1.000
80	K_{N1}	0.000	−0.291	−0.244	0.174	1.000	0.000	−0.274	−0.241	0.156	1.000	0.000	−0.111	−0.199	−0.024	1.000
	K_{N2}	−0.780	−0.209	0.386	1.022	1.702	−0.544	−0.159	0.262	0.755	1.329	−0.070	−0.075	−0.011	0.298	1.008
	K_{M1}	0.000	0.053	0.144	0.166	0.000	0.000	0.066	0.183	0.216	0.000	0.000	0.045	0.181	0.313	0.000
	K_{M2}	−0.103	−0.052	0.034	0.049	−0.122	−0.052	0.012	0.122	0.145	−0.079	0.000	0.044	0.171	0.278	−0.057
	$K_{\Delta R}$	−0.815	−0.218	0.394	1.034	1.707	−0.577	−0.167	0.269	0.767	1.334	−0.079	−0.082	−0.013	0.307	1.014
	K_ψ	1.968	1.986	2.053	2.160	2.230	1.005	1.029	1.135	1.314	1.431	−0.009	0.015	0.192	0.621	1.001
k_A	$\mu_D = 0.4$		$k_B =$ 159.434	$K_{\Delta h1} =$ 0.300	$K_{\Delta h2} = 7.471$		$\mu_D = 0.6$	$k_B =$ 159.151	$K_{\Delta h1} = 0.300$		$K_{\Delta h2} =$ 5.000	$\mu_D = 0.8$	$k_B =$ 158.869	$K_{\Delta h1} = 0.300$		$K_{\Delta h2} =$ 3.758
	Ω	0.000	0.250	0.500	0.750	1.000	0.000	0.250	0.500	0.750	1.000	0.000	0.250	0.500	0.750	1.000
160	K_{N1}	0.000	−0.307	−0.245	0.189	1.000	0.000	−0.307	−0.247	0.186	1.000	0.000	−0.306	−0.248	0.184	1.000
	K_{N2}	−2.466	−0.616	1.242	3.109	4.984	−1.645	−0.415	0.824	2.074	3.335	−1.227	−0.312	0.613	1.554	2.510
	K_{M1}	0.000	0.019	0.050	0.056	0.000	0.000	0.028	0.075	0.084	0.000	0.000	0.037	0.099	0.112	0.000
	K_{M2}	−0.497	−0.480	−0.450	−0.445	−0.502	−0.221	−0.193	−0.148	−0.140	−0.225	−0.123	−0.087	−0.026	−0.015	−0.128
	$K_{\Delta R}$	−2.484	−0.620	1.246	3.115	4.987	−1.662	−0.419	0.828	2.080	3.337	−1.245	−0.371	0.617	1.560	2.513
	K_ψ	18.607	18.618	18.635	18.656	18.673	8.269	8.279	8.301	8.333	8.355	4.625	4.635	4.667	4.718	4.751
k_A	$\mu_D = 1.2$		$k_B =$ 158.303	$K_{\Delta h1} =$ 0.300	$K_{\Delta h2} = 2.516$		$\mu_D = 1.6$	$k_B =$ 157.737	$K_{\Delta h1} = 0.300$		$K_{\Delta h2} =$ 1.903	$\mu_D = 3.2$	$k_B =$ 155.475	$K_{\Delta h1} = 0.298$		$K_{\Delta h2} =$ 1.086
	Ω	0.000	0.250	0.500	0.750	1.000	0.000	0.250	0.500	0.750	1.000	0.000	0.250	0.500	0.750	1.000
160	K_{N1}	0.000	−0.299	−0.247	0.177	1.000	0.000	−0.283	−0.245	0.161	1.000	0.000	−0.119	−0.207	−0.019	1.000
	K_{N2}	−0.796	−0.210	0.394	1.030	1.701	−0.557	−0.159	0.269	0.762	1.327	−0.074	−0.077	−0.009	0.307	1.006
	K_{M1}	0.000	0.054	0.146	0.167	0.000	0.000	0.068	0.186	0.217	0.000	0.000	0.048	0.188	0.317	0.000
	K_{M2}	−0.053	0.001	0.090	0.108	−0.061	−0.027	0.040	0.155	0.182	−0.039	0.000	0.047	0.183	0.300	−0.028
	$K_{\Delta R}$	−0.813	−0.214	0.397	1.036	1.704	−0.573	−0.163	0.272	0.768	1.330	−0.078	−0.081	−0.010	0.311	1.008
	K_ψ	1.983	1.998	2.062	2.166	2.233	1.014	1.036	1.142	1.318	1.431	−0.009	0.016	0.198	0.629	1.001

Case No., Loading	Formulas
5d. Uniform edge moment M_B at the small end 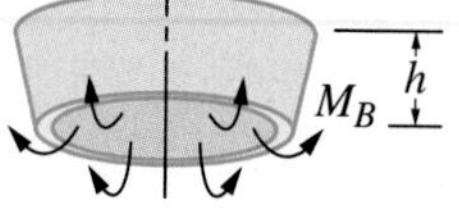	$N_1 = -M_B \dfrac{2\sqrt{2\beta}}{tk_B} K_{N1}, \quad N_2 = M_B \dfrac{\beta}{t} K_{N2}$ $M_1 = M_B K_{M1}, \quad M_2 = M_B \nu K_{M2}, \quad \Delta h = \dfrac{-M_B R_B \beta}{Et^2 \sin\alpha} K_{\Delta h1} - \dfrac{M_B \beta R_B \sin\alpha}{Et^2 \cos\alpha} K_{\Delta h2}$ $\Delta R = M_B \dfrac{\beta R_B}{Et^2} K_{\Delta R}, \quad \psi = -M_B \dfrac{2\sqrt{2}\beta^2 R_B}{Et^3 k_B \sin\alpha} K_\psi$ For $R_B/(t\cos\alpha) > 10$ and $k_B > 5$ and for $\nu = 0.3$, the following tables give the values of K at several locations along the shell $[\Omega = (R_A - R)/(R_A - R_B)]$.

(*Continued*)

TABLE 13.3 Formulas for Bending and Membrane Stresses and Deformations in Thin-Walled Pressure Vessels (*Continued*)

k_A																
k_A	$\mu_D = 0.4$		$k_B =$ 9.434	$K_{\Delta h1} =$ −0.018	$K_{\Delta h2} = -15.589$		$\mu_D = 0.6$	$k_B =$ 9.151	$K_{\Delta h1} = -0.027$		$K_{\Delta h2} =$ −10.293	$\mu_D = 0.8$	$k_B =$ 8.869	$K_{\Delta h1} = -0.034$		$K_{\Delta h2} =$ −7.029
	Ω	0.000	0.250	0.500	0.750	1.000	0.000	0.250	0.500	0.750	1.000	0.000	0.250	0.500	0.750	1.000
10	K_{N1}	0.000	−0.507	−0.712	−0.563	0.000	0.000	−0.465	−0.670	−0.546	0.000	0.000	−0.389	−0.578	−0.487	0.000
	K_{N2}	−7.020	−3.707	−0.170	3.622	7.702	−4.374	−2.378	−0.172	2.290	5.070	−2.800	−1.571	−0.164	1.486	3.469
	K_{M1}	0.000	0.187	0.461	0.758	1.000	0.000	0.160	0.443	0.761	1.000	0.000	0.141	0.425	0.757	1.000
	K_{M2}	−15.021	−15.396	−15.742	−16.134	−16.657	−6.220	−6.412	−6.550	−6.740	−7.110	−2.965	−3.052	−3.069	−3.143	−3.433
	$K_{\Delta R}$	−7.887	−4.100	−0.248	3.681	7.702	−5.223	−2.773	−0.257	2.348	5.070	−3.559	−1.936	−0.249	1.537	3.469
	K_{ψ}	18.558	18.722	18.917	19.149	19.416	7.922	8.031	8.183	8.390	8.651	3.896	3.973	4.101	4.305	4.582
k_A	$\mu_D = 1.2$		$k_B =$ 8.303	$K_{\Delta h1} =$ −0.043	$K_{\Delta h2} = -3.764$		$\mu_D = 1.6$	$k_B =$ 7.737	$K_{\Delta h1} = -0.049$		$K_{\Delta h2} =$ −2.352	$\mu_D = 3.2$	$k_B =$ 5.475	$K_{\Delta h1} = -0.068$		$K_{\Delta h2} =$ −0.927
	Ω	0.000	0.250	0.500	0.750	1.000	0.000	0.250	0.500	0.750	1.000	0.000	0.250	0.500	0.750	1.000
10	K_{N1}	0.000	−0.255	−0.407	−0.376	0.000	0.000	−0.161	−0.285	−0.301	0.000	0.000	−0.009	−0.051	−0.155	0.000
	K_{N2}	−1.271	−0.774	−0.153	0.690	1.920	−0.621	−0.429	−0.154	0.335	1.314	−0.002	−0.045	−0.107	−0.905	0.920
	K_{M1}	0.000	0.113	0.381	0.729	1.000	0.000	0.088	0.327	0.681	1.000	0.000	0.005	0.066	0.330	1.000
	K_{M2}	−0.862	−0.855	−0.771	−0.750	−1.004	−0.281	−0.245	−0.136	−0.092	−0.390	0.024	0.035	0.084	0.158	−0.517
	$K_{\Delta R}$	−1.843	−1.071	−0.239	0.725	1.920	−1.038	−0.672	−0.247	0.353	1.314	−0.007	−0.128	−0.248	−0.188	0.920
	K_{ψ}	1.210	1.254	1.366	1.592	1.939	0.423	0.451	0.558	0.814	1.254	−0.050	−0.052	−0.024	0.174	0.968
k_A	$\mu_D = 0.4$		$k_B =$ 19.434	$K_{\Delta h1} =$ −0.008	$K_{\Delta h2} = 27.998$		$\mu_D = 0.6$	$k_B =$ 19.151	$K_{\Delta h1} = 0.009$		$K_{\Delta h2} =$ −14.589	$\mu_D = 0.8$	$k_B =$ 18.869	$K_{\Delta h1} = 0.010$		$K_{\Delta h2} =$ 8.774
	Ω	0.000	0.250	0.500	0.750	1.000	0.000	0.250	0.500	0.750	1.000	0.000	0.250	0.500	0.750	1.000
20	K_{N1}	0.000	−0.979	−1.340	−1.031	0.000	0.000	−0.737	−1.022	−0.799	0.000	0.000	−0.567	−0.799	−0.636	0.000
	K_{N2}	−13.294	−6.833	−0.156	6.755	13.918	−6.729	−3.513	−0.130	3.448	7.251	−3.910	−2.081	−0.117	2.021	4.381
	K_{M1}	0.000	0.171	0.483	0.806	1.000	0.000	0.155	0.474	0.811	1.000	0.000	0.146	0.465	0.809	1.000
	K_{M2}	−14.236	−14.330	−14.303	−14.289	−14.426	−4.790	−4.771	−4.612	−4.465	−4.496	−2.074	−2.009	−1.798	−1.601	−1.594
	$K_{\Delta R}$	−14.079	−7.180	−0.221	6.809	13.918	−7.338	−3.787	−0.183	3.489	7.251	−4.393	−2.301	−0.162	2.054	4.381
	K_{ψ}	34.151	34.302	34.481	34.697	34.944	11.662	11.744	11.867	12.045	12.269	5.124	5.178	5.284	5.463	5.705
k_A	$\mu_D = 1.2$		$k_B =$ 18.303	$K_{\Delta h1} =$ −0.011	$K_{\Delta h2} = -4.165$		$\mu_D = 1.6$	$k_B =$ 17.737	$K_{\Delta h1} = -0.011$		$K_{\Delta h2} =$ −2.471	$\mu_D = 3.2$	$k_B =$ 15.475	$K_{\Delta h1} = -0.013$		$K_{\Delta h2} =$ −0.988
	Ω	0.000	0.250	0.500	0.750	1.000	0.000	0.250	0.500	0.750	1.000	0.000	0.250	0.500	0.750	1.000
20	K_{N1}	0.000	−0.362	−0.533	−0.448	0.000	0.000	−0.242	−0.385	−0.355	0.000	0.000	−0.024	−0.199	−0.259	0.000
	K_{N2}	−1.679	−0.948	−0.117	0.887	2.160	−0.835	−0.527	−0.136	0.452	1.410	0.004	−0.085	−0.175	−0.067	0.994
	K_{M1}	0.000	0.131	0.440	0.793	1.000	0.000	0.113	0.403	0.766	1.000	0.000	0.013	0.135	0.516	1.000
	K_{M2}	−0.571	−0.477	−0.241	−0.016	0.003	−0.189	−0.096	0.132	0.359	0.369	0.024	0.039	0.144	0.394	0.443
	$K_{\Delta R}$	−2.005	−1.105	−0.155	0.908	2.160	−1.061	−0.648	−0.175	0.464	1.410	0.007	−0.130	−0.242	−0.095	0.994
	K_{ψ}	1.454	1.486	1.593	1.810	2.127	0.496	0.520	0.635	0.898	1.305	−0.073	−0.073	−0.021	0.257	1.005
k_A	$\mu_D = 0.4$		$k_B =$ 39.434	$K_{\Delta h1} =$ −0.002	$K_{\Delta h2} = -34.665$		$\mu_D = 0.6$	$k_B =$ 39.151	$K_{\Delta h1} = -0.002$		$K_{\Delta h2} =$ −16.163	$\mu_D = 0.8$	$k_B =$ 38.869	$K_{\Delta h1} = -0.003$		$K_{\Delta h2} =$ −9.280
	Ω	0.000	0.250	0.500	0.750	1.000	0.000	0.250	0.500	0.750	1.000	0.000	0.250	0.500	0.750	1.000
40	K_{N1}	0.000	−1.256	−1.696	−1.289	0.000	0.000	−0.861	−1.172	−0.898	0.000	0.000	−0.644	−0.885	−0.685	0.000
	K_{N2}	−16.889	−8.564	−0.100	8.514	17.288	−7.752	−3.966	−0.080	3.926	8.071	−4.360	−2.259	−0.075	2.221	4.662
	K_{M1}	0.000	0.159	0.492	0.830	1.000	0.000	0.153	0.487	0.831	1.000	0.000	0.149	0.482	0.828	1.000
	K_{M2}	−9.047	−8.972	−8.731	−8.493	−8.432	−2.761	−2.648	−2.363	−2.082	−1.985	−1.157	−1.031	−0.734	−0.440	−0.335
	$K_{\Delta R}$	−17.377	−8.778	−0.139	8.548	17.288	−8.092	−4.116	−0.107	3.950	8.071	−4.617	−2.373	−0.097	2.238	4.662
	K_{ψ}	42.783	42.880	43.003	43.163	43.351	13.153	13.203	13.292	13.437	13.623	5.552	5.586	5.672	5.832	6.048

(*Continued*)

TABLE 13.3 Formulas for Bending and Membrane Stresses and Deformations in Thin-Walled Pressure Vessels (*Continued*)

k_A	$\mu_D = 1.2$		$k_B =$ 38.303	$K_{\Delta h1} =$ −0.003	$K_{\Delta h2} = -4.240$		$\mu_D = 1.6$	$k_B =$ 37.737	$K_{\Delta h1} = -0.003$		$K_{\Delta h2} =$ −2.476	$\mu_D = 3.2$	$k_B =$ 35.475	$K_{\Delta h1} = -0.003$		$K_{\Delta h2} =$ −0.992
	Ω	0.000	0.250	0.500	0.750	1.000	0.000	0.250	0.500	0.750	1.000	0.000	0.250	0.500	0.750	1.000
40	K_{N1}	0.000	−0.412	−0.584	−0.471	0.000	0.000	−0.283	−0.427	−0.372	0.000	0.000	−0.034	−0.156	−0.294	0.000
	K_{N2}	−1.854	−1.007	−0.090	0.961	2.218	−0.933	−0.562	−0.120	0.498	1.428	0.007	−0.107	−0.199	−0.042	1.001
	K_{M1}	0.000	0.140	0.466	0.818	1.000	0.000	0.127	0.437	0.797	1.000	0.000	0.018	0.171	0.583	1.000
	K_{M2}	−0.315	−0.187	0.110	0.408	0.514	−0.105	0.014	0.295	0.594	0.701	0.015	0.033	0.174	0.518	0.757
	$K_{\Delta R}$	−2.022	−1.085	−0.107	0.972	2.218	−1.048	−0.621	−0.138	0.505	1.428	0.009	−0.129	−0.231	−0.052	1.001
	K_ψ	1.535	1.560	1.660	1.870	2.170	0.521	0.544	0.660	0.923	1.313	−0.081	−0.079	−0.015	0.289	1.006
k_A	$\mu_D = 0.4$		$k_B =$ 79.434	$K_{\Delta h1} =$ −0.001	$K_{\Delta h2} = 36.799$		$\mu_D = 0.6$	$k_B =$ 79.151	$K_{\Delta h1} = -0.001$		$K_{\Delta h2} =$ −16.579	$\mu_D = 0.8$	$k_B =$ 78.869	$K_{\Delta h1} = -0.001$		$K_{\Delta h2} =$ −9.392
	Ω	0.000	0.250	0.500	0.750	1.000	0.000	0.250	0.500	0.750	1.000	0.000	0.250	0.500	0.750	1.000
80	K_{N1}	0.000	−1.356	−1.820	−1.374	0.000	0.000	−0.907	−1.222	−0.927	0.000	0.000	−0.675	−0.916	−0.700	0.000
	K_{N2}	−18.157	−9.146	−0.056	9.116	18.382	−8.105	−4.105	−0.048	4.081	8.299	−4.527	−2.314	−0.051	2.288	4.734
	K_{M1}	0.000	0.156	0.496	0.839	1.000	0.000	0.154	0.493	0.838	1.000	0.000	0.152	0.490	0.836	1.000
	K_{M2}	−4.864	−4.731	−4.417	−4.104	−3.976	−1.444	−1.301	−0.977	−0.653	−0.517	−0.601	−0.456	−0.130	0.195	0.333
	$K_{\Delta R}$	−18.417	−9.258	−0.076	9.134	18.382	−8.280	−4.181	−0.062	4.093	8.299	−4.658	−2.372	−0.062	2.297	4.734
	K_ψ	45.679	45.733	45.813	45.930	46.074	13.610	13.639	13.709	13.832	13.997	5.685	5.708	5.783	5.931	6.133
k_A	$\mu_D = 1.2$		$k_B =$ 78.303	$K_{\Delta h1} =$ −0.001	$K_{\Delta h2} = -4.244$		$\mu_D = 1.6$	$K_B =$ 77.737	$K_{\Delta h1} = -0.001$		$K_{\Delta h2} =$ −2.468	$\mu_D = 3.2$	$k_B =$ 75.475	$K_{\Delta h1} = -0.001$		$K_{\Delta h2} =$ −0.992
	Ω	0.000	0.250	0.500	0.750	1.000	0.000	0.250	0.500	0.750	1.000	0.000	0.250	0.500	0.750	1.000
80	K_{N1}	0.000	−0.435	−0.605	−0.478	−0.000	0.000	−0.303	−0.446	−0.378	0.000	0.000	−0.041	−0.175	−0.308	0.000
	K_{N2}	−1.929	−1.028	−0.075	0.990	2.230	−0.979	−0.575	−0.111	0.518	1.431	0.008	−0.117	−0.209	−0.028	1.001
	K_{M1}	0.000	0.146	0.478	0.828	1.000	0.000	0.134	0.452	0.811	1.000	0.000	0.020	0.188	0.611	1.000
	K_{M2}	−0.164	−0.023	0.297	0.623	0.761	−0.055	0.075	0.381	0.709	0.855	0.008	0.029	0.190	0.578	0.886
	$K_{\Delta R}$	−2.014	−1.066	−0.083	0.995	2.230	−1.037	−0.604	−0.199	0.521	1.431	0.009	−0.128	−0.224	−0.033	1.001
	K_ψ	1.564	1.585	1.682	1.887	2.178	0.531	0.552	0.670	0.931	1.314	−0.084	−0.081	−0.011	0.303	1.005
k_A	$\mu_D = 0.4$		$k_B =$ 159.434	$K_{\Delta h1} =$ −0.000	$K_{\Delta h2} = -37.352$		$\mu_D = 0.6$	$k_B =$ 159.151	$K_{\Delta h1} = -0.000$		$K_{\Delta h2} =$ −16.672	$\mu_D = 0.8$	$k_B =$ 158.869	$K_{\Delta h1} = -0.000$		$K_{\Delta h2} =$ −9.411
	Ω	0.000	0.250	0.500	0.750	1.000	0.000	0.250	0.500	0.750	1.000	0.000	0.250	0.500	0.750	1.000
160	K_{N1}	0.000	−1.388	−1.857	−1.398	0.000	0.000	−0.923	−1.239	−0.936	0.000	0.000	−0.688	−0.923	−0.705	0.000
	K_{N2}	−18.547	−9.310	−0.033	9.292	18.673	−8.228	−4.146	−0.032	4.130	8.355	−4.594	−2.333	−0.039	2.313	4.751
	K_{M1}	0.000	0.156	0.498	0.842	1.000	0.000	0.155	0.496	0.841	1.000	0.000	0.153	0.494	0.839	1.000
	K_{M2}	−2.484	−2.335	−2.000	−1.666	−1.518	−0.733	−0.581	−0.245	0.092	0.241	−0.305	−0.154	0.182	0.519	0.668
	$K_{\Delta R}$	−18.679	−9.367	−0.043	9.301	18.673	−8.316	−4.185	−0.039	4.136	8.355	−4.660	−2.361	−0.044	2.317	4.751
	K_ψ	46.497	46.527	46.584	46.677	46.796	13.745	13.764	13.823	13.935	14.089	5.729	5.746	5.816	5.957	6.153
k_A	$\mu_D = 1.2$		$k_B =$ 158.303	$K_{\Delta h1} =$ −0.000	$K_{\Delta h2} = -4.239$		$\mu_D = 1.6$	$k_B =$ 157.737	$K_{\Delta h1} = -0.000$		$K_{\Delta h2} =$ −2.461	$\mu_D = 3.2$	$k_B =$ 155.475	$K_{\Delta h1} = -0.000$		$K_{\Delta h2} =$ −0.991
	Ω	0.000	0.250	0.500	0.750	1.000	0.000	0.250	0.500	0.750	1.000	0.000	0.250	0.500	0.750	1.000
160	K_{N1}	0.000	−0.446	−0.614	−0.480	0.000	0.000	−0.313	−0.455	−0.381	0.000	0.000	−0.044	−0.184	−0.314	0.000
	K_{N2}	−1.964	−1.036	−0.067	1.002	2.233	−1.001	−0.581	−0.106	0.527	1.431	0.009	−0.122	−0.213	−0.022	1.001
	K_{M1}	0.000	0.148	0.484	0.833	1.000	0.000	0.137	0.460	0.817	1.000	0.000	0.022	0.197	0.624	1.000
	K_{M2}	−0.084	0.063	0.393	0.731	0.882	−0.028	0.107	0.424	0.767	0.929	0.004	0.026	0.198	0.608	0.945
	$K_{\Delta R}$	−2.006	−1.055	−0.071	1.005	2.233	−1.030	−0.596	−0.110	0.529	1.431	0.010	−0.128	−0.221	−0.024	1.001
	K_ψ	1.576	1.594	1.689	1.892	2.179	0.536	0.556	0.674	0.935	1.313	−0.085	−0.081	−0.008	0.310	1.005

(Continued)

TABLE 13.3 Formulas for Bending and Membrane Stresses and Deformations in Thin-Walled Pressure Vessels (*Continued*)

Case No., Loading		Formulas
6. Toroidal shells 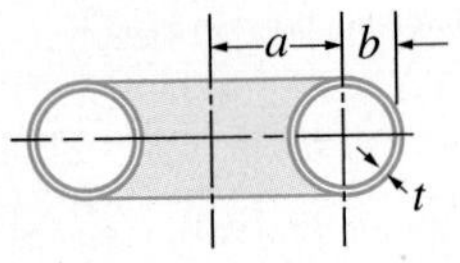 $\mu = \frac{b^2}{at}\sqrt{12(1-\nu^2)}$ $\frac{t}{b} < \frac{1}{10}$	6a. Split toroidal shell under axial load P (omega joint) 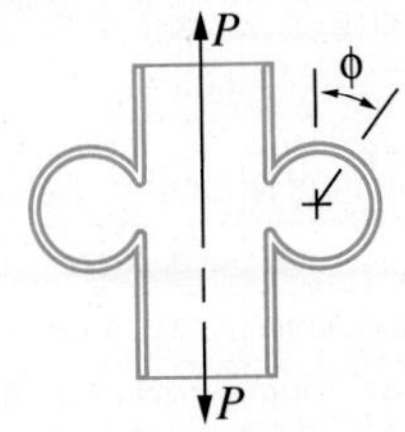 (Refs. 16 and 40)	For $4 < \mu < 40$, $\text{Stretch} = \frac{3.7Pb\sqrt{1-\nu^2}}{Et^2}$ $(\sigma_2)_{max} = \frac{2.15P}{2\pi at}\left[\frac{ab(1-\nu^2)}{t^2}\right]^{1/3}$ for $\phi = 0°, 180°$ $(\sigma_1')_{max} = \frac{2.99P}{2\pi at}\left[\frac{ab}{t^2\sqrt{1-\nu^2}}\right]^{1/3}$ Range* $\mu = 4, \phi = 50°$: $\mu = 40, \phi = 20°$ If $\mu < 4$, the following values for stretch should be used, where $\Delta = \frac{Pb^3}{2aD}$: $\begin{array}{c\|cccc} \mu & <1 & 1 & 2 & 3 \\ \hline \mathit{Stretch} & 1.00\Delta & 0.95\Delta & 0.80\Delta & 0.66\Delta \end{array}$ *Within range, ϕ is approximately linear versus $\log \mu$.
	6b. Corrugated tube under axial load P 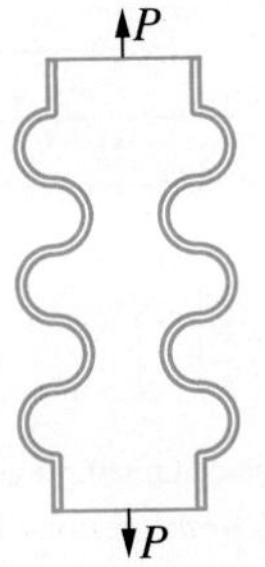	For $4 < \mu < 40$, $\text{Stretch} = \frac{0.577Pbn\sqrt{1-\nu^2}}{Et^2}$ where n is the number of semicircular corrugations (five shown in figure) $(\sigma_2)_{max} = \frac{0.925P}{2\pi at}\left[\frac{ab(1-\nu^2)}{t^2}\right]^{1/3}$ $(\sigma_1')_{max} = \frac{1.63P}{2\pi at}\left[\frac{ab}{t^2\sqrt{1-\nu^2}}\right]^{1/3}$ (Ref. 16) If $\mu < 4$, let $\Delta = \frac{Pb^3 n}{4aD}$ and use the tabulated values from case 6a. For U-shaped corrugations where a flat annular plate separates the inner and outer semicircles, see Ref. 41.
	6c. Corrugated tube under internal pressure, q. If internal pressure on the ends must be carried by the walls, calculate the end load and use case 6b in addition. (See Ref. 55 and Sec. 13.5 for a discussion of a possible instability due to internal pressure in a long bellows.)	For $4 < \mu < 40$, Stretch per semicircular corrugation $= \pm 2.45(1-\nu^2)^{1/3}\left(\frac{a}{t}\right)^{4/3}\left(\frac{b}{t}\right)^{1/3}\frac{bq}{E}$ Total stretch = 0 if there are an equal number of inner and outer corrugations $(\sigma_2)_{max} = 0.955q(1-\nu^2)^{1/6}\left(\frac{ab}{t^2}\right)^{2/3}$ $(\sigma_1')_{max} = 0.955q(1-\nu^2)^{-1/3}\left(\frac{ab}{t^2}\right)^{2/3}$ If $\mu < 1$, the stretch per semicircular corrugation $= \pm 3.28(1-\nu^2)\frac{b^4 q}{Et^3}$ For U-shaped corrugations, see Ref. 41.
7. Cylindrical shells with open ends 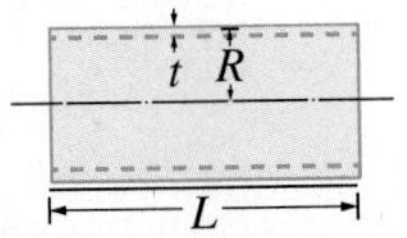	7a. Diametrically opposite and equal concentrated loads, P at mid-length 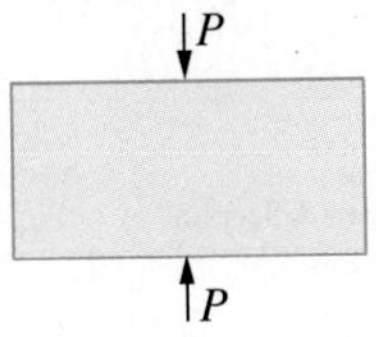	For $1 < L/R < 18$ and $R/t > 10$, Deflection under the load $= 6.5\frac{P}{Et}\left(\frac{R}{t}\right)^{3/2}\left(\frac{L}{R}\right)^{-3/4}$ For $L/R > 18$, the maximum stresses and deflections are approximately the same as for case 8a. For loads at the extreme ends, the maximum stresses are approximately four times as great as for loading at midlength. See Refs. 24 and 25.

(*Continued*)

TABLE 13.3 Formulas for Bending and Membrane Stresses and Deformations in Thin-Walled Pressure Vessels (*Continued*)

Case No., Loading		Formulas

8. Cylindrical shells with closed ends and end support

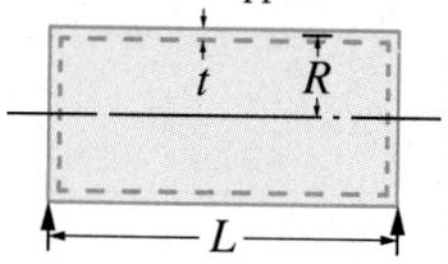

8a. Radial load P uniformly distributed over small area A, approximately square or round, located near midspan

Maximum stresses are circumferential stresses at center of loaded area and can be found from the following table. Values given are for $L/R = 8$ but may be used for L/R ratios between 3 and 40. [Coefficients adapted from Bjilaard (Refs. 22, 23, 28.)]

	A/R^2													
R/t	0.0004	0.0016	0.0036	0.0064	0.010	0.0144	0.0196	0.0256	0.0324	0.040	0.0576	0.090	0.160	0.25
	Values of $\sigma'_2\,(t^2/P)$													
300	1.475	1.11	0.906	0.780	0.678	0.600	0.522	0.450	0.390	0.348	0.264	0.186	0.120	0.078
100		1.44	1.20	1.044	0.918	0.840	0.750	0.666	0.600	0.540	0.444	0.342	0.240	0.180
50			1.44	1.254	1.11	1.005	0.900	0.840	0.756	0.720	0.600	0.480	0.360	0.264
15										0.990	0.888	0.780	0.600	0.468
	Values of $\sigma_2\,(Rt/P)$													
300	58	53.5	49	44.5	40	35.5	32	28	24	21	16	11	6	4
100		33.5	30.5	27.6	25	25.5	20	17.5	15	13	10	7	4.2	3.6
50					9.6	9	8.5	8.0	7.7	7.5	6.5	5.6	4.1	3.1
15										3.25	3.0	2.4	2.0	1.56

For A very small (nominal point loading) at point of load

$$\sigma_2 = \frac{0.4P}{t^2} \quad \sigma'_2 = \frac{2.4P}{t^2} \quad y = \frac{P}{Et}\left[0.48\left(\frac{L}{R}\right)^{1/2}\left(\frac{R}{t}\right)^{1.22}\right]$$

(Approximate empirical formulas which are based on tests of Refs. 2 and 19.)

For a more extensive presentation of Bjilaard's work in graphic form over an extended range of parameters, see Refs. 27 and 60–71.

8b. Center load, P lb, concentrated on a very short length $2b$

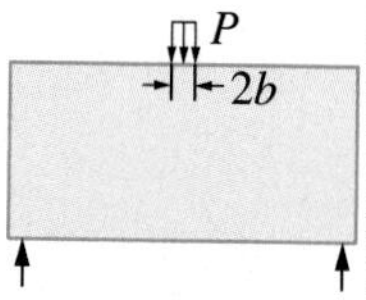

At the top center,

$$(\sigma_2)_{max} = -0.130BPR^{3/4}b^{-3/2}t^{-5/4}$$

$$(\sigma'_2)_{max} = -1.56B^{-1}PR^{1/4}b^{-1/2}t^{-7/4}$$

$$(\sigma_1)_{max} = -0.153B^3PR^{1/4}b^{-1/2}t^{-7/4}$$

$$\text{Deflection} = 0.0820B^5PR^{3/4}L^{1/2}t^{-9/4}E^{-1}$$

where $B = [12(1-\nu^2)]^{1/8}$ (Ref. 13)

8c. Uniform load, p lb/in, over entire length of top element

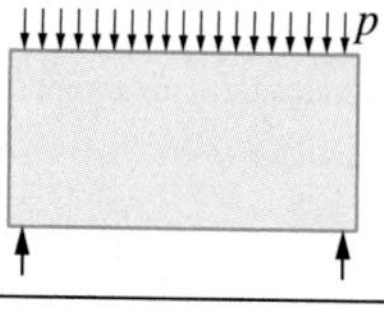

At the top center,

$$(\sigma_2)_{max} = -0.492BpR^{3/4}L^{-1/2}t^{-5/4}$$

$$(\sigma'_2)_{max} = -1.217B^{-1}pR^{1/4}L^{1/2}t^{-7/4}$$

$$(\sigma_1)_{max} = -0.1188B^3pR^{1/4}L^{1/2}t^{-7/4}$$

$$\text{Deflection} = 0.0305B^5pR^{3/4}L^{3/2}t^{-9/4}E^{-1}$$

where B is given in case 8b

Quarter-span deflection = 0.774 midspan deflection (Ref. 13)

TABLE 13.4 Formulas for Discontinuity Stresses and Deformations at the Junctions of Shells and Plates

Notation: R_A = radius of common circumference; ΔR_A is the radial deflection of the common circumference, positive outward; ψ_A is the rotation of the meridian at the common circumference, positive as indicated. The notation used in Tables 11.2 and 13.1–13.3 is retained where possible with added subscripts 1 and 2 used for left and right members, respectively, when needed for clarification. There are some exceptions in using the notation from the other tables when differences occur from one table to another.

1. Cylindrical shell connected to another cylindrical shell. Expressions are accurate if $R/t > 5$. E_1 and E_2 are the moduli of elasticity and ν_1 and ν_2 the Poisson's ratios for the left and right cylinders, respectively. See Table 13.2 for formulas for D_1 and λ_1. $R_A = R_1$, $b_1 = R_1 - t_1/2$, and $a_1 = R_1 + t_1/2$. Similar expressions hold for b_2, a_2, D_2, and λ_2.

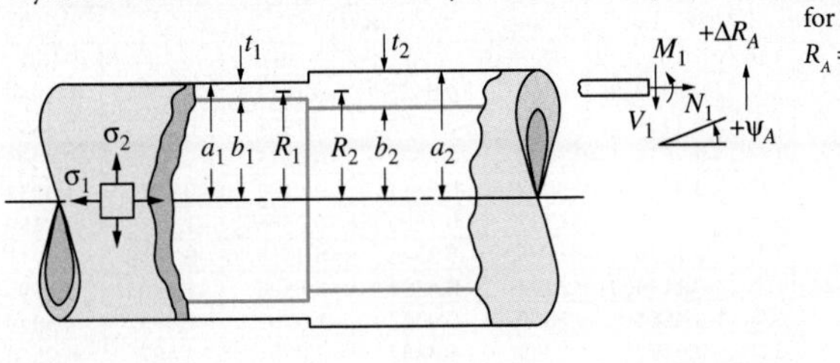

$$K_{V1} = \frac{LT_A C_{BB} - LT_B C_{AB}}{C_{AA} C_{BB} - C_{AB}^2}, \quad K_{M1} = \frac{LT_B C_{AA} - LT_A C_{AB}}{C_{AA} C_{BB} - C_{AB}^2}, \quad \left.\begin{matrix} LT_A = LT_{A1} + LT_{A2} + LT_{AC} \\ LT_B = LT_{B1} + TL_{B2} + LT_{BC} \end{matrix}\right\} \text{ See cases 1a–1d for these load terms.}$$

$$C_{AA} = C_{AA1} + C_{AA2}, \quad C_{AA1} = \frac{E_1}{2D_1\lambda_1^3}, \quad C_{AA2} = \frac{R_1 E_1}{2R_2 D_2 \lambda_2^3}$$

$$C_{AB} = C_{AB1} + C_{AB2}, \quad C_{AB1} = \frac{-E_1 t_1}{2D_1\lambda_1^2}, \quad C_{AB2} = \frac{R_1 E_1 t_1}{2R_2 D_2 \lambda_2^2}$$

$$C_{BB} = C_{BB1} + C_{BB2}, \quad C_{BB1} = \frac{E_1 t_1^2}{D_1 \lambda_1}, \quad C_{BB2} = \frac{R_1 E_1 t_1^2}{R_2 D_2 \lambda_2}$$

The stresses in the left cylinder at the junction are given by

$$\sigma_1 = \frac{N_1}{t_1}$$

$$\sigma_2 = \frac{\Delta R_A E_1}{R_A} + \nu_1 \sigma_1$$

$$\sigma_1' = \frac{-6M_1}{t_1^2}$$

$$\sigma_2' = \nu_1 \sigma_1'$$

Note: The use of joint load correction terms LT_{AC} and LT_{BC} depends upon the accuracy desired and the relative values of the thickness and the radii. Read Sec. 13.3 carefully. For thin-walled shells, $R/t > 10$, they can be neglected.

Loading and Case No.	Load Terms	Selected Values
1a. Internal* pressure q	$LT_{A1} = \frac{b_1 R_1}{t_1^2}$	For internal pressure, $b_1 = b_2$ (smooth internal wall), $E_1 = E_2$, $\nu_1 = \nu_2 = 0.3$, and for $R/t > 5$.

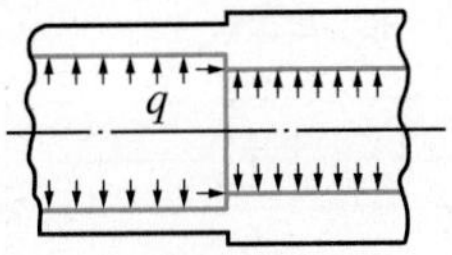

Note: There is no axial load on the left cylinder. A small axial load on the right cylinder balances any axial pressure on the joint. For an enclosed pressure vessel superpose an axial load $P = q\pi b_1^2$ using case 1b.

Load terms:

$$LT_{A1} = \frac{b_1 R_1}{t_1^2}$$

$$LT_{A2} = \frac{-b_2 R_2 E_1}{E_2 t_1 t_2}$$

$$LT_{AC} = \frac{E_1(b_1^2 - b_2^2)}{8t_1}\left(\frac{a_2 - b_1}{R_2 D_2 \lambda_2^2} - \frac{4\nu_2}{E_2 t_2}\right)$$

$$LT_{B1} = 0, \quad LT_{B2} = 0$$

$$LT_{BC} = E_1(b_1^2 - b_2^2)\frac{a_2 - b_1}{4R_2 D_2 \lambda_2}$$

At the junction of the two cylinders,

$$V_1 = q t_1 K_{V1}, \quad M_1 = q t_1^2 K_{M1}, \quad N_1 = 0$$

$$\Delta R_A = \frac{q t_1}{E_1}(LT_{A1} - K_{V1} C_{AA1} - K_{M1} C_{AB1})$$

$$\psi_A = \frac{q}{E_1}(K_{V1} C_{AB1} + K_{M1} C_{BB1})$$

Selected values:

$$\Delta R_A = \frac{qR_1^2}{E_1 t_1} K_{\Delta RA}, \quad \psi_A = \frac{qR_1}{E_1 t_1} K_{\psi A}, \quad \sigma_2 = \frac{qR_1}{t_1} K_{\sigma 2}$$

	t_2/t_1	R_1/t_1 = 10	15	20	30	50	100
K_{V1}	1.1	0.0542	0.0688	0.0808	0.1007	0.1318	0.1884
	1.2	0.1066	0.1353	0.1589	0.1981	0.2593	0.3705
	1.5	0.2574	0.3269	0.3843	0.4791	0.6273	0.8966
	2.0	0.4945	0.6286	0.7392	0.9220	1.2076	1.7264
	3.0		1.1351	1.3356	1.6667	2.1840	3.1231
K_{M1}	1.1	0.0065	0.0101	0.0137	0.0208	0.0352	0.0711
	1.2	0.0246	0.0382	0.0518	0.0790	0.1334	0.2695
	1.5	0.1295	0.2012	0.2730	0.4166	0.7038	1.4221
	2.0	0.3891	0.6050	0.8211	1.2535	2.1186	4.2815
	3.0		1.4312	1.9436	2.9691	5.0207	10.1505
$K_{\Delta RA}$	1.1	0.9080	0.9232	0.9308	0.9383	0.9444	0.9489
	1.2	0.8715	0.8853	0.8922	0.8991	0.9046	0.9087
	1.5	0.7835	0.7940	0.7992	0.8043	0.8084	0.8115
	2.0	0.6765	0.6827	0.6857	0.6887	0.6910	0.6927
	3.0		0.5285	0.5283	0.5281	0.5278	0.5275
$K_{\psi A}$	1.1	−0.1618	−0.2053	−0.2412	−0.3006	−0.3934	−0.5620
	1.2	−0.2862	−0.3633	−0.4269	−0.5321	−0.6965	−0.9952
	1.5	−0.5028	−0.6390	−0.7515	−0.9372	−1.2275	−1.7547
	2.0	−0.5889	−0.7501	−0.8831	−1.1026	−1.4454	−2.0676
	3.0		−0.6118	−0.7216	−0.9026	−1.1850	−1.6971
$K_{\sigma 2}$	1.1	0.9080	0.9232	0.9308	0.9383	0.9444	0.9489
	1.2	0.8715	0.8853	0.8922	0.8991	0.9046	0.9087
	1.5	0.7835	0.7940	0.7992	0.8043	0.8084	0.8115
	2.0	0.6765	0.6827	0.6857	0.6887	0.6910	0.6927
	3.0		0.5285	0.5283	0.5281	0.5278	0.5275

*For external pressure, substitute $-q$ for q, a_1 for b_1, b_2 for a_2, and a_2 for b_2 in the load terms.

(*Continued*)

TABLE 13.4 Formulas for Discontinuity Stresses and Deformations at the Junctions of Shells and Plates (*Continued*)

Loading and Case No.	Load Terms	Selected Values

1b. Axial load P

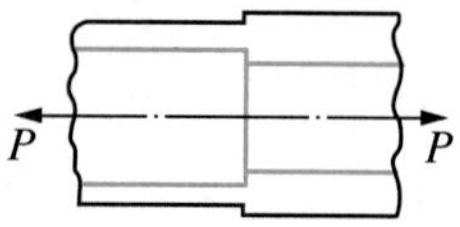

Load Terms

$$LT_{A1} = \frac{-\nu_1 R_1^2}{2t_1^2}, \qquad LT_{AC} = 0$$

$$LT_{A2} = \frac{E_1 R_1^2}{2t_1}\left(\frac{\nu_2}{E_2 t_2} - \frac{R_2 - R_1}{2R_2 D_2 \lambda_2^2}\right)$$

$$LT_{B1} = 0, \qquad LT_{BC} = 0$$

$$LT_{B2} = \frac{-(R_2 - R_1)R_1^2 E_1}{2R_2 D_2 \lambda_2}$$

At the junction of the two cylinders,

$$V_1 = \frac{Pt_1 K_{V1}}{\pi R_1^2}, \quad M_1 = \frac{Pt_1^2 K_{M1}}{\pi R_1^2}, \quad N_1 = \frac{P}{2\pi R_1}$$

$$\Delta R_A = \frac{Pt_1}{E_1 \pi R_1^2}(LT_{A1} - K_{V1}C_{AA1} - K_{M1}C_{AB1})$$

$$\psi_A = \frac{P}{E_1 \pi R_1^2}(K_{V1}C_{AB1} + K_{M1}C_{BB1})$$

Selected Values

For axial tension, $b_1 = b_2$ (smooth internal wall), $E_1 = E_2$, $\nu_1 = \nu_2 = 0.3$, and for $R/t > 5$.

$$\Delta R_A = \frac{P\nu_1}{2\pi E_1 t_1} K_{\Delta RA}, \qquad \psi_A = \frac{P\nu_1}{2\pi E_1 t_1^2} K_{\psi A}, \qquad \sigma_2 = \frac{P}{2\pi R_1 t_1} K_{\sigma 2}$$

	t_2/t_1	R_1/t_1 = 10	15	20	30	50	100
K_{v1}	1.1	−0.0583	−0.0715	−0.0825	−0.1011	−0.1306	−0.1847
	1.2	−0.1129	−0.1383	−0.1598	−0.1958	−0.2529	−0.3577
	1.5	−0.2517	−0.3089	−0.3571	−0.4378	−0.5657	−0.8005
	2.0	−0.4084	−0.5022	−0.5811	−0.7132	−0.9222	−1.3058
	3.0		−0.6732	−0.7800	−0.9586	−1.2411	−1.7589
K_{M1}	1.1	−0.1170	−0.1756	−0.2341	−0.3513	−0.5856	−1.1714
	1.2	−0.2194	−0.3294	−0.4394	−0.6594	−1.0995	−2.1995
	1.5	−0.4564	−0.6863	−0.9163	−1.3761	−2.2957	−4.5949
	2.0	−0.6892	−1.0389	−1.3887	−2.0883	−3.4876	−6.9860
	3.0		−1.2874	−1.7242	−2.5981	−4.3462	−8.7167
$K_{\Delta RA}$	1.1	−0.9416	−0.9416	−0.9416	−0.9416	−0.9416	−0.9416
	1.2	−0.8717	−0.8716	−0.8716	−0.8716	−0.8715	−0.8715
	1.5	−0.6414	−0.6410	−0.6408	−0.6406	−0.6405	−0.6404
	2.0	−0.3047	−0.3033	−0.3027	−0.3020	−0.3015	−0.3011
	3.0		0.0883	0.0900	0.0918	0.0931	0.0942
$K_{\psi A}$	1.1	−0.0810	−0.0662	−0.0573	−0.0468	−0.0363	−0.0257
	1.2	−0.1443	−0.1180	−0.1022	−0.0835	−0.0647	−0.0458
	1.5	−0.2630	−0.2154	−0.1869	−0.1528	−0.1185	−0.0839
	2.0	−0.3345	−0.2752	−0.2392	−0.1961	−0.1524	−0.1080
	3.0		−0.2663	−0.2326	−0.1915	−0.1494	−0.1062
$K_{\sigma 2}$	1.1	0.0175	0.0175	0.0175	0.0175	0.0175	0.0175
	1.2	0.0385	0.0385	0.0385	0.0385	0.0385	0.0385
	1.5	0.1076	0.1077	0.1077	0.1078	0.1079	0.1079
	2.0	0.2086	0.2090	0.2092	0.2094	0.2096	0.2097
	3.0		0.3265	0.3270	0.3275	0.3279	0.3283

1c. Hydrostatic internal* pressure q_1 at the junction for $x_1 > 3/\lambda_1$†

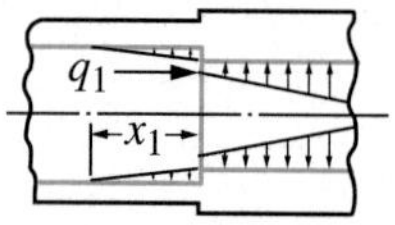

Note: There is no axial load on the left cylinder. A small axial load on the right cylinder balances any axial pressure on the joint.

Load Terms

$$LT_{A1} = \frac{b_1 R_1}{t_1^2}$$

$$LT_{A2} = \frac{-b_2 R_2 E_1}{E_2 t_1 t_2}$$

For LT_{AC} use the expression from case 1a.

$$LT_{B1} = \frac{-b_1 R_1}{x_1 t_1}, \qquad LT_{B2} = \frac{b_2 R_2 E_1}{x_1 E_2 t_2}$$

For LT_{BC} use the expression from case 1a.

At the junction of the two cylinders,

$$V_1 = q_1 t_1 K_{V1}, \qquad M_1 = q_1 t_1^2 K_{M1}, \qquad N_1 = 0$$

$$\Delta R_A = \frac{q_1 t_1}{E_1}(LT_{A1} - K_{V1}C_{AA1} - K_{M1}C_{AB1})$$

$$\psi_A = \frac{q_1}{E_1}(-LT_{B1} + K_{V1}C_{AB1} + K_{M1}C_{BB1})$$

Selected Values

For internal pressure, $b_1 = b_2$ (smooth internal wall), $E_1 = E_2$, $\nu_1 = \nu_2 = 0.3$, and for $R/t > 5$.

$$\Delta R_A = \frac{q_1 R_1^2}{E_1 t_1} K_{\Delta RA}, \qquad \psi_A = \frac{q_1 R_1}{E_1 t_1} K_{\psi A}, \qquad \sigma_2 = \frac{q_1 R_1}{t_1} K_{\sigma 2}$$

		R_1/t_1					
		10			20		
	t_2/t_1	x_1/R_1 = 1	2	4	x_1/R_1 = 1	2	4
K_{v1}	1.1	0.0536	0.0539	0.0541	0.0802	0.0805	0.0807
	1.2	0.1041	0.1054	0.1060	0.1564	0.1577	0.1583
	1.5	0.2445	0.2509	0.2542	0.3707	0.3775	0.3809
	2.0	0.4556	0.4751	0.4848	0.6982	0.7187	0.7290
	3.0				1.2385	1.2870	1.3113
K_{M1}	1.1	−0.0107	−0.0021	0.0022	−0.0119	0.0009	0.0073
	1.2	−0.0103	0.0071	0.0158	−0.0002	0.0258	0.0388
	1.5	0.0406	0.0850	0.1073	0.1407	0.2068	0.2399
	2.0	0.2152	0.3022	0.3456	0.5625	0.6918	0.7565
	3.0				1.4882	1.7159	1.8298
$K_{\Delta RA}$	1.1	0.9029	0.9055	0.9068	0.9269	0.9289	0.9298
	1.2	0.8619	0.8667	0.8691	0.8851	0.8886	0.8904
	1.5	0.7647	0.7741	0.7788	0.7852	0.7922	0.7957
	2.0	0.6507	0.6636	0.6701	0.6666	0.6761	0.6809
	3.0				0.5090	0.5187	0.5235
$K_{\psi A}$	1.1	0.7442	0.2912	0.0647	0.6875	0.2231	−0.0091
	1.2	0.5781	0.1460	−0.0701	0.4580	0.0155	−0.2057
	1.5	0.2511	−0.1259	−0.3143	0.0173	−0.3671	−0.5593
	2.0	0.0226	−0.2831	−0.4360	−0.2637	−0.5734	−0.7282
	3.0				−0.2905	−0.5060	−0.6138
$K_{\sigma 2}$	1.1	0.9029	0.9055	0.9068	0.9269	0.9289	0.9298
	1.2	0.8619	0.8667	0.8691	0.8851	0.8886	0.8904
	1.5	0.7647	0.7741	0.7788	0.7852	0.7922	0.7957
	2.0	0.6507	0.6636	0.6701	0.6666	0.6761	0.6809
	3.0				0.5090	0.5187	0.5235

*For external pressure, substitute $-q$ for q, a_1 for b_1, b_2 for a_2, and a_2 for b_2 in the load terms.

†If pressure increases right to left, substitute $-x_1$ for x_1 and verify that $|x_1| > 3/\lambda_2$.

(*Continued*)

TABLE 13.4 Formulas for Discontinuity Stresses and Deformations at the Junctions of Shells and Plates (*Continued*)

Loading and Case No.	Load Terms	Selected Values
1d. Rotation around the axis of symmetry at ω rad/s	$LT_{A1}=\frac{R_1^2}{t_1^2}$ $LT_{A2}=\frac{-\delta_2 R_2^3 E_1}{\delta_1 R_1 E_2 t_1^2}$ $LT_{AC}=0$ $LT_{B1}=0,\quad LT_{B2}=0$ $LT_{BC}=0$	For $b_1=b_2$ (smooth internal wall), $\delta_1=\delta_2$, $E_1=E_2$, $v_1=v_2=0.3$, and for $R/t>5$.

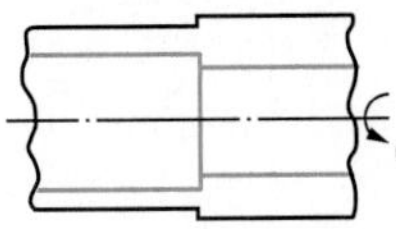

Note: δ = mass/unit volume

At the junction of the two cylinders,

$$V_1=\delta_1\omega^2 R_1 t_1^2 K_{V1},\qquad M_1=\delta_1\omega^2 R_1 t_1^3 K_{M1}$$

$$N_1=0$$

$$\Delta R_A=\frac{\delta_1\omega^2 R_1 t_1^2}{E_1}(LT_{A1}-K_{V1}C_{AA1}-K_{M1}C_{AB1})$$

$$\psi_A=\frac{\delta_1\omega^2 R_1 t_1}{E_1}(K_{V1}C_{AB1}+K_{M1}C_{BB1})$$

Selected values:

$$\Delta R_A=\frac{\delta_1\omega^2 R_1^3}{E_1}K_{\Delta RA},\qquad \psi_A=\frac{\delta_1\omega^2 R_1^2}{E_1}K_{\psi A},\qquad \sigma_2=\delta_1\omega^2 R_1^2 K_{\sigma 2}$$

	$\frac{t_2}{t_1}$	R_1/t_1 = 10	15	20	30	50	100
K_{V1}	1.1	-0.0100	-0.0081	-0.0070	-0.0057	-0.0044	-0.0031
	1.2	-0.0215	-0.0175	-0.0151	-0.0123	-0.0095	-0.0067
	1.5	-0.0658	-0.0534	-0.0461	-0.0375	-0.0289	-0.0204
	2.0	-0.1727	-0.1391	-0.1196	-0.0970	-0.0747	-0.0526
	3.0		-0.3893	-0.3322	-0.2673	-0.2046	-0.1434
K_{M1}	1.1	-0.0012	-0.0012	-0.0012	-0.0012	-0.0012	-0.0012
	1.2	-0.0049	-0.0049	-0.0049	-0.0049	-0.0049	-0.0049
	1.5	-0.0331	-0.0328	-0.0327	-0.0326	-0.0325	-0.0324
	2.0	-0.1359	-0.1339	-0.1328	-0.1318	-0.1310	-0.1304
	3.0		-0.4908	-0.4834	-0.4761	-0.4703	-0.4660
$K_{\Delta RA}$	1.1	1.0077	1.0051	1.0038	1.0026	1.0015	1.0008
	1.2	1.0158	1.0105	1.0079	1.0052	1.0031	1.0016
	1.5	1.0426	1.0282	1.0211	1.0140	1.0084	1.0042
	2.0	1.0955	1.0628	1.0468	1.0310	1.0185	1.0092
	3.0		1.1503	1.1111	1.0730	1.0433	1.0215
$K_{\psi A}$	1.1	0.0297	0.0242	0.0210	0.0171	0.0133	0.0094
	1.2	0.0577	0.0470	0.0406	0.0331	0.0256	0.0181
	1.5	0.1285	0.1043	0.0900	0.0733	0.0566	0.0400
	2.0	0.2057	0.1659	0.1429	0.1159	0.0894	0.0630
	3.0		0.2098	0.1795	0.1447	0.1110	0.0779
$K_{\sigma 2}$	1.1	1.0077	1.0051	1.0038	1.0026	1.0015	1.0008
	1.2	1.0158	1.0105	1.0079	1.0052	1.0031	1.0016
	1.5	1.0426	1.0282	1.0211	1.0140	1.0084	1.0042
	2.0	1.0955	1.0628	1.0468	1.0310	1.0185	1.0092
	3.0		1.1503	1.1111	1.0730	1.0433	1.0215

2. Cylindrical shell connected to a conical shell.* To ensure accuracy, evaluate k_A and the value of k in the cone at the position where $\mu=4$. The absolute values of k at both positions should be greater than 5. $R/(t_2\cos\alpha_2)$ should also be greater than 5 at both of these positions. E_1 and E_2 are the moduli of elasticity and ν_1 and ν_2 the Poisson's ratios for the cylinder and cone, respectively. See Table 13.2 for formulas for D_1 and λ_1. $b_1=R_1-t_1/2$ and $a_1=R_1+t_1/2$. $b_2=R_2-(t_2\cos\alpha_2)/2$ and $a_2=R_2+(t_2\cos\alpha_2)/2$, where R_2 is the mid-thickness radius of the cone at the junction. $R_A=R_1$. See Table 13.3, case 4, for formulas for k_A, β, μ, C_1, and the F functions.

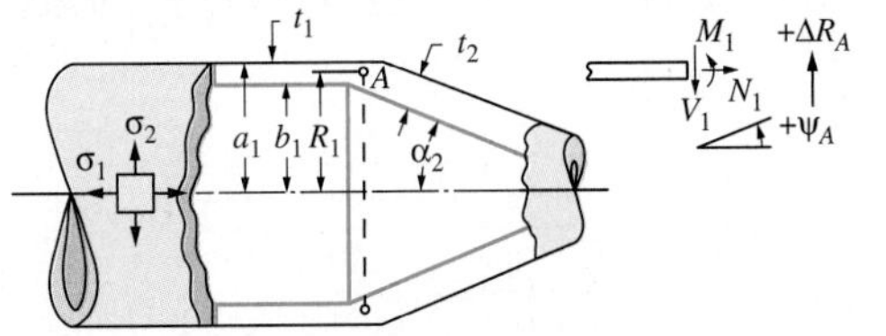

$$K_{V1}=\frac{LT_A C_{BB}-LT_B C_{AB}}{C_{AA}C_{BB}-C_{AB}^2},\qquad K_{M1}=\frac{LT_B C_{AA}-LT_A C_{AB}}{C_{AA}C_{BB}-C_{AB}^2}$$

$$\left.\begin{aligned}LT_A&=LT_{A1}+LT_{A2}+LT_{AC}\\ LT_B&=LT_{B1}+LT_{B2}+LT_{BC}\end{aligned}\right\}\ \text{See cases 2a–2d for these load terms.}$$

$$C_{AA}=C_{AA1}+C_{AA2},\qquad C_{AA1}=\frac{E_1}{2D_1\lambda_1^3},\qquad C_{AA2}=\frac{R_1E_1k_A\sin\alpha_2}{E_2t_2\sqrt{2}C_1}\left(F_{4A}-\frac{4\nu_2^2}{k_A^2}F_{2A}\right)$$

$$C_{AB}=C_{AB1}+C_{AB2},\qquad C_{AB1}=\frac{-E_1t_1}{2D_1\lambda_1^2},\qquad C_{AB2}=\frac{R_1E_1t_1\beta F_{7A}}{E_2t_2^2C_1}$$

$$C_{BB}=C_{BB1}+C_{BB2},\qquad C_{BB1}=\frac{E_1t_1^2}{D_1\lambda_1},\qquad C_{BB2}=\frac{R_1E_1t_1^2 2\sqrt{2}\beta^2F_{2A}}{E_2t_2^3k_AC_1\sin\alpha_2}$$

The stresses in the left cylinder at the junction are given by

$$\sigma_1=\frac{N_1}{t_1}$$

$$\sigma_2=\frac{\Delta R_A E_1}{R_A}+\nu_1\sigma_1$$

$$\sigma_1'=\frac{-6M_1}{t_1^2}$$

$$\sigma_2'=\nu_1\sigma_1'$$

Note: The use of joint load correction terms LT_{AC} and LT_{BC} depends upon the accuracy desired and the relative values of the thicknesses and the radii and the cone angle α. Read Sec. 13.4 carefully. For thin-walled shells, $R/t>10$ at the junction, they can be neglected.

*If the conical shell increases in radius away from the junction, substitute −α for α in all the formulas above and in those used from case 4, Table 13.3.

(*Continued*)

TABLE 13.4 Formulas for Discontinuity Stresses and Deformations at the Junctions of Shells and Plates (*Continued*)

Loading and Case No.	Load Terms	Selected Values
2a. Internal* pressure q		

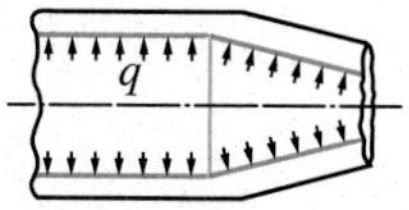

Note: There is no axial load on the cylinder. An axial load on the right end of the cone balances any axial component of the pressure on the cone and the joint. For an enclosed pressure vessel superpose an axial load $P = q\pi b_1^2$ using case 2b.

Load Terms

$$LT_{A1} = \frac{b_1 R_1}{t_1^2}$$

$$LT_{A2} = \frac{-b_2 E_1 R_2}{E_2 t_1 t_2 \cos\alpha_2}$$

$$LT_{AC} = \frac{t_2 \sin\alpha_2}{2t_1} C_{AA2} + \frac{t_2^2 - t_1^2}{8t_1^2} C_{AB2} + \frac{E_1 \nu_2 R_1 (t_1 - t_2 \cos\alpha_2)}{2E_2 t_1 t_2 \cos\alpha_2} \dagger$$

$$LT_{B1} = 0$$

$$LT_{B2} = \frac{-2E_1 b_2 \tan\alpha_2}{E_2 t_2 \cos\alpha_2}$$

$$LT_{BC} = \frac{t_2 \sin\alpha_2}{2t_1} C_{AB2} + \frac{t_2^2 - t_1^2}{8t_1^2} C_{BB2} + \frac{E_1 \sin\alpha_2}{E_2 t_2 \cos^2\alpha_2}(t_1 - t_2\cos\alpha_2)\dagger$$

At the junction of the cylinder and cone,

$$V_1 = qt_1 K_{V1}, \quad M_1 = qt_1^2 K_{M1}, \quad N_1 = 0$$

$$\Delta R_A = \frac{qt_1}{E_1}(LT_{A1} - K_{V1}C_{AA1} - K_{M1}C_{AB1})$$

$$\psi_A = \frac{q}{E_1}(K_{V1}C_{AB1} + K_{M1}C_{BB1})$$

Selected Values

For internal pressure, $R_1 = R_2$, $E_1 = E_2$, $\nu_1 = \nu_2 = 0.3$, and for $R/t > 5$, (*Note:* No correction terms are used.)

$$\Delta R_A = \frac{qR_1^2}{E_1 t_1} K_{\Delta RA}, \qquad \psi_1 = \frac{qR_1}{E_1 t_1} K_{\psi A}, \qquad \sigma_2 = \frac{qR_1}{t_1} K_{\sigma 2}$$

		R_1/t_1					
		20			40		
	$\frac{t_2}{t_1}$	α_2			α_2		
		−30	30	45	−30	30	45
K_{V1}	0.8	−0.3602	−0.3270	−0.5983	−0.5027	−0.4690	−0.8578
	1.0	−0.1377	−0.1419	−0.4023	−0.1954	−0.1994	−0.5633
	1.2	0.0561	0.0212	−0.2377	0.0725	0.0374	−0.3165
	1.5	0.3205	0.2531	−0.0083	0.4399	0.3715	0.0225
	2.0	0.7208	0.6267	0.3660	1.0004	0.9040	0.5628
K_{M1}	0.8	0.4218	−0.1915	−0.4167	0.6693	−0.2053	−0.4749
	1.0	0.3306	−0.3332	−0.6782	0.4730	−0.4756	−0.9638
	1.2	0.3562	−0.3330	−0.7724	0.5175	−0.4696	−1.1333
	1.5	0.5482	−0.1484	−0.6841	0.8999	−0.1011	−0.9487
	2.0	1.0724	0.4091	−0.1954	1.9581	0.9988	0.0084
$K_{\Delta RA}$	0.8	1.2517	1.1314	1.2501	1.2471	1.1612	1.2969
	1.0	1.1088	1.0015	1.0942	1.1060	1.0293	1.1368
	1.2	1.0016	0.9078	0.9840	1.0008	0.9335	1.0225
	1.5	0.8813	0.8050	0.8667	0.8830	0.8281	0.9000
	2.0	0.7378	0.6823	0.7323	0.7426	0.7026	0.7594
$K_{\psi A}$	0.8	1.9915	0.7170	1.1857	2.5602	1.2739	2.1967
	1.0	1.0831	−0.1641	0.0411	1.2812	0.0201	0.5668
	1.2	0.4913	−0.7028	−0.6817	0.4554	−0.7544	−0.4765
	1.5	−0.0178	−1.1184	−1.2720	−0.2449	−1.3635	−1.3488
	2.0	−0.3449	−1.2939	−1.5808	−0.6758	−1.6457	−1.8483
$K_{\sigma 2}$	0.8	1.2517	1.1314	1.2501	1.2471	1.1612	1.2969
	1.0	1.1088	1.0015	1.0942	1.1060	1.0293	1.1368
	1.2	1.0016	0.9078	0.9840	1.0008	0.9335	1.0225
	1.5	0.8813	0.8050	0.8667	0.8830	0.8281	0.9000
	2.0	0.7378	0.6823	0.7323	0.7426	0.7026	0.7594

*For external pressure, substitute $-q$ for q, a_1 for b_1, and a_2 for b_2 in the load terms.
†If α_2 approaches 0, use the correction term from case 1a.

(Continued)

TABLE 13.4 Formulas for Discontinuity Stresses and Deformations at the Junctions of Shells and Plates (*Continued*)

Loading and Case No.	Load Terms	Selected Values

2b. Axial load P

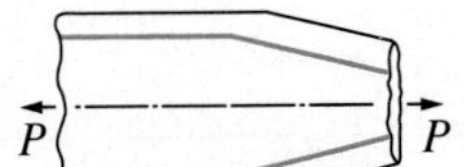

Load Terms

$$LT_{A1} = \frac{-\nu_1 R_1^2}{2t_1^2}$$

$$LT_{A2} = \frac{\nu_2 R_1^2 E_1}{2E_2 t_1 t_2 \cos\alpha_2} + \frac{R_1 C_{AA2}}{2t_1}\tan\alpha_2$$

$$LT_{AC} = 0$$

$$LT_{B1} = 0$$

$$LT_{B2} = \frac{E_1 R_1^2 \tan\alpha_2}{2E_2 R_2 t_2 \cos\alpha_2} + \frac{R_1 C_{AB2}}{2t_1}\tan\alpha_2$$

$$LT_{BC} = 0$$

At the junction of the cylinder and cone,

$$V_1 = \frac{Pt_1 K_{V1}}{\pi R_1^2}, \quad M_1 = \frac{Pt_1^2 K_{M1}}{\pi R_1^2}, \quad N_1 = \frac{P}{2\pi R_1}$$

$$\Delta R_A = \frac{Pt_1}{E_1 \pi R_1^2}(LT_{A1} - K_{V1}C_{AA1} - K_{M1}C_{AB1})$$

$$\psi_A = \frac{P}{E_1 \pi R_1^2}(K_{V1}C_{AB1} - K_{M1}C_{BB1})$$

Selected Values

For axial tension, $R_1 = R_2$, $E_1 = E_2$, $\nu_1 = \nu_2 = 0.3$, and for $R/t > 5$.

$$\Delta R_A = \frac{P\nu_1}{2\pi E_1 t_1}K_{\Delta RA}, \quad \psi_A = \frac{P\nu_1}{2\pi E_1 t_1^2}K_{\psi A}, \quad \sigma_2 = \frac{P}{2\pi R_1 t_1}K_{\sigma 2}$$

		R_1/t_1					
		20			40		
	$\frac{t_2}{t_1}$	α_2			α_2		
		−30	30	45	−30	30	45
K_{V1}	0.8	−2.9323	2.9605	4.8442	−5.8775	5.9175	9.7022
	1.0	−2.7990	2.7667	4.5259	−5.5886	5.5429	9.0767
	1.2	−2.6963	2.6095	4.2750	−5.3637	5.2411	8.5862
	1.5	−2.5522	2.3879	3.9263	−5.0493	4.8171	7.9088
	2.0	−2.3074	2.0205	3.3492	−4.5212	4.1161	6.7938
K_{M1}	0.8	−4.5283	4.5208	7.4287	−12.7003	12.6851	20.8134
	1.0	−4.9208	4.9206	8.0936	−13.8055	13.8053	22.6838
	1.2	−5.0832	5.0531	8.3299	−14.2489	14.1890	23.3674
	1.5	−5.0660	4.9455	8.1850	−14.1627	13.9229	23.0120
	2.0	−4.7100	4.3860	7.3095	−13.0797	12.4341	20.6655
$K_{\Delta RA}$	0.8	5.2495	−7.3660	−11.3820	7.9325	−10.0493	−15.8287
	1.0	4.3064	−6.1827	−9.4297	6.5410	−8.4173	−13.1035
	1.2	3.7337	−5.4346	−8.2078	5.6872	−7.3879	−11.3978
	1.5	3.2008	−4.7036	−7.0309	4.8827	−6.3855	−9.7578
	2.0	2.6544	−3.9120	−5.7838	4.0481	−5.3060	−8.0287
$K_{\psi A}$	0.8	0.3626	−0.3985	−0.6321	0.3938	−0.4193	−0.6840
	1.0	−0.0327	0.0683	0.1395	−0.0127	0.0378	0.0792
	1.2	−0.2488	0.3252	0.5655	−0.2359	0.2900	0.5024
	1.5	−0.3965	0.5013	0.8580	−0.3897	0.4639	0.7959
	2.0	−0.4408	0.5517	0.9392	−0.4381	0.5167	0.8847
$K_{\sigma 2}$	0.8	1.8748	−1.9098	−3.1146	2.6797	−2.7148	−4.4486
	1.0	1.5919	−1.5548	−2.5289	2.2623	−2.2252	−3.6310
	1.2	1.4201	−1.3304	−2.1623	2.0062	−1.9164	−3.1193
	1.5	1.2602	−1.1111	−1.8093	1.7648	−1.6156	−2.6273
	2.0	1.0963	−0.8736	−1.4351	1.5144	−1.2918	−2.1086

2c. Hydrostatic internal* pressure q_1 at the junction for $x_1 > 3/\lambda_1$.† If $x_1 < 3/\lambda_1$, the discontinuity in pressure gradient introduces small deformations at the junction.

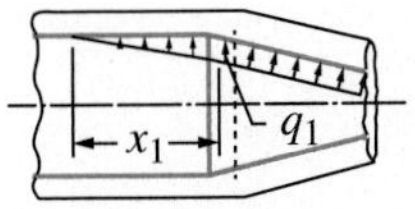

Note: There is no axial load on the cylinder. An axial load on the right end of the cone balances any axial component of pressure in the cone and the joint.

Load Terms

$$LT_{A1} = \frac{b_1 R_1}{t_1^2}$$

$$LT_{A2} = \frac{-b_2 R_2 E_1}{E_2 t_1 t_2 \cos\alpha_2}$$

For LT_{AC} use the expression from case 2a.

$$LT_{B1} = \frac{-b_1 R_1}{x_1 t_1}$$

$$LT_{B2} = \frac{E_1 b_2}{E_2 t_2 \cos\alpha_2}\left(\frac{R_2}{x_1} - 2\tan\alpha_2\right)$$

For LT_{AC} use the expression from case 2a.

At the junction of the cylinder and cone,

$$V_1 = q_1 t_1 K_{V1}, \quad M_1 = q_1 t_1^2 K_{M1}, \quad N_1 = 0$$

$$\Delta R_A = \frac{q_1 t_1}{E_1}(LT_{A1} - K_{V1}C_{AA1} - K_{M1}C_{AB1})$$

$$\psi_A = \frac{q_1}{E_1}(-LT_{B1} + K_{V1}C_{AB1} + K_{M1}C_{BB1})$$

*For external pressure, substitute $-q$ for q_1, a_1 for b_1 and a_2 for b_2 in the load terms.

†If pressure increases right to left, substitute $-x_1$ for x_1 and verify that $|x_1|$ is large enough to extend into the cone as far as the position where $|\mu| = 4$.

Selected Values

For internal pressure, $R_1 = R_2$, $x_1 = R_1$, $E_1 = E_2$, $\nu_1 = \nu_2 = 0.3$, and for $R/t > 5$. (*Note:* No correction terms are used.)

$$\Delta R_A = \frac{q_1 R_1^2}{E_1 t_1}K_{\Delta RA}, \quad \psi_A = \frac{q_1 R_1}{E_1 t_1}K_{\psi A}, \quad \sigma_2 = \frac{q_1 R_1}{t_1}K_{\sigma 2}$$

		R_1/t_1					
		20			40		
	$\frac{t_2}{t_1}$	α_2			α_2		
		−30	30	45	−30	30	45
K_{V1}	0.8	−0.3660	−0.3329	−0.6082	−0.5085	−0.4748	−0.8678
	1.0	−0.1377	−0.1418	−0.4020	−0.1954	−0.1994	−0.5631
	1.2	0.0555	0.0206	−0.2346	0.0719	0.0368	−0.3134
	1.5	0.3107	0.2429	−0.0109	0.4300	0.3614	0.0199
	2.0	0.6843	0.5890	0.3392	0.9639	0.8666	0.5363
K_{M1}	0.8	0.5047	−0.1081	−0.2814	0.7866	−0.0876	−0.2840
	1.0	0.3698	−0.2938	−0.5779	0.5285	−0.4200	−0.8224
	1.2	0.3442	−0.3451	−0.7175	0.5005	−0.4866	−1.0559
	1.5	0.4542	−0.2430	−0.7070	0.7670	−0.2347	−0.9811
	2.0	0.8455	0.1802	−0.3526	1.6371	0.6759	−0.2128
$K_{\Delta RA}$	0.8	1.2688	1.1485	1.2781	1.2592	1.1733	1.3168
	1.0	1.1153	1.0080	1.1106	1.1106	1.0338	1.1485
	1.2	1.0000	0.9062	0.9913	0.9996	0.9323	1.0277
	1.5	0.8715	0.7952	0.8645	0.8761	0.8212	0.8983
	2.0	0.7213	0.6662	0.7218	0.7310	0.6911	0.7519
$K_{\psi A}$	0.8	3.1432	1.8696	2.4504	3.7245	2.4390	3.4736
	1.0	2.1326	0.8856	1.2057	2.3432	1.0822	1.7437
	1.2	1.4455	0.2515	0.3872	1.4221	0.2123	0.6049
	1.5	0.8113	−0.2892	−0.3320	0.5967	−0.5218	−0.3962
	2.0	0.3197	−0.6291	−0.8157	0.0014	−0.9683	−1.0706
$K_{\sigma 2}$	0.8	1.2688	1.1485	1.2781	1.2592	1.1733	1.3168
	1.0	1.1153	1.0080	1.1106	1.1106	1.0338	1.1485
	1.2	1.0000	0.9062	0.9913	0.9996	0.9323	1.0277
	1.5	0.8715	0.7952	0.8645	0.8761	0.8212	0.8983
	2.0	0.7213	0.6662	0.7218	0.7310	0.6911	0.7519

(Continued)

TABLE 13.4 Formulas for Discontinuity Stresses and Deformations at the Junctions of Shells and Plates (*Continued*)

Loading and Case No.	Load Terms	Selected Values

2d. Rotation around the axis of symmetry at ω rad/s

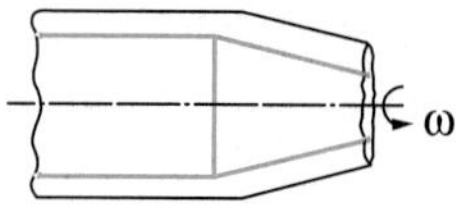

Note: δ = mass/unit volume

Load Terms

$$LT_{A1} = \frac{R_1^2}{t_1^2}$$

$$LT_{A2} = \frac{-\delta_2 R_2^3 E_1}{\delta_1 t_1^2 E_2 R_1}$$

$$LT_{AC} = 0$$

$$LT_{B1} = 0, \quad LT_{B2} = \frac{-\delta_2 R_2^2 E_1 (3+\nu_2)\tan\alpha_2}{\delta_1 E_2 t_1 R_1}$$

$$LT_{BC} = 0$$

At the junction of the cylinder and cone,

$$V_1 = \delta_1 \omega^2 R_1 t_1^2 K_{V1}, \quad M_1 = \delta_1 \omega^2 R_1 t_1^3 K_{M1}$$

$$N_1 = 0$$

$$\Delta R_A = \frac{\delta_1 \omega^2 R_1 t_1^2}{E_1}(LT_{A1} - K_{V1}C_{AA1} - K_{M1}C_{AB1})$$

$$\psi_A = \frac{\delta_1 \omega^2 R_1 t_1}{E_1}(K_{V1}C_{AB1} + K_{M1}C_{BB1})$$

Selected Values

For $R_1 = R_2$, $E_1 = E_2$, $\nu_1 = \nu_2 = 0.3$, $\delta_1 = \delta_2$, and for $R/t > 5$.

$$\Delta R_A = \frac{\delta_1 \omega^2 R_1^3}{E_1} K_{\Delta RA}, \quad \psi_A = \frac{\delta_1 \omega^2 R_1^2}{E_1} K_{\psi A}, \quad \sigma_2 = \delta_1 \omega^2 R_1^2 K_{\sigma 2}$$

		R_1/t_1 = 20			R_1/t_1 = 40		
	t_2/t_1	α_2 = −30	30	45	α_2 = −30	30	45
K_{v1}	0.8	−0.0248	0.0250	0.0425	−0.0249	0.0251	0.0428
	1.0	−0.0000	−0.0006	−0.0022	−0.0001	−0.0004	−0.0013
	1.2	0.0308	−0.0325	−0.0581	0.0308	−0.0320	−0.0565
	1.5	0.0816	−0.0849	−0.1507	0.0817	−0.0840	−0.1480
	2.0	0.1645	−0.1701	−0.3023	0.1649	−0.1687	−0.2979
K_{M1}	0.8	0.3565	−0.3583	−0.5815	0.5043	−0.5061	−0.8207
	1.0	0.4828	−0.4851	−0.7988	0.6829	−0.6852	−1.1269
	1.2	0.6056	−0.6089	−1.0149	0.8566	−0.8599	−1.4309
	1.5	0.7777	−0.7831	−1.3237	1.1000	−1.1053	−1.8646
	2.0	1.0229	−1.0320	−1.7708	1.4469	−1.4559	−2.4920
$K_{\Delta RA}$	0.8	1.0731	0.9264	0.8795	1.0518	0.9480	0.9148
	1.0	1.0798	0.9202	0.8693	1.0565	0.9435	0.9074
	1.2	1.0824	0.9181	0.8657	1.0582	0.9420	0.9047
	1.5	1.0816	0.9194	0.8679	1.0576	0.9428	0.9061
	2.0	1.0744	0.9273	0.8812	1.0525	0.9483	0.9152
$K_{\psi A}$	0.8	0.7590	−0.7633	−1.2450	0.7597	−0.7628	−1.2438
	1.0	0.9173	−0.9194	−1.5101	0.9176	−0.9192	−1.5094
	1.2	1.0488	−1.0494	−1.7360	1.0488	−1.0493	−1.7354
	1.5	1.2078	−1.2071	−2.0164	1.2074	−0.2070	−2.0156
	2.0	1.3995	−1.3984	−2.3649	1.3988	−1.3981	−2.3631
$K_{\sigma 2}$	0.8	1.0731	0.9264	0.8795	1.0518	0.9480	0.9148
	1.0	1.0798	0.9202	0.8693	1.0565	0.9435	0.9074
	1.2	1.0824	0.9181	0.8657	1.0582	0.9420	0.9047
	1.5	1.0816	0.9194	0.8679	1.0576	0.9428	0.9061
	2.0	1.0744	0.9273	0.8812	1.0525	0.9483	0.9152

3. Cylindrical shell connected to a spherical shell. To ensure accuracy $R/t > 5$ and the junction angle for the spherical shell must lie within the range $3/\beta_1 < \phi_2 < (\pi - 3/\beta_2)$. The spherical shell must also extend with no interruptions such as a second junction or a cutout, such that $\theta_2 > 3/\beta_2$. See the discussion on page 495. E_1 and E_2 are the moduli of elasticity and ν_1 and ν_2 the Poisson's ratios for the cylinder and sphere, respectively. See Table 13.2 for formulas for D_1 and λ_1. $b_1 = R_1 - t_1/2$ and $a_1 = R_1 + t_1/2$. See Table 13.3, case 1, for formulas for K_{12},* K_{22},* and β_2 for the spherical shell. $b_2 = R_2 - t_2/2$ and $a_2 = R_2 + t_2/2$. $R_A = R_1$ and normally $R_2 \sin\phi_2 = R_1$ but if $\phi_2 = 90°$ or is close to 90°, the midthickness radii at the junction may not be equal. Under this condition different correction terms will be used if necessary.

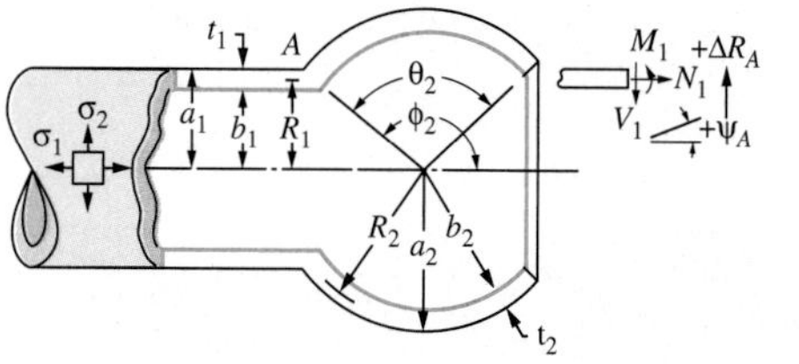

$$K_{V1} = \frac{LT_A C_{BB} - LT_B C_{AB}}{C_{AA}C_{BB} - C_{AB}^2}, \quad K_{M1} = \frac{LT_B C_{AA} - LT_A C_{AB}}{C_{AA}C_{BB} - C_{AB}^2}, \quad \left.\begin{aligned} LT_A &= LT_{A1} + LT_{A2} + LT_{AC} \\ LT_B &= LT_{B1} + LT_{B2} + LT_{BC} \end{aligned}\right\}$$ See cases 3a–3d for these load terms.

$$C_{AA} = C_{AA1} + C_{AA2}, \quad C_{AA1} = \frac{E_1}{2D_1\lambda_1^3}, \quad C_{AA2} = \frac{R_A E_1 \beta_2 \sin\phi_2}{E_2 t_2}\left(\frac{1}{K_{12}} + K_{22}\right)$$

$$C_{AB} = C_{AB1} + C_{AB2}, \quad C_{AB1} = \frac{-E_1 t_1}{2D_1\lambda_1^2}, \quad C_{AB2} = \frac{2E_1 R_A t_1 \beta_2^2}{E_2 R_2 t_2 K_{12}}$$

$$C_{BB} - C_{BB1} + C_{BB2}, \quad C_{BB1} = \frac{E_1 t_1^2}{D_1 \lambda_1}, \quad C_{BB2} = \frac{4E_1 R_A t_1^2 \beta_2^3}{R_2^2 E_2 t_2 K_{12} \sin\phi_2}$$

The stresses in the left cylinder at the junction are given by

$$\sigma_1 = \frac{N_1}{t_1}$$

$$\sigma_2 = \frac{\Delta R_A E_1}{R_A} + \nu_1 \sigma_1$$

$$\sigma_1' = \frac{-6M_1}{t_1^2}$$

$$\sigma_2' = \nu_1 \sigma_1'$$

Note: The use of joint load correction terms LT_{AC} and LT_{BC} depends upon the accuracy desired and the relative values of the thicknesses and the radii. Read Sec. 13.4 carefully. For thin-walled shells, $R/t > 10$, they can be neglected.

*The second subscript is added to refer to the right-hand shell. Evaluate K at the junction where $\omega = 0$.

(*Continued*)

TABLE 13.4 Formulas for Discontinuity Stresses and Deformations at the Junctions of Shells and Plates (*Continued*)

Loading and Case No.	Load Terms	Selected Values
3a. Internal* pressure q		

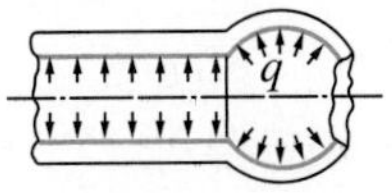

Note: There is no axial load on the cylinder. An axial load on the right end of the sphere balances any axial component of the pressure on the sphere and the joint. For an enclosed pressure vessel superpose an axial load $P = q\pi b_1^2$ using case 3b.

Load Terms

$$LT_{A1} = \frac{b_1 R_1}{t_1^2}$$

$$LT_{A2} = \frac{-b_2^2 E_1 \sin\phi_2}{E_2 t_1 t_1^2}$$

$$LT_{AC} = \frac{t_2 \cos\phi_2}{2t_1} C_{AA2} + \frac{t_2^2 - t_1^2}{8t_1^2} C_{AB2} + \frac{E_1 R_1 (1+\nu_2)(t_1 - t_2 \sin\phi_2)}{2E_2 t_1 t_2 \sin^2\phi_2} \dagger$$

$$LT_{B1} = 0, \quad LT_{B2} = 0$$

$$LT_{BC} = \frac{t_2 \cos\phi_2}{2t_1} C_{AB2} + \frac{t_2^2 - t_1^2}{8t_1^2} C_{BB2} + \frac{E_1 \cos\phi_2 (t_1 - t_2 \sin\phi_2)}{E_2 t_2 \sin^2\phi_2} \dagger$$

At the junction of the cylinders and sphere,

$$V_1 = q t_1 K_{V1}, \quad M_1 = q t_1^2 K_{M1}, \quad N_1 = 0$$

$$\Delta R_A = \frac{q t_1}{E_1}(LT_{A1} - K_{V1} C_{AA1} - K_{M1} C_{AB1})$$

$$\psi_A = \frac{q}{E_1}(K_{V1} C_{AB1} + K_{M1} C_{BB1})$$

Selected Values

For internal pressure, $E_1 = E_2$, $\nu_1 = \nu_2 = 0.3$, $R_2 \sin\phi_2 = R_1$, and for $R/t > 5$. (*Note:* No correction terms are used.)

$$\Delta R_A = \frac{qR_1^2}{E_1 t_1} K_{\Delta RA}, \quad \psi_A = \frac{qR_1}{E_1 t_1} K_{\psi A}, \quad \sigma_2 = \frac{qR_1}{t_1} K_{\sigma 2}$$

		R_1/t_1					
		10			20		
	$\frac{t_2}{t_1}$	ϕ_2 60	90	120	ϕ_2 60	90	120
K_{V1}	0.5	−0.5344	−0.3712	−0.5015	−0.7630	−0.5382	−0.7294
	0.8	−0.2216	−0.1062	−0.2115	−0.3299	−0.1676	−0.3192
	1.0	−0.0694	0.0292	−0.0668	−0.1190	0.0212	−0.1158
	1.2	0.0633	0.1517	0.0614	0.0650	0.1917	0.0636
	1.5	0.2472	0.3263	0.2410	0.3200	0.4344	0.3142
K_{M1}	0.5	0.2811	0.2058	0.2591	0.5660	0.4219	0.5344
	0.8	0.0560	0.0267	0.0483	0.1162	0.0597	0.1047
	1.0	0.0011	0.0000	−0.0010	0.0019	0.0000	−0.0018
	1.2	0.0130	0.0349	0.0148	0.0194	0.0624	0.0212
	1.5	0.1172	0.1636	0.1225	0.2169	0.3081	0.2238
$K_{\Delta RA}$	0.5	1.4774	1.3197	1.4433	1.5071	1.3541	1.4826
	0.8	1.1487	1.0452	1.1379	1.1838	1.0812	1.1758
	1.0	1.0068	0.9263	1.0040	1.0437	0.9628	1.0413
	1.2	0.9028	0.8382	0.9050	0.9408	0.8751	0.9420
	1.5	0.7878	0.7388	0.7946	0.8269	0.7762	0.8313
$K_{\psi A}$	0.5	2.5212	1.7793	2.3533	3.5965	2.5800	3.4255
	0.8	0.8827	0.4228	0.8287	1.3109	0.6674	1.2536
	1.0	0.2323	−0.0965	0.2180	0.3967	−0.0701	0.3792
	1.2	−0.1743	−0.4076	−0.1631	−0.1780	−0.5151	−0.1698
	1.5	−0.5021	−0.6387	−0.4671	−0.6453	−0.8504	−0.6132
$K_{\sigma 2}$	0.5	1.4774	1.3197	1.4433	1.5071	1.3541	1.4826
	0.8	1.1487	1.0452	1.1379	1.1838	1.0812	1.1758
	1.0	1.0068	0.9263	1.0040	1.0437	0.9628	1.0413
	1.2	0.9028	0.8382	0.9050	0.9408	0.8751	0.9420
	1.5	0.7878	0.7388	0.7946	0.8269	0.7762	0.8313

*For external pressure, substitute $-q$ for q, a_1 for b_1, b_2 for a_2, and a_2 for b_2 in the load terms.

†If $\phi_2 = 90°$ or is close to 90°, the following correction terms should be used:

$$LT_{AC} = \frac{b_1^2 - b_2^2}{4t_1^2}\left[\frac{a_2 - b_1}{R_1} C_{AB2} - \frac{2E_1 t_1 (1+\nu_2)}{E_2 t_2}\right], \quad LT_{BC} = \frac{b_1^2 - b_2^2}{4t_1^2} \frac{a_2 - b_1}{R_1} C_{BB2}$$

Loading and Case No.	Load Terms	Selected Values
3b. Axial load P		

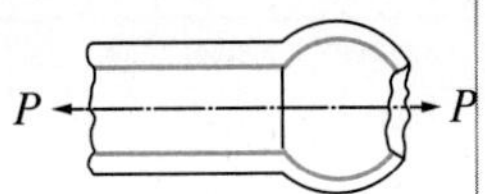

Load Terms

$$LT_{A1} = \frac{-\nu_1 R_1^2}{2t_1^2}$$

$$LT_{A2} = \frac{R_1^2 E_1 (1+\nu_2)}{2E_2 t_1 t_2 \sin\phi_2} + \frac{R_1 C_{AA2}}{2t_1 \tan\phi_2}$$

$$LT_{AC} = 0^*$$

$$LT_{B1} = 0$$

$$LT_{B2} = \frac{R_1 C_{AB2}}{2t_1 \tan\phi_2}$$

$$LT_{BC} = 0^*$$

At the junction of the cylinders and sphere,

$$V_1 = \frac{P t_1 K_{V1}}{\pi R_1^2}, \quad M_1 = \frac{P t_1^2 K_{M1}}{\pi R_1^2}, \quad N_1 = \frac{P}{2\pi R_1}$$

$$\Delta R_A = \frac{P t_1}{E_1 \pi R_1^2}(LT_{A1} - K_{V1} C_{AA1} - K_{M1} C_{AB1})$$

$$\psi_A = \frac{P}{E_1 \pi R_1^2}(K_{V1} C_{AB1} + K_{M1} C_{BB1})$$

Selected Values

For axial tension, $E_1 = E_2$, $\nu_1 = \nu_2 = 0.3$, $R_2 \sin\phi_2 = R_1$, and for $R/t > 5$.

$$\Delta R_A = \frac{P\nu_1}{2\pi E_1 t_1} K_{\Delta RA}, \quad \psi_A = \frac{P\nu_1}{2\pi E_1 t_1^2} K_{\psi A}, \quad \sigma_2 = \frac{P}{2\pi R_1 t_1} K_{\sigma 2}$$

		R_1/t_1					
		10			20		
	$\frac{t_2}{t_1}$	ϕ_2 60	90	120	ϕ_2 60	90	120
K_{V1}	0.5	2.2519	0.4487	−1.2194	4.1845	0.6346	−2.7251
	0.8	1.8774	0.3483	−1.0921	3.5154	0.4925	−2.4052
	1.0	1.7427	0.3075	−1.0535	3.2769	0.4349	−2.3023
	1.2	1.6409	0.2781	−1.0186	3.0945	0.3933	−2.2145
	1.5	1.5083	0.2435	−0.9620	2.8531	0.3444	−2.0804
K_{M1}	0.5	0.8275	−0.2487	−1.3712	2.5815	−0.4975	−3.6682
	0.8	1.4700	−0.0877	−1.6631	4.2478	−0.1754	−4.6337
	1.0	1.7061	−0.0000	−1.7064	4.8334	−0.0000	−4.8338
	1.2	1.8303	0.0640	−1.6870	5.1238	0.1280	−4.8367
	1.5	1.8775	0.1221	−1.5987	5.1991	0.2442	−4.6411
$K_{\Delta RA}$	0.5	−11.3815	−3.9800	2.5881	−14.1929	−3.9800	5.4031
	0.8	−7.9365	−3.0808	1.2550	−9.7933	−3.0808	3.1133
	1.0	−6.6866	−2.6667	0.9502	−8.2341	−2.6667	2.4986
	1.2	−5.8608	−2.3665	0.8041	−7.2151	−2.3665	2.1590
	1.5	−5.0386	−2.0509	0.6916	−6.2071	−2.0509	1.8605
$K_{\psi A}$	0.5	−3.4789	−1.4341	0.2306	−2.9746	−1.0140	0.6790
	0.8	−1.5032	−0.9243	−0.5727	−1.1826	−0.6536	−0.2848
	1.0	−0.7837	−0.6775	−0.7353	−0.5490	−0.4790	−0.5248
	1.2	−0.3368	−0.4982	−0.7774	−0.1642	−0.3532	−0.6233
	1.5	0.0397	−0.3178	−0.7440	0.1494	−0.2247	−0.6472
$K_{\sigma 2}$	0.5	−3.1144	−0.8940	1.0764	−3.9579	−0.8940	1.9209
	0.8	−2.0809	−0.6242	0.6765	−2.6380	−0.6242	1.2340
	1.0	−1.7060	−0.5000	0.5851	−2.1702	−0.5000	1.0496
	1.2	−1.4582	−0.4099	0.5412	−1.8645	−0.4099	0.9477
	1.5	−1.2116	−0.3153	0.5075	−1.5621	−0.3153	0.8581

*If $\phi_2 = 90°$ or is close to 90°, the following correction terms should be used:

$$LT_{AC} = \frac{-R_1(R_2 - R_1) C_{AB2}}{2t_1^2}, \quad LT_{BC} = \frac{-R_1(R_2 - R_1) C_{BB2}}{2t_1^2}$$

(*Continued*)

TABLE 13.4 Formulas for Discontinuity Stresses and Deformations at the Junctions of Shells and Plates (*Continued*)

Loading and Case No.	Load Terms	Selected Values
3c. Hydrostatic internal* pressure q_1 at the junction for $x_1 > 3/\lambda_1$.† If $x_1 < 3/\lambda_1$, the discontinuity in pressure gradient introduces small deformations at the junction. 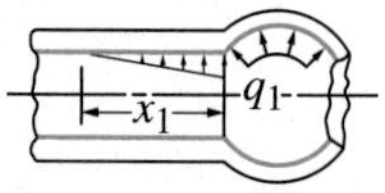 *Note:* There is no axial load on the cylinder. An axial load on the right end of the sphere balances the axial component of pressure in the sphere and on the joint.	$LT_{A1} = \frac{b_1 R_1}{t_1^2}$ $LT_{A2} = \frac{-b_2^2 E_1 \sin\phi_2}{E_2 t_1 t_2}$ For LT_{AC} use the expressions from case 3a. $LT_{B1} = \frac{-b_1 R_1}{x_1 t_1}$ $LT_{B2} = \frac{E_1 b_2 R_2 \sin\phi_2}{E_2 t_2 x_1}$ For LT_{BC} use the expressions from case 3a. At the junction of the cylinder and sphere, $V_1 = q_1 t_1 K_{V1}, \quad M_1 = q_1 t_1^2 K_{M1}, \quad N_1 = 0$ $\Delta R_A = \frac{q_1 t_1}{E_1}(LT_{A1} - K_{V1}C_{AA1} - K_{M1}C_{AB1})$ $\psi_A = \frac{q_1}{E_1}(-LT_{B1} + K_{V1}C_{AB1} + K_{M1}C_{BB1})$	For internal pressure, $x_1 = R_1$, $E_1 = E_2$, $\nu_1 = \nu_2 = 0.3$, $R_2 \sin\phi_2 = R_1$, and for $R/t > 5$. (*Note:* No correction terms are used.) $\Delta R_A = \frac{q_1 R_1^2}{E_1 t_1} K_{\Delta RA}, \quad \psi_A = \frac{q_1 R_1}{E_1 t_1} K_{\psi A}, \quad \sigma_2 = \frac{q_1 R_1}{t_1} K_{\sigma 2}$

Selected values (R_1/t_1 = 10 and 20; ϕ_2 = 60, 90, 120):

	t_2/t_1	R_1/t_1 = 10, ϕ_2 = 60	10, 90	10, 120	20, 60	20, 90	20, 120
K_{V1}	0.5	−0.5636	−0.3928	−0.5284	−0.7918	−0.5598	−0.7566
	0.8	−0.2279	−0.1095	−0.2169	−0.3360	−0.1710	−0.3247
	1.0	−0.0696	−0.0292	−0.0667	−0.1191	0.0212	−0.1157
	1.2	0.0628	0.1490	0.0607	0.0644	0.1890	0.0629
	1.5	0.2375	0.3119	0.2308	0.3101	0.4201	0.3041
K_{M1}	0.5	0.3642	0.2737	0.3430	0.6837	0.5180	0.6528
	0.8	0.1140	0.0615	0.1078	0.1986	0.1088	0.1885
	1.0	0.0285	0.0000	0.0271	0.0408	0.0000	0.0378
	1.2	0.0047	−0.0037	0.0062	0.0075	0.0078	0.0091
	1.5	0.0515	0.0653	0.0552	0.1236	0.1691	0.1289
$K_{\Delta RA}$	0.5	1.5286	1.3598	1.4929	1.5431	1.3824	1.5178
	0.8	1.1730	1.0593	1.1620	1.2010	1.0913	1.1928
	1.0	1.0160	0.9263	1.0131	1.0502	0.9628	1.0477
	1.2	0.9005	0.8276	0.9027	0.9392	0.8676	0.9403
	1.5	0.7740	0.7180	0.7806	0.8171	0.7615	0.8215
$K_{\psi A}$	0.5	3.7909	2.9832	3.6176	4.8904	3.8089	4.7155
	0.8	2.0095	1.4769	1.9564	2.4627	1.7466	2.4061
	1.0	1.2564	0.8535	1.2431	1.4459	0.9049	1.4291
	1.2	0.7551	0.4477	0.7659	0.7764	0.3652	0.7842
	1.5	0.3036	0.0949	0.3355	0.1849	−0.0919	0.2149
$K_{\sigma 2}$	0.5	1.5286	1.3598	1.4929	1.5431	1.3824	1.5178
	0.8	1.1730	1.0593	1.1620	1.2010	1.0913	1.1928
	1.0	1.0160	0.9263	1.0131	1.0502	0.9628	1.0477
	1.2	0.9005	0.8276	0.9027	0.9392	0.8676	0.9403
	1.5	0.7740	0.7180	0.7806	0.8171	0.7615	0.8215

*For external pressure, substitute $-q$ for q_1, a_1 for b_1, b_2 for a_2, and a_2 for b_2 in the load terms.

†If pressure increases right to left, substitute $-x_1$ for x_1 and verify that $|x_1|$ is large enough to extend into the sphere as far as the position where $\theta_2 = 3/\beta_2$.

(*Continued*)

TABLE 13.4 Formulas for Discontinuity Stresses and Deformations at the Junctions of Shells and Plates (*Continued*)

Loading and Case No.	Load Terms	Selected Values
3d. Rotation around the axis of symmetry at ω rad/s 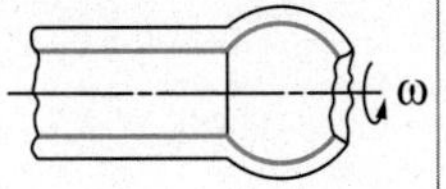 *Note:* δ = mass/unit volume	$LT_{A1} = \frac{R_1^2}{t_1^2}$ $LT_{A2} = \frac{-\delta_2 R_2^3 E_1 \sin^3 \phi_2}{\delta_1 R_1 E_2 t_1^2}$ $LT_{AC} = 0$ $LT_{B1} = 0$ $LT_{B2} = \frac{-\delta_2 R_2^2 E_1 (3+\nu_2) \sin\phi_2 \cos\phi_2}{\delta_1 R_1 E_2 t_1}$ $LT_{BC} = 0$ At the junction of the cylinders and sphere, $V_1 = \delta_1 \omega^2 R_1 t_1^2 K_{V1}, \quad M_1 = \delta_1 \omega^2 R_1 t_1^3 K_{M1}$ $N_1 = 0$ $\Delta R_A = \frac{\delta_1 \omega^2 R_1 t_1^2}{E_1}(LT_{A1} - K_{V1} C_{AA1} - K_{M1} C_{AB1})$ $\psi_A = \frac{\delta_1 \omega^2 R_1 t_1}{E_1}(K_{V1} C_{AB1} + K_{M1} C_{BB1})$	For $E_1 = E_2$, $\nu_1 = \nu_2 = 0.3$, $\delta_1 = \delta_2$, $R_2 \sin\phi_2 = R_1$, and for $R/t > 5$. $\Delta R_A = \frac{\delta_1 \omega^2 R_1^3}{E_1} K_{\Delta RA}, \quad \psi_A = \frac{\delta_1 \omega^2 R_1^2}{E_1} K_{\psi A}, \quad \sigma_2 = \delta_1 \omega^2 R_1^2 K_{\sigma 2}$

		R_1/t_1					
		10			20		
		ϕ_2			ϕ_2		
	$\frac{t_2}{t_1}$	60	90	120	60	90	120
K_{V1}	0.5	0.0425	0.0000	−0.0392	0.0420	0.0000	−0.0396
	0.8	0.0270	0.0000	−0.0233	0.0264	0.0000	−0.0238
	1.0	0.0020	0.0000	0.0018	0.0014	0.0000	0.0013
	1.2	−0.0292	0.0000	0.0330	−0.0298	0.0000	0.0325
	1.5	−0.0806	0.0000	0.0843	−0.0812	0.0000	0.0838
K_{M1}	0.5	−0.1209	0.0000	0.1220	−0.1712	0.0000	0.1723
	0.8	−0.2495	0.0000	0.2556	−0.3541	0.0000	0.3602
	1.0	−0.3371	0.0000	0.3465	−0.4788	0.0000	0.4881
	1.2	−0.4228	0.0000	0.4347	−0.6005	0.0000	0.6123
	1.5	−0.5437	0.0000	0.5576	−0.7719	0.0000	0.7858
$K_{\Delta RA}$	0.5	0.9255	1.0000	1.0722	0.9476	1.0000	1.0513
	0.8	0.8956	1.0000	1.1034	0.9263	1.0000	1.0732
	1.0	0.8870	1.0000	1.1130	0.9201	1.0000	1.0799
	1.2	0.8840	1.0000	1.1168	0.9179	1.0000	1.0825
	1.5	0.8859	1.0000	1.1157	0.9191	1.0000	1.0817
$K_{\psi A}$	0.5	−0.4653	0.0000	0.4573	−0.4639	0.0000	0.4583
	0.8	−0.7594	0.0000	0.7636	−0.7600	0.0000	0.7629
	1.0	−0.9122	0.0000	0.9248	−0.9140	0.0000	0.9229
	1.2	−1.0394	0.0000	1.0585	−1.0422	0.0000	1.0557
	1.5	−1.1943	0.0000	1.2195	−1.1980	0.0000	1.2158
$K_{\sigma 2}$	0.5	0.9255	1.0000	1.0722	0.9476	1.0000	1.0513
	0.8	0.8956	1.0000	1.1034	0.9263	1.0000	1.0732
	1.0	0.8870	1.0000	1.1130	0.9201	1.0000	1.0799
	1.2	0.8840	1.0000	1.1168	0.9179	1.0000	1.0825
	1.5	0.8859	1.0000	1.1157	0.9191	1.0000	1.0817

(*Continued*)

TABLE 13.4 Formulas for Discontinuity Stresses and Deformations at the Junctions of Shells and Plates (*Continued*)

4. Cylindrical shell connected to a circular plate. Expressions are accurate if $R_1/t_1 > 5$ and $R_1/t_2 > 4$. E_1 and E_2 are the moduli of elasticity and ν_1 and ν_2 the Poisson's ratios for the cylinder and plate, respectively. See Table 13.2 for formulas for D_1 and λ_1. $b_1 = R_1 - t_1/2$ and $a_1 = R_1 + t_1/2$. See Table 11.2 for the formula for D_2.

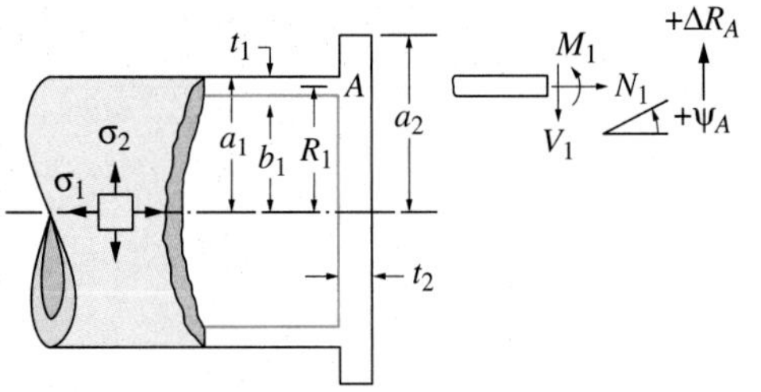

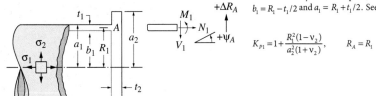

$$K_{P1} = 1 + \frac{R_1^2(1-\nu_2)}{a_2^2(1+\nu_2)}, \qquad R_A = R_1$$

$$K_{V1} = \frac{LT_A C_{BB} - LT_B C_{AB}}{C_{AA}C_{BB} - C_{AB}^2}, \qquad K_{M1} = \frac{LT_B C_{AA} - LT_A C_{AB}}{C_{AA}C_{BB} - C_{AB}^2}, \qquad \left.\begin{aligned} LT_A &= LT_{A1} + LT_{A2} + LT_{AC} \\ LT_B &= LT_{B1} + LT_{B2} + LT_{BC} \end{aligned}\right\}$$ See cases 4a–4d for these load terms.

$$C_{AA} = C_{AA1} + C_{AA2}, \qquad C_{AA1} = \frac{E_1}{2D_1\lambda_1^3}, \qquad C_{AA2} = \frac{E_1 t_2^2 R_1 K_{P1}}{6D_2}$$

$$C_{AB} = C_{AB1} + C_{AB2}, \qquad C_{AB1} = \frac{-E_1 t_1}{2D_1\lambda_1^2}, \qquad C_{AB2} = \frac{E_1 t_1 t_2 R_1 K_{P1}}{4D_2}$$

$$C_{BB} = C_{BB1} + C_{BB2}, \qquad C_{BB1} = \frac{E_1 t_1^2}{D_1\lambda_1}, \qquad C_{BB2} = \frac{E_1 t_1^2 R_1 K_{P1}}{2D_2}$$

The stresses in the left cylinder at the junction are given by

$$\sigma_1 = \frac{N_1}{t_1}$$

$$\sigma_2 = \frac{\Delta R_A E_1}{R_A} + \nu_1\sigma_1$$

$$\sigma_1' = \frac{-6M_1}{t_1^2}$$

$$\sigma_2' = \nu_1\sigma_1'$$

Note: The use of joint load correction terms LT_{AC} and LT_{BC} depends upon the accuracy desired and the relative values of the thicknesses and the radii. Read Sec. 13.4 carefully. For thin-walled shells, $R/t > 10$, they can be neglected.

Loading and Case No.	Load Terms	Selected Values

4a. Internal pressure* q

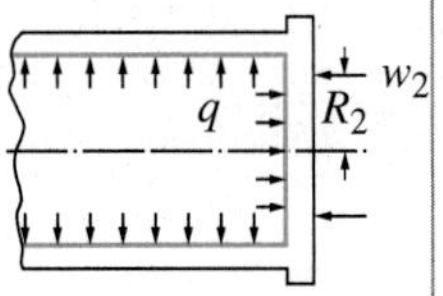

Note: There is no axial load on the cylinder. The axial load on the plate is reacted by the annular line load $w_2 = qb_1^2/(2R_2)$ at a radius R_2. For an enclosed pressure vessel superpose an axial load $P = q\pi b_1^2$ using case 4b.

Load Terms

$$LT_{A1} = \frac{b_1 R_1}{t_1^2}, \qquad LT_{AC} = 0$$

$$LT_{A2} = \frac{E_1 t_2 b_1^2}{32 D_2 t_1 R_1} K_{P2}$$

where

$$K_{P2} = \begin{cases} (2R_2^2 - b_1^2)K_{P1} & \text{for } R_2 \le R_1 \\ (2R_2^2 - b_1^2)K_{P1} - 2(R_2^2 - R_1^2) + 4R_1^2 \ln\dfrac{R_2}{R_1} & \text{for } R_2 \ge R_1 \end{cases}$$

$$LT_{B1} = 0, \qquad LT_{BC} = 0$$

$$LT_{B2} = \frac{E_1 b_1^2}{16 D_2 R_1} K_{P2}$$

At the junction of the cylinder and plate,

$$V_1 = qt_1 K_{V1}, \qquad M_1 = qt_1^2 K_{M1}, \qquad N_1 = 0$$

$$\Delta R_A = \frac{qt_1}{E_1}(LT_{A1} - K_{V1}C_{AA1} - K_{M1}C_{AB1})$$

$$\psi_A = \frac{q}{E_1}(K_{V1}C_{AB1} + K_{M1}C_{BB1})$$

Selected Values

For internal pressure, $E_1 = E_2$, $\nu_1 = \nu_2 = 0.3$, $a_2 = a_1$, $R_2 = 0.7R_1$, and for $R_1/t_1 > 5$ and $R_1/t_1 > 4$.

$$\Delta R_A = \frac{qR_1^2}{E_1 t_1} K_{\Delta RA}, \qquad \psi_A = \frac{qR_1}{E_1 t_1} K_{\psi A}, \qquad \sigma_2 = \frac{qR_1}{t_1} K_{\sigma 2}$$

	$\frac{t_2}{t_1}$	R_1/t_1 = 15	20	30	40	80	100
K_{v1}	1.5	1.5578	1.8088	2.1689	2.4003	2.5659	2.3411
	2.0	1.7087	1.9850	2.3934	2.6724	3.0336	2.9105
	2.5	1.8762	2.1839	2.6518	2.9896	3.5949	3.6011
	3.0	2.0287	2.3667	2.8931	3.2891	4.1422	4.2830
	4.0		2.6452	3.2629	3.7513	5.0100	5.3783
K_{M1}	1.5	1.1277	1.3975	1.6600	1.5275	−3.2234	−8.2255
	2.0	1.5904	2.0300	2.6646	2.9496	0.3139	−3.3803
	2.5	2.0582	2.6818	3.7244	4.4723	4.2328	2.0517
	3.0	2.4593	3.2474	4.6606	5.8356	7.8744	7.1711
	4.0		4.0612	6.0209	7.8344	13.3902	15.0391
$K_{\Delta RA}$	1.5	0.1811	0.1661	0.1482	0.1380	0.1231	0.1213
	2.0	0.1828	0.1693	0.1535	0.1449	0.1348	0.1350
	2.5	0.1747	0.1627	0.1489	0.1417	0.1353	0.1370
	3.0	0.1618	0.1510	0.1388	0.1327	0.1284	0.1309
	4.0		0.1254	0.1151	0.1099	0.1069	0.1093
$K_{\psi A}$	1.5	−2.6740	−3.3226	−4.5926	−5.8801	−11.5407	−14.7243
	2.0	−2.1579	−2.7034	−3.7762	−4.8690	−9.7264	−12.4897
	2.5	−1.6854	−2.1223	−2.9864	−3.8720	−7.8591	−10.1571
	3.0	−1.3096	−1.6521	−2.3314	−3.0303	−6.2088	−8.0614
	4.0		−1.0265	−1.4437	−1.8728	−3.8376	−4.9966
$K_{\sigma 2}$	1.5	0.1811	0.1661	0.1482	0.1380	0.1231	0.1213
	2.0	0.1828	0.1693	0.1535	0.1449	0.1348	0.1350
	2.5	0.1747	0.1627	0.1489	0.1417	0.1353	0.1370
	3.0	0.1618	0.1510	0.1388	0.1327	0.1284	0.1309
	4.0		0.1254	0.1151	0.1099	0.1069	0.1093

*For external pressure, substitute $-q$ for q and a_1 for b_1 in the load terms.

(*Continued*)

TABLE 13.4 Formulas for Discontinuity Stresses and Deformations at the Junctions of Shells and Plates (*Continued*)

Loading and Case No.	Load Terms	Selected Values

4b. Axial load P

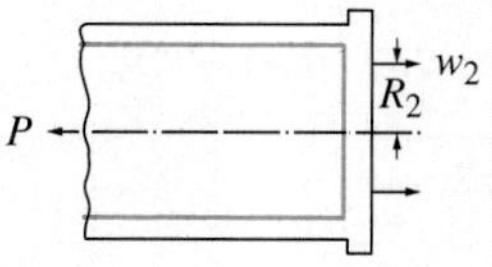

$$w_2 = \frac{P}{2\pi R_2}$$

Load Terms

$$LT_{A1} = \frac{-\nu_1 R_1^2}{2t_1^2}, \qquad LT_{AC} = 0$$

$$LT_{A2} = \frac{E_1 t_2 R_1^3}{16 D_2 t_1} K_{P2}$$

where

$$K_{P2} = \begin{cases} \left(1 - \dfrac{R_2^2}{R_1^2}\right) K_{P1} & \text{for } R_2 \le R_1 \\ -\dfrac{1-\nu_2}{1+\nu_2}\dfrac{R_2^2 - R_1^2}{a_2^2} - 2\ln\dfrac{R_2}{R_1} & \text{for } R_2 \ge R_1 \end{cases}$$

$$LT_{B1} = 0, \qquad LT_{BC} = 0$$

$$LT_{B2} = \frac{E_1 R_1^3}{8 D_2} K_{P2}$$

At the junction of the cylinder and plate,

$$V_1 = \frac{P t_1 K_{V1}}{\pi R_1^2}, \quad M_1 = \frac{P t_1^2 K_{M1}}{\pi R_1^2}, \quad N_1 = \frac{P}{2\pi R_1}$$

$$\Delta R_A = \frac{P t_1}{E_1 \pi R_1^2}(LT_{A1} - K_{V1}C_{AA1} - K_{M1}C_{AB1})$$

$$\psi_A = \frac{P}{E_1 \pi R_1^2}(K_{V1}C_{AB1} + K_{M1}C_{BB1})$$

Selected Values

For axial tension, $E_1 = E_2$, $\nu_1 = \nu_2 = 0.3$, $a_2 = a_1$, and for $R_1/t_1 > 5$ and $R_1/t_1 > 4$.

$$\Delta R_A = \frac{P\nu_1}{2\pi E_1 t_1} K_{\Delta RA}, \qquad \psi_A = \frac{P\nu_1}{2\pi E_1 t_1^2} K_{\psi A}, \qquad \sigma_2 = \frac{P}{2\pi R_1 t_1} K_{\sigma 2}$$

		R_2/t_1					
		0.8			0.9		
		R_1/t_1			R_1/t_1		
	$\frac{t_2}{t_1}$	15	20	30	15	20	30
K_{V1}	1.5	4.4625	7.2335	14.1255	2.2612	3.7073	7.3171
	2.0	3.4224	5.6661	11.3633	1.6969	2.8629	5.8397
	2.5	2.5276	4.2700	8.7738	1.2087	2.1078	4.4512
	3.0	1.8389	3.1711	6.6622	0.8311	1.5116	3.3167
	4.0		1.7432	3.8418		0.7339	1.7981
K_{M1}	1.5	12.0771	22.7118	54.8202	6.3440	11.9494	28.8827
	2.0	8.8772	17.0664	42.4231	4.6078	8.9085	22.2520
	2.5	6.3757	12.4854	31.7955	3.2429	6.4308	16.5534
	3.0	4.5632	9.0843	23.5956	2.2494	4.5854	12.1481
	4.0		4.9077	13.2133		2.3106	6.5578
$K_{\Delta RA}$	1.5	−3.0099	−3.7039	−4.9434	−1.6891	−2.0452	−2.6862
	2.0	−3.1072	−3.9154	−5.4038	−1.7418	−2.1590	−2.9323
	2.5	−2.8215	−3.6112	−5.1054	−1.5858	−1.9944	−2.7723
	3.0	−2.4354	−3.1463	−4.5192	−1.3741	−1.7421	−2.4573
	4.0		−2.2748	−3.3184		−1.2673	−1.8107
$K_{\psi A}$	1.5	5.2198	6.4135	8.5221	2.8636	3.4828	4.5818
	2.0	3.6278	4.5653	6.2776	1.9999	2.4873	3.3814
	2.5	2.5032	3.2024	4.5161	1.3863	1.7502	2.4369
	3.0	1.7479	2.2592	3.2405	0.9723	1.2385	1.7516
	4.0		1.1874	1.7331		0.6547	0.9399
$K_{\sigma 2}$	1.5	−0.6030	−0.8112	−1.1830	−0.2067	−0.3135	−0.5059
	2.0	−0.6322	−0.8746	−1.3211	−0.2225	−0.3477	−0.5797
	2.5	−0.5464	−0.7834	−1.2316	−0.1757	−0.2983	−0.5317
	3.0	−0.4306	−0.6439	−1.0558	−0.1122	−0.2226	−0.4372
	4.0		−0.3824	−0.6955		−0.0802	−0.2432

4c. Hydrostatic internal* pressure q_1 at the junction for $x_1 > 3/\lambda_1$.†

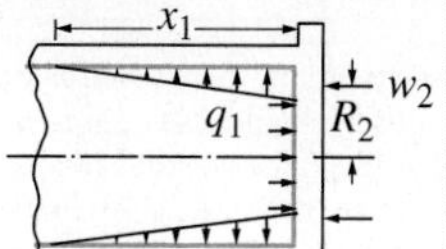

Note: There is no axial load on the cylinder. The axial load on the plate is reacted by the annular line load $w_2 = q_1(b_1^2/2R_2)$ at a radius R_2.

Load Terms

$$LT_{A1} = \frac{b_1 R_1}{t_1^2}, \qquad LT_{AC} = 0$$

$$LT_{A2} = \frac{E_1 t_2 b_1^2}{32 D_2 t_1 R_1} K_{P2}$$

where

$$K_{P2} = \begin{cases} (2R_2^2 - b_1^2)K_{P1} & \text{for } R_2 \le R_1 \\ (2R_2^2 - b_1^2)K_{P1} - 2(R_2^2 - R_1^2) + 4R_1^2 \ln\dfrac{R_2}{R_1} & \text{for } R_2 \ge R_1 \end{cases}$$

$$LT_{B1} = \frac{-b_1 R_1}{x_1 t_1}, \qquad LT_{BC} = 0$$

$$LT_{B2} = \frac{E_1 b_1^2}{16 D_2 R_1} K_{P2}$$

At the junction of the cylinder and plate,

$$V_1 = q_1 t_1 K_{V1}, \qquad M_1 = q_1 t_1^2 K_{M1}, \qquad N_1 = 0$$

$$\Delta R_A = \frac{q_1 t_1}{E_1}(LT_{A1} - K_{V1}C_{AA1} - K_{M1}C_{AB1})$$

$$\psi_A = \frac{q_1}{E_1}(-LT_{B1} + K_{V1}C_{AB1} + K_{M1}C_{BB1})$$

Selected Values

For internal pressure, $E_1 = E_2$, $\nu_1 = \nu_2 = 0.3$, $x_1 = R_1$, $a_2 = a_1$, $R_2 = 0.7R_1$, and for $R_1/t_1 > 5$ and $R_1/t_2 > 4$.

$$\Delta R_A = \frac{q_1 R_1^2}{E_1 t_1} K_{\Delta RA}, \qquad \psi_A = \frac{q_1 R_1}{E_1 t_1} K_{\psi A}, \qquad \sigma_2 = \frac{q_1 R_1}{t_1} K_{\sigma 2}$$

		R_1/t_1					
	$\frac{t_2}{t_1}$	15	20	30	40	80	100
K_{V1}	1.5	1.5304	1.7832	2.1457	2.3788	2.5486	2.3251
	2.0	1.6383	1.9170	2.3295	2.6119	2.9820	2.8620
	2.5	1.7652	2.0747	2.5466	2.8880	3.5041	3.5144
	3.0	1.8841	2.2228	2.7519	3.1509	4.0143	4.1593
	4.0		2.4528	3.0704	3.5599	4.8249	5.1962
K_{M1}	1.5	0.9101	1.1645	1.4064	1.2602	−3.5206	−8.5313
	2.0	1.2404	1.6448	2.2289	2.4780	−0.2425	−3.9630
	2.5	1.5946	2.1613	3.1188	3.8031	3.4026	1.1684
	3.0	1.9073	2.6197	3.9156	4.9999	6.7971	6.0099
	4.0		3.2916	5.0872	6.7688	11.9491	13.4589
$K_{\Delta RA}$	1.5	0.1513	0.1424	0.1311	0.1247	0.1158	0.1153
	2.0	0.1525	0.1448	0.1355	0.1305	0.1266	0.1283
	2.5	0.1463	0.1395	0.1316	0.1278	0.1271	0.1301
	3.0	0.1362	0.1300	0.1230	0.1198	0.1207	0.1243
	4.0		0.1089	0.1025	0.0997	0.1005	0.1039
$K_{\psi A}$	1.5	−2.0945	−2.7054	−3.9258	−5.1808	−10.7721	−13.9361
	2.0	−1.7263	−2.2355	−3.2576	−4.3151	−9.0908	−11.8293
	2.5	−1.3687	−1.7750	−2.5946	−3.4475	−7.3540	−9.6258
	3.0	−1.0757	−1.3940	−2.0371	−2.7086	−5.8157	−7.6438
	4.0		−0.8776	−1.2726	−1.6843	−3.6009	−4.7423
$K_{\sigma 2}$	1.5	0.1513	0.1424	0.1311	0.1247	0.1158	0.1153
	2.0	0.1525	0.1448	0.1355	0.1305	0.1266	0.1283
	2.5	0.1463	0.1395	0.1316	0.1278	0.1271	0.1301
	3.0	0.1362	0.1300	0.1230	0.1198	0.1207	0.1243
	4.0		0.1089	0.1025	0.0997	0.1005	0.1039

*For external pressure, substitute $-q$ for q_1 and a_1 for b_1 in the load terms.

†If pressure increases right to left, substitute $-x_1$ for x_1.

(*Continued*)

TABLE 13.4 Formulas for Discontinuity Stresses and Deformations at the Junctions of Shells and Plates (*Continued*)

Loading and Case No.	Load Terms	Selected Values

4d. Rotation around the axis of symmetry at ω rad/s

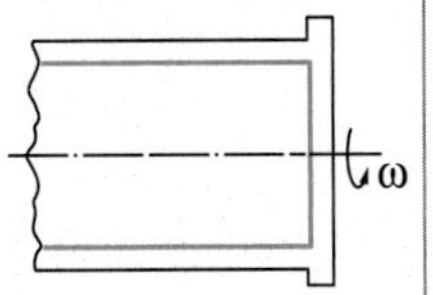

Note: δ = mass/unit volume

Load Terms

$LT_{A1} = \frac{R_1^2}{t_1^2}, \quad LT_{AC} = 0$

$$LT_{A2} = \frac{-E_1\delta_2 t_2^3}{96D_2\delta_1 t_1^2}\left[\frac{a_2^2(3+\nu_2)}{1+\nu_2} - R_1^2\right]$$

$LT_{B1} = 0$

$LT_{B2} = 0$

$LT_{BC} = 0$

At the junction of the cylinder and plate,

$V_1 = \delta_1\omega^2 R_1 t_1^2 K_{V1}, \quad M_1 = \delta_1\omega^2 R_1 t_1^3 K_{M1}$

$N_1 = 0$

$$\Delta R_A = \frac{\delta_1\omega^2 R_1 t_1^2}{E_1} \times (LT_{A1} - K_{V1}C_{AA1} - K_{M1}C_{AB1})$$

$$\psi_A = \frac{\delta_1\omega^2 R_1 t_1}{E_1}(K_{V1}C_{AB1} + K_{M1}C_{BB1})$$

Selected Values

For $E_1 = E_2$, $\nu_1 = \nu_2 = 0.3$, $\delta_1 = \delta_2$, $a_2 = a_1$, and for $R_1/t_1 > 5$ and $R_1/t_1 > 4$.

$$\Delta R_A = \frac{\delta_1\omega^2 R_1^3}{E_1}K_{\Delta RA}, \quad \psi_A = \frac{\delta_1\omega^2 R_1^2}{E_1}K_{\psi A}, \quad \sigma_2 = \delta_1\omega^2 R_1^2 K_{\sigma 2}$$

	t_2/t_1	R_1/t_1 = 15	20	30	40	80	100
K_{V1}	1.5	1.0683	1.2627	1.5887	1.8632	2.7105	3.0514
	2.0	1.2437	1.4585	1.8143	2.1105	3.0101	3.3674
	2.5	1.4257	1.6676	2.0656	2.3946	3.3815	3.7690
	3.0	1.5852	1.8538	2.2957	2.6604	3.7505	4.1763
	4.0		2.1297	2.6424	3.0676	4.3459	4.8472
K_{M1}	1.5	0.3416	0.4269	0.5778	0.7109	1.1434	1.3236
	2.0	0.8797	1.1304	1.5883	2.0037	3.4081	4.0109
	2.5	1.3881	1.8157	2.6188	3.3673	6.0009	7.1688
	3.0	1.8077	2.3920	3.5118	4.5776	8.4561	10.2267
	4.0		3.1982	4.7871	6.3383	12.2402	15.0458
$K_{\Delta RA}$	1.5	0.3661	0.3447	0.3180	0.3014	0.2682	0.2593
	2.0	0.3683	0.3483	0.3234	0.3076	0.2756	0.2668
	2.5	0.3595	0.3414	0.3189	0.3048	0.2759	0.2680
	3.0	0.3460	0.3295	0.3093	0.2968	0.2713	0.2643
	4.0		0.3042	0.2871	0.2767	0.2565	0.2511
$K_{\psi A}$	1.5	-2.7811	-3.3618	-4.3536	-5.2023	-7.8711	-8.9591
	2.0	-2.1801	-2.6725	-3.5321	-4.2829	-6.7099	-7.7203
	2.5	-1.6664	-2.0614	-2.7641	-3.3900	-5.4746	-6.3645
	3.0	-1.2732	-1.5822	-2.1392	-2.6428	-4.3620	-5.1128
	4.0		-0.9623	-1.3069	-1.6231	-2.7354	-3.2359
$K_{\sigma 2}$	1.5	0.3661	0.3447	0.3180	0.3014	0.2682	0.2593
	2.0	0.3683	0.3483	0.3234	0.3076	0.2756	0.2668
	2.5	0.3595	0.3414	0.3189	0.3048	0.2759	0.2680
	3.0	0.3460	0.3295	0.3093	0.2968	0.2713	0.2643
	4.0		0.3042	0.2871	0.2767	0.2565	0.2511

5. Conical shell connected to another conical shell.* To ensure accuracy, for each cone evaluate k_A and the value of k in that cone at the position where $\mu = 4$. The absolute values of k at all four positions should be greater than 5. $R/(t\cos\alpha)$ should also be greater than 5 at all these positions. E_1 and E_2 are the moduli of elasticity and ν_1 and ν_2 the Poisson's ratios for the left and right cones, respectively. $b_1 = R_1 - (t_1\cos\alpha_1)/2$, $a_1 = R_1 + (t_1\cos\alpha_1)/2$. $R_A = R_1$. Similar expressions are used for the right-hand cone. See Table 13.3, case 4, for formulas for k_A, β, μ, C_1, and the F functions for each of the two cones. Normally $R_2 = R_1$, but if $\alpha_1 + \alpha_2$ is close to zero, the midthickness radii may not be equal at the junction. Under this condition a different set of correction terms will be used if they are necessary. Note that rather than use an additional level of subscripting in the following equations, use has been made of subscripted parentheses or brackets to denote which cone the coefficients refer to.

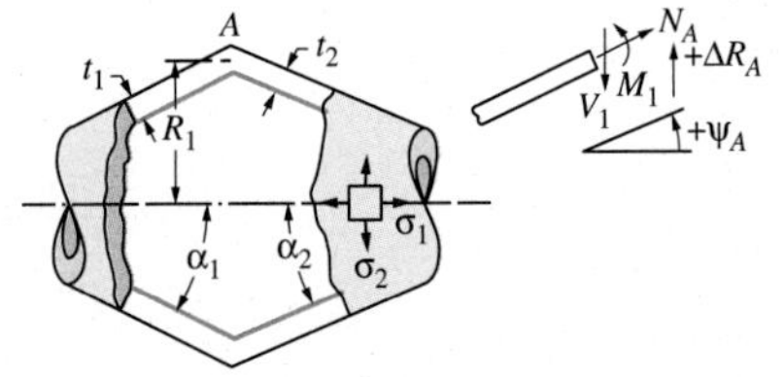

$$K_{V1} = \frac{LT_A C_{BB} - LT_B C_{AB}}{C_{AA}C_{BB} - C_{AB}^2}, \quad K_{M1} = \frac{LT_B C_{AA} - LT_A C_{AB}}{C_{AA}C_{BB} - C_{AB}^2}, \quad \left.\begin{matrix} LT_A = LT_{A1} + LT_{A2} + LT_{AC} \\ LT_B = LT_{B1} + LT_{B2} + LT_{BC} \end{matrix}\right\}$$ See cases 5a–5d for these load terms.

$$C_{AA} = C_{AA1} + C_{AA2}, \quad C_{AA1} = \frac{R_1\sin\alpha}{t_1\sqrt{2}}\left[\frac{k_A}{C_1}\left(F_{4A} - \frac{4\nu_1^2}{k_A^2}F_{2A}\right)\right], \quad C_{AA2} = \frac{R_1E_1\sin\alpha_2}{E_2t_2\sqrt{2}}\left[\frac{k_A}{C_1}\left(F_{4A} - \frac{4\nu_2^2}{k_A^2}F_{2A}\right)\right]_2$$

$$C_{AB} = C_{AB1} + C_{AB2}, \quad C_{AB1} = \frac{-R_1}{t_1}\left(\frac{\beta F_{7A}}{C_1}\right)_1, \quad C_{AB2} = \frac{R_1E_1t_1}{E_2t_2^2}\left(\frac{\beta F_{7A}}{C_1}\right)_2$$

$$C_{BB} = C_{BB1} + C_{BB2}, \quad C_{BB1} = \frac{R_1 2\sqrt{2}}{t_1\sin\alpha_1}\left(\frac{\beta^2F_{2A}}{k_AC_1}\right)_1, \quad C_{BB2} = \frac{R_1E_1t_1^2 2\sqrt{2}}{E_2t_2^3\sin\alpha_2}\left(\frac{\beta^2F_{2A}}{k_AC_1}\right)_2$$

The stresses in the left-hand cone at the junction are given by

$$\sigma_1 = \frac{N_1}{t_1}, \quad \sigma_2 = \frac{\Delta R_A E_1}{R_A} + \nu_1\sigma_1$$

$$\sigma_1' = \frac{-6M_1}{t_1^2}$$

$$\sigma_1' = \frac{-6V_1\sin\alpha_1}{t_1\beta_1}(1-\nu_1^2)\left(\frac{F_{7A}}{C_1}\right)_1 + \sigma_1'\left[\nu_1 + \frac{2\sqrt{2}}{k_AC_1}(1-\nu_1^2)F_{2A}\right]_1$$

**Note:* If either conical shell increases in radius away from the junction, substitute $-\alpha$ for α for that cone in all the appropriate formulas above and in those used from case 4, Table 13.3.

(*Continued*)

TABLE 13.4 Formulas for Discontinuity Stresses and Deformations at the Junctions of Shells and Plates (*Continued*)

Loading and Case No.	Load Terms	Selected Values

5a. Internal* pressure q

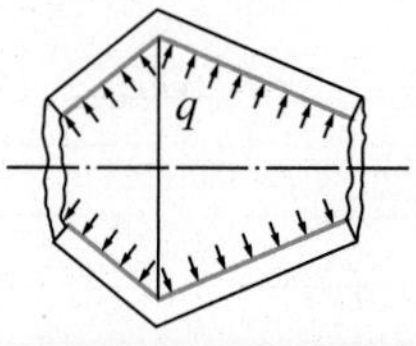

Note: There is no axial load on the junction. An axial load on the left end of the left cone balances any axial component of the pressure on the left cone, and an axial load on the right end of the right cone balances any axial component of the pressure on the cone and the joint. For an enclosed pressure vessel superpose an axial load $P = q\pi b_1^2$ using case 5b.

Load Terms

$$LT_{A1} = \frac{b_1 R_1}{t_1^2 \cos\alpha_1}$$

$$LT_{A2} = \frac{-b_2 R_2 E_1}{E_2 t_1 t_2 \cos\alpha_2}$$

$$LT_{AC} = \frac{t_2 \sin\alpha_2 + t_1 \sin\alpha_1}{2t_1} C_{AA2} + \frac{t_2^2 - t_1^2}{8t_1^2} C_{AB2} + \frac{E_1 R_1 \nu_2 (t_1 \cos\alpha_1 - t_2 \cos\alpha_2)}{2E_2 t_1 t_2 \cos\alpha_2} \dagger$$

$$LT_{B1} = \frac{-2b_1 \tan\alpha_1}{t_1 \cos\alpha_1}$$

$$LT_{B2} = \frac{-2E_1 b_2 \tan\alpha_2}{E_2 t_2 \cos\alpha_2}$$

$$LT_{BC} = \frac{t_2 \sin\alpha_2 + t_1 \sin\alpha_1}{2t_1} C_{AB2} + \frac{t_2^2 - t_1^2}{8t_1^2} C_{BB2} + \frac{E_1 \tan\alpha_2 (t_1 \cos\alpha_1 - t_2 \cos\alpha_2)}{E_2 t_2 \cos\alpha_2} \dagger$$

At the junction of the two cones,

$$V_1 = q t_1 K_{V1}, \quad M_1 = q t_1^2 K_{M1}, \quad N_1 = -V_1 \sin\alpha_1$$

$$\Delta R_A = \frac{q t_1}{E_1}(LT_{A1} - K_{V1} C_{AA1} - K_{M1} C_{AB1})$$

$$\psi_A = \frac{q}{E_1}(-LT_{B1} + K_{V1} C_{AB1} + K_{M1} C_{BB1})$$

Selected Values

For internal pressure, $E_1 = E_2$, $\nu_1 = \nu_2 = 0.3$, $t_1 = t_2$, $R_1 = R_2$, and for $R/t \cos\alpha > 5$. (*Note:* No correction terms are used.)

$$\Delta R_A = \frac{q R_1^2}{E_1 t_1} K_{\Delta RA}, \quad \psi_A = \frac{q R_1}{E_1 t_1} K_{\psi A}, \quad \sigma_2 = \frac{q R_1}{t_1} K_{\sigma 2}$$

		α_1					
		−30			15		
		R_1/t_1			R_1/t_1		
	α_2	10	20	50	10	20	50
K_{V1}	-30.0	0.0000	0.0000	0.0000	-0.0764	-0.1081	-0.1711
	-15.0	0.0746	0.1065	0.1696	0.0000	0.0000	0.0000
	15.0	0.0764	0.1081	0.1711	0.0000	0.0000	0.0000
	30.0	0.0000	0.0000	0.0000	-0.0800	-0.1114	-0.1742
	45.0	-0.1847	-0.2583	-0.4056	-0.2708	-0.3746	-0.5832
K_{M1}	-30.0	0.4408	0.6374	1.0216	0.1333	0.1924	0.3079
	-15.0	0.3200	0.4634	0.7433	0.0000	0.0000	0.0000
	15.0	0.1333	0.1924	0.3079	-0.1940	-0.2809	-0.4502
	30.0	0.0000	0.0000	0.0000	-0.3238	-0.4672	-0.7472
	45.0	-0.2441	-0.3482	-0.5529	-0.5603	-0.8025	-1.2770
$K_{\Delta RA}$	-30.0	1.2502	1.2351	1.2123	1.0889	1.1022	1.1062
	-15.0	1.1519	1.1479	1.1356	0.9853	1.0103	1.0253
	15.0	1.0889	1.1022	1.1062	0.9215	0.9640	0.9956
	30.0	1.1047	1.1297	1.1447	0.9414	0.9954	1.0377
	45.0	1.1689	1.2139	1.2477	1.0131	1.0868	1.1476
$K_{\psi A}$	-30.0	0.0000	0.0000	0.0000	1.1418	1.2683	1.4897
	-15.0	-0.5956	-0.7081	-0.9212	0.5280	0.5414	0.5494
	15.0	-1.1418	-1.2683	-1.4897	0.0000	0.0000	0.0000
	30.0	-1.2756	-1.3045	-1.3218	-0.0904	0.0078	0.2123
	45.0	-1.3724	-1.1627	-0.6949	-0.1067	0.2286	0.9146
$K_{\sigma 2}$	-30.0	1.2502	1.2351	1.2123	1.0894	1.1026	1.1065
	-15.0	1.1530	1.1487	1.1361	0.9853	1.0103	1.0253
	15.0	1.0900	1.1030	1.1067	0.9215	0.9640	0.9956
	30.0	1.1047	1.1297	1.1447	0.9421	0.9958	1.0380
	45.0	1.1661	1.2120	1.2465	1.0152	1.0882	1.1485

*For external pressure, substitute $-q$ for q, a_1 for b_1, b_2 for a_2, and a_2 for b_2 in the load terms.

†If $\alpha_1 + \alpha_2$ is zero or close to zero, the following correction terms should be used:

$$LT_{AC} = \frac{b_1^2 - b_2^2}{4t_1^2 \cos^2\alpha_2}\left(\frac{a_2 - b_1}{R_1} C_{AB2} - \frac{2\nu_2 E_1 t_1 \cos\alpha_2}{E_2 t_2}\right),$$

$$LT_{BC} = \frac{b_1^2 - b_2^2}{4t_1^2 \cos^2\alpha_2}\left(\frac{a_2 - b_1}{R_1} C_{BB2} - \frac{2E_1 t_1^2 \sin\alpha_2}{E_2 R_2 t_2}\right)$$

(*Continued*)

TABLE 13.4 Formulas for Discontinuity Stresses and Deformations at the Junctions of Shells and Plates (*Continued*)

Loading and Case No.	Load Terms	Selected Values
5b. Axial load P	$LT_{A1}=\frac{-\nu_1 R_1^2}{2t_1^2\cos\alpha_1}$	For axial tension, $E_1=E_2$, $\nu_1=\nu_2=0.3$, $t_1=t_2$, $R_1=R_2$, and for $R/t\cos\alpha>5$.

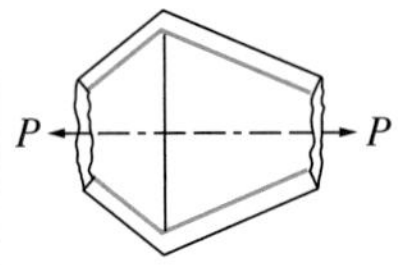

$$LT_{A1}=\frac{-\nu_1 R_1^2}{2t_1^2\cos\alpha_1}$$

$$LT_{A2}=\frac{\nu_2 R_1^2 E_1}{2E_2 t_1 t_2\cos\alpha_2}+\frac{R_1 C_{AA2}}{2t_1}(\tan\alpha_1+\tan\alpha_2)$$

$$LT_{AC}=0^*$$

$$LT_{B1}=\frac{R_1\tan\alpha_1}{2t_1\cos\alpha_1}$$

$$LT_{B2}=\frac{E_1 R_1^2\tan\alpha_2}{2E_2 R_2 t_2\cos\alpha_2}+\frac{R_1 C_{AB2}}{2t_1}(\tan\alpha_1+\tan\alpha_2)$$

$$LT_{BC}=0^*$$

At the junction of the two cones,

$$V_1=\frac{Pt_1 K_{V1}}{\pi R_1^2},\qquad M_1=\frac{Pt_1^2 K_{M1}}{\pi R_1^2}$$

$$N_1=\frac{P}{2\pi R_1\cos\alpha_1}-V_1\sin\alpha_1$$

$$\Delta R_A=\frac{Pt_1}{E_1\pi R_1^2}(LT_{A1}-K_{V1}C_{AA1}-K_{M1}C_{AB1})$$

$$\psi_A=\frac{P}{E_1\pi R_1^2}(-LT_{B1}+K_{V1}C_{AB1}+K_{M1}C_{BB1})$$

*If $\alpha_1+\alpha_2$ is zero or close to zero, the following correction terms should be used:

$$LT_{AC}=\frac{-R_1(R_2-R_1)C_{AB2}}{2t_1^2\cos^2\alpha_2},\qquad LT_{BC}=\frac{-R_1(R_2-R_1)C_{BB2}}{2t_1^2\cos^2\alpha_2}$$

Selected Values

For axial tension, $E_1=E_2$, $\nu_1=\nu_2=0.3$, $t_1=t_2$, $R_1=R_2$, and for $R/t\cos\alpha>5$.

$$\Delta R_A=\frac{P\nu_1}{2\pi E_1 t_1}K_{\Delta RA},\qquad \psi_A=\frac{P\nu_1}{2\pi E_1 t_1^2}K_{\psi A},\qquad \sigma_2=\frac{P}{2\pi R_1 t_1}K_{\sigma 2}$$

		α_1					
		−30			15		
		R_1/t_1			R_1/t_1		
	α_2	10	20	50	10	20	50
K_{V1}	−30.0	−2.8868	−5.7735	−14.4338	−0.7618	−1.5181	−3.7830
	−15.0	−2.1629	−4.3306	−10.8367	0.0000	0.0000	0.0000
	15.0	−0.7852	−1.5760	−3.9521	1.3397	2.6795	6.6987
	30.0	0.0000	0.0000	0.0000	2.0473	4.0996	10.2597
	45.0	0.9808	1.9747	4.9659	2.8992	5.8125	14.5630
K_{M1}	−30.0	−3.4070	−9.4911	−37.1735	−0.9420	−2.6184	−10.2418
	−15.0	−2.5628	−7.1386	−27.9573	0.0000	0.0000	0.0000
	15.0	−0.9420	−2.6184	−10.2418	1.6690	4.6507	18.2184
	30.0	0.0000	0.0000	0.0000	2.5628	7.1386	27.9573
	45.0	1.2012	3.3225	12.9563	3.6553	10.1672	39.7823
$K_{\Delta RA}$	−30.0	6.4309	9.5114	15.5525	0.8751	1.7262	3.3919
	−15.0	4.5065	6.8241	11.3678	−1.0353	−1.0353	−1.0353
	15.0	0.8751	1.7262	3.3919	−4.3681	−5.8778	−8.8384
	30.0	−1.1547	−1.1547	−1.1547	−6.1013	−8.4188	−12.9625
	45.0	−3.6360	−4.7187	−6.8292	−8.1516	−11.4538	−17.9219
$K_{\psi A}$	−30.0	0.0000	0.0000	0.0000	−0.1324	−0.0619	−0.0213
	−15.0	0.0362	0.0144	0.0028	−0.0925	−0.0462	−0.0185
	15.0	0.1324	0.0619	0.0213	0.0000	0.0000	0.0000
	30.0	0.2222	0.1111	0.0444	0.0713	0.0394	0.0187
	45.0	0.3819	0.2012	0.0885	0.1870	0.1045	0.0503
$K_{\sigma 2}$	−30.0	2.1891	3.1132	4.9256	0.5849	0.8402	1.3399
	−15.0	1.6335	2.3287	3.6917	0.0000	0.0000	0.0000
	15.0	0.5854	0.8406	1.3403	−1.0207	−1.4735	−2.3617
	30.0	0.0000	0.0000	0.0000	−1.5516	−2.2469	−3.6100
	45.0	−0.7150	−1.0396	−1.6725	−2.1799	−3.1707	−5.1112

(Continued)

TABLE 13.4 Formulas for Discontinuity Stresses and Deformations at the Junctions of Shells and Plates (*Continued*)

Loading and Case No.

5c. Hydrostatic internal* pressure q_1 at the junction if $|\mu| > 4$† at the position of zero pressure. If $|\mu| < 4$ at this position, the discontinuity in pressure gradient introduces deformations at the junction.

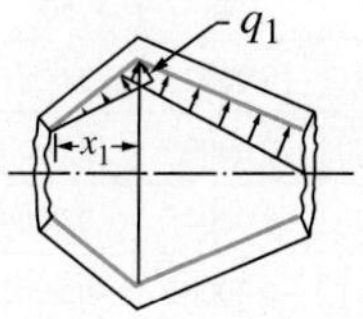

Note: There is no axial load on the junction. An axial load on the left end of the left cone balances any axial component of the pressure on the left cone, and an axial load on the right end of the right cone balances the axial component of the pressure on the right cone and on the joint.

Load Terms

$$LT_{A1} = \frac{b_1 R_1}{t_1^2 \cos\alpha_1}$$

$$LT_{A2} = \frac{-b_2 R_2 E_1}{E_2 t_1 t_2 \cos\alpha_2}$$

For LT_{AC} use the expressions from case 5a.

$$LT_{B1} = \frac{-b_1}{t_1 \cos\alpha_1}\left(\frac{R_1}{x_1} + 2 \tan\alpha_1\right)$$

$$LT_{B2} = \frac{E_1 b_2}{E_2 t_2 \cos\alpha_2}\left(\frac{R_2}{x_1} - 2 \tan\alpha_2\right)$$

For LT_{BC} use the expressions from case 5a.

At the junction of the two cones,

$$V_1 = q_1 t_1 K_{V1}, \quad M_1 = q_1 t_1^2 K_{M1}, \quad N_1 = -V_1 \sin\alpha_1$$

$$\Delta R_A = \frac{q_1 t_1}{E_1}(LT_{A1} - K_{V1}C_{AA1} - K_{M1}C_{AB1})$$

$$\psi_A = \frac{q_1}{E_1}(-LT_{B1} + K_{V1}C_{AB1} + K_{M1}C_{BB1})$$

Selected Values

For internal pressure, $x_1 = R_1$, $E_1 = E_2$, $\nu_1 = \nu_2 = 0.3$, $t_1 = t_2$, $R_1 = R_2$, and for $R/t \cos\alpha > 5$. (*Note:* No correction terms are used.)

$$\Delta R_A = \frac{q_1 R_1^2}{E_1 t_1} K_{\Delta RA}, \quad \psi_A = \frac{q_1 R_1}{E_1 t_1} K_{\psi A}, \quad \sigma_2 = \frac{q_1 R_1}{t_1} K_{\sigma 2}$$

		α_1					
		−30			15		
		R_1/t_1			R_1/t_1		
	α_2	10	20	50	10	20	50
K_{V1}	−30.0	0.0000	0.0000	0.0000	−0.0764	−0.1081	−0.1712
	−15.0	0.0746	0.1065	0.1696	0.0000	0.0000	0.0000
	15.0	0.0763	0.1081	0.1711	0.0000	0.0000	0.0000
	30.0	0.0000	0.0000	0.0000	−0.0799	−0.1114	−0.1742
	45.0	−0.1844	−0.2582	−0.4055	−0.2704	−0.3744	−0.5831
K_{M1}	−30.0	0.4408	0.6374	1.0216	0.1546	0.2224	0.3554
	−15.0	0.2988	0.4334	0.6958	0.0000	0.0000	0.0000
	15.0	0.1120	0.1623	0.2604	−0.1940	−0.2809	−0.4502
	30.0	0.0000	0.0000	0.0000	−0.3024	−0.4370	−0.6995
	45.0	−0.2010	−0.2875	−0.4572	−0.4955	−0.7113	−1.1335
$K_{\Delta RA}$	−30.0	1.2502	1.2351	1.2123	1.0959	1.1071	1.1093
	−15.0	1.1449	1.1429	1.1324	0.9853	1.0103	1.0253
	15.0	1.0818	1.0972	1.1031	0.9215	0.9640	0.9956
	30.0	1.1047	1.1297	1.1447	0.9484	1.0003	1.0408
	45.0	1.1829	1.2239	1.2540	1.0341	1.1016	1.1570
$K_{\psi A}$	−30.0	1.1047	1.1297	1.1447	2.1850	2.3365	2.5730
	−15.0	0.4478	0.3602	0.1621	1.5133	1.5517	1.5747
	15.0	−0.0986	−0.2000	−0.4064	0.9853	1.0103	1.0253
	30.0	−0.1709	−0.1748	−0.1771	0.9530	1.0762	1.2957
	45.0	−0.1440	0.0905	0.5732	1.0535	1.4138	2.1147
$K_{\sigma 2}$	−30.0	1.2502	1.2351	1.2123	1.0965	1.1075	1.1096
	−15.0	1.1460	1.1437	1.1329	0.9853	1.0103	1.0253
	15.0	1.0830	1.0980	1.1036	0.9215	0.9640	0.9956
	30.0	1.1047	1.1297	1.1447	0.9491	1.0008	1.0411
	45.0	1.1801	1.2219	1.2528	1.0362	1.1031	1.1579

*For external pressure, substitute $-q_1$ for q_1, a_1 for b_1, b_2 for a_2, and a_2 for b_2 in the load terms.

†If pressure increases right to left, substitute $-x_1$ for x_1 and verify that $|x_1|$ is large enough to extend into the right cone as far as the position where $|\mu| = 4$.

(*Continued*)

TABLE 13.4 Formulas for Discontinuity Stresses and Deformations at the Junctions of Shells and Plates (*Continued*)

Loading and Case No.	Load Terms	Selected Values

5d. Rotation around the axis of symmetry at ω rad/s

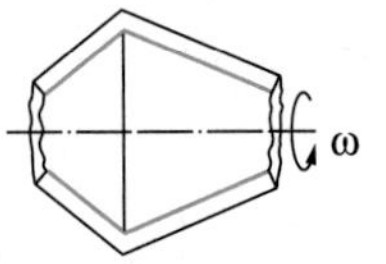

Note: δ = mass/unit volume

Load Terms

$$LT_{A1} = \frac{R_1^2}{t_1^2}$$

$$LT_{A2} = \frac{-\delta_2 R_2^3 E_1}{\delta_1 R_1 E_2 t_1^2}$$

$$LT_{AC} = 0$$

$$LT_{B1} = \frac{-R_1(3+\nu_1)\tan\alpha_1}{t_1}$$

$$LT_{B2} = \frac{-\delta_2 R_2^2 E_1(3+\nu_2)\tan\alpha_2}{\delta_1 E_2 R_1 t_1}$$

$$LT_{BC} = 0$$

At the junction of the two cones,

$$V_1 = \delta_1\omega^2 R_1 t_1^2 K_{V1}, \qquad N_1 = V_1 \sin\alpha_1$$

$$M_1 = \delta_1\omega^2 R_1 t_1^3 K_{M1}$$

$$\Delta R_A = \frac{\delta_1\omega^2 R_1 t_1^2}{E_1}(LT_{A1} - K_{V1}C_{AA1} - K_{M1}C_{AB1})$$

$$\psi_A = \frac{\delta_1\omega^2 R_1 t_1}{E_1}(-LT_{B1} + K_{V1}C_{AB1} + K_{M1}C_{BB1})$$

Selected Values

For $E_1 = E_2$, $\nu_1 = \nu_2 = 0.3$, $t_1 = t_2$, $R_1 = R_2$, $\delta_1 = \delta_2$, and for $R/t \cos\alpha > 5$.

$$\Delta R_A = \frac{\delta_1\omega^2 R_1^3}{E_1}K_{\Delta RA}, \qquad \psi_A = \frac{\delta_1\omega^2 R_1^2}{E_1}K_{\psi A}, \qquad \sigma_2 = \delta_1\omega^2 R_1^2 K_{\sigma 2}$$

		α_1					
		−30			15		
		R_1/t_1			R_1/t_1		
	α_2	10	20	50	10	20	50
K_{V1}	−30.0	0.0000	0.0000	0.0000	−0.0001	−0.0002	−0.0001
	−15.0	−0.0003	−0.0001	0.0000	0.0000	0.0000	0.0000
	15.0	0.0001	0.0002	0.0001	0.0000	0.0000	0.0000
	30.0	0.0000	0.0000	0.0000	−0.0011	−0.0006	−0.0003
	45.0	−0.0016	−0.0010	−0.0005	−0.0043	−0.0023	−0.0011
K_{M1}	−30.0	0.6583	0.9309	1.4726	0.1818	0.2569	0.4061
	−15.0	0.4951	0.7004	1.1079	0.0000	0.0000	0.0000
	15.0	0.1818	0.2569	0.4061	−0.3249	−0.4588	−0.7245
	30.0	0.0000	0.0000	0.0000	−0.5001	−0.7053	−1.1129
	45.0	−0.2315	−0.3260	−0.5141	−0.7149	−1.0061	−1.5851
$K_{\Delta RA}$	−30.0	1.2173	1.1539	1.0975	1.0599	1.0424	1.0268
	−15.0	1.1636	1.1158	1.0733	1.0000	1.0000	1.0000
	15.0	1.0599	1.0424	1.0268	0.8932	0.9245	0.9522
	30.0	1.0000	1.0000	1.0000	0.8364	0.8842	0.9267
	45.0	0.9248	0.9466	0.9662	0.7683	0.8356	0.8958
$K_{\psi A}$	−30.0	0.0000	0.0000	0.0000	1.3796	1.3799	1.3803
	−15.0	−0.4714	−0.4717	−0.4719	0.8842	0.8842	0.8842
	15.0	−1.3796	−1.3799	−1.3803	0.0000	0.0000	0.0000
	30.0	−1.9053	−1.9053	−1.9053	−0.4735	−0.4732	−0.4730
	45.0	−2.5700	−2.5693	−2.5686	−1.0477	−1.0473	−1.0466
$K_{\sigma 2}$	−30.0	1.2173	1.1539	1.0975	1.0599	1.0424	1.0268
	−15.0	1.1636	1.1158	1.0733	1.0000	1.0000	1.0000
	15.0	1.0599	1.0424	1.0268	0.8932	0.9245	0.9522
	30.0	1.0000	1.0000	1.0000	0.8364	0.8842	0.9267
	45.0	0.9248	0.9466	0.9662	0.7683	0.8356	0.8958

(*Continued*)

TABLE 13.4 Formulas for Discontinuity Stresses and Deformations at the Junctions of Shells and Plates (*Continued*)

6. Conical shell connected to a spherical shell.* To ensure accuracy, evaluate k_A and the value of k in the cone at the position where $\mu = 4$. The absolute values of k at both positions should be greater than 5. $R/(t_1 \cos \alpha_1)$ should also be greater than 5 at both these positions. The junction angle for the spherical shell must lie within the range $3/\beta_2 < \phi_2 < \pi - 3/\beta_2$. The spherical shell must also extend with no interruptions such as a second junction or a cutout, such that $\theta_2 > 3/\beta_2$. See the discussion on page 495. E_1 and E_2 are the moduli of elasticity and ν_1 and ν_2 the Poisson's ratios for the cone and the sphere, respectively. $b_1 = R_1 - (t_1 \cos\alpha_1)/2$, $a_1 = R_1 + (t_1 \cos\alpha_1)/2$, and $R_A = R_1$. See Table 13.3, case 4, for formulas for k_A, β_1, μ, C_1, and the F functions for the cone. See Table 13.3, case 1, for the formulas for K_{12},†, K_{22},† and β_2 for the spherical shell. $b_2 = R_2 - t_2/2$ and $a_2 = R_2 + t_2/2$. Normally $R_2 \sin\phi_2 = R_1$, but if $\phi_2 = \alpha_1 = 90°$ or is close to 90°, the midthickness radii at the junction may not be equal. Under this condition different correction terms will be used if necessary.

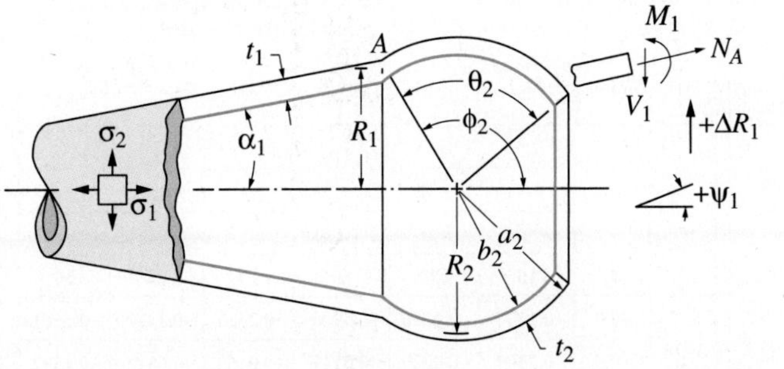

$$K_{V1} = \frac{LT_A C_{BB} - LT_B C_{AB}}{C_{AA} C_{BB} - C_{AB}^2}, \quad K_{M1} = \frac{LT_B C_{AA} - LT_A C_{AB}}{C_{AA} C_{BB} - C_{AB}^2}, \quad \left.\begin{matrix} LT_A = LT_{A1} + LT_{A2} + LT_{AC} \\ LT_B = LT_{B1} + LT_{B2} + LT_{BC} \end{matrix}\right\}$$ See cases 6a–6d for these load terms.

$$C_{AA} = C_{AA1} + C_{AA2}, \quad C_{AA1} = \frac{R_1 \sin\alpha_1}{t_1\sqrt{2}} \left[\frac{k_A}{C_1} \left(F_{4A} - \frac{4\nu_1^2}{k_A^2} F_{2A} \right) \right], \quad C_{AA2} = \frac{R_A E_1 \beta_2 \sin\phi_2}{E_2 t_2} \left(\frac{1}{K_{12}} + K_{22} \right)$$

$$C_{AB} = C_{AB1} + C_{AB2}, \quad C_{AB1} = \frac{-R_1}{t_1} \frac{\beta_1 F_{7A}}{C_1}, \quad C_{AB2} = \frac{2E_1 R_A t_1 \beta_2^2}{E_2 R_2 t_2 K_{12}}$$

$$C_{BB} = C_{BB1} + C_{BB2}, \quad C_{BB1} = \frac{R_1 2\sqrt{2}}{t_1 \sin\alpha_1} \frac{\beta_1^2 F_{2A}}{k_A C_1}, \quad C_{BB2} = \frac{4E_1 R_A t_1^2 \beta_2^3}{R_2^2 E_2 t_2 K_{12} \sin\phi_2}$$

The stresses in the left-hand cone at the junction are given by

$$\sigma_1 = \frac{N_1}{t_1}, \quad \sigma_2 = \frac{\Delta R_A E_1}{R_A} + \nu_1 \sigma_1$$

$$\sigma_1' = \frac{-6M_1}{t_1^2}, \quad \sigma_2' = \frac{-6V_1 \sin\alpha_1}{t_1 \beta_1} (1 - \nu_1^2) \frac{F_{7A}}{C_1} + \sigma_1' \left[\nu_1 + \frac{2\sqrt{2}}{k_A C_1} (1 - \nu_1^2) F_{2A} \right]$$

Note: The use of joint load correction terms LT_{AC} and LT_{BC} depends upon the accuracy desired and the relative values of the thicknesses and the radii. Read Sec. 13.4 carefully. For thin-walled shells, $R/t > 10$, they can be neglected.

*If the conical shell increases in radius away from the junction, substitute $-\alpha$ for α for the cone in all of the appropriate formulas above and in those used from case 4, Table 13.3.

†The second subscript is added to refer to the right-hand shell.

(*Continued*)

TABLE 13.4 Formulas for Discontinuity Stresses and Deformations at the Junctions of Shells and Plates (*Continued*)

Loading and Case No.	Load Terms	Selected Values

6a. Internal* pressure q

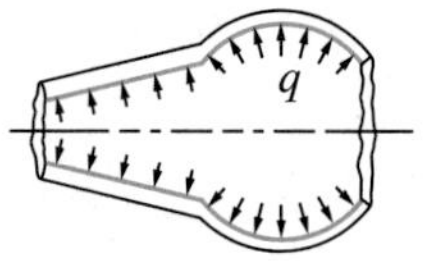

Note: There is no axial load on the junction. An axial load on the left end of the cone balances any axial component of the pressure on the cone, and an axial load on the right end of the sphere balances any axial component of the pressure on the sphere and on the joint. For an enclosed pressure vessel superpose an axial load $P = q\pi b_1^2$ using case 6b.

Load Terms

$$LT_{A1} = \frac{b_1 R_1}{t_1^2 \cos\alpha_1}$$

$$LT_{A2} = \frac{-b_2^2 E_1 \sin\phi_2}{E_2 t_1 t_2}$$

$$LT_{AC} = \frac{t_2 \cos\phi_2 + t_1 \sin\alpha_1}{2t_1} C_{AA2} + \frac{t_2^2 - t_1^2}{8t_1^2} C_{AB2} + \frac{E_1 R_1 (1+\nu_2)(t_1 \cos\alpha_1 - t_2 \sin\phi_2)}{2E_2 t_1 t_2 \sin\phi_2} + \dagger$$

$$LT_{B1} = \frac{-2b_1 \tan\alpha_1}{t_1 \cos\alpha_1}$$

$$LT_{B2} = 0$$

$$LT_{BC} = \frac{t_2 \cos\phi_2 + t_1 \sin\alpha_1}{2t_1} C_{AB2} + \frac{t_2^2 - t_1^2}{8t_1^2} C_{BB2} + \frac{E_1 \cos\phi_2 (t_1 \cos\alpha_1 - t_2 \sin\phi_2)}{E_2 t_2 \sin^2\phi_2} + \dagger$$

At the junction of the cone and sphere,

$$V_1 = q t_1 K_{V1}, \quad M_1 = q t_1^2 K_{M1}, \quad N_1 = -V_1 \sin\alpha_1$$

$$\Delta R_A = \frac{q t_1}{E_1}(LT_{A1} - K_{V1} C_{AA1} - K_{M1} C_{AB1})$$

$$\psi_A = \frac{q}{E_1}(-LT_{B1} + K_{V1} C_{AB1} + K_{M1} C_{BB1})$$

Selected Values

For internal pressure, $E_1 = E_2$, $\nu_1 = \nu_2 = 0.3$, $t_1 = t_2$, $R_2 \sin\phi_2 = R_1$, and for $R/t \cos\alpha_1 > 5$ and $R_2/t_2 > 5$. (*Note:* No correction terms are used.)

$$\Delta R_A = \frac{q R_1^2}{E_1 t_1} K_{\Delta RA}, \qquad \psi_A = \frac{q R_1}{E_1 t_1} K_{\psi A}, \qquad \sigma_2 = \frac{q R_1}{t_1} K_{\sigma 2}$$

		α_1					
		−30			15		
		R_1/t_1			R_1/t_1		
	ϕ_2	10	20	50	10	20	50
K_{V1}	45.0	−0.1504	−0.2338	−0.3900	−0.2313	−0.3470	−0.5661
	60.0	0.0303	0.0220	0.0141	−0.0468	−0.0877	−0.1592
	90.0	0.1266	0.1594	0.2327	0.0520	0.0528	0.0630
	105.0	0.1050	0.1287	0.1839	0.0296	0.0215	0.0138
	120.0	0.0316	0.0232	0.0150	−0.0460	−0.0861	−0.1570
K_{M1}	45.0	0.2084	0.3034	0.4889	−0.0811	−0.1182	−0.1908
	60.0	0.2172	0.3158	0.5081	−0.0920	−0.1335	−0.2146
	90.0	0.2285	0.3306	0.5301	−0.0978	−0.1416	−0.2270
	105.0	0.2288	0.3303	0.5285	−0.0973	−0.1408	−0.2255
	120.0	0.2240	0.3224	0.5146	−0.0960	−0.1388	−0.2221
$K_{\Delta RA}$	45.0	1.2913	1.3082	1.3113	1.1396	1.1841	1.2132
	60.0	1.1526	1.1698	1.1735	0.9917	1.0371	1.0675
	90.0	1.0809	1.0969	1.0997	0.9124	0.9576	0.9880
	105.0	1.0979	1.1137	1.1164	0.9301	0.9751	1.0055
	120.0	1.1538	1.1703	1.1736	0.9898	1.0354	1.0662
$K_{\psi A}$	45.0	−0.1762	0.0897	0.6014	1.0674	1.4565	2.1836
	60.0	−0.7470	−0.7309	−0.7111	0.4315	0.5731	0.8128
	90.0	−1.0321	−1.1548	−1.4058	0.0909	0.0946	0.0646
	105.0	−0.9601	−1.0540	−1.2466	0.1658	0.1993	0.2288
	120.0	−0.7317	−0.7212	−0.7058	0.4181	0.5574	0.7961
$K_{\sigma 2}$	45.0	1.2891	1.3064	1.3101	1.1414	1.1854	1.2140
	60.0	1.1531	1.1700	1.1735	0.9920	1.0375	1.0678
	90.0	1.0828	1.0981	1.1004	0.9120	0.9574	0.9879
	105.0	1.0995	1.1146	1.1169	0.9299	0.9750	1.0055
	120.0	1.1543	1.1705	1.1736	0.9901	1.0357	1.0665

*For external pressure substitude $-q$ for q, a_1 for b_1, b_2 for a_2, and a_2 for b_2 in the load terms.

†If $\phi_2 - \alpha_1 = 90°$ or is close to 90°, the following correction terms should be used:

$$LT_{AC} = \frac{b_1^2 - b_2^2 \sin^2\phi_2}{4t_1^2 \sin^2\phi_2}\left[\frac{a_2 \sin\phi_2 - b_1}{R_1} C_{AB2} - \frac{2E_1 t_1 (1+\nu_2)\sin\phi_2}{E_2 t_2}\right],$$

$$LT_{BC} = \frac{b_1^2 - b_2^2 \sin^2\phi_2}{4t_1^2 \sin^2\phi_2} \frac{a_2 \sin\phi_2 - b_1}{R_1} C_{BB2}$$

(*Continued*)

TABLE 13.4 Formulas for Discontinuity Stresses and Deformations at the Junctions of Shells and Plates (*Continued*)

Loading and Case No.	Load Terms	Selected Values
6b. Axial load P	see below	see below

6b. Axial load P

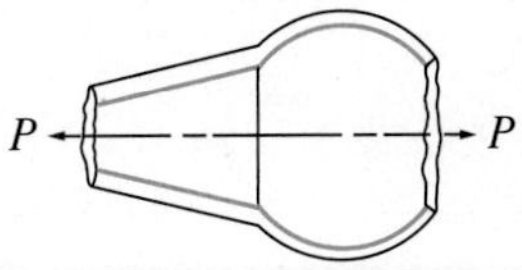

Load Terms

$$LT_{A1} = \frac{-\nu_1 R_1^2}{2t_1^2 \cos\alpha_1}$$

$$LT_{A2} = \frac{R_1^2 E_1(1+\nu_2)}{2E_2 t_1 t_2 \sin\phi_2} + \frac{R_1 C_{AA2}}{2t_1}\left(\tan\alpha_1 + \frac{1}{\tan\phi_2}\right)$$

$$LT_{AC} = 0$$

$$LT_{B1} = \frac{R_1 \tan\alpha_1}{2t_1 \cos\alpha_1}$$

$$LT_{B2} = \frac{R_1 C_{AB2}}{2t_1}\left(\tan\alpha_1 + \frac{1}{\tan\phi_2}\right)$$

$$LT_{BC} = 0$$

At the junction of the cone and sphere,

$$V_1 = \frac{Pt_1 K_{V1}}{\pi R_1^2}, \quad M_1 = \frac{Pt_1^2 K_{M1}}{\pi R_1^2}$$

$$N_1 = \frac{P}{2\pi R_1 \cos\alpha_1} - V_1 \sin\alpha_1$$

$$\Delta R_A = \frac{Pt_1}{E_1 \pi R_1^2}(LT_{A1} - K_{V1}C_{AA1} - K_{M1}C_{AB1})$$

$$\psi_A = \frac{P}{E_1 \pi R_1^2}(-LT_{B1} + K_{V1}C_{AB1} + K_{M1}C_{BB1})$$

Selected Values

For axial tension, $E_1 = E_2$, $\nu_1 = \nu_2 = 0.3$, $t_1 = t_2$, $R_2 \sin\phi_2 = R_1$, and $R/t \cos\alpha_1 > 5$ and $R_2/t_2 > 5$.

$$\Delta R_A = \frac{P\nu_1}{2\pi E_1 t_1} K_{\Delta RA}, \quad \psi_A = \frac{P\nu_1}{2\pi E_1 t_1^2} K_{\psi A}, \quad \sigma_2 = \frac{P}{2\pi R_1 t_1} K_{\sigma_2}$$

		α_1					
		−30			15		
		R_1/t_1			R_1/t_1		
	ϕ_2	10	20	50	10	20	50
K_{V1}	45.0	1.4637	2.6550	6.0381	3.3529	6.4470	15.5559
	60.0	0.3826	0.5404	0.8537	2.4117	4.6104	11.0611
	90.0	−1.1708	−2.5296	−6.7444	0.9917	1.7969	4.0811
	105.0	−1.8475	−3.8810	−10.1209	0.3241	0.4582	0.7244
	120.0	−2.5422	−5.2810	−13.6476	−0.4001	−1.0054	−2.9706
K_{M1}	45.0	1.0818	3.1542	12.6906	3.5439	10.0102	39.5349
	60.0	−0.0630	−0.0887	−0.1397	2.5015	7.0525	27.8219
	90.0	−1.7682	−4.9205	−19.2610	0.8411	2.3448	9.1872
	105.0	−2.5375	−7.1021	−27.8984	0.0271	0.0384	0.0608
	120.0	−3.3516	−9.4119	−37.0470	−0.8798	−2.5305	−10.1027
$K_{\Delta RA}$	45.0	−6.4214	−7.3936	−9.4058	−10.7636	−13.9818	−20.3748
	60.0	−3.2923	−3.2299	−3.1745	−8.1367	−10.4098	−14.9142
	90.0	1.0715	2.6784	5.8176	−4.3666	−5.1327	−6.6300
	105.0	2.9144	5.2194	9.7519	−2.6671	−2.6945	−2.7189
	120.0	4.7524	7.7965	13.8050	−0.8758	−0.0808	1.5350
$K_{\psi A}$	45.0	−0.9108	−0.6635	−0.4304	−1.0097	−0.6930	−0.4266
	60.0	−0.7412	−0.5450	−0.3563	−0.8384	−0.5768	−0.3558
	90.0	−0.6135	−0.4575	−0.3026	−0.7396	−0.5126	−0.3183
	105.0	−0.6090	−0.4562	−0.3028	−0.7536	−0.5247	−0.3272
	120.0	−0.6513	−0.4888	−0.3250	−0.8121	−0.5683	−0.3560
$K_{\sigma 2}$	45.0	−1.5361	−1.8318	−2.4391	−2.9706	−3.9340	−5.8502
	60.0	−0.6298	−0.6144	−0.6008	−2.1679	−2.8482	−4.1980
	90.0	0.6327	1.1120	2.0512	−1.0148	−1.2432	−1.6911
	105.0	1.1653	1.8540	3.2113	−0.4946	−0.5013	−0.5073
	120.0	1.6959	2.6062	4.4060	0.0540	0.2942	0.7803

(*Continued*)

TABLE 13.4 Formulas for Discontinuity Stresses and Deformations at the Junctions of Shells and Plates (*Continued*)

Loading and Case No.

6c. Hydrostatic internal* pressure q_1 at the junction when $|\mu| > 4$† at the position of zero pressure. If $|\mu| < 4$ at this position, the discontinuity in pressure gradient introduces small deformations at the junction.

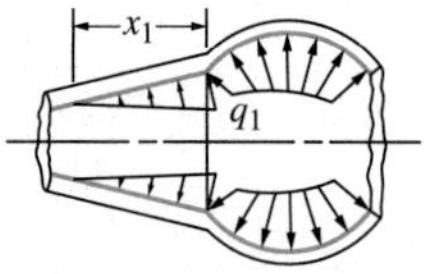

Note: There is no axial load on the junction. An axial load on the left end of the cone balances any axial component of the pressure on the cone, and an axial load on the right end of the sphere balances any axial component of the pressure on the sphere and on the joint.

Load Terms

$$LT_{A1} = \frac{b_1 R_1}{t_1^2 \cos\alpha_1}$$

$$LT_{A2} = \frac{-b_2^2 E_1 \sin\phi_2}{E_2 t_1 t_2}$$

For LT_{AC} use the expressions from case 6a.

$$LT_{B1} = \frac{-b_1}{t_1 \cos\alpha_1}\left(\frac{R_1}{x_1} + 2\tan\alpha_1\right)$$

$$LT_{B2} = \frac{E_1 b_2 R_2 \sin\phi_2}{E_2 t_2 x_1}$$

For LT_{BC} use the expressions from case 6a.

At the junction of the cone and sphere,

$$V_1 = q_1 t_1 K_{V1}, \quad M_1 = q_1 t_1^2 K_{M1}$$

$$N_1 = -V_1 \sin\alpha_1$$

$$\Delta R_A = \frac{q_1 t_1}{E_1}(LT_{A1} - K_{V1}C_{AA1} - K_{M1}C_{AB1})$$

$$\psi_A = \frac{q_1}{E_1}(-LT_{B1} + K_{V1}C_{AB1} + K_{M1}C_{BB1})$$

Selected Values

For internal pressure, $x_1 = R_1$, $E_1 = E_2$, $\nu_1 = \nu_2 = 0.3$, $t_1 = t_2$, $R_2 \sin\phi_2 = R_1$, and for $R/t\cos\alpha_1 > 5$ and $R_2/t_2 > 5$. (*Note:* No correction terms are used.)

$$\Delta R_A = \frac{q_1 R_1^2}{E_1 t_1} K_{\Delta RA}, \quad \psi_A = \frac{q_1 R_1}{E_1 t_1} K_{\psi A}, \quad \sigma_2 = \frac{q_1 R_1}{t_1} K_{\sigma_2}$$

		α_1					
		-30			15		
		R_1/t_1			R_1/t_1		
	ϕ_2	10	20	50	10	20	50
K_{V1}	45.0	-0.1508	-0.2341	-0.3902	-0.2320	-0.3475	-0.5664
	60.0	0.0303	0.0220	0.0141	-0.0469	-0.0878	-0.1593
	90.0	0.1266	0.1594	0.2327	0.0520	0.0528	0.0630
	105.0	0.1050	0.1287	0.1839	0.0296	0.0215	0.0138
	120.0	0.0316	0.0232	0.0150	-0.0459	-0.0860	-0.1569
K_{M1}	45.0	0.2501	0.3628	0.5833	-0.0185	-0.0292	-0.0492
	60.0	0.2172	0.3158	0.5081	-0.0710	-0.1037	-0.1673
	90.0	0.2008	0.2914	0.4681	-0.1043	-0.1508	-0.2416
	105.0	0.2074	0.3002	0.4809	-0.0973	-0.1408	-0.2255
	120.0	0.2240	0.3224	0.5146	-0.0744	-0.1084	-0.1743
$K_{\Delta RA}$	45.0	1.3054	1.3182	1.3176	1.1607	1.1990	1.2226
	60.0	1.1526	1.1698	1.1735	0.9987	1.0421	1.0707
	90.0	1.0718	1.0904	1.0956	0.9103	0.9560	0.9870
	105.0	1.0909	1.1087	1.1133	0.9301	0.9751	1.0055
	120.0	1.1538	1.1703	1.1736	0.9968	1.0403	1.0694
$K_{\psi A}$	45.0	1.0506	1.3418	1.8688	2.2254	2.6400	3.3827
	60.0	0.3577	0.3988	0.4336	1.4744	1.6411	1.8960
	90.0	-0.0076	-0.1053	-0.3413	1.0584	1.0871	1.0721
	105.0	0.0830	0.0141	-0.1634	1.1511	1.2096	1.2541
	120.0	0.3730	0.4085	0.4389	1.4618	1.6261	1.8797
$K_{\sigma 2}$	45.0	1.3031	1.3164	1.3164	1.1625	1.2004	1.2235
	60.0	1.1531	1.1700	1.1735	0.9990	1.0425	1.0709
	90.0	1.0737	1.0916	1.0963	0.9099	0.9558	0.9869
	105.0	1.0925	1.1097	1.1138	0.9299	0.9750	1.0055
	120.0	1.1543	1.1705	1.1736	0.9971	1.0407	1.0696

*For external pressure, substitute $-q_1$ for q_1, a_1 for b_1, b_2 for a_2, and a_2 for b_2 in load terms.

†If pressure increases right to left, substitute $-x_1$ for x_1 and verify that $|x_1|$ is large enough to extend into the sphere as far as the position where $\theta_2 = 3/\beta_2$.

(*Continued*)

TABLE 13.4 Formulas for Discontinuity Stresses and Deformations at the Junctions of Shells and Plates (*Continued*)

Loading and Case No.	Load Terms	Selected Values
6d. Rotation around the axis of symmetry at ω rad/s 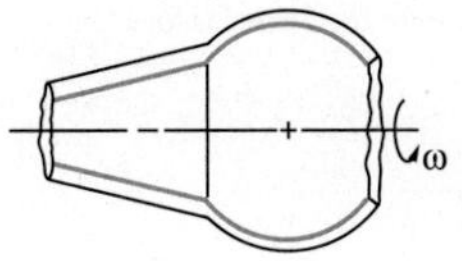 *Note*: δ = mass/unit volume	$LT_{A1} = \frac{R_1^2}{t_1^2}$ $LT_{A2} = \frac{-\delta_2 R_2^3 E_1 \sin^3\phi_2}{\delta_1 R_1 E_2 t_1^2}$ $LT_{AC} = 0$ $LT_{B1} = \frac{-R_1(3+\nu_1)\tan\alpha_1}{t_1}$ $LT_{B2} = \frac{-\delta_2 R_2^2 E_1(3+\nu_2)\sin\phi_2\cos\phi_2}{\delta_1 R_1 t_1 E_2}$ $LT_{BC} = 0$ At the junction of the cone and sphere, $V_1 = \delta_1\omega^2 R_1 t_1^2 K_{V1}, \quad M_1 = \delta_1\omega^2 R_1 t_1^3 K_{M1}$ $N_1 = -V_1 \sin\alpha_1$ $\Delta R_A = \frac{\delta_1\omega^2 R_1 t_1^2}{E_1}(LT_{A1} - K_{V1}C_{AA1} - K_{M1}C_{AB1})$ $\psi_A = \frac{\delta_1\omega^2 R_1 t_1}{E_1}(-LT_{B1} + K_{V1}C_{AB1} + K_{M1}C_{BB1})$	For $E_1 = E_2$, $\nu_1 = \nu_2 = 0.3$, $t_1 = t_2$, $\delta_1 = \delta_2$, $R_2 \sin\phi_2 = R_1$, and for $R/t\cos\alpha_1 > 5$ and $R_2/t_2 > 5$. $\Delta R_A = \frac{\delta_1\omega^2 R_1^3}{E_1}K_{\Delta RA}, \quad \psi_A = \frac{\delta_1\omega^2 R_1^2}{E_1}K_{\psi A}, \quad \sigma_2 = \delta_1\omega^2 R_1^2 K_{\sigma 2}$

		α_1					
		−30			15		
		R_1/t_1			R_1/t_1		
	ϕ_2	10	20	50	10	20	50
K_{V1}	45.0	0.0023	0.0015	0.0009	0.0078	0.0053	0.0033
	60.0	0.0000	0.0000	0.0000	0.0035	0.0024	0.0015
	90.0	−0.0001	0.0000	0.0001	0.0002	0.0001	0.0001
	105.0	0.0012	0.0010	0.0007	0.0000	0.0000	0.0000
	120.0	0.0034	0.0027	0.0019	0.0008	0.0006	0.0004
K_{M1}	45.0	−0.2240	−0.3189	−0.5073	−0.6913	−0.9836	−1.5636
	60.0	0.0000	0.0000	0.0000	−0.4906	−0.6961	−1.1039
	90.0	0.3414	0.4828	0.7636	−0.1635	−0.2310	−0.3651
	105.0	0.4988	0.7041	1.1117	0.0000	0.0000	0.0000
	120.0	0.6678	0.9408	1.4829	0.1845	0.2597	0.4090
$K_{\Delta RA}$	45.0	0.9243	0.9464	0.9661	0.7667	0.8350	0.8957
	60.0	1.0000	1.0000	1.0000	0.8360	0.8840	0.9267
	90.0	1.1127	1.0798	1.0505	0.9461	0.9619	0.9759
	105.0	1.1637	1.1159	1.0733	1.0000	1.0000	1.0000
	120.0	1.2177	1.1541	1.0975	1.0600	1.0424	1.0268
$K_{\psi A}$	45.0	−2.5613	−2.5631	−2.5646	−1.0229	−1.0292	−1.0349
	60.0	−1.9053	−1.9053	−1.9053	−0.4625	−0.4653	−0.4679
	90.0	−0.9169	−0.9173	−0.9177	0.4386	0.4385	0.4384
	105.0	−0.4656	−0.4676	−0.4694	0.8842	0.8842	0.8842
	120.0	0.0161	0.0113	0.0071	1.3838	1.3829	1.3821
$K_{\sigma 2}$	45.0	0.9243	0.9464	0.9661	0.7666	0.8350	0.8957
	60.0	1.0000	1.0000	1.0000	0.8360	0.8840	0.9267
	90.0	1.1127	1.0798	1.0505	0.9461	0.9619	0.9759
	105.0	1.1637	1.1159	1.0733	1.0000	1.0000	1.0000
	120.0	1.2177	1.1541	1.0975	1.0600	1.0424	1.0268

(*Continued*)

TABLE 13.4 Formulas for Discontinuity Stresses and Deformations at the Junctions of Shells and Plates (*Continued*)

7. Conical shell connected to a circular plate.* To ensure accuracy, evaluate k_A and the value of k in the cone at the position where $\mu = 4$. The absolute values of k at both positions should be greater than 5. $R/(t_1 \cos\alpha_1)$ should also be greater than 5 at both these positions. E_1 and E_2 are the moduli of elasticity and ν_1 and ν_2 the Poisson's ratios for the cone and the plate, respectively. $b_1 = R_1 - (t_1 \cos\alpha_1)/2$, $a_1 = R_1 + (t_1 \cos\alpha_1)/2$, and $R_A = R_1$. See Table 13.3, case 4, for formulas k_A, β_1, μ, C_1, and the F functions for the cone. See Table 11.2 for the formulas for D_2.

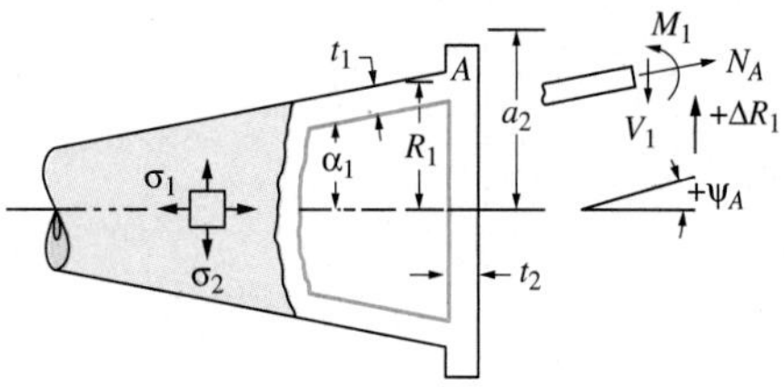

$$K_{P1} = 1 + \frac{R_1^2(1-\nu_2)}{a_2^2(1+\nu_2)}$$

$$K_{V1} = \frac{LT_A C_{BB} - LT_B C_{AB}}{C_{AA}C_{BB} - C_{AB}^2}, \qquad K_{M1} = \frac{LT_B C_{AA} - LT_A C_{AB}}{C_{AA}C_{BB} - C_{AB}^2}, \qquad \left.\begin{aligned} LT_A &= LT_{A1} + LT_{A2} + LT_{AC} \\ LT_B &= LT_{B1} + LT_{B2} + LT_{BC} \end{aligned}\right\} \text{ See cases 7a–7d for these load terms.}$$

$$C_{AA} = C_{AA1} + C_{AA2}, \qquad C_{AA1} = \frac{R_1 \sin\alpha_1}{t_1\sqrt{2}}\left[\frac{k_A}{C_1}\left(F_{4A} - \frac{4\nu_1^2}{k_A^2}F_{2A}\right)\right], \qquad C_{AA2} = \frac{E_1 t_2^2 R_1 K_{P1}}{6D_2}$$

$$C_{AB} = C_{AB1} + C_{AB2}, \qquad C_{AB1} = \frac{-R_1}{t_1}\left(\frac{\beta F_{7A}}{C_1}\right), \qquad C_{AB2} = \frac{E_1 t_1 t_2 R_1 K_{P1}}{4D_2}$$

$$C_{BB} = C_{BB1} + C_{BB2}, \qquad C_{BB1} = \frac{R_1 2\sqrt{2}}{t_1 \sin\alpha_1}\left(\frac{\beta^2 F_{2A}}{k_A C_1}\right)_1, \qquad C_{BB2} = \frac{E_1 t_1^2 R_1 K_{P1}}{2D_2}$$

The stresses in the left-hand cone at the junction are given by

$$\sigma_1 = \frac{N_1}{t_1}, \qquad \sigma_2 = \frac{\Delta R_A E_1}{R_A} + \nu_1\sigma_1$$

$$\sigma_1' = \frac{-6M_1}{t_1^2}, \qquad \sigma_2' = \frac{-6V_1 \sin\alpha_1}{t_1\beta_1}(1-\nu_1^2)\left(\frac{F_{7A}}{C_1}\right)_1 + \sigma_1'\left[\nu_1 + \frac{2\sqrt{2}}{k_A C_1}(1-\nu_1^2)F_{2A}\right]_1$$

Note: The use of joint load correction terms LT_{AC} and LT_{BC} depends upon the accuracy desired and the relative values of the thicknesses and the radii. Read Sec. 13.4 carefully. For thin-walled shells, $R/t > 10$, they can be neglected.

**Note:* If the conical shell increases in radius away from the junction, substitute $-\alpha$ for α for the cone in all of the appropriate formulas above and in those used from case 4, Table 13.3.

(*Continued*)

TABLE 13.4 Formulas for Discontinuity Stresses and Deformations at the Junctions of Shells and Plates (*Continued*)

Loading and Case No.

7a. Internal* pressure q

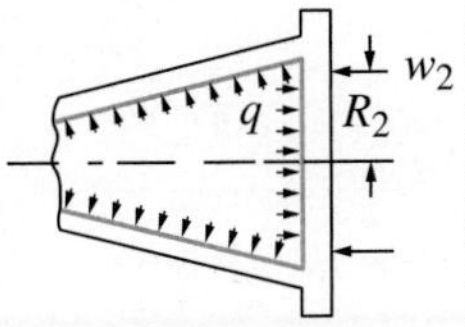

Note: There is no axial load on the junction. An axial load on the left end of the cone balances any axial component of the pressure on the cone, and an axial load on the plate is reacted by the annular line load $w_2 = qb_1^2/(2R_2)$ at a radius R_2. For an enclosed pressure vessel superpose an axial load $P = q\pi b_1^2$ using case 7b.

Load Terms

$$LT_{A1} = \frac{b_1 R_1}{t_1^2 \cos\alpha_1}, \qquad LT_{A2} = \frac{E_1 t_2 b_1^2}{32 D_2 t_1 R_1} K_{P2}$$

where

$$K_{P2} = \begin{cases} (2R_2^2 - b_1^2)K_{P1} & \text{for } R_2 \le R_1 \\ (2R_2^2 - b_1^2)K_{P1} - 2(R_2^2 - R_1^2) + 4R_1^2 \ln\dfrac{R_2}{R_1} & \text{for } R_2 \ge R_1 \end{cases}$$

$$LT_{AC} = \frac{E_1 b_1 t_2 \sin\alpha_1}{12 D_2}\left(t_2 - \frac{3t_1 \sin\alpha_1}{8}\right) K_{P1}$$

$$LT_{B1} = \frac{-2b_1 \tan\alpha_1}{t_1 \cos\alpha_1}$$

$$LT_{B2} = \frac{E_1 b_1^2}{16 D_2 R_1} K_{P2}$$

$$LT_{BC} = \frac{E_1 b_1 t_1 \sin\alpha_1}{8 D_2}\left(t_2 - \frac{t_1 \sin\alpha_1}{2}\right) K_{P1}$$

At the junction of the cone and plate,

$$V_1 = qt_1 K_{V1}, \qquad M_1 = qt_1^2 K_{M1}$$

$$N_1 = -V_1 \sin\alpha_1$$

$$\Delta R_A = \frac{qt_1}{E_1}(LT_{A1} - K_{V1}C_{AA1} - K_{M1}C_{AB1})$$

$$\psi_A = \frac{q}{E_1}(-LT_{B1} + K_{V1}C_{AB1} + K_{M1}C_{BB1})$$

Selected Values

For internal pressure, $E_1 = E_2$, $\nu_1 = \nu_2 = 0.3$, $a_2 = a_1$, $R_2 = 0.7R_1$, and for $R/t\cos\alpha_1 > 5$ and $R_2/t_2 > 4$. (*Note:* No correction terms are used.)

$$\Delta R_A = \frac{qR_1^2}{E_1 t_1} K_{\Delta RA}, \qquad \psi_A = \frac{qR_1}{E_1 t_1} K_{\psi A}, \qquad \sigma_2 = \frac{qR_1}{t_1} K_{\sigma 2}$$

		α_1					
		−30			30		
		R_1/t_1			R_1/t_1		
	t_2/t_1	15	30	50	15	30	50
K_{V1}	1.5	1.8080	2.5258	2.9931	1.8323	2.5642	3.0351
	2.0	2.0633	2.8778	3.4621	1.9988	2.8319	3.4256
	2.5	2.3269	3.2629	3.9945	2.1871	3.1396	3.8816
	3.0	2.5567	3.6105	4.4891	2.3608	3.4267	4.3139
	4.0		4.1282	5.2397		3.8687	4.9843
K_{M1}	1.5	1.3216	1.7736	0.9500	0.7803	1.1393	0.2492
	2.0	2.0621	3.2566	3.5429	1.2435	2.2323	2.3554
	2.5	2.7563	4.7394	6.2463	1.7200	3.3817	4.6146
	3.0	3.3251	6.0046	8.6287	2.1337	4.3961	6.6473
	4.0		7.7864	12.0641		5.8744	9.6442
$K_{\Delta RA}$	1.5	0.2625	0.2066	0.1777	0.1994	0.1699	0.1533
	2.0	0.2632	0.2130	0.1880	0.2011	0.1754	0.1622
	2.5	0.2458	0.2047	0.1837	0.1927	0.1701	0.1594
	3.0	0.2276	0.1891	0.1712	0.1793	0.1587	0.1496
	4.0		0.1546	0.1403		0.1321	0.1241
$K_{\psi A}$	1.5	−4.1463	−6.7069	−9.9970	−2.8696	−5.2149	−8.3411
	2.0	−3.2400	−5.3937	−8.1941	−2.3306	−4.2811	−6.9166
	2.5	−2.4705	−4.1900	−6.4582	−1.8320	−3.3853	−5.5109
	3.0	−1.8858	−3.2264	−5.0124	−1.4327	−2.6460	−4.3184
	4.0		−1.9618	−3.0515		−1.6459	−2.6691
$K_{\sigma 2}$	1.5	0.2806	0.2192	0.1867	0.1811	0.1570	0.1442
	2.0	0.2838	0.2274	0.1984	0.1811	0.1613	0.1519
	2.5	0.2717	0.2211	0.1957	0.1708	0.1544	0.1477
	3.0	0.2532	0.2072	0.1847	0.1557	0.1416	0.1366
	4.0		0.1753	0.1560		0.1127	0.1092

*For external pressure, substitute $-q$ for q and a_1 for b_1 in the load terms.

(*Continued*)

TABLE 13.4 Formulas for Discontinuity Stresses and Deformations at the Junctions of Shells and Plates (*Continued*)

Loading and Case No.	Load Terms
7b. Axial load P	$LT_{A1}=\dfrac{-\nu_1 R_1^2}{2t_1^2\cos\alpha_1}$, $LT_{AC}=0$

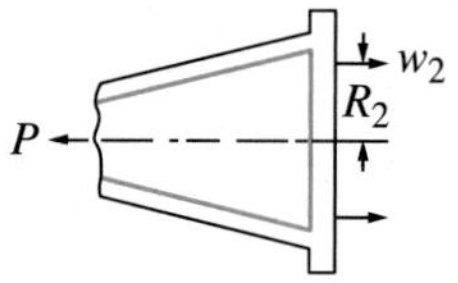

$$w_2=\frac{P}{2\pi R_2}$$

$$LT_{A2}=\frac{E_1 t_2 R_1^3}{16 D_2 t_1}K_{P2}+\frac{R_1 C_{AA2}}{2t_1}\tan\alpha_1$$

where

$$K_{P2}=\begin{cases}\left(1-\dfrac{R_2^2}{R_1^2}\right)K_{P1} & \text{for } R_2\le R_1\\[2ex] -\dfrac{1-\nu_2}{1+\nu_2}\dfrac{R_2^2-R_1^2}{a_2^2}-2\ln\dfrac{R_2}{R_1} & \text{for } R_2\ge R_1\end{cases}$$

$$LT_{B1}=\frac{R_1\tan\alpha_1}{2t_1\cos\alpha_1},\qquad LT_{BC}=0$$

$$LT_{B2}=\frac{E_1 R_1^3}{8D_2}K_{P2}+\frac{R_1 C_{AB2}}{2t_1}\tan\alpha_1$$

At the junction of the cone and plate,

$$V_1=\frac{Pt_1K_{V1}}{\pi R_1^2},\qquad M_1=\frac{Pt_1^2K_{M1}}{\pi R_1^2}$$

$$N_1=\frac{P}{2\pi R_1}\cos\alpha_1-V_1\sin\alpha_1$$

$$\Delta R_A=\frac{Pt_1}{E_1\pi R_1^2}(LT_{A1}-K_{V1}C_{AA1}-K_{M1}C_{AB1})$$

$$\psi_A=\frac{P}{E_1\pi R_1^2}(-LT_{B1}+K_{V1}C_{AB1}+K_{M1}C_{BB1})$$

Selected Values

For axial tension, $E_1=E_2$, $\nu_1=\nu_2=0.3$, $a_2=a_1$, $R_2=0.8R_1$, and for $R/t\cos\alpha_1>5$ and $R_1/t_2>4$.

$$\Delta R_A=\frac{P\nu_1}{2\pi E_1 t_1}K_{\Delta RA},\qquad \psi_A=\frac{P\nu_1}{2\pi E_1 t_1^2}K_{\psi A},\qquad \sigma_2=\frac{P}{2\pi R_1 t_1}K_{\sigma 2}$$

		α_1					
		−30			30		
		R_1/t_1			R_1/t_1		
	t_2/t_1	15	30	50	15	30	50
K_{V1}	1.5	3.5876	13.1267	31.6961	5.7802	16.6732	36.7478
	2.0	2.4577	10.0731	25.5337	4.6739	13.7167	30.7925
	2.5	1.5630	7.3736	19.6429	3.6712	10.8769	24.7596
	3.0	0.9203	5.2737	14.7796	2.8640	8.5120	19.5405
	4.0		2.6122	8.2677		5.2609	12.1583
K_{M1}	1.5	9.6696	48.9553	152.7700	13.8101	58.1862	169.2630
	2.0	6.3808	36.0554	118.6220	10.7213	46.0816	137.0810
	2.5	4.0214	25.6494	88.6891	8.1816	35.4643	107.1740
	3.0	2.4296	18.0038	65.2524	6.2589	27.1071	82.6250
	4.0		8.8324	35.4261		16.2264	49.6067
$K_{\Delta RA}$	1.5	−2.1662	−4.2130	−6.4572	−4.0595	−6.0789	−8.2689
	2.0	−2.2052	−4.6128	−7.3879	−4.1508	−6.5095	−9.1996
	2.5	−1.8764	−4.2328	−7.0820	−3.8563	−6.1865	−8.9613
	3.0	−1.4888	−3.6069	−6.2667	−3.4421	−5.5641	−8.1742
	4.0		−2.4320	−4.4893		−4.2647	−6.3131
$K_{\psi A}$	1.5	5.0355	8.6001	12.4137	5.8059	9.2022	12.8653
	2.0	3.2406	6.0555	9.2419	4.2033	6.8989	9.9584
	2.5	2.0765	4.1764	6.6773	3.0207	5.0584	7.4756
	3.0	1.3485	2.8815	4.7801	2.1953	3.7043	5.5547
	4.0		1.4330	2.5079		2.0669	3.1302
$K_{\sigma 2}$	1.5	−0.2317	−0.7862	−1.4006	−0.9870	−1.6440	−2.3547
	2.0	−0.2660	−0.9367	−1.7168	−0.9923	−1.7436	−2.5982
	2.5	−0.1853	−0.8497	−1.6603	−0.8839	−1.6183	−2.4905
	3.0	−0.0818	−0.6829	−1.4449	−0.7435	−1.4079	−2.2231
	4.0		−0.3571	−0.9508		−0.9856	−1.6205

(*Continued*)

TABLE 13.4 Formulas for Discontinuity Stresses and Deformations at the Junctions of Shells and Plates (*Continued*)

Loading and Case No.	Load Terms
7c. Hydrostatic internal* pressure q_1 at the junction when $\lvert\mu\rvert > 4$† at position of zero pressure. If $\lvert\mu\rvert < 4$ at this position, the discontinuity in pressure gradient introduces small deformations at the junction.	$LT_{A1} = \frac{b_1 R_1}{t_1^2 \cos\alpha_1}$ $LT_{A2} = \frac{E_1 t_2 b_1^2}{32 D_2 t_1 R_1}$ For K_{P2} use the expressions from case 7a. For LT_{AC} use the expressions from case 7a. $LT_{B1} = \frac{-b_1}{t_1 \cos\alpha_1}\left(\frac{R_1}{x_1} + 2\tan\alpha_1\right)$ $LT_{B2} = \frac{E_1 b_1^2}{16 D_2 R_1} K_{P2}$ For LT_{BC} use the expressions from case 7a.
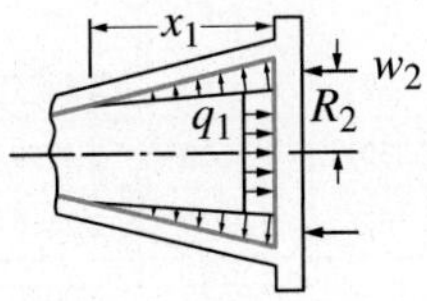 	At the junction of the cone and plate, $V_1 = q_1 t_1 K_{V1}, \quad M_1 = q_1 t_1^2 K_{M1}$ $N_1 = -V_1 \sin\alpha_1$ $\Delta R_A = \frac{q_1 t_1}{E_1}(LT_{A1} - K_{V1} C_{AA1} - K_{M1} C_{AB1})$ $\psi_A = \frac{q_1}{E_1}(-LT_{B1} + K_{V1} C_{AB1} + K_{M1} C_{BB1})$
Note: There is no axial load on the junction. An axial load on the left end of the cone balances any axial component of the pressure on the cone, and the axial load on the plate is reacted by the annular line load $w_2 = q_1 b_1^2/(2R_2)$ at a radius R_2.	

Selected Values

For internal pressure, $E_1 = E_2$, $\nu_1 = \nu_2 = 0.3$, $x_1 = R_1$, $a_2 = a_1$, $R_2 = 0.7R_1$, and for $R/t\cos\alpha_1 > 5$ and $R_1/t_2 > 4$. (*Note:* No correction terms are used.)

$$\Delta R_A = \frac{q_1 R_1^2}{E_1 t_1} K_{\Delta RA}, \qquad \psi_A = \frac{q_1 R_1}{E_1 t_1} K_{\psi A}, \qquad \sigma_2 = \frac{q_1 R_1}{t_1} K_{\sigma 2}$$

		α_1					
		−30			30		
		R_1/t_1			R_1/t_1		
	$\frac{t_2}{t_1}$	15	30	50	15	30	50
K_{V1}	1.5	1.7764	2.4985	2.9693	1.7997	2.5362	3.0107
	2.0	1.9832	2.8038	3.3944	1.9143	2.7550	3.3559
	2.5	2.2020	3.1423	3.8805	2.0538	3.0132	3.7634
	3.0	2.3954	3.4503	4.3338	2.1871	3.2577	4.1520
	4.0		3.9122	5.0243		3.6388	4.7582
K_{M1}	1.5	1.0786	1.4887	0.6370	0.5358	0.8534	−0.0645
	2.0	1.6780	2.7748	2.9876	0.8544	1.7465	1.7969
	2.5	2.2541	4.0782	5.4554	1.2072	2.7112	3.8157
	3.0	2.7328	5.1990	7.6390	1.5247	3.5749	5.6432
	4.0		6.7888	10.7995		4.8492	8.3528
$K_{\Delta RA}$	1.5	0.2290	0.1873	0.1652	0.1656	0.1505	0.1407
	2.0	0.2295	0.1929	0.1745	0.1667	0.1550	0.1486
	2.5	0.2173	0.1856	0.1706	0.1604	0.1505	0.1461
	3.0	0.1997	0.1718	0.1591	0.1501	0.1408	0.1372
	4.0		0.1410	0.1307		0.1179	0.1142
$K_{\psi A}$	1.5	−3.4953	−5.9545	−9.1782	−2.2131	−4.4588	−7.5197
	2.0	−2.7619	−4.8159	−7.5439	−1.8429	−3.6961	−6.2610
	2.5	−2.1234	−3.7579	−5.9594	−1.4738	−2.9444	−5.0049
	3.0	−1.6313	−2.9042	−4.6343	−1.1675	−2.3150	−3.9328
	4.0		−1.7760	−2.8303		−1.4529	−2.4417
$K_{\sigma 2}$	1.5	0.2467	0.1998	0.1741	0.1476	0.1379	0.1317
	2.0	0.2493	0.2069	0.1847	0.1475	0.1413	0.1385
	2.3	0.2393	0.2013	0.1822	0.1399	0.1355	0.1348
	3.0	0.2236	0.1890	0.1721	0.1282	0.1245	0.1248
	4.0		0.1606	0.1458		0.0997	0.0999

*For external pressure, substitute $-q_1$ for q_1 and a_1 for b_1 in the load terms.
†If pressure increases right to left, substitute $-x_1$ for x_1.

(*Continued*)

TABLE 13.4 Formulas for Discontinuity Stresses and Deformations at the Junctions of Shells and Plates (*Continued*)

Loading and Case No.	Load Terms	Selected Values
7d. Rotation around the axis of symmetry at ω rad/s		

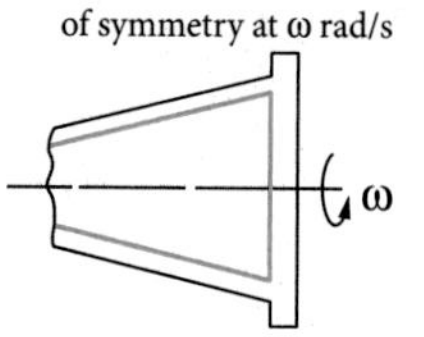

Note: d = mass/unit volume

Load Terms

$LT_{A1} = \frac{R_1^2}{t_1^2}$, $LT_{AC} = 0$

$LT_{A2} = \frac{-E_1\delta_2 t_2^3}{96D_2\delta_1 t_1^2}\left[\frac{a_2^3(3+\nu_2)}{1+\nu_2} - R_1^2\right]$

$LT_{B1} = \frac{-R_1(3+\nu_1)\tan\alpha_1}{t_1}$, $LT_{B2} = 0$

$LT_{BC} = 0$

At the junction of the cone and plate,

$V_1 = \delta_1\omega^2 R_1 t_1^2 K_{V1}$, $M_1 = \delta_1\omega^2 R_1 t_1^3 K_{M1}$

$N_1 = -V_1 \sin\alpha_1$

$\Delta R_A = \frac{\delta_1\omega^2 R_1 t_1^2}{E_1}(LT_{A1} - K_{V1}C_{AA1} - K_{M1}C_{AB1})$

$\psi_A = \frac{\delta_1\omega^2 R_1 t_1}{E_1}(-LT_{B1} + K_{V1}C_{AB1} + K_{M1}C_{BB1})$

Selected Values

For $E_1 = E_2$, $\nu_1 = \nu_2 = 0.3$, $\delta_1 = \delta_2$, $a_2 = a_1$, and for $R/t\cos\alpha_1 > 5$ and $R_1/t_2 > 4$.

$\Delta R_A = \frac{\delta_1\omega^2 R_1^3}{E_1}K_{\Delta RA}$, $\psi_A = \frac{\delta_1\omega^2 R_1^2}{E_1}K_{\psi A}$, $\sigma_2 = \delta_1\omega^2 R_1^2 K_{\sigma 2}$

		α_1					
		−30			30		
		R_1/t_1			R_1/t_1		
	$\frac{t_2}{t_1}$	15	30	50	15	30	50
K_{V1}	1.5	1.1636	1.7101	2.2561	1.1171	1.6863	2.2484
	2.0	1.4271	2.0268	2.6125	1.2301	1.8620	2.4739
	2.5	1.6871	2.3651	3.0170	1.3556	2.0639	2.7440
	3.0	1.9078	2.6661	3.3926	1.4704	2.2525	3.0036
	4.0		3.1083	3.9624		2.5427	3.4113
K_{M1}	1.5	0.7537	1.0646	1.3760	−0.0626	0.1247	0.3507
	2.0	1.5184	2.3995	3.3469	0.2520	0.8420	1.5667
	2.5	2.2034	3.7021	5.4006	0.5697	1.5965	2.9044
	3.0	2.7495	4.7976	7.2098	0.8428	2.2625	4.1251
	4.0		6.3200	9.8177		3.2332	5.9476
$K_{\Delta RA}$	1.5	0.4259	0.3543	0.3148	0.3206	0.2950	0.2765
	2.0	0.4268	0.3602	0.3226	0.3218	0.2986	0.2816
	2.5	0.4122	0.3529	0.3193	0.3162	0.2951	0.2799
	3.0	0.3922	0.3394	0.3099	0.3073	0.2876	0.2740
	4.0		0.3100	0.2864		0.2701	0.2585
$K_{\psi A}$	1.5	−3.9694	−5.7840	−7.5904	−1.8922	−3.4239	−5.0395
	2.0	−3.0329	−4.6016	−6.2200	−1.5261	−2.8111	−4.2173
	2.5	−2.2734	−3.5442	−4.9013	−1.1937	−2.2231	−3.3850
	3.0	−1.7119	−2.7097	−3.8033	−0.9300	−1.7377	−2.6689
	4.0		−1.6292	−2.3148		−1.0810	−1.6660
$K_{\sigma 2}$	1.5	0.4375	0.3629	0.3216	0.3094	0.2865	0.2697
	2.0	0.4410	0.3703	0.3304	0.3095	0.2893	0.2742
	2.5	0.4291	0.3647	0.3284	0.3026	0.2848	0.2717
	3.0	0.4113	0.3527	0.3200	0.2926	0.2764	0.2650
	4.0		0.3255	0.2983		0.2574	0.2483

(*Continued*)

TABLE 13.4 Formulas for Discontinuity Stresses and Deformations at the Junctions of Shells and Plates (*Continued*)

8. Spherical shell connected to another spherical shell. To ensure accuracy, $R/t > 5$ and the junction angles for each of the spherical shells must lie within the range $3/\beta < \phi < \pi - 3/\beta$. Each spherical shell must also extend with no interruptions such as a second junction or a cutout, such that $\theta > 3/\beta$. See the discussion on page 495. E_1 and E_2 are the moduli of elasticity and ν_1 and ν_2 the Poisson's ratios for the left and right spheres, respectively. $b_1 = R_1 - t_1/2$, $a_2 = R_1 + t_1/2$, and $R_A = R_1 \sin\phi_1$. See Table 13.3, case 1, for formulas K_{11},* K_{21},* and β_1 for the left-hand spherical shell. Similar expressions hold for b_2, a_2, β_2 and K_{12}* and K_{22}* for the right-hand sphere. Normally $R_2 \sin\phi_2 = R_A$, but if $\phi_1 + \phi_2 = 180°$ or close to 180°, the midthickness radii may not be equal at the junction. Under this condition a different set of correction terms will be used if necessary.

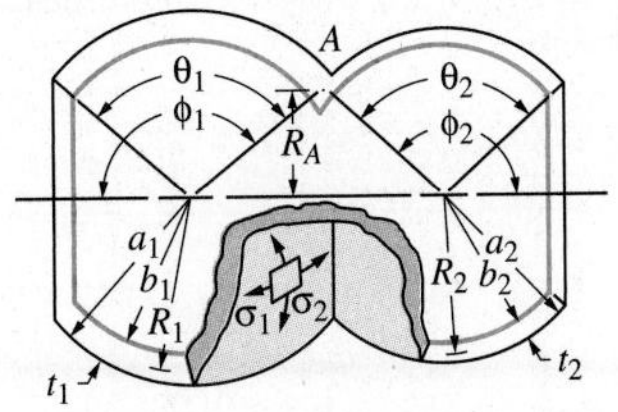

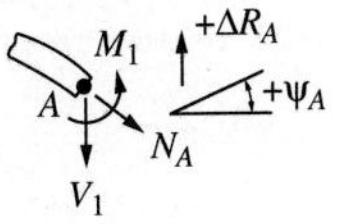

$$K_{V1} = \frac{LT_A C_{BB} - LT_B C_{AB}}{C_{AA}C_{BB} - C_{AB}^2}, \quad K_{M1} = \frac{LT_B C_{AA} - LT_A C_{AB}}{C_{AA}C_{BB} - C_{AB}^2},$$

$$\left.\begin{aligned} LT_A &= LT_{A1} + LT_{A2} + LT_{AC} \\ LT_B &= LT_{B1} + LT_{B2} + LT_{BC} \end{aligned}\right\}$$ See cases 8a–8d for these load terms.

$$C_{AA} = C_{AA1} + C_{AA2}, \quad C_{AA1} = \frac{R_1\beta_1 \sin^2\phi_1}{t_1}\left(\frac{1}{K_{11}} + K_{21}\right), \quad C_{AA2} = \frac{R_A E_1 \beta_2 \sin\phi_2}{E_2 t_2}\left(\frac{1}{K_{12}} + K_{22}\right)$$

$$C_{AB} = C_{AB1} + C_{AB2}, \quad C_{AB1} = \frac{-2\beta_1^2 \sin\phi_1}{K_{11}}, \quad C_{AB2} = \frac{2E_1 R_A t_1 \beta_2^2}{E_2 R_2 t_2 K_{12}}$$

$$C_{BB} = C_{BB1} + C_{BB2}, \quad C_{BB1} = \frac{4t_1\beta_1^3}{R_1 K_{11}}, \quad C_{BB2} = \frac{4E_1 R_A t_1^2 \beta_2^3}{R_2^2 E_2 t_2 K_{12} \sin\phi_2}$$

The stresses in the left sphere at the junction are given by

$$\sigma_1 = \frac{N_1}{t_1}$$

$$\sigma_2 = \frac{\Delta R_A E_1}{R_A} + \nu_1\sigma_1$$

$$\sigma_1' = \frac{-6M_1}{t_1^2}$$

$$\sigma_2' = \frac{V_1\beta_1^2\cos\phi_1}{K_{11}R_1} - \frac{6M_1}{t_1^2 K_{11}}\left(\nu_1 + \frac{1-\nu_1/2}{\beta_1\tan\phi_1}\right)$$

Note: The use of joint load correction terms LT_{AC} and LT_{BC} depends upon the accuracy desired and the relative values of the thicknesses and the radii. Read Sec. 13.4 carefully. For thin-walled shells, $R/t > 10$, they can be neglected.

*The second subscript refers to the left-hand (1) or right-hand (2) shell.

(*Continued*)

TABLE 13.4 Formulas for Discontinuity Stresses and Deformations at the Junctions of Shells and Plates (*Continued*)

Loading and Case No.

8a. Internal pressure* q

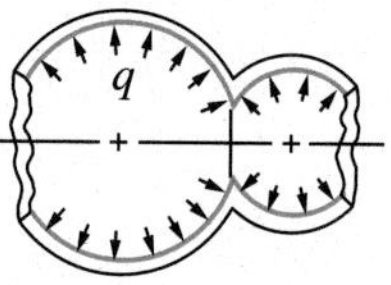

Note: There is no axial load on the junction. An axial load on the left end of the left sphere balances any axial component of the pressure on the left sphere, and an axial load on the right end of the right sphere balances any axial component of the pressure on the right sphere and on the joint. For an enclosed pressure vessel superpose an axial load $P = q\pi b_1^2 \sin^2\phi_1$ using case 8b.

Load Terms

$$LT_{A1} = \frac{b_1^2 \sin\phi_1}{t_1^2}$$

$$LT_{A2} = \frac{-b_2^2 E_1 \sin\phi_2}{E_2 t_1 t_2}$$

$$LT_{AC} = \frac{t_2\cos\phi_2 + t_1\cos\phi_1}{2t_1} C_{AA2} + \frac{t_2^2 - t_1^2}{8t_1^2} C_{AB2} + \frac{E_1 R_1 \nu_2 (t_1\sin\phi_1 - t_2\sin\phi_2)}{2E_2 t_1 t_2 \sin\phi_2} \dagger$$

$$LT_{B1} = 0, \quad LT_{B2} = 0$$

$$LT_{BC} = \frac{t_2\cos\phi_2 + t_1\cos\phi_1}{2t_1} C_{AB2} + \frac{t_2^2 - t_1^2}{8t_2^2} C_{BB2} + \frac{E_1 \cos\phi_2 (t_1\sin\phi_1 - t_2\sin\phi_2)}{E_2 t_2 \sin^2\phi_2} \dagger$$

At the junction of the two spheres,

$$V_1 = qt_1 K_{V1}, \quad M_1 = qt_1^2 K_{M1}$$

$$N_1 = -V_1\cos\phi_1$$

$$\Delta R_A = \frac{qt_1}{E_1}(LT_{A1} - K_{V1}C_{AA1} - K_{M1}C_{AB1})$$

$$\psi_A = \frac{q}{E_1}(-LT_{B1} + K_{V1}C_{AB1} + K_{M1}C_{BB1})$$

Selected Values

For internal pressure, $E_1 = E_2$, $\nu_1 = \nu_2 = 0.3$, $t_1 = t_2$, $R_1\sin\phi_1 = R_2\sin\phi_2$, and for $R/t > 5$. (*Note:* No correction terms are used.)

$$\Delta R_A = \frac{qR_1^2}{E_1 t_1} K_{\Delta RA}, \qquad \psi_A = \frac{qR_1}{E_1 t_1} K_{\psi A}, \qquad \sigma_2 = \frac{qR_1}{t_1} K_{\sigma 2}$$

		ϕ_1					
		75			120		
		R_1/t_1			R_1/t_1		
	ϕ_2	10	20	50	10	20	50
K_{V1}	45.0	−0.2610	−0.3653	−0.5718	−0.1689	−0.2386	−0.3765
	60.0	−0.0776	−0.1089	−0.1710	0.0000	0.0000	0.0000
	75.0	0.0000	0.0000	0.0000	0.0705	0.1003	0.1590
	90.0	0.0217	0.0306	0.0483	0.0896	0.1278	0.2032
	135.0	−0.2462	−0.3505	−0.5570	−0.1587	−0.2282	−0.3660
K_{M1}	45.0	0.0042	0.0059	0.0092	0.0063	0.0088	0.0139
	60.0	0.0006	0.0009	0.0014	0.0000	0.0000	0.0000
	75.0	0.0000	0.0000	0.0000	−0.0016	−0.0023	−0.0037
	90.0	0.0002	0.0002	0.0004	−0.0014	−0.0020	−0.0032
	135.0	−0.0074	−0.0105	−0.0168	−0.0012	−0.0018	−0.0029
$K_{\Delta RA}$	45.0	1.0672	1.1124	1.1395	0.8907	0.9305	0.9542
	60.0	0.9296	0.9760	1.0043	0.7816	0.8233	0.8488
	75.0	0.8717	0.9182	0.9467	0.7363	0.7784	0.8044
	90.0	0.8557	0.9021	0.9305	0.7242	0.7662	0.7921
	135.0	1.0524	1.1019	1.1329	0.8822	0.9244	0.9503
$K_{\psi A}$	45.0	0.8559	1.1883	1.8470	0.4864	0.6858	1.0805
	60.0	0.2526	0.3527	0.5506	0.0000	0.0000	0.0000
	75.0	0.0000	0.0000	0.0000	−0.2005	−0.2856	−0.4538
	90.0	−0.0696	−0.0982	−0.1546	−0.2530	−0.3621	−0.5780
	135.0	0.7762	1.1091	1.7683	0.4383	0.6369	1.0309
$K_{\sigma 2}$	45.0	1.1069	1.1530	1.1806	1.0260	1.0726	1.1007
	60.0	0.9630	1.0108	1.0400	0.9025	0.9506	0.9801
	75.0	0.9025	0.9506	0.9801	0.8512	0.8996	0.9293
	90.0	0.8857	0.9338	0.9632	0.8376	0.8857	0.9153
	135.0	1.0915	1.1421	1.1737	1.0162	1.0657	1.0963

*For external pressure, substitute $-q$ for q, a_1 for b_1, b_2 for a_2, and a_2 for b_2 in the load terms.

†If $\phi_1 + \phi_2 = 180°$ or is close to 180°, the following correction terms should be used:

$$LT_{AC} = \frac{b_1^2 - b_2^2}{4t_1^2}\left[\frac{a_2 - b_1}{R_1} C_{AB2} - \frac{2E_1 t_1}{E_2 t_2}(1+\nu_2)\sin\phi_2\right], \quad LT_{BC} = \frac{b_1^2 - b_2^2}{4t_1^2 R_1}(a_2 - b_1)\, C_{BB2}$$

(*Continued*)

TABLE 13.4 Formulas for Discontinuity Stresses and Deformations at the Junctions of Shells and Plates (*Continued*)

Loading and Case No.	Load Terms	Selected Values

8b. Axial load P

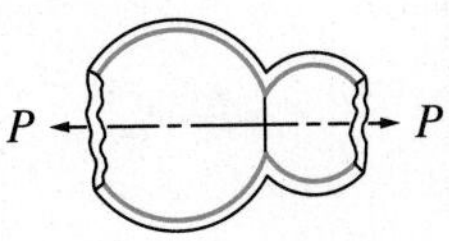

Load Terms

$$LT_{A1} = \frac{-R_1^2(1+\nu_1)}{2t_1^2 \sin\phi_1}$$

$$LT_{A2} = \frac{R_1^2 E_1(1+\nu_2)}{2E_2 t_1 t_2 \sin\phi_2} + \frac{R_1 C_{AA2}}{2t_1 \sin\phi_1}\left(\frac{1}{\tan\phi_1} + \frac{1}{\tan\phi_2}\right)$$

$$LT_{AC} = 0^*$$

$$LT_{B1} = 0$$

$$LT_{B2} = \frac{R_1 C_{AB2}}{2t_1 \sin\phi_1}\left(\frac{1}{\tan\phi_1} + \frac{1}{\tan\phi_2}\right)$$

$$LT_{BC} = 0^*$$

At the junction of the two spheres,

$$V_1 = \frac{Pt_1 K_{V1}}{\pi R_1^2}$$

$$M_1 = \frac{Pt_1^2 K_{M1}}{\pi R_1^2}$$

$$N_1 = \frac{P}{2\pi R_1 \sin^2\phi_1} - V_1\cos\phi_1$$

$$\Delta R_A = \frac{Pt_1}{E_1 \pi R_1^2}(LT_{A1} - K_{V1}C_{AA1} - K_{M1}C_{AB1})$$

$$\psi_A = \frac{P}{E_1 \pi R_1^2}(-LT_{B1} + K_{V1}C_{AB1} + K_{M1}C_{BB1})$$

*If $\phi_1 + \phi_2 = 180°$ or is close to 180°, the following correction terms should be used:

$$LT_{AC} = \frac{-R_1(R_2 - R_1)C_{AB2}}{2t_1^2 \sin^2\phi_2}, \quad LT_{BC} = \frac{-R_1(R_2 - R_1)C_{BB2}}{2t_1^2 \sin^2\phi_2}$$

Selected Values

For axial tension, $E_1 = E_2$, $\nu_1 = \nu_2 = 0.3$, $t_1 = t_2$, $R_1 \sin\phi_1 = R_2 \sin\phi_2$, and for $R/t > 5$.

$$\Delta R_A = \frac{P\nu_1}{2\pi E_1 t_1}K_{\Delta RA}, \qquad \psi_A = \frac{P\nu_1}{2\pi E_1 t_1^2}K_{\psi A}, \qquad \sigma_2 = \frac{P}{2\pi R_1 t_1}K_{\sigma 2}$$

		ϕ_1					
		75			120		
		R_1/t_1			R_1/t_1		
	ϕ_2	10	20	50	10	20	50
K_{V1}	45.0	3.1103	6.1700	15.3153	1.2155	2.3973	5.9187
	60.0	2.1514	4.2888	10.6917	0.0000	0.0000	0.0000
	75.0	1.3870	2.7740	6.9350	−0.9414	−1.8690	−4.6414
	90.0	0.6936	1.3907	3.4844	−1.7581	−3.4981	−8.7042
	135.0	−1.6776	−3.3955	−8.5790	−4.2542	−8.5470	−21.4561
K_{M1}	45.0	3.5944	10.1684	40.1986	1.2318	3.4870	13.7907
	60.0	2.5283	7.1514	28.2692	0.0000	0.0000	0.0000
	75.0	1.6480	4.6612	18.4250	−0.9769	−2.7638	−10.9267
	90.0	0.8311	2.3507	9.2922	−1.8388	−5.2015	−20.5626
	135.0	−2.0739	−5.8688	−23.2056	−4.6045	−13.0241	−51.4841
$K_{\Delta RA}$	45.0	−12.1536	−15.3196	−21.6051	−7.8717	−8.8421	−10.7727
	60.0	−9.6988	−11.9270	−16.3487	−5.0037	−5.0037	−5.0037
	75.0	−7.8081	−9.2607	−12.1428	−2.8247	−2.0535	−0.5219
	90.0	−6.1324	−6.8649	−8.3183	−0.9612	0.4913	3.3743
	135.0	−0.6421	1.1826	4.8086	4.4902	8.1282	15.3477
$K_{\psi A}$	45.0	−0.1836	−0.1273	−0.0791	−0.1098	−0.0073	−0.0487
	60.0	−0.0505	−0.0352	−0.0220	0.0000	0.0000	0.0000
	75.0	0.0000	0.0000	0.0000	0.0453	0.0322	0.0204
	90.0	0.0126	0.0089	0.0056	0.0594	0.0424	0.0271
	135.0	−0.1463	−0.1044	−0.0665	−0.1265	−0.0917	−0.0594
$K_{\sigma 2}$	45.0	−3.5015	−4.4844	−6.4362	−2.2904	−2.6270	−3.2962
	60.0	−2.7242	−3.4161	−4.7893	−1.3333	−1.3333	−1.3333
	75.0	−2.1251	−2.5762	−3.4713	−0.6067	−0.3394	0.1913
	90.0	−1.5939	−1.8214	−2.2728	0.0143	0.5177	1.5167
	135.0	0.1482	0.7152	1.8417	1.8278	3.0875	5.5879

(*Continued*)

TABLE 13.4 Formulas for Discontinuity Stresses and Deformations at the Junctions of Shells and Plates (*Continued*)

Loading and Case No.

8c. Hydrostatic internal* pressure q_1 at the junction where the angle to the position of zero pressure, $\theta_1 > 3/\beta_1$.† If $\theta_1 < 3/\beta_1$, the discontinuity in pressure gradient introduces small deformations at the junction.

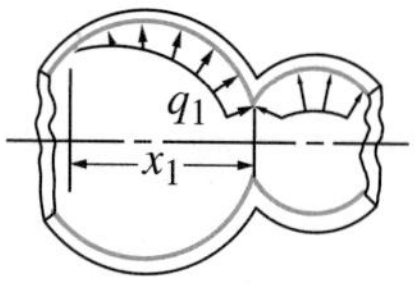

Note: There is no axial load on the junction. An axial load on the left end of the left sphere balances any axial component of the pressure on the left sphere, and an axial load on the right end of the right sphere balances any axial component of the pressure on the right sphere and on the joint.

Load Terms

$$LT_{A1} = \frac{b_1^2 \sin\phi_1}{t_1^2}$$

$$LT_{A2} = \frac{-b_2^2 E_1 \sin\phi_2}{E_2 t_1 t_2}$$

For LT_{AC} use the expressions from case 8a.

$$LT_{B1} = \frac{-b_1 R_1 \sin\phi_1}{x_1 t_1}$$

$$LT_{B2} = \frac{E_1 b_2 R_2 \sin\phi_2}{E_2 t_2 x_1}$$

For LT_{BC} use the expressions from case 8a.

At the junction of the two spheres,

$$V_1 = q_1 t_1 K_{V1}$$

$$M_1 = q_1 t_1^2 K_{M1}$$

$$N_1 = -V_1 \cos\phi_1$$

$$\Delta R_A = \frac{q_1 t_1}{E_1}(LT_{A1} - K_{V1}C_{AA1} - K_{M1}C_{AB1})$$

$$\psi_A = \frac{q_1}{E_1}(-LT_{B1} + K_{V1}C_{AB1} + K_{M1}C_{BB1})$$

Selected Values

For internal pressure, $x_1 = R_1$, $E_1 = E_2$, $\nu_1 = \nu_2 = 0.3$, $t_1 = t_2$, $R_1 \sin\phi_1 = R_2 \sin\phi_2$, and for $R/t > 5$. (*Note:* No correction terms are used.)

$$\Delta R_A = \frac{q_1 R_1^2}{E_1 t_1} K_{\Delta RA}, \qquad \psi_A = \frac{q_1 R_1}{E_1 t_1} K_{\psi A}, \qquad \sigma_2 = \frac{q_1 R_1}{t_1} K_{\sigma 2}$$

		ϕ_1 = 75			ϕ_1 = 120		
		R_1/t_1			R_1/t_1		
	ϕ_2	10	20	50	10	20	50
K_{V1}	45.0	−0.2614	−0.3656	−0.5720	−0.1695	−0.2390	−0.3768
	60.0	−0.0776	−0.1090	−0.1710	0.0000	0.0000	0.0000
	75.0	0.0000	0.0000	0.0000	0.0707	0.1004	0.1591
	90.0	0.0216	0.0306	0.0483	0.0898	0.1279	0.2033
	135.0	−0.2454	−0.3500	−0.5567	−0.1586	−0.2281	−0.3660
K_{M1}	45.0	0.0632	0.0900	0.1431	0.0403	0.0571	0.0905
	60.0	0.0204	0.0290	0.0461	0.0000	0.0000	0.0000
	75.0	0.0000	0.0000	0.0000	−0.0189	−0.0267	−0.0421
	90.0	−0.0060	−0.0085	−0.0134	−0.0241	−0.0339	−0.0535
	135.0	0.0543	0.0763	0.1199	0.0344	0.0481	0.0752
$K_{\Delta RA}$	45.0	1.0866	1.1261	1.1482	0.9006	0.9375	0.9586
	60.0	0.9361	0.9805	1.0072	0.7816	0.8233	0.8488
	75.0	0.8717	0.9182	0.9467	0.7314	0.7749	0.8022
	90.0	0.8537	0.9007	0.9296	0.7178	0.7617	0.7893
	135.0	1.0718	1.1156	1.1416	0.8920	0.9314	0.9547
$K_{\psi A}$	45.0	1.9353	2.2922	2.9657	1.3998	1.6213	2.0294
	60.0	1.2243	1.3485	1.5610	0.8227	0.8444	0.8574
	75.0	0.9176	0.9418	0.9563	0.5767	0.5131	0.3578
	90.0	0.8314	0.8270	0.7850	0.5101	0.4224	0.2194
	135.0	1.8593	2.2156	2.8887	1.3538	1.5740	1.9809
$K_{\sigma 2}$	45.0	1.1270	1.1672	1.1896	1.0374	1.0807	1.1058
	60.0	0.9697	1.0156	1.0430	0.9025	0.9506	0.9801
	75.0	0.9025	0.9506	0.9801	0.8456	0.8956	0.9268
	90.0	0.8837	0.9323	0.9623	0.8302	0.8805	0.9120
	135.0	1.1115	1.1563	1.1827	1.0276	1.0737	1.1014

*For external pressure, substitute $-q_1$ for q_1, a_1 for b_1, b_2 for a_2, and a_2 for b_2 in the load terms.

†If pressure increases right to left, substitute $-x_1$ for x_1 and verify that $|x_1|$ is large enough to extend into the right hand sphere as far as the position where $\theta_2 = 3/\beta_2$.

(*Continued*)

TABLE 13.4 Formulas for Discontinuity Stresses and Deformations at the Junctions of Shells and Plates (*Continued*)

Loading and Case No.

8d. Rotation around the axis of symmetry at ω rad/s

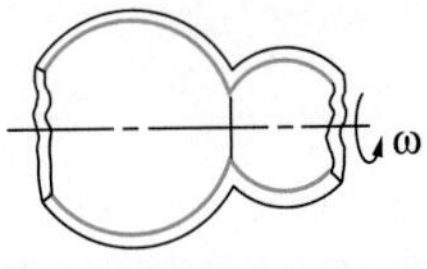

Note: δ = mass/unit volume

Load Terms

$$LT_{A1} = \frac{R_1^2 \sin^3 \phi_1}{t_1^2}$$

$$LT_{A2} = \frac{-\delta_2 R_2^3 E_1 \sin^3 \phi_2}{\delta_1 R_1 E_2 t_1^2}$$

$$LT_{AC} = 0$$

$$LT_{B1} = \frac{-R_1 \sin\phi_1 \cos\phi_1 (3+\nu_1)}{t_1}$$

$$LT_{B2} = \frac{-\delta_2 R_2^2 E_1 (3+\nu_2) \sin\phi_2 \cos\phi_2}{\delta_1 R_1 t_1 E_2}$$

$$LT_{BC} = 0$$

At the junction of the two sphere,

$$V_1 = \delta_1 \omega^2 R_1 t_1^2 K_{V1}$$

$$M_1 = \delta_1 \omega^2 R_1 t_1^3 K_{M1}$$

$$N_1 = -V_1 \cos\phi_1$$

$$\Delta R_A = \frac{\delta_1 \omega^2 R_1 t_1^2}{E_1}(LT_{A1} - K_{V1} C_{AA1} - K_{M1} C_{AB1})$$

$$\psi_A = \frac{\delta_1 \omega^2 R_1 t_1}{E_1}(-LT_{B1} + K_{V1} C_{AB1} + K_{M1} C_{BB1})$$

Selected Values

For $E_1 = E_2$, $\nu_1 = \nu_2 = 0.3$, $t_1 = t_2$, $R_1 \sin\phi_1 = R_2 \sin\phi_2$, $\delta_1 = \delta_2$, and $R/t > 5$.

$$\Delta R_A = \frac{\delta_1 \omega^2 R_1^3}{E_1} K_{\Delta RA}, \quad \psi_A = \frac{\delta_1 \omega^2 R_1^2}{E_1} K_{\psi A}, \quad \sigma_2 = \delta_1 \omega^2 R_1^2 K_{\sigma 2}$$

		ϕ_1					
		75			120		
		R_1/t_1			R_1/t_1		
	ϕ_2	10	20	50	10	20	50
K_{v1}	45.0	0.0047	0.0033	0.0020	0.0034	0.0024	0.0015
	60.0	0.0015	0.0010	0.0006	0.0000	0.0000	0.0000
	75.0	0.0000	0.0000	0.0000	-0.0014	-0.0010	-0.0006
	90.0	-0.0004	-0.0003	-0.0002	-0.0017	-0.0012	-0.0008
	135.0	0.0047	0.0034	0.0021	0.0025	0.0018	0.0012
K_{M1}	45.0	-0.6507	-0.9283	-1.4790	-0.1829	-0.2595	-0.4114
	60.0	-0.4616	-0.6568	-1.0441	0.0000	0.0000	0.0000
	75.0	-0.3030	-0.4302	-0.6826	0.1475	0.2082	0.3285
	90.0	-0.1538	-0.2180	-0.3453	0.2795	0.3937	0.6201
	135.0	0.3931	0.5534	0.8714	0.7152	1.0009	1.5678
$K_{\Delta RA}$	45.0	0.6872	0.7500	0.8056	0.5965	0.6121	0.6259
	60.0	0.7508	0.7949	0.8340	0.6495	0.6495	0.6495
	75.0	0.8032	0.8319	0.8574	0.6915	0.6792	0.6683
	90.0	0.8518	0.8663	0.8791	0.7284	0.7053	0.6848
	135.0	1.0249	0.9886	0.9565	0.8470	0.7891	0.7378
$K_{\psi A}$	45.0	-0.9617	-0.9655	-0.9689	-1.9161	-1.9188	-1.9213
	60.0	-0.4365	-0.4377	-0.4388	-1.4289	-1.4289	-1.4289
	75.0	0.0000	0.0000	0.0000	-1.0397	-1.0385	-1.0374
	90.0	0.4076	0.4080	0.4083	-0.6940	-0.6924	-0.6911
	135.0	1.8800	1.8754	1.8713	0.4324	0.4299	0.4278
$K_{\sigma 2}$	45.0	0.7115	0.7764	0.8340	0.6888	0.7068	0.7227
	60.0	0.7773	0.8229	0.8634	0.7500	0.7500	0.7500
	75.0	0.8315	0.8613	0.8876	0.7984	0.7842	0.7716
	90.0	0.8818	0.8968	0.9101	0.8411	0.8144	0.7907
	135.0	1.0610	1.0235	0.9902	0.9780	0.9112	0.8519

(*Continued*)

TABLE 13.4 Formulas for Discontinuity Stresses and Deformations at the Junctions of Shells and Plates (*Continued*)

9. Spherical shell connected to a circular plate. Expressions are accurate if $R_1/t_1 > 5$ and $R_1/t_1 > 4$. The junction angle for each of the spherical shells must lie within the range $3/\beta < \phi_1 < \pi - 3/\beta$. The spherical shell must also extend with no interruptions such as a second junction or a cutout, such that $\theta_1 > 3/\beta$. See the discussion on page 495. E_1 and E_2 are the moduli of elasticity and ν_1 and ν_2 the Poisson's ratios for the sphere and plate, respectively. $b_1 = R_1 - t_1/2$, $a_1 = R_1 + t_1/2$, and $\sin\phi_1 = R_A/R_1$. See Table 13.3, case 1, for formulas for K_1, K_2, and β for the spherical shell. See Table 11.2 for the formula for D_2.

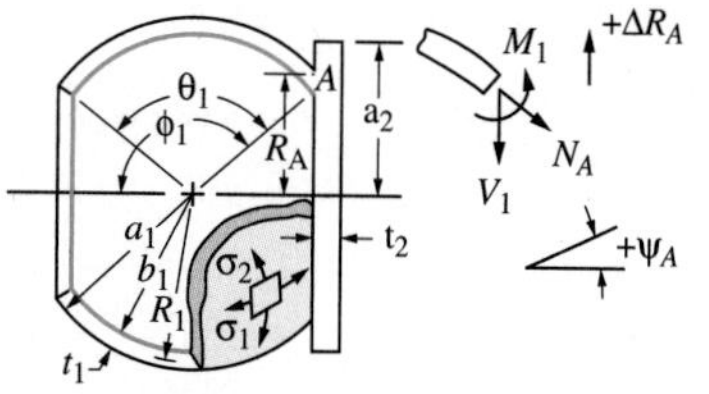

$$K_{P1} = 1 + \frac{R_A^2(1-\nu_2)}{a_2^2(1+\nu_2)}$$

$$K_{V1} = \frac{LT_A C_{BB} - LT_B C_{AB}}{C_{AA}C_{BB} - C_{AB}^2}, \quad K_{M1} = \frac{LT_B C_{AA} - LT_A C_{AB}}{C_{AA}C_{BB} - C_{AB}^2}, \quad \left.\begin{aligned} LT_A &= LT_{A1} + LT_{A2} + LT_{AC} \\ LT_B &= LT_{B1} + LT_{B2} + LT_{BC} \end{aligned}\right\} \text{See cases 9a–9d for these load terms.}$$

$$C_{AA} = C_{AA1} + C_{AA2}, \quad C_{AA1} = \frac{R_1\beta\sin^2\phi_1}{t_1}\left(\frac{1}{K_1} + K_2\right), \quad C_{AA2} = \frac{E_1 t_2^2 R_A K_{P1}}{6D_2}$$

$$C_{AB} = C_{AB1} + C_{AB2}, \quad C_{AB1} = \frac{-2\beta^2\sin\phi_1}{K_1}, \quad C_{AB2} = \frac{E_1 t_1 t_2 R_A K_{P1}}{4D_2}$$

$$C_{BB} = C_{BB1} + C_{BB2}, \quad C_{BB1} = \frac{4t_1\beta^3}{R_1 K_1}, \quad C_{BB2} = \frac{E_1 t_2^2 R_A K_{P1}}{2D_2}$$

The stresses in the left cylinder at the junction are given by

$$\sigma_1 = \frac{N_1}{t_1}$$

$$\sigma_2 = \frac{\Delta R_A E_1}{R_A} + \nu_1\sigma_1$$

$$\sigma_1' = \frac{-6M_1}{t_1^2}$$

$$\sigma_2' = \frac{V_1\beta^2\cos\phi_1}{K_1 R_1} - \frac{6M_1}{t_1^2 K_1}\left(\nu_1 + \frac{1-\nu_1/2}{\beta\tan\phi_1}\right)$$

Note: The use of joint load correction terms LT_{AC} and LT_{BC} depends upon the accuracy desired and the relative values of the thicknesses and the radii. Read Sec. 13.4 carefully. For thin-walled shells, $R/t > 10$, they can be neglected.

(*Continued*)

TABLE 13.4 Formulas for Discontinuity Stresses and Deformations at the Junctions of Shells and Plates (*Continued*)

Loading and Case No.	Load Terms	Selected Values

Loading and Case No.: 9a. Internal* pressure q

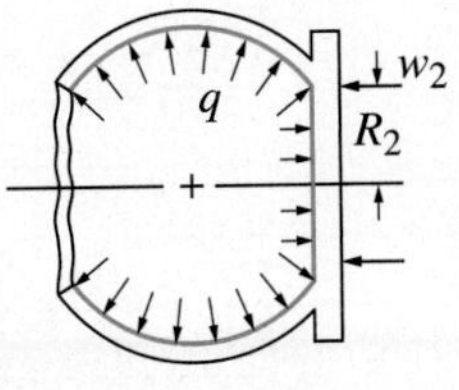

Note: There is no axial load on the junction. An axial load on the left end of the left sphere balances any axial component of the pressure on sphere, and the axial load on the plate is reacted by the annular line load $w_2 = qb_1^2 \sin^2\phi_1/(2R_2)$ at a radius R_2. For an enclosed pressure vessel, superpose an axial load $P = q\pi b_1^2 \sin^2\phi_1$ using case 9b.

Load Terms:

$$LT_{A1} = \frac{b_1^2 \sin\phi_1}{t_1^2}, \qquad LT_{A2} = \frac{E_1 t_2 b_1^2 \sin\phi_1}{32 D_2 t_1 R_1} K_{P2}$$

where

$$K_{P2} = \begin{cases} (2R_2^2 - b_1^2 \sin\phi_1)K_{P1} & \text{for } R_2 \le R_A \\ (2R_2^2 - b_1^2 \sin^2\phi_1)K_{P1} - 2(R_2^2 - R_A^2) + 4R_A^2 \ln\dfrac{R_2}{R_A} & \text{for } R_2 \ge R_A \end{cases}$$

$$LT_{AC} = \frac{E_1 b_1 t_2 \sin\phi_1 \cos\phi_1}{12 D_2}\left(t_2 - \frac{3t_1\cos\phi_1}{8}\right) K_{P1}$$

$$LT_{B1} = 0$$

$$LT_{B2} = \frac{E_1 b_1^2 \sin\phi_1}{16 D_2 R_1} K_{P2}$$

$$LT_{BC} = \frac{E_1 b_1 t_1 \sin\phi_1 \cos\phi_1}{8 D_2}\left(t_2 - \frac{t_1\cos\phi_1}{2}\right) K_{P1}$$

At the junction of the sphere and plate,

$$V_1 = qt_1 K_{V1}, \qquad M_1 = qt_1^2 K_{M1}$$

$$N_1 = -V_1 \cos\phi_1$$

$$\Delta R_A = \frac{qt_1}{E_1}(LT_{A1} - K_{V1}C_{AA1} - K_{M1}C_{AB1})$$

$$\psi_A = \frac{q}{E_1}(-LT_{B1} + K_{V1}C_{AB1} + K_{M1}C_{BB1})$$

Selected Values:

For internal pressure, $E_1 = E_2$, $\nu_1 = \nu_2 = 0.3$, $a_2 = a_1 \sin\phi_1$, $R_2 = 0.8a_2$, and for $R/t > 5$ and $R_A/t_2 > 4$. (*Note:* No correction terms are used.)

$$\Delta R_A = \frac{qR_1^2}{E_1 t_1} K_{\Delta RA}, \qquad \psi_A = \frac{qR_1}{E_1 t_1} K_{\psi A}, \qquad \sigma_2 = \frac{qR_1}{t_1} K_{\sigma 2}$$

		ϕ_1					
		60			120		
		R_1/t_1			R_1/t_1		
	t_2/t_1	15	30	50	15	30	50
K_{V1}	1.5	3.7234	8.1706	15.4808	3.4307	7.6706	14.6997
	2.0	3.4877	7.3413	13.6666	3.1733	6.8420	12.9155
	2.5	3.3436	6.6565	11.9903	3.0014	6.1506	11.2632
	3.0	3.2675	6.1516	10.6380	2.8981	5.6384	9.9338
	4.0		5.5526	8.8672		5.0264	8.2023
K_{M1}	1.5	5.2795	19.5856	53.6523	5.2058	19.4818	53.5198
	2.0	4.6874	16.5098	44.7285	4.4942	16.1774	44.2151
	2.5	4.3589	14.1907	37.1658	4.0654	13.6690	36.3338
	3.0	4.1955	12.5748	31.3788	3.8241	11.9177	30.3314
	4.0		10.7573	24.1731		9.9411	22.9142
$K_{\Delta RA}$	1.5	0.0508	-0.0690	-0.1820	0.0121	-0.1011	-0.2113
	2.0	0.0484	-0.0825	-0.2140	0.0121	-0.1124	-0.2414
	2.5	0.0535	-0.0722	-0.2040	0.0207	-0.0984	-0.2277
	3.0	0.0582	-0.0556	-0.1778	0.0290	-0.0780	-0.1975
	4.0		-0.0261	-0.1218		-0.0427	-0.1355
$K_{\psi A}$	1.5	0.9467	7.1118	20.4155	1.5646	8.1319	21.9534
	2.0	0.3077	4.6745	14.8148	0.7591	5.4222	15.9567
	2.5	-0.0078	3.0100	10.4710	0.3206	3.5412	11.2759
	3.0	-0.1516	1.9308	7.3488	0.0922	2.3109	7.9116
	4.0		0.8078	3.7130		1.0178	4.0052
$K_{\sigma 2}$	1.5	0.0214	-0.1205	-0.2566	0.0483	-0.0784	-0.1999
	2.0	0.0210	-0.1320	-0.2881	0.0457	-0.0956	-0.2400
	2.5	0.0283	-0.1167	-0.2716	0.0539	-0.0829	-0.2991
	3.0	0.0345	-0.0949	-0.2372	0.0624	-0.0619	-0.1982
	4.0		-0.0579	-0.1673		-0.0242	-0.1319

*For external pressure, substitute $-q$ for q and a_1 for b_1 in the load terms.

(*Continued*)

TABLE 13.4 Formulas for Discontinuity Stresses and Deformations at the Junctions of Shells and Plates (*Continued*)

Loading and Case No.	Load Terms	Selected Values
9b. Axial load P 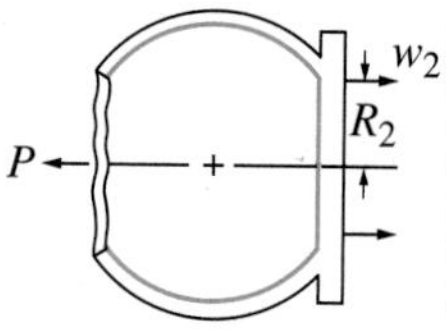 $w_2 = \dfrac{P}{2\pi R_2}$	$LT_{A1} = \dfrac{-R_1^2(1+\nu_1)}{2t_1^2 \sin\phi_1}, \quad LT_{AC} = 0$ $LT_{A2} = \dfrac{E_1 t_2 R_A R_1^2}{16 D_2 t_1} K_{P2} + \dfrac{R_1 C_{AA2}}{2t_1 \sin\phi_1} \dfrac{1}{\tan\phi_1}$ where $K_{P2} = \begin{cases} \left(1 - \dfrac{R_2^2}{R_A^2}\right) K_{P1} & \text{for } R_2 \le R_A \\ -\dfrac{1-\nu_2}{1+\nu_2} \dfrac{R_2^2 - R_A^2}{a_2^2} - 2\ln\dfrac{R_2}{R_A} & \text{for } R_2 \ge R_A \end{cases}$ $LT_{B1} = 0, \quad LT_{BC} = 0$ $LT_{B2} = \dfrac{E_1 R_A R_1^2}{8D_2} K_{P2} + \dfrac{R_1 C_{AB2}}{2t_1 \sin\phi_1} \dfrac{1}{\tan\phi_1}$ At the junction of the sphere and plate, $V_1 = \dfrac{P t_1 K_{V1}}{\pi R_1^2}, \quad M_1 = \dfrac{P t_1^2 K_{M1}}{\pi R_1^2}$ $N_1 = \dfrac{P}{2\pi R_1 \sin^2\phi_1} - V_1 \cos\phi_1$ $\Delta R_A = \dfrac{P t_1}{E_1 \pi R_1^2}(LT_{A1} - K_{V1} C_{AA1} - K_{M1} C_{AB1})$ $\psi_A = \dfrac{P}{E_1 \pi r_1^2}(K_{V1} C_{AB1} + K_{M1} C_{BB1})$	For axial tension, $E_1 = E_2$, $\nu_1 = \nu_2 = 0.3$, $a_2 = a_1 \sin\phi_1$, $R_2 = 0.8a_2$, and $R/t > 5$ and $R_A/t_2 > 4$. $\Delta R_A = \dfrac{P\nu_1}{2\pi E_1 t_1} K_{\Delta RA}, \quad \psi_A = \dfrac{P\nu_1}{2\pi E_1 t_1^2} K_{\psi A}, \quad \sigma_2 = \dfrac{P}{2\pi R_1 t_1} K_{\sigma 2}$ (see values below)

Selected values (ϕ_1):

		60			120		
		R_1/t_1			R_1/t_1		
	$\frac{t_2}{t_1}$	15	30	50	15	30	50
K_{V1}	1.5	4.6437	15.5843	36.3705	1.9491	10.9805	29.5359
	2.0	3.3547	12.2448	29.6890	0.7536	7.7075	22.8663
	2.5	2.1645	9.0429	22.9862	−0.2002	4.8369	16.5737
	3.0	1.1889	6.3731	17.2246	−0.8920	2.6161	11.4344
	4.0		2.6779	9.0973		−0.1930	4.6167
K_{M1}	1.5	11.8774	53.5919	160.3030	7.5277	43.6643	142.3870
	2.0	8.6537	41.2277	127.4680	4.2295	30.6183	107.6120
	2.5	5.9474	30.3930	97.2405	1.8532	20.2072	77.5990
	3.0	3.8585	21.8516	72.5898	0.2383	12.6155	54.3952
	4.0		10.6527	39.5330		3.5467	25.1934
$K_{\Delta RA}$	1.5	−4.4181	−6.0318	−7.9101	−2.3416	−4.0897	−6.0730
	2.0	−4.4852	−6.3806	−8.6849	−2.3351	−4.3814	−6.8196
	2.5	−4.1953	−6.0555	−8.4140	−2.0124	−3.9922	−6.4701
	3.0	−3.7890	−5.4660	−7.6662	−1.6431	−3.4017	−5.6919
	4.0		−4.2468	−5.9485		−2.3204	−4.0641
$K_{\psi A}$	1.5	5.8073	8.7035	11.9527	4.7495	7.9366	11.3940
	2.0	4.2685	6.5301	9.2075	3.0929	5.5605	8.4063
	2.5	3.1172	4.8036	6.8936	2.0145	3.8261	6.0299
	3.0	2.3029	3.5366	5.1207	1.3359	2.6412	4.2959
	4.0		2.0015	2.8983		1.3235	2.2456
$K_{\sigma 2}$	1.5	−1.2233	−1.8453	−2.5584	−0.3722	−0.9069	−1.5265
	2.0	−1.2208	−1.9328	−2.7867	−0.3938	−1.0407	−1.8252
	2.5	−1.0966	−1.7881	−2.6526	−0.3011	−0.9346	−1.7419
	3.0	−0.9363	−1.5572	−2.3590	−0.1870	−0.7522	−1.5031
	4.0		−1.0979	−1.7152		−0.4057	−0.9801

(*Continued*)

TABLE 13.4 Formulas for Discontinuity Stresses and Deformations at the Junctions of Shells and Plates (*Continued*)

Loading and Case No.

9c. Hydrostatic internal* pressure q_1 at the junction where the angle to the position of zero pressure, $\theta > 3/\beta$.† If $\theta < 3/\beta$, the discontinuity in pressure gradient introduces small deformations at the junction.

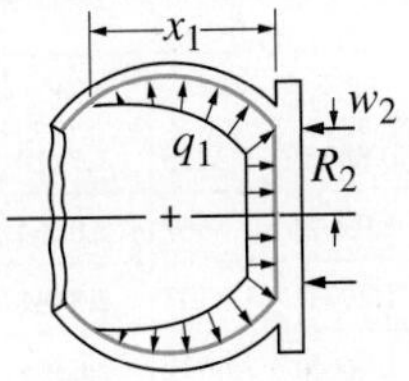

Note: There is no axial load on the junction. An axial load on the left end of the sphere balances any axial component of the pressure on the sphere, and the axial load on the plate is reacted by the annular line load $w_2 = q_1 b_1^2 \sin^2\phi_1/(2R_2)$ at a radius R_2.

Load Terms

$$LT_{A1} = \frac{b_1^2 \sin\phi_1}{t_1^2}$$

$$LT_{A2} = \frac{E_1 t_2 b_1^2 \sin\phi_1}{32 D_2 t_1 R_1} K_{P2}$$ For K_{P2} use the expressions from case 9a.

For LT_{BC} use the expressions from case 9a.

$$LT_{B1} = \frac{-b_1 R_1 \sin\phi_1}{x_1 t_1}$$

$$LT_{B2} = \frac{E_1 b_1^2 \sin\phi_1}{16 D_2 R_1} K_{P2}$$

For LT_{BC} use the expressions from case 9a.

At the junction of the sphere and plate,

$$V_1 = q_1 t_1 K_{V1}, \quad M_1 = q_1 t_1^2 K_{M1} \quad N_1 = -V_1 \cos\phi_1$$

$$\Delta R_A = \frac{q_1 t_1}{E_1}(LT_{A1} - K_{V1}C_{AA1} - K_{M1}C_{AB1})$$

$$\psi_A = \frac{q_1}{E_1}(-LT_{B1} + K_{V1}C_{AB1} + K_{M1}C_{BB1})$$

Selected Values

For internal pressure, $E_1 = E_2$, $\nu_1 = \nu_2 = 0.3$, $x_1 = R_1$, $a_2 = a_1 \sin\phi_1$, $R_2 = 0.8a_2$, and for $R/t > 5$ and $R_A/t_2 > 4$. (*Note:* No correction terms are used.)

$$\Delta R_A = \frac{q_1 R_1^2}{E_1 t_1} K_{\Delta RA}, \qquad \psi_A = \frac{q_1 R_1}{E_1 t_1} K_{\psi A}, \qquad \sigma_2 = \frac{q_1 R_1}{t_1} K_{\sigma 2}$$

		ϕ_1					
		60			120		
		R_1/t_1			R_1/t_1		
	t_2/t_1	15	30	50	15	30	50
K_{V1}	1.5	3.6924	8.1443	15.4580	3.4040	7.6470	14.6788
	2.0	3.4106	7.2712	13.6031	3.1050	6.7779	12.8563
	2.5	3.2236	6.5428	11.8838	2.8953	6.0468	11.1641
	3.0	3.1120	6.0007	10.4934	2.7617	5.5011	9.7996
	4.0		5.3486	8.6670		4.8425	8.0177
K_{M1}	1.5	5.0776	19.3466	53.3882	5.0013	19.2411	53.2544
	2.0	4.3702	16.1090	44.2642	4.1753	15.7756	43.7503
	2.5	3.9440	13.6425	36.5081	3.6524	13.1232	35.6786
	3.0	3.7048	11.9073	30.5580	3.3402	11.2576	29.5179
	4.0		9.9290	23.1258		9.1306	21.8848
$K_{\Delta RA}$	1.5	0.0262	−0.0831	−0.1912	−0.0122	−0.1152	−0.2205
	2.0	0.0235	−0.0973	−0.2239	−0.0122	−0.1270	−0.2512
	2.5	0.0302	−0.0864	−0.2137	−0.0016	−0.1122	−0.2371
	3.0	0.0371	−0.0685	−0.1867	0.0090	−0.0904	−0.2062
	4.0		−0.0364	−0.1290		−0.0525	−0.1424
$K_{\psi A}$	1.5	1.4212	7.6631	21.0176	2.0380	8.6826	22.5550
	2.0	0.6584	5.0979	15.2914	1.1033	5.8410	16.4299
	2.5	0.2494	3.3278	10.8366	0.5689	3.8521	11.6362
	3.0	0.0391	2.1689	7.6264	0.2736	2.5416	8.1831
	4.0		0.9467	3.8764		1.1502	4.1631
$K_{\sigma 2}$	1.5	−0.0067	−0.1367	−0.2671	0.0199	−0.0948	−0.2106
	2.0	−0.0070	−0.1487	−0.2994	0.0170	−0.1128	−0.2515
	2.5	0.0026	−0.1325	−0.2824	0.0271	−0.0993	−0.2403
	3.0	0.0117	−0.1091	−0.2470	0.0380	−0.0769	−0.2087
	4.0		−0.0688	−0.1749		−0.0364	−0.1404

*For external pressure, substitute $-q_1$ for q_1, and a_1 for b_1, in load terms.
†If pressure increases right to left, substitute $-x_1$ for x_1.

(*Continued*)

TABLE 13.4 Formulas for Discontinuity Stresses and Deformations at the Junctions of Shells and Plates (*Continued*)

Loading and Case No.	Load Terms
9d. Rotation around the axis of symmetry at ω rad/s 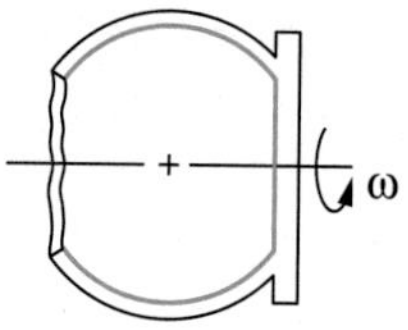 *Note*: δ = mass/unit volume	$LT_{A1}=\dfrac{R_1^2\sin^3\phi_1}{t_1^2}, \quad LT_{AC}=0$ $LT_{A2}=\dfrac{-E_1\delta_2t_2^3}{96D_2\delta_1t_1^2}\left[\dfrac{a_2^2(3+\nu_2)}{1+\nu_2}-R_A^2\right]$ $LT_{B1}=\dfrac{-R_1\sin\phi_1\cos\phi_1(3+\nu_1)}{t_1}, \quad LT_{B2}=0$ $LT_{BC}=0$ At the junction of the sphere and plate, $V_1=\delta_1\omega^2R_1t_1^2K_{V1}, \quad M_1=\delta_1\omega^2R_1t_1^3K_{M1} \quad N_1=-V_1\cos\phi_1$ $\Delta R_A=\dfrac{\delta_1\omega^2R_1t_1^2}{E_1}(LT_{A1}-K_{V1}C_{AA1}-K_{M1}C_{AB1})$ $\psi_A=\dfrac{\delta_1\omega^2R_1t_1}{E_1}(-LT_{B1}+K_{V1}C_{AB1}+K_{M1}C_{BB1})$

Selected Values

For $E_1=E_2$, $\nu_1=\nu_2=0.3$, $\delta_1=\delta_2$, $a_2=a_1\sin\phi_1$, and for $R/t>5$ and $R_A/t_2>4$.

$\Delta R_A=\dfrac{\delta_1\omega^2R_1^3}{E_1}K_{\Delta RA}$, $\quad \psi_A=\dfrac{\delta_1\omega^2R_1^2}{E_1}K_{\psi A}$, $\quad \sigma_2=\delta_1\omega^2R_1^2K_{\sigma 2}$

		ϕ_1					
		60			120		
		R_1/t_1			R_1/t_1		
	$\frac{t_2}{t_1}$	15	30	50	15	30	50
K_{V1}	1.5	0.8493	1.2962	1.7363	0.8885	1.3155	1.7419
	2.0	0.9373	1.4358	1.9173	1.0957	1.5667	2.0264
	2.5	1.0345	1.5946	2.1310	1.2971	1.8307	2.3441
	3.0	1.1233	1.7419	2.3346	1.4665	2.0631	2.6355
	4.0		1.9683	2.6525		2.4016	3.0728
K_{M1}	1.5	−0.0650	0.0730	0.2432	0.5903	0.8287	1.0685
	2.0	0.1543	0.5888	1.1308	1.1604	1.8275	2.5488
	2.5	0.3751	1.1257	2.0941	1.6617	2.7842	4.0628
	3.0	0.5652	1.5967	2.9649	2.0568	3.5782	5.3777
	4.0		2.2823	4.2570		4.6700	7.2494
$K_{\Delta RA}$	1.5	0.2286	0.2115	0.1992	0.3018	0.2528	0.2258
	2.0	0.2292	0.2135	0.2022	0.3013	0.2559	0.2304
	2.5	0.2256	0.2111	0.2009	0.2910	0.2504	0.2277
	3.0	0.2200	0.2062	0.1969	0.2774	0.2411	0.2211
	4.0		0.1949	0.1868		0.2215	0.2053
$K_{\psi A}$	1.5	−1.2055	−2.2272	−3.3080	−2.6485	−3.8683	−5.0838
	2.0	−0.9706	−1.8202	−2.7526	−2.0059	−3.0513	−4.1317
	2.5	−0.7594	−1.4356	−2.1999	−1.4946	−2.3348	−3.2333
	3.0	−0.5928	−1.1213	−1.7304	−1.1213	−1.7774	−2.4967
	4.0		−0.6988	−1.0794		−1.0645	−1.5120
$K_{\sigma 2}$	1.5	0.2554	0.2377	0.2248	0.3573	0.2984	0.2660
	2.0	0.2552	0.2394	0.2277	0.3589	0.3033	0.2722
	2.5	0.2501	0.2358	0.2255	0.3490	0.2983	0.2700
	3.0	0.2428	0.2294	0.2203	0.3350	0.2888	0.2632
	4.0		0.2153	0.2077		0.2678	0.2463

TABLE 13.5 Formulas for Thick-Walled Vessels Under Internal and External Loading

Notation: q = unit pressure (force per unit area); δ and δ_b = radial body forces (force per unit volume); a = outer radius; b = inner radius; σ_1, σ_2, and σ_3 are normal stresses in the longitudinal, circumferential, and radial directions, respectively (positive when tensile); E = modulus of elasticity; ν = Poisson's ratio. Δa, Δb, and Δl are the changes in the radii a and b and in the length l, respectively. ε_1 = unit normal strain in the longitudinal direction.

Case No., Form of Vessel	Case No., Manner of Loading	Formulas
1. Cylinderical disk or shell 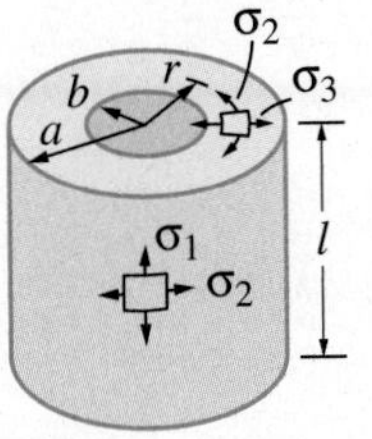	1a. Uniform internal radial pressure q, longitudinal pressure zero or externally balanced; for a disk or a shell	$\sigma_1 = 0$ $\sigma_2 = \dfrac{qb^2(a^2+r^2)}{r^2(a^2-b^2)}$, $(\sigma_2)_{max} = q\dfrac{a^2+b^2}{a^2-b^2}$, at $r=b$ $\sigma_3 = \dfrac{-qb^2(a^2-r^2)}{r^2(a^2-b^2)}$, $(\sigma_3)_{max} = -q$, at $r=b$ $\tau_{max} = \dfrac{\sigma_2-\sigma_3}{2} = q\dfrac{a^2}{a^2-b^2}$, at $r=b$ $\Delta a = \dfrac{q}{E}\dfrac{2ab^2}{a^2-b^2}$, $\Delta b = \dfrac{qb}{E}\left(\dfrac{a^2+b^2}{a^2-b^2}+\nu\right)$, $\Delta l = \dfrac{-q\nu l}{E}\dfrac{2b^2}{a^2-b^2}$
	1b. Uniform internal pressure q, in all directions; ends capped; for a disk or a shell	$\sigma_1 = \dfrac{qb^2}{a^2-b^2}$ [σ_2, σ_3, $(\sigma_2)_{max}$, $(\sigma_3)_{max}$, and τ_{max} are the same as for case 1a.] $\Delta a = \dfrac{qa}{E}\dfrac{b^2(2-\nu)}{a^2-b^2}$ $\Delta b = \dfrac{qb}{E}\dfrac{a^2(1+\nu)+b^2(1-2\nu)}{a^2-b^2}$ $\Delta l = \dfrac{ql}{E}\dfrac{b^2(1-2\nu)}{a^2-b^2}$
	1c. Uniform external radial pressure q, longitudinal pressure zero or externally balanced; for a disk or a shell	$\sigma_1 = 0$ $\sigma_2 = \dfrac{-qa^2(b^2+r^2)}{r^2(a^2-b^2)}$, $(\sigma_2)_{max} = \dfrac{-q2a^2}{a^2-b^2}$, at $r=b$ $\sigma_3 = \dfrac{-qa^2(r^2-b^2)}{r^2(a^2-b^2)}$, $(\sigma_3)_{max} = -q$, at $r=a$ $\tau_{max} = \dfrac{(\sigma_2)_{max}}{2} = \dfrac{qa^2}{a^2-b^2}$ at $r=b$ $\Delta a = \dfrac{-qa}{E}\left(\dfrac{a^2+b^2}{a^2-b^2}-\nu\right)$, $\Delta b = \dfrac{-q}{E}\dfrac{2a^2b}{a^2-b^2}$, $\Delta l = \dfrac{q\nu l}{E}\dfrac{2a^2}{a^2-b^2}$
	1d. Uniform external pressure q in all directions; ends capped; for a disk or a shell	$\sigma_1 = \dfrac{-qa^2}{a^2-b^2}$ [σ_2, σ_3, $(\sigma_2)_{max}$, $(\sigma_3)_{max}$, and τ_{max} are the same as for case 1c.] $\Delta a = \dfrac{-qa}{E}\dfrac{a^2(1-2\nu)+b^2(1+\nu)}{a^2-b^2}$, $\Delta b = \dfrac{-qb}{E}\dfrac{a^2(2-\nu)}{a^2-b^2}$ $\Delta l = \dfrac{-ql}{E}\dfrac{a^2(1-2\nu)}{a^2-b^2}$
	1e. Uniformly distributed radial body force δ acting outward throughout the wall; for a disk only	$\sigma_1 = 0$ $\sigma_2 = \dfrac{\delta(2+\nu)}{3(a+b)}\left[a^2+ab+b^2-(a+b)\left(\dfrac{1+2\nu}{2+\nu}\right)r+\dfrac{a^2b^2}{r_2}\right]$ $(\sigma_2)_{max} = \dfrac{\delta a^2}{3}\left[\dfrac{2(2+\nu)}{a+b}+\dfrac{b}{a^2}(1-\nu)\right]$ at $r=b$ $\sigma_3 = \dfrac{\delta(2+\nu)}{3(a+b)}\left[a^2+ab+b^2-(a+b)r-\dfrac{a^2b^2}{r^2}\right]$ (*Note:* $\sigma_3 = 0$ at both $r=b$ and $r=a$.) $\tau_{max} = \dfrac{(\sigma_2)_{max}}{2}$ at $r=b$ $\Delta a = \dfrac{\delta a^2}{3E}\left[1-\nu+\dfrac{2(2+\nu)b^2}{a(a+b)}\right]$, $\Delta b = \dfrac{\delta ab}{3E}\left[\dfrac{b}{a}(1-\nu)+\dfrac{2a(2+\nu)}{a+b}\right]$ $\varepsilon_1 = \dfrac{-\delta a\nu}{E}\left[\dfrac{2(a^2+ab+b^2)}{3a(a+b)}(2+\nu)-\dfrac{r}{a}(1+\nu)\right]$

(*Continued*)

TABLE 13.5 Formulas for Thick-Walled Vessels Under Internal and External Loading (*Continued*)

Case No., Form of Vessel	Case No., Manner of Loading	Formulas
	1f. Linearly varying radial body force from δ_b outward at $r=b$ to zero at $r=a$; for a disk only	$\sigma_1=0$ $\sigma_2=\delta_b\left[\frac{(7+5\nu)a^4-8(2+\nu)ab^3+3(3+\nu)b^4}{24(a-b)(a^2-b^2)}-\frac{(1+2\nu)a}{3(a-b)}r+\frac{1+3\nu}{8(a-b)}r^2+\frac{b^2a^2}{24r^2}\frac{(7+5\nu)a^2-8(2+\nu)ab+3(3+\nu)b^2}{(a-b)(a^2-b^2)}\right]$ $\sigma_3=\delta_b\left[\frac{(7+5\nu)a^4-8(2+\nu)ab^3+3(3+\nu)b^4}{24(a-b)(a^2-b^2)}-\frac{(2+\nu)a}{3(a-b)}r+\frac{(3+\nu)}{8(a-b)}r^2-\frac{b^2a^2}{24r^2}\frac{(7+5\nu)a-3(3+\nu)b}{a^2-b^2}\right]$ (*Note:* $\sigma_3=0$ at both $r=b$ and $r=a$.) $(\sigma_2)_{max}=\frac{\delta_b}{12}\frac{2a^4+(1+\nu)a^2(5a^2-12ab+6b^2)-(1-\nu)b^3(4a-3b)}{(a-b)(a^2-b^2)}$ at $r=b$ $\tau_{max}=\frac{(\sigma_2)_{max}}{2}$ at $r=b$ $\Delta a=\frac{\delta_b a}{12E}\frac{(1-\nu)a^4-8(2+\nu)ab^3+3(3+\nu)b^4+6(1+\nu)a^2b^2}{(a-b)(a^2-b^2)}$ $\Delta b=(\sigma_2)_{max}\frac{b}{E}$ $\varepsilon_1=\frac{-\delta_b\nu}{E}\left[\frac{(7+5\nu)a^4-8(2+\nu)ab^3+3(3+\nu)b^4}{12(a-b)(a^2-b^2)}-\frac{1+\nu}{a-b}\left(a-\frac{r}{2}\right)r\right]$
2. Spherical 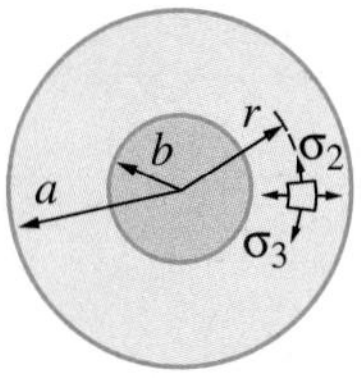 	2a. Uniform internal pressure q	$\sigma_1=\sigma_2=\frac{qb^3}{2r^3}\frac{a^3+2r^3}{a^3-b^3}$, $(\sigma_1)_{max}=(\sigma_2)_{max}=\frac{q}{2}\frac{a^3+2b^3}{a^3-b^3}$ at $r=b$ $\sigma_3=\frac{-qb^3}{r^3}\frac{a^3-r^3}{a^3-b^3}$, $(\sigma_3)_{max}=-q$ at $r=b$ $\tau_{max}=\frac{q3a^3}{4(a^3-b^3)}$ at $r=b$ The inner surface yields at $q=\frac{2\sigma_y}{3}\left(1-\frac{b^3}{a^3}\right)$ (Ref. 20) $\Delta a=\frac{qa}{E}\frac{3(1-\nu)b^3}{2(a^3-b^3)}$, $\Delta b=\frac{qb}{E}\left[\frac{(1-\nu)(a^3+2b^3)}{2(a^3-b^3)}+\nu\right]$ (Ref. 3)
	2b. Uniform external pressure q	$\sigma_1=\sigma_2=\frac{-qa^3}{2r^3}\frac{b^3+2r^3}{a^3-b^3}$, $(\sigma_1)_{max}=(\sigma_2)_{max}=\frac{-q3a^3}{2(a^3-b^3)}$ at $r=b$ $\sigma_3=\frac{-qa^3}{r^3}\frac{r^3-b^3}{a^3-b^3}$, $(\sigma_3)_{max}=-q$ at $r=a$ $\Delta a=\frac{-qa}{E}\left[\frac{(1-\nu)(b^3+2a^3)}{2(a^3-b^3)}-\nu\right]$, $\Delta b=\frac{-qb}{E}\frac{3(1-\nu)a^3}{2(a^3-b^3)}$ (Ref. 3)

13.9 REFERENCES

1. Southwell, R. V., "On the Collapse of Tubes by External Pressure," *Philos. Mag.*, 29: 67, 1915.
2. Roark, R. J., "The Strength and Stiffness of Cylindrical Shells under Concentrated Loading," *ASME J. Appl. Mech.*, 2(4): A-147, 1935.
3. Timoshenko, S., *Theory of Plates and Shells*, Engineering Societies Monograph, McGraw-Hill, 1940.
4. Schorer, H., "Design of Large Pipe Lines," *Trans. Am. Soc. Civil Eng.*, 98: 101, 1933.
5. Flügge, W., *Stresses in Shells*, Springer-Verlag, 1960.
6. Baker, E. H., L. Kovalevsky, and F. L. Rish, *Structural Analysis of Shells*, McGraw-Hill, 1972.
7. Saunders, H. E., and D. F. Windenburg, "Strength of Thin Cylindrical Shells under External Pressure," *Trans. ASME*, 53: 207, 1931.
8. Jasper, T. M., and J. W. W. Sullivan, "The Collapsing Strength of Steel Tubes," *Trans. ASME*, 53: 219, 1931.
9. American Society of Mechanical Engineers, Rules for Construction of Nuclear Power Plant Components, Sec. III; Rules for Construction of Pressure Vessels, Division 1, and Division 2, Sec. VIII; ASME Boiler and Pressure Vessel Code, 1971.
10. Langer, B. F., "Design-Stress Basis for Pressure Vessels," *Exp. Mech., J. Soc. Exp. Stress Anal.*, 11(1): 1971.
11. Hartenberg, R. S., "The Strength and Stiffness of Thin Cylindrical Shells on Saddle Supports," doctoral dissertation, University of Wisconsin, 1941.
12. Wilson, W. M., and E. D. Olson, "Tests on Cylindrical Shells," *Eng. Exp. Sta., Univ. Ill. Bull.* 331, 1941.
13. Odqvist, F. K. G., "Om Barverkan Vid Tunna Cylindriska Skal Ock Karlvaggar," *Proc. Roy. Swed. Inst. for Eng. Res.*, No. 164, 1942.
14. Hetényi, M., *Spherical Shells Subjected to Axial Symmetrical Bending*, vol. 5 of the Publications, International Association for Bridge and Structural Engineers, 1938.
15. Reissner, E., "Stresses and Small Displacements of Shallow Spherical Shells. II," *J. Math. and Phys.*, 25(4): 1947.
16. Clark, R. A., "On the Theory of Thin Elastic Toroidal Shells," *J. Math. and Phys.*, 29(3): 1950.
17. O'Brien, G. J., E. Wetterstrom, M. G. Dykhuizen, and R. G. Sturm, "Design Correlations for Cylindrical Pressure Vessels with Conical or Toriconical Heads," *Weld. Res. Suppl.*, 15(7): 336, 1950.
18. Osipova, L. N., and S. A. Tumarkin, *Tables for the Computation of Toroidal Shells*, P. Noordhoff, 1965 (English trans. M. D. Friedman).
19. Roark, R. J., "Stresses and Deflections in Thin Shells and Curved Plates Due to Concentrated and Variously Distributed Loading," *Natl. Adv. Comm. Aeron.*, Tech. Note 806, 1941.
20. Svensson, N. L., "The Bursting Pressure of Cylindrical and Spherical Vessels," *ASME J. Appl. Mech.*, 25(1): 1958.
21. Durelli, A. J., J. W. Dally, and S. Morse, "Experimental Study of Thin-Wall Pressure Vessels," *Proc. Soc. Exp. Stress Anal.*, 18(1): 1961.
22. Bjilaard, P. P., "Stresses from Local Loadings in Cylindrical Pressure Vessels," *Trans. ASME*, 77(6): 1955 (also in Ref. 28).
23. Bjilaard, P. P., "Stresses from Radial Loads in Cylindrical Pressure Vessels," *Weld. J.*, 33: December 1954 (also in Ref. 28).
24. Yuan, S. W., and L. Ting, "On Radial Deflections of a Cylinder Subjected to Equal and Opposite Concentrated Radial Loads," *ASME J. Appl. Mech.*, 24(6): 1957.
25. Ting, L., and S. W. Yuan, "On Radial Deflection of a Cylinder of Finite Length with Various End Conditions," *J. Aeron. Sci.*, 25: 1958.
26. Final Report, Purdue University Project, Design Division, Pressure Vessel Research Committee, Welding Research Council, 1952.
27. Wichman, K. R., A. G. Hopper, and J. L. Mershon, "Local Stresses in Spherical and Cylindrical Shells Due to External Loadings," *Weld. Res. Counc. Bull.* No. 107, August 1965.
28. von Kármán, Th., and Hsue-shen Tsien, "Pressure Vessel and Piping Design," ASME Collected Papers 1927–1959.
29. Galletly, G. D., "Edge Influence Coefficients for Toroidal Shells of Positive;" also "Negative Gaussian Curvature," *ASME J. Eng. Ind.*, 82: February 1960.
30. Wenk, E., Jr., and C. E. Taylor, "Analysis of Stresses at the Reinforced Intersection of Conical and Cylindrical Shells," U.S. Dept. of the Navy, David W. Taylor Model Basin, Rep. 826, March 1953.
31. Taylor, C. E., and E. Wenk, Jr., "Analysis of Stresses in the Conical Elements of Shell Structures," *Proc. 2d U.S. Natl. Congr. Appl. Mech.*, 1954.
32. Borg, M. F., "Observations of Stresses and Strains Near Intersections of Conical and Cylindrical Shells," U.S. Dept. of the Navy, David W. Taylor Model Basin, Rept. 911, March 1956.
33. Raetz, R. V., and J. G. Pulos, "A Procedure for Computing Stresses in a Conical Shell Near Ring Stiffeners or Reinforced Intersections," U.S. Dept of the Navy, David W. Taylor Model Basin, Rept. 1015, April 1958.

34. Narduzzi, E. D., and G. Welter, "High-Pressure Vessels Subjected to Static and Dynamic Loads," *Weld. J. Res. Suppl.*, 1954.
35. Dubuc, J., and G. Welter, "Investigation of Static and Fatigue Resistance of Model Pressure Vessels," *Weld. J. Res. Suppl.*, July 1956.
36. Kooistra, L. F., and M. M. Lemcoe, "Low Cycle Fatigue Research on Full-Size Pressure Vessels," *Weld. J. Res. Suppl.*, July 1962.
37. Weil, N. A., "Bursting Pressure and Safety Factors for Thin-Walled Vessels," *J. Franklin Inst.*, February 1958.
38. Brownell, L. E., and E. H. Young, *Process Equipment Design: Vessel Design*, John Wiley & Sons, 1959.
39. Faupel, J. H., "Yield and Bursting Characteristics of Heavy-Wall Cylinders," *Trans. ASME*, 78(5): 1956.
40. Dahl, N. C., "Toroidal-Shell Expansion Joints," *ASME J. Appl. Mech.*, 20: 1953.
41. Laupa, A., and N. A. Weil, "*Analysis of U-Shaped Expansion Joints*," *ASME J. Appl. Mech.*, 29(1): 1962.
42. Baker, B. R., and G. B. Cline, Jr., "Influence Coefficients for Thin Smooth Shells of Revolution Subjected to Symmetric Loads," *ASME J. Appl. Mech.*, 29(2): 1962.
43. Tsui, E. Y. W., and J. M. Massard, "Bending Behavior of Toroidal Shells," *Proc. Am. Soc. Civil Eng., J. Eng. Mech. Div.*, 94(2): 1968.
44. Kraus, H., *Thin Elastic Shells*, John Wiley & Sons, 1967.
45. Pflüger, A., *Elementary Statics of Shells*, 2nd ed., McGraw-Hill, 1961 (English trans., E. Galantay).
46. Gerdeen J. C., and F. W. Niedenfuhr, "Influence Numbers for Shallow Spherical Shells of Circular Ring Planform," *Proc. 8th Midwestern Mech. Conf.*, Development in Mechanics, vol. 2, part 2, Pergamon Press, 1963.
47. Zaremba, W. A., "Elastic Interactions at the Junction of an Assembly of Axisymmetric Shells," *J. Mech. Eng. Sci.*, 1(3): 1959.
48. Johns, R. H., and T. W. Orange, "Theoretical Elastic Stress Distributions Arising from Discontinuities and Edge Loads in Several Shell-Type Structures," *NASA Tech. Rept.* R-103, 1961.
49. Stanek, F. J., *Stress Analysis of Circular Plates and Cylindrical Shells*, Dorrance, 1970.
50. Blythe, W., and E. L. Kyser, "A Flugge-Vlasov Theory of Torsion for Thin Conical Shells," *ASME J. Appl. Mech.*, 31(3) 1964.
51. Payne, D. J., "Numerical Analysis of the Axisymmetric Bending of a Toroidal Shell," *J. Mech. Eng. Sci.*, 4(4): 1962.
52. Rossettos, J. N., and J. L. Sanders, Jr., "Toroidal Shells Under Internal Pressure in the Transition Range," *AIAA J.*, 3(10): 1965.
53. Jordan, P. F., "Stiffness of Thin Pressurized Shells of Revolution," *AIAA J.*, 3(5): 1965.
54. Haringx, J. A., "Instability of Thin-Walled Cylinders Subjected to Internal Pressure," *Phillips Res. Rept.* 7, 1952.
55. Haringx, J. A., "Instability of Bellows Subjected to Internal Pressure," *Phillips Res. Rept.* 7, 1952.
56. Seide, Paul, "The Effect of Pressure on the Bending Characteristics of an Actuator System," *ASME J. Appl. Mech.*, 27(3): 1960.
57. Chou, Seh-Ieh, and M. W. Johnson, Jr., "On the Finite Deformation of an Elastic Toroidal Membrane," *Proc. 10th Midwestern Mech. Conf.*, 1967.
58. Tsui, E. Y. W., "Analysis of Tapered Conical Shells," *Proc. 4th U.S. Natl. Congr. Appl. Mech.*, 1962.
59. Fischer, L., *Theory and Practice of Shell Structures*, Wilhelm Ernst, 1968.
60. Pao, Yen-Ching, "Influence Coefficients of Short Circular Cylindrical Shells with Varying Wall Thickness," *AIAA J.*, 6(8): 1968.
61. Turner, C. E., "Study of the Symmetrical Elastic Loading of Some Shells of Revolution, with Special Reference to Toroidal Elements," *J. Mech. Eng. Sci.*, 1(2) 1959.
62. Perrone, N., "Compendium of Structural Mechanics Computer Programs," *Computers and Structures*, 2(3): April 1972. (Also available as NTIS Paper N71-32026, April 1971.)
63. Bushnell, D., "Stress, Stability, and Vibration of Complex, Branched Shells of Revolution," *AIAA/ASME/SAE 14th Structures, Struct. Dynam. & Mater. Conf.*, Williamsburg, VA., March 1973.
64. Baltrukonis, J. H., "Influence Coefficients for Edge-Loaded Short, Thin, Conical Frustums," *ASME J. Appl. Mech.*, 26(2): 1959.
65. Taylor, C. E., "Simplification of the Analysis of Stress in Conical Shells," *Univ. Ill., TAM Rept. 385*, April 1974.
66. Cook, R. D., and W. C. Young, *Advanced Mechanics of Materials*, 2nd ed., Prentice-Hall, 1998.
67. Galletly, G. D., "A Simple Design Equation for Preventing Buckling in Fabricated Torispherical Shells under Internal Pressure," Trans. ASME, *J. Pressure Vessel Tech.*, 108(4): 1986.
68. Ranjan, G. V., and C. R. Steele, "Analysis of Torispherical Pressure Vessels," *ASCE J. Eng. Mech. Div.*, 102(EM4): 1976.
69. Forman, B. F., *Local Stresses in Pressure Vessels*, 2nd ed., Pressure Vessel Handbook Publishing, 1979.
70. Dodge, W. C., "Secondary Stress Indices for Integral Structural Attachments to Straight Pipe," *Weld. Res. Counc. Bull.* No. 198, September 1974.
71. Rodabaugh, E. C., W. G. Dodge, and S. E. Moore, "Stress Indices at Lug Supports on Piping Systems," *Weld. Res. Counc. Bull.* No. 198, September 1974.

72. Kitching, R., and B. Olsen, "Pressure Stresses at Discrete Supports on Spherical Shells," *Inst. Mech. Eng. J. Strain Anal.*, 2(4): 1967.
73. Evces, C. R., and J. M. O'Brien, "Stresses in Saddle-Supported Ductile-Iron Pipe," *J. Am. Water Works Assoc., Res. Tech.*, 76(11): 1984.
74. Harvey, J. F., *Theory and Design of Pressure Vessels*, Van Nostrand Reinhold, 1985.
75. Bednar, H. H., *Pressure Vessel Design Handbook*, 2nd ed., Van Nostrand Reinhold, 1986.
76. Calladine, C. R., *Theory of Shell Structures*, Cambridge University Press, 1983.
77. Stanley, P., and T. D. Campbell, "Very Thin Torispherical Pressure Vessel Ends under Internal Pressure: Strains, Deformations, and Buckling Behaviour," *Inst. Mech. Eng. J. Strain Anal.*, 16(3): 1981.
78. Kishida, M., and H. Ozawa, "Three-Dimensional Axisymmetric Elastic Stresses in Pressure Vessels with Torispherical Drumheads (Comparison of Elasticity, Photo-elasticity, and Shell Theory Solutions)," *Inst. Mech. Eng. J. Strain Anal.*, 20(3): 1985.
79. Cook, R. D., "Behavior of Cylindrical Tanks Whose Axes Are Horizontal," *Int. J. Thin-Walled Struct.*, 3(4): 1985.
80. Cheng, S., and T. Angsirikul, "Three-Dimensional Elasticity Solution and Edge Effects in a Spherical Dome," *ASME J. Appl. Mech.*, 44(4): 1977.
81. Holland, M., M. J. Lalor, and J. Walsh, "Principal Displacements in a Pressurized Elliptic Cylinder: Theoretical Predictions with Experimental Verification by Laser Interferometry," *Inst. Mech. Eng. J. Strain Anal.*, 9(3): 1974.
82. White, R. N., and C. G. Salmon (eds.), *Building Structural Design Handbook*, John Wiley & Sons, 1987.

CHAPTER 14

Bodies under Direct Bearing and Shear Stress

14.1 STRESS DUE TO PRESSURE BETWEEN ELASTIC BODIES

The stresses caused by the pressure between elastic bodies are of importance in connection with the design or investigation of ball and roller bearings, gears, trunnions, expansion rollers, track stresses, etc. Hertz (Ref. 1) developed the mathematical theory for the surface stresses and deformations produced by pressure between curved bodies, and the results of his analysis are supported by experiment. Formulas based on this theory give the maximum compressive stresses, which occur at the center of the surfaces of contact, but not the maximum shear stresses, which occur in the interiors of the compressed parts, nor the maximum tensile stress, which occurs at the boundary of the contact area and is normal thereto.

Both surface and subsurface stresses were studied by Belajef (Refs. 28 and 29), and some of his results are cited in Ref. 6. A tabulated summary of surface and subsurface stresses, greatly facilitating calculation, is given in Ref. 33. For a cylinder on a plane and for crossed cylinders, Thomas and Hoersch (Ref. 2) investigated mathematically surface compression and internal shear and checked the calculated value of the latter experimentally. The stresses due to the pressure of a sphere on a plate (Ref. 3) and of a cylinder on a plate (Ref. 4) have also been investigated by photoelasticity. The deformation and contact area for a ball in a race were measured by Whittemore and Petrenko (Ref. 8) and compared with the theoretical values. Additionally, investigations have considered the influence of tangential loading combined with normal loading (Refs. 35, 47–49, and 58).

In Table 14.1, formulas are given for the elastic stress and deformation produced by pressure between bodies of various forms, and for the dimensions of the circular, elliptical, or rectangular area of contact formed by the compressed surfaces. Except where otherwise indicated, these equations are based on Hertz's theory, which assumes the length of the cylinder and dimensions of the plate to be infinite. For a very short cylinder and for a plate having a width less than five or six times that of the contact area or a thickness less than five or six times the depth to the point of maximum shear stress, the actual stresses may vary considerably from the values indicated by the theory (see Refs. 4, 45, and 50). Tu (Ref. 50) discusses the stresses and deformations for a plate pressed between two identical elastic spheres with no friction; graphs are also presented. Pu and Hussain (Ref. 51) consider the unbonded contact between a flat plate and an elastic half-space when a normal load is applied to the plate. Graphs of the contact radii are presented for a concentrated load and two distributed loadings on circular areas.

Hertz (Ref. 1) based his work on the assumption that the contact area was small compared with the radius of the ball or cylinder; Goodman and Keer (Ref. 52) compare the work of Hertz with a solution which permits the contact area to be larger, such as the case when the negative radius of one surface is only very slightly larger (1.01–1) than the positive radius of the other. Cooper (Ref. 53) presents some reformulated hertzian coefficients in a more easily interpolated form and also points out some numerical errors in the coefficients originally published by Hertz. Dundurs and Stippes (Ref. 54) discuss the effect of Poisson's ratio on contact stress problems.

The use of the formulas of Table 14.1 is illustrated in the example at the end of this section. The general formula for case 4 can be used, as in the example, for any contact-stress problems involving any geometrically regular bodies except parallel cylinders, but for bearing calculations use should be made of charts such as those given in Refs. 33 and 34, which not only greatly facilitate calculations but provide for influences not taken into account in the formulas.

Because of the very small area involved in what initially approximates a point or line contact, contact stresses for even light loads are very high; but as the formulas show, the stresses do not increase in proportion to the loading. Furthermore, because of the facts that the stress is highly localized and triaxial, the actual stress intensity can be very high without producing apparent damage. To make use of the Hertz formulas for purposes of design or safe-load determination, it is necessary to know the relationship between theoretical stresses and likelihood of failure, whether from excessive deformation or fracture. In discussing this relationship, it is convenient to refer to the computed stress as the Hertz stress, whether the elastic range has been exceeded or not. Some of the available information showing the Hertz stress corresponding to loadings found to be safe and to loadings that produced excessive deformations or fracture may be summarized as follows.

Static or Near-Static Conditions

Cylinder

The American Railway Engineering Association gives as the allowable loading for a steel cylinder on a flat steel plate the formulas

$$p=\begin{cases}\dfrac{\sigma_y-13{,}000}{20{,}000}600d & \text{for } d<25 \text{ in}\\[2ex] \dfrac{\sigma_y-13{,}000}{20{,}000}3000\sqrt{d} & \text{for } 25<d<125 \text{ in}\end{cases}$$

Here (and in subsequent equations) p is the load per linear inch in pounds, d is the diameter of the cylinder in inches, and σ_y is the tensile yield point of the steel in the roller or plate, whichever is lower. If σ_y is taken as $32{,}000\ \text{lb/in}^2$, the Hertz stress corresponding to this loading is constant at $76{,}200\ \text{lb/in}^2$ for any diameter up to 25 in and decreases as $d^{-1/4}$ to 50,900 at $d=125$ in. See Ref. 10.

Wilson (Refs. 7, 11, and 32) carried out several series of static and slow-rolling tests on large rollers. From static tests on rollers of medium-grade cast steel having diameters of 120–720 in, he concluded that the load per linear inch required to produce appreciable permanent set could be represented by the empirical formula $p=500+110d$, provided the bearing plates were 3 in thick or more. He found that p increased with the axial length of the roller up to a length of 6 in, after which it remained practically constant (Ref. 32).

Slow-rolling tests (Ref. 11) undertaken to determine the load required to produce a permanent elongation or spread of 0.001 in/in in the bearing plate led to the empirical formula

$$p = (18{,}000 + 120d)\frac{\sigma_y - 13{,}000}{23{,}000}$$

for rollers with $d > 120$ in. Wilson's tests indicated that the average pressure on the area of contact required to produce set was greater for small rollers than for large rollers, and that there was little difference in bearing capacity under static and slow-rolling conditions, though the latter showed more tendency to produce surface deterioration.

Jensen (Ref. 4), making use of Wilson's test results and taking into account the three-dimensional aspect of the problem, proposed for the load-producing set the formula

$$p = \left(1 + \frac{1.78}{1 + d^2/800L^2}\right)\frac{\sigma_y^2 d\pi}{E}$$

where L is the length of the cylinder in inches and E is the modulus of elasticity in pounds per square inch. For values of the ratio d/L from 0.1 to 10, the corresponding Hertz stress ranges from $1.660\sigma_y$ to $1.72\sigma_y$.

Whittemore (Ref. 8) found that the elastic limit load for a flexible roller of hardened steel (tensile strength about 265,000 lb/in^2) tested between slightly hardened races corresponded to a Hertz stress of about 436,000 lb/in^2. The roller failed before the races.

Sphere

Tests reported in Whittemore and Petrenko (Ref. 8) gave, for balls 1, 1¼, and 1½ in in diameter, tested between straight races, Hertz stresses of 239,000, 232,000, and 212,000 lb/in^2, respectively, at loads producing a permanent strain of 0.0001. The balls were of steel having sclerescope hardness of 60–68, and the races were of approximately the same hardness. The critical strain usually occurred first in the races.

From the results of crushing tests of a sphere between two similar spheres, SKF derived the empirical formula $P = 1960(8d)^{1.75}$, where P is the crushing load in pounds and d is the diameter of the sphere in inches. The test spheres were made of steel believed to be of hardness 64–66 Rockwell C, and the formula corresponds to a Hertz stress of about $4{,}000{,}000 \times d^{-1/12}$.

Knife-Edge

Knife-edge pivots are widely used in scales and balances, and if accuracy is to be maintained, the bearing loads must not cause excessive deformation. It is impossible for a truly sharp edge to bear against a flat plane without suffering plastic deformation, and so pivots are not designed on the supposition that the contact stresses will be elastic; instead, the maximum load per inch consistent with the requisite degree of accuracy in weighing is determined by experience or by testing. In Wilson et al. (Ref. 9), the National Bureau of Standards is quoted as recommending that for heavy service the load per linear inch should not exceed 5000 lb/in for high-carbon steels or 6000 for special alloy steels; for light service the values can be increased to 6000 and 7000, respectively. In the tests described in Ref. 9, the maximum load that could be sustained without damage—the so-called *critical load*—was defined as the load per linear inch that produced an increase in the edge width of 0.0005 in or a sudden increase in the load rate of vertical deformation. The two methods gave about the same results when the bearing was harder than the pivot, as it should be for good operation. The conclusions drawn from the reported tests may be summarized as follows.

The bearing value of a knife-edge or pivot varies approximately with the wedge angle for angles of $30° - 120°$, the bearing value of a flat pivot varies approximately with the width of the edge for widths of 0.004–0.04 in, and the bearing value of pivots increases with the hardness for variations in hardness of 45–60 on the Rockwell *C* scale. Successive applications of a load less than the critical load will cause no plastic flow; the edge of a pivot originally sharp will increase in width with the load, but no further plastic deformation is produced by successive applications of the same or smaller loads. The application of a load greater than the critical load will widen the edge at the first application, but additional applications of the same load will not cause additional flow; the average unit pressure on 90° pivots having a hardness represented by Rockwell *C* numbers of 50–60 is about $400{,}000–500{,}000\ \text{lb/in}^2$ at the critical load. This critical unit pressure appears to be independent of the width of the edge but increases with the pivot angle and the hardness of the material (Ref. 9).

These tests and the quoted recommendations relate to applications involving heavy loads (thousands of pounds) and reasonable accuracy. For light loads and extreme accuracy, as in analytical balances, the pressures are limited to much smaller values. Thus, in Ref. 39, on the assumptions that an originally sharp edge indents the bearing and that the common surface becomes cylindrical, it is stated that the radius of the loaded edge must not exceed $0.25\ \mu\text{m}$ (approximately 0.00001 in) if satisfactory accuracy is to be attained, and that the corresponding loading would be about $35{,}000\ \text{lb/in}^2$ of contact area.

Dynamic Conditions

If the motion involved is a true rolling motion without any slip, then under conditions of slow motion (expansion rollers, bascules, etc.) the stress conditions are comparable with those produced by static loading. This is indicated by a comparison of the conclusions reached in Ref. 7, where the conditions are truly static, with those reached in Ref. 11, where there is a slow-rolling action. If there is even a slight amount of slip, however, the conditions are very much more severe and failure is likely to occur through mechanical wear. The only guide to proper design against wear is real or simulated service testing (Refs. 24, 41, and 46).

When the motion involved is at high speed and produces cyclic loading, as in ball and roller bearings, fatigue is an important consideration. A great many tests have been made to determine the fatigue properties of bearings, especially ball bearings, and such tests have been carried out to as many as 1 billion cycles and with Hertz stresses up to $750{,}000\ \text{lb/in}^2$ (Ref. 37). The number of cycles to damage (either spalling or excessive deformation) has been found to be inversely proportional to the cube of the load for point contact (balls) and to the fourth power for line contact; this would be inversely proportional to the ninth and eighth powers, respectively, of the Hertz stress. Styri (Ref. 40) found the cycles to failure to vary as the ninth power of the Hertz stress and was unable to establish a true endurance limit. Some of these tests show that ball bearings can run for a great number of cycles at very high stresses; for example, ½-in balls of SAE 52,100 steel (RC 63–64) withstood 17,500,000 cycles at a stress of $174{,}000\ \text{lb/in}^2$ before 10% failures occurred, and withstood 700,000,000 cycles at that stress before 90% failures occurred.

One difficulty in correlating different tests on bearings is the difference in criteria for judging damage; some experimenters have defined failure as a certain permanent deformation, others as visible surface damage through spalling. Palmgren (Ref. 36) states that a permanent deformation at any one contact point of rolling element and bearing ring combined equal to 0.001 times the diameter of the rolling element has no significant influence on the functioning of the bearing. In the tests of Ref. 37, spalling of the surface was taken as the sign of failure; this spalling generally originated on plates of maximum shear stress below the surface.

Large-diameter bearings, usually incorporating integral gearing, are heat-treated to produce a hardened case to resist wear and fatigue and a tough machinable core. Sague in Ref. 56 describes how high subsurface shear stresses have produced yielding in the core with subsequent failure of the case due to lack of support.

It is apparent from the foregoing discussion that the practical design of parts that sustain direct bearing must be based largely on experience since this alone affords a guide as to whether, at any given load and number of stress cycles, there is enough deformation or surface damage to interfere with proper functioning. The rated capacities of bearings and gears are furnished by the manufacturers, with proper allowance indicated for the conditions of service and recommendations as to proper lubrication (Ref. 38). Valid and helpful conclusions, however, can often be drawn from a comparison of service records with calculated stresses.

EXAMPLE A ball 1.50 in in diameter, in a race which has a diameter of 10 in and a groove radius of 0.80 in, is subjected to a load of 2000 lb. It is required to find the dimensions of the contact area, the combined deformation of ball and race at the contact, and the maximum compressive stress.

SOLUTION The formulas and table of case 4 (Table 14.1) are used. The race is taken as body 1 and the ball as body 2; hence $R_1 = -0.80$ in, $R_1' = -5$ in, and $R_2 = R_2' = 0.75$ in. Taking $E_1 = E_2 = 30{,}000{,}000$ lb/in^2 and $\nu_1 = \nu_2 = 0.3$, we have

$$C_E = \frac{2(1-0.09)}{30(10^6)} = 6.067(10^{-8})\ \text{in}^2/\text{lb}$$

$$K_D = \frac{1.5}{-1.25+1.33-0.20+1.33} = 1.233\ \text{in}$$

$$\cos\theta = \frac{1.233}{1.5}\sqrt{(-1.25+0.20)^2+0+0} = 0.863$$

From the table, by interpolation

$$\alpha = 2.710, \qquad \beta = 0.495, \qquad \lambda = 0.546$$

Then

$$c = 2.710\sqrt[3]{2000(1.233)(6.067)(10^{-8})} = 0.144\ \text{in}$$

$$d = 0.495\sqrt[3]{2000(1.233)(6.067)(10^{-8})} = 0.0263\ \text{in}$$

$$(\sigma_c)_{\max} = \frac{1.5(2000)}{0.144(0.0263)\pi} = 252{,}000\ \text{lb/in}^2$$

$$y = 0.546\sqrt[3]{\frac{2000^2(6.067^2)(10^{-16})}{1.233}} = 0.00125\ \text{in}$$

Therefore, the contact area is an ellipse with a major axis of 0.288 in and a minor axis of 0.0526 in.

14.2 RIVETS AND RIVETED JOINTS

Although the actual state of stress in a riveted joint is complex, it is customary—and experience shows it is permissible—to ignore such considerations as the stress concentration at the edges of rivet holes, unequal division of load among rivets, and nonuniform distribution of shear stress across the section of the rivet and of the bearing stress between rivet and plate. Simplifying assumptions are made, which may be summarized as follows: (1) The applied load is assumed to be transmitted entirely by the rivets, friction between the connected plates being ignored; (2) when the centroid of the rivet areas is on the line of action of the load, all the rivets of the joint are assumed to carry equal parts of the load if they are of the same size, or to be loaded proportionally to their respective section areas if they are of different sizes; (3) the shear stress is assumed to be uniformly distributed across the rivet section; (4) the bearing stress between plate and rivet is assumed to be uniformly distributed over an area equal to the rivet diameter times the plate thickness; (5) the stress in a tension member is assumed to be uniformly distributed over the net area; and (6) the stress in a compression member is assumed to be uniformly distributed over the gross area.

The design of riveted joints on the basis of these assumptions is the accepted practice, although none of them is strictly correct and methods of stress calculation that are supposedly more accurate have been proposed (Ref. 12).

Details of Design and Limitations

The possibility of secondary failure due to secondary causes, such as the shearing or tearing out of a plate between the rivet and the edge of a plate or between adjacent rivets, the bending or insufficient upsetting of long rivets, or tensile failure along a zigzag line when rivets are staggered, is guarded against in standard specifications (Ref. 13) by detailed rules for edge clearance, maximum grip of rivets, maximum pitch, and computing the net width of riveted parts. Provision is made for the use of high-strength bolts in place of rivets under certain circumstances (Ref. 42). Joints may be made by welding instead of riveting, but the use of welding in conjunction with riveting is not approved on new work; the division of the load as between the welds and the rivets would be indeterminate.

Tests on Riveted Joints

In general, tests on riveted joints show that although under working loads the stress conditions may be considerably at variance with the usual assumptions, the ultimate strength may be closely predicted by calculations based thereon. Some of the other conclusions drawn from such tests may be summarized as follows.

In either lap or double-strap butt joints in very wide plates, the unit tensile strength developed by the net section is greater than that developed by the plate itself when tested in full width and is practically equal to that developed by narrow tension specimens cut from the plate. The rivets in lap joints are as strong relative to undriven rivets tested in shear as are the rivets in butt joints. Lap joints bend sufficiently at stresses below the usual design stresses to cause opening of caulked joints (Ref. 14).

Although it is frequently specified that rivets shall not be used in tension, tests show that hot-driven buttonhead rivets develop a strength in direct tension greater than the strength of the rod from which they are made, and that they may be relied upon to develop this strength in every instance. Although the initial tension in such rivets due to cooling usually amounts to 70% or more of the yield strength, this initial tension does not reduce the ability of the rivets to resist an applied tensile load (see also Sec. 3.12). Unless a joint is subjected to reversals of primary load, the use of rivets in tension appears to be justified; but when the primary load producing shear in the rivets is reversed, the reduction in friction due to simultaneous rivet tension may permit slip to occur, with possible deleterious effects (Ref. 15).

With respect to the form of the rivet head, the rounded or buttonhead type is standard; but countersunk rivets are often used, and tests show that these develop the same ultimate strength, although they permit much more slip and deformation at working loads than do the buttonhead rivets (Ref. 16).

In designing riveted joints in very thin metals, especially the light alloys, it may be necessary to take into account factors that are not usually considered in ordinary structural-steel work, such as the radial stresses caused at the hole edges by closing pressure and the buckling of the plates under rivet pressure (Ref. 17).

Eccentric Loading

When the rivets of a joint are so arranged that the centroid G of the areas of the rivet group lies not on the line of action of the load but at a distance e therefrom, the load P can be replaced by an equal and parallel load P' acting through G and a couple Pe. The load on any one of the n rivets is then found by *vectorially* adding the load P/n due to P' and the load Q due to the couple Pe. This load Q acts normal to the line from G to the rivet and is given by the equation $Q = PeA_1r_1/J$, where A_1 is the area of the rivet in question, r_1 is its distance from G, and $J = \Sigma Ar^2$ for all the rivets of the group. When all rivets are of the same size, as is usually the case, the formula becomes $Q = Per_1/\Sigma r^2$. Charts and tables are available which greatly facilitate the labor of the calculation involved, and which make possible direct design of the joint without recourse to trial and error (Ref. 18). (The direct procedure, as outlined previously, is illustrated in the following example.)

The stiffness or resistance to angular displacement of a riveted joint determines the degree of fixity that should be assumed in the analysis of beams with riveted ends or of rectangular frames. Tests (Ref. 19) have shown that although joints made with wide gusset plates are practically rigid, joints made by simply riveting through clip angles are not even approximately so. A method of calculating the elastic constraint afforded by riveted joints of different types, based on an extensive series of tests, has been proposed by Rathbun (Ref. 20). Brombolich (Ref. 55) describes the use of a finite-element-analysis procedure to determine the effect of yielding, interference fits, and load sequencing on the stresses near fastener holes.

EXAMPLE Figure 14.1 represents a lap joint in which three 1-in rivets are used to connect a 15-in channel to a plate. The channel is loaded eccentrically as shown. It is required to determine the maximum shear stress in the rivets. (This is not, of course, a properly designed joint intended to develop the full strength of the channel. It represents a simple arrangement of rivets assumed for the purpose of illustrating the calculation of rivet stress due to a moment.)

SOLUTION The centroid of the rivet areas is found to be at G. The applied load is replaced by an equal load through G and a couple equal to

$$15{,}000 \times 5 = 75{,}000 \text{ lb-in}$$

as shown in Fig. 14.1(b). The distances r_1, r_2, and r_3 of rivets 1, 2, and 3, respectively, from G are as shown; the value of Σr^2 is 126. The loads on the rivets due to the couple of 75,000 lb-in are therefore

$$Q_1 = Q_2 = \frac{(75{,}000)(6.7)}{126} = 3990 \text{ lb}$$

$$Q_3 = \frac{(75{,}000)(6)}{126} = 3570 \text{ lb}$$

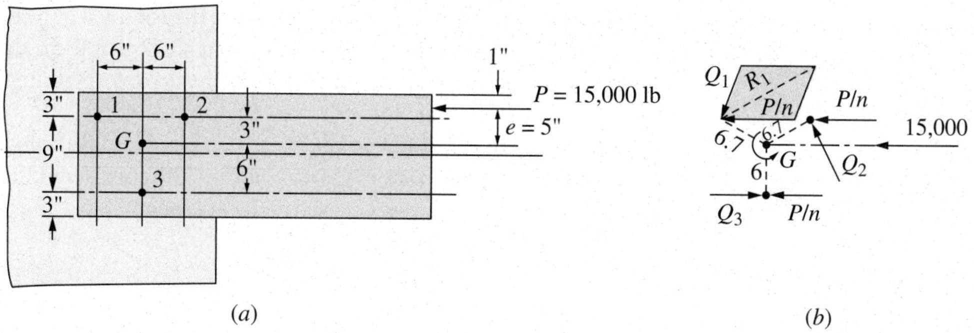

Figure 14.1

These loads act on the rivets in the directions shown. In addition, each rivet is subjected to a load in the direction of P' of $P/n = 5000$ lb. The resultant load on each rivet is then found by graphically (or algebraically) solving for the resultant of Q and P/n as shown. The resultant loads are $R_1 = R_2 = 7670$ lb; $R_3 = 1430$ lb. The maximum shear stress occurs in rivets 1 and 2 and is $\tau = 7670/0.785 = 9770$ lb/in^2.

14.3 MISCELLANEOUS CASES

In most instances, the stress in bodies subjected to direct shear or pressure is calculated on the basis of simplifying assumptions such as are made in analyzing a riveted joint. Design is based on rules justified by experience rather than exact theory, and a full discussion does not properly come within the scope of this book. However, a brief consideration of a number of cases is given here; a more complete treatment of these cases may be found in books on machine and structural design and in the references cited.

Pins and Bolts

These are designed on the basis of shear and bearing stress calculated in the same way as for rivets. In the case of pins bearing on wood, the allowable bearing stress must be reduced to provide for nonuniformity of pressure when the length of bolt is more than five or six times its diameter. When the pressure is inclined to the grain, the safe load is found by the formula

$$N = \frac{PQ}{P\sin^2\theta + Q\cos^2\theta}$$

where N is the safe load for the case in question, P is the safe load applied parallel to the grain, Q is the safe load applied transverse to the grain, and θ is the angle N makes with the direction of the grain (Ref. 21).

Hollow pins and rollers are thick-walled but can be analyzed as circular rings by the appropriate use of the formulas of Table 9.2. The loading is essentially as shown in Fig. 14.2, and the greatest circumferential stresses, which occur at points 1–4, may be found by the formula

$$\sigma = K\frac{2p}{\pi b}$$

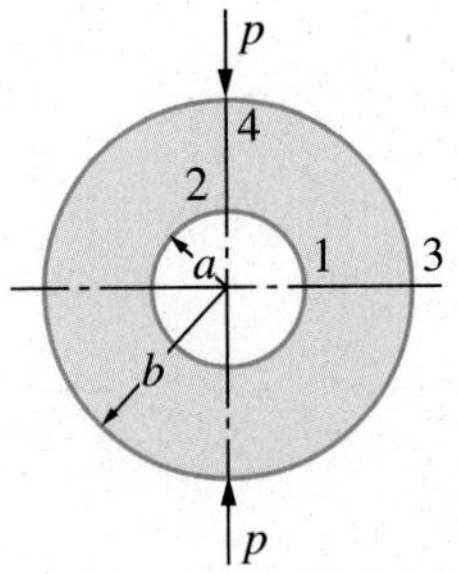

Figure 14.2

where p = load/unit length of the pin and the numerical coefficient K depends on the ratio a/b and has the following values [a plus sign for K indicates tensile stress and a minus sign compressive stress (Ref. 30)]:

	a/b							
Point	0	0.1	0.2	0.3	0.4	0.5	0.6	0.7
1	−5.0	−5.05	−5.30	−5.80	−7.00	−9.00	−12.9	−21.4
2	+3.0	+3.30	+3.80	+4.90	+7.00	+10.1	+16.0	+31.0
3	0	+0.06	+0.20	+1.0	+1.60	+3.0	+5.8	+13.1
4	+0.5	+0.40	0	−0.50	−1.60	−3.8	−8.4	−19.0

For changes in the mean vertical and horizontal diameters see case 1 in Table 9.2. Durelli and Lin in Ref. 59 have made extensive use of Nelson's equations for diametrically loaded hollow circular cylinders from Ref. 60 and present, in graphical form, stress factors and radial displacements at all angular positions along both inner and outer boundaries. Results are plotted for the radius ratio a/b from near zero to 0.92.

Gear Teeth

Gear teeth may be investigated by considering the tooth as a cantilever beam, the critical stress being the tensile bending stress at the base. This stress can be calculated by the modified Heywood formula for a very short cantilever beam (Sec. 8.10) or by a combination of the modified Lewis formula and stress concentration factor given for case 21 in Table 17.1 (see also Refs. 22–24). The allowable stress is reduced according to speed of operation by one of several empirical formulas (Ref. 24). Under certain conditions, the bearing stress between teeth may become important (especially as this stress affects wear), and this stress may be calculated by the formula for case 2b, Table 14.1. The total deformation of the tooth, the result of direct compression at the point of contact and of beam deflection and shear, may be calculated by the formula of case 2b and the methods of Sec. 8.1 (Ref. 23).

Keys

Keys are designed for a total shearing force $F = T/r$ (Fig. 14.3), where T represents the torque transmitted. The shear stress is assumed to be uniformly distributed over the horizontal section AB, and the bearing stress is assumed to be uniformly distributed over half

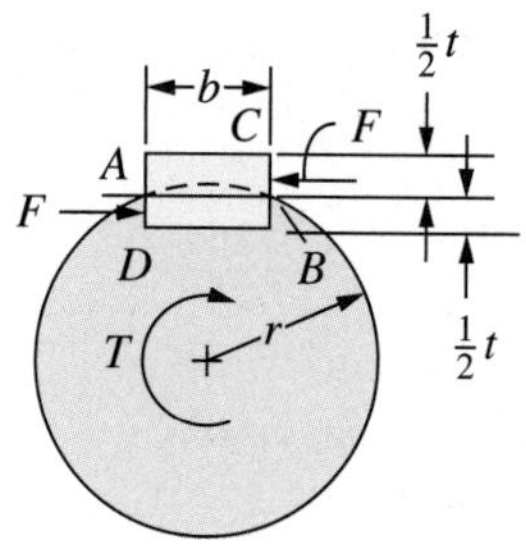

Figure 14.3

the face. These assumptions lead to the following formulas: $\tau = F/Lb$; $\sigma_b = 2F/tL$ on the sides; and $\sigma_b = 2Ft/b^2L$ on top and bottom. Here L is the length of the key; in conventional design $4b < L < 16b$. As usually made, $b \geq t$; hence the bearing stress on the sides is greater than that on the top and bottom.

Photoelastic analysis of the stresses in square keys shows that the shear stress is not uniform across the breadth b but is greatest at A and B, where it may reach a value two to four times the average value (Ref. 25). Undoubtedly the shear stress also varies in intensity along the length of the key. The bearing stresses on the surfaces of the key are also nonuniform, that on the sides being greatest near the common surface of shaft and hub, and that on the top and bottom being greatest near the corners C and D. When conservative working stresses are used, however, and the proportions of the key are such as have been found satisfactory in practice, the approximate methods of stress calculation that have been indicated result in satisfactory design.

Fillet Welds

These are successfully designed on the basis of uniform distribution of shear stress on the longitudinal section of least area, although analysis and tests show that there is considerable variation in the intensity of shear stress along the length of the fillet (Refs. 26 and 27). (Detailed recommendations for the design of welded structural joints are given in Ref. 13.)

Screwthreads

The strength of screwthreads is of great importance in the design of joints, where the load is transferred to a bolt or stud by a nut. A major consideration is the load distribution. The load is not transferred uniformly along the engaged thread length; both mathematical analysis and tests show that the maximum load per linear inch of thread, which occurs near the loaded face of the nut, is several times as great as the average over the engaged length. This ratio, called the *thread-load concentration factor* and denoted by H, is often 2, 3, or even 4 (Ref. 43). The maximum load per linear inch on a screwthread is therefore the total load divided by the helical length of the engaged screwthread times H. The maximum stress due to this loading can be computed by the Heywood–Kelley–Pedersen formula for a short cantilever, as given in Sec. 8.10. It is important to note that in some cases the values of k_f given in the literature are for loading through a nut, and so include H, while in other cases (as in rotating-beam tests) this influence is absent. Because of the combined effects of reduced area, nonuniform load distribution, and stress concentration, the efficiency of a bolted joint under reversed repeated loading is likely to be quite small. In Ref. 28 of Chap. 3, values from 18% (for a 60,000 lb/in^2 steel with rolled threads) to 6.6% (for a 200,000 lb/in^2 steel with machine-cut threads) are cited.

The design of bolted connections has received much study, and an extensive discussion and bibliography are given in Heywood (Chap. 3, Ref. 28) and in some of the papers of Ref. 32 of Chap. 3.

14.4 TABLE

TABLE 14.1 Formulas for Stress and Strain Due to Pressure on or between Elastic Bodies

Notation: P = total load; p = load per unit length; a = radius of circular contact area for case 1; b = width of rectangular contact area for case 2; c = major semiaxis and d = minor semiaxis of elliptical contact area for cases 3 and 4; y = relative motion of approach along the axis of loading of two points, one in each of the two contact bodies, remote from the contact zone; ν = Poisson's ratio; E = modulus of elasticity. Subscripts 1 and 2 refer to bodies 1 and 2, respectively. To simplify expressions let

$$C_E = \frac{1-\nu_1^2}{E_1} + \frac{1-\nu_2^2}{E_2}$$

Conditions and Case No.	Formulas	
1. Sphere 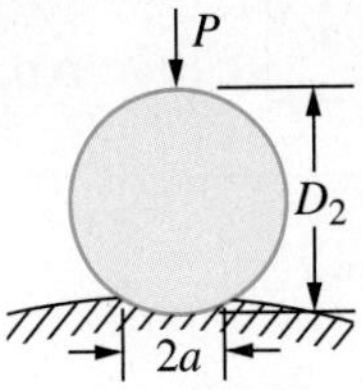	$a = 0.721\sqrt[3]{PK_DC_E}$ $(\sigma_c)_{max} = 1.5\dfrac{P}{\pi a^2} = 0.918\sqrt[3]{\dfrac{P}{K_D^2C_E^2}}$ $y = 1.040\sqrt[3]{\dfrac{P^2C_E^2}{K_D}}$	1a. Sphere on a plate 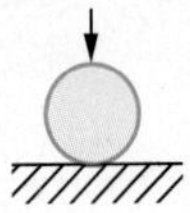 $K_D = D_2$
	If $E_1 = E_2 = E$ and $\nu_1 = \nu_2 = 0.3$, then $a = 0.881\sqrt[3]{\dfrac{PK_D}{E}}$ (Note: 50% of y occurs within a distance of 1.2 times the contact radius a and 90% within a distance $7a$ from the contact zone.) $(\sigma_c)_{max} = 0.616\sqrt[3]{\dfrac{PE^2}{K_D^2}}$ $y = 1.55\sqrt[3]{\dfrac{P^2}{E^2K_D}}$	1b. Sphere on a sphere 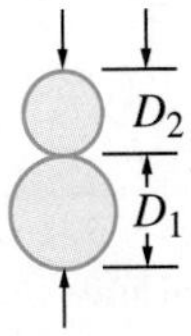 $K_D = \dfrac{D_1D_2}{D_1+D_2}$
	$(\sigma_t)_{max} \approx 0.133(\sigma_c)_{max}$ radially at the edge of contact area $\tau_{max} \approx \frac{1}{3}(\sigma_c)_{max}$ at a point on the load line a distance $a/2$ below the contact surface (Approximate stresses from Refs. 3 and 6) For graphs of subsurface stress variations, see Refs. 6 and 57.	1c. Sphere in a spherical socket 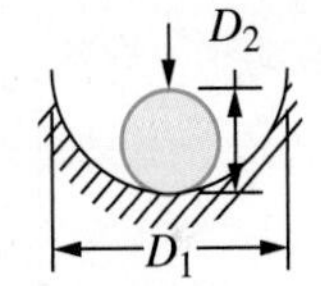 $K_D = \dfrac{D_1D_2}{D_1-D_2}$

(Continued)

TABLE 14.1 Formulas for Stress and Strain Due to Pressure on or between Elastic Bodies (*Continued*)

<table>
<tr><th>Conditions and Case No.</th><th colspan="2">Formulas</th></tr>
<tr>
<td rowspan="3">2. Cylinder of length L large as compared with D; p = load per unit length = P/L
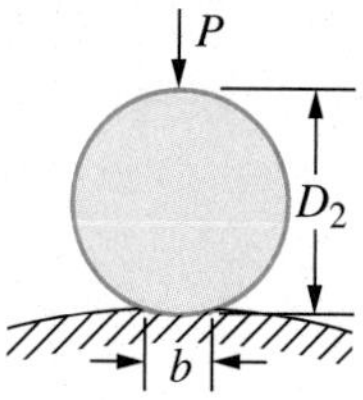</td>
<td rowspan="3">$b = 1.60\sqrt{pK_D C_E}$
$(\sigma_c)_{max} = 0.798\sqrt{\dfrac{p}{K_D C_E}}$
If $E_1 = E_2 = E$ and $\nu_1 = \nu_2 = 0.3$, then
$b = 2.15\sqrt{\dfrac{pK_D}{E}}$
$(\sigma_c)_{max} = 0.591\sqrt{\dfrac{pE}{K_D}}$
For a cylinder between two flat plates
$\Delta D_2 = \dfrac{4p(1-\nu^2)}{\pi E}\left(\dfrac{1}{3} + \ln\dfrac{2D}{b}\right)$ Refs. 5 and 44
For a cylinder on a cylinder the distance between centers is reduced by
$\dfrac{2p(1-\nu^2)}{\pi E}\left(\dfrac{2}{3} + \ln\dfrac{2D_1}{b} + \ln\dfrac{2D_2}{b}\right)$ Ref. 31
For graphs of subsurface stress variations, see Refs. 6 and 56.</td>
<td>2a. Cylinder on a flat plate
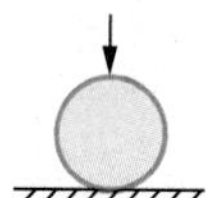
$K_D = D_2$
$\tau_{max} \approx \frac{1}{3}(\sigma_c)_{max}$ at a depth of $0.4b$ below the surface of the plane</td>
</tr>
<tr><td>2b. Cylinder on a cylinder
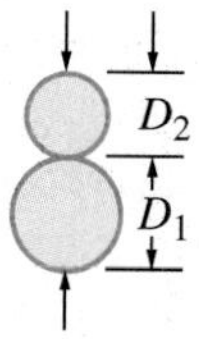
$K_D = \dfrac{D_1 D_2}{D_1 + D_2}$</td></tr>
<tr><td>2c. Cylinder in a cylindrical socket
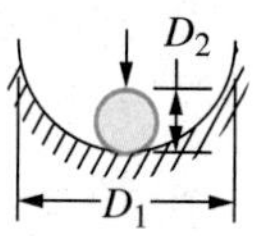
$K_D = \dfrac{D_1 D_2}{D_1 - D_2}$</td></tr>
<tr>
<td>3. Cylinder on a cylinder; axes at right angles
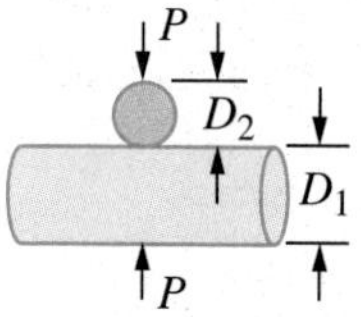</td>
<td colspan="2">$c = \alpha\sqrt[3]{PK_D C_E}$ $K_D = \dfrac{D_1 D_2}{D_1 + D_2}$ and α, β, and λ depend upon $\dfrac{D_1}{D_2}$ as shown
$d = \beta\sqrt[3]{PK_D C_E}$
$(\sigma_c)_{max} = \dfrac{1.5P}{\pi cd}$
$y = \lambda\sqrt[3]{\dfrac{P^2 C_E^2}{K_D}}$
$\tau_{max} = \dfrac{1}{3}(\sigma_c)_{max}$
<table>
<tr><th>D_1/D_2</th><th>1</th><th>1.5</th><th>2</th><th>3</th><th>4</th><th>6</th><th>10</th></tr>
<tr><td>α</td><td>0.908</td><td>1.045</td><td>1.158</td><td>1.350</td><td>1.505</td><td>1.767</td><td>2.175</td></tr>
<tr><td>β</td><td>0.908</td><td>0.799</td><td>0.732</td><td>0.651</td><td>0.602</td><td>0.544</td><td>0.481</td></tr>
<tr><td>γ</td><td>0.825</td><td>0.818</td><td>0.804</td><td>0.774</td><td>0.747</td><td>0.702</td><td>0.641</td></tr>
</table></td>
</tr>
<tr>
<td>4. General case of two bodies in contact; P = total load
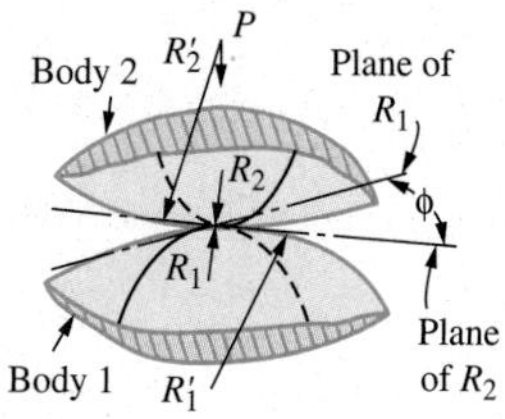</td>
<td colspan="2">At point of contact minimum and maximum radii of curvature are R_1 and R_1' for body 1, and R_2 and R_2' for body 2. Then $1/R_1$ and $1/R_1'$ are principal curvatures of body 1, and $1/R_2$ and $1/R_2'$ of body 2; and in each body the principal curvatures are mutually perpendicular. The radii are positive if the center of curvature lies within the given body, i.e., the surface is convex, and negative otherwise. The plane containing curvature $1/R_1$ in body 1 makes with the plane containing curvature $1/R_2$ in body 2 the angle ϕ. Then:
$c = \alpha\sqrt[3]{PK_D C_E}$ $d = \beta\sqrt[3]{PK_D C_E}$ $(\sigma_c)_{max} = \dfrac{1.5P}{\pi cd}$ and $y = \lambda\sqrt[3]{\dfrac{P^2 C_E^2}{K_D}}$ where $K_D = \dfrac{1.5}{1/R_1 + 1/R_2 + 1/R_1' + 1/R_2'}$
and α, β, and λ are given by the following table in which
$\cos\theta = \dfrac{K_D}{1.5}\sqrt{\left(\dfrac{1}{R_1} - \dfrac{1}{R_1'}\right)^2 + \left(\dfrac{1}{R_2} - \dfrac{1}{R_2'}\right)^2 + 2\left(\dfrac{1}{R_1} - \dfrac{1}{R_1'}\right)\left(\dfrac{1}{R_2} - \dfrac{1}{R_2'}\right)\cos 2\phi}$
<table>
<tr><th>$\cos\theta$</th><th>0.00</th><th>0.10</th><th>0.20</th><th>0.30</th><th>0.40</th><th>0.50</th><th>0.60</th><th>0.70</th><th>0.75</th><th>0.80</th><th>0.85</th><th>0.90</th><th>0.92</th><th>0.94</th><th>0.96</th><th>0.98</th><th>0.99</th></tr>
<tr><td>α</td><td>1.000</td><td>1.070</td><td>1.150</td><td>1.242</td><td>1.351</td><td>1.486</td><td>1.661</td><td>1.905</td><td>2.072</td><td>2.292</td><td>2.600</td><td>3.093</td><td>3.396</td><td>3.824</td><td>4.508</td><td>5.937</td><td>7.774</td></tr>
<tr><td>β</td><td>1.000</td><td>0.936</td><td>0.878</td><td>0.822</td><td>0.769</td><td>0.717</td><td>0.664</td><td>0.608</td><td>0.578</td><td>0.544</td><td>0.507</td><td>0.461</td><td>0.438</td><td>0.412</td><td>0.378</td><td>0.328</td><td>0.287</td></tr>
<tr><td>λ</td><td>0.750</td><td>0.748</td><td>0.743</td><td>0.734</td><td>0.721</td><td>0.703</td><td>0.678</td><td>0.644</td><td>0.622</td><td>0.594</td><td>0.559</td><td>0.510</td><td>0.484</td><td>0.452</td><td>0.410</td><td>0.345</td><td>0.288</td></tr>
</table>
(Ref. 8)</td>
</tr>
</table>

(*Continued*)

TABLE 14.1 Formulas for Stress and Strain Due to Pressure on or between Elastic Bodies (*Continued*)

Conditions and Case No.	Formulas
5. Rigid knife-edge across edge of semi-infinite plate; load $p = P/t$ where t is plate thickness 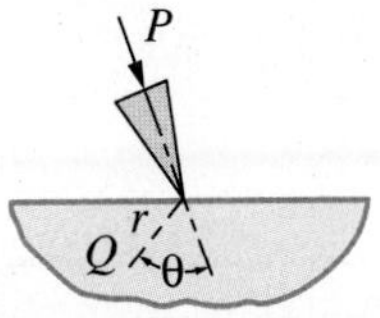	At any point Q, $\sigma_c = \dfrac{2p\cos\theta}{\pi r}$ in the direction of the radius r (Ref. 6)
6. Rigid block of width $2b$ across edge of semi-infinite plate; load $p = P/t$ where t is plate thickness 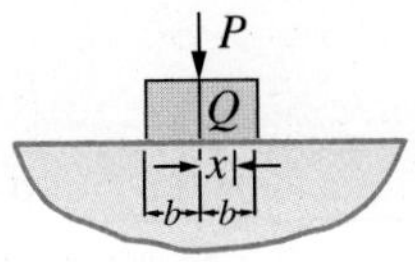	At any point Q on surface of contact, $\sigma_c = \dfrac{p}{\pi\sqrt{b^2 - x^2}}$ (For loading on block of finite width and influence of distance of load from corner see Ref. 45.) (Ref. 6)
7. Uniform pressure q over length L across edge of semi-infinite plate 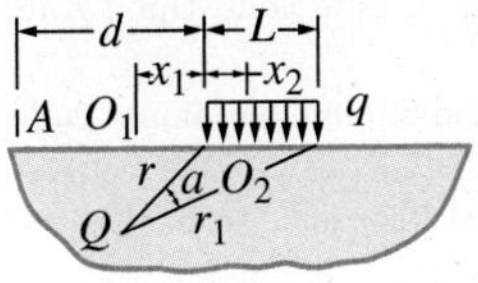	At any point O_1 outside loaded area, $y = \dfrac{2q}{\pi E}\left[(L + x_1)\ln\dfrac{d}{L + x_1} - x_1\ln\dfrac{d}{x_1}\right] + qL\dfrac{1-\nu}{\pi E}$ At any point O_2 inside loaded area, $y = \dfrac{2q}{\pi E}\left[(L - x_2)\ln\dfrac{d}{L - x_2} + x_2\ln\dfrac{d}{x_2}\right] + qL\dfrac{1-\nu}{\pi E}$ where y = deflection relative to a remote point A a distance d from edge of loaded area At any point Q, $\sigma_c = 0.318q(\alpha + \sin\alpha)$ $\tau = 0.318q\sin\alpha$ (Ref. 6)
8. Rigid cylindrical die of radius R on surface of semi-infinite body; total load P 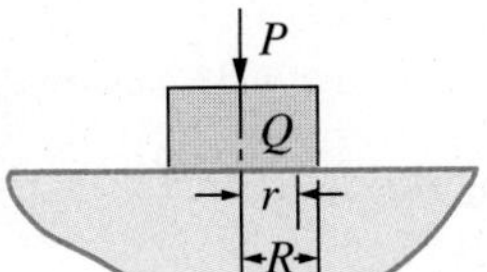	$y = \dfrac{P(1-\nu)^2}{2RE}$ At any point Q on surface of contact, $\sigma_c = \dfrac{P}{2\pi R\sqrt{R^2 - r^2}}$ $(\sigma_c)_{max} = \infty$ at edge (theoretically) $(\sigma_c)_{min} = \dfrac{P}{2\pi R^2}$ at center (Ref. 6)

(*Continued*)

TABLE 14.1 Formulas for Stress and Strain Due to Pressure on or between Elastic Bodies (*Continued*)

Conditions and Case No.	Formulas
9. Uniform pressure q over circular area of radius R on surface of semi-infinite body 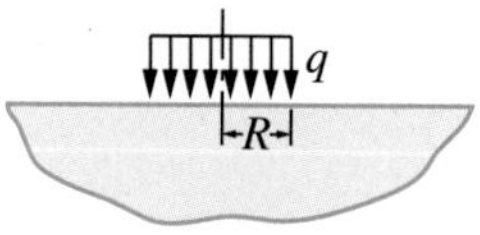	$y_{max} = \frac{2qR(1-\nu^2)}{E}$ at center y at edge $= \frac{4qR(1-\nu^2)}{\pi E}$ $\tau_{max} = 0.33q$ at point $0.638R$ below center of loaded area
10. Uniform pressure q over square area of sides $2b$ on the surface of semi-infinite body 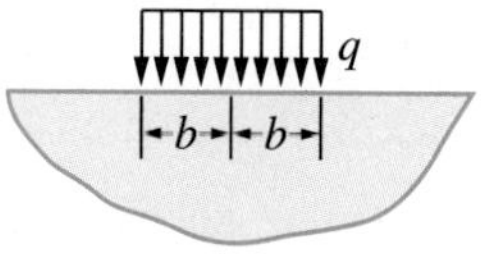	$y_{max} = \frac{2.24qb(1-\nu^2)}{E}$ at center $y = \frac{1.12qb(1-\nu^2)}{E}$ at corners $y_{average} = \frac{1.90qb(1-\nu^2)}{E}$ (Ref. 6)

14.5 REFERENCES

1. Hertz, H., *Gesammelte Werke*, vol. I, Leipzig, 1895.
2. Thomas, H. R., and V. A. Hoersch, "Stresses Due to the Pressure of One Elastic Solid Upon Another," *Eng. Exp. Sta. Univ. Ill.*, Bull. 212, 1930.
3. Oppel, G., "The Photoelastic Investigation of Three-Dimensional Stress and Strain Conditions," *Natl. Adv. Comm. Aeron.*, Tech. Memo. 824, 1937.
4. Jensen, V. P., "Stress Analysis by Means of Polarized Light with Special Reference to Bridge Rollers," *Bull. Assoc. State Eng. Soc.*, October 1936.
5. Föppl, A., *Technische Mechanik*, 4th ed., vol. 5, p. 350.
6. Timoshenko, S., and J. N. Goodier, *Theory of Elasticity*, 2nd ed., McGraw-Hill, 1951.
7. Wilson, W. M., "The Bearing Value of Rollers," *Eng. Exp. Sta. Univ. Ill.*, Bull. 263, 1934.
8. Whittemore, H. L., and S. N. Petrenko, "Friction and Carrying Capacity of Ball and Roller Bearings," *Tech. Paper Bur. Stand.*, No. 201, 1921.
9. Wilson,W. M., R. L. Moore, and F. P. Thomas, "Bearing Value of Pivots for Scales," *Eng. Exp. Sta. Univ. Ill.*, Bull. 242, 1932.
10. Manual of the American Railway Engineering Association, 1936.
11. Wilson, W. M., "Rolling Tests of Plates," *Eng. Exp. Sta. Univ. Ill.*, Bull. 191, 1929.
12. Hrenikoff, A., "Work of Rivets in Riveted Joints," *Trans. Am. Soc. Civil Eng.*, 99: 437, 1934.
13. Specifications for the Design, Fabrication, and Erection of Structural Steel for Buildings, and Commentary, American Institute of Steel Construction, 1969.
14. Wilson, W. M., J. Mather, and C. O. Harris, "Tests of Joints in Wide Plates," *Eng. Exp. Sta. Univ. Ill.*, Bull. 239, 1931.
15. Wilson, W. M., and W. A. Oliver, "Tension Tests of Rivets," *Eng. Exp. Sta. Univ. Ill.*, Bull. 210, 1930.
16. Kommers, J. B., "Comparative Tests of Button Head and Countersunk Riveted Joints," *Bull. Eng. Exp. Sta. Univ., Wis.*, 9(5): 1925.
17. Hilbes, W., "Riveted Joints in Thin Plates," *Natl. Adv. Comm. Aeron.*, Tech. Memo. 590.
18. Dubin, E. A., "Eccentric Riveted Connections," *Trans. Am. Soc. Civil Eng.*, 100: 1086, 1935.
19. Wilson, W. M., and H. F. Moore, "Tests to Determine the Rigidity of Riveted Joints of Steel Structures," *Eng. Exp. Sta. Univ. Ill.*, Bull. 104, 1917.
20. Rathbun, J. C., "Elastic Properties of Riveted Connections," *Trans. Am. Soc. Civil Eng.*, 101: 524, 1936.
21. *Wood Handbook*, Forest Products Laboratory, U.S. Dept. of Agriculture, 1987.
22. Baud, R. V., and R. E. Peterson, "Loads and Stress Cycles in Gear Teeth," *Mech. Eng.*, 51: 653, 1929.
23. Timoshenko, S., and R. V. Baud, "The Strength of Gear Teeth," *Mech. Eng.*, 48: 1108, 1926.

24. "Dynamic Loads on Gear Teeth," *ASME Res. Pub.*, 1931.
25. Solakian, A. G., and G. B. Karelitz, "Photoelastic Study of Shearing Stress in Keys and Keyways," *Trans. ASME*, 54(11): 97, 1932.
26. Troelsch, H. W. "Distributions of Shear in Welded Connections," *Trans. Am Soc. Civil Eng.*, 99: 409, 1934.
27. Report of Structural Steel Welding Committee of the American Bureau of Welding, 1931.
28. Belajef, N. M., "On the Problem of Contact Stresses," *Bull. Eng. Ways Commun.*, St. Petersburg, 1917.
29. Belajef, N. M., "Computation of Maximal Stresses Obtained from Formulas for Pressure in Bodies in Contact," *Bull. Eng. Ways Commun.*, Leningrad, 1929.
30. Horger, O. J., "Fatigue Tests of Some Manufactured Parts," *Proc. Soc. Exp. Stress Anal.*, 3(2): 135, 1946.
31. Radzimovsky, E. I., "Stress Distribution and Strength Conditions of Two Rolling Cylinders Pressed Together," *Eng. Exp. Sta. Univ. Ill.*, Bull. 408, 1953.
32. Wilson,W. M., "Tests on the Bearing Value of Large Rollers," *Univ. Ill., Eng. Exp. Sta.*, Bull. 162, 1927.
33. New Departure, Division of General Motors Corp., *Analysis of Stresses and Deflections*, Bristol, Conn., 1946.
34. Lundberg, G., and H. Sjovall, *Stress and Deformation in Elastic Contacts*, Institution of Theory of Elasticity and Strength of Materials, Chalmers University of Technology, Gothenburg, 1958.
35. Smith, J. O., and C. K. Lin, "Stresses Due to Tangential and Normal Loads on an Elastic Solid with Application to Some Contact Stress Problems," *ASME J. Appl. Mech.*, 20(2): 1953.
36. Palmgren, Arvid, *Ball and Roller Bearing Engineering*, 3rd ed., SKF Industries Inc., 1959.
37. Butler, R. H., H. R. Bear, and T. L. Carter, "Effect of Fiber Orientation on Ball Failure," *Natl. Adv. Comm. Aeron.*, Tech. Note 3933 (also Tech. Note 3930).
38. Selection of Bearings, Timken Roller Bearing Co.
39. Corwin, A. H.: *Techniques of Organic Chemistry*, 3rd ed., vol. 1, part 1, Interscience, 1959.
40. Styri, Haakon, "Fatigue Strength of Ball Bearing Races and Heat-Treated Steel Specimens," *Proc. ASTM*, vol. 51, 1951.
41. Burwell, J. T., Jr. (ed.), *Mechanical Wear*, American Society for Metals, 1950.
42. Specifications for Assembly of Structural Joints Using High Strength Steel Bolts, distributed by American Institute of Steel Construction; approved by Research Council on Riveted and Bolted Structural Joints of the Engineering Foundation; endorsed by American Institute of Steel Construction and Industrial Fasteners Institute.
43. Sopwith, D. C., "The Distribution of Load in Screw Threads," *Proc. Inst. Mech. Eng.*, vol. 159, 1948.
44. Lundberg, Gustaf, "Cylinder Compressed between Two Plane Bodies," reprint courtesy of SKF Industries Inc., 1949.
45. Hiltscher, R., and G. Florin, "Spalt- und Abreisszugspannungen in rechteckingen Scheiben, die durch eine Last in verschiedenem Abstand von einer Scheibenecke belastet sind," *Die Bautech.*, 12: 1963.
46. MacGregor, C. W. (ed.), *Handbook of Analytical Design for Wear*, IBM Corp., Plenum Press, 1964.
47. Hamilton. G. M., and L. E. Goodman, "The Stress Field Created by a Circular Sliding Contact," *ASME J. Appl. Mech.*, 33(2): 1966.
48. Goodman, L. E., "Contact Stress Analysis of Normally Loaded Rough Spheres," *ASME J. Appl. Mech.*, 29(3): 1962.
49. O'Connor, J. J., "Compliance Under a Small Torsional Couple of an Elastic Plate Pressed Between Two Identical Elastic Spheres," *ASME J. Appl. Mech.*, 33(2) 1966.
50. Tu, Yih-O, "A Numerical Solution for an Axially Symmetric Contact Problem," *ASME J. Appl. Mech.*, 34(2): 1967.
51. Pu, S. L., and M. A. Hussain, "Note on the Unbonded Contact Between Plates and an Elastic Half Space," *ASME J. Appl. Mech.*, 37(3): 1970.
52. Goodman, L. E., and L. M. Keer, "The Contact Stress Problem for an Elastic Sphere Indenting an Elastic Cavity," *Int. J. Solids Struct.*, 1: 1965.
53. Cooper, D. H., "Hertzian Contact-Stress Deformation Coefficients," *ASME J. Appl. Mech.*, 36(2): 1969.
54. Dundurs, J., and M. Stippes, "Role of Elastic Constants in Certain Contact Problems," *ASME J. Appl. Mech.*, 37(4): 1970.
55. Brombolich, L. J., "Elastic-Plastic Analysis of Stresses Near Fastener Holes," McDonnell Aircraft Co., MCAIR 73-002, January 1973.
56. Sague, J. E., "The Special Way Big Bearings Can Fail," *Mach. Des.*, September 1978.
57. Cook, R. D., and W. C. Young, *Advanced Mechanics of Materials*, 2nd ed., Prentice-Hall, 1999.
58. Shukla, A., and H. Nigam, "A Numerical-Experimental Analysis of the Contact Stress Problem," *Inst. Mech. Eng. J. Strain Anal.*, 30(4): 1985.
59. Durelli, A. J., and Y. H. Lin, "Stresses and Displacements on the Boundaries of Circular Rings Diametrically Loaded," *ASME J. Appl. Mech.*, 53(1): 1986.
60. Nelson, C. W., "Stresses and Displacements in a Hollow Circular Cylinder," Ph.D. thesis, University of Michigan, 1939.

CHAPTER 15

Elastic Stability

15.1 GENERAL CONSIDERATIONS

Failure through elastic instability has been discussed briefly in Sec. 3.13, where it was pointed out that it may occur when the bending or twisting effect of an applied load is proportional to the deformation it produces. In this chapter, formulas for the critical load or critical unit stress at which such failure occurs are given for a wide variety of members and conditions of loading.

Such formulas can be derived mathematically by integrating the differential equation of the elastic curve or by equating the strain energy of bending to the work done by the applied load in the corresponding displacement of its point of application, the form of the elastic curve being assumed when unknown. Of all possible forms of the curve, that which makes the critical load a minimum is the correct one; but almost any reasonable assumption (consistent with the boundary conditions) can be made without gross error resulting, and for this reason the strain-energy method is especially adapted to the approximate solution of difficult cases. A very thorough discussion of the general problem, with detailed solutions of many specified cases, is given in Timoshenko and Gere (Ref. 1), from which many of the formulas in this chapter are taken. Formulas for many cases are also given in Refs. 35 and 36; in addition Ref. 35 contains many graphs of numerically evaluated coefficients.

At one time, most of the problems involving elastic stability were of academic interest only since engineers were reluctant to use compression members so slender as to fail by buckling at elastic stresses and danger of corrosion interdicted the use of very thin material in exposed structures. The requirements for minimum-weight construction in the fields of aerospace and transportation, however, have given great impetus to the theoretical and experimental investigation of elastic stability and to the use of parts for which it is a governing design consideration.

There are certain definite advantages in lightweight construction, in which stability determines strength. One is that since elastic buckling may occur without damage, part of a structure—such as the skin of an airplane wing or web of a deep beam—may be used safely at loads that cause local buckling, and under these circumstances the resistance afforded by the buckled part is definitely known. Furthermore, members such as Euler columns may be loaded experimentally to their maximum capacity without damage or permanent deformation and subsequently incorporated in a structure.

15.2 BUCKLING OF BARS

In Table 15.1, formulas are given for the critical loads on columns, beams, and shafts. In general, the theoretical values are in good agreement with test results as long as the assumed conditions are reasonably well-satisfied. It is to be noted that even slight changes in the amount of end constraint have a marked effect on the critical loads, and therefore it is important that such constraint be closely estimated. Slight irregularities in form and small accidental eccentricities are less likely to be important in the case of columns than in the case of thin plates. For latticed columns or columns with tie plates, a reduced value of E may be used, calculated as shown in Sec. 12.3. Formulas for the elastic buckling of bars may be applied to conditions under which proportional limit is exceeded if a reduced value of E corresponding to the actual stress is used (Ref. 1), but the procedure requires a stress-strain diagram for the material and, in general, is not practical.

In Table 15.1, cases 1–3, the tabulated buckling coefficients are worked out for various combinations of concentrated and distributed axial loads. Tensile end loads are included so that the effect of axial end restraint under axial loading within the column length can be considered (see the example at the end of this section). Carter and Gere (Ref. 46) present graphs of buckling coefficients for columns with single tapers for various end conditions, cross sections, and degrees of taper. Culver and Preg (Ref. 47) investigate and tabulate buckling coefficients for singly tapered beam-columns in which the effect of torsion, including warping restraint, is considered for the case where the loading is by end moments in the stiffer principal plane.

Kitipornchai and Trahair describe (Ref. 55) the lateral stability of singly tapered cantilever and doubly tapered simple I-beams, including the effect of warping restraint; experimental results are favorably compared with numerical solutions. Morrison (Ref. 57) considers the effect of lateral restraint of the tensile flange of a beam under lateral buckling; example calculations are presented. Massey and McGuire (Ref. 54) present graphs of buckling coefficients for both stepped and tapered cantilever beams; good agreement with experiments is reported. Tables of lateral stability constants for laminated timber beams are presented in Fowler (Ref. 53) along with two design examples.

Clark and Hill (Ref. 52) derive a general expression for the lateral stability of unsymmetrical I-beams with boundary conditions based on both bending and warping supports; tables of coefficients as well as nomographs are presented. Anderson and Trahair (Ref. 56) present tabulated lateral buckling coefficients for uniformly loaded and endloaded cantilevers and center- and uniformly loaded simply supported beams having unsymmetric I-beam cross sections; favorable comparisons are made with extensive tests on cantilever beams.

The Southwell plot is a graph in which the lateral deflection of a column or any other linearly elastic member undergoing a manner of loading which will produce buckling is plotted versus the lateral deflection divided by the load; the slope of this line gives the critical load. For columns and some frameworks, significant deflections do occur within the range where small-deflection theory is applicable. If the initial imperfections are such that experimental readings of lateral deflection must be taken beyond the small-deflection region, then the Southwell procedure is not adequate. Roorda (Ref. 93) discusses the extension of this procedure into the nonlinear range.

Bimetallic Beams

Burgreen and Manitt (Ref. 48) and Burgreen and Regal (Ref. 49) discuss the analysis of bimetallic beams and point out some of the difficulties in predicting the *snap-through instability* of these beams under changes in temperature. The thermal expansion of the support structure is an important design factor.

Rings and Arches

Austin (Ref. 50) tabulates in-plane buckling coefficients for circular, parabolic, and catenary arches for pinned and fixed ends as well as for the three-hinged case; he considers cases where the cross section varies with the position in the span as well as the usual case of a uniform cross section. Uniform loads, unsymmetric distributed loads, and concentrated center loads are considered, and the stiffening effect of tying the arch to the girder with columns is also evaluated. (The discussion referenced with the paper gives an extensive bibliography of work on arch stability.)

A thin ring shrunk by cooling and inserted into a circular cavity usually will yield before buckling unless the radius/thickness ratio is very large and the elastic-limit stress is high. Chicurel (Ref. 51) derives approximate solutions to this problem when the effect of friction is considered. He suggests a conservative expression for the *no-friction* condition: $P_o/AE = 2.67(k/r)^{1.2}$, where P_o is the prebuckling hoop compressive force, A is the hoop cross-sectional area, E is the modulus of elasticity, k is the radius of gyration of the cross section, and r is the radius of the ring.

EXAMPLE A 4-in steel pipe is to be used as a column to carry 8000 lb of transformers centered axially on a platform 20 ft above the foundation. The factor of safety *FS* is to be determined for the following conditions, based on elastic buckling of the column:

(*a*) The platform is supported only by the pipe fixed at the foundation.

(*b*) A 3½-in steel pipe is to be slipped into the top of the 4-in pipe a distance of 4 in, welded in place, and extended 10 ft to the ceiling above, where it will extend through a close-fitting hole in a steel plate.

(*c*) This condition is the same as in (*b*) except that the 3½-in pipe will be welded solidly into a heavy steel girder passing 10 ft above the platform.

SOLUTION A 4-in steel pipe has a cross-sectional area of 3.174 in^2 and a bending moment of inertia of 7.233 in^4. For a 3½ -in pipe these are 2.68 in^2 and 4.788 in^4, respectively.

(*a*) This case is a column fixed at the bottom and free at the top with an end load only. In Table 15.1, case la, for $I_2/I_1 = 1.00$ and $P_2/P_1 = 0$, K_1 is given as 0.25. Therefore,

$$P_1' = 0.25\frac{\pi^2 30(10^6)(7.233)}{240^2} = 9295 \text{ lb}$$

$$FS = \frac{9295}{8000} = 1.162$$

(*b*) This case is a column fixed at the bottom and pinned at the top with a load at a distance of two-thirds the 30-ft length from the bottom: $I_1 = 4.788$ in^4, $I_2 = 7.233$ in^4, and $I_2/I_1 = 1.511$. In Table 15.1, case 2d, $E_2I_2/E_1I_1 = 1.5$, $P_1/P_2 = 0$, and $a/l = 2/3$, K_2 is given as 6.58. Therefore,

$$P_2' = 6.58\frac{\pi^2 30(10^6)(4.788)}{360^2} = 72{,}000 \text{ lb}$$

$$FS = \frac{72{,}000}{8000} = 9$$

(*c*) This case is a column fixed at both ends and subjected to an upward load on top and a downward load at the platform. The upward load depends to some extent on the stiffness of the girder to which the top is welded, and so we can only bracket the actual critical load. If we assume the girder is infinitely rigid and permits no vertical deflection of the top, the elongation of the upper 10 ft would equal the reduction in length of the lower 20 ft.

Equating these deformations gives

$$\frac{P_1(10)(12)}{2.68(30)(10^6)} = \frac{(P_2 - P_1)(20)(12)}{3.174(30)(10^6)} \qquad \text{or} \qquad P_1 = 0.628P_2$$

From Table 15.1, case 2e, for $E_2I_2/E_1I_1 = 1.5$ and $a/l = 2/3$, we find the following values of K_2 for the several values of P_1/P_2:

P_1/P_2	0	0.125	0.250	0.375	0.500
K_2	8.34	9.92	12.09	15.17	19.86

By extrapolation, for $P_1/P_2 = 0.628$, $K_2 = 26.5$.

If we assume the girder provides no vertical load but does prevent rotation of the top, $K_2 = 8.34$. Therefore, the value of P_2 ranges from 91,200 to 289,900 lb, and the factor of safety lies between 11.4 and 36.2. A reasonable estimate of the rotational and vertical stiffness of the girder will allow a good estimate to be made of the actual factor of safety from the values calculated.

15.3 BUCKLING OF FLAT AND CURVED PLATES

In Table 15.2, formulas are given for the critical loads and critical stresses on plates and thin-walled members. Because of the greater likelihood of serious geometrical irregularities and their greater relative effect, the critical stresses actually developed by such members usually fall short of the theoretical values by a wider margin than in the case of bars. The discrepancy is generally greater for pure compression (thin tubes under longitudinal compression or external pressure) than for tension and compression combined (thin tubes under torsion or flat plates under edge shear), and increases with the thinness of the material. The critical stress or load indicated by any one of the theoretical formulas should therefore be regarded as an upper limit, approached more or less closely according to the closeness with which the actual shape of the member approximates the geometrical form assumed. In Table 15.2, the approximate discrepancy to be expected between theory and experiment is indicated wherever the data available have made this possible.

Most of the theoretical analyses of the stability of plates and shells require a numerical evaluation of the resulting equations. Considering the variety of shapes and combinations of shapes as well as the multiplicity of boundary conditions and loading combinations, it is not possible in the limited space available to present anything like a comprehensive coverage of plate and shell buckling. As an alternative, Table 15.2 contains many of the simpler loadings and shapes. The following paragraphs and the References contain some, but by no means all, of the more easily acquired sources giving results in tabular or graphic form that can be applied directly to specific problems. See also Refs. 101–104, and 109–111.

Rectangular Plates

Stability coefficients for *orthotropic* rectangular plates with several combinations of boundary conditions and several ratios of the bending stiffnesses parallel to the sides of the plate are tabulated in Shuleshko (Ref. 60); these solutions were obtained by reducing the problem of plate buckling to that of an isotropic bar that is in a state of vibration and under tension. Srinivas and Rao (Ref. 63) evaluate the effect of shear deformation on the stability of simply supported rectangular plates under edge loads parallel to one side; the effect becomes noticeable for $h/b > 0.05$ and is greatest when the loading is parallel to the short side.

Skew Plates

Ashton (Ref. 61) and Durvasula (Ref. 64) consider the buckling of skew (parallelogram) plates under combinations of edge compression, edge tension, and edge shear. Since the loadings evaluated are generally parallel to orthogonal axes and not to both sets of the plate edges, we would not expect to find the particular case desired represented in the tables of coefficients; the general trend of results is informative.

Circular Plates

Vijayakumar and Joga Rao (Ref. 58) describe a technique for solving for the radial buckling loads on a *polar orthotropic annular plate.* They give graphs of stability coefficients for a wide range of rigidity ratios and for the several combinations of free, simply supported, and fixed inner and outer edges for the radius ratio (outer to inner) 2:1. Two loadings are presented: outer edge only under uniform compression and inner and outer edges under equal uniform compression.

Amon and Widera (Ref. 59) present graphs showing the effect of an edge beam on the stability of a circular plate of uniform thickness.

Sandwich Plates

There is a great amount of literature on the subject of sandwich construction. References 38 and 100 and the publications listed in Ref. 39 provide initial sources of information.

15.4 BUCKLING OF SHELLS

Baker, Kovalevsky, and Rish (Ref. 97) discuss the stability of unstiffened orthotropic composite, stiffened, and sandwich shells. They represent data based on theory and experiment which permit the designer to choose a loading or pressure with a 90% probability of no stability failure; the work is extensively referenced. For similar collected data see Refs. 41 and 42.

Stein (Ref. 95) discusses some comparisons of theory with experimentation in shell buckling. Rabinovich (Ref. 96) describes in some detail the work in structural mechanics, including shell stability, in the U.S.S.R. from 1917 to 1957.

In recent years, there have been increasing development and application of the finite-element method for the numerical solution of shell problems. Navaratna, Pian, and Witmer (Ref. 94) describe a finite-element method of solving axisymmetric shell problems where the element considered is either a conical frustum or a frustum with a curved meridian; examples are presented of cylinders with uniform or tapered walls under axial load, a truncated hemisphere under axial tension, and a conical shell under torsion. Bushnell (Ref. 99) presents a very general finite-element program for shell analysis, and Perrone (Ref. 98) gives a compendium of such programs. See also Refs. 101 to 108.

Cylindrical and Conical Shells

In general, experiments to determine the axial loads required to buckle cylindrical shells yield results that are between one-half and three-fourths of the classical buckling loads predicted by theory. The primary causes of these discrepancies are the deviations from a true cylindrical form in most manufactured vessels and the inability to accurately define the boundary conditions. Hoff (Refs. 67 and 68) shows that removing the in-plane shear stress at the boundary of a simply supported cylindrical shell under axial compression can reduce the theoretical buckling load by a factor of 2 from that predicted by the more usual boundary conditions associated with a simply supported edge. Baruch, Harari, and Singer (Ref. 84) find similar low-buckling loads for simply supported conical shells under axial load but for a different modification of the boundary support. Tani and Yamaki (Ref. 83) carry out further work on this problem, including the effect of clamped edges.

The random nature of manufacturing deviations leads to the use of the statistical approach, as mentioned previously (Ref. 97) and as Hausrath and Dittoe have done for conical shells (Ref. 77). Weingarten, Morgan, and Seide (Ref. 80) have developed empirical expressions for lower bounds of stability coefficients for cylindrical and conical shells under axial compression with references for the many data they present.

McComb, Zender, and Mikulas (Ref. 44) discuss the effects of internal pressure on the bending stability of very thin-walled cylindrical shells. Internal pressure has a stabilizing effect on axially and/or torsionally loaded cylindrical and conical shells. This subject is discussed in several references: Seide (Ref. 75), Weingarten (Ref. 76), and Weingarten, Morgan, and Seide (Ref. 82) for conical and cylindrical shells; Ref. 97 contains much information on this subject as well.

Axisymmetric snap-buckling of conical shells is discussed by Newman and Reiss (Ref. 73), which leads to the concept of the Belleville spring for the case of shallow shells. (See also Sec. 11.8.)

External pressure as a cause of buckling is examined by Singer (Ref. 72) for cones and by Newman and Reiss (Ref. 73) and Yao and Jenkins (Ref. 69) for elliptic cylinders. External pressure caused by pretensioned filament winding on cylinders is analyzed by Mikulas and Stein (Ref. 66); they point out that material compressibility in the thickness direction is important in this problem.

The combination of external pressure and axial loads on cylindrical and conical shells is very thoroughly examined and referenced by Radkowski (Ref. 79) and Weingarten and Seide (Ref. 81). The combined loading on orthotropic and stiffened conical shells is discussed by Singer (Ref. 74).

Attempts to manufacture nearly perfect shells in order to test the theoretical results have led to the construction of thin-walled shells by electroforming; Sendelbeck and Singer (Ref. 85) and Arbocz and Babcock (Ref. 91) describe the results of such tests.

A very thorough survey of buckling theory and experimentation for conical shells of constant thickness is presented by Seide (Ref. 78).

Spherical Shells

Experimental work is described by Loo and Evan-Iwanowski on the effect of a concentrated load at the apex of a spherical cap (Ref. 90) and the effect of multiple concentrated loads (Ref. 89). Carlson, Sendelbeck, and Hoff (Ref. 70) report on the experimental study of buckling of electroformed complete spherical shells; they report experimental critical pressures of up to 86% of those predicted by theory and the correlation of flaws with lower test pressures.

Burns (Ref. 92) describes tests of static and dynamic buckling of thin spherical caps due to external pressure; both elastic and plastic buckling are considered and evaluated in these tests. Wu and Cheng (Ref. 71) discuss in detail the buckling due to circumferential hoop compression which is developed when a truncated spherical shell is subjected to an axisymmetric tensile load.

Toroidal Shells

Stein and McElman (Ref. 86) derive nonlinear equations of equilibrium and buckling equations for segments of toroidal shells; segments that are symmetric with the equator are considered for both inner and outer diameters, as well as segments centered at the crown. Sobel and Flügge (Ref. 87) tabulate and graph the minimum buckling external pressures on full toroidal shells. Almroth, Sobel, and Hunter (Ref. 88) compare favorably the theory in Ref. 87 with experiments they performed.

Corrugated Tubes or Bellows

An instability can develop when a corrugated tube or bellows is subjected to an internal pressure with the ends partially or totally restrained against axial displacement. (This instability can also occur in very long cylindrical vessels under similar restraints.) For a discussion and an example of this effect, see Sec. 13.5.

15.5 TABLES

TABLE 15.1 Formulas for Elastic Stability of Bars, Rings, and Beams

Notation: P' = critical load (force); p' = critical unit load (force per unit length); T' = critical torque (force-length); M' = critical bending moment (force-length); E = modulus of elasticity (force per unit area); and I = moment of inertia of cross section about central axis perpendicular to plane of buckling.

Reference Number, Form of Bar, and Manner of Loading and Support

1a. Stepped straight bar under end load P_1 and intermediate load P_2; upper end free, lower end fixed. $P_1' = K_1 \frac{\pi^2 E_1 I_1}{l^2}$ where K_1 is tabulated below

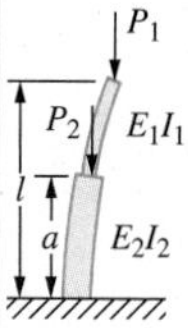

E_2I_2/E_1I_1	1.000					1.500					2.000				
P_2/P_1 \ a/l	1/6	1/3	½	2/3	5/6	1/6	1/3	½	2/3	5/6	1/6	1/3	½	2/3	5/6
0.0	0.250	0.250	0.250	0.250	0.250	0.279	0.312	0.342	0.364	0.373	0.296	0.354	0.419	0.471	0.496
0.5	0.249	0.243	0.228	0.208	0.187	0.279	0.306	0.317	0.306	0.279	0.296	0.350	0.393	0.399	0.372
1.0	0.248	0.237	0.210	0.177	0.148	0.278	0.299	0.295	0.261	0.223	0.296	0.345	0.370	0.345	0.296
2.0	0.246	0.222	0.178	0.136	0.105	0.277	0.286	0.256	0.203	0.158	0.295	0.335	0.326	0.267	0.210
4.0	0.242	0.195	0.134	0.092	0.066	0.274	0.261	0.197	0.138	0.099	0.294	0.314	0.257	0.184	0.132
8.0	0.234	0.153	0.088	0.056	0.038	0.269	0.216	0.132	0.084	0.057	0.290	0.266	0.174	0.112	0.076

1b. Stepped straight bar under end load P_1 and intermediate load P_2; both ends pinned. $P_1' = K_1 \frac{\pi^2 E_1 I_1}{l^2}$ where K_1 is tabulated below

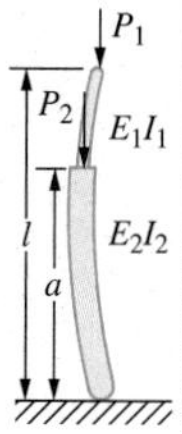

E_2I_2/E_1I_1	1.000					1.500					2.000				
P_2/P_1 \ a/l	1/6	1/3	½	2/3	5/6	1/6	1/3	½	2/3	5/6	1/6	1/3	½	2/3	5/6
0.0	1.000	1.000	1.000	1.000	1.000	1.010	1.065	1.180	1.357	1.479	1.014	1.098	1.297	1.633	1.940
0.5	0.863	0.806	0.797	0.789	0.740	0.876	0.872	0.967	1.091	1.098	0.884	0.908	1.069	1.339	1.452
1.0	0.753	0.672	0.663	0.646	0.584	0.769	0.736	0.814	0.908	0.870	0.776	0.769	0.908	1.126	1.153
2.0	0.594	0.501	0.493	0.473	0.410	0.612	0.557	0.615	0.676	0.613	0.621	0.587	0.694	0.850	0.814
4.0	0.412	0.331	0.325	0.307	0.256	0.429	0.373	0.412	0.442	0.383	0.438	0.397	0.470	0.566	0.511
8.0	0.254	0.197	0.193	0.180	0.147	0.267	0.225	0.248	0.261	0.220	0.272	0.240	0.284	0.336	0.292

1c. Stepped straight bar under end load P_1 and intermediate load P_2; upper end guided, lower end fixed. $P_1' = K_1 \frac{\pi^2 E_1 I_1}{l^2}$ where K_1 is tabulated below

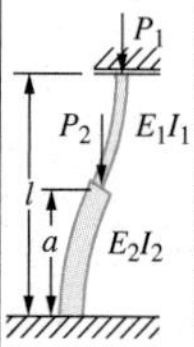

E_2I_2/E_1I_1	1.000					1.500					2.000				
P_2/P_1 \ a/l	1/6	1/3	½	2/3	5/6	1/6	1/3	½	2/3	5/6	1/6	1/3	½	2/3	5/6
0.0	1.000	1.000	1.000	1.000	1.000	1.113	1.208	1.237	1.241	1.309	1.184	1.367	1.452	1.461	1.565
0.5	0.986	0.904	0.792	0.711	0.672	1.105	1.117	1.000	0.897	0.885	1.177	1.288	1.192	1.063	1.063
1.0	0.972	0.817	0.650	0.549	0.507	1.094	1.026	0.830	0.697	0.669	1.171	1.206	1.000	0.832	0.805
2.0	0.937	0.671	0.472	0.377	0.339	1.073	0.872	0.612	0.482	0.449	1.156	1.047	0.745	0.578	0.542
4.0	0.865	0.480	0.304	0.231	0.204	1.024	0.642	0.397	0.297	0.270	1.126	0.794	0.486	0.358	0.327
8.0	0.714	0.299	0.176	0.130	0.114	0.910	0.406	0.232	0.169	0.151	1.042	0.511	0.284	0.203	0.182

(Continued)

TABLE 15.1 Formulas for Elastic Stability of Bars, Rings, and Beams (*Continued*)

Reference Number, Form of Bar, and Manner of Loading and Support

1d. Stepped straight bar under end load P_1 and intermediate load P_2; upper end pinned, lower end fixed. $P_1' = K_1 \frac{\pi^2 E_1 I_1}{l^2}$ where K_1 is tabulated below

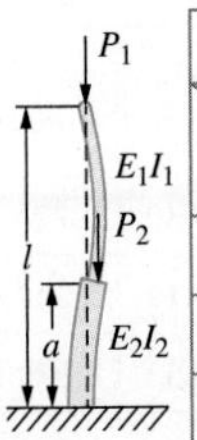

E_2I_2/E_1I_1	1.000					1.500					2.000				
a/l — P_2/P_1	1/6	1/3	½	2/3	5/6	1/6	1/3	½	2/3	5/6	1/6	1/3	½	2/3	5/6
0.0	2.046	2.046	2.046	2.046	2.046	2.241	2.289	2.338	2.602	2.976	2.369	2.503	2.550	2.983	3.838
0.5	1.994	1.814	1.711	1.700	1.590	2.208	2.071	1.991	2.217	2.344	2.344	2.286	2.196	2.570	3.066
1.0	1.938	1.613	1.464	1.450	1.290	2.167	1.869	1.727	1.915	1.918	2.313	2.088	1.915	2.250	2.525
2.0	1.820	1.300	1.130	1.111	0.933	2.076	1.535	1.355	1.506	1.390	2.250	1.742	1.518	1.796	1.844
4.0	1.570	0.918	0.773	0.753	0.594	1.874	1.107	0.941	1.042	0.891	2.097	1.277	1.065	1.270	1.184
8.0	1.147	0.569	0.469	0.454	0.343	1.459	0.697	0.582	0.643	0.514	1.727	0.812	0.664	0.796	0.686

1e. Stepped straight bar under end load P_1 and intermediate load P_2; both ends fixed. $P_1' = K_1 \frac{\pi^2 E_1 I_1}{l^2}$ where K_1 is tabulated below

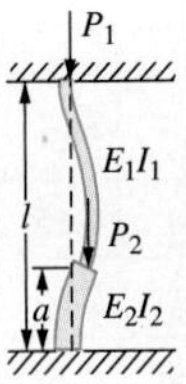

E_2I_2/E_1I_1	1.000					1.500					2.000				
a/l — P_2/P_1	1/6	1/3	½	2/3	5/6	1/6	1/3	½	2/3	5/6	1/6	1/3	½	2/3	5/6
0.0	4.000	4.000	4.000	4.000	4.000	4.389	4.456	4.757	5.359	5.462	4.657	4.836	5.230	6.477	6.838
0.5	3.795	3.298	3.193	3.052	2.749	4.235	3.756	3.873	4.194	3.795	4.545	4.133	4.301	5.208	4.787
1.0	3.572	2.779	2.647	2.443	2.094	4.065	3.211	3.254	3.411	2.900	4.418	3.568	3.648	4.297	3.671
2.0	3.119	2.091	1.971	1.734	1.414	3.679	2.459	2.459	2.452	1.968	4.109	2.766	2.782	3.136	2.496
4.0	2.365	1.388	1.297	1.088	0.857	2.921	1.659	1.649	1.555	1.195	3.411	1.882	1.885	2.008	1.523
8.0	1.528	0.826	0.769	0.623	0.479	1.943	1.000	0.992	0.893	0.671	2.334	1.138	1.141	1.158	0.854

2a. Stepped straight bar under end load P_1 and intermediate load P_2; upper end free, lower end fixed. $P_2' = K_2 \frac{\pi^2 E_1 I_1}{l^2}$ where K_2 is tabulated below

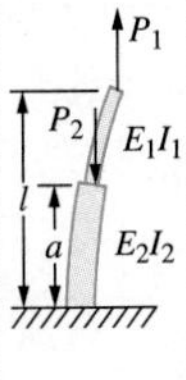

E_2I_2/E_1I_1	1.000					1.500					2.000				
a/l — P_1/P_2	1/6	1/3	½	2/3	5/6	1/6	1/3	½	2/3	5/6	1/6	1/3	½	2/3	5/6
0	9.00	2.25	1.00	0.56	0.36	13.50	3.38	1.50	0.84	0.54	18.00	4.50	2.00	1.13	0.72
0.125	15.55	3.75	1.48	0.74	0.44	21.87	5.36	2.19	1.11	0.65	27.98	6.92	2.89	1.48	0.87
0.250	21.33	5.30	2.19	1.03	0.55	29.51	7.36	3.13	1.53	0.82	37.30	9.31	4.02	2.02	1.10
0.375	29.02	7.25	3.13	1.52	0.74	39.89	9.97	4.37	2.21	1.10	50.10	12.52	5.52	2.86	1.46
0.500	40.50	10.12	4.46	2.31	1.08	55.66	13.92	6.16	3.28	1.60	69.73	17.43	7.73	4.18	2.12

(*Continued*)

TABLE 15.1 Formulas for Elastic Stability of Bars, Rings, and Beams (*Continued*)

Reference Number, Form of Bar, and Manner of Loading and Support

2b. Stepped straight bar under end load P_1 and intermediate load P_2; both ends pinned. $P_2' = K_2 \frac{\pi^2 E_1 I_1}{l^2}$ where K_2 is tabulated below

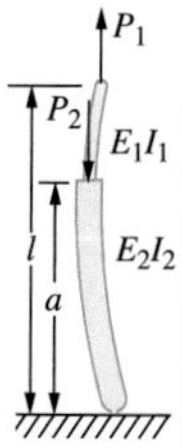

E_2I_2/E_1I_1	1.000					1.500					2.000				
P_1/P_2 \ a/l	1/6	1/3	½	2/3	5/6	1/6	1/3	½	2/3	5/6	1/6	1/3	½	2/3	5/6
0	2.60	1.94	1.89	1.73	1.36	2.77	2.24	2.47	2.54	2.04	2.86	2.41	2.89	3.30	2.72
0.125	3.51	2.49	2.43	2.14	1.62	3.81	2.93	3.26	3.18	2.43	3.98	3.21	3.89	4.19	3.24
0.250	5.03	3.41	3.32	2.77	1.99	5.63	4.15	4.64	4.15	2.99	5.99	4.65	5.75	5.52	3.98
0.375	7.71	5.16	4.96	3.76	2.55	8.98	6.61	7.26	5.63	3.82	9.80	7.67	9.45	7.50	5.09
0.500	12.87	9.13	8.00	5.36	3.48	15.72	12.55	12.00	7.96	5.18	17.71	15.45	16.00	10.54	6.87

2c. Stepped straight bar under tensile end load P_1 and intermediate load P_2; upper end guided, lower end fixed. $P_2' = K_2 \frac{\pi^2 E_1 I_1}{l^2}$ where K_2 is tabulated below

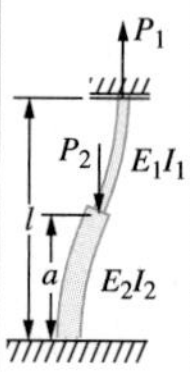

E_2I_2/E_1I_1	1.000					1.500					2.000				
P_1/P_2 \ a/l	1/6	1/3	½	2/3	5/6	1/6	1/3	½	2/3	5/6	1/6	1/3	½	2/3	5/6
0	10.40	3.08	1.67	1.19	1.03	14.92	4.23	2.21	1.55	1.37	19.43	5.37	2.73	1.88	1.65
0.125	15.57	4.03	2.03	1.40	1.18	21.87	5.57	2.71	1.82	1.57	27.98	7.07	3.36	2.21	1.90
0.250	21.33	5.37	2.54	1.67	1.38	29.52	7.40	3.42	2.20	1.84	37.32	9.34	4.26	2.68	2.24
0.375	29.02	7.26	3.31	2.08	1.67	39.90	9.97	4.50	2.76	2.24	50.13	12.53	5.61	3.39	2.73
0.500	40.51	10.12	4.53	2.72	2.10	55.69	13.91	6.21	3.66	2.84	69.76	17.43	7.76	4.52	3.47

2d. Stepped straight bar under tensile end load P_1 and intermediate load P_2; upper end pinned, lower end fixed. $P_2' = K_2 \frac{\pi^2 E_1 I_1}{l^2}$ where K_2 is tabulated below

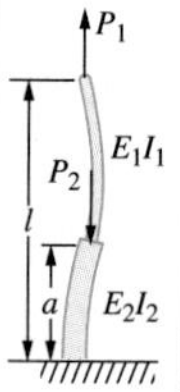

E_2I_2/E_1I_1	1.000					1.500					2.000				
P_1/P_2 \ a/l	1/6	1/3	½	2/3	5/6	1/6	1/3	½	2/3	5/6	1/6	1/3	½	2/3	5/6
0	13.96	5.87	4.80	4.53	3.24	18.66	7.33	6.04	6.58	4.86	23.26	8.64	6.98	8.40	6.48
0.125	20.21	7.93	6.50	5.84	3.91	27.12	10.12	8.43	8.71	5.86	33.71	12.06	9.92	11.51	7.81
0.250	28.58	11.35	9.64	7.68	4.87	38.32	14.82	13.13	11.50	7.27	47.36	17.85	15.96	15.30	9.65
0.375	41.15	17.64	15.82	10.15	6.26	55.43	23.67	23.44	14.96	9.30	68.40	28.88	30.65	19.66	12.28
0.500	62.90	31.73	23.78	13.58	8.42	85.64	44.28	34.97	19.80	12.40	106.27	55.33	45.81	25.83	16.27

(*Continued*)

TABLE 15.1 Formulas for Elastic Stability of Bars, Rings, and Beams (*Continued*)

Reference Number, Form of Bar, and Manner of Loading and Support

2e. Stepped straight bar under tensile end load P_1 and intermediate load P_2; both ends fixed. $P_2' = K_2 \frac{\pi^2 E_1 I_1}{l^2}$ where K_2 is tabulated below

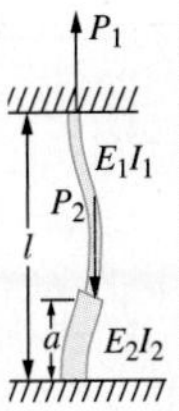

E_2I_2/E_1I_1	1.000					1.500					2.000				
P_1/P_2 \ a/l	1/6	1/3	½	2/3	5/6	1/6	1/3	½	2/3	5/6	1/6	1/3	½	2/3	5/6
0	16.19	8.11	7.54	5.79	4.34	21.06	9.93	9.89	8.34	6.09	25.75	11.44	11.55	10.87	7.78
0.125	21.83	10.37	9.62	6.86	5.00	28.74	12.93	13.03	9.92	7.05	35.28	15.06	15.55	12.96	9.01
0.250	30.02	14.09	12.86	8.34	5.91	39.81	17.99	18.36	12.09	8.35	48.88	21.25	22.98	15.79	10.69
0.375	42.72	20.99	17.62	10.47	7.19	57.14	27.66	26.02	15.17	10.20	70.23	33.29	34.36	19.79	13.11
0.500	64.94	36.57	24.02	13.70	9.16	86.23	50.39	35.09	19.86	13.07	102.53	61.71	45.87	25.86	16.85

3a. Uniform straight bar under end load P and a uniformly distributed load p over a lower portion of the length; several end conditions. $(pa)' = K\frac{\pi^2 EI}{l^2}$ where K is tabulated below (a negative value for P/pa means the end load is tensile)

End conditions	Upper end free, lower end fixed 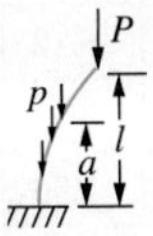				Both ends pinned 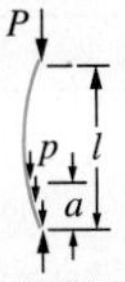				Upper end pinned, lower end fixed 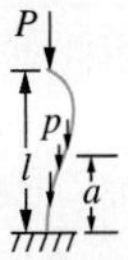				Both ends fixed 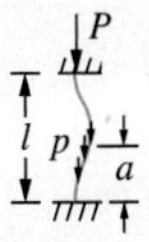			
P/pa \ a/l	¼	½	¾	1	¼	½	¾	1	¼	½	¾	1	¼	½	¾	1
−0.25		11.31	5.18	2.38	9.03	5.32	4.25	3.30		27.9	17.4	11.3		31.3	19.4	13.4
0.00	12.74	3.185	1.413	0.795	3.52	2.53	2.22	1.88	22.2	9.46	7.13	5.32	25.3	13.0	9.78	7.56
0.25	0.974	0.825	0.614	0.449	1.97	1.59	1.46	1.30	6.83	4.70	3.98	3.30	11.2	7.50	6.25	5.20
0.50	0.494	0.454	0.383	0.311	1.34	1.15	1.08	0.98	3.76	3.03	2.71	2.37	6.75	5.18	4.54	3.94
1.00	0.249	0.238	0.218	0.192	0.81	0.73	0.70	0.66	1.97	1.75	1.64	1.51	3.69	3.17	2.91	2.65

3b. Uniform straight bar under end load P and a uniformly distributed load p over an upper portion of the length; several end conditions. $(pa)' = K\frac{\pi^2 EI}{l^2}$ where K is tabulated below (a negative value for P/pa means the end load is tensile)

End conditions	Upper end free, lower end fixed 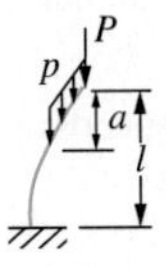			Both ends pinned 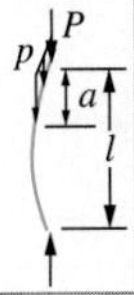			Upper end pinned, lower end fixed 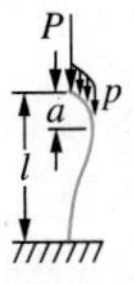			Both ends fixed 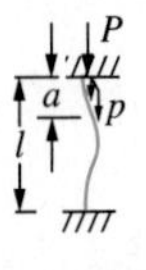		
P/pa \ a/l	¼	½	¾	¼	½	¾	¼	½	¾	¼	½	¾
−0.25	0.481	0.745	1.282	1.808	2.272	2.581	4.338	5.937	7.385	5.829	7.502	9.213
0.00	0.327	0.440	0.600	1.261	1.479	1.611	2.904	3.586	4.160	4.284	5.174	5.970
0.25	0.247	0.308	0.380	0.963	1.088	1.159	2.164	2.529	2.815	3.384	3.931	4.383
0.50	0.198	0.236	0.276	0.778	0.859	0.903	1.720	1.943	2.111	2.796	3.164	3.453
1.00	0.142	0.161	0.179	0.561	0.603	0.624	1.215	1.323	1.400	2.073	2.273	2.419

(*Continued*)

TABLE 15.1 Formulas for Elastic Stability of Bars, Rings, and Beams (*Continued*)

Reference Number, Form of Bar, and Manner of Loading and Support

3c. Stepped straight bar under end load P and a distributed load of maximum value p at the bottom linearly decreasing to zero at a distance a from the bottom, $(pa)' = K\dfrac{\pi^2 EI}{l^2}$ where K is tabulated below (a negative value for P/pa means the end load is tensile)

End conditions	Upper end free, lower end fixed 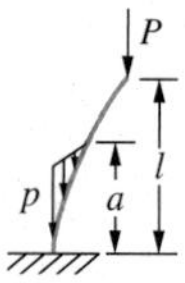				Both ends pinned 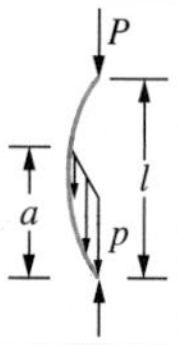				Upper end pinned, lower end fixed 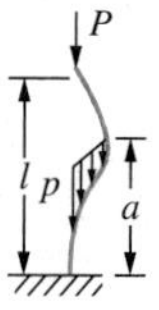				Both ends fixed 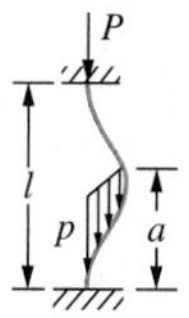			
P/pa \ a/l	¼	½	¾	1	¼	½	¾	1	¼	½	¾	1	¼	½	¾	1
−0.250						58.9	41.1	30.4								
−0.125			26.7	15.5	31.9	15.7	12.0	9.41			62.1	43.8		113.0	70.2	48.7
0.000	52.4	13.1	5.80	3.26	9.66	6.31	5.32	4.72		30.3	20.6	16.1		38.9	27.8	21.9
0.125	1.98	1.85	1.58	1.29	4.65	3.66	3.29	3.03	15.2	11.7	9.73	8.50	27.3	18.9	15.6	13.4
0.250	0.995	0.961	0.887	0.787	2.98	2.54	2.35	2.22	7.90	6.92	6.18	5.66	14.9	12.1	10.6	9.53
0.500	0.499	0.490	0.471	0.441	1.72	1.56	1.49	1.43	4.02	3.77	3.54	3.36	7.73	6.95	6.43	6.00
1.000	0.250	0.248	0.243	0.235	0.93	0.88	0.86	0.84	2.03	1.96	1.90	1.85	3.93	3.73	3.57	3.44

4. Uniform straight bar under end load P; both ends hinged and bar elastically supported by lateral pressure p proportional to deflection ($p = ky$, where k = lateral force per unit length per unit of deflection)

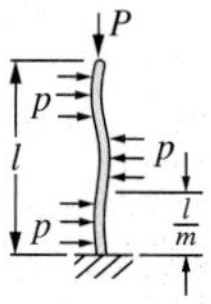

$$P' = \frac{\pi^2 EI}{l^2}\left(m^2 + \frac{kl^4}{m^2\pi^4 EI}\right)$$

where m represents the number of half-waves into which the bar buckles and is equal to the lowest integer greater than

$$\frac{1}{2}\left(\sqrt{1+\frac{4l^2}{\pi^2}\sqrt{\frac{k}{EI}}}-1\right)$$

(Ref. 1)

5. Uniform straight bar under end load P; both ends hinged and bar elastically supported by lateral pressure p proportional to deflection but where the constant of proportionality depends upon the direction of the deflection ($p = k_1 y$ for deflection toward the softer foundation; $p = k_2 y$ for deflection toward the harder foundation); these are also called unattached foundations

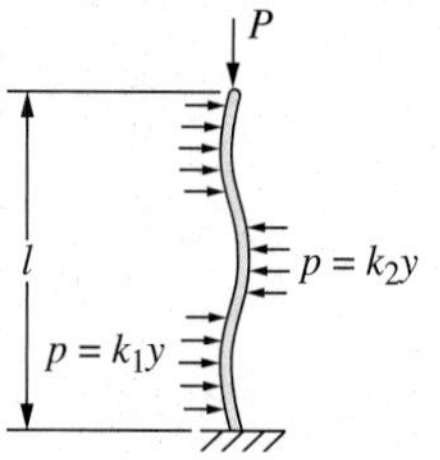

$$P' = \frac{\pi^2 EI}{l^2}\left(m^2 + \frac{k_2 l^4}{m^2\pi^4 EI}\phi^a\right) \quad \text{where } \phi = \frac{k_1}{k_2} \text{ and } \alpha \text{ depends upon } m \text{ as given below}$$

m	α
1	1
2	$1+\phi(0.23-0.017l^2\sqrt{k_2/EI})$
3	$0.75 - 0.56\phi$

This is an empirical expression which closely fits numerical solutions found in Ref. 45 and is valid only over the range $0 \le l^2\sqrt{k_2/EI} \le 120$. Solutions for P' are carried out for values of m = 1, 2, and 3, and the lowest one governs.

(*Continued*)

TABLE 15.1 Formulas for Elastic Stability of Bars, Rings, and Beams (*Continued*)

Reference Number, Form of Bar, and Manner of Loading and Support

6. Straight bar, middle portion uniform, end portions tapered and alike; end load; I = moment of inertia of cross section of middle portion; I_o = moment of inertia of end cross sections; I_x = moment of inertia of section x

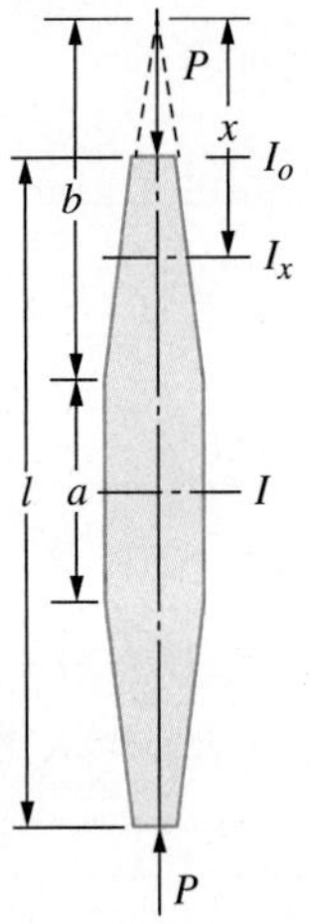

(For singly tapered columns see Ref. 46.)

6a. $I_x = I\frac{x}{b}$

for example, rectangular section tapering uniformly in width

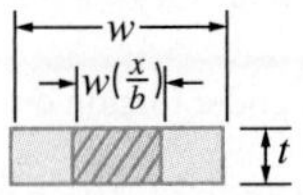

$P' = \frac{KEI}{l^2}$ where K depends on $\frac{I_0}{I}$ and $\frac{a}{l}$ and may be found from the following table:

I_o/I / a/l	K for ends hinged							K for ends fixed			
	0	0.01	0.10	0.2	0.4	0.6	0.8	0.2	0.4	0.6	0.8
0	5.78	5.87	6.48	7.01	7.86	8.61	9.27	20.36	26.16	31.04	35.40
0.2	7.04	7.11	7.58	7.99	8.59	9.12	9.53	22.36	27.80	32.20	36.00
0.4	8.35	8.40	8.63	8.90	9.19	9.55	9.68	23.42	28.96	32.92	36.36
0.6	9.36	9.40	9.46	9.73	9.70	9.76	9.82	25.44	30.20	33.80	36.84
0.8	9.80	9.80	9.82	9.82	9.83	9.85	9.86	29.00	33.08	35.80	37.84

(Ref. 5)

6b. $I_x = I\left(\frac{x}{b}\right)^2$

for example, section of four slender members latticed together

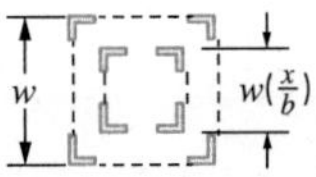

$P' = \frac{KEI}{l^2}$ where K may be found from the following table:

I_o/I / a/l	K for ends hinged							K for ends fixed			
	0	0.01	0.10	0.2	0.4	0.6	0.8	0.2	0.4	0.6	0.8
0	1.00	3.45	5.40	6.37	7.61	8.51	9.24	18.94	25.54	30.79	35.35
0.2	1.56	4.73	6.67	7.49	8.42	9.04	9.50	21.25	27.35	32.02	35.97
0.4	2.78	6.58	8.08	8.61	9.15	9.48	9.70	22.91	28.52	32.77	36.34
0.6	6.25	8.62	9.25	9.44	9.63	9.74	9.82	24.29	29.69	33.63	36.80
0.8	9.57	9.71	9.79	9.81	9.84	9.85	9.86	27.67	32.59	35.64	37.81

(Ref. 5)

6c. $I_x = I\left(\frac{x}{b}\right)^3$

for example, rectangular section tapering uniformly in thickness

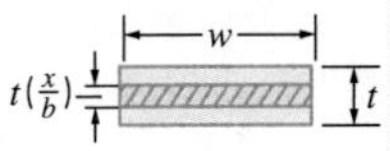

$P' = \frac{KEI}{l^2}$ where K may be found from the following table:

I_o/I / a/l	K for ends hinged						K for ends fixed			
	0.01	0.10	0.2	0.4	0.6	0.8	0.2	0.4	0.6	0.8
0	2.55	5.01	6.14	7.52	8.50	9.23	18.48	25.32	30.72	35.32
0.2	3.65	6.32	7.31	8.38	9.02	9.50	20.88	27.20	31.96	35.96
0.4	5.42	7.84	8.49	9.10	9.46	9.69	22.64	28.40	32.72	36.32
0.6	7.99	9.14	9.39	9.62	9.74	9.81	23.96	29.52	33.56	36.80
0.8	9.63	9.77	9.81	9.84	9.85	9.86	27.24	32.44	35.60	37.80

(Ref. 5)

6d. $I_x = I\left(\frac{x}{b}\right)^4$

for example, end portions pyramidal or conical

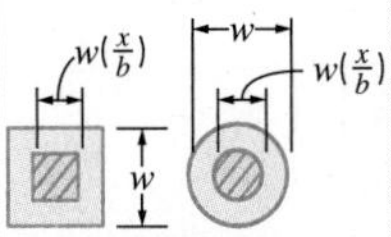

$P' = \frac{KEI}{l^2}$ where K may be found from the following table:

I_o/I / a/l	K for ends hinged						K for ends fixed			
	0.01	0.10	0.2	0.4	0.6	0.8	0.2	0.4	0.6	0.8
0	2.15	4.81	6.02	7.48	8.47	9.23	18.23	25.23	30.68	35.33
0.2	3.13	6.11	7.20	8.33	9.01	9.49	20.71	27.13	31.94	35.96
0.4	4.84	7.68	8.42	9.10	9.45	9.69	22.49	28.33	32.69	36.32
0.6	7.53	9.08	9.38	9.62	9.74	9.81	23.80	29.46	33.54	36.78
0.8	9.56	9.77	9.80	9.84	9.85	9.86	27.03	32.35	35.56	37.80

(Ref. 5)

(*Continued*)

TABLE 15.1 Formulas for Elastic Stability of Bars, Rings, and Beams (*Continued*)

<table>
<tr><th colspan="2">Reference Number, Form of Bar, and Manner of Loading and Support</th></tr>
<tr><td>7. Uniform straight bar under end loads P and end twisting couples T; cross section of bar has same I for all central axes; both ends hinged
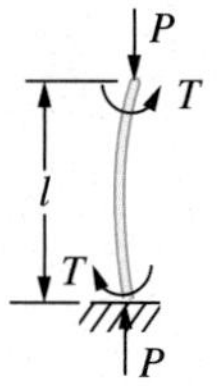</td>
<td>Critical combination of P and T is given by
$$\frac{T^2}{4(EI)^2}+\frac{P}{EI}=\frac{\pi^2}{l^2}$$ (Ref. 1)
If $P=0$, the formula gives critical twisting moment T' which, acting alone, would cause buckling.
If for a given value of T the formula gives a negative value for P, $T>T'$ and P represents tensile load required to prevent buckling.
For thin circular tube of diameter D and thickness t under torsion only, critical shear stress (Ref. 2)
$$\tau=\frac{\pi ED}{l(1-\nu)}\left(1-\frac{t}{D}+\frac{1}{3}\frac{t^2}{D^2}\right)$$ for helical buckling only (not for shell-type buckling in the thin wall)</td></tr>
<tr><td>8. Uniform circular ring under uniform radial pressure p lb/in; mean radius of ring r
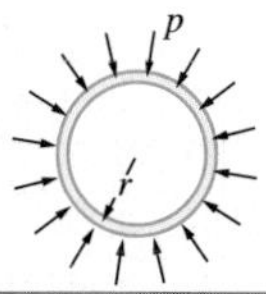</td>
<td>$$p'=\frac{3EI}{r^3}$$ (Ref. 1)</td></tr>
<tr><td>9. Uniform circular arch under uniform radial pressure p lb/in; mean radius r; ends hinged
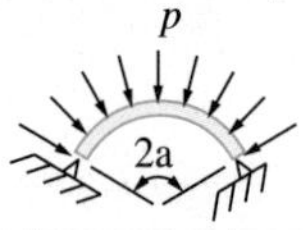</td>
<td>$$p'=\frac{EI}{r^3}\left(\frac{\pi^2}{a^2}-1\right)$$ (Ref. 1)
(For symmetrical arch of any form under central concentrated loading, see Ref. 40; for parabolic and catenary arches, see Ref. 50.)</td></tr>
<tr><td>10. Uniform circular arch under uniform radial pressure p lb/in; mean radius r; ends fixed
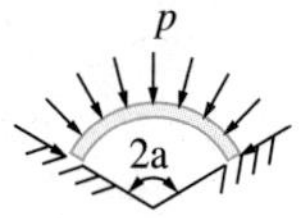</td>
<td>$$p'=\frac{EI}{r^3}(k^2-1)$$ (Ref. 1)
where k depends on α and is found by trial from the equation: $k \tan\alpha \cot kx = 1$ or from the following table:
<table>
<tr><td>α</td><td>15°</td><td>30°</td><td>45°</td><td>60°</td><td>75°</td><td>90°</td><td>120°</td><td>180°</td></tr>
<tr><td>k</td><td>17.2</td><td>8.62</td><td>5.80</td><td>4.37</td><td>3.50</td><td>3.00</td><td>2.36</td><td>2.00</td></tr>
</table>
(For parabolic and catenary arches, see Ref. 50.)</td></tr>
<tr><td>11. Straight uniform beam of narrow rectangular section under pure bending
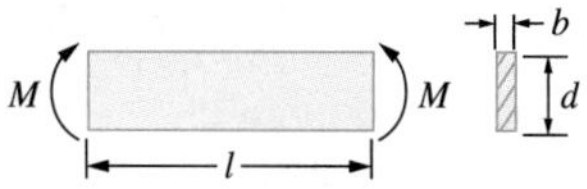</td>
<td>For ends held vertical but not fixed in horizontal plane:
$$M'=\frac{\pi b^3 d\sqrt{EG\left(1-0.63\frac{b}{d}\right)}}{6l}$$
For ends held vertical and fixed in horizontal plane:
$$M'=\frac{2\pi b^3 d\sqrt{EG\left(1-0.63\frac{b}{d}\right)}}{6l}$$ (Refs. 1, 3, 4)</td></tr>
<tr><td>12. Straight uniform cantilever beam of narrow rectangular section under end load applied at a point a distance a above (a positive) or below (a negative) centroid of section
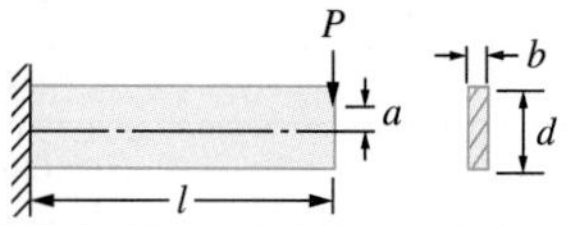</td>
<td>$$P'=\frac{0.669b^3d\sqrt{\left(1-0.63\frac{b}{d}\right)EG}}{l^2}\left[1-\frac{a}{2l}\sqrt{\frac{E}{G\left(1-0.63\frac{b}{d}\right)}}\right]$$
For a load W uniformly distributed along the beam, the critical load $W'=3P'$ (approximately).
(For tapered and stepped beams, see Ref. 54.)
(Refs. 1, 3, 4)</td></tr>
</table>

(*Continued*)

TABLE 15.1 Formulas for Elastic Stability of Bars, Rings, and Beams (*Continued*)

Reference Number, Form of Bar, and Manner of Loading and Support	
13. Straight uniform beam of narrow rectangular section under center load applied at a point a distance a above (a positive) or below (a negative) centroid of section; ends of beam simply supported and constrained against twisting 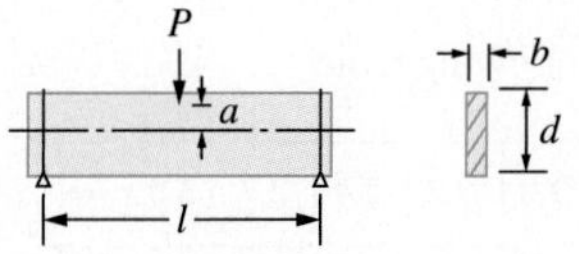	$P' = \frac{2.82b^3 d\sqrt{\left(1-0.63\frac{b}{d}\right)EG}}{l^2}\left[1-\frac{1.74a}{l}\sqrt{\frac{E}{G\left(1-0.63\frac{b}{d}\right)}}\right]$ For a uniformly distributed load, the critical load $W' = 1.67P'$ (approximately). If P is applied at an intermediate point, a distance C from one end, its critical value is practically the same as for central loading if $0.4l < C < 0.5l$; if $C < 0.4l$, the critical load is given approximately by multiplying the P' for central loading by $0.36 + 0.28\frac{l}{C}$. If the ends of the beam are fixed and the load P is applied at the centroid of the middle cross section, $P' = \frac{4.43b^3 d}{l^2}\sqrt{\left(1-0.63\frac{b}{d}\right)EG}$ (Refs. 1, 3, 4)
14. Straight uniform I-beam under pure bending; d = depth center to center of flange; ends constrained against twisting 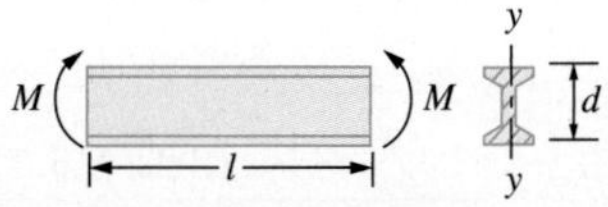	$M' = \frac{\pi\sqrt{EI_y KG}}{l}\sqrt{1+\pi^2\frac{I_f E d^2}{2KGl^2}}$ where I_y is the moment of inertia of the cross section about its vertical axis of symmetry, I_f is the moment of inertia of one flange about this axis, and KG is the torsional rigidity of the section (see Table 10.1, case 26). (For tapered I-beams, see Ref. 47.) (Refs. 1, 3)
15. Straight uniform cantilever beam of I-section under end load applied at centroid of cross section; d = depth center to center of flanges 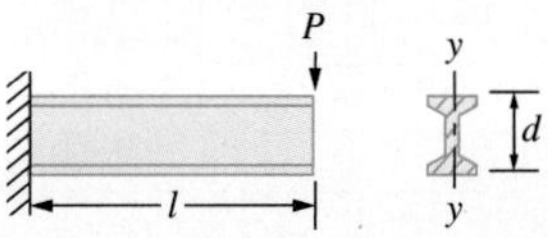	$P' = \frac{m\sqrt{EI_y KG}}{l^2}$ where m is approximately equal to $4.01 + 11.7\sqrt{\frac{I_f E d^2}{2KGl^2}}$ and I_y, I_f and KG have the same significance as in case 14. (For unsymmetric I-beams, see Refs. 52 and 56; for tapered I-beams, see Ref. 55.) (Refs. 1, 3)
16. Straight uniform I-beam loaded at centroid of middle section; ends simply supported and constrained against twisting 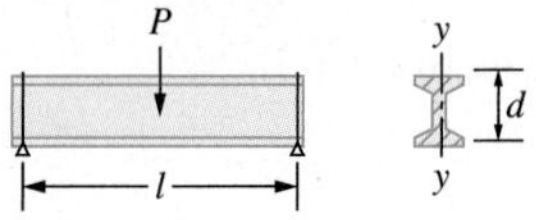	$P' = \frac{m\sqrt{EI_y KG}}{l^2}$ where m is approximately equal to $16.93 + 45\left(\frac{I_f E d^2}{2KGl^2}\right)^{0.8}$ and I_y, I_f and KG have the same significance as in case 14. (For unsymmetric I-beams, see Refs. 52 and 56; for tapered I-beams, see Ref. 55.) (Refs. 1, 3)

TABLE 15.2 Formulas for Elastic Stability of Plates and Shells

Notation: E = modulus of elasticity; ν = Poisson's ratio; and t = thickness for all plates and shells. All angles are in radians. Compression is positive; tension is negative. For the plates, the smaller width should be greater than 10 times the thickness unless otherwise specified.

Form of Plate or Shell and Manner of Loading	Manner of Support	Formulas for Critical Unit Compressive Stress σ', Unit Shear Stress τ', Load P', Bending Moment M', or Unit External Pressure q' at Which Elastic Buckling Occurs
1. Rectangular plate under equal uniform compression on two opposite edges b	1a. All edges simply supported	$\sigma' = K\frac{E}{1-\nu^2}\left(\frac{t}{b}\right)^2$ Here K depends on ratio $\frac{a}{b}$ and may be found from the following table (see below). (For unequal end compressions, see Ref. 33.) (Refs. 1, 6)
	1b. All edges clamped	$\sigma' = K\frac{E}{1-\nu^2}\left(\frac{t}{b}\right)^2$ (table below) (Refs. 1, 6, 7)
	1c. Edges b simply supported, edges a clamped	$\sigma' = K\frac{E}{1-\nu^2}\left(\frac{t}{b}\right)^2$ (table below) (Refs. 1, 6)
	1d. Edges b simply supported, one edge a simply supported, other edge a free	$\sigma' = K\frac{E}{1-\nu^2}\left(\frac{t}{b}\right)^2$ (table below) (Ref. 1)
	1e. Edges b simply supported, one edge a clamped, other edge a free	$\sigma' = K\frac{E}{1-\nu^2}\left(\frac{t}{b}\right)^2$ (table below) (Ref. 1)
	1f. Edges b clamped, edges a simply supported	$\sigma' = K\frac{E}{1-\nu^2}\left(\frac{t}{b}\right)^2$ (table below) (Ref. 1)

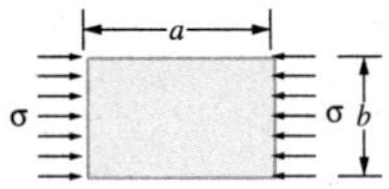

1a.

$\frac{a}{b}$	0.2	0.3	0.4	0.6	0.8	1.0	1.2	1.4	1.6	1.8	2.0	2.2	2.4	2.7	3.0	∞
K	22.2	10.9	6.92	4.23	3.45	3.29	3.40	3.68	3.45	3.32	3.29	3.32	3.40	3.32	3.29	3.29

1b.

$\frac{a}{b}$	1	2	3	∞
K	7.7	6.7	6.4	5.73

1c.

$\frac{a}{b}$	0.4	0.5	0.6	0.7	0.8	1.0	1.2	1.4	1.6	1.8	2.1	∞
K	7.76	6.32	5.80	5.76	6.00	6.32	5.80	5.76	6.00	5.80	5.76	5.73

1d.

b	0.5	1.0	1.2	1.4	1.6	1.8	2.0	2.5	3.0	4.0	5.0
K	3.62	1.18	0.934	0.784	0.687	0.622	0.574	0.502	0.464	0.425	0.416

1e.

$\frac{a}{b}$	1	1.1	1.2	1.3	1.4	1.5	1.6	1.7	1.8	1.9	2.0	2.2	2.4
K	1.40	1.28	1.21	1.16	1.12	1.10	1.09	1.09	1.10	1.12	1.14	1.19	1.21

1f.

$\frac{a}{b}$	0.6	0.8	1.0	1.2	1.4	1.6	1.7	1.8	2.0	2.5	3.0
K	11.0	7.18	5.54	4.80	4.48	4.39	4.39	4.26	3.99	3.72	3.63

(Continued)

TABLE 15.2 Formulas for Elastic Stability of Plates and Shells (*Continued*)

<table>
<tr><th>Form of Plate or Shell and Manner of Loading</th><th>Manner of Support</th><th>Formulas for Critical Unit Compressive Stress σ', Unit Shear Stress τ', Load P', Bending Moment M', or Unit External Pressure q' at Which Elastic Buckling Occurs</th></tr>
<tr><td rowspan="2">2. Rectangular plate under uniform compression (or tension) σ_x on edges b and uniform compression (or tension) σ_y on edges a
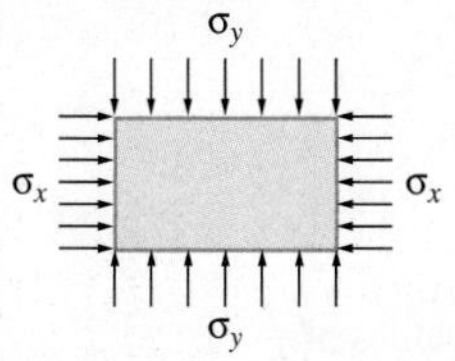
</td>
<td>2a. All edges simply supported</td>
<td>$$\sigma'_x \frac{m^2}{a^2} + \sigma'_y \frac{n^2}{b^2} = 0.823 \frac{E}{1-\nu^2} t^2 \left(\frac{m^2}{a^2} + \frac{n^2}{b^2} \right)^2$$
Here m and n signify the number of half-waves in the buckled plate in the x and y directions, respectively. To find σ'_y for a given σ_x, take $m = 1, n = 1$ if
$$C\left(1 - 4\frac{a^4}{b^4}\right) < \sigma_x < C\left(5 + 2\frac{a^2}{b^2}\right), \text{ where } C = \frac{0.823Et^2}{(1-\nu^2)a^2}.$$
If σ_x is too large to satisfy this inequality, take $n = 1$ and m to satisfy:
$$C\left(2m^2 - 2m + 1 + 2\frac{a^2}{b^2}\right) < \sigma_x < C\left(2m^2 + 2m + 1 + 2\frac{a^2}{b^2}\right).$$ If σ_x too small to satisfy the first inequality, take $m = 1$ and n to satisfy:
$$C\left[1 - n^2(n-1)^2 \frac{a^4}{b^4}\right] > \sigma_x > C\left[1 - n^2(n+1)^2 \frac{a^4}{b^4}\right]$$
(Refs. 1 ,6)</td></tr>
<tr><td>2b. All edges clamped</td>
<td>$$\sigma'_x + \frac{a^2}{b^2}\sigma'_y = 1.1 \frac{Et^2a^2}{1-\nu^2} \left(\frac{3}{a^4} + \frac{3}{b^4} + \frac{2}{a^2b^2} \right)$$
(This equation is approximate and is most accurate when the plate is nearly square and σ_x and σ_y nearly equal.)
(Ref. 1)</td></tr>
<tr><td>3. Rectangular plate under linearly varying stress on edges b (bending or bending combined with tension or compression)
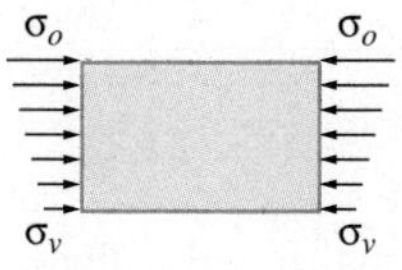
</td>
<td>3a. All edges simply supported</td>
<td>$$\sigma'_o = K \frac{E}{1-\nu^2} \left(\frac{t}{b} \right)^2$$
Here K depends on $\frac{a}{b}$ and on $\alpha = \frac{\sigma_o}{\sigma_o - \sigma_v}$ and may be found from the following table:
<table>
<tr><td></td><td>$\frac{a}{b} = 0.4$</td><td>0.5</td><td>0.6</td><td>0.667</td><td>0.75</td><td>0.8</td><td>0.9</td><td>1.0</td><td>1.5</td></tr>
<tr><td>$\alpha = 0.5$</td><td>$K = 23.9$</td><td>21.1</td><td>19.8</td><td>19.7</td><td>19.8</td><td>20.1</td><td>21.1</td><td>21.1</td><td>19.8</td></tr>
<tr><td>0.75</td><td>15.4</td><td></td><td>10.6</td><td></td><td>9.5</td><td>9.2</td><td></td><td>9.1</td><td>9.5</td></tr>
<tr><td>1.00</td><td>12.4</td><td></td><td>8.0</td><td></td><td>6.9</td><td>6.7</td><td></td><td>6.4</td><td>6.9</td></tr>
<tr><td>1.25</td><td>10.95</td><td></td><td>6.8</td><td></td><td>5.8</td><td>5.7</td><td></td><td>5.4</td><td>5.8</td></tr>
<tr><td>1.50</td><td>8.9</td><td></td><td>5.3</td><td></td><td>5.0</td><td>4.9</td><td></td><td>4.8</td><td>5.0</td></tr>
<tr><td>∞ pure compression)</td><td>6.92</td><td></td><td>4.23</td><td></td><td></td><td>3.45</td><td></td><td>3.29</td><td>3.57</td></tr>
</table>
(Refs. 1, 6)</td></tr>
</table>

(Continued)

TABLE 15.2 Formulas for Elastic Stability of Plates and Shells (*Continued*)

<table>
<tr><th>Form of Plate or Shell and Manner of Loading</th><th>Manner of Support</th><th>Formulas for Critical Unit Compressive Stress σ', Unit Shear Stress τ', Load P', Bending Moment M', or Unit External Pressure q' at Which Elastic Buckling Occurs</th></tr>
<tr><td rowspan="2">4. Rectangular plate under uniform shear on all edges
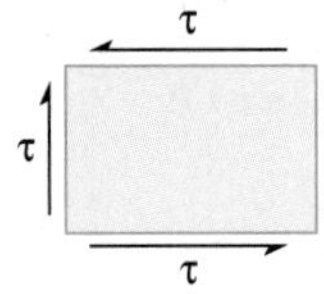
</td><td>4a. All edges simply supported</td><td>$$\tau' = K\frac{E}{1-\nu^2}\left(\frac{t}{b}\right)^2$$
<table>
<tr><td>$\frac{a}{b}$</td><td>1.0</td><td>1.2</td><td>1.4</td><td>1.5</td><td>1.6</td><td>1.8</td><td>2.0</td><td>2.5</td><td>3.0</td><td>∞</td></tr>
<tr><td>K</td><td>7.75</td><td>6.58</td><td>6.00</td><td>5.84</td><td>5.76</td><td>5.59</td><td>5.43</td><td>5.18</td><td>5.02</td><td>4.40</td></tr>
</table>
(Refs. 1, 6, 8, 22)</td></tr>
<tr><td>4b. All edges clamped</td><td>$$\tau' = K\frac{E}{1-\nu^2}\left(\frac{t}{b}\right)^2$$
<table>
<tr><td>$\frac{a}{b}$</td><td>1</td><td>2</td><td>∞</td></tr>
<tr><td>K</td><td>12.7</td><td>9.5</td><td>7.38</td></tr>
</table>
Test results indicate a value for K of about 4.1 for very large values of $\frac{a}{b}$. (For continuous panels, see Ref. 30.) (Ref. 9)</td></tr>
<tr><td rowspan="2">5. Rectangular plate under uniform shear on all edges; compression (or tension) σ_x on edges b; compression (or tension) σ_y on edges a; a/b very large
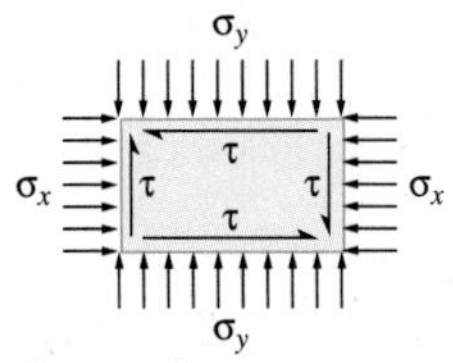
</td><td>5a. All edges simply supported</td><td>$$\tau' = \sqrt{C^2\left(2\sqrt{1-\frac{\sigma_y}{C}}+2-\frac{\sigma_x}{C}\right)\left(2\sqrt{1-\frac{\sigma_y}{C}}+6-\frac{\sigma_x}{C}\right)}$$
where $C = \frac{0.823}{1-\nu^2}\left(\frac{t}{b}\right)^2 E$
(Refs. 1, 6, 23, and 31)</td></tr>
<tr><td>5b. All edges clamped</td><td>$$\tau' = \sqrt{C^2\left(2.31\sqrt{4-\frac{\sigma_y}{C}}+\frac{4}{3}-\frac{\sigma_x}{C}\right)\left(2.31\sqrt{4-\frac{\sigma_y}{C}}+8-\frac{\sigma_x}{C}\right)}$$
where $C = \frac{0.823}{1-\nu^2}\left(\frac{t}{b}\right)^2 E$
(σ_x and σ_y are negative when tensile) (Ref. 6)</td></tr>
<tr><td>6. Rectangular plate under uniform shear on all edges and bending stresses on edges b
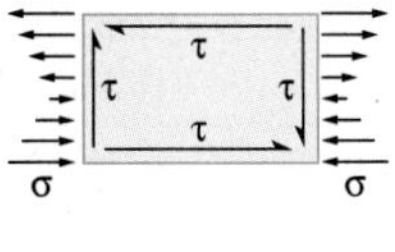
</td><td>6a. All edges simply supported</td><td>$$\sigma' = K\frac{E}{1-\nu^2}\left(\frac{t}{b}\right)^2$$
Here K depends on $\frac{\tau}{\tau'}$ (ratio of actual shear stress to shear stress that, acting alone, would be critical) and on $\frac{a}{b}$. K varies less than 10% for values $\frac{a}{b}$ from 0.5 to 1, and for $\frac{a}{b}=1$ is approximately as follows:
<table>
<tr><td>$\frac{\tau}{\tau'}$</td><td>0</td><td>0.2</td><td>0.3</td><td>0.4</td><td>0.5</td><td>0.6</td><td>0.7</td><td>0.8</td><td>0.9</td><td>1.0</td></tr>
<tr><td>K</td><td>21.1</td><td>20.4</td><td>19.6</td><td>18.5</td><td>17.7</td><td>16.0</td><td>14.0</td><td>11.9</td><td>8.20</td><td>0</td></tr>
</table>
(Refs. 1, 10)</td></tr>
</table>

(Continued)

TABLE 15.2 Formulas for Elastic Stability of Plates and Shells (*Continued*)

<table>
<tr><th>Form of Plate or Shell and Manner of Loading</th><th>Manner of Support</th><th>Formulas for Critical Unit Compressive Stress σ′, Unit Shear Stress τ′, Load P′, Bending Moment M′, or Unit External Pressure q′ at Which Elastic Buckling Occurs</th></tr>
<tr><td rowspan="2">7. Rectangular plate under concentrated center loads on two opposite edges
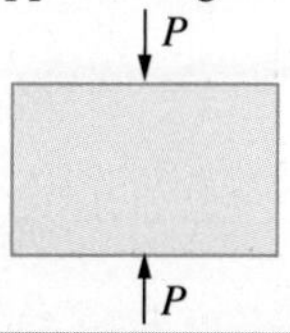
</td><td>7a. All edges simply supported</td><td>$P' = \frac{\pi}{3}\frac{Et^3}{(1-\nu^2)b}\quad\left(\text{for } \frac{a}{b} > 2\right)$ (Ref. 1)</td></tr>
<tr><td>7b. Edges b simply supported, edges a clamped</td><td>$P' = \frac{2\pi}{3}\frac{Et^3}{(1-\nu^2)b}\quad\left(\text{for } \frac{a}{b} > 2\right)$ (Ref. 1)</td></tr>
<tr><td>8. Rhombic plate under uniform compression on all edges
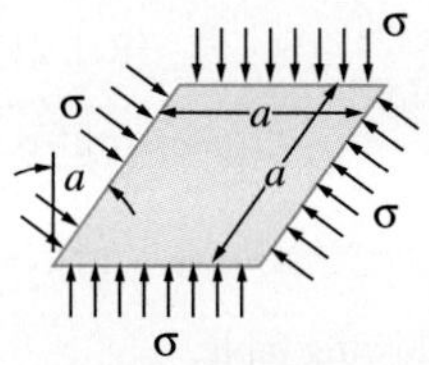
</td><td>8a. All edges simply supported</td><td>$\sigma' = K\frac{Et^2}{a^2(1-\nu^2)}$

<table>
<tr><td>α</td><td>0°</td><td>9°</td><td>18°</td><td>27°</td><td>36°</td><td>45°</td></tr>
<tr><td>K</td><td>1.645</td><td>1.678</td><td>1.783</td><td>1.983</td><td>2.338</td><td>2.898</td></tr>
</table>
(Ref. 65)</td></tr>
<tr><td>9. Polygon plate under uniform compression on all edges
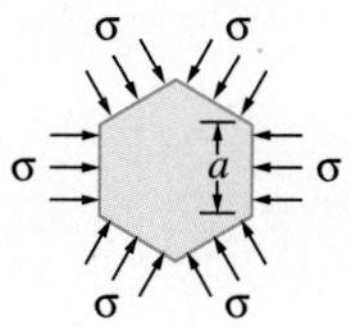

N = number of sides</td><td>9a. All edges simply supported</td><td>$\sigma' = K\frac{Et^2}{a^2(1-\nu^2)}$

<table>
<tr><td>N</td><td>3</td><td>4</td><td>5</td><td>6</td><td>7</td><td>8</td></tr>
<tr><td>K</td><td>4.393</td><td>1.645</td><td>0.916</td><td>0.597</td><td>0.422</td><td>0.312</td></tr>
</table>
(Ref. 65)</td></tr>
<tr><td rowspan="2">10. Parabolic and semielliptic plates under uniform compression on all the edges
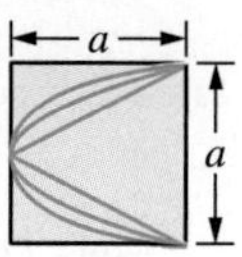
</td><td>10a. All edges simply supported</td><td rowspan="2">$\sigma' = K\frac{Et^2}{a^2(1-\nu^2)}$

where K is tabulated below for the several shapes and boundary conditions for $\nu = \frac{1}{3}$:

<table>
<tr><td></td><td>Square</td><td>Semiellipse</td><td>Parabola</td><td>Triangle</td></tr>
<tr><td>Simply supported</td><td>1.65</td><td>1.86</td><td>2.50</td><td>3.82</td></tr>
<tr><td>Fixed</td><td>4.36</td><td>5.57</td><td>7.22</td><td>10.60</td></tr>
</table>
(Ref. 62)</td></tr>
<tr><td>10b. All edges fixed</td></tr>
</table>

(*Continued*)

TABLE 15.2 Formulas for Elastic Stability of Plates and Shells (*Continued*)

<table>
<tr><th>Form of Plate or Shell and Manner of Loading</th><th>Manner of Support</th><th>Formulas for Critical Unit Compressive Stress σ', Unit Shear Stress τ', Load P', Bending Moment M', or Unit External Pressure q' at Which Elastic Buckling Occurs</th></tr>
<tr><td rowspan="2">11. Isotropic circular plate under uniform radial edge compression
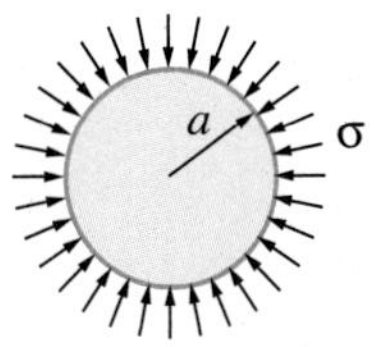
</td><td>11a. Edges simply supported</td><td>$\sigma' = K\frac{E}{1-\nu^2}\left(\frac{t}{a}\right)^2$
<table><tr><td>ν</td><td>0</td><td>0.1</td><td>0.2</td><td>0.3</td><td>0.4</td></tr><tr><td>K</td><td>0.282</td><td>0.306</td><td>0.328</td><td>0.350</td><td>0.370</td></tr></table>(Ref. 1)</td></tr>
<tr><td>11b. Edges clamped</td><td>$\sigma' = 1.22\frac{E}{1-\nu^2}\left(\frac{t}{a}\right)^2$ (Ref. 1)
For elliptical plate with major semiaxis a, minor semiaxis b, $\sigma' = K\frac{E}{1-\nu^2}\left(\frac{t}{b}\right)^2$, where K has values as follows:
<table><tr><td>$\frac{a}{b}$</td><td>1.0</td><td>1.1</td><td>1.2</td><td>1.3</td><td>2.0</td><td>5.0</td></tr><tr><td>K</td><td>1.22</td><td>1.13</td><td>1.06</td><td>1.01</td><td>0.92</td><td>0.94</td></tr></table>(Ref. 21)</td></tr>
<tr><td rowspan="2">12. Circular plate with concentric hole under uniform radial compression on outer edge
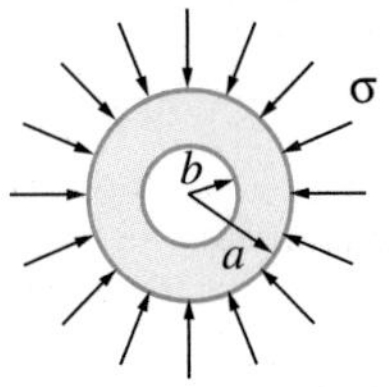

$\frac{a}{t} > 10$</td><td>12a. Outer edge simply supported, inner edge free</td><td>$\sigma' = K\frac{E}{1-\nu^2}\left(\frac{t}{a}\right)^2$
Here K depends on $\frac{b}{a}$ and is given approximately by the following table:
<table><tr><td>$\frac{b}{a}$</td><td>0</td><td>0.1</td><td>0.2</td><td>0.3</td><td>0.4</td><td>0.5</td><td>0.6</td><td>0.7</td><td>0.8</td><td>0.9</td></tr><tr><td>K</td><td>0.35</td><td>0.33</td><td>0.30</td><td>0.27</td><td>0.23</td><td>0.21</td><td>0.19</td><td>0.18</td><td>0.17</td><td>0.16</td></tr></table>(Ref. 1)</td></tr>
<tr><td>12b. Outer edge clamped, inner edge free</td><td>$\sigma' = K\frac{E}{1-\nu^2}\left(\frac{t}{a}\right)^2$
Here K depends on $\frac{b}{a}$ and is given approximately by the following table:
<table><tr><td>$\frac{b}{a}$</td><td>0</td><td>0.1</td><td>0.2</td><td>0.3</td><td>0.4</td><td>0.5</td></tr><tr><td>K</td><td>1.22</td><td>1.17</td><td>1.11</td><td>1.21</td><td>1.48</td><td>2.07</td></tr></table>(Ref. 1)</td></tr>
<tr><td>13. Curved panel under uniform compression on curved edges b (b = width of panel measured on arc; r = radius of curvature)
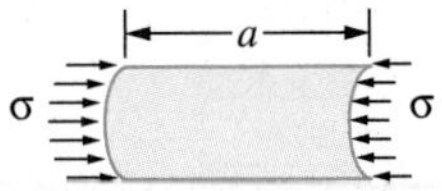

$\frac{b}{t} > 10$</td><td>13a. All edges simply supported</td><td>$\sigma' = \frac{1}{6}\frac{E}{1-\nu^2}\left[\sqrt{12(1-\nu^2)\left(\frac{t}{r}\right)^2 + \left(\frac{\pi t}{b}\right)^4} + \left(\frac{\pi t}{b}\right)^2\right]$
(*Note:* With a > b, the solution does not depend upon a.)
or $\sigma' = 0.6E\frac{t}{r}$ if $\frac{b}{r}$ (central angle of curve) is less than ½ and b and a are nearly equal.
(For compression combined with shear, see Refs. 28 and 34.)
(Refs. 1 and 6)</td></tr>
</table>

(*Continued*)

TABLE 15.2 Formulas for Elastic Stability of Plates and Shells (*Continued*)

Form of Plate or Shell and Manner of Loading	Manner of Support	Formulas for Critical Unit Compressive Stress σ', Unit Shear Stress τ', Load P', Bending Moment M', or Unit External Pressure q' at Which Elastic Buckling Occurs
14. Curved panel under uniform shear on all edges	14a. All edges simply supported	$\tau' = 0.1E\frac{t}{r} + 5E\left(\frac{t}{b}\right)^2$ (Refs. 6, 27, 29)
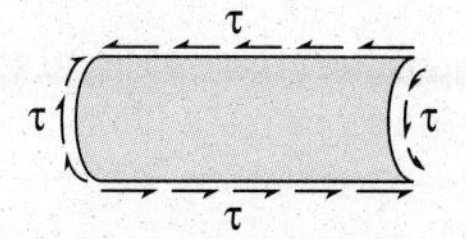 $b/t > 10$. See case 13 for b and r.	14b. All edges clamped	$\tau' = 0.1E\frac{t}{r} + 7.5E\left(\frac{t}{b}\right)^2$ (Ref. 6) Tests show $\tau' = 0.075E\frac{t}{r}$ for panels curved to form quadrant of a circle. (Ref. 11) (See also Refs. 27, 29)
15. Thin-walled circular tube under uniform longitudinal compression (radius of tube $= r$)  $\frac{r}{t} > 10$	15a. Ends not constrained	$\sigma' = \frac{1}{\sqrt{3}}\frac{E}{\sqrt{1-\nu^2}}\frac{t}{r}$ (Refs. 6, 12, 13, 24) Most accurate for very long tubes, but applicable if length is several times as great as $1.72\sqrt{rt}$, which is the length of a half-wave of buckling. Tests indicate an actual buckling strength of 40–60% of this theoretical value, or $\sigma' = 0.3Et/r$ approximately.
16. Thin-walled circular tube under a transverse bending moment M (radius of tube $= r$  $\frac{r}{t} > 10$	16a. No constraint	$M' = K\frac{E}{\sqrt{1-\nu^2}}rt^2$ Here the theoretical value of K for pure bending and long tubes is 0.99. The average value of K determined by tests is 1.14, and the minimum value is 0.72. Except for very short tubes, length effect is negligible and a small transverse shear produces no appreciable reduction in M'. A very short cylinder under transverse (beam) shear may fail by buckling at neutral axis when shear stress there reaches a value of about $1.25\tau'$ for case 17a. (Refs. 6, 14, 15)
17. Thin-walled circular tube under a twisting moment T that produces a uniform circumferential shear stress: $\tau = \frac{T}{2\pi r^2 t}$ (length of tube $= l$; radius of tube $= r$)	17a. Ends hinged, i.e., wall free to change angle with cross section, but circular section maintained	$\tau' = \frac{E}{1-\nu^2}\left(\frac{t}{l}\right)^2 (1.27 + \sqrt{9.64 + 0.466H^{1.5}})$ where $H = \sqrt{1-\nu^2}\frac{l^2}{tr}$ Tests indicate that the actual buckling stress is 60–75% of this theoretical value, with the majority of the data points nearer 75%. (Refs. 6, 16, 18, 25)
 $\frac{r}{t} > 10$	17b. Ends clamped, i.e., wall held perpendicular to cross section and circular section maintained	$\tau' \frac{E}{1-\nu^2}\left(\frac{t}{l}\right)^2 (-2.39 + \sqrt{96.9 + 0.605H^{1.5}})$ where H is given in part 17a. The statement in part a regarding actual buckling stress applies here as well. (Refs. 6, 16, 18, 25)

(*Continued*)

TABLE 15.2 Formulas for Elastic Stability of Plates and Shells (*Continued*)

Form of Plate or Shell and Manner of Loading	Manner of Support	Formulas for Critical Unit Compressive Stress σ', Unit Shear Stress τ', Load P', Bending Moment M', or Unit External Pressure q' at Which Elastic Buckling Occurs
18. Thin-walled circular tube under uniform longitudinal compression σ and uniform circumferential shear τ due to torsion (case 15 combined with case 17) 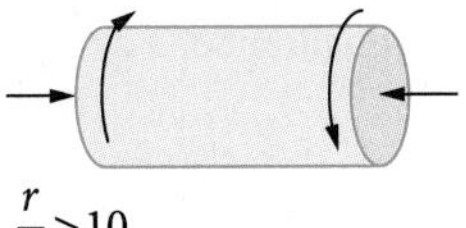 $\frac{r}{t}>10$	18a. Edges hinged as in case 17a 18b. Edges clamped as in case 17b	The equation $1-\frac{\sigma'}{\sigma'_o}=\left(\frac{\tau'}{\tau'_o}\right)^n$ holds, where σ' and τ' are the critical compressive and shear stresses for the combined loading, σ'_o is the critical compressive stress for the cylinder under compression alone (case 15), and τ'_o is the critical shear stress for the cylinder under torsion alone (case 17a or 17b according to end conditions). Tests indicate that n is approximately 3. If σ is tensile, then σ' should be considered negative. (Ref. 6) (See also Ref. 26. For square tube, see Ref. 32.)
19. Thin tube under uniform lateral external pressure (radius of tube $= r$) 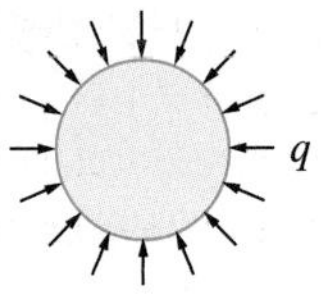 $\frac{r}{t}>10$	19a. Very long tube with free ends; length l	$q'=\frac{1}{4}\frac{E}{1-\nu^2}\frac{t^3}{r^3}$ Applicable when $l>4.90r\sqrt{\frac{r}{t}}$ (Ref. 19)
	19b. Short tube, of length l, ends held circular, but not otherwise constrained, or long tube held circular at intervals l	$q'=0.807\frac{Et^2}{lr}\sqrt[4]{\left(\frac{1}{1-\nu^2}\right)^3\frac{t^2}{r^2}}$ approximate formula (Ref. 19)
20. Thin tube with closed ends under uniform external pressure, lateral and longitudinal (length of tube $= l$; radius of tube $= r$) 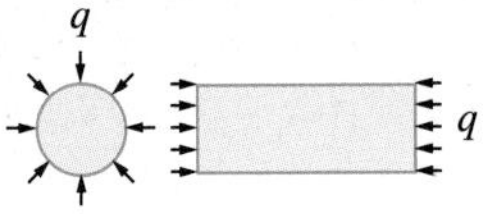 $\frac{r}{t}>10$	20a. Ends held circular	$q'=\frac{E\frac{t}{r}}{1+\frac{1}{2}\left(\frac{\pi r}{nl}\right)^2}\left\{\frac{1}{n^2\left[1+\left(\frac{nl}{\pi r}\right)^2\right]^2}+\frac{n^2t^2}{12r^2(1-\nu^2)}\left[1+\left(\frac{\pi r}{nl}\right)^2\right]^2\right\}$ (Refs. 19, 20) where $n=$ number of lobes formed by the tube in buckling. To determine q' for tubes of a given t/r, plot a group of curves, one curve for each integral value of n of 2 or more, with l/r as ordinates and q' as abscissa; that curve of the group which gives the least value of q' is then used to find the q' corresponding to a given l/r. If $60<\left(\frac{l}{r}\right)^2\left(\frac{r}{t}\right)<2.5\left(\frac{r}{t}\right)^2$, the critical pressure can be approximated by $q'=\frac{0.92E}{\left(\frac{l}{r}\right)\left(\frac{r}{t}\right)^{2.5}}$ (Ref. 81) For other approximations see ref. 109. Values of experimentally determined critical pressures range 20% above and below the theoretical values given by the expressions above. A recommended probable minimum critical pressure is $0.80q'$.

(*Continued*)

TABLE 15.2 Formulas for Elastic Stability of Plates and Shells (*Continued*)

<table>
<tr><th>Form of Plate or Shell and Manner of Loading</th><th>Manner of Support</th><th>Formulas for Critical Unit Compressive Stress σ′, Unit Shear Stress τ′, Load P′, Bending Moment M′, or Unit External Pressure q′ at Which Elastic Buckling Occurs</th></tr>
<tr><td rowspan="2">19. Thin tube under uniform lateral external pressure (radius of tube = r)
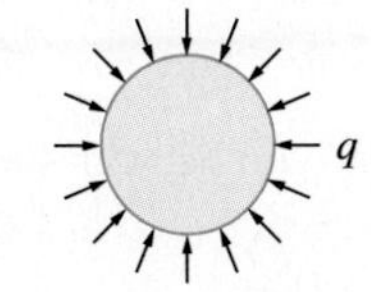

$\frac{r}{t} > 10$</td><td>19a. Very long tube with free ends; length l</td><td>$q' = \frac{1}{4}\frac{E}{1-\nu^2}\frac{t^3}{r^3}$
Applicable when $l > 4.90r\sqrt{\frac{r}{t}}$ (Ref. 19)</td></tr>
<tr><td>19b. Short tube, of length l, ends held circular, but not otherwise constrained, or long tube held circular at intervals l</td><td>$q' = 0.807\frac{Et^2}{lr}\sqrt[4]{\left(\frac{1}{1-\nu^2}\right)^3\frac{t^2}{r^2}}$ approximate formula (Ref. 19)</td></tr>
<tr><td>20. Thin tube with closed ends under uniform external pressure, lateral and longitudinal (length of tube = l; radius of tube = r)
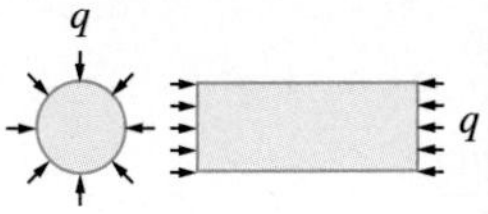

$\frac{r}{t} > 10$</td><td>20a. Ends held circular</td><td>$q' = \frac{E\frac{t}{r}}{1+\frac{1}{2}\left(\frac{\pi r}{nl}\right)^2}\left\{\frac{1}{n^2\left[1+\left(\frac{nl}{\pi r}\right)^2\right]^2} + \frac{n^2t^2}{12r^2(1-\nu^2)}\left[1+\left(\frac{\pi r}{nl}\right)^2\right]^2\right\}$
(Refs. 19, 20)
where n = number of lobes formed by the tube in buckling. To determine q' for tubes of a given t/r, plot a group of curves, one curve for each integral value of n of 2 or more, with l/r as ordinates and q' as abscissa; that curve of the group which gives the least value of q' is then used to find the q' corresponding to a given l/r. If $60 < \left(\frac{l}{r}\right)^2\left(\frac{r}{t}\right) < 2.5\left(\frac{r}{t}\right)^2$, the critical pressure can be approximated by $q' = \frac{0.92E}{\left(\frac{l}{r}\right)\left(\frac{r}{t}\right)^{2.5}}$ (Ref. 81)
For other approximations see Ref. 109.
Values of experimentally determined critical pressures range 20% above and below the theoretical values given by the expressions above. A recommended probable minimum critical pressure is $0.80q'$.</td></tr>
<tr><td rowspan="2">21. Curved panel under uniform radial pressure (radius of curvature r, central angle 2α, when $2\alpha = \text{arc } AB/r$)
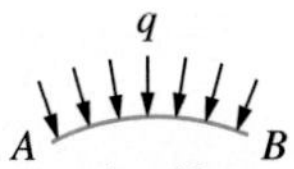

$r/t > 10$</td><td>21a. Curved edges free, straight edges at A and B simply supported (i.e., hinged)</td><td>$q' = \frac{Et^3\left(\frac{\pi^2}{\alpha^2}-1\right)}{12r^3(1-\nu^2)}$ (Ref. 1)</td></tr>
<tr><td>21b. Curved edges free, straight edges at A and B clamped</td><td>Here k is found from the equation $k \tan\alpha \cot k\alpha = 1$ and has the following values:
$q' = \frac{Et^3(k^2-1)}{12r^3(1-\nu^2)}$
<table><tr><td>α</td><td>15°</td><td>30°</td><td>60°</td><td>90°</td><td>120°</td><td>150°</td><td>180°</td></tr><tr><td>k</td><td>17.2</td><td>8.62</td><td>4.37</td><td>3.0</td><td>2.36</td><td>2.07</td><td>2.0</td></tr></table>(Ref. 1)</td></tr>
</table>

(Continued)

TABLE 15.2 Formulas for Elastic Stability of Plates and Shells (*Continued*)

Form of Plate or Shell and Manner of Loading	Manner of Support	Formulas for Critical Unit Compressive Stress σ', Unit Shear Stress τ', Load P', Bending Moment M', or Unit External Pressure q' at Which Elastic Buckling Occurs
22. Thin sphere under uniform external pressure (radius of sphere $= r$) 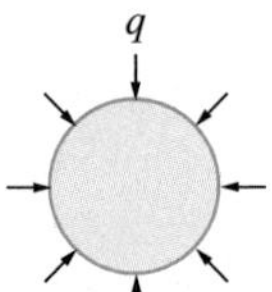 $r/t > 10$	22a. No constraint	$q' = \frac{2Et^2}{r^2\sqrt{3(1-\nu^2)}}$ (for ideal case) (Refs. 1, 37) $q' = \frac{0.365Et^2}{r^2}$ (probable actual minimum q') For spherical cap, half-central angle ϕ between 20 and 60°, R/t between 400 and 2000, $q' = [1 - 0.00875(\phi° - 20°)]\left(1 - 0.000175\frac{R}{t}\right)(0.3E)\left(\frac{t}{R}\right)^2$ (Empirical formula, Ref. 43)
23. Thin truncated conical shell with closed ends under external pressure (both lateral and longitudinal pressure) 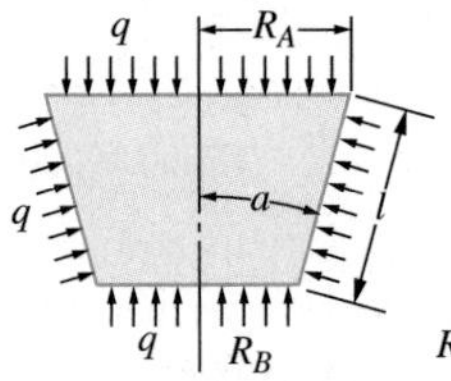 $R_B/t > 10$	23a. Ends held circular	q' can be found from the formula of case 20a if the slant length of the cone is substituted for the length of the cylinder and if the average radius of curvature of the wall of the cone normal to the meridian $(R_A + R_B)/(2\cos\alpha)$ is substituted for the radius of the cylinder. The same recommendation of a probable minimum critical pressure of $0.8q'$ is made from the examination of experimental data for cones. (Refs. 78, 81)
24. Thin truncated conical shell under axial load 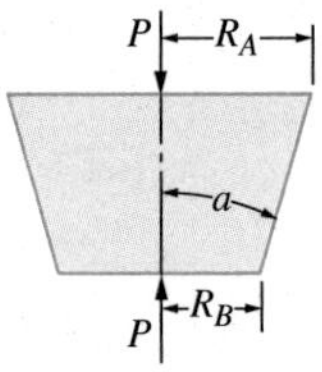 $R_B/t > 10$	24a. Ends held circular	$P' = \frac{2\pi Et^2\cos^2\alpha}{\sqrt{3(1-\nu^2)}}$ (theoretical) Tests indicate an actual buckling strength of 40–60% of the above theoretical value, or $P' = 0.3(2\pi Et^2\cos^2\alpha)$ approximately. (Ref. 78) In Ref. 77 it is stated that $P' = 0.277(2\pi Et^2\cos^2\alpha)$ will give 95% confidence in at least 90% of the cones carrying more than this critical load. This is based on 170 tests.
25. Thin truncated conical shell under combined axial load and internal pressure 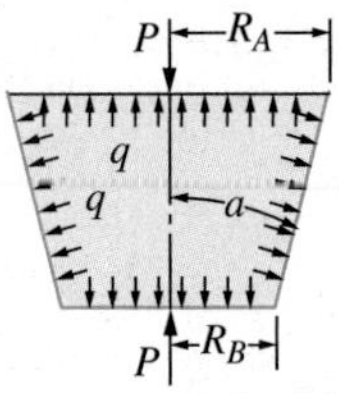 $R_B/t > 10$	25a. Ends held circular	$P' - q\pi R_B^2 = K_A 2\pi Et^2\cos^2\alpha$ The probable minimum values of K_A are tabulated for several values of $K_p = \frac{q}{E}\left(\frac{R_B}{t\cos\alpha}\right)^2$. $k_B = 2\left[\frac{12(1-\nu^2)R_B^2}{t^2\tan^2\alpha\sin^2\alpha}\right]^{1/4}$ K_p: 0.00, 0.25, 0.50, 1.00, 1.50, 2.00, 3.00 for $k_B \le 150$: 0.30, 0.52, 0.60, 0.68, 0.73, 0.76, 0.80 for $k_B > 150$: 0.20, 0.36, 0.48, 0.60, 0.64, 0.66, 0.69 (Ref. 78)

(*Continued*)

TABLE 15.2 Formulas for Elastic Stability of Plates and Shells (*Continued*)

Form of Plate or Shell and Manner of Loading	Manner of Support	Formulas for Critical Unit Compressive Stress σ', Unit Shear Stress τ', Load P', Bending Moment M', or Unit External Pressure q' at Which Elastic Buckling Occurs
26. Thin truncated conical shell under combined axial load and external pressure 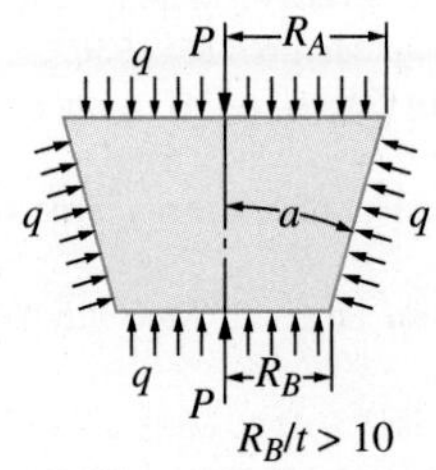	26a. Ends held circular	The following conservative interaction formula may be used for design. It is applicable equally to theoretical values or to minimum probable values of critical load and pressure. $$\frac{P'}{P'_{\text{case 24}}}+\frac{q'}{q'_{\text{case 23}}}=1$$ This expression can be used for cylinders if the angle α is set equal to zero or use is made of cases 15 and 20. For small values of $P'/P'_{\text{case 24}}$ the external pressure required to collapse the shell is greater than that required to initiate buckling. See Ref. 78.
27. Thin truncated conical shell under torsion 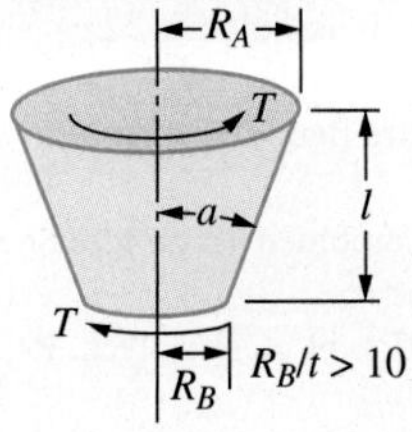	27a. Ends held circular	Let $T=\tau' 2\pi r_e^2 t$ and for τ' use the formulas for thin-walled circular tubes, case 17, substituting for the radius r of the tube the equivalent radius r_e, where $r_e = R_B \cos\alpha \left\{1+\left[\frac{1}{2}\left(1+\frac{R_A}{R_B}\right)\right]^{1/2} - \left[\frac{1}{2}\left(1+\frac{R_A}{R_B}\right)\right]^{-1/2}\right\}$. l and t remain the axial length and wall thickness, respectively. (Ref. 17)

15.6 REFERENCES

1. Timoshenko, S. P., and J. M. Gere, *Theory of Elastic Stability*, 2nd ed., McGraw-Hill, 1961.
2. Schwerin, E., "Die Torsionstabilität des dünnwandigen Rohres," *Z. angew. Math. Mech.*, 5(3): 235, 1925.
3. Trayer, G. W., and H. W. March, "Elastic Instability of Members Having Sections Common in Aircraft Construction," *Natl. Adv. Comm. Aeron.*, Rept. 382, 1931.
4. Dumont, C., and H. N. Hill, "The Lateral Instability of Deep Rectangular Beams," *Natl. Adv. Comm. Aeron.*, Tech. Note 601, 1937.
5. Dinnik, A., "Design of Columns of Varying Cross-Section," *Trans. ASME*, 54(18): 165, 1932.
6. Heck, O. S., and H. Ebner, "Methods and Formulas for Calculating the Strength of Plate and Shell Construction as Used in Airplane Design," *Natl. Adv. Comm. Aeron.*, Tech, Memo. 785, 1936.
7. Maulbetsch, J. L., "Buckling of Compressed Rectangular Plates with Built-In Edges," *ASME J. Appl. Mech.*, 4(2), 1937.
8. Southwell, R. V., and S. W. Skan, "On the Stability under Shearing Forces of a Flat Elastic Strip," *Proc. R. Soc. Lond.*, Ser. A., 105: 582, 1924.
9. Bollenrath, F., "Wrinkling Phenomena of Thin Flat Plates Subjected to Shear Stresses," *Natl. Adv. Comm. Aeron.*, Tech. Memo. 601, 1931.
10. Way, S., "Stability of Rectangular Plates under Shear and Bending Forces," *ASME J. Appl. Mech.*, 3(4): 1936.
11. Smith, G. M., "Strength in Shear of Thin Curved Sheets of Alclad," *Natl. Adv. Comm. Aeron.*, Tech. Note 343, 1930.
12. Lundquist, E. E., "Strength Tests of Thin-Walled Duralumin Cylinders in Compression," *Natl. Adv. Comm. Aeron.*, Rept. 473, 1933.
13. Wilson, W. M., and N. M. Newmark, "The Strength of Thin Cylindrical Shells as Columns," *Eng. Exp. Sta. Univ. Ill.*, Bull. 255, 1933.
14. Lundquist, E. E., "Strength Tests of Thin-Walled Duralumin Cylinders in Pure Bending," *Natl. Adv. Comm. Aeron.*, Tech. Note 479, 1933.
15. Lundquist, E. E., "Strength Tests of Thin-Walled Duralumin Cylinders in Combined Transverse Shear and Bending," *Natl. Adv. Comm. Aeron.*, Tech. Note 523, 1935.
16. Donnell, L. H., "Stability of Thin-Walled Tubes under Torsion," *Natl. Adv. Comm. Aeron.*, Tech. Rept. 479, 1933.
17. Seide, P., "On the Buckling of Truncated Conical Shells in Torsion," *ASME, J. Appl. Mech.*, 29(2): 1962.
18. Ebner, H., "Strength of Shell Bodies—Theory and Practice," *Natl. Adv. Comm. Aeron.*, Tech. Memo. 838, 1937.
19. Saunders, H. E., and D. F. Windenberg, "Strength of Thin Cylindrical Shells under External Pressure," *Trans. ASME*, 53(15): 207, 1931.
20. von Mises, R., "Der kritische Aussendruck zylindrischer Rohre," *Z. Ver Dtsch. Ing.*, 58: 750, 1914.
21. Woinowsky-Krieger, S., "The Stability of a Clamped Elliptic Plate under Uniform Compression," *ASME J. Appl. Mech.*, 4(4): 1937.
22. Stein, M., and J. Neff, "Buckling Stresses in Simply Supported Rectangular Flat Plates in Shear," *Natl. Adv. Comm. Aeron.*, Tech. Note 1222, 1947.
23. Batdorf, S. B., and M. Stein, "Critical Combinations of Shear and Direct Stress for Simply Supported Rectangular Flat Plates," *Natl. Adv. Comm. Aeron.*, Tech. Note 1223, 1947.
24. Batdorf, S. B., M. Schildcrout, and M. Stein, "Critical Stress of Thin-Walled Cylinders in Axial Compression," *Natl. Adv. Comm. Aeron.*, Tech. Note 1343, 1947.
25. Batdorf, S. B., M. Stein, and M. Schildcrout, "Critical Stress of Thin-Walled Cylinders in Torsion," *Natl. Adv. Comm. Aeron.*, Tech. Note 1344, 1947.
26. Batdorf, S. B., M. Stein, and M. Schildcrout, "Critical Combination of Torsion and Direct Axial Stress for Thin-Walled Cylinders," *Natl. Adv. Comm. Aeron.*, Tech. Note 1345, 1947.
27. Batdorf, S. B., M. Schildcrout, and M. Stein, "Critical Shear Stress of Long Plates with Transverse Curvature," *Natl. Adv. Comm. Aeron.*, Tech. Note 1346, 1947.
28. Batdorf, S. B., M. Schildcrout, and M. Stein, "Critical Combinations of Shear and Longitudinal Direct Stress for Long Plates with Transverse Curvature," *Natl. Adv. Comm. Aeron.*, Tech. Note 1347, 1947.
29. Batdorf, S. B., M. Stein, and M. Schildcrout, "Critical Shear Stress of Curved Rectangular Panels," *Natl. Adv. Comm. Aeron.*, Tech. Note 1348, 1947.
30. Budiansky, B., R. W. Connor, and M. Stein: Buckling in Shear of Continuous Flat Plates, Natl. Adv. Comm. Aeron., Tech. Note 1565, 1948.
31. Peters, R. W., "Buckling Tests of Flat Rectangular Plates under Combined Shear and Longitudinal Compression," *Natl. Adv. Comm. Aeron.*, Tech. Note 1750, 1948.
32. Budiansky, B., M. Stein, and A. C. Gilbert, "Buckling of a Long Square Tube in Torsion and Compression," *Natl. Adv. Comm. Aeron.*, Tech. Note 1751, 1948.

33. Libove, C., S. Ferdman, and J. G. Reusch: Elastic Buckling of a Simply Supported Plate under a Compressive Stress that Varies Linearly in the Direction of Loading, Natl. Adv. Comm. Aeron., Tech. Note 1891, 1949.
34. Schildcrout, M., and M. Stein: Critical Combinations of Shear and Direct Axial Stress for Curved Rectangular Panels, Natl. Adv. Comm. Aeron., Tech. Note 1928, 1949.
35. Pflüger, A., *Stabilitätsprobleme der Elastostatik*, Springer-Verlag, 1964.
36. Gerard, G., and H. Becker, "Handbook of Structural Stability," *Natl. Adv. Comm. Aeron.*, Tech. Notes 3781–3786 inclusive, and D163, 1957–1959.
37. von Kármán, Th., and Hsue-shen Tsien, "The Buckling of Spherical Shells by External Pressure," *Pressure Vessel and Piping Design*, ASME Collected Papers 1927–1959.
38. Cheng, Shun, "On the Theory of Bending of Sandwich Plates," *Proc. 4th U.S. Natl. Congr. Appl. Mech.*, 1962.
39. U.S. Forest Products Laboratory, "List of Publications on Structural Sandwich, Plastic Laminates, and Wood-Base Aircraft Components," 1962.
40. Lind, N. C., "Elastic Buckling of Symmetrical Arches, Univ. Ill.," *Eng. Exp. Sta. Tech. Rept. 3*, 1962.
41. Goodier, J. N., and N. J. Hoff (eds.), "Structural Mechanics," *Proc. 1st Symp. Nav. Struct. Mech.*, Pergamon Press, 1960.
42. "Collected Papers on Instability of Shell Structures," *Natl. Aeron. Space Admin.*, Tech. Note D-1510, 1962.
43. Kloppel, K., and O. Jungbluth, "Beitrag zum Durchschlagproblem dünnwandiger Kugelschalen," *Der Stahlbau*, 1953.
44. McComb, H. G. Jr., G. W. Zender, and M. M. Mikulas, Jr., "The Membrane Approach to Bending Instability of Pressurized Cylindrical Shells (in Ref. 42)," p. 229.
45. Burkhard, A., and W. Young, "Buckling of a Simply-Supported Beam between Two Unattached Elastic Foundations," *AIAA J.*, 11(3): 1973.
46. Gere, J. M., and W. O. Carter, "Critical Buckling Loads for Tapered Columns," *Trans. Am. Soc. Civil Eng.*, 128(2): 1963.
47. Culver, C. G., and S. M. Preg, Jr., "Elastic Stability of Tapered Beam-Columns," *Proc. Am. Soc. Civil Eng.*, 94(ST2): 1968.
48. Burgreen, D., and P. J. Manitt, "Thermal Buckling of a Bimetallic Beam," *Proc. Am. Soc. Civil Eng.*, 95(EM2): 1969.
49. Burgreen, D., and D. Regal, "Higher Mode Buckling of Bimetallic Beam," *Proc. Am. Soc. Civil Eng.*, 97(EM4): 1971.
50. Austin, W. J., "In-Plane Bending and Buckling of Arches," *Proc. Am. Soc. Civil Eng.*, 97(ST5): May 1971. Discussion by R. Schmidt, D. A. DaDeppo, and K. Forrester: ibid., 98(ST1): 1972.
51. Chicurel, R., "Shrink Buckling of Thin Circular Rings," *ASME J. Appl. Mech.*, 35(3): 1968.
52. Clark, J. W., and H. N. Hill, "Lateral Buckling of Beams," *Proc. Am. Soc. Civil Eng.*, 86(ST7): 1960.
53. Fowler, D. W., "Design of Laterally Unsupported Timber Beams," *Proc. Am. Soc. Civil Eng.*, 97(ST3): 1971.
54. Massey, C., and P. J. McGuire, "Lateral Stability of Nonuniform Cantilevers," *Proc. Am. Soc. Civil Eng.*, 97(EM3): 1971.
55. Kitipornchai. S., and N. S. Trahair, "Elastic Stability of Tapered I-Beams," *Proc. Am. Soc. Civil Eng.*, 98(ST3): 1972.
56. Anderson, J. M., and N. S. Trahair, "Stability of Monosymmetric Beams and Cantilevers," *Proc. Am. Soc. Civil Eng.*, 98(ST1): 1972.
57. Morrison, T. G., "Lateral Buckling of Constrained Beams," *Proc. Am. Soc. Civil Eng.*, 98(ST3): 1972.
58. Vijayakumar, K., and C. V. Joga Rao, "Buckling of Polar Orthotropic Annular Plates," *Proc. Am. Soc. Civil Eng.*, 97(EM3): 1971.
59. Amon, R., and O. E. Widera, "Stability of Edge-Reinforced Circular Plate," *Proc. Am. Soc. Civil Eng.*, 97(EM5): 1971.
60. Shuleshko, P., "Solution of Buckling Problems by Reduction Method," *Proc. Am. Soc. Civil Eng.*, 90(EM3), 1964.
61. Ashton, J. E., "Stability of Clamped Skew Plates Under Combined Loads," *ASME J. Appl. Mech.*, 36(1): 1969.
62. Robinson, N. I., "Buckling of Parabolic and Semi-Elliptic Plates," *AIAA J.*, 7(6): 1969.
63. Srinivas, S., and A. K. Rao, "Buckling of Thick Rectangular Plates," *AIAA J.*, 7(8): 1969.
64. Durvasula, S., "Buckling of Clamped Skew Plates," *AIAA J.*, 8(1): 1970.
65. Roberts, S. B., "Buckling and Vibrations of Polygonal and Rhombic Plates," *Proc. Am. Soc. Civil Eng.*, 97(EM2): AIAA J., 3(3): 1965.
66. Mikulas, M. M., Jr., and M. Stein: Buckling of a Cylindrical Shell Loaded by a Pre-Tensioned Filament Winding, AIAA J., vol. 3, no. 3, 1965.
67. Hoff, N. J., "Low Buckling Stresses of Axially Compressed Circular Cylindrical Shells of Finite Length," *ASME J. Appl. Mech.*, 32(3): 1965.
68. Hoff, N. J., and L. W. Rehfield, "Buckling of Axially Compressed Circular Cylindrical Shells at Stresses Smaller Than the Classical Critical Value," *ASME J. Appl. Mech.*, 32(3): 1965.

69. Yao, J. C., and W. C. Jenkins, "Buckling of Elliptic Cylinders under Normal Pressure," *AIAA J.*, 8(1): 1970.
70. Carlson, R. L., R. L. Sendelbeck, and N. J. Hoff, "Experimental Studies of the Buckling of Complete Spherical Shells, Experimental Mechanics," *J. Soc. Exp. Stress Anal.*, 7(7): 1967.
71. Wu, M. T., and Shun Cheng, "Nonlinear Asymmetric Buckling of Truncated Spherical Shells," *ASME J. Appl. Mech.*, 37(3): 1970.
72. Singer, J., "Buckling of Circular Cortical Shells under Axisymmetrical External Pressure," *J. Mech. Eng. Sci.*, 3(4): 1961.
73. Newman, M., and E. L. Reiss, "Axisymmetric Snap Buckling of Conical Shells (in Ref. 42)," p. 45.
74. Singer, J., "Buckling of Orthotropic and Stiffened Conical Shells (in Ref. 42)," p. 463.
75. Seide, P., "On the Stability of Internally Pressurized Conical Shells under Axial Compression," *Proc. 4th U.S. Natl. Cong. Appl. Mech.*, June 1962.
76. Weingarten, V. I., "Stability of Internally Pressurized Conical Shells under Torsion," *AIAA J.*, 2(10): 1964.
77. Hausrath, A. H., and F. A. Dittoe, "Development of Design Strength Levels for the Elastic Stability of Monocoque Cones under Axial Compression (in Ref. 42)," p. 45.
78. Seide, P., "A Survey of Buckling Theory and Experiment for Circular Conical Shells of constant Thickness (in Ref. 42)," p. 401.
79. Radkowski, P. P., "Elastic Instability of Conical Shells under Combined Loading (in Ref. 42)," p. 427.
80. Weingarten, V. I., E. J. Morgan, and P. Seide, "Elastic Stability of Thin-Walled Cylindrical and Conical Shells under Axial Compression," *AIAA J.*, 3(3): 1965.
81. Weingarten, V. I., and P. Seide, "Elastic Stability of Thin-Walled Cylindrical and Conical Shells under Combined External Pressure and Axial Compression," *AIAA J.*, 3(5): 1965.
82. Weingarten, V. I., E. J. Morgan, and P. Seide, "Elastic Stability of Thin-Walled Cylindrical and Conical Shells under Combined Internal Pressure and Axial Compression," *AIAA J.*, 3(6): 1965.
83. Tani, J., and N. Yamaki, "Buckling of Truncated Conical Shells under Axial Compression," *AIAA J.*, 8(3): 1970.
84. Baruch, M., O. Harari, and J. Singer, "Low Buckling Loads of Axially Compressed Conical Shells," *ASME J. Appl. Mech.*, 37(2): 1970.
85. Sendelbeck, R. L., and J. Singer, "Further Experimental Studies of Buckling of Electroformed Conical Shells," *AIAA J.*, 8(8): 1970.
86. Stein, M., and J. A. McElman, "Buckling of Segments of Toroidal Shells," *AIAA J.*, 3(9): 1965.
87. Sobel, L. H., and W. Fluugge, "Stability of Toroidal Shells under Uniform External Pressure," *AIAA J.*, 5(3): 1967.
88. Almroth, B. O., L. H. Sobel, and A. R. Hunter, "An Experimental Investigation of the Buckling of Toroidal Shells," *AIAA J.*, 7(11): 1969.
89. Loo, Ta-Cheng, and R. M. Evan-Iwanowski, "Interaction of Critical Pressures and Critical Concentrated Loads Acting on Shallow Spherical Shells," *ASME J. Appl. Mech.*, 33(3): 1966.
90. Loo, Ta-Cheng, and R. M. Evan-Iwanowski, "Experiments on Stability on Spherical Caps," *Proc. Am. Soc. Civil Eng.*, 90(EM3): 1964.
91. Arbocz, J., and C. D. Babcock, Jr., "The Effect of General Imperfections on the Buckling of Cylindrical Shells," *ASME J. Appl. Mech.*, 36(1): 1969.
92. Burns, J. J. Jr., "Experimental Buckling of Thin Shells of Revolution," *Proc. Am. Soc. Civil Eng.*, 90(EM3): 1964.
93. Roorda, J., "Some Thoughts on the Southwell Plot," *Proc. Am. Soc. Civil Eng.*, 93(EM6): 1967.
94. Navaratna, D. R., T. H. H. Pian, and E. A. Witmer, "Stability Analysis of Shells of Revolution by the Finite-Element Method," *AIAA J.*, 6(2): 1968.
95. Stein, M., "Some Recent Advances in the Investigation of Shell Buckling," *AIAA J.*, 6(12): 1968.
96. Rabinovich, I. M. (ed.), *Structural Mechanics in the U.S.S.R. 1917–1957*, Pergamon Press, 1960 (English transl. edited by G. Herrmann).
97. Baker, E. H., L. Kovalevsky, and F. L. Rish, *Structural Analysis of Shells*, McGraw-Hill, 1972.
98. Perrone, N., "Compendium of Structural Mechanics Computer Programs," *Comput. & Struct.*, 2(3): April 1972. (Available from NTIS as N71-32026, April 1971.)
99. Bushnell, D., "Stress, Stability, and Vibration of Complex, Branched Shells of Revolution," *AIAA/ASME/SAE 14th Struct. Dynam. & Mater. Conf.*, March, 1973.
100. *Structural Sandwich Composites*, MIL-HDBK-23, U.S. Dept. of Defense, 1968.
101. Allen, H. G., and P. S. Bulson, *Background to Buckling*, McGraw-Hill, 1980.
102. Thompson, J. M. T., and G. W. Hunt (eds.), *Collapse: The Buckling of Structures in Theory and Practice*, IUTAM/Cambridge University Press, 1983.
103. Narayanan, R. (ed.), *Axially Compressed Structures: Stability and Strength*, Elsevier Science, 1982.
104. Brush, D. O., and B. O. Almroth, *Buckling of Bars, Plates, and Shells*, McGraw-Hill, 1975.

105. Kollár, L., and E. Dulácska, *Buckling of Shells for Engineers*, English transl. edited by G. R. Thompson, John Wiley & Sons, 1984.
106. Yamaki, N., *Elastic Stability of Circular Cylindrical Shells*, Elsevier Science, 1984.
107. *Collapse Analysis of Structures*, Pressure Vessels and Piping Division, ASME, PVP, vol. 84, 1984.
108. Bushnell, D., *Computerized Buckling Analysis of Shells*, Kluwer, 1985.
109. Johnston, B. G. (ed.), *Guide to Stability Design Criteria for Metal Structures*, 3rd ed., Structural Stability Research Council, John Wiley & Sons, 1976.
110. White, R. N., and C. G. Salmon (eds.), *Building Structural Design Handbook*, John Wiley & Sons, 1987.
111. American Institute of Steel Construction, *Manual of Steel Construction—Load and Resistance Factor Design*, 1st ed., 1986.

CHAPTER 16

Dynamic and Temperature Stresses

16.1 DYNAMIC LOADINGS; GENERAL CONDITIONS

Dynamic loading was defined in Chap. 3 as any loading during which the parts of the body cannot be considered to be in static equilibrium. It was further pointed out that two kinds of dynamic loadings can be distinguished as: (1) that in which the body has imposed upon it a particular kind of motion involving known accelerations, and (2) impact, of which sudden loading may be considered a special case. In the following sections, specific cases of each kind of dynamic loading will be considered.

16.2 BODY IN A KNOWN STATE OF MOTION

The acceleration a of each particle of mass dm being known, the effective force on each particle is $dm \times a$, directed like a. If to each particle a force equal and opposite to the effective force were applied, equilibrium would result. If then such reversed effective forces are assumed to be applied to all the constituent particles of the body, the body may be regarded as being in equilibrium under these forces and the actual forces (loads and reactions) that act upon it, and the resulting stresses can be found exactly as for a body at rest. The reversed effective forces are *imaginary* forces exerted *on* the particles but are equal to and directed like the actual reactions the particles exert on whatever gives them their acceleration, that is, in general, on the rest of the body. Since these reactions are due to the inertia of the particles, they are called *inertia forces*, and the body may be thought of as loaded by these inertia forces. Similarly, any attached mass will exert on a body inertia forces equal and opposite to the forces which the body has to exert on the attached mass to accelerate it.

The results of applying this method of analysis to a number of more or less typical problems are given below. In all cases, in finding the accelerations of the particles, it has been assumed that the effect of deformation could be ignored, that is, the acceleration of each particle has been found as though the body were rigid. For convenience, stresses, bending moments, and shears due to inertia forces only are called *inertia* stresses, moments, and shears, respectively; they are calculated as though the body were outside the field of gravitation. Stresses, moments, and shears due to balanced forces (including gravity) may be superimposed thereon. The gravitational acceleration constant g depends upon the units used for the imposed acceleration.

1. A slender uniform rod of weight W, length L, section area A, and modulus of elasticity E is given a motion of translation with an acceleration of a parallel to its axis by a pull (push) applied at one end. The maximum tensile (compressive) stress occurs at the loaded end and is $\sigma = Wa/gA$. The elongation (shortening) due to the inertia stresses is

$$e = \frac{1}{2}\frac{W}{g}\frac{aL}{AE}$$

2. The rod described in problem 1 is given a motion of translation with an acceleration of a normal to its axis by forces applied at each end. The maximum inertia bending moment occurs at the middle of the bar and is $M = 1/8\ WaL/g$. The maximum inertia vertical (transverse) shear occurs at the ends and is $V = ½\ Wa/g$.

3. The rod described in problem 1 is made to rotate about an axis through one end normal to its length at a uniform angular velocity of ω rad/s. The maximum tensile inertia stress occurs at the pinned end and is

$$\sigma = \frac{1}{2}\frac{W}{g}\frac{L\omega^2}{A}$$

The elongation due to inertia stress is

$$e = \frac{1}{3}\frac{W}{g}\frac{L^2\omega^2}{AE}$$

4. The rod described in problem 1 is pinned at the lower end and allowed to swing down under the action of gravity from an initially vertical position. When the rod reaches a position where it makes an angle θ with the vertical, it is subjected to a positive bending moment (owing to its weight and the inertia forces) which has its maximum value at a section a distance 1/3 L from the pinned end. This maximum value is $M = 1/27\ WL \sin\theta$. The maximum positive inertia shear occurs at the pinned end and is $V = ¼\ W \sin\theta$. The maximum negative inertia shear occurs at a section a distance 2/3 L from the pinned end and is $V = -1/12\ W \sin\theta$. The axial force at any section x in from the pinned end is given by

$$H = \frac{3W}{2}\left(1 - \frac{x^2}{L^2}\right) - \frac{W\cos\theta}{2}\left(5 - 2\frac{x}{L} - 3\frac{x^2}{L^2}\right)$$

and becomes tensile near the free end when θ exceeds 41.4°. (This case represents approximately the conditions existing when a chimney or other slender structure topples over, and the bending moment M explains the tendency of such a structure to break near the one-third point while falling.)

5. The rod described in problem 1 is pinned at the lower end and, while in the vertical position, has imposed upon its lower end a horizontal acceleration of a. The maximum inertia bending moment occurs at a section a distance $\frac{1}{3}L$ from the lower end and is $M = \frac{1}{27}Wla/g$. The maximum inertia shear is in the direction of the acceleration, is at the lower end, and is $V = ¼\ Wa/g$. The maximum inertia shear in the opposite direction occurs at a section a distance $\frac{2}{3}L$ from the lower end and is $V = \frac{1}{12}Wa/g$. (This case represents approximately the conditions existing when a chimney or other slender structure without anchorage is subjected to an earthquake shock.)

6. A uniform circular ring of mean radius R and weight per unit volume δ, having a thickness in the plane of curvature that is very small compared with R, rotates about its own axis with a uniform angular velocity of ω rad/s. The ring is subjected to a uniform tangential inertial stress

$$\sigma = \frac{\delta R^2 \omega^2}{g}$$

7. A solid homogeneous circular disk of uniform thickness (or a solid cylinder) of radius R, Poisson's ratio ν, and weight per unit volume δ rotates about its own axis with a uniform angular velocity of ω rad/s. At any point a distance r from the center there is a radial tensile inertial stress

$$\sigma_r = \frac{1}{8}\frac{\delta\omega^2}{g}[(3+\nu)(R^2 - r^2)] \tag{16.2-1}$$

and a tangential tensile inertia stress

$$\sigma_t = \frac{1}{8}\frac{\delta\omega^2}{g}[(3+\nu)R^2 - (1+3\nu)r^2)] \tag{16.2-2}$$

The maximum radial stress and maximum tangential stress are equal, occur at the center, and are

$$(\sigma_r)_{\max} = (\sigma_t)_{\max} = \frac{1}{8}\frac{\delta\omega^2}{g}(3+\nu)R^2 \tag{16.2-3}$$

8. A homogeneous annular disk of uniform thickness outer radius R and weight per unit volume δ, with a central hole of radius R_0, rotates about its own axis with a uniform angular velocity of ω rad/s. At any point a distance r from the center there is a radial tensile inertia stress

$$\sigma_r = \frac{3+\nu}{8}\frac{\delta\omega^2}{g}\left(R^2 + R_0^2 - \frac{R^2R_0^2}{r^2} - r^2\right) \tag{16.2-4}$$

and a tangential tensile inertia stress

$$\sigma_t = \frac{1}{8}\frac{\delta\omega^2}{g}\left[(3+\nu)\left(R^2 + R_0^2 + \frac{R^2R_0^2}{r^2}\right) - (1+3\nu)r^2\right] \tag{16.2-5}$$

The maximum radial stress occurs at $r = \sqrt{RR_0}$ and is

$$(\sigma_r)_{\max} = \frac{3+\nu}{8}\frac{\delta\omega^2}{g}(R - R_0)^2 \tag{16.2-6}$$

and the maximum tangential stress occurs at the perimeter of the hole and is

$$(\sigma_t)_{\max} = \frac{1}{4}\frac{\delta\omega^2}{g}\left[(3+\nu)R^2 + (1-\nu)R_0^2\right] \tag{16.2-7}$$

The change in the outer radius is

$$\Delta R = \frac{1}{4}\frac{\delta\omega^2}{g}\frac{R}{E}\left[(1-\nu)R^2 + (3+\nu)R_0^2\right] \tag{16.2-8}$$

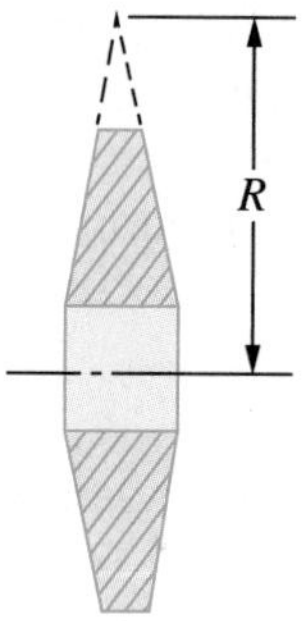

Figure 16.1

and the change in the inner radius is

$$\Delta R_0 = \frac{1}{4}\frac{\delta\omega^2}{g}\frac{R_0}{E}\left[(3+\nu)R^2 + (1-\nu)R_0^2\right] \qquad (16.2\text{-}9)$$

If there are radial pressures or pulls distributed uniformly along either the inner or outer perimeter of the disk, such as a radial pressure from the shaft or a centrifugal pull from parts attached to the rim, the stresses due thereto can be found by the formula for thick cylinders (Table 13.5) and superimposed upon the inertia stresses given by the preceding formulas.

9. A homogeneous circular disk of conical section (Fig. 16.1) of density δ lb/in^3 rotates about its own axis with a uniform angular velocity of N rpm. At any point a distance r from the center, the tensile inertia stresses σ_r and σ_t are given, in lb/in^2, by

$$\sigma_r = TK_r + Ap_1 + Bp_2 \qquad (16.2\text{-}10)$$

$$\sigma_t = TK_t + Aq_1 + Bq_2 \qquad (16.2\text{-}11)$$

where $T = 0.0000282R^2N^2\delta$ (or for steel, $T = 0.0000082R^2N^2$); K_r, K_t, p_1, p_2, q_1, and q_2 are given in the following table; and A and B are constants which may be found by setting σ_r equal to its known or assumed values at the outer perimeter and solving the resulting equations simultaneously for A and B, as in the example on pages 637 and 638. [See papers by Hodkinson and Rushing (Refs. 1 and 2) from which Eqs. (16.2-8) and (16.2-9) and the tabulated coefficients are taken.]

10. A homogeneous circular disk of hyperbolic section (Fig. 16.2) of density δ lb/in^3 rotates about its own axis with uniform angular velocity ω rad/s. The equation $t = cr^a$ defines the section, where if t_1, thickness at radius r_1 and t_2 at radius r_2,

$$a = \frac{\ln(t_1/t_2)}{\ln(r_1/r_2)}$$

and

$$\ln c = \ln t_1 - a\ln r_1 = \ln t_2 - a\ln r_2$$

Tabulated Values of Coefficients

r/R	K_r	K_t	p_1	q_1	p_2	q_2
0.00	0.1655	0.1655	1.435	1.435	∞	∞
0.05	0.1709	0.1695	1.475	1.497	−273.400	288.600
0.10	0.1753	0.1725	1.559	1.518	−66.620	77.280
0.15	0.1782	0.1749	1.627	1.565	−28.680	36.550
0.20	0.1794	0.1763	1.707	1.617	−15.540	21.910
0.25	0.1784	0.1773	1.796	1.674	−9.553	14.880
0.30	0.1761	0.1767	1.898	1.738	−6.371	10.890
0.35	0.1734	0.1757	2.015	1.809	−4.387	8.531
0.40	0.1694	0.1739	2.151	1.890	−3.158	6.915
0.45	0.1635	0.1712	2.311	1.983	−2.328	5.788
0.50	0.1560	0.1675	2.501	2.090	−1.743	4.944
0.55	0.1465	0.1633	2.733	2.217	−1.309	4.301
0.60	0.1355	0.1579	3.021	2.369	−0.9988	3.816
0.65	0.1229	0.1525	3.390	2.556	−0.7523	3.419
0.70	0.1094	0.1445	3.860	2.794	−0.5670	3.102
0.75	0.0956	0.1370	4.559	3.111	−0.4161	2.835
0.80	0.0805	0.1286	5.563	3.557	−0.2971	2.614
0.85	0.0634	0.1193	7.263	4.276	−0.1995	2.421
0.90	0.0442	0.1100	10.620	5.554	−0.1203	2.263
0.95	0.0231	0.0976	20.645	8.890	−0.0555	2.140
1.00	0.0000	0.0840	∞	∞	−0.0000	2.051

(For taper toward the rim, a is negative; and for uniform t, $a = 0$.) At any point a distance r in from the center the tensile inertia stresses σ_r and σ_t, in lb/in², are

$$\sigma_r = \frac{E}{1-\nu^2}[(3+\nu)Fr^2 + (m_1+\nu)Ar^{m_1-1} + (m_2+\nu)Br^{m_2-1}] \tag{16.2-12}$$

$$\sigma_t = \frac{E}{1-\nu^2}[(1+3\nu)Fr^2 + (1+m_1\nu)Ar^{m_1-1} + (1+m_2\nu)Br^{m_2-1}] \tag{16.2-13}$$

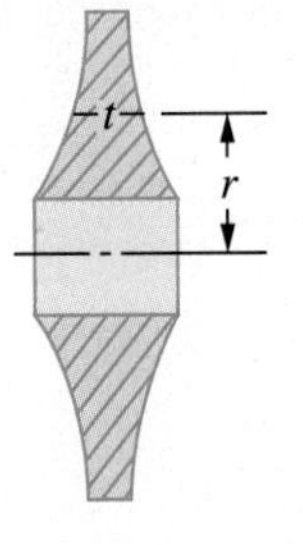

Figure 16.2

$$\text{where } F = \frac{-(1-\nu^2)\delta\omega^2/386.4}{E[8+(3+\nu)a]}$$

$$m_1 = -\frac{a}{2} - \sqrt{\frac{a^2}{4} - a\nu + 1}$$

$$m_2 = -\frac{a}{2} + \sqrt{\frac{a^2}{4} - a\nu + 1}$$

A and *B* are constants, found by settings σ_r equal to its known or assumed values at the inner and outer perimeters and solving the two resulting equations simultaneously for *A* and *B*. [Equations (16.2-12) and (16.2-13) are taken from Stodola (Ref. 3) with some changes in notation.]

11. A homogeneous circular disk with section bounded by curves and straight lines (Fig. 16.3) rotates about its own axis with a uniform angular velocity *N* rpm. The disk is imagined divided into annular rings of such width that each ring can be regarded as having a section with hyperbolic outline, as in problem 10. For each ring, *a* is calculated by the formulas of problem 10, using the inner and outer radii and the corresponding thicknesses. Then, if r_1 and r_2 represent, respectively, the inner and outer radii of any ring, the tangential stresses σ_{t_1} and σ_{t_2} at the inner and outer boundaries of the ring are related to the corresponding radial stresses σ_{r_1} and σ_{r_2}, in lb/in², as follows:

$$\sigma_{t_1} = Ar_2^2 - B\sigma_{r_1} + C\sigma_{r_2} \tag{16.2-14}$$

$$\sigma_{t_2} = Dr_2^2 - E\sigma_{r_1} + F\sigma_{r_2} \tag{16.2-15}$$

$$\text{where } B = \frac{m_2 K^{m_1-1} - m_1 K^{m_2-1}}{K^{m_2-1} - K^{m_1-1}}$$

$$K = \frac{r_1}{r_2}$$

$$E = -\frac{m_2 - m_1}{K^{m_2-1} - K^{m_1-1}}$$

$$C = \frac{E}{K^{a+2}}$$

$$F = B + a$$

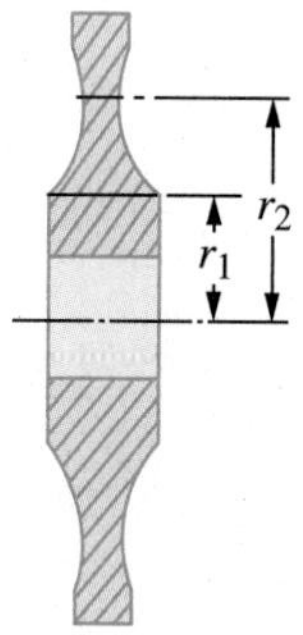

Figure 16.3

$$A = -\frac{7.956(N/1000)^2}{8+3.3a}[1.9K^2 + 3.3(K^2B - C)]$$

$$D = -\frac{7.956(N/1000)^2}{8+3.3a}[1.9 + 3.3(K^2E - F)]$$

$$m_1 = -\frac{a}{2} - \sqrt{\frac{a^2}{4} - 0.3a + 1}$$

$$m_2 = -\frac{a}{2} + \sqrt{\frac{a^2}{4} - 0.3a + 1}$$

The preceding formulas, which are given by Loewenstein (Ref. 4), are directly applicable to steel, for which the values $\nu = 0.3$ and $\delta = 0.28$ lb/in^3 have been assumed.

Two values of σ_r are known or can be assumed, viz. the values at the inner and outer perimeters of the disk. Then, by setting the tangential stress at the outer boundary of each ring equal to the tangential stress at the inner boundary of the adjacent larger ring, one equation in σ_r will be obtained for each common ring boundary. In this case, the modulus of elasticity is the same for adjacent rings and the radial stress σ_r at the boundary is common to both rings, and so the tangential stresses can be equated instead of the tangential strains [Eq. (16.2-15) for the smaller ring equals Eq. (16.2-14) for the larger ring]. Therefore there are as many equations as there are unknown boundary radial stresses, and hence the radial stress at each boundary can be found. The tangential stresses can then be found by Eqs. (16.2-14) and (16.2-15), and then the stresses at any point in a ring can be found by using, in Eq. (16.2-14), the known values of σ_{r_1} and σ_{t_1} and substituting for σ_{r_2} the unknown radial stress σ_r, and for r_2 the corresponding radius r.

A fact of importance with reference to turbine disks or other rotating bodies is that geometrically similar disks of different sizes will be equally stressed at corresponding points when running at the same *peripheral* velocity. Furthermore, for any given peripheral velocity, the axial and radial dimensions of a rotating body may be changed independently of each other and in any ratio without affecting the stresses at similarly situated points.

EXAMPLE The conical steel disk shown in section in Fig. 16.4 rotates at 2500 rpm. To its rim, it has attached buckets whose aggregate mass amounts to $w = 0.75$ lb/linear in of rim; this mass may be considered to be centered 30 in from the axis. It is desired to determine the stresses at a point 7 in from the axis.

SOLUTION From the dimensions of the section, R is found to be 28 in. The values of r/R for the inner and outer perimeters and for the circumference $r = 7$ are calculated, and the corresponding coefficients K_r, K_t, etc., are determined from the table on page 635 by graphic interpolation. The results are tabulated here for convenience:

	r/R	K_r	K_t	p_1	q_1	p_2	q_2
Inner rim	0.143	0.1780	0.1747	1.616	1.558	−32.5	40.5
Outer rim	0.714	0.1055	0.1425	4.056	2.883	−0.534	3.027
$r = 7$ in	0.25	0.1784	0.1773	1.796	1.674	−9.553	14.88

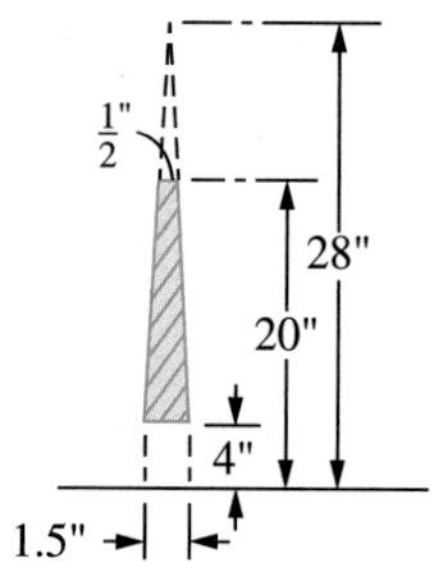

Figure 16.4

The attached mass exerts on the rim outward inertia forces which will be assumed to be uniformly distributed; the amount of force per linear inch is

$$p=\frac{w}{g}\omega^2 r=\frac{0.75}{386.4}(261.5^2)(30)=3980\ \text{lb/linear in}$$

Therefore at the outer rim $\sigma_r = 7960\ \text{lb/in}^2$.

It is usual to design the shrink fit so that in operation the hub pressure is a few hundred pounds per square inch; it will be assumed that the radial stress at the inner rim $\sigma_r = -700\ \text{lb/in}^2$. The value of $T = 0.000008(28^2)(2500^2) = 39{,}200$. Having two values of σ_r, Eq. (16.2-10) can now be written

$$-700=(39{,}200)(0.1780)+A(1.616)+B(-32.5) \qquad \text{(inner rim)}$$

$$7960=(39{,}200)(0.1055)+A(4.056)+B(-0.534) \qquad \text{(outer rim)}$$

The solution gives

$$A=973, \qquad B=285$$

The stresses at $r = 7$ are now found by Eqs. (16.2-10) and (16.2-11) to be

$$\sigma_r=(39{,}200)(0.1784)+(973)(1.796)+(285)(-9.553)=6020\ \text{lb/in}^2$$

$$\sigma_t=(39{,}200)(0.1773)+(973)(1.674)+(285)(14.88)=12{,}825\ \text{lb/in}^2$$

Bursting Speed

The formulas given above for stresses in rotating disks presuppose *elastic* conditions; when the elastic limit is exceeded, plastic yielding tends to equalize the stress intensity along a diametral plane. Because of this, the average stress σ_a on such a plane is perhaps a better criterion of margin of safety against bursting than is the maximum stress computed for elastic conditions. For a solid disk of uniform thickness (case 7),

$$\sigma_a=\frac{\delta\omega^2 R^2}{3g}$$

For a pierced disk (case 8),

$$\sigma_a=\frac{\delta\omega^2\left(R^3-R_0^3\right)}{3g(R-R_0)}$$

Tests (Refs. 12 and 13) have shown that for some materials, rupture occurs in both solid and pierced disks when σ_a, computed for the original dimensions, becomes equal to the ultimate tensile strength of the material as determined by a conventional test. On the other hand, some materials fail at values of σ_a as low as 61.5% of the ultimate strength, and the lowest values have been observed in tests of solid disks. The ratio of σ_a at failure to the ultimate strength does not appear to be related in any consistent way to the ductility of the material; it seems probable that it depends on the form of the stress-strain diagram. In none of the tests reported did the weakening effect of a central hole prove to be nearly as great as the formulas for elastic stress would seem to indicate.

16.3 IMPACT AND SUDDEN LOADING

When a force is suddenly applied to an elastic body (as by a blow), a wave of stress is propagated, which travels through the body with a velocity

$$V = \sqrt{\frac{gE}{\delta}} \tag{16.3-1}$$

where E is the modulus of elasticity of the material and δ is the weight of the material per unit volume.

Bar with Free Ends

When one end of an unsupported uniform elastic bar is subjected to longitudinal impact from a rigid body moving with velocity v, a wave of compressive stress of intensity

$$\sigma = \frac{v}{V}E = v\sqrt{\frac{\delta E}{g}} \tag{16.3-2}$$

is propagated. The intensity of stress is seen to be independent of the mass of the moving body, but the length of the stressed zone, or volume of material simultaneously subjected to this stress, does depend on the mass of the moving body. If this mass is infinite (or very large compared with that of the bar), the wave of compression is reflected back from the free end of the bar as a wave of tension and returns to the struck end after a period $t_1 = 2L/V$s, where L is the length of the bar and the period t_1 is the duration of contact between bar and body.

If the impinging body is very large compared with the bar (so that its mass may be considered infinite), the bar, after breaking contact, moves with a velocity $2v$ in the direction of the impact and is free of stress. If the mass of the impinging body is μ times the mass of the bar, the average velocity of the bar after contact is broken is

$$\mu v(1 - e^{-2/\mu})$$

and it is left vibrating with a stress of intensity

$$\sigma = \frac{v}{V}Ee^{-\beta t_1}$$

where $\beta = A\sqrt{\delta E g}/W$, A being the section area of the bar and W the weight of the moving body.

Bar with One End Fixed

If one end of a bar is fixed, the wave of compressive stress resulting from impact on the free end is reflected back unchanged from the fixed end and combines with advancing waves to produce a maximum stress very nearly equal to

$$\sigma_{max} = \frac{v}{V}E\left(1+\sqrt{\mu+\frac{2}{3}}\right) \qquad (16.3\text{-}3)$$

where, as before, μ denotes the ratio of the mass of the moving body to the mass of the bar. The total time of contact is approximately

$$t_1 = \frac{L}{V}\left[\pi\sqrt{\mu+\frac{1}{2}}-\frac{1}{2}\right] \text{s}$$

[The above formulas are taken from the paper by Donnell (Ref. 5); see also Ref. 17.]

Sudden Loading

If a dead load is suddenly transferred to the free end of a bar, the other end being fixed, the resulting state of stress is characterized by waves, as in the case of impact. The space-average value of the pull exerted by the bar on the load is not half the maximum tension, as is usually assumed, but is somewhat greater than that, and therefore the maximum stress that results from sudden loading is somewhat less than twice that which results from static loading. Love (Ref. 6) shows that if μ (the ratio of the mass of the load to that of the bar) is 1, sudden loading causes 1.63 times as much stress as static loading; for $\mu = 2$, the ratio is 1.68; for $\mu = 4$, it is 1.84; and it approaches 2 as a limit as μ increases. It can be seen that the ordinary assumption that sudden loading causes twice as much stress and deflection as static loading is always a safe one to make.

Moving Load on Beam

If a constant *force* moves at uniform speed across a beam with simply supported ends, the maximum deflection produced exceeds the static deflection that the same force would produce. If v represents the velocity of the force, l the span, and ω the lowest natural vibration frequency of the (unloaded) beam, then theoretically the maximum value of the ratio of dynamic to static deflection is 1.74; it occurs for $v = \omega l/1.64\pi$ and at the instant when the force has progressed at a distance 0:757l along the span (Refs. 15 and 16).

If a constant *mass W* moves across a simple beam of relatively negligible mass, then the maximum ratio of dynamic to static deflection is $[1+(v^2/g)(Wl/3EI)]$ (see Ref. 30). (Note that consistent units must be used in the preceding equations.)

16.4 IMPACT AND SUDDEN LOADING; APPROXIMATE FORMULAS

If it is assumed that the stresses due to impact are distributed throughout any elastic body exactly as in the case of static loading, then it can be shown that the vertical deformation d_i and stress σ_i produced in any such body (bar, beam, truss, etc.) by the vertical impact of a body falling from a height of h are greater than the deformation d and stress σ produced by the weight of the same body applied as a static load in the ratio

$$\frac{d_i}{d} = \frac{\sigma_i}{\sigma} = 1+\sqrt{1+2\frac{h}{d}} \qquad (16.4\text{-}1)$$

If $h = 0$, we have the case of sudden loading, and $d_i/d = \sigma_i/\sigma = 2$, as is usually assumed.

If the impact is horizontal instead of vertical, the impact deformation and stress are given by

$$\frac{d_i}{d} = \frac{\sigma_i}{\sigma} = \sqrt{\frac{v^2}{gd}} \qquad (16.4\text{-}2)$$

where, as before, d is the deformation the weight of the moving body would produce if applied as a static load in the direction of the velocity and v is the velocity of impact.

Energy Losses

The above approximate formulas are derived on the assumptions that impact strains the elastic body in the same way (though not in the same degree) as static loading and that all the kinetic energy of the moving body is expended in producing this strain. Actually, on impact, some kinetic energy is dissipated; and this loss, which can be found by equating the momentum of the entire system before and after impact, is most conveniently taken into account by multiplying the available energy (measured by h or by v^2) by a factor K, the value of which is as follows for a number of simple cases involving members of uniform section:

1. A moving body of mass M strikes axially one end of a bar of mass M_1, the other end of which is fixed. Then

$$K = \frac{1 + \frac{1}{3}\frac{M_1}{M}}{\left(1 + \frac{1}{2}\frac{M_1}{M}\right)^2}$$

   If there is a body of mass M_2 attached to the struck end of the bar, then

$$K = \frac{1 + \frac{1}{3}\frac{M_1}{M} + \frac{M_2}{M}}{\left(1 + \frac{1}{2}\frac{M_1}{M} + \frac{M_2}{M}\right)^2}$$

2. A moving body of mass M strikes transversely the center of a simple beam of mass M_1. Then

$$K = \frac{1 + \frac{17}{25}\frac{M_1}{M}}{\left(1 + \frac{5}{8}\frac{M_1}{M}\right)^2}$$

   If there is a body of mass M_2 attached to the beam at its center, then

$$K = \frac{1 + \frac{17}{35}\frac{M_1}{M} + \frac{M_2}{M}}{\left(1 + \frac{5}{8}\frac{M_1}{M} + \frac{M_2}{M}\right)^2}$$

3. A moving body of mass M strikes transversely the end of a cantilever beam of mass M_1. Then

$$K = \frac{1 + \frac{33}{140}\frac{M_1}{M}}{\left(1 + \frac{3}{8}\frac{M_1}{M}\right)^2}$$

If there is a body of mass M_2 attached to the beam at the struck end, then

$$K = \frac{1 + \frac{33}{140}\frac{M_1}{M} + \frac{M_2}{M}}{\left(1 + \frac{3}{8}\frac{M_1}{M} + \frac{M_2}{M}\right)^2}$$

4. A moving body of mass M strikes transversely the center of a beam with fixed ends and of mass M_1. Then

$$K = \frac{1 + \frac{13}{35}\frac{M_1}{M}}{\left(1 + \frac{1}{2}\frac{M_1}{M}\right)^2}$$

If there is a body of mass M_2 attached to the beam at the center, then

$$K = \frac{1 + \frac{13}{35}\frac{M_1}{M} + \frac{M_2}{M}}{\left(1 + \frac{1}{2}\frac{M_1}{M} + \frac{M_2}{M}\right)^2}$$

16.5 REMARKS ON STRESS DUE TO IMPACT

It is improbable that in any actual case of impact the stresses can be calculated accurately by any of the methods or formulas given previously. Equation (16.3-3), for instance, is supposedly very nearly precise if the conditions assumed are realized, but those conditions—perfect elasticity of the bar, rigidity of the moving body, and simultaneous contact of the moving body with all points on the end of the rod—are obviously unattainable. On the one hand, the damping of the initial stress wave by elastic hysteresis in the bar and the diminution of the intensity of that stress wave by the cushioning effect of the actually nonrigid moving body would serve to make the actual maximum stress less than the theoretical value; on the other hand, uneven contact between the moving body and the bar would tend to make the stress conditions nonuniform across the section and would probably increase the maximum stress.

The formulas given in Sec. 16.4 are based upon an admittedly false assumption, viz. that the distribution of stress and strain under impact loading is the same as under static loading. It is known, for instance, that the elastic curve of a beam under impact is different from that under static loading. Such a difference exists in any case, but it is less marked for low than for high velocities of impact, and Eqs. (16.4-1) and (16.4-2) probably give reasonably accurate values for the deformation and stress (especially the deformation) resulting from the impact of a relatively heavy body moving at low velocity. The lenitive effect of the inertia of the body struck and of attached bodies, as expressed by K, is greatest when the masses of these parts are large compared with that of the moving body. When this is the case, impact can be serious only if the velocity is relatively high, and under such circumstances the formulas probably give only a rough indication of the actual stresses and deformations to be expected. (See Ref. 18.)

16.6 VIBRATION

A very important type of dynamic loading occurs when an elastic body vibrates under the influence of a periodic load. This occurs whenever a rotating or reciprocating mass is unbalanced and also under certain conditions of fluid flow. The most serious situation arises when the impulse synchronizes (or nearly synchronizes) with the natural period of vibration, and it is of the utmost importance to guard against this condition of resonance (or near resonance). There is always some resistance to vibration, whether natural or introduced; this is called damping and tends to prevent vibrations of excessive amplitude. In the absence of effective damping, the amplitude for near-resonance vibration will much exceed the deflection that would be produced by the same force under static conditions. Obviously, it is necessary to know at least approximately the natural period of vibration of a member in order to guard against resonance. Thomson and Dahleh (Ref. 19) describe in detail analytical and numerical techniques for determining resonant frequencies for systems with single and multiple degrees of freedom; they also describe methods and give numerous examples for torsional and lateral vibrations of rods and beams. Huang (Ref. 22) has tabulated the first 5 resonant frequencies as well as deflections, slopes, bending moments, and shearing forces for each frequency at intervals of 0.021 for uniform beams; these are available for 6 combinations of boundary conditions. He has also included the first 5 resonant frequencies for all combinations of 7 different amounts of correction for rotary inertia and 10 different amounts of correction for lateral shear deflection; many mode shapes for these corrections are also included. In Table 16.1, the resonant frequencies and nodal locations are listed for several boundary conditions with no corrections for rotary inertia or shear deflection (Corrections for rotary inertia and shear deflection have a relatively small effect on the fundamental frequency but a proportionally greater effect on the higher modes.). Leissa (Ref. 20) has compiled, compared, and in some cases extended most of the known work on the vibration of plates; where possible, mode shapes are given in addition to the many resonant frequencies. Table 16.1 lists only a very few simple cases. Similarly, Leissa (Ref. 21) has done an excellent job of reporting the known work on the vibration of shells. Since, in general, this work must involve three additional variables—the thickness/radius ratio, length/radius ratio, and Poisson's ratio—no results are included here. Blevins (Ref. 24) gives excellent coverage, by both formulas and tables, of resonant frequencies and mode shapes for cables, straight and curved beams, rings, plates, and shells. A simple but close approximation for the fundamental frequency of a uniform thin plate of arbitrary shape having any combination of fixed, partially fixed, or simply supported boundaries is given by Jones in Ref. 23. The equation

$$f = \frac{1.2769}{2\pi}\sqrt{\frac{g}{\delta_{max}}}$$

is based on his work where δ_{max} is the maximum static deflection produced by the weight of the plate and any uniformly distributed mass attached to the plate and vibrating with it. It is based on the expression for the fundamental frequency of a clamped elliptical plate, but, as Jones points out with several examples of triangular, rectangular, and circular plates having

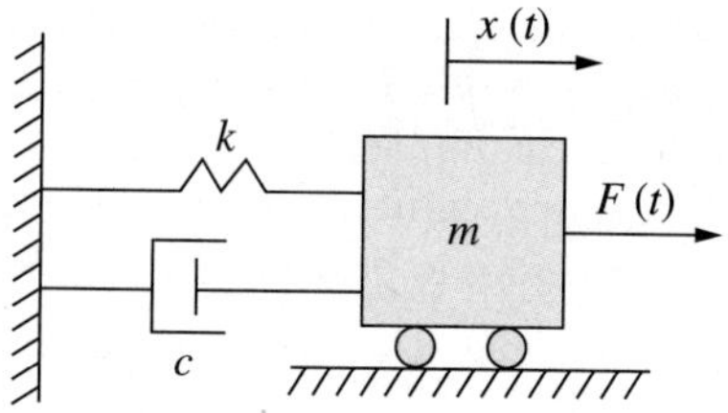

Figure 16.5 Single degree, viscous damping, vibration system.

various combinations of boundary conditions, it should hold equally well for all uniform plates having no free boundaries. In the 16 examples he presents, the maximum error in frequency is about 3%.

Fundamentals of Vibration

Vibration in general is a periodic or oscillatory motion of a mechanical system. Depending on the existence of an external load, vibration of a mechanical system can be free vibration or forced vibration, where in either case the system can be considered damped or undamped. The linear dynamic behavior of many engineering systems can be approximated with good accuracy by the "mass-damper-spring model" as shown in Fig. 16.5.

The differential equation of motion of the system is

$$m\ddot{x}(t)+c\dot{x}(t)+kx(t)=F(t) \tag{16.6-1}$$

where m is the mass, c is the damping coefficient, and k is the spring constant.

Free Vibration of Undamped Systems

Assuming zero damping and external forces and dividing Eq. (16.6-1) by m we obtain

$$\ddot{x}+\omega_n^2 x=0 \qquad \omega_n=\sqrt{k/m} \tag{16.6-2}$$

In this case, the vibration is caused by the initial excitations alone. If the initial conditions are $x(0)=x_0$ and $\dot{x}(0)=v_0$, where x_0 and v_0 are the initial displacement and velocity, respectively, then the solution of Eq. (16.6-2) is

$$x(t)=A\cos(\omega_n t-\phi) \tag{16.6-3}$$

which represents simple sinusoidal, or simple harmonic, oscillation with the amplitude $A = [x_0^2+(v_0/\omega_n)^2]^{1/2}$, the phase angle $\phi=\tan^{-1}[v_0/(x_0\omega_n)]$. The natural frequency ω_n is given by

$$\omega_n=\sqrt{k/m}\ \text{rad/s} \tag{16.6-4}$$

The time necessary to complete one cycle of motion defines the period, as

$$T-2\pi/\omega_n\ \text{seconds} \tag{16.6-5}$$

The reciprocal of the period provides another definition of the natural frequency, namely,

$$f_n=\frac{1}{T}=\frac{\omega_n}{2\pi}\ \text{Hz}$$

where Hz denotes hertz [1 Hz = 1 cycle per second (cps)].

Free Vibration of a Damped System

Consider the system of Fig. 16.5 when the external force is zero [$F(t) = 0$]; the equation is reduced to

$$\ddot{x}(t)+2\zeta\omega_n\dot{x}(t)+\omega_n^2 x(t)=0 \tag{16.6-6}$$

where $\zeta = c/2m\omega_n$ is the damping factor, a nondimensional quantity. The form of the motion depends on ζ. The most important case is that in which $0 < \zeta < 1$.

The solution of the equation of motion is

$$x(t)=Ae^{-\zeta\omega_n t}\cos(\omega_d t-\phi) \tag{16.6-7}$$

where $\omega_d = (1-\zeta^2)^{1/2}\omega_n$ is the frequency of damped free vibration and

$$T = 2\pi/\omega_d$$

is the period of damped oscillation. The amplitude and phase angle depend on the initial displacement and velocity, as follows:

If $\varsigma < 1$, then the system is decaying oscillation and the term $Ae^{-\zeta\omega_n t}$ is a time-dependent amplitude, and it is decreasing as the time increases. In this case, the system is called underdamped. A typical decaying oscillation of an underdamped system is shown in Fig. 16.6. The ratio of two displacements x_1 and x_2 separated by a complete period T is

$$\frac{x_1}{x_2}=\frac{Ae^{-\zeta\omega_n t_1}\cos(\omega_d t_1-\phi)}{Ae^{-\zeta\omega_n(t_1+T)}\cos[\omega_d(t_1+T)-\phi]}=e^{\zeta\omega_n T} \tag{16.6-8}$$

where $\cos[\omega_d(t_1+T)-\phi]=\cos(\omega_d t_1-\phi+2\pi)=\cos(\omega_d t_1-\phi)$.

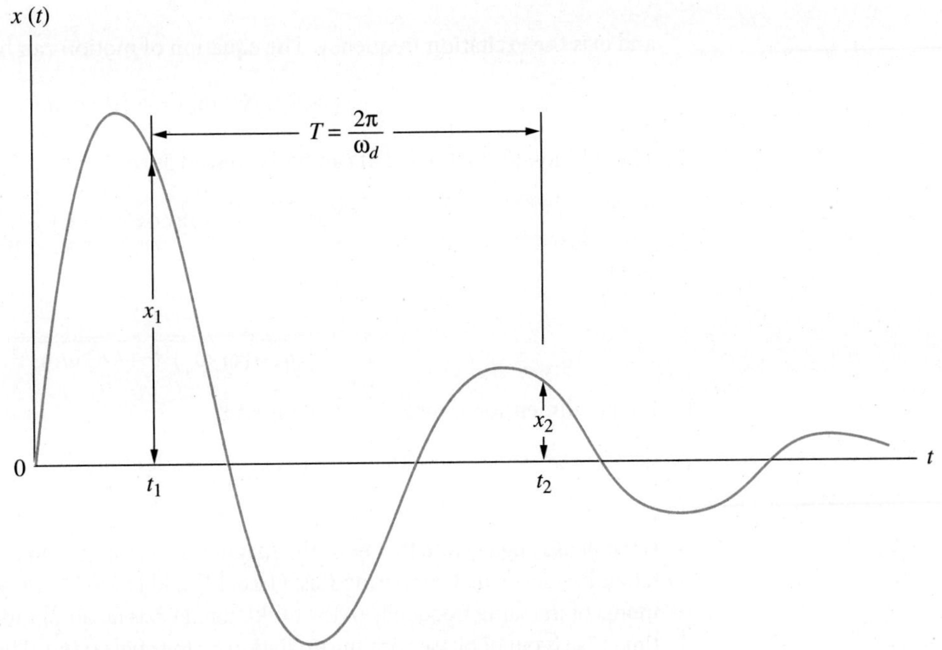

Figure 16.6 Decaying oscillation of an underdamped system.

Equation (16.6-8) yields the logarithmic decrement

$$\delta = \ln\frac{x_1}{x_2} = \zeta\omega_n T = \frac{2\pi\zeta}{\sqrt{1-\zeta^2}} \tag{16.6-9}$$

which can be used to obtain the damping factor

$$\zeta = \frac{\delta}{\sqrt{(2\pi)^2 + \delta^2}} \tag{16.6-10}$$

For small damping, the logarithmic decrement is also small, and the damping factor can be approximated by

$$\zeta \approx \frac{\delta}{2\pi} \tag{16.6-11}$$

The case $\zeta = 1$ represents "Critical Damping" and the critical damping coefficient is $C_c = 2m\omega_n$.

When $\zeta \geq 1$ the solution of the equation represents "A periodic Decay." Note that $\zeta = 1$ is the borderline between oscillatory decay and a periodic decay.

Forced Vibration of a Damped System

Consider a steady state loading of the system of Fig. 16.5, which the excitation force $F(t)$ is harmonic in the form of

$$F(t) = F_0 \cos\omega t = k\,A\cos\omega t$$

where F_0 is the force amplitude,

$$A = F_0/k$$

and ω is the excitation frequency. The equation of motion can be reduced to

$$\ddot{x} + 2\zeta\omega_n\dot{x} + \omega_n^2 x = \omega_n^2 A\cos\omega t \tag{16.6-12}$$

The solution of Eq. (16.6-12) can be expressed as

$$x(t) = A|G(\omega)|\cos(\omega t - \phi) \tag{16.6-13}$$

where

$$G(\omega) = \frac{1}{\sqrt{[1-(\omega/\omega_n)^2]^2 + (2\zeta\omega/\omega_n)^2}} \tag{16.6-14}$$

is a nondimensional magnitude factor and

$$\phi(\omega) = \tan^{-1}\frac{2\zeta\omega/\omega_n}{1-(\omega/\omega_n)^2} \tag{16.6-15}$$

is the phase angle. Note that both the magnitude factor and phase angle depend on the excitation frequency ω. Comparing Eqs. (16.6-12) and (16.6-13) we see that the response is harmonic of the same frequency of the excitation, ω, has an amplitude $AG(\omega) = (F_0/k)\,G(\omega)$, and time wise is out of phase with the excitation by the angle $\phi\,(\omega)$. The variation of the amplitude ratio $G(\omega)$ and the phase angle with respect to the ratio of ω/ω_n are plotted in Fig. 16.7, in

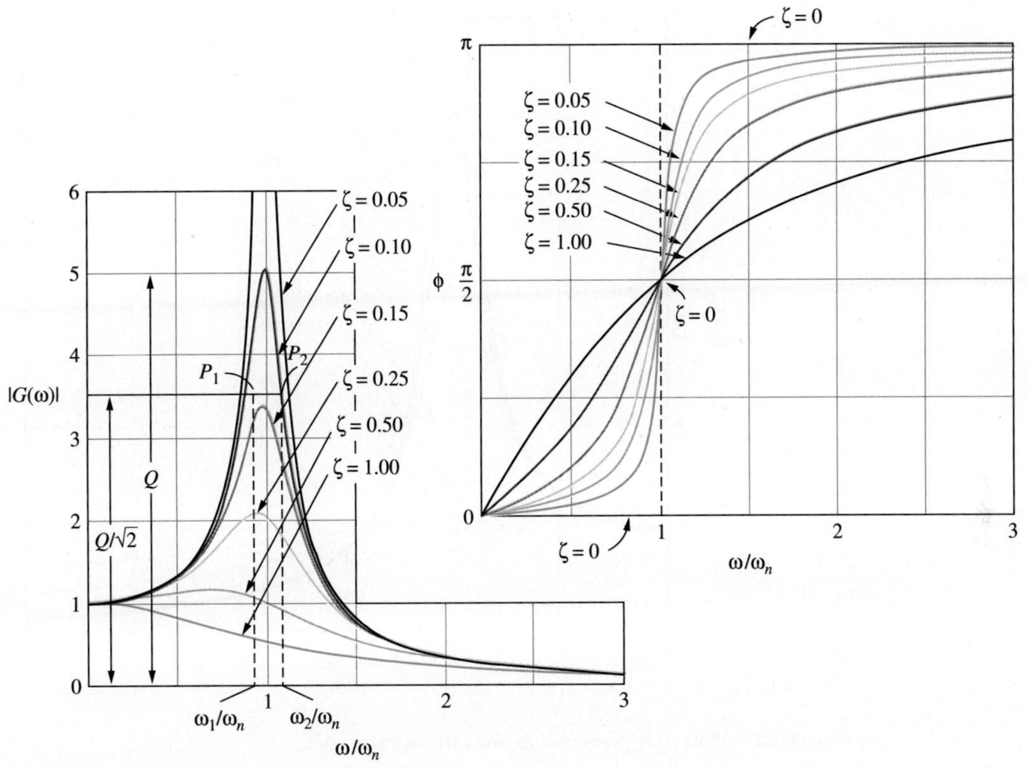

Figure 16.7 Variation of $G(\omega)$ and the phase angle with respect to the ratio of ω/ω_n for different values of damping ratio.

which the following observations can be made.

1. For an undamped system $\zeta = 0$, the response of the system is

$$x(t)=\frac{F_0/k}{1-(\omega/\omega_n)^2}\cos\omega t$$

2. The damping reduces the amplitude ratio $G(\omega)$ for all values of the forcing frequency.
3. The maximum value of the amplitude ratio is at $\frac{1}{2}\zeta(1-\zeta^2)^{1/2}$.
4. For a small value of ζ, the system experiences large-amplitude vibration, a condition known as "resonance."
5. For a small value of ζ, the peaks of the amplitude ratio occur approximately at $\omega/\omega_n = 1$ and have an approximate value of $\frac{1}{2}\zeta$. In such cases, the phase angle tends to 90°.
6. For $\omega/\omega_n < 1$, the displacement is in the same direction as the force, so that the phase angle is zero; the response is in phase with the excitation. For $\omega/\omega_n > 1$, the displacement is in the direction opposite to the force, so that the phase angle is 180° out of phase with the excitation. Finally, when $\omega = \omega_n$, the response is

$$x(t)=\frac{F_0}{2k}\omega_n t\sin\omega_n t$$

This is typical of the resonance condition, when the response increases without bounds as time increases.

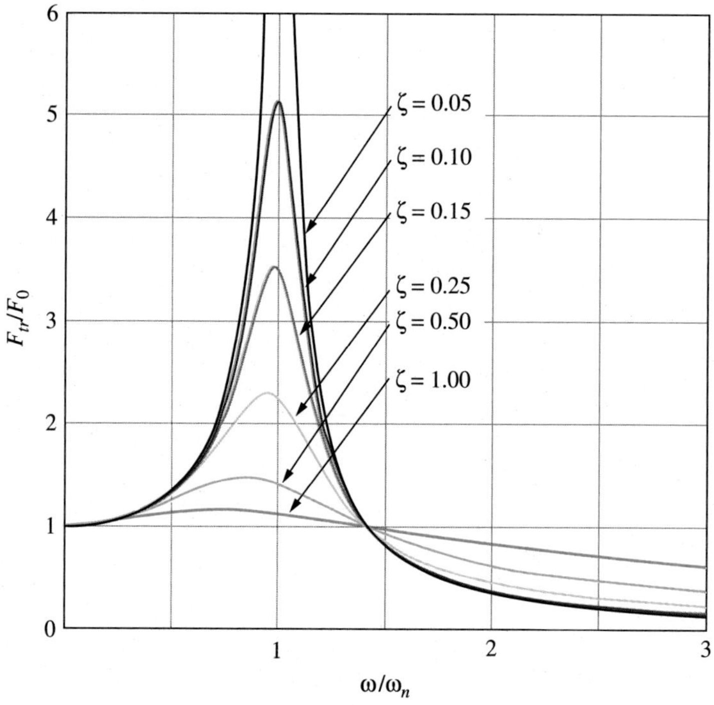

Figure 16.8 Plots F_{tr}/F_0 versus ω/ω_n for various values of ζ.

Vibration Isolation

A supporting structure of a machine may experience excessive vibration if the excitation frequency of the machine nearly coincides with the natural frequency of the structure. Therefore, the force transmitted to the base or foundation of a vibration system is of great interest, since it could cause damage to the supporting structure.

Let the magnitude of the harmonic excitation be $F_0 = kA$; the measure of the force transmitted to the base is then

$$T = \frac{F_{tr}}{F_0} = G\sqrt{1+(2\zeta\omega/\omega_n)^2}$$
$$= \sqrt{\frac{1+(2\zeta\omega/\omega_n)^2}{[1-(\omega/\omega_n)^2]^2+(2\zeta\omega/\omega_n)^2}} \tag{16.6-16}$$

which represents a nondimensional ratio called transmissibility. Figure 16.8 reveals that for $\omega/\omega_n > \sqrt{2}$ the transmissibility is less than one as ω/ω_n increases. Therefore, to have an effective vibration isolator, the natural frequency of the system must be much smaller than the excitation frequency.

16.7 TEMPERATURE STRESSES

Whenever the expansion or contraction that would normally result from the heating or cooling of a body is prevented, stresses are developed that are called *thermal*, or *temperature, stresses.* It is convenient to distinguish between two different sets of circumstances under which thermal stresses occur: (1) The form of the body and the temperature conditions are such that there would be no stresses except for the *constraint of external forces;* in any such case, the stresses may be found by determining the shape and dimensions the body would assume if unconstrained and then calculating the stresses produced by forcing it back to its original shape and dimensions (see Sec. 7.2, Example 2). (2) The form of the body and the temperature conditions are such that stresses are produced in the *absence of external constraint* solely because of the incompatibility of the natural expansions or contractions of the different parts of the body.

A number of representative examples of each type of thermal stress will now be considered.* In all instances, the modulus of elasticity E and the coefficient of thermal expansion γ are assumed to be constant for the temperature range involved, and the increment or difference in temperature ΔT is assumed to be positive; when ΔT is negative, the stress produced is of the opposite kind. Also, it is assumed that the compressive stresses produced do not produce buckling and that yielding does not occur; if either buckling or yielding is indicated by the stress levels found, then the solution must be modified by appropriate methods discussed in previous chapters.

Stresses Due to External Constraint

1. A uniform straight bar is subjected to a temperature change ΔT throughout while held at the ends; the resulting unit stress is $\Delta T\gamma E$ (compression). (For other conditions of end restraint see Table 8.2 cases 1q–12q.)
2. A uniform flat plate is subjected to a temperature change ΔT throughout while held at the edges; the resulting unit stress is $\Delta T\gamma E/(1-\nu)$ (compression).
3. A solid body of any form is subjected to a temperature change ΔT throughout while held to the same form and volume; the resulting stress is $\Delta T\gamma E/(1-2\nu)$ (compression).
4. A uniform bar of rectangular section has one face at a uniform temperature T and the opposite face at a uniform temperature $T+\Delta T$ with the temperature gradient between these faces being linear. The bar would normally curve in the arc of the circle of radius $d/\Delta T\gamma$, where d is the distance between the hot and cold faces. If the ends are fixed, the bar will be held straight by end couples $EI\Delta T\gamma/d$ and the maximum resulting bending stress will be $\frac{1}{2}\Delta T\gamma E$ (compression on the hot face; tension on the cold face). (For many other conditions of end restraint and partial heating, see Table 8.1, cases 6a–6f; Table 8.2, cases 1r–12r; Table 8.5, case 7; Table 8.6, case 7; Table 8.8, cases 6a–6f; and Table 8.9 cases 6a–6f.)
5. A flat plate of uniform thickness t and of any shape has one face at a uniform temperature T and the other face at a uniform temperature $T+\Delta T$, the temperature gradient between the faces being linear. The plate would normally assume a spherical curvature with radius $t/(\Delta T\gamma)$. If the edges are fixed, the plate will be held flat by uniform edge moments and the maximum resulting bending stress will be $\frac{1}{2}\Delta T\gamma E/(1-\nu)$ (compression on the hot face; tension on the cold face). (For many other conditions of edge restraint and axisymmetric partial heating, see Table 11.2, cases 8a–8h; a more general treatment of the solid circular plate is given in Table 11.2, case 15.)
6. If the plate described in case 5 is circular, no stress is produced by supporting the edges in a direction normal to the plane of the plate.

*Most of the formulas given here are taken from the papers by Goodier (Refs. 7 and 14), Maulbetsch (Ref. 8), and Kent (Ref. 9).

7. If the plate described in case 5 has the shape of an equilateral triangle of altitude a (sides $2a/\sqrt{3}$) and the edges are rigidly supported so as to be held in a plane, the supporting reactions will consist of a uniform load $1/8\,\Delta T\gamma Et^2/a$ per unit length along each edge against the hot face and a concentrated load $\sqrt{3}\Delta T\gamma Et^2/12$ at each corner against the cold face. The maximum resulting bending stress is $\frac{3}{4}\Delta T\gamma E$ at the corners (compression on the hot face; tension on the cold face). There are also high shear stresses near the corners (Ref. 8).
8. If the plate described in case 5 is square, no simple formula is available for the reactions necessary to hold the edges in their original plane. The maximum bending stress occurs near the edges, and its value approaches $\frac{1}{2}\Delta T\gamma E$. There are also high shear stresses near the corners (Ref. 8).

Stresses Due to Internal Constraint

9. Part or all of the surface of a solid body is suddenly subjected to a temperature change ΔT; a compressive stress $\Delta T\gamma E/(1-\nu)$ is developed in the surface layer of the heated part (Ref. 7).
10. A thin circular disk at uniform temperature has the temperature changed ΔT throughout a comparatively small central circular portion of radius a. Within the heated part there are radial and tangential compressive stresses $\sigma_r = \sigma_t = \frac{1}{2}\Delta T\gamma E$. At points outside the heated part, a distance r from the center of the disk but still close to the central portion, the stresses are $\sigma_r = \frac{1}{2}\Delta T\gamma Ea^2/r^2$ (compression) and $\sigma_t = \frac{1}{2}\Delta T\gamma Ea^2/r^2$ (tension); at the edge of the heated portion, there is a maximum shear stress $\frac{1}{2}\Delta T\gamma E$ (Ref. 7).
11. If the disk of case 10 is heated uniformly throughout a small central portion of elliptical instead of circular outline, the maximum stress is the tangential stress at the ends of the ellipse and is $\sigma_t = \Delta T\gamma E/[1+(b/a)]$, where a is the major and b the minor semi-axis of the ellipse (Ref. 7).
12. If the disk of case 10 is heated symmetrically about its center and uniformly throughout its thickness so that the temperature is a function of the distance r from the center only, the radial and tangential stresses at any point a distance r_1 from the center are

$$\sigma_{r_1} = \gamma E\left(\frac{1}{R^2}\int_0^R Tr\,dr - \frac{1}{r_1^2}\int_0^{r_1} Tr\,dr\right)$$

$$\sigma_{t_1} = \gamma E\left(-T + \frac{1}{R^2}\int_0^R Tr\,dr + \frac{1}{r_1^2}\int_0^{r_1} Tr\,dr\right)$$

where R is the radius of the disk and T is the temperature at any point a distance r from the center minus the temperature of the coldest part of the disk. [In the preceding expressions, the negative sign denotes compressive stress (Ref. 7).]
13. A rectangular plate or strip $ABCD$ (Fig. 16.9) is heated along a transverse line FG uniformly throughout the thickness and across the width so that the temperature varies only along the length with x. At FG the temperature is T_1; the minimum temperature in the plate is T_0. At any point along the edges of the strip where the temperature is T, a tensile stress $\sigma_x = E\gamma(T - T_0)$ is developed; this stress has its maximum value at F and G, where it becomes $E\gamma(T_1 - T_0)$. Halfway between F and G, a compressive stress σ_y of equal intensity is developed along line FG (Ref. 7).

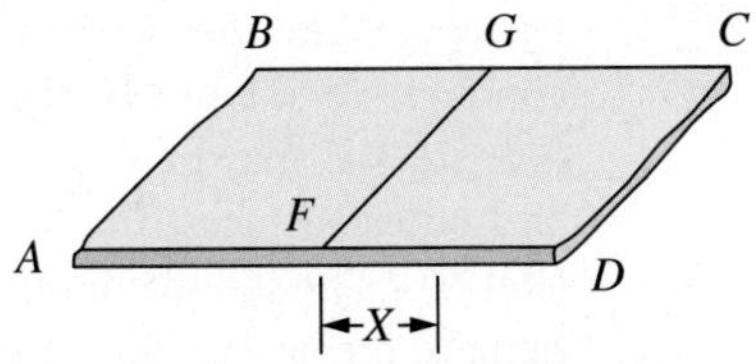

Figure 16.9

14. The plate of case 13 is heated as described except that the lower face of the plate is cooler than the upper face, the maximum temperature there being T_2 and the temperature gradient through the thickness being linear. The maximum tensile stress at F and G is (see Ref. 7)

$$\sigma_x = \frac{1}{2} E\gamma \left[T_1 + T_2 - 2T_0 + \frac{1-\nu}{3+\nu}(T_1 - T_2) \right]$$

15. A long hollow cylinder with thin walls has the outer surface at the uniform temperature T and the inner surface at the uniform temperature $T + \Delta T$. The temperature gradient through the thickness is linear. At points remote from the ends, the maximum circumferential stress is $\frac{1}{2}\Delta T\gamma E/(1-\nu)$ (compression at the inner surface; tension at the outer surface) and the longitudinal stress is $\frac{1}{2}\Delta T\gamma E/(1-\nu)$ (compression at the inside; tension at the outside). (These formulas apply to a thin tube of any cross section.) At the ends, if these are free, the maximum tensile stress in a tube of circular section is about 25% greater than the value given by the formula (Ref. 7).

16. A hollow cylinder with thick walls of inner radius b and outer radius c has the outer surface at the uniform temperature T and the inner surface at the uniform temperature $T + \Delta T$. After steady-state heat flow is established, the temperature decreases logarithmically with r and then the maximum stresses, which are circumferential and which occur at the inner and outer surfaces, are

(Outer surface)

$$\sigma_t = \frac{\Delta T\gamma E}{2(1-\nu)\ln(c/b)}\left(1 - \frac{2b^2}{c^2 - b^2}\ln\frac{c}{b}\right) \qquad \text{tension}$$

(Inner surface)

$$\sigma_t = \frac{\Delta T\gamma E}{2(1-\nu)\ln(c/b)}\left(1 - \frac{2c^2}{c^2 - b^2}\ln\frac{c}{b}\right) \qquad \text{compression}$$

At the inner and outer surfaces, the longitudinal stresses are equal to the tangential stresses (Ref. 7).

17. If the thick tube of case 16 has the temperature of the outer surface raised at the uniform rate of $m°/\text{s}$, then, after a steady rate of temperature rise has been reached throughout, the maximum tangential stresses are

(Outer surface)

$$\sigma_t = \frac{E\gamma m}{8A(1-\nu)}\left(3b^2 - c^2 - \frac{4b^4}{c^2 - b^2}\ln\frac{c}{b}\right) \qquad \text{compression}$$

(Inner surface)

$$\sigma_t = \frac{E\gamma m}{8A(1-\nu)}\left(b^2 + c^2 - \frac{4b^2c^2}{c^2 - b^2}\ln\frac{c}{b}\right) \qquad \text{tension}$$

where A is the coefficient of thermal diffusivity equal to the coefficient of thermal conductivity divided by the product of density of the material and its specific heat. (For steel, A may be taken as 0.027 in^2/s at moderate temperatures.) [At the inner and outer surfaces, the longitudinal stresses are equal to the tangential stresses (Ref. 9).] The stated conditions in case 17 as well as those in cases 19 to 21 are difficult to create in a short time except for small parts heated or cooled in liquids.

18. A solid rod of circular section is heated or cooled symmetrically with respect to its axis, the condition being uniform along the length, so that the temperature is a function of r (the distance from the axis) only. The stresses are equal to those given by the formulas for case 12 divided by $1 - \nu$ (Ref. 7).

19. If the solid rod of case 18 has the temperature of its convex surface raised at the uniform rate of $m°/s$, then, after a steady rate of temperature rise has been reached throughout, the radial, tangential, and longitudinal stresses at any point a distance r from the center are

$$\sigma_r = \frac{E\gamma m}{1-\nu}\frac{c^2 - r^2}{16A}$$

$$\sigma_t = \frac{E\gamma m}{1-\nu}\frac{c_2 - 3r^2}{16A}$$

$$\sigma_x = \frac{E\gamma m}{1-\nu}\frac{c^2 - 2r^2}{8A}$$

Here A has the same meaning as in case 17 and c is the radius of the shaft. [A negative result indicates compression, a positive result tension (Ref. 9).]

20. A solid sphere of radius c is considered instead of a solid cylinder but with all other conditions kept the same as in case 19. The radial and tangential stresses produced at any point a distance r from the center are

$$\sigma_r = \frac{E\gamma m}{15A(1-\nu)}(c^2 - r^2)$$

$$\sigma_t = \frac{E\gamma m}{15A(1-\nu)}(c^2 - 2r^2)$$

[A negative result indicates compression, a positive result tension (Ref. 9).]

21. If the sphere is hollow, with outer radius c and inner radius b, and with all other conditions kept as stated in case 17, the stresses at any point are

$$\sigma_r = \frac{E\gamma m}{15A(1-\nu)}\left(-r^2 - \frac{5b^3}{r} + \phi - \psi\right)$$

$$\sigma_t = \frac{E\gamma m}{15A(1-\nu)}\left(-2r^2 - \frac{5b^3}{2r} + \phi + \frac{\psi}{2}\right)$$

where

$$\phi = \frac{c^5 + 5c^2b^3 - 6b^5}{c^3 - b^3}$$

$$\psi = \frac{c^5b^3 - 6c^3b^5 + 5c^2b^6}{r^3(c^3 - b^3)}$$

[A negative result indicates compression, a positive result tension (Ref. 9).]

Other problems involving thermal stress, the solutions of which cannot be expressed by simple formulas, are considered in the references cited above and in Refs. 3, 10, and 25 to 29; charts for the solution of thermal stresses in tubes are given in Ref. 11. Derivations for many of the thermal loadings shown above along with thermal loadings on many other examples of bars, rings, plates, and cylindrical and spherical shells are given in Ref. 28.

16.8 TABLES

TABLE 16.1 Natural Frequencies of Vibration for Continuous Members

Notation: f = natural frequency (cycles per second); K_n = constant where n refers to the mode of vibration; g = gravitational acceleration (units consistent with length dimension); E = modulus of elasticity; I = area moment of inertia; $D = Et^3/12(1-\nu^2)$

<table>
<tr><th>Case No. and Description</th><th></th><th>Natural Frequencies</th></tr>
<tr><td rowspan="3">1. Uniform beam; both ends simply supported</td><td>1a. Center load W, beam weight negligible</td><td>$f_1 = \frac{6.93}{2\pi}\sqrt{\frac{EIg}{Wl^3}}$</td></tr>
<tr><td>1b. Uniform load w per unit length including beam weight</td><td>$f_n = \frac{K_n}{2\pi}\sqrt{\frac{EIg}{wl^4}}$
<table>
<tr><th>Mode</th><th>K_n</th><th colspan="6">Nodal position/l</th></tr>
<tr><td>1</td><td>9.87</td><td>0.0</td><td>1.0</td><td></td><td></td><td></td><td></td></tr>
<tr><td>2</td><td>39.5</td><td>0.0</td><td>0.50</td><td>1.00</td><td></td><td></td><td></td></tr>
<tr><td>3</td><td>88.8</td><td>0.0</td><td>0.33</td><td>0.67</td><td>1.00</td><td></td><td></td></tr>
<tr><td>4</td><td>158</td><td>0.0</td><td>0.25</td><td>0.50</td><td>0.75</td><td>1.00</td><td></td></tr>
<tr><td>5</td><td>247</td><td>0.0</td><td>0.20</td><td>0.40</td><td>0.60</td><td>0.80</td><td>1.00</td></tr>
</table>
Ref. 22</td></tr>
<tr><td>1c. Uniform load w per unit length plus a center load W</td><td>$f_1 = \frac{6.93}{2\pi}\sqrt{\frac{EIg}{Wl^3 + 0.486wl^4}}$ approximately</td></tr>
<tr><td rowspan="3">2. Uniform beam; both ends fixed</td><td>2a. Center load W, beam weight negligible</td><td>$f_1 = \frac{13.86}{2\pi}\sqrt{\frac{EIg}{Wl^3}}$</td></tr>
<tr><td>2b. Uniform load w per unit length including beam weight</td><td>$f_n = \frac{K_n}{2\pi}\sqrt{\frac{EIg}{wl^4}}$
<table>
<tr><th>Mode</th><th>K_n</th><th colspan="6">Nodal position/l</th></tr>
<tr><td>1</td><td>22.4</td><td>0.0</td><td>1.0</td><td></td><td></td><td></td><td></td></tr>
<tr><td>2</td><td>61.7</td><td>0.0</td><td>0.50</td><td>1.00</td><td></td><td></td><td></td></tr>
<tr><td>3</td><td>121</td><td>0.0</td><td>0.36</td><td>0.64</td><td>1.00</td><td></td><td></td></tr>
<tr><td>4</td><td>200</td><td>0.0</td><td>0.28</td><td>0.50</td><td>0.72</td><td>1.00</td><td></td></tr>
<tr><td>5</td><td>299</td><td>0.0</td><td>0.23</td><td>0.41</td><td>0.59</td><td>0.77</td><td>1.00</td></tr>
</table>
Ref. 22</td></tr>
<tr><td>2c. Uniform load w per unit length plus a center load W</td><td>$f_1 = \frac{13.86}{2\pi}\sqrt{\frac{EIg}{Wl^3 + 0.383wl^4}}$ approximately</td></tr>
<tr><td rowspan="3">3. Uniform beam; left end fixed, right end free (cantilever)</td><td>3a. Right end load W, beam weight negligible</td><td>$f_1 = \frac{1.732}{2\pi}\sqrt{\frac{EIg}{Wl^3}}$</td></tr>
<tr><td>3b. Uniform load w per unit length including beam weight</td><td>$f_n = \frac{K_n}{2\pi}\sqrt{\frac{EIg}{wl^4}}$
<table>
<tr><th>Mode</th><th>K_n</th><th colspan="5">Nodal position/l</th></tr>
<tr><td>1</td><td>3.52</td><td>0.0</td><td></td><td></td><td></td><td></td></tr>
<tr><td>2</td><td>22.0</td><td>0.0</td><td>0.783</td><td></td><td></td><td></td></tr>
<tr><td>3</td><td>61.7</td><td>0.0</td><td>0.504</td><td>0.868</td><td></td><td></td></tr>
<tr><td>4</td><td>121</td><td>0.0</td><td>0.358</td><td>0.644</td><td>0.905</td><td></td></tr>
<tr><td>5</td><td>200</td><td>0.0</td><td>0.279</td><td>0.500</td><td>0.723</td><td>0.926</td></tr>
</table>
Ref. 22</td></tr>
<tr><td>3c. Uniform load w per unit length plus an end load W</td><td>$f_1 = \frac{1.732}{2\pi}\sqrt{\frac{EIg}{Wl^3 + 0.236wl^4}}$ approximately</td></tr>
</table>

(Continued)

TABLE 16.1 Natural Frequencies of Vibration for Continuous Members (*Continued*)

<table>
<tr><th>Case No. and Description</th><th></th><th>Natural Frequencies</th></tr>
<tr><td>4. Uniform beam; both ends free</td><td>4a. Uniform load w per unit length including beam weight</td><td>

$f_n = \frac{K_n}{2\pi}\sqrt{\frac{EIg}{wl^4}}$

<table>
<tr><th>Mode</th><th>K_n</th><th colspan="6">Nodal position/l</th></tr>
<tr><td>1</td><td>22.4</td><td>0.224</td><td>0.776</td><td></td><td></td><td></td><td></td></tr>
<tr><td>2</td><td>61.7</td><td>0.132</td><td>0.500</td><td>0.868</td><td></td><td></td><td></td></tr>
<tr><td>3</td><td>121</td><td>0.095</td><td>0.356</td><td>0.644</td><td>0.905</td><td></td><td></td></tr>
<tr><td>4</td><td>200</td><td>0.074</td><td>0.277</td><td>0.500</td><td>0.723</td><td>0.926</td><td></td></tr>
<tr><td>5</td><td>299</td><td>0.060</td><td>0.226</td><td>0.409</td><td>0.591</td><td>0.774</td><td>0.940</td></tr>
</table>

Ref. 22</td></tr>
<tr><td>5. Uniform beam; left end fixed, right end hinged</td><td>5a. Uniform load w per unit length including beam weight</td><td>

$f_n = \frac{K_n}{2\pi}\sqrt{\frac{EIg}{wl^4}}$

<table>
<tr><th>Mode</th><th>K_n</th><th colspan="6">Nodal position/l</th></tr>
<tr><td>1</td><td>15.4</td><td>0.0</td><td>1.000</td><td></td><td></td><td></td><td></td></tr>
<tr><td>2</td><td>50.0</td><td>0.0</td><td>0.557</td><td>1.000</td><td></td><td></td><td></td></tr>
<tr><td>3</td><td>104</td><td>0.0</td><td>0.386</td><td>0.692</td><td>1.000</td><td></td><td></td></tr>
<tr><td>4</td><td>178</td><td>0.0</td><td>0.295</td><td>0.529</td><td>0.765</td><td>1.000</td><td></td></tr>
<tr><td>5</td><td>272</td><td>0.0</td><td>0.239</td><td>0.428</td><td>0.619</td><td>0.810</td><td>1.000</td></tr>
</table>

Ref. 22</td></tr>
<tr><td>6. Uniform beam; left end hinged, right end free</td><td>6a. Uniform load w per unit length including beam weight</td><td>

$f_n = \frac{K_n}{2\pi}\sqrt{\frac{EIg}{wl^4}}$

<table>
<tr><th>Mode</th><th>K_n</th><th colspan="6">Nodal position/l</th></tr>
<tr><td>1</td><td>15.4</td><td>0.0</td><td>0.736</td><td></td><td></td><td></td><td></td></tr>
<tr><td>2</td><td>50.0</td><td>0.0</td><td>0.446</td><td>0.853</td><td></td><td></td><td></td></tr>
<tr><td>3</td><td>104</td><td>0.0</td><td>0.308</td><td>0.617</td><td>0.898</td><td></td><td></td></tr>
<tr><td>4</td><td>178</td><td>0.0</td><td>0.235</td><td>0.471</td><td>0.707</td><td>0.922</td><td></td></tr>
<tr><td>5</td><td>272</td><td>0.0</td><td>0.190</td><td>0.381</td><td>0.571</td><td>0.763</td><td>0.937</td></tr>
</table>

Ref. 22</td></tr>
<tr><td rowspan="3">7. Uniform bar or spring vibrating along its longitudinal axis; upper end fixed, lower end free</td><td>7a. Weight W at lower end, bar weight negligible</td><td>

$f_1 = \frac{1}{2\pi}\sqrt{\frac{kg}{W}}$ for a spring where k is the spring constant

$f_1 = \frac{1}{2\pi}\sqrt{\frac{AEg}{Wl}}$ for a bar where A is the area, l the length, and E the modulus</td></tr>
<tr><td>7b. Uniform load w per unit length including bar weight</td><td>

$f_n = \frac{K_n}{2\pi}\sqrt{\frac{AEg}{wl^2}}$ where $K_1 = 1.57$ $K_2 = 4.71$ $K_3 = 7.85$</td></tr>
<tr><td>7c. Uniform load w per unit length plus a load W at the lower end</td><td>

$f_1 = \frac{1}{2\pi}\sqrt{\frac{kg}{W + wl/3}}$ approximately for a spring where k is the spring constant

$f_1 = \frac{1}{2\pi}\sqrt{\frac{AEg}{Wl + wl^2/3}}$ approximately for a bar where A is the area</td></tr>
<tr><td rowspan="3">8. Uniform shaft or bar in torsional vibration; one end fixed, the other end free</td><td>8a. Concentrated end mass of J mass moment of inertia, shaft weight negligible</td><td>

$f_1 = \frac{1}{2\pi}\sqrt{\frac{GK}{Jl}}$ G is the shear modulus of elasticity and K is the torsional stiffness constant (see Chap. 10)</td></tr>
<tr><td>8b. Uniform distribution of mass moment of inertia along shaft; J_s = total distributed mass moment of inertia</td><td>

$f_n = \frac{K_n}{2\pi}\sqrt{\frac{GK}{J_s l}}$ where $K_1 = 1.57$ $K_2 = 4.71$ $K_3 = 7.85$</td></tr>
<tr><td>8c. Uniformly distributed inertia plus a concentrated end mass</td><td>

$f_1 = \frac{1}{2\pi}\sqrt{\frac{GK}{(J + J_s/3)l}}$ approximately</td></tr>
<tr><td>9. String vibrating laterally under a tension T with both ends fixed</td><td>9a. Uniform load w per unit length including own weight</td><td>

$f_n = \frac{K_n}{2\pi}\sqrt{\frac{Tg}{wl^2}}$ where $K_1 = \pi$ $K_2 = 2\pi$ $K_3 = 3\pi$</td></tr>
</table>

(*Continued*)

TABLE 16.1 Natural Frequencies of Vibration for Continuous Members (*Continued*)

<table>
<tr><th>Case No. and Description</th><th></th><th>Natural Frequencies</th></tr>
<tr><td>10. Circular flat plate of uniform thickness t and radius r; edge fixed</td><td>10a. Uniform load w per unit area including own weight</td><td>$f = \frac{K_n}{2\pi}\sqrt{\frac{Dg}{wr^4}}$ where $K_1 = 10.2$ fundamental
$K_2 = 21.3$ one nodal diameter
$K_3 = 34.9$ two nodal diameters
$K_4 = 39.8$ one nodal circle
Ref. 20</td></tr>
<tr><td>11. Circular flat plate of uniform thickness t and radius r; edge simply supported</td><td>11a. Uniform load w per unit area including own weight; $\nu = 0.3$</td><td>$f = \frac{K_n}{2\pi}\sqrt{\frac{Dg}{wr^4}}$ where $K_1 = 4.99$ fundamental
$K_2 = 13.9$ one nodal diameter
$K_3 = 25.7$ two nodal diameters
$K_4 = 29.8$ one nodal circle
Ref. 20</td></tr>
<tr><td>12. Circular flat plate of uniform thickness t and radius r; edge free</td><td>12a. Uniform load w per unit area including own weight; $\nu = 0.3$</td><td>$f = \frac{K_n}{2\pi}\sqrt{\frac{Dg}{wr^4}}$ where $K_1 = 5.25$ two nodal diameters
$K_2 = 9.08$ one nodal circle
$K_3 = 12.2$ three nodal diameters
$K_4 = 20.5$ one nodal diameter and one nodal circle
Ref. 20</td></tr>
<tr><td>13. Circular flat plate of uniform thickness t and radius r; edge simply supported with an additional edge constraining moment $M = \beta\psi$ per unit circumference where ψ is the edge rotation</td><td>13a. Uniform load w per unit area including own weight; $\nu = 0.3$</td><td>$f = \frac{K_n}{2\pi}\sqrt{\frac{Dg}{wr^4}}$ where K_n is tabulated for various degrees of edge stiffness in the form of $\beta r/D$.
<table>
<tr><th></th><th colspan="4">K_n</th></tr>
<tr><th>$\beta r/D$</th><th>Fundamental</th><th>1 Nodal Diameter</th><th>2 Nodal Diameters</th><th>1 Nodal Circle</th></tr>
<tr><td>∞</td><td>10.2</td><td>21.2</td><td>34.8</td><td>39.7</td></tr>
<tr><td>1</td><td>10.2</td><td>21.2</td><td>34.8</td><td>39.7</td></tr>
<tr><td>0.1</td><td>10.0</td><td>20.9</td><td>34.2</td><td>39.1</td></tr>
<tr><td>0.01</td><td>8.76</td><td>18.6</td><td>30.8</td><td>35.2</td></tr>
<tr><td>0.001</td><td>6.05</td><td>15.0</td><td>26.7</td><td>30.8</td></tr>
<tr><td>0</td><td>4.93</td><td>13.9</td><td>25.6</td><td>29.7</td></tr>
</table>
Ref. 20</td></tr>
<tr><td>14. Elliptical flat plate of major radius a, minor radius b, and thickness t; edge fixed</td><td>14a. Uniform load w per unit area including own weight</td><td>$f = \frac{K_1}{2\pi}\sqrt{\frac{Dg}{wa^4}}$ where K_1 is tabulated for various ratios of a/b
<table>
<tr><td>a/b</td><td>1.0</td><td>1.1</td><td>1.2</td><td>1.5</td><td>2.0</td><td>3.0</td></tr>
<tr><td>K_1</td><td>10.2</td><td>11.3</td><td>12.6</td><td>17.0</td><td>27.8</td><td>57.0</td></tr>
</table>
Ref. 20</td></tr>
<tr><td>15. Rectangular flat plate with short edge a, long edge b, and thickness t; all edges fixed</td><td>15a. Uniform load w per unit area including own weight</td><td>$f = \frac{K_1}{2\pi}\sqrt{\frac{Dg}{wa^4}}$ where K_1 is tabulated for various ratios of a/b
<table>
<tr><td>a/b</td><td>1</td><td>0.9</td><td>0.8</td><td>0.6</td><td>0.4</td><td>0.2</td><td>0</td></tr>
<tr><td>K_1</td><td>36.0</td><td>32.7</td><td>29.9</td><td>25.9</td><td>23.6</td><td>22.6</td><td>22.4</td></tr>
</table>
Ref. 20</td></tr>
<tr><td>16. Rectangular flat plate with short edge a, long edge b, and thickness t; all edges simply supported</td><td>16a. Uniform load w per unit area including own weight</td><td>$f = \frac{K_n}{2\pi}\sqrt{\frac{Dg}{wa^4}}$ where $K_n = \pi^2\left[m_a^2 + \left(\frac{a}{b}\right)^2 m_b^2\right]$
<table>
<tr><td></td><td>a/b</td><td>1.0</td><td>0.8</td><td>0.6</td><td>0.4</td><td>0.2</td><td>0.0</td></tr>
<tr><td>$(m_a = 1, m_b = 1)$</td><td>K_1</td><td>19.7</td><td>16.2</td><td>13.4</td><td>11.5</td><td>10.3</td><td>9.87</td></tr>
<tr><td>$(m_a = 1, m_b = 2)$</td><td>K_2</td><td>49.3</td><td>35.1</td><td>24.1</td><td>16.2</td><td>11.5</td><td></td></tr>
<tr><td>$(m_a = 2, m_b = 1)$</td><td>K_3</td><td>49.3</td><td>45.8</td><td></td><td></td><td></td><td></td></tr>
<tr><td>$(m_a = 1, m_b = 3)$</td><td>K_4</td><td></td><td></td><td>41.9</td><td>24.1</td><td>13.4</td><td></td></tr>
</table>
Ref. 20</td></tr>
<tr><td>17. Rectangular flat plate with two edges a fixed, one edge b fixed, and one edge b simply supported</td><td>17a. Uniform load w per unit area including own weight</td><td>$f = \frac{K_1}{2\pi}\sqrt{\frac{Dg}{wa^4}}$ where K_1 is tabulated for various ratios of a/b
<table>
<tr><td>a/b</td><td>3.0</td><td>2.0</td><td>1.6</td><td>1.2</td><td>1.0</td><td>0.8</td><td>0.6</td><td>0.4</td><td>0.2</td><td>0</td></tr>
<tr><td>K_1</td><td>213</td><td>99</td><td>67</td><td>42.4</td><td>33.1</td><td>25.9</td><td>20.8</td><td>17.8</td><td>16.2</td><td>15.8</td></tr>
</table>
Ref. 22</td></tr>
</table>

TABLE 16.2 Natural Frequencies of Vibration of Various Systems

Natural Frequencies of Simple Translational Systems		
System	$\omega^2 =$	Remarks
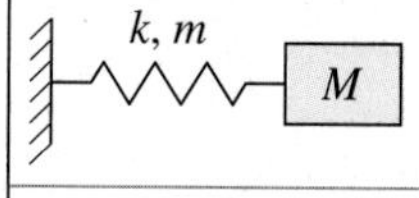	$\frac{k}{M+0.33\,m}$	
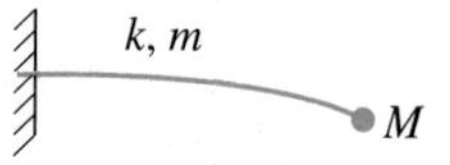	$\frac{k}{M+0.23\,m}$	Point mass on cantilever beam
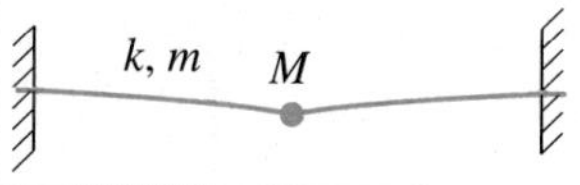	$\frac{k}{M+0.375\,m}$	Point mass at center of clamped beam; See Table 4.9c for k (applies to the three beam rows)
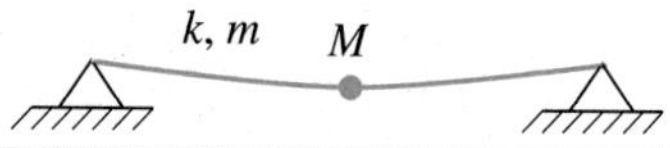	$\frac{k}{M+0.50\,m}$	Point mass at center of simply supported beam
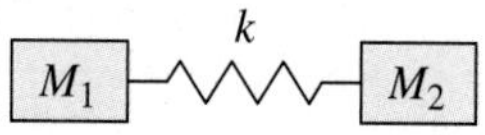	$k\left[\frac{1}{M_1}+\frac{1}{M_2}\right]$	
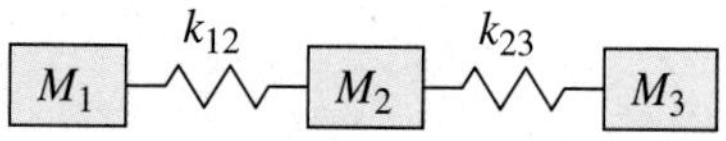		$2B=\frac{k_{12}}{M_1}+\frac{k_{23}}{M_3}+\frac{k_{12}+k_{23}}{M_2}$ $C=\frac{k_{12}+k_{23}}{M_1M_2M_3}(M_1+M_2+M_3)$
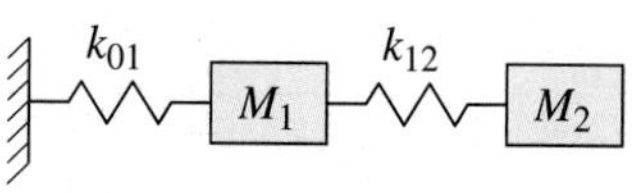	$B\pm\sqrt{B^2-C}$ (applies to the three rows)	$2B=\frac{k_{01}+k_{12}}{M_1}+\frac{k_{12}}{M_2}$ $C=\frac{k_{01}k_{12}}{M_1M_2}$
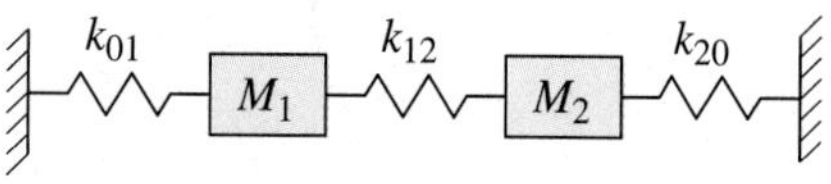		$2B=\frac{k_{01}+k_{12}}{M_1}+\frac{k_{12}+k_{20}}{M_2}$ $C=\frac{1}{M_1M_2}(k_{01}k_{12}+k_{12}k_{20}+k_{20}k_{01})$
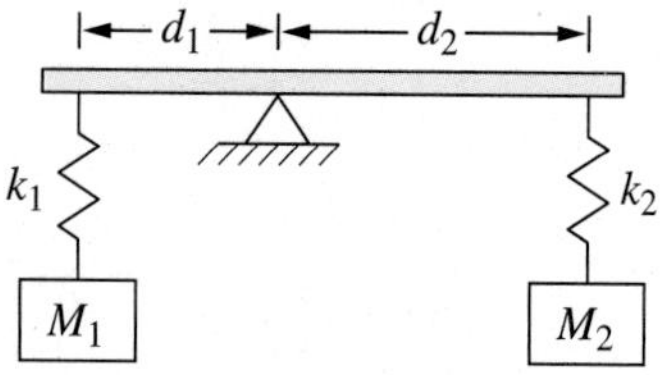	$\dfrac{\frac{1}{M_1}+\frac{1}{R^2M_2}}{\frac{1}{k_1}+\frac{1}{R^2k_2}}$	Inertia of lever negligible $R=\frac{d_2}{d_1}$
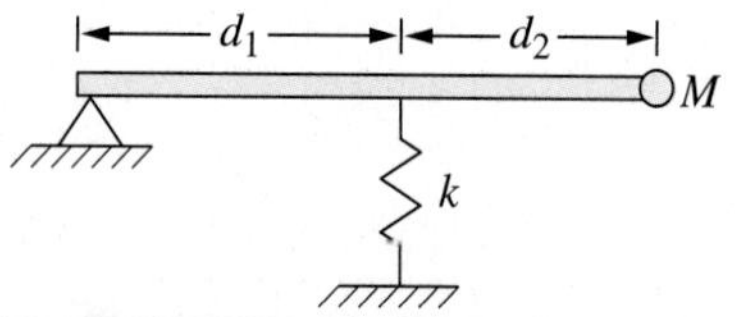	$\frac{k}{M}\left(\frac{d_1}{d_1+d_2}\right)^2$	Inertia of lever negligible

Symbols:
ω = circular natural frequency = $2\pi f$
M = mass
m = total mass of spring element
k = spring constant

(*Continued*)

TABLE 16.2 Natural Frequencies of Vibration of Various Systems (*Continued*)

Natural Frequencies of Simple Torsional Systems		
System	$\omega^2 =$	Remarks
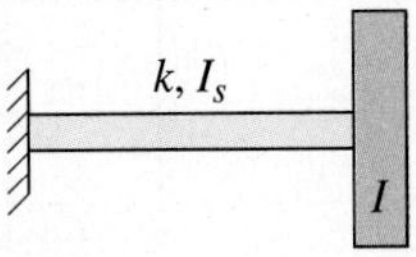	$\dfrac{k}{I + I_s/3}$	
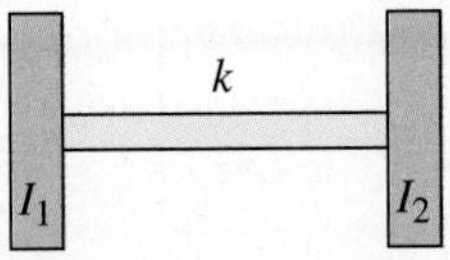	$k\left(\dfrac{1}{I_1} + \dfrac{1}{I_2}\right)$	
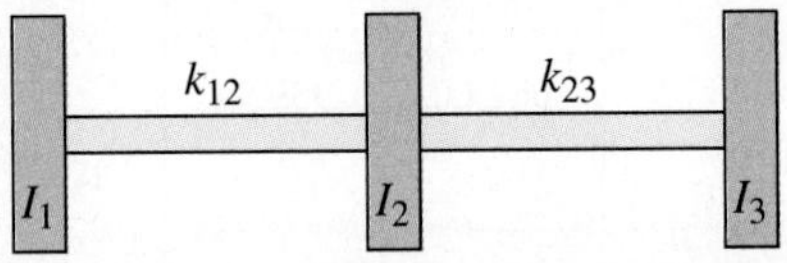	$B \pm \sqrt{B^2 - C}$	$2B = \dfrac{k_{12}}{I_1} + \dfrac{k_{23}}{I_3} + \dfrac{k_{12} + k_{23}}{I_2}$ $C = \dfrac{k_{12}k_{23}}{I_1 I_2 I_3}(I_1 + I_2 + I_3)$
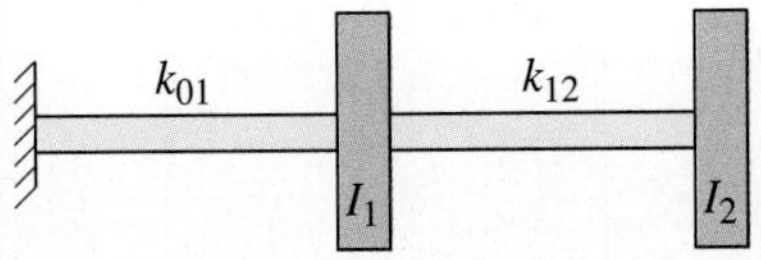		$2B = \dfrac{k_{01} + k_{12}}{I_1} + \dfrac{k_{12}}{I_2}$ $C = \dfrac{k_{01}k_{12}}{I_1 I_2}$
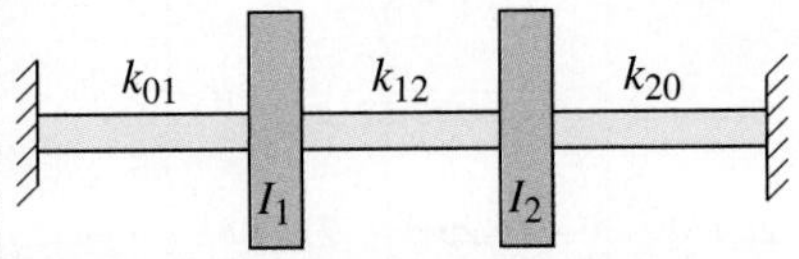		$2B = \dfrac{k_{01} + k_{12}}{I_1} + \dfrac{k_{12} + k_{20}}{I_2}$ $C = \dfrac{1}{I_1 I_2}(k_{01}k_{12} + k_{12}k_{20} + k_{20}k_{01})$
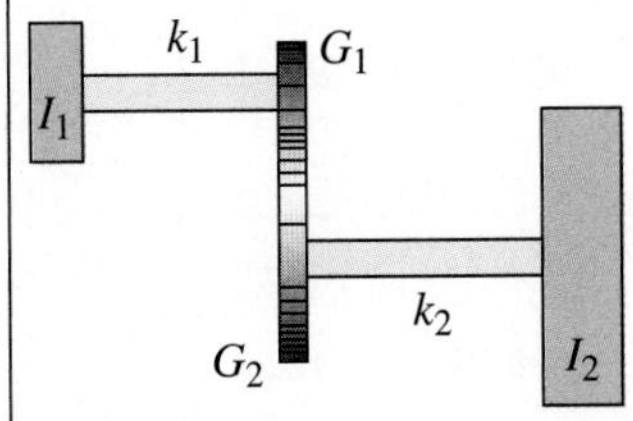	$\dfrac{\dfrac{1}{I_1} + \dfrac{1}{R^2 I_2}}{\dfrac{1}{k_1} + \dfrac{1}{R^2 k_2}}$	Inertia of gears G_1, G_2 assumed negligible $R = \dfrac{\text{number of teeth on gear 1}}{\text{number of teeth on gear 2}}$ $= \dfrac{\text{rpm of shaft 2}}{\text{rpm of shaft 1}}$

Symbols:
ω = circular natural frequency = $2\pi f$
k = torsional spring constant
I = polar mass moment of inertia
I_s = polar mass moment of inertia of entire shaft

(*Continued*)

TABLE 16.2 Natural Frequencies of Vibration of Various Systems (*Continued*)

Natural Frequencies of Beams in Flexure

$$f_n = C_n \frac{r}{L^2} \times 10^4 \times K_m$$

f_n = nth natural frequency, hertz

C_n = frequency constant listed in these tables

r = radius of gyration of cross section = $\sqrt{I/A}$, inches*

L = beam length, inches*

K_m = material constant (Table 16.4) = 1.00 for steel

Uniform-section beams	n				
	1	2	3	4	5
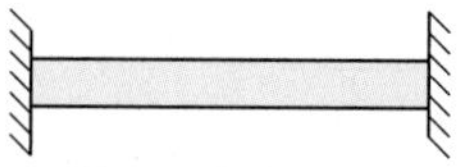 Clamped-clamped  Free-free	71.95	198.29	388.73	642.60	959.94
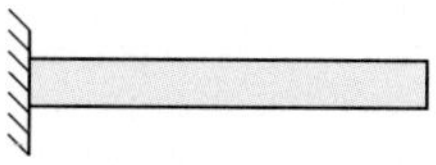 Clamped-free	11.30	70.85	198.30	388.73	642.60
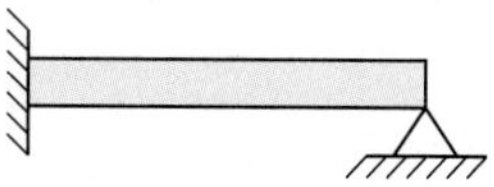 Clamped-hinged 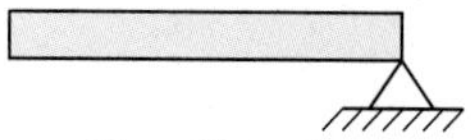 Free-hinged	49.57	160.65	335.17	573.20	874.65
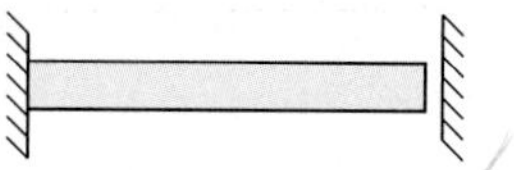 Clamped-guided 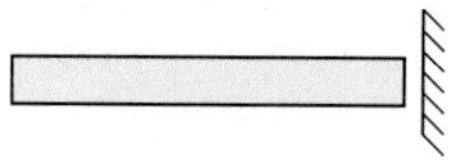 Free-guided	17.98	97.18	239.98	446.25	715.98
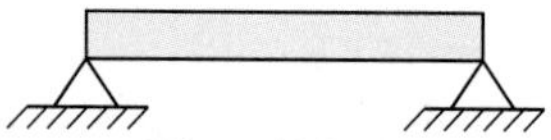 Hinged-hinged 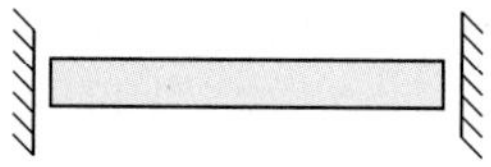 Guided-guided	31.73	126.93	285.60	507.73	793.33
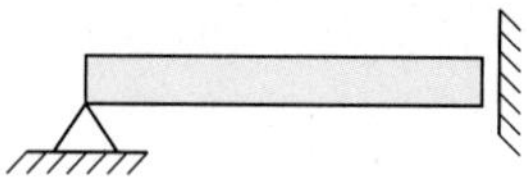 Hinged-guided	7.93	71.40	198.33	388.73	642.60

*For r and L in centimeters, use C_n values listed in table multiplied by 2.54.

(*Continued*)

TABLE 16.2 Natural Frequencies of Vibration of Various Systems (*Continued*)

Modal Properties for Flexural Vibrations of Uniform Beams

Natural frequency $\omega_n = 2\pi f_n = k_n^2 c_L r = k_n^2\sqrt{EI/\rho A}$

Wavelength $\lambda_n = 2\pi/k_n$

Wave velocity $c_n = \omega_n/k_n = k_n c_L r = k_n\sqrt{EI/\rho A} = \sqrt{\omega_n c_L r}$

For mode shapes $\phi_n(x)$ listed below, $\Phi_n = (1/L)\int_0^L \phi_n^2\,dx = 1$

Boundary Conditions					Roots of Frequency Equation					
$x=0$	$x=L$	Mode Shape $\phi_n(x)$	σ_n	Frequency Equation	k_1L	k_2L	k_3L	k_4L	k_5L	$k_nL,\ n>5$
Pinned	Pinned	$\sqrt{2}\sin(k_nx)$		$\sin(k_nL)=0$	π	2π	3π	4π	5π	$n\pi$
Clamped	Clamped	$\cosh(k_nx)-\cos(k_nx)$ $-\sigma_n[\sinh(k_nx)-\sin(k_nx)]$	$\dfrac{\cosh(k_nL)-\cos(k_nL)}{\sinh(k_nL)-\sin(k_nL)}$	$\cos(k_nL)\cosh(k_nL)=1$	4.73	7.85	11.00	14.14	17.29	$\approx\dfrac{2n+1}{2}\pi$
Free	Free	$\cosh(k_nx)+\cos(k_nx)$ $-\sigma_n[\sinh(k_nx)+\sin(k_nx)]$								
Clamped	Free	$\cosh(k_nx)-\cos(k_nx)$ $-\sigma_n[\sinh(k_nx)-\sin(k_nx)]$	$\dfrac{\sinh(k_nL)-\sin(k_nL)}{\cosh(k_nL)+\cos(k_nL)}$	$\cos(k_nL)\cosh(k_nL)=-1$	1.875	4.69	7.85	11.00	14.14	$\approx\dfrac{2n-1}{2}\pi$
Clamped	Pinned	$\cosh(k_nx)-\cos(k_nx)$ $-\sigma_n[\sinh(k_nx)-\sin(k_nx)]$	$\cot(k_nL)$	$\tan(k_nL)=\tanh(k_nL)$	3.93	7.07	10.21	13.35	16.49	$\approx\dfrac{4n+1}{4}\pi$
Free	Pinned	$\cosh(k_nx)+\cos(k_nx)$ $-\sigma_n[\sinh(k_nx)-\sin(k_nx)]$								

Symbols:
k_n = wave number
$c_L = \sqrt{E/\rho}$ = longitudinal wave velocity
$r = \sqrt{I/A}$ = radius of gyration
E = Young's modulus
A = cross-section area
L = beam length
ρ = material density
I = moment of inertia of A

(*Continued*)

TABLE 16.2 Natural Frequencies of Vibration of Various Systems (*Continued*)

Natural Frequencies of Uniform Beams on Multiple Equally Spaced Supports

$$f_n = C_n \frac{r}{L^2} \times 10^4 \times K_m$$

f_n = nth natural frequency, hertz

C_n = frequency constant listed in these tables

r = radius of gyration of cross section = $\sqrt{I/A}$, inches*

L = beam length, inches*

K_m = material constant (Table 16.4) = 1.00 for steel

	Number of spans	n = 1	2	3	4	5
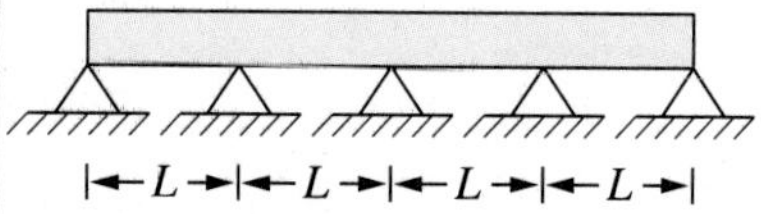 Ends simply supported	1	31.73	126.94	285.61	507.76	793.37
	2	31.73	49.59	126.94	160.66	285.61
	3	31.73	40.52	59.56	126.94	143.98
	4	31.73	37.02	49.59	63.99	126.94
	5	31.73	34.99	44.19	55.29	66.72
	6	31.73	34.32	40.52	49.59	59.56
	7	31.73	33.67	38.40	45.70	53.63
	8	31.73	33.02	37.02	42.70	49.59
	9	31.73	33.02	35.66	40.52	46.46
	10	31.73	33.02	34.99	39.10	44.19
	11	31.73	32.37	34.32	37.70	41.97
	12	31.73	32.37	34.32	37.02	40.52
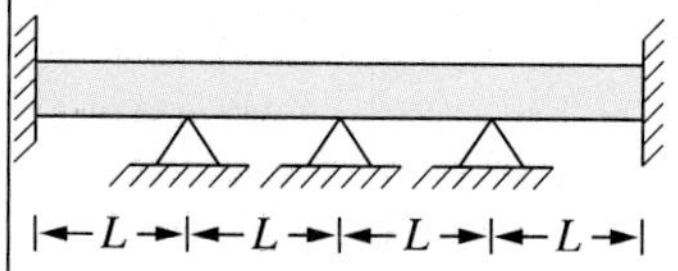 Ends clamped	1	72.36	198.34	388.75	642.63	959.98
	2	49.59	72.36	160.66	198.34	335.20
	3	40.52	59.56	72.36	143.98	178.25
	4	37.02	49.59	63.99	72.36	137.30
	5	34.99	44.19	55.29	66.72	72.36
	6	34.32	40.52	49.59	59.56	67.65
	7	33.67	38.40	45.70	53.63	62.20
	8	33.02	37.02	42.70	49.59	56.98
	9	33.02	35.66	40.52	46.46	52.81
	10	33.02	34.99	39.10	44.19	49.59
	11	32.37	34.32	37.70	41.97	47.23
	12	32.37	34.32	37.02	40.52	44.94
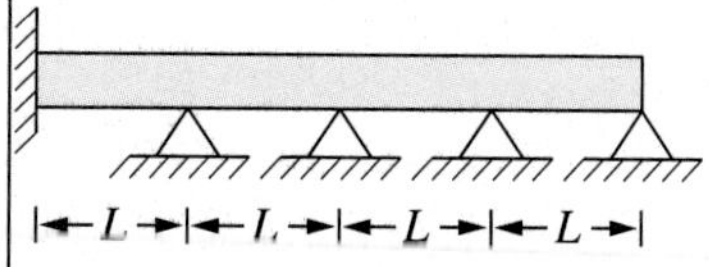 Ends clamped-supported	1	49.59	160.66	335.2	573.21	874.69
	2	37.02	63.99	137.30	185.85	301.05
	3	34.32	49.59	67.65	132.07	160.66
	4	33.02	42.70	56.98	69.51	129.49
	5	33.02	39.10	49.59	61.31	70.45
	6	32.37	37.02	44.94	54.46	63.99
	7	32.37	35.66	41.97	49.59	57.84
	8	31.73	34.99	39.81	45.70	53.63
	9	31.73	34.32	38.40	43.44	49.59
	10	31.73	33.67	37.02	41.24	46.46
	11	31.73	33.67	36.33	39.81	44.19
	12	31.73	33.02	35.66	39.10	42.70

*For r and L in centimeters, use C_n values multiplied by 2.54.

(*Continued*)

TABLE 16.2 Natural Frequencies of Vibration of Various Systems (*Continued*)

Natural Frequencies of Cantilever Plates

$$f_n = C_n \frac{h}{a^2} \times 10^4 \times K_m$$

f_n = nth natural frequency, hertz
C_n = frequency constant listed in table
h = plate thickness, inches*
a = plate dimension, as shown inches*
K_m = material constant (Table 16.4) = 1.00 for steel

Boundary conditions *F* = free *C* = clamped *S* = simply supported		*n*				
		1	2	3	4	5
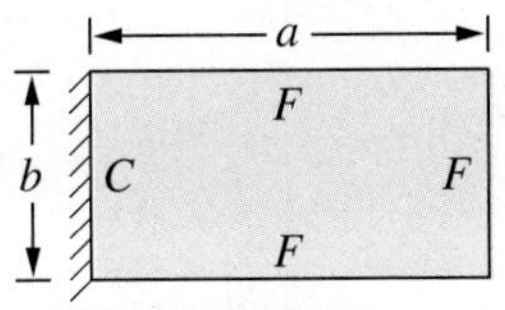	$a/b = ½$	3.41	5.23	9.98	21.36	24.18
	$a/b = 1$	3.40	8.32	26.71	20.86	30.32
	$a/b = 2$	3.38	14.52	91.92	21.02	47.39
	$a/b = 5$	3.36	33.79	548.60	20.94	103.3
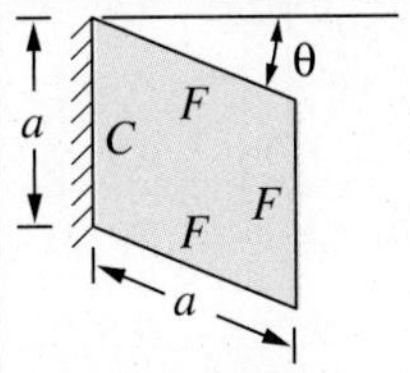	$\theta = 15°$	3.50	8.63			
	$\theta = 30°$	3.85	9.91			
	$\theta = 45°$	4.69	13.38			
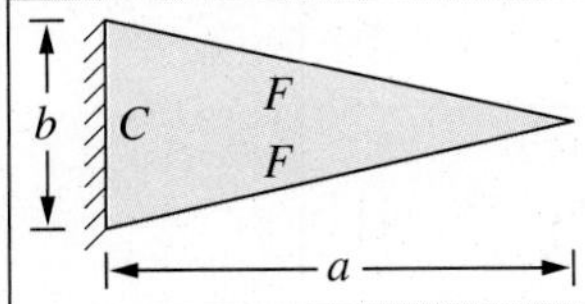	$a/b = 2$	6.7	28.6	56.6	137	
	$a/b = 4$	6.6	28.5	83.5	240	
	$a/b = 8$	6.6	28.4	146.1	457	
	$a/b = 14$	6.6	28.4	246	790	
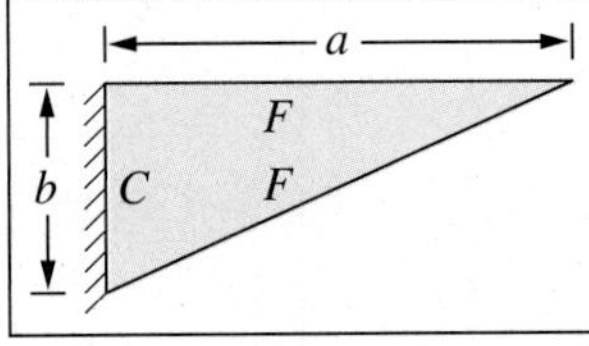	$a/b = 2$	5.5	23.6			
	$a/b = 4$	6.2	26.7			
	$a/b = 7$	6.4	28.1			

*For h and a in centimeters, use given C_n values multiplied by 2.54.

(*Continued*)

TABLE 16.2 Natural Frequencies of Vibration of Various Systems (*Continued*)

Natural Frequencies of Circular Plates

$$f = C\frac{h}{r^2} \times 10^4 \times K_m$$

f = natural frequency, hertz

C = frequency constant listed in table

h = plate thickness, inches*

r = plate radius, inches*

K_m = material constant (Table 16.4) = 1.00 for steel

Boundary conditions	Number of nodal circles	Number of Nodal Diameters: 0	1	2	3
Clamped at circumference 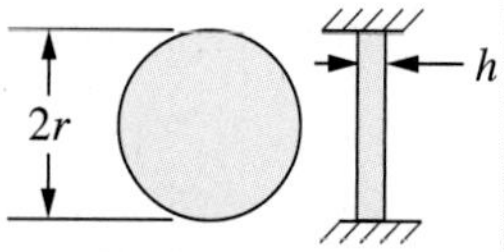	0	9.94	20.67	33.91	49.61
	1	38.66	59.12	82.22	107.9
	2	86.61	116.8	149.5	185.0
	3	153.8	193.5	235.9	281.1
Free 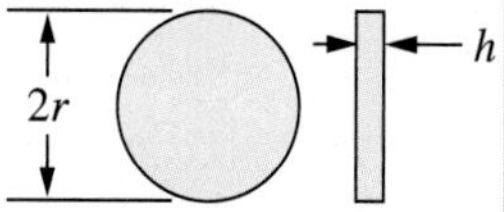	0		19.94	5.11	11.89
	1	8.83	58.19	34.27	51.43
	2	37.47	115.68	81.6	108.2
	3	85.35		149.7	186.1
Clamped at center; free at circumference 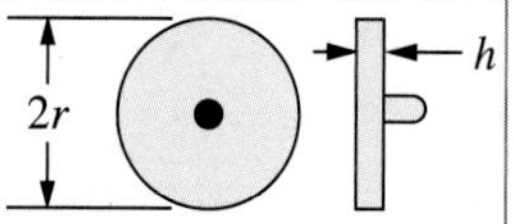	0	3.65			
	1	20.33			
	2	59.49			
	3	117.2			
Simply supported at circumference 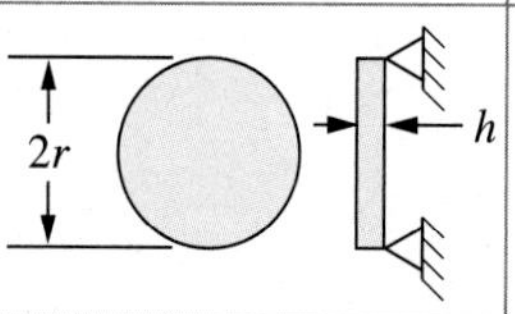	0	4.84	13.55	24.93	
	1	28.93	47.16	68.18	
	2	72.13	99.93	130.6	
	3	134.4	171.9	212.1	
Clamped at center; simply supported circumference 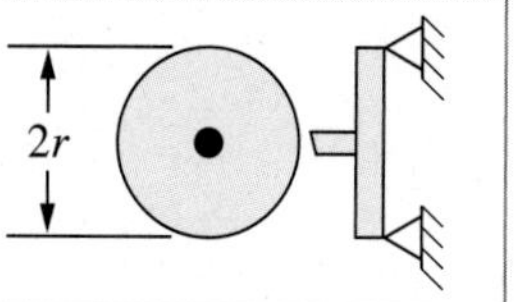	0	14.4			
	1	48.0			
Clamped at center and at circumference 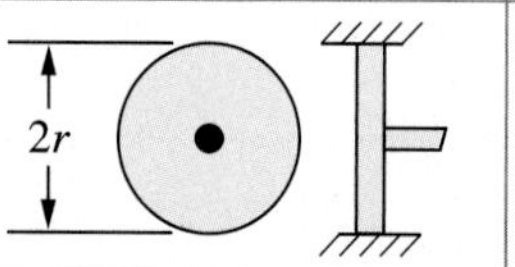	0	22.1			
	1	60.2			

Simply supported at radius a; free at circumference 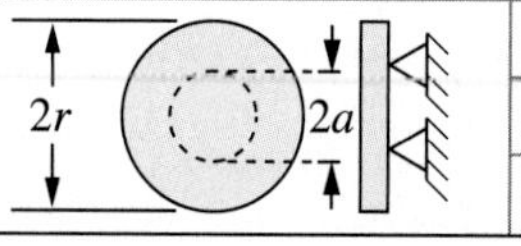

Fundamental Mode						
a/r	0	0.2	0.4	0.6	0.8	1.0
C	3.64	4.4	6.5	8.6	7.2	4.84

*For h and r in centimeters, use given C_n values multiplied by 2.54.

(*Continued*)

TABLE 16.2 Natural Frequencies of Vibration of Various Systems (*Continued*)

Natural Frequencies of Circular Membership

$$f = C\sqrt{T/h\rho r^2}$$

f = natural frequency, hertz
C = frequency constant listed in table
r = membrane radius, inches*
h = membrane thickness, inches*
T = tension at circumference, lb_f/in*
ρ = material density, lb_m/in^3 or $lb_f \cdot s^2/in^4$*

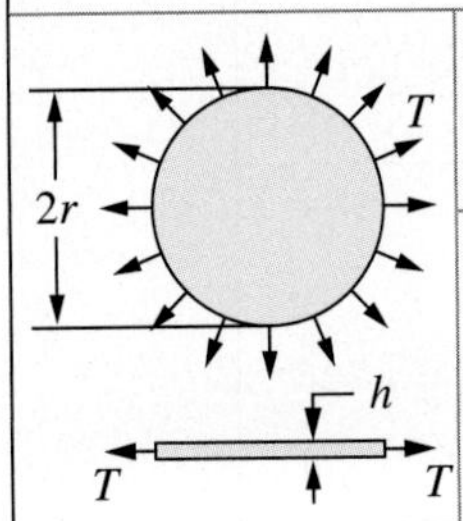

Number of Nodal Circles	Number of Nodal Diameters			
	1	2	3	4
1	0.383	0.610	0.817	1.015
2	0.887	1.116	1.340	1.553
3	1.377	1.619	1.849	2.071
4	1.877	2.120	2.355	2.582

Natural Frequencies of Cylindrical Shells

f_{ij} = natural frequency for mode shape ij, hertz
h = shell thickness, inches†
L = length, inches†
i = number of circumferential waves in mode shape
j = number of axial half-waves in mode shape
c_L = longitudinal wavespeed in shell material, in/s†
ν = Poission's ratio, dimensionless

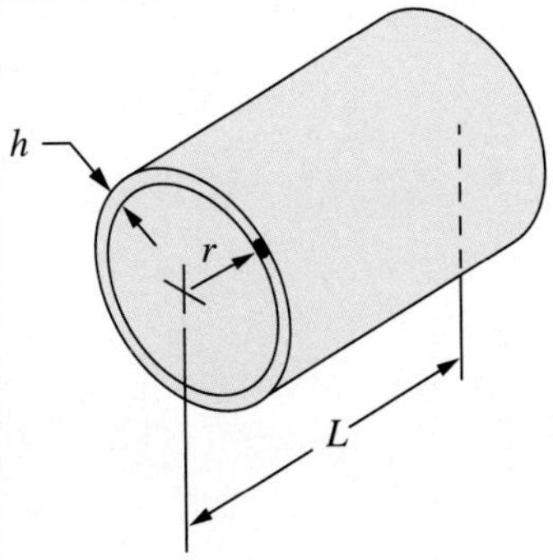

$$f_{ij} = \frac{\lambda_{ij}}{2\pi r}\frac{c_L}{\sqrt{1-\nu^2}}$$

λ_{ij}'s for different boundary conditions are given in the next table

*Any set of consistant units may be used that causes all units except s to cancel in the square root.
†Any other consistent set of units may be used. For example, h, r, L, in centimeters and c_L in centimeters per second.

(*Continued*)

TABLE 16.2 Natural Frequencies of Vibration of Various Systems (*Continued*)

Natural Frequencies of Cylindrical Shells (λ_{ij}) with Different Modes and Boundary Conditions		
Infinitely Long		
Axial modes:	$\lambda_{ij}=i\sqrt{\frac{1-\nu}{2}}$,	$i=1, 2, 3, \ldots$
Radial (extensional) modes:	$\lambda_{ij}=\sqrt{1+i^2}$,	$i=0, 1, 2, \ldots$
	Note: $i=0$ corresponding to breathing mode	
Radial-circumferential flexural modes:	$\lambda_{ij}=\frac{h/r}{\sqrt{12}}\frac{i(i^2-1)}{\sqrt{i^2+1}}$,	$i=2, 3, 4, \ldots$
Finite Length, Simply Supported Edges without Axial Constraint		
Torsional modes:	$\lambda_{ij}=\frac{j\pi r}{L}\sqrt{\frac{1-\nu}{2}}$	$i=0$; $j=1, 2, 3$
Axial modes*:	$\lambda_{ij}=\frac{j\pi r}{L}\sqrt{1-\nu^2}$	$i=0$; $j=1, 2, 3$
Radial modes*:	$\lambda_{ij}=1$	$i=0$; $j=1, 2, 3$
Bending modes*:	$\lambda_{ij}=\left(\frac{j\pi r}{L}\right)^2\sqrt{\frac{1-\nu}{2}}$	$i=1$; $j=1, 2, 3$
Radial-axial modes:	$\lambda_{ij}=\frac{\sqrt{(1-\nu^2)\beta_j^4+\frac{h^2}{12r^2}(i^2+\beta_j^2)^4}}{i^2+\beta_j^2}$	$\beta_j=j\pi r/L$; $i=2, 3, 4, \ldots$; $j=1, 2, 3, \ldots$

*Values given here apply for long shells, for which $L>8jr$.

TABLE 16.3 Spring Constants

Spring Constants of Beams		
	Translational (Force F)	Rotational (Moment M)
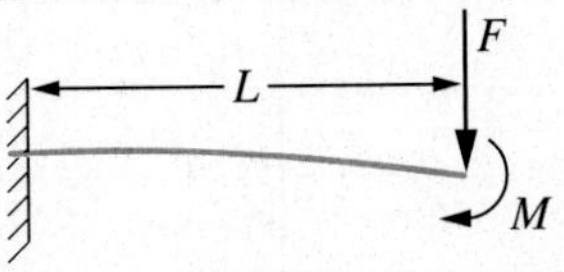	$k=\frac{3EI}{L^3}$	$k_r=\frac{EI}{L}$
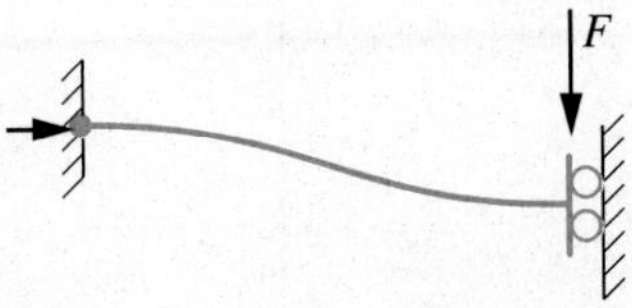	$k=\frac{12EI}{L^3}$	
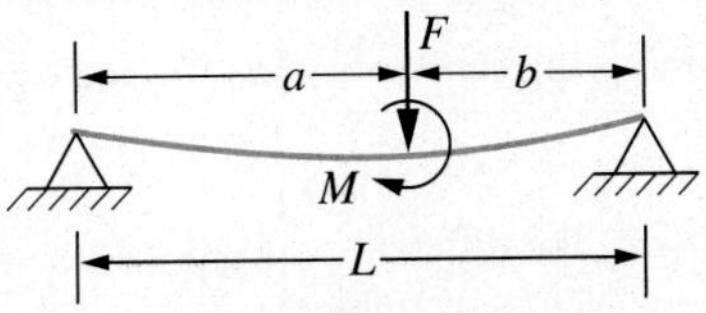	$k=\frac{3EIL}{(ab)^2}$ For $a=b$ $k=\frac{48EI}{L^3}$	$k_r=\frac{3EIL}{L^2-3ab}$ For $a=b$ $k_r=\frac{12EI}{L}$
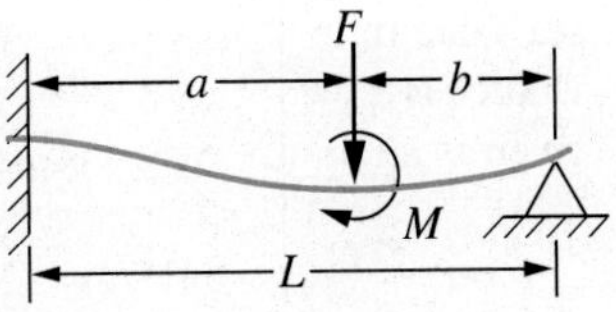	$k=\frac{12EIL^3}{a^3b^2(3L+b)}$ For $a=b$ $k=\frac{768EI}{7L^3}$	$k_r=\frac{4EIL^3/b}{4L^3-3b(L+a)^2}$ For $a=b$ $k_r=\frac{64EI}{5L}$
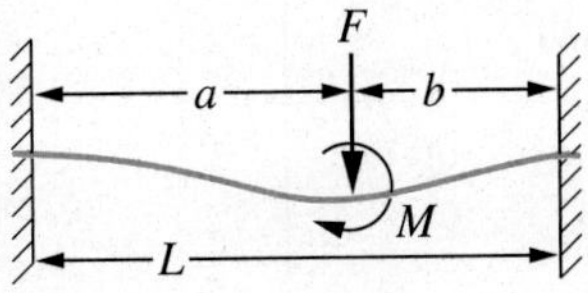	$k=\frac{3EIL^3}{(ab)^3}$ For $a=b$ $k=\frac{192EI}{L^3}$	$k_r=\frac{EIL^3}{ab(L^2-3ab)}$ For $a=b$ $k_r=\frac{16EI}{L}$
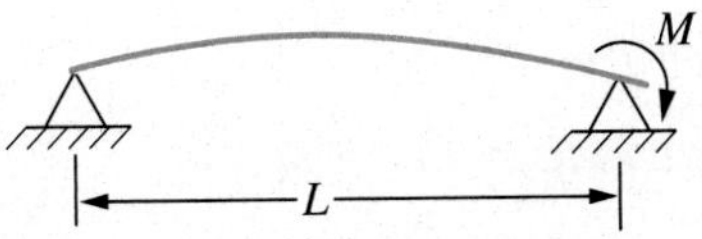		$k_r=\frac{3EI}{L}$
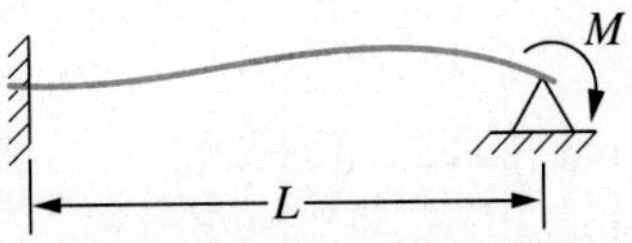		$k_r=\frac{4EI}{L}$

Symbols:

E = modulus of elasticity
I = centroidal moment of inertia of cross section
F = force
M = bending moment
$k=F/\delta$, translational spring constant
$k_r=M/\theta$, rotational spring constant
δ = lateral deflection at F
θ = flexural angle at M

(Continued)

TABLE 16.3 Spring Constants (*Continued*)

Torsion Springs

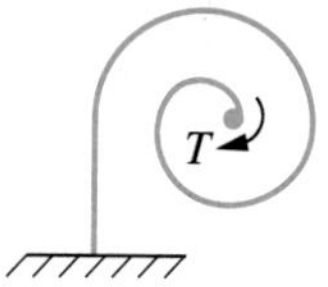

Spiral spring

$k_r = T/\phi = EI/L$

E = modulus of elasticity

L = total spring length

I = moment of inertia of cross section

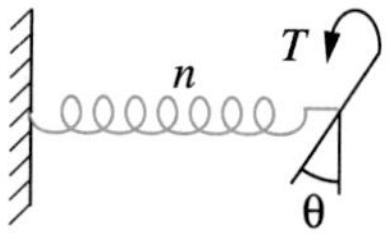

Helical spring

$$\frac{T}{\theta} = \frac{Ed^4}{64nD} = k_r$$

n is the number of coils.

See also the next table for more spring constants.

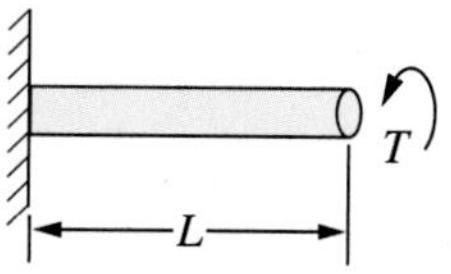

Uniform shaft

$k_r = GJ/L$

G = shear modulus

L = length

J = torsional constant $= \frac{1}{2}\pi r^4$

For different cross sections of the shaft see Table 10.1 where J is given as K

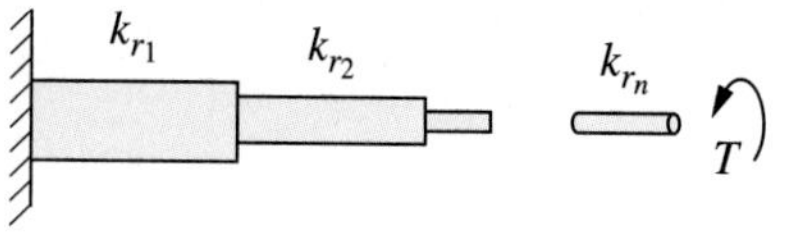

Stepped shaft

$1/k_r = 1/k_{r_1} + 1/k_{r_2} + \cdots 1/k_{r_n}$

$k_{r_i} = k_r$ of jth uniform part by itself

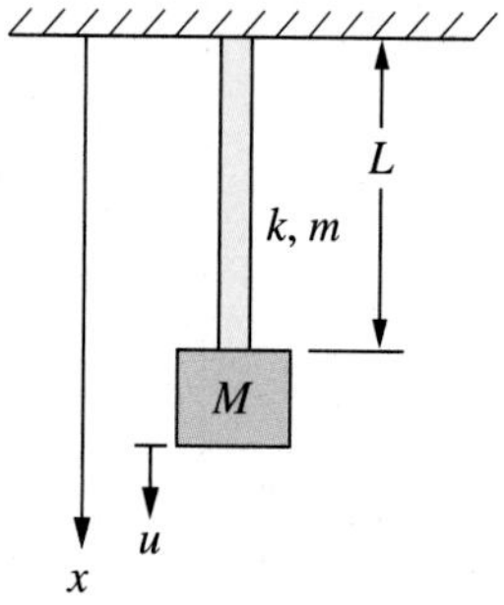

Mass on Rod or Spring of Static Stiffness k, Spring Total Mass m Uniformly Distributed

Frequency equation:

$$(M/m)(\omega\sqrt{m/k}) = \cot\,(\omega\sqrt{m/k})$$

$$\phi_n(x) = \sin\,(\omega_n x/L\sqrt{m/k})$$

Weighting function $\mu(x) = m/L + M\delta(x - L)$

M/m	$\omega_1\sqrt{m/k}$	$\omega_2\sqrt{m/k}$	$\omega_3\sqrt{m/k}$	$\omega_4\sqrt{m/k}$	$\omega_5\sqrt{m/k}$
0*	$\pi/2$	$3\pi/2$	$5\pi/2$	$7\pi/2$	$9\pi/2$
½	1.077	3.644	6.579	9.630	12.772
1	1.00063	3.435	6.437	9.426	12.645
∞†	0	π	2π	3π	4π

*Corresponds to spring free at $x = L$.
†Corresponds to spring clamped at $x = L$.

(*Continued*)

TABLE 16.3 Spring Constants (*Continued*)

Spring Constants of Round-Wire Helical Springs		
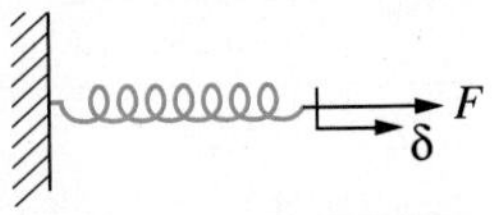	Axial loading	$\frac{F}{\delta} = k = \frac{Gd^4}{8nD^3}$
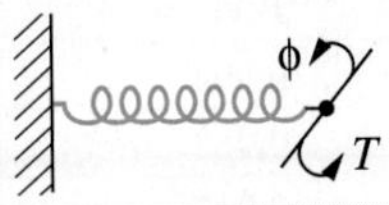	Torsion	$\frac{T}{\phi} = k_r = \frac{Ed^4}{64nD}$
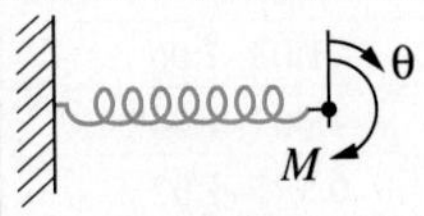	Bending	$\frac{M}{\theta} = k_b = \frac{2}{2+\nu} k_r$

Symbols:

F = axial force
δ = axial deflection
T = torque
ϕ = torsion angle
M = bending moment
θ = flexural angle

G = shear modulus
E = elastic modulus
ν = Poisson's ratio
d = wire diameter
D = mean coil diameter
n = number of coils

TABLE 16.4 Longitudinal Wavespeed and K_m for Engineering Materials

Material	Temperature, °C	Longitudinal Wavespeed $c_L = \sqrt{E/\rho}$ 10^5 in/s*	$K_m = \frac{(c_L)_{\text{material}}}{(c_L)_{\text{structural steel}}}$
Metals			
Aluminum	20	2.00	1.04
Beryllium	20	4.96	2.57
Brass, bronze	20	1.38–1.57	0.715–0.813
Copper	20	1.34	0.694
Copper	100	1.21	0.627
Copper	200	1.16	0.601
Cupro-nickel	20	1.47–1.61	0.762–0.834
Iron, cast	20	1.04–1.64	0.539–0.850
Lead	20	0.49	0.25
Magnesium	20	2.00	1.04
Monel metal	20	1.76	0.912
Nickel	20	1.24–1.90	0.642–0.984
Silver	20–100	1.03	0.534
Tin	20	0.98	0.51
Titanium	20	1.96–2.00	1.02–1.04
Titanium	90	1.90–1.96	0.984–1.02
Titanium	200	1.84–1.88	0.953–0.974
Titanium	325	1.68–1.75	0.870–0.907
Titanium	400	1.57–1.68	0.813–0.870
Zinc	20	1.46	0.756
Steel, typical	20	2.00	1.04

*To convert to centimeters per second, multiply by 2.54

(*Continued*)

TABLE 16.4 Longitudinal Wavespeed and K_m for Engineering Materials (*Continued*)

Material	Temperature, °C	Longitudinal Wavespeed $c_L = \sqrt{E/\rho}$ 10^5 in/s*	$K_m = \frac{(c_L)_{\text{material}}}{(c_L)_{\text{structural steel}}}$
Steel, structural	−200	2.00	1.04
	−100	1.97	1.02
	25	1.93	1.00
	150	1.92	0.994
	300	1.86	0.963
	400	1.77	0.917
Steel, stainless	−200	1.98–2.04	1.03–1.06
	−100	1.94–2.02	1.005–1.047
	25	1.92–2.06	0.995–1.07
	150	1.89–2.02	0.979–1.05
	300	1.84–1.98	0.953–0.979
	400	1.76–1.94	0.912–1.005
	600	1.70–1.87	0.881–0.969
	800	1.57–1.76	0.813–0.912
Aluminum Alloys	−200	2.08–2.13	1.078–1.104
	−100	2.01–2.06	1.04–1.07
	25	1.97–2.02	1.02–1.05
	200	1.79–1.87	0.927–0.969
Plastics†			
Cellulose acetate	25	0.40	0.21
Methyl methacrylate	25	0.60	0.31
Nylon 6, 6	25	0.54	0.78
Vinyl chloride	25	0.26	0.13
Polyethylene	25	0.41	0.21
Polypropylene	25	0.42	0.22
Polystyrene	25	0.68	0.35
Teflon	25	0.17	0.088
Epoxy resin	25	0.61	0.32
Phenolic resin	25	0.57	0.30
Polyester resin	25	0.61	0.32
Glass-fiber-epoxy laminate	25	1.72	0.89
Glass-fiber-phenolic laminate	25	1.44	0.75
Glass-fiber-polyester laminate	25	1.25	0.65
Other			
Concrete	25	1.44–1.80	0.75–0.93
Cork	25	0.20	0.104
Glass	25	1.97–2.36	1.02–1.22
Granite	25	2.36	1.22
Marble	25	1.50	0.78
Plywood	25	0.84	0.44
Woods, along fibers		1.32–1.92	0.68–0.99
Woods, across fibers		0.48–0.54	0.25–0.28

*To convert to centimeters per second, multiply by 2.54

†Values given here are typical. Wide variations may occur with composition, temperature, and frequency.

16.9 REFERENCES

1. Hodkinson, B., "Rotating Discs of Conical Profile," *Engineering*, 115: 1, 1923.
2. Rushing, F. C., "Determination of Stresses in Rotating Disks of Conical Profile," *Trans. ASME*, 53: 91, 1931.
3. Stodola, A., *Steam and Gas Turbines*, 6th ed., McGraw-Hill, 1927 (transl. by L. C. Loewenstein).
4. Loewenstein, L. C., *Marks' Mechanical Engineers' Handbook*, McGraw-Hill, 1930.
5. Donnell, L. H., "Longitudinal Wave Transmission and Impact," *Trans. ASME*, 52(1): 153, 1930.
6. Love, A. E. H., *Mathematical Theory of Elasticity*, 2nd ed., Cambridge University Press, 1906.
7. Goodier, J. N., "Thermal Stress," *ASME J. Appl. Mech.*, 4(1): 1937.
8. Maulbetsch, J. L., "Thermal Stresses in Plates," *ASME J. Appl. Mech.*, 2(4): 1935.
9. Kent, C. H., "Thermal Stresses in Spheres and Cylinders Produced by Temperatures Varying with Time," *Trans. ASME*, 54(18): 185, 1932.
10. Timoshenko, S., *Theory of Elasticity*, McGraw-Hill, 1934.
11. Barker, L. H., "The Calculation of Temperature Stresses in Tubes," *Engineering*, 124: 443, 1927.
12. Robinson, E. L., "Bursting Tests of Steam-turbine Disk Wheels," *Trans. ASME*, 66(5): 373, 1944.
13. Holms, A. G., and J. E. Jenkins, "Effect of Strength and Ductility on Burst Characteristics of Rotating Disks," *Natl. Adv. Comm. Aeron.*, Tech. Note 1667, 1948.
14. Goodier, J. N., "Thermal Stress and Deformation," *ASME J. Appl. Mech.*, 24(3): 1957.
15. Eichmann, E. S., "Note on the Maximum Effect of a Moving Force on a Simple Beam," *ASME J. Appl. Mech.*, 20(4): 1953.
16. Ayre, R. S., L. S. Jacobsen, and C. S. Hsu, "Transverse Vibration of One- and Two-span Beams under Moving Mass-Load," *Proc. 1st U.S. Nail. Congr. Appl. Mech.*, 1952.
17. Burr, A. H., "Longitudinal and Torsional Impact in a Uniform Bar with a Rigid Body at One End," *ASME J. Appl. Mech.*, 17(2): 1950.
18. Schwieger, H., "A Simple Calculation of the Transverse Impact on Beams and Its Experimental Verification," *J. Soc. Exp. Mech.*, 5(11): 1965.
19. Thomson, W. T., and M. D. Dahleh, *Theory of Vibrations with Applications*, 5th ed., Prentice-Hall, 1998.
20. Leissa, A. W., Vibration of Plates, NASA SP-160, National Aeronautics and Space Administration, 1969.
21. Leissa, A. W., Vibration of Shells, NASA SP-288, National Aeronautics and Space Administration, 1973.
22. Huang, T. C., "Eigenvalues and Modifying Quotients of Vibration of Beams, and Eigenfunctions of Vibration of Beams," *Univ. Wis. Eng. Exp. Sta.*, Repts. Nos. 25 and 26, 1964.
23. Jones, R., "An Approximate Expression for the Fundamental Frequency of Vibration of Elastic Plates," *J. Sound Vib.*, 38(4): 1975.
24. Blevins, R. D., *Formulas for Natural Frequency and Mode Shape*, Van Nostrand Reinhold, 1979.
25. Fridman, Y. B. (ed.), *Strength and Deformation in Nonuniform Temperature Fields*, transl. from the Russian, Consultants Bureau, 1964.
26. Johns, D. J., *Thermal Stress Analyses*, Pergamon Press, 1965.
27. Boley, B. A., and J. H. Weiner, *Theory of Thermal Stresses*, John Wiley & Sons, 1960.
28. Burgreen, D., *Elements of Thermal Stress Analysis*, C. P. Press, 1971.
29. Nowacki, W., *Thermoelasticity*, 2nd ed., English transl. by H. Zorski, Pergamon Press, 1986.
30. Timoshenko, S., *Vibration Problems in Engineering*, Van Nostrand, 1955.
31. Rothbart, H. A., and T. H. Brown Jr., *Mechanical Design Handbook*, 2nd ed., McGraw-Hill, 2006.
32. Hoffman, J. D., *Numerical Methods for Engineers and Scientists*, McGraw-Hill, 2001.
33. Hall, A. S., A. R. Holowenko, and H. G. Laughlin, *Schaum's Outlines Machine Design*, McGraw-Hill, 1961.
34. Barton, M. V. (ed.), *Shock and Structural Response*, American Society of Mechanical Engineers, New York, 1960.

CHAPTER 17

Stress Concentration

When a large stress gradient occurs in a small, localized area of a structure, the high stress is referred to as a *stress concentration*. Near changes in geometry of a loaded structure, the *flow* of stress is interfered with, causing high stress gradients where the maximum stress and strain may greatly exceed the average or nominal values based on simple calculations. Contact stresses, as discussed in Chap. 14, also exhibit, near the point of contact, high stress gradients, which subside quickly as one moves away from the contact area. Thus, the two most common occurrences of stress concentrations are due to (1) discontinuities in continuum and (2) contact forces. Discontinuities in continuum include changes in geometry and material properties. This chapter is devoted to geometric changes.

Rapid geometry changes disrupt the smooth flow of stresses through the structure between load application areas. Plates in tension or bending with holes, notches, steps, etc., are simple examples involving direct normal stresses. Shafts in tension, bending, and torsion, with holes, notches, steps, keyways, etc., are simple examples involving direct and bending normal stresses and torsional shear stresses. More complicated geometries must be analyzed either by experimental or numerical techniques such as the finite element method. Other, less obvious, geometry changes include rough surface finishes and external and internal cracks.

Changes in material properties are discussed in Chap. 7 and demonstrated in an example where a change in modulus of elasticity drastically changed the stress distribution. Changes in material properties can occur both at macroscopic and microscopic levels which include alloy formulation, grain size and orientation, foreign materials, etc.

17.1 STATIC STRESS AND STRAIN CONCENTRATION FACTORS

Consider the plate shown in Fig. 17.1, loaded in tension by a force per unit area, σ. Although not drawn to scale, consider that the outer dimensions of the plate are infinite compared with the diameter of the hole, $2a$. It can be shown, from linear elasticity, that the tangential stress throughout the plate is given by (see Ref. 60)

$$\sigma_\theta = \frac{\sigma}{2}\left[1+\frac{a^2}{r^2}-\left(1+3\frac{a^4}{r^4}\right)\cos 2\theta\right] \tag{17.1-1}$$

The maximum stress is $\sigma_\theta = 3\sigma$ at $r = a$ and $\theta = \pm 90°$. Figure 17.2 shows how the tangential stress varies along the x and y axes of the plate. For the top (and bottom) of the hole, we see the stress gradient is extremely large compared with the nominal stress, and hence the term *stress concentration* applies. Along the surface of the hole, the tangential stress is $-\sigma$ at $\theta = 0°$ and 180°, and increases, as θ increases, to 3σ at $\theta = 90°$ and 270°.

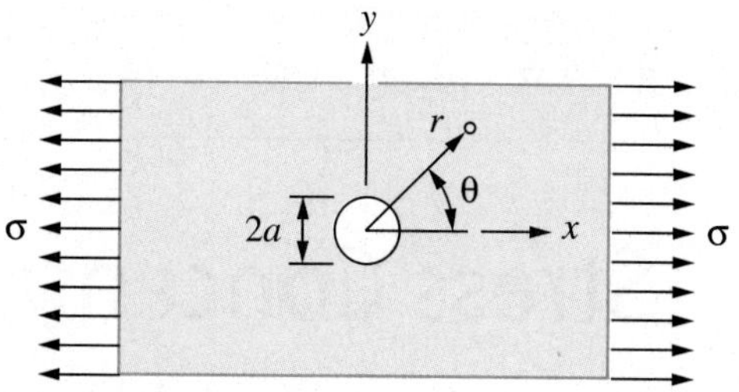

Figure 17.1 Circular hole in a plate loaded in tension.

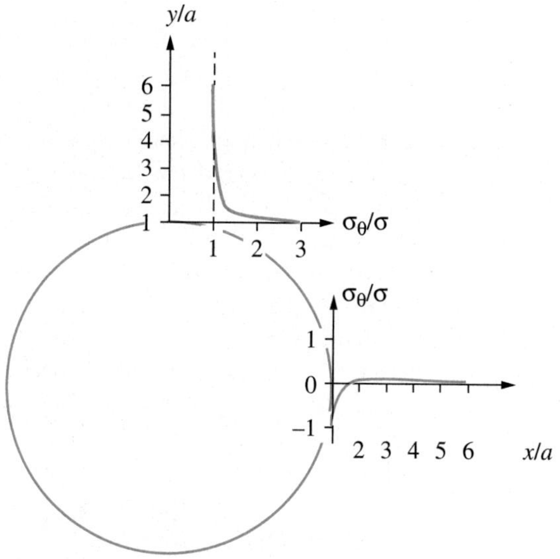

Figure 17.2 Tangential stress distribution for $\theta = 0°$ and $90°$.

The *static stress concentration factor* in the elastic range, K_t, is defined as the ratio of the maximum stress, σ_{max}, to the nominal stress, σ_{nom}. That is,

$$K_t = \frac{\sigma_{max}}{\sigma_{nom}} \tag{17.1-2}$$

For the infinite plate containing a hole and loaded in tension, $\sigma_{nom} = \sigma$, $\sigma_{max} = 3\sigma$, and thus $K_t = 3$.*

The analysis of the plate in tension with a hole just given is for a very wide plate (infinite in the limit). As the width of the plate decreases, the maximum stress becomes less than three times the nominal stress at the zone containing the hole. Figure 17.3(*a*) shows a plate of thickness $t = 0.125$ in, width $D = 1.50$ in, with a hole of diameter $2r = 0.50$ in, and an applied uniform stress of $\sigma_0 = 320$ psi.

A photoelastic† model is shown in Fig. 17.3(*b*). From a photoelastic analysis, the stresses at points *a*, *b*, and *c* are found to be

zone *A*–*A*: $\sigma_a = 320$ psi

zone *B*–*B*: $\sigma_b = 280$ psi, $\sigma_c = 1130$ psi

*See Case 7a of Table 17.1. As $2a/D \to 0$, $K_t \to 3.00$.

†Photoelasticity is discussed at some length in Ref. 60.

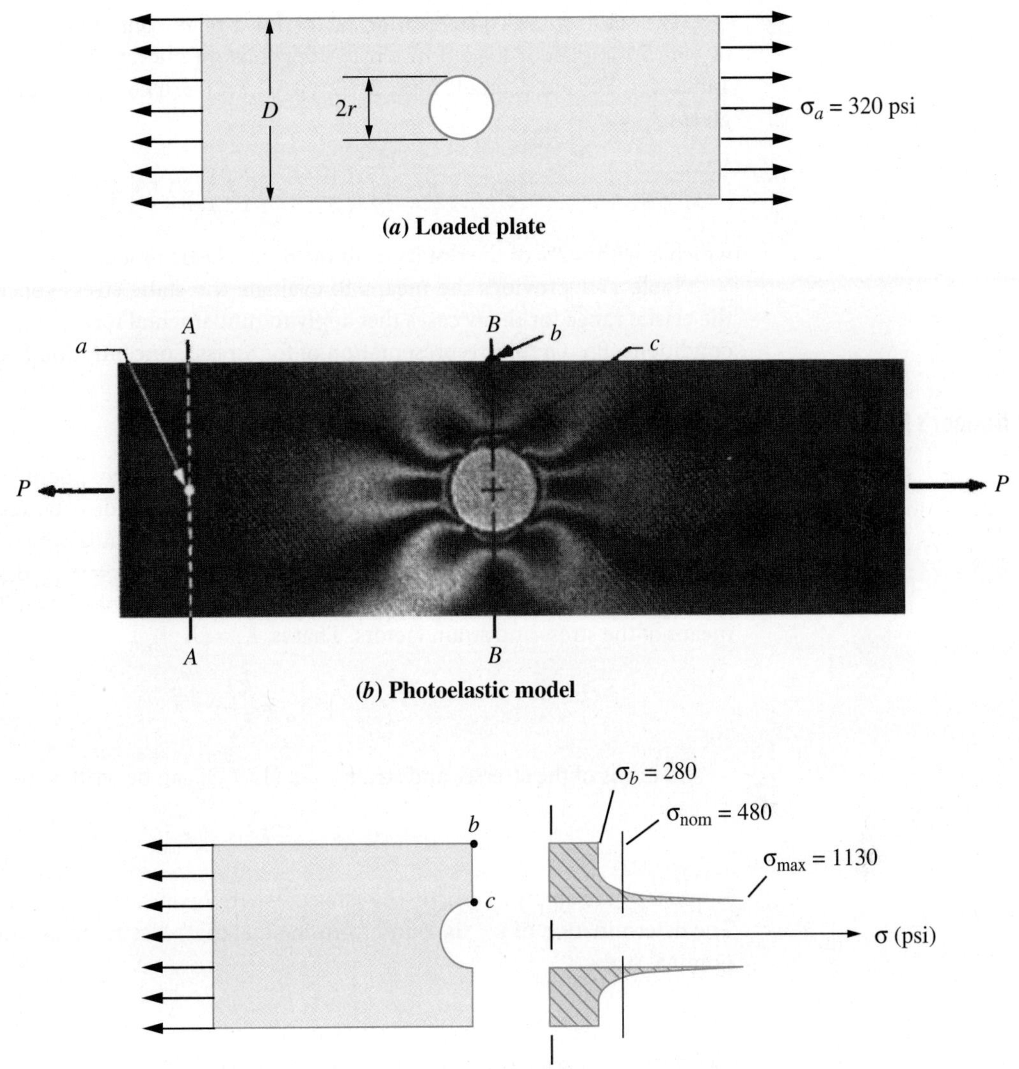

(a) Loaded plate

(b) Photoelastic model

(c) Stress distribution

Figure 17.3 Stress distribution for a plate in tension containing a centrally located hole.

The nominal stress in zone *B–B* is

$$\sigma_{nom} = \frac{D}{D-2r}\sigma_0 = \frac{1.50}{1.50-0.5}320 = 480 \text{ psi}$$

If the stress was uniform from b to c, the stress would be 480 psi. However, the photoelastic analysis shows the stress to be nonuniform, ranging from 280 psi at b to a maximum stress at c of 1130 psi. Thus, for this example, the stress concentration factor is found to be

$$K_t = \frac{\sigma_{max}}{\sigma_{nom}} = \frac{1130}{480} = 2.35$$

The static stress concentration factor for a plate containing a centrally located hole in which the plate is loaded in tension depends on the ratio $2r/D$ as given for case 7a of Table 17.1. For our example here, $2r/D = 0.5/1.5 = 1/3$. The equation for K_t from Table 17.1 gives

$$K_t = 3.00 - 3.13\left(\frac{1}{3}\right) + 3.66\left(\frac{1}{3}\right)^2 - 1.53\left(\frac{1}{3}\right)^3 = 2.31$$

which is within 2% of the results from the photoelastic model.

Table 17.1 provides the means to evaluate the static stress concentration factors in the elastic range for many cases that apply to fundamental forms of geometry and loading conditions. For Graphic representation of for Stress Concentration Factors see Table 17.2.

Neuber's Formula for Nonlinear Material Behavior

If the load on a structure exceeds the value for which the maximum stress at a stress concentration equals the elastic limit of the material, the stress distribution changes from that within the elastic range. Neuber (Ref. 61) presented a formula which includes stress and strain. Defining an effective stress concentration factor, $K_\sigma = \sigma_{max}/\sigma_{nom}$, and an effective strain concentration factor, $K_\varepsilon = \varepsilon_{max}/\varepsilon_{nom}$, Neuber established that K_t is the geometric means of the stress and strain factors. That is, $K_t = (K_\sigma K_\varepsilon)^{1/2}$, or

$$K_\sigma = \frac{K_t^2}{K_\varepsilon} \tag{17.1-3}$$

In terms of the stresses and strains, Eq. (17.1-3) can be written as

$$\sigma_{max}\varepsilon_{max} = K_t^2 \sigma_{nom}\varepsilon_{nom} \tag{17.1-4}$$

K_t and σ_{nom} are obtained exactly the same as when the max stress is within the elastic range. The determination of ε_{nom} is found from the material's elastic stress–strain curve using the nominal stress.

EXAMPLE A circular shaft with a square shoulder and fillet is undergoing bending (case 17b of Table 17.1). A bending moment of 500 N-m is being transmitted at the fillet section. For the shaft, $D = 50$ mm, $h = 9$ mm, and $r = 3$ mm. The stress–strain data for the shaft material is tabulated below and plotted in Fig. 17.4. Determine the maximum stress in the shaft.

ε, 10^{-5}	0	25	50	75	100	125	150	175	200	225	250	275	300	325	350	375	400
σ, (MPa)	0	50	100	150	200	235	252	263	267	272	276	279	282	285	287	289	290

SOLUTION From the given dimensions, $h/r = 9/3 = 3$. From case 17b of Table 17.1,

$$C_1 = 1.225 + 0.831\sqrt{3} - 0.010(3) = 2.634$$
$$C_2 = -3.790 + 0.958\sqrt{3} - 0.257(3) = -2.902$$
$$C_3 = 7.374 - 4.834\sqrt{3} + 0.862(3) = 1.587$$
$$C_4 = -3.809 + 3.046 - 0.595(3) = -0.3182$$

With $2h/D = 18/50 = 0.36$,

$$K_t = 2.634 - 2.902(0.36) + 1.587(0.36)^2 - 0.3182(0.36)^3 = 1.780$$

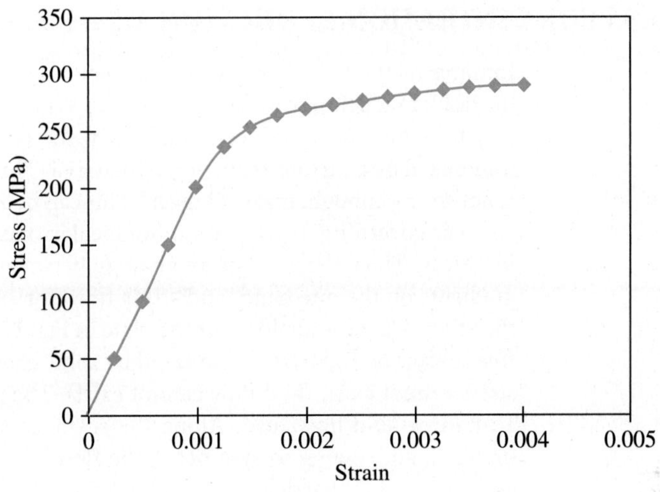

Figure 17.4

The nominal stress at the minor radius of the step shaft is

$$\sigma_{nom} = \frac{32M}{\pi(D-2h)^3} = \frac{32(500)}{\pi[50-2(9)]^3(10^{-3})^3} = 155.4(10^6)\ \text{N/m}^2 = 155.4\ \text{MPa}$$

If σ_{max} is in the elastic range, then

$$\sigma_{max} = K_t\sigma_{nom} = 1.780(155.4) = 276.6\ \text{MPa}$$

However, as one can see from the stress–strain plot that this exceeds the elastic limit of 200 MPa. Thus, σ_{max} must be determined from Neuber's equation.

The modulus of elasticity in the elastic range of the material is $E = 20$ GPa. Thus, the nominal strain is found to be $\varepsilon_{nom} = \sigma_{nom}/E = 155.4(10^6)/20(10^9) = 77.7(10^{-5})$. Thus,

$$K_t^2\sigma_{nom}\varepsilon_{nom} = (1.780)^2(155.4)(77.7)(10^{-5}) = 0.3826\ \text{MPa}$$

From the tabulated data, the product $\sigma\,\varepsilon$ can be tabulated as a function of σ. This results in the following:

σ (MPa)	0	50	100	150	200	235	252	263	267
$\sigma\,\varepsilon$ (MPa)	0	0.0125	0.05	0.1125	0.2	0.29375	0.378	0.46025	0.534

σ (MPa)	272	276	279	282	285	287	289	290
$\sigma\varepsilon$ (MPa)	0.612	0.69	0.76725	0.846	0.92625	1.0045	1.08375	1.16

Since, based on Eq. (17.1-4), we are looking for the value of $\sigma_{max}\varepsilon_{max} = 0.3826$, we will interpolate $\sigma\,\varepsilon$ between 0.378 and 0.46025. Thus,

$$\frac{\sigma_{max} - 252}{0.3826 - 0.378} = \frac{263 - 252}{0.46025 - 0.378}$$

This yields $\sigma_{max} = 252.6$ MPa.

For dynamic problems where loading is cycling, the fatigue stress concentration factor is more appropriate to use. See Sec. 3.10 for a discussion of this.

17.2 STRESS CONCENTRATION REDUCTION METHODS

Intuitive methods such as the *flow analogy* are sometimes helpful to the analyst faced with the task of reducing stress concentrations. When dealing with a situation where it is necessary to reduce the cross section abruptly, the resulting stress concentration can often be minimized by a further reduction of material. This is contrary to the common advice "if it is not strong enough, make it bigger." This can be explained by examining the flow analogy.

The governing field equations for ideal irrotational fluid flow are quite similar to those for stress. Thus, there exists an analogy between fluid flow lines, velocity, and pressure gradients on the one hand, and stress trajectories, magnitudes, and principal stresses on the other. The flow analogy for the plate in Fig. 17.3 is shown in Fig. 17.5(a), where stress-free surface boundaries are replaced by solid-channel boundaries for the fluid (wherever stress cannot exist, fluid flow cannot exist). The uniformly applied loads are replaced by a uniform fluid flow field. Along the entrance section A–A of Fig. 17.5(a), the flow is uniform, and, owing to symmetry, the flow is uniform at the exit of the channel. However, as the fluid particles approach section B–B, the streamlines need to adjust to move around the circular obstacle. In order to accomplish this, particles close to streamline 1

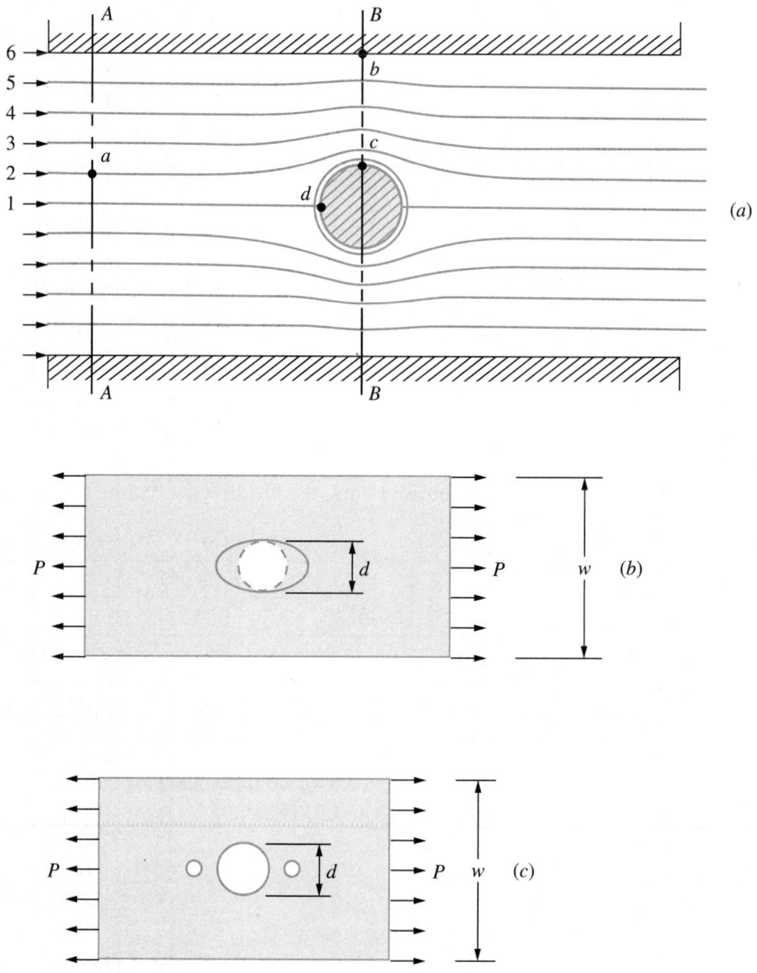

Figure 17.5 Stress–flow analogy.

must make the greatest adjustment and must accelerate until they reach section *B–B*, where they reach maximum velocity, and then decelerate to their original uniform velocity some distance from *B–B*. Thus, the velocity at point *c* is the maximum. The compaction of the streamlines at *c* will lead to the development of a pressure gradient, which will actually cause the velocity of point *b* to be less than that of the incoming velocity of streamline 6 at *A–A*. Note also that when a particle on streamline 1 reaches point *d*, the particle theoretically takes on a velocity perpendicular to the net flow. This analogy agrees with that of the plate loaded in tension with a centrally located hole. The stress is a maximum at the edge of the hole corresponding to point *c* in Fig. 17.5(*a*). The stress in the plate corresponding to point *b* is lower than the applied stress, and for point *d* the stress in the plate is compressive perpendicular to the axial direction.

This analogy can be used to suggest improvements to reduce stress concentrations. For example, for the plate with the hole, the hole can be elongated to an ellipse as shown in Fig. 17.5(*b*), which will improve the flow transition into section *B–B* (note that this is a reduction of material). An ellipse, however, is not a practical solution, but it can effectively be approximated by drilling two smaller relief holes in line and in close proximity to the original hole as shown in Fig. 17.5(*c*). The material between the holes, provided the holes are close, will be a stagnation area where the flow (stresses) will be low. Consequently, the configuration acts much like that of an elliptical hole.

At first, this might not seem to make sense, since this is a reduction of more material—and if one hole weakens the part, obviously more holes will make things worse. One must keep in mind that the first hole increased the stress in two ways: (1) by reducing the cross-sectional area and (2) by changing the shape of the stress distribution. The two additional holes in Fig. 17.5(*c*) do not change the area reduction unless they are larger than the original hole. However, as stated, the additional holes will improve the flow transition, and consequently reduce the stress concentration. Another way of improving the plate with the hole is to elongate the hole in the axial direction to a slot.

Some other examples of situations where stress concentrations occur and possible methods of improvements are given in Fig. 17.6. Note that in each case, improvement is made by reducing material.

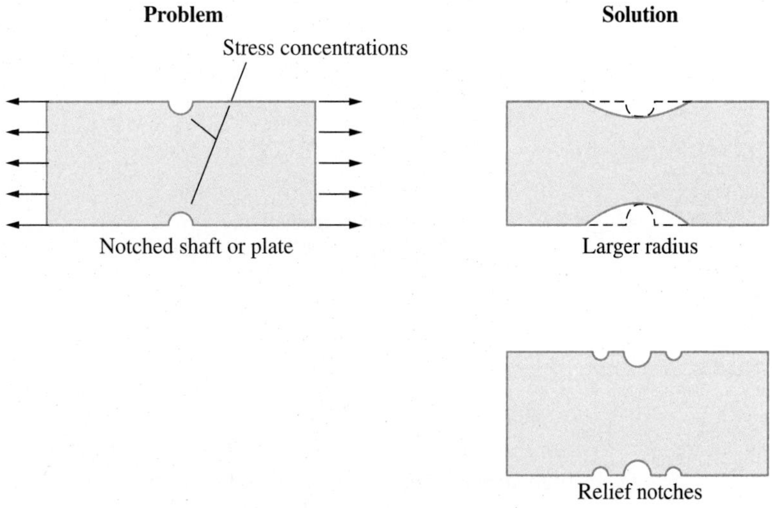

Figure 17.6 Stress concentration reductions.

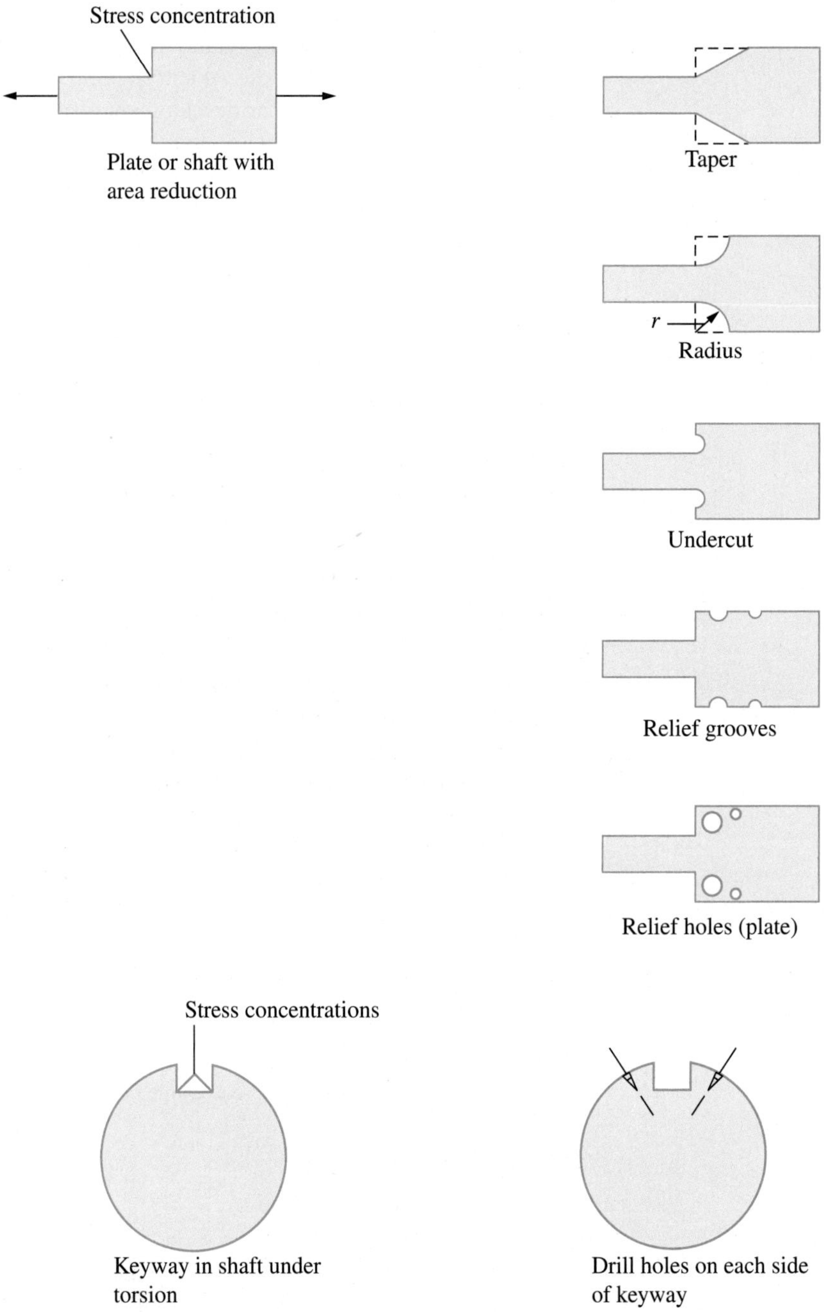

Figure 17.6 Stress concentration reductions. (*Continued*)

This is not a hard and fast rule, however, most reductions in high-stress concentrations are made by removing material from adjacent low-stressed areas. This "draws" the high stresses away from the stress concentration area toward the low-stressed area, which decreases the stress in the high-stressed areas.

17.3 TABLES

TABLE 17.1 Stress Concentration Factors for Elastic Stress (K_t)

<table>
<tr><th>Type of Form Irregularity or Stress Raiser</th><th>Stress Condition and Manner of Loading</th><th>Stress Concentration Factor K_t for Various Dimensions</th></tr>
<tr><td rowspan="3">1. Two U-notches in a member of rectangular section
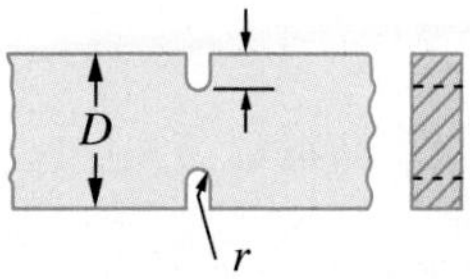
</td>
<td>1a. Elastic stress, axial tension</td>
<td>$$K_t = C_1 + C_2\left(\frac{2h}{D}\right) + C_3\left(\frac{2h}{D}\right)^2 + C_4\left(\frac{2h}{D}\right)^3$$
where
<table>
<tr><th></th><th>$0.1 \le h/r \le 2.0$</th><th>$2.0 \le h/r \le 50.0$</th></tr>
<tr><td>C_1</td><td>$0.850 + 2.628\sqrt{h/r} - 0.413h/r$</td><td>$0.833 + 2.069\sqrt{h/r} - 0.009h/r$</td></tr>
<tr><td>C_2</td><td>$-1.119 - 4.826\sqrt{h/r} + 2.575h/r$</td><td>$2.732 - 4.157\sqrt{h/r} + 0.176h/r$</td></tr>
<tr><td>C_3</td><td>$3.563 - 0.514\sqrt{h/r} - 2.402h/r$</td><td>$-8.859 + 5.327\sqrt{h/r} - 0.320h/r$</td></tr>
<tr><td>C_4</td><td>$-2.294 + 2.713\sqrt{h/r} + 0.240h/r$</td><td>$6.294 - 3.239\sqrt{h/r} + 0.154h/r$</td></tr>
</table>
(Refs. 1–10)
For the semicircular notch ($h/r = 1$)
$$K_t = 3.065 - 3.370\left(\frac{2h}{D}\right) + 0.647\left(\frac{2h}{D}\right)^2 + 0.658\left(\frac{2h}{D}\right)^3$$</td></tr>
<tr><td>1b. Elastic stress, in-plane bending
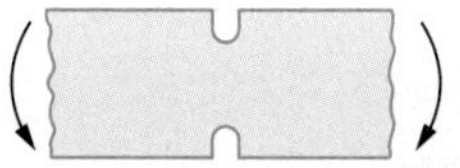</td>
<td>$$K_t = C_1 + C_2\left(\frac{2h}{D}\right) + C_3\left(\frac{2h}{D}\right)^2 + C_4\left(\frac{2h}{D}\right)^3$$
where
<table>
<tr><th></th><th>$0.25 \le h/r \le 2.0$</th><th>$2.0 \le h/r \le 50.0$</th></tr>
<tr><td>C_1</td><td>$0.723 + 2.845\sqrt{h/r} - 0.504h/r$</td><td>$0.833 + 2.069\sqrt{h/r} - 0.009h/r$</td></tr>
<tr><td>C_2</td><td>$-1.836 - 5.746\sqrt{h/r} + 1.314h/r$</td><td>$0.024 - 5.383\sqrt{h/r} + 0.126h/r$</td></tr>
<tr><td>C_3</td><td>$7.254 - 1.885\sqrt{h/r} + 1.646h/r$</td><td>$-0.856 + 6.460\sqrt{h/r} - 0.199h/r$</td></tr>
<tr><td>C_4</td><td>$-5.140 + 4.785\sqrt{h/r} - 2.456h/r$</td><td>$0.999 - 3.146\sqrt{h/r} + 0.082h/r$</td></tr>
</table>
For the semicircular notch ($h/r = 1$) (Refs. 1, 3, 8, 11, and 12)
$$K_t = 3.065 - 6.269\left(\frac{2h}{D}\right) + 7.015\left(\frac{2h}{D}\right)^2 - 2.812\left(\frac{2h}{D}\right)^3$$</td></tr>
<tr><td>1c. Elastic stress, out-of-plane bending
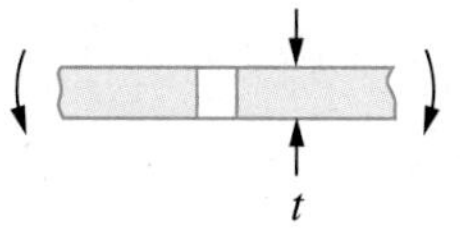
</td>
<td>$$K_t = C_1 + C_2\left(\frac{2h}{D}\right) + C_3\left(\frac{2h}{D}\right)^2 + C_4\left(\frac{2h}{D}\right)^3$$
where for $0.25 \le h/r \le 4.0$ and h/t is large
$C_1 = 1.031 + 0.831\sqrt{h/r} + 0.014h/r$
$C_2 = -1.227 - 1.646\sqrt{h/r} + 0.117h/r$
$C_3 = 3.337 - 0.750\sqrt{h/r} + 0.469h/r$
$C_4 = -2.141 + 1.566\sqrt{h/r} - 0.600h/r$
For the semicircular notch ($h/r = 1$) (Refs. 1, 9, 13, and 14)
$$K_t = 1.876 - 2.756\left(\frac{2h}{D}\right) + 3.056\left(\frac{2h}{D}\right)^2 - 1.175\left(\frac{2h}{D}\right)^3$$</td></tr>
</table>

The elastic stress concentration factor K_t is the ratio of the maximum stress in the stress raiser to the nominal stress computed by the ordinary mechanics-of-materials formulas, using the dimensions of the net cross section unless defined otherwise in specific cases.

For those data presented in the form of equations, the equations have been developed to fit as closely as possible the many data points given in the literature referenced in each case. Over the majority of the ranges specified for the variables, the curves fit the data points with much less than a 5% error.

It is not possible to tabulate all the available values of stress concentration factors found in the literature, but the following list of topics and sources of data will be helpful. All the following references have extensive bibliographies.

Fatigue stress concentration factors (K_f); see the 4th edition of this book, Ref. 23.

Stress concentration factors for rupture (K_r); see the 4th edition of this book, Ref. 23.

Stress concentration factors pertaining to odd-shaped holes in plates, multiple holes arranged in various patterns, and reinforced holes under multiple loads; see Refs. 1 and 24.

Stress concentration around holes in pressure vessels; see Ref. 1.

For a discussion of the effect of stress concentration on the response of machine elements and structures, see Ref. 25.

(Continued)

TABLE 17.1 Stress Concentration Factors for Elastic Stress (K_t) (*Continued*)

Type of Form Irregularity or Stress Raiser	Stress Condition and Manner of Loading	Stress Concentration Factor K_t for Various Dimensions
2. Two V-notches in a member of rectangular section 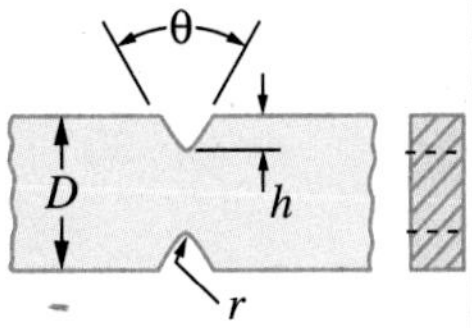 	2a. Elastic stress, axial tension	The stress concentration factor for the V-notch, K_{θ}, is the smaller of the values $$K_{t\theta} = K_{tu}$$ or $$K_{t\theta} = 1.11K_{tu} - \left[0.0275 + 0.000145\theta + 0.0164\left(\frac{\theta}{120}\right)^8\right]K_{tu}^2 \quad \text{for } \frac{2h}{D} = 0.40 \text{ and } \theta \le 120°$$ or $$K_{t\theta} = 1.11K_{tu} - \left[0.0275 + 0.00042\theta + 0.0075\left(\frac{\theta}{120}\right)^8\right]K_{tu}^2 \quad \text{for } \frac{2h}{D} = 0.667 \text{ and } \theta \le 120°$$ where K_{tu} is the stress concentration factor for a U-notch, case 1a, when the dimensions h, r, and D are the same as for the V-notch and θ is the notch angle in degrees. (Refs. 1 and 15)
3. One U-notch in a member of rectangular section 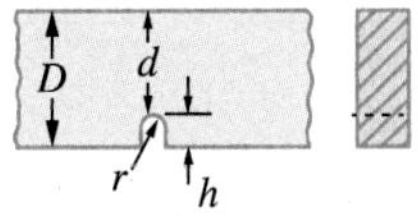 	3a. Elastic stress, axial tension 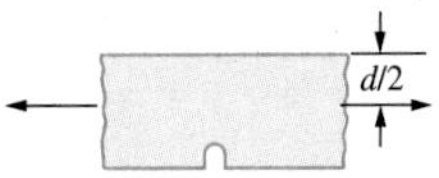 	$$K_t = C_1 + C_2\left(\frac{h}{D}\right) + C_3\left(\frac{h}{D}\right)^2 + C_4\left(\frac{h}{D}\right)^3$$ where for $0.5 \le h/r \le 4.0$ $C_1 = 0.721 + 2.394\sqrt{h/r} - 0.127h/r$ $C_2 = 1.978 - 11.489\sqrt{h/r} + 2.211h/r$ $C_3 = -4.413 + 18.751\sqrt{h/r} - 4.596h/r$ $C_4 = 2.714 - 9.655\sqrt{h/r} + 2.512h/r$ For the semicircular notch ($h/r = 1$) (Refs. 16 and 17) $$K_t = 2.988 - 7.300\left(\frac{h}{D}\right) + 9.742\left(\frac{h}{D}\right)^2 - 4.429\left(\frac{h}{D}\right)^3$$
	3b. Elastic stress, in-plane bending 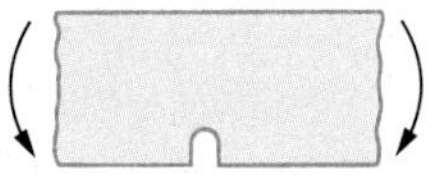	$$K_t = C_1 + C_2\left(\frac{h}{D}\right) + C_3\left(\frac{h}{D}\right)^2 + C_4\left(\frac{h}{D}\right)^3$$ where for $0.5 \le h/r \le 4.0$ $C_1 = 0.721 + 2.394\sqrt{h/r} - 0.127h/r$ $C_2 = -0.426 - 8.827\sqrt{h/r} + 1.518h/r$ $C_3 = 2.161 + 10.968\sqrt{h/r} - 2.455h/r$ $C_4 = -1.456 - 4.535\sqrt{h/r} + 1.064h/r$ For the semicircular notch ($h/r = 1$) (Refs. 17 and 18) $$K_t = 2.988 - 7.735\left(\frac{h}{D}\right) + 10.674\left(\frac{h}{D}\right)^2 - 4.927\left(\frac{h}{D}\right)^3$$
4. One V-notch in a member of rectangular section 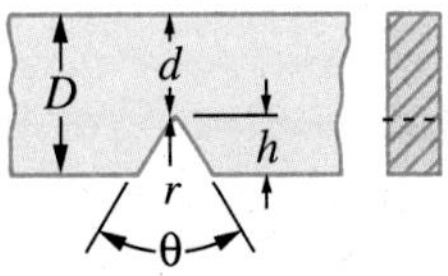 	4b. Elastic stress, in-plane bending 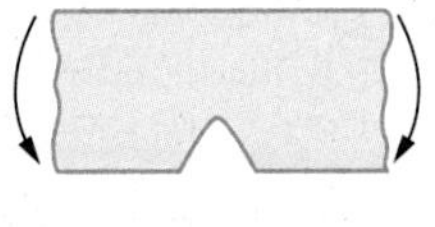	The stress concentration factor for the V-notch, $K_{t\theta}$, is the smaller of the values $$K_{t\theta} = K_{tu}$$ or $$K_{t\theta} = 1.11K_{tu} - \left[0.0275 + 0.1125\left(\frac{\theta}{150}\right)^4\right]K_{tu}^2 \quad \text{for } \theta \le 150°$$ where K_{tu} is the stress concentration factor for a U-notch, case 3b, when the dimensions h, r and D are the same as for the V-notch and θ is the notch angle in degrees. (Ref. 18)

(*Continued*)

TABLE 17.1 Stress Concentration Factors for Elastic Stress (K_t) (*Continued*)

<table>
<tr><th>Type of Form Irregularity or Stress Raiser</th><th>Stress Condition and Manner of Loading</th><th>Stress Concentration Factor K_t for Various Dimensions</th></tr>
<tr><td rowspan="2">5. Square shoulder with fillet in a member of rectangular section
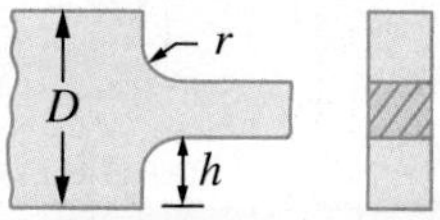
</td>
<td>5a. Elastic stress, axial tension
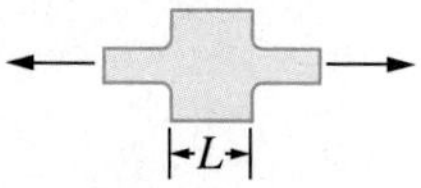
</td>
<td>$$K_t = C_1 + C_2\left(\frac{2h}{D}\right) + C_3\left(\frac{2h}{D}\right)^2 + C_4\left(\frac{2h}{D}\right)^3$$
where $\frac{L}{D} > \frac{3}{[r/(D-2h)]^{1/4}}$ and where
<table>
<tr><th></th><th>$0.1 \le h/r \le 2.0$</th><th>$2.0 \le h/r \le 20.0$</th></tr>
<tr><td>C_1</td><td>$1.007 + 1.000\sqrt{h/r} - 0.031h/r$</td><td>$1.042 + 0.982\sqrt{h/r} - 0.036h/r$</td></tr>
<tr><td>C_2</td><td>$-0.114 - 0.585\sqrt{h/r} + 0.314h/r$</td><td>$-0.074 - 0.156\sqrt{h/r} - 0.010h/r$</td></tr>
<tr><td>C_3</td><td>$0.241 - 0.992\sqrt{h/r} - 0.271h/r$</td><td>$-3.418 + 1.220\sqrt{h/r} - 0.005h/r$</td></tr>
<tr><td>C_4</td><td>$-0.134 + 0.577\sqrt{h/r} - 0.012h/r$</td><td>$3.450 - 2.046\sqrt{h/r} + 0.051h/r$</td></tr>
</table>
(Refs. 1, 8, 11, and 19)
For cases where $\frac{L}{D} < \frac{3}{[r/(D-2h)]^{1/4}}$ see Refs. 1, 21, and 22.</td></tr>
<tr><td>5b. Elastic stress, in-plane bending
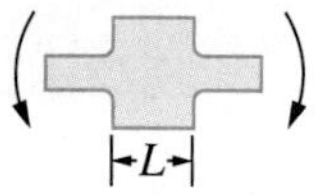
</td>
<td>$$K_t = C_1 + C_2\left(\frac{2h}{D}\right) + C_3\left(\frac{2h}{D}\right)^2 + C_4\left(\frac{2h}{D}\right)^3$$
where $\frac{L}{D} > \frac{0.8}{[r/(D-2h)]^{1/4}}$ and where
<table>
<tr><th></th><th>$0.1 \le h/r \le 2.0$</th><th>$2.0 \le h/r \le 20.0$</th></tr>
<tr><td>C_1</td><td>$1.007 + 1.000\sqrt{h/r} - 0.031h/r$</td><td>$1.042 + 0.982\sqrt{h/r} - 0.036h/r$</td></tr>
<tr><td>C_2</td><td>$-0.270 - 2.404\sqrt{h/r} + 0.749h/r$</td><td>$-3.599 + 1.619\sqrt{h/r} - 0.431h/r$</td></tr>
<tr><td>C_3</td><td>$0.677 + 1.133\sqrt{h/r} - 0.904h/r$</td><td>$6.084 - 5.607\sqrt{h/r} + 1.158h/r$</td></tr>
<tr><td>C_4</td><td>$-0.414 + 0.271\sqrt{h/r} + 0.186h/r$</td><td>$-2.527 + 3.006\sqrt{h/r} - 0.691h/r$</td></tr>
</table>
(Refs. 1, 11, and 20)
For cases where $\frac{L}{D} < \frac{0.8}{[r/(D-2h)]^{1/4}}$ see Refs. 1 and 20.</td></tr>
<tr><td rowspan="2">6. Circular hole in an infinite plate
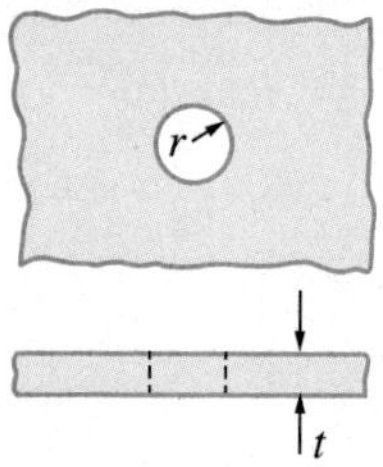
</td>
<td>6a. Elastic stress, in-plane normal stress
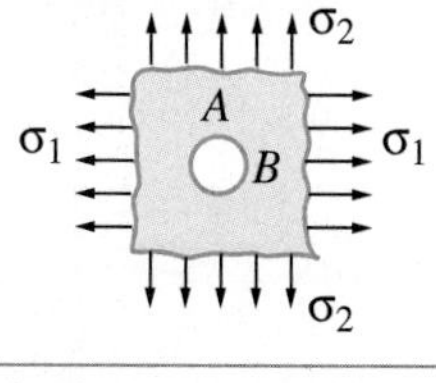
</td>
<td>(a1) Uniaxial stress, $\sigma_2 = 0$
$\sigma_A = 3\sigma_1 \quad \sigma_B = -\sigma_1$
(a2) Biaxial stress, $\sigma_2 = \sigma_1$
$\sigma_A = \sigma_B = 2\sigma_1$
(a3) Biaxial stress, $\sigma_2 = -\sigma_1$ (pure shear)
$\sigma_A = -\sigma_B = 4\sigma_1$</td></tr>
<tr><td>6b. Elastic stress, out-of-plane bending
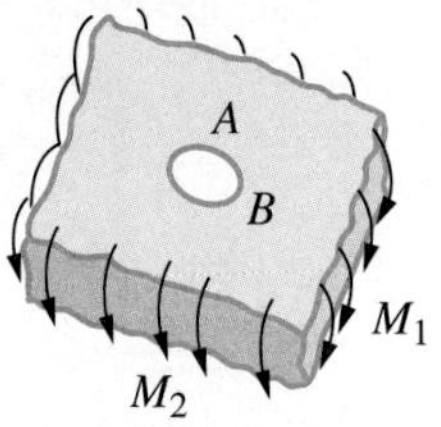

Note: M_1 and M_2 are unit moments (force-length per unit length).</td>
<td>(b1) Simple bending, $M_2 = 0$
$$\sigma_A = K_t \frac{6M_1}{t^2}$$
where $K_t = 1.79 + \frac{0.25}{0.39 + (2r/t)} + \frac{0.81}{1 + (2r/t)^2} - \frac{0.26}{1 + (2r/t)^3}$
(b2) Cylindrical bending, $M_2 = \nu M_1$
$$\sigma_A = K_t \frac{6M_1}{t^2}$$
where $K_t = 1.85 + \frac{0.509}{0.70 + (2r/t)} - \frac{0.214}{1 + (2r/t)^2} + \frac{0.335}{1 + (2r/t)^3}$ for $\nu = 0.3$
(b3) Isotropic bending, $M_2 = M_1$
$$\sigma_A = K_t \frac{6M_1}{t^2}$$
where $K_t = 2$ (independent of r/t)
(Refs. 1 and 26–30)</td></tr>
</table>

(*Continued*)

TABLE 17.1 Stress Concentration Factors for Elastic Stress (K_t) (*Continued*)

Type of Form Irregularity or Stress Raiser	Stress Condition and Manner of Loading	Stress Concentration Factor K_t for Various Dimensions
7. Central circular hole in a member of rectangular cross section 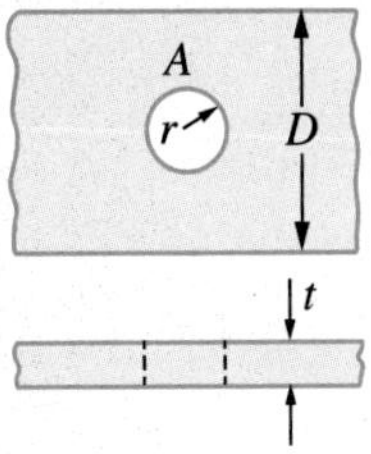	7a. Elastic stress, axial tension 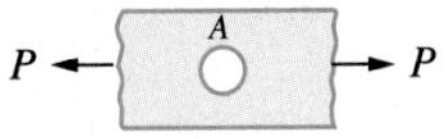	$\sigma_{max} = \sigma_A = K_t\sigma_{nom}$ where $\sigma_{nom} = \dfrac{P}{t(D-2r)}$ $K_t = 3.00 - 3.13\left(\dfrac{2r}{D}\right) + 3.66\left(\dfrac{2r}{D}\right)^2 - 1.53\left(\dfrac{2r}{D}\right)^3$ (Refs. 5 and 25)
	7b. Elastic stress, in-plane bending 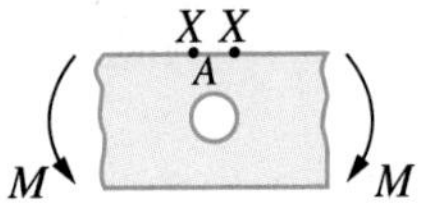	The maximum stress at the edge of the hole is $\sigma_A = K_t\sigma_{nom}$ where $\sigma_{nom} = \dfrac{12Mr}{t[D^3-(2r)^3]}$ (at the edge of the hole) $K_t = 2$ (independent of r/D) The maximum stress at the edge of the plate is not directly above the hole but is found a short distance away in either side, points X. $\sigma_X = \sigma_{nom}$ where $\sigma_{nom} = \dfrac{6MD}{t[D^3-(2r)^3]}$ (at the edge of the plate) (Refs. 1, 25, and 31)
	7c. Elastic stress, out-of-plane bending 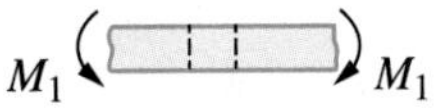 *Note*: see case 6b for interpretation of M_2.	(c1) Simple bending, $M_2 = 0$ $\sigma_{max} = \sigma_A = K_t\dfrac{6M_1}{t^2(D-2r)}$ where $K_t = \left[1.79 + \dfrac{0.25}{0.39+(2r/t)} + \dfrac{0.81}{1+(2r/t)^2} - \dfrac{0.26}{1+(2r/t)^3}\right]\left[1 - 1.04\left(\dfrac{2r}{D}\right) + 1.22\left(\dfrac{2r}{D}\right)^2\right]$ for $\dfrac{2r}{D} < 0.3$ (c2) Cylindrical bending (plate action), $M_2 = \nu M_1$ $\sigma_{max} = \sigma_A = K_t\dfrac{6M_1}{t^2(D-2r)}$ where $K_t = \left[1.85 + \dfrac{0.509}{0.70+(2r/t)} - \dfrac{0.214}{1+(2r/t)^2} + \dfrac{0.335}{1+(2r/t)^3}\right]$ $\times\left[1 - 1.04\left(\dfrac{2r}{D}\right) + 1.22\left(\dfrac{2r}{D}\right)^2\right]$ for $\dfrac{2r}{D} < 0.3$ and $\nu = 0.3$ (Refs. 1 and 27–29)
8. Off-center circular hole in a member of rectangular cross section 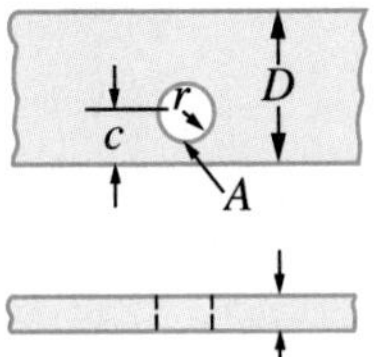	8a. Elastic stress, axial tension 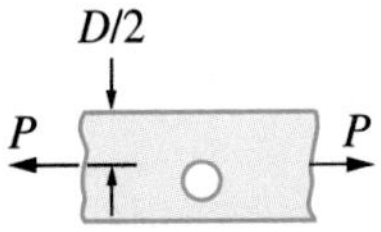	$\sigma_{max} = \sigma_A = K_t\sigma_{nom}$ where $\sigma_{nom} = \dfrac{P}{Dt}\dfrac{\sqrt{1-(r/c)^2}}{1-(r/c)}\dfrac{1-(c/D)}{1-(c/D)[2-\sqrt{1-(r/c)^2}]}$ $K_t = 3.00 - 3.13\left(\dfrac{r}{c}\right) + 3.66\left(\dfrac{r}{c}\right)^2 - 1.53\left(\dfrac{r}{c}\right)^3$ (Refs. 1 and 32)
	8b. Elastic stress, in-plane bending 	$\sigma_{max} = \sigma_A = K_t\sigma_{nom}$ where $\sigma_{nom} = \dfrac{12M}{tD^3}\left(\dfrac{D}{2} - c + r\right)$ and $K_t = 3.0$ if $r/c < 0.05$ or $K_t = C_1 + C_2\left(\dfrac{2c}{D}\right) + C_3\left(\dfrac{2c}{D}\right)^2 + C_4\left(\dfrac{2c}{D}\right)^3$ where for $0.05 \le r/c \le 0.5$ $C_1 = 3.022 - 0.422r/c + 3.556(r/c)^2$ $C_2 = -0.569 + 2.664r/c - 4.397(r/c)^2$ $C_3 = 3.138 - 18.367r/c + 28.093(r/c)^2$ $C_4 = -3.591 + 16.125r/c - 27.252(r/c)^2$ (Ref. 31)

(*Continued*)

TABLE 17.1 Stress Concentration Factors for Elastic Stress (K_t) (*Continued*)

Type of Form Irregularity or Stress Raiser	Stress Condition and Manner of Loading	Stress Concentration Factor K_t for Various Dimensions
9. Elliptical hole in an infinite plate 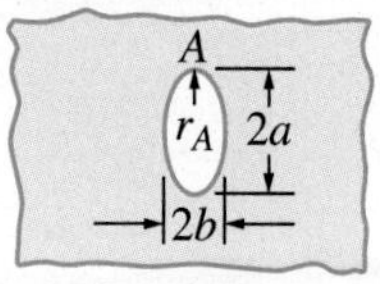 $r_A = \frac{b^2}{a}$	9a. Elastic stress, in-plane normal stress 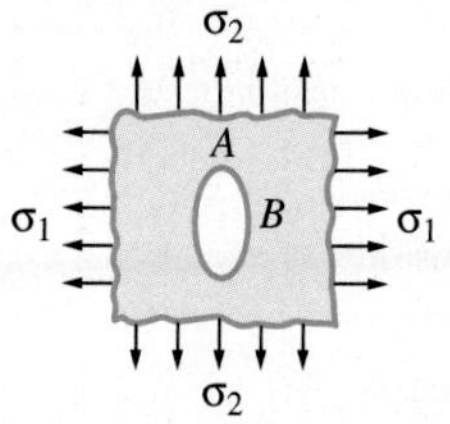	(a1) Uniaxial stress, $\sigma_2 = 0$ $\sigma_A = \left(1+\frac{2a}{b}\right)\sigma_1$ or $\sigma_A = \left(1+2\sqrt{\frac{a}{r_A}}\right)\sigma_1$ $\sigma_B = -\sigma_1$ (a2) Biaxial stress, $\sigma_2 = \sigma_1$ $\sigma_A = 2\frac{a}{b}\sigma_1$ $\sigma_B = 2\frac{b}{a}\sigma_1$ (a3) Biaxial stress, $\sigma_2 = -\sigma_1$ $\sigma_A = 2\left(1+\frac{a}{b}\right)\sigma_1$ $\sigma_B = -2\left(1+\frac{b}{a}\right)\sigma_1$ This stress condition would also be created by pure shear inclined at 45° to the axes of the ellipse.
10. Central elliptical hole in a member of rectangular cross section 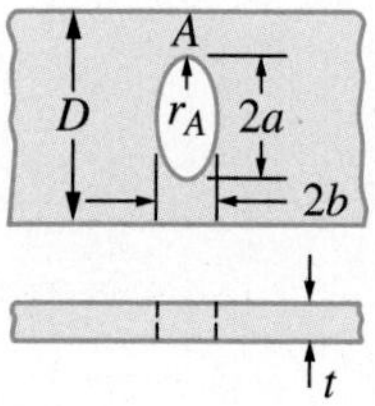 $r_A = \frac{b^2}{a}$	10a. Elastic stress, axial tension 	$\sigma_{max} = \sigma_A = K_t\sigma_{nom}$ where $\sigma_{nom} = \frac{P}{t(D-2a)}$ $K_t = C_1 + C_2\left(\frac{2a}{D}\right) + C_3\left(\frac{2a}{D}\right)^2 + C_4\left(\frac{2a}{D}\right)^3$ where for $0.5 \le a/b \le 10.0$ $C_1 = 1.000 + 0.000\sqrt{a/b} + 2.000a/b$ $C_2 = -0.351 - 0.021\sqrt{a/b} - 2.483a/b$ $C_3 = 3.621 - 5.183\sqrt{a/b} + 4.494a/b$ $C_4 = -2.270 + 5.204\sqrt{a/b} - 4.011a/b$ (Refs. 33–37)
	10b. Elastic stress, in-plane bending 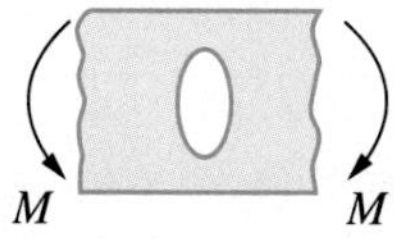	The maximum stress at the edge of the hole is $\sigma_A = K_t\sigma_{nom}$ where $\sigma_{nom} = \frac{12Ma}{t[D^3-(2a)^3]}$ (at the edge of the hole) $K_t = C_1 + C_2\left(\frac{2a}{D}\right) + C_3\left(\frac{2a}{D}\right)^2$ where for $1.0 \le a/b \le 2.0$ and $0.4 \le 2a/D \le 1.0$ $C_1 = 3.465 - 3.739\sqrt{a/b} + 2.274a/b$ $C_2 = -3.841 + 5.582\sqrt{a/b} - 1.741a/b$ $C_3 = 2.376 - 1.843\sqrt{a/b} - 0.534a/b$ (Refs. 1, 36, and 37)
11. Off-center elliptical hole in a member of rectangular cross section 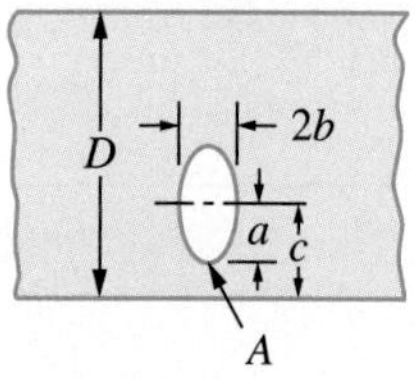	11a. Elastic stress, axial tension 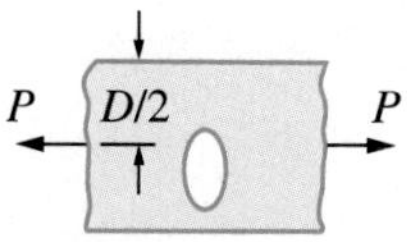	$\sigma_{max} = \sigma_A = K_t\sigma_{nom}$ The expression for σ_{nom} from case 8a can be used by substituting a/c for r/c. Use the expression for K_t from case 10a by substituting a/c for $2a/D$.

(*Continued*)

TABLE 17.1 Stress Concentration Factors for Elastic Stress (K_t) (*Continued*)

<table>
<tr><th>Type of Form Irregularity or Stress Raiser</th><th>Stress Condition and Manner of Loading</th><th>Stress Concentration Factor K_t for Various Dimensions</th></tr>
<tr><td>12. Rectangular hole with round corners in an infinite plate
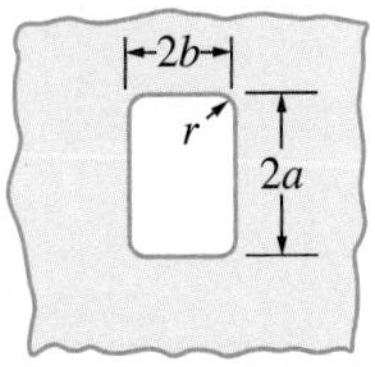
</td><td>12a. Elastic stress, axial tension
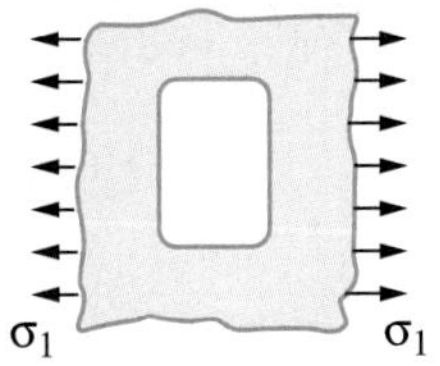
</td><td>$$\sigma_{max} = K_t\sigma_1 \quad \text{and} \quad K_t = C_1 + C_2\left(\frac{b}{a}\right) + C_3\left(\frac{b}{a}\right)^2 + C_4\left(\frac{b}{a}\right)^3$$
where for $\le 0.2 \le r/b \le 1.0$ and $0.3 \le b/a \le 1.0$
$C_1 = 14.815 - 15.774\sqrt{r/b} + 8.149r/b$
$C_2 = -11.201 - 9.750\sqrt{r/b} + 9.600r/b$
$C_3 = 0.202 + 38.622\sqrt{r/b} - 27.374r/b$
$C_4 = 3.232 - 23.002\sqrt{r/b} + 15.482r/b$
(Refs. 38–40)</td></tr>
<tr><td rowspan="2">13. Lateral slot with circular ends in a member of rectangular section
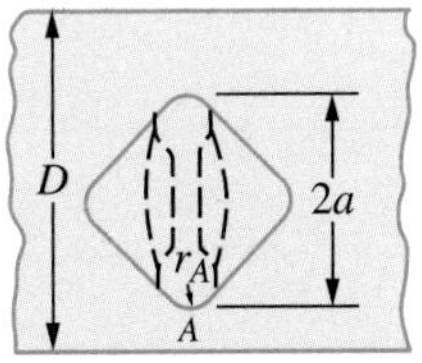

The equivalent ellipse has a width $2b_{eq}$ where $b_{eq} = \sqrt{r_A a}$</td><td>13a. Elastic stress, axial tension</td><td>A very close approximation to the maximum stress σ_A can be obtained by using the maximum stress for the given loading with the actual slot replaced by an ellipse having the same overall dimension normal to the loading direction, $2a$, and the same end radius r_A. See cases 9a, 10a, and 11a.</td></tr>
<tr><td>13b. Elastic stress, in-plane bending</td><td>As above, but see case 10b.</td></tr>
<tr><td>14. Reinforced circular hole in a wide plate
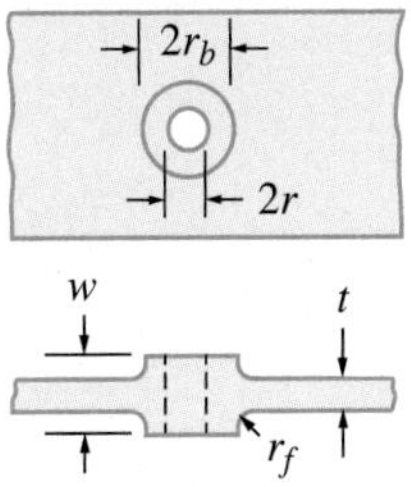
</td><td>14a. Elastic stress, axial tension
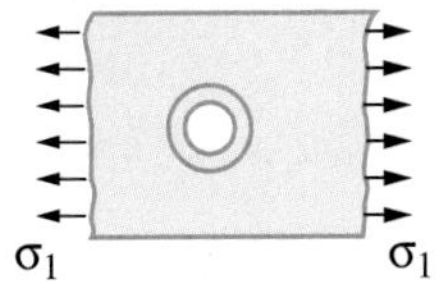
</td><td>$$\sigma_{max} = K_t\sigma_1 \text{ where for } r_f \ge 0.6t \text{ and } w \ge 3t$$
$$K_t = 1.0 + \frac{1.66}{1+A} - \frac{2.182}{(1+A)^2} + \frac{2.521}{(1+A)^3}$$
A is the ratio of the transverse area of the added reinforcement to the transverse area of the hole:
$$A = \frac{(r_b - r)(w-t) + 0.429r_f^2}{rt}$$
(Ref. 41)</td></tr>
<tr><td>15. U-notch in a circular shaft
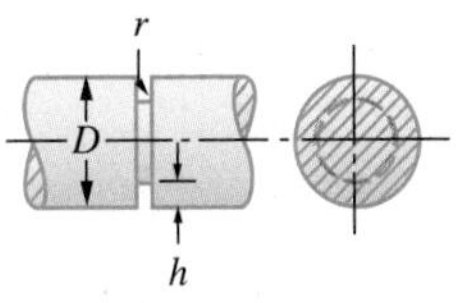
</td><td>15a. Elastic stress, axial tension</td><td>$$\sigma_{max} = K_t\frac{4P}{\pi(D-2h)^2} \text{ where } K_t = C_1 + C_2\left(\frac{2h}{D}\right) + C_3\left(\frac{2h}{D}\right)^2 + C_4\left(\frac{2h}{D}\right)^3$$
where
<table><tr><th></th><th>$0.25 \le h/r \le 2.0$</th><th>$2.0 \le h/r \le 50.0$</th></tr><tr><td>C_1</td><td>$0.455 + 3.354\sqrt{h/r} - 0.769h/r$</td><td>$0.935 + 1.922\sqrt{h/r} + 0.004h/r$</td></tr><tr><td>$C_2$</td><td>$3.129 - 15.955\sqrt{h/r} + 7.404h/r$</td><td>$0.537 - 3.708\sqrt{h/r} + 0.040h/r$</td></tr><tr><td>$C_3$</td><td>$-6.909 + 29.286\sqrt{h/r} - 16.104h/r$</td><td>$-2.538 + 3.438\sqrt{h/r} - 0.012h/r$</td></tr><tr><td>$C_4$</td><td>$4.325 - 16.685\sqrt{h/r} + 9.469h/r$</td><td>$2.066 - 1.652\sqrt{h/r} - 0.031h/r$</td></tr></table>
For the semicircular notch ($h/r = 1$) (Refs. 1, 9, and 42)
$$K_t = 3.04 - 5.42\left(\frac{2h}{D}\right) + 6.27\left(\frac{2h}{D}\right)^2 - 2.89\left(\frac{2h}{D}\right)^3$$</td></tr>
</table>

(*Continued*)

TABLE 17.1 Stress Concentration Factors for Elastic Stress (K_t) (*Continued*)

<table>
<tr><th>Type of Form Irregularity or Stress Raiser</th><th>Stress Condition and Manner of Loading</th><th>Stress Concentration Factor K_t for Various Dimensions</th></tr>
<tr><td></td><td>15b. Elastic stress, bending</td><td>

$$\sigma_{max} = K_t \frac{32M}{\pi(D-2h)^3} \text{ where } K_t = C_1 + C_2\left(\frac{2h}{D}\right) + C_3\left(\frac{2h}{D}\right)^2 + C_4\left(\frac{2h}{D}\right)^3$$

where

<table>
<tr><th></th><th>$0.25 \le h/r \le 2.0$</th><th>$2.0 \le h/r \le 50.0$</th></tr>
<tr><td>C_1</td><td>$0.455 + 3.354\sqrt{h/r} - 0.769h/r$</td><td>$0.935 + 1.922\sqrt{h/r} + 0.004h/r$</td></tr>
<tr><td>C_2</td><td>$0.891 - 12.721\sqrt{h/r} + 4.593h/r$</td><td>$-0.552 - 5.327\sqrt{h/r} + 0.086h/r$</td></tr>
<tr><td>C_3</td><td>$0.286 + 15.481\sqrt{h/r} - 6.392h/r$</td><td>$0.754 + 6.281\sqrt{h/r} - 0.121h/r$</td></tr>
<tr><td>C_4</td><td>$-0.632 - 6.115\sqrt{h/r} + 2.568h/r$</td><td>$-0.138 - 2.876\sqrt{h/r} + 0.031h/r$</td></tr>
</table>

For the semicircular notch ($h/r = 1$) (Refs. 1 and 9)

$$K_t = 3.04 - 7.236\left(\frac{2h}{D}\right) + 9.375\left(\frac{2h}{D}\right)^2 - 4.179\left(\frac{2h}{D}\right)^3$$

</td></tr>
<tr><td></td><td>15c. Elastic stress, torsion</td><td>

$$\sigma_{max} = K_t \frac{16T}{\pi(D-2h)^3} \text{ where } K_t = C_1 + C_2\left(\frac{2h}{D}\right) + C_3\left(\frac{2h}{D}\right)^2 + C_4\left(\frac{2h}{D}\right)^3$$

where

<table>
<tr><th></th><th>$0.25 \le h/r \le 2.0$</th><th>$2.0 \le h/r \le 50.0$</th></tr>
<tr><td>C_1</td><td>$1.245 + 0.264\sqrt{h/r} + 0.491h/r$</td><td>$1.651 + 0.614\sqrt{h/r} + 0.040h/r$</td></tr>
<tr><td>C_2</td><td>$-3.030 + 3.269\sqrt{h/r} - 3.633h/r$</td><td>$-4.794 - 0.314\sqrt{h/r} - 0.217h/r$</td></tr>
<tr><td>C_3</td><td>$7.199 - 11.286\sqrt{h/r} + 8.318h/r$</td><td>$8.457 - 0.962\sqrt{h/r} + 0.389h/r$</td></tr>
<tr><td>C_4</td><td>$-4.414 + 7.753\sqrt{h/r} - 5.176h/r$</td><td>$-4.314 + 0.662\sqrt{h/r} - 0.212h/r$</td></tr>
</table>

For the semicircular notch ($h/r = 1$) (Refs. 1, 9, and 43–46)

$$K_t = 2.000 - 3.394\left(\frac{2h}{D}\right) + 4.231\left(\frac{2h}{D}\right)^2 - 1.837\left(\frac{2h}{D}\right)^3$$

</td></tr>
<tr><td>16. V-notch in a circular shaft

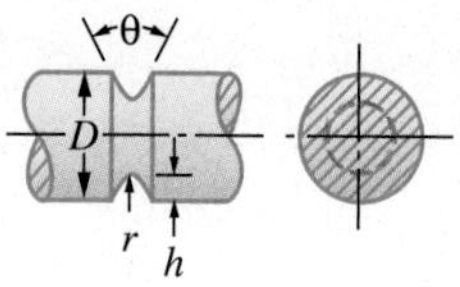

</td><td>16c. Elastic stress, torsion</td><td>

The stress concentration factor for the V-notch, $K_{t\theta}$, is the smaller of the values

$$K_{t\theta} = K_{tu} \quad \text{or} \quad K_{t\theta} = 1.065K_{tu} - \left[0.022 + 0.137\left(\frac{\theta}{135}\right)^2\right](K_{tu} - 1)K_{tu}$$

for $\frac{r}{D-2h} \le 0.01$ and $\theta \le 135°$ where K_{tu} is the stress concentration factor for a U-notch, case 15c, when the dimensions h, r, and D are the same as for the V-notch and θ is the notch angle in degrees.

(Refs. 1 and 44)

</td></tr>
<tr><td>17. Square shoulder with fillet in circular shaft

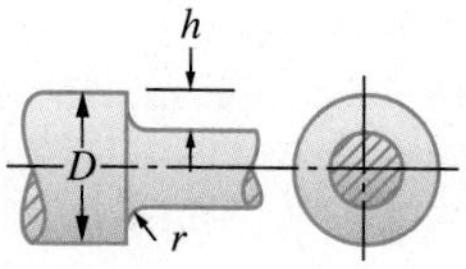

</td><td>17a. Elastic stress, axial tension</td><td>

$$\sigma_{max} = K_t \frac{4P}{\pi(D-2h)^2} \text{ where } K_t = C_1 + C_2\left(\frac{2h}{D}\right) + C_3\left(\frac{2h}{D}\right)^2 + C_4\left(\frac{2h}{D}\right)^3$$

where

<table>
<tr><th></th><th>$0.25 \le h/r \le 2.0$</th><th>$2.0 \le h/r \le 20.0$</th></tr>
<tr><td>C_1</td><td>$0.927 + 1.149\sqrt{h/r} - 0.086h/r$</td><td>$1.225 + 0.831\sqrt{h/r} - 0.010h/r$</td></tr>
<tr><td>C_2</td><td>$0.011 - 3.029\sqrt{h/r} + 0.948h/r$</td><td>$-1.831 - 0.318\sqrt{h/r} - 0.049h/r$</td></tr>
<tr><td>C_3</td><td>$-0.304 + 3.979\sqrt{h/r} - 1.737h/r$</td><td>$2.236 - 0.522\sqrt{h/r} + 0.176h/r$</td></tr>
<tr><td>C_4</td><td>$0.366 - 2.098\sqrt{h/r} + 0.875h/r$</td><td>$-0.630 + 0.009\sqrt{h/r} - 0.117h/r$</td></tr>
</table>

(Refs. 1, 19, and 47)

</td></tr>
</table>

(Continued)

TABLE 17.1 Stress Concentration Factors for Elastic Stress (K_t) (*Continued*)

<table>
<tr><th>Type of Form Irregularity or Stress Raiser</th><th>Stress Condition and Manner of Loading</th><th>Stress Concentration Factor K_t for Various Dimensions</th></tr>
<tr><td></td><td>17b. Elastic stress, bending</td><td>$\sigma_{max} = K_t \frac{32M}{\pi(D-2h)^3}$ where $K_t = C_1 + C_2\left(\frac{2h}{D}\right) + C_3\left(\frac{2h}{D}\right)^2 + C_4\left(\frac{2h}{D}\right)^3$
where
<table>
<tr><th></th><th>$0.25 \le h/r \le 2.0$</th><th>$2.0 \le h/r \le 20.0$</th></tr>
<tr><td>C_1</td><td>$0.927 + 1.149\sqrt{h/r} - 0.086h/r$</td><td>$1.225 + 0.831\sqrt{h/r} - 0.010h/r$</td></tr>
<tr><td>C_2</td><td>$0.015 - 3.281\sqrt{h/r} + 0.837h/r$</td><td>$-3.790 + 0.958\sqrt{h/r} - 0.257h/r$</td></tr>
<tr><td>C_3</td><td>$0.847 + 1.716\sqrt{h/r} - 0.506h/r$</td><td>$7.374 - 4.834\sqrt{h/r} + 0.862h/r$</td></tr>
<tr><td>C_4</td><td>$-0.790 + 0.417\sqrt{h/r} - 0.246h/r$</td><td>$-3.809 + 3.046\sqrt{h/r} - 0.595h/r$</td></tr>
</table>
(Refs. 1, 20, and 48)</td></tr>
<tr><td></td><td>17c. Elastic stress, torsion</td><td>$\sigma_{max} = K_t \frac{16T}{\pi(D-2h)^3}$ where $K_t = C_1 + C_2\left(\frac{2h}{D}\right) + C_3\left(\frac{2h}{D}\right)^2 + C_4\left(\frac{2h}{D}\right)^3$
where for $0.25 \le h/r \le 4.0$
$C_1 = 0.953 + 0.680\sqrt{h/r} - 0.053h/r$
$C_2 = -0.493 - 1.820\sqrt{h/r} + 0.517h/r$
$C_3 = 1.621 + 0.908\sqrt{h/r} - 0.529h/r$
$C_4 = -1.081 + 0.232\sqrt{h/r} + 0.065h/r$
(Refs. 1, 19, 46, 49, and 50)</td></tr>
<tr><td>18. Radial hole in a hollow or solid circular shaft
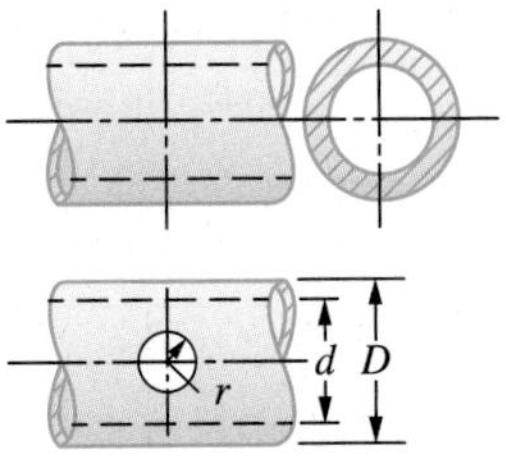

for a solid shaft $d = 0$</td><td>18a. Elastic stress, axial tension</td><td>$\sigma_{max} = K_t \frac{4P}{\pi(D^2 - d^2)}$ where $K_t = C_1 + C_2\left(\frac{2r}{D}\right) + C_3\left(\frac{2r}{D}\right)^2 + C_4\left(\frac{2r}{D}\right)^3$
and where for $d/D \le 0.9$ and $2r/D \le 0.45$
$C_1 = 3.000$
$C_2 = 2.773 + 1.529d/D - 4.379(d/D)^2$
$C_3 = -0.421 - 12.782d/D + 22.781(d/D)^2$
$C_4 = 16.841 + 16.678d/D - 40.007(d/D)^2$
(Refs. 1, 43, 51, and 52)</td></tr>
<tr><td></td><td>18b. Elastic stress, bending when hole is farthest from bending axis</td><td>$\sigma_{max} = K_t \frac{32MD}{\pi(D^4 - d^4)}$ where $K_t = C_1 + C_2\left(\frac{2r}{D}\right) + C_3\left(\frac{2r}{D}\right)^2 + C_4\left(\frac{2r}{D}\right)^3$
and where for $d/D \le 0.9$ and $2r/D \le 0.3$
$C_1 = 3.000$
$C_2 = -6.690 - 1.620d/D + 4.432(d/D)^2$
$C_3 = 44.739 + 10.724d/D - 19.927(d/D)^2$
$C_4 = -53.307 - 25.998d/D + 43.258(d/D)^2$
(Refs. 1 and 51–54)</td></tr>
<tr><td></td><td>18c. Elastic stress, torsional loading</td><td>$\sigma_{max} = K_t \frac{16TD}{\pi(D^4 - d^4)}$ where $K_t = C_1 + C_2\left(\frac{2r}{D}\right) + C_3\left(\frac{2r}{D}\right)^2 + C_4\left(\frac{2r}{D}\right)^3$
and where for $d/D \le 0.9$ and $2r/D \le 0.4$
$C_1 = 4.000$
$C_2 = -6.793 + 1.133d/D - 0.126(d/D)^2$
$C_3 = 38.382 - 7.242d/D + 6.495(d/D)^2$
$C_4 = -44.576 - 7.428d/D + 58.656(d/D)^2$
(Refs. 1 and 51–53)</td></tr>
</table>

(*Continued*)

TABLE 17.1 Stress Concentration Factors for Elastic Stress (K_t) (*Continued*)

Type of Form Irregularity or Stress Raiser	Stress Condition and Manner of Loading	Stress Concentration Factor K_t for Various Dimensions
19. Multiple U-notches in a member of rectangular section 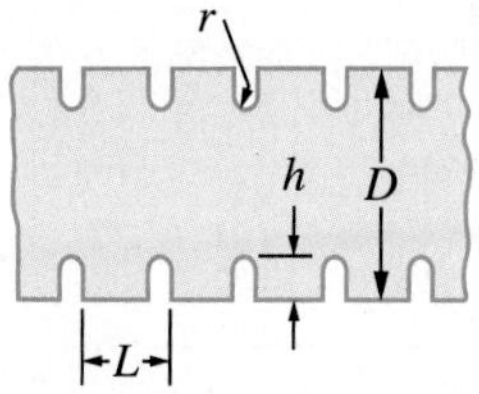	19a. Elastic stress, axial tension, semicircular notches only, i.e., $h = r$	The stress concentration factor for the multiple semicircular U-notches, K_{tm}, is the *smaller* of the values $$K_{tm} = K_{tu}$$ or $$K_{tm}=\left\{1.1-\left[0.88-1.68\left(\frac{2r}{D}\right)\right]\frac{2r}{L}+\left[1.3\left(0.5-\frac{2r}{D}\right)^2\right]\left(\frac{2r}{L}\right)^3\right\}K_{tu} \quad \text{for } \frac{2r}{L}<1$$ where K_{tu} is the stress concentration factor for a single pair of semicircular U-notches, case 1a. (Refs. 1, 55, and 56)
20. Infinite row of circular holes in an infinite plate 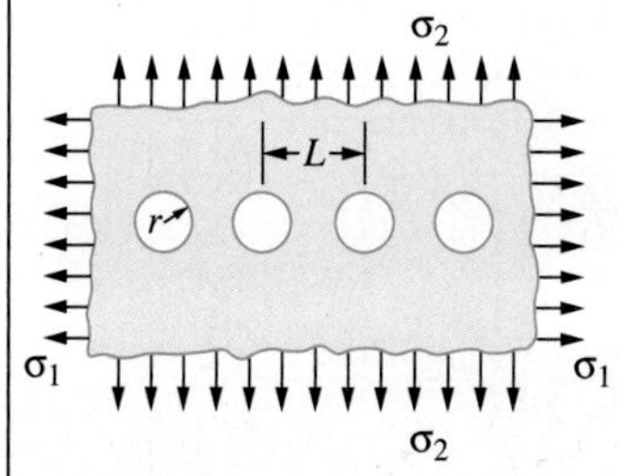	20a. Elastic stress, axial tension parallel to the row of holes	$$\sigma_{max} = K_t\sigma_1, \quad \sigma_2 = 0$$ where $K_t = 3.0-1.061\left(\frac{2r}{L}\right)-2.136\left(\frac{2r}{L}\right)^2+1.877\left(\frac{2r}{L}\right)^3$ (Refs. 1 and 57)
	20b. Elastic stress, axial tension normal to the row of holes	$$\sigma_{max} = K_t\sigma_2, \quad \sigma_1 = 0$$ where $K_t = 3.0-3.057\left(\frac{2r}{L}\right)+0.214\left(\frac{2r}{L}\right)^2+0.843\left(\frac{2r}{L}\right)^3$ (Refs. 1 and 57)
21. Gear tooth 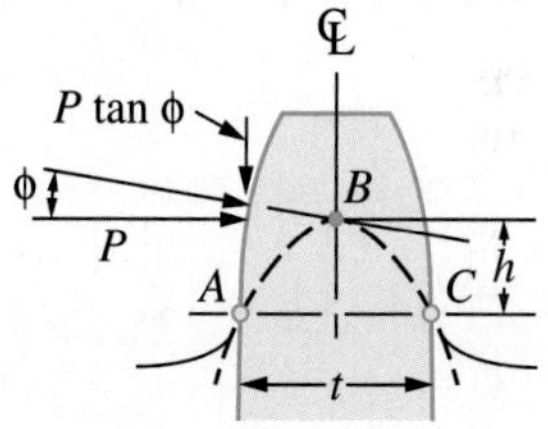 *A* and *C* are points of tangency of the inscribed parabola *ABC* with tooth profile, b = tooth width normal to plane of figure, r = minimum radius of tooth fillet.	Elastic stress, bending plus some compression	For 14.5° pressure angle: $K_t = 0.22+\left(\frac{t}{r}\right)^{0.2}\left(\frac{t}{h}\right)^{0.4}$ For 20° pressure angle: $K_t = 0.18+\left(\frac{t}{r}\right)^{0.15}\left(\frac{t}{h}\right)^{0.45}$ $K_t = [(\sigma_t)_{max}$ by photoelastic analysis$]/\left[\text{calculated } (\sigma_t)_{max} = \frac{6Ph}{bt^2}-\frac{P\tan\phi}{bt}\right]$ (Alternatively, the maximum stress can be found by the formula for a short cantilever, pages 187 and 188.) (Ref. 58)
22. Square or filleted corner in tension 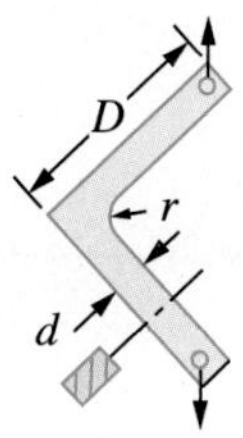	Elastic stress	$\frac{D}{d} = 5.5$ r/d: 0.125, 0.15, 0.20, 0.25, 0.30, 0.40, 0.50, 0.70, 1.00 K_t: 2.50, 2.30, 2.03, 1.88, 1.70, 1.53, 1.40, 1.26, 1.20 (Ref. 17)

(*Continued*)

TABLE 17.1 Stress Concentration Factors for Elastic Stress (K_t) (*Continued*)

Type of Form Irregularity or Stress Raiser	Stress Condition and Manner of Loading	Stress Concentration Factor K_t for Various Dimensions
23. U-shaped member 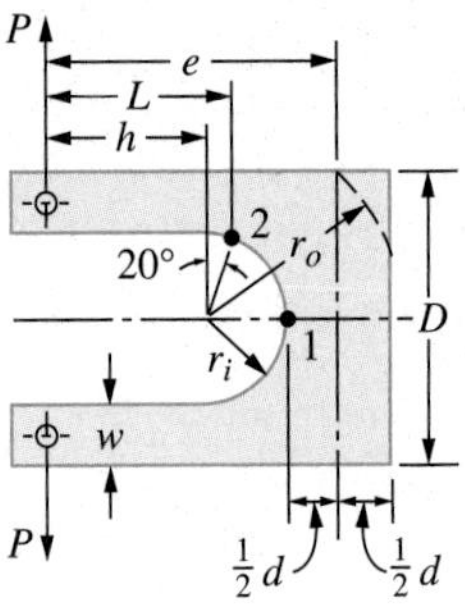	Elastic stress, as shown	K_{t1} is the ratio of actual to nominal bending stress at point 1, and K_{t2} is this ratio at point 2. Nominal bending stress = Pey/I at point 1 and PLy/I at point 2, where I/y = section modulus at the section in question.

Outer corners	Dimension Ratios and Values of K_t					
	$\frac{e}{r_i}=\frac{e}{w}=\frac{e}{d}$		K_{t1}		K_{t2}	
Square	4.5		1.24		1.24	
	3.5		1.20		1.24	
	2.5		1.30		1.20	
	1.5		1.24		1.61	
	$\frac{e}{2r_i}=\frac{e}{2w}=\frac{e}{d}$					
Square	2.5		1.50		12.9	
	2.0		1.52		1.33	
	1.5		1.53		1.22	
	1.0		1.46		1.75	
	$\frac{d}{r_i}=\frac{d}{w}$		$h=\frac{3}{4}D$		$h=\frac{1}{4}D$	
			K_{t1}	K_{t2}	K_{t1}	K_{t2}
Square	2.0		1.50	1.29	1.53	1.22
	1.5		1.34	1.10	1.37	1.40
	1.25		1.29	1.23	1.33	1.41
	1.0		1.24	1.24	1.30	1.20
	0.75		1.21	1.10	1.24	1.22
	$\frac{r_0}{r_i}=\frac{r_0}{d}=\frac{r_0}{w}$					
Rounded to radius r_0	2.75		1.24	1.24	1.30	1.20
	2.37		1.18	1.21	1.18	1.22
	2.12		1.16	1.22	1.21	1.31
	2.0		1.27	1.42	1.31	1.56
	$\frac{d}{r_i}=\frac{w}{r_i}$					
Square	7.0		2.29	1.93	2.38	2.38
	3.0		1.72	1.59	1.76	1.62
	1.67		1.49	1.37	1.41	1.52
	1.0		1.24	1.24	1.30	1.20
	$\frac{d}{r_i}$	$\frac{d}{w}$				
Square	5	1.67	2.33	1.73	2.32	2.00
	3	1.80	1.82	1.30	1.75	1.56
	2	2.0	1.50	1.29	1.53	1.22

(Ref. 59)

Type of Form Irregularity or Stress Raiser	Stress Condition and Manner of Loading	Stress Concentration Factor K_t for Various Dimensions
24. Splined shaft torsion Splined shaft	Torsion 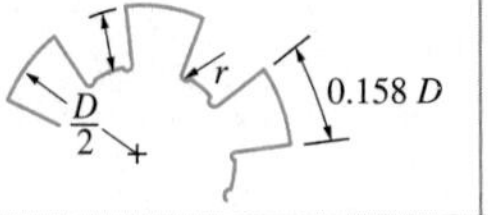	For an eight-tooth spline $K_{tS}=\tau_{max}/\tau,\quad \tau=16T/\pi D^3$ For $0.01 \le r/D \le 0.04$ $K_{tS}=6.083-14.775\left(\frac{10r}{D}\right)+18.250\left(\frac{10r}{D}\right)^2$

25. Torsion of angles and box sections

Angles and box sections

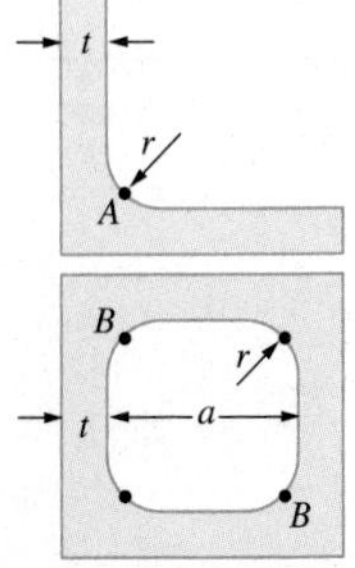

(1) For angle section:

$$\tau_{max}=\tau_A=K_t\tau$$

$$K_t=6.554-16.077\sqrt{\frac{r}{t}}+16.987\left(\frac{r}{t}\right)-5.886\sqrt{\frac{r}{t}}\left(\frac{r}{t}\right)$$

where $0.1 \le r/t \le 1.4$

(2) For box section:

$$\tau_{max}=\tau_B=K_t\tau$$

$$K_t=3.962-7.359\left(\frac{r}{t}\right)+6.801\left(\frac{r}{t}\right)^2-2.153\left(\frac{r}{t}\right)^3$$

where a is 15–20 times larger than t,
$0.2 \le r/t \le 1.4$

(*Continued*)

TABLE 17.2 Graphs for Stress Concentration Factors

(a)

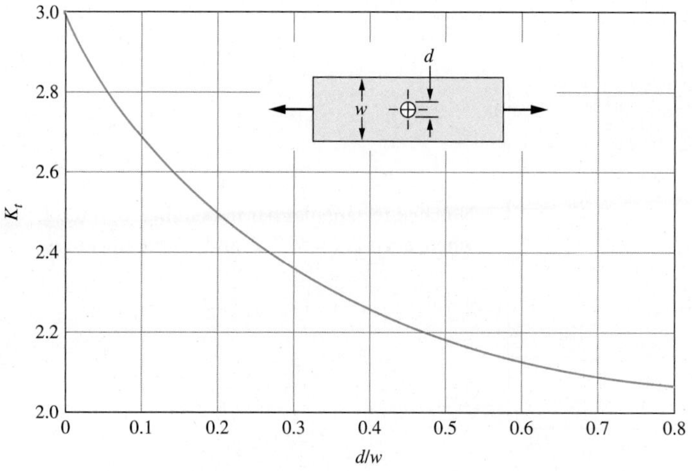

Bar in tension or simple compression with a transverse hole. $\sigma_0 = F/A$, where $A = (w - d)t$ and t is the thickness.

(b)

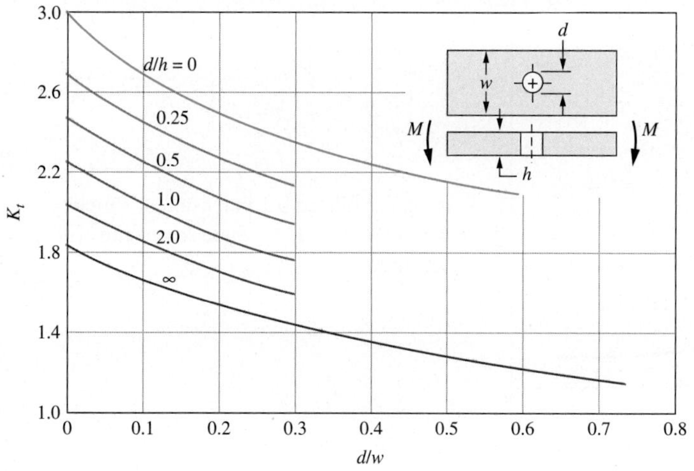

Rectangular bar with a transverse hole in bending. $\sigma_0 = Mc/I$, where $I = (w - d)h^3/12$.

(c)

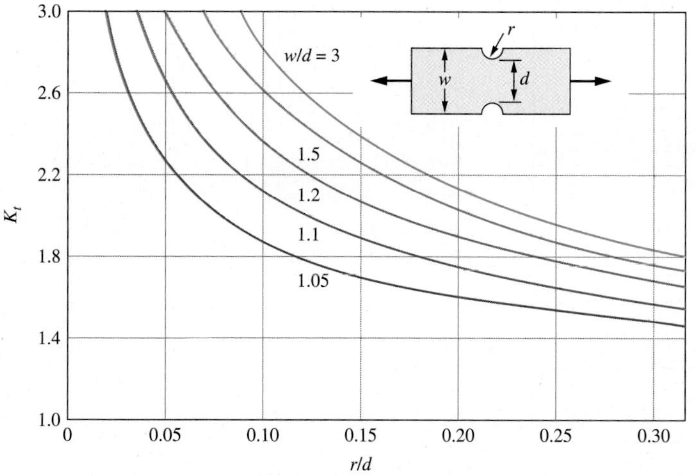

Notched rectangular bar in tension or simple compression. $\sigma_0 = F/A$, where $A = dt$ and t is the thickness.

(Continued)

TABLE 17.2 Graphs for Stress Concentration Factors (*Continued*)

(d)

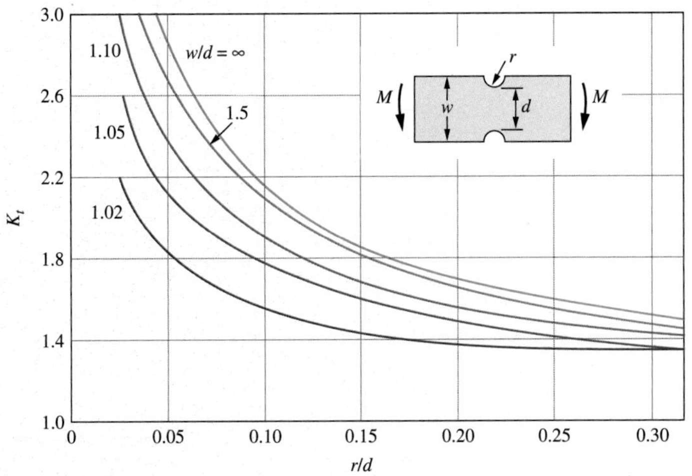

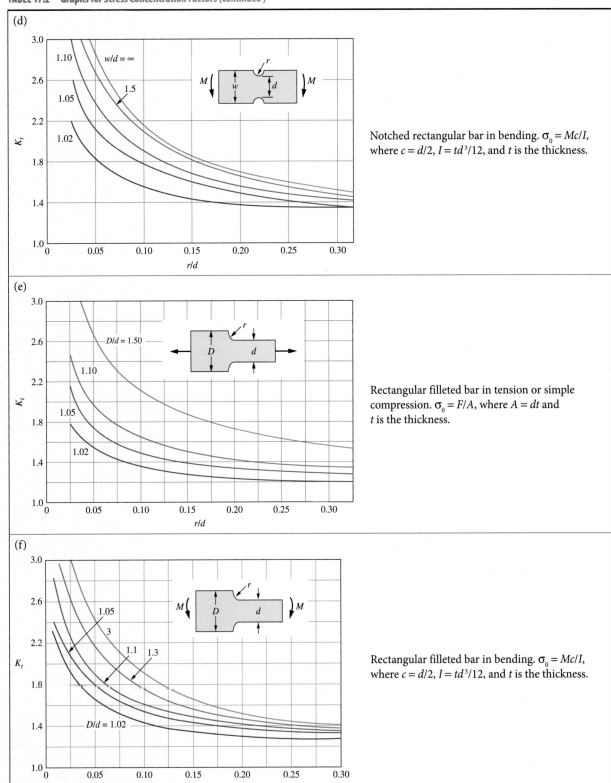

Notched rectangular bar in bending. $\sigma_0 = Mc/I$, where $c = d/2$, $I = td^3/12$, and t is the thickness.

(e)

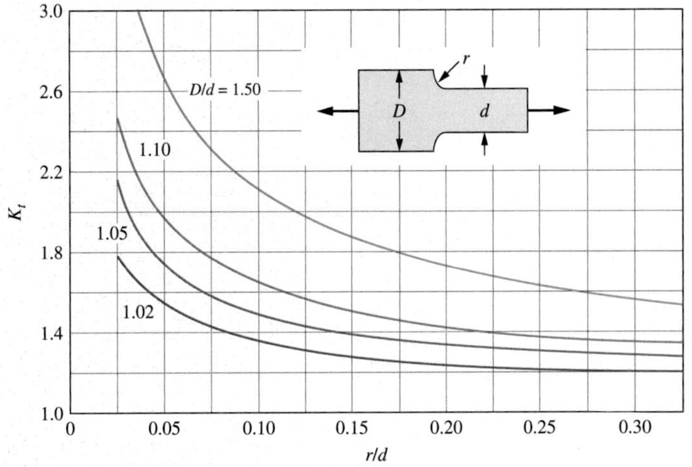

Rectangular filleted bar in tension or simple compression. $\sigma_0 = F/A$, where $A = dt$ and t is the thickness.

(f)

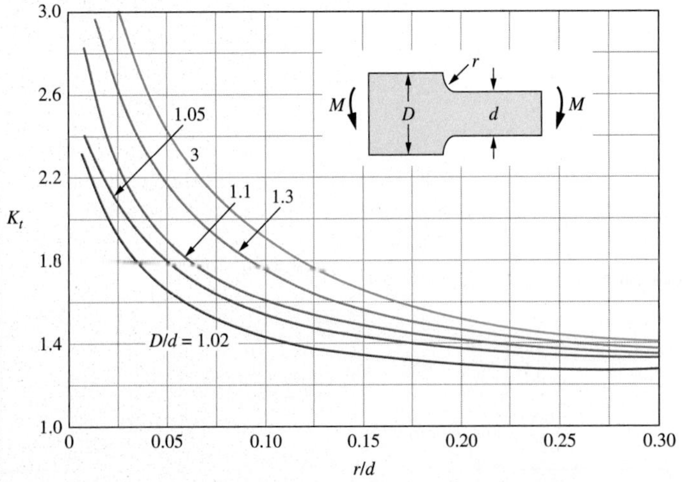

Rectangular filleted bar in bending. $\sigma_0 = Mc/I$, where $c = d/2$, $I = td^3/12$, and t is the thickness.

(Continued)

TABLE 17.2 Graphs for Stress Concentration Factors (*Continued*)

(g)

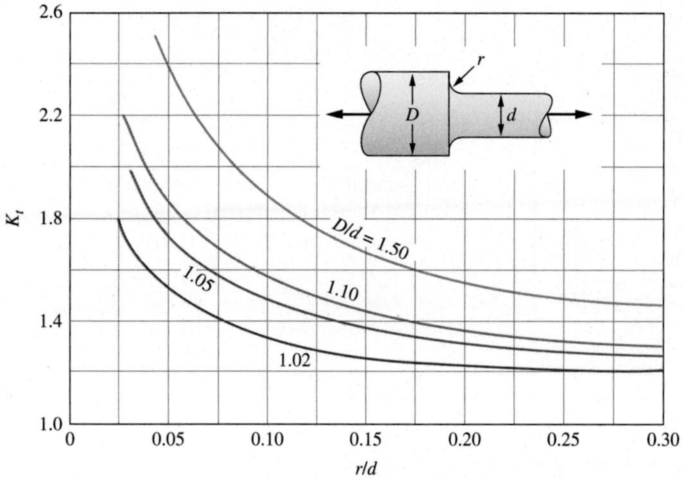

Round shaft with shoulder fillet in tension.
$\sigma_0 = F/A$, where $A = \pi d^2/4$.

(h)

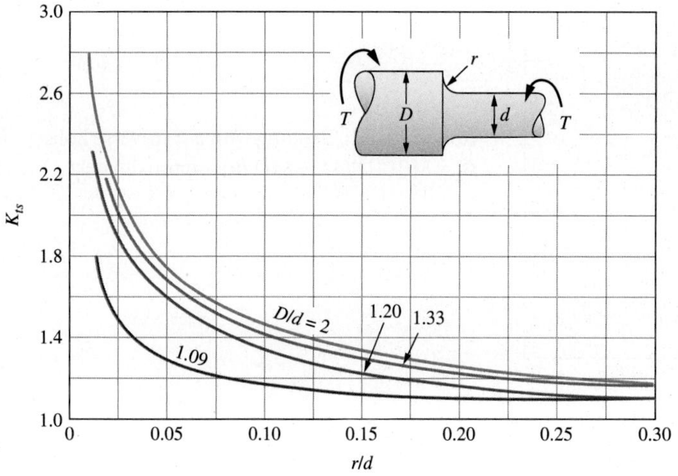

Round shaft with shoulder fillet in torsion.
$\tau_0 = Tc/J$, where $c = d/2$ and $J = \pi d^4/32$.

(i)

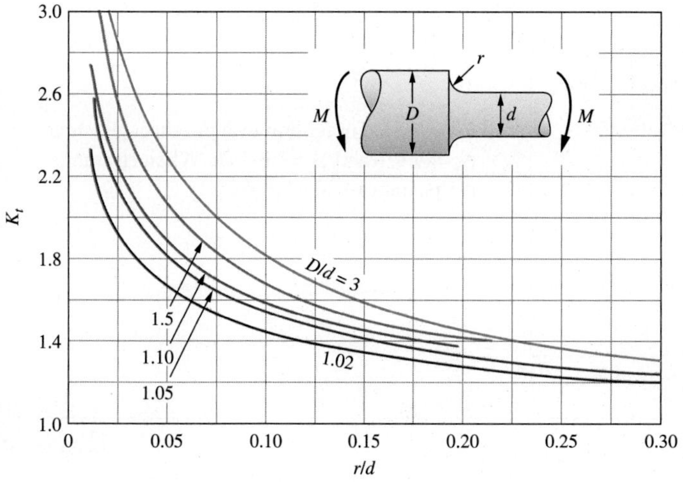

Round shaft with shoulder fillet in bending.
$\sigma_0 = Mc/I$, where $c = d/2$ and $I = \pi d^4/64$.

(*Continued*)

TABLE 17.2 Graphs for Stress Concentration Factors (*Continued*)

(j)

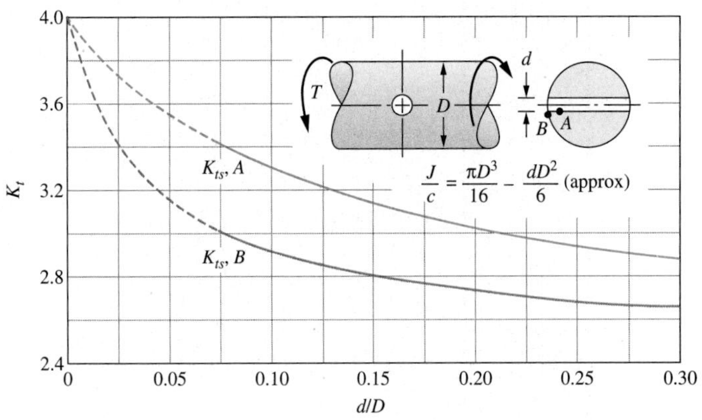

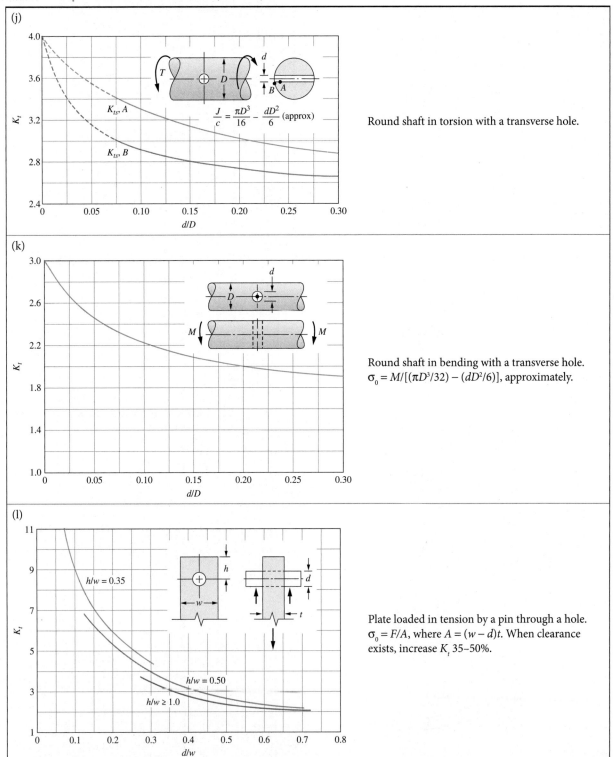

Round shaft in torsion with a transverse hole.

(k)

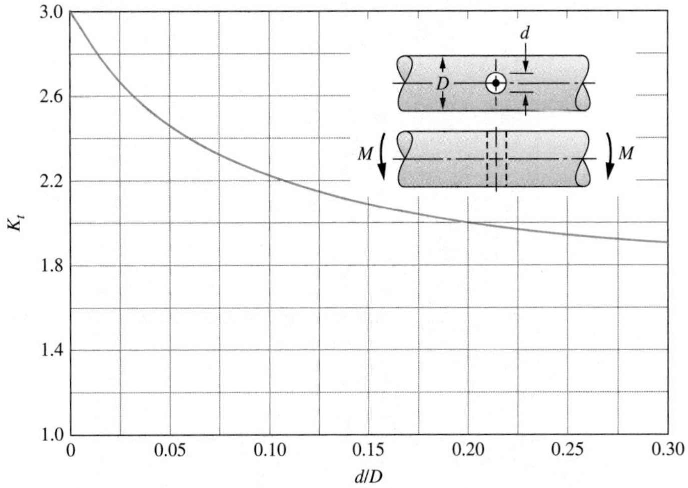

Round shaft in bending with a transverse hole. $\sigma_0 = M/[(\pi D^3/32) - (dD^2/6)]$, approximately.

(l)

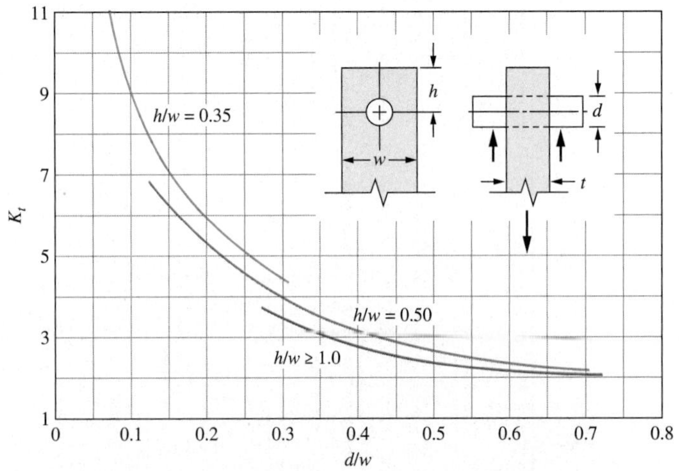

Plate loaded in tension by a pin through a hole. $\sigma_0 = F/A$, where $A = (w - d)t$. When clearance exists, increase K_t 35–50%.

(*Continued*)

TABLE 17.2 Graphs for Stress Concentration Factors (*Continued*)

(m)

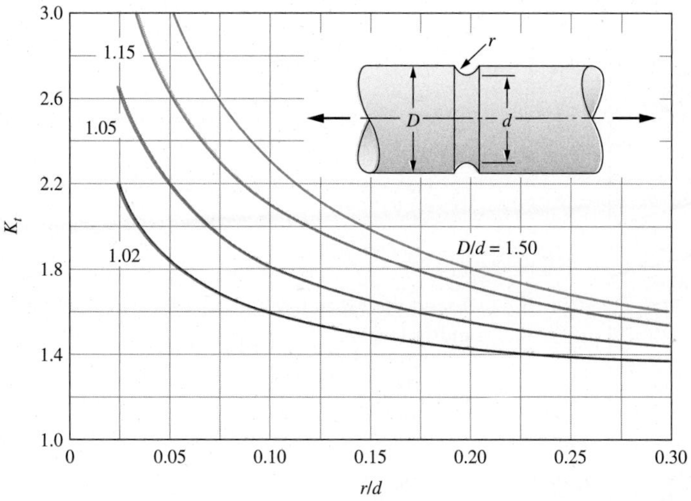

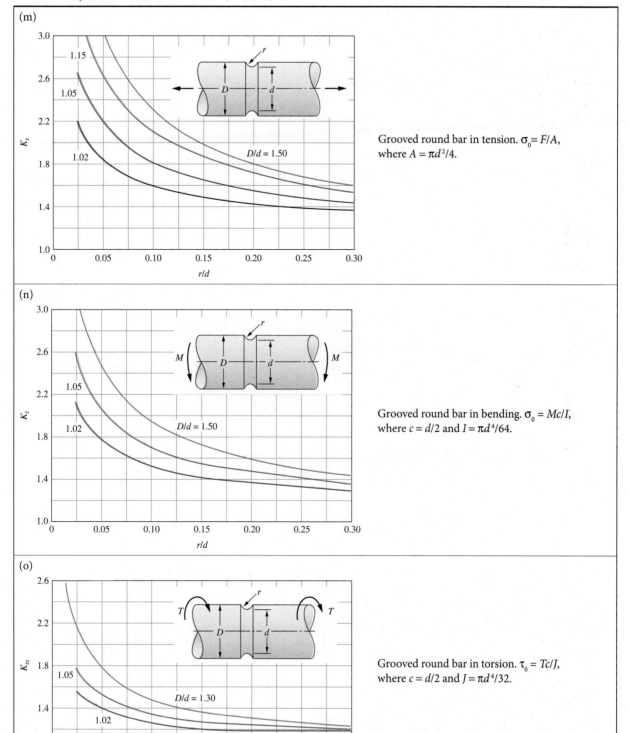

Grooved round bar in tension. $\sigma_0 = F/A$, where $A = \pi d^2/4$.

(n)

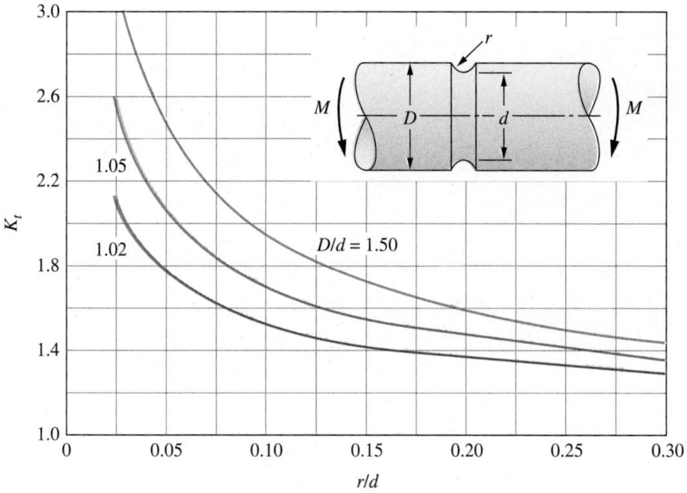

Grooved round bar in bending. $\sigma_0 = Mc/I$, where $c = d/2$ and $I = \pi d^4/64$.

(o)

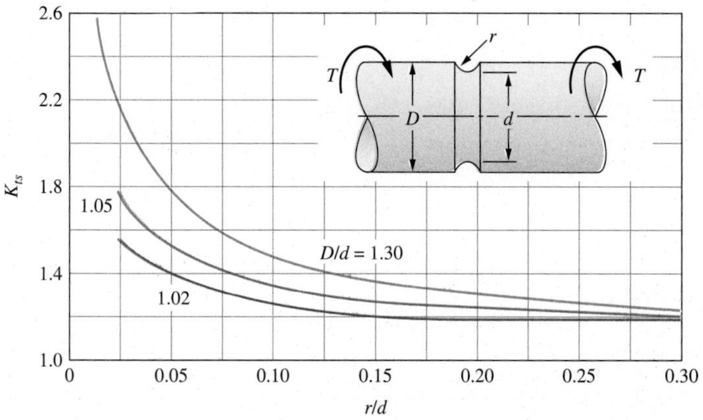

Grooved round bar in torsion. $\tau_0 = Tc/J$, where $c = d/2$ and $J = \pi d^4/32$.

(*Continued*)

TABLE 17.2 Graphs for Stress Concentration Factors (*Continued*)

(p)

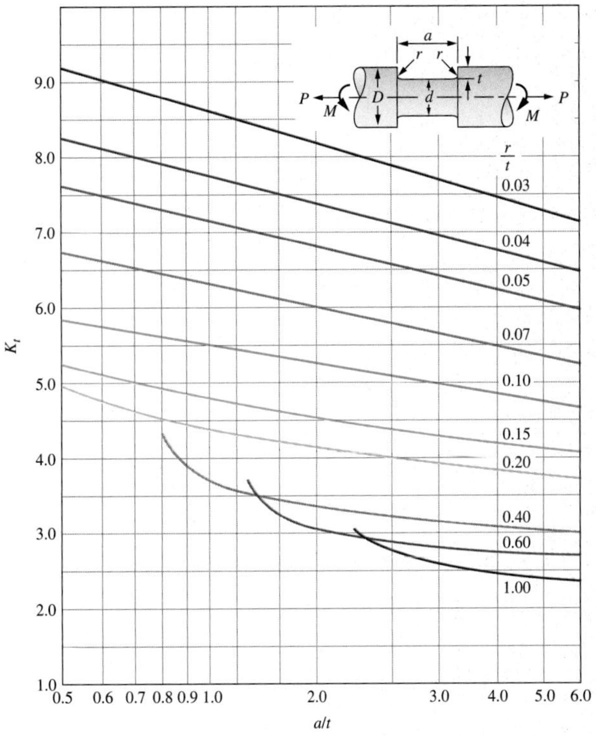

Round shaft with flat-bottom groove in bending and/or tension.

$$\sigma_0 = \frac{4P}{\pi d^2} + \frac{32M}{\pi d^3}$$

(q)

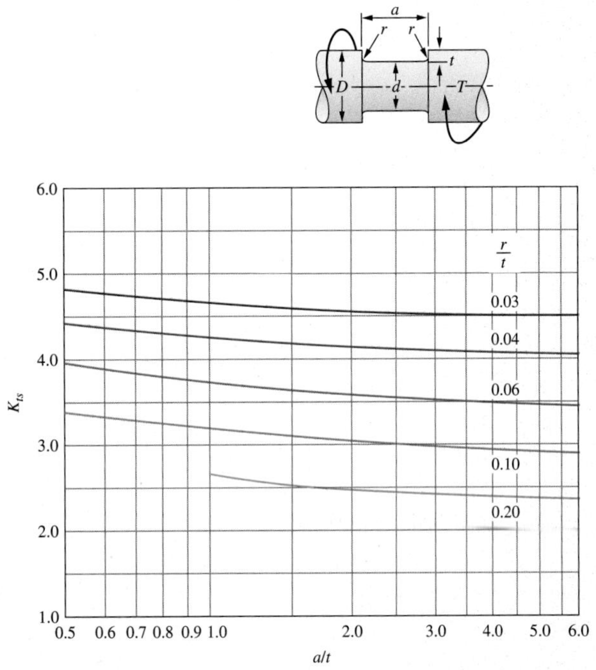

Round shaft with flat-bottom groove in torsion.

$$\tau_0 = \frac{16T}{\pi d^3}$$

Factors from R. Peterson, "Design Factors for Stress Concentration," *Machine Design*, Vol. 23, No. 2, 1951, p. 169. Reprinted with permission from Machine Design, a Penton Media Inc. Publication.

17.4 REFERENCES

1. Peterson, R. E., *Stress Concentration Factors*, John Wiley & Sons, 1974.
2. Isida, M., "On the Tension of the Strip with Semicircular Notches," *Trans. Jap. Soc. Mech. Eng.*, 19(83): 1953.
3. Ling, C-B., "On Stress-Concentration Factor in a Notched Strip," *ASME J. Appl. Mech.*, 35(4): 1968.
4. Flynn, P. D., and A. A. Roll, "A Comparison of Stress-Concentration Factors in Hyperbolic and U-shaped Grooves," *Exp. Mech., J. Soc. Exp. Stress Anal.*, 7(no.): 1967.
5. Flynn, P. D., "Photoelastic Comparison of Stress Concentrations Due to Semicircular Grooves and a Circular Hole in a Tension Bar," *ASME J. Appl. Mech.*, 36(4): December 1969.
6. Slot, T., and D. F. Mowbray, "A Note on Stress-Concentration Factors for Symmetric U-shaped Notches in Tension Strips," *ASME J. Appl. Mech.*, 36(4): 1969.
7. Kikukawa, M., "Factors of Stress Concentration for Notched Bars under Tension and Bending," *Proc. 10th Int. Congr. Appl. Mech.*, Stresa, Italy, 1960.
8. Frocht, M., "Factors of Stress Concentration Photoelastically Determined," *ASME J. Appl. Mech.*, 2(2): 1935.
9. Neuber, H., "Notch Stress Theory," Tech. Rep. AFML-TR-65-225, July 1965.
10. Baratta, F. I., "Comparison of Various Formulae and Experimental Stress-Concentration Factors for Symmetrical U-notched Plates," *J. Strain Anal.*, 7(2): 1972.
11. Wilson, I. H., and D. J. White, "Stress-Concentration Factors for Shoulder Fillets and Grooves in Plates," *J. Strain Anal.*, 8(1): 1973.
12. Ling, C-B., "On the Stresses in a Notched Strip," *ASME J. Appl. Mech.*, 19(2): 1952.
13. Lee, G. H., "The Influence of Hyperbolic Notches on the Transverse Flexure of Elastic Plates," *Trans. ASME*, 62: 1940.
14. Shioya, S., "The Effect of Square and Triangular Notches with Fillets on the Transverse Flexure of Semi-Infinite Plates," *Z. angew. Math. Mech.*, 39: 1959.
15. Appl, F. J., and D. R. Koerner, "Stress Concentration Factors for U-shaped, Hyperbolic and Rounded V-shaped Notches," ASME Pap. 69-DE-2, Engineering Society Library, United Engineering Center, New York, 1969.
16. Cole, A. G., and A. F. C. Brown, "Photoelastic Determination of Stress Concentration Factors Caused by a Single U-Notch on One Side of a Plate in Tension," *J. Roy. Aeronaut. Soc.*, 62: 1958.
17. Roark, R. J., R. S. Hartenberg, and R. Z. Williams, "Influence of Form and Scale on Strength," *Univ. Wis. Eng. Exp. Sta. Bull.*, 1938.
18. Leven, M. M., and M. M. Frocht, "Stress Concentration Factors for a Single Notch in a Flat Bar in Pure and Central Bending," *Proc. Soc. Exp. Stress Anal.*, 11(2): 1953.
19. Fessler, H., C. C. Rogers, and P. Stanley, "Shouldered Plates and Shafts in Tension and Torsion," *J. Strain Anal.*, 4(3): 1969.
20. Hartman, J. B., and M. M. Leven, "Factors of Stress Concentration for the Bending Case of Fillets in Flat Bars and Shafts with Central Enlarged Section," *Proc. Soc. Exp. Stress Anal.*, 19(1): 1951.
21. Kumagai, K., and H. Shimada, "The Stress Concentration Produced by a Projection under Tensile Load," *Bull. Jap. Soc. Mech. Eng.*, 11: 1968.
22. Derecho, A. T., and W. H. Munse, "Stress Concentration at External Notches in Members Subjected to Axial Loading," *Univ. Ill. Eng. Exp. Sta.* Bull. 494, 1968.
23. Roark, R. J., *Formulas for Stress and Strain*, 4th ed., McGraw-Hill, 1965.
24. Savin, G. N., *Stress Concentration Around Holes*, Pergamon Press, 1961.
25. Heywood, R. B., *Designing by Photoelasticity*, Chapman & Hall, 1952.
26. Goodier, J. N., "Influence of Circular and Elliptical Holes on Transverse Flexure of Elastic Plates," *Phil. Mag.*, 22: 1936.
27. Goodier, J. N., and G. H. Lee, "An Extension of the Photoelastic Method of Stress Measurement to Plates in Transverse Bending," *Trans. ASME*, 63: 1941.
28. Drucker, D. C., "The Photoelastic Analysis of Transverse Bending of Plates in the Standard Transmission Polariscope," *Trans. ASME*, 64: 1942.
29. Dumont, C., "Stress Concentration Around an Open Circular Hole in a Plate Subjected to Bending Normal to the Plane of the Plate," NACA Tech. Note 740, 1939.
30. Reissner, E., "The Effect of Transverse Shear Deformation on the Bending of Elastic Plates," Trans. ASME, 67: 1945.
31. Isida, M., "On the Bending of an Infinite Strip with an Eccentric Circular Hole," *Proc. 2nd Jap. Congr. Appl. Mech.*, 1952.
32. Sjöström, S., "On the Stresses at the Edge of an Eccentrically Located Circular Hole on a Strip under Tension," Aeronaut. Res. Inst. Rept. 36, Sweden, 1950.
33. Isida, M., "On the Tension of a Strip with a Central Elliptic Hole," *Trans. Jap. Soc. Mech. Eng.*, 21: 1955.
34. Durelli, A. J., V. J. Parks, and H. C. Feng, "Stresses Around an Elliptical Hole in a Finite Plate Subjected to Axial Loading," *ASME J. App 1. Mech.*, 33(1): 1966.

35. Jones, N., and D. Hozos, "A Study of the Stresses Around Elliptical Holes in Flat Plates," *ASME J. Eng. Ind.*, 93(2): 1971.
36. Isida, M., "Form Factors of a Strip with an Elliptic Hole in Tension and Bending," *Sci. Pap. Fac. Eng.*, Tokushima Univ., 4: 1953.
37. Frocht, M. M., and M. M. Leven, "Factors of Stress Concentration for Slotted Bars in Tension and Bending," *ASME J. Appl. Mech.*, 18(1): 1951.
38. Brock, J. S., "The Stresses Around Square Holes with Rounded Corners," *J. Ship Res.*, October 1958.
39. Sobey, A. J., "Stress Concentration Factors for Rounded Rectangular Holes in Infinite Sheets," Aeronaut. Res. Counc. R&M 3407, Her Majesty's Stationery Office, 1963.
40. Heller, S. R., J. S. Brock, and R. Bart, "The Stresses Around a Rectangular Opening with Rounded Corners in a Uniformly Loaded Plate," *Proc. U.S. Nat. Congr. Appl. Mech.*, 1958.
41. Seika, M., and A. Amano, "The Maximum Stress in a Wide Plate with a Reinforced Circular Hole under Uniaxial Tension—Effects of a Boss with Fillet," *ASME J. Appl. Mech.*, 34(1): 1967.
42. Cheng, Y. F., "Stress at Notch Root of Shafts under Axially Symmetric Loading," *Exp. Mech., J. Soc. Exp. Stress Anal.*, 10(12): 1970.
43. Leven, M. M., "Quantitative Three-Dimensional Photoelasticity," *Proc. Soc. Exp. Stress Anal.*, 12(2): 1955.
44. Rushton, K. R., "Stress Concentrations Arising in the Torsion of Grooved Shafts," *J. Mech. Sci.*, 9: 1967.
45. Hamada, M., and H. Kitagawa, "Elastic Torsion of Circumferentially Grooved Shafts," *Bull. Jap. Soc. Mech. Eng.*, 11: 1968.
46. Matthews, G. J., and C. J. Hooke, "Solution of Axisymmetric Torsion Problems by Point Matching," *J. Strain Anal.*, 6: 1971.
47. Allison, I. M., "The Elastic Concentration Factors in Shouldered Shafts, Part III: Shafts Subjected to Axial Load," *Aeronaut. Q.*, 13: 1962.
48. Allison, I. M., "The Elastic Concentration Factors in Shouldered Shafts, Part II: Shafts Subjected to Bending," *Aeronaut. Q.*, 12: 1961.
49. Allison, I. M., "The Elastic Concentration Factors in Shouldered Shafts," *Aeronaut. Q.*, 12: 1961.
50. Rushton, K. R., "Elastic Stress Concentrations for the Torsion of Hollow Shouldered Shafts Determined by an Electrical Analogue," *Aeronaut. Q.*, 15: 1964.
51. Jessop, H. T., C. Snell, and I. M. Allison, "The Stress Concentration Factors in Cylindrical Tubes with Transverse Circular Holes," *Aeronaut. Q.*, 10: 1959.
52. British Engineering Science Data, 65004, Engineering Science Data Unit, London, 1965.
53. Thum, A., and W. Kirmser, "Uberlagerte Wechselbeanspruchungen, ihre Erzeugung und ihr Einfluss auf die Dauerbarkeit und Spannungsausbildung quergebohrten Wellen," VDI-Forschungsh. 419, vol. 14b, 1943.
54. Fessler, H., and E. A. Roberts, "Bending Stresses in a Shaft with a Transverse Hole," *Selected Papers on Stress Analyses, Stress Analysis Conference*, Delft, 1958, Reinhold, 1961.
55. Atsumi, A., "Stress Concentrations in a Strip under Tension and Containing an Infinite Row of Semicircular Notches," *Q. J. Mech. Appl. Math.*, 11(4): 1958.
56. Durelli, A. J., R. L. Lake, and E. Phillips, "Stress Concentrations Produced by Multiple Semi-circular Notches in Infinite Plates under Uniaxial State of Stress," *Proc. Soc. Exp. Stress Anal.*, 10(1): 1952.
57. Schulz, K. J., "On the State of Stress in Perforated Strips and Plates," *Proc. Neth. Roy. Acad. Sci.*, 45–48: 1942–1945.
58. Dolan, T. J., and E. L. Broghamer, "A Photo-elastic Study of Stresses in Gear Tooth Fillets," *Univ. Ill. Eng. Exp. Sta. Bull. 335*, 1942.
59. Mantle, J. B., and T. J. Dolan, "A Photoelastic Study of Stresses in U-shaped Members," *Proc. Soc. Exp. Stress Anal.*, 6(1): 1948.
60. Budynas, R. G., *Advanced Strength and Applied Stress Analysis*, 2nd ed., McGraw-Hill, 1999.
61. Neuber, H., "Theory of Stress Concentration for Shear Strained Prismatic Bodies with Nonlinear Stress–Strain Law," *J. Appl. Mech.*, Series E, 28(4): 544–550, 1961.
62. Pilkey, Walter D., *Peterson's Stress Concentration Factors*, 2nd ed., 1997.
63. Budynas, R., and K. J. Nisbett, *Shigley's Mechanical Engineering Design*, 9th ed., McGraw-Hill, 2011.

CHAPTER 18

Fatigue and Fracture

Failure due to fatigue and fracture is the primary threat to the integrity, safety, and performance of nearly all highly stressed mechanical structures and machineries. It can have major negative consequences, including serious injury or loss of life, severe environmental damage, and substantial economic loss. The fatigue failure occurs when a material is subjected to repeated loading and unloading that are above a certain threshold. For example, the elements of the structures in spacecraft, gas turbine, automobiles, pipelines and pressure vessels, and ships that are subjected to cyclic repeated loading are susceptible to fatigue and/or fracture failure. Upon localized fatigue failure of a material, a flaw or microscopic crack will initiate and grow to a critical size, and the structure will suddenly fracture. The shape of the structure will significantly affect the fatigue life.

In this chapter, the phenomenon of fatigue and the S-N curve are discussed. The discussion on fracture mechanics, including different modes of fracture and the stress intensity factors of different orientation and geometry of a crack, are then presented.

18.1 FATIGUE IN MATERIALS

When machine elements or structural elements are subjected to time-varying loads, rather than static loads, they fail typically at stress levels significantly lower than the yield strengths of the materials. This phenomenon, which is called "fatigue failure," was first recognized by Rankine who in 1843 published a paper entitled "On the Causes of Unexpected Breakage of Journal of Railway Axles" in which he hypothesized that the material had "crystallized" and become brittle due to the fluctuation stress. Fatigue failure of materials has become progressively more prevalent as technology has developed a greater amount of equipment, such as automobiles, aircraft, compressors, pumps, turbines, etc., which are subjected to repeated loading and vibration. Today it is often stated that fatigue accounts for a significant percent of all service failures due to mechanical causes.

Fatigue failures occur in many different loading conditions such as axial, bending, and torsion. When the temperature of the cyclically loaded component also fluctuates, thermo-mechanical fatigue is induced. The repeated application of loads in conjunction with sliding and rolling contact between materials produces sliding contact fatigue and rolling contact fatigue, respectively, while fretting fatigue occurs as a result of pulsating stresses along with oscillatory relative motion and frictional sliding between surfaces.

At microscale, metals are not homogeneous and isotropic, rather they have discontinuities at the grain boundaries that are induced in the manufacturing or fabrication process. When metals are subjected to cyclic load for a long time, local yielding occurs at a notch or other stress concentration areas. The material yielding causes distortion and creates a slip band along the crystal boundaries of the material. As the stress cycles increase, additional slip bands occur and coalesce into micro-cracks. Once a micro-crack is established, it creates further stress concentration larger than the original notch.

18.2 FATIGUE TESTING

Fatigue characteristics of materials and stress states are generally determined through testing. In fatigue testing, a specimen is subjected to periodically varying constant amplitude stresses. The applied stresses may alternate between equal positive and negative values, from zero to maximum positive or negative values. The most common loading is alternate tension and compression of equal numerical values obtained by rotating a smooth cylindrical specimen while under a bending load. The cyclic load is repeated until the material fails. A series of fatigue tests are made on a number of specimens of the material at different stress levels. The fatigue strength $S_f(N)$ of each specimen, that is, the stress level that a material can endure for N cycles, is then plotted against the number of cycles. This stress-cycle diagram is called an *S-N diagram*. The S-N diagram is generally plotted as stress versus the logarithm of the number of cycles as shown in Fig. 18.1. For ferrous materials, the diagram shows a relatively sharp bend in the curve as the number of cycles increases. The stress level at which the material can withstand an infinite number of cycles is called the *endurance limit*. The endurance limit may be established for most steels between 2 and 10 (most common at 7) million cycles. That is, in most steels the endurance limit is observed as a horizontal line on the S-N curve. Nonferrous metals usually show no clearly defined fatigue limit. The fatigue strength of a part reduces if it has surface defects such as roughness or scratches and notches.

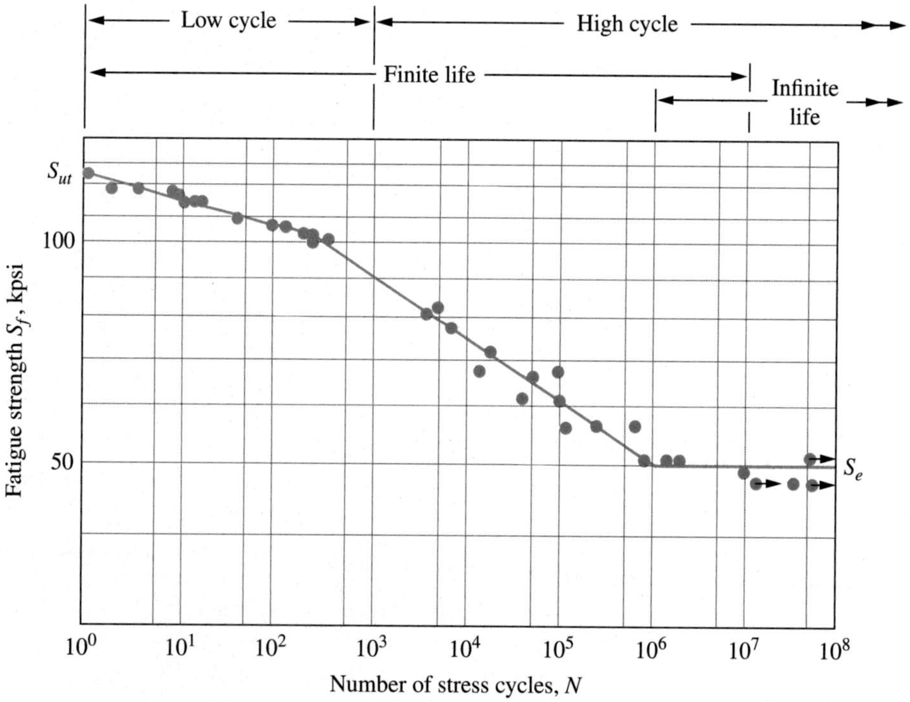

Figure 18.1 Completely reversed cyclic stress, UNS G41200 steel (Ref. 6).

Fatigue loading in some rotating machines often takes the form of a sinusoidal periodical pattern. The cycle of these patterns exhibit a maximum stress σ_{max} and a minimum stress σ_{min}, where the mean stress σ_m and the amplitude stress σ_a are

$$\sigma_m = (\sigma_{max} + \sigma_{min})/2 \qquad \text{and} \qquad \sigma_a = (\sigma_{max} - \sigma_{min})/2, \text{ respectively.}$$

The mean stress σ_m has a profound influence on the limit of amplitude stress σ_a. Several empirical formulas and graphical methods such as the "Soderberg line" and "modified Goodman diagram" have been developed to show the influence of the mean stress on the amplitude stress for failure (Ref. 11). A simple but conservative approach is to plot the amplitude stress σ_a as a function of the mean stress σ_m, as shown in Fig. 8.2. At zero mean stress, that is, completely reversed stress, the amplitude stress σ_a becomes the fatigue limit or the endurance limit σ_e. As the value of the mean stress σ_m increases the failure occurs at the value of the amplitude stress that is less than the endurance limit, that is, σ_a decreases. Yielding will occur if the mean stress exceeds the yield stress σ_y, and this establishes the extreme right-hand point of the Fig. 18.2. A straight line is drawn between these two points, known as Soderberg line. The coordinates of any other point on this line are values of σ_m and σ_a, which may produce failure. This line is modified by Goodman, where instead of the yield stress σ_y the ultimate stress σ_{ult} of the material is used.

The criterion equation for the Soderberg line is

$$\frac{\sigma_a}{\sigma_e} + \frac{\sigma_m}{\sigma_y} = 1$$

Similarly the criterion equation for the modified Goodman line is

$$\frac{\sigma_a}{\sigma_e} + \frac{\sigma_m}{\sigma_{ult}} = 1$$

A common fatigue loading is reverse bending, where the maximum stress is equal to the minimum stress, that is, the amplitude stress σ_a is zero. The typical approximate fatigue limits for reverse bending for various metals is shown in Table 18.1. The endurance limit σ_e for steel materials, when $\sigma_a = 0$, is approximately equal to 0.504 of the tensile strength of the steel. Note that the endurance limit could vary depending on the manufacturing methods, surface condition (roughness), size, and temperature of the loading. Particularly surface defects such as roughness or scratches, and notches all reduce the fatigue strength of a part. With a notch of prescribed geometric form and known concentration factor, the reduction

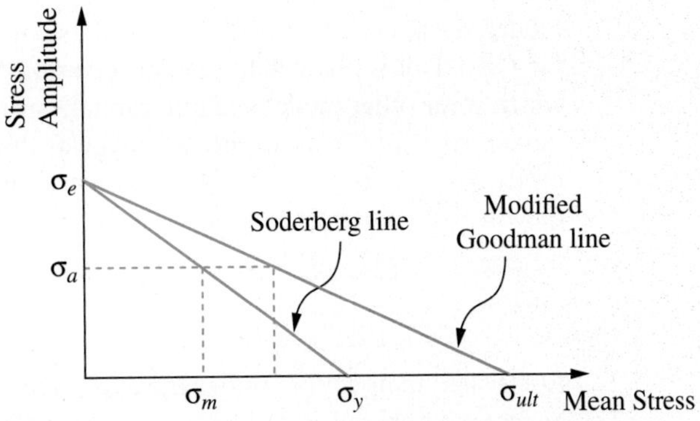

Figure 18.2 Variation of amplitude stress on the mean stress.

TABLE 18.1 Typical Fatigue Limits for Reverse Bending Load

Metal	Tensile Strength, 1000 lb/in^2	Fatigue Limit, 1000 lb/in^2	Metal	Tensile Strength, 1000 lb/in^2	Fatigue Limit, 1000 lb/in^2
Cast iron	20–50	6–18	Copper	32–50	12–17
Malleable iron	50	24	Monel	70–120	20–50
Cast steel	60–80	24–32	Phosphor bronze	55	12
Armco iron	44	24	Tobin bronze, hard	65	21
Plain carbon steels	60–150	25–75	Cast aluminum alloys	18–40	6–11
SAE 6150, heat-treated	200	80	Wrought aluminum alloys	25–70	8–18
Nitralloy	125	80	Magnesium alloys	20–45	7–17
Brasses, various	25–75	7–20	Molybdenum, as cast	98	45
Zirconium crystal bar	52	16–18	Titanium (Ti-75A)	91	45

in strength is appreciably less than would be called for by the concentration factor itself, but the various metals differ widely in their susceptibility to the effect of roughness and concentrations, or notch sensitivity.

18.3 FATIGUE AND CRACK GROWTH

As the material of a machine part is subjected to repeated loads, physical changes and deterioration of the material occur. As indicated in Sec. 18.1, during the cyclic loading small cracks or flaws are generated. These flaws are likely to appear on the surface of the part, that is, fatigue failures generally originate at a surface (except for contact problems). Following nucleation of the crack, it grows during the crack-propagation stage. Eventually the crack becomes large enough and propagates through the material and the part fails. Depending on the material, the failure can be a ductile rupture or a brittle fracture. The rate of crack growth in the crack-propagation stage can be accurately quantified by fracture mechanics methods. The rate of crack growth per cycle can generally be expressed as

$$da/dN = C(\Delta K_{\mathrm{I}})^m \tag{18.3-1}$$

where a is the length of the crack, C and m are constants for a particular material, and ΔK_{I} is the range of mode I stress intensity factor per cycle. Note that the various fracture modes, the stress intensity factor K_{I} and its critical value K_{IC} are discussed in Sec. 18.5. Using Eq. (18.3-1) it is possible to predict the number of cycles for the crack to grow to a size at which some other mode of failure can take over. Values of the constants C and m are determined from specimens of the same type as those used for determination of K_{IC} (known as critical stress intensity factor) but are instrumental for accurate measurement of slow crack growth. The values of C and m are given in Table 18.2.

18.4 CREEP

At elevated temperature under the application of a constant load, metallic and nonmetallic materials exhibit a gradual flow or creep. The deformation that can be permitted in the satisfactory operation of most high-temperature equipment is limited. In metals, creep is a

TABLE 18.2 Values of Factor *C* and Exponent *m* in Eq. (18.3-1)

Material	$C, \dfrac{m/\text{cycle}}{(\text{MPa}\sqrt{\text{m}})^m}$	m
Ferritic-pearlitic steels	6.89 (10^{-12})	3.00
Martensitic steels	1.36 (10^{-10})	2.25
Austenitic stainless steels	5.61 (10^{-12})	3.25
Al 7075-T651	2.33 (10^{-11})	2.88
Al 7075-T73	1.49 (10^{-11})	3.32
Al 7076-T6	5.01 (10^{-10})	3.00
Man-Ten Steel	1.62 (10^{-11})	3.11
USS-T1 Steel	2.51 (10^{-11})	3.08

From J. M. Barsom and S. T. Rolfe, "Fatigue and Fracture Control in Structure," 2nd ed., Prentice Hall, Upper Saddle River, NJ, 1987, pp. 288–291. Copyright ASTM International. Reprinted with permission.

plastic deformation caused by slip occurring along crystallographic directions in the individual crystals, together with some flow of the grain-boundary material. Upon removing the load, a small fraction of this plastic deformation is recovered with time; however, most of the flow is nonrecoverable, the deformation is permanent.

The *long-time creep test* is conducted by applying a dead weight to one end of a lever system, the other end being attached to the specimen surrounded by a furnace and held at constant temperature. The axial deformation is read periodically throughout the test and a curve is plotted of the strain ε_0 as a function of time t. Based on 1000-h tests (stresses in 1000 psi and for a given creep rate and at a different temperature), the generated stresses in the metals (in psi) are given in Table 18.3.

18.5 FRACTURE MECHANICS

Almost all engineering materials used in machines and structures have imperfections such as infinitesimal flaws and cracks which reduce the load-carrying capacity of the structure. That is, materials sometimes fail by rapid fracture at a stress much below their strength level as determined in static tests. These catastrophic and brittle failures originate at preexisting stress-concentrating flaws which may be inherent in a material. In fact, the term *brittle fracture* generally implies the separation of a component into two parts without excessive deformation.

The first successful analysis of a fracture problem is due to A. A. Griffith who in 1920 (Ref. 1) postulated that an existing crack rapidly propagates when the strain energy released from the stressed body equals or exceeds that required to create the surfaces of the crack. After World War II, when some catastrophic failures such as bridges and pressure vessels were attributed to fracture failures, attention of researchers was drawn to growth of a crack. In the middle of 1950s, Griffith's concept was considerably expanded by G. R. Irwin (Ref. 2), who showed that the energy approach is equivalent to a stress intensity factor (K) approach. That is, fracture occurs when a critical stress distribution ahead of the crack tip reaches a critical value of stress intensity factor K_c, or in terms of energy as a critical value G_c (K_c and G_c are explained in Secs. 18.6 and 18.8, respectively), then the crack propagates.

TABLE 18.3 Stresses in Metals at High Temperature for Two Given Creep Rates

	Creep Rate 0.1% per 1000 h					Creep Rate 0.01% per 1000 h				
Material Temp, °F	800	900	1000	1100	1200	800	900	1000	1100	1200
Wrought steels:										
SAE 1015	17–27	11–18	3–12	2–7	1	10–18	6–14	3–8	1	
0.20 C, 0.50 Mo	26–33	18–25	9–16	2–6	1–2	16–24	11–22	4–12	2	1
0.10–0.25 C, 4–6 Cr + Mo	22	15–18	9–11	3–6	2–3	14–17	11–15	4–7	2–3	1–2
SAE 4140	27–33	20–25	7–15	4–7	1–2	19–28	12–19	3–8	2–4	1
SAE 1030–1045	8–25	5–15	5	2	1	5–15	3–7	2–4	1	
Commercially pure iron	7		4		3	5		2		
0.15 C, 1–2.5 Cr, 0.50 Mo	25–35	18–28	8–20	6–8	3–4	20–30	12–18	3–12	2–5	1–2
SAE 4340	20–40	15–30	2–12	1–3		8–20		1–6		
SAE X3140	7–10		5–4			3–8		1–2		
0.20 C, 4–6 Cr	30	10–20	7–10	1				3–5		
0.25 C, 4–6 Cr + W	30	10–15	4–10	2–8			6–11	2–7		
0.16 C, 1.2 Cu		18	10–15	3	1		10–18	7–12		
0.20 C, 1 Mo	35	27	12			25	12	6		
0.10–0.40 C, 0.2–0.5 Mo, 1–2 Mn	30–40	12–20	4–14			25–28	8–15	2–8		0.5
SAE 2340	7–12	5	2							
SAE 6140	30	12	4			7	6	1		
SAE 7240	30	21	6–15	2		30	11	3–9	1	
Cr + Va + W, various	20–70	14–30	5–15			18–50	8–18	2–13		
Temp, °F	1100	1200	1300	1400	1500	1000	1100	1200	1300	1400
Wrought chrome-nickel steels:										
18–8	10–18	5–11	3–10	2–5	2.5	11–16	5–12	2–10		1–2
10–25 Cr, 10–30 Ni	10–20	5–15	3–10	2–5			6–15	3–10	2–8	1–3
Temp, °F	800	900	1000	1100	1200	800	900	1000	1100	1200
Cast steels:										
0.20–0.40 C	10–20	5–10	3			8–15		1		
0.10–0.30 C, 0.5–1 Mo	28	20–30	6–12	2		20	10–15	2–5		
0.15–0.30 C, 4–6 Cr + Mo	25–30	15–25	8–15	8		20–25	9–15	2–7	2	
18–8			20–25	15	10			20	15	8
Cast iron	20	8	4			10		2		
Cr-Ni cast iron			9					3		

These values were shown to be material properties and provided the basis for development of the discipline of linear elastic fracture mechanics (LEFM).

In linear elastic fracture mechanics (LEFM) the material is assumed to be isotropic and linear elastic. Based on the assumption, the stress field near the crack tip is calculated using the theory of elasticity. LEFM is valid only when the inelastic deformation at the crack tip is small compared to the size of the crack, what is so called "small-scale yielding." If large zones of plastic deformation develop before the crack grows, *elastic-plastic* fracture mechanics (EPFM) must be used.

18.6 THE STRESS INTENSITY FACTOR

Consider an elliptical hole of major axis $2a$ and minor axis $2b$ in an infinite plane that is subjected to a uniaxial tensile stress $\sigma_y = \sigma$, as shown in Fig. 18.3. The maximum stress (stress concentration) occurs at (+/– a, 0) and is given by, $\sigma_y = \sigma\,(1 + 2a/b)$. If b is decreased such that $a >> b$, then the elliptical hole becomes a crack of length $2a$ (Fig. 18.4a). Utilizing the elastic solution, the tensile stress in x and y directions, and the shear stress in the x-y plane, at the tip of the crack, can be calculated as (Fig. 18.4b)

$$\sigma_x = \sigma\frac{\sqrt{a}}{\sqrt{2r}}\cos\frac{\theta}{2}\left(1-\sin\frac{\theta}{2}\sin\frac{3\theta}{2}\right),\quad \sigma_y = \sigma\frac{\sqrt{a}}{\sqrt{2r}}\cos\frac{\theta}{2}\left(1+\sin\frac{\theta}{2}\sin\frac{3\theta}{2}\right) \tag{18.6-1}$$

$$\tau_{xy} = \sigma\frac{\sqrt{a}}{\sqrt{2r}}\sin\frac{\theta}{2}\cos\frac{\theta}{2}\cos\frac{3\theta}{2}$$

where σ is the uniaxial tensile stress in the y direction, and r and θ are the cylindrical polar coordinates of a point with respect to the crack tip. Note that σ_z is zero for plane stress and is equal to $\nu(\sigma_x + \sigma_y)$ for plane strain. Equation (18.6-1) reveals that as r approaches zero all the stresses approach infinity. That is, there is a singularity at the crack tip. Because of the singularity, the usual stress concentration approach is not valid.

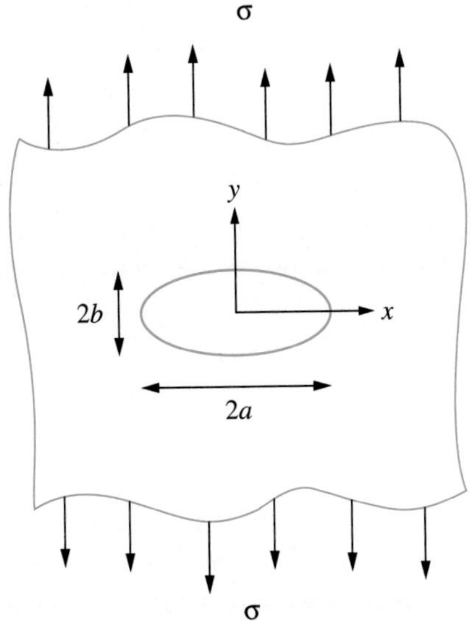

Figure 18.3 Crack 4 infinite plane.

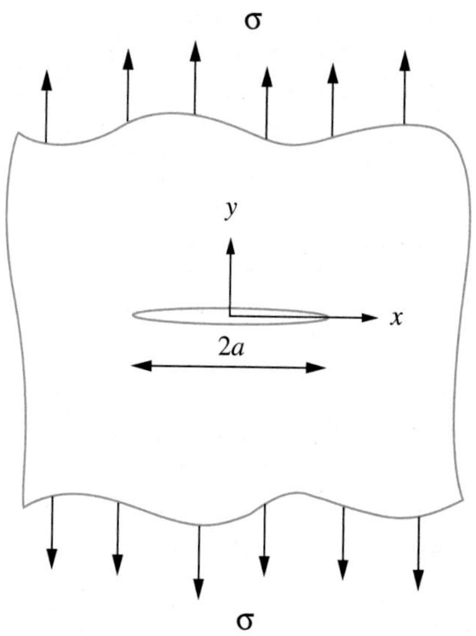

Figure 18.4a An infinite plate and a horizontal crack of 2*a*.

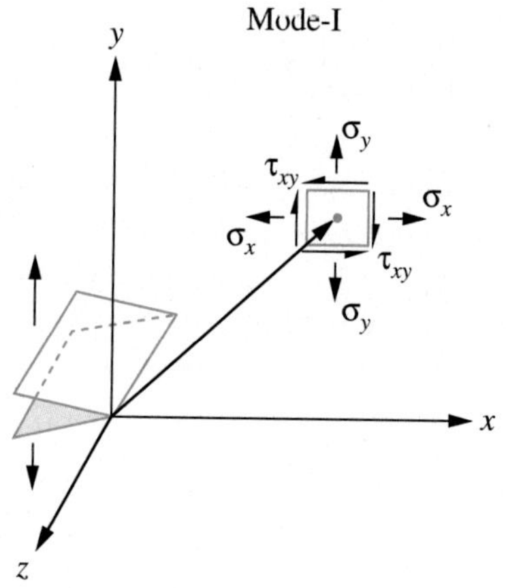

Figure 18.4b Stresses near the crack tip.

However, the intensity of the stress near the crack tip, where $\theta = 0$, is introduced by a factor K, known as *stress intensity factor,* as

$$K = \sigma\sqrt{2\pi r} \tag{18.6-2}$$

Substituting σ_y at the tip with $\theta = 0$ into Eq. (18.6-2), we have

$$K = (\sigma\sqrt{a}/\sqrt{2r})\sqrt{2\pi r} = \sigma\sqrt{\pi a}$$

Depending on the shape and size of the crack and the geometry of the specimen, a constant C_0 is introduced. Therefore, in a more general form, the *stress intensity factor*, K, with the units of $\text{MPa}\sqrt{\text{m}}$ or $\text{ksi}\sqrt{\text{in}}$, is given as

$$K = C_0\,\sigma\sqrt{\pi a} \tag{18.6-3}$$

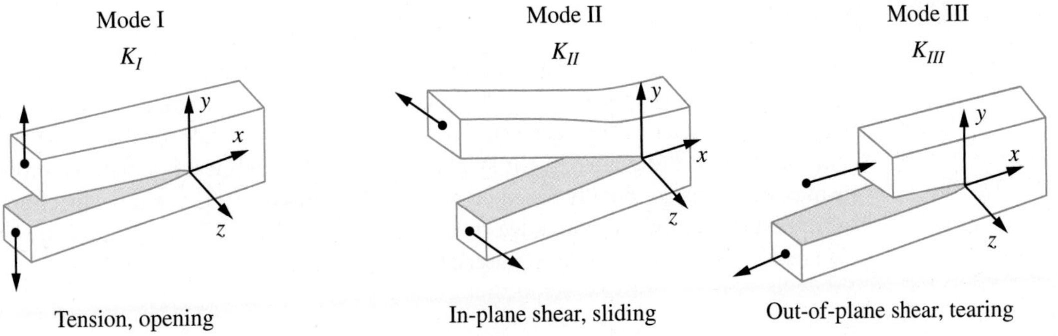

Figure 18.5 Three modes of crack surface displacements.

Generally there are three modes of crack growth, namely, opening, sliding, and tearing, called modes I, II, and III, respectively. These modes describe different crack propagation and surface displacement, as shown in Fig. 18.5. The stress intensity factors associated with the three modes are K_{I}, K_{II}, and K_{III}. The geometry and loading of an infinite plane corresponding to the three modes are shown in Fig. 18.6. In a particular geometry and loading, a crack may propagate in a particular mode or in a combination of these modes.

Therefore, based on a mode I fracture, which is the most common load type encountered in engineering design, the equation (18.6-1) can be written as

$$\sigma_x = \frac{K_{\mathrm{I}}}{\sqrt{2\pi r}} \cos\left(\frac{\theta}{2}\right)\left[1 - \sin\left(\frac{\theta}{2}\right)\sin\left(\frac{3\theta}{2}\right)\right], \quad \sigma_y = \frac{K_{\mathrm{I}}}{\sqrt{2\pi r}} \cos\left(\frac{\theta}{2}\right)\left[1 + \sin\left(\frac{\theta}{2}\right)\sin\left(\frac{3\theta}{2}\right)\right]$$

$$\tau_{xy} = \frac{K_{\mathrm{I}}}{\sqrt{2\pi r}} \sin\left(\frac{\theta}{2}\right)\cos\left(\frac{\theta}{2}\right)\cos\left(\frac{3\theta}{2}\right) \tag{18.6-4}$$

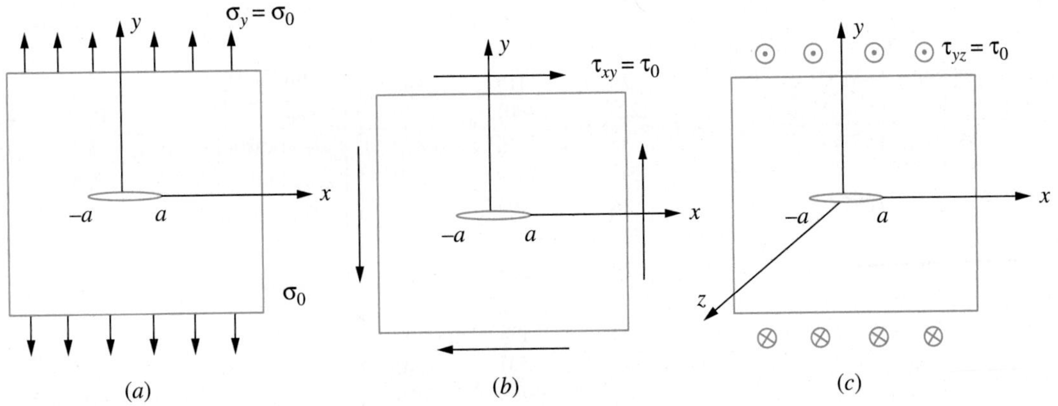

Figure 18.6 Examples for three basic modes of crack tip stress fields.

18.7 FRACTURE TOUGHNESS

The stress intensity factor is a function of the applied load and the geometry of the crack. It has been shown experimentally that as the applied load is increased, at a critical value of stress intensity factor the crack starts to propagate. This critical value is known as *fracture toughness* and is designated by the symbol K_c. Fracture toughness is a material property and is measured experimentally based on an ASTM standard E399 specimen, by either a beam or a tension member with an edge crack at the root of the notch. The critical stress intensity factors K_{Ic}'s for a few common engineering materials are listed in Table 18.4.

TABLE 18.4 Fracture Toughness of Some Materials

Fracture Toughness and Corresponding Tensile Properties for Representative Metals at Room Temperature

Material	Toughness K_{Ic} MPa$\sqrt{m}$ (ksi$\sqrt{in}$)	Yield σ_{ys} MPa (ksi)	Ultimate σ_u MPa (ksi)
(a) Steels			
AISI 1144	66 (60)	540 (78)	840 (122)
ASTM A470-8 (Cr-Mo-V)	60 (55)	620 (90)	780 (113)
ASTM A517-F	187 (170)	760 (110)	830 (121)
AISI 4130	110 (100)	1090 (158)	1150 (167)
18-Ni maraging air melted	123 (112)	1310 (190)	1350 (196)
18-Ni maraging vacuum melted	176 (160)	1290 (187)	1345 (195)
300-M 650°C temper	152 (138)	1070 (156)	1190 (172)
300-M 300°C temper	65 (59)	1740 (252)	2010 (291)
(b) Aluminum and Titanium Alloys (L-T Orientation)			
2014-T651	24 (22)	415 (60)	485 (70)
2024-T351	34 (31)	325 (47)	470 (68)
2219-T851	36 (33)	350 (51)	455 (66)
7075-T651	29 (26)	505 (73)	570 (83)
7475-T7351	52 (47)	435 (63)	505 (73)
Ti-6Al-4V annealed	66 (60)	925 (134)	1000 (145)

Material Polymers	K_{ic} MPa$\sqrt{m}$	K_{ic} (ksi$\sqrt{in}$)
ABS	3.0	(2.7)
Acrylic	1.8	(1.6)
Epoxy	0.6	(0.55)
PC	2.2	(2.0)
PET	5.0	(4.6)
Polyester	0.6	(0.55)
PS	1.15	(1.05)
PVC	2.4	(2.2)
PVC rubber mod.	3.35	(3.05)

Material Ceramics	K_{ic} MPa$\sqrt{m}$	K_{ic} (ksi$\sqrt{in}$)
Soda-lime glass	0.76	(0.69)
Magnesia, MgO	2.9	(2.6)
Alumina, Al_2O_3	4.0	(3.6)
Al_2O_3, 15% ZrO_2	10	(9.1)
Silicon carbide SiC	3.7	(3.4)
Silicon nitride Si_3N_4	5.6	(5.1)
Dolomitic limestone	1.30	(1.18)
Westerly granite	0.89	(0.81)
Concrete	1.19	(1.08)

For some structural materials such as high-strength steels, titanium and aluminum, the fracture toughness slightly changes with the temperature. However, the resistance of steel to fracture decreases as the temperature decreases and as the loading rate increases.

The value of the stress intensity factor, K, is a function of the applied stress, the size and the position of the crack, as well as the geometry of the solid piece where the cracks are detected. Tables in Sec. 18.10 provides the stress intensity factors for different modes, geometry, and loading of materials.

To avoid crack propagation, when designing for a safe crack length a or for the applied stress σ in a geometry, we must have

$$K < K_c$$

Note that the crack size a, the magnitude of the stress σ, and the fracture toughness are primary factors in controlling the brittle fracture failure of a part. To have a nonpropagating crack, the applied stress should create a stress intensity factor K which is less than the fracture toughness of the material. That is, the failure in a part occurs when the stress intensity factor reaches the value of the fracture toughness. Therefore, in design, the factor of safety for fracture mechanics is given by

$$n = K_c/K$$

EXAMPLE Consider a semi-infinite plate made of aluminum alloy 2024, having an edge crack of length 15 mm. The plate is subjected to an in-plane tension σ_0, shown in Fig. 18.7. Estimate the maximum stress σ_0 that can be applied without causing a sudden fracture in the plate; use the factor of safety of 2.5.

For the geometry shown (from case 12 of Table 18.5), we have $K_I = 1.122\sigma_o\sqrt{\pi a}$.

From Table 18.4, the fracture toughness, K_{Ic}, of the material, Al-Ti 2024, is $K_{Ic} = 31$ MPa. Using the factor of safety of 2.5, the allowable fracture toughness is

$$K_{\text{Ic(allowable)}} = 31/2.5 = 12.4 \text{ MPa}$$

Or, $\sigma_0 = K_{\text{Ic(allowable)}}/1.122\sqrt{\pi a}$. Using crack length of $a = 15$ mm, the allowable tensile stress σ_0 is

$$\sigma_0 = 12.4\sqrt{1000}/\left[1.122\sqrt{\pi(15)}\right] = 50.9 \text{ MPa}$$

To prevent crack growth, σ_0 should be less than 50.9 MPa.

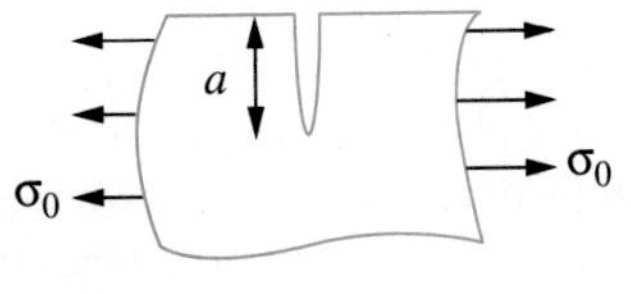

Figure 18.7 Edge crack in a semi-infinite plate.

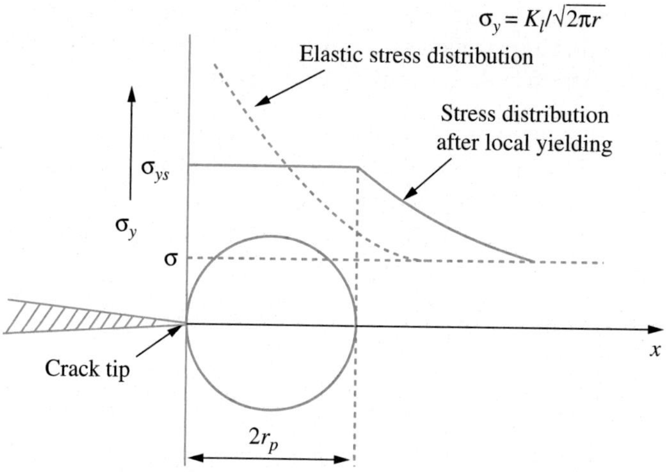

Figure 18.8 A circular plastic zone at the crack tip.

18.8 CRACK TIP PLASTICITY

The linear fracture mechanics method assumes no plastic region at the crack tip and thus there is a stress singularity at the tip of the crack. However, since structural materials deform plastically above the yield stress, in reality, there will be a plastic zone surrounding the crack tip. Tada, Paris, and Irwin (Ref. 3) considered a circular plastic zone with a diameter of r_p to exist at the crack tip under tensile loading (Fig. 18.8). The radius of the plastic zone for plane stress is

$$r_p = \cos^2\frac{\theta}{2}\left(1+3\sin^2\frac{\theta}{2}\right)\left(\frac{1}{2\pi}\right)\left(\frac{K_\text{I}}{\sigma_{ys}}\right)^2$$

At $\theta = 0$, for plane stress the radius of plastic zone is

$$r_p = \frac{1}{2\pi}\left(\frac{K_\text{I}}{\sigma_{ys}}\right)^2 \qquad (18.8\text{-}1)$$

for plane strain it is

$$r_p = \frac{1}{6\pi}\left(\frac{K_\text{I}}{\sigma_{ys}}\right)^2 \qquad (18.8\text{-}2)$$

where σ_{ys} is the yield stress in y direction.

In reality, experimental studies have shown that the plastic zone size and shape at the crack tip of a plate that is subjected to an in-plane tensile stress mode I varies through the thickness of the plate. Generally, depending on the thickness of the plate, there are two distinctive plastic zones—plane stress (zone 1) at the surface and plane strain (zone 2) in the middle of the plane, as shown in Fig. 18.9. In plane stress the size of the plastic zone is large and is in the order of the plate thickness. However, in plane strain, the size of the plastic zone (zone 1) is much smaller than the plate thickness and the second zone will dominate.

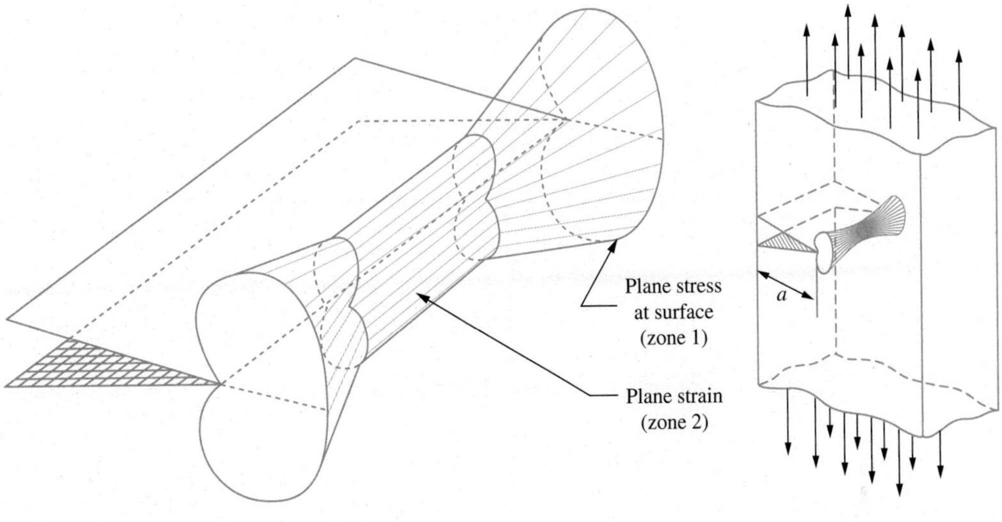

Figure 18.9 Plastic zones at the crack tip.

18.9 THE ENERGY BALANCE APPROACH OF FRACTURE

In linear fracture mechanics, in addition to the elastic stress field (stress intensity factor) approach, another method called *energy balance approach* is also used. Consider a crack of length $2a$ in an elastic plate with unit thickness (Fig. 18.4a). Then the total energy, U, of the elastic plate is

$$U = U_0 + U_a + U_\gamma - F \tag{18.9-1}$$

where U_0 is the elastic energy of the loaded uncracked plate, U_a is the change in the elastic strain energy caused by introducing the crack in the plate, U_γ is the change in elastic surface energy caused by the formation of the crack surfaces, and F is work performed by external forces.

Using the stress field in the vicinity of the crack and Hooke's law, the opening (vertical displacement) of the crack at a point x can be written as

$$d = (2\sigma/E)\sqrt{a^2 - x^2} \tag{18.9-2}$$

Multiplying the opening d by σ and integrating it over the total crack length, the change in the elastic strain energy caused by introducing the crack in the plate is

$$\begin{aligned} U_a &= (\pi\sigma^2 a^2)/E && \text{for plane stress, and} \\ U_a &= (1-\nu^2)(\pi\sigma^2 a^2)/E && \text{for plane strain} \end{aligned} \tag{18.9-3}$$

Note that Eq. (18.9-3) was developed by Griffith, who used the Inglis stress analysis.

Consider now that the stress field σ is increased such that the crack starts to propagate (i.e., crack growth) Δa. The elastic energy of the plate is released as the crack growth continues. Therefore, for infinitesimal crack extension, the elastic *energy release rate,* G, is

$$G = dU/da$$

Using equations (18.9-3), the elastic *energy release rate* is

$$G = dU/da = (\pi\sigma^2 a)/E \qquad \text{for plane stress, and}$$
$$G = dU/da = (1 - \nu^2)\,(\pi\sigma^2 a)/E \qquad \text{for plane strain} \tag{18.9-4}$$

Using the mode I stress intensity factor K_I, we have

$$G = K_I^2/E \qquad \text{for plane stress, and}$$
$$G = (1 - \nu^2)\,K_I^2/E \qquad \text{for plane strain} \tag{18.9-5}$$

Similarly, for a mode II fracture we have

$$G_{II} = K_{II}^2/E \qquad \text{for plane stress, and}$$
$$G_{II} = (1 - \nu^2)\,K_{II}^2/E \qquad \text{for plane strain} \tag{18.9-6}$$

Finally, the *energy release rate* for the case when all three modes of fracture are present is

$$G = [(K_I^2 + K_{II}^2) + (1 + \nu)K_{III}^2]/E \qquad \text{for plane stress, and}$$
$$G = (1 - \nu^2)\,[(K_I^2 + K_{II}^2) + (1 + \nu)K_{III}^2]/E \qquad \text{for plane strain} \tag{18.9-7}$$

The critical value of G is known as R, or the *critical energy release rate*. The fracture will be unstable and will start to propagate if

$$dG/da \geq dR/da \tag{18.9-8}$$

18.10 THE *J* INTEGRAL

In the theory of linear elastic fracture mechanics, where the plastic zone at the crack tip is relatively small, the quantities G and K are used. However, if the plastic zone at the crack tip is not negligible and must be considered, the J integral is used. Rice (Ref. 5) introduced a contour integral J taken about the crack tip as

$$J = \int_{\Gamma} \left(U_0\,dy - t\frac{\partial u}{\partial x}\,ds \right) \tag{18.10-1}$$

where U_0 is the strain energy density, x, y are the coordinate directions, $t = n \cdot \sigma$ is the traction vector, n is the normal to the curve Γ, σ is the Cauchy stress tensor, and u is the displacement vector, Fig. 18.10. Typical units for the J integral are MN/m (meganewton per meter). Note that the J integral is path independent, that is, Γ could be any curve. Note also that if there is no crack, the J integral vanishes along any closed contour. For example, in Fig. 18.11, the J integral for the path of $\Gamma = (\Gamma_1 + AB + \Gamma_2 + CD)$ is zero.

The critical value of J integral is denoted as J_c. Similar expression for the energy release rate is written for the J integral as

$$dJ/da \geq dJ_c/da \tag{18.10-2}$$

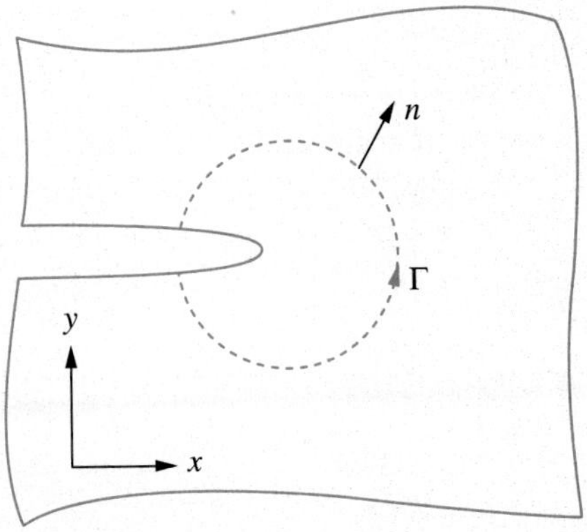

Figure 18.10 J-integral at the crack tip.

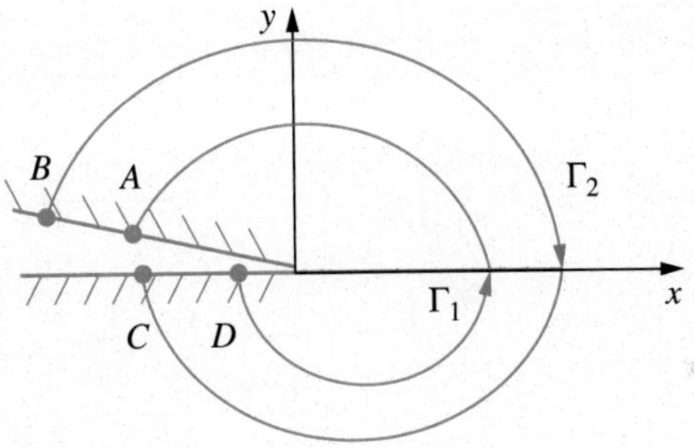

Figure 18. 11 Two contours of J-integral at the crack tip.

For mode I the J at initiation of crack propagation is given by ASTM standard E813. If there is a limited yielding (very small) at the crack tip then J and G are equal to

$$J_I = G_I = K_I^2/E \quad \text{for plane stress, and}$$
$$J_I = G_I = (1 - \nu^2)\, K_I^2/E \quad \text{for plane strain} \tag{18.10-3}$$

Similarly, for a mode II fracture we have

$$J_{II} = G_{II} = K_{II}^2/E \quad \text{for plane stress, and}$$
$$J_{II} = G_{II} = (1 - \nu^2)\, K_{II}^2/E \quad \text{for plane strain} \tag{18.10-4}$$

Finally, for the case when all three modes of fracture are present

$$J_{II} = G = [(K_I^2 + K_{II}^2) + (1 + \nu)K_{III}^2]/E \quad \text{for plane stress, and}$$
$$J_{II} = G = (1 - \nu^2)\,[(K_I^2 + K_{II}^2) + (1 + \nu)K_{III}^2]/E \quad \text{for plane strain} \tag{18.10-5}$$

18.11 TABLES

Following is a useful list of additional stress intensity factors for some simple crack geometries. For plane regions under in-plane loading conditions the results are valid for both plane stress and plane strain cases.* For more tables of stress intensity factors see Ref. 9.

Table 18.5

1

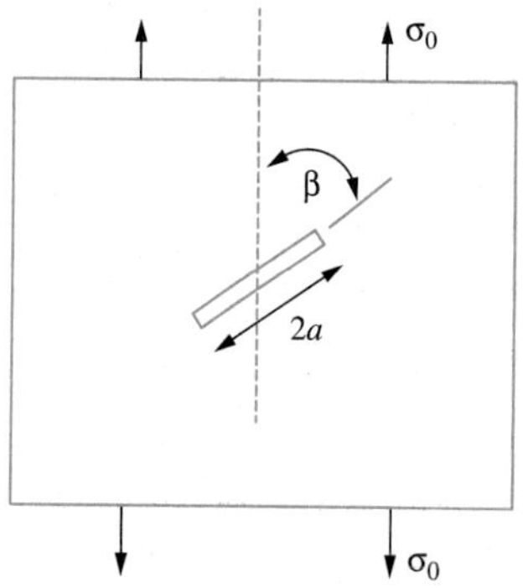

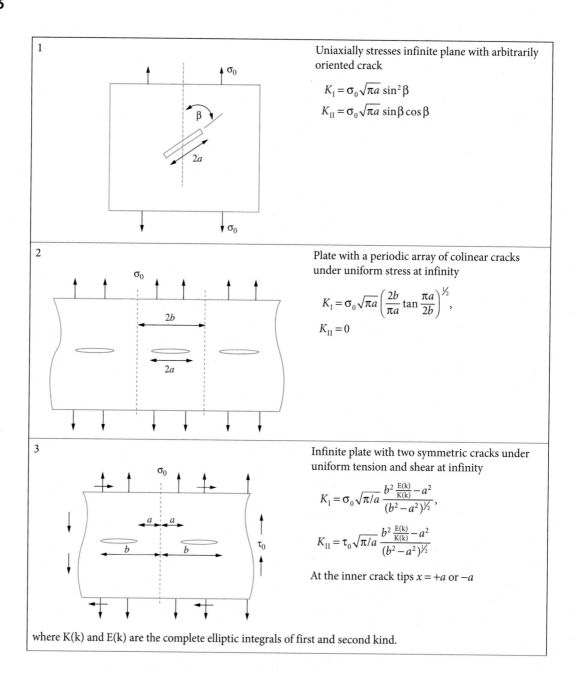

Uniaxially stresses infinite plane with arbitrarily oriented crack

$$K_I = \sigma_0\sqrt{\pi a}\,\sin^2\beta$$

$$K_{II} = \sigma_0\sqrt{\pi a}\,\sin\beta\cos\beta$$

2

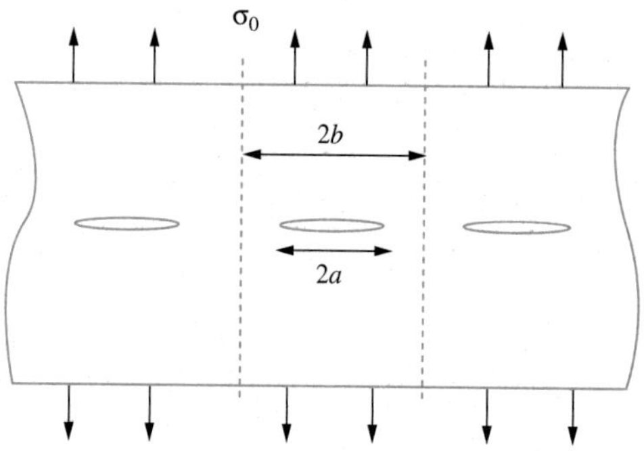

Plate with a periodic array of colinear cracks under uniform stress at infinity

$$K_I = \sigma_0\sqrt{\pi a}\left(\frac{2b}{\pi a}\tan\frac{\pi a}{2b}\right)^{1/2},$$

$$K_{II} = 0$$

3

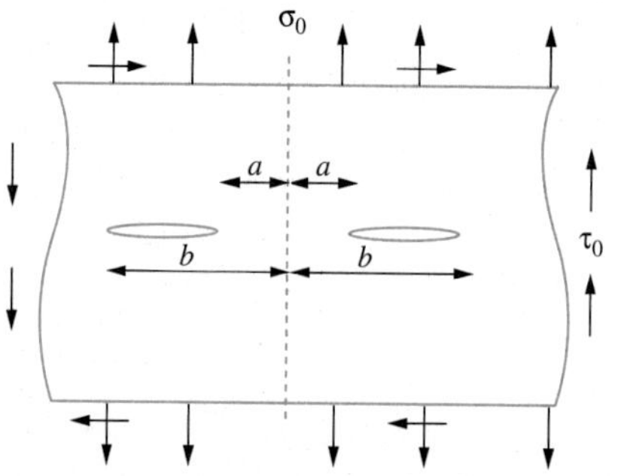

Infinite plate with two symmetric cracks under uniform tension and shear at infinity

$$K_I = \sigma_0\sqrt{\pi/a}\,\frac{b^2\frac{E(k)}{K(k)} - a^2}{(b^2 - a^2)^{1/2}},$$

$$K_{II} = \tau_0\sqrt{\pi/a}\,\frac{b^2\frac{E(k)}{K(k)} - a^2}{(b^2 - a^2)^{1/2}}$$

At the inner crack tips $x = +a$ or $-a$

where K(k) and E(k) are the complete elliptic integrals of first and second kind.

*Items 1 to 19 in the table are from Prof. F. Erdogan's lector notes, with permission, [Ref. 8].

4 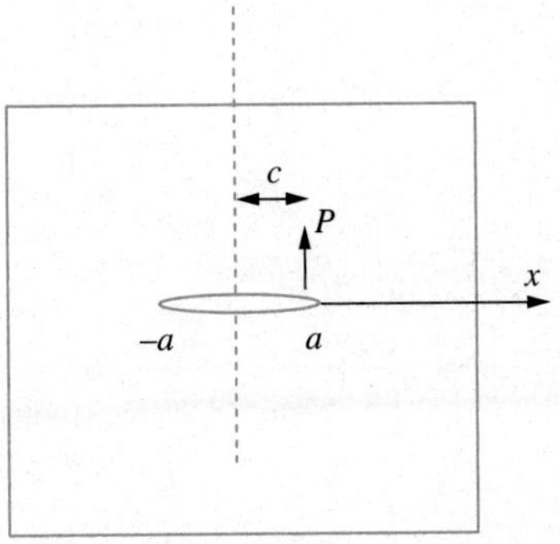	Infinite plate with a concentrated normal force acting on the crack surface $K_{\mathrm{I}}(a)=\frac{P}{2\sqrt{\pi a}}\sqrt{\frac{a+c}{a-c}}$ $K_{\mathrm{II}}(a)=-\frac{\kappa-1}{\kappa+1}\frac{P}{2\sqrt{\pi a}}$ $\kappa=3-\nu$ for plane strain $\kappa=\frac{3-\nu}{1+\nu}$ for plane stress, and ν = Poisson's ratio
5 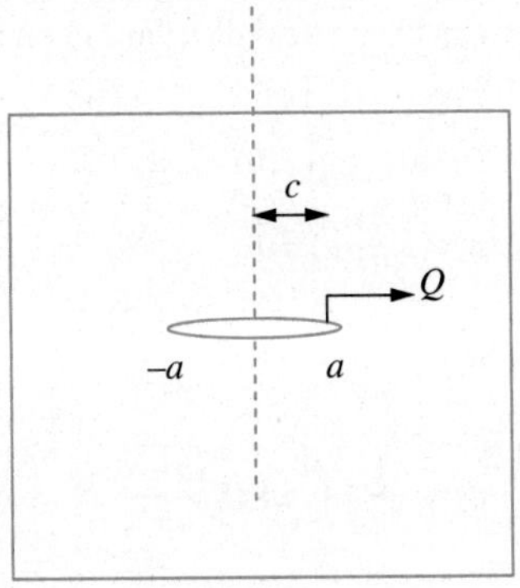	Concentrated shear force $K_{\mathrm{I}}(a)=-K_{\mathrm{I}}(-a)=\frac{\kappa-1}{\kappa+1}\frac{Q}{2\sqrt{\pi a}}$ $K_{\mathrm{II}}(a)=\frac{Q}{2\sqrt{\pi a}}\sqrt{\frac{a+c}{a-c}},$ $\kappa=3-\nu$ for plane strain $\kappa=\frac{3-\nu}{1+\nu}$ for plane stress, and ν = Poisson's ratio
6 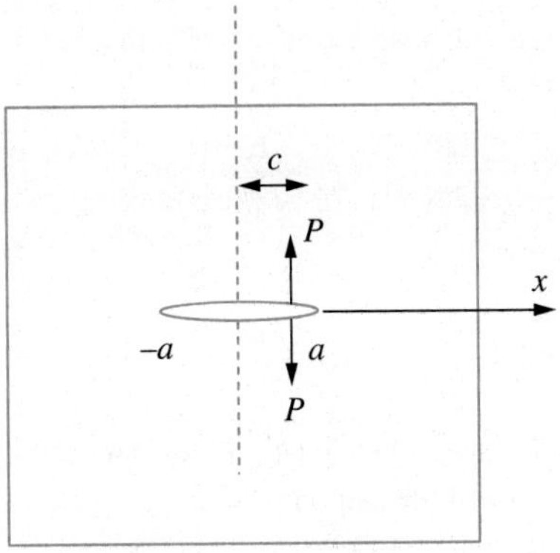	Equal and opposite concentrated wedge forces $K_{\mathrm{I}}(a)=\frac{P}{\sqrt{\pi a}}\sqrt{\frac{a+c}{a-c}},$ $K_{\mathrm{II}}(a)=K_{\mathrm{II}}(-a)=0,$ $K_{\mathrm{I}}(-a)=\frac{P}{\sqrt{\pi a}}\sqrt{\frac{a-c}{a+c}}$
7 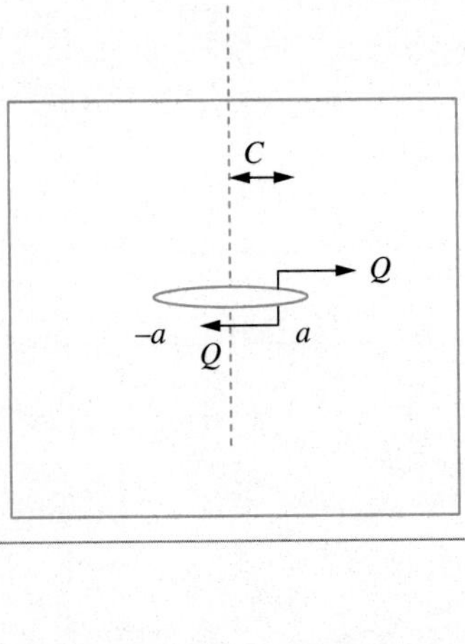	Equal and opposite concentrated shear loads on the crack surface $K_{\mathrm{I}}(a)=0=K_{\mathrm{I}}(-a)$ $K_{\mathrm{II}}(a)=\frac{Q}{\sqrt{\pi a}}\sqrt{\frac{a+c}{a-c}},\quad K_{\mathrm{II}}(-a)=\frac{Q}{\sqrt{\pi a}}\sqrt{\frac{a-c}{a+c}}$

8

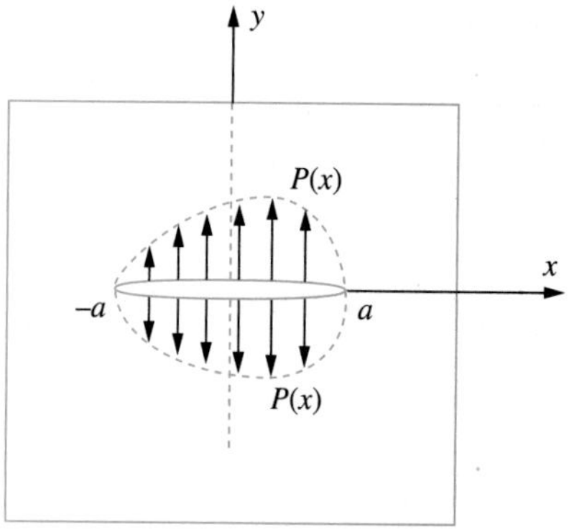

Arbitrary pressure distribution on the crack surface

$$\sigma_y(x,0) = -p(x), \quad \tau_{xy}(x,0) = 0, \quad -a < x < a$$

$$K_{\text{I}}(a) = \frac{1}{\sqrt{\pi a}} \int_{-a}^{a} p(x) \sqrt{\frac{a+x}{a-x}}\, dx,$$

$$K_{\text{II}}(a) = K_{\text{II}}(-a) = 0$$

$$K_{\text{I}}(-a) = \frac{1}{\sqrt{\pi a}} \int_{-a}^{a} p(x) \sqrt{\frac{a-x}{a+x}}\, dx$$

9

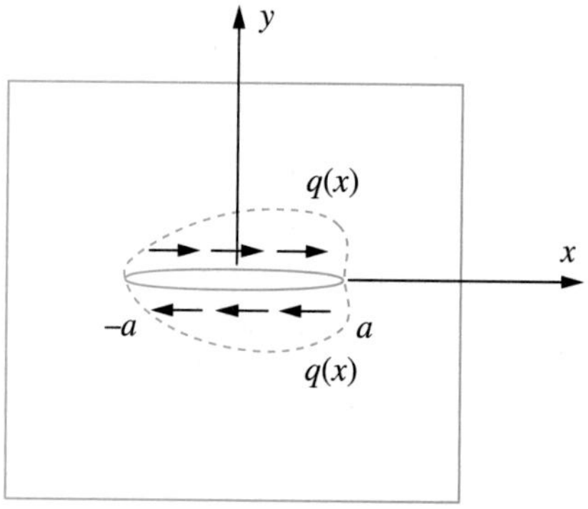

Arbitrary shear stress distribution on the crack surface

$$\sigma_y(x,0) = 0, \quad \tau_{xy}(x,0) = -q(x)$$

$$K_I(a) = K_I(-a) = 0$$

$$K_{\text{II}}(a) = \frac{1}{\sqrt{\pi a}} \int_{-a}^{a} q(x) \sqrt{\frac{a+x}{a-x}}\, dx,$$

$$K_{\text{II}}(-a) = \frac{1}{\sqrt{\pi a}} \int_{-a}^{a} q(x) \sqrt{\frac{a-x}{a+x}}\, dx$$

10

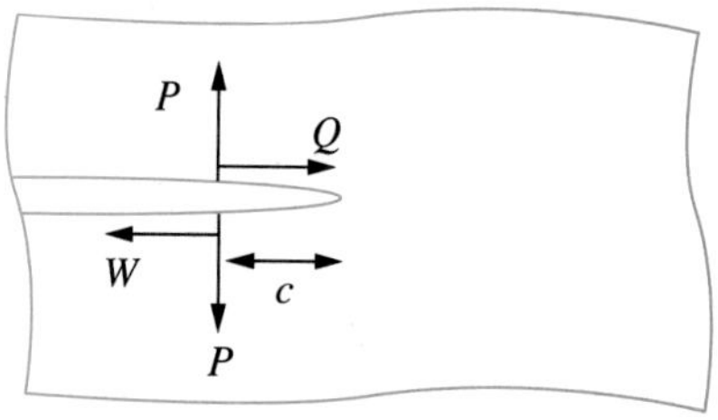

Wedge-loaded infinite plate with a semi-infinite crack

$$K_{\text{I}} = \frac{2P}{\sqrt{2\pi c}}, \quad K_{\text{II}} = \frac{2Q}{\sqrt{2\pi c}}$$

11

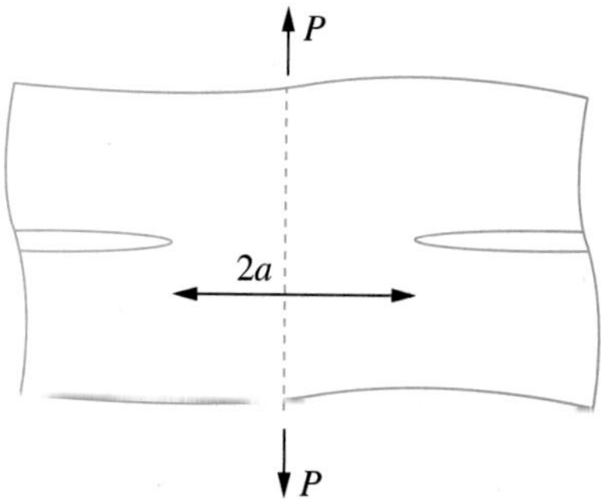

Symmetrically loaded infinite plate with two semi-infinite edge cracks

$$K_{\text{I}} = \frac{P}{\sqrt{\pi a}}, \quad K_{\text{II}} = 0$$

12

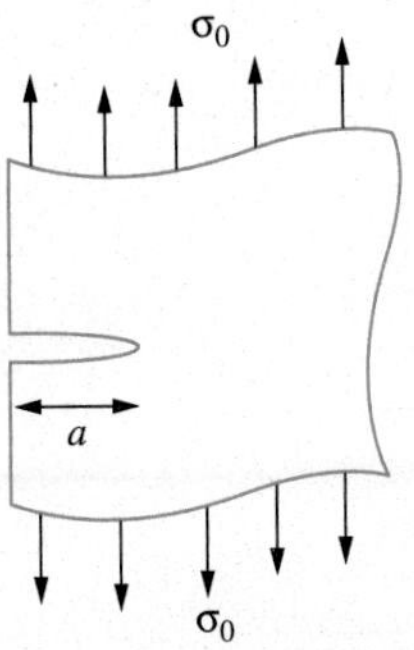

Uniformly loaded semi-infinite plate with an edge crack

$K_I = 1.122\sigma_0\sqrt{\pi a}, \quad K_{II} = 0$

13

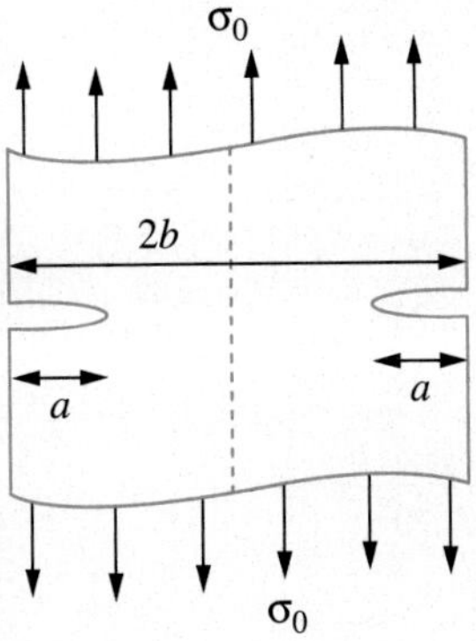

Uniformly loaded strip with symmetric edge cracks

$\frac{a}{b}$	$\frac{K_I}{\sigma_0\sqrt{\pi a}}$
0.9	2.084
0.8	1.539
0.7	1.325
0.6	1.212
0.5	1.155
0.4	1.137
0.3	1.129
0.2	1.125
0.1	1.123
0.001	1.122

$a/b < 0.7$ a good approximation is

$$K_I = \sigma_0\sqrt{\pi a}\ f(a/b)$$

$$f(a/b) = \frac{1}{\sqrt{\pi}}\left[1.98 + 0.36\frac{a}{b} - 2.12\frac{a^2}{b^2} + 3.42\frac{a^3}{b^3}\right]$$

14

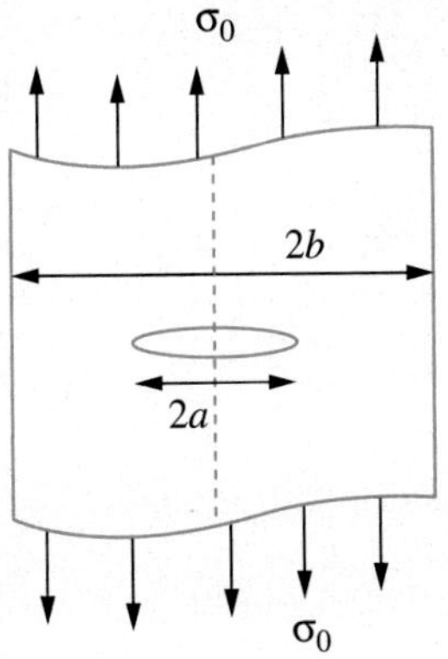

Uniformly loaded strip with a symmetric internal crack

$\frac{a}{b}$	$\frac{K_I}{\sigma_0\sqrt{\pi a}}$	$\frac{1}{\sqrt{\cos\frac{\pi a}{2b}}}$
0.1	1.0060	1.0062
0.2	1.0246	1.0254
0.3	1.0577	1.0594
0.4	1.1094	1.1118
0.5	1.1867	1.1892
0.6	1.3033	1.3043
0.7	1.4884	1.4841
0.8	1.8169	1.7989
0.9	2.585	2.5283
0.95	4.252	3.5701

Good approximation provided $a/b < 0.7$

$$K_I = \sigma_0\sqrt{\pi a}/\sqrt{\cos(\pi a/2b)}$$

15

Infinite plate containing a circular hole with single or double edge cracks under uniaxial or biaxial stress at infinity

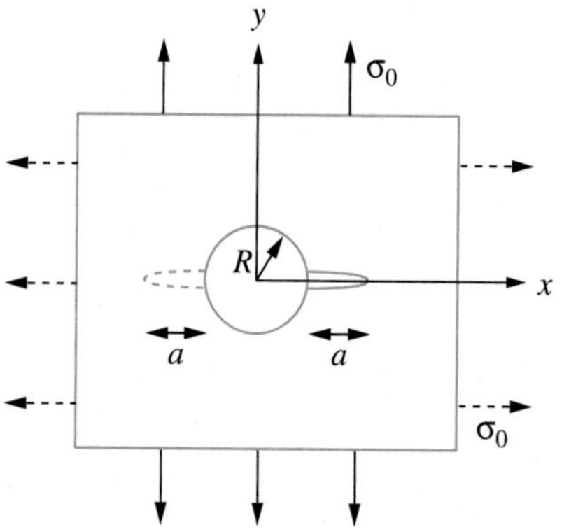

	$K_I/\sigma_0\sqrt{\pi a}$			
	One Crack		Two Cracks	
a/R	$\sigma_y=\sigma_0,\ \sigma_x=0$	$\sigma_y=\sigma_x=\sigma_0$	$\sigma_y=\sigma_0,\ \sigma_x=0$	$\sigma_y=\sigma_x=\sigma_0$
$\to 0$	$\to 3.366$	$\to 2.244$	$\to 3.366$	$\to 2.244$
0.1	2.73	1.98	2.73	1.98
0.2	2.30	1.82	2.41	1.83
0.3	2.04	1.67	2.15	1.70
0.4	1.86	1.58	1.96	1.61
0.5	1.73	1.49	1.83	1.57
0.6	1.64	1.42	1.71	1.52
0.8	1.47	1.32	1.58	1.43
1.0	1.37	1.22	1.45	1.38
1.5	1.18	1.06	1.29	1.26
2.0	1.06	1.01	1.21	1.20
3.0	0.94	0.93	1.14	1.13
5.0	0.81	0.81	1.07	1.06
10.1	0.75	0.75	1.03	1.03
$\to \infty$	$\to 0.707$	$\to 0.707$	$\to 1.00$	$\to 1.00$

16

Uniformly loaded strip with an edge crack

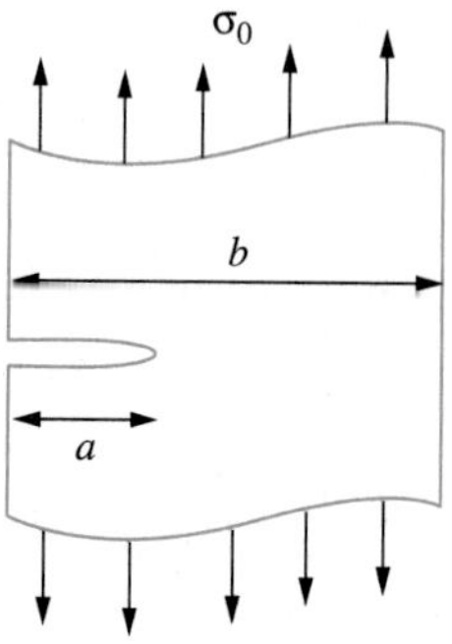

$$K_I=\sigma_0\sqrt{\pi a}\ f(a/b)$$

for

$$a/b \le 0.6$$

$$f(a/b)\cong\frac{1}{\sqrt{\pi}}\left[1.99-0.41\frac{a}{b}+18.70\frac{a^2}{b^2}-38.48\frac{a^3}{b^3}+53.85\frac{a^4}{b^4}\right]$$

17

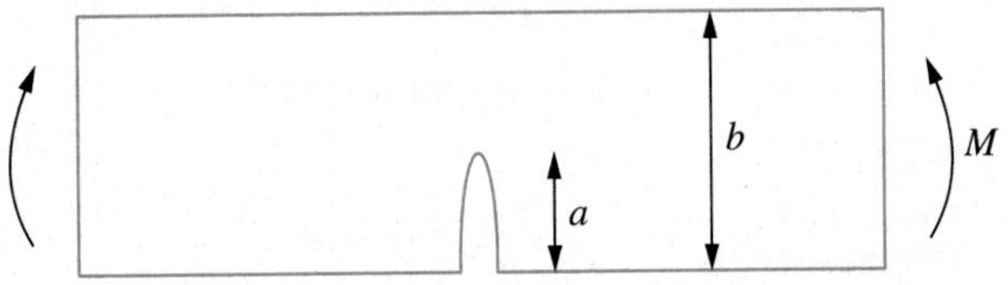

(a)

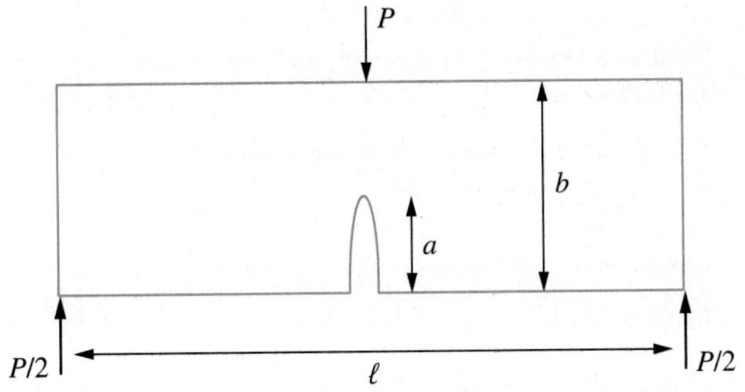

(b)

Edge-cracked strip under bending

$$K_{\mathrm{I}} = \sigma_0 \sqrt{\pi a}\, f(a/b)$$

(a)

$$\sigma_0 = \frac{6\mathrm{M}}{\mathrm{Bb}^2}$$

where B is the thickness and for $a/b \le 0.6$ $f(a/b)$ is approximated by

$$f(a/b) \cong \frac{1}{\sqrt{\pi}} \sum_{n=0}^{4} A_n (a/b)^n$$

(b) constants A_n have the following values:

$$\sigma_0 = \frac{6}{\mathrm{Bb}^2}\frac{Pl}{4}$$

		A_0	A_1	A_2	A_3	A_4
Pure bending (a)		1.99	−2.47	12.97	−23.17	24.80
Three-point bending (b)	$l/b = 8$	1.96	−2.75	13.66	−23.98	25.22
	$l/b = 4$	1.93	−3.07	14.53	−25.11	25.80

18

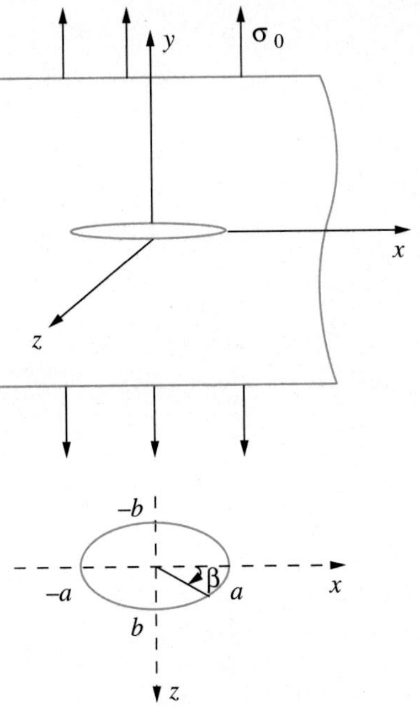

Flat elliptic crack in an infinite solid stressed uniformly perpendicular to the crack plane ($a > b$)

$$K_{\mathrm{I}} = \frac{\sigma_0\sqrt{\pi b}}{\mathrm{E(k)}}\left(\sin^2\beta + \frac{b^2}{a^2}\cos^2\beta\right)^{1/4}$$

where E(k) is the complete elliptic integral of the second kind and is given by

$$\mathrm{E(k)} = \int_0^{\pi/2} \sqrt{1 - \mathrm{k}^2 \sin^2\theta}\, d\theta, \quad \mathrm{k}^2 = (a^2 - b^2)/a^2$$

19

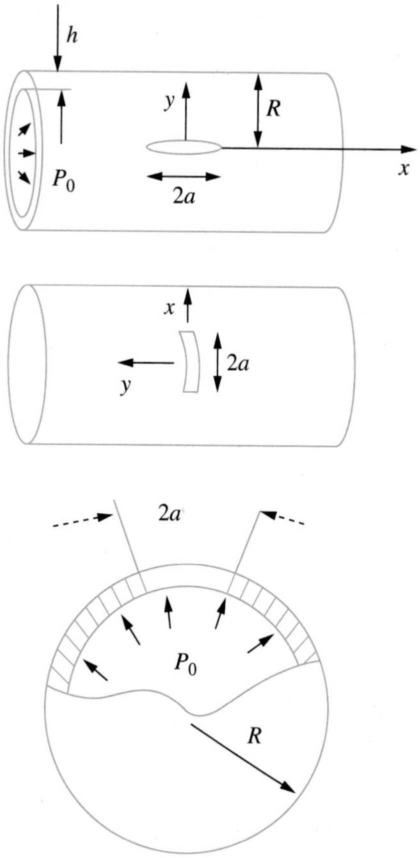

Cracked cylindrical and spherical shells

Membrane component: $k_I^m = \sigma_0\sqrt{\pi a}\, A_m,\quad k_{II}^m = 0$

Bending component: (for the stress on outer surface) $k_I^b = \sigma_0\sqrt{\pi a}\, A_b,\quad k_{II}^b = 0$

The constants A_m and A_b are given in Figures A and B
The shell parameter λ is defined by

$$\lambda = \left[12(1-\nu^2)\right]^{1/4} a/\sqrt{Rh}$$

$2a$, R, and H are the crack length, shell radius, and shell thickness

Cylindrical and spherical shells under skew-symmetric loading (torsion)

Membrane component: $k_I^m = 0,\quad k_{II}^m = \tau_0\sqrt{\pi a}\, C_m$

Bending component: $k_I^b = 0,\quad k_{II}^b = \tau_0\sqrt{\pi a}\, C_b$

where τ_0 is the membrane component of the shear stress away from the crack region and the constants C_m and C_b are given in Figs. C and D

$$\sigma_0 = \frac{P_0 R}{h}$$

where $\sigma_0 = P_0 R/h$ is the membrane stress away from the crack region

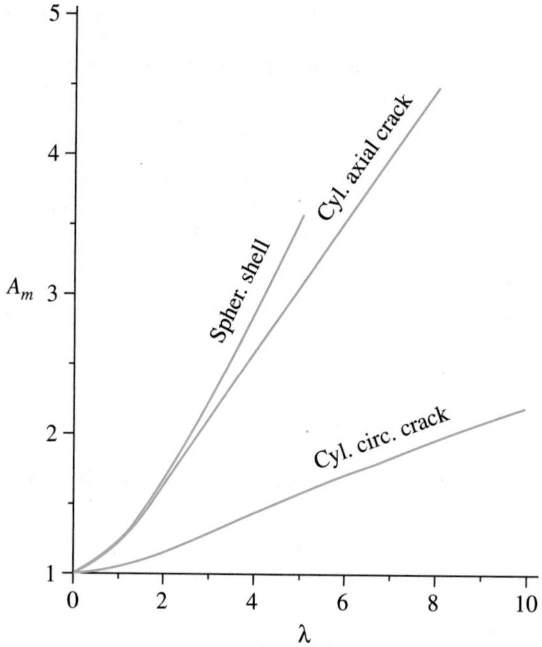

Figure A Membrane component of the stress intensity factor ratio for pressurized shells.

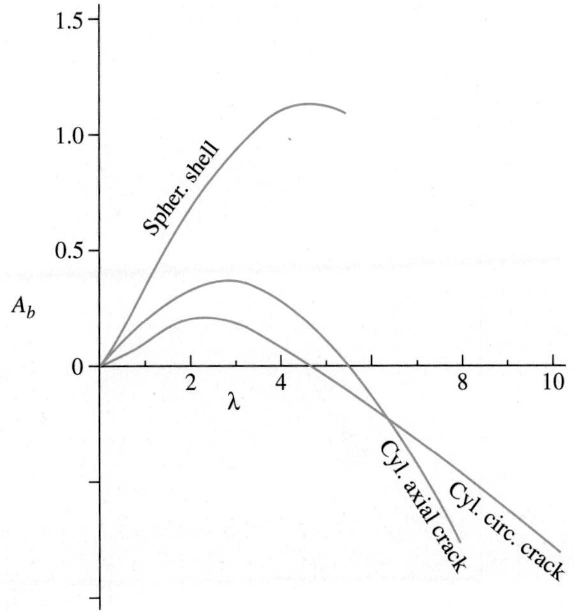

Figure B Bending component of the stress intensity factor ratio for pressurized shells.

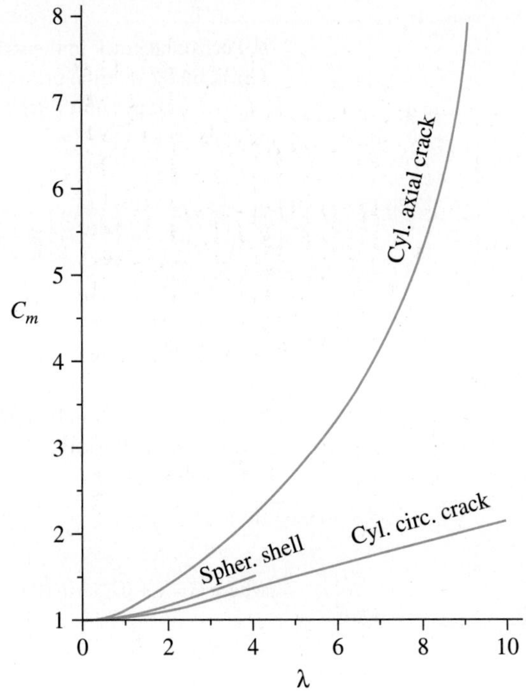

Figure C Membrane component of the stress intensity factor ratio for shells under skew symmetric loading (or under torsion).

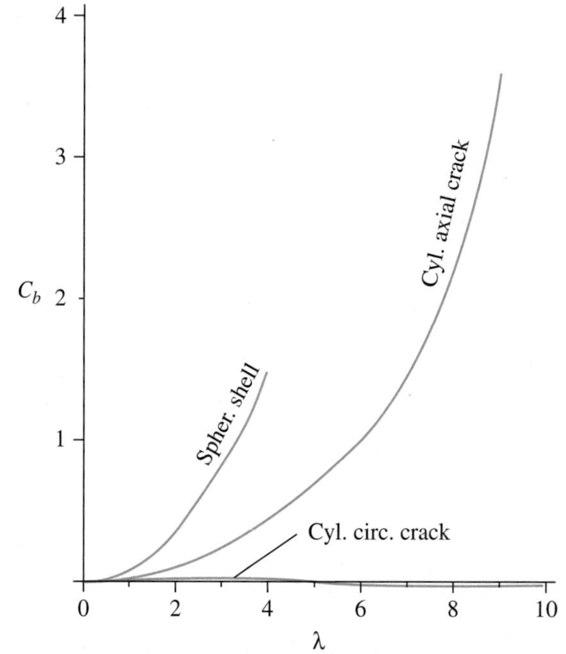

Figure D Bending component of the stress intensity factor ratio for shells under skew symmetric loading.

20

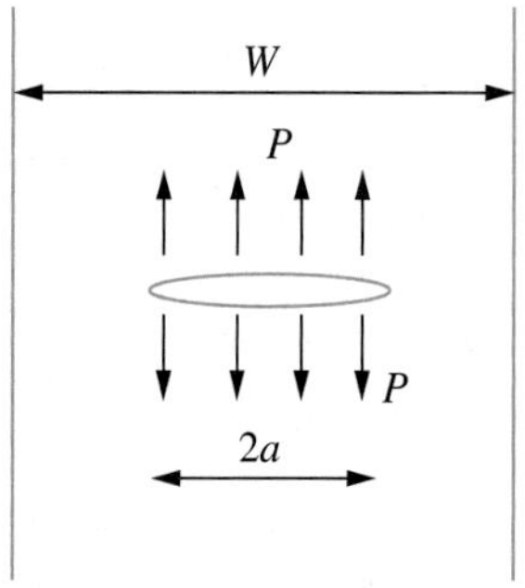

A horizontal crack in the strip of material.
Crack under internal pressure
$K_{\mathrm{I}} = CP\sqrt{\pi a}$ per unit thickness
where $C = 1 + 0.256(a/W) - 1.152(a/W)^2 + 12.200(a/W)^3$

21

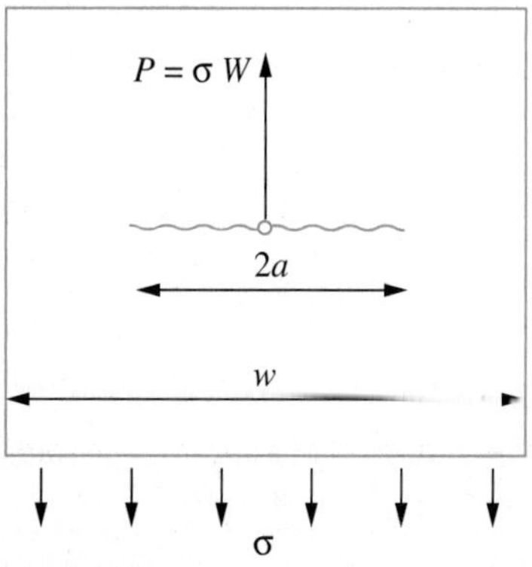

Cracks growing from both sides of a loaded hole, where the hole is small with respect to the crack.

$$K_{\mathrm{I}} = C\left(\frac{\sigma\sqrt{\pi a}}{2} + \frac{P}{2\sqrt{\pi a}}\right) \text{ per unit thickness}$$

where $C = 1 + 0.256(a/W) - 1.152(a/W)^2 + 12.200(a/W)^3$

Formulas of Items 20 and 21 are from Ref. 7.

22

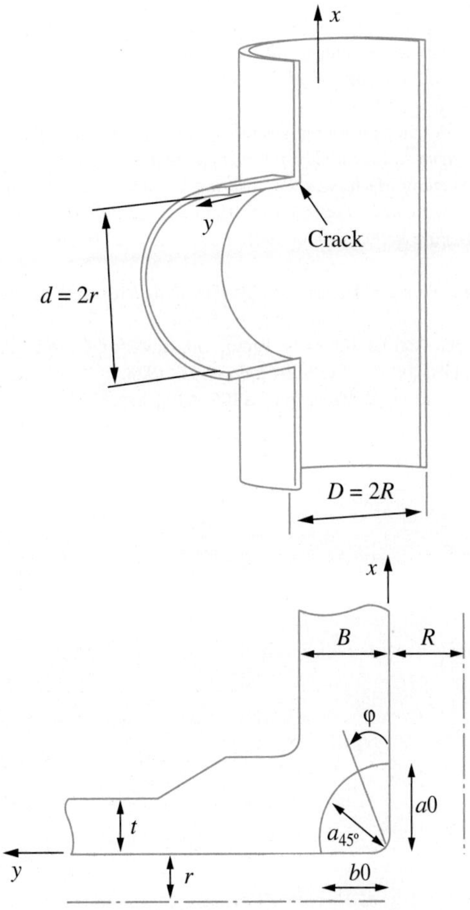

Corner crack in a longitudinal section of a pipe vessel intersection in a pressure vessel.

$$K_{\mathrm{I}}/\left[\sigma_H\sqrt{\pi a}\right]=F_m\left\{1+\left(\frac{rt}{RB}\right)^{1/2}\right\}$$

where σ_H is the hoop stress in the vessel wall

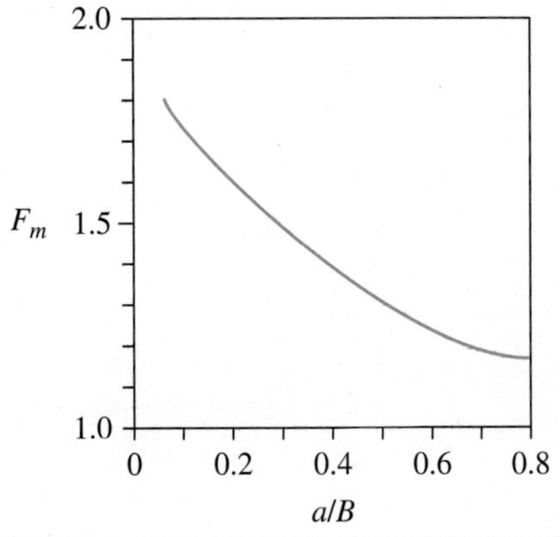

Correction factor for a corner crack in a longitudinal section of a pipe vessel intersection in a pressure vessel.*

*See Ref. 10.

18.12 REFERENCES

1. Griffith, A. A., "The Phenomena of Rupture and Flow in Solids," *Trans. Roy. Soc. Lond.*, Vol. A-221, 1920.
2. Irwin, G. R., *Fracture, Encyclopida of Physics*, ed. Fluge, Springer Verlag, pp. 551–589, 1958, Berlin.
3. Tada, H., P. C. Paris, and G. R. Irwin, *The Stress Analysis of Cracks Handbooks*, St. Louis, MO: Paris Production, 1973.
4. The 7th Joint DoD/FAA/NASA Conference on Aging Aircraft, 8–11 September, 2003, New Orleans, LA.
5. Manson, S. S., and G. R. Halford, *Fatigue and Durability of Structural Materials*, 2006.
6. Budynas, R. G., and J. K. Nisbett, *Shigley's Mechanical Engineering Design*, 9th ed., McGraw-Hill, 2011.
7. Ewalds, H. L., and R. J. H. Wanhill, *Fracture Mechanics*, UK: Edward Arnold Publishers Ltd, 1984.
8. Erdogan, F., "Fracture Mechanics," *Lecture Notes*, Dept. of Mechanical Eng. and Mechanics, Lehigh University, 1976.
9. Cherepanov, G. P., *Mechanics of Brittle Fracture*, Translation edited by R. deWit and W. Cooley, McGraw-Hill, 1979.
10. Mohamed, M. A., and J. Schroeder, "Stress Intensity Factor Solution for Crotch-Corner Cracks of Tee-Intersections of Cylindrical Shells," *Int. J. Fract.*, 14(6): 605–621, 1978.
11. Soderberg, C. R., "Working Stresses," *Jour. Appl. Mech.*, 2(1): A106–A108, Sep. 1935.

CHAPTER 19

Stresses in Fasteners, Joints, and Gears

Two or more separate structural elements are commonly connected using fastening elements such as welds, rivets, bolts, etc. In many fabricated structures, joints are often critically important in affecting the durability, reliability, and safety of the structure. It is important to design these joints such that no separation of the structural elements or failures of the fasteners occur. In this chapter, the analyses of welds, rivets, and bolts subjected to tension, bending, and shearing loads are presented.

In Secs. 19.5 and 19.6, reference is made only to rivets. However, the analyses presented apply to bolts as well.

19.1 WELDING

Welding is a fabrication process where two or more pieces of base metal are joined through melting the workpieces and adding a filler material to form a pool of molten material. Welded connections represent a large group of fabricated metal workpieces. Unlike soldering and brazing, which involve melting a lower-melting-point material between the workpieces to form a bond between them, without melting the workpieces, welding is a complex process involving heat and liquid-metal transfer. There are many different welding processes. Examples include electric arc welding, spot welding (applying an electric current through the workpieces), gas flame welding, laser welding, electron-beam welding, ultrasonic welding, and friction or friction stir welding. These welding processes can be performed as manual, semi-automatic, and automatic processes. Robots can be designed to be used in industrial settings, and new welding methods are being developed to improve the quality and properties of the welded joints. While many welding applications are done in factories and repair shops, some welding processes are performed in underwater and in vacuums.

There are five types of configurations of joining (relative positions) two pieces; butt, tee, corner, lap, and edge, as shown in Fig. 19.1. While there are many forms and shapes of the welds, they fall into three categories: fillet welds, groove welds, and plug or slot welds. Fillet welds have a triangular cross section and are applied to the surface of the materials they join. Groove welds could be complete or partial, penetrating through the thickness of the workpieces. As shown in Fig. 19.2, they are in the form of square, bevel, V-groove, J-groove, and U-groove. In the plug or slot welds, the two pieces of materials are joined through one or more plug(s) or slot(s) that are prefabricated in one of the parts.

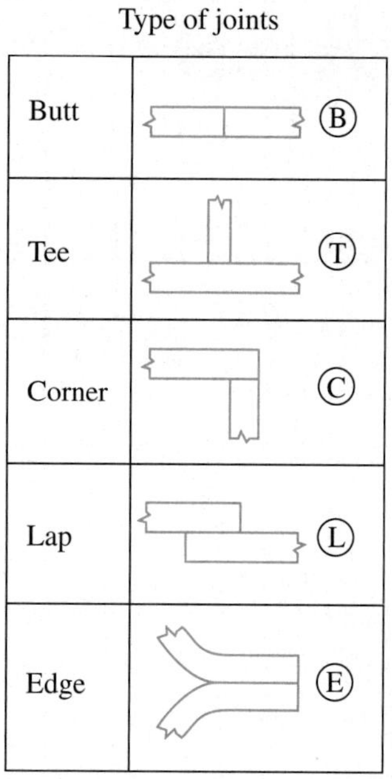

Figure 19.1 Types of joints.

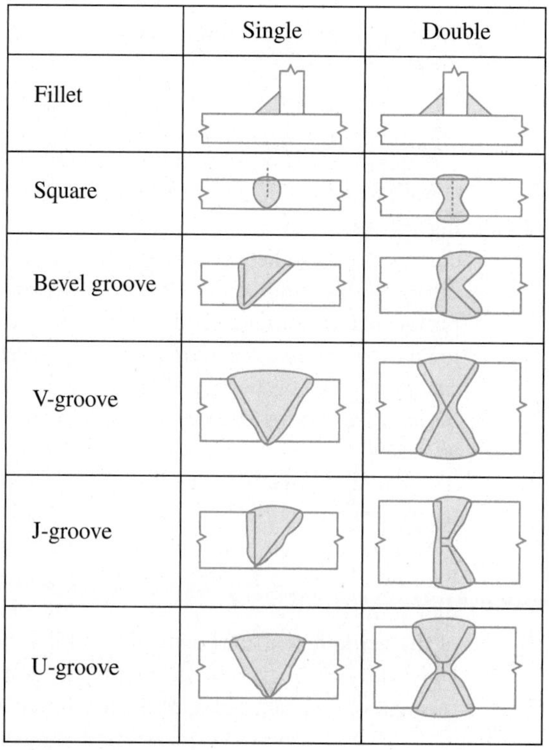

Figure 19.2 Types of welds.

Weld symbol	Description
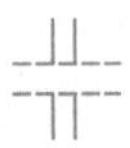	Square
	Scarf
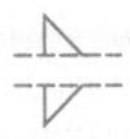	Fillet
	V
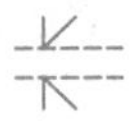	Bevel
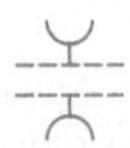	U
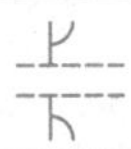	J
	V flare
	Bevel flare

Weld symbol	Description
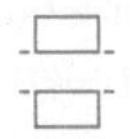	Plug
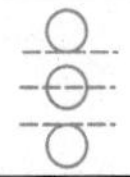	Spot
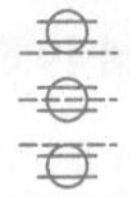	Seam
	Backing
	Surfacing
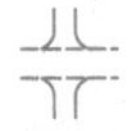	Flange edge
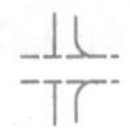	Flange corner
	All around
	Flush
	Convex
	Concave

Figure 19.3 Symbols for welds.

Weld symbols are used to describe the types and forms of the weld. They generally have an arrow-shaped form where the symbols of Fig. 19.3 are used on the arrow to describe the type of weld. For further information see the standards provided by the American Welding Society (AWS).

19.2 ANALYSIS OF WELDED JOINTS

There are three dimensions associated with the welded joints: the throat h, the leg w, and the length of the weld, L. With exception of the length, these dimensions are shown in Fig. 19.4 for different types of welds. In fillet welds, the throat, h, is related to the leg, w, as $h = w \cos 45°$ or $h = 0.707\, w$. In the following subsections, the stresses in welds under different loading are presented.

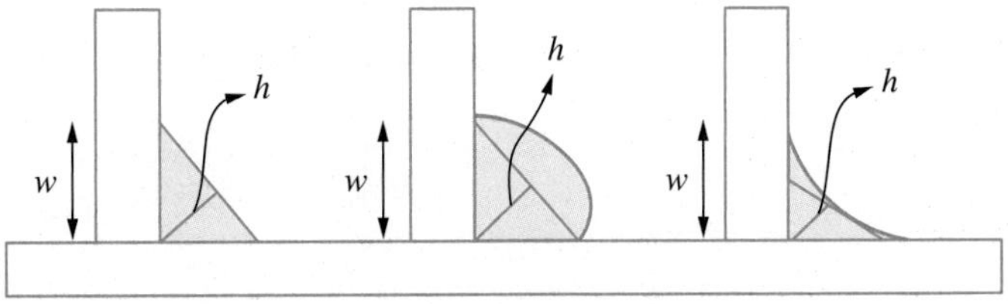

Figure 19.4 Throat h and leg w for different types of fillet welds.

A. Tensile or Compressive Load (Fig. 19.5a)

For a butt, V-groove, or fillet weld loaded by a tensile or compressive force F that is going through the centroid of the weld, the average normal stress is:

$$\sigma = \frac{F}{hL} \tag{19.2-1}$$

B. Shear Load (Figs. 19.5b and 19.6)

For a butt, V-groove, or fillet weld loaded by a shear force F that is going through the centroid of the weld(s) and does not produce any torsion, the average shear stress is

$$\tau = \frac{\text{Force}}{\text{Weld throat area}} = \frac{F}{0.707\, wL} \tag{19.2-2}$$

C. Torsion (Fig. 19.7)

For a weld joint subjected to a shear force F that is not going through the center of the weld, the weld is subjected to torsion. In this case two shearing stresses, *primary* and *secondary shear stresses,* are produced. The primary shear stress due to the overall shear on the length of the weld that is uniformly distributed on the weld is

$$\tau_p = \frac{F}{hL} \tag{19.2-3}$$

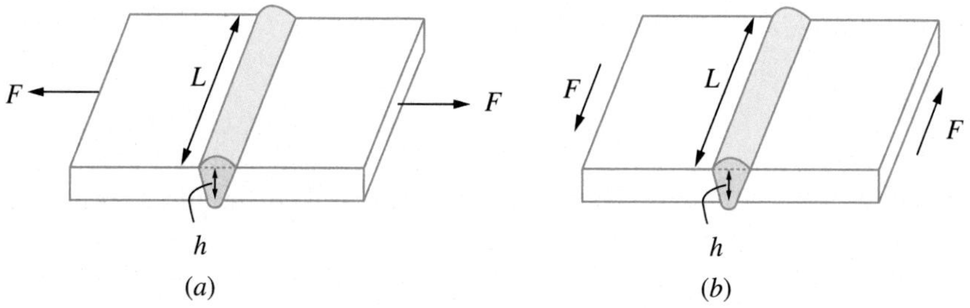

Figure 19.5 (*a*) Tensile and (*b*) shear load of V-groove welds.

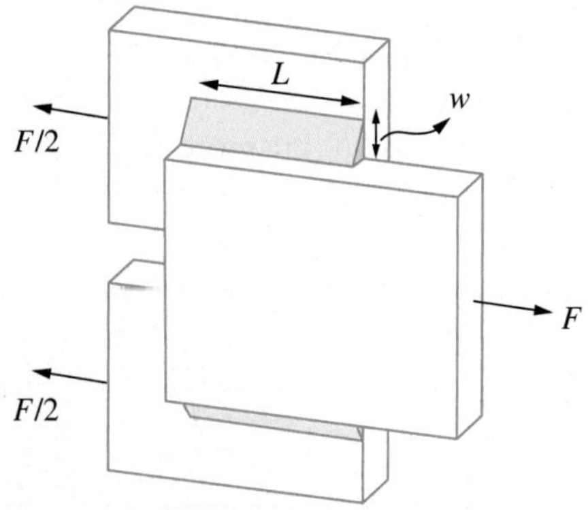

Figure 19.6 Parallel fillet weld under shear load.

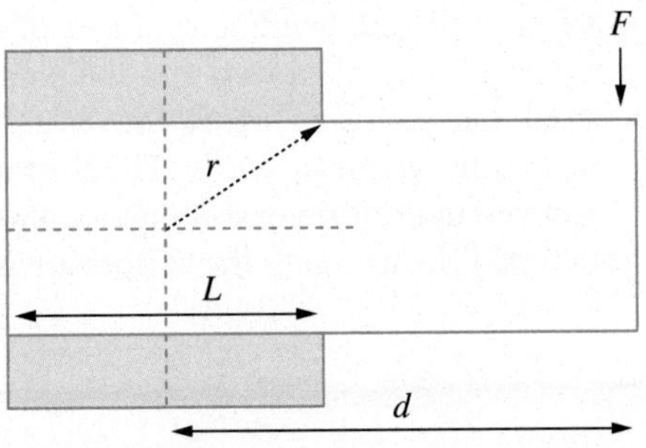

Figure 19.7 Torsional shear loading of the welds.

The secondary shear, or the shear due to the torsion, $T = Fd$, at a point on the weld(s), is

$$\tau_s = \frac{Tr}{J} \tag{19.2-4}$$

where r is the distance between the centroid and the point where the shear stress is measured. The maximum secondary shear stress occurs at farthest distance from the centroid. The term J is the polar moment of inertia of the weld system. Formulas for J of different configurations and welding systems are given in Table 19.4.

The total shear stress is the vectorial summation of the two stresses as

$$\tau_{\text{total}} = \frac{F}{hL} + \frac{Fdr}{J} \tag{19.2-5}$$

D. Bending (Fig. 19.8)

In many cases the weld is subjected to out-of-plane loading which causes bending stress in the welds. Similar to torsion case C, the weld system is subjected to a primary shear and the secondary bending shear at the throat of the welds. That is, the primary shear is

$$\tau_p = \frac{F}{hL} \tag{19.2-6}$$

and the secondary shear stress due to bending is

$$\tau_s = \frac{Mc}{I} \tag{19.2-7}$$

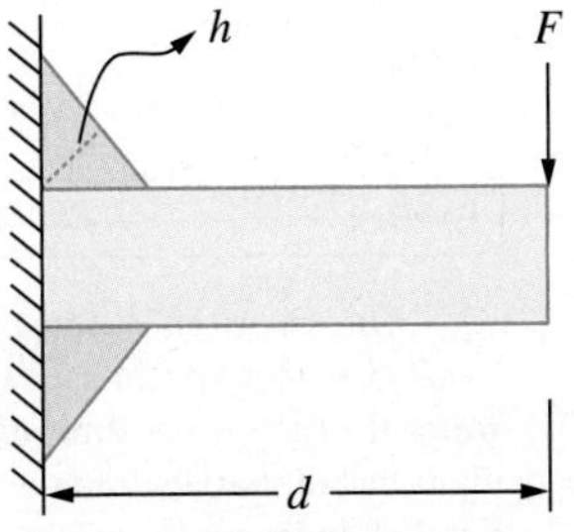

Figure 19.8 Cantilever welded to a support by fillet welds at top and bottom.

where M is the bending moment of the force F ($M = Fd$) (with d being the distance between the applied load and the plane of the welds), c is the distance between the centroid and the point where the shear stress is measured, and I is the moment of inertia of the system of welds. The formulas for I for different configurations and welding systems, in the form of section modulus S_w, are given in Table 19.4. The total shear stress in the weld in this case is the vector sum of the two shear stresses as

$$\tau_{\text{total}} = \sqrt{\left(\frac{F}{hL}\right)^2 + \left(\frac{Mc}{I}\right)^2} \tag{19.2-8}$$

19.3 STRENGTH OF WELDED JOINTS

There are many electrodes for various welding applications. The American Welding Society (AWS) provides specific codes for electrodes (see Table 19.1). The numbering starts with the letter E and is followed by four to five digits. The first two digits represent the approximate tensile strength of the weld. The last two or three digits refer to variables in the welding technique, such as current supply. For more information see the AWS specifications.

In groove welds where two parts of different thickness are to be welded, there is a partial joint penetration at the weld site. The minimum allowable sizes for fillet welds are given in Table 19.2. Generally, the minimum fillet weld size is governed by the thicker material; however, the minimum size of the weld should not exceed the thickness of the thinner material unless it is required by the calculated stress.

Table 19.1 reveals that for common structural materials, the weld material is often stronger than the welded materials. The allowable shear stress and unit load on fillet weld and the more common fillet sizes are given in Table 19.3. These values are for equal fillet welds where the effective throat h is equal to $0.707w$. This table provides the allowable unit force per lineal inch of weld made with a particular electrode type. For example, the allowable unit force per lineal inch weld for a 7/8 in fillet weld made with an E80 electrode is

$$f = 0.707\,w\tau = 0.707\,w(24.0) = 16.97\,w = 16.97\,(7/8) = 14.85 \text{ kips/lin in}$$

When a weld is treated as a line, the formulas are simplified and are shown in Table 19.4.

TABLE 19.1 Minimum Weld Metal Properties

AWS Electrode Number*	Tensile Strength kpsi (MPa)	Yield Strength kpsi (MPa)	Percent Elongation
E60xx	62 (427)	50 (345)	17–25
E70xx	70 (482)	57 (393)	22
E80xx	80 (551)	67 (462)	19
E90xx	90 (620)	77 (531)	14–17
E100xx	100 (689)	87 (600)	13–16
E120xx	120 (827)	107 (737)	14

*The American Welding Society (AWS) specification code numbering system for electrodes. This system uses an E prefixed to a four- or five-digit numbering system in which the first two or three digits designate the approximate tensile strength. The last digits include variables in the welding technique, such as current supply. The next-to-last digit indicates the welding position, as, for example, flat, or vertical, or overhead. The complete set of specifications may be obtained from the AWS upon request.

TABLE 19.2 Minimum Fillet Weld Size *w*

Material Thickness of Thicker Part Joined, in	*w*, in
*To ¼ incl.	1/8
Over ¼ to ½	3/16
Over ½ to ¾	¼
†Over ¾ to 1½	5/16
Over 1½ to 2¼	3/8
Over 2¼ to 6	½
Over 6	5/8

Not to exceed the thickness of the thinner part.
*Minimum size for bridge application does not go below 3/16 in.
†For minimum fillet weld size, table does not go above 5/16-in fillet weld for over ¾-in material.

TABLE 19.3 Allowable Loads for Various Sizes of Fillet Welds

	Strength Level of Weld Metal (EXX)						
	60*	70*	80	90*	100	110*	120
	Allowable Shear on Throat, ksi (1000 lb/in^2), of Fillet Weld						
$\tau =$	18.0	21.0	24.0	27.0	30.0	33.0	36.0
	Allowable Unit Force on Fillet Weld, kips/lin in						
$f =$	12.73w	14.85w	16.97w	19.09w	21.21w	23.33w	25.45w
Leg Size *w*, in	Allowable Unit Force for Various Size of Fillet Welds, kips/lin in						
1	12.73	14.85	16.97	19.09	21.21	23.33	25.45
7/8	11.14	12.99	14.85	16.70	18.57	20.41	22.27
¾	9.55	11.14	12.73	14.32	15.92	17.50	19.09
5/8	7.96	9.28	10.61	11.93	13.27	14.58	15.91
½	6.37	7.42	8.48	9.54	10.61	11.67	12.73
7/16	5.57	6.50	7.42	8.35	9.28	10.21	11.14
3/8	4.77	5.57	6.36	7.16	7.95	8.75	9.54
5/16	3.98	4.64	5.30	5.97	6.63	7.29	7.95
¼	3.18	3.71	4.24	4.77	5.30	5.83	6.36
3/16	2.39	2.78	3.18	3.58	3.98	4.38	4.77
1/8	1.59	1.86	2.12	2.39	2.65	2.92	3.18
1/16	0.795	0.930	1.06	1.19	1.33	1.46	1.59

*Fillet welds actually tested by the joint AISC-AWS Task Committee.

EXAMPLE Find the fillet weld size for the connection shown.

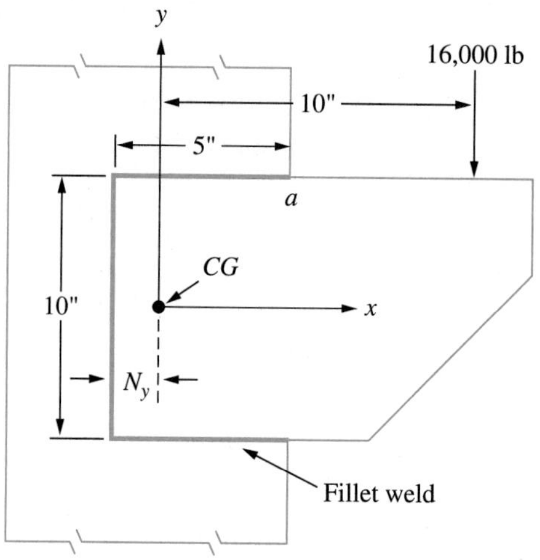

SOLUTION Properties of weld treated as a line (from Table 19.4)

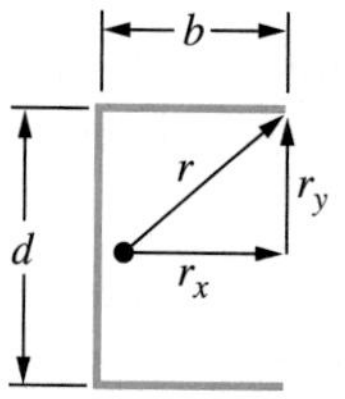

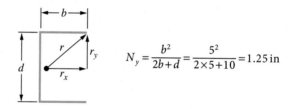

$$N_y = \frac{b^2}{2b+d} = \frac{5^2}{2\times5+10} = 1.25\,\text{in}$$

$$J_w = \frac{(2b+d)^3}{12} - \frac{b^2(b+d)^2}{(2b+d)} = \frac{[2(5)+10]^3}{12} - \frac{5^2(5+10)^2}{[2(5)+10]} = 385.4\,\text{in}^3$$

$$A_w = 2b+d = 2(5)+10 = 20\,\text{in}$$

The vertical and horizontal unit forces due to torsion on the weld are found separately. The maximum shear stress occurs at point *a*. There, the secondary unit force in the vertical direction is

$$f_{sy} = \frac{Tr_x}{J} = \frac{16{,}000(10)3.75}{385.4} = 1557\,\text{lb/in}$$

The secondary unit force in the horizontal direction is

$$f_{sx} = \frac{Tr_y}{J} = \frac{16{,}000(10)5}{385.4} = 2076\,\text{lb/in}$$

The direct (primary) vertical unit force is

$$f_p = \frac{F}{A_w} = \frac{16{,}000}{20} = 800\,\text{lb/in}$$

The total vertical unit force is $f_{syt} = 1557 + 800 = 2357$ lb/in. The total resultant unit force is

$$f_{\text{total}} = \sqrt{(2357)^2 + (2076)^2} = 3141\,\text{lb/in}$$

Assume an E60 electrode. From Table 19.3, $f = 12.73\,(10^3)\,w$. Then

$w = 3141/12.73\,(10^3) = 0.247$ in   or   ¼ in leg fillet

TABLE 19.4 Treating a Weld as a Line

Type of Loading		Standard Design Formula Stress, lb/in²	Treating the Weld as a Line Force, lb/in
Primary welds transmit entire load at this point			
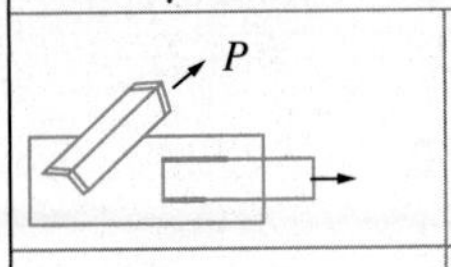	Tension or compression	$\sigma = \frac{P}{A}$	$f = \frac{P}{A_w}$
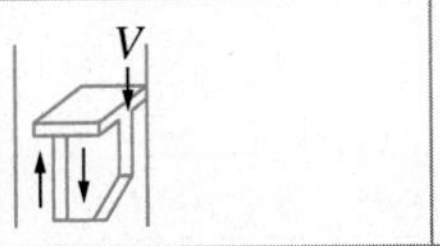	Vertical shear	$\sigma = \frac{V}{A}$	$f = \frac{V}{A_w}$
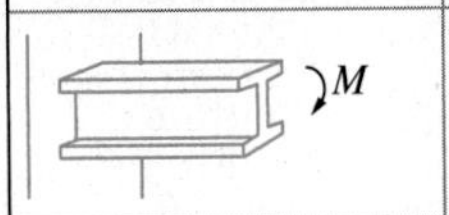	Bending	$\sigma = \frac{M}{S}$	$f = \frac{M}{S_w}$
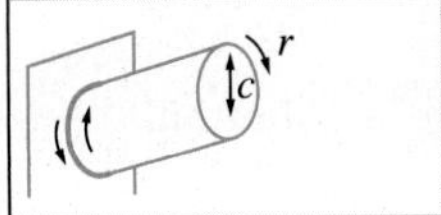	Twisting	$\tau = \frac{Tc}{J}$	$f = \frac{Tc}{J_w}$
Secondary welds hold section together—low stress			
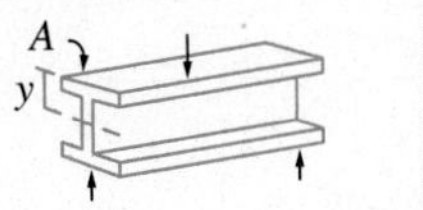	Horizontal shear	$\tau = \frac{VA_y}{h}$	$f = \frac{VA_y}{In}$
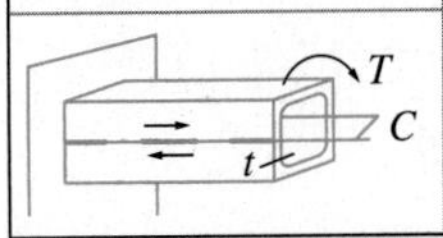	Torsion horizontal shear*	$\tau = \frac{T}{2A_t}$	$f = \frac{T}{2A}$

Outline of Welded Joint: b = Width	Outline of Welded Joint: d = Depth	Bending (About Horizontal Axis $x-x$), in²	Twisting, in³
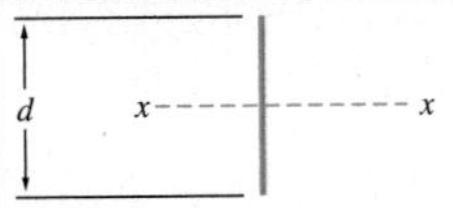		$S_w = \frac{d^2}{6}$	$J_w = \frac{d^3}{12}$
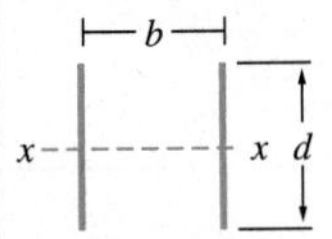		$S_w = \frac{d^2}{3}$	$J_w = \frac{d(3b^2+d^2)}{6}$
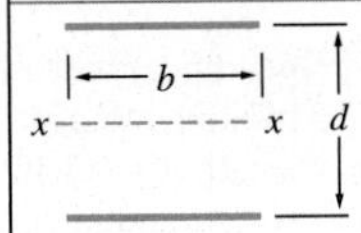		$S_w = bd$	$J_w = \frac{b^3+3bd^2}{6}$
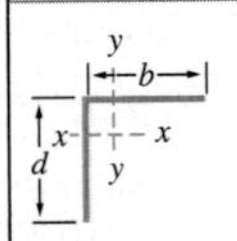	$N_y = \frac{b^2}{2(b+d)}$ $N_x = \frac{d^2}{2(b+d)}$	$S_w = \frac{4bd+d^2}{6} = \frac{d^2(4b+d)}{6(2b+d)}$ top bottom	$J_w = \frac{(b+d)^4-6b^2d^2}{12(b+d)}$
$N_y = \frac{b^2}{2b+d}$	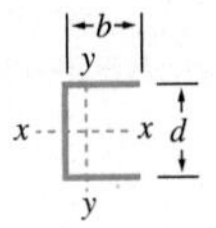	$S_w = bd + \frac{d^2}{6}$	$J_w = \frac{(2b+d)^3}{12} - \frac{b^2(b+d)^2}{2b+d}$

(Continued)

Table 19.4 Treating a Weld as a Line (*Continued*)

Outline of Welded Joint (b = Width, d = Depth)	Bending (About Horizontal Axis $x-x$), in²	Twisting, in³
$N_x = \frac{d^2}{b+2d}$ 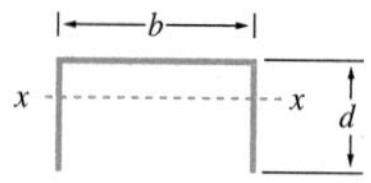	$S_w = \frac{2bd+d^2}{3} = \frac{d^2(2b+d)}{3(b+d)}$ top bottom	$J_w = \frac{(b+2d)^3}{12} - \frac{d^2(b+d)^2}{b+2d}$
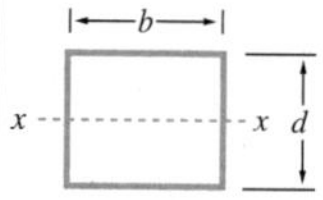	$S_w = bd + \frac{d^2}{3}$	$J_w = \frac{(b+d)^3}{6}$
$N_y = \frac{d^2}{b+2d}$ 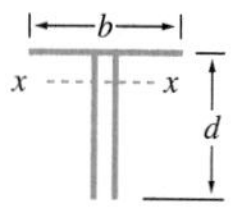	$S_w = \frac{2bd+d^2}{3} = \frac{d^2(2b+d)}{3(b+d)}$ top bottom	$J_w = \frac{(b+2d)^3}{12} - \frac{d^2(b+d)^2}{b+2d}$
$N_y = \frac{d^2}{2(b+d)}$ 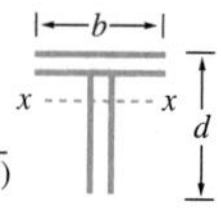	$S_w = \frac{4bd+d^2}{3} = \frac{4bd^2+d^3}{6b+3d}$ top bottom	$J_w = \frac{d^3(4b+d)}{6(b+d)} + \frac{b^3}{6}$
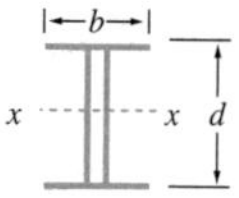	$S_w = bd + \frac{d^2}{3}$	$J_w = \frac{b^3+3bd^2+d^3}{6}$
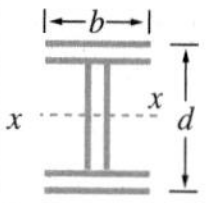	$S_w = 2bd + \frac{d^2}{3}$	$J_w = \frac{2b^3+6bd^2+d^3}{6}$
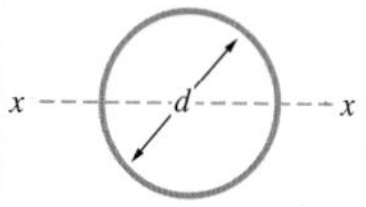	$S_w = \frac{\pi d^2}{4}$	$J_w = \frac{\pi d^3}{4}$
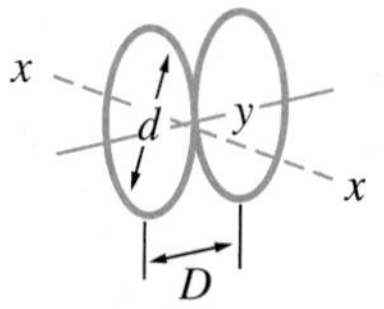	$I_w = \frac{\pi d}{2}\left(D^2 + \frac{d^2}{2}\right)$ $S_w = \frac{I_w}{c}$ where $c = \frac{\sqrt{D^2+d^2}}{2}$	

*Applies to closed tubular section only.

b = width of connection, in
d = depth of connection, in
A = area of flange material held by welds in horizontal shear, in²
y = distance between center of gravity of flange material and N.A. of whole section, in
I = moment of inertia of whole section, in⁴
c = distance of outer fiber, in
l = thickness of plate, in
J = polar moment of inertia of section, in⁴
P = tensile or compressive load, lb
V = vertical shear load, lb
M = bending moment, in·lb
T = twisting moment, in·lb
A_w = length of weld, in
S_w = section modulus of weld, in²
J_w = polar moment of inertia of weld, in³
N_x = distance from x axis to face
N_y = distance from y axis to face
S = section modulus, in³
f = force in standard design formula when weld is treated as a line, lb/in
n = number of welds

19.4 RIVETED AND BOLTED JOINTS

Rivets and bolts are used for connecting members. However, rivets are considered permanent mechanical fasteners while bolts are removable. A rivet consists of a smooth cylindrical shaft with a head on one end and is usually made from a soft grade of steel that does not become brittle when heated or hammered. After a rivet is placed inside a pre-drilled hole in connecting members, the opposite side of the head (known as a buck-tail) is hammered with a pneumatic riveting gun so that it expands to about 1.5 times the original shaft diameter of the rivet, holding the rivet in place, thereby joining the members.

While bolts and screws are generally used for tension loads, rivets are more capable of supporting shear loads (loads perpendicular to the axis of the shaft). A typical single shear rivet joint is shown in Fig. 19.9. A double (two-sided) shear rivet connects three members, as shown in Fig. 19.10.

19.5 SHEARING AND FAILURE MODES IN RIVETED JOINTS

There are several modes of failure in riveted joints. In all the following cases, to ensure that no failure occurs, the calculated shear stress or the bearing stress should be less than that of the corresponding material strength with the consideration of the factor of safety.

A. Shearing Mode of the Rivet

In this mode the rivet material is subjected to a shearing force across its cross section. The average shearing stress in the rivet τ is

$$\tau = \frac{F}{nA} = \frac{4F}{n\pi d^2} \tag{19.5-1}$$

where F is the shearing force, A is the cross section of the rivet ($A = \pi d^2/4$), d is the diameter of the rivet, and n is the number of shearing surfaces. For a single rivet in double shear, $n = 2$. If there is more than one rivet in the connecting joints then n is the total number of shearing surfaces in the rivets. For example, for the case of Fig. 19.9, $n = 1$, and for the case of Fig. 19.10, $n = 2$.

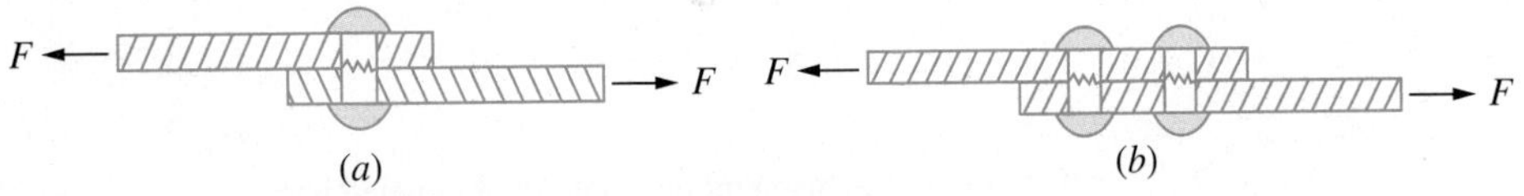

Figure 19.9 (*a*) Single rivet in shear. (*b*) Two rivets in shear.

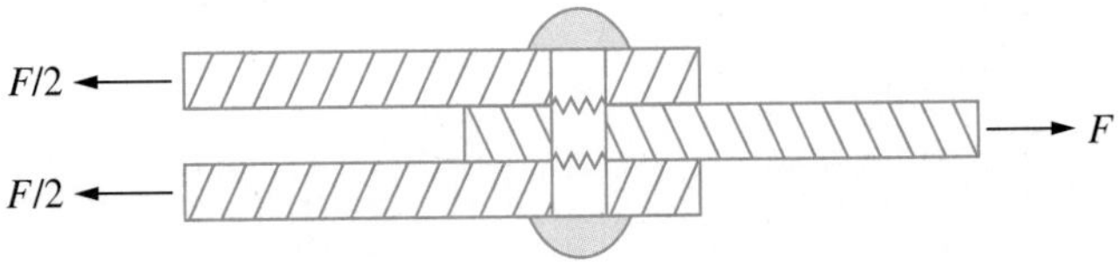

Figure 19.10 Double shear rivet.

B. Tension Mode of the Member

If h in Fig. 19.11 is sufficiently small the member may fail under tension. The tensile stress in the member with one rivet is

$$\sigma = \frac{F}{2ht} \tag{19.5-2}$$

C. Tearing Mode of the Member

If the rivet is placed very close to the edge of the member, it could tear or shear the member, as shown in Fig. 19.12. The shearing stress of the member with one rivet is

$$\tau_p = \frac{F}{2Lt} \tag{19.5-3}$$

D. Compression or Bearing Mode of Members and Rivets

When the member material is softer than that of the rivet, or the rivet is located close to the edge of the member, then the rivet could crush the member (Fig. 19.13). In this case the compressive bearing stress of the member σ_{br} is

$$\sigma_{br} = \frac{F}{td} \tag{19.5-4}$$

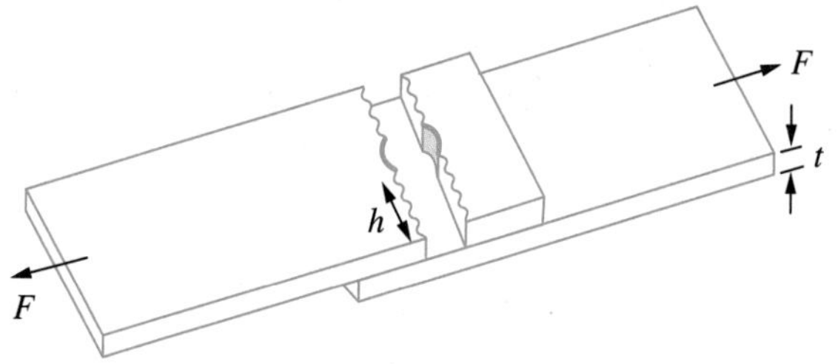

Figure 19.11 Tension mode of failure of the member (tension on net section).

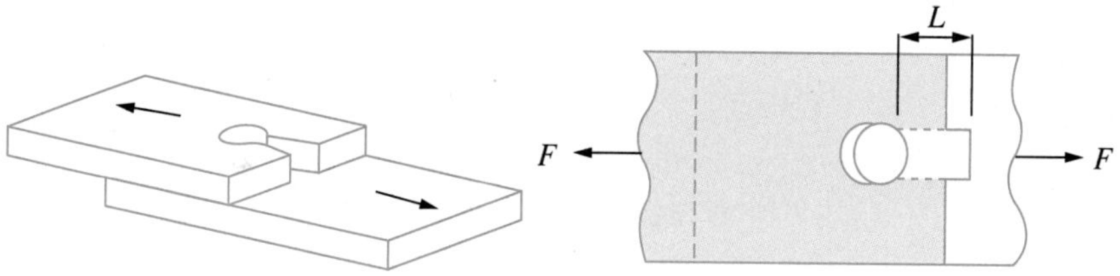

Figure 19.12 Tearing mode of failure of the member.

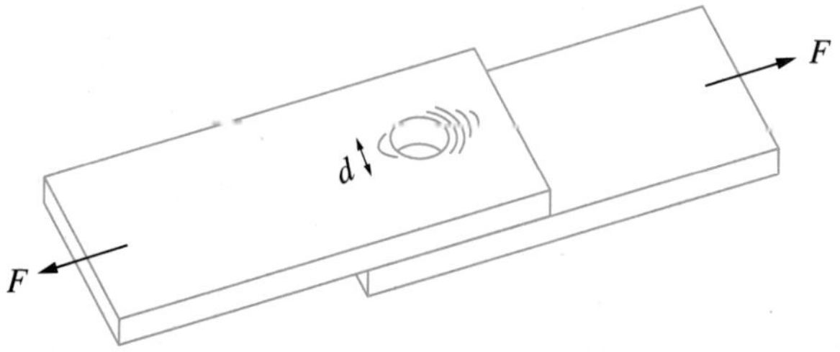

Figure 19.13 Compression or beading modes of failure.

where t is the thickness of the member and d is the diameter of the rivet. On the other hand, if the material of the rivet is much softer than that of the member, then the rivet itself could fail and plastically deform under the bearing stress before it shears off. In this case the same formula Eq. (19.5-4) is used except that the σ_{br} should be compared with the bearing stress of the rivet material.

19.6 ECCENTRIC LOADING OF RIVETED JOINTS

In the previous section, Sec. 19.5, the resultant external force of the members passes through the centroid of the cross-sectional area of the rivet or the areas of a group of the rivets or fasteners, i.e., the rivets are *concentrically loaded.* However, a group of fasteners could be subjected to a system of load where the resultant of the forces does not pass through the centroid of the fasteners, i.e., the groups of the fasteners are *eccentrically loaded.* In concentrically loaded fasteners, the load is assumed to be uniformly distributed among the fasteners. For an eccentrically loaded joint the force is replaced by an equal force at the centroid of the fasteners and a moment equal in magnitude to the force times its distance from the centroid. The shearing force due to the moment is resisted by each fastener according to its cross-sectional area and distance from the centroid of the fastener group. The analysis in this section assumes that the cross-section areas of the rivets in the rivet group are equal. For unequal areas, see Example 7.8-1 of Ref. 4.

Consider a group of n rivets at a connecting joint where an eccentric load F is applied at a distance d from the centroid of the rivet group (Fig. 19.14). To calculate the stresses in each rivet in an *eccentrically loaded* rivet group, first the centroid of the group $(\overline{X},\overline{Y})$ is calculated as

$$\overline{X}=\frac{\sum_1^n A_i X_i}{\sum_1^n A_i} \qquad \overline{Y}=\frac{\sum_1^n A_i Y_i}{\sum_1^n A_i} \tag{19.6-1}$$

where A_i is the area of the ith rivet and n is the number of rivets. Then the force F is moved to the centroid and the primary shear stress that is equally distributed to all rivets is calculated as

$$F_p = F/n \tag{19.6-2}$$

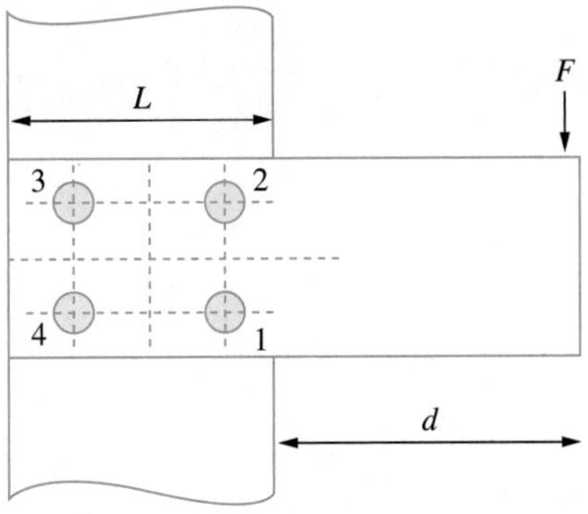

Figure 19.14 Rivet joints with eccentric loading.

The secondary shear force F_s in each rivet that is due to the moment, $M = F(d + L/2)$, satisfies the moment equilibrium as

$$M = F_{s1}r_1 + F_{s2}r_2 + F_{s3}r_3 + \cdots \tag{19.6-3}$$

However, similar to a force on a lever arm, and assuming the areas of the rivets are equal, each rivet is subjected to a force directly proportional to its distance from the group centroid. Thus, the farthest rivet from the centroid takes the greatest force and the nearest rivet takes the smallest. This results in,

$$\frac{F_{s1}}{r_1} = \frac{F_{s2}}{r_2} = \frac{F_{s3}}{r_3} = \cdots \tag{19.6-4}$$

Solving Eqs. (19.6-3) and (19.6-4) simultaneously, the secondary shear force of each rivet is

$$F_{si} = \frac{Mr_i}{r_1^2 + r_2^2 + r_3^2 + \cdots} \tag{19.6-5}$$

The primary and secondary shear forces F_p and F_s should be added vectorially as shown in Fig. 19.15. Then, the total shear force, F_{ti}, is

$$\vec{F}_{ti} = \vec{F}_{pi} + \vec{F}_{si} \tag{19.6-6}$$

The shear stress in each rivet is

$$\tau = \frac{4F_{ti}}{\pi d^2} \tag{19.6-7}$$

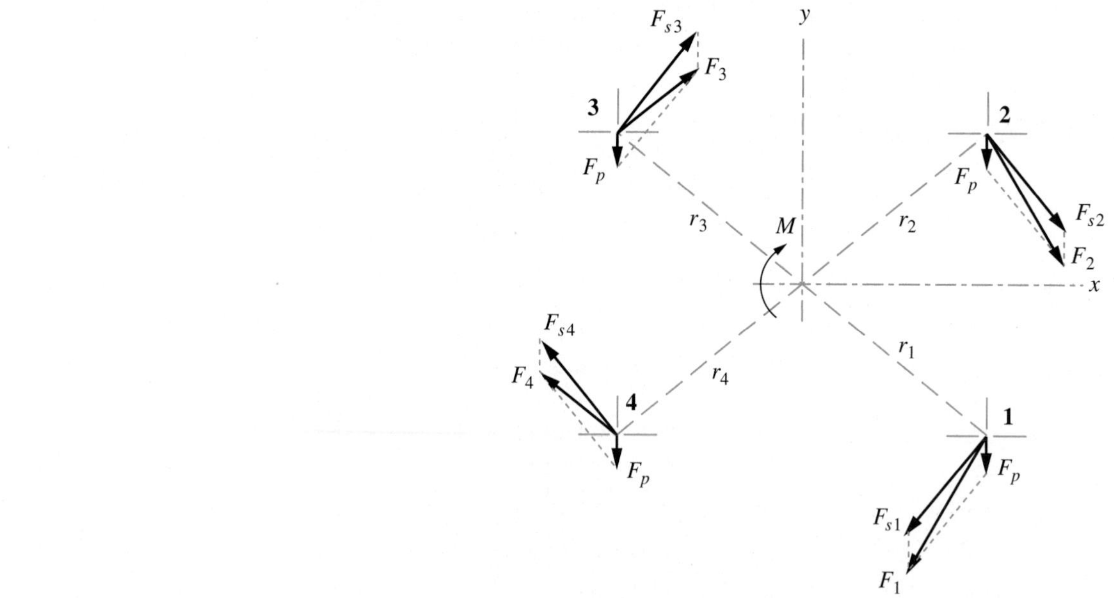

Figure 19.15 Free body diagram of primary and secondary shear forces on the rivets.

EXAMPLE Considering the shear failure of the rivets only, determine the maximum allowable force F that can be applied to the rivet group shown below. The force passes directly through the center of rivet B. The diameters of the rivets are 7/8 in and the allowable shear stress is 16 kpsi.

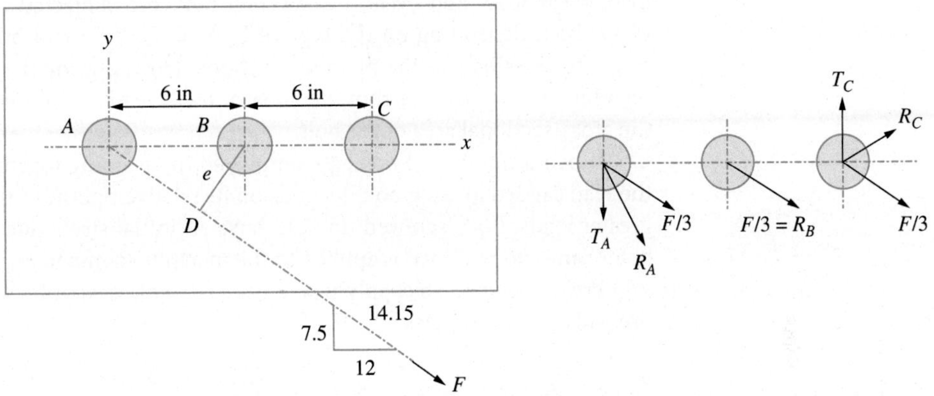

SOLUTION By similar triangles, the eccentricity e can be found by considering triangle ABD. Thus

$$\frac{e}{6} = \frac{7.5}{14.15} \quad \Rightarrow \quad e = 3.18 \text{ in}$$

Due to symmetry, the centroid of the rivet group is on the center of rivet B. The moment at the centroid due to the force is $M = Fe = 3.18F$.

The primary and secondary shear forces of each rivet are shown in the figure on the right. The secondary shear forces in rivets A and C can be found with Eq. (19.6-5) using the rivet diameters instead of radii. Thus

$$\tau_A = \tau_C = \frac{Md_A}{d_A^2 + d_C^2} = \frac{3.18F(7/8)}{(7/8)^2 + (7/8)^2} = 1.82F$$

The primary shear force for each rivet is $F/3$. Therefore, the vertical (y) and horizontal (x) components of the shearing force on rivet A are

$$\left(\sum F_A\right)_y = 1.82F + \left(\frac{7.5}{14.2}\right)F/3 = 1.99F$$

$$\left(\sum F_A\right)_x = \left(\frac{12}{14.15}\right)F/3 = 0.282F$$

The resultant shear force on A is

$$R_A = \sqrt{(1.99F)^2 + (0.282F)^2} = 2.009F$$

The shear stress is $\tau = R_A/A$. Rearranging, $R_A = \tau A$. Thus

$$2.009F = 16(10)^3 \frac{\pi}{4}\left(\frac{7}{8}\right)^2$$

Solving yields $F = 4788$ lb.

19.7 BOLT STRENGTH AND DESIGN

Bolts are employed in fastening together two or more members of a machine or a structure using nuts/washers or threaded members. They are subjected to preload torque, tension load, shear load, and cyclic load. When bolts are subjected to shear, the shear strength of the bolt, depending on the type of joint and number of bolts are calculated the same as rivets described in the previous sections. However, for the shear loading of bolts it is recommended to use the root diameter of the thread (or minor diameter of the thread) for the cross-sectional area of the bolt.

When a number of bolts are employed in fastening together two parts of a machine, the load carried by each bolt depends on its relative tightness, the tighter bolts carrying the greater loads. Bolts screwed up tight have an initial stress due to the tightening (preload) before any external load is applied to the machine member.

Preloading bolts by applying a torque is highly desirable. When a bolt is subjected to a preload of F_i it elongates δ by the formula:

$$\delta = F_i L/(AE) \tag{19.7-1}$$

where L is the length of the screw, A is the cross-sectional area, and E is the elastic moduli of the bolt. Before preloading a bolt it is snug-tightened, where a person uses an ordinary wrench with some initial torque. Once the snug tight is attained, all additional turning creates preloading tension in the bolt. A torque wrench that has a built-in dial to indicate the proper torque is used to apply specific torque that develops desired preloading F_i. The torque T needed to create the preload F_i is given as

$$T = \frac{F_i d_m}{2}\left(\frac{l + \pi f d_m \sec\alpha}{\pi d_m - f l \sec\alpha}\right) + \frac{F_i f_c d_c}{2} \tag{19.7-2}$$

where, d_m is the mean diameter, l is the pitch length, f is the coefficient of friction, and α is the half of the thread angle. The second term is the torque associated with the contact of the surface of the bolt head or a nut with a member or a washer, where f_c is the coefficient of friction between the bolt head or a nut and the washer/member, and d_c is the medium diameter of the bolt head or the nut. The coefficient of friction depends on the surface smoothness and if there is a lubricant present. On the average, both f and f_c are equal to 0.15.

Equation (19.7-2) could be simplified as

$$T = K F_i d \tag{19.7-3}$$

where K is approximately between 0.12 and 0.2.

In cases where bolts are subjected to cyclic loading, an increase in the initial tightening load decreases the operating stress range. In certain applications it is customary to fix the tightening load as a fraction of the yield-point load of the bolt. When the cyclic load is in the form of a sinusoidal periodical pattern, the mean stress σ_m and the amplitude stress σ_a of the bolt need to be determined based on the fatigue failure criterion. The criterion equation for the Soderberg line is

$$\frac{\sigma_a}{\sigma_e} + \frac{\sigma_m}{\sigma_y} = 1 \tag{19.7-4}$$

Similarly the criterion equation for the modified Goodman line is

$$\frac{\sigma_a}{\sigma_e} + \frac{\sigma_m}{\sigma_{\text{ult}}} = 1 \tag{19.7-5}$$

where σ_y and σ_{ult} are bolt's yield and ultimate stresses and

$$\sigma_m = \frac{(\sigma_{max} + \sigma_{min})}{2} \quad \text{and} \quad \sigma_a = \frac{\sigma_{max} - \sigma_{min}}{2} \tag{19.7-6}$$

where σ_{max} is the maximum bolt stress and σ_{min} is the minimum bolt stress.

19.8 GEARING AND GEAR STRESS

Gear systems as means of motion and power transmission between shafts with change in rotational speed are classified as spur, helical, worm, bevel, and hypoid bevel. The shaft arrangement for these gear systems could be parallel, under an angle, intersecting or offset between the shafts. Every gear has a pitch circle. Pitch circle is the imaginary circle associated with each gear that that is tangent to the pitch circle of a mating gear and rolls without slippage with the pitch circle of the mating gear. Diametral pitch P_d is the ratio of number of teeth in the gear to the diameter of the pitch circle D measured in inches, $P_d = N/D$. Circular pitch p is the linear measure in inches along the pitch circle between corresponding points of adjacent teeth. From these definitions, $P_d p = \pi$. The base pitch p_b is the distance along the line of action between successive involute tooth surfaces. The base and circular pitches are related as $p_b = p \cos \Phi$, where for all gear type the pressure angle Φ is the acute angle between the common normal to the profiles at the contact point and the common pitch plane. For more detailed kinematics of the gear systems see Marks' Handbook (Ref. 5).

Gear teeth fail in two classical manners: tooth breakage and surface fatigue pitting. Small and lightly loaded gears are designed primarily for tooth-bending beam strength. However, larger gears usually used for power transmission are designed for both strengths and with surface durability that is often more critical. Expressions for calculating the beam and surface stresses started with the Lewis-Buckingham formulas and now extend to the latest AGMA (American Gear Manufacturer Association) formulas. A tooth layout of Lewis formula is shown in Fig. 19.16.

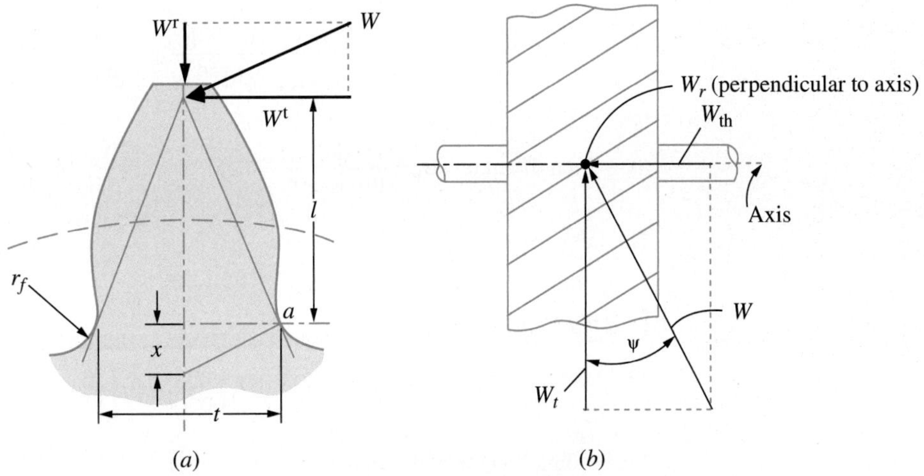

Figure 19.16 (*a*) Layout of forces on a spur gear tooth strength for Lewis formula, (*b*) forces on helical gear.

The maximum bending stress σ_t is

$$\sigma_t = \frac{W_t P_d}{FY} \tag{19.8-1}$$

where W_t is the tangential load (lb), F = face width (in), Y = Lewis form factor, and P_d = diametral pitch (in^{-1}). When only bending of the tooth is considered, the form factor Y is derived from the layout as $Y = 2xP_d/3$, where from Fig. 19.16a, $x = t^2/4l$. The value of Y varies with tooth design (form and pressure angle) and number of teeth, see ANSI/AGMA 2110-C95 standard for the details. In the case of a helical gear tooth, there is a thrust force W_{th} in the axial direction that arises and must be considered as a component of bearing load, see Fig. 19.16b.

In cases where a pair of gears is driven at moderate to high speed, Carl G. Barth (Ref. 8). modified the Lewis formula by multiplying the formula by velocity factor as

$$\frac{(1200+V)}{1200} \tag{19.8-2}$$

Thus for cut or milled profile, the stress is

$$\sigma_t = \frac{W_t P_d}{FY}\frac{(1200+V)}{1200} \tag{19.8-3}$$

and for cast iron, cast profile, the stress is

$$\sigma_t = \frac{W_t P_d}{FY}\frac{(600+V)}{600} \tag{19.8-4}$$

where V in the bracket has the unit of ft/min.

The strength of a gear tooth is evaluated based on the bending stress and contact stress (pitting resistance). Considering many factors, the bending and contact stresses formulas for spur and helical gears are given by ANSI/AGMA 2001-D04, AGMA 908-B89 (Refs. 6, 7).

EXAMPLE A spur gear has 50 teeth and pressure angle of 20°. The gear diametral pitch is 8 per inch and has the face of 0.5 in. It transmits 4.5 hp at 1200 rpm. Determine the tooth bending stress.

Torque transmitted is

$$T = \frac{4.5\,(\text{hp})\times 550\left(\text{ft}\cdot\frac{\text{lb/s}}{\text{hp}}\right)\times 12\left(\frac{\text{in}}{\text{ft}}\right)}{1200\,(\text{rpm})\times 2\pi\left(\frac{\text{rad}}{\text{rev}}\right)\times\left(\frac{1}{60}\frac{\text{min}}{\text{s}}\right)} = 236.4 \text{ lb·in}$$

$$\text{The pitch diameter, } D_p = \frac{\text{Teeth}}{\text{pitch}} = \frac{50}{8} = 6.25$$

$$\text{The tangential load, } W_t = \frac{2T}{D_p} = 75.6$$

Pitch line velocity,

$$V = 1200\,(\text{rpm})\times \pi D_p\left(\frac{\text{in}}{\text{rev}}\right)\times\frac{1}{12}\left(\frac{\text{ft}}{\text{in}}\right) = 1962.5\frac{\text{ft}}{\text{min}}$$

The bending strength of the tooth is

$$\sigma_t = \frac{W_t P_d}{FY}\cdot\frac{1200+V}{1200} = \frac{(75.6\times 8)(1200+1962.5)}{0.5\times 0.409\times 1200} = 7794.1 \text{ psi}$$

19.9 REFERENCES

1. Avallone, E. A., T. Baumeister, and A. M. Sadegh, *Marks' Standard Handbook for Mechanical Engineers*, 11th ed., McGraw Hill, 2007.
2. Budynas, R., and K. J. Nisbett, *Shigley's Mechanical Engineering Design*, 9th ed., McGraw-Hill, 2011.
3. Deutschman, A. D., W. J. Michels, and C. E. Wilson, *Machine Design*, Macmillan Publishing Co., Inc., 1975.
4. Budynas, R. G.: *Advanced Strength and Applied Stress Analysis*, 2nd ed., McGraw-Hill, 1999.
5. Sadegh, A. M., and W. M. Worek, *Marks' Standard Handbook for Mechanical Engineers*, 12th ed., McGraw Hill, 2018.
6. ANSI/AGMA 2001-D04, *Fundamental Rating Factors and Calculation Methods for Involute, Spur, and Helical Gear Teeth*, (available in English and metric units).
7. AGMA 908-B89, *Geometry Factors for Determining the Pitting Resistance and Bending Strength of Spur, Helical, and Herringbone Gear Teeth*. These standards have replaced AGMA 218.01 with improved formulas and details.
8. Barth, C. G., *The Transmission of Power by Leather Belting*. Transactions of the American Society of Mechanical Engineers 31 (1909).

CHAPTER 20

Composite Materials

This chapter deals with the mechanical behavior of composite materials that are formed from two or more dissimilar materials, on a macroscopic scale, where each material is continuous and homogeneous. Common composite materials generally consist of stiff and strong fibers, such as glass, carbon, or aramid, embedded in a softer material, such as a thermosetting or thermoplastic polymer, known as the matrix material, or simply, the *matrix*. Because of their strength, the fibers carry the load while the matrix keeps the fibers aligned, transfer the load from fiber to fiber, and protect the fibers from direct mechanical contact. Composite materials are ideal for structures where high strength-to-weight and stiffness-to-weight ratios are necessary. For example, composite materials are used in aircraft and spacecraft due to their weight sensitive structures.

20.1 COMPOSITE MATERIALS CLASSIFICATIONS AND COMPONENTS

There are four classifications of composite materials with different characteristics.

1. Fibrous composite materials, where long, strong, and stiff fibers, oriented in a specific direction, or woven fibers/fabrics are embedded in a softer matrix material, as shown in Fig. 20.1.
2. Laminated composite materials, where layers of various homogeneous or composite materials are sandwiched and are bonded together, as shown in Fig. 20.2.
3. Particulate composite materials, where particles of different size and shape are embedded in a matrix material, as shown in Fig. 20.3.
4. Combinations of some or all of the first three types.

Fibers are characterized by their high length-to-diameter ratio and their near-crystal-sized diameter. Because of these characteristics, long fibers are intrinsically stronger and stiffer than the same material in bulk form. Table 20.1 shows the density, strength, and stiffness of a few selected fiber materials with respect to their density. On the other hand, short fibers, known as whiskers, are also used as particulates in composite materials which essentially have the same near-crystal-sized diameter as fibers. Generally, the length-to-diameter ratio of the short fibers is in the order of hundreds. Short fibers have even higher mechanical properties than long fibers because they are manufactured by crystallization on a very small scale resulting in a nearly perfect alignment of crystals.

Matrix materials are usually polymers; however, metals, carbons, or ceramics are used in special composites. Major classes of structural polymers are thermoplastic, thermosets, and rubbers. A typical organic epoxy matrix material such as Narmco 2387 has tensile

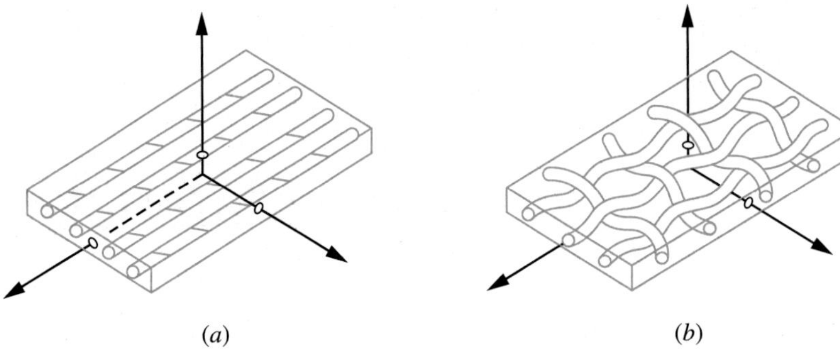

Figure 20.1 A layer (lamina) of fiber reinforced composite materials: (*a*) unidirectional fibers, (*b*) woven fibers/fabric.

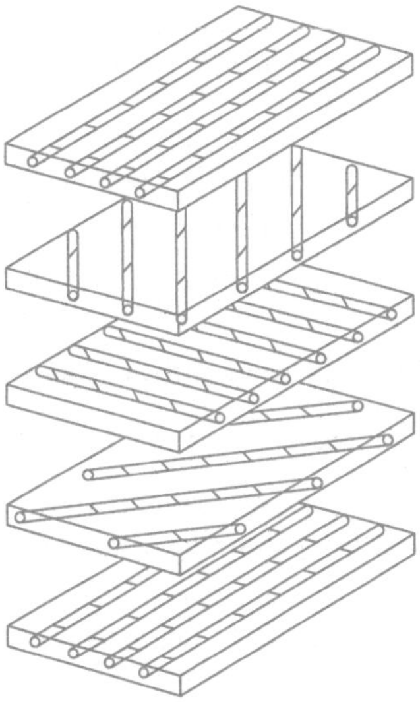

Figure 20.2 Exploded view of laminated composite materials.

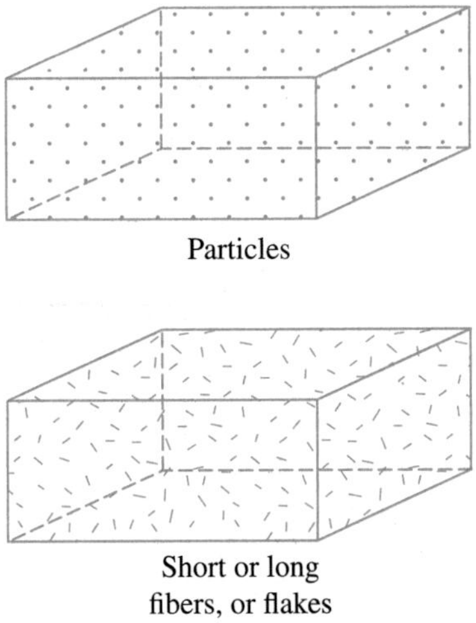

Figure 20.3 Particulate composite materials.

strength of 4200 psi and modulus of elasticity of $0.49(10^6)$ psi. Metal matrices are generally made by pouring molten metal around an in-place fiber system by diffusion bonding or by heating and vacuum infiltration. Ceramic matrices are generally casted or vapor deposited around in-place fiber systems. Table 20.2 describes the characteristics of some common fiber and matrix materials used in a composite.

Laminated composite materials consist of at least two layers of different composite or homogeneous materials that are bonded together. For example, two or more sheets of composite materials each containing a fiber orientation system can be bonded together and form a laminated composite. Also, two or more sheets of base materials such as glass, aluminum, plastic, or wood, etc., can be bonded together and form a new composite. If two different metals that usually have significantly different coefficients of thermal expansion are bonded

TABLE 20.1 Density, Strength, and Stiffness of Selected Fibers

	Fiber and Wire Properties*				
Fiber or Wire	Density, ρ lb/in^3 (kN/m^3)	Tensile Strength, S 10^3 lb/in^2 (GN/m^2)	S/ρ 10^5 in (km)	Tensile Stiffness, E 10^6 lb/in^2 (GN/m^2)	E/ρ 10^7 in (Mm)
Aluminum	.097 (26.3)	90 (.62)	9 (24)	10.6 (73)	11 (2.8)
Titanium	.170 (46.1)	280 (1.9)	16 (41)	16.7 (115)	10 (2.5)
Steel	.282 (76.6)	600 (4.1)	21 (54)	30 (207)	11 (2.7)
E-Glass	.092 (25.0)	500 (3.4)	54 (136)	10.5 (72)	11 (2.9)
S-Glass	.090 (24.4)	700 (4.8)	78 (197)	12.5 (86)	14 (3.5)
Carbon	.051 (13.8)	250 (1.7)	49 (123)	27 (190)	53 (14)
Beryllium	.067 (18.2)	250 (1.7)	37 (93)	44 (300)	66 (16)
Boron	.093 (25.2)	500 (3.4)	54 (137)	60 (400)	65 (16)
Graphite	.051 (13.8)	250 (1.7)	49 (123)	37 (250)	72 (18)

*Adapted from Dietz [1-1]

TABLE 20.2 Fiber and Matrix Materials Used in a Composite

Material	Characteristics
Fibers	
Glass	High strength, low stiffness, high density, lowest cost; E (calcium aluminoborosilicate) and S (magnesia-aluminosilicate) types commonly used.
Graphite	Available as high-modulus or high-strength; low cost; less dense than glass.
Boron	High strength and stiffness; highest density, highest cost; has tungsten filament at its center.
Aramids (Kevlar)	Highest strength-to-weight ratio of all fibers; high cost.
Other fibers	Nylon, silicon carbide, silicon nitride, aluminum oxide, boron carbide, boron nitride, tantalum carbide, steel, tungsten, molybdenum.
Matrix materials	
Thermosets	Epoxy and polyester, with the former most commonly used; others are phenolics, fluorocarbons, polyethersulfone, silicon, and polyimides.
Thermoplastics	Polyetheretherketone; tougher than thermosets but lower resistance to temperature.
Metals	Aluminum, aluminum-lithium, magnesium, and titanium; fibers are graphite, aluminum oxide, silicon carbide, and boron.
Ceramics	Silicon carbide, silicon nitride, aluminum oxide, and mullite; fibers are various ceramics.

together, they create what is known as a "*bimaterial*." Cantilever forms of bimaterials are used in thermostats and temperature sensing devices.

When particles of one or more materials are suspended in a matrix of another material, a composite material known as a particulate composite is formed. The particles or the matrix could be metallic or non-metallic. Examples of these composite materials are concrete where sand/gravel is mixed with cement as matrix and cermet where ceramic is used in a metal matrix as well as many others.

20.2 MECHANICS OF COMPOSITE MATERIALS

Each material of a composite is generally uniform, or isotropic, in nature. However, due to the discontinuous nature of the composite, the material properties of the composite can vary with both position and direction and therefore are anisotropic. For example, an epoxy resin reinforced with continuous graphite fibers will have very high strength and stiffness properties in the direction of the fibers, but very low properties normal to or transverse to the fibers.

Fiber reinforced composite properties depend not only on the properties of each constituent, but also on the quantity of each constituent in the composite material. A key measure of the quantity of a constituent is the *volume fraction* of that constituent in a given volume of composite material. There are models for predicting the properties of a composite material based on the properties and volume fraction of each constituent, in particular, the volume fraction of fibers and the volume fraction of matrix material. These models are referred to as *micromechanical models* where a layer of composite material is assumed to exhibit combined effects of the composing materials that are uniformly distributed throughout the volume of the layer (lamina). This process is also called homogenization where the mechanical properties of the matrix and the fiber lose their identities and a layer is treated as a single material with a single set of properties.

In laminate composites where several layers are bonded together, each layer interacts with the surrounding layers. The study of the individual layers in a laminate and how they interact is often referred to as *macromechanics*.

20.3 MACROMECHANICS OF A LAYER (LAMINA)

In this section we address the "apparent" properties of a lamina which is large enough such that the details of the fibers and the constituent of the lamina are not dealt with. That is, a single layer unidirectional fiber reinforced composite material is often represented by idealized packing, as shown in Fig. 20.4. For this composite material, it is convenient to use an orthogonal coordinate system that has one axis aligned with the direction of the fiber reinforcement, known as the longitudinal axis; the transverse axis 2, that lies in the plane of the layer, perpendicular to the longitudinal axis; and a third axis 3 through the thickness of the layer. Such a coordinate system is referred to as the *principal material coordinate system* (or often as the 1-2-3 coordinate system). In this case, the 1 direction is the fiber direction, while the 2 and 3 directions are transverse directions, often referred to as the matrix directions.

Unidirectional fiber reinforced composite materials have two orthogonal planes of material property symmetry, that is, longitudinal and transverse (1 and 2) directions, that are mutually orthogonal to the third plane. Therefore, the stress-strain relations,

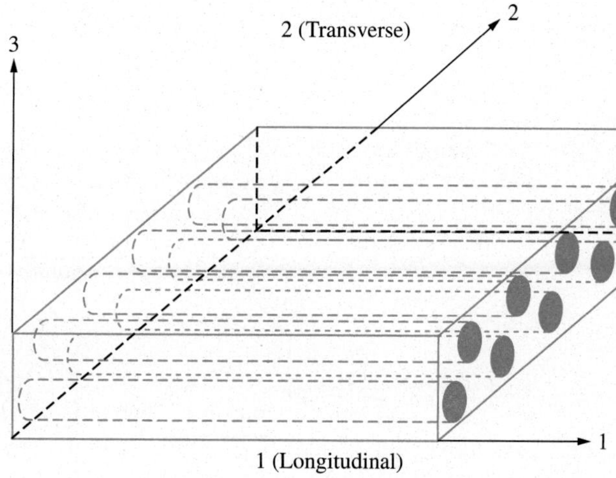

Figure 20.4 Principal material coordinate system of a lamina.

the generalized Hooke's law, in coordinates aligned with principal material directions are written in terms of the *stiffness matrix, C,* as

$$\begin{Bmatrix} \sigma_1 \\ \sigma_2 \\ \sigma_3 \\ \tau_{23} \\ \tau_{13} \\ \tau_{12} \end{Bmatrix} = \begin{bmatrix} C_{11} & C_{12} & C_{13} & 0 & 0 & 0 \\ C_{21} & C_{22} & C_{23} & 0 & 0 & 0 \\ C_{31} & C_{32} & C_{33} & 0 & 0 & 0 \\ 0 & 0 & 0 & C_{44} & 0 & 0 \\ 0 & 0 & 0 & 0 & C_{55} & 0 \\ 0 & 0 & 0 & 0 & 0 & C_{66} \end{bmatrix} \begin{Bmatrix} \varepsilon_1 \\ \varepsilon_2 \\ \varepsilon_3 \\ \gamma_{23} \\ \gamma_{13} \\ \gamma_{12} \end{Bmatrix} \tag{20.3-1}$$

The form of the stress-strain relations in Eq. (20.3-1) defines *orthotropic* material behavior. Such behavior is characterized by: (1) a decoupling of shear and normal responses, resulting in many of the off-diagonal terms being zero; and (2) different material properties in the three mutually perpendicular directions. The stress-strain relations, in terms of the *compliance matrix, S,* are then given as

$$\begin{Bmatrix} \varepsilon_1 \\ \varepsilon_2 \\ \varepsilon_3 \\ \gamma_{23} \\ \gamma_{13} \\ \gamma_{12} \end{Bmatrix} = \begin{bmatrix} S_{11} & S_{12} & S_{13} & 0 & 0 & 0 \\ S_{21} & S_{22} & S_{23} & 0 & 0 & 0 \\ S_{31} & S_{32} & S_{33} & 0 & 0 & 0 \\ 0 & 0 & 0 & S_{44} & 0 & 0 \\ 0 & 0 & 0 & 0 & S_{55} & 0 \\ 0 & 0 & 0 & 0 & 0 & S_{66} \end{bmatrix} \begin{Bmatrix} \sigma_1 \\ \sigma_2 \\ \sigma_3 \\ \tau_{23} \\ \tau_{13} \\ \tau_{12} \end{Bmatrix} \tag{20.3-2}$$

Consider a small element in the composite material of Fig. 20.4. The stresses on this element are shown in Fig. 20.5.

The σ_1, σ_2, and σ_3 are normal stresses in the principal material directions, τ_{12} is the in-plane shear stress, and τ_{13} and τ_{23} are the through-thickness shear stresses. The

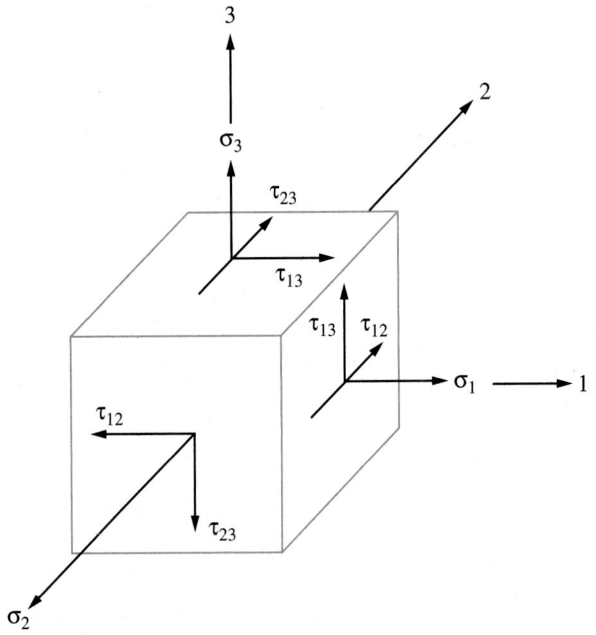

Figure 20.5 The stresses on the element.

corresponding normal strains are given as ε_1, ε_2, and ε_3. The engineering shear strains are given as γ_{12}, γ_{13}, and γ_{23}. The strain-stress relations for orthotropic materials are written as

$$\begin{Bmatrix} \varepsilon_1 \\ \varepsilon_2 \\ \varepsilon_3 \\ \gamma_{23} \\ \gamma_{13} \\ \gamma_{12} \end{Bmatrix} = \begin{bmatrix} \frac{1}{E_1} & -\frac{\nu_{21}}{E_2} & -\frac{\nu_{31}}{E_3} & 0 & 0 & 0 \\ -\frac{\nu_{12}}{E_1} & \frac{1}{E_2} & -\frac{\nu_{32}}{E_3} & 0 & 0 & 0 \\ -\frac{\nu_{13}}{E_1} & -\frac{\nu_{23}}{E_2} & \frac{1}{E_3} & 0 & 0 & 0 \\ 0 & 0 & 0 & \frac{1}{G_{23}} & 0 & 0 \\ 0 & 0 & 0 & 0 & \frac{1}{G_{13}} & 0 \\ 0 & 0 & 0 & 0 & 0 & \frac{1}{G_{12}} \end{bmatrix} \begin{Bmatrix} \sigma_1 \\ \sigma_2 \\ \sigma_3 \\ \tau_{23} \\ \tau_{13} \\ \tau_{12} \end{Bmatrix} \tag{20.3-3}$$

where E_1 is the elastic modulus in the fiber 1-direction, E_2 is the elastic modulus in the transverse 2-direction, and E_3 is the elastic modulus in the transverse 3-direction. The Poisson's ratios ν_{ij} are given so that $-(\nu_{ij}/E_i)\sigma_i$ is the strain in the j-direction due to a stress applied in the i-direction. The three shear moduli are G_{23}, G_{13}, and G_{12}. It can be shown that the compliance matrix, S, and stiffness matrix, C, are symmetric, i.e., $S_{ij} = S_{ji}$, and $C_{ij} = C_{ji}$. For that to be true, the following relations must hold:

$$\frac{\nu_{12}}{E_1} = \frac{\nu_{21}}{E_2} \qquad \frac{\nu_{13}}{E_1} = \frac{\nu_{31}}{E_3} \qquad \frac{\nu_{23}}{E_2} = \frac{\nu_{32}}{E_3} \tag{20.3-4}$$

Therefore, there are only nine independent engineering properties that are needed to describe the elastic response of a layer of composite material.

Composite materials are usually in a layer or planar shape and are less often in a bulk form. Therefore, it may be assumed that the composite is in a state of *plane stress*. That is, a state where three of the stress components in a principal direction are so much smaller than the other stresses that they may be assumed to be equal to zero for subsequent stress analysis. As an example, consider a plate in which the thickness direction, which is aligned with the 3 direction, is much less than the in-plane dimensions. The out-of-plane stress components σ_3, τ_{23}, and τ_{13} will then be much smaller than the in-plane stress components σ_1, σ_2, and τ_{12} and are taken to be identically zero. As a result, Eq. (20.3-3) can be reduced to:

$$\begin{Bmatrix} \varepsilon_1 \\ \varepsilon_2 \\ \gamma_{12} \end{Bmatrix} = \begin{bmatrix} \dfrac{1}{E_1} & -\dfrac{\nu_{12}}{E_1} & 0 \\ -\dfrac{\nu_{12}}{E_1} & \dfrac{1}{E_2} & 0 \\ 0 & 0 & \dfrac{1}{G_{12}} \end{bmatrix} \begin{Bmatrix} \sigma_1 \\ \sigma_2 \\ \tau_{12} \end{Bmatrix} \tag{20.3-5}$$

Note that the plane-stress assumption can lead to inaccuracies at the free-edges of a layer where delamination of laminates is often initiated. It is important to remember that while the normal stress σ_3 is equal to zero for the plane-stress condition, the corresponding normal strain ε_3 is non-zero and is given by

$$\varepsilon_3 = -\frac{\nu_{13}}{E_1}\sigma_1 - \frac{\nu_{23}}{E_2}\sigma_2 \tag{20.3-6}$$

When the composite material is subjected to a temperature change ΔT, and sometimes to a moisture absorption change ΔM, then the strain-stress relations of Eq. (20.3-5) and (20.3-6) are modified as

$$\begin{Bmatrix} \varepsilon_1 \\ \varepsilon_2 \\ \gamma_{12} \end{Bmatrix} = \begin{bmatrix} \dfrac{1}{E_1} & -\dfrac{\nu_{12}}{E_1} & 0 \\ -\dfrac{\nu_{12}}{E_1} & \dfrac{1}{E_2} & 0 \\ 0 & 0 & \dfrac{1}{G_{12}} \end{bmatrix} \begin{Bmatrix} \sigma_1 \\ \sigma_2 \\ \tau_{12} \end{Bmatrix} + \begin{Bmatrix} \alpha_1 \\ \alpha_2 \\ 0 \end{Bmatrix} \Delta T + \begin{Bmatrix} \beta_1 \\ \beta_2 \\ 0 \end{Bmatrix} \Delta M \tag{20.3-7}$$

and

$$\varepsilon_3 = -\frac{\nu_{13}}{E_1}\sigma_1 - \frac{\nu_{23}}{E_2}\sigma_2 + \alpha_3 \Delta T + \beta_3 \Delta M \tag{20.3-8}$$

where α_1, α_2, and α_3 are coefficients of thermal expansion; β_1, β_2, and β_3 are coefficients of moisture expansion; ΔT is the change in temperature; and ΔM is the change in moisture content within the material from a reference moisture content.

20.4 MICROMECHANICS OF A LAYER (LAMINA)

In this section the basic relationship of the composite material properties to the properties of its constituents is presented. The objective of the micromechanics approach is to determine the elastic moduli or stiffnesses of a composite material in terms of the elastic

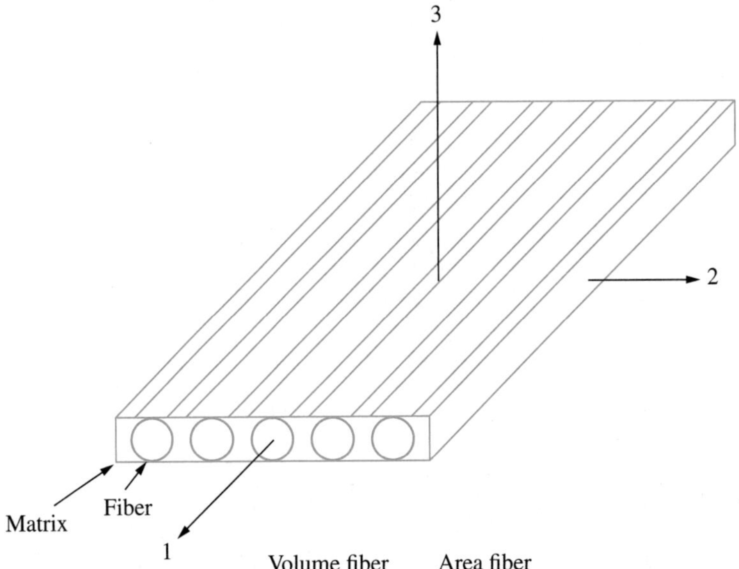

$$V_f = \frac{\text{Volume fiber}}{\text{Total volume}} = \frac{\text{Area fiber}}{\text{Total area}}$$

Figure 20.6 Representative volume element.

moduli of its constituent materials. The solutions of the micromechanics of composite materials can be obtained either by the mechanics of materials method or the theory of elasticity. The mechanics of material method, which is discussed here, employs simplified assumptions regarding the hypothesized behavior of the composite materials.

The basic starting point for the mechanics of materials model is the representative volume element shown in Fig. 20.6, where the elastic moduli of the fiber and the matrix in longitudinal 1-direction are E_1^f and E_1^m (superscripts f and m are for the fiber and matrix), respectively. Therefore, the fiber-direction modulus of the composite, E_1, can be developed by assuming that when a load is applied in the 1-direction, the strains in the fiber and matrix are the same. This *iso-strain* assumption leads to the result of

$$E_1 = E_1^f V_f + E_1^m V_m \qquad (20.4\text{-}1)$$

where V_f is the fiber volume fraction and is

$$V_f = \frac{\text{Volume of fibers}}{\text{Total volume of composite material}}$$

V_m is the matrix volume fraction and is

$$V_m = \frac{\text{Volume of matrix}}{\text{Total volume of composite material}}$$

Note that $V_f + V_m = 1$. Therefore, Eq. (20.4-1) can be rewritten as

$$E_1 = V_f E_1^f + (1 - V_f) E_1^m \qquad (20.4\text{-}2)$$

Equation (20.4-2) is often referred to as the *rule-of-mixtures* for apparent Young's modulus of the composite material in the direction of the fibers. For a unidirectional fiber composite, the fiber and matrix volume fractions are reduced to

$$V_f = A_f / A_t \qquad \text{and} \qquad V_m = A_m / A_t$$

where A_f is the area of fibers, A_m is the area of matrix, and A_t is the total cross-sectional area of the composite.

In an iso-strain model the estimated Poisson's ratio of the composite material leads to the following expression,

$$\nu_{12} = V_f \nu_{12}^f + (1 - V_f)\nu_{12}^m \tag{20.4-3}$$

where ν_{12}^f is the fiber Poisson's ratio and ν_{12}^m is the Poisson's ratio of the matrix.

The apparent modulus of elasticity of the unidirectional fiber composite material in the transverse 2-direction (perpendicular to the fibers direction) is considered next. As shown in Fig. 20.6, unlike the 1-direction, the iso-strain assumption is not appropriate for estimating the apparent modulus E_2 in 2-direction. However, it is reasonable to assume that when a transverse load is applied, the stress σ_2 is the same in the fiber and the matrix. This *iso-stress assumption* results in an estimate for E_2 of the composite given by

$$E_2 = \frac{E_2^f E_2^m}{V_m E_2^f + V_f E_2^m} \tag{20.4-4}$$

where E_2^f and E_2^m are the fiber and matrix transverse modulus values, respectively.

Equation (20.4-4) can be written in terms of V_f only as

$$\frac{1}{E_2} = \frac{V_f}{E_2^f} + \frac{(1 - V_f)}{E_2^m} \tag{20.4-5}$$

Note that the matrix and fiber materials are usually isotropic and thus, $E_1^f = E_2^f$ and $E_1^m = E_2^m$.

The in-plane shear modulus of a lamina, G_{12} is determined using the mechanics of materials method by presuming that the shear stresses on the fiber and on the matrix are the same, the *iso-stress assumption*. Note that the shear deformations of the fiber and the matrix are not the same. The shear modulus of the composite material can be written as

$$G_{12} = \frac{G_{12}^f G_{12}^m}{V_m G_{12}^f + V_f G_{12}^m} \tag{20.4-6}$$

Equation (20.4-6) can be written in terms of V_f only as,

$$\frac{1}{G_{12}} = \frac{V_f}{G_{12}^f} + \frac{(1 - V_f)}{G_{12}^m} \tag{20.4-7}$$

The mechanical properties of four commonly used, unidirectional reinforced composite materials, that is, glass-epoxy, boron-epoxy, graphite-epoxy, and kevlor-epoxy, are given in Table 20.3. Note, X_t and X_c are the strengths of the composite in tension and compression in the 1-direction, respectively, Y_t and Y_c are the strengths of the composite in tension and compression in the 2-direction, respectively, and S is the strength in shear.

TABLE 20.3 The Mechanical Properties of Unidirectional Reinforced Composite Materials

	Unidirectionally Reinforced Composite Material			
	Glass-Epoxy		Boron-Epoxy	
Property	GPA/MPa	psi	GPA/MPa	psi
E_1	54 GPa	7.8×10^6 psi	207 GPa	30×10^6 psi
E_2	18 GPa	2.6×10^6 psi	21 GPa	3×10^6 psi
ν_{12}	0.25	0.25	0.3	0.3
G_{12}	9 GPa	1.3×10^6 psi	7 GPa	1×10^6 psi
X_t	1035 MPa	150×10^3 psi	1380 MPa	200×10^3 psi
Y_t	28 MPa	4×10^3 psi	83 MPa	12×10^3 psi
S	41 MPa	6×10^3 psi	124 MPa	18×10^3 psi
X_c	1035 MPa	150×10^3 psi	2760 MPa	400×10^3 psi
Y_c	138 MPa	20×10^3 psi	276 MPa	40×10^3 psi
	Graphite-Epoxy		Kevlar®-Epoxy	
Property	GPA/MPa	psi	GPA/MPa	psi
E_1	207 GPa	30×10^6 psi	76 GPa	11×10^6 psi
E_2	5 GPa	0.75×10^6 psi	5.5 GPa	0.8×10^6 psi
ν_{12}	0.25	0.25	0.34	0.34
G_{12}	2.6 GPa	0.375×10^6 psi	2.1 GPa	0.3×10^6 psi
X_t	1035 MPa	150×10^3 psi	1380 MPa	200×10^3 psi
Y_t	41 MPa	6×10^3 psi	28 MPa	4×10^3 psi
S	69 MPa	10×10^3 psi	44 MPa	6.4×10^3 psi
X_c	689 MPa	100×10^3 psi	276 MPa	40×10^3 psi
Y_c	117 MPa	17×10^3 psi	138 MPa	20×10^3 psi

20.5 FAILURE CRITERION FOR A LAYER (LAMINA)

There are several failure criteria proposed to predict the failure of a unidirectional fiber composite material. These can be divided into two main groups, that of yield and a failure hypothesis.

1. **Yield criterion** These criterions address the yielding and not the failure modes of composite materials. That is, these criterions propose a polynomial and tensorial criteria, using mathematical expressions to describe the failure surface as a function of the material strengths. Generally, these expressions are based on the process of adjusting (fitting) an expression to a curve obtained by experimental tests. The most general polynomial failure criterion for composite materials of this group is the *Tensor Polynomial Criterion* proposed by Tsai and Wu (Ref. 1). For brevity, the derivations of these criterions are not presented here and the readers are referred to the references at the end of this chapter.

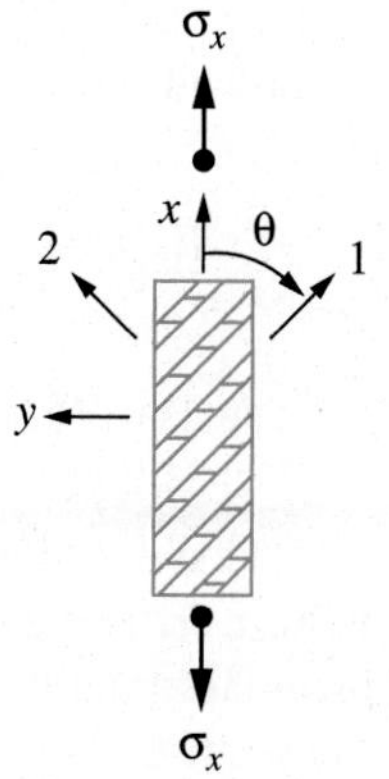

Figure 20.7 Off-axis loading of a unidirectional fiber composite lamina.

a. **Tsai-Hill Failure Criterion** Consider the unidirectional fiber composite lamina of Fig. 20.7 that is subjected to an off-axis loading of σ_x. The plane stresses σ_1, σ_2, and τ_{12} (the stresses in the principal material directions) are

$$\sigma_1 = \sigma_x \cos^2\theta$$

$$\sigma_2 = \sigma_x \sin^2\theta$$

$$\tau_{12} = -\sigma_x \cos\theta \sin\theta$$

For plane stress in the 1-2 plane of a unidirectional lamina (a layer) with fibers in the 1-direction, where $\sigma_3 = \tau_{13} = \tau_{23} = 0$, Tsai-Hill proposed the following governing failure criterion:

$$\frac{\sigma_1^2}{X^2} - \frac{\sigma_1\sigma_2}{X^2} + \frac{\sigma_2^2}{Y^2} + \frac{\tau_{12}^2}{S^2} = 1 \qquad (20.5\text{-}1)$$

where X and Y are the strength of the composite material in the principal material 1 and 2 directions, and S is the shear strength in the 1-2 coordinates. Depending on the sign of the σ_1 and σ_2, the strength in tension, X_t and Y_t, or compression, X_c and Y_c, must be used.

b. **Tsai-Wu Failure Criterion** Tsai and Wu postulated that a failure surface in six-dimensional stress space exists in the form of

$$F_i\sigma_i + F_{ij}\sigma_i\sigma_j = 1 \qquad i, j = 1, \ldots, 6 \qquad (20.5\text{-}2)$$

where F_i and F_{ij} are strength tensors of the second and fourth rank tensor, respectively. Equation (20.5-2) is complicated, however; the simplified version of the equation for a unidirectional lamina (a layer) with fibers in the 1-direction, similar to Eq. (20.5-1), can be written as

$$\frac{\sigma_1^2}{X^2} + 2F_{12}\sigma_1\sigma_2 + \frac{\sigma_2^2}{Y^2} + \frac{\tau_{12}^2}{S^2} = 1 \qquad (20.5\text{-}3)$$

To calculate F_{12} we can impose a state of biaxial tension described as $\sigma_1 = \sigma_2 = \sigma$ and all other stresses are zero. Then F_{12} is

$$F_{12} = \frac{1}{2\sigma^2}\left[1 - \left[\frac{1}{X_t} + \frac{1}{X_c} + \frac{1}{Y_t} + \frac{1}{Y_c}\right]\sigma + \left[\frac{1}{X_t X_c} + \frac{1}{Y_t Y_c}\right]\sigma^2\right] \tag{20.5-4}$$

The Tsai-Wu failure criterion fits closer to the experimental results than the Tsai-Hill criterion; however, to determine F_{12} one needs to perform a rather difficult biaxial test of the composite material.

2. **Failure criteria associated with failure modes** These criteria consider different material strength failure modes of the constituents, that is, the fiber and the matrix. These criteria have the advantage of being able to predict failure modes, such as fiber fracture, transverse matrix cracking, and shear matrix cracking.

   a. **Maximum stress criterion** This criterion considers that the composite material fails when the stress exceeds the respective allowable stress of the constituents. That is, for tensile stresses,

$$\sigma_1 > X_t \qquad \text{and} \qquad \sigma_2 > Y_t$$

   and for compressive stresses,

$$\sigma_1 < X_c \qquad \text{and} \qquad \sigma_2 < Y_c$$

   and for shear

$$|\tau_{12}| > S \tag{20.5-5}$$

   There is no interaction between modes of failure in this criterion. That is, there are actually five sub-criteria and five failure mechanisms.

   b. **Maximum strain criterion** This criterion considers that the composite fails when the strain exceeds the respective allowable strain of the constituents. That is, for tensile stresses,

$$\varepsilon_1 > X_{\varepsilon t} \qquad \text{and} \qquad \varepsilon_2 > Y_{\varepsilon t}$$

   and for compressive stresses,

$$\varepsilon_1 < X_{\varepsilon c} \qquad \text{and} \qquad \varepsilon_2 < Y_{\varepsilon c}$$

   and for shear

$$|\gamma_{12}| > S_\varepsilon \tag{20.5-6}$$

   where $X_{\varepsilon t}$, $X_{\varepsilon c}$ and $Y_{\varepsilon t}$, $Y_{\varepsilon c}$ are the allowable strains in the 1 and 2 directions, respectively (t for tension and c for compression), and S_ε is the allowable shear strain.

   c. **Hashin-Rotem criterion** Unlike the maximum stress and strain criterion, the Hashin-Rotem (Ref. 15) criterion takes into account interactions between the stresses and strains acting on a lamina. It involves two failure mechanisms, one associated with fiber failure and the other with matrix failure, distinguishing between tension and compression modes. That is, for

Fiber failure in tension: $(\sigma_1 > 0)$

$$\sigma_1 = \sigma_{1t}^u$$

Fiber failure in compression: $(\sigma_1 < 0)$

$$-\sigma_1 = \sigma_{1c}^u$$

Matrix failure in tension: $(\sigma_2 > 0)$

$$\left(\frac{\sigma_2}{\sigma_{2t}^u}\right)^2 + \left(\frac{\sigma_{12}}{\sigma_{12}^u}\right)^2 = 1$$

Matrix failure in compression: $(\sigma_2 < 0)$

$$\left(\frac{\sigma_2}{\sigma_{2c}^u}\right)^2 + \left(\frac{\sigma_{12}}{\sigma_{12}^u}\right)^2 = 1 \tag{20.5-7}$$

where t and c indicate tension and compression, respectively, and the superscript u is for ultimate stress.

d. **Hashin Criterion** Hashin (Ref. 3) later proposed a failure criterion for fibrous composites under a three-dimensional state of stress. The method uses the effect of the shear stress in the tensile fiber failure mode. For the matrix failure mode, a quadratic approach is used since a linear criterion underestimates the material strength and a polynomial of higher degree than quadratic would be too complicated. Therefore, the criterion states for

Fiber failure in tension: $(\sigma_1 > 0)$

$$\left(\frac{\sigma_1}{\sigma_{1t}^u}\right)^2 + \frac{\sigma_{12}^2 + \sigma_{13}^2}{(\sigma_{12}^u)^2} = 1 \qquad \text{or} \qquad \sigma_1 = \sigma_{1t}^u$$

Fiber failure in compression: $(\sigma_1 < 0)$

$$-\sigma_1 = \sigma_{1c}^u$$

Matrix failure in tension: $((\sigma_2 + \sigma_3) > 0)$

$$\left(\frac{\sigma_2 + \sigma_3}{\sigma_{2t}^u}\right)^2 + \frac{\sigma_{23}^2 - \sigma_2\sigma_3}{(\sigma_{23}^u)^2} + \frac{\sigma_{12}^2 + \sigma_{13}^2}{(\sigma_{12}^u)^2} = 1$$

Matrix failure in compression: $((\sigma_2 + \sigma_3) < 0)$

$$\left[\left(\frac{\sigma_{2c}^u}{2\sigma_{23}^u}\right)^2 - 1\right]\frac{\sigma_2 + \sigma_3}{\sigma_{2c}^u} + \left(\frac{\sigma_2 + \sigma_3}{2\sigma_{23}^u}\right)^2 + \frac{\sigma_{23}^2 - \sigma_2\sigma_3}{(\sigma_{23}^u)^2} + \frac{\sigma_{12}^2 + \sigma_{13}^2}{(\sigma_{12}^u)^2} = 1 \tag{20.5-8}$$

where $\sigma_1^u, \sigma_2^u, \sigma_3^u$: are the ultimate normal strengths of the lamina in the 1, 2, and 3 directions, and $\sigma_{23}^u, \sigma_{13}^u, \sigma_{12}^u$ are the ultimate shear strengths of the material in the 23, 31, and 12 planes.

20.6 MACROMECHANICS OF A LAMINATE

A laminate consists of two or more laminae (layers) bonded together to create an integral structural element. An important part of the description of a laminate is the specification of its *stacking sequence* which is the fiber orientation of all the layers through the thickness of a laminate. Figure 20.8 shows the exploded view of a laminate and the *laminate coordinate system.* The stacking sequence lists the layer fiber orientations relative to the $+x$ axis of the laminate coordinate system, starting with the layer at the negative-most z-position. For example, for regular (equal thickness layers) laminates of Fig. 20.8 a list of the layers and their orientation with respect to the $+x$ axis is sufficient and can be shown as $[+90/+30/0/-30/-90]_T$, where T signifies that the total laminate has been described. Note that only the angles of the principal material coordinates of each layer need to be given. When the thickness of the layers are not the same, a notation of layer thickness, t, must be appended to the previous notation. For example, $[+30_t/+30_{2t}/-45_{4t}]_T$ indicates that the thickness of the second and third layers are twice and four times the first layer, respectively. When different materials and thicknesses are used in a laminate, then it is necessary to specify the material properties and thickness of each layer to completely describe a laminate.

For symmetric laminates with respect to the xy plane, the symbol S is used. For example, [0/30/45/45/30/0] can be written in shorthand notation as $[0/30/45]_S$. Other shorthand notation can be used, such as $[(\pm 30/0)_3/(90/0)_2]_S$ to describe a 26-layer symmetric laminate which has two sub-sequences that repeat, one three times, the other twice. The notation $[\pm 30]_{2S}$ is interpreted to mean $[(\pm 30)_2]_S$.

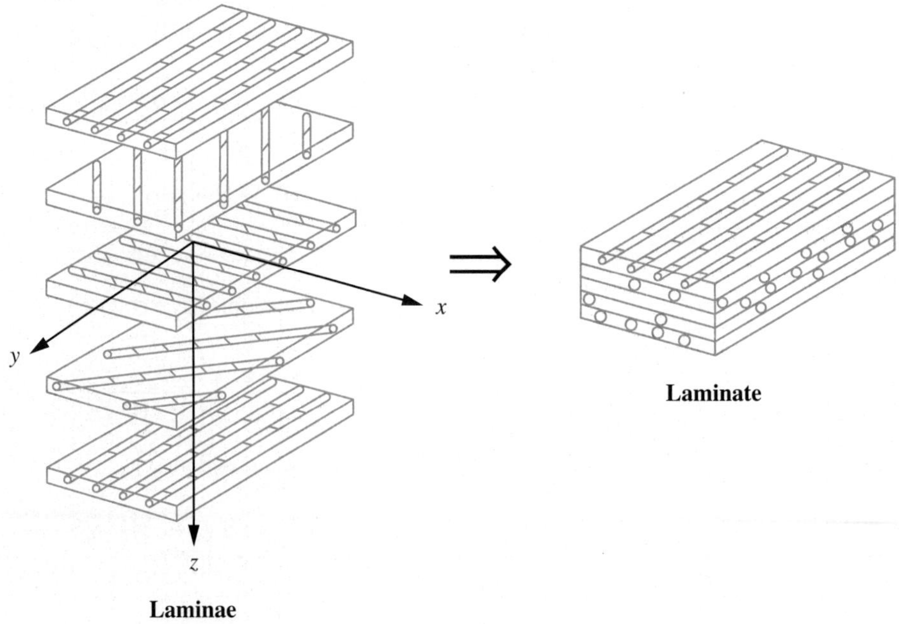

Figure 20.8 Exploded view of a laminate and the laminate coordinate system.

20.7 CLASSICAL LAMINATION THEORY

For the *classical lamination theory*, a layer is assumed to be in a state of plane stress. From Eq. (20.3-5), the stress-strain relations for a lamina can be written as

$$\begin{Bmatrix} \sigma_1 \\ \sigma_2 \\ \tau_{12} \end{Bmatrix} = \begin{pmatrix} Q_{11} & Q_{12} & 0 \\ Q_{12} & Q_{22} & 0 \\ 0 & 0 & Q_{66} \end{pmatrix} \begin{Bmatrix} \varepsilon_1 \\ \varepsilon_2 \\ \gamma_{12} \end{Bmatrix} \tag{20.7-1}$$

where the Q matrix is the *reduced stiffness matrix* having the element of

$$Q_{11} = \frac{E_1}{1-\nu_{12}\nu_{21}} \qquad Q_{12} = \frac{\nu_{12}E_2}{1-\nu_{12}\nu_{21}} \qquad Q_{22} = \frac{E_2}{1-\nu_{12}\nu_{21}} \qquad Q_{66} = G_{12}$$

For a temperature change of the composite material, Eq. (20.7-1) can be rewritten as

$$\begin{Bmatrix} \sigma_1 \\ \sigma_2 \\ \tau_{12} \end{Bmatrix} = \begin{pmatrix} Q_{11} & Q_{12} & 0 \\ Q_{12} & Q_{22} & 0 \\ 0 & 0 & Q_{66} \end{pmatrix} \begin{Bmatrix} \varepsilon_1 - \alpha_1 \Delta T \\ \varepsilon_2 - \alpha_2 \Delta T \\ \gamma_{12} \end{Bmatrix} \tag{20.7-2}$$

Note that in nearly all cases of composites, $\nu_{12} \neq \nu_{21}$. Since there can be many fiber orientations within a laminate, each orientation having its own principal material coordinate system, it is more convenient to describe laminate behavior in terms of one coordinate system, namely the laminate, or *xyz*, coordinate system shown in Fig. 20.8. The stress-strain relations of Eq. (20.7-1) can be transformed to the laminate *xyz* coordinate system as

$$\begin{Bmatrix} \sigma_x \\ \sigma_y \\ \tau_{xy} \end{Bmatrix} = [T]^{-1} \begin{Bmatrix} \sigma_1 \\ \sigma_2 \\ \tau_{12} \end{Bmatrix}, \quad \begin{Bmatrix} \varepsilon_x \\ \varepsilon_y \\ \gamma_{xy} \end{Bmatrix} = [T]^{-1} \begin{Bmatrix} \varepsilon_1 \\ \varepsilon_2 \\ \gamma_{12} \end{Bmatrix}, \quad \begin{Bmatrix} \alpha_x \\ \alpha_y \\ \alpha_{xy} \end{Bmatrix} \Delta T = [T]^{-1} \begin{Bmatrix} \alpha_1 \\ \alpha_2 \\ 0 \end{Bmatrix} \Delta T \tag{20.7-3}$$

where the transformation matrix $[T]$ is written as

$$[T] = \begin{bmatrix} \cos^2\theta & \sin^2\theta & 2\sin\theta\cos\theta \\ \sin^2\theta & \cos^2\theta & -2\sin\theta\cos\theta \\ -\sin\theta\cos\theta & \sin\theta\cos\theta & \cos^2\theta - \sin^2\theta \end{bmatrix}$$

The transformed stress-strain relations of the kth layer of a multilayered laminate can be written as

$$\{\sigma\}_k = [\bar{Q}]_k \{\varepsilon\}_k$$

$$\begin{Bmatrix} \sigma_x \\ \sigma_y \\ \tau_{xy} \end{Bmatrix} = \begin{pmatrix} \bar{Q}_{11} & \bar{Q}_{12} & \bar{Q}_{16} \\ \bar{Q}_{12} & \bar{Q}_{22} & \bar{Q}_{26} \\ \bar{Q}_{16} & \bar{Q}_{26} & \bar{Q}_{66} \end{pmatrix} \begin{Bmatrix} \varepsilon_x \\ \varepsilon_y \\ \gamma_{xy} \end{Bmatrix} \tag{20.7-4}$$

where the $\bar{Q}$ matrix is referred to as the *transformed reduced stiffness matrix*, given by $[\bar{Q}]=[T]^{-1}[Q][T]^{-T}$. The $\bar{Q}$ matrix terms are

$$\begin{aligned}
\bar{Q}_{11} &= Q_{11}\cos^4\theta + 2(Q_{12}+2Q_{66})\sin^2\theta\cos^2\theta + Q_{22}\sin^4\theta \\
\bar{Q}_{12} &= (Q_{11}+Q_{22}-4Q_{66})\sin^2\theta\cos^2\theta + Q_{12}(\sin^4\theta+\cos^4\theta) \\
\bar{Q}_{22} &= Q_{11}\sin^4\theta + 2(Q_{12}+2Q_{66})\sin^2\theta\cos^2\theta + Q_{22}\cos^4\theta \\
\bar{Q}_{16} &= (Q_{11}-Q_{12}-2Q_{66})\sin\theta\cos^3\theta + (Q_{12}-Q_{22}+2Q_{66})\sin^3\theta\cos\theta \\
\bar{Q}_{26} &= (Q_{11}-Q_{12}-2Q_{66})\sin^3\theta\cos\theta + (Q_{12}-Q_{22}+2Q_{66})\sin\theta\cos^3\theta \\
\bar{Q}_{66} &= (Q_{11}+Q_{22}-2Q_{12}-2Q_{66})\sin^2\theta\cos^2\theta + Q_{66}(\sin^4\theta + \cos^4\theta)
\end{aligned} \tag{20.7-5}$$

Note that in the $\bar{Q}$ matrix the shear stress is related to the two normal strains, and the two normal stresses are related to the shear strain. There are no zeros in the $\bar{Q}$ matrix as there are with the Q matrix. Therefore, the behavior represented by Eq. (20.7-5) is referred to as *anisotropic* behavior.

20.8 MACROMECHANICS OF A LAMINATE: STRESS AND STRAIN IN A LAMINATE

In the analysis of laminates it is assumed that all the layers of a laminate are perfectly bonded and that the laminate is a thin plate and its deformation is small. It is also assumed that a normal to the middle surface of the laminate remains perpendicular to the middle surface.

The *Kirchhoff hypothesis* is employed in which the strains ε_x, ε_y, and γ_{xy} vary linearly through the thickness of the laminate. Following the geometry of the cross section shown in Fig. 20.9, the displacements in the *x*, *y*, and *z* directions of a point *P* at location *x*, *y*, *z* within the laminate are

$$\begin{aligned}
u(x,y,z) &= u^o(x,y) - z\frac{\partial w^o(x,y)}{\partial x} \\
v(x,y,z) &= v^o(x,y) - z\frac{\partial w^o(x,y)}{\partial y} \\
w(x,y,z) &= w^o(x,y)
\end{aligned} \tag{20.8-1}$$

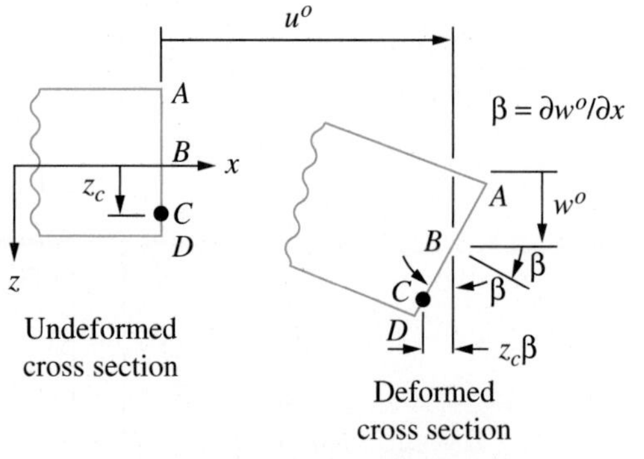

Figure 20.9 Undeformed and deformed cross section of a laminate.

where the superscript, 0, denotes the displacements of corresponding point P^0 on the *geometric mid-plane* of the laminate, and where $\beta = \partial w^0/\partial x$, as shown in Fig. 20.9.

The strains at any location within the laminate can be written in terms of the displacements as

$$\begin{aligned}
\varepsilon_x &= \frac{\partial u}{\partial x} = \frac{\partial u^o}{\partial x} - z\frac{\partial^2 w^o}{\partial x^2} = \varepsilon_x^o + z\kappa_x^o \\
\varepsilon_y &= \frac{\partial v}{\partial y} = \frac{\partial v^o}{\partial y} - z\frac{\partial^2 w^o}{\partial y^2} = \varepsilon_y^o + z\kappa_y^o \\
\gamma_{xy} &= \frac{\partial u}{\partial y} + \frac{\partial v}{\partial x} = \frac{\partial u^o}{\partial y} + \frac{\partial v^o}{\partial x} - 2z\frac{\partial^2 w^o}{\partial x \partial y} = \gamma_{xy}^o + z\kappa_{xy}^o
\end{aligned} \tag{20.8-2}$$

where

$$\begin{Bmatrix} \varepsilon_x^o \\ \varepsilon_y^o \\ \gamma_{xy}^o \end{Bmatrix} = \begin{Bmatrix} \dfrac{\partial u^o}{\partial x} \\ \dfrac{\partial v^o}{\partial y} \\ \dfrac{\partial u^o}{\partial y} + \dfrac{\partial v^o}{\partial x} \end{Bmatrix} \qquad \begin{Bmatrix} \kappa_x^o \\ \kappa_y^o \\ \kappa_{xy}^o \end{Bmatrix} = \begin{Bmatrix} -\dfrac{\partial^2 w^o}{\partial x^2} \\ -\dfrac{\partial^2 w^o}{\partial y^2} \\ -2\dfrac{\partial^2 w^o}{\partial x \partial y} \end{Bmatrix} \tag{20.8-3}$$

The quantities ε_x^o, ε_y^o, and γ_{xy}^o are referred to as the *mid-plane* or *reference surface strains*. The quantities κ_x^o, κ_y^o, and κ_{xy}^o are the *mid-plane* or *reference surface curvatures* and are the bending reference surface curvatures in the *x*, *y* directions and the twist curvature, respectively. Substituting Eq. (20.8-2) into the stress-strain relations of Eq. (20.7-4), the stresses in the *k*th layer can be expressed in terms of the laminate middle-surface strain and curvatures as

$$\begin{Bmatrix} \sigma_x \\ \sigma_y \\ \tau_{xy} \end{Bmatrix} = \begin{pmatrix} \bar{Q}_{11} & \bar{Q}_{12} & \bar{Q}_{16} \\ \bar{Q}_{12} & \bar{Q}_{22} & \bar{Q}_{26} \\ \bar{Q}_{16} & \bar{Q}_{26} & \bar{Q}_{66} \end{pmatrix} \left(\begin{Bmatrix} \varepsilon_x \\ \varepsilon_y \\ \gamma_x \end{Bmatrix} + z \begin{Bmatrix} \kappa_x^o \\ \kappa_y^o \\ \kappa_{xy}^o \end{Bmatrix} \right) \tag{20.8-4}$$

Next, the relationships between the *loads* and the mid-plane strains and curvatures are developed. The resultant forces and moments acting on a laminate are obtained by integration of the stresses in each layer or lamina through the laminate thickness, as follows. Let

$$\begin{aligned}
N_x &= \int_{-t/2}^{t/2} \sigma_x \, dz \qquad N_y = \int_{-t/2}^{t/2} \sigma_y \, dz \qquad N_{xy} = \int_{-t/2}^{t/2} \tau_{xy} \, dz \\
M_x &= \int_{-t/2}^{t/2} \sigma_x z \, dz \qquad M_y = \int_{-t/2}^{t/2} \sigma_y z \, dz \qquad M_{xy} = \int_{-t/2}^{t/2} \tau_{xy} z \, dz
\end{aligned} \tag{20.8-5}$$

where t is the thickness of the laminate, the N terms are the *force resultants per unit length,* and the M terms are the *moment resultants per unit length,* as shown in Fig. 20.10.

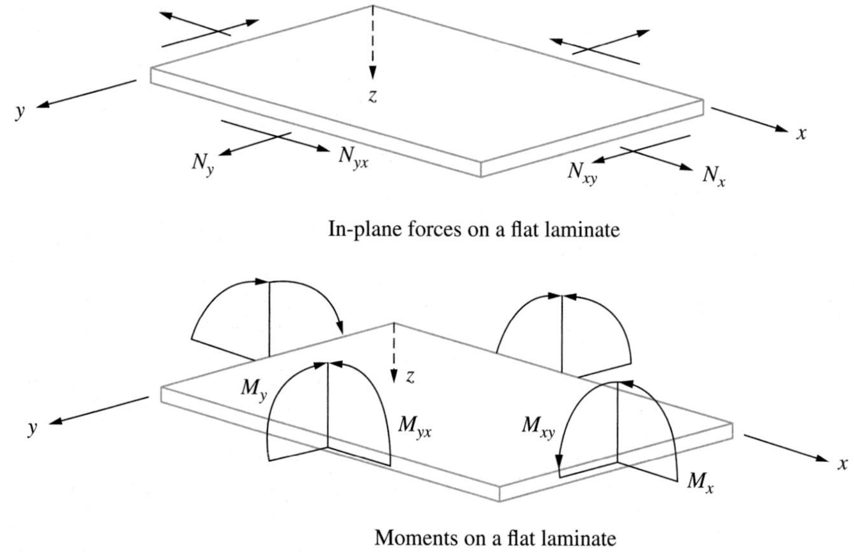

Figure 20.10 In-plane forces and moments on a laminate.

If the three components of stress from Eq. (20.8-4) are substituted into the integrands of the definitions of the six stress resultants from Eq. (20.8-5), integration with respect to z results in

$$\begin{bmatrix} N_x \\ N_y \\ N_{xy} \end{bmatrix} = \sum_{k=1}^{n} \begin{bmatrix} \bar{Q}_{11} & \bar{Q}_{12} & \bar{Q}_{16} \\ \bar{Q}_{12} & \bar{Q}_{22} & \bar{Q}_{26} \\ \bar{Q}_{16} & \bar{Q}_{26} & \bar{Q}_{66} \end{bmatrix}_k \left[\int_{z_{k-1}}^{z_k} \begin{bmatrix} \varepsilon_x^o \\ \varepsilon_y^o \\ \gamma_{xy}^o \end{bmatrix} dz + \int_{z_{k-1}}^{z_k} \begin{bmatrix} \kappa_x \\ \kappa_y \\ \kappa_{xy} \end{bmatrix} z\,dz \right]$$

$$\begin{bmatrix} M_x \\ M_y \\ M_{xy} \end{bmatrix} = \sum_{k=1}^{n} \begin{bmatrix} \bar{Q}_{11} & \bar{Q}_{12} & \bar{Q}_{16} \\ \bar{Q}_{12} & \bar{Q}_{22} & \bar{Q}_{26} \\ \bar{Q}_{16} & \bar{Q}_{26} & \bar{Q}_{66} \end{bmatrix}_k \left[\int_{z_{k-1}}^{z_k} \begin{bmatrix} \varepsilon_x^o \\ \varepsilon_y^o \\ \gamma_{xy}^o \end{bmatrix} z\,dz + \int_{z_{k-1}}^{z_k} \begin{bmatrix} \kappa_x \\ \kappa_y \\ \kappa_{xy} \end{bmatrix} z^2\,dz \right] \tag{20.8-6}$$

where z_k and z_{k-1} are defined in the basic laminate geometry of Fig. 20.11.

Note that the strains ε_x^o, ε_y^o, and γ_{xy}^o and the curvature κ_x^o, κ_y^o, and κ_{xy}^o are the middle-surface values and are not functions of z, and, thus, they are removed from the integration. Equation (20.8-6) can be written as

$$\begin{bmatrix} N_x \\ N_y \\ N_{xy} \end{bmatrix} = \begin{bmatrix} A_{11} & A_{12} & A_{16} \\ A_{12} & A_{22} & A_{26} \\ A_{16} & A_{26} & A_{66} \end{bmatrix} \begin{bmatrix} \varepsilon_x^o \\ \varepsilon_y^o \\ \gamma_{xy}^o \end{bmatrix} + \begin{bmatrix} B_{11} & B_{12} & B_{16} \\ B_{12} & B_{22} & B_{26} \\ B_{16} & B_{26} & B_{66} \end{bmatrix} \begin{bmatrix} \kappa_x \\ \kappa_y \\ \kappa_{xy} \end{bmatrix}$$

$$\begin{bmatrix} M_x \\ M_y \\ M_{xy} \end{bmatrix} = \begin{bmatrix} B_{11} & B_{12} & B_{16} \\ B_{12} & B_{22} & B_{26} \\ B_{16} & B_{26} & B_{66} \end{bmatrix} \begin{bmatrix} \varepsilon_x^o \\ \varepsilon_y^o \\ \gamma_{xy}^o \end{bmatrix} + \begin{bmatrix} D_{11} & D_{12} & D_{16} \\ D_{12} & D_{22} & D_{26} \\ D_{16} & D_{26} & D_{66} \end{bmatrix} \begin{bmatrix} \kappa_x \\ \kappa_y \\ \kappa_{xy} \end{bmatrix} \tag{20.8-7}$$

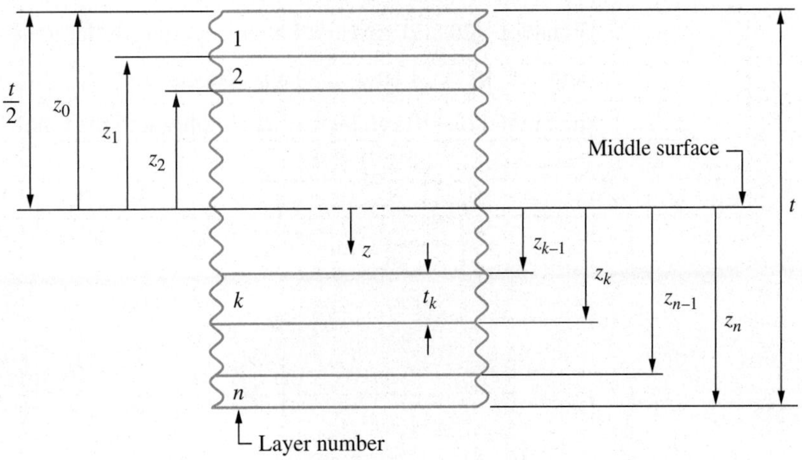

Geometry of an N-layered laminate

Figure 20.11 Coordinates and numbering of each layer.

where,

$$A_{ij}=\sum_{k=1}^{n}\bar{Q}_{ij_k}\left(z_k-z_{k-1}\right)\qquad B_{ij}=\frac{1}{2}\sum_{k=1}^{n}\bar{Q}_{ij_k}\left(z_k^2-z_{k-1}^2\right)\qquad D_{ij}=\frac{1}{3}\sum_{k=1}^{n}\bar{Q}_{ij_k}\left(z_k^3-z_{k-1}^3\right)\tag{20.8-8}$$

The elements of the *ABD* matrices needs to be discussed. The terms of the *A* matrix involve only the layer thicknesses, i.e., $z_k - z_{k-1} = t_k$, the thickness of the *k*th layer, while terms of the *B* and *D* matrices involve layer thicknesses and locations. Physically, the terms of the *A* matrix are the in-plane stiffnesses of the laminate, or extensional stiffnesses. The terms of the *D* matrix are the bending stiffnesses, and the terms of the *B* matrix are unique to composite materials and are referred to as the *bending-stretching stiffnesses.* This nomenclature refers to the fact that through these terms, bending effects, namely, the curvatures, κ_x^o, κ_y^o, and κ_{xy}^o, are coupled to in-plane force resultants, N_x, N_y, and N_{xy}, and conversely, the in-plane strains, ε_x^o, ε_y^o, and γ_{xy}^o, are coupled to the bending moment resultants, M_x, M_y, and M_{xy}.

When a laminate has symmetrical layers *(symmetric laminates)* all elements of the *B* matrix are zero. A symmetric laminate is defined as follows: If for every layer at a specific location with a specific thickness, material properties, and fiber orientation located on one side of the geometric mid-plane, there is an identical layer at the mirror-image location on the other side of the geometric mid-plane, then the laminate is said to be symmetric.

For so-called *balanced laminates*, A_{16}, A_{26}, and $\hat{N}_{xy}^T$ (the thermal resultant shear force) are zero. A laminate is said to be *balanced* if every layer of a specific thickness, material properties, and fiber orientation, has an identical layer with the *opposite* fiber orientation *somewhere* within the laminate. Symmetric and balanced laminates are very common and the relations of Eq. (20.8-7) simplifies considerably to

$$\begin{Bmatrix} N_x \\ N_y \\ N_{xy} \\ M_x \\ M_y \\ M_{xy} \end{Bmatrix} = \begin{bmatrix} A_{11} & A_{12} & 0 & 0 & 0 & 0 \\ A_{12} & A_{22} & 0 & 0 & 0 & 0 \\ 0 & 0 & A_{66} & 0 & 0 & 0 \\ 0 & 0 & 0 & D_{11} & D_{12} & D_{16} \\ 0 & 0 & 0 & D_{12} & D_{22} & D_{26} \\ 0 & 0 & 0 & D_{16} & D_{26} & D_{66} \end{bmatrix} \begin{Bmatrix} \varepsilon_x^o \\ \varepsilon_y^o \\ \gamma_{xy}^o \\ \kappa_x^o \\ \kappa_y^o \\ \kappa_{xy} \end{Bmatrix}\tag{20.8-9}$$

Equation (20.8-9) reveals that for symmetric laminates, bending effects $\left(M_x, M_y, M_{xy}\right.$ and $\left.\kappa_x^o, \kappa_y^o, \kappa_{xy}^o\right)$ and in-plane effects $\left(N_x, N_y, N_{xy} \text{ and } \varepsilon_x^o, \varepsilon_y^o, \gamma_{xy}^o\right)$ are decoupled. Equation (20.8-9) can be further simplified to the following equations:

$$\begin{Bmatrix} N_x \\ N_y \end{Bmatrix} = \begin{bmatrix} A_{11} & A_{12} \\ A_{12} & A_{22} \end{bmatrix} \begin{Bmatrix} \varepsilon_x^o \\ \varepsilon_y^o \end{Bmatrix} \tag{20.8-10}$$

$$N_{xy} = A_{66}\gamma_{xy}^o \tag{20.8-11}$$

$$\begin{Bmatrix} M_x \\ M_y \end{Bmatrix} = \begin{bmatrix} D_{11} & D_{12} \\ D_{12} & D_{22} \end{bmatrix} \begin{Bmatrix} \kappa_x^o \\ \kappa_y^o \end{Bmatrix} \tag{20.8-12}$$

$$M_{xy} = D_{66}\kappa_{xy}^o \tag{20.8-13}$$

Cross-ply laminates are laminates constructed of layers with fiber orientations of only 0 or 90°. For such laminates, because $\bar{Q}_{16}$ and $\bar{Q}_{26}$ are zero for every layer, all the terms in Eq. (20.8-7) with subscripts 16 and 26 are zero. For a symmetric cross-ply laminate, the relations of Eq. (20.8-7) simplify similar to Eqs. (20.8-10) to (20.8-13). In a case where the temperature of a *balanced laminate or cross-ply laminate* changes, the Eqs. (20.8-10) to (20.8-13) can be modified as

$$\begin{Bmatrix} N_x \\ N_y \end{Bmatrix} = \begin{bmatrix} A_{11} & A_{12} \\ A_{12} & A_{22} \end{bmatrix} \begin{Bmatrix} \varepsilon_x^o \\ \varepsilon_y^o \end{Bmatrix} - \begin{Bmatrix} \hat{N}_x^T \\ \hat{N}_y^T \end{Bmatrix} \Delta T \tag{20.8-14}$$

$$N_{xy} = A_{66}\gamma_{xy}^o \tag{20.8-15}$$

$$\begin{Bmatrix} M_x \\ M_y \end{Bmatrix} = \begin{bmatrix} D_{11} & D_{12} \\ D_{12} & D_{22} \end{bmatrix} \begin{Bmatrix} \kappa_x^o \\ \kappa_y^o \end{Bmatrix} \tag{20.8-16}$$

$$M_{xy} = D_{66}\kappa_{xy}^o \tag{20.8-17}$$

where $\hat{N}_i^T$ is the thermal resultant force.

Note that $\hat{N}_i^T$ is the *unit effective thermal force resultant* which is only a function of material properties and the laminate geometry and is defined as

$$\hat{N}_x^T = \int_{-t/2}^{t/2} \left\{\bar{Q}_{11}\alpha_x + \bar{Q}_{12}\alpha_y + \bar{Q}_{16}\alpha_{xy}\right\} dz$$

Note also that if the temperature is not uniform through the thickness of the laminate, then Eqs. (20.8-14) to (20.8-17) do not apply.

EXAMPLE 1 A composite layer is made of a glass-fiber-reinforced polymer having a fiber volume fraction of 60%. Determine the mechanical properties of the composite layer if the mechanical properties of the glass fiber are $E_1 = 73.1\,\text{GPa}$, $E_2 = 73.1\,\text{GPa}$, $G_{12} = 30.0\,\text{GPa}$, $\nu_{12} = 0.22$, $\alpha_1 = 5.04/°\text{C}$, and $\alpha_2 = 5.04/°\text{C}$, and the mechanical properties of the Polymer matrix are $E_1 = 4.62\,\text{GPa}$, $E_2 = 4.62\,\text{GPa}$, $G_{12} = 1.699\,\text{GPa}$, $\nu_{12} = 0.36$, $\alpha_1 = 41.4 \times 10^{-6}/°\text{C}$, and $\alpha_2 = 41.4 \times 10^{-6}/°\text{C}$.

SOLUTION We use the *rule-of-mixtures* to find the E_1, E_2, G_{12}, and ν_{12} of the layer.

For E_1, we have

$$E_1 = V_f E_1^f + (1-V_f)E_1^m$$
$$= (0.60)(73.1) + (1-0.60)(4.62)$$
$$= 45.7 \text{ GPa}$$

For ν_{12},

$$\nu_{12} = V_f \nu_{12}^f + (1-V_f)\nu_{12}^m$$
$$= (0.60)(0.22) + (1-0.60)(0.36)$$
$$= 0.276$$

For E_2,

$$\frac{1}{E_2} = \frac{V_f}{E_2^f} + \frac{(1-V_f)}{E_2^m}$$
$$= \frac{0.60}{73.1} + \frac{(1-0.60)}{4.62}$$
$$= 0.09478$$
$$E_2 = \frac{1}{0.09478} = 10.55 \text{ GPa}$$

For G_{12},

$$\frac{1}{G_{12}} = \frac{V_f}{G_{12}^f} + \frac{(1-V_f)}{G_{12}^m}$$
$$= \frac{0.60}{30.0} + \frac{(1-0.60)}{1.699}$$
$$= 0.255$$
$$G_{12} = 3.91 \text{ GPa}$$

EXAMPLE 2 Consider a glass-fiber-reinforced polymer composite layer having the mechanical properties of $E_1 = 45.7$ GPa, $E_2 = 16.07$ GPa, $\nu_{12} = 0.27$, $\nu_{21} = 0.0970$, $G_{12} = 5.62$ GPa, $\alpha_1 = 6.51 \times 10^6/°C$, and $\alpha_2 = 24.4 \times 10^6/°C$. Determine the numerical values of the reduced stiffness matrix Q for the glass-fiber-reinforced composite.

SOLUTION

$$Q_{11} = \frac{E_1}{1-\nu_{12}\nu_{21}} = \frac{45.7\times 10^9}{1-(0.276)(0.0970)} = 47.0\times 10^9 = 47.0 \text{ GPa}$$

$$Q_{12} = \frac{\nu_{12}E_2}{1-\nu_{12}\nu_{21}} = \frac{(0.276)(16.07\times 10^9)}{1-(0.276)(0.0970)} = 4.56 \text{ GPa}$$

$$Q_{22} = \frac{E_2}{1-\nu_{12}\nu_{21}} = \frac{16.07\times 10^9}{1-(0.276)(0.0970)} = 16.51 \text{ GPa}$$

$$[Q] = \begin{pmatrix} Q_{11} & Q_{12} & 0 \\ Q_{12} & Q_{22} & 0 \\ 0 & 0 & Q_{66} \end{pmatrix} = \begin{pmatrix} 47.0 & 4.56 & 0 \\ 4.56 & 16.51 & 0 \\ 0 & 0 & 5.62 \end{pmatrix} \times 10^9 = \begin{pmatrix} 47.0 & 4.56 & 0 \\ 4.56 & 16.51 & 0 \\ 0 & 0 & 5.62 \end{pmatrix} \text{GPa}$$

EXAMPLE 3 The fiber reinforced composite of Example 2 is rotated +30° with respect to the x axis and is subjected to a normal stress of $\sigma_x = 50$ MPa. Determine the strains of this composite. Assume the Q's of Example 2.

SOLUTION From Eq. (20.7-4) we have

$$\begin{Bmatrix} \sigma_x \\ \sigma_y \\ \tau_{xy} \end{Bmatrix} = \begin{Bmatrix} 50\times10^6 \\ 0 \\ 0 \end{Bmatrix} = \begin{pmatrix} \bar{Q}_{11}(30°) & \bar{Q}_{12}(30°) & \bar{Q}_{16}(30°) \\ \bar{Q}_{12}(30°) & \bar{Q}_{22}(30°) & \bar{Q}_{26}(30°) \\ \bar{Q}_{16}(30°) & \bar{Q}_{26}(30°) & \bar{Q}_{66}(30°) \end{pmatrix} \begin{Bmatrix} \varepsilon_x \\ \varepsilon_y \\ \gamma_{xy} \end{Bmatrix}$$

where the $\bar{Q}'s$ are computed using Eq. (20.7-5) with $\theta = 30°$,

$$\begin{pmatrix} \bar{Q}_{11}(30°) & \bar{Q}_{12}(30°) & \bar{Q}_{16}(30°) \\ \bar{Q}_{12}(30°) & \bar{Q}_{22}(30°) & \bar{Q}_{26}(30°) \\ \bar{Q}_{16}(30°) & \bar{Q}_{26}(30°) & \bar{Q}_{66}(30°) \end{pmatrix} = \begin{pmatrix} 31.29 & 10.54 & 10.05 \\ 10.54 & 18.15 & 3.14 \\ 10.05 & 3.14 & 11.60 \end{pmatrix} \times 10^9 \text{ N/m}^2$$

resulting in

$$\begin{pmatrix} 33.4 & 10.54 & 10.05 \\ 10.54 & 18.15 & 3.14 \\ 10.05 & 3.14 & 11.60 \end{pmatrix} \times 10^9 \begin{Bmatrix} \varepsilon_x \\ \varepsilon_y \\ \gamma_{xy} \end{Bmatrix} = \begin{Bmatrix} 50\times10^6 \\ 0 \\ 0 \end{Bmatrix}$$

Solving for the strains yields

$$\begin{Bmatrix} \varepsilon_x \\ \varepsilon_y \\ \gamma_{xy} \end{Bmatrix} = \begin{Bmatrix} 2370 \\ -1069 \\ -1760 \end{Bmatrix} \times 10^{-6}$$

EXAMPLE 4 Consider a laminate having layers of $[\pm45/0/90]_S$. Each layer is a glass-fiber-reinforced polymer composite of Example 2 with thickness of 0.125 mm. Determine the A, B, and D matrices and the resultant normal force when the temperature change is −100°C.

SOLUTION Since the laminate is symmetric, the B matrix is zero. Because there is a −45° layer for each +45° layer, the laminate is balanced. Neither the 0° nor 90° layers contribute to any 16 terms or the unit effective thermal stress resultants.

$$\begin{bmatrix} A_{11} & A_{12} & A_{16} \\ A_{12} & A_{22} & A_{26} \\ A_{16} & A_{26} & A_{66} \end{bmatrix} = \begin{bmatrix} 27.8 & 8.55 & 0 \\ 8.55 & 27.8 & 0 \\ 0 & 0 & 9.60 \end{bmatrix} \text{MN/m}$$

$$\begin{bmatrix} D_{11} & D_{12} & D_{16} \\ D_{12} & D_{22} & D_{26} \\ D_{16} & D_{26} & D_{66} \end{bmatrix} = \begin{bmatrix} 2.18 & 0.961 & 0.1784 \\ 0.961 & 1.945 & 0.1784 \\ 0.1784 & 0.1784 & 1.050 \end{bmatrix} \text{N-m}$$

$$\begin{Bmatrix} \hat{N}_x^T \\ \hat{N}_y^T \end{Bmatrix} = \begin{Bmatrix} 425 \\ 425 \end{Bmatrix} \text{N/m °C}$$

20.9 INVERSION OF STIFFNESS EQUATION IN A LAMINATE

For engineering applications of composite laminates, usually the force and moment resultants and temperature changes are known for a laminate. It may be of interest to compute the strains and stresses in individual layers or throughout the entire thickness of the laminate. Therefore, we invert the relations previously discussed in order to calculate the strains and curvatures of the laminate. Rewriting Eq. (20.8-7) in a condensed form as

$$\begin{bmatrix} N \\ \hline M \end{bmatrix} = \left[\begin{array}{c|c} A & B \\ \hline B & D \end{array}\right] \begin{bmatrix} \varepsilon^o \\ \hline \kappa^o \end{bmatrix} \tag{20.9-1}$$

the inverse relation is

$$\begin{bmatrix} \varepsilon_x^o \\ \varepsilon_y^o \\ \gamma_{xy}^o \\ \kappa_x^o \\ \kappa_y^o \\ \kappa_{xy}^o \end{bmatrix} = \begin{bmatrix} a_{11} & a_{12} & a_{16} & b_{11} & b_{12} & b_{16} \\ a_{12} & a_{22} & a_{26} & b_{21} & b_{22} & b_{26} \\ a_{16} & a_{26} & a_{66} & b_{61} & b_{62} & b_{66} \\ b_{11} & b_{21} & b_{61} & d_{11} & d_{12} & d_{16} \\ b_{12} & b_{22} & b_{62} & d_{12} & d_{22} & d_{26} \\ b_{16} & b_{26} & b_{66} & d_{16} & d_{26} & d_{66} \end{bmatrix} \begin{Bmatrix} N_x \\ N_y \\ N_{xy} \\ M_x \\ M_y \\ M_{xy} \end{Bmatrix} \tag{20.9-2}$$

where

$$\begin{bmatrix} a_{11} & a_{12} & a_{16} & b_{11} & b_{12} & b_{16} \\ a_{12} & a_{22} & a_{26} & b_{21} & b_{22} & b_{26} \\ a_{16} & a_{26} & a_{66} & b_{61} & b_{62} & b_{66} \\ b_{11} & b_{21} & b_{61} & d_{11} & d_{12} & d_{16} \\ b_{12} & b_{22} & b_{62} & d_{12} & d_{22} & d_{26} \\ b_{16} & b_{26} & b_{66} & d_{16} & d_{26} & d_{66} \end{bmatrix} = \begin{bmatrix} A_{11} & A_{12} & A_{16} & B_{11} & B_{12} & B_{16} \\ A_{12} & A_{22} & A_{26} & B_{12} & B_{22} & B_{26} \\ A_{16} & A_{26} & A_{66} & B_{16} & B_{26} & B_{66} \\ B_{11} & B_{12} & B_{16} & D_{11} & D_{12} & D_{16} \\ B_{12} & B_{22} & B_{26} & D_{12} & D_{22} & D_{26} \\ B_{16} & B_{26} & B_{66} & D_{16} & D_{26} & D_{66} \end{bmatrix}^{-1} \tag{20.9-3}$$

For a symmetric balanced laminate, the inverse equations reduces to

$$\begin{Bmatrix} \varepsilon_x^o \\ \varepsilon_y^o \end{Bmatrix} = \begin{bmatrix} a_{11} & a_{12} \\ a_{12} & a_{22} \end{bmatrix} \begin{Bmatrix} N_x \\ N_y \end{Bmatrix} \tag{20.9-4}$$

And

$$\gamma_{xy}^o = a_{66} N_{xy} \tag{20.9-5}$$

$$\begin{Bmatrix} \kappa_x^o \\ \kappa_y^o \\ \kappa_{xy}^o \end{Bmatrix} = \begin{bmatrix} d_{11} & d_{12} & d_{16} \\ d_{12} & d_{22} & d_{26} \\ d_{16} & d_{26} & d_{66} \end{bmatrix} \begin{Bmatrix} M_x \\ M_y \\ M_{xy} \end{Bmatrix} = [d] \begin{Bmatrix} M_x \\ M_y \\ M_{xy} \end{Bmatrix} \tag{20.9-6}$$

where

$$\begin{bmatrix} a_{11} & a_{12} & 0 \\ a_{12} & a_{22} & 0 \\ 0 & 0 & a_{66} \end{bmatrix} = \begin{bmatrix} A_{11} & A_{12} & 0 \\ A_{12} & A_{22} & 0 \\ 0 & 0 & A_{66} \end{bmatrix}^{-1} \quad \begin{bmatrix} d_{11} & d_{12} & d_{16} \\ d_{12} & d_{22} & d_{26} \\ d_{16} & d_{26} & d_{66} \end{bmatrix} = \begin{bmatrix} D_{11} & D_{12} & D_{16} \\ D_{12} & D_{22} & D_{26} \\ D_{16} & D_{26} & D_{66} \end{bmatrix}^{-1} \tag{20.9-7}$$

EXAMPLE 5 Determine the $[a]$ and $[d]$ matrixes for Example 4.

SOLUTION Inverting the problem of Example 4 for the laminate of $[\pm 45/0/90]_s$, we have

$$\begin{bmatrix} a_{11} & a_{12} & a_{16} \\ a_{12} & a_{22} & a_{26} \\ a_{16} & a_{26} & a_{66} \end{bmatrix} = \begin{bmatrix} 39.8 & -12.26 & 0 \\ -12.26 & 39.8 & 0 \\ 0 & 0 & 104.1 \end{bmatrix} \text{m(GN)}^{-1}$$

$$\begin{bmatrix} d_{11} & d_{12} & d_{16} \\ d_{12} & d_{22} & d_{26} \\ d_{16} & d_{26} & d_{66} \end{bmatrix} = \begin{bmatrix} 0.588 & -0.286 & -0.0514 \\ -0.286 & 0.662 & -0.0638 \\ -0.0514 & -0.0638 & 0.972 \end{bmatrix} (\text{N-m})^{-1}$$

20.10 EXAMPLE OF STRESSES AND STRAINS IN A LAMINATE

When a laminate is utilized as a structural element, its material properties, fiber orientation, and z-coordinates of each layer of the laminate are known. This information would lead to the calculation of the numerical values for the elements of the A, B, and D matrices. Knowing the magnitude and direction of the applied loads and the dimensions of the element, the force and moment resultants can be determined. The strains and curvatures of the reference surface of the laminate can be computed using the various inverse relations of the previous section, depending on the particular laminate. These calculations are best executed by programming a calculator or computer to deal with the many algebraic relations, where hand calculations may lead to some errors. In this section, several examples that demonstrate the methodology of the stress and strain calculations of a laminate are presented.

EXAMPLE 6 Consider a flat eight-layer $[\pm 45/0/90]_s$ quasi-isotropic laminate 0.1 by 0.4 m loaded by a 0.7 N·m moment, as shown in Fig. 20.12. The temperature change of the laminate is −100°C. Use the material properties of Example 4 and determine the through-thickness distribution of the stresses in the principal material coordinate system of the laminate.

SOLUTION The symmetric and balanced nature of the laminate, along with the fact that only a single moment is applied, simplifies the computations considerably. If it is assumed that the

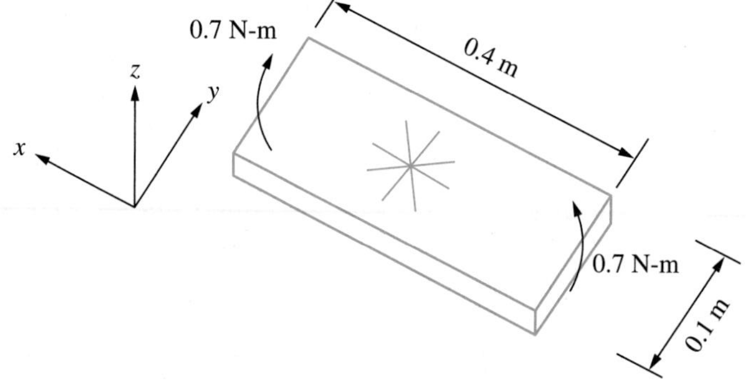

Figure 20.12 Laminate subjected to bending moment.

applied moment is uniformly distributed along the opposite edges, the resultant unit moment is given by

$$M_x = -0.7/0.1 = -7 \text{ N}\cdot\text{m/m}$$

The values of M_y and M_{xy} are zero. There are no unit equivalent thermal moment resultants, and the unit equivalent thermal force resultants are given by Example 4 as

$$\begin{Bmatrix} \hat{N}_x^T \\ \hat{N}_y^T \\ \hat{N}_{xy}^T \end{Bmatrix} = \begin{Bmatrix} 425 \\ 425 \\ 0 \end{Bmatrix} \text{ N/m/°C}$$

The inverse matrices of A and D are

$$\begin{bmatrix} a_{11} & a_{12} & a_{16} \\ a_{12} & a_{22} & a_{26} \\ a_{16} & a_{26} & a_{66} \end{bmatrix} = \begin{bmatrix} 39.8 & -12.26 & 0 \\ -12.26 & 39.8 & 0 \\ 0 & 0 & 104.1 \end{bmatrix} \text{m(GN)}^{-1}$$

$$\begin{bmatrix} d_{11} & d_{12} & d_{16} \\ d_{12} & d_{22} & d_{26} \\ d_{16} & d_{26} & d_{66} \end{bmatrix} = \begin{bmatrix} 0.588 & -0.286 & -0.0514 \\ -0.286 & 0.662 & -0.0638 \\ -0.0514 & -0.0638 & 0.972 \end{bmatrix} (\text{N-m})^{-1}$$

The reference surface strains are due only to the equivalent thermal force resultants and are computed as

$$\begin{Bmatrix} \varepsilon_x^o \\ \varepsilon_y^o \\ \gamma_{xy}^o \end{Bmatrix} = \begin{bmatrix} a_{11} & a_{12} & 0 \\ a_{12} & a_{22} & 0 \\ 0 & 0 & a_{66} \end{bmatrix} \begin{Bmatrix} \hat{N}_x^T \Delta T \\ \hat{N}_y^T \Delta T \\ \hat{N}_{xy}^T \Delta T \end{Bmatrix} = \begin{bmatrix} 39.8 & -12.26 & 0 \\ -12.26 & 39.8 & 0 \\ 0 & 0 & 104.1 \end{bmatrix} (10^{-9}) \begin{Bmatrix} 425(-100) \\ 425(-100) \\ 0 \end{Bmatrix}$$

$$= \begin{Bmatrix} -1171 \\ -1171 \\ 0 \end{Bmatrix} (10^{-6})$$

The reference surface curvatures are due to the applied moment and are computed as

$$\begin{Bmatrix} \kappa_x^o \\ \kappa_y^o \\ \kappa_{xy}^o \end{Bmatrix} = \begin{Bmatrix} d_{11} & d_{12} & d_{16} \\ d_{12} & d_{22} & d_{26} \\ d_{16} & d_{26} & d_{66} \end{Bmatrix} \begin{Bmatrix} M_x \\ M_y \\ M_{xy} \end{Bmatrix} = \begin{Bmatrix} 0.588 & -0.286 & -0.0514 \\ -0.286 & 0.662 & -0.0638 \\ -0.0514 & -0.0638 & 0.972 \end{Bmatrix} \begin{Bmatrix} -7 \\ 0 \\ 0 \end{Bmatrix} = \begin{Bmatrix} -4.12 \\ 2.00 \\ 0.360 \end{Bmatrix} \text{m}^{-1}$$

The strain as a function of z in the laminate coordinate system is

$$\begin{Bmatrix} \varepsilon_x \\ \varepsilon_y \\ \gamma_{xy} \end{Bmatrix} = \begin{Bmatrix} \varepsilon_x^o \\ \varepsilon_y^o \\ \gamma_{xy}^o \end{Bmatrix} + z \begin{Bmatrix} \kappa_x^o \\ \kappa_y^o \\ \kappa_{xy}^o \end{Bmatrix} = \begin{Bmatrix} -1171 \\ -1171 \\ 0 \end{Bmatrix} (10^{-6}) + z \begin{Bmatrix} -4.12 \\ 2.00 \\ 0.360 \end{Bmatrix}$$

These relations are valid for the full range of z, $-0.000500 \text{ m} \leq z \leq +0.000500$ m. Transforming these strains to the principal material coordinate system for the various fiber orientations gives,

for $-0.000500 \text{ m} \leq z \leq -0.000375$ m and $0.000375 \text{ m} \leq z \leq 0.000500$ m

$$\begin{Bmatrix} \varepsilon_1(45°) \\ \varepsilon_2(45°) \\ \gamma_{12}(45°) \end{Bmatrix} = [T(45°)] \begin{Bmatrix} \varepsilon_x \\ \varepsilon_y \\ \gamma_{xy} \end{Bmatrix} = \begin{bmatrix} \frac{1}{2} & \frac{1}{2} & 1 \\ \frac{1}{2} & \frac{1}{2} & -1 \\ -\frac{1}{2} & \frac{1}{2} & 0 \end{bmatrix} \begin{Bmatrix} -1171(10^{-6}) - 4.12z \\ -1171(10^{-6}) + 2.00z \\ 0.180z \end{Bmatrix} = \begin{Bmatrix} -1171(10^{-6}) - 0.878z \\ -1171(10^{-6}) - 1.238z \\ 3.06z \end{Bmatrix}$$

for $-0.000375\text{ m} \le z \le -0.000250\text{ m}$ and $0.000250\text{ m} \le z \le 0.000375\text{ m}$

$$\begin{Bmatrix} \varepsilon_1(-45°) \\ \varepsilon_2(-45°) \\ \gamma_{12}(-45°) \end{Bmatrix} = \begin{Bmatrix} -1171(10^{-6}) - 1.238z \\ -1171(10^{-6}) - 0.878z \\ -3.06z \end{Bmatrix}$$

for $-0.000250\text{ m} \le z \le -0.000125\text{ m}$ and $0.000125\text{ m} \le z \le 0.000250\text{ m}$

$$\begin{Bmatrix} \varepsilon_1(0°) \\ \varepsilon_2(0°) \\ \gamma_{12}(0°) \end{Bmatrix} = \begin{Bmatrix} -1171(10^{-6}) - 4.12z \\ -1171(10^{-6}) + 2.00z \\ 0.1799z \end{Bmatrix}$$

for $-0.000125\text{ m} \le z \le 0.000125\text{ m}$

$$\begin{Bmatrix} \varepsilon_1(90°) \\ \varepsilon_2(90°) \\ \gamma_{12}(90°) \end{Bmatrix} = \begin{Bmatrix} -1171(10^{-6}) + 2.00z \\ -1171(10^{-6}) - 4.12z \\ -0.1799z \end{Bmatrix}$$

Note each equation is valid for two ranges of z, depending on the z-location of the layers with the different fiber orientations. The stresses in the various layers in the principal material coordinate system can be computed directly from

$$\begin{Bmatrix} \sigma_1 \\ \sigma_2 \\ \tau_{12} \end{Bmatrix} = \begin{bmatrix} Q_{11} & Q_{12} & 0 \\ Q_{12} & Q_{22} & 0 \\ 0 & 0 & Q_{66} \end{bmatrix} \begin{Bmatrix} \varepsilon_1 - \alpha_1 \Delta T \\ \varepsilon_2 - \alpha_2 \Delta T \\ \gamma_{12} \end{Bmatrix}$$

Again there is a range of z for each relation, depending on the layer. The stresses in the principal material coordinate system are

for $-0.000500\text{ m} \le z \le -0.000375\text{ m}$ and $0.000375\text{ m} \le z \le 0.000500\text{ m}$

$$\begin{Bmatrix} \sigma_1(45°) \\ \sigma_2(45°) \\ \tau_{12}(45°) \end{Bmatrix} = \begin{Bmatrix} Q_{11} & Q_{12} & 0 \\ Q_{12} & Q_{22} & 0 \\ 0 & 0 & Q_{66} \end{Bmatrix} \begin{Bmatrix} \varepsilon_1(45°) - \alpha_1 \Delta T \\ \varepsilon_2(45°) - \alpha_2 \Delta T \\ \gamma_{12}(45°) \end{Bmatrix}$$

$$\begin{Bmatrix} \sigma_1(45°) \\ \sigma_2(45°) \\ \tau_{12}(45°) \end{Bmatrix} = \begin{bmatrix} 47.0 & 4.56 & 0 \\ 4.56 & 16.51 & 0 \\ 0 & 0 & 5.62 \end{bmatrix} \times 10^9 \begin{Bmatrix} -1171(10^{-6}) - 0.878z - 6.51(10^{-6})(-100) \\ -1171(10^{-6}) - 1.238z - 24.4(10^{-6})(-100) \\ 6.12z \end{Bmatrix}$$

$$\begin{Bmatrix} \sigma_1(45°) \\ \sigma_2(45°) \\ \tau_{12}(45°) \end{Bmatrix} = \begin{Bmatrix} -18.62 - 46{,}900z \\ 18.62 - 24{,}400z \\ 34{,}400z \end{Bmatrix} \text{MPa}$$

for $-0.000375\text{ m} \le z \le -0.000250\text{ m}$ and $0.000250\text{ m} \le z \le 0.000375\text{ m}$

$$\begin{Bmatrix} \sigma_1(-45°) \\ \sigma_2(-45°) \\ \tau_{12}(-45°) \end{Bmatrix} = \begin{Bmatrix} -18.62 - 62{,}100z \\ 18.62 - 20{,}100z \\ -34{,}400z \end{Bmatrix} \text{MPa}$$

for $-0.000250\text{ m} \le z \le -0.000125\text{ m}$ and $0.000125\text{ m} \le z \le 0.000250\text{ m}$

$$\begin{Bmatrix} \sigma_1(0°) \\ \sigma_2(0°) \\ \tau_{12}(0°) \end{Bmatrix} = \begin{Bmatrix} -18.62 - 184{,}300z \\ 18.62 + 14{,}310z \\ 2{,}020z \end{Bmatrix} \text{MPa}$$

for $-0.000150\ \text{m} \le z \le 0.000150\ \text{m}$

$$\begin{Bmatrix} \sigma_1(90°) \\ \sigma_2(90°) \\ \tau_{12}(90°) \end{Bmatrix} = \begin{Bmatrix} -18.62+75{,}300z \\ 18.62-58{,}900z \\ -2{,}020z \end{Bmatrix} \text{MPa}$$

The results show that the largest tensile and compressive stresses in the principal material coordinate system due to the applied moment occur in layers 3 and 6, respectively, not in the outer layers. This is very much unlike the standard linear stress distribution due to bending of a beam where the largest magnitudes occur at $z = \pm t/2$. This characteristic of layered materials is often overlooked.

20.11 STRENGTH AND FAILURE ANALYSES OF LAMINATE

The failure criteria previously described (Sec. 20.3) deal with the failure of a lamina (a layer). Failure mechanisms in laminates are more complicated than those in a unidirectional composite under in-plane loading. This is due to the fact that multiple modes of failure occur and because the strength of a composite material in tension is considerably different than the strength in compression. A composite can fail due to excess stress in the fiber direction, and the strength in the fiber direction in tension is greater than the strength in compression. Alternatively, a composite material can fail because of excess stress perpendicular to the fibers. The strength perpendicular to the fibers in tension is considerably less than the strength in compression. And finally, a composite can fail due to excessive shear stress. The strength in shear is considerably less than the strength in tension in the fiber direction, although the strength in shear does not depend on the sign of the shear stress. These issues are further complicated by the fact that when the fibers are oriented at an angle relative to the loading direction, there is a stress component in the fiber direction, a component perpendicular to the fiber direction, and a shear stress. There are a number of theories of failure of laminates; however, no one theory, or criterion, works perfectly for all materials in all situations.

Damage accumulation is an important failure mode in laminated composites. New damage mechanisms, such as delamination, and complex interactions between intralaminar and interlaminar damage mechanisms may occur in a laminate. The effects of delamination are usually treated separately from intralaminar damage mechanisms, although recent work has taken into consideration all the damage mechanisms in the failure analysis of a skin-stiffener composite structure (Ref. 23). Experimental evidence (Ref. 24) has shown that the failure in a laminated composite is very often progressive in nature, occurring by a process of damage accumulation. Therefore, the progressive loss of lamina stiffness must be taken into account as a function of the type of damage predicted.

Delamination is one of the predominant forms of failure in laminated composites and is due to the lack of reinforcement in the thickness direction. Delamination as a result of impact or a manufacturing defect can cause significant reductions in the compressive load-carrying capacity and bending stiffness of a structure. The analysis of delamination is commonly divided into the study of the initiation and the propagation of an already initiated area. Delamination initiation analysis is usually based on stresses and use of criteria such as the quadratic interaction of the interlaminar stresses in conjunction with a characteristic distance (Ref. 30). This distance is a function of specimen geometry and material properties, and its determination always requires extensive testing. Delamination propagation, on the other hand, is usually predicted using fracture mechanics.

The fracture mechanics approach have been successfully utilized to predict laminates fracture failure. This type of approach has been successfully used to predict laminate failure in the presence of stress concentrations, and it can accurately simulate hole size effects in laminates (characterized by a strength decrease for larger hole sizes in laminates without finite width effects). Methods based on fracture mechanics require more experimental information than the method previously described. However, since virtually all composite structures contain stress concentrations, e.g., joints, it is considered that methods based on fracture mechanics should also be investigated, for example see Whitney and Nuismer's (Ref. 28) failure criterion for unloaded holes.

A practical failure theory of laminates that is easy to implement is the *maximum stress failure criterion* explained in Sec. 20.5. To review the section, there are five possible modes of failure in the maximum stress failure criteria. Failures are due to: tension in the fiber direction, compression in the fiber direction, tension perpendicular to the fiber direction, compression perpendicular to the fiber direction, and matrix in shear. A known level of stress failure for each mode is required. Failure in tension in the fiber direction is a direct result of fibers fracturing or otherwise breaking. Failure in compression in the fiber direction is due to the kinking of the fibers, generally due to the lack of support of the matrix material surrounding the fibers and the development of shear, or kink, bands within the fiber. Either alone or in combination, failure in tension perpendicular to the fibers is due to failures of the fibers, the matrix material between the fibers, and/or the bond between the matrix material and the fibers. Failure in compression perpendicular to the fibers is generally due to crushing of the matrix material. Finally, shear can cause failures in the fibers, the matrix material between the fibers, and/or the bond between the matrix material and the fibers.

The five sub-criteria and five failure mechanisms of this criterion are

1. $\sigma_1 > \sigma_1^t$ = tensile failure stress in the 1-direction
2. $\sigma_2 > \sigma_2^t$ = tensile failure stress in the 2-direction
3. $\sigma_1 < \sigma_1^c$ = compression failure stress in the 1-direction (σ_1 being a negative number)
4. $\sigma_2 < \sigma_2^c$ = compression failure stress in the 2-direction (σ_2 being a negative number)
5. $|\tau_{12}| > \tau_{12}^f$ = shear failure stress in 1–2 plane (a positive number) (20.11-1)

The maximum stress failure criterion states that failure occurs if any of the five inequalities of (20.11-1) are satisfied.

EXAMPLE 7 Determine the internal pressure p of a thin-walled cylindrical pressure vessel with radius $R = 0.25$ m when failure occur. The cylinder is constructed using eight-layers of glass-fiber-reinforced material and a stacking sequence of $[\pm 60/0/90]_s$. Neglect the end cap effect. The change in temperature due to curing is −100°C. In what layer, or layers, does the first failure occur, and what is the mode of failure?

SOLUTION In Fig. 20.13, the net force in the y direction from p is $2pRDl$, where Dl is the length in the x direction. This is balanced by the internal force $2N_q Dl$ yielding $N_q = pR$. In the x direction, the end cap force ppR^2 is balanced by $2p\,N_x R$ resulting in $N_x = pR/2$.

or

$$\begin{Bmatrix} N_x \\ N_\theta \\ N_{x\theta} \end{Bmatrix} = \begin{Bmatrix} \frac{1}{2} \\ 1 \\ 0 \end{Bmatrix} pR$$

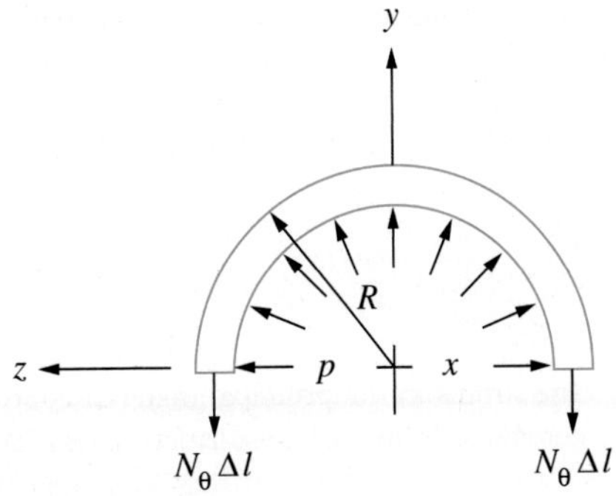

Figure 20.13 Forces acting on section of the cylinder.

Due to the internal pressure, these are the only two non-zero stress resultants acting within the cylinder wall away from the end caps. There are no moment resultants. There are, of course, equivalent thermal force resultants from the temperature change of −100°C.

Following the procedure of Example 4, for composite $[\pm 60/0/90]_s$ we have

$$\begin{bmatrix} A_{11} & A_{12} & A_{16} \\ A_{12} & A_{22} & A_{26} \\ A_{16} & A_{26} & A_{66} \end{bmatrix} = \begin{bmatrix} 24.9 & 7.55 & 0 \\ 7.55 & 32.6 & 0 \\ 0 & 0 & 8.61 \end{bmatrix} \text{MN/m}$$

$$\begin{bmatrix} D_{11} & D_{12} & D_{16} \\ D_{12} & D_{22} & D_{26} \\ D_{16} & D_{26} & D_{66} \end{bmatrix} = \begin{bmatrix} 1.773 & 0.816 & 0.0736 \\ 0.816 & 2.65 & 0.235 \\ 0.0736 & 0.235 & 0.904 \end{bmatrix} \text{N-m}$$

$$\begin{Bmatrix} \hat{N}_x^T \\ \hat{N}_y^T \end{Bmatrix} = \begin{Bmatrix} 427 \\ 423 \end{Bmatrix} \text{N/m °C}$$

Or

$$\begin{Bmatrix} \hat{N}_x^T \\ \hat{N}_\theta^T \\ 0 \end{Bmatrix} = \begin{Bmatrix} 427 \\ 423 \\ 0 \end{Bmatrix} \text{N/m/°C}$$

The inverse matrix A and D are

$$\begin{bmatrix} a_{11} & a_{12} & a_{16} \\ a_{12} & a_{22} & a_{26} \\ a_{16} & a_{26} & a_{66} \end{bmatrix} = \begin{bmatrix} 43.1 & -10.00 & 0 \\ -10.00 & 33.0 & 0 \\ 0 & 0 & 116.2 \end{bmatrix} \text{m(GN)}^{-1}$$

$$\begin{bmatrix} d_{11} & d_{12} & d_{16} \\ d_{12} & d_{22} & d_{26} \\ d_{16} & d_{26} & d_{66} \end{bmatrix} = \begin{bmatrix} 0.657 & -0.203 & -0.000718 \\ -0.203 & 0.450 & -0.1006 \\ -0.000718 & -0.1006 & 1.132 \end{bmatrix} (\text{N-m})^{-1}$$

Therefore the strains are

$$\begin{Bmatrix} \varepsilon_x^o \\ \varepsilon_y^o \\ \gamma_{xy}^o \end{Bmatrix} = [a] \begin{Bmatrix} N_x + \hat{N}_x^T \Delta T \\ N_y + \hat{N}_y^T \Delta T \\ N_{xy} + \hat{N}_{xy}^T \Delta T \end{Bmatrix} = \begin{bmatrix} 43.1 & -10.00 & 0 \\ -10.00 & 33.0 & 0 \\ 0 & 0 & 116.2 \end{bmatrix} \times 10^{-9} \begin{Bmatrix} \frac{1}{2}pR + 427(-100) \\ pR + 423(-100) \\ 0 \end{Bmatrix}$$

For convenience, the pressure p in Pascals is converted to atmospheres using the relation

$$p = 101.4\times10^3 p_a$$

where p_a is the internal pressure in atmospheres. The reference surface strains become

$$\begin{Bmatrix} \varepsilon_x^o \\ \varepsilon_y^o \\ \gamma_{xy}^o \end{Bmatrix} = \begin{Bmatrix} -1418+1172p_aR \\ -970+2840p_aR \\ 0 \end{Bmatrix} \times 10^{-6}$$

The strains in the principal material coordinate system for the layers with 60° fiber orientation are computed from the transformation relation as

$$\begin{Bmatrix} \varepsilon_1(60°) \\ \varepsilon_2(60°) \\ \gamma_{12}(60°) \end{Bmatrix} = [T(60°)] \begin{Bmatrix} \varepsilon_x^o \\ \varepsilon_y^o \\ \gamma_{xy}^0 \end{Bmatrix} = \begin{bmatrix} \frac{1}{4} & \frac{3}{4} & \frac{\sqrt{3}}{2} \\ \frac{3}{4} & \frac{1}{4} & -\frac{\sqrt{3}}{2} \\ -\frac{\sqrt{3}}{4} & \frac{\sqrt{3}}{4} & -\frac{1}{2} \end{bmatrix} \begin{Bmatrix} -1418+1172p_aR \\ -970+2840p_aR \\ 0 \end{Bmatrix} \times 10^{-6}$$

Carrying out the algebra and repeating the calculations for the other layer orientations gives

$$\begin{Bmatrix} \varepsilon_1(60°) \\ \varepsilon_2(60°) \\ \gamma_{12}(60°) \end{Bmatrix} = \begin{Bmatrix} -1082+2420p_aR \\ -1306+1589p_aR \\ 388+1446p_aR \end{Bmatrix} \times 10^{-6} \qquad \begin{Bmatrix} \varepsilon_1(-60°) \\ \varepsilon_2(-60°) \\ \gamma_{12}(-60°) \end{Bmatrix} = \begin{Bmatrix} -1082+2420p_aR \\ -1306+1589p_aR \\ -388-1446p_aR \end{Bmatrix} \times 10^{-6}$$

$$\begin{Bmatrix} \varepsilon_1(0°) \\ \varepsilon_2(0°) \\ \gamma_{12}(0°) \end{Bmatrix} = \begin{Bmatrix} -1418+1172p_aR \\ -970+2840p_aR \\ 0 \end{Bmatrix} \times 10^{-6} \qquad \begin{Bmatrix} \varepsilon_1(90°) \\ \varepsilon_2(90°) \\ \gamma_{12}(90°) \end{Bmatrix} = \begin{Bmatrix} -970+2840p_aR \\ -1418+1172p_aR \\ 0 \end{Bmatrix} \times 10^{-6}$$

The stresses in the various layers can be computed using the above strains as follows:

$$\begin{Bmatrix} \sigma_1(60°) \\ \sigma_2(60°) \\ \tau_{12}(60°) \end{Bmatrix} = [Q] \begin{Bmatrix} \varepsilon_1(60°) - \alpha_1\Delta T \\ \varepsilon_2(60°) - \alpha_2\Delta T \\ \gamma_{12}(60°) \end{Bmatrix}$$

$$= \begin{bmatrix} 47.0 & 4.56 & 0 \\ 4.56 & 16.51 & 0 \\ 0 & 0 & 5.62 \end{bmatrix} \times 10^9 \begin{Bmatrix} -1082+2420p_aR-6.51(-100) \\ -1306+1589p_aR-24.4(-100) \\ 388+1446p_aR \end{Bmatrix} \times 10^{-6}$$

Carrying out the algebra, substituting the radius $R = 0.25$ m, and performing similar calculations for the other three layer angles results in

$$\begin{Bmatrix} \sigma_1(60°) \\ \sigma_2(60°) \\ \tau_{12}(60°) \end{Bmatrix} = \begin{Bmatrix} -15.08+30.3p_a \\ 16.78+9.32p_a \\ 2.18+2.03p_a \end{Bmatrix} \text{ MPa} \qquad \begin{Bmatrix} \sigma_1(-60°) \\ \sigma_2(-60°) \\ \tau_{12}(-60°) \end{Bmatrix} = \begin{Bmatrix} -15.08+30.3p_a \\ 16.78+9.32p_a \\ -2.18-2.03p_a \end{Bmatrix} \text{ MPa}$$

$$\begin{Bmatrix} \sigma_1(0°) \\ \sigma_2(0°) \\ \tau_{12}(0°) \end{Bmatrix} = \begin{Bmatrix} -29.3+17.00p_a \\ 20.8+13.07p_a \\ 0 \end{Bmatrix} \text{ MPa} \qquad \begin{Bmatrix} \sigma_1(90°) \\ \sigma_2(90°) \\ \tau_{12}(90°) \end{Bmatrix} = \begin{Bmatrix} -10.34+34.7p_a \\ 15.45+8.08p_a \\ 0 \end{Bmatrix} \text{ MPa}$$

We now determine p_a using Eq. (20.9-7). That is, the stresses in each layer are equated to the failure levels for each failure mode. As the pressure increases from zero, we select the smallest positive value of the internal pressure that first causes failure. The failure conditions are

$$\begin{Bmatrix} \sigma_1 \\ \sigma_2 \\ \tau_{12} \end{Bmatrix} = \begin{Bmatrix} \sigma_1^T \\ \sigma_2^T \\ \tau_{12}^F \end{Bmatrix} = \begin{Bmatrix} 1000 \\ 30 \\ 70 \end{Bmatrix} \text{ MPa} \quad \text{and} \quad \begin{Bmatrix} \sigma_1 \\ \sigma_2 \\ \tau_{12} \end{Bmatrix} = \begin{Bmatrix} \sigma_1^C \\ \sigma_2^C \\ -\tau_{12}^F \end{Bmatrix} = \begin{Bmatrix} -600 \\ -120 \\ -70 \end{Bmatrix} \text{ MPa}$$

For the 60° layers, using MPa, the above six failure equations are

$$\begin{Bmatrix} -15.08+30.3p_a \\ 16.78+9.32p_a \\ 2.18+2.03p_a \end{Bmatrix} = \begin{Bmatrix} 1000 \\ 30 \\ 70 \end{Bmatrix} \quad \text{and} \quad \begin{Bmatrix} -15.08+30.3p_a \\ 16.78+9.32p_a \\ 2.18+2.03p_a \end{Bmatrix} = \begin{Bmatrix} -600 \\ -120 \\ -70 \end{Bmatrix}$$

For the –60° layers, the six failure equations are

$$\begin{Bmatrix} -15.08+30.3p_a \\ 16.78+9.32p_a \\ -2.18-2.03p_a \end{Bmatrix} = \begin{Bmatrix} 1000 \\ 30 \\ 70 \end{Bmatrix} \quad \text{and} \quad \begin{Bmatrix} -15.08+30.3p_a \\ 16.78+9.32p_a \\ -2.18-2.03p_a \end{Bmatrix} = \begin{Bmatrix} -600 \\ -120 \\ -70 \end{Bmatrix}$$

For the 0° layers, the six failure equations are

$$\begin{Bmatrix} -29.3+17.00p_a \\ 20.8+13.07p_a \\ 0 \end{Bmatrix} = \begin{Bmatrix} 1000 \\ 30 \\ 70 \end{Bmatrix} \quad \text{and} \quad \begin{Bmatrix} -29.3+17.00p_a \\ 20.8+13.07p_a \\ 0 \end{Bmatrix} = \begin{Bmatrix} -600 \\ -120 \\ -70 \end{Bmatrix}$$

For the 90° layers, the six failure equations are

$$\begin{Bmatrix} -10.34+34.7p_a \\ 15.45+8.08p_a \\ 0 \end{Bmatrix} = \begin{Bmatrix} 1000 \\ 30 \\ 70 \end{Bmatrix} \quad \text{and} \quad \begin{Bmatrix} -10.34+34.7p_a \\ 15.45+8.08p_a \\ 0 \end{Bmatrix} = \begin{Bmatrix} -600 \\ -120 \\ -70 \end{Bmatrix}$$

The results indicate that there are 10 positive values of p_a, 10 negative values of p_a, and 4 values of infinity. The infinite values come from the two shear equations for the 0° layers and the two shear equations for the 90° layers. From these multiple values of p_a, the lowest level of internal pressure is $p_a = 0.704$ atmospheres. This value is obtained from the solution of the second equation for the positive failure stress levels for the 0° layer, i.e., $20.8 + 13.07p_a = 30$. At $p_a = 0.704$ atmospheres, cracks parallel to the fibers occur in the 0° layers. If matrix cracking is to be avoided, only pressures below this level are allowed. With this pressure the failure mode is tensile. However, a negative pressure of magnitude 10.78 atmospheres is obtained from the solution of the second equation for negative failure stress levels for the 0° layers, i.e., $20.8 + 13.07p_a = -120$. This value is the negative value of p_a with the lowest magnitude. That is, an *external* pressure of 10.78 atmospheres will cause failure of the 0° layers due to excessive compressive stress perpendicular to the fibers, assuming that the cylinder has not buckled due to the external pressure.

20.12 COMPOSITE SANDWICH STRUCTURES

A sandwich structure composed of two thin, stiff and strong sheets, as outer faces (sheets), or skins and a weaker core that could be a web-like structure sandwich between the faces. The sheets must be adhesively bonded to the core in order to transfer the applied loads between the constituents. Once the face sheets and core are combined into a single structure or panel, they create a structure with high stiffness, strength, and low in weight. The main components in a sandwich structure were shown in Fig. 20.14. The use of composite materials and composite sandwich structures are greatly implemented in the civil and aerospace design of structural components. They are many times the preferred structures due to their low weight and high weight-to-stiffness ratio. Sandwich structures are also capable

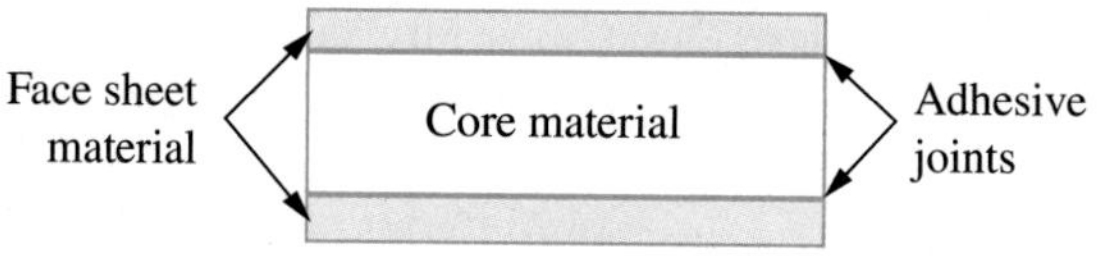

Figure 20.14 Sandwich structure.

of sustaining higher bending and shear loads than regular laminated composites. Impact resistance and energy absorption of sandwich structures make them suitable for applications in which crashing may occur.

Face sheets are generally strong and thick enough capable of sustaining bending loads while the core material is generally subjected to shear loads (Ref. 50). Note the adhesive must be able to carry shear stress from the face sheets to the core. Sandwich plates can be treated as a laminated composite, where the core is considered as another ply. Therefore, classical laminated plates theory can be used to determine the corresponding matrices A_{ij}, B_{ij}, and D_{ij}. In sandwich plates A_{ij} matrix does not change and B_{ij} matrix exists if the layers are not symmetric; however, D_{ij} matrix changes significantly as

$$D_{ij} = 2(D_{ij})_f + 2(A_{ij})_f \left[\frac{t_c + t_f}{2}\right]^2 \tag{20.12-1}$$

When the sandwich structure is subjected to bending, the one side of the face sheets is under tension while the other is under compression. Consider a sandwich plate is subjected to combination of axial load N and shear load V and a bending moment M, having a width of b, as shown in Fig. 20.15.

The axial and shear loads on the sandwich structure are equally divided between the two face sheets. However, the moment is divided by the width, $t_c + t_f$. Thus the average axial force due to the moment is

$$N_m = \frac{M}{t_c + t_f} \tag{20.12-2}$$

where t_c is the thickness of the core and t_f is the thickness of the face sheet.

The bending stiffness EI of the sandwich plate is

$$EI = E_f\left[\frac{bt_f^3}{12}\right] + A_f d_f^2 \tag{20.12-3}$$

where E_f is the modulus of face sheet, b and t_f are the width and thickness of the face sheet, respectively, and d_f is the distance of the face sheet from the center of the sandwich.

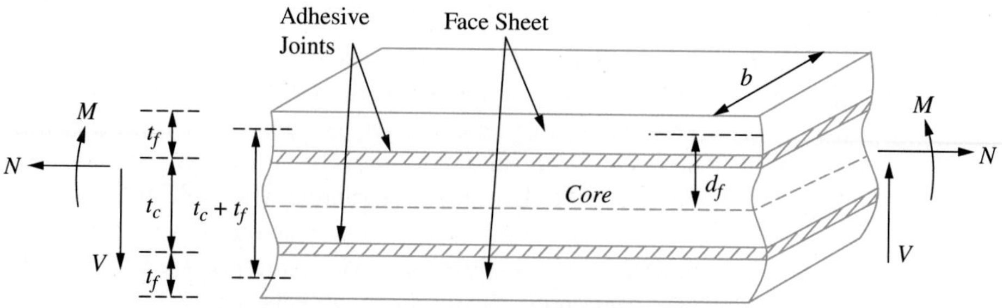

Figure 20.15 Sandwich structure subjected to axial, shear, and bending loads.

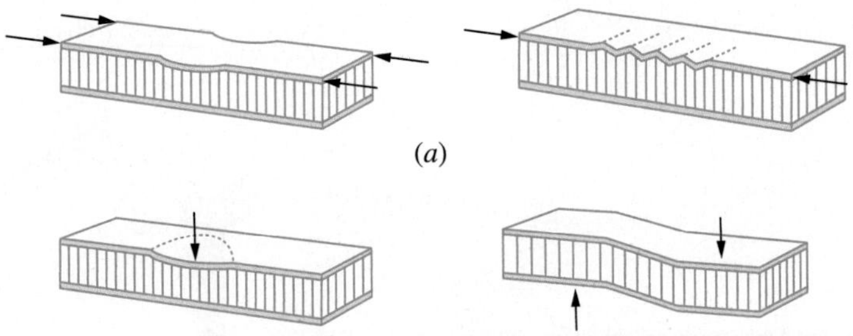

Figure 20.16 (*a*) Buckling loads—face sheet buckling failure mode due to the axial loads. (*b*) Shear loads—core buckling failure due to the shear and concentrated transverse loads.

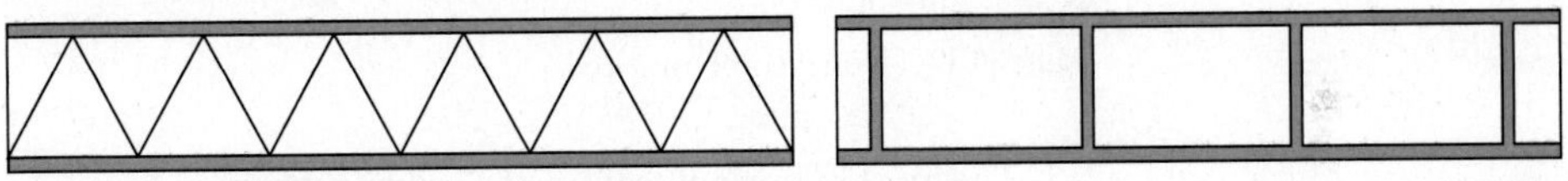

Figure 20.17 Corrugated and web cores.

It is important to note that sandwich plates have a tendency of buckling under excessive axial load, similar to the buckling of laminate composites, except the transvers shear make a difference. That is Kirchhoff hypothesis, which is plane remain perpendicular to the mid plane is no longer valid. Some of the buckling modes, namely, face sheet failure buckling and core buckling failure of the sandwich composites are shown in Fig. 20.16.

The core of the sandwich composite could be comprised of four classes of materials: (1) solid core or foam, (2) corrugated or truss core, (3) web core, and (4) honeycomb core. Figure 20.14 depicted the solid or foam type core and corrugated or truss cores and web core are shown in Figs. 20.16 and 20.17. Honeycomb cores are discussed in the next section.

20.13 COMPOSITE CELLULAR STRUCTURES

A cellular solid is made up of an interconnected network of solid struts or plates, which form the edges and faces of cells. The simplest form of cellular structure is a two-dimensional array of polygons, which pack to fill a plane area like the hexagonal cells known as honeycomb structures, Fig. 20.18.

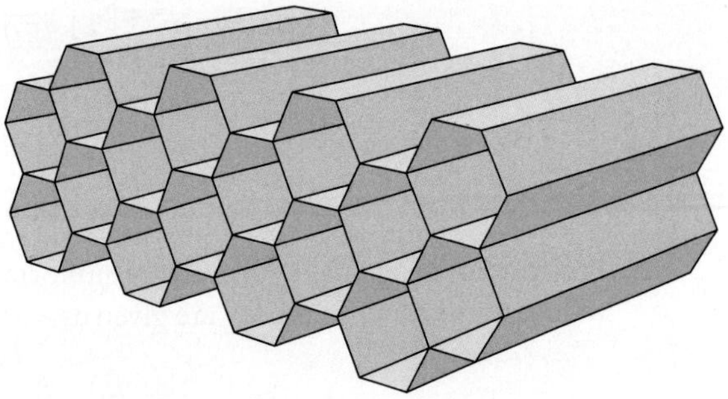

Figure 20.18 Honeycomb structures.

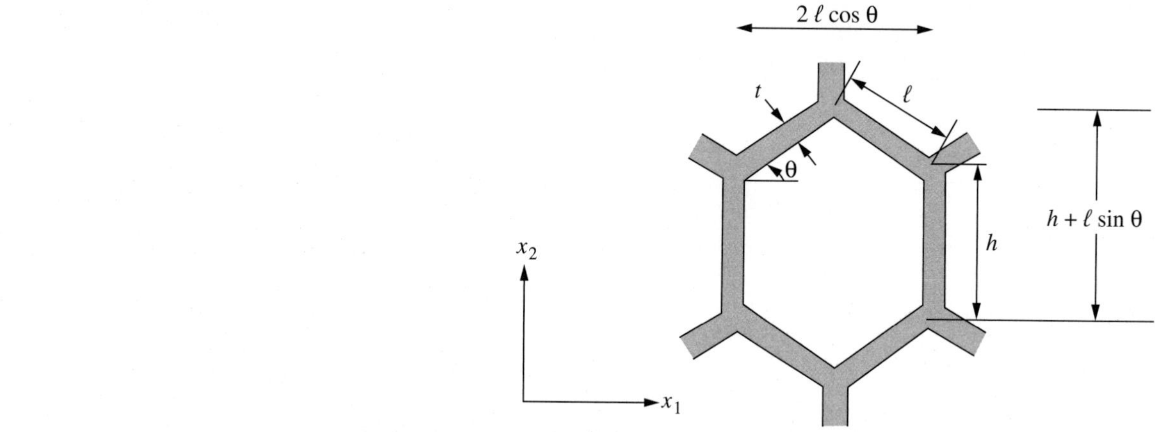

Figure 20.19 Honeycomb unit cell.

The most important feature of a cellular structure is its relative density $\rho*/\rho_s$, where $\rho*$ is the density of the cellular material, and ρ_s represents the density of the solid from which the cells are made. The thickness of the cell walls is determined by the value of the relative density. The higher the relative density, the greater the thickness of the cell walls.

In many honeycomb structures the length of cell walls are not equal and thus the angle between the walls are not 120°. In general, honeycombs do not have equal length cell walls. The in-plane properties of these honeycombs are anisotropic and the structure has four elastic constants (E_1, E_2, G_{12}, and ν_{12}) and two values for the yield stresses in two directions (σ_{y1} and σ_{y2}). A general honeycomb cell is shown in Fig. 20.19.

Through simple geometry the relative density is given:

$$\frac{\rho*}{\rho_s} = \frac{\frac{t}{l}\left(\frac{h}{l}+2\right)}{2\cos\theta\left(\frac{h}{l}+\sin\theta\right)} \tag{20.13-1}$$

For a special case when $\theta = 30°$ and $h = l$, the relative density reduces to

$$\frac{\rho*}{\rho_s} = \frac{2}{\sqrt{3}}\frac{t}{l} \tag{20.13-2}$$

When the cell walls are treated as beams of length l, thickness t, depth b, and Young's modulus E_s, then the elastic moduli in the direction X_1 and X_2 are given as

$$\frac{E_1^*}{E_s} = \left(\frac{t}{l}\right)^3 \frac{\cos\theta}{(h/l+\sin\theta)\sin^2\theta} \tag{20.13-3}$$

and,

$$\frac{E_2^*}{E_s} = \left(\frac{t}{l}\right)^3 \frac{(h/l+\sin\theta)}{\cos^3\theta} \tag{20.13-4}$$

For regular hexagonal with walls of uniform thickness, both Young's moduli, E_1^* and E_2^*, reduce to the same value and are given as

$$\frac{E_1^*}{E_s} = \frac{E_2^*}{E_s} = 2.3\left(\frac{t}{l}\right)^3 \tag{20.13-5}$$

20.14 TABLES

TABLE 20.4 Fibers and Matrix Materials and Their Applications

Fiber	Matrix	Applications
Graphite	Aluminum	Satellite, missile, and helicopter structures
	Magnesium	Space and satellite structures
	Lead	Storage-battery plates
	Copper	Electrical contacts and bearings
Boron	Aluminum	Compressor blades and structural supports
	Magnesium	Antenna structures
	Titanium	Jet-engine fan blades
Alumina	Aluminum	Superconductor restraints in fission power reactors
	Lead	Storage-battery plates
	Magnesium	Helicopter transmission structures
Silicon carbide	Aluminum, titanium	High-temperature structures
	Superalloy (cobalt-base)	High-temperature engine components
Molybdenum, tungsten	Superalloy	High-temperature engine components

TABLE 20.5 Properties of Key Reinforcing Fibers

Fiber	Density g/cm^3 (lb/in^3)	Axial Modulus GPa (Msi)	Tensile Strength MPa (Ksi)	Axial Coefficient of Thermal Expansion ppm/K (ppm/F)	Axial Thermal Conductivity W/mK
E-glass	2.6 (0.094)	70 (10)	2000 (300)	5 (2.8)	0.9
HS glass	2.5 (0.090)	83 (12)	4200 (650)	4.1 (2.3)	0.9
Aramid	1.4 (0.052)	124 (18)	3200 (500)	−5.2 (−2.9)	0.04
Boron	2.6 (0.094)	400 (58)	3600 (520)	4.5 (2.5)	—
SM carbon (PAN)	1.7 (0.061)	235 (34)	3200 (500)	−0.5 (−0.3)	9
UHM carbon (PAN)	1.9 (0.069)	590 (86)	3800 (550)	−1 (−0.6)	18
UHS carbon (PAN)	1.8 (0.065)	290 (42)	7000 (1000)	−1.5 (−0.8)	160
UHM carbon (pitch)	2.2 (0.079)	895 (130)	2200 (320)	−1.6 (−0.9)	640
UHK carbon (pitch)	2.2 (0.079)	830 (120)	2200 (320)	−1.6 (−0.9)	1100
SiC monofilament	3.0 (0.11)	400 (58)	3600 (520)	4.9 (2.7)	—
SiC multifilament	3.0 (0.11)	400 (58)	3100 (450)	—	—
Si-C-O	2.6 (0.094)	190 (28)	2900 (430)	3.9 (2.2)	1.4
Si-Ti-C-O	2.4 (0.087)	190 (27)	3300 (470)	3.1 (1.7)	—
Aluminum oxide	3.9 (0.14)	370 (54)	1900 (280)	7.9 (4.4)	—
High-density polyethylene	0.97 (0.035)	172 (25)	3000 (440)	—	—

TABLE 20.6 Properties of Selected Thermosetting and Thermoplastic Matrices

	Density g/cm^3 (lb/in^3)	Modulus GPa (Msi)	Tensile Strength MPa (Ksi)	Elongation to Break (%)	Thermal Conductivity W/mK	Coefficient of Thermal Expansion ppm/K (ppm/F)
Epoxy (1)	1.1–1.4 (0.040–0.050)	3–6 (0.43–0.88)	35–100 (5–15)	1–6	0.1	60 (33)
Thermosetting polyester (1)	1.2–1.5 (0.043–0.054)	2–4.5 (0.29–0.65)	40–90 (6–13)	2	0.2	100–200 (56–110)
Polypropylene (2)	0.90 (0.032)	1–4 (0.15–0.58)	25–38 (4–6)	>300	0.2	110 (61)
Nylon 6-6 (2)	1.14 (0.041)	1.4–2.8 (0.20–0.41)	60–75 (9–11)	40–80	0.2	90 (50)
Polycarbonate (2)	1.06–1.20 (0.038–0.043)	2.2–2.4 (0.32–0.35)	45–70 (7–10)	50–100	0.2	70 (39)
Polysulfone (2)	1.25 (0.045)	2.2 (0.32)	76 (11)	50–100	—	56 (31)
Polyetherimide (2)	1.27 (0.046)	3.3 (0.48)	110 (16)	60	—	62 (34)
Polyamideimide (2)	1.4 (0.050)	4.8 (0.7)	190 (28)	17	—	63 (35)
Polyphenylene sulfide (2)	1.36 (0.049)	3.8 (0.55)	65 (10)	4	—	54 (30)
Polyether etherketone (2)	1.26–1.32 (0.046–0.048)	3.6 (0.52)	93 (13)	50	—	47 (26)

(1) Thermoset, (2) Thermoplastic.

TABLE 20.7 Effect of Fiber Form and Volume Fraction on Mechanical Properties of E-Glass-Reinforced Polyester

	Bulk Molding Compound	Sheet Molding Compound	Chopped Strand Mat	Woven Roving	Unidirectional Axial	Unidirectional Transverse
Glass content (wt %)	20	30	30	50	70	70
Tensile modulus GPa (Msi)	9 (1.3)	13 (1.9)	7.7 (1.1)	16 (2.3)	42 (6.1)	12 (1.7)
Tensile strength MPa (Ksi)	45 (6.5)	85 (12)	95 (14)	250 (36)	750 (110)	50 (7)

TABLE 20.8 Mechanical Properties of Selected Unidirectional Polymer Matrix Composites

Fiber	Axial Modulus GPa (Msi)	Transverse Modulus GPa (Msi)	In-plane Shear Modulus GPa (Msi)	Poisson's Ratio	Axial Tensile Strength MPa (Ksi)	Transverse Tensile Strength MPa (Ksi)	Axial Compressive Strength MPa (Ksi)	Transverse Compressive Strength MPa (Ksi)	In-plane Shear Strength MPa (Ksi)
E-glass	45 (6.5)	12 (1.8)	5.5 (0.8)	0.28	1020 (150)	40 (7)	620 (90)	140 (20)	70 (10)
Aramid	76 (11)	5.5 (0.8)	2.1 (0.3)	0.34	1240 (180)	30 (4.3)	280 (40)	140 (20)	60 (9)
Boron	210 (30)	19 (2.7)	4.8 (0.7)	0.25	1240 (180)	70 (10)	3310 (480)	280 (40)	90 (13)
SM carbon (PAN)	145 (21)	10 (1.5)	4.1 (0.6)	0.25	1520 (220)	41 (6)	1380 (200)	170 (25)	80 (12)
UHS carbon (PAN)	170 (25)	10 (1.5)	4.1 (0.6)	0.25	3530 (510)	41 (6)	1380 (200)	170 (25)	80 (12)
UHM carbon (PAN)	310 (45)	9 (1.3)	4.1 (0.6)	0.20	1380 (200)	41 (6)	760 (110)	170 (25)	80 (12)
UHM carbon (pitch)	480 (70)	9 (1.3)	4.1 (0.6)	0.25	900 (130)	20 (3)	280 (40)	100 (15)	41 (6)
UHK carbon (pitch)	480 (70)	9 (1.3)	4.1 (0.6)	0.25	900 (130)	20 (3)	280 (40)	100 (15)	41 (6)

TABLE 20.9 Mechanical Properties of Selected Quasi-Isotropic Polymer Matrix Composites

Fiber	Axial Modulus GPa (Msi)	Transverse Modulus GPa (Msi)	In-plane Shear Modulus GPa (Msi)	Poisson's Ratio	Axial Tensile Strength MPa (Ksi)	Transverse Tensile Strength MPa (Ksi)	Axial Compressive Strength MPa (Ksi)	Transverse Compressive Strength MPa (Ksi)	In-plane Shear Strength MPa (Ksi)
E-glass	23 (3.4)	23 (3.4)	9.0 (1.3)	0.28	550 (80)	550 (80)	330 (48)	330 (48)	250 (37)
Aramid	29 (4.2)	29 (4.2)	11 (1.6)	0.32	460 (67)	460 (67)	190 (28)	190 (28)	65 (9.4)
Boron	80 (11.6)	80 (11.6)	30 (4.3)	0.33	480 (69)	480 (69)	1100 (160)	1100 (160)	360 (52)
SM carbon (PAN)	54 (7.8)	54 (7.8)	21 (3.0)	0.31	580 (84)	580 (84)	580 (84)	580 (84)	410 (59)
UHS carbon (PAN)	63 (9.1)	63 (9.1)	21 (3.0)	0.31	1350 (200)	1350 (200)	580 (84)	580 (84)	410 (59)
UHM carbon (PAN)	110 (16)	110 (16)	41 (6.0)	0.32	490 (71)	490 (71)	270 (39)	70 (39)	205 (30)
UHM carbon (pitch)	165 (24)	165 (24)	63 (9.2)	0.32	310 (45)	310 (45)	96 (14)	96 (14)	73 (11)
UHK carbon (pitch)	165 (24)	165 (24)	63 (9.2)	0.32	310 (45)	310 (45)	96 (14)	96 (14)	73 (11)

TABLE 20.10 Fracture Toughness of Structural Alloys, Monolithic Ceramics, and Ceramic Matrix Composites

Matrix	Reinforcement	Fracture Toughness MPa $m^{1/2}$
Aluminum	none	30–45
Steel	none	40–65[a]
Alumina	none	3–5
Silicon carbide	none	3–4
Alumina	Zirconia particles[b]	6–15
Alumina	Silicon carbide whiskers	5–10
Silicon carbide	Continuous silicon carbide fibers	25–30

[a]The toughness of some alloys can be much higher.
[b]Transformation-toughened.

TABLE 20.11 Mechanical Properties of Selected Unidirectional Continuous Fiber-Reinforced Metal Matrix Composites

Fiber	Matrix	Density g/cm³ (lb/in³)	Axial Modulus GPa (Msi)	Transverse Modulus GPa (Msi)	Axial Tensile Strength MPa (Ksi)	Transverse Tensile Strength MPa (Ksi)	Axial Compressive Strength MPa (Ksi)
UHM carbon (pitch)	Aluminum	2.4 (0.090)	450 (65)	15 (5)	690 (100)	15 (5)	340 (50)
Boron	Aluminum	2.6 (0.095)	210 (30)	140 (20)	1240 (180)	140 (20)	1720 (250)
Alumina	Aluminum	3.2 (0.12)	240 (35)	130 (19)	1700 (250)	120 (17)	1800 (260)
Silicon carbide	Titanium	3.6 (0.13)	260 (38)	170 (25)	1700 (250)	340 (50)	2760 (400)

TABLE 20.12 Mechanical Properties of Silicon Carbide Particle-Reinforced Aluminum

Property	Aluminum (6061-T6)	Titanium (6AI-4V)	Steel (4340)	Composite Particle Volume Fraction		
				25	55	70
Modulus, GPa (Msi)	69 (10)	113 (16.5)	200 (29)	114 (17)	186 (27)	265 (38)
Tensile yield strength, MPa (Ksi)	275 (40)	1000 (145)	1480 (215)	400 (58)	495 (72)	225 (33)
Tensile ultimate strength, MPa (Ksi)	310 (45)	1100 (160)	1790 (260)	485 (70)	530 (77)	225 (33)
Elongation, (%)	15	5	10	3.8	0.6	0.1
Density, g/cm³ (lb/in³)	2.77 (0.10)	4.43 (0.16)	7.76 (0.28)	2.88 (0.104)	2.96 (0.107)	3.00 (0.108)
Specific modulus, GPa	5	26	26	40	63	88

TABLE 20.13 Physical Properties of Selected Unidirectional Composites and Monolithic Metals

Matrix	Reinforcement	V/O %	Density g/cm^3 (lb/in^3)	Axial Coefficient of Thermal Expansion ppm*/K (ppm*/F)	Axial Thermal Conductivity W/mK (BTU/h·ft·F)	Transverse Thermal Conductivity W/mK (BTU/h·ft·F)	Specific Axial Thermal Conductivity W/mK (BTU/h·ft·F)
Aluminum (6063)	—	—	2.7 (0.098)	23 (13)	218 (126)	218 (126)	81
Copper	—	—	8.9 (0.32)	17 (9.8)	400 (230)	400 (230)	45
Epoxy	UHK carbon fibers	60	1.8 (0.065)	−1.2 (−0.7)	660 (380)	2 (1.1)	370
Aluminum	UHK carbon fibers	50	2.45 (0.088)	−0.5 (−0.3)	660 (380)	50 (29)	110
Copper	UHK carbon fibers	50	5.55 (0.20)	−0.5 (−0.3)	745 (430)	140 (81)	130
Carbon	UHK carbon fibers	40	1.85 (0.067)	−1.5 (−0.8)	740 (430)	45 (26)	400

*ppm is part per million.

TABLE 20.14 Physical Properties of Isotropic and Quasi-Isotropic Composites and Monolithic Materials Used in Electronic Packaging

Matrix	Reinforcement	V/O %	Density g/cm^3 (lb/in^3)	Coefficient of Thermal Expansion ppm*/K (ppm*/F)	Thermal Conductivity W/mK (BTU/h·ft·F)	Specific Thermal Conductivity W/mK
Aluminum (6063)	—	—	2.7 (0.098)	23 (13)	218 (126)	81
Copper	—	—	8.9 (0.32)	17 (9.8)	400 (230)	45
Beryllium	—	—	1.86 (0.067)	13 (7.2)	150 (87)	81
Magnesium	—	—	1.80 (0.065)	25 (14)	54 (31)	12
Titanium	—	—	4.4 (0.16)	9.5 (5.3)	16 (9.5)	4
Stainless steel (304)	—	—	8.0 (0.29)	17 (9.6)	16 (9.4)	2
Molybdenum	—	—	10.2 (0.37)	5.0 (2.8)	140 (80)	14
Tungsten	—	—	19.3 (0.695)	4.5 (2.5)	180 (104)	9
Invar	—	—	8.0 (0.29)	1.6 (0.9)	10 (6)	1
Kovar	—	—	8.3 (0.30)	5.9 (3.2)	17 (10)	2
Alumina (99% pure)	—	—	3.9 (0.141)	6.7 (3.7)	20 (12)	5
Beryllia	—	—	2.9 (0.105)	6.7 (3.7)	250 (145)	86
Aluminum nitride	—	—	3.2 (0.116)	4.5 (2.5)	250 (145)	78
Silicon	—	—	2.3 (0.084)	4.1 (2.3)	150 (87)	65
Gallium arsenide	—	—	5.3 (0.19)	5.8 (3.2)	44 (25)	8
Diamond	—	—	3.5 (0.13)	1.0 (0.6)	2000 (1160)	570
Pyrolitic graphite	—	—	2.3 (0.083)	−1 (−0.6)	1700 (980)	750
Aluminum-silicon	—	—	2.5 (0.091)	13.5 (7.5)	126 (73)	50
Beryllium-aluminum	—	—	2.1 (0.076)	13.9 (7.7)	210 (121)	100
Copper-tungsten (10/90)	—	—	17 (0.61)	6.5 (3.6)	209 (121)	12
Copper-molybdenum (15/85)	—	—	10 (0.36)	6.6 (3.7)	184 (106)	18
Aluminum	SiC particles	70	3.0 (0.108)	6.5 (3.6)	190 (110)	63
Beryllium	BeO particles	60	2.6 (0.094)	6.1 (3.4)	240 (139)	92
Copper	Diamond particles	55	5.9 (0.21)	5.8 (3.2)	420 (243)	71
Epoxy	UHK carbon fibers	60	1.8 (0.065)	−0.7 (−0.4)	330 (191)	183
Aluminum	UHK carbon fibers	26	2.6 (0.094)	6.5 (3.6)	290 (168)	112
Copper	UHK carbon fibers	26	7.2 (0.26)	6.5 (3.6)	400 (230)	56
Carbon	UHK carbon fibers	40	1.8 (0.065)	−1 (−0.6)	360 (208)	195

*ppm is part per million.

TABLE 20.15 Physical Properties of Selected Unidirectional Polymer Matrix Composites

Fiber	Density g/cm³ (lb/in³)	Axial CTE 10^{-6}/K (10^{-6}/F)	Transverse CTE* 10^{-6}/K (10^{-6}/F)	Axial Thermal Conductivity W/mK (BTU/h·ft·F)	Transverse Thermal Conductivity W/mK (BTU/h·ft·F)
E-glass	2.1 (0.075)	6.3 (3.5)	22 (12)	1.2 (0.7)	0.6 (0.3)
Aramid	1.38 (0.050)	−4.0 (−2.2)	58 (32)	1.7 (1.0)	0.1 (0.08)
Boron	2.0 (0.073)	4.5 (2.5)	23 (13)	2.2 (1.3)	0.7 (0.4)
SM carbon (PAN)	1.58 (0.057)	0.9 (0.5)	27 (15)	5 (3)	0.5 (0.3)
UHS carbon (PAN)	1.61 (0.058)	0.5 (0.3)	27 (15)	10 (6)	0.5 (0.3)
UHM carbon (PAN)	1.66 (0.060)	−0.9 (−0.5)	40 (22)	45 (26)	0.5 (0.3)
UHM carbon (pitch)	1.80 (0.065)	−1.1 (−0.6)	27 (15)	380 (220)	10 (6)
UHK carbon (pitch)	1.80 (0.065)	−1.1 (−0.6)	27 (15)	660 (380)	10 (6)

*CTE is Coefficient of Thermal Expansion.

TABLE 20.16 Physical Properties of Selected Quasi-Isotropic Polymer Matrix Composites

Fiber	Density g/cm³ (lb/in³)	Axial CTE 10^{-6}/K (10^{-6}/F)	Transverse CTE* 10^{-6}/K (10^{-6}/F)	Axial Thermal Conductivity W/mK (BTU/h·ft·F)	Transverse Thermal Conductivity W/mK (BTU/h·ft·F)
E-glass	2.1 (0.075)	10 (5.6)	10 (5.6)	0.9 (0.5)	0.9 (0.5)
Aramid	1.38 (0.050)	1.4 (0.8)	1.4 (0.8)	0.9 (0.5)	0.9 (0.5)
Boron	2.0 (0.073)	6.5 (3.6)	6.5 (3.6)	1.4 (0.8)	1.4 (0.8)
SM carbon (PAN)	1.58 (0.057)	3.1 (1.7)	3.1 (1.7)	2.8 (1.6)	2.8 (1.6)
UHS carbon (PAN)	1.61 (0.058)	2.3 (1.3)	2.3 (1.3)	6 (3)	6 (3)
UHM carbon (PAN)	1.66 (0.060)	0.4 (0.2)	0.4 (0.2)	23 (13)	23 (13)
UHM carbon (pitch)	1.80 (0.065)	−0.4 (−0.2)	−0.4 (−0.2)	195 (113)	195 (113)
UHK carbon (pitch)	1.80 (0.065)	−0.4 (−0.2)	−0.4 (−0.2)	335 (195)	335 (195)

*CTE is Coefficient of Thermal Expansion.

TABLE 20.17 Physical Properties of Silicon Carbide Particle-Reinforced Aluminum

Property	Aluminum (6061-T6)	Titanium (6Al-4V)	Steel (4340)	Composite Particle Volume Fraction		
				25	55	70
CTE, 10^{-6}/K (10^{-6}/F)	23 (13)	9.5 (5.3)	12 (6.6)	16.4 (9.1)	10.4 (5.8)	6.2 (3.4)
Thermal Conductivity W/m·K (BTU/h·ft·F)	218 (126)	16 (9.5)	17 (9.4)	160–220 (92–126)	160–220 (92–126)	160–220 (92–126)
Density, g/cm³ (lb/in³)	2.77 (0.10)	4.43 (0.16)	7.76 (0.28)	2.88 (0.104)	2.96 (0.107)	3.00 (0.108)

TABLE 20.18 Mechanical Properties of Graphite-Polymer Composite Material with Different Volume Fraction

	Fiber Volume Fraction							
	0.0	0.1	0.2	0.3	0.4	0.5	0.6	0.7
E_1 (GPa)	4.62	27.5	50.3	73.1	96.0	118.8	141.6	164.5
ν_{12}	0.360	0.344	0.328	0.312	0.296	0.280	0.264	0.248
E_2 (GPa)	4.62	5.36	6.13	6.99	7.99	9.18	10.63	12.45
G_{12} (GPa)	1.70	1.91	2.16	2.46	2.82	3.26	3.80	4.52
α_1(/°C) $\times 10^6$	41.4	5.89	2.61	1.38	0.73	0.32	7.47×10^{-2}	−0.16
α_2(/°C) $\times 10^6$	41.4	49.3	45.55	41.15	36.65	32.15	27.6	23.15

TABLE 20.19 Mechanical Properties of Glass-Polymer Composite Material with Different Volume Fraction

	Fiber Volume Fraction							
	0.0	0.1	0.2	0.3	0.4	0.5	0.6	0.7
E_1 (GPa)	4.62	11.47	18.32	25.2	32.0	38.9	45.7	52.6
ν_{12}	0.360	0.346	0.332	0.318	0.304	0.290	0.276	0.262
E_2 (GPa)	4.62	5.74	6.88	8.23	9.90	12.07	15.01	19.27
G_{12} (GPa)	1.70	1.98	2.32	2.75	3.30	4.03	5.06	6.61
α_1(/°C) $\times 10^6$	41.4	18.39	12.565	9.885	8.345	7.33	6.615	6.08
α_2(/°C) $\times 10^6$	41.4	44.45	41.25	37	32.45	27.75	23.1	18.5

20.15 REFERENCES

1. Tsai, S. W., and E. M. Wu, "A General Theory of Strength for Anisotropic Materials," *J. Compos. Mater.*, 5: 58–80, 1971.
2. Tsai, S. W., *Strength Characteristics of Composite Materials*, NASA CR-224, 1965.
3. Hashin, Z., "Failure Criteria for Unidirectional Fiber Composites," *J. App. Mech.*, 47: 329–334, 1980.
4. Rowlands, R. E., "Strength (Failure) Theories and Their Experimental Correlation," *Handbook of Composites, 3-Failure Mechanics of Composites*, Sih, G. C., and Skuda, A. M. (eds.), Elsevier, Amsterdam, 1985.
5. Hart-Smith, L. J., "Predictions of the Original and Truncated Maximum-Strain Failure Models for Certain Fibrous Composite Laminates," *Compos. Sci. Tech.*, 58: 1151–1179, 1998.
6. Rotem, A., "Prediction of Laminate Failure with the Rotem Failure Criterion," *Compos. Sci. Tech.*, 58: 1083–1094, 1998.
7. Cristensen, R. M., "Stress Based Yield/Failure Criteria for Fiber Composites." 12th International Conference on Composite Materials, Paris: 1999.
8. Puck, A., and H. Schürmann, "Failure Analysis of FRP Laminates by Means of Physically Based Phenomenological Models," *Compos. Sci. Tech.*, 58: 104510–67, 1998.
9. Hinton, M. J. and P. D. Soden, "Predicting Failure in Composite Laminates: The Background to the Exercise." *Compos. Sci. Tech.*, 58: 1001–1010, 1998.
10. Hinton, M. J., A. S. Kaddour, and P. D. Soden, "Predicting Failure in Fibre Composites: Lessons Learned from the Word-Wide Failure Exercise." 13th International Conference on Composite Materials. 2001; Beijing, China. China: 2001: Paper 1198.
11. Sun, C. T., B. J. Quinn, J. Tao, and D. W. Oplinger, "Comparative Evaluation of Failure Analysis Methods for Composite Laminates," NASA, DOT/FAA/AR-95/109, 1996.
12. Azzi, V. D., and S. W. Tsai: "Anisotropic Strength of Composites", Experimental Mechanics, September 1965, pp. 283–288.
13. Hoffman, O., "The Brittle Strength of Orthotropic Materials," *J. Compos. Mater.*, 1: 200–206, 1967.
14. Chamis, C. C., "Failure Criteria for Filamentary Composites," Composite Materials: Testing and Design, STP 460, ASTM, Philadelphia, pp. 336–351, 1969.

15. Hashin, Z., and A. Rotem, "A Fatigue Failure Criterion for Fibre Reinforced Materials," *J. Compos. Mater.*, 7: 448–464, 1973.
16. Puck, A., *Festigkeitsanalyse von Faser-Matrix-Laminaten, Modelle für die Praxis*, Hanser, 1995.
17. Cuntze, R. G., "Progressive Failure of 3-D-Stresses Laminates: Multiple Nonlinearity Treated by the Failure Mode Concept." Cardon FRV. Recent Developments in Durability Analysis of Composite Systems; Brussels, Belgium, pp. 3–27, July 1999.
18. Yamada, S. E., and C. T. Sun, "Analysis of Laminate Strength and its Distribution," *J. Compos. Mater.*, 12: 275–284, 1978.
19. Kropp J., and Michaeli, W., "Dimensioning of Thick Laminates Using New IFF Strength Criteria and Some Experiments for their Verification," *Proceedings of the ESA-ESTEC Conference*, 305–312, 1996.
20. Kroll, L., and W. Hufenbach, "Physically Based Failure Criteria for Dimensioning of Thickwalled Laminates," *App. Compos. Mater.*, 4: 321–332, 1997.
21. Zinoviev, P. A., S. V. Griogoriev, O. V. Lrdedeva, and L. P. Tairova, "The Strength of Multilayered Composites under a Plane-Stress State," *Compos. Sci. Tech.*, 58: 1209–1214, 1998.
22. Gosse, J. H., "Strain Invariant Failure Criteria for Polymers in Composite Materials," *42nd AIAA/ASME/ASCE/AHS/ASC Structures, Structural Dynamics and Materials Conference*. Seattle, WA.
23. Dávila, C. G., P. P. Camanho, and M. F. Moura, "Mixed-Mode Decohesion Elements for Analyses with Progressive Delamination." *42nd AIAA/ASME/ASCE/AHS/ASC Structures, Structural Dynamics and Materials Conference*, Seattle, WA, April 2001.
24. Camanho, P. P., S. Bowron, and F. L. Matthews, "Failure Mechanisms in Bolted CFRP," *J. Reinf. Plast. Comp.*, 17: 205–233, 1998.
25. Camanho, P. P., and F. L. Matthews, "A Progressive Damage Model for Mechanically Fastened Joints in Composite Laminates," *J. Compos. Mater.*, 33: 2248–2280, 1999.
26. Shahid, I. S., and F. K. Chang, "An Accumulative Damage Model for Tensile and Shear Failures of Laminated Composite Plates," *J. Compos. Mater.*, 29: 926–981, 1995.
27. Camanho, P. P., *Application of Numerical Methods to the Strength Prediction of Mechanically Fastened Joints in Composite Laminates*, Imperial College of Science, Technology and Medicine, University of London, U.K., 1999.
28. Whitney, J. M., and R. J. Nuismer, "Stress Fracture Criteria for Laminated Composites Containing Stress Concentrations," *J. Compos. Mater.*, 8: 253–265, 1974.
29. Aronsson, C. G., "Strength of Carbon/Epoxy Laminates With Countersunk Hole," *Compos. Struct.*, 24: 283–289, 1993.
30. Camanho, P. P., and F. L. Matthews, "Delamination Onset Prediction in Mechanically Fastened Joints in Composite Laminates," *J. Compos. Mater.*, 33: 906–927, 1999.
31. Krueger, R., and T. K. O'Brien, "A Shell/3D Modeling Technique for the Analysis of Delaminated Composite Laminates." *Composites-Part A*, 32: 25–44, 2001.
32. Camanho, P. P., C. G. Dávila, and D. R. Ambur, "Numerical Simulation of Delamination Growth in Composite Materials," NASA-TP-2001-211041, National Aeronautics and Space Administration, U.S.A., 2001.
33. Nuismer, R. J., and S. C. Tan, "Constitutive Relations of a Cracked Composite Lamina," *J. Compos. Mater.*, 22: 306–321, 1988.
34. Sun, C.T., and J. Tao, "Prediction of Failure Envelope and Stress/Strain Behaviour of Composite Laminates," *Compos. Sci. Tech.*, 58: 1125–1136, 1998.
35. Sun, C. T., "Strength Analysis of Unidirectional Composites and Laminates." In: *Comprehensive Composite Materials*. A Kelly, C. Zweben (eds.), Pergamon Press, Oxford, 2000.
36. Hyer, M. W., *Stress Analysis of Fiber-Reinforced Composite Materials*, WCB/McGraw-Hill, New York, NY, 1998.
37. Jones, R. M., *Mechanics of Composite Materials*, 2nd ed., Taylor and Francis, Philadelphia, PA, 1999.
38. Herakovich, C.T., *Mechanics of Fibrous Composites*, John Wiley and Sons, New York, NY, 1998.
39. Daniel, I. M., and O. Ishai, *Engineering Mechanics of Composite Materials*, Oxford University Press, New York, NY, 1994.
40. Gibson, R. F., *Principles of Composite Material Mechanics*, McGraw-Hill, New York, NY, 1994.
41. Swanson, S. R., *Introduction to Design and Analysis with Advanced Composite Materials*, Prentice-Hall, Upper Saddle River, NJ, 1997.
42. Vinson, J. R., and R. L. Sierakowski, *The Behavior of Structures Composed of Composite Materials*, 2nd ed., Martinus Nijhoff Publishers, Boston, MA, 2002.
43. Wolff, E. G., *Introduction to the Dimensional Stability of Composite Materials*, DEStech Publications, Inc., Lancaster, PA, 2004.
44. Reifsnider, K. L., and S. W. Case, *Damage Tolerance and Durability of Material Systems*, John Wiley & Sons, New York, NY, 2002.

45. Hinton, M. J., P. D. Soden, and A. S. Kaddour, *Failure Criteria in Fibre-Reinforced-Polymer Composites*, Elsevier Publishers, 2004.
46. Anonymous, *The Composite Materials Handbook MIL-17, ASTM*, West Conshohocken, PA.
47. Callister, W. D., *Material Science and Engineering: An Introduction*, John Wiley and Sons, 2007.
48. *Hexweb Honeycomb Sandwich Design Technology*, Hexcel. Hexcel, December 2000. Web. 17 December 2015.
49. Jones, R. M., *Mechanics of Composite Materials*, Taylor and Francis, New York, NY, 1999.
50. Nguyen, M. Q., et al., "Simulation of impact on sandwich structures," *Compos. Struct.*, 67: 217–227, 2005.
51. Sezgin, F. E., *Mechanical Behavior and Modeling of Honeycomb Core Laminated Fiber/Polymer Sandwich Structures*, Izmir Institute of Technology. 2008. Web.
52. Vinson, J. R., *The Behavior of Sandwich Structures of Isotropic and Composite Materials*, Technomic, Basel, Switzerland, 1999.
53. Zureick, A., B. Shi, and E. Munley, "Fiber Reinforced Polymeric Bridge Decks," *Struct. Eng. Rev.*, 7(3): 257–266, 1995.
54. Kassapoglou, C., *Design and Analysis of Composite Structures with Application to Aerospace Structures*, AIAA Educational Series John Wiley, 2010.

CHAPTER 21

Solid Biomechanics

21.1 INTRODUCTION

Biomechanics is the study of mechanical laws and their application to living organisms, especially the human body. In short, biomechanics is the application of mechanics in biology and physiology. That is, utilizing laws of physics to study and to analyze the anatomical and functional aspects of living organisms. Solid biomechanics, however, specifically deals with the internal and external forces acting on the locomotor system of the human body and the effects produced by these forces. While the conception of biomechanics, in limited areas of medicine, started in the late nineteenth century, it was not until the late twentieth century that some researchers employed principles of mechanics to explain some phenomenon in human physiology.

Biomechanics has been involved in virtually every facet of modern medical science and technology. Biomechanics has helped in solving all kinds of clinical problems associated with virtually all human organs from the cardiovascular system to prosthetic devices. Even the molecular biology field, which may seem far from biomechanics, utilizes physics principles to understand the mechanics of the formation, design, function, and production of molecules. Orthopedics has profoundly relied on biomechanics research to treat patients with musculoskeletal problems. Orthopedic surgeons employ biomechanics as their clinical tools. Rehabilitation and healing are intimately related to the stress and strain in the tissues. Due to its economic impact, injury biomechanics in automotive accidents and other trauma is becoming more important to modern society.

Similar to an engineering approach, biomechanics problems are generally posed through the governing equation and fundamental laws of physics when the geometry, anatomy, and configuration of an organ are known. Physiological experiments including animals (in vivo or in vitro) are needed to verify or to compare with the analytical/numerical solutions. In this chapter, anatomy, material characteristics, and biomechanics of several generic types of human tissues such as bone, muscles, ligaments, etc., will be discussed. Prostheses and biomaterials will then be discussed. The body spatial coordinate system, used and frequently referred to in this chapter, is three orthogonal planes, namely, the sagittal plane (a vertical plane passes through the anterior and posterior of the body), the transverse plane (a horizontal plane), and the frontal plane (a vertical plane that passes through the left and right sides).

21.2 BIOMECHANICS OF BONE

Bones are the main load-bearing tissue in the skeletal system, the purpose of which is to protect internal organs, to provide rigid links and muscle attachment sites, and to facilitate muscle action and body movement. There are 206 bones in the human body that are categorized into four groups according to their general shapes and functions: short bones, which provide gliding motion and serve as shock absorbers (e.g., carpals); flat bones, which provide protection for underlying organs (e.g., scapulae or shoulder blades); irregular bones, which have different shapes to fulfill specific functions in the body (e.g., vertebrae in the spine); and long bones, which provide movements for upper and lower extremities (Fig. 21.1).

After the enamel of the teeth, bone is the second hardest structure of the body. Bone is a highly vascular tissue with excellent remodeling capability. That is, it can repair itself, alter its properties, and change its geometry in response to changes in mechanical loading. For example, when bone is not subjected to mechanical loading, its density changes, or when it is subjected to new and repeated mechanical loads, its shape, architecture, and density change.

Bone is composed of inorganic and organic phases and water. Based on the weight, bone is approximately 60% inorganic, 30% organic, and 10% water. On a volume basis, bone is approximately 40% inorganic, 35% organic, and 25% water. The inorganic portion of the bone, which is in the form of small ceramic crystalline-type minerals, consists of calcium

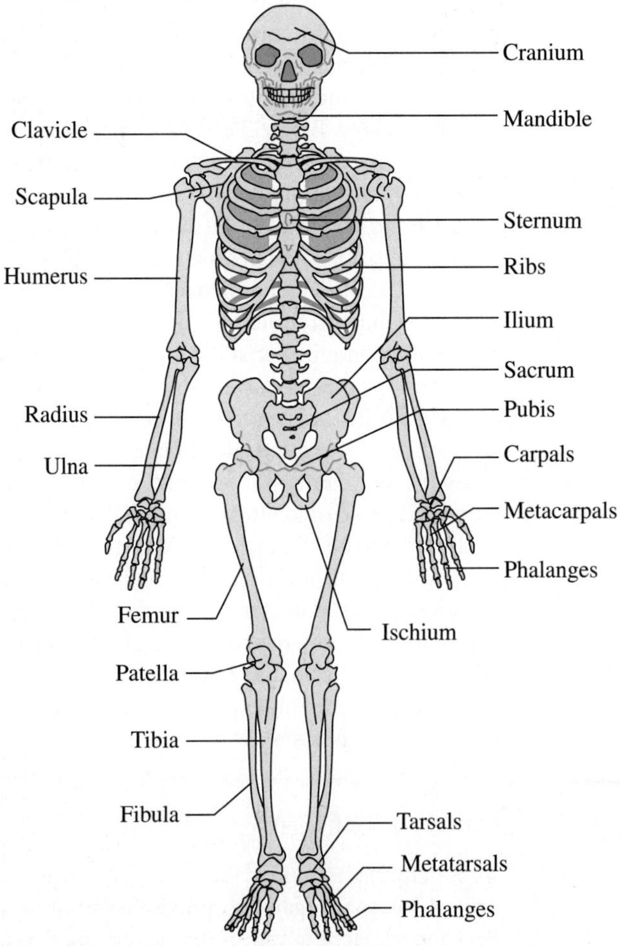

Figure 21.1 The human skeleton.

and phosphate, resembling synthetic hydroxyapatite [$Ca_{10}(PO_4)_6(OH)_2$]. The organic phase of the bone consists primarily of type I collagen (90% by weight), some other collagen type III and IV, and a variety of noncollagenous proteins.

Bone Structure

Macroscopically, there are two types of bone, cortical or compact bone, and cancellous or trabecular bone. Cortical bone forms the outer shell or cortex of the bone and has a dense structure, while cancellous bone is within this shell and is composed of thin plates, or rods (known as trabeculae), in a mesh structure, as shown in the femoral bone of Fig. 21.2. The interstices between the trabeculae are filled with red marrow. The porosity of cortical bone ranges from 5 to 30% and cancellous (trabecular) bone ranges from 30 to 90%. Except for the joint surfaces, which are covered with articular cartilage, a dense fibrous membrane called the periosteum surrounds all bones.

Microscopically, the osteon or haversian system is the main structural unit of bone. At the center of each osteon is a small channel, called the haversian canal, which contains blood vessels and nerve fibers. The osteon itself consists of a concentric series of layers

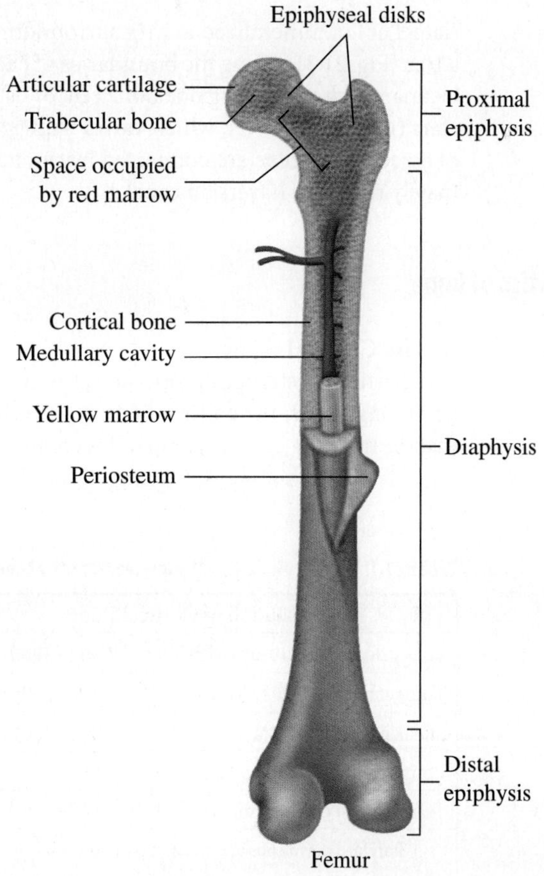

Figure 21.2 Femur, showing cortical and trabecular bones. *Adapted from Hole's Human Anatomy and Physiology (Ref. 26).*

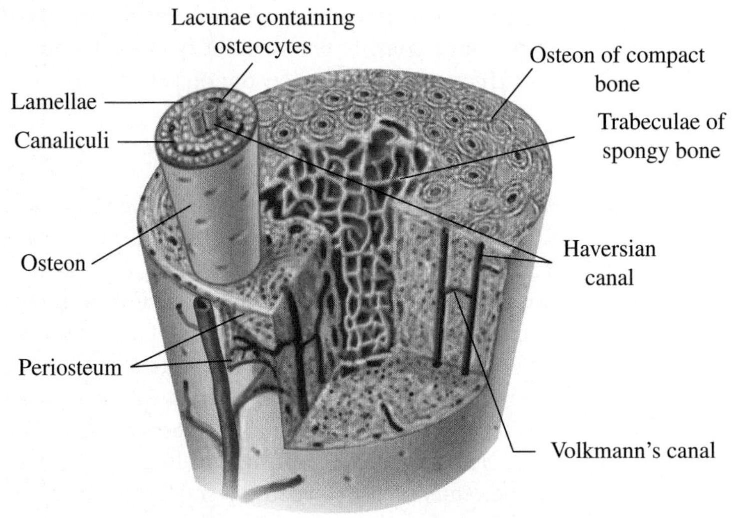

Figure 21.3 Diagram of a sector of the shaft of a long bone showing structural details of cortical bone.

(lamellae) of mineralized matrix surrounding the central canal, similar to growth rings on a tree (Fig. 21.3). Along the boundaries of each layer, or lamella, are small cavities known as lacunae, each containing one bone cell, or osteocyte. Osteon is typically about 200 micrometers (μm) in diameter, which varies with age, and has a length of 1 to 3 mm. The lacunae of adjacent lamellae are connected by numerous small channels, or canaliculi, which ultimately reach the haversian canal.

Mechanical Properties of Bone

A. The Cortical Bone

The elastic and strength properties of human cortical bone are anisotropic. In the diaphysis of long bones, the cortical bone is both stronger and stiffer in the longitudinal direction than in radial or circumferential directions (Table 21.1). Cortical bones are also stronger in

TABLE 21.1 Elastic Muduli and Ultimate Strength of Cortical Bone, from Ref. 51

Elastic Muduli of Cortical Bone	
Longitudinal modulus (MPa)	17,900 (3900)*
Transverse modulus (MPa)	10,100 (2400)
Shear modulus (MPa)	3300 (400)
Longitudinal Poisson's ratio	0.40 (0.16)
Transverse Poisson's ratio	0.62 (0.26)

*Standard deviations are given in parentheses.

Ultimate Strength of Cortical Bone	
Longitudinal (MPa)	
Tension	135 (15.6)*
Compression	205 (17.3)
Transverse (MPa)	
Tension	53 (10.7)
Compression	131 (20.7)
Shear (MPa)	65 (4.0)

*Standard deviations are given in parentheses.

compression than in tension (see Table 21.1). The mechanical properties of human cortical bone change (decrease) as the age increases. While the modulus does not reduce much, if at all, the strength is reduced at a rate of about 2% per decade.

As with all biological materials, bone is viscoelastic and nonlinear. It has been shown that cortical bone is moderately strain-rate dependent. That is, for over a six order of magnitude increase in strain rate, the modulus only changes by a factor of 2, and the strength by a factor of 3 (Fig. 21.4). The majority of physiological activities occur in the range of 0.01 to 1.0% strain per second where the changes in bone strength are negligible.

In some simple failure analyses of bone, cortical bone has been considered a simple isotropic materials and the von Mises criterion has been employed. This approach is not capable of describing multiaxial failure of cortical bone. However, the Tsai-Wu criterion, see Sec. 20.5, commonly used for composite materials, has been applied to cortical bone using the transversely isotropic and orthotropic properties of bone.

B. Trabecular Bone

The modulus of elasticity and the strength of trabecular bone vary depending on the anatomic site, loading direction, loading mode, and age and health of the individual. In compression, the anisotropy of trabecular bone, both modulus and strength, decrease with age, falling approximately 10% per decade (Fig. 21.5). The trabecular bone is stronger in compression and is weaker in shear. Both the modulus and strength of trabecular bone depend heavily on the apparent density, yet they depend on the anatomic site. Note that the density of bone is the ratio of mass to volume of the actual bone tissue; however, the apparent density is defined as the ratio of the mass of the bone tissue to the bulk volume of the specimen, including the volume associated with the vascular channels. The density of the trabecular bone also decreases with age. The apparent density of cortical bone is about 1.85 g/cm^3, which does not vary much with anatomic sites or species. However, the apparent density of trabecular bone is as low as 0.10 g/cm^3 for the spine (vertebra) (Kopperdahl and Keaveny, 1998), about 0.3 g/cm^3 for the human tibia (Linde et. al., 1989), and up to 0.6 g/cm^3 for the load-bearing portion of the proximal femur (Morgan and Keaveny, 2001).

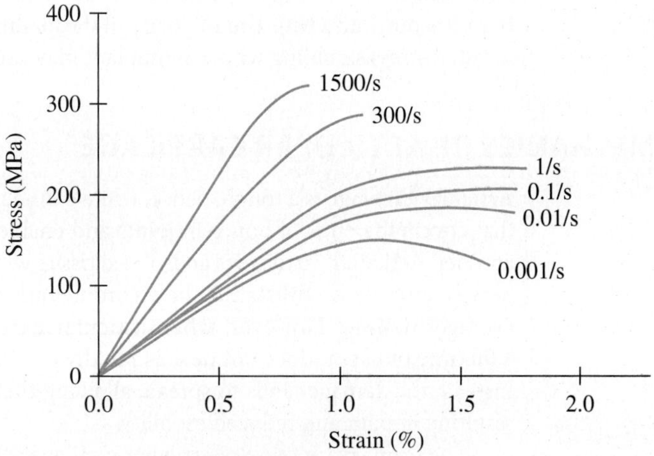

Figure 21.4 Strain–rate sensitivity of cortical bone for longitudinal tensile loading. *Adapted from Ref. 37.*

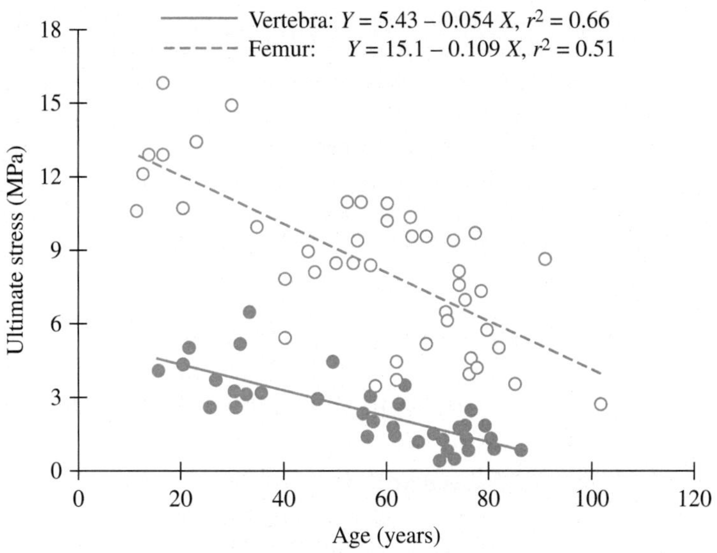

Figure 21.5 Variation of ultimate stress of trabecular bone (from the human vertebra and femur) with age. *Adapted from Refs. 36 and 41.*

Bone Remodeling

Bone, as a living tissue, has the ability to remodel, by altering its size, shape, architecture, density, and structure, to accommodate the mechanical loadings placed on it. According to Wolff's law, bone remodeling is a function of the magnitude and direction of the mechanical stresses that act on the bones (Wolff, 1892). In other words, Wolff's law states that bone is laid down (deposition) where needed and is resorbed where not needed. Bone remodeling particularly occurs around implants and prostheses where the mechanical stresses to the bone have altered, Sadegh et al. (1993), Cowin and Sadegh (1992), and Luo, Cowin, and Sadegh (1995). As an example of bone remodeling, consider a bone fracture repair process where an implant is firmly attached to the bone to stabilize the fracture site. After the fracture has healed, if the implant is not extracted, the strength and stiffness of the bone may be diminished. This is due to the fact that the material of implants is about 10 times stiffer than the bone and thus the majority of the loading will be carried out by the implant. This is called stress shielding where an implant may cause bone resorbtion.

21.3 BIOMECHANICS OF ARTICULAR CARTILAGE

Articular cartilage is a tough, dense, white connective elastic tissue, about 1 to 5 mm thick, that covers the ends of bones in joints and enables the bones to move smoothly over one another. Articular cartilage is an isolated tissue without blood vessels, lymph channels, and nerves, and yet it withstands the rigorous joint environment without failing during an average lifetime. However, when articular cartilage is damaged through an injury or a lifetime of use, it does not heal as rapidly or effectively as the other tissues in the body. Instead, the damage tends to spread, allowing the bones to rub directly against each other resulting in pain and reduced mobility.

The primary function of articular cartilage is to distribute joint loads over a wide area, reducing the stress concentration on the joint. Its secondary function is to provide a contacting surface with minimal friction and wear.

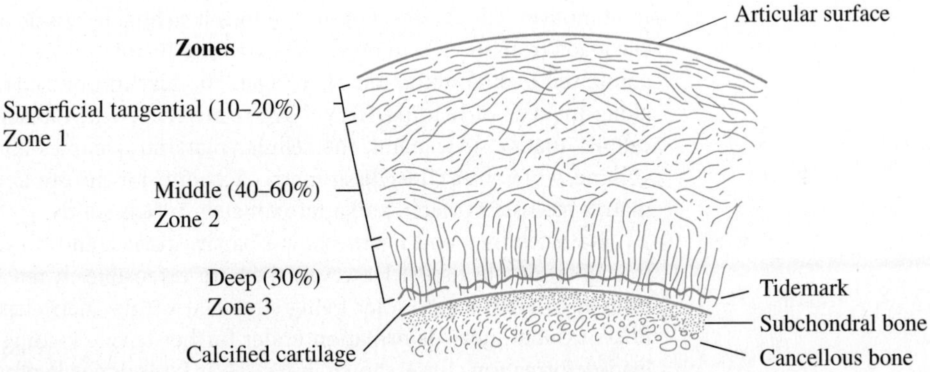

Figure 21.6 The arrangement of the collagen fibrils of chondrocytes through the thickness of cartilage, with three distinct zones: surface tangential, middle, and deep. *Adapted from Nordin and Frankel (Ref. 47).*

TABLE 21.2 Representative Properties of the Human Articular Cartilage Taken from the Lateral Condyle of the Femur, from Ref. 35

Property	Value
Poisson's ratio	0.10
Compressive modulus, MPa	0.70
Permeability coefficient, $m^4/N{\cdot}s$	1.18×10^{-15}

Chondrocytes are mature cartilage cells. They are sparsely distributed and account for less than 10% of the tissue's volume. As shown in Fig. 21.6, chondrocytes are arranged in three zones through the thickness of cartilage. The first zone is close to the articular surface where chondrocytes's long axes are parallel to the articular surface. In the second, middle zone, the chondrocytes are round and randomly distributed. In the third deep zone the chondrocytes are arranged in columnar fashion perpendicular to the articular surface.

The organic matrix of cartilage is composed of a dense network of fine collagen fibrils enmeshed in a concentrated solution of proteoglycans (PGs). The collagen content of cartilage tissue ranges from 10 to 30% by net weight and the PG content from 3 to 10% by net weight; the remaining 60 to 87% is water, inorganic salts, and small amounts of other matrix proteins, glycoproteins, and lipids.

Cartilage is predominantly loaded in compression and is biphasic (solid-fluid) viscoelastic in nature. When cartilage is compressively loaded, the water within the proteoglycan matrix is extruded. The stiffness of cartilage is a function of its permeability, and the fluid pressure within the matrix supports approximately 20 times more load than the underlying material during physiological loading, Mankin et. al. 1994. Cartilage material properties are given in Table 21.2.

21.4 BIOMECHANICS OF TENDONS AND LIGAMENTS

Tendons and ligaments, along with the joint capsule, surround, connect, and stabilize the joints of the skeletal system. Ligaments and joint capsules, which connect bone to other bones, provide stability to the joints, by guiding the joint motion, and prevent excessive joint motion. The function of the tendons is to attach muscle to bone and to transmit tensile loads from muscle to bone. Tendons and ligaments are sparsely vascularized and are composed of a combination of collagen and elastin fibers arranged primarily in parallel along the axis of loading. They have relatively few fibroblast cells and abundant extracellular matrix. In general, the cellular material occupies about 20% of the total tissue volume, while the extracellular matrix accounts for the remaining 80%. About 70% of the matrix consists of water, and approximately 30% is solids.

Both tendons and ligaments are passive tissues and do not have the ability to contract like muscle tissue. However, they are extensible in tensile loading and will return to their original length after being stretched within their elastic limits. The straightened fibers exhibit linear deformation under further tensile loading, resulting in the nonlinear load-deformation curve, shown in Fig. 21.7. Typical tensile properties of ligament and tendon are shown in Table 21.3. Tendons are generally stiffer than ligament due to the higher concentration of collagen. Tendons and ligaments are highly viscoelastic and strain rate dependent; that is, they can tear or rupture at lower extensions when loaded at high rates.

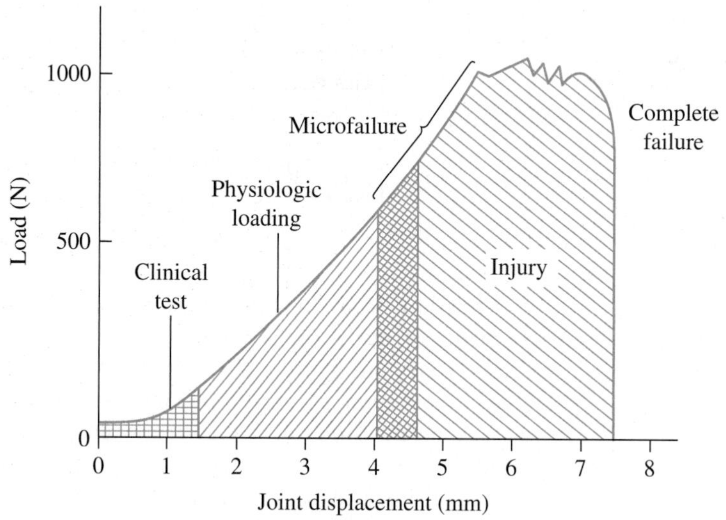

Figure 21.7 Typical load-deformation of collagenous soft tissue such as ligament and tendon. *Adapted from Kutz (Ref. 31).*

TABLE 21.3 Representative Properties of Human Tendon and Ligament under Tensile Loading, from Refs. 31 and 61

Tissue	Property	Value
Tendon	Elastic modulus, GPa	1.2–1.8
	Ultimate strength, MPa	50–105
	Ultimate strain, %	9–35
	Energy absorbed during elastic deformation	4–10% per cycle
Ligament	Stiffness, N/mm	150
	Ultimate load, N	368
	Energy absorbed to failure, N-mm	1330

21.5 BIOMECHANICS OF MUSCLES

Skeletal muscle is the most abundant tissue in the human body, accounting for 40 to 45% of the total body weight. There are 430 skeletal muscles, found in pairs on the right and left sides of the body. There are three types of muscles: (1) cardiac muscles that compose the heart; (2) smooth or involuntary muscles, which line the hollow internal organs; and (3) skeletal or voluntary muscles, which are attached to the skeleton through the tendons and cause them to move.

Muscles are molecular machines that convert chemical energy into a force (mechanical energy). The structural unit of skeletal muscle is the fiber, a long cylindrical cell with hundreds of nuclei, ranging from about 10 to 100 microns in thickness and about 1–30 cm in length. Individual muscle fibers are connected together by three levels of collagenous tissue: endomysium, which surrounds individual muscle fibers; perimysium, which collects bundles of fibers into fascicles; and epimysium, which encloses the entire muscle belly (Fig. 21.8).

The tendons and the connective tissues in and around the muscle belly are visco-elastic structures that help determine the mechanical characteristics of whole muscle during contraction and passive extension. In simulating musculotendon actuation, Hill (1970) showed that the tendons represent a spring-like elastic component located in series with the contractile-element (CE). This element is in parallel with parallel-elastic-element (PEE) and in series with series-elastic-element (SEE) (Fig. 21.9).

The mechanical response of muscle fibers is known as twitching. Some muscle fibers, with small mass, contract with a speed of only 10 msec, while others with larger mass may take 100 msec or longer. During muscle contraction, a tension load is exerted on the bone to overcome the external load of the body. The muscle tension is generally applied at a distance from the center of rotation of a joint in order to create a moment (torque) on the involved joint.

There are three types of muscle contraction. First, when muscles develop sufficient tension to overcome the resistance of the body segment, the muscles shorten and cause joint

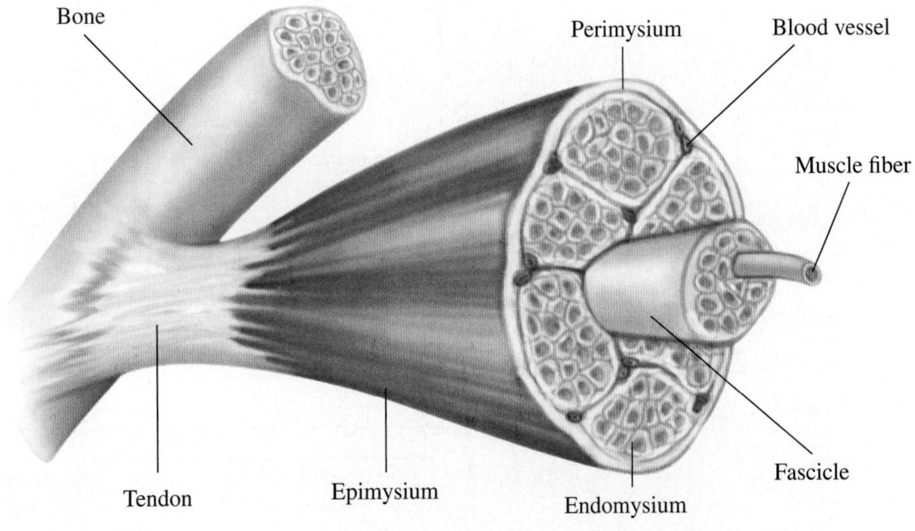

Figure 21.8 Structure of a skeletal muscle.

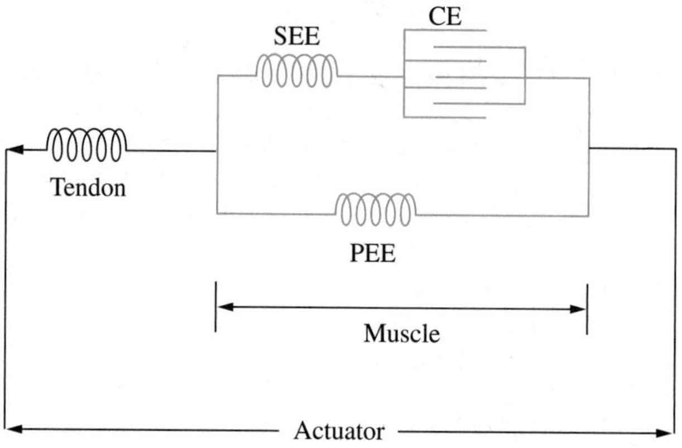

Figure 21.9 Schematic diagram of a model commonly used to simulate musculotendon actuation. *Modified from Refs. 49 and 62.*

movement in the same direction as the net torque generated by the muscles. These contractions are called *concentrics*. Second, when muscle tension is developed but no change in muscle length occurs, the contraction is said to be *isometric*. In this case, due to the developed contraction, the diameter of the muscles is increased. Third, when a muscle cannot develop sufficient tension and the opposing joint torque exceeds that produced by the tension, the muscle progressively lengthens instead of shortening. In this case the muscle is said to contract *eccentrically*. Note that concentric and eccentric contractions involve dynamic work, in which the muscle moves a joint or controls the movement. Whereas isometric contractions involve static work, in which the joint position is maintained (Fig. 21.10).

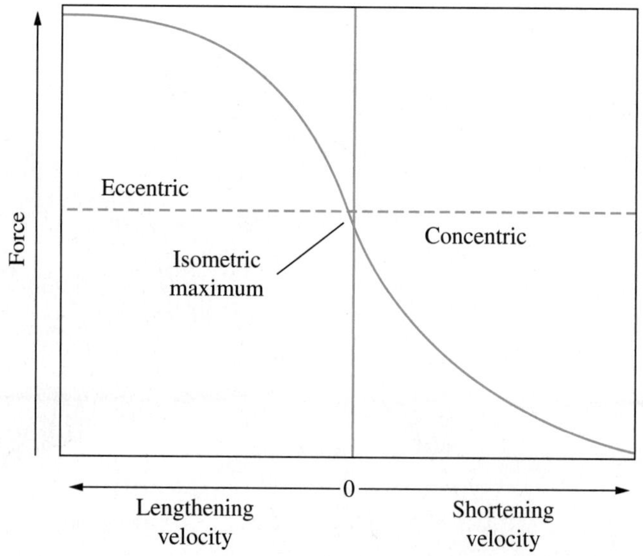

Figure 21.10 The three types of muscle contraction shown schematically.

21.6 BIOMECHANICS OF JOINTS

Anatomically, there are three types of joints: immovable joints such as sutures of the skull; slightly movable joints such as cartilaginous joints (discs) where applied forces are attenuated to the adjacent bones; and, finally, freely moveable (synovial) joints such as articular joints in upper and lower extremities. In the following sections the biomechanics of articular joints in upper and lower extremities is discussed.

21.7 BIOMECHANICS OF THE KNEE

The human knee is the largest and one of the most complex joints in the body. The knee joint is situated between the body's two longest bones, the femur and tibia, and bears tremendous loads. The fact that the knee joint provides the mobility required for locomotor activities makes it one of the highest weight-bearing joints of the body. Anatomically, the knee joint includes (a) two condylar articulations of the tibiofemoral, with a third articulation being the patellofemoral joint; (b) the medial and lateral menisci cartilaginous discs located between the tibial and femoral condyles; (c) the articular capsule, which is a double-layered membrane that surrounds the joint; (d) the synovial fluid, which is clear, slightly yellow liquid that provides lubrication inside the articular capsule at the joint; and (e) the patella, also known as the knee cap (Fig. 21.11). In addition, there are several ligaments in the knee joint; the two important ones are the anterior cruciate ligament (ACL) and the posterior cruciate ligament (PCL) (see Fig. 21.11).

Knee Kinematics

Motion of the knee joint takes place in all three planes. However, the range of motion is greatest in the sagittal plane. Motion in this plane is from full extension to full flexion of the knee, which is approximately 140 degrees. When the knee is fully extended, its rotation

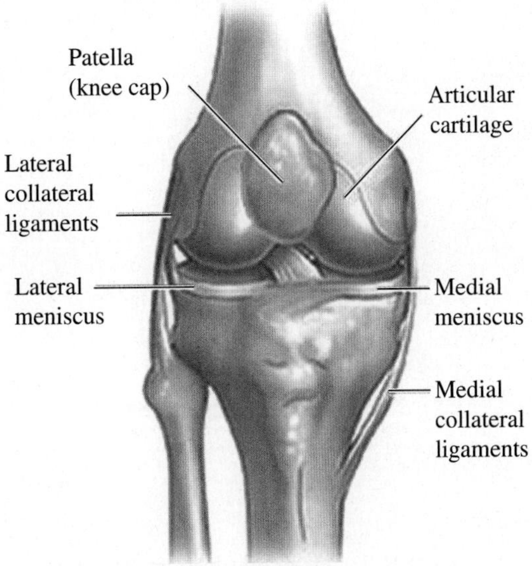

Figure 21.11 Anatomy of the right knee.

is almost completely restricted by the interlocking of the femoral and tibial condyles. However, when the knee reaches about 90 degrees flexion, the external rotation ranges from zero to approximately 45 degrees. During flexion/extension, in addition to the rolling contact between the femoral and tibial condyles, there is tangential gliding between the two condyles as well. Also, in flexion-extension motion, the patella glides inferiorly and superiorly against the distal end of the femur in the vertical direction. The anterior and posterior cruciate ligaments (ACL and PCL) limit the forward and backward sliding of the femur on the tibial plateaus during knee flexion and extension, and also limit knee hyperextension. They both play an important role in the stability of the knee.

Knee Loads

In flexing the knee, the three hamstring muscles are the primary flexors acting on the knee. However, during extension, the quadricep muscles provide the primary force for the knee. During daily activities, the tibiofemoral joint is subjected to compression and shear forces. Specifically, during the stance phase of a gait, the compressive force at the tibiofemoral joint can exceed three times the body weight, or during stair climbing, the compressive force is approximately four times the body weight. Figure 21.12 shows the static load diagram of the knee during stair climbing. Figure 21.13 shows the joint reaction force for one gait cycle, in terms of the body weight transmitted through the tibial plateau during walking. During the gait cycle, the joint reaction force shifts from the medial to the lateral tibial plateau. In the

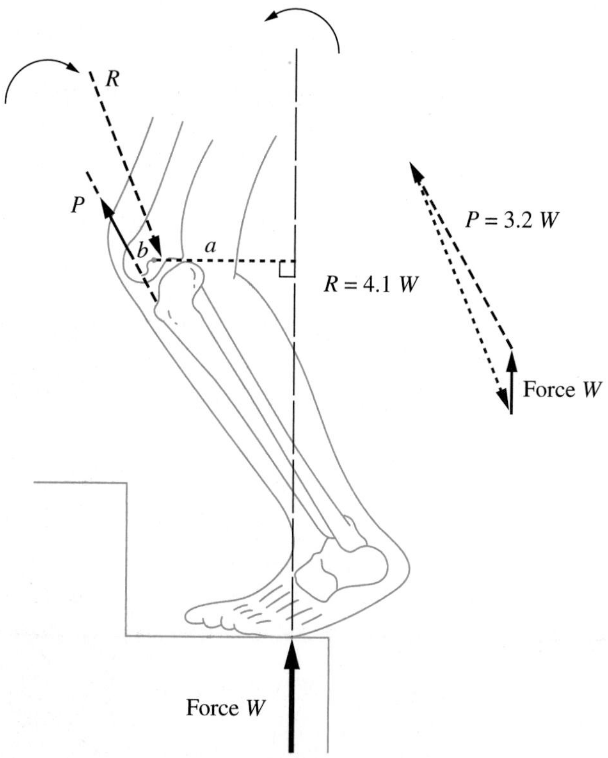

Figure 21.12 Free-body diagram of the knee during stair climbing. *Adapted from Nordin and Frankel (Ref. 47).*

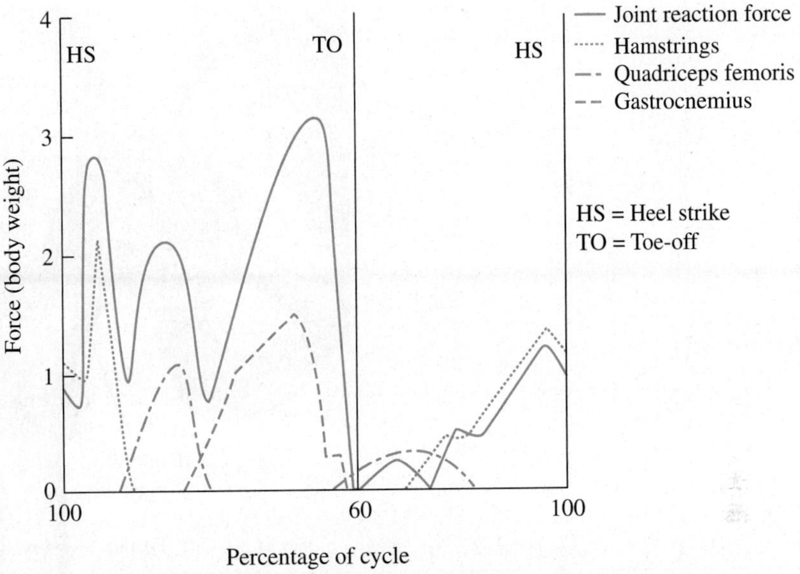

Figure 21.13 The knee reaction force in terms of body weight transmitted through the tibial plateau during walking in one cycle. *Adapted from Morrison (Ref. 39).*

stance phase, when the force reaches its peak value, the medial tibial plateau carries most of the load. In the swing phase, when the force is minimal, the load is carried by the lateral tibial plateau. The contact area of the medial tibial plateau is approximately 50% larger than that of the lateral tibial plateau.

Although the tibial plateaus are the main load-bearing structures in the knee, the menisci provide force absorption and distribute loads at the tibiofemoral joint over a larger area. This reduces the magnitude of contact stress at the joint. The joint cartilages and the ligaments also contribute to load bearing.

Biomechanically, the patella serves two important functions. First, it aids knee extension by producing anterior displacement of the quadricep tendons throughout the entire range of motion, thereby lengthening the lever arm of the muscle force. Second, it allows wider distribution of compressive stress on the femur by increasing the area of contact. During the normal walking gait, the compressive force at the patellofemoral joint is about half of the body weight. During stair climbing, it increases to over three times the body weight (3.2 w), as shown in Fig. 21.12.

21.8 BIOMECHANICS OF THE HIP

The hip joint is a ball-and-socket joint and is one of the largest and most stable joints (less likely to be dislocated) and with a great deal of mobility. Anatomically, the hip joint is composed of the head of the femur, which forms approximately two-thirds of a sphere, and the acetabulum of the pelvis, which is a socket (Fig. 21.14).

The acetabular surface and the head are covered with articular cartilages. The femur is a major weight-bearing bone and is the longest, largest, and strongest bone in the body. The femoral neck is its weakest component and is primarily composed of trabecular bone. The hip-joint cartilage covers both articulating surfaces. Several large and strong ligaments provide the stability of the hip.

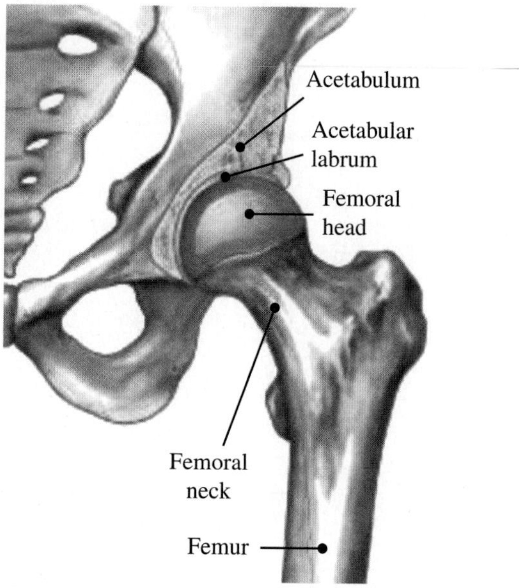

Figure 21.14 Anatomy of the hip.

Hip Kinematics and Loads

The hip joint, due to its ball and socket nature, moves in all three planes. In the sagittal plane it flexes or extends. In the transverse plane it rotates internally and externally, and in the frontal plane it has abduction and adduction motions. The maximum range of motion is in the sagittal plane where the range of flexion is from zero to approximately 140 degrees and the range of extension is from zero to 15 degrees. The range of abduction is from zero to 30 degrees, whereas that of adduction is from zero to 25 degrees. Its external rotation ranges from zero to 90 degrees, and its internal rotation ranges from zero to 70 degrees when the hip is flexed.

In terms of loading during flexion, six muscles crossing the joint anteriorly are primarily responsible for the motion. However, during extension, the three hamstring muscles and the gluteus provide the motion. These muscles are the knee flexors as well. The hip is a major weight-bearing joint. During upright standing on both feet, the body weight is evenly distributed to each of the hip joints. To calculate the forces and torques in each joint or muscle of the body, the dimensions and length of each body segment are needed. The length between each joint of the human body varies with body build, sex, and racial origin. However, an average set of segment lengths expressed as a percentage of body height was prepared by Drillis and Contini (1966) and is shown in Fig. 21.15. To determine the mechanism of balance during quiet standing, Winter et. al. (1998) modeled the body in 14 segments with 21 markers, as shown in Fig. 21.16. The location of the markers are shown, and the accompanying table gives the definition of each of the 14 segments, along with the mass fraction of each segment.

During a single-leg stance, the force of the abductor muscle is calculated as follows. The weight of the body above the hip joint is approximately equal to 5/6 of the total body weight. The moment arising from the weight above the hip joint must be balanced by a moment occurring from the force of the abductor muscles, as shown in Fig. 21.17.

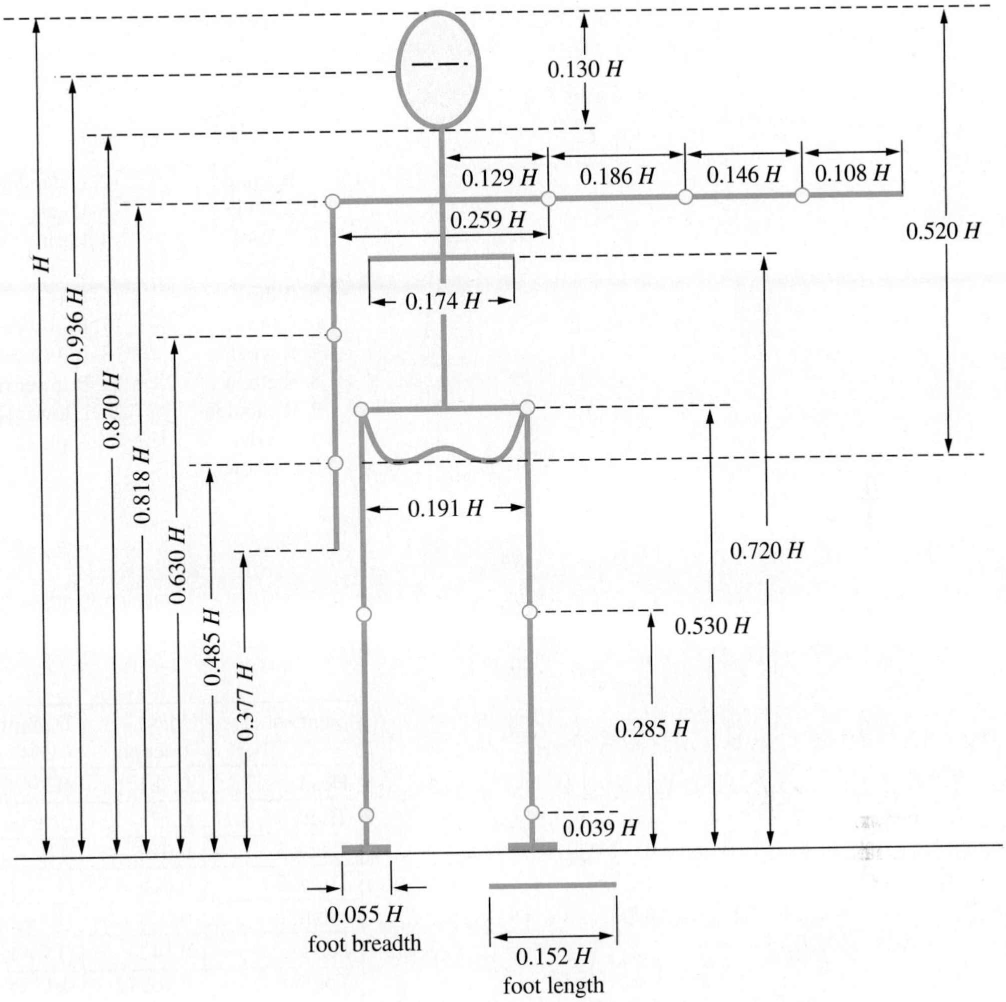

Figure 21.15 Body segment length expressed as a fraction of body height H. *Adapted from Drillis and Contini (Ref. 15).*

A simple balance of moment with respect to the center of the femoral head reveals that the abductor muscle force is $F = [(5/6)\ Wb]/c$, where c is the radius of the ball (femoral head) and b is the distance between the center of the ball and the CG of mass of the body, without the right hip (minus the one leg). Substitution of the values b and c leads to the abductor muscle force, which is about two times the body weight. The reaction force of the hip joint during a single-leg stance is approximately three times the body weight. This magnitude varies as the position of the upper body changes.

During dynamic activities, several investigators (Paul 1967, Andriacchi et al. 1980, and Bergmann et. al 1997 and 2001) have studied the loads on the hip joint. Figure 21.18 shows the hip joint reaction force in units of body weight during a one-gait walking cycle). In both men and women, the largest peak of about seven times body weight is reached just before toe-off. An increase in gait velocity increases the magnitude of the hip-joint reaction force in both swing and stance phases.

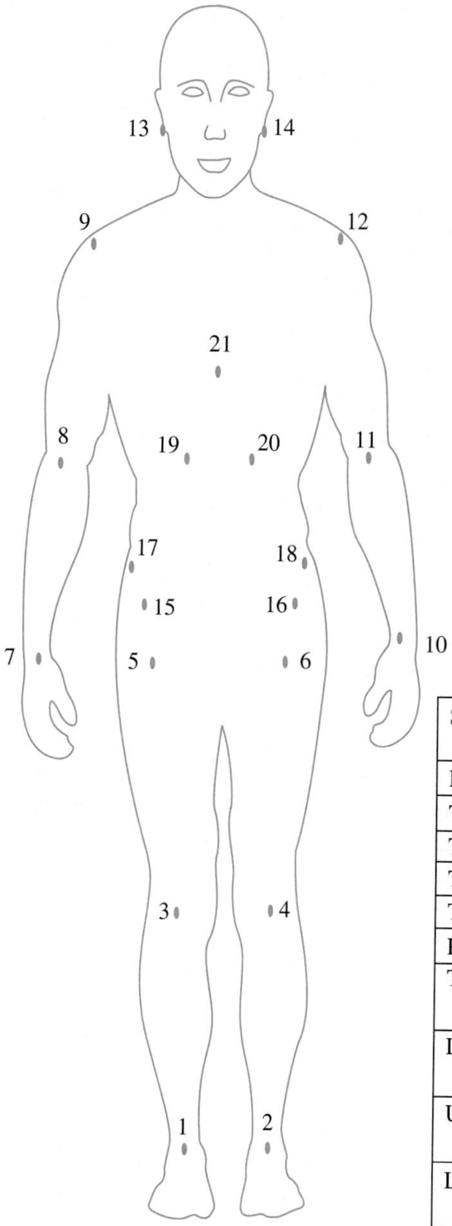

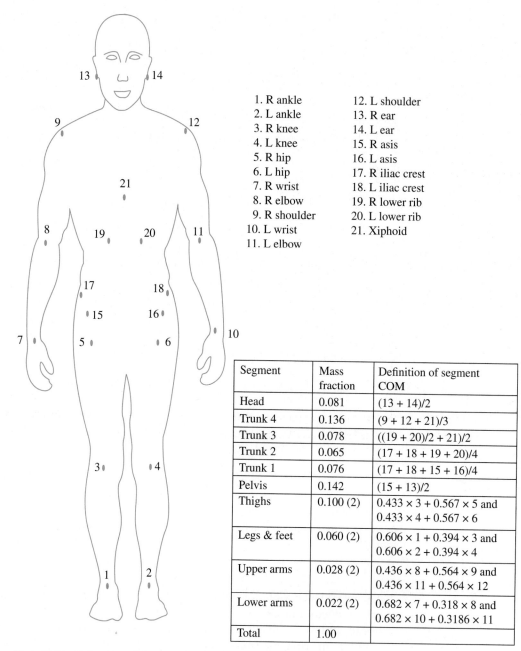

1. R ankle
2. L ankle
3. R knee
4. L knee
5. R hip
6. L hip
7. R wrist
8. R elbow
9. R shoulder
10. L wrist
11. L elbow
12. L shoulder
13. R ear
14. L ear
15. R asis
16. L asis
17. R iliac crest
18. L iliac crest
19. R lower rib
20. L lower rib
21. Xiphoid

Segment	Mass fraction	Definition of segment COM
Head	0.081	(13 + 14)/2
Trunk 4	0.136	(9 + 12 + 21)/3
Trunk 3	0.078	((19 + 20)/2 + 21)/2
Trunk 2	0.065	(17 + 18 + 19 + 20)/4
Trunk 1	0.076	(17 + 18 + 15 + 16)/4
Pelvis	0.142	(15 + 13)/2
Thighs	0.100 (2)	0.433 × 3 + 0.567 × 5 and 0.433 × 4 + 0.567 × 6
Legs & feet	0.060 (2)	0.606 × 1 + 0.394 × 3 and 0.606 × 2 + 0.394 × 4
Upper arms	0.028 (2)	0.436 × 8 + 0.564 × 9 and 0.436 × 11 + 0.564 × 12
Lower arms	0.022 (2)	0.682 × 7 + 0.318 × 8 and 0.682 × 10 + 0.3186 × 11
Total	1.00	

Figure 21.16 Estimation of the mass fraction and the center of mass (COM) of 14 segments of the body with 21 markers. *Adapted from Winter, et. al. (Ref. 58).*

21.9 BIOMECHANICS OF THE SPINE

The human spine is a complex and functionally significant structure of the human body. The principal function of the spine is to protect the spinal cord and transfer loads from the upper body, head, and trunk to the lower extremities. The spine is capable of motion in all three planes through its 24 vertebrae which are articulated by intervertebral discs. The discs

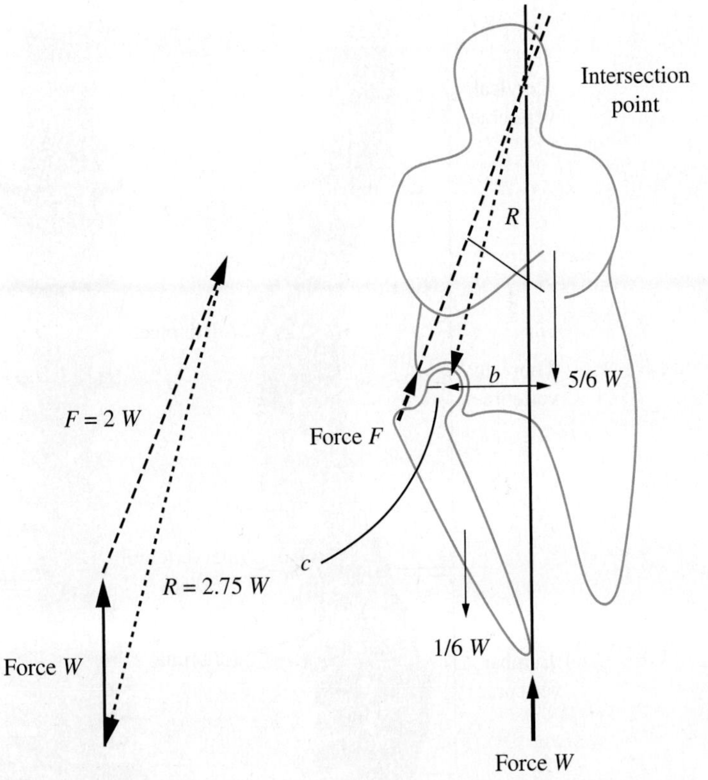

Figure 21.17 Free-body diagram of the hip joint during a single-leg stance. The abductor muscle counterbalances the moment arising from gravitational force of the body. *Adapted from Nordin and Frankel (Ref. 47).*

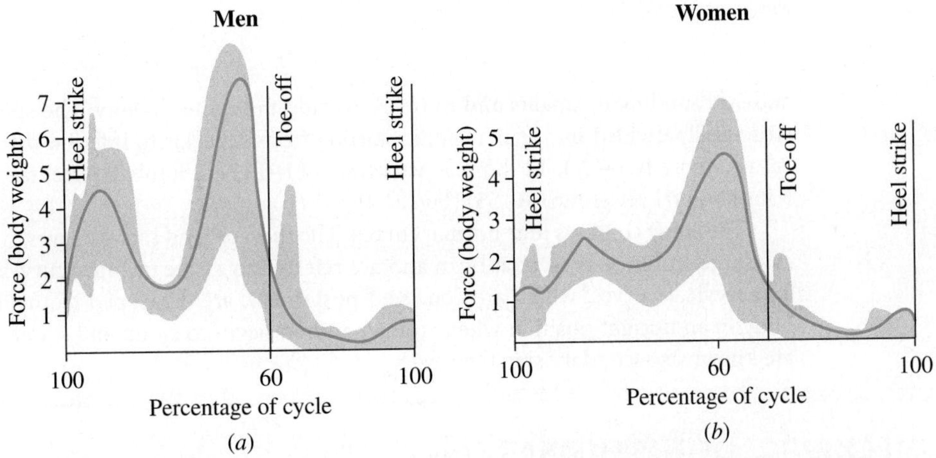

Figure 21.18 Hip-joint reaction forces in units of body weight during a one-gait walking cycle. Shaded area indicates variations among subjects for (*a*) men and (*b*) women. *Adapted from Paul (Ref. 50).*

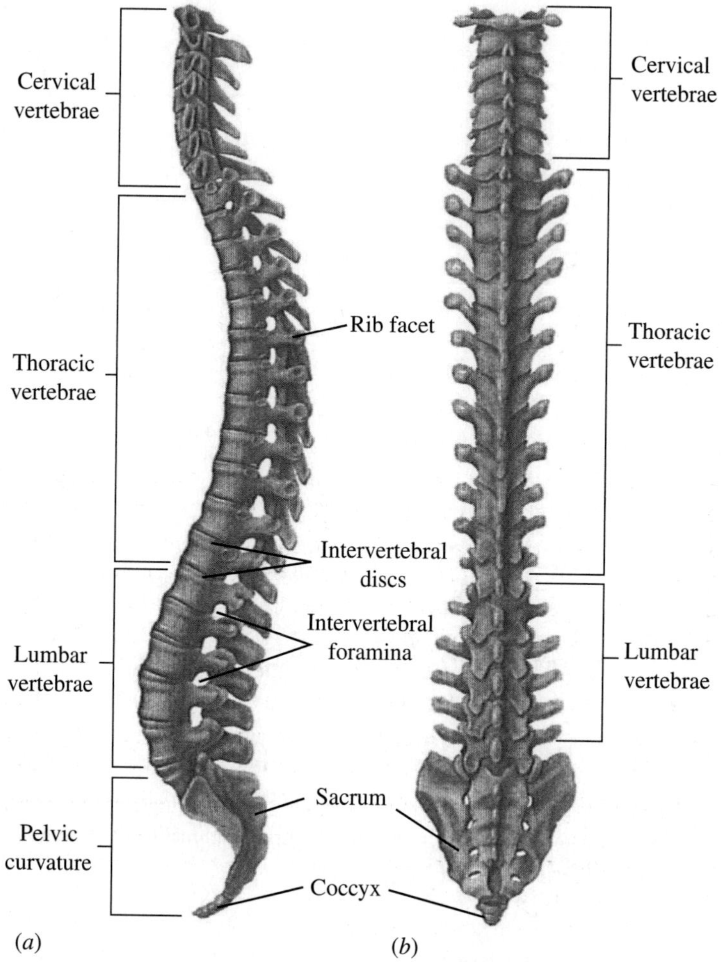

Figure 21.19 The spine (*a*) the lateral view and (*b*) the posterior view. *Hole's Human Anatomy and Physiology (Ref. 26).*

and surrounding ligaments and muscles provide the spine stability. The spinal column is structurally divided into four regions. Starting from superior to inferior, there are 7 cervical vertebrae (C1–C7), 12 thoracic vertebrae (T1–T12), 5 lumbar vertebrae (L1–L5), and 5 fused sacral vertebrae (S1–S5) (Fig. 21.19).

The spine contains four normal curves. The thoracic and sacral curves, which are concave anteriorly, are present at birth and are referred to as the primary curves. The lumbar and cervical curves, which are concaved posteriorly, are developed from supporting the body in an upright position when young children begin to sit up and stand. These curves are known as secondary spinal curves.

21.10 BIOMECHANICS OF THE LUMBAR SPINE

The vertebral bodies are designed to bear mainly compressive loads. Anatomically, the vertebral bodies of the lumbar region are thicker and wider than those in the thoracic and cervical regions. The intervertebral discs in adults account for approximately one-fourth of the

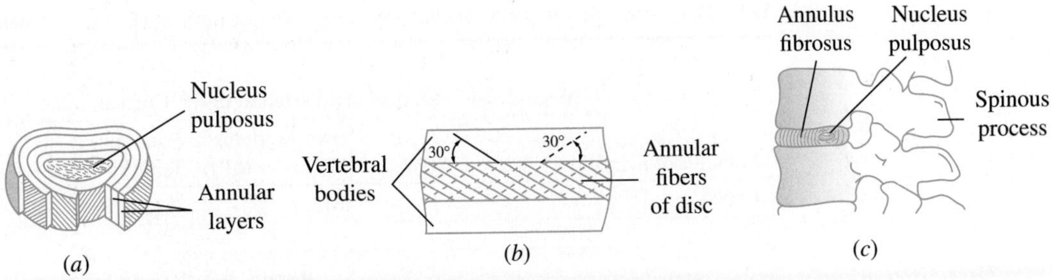

Figure 21.20 Schematic drawings of the intervertebral disc. (*a*) concentric layers of the annulus fibrosus, (*b*) orientation of the annulus fibrosus, and (*c*) the lateral view of the lumbar vertebral motion segment. [(*a*) *and* (*b*) *Adapted from Nordin and Frankel, 1989. By permission.* (*c*) *Adapted from Hall, 2003.*]

height of the spine. The intervertebral disc bears and distributes loads and restrains excessive motion. The intervertebral disc is composed of two functional structures: a thick outer ring composed of fibrous cartilage called annulus fibrosus, and a central gelatinous material known as the nucleus pulposus, or nucleus (Fig. 21.20). The annulus surrounds the nucleus pulposus. The annulus fibrosus, composed of fibrocartilage, which is a crisscross arrangement of the coarse collagen fiber bundles within the fibrocartilage. This allows the annulus fibrosus to withstand the high bending and torsional loads. The nucleus pulposus lies directly in the center of every disc except those in the lumbar segment, where it has a slightly posterior position, as shown in Fig. 21.20c. The vertebral body comprises a thin outer cortical shell and the encompassed cancellous or trabecular bone. For an adult of 45 years or younger, the cortical shell and the cancellous bone, with the ratio of 45:55, share the load transmitted through the vertebral body, 45 years or younger, Goal and Weinstein (Ref. 19). For subjects over 40 years, the cortical bone shell transmits about 65% of the total load exerted on the body, due to an increase in the porosity of the cancellous bone. Table 21.4 shows the mechanical properties of the cortical shell and the cancellous bone in the vertebral body. The mechanical properties of ligaments surrounding the vertebrae are given in Table 21.5.

Spine Kinematics

The ranges of motion of the individual motion-segments of the spine have been measured by investigators. The representative values of the relative amount of motion at different levels of the spine are presented in Fig. 21.21, White and Panjabi (Ref. 55). The range of flexion and extension is approximately 4 degrees in the upper thoracic motion segments, about 6 degrees in the middle thoracic region, and about 12 degrees in the two lower thoracic segments. This range progressively increases in the lumbar motion segments, reaching a maximum of 20 degrees at the lumbosacral level. The range of motion is strongly age-dependent, decreasing by about 50% from youth to old age. The first 50–60 degrees of spine flexion occurs in the lumbar spine, mainly in the lower motion segments. The thoracic spine contributes little to flexion of the total spine because of the oblique orientation of the facets. Significant axial rotation occurs at the thoracic and lumbosacral levels.

Spinal Load

During daily activities, the disc is subjected to a combination of compression, bending, and torsional loads. While the facets appear to function as a guide for the relative movement of adjacent vertebrae, they in fact carry about 30% of the total load when the spine is

TABLE 21.4 Mechanical Properties of a Vertebral Body, A: Cortical Bone, and B: Trabecular Bone, from Ref. 20

A			
Mechanical Properties of a Vertebral Body (Cortical Bone)			
Modulus of Elasticity (N/mm^2 or MPa)	Shear Modulus (N/mm^2 or MPa)	Poisson's Ratio (ν)	References
12,000	4615.0	0.30	63

B				
Mechanical Properties of a Vertebral Body (Spongy or Cancellous Bone)				
	Male	Female		
Tensile strength (N/mm^2 or MPa)	—	—	—	1.18
Compressive strength (N/mm^2 of MPa)	4.6(0.3)[a] (0.2–10.5)[b]	2.7(0.2) (0.3–7.0)	—	1.37–1.86
Compression at rupture (%)	9.5(0.4) (5.3–14.4)	9.0(0.6) (3.2–14.7)	—	2.5
Limit of proportionality (N/mm^2 or MPa)	4.0(0.1) (0.1–9.7)	2.2(0.1) (0.2–6.0)	—	
Compression at the limit of proportionality (%)	6.7(0.2) (4.1–8.6)	6.1(0.4) 2.6–10.6	—	
Modulus of elasticity (N/mm^2 or MPa)	55.6(0.7) (1.1–139.1)	35.1(0.6) (5.2–103.6)	100	68.7–88.3
Shear modulus (N/mm^2 or MPa)			41	
Poission's ratio (ν)			0.2	0.14
References	64	64	63	65

[a]Mean (SD).
[b]Range.

TABLE 21.5 Mechanical Properties of Spinal Elements, from Ref. 20

Property	Ligament Parameter	ALL	TL	CL	LF	ISL	SSL
Structural	X-area (mm^2)	63.7 (11.3)	3.0 (0.4)	20.9 —	56.0 (7.0)	75.7 (41.3)	30.1 (8.8)
	Load at failure (N)	177.0 (30.0)	15.0 (1.0)	191.5 (58.2)	87.0 (24.7)	90.5 (44.5)	118.3 (68.9)
	Stiffness (N/mm)	164.0 (10.3)	3.7 (0.9)	49.7 (21.9)	(66.5) (43.5)	41.5 (1.5)	55.1 (45.9)
Material	Stress at failure (MPa)	1.5 (0.1)	3.6 (1.3)	(3.0) —	1.3 (0.1)	1.3 (0.1)	3.6 (2.2)
	Strain at failure (%)	8.2 (0.2)	17.1 (1.2)	9.0 (0.8)	13.0 (1.0)	7.8 (1.4)	7.8 (0.8)
	Young's modulus (MPa)	20.1 (2.2)	58.7 (26.5)	32.9 —	19.5 (0.5)	11.6 (3.5)	62.7 (37.5)

ALL: anterior longitudinal ligament; CL: capsular ligament; ISL: interspinous ligament; LF: ligament flavum; SSL: supraspinous ligament; TL: transverse ligament

Note: numbers in parentheses are SD.

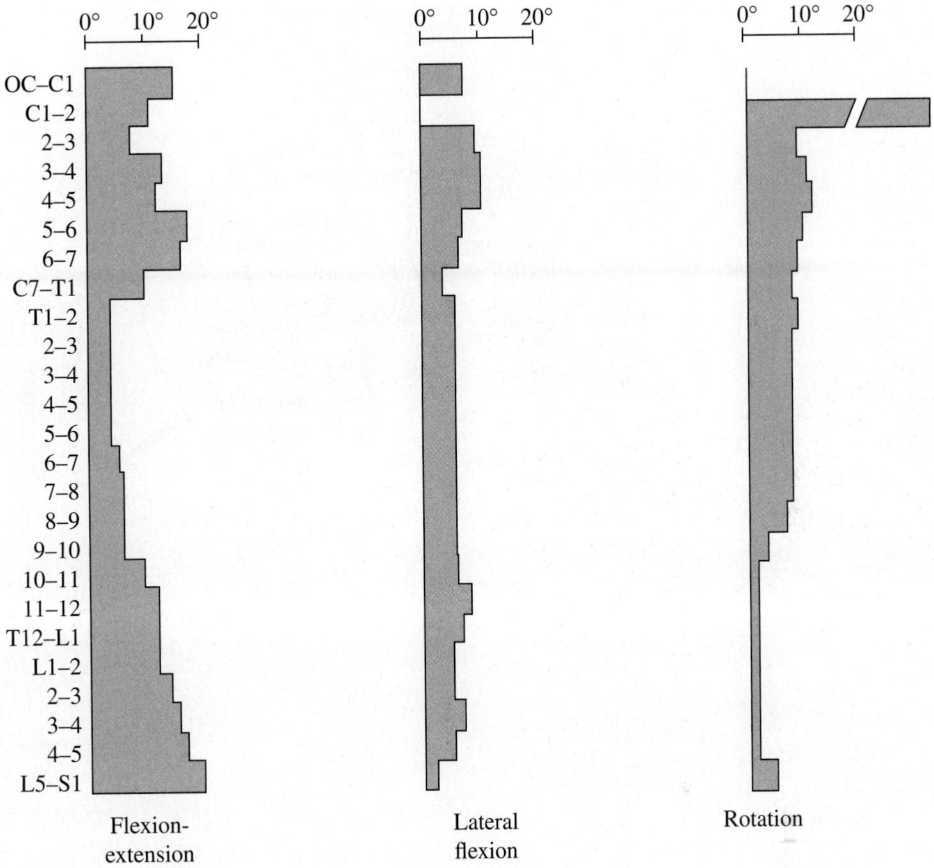

Figure 21.21 Representative values for type and range of motion at different levels of spine. *Adapted from White and Panjabi (Ref. 55).*

hyperextended, El-Bohy and King et al. (Ref. 16). The vertebral arches and intervertebral joints play an important role in resisting shear forces. The abdominal muscles and the vertebral portion of the psoas muscle initiate flexion. The weight of the upper body produces further flexion.

When standing, the center of gravity of the total body is anterior to the spinal column (passes through L5), which makes the postural muscles active. That is, the spine is under a constant forward bending moment. As the body continues to flex, the moment arm of the trunk, arm, head, and neck increases leading to an increase in tension in the back extensor muscles. Figure 21.22 illustrates the back muscle tension, with a lever arm approximately 6 cm long countering the torque created by the weight of the body and the external load. When the body is flexed (forward inclination), the spine makes the disc bulge on the concave side of the spine and retracts on the convex side. That is, the disc protrudes anteriorly and is retracted posteriorly.

When in a standing position a disc is subjected to a compressive load. Compressive stress predominates in the inner portion of the disc, and the nucleus pulposus, whereas tensile stress predominates in the annulus fibrosus. While lifting an object, the increase in abdominal pressure contributes to the stiffness of the lumbar spine and thus helps in lifting. While in a sitting position, the loads on the lumbar spine are less if the back is supported, since part of the weight of the upper body is supported by the backrest. The trunk muscles

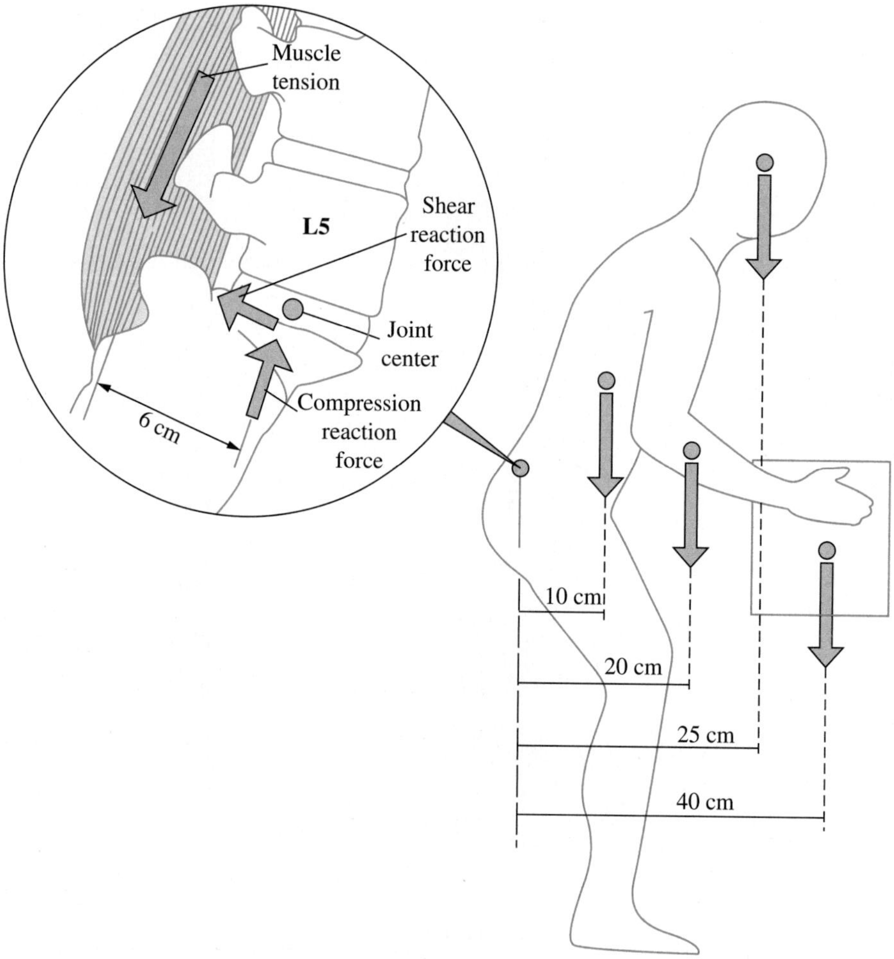

Figure 21.22 The free body diagram of the spine and the body loads when a subject lifts an external load. The back muscle with a moment of arm of approximately 6 cm must support the moment produced by the body and the load. *Adapted from Hall (Ref. 22).*

involved in motion contribute to the stability of the spine and maintaining the posture in an upright position.

During slow walking, the increase in loads on the lumbar spine is relatively modest. However, the loads increase with an increase in walking speed. In his study, Cappozzo (Ref. 9) showed that the loads were maximal around toe-off and increased approximately linearly with walking speed. Figure 21.23 reveals the axial load on the L3-L4 motion segment in terms of body weight during walking at four speeds. The solid horizontal line in the figure shows the upper body weight (UBW) and the maximum load occurred about the left and right heel strikes (LHS and RHS, respectively).

21.11 BIOMECHANICS OF THE CERVICAL SPINE

The cervical spine consists of seven vertebrae. The first two vertebrae, C1 and C2, are atypical, each having a unique structure and role, and the rest of the vertebrae, C3–C7, have similar structures and functions to that of the thoracic and lumbar spines.

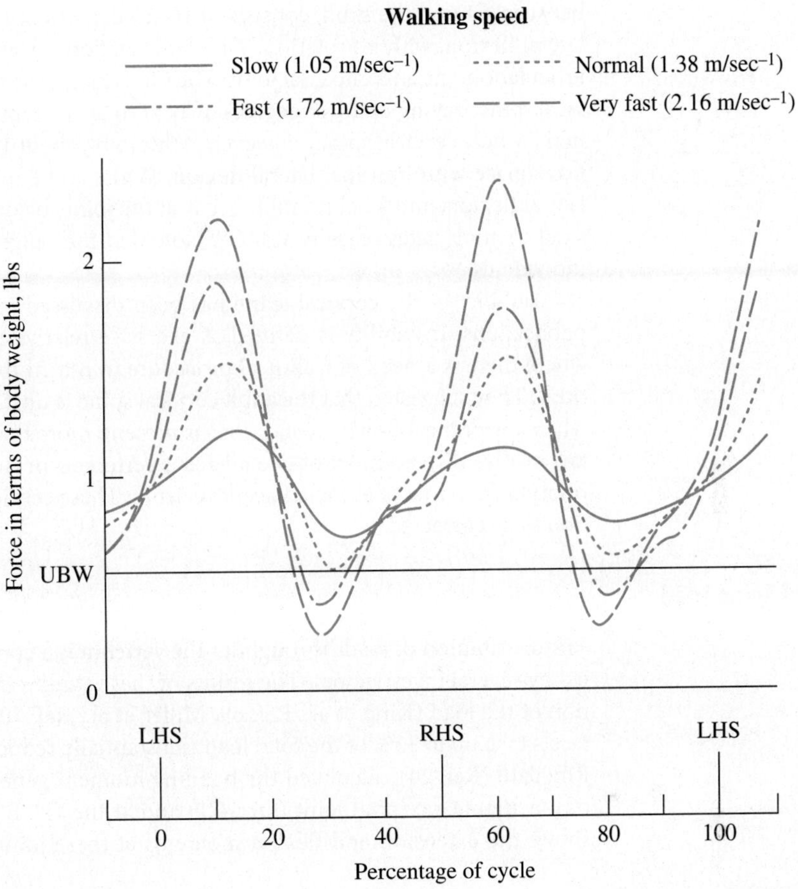

Figure 21.23 Variation of axial load on L3-L4 motion segment based on the body weight during walking at four speeds. The horizontal line represents the upper body weight (UBW), which represents the gravitational component of the load. *Adapted from Nordin and Frankel (Ref. 47).*

The first vertebra in the cervical spine is called the *atlas*, which has no vertebral body, and is composed of a ring within which an oval fossa articulates anteriorly with the dens of C2. The superior facets of C1, which form the base of the atlanto-occipital joint, bear the weight of the skull. The second vertebra is called *axis* and is composed of dens, superior facets, pedicle, and spinous process. The dens is an articular process that protrudes superiorly from the vertebral body and around which C1 rotates. This configuration allows significant rotation of the head.

Each of the five typical cervical vertebrae, C3–C7, has a body that is elliptical with a transversely concave upper surface, two pedicles, two laminae, and a spinous process. The intervertebral disc, similar to the lumbar disc, consists of a central gelatinous mass known as the nucleus pulposus, surrounded by a tough outer covering, the annulus fibrosus.

Cervical Spine Kinematics

The cervical spine is the most mobile region of the spine. The range of motion of the cervical spine in flexion-extension is about 145 degrees, in axial rotation about 180 degrees, and in lateral bending about 90 degrees. Specifically, motion at the atlanto-occipital articulation, between C1 and the skull, consists of 10–15 degrees of flexion-extension and 8 degrees of lateral flexion, with no rotation. The skull rotation is transferred into motion at the C1-C2 articulation, the antlantoaxial joint. The C1-C2 joint is the most mobile segment of the cervical spine, having up to 47 degrees of axial rotation, representing 50% of the axial rotation of the whole cervical spine. However, at this joint, about 10 degrees of flexion and extension takes place with minimal lateral flexion, White and Panjabi (Ref. 55). Flexion, extension, lateral flexion, and axial rotation occur at the joints below C2. A total of +/– 45 degrees of axial rotation takes place in C3–C7. Note that the range of motions varies widely among individuals.

Stability of the cervical spine has been discussed extensively in White and Panjabi's publications. Instability is defined as excessive relative movements of adjacent vertebrae where there is a need of a clinical procedure to return the vertebrae to their normal positions. They suggested that the adult cervical spine is unstable, or on the brink of instability, when any of the following conditions is present: more than 3.5 mm of transverse (horizontal) relative displacement of two adjacent vertebrae, or more than 11 degrees difference in rotation from that of either adjacent vertebra. These cervical conditions need clinical attention to be corrected.

Cervical Spine Load

The distribution of loads throughout the vertebra is a complex issue, which has been studied by several investigators. The results of these studies indicate that the facets bear a portion of the load (King et al., Ref. 69, Miller et al., Ref. 70). In particular, in extension, the facets take up to 33% of the total load, substantially reducing the load of the discs. Harms-Ringdahl (Ref. 24) calculated the bending moment generated around the axes of motion of the atlanto-occipital joint (Occ-C1) and at the C7-T1 motion segments. Figure 21.24 shows the extension and flexion moments at these joints in a variety of head positions.

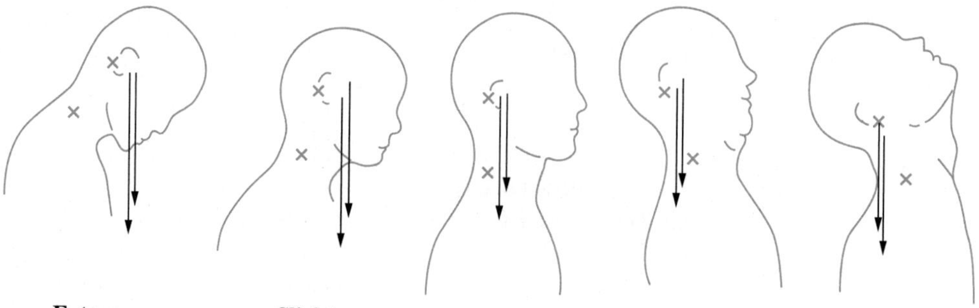

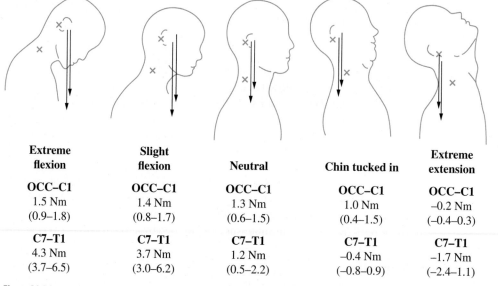

Extreme flexion	Slight flexion	Neutral	Chin tucked in	Extreme extension
OCC–C1	**OCC–C1**	**OCC–C1**	**OCC–C1**	**OCC–C1**
1.5 Nm	1.4 Nm	1.3 Nm	1.0 Nm	–0.2 Nm
(0.9–1.8)	(0.8–1.7)	(0.6–1.5)	(0.4–1.5)	(–0.4–0.3)
C7–T1	**C7–T1**	**C7–T1**	**C7–T1**	**C7–T1**
4.3 Nm	3.7 Nm	1.2 Nm	–0.4 Nm	–1.7 Nm
(3.7–6.5)	(3.0–6.2)	(0.5–2.2)	(–0.8–0.9)	(–2.4–1.1)

Figure 21.24 Extension-flexion moments at atlanto-occipital (OCC-C1) joint and C7-T1 motion segment (marked with X's) for five positions of the head. The arrows represent the force vectors produced by the weight of the head. *Adapted from Harms-Ringdahl (Ref. 24).*

During extreme flexion, the maximum bending moment in the neck occurs at the C7-T1 joint, while the moment at the Occ-C1 joint is relatively high. As the head moves to its natural position, the moment at the Occ-C1 joint slightly decreases while the moment at C7-T1 significantly decreases so that both joints carry approximately an equal amount of load. As the head continues to rotate and approaches the extreme extension, the moment on the Occ-C1 is further reduced while that of C7-T1 increases in the negative direction.

The head and neck, a large mobile ball atop a column, are particularly vulnerable to dynamic injuries. Excessive loading and motion of the head can easily create stresses and bending moments that exceed the strength of the structure, which leads to subluxation and the destabilization of the neck.

21.12 BIOMECHANICS OF THE SHOULDER

The shoulder, shown in Fig. 21.25, is the most complex joint in the human body. It consists of four major separate articulations as described below.

1. The sternoclavicular joint is located at the proximal end of the clavicle articulates with the clavicular notch of the sternum along with the cartilage of the first rib.
2. The acromioclavicular joint is located at the acromion process of the scapula (shoulder blade) articulates with the distal end of the clavicles.

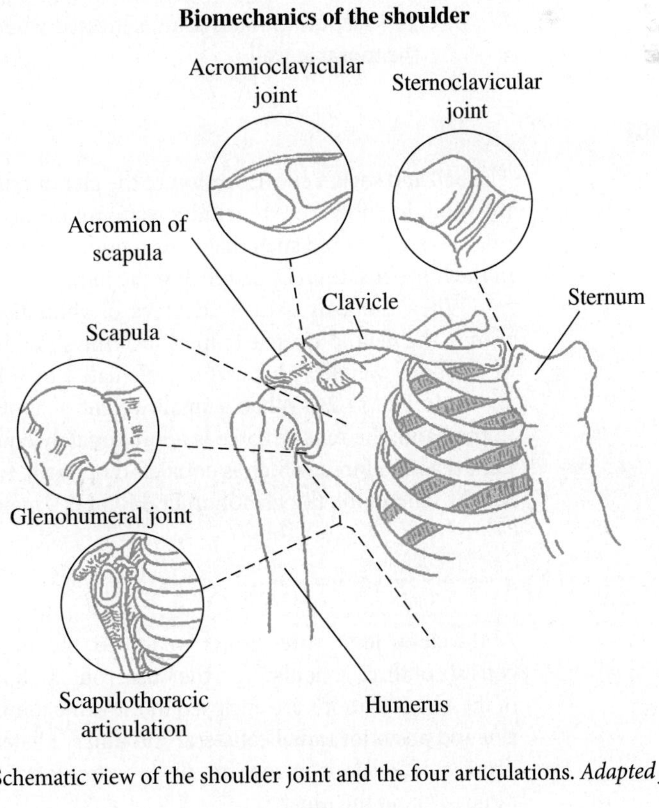

Figure 21.25 Schematic view of the shoulder joint and the four articulations. *Adapted from DePalma (Ref. 14).*

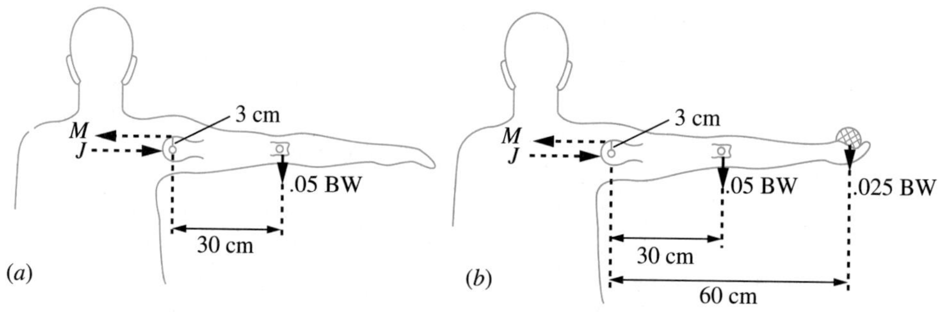

Figure 21.26 Free body diagram of the shoulder joint, showing the reaction force and the deltoid muscle force. A: without the load, and B: with a load in the hand. *Adapted from Nordin and Frankel (Ref. 47).*

3. The glenohumeral joint is located at the hemispherical head of the humerus articulates with the glenoid fossa of the scapula. This is a ball-and-socket joint configuration and is the most freely moving joint in the human body, enabling flexion, extension, hyperextension, abduction, adduction, horizontal abduction and adduction, and the medial and lateral rotation of the humerus. There are four major muscles on the posteriori, superior, and anterior sides of the joint; providing the joint movements, they are called rotator cuff muscles. The capsule of the joint is surrounded and is stabilized by several ligaments and the tendons of four muscles, as well as the biceps. Negative pressure within the capsule of the glenohumeral joint also helps to stabilize the joint.
4. The scapulothoracis joint is located where the anterior scapula is articulated with the thoracic wall.

Shoulder Loads

The ball and socket configuration of the glenohumeral joint provides mainly the rotation of the shoulder; however, it includes some rolling and gliding motion as well. The rotator cuff muscles are oriented such that their tendons and their muscle masses may push on the head of the humerus, thereby stabilizing the joint.

When the arm is in 90 degrees of abduction (extended horizontally in the frontal plane), the deltoid muscle (superior of the shoulder joint) provides a tension of about half of the body weight. This can be calculated by a simple static equilibrium of the joint as shown in Fig. 21.26. When a small weight of about 0.025 times of the body weight is held in the hand, the muscle force is approximately equal to the body weight. In these cases the joint reaction force, which is equal and opposite to the muscle force, is as high as the body weight. Therefore, the glenohumeral joint is a major load-bearing joint.

21.13 BIOMECHANICS OF THE ELBOW

At the elbow joint, three bones are joined, the humorous, ulna, and radial. The elbow joint consists of three articulations, the humeroulnar, humeroradial, and proximal radioulnar. All of these articulations are enclosed in the same joint capsule, which is reinforced by the anterior and posterior radial collateral and ulnar collateral ligaments. The elbow joint allows two degrees of freedom in motion, flexio-extension and pronation-supination (rotation of the radius around the ulna).

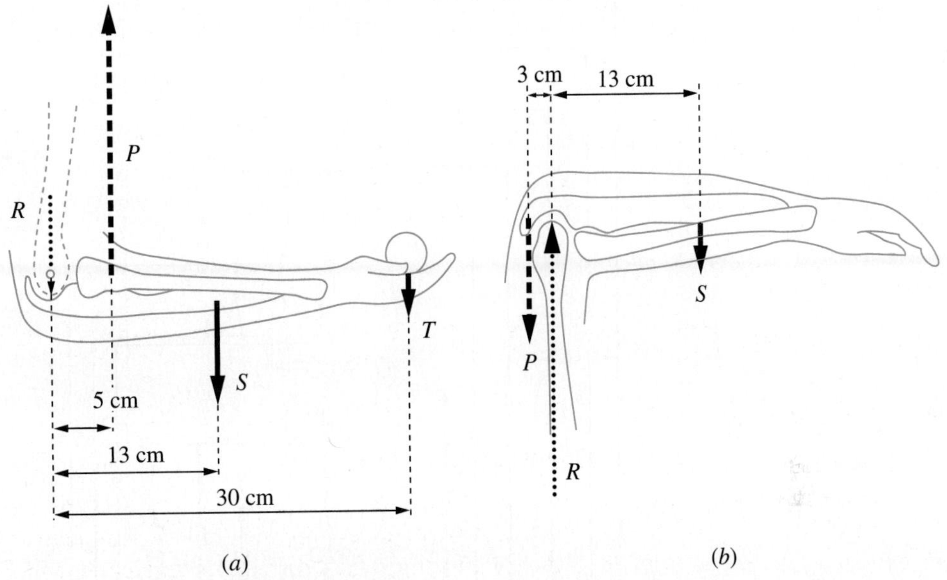

Figure 21.27 Free body diagram of the elbow joint. (*a*) showing the reaction force in flexion and the tendons of the biceps and the brachialis muscles forces, and (*b*) the tendon of the extensor (triceps) muscle. *Adapted from Nordin and Frankel (Ref. 47).*

Elbow Loading

The elbow is not considered a weight-bearing joint; however, it is regularly subjected to large loads during our daily activities. The two simple diagrams in Fig. 21.27 show the muscle forces and joint reaction forces, which could be calculated by a simple static equilibrium. The research has shown that the compressive loads during activities such as dressing and eating could reach as high as 300 N (67 lb). When rising from a chair and the body is supported by the arms, the load could reach as high as 1700 N (382 lb). During push-up exercises the peak forces on the elbow could reach as high as 45% of the body weight.

21.14 HUMAN FACTORS IN DESIGN

When interacting with machines and devices, the human musculoskeletal system often has to provide forces to power a product or actuate a controller. The designs of machines and devices are influenced by the human's ability to supply the force and to interact with the machine. The interface between humans and machines requires that humans sense the state of the device and be able to control it. That is, a product must be comfortable for a person to use. Thus, products and devices must be designed with the human factors in mind. The geometric properties of humans, that is, their height, reach, seating requirements, the size of the holes they can fit through, and the forces that they can exert in specific directions with their body segments are called *Anthropometric data*. The armed forces in the Military Standard MIL-STD-1472 have collected the majority of these data. A sample of these data is shown in Fig. 21.28 and Table 21.6. In addition, occupational biomechanics researchers have studied measures of the human ability to perform various activities. Figure 21.29 shows the average human strength for differing body positions. For example, in a standing position, the push and pull force of the hand in the position slightly lower than the torso region is higher than

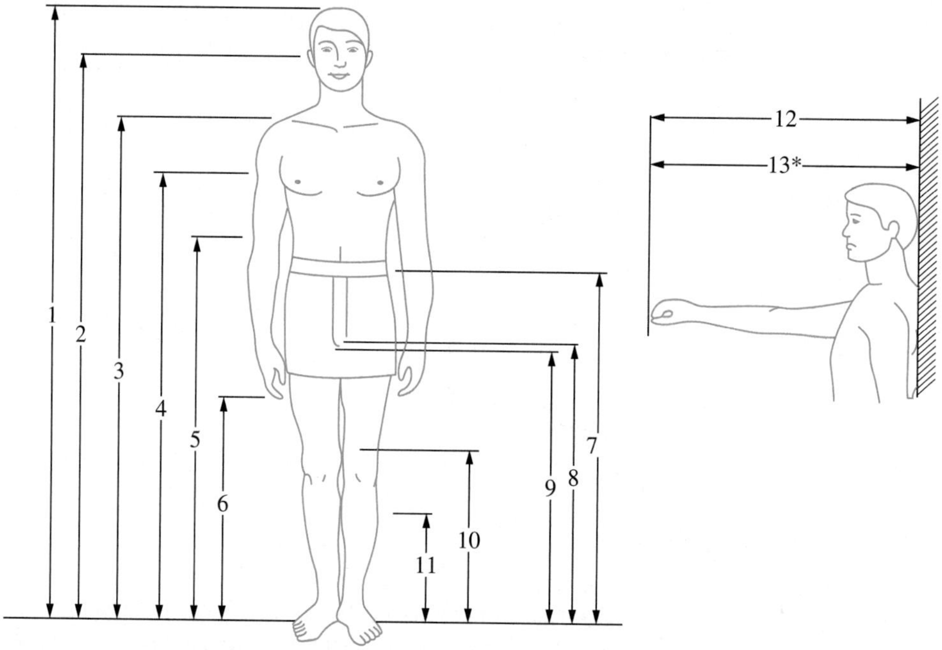

Figure 21.28 Anthropometric man (from MIL-STD-1472D).

TABLE 21.6 Anthropometric Data from Dreyfuss (Ref. 21)

		Percentile Values											
		5th Percentile						95th Percentile					
		Ground Troops		Aviators		Women		Ground Troops		Aviators		Women	
Weight kg (lb)		65.5 (122.4)		60.4 (133.1)		46.6 (102.3)		91.6 (201.9)		96.0 (211.6)		74.5 (164.3)	
		Standing body dimensions											
		cm	in	cm	in	cm	in	cm	in	cm	in	cm	in
1.	Stature	162.8	64.1	164.2	64.6	152.4	60.0	185.6	73.1	187.7	73.9	174.1	68.5
2.	Eye height (standing)	151.1	59.5	152.1	59.9	140.9	55.5	173.3	68.2	175.2	69.0	162.2	63.9
3.	Shoulder (acromial) height	133.6	52.6	133.3	52.5	123.0	48.4	154.2	60.7	154.8	60.9	143.7	56.6
4.	Chest (nipple) height*	117.9	46.4	120.8	47.5	109.3	43.0	136.5	53.7	138.5	54.5	127.6	50.3
5.	Elbow (radiate) height	101.0	39.8	104.8	41.3	94.9	37.4	117.8	46.4	120.0	47.2	110.7	43.6
6.	Fingertip (dactylion) height			61.5	24.2					73.2	28.8		
7.	Waist height	96.6	38.0	97.6	38.4	93.1	36.6	115.2	45.3	115.1	45.3	110.3	43.4
8.	Crotch height	76.3	30.0	74.7	29.4	68.1	26.8	91.8	36.1	92.0	36.2	83.9	33.0
9.	Gluteal furrow height	73.3	28.8	74.6	29.4	66.4	26.2	87.7	34.5	88.1	34.7	81.0	31.9
10.	Kneecap height	47.5	18.7	46.8	18.4	43.8	17.2	58.6	23.1	57.8	22.8	52.5	20.7
11.	Calf height	31.1	12.2	30.9	12.2	29.0	11.4	40.6	16.0	39.3	15.5	36.6	14.4
12.	Functional reach	72.6	28.6	73.1	28.8	64.0	25.2	90.9	35.8	87.0	34.3	80.4	31.7
13.	Functional reach, extended	84.2	33.2	82.3	32.4	73.5	28.9	101.2	39.8	97.3	38.3	92.7	36.5

*Bustpoint height for women.

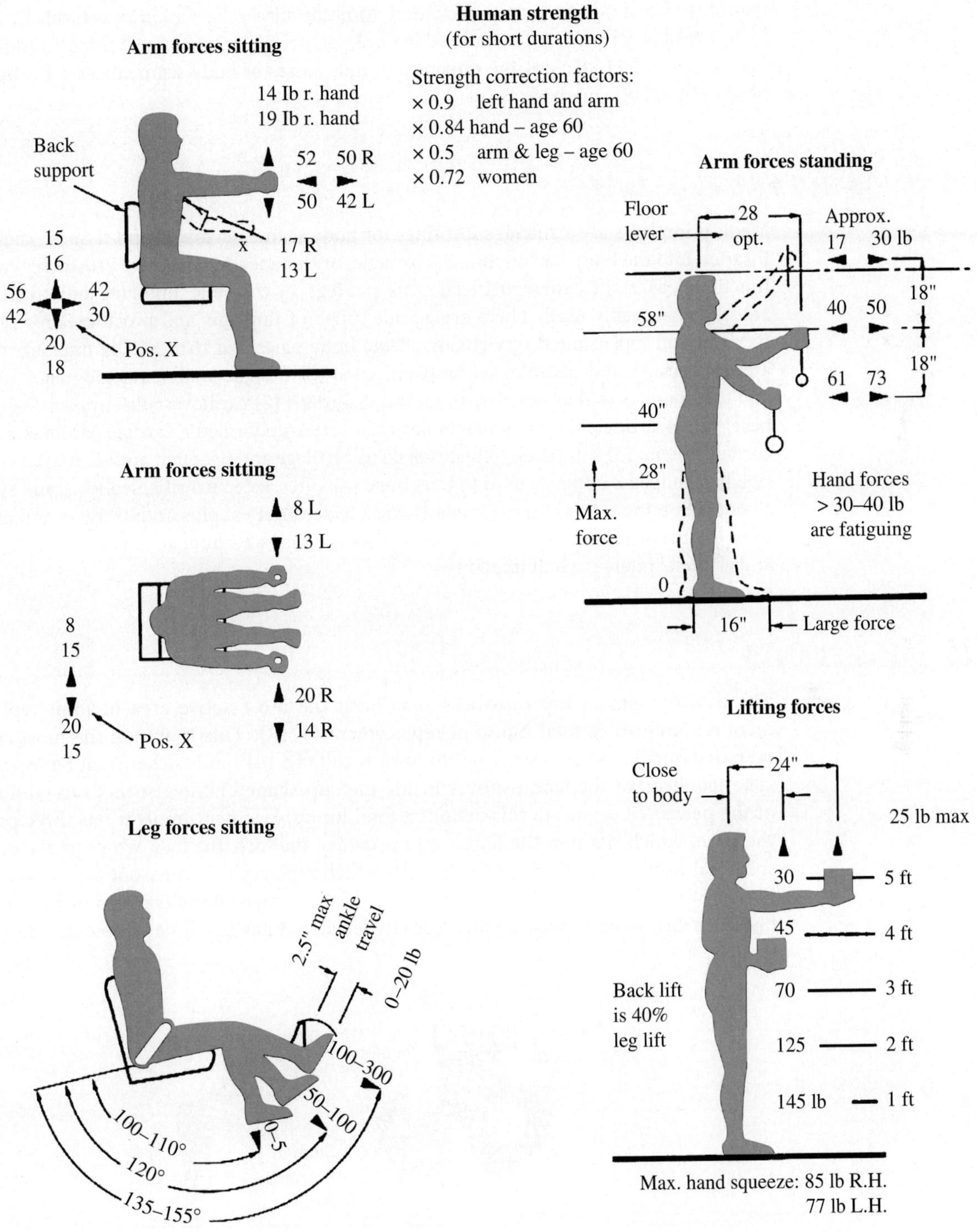

Figure 21.29 Average human strength for different tasks. *Adapted from Dreyfuss (Ref. 21).*

that of the hand position at the neck level. More detailed information is available in MIL-HDBK (Military Handbook) 759A and in books on occupational biomechanics. In addition, Figs. 21.15 and 21.16 show the dimensions and masses of body segments as a fraction of height H and body mass M.

21.15 IMPLANTS AND PROSTHESES

Medical implants are artificial substitutes for body segments, joints, and tissues. They are inserted into the body for functional, cosmetic, or therapeutic purposes. Prostheses can be functional, as in the case of artificial arms and legs, or cosmetic. Implants and prostheses are interchangeably used. There are a wide range of implants and prostheses for repair, fixation, and replacement of virtually all the body parts and tissues. The implants types are: (1) sensory and neurological implants used for disorders affecting the major senses and the brain, as well as other neurological disorders; (2) cardiovascular implants such as heart valves, artificial heart, implantable cardioverter-defibrillator, cardiac pacemaker, and coronary stent; (3) soft tissue implants such as cartilage and ligament and tendons; (4) and finally orthopedic implants used to treat bone fractures, osteoarthritis, scoliosis, and spinal stenosis. Specific forms of these implants are a wide variety of pins, rods, screws, and plates used to anchor fractured bones and spacer for cervical and lumbar intervertebral bodies and artificial intervertebral implants.

21.16 HIP IMPLANTS

The development of hip prostheses has been the most active area of joint replacement research since total hip joint replacement (THR). This is among the most common orthopedic procedures. The hip joint is called a ball-and-socket joint because the spherical head of the femur moves inside the cup-shaped hollow socket (acetabulum) of the pelvis. To duplicate this action, a total hip replacement implant has three parts: the stem, which fits into the femur and provides stability; the ball, which replaces the spherical head of the femur; and the cup, which replaces the worn-out hip socket, see Fig. 21.30. Depending on the manufacturer, there are many designs of hip implant parts based on age, weight, bone quality, activity level, and health of the recipient. The stem

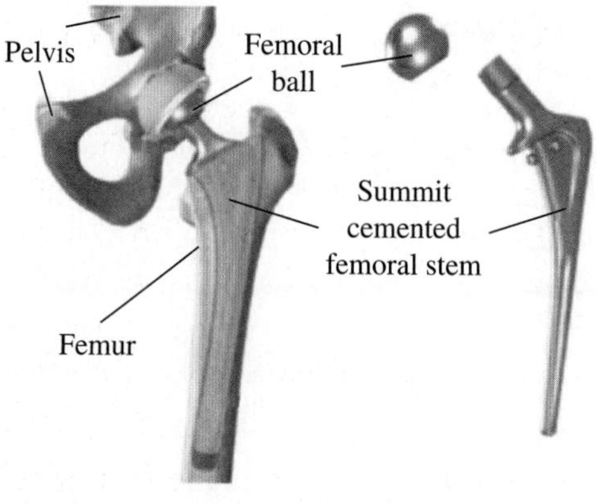

Figure 21.30 Total hip implant.

portions of most hip implants are made of titanium or cobalt/chromium based alloys. They come in different shapes and degrees of roughness. Cobalt/chromium based alloys or ceramic materials (aluminum oxide or zirconium oxide) are used in making the ball portions, which are polished smooth to allow easy rotation within the prosthetic socket. The acetabular socket can be made of metal, with a liner made of ultrahigh molecular weight polyethylene, UHMWPE, or a combination of polyethylene backed by metal. All together, these components weigh between 14 and 18 oz depending on the size needed. The chance of a hip replacement lasting 20 years is about 80%.

Implant Fixation

To hold the femoral and acetabular components in place, two methods—cemented and cementless (bone ingrowth or Osseointegration)—are used. The cement, which is an acrylic polymer called polymethylmethacrylate (PMMA), is used for older patients. However, the cementless method is used for younger patients (under 50 years of age) who are active and have good bone quality. Cemented fixation relies on a stable interface between the prosthesis and the cement and a solid mechanical bond between the cement and the bone.

Cementless implants, where the implant is directly attached to bone without the use of cement, were introduced in the 1980s. The surface topography of the implant is textured or has porous regions, and the implant is generally coated with hydroxyapatite so that the new bone actually grows into the surface of the implant. In general, these designs are larger and longer than those used with cement. A surgeon's accuracy in placing the implant plays an important role in the stability of the implant since the channel must match the shape of the implant precisely. The intimate contact between the component and bone is crucial to permit bone ingrowth. New bone growth cannot bridge gaps larger than 1 mm to 2 mm. Hybrid total hip joint replacement (THR), which was introduced in the early 1980s, has one component, usually the acetabular socket, inserted without cement, and the other component, usually the femoral stem, inserted with cement. A hybrid hip takes advantage of the excellent track record of cementless hip sockets and cemented stems.

Implant Lossening

Hip implants may loosen due to several factors: (1) Cracks (fatigue fractures) in the cement that occur over time can cause the prosthetic stem to loosen and become unstable. This is more often the case with patients who are very active or very heavy. (2) The action of the metal ball against the polyethylene cup of the acetabular component creates polyethylene wear debris. The microscopic debris particles are absorbed by cells around the joint and initiate an inflammatory response from the body, which tries to remove them. (3) Micromotion between the implant and the bone prevents bone ingrowth into the implant.

The hip implants could be made of two or three types of materials, namely, metal, ceramic, and polyethylene. The crucial part of the implant is the contact between the ball and the acetabular liner. The ball could be made of metal or ceramic and the liner could be made of ceramic, metal, or polyethylene (UHMWPE). The metal and plastic implants wear at a rate of about 0.1 mm each year. Metal-on-metal implants wear at a rate of about 0.01 mm each year, about 10 times slower than metal and plastic. There are also concerns about the wear debris that is generated from the metal-on-metal implants. Ceramics are the most resistant to wear of all available hip replacement implants. However, due to its low fracture toughness, ceramic implants can break inside the body.

21.17 KNEE IMPLANTS

The total knee arthroplasty (TKA) or replacement implants are more complex because of the articulation of three bones, the distal femur, proximal tibia, and patella (the kneecap), where the surfaces actually roll and glide as the knee bends. There are many knee implant designs available having the following major components (Fig. 21.31):

1. Femoral component: The metal femoral component curves around the distal end of the femur and has an interior groove so the kneecap can move up and down smoothly against the bone as the knee bends and straightens. In some designs a smaller piece may be used (unicompartmental knee replacement) to resurface just the medial or lateral part of the condyle.
2. Tibial component: The tibial component is a flat metal platform.
3. Polyethylene cushion: The cushion may be part of the platform (fixed) or separate (mobile).
4. Patellar component: The patellar component is a dome-shaped piece of polyethylene that duplicates the shape of the kneecap anchored to a flat metal plate.

The metal parts of the implant are made of titanium or cobalt/chromium based alloys. The plastic parts are made of ultrahigh molecular weight polyethylene (UHMWPE). Altogether, the components weigh between 15 and 20 oz, depending on the size selected. Due to its continuous sliding and rolling, knee implant wear is a major problem where microscopic particulate debris are generated. Similar to the hip implants, the knee implants may be "cemented," "cementless," or "hybrid," depending on the type of fixation used to hold the implant in place.

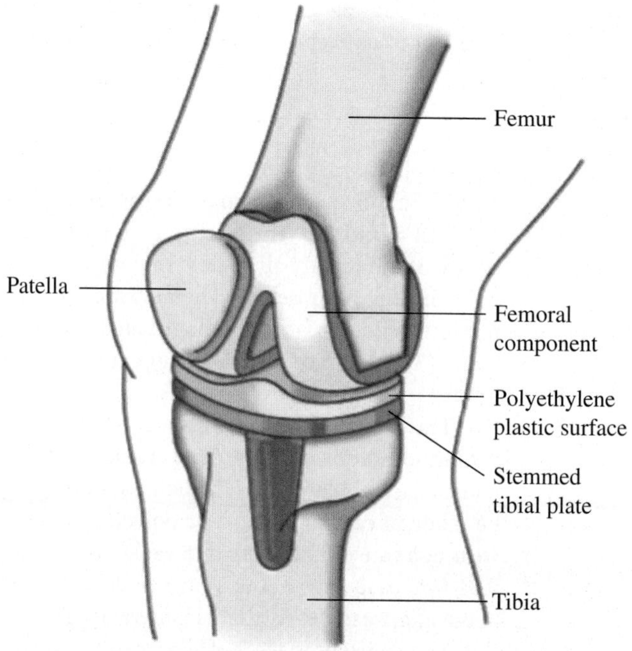

Figure 21.31 The components of the total knee arthroplasty implant.

21.18 OTHER IMPLANTS

The majority of orthopedic implants are made of metals. There is considerable mismatch between the metal and bone properties, which leads to stress shielding and bone resorbtion. New composite implants or graded materials are being developed to closely match the bone properties, thereby reducing the risk of bone resorbtion.

There are many other orthopedic implants and prostheses for virtually every joint in the body as well as plates and rods for bone repair and fracture fixations. In addition, there are many types of screws for different locations and variety of bones. For these topics the reader is referred to specialized orthopedic books and literature. Some of the implants are described below.

Artificial Intervertebral Disc

In cases of degenerative intervertebral disc disease or severe intervertebral disc herniation, a spinal disc fusion has been a routine surgical approach. However, recently a variety of artificial intervertebral discs, as total disc replacement or spinal arthroplasty, have been investigated and tested as an alternative approach to spinal fusion. Combinations of metal cobalt-chrome (as the end plates) and medical grade plastic (polyethylene) similar to joint replacements in the knee and hip have been used to create the biological range of motion and of the disc at cervical or lumbar spine. The design varies depending on the manufacturer and the anatomical site, i.e., C1 -C7, T1 -T12, or L1 -L5 levels.

Dental Implants

Dental implants have been widely used in recent years. There are two categories of dental implants: endosteal implants, which enter the bone tissue; and subperiosteal systems, which contact the exterior bone surfaces (Lemons, 1988). However, endosteal implants where the implant body is screwed into the jaw bone is more common. Once the jaw bone is osteointegrated to the screw, an abutment is placed on top of the screw. The crown then is installed on the abutment. The types and designs of dental implants availed are numerous. The surface of the threads of the screw may be covered by hydroxyapatite for a better bone integration.

Heart Valves

Mechanical or tissue prosthetic heart valves are used to replace the natural heart valves (miteral, tricuspid, aortic, and pulmonic) when these no longer perform their normal function because of disease (Gidden et al., 1993). Mechanical valves are of two types, caged ball and tilting disk. The materials most widely used are silicone elastomer, cobalt-chrome-based alloys, titanium, and pyrolytic carbon.

Stents

A stent is a device made of inert materials and designed to serve as a temporary or permanent internal scaffold to maintain or increase the lumen of a vessel. For further information see Schatz (Ref. 53).

21.19 BIOMATERIALS

By definition, a biomaterial is a biocompatible material suitable for inclusion, augmentation or replacement of the function of bodily tissues or organs. Biomaterials science is a multidisciplinary field that brings together researchers from diverse academic backgrounds, such as bioengineering, chemistry, chemical engineering, electrical engineering, mechanical engineering, materials science, biology, microbiology, and medicine. Biomaterials can be synthetic or natural, solid, and sometimes liquid. Solid biomaterials can be metals, ceramics, polymers, glasses, carbons, and composite materials. Such materials are used as molded or machined parts, coatings, fibers, films, foams, and fabrics. Biomaterials are generally integrated into medical devices and implants and are rarely used on their own.

Historically, in the early 1900s bone plates were successfully implanted to stabilize bone fractures and to accelerate their healing, and since the 1960s, blood vessel replacement and artificial heart valves and joint replacements were used. Since then, with the advent of new materials and prostheses, the field of biomaterials has rapidly expanded and saved many lives.

Key factors in the design and use of biomaterials are biocompatibility, toxicology, biofunctionality, and, to a lesser extent, the availability of biomaterials. Except for their brittle characteristics, ceramic materials are ideal candidates as biomaterials. Most biomaterials and medical devices perform satisfactorily, improving the quality of life for the recipient or saving lives; however, they have exhibited some failures. This is due to the fact that manufactured materials and devices have a failure rate. Other factors such as genetic differences, gender, body chemistries, living environment and physical activity, as well as a physician's "talent" for implanting devices may contribute to the success or failure of the biomaterials and the implants.

Biocompatibility

The understanding and measurement of biocompatibility of biomaterials is not fully developed. It has been accepted that no foreign material placed within a living body is completely compatible. Only those materials (natural tissues) manufactured by the body itself (autogenous) are compatible, and any other substance that is recognized as foreign initiates some type of reaction (host-tissue response). Natural material that is obtained from an individual and that will be implanted into the same person is called **autograft**. However, if the donor is a different individual, the material is called **allograft**.

Prostheses and Implant Materials

A wide range of materials are used in the design and manufacturing of prostheses and implants. These materials are grouped as metals, polyethylene, ceramics, bioresorbables, bioactive, and bone cements, or a combination of them. All these materials must possess strength and biocompatibility to withstand the forces and stresses imposed in the patient's body and to not be rejected by the body.

A. Metals

There are four types of biocompatible metals. (1) *Cobalt-Chrome alloys* where the base metals are cobalt (> 34%) and chrome (> 19%) mixed with smaller quantities of other metals, even nickel. (2) *Titanium alloys* where the base metal is titanium; used commercially with (ca 4%) Alumina. (3) *Stainless Steel alloys* where the base metal is iron (> 58%), mixed with larger quantities of chrome and nickel and some other metals. (4) *Porous tantalum or Spongy metal* which is composed of microscopic trabecular-type structure similar to trabecular bone. Thus, its mechanical characteristics come very close to the mechanical characteristics of the spongious bone itself. It is thus used mainly in reconstructive procedures where it replaces the lost bone and as the porous coating of implants that allow greater ingrowth of bone tissue.

B. Polyethylene

The most common form of polyethylene is ultra-high molecular-weight polyethylene (UHMWPE) that is used for bearing surfaces in total joint prostheses and articular joints. There is a growing body of evidence showing that wear particles worn off the surface of a UHMWPE component may cause a failure of the whole joint.

C. Ceramics

Ceramic materials used in prostheses are zirconium oxides. Such ceramics are the most chemically and biologically inert of all materials. They are also stiff, strong, and hard. The main disadvantage of medical ceramic materials is their low fracture toughness or fragility. When the stress acting on medical ceramic materials exceeds a certain limit, ceramic materials burst exploding into many splinters.

There are two kinds of ceramic materials that are used in implants, alumina ceramic and zirconia ceramic. Zirconia ceramic is one of the highest-strength ceramics suitable for medical use. It is two to three times stronger than alumina ceramic. Particularly, femoral balls that are made out of this material have a smaller diameter than those made out of alumina ceramic. The surface of the zirconia ball can be made smoother than that of alumina ceramic and the wear produced by the zirconia ball coupled with polyethylene cup is only half as large as the wear produced by an alumina ceramic ball in identical coupling. Recent studies demonstrate that zirconia ceramic ages in the body's temperature and the surface of the zirconia ball's surface roughens. In non-cement (cementless) or osseointegrating implants a coating material known as synthetic hydroxyapatite [$Ca_{10}(PO_4)_6(OH)_2$] or HA is utilized. The HA facilitates the bone ingrowth and interlocks of bone with implant.

D. Bioresorbable

Bioresorbables are the materials that, upon placement within the human body, start to dissolve (resorb) and slowly are replaced by advancing tissue (such as bone). Common examples of bioresorbable materials are tricalcium phosphate [$Ca_3(PO_4)_2$] and polylactic–polyglycolic acid copolymers. Calcium oxide, calcium carbonate, and gypsum are other common materials that have been utilized during the last three decades.

E. Bioactive

Bioactives are the materials that, upon placement in the human body, interact with the surrounding bone and, in some cases, even soft tissue. This occurs through a time-dependent kinetic modification of the surface, triggered by their implantation within the living bone. An ion-exchange reaction between the bioactive implant and surrounding body fluids results in the formation of a biologically active carbonate apatite layer on the implant. For example, coating an implant with synthetic hydroxyapatite [$Ca_{10}(PO_4)_6(OH)_2$] triggers bone ingrowth.

F. Bone Cements

Bone cement is a filler that is used for fixation of the artificial joints to the bones. Contrary to general perception, bone cement is not a glue, rather it is grout that fills the space between the implant and the bone. Bone cement is a compound consisting of 90% of polymethylmetacrylate (PMM); the remaining 10% is mainly crystals of barium sulfate or zirconium oxide that make the resulting product radio-opaque. Bone cements are produced manually, moments before use, by mixing a white polymer powder of PMM (PolyMethylMetacrylate) with a monomer fluid. This mixture polymerizes to a hard and brittle substance within ten minutes. The mixing process could have an effect on the quality of the cement. Studies have demonstrated that the use of low viscosity cements in the surgery of total hip replacements produced more failures than the use of conventional doughy products.

The following subsections deal with issues associated with the tissue response to biomaterials and implants.

G. Bioinert

Any material placed in the human body that has minimal interaction with its surrounding tissue is called bioinert. For example, stainless steel, titanium, alumina, partially stabilised zirconia, and ultra-high molecular weight polyethylene (UHMWPE) (to some degree) are bioinert. A fibrous capsule that might form around bioinert implants may compromise the biofunctionality of bioinert materials. In addition, the tissue remodeling around the implant (bone ingrowth or architectural changes such as resorption or depositon of bone) may alter the functionality of the implant.

H. Toxicity

Toxicology in biomaterials has evolved into a sophisticated science that deals with the substances that migrate out of biomaterials. For example, for polymers, many low molecular weights have shown some level of physiologic activity and cell toxicity. It is reasonable to say that a biomaterial should not give off anything from its mass unless it is specifically designed to do so, for example, cancer drug delivery systems used to destroy specific cells.

21.20 TABLES

TABLE 21.7 Significant Physical Properties of Different Biomaterials

Material	Density (gcm^{-3})	Ultimate Tensile Stress (MPa)	Compressive Strength (MPa)	E (GPa)	Fracture Toughness K_{Ic} ($MPa\ m^{1/2}$)	Hardness (Knoop)	α (ppm/°C)	Fracture Surface Energy (J/m^2)	Poisson's Ratio	k ($Wm^{-1}K^{-1}$)
HA	3.1	40–300	300–900	80–120	0.6–1.0	400–4,500	11	2.3–20	0.28	
TCP	3.14	40–120	450–650	90–120	1.20		14–15	6.3–8.1		
Bioglass	1.8–2.9	20–350	800–1200	40–140	–2	4,000–5,000	0–14	14–50	0.21–0.24	1.5–3.6
A-W glass ceramic	3.07	215	1080	118	2					
SiO_2 glass	2.2	70–120		–70	0.7–0.8	7,000–7,500	0.6	3.5–4.6	0.17	1.5
Al_2O_3	3.85–3.99	270–500	3000–5000	380–410	3–6	15,000–20,000	6–9	7.6–30	0.27	30
PSZ	5.6–5.89	500–650	1850	195–210	5–8	–17,000	9.8	160–350	0.27	4.11
Si_3N_4	3.18	600–850	500–2500	300–320	3.5–8.0	–22,000	3.2	20–100	0.27	10–25
SiC	3.10–3.21	250–600	–650	350–450	3–6	–27,000	4.3–5.5	22–40	0.24	100–150
Graphite	1.5–2.25	5.6–25	35–80	3.5–12	1.9–3.5		1–3	–500	0.3	120–180
LTI-ULTI	1.5–2.2	200–700	330–360	25–40			1–10		0.3	2.5–420
Carbon fiber	1.5–1.8	400–5000	330–360	200–700						
Glassy carbon	1.4–1.6	150–250	–690	25–40		8,200	2.2–3.2			
PE	0.9–1.0	0.5–65		0.1–1.0	0.4–4.0	170	11–22	500–8,000	0.4	0.3–0.5
PMMA	1.2	60–70	–80	3.5	1.5	160	5–8.1	300–400		0.20
Ti	4.52	345	250–600	117	60	1,800–2,600	8.7–10.1	–15,000	0.31	
Ti/Al/V alloys	4.4	780–1050	450–1850	110	40–70	3,200–3,600	8.7–9.8	–10,000	0.34	
Ti/Al/Nb/Ta alloys	4.4–4.8	840–1010		105	50–80			–17,000	0.32	
Vitalliumstellite alloys (Co-Cr-Mn)	7.8–8.2	400–1030	480–600	230	120–160	3,000	15.6–17.0	–5,000	0.30	
Low C steel Fe-Cr-Ni alloys	7.8–8.2	540–4000	1000–4000	200	55–95	1,200–9,000	16.0–19.0	–50,000	0.20–0.33	46

TABLE 21.8 Compositions of Surface-Active Glasses and Glass Ceramics in Weight Percent, Data from Ref. 66

Material	SiO_2	P_2O_5	CaO	Na_2O	Others
45S5	45.0	6.0	24.5	24.5	—
45S5-F	45.0	6.0	12.25	24.5	Caf_2, 12.15
45S5-B5	40.0	6.0	24.5	24.5	B_2O_3, 5
45S5-OP	45.0	0	24.5	30.5	—
45S5-M	48.3	6.4	—	26.4	MgO, 18.5
Alkali-rich glass-ceramic	42–47	5–7.5	20–25	20–25	—
Ceravital	40–50	10–15	30–35	5–10	K_2, 0.5–3.0 MgO, 2.5–5.0
Composition C	42.4	11.2	22.0	24.4	—

TABLE 21.9 Properties of Polyethylene, Data from ASTM F 648, and Ref. 67

	High Density	Low Density	UHMWPE
Molecular weight (g/mol)	34×10^3	5×10^5	2×10^6
Density (g/ml)	0.90–0.92	0.92–0.96	0.93–0.944
Tensile strength (MPa)	7.6	23–40	3
Elongation (%)	150	400–500	200–250
Modulus of elasticity (MPa)	96–260	410–1240	—

TABLE 21.10 Chemical and Mechanical Properties of Alumina for Implants, Data from Ref. 18

Property	Value
Chemical	
Al_2O_3	99.7%
MgO	0.23% (max)
SiO_2 and alkali metal oxides	0.1%
Physical and mechanical	
Density	3.4 g cm^3
Grain size	4 μm
Fracture strength (compression)	580.0 ksi or 4000.0 N mm^{-2}
Flexure strength (MOR)	58.0 ksi or 400.0 N mm^{-2}
Young's modulus	55.000 ksi or 380,000 N mm^2
Coefficient of friction: water lubricant	0.05
Wear rate (50 lb/in^2; 400 mm sec^{-1})	3×10^{-9} mg/mm

TABLE 21.11 Mechanical Properties of Cortical Bone, 316L Stainless Steel, Cobalt–Chromium Alloy, Titanium, and Titanium-6-Aluminum-4-Vanadium

Material	Young's Modulus (GPa)	Compressive Strength (GPa)	Tensile Strength (GPa)
Bone			
(wet at low strain rate)	15.2	0.15	0.090
(wet at high strain rate)	40.7	0.27–0.40	—
316L stainless steel	193	—	0.54
Co–Cr (cast)	214	—	0.48
Ti			
0% porosity	110	—	0.40
40% porosity	24	—	0.076
Ti–6AI–4V			
0% porosity	124	—	0.94
40% porosity	27	—	0.14

TABLE 21.12 Properties of Bones at Different Age

Property	Age (Years) 10–20	20–30	30–40	40–50	50–60	60–70	70–80
Ultimate strength (MPa)							
Tension	114	123	120	112	93	86	86
Compression	—	167	167	161	155	145	—
Bending	151	173	173	162	154	139	139
Torsion	—	57	57	52	52	49	49
Ultimate strain (%)							
Tension	1.5	1.4	1.4	1.3	1.3	1.3	1.3
Compression	—	1.9	1.8	1.8	1.8	1.8	—
Torsion	—	2.8	2.8	2.5	2.5	2.7	2.7

TABLE 21.13 Typical Mechanical Properties of Polymer-Carbon Composites (Three-Point Bending)

Polymer	Ultimate Strength (MPa)	Modulus (GPa)
PMMA	772	55
Polysulfone	938	76
Epoxy		
Stycast	535	30
Hysol	207	24
Polyurethane	289	18

TABLE 21.14 Mechanical Properties of Some Degradable Polymers[a]

Polymer	Glass Transition (°C)	Melting Temperature (°C)	Tensile Strength (MPa)	Tensile Modulus (MPa)	Flexural Modulus (MPa)	Elongation Yield (%)	Elongation Break (%)
Poly(glycolic acid) (MW: 50,000)	35	210	n/a	n/a	n/a	n/a	n/a
Poly(lactic acids)							
L-PLA (MW: 50,000)	54	170	28	1200	1400	3.7	6.0
L-PLA (MW: 100,000)	58	159	50	2700	3000	2.6	3.3
L-PLA (MW: 300,000)	59	178	48	3000	3250	1.8	2.0
D,L-PLA (MW: 20,000)	50	—	n/a	n/a	n/a	n/a	n/a
D,L-PLA (MW: 107,000)	51	—	29	1900	1950	4.0	6.0
D,L-PLA (MW: 550,000)	53	—	35	2400	2350	3.5	5.0
Poly(β-hydroxybutyrate) (MW: 422,000)	1	171	36	2500	2850	2.2	2.5
Poly(ε-caprolactone) (MW: 44,000)	−62	57	16	400	500	7.0	80

(Continued)

TABLE 21.14 Mechanical Properties of Some Degradable Polymers[a] (*Continued*)

Polymer	Glass Transition (°C)	Melting Temperature (°C)	Tensile Strength (MPa)	Tensile Modulus (MPa)	Flexural Modulus (MPa)	Elongation	
						Yield (%)	Break (%)
Polyanhydrides[b]							
Poly(SA-HDA anhydride) (MW: 142,000)	n/a	49	4	45	n/a	14	85
Poly(ortho esters)[c]							
DETOSU: t-CDM: 1, 6-HD (MW: 99,700)	55	—	20	820	950	4.1	220
Polyiminocarbonates[d]							
Poly(BPA iminocarbonate) (MW: 105,000)	69	—	50	2150	2400	3.5	4.0
Poly(DTH iminocarbonate) (MW: 103,000)	55	—	40	1630	n/a	3.5	7.0

[a]Based on data published by Engelberg and Kohn (Ref. 68). n/a = not available, (—) = not applicable.
[b]A 1 : 1 copolymer of sebacic acid (SA) and hexadecanedioic acid (HDA) was selected as a specific example.
[c]A 100 : 35 : 65 copolymer of 3, 9-bis(ethylidene 2, 4, 8, 10-tetraoxaspiro[5, 5] undecane) (DETOSU), *trans*-cyclohexane dimethanol (t-CDM), and 1, 6-hexanediol (1, 6-HD) was selected as a specific example.
[d]BPA: Bisphenol A; DTH: desaminotyrosyl-tyrosine hexyl ester.

TABLE 21.15 Representative Mechanical Properties of Commercial Sutures

Suture Type	St. Pull (MPa)	Kt. Pull (MPa)	Elongation to Break (%)	Subjective Flexibility
Natural materials				
Catgut	370	160	25	Stiff
Silk	470	265	21	Very supple
Synthetic absorbable				
Poly(glycolic acid)	840	480	22	Supple
Poly(glycolide-co-lactide)	740	350	22	Supple
Poly(*p*-dioxanone)	505	290	34	Mod. stiff
Poly(glycolide-co-trimethylene carbonate)	575	380	32	Mod. stiff
Synthetic nonabsorbable				
Poly(butylene terephthalate)	520	340	20	Supple
Poly(ethylene terephthalate)	735	345	25	Supple
Poly[p(tetramethylene ether) terephthalate-co-tetramethylene terephthalate]	515	330	34	Supple
Polypropylene	435	300	43	Stiff
Nylon 66	585	315	41	Stiff
Steel	660	565	45	Rigid

21.21 REFERENCES

1. Andriacchi, T., et al., "A Study of Lower-limb Mechanics During Stair-climbing," *J. Bone Joint Surg.*, 62A: 749–757, 1980.
2. Aydin T., *Human Body Dynamics*, Springer Verlag, 2000.
3. Berger, S. A., W. Goldsmith, and E. R. Lewis, *Introduction to Bioengineering*, Oxford Press, 2000.
4. Bergmann, G., A. Griachen, and A. Rohlamann, "Hip Joint Forces During Load Carrying," *Clin. Orthop.*, 335: 190, 1997.
5. Bergmann, G., et al., "Hip Contact Forces and Gait Patterns from Routine Activities," *J. Biomech.*, 34: 859, 2001.
6. Bogduk, N., and S. Mercer, "Biomechanics of the Cervical Spine, I. Normal Kinematics," *Clin. Biomech.*, 15: 633, 2000.
7. Brault, J. R., G. P. Siegmund, and J. B. Wheeler, "Cervical Muscle Response During Whiplash: Evidence of a Lengthening Muscle Contraction," *Clinical Biomech.*, 15: 426, 2000.
8. Ratner, B., A. Hoffman, F. Schoen, and J. Lemons, *Biomaterials Science: An Introduction to Materials in Medicine*, Academic Press, 1996.
9. Cappozzo, A., "Compressive Load in the Lumbar Vertebral Column During Normal Level Walking," *J. Orthop. Res.*, 1: 292, 1984.
10. Chaffin, D., A. Gurnnar, and M. Barnard, *Occupational Biomechanics*, John Wiley, 1999.
11. Cholewicki, J., K. Juluru, and S. M. McGill, "Intra-Abdominal Pressure Mechanism for Stabilizing the Lumbar Spine," *J. Biomech.*, 32: 13, 1999.
12. Cowin, S., A. Sadegh, and G. Luo, "An Evolutionary Wolff's Law for Trabecular Architecture," *J. Biomech. Eng.*, 116: 129–136, 1992.
13. Davis, K. G., and W. S. Marras, "The Effect of Motion on Trunk Biomechanics," *Clin. Biomech*, 15: 703, 2000.
14. DePalma, A. F., *Biomechanics of the Shoulder, in "Surgery of the Shoulder,"* 3rd ed., J. B. Lippincott, Philadelphia, pp. 65–85, 1983.
15. Drillis, R., and R. Contini, *Body Segment Parameters*, Rep. 1163-03, Office of Vocational Rehabilitation, Dept. of Health, Education, and Welfare, New York, 1966.
16. El-Bohy, A., and A. King, "Intervertebral Disc and Facet Contact Pressure in Axial Torsion," in Adv. In Bioeng., edited Lantz and King, ASME, pp. 26–27, 1986.
17. Fung, Y. C., *Biomechanics of Mechanical Properties of Living Tissues*, Springer Verlag, 1993.
18. Gibson, L., and M. Ashby, *Cellular Solids: Structures and Properties*, 2nd ed, Pergamon Press, Oxford, UK, 1984.
19. Gidden, D. P., A. P. Yagonathan, and F. J. Schoen, "Prosthetic Cardiac Valves," Special supplement to Vol. 2 No. 3, *Cardiovascular Pathology*, pp. 167S–177S, 1993.
20. Vijay, G., and J. Weinstein, *Biomechanics of the Spine: Clinical and Surgical Perspective*, CRC Press, 2000.
21. Dreyfuss, H., *The Measure of Man: Human Factors in Design*, Whitney Library of Design, New York, 1967. This is a loose-leaf book of 30 anthropometric and biomedical charts, a classic.
22. Hall, S. J., *Basic Biomechanics*, McGraw Hill, 4th ed., 2003.
23. Hall, S. J., "Effect of Attempted Lifting Speed on Forces and Torque Exerted on the Lumbar Spine," *Med Sci Sports Exerc.*, 17:440, 1985.
24. Harms-Ringdahl, K., *On Assessment of Shoulder Exercise and Load-elicited Pain in the Cervical Spine*, Thesis, Karolinska Institute, University of Stockholm, 1986.
25. Hodges, P. W., A. G. Cresswell, K. Daggfeldt, and A. Thorstensson, "In Vivo Measurement of the Effect of Intra-Abdominal Pressure on the Human Spine," *J. Biomech.* 34:347, 2001.
26. *Hole's Human Anatomy and Physiology*, McGraw-Hill, 1996.
27. *Human Factors Engineer Design Criterion for Military Systems, Equipment, and Facilities*, MIL–STD 1472 D.
28. *Human Factors Engineer Design for Army Material*, MIL-HDBK 759A, almost 700 pages of information on all aspects of human factors.
29. Imwold, D., and J. Parker, *Anatomica: The Complete Home Medical Reference*, Global Book Publishing, 2000.
30. Kopperdahl, D., and T. Keaveny, *Yield Strain Behavior of Trabecular Bone, J. of Biomechanics*, 31 (7): 601–608, 1998.
31. Myer, K., ed, *Standard Handbook of Biomedical Engineering & Design*, McGraw Hill, 2002.
32. Lemons, J. E., "Dental Implant Retrieval Analyses," *J. Dent. Ed.*, 52: 748–757, 1988.
33. Linde, F., I. Hvid, and B. Pongsoipetch, "Energy Absorptive Properties of Human Trabecular Bone Specimens During Axial Compression," *J. Orthop. Res.*, 7(3): 432–439, 1989.
34. Luo, G., S. Cowin, A. Sadegh, and Y. Arramon, "Implementation of Strain Rate as a Bone Remodeling Stimulus," *J. Biomech Eng.*, 117: 329–338, 1995.
35. Mankin, H., V. C. Mow, J. Buckwalter, J. Iannotti, and A. Ratcliffe, "Form and Function of Articular Cartilage, in S. R. Simon (ed.) Orthopedic Basic Science," *Amer. Acad. of Orthop. Surgeons*, Chicago, 1994.

36. McCalden, R., J. McGeough, and C. Court-Brown, "Age-related Changes in the Compressive Strength of Cancellous Bone: The Relative Importance of Changes in Density and Trabecular Architecture," *J. Bone Joint Surg.*, 79 A(3): 421–427, 1997.
37. McElhaney, J., and E. Byars, "Dynamic Response of Biological Materials," *Proc. Amer. Soc. Mech. Eng.*, ASME 65-WA/HUF-9:8, Chicago, 1965.
38. Morgan, E., and T. Keaveny, "Dependence of Yield Strain of Human Trabecular Bone on Anatomic Site," *J. Biomechanics*, 34(5): 569–577, 2001.
39. Morrison, J., "The Mechanics of the Knee Joint in Relation to Normal Walking," *J. Biomech*, 3: 164–171, 1970.
40. Mosekilde, L., "Normal Vertebral Body Size and Compressive Strength: Relations to Age and to Vertebral and Iliac Trabecular Bone Compressive Strength," *Bone*, 7: 207–212, 1986.
41. Mosekilde, L., and C. Danielsen, "Biomechanical Competence of Vertebral Trabecular Bone in Relation to Ash Density and Age in Normal Individuals," *Bone*, 8(2): 79–85, 1987.
42. Mow, V. C., W. M. Lai, and I. Redler, "Some Surface Characteristics of Articular Cartilage, I: A Scanning Electron Microscopy Study and a Theoretical Model for the Dynamic Interaction of Synovial Fluid and Articular Cartilage," *J. Biomech.*, 7: 449–456, 1974.
43. Muray, M., "Gait as a Total Pattern of Movement," *Am. J. Phys. Med.*, 46: 290, 1967.
44. Nachemson, A., G. B. Anderson, and A. B. Schultz, "Valsalva Manoeuvre Biomechanics: Effect on Lumbar Trunk Loads of Elevated Intra-Abdominal Pressure," *Spine*, 11: 476, 1986.
45. Nachemson, A., "Towards a Better Understanding of Back Pain: A Review of the Mechanics of the Lumbar Disc," *Rheumatol. Rehabil.*, 14: 129, 1975.
46. Nahum, A., and J. Melvin, *Accidental Injury, Biomechanics and Prevention*, Springer Verlag, 1993.
47. Nordin, M., and V. Frankel, "*Basic Biomechanics of the Musculoskeletal System*," 2nd ed. Williams and Williams, 1989.
48. Noyes, F. R., "Fuctional Properties of Knee Ligaments and Alteration Induced by Immobilization," *Clin, Orthop.*, 123: 210, 1977.
49. Pandy, M., F. Zajac, E. Sim, and W. Levine, An Optimal Control Model for Maximum Height Human Jumping," *J. Biomech.*, 23: 1185–1198, 1990.
50. Paul, J., *Forces at the Human Hip Joint*, PhD Thesis, University of Chicago, 1967.
51. Reilly, D., and A. Burstein, "The Elastic and Ultimate Properties of Compact Bone Tissue," *J. Biomech.*, 8: 393–405, 1975.
52. Sadegh, A., G. Luo, and S. Cowin, "Bone Ingrowth: An Application of the Boundary Element Method to Bone Remodeling at the Implant Interface," *J. Biomech.*, 26(2): 167–182, 1993.
53. Schatz, R. A., "A View of Vascular Stents," *J. Am. Coll. Cardiol.*, 13: 445–447, 1989.
54. Ullman, D., *The Mechanical Design Process*, McGraw Hill, 1997.
55. White, A., and M. Panjabi, *Clinical Biomechanics of the Spine*, J. B. Lippincott, Philadelphia, 1978.
56. White, A., et al., "Biomechanical Analysis of Clinical Stability in the Cervical Spine," *Clin. Orthop.*, 109: 85, 1975.
57. Williams, P., and R. Warwick, *Gray's Anatomy*, 36th ed., Churchill Livingston, Edinburgh, pp. 506–515, 1980.
58. Winter, D., A. Patla, F. Prince, M. Ishac, and Gielo-Perczak, "Stiffness Control of Balance in Quiet Standing," *J. Neurophysiology*, 80: 1211–1221, 1998.
59. Winter D., *Biomechanics and Motor Control of Human Movement*, John Wiley, 2005.
60. Woo, S. L., G. Livesay, T. Runco, and E. Young, *Structure and Function of Tendons and Ligaments*, in V. C. Mow and W. C. Hayes (eds.), "Basic Orthop. Biomech.," Lippincott-Raven, Philadelphia, 1997.
61. Woo, S. L., K. An, S. Arnoczky, J. Wayne, D. Fithian, and B. Myers, *Anatomy, Biology and Biomechanics of Tendon, Ligament and Meniscus*, in S. R. Simon (ed.), Ortho. Basic Science, Amer. Acad. of Orthop. Surgeons, Chicago, 1994.
62. Zajac, F., and M. Gordon, "Determining Muscle's Force and Action in Multi-Articular Movement," *Exerc. Sport Sci. Rev.*, 17: 187–230, 1989.
63. Shirazi-Adl, A., S. C. Shrivastca, and A. M. Ahmed, "Stress Analysis of Lumbar Disc-body Unit in Compression," *Spine*, 9: 120, 1984.
64. Lindhal, O., "Mechanical Properties of Dried Defatted Spongy Bone," *Acta Orthop. Scand.*, 47: 11, 1976.
65. Yamada, H., *Strength of Biological Materials*, in Evans, F. G. (ed.), Williams and Wilkins, Baltimore, 1970.
66. Hench, L. L., and E. C. Ethridge, *Biomaterials: An Interfacial Approach*, Academic Press, New York, 1982.
67. Park, J. B., *Biomaterials Science and Engineering*, Plenum Publications, New York, 1984.
68. Engelberg, I., and J. Kohn, "Physico-Mechanical Properties of Degradable Polymers Used in Medical Applications: A Comparative Study," *Biomaterials*, 12: 292–304, 1991.
69. Kin, A. I., P. Prasad, and C. L. Ewing, "Mechanism of Spinal Injury Due to Caudocephalad Acceleration," *Orthop. Clin. North Am.*, 6: 19, 1975.
70. Miller, M. D., et al., "Significant New Observations on Cervical Spine Trauma," *Am. J. Roentgenol.*, 130: 659, 1978.

21.22 GLOSSARY

Abduction Motion away from the body midline

Adduction Motion toward the body midline

Anterior Toward the front of the body

Distal Away from the trunk

Inferior Farther away from the head

In vitro In an artificial environment outside the living organism

In vivo Within a living organism

Lateral Away from the midline of the body

Medial Toward the midline of the body

Posterior Toward the back of the body

Proximal Closer to the trunk

Range of motion The range of translation and rotation of a joint for each of its six degrees of freedom

Superior Closer to the head

APPENDIX A

Properties of a Plane Area

Because of their importance in connection with the analysis of bending and torsion, certain relations for the *second-area moments,* commonly referred to as *moments of inertia,* are indicated in the following paragraphs. The equations given are in reference to Fig. A.1, and the notation is as follows:

A: area of the section

X, Y: rectangular axes in the plane of the section at arbitrary point O

x, y: rectangular axes in the plane of the section parallel to X, Y, respectively with origin at the centroid, C, of the section

z: polar axis through C

x', y': rectangular axes in the plane of the section, with origin at C, inclined at a counterclockwise angle θ from x, y

1, 2: principal axes at C inclined at a counterclockwise angle θ_p from x, y

r: the distance from C to the dA element, $r = \sqrt{x^2 + y^2}$

By definition,

Moments of inertia: $I_x = \int_A y^2\, dA, \qquad I_y = \int_A x^2\, dA$

Polar moment of inertia:

$$I_z = J = \int_A r^2\, dA = I_x + I_y = I_{x'} + I_{y'} = I_1 + I_2$$

Product of inertia: $I_{xy} = \int_A xy\, dA$

Radii of gyration: $k_x = \sqrt{I_x / A}, \qquad k_y = \sqrt{I_y / A}$

Parallel axis theorem:

$$I_X = I_x + Ay_c^2, \qquad I_Y = I_y + Ax_c^2, \qquad I_{XY} = I_{xy} + Ax_c y_c$$

Transformation equations:

$$I_{x'} = I_x \cos^2\theta + I_y \sin^2\theta - I_{xy} \sin 2\theta$$

$$I_{y'} = I_x \sin^2\theta + I_y \cos^2\theta + I_{xy} \sin 2\theta$$

$$I_{x'y'} = \frac{1}{2}(I_x - I_y)\sin 2\theta + I_{xy} \cos 2\theta$$

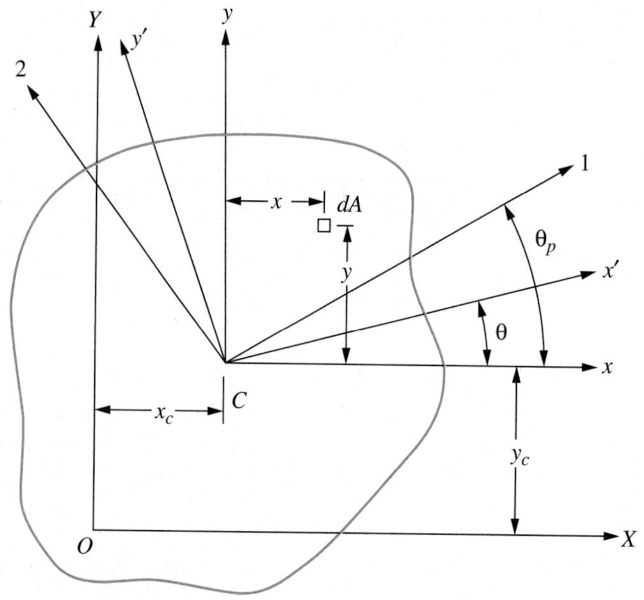

Figure A.1 Plane area.

Principal moments of inertia and directions:

$$I_{1,2} = \frac{1}{2}\left[(I_x + I_y) \pm \sqrt{(I_y - I_x)^2 + 4I_{xy}^2}\right], \qquad I_{12} = 0,$$

$$\theta_p = \frac{1}{2}\tan^{-1}\left(\frac{2I_{xy}}{I_y - I_x}\right)$$

Upon the determination of the two principal moments of inertia, I_1 and I_2, two angles, 90° apart, can be solved for from the equation for θ_p. It may be obvious which angle corresponds to which principal moment of inertia. If not, one of the angles must be substituted into the equations $I_{x'}$ and $I_{y'}$ which will again yield the principal moments of inertia but also their orientation.

Note, if either one of the xy axes is an axis of symmetry, $I_{xy} = 0$, with I_x and I_y being the principal moments of inertia of the section.

If $I_1 = I_2$ for a set of principal axes through a point, it follows that the moments of inertia for all $x'y'$ axes through that point, in the same plane, are equal and $I_{x'y'} = 0$ regardless of θ. Thus the moment of inertia of a square, an equilateral triangle, or any section having two or more axes of identical symmetry is the same for any central axis.

The moment of inertia and radius of gyration of a section with respect to a centroidal axis are less than for any other axis parallel thereto.

The moment of inertia of a composite section (one regarded as made up of rectangles, triangles, circular segments, etc.) about an axis is equal to the sum of the moments of inertia of each component part about that axis. Voids are taken into account by subtracting the moment of inertia of the void area.

Expressions for the area, distances of centroids from edges, moments of inertia, and radii of gyration are given in Table A.1 for a number of representative sections. The moments of products of inertia for composite areas can be found by addition; the centroids of composite areas can be found by using the relation that the statical moment about any line of the entire area is equal to the sum of the statical moments of its component parts.

Although properties of structural sections—wide-flange beams, channels, angles, etc.—are given in structural handbooks, formulas are included in Table A.1 for similar sections. These are applicable to sections having a wider range of web and flange thicknesses than normally found in the rolled or extruded sections included in the handbooks.

Plastic or ultimate strength design is discussed in Secs. 8.15 and 8.16, and the use of this technique requires the value of the fully plastic bending moment—the product of the yield strength of a ductile material and the plastic section modulus Z. The last column in Table A.1 gives for many of the sections the value or an expression for Z and the location of the neutral axis under fully plastic pure bending. This neutral axis does not, in general, pass through the centroid, but instead divides the section into equal areas in tension and compression.

TABLE A.1 Properties of Sections

Notation: A = area (length)^2; y = distance to extreme fiber (length); I = moment of inertia (length^4); r = radius of gyration (length); Z = plastic section modulus (length^3); SF = shape factor. See Sec. 8.15 for applications of Z and SF.

Form of Section	Area and Distances from Centroid to Extremities	Moments and Products of Inertia and Radii of Gyration about Central Axes	Plastic Section Moduli, Shape Factors, and Locations of Plastic Neutral Axes
1. Square 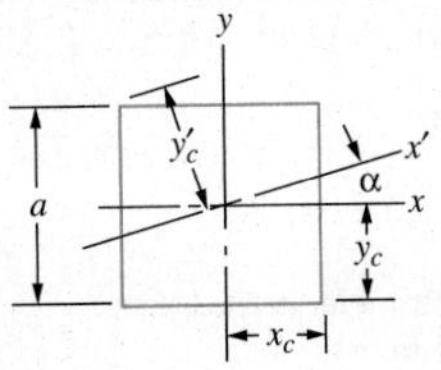	$A=a^2$ $y_c=x_c=\frac{a}{2}$ $y'_c=0.707a\cos\left(\frac{\pi}{4}-\alpha\right)$	$I_x=I_y=I'_x=\frac{1}{12}a^4$ $r_x=r_y=r'_x=0.2887a$	$Z_x=Z_y=0.25a^3$ $SF_x=SF_y=1.5$
2. Rectangle 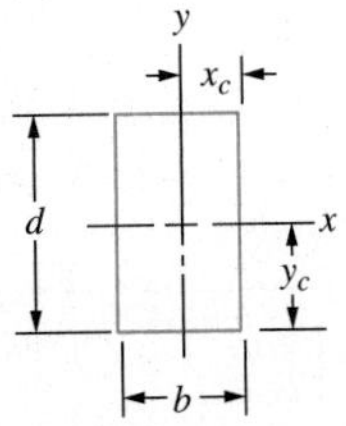	$A=bd$ $y_c=\frac{d}{2}$ $x_c=\frac{b}{2}$	$I_x=\frac{1}{12}bd^3$ $I_y=\frac{1}{12}db^3$ $I_x>I_y$ if $d>b$ $r_x=0.2887d$ $r_y=0.2887b$	$Z_x=0.25bd^2$ $Z_y=0.25db^2$ $SF_x=SF_y=1.5$
3. Hollow rectangle 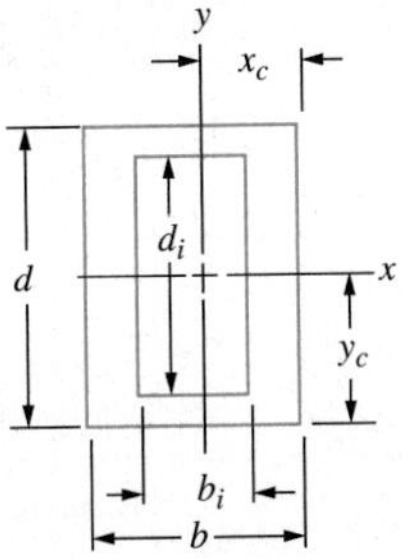	$A=bd-b_id_i$ $y_c=\frac{d}{2}$ $x_c=\frac{b}{2}$	$I_x=\frac{bd^3-b_id_i^3}{12}$ $I_y=\frac{db^3-d_ib_i^3}{12}$ $r_x=\left(\frac{I_x}{A}\right)^{1/2}$ $r_y=\left(\frac{I_y}{A}\right)^{1/2}$	$Z_x=\frac{bd^2-b_id_i^2}{4}$ $SF_x=\frac{Z_xd}{2I_x}$ $Z_y=\frac{db^2-d_ib_i^2}{4}$ $SF_y=\frac{Z_yb}{2I_y}$

(Continued)

TABLE A.1 Properties of Sections (*Continued*)

Form of Section	Area and Distances from Centroid to Extremities	Moments and Products of Inertia and Radii of Gyration about Central Axes	Plastic Section Moduli, Shape Factors, and Locations of Plastic Neutral Axes

4. Tee section
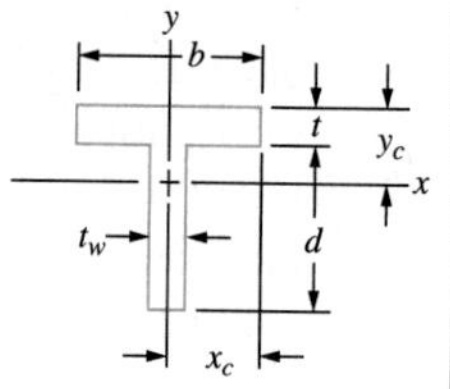
 	
	$A = tb + t_w d$
	$y_c = \frac{bt^2 + t_w d(2t+d)}{2(tb + t_w d)}$
	$x_c = \frac{b}{2}$
		$I_x = \frac{b}{3}(d+t)^3 - \frac{d^3}{3}(b - t_w) - A(d+t-y_c)^2$
		$I_y = \frac{tb^3}{12} + \frac{dt_w^3}{12}$
		$r_x = \left(\frac{I_x}{A}\right)^{1/2}$
		$r_y = \left(\frac{I_y}{A}\right)^{1/2}$
			If $t_w d \geq bt$, then
			$Z_x = \frac{d^2 t_w}{4} - \frac{b^2 t^2}{4t_w} + \frac{bt(d+t)}{2}$
			Neutral axis x is located a distance $(bt/t_w + d)/2$ from the bottom.
			If $t_w d \leq bt$, then
			$Z_x = \frac{t^2 b}{4} + \frac{t_w d(t + d - t_w d/2b)}{2}$
			Neutral axis x is located a distance $(t_w d/b + t)/2$ from the top.
			$SF_x = \frac{Z_x(d+t-y_c)}{I_1}$
			$Z_y = \frac{b^2 t + t_w^2 d}{4}$
			$SF_y = \frac{Z_y b}{2I_y}$

5. Channel section
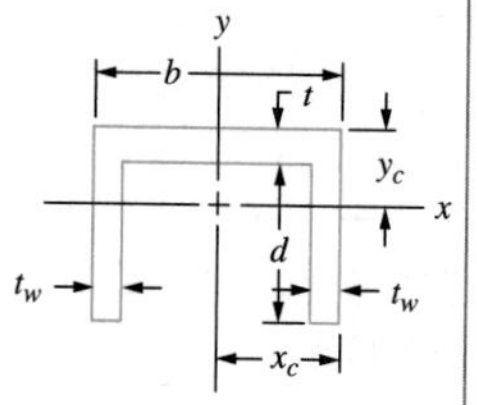
 	
	$A = tb + 2t_w d$
	$y_c = \frac{bt^2 + 2t_w d(2t+d)}{2(tb + 2t_w d)}$
	$x_c = \frac{b}{2}$
		$I_x = \frac{b}{3}(d+t)^3 - \frac{d^3}{3}(b - 2t_w) - A(d+t-y_c)^2$
		$I_y = \frac{(d+t)b^3}{12} - \frac{d(b-2t_w)^3}{12}$
		$r_x = \left(\frac{I_x}{A}\right)^{1/2}$
		$r_y = \left(\frac{I_y}{A}\right)^{1/2}$
			If $2t_w d \geq bt$, then
			$Z_x = \frac{d^2 t_w}{2} - \frac{b^2 t^2}{8t_w} + \frac{bt(d+t)}{2}$
			Neutral axis x is located a distance $(bt/2t_w + d)/2$ from the bottom.
			If $2t_w d \leq bt$, then
			$Z_x = \frac{t^2 b}{4} + t_w d\left(t + d - \frac{t_w d}{b}\right)$
			Neutral axis x is located a distance $t_w d/b + t/2$ from the top.
			$SF_x = \frac{Z_x(d+t-y_c)}{I_x}$
			$Z_y = \frac{b^2 t}{4} + t_w d(b - t_w)$
			$SF_y = \frac{Z_y b}{2I_y}$

6. Wide-flange beam with equal flanges 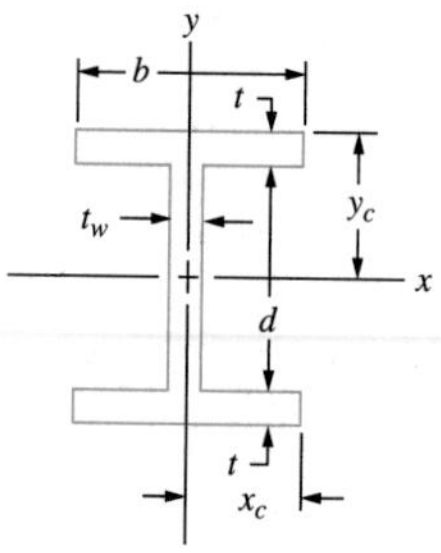 	$A = 2bt + t_w d$ $y_c = \frac{d}{2} + t$ $x_c = \frac{b}{2}$	$I_x = \frac{b(d+2t)^3}{12} - \frac{(b-t_w)d^3}{12}$ $I_y = \frac{b^3 t}{6} + \frac{t_w^3 d}{12}$ $r_x = \left(\frac{I_x}{A}\right)^{1/2}$ $r_y = \left(\frac{I_y}{A}\right)^{1/2}$	$Z_x = \frac{t_w d^2}{4} + bt(d+t)$ $SF_x = \frac{Z_x y_c}{I_x}$ $Z_y = \frac{b^2 t}{2} + \frac{t_w^2 d}{4}$ $SF_y = \frac{Z_y x_c}{I_y}$

(*Continued*)

TABLE A.1 Properties of Sections (*Continued*)

Form of Section	Area and Distances from Centroid to Extremities	Moments and Products of Inertia and Radii of Gyration about Central Axes	Plastic Section Moduli, Shape Factors, and Locations of Plastic Neutral Axes
7. Equal-legged angle 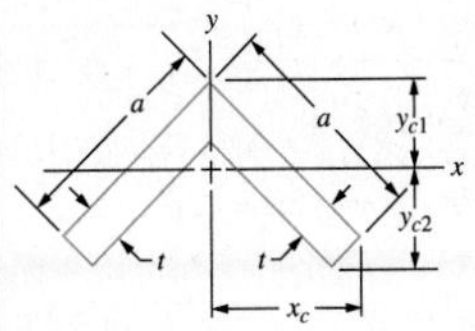	$A = t(2a-t)$ $y_{c1} = \dfrac{0.7071(a^2+at-t^2)}{2a-t}$ $y_{c2} = \dfrac{0.7071a^2}{2a-t}$ $x_c = 0.7071a$	$I_x = \dfrac{a^4-b^4}{12} - \dfrac{0.5ta^2b^2}{a+b}$ $I_y = \dfrac{a^4-b^4}{12}$ where $b = a-t$ $r_x = \left(\dfrac{I_x}{A}\right)^{1/2}$ $r_y = \left(\dfrac{I_y}{A}\right)^{1/2}$	Let y_p be the vertical distance from the top corner to the plastic neutral axis. If $t/a \geq 0.40$, then $y_p = a\left[\dfrac{t}{a} - \dfrac{(t/a)^2}{2}\right]^{1/2}$ $Z_x = A(y_{c1} - 0.6667y_p)$ If $t/a \leq 0.4$, then $y_p = 0.3536(a+1.5t)$ $Z_x = Ay_{c1} - 2.8284y_p^2 t + 1.8856t^3$
8. Unequal-legged angle 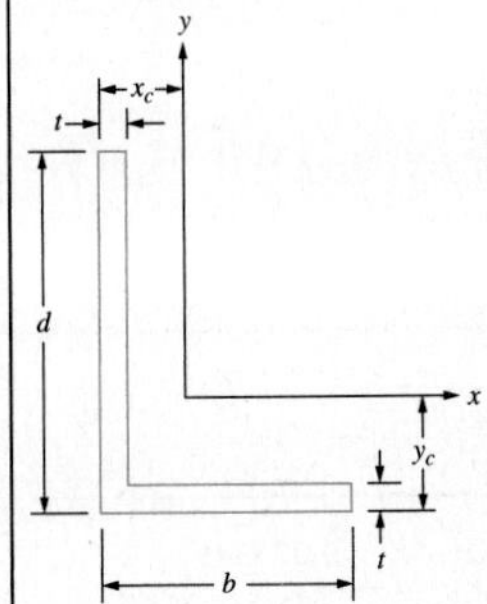	$A = t(b+d-t)$ $x_c = \dfrac{b^2+dt-t^2}{2(b+d-t)}$ $y_c = \dfrac{d^2+bt-t^2}{2(b+d-t)}$	$I_x = \dfrac{1}{3}\left[bd^3-(b-t)(d-t)^3\right] - A(d-y_c)^2$ $I_y = \dfrac{1}{3}\left[db^3-(d-t)(b-t)^3\right] - A(b-x_c)^2$ $I_{xy} = \dfrac{1}{4}\left[b^2d^2-(b-t)^2(d-t)^2\right] - A(b-x_c)(d-y_c)$ $r_x = \left(\dfrac{I_x}{A}\right)^{1/2}$ $r_y = \left(\dfrac{I_y}{A}\right)^{1/2}$	
9. Equilateral triangle 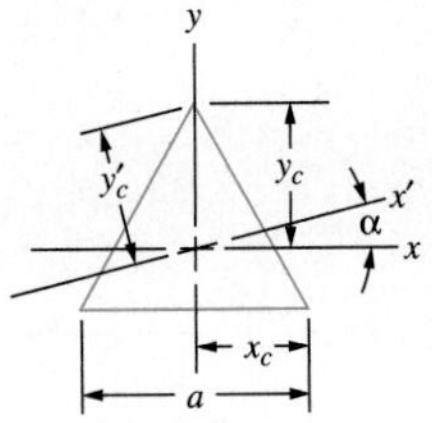	$A = 0.4330a^2$ $y_c = 0.5774a$ $x_c = 0.5000a$ $y_c' = 0.5774a\cos\alpha$	$I_x = I_y = I_{x'} = 0.01804a^4$ $r_x = r_y = r_{x'} = 0.2041a$	$Z_x = 0.0732a^3$, $Z_y = 0.0722a^3$ $SF_x = 2.343$, $SF_y = 2.000$ Neutral axis x is $0.2537a$ from the base.
10. Isosceles triangle 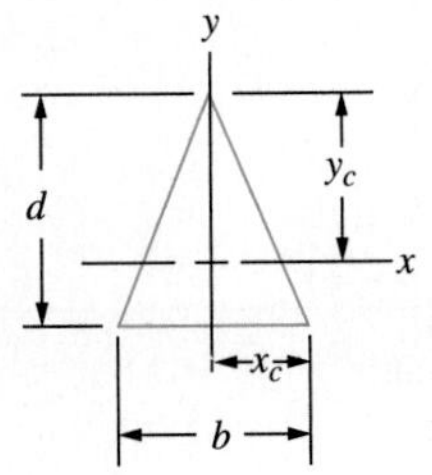	$A = \dfrac{bd}{2}$ $y_c = \dfrac{2}{3}d$ $x_c = \dfrac{b}{2}$	$I_x = \dfrac{1}{36}bd^3$ $I_y = \dfrac{1}{48}db^3$ $I_x > I_y$ if $d > 0.866b$ $r_x = 0.2357d$ $r_y = 0.2041b$	$Z_x = 0.097bd^2$, $Z_y = 0.0833db^2$ $SF_x = 2.343$, $SF_y = 2.000$ Neutral axis x is $0.299d$ from the base.

(Continued)

TABLE A.1 Properties of Sections (*Continued*)

Form of Section	Area and Distances from Centroid to Extremities	Moments and Products of Inertia and Radii of Gyration about Central Axes	Plastic Section Moduli, Shape Factors, and Locations of Plastic Neutral Axes
11. Triangle 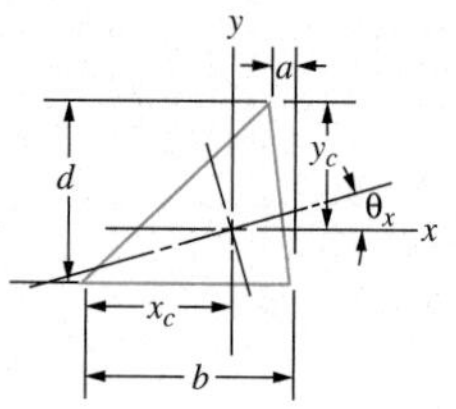	$A=\frac{bd}{2}$ $y_c=\frac{2}{3}d$ $x_c=\frac{2}{3}b-\frac{1}{3}a$	$I_x=\frac{1}{36}bd^3$ $I_y=\frac{1}{36}bd(b^2-ab+a^2)$ $I_{xy}=\frac{1}{72}bd^2(b-2a)$ $\theta_x=\frac{1}{2}\tan^{-1}\frac{d(b-2a)}{b^2-ab+a^2-d^2}$ $r_x=0.2357d$ $r_y=0.2357\sqrt{b^2-ab+a^2}$	
12. Parallelogram 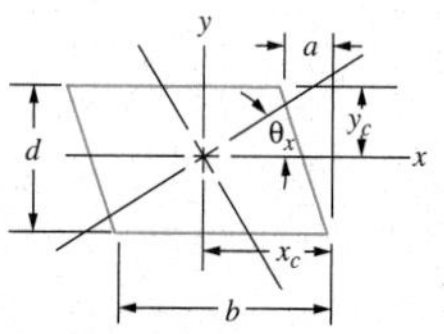	$A=bd$ $y_c=\frac{d}{2}$ $x_c=\frac{1}{2}(b+a)$	$I_x=\frac{1}{12}bd^3$ $I_y=\frac{1}{12}bd(b^2+a^2)$ $I_{xy}=-\frac{1}{12}abd^2$ $\theta_x=\frac{1}{2}\tan^{-1}\frac{-2ab}{b^2+a^2-d^2}$ $r_x=0.2887d$ $r_y=0.2887\sqrt{b^2+a^2}$	
13. Diamond 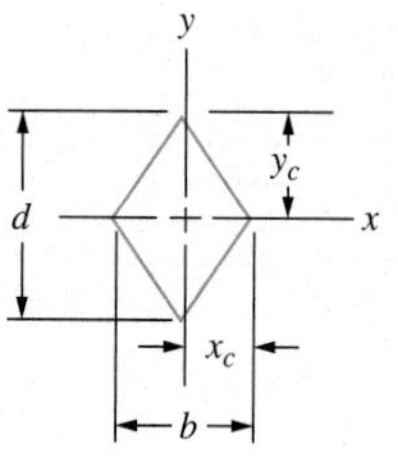	$A=\frac{bd}{2}$ $y_c=\frac{d}{2}$ $x_c=\frac{b}{2}$	$I_x=\frac{1}{48}bd^3$ $I_y=\frac{1}{48}db^3$ $r_x=0.2041d$ $r_y=0.2041b$	$Z_x=0.0833bd^2$, $Z_y=0.0833db^2$ $SF_x=SF_y=2.000$
14. Trapezoid 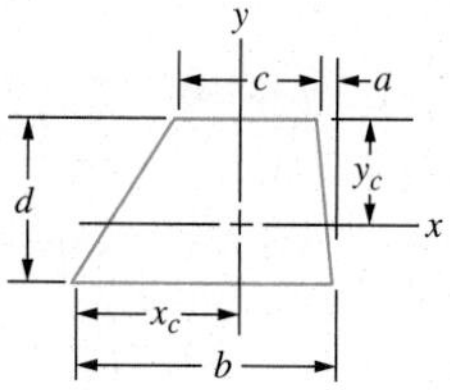	$A=\frac{d}{2}(b+c)$ $y_c=\frac{d}{3}\frac{2b+c}{b+c}$ $x_c=\frac{2b^2+2bc-ab-2ac-c^2}{3(b+c)}$	$I_x=\frac{d^3}{36}\frac{b^2+4bc+c^2}{b+c}$ $I_y=\frac{d}{36(b+c)}[b^4+c^4+2bc(b^2+c^2)$ $-a(b^3+3b^2c-3bc^2-c^3)+a^2(b^2+4bc+c^2)]$ $I_{xy}=\frac{d^2}{72(b+c)}[c(3b^2-3bc-c^2)+b^3-a(2b^2+8bc+2c^2)]$	
15. Solid circle 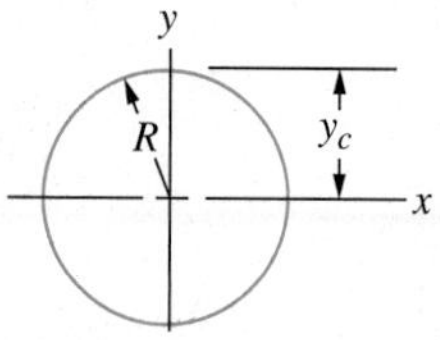	$A=\pi R^2$ $y_c=R$	$I_x=I_y=\frac{\pi}{4}R^4$ $r_x=r_y=\frac{R}{2}$	$Z_x=Z_y=1.333R^3$ $SF_x=1.698$

(*Continued*)

TABLE A.1 Properties of Sections (*Continued*)

Form of Section	Area and Distances from Centroid to Extremities	Moments and Products of Inertia and Radii of Gyration about Central Axes	Plastic Section Moduli, Shape Factors, and Locations of Plastic Neutral Axes
16. Hollow circle 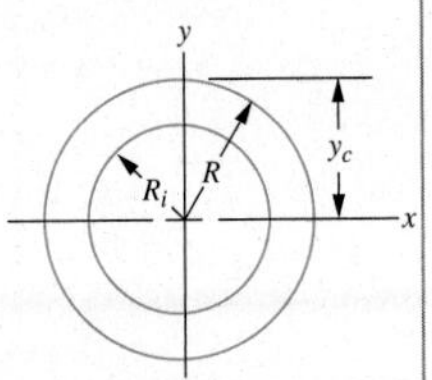	$A=\pi(R^2-R_i^2)$ $y_c=R$	$I_x=I_y=\frac{\pi}{4}(R^4-R_i^4)$ $r_x=r_y=\frac{1}{2}\sqrt{R^2+R_i^2}$	$Z_x=Z_y=1.333(R^3-R_i^3)$ $SF_x=1.698\frac{R^4-R_i^3R}{R^4-R_i^4}$
17. Very thin annulus 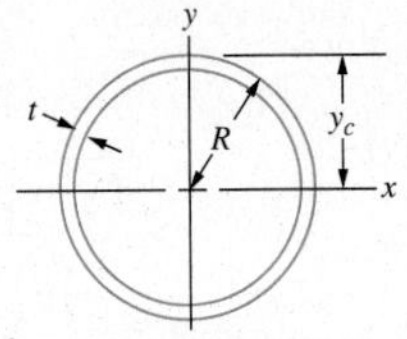	$A=2\pi Rt$ $y_c=R$	$I_x=I_y=\pi R^3t$ $r_x=r_y=0.707R$	$Z_x=Z_y=4R^2t$ $SF_x=SF_y=\frac{4}{\pi}$

18. Sector of solid circle 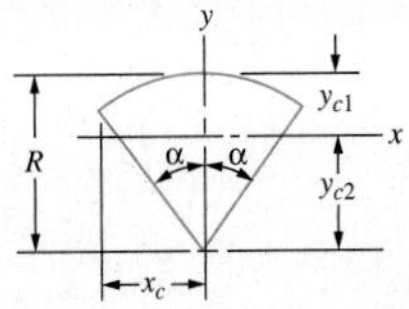	$A=\alpha R^2$
	$y_{c1}=R\left(1-\frac{2\sin\alpha}{3\alpha}\right)$
	$y_{c2}=\frac{2R\sin\alpha}{3\alpha}$
	$x_c=R\sin\alpha$
		$I_x=\frac{R^4}{4}\left(\alpha+\sin\alpha\cos\alpha-\frac{16\sin^2\alpha}{9\alpha}\right)$
		$I_y=\frac{R^4}{4}(\alpha-\sin\alpha\cos\alpha)$
		(*Note:* If α is small, $\alpha-\sin\alpha\cos\alpha=\frac{2}{3}\alpha^3-\frac{2}{15}\alpha^5$)
		$r_x=\frac{R}{2}\sqrt{1+\frac{\sin\alpha\cos\alpha}{\alpha}-\frac{16\sin^2\alpha}{9\alpha^2}}$
		$r_y=\frac{R}{2}\sqrt{1-\frac{\sin\alpha\cos\alpha}{\alpha}}$
			If $\alpha\le 54.3°$, then
			$Z_x=0.6667R^3\left[\sin\alpha-\left(\frac{\alpha^3}{2\tan\alpha}\right)^{1/2}\right]$
			Neutral axis x is located a distance $R(0.5\alpha/\tan\alpha)^{1/2}$ from the vertex.
			If $\alpha\ge 54.3°$, then
			$Z_x=0.6667R^3(2\sin^3\alpha_1-\sin\alpha)$ where the expression $2\alpha_1-\sin 2\alpha_1=\alpha$ is solved for the value of α_1.
			Neutral axis x is located a distance $R\cos\alpha_1$ from the vertex.
			If $\alpha\le 73.09°$, then $SF_x=\frac{Z_x y_{c2}}{I_x}$
			If $73.09°\le\alpha\le 90°$, then $SF_x=\frac{Z_x y_{c1}}{I_x}$
			$Z_y=0.6667R^3(1-\cos\alpha)$
			If $\alpha\le 90°$, then
			$SF_y=2.6667\sin\alpha\frac{1-\cos\alpha}{\alpha-\sin\alpha\cos\alpha}$
			If $\alpha\ge 90°$, then
			$SF_y=2.6667\frac{1-\cos\alpha}{\alpha-\sin\alpha\cos\alpha}$

19. Segment of solid circle (*Note:* If $\alpha\le\pi/4$, use expressions from case 17) 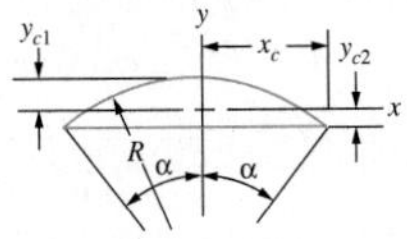	$A=R^2(\alpha-\sin\alpha\cos\alpha)$ $y_{c1}=R\left[1-\frac{2\sin^3\alpha}{3(\alpha-\sin\alpha\cos\alpha)}\right]$ $y_{c2}=R\left[\frac{2\sin^3\alpha}{3(\alpha-\sin\alpha\cos\alpha)}-\cos\alpha\right]$ $x_c=R\sin\alpha$	$I_x=\frac{R^4}{4}\left[\alpha-\sin\alpha\cos\alpha+2\sin^3\alpha\cos\alpha-\frac{16\sin^6\alpha}{9(\alpha-\sin\alpha\cos\alpha)}\right]$ $I_y=\frac{R^4}{12}(3\alpha-3\sin\alpha\cos\alpha-2\sin^3\alpha\cos\alpha)$ $r_x=\frac{R}{2}\sqrt{1+\frac{2\sin^3\alpha\cos\alpha}{\alpha-\sin\alpha\cos\alpha}-\frac{16\sin^6\alpha}{9(\alpha-\sin\alpha\cos\alpha)^2}}$ $r_y=\frac{R}{2}\sqrt{1-\frac{2\sin^3\alpha\cos\alpha}{3(\alpha-\sin\alpha\cos\alpha)}}$	

(*Continued*)

TABLE A.1 Properties of Sections (*Continued*)

Form of Section	Area and Distances from Centroid to Extremities	Moments and Products of Inertia and Radii of Gyration about Central Axes	Plastic Section Moduli, Shape Factors, and Locations of Plastic Neutral Axes
20. Segment of solid circle (*Note:* Do not use if $\alpha > \pi/4$) 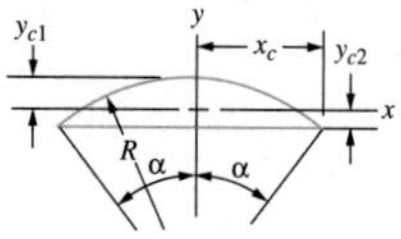	$A=\frac{2}{3}R^2\alpha^3(1-0.2\alpha^2+0.019\alpha^4)$ $y_{c1}=0.3R\alpha^2(1-0.0976\alpha^2+0.0028\alpha^4)$ $y_{c2}=0.2R\alpha^2(1-0.0619\alpha^2+0.0027\alpha^4)$ $x_c=R\alpha(1-0.1667\alpha^2+0.0083\alpha^4)$	$I_x=0.01143R^4\alpha^7(1-0.3491\alpha^2+0.0450\alpha^4)$ $I_y=0.1333R^4\alpha^5(1-0.4762\alpha^2+0.1111\alpha^4)$ $r_x=0.1309R\alpha^2(1-0.0745\alpha^2)$ $r_y=0.4472R\alpha(1-0.1381\alpha^2+0.0184\alpha^4)$	
21. Sector of hollow circle 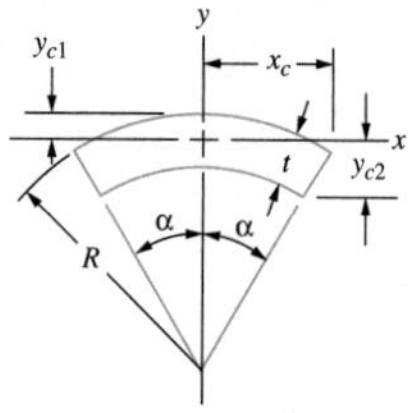 (*Note:* If t/R is small, α can exceed π to form an overlapped annulus)	$A=\alpha t(2R-t)$ $y_{c1}=R\left[1-\frac{2\sin\alpha}{3\alpha}\left(1-\frac{t}{R}+\frac{1}{2-t/R}\right)\right]$ $y_{c2}=R\left[\frac{2\sin\alpha}{3\alpha(2-t/R)}+\left(1-\frac{t}{R}\right)\frac{2\sin\alpha-3\alpha\cos\alpha}{3\alpha}\right]$ $x_c=R\sin\alpha$	$I_x=R^3t\left[\left(1-\frac{3t}{2R}+\frac{t^2}{R^2}-\frac{t^3}{4R^3}\right)\times\left(\alpha+\sin\alpha\cos\alpha-\frac{2\sin^2\alpha}{\alpha}\right)+\frac{t^2\sin^2\alpha}{3R^2\alpha(2-t/R)}\left(1-\frac{t}{R}+\frac{t^2}{6R^2}\right)\right]$ $I_y=R^3t\left(1-\frac{3t}{2R}+\frac{t^2}{R^2}-\frac{t^3}{4R^3}\right)(\alpha-\sin\alpha\cos\alpha)$ $r_x=\sqrt{\frac{I_x}{A}},\quad r_y=\sqrt{\frac{I_y}{A}}$	
	Note: If α is small $\frac{\sin\alpha}{\alpha}=1-\frac{\alpha^2}{6}+\frac{\alpha^4}{120},\ \alpha-\sin\alpha\cos\alpha=\frac{2}{3}\alpha^3\left(1-\frac{\alpha^2}{5}+\frac{2\alpha^4}{105}\right),\ \frac{\sin^2\alpha}{\alpha}=\alpha\left(1-\frac{\alpha^2}{3}+\frac{2\alpha^4}{45}\right)$ $\cos=1-\frac{\alpha^2}{2}+\frac{\alpha^4}{24},\ \alpha+\sin\alpha\cos\alpha-\frac{2\sin^2\alpha}{\alpha}=\frac{2\alpha^5}{45}\left(1-\frac{\alpha^2}{7}+\frac{\alpha^4}{105}\right)$		
22. Solid semicircle 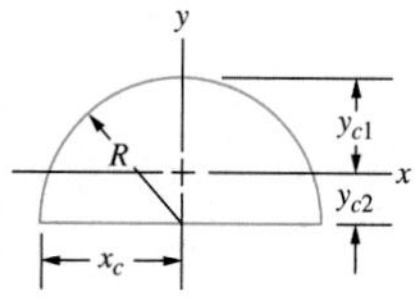	$A=\frac{\pi}{2}R^2$ $y_{c1}=0.5756R$ $y_{c2}=0.4244R$ $x_c=R$	$I_x=0.1098R^4$ $I_y=\frac{\pi}{8}R^4$ $r_x=0.2643R$ $r_y=\frac{R}{2}$	$Z_x=0.3540R^3, Z_y=0.6667R^3$ $SF_x=1.856, SF_y=1.698$ Plastic neutral axis x is located a distance $0.4040R$ from the base.
23. Hollow semicircle 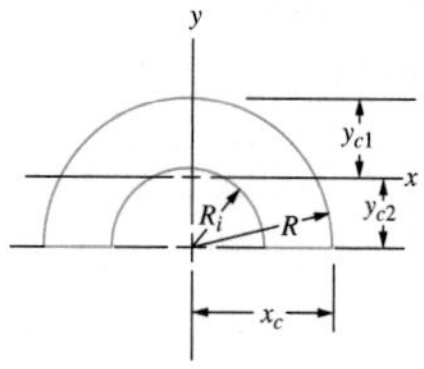 *Note:* $b=\frac{R+R_i}{2}$ $t=R+R_i$	$A=\frac{\pi}{2}(R^2-R_i^2)$ $y_{c2}=\frac{4}{3\pi}\frac{R^3-R_i^3}{R^2-R_i^2}$ or $y_{c2}=\frac{2b}{\pi}\left[1+\frac{(t/b)^2}{12}\right]$ $y_{c1}=R-y_{c2}$ $x_c=R$	$I_x=\frac{\pi}{8}(R^4-R_i^4)-\frac{8}{9\pi}\frac{(R^3-R_i^3)^2}{R^2-R_i^2}$ or $I_x=0.2976tb^3+0.1805bt^3-\frac{0.00884t^5}{b}$ $I_y=\frac{\pi}{8}(R^4-R_i^4)$ or $I_y=1.5708b^3t+0.3927bt^3$	Let y_p be the vertical distance from the bottom to the plastic neutral axis. $y_p=(0.7071-0.2716C-0.4299C^2+0.3983C^3)R$ $Z_x=(0.8284-0.9140C+0.7245C^2-0.2850C^3)R^2t$ where $C=t/R$ $Z_y=0.6667(R^3-R_i^3)$

(*Continued*)

TABLE A.1 Properties of Sections (*Continued*)

Form of Section	Area and Distances from Centroid to Extremities	Moments and Products of Inertia and Radii of Gyration about Central Axes	Plastic Section Moduli, Shape Factors, and Locations of Plastic Neutral Axes
24. Solid ellipse 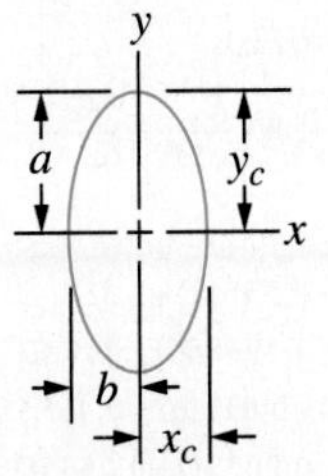	$A=\pi ab$ $y_c=a$ $x_c=b$	$I_x=\frac{\pi}{4}ba^3$ $I_y=\frac{\pi}{4}ab^3$ $r_x=\frac{a}{2}$ $r_y=\frac{b}{2}$	$Z_x=1.333a^2b,\ Z_y=1.3333b^2a$ $SF_x=SF_y=1.698$
25. Hollow ellipse 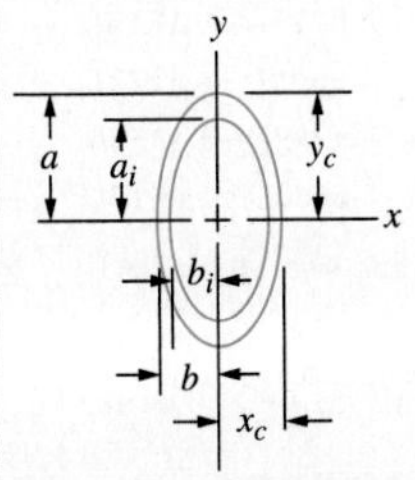	$A=\pi(ab-a_ib_i)$ $y_c=a$ $x_c=b$	$I_x=\frac{\pi}{4}(ba^3-b_ia_i^3)$ $I_y=\frac{\pi}{4}(ab^3-a_ib_i^3)$ $r_x=\frac{1}{2}\sqrt{\frac{ba^3-b_ia_i^3}{ab-a_ib_i}}$ $r_y=\frac{1}{2}\sqrt{\frac{ab^3-a_ib_i^3}{ab-a_ib_i}}$	$Z_x=1.333(a^2b-a_i^2b_i)$ $Z_y=1.333(b^2a-b_i^2a_i)$ $SF_x=1.698\frac{a^3b-a_i^2b_ia}{a^3b-a_i^3b_i}$ $SF_y=1.698\frac{b^3a-b_i^2a_ib}{b^3a-b_i^3a_i}$
Note: For this case the inner and outer perimeters are both ellipses and the wall thickness is not constant. For a cross section with a constant wall thickness see case 26.			
26. Hollow ellipse with constant wall thickness t. The midthickness perimeter is an ellipse (shown dashed). $0.2<a/b<5$ 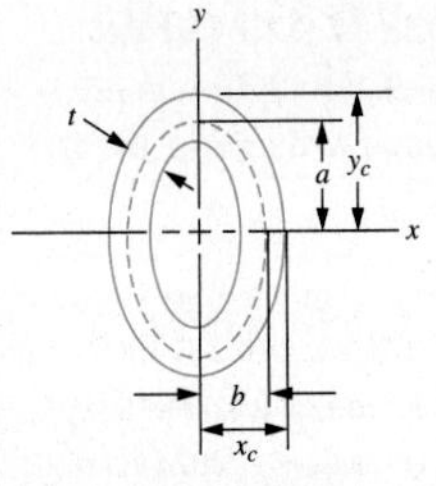 See the note on maximum wall thickness in case 27.	$A=\pi t(a+b)\left[1+K_1\left(\frac{a-b}{a+b}\right)^2\right]$ where $K_1=0.2464+0.002222\left(\frac{a}{b}+\frac{b}{a}\right)$ $y_c=a+\frac{t}{2}$ $x_c=b+\frac{t}{2}$	$I_x=\frac{\pi}{4}ta^2(a+3b)\left[1+K_2\left(\frac{a-b}{a+b}\right)^2\right]$ $+\frac{\pi}{16}t^3(3a+b)\left[1+K_3\left(\frac{a-b}{a+b}\right)^2\right]$ where $K_2=0.1349+0.1279\frac{a}{b}-0.01284\left(\frac{a}{b}\right)^2$ $K_3=0.1349+0.1279\frac{b}{a}-0.01284\left(\frac{b}{a}\right)^2$ For I_y interchange a and b in the expressions for I_x, K_2, and K_3	$Z_x=1.3333ta(a+2b)\left[1+K_4\left(\frac{a-b}{a+b}\right)^2\right]+\frac{t^3}{3}$ where $K_4=0.1835+0.895\frac{a}{b}-0.00978\left(\frac{a}{b}\right)^2$ For Z_y interchange a and b in the expression for Z_x and K_4.

(*Continued*)

TABLE A.1 Properties of Sections (*Continued*)

Form of Section	Area and Distances from Centroid to Extremities	Moments and Products of Inertia and Radii of Gyration about Central Axes	Plastic Section Moduli, Shape Factors, and Locations of Plastic Neutral Axes
27. Hollow semiellipse with constant wall thickness *t*. The midthickness perimeter is an ellipse (shown dashed). $0.2<a/b<5$ 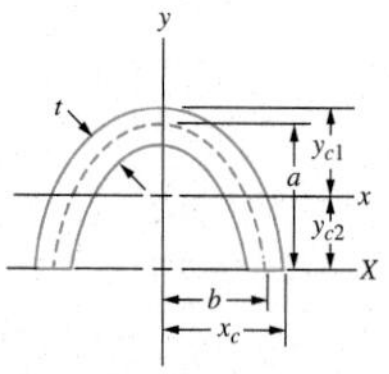 *Note:* There is a limit on the maximum wall thickness allowed in this case. Cusps will form in the perimeter at the ends of the major axis if this maximum is exceeded. If $\frac{a}{b}\le 1$, then $t_{max}=\frac{2a^2}{b}$ If $\frac{a}{b}\ge 1$, then $t_{max}=\frac{2b^2}{a}$	$A=\frac{\pi}{2}t(a+b)\left[1+K_1\left(\frac{a-b}{a+b}\right)^2\right]$ where $K_1=0.2464+0.002222\left(\frac{a}{b}+\frac{b}{a}\right)$ $y_{c2}=\frac{2a}{\pi}K_2+\frac{t^2}{6\pi a}K_3$ where $K_2=1-0.3314C+0.0136C^2+0.1097C^3$ $K_3=1+0.9929C-0.2287C^2-0.2193C^3$ Using $C=\frac{a-b}{a+b}$ $y_{c1}=a+\frac{t}{2}-y_{c2}$ $x_c=b+\frac{t}{2}$	$I_x=\frac{\pi}{8}ta^2(a+3b)\left[1+K_4\left(\frac{a-b}{a+b}\right)^2\right]$ $+\frac{\pi}{32}t^3(3a+b)\left[1+K_5\left(\frac{a-b}{a+b}\right)^2\right]$ where $K_4=0.1349+0.1279\frac{a}{b}-0.01284\left(\frac{a}{b}\right)^2$ $K_5=0.1349+0.1279\frac{b}{a}-0.01284\left(\frac{b}{a}\right)^2$ $I_X=I_x-Ay_{c2}^2$ For I_y use one-half the value for I_y in case 23.	Let y_p be the vertical distance from the bottom to the plastic neutral axis. $y_p=\left[C_1+\frac{C_2}{a/b}+\frac{C_3}{(a/b)^2}+\frac{C_4}{(a/b)^3}\right]a$ where if $0.25<a/b\le 1$, then $C_1=0.5067-0.5588D+1.3820D^2$ $C_2=0.3731+0.1938D-1.4078D^2$ $C_3=-0.1400+0.0179D+0.4885D^2$ $C_4=0.0170-0.0079D-0.0565D^2$ or if $1\le a/b<4$, then $C_1=0.4829+0.0725D-0.1815D^2$ $C_2=0.1957-0.6608D+1.4222D^2$ $C_3=0.0203+1.8999D-3.4356D^2$ $C_4=0.0578-1.6666D+2.6012D^2$ where $D=t/t_{max}$ and where $0.2<D\le 1$ $Z_x=\left[C_5+\frac{C_6}{a/b}+\frac{C_7}{(a/b)^2}+\frac{C_8}{(a/b)^3}\right]4a^2t$ where if $0.25<a/b\le 1$, then $C_5=-0.0292+0.3749D^{1/2}+0.0578D$ $C_6=0.3674-0.8531D^{1/2}+0.3882D$ $C_7=-0.1218+0.3563D^{1/2}-0.1803D$ $C_8=0.0154-0.0448D^{1/2}+0.0233D$ or if $1\le a/b<4$, then $C_5=0.2241-0.3922D^{1/2}+0.2960D$ $C_6=-0.6637+2.7357D^{1/2}-2.0482D$ $C_7=1.5211-5.3864D^{1/2}+3.9286D$ $C_8=-0.8498+2.8763D^{1/2}+1.8874D$ For Z_y use one-half the value for Z_y in case 23.
28. Regular polygon with *n* sides 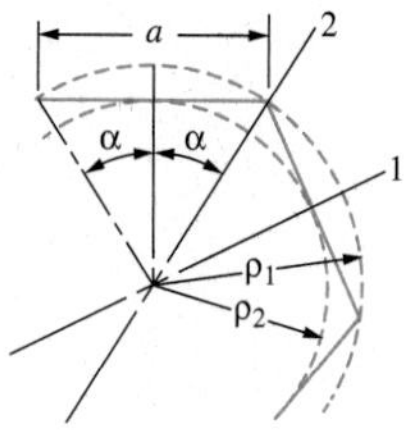 	$A=\frac{a^2n}{4\tan\alpha}$ $\rho_1=\frac{a}{2\sin\alpha}$ $\rho_2=\frac{\alpha}{2\tan\alpha}$ If *n* is odd $y_1=y_2=\rho_1\cos\left[\alpha\left(\frac{n+1}{2}\right)-\frac{\pi}{2}\right]$ If *n*/2 is odd $y_1=\rho_1,\ y_2=\rho_2$ If *n*/2 is even $y_1=\rho_2,\ y_2=\rho_1$	$I_1=I_2=\frac{1}{24}A(6\rho_1^2-a^2)$ $r_1=r_2=\sqrt{\frac{1}{24}(6\rho_1^2-a^2)}$	For *n* = 3, see case 9. For *n* = 4, see cases 1 and 13. For *n* = 5, $Z_1=Z_2=0.8825\rho_1^3$. For an axis perpendicular to axis 1, $Z=0.8838\rho_1^3$. The location of this axis is 0.7007*a* from that side which is perpendicular to axis 1. For $n\ge 6$, use the following expression for a neutral axis of any inclination: $Z=\rho_1^3\left[1.333-13.908\left(\frac{1}{n}\right)^2+12.528\left(\frac{1}{n}\right)^3\right]$

(*Continued*)

TABLE A.1 Properties of Sections (*Continued*)

Form of Section	Area and Distances from Centroid to Extremities	Moments and Products of Inertia and Radii of Gyration about Central Axes	Plastic Section Moduli, Shape Factors, and Locations of Plastic Neutral Axes
29. Hollow regular polygon with n sides 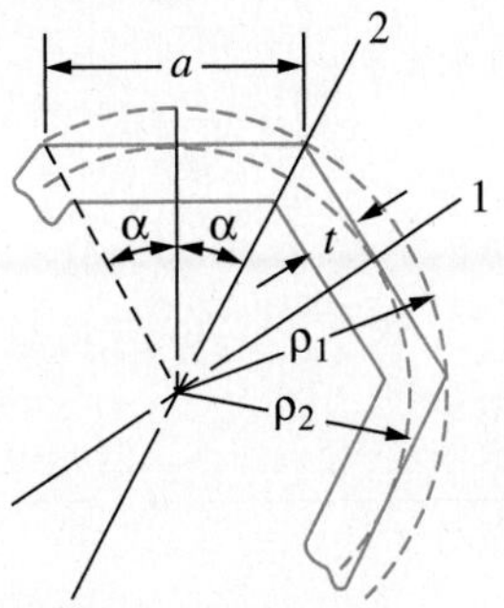	$A = nat\left(1 - \frac{t\tan\alpha}{a}\right)$ $\rho_1 = \frac{a}{2\sin\alpha}$ $\rho_2 = \frac{\alpha}{2\tan\alpha}$ If n is odd $y_1 = y_2 = \rho_1 \cos\left(\alpha\frac{n+1}{2} - \frac{\pi}{2}\right)$ If $n/2$ is odd $y_1 = \rho_1, \quad y_2 = \rho_2$ If $n/2$ is even $y_1 = \rho_2, \quad y_2 = \rho_1$	$I_1 = I_2 = \frac{na^3t}{8}\left(\frac{1}{3} + \frac{1}{\tan^2\alpha}\right)$ $\times\left[1 - 3\frac{t\tan\alpha}{a} + 4\left(\frac{t\tan\alpha}{a}\right)^2 - 2\left(\frac{t\tan\alpha}{a}\right)^3\right]$ $r_1 = r_2 = \frac{a}{\sqrt{8}}$ $\times\sqrt{\left(\frac{1}{3}\right) + \frac{1}{\tan^2\alpha}\left[1 - 2\frac{t\tan\alpha}{a} + 2\left(\frac{t\tan\alpha}{a}\right)^2\right]}$	
30. Hollow square 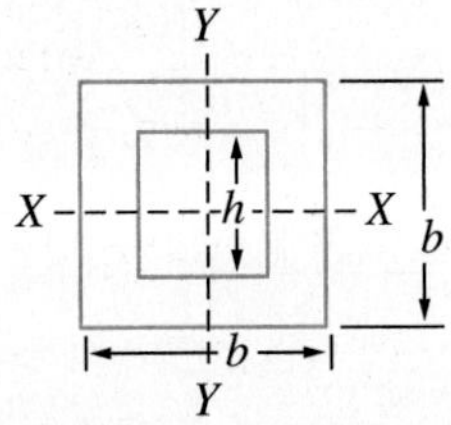	$A = b^2 - h^2$ $y_c = \frac{b}{2}$	$I_{xx} = I_{yy} = \frac{b^4 - h^4}{12}$ $r_x = 0.289\sqrt{b^2 + h^2}$	$Z_{xx} = Z_{yy} = \frac{b^4 - h^4}{6b}$
31. H-section 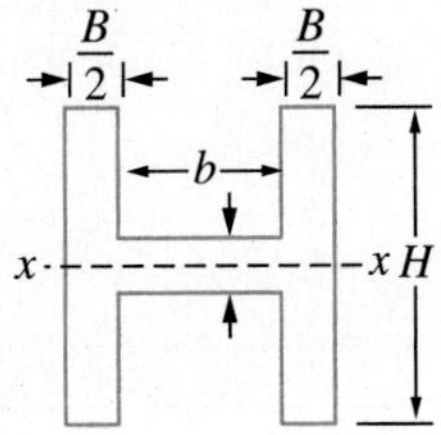	$A = BH + bh$ $y_c = \frac{H}{2}$	$I_{xx} = \frac{BH^3 + bh^3}{12}$ $r_x = 0.289\sqrt{\frac{BH^3 + bh^3}{BH + bh}}$	$Z_{xx} = \frac{BH^3 + bh^3}{6H}$
32. Cross section 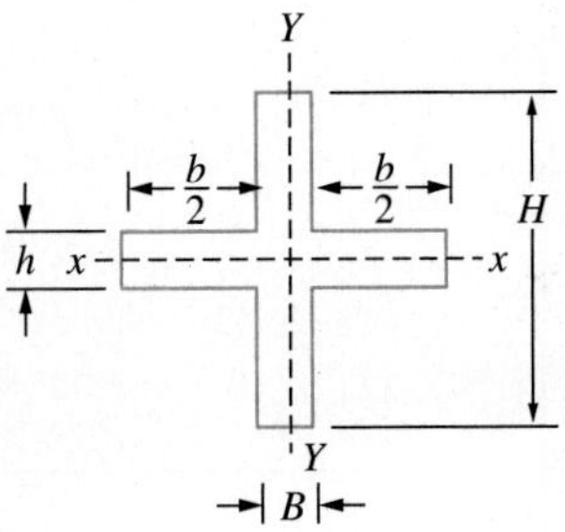	$A = BH + bh$ $y_c = \frac{H}{2}$	$I_{xx} = \frac{Bh^3 + bh^3}{12}$ $r_x = 0.289\sqrt{\frac{BH^3 + bh^3}{BH + bh}}$	$Z_{xx} = \frac{BH^3 + bh^3}{6H}$

(*Continued*)

TABLE A.1 Properties of Sections (*Continued*)

Form of Section	Area and Distances from Centroid to Extremities	Moments and Products of Inertia and Radii of Gyration about Central Axes	Plastic Section Moduli, Shape Factors, and Locations of Plastic Neutral Axes
33. Right angle triangle 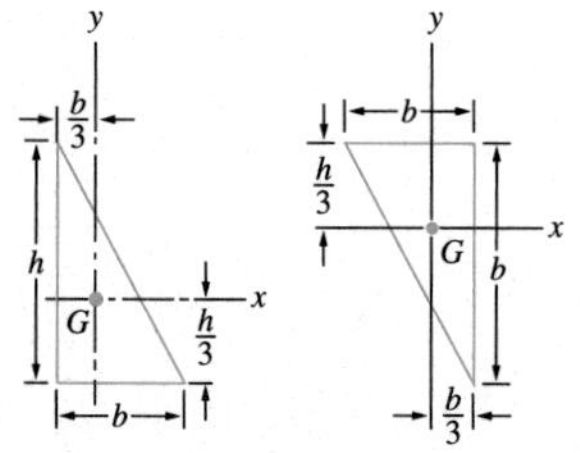	$A=\frac{bh}{2}$	$I_x=\frac{bh^3}{36}$ $I_y=\frac{b^3h}{36}$ $I_{xy}=\frac{-b^2h^2}{72}$	
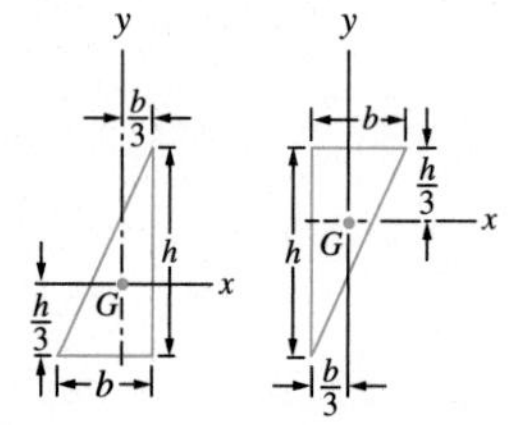	$A=\frac{bh}{2}$	$I_x=\frac{bh^3}{36}$ $I_y=\frac{b^3h}{36}$ $I_{xy}=\frac{b^2h^2}{72}$	
34. Z section 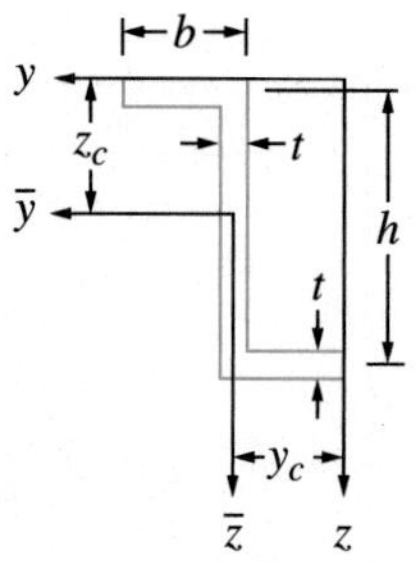	$H=h+t$ $B=b+\frac{1}{2}t$ $C=b-\frac{1}{2}t$ $A=t(h+2b)$ $y_c=b_x z_c=\frac{1}{2}(h+t)$	$I_y=\frac{1}{12}[BH^3-C(H-2t)^3]$ $I_z=\frac{1}{12}[H(B+C)^3-2hC^3-6B^2hC]$ $I_{yz}=-\frac{1}{2}htb^2$ $J_x=I_y+I_z$	
35. 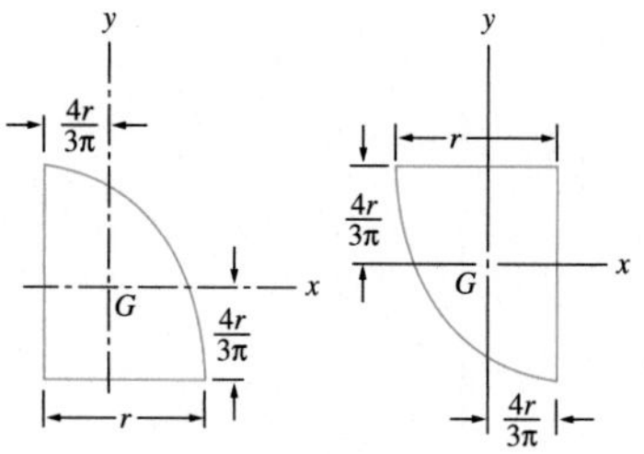	$A=\frac{\pi r^2}{4}$	$I_x=I_y=r^4\left(\frac{\pi}{16}-\frac{4}{9\pi}\right)$ $I_{xy}=r^4\left(\frac{1}{8}-\frac{4}{9\pi}\right)$	
36. 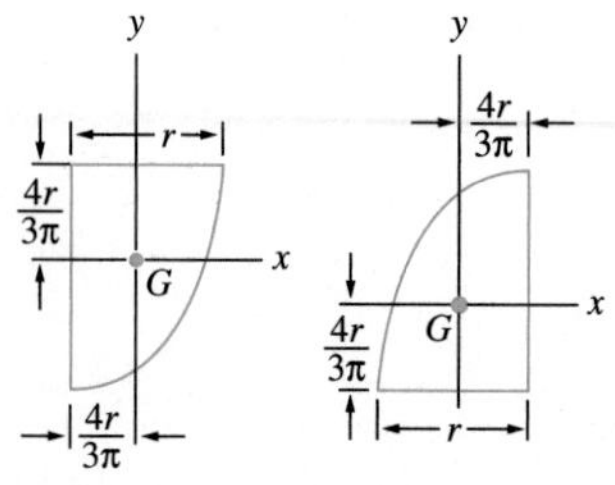	$A=\frac{\pi r^2}{4}$	$I_x=I_y=r^4\left(\frac{\pi}{16}-\frac{4}{9\pi}\right)$ $I_{xy}=r^4\left(\frac{4}{9\pi}-\frac{1}{8}\right)$	

(Continued)

TABLE A.1 Properties of Sections (*Continued*)

Form of Section	Area and Distances from Centroid to Extremities	Moments and Products of Inertia and Radii of Gyration about Central Axes	Plastic Section Moduli, Shape Factors, and Locations of Plastic Neutral Axes
37. Parabola 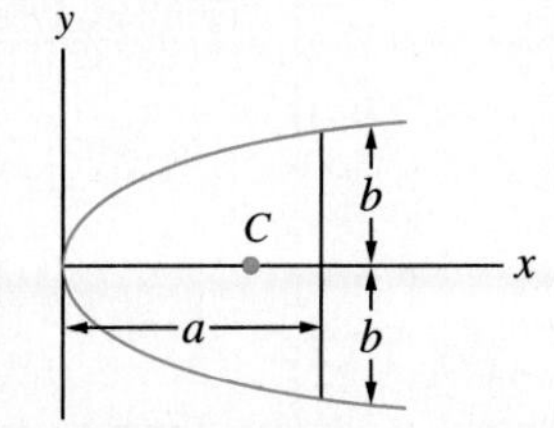	$A = 4ab/3$ $x_c = 3a/5$ $y_c = 0$	$I_{x_c} = I_x = 4ab^3/15$ $r_{x_c}^2 = r_x^2 = b^2/5$ $I_{y_c} = 16a^3b/175$ $r_{y_c}^2 = 12a^2/175$ $I_y = 4a^3b/7$ $r_y^2 = 3a^2/7$	
38. Half a parabola 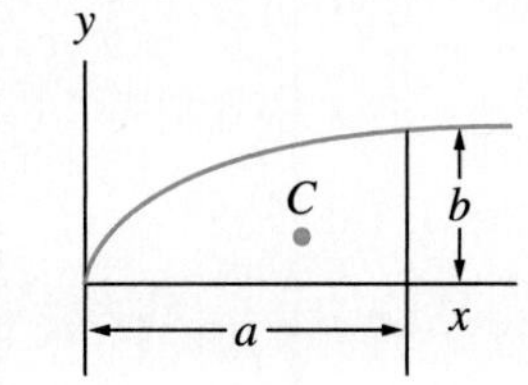	$A = 2ab/3$ $x_c = 3a/5$ $y_c = 3b/8$	$I_x = 2ab^3/15$ $r_x^2 = b^2/5$ $I_y = 2ba^3/7$ $r_y^2 = 3a^2/7$	
39. n^{th} degree parabola 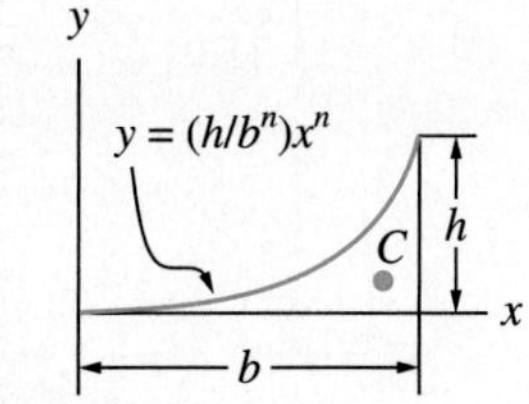	$A = bh/(n+1)$ $x_c = \frac{n+1}{n+2}b$ $y_c = \frac{h}{2}\frac{n+1}{2n+1}$	$I_x = \frac{bh^3}{3(3n+1)}$ $r_x^2 = \frac{h^2(n+1)}{3(3n+1)}$ $I_y = \frac{hb^3}{n+3}$ $r_y^2 = \frac{n+1}{n+3}b^2$	
40. n^{th} degree parabola 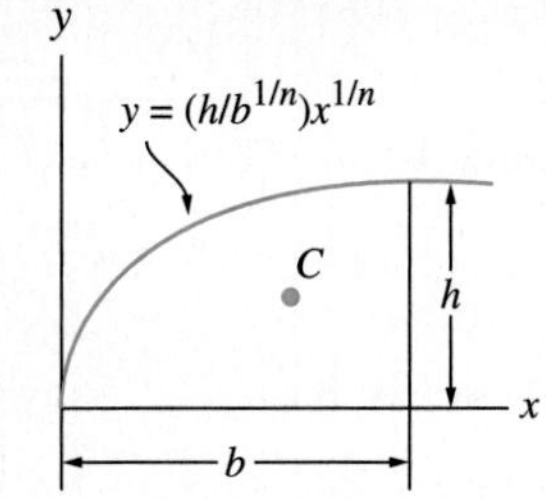	$A = \frac{n}{n+1}bh$ $x_c = \frac{n+1}{2n+1}b$ $y_c = \frac{n+1}{2(n+2)}h$	$I_x = \frac{n}{3(n+3)}bh^3$ $r_x^2 = \frac{n+1}{3(n+1)}h^2$ $I_y = \frac{n}{3n+1}b^3h$ $r_y^2 = \frac{n+1}{3n+1}b^2$	

TABLE A.2 Moment of Inertia of Sections

Section	$\bar{y}$	Moment of Inertia I_y	Section	$\bar{y}$	Moment of Inertia I_y
1. 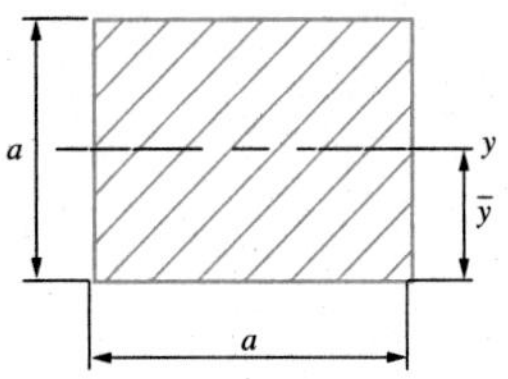	$\frac{a}{2}$	$\frac{a^4}{12}$	5. 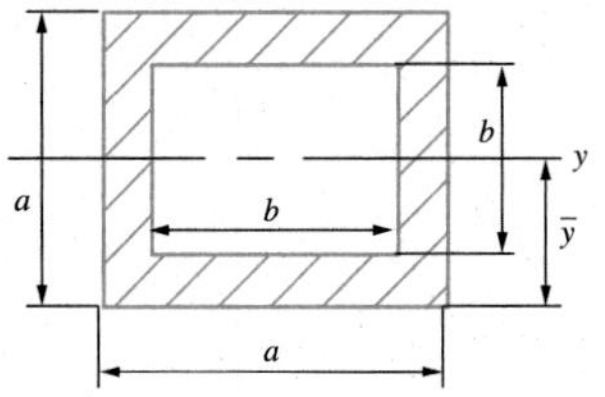	$\frac{a}{2}$	$\frac{a^4-b^4}{12}$
2. 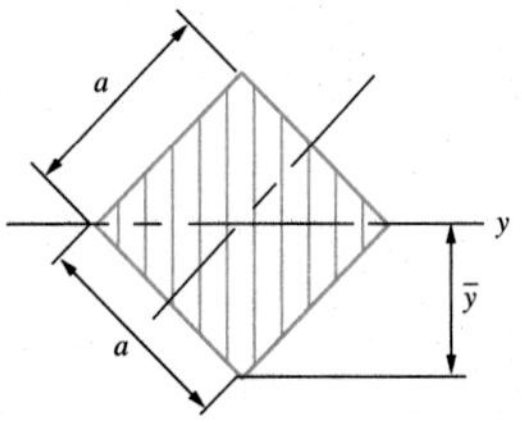	$\frac{a}{\sqrt{2}}$	$\frac{a^4}{12}$	6. 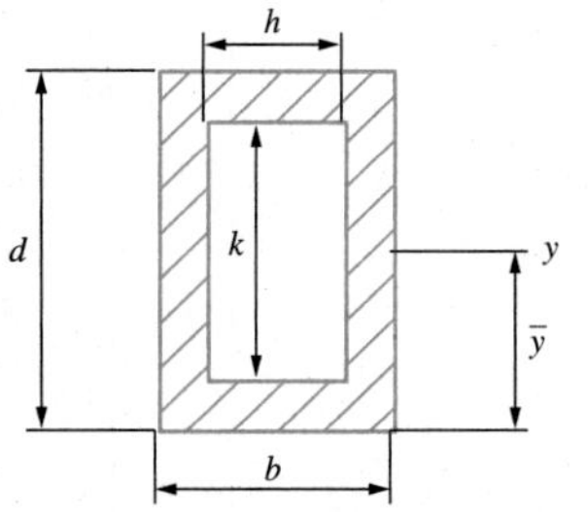	$\frac{d}{2}$	$\frac{bd^3-hk^3}{12}$
3. 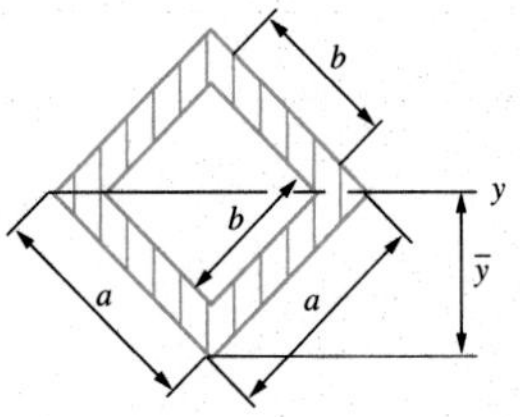	$\frac{a}{\sqrt{2}}$	$\frac{a^4-b^4}{12}$	7. 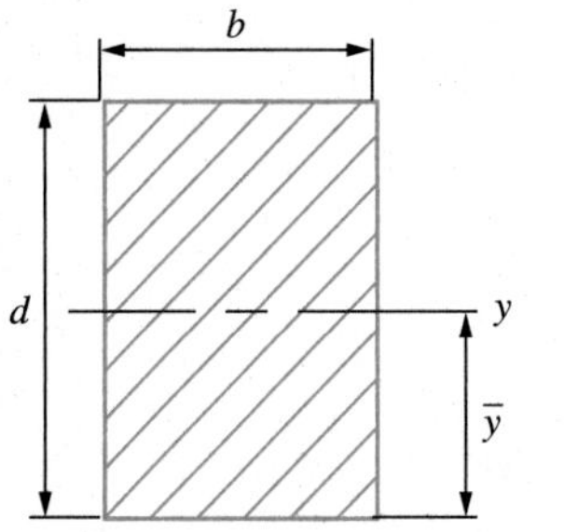	$\frac{d}{2}$	$\frac{bd^3}{12}$
4. 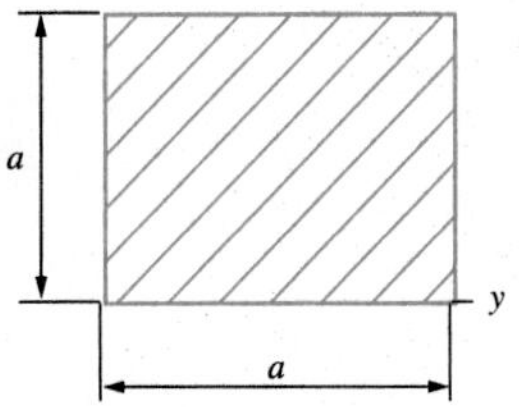		$\frac{a^4}{3}$	8. 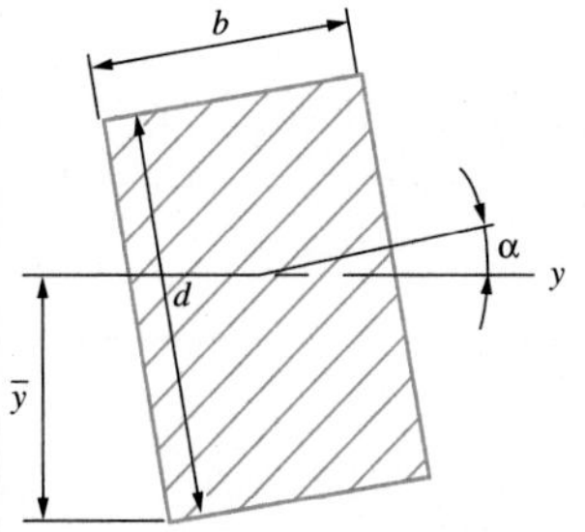	$\frac{1}{2}(d\cos\alpha+b\sin\alpha)$	$\frac{bd}{12}(d^2\cos^2\alpha + b^2\sin^2\alpha)$

(Continued)

TABLE A.2 Moment of Inertia of Sections (*Continued*)

Section	$\bar{y}$	Moment of Inertia I_y	Section	$\bar{y}$	Moment of Inertia I_y
9. 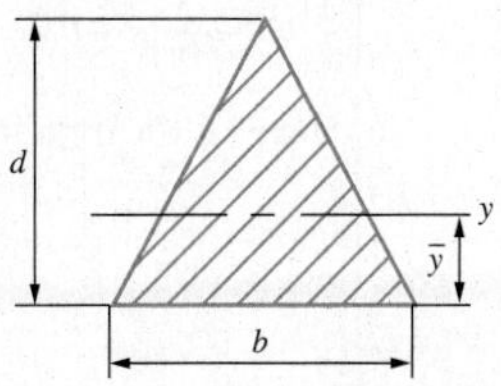	$\frac{1}{3}d$	$\frac{bd^3}{36}$	13. 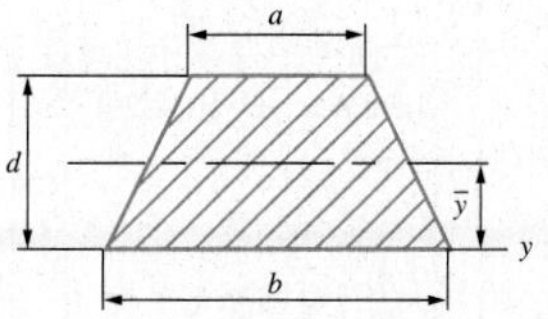	$\frac{d(2a+b)}{3(a+b)}$	$\frac{d^3(a^2+4ab+b^2)}{36(a+b)}$
10. 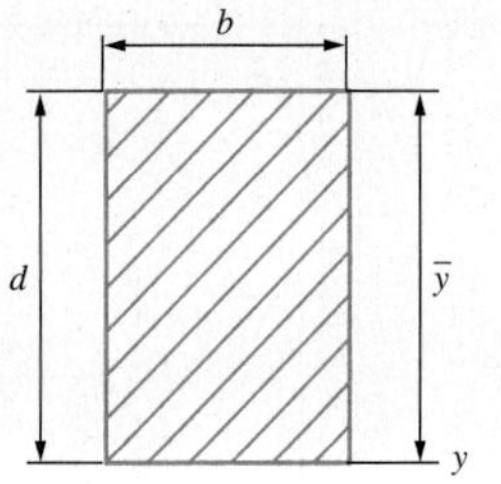	d	$\frac{bd^3}{3}$	14. 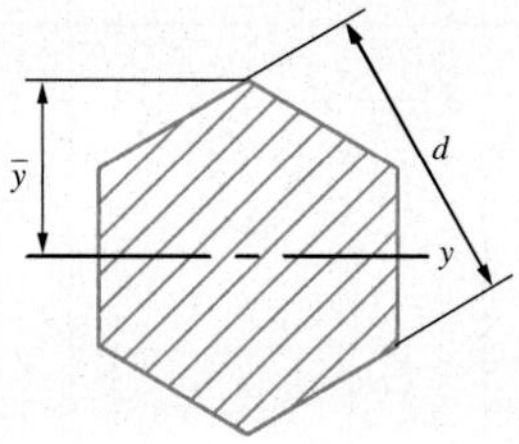	$\frac{d}{2\cos 30°}$	$\frac{A}{12}\left[\frac{d^2(1+2\cos^2 30°)}{4\cos^2 30°}\right]$ where $A = (3d^2 \tan 30)/2$
11. 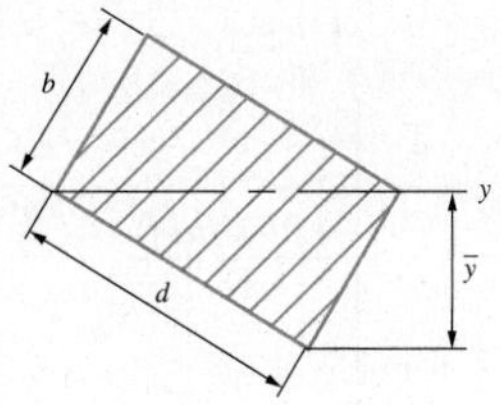	$\frac{bd}{\sqrt{b^2+d^2}}$	$\frac{b^3d^3}{6(b^2+d^2)}$	15. 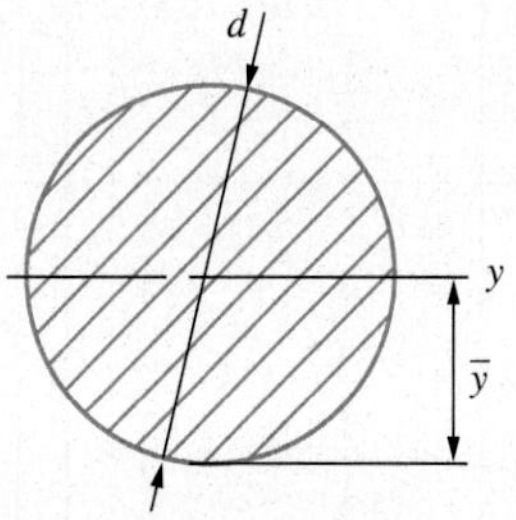	$\frac{d}{2}$	$\frac{\pi d^4}{64}$
12. 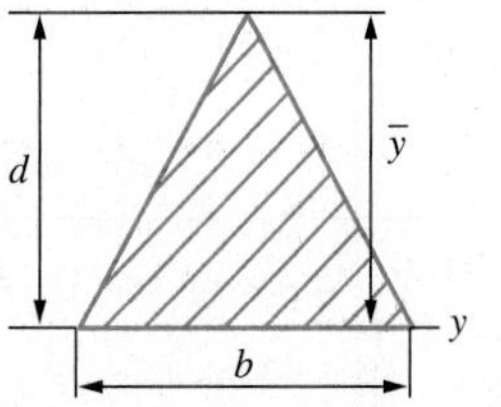	d	$\frac{bd^3}{12}$	16. 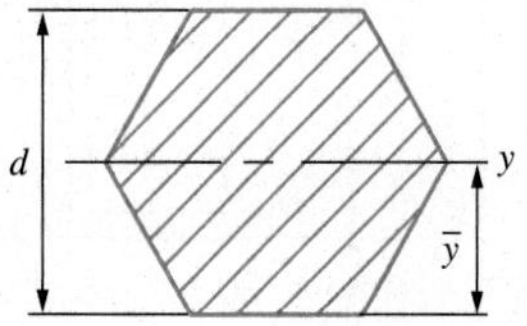	$\frac{d}{2}$	$\frac{A}{12}\left[\frac{d^2(1+2\cos^2 30°)}{4\cos^2 30°}\right]$ where $A = (3d^2 \tan$ $30)/2$

(*Continued*)

TABLE A.2 Moment of Inertia of Sections (*Continued*)

Section	$\bar{y}$	Moment of Inertia I_y	Section	$\bar{y}$	Moment of Inertia I_y
17. 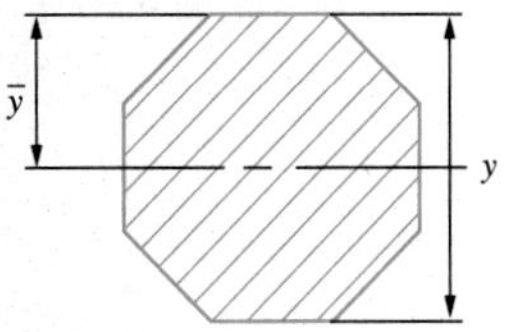	$\frac{d}{2}$	$\frac{A}{12}\left[\frac{d^2(1+2\cos^2 22\frac{1}{2}^\circ)}{4\cos^2 22\frac{1}{2}^\circ}\right]$ where $A = 2d^2(\tan 22\frac{1}{2})$	21. 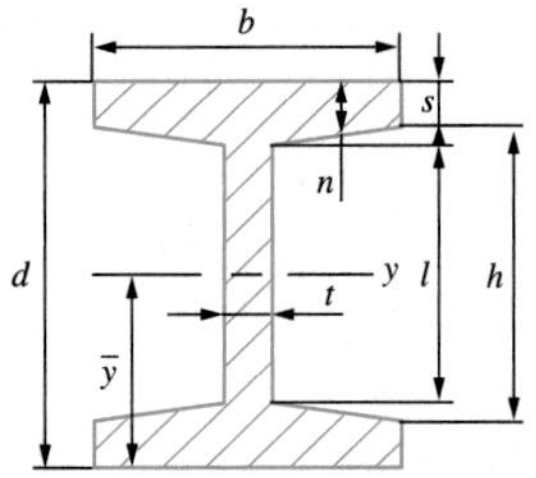	$\frac{d}{2}$	$\frac{1}{2}\left[bd^3 - \frac{1}{4g}(h^4 - l^4)\right]$ Slope $= g = (h-1)/(b-t)$
18. 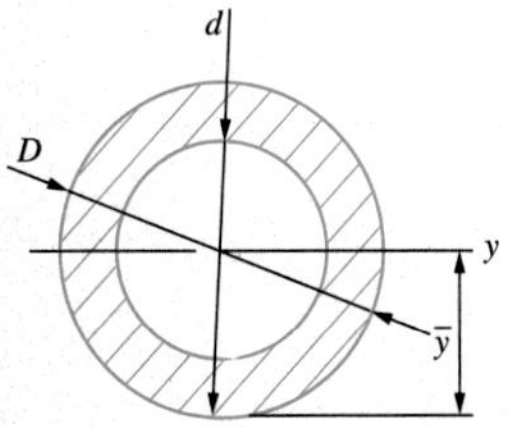	$\frac{D}{2}$	$\frac{\pi(D^4-d^4)}{64}$	22. 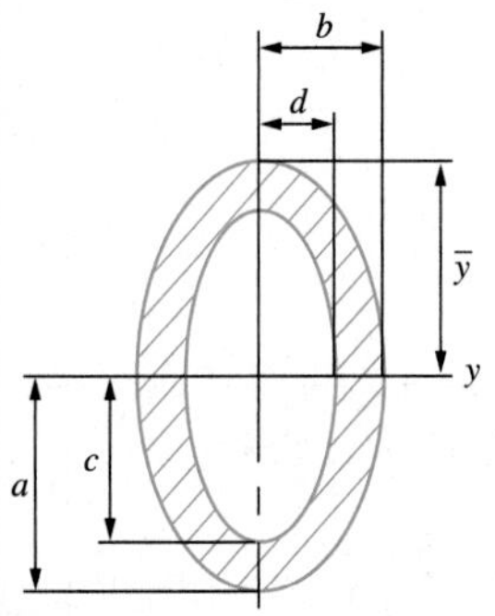	a	$\frac{\pi}{4}(a^3b-c^3d)$
19. 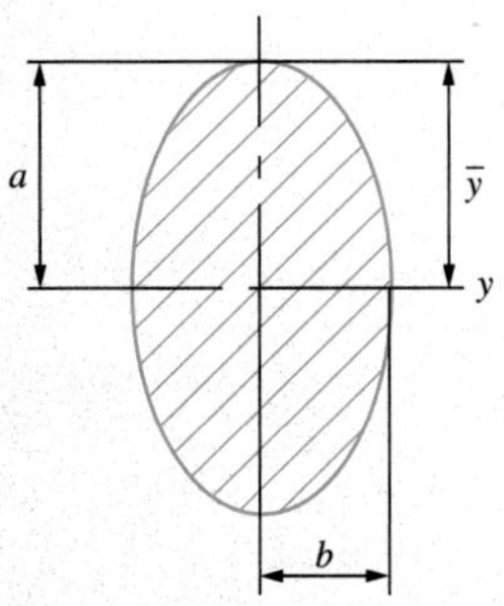	a	$\frac{\pi a^3 b}{4}$	23. 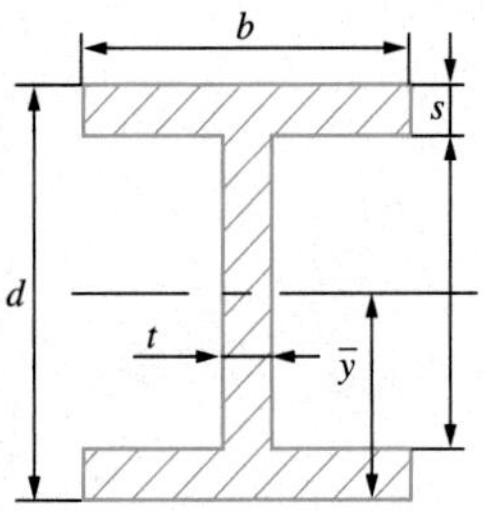	$\frac{d}{2}$	$\frac{bd^3-h^3(b-t)}{12}$
20. 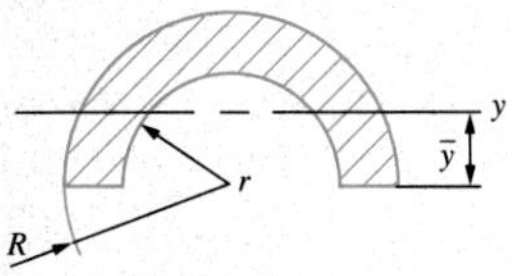	$\frac{4(R^3-r^3)}{3\pi(R^2-r^2)}$	$0.1098(R^4-r^4) - \frac{0.283R^2r^2(R-r)}{R+r}$	24. 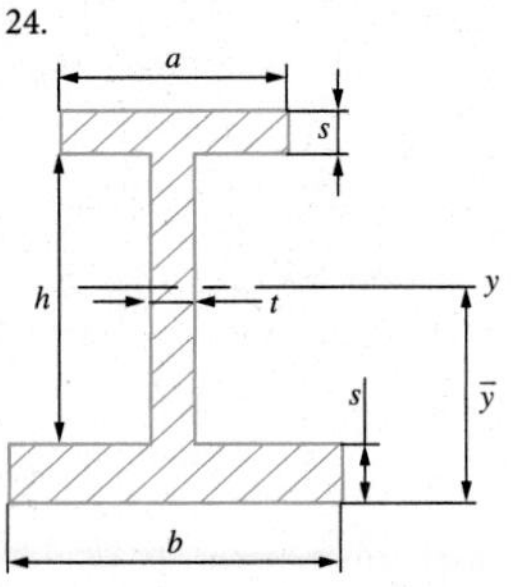	$d-[td^2+s^2(b-t)$ $+s(a-t)(2d-s)]$ $\div 2A$ where $A = bs + ht + as$	$\frac{1}{3}[b(d-y)^3+ay^3$ $-(b-t)(d-y-s)^3$ $-(a-t)(y-s)^3]$

(*Continued*)

TABLE A.2 Moment of Inertia of Sections (*Continued*)

Section	$\bar{y}$	Moment of Inertia I_y	Section	$\bar{y}$	Moment of Inertia I_y
25. 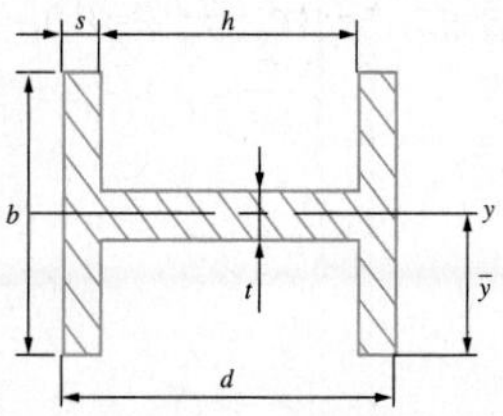	$\frac{b}{2}$	$\frac{2sb^3+ht^3}{12}$	29. 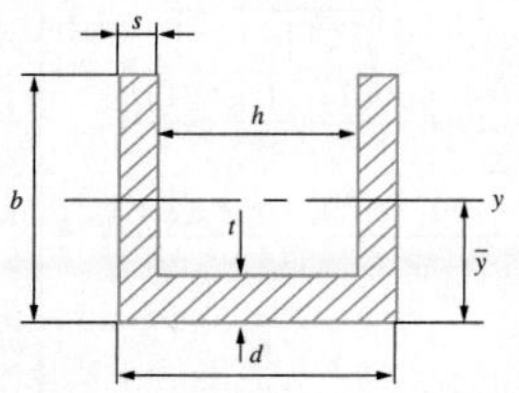	$b-\frac{2b^2s+ht^2}{2bd-2h(b-t)}$	$\frac{2sb^3+ht^3}{3}-A(b-y)^2$ where $A = bd - h(b - t)$
26. 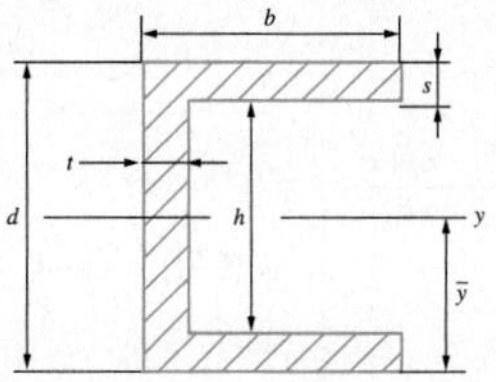	$\frac{d}{2}$	$\frac{bd^3-h^3(b-t)}{12}$	30. 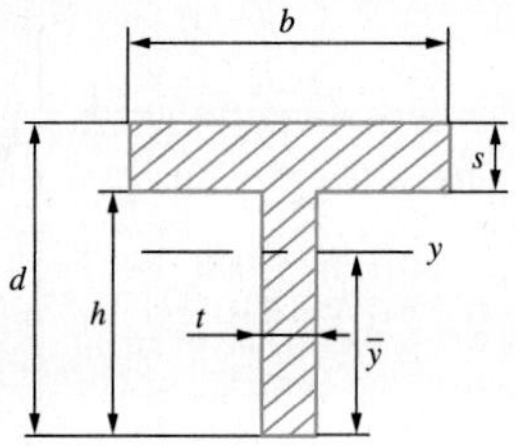	$d-\frac{d^2t+s^2(b-t)}{2(bs+ht)}$	$\frac{1}{3}[ty^3+b(d-y)^3$ $-(b-t)(d-y-s)^3]$
27. 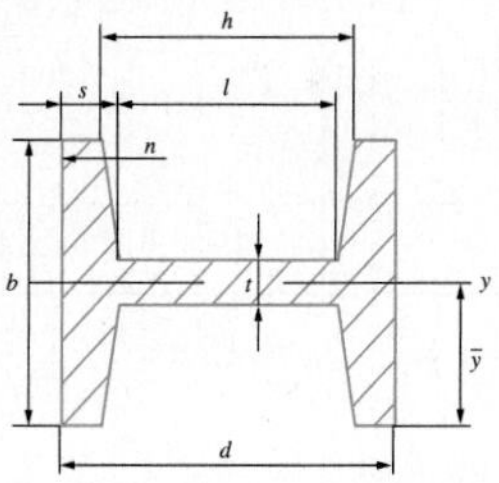	$\frac{b}{2}$	$\frac{1}{12}\Big[b^3(d-h)+lt^3$ $+\frac{g}{4}(b^4-t^4)\Big]$ $g=$ flange slope $=(h-l)(b-t)$	31. 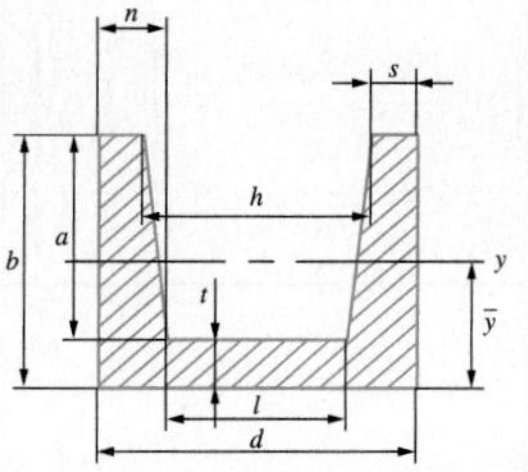	$b-\Big[b^2s+\frac{ht^2}{2}$ $+\frac{g}{3}(b-t)^2$ $\times(b+2t)\Big]\div A$ $g=$ slope of flange $=\frac{h-l}{2(b-t)}$ where $A =$ $dt + 2a(s + n)$	$\frac{1}{3}\Big[2sb^3+lt^3+\frac{g}{2}(b^4-t^4)\Big]$ $-A(b-y)^2$ $=\frac{h-l}{2(b-t)}$
28. 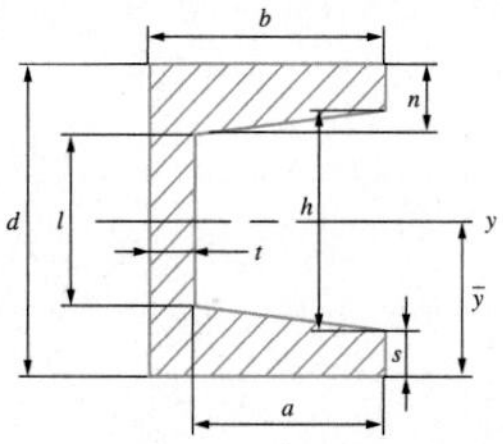	$\frac{d}{2}$	$\frac{1}{12}\Big[b^3d^3-\frac{1}{8g}(h^4-l^4)\Big]$ $g=$ flange slope (Standard Channels) $=\frac{h-l}{2(b-t)}$	32. 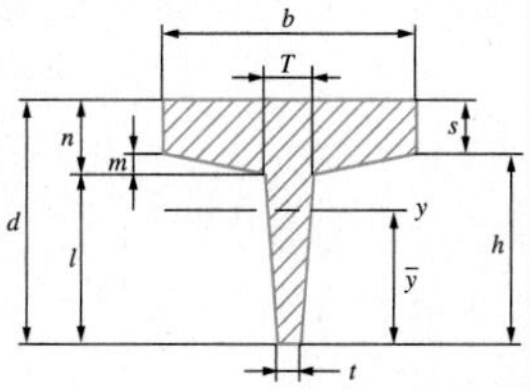	$d-[3s^2(b-T)$ $+2am(m+3s)+3Td^2$ $-l(T-t)(3d-l)]\div 6A$ where $A = l(T + t)/2$ $+ Tn + a(s + n)$	$\frac{1}{12}[l^3(T+3t)+4bn^3-2am^3]$ $-A(d-y-n)^2$

(*Continued*)

TABLE A.2 Moment of Inertia of Sections (*Continued*)

Section	$\bar{y}$	Moment of Inertia I_y	Section	$\bar{y}$	Moment of Inertia I_y
33. 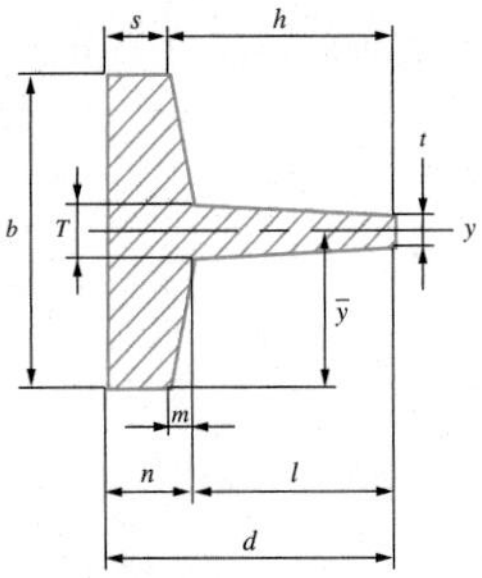	$\frac{b}{2}$	$\frac{sb^3+mT^3+lt^3}{12}+\frac{am[2a^2+(2a+3T)^2]}{36}+\frac{l(T-t)[(T-t)^2+2(T+2t)^2]}{144}$	35. 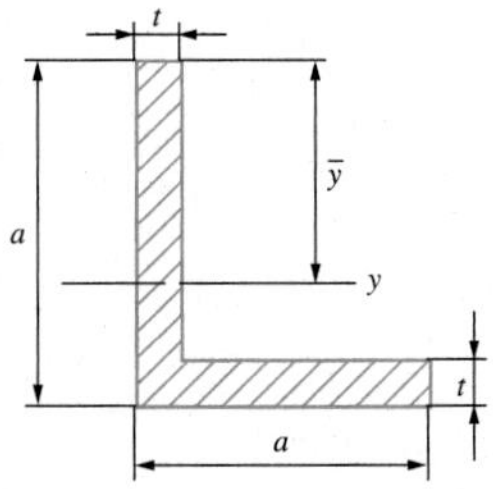	$a-\frac{a^2+at-t^2}{2(2a-t)}$	$\frac{1}{3}[ty^3+a(a-y)^3-(a-t)(a-y-t)^3]$
34. 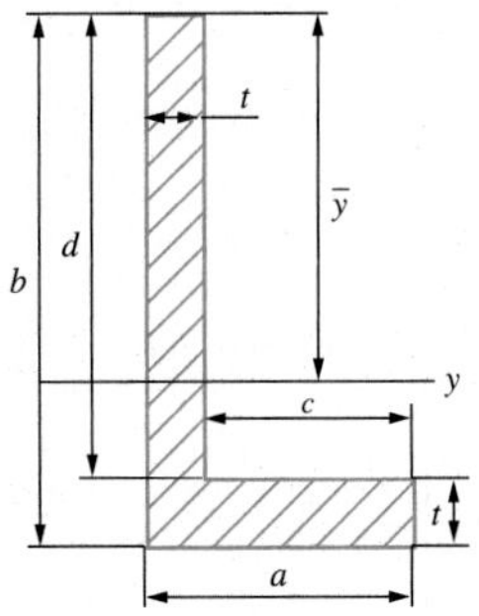	$b-\frac{t(2d+a)+d^2}{2(d+a)}$	$\frac{1}{3}[ty^3+a(b-y)^3-(a-t)(b-y-t)^3]$	36. 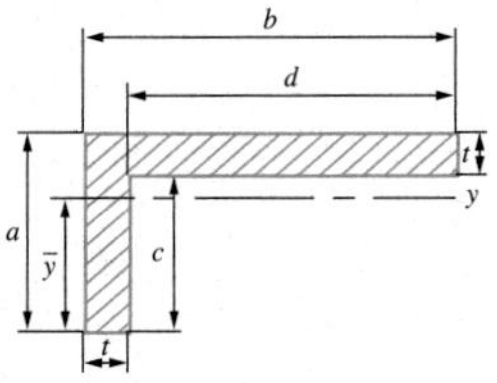	$a-\frac{t(2c+b)+c^2}{2(c+b)}$	$\frac{1}{3}[ty^3+b(a-y)^3-(b-t)(a-y-t)^3]$

TABLE A.3 Moment of Inertia of Uniform Objects

Axis of Rotation	Moment of Inertia	Axis of Rotation	Moment of Inertia
1. 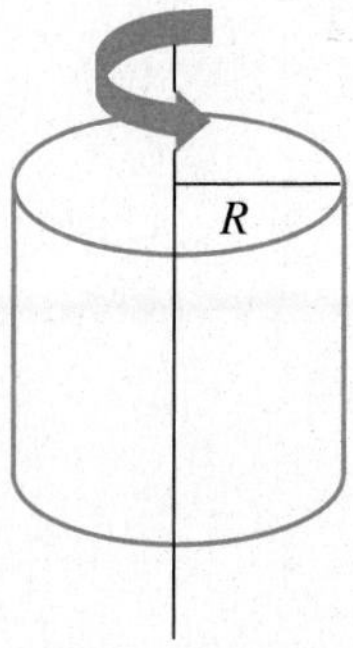 Solid Cylinder Central Axis of Cylinder	$\frac{1}{2}MR^2$	4. 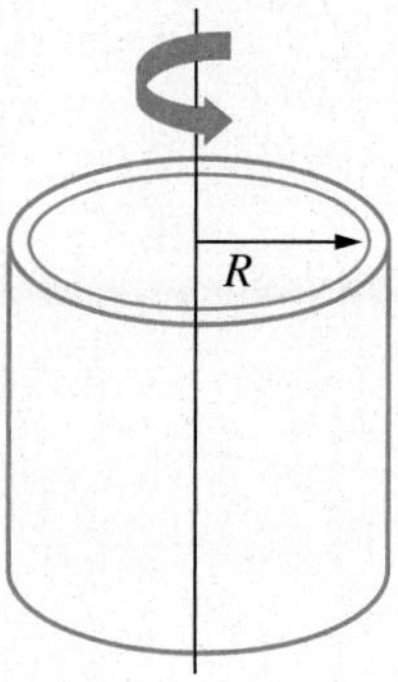 Hollow Cylinder Central Axis	MR^2
2. 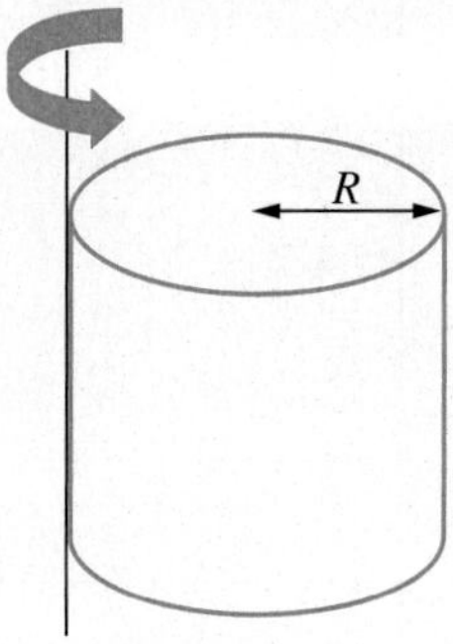 Solid Cylinder Axis on Surface	$\frac{3}{2}MR^2$	5. 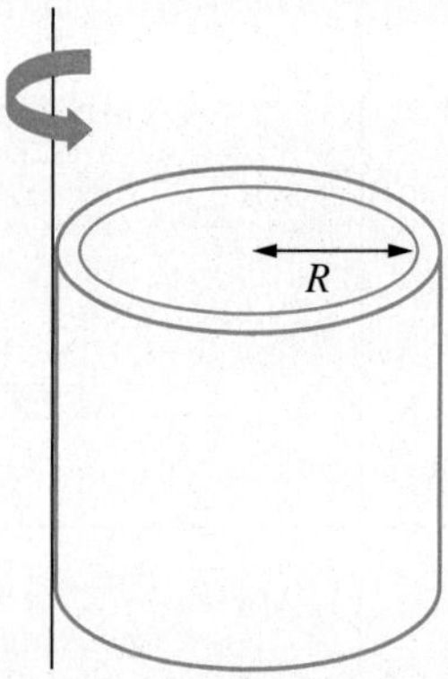 Hollow Cylinder Axis on Surface	$2MR^2$
3. 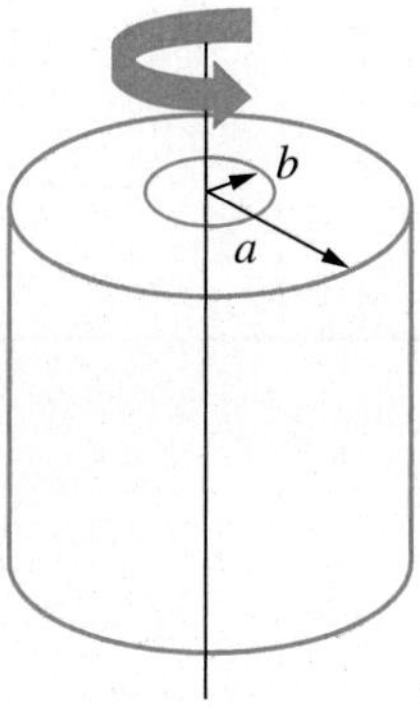 Cylindrical Shell Axis at Center	$\frac{1}{2}M(a^2+b^2)$	6. 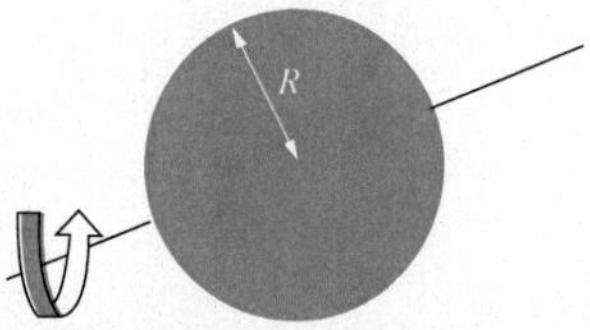 Solid Sphere Axis at Center	$\frac{2}{5}MR^2$

(*Continued*)

TABLE A.3 Moment of Inertia of Uniform Objects (*Continued*)

Axis of Rotation	Moment of Inertia	Axis of Rotation	Moment of Inertia
7. 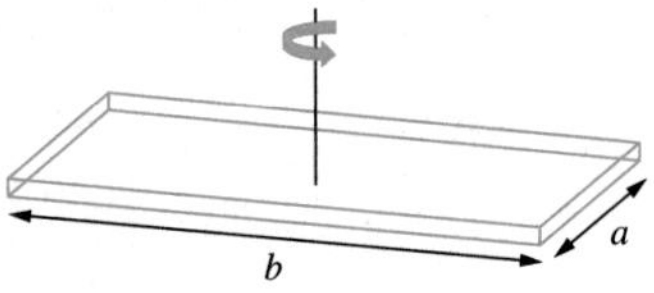 Rectangular Plate Axis through Center	$\frac{1}{2}M(a^2+b^2)$	10. 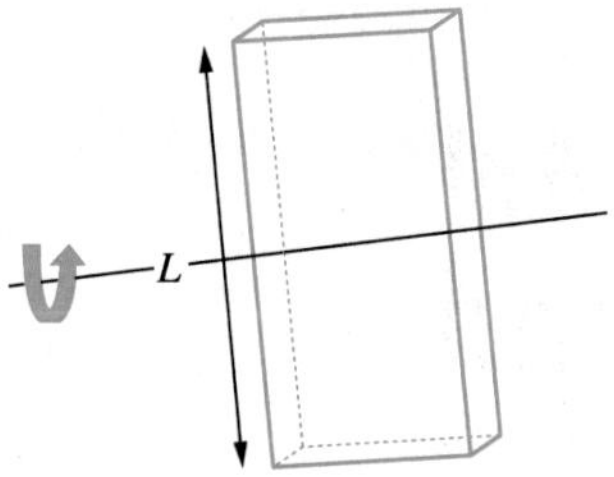 Rectangular Plate Axis through Center in Plane of Plate	$\frac{1}{12}ML^2$
8. 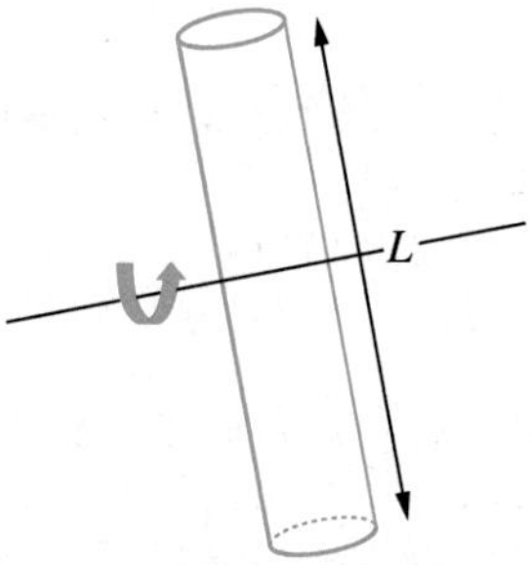 Thin Rod Axis through Mid Point	$\frac{1}{12}ML^2$	11. 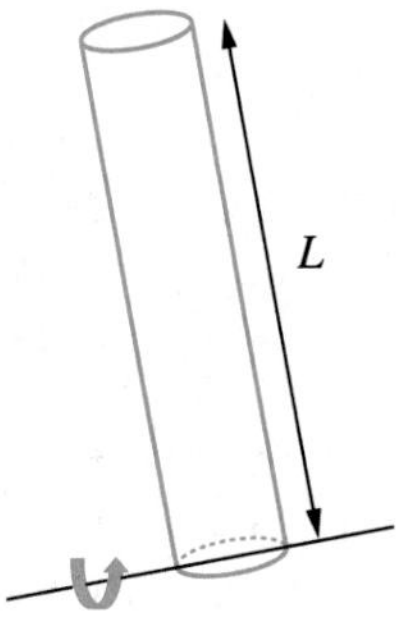 Thin Rod Axis at One End	$\frac{1}{3}ML^2$
9. 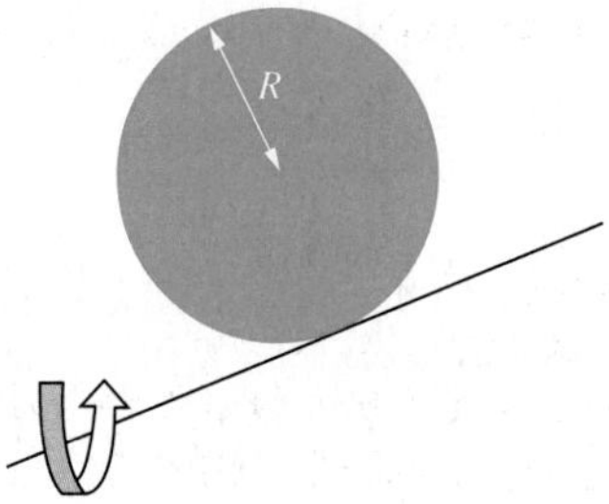 Solid Sphere Axis on Surface	$\frac{7}{5}MR^2$		

APPENDIX B

Mathematical Formulas and Matrices

ALGEBRAIC IDENTITIES

$$(a \pm b)^2 = a^2 \pm 2ab + b^2$$

$$(a \pm b)^3 = a^3 \pm 3a^2b + 3ab^2 \pm b^3$$

$$(a \pm b)^4 = a^4 \pm 4a^3b + 6a^2b^2 \pm 4ab^3 + b^4$$

$$(a \pm b)^n = \sum_{k=0}^{n} \binom{n}{k} a^k (\pm b)^{n-k} \quad \text{where} \binom{n}{k} = \frac{n!}{k!(n-k)!}$$

$$a^2 + b^2 = (a + bi)(a - bi)$$

$$a^4 + b^4 = (a^2 + \sqrt{2}ab + b^2)(a^2 - \sqrt{2}ab + b^2)$$

$$a^2 - b^2 = (a - b)(a + b)$$

$$a^3 - b^3 = (a - b)(a^2 + ab + b^2)$$

$$a^n - b^n = (a - b)(a^{n-1} + a^{n-2}b + \cdots + ab^{n-2} + b^{n-1})$$

$$(a + b + c)^2 = a^2 + b^2 + c^2 + 2ab + 2ac + 2bc$$

$$(a + b + c)^3 = a^3 + b^3 + c^3 + 3(a^2b + ab^2 + a^2c + ac^2 + b^2c + bc^2) + 6abc$$

$$a^n + b^n = (a + b)(a^{n-1} + (-b)a^{n-2} + (-b)^2 a^{n-3} + \cdots + (-b)^{n-1}) \qquad n \text{ odd}$$

TRIGONOMETRIC AND HYPERBOLIC IDENTITIES

$$\sin x = \frac{1}{2i}(e^{ix} - e^{-ix}) = -i \sinh(ix) \qquad \sinh x = \frac{1}{2}(e^{x} - e^{-x}) = -i \sin(ix)$$

$$\cos x = \frac{1}{2}(e^{ix} + e^{-ix}) = \cosh(ix) \qquad \cosh x = \frac{1}{2}(e^{x} + e^{-x}) = \cos(ix)$$

$$\sin(x \pm y) = \sin x \cos y \pm \cos x \sin y \qquad \sinh(x \pm y) = \sinh x \cosh y \pm \cosh x \sinh y$$

$$\cos(x \pm y) = \cos x \cos y \mp \sin x \sin y \qquad \cosh(x \pm y) = \cosh x \cosh y \pm \sinh x \sinh y$$

$$\sin x \pm \sin y = 2 \sin\frac{x \pm y}{2}\cos\frac{x \mp y}{2} \qquad \sinh x \pm \sinh y = 2 \sinh\frac{x \pm y}{2}\cosh\frac{x \mp y}{2}$$

$$\cos x + \cos y = 2\cos\frac{x+y}{2}\cos\frac{x-y}{2} \qquad \cosh x + \cosh y = 2\cosh\frac{x+y}{2}\cosh\frac{x-y}{2}$$

$$\cos x - \cos y = -2\sin\frac{x+y}{2}\sin\frac{x-y}{2} \qquad \cosh x - \cosh y = 2\sinh\frac{x+y}{2}\sinh\frac{x-y}{2}$$

$$\sin x \cos y = \frac{1}{2}[\sin(x+y) + \sin(x-y)] \qquad \sinh x \cosh y = \frac{1}{2}[\sinh(x+y) + \sinh(x-y)]$$

$$\cos x \cos y = \frac{1}{2}[\cos(x+y) + \cos(x-y)] \qquad \cosh x \cosh y = \frac{1}{2}[\cosh(x+y) + \cosh(x-y)]$$

$$\sin x \sin y = -\frac{1}{2}[\cos(x+y) - \cos(x-y)] \qquad \sinh x \sinh y = \frac{1}{2}[\cosh(x+y) - \cosh(x-y)]$$

$$\sin(2x) = 2 \sin x \cos x \qquad \sinh(2x) = 2 \sinh x \cosh x$$

$$\cos(2x) = \cos^2 x - \sin^2 x \qquad \cosh(2x) = \cosh^2 x + \sinh^2 x$$

$$\sin^2 x = \frac{1}{2}[1 - \cos(2x)] \qquad \sinh^2 x = \frac{1}{2}[\cosh(2x) - 1]$$

$$\cos^2 x = \frac{1}{2}[1 + \cos(2x)] \qquad \cosh^2 x = \frac{1}{2}[\cosh(2x) + 1]$$

$$\sin\left(\frac{\pi}{2} - x\right) = \cos x \qquad \sin(\pi - x) = \sin x \qquad \cos(\pi - x) = -\cos x$$

COMPLEX RELATIONSHIPS

$$e^{ix} = \cos x + i \sin x \qquad \sinh x = \frac{e^x - e^{-x}}{2}$$

$$\sin x = \frac{e^{ix} - e^{-ix}}{2i} \qquad \cosh x = \frac{e^x + e^{-x}}{2}$$

$$\cos x = \frac{e^{ix} + e^{-ix}}{2} \qquad \sin ix = i \sinh x$$

$$(\cos x + i \sin x)^n = \cos nx + i \sin nx \qquad \cos ix = \cosh x$$

$$\sin nx = \mathrm{Im}\{(\cos x + i \sin x)^n\} \qquad \sinh ix = i \sin x$$

$$\cos nx = \mathrm{Re}\{(\cos x + i \sin x)^n\} \qquad \cosh ix = \cos x$$

DERIVATIVES OF ELEMENTARY FUNCTIONS

General Rules of Differentiation

Consider $y = f(x)$ is a function then the derivative of y can be represented by

$$y' = \frac{dy}{dx}$$

Rule 1 : $\frac{dy}{dx}c = 0$; (c = constant)

Rule 2 : $\frac{dy}{dx}\,x^n = nx^{n-1}$;

Rule 3 : $\frac{dy}{dx}\,c\,f(x) = c\,f'(x)$;

Rule 4 : $f(x) = h(x) + g(x)$ then $\frac{df}{dx} = \frac{dh}{dx} + \frac{dg}{dx}$; similar with subtraction

Rule 5 : $y = u(x)v(x)$ then $\frac{dy}{dx} = u\left(\frac{dv}{dx}\right) + v\left(\frac{du}{dx}\right)$;

Rule 6 : $f(x) = \frac{u(x)}{v(x)}$ then $\frac{df}{dx} = \frac{v\left(\frac{du}{dx}\right) - u\left(\frac{dv}{dx}\right)}{(v)(v)}$;

Rule 7 : If $y = f(u)$ and $u = g(x)$ then $\frac{dy}{dx} = \frac{dy}{du}\cdot\frac{du}{dx}$;

$$\frac{d}{dx}\sin u = \cos u\frac{du}{dx} \qquad \frac{d}{dx}\cot u = -\csc^2 u\frac{du}{dx}$$

$$\frac{d}{dx}\cos u = -\sin u\frac{du}{dx} \qquad \frac{d}{dx}\sec u = \sec u\tan u\frac{du}{dx}$$

$$\frac{d}{dx}\tan u = \sec^2 u\frac{du}{dx} \qquad \frac{d}{dx}\csc u = -\csc u\cot u\frac{du}{dx}$$

$$\frac{d}{dx}\sin^{-1} u = \frac{1}{\sqrt{1-u^2}}\frac{du}{dx} \quad \left[-\frac{\pi}{2} < \sin^{-1} u < \frac{\pi}{2}\right]$$

$$\frac{d}{dx}\cos^{-1} u = \frac{-1}{\sqrt{1-u^2}}\frac{du}{dx} \quad \left[0 < \cos^{-1} u < \pi\right]$$

$$\frac{d}{dx}\tan^{-1} u = \frac{1}{1+u^2}\frac{du}{dx} \quad \left[-\frac{\pi}{2} < \tan^{-1} u < \frac{\pi}{2}\right]$$

$$\frac{d}{dx}\cot^{-1} u = \frac{-1}{1+u^2}\frac{du}{dx} \quad \left[0 < \cot^{-1} u < \pi\right]$$

$$\frac{d}{dx}\sec^{-1} u = \frac{1}{|u|\sqrt{u^2-1}}\frac{du}{dx} = \frac{\pm 1}{u\sqrt{u^2-1}}\frac{du}{dx} \quad \begin{bmatrix} +\text{ if } 0 < \sec^{-1} u < \pi/2 \\ -\text{ if } \pi/2 < \sec^{-1} u < \pi \end{bmatrix}$$

$$\frac{d}{dx}\csc^{-1} u = \frac{-1}{|u|\sqrt{u^2-1}}\frac{du}{dx} = \frac{\mp 1}{u\sqrt{u^2-1}}\frac{du}{dx} \quad \begin{bmatrix} -\text{ if } 0 < \csc^{-1} u < \pi/2 \\ +\text{ if } -\pi/2 < \csc^{-1} u < 0 \end{bmatrix}$$

$$\frac{d}{dx}\log_a u = \frac{\log_a e}{u}\frac{du}{dx} \qquad a \neq 0, 1$$

$$\frac{d}{dx}\ln u = \frac{d}{dx}\log_e u = \frac{1}{u}\frac{du}{dx}$$

$$\frac{d}{dx}a^u = a^u \ln a \frac{du}{dx}$$

$$\frac{d}{dx}e^u = e^u \frac{du}{dx}$$

$$\frac{d}{dx}u^v = \frac{d}{dx}e^{v \ln u} = e^{v \ln u}\frac{d}{dx}[v \ln u] = vu^{v-1}\frac{du}{dx} + u^v \ln u \frac{dv}{dx}$$

$$\frac{d}{dx}\sinh u = \cosh u \frac{du}{dx} \qquad \frac{d}{dx}\coth u = -\operatorname{csch}^2 u \frac{du}{dx}$$

$$\frac{d}{dx}\cosh u = \sinh u \frac{du}{dx} \qquad \frac{d}{dx}\operatorname{sech} u = -\operatorname{sech} u \tanh u \frac{du}{dx}$$

$$\frac{d}{dx}\tanh u = \operatorname{sech}^2 u \frac{du}{dx} \qquad \frac{d}{dx}\operatorname{csch} u = -\operatorname{csch} u \coth u \frac{du}{dx}$$

$$\frac{d}{dx}\sinh^{-1} u = \frac{1}{\sqrt{u^2+1}}\frac{du}{dx}$$

$$\frac{d}{dx}\cosh^{-1} u = \frac{\pm 1}{\sqrt{u^2-1}}\frac{du}{dx} \qquad \begin{bmatrix} +\text{if } \cosh^{-1} u > 0, u > 1 \\ -\text{if } \cosh^{-1} u < 0, u > 1 \end{bmatrix}$$

$$\frac{d}{dx}\tanh^{-1} u = \frac{1}{1-u^2}\frac{du}{dx} \qquad [-1 < u < 1]$$

$$\frac{d}{dx}\coth^{-1} u = \frac{1}{1-u^2}\frac{du}{dx} \qquad [u > 1 \text{ or } u < -1]$$

$$\frac{d}{dx}\operatorname{sech}^{-1} u = \frac{\mp 1}{u\sqrt{1-u^2}}\frac{du}{dx} \qquad \begin{bmatrix} -\text{if } \operatorname{sech}^{-1} u > 0, 0 < u < 1 \\ +\text{if } \operatorname{sech}^{-1} u < 0, 0 < u < 1 \end{bmatrix}$$

$$\frac{d}{dx}\operatorname{csch}^{-1} u = \frac{-1}{|u|\sqrt{1+u^2}}\frac{du}{dx} = \frac{\mp 1}{u\sqrt{1+u^2}}\frac{du}{dx} \qquad [-\text{if } u > 0, + \text{ if } u < 0]$$

The second, third, and higher derivatives are defined as follows:

$$\text{Second derivative} = \frac{d}{dx}\left(\frac{dy}{dx}\right) = \frac{d^2 y}{dx^2} = f''(x) = y''$$

$$\text{Third derivative} = \frac{d}{dx}\left(\frac{d^2 y}{dx^2}\right) = \frac{d^3 y}{dx^3} = f'''(x) = y'''$$

$$n\text{th derivative} = \frac{d}{dx}\left(\frac{d^{n-1} y}{dx^{n-1}}\right) = \frac{d^n y}{dx^n} = f^{(n)}(x) = y^{(n)}$$

PARTIAL DERIVATIVES

Consider a function $f(x, y)$ then the partial derivative of $f(x, y)$ with respect to x, and y as constant

$$\frac{\partial f}{\partial x} = \lim_{\Delta x \to 0} \frac{f(x+\Delta x, y) - f(x, y)}{\Delta x}$$

Similarly, considering x as constant we have

$$\frac{\partial f}{\partial y} = \lim_{\Delta y \to 0} \frac{f(x, y+\Delta y) - f(x, y)}{\Delta y}$$

Partial derivatives of higher order

$$\frac{\partial^2 f}{\partial x^2} = \frac{\partial}{\partial x}\left(\frac{\partial f}{\partial x}\right), \quad \frac{\partial^2 f}{\partial y^2} = \frac{\partial}{\partial y}\left(\frac{\partial f}{\partial y}\right)$$

$$\frac{\partial^2 f}{\partial x \partial y} = \frac{\partial}{\partial x}\left(\frac{\partial f}{\partial y}\right), \quad \frac{\partial^2 f}{\partial y \partial x} = \frac{\partial}{\partial y}\left(\frac{\partial f}{\partial x}\right)$$

BASIC DIFFERENTIAL EQUATIONS AND SOLUTIONS

Differential Equation	Solution
Separation of variables	
$f_1(x)g_1(y)dx + f_2(x)g_2(y)dy = 0$	$\int \frac{f_1(x)}{f_2(x)}dx + \int \frac{g_2(y)}{g_1(y)}dy = c$
Linear first order equation	
$\frac{dy}{dx} + P(x)y = Q(x)$	$ye^{\int P dx} = \int Qe^{\int P dx}\,dx + c$
Bernoulli's equation	
$\frac{dy}{dx} + P(x)y = Q(x)y^n$	$ve^{(1-n)\int P dx} = (1-n)\int Qe^{(1-n)\int P dx}dx - c$ where $v = y^{1-n}$. If $n = 1$, the solution is $\ln y = \int (Q-P)dx + c$
Exact equation	
$M(x, y)dx + N(x, y)dy = 0$ where $\partial M/\partial y = \partial N/\partial x$	$\int M\,\partial x + \int\left(N - \frac{\partial}{\partial y}\int M\,\partial x\right)dy = c$ where ∂x indicates that the integration is to be performed with respect to x keeping y constant.
Homogeneous equation	
$\frac{dy}{dx} = F\left(\frac{y}{x}\right)$	$\ln x = \int \frac{dv}{F(v)-v} + c$ where $v = y/x$. If $F(v) = v$, the solution is $y = cx$.

(*Continued*)

Differential Equation	Solution
$y\,F(xy)dx+x\,G(xy)dy=0$	$\ln x=\int\frac{G(v)dv}{v\{G(v)-F(v)\}}+c$ where $v=xy$. If $G(v)-F(v)$, the solution is $xy=c$.
Linear, homogeneous second order equation $\frac{d^2y}{dx^2}+a\frac{dy}{dx}+by=0$ a, b are real constants.	Let m_1, m_2 be the roots of $m^2+am+b=0$. Then there are 3 cases. Case 1. m_1, m_2 real and distinct: $y=c_1e^{m_1x}-c_2e^{m_2x}$ Case 2. m_1, m_2 real and equal: $y=c_1e^{m_1x}+c_2xe^{m_1x}$ Case 3. $m_1=p+qi$, $m_2=p-qi$: $y=e^{px}(c_1\cos qx-c_2\sin qx)$ where $p=-a/2$, $q=\sqrt{b-a^2/4}$.
Linear, nonhomogeneous second order equation $\frac{d^2y}{dx^2}+a\frac{dy}{dx}+by=R(x)$ a, b are real constants.	There are 3 cases corresponding to those of entry above. Case 1. $y=c_1e^{m_1x}+c_2e^{m_2x}$ $+\frac{e^{m_1x}}{m_1-m_2}\int c^{-m_1x}R(x)dx$ $+\frac{e^{m_2x}}{m_2-m_1}\int e^{-m_2x}R(x)dx$ Case 2. $y=c_1e^{m_1x}+c_2xe^{m_1x}$ $+\,xe^{m_1x}\int e^{m_1x}R(x)dx$ $-e^{m_1x}\int xe^{-m_1x}R(x)dx$ Case 3. $y=e^{px}(c_1\cos qx+c_2\sin qx)$ $+\frac{e^{px}\sin qx}{q}\int e^{px}R(x)\cos qx\,dx$ $-\frac{e^{px}\cos qx}{q}\int e^{-px}R(x)\sin qx\,dx$
Euler or Cauchy equation $x^2\frac{d^2y}{dx^2}+ax\frac{dy}{dx}+by=S(x)$	Putting $x=e^t$, the equation becomes $\frac{d^2y}{dt^2}+(a-1)\frac{dy}{dt}+by=S(e^t)$ and can then be solved as in entries above.
Bessel's equation $x^2\frac{d^2y}{dx^2}+x\frac{dy}{dx}+(\lambda^2x^2-n^2)y=0$	$y=c_1J_n(\lambda x)+c_2Y_n(x)$

(Continued)

Differential Equation	Solution
Transformed Bessel's equation	
$x^2\dfrac{d^2y}{dx^2}+(2p+1)x\dfrac{dy}{dx}+(\alpha^2x^{2r}-\beta^2)y=0$	$y=x^{-p}\left\{c_1J_{q/r}\left(\dfrac{\alpha}{r}x^r\right)+c_2Y_{q/r}\left(\dfrac{\alpha}{r}x^r\right)\right\}$ where $q=\sqrt{p^2-\beta^2}$.
Legendre's equation	
$(1-x^2)\dfrac{d^2y}{dx^2}-2x\dfrac{dy}{dx}+n(n+1)y=0$	$y=c_1P_n(x)+c_2Q_n(x)$

STANDARD INTEGRALS

1. $\int cu\,dx=c\int u\,dx,$
2. $\int (u+v)dx=\int u\,dx+\int v\,dx,$
3. $\int x^n dx=\dfrac{1}{n+1}x^{n+1}, \qquad n\neq -1,$
4. $\int \dfrac{1}{x}dx=\ln x,$
5. $\int e^x dx=e^x,$
6. $\int \dfrac{dx}{1+x^2}=\arctan x,$
7. $\int u\dfrac{dv}{dx}dx=uv-\int v\dfrac{du}{dx}dx,$
8. $\int \sin x\,dx=-\cos x,$
9. $\int \cos x\,dx=\sin x,$
10. $\int \tan x\,dx=-\ln|\cos x|,$
11. $\int \cot x\,dx=\ln|\cos x|,$
12. $\int \sec x\,dx=\ln|\sec x+\tan x|,$
13. $\int \csc x\,dx=\ln|\csc x+\cot x|,$
14. $\int \arcsin\dfrac{x}{a}dx=\arcsin\dfrac{x}{a}+\sqrt{a^2-x^2}, \qquad a>0,$
15. $\int \arccos\dfrac{x}{a}dx=\arccos\dfrac{x}{a}-\sqrt{a^2-x^2}, \qquad a>0,$
16. $\int \arctan\dfrac{x}{a}dx=x\arctan\dfrac{x}{a}-\dfrac{a}{2}\ln(a^2+x^2), \qquad a>0,$

17. $\int \sin^2(ax)dx = \frac{1}{2a}(ax - \sin(ax)\cos(ax)),$

18. $\int \cos^2(ax)dx = \frac{1}{2a}(ax + \sin(ax)\cos(ax)),$

19. $\int \sec^2 x\,dx = \tan x,$

20. $\int \csc^2 x\,dx = -\cot x,$

21. $\int \sin^n x\,dx = -\frac{\sin^{n-1} x \cos x}{n} + \frac{n-1}{n}\int \sin^{n-2} x\,dx,$

22. $\int \cos^n x\,dx = \frac{\cos^{n-1} x \sin x}{n} + \frac{n-1}{n}\int \cos^{n-2} x\,dx,$

23. $\int \tan^n x\,dx = \frac{\tan^{n-1} x}{n-1} - \int \tan^{n-2} x\,dx, \qquad n \neq 1,$

24. $\int \cot^n x\,dx = -\frac{\cot^{n-1} x}{n-1} - \int \cot^{n-2} x\,dx, \qquad n \neq 1,$

25. $\int \sec^n x\,dx = \frac{\tan x \sec^{n-1} x}{n-1} + \frac{n-2}{n-1}\int \sec^{n-2} x\,dx, \qquad n \neq 1,$

26. $\int \csc^n x\,dx = -\frac{\cot x \csc^{n-1} x}{n-1} + \frac{n-2}{n-1}\int \csc^{n-2} x\,dx, \qquad n \neq 1,$

27. $\int \sinh x\,dx = \cosh x,$

28. $\int \cosh x\,dx = \sinh x,$

29. $\int \tanh x\,dx = \ln|\cosh x|,$

30. $\int \coth x\,dx = \ln|\sinh x|,$

31. $\int \operatorname{sech} x\,dx = \arctan \sinh x,$

32. $\int \operatorname{csch} x\,dx = \ln\left|\tanh\frac{x}{2}\right|,$

33. $\int \sinh^2 x\,dx = \frac{1}{4}\sinh(2x) - \frac{1}{2}x,$

34. $\int \cosh^2 x\,dx = \frac{1}{4}\sinh(2x) + \frac{1}{2}x,$

35. $\int \operatorname{sech}^2 x\,dx = \tanh x,$

36. $\int \operatorname{arcsinh}\frac{x}{a}dx = x\operatorname{arcsinh}\frac{x}{a} - \sqrt{x^2 + a^2}, \qquad a > 0,$

37. $\int \operatorname{arctanh}\frac{x}{a}dx = x\operatorname{arctanh}\frac{x}{a} + \frac{a}{2}\ln|a^2 - x^2|,$

38. $$\int \operatorname{arccosh}\frac{x}{a}\,dx = \begin{cases} x\operatorname{arccosh}\dfrac{x}{a} - \sqrt{x^2+a^2}, & \text{if } \operatorname{arccosh}\dfrac{x}{a} > 0 \text{ and } a > 0, \\ x\operatorname{arccosh}\dfrac{x}{a} + \sqrt{x^2+a^2}, & \text{if } \operatorname{arccosh}\dfrac{x}{a} < 0 \text{ and } a > 0, \end{cases}$$

39. $$\int \frac{dx}{\sqrt{a^2+x^2}} = \ln\left(x+\sqrt{a^2+x^2}\right), \qquad a>0,$$

40. $$\int \frac{dx}{a^2+x^2} = \frac{1}{a}\arctan\frac{x}{a}, \qquad a>0,$$

41. $$\int \sqrt{a^2-x^2}\,dx = \frac{x}{2}\sqrt{a^2-x^2} + \frac{a^2}{2}\arcsin\frac{x}{a}, \qquad a>0,$$

42. $$\int (a^2-x^2)^{3/2}\,dx = \frac{x}{8}(5a^2-2x^2)\sqrt{a^2-x^2} + \frac{3a^4}{8}\arcsin\frac{x}{a}, \qquad a>0,$$

43. $$\int \frac{dx}{\sqrt{a^2-x^2}} = \arcsin\frac{x}{a}, \qquad a>0,$$

44. $$\int \frac{dx}{a^2-x^2} = \frac{1}{2a}\ln\left|\frac{a+x}{a-x}\right|,$$

45. $$\int \frac{dx}{(a^2-x^2)^{3/2}} = \frac{x}{a^2\sqrt{a^2-x^2}},$$

46. $$\int \sqrt{a^2\pm x^2}\,dx = \frac{x}{a}\sqrt{a^2\pm x^2} \pm \frac{a^2}{2}\ln\left|x+\sqrt{a^2\pm x^2}\right|,$$

47. $$\int \frac{dx}{\sqrt{x^2-a^2}} = \ln\left|x+\sqrt{x^2-a^2}\right|, \qquad a>0,$$

48. $$\int \frac{dx}{ax^2+bx} = \frac{1}{a}\ln\left|\frac{x}{a+bx}\right|,$$

49. $$\int x\sqrt{a+bx}\,dx = \frac{2(3bx-2a)(a+bx)^{3/2}}{15b^2},$$

50. $$\int \frac{\sqrt{a+bx}}{x}\,dx = 2\sqrt{a+bx} + a\int \frac{1}{x\sqrt{a+bx}}\,dx,$$

51. $$\int \frac{x}{\sqrt{a+bx}}\,dx = \frac{1}{\sqrt{2}}\ln\left|\frac{\sqrt{a+bx}-\sqrt{a}}{\sqrt{a+bx}+\sqrt{a}}\right|, \qquad a>0,$$

52. $$\int \frac{\sqrt{a^2-x^2}}{x}\,dx = \sqrt{a^2-x^2} - a\ln\left|\frac{a+\sqrt{a^2-x^2}}{x}\right|,$$

53. $$\int x\sqrt{a^2-x^2}\,dx = -\frac{1}{3}(a^2-x^2)^{3/2},$$

54. $$\int x^2\sqrt{a^2-x^2}\,dx = \frac{x}{8}(2x^2-a^2)\sqrt{a^2-x^2} + \frac{a^4}{8}\arcsin\frac{x}{a}, \qquad a>0,$$

55. $$\int \frac{dx}{\sqrt{a^2-x^2}} = -\frac{1}{a}\ln\left|\frac{a+\sqrt{a^2-x^2}}{x}\right|,$$

56. $\int \frac{x\,dx}{\sqrt{a^2-x^2}} = -\sqrt{a^2-x^2},$

57. $\int \frac{x^2\,dx}{\sqrt{a^2-x^2}} = -\frac{x}{2}\sqrt{a^2-x^2} + \frac{a^2}{2}\arcsin\frac{x}{a}, \qquad a>0,$

58. $\int \frac{\sqrt{a^2+x^2}}{x}dx = \sqrt{a^2+x^2} - a\ln\left|\frac{a+\sqrt{a^2+x^2}}{x}\right|,$

59. $\int \frac{\sqrt{x^2-a^2}}{x}dx = \sqrt{x^2-a^2} - a\arccos\frac{a}{|x|}, \qquad a>0,$

60. $\int x\sqrt{x^2 \pm a^2}\,dx = \frac{1}{3}(x^2 \pm a^2)^{3/2},$

61. $\int \frac{dx}{x\sqrt{x^2+a^2}} = \frac{1}{a}\ln\left|\frac{x}{a+\sqrt{a^2+x^2}}\right|,$

62. $\int \frac{dx}{x\sqrt{x^2-a^2}} = \frac{1}{a}\arccos\frac{a}{|x|}, \qquad a>0,$

63. $\int \frac{dx}{x^2\sqrt{x^2 \pm a^2}} = \mp\frac{\sqrt{x^2 \pm a^2}}{a^2x},$

64. $\int \frac{x\,dx}{\sqrt{x^2 \pm a^2}} = \sqrt{x^2 \pm a^2},$

65. $\int \frac{\sqrt{x^2 \pm a^2}}{x^4}dx = \mp\frac{(x^2+a^2)^{3/2}}{3a^2x^3},$

66. $$\int \frac{dx}{ax^2+bx+c} = \begin{cases} \frac{1}{\sqrt{b^2-4ac}}\ln\left|\frac{2ax+b-\sqrt{b^2-4ac}}{2ax+b+\sqrt{b^2-4ac}}\right|, & \text{if } b^2>4ac, \\ \frac{2}{\sqrt{4ac-b^2}}\arctan\frac{2ax+b}{\sqrt{4ac-b^2}}, & \text{if } b^2<4ac, \end{cases}$$

67. $$\int \frac{dx}{\sqrt{ax^2+bx+c}} = \begin{cases} \frac{1}{\sqrt{a}}\ln\left|2ax+b+2\sqrt{a}\sqrt{ax^2+bx+c}\right|, & \text{if } a>0, \\ \frac{1}{\sqrt{-a}}\arcsin\frac{-2ax-b}{\sqrt{b^2-4ac}}, & \text{if } a<0, \end{cases}$$

68. $$\int \sqrt{ax^2+bx+c}\,dx = \frac{2ax+b}{4a}\sqrt{ax^2+bx+c} + \frac{4ax-b^2}{8a}\int\frac{dx}{\sqrt{ax^2+bx+c}},$$

69. $$\int \frac{x\,dx}{\sqrt{ax^2+bx+c}} = \frac{\sqrt{ax^2+bx+c}}{a} - \frac{b}{2a}\int\frac{dx}{\sqrt{ax^2+bx+c}},$$

70. $$\int \frac{dx}{x\sqrt{ax^2+bx+c}} = \begin{cases} \frac{-1}{\sqrt{c}}\ln\left|\frac{2\sqrt{c}\sqrt{ax^2+bx+c}+bx+2c}{x}\right|, & \text{if } c>0, \\ \frac{1}{\sqrt{-c}}\arcsin\frac{bx+2c}{|x|\sqrt{b^2-4ac}}, & \text{if } c<0, \end{cases}$$

71. $\int x^3\sqrt{x^2+a^2}\,dx=\left(\frac{1}{3}x^2-\frac{2}{15}a^2\right)(x^2+a^2)^{3/2},$

72. $\int x^n \sin(ax)dx=-\frac{1}{a}x^n\cos(ax)+\frac{n}{a}\int x^{n-1}\cos(ax)dx,$

73. $\int x^n \cos(ax)dx=\frac{1}{a}x^n\sin(ax)-\frac{n}{a}\int x^{n-1}\sin(ax)dx,$

74. $\int x^n e^{ax}\,dx=\frac{x^n e^{ax}}{a}-\frac{n}{a}\int x^{n-1}e^{ax}dx,$

75. $\int x^n \ln(ax)dx=x^{n+1}\left(\frac{\ln(ax)}{n+1}-\frac{1}{(n+1)^2}\right),$

76. $\int x^n(\ln ax)^m\,dx=\frac{x^{n+1}}{n+1}(\ln ax)^m-\frac{m}{n+1}\int x^n(\ln ax)^{m-1}dx.$

SERIES

Powers of natural numbers

$$\sum_{k=1}^{n}k=\frac{1}{2}n(n+1);\ \sum_{k=1}^{n}k^2=\frac{1}{6}n(n+1)(2n+1);\ \sum_{k=1}^{n}k^3=\frac{1}{4}n^2(n+1)^2$$

Arithmetic $\quad S_n=\sum_{k=0}^{n-1}(a+kd)=\frac{n}{2}\{2a+(n-1)d\}$

Geometric (convergent for $-1<r<1$)

$$S_n=\sum_{k=0}^{n-1}ar^k=\frac{a(1-r^n)}{1-r},\ S_\infty=\frac{a}{1-r}$$

Binomial (convergent for $|x|<1$)

$$(1+x)^n=1+nx+\frac{n!}{(n-2)!2!}x^2+\cdots+\frac{n!}{(n-r)!r!}x^r+\cdots$$

where $\frac{n!}{(n-r)!r!}=\frac{n(n-1)(n-2)\ldots(n-r+1)}{r!}$

Maclaurin series

$$f(x)=f(0)+xf'(0)+\frac{x^2}{2!}f''(0)+\cdots+\frac{x^k}{k!}f^{(k)}(0)+R_{k+1}$$

where $R_{k+1}=\frac{x^{k+1}}{(k+1)!}f^{(k+1)}(\theta x),\ 0<\theta<1$

Taylor series

$$f(a+h)=f(a)+hf'(a)+\frac{h^2}{2!}f''(a)+\cdots+\frac{h^k}{k!}f^{(k)}(a)+R_{k+1}$$

$$\text{where } R_{k+1} = \frac{h^{k+1}}{(k+1)!} f^{(k+1)}(a+\theta h),\ 0<\theta<1$$

OR

$$f(x) = f(x_0) + (x-x_0)f'(x_0) + \frac{(x-x_0)^2}{2!} f''(x_0) + \cdots + \frac{(x-x_0)^k}{k!} f^{(k)}(x_0) + R_{k+1}$$

$$\text{where } R_{k+1} = \frac{(x-x_0)^{k+1}}{(k+1)!} f^{(k+1)}(x_0 + (x-x_0)\theta),\ 0<\theta<1$$

Special Power Series

$$e^x = 1 + x + \frac{x^2}{2!} + \frac{x^3}{3!} + \cdots + \frac{x^r}{r!} + \cdots \qquad (\text{all } x)$$

$$\sin x = x - \frac{x^3}{3!} + \frac{x^5}{5!} - \frac{x^7}{7!} + \cdots + \frac{(-1)^r x^{2r+1}}{(2x+1)!} + \cdots \qquad (\text{all } x)$$

$$\cos x = 1 - \frac{x^2}{2!} + \frac{x^4}{4!} - \frac{x^6}{6!} + \cdots + \frac{(-1)^r x^{2r}}{(2r)!} + \cdots \qquad (\text{all } x)$$

$$\tan x = x + \frac{x^3}{3} + \frac{2x^5}{15} + \frac{17x^7}{315} + \cdots \qquad \left(|x| < \frac{\pi}{2}\right)$$

$$\sin^{-1} x = x + \frac{1}{2}\frac{x^3}{3} + \frac{1.3}{2.4}\frac{x^5}{5} + \frac{1.3.5}{2.4.6}\frac{x^7}{7} + \cdots + \frac{1.3.5....(2n-1)}{2.4.6....(2n)}\frac{x^{2n+1}}{2n+1} + \cdots \qquad (|x| < 1)$$

$$\tan^{-1} x = x - \frac{x^3}{3} + \frac{x^5}{5} - \frac{x^7}{7} + \cdots + (-1)^n \frac{x^{2n+1}}{2n+1} + \cdots \qquad (|x| < 1)$$

$$\ell n(1+x) = x - \frac{x^2}{2} + \frac{x^3}{3} - \frac{x^4}{4} + \cdots + (-1)^{n+1}\frac{x^n}{n} + \cdots \qquad (-1 < x \le 1)$$

$$\sinh x = x + \frac{x^3}{3!} + \frac{x^5}{5!} + \frac{x^7}{7!} + \cdots + \frac{x^{2n+1}}{(2n+1)!} + \cdots \qquad (\text{all } x)$$

$$\cosh x = 1 + \frac{x^2}{2!} + \frac{x^4}{4!} + \frac{x^6}{6!} + \cdots + \frac{x^{2n}}{(2n)!} + \cdots \qquad (\text{all } x)$$

$$\tanh x = x - \frac{x^3}{3} + \frac{2x^5}{15} - \frac{17x^7}{315} + \cdots \qquad \left(|x| < \frac{\pi}{2}\right)$$

$$\sinh^{-1} x = x - \frac{1}{2}\frac{x^3}{3} + \frac{1.3}{2.4}\frac{x^5}{5} - \frac{1.3.5}{2.4.6}\frac{x^7}{7} + \cdots + (-1)^n \frac{1.3.5...(2n-1)}{2.4.6...2n}\frac{x^{2n+1}}{2n+1} + \cdots \qquad (|x| < 1)$$

$$\tanh^{-1} x = x + \frac{x^3}{3} + \frac{x^5}{5} + \frac{x^7}{7} + \cdots \frac{x^{2n+1}}{2n+1} + \cdots \qquad (|x| < 1)$$

EQUATIONS OF BASIC GEOMETRIES

Straight Line

- General equation is $Ax + By + C = 0$
- Standard form of the equation is $y = mx + b$, *slope intercept form*
- *Point-slope* form is $y - y_1 = m(x - x_1)$ slope, $m = (y_2 - y_1)/(x_2 - x_1)$
- Angle between lines with slopes m_1 and m_2 is $\alpha = \arctan\,[(m_2 - m_1)/(1 + m_2 \cdot m_1)]$
- $m_1 = -1/m_2$ condition for perpendicular lines
- Distance between two points is $d = \sqrt{(y_2 - y_1)^2 + (x_2 - x_1)^2}$
- *Quadratic equation* $ax^2 + bx + c = 0$ the roots of the equation are x = Roots = $\dfrac{-b \pm \sqrt{b^2 - 4ac}}{2a}$

Circle

General equation is

$$(x - h)^2 + (y - k)^2 = r^2$$

where the center is at (h, k) and the radius is

$$r = \sqrt{(x - h)^2 + (y - k)^2}$$

Conic Section

The general form of the conic section equation is

$$Ax^2 + Bxy + Cy^2 + Dx + Ey + F = 0$$

where not both A and C are zero.

If $B^2 - 4AC < 0$, an *ellipse* is defined.
If $B^2 - 4AC > 0$, a *hyperbola* is defined.
If $B^2 - 4AC = 0$, the conic is a *parabola*.

If $A = C$ and $B = 0$, a *circle* is defined.
If $A = B = C = 0$, a *straight line* is defined.

Complex Numbers

Definition $i = \sqrt{-1}$

$$(a + ib) + (c + id) = (a + c) + i(b + d)$$

$$(a + ib) - (c + id) = (a - c) + i(b - d)$$

$$(a + ib)(c + id) = (ac - bd) + i(ad + bc)$$

$$\frac{a + ib}{c + id} = \frac{(a + ib)(c - id)}{(c + id)(c - id)} = \frac{(ac + bd) + i(bc - ad)}{c^2 + d^2}$$

Polar Coordinates

$$x = r\cos\theta;\ y = r\sin\theta;\ \theta = \arctan(y/x)$$

$$r = |x + iy| = \sqrt{x^2 + y^2}$$

$$x + iy = r(\cos\theta + i\sin\theta) = re^{i\theta}$$

$$[r_1(\cos\theta_1 + i\sin\theta_1)][r_2(\cos\theta_2 + i\sin\theta_2)] = r_1r_2[\cos(\theta_1 + \theta_2) + i\sin(\theta_1 + \theta_2)]$$

$$(x + iy)^n = [r(\cos\theta + i\sin\theta)]^n$$

$$= r^n(\cos n\theta + i\sin n\theta)$$

$$\frac{r_1(\cos\theta_1 + i\sin\theta_1)}{r_2(\cos\theta_2 + i\sin\theta_2)} = \frac{r_1}{r_2}[\cos(\theta_1 - \theta_2) + i\sin(\theta_1 - \theta_2)]$$

VECTOR FORMULAS

Scalar product $\mathbf{a}\cdot\mathbf{b} = ab\cos\theta = a_1b_1 + a_2b_2 + a_3b_3$

Vector product $\mathbf{a}\times\mathbf{b} = ab\sin\theta\ \hat{\mathrm{n}} = \begin{vmatrix} \mathbf{i} & \mathbf{j} & \mathbf{k} \\ a_1 & a_2 & a_3 \\ b_1 & b_2 & b_3 \end{vmatrix}$

$$= (a_2b_3 - a_3b_2)\mathbf{i} + (a_3b_1 - a_1b_3)\mathbf{j} + (a_1b_2 - a_2b_1)\mathbf{k}$$

Triple products

$$(\mathbf{a}\times\mathbf{b})\cdot\mathbf{c} = \mathbf{a}\cdot(\mathbf{b}\times\mathbf{c}) = \begin{vmatrix} a_1 & a_2 & a_3 \\ b_1 & b_2 & b_3 \\ c_1 & c_2 & c_3 \end{vmatrix}$$

$$\mathbf{a}\times(\mathbf{b}\times\mathbf{c}) = (\mathbf{a}\cdot\mathbf{c})\mathbf{b} - (\mathbf{a}\cdot\mathbf{b})\mathbf{c}$$

Vector Calculus

$$\nabla \equiv \left(\frac{\partial}{\partial x}, \frac{\partial}{\partial y}, \frac{\partial}{\partial z}\right)$$

grad $\phi \equiv \nabla\phi$,

div $\mathbf{A} \equiv \nabla\cdot\mathbf{A}$,

curl $\mathbf{A} \equiv \nabla\times\mathbf{A}$

div grad $\phi \equiv \nabla\cdot(\nabla\phi) \equiv \nabla^2\phi$ (for scalars only)

div curl $\mathbf{A} = 0$ curl grad $\phi \equiv 0$

$\nabla^2\mathbf{A}$ = grad div $\mathbf{A}$ – curl curl $\mathbf{A}$

$$\nabla(\alpha\beta) = \alpha\nabla\beta + \beta\nabla\alpha$$

$$\text{div}\,(\alpha\mathbf{A}) = \alpha\,\text{div}\,\mathbf{A} + \mathbf{A}\cdot(\nabla\alpha)$$

$$\text{curl}\,(\alpha\mathbf{A}) = \alpha\,\text{curl}\,\mathbf{A} - \mathbf{A}\times(\nabla\alpha)$$

$$\text{div}\,(\mathbf{A}\times\mathbf{B}) = \mathbf{B}\cdot\text{curl}\,\mathbf{A} - \mathbf{A}\cdot\text{curl}\,\mathbf{B}$$

$$\text{curl}\,(\mathbf{A}\times\mathbf{B}) = \mathbf{A}\,\text{div}\,\mathbf{B} - \mathbf{B}\,\text{div}\,\mathbf{A} + (\mathbf{B}\cdot\nabla)\mathbf{A} - (\mathbf{A}\cdot\nabla)\mathbf{B}$$

$$\text{grad}\,(\mathbf{A}\cdot\mathbf{B}) = \mathbf{A}\times\text{curl}\,\mathbf{B} + \mathbf{B}\times\text{curl}\,\mathbf{A} + (\mathbf{A}\cdot\nabla)\mathbf{B} + (\mathbf{B}\cdot\nabla)\mathbf{A}$$

Integral Theorems

Divergence theorem

$$\int_{\text{surface}} \mathbf{A}\cdot d\mathbf{S} = \int_{\text{volume}} \text{div}\,\mathbf{A}\;dV$$

Stokes' theorem

$$\int_{\text{surface}} (\text{curl}\,\mathbf{A})\cdot d\mathbf{S} = \oint_{\text{contour}} \mathbf{A}\cdot d\mathbf{r}$$

Green's theorems

$$\int_{\text{volume}} (\psi\nabla^2\phi - \phi\nabla^2\psi)\,dV = \int_{\text{surface}} \left(\psi\frac{\partial\phi}{\partial n} - \phi\frac{\partial\psi}{\partial n}\right)|d\mathbf{S}|$$

$$\int_{\text{volume}} \{\psi\nabla^2\phi + (\nabla\phi)(\nabla\psi)\}\,dV = \int_{\text{surface}} \psi\frac{\partial\phi}{\partial n}|d\mathbf{S}|$$

where

$$d\mathbf{S} = \hat{\text{n}}|d\mathbf{S}|$$

Green's theorem in the plane

$$\oint (P\,dx + Q\,dy) = \iint\left(\frac{\partial Q}{\partial x} - \frac{\partial P}{\partial y}\right)dx\,dy$$

GEOMETRIC PROPERTIES

Rectangle

Area $= ab$

Perimeter $= 2a + 2b$

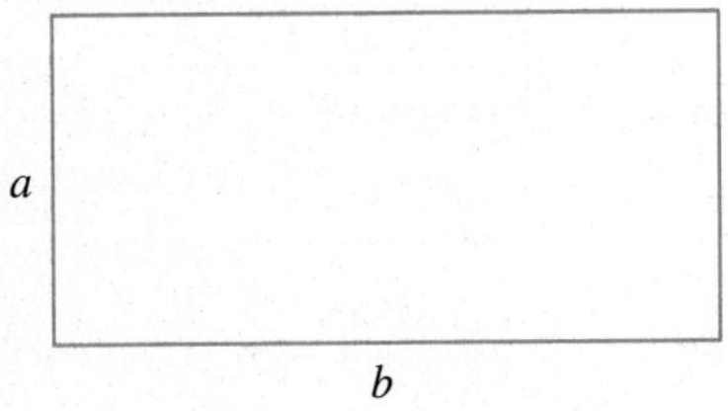

Parallelogram

Area $= bh = ab\sin\theta$

Perimeter $= 2a + 2b$

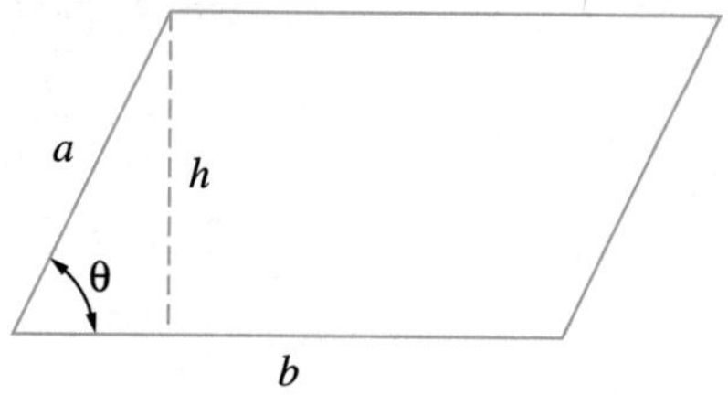

Triangle

$$\text{Area} = \frac{1}{2}bh = \frac{1}{2}ab\sin\theta$$

$$= \sqrt{s(s-a)(s-b)(s-c)}$$

where $s = \frac{1}{2}(a+b+c) =$ semiperimeter

Perimeter $= a + b + c$

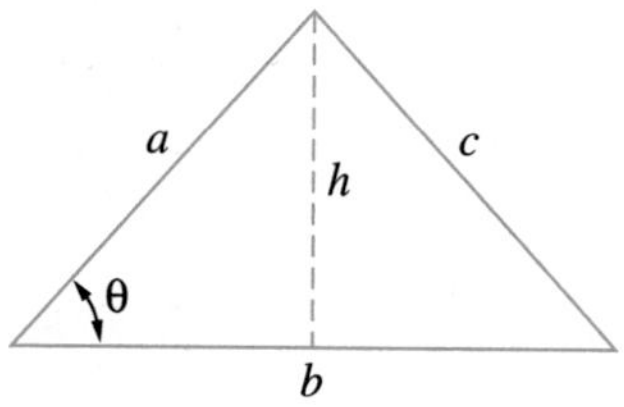

Trapezoid

$$\text{Area} = \frac{1}{2}h(a+b)$$

$$\text{Perimeter} = a + b + h\left(\frac{1}{\sin\theta} + \frac{1}{\sin\phi}\right)$$

$$= a + b + h(\csc\theta + \csc\phi)$$

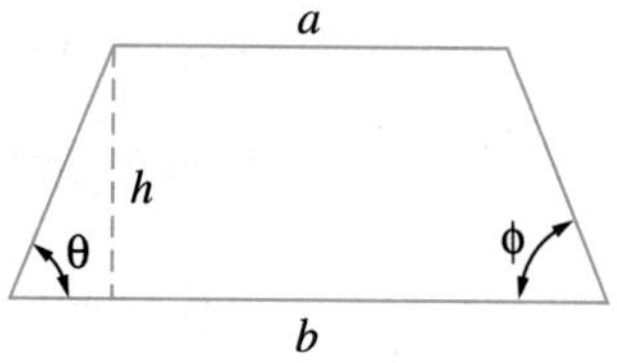

Circle

Area $= \pi r^2$

Perimeter $= 2\pi r$

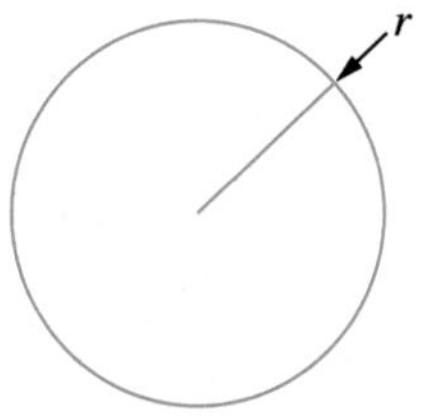

Polygon of *n* Sides

$$\text{Area} = \frac{1}{4}nb^2\cot\frac{\pi}{n} = \frac{1}{4}nb^2\frac{\cos(\pi/n)}{\sin(\pi/n)}$$

Perimeter $= nb$

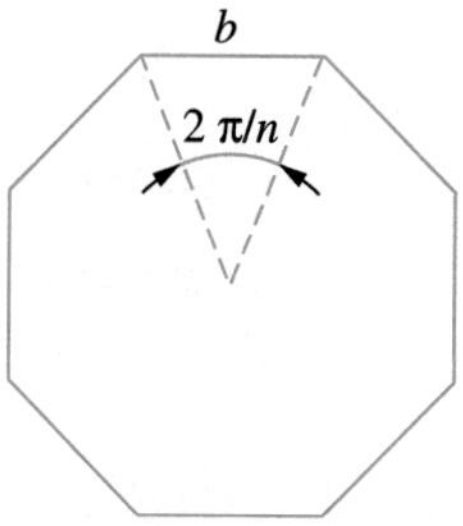

Sector of a Circle

$$\text{Area} = \frac{1}{2}r^2\theta \qquad [\theta \text{ in radians}]$$

Arc length $s = r\theta$

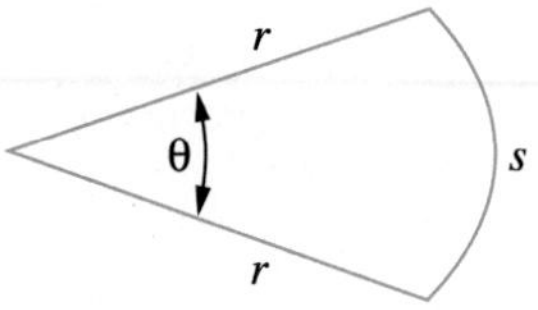

Circle Inscribed in a Triangle

$$r = \frac{\sqrt{s(s-a)(s-b)(s-c)}}{s}$$

where $s = \frac{1}{2}(a+b+c) =$ semiperimeter

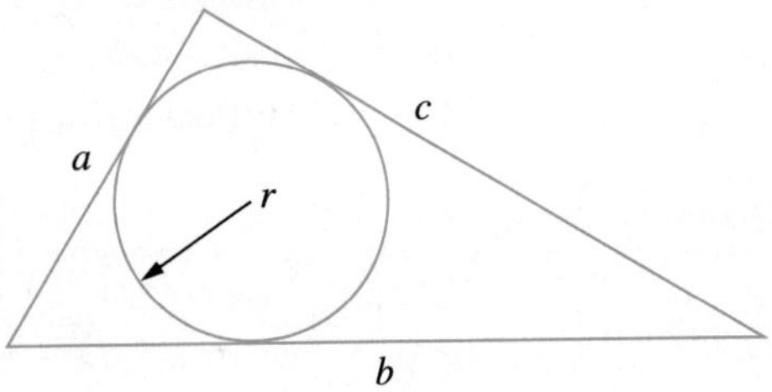

Circle Circumscribing a Triangle

$$R = \frac{abc}{4\sqrt{s(s-a)(s-b)(s-c)}}$$

where $s = \frac{1}{2}(a+b+c) =$ semiperimeter

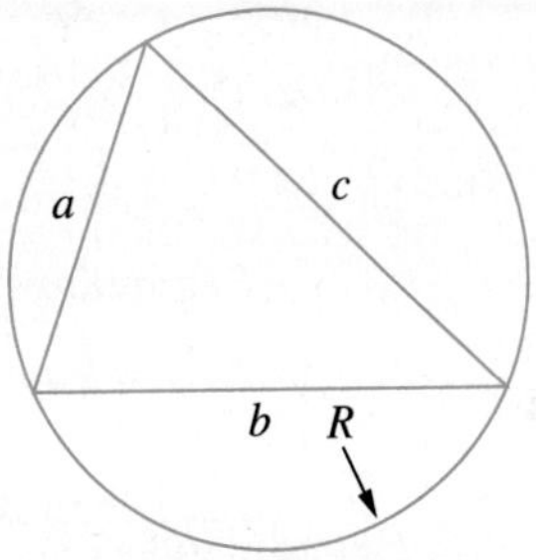

Polygon Inscribed in a Circle

$$\text{Area} = \frac{1}{2}nr^2 \sin\frac{2\pi}{n} = \frac{1}{2}nr^2 \sin\frac{360°}{n}$$

$$\text{Perimeter} = 2nr\sin\frac{\pi}{n} = 2nr\sin\frac{180°}{n}$$

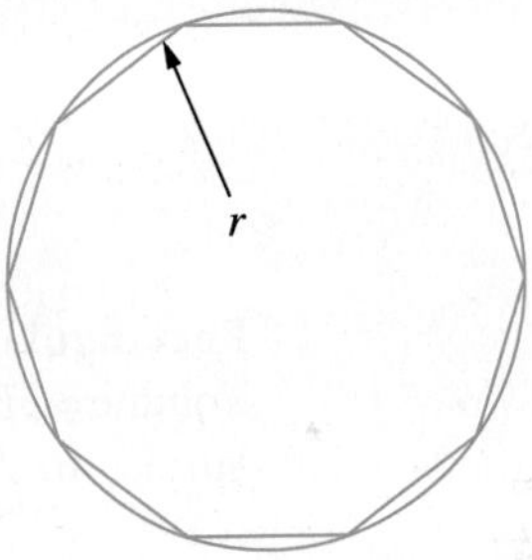

Polygon Circumscribing a Circle

$$\text{Area} = nr^2 \tan\frac{\pi}{n} = nr^2 \tan\frac{180°}{n}$$

$$\text{Perimeter} = 2nr\tan\frac{\pi}{n} = 2nr\tan\frac{180°}{n}$$

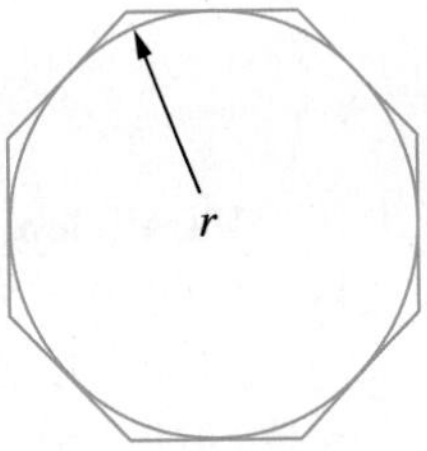

Segment of a Circle

Area of shaded part $= \frac{1}{2}r^2(\theta - \sin\theta)$

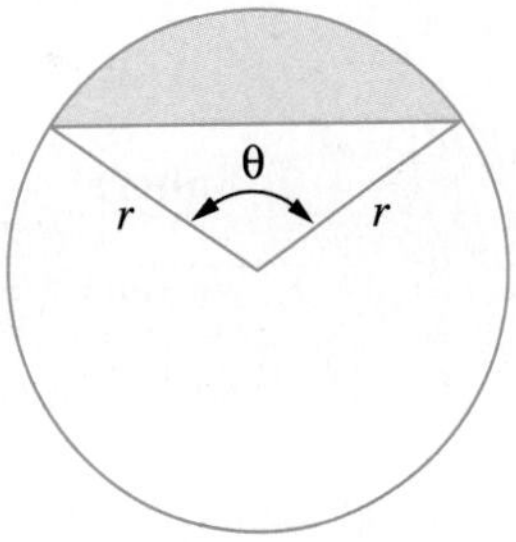

Ellipse

Area $= \pi ab$

Perimeter $= 4a\int_0^{\pi/2} \sqrt{1-k^2 \sin^2\theta}\, d\theta$

$= 2\pi\sqrt{\frac{1}{2}(a^2+b^2)}$ [approximately]

where $k = \sqrt{a^2-b^2}/a$.

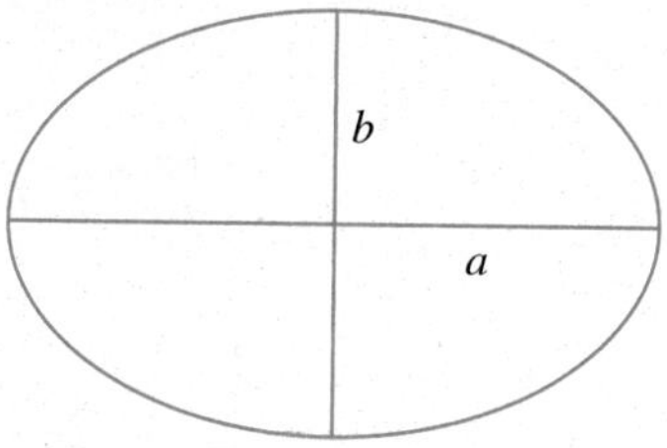

Segment of a Parabola

Area $= \frac{2}{3}ab$

Arc length $ABC = \frac{1}{2}\sqrt{b^2+16a^2} + \frac{b^2}{8a}\ln\left(\frac{4a+\sqrt{b^2+16a^2}}{b}\right)$

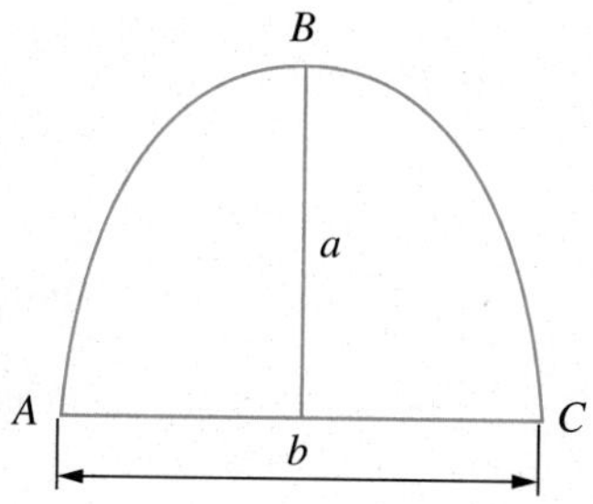

Rectangular Parallelepiped

Volume $= abc$

Surface area $= 2(ab+ac+bc)$

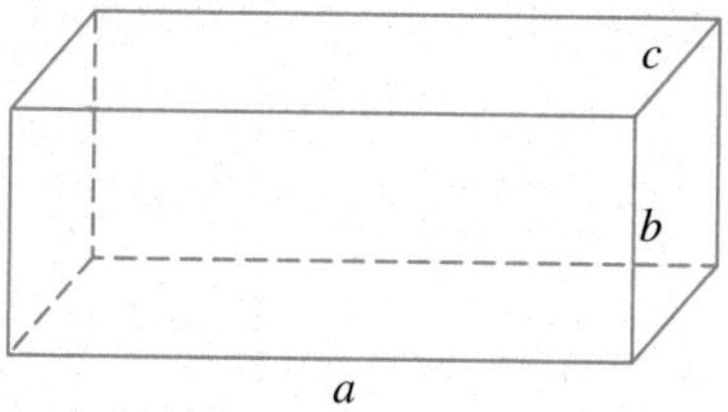

Parallelepiped of Area A

Volume $= Ah = abc\sin\theta$

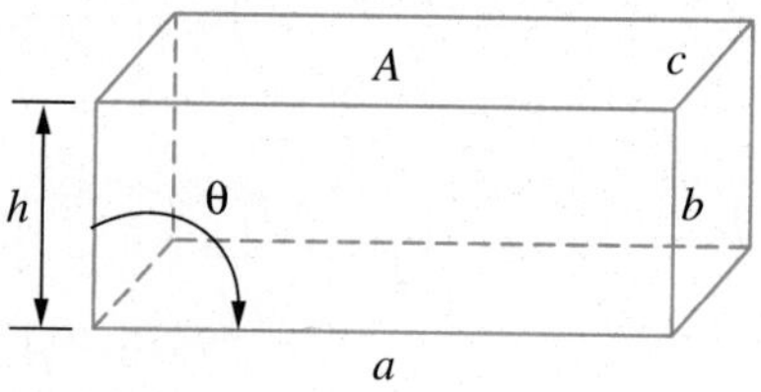

Sphere

Volume $= \frac{4}{3}\pi r^3$

Surface area $= 4\pi r^2$

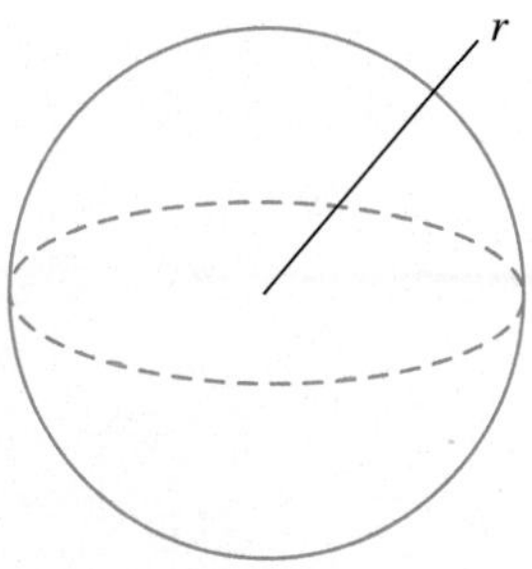

Right Circular Cylinder

Volume $= \pi r^2 h$

Lateral surface area $= 2\pi rh$

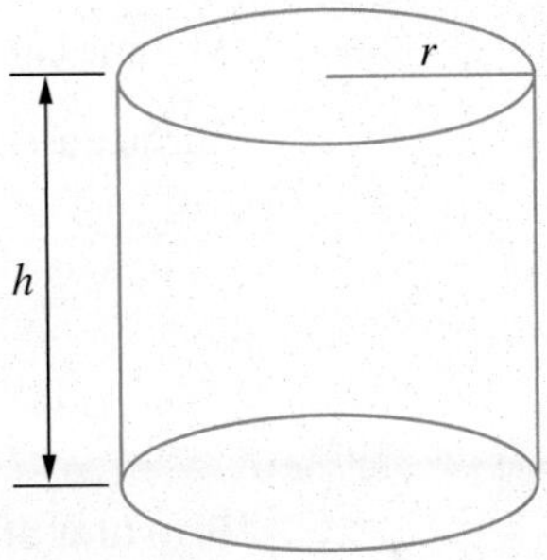

Circular Cylinder

Volume $= \pi r^2 h = \pi r^2 l \sin\theta$

Lateral surface area $= 2\pi rl = \dfrac{2\pi rh}{\sin\theta} = 2\pi rh\csc\theta$

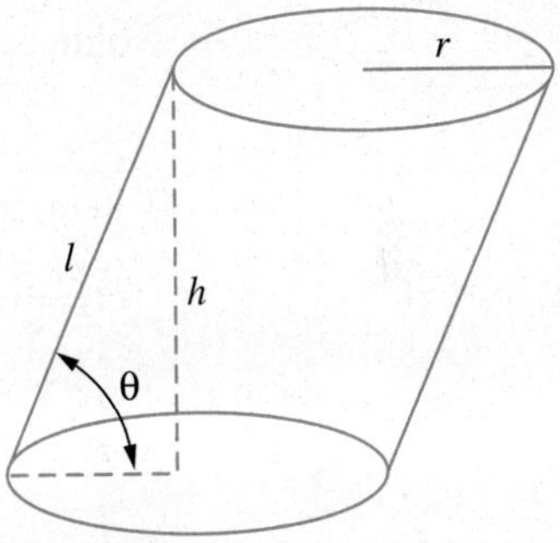

Cylinder of Cross-Sectional Area A

Volume $= Ah = Al\sin\theta$

Lateral surface area $= pl = \dfrac{ph}{\sin\theta} = ph\csc\theta$

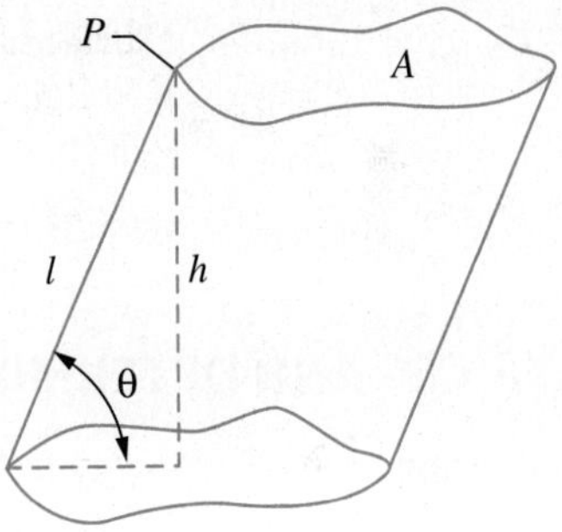

Right Circular Cone

Volume $= \dfrac{1}{3}\pi r^2 h$

Lateral surface area $= \pi r\sqrt{r^2 + h^2} = \pi rl$

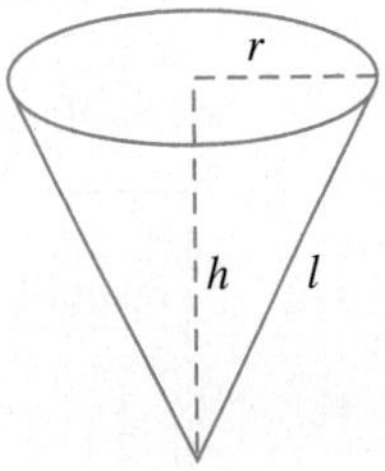

Pyramid

Volume $= \dfrac{1}{3}Ah$

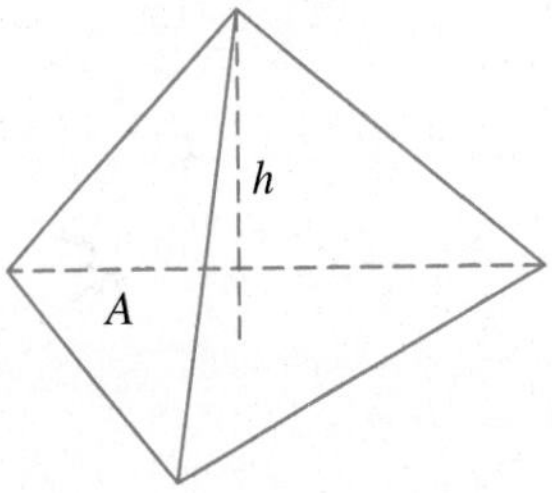

Spherical Cap

Volume (shaded in figure) $= \frac{1}{3}\pi h^2(3r - h)$

Surface area $= 2\pi rh$

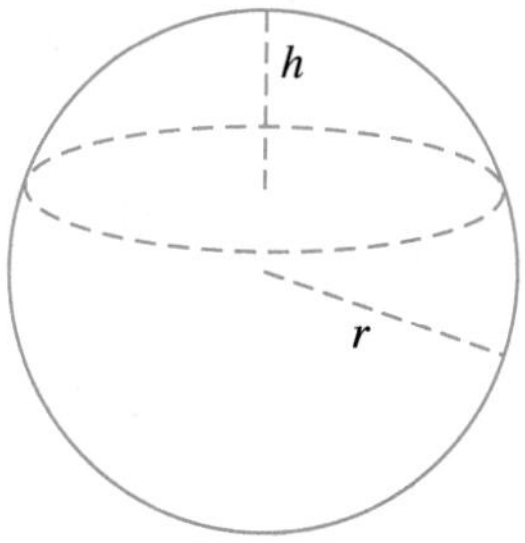

Frustum of Right Circular Cone

Volume $= \frac{1}{3}\pi h(a^2 + ab + b^2)$

Lateral surface area $= \pi(a+b)\sqrt{h^2 + (b-a)^2}$

$= \pi(a+b)l$

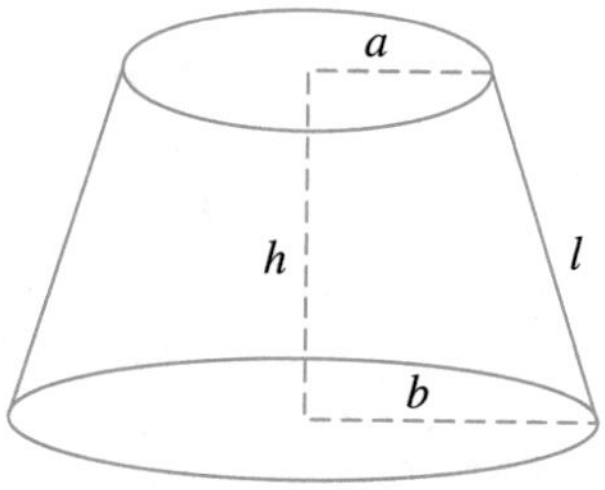

Torus

Volume $= \frac{1}{4}\pi^2(a+b)(b-a)^2$

Surface area $= \pi^2(b^2 - a^2)$

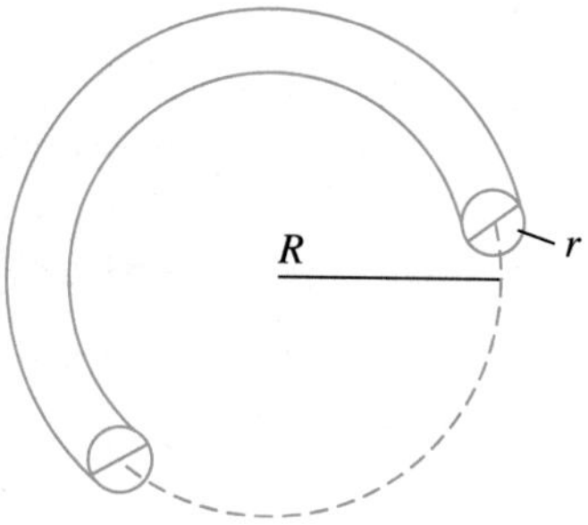

MATRICES AND DETERMINANTS

Introduction

A matrix is defined as an ordered rectangular array of numbers. They can be used to represent systems of linear equations, as will be explained below.

Examples of different types of matrices:

Symmetric	Diagonal	Upper Triangular	Lower Triangular	Zero	Identity
$\begin{bmatrix} 1 & 2 & 3 \\ 2 & 0 & -5 \\ 3 & -5 & 6 \end{bmatrix}$	$\begin{bmatrix} 1 & 0 & 0 \\ 0 & 4 & 0 \\ 0 & 0 & 6 \end{bmatrix}$	$\begin{bmatrix} 1 & 2 & 3 \\ 0 & 7 & -5 \\ 0 & 0 & -4 \end{bmatrix}$	$\begin{bmatrix} 1 & 0 & 0 \\ -4 & 7 & 0 \\ 12 & 5 & 3 \end{bmatrix}$	$\begin{bmatrix} 0 & 0 & 0 \\ 0 & 0 & 0 \\ 0 & 0 & 0 \end{bmatrix}$	$\begin{bmatrix} 1 & 0 & 0 \\ 0 & 1 & 0 \\ 0 & 0 & 1 \end{bmatrix}$

A fully expanded $m \times n$ matrix A is expressed as:

$$A = \begin{pmatrix} a_{11} & a_{12} & \cdots & a_{1n} \\ a_{21} & a_{22} & \cdots & a_{2n} \\ \vdots & \vdots & \ddots & \vdots \\ a_{m1} & a_{m2} & \cdots & a_{mn} \end{pmatrix}$$

or in a more compact form (indicial notation): $A = a_{ij}$

Matrix Addition and Subtraction

Two matrices A and B can be added or subtracted if and only if their dimensions are the same (i.e., both matrices have the identical amount of rows and columns). For example:

$$A=\begin{pmatrix}1 & 2 & 3\\ 1 & 0 & 2\end{pmatrix} \text{ and } B=\begin{pmatrix}2 & 1 & 2\\ 1 & 0 & 3\end{pmatrix}$$

Addition

If A and B above are matrices of the same type then the sum is found by adding the corresponding elements $a_{ij} + b_{ij}$.

EXAMPLE:

$$A+B=\begin{pmatrix}1 & 2 & 3\\ 1 & 0 & 2\end{pmatrix}+\begin{pmatrix}2 & 1 & 2\\ 1 & 0 & 3\end{pmatrix}=\begin{pmatrix}3 & 3 & 5\\ 2 & 0 & 5\end{pmatrix}$$

Subtraction

If A and B are matrices of the same type then the subtraction is found by subtracting the corresponding elements $a_{ij} - b_{ij}$.

EXAMPLE:

$$A-B=\begin{pmatrix}1 & 2 & 3\\ 1 & 0 & 2\end{pmatrix}-\begin{pmatrix}2 & 1 & 2\\ 1 & 0 & 3\end{pmatrix}=\begin{pmatrix}-1 & 1 & 1\\ 0 & 0 & -1\end{pmatrix}$$

Matrix Multiplication

When the number of columns of the first matrix is the same as the number of rows in the second matrix then matrix multiplication can be performed.

EXAMPLE 1:

$$\begin{pmatrix}a & b\\ c & d\end{pmatrix}\begin{pmatrix}e & f\\ g & h\end{pmatrix}=\begin{pmatrix}(ae+bg) & (af+bh)\\ (ce+dg) & (cf+dh)\end{pmatrix}$$

EXAMPLE 2:

$$\begin{pmatrix}a & b & c\\ d & e & f\\ g & h & i\end{pmatrix}\begin{pmatrix}j & k & l\\ m & n & o\\ p & q & r\end{pmatrix}=\begin{pmatrix}(aj+bm+cp) & (ak+bn+cq) & (al+bo+cr)\\ (dj+em+fp) & (dk+en+fq) & (dl+eo+fr)\\ (gj+hm+ip) & (gk+hn+iq) & (gl+ho+ir)\end{pmatrix}$$

Consider matrice A $(m \times n)$ and B $(n \times p)$, then the product of A and B is the matrix C, which has dimensions $m \times p$. The ij^{th} element of matrix C is found by multiplying the entries of the ith row of A with the corresponding entries in the jth column of B and summing the n terms. The elements of C are

$$
\begin{aligned}
c_{11} &= a_{11}b_{11} + a_{12}b_{21} + \cdots + a_{1n}b_{n1} = \sum_{j-1}^{n} a_{1j}b_{j1} \\
c_{12} &= a_{11}b_{12} + a_{12}b_{22} + \cdots + a_{1n}b_{n2} \\
\vdots \quad & \quad \vdots \qquad \vdots \qquad\qquad \vdots \\
c_{mp} &= a_{m1}b_{1p} + a_{m2}b_{2p} + \cdots + a_{mn}b_{np}
\end{aligned}
$$

Note: That $A \times B$ is not the same as $B \times A$.

Transpose of Matrices

The transpose of a matrix is found by exchanging rows for columns, that is, matrix $A = (a_{ij})$ and the transpose of A is:

$A^T = (a_{ji})$ where j is the column number and i is the row number of matrix A.

For example, the transpose of a matrix would be:

$$
A = \begin{pmatrix} 5 & 2 & 3 \\ 4 & 7 & 1 \\ 8 & 5 & 9 \end{pmatrix} \qquad A^T = \begin{pmatrix} 5 & 4 & 8 \\ 2 & 7 & 5 \\ 3 & 1 & 9 \end{pmatrix}
$$

In the case of a square matrix $(m = n)$, the transpose can be used to check if a matrix is symmetric. For a symmetric matrix $A = A^T$:

$$
A = \begin{pmatrix} 1 & 2 \\ 2 & 3 \end{pmatrix} = A^T = \begin{pmatrix} 1 & 2 \\ 2 & 3 \end{pmatrix} = A
$$

The Determinant of a Matrix

Determinants play an important role in finding the inverse of a matrix and also in solving systems of linear equations. In the following we assume we have a square matrix $(m = n)$. The determinant of a matrix A will be denoted by $\det(A)$ or $|A|$. Firstly the determinant of a 2×2 and 3×3 matrix will be introduced, then the $n \times n$ case will be shown.

Determinant of a 2 × 2 Matrix

Assuming A is an arbitrary 2×2 matrix, where the elements are given by:

$$
A = \begin{pmatrix} a_{11} & a_{12} \\ a_{21} & a_{22} \end{pmatrix}
$$

then the determinant of this matrix is as follows:

$$
\det(A) = |A| = \begin{vmatrix} a_{11} & a_{12} \\ a_{21} & a_{22} \end{vmatrix} = a_{11}a_{22} - a_{21}a_{12}
$$

Determinant of a 3 × 3 Matrix
Consider a 3 × 3 matrix A as

$$A = \begin{pmatrix} a_{11} & a_{12} & a_{13} \\ a_{21} & a_{22} & a_{23} \\ a_{31} & a_{32} & a_{33} \end{pmatrix}$$

then the determinant of this matrix is as follows:

$$\det(A) = |A| = \begin{vmatrix} a_{11} & a_{12} & a_{13} \\ a_{21} & a_{22} & a_{23} \\ a_{31} & a_{32} & a_{33} \end{vmatrix} = a_{11}\begin{vmatrix} a_{22} & a_{23} \\ a_{32} & a_{33} \end{vmatrix} - a_{12}\begin{vmatrix} a_{21} & a_{23} \\ a_{31} & a_{33} \end{vmatrix} + a_{13}\begin{vmatrix} a_{21} & a_{22} \\ a_{31} & a_{32} \end{vmatrix}$$

Determinant of a n × n Matrix
For the general case, where A is an $n \times n$ matrix, the determinant is given by:

$$\det(A) = |A| = a_{11}\alpha_{11} + a_{12}\alpha_{12} + \cdots + a_{1n}\alpha_{1n}$$

where the coefficients α_{ij} are given by the relation

$$\alpha_{ij} = (-1)^{i+j}\beta_{ij}$$

where β_{ij} is the determinant of the $(n-1) \times (n-1)$ matrix that is obtained by deleting row i and column j. This coefficient α_{ij} is also called the cofactor of a_{ij}.

The Inverse of a Matrix

Assuming we have a square matrix A, which is non-singular (i.e., $\det(A)$ does not equal to zero), then there exists an $n \times n$ matrix A^{-1} which is called the inverse of A, such that this property holds:

$$AA^{-1} = A^{-1}A = I$$

where I is the identity matrix.

The Inverse of a 2 × 2 Matrix
Take for example an arbitrary 2 × 2 matrix A whose determinant $(ad - bc)$ is not equal to zero

$$A = \begin{pmatrix} a & b \\ c & d \end{pmatrix}$$

where a, b, c, d are numbers, the inverse is

$$A^{-1} = \begin{pmatrix} a & b \\ c & d \end{pmatrix}^{-1} = \frac{1}{ad - bc}\begin{pmatrix} d & -b \\ -c & a \end{pmatrix}$$

The Inverse of a n × n Matrix
The inverse of a general $n \times n$ matrix A can be found by using the following equation:

$$A^{-1} = \frac{adj(A)}{\det(A)}$$

where the adj(A) denotes the adjoint (or adjugate) of a matrix. It can be calculated by the following method:

- Given the $n \times n$ matrix A, define $B = (b_{ij})$ to be the matrix whose coefficients are found by taking the determinant of the $(n-1) \times (n-1)$ matrix obtained by deleting the ith row and j^{th} column of A. The terms of B (i.e., $B = b_{ij}$) are known as the cofactors of A.
- And define the matrix C, where

$$c_{ij} = (-1)^{i+j} b_{ij}$$

- The transpose of C (i.e., C^T) is called the adjoint of matrix A.

Lastly, to find the inverse of A divide the matrix C^T by the determinant of A to give its inverse.

Solving Systems of Equations Using Matrices

A system of linear equations is a set of equations with n equations and n unknowns, and is of the form of

$$\begin{aligned}
a_{11}x_1 + a_{12}x_2 + \cdots + a_{1n}x_n &= b_1 \\
a_{21}x_1 + a_{22}x_2 + \cdots + a_{2n}x_n &= b_2 \\
&\vdots \\
a_{n1}x_1 + a_{n2}x_2 + \cdots + a_{nn}x_n &= b_n
\end{aligned}$$

The unknowns are denoted by $x_1, x_2, \ldots, x_n$ and the coefficients (a's and b's above) are assumed to be given. In matrix form the system of equations above can be written as:

$$\begin{pmatrix} a_{11} & a_{12} & \cdots & a_{1n} \\ a_{21} & a_{22} & \cdots & a_{2n} \\ \vdots & \vdots & \ddots & \vdots \\ a_{n1} & a_{n2} & \cdots & a_{nn} \end{pmatrix} \begin{pmatrix} x_1 \\ x_2 \\ \vdots \\ x_n \end{pmatrix} = \begin{pmatrix} b_1 \\ b_2 \\ \vdots \\ b_n \end{pmatrix}$$

A simplified way of writing the above is like this: $Ax = b$

After looking at this we will now look at two methods used to solve matrices:

- Inverse matrix method
- Cramer's rule

Inverse Matrix Method

The inverse matrix method uses the inverse of a matrix to help solve a system of equations, such as the above $Ax = b$. Pre-multiplying both sides of this equation by A^{-1} gives

$$A^{-1}(Ax) = A^{-1}b$$

$$(A^{-1}A)x = A^{-1}b$$

or alternatively this gives

$$x = A^{-1}b$$

So by calculating the inverse of the matrix and multiplying this by the vector b we can find the solution to the system of equations directly. And from earlier we found that the inverse is given by

$$A^{-1} = \frac{adj(A)}{\det(A)}$$

From the above, it is clear that the existence of a solution depends on the value of the determinant of A. There are three cases:

1. If the $\det(A)$ does not equal zero then solutions exist using $x = A^{-1}b$.
2. If the $\det(A)$ is zero and $b = 0$ then the solution will neither be unique nor exist.
3. If the $\det(A)$ is zero and $b = 0$ then the solution can be $x = 0$, but as in point above (2) it will neither be unique nor will it exist.

Looking at two equations we might have that

$$ax + by = c$$

$$dx + ey = f$$

Written in matrix form would look like

$$\begin{pmatrix} a & b \\ d & e \end{pmatrix}\begin{pmatrix} x \\ y \end{pmatrix} = \begin{pmatrix} c \\ f \end{pmatrix}$$

and by rearranging we would get that the solution would look like

$$\begin{pmatrix} x \\ y \end{pmatrix} = \begin{pmatrix} a & b \\ d & e \end{pmatrix}^{-1}\begin{pmatrix} c \\ f \end{pmatrix}$$

Similarly for three simultaneous equations we would have:

$$a_{11}x + a_{12}y + a_{13}z = b_1$$

$$a_{21}x + a_{22}y + a_{23}z = b_2$$

$$a_{31}x + a_{32}y + a_{33}z = b_3$$

Written in matrix form would look like

$$\begin{pmatrix} a_{11} & a_{12} & a_{13} \\ a_{21} & a_{22} & a_{23} \\ a_{31} & a_{32} & a_{33} \end{pmatrix}\begin{pmatrix} x \\ y \\ z \end{pmatrix} = \begin{pmatrix} b_1 \\ b_2 \\ b_3 \end{pmatrix}$$

and by rearranging we would get that the solution would look like

$$\begin{pmatrix} x \\ y \\ z \end{pmatrix} = \begin{pmatrix} a_{11} & a_{12} & a_{13} \\ a_{21} & a_{22} & a_{23} \\ a_{31} & a_{32} & a_{33} \end{pmatrix}^{-1}\begin{pmatrix} b_1 \\ b_2 \\ b_3 \end{pmatrix}$$

Cramer's Rule

Cramer's rule uses a method of determinants to solve systems of equations. Starting with the equation below:

$$a_{11}x_1 + a_{12}x_2 + \cdots + a_{1n}x_n = b_1$$

$$a_{21}x_1 + a_{22}x_2 + \cdots + a_{2n}x_n = b_2$$

$$\vdots$$

$$a_{n1}x_1 + a_{n2}x_2 + \cdots + a_{nn}x_n = b_n$$

The first term x_1 above can be found by replacing the first column of A with $(b_1 \quad b_2 \quad \ldots \quad b_n)^T$. Doing this we obtain:

$$x_1 = \frac{1}{|A|}\begin{vmatrix} b_1 & a_{12} & a_{13} & \cdots & a_{1n} \\ b_2 & a_{22} & a_{23} & \cdots & a_{2n} \\ \vdots & \vdots & \vdots & \ddots & \vdots \\ b_n & a_{n2} & a_{n3} & \cdots & a_{nn} \end{vmatrix}$$

Similarly, for the general case for solving x_r we replace the rth column of A with $(b_1 \quad b_2 \quad \ldots \quad b_n)^T$ and expand the determinant.

This method of using determinants can be applied to solve systems of linear equations. We will illustrate this for solving two simultaneous equations in x and y and three equations with three unknowns x, y, and z.

Two Simultaneous Equations in x and y

$$ax + by = p$$

$$cx + dy = q$$

To solve, use the following:

$$x = \frac{\text{Det}\left(\begin{bmatrix} p & b \\ q & d \end{bmatrix}\right)}{\text{Det}\left(\begin{bmatrix} a & b \\ c & d \end{bmatrix}\right)} \quad \text{and} \quad y = \frac{\text{Det}\left(\begin{bmatrix} a & p \\ c & q \end{bmatrix}\right)}{\text{Det}\left(\begin{bmatrix} a & b \\ c & d \end{bmatrix}\right)}$$

or simplified:

$$x = \frac{pd - bq}{ad - bc} \quad \text{and} \quad y = \frac{aq - cp}{ad - bc}$$

Three Simultaneous Equations in x, y and z

$$ax + by + cz = p$$

$$dx + ey + fz = q$$

$$gx + hy + iz = r$$

The solution is

$$x = \frac{\text{Det}\left(\begin{bmatrix} p & b & c \\ q & e & f \\ r & h & i \end{bmatrix}\right)}{\text{Det}\left(\begin{bmatrix} a & b & c \\ d & e & f \\ g & h & i \end{bmatrix}\right)}, \quad y = \frac{\text{Det}\left(\begin{bmatrix} a & p & c \\ d & q & f \\ g & r & i \end{bmatrix}\right)}{\text{Det}\left(\begin{bmatrix} a & b & c \\ d & e & f \\ g & h & i \end{bmatrix}\right)}, \quad z = \frac{\text{Det}\left(\begin{bmatrix} a & b & p \\ d & e & q \\ g & h & r \end{bmatrix}\right)}{\text{Det}\left(\begin{bmatrix} a & b & c \\ d & e & f \\ g & h & i \end{bmatrix}\right)}$$

EIGEN VALUES

Given a linear transformation A, a non-zero vector $\mathbf{x}$ is defined to be an *eigenvector* of the transformation if it satisfies the eigenvalue equation, defined as

$$A\mathbf{x} = \lambda\mathbf{x}$$

for some scalar λ. In this situation, the scalar λ is called an *eigenvalue* of A corresponding to the eigenvector $\mathbf{x}$.

The matrix representation is

$$\begin{bmatrix} a_{11} & a_{12} \\ a_{21} & a_{22} \end{bmatrix}\begin{bmatrix} x \\ y \end{bmatrix} = \lambda\begin{bmatrix} x \\ y \end{bmatrix}$$

Computation of Eigenvalues, and the Characteristic Equation

When a transformation is represented by a square matrix A, the eigenvalue equation can be expressed as $A\mathbf{x} - \lambda I\mathbf{x} = \mathbf{0}$.

This can be rearranged to

$$(A - \lambda I)\mathbf{x} = 0$$

If there exists an *inverse*

$$(A - \lambda I)^{-1}$$

then both sides can be left-multiplied by the inverse to obtain the trivial solution: $\mathbf{x} = \mathbf{0}$. For a non-trivial solution, the determinant must be equal to zero:

$$\det(A - \lambda I) = 0$$

The determinant requirement is called the *characteristic equation* of A, and the left-hand side is called the *characteristic polynomial.* When expanded, this gives a polynomial equation for λ.

REFERENCES

Adams, R. A., *Calculus: A Complete Course*, 5th ed., Pearson Education Ltd., 2003.

Jeffrey, A., and H. H. Dai, *Handbook of Mathematical Formulas and Integrals*, 4th ed., Academic Press, 2008.

Spiegel, M. R., *Schaum's Mathematical Handbook of Formulas and Tables*, McGraw-Hill, 1998.

Zwillinger, D., *CRC Standard Mathematical Tables and Formulae*, 31st ed., Chapman and Hall/CRC, 2003.

APPENDIX C

Glossary

The definitions given here apply to the terminology used throughout this book. Some of the terms may be defined differently by other authors; when this is the case, alternative terminology is noted. When two or more terms with identical or similar meaning are in general acceptance, they are given in the order of preference of the current writers.

Allowable stress (working stress) If a member is so designed that the maximum stress as calculated for the expected conditions of service is less than some limiting value, the member will have a proper margin of security against damage or failure. This limiting value is the allowable stress subject to the material and condition of service in question. The allowable stress is made less than the damaging stress because of uncertainty as to the conditions of service, nonuniformity of material, and inaccuracy of the stress analysis (see Ref. 1). The margin between the allowable stress and the damaging stress may be reduced in proportion to the certainty with which the conditions of the service are known, the intrinsic reliability of the material, the accuracy with which the stress produced by the loading can be calculated, and the degree to which failure is unattended by danger or loss. (Compare with *Damaging stress; Factor of safety; Margin of safety*. See Refs. 1–3.)

Apparent elastic limit (useful limit point) The stress at which the rate of change of strain with respect to stress is 50% greater than at zero stress. It is more definitely determinable from the stress–strain diagram than is the proportional limit, and is useful for comparing materials of the same general class. (Compare with *Elastic limit; Proportional limit; Yield point, Yield strength.*)

Apparent stress The stress corresponding to a given unit strain on the assumption of uniaxial elastic stress. It is calculated by multiplying the unit strain by the modulus of elasticity, and may differ from the true stress because the effect of the transverse stresses is not taken into account.

Bending moment Reference is to a simple straight beam, assumed for convenience to be horizontal and loaded and supported by forces, all of which lie in a vertical plane. The bending moment at any section of the beam is the moment of all forces that act on the beam to the left (or right) of that section, taken about the horizontal axis in the plane of the section. When considering the moment at the section due to the forces to the left of the section, the bending moment is positive when counterclockwise and negative when clockwise. The reverse is true when considering the moment due to forces to the right of the section. Thus, a positive bending moment bends the beam such that the beam deforms concave upward, and a negative bending moment bends it concave downward. The *bending moment equation* is an expression for the bending moment at any section in terms of x, the distance along the longitudinal axis of the beam to the section measured from an origin, usually taken to be the left end of the beam.

Bending moments as applied to straight beams in two-plane symmetric or unsymmetric bending, curved beams, or plates are a bit more involved and are discussed in the appropriate sections of this book.

Bending stress (flexural stress) The tensile and compressive stress transmitted in a beam or plate that arises from the bending moment. (See also *Flexure equation.*)

Boundary conditions As used in structural analysis, the term usually refers to the condition of stress, displacement, or slope at the ends or edges of a member, where these conditions are apparent from the circumstances of the problem. For example, given a beam with fixed ends, the zero displacement and slope at each end are boundary conditions. For a plate with a freely supported edge, the zero-stress state is a boundary condition.

Brittle fracture The tensile failure of a material with negligible plastic deformation. The material can inherently be a brittle material in its normal state such as glass, masonry, ceramic, cast iron, or high strength high-carbon steel (see Sec. 3.7); or it can be a material normally considered ductile which contains imperfections exceeding specific limits, or in a low-temperature environment, or undergoing high strain rates, or any combination thereof.

Bulk modulus of elasticity The ratio of a tensile or compressive stress, triaxial and equal in all directions (e.g., hydrostatic pressure) to the relative change it produces in volume.

Central axis (centroidal axis) A central axis of a line, area, or volume is one that passes through the *centroid*; in the case of an area, it is understood to lie in the plane of the area unless stated otherwise. When taken normal to the plane of the area, it is called the *central polar axis*.

Centroid of an area That point in the plane of an area where the moment of the area is zero about any axis. The centroid coincides with the center of gravity in the plane of an infinitely thin homogeneous uniform plate.

Corrosion fatigue Fatigue aggravated by corrosion, as in parts repeatedly stressed while exposed to a corrosive environment.

Creep Continuous increase in deformation under constant or decreasing stress. The term is ordinarily used with reference to the behavior of metals under tension at elevated temperatures. The similar yielding of a material under compressive stress is called *plastic flow*, or *flow*. Creep at atmospheric temperature due to sustained elastic stress is sometimes called *drift*, or *elastic drift*. (See also *Relaxation*.)

Damaging stress The least unit stress of a given kind and for a given material and condition of service that will render a member unfit for service before the end of its useful life. It may do this by excessive deformation, by excessive yielding or creep, or through fatigue cracking, excessive strain hardening, or rupture.

Damping capacity The amount of energy dissipated into heat per unit of total strain energy present at maximum strain for a complete cycle. (See Ref. 4.)

Deformation Change in the shape or dimensions of a body produced by stress. *Elongation* is often used for tensile deformation, *compression* or *shortening* for compressive deformation, and *distortion* for shear deformation. *Elastic deformation* is deformation that invariably disappears upon removal of stress, whereas *permanent deformation* is that which remains after the removal of stress. (Compare with *Set*.)

Eccentricity A load or component of a load normal to a given cross section of a member is eccentric with respect to that section if it does not act through the centroid. The perpendicular distance from the line of action of the load to the central polar axis is the eccentricity with respect to that axis.

Elastic Capable of sustaining stress without permanent deformation; the term is also used to denote conformity to the law of stress– strain proportionality (Hooke's law). An elastic stress or strain is a stress or strain within the elastic limit.

Elastic axis The elastic axis of a beam is the line, lengthwise of the beam, along which transverse loads must be applied to avoid torsion of the beam at any section. Strictly speaking, no such line exists except for a few conditions of loading. Usually the elastic axis is assumed to be the line through the elastic center of every section. The term is most often used with reference to an airplane wing of either the shell or multiple spar type. (Compare with *Torsional center*; *Flexural center*; *Elastic center*. See Ref. 5.)

Elastic center The elastic center of a given section of a beam is that point in the plane of the section lying midway between the shear center and center of twist of that section. The three points may be identical—which is the normal assumption. (Compare with *Shear center*; *Torsional center*; *Elastic axis*. See Refs. 5 and 6.)

Elastic curve The curve assumed by the longitudinal axis of an initially straight beam or column in bending where the stress is within the elastic limit.

Elastic instability (buckling) Unstable local or global elastic deformations caused by compressive stresses in members with large length to lateral dimensions. (See *Slenderness ratio*.)

Elastic limit The least stress that will cause permanent set. (Compare with *Proportional limit*; *Apparent elastic limit, Yield point*; *Yield strength*. See Sec. 3.2 and Ref. 7.)

Elastic, perfectly plastic material A model that represents the stress–strain curve of a material as linear from zero stress and strain to the elastic limit. Beyond the elastic limit, the stress remains constant with strain.

Elastic ratio The ratio of the elastic limit to the ultimate strength.

Ellipsoid of strain An ellipsoid that represents the state of strain at any given point in a body. It has the shape assumed under stress by a sphere centered at the point in question (Ref. 8).

Ellipsoid of stress An ellipsoid that represents the state of stress at any given point in a body; its semi-axes are vectors representing the principal stresses at the point, and any radius vector represents the resultant stress on a particular plane through the point. For a condition of plane stress, where one of the principal stresses is zero, the ellipsoid becomes the *ellipse of stress* (see Ref. 9).

Endurance limit (fatigue strength) The maximum stress amplitude of a purely reversing stress that can be applied to a material an indefinitely large number of cycles without producing fracture (see Sec. 3.8).

Endurance ratio Ratio of the endurance limit to the ultimate static tensile strength.

Endurance strength The maximum stress amplitude of a purely reversing stress that can be applied to a material for a specific number of cycles without producing fracture. (Compare with *Endurance limit*.)

Energy of rupture (modulus of toughness) The work done per unit volume in producing fracture. It is not practicable to establish a specific energy of rupture value for a given material, because the result obtained depends upon the form and proportions of the test specimen and the manner of loading. As determined by similar tests on similar specimens, the energy of rupture affords a criterion for comparing the toughness of different materials.

Equivalent bending moment A bending moment that, acting alone, would produce in a circular shaft a normal (tensile or compressive) stress of the same magnitude as the maximum normal stress produced by a given bending moment and a given twisting moment acting simultaneously.

Equivalent twisting moment A twisting moment that, acting alone, would produce in a circular shaft a shear stress of the same magnitude as the maximum shear stress produced by a given twisting moment and a given bending moment acting simultaneously.

Factor of safety The intent of the factor of safety is to provide a safeguard to failure. The term usually refers to the ratio of the load that would cause failure of a member or structure to the load that is imposed upon it in service. The term may also be used to represent the ratio of the failure to service value of speed, deflection, temperature variation, or other stress-producing quantities. (Compare with *Allowable stress*; *Margin of safety*.)

Fatigue The fracture of a material under many repetitions of a stress at a level considerably less than the ultimate strength of the material.

Fatigue strength See *Endurance limit.*

Fixed (clamped) A support condition at the end of a beam or column or at the edge of a plate or shell that prevents *rotation and transverse displacement* of the edge of the neutral surface but permits *longitudinal displacement.* (Compare *Guided; Held; Simply supported.*)

Flexural center See *Shear center.*

Flexural rigidity (beam, plate) A measure of the resistance of the bending deformation of a beam or plate. For a beam, the flexural rigidity is given by EI; whereas for a plate of thickness t, it is given by $Et^3/[12(1-\nu)]$.

Flexure equation The equation for tensile and compressive stresses in beams undergoing bending, given by $\sigma = Mc/I$.

Form factor The term is applied to several situations pertaining to beams:

(1) Given a beam section of a given shape, the form factor is the ratio of the modulus of rupture of a beam having that particular section to the modulus of rupture of a beam otherwise similar but having a section adopted as a standard. This standard section is usually taken as rectangular or square; for wood it is a 2 in by 2 in square with edges horizontal and vertical (see Secs. 3.11 and 8.15).

(2) For the shear deflection of a beam due to transverse loading, the form factor is a correction factor that is the ratio of the actual shear deflection to the shear deflection calculated on the assumption of a uniform shear stress across the section (see Sec. 8.10).

(3) For a given maximum fiber stress within the elastic limit, the form factor is the ratio of the actual resisting moment of a wide-flanged beam to the resisting moment the beam would develop if the fiber stress were uniformly distributed across the entire width of the flanges. So used, the term expresses the strength-reducing effort of shear lag.

Fracture toughness, KIc A material property which describes the ability of a material containing a crack to resist fracture. Has the units of MPa-m1/2. See Chap. 19 for more details.

Fretting fatigue (chafing fatigue) Fatigue aggravated by surface rubbing, as in shafts with press-fitted collars.

Guided A support condition at the end of a beam or column or at the edge of a plate or shell that prevents *rotation* of the edge of the neutral surface in the plane of bending but permits *longitudinal and transverse displacement.* (Compare with *Fixed; Held; Simply supported.*)

Held A support condition at the end of a beam or column or at the edge of a plate or shell that prevents *longitudinal and transverse displacement* of the edge of the neutral surface but permits *rotation* in the plane of bending. (Compare with *Fixed; Guided; Simply supported.*)

Hertzian stress (contact stress) Stress caused by the pressure between elastic bodies in contact.

Hysteresis The dissipation of energy as heat during a stress cycle of a member.

Influence line Usually pertaining to a particular section of a beam, an influence line is a curve drawn so that its ordinate at any point represents the value of the reaction, vertical shear, bending moment, or deflection produced at the particular section by a unit load applied at the point where the ordinate is measured. An influence line may be used to show the effect of load position on any quantity dependent thereon, such as the stress in a given truss member, the deflection of a truss, or the twisting moment in a shaft.

Isoclinic A line (in a stressed body) at all points on which the corresponding principal stresses have the same direction.

Isotropic Having the same properties in all directions. In discussions pertaining to strength of materials, isotropic usually means having the same strength and elastic properties (modulus of elasticity, modulus of rigidity, and Poisson's ratio) in all directions.

Kern (kernal) Reference is to some particular section of a member. The kern is that area in the plane of a section through which the line of action of a force must pass if that force is to produce, at all points in the given section, the same kind of normal stress, that is, tension throughout or compression throughout.

Limit load The fictitious theoretical load that the cross section of a member made of an elastic, perfectly plastic material reaches when the entire section goes into the plastic range.

Lüder's lines See *Slip lines*.

Margin of safety As used in aeronautical design, margin of safety is the percentage by which the ultimate strength of a member exceeds the *design load*. The *design load* is the applied load, or maximum probable load, multiplied by a specified factor of safety. [The use of the terms *margin of safety* and *design load* in this sense is practically restricted to aeronautical engineering (see Ref. 11).]

Member Any single part or element of a machine or structure, such as a beam, column, shaft, etc.

Modulus of elasticity, *E* (Young's modulus) The rate of change of normal stress, σ, to normal strain, ε, for the condition of uniaxial stress within the proportional limit of a given material. For most, but not all, materials, the modulus of elasticity is the same for tension and compression. For nonisotropic materials such as wood, it is necessary to distinguish between the moduli of elasticity in different directions.

Modulus of resilience The strain energy per unit volume absorbed up to the elastic limit under conditions of uniform uniaxial stress.

Modulus of rigidity, *G* (modulus of elasticity in shear) The rate of change of shear stress, τ, with respect to shear strain, γ, within the proportional limit of a given material. For nonisotropic materials such as wood, it is necessary to distinguish between the moduli of rigidity in different directions.

Modulus of rupture in bending (computed ultimate bending strength) The fictitious normal stress in the extreme fiber of a beam computed by the flexure equation $\sigma = M_R c/I$, where M_R is the bending moment that causes rupture.

Modulus of rupture in torsion (computed ultimate torsional strength) The fictitious shear stress at the outer radius of a circular shaft computed by the torsion equation $\tau = T_R r/J$, where T_R is the torsional moment that causes rupture.

Moment of an area (first moment of an area) With respect to an axis within the plane of an area, the sum of the products obtained by multiplying each element of the area dA by its distance, y, from the axis: it is therefore the quantity $\int y\,dA$.

Moment of inertia of an area (second moment of an area) With respect to an axis x within the xy plane of an area, the sum of the products obtained by multiplying each element of the area dA by the square of the distance y from the x axis: it is thus the quantity $I_x = \int y^2\,dA$ (see App. A).

Neutral axis The line of zero fiber stress in any given section of a member subject to bending; it is the line formed by the intersection of the neutral surface and the section.

Neutral surface The longitudinal surface of zero fiber stress in a member subject to bending; it contains the neutral axis of every section.

Notch-sensitivity factor Used to compare the *stress concentration factor* K_t and *fatigue-strength reduction factor* K_f. The notch-sensitivity factor q is commonly defined as the ratio $(K_f - 1)/(K_t - 1)$, and varies from 0, for some soft ductile materials, to 1, for some hard brittle materials.

Plane strain A condition where the normal and shear strains in a particular direction are zero; for example, $\varepsilon_z = \gamma_{zx} = \gamma_{zy} = 0$.

Plane stress A condition where the normal and shear stresses in a particular direction are zero; for example, $\sigma_z = \tau_{zx} = \tau_{zy} = 0$.

Plastic moment; plastic hinge; plastic section modulus The maximum hypothetical bending moment for which the stresses in *all* fibers of a section of a ductile member in bending reach the lower yield point σ_y is called the *plastic moment*, M_p. Under this condition the section cannot accommodate any additional load, and a *plastic hinge* is said to form. The *section modulus* Z_p is defined as M_p/σ_y.

Plasticity The property of sustaining appreciable permanent deformation without rupture. The term is also used to denote the property of yielding or flowing under steady load (Ref. 13).

Poisson's ratio, ν The ratio of lateral to longitudinal strain under the condition of uniform and uniaxial longitudinal stress within the proportional limit.

Polar moment of inertia With respect to an axis normal to the plane of an area, the sum of the products obtained by multiplying each element of the area dA by the square of the distance r from the axis; it is thus the quantity $\int r^2\, dA$ (see App. A).

Principal axes of inertia The two mutually perpendicular axes in the plane of an area, centered at the centroid of the area, with moments of inertia that are maximum and minimum (see App. A).

Principal axes of stress The three mutually perpendicular axes at a specific point within a solid where the state of stress on each surface normal to the axes contains a tensile or compressive stress and zero shear stress.

Principal moment of inertia The moment of inertia of an area about a principal axis of inertia (see App. A).

Principal stresses The tensile or compressive stresses acting along the principal axes of stress.

Product of inertia of an area With respect to a pair of xy rectangular axes in the plane of an area, the sum of the products obtained by multiplying each element of area dA by the coordinates with respect to these axes; that is, $\int xy\, dA$ (see App. A). The product of inertia relative to the principal axes of inertia is zero.

Proof stress Pertaining to acceptance tests of metals, a specified tensile stress that must be sustained without deformation in excess of a specified amount.

Proportional limit The greatest stress that a material can sustain without deviating from the law of stress–strain proportionality (*Hookes' law*). (Compare *Apparent elastic limit*; *Elastic limit*; *Yield point*; *Yield strength*. See Sec. 3.2 and Ref. 8.)

Radius of gyration, k The *radius of gyration* of an area with respect to a given axis is the square root of the quantity obtained by dividing the moment of inertia of the area I with respect to that axis by the area A; that is, $k = \sqrt{I/A}$ (see App. A).

Reduction of area The difference between the cross-sectional areas of a tensile specimen at the section of rupture before loading and after rupture.

Relaxation The reduction in stress when the deformation is maintained constant. (Compare with *Creep*.)

Rupture factor Used in reference to brittle materials, that is, materials in which failure occurs through tensile rupture rather than excessive deformation. For a member of given form, size, and material, loaded and supported in a given manner, the *rupture factor* is the ratio of the fictitious maximum tensile stress at failure, as calculated by the appropriate formula for elastic stress, to the ultimate tensile strength of the material, as determined by a conventional tension test (Sec. 3.11).

Saint-Venant's principle If a load distribution is replaced by a statically equivalent force system, the distribution of stress throughout the body is possibly altered *only* near the regions of load application.

Section modulus (section factor), *S* Pertaining to the cross section of a beam, the *section modulus* with respect to either principal axis of inertia is the moment of inertia with respect to that axis, I, divided by the distance from that axis to the most remote point of the section, c; that is, $S = I/c$. (Compare with *Plastic section modulus.*)

Set (permanent deformation) Strain remaining after the removal of the applied loading.

Shakedown load (stabilizing load) The maximum load that can be applied to a beam or rigid frame and upon removal leave residual moments such that subsequent applications of the same or a smaller load will cause only elastic stresses.

Shape factor The ratio of the plastic section modulus to the elastic section modulus.

Shear center (flexural center) With reference to a beam, the shear center of any section is that point in the plane of the section through which a transverse load, applied at the section, must act to produce bending deflection only and no twist of the section. (Compare with *Torsional center*; *Elastic center*; *Elastic axis*. See Refs. 5 and 10.)

Shear lag Because of shear strain, the longitudinal tensile or compressive bending stresses in wide beam flanges decrease with the distance from the web(s), and this stress reduction is called *shear lag.*

Simply supported A support condition at the end of a beam or column or at the edge of a plate or shell that prevents *transverse displacement* of the edge of the neutral surface but permits *rotation and longitudinal displacement.* (Compare with *Fixed*, *Guided*; *Held.*)

Singularity functions A class of mathematical functions that can be used to describe discontinuous behavior using one equation. *Singularity functions* are commonly employed to represent shear forces, bending moments, slopes, and deformations as functions of position for discontinuous loading of beams, plates, and shells. The functions are written using bracket notation as $F_n = \langle x - a \rangle^n$, where $F_n = 0$ for $x \leq a$, and $F_n = (x = a)^n$ for $x > a$. (See Ref. 12.)

Slenderness ratio The ratio of length of a uniform column to the minimum radius of gyration of the cross section.

Slip lines (Lüder's lines) Lines that appear on the polished surface of a crystal or crystalline body that has been stressed beyond the elastic limit. They represent the intersection of the surface by planes on which shear stress has produced plastic slip (see Sec. 3.5 and Ref. 13).

Strain Any forced change in the dimensions and/or shape of an elastic element. A stretch is a *tensile strain*; a shortening is a *compressive strain*; and an angular distortion is a *shear strain.*

Strain concentration factor Localized peak strains develop in the presence of stress raisers. The strain concentration factor is the ratio of the *localized maximum strain* at a given location to the *nominal average strain* at that location. The nominal average strain is computed from the average stress and a knowledge of the stress–strain behavior of the material. In a situation where all stresses and strains are elastic, the stress and strain concentration factors are equal. (Compare with *Stress concentration factor.*)

Strain energy Mechanical energy stored in a stressed material. Stress within the elastic limit is implied where the strain energy is equal to the work done by the external forces in producing the stress and is recoverable.

Strain rosette At any point on the surface of a stressed body, strains measured along each of three intersecting gage lines make the calculation of the principal stresses possible. The gage lines and the corresponding strains are called *strain rosettes.*

Strength Typically refers to a particular limiting value of stress for which a material ceases to behave according to some prescribed function. [Compare *Endurance limit (fatigue strength)*; *Endurance strength*; *Ultimate strength*; *Yield strength.*]

Stress Internal force per unit area exerted on a specified surface. When the force is tangential to the surface, the stress is called a *shear stress*; when the force is normal to the surface, the stress is called a *normal stress*; when the normal stress is directed toward the surface, it is called a *compressive stress*; and when the normal stress is directed away from the surface, it is called a *tensile stress*.

Stress concentration factor, K_t Irregularities of form such as holes, screw threads, notches, and sharp shoulders, that, when present in a beam, shaft, or other member subject to loading, may produce high localized stresses. This phenomenon is called a *stress concentration*, and the form irregularities that cause it are called *stress raisers*. For the particular type of stress raiser in question, the ratio of the true maximum stress to the nominal stress calculated by the ordinary formulas of mechanics (P/A, Mc/I, Tc/J, etc.) is the stress concentration factor. The nominal stress calculation is based on the net section properties at the location of the stress raiser ignoring the redistribution of stress caused by the form irregularity. (See Sec. 3.10.)

Stress intensity factor, K A term employed in fracture mechanics to describe the elastic stress field surrounding a crack tip caused by a remote load. The magnitude of K depends on the sample geometry, and the size and location of the crack. There are three modes associated with the stress intensity factor; Mode I, KI, or opening mode; Mode II, KII ,or sliding mode; and Mode III, KIII, or tearing mode. See Chap. 19 for more details.

Stress trajectory (isostatic) A line (in a stressed body) tangent to the direction of one of the principal stresses at every point through which it passes.

Superposition, principle of With certain exceptions, the effect of a given combined loading on a structure may be resolved by determining the effects of each load separately and adding the results algebraically. The principle may be applied provided: (1) each effect is linearly related to the load that produces it, (2) a load does not create a condition which affects the result of another load, and (3) the deformations resulting from any specific load are not large enough to appreciably alter the geometric relations of the parts of the structural system. (See Sec. 4.2.)

Torsional center (center of twist) If a twisting couple is applied at a given section of a straight member, that section rotates about some point in its plane. This point, which does not move when the member twists, is the torsional center of that section. (See Refs. 5 and 6.)

Torsional moment (torque, twisting moment) At any section of a member, the moment of all forces that act on the member to the left (or right) of that section, taken about a polar axis through the *flexural center* of that section. For sections that are symmetrical about each principal axis, the flexural center coincides with the centroid (see Refs. 6 and 10).

Transformations of stress or strain Conversions of stress or strain at a point from one three-dimensional coordinate system to another. (See Secs. 2.3 and 2.4.)

Transverse shear force (vertical shear) Reference is to a simple straight beam, assumed for convenience to be horizontal and loaded and supported by forces, all of which lie in a vertical plane. The transverse shear force at any section of the beam is the vertical component of all forces that act on the beam to the left (or right) of that section. The *shear force equation* is an expression for the transverse shear at any section in terms of x, the distance to that section measured from a chosen origin, usually taken from the left of the beam.

Tresca stress Based on the failure mode of a ductile material being due to shear stress, the *Tresca stress* is a single shear stress value, which is equivalent to an actual combined state of stress.

True strain The summation (integral) of each infinitesimal elongation ΔL of successive values of a specific gage length L divided by that length. It is equal to $\int_{L_0}^{L} (dL/L) = \log_e (L/L_0) = \log_e (1 + \varepsilon)$, where L_0 is the original gage length and ε is the normal strain as ordinarily defined (Ref. 14).

True stress For an axially loaded bar, the force divided by the actual cross-sectional area undergoing loading. It differs from the *engineering stress* defined in terms of the original area.

Ultimate elongation The percentage of permanent deformation remaining after tensile rupture (measured over an arbitrary length including the section of rupture).

Ultimate strength The ultimate strength of a material in uniaxial tension or compression, or pure shear, respectively, is the maximum tensile, compressive, or shear stress that the material can sustain calculated on the basis of the greatest load achieved prior to fracture and the original unstrained dimensions.

von Mises stress Based on the failure mode of a ductile material being due to distortional energy caused by a stress state, the *von Mises stress* is a single normal stress value, which is equivalent to an actual combined state of stress.

Yield point The stress at which the strain increases without an increase in stress. For some purposes, it is important to distinguish between *upper* and *lower* yield points. When they occur, the upper yield point is reached first and is a maxima that is followed by the lower yield point, a minima. Only a few materials exhibit a true yield point. For other materials the term is sometimes used synonymously with yield strength. (Compare *Apparent elastic limit*; *Elastic limit*; *Proportional limit*; *Yield strength*. See Ref. 7.)

Yield strength The stress at which a material exhibits a specified permanent deformation or set. This stress is usually determined by the offset method, where the strain departs from the linear portion of the actual stress–strain diagram by an offset unit strain of 0.002. (See Ref. 7.)

REFERENCES

1. Soderberg, C. R., "Working Stresses," ASME Paper A-106, *J. Appl. Mech.*, 2(3): 1935.
2. "Unit Stress in Structural Materials (Symposium)," *Trans. Am. Soc. Civil Eng.*, 91(388): 1927.
3. Johnson, R. C., "Predicting Part Failures," *Mach. Des.*, 37(1): 137–142, 1965; no. 2, 157–162, 1965.
4. von Heydenkamph, G. S., "Damping Capacity of Materials," *Proc. ASTM*, vol. 21, part II, p. 157, 1931.
5. Kuhn, P., "Remarks on the Elastic Axis of Shell Wings," *Nat. Adv. Comm. Aeron.*, Tech. Note 562, 1936.
6. Schwalbe, W. L., "The Center of Torsion for Angle and Channel Sections," *Trans. ASME*, Paper APM-54-11, 54(1): 1932.
7. "Tentative Definitions of Terms Relating to Methods of Testing," *Proc. ASTM*, vol. 35, part I, p. 1315, 1935.
8. Morley, S., *Strength of Materials*, 5th ed., Longmans, Green, 1919.
9. Timoshenko, S., *Theory of Elasticity*, 3rd ed., McGraw-Hill, 1970.
10. Griffith, A. A., and G. I. Taylor, "The Problems of Flexure and Its Solution by the Soap Film Method," *Tech Rep. Adv. Comm. Aeron. (British)*, Reports and Memoranda no. 399, p. 950, 1917.
11. "Airworthiness Requirements for Aircraft," *Aeron. Bull. 7-A*, U.S. Dept. of Commerce, 1934.
12. Budynas, R. G., *'Advanced Strength and Applied Stress Analysis,'* WCB/McGraw-Hill, 1999.
13. Nadai, A., *Plasticity*, McGraw-Hill, 1931.
14. Freudenthal, A. M., *The Inelastic Behavior of Engineering Materials and Structures*, John Wiley & Sons, 1950.

Index

Note: Figures are indicated by *f*; Tables are indicated by *t*.